The Natural History
of
Canadian Mammals

The Natural History of Canadian Mammals

DONNA NAUGHTON

COLOUR ART: PAUL GERAGHTY,
JULIUS CSOTONYI, AND BRENDA CARTER
LINE ART: DONNA NAUGHTON,
MICHELINE BEAULIEU-BOUCHARD,
AND ALAN McDONALD

CANADIAN MUSEUM OF NATURE
AND
UNIVERSITY OF TORONTO PRESS
TORONTO   BUFFALO   LONDON

© Canadian Museum of Nature 2012
Published by University of Toronto Press
Toronto Buffalo London
www.utppublishing.com
Printed in Canada

ISBN 978-1-4426-4483-0

Printed on acid-free paper
Jacket and interior design by Linda Gustafson / Counterpunch Inc.

Library and Archives Canada Cataloguing in Publication
Naughton, Donna
The natural history of Canadian mammals / Donna Naughton.
Includes bibliographical references and index.
Co-published by: Canadian Museum of Nature.
ISBN 978-1-4426-4483-0
1. Mammals – Canada – Identification.   2. Natural history – Canada.   I.
Canadian Museum of Nature.   II. Title.
QL721.N386 2012   599.0971   C2012-902669-7

PHOTO CREDITS
Page i: Neal Weisenberg
Page ii: Al Parker / iStockphoto
Page 536: Louise Devenney

University of Toronto Press acknowledges the Royal Canadian Geographical
Society and *Canadian Geographic* for the use of images submitted as part of
their annual Wildlife Photography Competition.

University of Toronto Press acknowledges the financial assistance
to its publishing program of the Canada Council for the Arts
and the Ontario Arts Council.

University of Toronto Press acknowledges the financial support of
the Government of Canada through the Canada Book Fund for its
publishing activities.

Canadä

Canada Council   Conseil des arts
for the Arts   du Canada

ONTARIO ARTS COUNCIL
CONSEIL DES ARTS DE L'ONTARIO

To Stewart D. MacDonald, retired Curator of Vertebrate Ethology, National Museum of Natural Sciences of Canada (now the Canadian Museum of Nature), whose photographic artistry, keen powers of observation, and generosity of spirit continue to be an inspiration. His faith and support over these many years are gratefully acknowledged and appreciated.

Contents

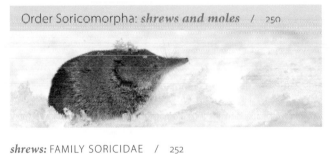

Order Chiroptera: *bats* / 306

Order Carnivora: *carnivores* / 356

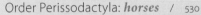

Order Perissodactyla: *horses* / 530

Order Artiodactyla: *deer, bison, sheep, and other even-toed ungulates* / 538

Order Cetacea: *whales, dolphins, and porpoises* / 600

Domestic mammals / 702

Preface

In its most basic definition, a mammal is a warm-blooded vertebrate with hair and mammary glands. We humans, by our very nature, relate in a primal, visceral way to our fellow mammals. We see baby mammals as cute and endearing, we ascribe human attributes to the behaviour of other mammals, and we are excited by sightings of wild mammals. Mammals were our earliest domesticated species, our major prey, and our most feared predators. It could be said that we are pre-programmed to notice other mammals. While hunting and trapping are still significant activities in North America, more and more people are taking an interest in watching and photographing wild mammals. In all of these situations, the ability to identify a creature and its attributes is necessary to make the leap to a real understanding and appreciation of it. At a time when the natural environment is more threatened by human activity than ever before, the decisions we make in the next few decades could be the key to species and habitat survival.

This volume is a comprehensive source of information but is by no means all inclusive. Most of the literature that was researched and cited is dated from 1990 to the present. The idea for this book originated in the mid 1980s with the late Dr C.G. van Zyll de Jong, then Curator of Mammals for the National Museum of Natural Sciences (now called the Canadian Museum of Nature). He worked with Brenda Carter to produce the plates on the opossum and the shrews and moles. These were first published in volume 1 of *The Handbook of Canadian Mammals*. Dr van Zyll de Jong and the artist Paul Geraghty worked for over a decade to produce the majority of the remaining colour illustrations. The plates on bats were first published in volume 2 of *The Handbook of Canadian Mammals* (the last in the series); the remaining plates were not published until now. Dr van Zyll de Jong's idea for a comprehensive book featuring all of the art was not realized before his retirement in 1993 and his untimely death a few years later. The museum revived the project in 2001, thanks to the diligence and resourcefulness of Wendy McPeake and the participation of the publisher, University of Toronto Press.

Introduction

It does no harm to the mystery to know a little about it.
Richard Feynman

The main purpose of this book is to inform and encourage the observation, appreciation, and understanding of Canada's mammals. Identification is often the precursor to understanding; therefore much of the volume attempts to help the reader to distinguish one species from another.

ORGANIZATION

Species arrangement conforms to the evolutionary or phylogenetic organization favoured by mammalogists today. Similar species are grouped together and these groupings are presented in order of evolutionary advancement as it is currently understood. The preliminary groups are considered more primitive or older while the later groups are considered more advanced or more recent in origin. This organization is dynamic, as it is constantly being reviewed and updated as we gain more understanding and knowledge, and will undoubtedly continue to change over time.

The group or class of animals known as Mammalia is separated into 29 subgroups or orders (there are an additional 19 extinct orders). Each order is further subdivided into families. Within each family are the building blocks of scientific naming – the genus and species. This double naming system, known as binomial nomenclature, was created by the Swedish botanist Karl von Linné (commonly called Linnaeus) in 1758. Each two-part scientific name is a unique identifier, which, by agreed international convention, cannot be the same as any other. For example, in the case of the Bighorn Sheep, the genus is *Ovis,* and the trivial name or epithet is *canadensis.* Together they create the species name, *Ovis canadensis.* Other similar sheep share the genus name of *Ovis,* but no other species within that genus is called *Ovis canadensis.* The first word in a species name is capitalized while the second is not, and both are usually italicized.

For the purpose of this volume, the genera within each family are presented alphabetically by their scientific name, as are the species within each genus. Both the common and scientific names are included in the index for easy reference.

WHAT IS A SPECIES?

In the past, biologists defined a species as a group of physically distinct animals that mate in the wild and produce fertile offspring. Two groups whose ranges overlap, but do not interbreed, are deemed to be different species. Mating between two such species in the artificial environment of captivity could produce fertile offspring, but in nature they remain reproductively isolated. This "biological species concept" works well in cases where the two species in question have overlapping distributions.

However, in the case where very similar species are geographically isolated, there is no adequate method of testing the wild fertility possibilities. Furthermore, biologists consider that spatial isolation over an extended period is essential for two populations of the same species to gradually become different enough to be considered unique species. So, depending on how long the two populations have been separated and how much genetic divergence has occurred, they may well be on the road to becoming distinct. The difficulty arises in trying to establish where to draw the line. How much change is enough to call it a different species? Is it still only distinct enough to be considered a subspecies? The recent "phylogenetic species concept" uses DNA to help determine the extent of genetic differences. Recent advances in DNA testing have refined the procedure, reducing the cost and the amount of tissue needed, so that it is now becoming an attractive and effective research tool that may eventually help us to address this dilemma. DNA evidence is especially useful when it supports measurable physical differences. This still begs the question: How do we deal with two populations that are distinct genetically, but indistinguishable physically? The Pacific and Atlantic Right Whales are an example. We have been unable to find any measurable external or skeletal differences between the two species. In this case, the identification is simplified by the geographical reality that one species inhabits the Pacific Ocean and one the Atlantic Ocean and that there is no possibility of intermingling. However, museum specimens from unknown localities cannot be identified without a DNA analysis. Nevertheless, the compelling DNA evidence has swayed the majority of scientists to accept the two species as unique.

As our understanding of how to define a species increases, there will be changes in how we classify species. For the purposes of this volume, recognition of species follows the standard set by Wilson and Reader (2005), with a few exceptions. The Heather Vole (*Phenacomys intermedius*) has recently been split by some researchers into an Eastern (*P. ungava*) and a Western (*P. intermedius*) form. Unfortunately, there appears to be no recent genetic work to support this split and there are no known morphological characteristics that reliably separate the two forms. I have chosen to follow Nagorsen (2005), Verts and Carraway (1998), and Smith (1993) in preserving the single species until further definitive work is conducted to support the separation. Very recent genetic work on the Long-tailed Shrew and the Gaspé Shrew has confirmed, as has long been suspected, that the two species are actually the same species (Shafer and Stewart, 2007) and that the two forms are end results of a clinal gradient from a larger southern form, formerly called the Long-tailed Shrew (*Sorex dispar*), to a smaller northern form, formerly called the Gaspé Shrew (*Sorex gaspensis*). The newly merged species is called the Long-tailed Shrew (*Sorex dispar*), as that name is older and hence has precedence. A new species of North American shrew, *Sorex rohweri*, the Olympic Shrew, was described in 2007 and has been included. This species was formerly considered part of the Cinereous Shrew (*Sorex cinereus*) species complex.

WHICH SPECIES TO INCLUDE?

The selection of species to include in this book was based on two principles:

1. Those that have in recent times had a viable, naturally occurring wild population in Canada, its continental islands, or in the marine waters of its continental shelf. This includes all native species including humans and any recently extinct or extirpated species.

2. Species introduced into Canada by humans – whether intentionally or unintentionally – which currently have established self-sustaining, long-lasting wild populations (such as the Norway Rat and the Domestic Horse), but excluding those other escaped or introduced species whose ongoing presence in the wild is not assured. Most of our domesticated species do not fit these criteria for inclusion (with the exception of the horse), but for the sake of comparison and completeness, a special section is provided to assist in the identification of skull remains and tracks of Domestic Dogs, Domestic Cats, Cattle, Pigs, Sheep, and Goats.

Two hundred and fifteen mammal species in 10 different orders are included. Each species account includes a colour illustration (or several if there are significant differences between genders or in pelt colours), a distribution map, and a series of skull illustrations along with the text. When appropriate, diagrams of tracks of terrestrial species are also included. Four exceptions are made. One is for humans, as there are innumerable other texts containing this information. The second is for the Big Free-tailed Bat. This species appears without a colour illustration (a black and white sketch is provided). Although a specimen was collected in British Columbia in 1938, it represents the only known Canadian occurrence of a normally much more southerly bat, and the species is very unlikely to be encountered again in Canada. The third is the account for the recently extinct Sea Mink, which appears without colour or skull illustrations as this information is unavailable. The fourth exception is the North Pacific Right Whale, which appears without a colour or skull illustration as these are provided in the North Atlantic Right Whale account and the two are indistinguishable. Each order is introduced with a description of what is unique and distinctive about the group. This text is followed by a short discussion of the families within the order and their unique characteristics.

DISTRIBUTION MAPS

Eighteen to twenty thousand years ago, most of Canada was covered by ice up to three kilometres thick. This was the peak of the last major glaciation, called the Wisconsinan Glaciation. There were only a few areas that were not ice-covered. These very important refugia provided a place for many species of plants and animals to survive the prolonged cold period. The mammals that survived the glaciation did so either in these small refugia within the ice sheet or in the much larger refugium of the southern portion of North America. By about 10,000 years ago, the ice was mostly melted and the stage was thus set for a major recolonization, which in some minor ways, continues today.

Distribution is an important key to understanding many things about a species. In order to place each species within a global context, the maps show the entire worldwide distribution, with additional inserts to highlight the Canadian portions if necessary.

Bear in mind that while the range maps appear to indicate precisely where a species can be found, the story is rather more complex. Distribution is dynamic for mammals. As already mentioned, some species are still slowly moving back into areas previously abandoned during the last great glaciation. Other species are reacting to human activity and habitat change due to urbanization, agriculture, deforestation, and reforestation. Occurrence within the range may be spotty or seasonal, with numbers abundant in some areas and uncommon or occasional in others. Many mammals, especially bats and whales, are very mobile and are often found outside their usual range. Individuals of even the more sedentary species may display considerable extralimital movement, as in the case of individuals searching for food during a period of shortage or young adults looking for a mate or an unoccupied home range. Furthermore, many mammal species in temperate regions undergo significant migrations twice a year, during which misplaced individuals occasionally turn up in unusual places owing to factors such as weather, disease, or just loss of direction. The boundary of a range is usually an educated guess based on known habitat requirements and sometimes a few or a single specimen. Often the peripheries of the range provide less than optimum habitat and this is reflected in low numbers of individuals, or the possibility that the species is found there only during very favourable years. We do not know all there is to know about where mammals are found, so as more information is obtained the ranges will change accordingly. Finally, and most significantly, mammal distribution is based on the distinct habitat requirements of each species. As climate changes, we can expect habitats to change. Mammals may be forced to move from previously hospitable areas or may find that formerly unacceptable areas are now appropriate for their needs. In response to rapid habitat change, some species could become threatened, resulting in serious reductions in their population sizes even to the point of extinction, while other species may find more appropriate habitat available to them and have the capacity to extend their range.

At best, the distribution maps should be viewed as *likely* indicators of species presence and, at worst, as *suggested* indicators of species occurrence. Different colours are used to indicate specific information, such as former and current distribution or to highlight areas of introduction. A descriptive legend for each map with multiple colours explains the significance of the colours.

SKULL ILLUSTRATIONS

Much of the scholarly study and species identification of mammals centres on the skull and teeth. In order to provide a simulation of three dimensions, the skull of each species is drawn in dorsal, ventral, and lateral views with the lower jaw included with the lateral view. Species that display dramatic differences in skull shape or size between the male and female will have skulls of both genders illustrated. In most cases, the drawings reflect a single specimen; the catalogue number and source museum are included in the caption. Occasionally, when there is no complete adult skull available, the

drawing is necessarily a composite. The caption in these cases may include more than one catalogue number.

The skull illustrations are necessarily simplified. Important identification criteria are incorporated, but much that is unnecessary or irrelevant is left out. Sutures can be very prominent on some skulls; however, as they are rarely diagnostic and change with the age of the animal, a conscious decision has been made not to draw them unless they are diagnostic. Figure 1 shows a skull with the sutures included for the purpose of identifying the major individual skull bones.

DENTAL FORMULA

Most mammal teeth can be separated into incisors, canines, premolars, and molars, such as in the Cougar diagram (Figure 1), but

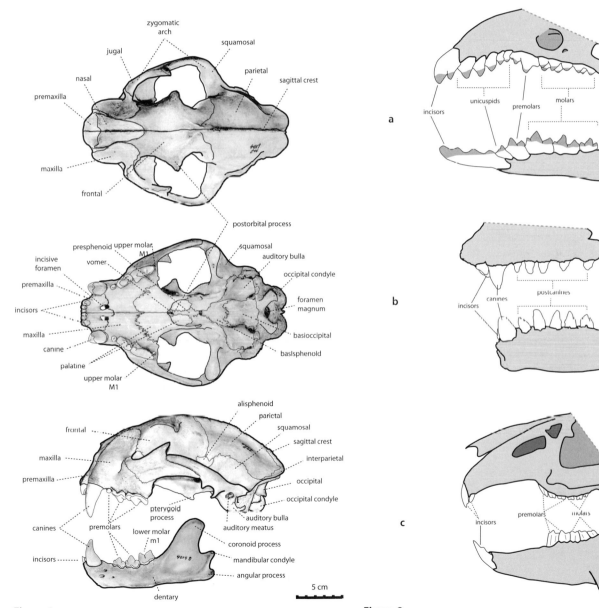

Figure 1
Skull of an adult male Cougar with the teeth and primary bones identified.
(from the collection of the Canadian Museum of Nature CMNMA 4019)

Figure 2
Toothrows of various mammals:
a) Arctic Shrew, b) male Grey Seal, c) Snowshoe Hare.

this is not always the case. Figure 2 illustrates some of the divergent tooth patterns found in Canadian mammals. For example, shrews have highly modified and characteristic incisors that protrude forward and are used like pinchers. The remaining incisors, canines, and some of the premolars are modified into simple, single-cusped teeth behind the incisors called unicuspids. Seal teeth behind the canines are all similar and cannot be separated into recognizable premolars and molars, so all are simply called postcanine teeth. A standard dental formula is composed of a single upper and lower toothrow so the teeth can be added up and then multiplied by two (thereby including the other sides of the jaws), to arrive at a total number of teeth in the mouth of each individual. Teeth in the upper jaw are denoted by upper-case letters and those in the lower jaw by lower-case, respectively. For example, M1 means the first upper molar, while p2 signifies the second lower premolar; I1 indicates the first upper incisor, while c1 indicates the first lower canine.

Some species, such as bears, may have variable numbers of certain small, and for them, unnecessary teeth, in this case, premolars.

IDENTIFICATION

It is not always possible to identify the mammal you have just seen, especially from afar. Many of the smaller mammals must be observed close at hand in order to identify them, and even then it is not always easy; often teeth or skull characteristics are needed for a definitive identification. Thus, diagnostic dental and cranial characteristics required for identification of these cryptic species are illustrated as necessary and similar species are listed in the text. Even in the larger species, the flash of brown you were lucky enough to see between the trees, or the dorsal fin of a whale in choppy seas, will not always provide enough information for identification.

Many mammals are furtive and nocturnal, and often the only indicator of their presence in an area is what they leave behind. Scat, tracks, and other signs offer clues about their presence and activity. These signs are to the mammal watcher what bird song is to the bird watcher. A clear sighting of the often elusive mammal is not always necessary to know that it is around.

Figure 3 shows the aspects of a mammal track that are mentioned in the text. Species that commonly leave tracks have track diagrams included in their accounts. These illustrate a left hind and left front track in detail and provide one or two of the more common track patterns. Most trackline illustrations in this volume begin with the left front foot and the direction of travel is towards the top of the page. Front and hind track diagrams are artificially filled with different colours to aid in discerning the differences and placement of each. The dimensions provided for a front and hind track include the maximum length where the heel pad registers and the maximum width where the toes are spread. The minimum length provided is typically the length of a track when the heel pad (on either front or hind, but especially the hind foot) does not strike the ground. Figure 4 explains the basics of stride measurement.

The treatment of tracking herein barely scratches the surface of this complex subject as it provides only the bare minimum of information in simplified form. Readers interested in further study of this interesting field are encouraged to consult a volume that specializes in the subject, several of which are mentioned in the reference list at the end of the Introduction.

A series of colour identification plates follows this Introduction. These illustrate the Canadian representatives within each group in relative size to each other for ease of comparison. Accompanying page numbers direct the reader to the full species account. Major species variations are included on the identification plates.

NAMES

English and French common names are listed at the beginning of each species account. Often there is more than one common name, in which case the alternatives are listed. Many mammalogists prefer not to capitalize the common name as it is not an "official name," as is the scientific version. The usual exception occurs when the common name includes a proper name, as in the case of Townsend's mole. However, a lack of capitalization, especially when dealing with descriptive names such as the Southern Flying Squirrel or the Black Bear, could easily lead to confusion.

In this volume, for the sake of clarity, a common name that refers to a distinct species of flora or fauna is capitalized, acknowledging that the full common name is a proper noun. The single exception is the common name for the human, which is generally not capitalized. If a common name refers to two or more possible species, it is not capitalized (e.g., dolphins, deer, mice).

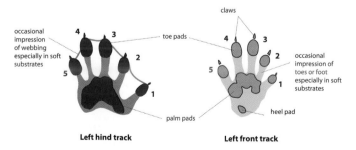

Figure 3
Track morphology. Hind and front tracks of a Northern River Otter illustrating the primary components of a track. Numbers refer to the toe formula with toe 1 being the inside toe (either the thumb or big toe).

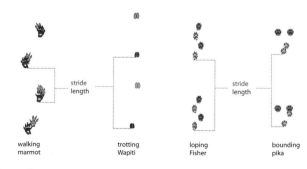

Figure 4
Stride length. Normally measured from the front to front or back to back of consecutive hind tracks (walking and trotting) or from back to back of a group of tracks (loping and bounding).

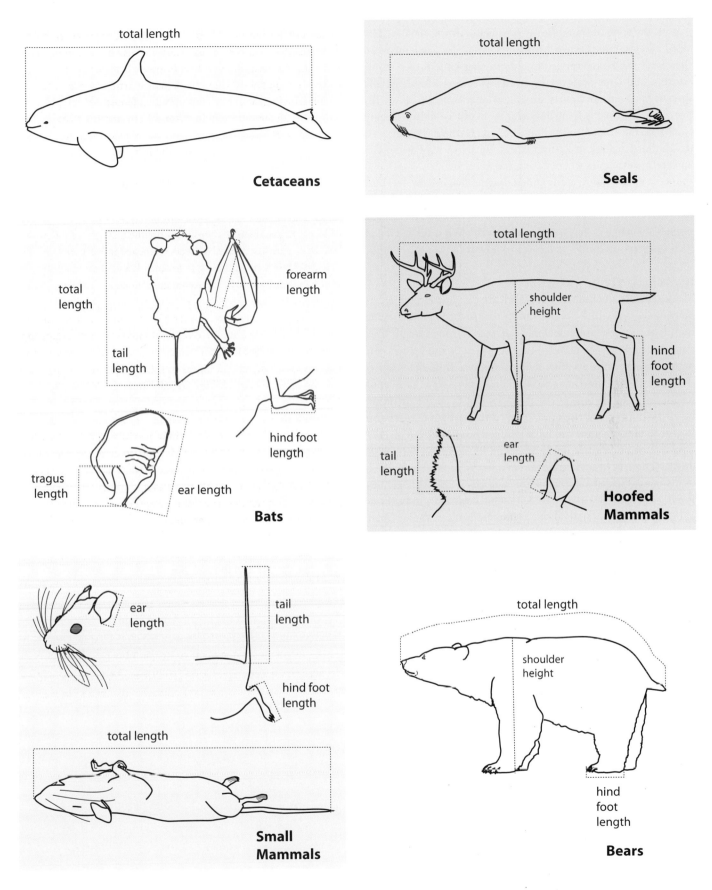

Figure 5
Standard measurements for various body types.

With the application of these principles, a listing such as short-tailed shrews, Little Brown Myotis, muskrats, humans, pikas, and Grey Seals can be interpreted to indicate that Little Brown Myotis, humans, and Grey Seals are individual species, while short-tailed shrews, muskrats, and pikas are groups of species. In cases where the common name is spelled differently in other English-speaking countries, the Canadian spelling is used. Harbour Porpoise versus Harbor Porpoise or Eastern Grey Squirrel versus Eastern Gray Squirrel are examples where this choice was made.

The scientific name is the unique name assigned by the International Commission on Zoological Nomenclature to identify the species. Usually these names are variations on Latin or Greek. Occasionally the name is changed as more information is discovered. In such cases, recent former names are also listed.

MEASUREMENTS

Measurements and how they are taken are explained in the accompanying figures, which cover a variety of standardized mammalian body types. One measurement that can be confusing is length. There are at least two means of measuring body length and readers comparing data provided in this volume with those from other sources should verify that the data were taken in the same way. For example, length has been defined as the measurement from the *tip* of nose to the *tip* of the tail vertebrae and called total length (TL), or from the *tip* of nose to the *base* of tail and called head-body length (HB). This volume utilizes the North American Museum standard, which defines length (TL) as the distance from the tip of the nose to the tip of the tail vertebrae with the backbone relatively straight. The term shoulder height (SH) is used to define the distance from the height of the back at the shoulder to the bottom of the foot. Measurements for bats include two not used for other mammals: the forearm length and the tragus length if present in the species. All measurements and weights are provided in metric units. A conversion guide is provided below.

Area
1 hectare (ha) = 2.47 acres
1 square kilometre (km²) = 0.86 square miles
1 km² = 100 ha

Mass (weight)
1 gram (g) = 0.035 ounces
1 kilogram (kg) = 2.204 pounds
1 metric tonne = 2204 pounds

Distance
1 kilometre (km) = 0.62 miles
1 metre (m) = 3.28 feet
1 m = 1.09 yards
1 centimetre (cm) = 0.39 inches
1 millimetre (mm) = 0.039 inches

ABBREVIATIONS

AMNH	American Museum of Natural History
CMN	Canadian Museum of Nature
CMNMA	Canadian Museum of Nature Mammal Collection
COSEWIC	Committee on the Status of Endangered Wildlife in Canada
IUCN	International Union for Conservation of Nature
NBM	New Brunswick Museum
NMNH	National Museum of Natural History (Smithsonian)
RAM	Royal Alberta Museum
ROM	Royal Ontario Museum
RSM	Royal Saskatchewan Museum
USNM	United States National Museum (Smithsonian)

SELECTED REFERENCES

Elbroch, M. 2003. *Mammal Tracks and Signs.* Stackpole Books, Mechanicsburg, PA.

Lowrey, J.C. 2006. *The Tracker's Field Guide.* FalconGuide, Globe Pequot Press, Guilford, CT.

Murie, O.J., and Elbroch, M. 2005. *A Field Guide to Animal Tracks.* Peterson Field Guide Series, 3rd ed. Houghton Mifflin, New York, NY.

McKenna, M.C., and Bell, S.G. 1997. *Classification of Mammals: Above the Species Level.* Columbia University Press, New York, NY.

Wilson, D.E., and Reeder, D.M., eds. 2005. *Mammal Species of the World: A Taxonomic and Geographic Reference.* Vols. 1 & 2. Johns Hopkins University Press, Baltimore, MD.

Identification Plates

Each plate illustrates a group of species at the same scale to convey relative size. Similar animals are displayed in a group for ease of comparison. Individual species are cross-referenced to the accounts in the text.

North American Opossum, Mountain Beaver, and squirrels

Woodchuck
page 33

North American
Opossum
Page 3

Mountain Beaver
page 12

Hoary Marmot
page 27

Vancouver Island
Marmot
page 36

Yellow-bellied
Marmot
page 30

Northern Flying
Squirrel page 20

Eastern Grey Squirrel
page 39

Eastern Fox Squirrel
page 44

Southern Flying
Squirrel
page 23

All are approximately 19% of actual size

Squirrels continued

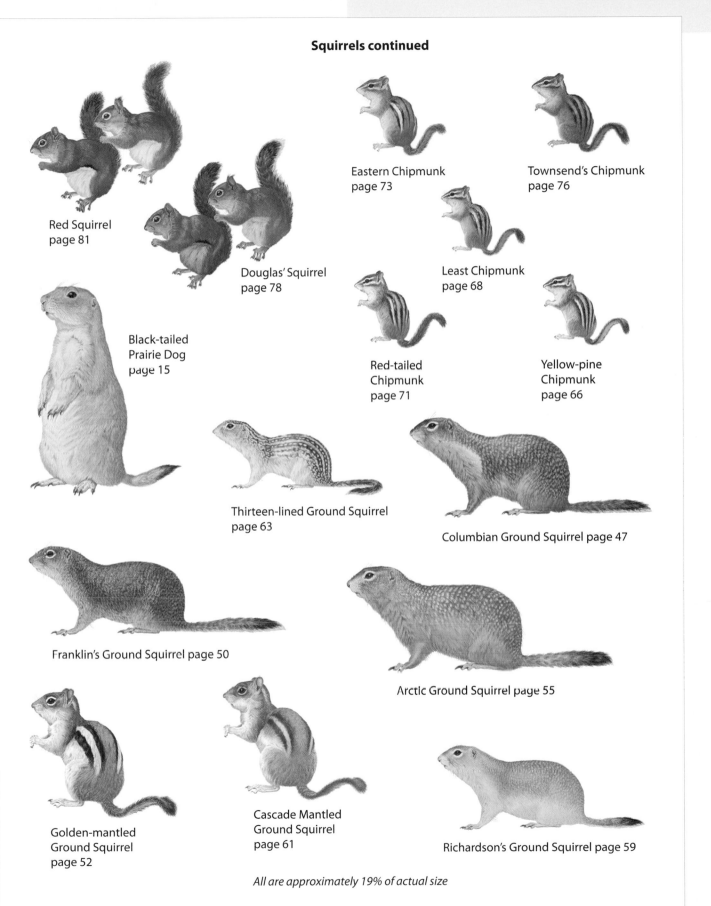

Red Squirrel
page 81

Douglas' Squirrel
page 78

Eastern Chipmunk
page 73

Townsend's Chipmunk
page 76

Least Chipmunk
page 68

Red-tailed
Chipmunk
page 71

Yellow-pine
Chipmunk
page 66

Black-tailed
Prairie Dog
page 15

Thirteen-lined Ground Squirrel
page 63

Columbian Ground Squirrel page 47

Franklin's Ground Squirrel page 50

Arctic Ground Squirrel page 55

Golden-mantled
Ground Squirrel
page 52

Cascade Mantled
Ground Squirrel
page 61

Richardson's Ground Squirrel page 59

All are approximately 19% of actual size

Large rodents

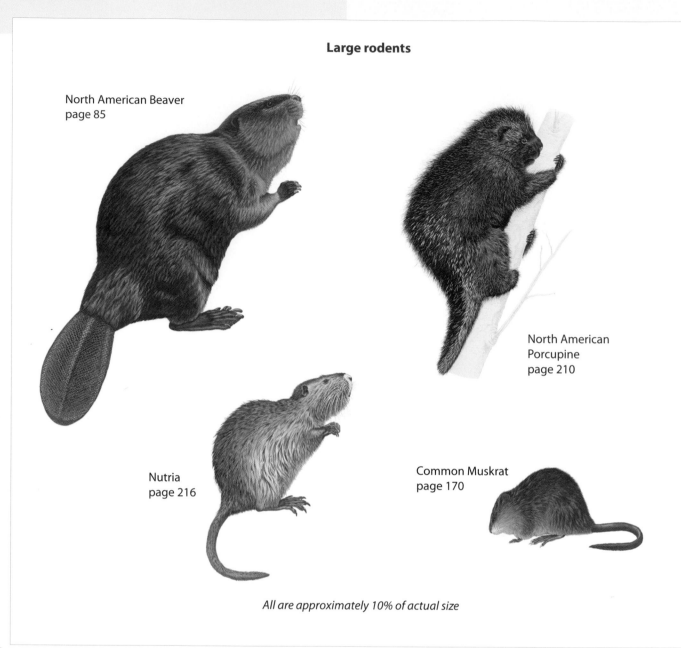

North American Beaver
page 85

North American
Porcupine
page 210

Nutria
page 216

Common Muskrat
page 170

All are approximately 10% of actual size

Pocket gophers, kangaroo rat, pocket mice, and jumping mice

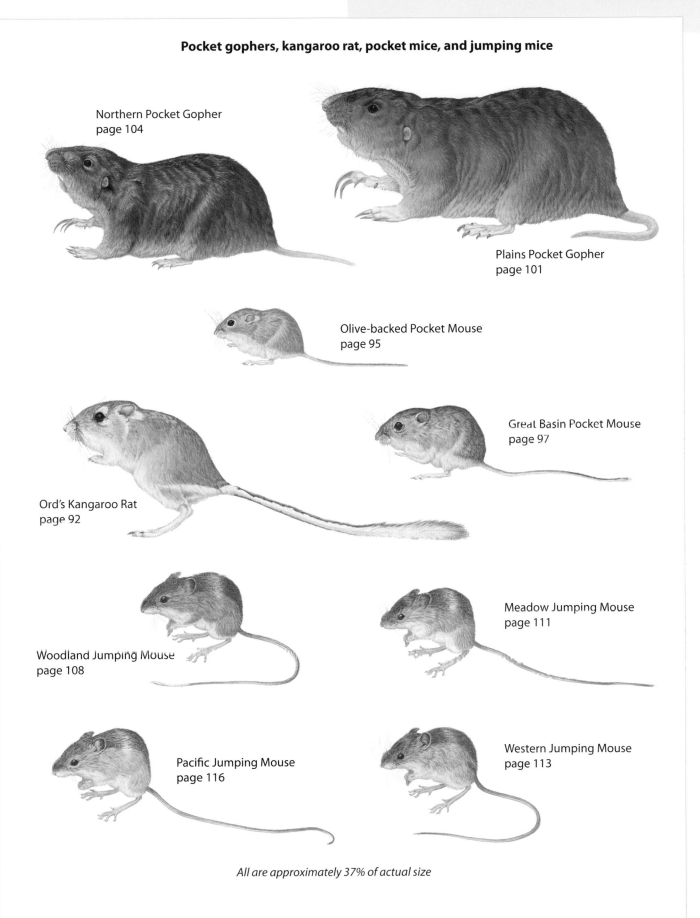

Northern Pocket Gopher
page 104

Plains Pocket Gopher
page 101

Olive-backed Pocket Mouse
page 95

Ord's Kangaroo Rat
page 92

Great Basin Pocket Mouse
page 97

Woodland Jumping Mouse
page 108

Meadow Jumping Mouse
page 111

Pacific Jumping Mouse
page 116

Western Jumping Mouse
page 113

All are approximately 37% of actual size

Voles

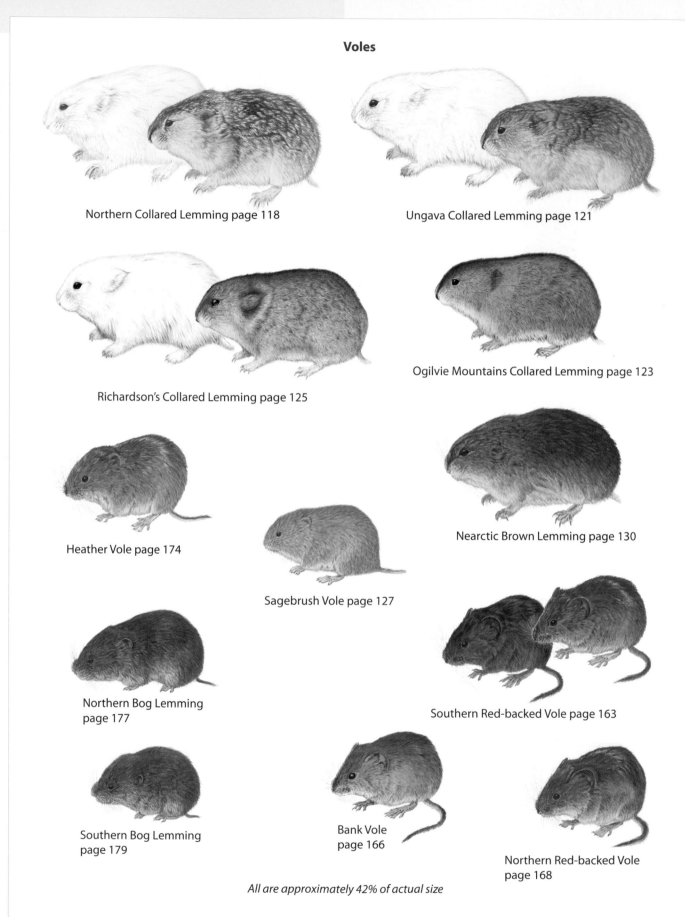

Northern Collared Lemming page 118

Ungava Collared Lemming page 121

Richardson's Collared Lemming page 125

Ogilvie Mountains Collared Lemming page 123

Heather Vole page 174

Nearctic Brown Lemming page 130

Sagebrush Vole page 127

Northern Bog Lemming page 177

Southern Red-backed Vole page 163

Southern Bog Lemming page 179

Bank Vole page 166

Northern Red-backed Vole page 168

All are approximately 42% of actual size

Voles continued

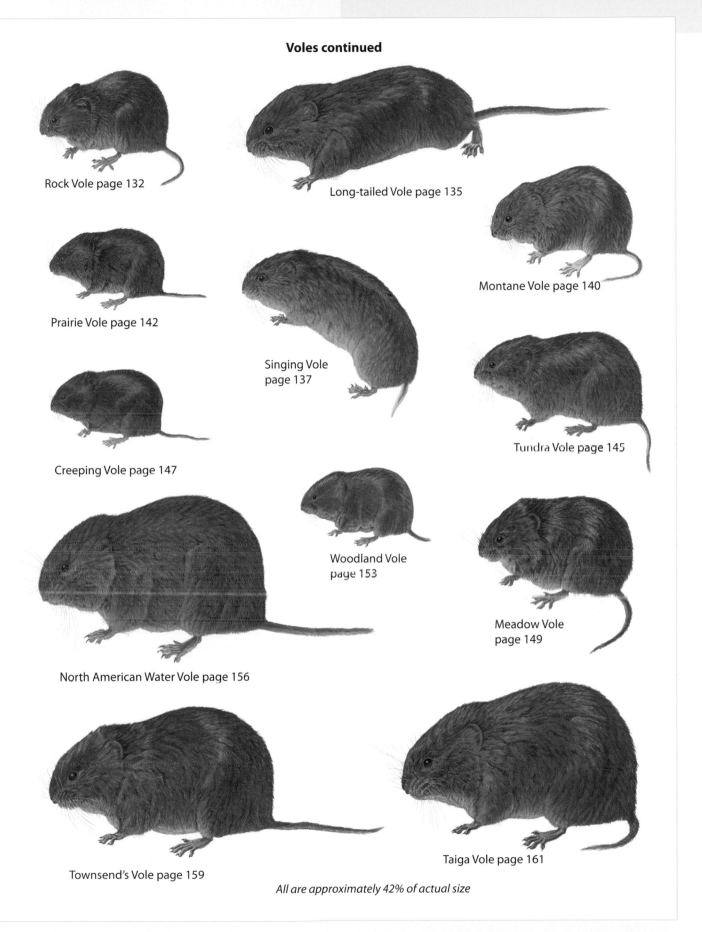

Rock Vole page 132

Long-tailed Vole page 135

Montane Vole page 140

Prairie Vole page 142

Singing Vole page 137

Tundra Vole page 145

Creeping Vole page 147

Woodland Vole page 153

Meadow Vole page 149

North American Water Vole page 156

Townsend's Vole page 159

Taiga Vole page 161

All are approximately 42% of actual size

Rats and mice

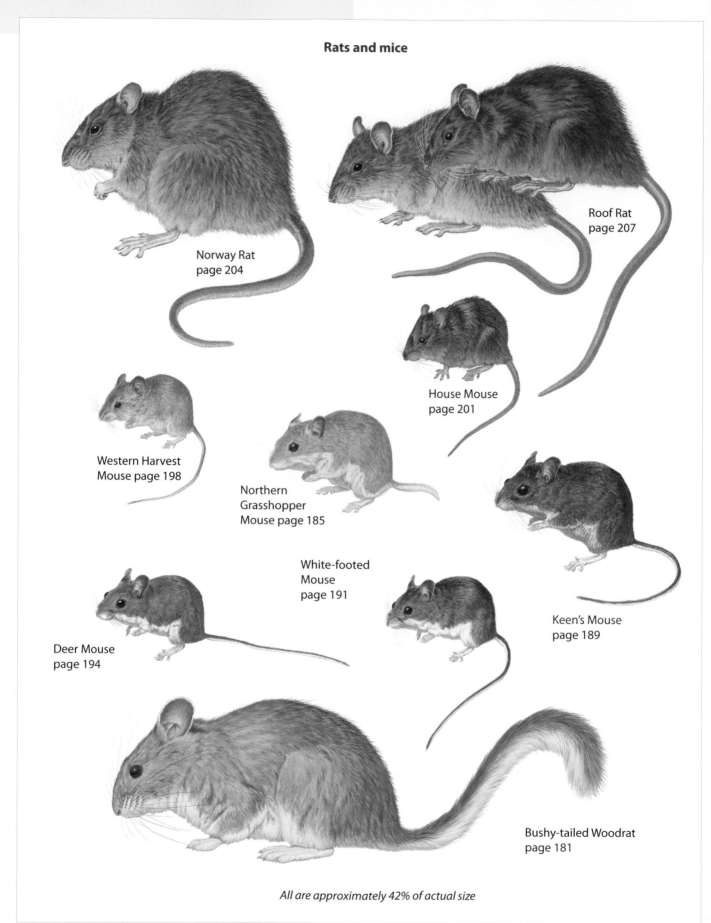

Norway Rat
page 204

Roof Rat
page 207

House Mouse
page 201

Western Harvest
Mouse page 198

Northern
Grasshopper
Mouse page 185

White-footed
Mouse
page 191

Keen's Mouse
page 189

Deer Mouse
page 194

Bushy-tailed Woodrat
page 181

All are approximately 42% of actual size

Pikas, hares, and rabbits

Collared Pika
page 222

American Pika
page 225

Eastern Cottontail
page 243

Mountain Cottontail
page 247

European Rabbit
page 241

Snowshoe Hare
page 228

Arctic Hare
page 232

European Hare
page 235

White-tailed Jackrabbit
page 238

All are approximately 14% of actual size

Shrews

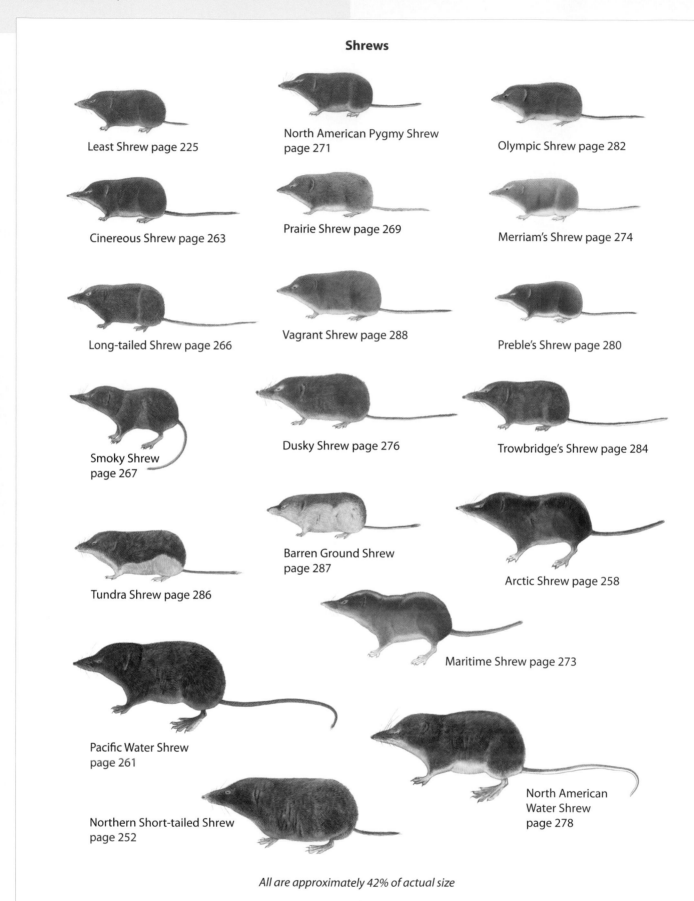

Least Shrew page 225

North American Pygmy Shrew page 271

Olympic Shrew page 282

Cinereous Shrew page 263

Prairie Shrew page 269

Merriam's Shrew page 274

Long-tailed Shrew page 266

Vagrant Shrew page 288

Preble's Shrew page 280

Smoky Shrew page 267

Dusky Shrew page 276

Trowbridge's Shrew page 284

Tundra Shrew page 286

Barren Ground Shrew page 287

Arctic Shrew page 258

Maritime Shrew page 273

Pacific Water Shrew page 261

Northern Short-tailed Shrew page 252

North American Water Shrew page 278

All are approximately 42% of actual size

Moles

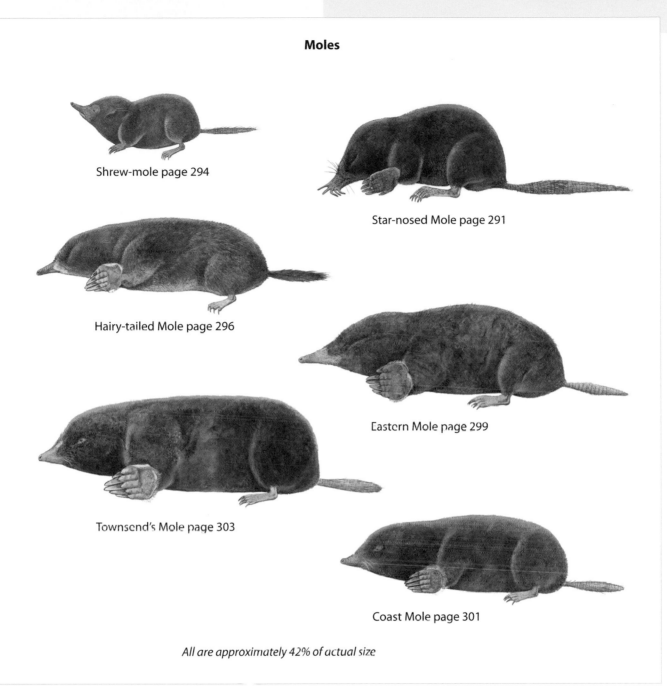

Shrew-mole page 294

Star-nosed Mole page 291

Hairy-tailed Mole page 296

Eastern Mole page 299

Townsend's Mole page 303

Coast Mole page 301

All are approximately 42% of actual size

Bats

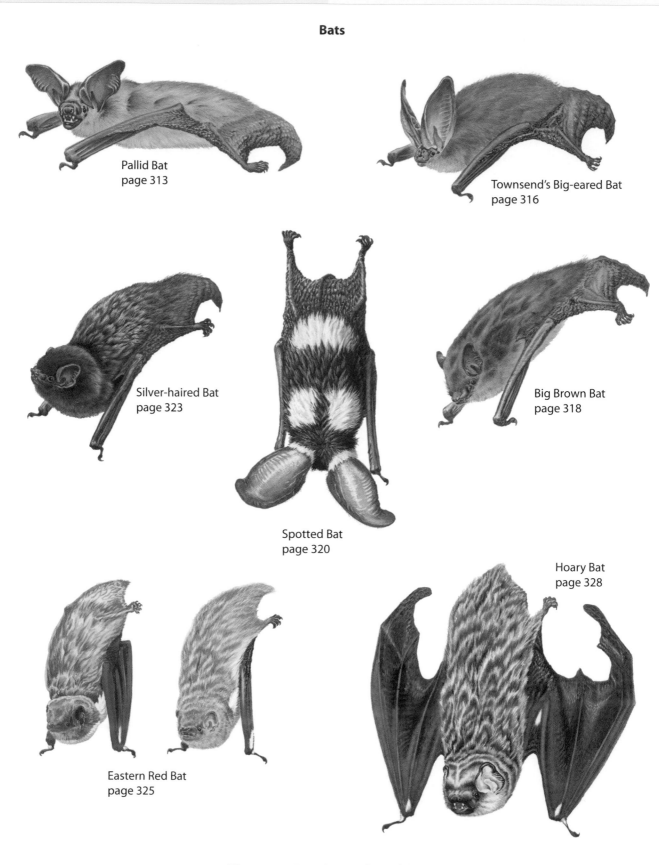

Pallid Bat
page 313

Townsend's Big-eared Bat
page 316

Silver-haired Bat
page 323

Spotted Bat
page 320

Big Brown Bat
page 318

Hoary Bat
page 328

Eastern Red Bat
page 325

All are approximately 60% of actual size

Bats continued

California Myotis
page 330

Western Small-footed Myotis
page 332

Long-eared Myotis
page 334

Little Brown Myotis
page 340

Eastern Small-footed Myotis
page 338

Keen's Myotis
page 336

Northern Myotis
page 343

Fringed Myotis
page 345

Long-legged Myotis
page 347

Yuma Myotis
page 349

Evening Bat
page 351

Eastern Pipistrelle
page 353

All are approximately 60% of actual size

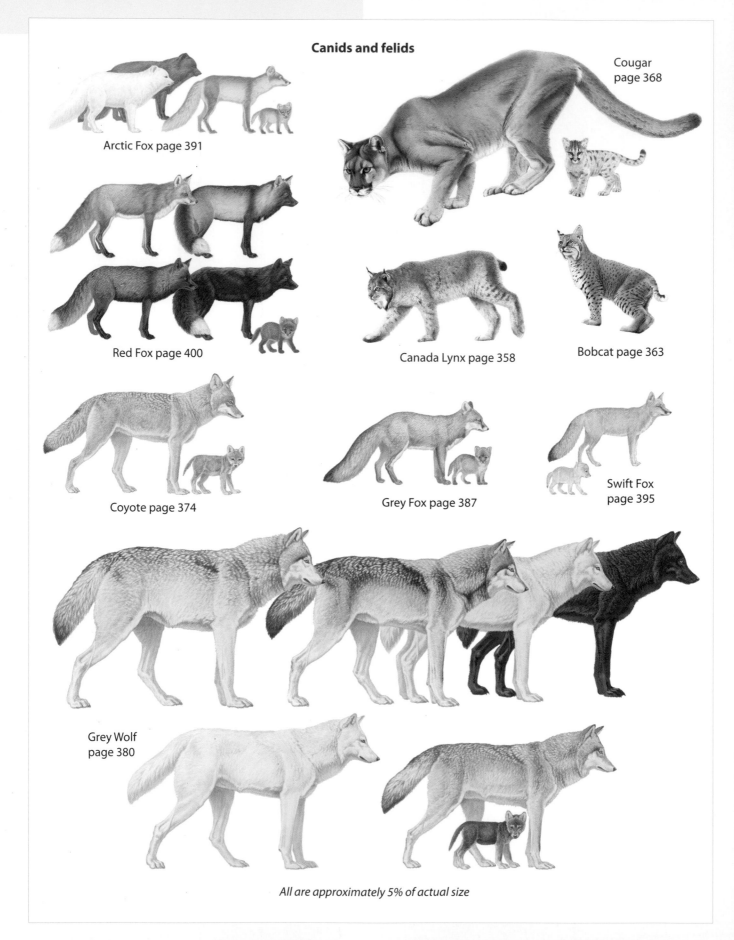

Canids and felids

Cougar page 368

Arctic Fox page 391

Red Fox page 400

Canada Lynx page 358

Bobcat page 363

Coyote page 374

Grey Fox page 387

Swift Fox page 395

Grey Wolf page 380

All are approximately 5% of actual size

Bears

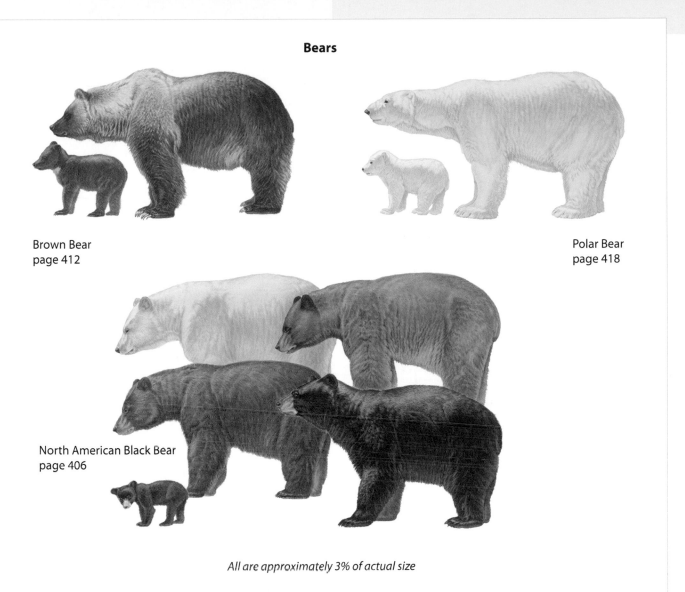

Brown Bear
page 412

Polar Bear
page 418

North American Black Bear
page 406

All are approximately 3% of actual size

Walrus and seals

Walrus
page 435

California Sea Lion
page 432

Steller Sea Lion
page 428

Northern Fur Seal
page 424

Harp Seal
page 453

Ringed Seal
page 463

Hooded Seal
page 440

Northern Elephant Seal
page 449

Bearded Seal
page 443

Grey Seal
page 446

Spotted Seal
page 456

Harbour Seal
page 458

All are approximately 2% of actual size

Raccoon, weasels, and skunks

Northern
Raccoon
page 523

Fisher
page 485

Sea Otter page 466

Northern
River Otter
page 476

American
Marten
page 480

American Mink
page 507

Ermine
page 489

Least Weasel
page 502

Long-tailed
Weasel page 494

Striped Skunk
page 515

Western Spotted
Skunk page 519

Black-footed Ferret
page 497

Wolverine
page 471

American Badger page 511

All are approximately 9% of actual size

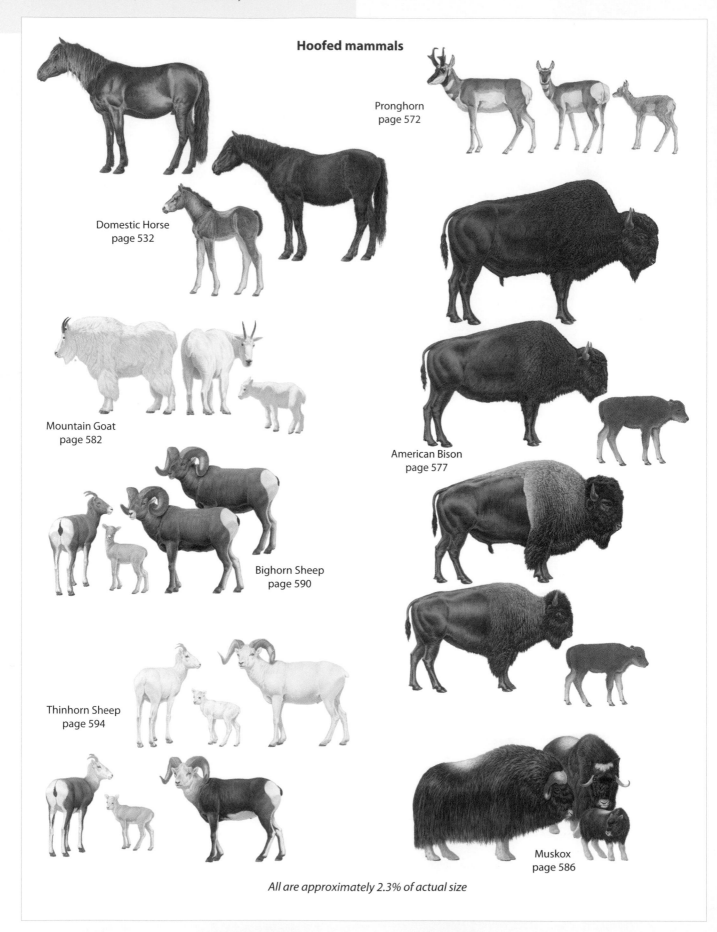

Hoofed mammals

Pronghorn
page 572

Domestic Horse
page 532

Mountain Goat
page 582

American Bison
page 577

Bighorn Sheep
page 590

Thinhorn Sheep
page 594

Muskox
page 586

All are approximately 2.3% of actual size

Hoofed mammals continued

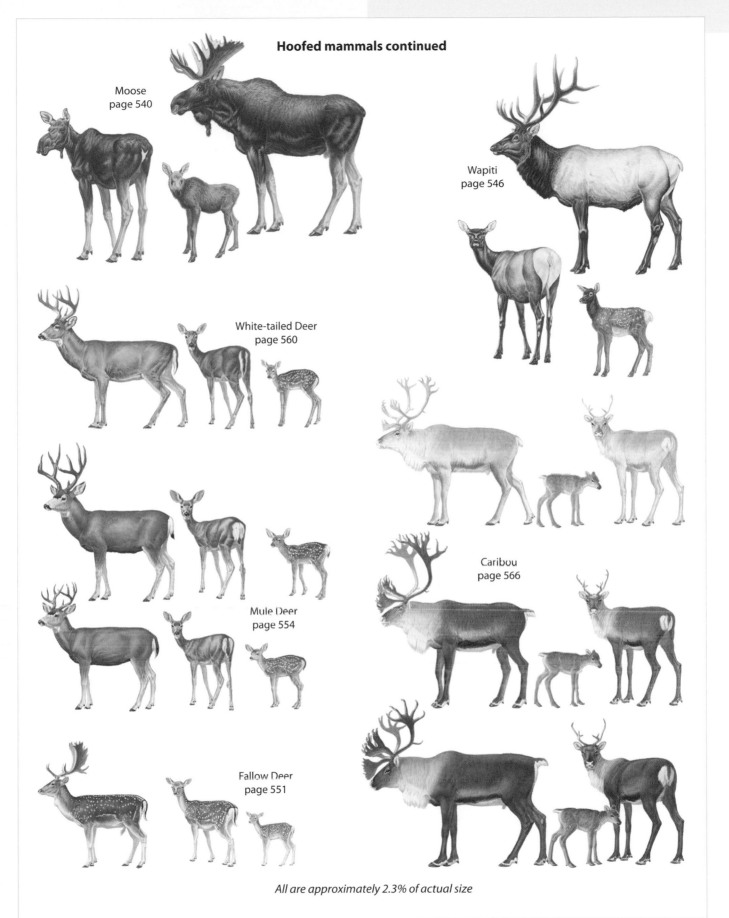

Moose
page 540

Wapiti
page 546

White-tailed Deer
page 560

Mule Deer
page 554

Caribou
page 566

Fallow Deer
page 551

All are approximately 2.3% of actual size

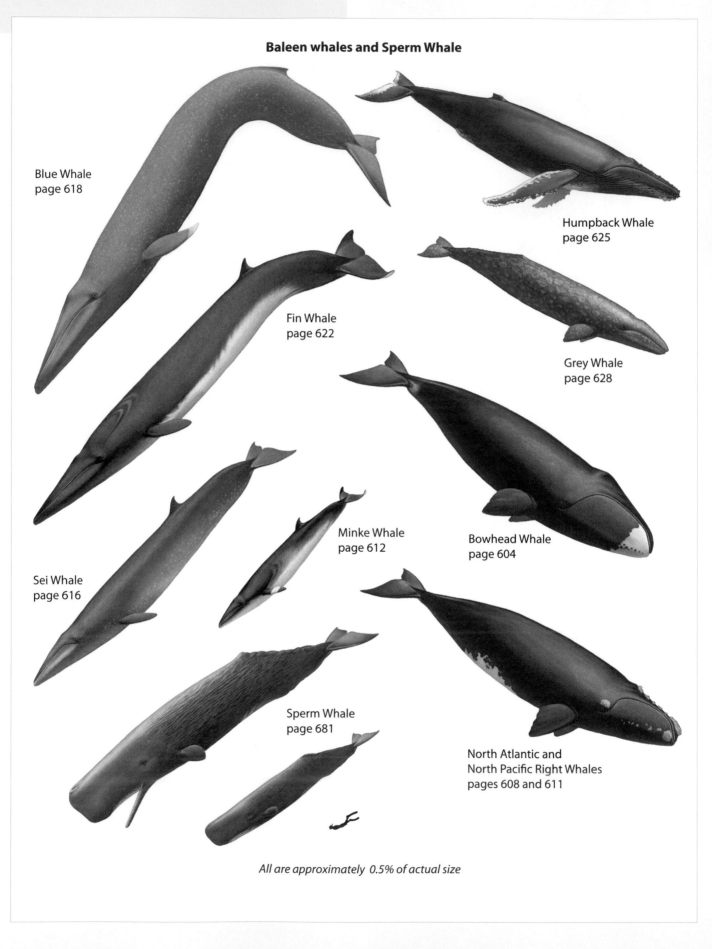

Baleen whales and Sperm Whale

Blue Whale
page 618

Humpback Whale
page 625

Fin Whale
page 622

Grey Whale
page 628

Sei Whale
page 616

Minke Whale
page 612

Bowhead Whale
page 604

Sperm Whale
page 681

North Atlantic and
North Pacific Right Whales
pages 608 and 611

All are approximately 0.5% of actual size

Smaller toothed whales

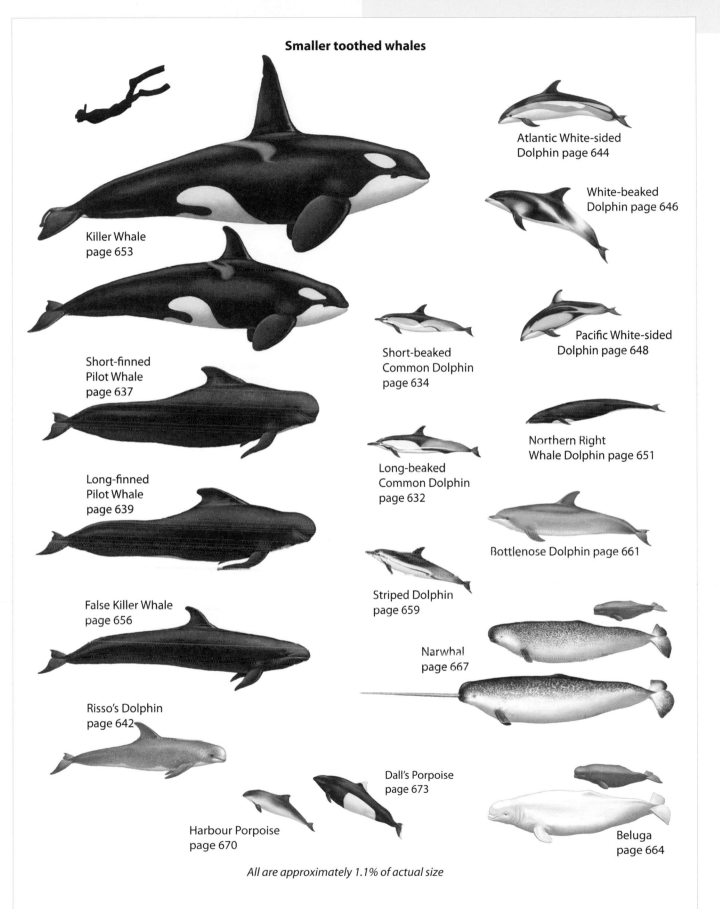

Killer Whale
page 653

Atlantic White-sided
Dolphin page 644

White-beaked
Dolphin page 646

Short-finned
Pilot Whale
page 637

Short-beaked
Common Dolphin
page 634

Pacific White-sided
Dolphin page 648

Long-finned
Pilot Whale
page 639

Long-beaked
Common Dolphin
page 632

Northern Right
Whale Dolphin page 651

False Killer Whale
page 656

Bottlenose Dolphin page 661

Striped Dolphin
page 659

Narwhal
page 667

Risso's Dolphin
page 642

Dall's Porpoise
page 673

Harbour Porpoise
page 670

Beluga
page 664

All are approximately 1.1% of actual size

Smaller toothed whales continued

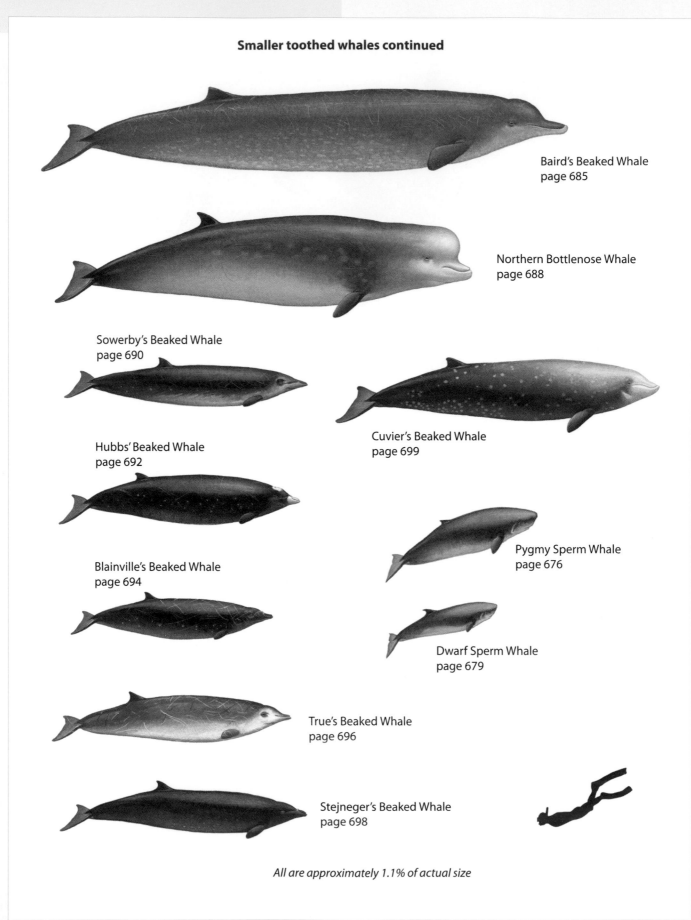

Baird's Beaked Whale
page 685

Northern Bottlenose Whale
page 688

Sowerby's Beaked Whale
page 690

Hubbs' Beaked Whale
page 692

Cuvier's Beaked Whale
page 699

Blainville's Beaked Whale
page 694

Pygmy Sperm Whale
page 676

Dwarf Sperm Whale
page 679

True's Beaked Whale
page 696

Stejneger's Beaked Whale
page 698

All are approximately 1.1% of actual size

The Natural History
of
Canadian Mammals

Order Didelphimorphia
New World opossums

North American Opossum (*Didelphis virginiana*)

Photo: Tammy Wolfe / iStockphoto

The group of animals of the order Didelphimorphia was formerly combined with all the other marsupials into the order Marsupialia, the "pouched" mammals. Recent taxonomic research strongly suggests that there are seven valid orders represented within the former "super" order. Didelphimorphia is the only one of these with representatives in North America, and only one

hardy species is found as far north as Canada. There are 17 genera and 87 species in the Didelphimorphia, and all are found only in the Western Hemisphere – North, Central, and South America – and most in South America. *Opossum* is the name used for the pouched mammals in the New World, and *possum* is used for similar small marsupials in Australia.

FOSSIL RECORD

The order is believed to have originated in South America. It is considered to be a very old lineage of mammals and quite primitive in many ways. However, it has also proven to be highly successful and has existed in North America for about 100 million years, since the middle Cretaceous Period.

ANATOMY

Most female marsupials, although not all, have pouches used to carry the young. One of the primitive characteristics of most marsupials, and of all those in this order, is that they do not possess a true placenta. As a result, the fetus can only achieve a limited development inside the mother. Members of this order have the shortest gestation periods among the mammals – as short as 10–12 days. The young are born in a premature state with the exception of well-developed front legs, often with claws. These legs are used to crawl from the exterior opening of the birth canal into the pouch. Once there, the young attach to a nipple and continue their development. The breathing and swallowing passages are

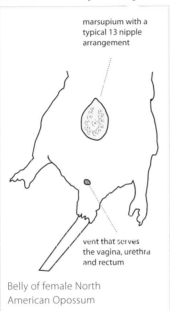

marsupium with a typical 13 nipple arrangement

vent that serves the vagina, urethra and rectum

Belly of female North American Opossum

distinct in very young opossums. They can breathe and suck at the same time. As the pups mature, this characteristic reverts to the normal mammalian condition of a shared passage for air and food. After such a short gestation the infants stay with their mother for an extended period. Newborns are unable to regulate their body temperature and rely on the heat produced by their mother to keep warm. Baby opossums have an unusual ability to tolerate high levels of carbon dioxide in the air they breathe, very likely an adaptation to pouch life.

Some of the other primitive features displayed by this order include a narrow braincase with a small brain and more than normal numbers of teeth. Only the third premolar is deciduous; all of the other teeth grow in as permanent teeth.

ECOLOGY

The majority of Didelphimorphia mammals are tropical and subtropical in habitat. Most are nocturnal or crepuscular and solitary. They are omnivorous or carnivorous and are frequently blamed for crop depredations. Most species in this order are terrestrial, but all have arboreal capabilities, and one is semiaquatic.

FAMILY DIDELPHIDAE
New World opossums

The Didelphidae family of marsupials comprises animals ranging from mouse size to the largest, the North American Opossum. All have short legs; prominent ears; a long, narrow snout; a naked, scaled, varyingly prehensile tail; an opposable hallux (big toe) for grasping; and an abdominal pouch for transporting young. Canine teeth are large, and adult males are larger than females.

North American Opossum
also called Virginia Opossum, American Opossum, American Possum

FRENCH NAME: **Opossum de Virginie**
SCIENTIFIC NAME: *Didelphis virginiana*

The North American Opossum is the only New World marsupial that reaches North America. None of the others extend northward beyond Central America. The North American Opossum is not a living fossil, as is often reported; it is a relatively new species, first appearing in the fossil record about 120,000 years ago, during the late Pleistocene epoch. Laboratory tests show that opossums, often thought to be dull witted, are actually smarter than Domestic Dogs and similar to pigs in intelligence.

DESCRIPTION

About the size of a Domestic Cat, the North American Opossum has shorter legs, a much longer, pointed muzzle, and an equally long, but rat-like, prehensile tail. Although it often looks pale grey or even whitish from a distance, the body colour of this scruffy, northern marsupial is actually a grizzled mix of pale underfur and long, white guard hairs, some of which are tipped with greyish brown or black. The all-white guard hairs are usually longer than the dark-tipped guard hairs, giving the animal a haloed appearance when backlit. The amount of dark-tipped hairs varies, and some animals are much darker than others. The base of the tail is well-furred, but the remainder is sparsely furred. The legs, the tops of the feet, and about the basal third of the tail are black. The remainder of the scaly tail and the toes are pinkish or white. The nose is hairless and pink. Long whiskers on the muzzle and cheeks are mixed black and white. The throat, cheeks, and belly are whitish, but parts of the chest, cheeks, and throat may be stained

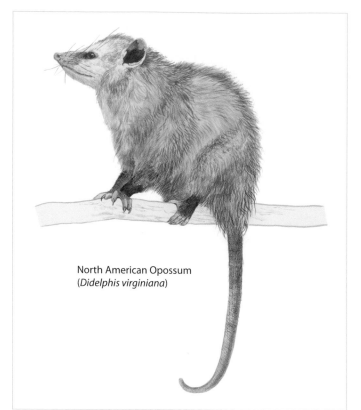

North American Opossum
(*Didelphis virginiana*)

The following measurements apply to animals of reproductive age from the northern part of the range. Males are considerably heavier than females. Opossums continue to grow throughout their life, so older animals are larger and heavier, but the high energy demands of reproduction result in a smaller overall size and a lower weight in females.

Total length: 635–883 mm; tail length: 216–358 mm; hind foot length: 57–78 mm; ear length: 44–62 mm; males' weight: 1.5–4.6 kg (average around 3.5 kg); females' weight: 1.3–3.2 kg (average around 2.5 kg). Newborns: weight about 0.13–0.2 g.

RANGE

North American Opossums occur from southern Ontario and southern British Columbia over much of the United States and Mexico to northern Costa Rica. After the retreat of the last ice sheet (between 15,000 and 10,000 years ago), they quickly reoccupied large parts of the United States and continue to expand their range, often aided by human activities. When Europeans first arrived in North America, opossums ranged as far north as West Virginia and Ohio. By the 1850s the first opossums were detected in eastern Canada along the shore of Lake Erie in southwestern Ontario. Their northward expansion is governed primarily by the severity of winters. Critical for opossum survival is the lack of extended periods of –20°C temperatures and standing snow depths of less than 28 cm. A milder winter permits the animals to survive, and the following summer they may penetrate further north. A few mild winters in succession encourages this expansion. A particularly cold winter will kill all the animals in the area. Juveniles, being smaller, are especially affected by cold weather. A general warming trend over the last hundred years has encouraged this marsupial's gradual northward expansion. Enhancement of this trend as a result of global climate change could see North American Opossums penetrating further north than ever before. Introduced into Washington, Oregon, and California in the 1910s and 1920s, the North American Opossum then spread into British Columbia (Crescent Beach) by 1949. These successful introductions have allowed populations to become established all along the west coast from southern California to southwestern British Columbia. All North American Opossums living west of the Rocky Mountains are the result of introductions. In British Columbia populations occur on the mainland in the lower Fraser Valley. An island population exists on Hornby Island off the east coast of Vancouver Island, following a successful introduction in 1986. Although reports in the early 1990s suggested that North American Opossums had reached southern Vancouver Island, no further sightings have been reported since.

ABUNDANCE

Sparse in the far northern limits of their range where numbers fluctuate dramatically as a result of weather conditions, North American Opossums are otherwise common to abundant in most of the remainder of their range. In suitable habitat they can reach densities of about one per 0.6 ha, but most populations occur at densities of one per 8–15 ha. Opossum numbers follow a very typical mammalian pattern; they are lowest in late winter following the cold season

yellowish orange on adult males, owing to a secretion from glands on the sternum. Dark eyes are often surrounded by dark fur, and the top of the head commonly has a somewhat darker line down the centre. Leathery, hairless, black ears are large and rounded and usually tipped with white or pink. A melanistic or black-phase individual typically has white cheeks, lips, toes, and tail tip and is grizzled greyish black on the rest of its body. This coloration is more common in the southern and coastal parts of the range, while the light phase is the most common colour in the north. Albino (with red eyes and pink ears) and white animals (with dark eyes and ears) have been reported but are rare. Females develop a fur-lined abdominal pouch (marsupium) during the breeding season, which encloses the teats and opens towards the front. Teats are normally arranged in an arc composed of 12 teats, with an additional teat placed centrally. The number of teats can vary from 9 to 17. Opossums in regions that experience bitter winter temperatures commonly lose toes and the tips of their ears and tail to frostbite. The dental formula is incisors 5/4, canines 1/1, premolars 3/3, and molars 4/4, for a total of 50 teeth. Older individuals develop prominent sagittal crests. Eyeshine is bright orange red.

SIMILAR SPECIES

The North American Opossum's pinkish nose, toes, and tail tip are distinctive. No other North American mammal resembles an opossum except possibly the much smaller and darker Brown Rat. The North American Opossum's skull is equally distinctive for the combination of tiny braincase, robust canines, and high tooth number.

Distribution of the North American Opossum (*Didelphis virginiana*)

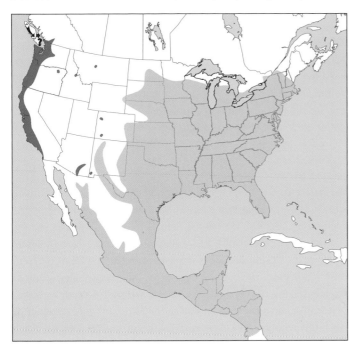

 introduced

attrition and highest in early autumn as the young of the year are recruited into the population.

ECOLOGY

In the heart of their native range North American Opossums are most abundant in deciduous woodlands that are bisected by streams and interspersed with agricultural lands. However, many other habitats from sea level to over 300 feet are occupied. Suburban and urban gardens and parks have proven quite acceptable, provided that food and water are available. In the lower Fraser Valley of British Columbia, where they are an introduced species, North American Opossums occupy similar habitats, including densely urban areas such as Vancouver. They are capable swimmers, even able to swim under water. As their fur gets wet and heavy, they swim with only the nose above water. North American Opossums are slow and careful but very proficient climbers, using their prehensile tails for balance and as a fifth limb. They commonly climb trees to forage, escape predators, and seek tree cavities or bird or squirrel nests to use as day resting dens. They do not hibernate and are largely nocturnal, although they sometimes forage during the daytime in winter during warmer weather following a prolonged cold spell. Opossums are most active between sunset and midnight.

Dens are created in hollow logs, rock and wood piles, underground burrows, or abandoned, but usually refurbished, squirrel dreys and larger bird nests, such as those of crows or raptors. Culverts and drains as well as basements may serve in urban areas. Opossums do not dig their own burrows; instead they make use of those of a suitable size that they find abandoned. Those dug by foxes, skunks, Northern Raccoons, and Woodchucks are preferred. North American

Opossums cope with periods of below-freezing cold by remaining curled up in their well-insulated winter dens. Underground burrows are especially important for individuals that are overwintering in cold regions. Opossums show no sign of territoriality. The home ranges of adults and juveniles of both sexes can overlap extensively. Adult males typically occupy larger home ranges than do adult females. Juveniles begin with a relatively small home range, which is slowly expanded as they gain experience and size. As North American Opossums are often semi-nomadic, the boundaries of their home ranges can shift over time. Furthermore, males may double their home range during the breeding season when they visit several females regularly to assess their breeding condition. Estimates vary from 5 ha to 114 ha and are largely determined by sex and by food and water availability. During their nightly foraging, adults may travel from 0.1 km to more than 4.5 km, depending on food availability.

Apart from maternal females, particularly those with young in the weaning phase, most North American Opossums change their warm-weather den sites often. The dispersal of juveniles occurs either in autumn or, more commonly, in early spring at the start of the breeding season, and they can travel several kilometres in a single night. Female juveniles tend to remain near the area in which they were born, while juvenile males disperse greater distances. Adults, even females with pouch young, will also disperse long distances at times (several kilometres). This ability to advance into new areas has allowed North American Opossums to expand their range and to repopulate regions where hunting or bad weather had decimated former populations. Maximum lifespan in the wild is likely around three years, but few live to be even two years old. In captivity they may live more than four years. Known predators include foxes, Coyotes, the larger owls, and Domestic Dogs and Cats. Many are killed by motor vehicles.

DIET

North American Opossums are highly adaptable, opportunistic omnivores that are willing to eat a wide variety of animal and plant material. Their diet is largely determined by what is abundantly available, and they will readily include fungi, carrion, and garbage. The bulk of their diet is composed of insects, earthworms, slugs and other invertebrates, fruit and berries, nuts, tubers, shoots, garden vegetables, mice, shrews, moles, bird eggs and nestlings, frogs, toads, salamanders, and snakes, including poisonous snakes. North American Opossums are unharmed by the venom from North American rattlesnakes and so can safely prey on these snakes. Most of the birds and mammals are consumed as carrion. During periods of food deprivation, as may occur during winter, North American Opossums will cannibalize others of their species. Their fearsome reputation as voracious raiders of domestic fowl is unwarranted, although they will kill chickens and wild and domesticated ducks and will raid their nests. North American Opossums are one of the few mammals that will readily consume shrews and moles, apparently not at all discouraged by their rank odour.

REPRODUCTION

Apart from the large number of teeth, the most distinctive traits of marsupials are related to their reproductive anatomy. Males have

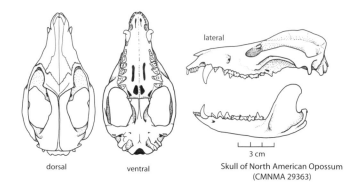

lateral

3 cm

dorsal   ventral

Skull of North American Opossum
(CMNMA 29363)

a forked penis. Their testes are permanently descended and lie in a scrotum that is ahead of the penis. Females have two uteri and secondary vaginas connected to a single primary vagina. Marsupial young are born in a relatively underdeveloped condition. Newborn North American Opossums are hairless and pink, with an incomplete nervous system, rudimentary hind limbs and tail, and non-functional eyes and ears. Their digestive, circulatory, and respiratory systems are functional but not fully formed.

The breeding season of North American Opossums in southern populations may begin in December or January, and in northern parts of the range it begins in January or February or even later. Breeding ends by October or November in the north and can last almost year-round in the south. After a very short gestation of 12–13 days, one to 21 still-embryonic-looking young are born. Newborns are about the size of a Honey Bee. The size of a litter that survives to weaning is largely determined by the number of neonates that reach the pouch and are able to locate an unoccupied teat. Most North American Opossums have 13 teats, so surviving litters that are larger than 13 are exceptional and are only possible if the female is among the minority that has more than 13 teats. The average number of young from a litter that survives to weaning is six to nine and is largest in the northern parts of the range. Females in the north (including populations in British Columbia) tend to wean one or, occasionally, two litters annually. A few minutes before parturition the female assumes a hunched sitting position on her rump, with her tail in front of her. This places the birth canal as close as possible to the pouch and shortens the distance that the young have to climb to about 40–50 mm. The neonates have relatively well-developed forelegs with fingers and deciduous claws, which they use to grasp their mother's belly fur and climb upwards from the birth canal into the pouch. There they search for and then cling to a nipple. Those young that do not reach the pouch, or that get there but cannot find an unoccupied nipple, die.

During most of the time that the young spend in the pouch they are attached to a nipple. The nipple swells soon after they begin suckling, firmly attaching them to their mother until they have grown sufficiently to open their mouths and release it. The nipple gradually lengthens while the young suckle, until the stalk is about 35 mm in length and the tethered young can move freely within the pouch. Around 60–70 days old, they begin to temporarily detach and crawl around outside the pouch and may be carried, clinging to their

mother's fur. Opossums suckle for close to 100 days. By about 80 days old they are left in the den while their mother forages. Weaning begins at around 87 days old as the young begin to take solid food, and by 92–107 days old they are fully weaned. Females will cycle back into oestrus within two to eight days, unless they are stimulated to continue lactating by the presence of at least two suckling young. Juveniles become sexually mature, at the earliest, around six months old, but few females in the north, if any, are mature enough in their first year to enter oestrus before the end of the breeding season. Most reproduce for the first time the following spring.

BEHAVIOUR

Adults are largely solitary and semi-nomadic, using a series of den sites in their shifting home range. They generally attempt to avoid each other, although occasional reports of two individuals sharing the same den have been noted. Female Opossums may be tolerant of other females unless one is in oestrus. In that situation the non-oestrous female will become aggressive. Although males are non-aggressive towards females, the female will not accept the male's proximity unless she is in oestrus, and has no difficulty driving the often much larger male away. Male-male encounters typically result in open-mouthed threats and hissing, growling, or screeching. Unless one individual withdraws, the confrontation escalates into a "fighting dance" dominance display as each animal faces the other and rhythmically rocks its head and shoulders from side to side. If this display does not lead to a retreat, the animals will do battle, slashing and biting until one withdraws or is killed.

North American Opossums utilize scent marking of various sorts to communicate with other Opossums. Scent marking a site likely advertises a North American Opossum's presence, sex, reproductive status, and perhaps even dominance rank. Saliva marking is a common form of scent marking whereby an individual licks an object, then rubs it with the side of its head. Although males saliva mark more frequently than do females, both sexes mark most frequently during the breeding season. Anogenital dragging is another type of North American Opossum scent-marking behaviour, as is sternal (chest) gland dragging by males. Males begin visiting a pre-oestrous female at least 10 days before she comes into heat. They follow her, making a continuous clicking call, until she drives them away. By the time she reaches oestrus, she may be followed by several males all vying for access. Generally the largest and heaviest male is dominant, and when she is receptive, she will allow him to mount, which he does by placing all of his weight onto her back. This results in the collapse of the pair almost invariably onto their right sides. Copulation lasts for about 15–20 minutes, and repeat attempts by the male are rebuffed. Males typically lose weight during the breeding season as they spend considerable energy looking for mates.

If approached too closely by a potential predator, a North American Opossum may freeze or crouch, hiss, and growl, defecate, flatulate, and excrete a foul-smelling greenish fluid from its anal glands in an effort to discourage its attacker. Death feigning ("playing possum") is a well-known defensive strategy of opossums. This catatonic state is typically induced when the individual is placed in a dangerous, inescapable situation and usually occurs after it is

Walk

Left hind track
Length: 3.0–7.0 cm
Width: 3.8–7.8 cm

Left front track
Length: 2.5–7.5 cm
Width: 3.2–6.3 cm

North American Opossum

suddenly grabbed or bitten by the aggressor. The North American Opossum falls onto its side, the corners of the mouth are drawn back, the tongue lolls, and copious saliva is drooled from the slightly open mouth. Its eyes usually remain slitted open, and it may defecate and express its anal glands. Although it will not respond overtly to being poked, it remains aware and can recover instantly once it feels safe.

Nest building usually consists of lining a pre-existing cavity with dry leaves and grasses. Nesting material is gathered in the mouth and then passed between the front legs and placed on the tail, which has been curved forward between the hind legs to receive it. The hind feet tread on the material to pack it down, and then the tail curls around it. This process may be repeated several times until the load is sufficient. The nest material is carried back to the den in the coiled tail, which is held out behind the body. Warm-weather nests may have loose and scattered nesting material, but cold-weather nests are dense and tightly packed for insulation.

VOCALIZATIONS

The North American Opossum produces four basic vocalizations: clicks, hisses, growls, and screeches. Females call their ambulatory sucklings with a clicking call. The same call may be used by males pursuing an oestrous female and by both young and adult animals of both sexes during aggressive encounters. Hisses, growls, and screeches, in that order, indicate an increasing level of intensity during aggressive or threatening encounters.

SIGNS

The North American Opossum walks with a clumsy-looking, waddling gait, with its tail held stiffly out at slightly above body level. Opossums tend to walk while foraging, but they will trot when crossing open ground and will bound to escape a predator or to pursue prey. Tail drag may rarely occur at a walk but not at faster paces. Both front and hind feet have five toes. All are clawed except the first toe on each hind foot (the big toe or hallux), which is thumb-like and opposable, and all claws may or may not register in a track. Tracks are distinctive because all five toes on the front feet radiate in a star pattern, and on the very asymmetrical hind feet the second to fifth toes clump together and the opposable big toe angles down at least 100° from the other toes (see the accompanying track schematic). Tracks in snow are possible in northern parts of the range, and this short-legged marsupial tends to leave long foot-drag marks between the tracks. Tail drag may or may not be evident depending on the snow depth. Scat is highly variable depending on the diet (1.0–2.9 cm in diameter and 2.5–11.4 cm in length) and is uncommonly found in the wild.

REFERENCES

Dobbyn 1994; Elbroch 2003; Fish 1993; Gardner and Sunquist 2003; Gehrt et al. 2006; Gipson and Kamler 2001; Harder et al. 1993; Holmes, D.J., 1990, 1991, 1992; Hossler et al. 1994; Jameson and Peeters 2004; Kurta 1995; Ladine 1997; Long, C.A., 2008; Nagorsen 1996; Nixon et al. 1994; Peterson, R.L., 1966; Prescott and Richard 2004; Reid, F.A., 1997, 2006; Ryser 1992; Schmidly 2004; van Zyll de Jong 1983a; Verts and Carraway 1998.

Order Primates
humans

Human *(Homo sapiens)*

Photo: Paul McKinnon

The order Primates includes 15 families, 69 genera, and 376 species, of which only one is native to North America: *Homo sapiens,* humans. There are many volumes devoted exclusively to the history, behaviour, and evolution of our species, and while it is not the intention of this book to treat humans in the same degree of detail as other Canadian mammals, still it needs to be recognized that we are valid faunal components, in many ways pivotal to the future of the mammalian fauna of Canada.

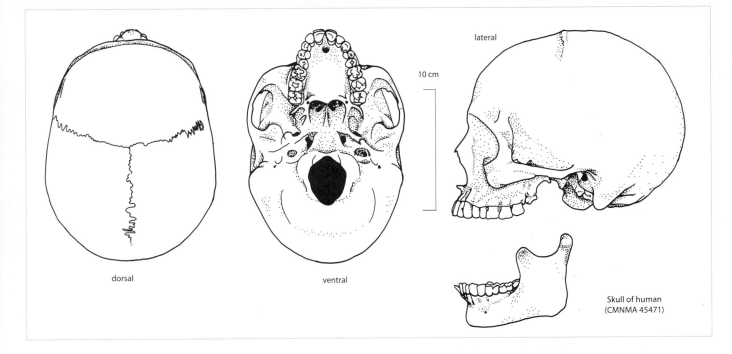

dorsal

ventral

lateral

10 cm

Skull of human
(CMNMA 45471)

FOSSIL RECORD

The first unequivocal fossil primate is 55 million years old.

HISTORY IN NORTH AMERICA

Based on the present state of our knowledge, humans arrived on the North American continent at least 14,000 years ago, and possibly more than 30,000 years ago. The first humans to reach North America may have walked across the Bering Isthmus, which has formed a solid connection between North America and Asia at various times in prehistory, but an increasingly popular hypothesis suggests that they came by boat along the coast from eastern Asia. There have been several waves of immigration since, either over more recent land connections or more commonly by boat. *Homo sapiens*, unlike the other species in the order Primates, is extremely adaptable and is found on all of the continents and major islands. Most of the other species in the order are tropical, subtropical forest, or warm savannah species.

ECOLOGY

There are two kinds of mammals whose impact on their environment is so massive as to be clearly visible from space with the naked eye. The first is the North American Beaver, and the other is a primate – *Homo sapiens* – us!

It took modern humans more than a million years to achieve a global population of one billion in the year 1830, a mere 100 years to double that to two billion by 1930, and only 45 years to double that to four billion by 1975. We are projected to reach eight billion by the year 2024. Our species has had more impact on this planet than have all of the other mammal species combined. By the year 2000 almost one-half of the land surface on the globe had been transformed by human action, and the average rate of plant and animal extinctions has risen by a factor of almost 1000 times since humans achieved dominance.

Order Rodentia
rodents

North American Beaver (*Castor canadensis*)   Photo: Paul Horsley

This group is the most numerous, diverse, and widespread of all the mammals. With 33 families, 481 genera, and 2277 species of living rodents, this is clearly the most successful mammalian order. Close to 40% of all mammal species are rodents. They can be found on all the continental land masses except Antarctica, on most of the larger islands, and on many of the smaller islands. Canada supports 9 families, 31 genera, and 69 species of rodents.

FOSSIL RECORD

Early rodents began to appear in the fossil record in central Asia during the late Palaeocene Period, about 60 million years ago.

ANATOMY

Often called the gnawing mammals, rodents are characterized by paired upper and lower incisor teeth at the front of their jaws. In many groups, these teeth grow at a slow rate throughout the life of the animal and thereby are never worn down. Due to the placement and structure of these teeth, constant gnawing grinds the lower pair against the upper pair, serving to continually sharpen the chisel-like leading edge. When natural chewing activities do not provide suffi-cient wear, the animals will grind their incisors against each other. A malocclusion – where the opposing tooth doesn't line up properly – usually caused either by malformation or accident, can result in insufficient tooth wear, as illustrated in the accompanying figure. As unopposed growth proceeds, the resulting abnormal tooth has been known to curl back into the skull and cause death. Rodents have a soft flap of lip that can be tucked back into the gap between the incisors and the cheek teeth. When an animal does this, it effect-ively separates the cutting incisors from the mouth. It can then chew hard, messy bark or nut shells or even use its teeth for digging with-out filling its mouth with debris.

To help digest their mainly herbaceous diet, most rodents have a caecum, an expanded area in the intestine where fermentation takes place. Bacteria, protozoans, and yeasts occupy this pouch, where they break down some of the cellulose delivered to them via the digestive tract.

Interestingly, there are no large modern rodents. The largest rodent alive today is a South American species called the Capybara (*Hydrochaeris hydrochaeris*), a semiaquatic animal about the size of a small pig (around 25–75 kg). A few extinct species achieved greater size, for example, the North American Giant Beaver (*Casteroides ohioensis*) reached the size of an adult black bear. Now our largest Canadian rodent species is the North American Beaver, followed closely by the North American Porcupine.

ECOLOGY

Rodents occupy a wide variety of ecological niches. There are fosso-rial, terrestrial, arboreal, and semiaquatic rodents. There are rodents that hop on two legs (bipedal), that glide from tree to tree on out-stretched legs, that climb with ease, that can dig through the ground faster than a man can shovel, as well as the more usual terrestrial forms. They are mainly herbivorous and granivorous, although some groups are secondarily insectivorous and many are opportun-istic meat-eaters, even cannibalistic at times. Rodents, particularly porcupines, squirrels, and some mice, are nature's recyclers of bone and discarded antler. They crave the calcium and will nibble on any sources they come across.

With a few exceptions, rodents specialize in achieving a high reproductive potential. Large litters, frequent pregnancies, rapid development of the young, and early fecundity create the possibility for such fantastic scenarios as a pair of House Mice being capable of having more than a million descendents in the space of a single year

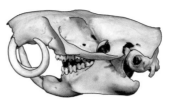
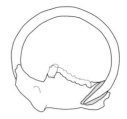

Woodchuck skull with severe incisor malocclusions (from the private collection of David Campbell).

Lingual view of right mandible of a North American Beaver with a severe malocclusion of the incisor. The tooth grew along a curved path until the tip penetrated the back of the dentary bone and likely the skull in the process. (Adapted from photo by David Campbell, of a specimen at the Royal Saskatchewan Museum.)

(assuming that all survive). Fortunately, these prodigious numbers are tempered by realities such as weather, predation, disease, food and water shortages, parasites, accidents, and competition for avail-able living space and mates. Nevertheless, rodents are admirably suited to rapidly exploit new opportunities.

Rodents have evolved a wide variety of social behaviours and societies. These vary from largely solitary lifestyles, to loose group-ings, to cooperative social units. Naked Mole Rats (*Heterocephalus glaber*) from Africa have taken the social unit to the extreme. Their society is a linear matriarchy similar to the social organization of Honey Bees. There is a principal breeding female and a few breeding males. All the other animals in the colony, both male and female, are stunted in their growth and do not develop breeding potential. They become the "workers" who are pushed around by the larger, more dominant "breeders" to keep them working on the colonies' business.

CLASSIFICATION

There is still considerable controversy about rodent classification. This group is in the midst of a major and rapid diversification. The systematics is complex and confusing, and frequently changes as more information is gathered.

FAMILY APLODONTIIDAE

mountain beavers

This family consists of a single living species, which is considered the most primitive living rodent.

Mountain Beaver
(*Aplodontia rufa*)

Mountain Beaver

also called Sewellel, Aplodontia

FRENCH NAME: **Aplodonte**
SCIENTIFIC NAME: *Aplodontia rufa*

This unique and seldom seen rodent is found only in the moist forests of western North America. It spends most of its life in underground burrows, venturing out to clip vegetation and haul it to a drying pile near the burrow entrance. Notably unbeaver-like, it can climb trees and more closely resembles a tail-less muskrat or a dark woodchuck.

DESCRIPTION

Mountain Beavers are short-legged, stocky rodents with dark-reddish to blackish-brown coats that are slightly lighter on the underside. Many have a small, whitish spot at the base of each ear and variable-sized whitish patches on their underside. Eyes and ears are small, but sensory whiskers on the muzzle and sides of the face are well developed. Claws on the front feet are long, with the exception of the pollex (thumb), which has a very short, nail-like claw and is partly opposable. Their furred tails are very short and often not visible. They undertake a single moult in autumn and maintain a woolly undercoat throughout the year. Pregnant and lactating females develop patches of coarse black hairs around the nipples. Mountain Beaver cheek teeth are distinctive due to the presence of a spine-like process, which projects to the outside on the upper teeth and to the inside on the lower teeth. Enamel on the front surface of the upper and lower incisors is orange in colour. The dental formula is incisors 1/1, canines 0/0, premolars 2/1, and molars 3/3, for a total of 22 teeth.

SIMILAR SPECIES

Muskrats are similarly sized; however, their laterally flattened, blackish tail is naked-looking and prominent, and their coat is glossy and long. Some Woodchucks may have similar dull, dark-brown pelage, but the presence of an obvious, thick, furred tail distinguishes them from Mountain Beavers. Young marmots may be mistaken for Mountain Beavers, but they, like Woodchucks, have distinct and furry tails, and short claws on the front feet.

SIZE

Males and females are similarly sized.
Total length: 260–470 mm; tail length: 17–47 mm; hind foot length: 48–68 mm; ear length: 16–22 mm.
Weight: 670–1300 g; newborns, uncertain and likely varies with litter size and maternal health; neonates 16 to 48 hours old, born to captive females, weighed 18–27 g.

RANGE

This is an endemic western North American species found only within, and to the west of, the Cascade and Sierra Nevada mountain ranges. The Fraser River, in southwestern British Columbia, appears to be an effective northern barrier to their dispersal. Although Mountain Beavers were formerly found as far west as Langley on the southern side of the Fraser River Valley, recent surveys suggest that they no longer occur in the western Fraser Valley regions, likely due to urbanization. The most southerly record is a recent discovery in Inyo County, California.

ABUNDANCE

Mountain Beavers in British Columbia are considered by COSEWIC to be in the "Special Concern" category, which means that the populations are vulnerable. The range of Mountain Beavers in British Columbia appears to have shrunk in the lower Fraser River Valley due to habitat lost to urbanization and agriculture. Furthermore, the species is susceptible to certain forms of intensive forestry operations throughout its Canadian distribution (see the "Ecology" section following). Population size is unknown. Provincially, the northern subspecies in British Columbia is considered to be "Vulnerable," while the southern subspecies found in the southern Fraser River Valley is considered "Critically Imperilled."

ECOLOGY

Mountain Beavers most commonly occur at lower elevations in humid, open-canopy forests with a dense understory of herbaceous

Distribution of the Mountain Beaver (*Aplodontia rufa*)

possibly extirpated

plants are retrieved from the pile and taken underground when they are partly dried and are then consumed immediately, as they would otherwise rot in the high humidity of the burrow. Some more rot-resistant plants, such as ferns or conifer twigs, are stockpiled in autumn into plugged chambers in the walls of burrows or feeding chambers for use during cold weather. Mountain Beavers do not hibernate and cannot store sufficient food to last all winter, so they must continue to forage, even under the snow, to find fresh food. They are capable of climbing trees and will do so to chew the bark or to clip twigs. Bark chewing and bud clipping of woody plants tend to be winter-foraging techniques. Home ranges vary in size from 0.02 to 4.0 ha, largely as a result of habitat quality, but their size is similar for adults of either sex. Population density rarely exceeds four per ha. Nest chambers are 50–60 cm in diameter, 30–40 cm high, and tightly packed with vegetation. The animal constructs a small, central nesting cavity in this ball of vegetation. In warmer regions or seasons, the vegetation may be trampled into a flat platform upon which the animal rests. Highest population densities are found around seeps and banks of small streams, especially where the undergrowth is dense, and where there are nearby open areas. Logging operations that trample and compact the soil in these wet areas can be very deleterious to Mountain Beaver survival. Even a 5 m machinery-free buffer zone in the vicinity of water will allow populations to survive the harvesting of the rest of the forest. Mountain Beavers will enter the water and can swim well. Juveniles disperse in the autumn and become especially vulnerable to Coyote and raptors if they remain above ground for long. They may be forced to travel more than 500 m to locate an empty territory. Should it survive its first year, a Mountain Beaver may live up to six years.

DIET

Strictly vegetarian, Mountain Beavers eat a variety of forbs, grasses, and woody plants, including some that are difficult to digest or are toxic and hence are avoided by other animals. Ferns, horsetails, rhododendrons, and stinging nettles fit into this latter class. Diet varies seasonally as plants become available. Mountain Beavers do not venture farther than about 50 m from their burrows while foraging. Their ability to climb saplings to harvest the terminal twigs makes them a pest in some reforested areas. During winter, they readily climb and chew areas of bark from firs, spruces, and other conifers as well as Vine and Bigleaf Maples and the previously mentioned rhododendrons. In Canada, Narrow-leafed Fireweed is a preferred food where available. They are hindgut fermenters and reingest some faecal material. As they defecate, they sit back on their

growth and an abundance of woody debris. They are largely nocturnal, but will venture above ground during the day. They are fossorial and spend about 22 or more of every 24 hours underground in their elaborate burrow system, which has multiple openings and may be up to 2 m deep. Typically, a large central feeding chamber forms a hub from which the toilet chambers, nesting chambers, and tunnels radiate. Tunnels usually extend into wet areas with abundant herbaceous vegetation and may have soggy footing, but the nest and food chambers are located above the water table. Sometimes, tent-like structures of sticks topped with leafy twigs are constructed, presumably to conceal a principal burrow entrance or perhaps to divert rainwater. During the short above-ground outings, vegetation is clipped and carried to a heap near the burrow entrance. Most

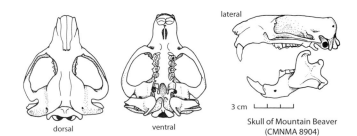

lateral

3 cm

Skull of Mountain Beaver
(CMNMA 8904)

dorsal   ventral

rump and bending their head downwards, receive the pellets into their mouth. Large, soft, tube-shaped caecal pellets are consumed and the much smaller, harder-type pellets are tossed into a latrine with a flick of the head. Since their kidneys are fairly primitive and inefficient, Mountain Beavers must have access to water and a cool, humid burrow system.

REPRODUCTION

Mountain Beaver females have a single oestrus period each year, and there is considerable synchrony among the breeding females of any population. Onset of oestrus varies by location, but not apparently by latitude. Reproductive activity begins in November or December when the male testes enlarge and become semi-scrotal (testes do not descend fully into a scrotum in this species). Mating takes place from early January to early April. A litter of 1 to 6 (most commonly 2–3) young are born in the underground nest about 30 days later. Newborns are naked and toothless, with closed ears and eyes and incompletely developed hind feet. They suckle for about six to eight weeks and begin taking solid foot shortly before weaning. Although yearling females will breed, it is thought that they do not conceive for the first time until they are two years old, although there is some evidence that especially well-fed yearlings can conceive. Yearling males may produce viable sperm in their first breeding season, especially if their body weight is high, but most do not begin breeding until their second year.

BEHAVIOUR

Apart from mothers with offspring and during mating season, Mountain Beavers are solitary. Home ranges frequently overlap, especially in lush habitats, and interconnecting burrows may be shared, but the feeding, nest, and toilet chambers are protected by the proprietor and other Mountain Beavers are aggressively ejected. Above-ground aggression is not known, as feeding areas are not protected and nearby Mountain Beavers are ignored. All adults scent-mark their burrows by rubbing sebaceous glands around their lips against the burrow walls. They also rub their feet over their face, which likely transfers the scent from there to the ground. These rodents will also produce a distinctive white secretion from their eyes in some stressful situations. This is accompanied by tooth clicking and agitation, but as the eyes fill, the animal becomes still. As the condition has only been seen upon contact with humans, any other occurrences in more natural situations are unknown. When eating, Mountain Beavers recline on their rump with their hind feet sticking out in front and grasp the plants using their semi-opposable pollex (thumb).

VOCALIZATIONS

Generally quiet, Mountain Beavers will squeal if captured or when fighting, and moan or whine softly at times. Tooth grinding is also used for communication.

SIGNS

Burrows are slightly oval, 15–20 cm wide and 13–18 cm high. A main entrance may be hidden with a pile of sticks, especially on an exposed slope, and not all burrows are marked by waste soil at their

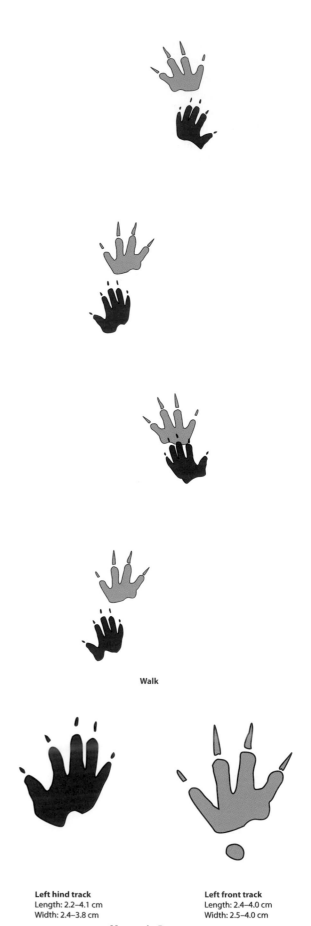

Walk

Left hind track
Length: 2.2–4.1 cm
Width: 2.4–3.8 cm

Left front track
Length: 2.4–4.0 cm
Width: 2.5–4.0 cm

Mountain Beaver

entrance. Regions with shallow burrows can produce treacherous footing, as the roof of the burrow may collapse under the weight of a hiker. The easiest way to determine the presence of Mountain Beavers is to look in summer for the clipped piles of fresh or wilting vegetation outside a burrow entrance. Since these are left outside only until partly dried, they typically indicate an active beaver site. During housecleaning, Mountain Beavers will discard old nesting material or vegetation that is no longer fit to eat and this is often left strewn around the entrance. Usually where there is one Mountain Beaver, there are others nearby. Caecal pellets are reingested and most hard pellets are deposited into underground latrines. Rarely, a single hard scat may be found above ground. These are oval, 0.5–0.8 cm in width and 1.3–1.7 cm in length. Clipped twigs harvested by Mountain Beavers display a characteristic sharp angle (less than 45°) and a surface texture best described as "stepped," as each bite of the incisors leaves a distinct groove. Most twigs are clipped around 15–25 cm from the ground, about chin height on a Mountain Beaver in either a quadrupedal or bipedal stance. Higher clipping may occur if the branches were bent to the ground by snow load, or if the tree was large enough to support the animal's weight as it climbed. Repetitive use of the same trail typically produces a clearly trampled or packed dirt runway that begins at a burrow entrance and ends at a feeding site. Front tracks usually show the massive claws that can make the toes, especially 3 and 4, look elongated. The first toe (thumb) is very short, lacks a claw, and rarely leaves a mark in the track. The hind track looks very similar to a human handprint. The short claws may not register. The first toe is smallest and often leaves a pointed track. Toes 2, 3, and 4 are about equal in length and register close together. Toe 5 points off to the side and is intermediate in size. The heel portion of either foot may or may not register in a track.

REFERENCES

Arjo et al. 2007; Bennett, B., 1991; Carraway and Verts 1993; Elbroch 2003; Feldhamer et al. 2003; Gyug 2000; Hackmann et al. 1990; Karban et al. 2007; Nagorsen 2005a; Nolte et al. 1993.

FAMILY SCIURIDAE
squirrels and marmots

Sciurids are commonly large-eyed mammals with bushy tails. They tend to sit up on their haunches to manipulate food with their forepaws and many have cheek pouches to transport food. The thumb is highly reduced, so there are, effectively, four toes on the front feet while the hind feet have five. Most sciurids are diurnal, the flying squirrels being notable exceptions. Some are arboreal (tree squirrels), some fossorial (marmots), and some are both (chipmunks).

Black-tailed Prairie Dog

FRENCH NAME: **Chien-de-prairie à queue noire**
SCIENTIFIC NAME: *Cynomys ludovicianus*

These buff-coloured, highly social rodents live in extensive colonies with intricately connected burrow systems. The large mounds of waste soil at the mouths of burrows make the colonies easy to spot. Vegetation around the colony is carefully clipped to ensure a wide field of view.

DESCRIPTION

In general, Black-tailed Prairie Dogs are sandy brown to reddish brown above and whitish below. Summer coats are darker, with more black hairs than on winter coats. Albinos are rare, but have been reported. Eyes, claws, whiskers, and the last third of the tail are black. Prairie dogs moult twice a year. Their summer coat is light, with almost no underfur. Their winter coat has a thick, warm underfur. Spring moult begins on the underside, proceeds to the dorsal side, and then moves from the face to the rear. Non-reproductive animals moult early in the spring, but moult in reproductive females may be delayed until June. The pre-winter moult is more synchronous. Individuals begin in late August or early September and complete their winter coat in about two weeks. Curiously, the progression of the autumn moult is the reverse of the spring moult. Young of the year moult several times during their first summer before growing the winter pelage. The dental formula is incisors 1/1, canines 0/0, premolars 2/1, and molars 3/3, for a total of 22 teeth. Black-tailed Prairie Dogs appear to have colour vision in the blue, green, and yellow wavelengths. Three anal glands produce

Black-tailed Prairie Dog
(*Cynomys ludovicianus*)

RANGE

Prairie Dogs occur only in North America. There are five species, of which the Black-tailed Prairie Dog has the largest range and is the only one whose range extends into Canada. It reaches the northern limit of its distribution in southwestern Saskatchewan. The only remaining Canadian populations are in the Frenchman River valley in the Grasslands National Park area near the village of Val Marie, Saskatchewan. The first prairie-dog town was discovered in Canada in 1927 about 10 km north of Val Marie. During a warm-dry period, 4000–6000 years ago, it is possible that this species also occurred in southern Alberta and Manitoba.

ABUNDANCE

Black-tailed Prairie Dogs first appeared on an endangered species list in 1974 following decades of shooting and poisoning by farmers and ranchers. Millions were killed in the 1900s and this continues to a more limited degree even today. The conversion of grasslands to croplands during the agricultural boom of the 1900s further compromised the populations. The small region of prairie-dog occupation in Canada is not located on arable land, so the animals there largely escaped this persecution. Once widespread and numbering in the billions, Black-tailed Prairie Dogs now occur in isolated pockets, occupying about 2% or less of their former range. There are no modern estimates of population size. With protection, some of the remaining colonies have expanded and prairie dog numbers have increased over the past 20 years. The Black-tailed Prairie Dog is not

a skunky-smelling secretion that serves to identify individuals and is used to mark territories and announce territorial disputes.

SIMILAR SPECIES

Black-tailed Prairie Dogs look like portly, oversized ground squirrels. Ground Squirrels (genus *Spermophilus*) are smaller, with a light-coloured ring around each eye. Most have more colourful or darker pelage than Black-tailed Prairie Dogs. Richardson's Ground Squirrels are the most similar, as they are light coloured and have a black tail tip. In addition to having an eye ring and a smaller size, their pelage colour is more grey-yellow than the cinnamon-buff coloration of prairie dogs.

SIZE

The following measurements apply to adult animals from the northern United States and Canada, as there are insufficient data available solely from the Canadian range. Body weight varies over the year. All age classes are lightest in spring and heaviest in October.

Males are about 5%–15% heavier than females, except about a month after the breeding season, when exhausted males are still thin and females are heavily pregnant.

Total length: 355–415 mm; tail length: 80–115 mm; hind foot length: 57.5–64.5 mm.

Weight: males, 437–1390 g; females, 406–1149 g; newborns, about 15 g.

Distribution of the Black-tailed Prairie Dog (*Cynomys ludovicianus*)

included on the US endangered species list, although it is being considered for "Threatened" status. In Canada, the species is designated by COSEWIC as "Vulnerable." At present there are some 25 colonies in and around Grasslands National Park, ranging from 0.6 ha to 170.0 ha in size. It is estimated that the colonies contain as many as 17,000–23,000 animals. They are slowly increasing in size and in number. Occupying less than 690 ha in 1985, Canadian colonies had grown to fill almost 1050 ha by 2004. It is hoped that the dog towns are sufficiently large and close enough together now to support a reintroduction of Black-footed Ferrets. This reintroduction began in 2009. Before European development in the prairies, prairie dog numbers were held in check by native predators. As these predators declined with agricultural advancement, prairie dog numbers blossomed and they did become a threat to crops in some locations. Past eradication efforts were undertaken mainly to protect domestic horses and cattle from breaking their legs by falling into the burrows and because ranchers felt that the prairie dogs were competing with livestock for food. But leg fractures attributable to burrows are rare and Pronghorns and livestock prefer to graze in dog towns because the forage is lush and more nutritious due to the enrichment of scat and the clipping undertaken by the prairie dogs that stimulates plant growth. Furthermore, many of the plant species preferred by the prairie dogs are avoided by livestock and visa versa. Only an estimated 4%–7% of plants are consumed by both prairie dogs and livestock. Prairie dogs and Bison coexisted for thousands of years to mutual benefit. Since cattle and Bison have similar diets, cattle and prairie dogs should be able to coexist.

ECOLOGY

Black-tailed Prairie Dogs inhabit short- to mixed-grass prairies. They are semi-fossorial, constructing and inhabiting extensive underground burrow systems that are used for sleeping, for raising young, and for protection from weather and predators. Burrows are typically around 5–10 m long and 2–3 m deep, but may be up to 33 m long and 5 m deep. Three types of burrow entrances are created. Scattered primarily around the periphery of the territory, the entrances are hidden and small, with no throw mound of waste soil. They lead to short burrows and are used chiefly for refuge during predation attempts. Within the core of the group territory, the burrow entrances are dug through the middle of either a shallow (0.2–0.3 m high) dome-shaped mound or a higher (up to 1 m) cone-shaped mound and both of these entrance types lead to more complex sleeping, hibernation, and nursery burrow systems. Entrances penetrating into the middle of a mound are raised and thereby protected from flooding. The mounds are used as perches to provide higher vantage points and they direct air into the entrance, which may increase the rate of air flowing through the tunnels. The cone-shaped mounds, called rim mounds or rim craters, are diagnostic of the Black-tailed Prairie Dogs and the endangered and closely related Mexican Prairie Dog (*Cynomys mexicanus*). All the other species of prairie dogs construct only dome mounds. Extent or density of burrows is not a useful indicator of colony size or density, as many burrows may be empty due to disease or emigration, and long-time colonies may contain many unused burrows.

Furthermore, the number of burrow entrances in a colony fluctuates widely depending on soil type and ease of digging. Colony density varies from less than 10 adults to over 35 adults per hectare.

A diurnal species, Black-tailed Prairie Dogs typically forage during the first few hours after sunrise and spend the remainder of the day sun bathing, dust bathing, grooming, socializing, and conducting burrow repairs or renovations. During hot days, they may disappear into the burrows for up to half an hour at a time, presumably to cool off. Black-tailed Prairie Dogs do not hibernate in the true sense of the word. They may appear above ground during any winter month, especially if temperatures moderate. Black-tailed Prairie Dogs in Canada exhibit more extensive and predictable bouts of torpor than they do elsewhere in the range. Periods of torpor occur throughout the cold season and some may last for several weeks during severe weather conditions. Their body temperature drops only a few degrees during short periods of torpor or may drop to about half of their normal temperature during extended periods. Since these squirrels do not store food for the winter, they rely instead on their fat reserves and whatever food they can find above the snow. Sylvatic Plague (also called Black or Bubonic Plague) was accidentally introduced into the New World through several western US ports in the early 1900s. It has since become established in group-living rodents such as prairie dogs and ground squirrels west of the Mississippi River. These rodents are especially susceptible to this disease, as it is new to them and they lack resistance. Living as they do in large, dense colonies, disease transmission is rapid. Outbreaks commonly wipe out whole colonies, and plague is the only disease known to produce such extensive die-offs in Black-tailed Prairie Dogs. This disease reached Canadian populations in 2009 and its long-term effect is still undetermined. Female Black-tailed Prairie Dogs may live eight to nine years, but few males live more than four years. Their list of predators is extensive and includes Coyotes, American Badgers, Long-tailed Weasels, large raptors, Prairie Rattlesnakes, Gopher Snakes, Swift Foxes, Red Foxes, Bobcats, and Black-footed Ferrets.

Prairie dogs are considered keystone species in the prairie grasslands. Their activity turns, aerates, and fertilizes the often-dry, hard-packed soil. Black-tailed Prairie Dogs carefully clip all the tall vegetation that grows in and around their colonies, presumably to permit a wider field of view for visual communication between the inhabitants and as a defence against predators. Clipping stimulates the grasses to continue growing in a younger and more nutritious state and it prevents the ever-increasing creep of shrubs such as sage and creosote onto the grasslands. Their burrows and the rich grassland they create within their towns create homes and microclimates for over 200 associated animal species of which at least 9 (including the endangered Black-footed Ferret) are dependent on prairie-dog towns for their survival. Other species associated with, but not dependent on, prairie dogs are similarly endangered, including the Swift Fox, Mountain Plover, and Burrowing Owl.

DIET

Black-tailed Prairie Dogs are primarily herbivorous, although they occasionally eat insects. They selectively consume a variety of

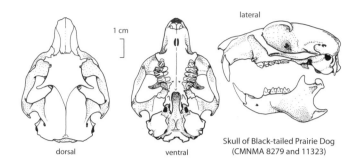

1 cm

lateral

dorsal   ventral

Skull of Black-tailed Prairie Dog
(CMNMA 8279 and 11323)

grasses and forbs as they become seasonally available. There are a number of plants that they do not eat, such as threeawn, Prairie Dog Weed, and horseweed, and these may become common within the dog towns. In fact, Prairie Dog Weed is rarely found outside of dog towns. During the winter, their diet is more limited and includes prickly pear cactus, thistles, dried grasses, and underground roots of various species. Black-tailed Prairie Dogs are especially inclined towards infanticide and cannibalism (see the "Behaviour" section below for more information).

REPRODUCTION

Most prairie dogs breed for the first time as two-year-olds. A few become sexually mature as yearlings (mostly females – about 35% or less) and some delay until they are three years old (mostly males). Onset of breeding varies with latitude. The breeding season for Canadian animals occurs over about two to three weeks between mid-March and mid-April, but may be as early as January for prairie dogs at the southern limits of the range. Each female is receptive for less than a day and produces a single litter each year. Yearling females that do mate appear to enter oestrus later than more-mature females. Mating usually takes place underground, and the vast majority of females mate with the harem male of their social unit (called a coterie). Gestation lasts 33–38 days and the hairless, blind pups are born in an underground nest. Litter size may be one to eight young, with an average of three that usually survive to weaning age. The pups develop relatively slowly. Their eyes open at 33–37 days old, and they begin walking soon after that. They are suckled and cared for by their mother, who may also bring them vegetation to supplement her milk as they become older. In Saskatchewan, newly emergent young of about six weeks old (40–43 days) begin appearing in late June and all are above ground by July. The pups then forage and mostly eat vegetation; however, weaning is gradual and their diet is usually supplemented for another week or two with milk dispensed freely by any adult female in the group. Sometimes all the pups in a coterie will group together and even attempt to nurse from a single female. This behaviour may lead to overestimations of litter size by casual observers. Juvenile males disperse from their natal coterie before they become sexually mature, typically in May or June of the year following their birth. They usually settle in another coterie in the same colony, but may also disperse to other more distant colonies up to 5 km away. Mortality of emigrating juveniles is understandably high, as they run a gauntlet of predators without the protection of a secure burrow. Females tend to remain in their natal coterie when they become sexually mature. Adult males are nomadic and rarely remain in their coterie longer than two years, which prevents them from mating with their daughters. Females can recognize kin and are unwilling to mate with close relatives. They will not come into oestrus or will seek copulations with neighbours rather than risk mating with their own resident male if he is a father, brother, or son. They will, however, mate with more distant relatives, such as cousins. Ultimately, a small isolated colony lacks genetic diversity, making the current state of Black-tailed Prairie Dog distribution (scattered colonies, many of them small) a conservation concern.

BEHAVIOUR

Black-tailed Prairie Dogs are highly social animals, which makes them fascinating to observe. Each dog town is made up of an assortment of territorial social units called coteries. A coterie typically comprises one adult male, three to four adult females, and their non-reproductive young. The females, unlike the more nomadic males, usually spend their whole lives within their natal coterie territory and are usually close kin. Interactions between coterie members are generally amiable, as animals groom each other, play, and perform mouth-to-mouth greetings that look like kissing. They also engage in self-grooming activities such as scratching fleas and dust bathing. Kicking, shoving, and packing dirt to reconstruct burrows and mounds are common, often communal activities, especially when the earth is damp. Prairie dogs spend a considerable amount of time resting on a mound watching the ground and sky for approaching predators. They call loudly when a threat is detected to warn kin and distant neighbours alike. Burrows within a coterie are available to all the members and are commonly shared at night, especially during cool weather. Exceptions occur in two situations. Occasionally two breeding males (usually brothers) will subdivide a large coterie and forcibly exclude the other during the short breeding season. Pregnant females and those with pups will temporarily and aggressively defend their chosen nursery burrow from the rest of the coterie until their young are weaned. During an encounter between animals from different coteries, each hastens to defend its respective territory. Disputes begin with tooth chattering, staring, tail fluffing, and sniffing each other's anus, and may advance to bluff charges, chases, and fights, especially between harem masters during the breeding season. A startling peculiarity of Black-tailed Prairie Dog society is the high degree of juvenile mortality due to infanticide and subsequent cannibalism. This is perpetrated by adults of either gender, but usually either by adult males that are new immigrants into a coterie or, more frequently, by closely related adult females who are members of the coterie. In fact, the major cause of juvenile mortality is infanticide and it affects about 40%–50% of litters. It is unclear why this behaviour is so prevalent. It is theorized that females are: stressed to produce milk for their own litters and may be looking for the protein; removing future competitors to themselves or their offspring; defending their own and their litter's foraging territory; protecting their offspring from the other female (when that female stops lactating she will become non-aggressive towards other litters); or perhaps attempting to procure a helper in

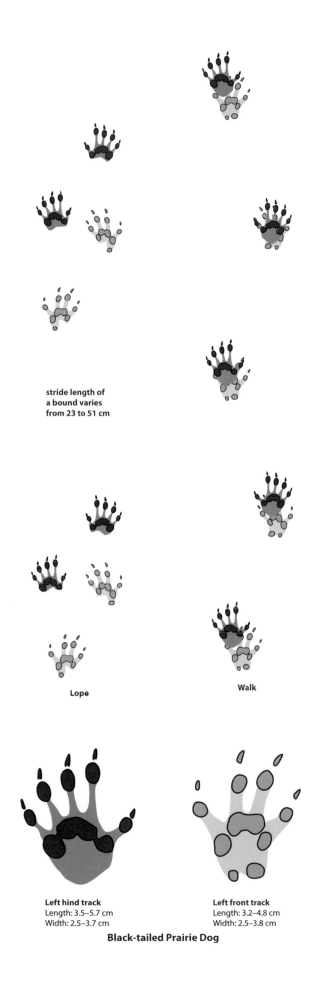

**stride length of
a bound varies
from 23 to 51 cm**

Lope

Walk

Left hind track
Length: 3.5–5.7 cm
Width: 2.5–3.7 cm

Left front track
Length: 3.2–4.8 cm
Width: 2.5–3.8 cm

Black-tailed Prairie Dog

raising her litter (adult females that have recently lost a litter make better helpers).

Although resident males have nothing to do with the rearing of young, they do play a role after the young have emerged from the natal burrow and are spending their time playing and foraging above ground. Their mother is distracted by her drive to gain weight before autumn, so the males are more likely to defend the young from infanticidal neighbours. Large males are most likely to secure breeding territories and consequently sire more offspring. Oestrous females prefer to breed with larger males, so some young males must wait until their third year to achieve the size necessary to gain a territory and breed for the first time. Should a rattlesnake or Gopher Snake take up residence in a burrow, the prairie dogs are quick to plug it in an effort to trap the predator within.

VOCALIZATIONS

Black-tailed Prairie Dogs have a varied vocabulary of at least 12 different barks and calls. They use a comical "yip-jump display" to emit a territorial call. Stretching nearly vertically, they fling their fore feet upwards and call. Typically, one yip-jumping prairie dog will spark another, so that a chain reaction might have several dozen prairie dogs calling. Yip-jump calls are sometimes used to advertise the presence of a predatory snake. Apart from this snake alarm call, and unlike many colonial rodents that have distinctly different alarm calls to identify an aerial from a terrestrial predator, Black-tailed Prairie Dogs use the same basic alarm call for all other predators despite their otherwise extensive vocal repertoire. They appear to make minor variations to the basic alarm call to impart information about the degree of danger and even the individual identity of the predator. For example, they can discern the differences between humans and will then assess the relative degree of danger each presents to the colony based on their current level of threat and the actions they took on previous visits. Alarm calls consistently reflect this information. Such calls (which are similar to a dog bark) are also contagious, and many animals will take up a call, to the obvious frustration of many a potential predator. The similarity of their commonly heard bark call to that of a Domestic Dog is why they are called prairie dogs. Males produce an often-lengthy, repetitious mating call after copulation. Other calls have been described as a clitter bark, used in defence, muffled bark, chirr used during disputes, raspy purr, chuckle, snarl, growl, scream uttered in pain or upon capture, and tooth chattering, which is used as a threat.

SIGNS

The throw mounds and short turf around the colony render Black-tailed Prairie Dog towns visible from a distance. Burrow entrances are roughly circular, and 10–28 cm in diameter. Major burrows may have up to six entrances. Most burrow construction and renovation occurs during the summer. The three types of burrow entrances are described in the "Ecology" section above. The front feet have the classic rodent arrangement of five toes, with the pollex (thumb) being vestigial and not registering in a track. Toes 2 and 5 angle to the sides and toes 3 and 4 point forward. The hind feet also have five toes. Toes 1 and 5 angle to the sides and toes 2, 3, and 4 register close together, aiming forward. The front track typically shows three distinct palm pads and two heel pads, while the hind foot track shows four semi-distinct pads. Claws are prominent on both front and hind feet. Black-tailed Prairie Dogs use three principal gaits. They mostly walk, but may lope or bound when travelling faster. Tracks may be found all year round, even in snow. Runs and trails begin and end at a burrow. Drag marks between tracks are a common characteristic of snow tracks left by these short-legged squirrels. Dust baths generally measure 22–35 cm long by 10–20 cm wide and are simple shallow pits often located on the throw mound. Scat usually litters the ground around a burrow entrance. It has either been flung out during housecleaning or has been deposited near the security of the burrow. Since burrows may be occupied for decades or longer, the fertilization value of the scat cannot be underestimated. Scat shape varies with the moisture content of the diet. Individual hard pellets tend to be blunt at one end and somewhat pointed at the other. Softer pellets may be held together with strands of fibre and form into sometimes lengthy chains. Scat varies from 0.5 to 1.1 cm in diameter and 1.3 to 5.0 cm in length.

REFERENCES
Cully and Williams 2001; Dobson, F.S., et al. 1997, 2004; Elbroch 2003; Frederiksen and Slobodchikoff 2007; Hoogland 1995, 1996, 2001; Kotliar et al. 1999; Lehmer et al. 2001; Lomolino and Smith 2001; Seabloom and Theisen 1990; Sidle et al. 2001; Smith, W.J., et al. 1977.

Northern Flying Squirrel

FRENCH NAME: **Grand polatouche**

SCIENTIFIC NAME: *Glaucomys sabrinus*

Despite its name, this tree squirrel cannot truly fly; but it is a master at gliding.

DESCRIPTION
The Northern Flying Squirrel is a medium-sized tree squirrel with a thick, soft pelage, large, dark eyes, large, pinkish ears, and a wide, flattened tail. Its most distinct characteristic, which it shares with the Southern Flying Squirrel, is the furred flap of skin (called the patagium) along each side from wrist to ankle, which, when the legs are extended, forms the gliding surface. Coloration varies somewhat with location. The back and sides may be cinnamon brown to ashy brown. The underparts are creamy white to buff to grey. Even when the tip of each belly hair is white, the base of the hair is dark, which dims the brightness of the white and provides a greyish to brownish undertone. The tail varies from light ashy brown to reddish brown to almost black on the upper surface with a dark tip, and the underside varies from pale grey or cinnamon to nearly black. The dark tail tip shows on the underside as well as the top of the tail. The dental formula is incisors 1/1, canines 0/0, premolars 2/1, and molars 3/3, for a total of 22 teeth.

Northern Flying Squirrel (*Glaucomys sabrinus*)

A cartilaginous rod several centimetres long juts out from each wrist to provide additional support to the front edge while gliding. It allows a wider spread of the patagium and is used for steering. Northern Flying Squirrels have a single moult each year in early summer.

SIMILAR SPECIES
Nocturnal habits, large eyes, soft, silky fur, and a noticeably flattened tail set this squirrel apart from all but the Southern Flying Squirrel. Apart from size, these two species are very similar. The Northern Flying Squirrel is much larger, but this may be difficult to assess if you can only see one animal. The belly fur of Northern Flying Squirrels is sooty to buffy to creamy white, but each hair is grey or grey-brown at its base, which imparts a darker undertone; by contrast, Southern Flying Squirrel belly hairs are creamy white right to the base of each hair. The colours of upper parts of a Northern Flying Squirrel are often brighter and their tail may have a darker tip. The dark side stripe usually quite evident on the Southern Flying Squirrel is often less noticeable on the Northern Flying Squirrel, as the upper fur is darker, and so the contrast is less distinct. Although the two species do co-occur in overlapping parts of the ranges where the forests are mixed, Southern Flying Squirrels predominate in deciduous habitats and Northern Flying Squirrels in coniferous and mixed forests. Recent evidence indicates that the two species can interbreed and produce hybrid offspring that are intermediate in size and display traits of both parents. Apparently global warming is permitting the smaller Southern Flying Squirrel to penetrate northwards into the range of the larger Northern Flying Squirrel, and this is where the hybridization is occurring.

SIZE
The following measurements apply to adult animals from the Canadian portion of the range. Adult males and females are similar in size, but there is considerable regional variation. The largest animals are found in the west.
Total length: 275–358 mm; tail length: 120–164 mm; hind foot length: 36–50 mm; ear length: 16–26 mm.
Weight: 99.5–207.5 g; newborns, 5–6 g.

RANGE
This is a widely distributed squirrel occurring in much of Canada and Alaska and in cool, boreal/montane habitats in the conterminous 48

states. The northern limit of distribution coincides generally with the treeline. Northern Flying Squirrels are found on many islands such as Cape Breton Island, Manitoulin Island, Prince Edward Island, Prince of Wales, Wrangell, Dall, Heceta, Revillagigedo, and Kosciusko Islands in southeast Alaska, and several islands off the coast of British Columbia, including Campbell, Princess Royal, Cortes, Stuart, and Quadra Islands. How the species reached these islands poses an interesting zoogeographical question, especially since they have not been found on Vancouver Island despite the mere 700 metres between Quadra Island and Vancouver Island.

ABUNDANCE

The shy and nocturnal habits of the Northern Flying Squirrel make it appear scarce or even rare. This is not the case over most of the range, although it is usually well dispersed and seldom abundant. Populations are often common in well-forested areas in cities and towns without residents knowing of their presence.

ECOLOGY

Although commonly associated with boreal and montane coniferous forests, these squirrels are also found in mixed deciduous/coniferous forests and occasionally in purely deciduous habitats. They are almost entirely nocturnal, with peaks of activity for a few hours after sunset and before sunrise. During very cold weather, they may reduce the amount of outside activity and remain in their nest for longer periods. Northern Flying Squirrels are not known to enter torpor. They remain active all year and appear impervious even to very cold temperatures. Spherical nests similar to Red Squirrel dreys, but constructed with finer materials (dried grasses, mosses, shredded bark, and lichens), may be constructed for summer use, but most often they take over and refurbish abandoned nests of other squirrels and birds. They require a warm nest in a tree cavity for the winter. Well-insulated winter nests are frequently created

Distribution of the Northern Flying Squirrel (*Glaucomys sabrinus*)

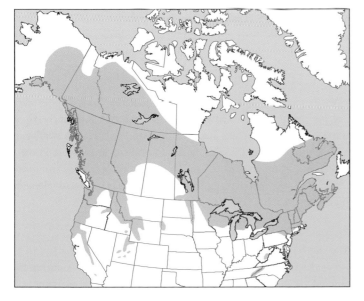

almost entirely of lichens, which may have the secondary function of providing food during extended periods of inclement weather. Winter nest-building material often includes shredded tree bark from species with frayed bark, such as cedar. There are reports of subterranean nest sites in some regions, but such usage appears to be rare. Northern Flying Squirrels are capable of gliding for 65 m, if they begin high in a tall tree, but normal glides are in the 20 m or less range. Wing loading is around 50 newtons/m², about the same as is recommended for hang-gliders. The flattened tail serves as a rudder and also generates around 20%–30% of the lift. Their glide ratio is about 2:1, which means that a glide of 40 m would require a drop of 20 m, assuming no influence from wind or need to bank to avoid obstacles. Air speed varies from 6.3–8.1 m/s. The average glide angle is 36°–44°. The path of an actual glide is not straight, as the squirrel drops steeply for 1.5–4.0 m to gain velocity, then levels out to increase the flight distance, and finally increases the angle of attack to gain altitude and slow for a landing. They typically try to land just as their velocity drops to near zero, so the touchdown is feather soft, then they zip around or up the trunk to avoid any predator that may have heard or seen the landing. Manoeuvrability is so fine-tuned that a flying squirrel is able to corkscrew through the air to land almost immediately below the take-off point. Understandably, flying squirrels resort to regular quadrupedal locomotion during inclement weather such as high winds, heavy rain, or fog.

The size of a Northern Flying Squirrel's home range differs between the sexes, with adult males occupying ranges as much as 40% larger than those of adult females. Home ranges vary from 0.8 ha to several km² depending on gender and food abundance. Densities vary from 0.1 to 3.5 squirrels per ha and are highest in old-growth forests. Juveniles may remain near their natal area, or move several kilometres away. Northern Flying Squirrels are considered to be a "keystone" species, crucial to their environment due to their feeding activities, which disperse tree seeds and the spores of symbiotic fungi throughout the forest. The hardy spores of truffles are unaffected by transmission through the gut of the squirrel and are scattered in their droppings along with nitrogen-fixing bacteria. The fungi's mycorrhizae help the trees absorb nutrients and water and are essential to the health of the forest. Because of their largely nocturnal activity, Northern Flying Squirrels are especially vulnerable to large owls such as the Spotted, Barn, Barred, and Great Horned Owls. There has been a flurry of new work done in the last 20 years on western populations of these squirrels, as it was discovered that they are a favoured prey of the endangered Spotted Owl. A Spotted Owl in Oregon is estimated to consume 260 Northern Flying Squirrels over the course of a year, which constitutes about 50%–60% of their diet. Mammalian predators include arboreal carnivores such as the Ermine, Long-tailed Weasel, and American Marten as well as ground predators such as the Bobcat, Canada Lynx, Grey Wolf, Red and Grey Foxes, Coyote, Cougar, and Domestic Cat. Like some low-flying bats, Northern Flying Squirrels occasionally die when their patagium becomes tangled in barbed-wire fencing or the body is punctured by a sharp branch or bud. The maximum lifespan in the wild is thought to be around seven years, but most squirrels live less than three to four years. Maximum running speed is around 13 km/h.

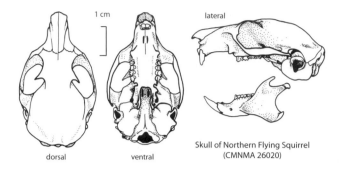

1 cm

dorsal   ventral   lateral

Skull of Northern Flying Squirrel
(CMNMA 26020)

DIET

Northern Flying Squirrels have a wide and eclectic diet that varies across the country as foods are available. They mainly eat fungi and lichen along with the seeds and nuts of coniferous and deciduous trees, but readily supplement these choices with fruit, catkins, flowers, tree sap and buds, insects and other invertebrates, small roosting birds, bird eggs and nestlings, small mammals, and carrion. They willingly come year-round at night to a bird feeder for suet or sunflower seeds. The favourite type of fungus they eat is the group of underground species collectively called truffles (hypogeous mycorrhizal fungi). Fruiting bodies of truffles do not reach above the soil surface. These fungi have evolved to depend on animal vectors to disperse their spores. The aromatic compounds produced by the truffles attract the animal's attention. Truffles are most abundant in moist forests both in the west and in the east. Northern Flying Squirrels find and consume over 40 different species, leaving distinctive little pits in the ground from their efforts. In coastal regions of British Columbia, where there is little snow, fungi, especially truffles, are abundant, and form a major part of the diet year round. In other areas with more snow and many tree-hanging lichens, the lichens make up the bulk of the winter diet. In northerly areas with snow, but few hanging lichens, the squirrels eat a wider variety of foods including dormant insects, seeds, tree bark and buds, and a large part of their diet appears to be fungi. Although some more southwestern squirrels are reported to cache small amounts of fungi for winter consumption, those in the north appear to get much of theirs by stealing from Red Squirrel caches. These are pilfered at night when the hard-working proprietor is sleeping. Most Northern Flying Squirrels drink free-standing water if it is available or fill most of their water requirements from the food they eat (fungi are 70%–80% water). There are regions of dry forests with little fungi and no standing water that still support this species. How the squirrels adapt to these situations remains unknown.

REPRODUCTION

The breeding season occurs from mid-March to early June. Following a gestation period of 37–42 days, a litter of two to four young on average (range 1–6) is born in a tree-cavity nest or in dreys. First-time mothers tend to have smaller litters than mature females. Females likely have a single litter each year, although there may be some evidence to suggest a second litter in warm regions, which probably does not occur in Canada. Newborns are pink and

helpless, with ears and eyes closed over, and toes fused together. By day 6 their toes separate. Their teeth begin erupting by 20–26 days old. Their ears open by day 24, and their eyes around day 35, when they are covered in a fuzzy coat of hair. They can walk, and begin to venture out of the nest at about 40–50 days old. Solid food is introduced into their diet at this time, and about two to three weeks later they are weaned. The pups usually remain with their mother for several more months, and may spend their first winter with her, or she might abandon the nest to them and find a solitary one for herself. Most juveniles become reproductively active in the breeding season following their birth (at 8–11 months old), but some may not breed until the following spring (at 20–23 months old).

BEHAVIOUR

The smaller Southern Flying Squirrel is more aggressive than the larger Northern Flying Squirrel (much like the situation of the Red Squirrel and the Grey Squirrel) and is usually dominant over its larger cousin. Among groups of Northern Flying Squirrels, the adults usually dominate the juveniles, but apart from the occasional short dispute over food, few altercations are noted. Northern Flying Squirrels often nest alone in the winter, but groups of up to five or possibly more animals of both genders may occupy the same nest cavity, likely as an energy-conservation measure. Patterns of den sharing suggest that kinship and social contact may be involved to some extent. Pregnant and nursing females defend a nest site and exclude other squirrels. Through studies using radio collars, we know that Northern Flying Squirrels move from nest to nest on a fairly regular basis. They use both dreys and tree-cavity sites through spring, summer, and autumn, and use cavities only during winter. In regions with high rainfall they use cavity nests most often, even during warm weather, probably to escape the rain. Larger, older trees are preferred for nesting. These squirrels use olfactory clues to detect truffles hidden underground. They also learn to look for truffles in the most likely locations, such as around decaying logs, and it is probable that they remember the locations where they have found truffles before and check the following year in the same places. During the breeding season, males travel long distances (sometimes > 5 km) in search of oestrous females. They expend a tremendous amount of energy during this period and lose body weight as a result.

VOCALIZATIONS

Northern Flying Squirrels are not particularly vocal. While mating, females produce a "chirring" call and the males produce a long nasal "whine" call. When disturbed, or while in a dispute with one another, they produce a faint clucking sound, or high-pitched chittering calls, and may accompany this sound with vigorous stomping of their hind feet and lashing of their tail. Their primary call is a soft, high-pitched chirp.

SIGNS

The smaller front feet have four toes, the outside two (toes 2 and 5) angle away from the foot, while the middle toes (toes 3 and 4) point straight ahead. Toe 1 (the thumb) is vestigial and rarely registers in a

stride length
of a bound
can vary from
15 to 86 cm and
occasionally
farther

**Typical
bound**

**Bound
variation**

Left hind track
Length: 3.2–4.8 cm
Width: 1.6–2.5 cm

Left front track
Length: 1.7–3.2 cm
Width: 1.3–1.9 cm

Northern Flying Squirrel

track. Hind feet have five toes, with the middle three aiming forward in a cluster and the outside two angling away from the line of the track. Claws may or may not register in the track. Although the size of tracks overlaps with those of Red Squirrels, there is considerably more fur between the toes in a flying squirrel's foot, causing the track to appear both larger and less defined. Northern Flying Squirrels normally bound when travelling. Their track pattern has been described as "boxy," as the front feet impressions typically register directly behind those of the hind feet (see track illustration). The landing pattern left when a gliding squirrel touches down is called a "sitzmark." It is most commonly seen in snow and may be 60 cm or more in length. (Sitzmarks are uncommon, as most squirrels take off and land in trees, then climb down the tree to the ground.) The placement of this mark in an otherwise clear snowy spot with a trail leading away from it leaves little doubt to its origin. Scat is about 0.2–0.5 cm in diameter and 0.3–1.0 cm in length and is typically bluntly oval in shape and dark brown or black in colour. Dreys are up to about 40 cm in diameter.

REFERENCES

Carey et al. 1997, 2002; Cotton and Parker 2000a, 2000b; Currah et al. 2000; Elbroch 2003; Ferron 1983; Hackett and Pagels 2004; Hayward and Rosentreter 1994; Holloway and Malcolm 2007; Martin, K.J., and Anthony 1999; McMaster 2002; Menzel, J.M., and Ford 2004; Nagorsen 2005a; Nero 1993; Patterson, J.E.H., et al. 2007; Payne, S., and Longland 2001; Payne, S., et al. 2002; Scheibe et al. 2006; Smith, H.C., 1993; Vernes 2001; Vernes et al. 2004; Villa et al. 1999; Wells-Gosling 1985; Wells-Gosling and Heaney 1984; Witt 1992.

Southern Flying Squirrel

FRENCH NAME: **Petit polatouche**
SCIENTIFIC NAME: *Glaucomys volans*

Master gliders, a Southern Flying Squirrel, like its larger cousin the Northern Flying Squirrel, can glide for over 50 m from tree to tree. See the "Ecology" section of the Northern Flying Squirrel account for a description of the aerodynamics of their flight.

DESCRIPTION

Southern Flying Squirrels have large, dark eyes, soft, silky fur, relatively large pinkish to greyish ears, and a long, flattened tail. Their belly and throat are creamy white and the hairs are white all the way to their base. Sometimes older animals become stained a rusty orange colour on the edges of their belly and under the hind legs, but the centre portion of the belly and chest remains clear white. Fur colour on their back and sides is variable depending on location. Tips of the hairs may be ashy brown, cinnamon brown, or yellowish brown, but always with a dark-grey base. Most Canadian Southern Flying Squirrels have yellowish to ashy-brown upper parts. A

Southern Flying Squirrel (*Glaucomys volans*)

ABUNDANCE

It is difficult to assess the size of Southern Flying Squirrel populations as they are shy, nocturnal animals. Furthermore, they exhibit a wide annual fluctuation in numbers and despite the large distribution shown on the range map, actually occur in scattered pockets within that area. Given the decline in woodlot coverage in the Great Lakes region, it is likely that the population there is declining, but recent survey work has shown that the distribution in Ontario is more widespread than was previously understood. The disjunct population in Nova Scotia may be threatened by the conversion of the natural mixed hardwood stands into conifer plantations following logging efforts. Ontario considers Southern Flying Squirrels to be of "Special Concern," and the population in Nova Scotia is considered by the province to be "Sensitive to Human Activities or Natural Events." Federally the species is considered to be "Not at Risk."

ECOLOGY

Southern Flying Squirrels occupy a variety of habitats from mixed coniferous/deciduous forests in the north to pine forests in the southeast and dry deciduous woodlands and oak savanna in the far south. They are most common in mature deciduous forests. Their population size is highly variable and dependent on food abundance. Winter survival is reduced and litters are smaller in the year following an autumn of poor mast production. These small squirrels are almost entirely nocturnal, with peaks of activity for a few hours after sunset and before sunrise, and so are active during the coolest time in each 24-hour period. During very cold weather, they reduce the amount of outside activity and remain in their nest for longer periods. Occasional torpor has been reported, usually associated with cold temperatures and lack of food. Body temperature may drop by over 15°C for as long as four

distinct dark stripe runs along each side from elbow to knee, separating the dark back and the pale underbelly. Tails are pale buff below and brownish above. Their most distinctive characteristic, which they share with Northern Flying Squirrels, is the flying membrane (patagium), which stretches from wrist to ankle on each side. The dental formula is incisors 1/1, canines 0/0, premolars 2/1, and molars 3/3, for a total of 22 teeth. Juveniles begin to moult their natal pelage when they are around 12 weeks old. This moult takes about three to four weeks and travels from the head to the tail, which is transformed last. Adults undergo a partial moult in early summer. Males and non-reproductive females complete it first and lactating females follow, after their litter is three to six weeks old. This partial moult begins at the nose and ends at the neck, so that only the fur on the head is renewed. A full moult begins in late summer. This moult starts again at the head and proceeds rearward and includes the tail. Females with a summer litter moult last, beginning as the litter is close to being weaned.

SIMILAR SPECIES

The only other squirrel likely to be mistaken for a Southern Flying Squirrel is a Northern Flying Squirrel. See the "Similar Species" section of the Northern Flying Squirrel account for a discussion of the distinctions.

SIZE

The following measurements apply to adult animals from the Canadian portion of the range. Adult males and females are similar in size.
Total length: 180–260 mm; tail length: 80–120 mm; hind foot length: 29–34 mm; ear length: 15–21 mm.
Weight: 26–67 g; newborns, 3–5 g.

RANGE

Southern Flying Squirrels occur across eastern North America and in several isolated locations in the mountains of Mexico and Central America. They reach the northern limit of their distribution in southern and eastern Ontario and southwestern Quebec. There is a disjunct population in south-central Nova Scotia that became isolated from the populations in Ontario and Maine by loss of the intermediate populations, likely due to climatic fluctuations that began around 7000 years ago. The barrier on the western edge of the range is the treeless Central Plains.

Distribution of the Southern Flying Squirrel (*Glaucomys volans*)

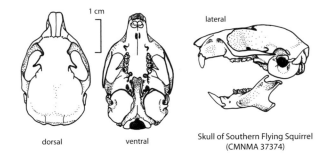

1 cm

dorsal   ventral   lateral

Skull of Southern Flying Squirrel
(CMNMA 37374)

weeks at a time as the animal curls in a nest, but prolonged periods of true hibernation do not occur. Nest sites are scattered throughout the home range and are used by any of the group in time of need, and all the squirrels are well acquainted with their location. Abandoned dreys (outside nests made of twigs, shredded bark, and dried grasses) constructed by larger squirrels and old bird nests are commonly renovated for flying squirrels' use. In northern parts of their range, including Canada, dreys are only used in summer, and even then cavity nests are favoured. Tree-cavity sites are in high demand by many animal species, especially Red Squirrels. Although the flying squirrel will box with its front feet to prevent the entrance of another animal, a persistent Red Squirrel can usually expel the considerably smaller Southern Flying Squirrel. Occasionally an underground nest is used. Both regular and nursery nests are frequently abandoned after they become fouled or infested with parasites, typically fleas, so that each squirrel may reside intermittently in several different locations. Southern Flying Squirrels carry a parasite that is not lethal to them, but which, when transmitted to Northern Flying Squirrels, can cause death. The two species are able to coexist in many parts of their range, especially in colder regions where the parasite is less prevalent, but the actions of this parasite may make the competition between the two somewhat more equal, and balance the size advantage of the Northern Flying Squirrel. The home ranges of these highly mobile animals can be quite large, depending on food availability. A study in Arkansas describes average ranges of 3.8 ha for females and 7.8 ha for males. Other studies found ranges of 0.4–6.4 ha for females and 0.5–19.0 ha for males. Their nocturnal lifestyle makes them susceptible to predation by larger owls such as Barred, Great Horned, Long-eared, Spotted, and Barn owls. Time spent on the ground makes them vulnerable to foxes, Coyotes, Bobcats, Canada Lynx, skunks, Northern Raccoons, Domestic Cats, and Grey Wolves. Arboreal predators, including Ermine, Long-tailed Weasels, and American Martens, will pursue them through the trees. Captive animals may live to be 13 years old, but few wild squirrels live more than 4 to 5 years.

DIET

The diet is diverse and variable, depending on local abundance and availability. Nuts, seeds, flowers, fruit and berries, tree buds, bark and sap, insects, small birds, bird eggs and nestlings, and carrion are all eaten. In Nova Scotia, these squirrels are reputed to consume fungi year-round, but fungi are rarely eaten elsewhere in the range. During late summer and autumn, large food items such as nuts are scatter-hoarded (single nuts hidden or buried in many different locations) and smaller nuts, like beech nuts, may be larder-hoarded

(many items placed in one location) if an appropriate tree cavity is available. These caches are consumed during the winter when food is otherwise unavailable or scarce. Most water requirements are met through their food and by drinking dew or eating snow.

REPRODUCTION

The breeding season in any particular area is relatively short (10–14 days) and may vary from year to year, depending on inclement weather, which can cause delays. Across the species distribution, mating occurs from late February to mid-April, and then again in mid-June to early August. Testes of mature males descend from the abdomen and become pendant in the scrotum in late January or early February and are retracted back up into the abdomen in late August. Two litters per year are possible if the females are well fed. It is suspected that in many regions females do not achieve their reproductive potential and a single litter each year is more common. Spring litters are normally smaller and lighter than summer litters. Following about a 40-day gestation period, an average litter of 3–4 (range 1–7) young are born in a tree cavity or drey. Yearling females producing their first litter tend to have a small litter, commonly only one pup. Females may come into oestrus again 60–90 days following the birth of a spring litter. Pups are born essentially hairless (except for short vibrissae), and with their eyes and ears closed and their toes fused. Females crouch over their litter for at least the first few days, blanketing them with her patagium and leaving the nest only for short periods, if at all. When she does leave, the mother carefully covers them with nest material to keep them warm. At two weeks old, the pup's toes are separate and their teeth are beginning to erupt. They are fully furred by three weeks old and their eyes open at around four weeks. By four weeks old they are sampling solid food and becoming increasingly active inside the nest. At five weeks they begin to cautiously venture outside the nest, gradually exploring and becoming stronger and more agile and confident. They begin gliding at around seven weeks old. Weaning begins gradually around six weeks of age and is completed in a few weeks. After that, the young follow their mother on her foraging trips. Juveniles typically breed for the first time as yearlings at ten to eleven months old.

BEHAVIOUR

Southern Flying Squirrels often share nests, especially in the winter to conserve body heat. Larger aggregations form as temperatures drop. The largest number on record is 50 squirrels in one tree cavity, and squirrel balls of one or two dozen are not uncommon. The aggregations disband in spring as the temperatures moderate, but reassemble in mid-June for a month or so, suggesting that there is a mating component to the assembly; perhaps to gather potential mates and to allow the males a chance to determine the level of oestrus of each female. After breeding, females become increasingly solitary as their pregnancy advances. A week or two before parturition, the female isolates herself in a drey or a tree cavity, which she prepares by filling with soft nesting material. A pregnant females becomes quite aggressive. No other flying squirrel is permitted near the chosen nursery site. She will stamp her hind feet and lunge at an intruder, clearly threatening a full attack if necessary. She also will become the "boss" of a feeder should

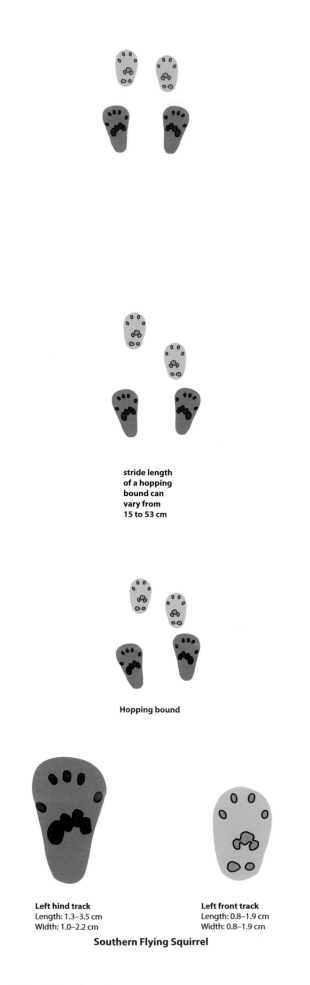

stride length of a hopping bound can vary from 15 to 53 cm

Hopping bound

Left hind track
Length: 1.3–3.5 cm
Width: 1.0–2.2 cm

Left front track
Length: 0.8–1.9 cm
Width: 0.8–1.9 cm

Southern Flying Squirrel

one be nearby. All other squirrels may eat only with her tacit approval, usually when she is either busy eating herself or is no longer hungry. There are no other indications that males or non-reproductive females claim or defend a territory. Maternal females maintain several possible nursery nests and move their pups should the first nest be disturbed or become fouled or heavily parasitized. They carry the pups one at a time by grasping their belly fur. The young instinctively curl around under her throat and cling to the back of her neck if they are older than about three weeks. She will even glide while carrying a pup. A diligent and careful mother, she will not stop returning to the former nest until all the young are moved and she has gone on her last trip to find an empty nest. Pups usually disperse from the nursery nest only when their mother forces them out. If they are the product of a spring litter, she may push them out as she nears parturition with her summer litter, some time in July. A summer litter may remain with their mother through their first winter. Southern Flying Squirrels memorize their routes and will commonly launch into a glide without looking at their possible landing. A favourite landing tree that falls or is cut down may result in several missed landings before the squirrel learns that it is not there anymore. Southern Flying Squirrels are quite wary, commonly running around to the back of the tree after a landing to foil a potential predator and instantly responding to another flying squirrel's alarm call. Males detect a female coming into heat, probably by odour. They pursue her, vibrating their hind end up and down and kicking backwards with their hind feet. Females may squeal and retreat if they are not yet receptive. The male remains nearby, persistently testing her readiness until she accepts his advances. The breeding behaviour of Southern Flying Squirrels is poorly understood and has mainly been deduced from captive animals. We do not know if the female will accept more than one mate or whether there is any competition between the males. Southern Flying Squirrels are often described as "lively" during the dark hours, as they are constantly in motion, especially in late summer and autumn as they gather food for the winter. The opposite is true during the daytime, when they rest and sleep in their nest.

VOCALIZATIONS

Pups produce a squeaking distress call and will squawk loudly if hurt. Their distress call is very high-pitched and may be partly ultrasonic. Squeaking pups removed from a nest can draw a maternal female to brave almost any threat to retrieve them. A squeal is the adult's distress call. Females will squeal at a male if he approaches her before she is ready to mate. Adult squirrels will squeal if their nest is invaded or if they are injured or captured. The typical flying-squirrel vocalization is a "chip" that sounds like that of a small bird. A louder version is also produced that appears to be an alarm call. The usual response after hearing this alarm call is to run up a tree and freeze.

SIGNS

Southern Flying Squirrels commonly use a tree cavity that is unsuitable for nesting as a latrine. Long cavities may develop a thick humus layer (over 0.5 m thick) or spill out if the cavity has an open bottom. Other cavities are used as a temporary shelter while the squirrels consume a large meal like an acorn or hickory nut. The remnants of the nut casings pile up in the cavity, sometimes filling it. Southern Flying Squirrel tracks are smaller than those of Northern Flying Squirrels, but otherwise are very similar. However, the Southern Flying Squirrels most commonly use a hopping gait rather than the typical bound of the Northern Flying Squirrel. Dreys are about 30 cm in diameter and are spherical.

REFERENCES

Bowman et al. 2005; Elbroch 2003; Holloway and Malcolm 2007; Lavers et al. 2006; Merritt et al. 2001; Petersen, S.D., and Stewart 2006; Rezendes 1992; Sollberger 1940, 1943; Stapp 1992; Stapp and Mautz 1991; Stapp et al. 1991; Stone, K.D., et al. 1997; Taulman 1998; Taulman and Seaman 2000; Wells-Gosling 1985; Winterrowd et al. 2005.

Hoary Marmot (*Marmota caligata*)

Hoary Marmot

also called Rockchuck

FRENCH NAME: **Marmotte des Rocheuses**

SCIENTIFIC NAME: *Marmota caligata*

This is our largest marmot. It is associated with rocky talus slopes in the western mountain ranges, where boulders shelter its burrow entrances.

DESCRIPTION

Hairs on the back and shoulders are white-rooted with black tips and on the rump are creamy with more extensive black or brown tips. The species is named for the "mantle" of white hairs over the shoulders, where the black tips on the hairs are either not present or very narrow. Fur on the throat, chest, and belly is yellowish white to yellowish grey and sometimes light brown. Around the nose and on the top of the snout between the eyes the fur is white. The feet, ears, eyes, and top of the head are blackish and the tail and cheeks are brown to light brown. Hoary Marmots have the typical bushy, short, blunt marmot-type tail. They are strong, stocky animals with short legs. The general impression upon viewing this animal in the wild is of a light-grey marmot with a dark band across its snout and a dark tail. Dark brown or black (melanistic) individuals occur from time to time in some northern populations, and significant variation of colour and extent of white markings exists even within colonies, such that some animals may be individually recognizable. Hoary Marmots undergo a single moult each summer, with adult males and non-reproductive adult females completing the moult first, followed by yearlings and two-year-olds, then by females with litters, and finally by young of the year. The dental formula is incisors 1/1, canines 0/0, premolars 2/1, and molars 3/3, for a total of 22 teeth.

SIMILAR SPECIES

Woodchucks are smaller, typically darker, and do not occur in the western mountains. Yellow-bellied Marmots are also smaller and darker in overall appearance and their distribution overlaps that of Hoary Marmots in southern British Columbia and extreme southwestern Alberta. Apart from the size difference, Yellow-bellied Marmots have a yellowish-orange throat, front legs, and belly, and their forehead and feet are brown.

SIZE

Males are somewhat larger than females. The following measurements apply to adult animals within the Canadian portion of the range. Weight fluctuates dramatically through the year, with a low in spring and a peak in late summer just before hibernation.
Total length: 630–800 mm; tail length: 170–252 mm; hind foot length: 87–114 mm; ear length: 19–38 mm.
Weight: 3283–6804 g (top weight may be as much as 9000 g); newborns, unknown.

RANGE

Hoary Marmots occur only in western North America in mountainous regions from central Alaska to Washington State, and eastward as far as the border of Yukon and the Northwest Territories in the north to western Alberta and western Montana in the south.

ABUNDANCE

These conspicuous marmots are considered to be of no conservation concern in British Columbia, where the bulk of their Canadian range occurs.

ECOLOGY

Hoary Marmots are found at high elevations (except in Alaska, where they may reach the coastal foothills), where they occupy alpine meadows, open forests, and vegetated rock fields, always near scree slopes with large boulders. They are capable diggers with strong legs and large claws, which they use to construct burrow systems, using the boulders for shelter and protection. They dig with their front feet and kick the loose soil behind them with their hind feet. They may then turn and push the pile of soil with their front paws or chest. Small rocks are carried out of the hole in their teeth to be dumped in the throw mound. Three different types of burrow

Distribution of the Hoary Marmot (*Marmota caligata*)

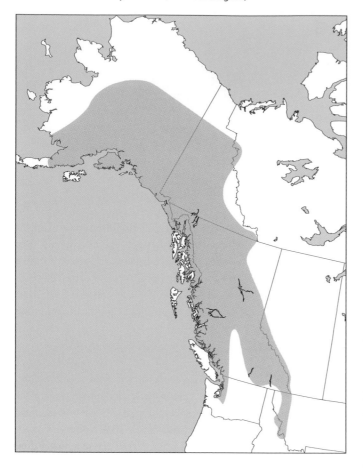

Marmots may live in the wild, but females can reproduce in their ninth year. The expected lifespan may be 13–14 or more years.

DIET

Hoary Marmots are herbivores whose diet includes grasses, sedges, forbs, willows, lichens, and mosses. Alpine forbs such as lupines, betony, Yellow Glacier Lily, Western Anemone, paintbrush, False Indian Hellebore, vetches, and fleabanes are important and often preferred foods. Snow is occasionally eaten, but Hoary Marmots rarely drink, as they satisfy most, if not all, of their water requirements from the vegetation they consume.

REPRODUCTION

Breeding occurs in May soon after the animals emerge from hibernation. The mating system appears to be facultative. In other words, females in small colonies with only one adult male available have no choice of sire for their litters and are by circumstances monogamous. Females with more choice, either within or between social groups, may be polyandrous. Adult females produce a single litter in a year and commonly take a year's hiatus between litters. In some colonies, this hiatus can extend to three or even four years. Juveniles breed for the first time as three-year-olds. An average litter is 3 to 4 young, produced following a pregnancy of about 30 days. Births occur in an underground nesting chamber and the young are born helpless and essentially hairless. They grow quickly and weaning occurs 25–35 days later, shortly after they first appear above ground.

BEHAVIOUR

As these marmots are colonial they have many interactions with each other and have developed an extensive repertoire of social behaviours. Individuals moving about on the trails within the colony territory greet each other and check identity by nose-to-nose sniffing, followed by sniffing in the region of the facial glands. This may sometimes develop into a wrestling match, where the two animals rear onto their hind legs, thrust their chins in the air, and push at each other's chest with their front feet. Such a wrestling match can result in both animals rolling together down the hill. A squeal from either participant will cause a pause and sometimes cessation of the action. This behaviour appears to be play and is non-aggressive and enjoyed by all members of the colony. All adults regularly rub their cheek glands onto rocks and other physical obstructions within the colony's territory and especially near burrow entrances

are produced. Hibernacula may be separate burrows from those used over the summer and are typically located on an exposed ridge with eastern to southern exposure, where the snow does not build into a deep drift and where it melts early. Typically, each colony has only one hibernation burrow. Placement is crucial for the colony's survival as the hibernacula must be adjacent to a reliable early spring food supply. Two other types of burrows – summer sleeping burrows and refuge burrows – are more numerous and scattered throughout the colony. Sleeping and hibernation burrows are more extensive, typically over 3.5 m in length, and are distinguished from the refuge burrows by a throw mound of loose soil at their entrance. Refuge burrows are short (less than 1.5 m long) and used mainly as a temporary bolt hole. Like all the marmots, Hoary Marmots hibernate. Since the whole colony often hibernates together, all age classes enter hibernation at about the same time, usually in early to mid-September. The colony emerges from hibernation over the course of a few days in early to mid-May, following about eight months of torpor. Hoary Marmots are diurnal, with activity peaks in early morning and evening, and except on hot days the midday lull is commonly spent resting, grooming, and sunbathing above ground. Predators include mammals such as the Grey Wolf, Coyote, Brown Bear, and Red Fox. Among the raptors, only a Golden Eagle is large enough to take a juvenile marmot. We do not know how long Hoary

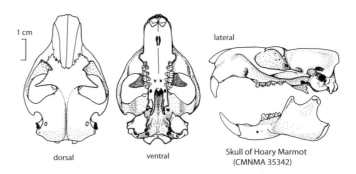

1 cm

lateral

dorsal

ventral

Skull of Hoary Marmot
(CMNMA 35342)

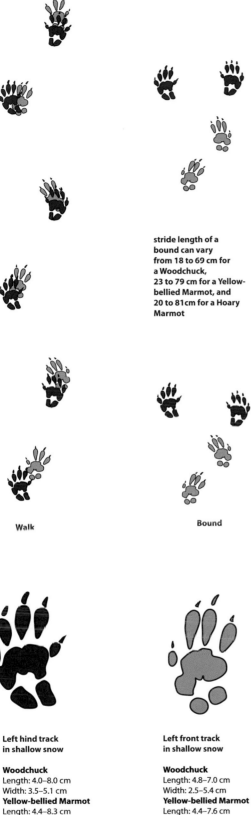

stride length of a
bound can vary
from 18 to 69 cm for
a Woodchuck,
23 to 79 cm for a Yellow-
bellied Marmot, and
20 to 81 cm for a Hoary
Marmot

Walk

Bound

**Left hind track
in shallow snow**

Woodchuck
Length: 4.0–8.0 cm
Width: 3.5–5.1 cm
Yellow-bellied Marmot
Length: 4.4–8.3 cm
Width: 4.1–6.4 cm
Hoary Marmot
Length: 5.1–8.6 cm
Width: 4.4–6.4 cm

**Left front track
in shallow snow**

Woodchuck
Length: 4.8–7.0 cm
Width: 2.5–5.4 cm
Yellow-bellied Marmot
Length: 4.4–7.6 cm
Width: 3.8–6.0 cm
Hoary Marmot
Length: 5.4–8.0 cm
Width: 3.2–6.7 cm

**Hoary and Yellow-bellied
Marmots and Woodchuck**

to scent-mark the area. The adult male of the colony marks more extensively around the periphery of the territory, usually on a regular, if not daily, basis. Non-members of the colony are chased from the territory by the first adult that detects the intrusion.

Within a colony there is little territorial behaviour exhibited, except by females with litters, who will defend their nesting burrow from the other marmots. Apart from her burrow, all the animals in the colony amiably share the burrows and available food. Females can be very protective of young. One female, sunning with two recently emerged young, was seen to push them off the rock they were sharing to remove them from the path of a suddenly approaching raptor. Most of the chases within a colony involve a female protecting her nesting burrow or newly emerged young from other colony members. A colony is typically composed of a male and two adult females and their offspring up to two to three years old, but may be much larger, including several adult males and many adult females and associated young. Usually two-year-olds of both sexes disperse from their natal territory when their mother has a new litter, but if she does not, they may remain another year. Hoary Marmots rarely venture away from their colony's territory, dispersing juveniles being the main exception. The adult females within a colony develop a dominance hierarchy, with one being more likely to breed earlier and protect her burrow more effectively from the other.

Hoary Marmot movements are usually leisurely as they amble around foraging, keeping their body low and close to the ground, stopping to check for danger or to graze. They frequently squat onto their hind legs to leave the front feet free to pull a flower down or manipulate a stem or leaf prior to consumption. While moving about, each marmot stops frequently to look around and then quickly flips its fluffed-up tail over its back and then down as it moves on. This visual signal appears to inform other animals in the colony of the individual's whereabouts. Hoary Marmots will break into a gallop, especially if a predator is approaching and they are hurrying back to the safety of a burrow. Often as the adults retire for the day, they will enter a sleeping burrow with a mouthful of fresh bedding material to replace that used previously. As hibernation approaches, they begin sleeping and taking fresh bedding into the hibernation burrow, appearing above ground for shorter periods; all the colony members begin foraging together more regularly.

VOCALIZATIONS

The most commonly heard call produced by the Hoary Marmot is a loud, high-pitched alarm whistle. This vocalization is usually produced while the animal is sitting upright, balanced on its hind legs, and originates from air vibrating in the throat rather than the lips. A single call is the norm, but sitting stiffly on an entrance mound, a Hoary Marmot watching a potential predator (including a human) may perform an alarm-whistle sequence that increases in repetition rate and diminishes in volume as the predator approaches. Individuals may squeal when wrestling, playing, or fighting. Growls and barking growls accompany play, fights, or chases. Hoary Marmots will sometimes emit a low-volume "chuck" sound when they are uneasy or aroused.

SIGNS

Toe 1 (the thumb) on the forefoot is rudimentary and does not register in a track. Heel pads on the front feet are roughly circular. Hind feet have five toes. All the marmots in Canada (Hoary, Yellow-bellied, Vancouver Island, and Woodchuck) have very similar tracks and track patterns that vary mainly by size (see accompanying illustration). Tracks in snow are possible, especially in spring. Hoary Marmots deposit some of their faeces in above-ground latrine sites, so their scat is seen more frequently than other marmots'. Scat ranges from 10–24 mm in diameter and 38–90 mm long. Fresh scat is shiny black, but weathering roughens the surface and mutes the colour. Pellets are often blunt on one end and finely pointed at the other. Chains of pellets connected by a thin thread of faecal material can occur that combine the length of many individual pellets. Latrine sites are commonly found near well-used perches and under overhangs. Burrow entrances may be 15–23 cm high by 15–31 cm wide. Hibernation and sleeping burrows commonly have throw mounds in front of them, while refuge burrows (bolt holes) are less noticeable, typically with smaller openings and no throw mound. Runs often radiate from a throw mound towards a foraging area or favourite perch. Because Hoary Marmots are the most gregarious of the marmots found in Canada, the burrow entrances of a colony will typically be fairly close together.

REFERENCES

Armitage 2003; Barash 1989; Blumstein 1999; Gray 1967, 1975; Kyle et al. 2007b; Nagorsen 2005a; Slough and Jung 2007; Taulman 1990a, 1990b.

Yellow-bellied Marmot

FRENCH NAME: **Marmotte à ventre jaune**

SCIENTIFIC NAME: *Marmota flaviventris*

These colourful marmots are about the same size as their close relative, the Woodchuck, but unlike Woodchucks are usually found living in colonies. They also enjoy the distinction of being the longest studied marmots – a group of colonies in Colorado have been closely monitored since 1962.

DESCRIPTION

Yellow-bellied Marmots are small marmots with a dark head and reddish-brown fur on their backs that is grizzled with tan, light grey, and black. There is a patch of white fur around the dark nose and a tan or white patch on the snout in front of the eyes. The distinguishing feature, and the reason for both its common and scientific names, is the bright yellowish fur on the throat, sides of the neck, and belly. The tail is short, with a mixture of buffy, reddish, and black hairs that look reddish brown from a distance. Feet are

Yellow-bellied Marmot (*Marmota flaviventris*)

light brown to yellowish brown. Melanistic animals are rare. Yellow-bellied Marmots undergo a single annual moult each year, some time over the summer, led by adult males, followed by non-reproductive females, then yearlings, then reproductive females, and finally young of the year. Usually all but the juveniles are complete by midsummer. The dental formula is incisors 1/1, canines 0/0, premolars 2/1, and molars 3/3, for a total of 22 teeth.

SIMILAR SPECIES

Only the Woodchuck and Hoary Marmot occur within the limited Canadian distribution of Yellow-bellied Marmots. Woodchucks are darker overall without white markings on their face and have black or dark brown feet and tail. They also have a reddish to brownish throat and belly and usually tan-coloured fur along the sides of the neck. Hoary Marmots are considerably larger, greyish overall with black feet and a white throat and belly.

SIZE

Males are somewhat larger than females. The following measurements apply to adult animals within the Canadian portion of the range. Weight fluctuates dramatically through the year, with a low in early spring and a peak in early autumn.

Total length: 520–670 mm; tail length: 125–184 mm; hind foot length: 60–89 mm; ear length: 25–37 mm.

Weight: 2380–3900 g; newborns, around 34 g.

RANGE

These marmots occur from south-central British Columbia and southern Alberta, southwards to central California and New Mexico. They are widely distributed in the western United States, but become isolated on "montane" islands in southern portions of their range. Yellow-bellied Marmots are expanding slowly northwards in Alberta.

ABUNDANCE

Colony size varies from year to year based primarily on weather and its effect on reproductive success. The British Columbia government does not consider the population of Yellow-bellied Marmots in the province to be of conservation concern. In Alberta, the

Distribution of the Yellow-bellied Marmot (*Marmota flaviventris*)

species is common enough in some areas to be an agricultural pest, but overall it is considered uncommon, although numbers appear to be increasing.

ECOLOGY

In British Columbia and Alberta, Yellow-bellied Marmots live at lower elevations on mountain sides or on the edges of valleys. They inhabit vegetated talus slopes or rock outcrops alongside meadows. Large boulders appear to be an essential component for a good marmot habitat. As the range becomes more southerly, Yellow-bellied Marmots are found higher in the mountains, until in the far southern portions of their range they are found only at high elevations. Like all marmots, they are diurnal, with peaks of activity in the morning and early evening, except in early spring when they are most active in the warmest part of the day. They avoid hot summer afternoons by retreating to their cool burrows, which are typically located near or under boulders or trees. There are three types of burrows constructed and used by Yellow-bellied Marmots. Refuge burrows are simple and short, scattered throughout the home territory and are used as bolt holes to escape predators. Home and hibernation burrows are more extensive with several entrances, often several short blind passages, and a nest chamber, and may be up to 4.5 m long. Construction of both of these burrow types creates a characteristic throw mound at their entrance. A home burrow may

also be used for hibernation or a hibernation burrow may be single use and located some distance from the home burrows.

Marmots seal themselves into their hibernation chamber with a mixture of soil and faeces as a protection from small predators. Yellow-bellied Marmots are true hibernators as their body temperature, heart rate, respiration rate, and physiological processes slow to very low levels. They tend to hibernate in social groups; adult males typically hibernate with one or more adult females from their harem, and mothers usually hibernate with their offspring. They begin hibernating in early to mid-September in most parts of their range. Over the course of about two weeks, adult males depart first, followed by non-reproductive females, then reproductive females, and finally juveniles. Spring emergence is determined by elevation and local temperature. Again, over about a two-week period, adult males emerge first, followed by adult females, then by yearling males, and finally by yearling females. An early spring is usually beneficial, as the young are born earlier and have more time to accumulate body fat before hibernating. The heavier a juvenile is when it enters hibernation, the more likely it is to survive until spring. This is why juveniles are last to hibernate, needing every opportunity to accumulate body fat. Conversely, a late spring can stress the reserves of even the adult animals, to the point that no females retain sufficient fat to allow them to reproduce that year. In the colder Canadian range, the marmots emerge in late April to mid-May, but it is thought that in south-central British Columbia, where the climate is warmer, they may emerge up to a month earlier. In the central United States, Yellow-bellied Marmots are emerging more than a month earlier than they did in the early 1980s, apparently in response to warmer spring temperatures due to climate change.

Although the sex ratio at birth is equal, females soon come to outnumber males. All young males disperse, but only some of the females disperse. Females are thus exposed to fewer hazards. Also, females are longer-lived. Although not proven, males may be more susceptible to stress-related deaths. Males live for a maximum of about 9 years, but females are known to live to be as old as 15 years. Predation and hibernation mortality are the two main mortality factors. Predators include the Coyote, American Marten, American Black Bear, American Badger, Long-tailed Weasel, and Golden Eagle.

DIET

Yellow-bellied Marmots are generalist herbivores with food choices largely based on availability, nutritional quality, and lack of secondary compounds that make the plant distasteful, poisonous, or poorly digestible. Grasses and forbs make up the majority of food eaten, with the addition of seeds in late summer. Alfalfa and clover are particular favourites. These marmots will climb trees to forage on leaves. Occasional reports of cannibalism, infanticide, and consumption of indefensible young of other species, or of carrion, reflect the general rodent propensity to infrequently consume protein from animal sources. Although they will drink free-standing water, most of their water requirements are satisfied by the plants that they eat. During spells of hot weather, Yellow-bellied Marmots

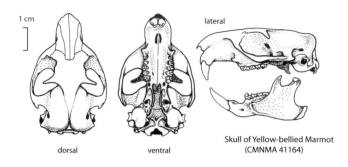

1 cm

lateral

dorsal   ventral

Skull of Yellow-bellied Marmot
(CMNMA 41164)

retreat to their burrows and enter periods of torpor to wait out the heat and to conserve water.

REPRODUCTION

A small percentage of Yellow-bellied Marmots reproduce as two-year-olds, but the vast majority delay first reproduction until they are three or older, sometimes as old as six years. These young females, although physically capable, are socially suppressed and prevented from breeding by older, more-dominant females, even if those females are their mothers. Breeding occurs within two weeks of spring emergence, and following a gestation of around 30 days, 3–8 young (average 4) are born in an underground nest chamber. The sex ratio of the offspring is equal. Growth of young is rapid and they appear above ground at about 30 days old and are weaned soon after. Adult females produce a litter on average every two years, although they are capable of breeding every year if well nourished. Reproductive senescence appears to occur in females beyond the age of nine years.

BEHAVIOUR

Yellow-bellied Marmots are variably colonial, depending on the size of the appropriate habitat. Large habitat patches typically support large colonies, while small areas of habitat or patchy habitat support smaller ones, made up of smaller family groups. The structural complexity of social groups falls along a continuum from a single marmot to a solitary female and her current year's offspring, to a male and female and their current year's offspring, to a male and several females and their offspring, to a large colony made up of a patchwork of groups of 1 to 3 males with their harems of up to 12 females and numerous offspring. The adult male establishes a territory of around 1 ha on average (range, 0.06–47.5 ha), which he rigorously defends from other males. Most of the females in a harem are of the same matriline and access to foraging areas is based on kinship, as only related females and their offspring will share a foraging patch. Adult females that are closely related tend to retain amicable relationships, but once the degree of relatedness drops, the group members becomes more antagonistic towards each other and may split up. Behaviour towards non-related females living within a male's territory is predominantly antagonistic, just as it would be if the females lived on separate territories. Solitary animals and very small family groups settle around the periphery of a large colony, where they are more vulnerable to predation. These are often young animals trying to establish themselves. Should a harem master

within the colony die or leave his territory, one of these neighbouring males is likely to take his place. All age classes of Yellow-bellied Marmots scent-mark their territory using secretions from their cheek glands, but adults and especially harem masters are more persistent and consistent markers. They rub their cheeks on boulders or other obstructions or even on the soil by turning their head and rubbing their cheek against the ground. These markings are thought to identify the territory against intruders. They also help juveniles delineate their safe zone and prevent them from wandering onto an aggressive neighbour's domain. Burrow location and occupancy is also advertised by scent marking. Greeting behaviour, conducted by members of the colony as they meet each other, involves a lengthy sniff at the cheek glands and often tail flagging (flipping the tail up and down). This may not always be just an amicable behaviour; it may lead to chases or be the prelude to sexual activity. Males are the territory holders and they vigorously defend their home range from rival males. Reproductive females defend their nesting burrow from other marmots and all females react antagonistically towards non-related females. Chases are most common, but fights may occur, although injuries are rare.

About 25% of each 24-hour period is spent above ground. Aboveground activities include foraging, resting, sunbathing, mating, and interacting with members of the family group or neighbours. Vigilance or alert behaviour is part of everyday life as foraging, resting, and otherwise active animals regularly lift their heads to check their surroundings. Females with young are especially vigilant. Yellow-bellied Marmots commonly climb trees to gain a wider view. Adult males are aggressive, especially towards other males more than about one year old, even their own sons, so most juveniles disperse as yearlings. A few of the young females get recruited into the harem and stay. The dispersing youngsters are very vulnerable, as they may travel over 15 km in search of a suitable unoccupied location, and, without access to a refuge burrow during that time, they run a high risk of predation. If there are regions on their natal territory where the juveniles can avoid their aggressive sire, and their mother does not produce a consecutive litter, the juveniles may stay until they are two-year-olds or sometimes even three-year-olds. They spend at least one winter hibernating with their mothers.

VOCALIZATIONS

Yellow-bellied Marmots whistle, "chuck," trill, scream, and chatter their teeth. Whistles are categorized into six different subtypes and are usually used to signal alarm or a perceived threat to the colony. Although Yellow-bellied Marmots were formerly thought to use different alarm calls to signify different types of predator, recent studies indicate that the calls vary to indicate the degree of risk involved, rather than the predator type. The most common alarm call is a brief, single-note whistle that may be repeated many times. Both juveniles and adults will produce this call, but those from small pups are noticeably higher pitched. These marmots also recognize and respond appropriately to alarm calls made by other animals within their environment. If a nearby Golden-mantled Ground Squirrel or Least Chipmunk sounds an alarm call, all the Yellow-bellied Marmots within hearing range will look up and then take appropriate action. "Chucks" are low-volume alarm calls usually produced when the animal is only mildly alarmed. Trills begin with an alarm whistle. The marmot then quickly increases the whistle rate and modulates the frequency. Trills occur during normal predator-detection situations and especially if the marmot is being pursued, or as it disappears into its burrow during pursuit. Occasionally marmots trill in social situations, particularly when being pursued by a rival. Screams are responses to fear or excitement and tooth chattering is used to threaten.

SIGNS

Yellow-bellied Marmots, unlike other marmots, are not particular about where they deposit scat. It may be found along travel routes and near burrows throughout a colony, but not concentrated in a latrine area. Fresh scat is initially a shiny black cylinder, often with one blunt end and the other sharply pointed, about 1–2 cm in diameter and 3.5–5 cm in length. The scat becomes frosted looking and brownish as it weathers. Scat from succulent vegetation sometimes forms into chains with a slender thread of material holding one scat to the next. Burrows with throw mounds in front have entrances 15–23 cm tall and 15–28 cm in width. Entrances to refuge burrows typically do not have a throw mound and are generally smaller. Runs commonly radiate from burrow entrances to feeding areas or favourite viewing or sunning perches. See the "Signs" section of the Hoary Marmot account for information and illustrations of marmot tracks.

REFERENCES

Armitage 1991, 2003; Barash 1989; Blumstein and Armitage 1997; Blumstein et al. 2004; Borrego et al. 2008; Brady and Armitage 1999; Elbroch 2003; Frase and Hoffman 1980; Inouye et al. 2000; Lenihan and Van Vuren 1996; Nagorsen 2005a; Oli and Armitage 2003; Petterson 1992; Salsbury and Armitage 1994; Schwartz, O.A., and Armitage 2005; Shriner 1998; Stallman and Holmes 2002; Van Vuren and Armitage 1991.

Woodchuck (*Marmota monax*)

Woodchuck

also called Groundhog

FRENCH NAME. **Marmotte commune**

SCIENTIFIC NAME: *Marmota monax*

This rodent is the smallest of our marmots and the most common, widespread, and best known. Sometimes a serious pest of gardens and crops, it is fortunately fairly easy to live-trap, using a strawberry, chunk of apple, or peanut butter mixed with rolled oats for bait.

DESCRIPTION

Most Woodchucks in eastern North America have black feet and tail, a dark nose, often with a somewhat lighter region of fur around it, and a darker face and top of the head. Neck, back, and belly are rusty brown, often with some grizzled grey hairs on the shoulders and back. Animals in the west are more brownish (less rufous), and Woodchucks in Alaska have pinkish feet. Ears are short and rounded at the tip and eyes are small and dark and placed high along the top line of the skull, about midway between the ears and nose. The ears can be compressed to block the auditory canal when the animals are swimming or digging. Colour variations such as black, dark brown, blond, and even white are not uncommon. Woodchucks undergo a single annual moult that takes place over the course of about a month between May and August, depending on the individual's age – young of the year and reproductive females moult late, adult males and non-reproductive females moult early, and yearlings some time between. Moulting begins at the tail and at the nose simultaneously, but proceeds more quickly from back to front. Consequently, the last area to grow new hair is typically over the shoulders; however, moult patterns are highly variable. The dental formula is incisors 1/1, canines 0/0, premolars 2/1, and molars 3/3, for a total of 22 teeth.

SIMILAR SPECIES

Woodchucks may be mistaken for other marmots such as Hoary or Yellow-bellied Marmots in regions where they co-occur. The Hoary Marmot has a paler, grizzled greyish white coat and a large white patch on its forehead. The range of Yellow-bellied Marmots is mostly separate from Woodchucks, except in the upper Fraser Valley and Chilcotin regions of British Columbia. Yellow-bellied Marmots have dark cheeks and extensive white patches on the snout and sides of the neck. In British Columbia, young Woodchucks may be mistaken for Mountain Beavers, which are found only in coastal regions outside the range of Woodchucks. In addition, Mountain Beavers have longer claws and a much shorter, stubby tail.

SIZE

Males are somewhat larger than females. The following measurements apply to adult animals within the Canadian portion of the range. Weight fluctuates dramatically through the year, with a low in early spring and a peak in early autumn.

Distribution of the Woodchuck (*Marmota monax*)

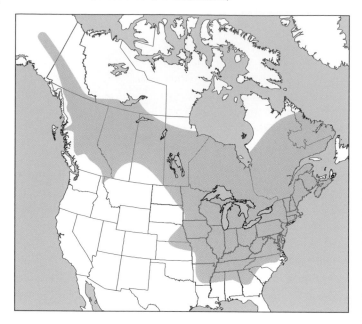

Total length: 388–654 mm; tail length: 95–180 mm; hind foot length: 58–89 mm; ear length: 20–37 mm.
Weight: 1630–4310 g; newborns, 23–33 g.

RANGE

Woodchucks occur from central Alaska to northern Quebec and Labrador and southwards to Georgia and Arkansas. They are scarce or absent in the prairies, although they may be seen in parklands around the periphery and sometimes far up wooded river valleys that penetrate into the prairies. Although they have not dispersed to larger maritime islands such as Newfoundland, Prince Edward Island, or Anticosti Island, they are found on Manitoulin Island in Lake Huron and have recently reached Cape Breton Island, no doubt due to the causeway.

ABUNDANCE

This species is common and frequently plentiful throughout much of its range, especially those portions near human habitation and agriculture. It is most rare in its original forest habitats (see the "Ecology" section following for more information).

ECOLOGY

Before European settlement in North America, Woodchucks were uncommon in forests and forest clearings. Deforestation for cultivation and human habitation has greatly benefited this species. The more northerly populations are sparse and probably reflect the pre-seventeenth-century ecology of the species. Estimates of population density range from 0.01 per ha in northern forests or poor habitats to 4.8 animals per ha in rich farmland. Typically, Woodchucks occupy low-elevation woodland-field edges, preferring to forage in the fields and hibernate in the woodlands or fence rows. They are generally diurnal, with activity peaks around dusk and dawn. This

pattern is altered somewhat in early spring when they first emerge from hibernation, as the males begin to explore the burrows of neighbours in search of receptive females and commonly do so at night. Some Woodchucks in the far southern portions of the range are reported to remain active through the winter, but most, and certainly all in Canada, spend the winter in true hibernation. Their normal 37°C body temperature declines to as low as 2°C–4°C and their heart rate drops from an average of 130 down to as low as 5 beats per minute. A breath may be taken as infrequently as once every six minutes. Males maintain a higher body temperature during hibernation than do females. In the east most Woodchucks enter hibernation in September to mid-November, with juveniles departing last. Adults acquire about a 2 cm-thick layer of fat by early autumn as a prelude to hibernation. Apparently a high level of body fat is one of the cues to begin hibernation and the leaner juveniles need more time to acquire this essential fat, since much of their energy resources have been channelled into growth. Hibernation initially consists of four- to ten-day bouts of torpor followed by one to five days of arousal. As the season progresses, the length and frequency of the arousal periods decline and the length of torpor increases. Timing of emergence varies with latitude and local weather patterns, but is usually from late February (British Columbia) to March or April (Ontario and Quebec). Adult males emerge first, around three to four weeks before the females and juveniles appear. Following initial emergence, the animals will retreat back to their burrows to enter short periods of torpor during cold spells. Adult Woodchucks lose 25%–47% of their body weight during hibernation, but juveniles may lose a staggering 50% of their body weight over their first winter. It is no wonder that many do not live to emerge from hibernation.

The Woodchucks' sharp claws and stocky bodies make them superb diggers. They excavate with their front feet and use the hind feet to throw the dirt backwards. Sharp incisors make short work of obstructing roots. Their burrow systems vary from simple, usually new burrows by young animals, to quite complex for older Woodchucks or burrows that have been used by several generations. A burrow may have a single entrance or up to a dozen. Main entrances are marked by the throw pile of fresh dirt, while other bolt holes are

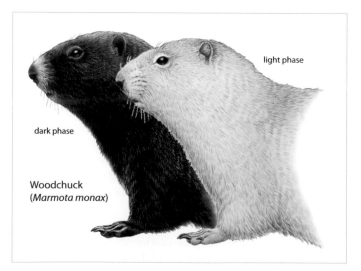

light phase

dark phase

Woodchuck
(*Marmota monax*)

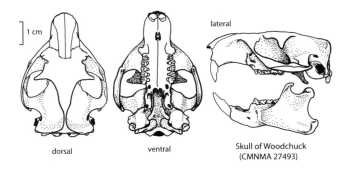

] 1 cm

dorsal   ventral   lateral

Skull of Woodchuck
(CMNMA 27493)

hidden and smaller, without the obvious mound of dirt in front of them. Burrows may be up to 9 m long and more than 1 m deep. Most contain a nesting chamber and a latrine, with the rest of the tunnels leading to main entrances or bolt holes. Summer may be spent in a burrow or series of burrows near prime foraging areas, while the winter burrow may be deeper and in a more protected site. A nesting chamber used for hibernation is usually higher than tunnel-level, lined with leaves and dried grasses, and is walled off from the rest of the burrow as a protection from small predators such as weasels that may enter while the Woodchuck is dormant and defenceless. Many species like rabbits, North American Opossums, skunks, mice, squirrels, and even the occasional Northern Raccoon and fox use Woodchuck burrows for shelter, while others such as insects, snakes, and lizards use them as handy hibernating sites, even if the Woodchuck is still in residence. Many old burrows are remodelled by foxes, Coyotes, or Grey Wolves for their own use. During their daily foraging, most Woodchucks stay within 30 m of their main burrow entrance. Dispersing juveniles and breeding males are notable exceptions. The size of home ranges varies with location and food availability, and is not always stable over the life of the animal. Adult males maintain ranges that are larger and overlap those of several females. Woodchucks are fastidious creatures that bury their faeces in an underground latrine or dig them into the throw mound. They are remarkably resilient to injury and unless killed outright are often able to run or drag themselves back to their burrows, where they may recover from even severe injuries. Woodchucks are capable swimmers and proficient climbers, commonly spotted resting or sunning on fence posts and easily climbing 6 m or more into treetops. Despite their agility going up, the descent is not nearly as graceful, as they must slowly back down. Swimming or climbing a tree may allow a Woodchuck to evade a predator if a bolt hole is not handy. Top speed is thought to be around 15 km/h for a limited distance. The oldest captive Woodchuck survived for nine years, eight months; however, life expectancy is the wild is typically four to five years. The oldest known wild Woodchuck lived 6.5 years. Predators include the Domestic Dog, Red Fox, Coyote, raptors, Bobcat, American Mink and other weasels, and of course humans. Death by motor vehicle impact is common and thousands are shot each year.

DIET

Woodchucks are classified as generalist herbivores. They eat a wide variety of plants depending upon availability and season. Alfalfa and clover are favourites, but grasses including grain crops, leaves, bark, and fruit of shrubs and trees, and many forbs are commonly eaten. Even June bugs, snails, and grasshoppers are consumed at times. Most of their water requirements are satisfied from dew and succulent vegetation, but they will occasionally drink if open water is available. It is possible, during periods of drought, for Woodchucks to allow their core body temperature to fluctuate somewhat in synchrony with the ambient temperatures, presumably to conserve energy and water.

REPRODUCTION

This species has a single oestrous period each year. Mating takes place in spring shortly after the females emerge from hibernation. Some yearling females breed in the spring following their birth, but most wait until they are two-year-olds to produce their first litter. Litters produced late in the season are most commonly those of yearling females. Following a pregnancy of about 30 days, a litter of 1–9 (average 3–4) helpless, pink young is born in an underground nest chamber. Females are thought to stand while nursing. By around four weeks old, the young are fully haired, with eyes and ears open, and are beginning to eat solid food delivered by their mother. At around five to six weeks old they first venture out of the burrow to feed and frolic near the entrance, and are weaned shortly after that. Male juveniles normally disperse from their mother's territory at about two months of age, but some female young may remain with her over their first winter, especially if there are few empty territories nearby.

BEHAVIOUR

Sociality among Woodchucks tends to be variable between populations. Some are strictly solitary except during the mating period and while raising young. Others form loose social groups comprising an adult male and one or two female kin groups, each consisting of an adult female, possibly a yearling (usually female) from last year's litter, and the young of the most recent litter. Adult males may be tolerated near gravid females or even infants in some populations, but not in others, so the parental role of the male varies, although it remains limited in either case. For the most part Woodchucks hibernate alone, but sometimes a female will share her hibernation chamber with young from her last litter, and occasionally two males have been found hibernating together. After ducking into a burrow to avoid a potential predator, these curious rodents will commonly peek out or re-emerge to observe the progress of the intruder. Oral glands in the corners of the mouth excrete a pungent secretion that is used to scent-mark their territory. Anal glands disseminate a characteristic musky odour, but the secretions cannot be sprayed as in skunks. Although the purpose of this scent is unclear, it appears to have a social communication function, as it has been noted just before the animals appear above ground to forage.

As explained already, females emerge from hibernation later than males, and when they do, they are already in oestrus. In early spring, adult males spend much of their time checking for signs of activity at neighbouring burrows, hoping to discover a newly emerged receptive female. When two males meet, a battle often ensues, and many males carry scars from these encounters. Most copulations occur near the mouth of the female's burrow and last for three to fourteen minutes. Multiple copulations with more than one male are likely, but no female will permit a male to mount if she is not in breeding condition. Typically, Woodchucks sit upright on their haunches when eating, to free their forefeet to manipulate the food. They will commonly reach up, pull the top of the plant down and eat it, leaving the bottom of the stem and some lower leaves intact. Smaller plants are yanked free and consumed completely.

VOCALIZATIONS

The loud, sharp alarm whistle is the most commonly heard call and it may be taken up and repeated by neighbours. Another call includes a sharp initial shriek followed by a diminishing series of chattering or warble notes. Although the purpose of this second call is unknown, it seems to play a more social function. During spring when fights between males occur, Woodchucks may be heard to scream in pain when bitten, especially if the aggressor bites with its large incisors and hangs on to its hapless opponent. Woodchucks grind their teeth together at times to denote anger or fear. This activity produces a low-volume, but distinct grating sound likely heard only by nearby animals. A cornered Woodchuck will grind and clatter its teeth at a predator and make charging movements towards it, apparently not at all intimidated by a much larger opponent.

SIGNS

Front feet have four toes (a tiny vestigial thumb, too small to register in a track, rests beside the heel pad), while hind feet have five. The track may superficially resemble that of a small Northern Raccoon due to the similar long toes; however, Northern Raccoons have five toes on both front and hind feet. Tracks in snow are possible, especially in early spring after a late snowfall. See the "Signs" section of the Hoary Marmot account for information and illustrations of marmot tracks. The blackish scat is rarely seen as it is usually deposited underground or buried in the throw mound. Burrow entrances vary depending on the size of the creator (13–30 cm in diameter), but inevitably the tunnel behind the entrance is narrower than the entrance. A large entrance and the turn-around area immediately behind it will likely be backed by a tunnel only 10–12 cm in diameter. Throw mounds can be as much as 0.6 m high and serve as sight and sunning posts. Occupied burrow systems have fresh dirt in the throw mound, as Woodchucks are inveterate diggers and burrow renovators. Repeated use may create trampled runways radiating from burrows to feeding areas. These runways are typically 10–15 cm wide and in lush growth may be canopied with overarching vegetation. Bite marks of Woodchucks are larger than those of rabbits and small rodents, due to the larger tooth size. Nipped stems commonly, but not always, show a 45° angle cut.

REFERENCES

Armitage 2003; Barash 1989; Elbroch 2003; Ferron 1996; Ferron and Ouellet 1991; Hamilton, W.J., 1934; Jones, J.K., et al. 1978; Kwiecinski 1998; Lloyd 1972; Meier 1992; Nagorsen 2005a; Rezendes 1992; Sinda Hikin et al. 1992; Slough and Jung 2007; Swihart and Picone 1991; Wilson, G.M., and Choate 1996; Woodward, S.M., 1990; Youngman 1975; Zervanos and Salsbury 2003.

Vancouver Island Marmot

FRENCH NAME: **Marmotte de l'île Vancouver**
SCIENTIFIC NAME: *Marmota vancouverensis*

These marmots are endemic to Canada and found only on Vancouver Island. They are the most endangered mammals in Canada and perhaps in the world. A survey in 2006 found only 35 remaining in the wild. A partnership of conservation organizations is conducting a captive breeding and release program to ensure the species' survival.

DESCRIPTION

Vancouver Island Marmots are large marmots, second in size only to Hoary Marmots. They exhibit the typical marmot features of a thickset body, short legs, and short, blunt, densely furred tail. Their pelage is primarily a rich dark-chocolate brown, which fades to a lighter brown by the time the summer moult takes place. Single white hairs may be scattered throughout the dark hairs. A large patch of white fur surrounds the dark nose and many animals have a small spot of white on their forehead. The chest and abdomen commonly has an irregular mottling of white hairs as well. Vancouver Island Marmots undergo a single moult each summer, with adult males and non-reproductive adult females completing the moult first, followed by yearlings and two-year-olds, then by reproductive females after their litter is weaned, and finally by young of the year. Incomplete moulting, which is common, may leave an animal looking mottled as patches of the paler old hairs and dark new hairs occur randomly over the body. The dental formula is incisors 1/1, canines 0/0, premolars 2/1, and molars 3/3, for a total of 22 teeth.

SIMILAR SPECIES

Vancouver Island Marmots closely resemble a melanistic Hoary Marmot, but their identification is uncomplicated, because there are no other marmot species found on Vancouver Island.

SIZE

Males are somewhat larger than females. Weight fluctuates dramatically through the year, with a low in spring and a peak in late summer just before hibernation.
Total length: 580–750 mm; tail length: 162–300 mm; hind foot length: 80–105 mm; ear length: 24–35 mm.
Weight: 3500–6900 g.
Newborns: Weight remains unknown. Despite captive breeding, newborns are so precious that parturient females are not disturbed for fear they will abandon their litter.

Vancouver Island Marmot (*Marmota vancouverensis*)

growth in the higher elevations and thereby reduce marmot habitat, but high-altitude clearcut logging has had the greatest obvious impact. This practice temporarily increases marmot habitat and thus encourages dispersing young to settle nearby, rather than travel to the neighbouring mountain. Unfortunately, this habitat is transient and as the vegetation regrows, the resident marmots become more vulnerable to predation, and thus reproduction rates are lower in clearcut colonies than in natural sites. Due to the drastic decline of Vancouver Island Marmots over the past 25 years, from 300–350 animals in 30 colonies in the 1980s to 50 animals in only 6 colonies in 2004, and 35 animals in 2006, this species is considered to be the most endangered mammal in the world. Clearly in danger of extinction, these marmots have been the focus of intensive scrutiny and efforts on their behalf. Four captive-breeding facilities have been established: at the Toronto Zoo, the Calgary Zoo, the Mountain View Conservation and Breeding Centre in Langley, BC, and the Tony Barrett Mount Washington Marmot Recovery Centre on Vancouver Island. Some success has already been achieved, with pups born each year since the year 2000. By 2004 there were 93 captive animals and 41 pups had been born. Reintroductions have already begun and numbers in the wild have begun to climb. There were an estimated 300–350 wild marmots in 2011. Whether these efforts will be sufficient to restore this animal to a healthy wild population remains to be seen. Disease and predation remain major obstacles to the species' survival. Vancouver Island Marmots were one of the first species to be listed as "Endangered" by the newly founded Committee on the Status of Endangered Species in Canada (COSEWIC) in 1979.

ECOLOGY

Most Vancouver Island Marmots live at elevations between 900 and 1500 m in clearcuts, alpine meadows, or avalanche bowls with a southerly to westerly exposure, numerous boulders, and rocky

RANGE

Vancouver Island Marmots were formerly more widespread through the middle of Vancouver Island, with most colonies along the east side of the central spine of mountain tops and many records of solitary dispersing animals in unlikely locations. The current range is greatly reduced to two disjunct portions. The core of the population is found in the Nanaimo Lakes area northwest of Lake Cowichan. Two isolated colonies continue to hang on almost 100 km northeast on Mount Washington. This is one of only five mammal species whose distribution occurs wholly within Canada. (The others are Ungava Collared Lemming, Ogilvie Mountains Collared Lemming, Richardson's Collared Lemming, and Maritime Shrew).

ABUNDANCE

Naturally rare because of their patchy and limited habitat, reasons for the Vancouver Island Marmots' historic and modern declines are not entirely clear. A warm period between 10,000 and 7000 years ago may account for their disappearance from lower-elevation locations, and recent warm trends may encourage tree

Distribution of the Vancouver Island Marmot (*Marmota vancouverensis*)

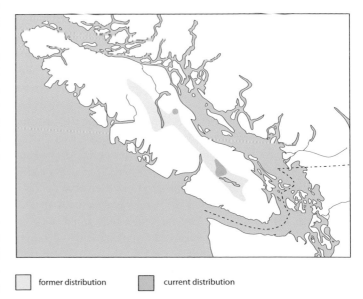

former distribution   current distribution

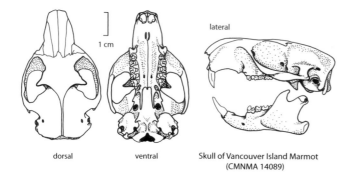

dorsal   ventral   Skull of Vancouver Island Marmot
(CMNMA 14089)

outcrops and friable soil to support burrow construction. Habitat of this sort is patchy and colonies are usually small (a few hectares) and scattered. A colony may be composed of one or more family groups – each of which consist of an adult male and one to three adult females, yearlings from the previous year, and juveniles from the current year. Most family groups are composed of only two adults, a male and a female, and most colonies have fewer than five adults. Juveniles remain in their natal colony until they are two or perhaps three years old, at which time most males and some females disperse. The remaining females are recruited into their natal colonies and the few remaining males either take over the family territory after the death of their sire, or settle in a neighbouring territory. Dispersing marmots typically travel long distances, such as down into the valley and up the next mountain to find a colony or a new region into which to settle. Journeys of 11 km have been recorded for ear-tagged individuals and early records of solitary animals wandering into urban or agricultural areas suggest that distances of more than 25 km may have been traversed. A single individual or a colony that establishes in nearby clearcuts is generally successful for a few years, but fails to thrive over the longer term. An eight-year study of clearcut colonies found only one juvenile survived to be recruited into the colony and then reproduce, and no adult females lived long enough to produce a second litter.

Vancouver Island Marmots dig three types of burrows. Refuge burrows are fairly numerous. They are short, simple burrows used as bolt holes. Home burrows are the most complex, with several entrances and long tunnels that terminate in a nesting chamber, usually located below a large rock for protection. These burrows are occupied during the summer and are used by all the family members, including birthing mothers. Family groups have one or more home burrows. Hibernation burrows are mainly single-purpose burrows, presumably dug down below the frost line, and used by the whole family group over the winter. Usually each family group has only one hibernation burrow. At the elevations occupied by Vancouver Island Marmots, winters are characterized by heavy snowfall, and hibernation burrows are located to take advantage of deep drifts that insulate and moderate the burrow temperatures. Burrow systems are commonly reused by the same family group and their descendents for decades. Several are known to have been occupied for over 30 years.

Like other marmots, these animals are mainly diurnal, with peaks of activity in early morning and late afternoon during summer, and in the warmer afternoons during spring. Hot summer afternoons are spent in the cooler comfort of their burrows. The cold period (seven to eight months per year) is spent in hibernation, wherein their body temperature, metabolism, pulse rate, and respiration decline considerably. In preparation for hibernation, the animals accumulate body fat that fuels their reduced activity for the whole winter. Most adults lose 26%–30% of their body weight over the course of the winter. Juveniles may lose up to 50%. Vancouver Island Marmots begin hibernation in early October and the entire family group shares the hibernation chamber after plugging the burrow entrance with grass and mud to protect themselves from predation. The latest date of a marmot sighting is mid-October. They emerge from hibernacula the following spring in late April or early May. The earliest marmot sighting in spring is mid-April. Male survivorship is lower than that of females, so despite a 1:1 birth ratio, adult females outnumber adult males. Predation is the major cause of death of adult marmots and is most intense in the clearcut colonies during late summer when the vegetation growth disrupts the marmot's sight lines and permits predators to approach undetected. Known predators include Golden Eagles, Cougars, and Grey Wolves, and at least 75% of mortality is caused by predation. Hibernation mortality, often caused by disease, is an additional significant cause of mortality, especially of young during their first winter. Approximately 54% of pups in natural sites survive their first winter, but only 43% in the clearcut colonies do so. Maximum known lifespan of both males and females in the wild is 10 years. In captivity, one female has lived to be 14 years old and a 10-year-old female was still actively reproducing.

DIET

The diet of Vancouver Island Marmots is primarily determined by availability. In early spring, they consume the first plants to emerge, usually grasses, sedges, some ferns, and lichens. By summer, their main diet is a wide variety of forbs, particularly lupines and Woolly Eriophyllum. Over 50 species of grasses and forbs are eaten by Vancouver Island Marmots. Clearcut colonies consume very different plants, as most that grow in alpine and subalpine sites are not found in clearcuts. These marmots thrive on their new diet and actually grow to larger size than those in the natural colonies, but their food choices have not been studied.

REPRODUCTION

Mating occurs below ground within about two weeks of spring emergence. Most females are four years old when they produce their first litter (range three to five years old). Well-nourished, captive-born females commonly breed before they are four years old. Litters are produced typically every two years, although it is possible for well-nourished females to breed in consecutive years. This does occur in the wild but is more common in captive-bred females. An interval of more than one year between litters is also possible. Litter size averages three pups (range 2–5). The gestation period is thought to be 30–32 days. Most young emerge from the nesting burrows for the first time during the first week of July and are weaned soon after.

BEHAVIOUR

These are very social animals. The most common social behaviour observed is the greeting. This nose-to-nose or nose-to-cheek ceremony occurs between all age classes and both genders, whenever one marmot encounters another. This allows individuals to verify identity by sniffing the distinctive odour of each other's cheek glands. It also likely has a dominant-subordinate recognition component, as most initiators of greeting ceremonies are subordinate animals. In family groups with multiple adult females, the females are usually related (typically mother and daughter), with the senior female being dominant. A typical dominance hierarchy within a family group in decreasing order of dominance is adult male, adult females (from oldest to youngest), two-year-old females, yearlings. All adult marmots are territorial, scent marking their territory regularly with secretions from their cheek glands by rubbing on burrow entrances, projecting rocks, or even the ground. Both sexes are active in defence of the family territory from intruders. Play is common among Vancouver Island Marmots. Most play (also called play-fighting) involves wrestling, upright pushing matches, lunging, or chasing. Play-fighting between adults may deteriorate into more aggressive behaviours. Typically, only dominant animals will lunge and chase while subordinate animals avoid these actions by running away. Vancouver Island Marmots are vigilant while above ground, regularly rising to sit on their hind legs to check out the area around the colony for predators. Commonly a resting adult perched on a rocky outcrop will take on sentinel duty for the family group.

VOCALIZATIONS

Vancouver Island Marmots produce two loud, distinctive alarm calls that can be heard easily throughout the whole colony. A short, sharp call indicates the approach of an aerial predator such as an eagle. A long alarm call indicates that the danger is from a terrestrial predator such as a Grey Wolf or Cougar. Both of these calls are short-bandwidth signals, which can make the source animal very difficult to locate unless you can actually see it vocalizing. A softer "kee-aw" call probably indicates anxiety or uncertainty or a low level of alarm. Tooth chattering is used to express threat. During social interactions, a dominant animal will sometimes growl and the subordinate animal will reply with a submissive whine. Pups playing will utter shrieks and hisses, as will fighting adults.

SIGNS

Home burrows and hibernation burrows are extensive enough to create a throw mound of waste soil at their main entrance. Home burrows may have several entrances, not all of which have a throw mound. Refuge burrows are short with smaller entrances and typically do not have throw mounds. Burrow entrance sizes have not been recorded in the literature, but are likely very similar to those of Hoary Marmots. Like the closely related Hoary Marmot, Vancouver Island Marmots deposit their scat into latrines both under and above ground. The above-ground latrines are usually in rock crevices or near favourite day perches. Fresh scat is shiny and black, often blunt at one end and sharply tapered at the other. During midsummer, when vegetation is lush, droppings may be connected into "chains" by thin strands of faecal matter. Weathered scat takes on a frosted appearance and fades to lighter shades of brown. Pellet size is similar to the Hoary Marmot's. Tracks in snow are quite possible in spring. Tracks and track patterns of Vancouver Island Marmots are largely unreported, but as the animals are similar in size and weight to Hoary Marmots, it is likely that their tracks are similar. See the "Signs" section of the Hoary Marmot account for information and illustrations of marmot tracks.

REFERENCES

Armitage 2003; Bryant 1996, 2005; Bryant and Janz 1996; Bryant and Page 2005; Casimir et al. 2007; Heard 1977; Martell and Milko 1986; Nagorsen 1987a, 2005a.

Eastern Grey Squirrel

also called Black Squirrel

FRENCH NAME: **Écureuil gris**

SCIENTIFIC NAME: *Sciurus carolinensis*

If you live in the more densely populated regions of Canada, you likely know this animal. It is widespread, often introduced, and especially common in urban parklands, where its predators are few.

DESCRIPTION

This squirrel is quite variable in colour. The two main phases are shown. The melanistic (black) phase is most commonly seen in northerly parts of the range in Ontario, where there may be some thermal advantage to the coloration during winter. The melanistic colouring may also just be due to chance (i.e., genetic drift), a not-uncommon occurrence in animal populations at the edge of the range. Grey is still the predominant colour in most parts of the range. Albinos and white animals are rare, but have been reported, as has a most attractive, but uncommon red phase. The typical coloration is grizzled grey or grey-brown above with a white underbelly and

Eastern Grey Squirrel
(*Sciurus carolinensis*)

grey phase

black phase

Distribution of the Eastern Grey Squirrel (*Sciurus carolinensis*)

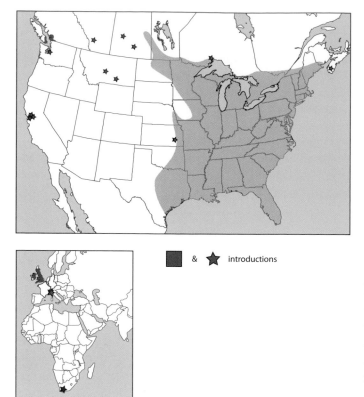

  ▇   & ★   introductions

and scruffy looking. The autumn moult begins in September and starts around the flanks, proceeding forward, but with a less-noticeable demarcation than that of the spring moult. The tail is moulted only once a year, in about July. Spring moult of reproductive females is delayed until their offspring are weaned and about this same time the young of the year begin their first moult. Eastern Grey Squirrels have keen eyesight and can see colours, especially in the red-green range, which is no doubt helpful as they forage for food.

SIMILAR SPECIES

In Canada the most similar squirrel is the Eastern Fox Squirrel, which is larger than the Eastern Grey and is more orange to yellow brown on the belly and throat. The underside of the tail is reddish and the back is an olive-brown, often with a yellowish cast. The ranges of the two species overlap in Manitoba, Saskatchewan, western Ontario, and on Pelee Island. Eastern Fox Squirrels also have a dark phase that can be very similar to but larger than the dark phase of Eastern Grey Squirrels. Skulls of the two species are usually distinguished simply by a tooth count. Most, but not all, Eastern Grey Squirrels have a small premolar that rests at the beginning of the upper toothrow. This pair of teeth is missing in the Eastern Fox Squirrel. The range of Red Squirrels overlaps that of Eastern Grey Squirrels; however, their smaller size and rufous pelage make them fairly easy to differentiate.

SIZE

The following measurements apply to adult animals from the Canadian portion of the range. Adult males and females are similar in size. Body weight fluctuates over the year. It is highest in late autumn and lowest in spring or early summer. It appears that with increased access to year-round bird feeders, the maximum weight of this species has risen. In 1966, R.L. Peterson in his *Mammals of Eastern Canada* listed the heaviest squirrel as 690 g.

Total length: 430–540 mm; tail length: 200–255 mm; hind foot length: 60–75 mm; ear length: 25–38 mm.

Weight: 340–831 g; newborns, 13–18 g.

RANGE

These tree squirrels occur naturally in southeastern North America, from southern New Brunswick to southern Saskatchewan, and southwards to the Gulf of Mexico. They invaded Manitoba in the 1930s, possibly due to agricultural changes created by European settlement and the availability of crops for food and buildings and shelter belts for nesting. They have recently appeared in neighbouring Saskatchewan. Eastern Grey Squirrels have been successfully introduced into many parts of the world, including Vancouver Island, Vancouver, Calgary, Regina, Saskatoon, Winnipeg, near Nipigon in northern Ontario, and into southern Nova Scotia and several locations in the western United States. The repeated Nova Scotia introductions apparently failed, but occasional individuals continue to be reported from urban centres and in the central Annapolis Valley. Introductions into Britain, Scotland, Ireland, Italy, and South Africa in the late 1800s and early 1900s have been highly successful. The species is widespread now in the British Isles and increasing

throat. Varying amounts of yellow-brown occur usually around the face and flanks, down the midline of the back, and on the underside of the tail. The winter pelage is typically greyer in appearance than the summer coat. The tail is long and bushy and somewhat flattened. The upper surface is usually grizzled grey with a light-coloured rim. In summer pelage, the grey phase animals have a noticeable light ring around their eye, which is muted in winter pelage. Many develop a noticeable light patch at the rear base of the ears during winter as well. The melanistic phase may not be entirely black. Some dark squirrels have reddish tails or appear more brown than black in strong sunlight. Litters may contain more than one colour phase. Adaptations for an arboreal lifestyle include: claws that are recurved and prominent on all four feet and grow at the end of elongated digits; ankle joints on the hind limbs that are unusually flexible and rotate 180° to allow the squirrel to climb down a tree head-first while still retaining a claw-hold; and a long bushy tail that can act as a rudder when moving through the treetops. The dental formula is incisors 1/1, canines 0/0, premolars 2/1, and molars 3/3, for a total of 22 teeth. The incisors on the upper and lower jaws are open-rooted and grow continuously through the life of the animal. They are kept sharp by constant use, with the front layer of enamel wearing more slowly than the rear dentine.

Eastern Grey Squirrels moult twice each year. The spring moult begins in March, starting on the head and proceeding back along a fairly distinct moult line that bisects the body. An animal in mid-moult may present a strange appearance, as the new coat is short and sleek, while the remaining winter coat is thick

rapidly in Italy. It is considered a pest in Europe, as it appears to be displacing the native Red Squirrel (*Sciurus vulgaris*).

ABUNDANCE

Eastern Grey Squirrels are generally common, although density can vary from year to year depending on weather and annual tree production of seeds and nuts. Densities as high as 21 per ha can occur in productive urban woodlands, but continuous woodlands typically support densities of fewer than three per ha. Although these squirrels are extensively hunted in the southern United States, they are rarely or lightly hunted in Canada.

ECOLOGY

Eastern Grey Squirrels are most common in mature deciduous woodlands with a diversity of nut- and seed-bearing trees and shrubs, especially in urban and suburban parkland settings where regular access to bird feeders during the winter enhances survival. Availability of suitably sized trees that provide the all-important cavity for a winter den is crucial for these squirrels' survival. Alternatively, in urban locations, the attic of a home is an acceptable substitute. With a well-insulated winter nest, Eastern Grey Squirrels survive the cold by allowing their basal body temperature of 36.4°C–38.7°C to decline by 1°C–4°C and by boosting their metabolic rate to produce more internal heat. As long as they have sufficient food to fuel this heat production, they can endure the coldest winter temperatures. While they may choose to continue using a tree cavity during the summer, most prefer a leaf nest (also called a drey) in warm weather. These often ragged-looking dried leaf clumps are built onto a twig platform on a tree branch. The outer shell is made of twigs and leaves and the central cavity is lined with shredded vegetation. A well-constructed drey made by an experienced adult is waterproof, while at the same time well ventilated, but those constructed by inexperienced juveniles may be so flimsy as to disintegrate in a brisk wind. Only the very best will remain intact through a winter. Eastern Grey Squirrels are diurnal, with summer activity peaks in the morning and late afternoon to avoid the midday heat. They remain active all winter and are most seen at midday when the weather is warmest. During very cold weather, the squirrels often choose to remain in their nests, sometimes for days at a time, to avoid the elements. Home ranges vary from 0.5 to more than 20 ha, but average less than 5 ha. Ranges of adult males are almost twice the size of those of adult females. There is extensive overlap of home ranges.

A poor mast year (few seeds and nuts produced by trees and shrubs) creates food shortages during the following cold season and increases winter mortality, particularly of juveniles. Furthermore, it negatively affects the reproductive potential of females the following spring, as litters are smaller or may not occur at all. Populations quickly crash after even a single poor mast crop and especially after recurring years of low mast production. Since mast-producing trees maintain a variable cycle of high-mast production, the squirrel populations are unable to sustain such high densities as to prevent germination and regrowth. Eastern Grey Squirrels contract, and sometimes die of, mange and severe cases of ringworm, but rabies is rare. While very agile as they climb up and down and leap from tree to tree, falls and even fatal falls do happen on occasion. The maximum lifespan for wild females is 12–13 years and for wild males is 9. A captive female lived more than 20 years. Eastern Grey Squirrels spend considerable time on the ground, especially during autumn while food gathering. Although they pause often and are relatively vigilant, they are vulnerable to aerial, arboreal, and terrestrial predators including rattlesnakes, raptors, weasels, foxes, Bobcats, Coyotes, Grey Wolves, and Domestic Dogs and Cats. These squirrels are capable swimmers and will undertake river and lake crossings, which explains how they occasionally are found in the gut of large fish. They can bound along the ground at speeds up to 27 km/h and are remarkably quick in the canopy.

DIET

Eastern Grey Squirrels feed heavily on seeds and nuts, flowers, and buds of oak, hickory, beech, walnut, pecan, and maple. They also consume fruits, conifer seeds, flowers, buds, catkins, and bark of many other plants as well as insects, fungi, crops such as wheat and corn, frogs, and bird eggs, nestlings, and occasionally adult birds. They are quick to chew on any bones or dropped antlers they find, likely for the calcium. Cannibalism has been reported. They cannot digest cellulose, so they must eat the higher-calorie seeds, fruit, and nuts rather than leaves or stems. Consumption of buds, catkins, bark, and sap is reserved for times of high nutritional stress when they are desperate. Acorns are a favourite food of Eastern Grey Squirrels. Oaks can be generally classified into one of two types – white oak and red oak. Their acorns sprout at different times, white in autumn and red in spring. Once sprouted, the nutritional value of the acorn diminishes. The squirrels react to this biological fact by eating white oak–type acorns immediately before they sprout and caching the longer-lasting red oak–type acorns for consumption over the winter. Or they nip the embryos off before caching the white-type acorns. The cycle of an Eastern Grey Squirrel's year, like most northern mammals, revolves around the cold season when food is in short supply. These tree squirrels are classic scatter-hoarders. Come late summer and autumn, they gather and hide small amounts of food in multiple places to guarantee that at least some will escape the attention of other animals, including other Eastern Grey Squirrels. They usually dig a small hole on the forest floor and cover it over after placing a single nut inside. This food may then be used to supplement whatever can be found in winter. Returning to a food cache is probably a matter of spatial memory and odour, but the

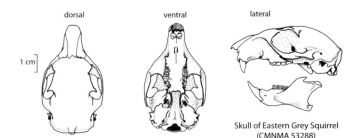

dorsal   ventral   lateral

1 cm

Skull of Eastern Grey Squirrel
(CMNMA 53288)

extent to which Eastern Grey Squirrels remember and relocate their stored food and then manage these stores remains under investigation and is particularly difficult to study. About 15% are forgotten or unused and may sprout the next spring. The extent and diversity of nut- and seed-bearing hardwood trees in North America owes much to these busy animals. Consumption of food typically takes place as the squirrel sits up on its haunches and uses its front paws to manipulate the food item.

REPRODUCTION

Both females and males in the northern portion of the range typically breed for the first time as yearlings. Farther south, such as in North Carolina, an early-litter female may mature quickly enough to produce a late litter in the same year as her birth. The breeding season is stimulated by increasing hours of daylight and in eastern Canada normally occurs in late winter (February), with a possible second season in mid-summer (July) some years. In the northern parts of the range, few females are able to produce a second litter unless the weather is exceptionally favourable and food is abundant. The timing of breeding means that the males must have fully descended testes (normally they reside in the abdominal cavity) and that females are pregnant during the most difficult part of the winter. Each female is in oestrus for about eight hours and may attract numerous males who will have to compete for first access. Females will mate with several males, but part of a male's ejaculate coagulates to form a temporary gelatinous plug in the vagina of the female, effectively reducing the success rate of later matings and ensuring that the first mating fertilizes most of her eggs. At least this is the case unless the female, while grooming her genitalia after mating, is able to reach the plug and remove it. Gestation lasts about 44 days, so young are born in March or early April in Canadian populations, even in the warm habitats of southern British Columbia. Average litter size is two to three (range 1–8) and is highest when food is plentiful and lowest when food is scarce and embryos may be reabsorbed. Young are born in a leaf nest, a nest in a tree cavity, or sometimes in an attic. At birth they are hairless, except for a few short whiskers, and their eyes and ears are closed over. Their heads and feet appear disproportionately large at this stage. They are cared for solely by their mother. Nestlings have a curious, but fortunately uncommon, propensity for getting their tails tangled. This can occur between two or more siblings and may extend to involve the whole litter. Likely some sticky substance such as tree sap or faeces begins the deadly process and their own squirming compounds the problem. In less severe cases, the entangled youngsters may be able to disengage and survive at the cost of a broken, deadened, or even missing tail. Weaning begins at about seven weeks old and is usually complete by the time the young are around ten weeks old. Juveniles newly out of the nest are poorly coordinated and fearless, making them very vulnerable to predation for a few days. The juveniles remain with their mother for a month or two after weaning, but gradually spend more time foraging on their own. They may disperse any time after weaning, but most remain on her territory until they are yearlings, often sharing her winter den. They are indistinguishable from adults by eight to nine months old.

BEHAVIOUR

Eastern Grey Squirrels are not highly territorial. Neighbours are often related or at least well known and are amiably tolerated. This same tolerance does not extend to potential immigrants, which are chased away and may be bitten if persistent. During the breeding season, yearlings must establish a territory of their own, and this activity can lead to displays of aggression even by relatives. Some yearlings, especially males, may disperse up to several kilometres away, but most females remain nearby and easily insinuate themselves into their mother's group. A dominance hierarchy develops within each group, with adult males dominant over adult females and adults dominant over juveniles.

During the breeding season a second dominance hierarchy is worked out, this time among the group of adult males that gather around each pre-oestrous female. Typically, the older males (three years or older) become dominant. Battles for male dominance may be bloody, with ears and tails being the main targets. The highest-ranked male may then achieve the important first access. Females sometimes foil this dominance hierarchy by breaking away from the group and hiding. At this point, it is every squirrel for himself and the first one that finds her has a mating opportunity. Frustrated males in search of such a female will chase almost anything that moves, including birds, rabbits, and even baseballs. Should she actually elude all the males, she merely gives a chattering call to attract their attention. Males become active early in the day during mating season, often while light levels and temperatures are still low. They travel around searching for females approaching oestrus. Each female they encounter is cautiously sniffed around the rear. Most females are quick to rebuff such efforts and may even nip. Males can apparently detect the odour of approaching oestrus, as they will persistently follow a female who is five days or less away from her eight hours of receptivity and will arrive expectantly at her nest every morning to await her exit. By the time she is receptive, she may have gathered a squabbling entourage of several males (up to 34 were recorded in Texas following one hapless female). Males are drawn from as far away as 850 m to join a mating chase. They expend so much energy hunting and competing for females during the short breeding season that their weight plummets, sometimes to the extent that their survival is threatened.

Eastern Grey Squirrels use visual, auditory, and olfactory signals to communicate with each other. Auditory communication is dealt with in the "Vocalization" section following. Visual signals include piloerection (erecting the hair), tail flagging, hind-foot stomping, posture, and even simple staring. Olfactory signals are mainly used to mark territory. Drops of urine are purposefully deposited on the sheltered side of tree trunks, deadfall, or boulders in an effort to preserve the scent mark for greater lengths of time. Additional markings are left by rubbing the oral glands at the corner of the mouth onto the substrate. These squirrels commonly nest communally on cold nights, thereby conserving heat. The sexes tend to segregate at this time. Unrelated males may share accommodation, but most females will share only with related females or with their young of the year. Communally nesting females form stable groups that may continue nesting together even during the warm season, although

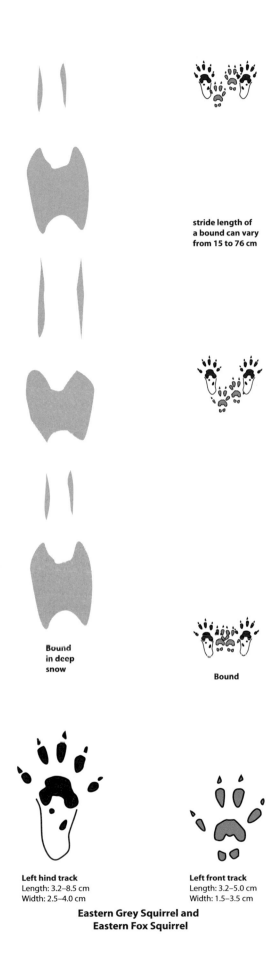

stride length of a bound can vary from 15 to 76 cm

Bound in deep snow

Bound

Left hind track
Length: 3.2–8.5 cm
Width: 2.5–4.0 cm

Left front track
Length: 3.2–5.0 cm
Width: 1.5–3.5 cm

Eastern Grey Squirrel and Eastern Fox Squirrel

females nearing labour separate from the group to produce their litters alone. During warm weather the squirrels may rest during the day inside a well-ventilated drey or stretched out on their belly on a tree limb where they can keep an eye out for predators. They may also be seen, motionless on a branch, soaking up some sun on a cold winter day. Eastern Grey Squirrels are capable of duplicitous behaviour, one of the few species outside the primate group to display such behaviour. If a squirrel is being watched by another squirrel while burying a nut, it will pretend to dig a hole and cover the stash, then will leave with the nut hidden in its mouth. Eastern Grey Squirrels possess a strong homing ability and may find their way back to their own territory after being displaced as far away as 4.5 km. The maximum known dispersal movement of a juvenile is 100 km. Nursing females commonly move their litter over the course of the lactation period, especially if they have been disturbed. Each nestling is moved individually. She grabs the loose skin of its chest or abdomen in her mouth and the infant curls around and clings to her head and neck.

VOCALIZATIONS

Usage of alarm calls by this species is quite sophisticated. Four calls are used, often repeated and sometimes in combination. A series of "kuk" calls indicates mild alarm, a "quaa" call is just a drawn-out "kuk" call and indicates moderate alarm, a moan is used to indicate startlement or mild distress, and a low-volume buzz call emitted through the nose indicates moderate alarm. These calls are usually accompanied by synchronous tail flicking. A scream is emitted under severe distress, such as capture. Aggressive encounters are generally accompanied by escalating amounts of tooth chattering, hind-foot stomping, and tail flagging, which reflect the level of the conflict. Males, when pursuing a female during the breeding season, emit a sneeze-like mating call that may be rapidly repeated several times. Females involved in mating chases produce a high-pitched whine. An infant's distress call varies with age. Between about 4 and 38 days old, they squeak loudly and can be heard for a considerable distance. Beyond 38 days, they emit the usual adult-type scream distress call. Before 4 days old, they are insufficiently developed to produce any but very soft sounds. Between 20 and 30 days old, the infants begin to produce a low-volume growl if handled. Suckling nestlings produce a raspy "muk-muk" sound through their noses as they nurse. They also create a "lip-smacking" sound as they search for, or if they lose contact with, the nipple. Both these sounds are classified as infant solicitation calls.

SIGNS

Eastern Grey Squirrels use two gaits: they walk and bound. Their front feet have the typical rodent shape, with four toes and a vestigial thumb that does not register in a track. Toes 2 and 5, the outside toes, angle outwards, and toes 3 and 4 point forward. The hind foot has five toes. The two outside toes, 1 and 5, angle out from the foot, and toes 3, 4, and 5 cluster together aiming forward. Claws do not always register in a track. When bounding in deep fluffy snow, the hind feet land almost on top of the front foot tracks, leaving the impression of only two tracks, rather than the expected four. Such a trail frequently has long drag marks leading into the track. A trackway of a bounding Eastern Grey Squirrel is typically 10–15 cm wide. There is considerable size overlap between the tracks of Eastern Grey Squirrels and Eastern Fox Squirrels. During warm weather Eastern Grey Squirrels often shelter in leafy dreys that are usually 30–48 cm in diameter. For more information on dreys see the "Ecology" section above. Most scat is oval and smooth and variable in size. It is usually a solid to semi-solid pellet 0.3–0.6 cm in diameter and 0.4–1.0 cm in length. Scattered randomly throughout the home range, it is most commonly found in heavily used areas. Scat deposited onto wet snow quickly absorbs water and can swell to several times its original dimensions, sometimes even bursting to create a small brownish smudge on the snow. Although Eastern Grey Squirrel scat is generally larger than scat from Red Squirrels, there is considerable overlap and distinguishing one from the other is often impossible. An acorn consumed by an Eastern Grey Squirrel is typically eaten after first peeling off the shell. A green shell is peeled in wide strips to expose the nut meat. If dry (having been stored for winter consumption), the nut is more easily cracked opened and the strips are even wider, approaching half the nut at times. Due in large part to their size, Eastern Grey Squirrels are capable of opening even the hardest and largest nut, and they usually break it into small fragments to ensure they reach all the meat. This is especially true of hickories, butternuts, and walnuts, whose flesh is compartmentalized. When eating coniferous cones they usually carry them up into a tree before peeling the bracts to reveal the seeds. The discarded bracts fall to the ground over a wide area below the tree. This pattern is quite unlike the midden of a Red Squirrel, where the animal sits on a stump, rock, or low branch to perform this activity and all the bracts and the stripped cone core are left in an untidy pile beside their perch.

REFERENCES

Eason 1998; Elbroch 2003; Gonzales 2005; Jacobs, L.F., and Liman 1991; Jannett et al. 2007; Koprowski 1992a, 1992b, 1994a, 1996; Lishak 1982a, 1982b, 1984; MacDonald, I.M.V., 1992; Nagorsen 2005a; Pereira et al. 2002; Peterson 1966; Rezendes 1992; Steele and Koprowski 2002; Wrigley 1979.

Eastern Fox Squirrel

FRENCH NAME: **Écureuil fauve**

SCIENTIFIC NAME: *Sciurus niger*

A newcomer to Canada, this large colourful squirrel is expanding its range in the west by moving north and westwards from North Dakota into Manitoba and Saskatchewan. Thanks to human activities such as shelter-belt creation, corn farming, and construction or abandonment of outbuildings that provide shelter, this squirrel is finding the habitat changes to its liking. An introduced population

Eastern Fox Squirrel
(*Sciurus niger*)

has existed on Pelee Island, the southernmost part of Canada, since the 1890s and an introduced population in northern Washington State has recently expanded into southern British Columbia.

DESCRIPTION

This squirrel is larger (20%–40% heavier) than the similar Eastern Grey Squirrel. Its pelage colour is variable, but only one colour phase is represented in the Canadian populations. These squirrels are grizzled grey to olive grey on the back, depending on how much ochre is mixed with the grey. Underparts and feet are yellowish ochre or orange and the underside of the tail is reddish ochre. The tail is bordered by a black and then an orange to yellowish fringe of hairs, and its upper surface is rusty brown interspersed with black hairs. A light border around each eye is yellowish and tends to blend with the surrounding fur. The ears are yellowish brown on the inside and may have a pale orange or tan patch on the backside near the base. There is potential for a melanistic (black) colour-phase animal to appear (this form is common in the southern United States), but to date none has been reported in Canada. The Eastern Fox Squirrel is well adapted for an arboreal lifestyle. Its claws are relatively large and recurved to ensure a good grip. Ankle joints on the hind limbs are unusually flexible and allow a rotation of 180° to permit the squirrel to climb down a tree head first while still retaining a claw-hold. The long bushy tail assists balance when moving through the treetops and acts as a sunshade during hot weather. The dental formula is incisors 1/1, canines 0/0, premolars 1/1, and molars 3/3, for a total of 20 teeth. Two yearly moults occur. The summer coat is made up only of guard hairs with no underfur. This pelage is shed in autumn starting at the flanks and proceeding forward. The winter coat, which is composed of both guard hairs and a dense underfur, is shed in spring. This moult travels from the head rearward in a noticeable moult line that bisects the body. A spring animal in mid-moult can present a strange appearance with sleek new summer fur on the front of the body, contrasting with the scruffy remains of the thick winter fur on the rear. The transition from summer to winter coat is less noticeable. The tail is moulted once each year in July or August. Spring moult in lactating females is delayed until the young are weaned. Young of the year moult into their first adult coat at 75–90 days old, soon after they are weaned.

SIMILAR SPECIES

Although smaller, Eastern Grey Squirrels are similar to Eastern Fox Squirrels and their distributions overlap in Ontario, Manitoba, and Saskatchewan. The colour of the belly fur is usually the best distinguishing characteristic, as the white or pale grey (occasionally with brownish patches) of the Eastern Grey Squirrels is quite different from the typical buffy orange of the Eastern Fox Squirrel. Furthermore, the underside of an Eastern Grey Squirrel tail is grey to yellow-brown, while the same area on an Eastern Fox Squirrel is orange. The circle around the eye is brighter white (during summer) on Eastern Grey Squirrels. Skulls can usually be told apart by the presence of a small peg-like premolar at the beginning of the row of upper cheek teeth. This small tooth is usually present in Eastern Grey Squirrels and is always absent in Eastern Fox Squirrels. The much smaller Red and Douglas Squirrels are closely associated with mixed coniferous forests and are not found in the largely deciduous forests frequented by Eastern Fox Squirrels, except in southern Manitoba, where Fox, Grey, and Red Squirrels may occur in the same area and habitats.

SIZE

The following measurements apply to adult animals from the northern portion of the range, but not only from Canada, as there are so few Canadian specimens. Adult males and females are similar in size. Body weight fluctuates over the year and is usually highest in late autumn and lowest in spring or early summer.
Total length: 450–698 mm; tail length: 210–330 mm; hind foot length: 65–82 mm; ear length: 26–31 mm.
Weight: 529–1361 g; newborns, 13–18 g.

RANGE

Before 1970, Eastern Fox Squirrels were only found in Canada on Pelee Island, where they had been successfully introduced in the 1890s. In the early 1970s they began moving from North Dakota into southern Manitoba, where they have extended to the northern edge of the mixed forest. The first Saskatchewan records began to appear in the early 1980s, and the species reached Kutawagan Lake, in the east-central part of the province, by 2008. The ability of these squirrels to be comfortable on the ground, at some distance from trees, has allowed them to penetrate prairie grasslands along treed river corridors. The population in the southern Okanagan valley appeared in the mid-1980s around Osoyoos and has now penetrated as far north as Okanagan Falls. Biologists suspect that these squirrels are derived from an introduced population in Okanagan County, Washington. There is every expectation that this population of Eastern Fox Squirrels will spread in southern British

Columbia as much suitable habitat awaits them, especially in urban areas such as Penticton, Kelowna, and Vernon.

ABUNDANCE

Population densities are not high anywhere in the Canadian range, but numbers fluctuate in association with annual mast (seeds and nuts of trees and shrubs) production. Squirrel numbers may decline precipitously after a poor mast year. In the midwestern United States, population densities may reach 12 animals per hectare, but throughout most of the range, densities are commonly less than 3.5/ha and often less than 0.1/ha.

ECOLOGY

Less dependent on proximity to trees than their close cousin, the Eastern Grey Squirrel, Eastern Fox Squirrels are bolder and may forage several kilometres from the nearest tree. They are adapted to living in open forests and edge regions with sparse vegetation and find urban parks, orchards, and nut groves particularly favourable habitats. Much of their time is spent on the ground. Viable populations can exist in fingers of woodland or even in hedgerows and fencerows of small trees that penetrate into prairie grasslands. Few studies have been undertaken on Canadian populations, so most of the information here is based on work done in the United States. Eastern Fox Squirrels do not hibernate, so are active all winter. They are strictly diurnal, with periods of high activity, usually for a few hours, after sunrise and before sunset during warm periods and at midday during cold periods. This midday pattern is broken during the late-winter breeding period, as the reproductive males are active by dawn. The early breeding season places tremendous strain on the males, who must use up their precious energy to locate and compete with each other for females, at a time of year when food is scarce. The females in turn are pregnant and rearing their young in the

Distribution of the Eastern Fox Squirrel (*Sciurus niger*)

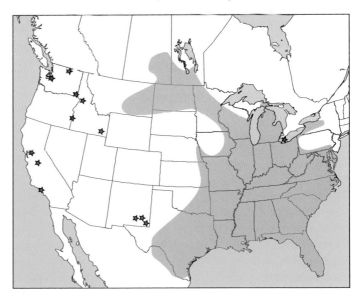

⭐ introductions

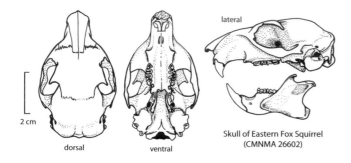

2 cm

dorsal   ventral

lateral

Skull of Eastern Fox Squirrel
(CMNMA 26602)

early spring when food is only slightly more available. This timing does ensure, however, that the young become independent when food is abundant. During warm months, Eastern Fox Squirrels construct and occupy leaf nests or dreys. These are built on a twig platform high in a tree. The exterior is made up of twigs and leaves and the nest chamber in the centre is lined with shredded vegetation. Cold months are usually spent in a tree cavity or abandoned building in a nest similarly lined with shredded and dry vegetation. Young are reared in both tree cavities and dreys. Females may live 12–13 years and remain reproductively active until death, but few males survive beyond 8 years. Predators include rattlesnakes, Black Rat Snakes, raptors, owls, weasels, Northern Raccoons, Red Foxes, Grey Foxes, Bobcats, Grey Wolves, Coyotes, and Domestic Cats and Dogs. Eastern Fox Squirrels can gallop at speeds of 24–25 km/h. They are susceptible to mange and West Nile Virus, in addition to many parasites.

DIET

These large tree squirrels eat mainly nuts and seeds of deciduous trees, but will also consume flowers, tree buds, fruits, fungi, invertebrates, and agricultural crops such as corn, wheat, oats, sorghum, and soybeans. Aside from insects, some animal protein is ingested in the form of dead fish, birds, and bird eggs. Cannibalism has been reported. Buds and flowers are mainly eaten during winter and spring when other foods are not available and the squirrels are very hungry. They are scatter-hoarders, that is, they cache small amounts of food in numerous places for later winter consumption. As the nuts ripen in autumn, they are gathered and buried (usually one at a time) in a shallow hole in the soil. Human plantings of oak, elm, and maple, as well as agricultural crops and bird feeders encourage the expansion of this species. Eastern Fox Squirrels also play a role in the woodland succession of grasslands, as some of the nuts they bury in the grasslands that are not consumed may go on to sprout.

REPRODUCTION

This is a widespread species reaching the northern limit of its range in Canada. Particulars of reproduction in the southern regions differ from the following data, which refers to the populations in Canada, excluding the small British Columbia population, for which reproductive details are unknown. Females are at least a year old before they become sexually mature and most do not become pregnant for the first time until they are two. The percentage of adult females that

breed each year varies from none to more than 90% and is directly related to food availability. Well-fed females will breed, while malnourished females do not. Adult males are in reproductive readiness when their testes are pendant in the scrotum. This is usually the case from November through August; otherwise, the testes shrink and retreat into the abdominal cavity. Northern populations have a single annual breeding season between December and February, depending on the location. Mating in warmer regions (such as Pelee Island and likely British Columbia) may occur earlier in the specified period, while in colder regions (such as Manitoba and Saskatchewan) it occurs towards the end. Each reproductive female has a single eight-hour period of oestrus each year. Part of the male ejaculate coagulates into a temporary copulatory plug that blocks further sperm from entering the reproductive track of the female for a few hours. Unless she can physically remove this plug, subsequent matings are generally unproductive and her first mate likely fertilizes most of her litter. Gestation lasts 44–45 days. Average litter size is two to three (range 1–7). Newborns are pink and hairless apart from a few short whiskers and their eyes and ears are closed over. Ears open at three weeks old and eyes open at five weeks. Gradual weaning begins at eight weeks old and is completed about a month later at twelve weeks.

BEHAVIOUR

Eastern Fox Squirrels are ordinarily non-territorial, except for pregnant and lactating females, which defend a core area of their home range against other squirrels. Otherwise, there is considerable overlap of home ranges. Home range size varies, depending on food abundance, from 0.9 to 17.2 ha for females and 1.5 to 42.8 ha for males. Within each extended group of squirrels, a dominance hierarchy is established. Males dominate females and females dominate juveniles. During the breeding season this hierarchy is refined among the males who compete, posture, and fight until they establish a linear hierarchy, which then permits the dominant males proximity to a receptive female and therefore greater chances of mating. A female does manage to foil this hierarchy on occasion by breaking away from the mob and hiding. She will copulate with the first male to find her, and in this manner, the younger, less dominant males occasionally have a mating opportunity. Age is the primary factor that determines male dominance. Males gather on the home range of a female on her day of oestrus, forming a squabbling mob. They chase her and manoeuvre among themselves for the day, then move on to look for another oestrous female. Females mate with several males.

Typically, the young will remain on their mother's territory until autumn, at which time they disperse to establish territories of their own, although rarely some may nest with her during their first winter. If so, they disperse the following spring, as their mothers will not tolerate them nearby during the breeding season and thereafter when she is pregnant. The farthest-known dispersal distance is 64.4 km. Apart from the suckling pups and an occasional juvenile sharing its mothers nest, most Eastern Fox Squirrels nest alone even during winter. Although Eastern Fox Squirrel males will join in the chase of an oestrous female Eastern Grey Squirrel, copulations of mixed-species pairs have not been observed and hybrids are unknown. Eastern Fox Squirrels use body and tail posture, scent, and sounds to communicate. Sounds are dealt with in the following section. Threat is expressed with a stare and then an upright posture, with the tail raised and flicking over the back. Hind-foot stomping and tooth chattering may also accompany an aggressive encounter and increase in intensity as the threat level increases. This may be followed by a charge by the aggressor and a chase or fight may ensue. Submission is signalled by retreat with a lowered tail position. Alarm calls are frequently combined with the visual signal of a flicking tail. Scent marking is a commonly used signal by both males and females. Urine is deposited in protected locations, such as under a log or on the leeward side of a boulder or large tree. These traditional sites are created and visited by all the local squirrels. Oral marking by rubbing the gland in the corner of the mouth onto the substrate is another method of marking. Both genders mark regularly used pathways in this manner, but only males mark traditional oral-marking sites.

VOCALIZATIONS

A series of barks is the most commonly heard vocalization. These are used to indicate alarm or distress. A chatter bark is used when the animals are startled or during an aggressive chase. A restrained animal may groan. A squirrel in severe distress, such as when captured, may emit a two-part scream. Males in mating chases will grunt and squeal. Females in mating chases produce a high-pitched whine. Chattering teeth signifies aggression.

SIGNS

Eastern Fox Squirrels use two gaits. They walk and bound when foraging and bound when travelling. Tracks and track patterns of this squirrel may be larger, but are very similar to those of the Eastern Grey Squirrel. There is considerable track-size overlap between the two species. See the "Signs" section of the Eastern Grey Squirrel account for further information. When Eastern Fox Squirrels are bounding, their trackways are typically 11–18 cm wide. They will wander much farther from the safety of trees than will the Eastern Grey Squirrel. Their leafy dreys are similar in size to those made by Eastern Grey Squirrels (30–48 cm in diameter). The openings of Eastern Fox Squirrel tree dens are usually circular and 7–9 cm in diameter. The occasional oval openings have also been found, in which case the squirrels typically have used only one end. These squirrels combat the efforts of the trees to heal over the opening by regularly chewing at the scar tissue.

REFERENCES

Baumgartner 1939; Elbroch 2003; Fairbanks and Koprowski 1992; Kiupel et al. 2002; Koprowski 1993, 1994b, 1996; Rezendes 1992; Smith, A.R., 2009; Steele and Koprowski 2002; Wrigley 1979; Wrigley et al. 1973, 1991.

Columbian Ground Squirrel

FRENCH NAME: **Spermophile de Columbia**
SCIENTIFIC NAME: *Spermophilus columbianus*

Due to its colonial nature, diurnal habits, and upright vigilance posture, this is one of the most conspicuous mammals in mountain clearings and meadows.

DESCRIPTION

Fur around the mouth is tawny to creamy white and the eye ring is white to buff. Cheeks are dark to medium grey. The top of the head and snout are dark reddish orange. The back is dark grey washed with brownish yellow, particularly over the shoulders, and covered with lateral rows of small yellow-brown to greyish spots. Throat, belly, flanks, and feet are reddish orange and claws are long and black. The top of the tail is brownish black and the underside is a mix of white and black hairs. There is an indistinct black tip on the tail. Adult males develop a black pigmentation of the skin of the scrotum. Moult occurs once each year over the summer. Individual hairs are shed and replaced throughout the body and there is no obvious pattern to indicate that moult is taking place. Males and non-reproductive females moult in early summer, but moult is delayed in reproductive females until after the pups are weaned. Young of the year moult last, late in the summer. The dental formula is incisors 1/1, canines 0/0, premolars 2/1, and molars 3/3, for a total of 22 teeth.

SIMILAR SPECIES

The only similarly sized and coloured ground squirrel is the Arctic Ground Squirrel, whose range does not overlap that of the Columbian Ground Squirrel. There is a distance of about 300 km between the most southerly Arctic Ground Squirrels and the most northerly Columbian Ground Squirrels. Arctic Ground Squirrels are generally paler, with less contrast between the dorsal and ventral coloration, and their spots are whiter, especially over the shoulders. The cheek region of Arctic Ground Squirrels is brownish orange rather than dark grey like Columbian Ground Squirrels.

SIZE

The following measurements apply to adult animals within the Canadian portion of the range. Body weight varies over the year. Adult males are lightest three to four weeks after emergence in spring and

Columbian Ground Squirrel
(*Spermophilus columbianus*)

heaviest in late summer before hibernation. Adult females are lightest upon emergence and heaviest in early August just before hibernation. Males and females are about equal in size.

Total length: 280–395 mm; tail length: 75–124 mm; hind foot length: 42–59 mm; ear length: 10–27 mm.

Weight: 195.0–820.6 g; newborns, 7–8 g.

RANGE

This squirrel is associated with the western mountains and is found in southeastern British Columbia and southwestern Alberta and southwards to western Oregon and Washington, northern Idaho, and northwestern Montana. Their distribution is patchy and discontinuous. There is some evidence that the range of this species has expanded westwards in British Columbia in the Cascades Mountains. In the 1950s it was confined to the eastern slopes. Populations are now found on the western slopes as far west as Allison Pass in Manning Provincial Park and at Tulameen, west of Princeton.

ABUNDANCE

Columbian Ground Squirrels are locally common in suitable habitat throughout their range.

ECOLOGY

These social squirrels live in colonies in alpine and subalpine meadows, open forests, rangelands, and agricultural lands, from the higher elevations to the valleys. Habitat prerequisites are sufficient food and deep soil suitable for burrowing that is not prone

Distribution of the Columbian Ground Squirrel (*Spermophilus columbianus*)

to spring flooding. Colonies of over 1000 individuals are known from particularly favourable sites. Population densities in some Canadian colonies in the Alberta foothills may be as high as 70 animals per hectare. More commonly the population density is 10–20 animals per ha. The social structure of the colony is made up of dominant adult males called "resident" males occupying large home ranges of 3000–6000 m² that overlap or encompass the much smaller homes ranges (200–700 m²) of one or more females. Juveniles occupy their mother's territory until dispersing as yearlings. Columbian Ground Squirrels are diurnal. In spring, they are most active in the afternoons during the warmer temperatures. During summer, they commonly remain in their cool burrows through the heat of the day and are most active in early morning and evening. They hibernate for seven to nine months each year depending on the elevation and latitude, and hence the severity and length of winter. Adult males and non-reproductive adult females enter hibernation first, followed by reproductive females, then yearlings, and finally young of the year. Almost all Canadian Columbian Ground Squirrels are hibernating by late August and some males as early as mid-July. Adult males are the first to emerge from hibernation, followed about a week or two later by adult females and finally by yearlings. Most animals emerge between mid-April and mid-May. Mortality is high for pups during their first winter. The most common cause for these deaths is late birth the previous spring and insufficient time to acquire the necessary fat stores to sustain them through the winter. The highest-quality forage occurs early in the season and typically begins to desiccate in July. Pups weaned later than early July may miss out on this bounty. Most juveniles disperse as yearlings in the summer following their first hibernation. More males than females leave their natal colony, as some of the young females remain to take over empty burrows or to occupy their mother's burrow, while their mother moves to a nearby site. Dispersing juveniles may travel up to 8.5 km from their birth colony to settle in another active colony.

Four types of burrows are constructed. Home burrows, also called summer burrows, are occupied by generations of squirrels and may be extended or enhanced many times to become quite extensive. Commonly, a hibernation burrow connects with this system, as may several refuge burrows and brood burrows. Hibernation burrows typically contain a spherical nesting chamber and a smaller latrine area nearby, which are at a level such that tunnels below would divert any water entering the burrow system. Access to an entrance or the rest of the burrow system is blocked by an earthen plug within 1–2 m of the hibernacula, when the squirrel is in torpor. Refuge burrows scattered throughout the territory have small unobtrusive openings and are used as temporary bolt holes. Some are discrete, while others connect to the home burrow system. Brood burrows tend to be small and inconspicuous as well, with no throw mound of waste soil at their entrance, and are usually located on the periphery of the home burrow system. Females commonly plug the external access to their brood burrow. Although typically occurring in clearings and meadows, those Columbian Ground Squirrels living near taller vegetation will easily climb trees and shrubs to forage or escape predators. Female Columbian Ground

Squirrels may live to be 10–12 years old, but few males live longer than 6 years. Predators include bears, especially Brown Bears, as well as Coyotes, American Martens, American Badgers, Canada Lynx, Bobcats, weasels, Cougars, and a variety of raptors, including Golden Eagles, Red-tailed Hawks, and Northern Goshawks. Smaller weasels, such as Long-tailed Weasels and Ermine, are likely only dangerous to unprotected young of the year, as they have been seen attacking juveniles and being driven away by other family members. The Western Rattlesnake in south-central British Columbia may also present a threat to young ground squirrels.

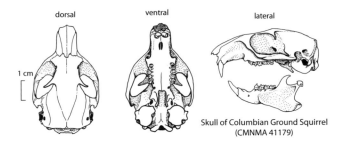

dorsal   ventral   lateral

1 cm

Skull of Columbian Ground Squirrel
(CMNMA 41179)

DIET

Primarily herbivorous, Columbian Ground Squirrels consume a variety of seeds, flowers, fruit, berries, bulbs, and leaves of forbs and grasses. Clover and seeds of many grain crops are avidly eaten when available, to the extent that they may be considered agricultural pests in places. Insects, small mammals, and carrion are sometimes consumed, including the dead bodies of colony members. Infanticide has been observed.

REPRODUCTION

The onset of breeding varies somewhat from year to year, depending on the local weather conditions and it also varies by several weeks between the low-elevation locations and those higher up, where the snow melt is delayed. The breeding season lasts about three weeks and begins as the first adult males appear above ground in spring and establish a territory. Adult females begin emerging one to two weeks later. Females breed about four days after they arouse from hibernation and are receptive for about four hours. Mating occurs both above and below ground. Part of the male's ejaculate commonly coagulates into a "copulatory plug" in the female's vagina, which prevents insemination by rival males for at least a few hours. Receptive females typically mate with several males, but the bulk of the offspring (65%) are sired during the first mating, which is normally with the resident territorial male. Most adult females produce a single litter each year, but some take a one-year hiatus between litters. These females gain more weight over the summer that they are not raising a litter and are more fit the following summer to produce healthy pups that survive their all-important first winter. Young are born in May or June after a gestation of 24–27 days. A litter consists of two to seven pups, but rarely do more than three survive to weaning. Born naked and blind, the young grow rapidly. Pups emerge from their burrows for the first time in mid-June to mid-July at around 27 days old and are weaned soon thereafter. Columbian Ground Squirrels reach their adult body weight in their fourth summer. Although yearlings may occasionally breed in lowland populations, few animals in high elevations are sexually mature at that age unless they were born during an especially productive year. Most females in Canada begin breeding as two-year-olds and most males at three years old.

BEHAVIOUR

Each animal, except females with their young, occupies a solitary burrow and all, including juveniles, hibernate alone. Juveniles may share a burrow system with their mothers, but not the same hibernation chamber. In spring, newly emerged adult males immediately set about establishing and defending a territory. Boundary disputes leading to chases, flank-to-flank shoving matches, or furious battles are most frequent in spring during the breeding season and diminish later in the summer. Older males commonly carry scars from these encounters. Females defend their much smaller territory from other ground squirrels, but particularly protect the burrows where their infants are nesting, especially from other females, who are likely to be the main perpetrators of infanticide. Before giving birth, female Columbian Ground Squirrels often create nests in several burrows within their territory before finally choosing one. Relations between adults of different sexes are usually antagonistic, even between the resident male and the females that live within his territory, except during the short breeding season. Closely related females (mother and daughter or sisters) may be less defensive towards each other.

Columbian Ground Squirrels use body posture, tail flicking, vocalizations, and scent in communication. An engaging behaviour regularly performed throughout a colony is a greeting ceremony called "kissing." Two animals with heads tilted touch noses and mouths for a few seconds, then may proceed to other social behaviours such as mating, dominance displays, or a chase. An individual's identity and possibly condition can apparently be determined by the odour of secretions from glands at the corners of the mouth. All adult squirrels, but especially females, become intolerant of yearlings, particularly yearling males, commonly chasing them into lesser-quality terrain away from the colony core. Most emigration of yearlings occurs in June and July, about the time that adult aggression towards them reaches its peak. Columbian Ground Squirrels are typically quite vigilant and adults regularly raise their heads to look around or rise on their haunches to scan their surroundings. If grooming, sunbathing, or resting above ground they usually choose a high perch where they can keep watch at the same time. Females with above-ground young, which are typically playing and often unaware, are most watchful. It appears that some males store food in autumn for consumption in early spring, but this is clearly not done by all males.

VOCALIZATIONS

There is no evidence that Columbian Ground Squirrels use different alarm calls to distinguish between aerial or terrestrial predators. The generalized alarm call is a series of loud chirps typically uttered as the animal rises on its hind legs and flicks its tail. Softer chirps or chirrs from the burrow entrance may indicate a lesser degree of alarm. They do recognize and respond appropriately to a marmot alarm call. When an alarm is sounded, those animals close to a burrow entrance seek a nearby high spot or rise on their haunches to scan the environment, while those at a distance from their burrows scurry back. Columbian Ground Squirrels will squeal loudly when injured. Tooth chattering is used to indicate threat.

SIGNS

Since these squirrels may become active before the snow disappears, tracks and runs in the snow are possible. Columbian Ground Squirrels adopt two main gaits. They usually bound while travelling and walk when foraging. They can jump as far as two metres between bounds when clearing an obstacle. See the illustration and description of tracks in the "Signs" section of the Arctic Ground Squirrel account, as Columbian Ground Squirrel tracks are similar. Burrow entrances vary in size depending on their use. Refuge burrows may be only 6–9 cm in diameter, but hibernation and summer or home burrows may be up to 20 cm wide by 13 cm high. Typically, refuge burrows have no noticeable throw mound in front of their entrance. Runs extend between burrow entrances and radiate to foraging sites or favourite perches. Depending on the water content of their diet, scat may be hard and dry or softer and clumped, but like all ground-squirrel scat, typically has a twisted pointy end. Scat is commonly found near burrow entrances and beside rock piles that are favoured surveillance perches. See the Arctic Ground Squirrel account for dimensions and a description of the scat.

REFERENCES

Bennett, R.P., 1999; Betts 1992; Dobson, F.S., and Oli 2001; Elliott and Flinders 1991; Festa-Bianchet and King 1984; Hansen, R.M., 1954; Harestad 1990; Harris, M.A., and Murie 1984; MacHutchon and Harestad 1990; McLean 1978; Murie and Harris 1988; Nagorsen 2005a; Neuhaus 2000a, 2000b; Neuhaus and Pelletier 2001; Smith, H.C., 1993; Stevens, S.D., 1998; Stevens, S.D., et al. 1997; Verts and Carraway 1998; Young 1990.

Franklin's Ground Squirrel

also called Grey Gopher

FRENCH NAME: **Spermophile de Franklin**

SCIENTIFIC NAME: *Spermophilus franklinii*

These ground squirrels are usually shy and secretive, preferring to stay under cover of protecting foliage. They commonly produce their musical bird-like alarm trill from a patch of dense vegetation, so you will likely hear one long before you see it. This species has not

Franklin's Ground Squirrel
(*Spermophilus franklinii*)

been well studied and there are many aspects of its natural history that are poorly understood.

DESCRIPTION

Franklin's Ground Squirrels are generally dark with a brown to brownish-yellow wash over their back, rump, and sides, liberally mixed with black hairs providing a speckled appearance. They have a white eye ring and a grey belly and feet. Tail fur is grizzled grey to grey-brown above and grizzled grey below with a very indistinct dark tip. There is a faint suggestion of a pale rim of hairs around the edge of the tail. The body of this squirrel is longer, making it appear slimmer than other ground squirrels. Like all the ground squirrels, their claws, both fore and hind, are prominent and dark and their ears are short and tight to their head. They moult once each year, during the summer, in a diffuse pattern where individual hairs are replaced throughout the coat and there is no overt indication of moult. The dental formula is incisors 1/1, canines 0/0, premolars 2/1, and molars 3/3, for a total of 22 teeth.

SIMILAR SPECIES

Despite its kinship to other ground squirrels, this species is most easily mistaken for an Eastern Grey Squirrel due to its fur colour, relatively long tail, and long slim body. Eastern Grey Squirrels are tree squirrels with larger ears, longer, fuller tails with longer tail hairs, and a back colour that is usually a clear grizzled grey without any brown or yellow-brown tint. The black form of the Eastern Grey Squirrel is not common in the area where the ranges of the two species overlap. All the somewhat similar ground squirrels are either lighter (Richardson's), smaller (Thirteen-lined), or rufous-bellied (Columbian).

SIZE

The following measurements apply to adult animals within the Canadian portion of the range. Body weight varies over the year. Adult males are lightest three to four weeks after emergence in spring and heaviest just before hibernation. Adult females are lightest upon emergence and heaviest in mid-July just before hibernation. During spring and early summer, adult males and adult females are similar in their body size and weight. Males only become heavier than females in the summer, as they begin storing fat for hibernation and the females are still lactating.

Males: total length, 329–412 mm; tail length: 113–153 mm; hind foot length: 51–58 mm; ear length: 14–15 mm.

Females: total length, 330–400 mm; tail length: 80–149 mm; hind foot length: 50–57 mm; ear length: 15–19 mm.

Weight: males, 314–711 g; females, 336–484 g; newborns, 7–9 g, with pups in larger litters being lighter than those from small litters.

RANGE

Franklin's Ground Squirrels occur in the Canadian prairies from central Alberta across central Saskatchewan to southern Manitoba. South of the border in the United States, they are found from the western side of Lake Michigan to northern Kansas and northwards to North Dakota.

ABUNDANCE

There is evidence, especially in the eastern parts of their range, that these squirrels are becoming uncommon and in some places disappearing. This may reflect a loss of prairie habitat and an increase in woodlands in those regions. Franklin's Ground Squirrels tend to be scattered within their range at the best of times and are rarely common, except locally in some years. Indications of cyclic population fluctuations, which may be four to ten years between peaks, are reported in some parts of the range in the United States. Franklin's Ground Squirrels are usually considered agricultural pests and are regularly poisoned or shot. Once extirpated from an area, they are slow to recolonize, as dispersing juveniles look for active colonies into which to settle, rather than occupying suitable, but unpopulated alternatives. Franklin's Ground Squirrels have become the target of predator-control programs where hunters believe they are predating too many duck nests.

ECOLOGY

Franklin's Ground Squirrels are strictly diurnal. They inhabit tall-grass habitats, which provide them with dense cover, but are not so tall that they cannot see over them when standing or sitting up. Colonies are commonly found at forest-prairie edges, in shrubby grasslands, and around marshes and sloughs. They may also be found along highways and rail lines where the grasses remain uncut. There are three prerequisites for colony location – a site with easy-to-dig soil, not subject to spring flooding or ploughing, and with suitable vegetation. Burrow systems are often dug into steep slopes, which avoids flooding. Burrows can be extensive, with many branches and chambers. Tunnel diameter is 8–10 cm, while chambers may be 30 × 25 × 20 cm in size. Tunnels extend to 2 m below the surface. Burrow entrances are generally in groups of three or four, or more, not far apart and evidently connected below ground. A brush heap, log, dense thatch of grasses, or stone pile commonly provides protection for the entrance. The main access hole has a large mound of waste soil in front of it, but most of the other burrow accesses are considerably less conspicuous. These smaller unobtrusive entrances are scattered throughout the colony and are used mainly as bolt holes in an emergency. Colonies are typically small and loosely knit and often go unnoticed in the thickly vegetated habitats that they prefer. There are 1.3 to 2.5 adults per ha in those Canadian colonies so far measured. The sex ratio of adults in this species is close to 1:1, which represents significantly more adult males than is found in other ground squirrel species in Canada and may reflect differences in reproductive efforts or social organization. These squirrels can climb the small shrubs and trees often found within their chosen habitat.

Franklin's Ground Squirrels escape the harsh winters by hibernating in underground chambers, which are well insulated with dried vegetation. Several may hibernate together in the same burrow. The adults begin to hibernate early, usually as a reaction to seasonal dry weather that results in desiccation of their food plants. The juveniles need more time to accumulate winter fat, and so must remain above ground longer to continue foraging. Adult males enter hibernation in July, adult females about two weeks later in late July to mid-August, and juveniles begin hibernating in September or early October. Adult males are the first to emerge in spring. They may be seen above ground in early April. Adult females emerge about two weeks later and yearlings follow a few months later, in early to mid-July. Time of emergence varies with local temperatures and snow loads. Emergence and departure into hibernation is about two to four weeks later in northern Alberta than it is in southern Manitoba. The species has been reported to store grain in the hibernation chambers, but it is unknown whether, like Richardson's Ground Squirrels, only the adult males do so. It is likely that these males do cache food for consumption in early spring, as they emerge from their hibernation burrows with their testes fully scrotal and filled with viable sperm. This level of breeding readiness is only possible if the animals had aroused from torpor some time earlier and maintained a normal body temperature for a week or more. Since these same males appear above ground in good physical condition, it is likely they were eating something during that time. The situation is less clear for females, but if they are like other ground squirrels, they probably do not collect a cache of seeds in late summer, relying instead on their body fat to sustain them through hibernation.

Distribution of the Franklin's Ground Squirrel (*Spermophilus franklinii*)

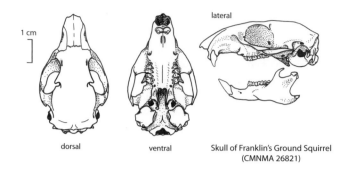

dorsal   ventral   Skull of Franklin's Ground Squirrel
(CMNMA 26821)

1 cm

lateral

Juveniles are thought to disperse in either spring or late summer. Males disperse farther than females and dispersal distances of 1.0–3.6 km have been reported. American Badger and Red-tailed Hawk are the main predators of Franklin's Ground Squirrels and others include Coyote, Red Foxes, weasels, skunks, snakes, and other raptors.

DIET

These ground squirrels are quite omnivorous. They eat primarily vegetation (leaves, bulbs, roots, fruits, flowers, nuts, and seeds), but when available will catch and consume insects, small mammals, frogs, and birds and bird eggs. In some marsh locations they have become significant nest predators of eggs and ducklings and will even take a mature duck. Like other ground squirrels, they also eat the dead of their own species. Franklin's Ground Squirrels are fond of most grain crops and will pull the seed head down in order to eat it rather than cut the stem, as do Thirteen-lined and Richardson's Ground Squirrels. Garden vegetables are also relished when available.

REPRODUCTION

The breeding season begins soon after the females emerge from hibernation and is usually complete by early June. Healthy, well-nourished females are capable of producing a single litter each year, but in some colonies between 25% and 50% of adult females do not reproduce in any given year. Gestation takes 28 days and the pups are born pink, deaf, blind, and with their toes fused together. Litter size ranges from two to thirteen, with an average of seven to nine pups. Growth is rapid and pups appear above ground about 26–28 days after their birth. They begin eating solid food when they come to the surface and are weaned at 28–30 days old. In those Canadian populations that have been studied, it is thought that the vast majority of yearlings do not reproduce in the summer after their birth, waiting instead until the following year, but this may be a factor of nutrition rather than age, so that well-fed yearlings may be capable of reproduction.

BEHAVIOUR

Franklin's Ground Squirrels are highly secretive and typically live alone, in pairs, or in family groups within a scattered and diffuse colony. They are said to be the least social of all the ground squirrels. They have not been seen to engage in the nose-to-nose greeting used by many other ground squirrels. They are also rarely seen scent marking, even though they do have scent glands in the corners of their mouth, on their backs, and beside their anus. A musky smell is evident around a colony during the breeding season, indicating that scent does play a role, perhaps in courtship or territoriality. Individuals rarely contact one another and appear to avoid interacting. If they do engage another ground squirrel, outside of a male-female connection during breeding season, their behaviour is typically threatening as they growl and posture and occasionally fight. Unlike most of the other ground squirrels, Franklin's have developed a method of puncturing a duck egg in order to consume the interior. They curl around the egg and thrust it against the upper incisors with their hind legs. This species often becomes bold around campgrounds, where it will approach people closely for food handouts.

VOCALIZATIONS

Their alarm call is a long, bubbling trill that is almost bird-like. Squeals are produced in fear or pain. Growls are used as a threat. Young in the nest emit a high-frequency peeping noise if disturbed.

SIGNS

Burrow entrances are generally concealed in brush or weed patches, from which well-worn trails or runways radiate to other burrows or feeding grounds. There is a conspicuous throw mound in front of a main entrance (10–12 cm in diameter) but bolt holes have no throw mound and tend to have smaller openings (8 cm). A duck egg that has been predated by a Franklin's Ground Squirrel is usually less than half broken, the opening has small chips still attached to the inner membrane, and sometimes tooth marks are evident. Some scat is deposited in an underground latrine, but usually there is some present on the surface as well. See the Arctic Ground Squirrel "Signs" section for descriptions of scat and tracks.

REFERENCES

Haberman and Fleharty 1971; Iverson and Turner 1972; Jannett et al. 2007; Johnson, S.A., and Choromanski-Norris 1992; Kivett et al. 1976; Martin, J.M., and Heske 2005; Martin, J.M., et al. 2003; Murie, J.O., 1973; Ostroff and Finck 2003; Pergams and Nyberg 2001; Sargeant et al. 1987; Sowls 1948.

Golden-mantled Ground Squirrel

FRENCH NAME: **Spermophile à mante dorée**
SCIENTIFIC NAME: *Spermophilus lateralis*

Golden-mantled Ground Squirrels and the slightly smaller Cascade Mantled Ground Squirrels look like large chipmunks. Both of these montane ground squirrels have a single white stripe on each side, which stops at the shoulder and does not reach the face.

Golden-mantled Ground Squirrel
(*Spermophilus lateralis*)

DESCRIPTION

The striking white stripe on each side, bracketed by two distinct black bands, coupled with a golden russet hood over the neck and shoulders, distinguish this species. The eye ring is white, the back is grey brown, and the belly and chin are creamy white. The sides of the body below the stripe may be greyish to yellowish white. The top of the tail is brown in the centre with a blackish margin. Underneath it is reddish to yellowish brown in the centre, edged again with black. The Golden-mantled Ground Squirrel has a single annual moult that occurs in early summer for the adult males, following weaning for the adult females, and soon after weaning for the young of the year. Moult begins at the nose and proceeds backwards and down the sides. A fresh coat is noticeably brighter than an old coat, which fades with time. The dental formula is incisors 1/1, canines 0/0, premolars 2/1, and molars 3/3, for a total of 22 teeth.

SIMILAR SPECIES

The most similar species to the Golden-mantled Ground Squirrel is the closely related Cascade Mantled Ground Squirrel. In general, the colours on the Cascade Mantled Ground Squirrel are more muted. The white stripe is less bright and distinct, and the black borders are more diffuse. The hood is duller and more brownish red than yellowish red. The ranges of the two species do not overlap. Golden-mantled Ground Squirrels also closely resemble large chipmunks, which have multiple light stripes that continue onto the head and through the eye. Chipmunks carry their tails stiffly vertical when they run.

SIZE

The following measurements apply to adult animals within the Canadian portion of the range. Body weight varies over the year. Adult males are lightest three to four weeks after emergence in spring and heaviest just before hibernation. Adult females are lightest upon emergence and heaviest just before hibernation. Adults of both genders are similarly sized.

Total length: 171–360 mm; tail length: 60–132 mm; hind foot length: 35–55 mm; ear length: 12–24 mm.

Weight: 152–378 g; newborns, around 6 g.

RANGE

This squirrel occurs throughout the western mountains from New Mexico and California to Alberta and British Columbia.

ABUNDANCE

The Golden-mantled Ground Squirrel is relatively common in appropriate habitats.

ECOLOGY

Golden-mantled Ground Squirrels are most abundant on talus slopes, boulder fields, and rock-slide regions bordering alpine and subalpine meadows and grasslands. They are fairly versatile in their environmental requirements and can also exploit open forests, clear-cuts, rocky railway beds, recently burned forest, and even some lower-elevation sites such as the open Ponderosa Pine forests around Okanagan Lake in southern British Columbia. Common features of all their habitats are boulders or stumps for viewing platforms and rocky crevices or deadfall for protection from predators with nearby open areas with sufficient food. The severe winters that occur in their range at those elevations are avoided by hibernating for six or seven months. Animals at higher elevations hibernate longer than those at lower elevations. While hibernating, these squirrels curl up with their nose near the base of their tail. The body temperature approaches the ambient temperature in the burrow and they breathe about once each minute. Males emerge from hibernation first, typically in early April to late May, and females several weeks later. At higher elevations, adults begin retreating into their hibernation burrows in August, with males going first. Young of the year are last to hibernate, but all of them are underground by late October. At lower elevations, some juveniles may remain above ground into November. During hibernation, the animals periodically arouse to move around for a short time before falling into torpor again.

These diurnal squirrels are most active above ground in early morning and late afternoon, avoiding the heat of the day underground. The sex ratio of pups is 1:1, but adult females outnumber adult males, possibly due to the greater risk of exposure to predators by the more mobile males, particularly during the breeding season. Golden-mantled Ground Squirrels confine most of their activity to scurrying over and around obstacles at ground level, but they are capable climbers, and may climb up to 10 m into shrubs and trees, especially when harvesting pine seeds. Burrows are usually constructed under the protection of boulders and tree roots, and even under or beside cabins. Burrow entrances are typically well hidden by rocks, logs, or vegetation, and rarely does this squirrel leave a mound or even any waste soil to mark a burrow entrance. Main burrows are 20–90 cm deep and may be up to 180 cm long, with several side passages and more than one entrance. Usually one or more of these side passages will terminate in a nest chamber filled with dry vegetation or

Distribution of the Golden-mantled Ground Squirrel (*Spermophilus lateralis*)

no indications that the animals store food. Perhaps populations in some locations, such as those at higher elevations or higher latitudes cache food, while those at lower elevations or lower latitudes do not. Much of their water requirement is supplied by succulent vegetation or dew, but free water will be readily consumed, especially during hot weather. Easily tamed and willing to eat almost anything, these engaging squirrels quickly learn to beg for handouts.

REPRODUCTION

Almost all males are in breeding condition when they emerge from their hibernation burrows in the spring. This likely means that they arouse from torpor underground and maintain a normal body temperature for some time before they emerge, which permits testes development and sperm formation. As the testes enlarge, they drop into the scrotum from an abdominal position and the skin of the scrotum blackens. Testes regress rapidly after the breeding season and assume an abdominal position again by June. Females emerge two to three weeks after the adult males and become sexually receptive soon after that. The breeding season lasts a few weeks. The gestation period is about 27–28 days and the average litter consists of 5 pups (range 2–8). One litter is produced per year, but should a female lose her litter early, she may come into heat again. Most births occur in May and June. At birth the pups are deaf, blind, pink, and almost hairless and their toes are fused together on each foot. Growth is rapid. Their ears open between 14 and 20 days old. By 20 days old the toes are distinct, the lower incisors have erupted, and most of the skin is furred. Their eyes open between 21 and 30 days. Young begin eating solid food at around 29–35 days and are weaned soon after that. Both sexes achieve sexual maturity in their second spring, but despite this, many yearlings and even two-year-olds in the higher elevation populations fail to breed.

BEHAVIOUR

Golden-mantled Ground Squirrels are generally solitary, apart from the short breeding season and while females are rearing pups. Although quite solicitous while the infants are helpless, maternal females become increasingly aloof and intolerant once their pups are above ground and eating solid food, so the young disperse within two to three weeks of weaning. Pups newly emerged from the nest stick together and play frequently. As they become more confident over the next few days, they begin to forage and play farther from the burrow. Eventually they strike out to find a territory of

shredded bark, but some chambers are simple swellings of the tunnels themselves. Shorter, more rudimentary burrows (called refuge burrows or bolt holes) are scattered throughout the territory for use in emergencies. Quick and agile, these ground squirrels will forage at some distance from a burrow entrance, counting on their top running speed of 11–12 km/h to get them to safety if a predator appears. Maximum lifespan of the Golden-mantled Ground Squirrel in the wild is at least seven years. Predators include raptors, Coyotes, Bobcats, skunks, and weasels.

DIET

Like other ground squirrels, this species is omnivorous. The bulk of their diet is made up of seasonally obtainable leaves, stems, seeds, nuts, flowers, fruit, and berries of a wide range of grasses, forbs, shrubs, and trees. This plant consumption is supplemented by a variety of fungi (e.g., truffles), invertebrates, small vertebrates, and carrion when available. Golden-mantled Ground Squirrels will catch and kill small birds, mammals, and reptiles and plunder the nests of small birds for their eggs and nestlings. They will also consume dead of their own species. Some researches claim that they gather and store food in late summer and autumn. Most is cached, in or near their main burrow, but smaller or temporary caches may be scattered throughout the territory. Other researchers have found

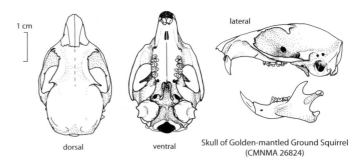

Skull of Golden-mantled Ground Squirrel (CMNMA 26824)

dorsal   ventral   lateral

1 cm

their own. Individuals are territorial, and in one study, home ranges varied in size from 0.4 to 4.1 ha. Threat displays, fights, and chases between neighbours are common. Golden-mantled Ground Squirrels do not usually assume the vigilant and upright "picket pin" posture so typical of many other ground squirrels. They tend rather to remain on all fours or to sit on their haunches (see the illustration) to free the forepaws for food manipulation. They will casually hunch beside a tall plant and using a hand-over-hand technique they pull the head of the plant down until they can reach the seeds or flowers. Forepaws are used to stuff seeds into the cheek pouches, or grasses for nesting material into the mouth. Elevated positions on stumps or boulders are used as lookout perches and for basking in the heat of the sun. Dust baths are a frequent activity and are usually followed by a bout of grooming.

VOCALIZATIONS

This animal tends to be relatively quiet and rather secretive. Very young pups in the underground nest squeak if disturbed or excited. At around 15 days old, this sound changes to a chucking call. Adults use a variety of calls, many of which are poorly described and have not been well studied. These calls, and likely others as yet unidentified, are used to communicate fear, aggression, anxiety, and alarm. Some calls are accompanied by flicks of the tail to add a visual dimension to the signal. A high-note alarm call warns of an aerial predator. A high-pitched trill warns of terrestrial predators. Both these calls have high-frequency components that are into the ultrasonic range and beyond human hearing. Golden-mantled Ground Squirrels do respond appropriately to at least one other species' alarm call. When a Yellow-bellied Marmot sounds an alarm, nearby Golden-mantled Ground Squirrels stop to take a look or run for cover.

SIGNS

Burrow entrances are small, typically 5–9 cm in diameter, and rarely have any waste soil at their well-hidden entrances. Often too cryptic to be found, they are sometimes easier to detect by looking for the runs (7–9 cm in width) that radiate from burrow entrances to foraging sites or other burrows. (See the track discussion in the "Signs" section of the Arctic Ground Squirrels account for information on ground squirrel track patterns.) Golden-mantled Ground Squirrels use three gaits. Most commonly, they bound when moving above ground and slow to a walk when foraging. They occasionally lope when moving across open spaces. Their front track is 2.2–2.5 cm long and 1.3–1.4 cm wide. Their hind track is usually 2.2 cm in length and 1.6–1.7 cm wide. Dust-bath craters are shallow pits about 18 cm in length and 9–10 cm in width. Golden-mantled Ground Squirrels do not litter their burrow entrances with scat, as do many other species of ground squirrels. They use the droppings and urine to mark their territory. Look for scat on favourite perches and along runs. They also create a shallow scrape near their burrow entrances into which they urinate and defecate. These scrapes can completely encircle a well-used burrow. It is not known whether both sexes are responsible for these scrapes or whether males alone create them, as is the case with Arctic Ground Squirrels. Scat consistency varies with the amount of moisture in the diet. Scat is typically oblong (0.5–0.8 cm in diameter by 0.5–1.3 cm long), but tends to be irregular in shape and often has one blunt and one pointed end.

REFERENCES

Bartels and Thompson 1993; Bihr and Smith 1998; Eiler and Banack 2004; McKenzie 1990; Nagorsen 2005a; Shriner 1998; Smith, H.C., 1993; Smith, R.J., 1995; Verts and Carraway 1998.

Arctic Ground Squirrel

FRENCH NAME: **Spermophile arctique**
SCIENTIFIC NAME: *Spermophilus parryii*

This is the most northerly of the ground squirrels and is easy to identify, as it is the only ground squirrel that lives in northern alpine and barren-ground tundra. Because of the short growing season and long winters at those latitudes, this squirrel has the longest hibernation period of any mammal.

DESCRIPTION

Fur on the forehead and top of the snout is a bright rufous colour. Fur on the back is greyish washed with brown or reddish brown and lightly marked with small, regularly spaced tan or whitish spots. The eye ring and fur around the chin is pale yellow to yellowish-grey. Claws are black and they show up well against the pale yellowish-orange feet. The upper part of the tail is brown, while the underside is reddish and it has a black tip. The throat and underparts are a bright tawny orange to buffy orange. Melanistic, white, and unusually coloured individuals are rare, but most frequent in some more northerly populations. Arctic Ground Squirrels moult once each year, during the summer, in a diffuse pattern where individual hairs are replaced throughout the coat and there is no overt indication of moult. The dental formula is incisors 1/1, canines 0/0, premolars 2/1, and molars 3/3, for a total of 22 teeth.

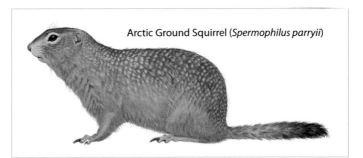

Arctic Ground Squirrel (*Spermophilus parryii*)

SIMILAR SPECIES

Columbian Ground Squirrels are quite similar to Arctic Ground Squirrels, but their ranges do not overlap. Columbian Ground Squirrels occur in Canada only in southeastern British Columbia and southwestern Alberta.

SIZE

The following measurements apply to adult animals within the Canadian portion of the range. Body weight varies over the year. Adult males are lightest three to four weeks after emergence in spring and heaviest in autumn before hibernation. Adult females are lightest upon emergence and heaviest in early August just before hibernation. Males are slightly longer than females.

Total length: 316–495 mm; tail length: 65–153 mm; hind foot length: 42–70 mm; ear length: 9–18 mm.

Weight: 355–1325 g; newborns, 11–14 g.

RANGE

This ground squirrel occurs across northern North America and into western Siberia. In North America, it occurs as far north as the north coast of Alaska and Yukon, and to the limit of mainland Northwest Territories and Nunavut. Southerly distributional limits are northern Manitoba at Hudson Bay across to northwestern British Columbia.

ABUNDANCE

Arctic Ground Squirrels are locally common in patches of favourable habitats throughout their range, especially on well-drained gravel ridges. Population size of colonies, especially those in semi-treeded areas, often fluctuates from year to year.

ECOLOGY

Arctic Ground Squirrels are a diurnal, colonial, Arctic-adapted species found in northern habitats that provide a deep soil with few rocks and permafrost that is at least a metre down for easy digging of burrows, and where there is sufficient food and no seasonal flooding. Most occur on tundra or in alpine meadows above the treeline, but there are colonies in clearings in the boreal forest, especially in the region of stunted, wind-scoured spruces along the transition zone where the boreal forest gradually accedes to tundra. Such locations often prove to be population sinks (less than optimum habitat where surplus animals can survive for a while, especially in good years, but where long-term reproductive success is usually limited). Unlike most ground squirrels, Arctic Ground Squirrels emerge from hibernation while the climate is still harsh, and before the short growing season has begun. Mature males are the first to emerge in spring. They must be reproductively capable when they emerge, so they awaken from their winter torpor one to four weeks before leaving their burrow. During this time they remain sequestered, eating from their larder of stored food gathered the previous summer, and regain all the weight they lost during hibernation. When they do arrive above ground in about mid-April to mid-May, they are fit and ready for the rigours of breeding season. They then essentially fast for the next two to three weeks and lose around 21% of their body

Global distribution of the Arctic Ground Squirrel (*Spermophilus parryii*)

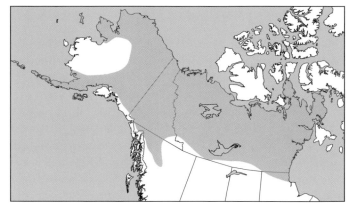

expanded view of the North American range

weight as they battle other males to establish a territory and then search for newly emerged mature females with which to mate. More than half of the adult males disappear soon after the breeding season and it is thought they die due to the stresses of the period. Females do not store food in their burrows, but depend on their substantial fat stores from the previous summer to sustain them during hibernation. They lose about 30% of their body weight over the winter, but still have energy in reserve to begin gestation before the growing season is truly under way. Reproductive females begin emerging about 10–14 days after the males and the non-reproductive yearlings may emerge as late as early June. Autumn departure into hibernation burrows is also staggered, with adult females going first in late July or early August, followed by adult males in September. Juveniles of both genders begin hibernation between mid-August and mid-September. In Arctic regions, most individuals in a colony hibernate for eight to ten months. In the subarctic parts of Yukon and British Columbia, where winters are less prolonged and severe, the hibernation period is more likely seven to eight months. Juveniles typically disperse in July and more males leave than do females, since many females remain in empty burrows in their natal colony. Some juveniles are known to travel over 3.5 km. Dispersing ground squirrels run a gauntlet of predators and only 25%–40% survive the trip to a new colony. Young males tend to travel farther than young females, and since the risk of predation increases with distance, more of them die.

Arctic Ground Squirrels dig two types of burrows. Home or resident burrows are extensive, with a throw mound at the entrance and multiple openings (up to 15), and are used for shelter, sleeping, hibernation, and raising young. They typically contain several metres of tunnels 10–15 cm wide, with at least one nesting chamber lined with dry vegetation and fur, and in the case of adult males, an area to store food. Each home range also contains a number of refuge burrows, which are short and meant to provide temporary shelter. These burrows usually have smaller entrances and no throw mound. In southerly reaches of their range, where they may live in clearings in boreal forests, Arctic Ground Squirrels will climb into trees to heights of up to 6 m to escape predators if they find themselves too far from the security of a burrow. Females are known to nest in woodpecker holes, and all age classes, except pups, will forage in trees and shrubs for shoots, leaves, berries, and flowers. Stumps and trees are also used to provide surveillance perches to watch for predators. Colonies in alpine and boreal forests are heavily predated, especially in years of low Snowshoe Hare abundance, when the predators are forced to hunt alternative prey. With more plant growth in that ecozone, the only ground squirrel colonies that survive this period of severe predation pressure are those on sloped sites with high visibility. Alpine and boreal-forest colony size varies considerably depending on the level of predation, whereas colonies on the tundra tend to have more stable populations. Arctic Ground Squirrels are highly sought after prey. They are hunted by Brown Bears, Coyotes, Red and Arctic Foxes, Canada Lynx, weasels, and an assortment of raptors. Adult squirrels have been seen chasing Ermine away from their burrow. Clearly, these weasels are too small to capture adults and are likely only a threat to unprotected pups.

DIET

Food includes shoots, leaves, berries, and seeds of a variety of forbs and shrubs, as well as grasses, horsetails, and fungi. Although primarily herbivorous, Arctic Ground Squirrels will supplement their diet occasionally with carrion, insects, small rodents, and even the dead of their own species. During July and August, males spend considerable time carrying seeds, leaves, grasses, and sedges into their burrows to consume in early spring.

REPRODUCTION

Females have a single litter each year and are receptive to mating for approximately four hours. They breed about three to four days after emerging from hibernation and normally first with the resident male whose territory overlaps theirs. Although they will often then proceed to mate with other males, 90% or more of the pups are sired from the first mating. This may be because a portion of the male ejaculate forms a coagulated mass called a "copulatory plug" in the female's vagina, which effectively blocks it for a time, and thereby hinders later sperm from reaching the fertile eggs. The gestation period is around 25 days and the litter of 4 to 7 young is typically born in an underground nest chamber in mid- to late May. The pups are hairless and blind at birth, but develop rapidly. They appear above ground about 30 days later, around mid- to late June and are weaned within a week or two after that. About 40%

of yearling males are sexually mature and the remainder become mature in their second year. Most females are sexually mature as yearlings, in the spring following their birth.

BEHAVIOUR

Arctic Ground Squirrels live in loose communities where each animal occupies its own burrow system. Home ranges of males may be 3–20 ha in size, while those of females are 2.3–3.1 ha. Size of the home range is directly dependent on food availability for females and food and mate availability for males. When the adult males first emerge from hibernation, they immediately establish a territory and then aggressively attempt to defend it from rival males. Calls, chases, and fights are common in spring as ownership is disputed and territorial boundaries shift. Each male's territory is selected to encompass or overlap the home ranges of several females. Females are less territorial, although they will defend the burrow containing their litter. Their home ranges commonly overlap, not only with the dominant male, but also with neighbouring females that are related, typically mothers and daughters. Mating mostly occurs underground. Following mating, some males will escort the female for a time (average 33 minutes, but up to 90 minutes), both preventing her from leaving the area and aggressively denying access by other males. Males that escort females after mating have always proved to be the first to mate with that female, and males of subsequent matings did not escort after copulation. The breeding territories break down in late summer, but small areas around the hibernation burrows are still defended. Males of this species cache food in summer. A hibernating ground squirrel plugs the entrance of its hibernation burrow to prevent predation while it is in torpor. Like other ground squirrels, these animals tend to rise on their hind legs to scan their surroundings and to sound an alarm call.

VOCALIZATIONS

A loud shrill whistle warns of aerial predators and a chattering "chick-chick-chick" alarm call warns of terrestrial dangers. Receptive females produce a distinctive bird-like chirp during courtship. Courting males utter a louder chirping call, commonly repeated many times, which is distinct from that produced by females. Territorial males use yet another loud chirping call to advertise their presence and to challenge possible rivals. An injured animal may squeal.

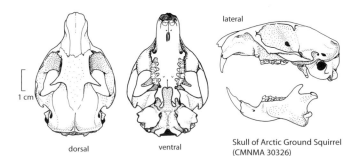

dorsal   ventral   Skull of Arctic Ground Squirrel
(CMNMA 30326)

1 cm

stride length
of a bound
can vary from
10 to 55 cm with
the larger ground
squirrels taking
the longest strides

Bound

Left hind track
Length: 2.4–4.6 cm
Width: 1.6–3.0 cm

**Arctic
Ground Squirrel**

Left front track
Length: 2.7–4.3 cm
Width: 1.3–3.7 cm

SIGNS

Tundra colonies can be identified by the small rises caused by the waste soil in the throw mounds. Colonies in the alpine and boreal forest are harder to see from a distance, but tend to be in clearings and forest edges where there is a better line of sight for spotting predators. As you approach an active colony, the terrestrial predator alarm call will be sounded. Since they are active long before the snow disappears, tracks and runs in the snow are possible. All the ground squirrels produce similar tracks, although the size and stride lengths may vary. The front feet have five toes, but the pollex or thumb is vestigial and rarely leaves an impression in a track. The hind feet have five toes. Toes 1 and 5 angle out to the side and the middle three toes register close together pointing forward. Claw impressions from both front and rear feet are prominent and longer than those of tree squirrels. In a typically bounding gait track pattern (as illustrated) the front feet registration is offset behind the hind feet. Most tree squirrel bound patterns leave an impression of the front feet that are roughly side by side. The exceptions are some track patterns of Red and Douglas' Squirrels, which may display the offset pattern, similar to that of smaller ground squirrels. Arctic Ground Squirrels have two principal gaits. They bound when they are travelling and walk while foraging.

Burrow entrances vary in size depending on their use. Refuge burrows may be only 9 cm in diameter, but hibernation and summer or home burrows may be up to 20 cm wide by 13 cm high. (See in the "Ecology" section above for more information on burrows.) Runs radiate from burrow entrances to foraging sites or favourite perches. Depending on the water content of the squirrels' diet, scat may be hard and dry, or softer and clumped. Both forms tend to be irregular in shape and often have one blunt and one pointed end. Hard scat measures 0.4–0.6 cm in diameter and 0.8–1.0 cm in length. Soft scat may be 1.0–1.3 cm in diameter and 1–2 cm in length. Scat is commonly found near burrow entrances and males may deposit scat into the middle of scrapes made to mark their territory. These scrapes are shallow pits quickly created with a few scratches of the front paws to clear the ground surface. Arctic Ground Squirrel territorial males will also dig a larger, deeper pit to mark the location of a rival's defeat. Although not a sign left by the ground squirrels, Brown Bear excavations of Arctic Ground Squirrel burrows can leave great holes in the ground marking the location of former burrows.

REFERENCES

Boonstra et al. 1990, 2001; Buck and Barnes 1999; Byrom and Krebs 1999; Elbroch 2003; Gillis et al. 2005a, 2005b; Hansen, R.M., 1954; Hubbs and Boonstra 1998; Hubbs et al. 1996; Karels and Boonstra 1999; Lacey, E.A., et al. 1997; Murie, J.O., and McLean 1980; Nagorsen 2005a; Simpson 1990, 1994; Wrigley 1974.

Richardson's Ground Squirrel
also called Prairie Gopher, Gopher, Flickertail

FRENCH NAME: **Spermophile de Richardson**

SCIENTIFIC NAME: *Spermophilus richardsonii*

This species, along with Thirteen-lined, Franklin's, and Columbian ground squirrels are called gophers by people living within their western range. All three species have long been considered pests by most prairie farmers and ranchers, partly due to their crop depredations and partly because their burrow entrances are thought to pose a potential danger to livestock.

DESCRIPTION

The forehead and top of the snout are a pale rusty orange to buffy grey. The eye ring is creamy white, as is the region around the nose and mouth. Cheeks are a pale, yellowish grey. The back is buffy grey with a flecked appearance caused by a speckling of dark hairs. The throat, belly, legs, and feet are sandy to yellowish buff and are lighter than the back. The tail is buffy grey above with an indistinct black tip and is buff or light grey below with a buff-coloured edge, sometimes visible from above. Richardson's Ground Squirrels moult once each year during the summer, in a diffuse pattern where individual hairs are replaced throughout the coat and there is no overt indication of moult. The dental formula is incisors 1/1, canines 0/0, premolars 2/1, and molars 3/3, for a total of 22 teeth.

SIMILAR SPECIES

Other ground squirrels that might be mistaken for the Richardson's Ground Squirrel dig similar-looking burrows, but are quite different in appearance. A Franklin's Ground Squirrel is about the same size as a Richardson's, but is considerably darker, has a longer tail and coarser hair. Black-tailed Prairie Dogs are much larger and paler than Richardson's Ground Squirrels. They have a creamy white belly and a tan-coloured back, and the black tip on their tail is very distinct. Thirteen-lined Ground Squirrels are smaller and slimmer and can be distinguished by the pattern of dark lines bracketed by pale lines and light spots running longitudinally along the back.

SIZE

The following measurements apply to adult animals within the Canadian portion of the range. Body weight varies over the year. Adult

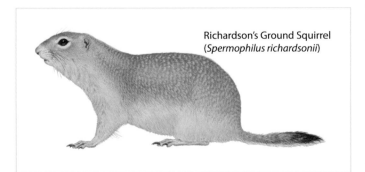

Richardson's Ground Squirrel
(*Spermophilus richardsonii*)

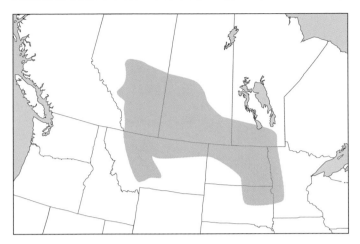

males are at their lightest three to four weeks after emergence in spring and are heaviest in June before hibernation. Adult females are lightest upon emergence and heaviest in early July, just before hibernation. Males are larger than females.

Total length, males: 279–337 mm; tail length: 65–92 mm; hind foot length: 41–50 mm; ear length: 10–15 mm.

Total length, females: 264–318 mm; tail length: 55–82 mm; hind foot length: 42–46 mm; ear length: 12–14 mm.

Weight: males (pre-hibernation) 440–745 g, (post-hibernation) 271–500 g; females (pre-hibernation) 400–475 g, (post-hibernation) 175–360 g.

Newborns: weight 4.4–9.6 g.

RANGE

Richardson's Ground Squirrels are animals of the northern plains. They occur from southern Alberta to southern Manitoba and southwards to western Wisconsin and southwestern Montana.

ABUNDANCE

These ground squirrels are considered common in short-grass prairie throughout their range and may be locally abundant, although in many agricultural regions they are sufficiently controlled that numbers may be low. A well-studied colony in Alberta occupied an area of around 18–19 ha, and the population size fluctuated between 7 and 37 adults per ha.

ECOLOGY

This colonial ground squirrel lives in open grasslands, including rangelands, ditches, hay fields, and the edges of grain fields in the western prairies. They prefer areas where their burrows are not prone to spring flooding or ploughing, and where their view is not obscured by tall grasses or shrubs. Most colonies are separated by agricultural lands that are unsuitable for occupation. The Richardson's Ground Squirrel is diurnal, retreating to the security of its burrows for the night. During the heat of the summer, it ventures out only in early mornings and evenings, remaining in its cool burrow during the midday heat. It avoids the rigours of winter by

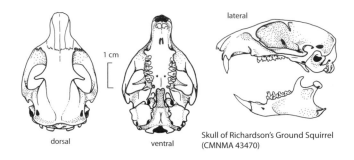

dorsal   ventral   lateral

Skull of Richardson's Ground Squirrel
(CMNMA 43470)

hibernating underground. Juvenile males hibernate for about four months, juvenile females for about six months, and adults for about eight months. Departure into hibernation depends on latitude, but also on rainfall. A dry summer may trigger an early departure. Adult males begin hibernating first, in mid-June to early July; adult females follow in mid-July to early August; juvenile females by late August or early September; and juvenile males by mid-October. Emergence from hibernation depends on local temperatures. In Manitoba, Saskatchewan, and central Alberta and the southern foothills, these squirrels begin emerging in late March. In other parts of southern Alberta they begin emerging about a month earlier. Males emerge first, with females following about two to three weeks later. Emergence may be delayed by a heavy snowfall or by unseasonably cold temperatures. Most males arouse from hibernation one to three weeks before emerging from their hibernacula. During that time, they eat from their larder of stored seeds and regain most of the weight they lost during hibernation. Once awake, their elevated body temperature allows the testes to begin producing sperm in readiness for breeding. When they finally emerge, they are in full breeding condition, with their enlarged testes descended into a scrotum of darkly pigmented skin. During the rest of the year, the testes are abdominal and the scrotal skin is pale pink.

Each ground squirrel hibernates alone, in a burrow used only for hibernation and rarely reused. Multipurpose burrows are occupied during the active period, for sleeping, birthing, and raising litters. Burrows may be as long as 15 m and 1–2 m below the ground surface. Sleeping, hibernation, and birthing chambers are 15–23 cm in diameter and are lined with dried grasses to form a comfortable and insulated nest. Although neonates and juveniles are about equally divided between males and females, among adult Richardson's Ground Squirrels, females outnumber males by more than 3:1. This is mainly due to two factors. Juveniles disperse from their natal colony about June or July when they are 8–12 weeks old and this dispersal is sex-biased, as more males leave than do females. Dispersing juveniles are very vulnerable to predation and many do not make it to neighbouring colonies where they might claim an empty burrow. In addition, the rigours of the breeding season cause extensive mortality among the adult males. About 45% of them disappear during, or right after, the breeding season. Males rarely live to three to four years old, while females may live five to six years. Richardson's Ground Squirrels are favoured prey of many predators. The Long-tailed Weasel, American Badger, Coyote, Red Fox, American Mink, Prairie Rattlesnake, skunks, and many raptors hunt them, as do the

Domestic Dog and Cat when the opportunity presents itself. Even Great Blue Herons have been seen catching and eating Richardson's Ground Squirrels. The smaller predators, such as Long-tailed Weasels and snakes, are more of a threat to juveniles and young still in the nest than they are to adults.

DIET

Richardson's Ground Squirrels are primarily herbivorous. They eat leaves, flowers, and seeds of a variety of grasses and forbs. Many males, especially those in the northern parts of the range, provision their hibernacula with a cache of seeds for consumption in the early spring before emerging. These squirrels will feed on road-killed ground squirrels, sometimes resulting in their own death by a subsequent motor vehicle. These ground squirrels learn to consume small eggs of prairie passerines and already damaged larger eggs, but are unable to penetrate the shells of eggs larger than quail.

REPRODUCTION

Breeding takes place soon after the females emerge from hibernation. Females are receptive for about two to three hours, usually in late afternoon within two to four days of emergence. Most will mate with several males, and about 80% of litters are sired by more than one male. Females have a single litter annually, beginning when they are a year old. Mating usually occurs underground. The gestation period is 23 days. An average litter is six to eight pups (range 3–13). Infants are born pink, more or less hairless, and deaf and blind. They develop rapidly, but only become ambulatory a few days before appearing above ground, at about 28–30 days old. Juveniles begin to eat vegetation as soon as they appear above ground, and they are weaned shortly afterwards. Juvenile males delay their first departure into hibernation in order to continue growing to their full adult size. They are then developed sufficiently to engage in the breeding season the following spring. Juvenile females depart for their first hibernation at least two months earlier than the juvenile males and complete their growth to adult size the following spring, while pregnant with their first litter. Amazingly, the yearling females are able to complete their own growth and still produce litters of equal size and weight to those of mature females.

BEHAVIOUR

Colonies are composed of a number of clustered matrilines where closely related females live and share a territory. Adult males occupy more transitory territories, which fluctuate widely during the breeding season. If receptive females are sparse or widespread, males may abandon territory defence altogether for the length of the breeding period. Related females commonly share an overnight sleeping chamber for a while after emerging from hibernation, but shortly before parturition the pregnant females relocate to separate birthing burrows and sleep alone until their young are born. Male-male aggression is highest during the breeding season, when chases and fights are common as males vie for access to receptive females. Loss of weight is typical for males during this time, as they are often too busy to forage, and injuries are frequent. Chases are the predominant antagonistic behaviour, as the pursuing male attempts to inflict injury on the rump of his opponent. Two males squaring off frequently present the side of their bodies to each other with their backs humped and the fur on their tails erected. This is called a "lateral display" and is meant to communicate intent to fight, but is also used as a tactic to allow the combatants to protect vulnerable parts of their bodies from injury, and to place themselves into a good position to counterattack. At the peak of the breeding season, males divide their time between looking for, and then guarding, receptive females, and fighting with other males with the same purpose. To check if a female is entering oestrus, a male repeatedly approaches and retreats while flicking his vertically held tail. She may in turn respond aggressively and pounce on him, push him away with her forepaws, or avoid contact and hide. If she accepts his proximity, he will advance further to sniff her anogenital region, frequently by approaching her from the front and pushing under her chest and belly. The lighter female is commonly flipped over or onto her side by this tactic. Like most of the ground squirrels, Richardson's Ground Squirrels regularly engage in a charming nose-to-nose greeting ceremony as they meet neighbours and kin. Usually this remains a peaceful encounter, but occasionally it will evolve into a chase if one animal takes exception to the other.

VOCALIZATIONS

Richardson's Ground Squirrels produce and often repeat three principal alarm calls. A short chirp warns of aerial predators, while a long whistle warns of terrestrial predators. Typically while vocalizing, the squirrels open their mouths and flick their tails in unison with the sound production. At around 90 decibels, these sounds may be heard up to a kilometre away. The third alarm call, produced about 10% of the time, is a highly directional ultrasonic call inaudible to humans. It is accompanied by the same tail flicking as the other alarm calls, but we can only hear a faint air rushing sound and hence have dubbed the call the whisper call. In the 48 kHz range, this quiet call travels only a short distance. The low volume and directionality of the signal allows the squirrels to warn nearby relatives without revealing their hidden presence to the predator. Repetition of alarm calls suggests the degree of threat, as more repetitions are uttered if the threat is closer. Other individuals hearing the calls appear to be able to identify the caller and thereby know if the call was produced by a nearby or distant squirrel, or perhaps by an unreliable caller, and to govern their actions accordingly. Tooth chattering is used as a threat. Squeals are emitted by both juveniles and adults in fear or pain. Chirrs are of a lower volume and may follow a chirp or whistle. Males produce a unique call, reminiscent of the aerial predator call, just before or just after they mate.

SIGNS

Since the squirrels are active long before the snow disappears, tracks and runs in the snow are possible. The front foot has five toes, but the pollex or thumb is vestigial and rarely leaves an impression in a track. The hind foot has five toes. Toes 1 and 5 angle out to the side and the middle three toes register close together pointing forward. Claw impressions from both front and rear feet are prominent. Richardson's Ground Squirrels have two principal gaits. They bound

when they are travelling and walk while foraging. Burrow entrances vary in size, depending on their use, and vary from 9 cm to around 16 cm in diameter. Runs radiate from burrow entrances. Depending on the water content of their diet, scat may be hard and dry or softer and clumped. Scat is commonly found near burrow entrances. See the track illustrations in the "Signs" section of the Arctic Ground Squirrel, which are larger but otherwise similar.

REFERENCES

Broussard et al. 2006; Hare 1998; Hare and Atkins 2001; Hare et al. 2004; Koeppl et al. 1978; Michener 1998, 2002, 2005; Michener and Koeppl 1985; Michener and McLean 1996; Pellis et al. 1996; Smith, H.C., 1993; Sykes 1996; van Staaden et al. 1996; Warkentin et al. 2001; Wilson, D.R., and Hare 2004, 2006.

Cascade Mantled Ground Squirrel

also called Cascade Golden-mantled Ground Squirrel

FRENCH NAME: **Spermophile des Cascades**

SCIENTIFIC NAME: *Spermophilus saturatus*

This species and the closely related Golden-mantled Ground Squirrel are often mistaken for large chipmunks. Unlike chipmunks, they have only one white stripe on each side, which ends at the shoulder and does not reach the head.

DESCRIPTION

The top and sides of the head and shoulders have a poorly defined mantle of brownish to russet hairs. Some individuals lack the mantle. A buff-coloured ring outlines each eye. The back is dark greyish

Cascade Mantled Ground Squirrel
(*Spermophilus saturatus*)

brown, sometimes with a rufous wash over the rump. Each side has a muted white or tan stripe, bordered below and usually above by poorly defined black stripes. The lower black stripe is darker and the upper stripe may be so reduced as to be non-existent. The throat, belly, and feet are buff or pale brown. The underside of the tail is yellowish brown to light orange and the top is blackish with an indistinct fringe of buff-coloured hairs. Cascade Mantled Ground Squirrels have a single annual moult that occurs in early summer for the adult males, following weaning for the adult females, and soon after weaning for the young of the year. Moult begins at the nose and proceeds backwards and down the sides. A fresh coat is noticeably brighter than an old coat, which fades with time. The dental formula is incisors 1/1, canines 0/0, premolars 2/1, and molars 3/3, for a total of 22 teeth.

SIMILAR SPECIES

The most similar species to the Cascade Mantled Ground Squirrel is the closely related, but slightly larger, Golden-mantled Ground Squirrel. These sister species were long thought to be subspecies of a single species. They can be difficult to tell apart, but their ranges do not overlap. Golden-mantled Ground Squirrels have more distinct markings and brighter colours. The boundary between the white stripe and its black borders is very sharp and the mantle on the head and shoulders is usually a more intense tawny to rufous colour. Chipmunks are smaller and have multiple light stripes that continue onto the head and through the eye. They also carry their tails stiffly vertical when they run.

SIZE

The following measurements apply to adult animals within the Canadian portion of the range. Body weight varies over the year. Adult males are lightest three to four weeks after emergence in spring and heaviest just before hibernation. Adult females are lightest upon emergence and heaviest just before hibernation. Adults of both sexes are similarly sized. There are three size classes of Cascade Mantled Ground Squirrel. Adults weigh the most, followed by yearlings and then juveniles.

Total length: 253–320 mm; tail length: 89–115 mm; hind foot length: 40–53 mm; ear length: 21–23 mm.
Weight: 162–340 g; newborns, around 6 g.

RANGE

Cascade Mantled Ground Squirrels have a limited range and are found only in the Cascade Mountains of south-central British Columbia and central Washington State. They occur as far north in British Columbia as Iron Mountain near Merritt, as far west as the Chilliwack River, and as far east as the Similkameen River. It is possible that the range in British Columbia, especially in the east and north, is shrinking, but evidence at this point is scanty.

ABUNDANCE

Cascade Mantled Ground Squirrels are considered uncommon in British Columbia. They are listed as "Not at Risk" by COSEWIC. A well-studied population in Washington State was estimated to have a density of 4.5 or more animals per ha in open meadows and 2 per ha in forests.

ECOLOGY

In the Canadian portion of their range, Cascade Mantled Ground Squirrels occur from 700 to 2500 m elevation and are most frequent between 1000 and 2000 m. Versatile in their habitat requirements, they live on talus slopes, in open meadows, in both open and dense coniferous forests, and in clearcuts and rocky railroad embankments. Cascade Mantled Ground Squirrels are diurnal. Like all the ground squirrels, they dig burrows that provide shelter and protection and are used for refuge, sleeping, raising young, and hibernation. Burrow entrances are well hidden and located under boulders, logs, trees, and stumps to discourage predators from attempting to dig them out. Each burrow typically has several entrances and tunnels that lead at least a metre underground to a nest chamber. The cup-like nest is composed of dried vegetation lined with fresh vegetation. These squirrels hibernate for six to eight months each year. They begin to disappear in mid-August. Adults and yearlings are first to go underground, followed up to 45 days later by the young

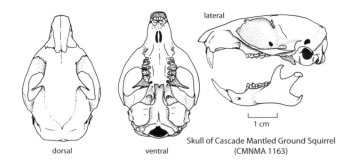

dorsal ventral lateral

Skull of Cascade Mantled Ground Squirrel
(CMNMA 1163)

of the year. The juveniles need more time to accumulate the necessary body fat to survive hibernation. All have entered into winter torpor by late September. They begin to appear above ground in early April (later at higher elevations), with the adult males emerging first, about a week before the adult females. Yearling females emerge one to two weeks after the adult females and yearling males about one week after that. Timing of emergence varies from year to year depending on the local weather and may be as late as mid-May. The sex ratio of Cascade Mantled Ground Squirrels is essentially 1:1 from birth through adulthood. Cascade Mantled Ground Squirrels are capable of bounding at speeds of up to about 21 km/h and can climb up to 5 m into trees and bushes, although most foraging occurs on the ground. Predators include raptors, Coyote, Red Fox, Long-tailed Weasel, American Marten, and Bobcat.

DIET

Cascade Mantled Ground Squirrels are essentially herbivorous and consume mainly fungi and stems and leaves of a variety of plants. They also eat some flowers, berries, seeds, and bark. Plant material is the principal food in spring and early summer. As it becomes drier in late summer and the plants desiccate, they switch their diet to include more fungi, especially truffles. Their fungal consumption may contribute greatly to spore dispersion, as the spores pass unharmed through the digestive tract and are viable when deposited in the scat. They will eat animal protein, usually in the form of carrion, and are known to feed on animals killed on the roadways, including their own species.

REPRODUCTION

Adult males emerge from their hibernation burrows in full reproductive condition. Their testes, which had shrunken and were abdominal since the previous summer, are enlarged, fully descended into the darkened scrotum, and contain motile sperm. Females mate within a few days of emerging from hibernation. The breeding period lasts about two weeks, during which time all the reproductive females have emerged and been bred. In Washington State, this takes place in late April. To date there are no breeding studies undertaken on Canadian individuals. The young are born in an underground burrow after a gestation period of about 28 days. Most are born in the latter half of May. Four pups make up the average litter (range 1–5). The pups are born pink and essentially hairless. After about 36 days underground being suckled and cared for by their mothers, they are old enough to venture outside and begin eating solid food. Weaning then occurs gradually over the course of at least a week. Each adult female is capable of producing one litter per year and most females (54%–100%) begin breeding as yearlings. The age of first reproduction of males is variable and most do not begin to breed until they are two years old. Those that begin breeding as yearlings were the best nourished and at least 50 g heavier going into hibernation the previous autumn than non-breeders of their age class.

BEHAVIOUR

The behaviour and social system of this species has not been closely studied. It is assumed to be essentially solitary, apart from the brief breeding period and while females are rearing young. When the adult males first emerge from hibernation, they spend much of their time searching for receptive females and will travel extensively in this pursuit. During the rest of the active season, Cascade Mantled Ground Squirrels spend about seven hours each day above ground. The bulk of this time is spent sunbathing or resting, foraging, and grooming. Both male and female juveniles disperse from their natal burrow soon after they are weaned. Dispersal distances vary with the terrain; they are greatest in open habitats and shortest in forested habitats. Males tend to disperse farther than females regardless of habitat type. Like Golden-mantled Ground Squirrels, Cascade Mantled Ground Squirrels are readily habituated to human presence and quickly learn to approach for a handout.

VOCALIZATIONS

The calls of Cascade Mantled Ground Squirrels have not been well studied. It is not known whether they produce distinct alarm calls for aerial and terrestrial predators, but this is probable. In a study examining their alarm call towards a terrestrial predator, the call is described as a trill, parts of which extend into the ultrasonic range beyond human hearing.

SIGNS

Cascade Mantled Ground Squirrels use three gaits: walk, trot, and, rarely, a lope. Their burrow entrances are cryptic with no throw mound of waste soil at their entrances. Otherwise, few indications of tracks or signs are recorded in the literature, although it is supposed that they are similar to those of Golden-mantled Ground Squirrels. See the "Signs" section of the Golden-mantled Ground Squirrel account for more information and the "Signs" section of the Arctic Ground Squirrel for illustrations of ground squirrel tracks.

REFERENCES

Cork and Kenagy 1989; Eiler and Banack 2004; Elbroch 2003; Kenagy and Barnes 1988; Leung and Cheng 1997; Nagorsen 2005a; Trombulak 1987, 1988.

Thirteen-lined Ground Squirrel
also called Thirteen-striped Ground Squirrel,
Gopher, Striped Gopher

FRENCH NAME: **Spermophile rayé**
SCIENTIFIC NAME: *Spermophilus tridecemlineatus*

Thirteen-lined Ground Squirrels are the smallest and quickest moving of our ground squirrels. Their striking coloration, with alternating lateral dark and pale stripes, is distinct, making this animal easy to identify, and gives it both its common and scientific names.

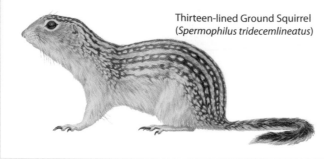

Thirteen-lined Ground Squirrel
(*Spermophilus tridecemlineatus*)

DESCRIPTION

The seven conspicuous, dark lines running parallel to the body are bracketed by six pale lines. Within each of the dark lines is a central row of spots, which may remain distinct and square shaped, or become irregular in shape and even merge into a continuous chain with narrow strands of pale-coloured fur between. The lines of colour run from the crown and sides of the neck to the base of the tail. Fur on the underside is yellowish white and the eye ring is a creamy white to buff. The relatively long (about half the body length) pencil-thin tail is well furred. It is tawny in the centre, both above and below, with a margin of black hairs edged with a thin fringe of buff-coloured hairs. The dental formula is incisors 1/1, canines 0/0, premolars 2/1, and molars 3/3, for a total of 22 teeth.

SIMILAR SPECIES

All similar ground squirrels are considerably larger and do not have the distinct lines on their backs and sides.

SIZE

The following measurements apply to adult animals within the Canadian portion of the range. Body weight varies over the year. Adult males are lightest three to four weeks after emergence in spring and heaviest just before hibernation. Adult females are lightest upon emergence and heaviest in mid-July just before hibernation. Adults of both genders are similar in size.
Total length: 244–305 mm; tail length: 70–113 mm; hind foot length: 34–46 mm; ear length: 7 mm.
Weight: 140–202 g; newborns, around 6 g.

RANGE

Thirteen-lined Ground Squirrels are the most widespread of the ground squirrels. They occur from the southern Great Lakes to central Alberta and all the way south to the Gulf Coast of Texas. During the last 200 years, this squirrel has expanded its range eastwards in the Great Lakes region, especially in Michigan and Wisconsin, as settlement and agricultural practices have resulted in the removal of trees and draining of wetlands. There is a wide area in south-central Alberta where the species is not found.

ABUNDANCE

Thirteen-lined Ground Squirrels are generally sporadic in their occurrence. They may be common at times in some areas and scarce or absent at times in others. Although aggregations may exist intermittently, these squirrels are essentially solitary and consequently appear in low densities and scattered locations. Population densities may approach four animals per ha during favourable years.

ECOLOGY

Thirteen-lined Ground Squirrels prefer to live in short-grass regions where they can see over the vegetation without having to stand upright on their hind legs. Wet areas and floodplains are avoided. Heavily grazed rangelands and mowed grasses on roadsides, golf courses, and lawns are preferred, along with the traditional (but now rare) regions of native short-grass prairie habitat. Sometimes, especially farther south in their range, they may live in dry open scrublands, where their view is impeded but more cover is available. Strictly diurnal, these squirrels like heat and spend the warmer portion of most, except the hottest, summer days foraging above ground. They are imperfectly homeothermic, as their body temperature drops about five degrees during the night while they sleep in their cool burrows.

Burrows are dug for protection, hibernation, and rearing young. They descend steeply at first, then level off somewhere between 10 cm and 120 cm below the surface. Often, more than one burrow is excavated and burrows may have two or more entrances. Some

Distribution of the Thirteen-lined Ground Squirrel
(*Spermophilus tridecemlineatus*)

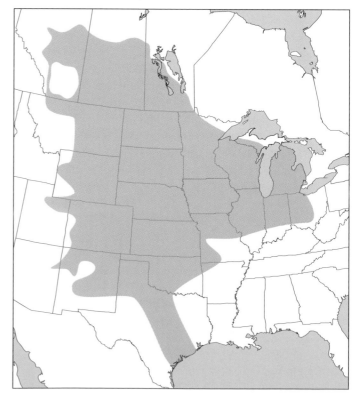

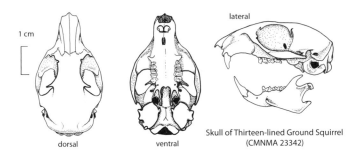

1 cm

dorsal ventral lateral

Skull of Thirteen-lined Ground Squirrel (CMNMA 23342)

burrows are used as bolt holes and these tend to be short, while others are longer and contain slightly oval chambers for nesting and food storage. Juveniles dig short simple burrows at first, but usually have an adequate hibernation burrow dug by early September. During hibernation, an earth plug blocks the burrow's entrance. The ratio of males to females in this species remains at about 1:1 from birth through adulthood. Adult males are the first to emerge from hibernation in April and they retire to their hibernation nest as early as late July or August, even though food remains abundant. Females emerge about one to two weeks after the males and depart to hibernate in August. Juveniles begin hibernating in September or October and emerge at the same time as adults of their gender.

These squirrels are capable swimmers. Thirteen-lined Ground Squirrels spend much of their vigilance energy monitoring the sky, as their principal daytime predators are hawks. Badgers are another serious predator, digging up burrows and taking both adults and young, mostly at night. Snakes, skunks, weasels, Coyote, foxes, crows, Domestic Dogs and Cats are also predators of Thirteen-lined Ground Squirrels.

DIET

These squirrels are omnivorous. Insects such as caterpillars, grasshoppers, and beetles, small vertebrates, and small bird eggs and nestlings as well as carrion are readily consumed. They will even eat the dead of their own species and infanticide does occur. The bulk of their diet is made up of leaves and seeds of various grasses and forbs with the addition of bulbs, nuts, and tubers when available. The fruit of chokecherries is also eaten and these squirrels can climb to reach this favourite treat. They relish corn, wheat, and oats and can do considerable damage to the edges of grain crops by clipping stems to reach leaves and seeds and by digging up newly planted seed. Garden vegetables are also eaten when obtainable. Most water requirements are met by eating succulent vegetation or drinking dew or rainwater. During periods of high heat, free water will be visited daily if available.

REPRODUCTION

Males are capable of reproduction as soon as, or soon after, they emerge from their hibernation burrow. Females become receptive about four or five days after they emerge from hibernation. The breeding season may be prolonged if unseasonably cold weather occurs after some initial warm weather, because female receptivity appears to be adversely affected by low temperatures. It normally lasts about two to three weeks and is over by early May.

This squirrel is polygamous as both females and males commonly mate with more than one partner. Each female's first mate sires roughly 75% of her offspring that year. Each male leaves behind a "copulatory plug" in her vagina, which is formed by coagulation of part of his ejaculate. This temporary plug blocks the success of later matings at least for a few hours. The gestation period is 27–28 days and most births in Canada occur in late May or June. Litter size ranges from three to thirteen young, with an average of six to eight. Older females have larger litters than do yearlings. Since each female has only ten nipples, there is often some early mortality in the larger litters. Although females in the southern portions of the range may have two litters some years, all Thirteen-lined Ground Squirrels in Canada have a single litter per year. The pups are born deaf, blind, and largely hairless, but develop rapidly and emerge from their natal burrow at around 28 days old. They are fully weaned shortly after that. Yearlings are sexually mature, but both sexes in their first breeding season become fertile somewhat later than do older animals. Late litters are usually produced by these young breeders or sometimes by older females that lost their first litter early and undergo a second heat.

BEHAVIOUR

Thirteen-lined Ground Squirrels are generally considered to be asocial. Adults are solitary except briefly during the breeding season and while mothers are rearing young. Mothers appear to avoid their above-ground young and rebuff most of their efforts to initiate interactions. Newly emerged litters stick together in a tight group for several days, but gradually they begin to forage independently and move progressively farther from their natal burrow. They disperse soon after this and most take up a territory nearby. Adult males have been observed chasing females away from their nests, then entering to take and eat the pups. This type of behaviour may be common. In the laboratory, males seem to recognize their own offspring, so perhaps this infanticide is perpetrated by males that are not the sire of the litter. During the breeding season, males spend much of their time above ground searching for females that are coming into oestrus. They can recognize and remember the location of such females and will check on them repeatedly until they are receptive. Males typically copulate several times with a female before departing in search of other mates. While searching for prospective mates, the males encounter each other frequently and chases and fights are common. Few wounds are inflicted on the head and neck during the wrestling fights. Most wounded animals are bitten on the hind end, likely during chases.

Both sexes of Thirteen-lined Ground Squirrels hoard seeds in their burrows. They transport the food in their large cheek pouches and store some of it in their hibernation nest. Additional hoards may be cached in other locations, sometimes even above ground. It is uncertain when this food is eaten. It may be consumed over the winter during periods of arousal from torpor, or more likely is eaten just before, or shortly after, emergence. It may also be utilized to minimize weight loss during the reproductive period the following spring. Squirrels gather dry grasses for use as nesting material by shoving it crosswise into their mouths with their front feet until they are almost obscured by the load. Navigating to and down their hole can then be quite a challenge. Dust bathing is a regular occurrence.

VOCALIZATIONS

The alarm call is a high-pitched trill that is ventriloquistic in its properties. A chattering series of notes may be heard from inside the burrow after the squirrel has taken refuge.

SIGNS

Adult burrow entrances are 5–9 cm tall and 4.5–9.0 cm wide, and unlike those of other ground squirrels, usually do not have a throw mound of waste soil at their mouth. Burrows created by juveniles have smaller entrances, usually 3–4 cm in diameter. Burrow entrances are often hidden in a clump of vegetation and all runs begin and end at a burrow entrance. Runs are generally 7–13 cm wide and are widest near the mouth of the burrow. Tracks are similar to those of other ground squirrels, but smaller (see the "Signs" section in the Arctic Ground Squirrel account for track illustrations). Front tracks are 2.0–3.5 cm long and 1.7–3.0 cm wide. Hind tracks are 2.2–3.0 cm long by 1.7–3.0 cm wide. Three gaits are used. Most common is the bound, their fastest pace. They walk while foraging. Occasionally a lope is used to cross open spaces at a medium pace. Tracks in snow are possible if there is a late snowfall in spring, after the animals have emerged from hibernation.

REFERENCES

Criddle 1939; Elbroch 2003; Livoreil and Baudoin 1996; Raynor and Armitage 1991; Schwagmeyer 1995; Schwagmeyer and Parker 1990; Smith, H.C., 1993; Streubel and Fitzgerald 1978; Vestal 1991; Vispo and Bakken 1993.

Yellow-pine Chipmunk

also called Yellow Pine Chipmunk, Northwestern Chipmunk

FRENCH NAME: **Tamia amène**

SCIENTIFIC NAME: *Tamias amoenus*

Chipmunks originated in the New World and are found in North America and northern Asia. There are 22 species in this group of squirrels, only 5 of which occur in Canada.

DESCRIPTION

These are small chipmunks renowned for the variability of their coats. The basic pattern of five dark stripes and four pale ones on the back and sides, and three dark stripes and two pale ones on the face, remains consistent throughout. Typically the colours are bright, with yellowish-brown sides and a buff or tawny belly. The underside of the tail is yellow to orange. Coastal forms are darker brown on the sides and belly, while dry interior British Columbia forms are paler grey to brown, sometimes with an almost white belly. The length of

Yellow-pine Chipmunk
(*Tamias amoenus*)

the tail is about 74% of the head-body length. Chipmunks undergo two moults each year, one in spring and one in autumn. The winter pelage is slightly duller, but otherwise similar to the summer coat. The spring moult begins at mid-back and proceeds forward, then back and finally under the belly and legs. The dental formula is incisors 1/1, canines 0/0, premolars 2/1, and molars 3/3, for a total of 22 teeth.

SIMILAR SPECIES

The striped body of this species makes it easily identifiable as a chipmunk, and only mistaken for other small squirrels with similar pelt patterns. Townsend's Chipmunk is larger, darker, has duller side stripes and the pale stripes fade out as they progress towards the tail. The Cascade Mantled and Golden-mantled Ground Squirrels are considerably larger than the Yellow-pine Chipmunk and have only one white stripe on each side. They also lack the characteristic striping found on the face of chipmunks. Yellow-pine Chipmunks are highly variable in coat colour and may converge in appearance with Red-tailed Chipmunks in the Selkirk Mountains of south-central British Columbia. Shape and size of genital bones is the only sure way to distinguish the two in that region. In the other area where both species occur, the southern Rocky Mountains, the underside of the Red-tailed Chipmunk's longer tail is a deep rust-red while the undertail of the Yellow-pine Chipmunk is usually lighter. Additionally, the body of the Red-tailed Chipmunk is larger and their colouring tends to be less bright. In northern North America, the chipmunk most similar to a Yellow-pine Chipmunk is a Least Chipmunk. Both species are similar in size (although the Least Chipmunk is slightly smaller) and coat colouring for both is equally variable. In general, Yellow-pine Chipmunks have a brighter coat, but this can be difficult to compare if only one species is in view. In the southern Rocky Mountains, Least Chipmunks are smaller and have a white or light grey belly. In central British Columbia, Yellow-pine Chipmunks may also have a light belly, so distinguishing between the two is again a matter of genital bone differences. As these tiny bones must be removed from the animals and specially

Distribution of the Yellow-pine Chipmunk (*Tamias amoenus*)

cleared and stained, the differences are hardly useful in the field with a living chipmunk.

SIZE
Females are about 5% larger in the body than males. Tail, ear, and hind foot lengths are about the same for both sexes. The following measurements apply to adult animals within the Canadian portion of the range.
Total length: 195–230 mm; tail length: 66–110 mm; hind foot length: 26–34 mm; ear length: 11–21 mm.
Weight: 42–89 g; newborns, about 1.7–2.7 g.

RANGE
These chipmunks occur in the mountains of southwestern Alberta and in most of southern British Columbia. The distribution extends southwards to northern California and westwards to western Montana and Wyoming.

ABUNDANCE
Yellow-pine Chipmunks vary in population density depending on habitat quality. They are common to rare in different parts of their range.

ECOLOGY
Yellow-pine Chipmunks occur in a wide variety of dry, open forests and are most abundant where there are rock outcrops and dense understorey vegetation. They create underground burrows with only a single entrance, a long (up to 3 m) tunnel, which usually has at least one widened turn-around section, and a combined nest and storage chamber at the end. Underground nest chambers are an average 16.5 cm in diameter and are usually used only in winter. The winter nest is constructed on top of the stored food. Summer dens could be in stumps or logs or high in the branches of a tree. These can be quite large (up to 36 cm in diameter) with about a 4 cm access hole into a central chamber. Old bird nests may supply the base of a Yellow-pine Chipmunk's tree nest (drey). Dried grasses and sedges are commonly used to construct nests and feathers, lichens, thistledown, and more grasses are used for lining. Boulder piles and talus are favoured areas for burrow creation, as entrances can be placed in natural crevices in the rocks for protection. These chipmunks are active during the daytime. They emerge from their dens around sunrise and with the exception of a midday rest during hot days (perhaps as long as from 9 am to 5 pm on very hot days), they are active until dusk. Activity on cool or cloudy days may continue throughout the day. During cool weather the chipmunks will appear above ground only during the warmest part of the day. They remain in their underground burrows throughout the winter (close to six months). Since they do not increase their body stores of fat before the cold season, they enter short periods of hibernation (five to seven days), then arouse to feed from their winter stockpile of food. Food supply and weather influence the onset of hibernation, which can vary from year to year and even between individuals. Some animals may remain active for a while, even after the rest have gone into hibernation. Home ranges remain stable, with the same animals occupying the same summer territory year after year. Size of this territory varies from around 0.8–1.6 ha, with those of males generally larger than those of females. Principal predators are hawks, weasels, Bobcat, Coyote, American Badger, Red Fox, and rattlesnakes. Juveniles newly emerged above ground are curious and fearless and are especially vulnerable to predation. Maximum recorded lifespan of a wild Yellow-pine Chipmunk is five years and two months. Yellow-pine Chipmunks can be infected with Colorado Tick Fever and Rocky Mountain Spotted Fever and thereby act as reservoirs for these tick-spread diseases.

DIET
Yellow-pine Chipmunks consume an eclectic omnivorous diet. Seeds of shrubs, cone-bearing conifers, and grasses form the bulk of their winter stores, but during the rest of the year they also eat bulbs and tubers, fruit, flowers, buds, green vegetation, fungi, insects, bird eggs, and even small animals such as fledglings and juvenile small mammals.

REPRODUCTION
Females come into oestrus for only one day within about a week of emerging from hibernation. In Canada, females mate in late April or early May. Farther south these dates may be advanced by a month or more. Following a gestation of about 30 days, an average litter of

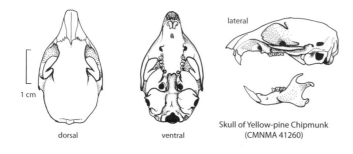

1 cm

dorsal   ventral   Skull of Yellow-pine Chipmunk
(CMNMA 41260)

4 (range 2–8) is born in an underground nest. These neonates are essentially hairless (except for short vibrissae on their head), deaf, and blind. By 20 days old they are fully haired, by 22 days old their ear canals open, and by 33 days old their eyes open. They appear above ground at around six weeks old and are weaned by about eight weeks. Most juveniles breed for the first time the following year (11 months old) although some may not become fertile until the next year (23 months old). One litter is produced each year. Yearling females have a smaller litter than more mature females. It appears that females expel their offspring soon after they are weaned, forcing them to find an unoccupied territory of their own before their first winter.

BEHAVIOUR

These small squirrels are active and nervous. Movements are quick and appear jerky. Even when climbing, they will dart up a metre or two and then stop to survey for danger. Watching the perpendicularly held tail disappearing into the undergrowth is often the only view one gets of these vigilant creatures. As with all chipmunks, however, time, patience, and a food reward may encourage them to be less timid with humans. Yellow-pine Chipmunks frequently dust-bathe to rid their coat of excess oils, and after eating they sit on their haunches to clean their faces with their front paws, much as a cat does. When resting they curl into a ball with the tail around the body and cover their face with their front paws. The tail is usually active, swishing gently from side-to-side when the animal is contented, or flicking up and down or more energetically from side-to-side if the animal is excited or calling. The tail is held stiffly, either vertically or horizontally, as the animal runs.

Yellow-pine Chipmunks scatter-hoard much of their food. They gather it into their cheek pouches and carry it to a variety of hiding places, usually as temporary storage before moving it to the main underground winter-storage chamber. While it is clear that they do not utilize all their caches (some seeds sprout the following spring), still they do retrieve a high percentage of their hidden food and appear to use nearby landmarks to remember the cache location. They also utilize different caching strategies depending on the value of the food in order to foil raiding neighbours. Preferred seeds, either due to their large size or their high food value, are stashed farther away from their source to make cache discovery more difficult, and fewer are placed in each cache to mitigate their loss if the hoard should be discovered. Despite the chance other Yellow-pine Chipmunks may steal their hidden caches, squabbles and chases at territorial boundaries are rare, but the immediate vicinity of the den with its crucial stockpile

of winter food may be vigorously defended. Interactions with other chipmunk species are not always so benign. Yellow-pine Chipmunks are behaviourally dominant over Least Chipmunks and commonly exclude them from preferred habitats. Pre-oestrous females call to advertize their condition to surrounding males. Several males (two to six) gather near her den and pursue her on her oestrous day. She may copulate with one or more during this "mating chase."

VOCALIZATIONS

Yellow-pine Chipmunks make at least 10 different vocalizations, varying from faint squeaks to loud squeals, in a variety of social interactions. Some may be repeated again and again and others are produced just once. Some express anger (a throaty growl or a buzzing sound), some express fear or distress (squeaks, squeals, long quavering notes, a loud descending series of five to six notes, a sharp pair of notes repeated), and others express alarm or excitement (a sharp "kwst" sound, a series of four to six "ks-ks" sounds, a "kyuk" sound). Nursing young emit faint squeaks when handled or when their mother stirs. Pre-oestrous females squeal loudly to attract males.

SIGNS

See the Eastern Chipmunk account for a discussion of chipmunk sign and sketches of tracks. Ground burrow entrances of Yellow-pine Chipmunks are 3–4 cm in diameter.

REFERENCES

Broadbooks 1958; Demboski and Sullivan 2003; Jameson and Peeters 2004; Nagorsen 2002, 2005a; Schultze-Hostedde and Millar 2002; Sutton, D.A., 1992; Vander Wall 1991, 1995; Verts and Carraway 1998.

Least Chipmunk

FRENCH NAME: **Tamia mineur**

SCIENTIFIC NAME: *Tamias minimus*

As its name indicates, this is the smallest of our five chipmunk species. It is also the most widespread.

DESCRIPTION

This chipmunk generally has a greyish rump, five dark and four light stripes on its back and sides, and three dark and two light stripes on its face. The middle dark dorsal stripe runs all the way to the base of the tail and the others may or may not quite reach the tail. The lower pair of light stripes are whitish, while the upper pair are greyish to brownish in colour. The underside of the tail is pale yellowish to yellowish orange, as are the sides of the body below the stripes and the tops of the hind feet. The belly and throat are white to greyish white. Darker individuals do rarely occur and melanism is most common in northerly populations. Populations in the

Least Chipmunk (*Tamias minimus*)

Rocky Mountains of southern Alberta and British Columbia tend to be smaller and paler. There are two annual moults. The summer coat replaces the winter pelage in June and July and the winter coat is acquired by late September or early October. The winter pelage is slightly paler than the summer, but otherwise similar. The dental formula is incisors 1/1, canines 0/0, premolars 2/1, and molars 3/3, for a total of 22 teeth.

SIMILAR SPECIES

The Least Chipmunk is the smallest chipmunk in Canada; the size difference is only noticeable when two species are visible for comparison. This means that other characteristics, some equally subtle, must be studied and even then a satisfactory identification may not be possible. This species overlaps with three other species of chipmunks. In the west, Red-tailed Chipmunks have a bright reddish-orange undertail, while Least Chipmunks in the same area have a yellowish undertail colouring. Additionally, Red-tailed Chipmunks are larger, with a longer tail and more orange on their sides, especially in summer. Yellow-pine Chipmunks are very easily confused with Least Chipmunks. In most areas their belly fur is buffy, unlike the clear white of the Least Chipmunk and they are slightly larger. These traits are difficult to distinguish in the field. In central British Columbia, Yellow-pine Chipmunks are generally paler and their belly fur becomes white, so positive identification cannot be achieved in the field. A detailed laboratory study of the genital bones (baculum for males and baubelum for females) is necessary to distinguish these two species from that region. In the east, the only other chipmunk present is the Eastern Chipmunk, which is readily distinguished from Least Chipmunks. It has a reddish rump, stripes that terminate well before reaching the tail, and only three dark and two light stripes. The presence of the peg-like upper premolar (P3) distinguishes all the other North American chipmunks from the Eastern Chipmunk, which lacks a P3. The much larger Golden-mantled Ground Squirrel looks superficially like a chipmunk as it has a single wide white stripe on each side bracketed by two dark stripes. It lacks the facial stripes of chipmunks and has a white ring around each eye.

SIZE

Females are slightly larger than males by 4%–5% in body length. The following measurements apply to adult animals within the Canadian portion of the range.
Total length: 172–228 mm; tail length: 69–105 mm; hind foot length: 27–38 mm; ear length: 10–19 mm.
Weight: 31–63 g; newborns, about 2.3 g.

RANGE

Least Chipmunks occur from central Yukon, across the western provinces to central Ontario, and down through the western states to California.

ABUNDANCE

Their population density is spotty, as there may not be any Least Chipmunks in poor or even suitable habitats, or as many as 22 per ha in optimum habitat during favourable years.

ECOLOGY

Least Chipmunks do not compete successfully against any other of our Canadian chipmunk species. Wherever their range overlaps with another species, they are excluded from forested regions and instead inhabit the more-harsh habitats where their smaller size and physiological adaptations give them a competitive edge. They are frequently found in rocky, alpine habitats in British Columbia for these reasons. In the northern part of the province and across the prairies where they have no other competing chipmunk species, Least Chipmunks occur in recently burned areas or in clearcuts

Distribution of the Least Chipmunk (*Tamias minimus*)

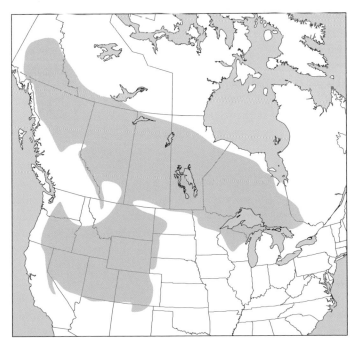

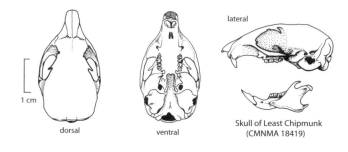

lateral

Skull of Least Chipmunk
(CMNMA 18419)

dorsal

ventral

1 cm

in the boreal forest and in open forests or wooded parklands, and shrubby savannah. They are active during daylight hours, but retire to their burrows during hot afternoons to avoid the midday heat. On cool days, they may not become active until the warmer part of the afternoon and they stay below ground on stormy days. They commonly forage in trees, but rarely build nests above ground. When they do, it is in a pre-existing cavity rather than in a typical leafy squirrel drey. The usual underground burrow has one or several entrances that lead through about a metre-long tunnel to a nest/storage chamber that is around 16 cm long and 14 cm high and is 14–50 cm below the ground surface. A nest constructed of dried grasses lined with feathers, fur, willow or poplar fluff, or soft shredded bark rests on the food cache and fills the chamber. A small opening in the nest ball provides access to the central space where the chipmunk rests. Least Chipmunks do not accumulate fat in autumn. Instead they use a combination of stored food and torpor to survive the cold period. They spend the winter curled up in their underground nest, waking up every three to five days days to eat from their food cache. While dormant, their body temperature drops, as does their heart rate. They go below ground in late September to late October as cold weather sets in. Adult males precede adult females and juveniles into dormancy. In spring, the males usually emerge first, followed about a week later by the females. Timing of emergence depends on location and local weather and may vary from year to year. Typically, in the Canadian portions of the range this occurs in late March to early May. Principal predators include raptors, weasels, and foxes. Domestic Cats may be major predators in some areas. Maximum known longevity is just over six years for a captive. A Least Chipmunk has been recorded running at a top speed of 7.7 km/h when pursued by a researcher. These animals can swim well, but enter the water reluctantly.

DIET

The summer diet is composed of seeds, insects, leaves, fruits, flowers, fungi, and sometimes bird eggs and nestlings. Their winter diet is primarily stored seeds. These arid-adapted squirrels can gain much of their water requirements from their food, but will drink regularly if water is available.

REPRODUCTION

If born during a year with good weather and abundant food, as many as 90% of juveniles will reproduce for the first time the following spring as yearlings. Following poor years, this percentage may drop

to 15%, with the majority needing another year to achieve sexual maturity. Mature females enter oestrus within a week of emerging in the spring. After a gestation period of 28–30 days, a litter of 4–7 young (average 4–5) is born. In northern parts of the range, most births occur in late May and June. In Canada, Least Chipmunks produce a single litter each year, but farther south a second litter may occur some years. Should a female lose her litter shortly after birth, she may go into oestrus a second time and then produce a late litter. The young are born helpless and mostly hairless. They grow quickly and are weaned around seven to eight weeks old, at which time they are almost adult-sized.

BEHAVIOUR

Normally shy, Least Chipmunks may become habituated to humans and bold to the point of raiding unguarded food at campsites and generally making nuisances of themselves as they explore all the baggage. These animals are agile and can climb and perch in trees with ease and dexterity. Summer nests may be constructed inside cavities appropriated from woodpeckers or in rotting logs, but many continue to use their underground nest year-round. Least Chipmunks run with their tails held vertically and will nervously flick them up and down even when standing still. The tail commonly twitches in rhythm with vocalizations. When sneaking slowly away to prevent being seen, they carry their tails horizontally. In late summer, these chipmunks enter a period of single-minded activity as they busily gather winter food. Seeds are carried in cheek pouches and usually scatter-hoarded in small caches around their territory for later retrieval into an underground storage site. This species is the only known rodent to deliberately use urine to mark a buried cache of seeds, typically after it has been emptied. This marking of a depleted food source may reinforce the memory of the creator and prevents further effort being expended looking for the emptied cache. The main winter cache may be built into the underground winter nest (as the lowest layer) or stored in a separate food chamber near the nest. Dust bathing is a regular occurrence, as are sessions of grooming by scratching or by licking the forepaws and rubbing them vigorously over the head and body. Newly weaned young are encouraged by their mothers to set off to find their own territory.

VOCALIZATIONS

These squirrels produce calls that can be ventriloquistic, making it difficult for predators to locate their position. A variety of alarm calls (at least six) are produced that communicate escalating levels of anxiety, including the commonly heard chattering "tuk-tuk-tuk" – a general alarm call. Interestingly, a whispered alarm call is part of the repertoire for those special occasions when a louder call may attract too much attention. A low-volume "kek-kek-kek" is used as a contact call to neighbours or offspring. Each call is punctuated with a corresponding tail twitch that becomes more dramatic as the animal becomes more excited.

SIGNS

Burrow entrances and the underground portion of the burrow, including the nest chamber, are commonly positioned beneath a boulder to protect and hide them. Entrance diameter is around 3–4 cm and burrow tunnels are slightly larger, at 3.5–4.5 cm in diameter. See in the "Signs" section of the Eastern Chipmunk account for a discussion of chipmunk signs that generally apply to all five Canadian species.

REFERENCES

Bihr and Smith 1998; Devenport et al. 1999; Jameson and Peeters 2004; Nagorsen 2002, 2005a; Slough and Jung 2007; Smith, R.J., 1995; Verts and Carraway 1998, 2001.

Red-tailed Chipmunk

FRENCH NAME: **Tamia à queue rousse**
SCIENTIFIC NAME: *Tamias ruficaudus*

These colourful chipmunks reach the northern limits of their distribution in southern British Columbia and Alberta.

DESCRIPTION

The pelage of this species is generally bright rufous on the shoulders and sides, dark greyish on the rump, and light to dark greyish or buffy brown on the belly. The underside of the tail is bright reddish orange. Five dark stripes run along the back from the shoulder to the rump in the typical chipmunk pattern. There are four lighter stripes bracketed between the dark stripes. The lowest pair is creamy white, while the upper pair is light grey. The belly, throat, chin, and lower cheeks are white. Ears are dark with a broad band of buffy white along the trailing edge. The tops of the hind feet are tawny or rufous. The dental formula is incisors 1/1, canines 0/0, premolars 2/1, and molars 3/3, for a total of 22 teeth.

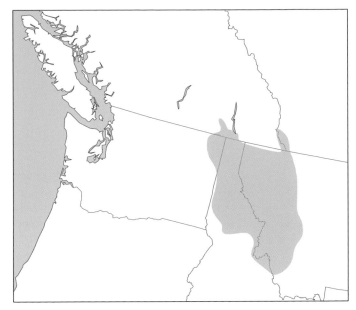

Distribution of the Red-tailed Chipmunk (*Tamias ruficaudus*)

Red-tailed Chipmunk (*Tamias ruficaudus*)

SIMILAR SPECIES

In areas of British Columbia and Alberta, where the range of the Least Chipmunk overlaps that of the Red-tailed Chipmunk, the two can be distinguished by both size and colouring. The Least Chipmunk is smaller and the fur colour on its sides is considerably paler and less rufous, as is the colour of the underside of the tail. It also tends to occur at higher elevations. Where the Rocky Mountain population of Red-tailed Chipmunks overlaps with the Yellow-pine Chipmunk identification is more difficult. Although the Red-tailed Chipmunk is more reddish on the sides and under the tail, the Yellow-pine Chipmunk is sufficiently variable that there is colour overlap between the two species, making field identification uncertain. Generally Yellow-pine Chipmunks occur at lower elevations. See the discussion of "Similar Species" in the Yellow-pine Chipmunk account for further information. Cascade Mantled Ground Squirrels are superficially similar, but have only a single light blaze on each side, no facial stripes, and are much larger.

SIZE

Males average about 3% smaller than females. This size difference is most notable in the body length and weight measurements. The following measurements apply to adult animals within the Canadian portion of the range.
Total length: 197–237 mm; tail length: 85–115 mm; hind foot length: 30–36 mm; ear length: 13–19 mm.
Weight: 44–79 g; newborns, about 2.3 g.

RANGE

This chipmunk reaches Canada in only two small areas: along the extreme southern border of British Columbia and Alberta, and in southeastern British Columbia in the southern Selkirk Mountains. The bulk of the range is farther south in northeastern Washington, northern Idaho, and western Montana.

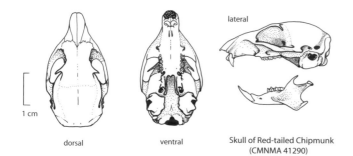

dorsal   ventral

1 cm

Skull of Red-tailed Chipmunk
(CMNMA 41290)

ABUNDANCE

Both populations of Red-tailed Chipmunk in British Columbia are listed by the British Columbia government as of conservation concern, mainly due to their limited range in the province. The population along the border with Alberta is on the "Red List," while the population in the Selkirk Mountains is on the "Blue List." Neither population is clearly threatened at this time. The Alberta population is currently listed on the provincial "Blue List" as a species that may be at risk due to its limited distribution and possible threats to its habitat. Red-tailed Chipmunks can be locally abundant in Idaho and Montana, but are generally more uncommon in Washington, British Columbia, and Alberta. Population density in Montana ranges from 4.2 to 7.0 chipmunks per ha.

ECOLOGY

Like all chipmunks, these small squirrels are diurnal. Red-tailed Chipmunks live in relatively moist coniferous and mixed forests and are most common in semi-open edge habitat where shrubby growth and woody debris is abundant. Where other chipmunks co-occur (border of British Columbia and Alberta), the species are roughly separated by elevation: Red-tailed Chipmunks occupying the intermediate heights, with Least Chipmunks above (over 2000 m) and Yellow-pine Chipmunks below (less than 1850 m). Both Least and Yellow-pine Chipmunks are usually found in drier habitat than that preferred by the Red-tailed Chipmunk. In the Selkirk Mountains of southeastern British Columbia, where Least Chipmunks are absent and Yellow-pine Chipmunks are rare, Red-tailed Chipmunks occupy all the elevation zones. Males have larger territories than females. Red-tailed Chipmunks are agile climbers and commonly forage and nest high in trees and shrubs. Tree nests, typically large, oval constructions (18–100 cm × 18–45 cm) of dried vegetation, tucked into dense growth beside the trunk, are commonly used in summer, and underground nests are occupied in winter and spring. Winter and summer nests are typically made of dried grasses and the hollow, central chamber is lined with lichens and feathers. Cavities in trees and in rock crevices may also be used in summer, especially in regions where large trees are scarce.

Young are usually born in an underground nest in the Canadian parts of the range, and are frequently moved by their mother to a tree nest as they become mobile. The burrow system is simple, typically about a metre long and about a half-metre below the surface, with a nest chamber at the terminus. Food is gathered in late summer and autumn using the extensible cheek pouches, and stored in the underground nest chamber. Juveniles are last to enter hibernation in autumn. Sightings begin to diminish in late September, and by early to mid-November, even the juveniles have gone underground for the winter, where they go into torpor for several days and then wake to eat from their food cache before entering torpor again. They may venture above ground during warm periods over the winter. Males emerge first in the spring. There are no direct reports of predation, but likely predators that occur within their range include the small weasels, Coyote, Red Fox, Bobcat, American Crow, and Common Raven and several species of diurnal raptors. Greatest-known age of a wild Red-tailed Chipmunk is eight years, but most (90% or more), live less than three years.

DIET

Seeds, berries, fruit, leaves and buds, insects, and flowers are eaten in spring and summer. Seeds, which are stored underground in late summer and autumn, are the principal winter food.

REPRODUCTION

There is sparse information concerning reproduction of this species in Canada. The following is based on studies conducted in Montana. Breeding occurs soon after emergence, which is directly correlated with elevation. Populations at lower elevations breed in late April and early May, while those at higher elevations breed seven to ten days later. This species has only one litter per year. Length of gestation is unknown, but thought to be around 31 days. Most births occur in late May to late June. The young are born pink and hairless, and usually in underground nest chambers. A litter consists of three to six young. They nurse for about five to six weeks and appear above ground at about six to seven weeks old. Yearling females will breed, but few conceive, and if they do, they produce smaller litters than do mature females.

BEHAVIOUR

Red-tailed Chipmunks frequently sand-bathe. They rub their belly and chin into the sand and roll from side to side with every indication of enjoyment. They practise two types of hoarding behaviour: scatter-hoarding (caching small amounts of food in numerous places) and larder-hoarding (gathering a pile of food in one place, usually the underground burrow). To hide a food item, or several food items, in the scatter-hoarding process, the animal uses its forefeet to dig a shallow hole into which it pushes its head, with the food in its teeth. After thrusting the food to the bottom of the hole several times, it will deposit it into the pit and then, again using its forefeet, scrape dirt and debris into the hole and tamp it down.

VOCALIZATIONS

Little information is available concerning the vocalizations of this species, apart from the knowledge that at least three different calls are produced: a scolding alarm call, a trill, and a bell-like call.

SIGNS
See the "Signs" section of the Eastern Chipmunk account for a generalized discussion that applies to all chipmunks.

REFERENCES
Bennett, R. 1999; Best 1993; Nagorsen 2002, 2005a; Smith, H.C., 1993.

Eastern Chipmunk

FRENCH NAME: **Tamia rayé,** *also* **Petit Suisse**

SCIENTIFIC NAME: *Tamias striatus*

This is the largest of the Canadian chipmunks and the only one found in much of eastern North America. Its well known "chipping" call is one of the most recognizable in the eastern forests. Vigilance, small size, and camouflage ensure that you will usually hear one before seeing it.

DESCRIPTION
The Eastern Chipmunk has a single white stripe on each side, bracketed by two dark stripes. An additional thin dark stripe runs down the middle of the back. The fur colour on the back is frosted grey brown to dark rufous. All the stripes, both light and dark, terminate well before the tail. The rump is bright rufous and this colour often extends onto the top of the tail for a short distance. The underside of the tail is similarly reddish orange with a dark-grey border. The sides of the body below the stripe pattern are light rufous and this colour gradually fades into the whitish belly and throat. The tops of the hind feet are buffy brown. The face pattern is typical of all chipmunks, two light and three dark stripes with the middle dark stripe running through the eye. These facial markings are not as bright and distinct

Eastern Chipmunk (*Tamias striatus*)

as on other chipmunks. Two moults occur each year and the winter pelage is slightly paler than the summer. The winter coat is shed in July or August and the summer coat is shed in October or November. Animals at the northern edges of the range (which includes animals in Canada) are paler and less reddish than their more southern counterparts. Albino (with pink eyes), white (with dark eyes), and melanistic (dark or black) individuals are rare, but are seen from time to time and may become numerous within a localized population for a period of time. The dental formula is incisors 1/1, canines 0/0, premolars 1/1, and molars 3/3, for a total of 20 teeth.

SIMILAR SPECIES
The Eastern Chipmunk with its single light blaze on each side is unlikely to be mistaken for any other mammal throughout most of its range, and even the odd-coloured individuals retain the perky attitude, tail carriage, and distinctive calls of the species. In the western half of its range, the Eastern Chipmunk overlaps the distribution of the Least Chipmunk. Least Chipmunks are smaller, have a greyish rump, and two pale stripes on each side.

SIZE
Unlike the other four species of chipmunks in Canada, whose females are larger, male and female Eastern Chipmunks are similarly sized. The following measurements apply to adult animals within the Canadian portion of the range.
Total length: 225–299 mm; tail length: 65–115 mm; hind foot length: 32–40 mm; ear length: 14–20 mm.
Weight: 75–115 g; newborns, 2.5–5.0 g.

RANGE
This chipmunk occurs from south of Lake Manitoba to the southern tip of James Bay, east to Quebec's North Shore and Cape Breton, Nova Scotia. The southern extent of its range is into the Gulf States of the United States. Eastern Chipmunks have reached some of the major islands such as Manitoulin Island, Cape Breton Island, and Prince Edward Island, but not Anticosti Island or Newfoundland. They were introduced into Newfoundland in the 1960s and are still fairly common at the introduction sites (Sir Richard Squires Park, Barachois Pond Provincial Park, and Butterpot Provincial Park), but have not spread extensively.

ABUNDANCE
Eastern Chipmunks can be locally abundant throughout their range, but populations are subject to severe declines due to fluctuations of food and weather.

ECOLOGY
This chipmunk is well adapted to mature eastern deciduous forests, but can also take advantage of clearcuts, small woodlots, shrubby edges of fields, and residential areas. Individual home ranges vary by habitat and may be as small as 100 m² in a rich environment, to more than a hectare in a poor one. Males generally occupy larger home ranges than do females. Their burrows may be simple, but can be elaborate, with many plugged tunnels, and are often located on a well-drained

Distribution of Eastern Chipmunk (*Tamias striatus*)

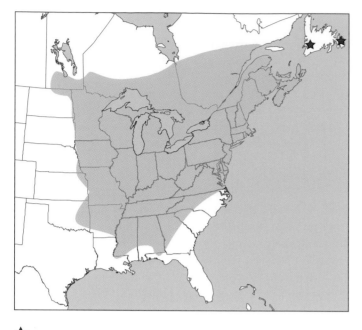

 introduced

slope. Simple burrows are just an entrance connected to a short dead-end tunnel used as an escape hole. More complex burrows contain one or more food storage chambers, and various other chambers for nesting, latrine use, and debris storage. Nest chambers are floored with a thick pile of crushed or shredded leaves about 30 cm in diameter. The access tunnel drops almost straight down at the entrance and then levels off within a metre. Total tunnel length can be more than 10 m, and most are not deeper than about a metre below the surface. Typically, each burrow system has a single active entrance, but may have several plugged entrances that may be temporarily opened as "work holes" to allow safe access to particular foraging areas. Eastern Chipmunks usually occupy the same burrow system for their whole life, after which it may be inherited by a descendant or a juvenile from a nearby litter. Over time, burrow systems are renovated and enlarged, with many side tunnels added to the main one.

In late summer, Eastern Chipmunks begin hoarding food for winter use. Like all chipmunks, they have large cheek pouches that they stuff with seeds, nuts, or acorns to carry to a cache site. Full cheek pouches can virtually double the size of their head. Eastern Chipmunks assemble food for the winter in two different ways. They scatter-hoard (many small caches spread around their territory) and they larder-hoard (one major cache stored usually in an underground chamber within their burrow). The occurrence of one or the other hoarding strategy depends on many factors, including the animal's age (younger animals scatter-hoard more often); their reproductive state (females with infants scatter-hoard more often); whether the animal has a secure burrow and territory (if so, it is more likely to larder-hoard); and lastly, whether it is sharing a food patch with neighbours (chipmunks feeling harassed will scatter-hoard more often). Eventually, most of the various smaller caches are gathered into the underground burrow for the winter. These caches, if

not entirely used up over the winter, may augment spring and summer diets the following year. Population density is governed in large part by food abundance, as the year following a sizable acorn crop is typically a year of high chipmunk numbers; the reverse also occurs. When food is in short supply, chipmunks have higher winter mortality and lower reproductive rates, so abundance drops. Average population density in suitable habitat varies from 10 to 22 per ha.

Eastern Chipmunks spend the winter underground in intermittent hibernation, waking every four to six days to eat from their larder. Date of onset and departure from hibernation varies with weather. Most enter hibernation before snowfall (early to mid-November) and emerge in late March or early April. Some individuals may dig through the snow to appear above ground on mild winter days, especially if a ready food source where they can scrounge a meal, such as a bird feeder, is nearby. Juveniles remain active later in autumn than do adults. Eastern Chipmunks are preyed upon by a wide variety of predators including snakes, hawks, foxes, weasels, Coyotes, Bobcats, and Domestic Dogs and Cats. Eastern Chipmunks are relatively long-lived for their size – as long as eight years in captivity – but survival in the wild is considerably shorter (up to five years), and few live to be even two years old.

DIET

Seeds, nuts, acorns, and fungi are the principal winter foods. Fruit, berries, leaves, flowers, and buds are added over the summer. This species exhibits significant carnivorous tendencies and consumes insects, earthworms, slugs, small frogs and snakes, birds and bird eggs, and small mammals.

REPRODUCTION

Female Eastern Chipmunks have two oestrous cycles each year, the first in late March and early April and the second in late July and early August. The more-northerly populations may only have time for one mating period each year. Most mature females are in breeding condition upon emergence and normally will mate within a week. A female may mate 10–30 times within about a 6–7 hour receptive period, not necessarily with the same male. Following a gestation of around 31–32 days, a litter averaging 4–5 (range 1–8) young are born in an underground nest. They are pink, blind, and almost hairless. Development is rapid and the litter emerges at five to seven weeks old and the young are weaned at about that same time. They remain in their mothers' burrow until they are two to three months old and reach adult size and dentition. Males rarely become sexually mature until after their first winter. Less than around 20% of

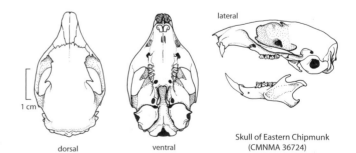

Skull of Eastern Chipmunk
(CMNMA 36724)

dorsal   ventral

females will breed for the first time at their first opportunity (i.e., in the fall). The majority delay their first litter until the next opportunity. For a spring-born female, this means she will miss the fall season in the year she is born and instead produce a spring litter the following year. A fall-born female will miss the next spring season and produce a fall litter instead. From then on, a female will enter oestrus twice each year provided her body weight is sufficient. Should her weight be too low, she will not enter oestrus.

BEHAVIOUR

Eastern Chipmunks rely heavily on vision and sound to detect predators. They are active during daylight, usually waiting until the dew is off the grass before rising. During hot days, they tend to limit their activities to shady locations and to early mornings and late evenings when the temperatures moderate. Logs, stumps, and rocks that provide a wide view are favoured perches. While foraging and travelling, chipmunks regularly stop to look around, since their survival depends on their vigilance. They also pay careful attention to the alarm calls of neighbours. While normally very shy, this small squirrel can be habituated to a non threatening human presence and may even become tame enough to take food from a hand. During the mating season, females tend to remain on their territory, whereas males spend much of their time exploring off their territories as they search for receptive females. Males congregate on the site of an oestrous female, sometimes as far as 400 m from their own burrow. If only a few males are present, they will sort out a dominance hierarchy that allows the dominant male mating access. If several males are present, the dominant male could spend so much time chasing his rivals that he will not have a chance to mate. During dominance battles, the males vocalize, wave their upright tails from side to side, chase each other, and fight.

Eastern Chipmunks are largely solitary, except during mating season and while rearing young. Adults defend the core of their home range around their burrow system by verbally announcing their presence, and if that is insufficient, they chase interlopers away. Scuffles may occur if the pursuer catches the interloper. The peripheries of home ranges may overlap broadly with those of neighbours. Once the juveniles have emerged from the burrow and are weaned, their mother's tolerance for their proximity begins to diminish, although she is more tolerant of the presence of daughters than of sons. Typically, the young males disperse farther away from their natal territory than do young females. Young females may occupy a territory neighbouring their mother, which may even overlap with hers, but young males almost never live that close. Clearly this results in a reduced likelihood of kin matings. These small squirrels are competent, but not daring climbers and they sometimes forage in trees and shrubs. They typically run with their tail held stiffly vertical.

VOCALIZATIONS

Eastern Chipmunks produce three different calls to indicate alarm at the approach of a potential predator: "chucks," "chips," and "trills." The first two are given repeatedly from a stationary position while a threat is in view and often for a while afterwards. Chucks are of a lower frequency and are associated with an aerial predator, while higher-pitched chips indicate a terrestrial predator. Several

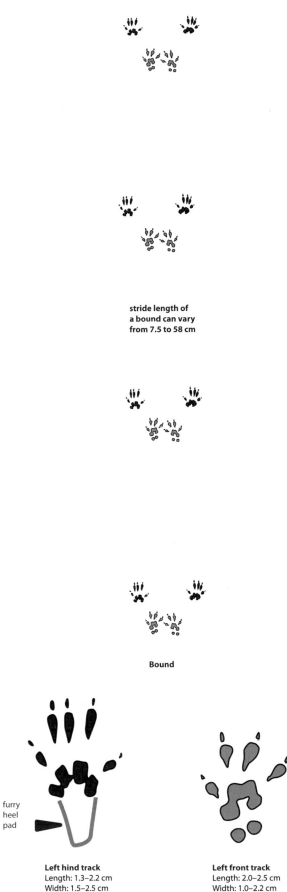

stride length of
a bound can vary
from 7.5 to 58 cm

Bound

furry
heel
pad

Left hind track
Length: 1.3–2.2 cm
Width: 1.5–2.5 cm

Left front track
Length: 2.0–2.5 cm
Width: 1.0–2.2 cm

Chipmunks

neighbours may join the originator in a chorus of chucks or chips that goes on for many minutes. A trill is a rapid series of notes usually only emitted once, as a chipmunk evades a pursuing predator of either type, and is thought to warn nearby relatives that a hungry predator has just missed and will be looking for another opportunity.

SIGNS

Eastern Chipmunk burrow entrances are 4–5 cm diameter clean holes (3–5 cm diameter in the case of other chipmunks), without debris or a fan of discarded soil in front of them. Likely they carry the waste soil out a back entrance and scatter it around before plugging that access, which may be opened again during the next renovation. Often the tunnel slopes sharply down from the entrance before straightening out and continuing. All chipmunk tracks are similar, and even with the size range in the group cannot be confidently ascribed to a species, unless there are no other chipmunk species sharing the range. Some Eastern Chipmunks may remain active after the first snowfall, or venture above ground briefly during the winter (especially during warm spells), so tracks in the snow are possible. The front feet have four functional toes (the pollex or thumb is vestigial, with a nail rather than a claw, and rarely registers in a track), while the hind feet have five longer toes. Foot placement tends to be very similar to that of other squirrels, but smaller. Trail widths are 6.0–8.5 cm from outside edge to outside edge. Chipmunk scat varies depending on the water content of the food. It is commonly 2–7 mm long (average 4.5 mm) and 2–4 mm in diameter and varies from oblong to almost round, often with a rough surface texture. Summer scat is usually slightly soft, sticky, and dark, often forming clumps or linked series. Winter scat tends to be harder and paler. Much of a chipmunk's scat is deposited in its burrow system where it is not visible, but some may be left near feeding sites or perches. Signs of chipmunk foraging can be difficult to distinguish from that of Red or Douglas' Squirrels as both eat acorns, nuts, and conifer cones. Chipmunks will peel thin strips off the shell of thin-shelled acorns like those of White Oak, and will open only half of a heavy-shelled acorn like those of Red Oak, and pull the meat out the open side. Cone scales are peeled even from the smallest cones (like hemlock or spruce). Chipmunks commonly leave small piles of debris on stumps, logs, rocks, or other elevated locations where they can eat and command a wide view at the same time.

REFERENCES

Baack and Switzer 2000; Burke da Silva et al. 1994, 2002; Clarke and Kramer 1994; Dobbyn 1994; Elbroch 2003; French, A.R., 2000; Kurta 1995; Loew 1999; Mahan and Yahner 1996; Rezendes 1992; Sayner 1982; Svendsen and White 1997; Wolff 1996.

Townsend's Chipmunk

FRENCH NAME: **Tamia de Townsend**
SCIENTIFIC NAME: *Tamias townsendii*

The Townsend's Chipmunk is the largest of the western chipmunks. Although this species may not hibernate in more southerly parts of its range, all the individuals in Canada do, spending four to five months underground each winter. Shy by nature, Townsend's Chipmunks are more often heard than seen.

DESCRIPTION

The Townsend's Chipmunk displays the typical chipmunk coloration on their back and sides, with five dark stripes alternating with four paler ones. The lower pair of paler stripes may be brighter than the upper pair, but all are greyish or cinnamon rather than white. None of the stripes reach the root of the tail, but the middle dark one is longest. The shoulders and lower sides are a greyish tawny brown and the rump is dark grey. The belly and throat are whitish to greyish and there is a fairly prominent light patch on the back of each ear along the trailing edge. The top of the tail is a grizzled grey black and the underside is bright orange. The tops of the hind feet are greyish. In general the pelage of this chipmunk is darker and duller than that of other chipmunks. There are two annual moults, one in spring (beginning in May) and the other in late summer and autumn (beginning in August). The winter pelage is darker and more olivaceous than the summer coat. The dental formula is incisors 1/1, canines 0/0, premolars 2/1, and molars 3/3, for a total of 22 teeth.

SIMILAR SPECIES

The only other chipmunks that co-occur with Townsend's Chipmunks are Yellow-pine Chipmunks, which are smaller and brighter in coloration with more distinct side stripes. Cascade Mantled Ground Squirrels, which may occur along the eastern border of the Townsend's range, are superficially similar. They are considerably

Townsend's Chipmunk
(*Tamias townsendii*)

Distribution of the Townsend's Chipmunk (*Tamias townsendii*)

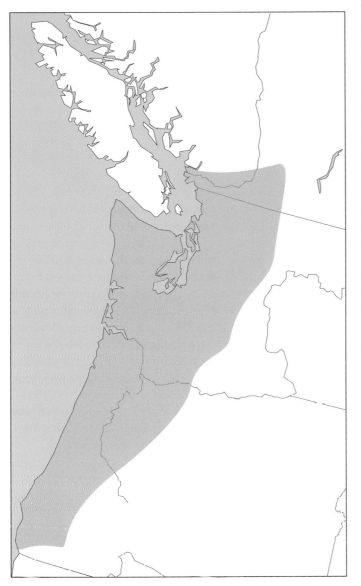

Their secretive and shy nature makes them difficult to survey; however, Townsend's Chipmunks are thought to be well established and common in suitable habitat within their limited range in British Columbia.

ECOLOGY
In British Columbia, these chipmunks live in coastal and upland forests from sea level to 1800 m. Coniferous, deciduous, or mixed forests are all occupied, especially if there is a thick understorey of shrubs and ground cover. Forestry practices and disturbances such as fires, which reduce the canopy and stimulate undergrowth, can encourage Townsend's Chipmunk populations and densities as high as 24.8 per ha have been reported from favourable habitats in the United States. There is only one estimate available for Canadian population density. A study in the Fraser River Valley found that in late summer, when the density was highest, there were 2 chipmunks per ha at the study site. Home range size varies with habitat. Mountain populations in Washington State have home ranges of around 0.6–0.8 ha. Like all chipmunks, Townsend's Chipmunks are diurnal. They are most active around midday. Like other chipmunks, they construct underground burrows, but little is known of their structure or complexity. Canadian animals, and indeed many in the United States, spend the bulk of winter in intermittent hibernation in their underground nest. They rouse during this period to eat from their food cache and then sink into dormancy again. Most disappear underground in early to mid-November and emerge from hibernation in late March to early April. They may appear above ground for a short time in the winter during warm periods. Chipmunks in warmer parts of the range may remain active throughout the winter. Males emerge earlier than females by up to a month and juveniles remain above ground for a month or more, after the majority of the adults have disappeared into their burrows in autumn. Chipmunks are attractive prey for many diurnal predators and Townsend's Chipmunks are vulnerable to weasels, skunks, hawks, Coyotes, Bobcats, Northern Raccoons, and Domestic Cats and Dogs. Wild Townsend's Chipmunks can live seven years and perhaps longer.

DIET
During spring, summer, and autumn, Townsend's Chipmunks consume a wide variety of seeds, nuts, acorns, fungi, berries, fruit, roots, bulbs, leaves, and flowers, and small animals such as insects, birds, and bird eggs. Because of their taste for underground fungi (such as truffles), Townsend's Chipmunks are thought to be important dispersers of mycorrhizal fungi, which are so vital to tree health, as the spores are resistant to digestive juices and are scattered in the faecal pellets. Winter food is made up of items that preserve well in underground storage: principally seeds, nuts, and acorns.

REPRODUCTION
Females come into oestrus soon after emerging from hibernation and most breed in mid-April to early May. Lowland females emerge earlier and may come into season earlier than those in upland

larger than any of the chipmunks, have only a single light blaze on each side, and lack the facial stripes that are so characteristic of the chipmunks.

SIZE
Males are around 3%–4% smaller than females, mainly in total length and weight measurements. The following measurements apply to adult animals within the Canadian portion of the range.
Total length: 234–286 mm; tail length: 103–130 mm; hind foot length: 32–40 mm; ear length: 17–23 mm.
Weight: 65–89 g; newborns, 3–4 g.

RANGE
Townsend's Chipmunks occur in coastal lowlands and mountains from southern Oregon to southwestern British Columbia. Formerly thought to occur on southern Vancouver Island, these early reports have been the result of several unsuccessful introduction attempts.

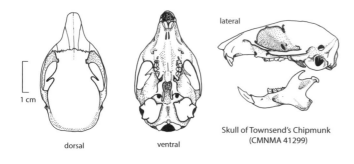

lateral

Skull of Townsend's Chipmunk
(CMNMA 41299)

1 cm

dorsal

ventral

populations. Gestation is thought to last around 30 days. Females in this species are monoestrus and so have a single litter each year. Range of litter size is two to seven young, with an average of four. The young are born blind, deaf, pink, and hairless but for vibrissae. They grow rapidly, begin eating solid food in their fifth week, and are weaned at six to seven weeks old. In early July, when the young first appear above ground, they are about one-half the size of the adults. Full adult dentition is attained by around 90 days old. Sexual maturity for most youngsters is reached by the spring following their birth, but 30%–40% do not breed for the first time until the following year.

BEHAVIOUR

These chipmunks are typically solitary and interact infrequently. When they do, they tend to be antagonistic, but less so to kin than to strangers. They are also capable of recognizing a close relative and thereby avoid inbreeding. Generally shy, they tend to remain undercover or in the shadows. They are most easily seen in late summer and autumn when food gathering activities consume most of their attention and vigilance is reduced. These chipmunks are less nervous and jerky in their movements than are other species, but they still carry their tails stiffly erect as they run from danger. Although they are excellent climbers and spend considerable time foraging high in shrubs and trees, no tree nests have been reported in Canada, though there are some records of tree nests in California.

VOCALIZATIONS

One of their alarm calls is very bird-like. They also produce a "chuck" alarm call similar to that of other chipmunks. Adults and juveniles of both sexes produce these alarm calls, which function to warn kin and neighbours of approaching predators. It is possible, as with Eastern Chipmunks, that each of these alarm calls warns of a different type of predator, aerial or terrestrial.

SIGNS

See the Eastern Chipmunk account for a discussion of chipmunk sign and sketches of tracks.

REFERENCES

Carey 1991; Carey et al. 2002; Fuller and Blaustein 1990; Harestad 1991; Hayes, J.P., et al. 1995; Kenagy and Barnes 1988; Nagorsen 2005a; Sutton, D.A., 1993; Verts and Carraway 1998.

Douglas' Squirrel
also called Chickaree

FRENCH NAME: **Écureuil de Douglas**
SCIENTIFIC NAME: *Tamiasciurus douglasii*

Energetic and noisy, these squirrels are quick to make their presence known by loudly calling an alarm upon the approach of a potential predator. Recent genetic evidence calls the distinction of this species from Red Squirrels into question and perhaps this population will be reduced to subspecific status.

DESCRIPTION

Douglas' Squirrels are olive brown to chestnut brown on the back, with brown to olive grey sides and yellow to orange underparts. The light ring around the eye is buff to light orange. The summer coat is browner and the underparts are more orange. A black side stripe between the belly fur and the back fur is typically prominent in summer. The winter coat is more olive grey, but often with a reddish-brown region along the spine. The black side stripe is muted or absent on the winter coat. Ear tufts are noticeable on the winter pelage. Tails vary in colour from rusty brown to yellowish black with an indistinct black edge and buffy white on the hair tips. The dental formula is incisors 1/1, canines 0/0, premolars 1-2/1, and molars 3/3, for a total of 20 or, rarely, 22 teeth. The incisors on the upper and lower jaws are open-rooted and grow continuously through the life of the animal. They are kept sharp by constant use. These squirrels moult twice annually. The spring moult takes place sometime in May or June, and takes about a month, beginning at the head and proceeding rearwards. The autumn moult occurs between late August and early October, beginning at the tail and

Douglas' Squirrel (*Tamiasciurus douglasii*)

summer

winter

proceeding forward. The tail hairs are probably moulted once each year in midsummer.

SIMILAR SPECIES

Red Squirrels are similar in size, behaviour, and physical characteristics to Douglas' Squirrels. They usually have white or light-grey underparts, but some have a faint yellowish tinge to their belly fur. The tips of their tail hairs are orange to reddish. Douglas' Squirrels typically have a yellow to orange belly and throat especially in summer, and the tips of the tail hairs are buff to pale yellow. In some parts of British Columbia (Manning Provincial Park and Fraser River Canyon areas), distinguishing the two is very difficult, as there appear to be individuals of intermediate appearance, perhaps suggesting a zone of hybridization. An experienced observer can learn to distinguish the vocalizations of the two species. Their alarm calls are especially distinct. That of Douglas' Squirrels has longer notes. Their rattle call also has longer notes spaced farther apart than those of Red Squirrels. Calls of intermediate animals in the possible hybridization zone are also intermediate.

SIZE

The following measurements apply to adult animals from the Canadian portion of the range. Adult males and females are similar in size.

Total length: 271–351 mm; tail length: 90–150 mm; hind foot length: 43–57 mm; ear length: 20–31 mm.

Weight: 157–270 g; newborns, unknown, but probably similar to that of Red Squirrels (6–8 g).

RANGE

Douglas' Squirrels occur in coniferous forests in western North America from southern British Columbia to California. In British Columbia, they are found as far north as the south side of Owikeno Lake in Rivers Inlet to at least as far east as the Pemberton Valley. The eastern limits of distribution in British Columbia are poorly delineated and require further surveying. They do not occur on Vancouver Island, being replaced there by Red Squirrels, but they are resident on Bowen and Gambier Islands in Howe Sound and on Cortes, East Redonda, Stuart, and Subtle Islands between central Vancouver Island and the mainland. Although Red Squirrels and Douglas' Squirrels tend not to occur together, the eastern border of the Douglas' Squirrel distribution does overlap with Red Squirrels.

ABUNDANCE

Douglas' Squirrels are relatively common in suitable habitat, but their population density fluctuates depending on food abundance.

ECOLOGY

Douglas' Squirrels inhabit coastal lowlands and mountain ranges in coniferous or mixed coniferous/deciduous forests. They tend to associate with stands of Douglas-fir, fir, Western Hemlock, and spruce. They are diurnal and remain active all year. During warm months their activity peaks in the morning and again in late afternoon, and during cold months they are most active at midday

Distribution of the Douglas' Squirrel (*Tamiasciurus douglasii*)

when it is warmest. During very cold days or heavy rain, they may forgo outside excursions altogether and remain in their nest. It is thought that, like the closely related Red Squirrels, Douglas' Squirrels manage to survive cold weather despite their small size and lack of body fat by maintaining a high internal temperature and a high metabolism. The price for this energy expenditure is their dependence on abundant food during the winter, a time when food is scarce. To overcome this problem they are highly territorial, ensuring the trees that usually provide their food, and the caches of cones they gather in summer and autumn to eat during the winter, are protected for their own exclusive use. The larder may be

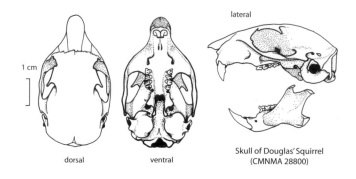

1 cm

dorsal   ventral

lateral

Skull of Douglas' Squirrel
(CMNMA 28800)

large enough to feed the animal for more than one, or even two, seasons. They also must maintain a sheltered, well-insulated nest. This squirrel builds dreys (leaf nests with a twig support insulated with moss, lichen, or shredded bark), uses tree cavities, and creates underground burrows. The density of Douglas' Squirrels is directly related to territory size, which is in turn inversely related to food abundance. Numbers may plummet following a poor cone crop or climb following a bumper crop. Likely due to their more abundant cone supply, Douglas' Squirrel winter densities are highest in old-growth forests. In fir forests, population densities have been found to be 0.2 squirrels per ha on average, while in hemlock forests there are < 0.01–0.5/ha. In a mixed forest of 40–50-year-old Western Hemlock, Western Red Cedar, and Douglas-fir in British Columbia, densities varied from 0.2 to 0.9 per ha. Northern Flying Squirrels and Townsend's Chipmunks may be seed and fungi competitors, as Douglas' Squirrels are more abundant in regions where these squirrels occur in low numbers. Furthermore, Douglas' Squirrels will defend their territory from these species as well as from conspecifics particularly in autumn, when they are gathering and protecting their cache. Feeding by Douglas' Squirrel may result in serious damage to forests, especially following a poor cone crop, when the squirrels are starving and resort to less preferred foods such as shoots, buds, and cambium (nutritious inner bark) of conifers. There is no information available concerning the potential lifespan of this species, but it is likely in the three-to-five-year range. Predators include raptors such as Cooper's Hawk, Northern Goshawk, Great Horned Owl, and Spotted Owl, and mammals such as Ermine and American Marten.

DIET

The food of Douglas' Squirrels is composed mainly of conifer seeds, fungi, deciduous seeds and nuts, and pine cambium. They also eat leaves, flowers, insects, bones (mainly by juveniles and pregnant or nursing females), birds, and bird eggs. Free water is rarely available to these squirrels, other than dew and rain, so most of their water requirements are achieved through their food. Some fungi are 90%–95% water.

REPRODUCTION

The breeding season in British Columbia lasts for four or five months. Males are in reproductive condition from March to July. A reproductively active male has pendant scrotal testes. During the remainder of the year, the testes shrink and regress to an abdominal position. Nursing females may be found from May to September. A female is in oestrus for only eight hours, during which time she will mate several times, possibly with multiple males. There are two peaks in the breeding season, one in early spring and another in early summer. Individual females may produce two litters per year if food is abundant, but following a hard winter they may breed late or not at all. Likely in Canada, few females realize their full reproductive potential in any given year. Following a gestation of about 35 days, 1–8 (average 4–6) young are born. Litters are smaller in years following a poor cone crop. The young squirrels are weaned at around 64–78 days old. Otherwise, little is known of their growth and development. Juveniles probably begin breeding as yearlings, provided they have had sufficient food through their first winter on their own.

BEHAVIOUR

Like the closely related Red Squirrel, the Douglas' Squirrel is highly territorial. Most disputes are resolved at the vocalization stage (see the "Vocalizations" section below), although some may lead to a chase. Rarely do the protagonists actually physically fight, unless it is late autumn and a food cache is at stake. A territory is essential to survival and juveniles and vagrants are quick to lay claim to an abandoned one. Even a nearby adult already with a territory may switch over to a more preferred site if it becomes vacant. Nevertheless, during years of extreme food shortage, squirrels will abandon their territories and travel in search of food. Normally, each adult squirrel defends an exclusive territory and will permit no intrusions. The only exception occurs during the breeding season, when males roam widely to investigate the reproductive readiness of females. A female will allow males to enter her territory only for the duration of the eight-hour period during which she is receptive. The males compete with each other to determine which is dominant and only the dominant male stays close, occasionally making tail-waving cautious approaches when she is still, and chasing her when she runs. As he remains near her, he continues to aggressively call to the subordinate males waiting on the periphery. Sometimes he leaves her to chase one of these males who has dared to approach too closely. The female usually feeds and waits for him to return. But females do not always accept this hierarchy, sometimes choosing to avoid the dominants and allow subordinate males a mating opportunity. Mating occurs when the female stops rejecting a male's advances and most copulations take place on the ground or on low branches. The pair must be alert for attacks by other males wishing to interrupt their activity. It is likely that male dominance in these cases is at least partly determined by proximity to the male's territory. A male from an adjacent territory is more likely to become the dominant male. Douglas' Squirrels are larder-hoarders, gathering one or a few caches of conifer cones for winter consumption. The caches are also called middens, as they are composed partly of cone debris from previous feeding. This debris is crucial to maintaining a moist environment, to prevent the intact cones from drying out and opening. The dispersal of juveniles is poorly understood, but it is likely that, like Red Squirrels, most settle near where they were born. The Douglas' Squirrel has odoriferous glands in the corners of the mouth that both sexes use to mark frequently used paths.

VOCALIZATIONS

Douglas' Squirrel and Red Squirrel call repertoires and usages are very similar (see in the "Similar Species" section above). Both are highly vocal, using calls to advertise and defend territory ownership. The "rattle" call announces territory occupancy and is also used as a warning if a neighbour is spotted. The "screech" call is louder and often follows the rattle call if the potential intruder actually enters the territory. "Growl" calls are used during a fight by dominant males in an effort to deter subordinates during courtship, and by females to deter an overly ardent male. A squirrel in a live trap may growl as it is approached. "Buzz" calls are produced by males to the female during mating chases, when juveniles interact with adults, or when two subordinate males approach one another. This call appears to indicate non-aggressive intent. Two types of alarm calls are used: the "chirp" call warns of an aerial predator and the "bark" call warns of a terrestrial predator. Apparently, other nearby squirrels, such as chipmunks and ground squirrels, respond to these alarm calls just as Douglas' Squirrels do. Squirrels of any age or gender may scream if injured or captured.

SIGNS

Tracks and trackways, habits, and signs are indistinguishable from those of Red Squirrels. See the "Signs" section of the Red Squirrel account for a discussion and illustrations.

REFERENCES

Arbogast et al. 2001; Buchanan et al. 1990; Elbroch 2003; Koford 1982; Nagorsen 2005a; Shaw and Flick 2002; Smith, C.C., 1968; Steele 1999; Steele and Koprowski 2002.

Red Squirrel

also called Pine Squirrel, North American Red Squirrel

FRENCH NAME: **Écureuil roux**
SCIENTIFIC NAME: *Tamiasciurus hudsonicus*

Aggressive and bold, these small denizens of our northern forests make their presence known with loud alarm calls, which alert other individuals that a potential predator has been seen. They also create some of the largest cone larders of any squirrel in North America.

DESCRIPTION

As their common name indicates, these small squirrels often have a reddish-coloured back and sides. The colour does vary from olive grey to rusty brown or often reddish on the midline of the back,

Red Squirrel (*Tamiasciurus hudsonicus*)

summer

winter

with olive-grey sides. The belly and ring around the eye are usually white, but sometimes slightly yellow or greyish. The tail colour varies from yellowish grey to rusty red and is often grizzled, with white or light buff tips on the hairs. The edges and often the tip of the tail are black and some animals have a black strip down the centre of the upper surface of the tail. The summer coat is sleeker and browner than the winter coat, with a more obvious black blaze on the side between the belly fur and the back fur. The winter coat has longer tufts on the tips of the ears, is often redder than the summer coat, and has a more muted black blaze. Rare melanistic animals have been reported from the Maritimes, south-central British Columbia, and southern Yukon. White and partial-white individuals occur occasionally in Alberta and British Columbia. The dental formula is incisors 1/1, canines 0/0, premolars 1–2/1, and molars 3/3, for a total of 20 or, rarely, 22 teeth. These squirrels moult twice annually. The spring moult takes place some time from late March through August, with the females and young of the year moulting last after weaning. This moult takes about a month, beginning at the head and proceeding rearwards. The summer pelt is made up only of guard hairs and lacks the insulating underfur. The autumn moult occurs between late August and early December, beginning at the tail and proceeding forward. The winter pelt is made up of thick underfur as well as guard hairs. The tail hairs are moulted once each year in midsummer.

SIMILAR SPECIES

The closely related Douglas' Squirrel is the most similar species and in some parts of British Columbia (Manning Provincial Park and Fraser River Canyon areas) distinguishing the two is very difficult as there appear to be individuals of intermediate appearance, perhaps suggesting a zone of hybridization. In general Douglas' Squirrels

Distribution of the Red Squirrel (*Tamiasciurus hudsonicus*)

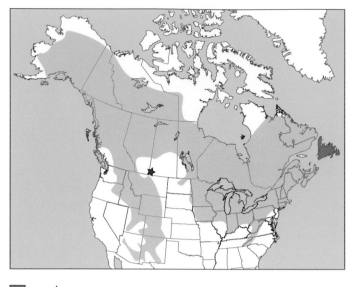

■ & ★ introductions

are distinguished by their orange to rusty-yellow belly and their tail hairs, which often terminate with pale-yellow tips. The larger Grey Squirrels are usually fairly easy to distinguish by colour and size. They also have a longer, bushier tail.

SIZE

The following measurements apply to adult animals from the Canadian portion of the range. Adult males and females are similar in size.

Total length: 282–402 mm; tail length: 81–187 mm; hind foot length: 23–57 mm; ear length: 11–29 mm.
Weight: 197–298 g; newborns, 6–8 g.

RANGE

These tree squirrels usually occur in the boreal and mixed coniferous-deciduous forests of North America from Alaska to the Maritimes and as far south as Texas. Along the edge of the prairies in southern Manitoba this species can be found in aspen-oak forests and savannah, and deciduous riparian forests. Their range is commonly fragmented, especially in the western United States, where they occupy the subalpine zone of the mountains. This zone retreats farther and farther up the mountainside as the mountains become closer to the equator. Red Squirrels are found on many offshore islands, including Vancouver Island and many of the associated islands, as well as numerous coastal islands along the British Columbia coast. Interestingly, Red Squirrels do not occur in the Lower Mainland of British Columbia adjacent to Vancouver Island, which is occupied instead by Douglas' Squirrels. Red Squirrels from Vancouver Island have been introduced and are now established on many of the Queen Charlotte Islands. They were introduced into Newfoundland twice: Northern Peninsula in 1963 and Notre Dame Bay in 1964, and have since expanded to occupy all the Newfoundland forests. There is an introduced population in the Cypress Hills of southern Saskatchewan and Alberta.

ABUNDANCE

Red Squirrels are generally common in their preferred habitats and common to rare in marginal habitats. Density depends on food availability. A poor cone year can drastically reduce the population, as the ensuing winter's mortality is high for both juveniles and adults and reproductive productivity the next spring is diminished. Climate change in northern forests has resulted in an 18-day advancement of breeding in a Yukon forest over the past ten years. This clearly shows that Red Squirrels are adaptable to some environmental variation and may be able to cope with warmer weather resulting from global warming.

ECOLOGY

Red Squirrels remain active all year and are primarily diurnal. During the frenzy of food gathering that takes place in autumn, some may continue harvesting cones until well after dark. Activity peaks for a few hours after sunrise and before sunset in warm weather and at midday during the cold months. They prefer to live in dense boreal or montane coniferous forests, which provide abundant conifer seeds and fungi and have interlocking branches for arboreal travel while the squirrels are foraging and for predator avoidance. Less-optimal habitat colonized by Red Squirrels includes hedgerows, mixed deciduous/coniferous forests, old orchards, planted stands of pine and spruce, and urban backyards. Despite their small size, they are the most northerly of the North American tree squirrels. They compensate for their low-fat insulation and small size, and therefore high heat loss, by maintaining a high body temperature and metabolism. Sustaining this level of energy output requires three things: sufficient food, dense underfur, and a well-insulated winter shelter. Red Squirrels build leaf nests (dreys), use natural cavities in trees, logs, and rocks, and create underground burrows for shelter. All may make suitable winter locations, but leaf nests are only adequate in warmer climates. Winter nesting usually occurs in tree cavities, and nests are constructed of shredded bark and other dried vegetation. Most burrows require minimal digging, as these squirrels will use an existing hole, especially if it gives them access to an underground tree-root system with its natural air spaces. Their globular dreys are constructed 2–20 m up near the trunk of a conifer and are made of twigs, leaves, mosses, dried grasses, and lichen, with an inner lining of shredded bark, leaves, fur, and feathers. The entrance is on the side. Placement of the drey is inevitably near the main cache site.

Red Squirrels have been classified in the past as classic larder-hoarders, with one main cache site where they store the bulk of their winter food supply. A large cache such as this may occupy 50–154 m² and reach over 1.2 m deep and is likely the accumulation of generations of squirrels that occupied the territory. Such a cache typically lies on the forest floor beneath a tree. The larder or cache may also be called a midden, as it is composed of both intact cones with their seeds still enclosed and debris from feeding. Cone debris accumulating beneath favourite feeding sites eventually creates a decomposing, moist environment, appropriate for intact cone storage. High moisture keeps the cones from drying and opening. Thus, the large caches are composed of both food cones and debris. In addition to, and sometimes instead of, the major stockpile, many Red Squirrels, especially in eastern North America, have a number of smaller caches stashed around their territory, including in an underground burrow if they have one large enough. Disturbed habitats due to logging or fire may favour such scatter-hoarding rather than the more traditional larder-hoarding. Beginning in late summer and continuing into autumn, mature cones are harvested directly from the tree or off the ground before they dry and open to allow seed dispersal.

Population density across North America ranges from 0.2 to 4.5 squirrels per ha and is highest in mixed old-growth forests and lowest in spruce-dominated forests. Population size fluctuates dramatically, depending on yearly cone production. Somehow the squirrels can detect months ahead when the trees are about to have a bumper cone crop. Litter size increases in the summer before such a crop and the young born have a higher chance of surviving their first winter, because of the abundant food supply. Mortality is normally high, as their high metabolism requires a plentiful food supply and many predators hunt Red Squirrels. On average, more than 60% die in their first year, 80% in their second year, and 90% in their third year. Maximum longevity in the wild is ten years, but few live more than five years. It is possible that females live shorter lives than males, as there appear to be slightly more adult males than females in most populations. Red Squirrels are capable swimmers known to swim up to 2 km to islands in lakes. Predators include raptors, owls, snakes, American Martens, Fishers, Bobcats, Red Foxes, Canada Lynx, Long-tailed Weasels, Ermine, American Mink, and Domestic Dogs and Cats.

DIET

Red Squirrels specialize in eating seeds of coniferous trees, but will eat other food items such as nuts and seeds of deciduous trees such as oak, beech, walnut, maple, and poplar. Fungi are an important summer food and are also dried and stored for winter consumption. Additional foods, consumed in times of food shortage, or when opportunity permits, include tree buds, flowers, fruit and berries, bark, tree sap (usually in spring), insects including many that are forest pests, birds and bird eggs, and small infant mammals (such as rabbits and other squirrels). Cannibalism has not been reported in this species. Red Squirrels have very efficient kidneys, so most of their water requirements are supplied in their food, and they rarely drink free-standing water or eat snow, even when it is available.

REPRODUCTION

Breeding season occurs during February and March in most parts of Canada. Females are in oestrus for eight hours. Most produce a single litter each year in spring (March–May), but during exceptional cone years with abundant food, some females, in warmer regions, may have a second heat in late June or July and produce a second litter in late July to early September. In Canada, evidence of a few females having a second litter has been detected in areas like southern Quebec and Vancouver Island. Many females fail to breed at all in the spring following a year of low cone production. Gestation is thought to last for 31–35 days. Litter size ranges from one to eight pups, but averages two to four. Newborns are hairless but for a few short whiskers, and their ears and eyes are unopened. Their ears open at around 18 days old and their eyes at about 27–35 days. They grow quickly and are fully furred by 40 days. They begin venturing out of the nest at six to seven weeks old and are weaned by eight weeks. They become fully independent of their mother and disperse from her territory at about 10–14 weeks. Both males and females may begin breeding as yearlings, if they are well nourished. In populations occupying less than optimum habitat, most squirrels begin reproducing as two-year-olds. Males are reproductively ready when their testes are enlarged and pendant in the scrotum. In regions with a single breeding season, this is typically the case from February to June. In regions where a second season may occur, the males' testes are scrotal from February to August. The remainder of the year, the testes shrink and regress into a more protected abdominal position.

BEHAVIOUR

These aggressively territorial squirrels maintain an exclusive and roughly circular feeding territory that centres on their main cone cache. Since survival depends upon a ready food supply during the winter, territorial ownership is essential. Territory holders repel trespassers by loudly advertising their presence by calling (see the "Vocalizations" section below), by threat displays, and by attacking or chasing intruders away. Territorial disputes rarely lead to battles in which one or both animals are seriously injured, but both participants reduce their predator vigilance during these chases and may then be more vulnerable, especially since the noise may attract the attention of a nearby predator. Regular patrols of the boundaries are required to maintain ownership, and despite their vigilance, many territory owners are challenged and lose. Only during the breeding season, when males travel around seeking oestrous females, will the females permit invasion of their territory. Closely related neighbours are sometimes treated somewhat differently. Nearby sisters may share a nest during very cold nights, but each returns to her own territory during the day.

Territory size varies with food availability. Pilfering from other squirrels' larders is a behaviour apparently practised by most Red Squirrels in some areas. Approximately 25% of the food they eat is stolen, while they lose about the same amount from their own cache. Some Red Squirrels appear to specialize in pilfering and will steal

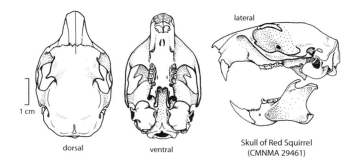

lateral

1 cm

dorsal   ventral

Skull of Red Squirrel
(CMNMA 29461)

stride length
of a bound
can vary from
10 to 64 cm and
occasionally
farther

Bound

Left hind track
Length: 2.2–5.7 cm
Width: 1.6–2.9 cm

Left front track
Length: 2.5–3.2 cm
Width: 1.4–2.5 cm

Red Squirrel

almost all their food. Many of the chases witnessed outside the breeding season may involve territory holders evicting thieving neighbours. In central Alberta, territories are 0.24–0.35 ha, in southern British Columbia they are 0.62–1.18 ha, and they average 0.4 ha in New Brunswick. In a poor mast year the territories grow as the squirrels search farther afield for food. Squirrels will also make occasional forays of several hundred metres, likely to find, or check on the ripening of, seasonally available foods. After they are weaned and before they occupy their own territory, most juveniles sleep on their mother's territory and make daily exploratory forays of up to 900 m beyond it. Both juveniles and subordinates are quick to recognize a vacant territory and rapidly move to establish occupancy. Although a dispersal distance of over 5 km has been recorded, most juveniles establish a territory near where they were born. Commonly, an adult female will relinquish her territory to one of her litter and establish a new adjacent territory or take over a nearby vacant territory for herself, or she might allow a juvenile to settle on part of her territory while she expands the remainder. Territories change ownership regularly, but perimeters of territories are remarkably stable through many generations of territory holders. Short-term mate pairings may occur, but most adult Red Squirrels are solitary. Young Red Squirrels improve their foraging and food-handling effectiveness by watching their mothers. Adults of both sexes use scent marking more or less equally. Oral glands in the corners of the mouth are rubbed on familiar spots in the territory, such as on resting, grooming, and feeding sites. This behaviour appears to be mainly for self-orientation, to reassure and maintain familiarity, rather than to mark ownership. During the breeding season, males travel around, checking on the condition of females in their area primarily by their odour. As many as 10 or more males may gather during her one-day period of oestrous. These males work out a dominance hierarchy, as the dominant male will call and aggressively chase the others away. Battles do occur and blood may be shed in the process. The female usually breeds with the dominant male, although this dominance often changes during the day, and she may eventually mate with multiple males. Copulation is instigated by the female as she assumes a breeding posture with tail over her back and tipped to one side.

VOCALIZATIONS

Adult Red Squirrels produce six recognizable calls. The "rattle" call is mainly used to advertise ownership of a territory and is emitted as a preventative measure when no other squirrels are around or sometimes if a potential intruder is visible. It carries for up to 130 m, lasts about 10 seconds, and seems to mean "keep out." Some males will rattle upon approaching a female during the breeding season. In this context, it appears to identify his presence and possibly his intent to approach. The rattle call is also used by males during courtship. The "screech" call, another territorial call, usually indicates that the squirrel is willing to aggressively defend its territory from an intruder. It is a series of high-pitched "tsew" sounds. Typically only the territory holder will screech. A screech call often follows a rattle call. The "growl" call usually occurs in complex social situations such as territorial disputes, when a female's young are too persistent in their demands to suckle or play or, as often occurs during the breeding season, when a female receives relentless unwelcome advances from an ardent male. A "buzz" call is uncommonly heard. A male sometimes buzzes during a mating chase or when he enters the territory of a female. Juveniles will buzz if approached by an aggressive adult. Use of this call typically occurs when one squirrel approaches another with non-aggressive intent. Red Squirrels may also produce two different alarm calls when they detect a potential predator. A short, high-frequency "seet" call is said to warn of an aerial predator, and a louder "bark" call warns of a terrestrial predator. The bark call is the most commonly heard Red Squirrel vocalization and may continue for only a few seconds, or for almost an hour. It is typically made while flicking the tail and stamping the feet. Social usage of the bark call and its variability need further study. Red Squirrels sometimes use the bark call during territorial disputes or when inside, defending their den from an aggressor. In addition, all ages of Red Squirrels will scream when injured or captured.

SIGNS

Red Squirrel tracks and trails are indistinguishable from those of the Douglas' Squirrel. Both exhibit the typical rodent tracks. The smaller front feet have four toes and the outside two (toes 2 and 5) angle away from the foot, while the middle toes (3 and 4) point straight ahead. Toe 1 (the thumb) is vestigial and rarely registers in a track. The hind feet have five toes, with the middle three aiming forward in a cluster and the outside two toes angling away from the line of the track. Both species almost always bound while travelling on the ground. In deep fluffy snow, bounds commonly leave a single pair of tracks, as the hind tracks register directly on top of the front tracks. Such a trail is generally connected by drag marks, as the hind legs skim through the snow both into and out of the track mark. Red Squirrels also make tunnels in the snow. Snow tunnels and ground-burrow entrances are about 6–8 cm in diameter. A ground burrow used as a cache site is betrayed by the pile of cone debris near the entrance, as these squirrels rarely travel far from the cache to eat. Red Squirrel scat may be dropped randomly around the territory, but is most frequently found on high-traffic routes such as a fallen log, a favourite stump, or on a high point such as a burl. A fungus wedged into the crotch of a tree branch to dry is a sure sign of a Red Squirrel or Douglas' Squirrel. When eating from their cache, Red and Douglas' Squirrels sit on their haunches, usually close to the ground on a low branch, stump, or log. They manipulate the cone, turning it systematically with their forefeet as they chew the bracts off to reveal the seeds. The discarded debris piles up below them. Since they rarely travel far from their cache to feed, the debris pile, besides identifying the feeding site, suggests the nearby location of a cache.

REFERENCES

Becker et al. 1998; Boutin et al. 1993; Bovet 1990; Elbroch 2003; Ferron 1983; Gerhardt, F., 2005; Greene and Meagher 1998; Hurly and Robertson 1990; Lair 1990; Larsen and Boutin 1994; McAdam and Boutin 2004; Nagorsen 2005a; Norment et al. 1999; Patterson, J.E.H., et al. 2007; Pauli 2005; Pretzlaw et al. 2006; Price and Boutin 1993; Réale et al. 2003; Smith, C.C., 1968; Smith, H.C., 1993; Steele and Koprowski 2002; Stuart-Smith and Boutin 1994; Sun 1997; Vernes et al. 2004; Wheatley et al. 2002; Yamamoto et al. 2001.

FAMILY CASTORIDAE
beavers

There are two living species of beavers, one in Europe and Asia *(Castor fiber)* and the other in North America *(Castor canadensis)*. Both are temperate and boreal forest inhabitants. They are very similar in appearance and lifestyle, but differ in some skull characteristics and behaviours. North American Beavers have been introduced into parts of Europe and are now out-competing the native beavers in some areas. Eurasian Beavers have never been introduced into North America. Successful hybridization between the two species does not appear to occur.

North American Beaver
also called Canadian Beaver

FRENCH NAME: **Castor du Canada**
SCIENTIFIC NAME: *Castor canadensis*

Long considered a fitting symbol of Canadian ingenuity and hard work, the North American Beaver has graced our coinage, fed and clothed us, and provided the impetus for the exploration of the New World. Beavers, like humans, can leave a presence visible from space. A large dam in an inaccessible part of Wood Buffalo National Park, Alberta, is estimated to be 850 m long. It was discovered through the study of satellite images.

DESCRIPTION

The North American Beaver is the largest rodent on the continent and the second largest in the world. Only the South American Capybara is larger. Beavers have short legs, large heads, large, webbed hind feet, and distinctive, broad, horizontally flattened, scaled tails. They are semiaquatic and display many adaptations that suit them to that lifestyle. Their eyes, ears, and nose are located along the upper plane of the head, so are above the waterline as the animal floats with the rest of its body hidden below. Well-furred lips can be closed behind the incisors so waste chips do not fall into the mouth and sticks can be carried and even gnawed while underwater without the risk of the mouth filling with water. Breathing occurs only through the nostrils, so food can be swallowed underwater without fear of choking. Clear nictitating membranes slide over the eyeballs like goggles, and flaps can be closed over the ear and nostril

North American Beaver (*Castor canadensis*)

openings when diving. Large hind feet are heavily webbed and provide all the animal's underwater propulsion. Special grooming claws on the inside two toes of each hind foot are slightly serrated, with double nails to "comb" the fur, and are especially useful in squeezing water out. The claw on the second toe is most developed for this purpose. Two oil glands behind the anus continuously secrete an oily substance that each animal spreads over its fur with its front feet to make it waterproof. Their double coat is composed of two kinds of hairs: long, flattened guard hairs and dense, shorter, curly, grey underfur. The underfur provides insulation and traps a layer of air, preventing water from touching the skin, and the guard hairs protect the underfur and direct water away from the animal. Typical fur colour is dark brown with reddish highlights, but can vary from black through dark brown, reddish brown to pale brown, and occasionally silver. Albino and white or part-white animals are rare. The base of the tail is heavily furred, but the remaining two-thirds, which forms the "paddle," has only a few coarse hairs between some of the scales. This flattened portion of the tail is almost black in young beavers, but lightens to brown with age. Castor glands, which produce a yellowish, pasty, odoriferous secretion called castoreum, are located on either side of the anus (see in the "Behaviour" and "Signs" sections below for more information). On a mature beaver, each of the paired castor glands is around 13 cm in circumference (about 5 cm in diameter). Beavers have keen senses of sight, hearing, touch, and smell. Their dental formula is incisors 1/1, canines 0/0, premolars 1/1, and molars 3/3, for a total of 20 teeth. Incisors are large and ever growing, and both upper and lower incisors are heavily pigmented with orange on their front surface, caused by iron in the enamel. It hardens the area so that the softer material behind wears away faster. This results in the tooth's chisel shape, which is so effective in cutting wood. See the "Signs" section for an explanation of how beavers cut wood. Kits are born with white incisors, which gradually become orange. One annual moult occurs during the summer.

SIMILAR SPECIES

A beaver's large size and broad flat tail are distinctive. Other semi-aquatic rodents such as muskrats and the introduced Nutria may be mistaken for small beavers, but their tails are more rodent-like, and in the case of the Common Muskrat, somewhat compressed laterally, in the opposite direction to a beaver's tail. Muskrats construct much smaller lodges than do beavers and they are typically composed of softer vegetation, such as cattail and bulrush stems, rather than woody branches. See the figure in the account of the North American Porcupine that shows the differences between the skulls of the North American Beaver and the North American Porcupine.

SIZE

Although large adult males tend to be larger and heavier than large adult females, the difference is not normally detectable in the field, especially since there is considerable overlap in the size ranges of the genders. The following measurements apply to adults from Canada. Weight varies with the season: adults are heaviest in late autumn and lightest in early spring. Beavers may continue to grow (albeit at a much reduced rate) after reaching physical maturity, so the largest animals tend also to be the oldest.

Total length, males: 98–126 cm; tail length: 28.3–53.0 cm; hind foot length: 16–20 cm; ear length: 3.2–4.2 cm.

Total length, females: 95–117 cm; tail length: 20.6–51.0 cm; hind foot length: 16.0–19.3 cm; ear length: 3.1–3.7 cm.

Tail width varies from 9–20 cm and is widest in mature adults. An adult's tail is about 2.5 cm thick at the mid-base, its thickest point. Weight: males, 15.9–35.3 kg; females, 21.1–23.6 kg; newborns, about 500 g.

RANGE

North American Beavers occupy a widespread range that includes most of North America, as well as many inshore islands. Anticosti Island and Queen Charlotte Islands (Haida Gwaii) are sites of introductions. Successful introductions have also occurred in the Old World (Finland, Kamchatka, and Poland) as well as in South America (southern Argentina). As they are reliant on tree bark for winter food, beavers do not extend far beyond the treeline.

ABUNDANCE

Estimated by E.T. Seton in 1929 at 60–400 million before European settlement in North America, beavers were very nearly exterminated by fur trapping in the 1700s and 1800s. Even populations in the eastern United States were trapped out before 1900. The primary market for this fur was the felt-hat trade in Europe. Large companies like the Hudson Bay Company employed European as well as native trappers, who continually moved to new locations to harvest previously untrapped populations. They left so few beavers behind them that the remaining animals were unable to support a resurgence of the species. Following public outcry, federal legislators in Canada and the United States finally put a stop to the unregulated trapping in the

Distribution of the North American Beaver (*Castor canadensis*)

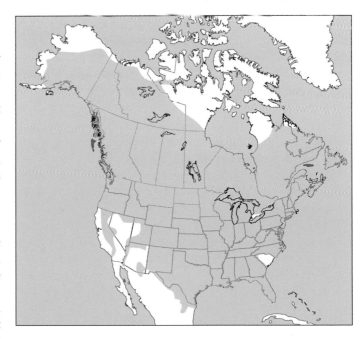

 introduced

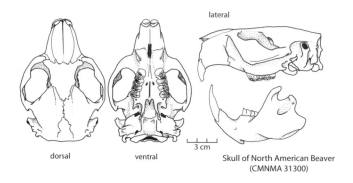

lateral

dorsal   ventral

3 cm

Skull of North American Beaver
(CMNMA 31300)

early 1900s. This was followed by a massive continent-wide reintro-duction period where the gene pools of many depleted populations were completely swamped by transplanted newcomers. Since then, North American Beavers have made a remarkable recovery, and thanks to both accidental and intentional introductions, now occupy a wider range than they did historically. Carefully regulated trapping continues in many jurisdictions. Roughly estimated in 1988 to be 6–12 million strong, beavers have proved to be both adaptable (to human landscape modifications) and tenacious. They are considered common in many regions and are numerous enough in places to be thought of as pests, usually due to widespread flooding that some-times results from their dam building, but also to their downing of valuable ornamental trees. Beavers are uncommon in open prairie, arid regions, and montane habitats.

ECOLOGY

North American Beavers are semiaquatic and rely on deep water for escape and protection from predators. If a lake or river is unavail-able, a beaver will, by the judicious placement of sticks, mud, and stones, dam a trickle of water, forcing it to back up. The impounded pond must be deep enough that it does not freeze to the bottom in winter. Amazingly, most dams are created by a pair of two- to three-year-old, newly paired beavers on a newly claimed territory. Years later, when all the surrounding food trees have been felled and eaten, the beavers will abandon the pond. Eventually the unmaintained dam will leak or break and a swamp or marsh will result. As this fills with silt or the impounded water drains completely, a meadow forms, and eventually the area reverts to forest with a stream run-ning through it, ready for another beaver family. While the pond is in existence, it is home to many other animals from fish and frogs to birds, insects, and other mammals. The dam and pond hold water long into a drought, providing essential moisture to the surrounding plants and animals. During floods, the dams slow down the water flow, preventing erosion and flooding.

An established colony of North American Beavers typically consists of a family group of six or seven animals sharing a lodge. The oldest and largest beavers are the primary breeding pair and the others are yearlings from the previous litter and kits of the cur-rent year. The summer home range of a colony depends on the food supply, as animals will range farther afield when food is scarce. In general, a home range varies from around 0.5–43 ha. During win-ter, the range shrinks to the immediate vicinity of the lodge. While

swimming leisurely, a beaver will push with both hind feet in uni-son and leave them stretched out behind during the glide portion of each stroke. The tail is engaged as a rudder during manoeuvring or to offset drag when hauling branches. The tightly fisted front feet are pressed to the chest to improve streamlining, unless they are needed to carry food or manipulate an obstacle. When swimming at speed, the beaver's hind feet stroke alternately. Beavers can swim as fast as 5 km/h. During the normal course of their underwater activity, beavers remain submerged for 3–5 minutes per dive, but are capable of holding their breath for as long as 15 minutes if pressed, and they can travel up to 0.8 km underwater between breaths. Beavers have large, efficient lungs that can exchange at least 75% of their capacity during each breath. Humans typically exchange 15%–20%.

Although they live in fresh water, beavers can navigate salt water and do so to reach islands. Their lodge or burrow is essential for their winter survival. Temperatures in a lodge are considerably warmer in winter than the ambient level, rarely dropping below freezing even when outside temperatures drop below −40°C. To conserve energy during the cold season, beavers typically huddle together in their lodge and spend minimum time in the water. Largely crepuscular and nocturnal, beavers shelter in their lodges during the daytime. This schedule is often abandoned during winter when the water is ice-covered, which suggests that their choice of activity period may be a predator-avoidance strategy. Dispersal is normally undertaken by subadults (2–3 years old) in April–June and dispersal distan-ces vary widely depending on available habitat. Overland travel of several kilometres is not uncommon and waterways travel of more than 10 km is possible. Beavers are sometimes killed by falling trees that they are in the process of felling, but recent work indicates that they have at least some control over the direction a tree falls. Pri-mary predators are humans, Grey Wolves, and Coyotes. Wolverines, Cougars, Bobcats, Canada Lynx, and bears take some adults, and American Mink and Northern River Otters may kill kits. Although 21-year-old wild beavers have been reported, few are thought to live beyond 10 years.

DIET

Beavers are primarily herbivorous and will eat a wide variety of plants as befits such a widespread species. During summer, the bulk of their diet is composed of forbs, aquatic plants, and mushrooms along with some leaves and bark of woody plants. During winter, they eat twigs and bark of most woody plants that grow near water. The majority of the trees harvested are poplars and willows, if they are available, although beavers are quite capable of flourishing with-out these favourites. They produce two types of scat and engage in coprophagy (eating of scat) by carefully consuming the soft green scats that are the product of fermentation in their hind gut (cae-cum), where symbiotic bacteria digest the cellulose in the cell walls to release the cell contents. By reingesting this material they give their digestive tracts another chance to absorb more nutrients. About 32% of the cellulose ingested is assimilated through digestion, and it takes at least 60 hours for food to travel through the gut. Win-ter food is gathered during autumn until freeze-up. The branches and stems of freshly felled trees are dragged down to the bottom of the pond near the lodge, where they are left anchored in the mud. When the water is ice-covered, the beavers feed from this underwater cache, but will also go ashore to harvest fresh food if they can break through the ice. The stored cache is crucial for northern beavers, as it permits them not only to survive when the pond is ice-covered and they are unable to forage, but also sustains them during the winter breeding and gestation period. The size of the cache increases in regions with longer winters, typically the farther north one goes. A northern cache may contain a tonne or more of wood. Beavers in more moderate climates, where ice does not form, do not store food. In the Pacific Northwest they have been observed feeding on freshly dead salmon carcasses.

REPRODUCTION

Beavers are monogamous. Pair bonds typically persist for life and new mates are found only upon the death of a partner. Breeding occurs during the winter, usually between December and March. Latitude and climate affects the length of the breeding season, which is shorter (and later) in colder, more northern locations and longer (and earlier) in warmer, more southerly regions. Mating takes place in the water. Female beavers produce one litter annually. Gestation lasts 98–111 days (typically about 105 days) and 1–9 (average 2–4) young are born from March to early June. Heavier females produce larger litters and first-time mothers typically produce small ones. An adult male may leave the lodge a few days before the kits are born to take up temporary residence in a bank burrow, returning after his mate has whelped. The kits are precocial, being born fully furred and able to see and hear. Their incisors are just erupting or are already partly erupted. Capable of swimming as early as four days old, they can dive and remain submerged by two months old. They begin grooming the water-repelling secretions of their anal glands onto their fur when they are around three to four weeks old, making themselves fully water-repellent by the time they are five to eight weeks old. By three months old, they can contribute to the construction projects of the colony, as they are sufficiently coordinated to walk on their hind legs while carrying sticks or mud with their front feet. Kits suckle for about 45–50 days, but begin taking solid food any time between one and four weeks old, and thereafter gradually become less reliant on milk. Young beavers reach sexual maturity between the ages of one and a half and three years old, with well-nourished juveniles achieving sexual maturity earlier than those more poorly nourished. They weigh from 10–16 kg at this time. Beavers achieve physical maturity at around four years old.

BEHAVIOUR

Highly social within the family group, all the members cooperate to maintain dams and lodges and to create their winter food cache. Kits newly out of the lodge swim alongside their parents, sometimes clinging to their fur or riding on their backs and tails. Yearlings from the previous litter help to care for and play with the new kits, learning parenting skills that will help them when they have litters of their own. At around two years old, following their second winter in their parents' lodge, the juveniles are encouraged to set out to seek a mate and establish their own territory. Most travel downstream a few kilometres, but

some may relocate at remarkable distances. The longest dispersal distance on record is 390 km. Reports of non-breeding adults in some colonies suggest that all subadults may not disperse, especially when the population density is high. Each colony scent-marks and defends its territory, preventing strangers from settling. Territorial beavers, especially males, patrol their turf daily and are prompt in reapplying their scent markings in spring. Usually this odour is sufficient to discourage a wandering two-year-old. Scent-marking behaviour is most frequent in spring and early summer, the peak period for dispersal. Beavers rarely fight, as even the most stubborn juvenile can clearly see that the resident is considerably larger. Recent evidence suggests that beavers use anal-gland secretions to identify relatedness. A beaver that settled near its colony of origin would not accept a mate from the same colony, even if it had never met that animal before, as would likely be the case with younger siblings. It might, however, allow them temporary shelter.

Some colonies may maintain two lodges and move between them, perhaps to reduce the parasite load in each. One lodge is typically used during summer, the other in winter. The lodge to which they make repairs in late summer is the one in which they will winter. Beavers generally feed on the downwind side of their pond to ensure that they have a clear path to the safety of the water should they detect an approaching predator. Tail slapping is a commonly used communication behaviour that expresses alarm. A floating animal slaps its tail against the water surface, creating a loud, sometimes startling splash, then dives. It tells a potential predator that it has been spotted, at the same time warning its family of danger. Tail slaps by young beavers are generally ignored, while the louder slap of an adult usually results in rapid defensive action. Animals on shore bound for the water and those in the shallows head for deeper water. Secretions from both the oil and castor glands are used for scent marking of territory. Scent is deposited on heaps of wet plants and mud that are pulled from the water's edge. Over the years these mounds can become fairly large as more debris is added (see the "Signs" section). The number of scent mounds in a beaver family's territory varies, depending on population density. Isolated colonies with no near neighbours may not produce scent mounds at all. The paired castor and oil glands empty into the cloaca, a pouch at the end of the digestive and urinary system. As the beaver urinates on a scent mound some of the secretions in the cloaca are mixed with the urine. A beaver can also intentionally deposit larger amounts of castoreum by dragging the cloaca across the scent mound. The odour of this material is thought to be individually distinctive, so a family group will immediately recognize the deposit of a stranger. As it gradually washes into the water, its scent may be detectable by other beavers for some distance downstream.

When towing sticks, beavers always haul from the butt end, as this is hydrodynamically more efficient and the branches are less likely to snag in the forest. They also try to use the current to help them whenever possible. During dam and lodge repairs, beavers commonly use the peeled logs and branches from the previous year's winter food cache, thereby making space for the new food stores. They will also fell non-food trees such as conifers for construction purposes. Once the sticks are woven into the structure, the animals collect mud from the bottom of the pond, which they hold under their chin with their front paws. This is carried to the repair site, often with difficulty, as the animal must balance on its hind legs and cannot see where it is placing its feet. The mud is patted into place using the front paws. Lodges are encrusted with mud all around their base and up to almost the top, which is left uncoated to allow air to filter into the living chamber. As this mud freezes, it creates an almost impenetrable barrier to hopeful predators using the ice to reach the lodge.

VOCALIZATIONS

Although some other vocalizations are produced in the socially charged environment inside the lodge, only three are normally produced outside where we are likely to hear them. "Whining" is the most common and is used by all ages and genders. Kits are most likely to whine, especially when begging for food or worried that the stick they are chewing on may be taken. Whining may also be used as a signal to initiate play or mutual grooming. "Hissing" is normally produced by adults when angry, for example, a territorial male checking a scent post may hiss when he detects a stranger's odour. "Growling" indicates a clear aggressive intent and might be heard during territorial disputes, fights, or if cornered by a human or predator. Another signal of aggression, which is not strictly a vocalization, is tooth chattering.

SIGNS

Apart from humans, no other mammals have more impact upon their environment than do beavers. The most commonly associated sign is their dam, constructed of sticks, stones, and mud. These vary in length and height depending on the terrain and water source. A beaver begins a dam by anchoring large sticks in the current with mud and rocks with one end pointing upstream. This provides the base for more sticks cemented with mud. Most stems used are in the 1.5–3.5 cm diameter range. Dams have been reported that were over 1.2 km long (this one created a large lake that housed 40 lodges) and a large dam may be 3–4 m high. The highest recorded dam is 5.5 m high (built in a steep-walled canyon). Beavers sometimes create a stepped series of dams, each of which backs water up against the upper dam, reducing the pressure on it. Long-term accommodation is created by the construction of one of two types of lodges depending on whether it is built in the middle of a pond or on the shore. Shore lodges begin with a burrow dug into the soil on the side of a waterway (a bank burrow) containing the living quarters. This burrow is then fortified at the underwater entrance(s) and on the ground surface above with sticks and mud. Pond lodges are usually an unruly looking heap of mud and sticks rising out of the water. Typically around 5.0–7.5 m in diameter, they have a hollow centre above the water level, that creates the living space. The top is typically 2–3 m above the bottom of the pond, and at least 1.2 m above the water level especially if it is an old lodge that has been added to over the years. The largest recorded lodge was 12 m across and almost 5 m high. Again, access is through one or more underwater entrances. The hollow core of a lodge slopes up from the water level, leaving space for a wet animal to feed or to shake and groom out

stride length of a bound can vary from 25 to 80 cm

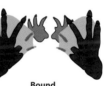

Bound

Direct register walk

Left hind track
Length: 12.0–18.0 cm
Width: 8.0–13.5 cm

Left front track
Length: 6.5–10.0 cm
Width: 5.5–9.0 cm

Beaver

surplus water before entering a more level dry-nest area, which is floored with thinly shredded strips of wood (like excelsior) for insulation and comfort. The inside curving walls with protruding sticks are plastered with adobe-like dried mud. Typically, a few air passages near the apex are left unplastered. An active lodge in winter will likely have some hoar frost around the outer apex. The beavers commonly create short-term and temporary accommodation by digging a bank burrow with the usual underwater access, but no fortification. These may be used intermittently or at specific times of the year (for example, by an adult male for the few days before and following the birth of the litter). Beavers may dig water-filled ditches that connect to the pond proper, usually after they have exhausted the immediately available supply of trees and must travel farther from the water to find food. These canals are only feasible on relatively flat land. They are dug as straight as possible to avoid snagging the branches of a load and are just big enough, usually around 45–60 cm wide and deep, to provide a margin of safety and to allow the animals to float their timber from the collecting site. The longest canal so far reported was 228 m long and about 1 m deep. Trails are quite common around a pond, even though beavers spend most of their time in the water. Runs that lead from the water to prime foraging or scent-marking locations tend to be most heavily used. Good beaver tracks can be surprisingly hard to find. The quality of footprints depends primarily on how much tail drag occurs. Beavers can walk on their hind legs for short distances as they carry sticks or mud. In these cases, the tail drag commonly obliterates the tracks, as the animal drags and presses its tail into the ground for balance. Even during a regular four-footed walk, some tail drag may occur and the front tracks are frequently trampled by an overstepping and larger hind foot. If they are not dragging their tails, the beavers are dragging a tree branch, which further compromises the tracks. Often the only tracks left by beavers are made by the branches and twigs they are carrying, which sweep the often muddy ground in the runs. When threatened or alarmed, a beaver drops what it is carrying and bounds towards the water. In this gait the tail is elevated and the overstepping hind feet register well beyond the daintier front tracks, so all four tracks are left relatively clearly. The small front feet are not webbed. Each toe is elongated, except the pollex (thumb), which is clawed but highly reduced and often fails to leave a mark. The other toes typically leave long-fingered, distinctly clawed impressions. The considerably larger, webbed rear track is five-toed and normally leaves a heel impression, but not all the five toes may register. Often toes 1 and 2 are not obvious in the track. Hind claws are large, but not as curved as those on the front feet, so they do not always register, especially if the ground surface is hard. Winter food caches should be visible in the water near an active lodge, typically looking like a pile of drowned woody debris. Scent-marking mounds occur on the shore near the pond edge. Older mounds that have been used and added to over years may reach 0.5 m or more in height. Castoreum, which is commonly deposited on these mounds, is said by some to smell like a horse barn or a well-used saddle, while others describe it as sweet smelling. When fresh it may be detected by a human as far away as 9–10 m if the wind is blowing from the right direction. Castoreum is widely used as a perfume base, and also as bait by trappers, as no beaver can pass by the scent of a stranger without examining it. Beavers produce two different kinds of scat, one of which is reingested (see the "Diet" section). The other is typically deposited in the water, but may be found in shallow clear water or when water levels have dropped. Summer scat forms large coarsely textured pellets or tubes, around 2–4 cm in diameter and 3.0–7.5 cm in length, composed of small wood chips cemented with sawdust. Winter scat is rarely seen. It is so dry it forms no shape, quickly disintegrating into a pile of powdered sawdust.

When felling a tree, beavers tip their head sideways at about a 45° angle to make their first cut. They set their upper incisors and pull their lower incisors up using their powerful lower jaw muscles. Typically, they engage only one opposing pair of incisors for each cut. Then they turn their head in the other direction and moving down the tree make an opposing cut at the opposite 45° angle. The resulting cuts are at 90° to each other, the most efficient wood-chopping angle. The width of each incisor groove is greater than 5 mm. A beaver may cut all around the tree or, if the tree is on a slope, will usually concentrate its efforts on the upslope side. As the tree begins to go down, the beaver is alerted by the popping of wood fibres and makes a dash for the water. Should the tree fall towards the water, the beaver may run under it and be crushed. Fortunately, this is a rare occurrence. Winter eaten trees are not usually felled; instead, they are stripped of their bark up as high as the beavers can reach standing on the snow.

REFERENCES

Barnes and Mallik 1996; Chubbs and Phillips 1994; DeStefano et al. 2006; Doucet et al. 1994; Dyck, A.P., and MacArthur 1993; Elbroch 2003; Elsey and Kinler 1996; Gallant et al. 2004; Gleason et al. 2005; Halley and Rosell 2003; Jenkins and Busher 1979; Nagorsen 2005a; Norment et al. 1999; Price et al. 2005; Reynolds, P.S., 1993; Rosell et al. 2005; Rue 2002; Samways et al. 2004; Schulte 1998; Seton 1929; Smith, D.W., and Jenkins 1997; Smith, H.C., 1993; Sun and Müller-Schwarze 1997; Sun et al. 2000; Van Deelen and Pletscher 1996; Wheatley 1997.

FAMILY HETEROMYIDAE

kangaroo rats and pocket mice

Heteromyids are arid-adapted rodents that can exist with little or no free water intake. They are nocturnal, sheltering in cool, underground burrows during the heat of the day. Most have enlarged hind legs and are prodigious jumpers. One Canadian species, Ord's Kangaroo Rat, is strongly bipedal.

Ord's Kangaroo Rat

FRENCH NAME: **Rat-kangourou d'Ord**

SCIENTIFIC NAME: *Dipodomys ordii*

Basically bipedal, these beautiful rodents are named for their resemblance to much larger, but similarly bipedal, kangaroos. Their long, banner-like tail helps them to balance on two hind legs. They are members of the family Heteromyidae, which includes 6 genera and 57 species, all of which are endemic to arid regions of western North America and northern South America.

DESCRIPTION

The disproportionately large hind feet provide most of the propulsion for this saltatorial rodent. The small front legs are used mainly for food manipulation and digging, but do touch the ground to support slow quadrupedal locomotion, such as when the rodent is foraging or moving in the burrows. The subspecies that occurs in Canada is grizzled greyish black on the midline of its back and brownish orange below that. The brown fades to yellowish on the sides and the belly, throat, snout, and feet. The base and sides of the tail are also white. A bold, tapered white stripe slashes from the base of the tail across the mid portion of each hind leg. There are white patches above each eye and around and below the ears. The large eyes, nose pad, claws, and hairs on the upper and lower tail surfaces are black, as is a narrow band around the base of the muzzle. The long hairs in the tuft at the tip of the tail are grey. The tail is well furred and noticeably longer than the body. Ord's Kangaroo Rats have a relatively large head and large, external cheek pouches that are furred on the inside and extend from the mouth to the shoulders. Their skull is unusual, with highly inflated auditory bulla, causing it to have a distinctly triangular appearance when viewed from above or below. As might be expected, their hearing is inordinately keen. Upper incisors are grooved on the front surface. The dental formula is incisors 1/1, canines 0/0, premolars 1/1, and molars 3/3, for a total of 20 teeth.

SIMILAR SPECIES

There are no other kangaroo rats in Canada to complicate the identification of this distinctive species and no other Canadian mammals display the combination of colourful pelage, long tail, large head, and long hind legs. Kangaroo rat burrows might be confused with those of Northern Pocket Gophers, as both favour sandy soils, but burrows of gophers typically terminate in a peaked pile of loose soil with a plugged entrance, while those of kangaroo rats end in a burrow entrance, often with a fan of discarded soil radiating from the hole.

SIZE

The following measurements apply to adult animals in the Canadian part of the range.

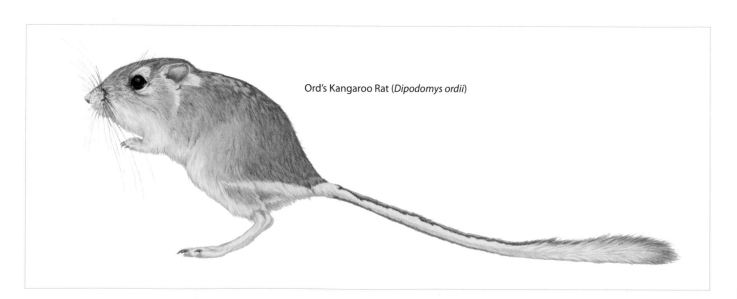

Ord's Kangaroo Rat (*Dipodomys ordii*)

Distribution of the Ord's Kangaroo Rat (*Dipodomys ordii*)

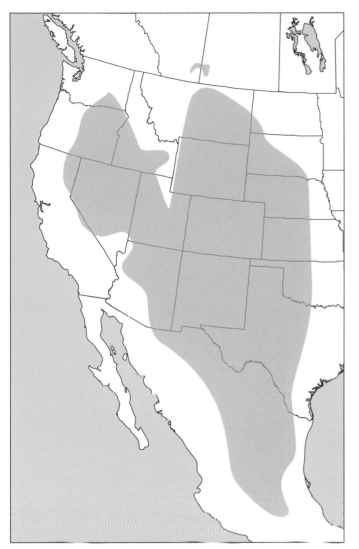

Total length: 233–282 mm; tail length: 131–167 mm; hind foot length: 39–45 mm; ear length: 10–19 mm (average 13–15 mm). Weight: 36–96 g (average 70 g); newborns, about 7 g.

RANGE

One of 21 species of kangaroo rats (genus *Dipodomys*), Ord's Kangaroo Rat is the only one that extends into Canada. The species is at the northern limit of its distribution in southern Alberta and Saskatchewan, but occurs southward to Mexico (southern Hidalgo State), eastward to central Texas, and westward to central Oregon. The Canadian populations are separated (disjunct) from the American by more than 270 km. It also appears that the Alberta population is now disjunct from that in Saskatchewan.

ABUNDANCE

These animals are uncommon in Canada. Their distribution is very localized and population size varies seasonally and regionally and appears to be declining. Ord's Kangaroo Rats are considered "Endangered" in Canada by COSEWIC and may be in imminent risk of extinction. Rough estimates of population size suggest that the entire Canadian population may number up to 7500 animals in the autumn, but likely drops to as few as 580–750 following a severe winter. Principal reasons for the decline of this species in Canada is the loss of habitat (sandy areas) by removal of primary grazers such as American Bison and Wapiti, reduction by fires, farming practices that reduce erosion and stabilize sand dunes, and increasing human habitation and road building. On the military base CFB Suffield, which encompasses around 13% of the Canadian range, efforts are being made to conduct military exercises in an environmentally sustainable manner. Grass fires are relatively common during live munitions practice and Wapiti were introduced in 1997 and 1998 in an effort to further promote kangaroo rat habitat. While normal military activity continues around half of the kangaroo rat habitat on the base, efforts are made to remain at least 250 m from active dens. The remainder of the rat habitat is protected as a National Wildlife Area, where access is restricted.

ECOLOGY

Primary habitats include deserts, arid and semi-arid grasslands, and dry scrublands. Sandy soils are necessary to support the rats' hopping locomotion and extensive burrowing. Ord's Kangaroo Rats create lengthy underground burrow systems for several purposes: to buffer seasonal and daily temperature fluctuations; to provide shelter; to provide predator protection; to preserve moisture and to protect and humidify their winter food cache. Home ranges cover about 4000–10,500 m², but the main core of the home range tends to be 1500–2500 m². The rats remain active all year and do not hibernate, although they do go into a daily torpor during the winter to conserve energy. Canadian populations are the only Ord's Kangaroo Rats currently known to use torpor. Ord's Kangaroo Rats are well adapted to an arid environment. Their nasal passages extract moisture from humid air by condensation, thereby diminishing water loss through respiration. In addition, their kidneys are highly efficient and can concentrate urine, further conserving water. They forage during the cooler night, but are less active in bright moonlight. Animals in captivity have survived longer than seven years, but their potential lifespan in the wild in Canada is considerably less, as few actually survive our cold northern winters (only about 10%–20% live to see the following spring depending on the severity of the winter). Populations are, hence, smallest in spring and largest in autumn. The oldest known

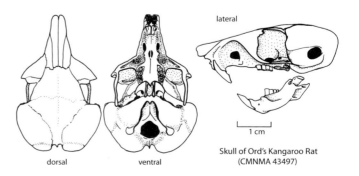

lateral

Skull of Ord's Kangaroo Rat (CMNMA 43497)

dorsal   ventral

1 cm

wild animal caught in Canada was four years old. Winter starvation is the dominant mortality factor among Canadian animals. Other causes of mortality include parasites (mainly botfly), motor vehicle strikes, crushing in burrows by heavy equipment, and predation by such animals as owls, snakes, Coyotes, Swift Foxes, Red Foxes, American Badgers, weasels, and skunks. Domestic Cats and Dogs may also predate kangaroo rats if they are nearby. Ord's Kangaroo Rats have very sensitive hearing and can detect the approach of an owl or the strike of a snake. They also see well in dim light.

DIET

Seeds of grasses and forbs constitute the primary food for these granivorous rodents. During summer, they will also consume the leaves, stems, and roots of fresh plants and supplement their diet with arthropods, primarily grasshoppers and beetles. All their water requirements are satisfied by the metabolic by-products of digestion. Although wild kangaroo rats do not require free water, if they do drink, they scoop it into their mouth with their front feet. During the warm season, they gather seeds in their cheek pouches and store them in an underground larder to eat over the winter when food is scarce. It is estimated that the contents of two full cheek pouches supplies slightly more than the necessary food for one day. Infanticide and cannibalism have been reported.

REPRODUCTION

Canadian kangaroo rats breed during the snow-free period from early spring to early autumn. Mating occurs above ground. Pregnancy lasts from 28–32 days. Litter size varies from one to six, with three on average. Young are born hairless, deaf, and blind and are suckled in an underground nest burrow for their first two to three weeks. Female juveniles typically breed for the first time around 3–4 months old, but well-nourished youngsters and most Canadian females begin breeding around 47 days old. This reproductive precocity helps counteract the severe winter mortality that occurs in Canada. Young males take a little longer to reach reproductive maturity. They become fertile around 61–83 days old. Adult females in Canada may produce up to four litters per year and an early-born female may raise two litters in her first year. Populations farther south average two litters per year and juveniles often do not raise any litters in their first year.

BEHAVIOUR

Ord's Kangaroo Rats are primarily nocturnal, although they may occasionally be seen above ground during daylight, especially if their burrow is invaded by a predator or crushed by farming equipment. They avoid the desiccating heat of the day by staying in their cool, humid burrows and forage during the cooler nighttime. Unlike other kangaroo rats, Ord's are largely solitary, aggressively defending their burrow system from all other rats, except during the breeding season, when females relax their aggression to allow a male's approach. Courtship appears to consist mainly of sometimes lengthy chases where the pursuer may be either male or female. Battles between males for access to an oestrous female may involve kicking with the hind feet, biting, and clawing with the fore feet. It is thought that Ord's Kangaroo Rats, like other kangaroo rats, foot-drum. This is a vibration-producing action

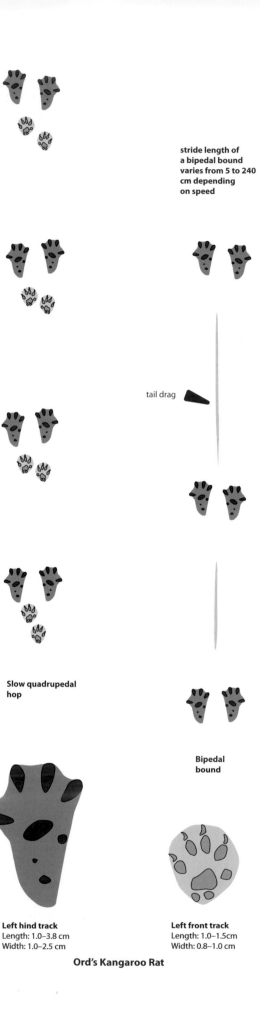

stride length of a bipedal bound varies from 5 to 240 cm depending on speed

tail drag

Slow quadrupedal hop

Bipedal bound

Left hind track
Length: 1.0–3.8 cm
Width: 1.0–2.5 cm

Left front track
Length: 1.0–1.5cm
Width: 0.8–1.0 cm

Ord's Kangaroo Rat

thought to alert snakes to the fact that the prey has detected them and they might as well leave. Another anti-predator strategy adopted by Ord's Kangaroo Rats is their zigzagging gait when they are above ground. This erratic bounce makes them difficult to catch. Grooming activities include licking the fur, as well as sand bathing and sliding on their sides and bellies in the sand, propelled by strong kicks of the hind legs. Dispersal of the species is primarily carried out by juveniles. They leave their mother's burrow at around four to six weeks old and although most travel less than 500 m to establish their own burrow system, some travel considerably farther (up to 10 km). Ord's Kangaroo Rats are capable of swimming, although the opportunity to do so is probably rare. The use of the secretions from a scent gland between the shoulders of both female and male adults is unknown, but a marking function is likely.

VOCALIZATIONS

Largely silent, Ord's Kangaroo Rats do chatter their teeth at each other during agonistic interactions, probably as a warning. A loser in such an encounter leaves the field making squeaky snorts and a faint chuckling call. Newborns and juveniles up to about two weeks old squeak to attract their mother's attention and encourage her to suckle them. Premature or ill neonates that cannot vocalize are ignored and may be eaten by their mother.

SIGNS

Burrow entrances are roughly oval, 3.7–7.3 cm tall, and 5.7–13.2 cm wide. They are often plugged during hot days, but the plugs may be located within the burrow access tunnel and not apparent from the surface. Active burrows are most noticeable in early morning, when the moist soil plugs that were kicked out the entrance overnight contrast with the dryer surface soil. Scat is most commonly seen near burrow entrances, but may also be found scattered around areas within the home range where heavy foraging occurs. Much of the scat is deposited below ground in a latrine chamber. Oval in shape with a smooth or slightly rough surface, scat normally measures 2–3 mm in diameter and 6–13 mm in length. Tracks are rarely seen in snow, as the animals usually remain in their burrows when the ground is snow-covered, but sand tracks are common. Ord's Kangaroo Rats have five toes on each hind foot, but the clawed hallux (big toe) is vestigial and does not usually register in a track. The five-toed front feet may touch the ground if the animal is travelling at slow speed using a quadrupedal gait. The more usual gait is bipedal. The full heel of the hind foot may register at slow to moderate speeds, but only the toes register during faster paces. Splay of the hind toes varies, wider on soft substrate or at high speeds, tighter at slower speeds and on a firmer substrate. A splayed-toed hind track resembles that of a Snowshoe Hare in miniature. Tail-drag marks are common. Distance between the bounces also varies with speed. A leisurely gait may produce tracks only 5 cm apart, while the tracks of a fleeing animal may be up to 2.4 m apart. Conspicuous runs or trackways may develop from the burrow entrance to favoured feeding areas. Often elongated remains of dust baths are usually cupped in the centre and typically found in sand near the burrow entrance.

REFERENCES

Alexander and Riddle 2005; Baron 1979; Elbroch 2003; Garrison and Best 1990; Gummer and Bender 2006; Gummer and Robertson 2003; Kilburn 1997; Lowery 2006; Smith, H.C., 1993; Verts and Carraway 1998.

Olive-backed Pocket Mouse

FRENCH NAME: **Souris-à-abajoues des Plains**
SCIENTIFIC NAME: *Perognathus fasciatus*

This diminutive mouse is one of the smallest in Canada. Its long tail and relatively large hind feet, along with a creamy yellow lateral band running from nose to tail, help distinguish it. Large fur-lined cheek pouches or "pockets" are used to carry food to underground larders.

DESCRIPTION

The Olive-backed Pocket Mouse has a large head with large dark eyes and small ears. A buffy yellowish band separates the white underside from the darker olive-coloured fur on the back. Black hairs on the back are usually scattered, but sometimes form a darker band along the spine. Melanistic animals have been reported. Juveniles have a slate-grey back. This pocket mouse has light rings around its eyes and a light patch directly behind each ear. The indistinctly bicoloured tail is about the same length as the body, but often looks longer, as the animal generally compresses itself by crouching over its hind legs to free up the fore feet for food handling. Hind feet are large and front feet are small. Eye shine is pale amber. Their fur-lined cheek pouches extend from the mouth to the neck, with external openings beside the mouth. The dental formula is incisors 1/1, canines 0/0, premolars 1/1, and molars 3/3, for a total of 20 teeth. The forward face of each upper incisor is shallowly grooved.

SIMILAR SPECIES

An Olive-backed Pocket Mouse could easily be mistaken for a jumping mouse, as their movements are similar. The only jumping mouse that shares its range is the Western Jumping Mouse. This mouse is larger, with a very long bicoloured tail, a broad olive-brown stripe along its backbone, and yellowish-orange sides. It occurs in heavily vegetated, moist meadows and near streams, not in the dry grassland habitat of the pocket mouse. The Western Harvest mouse is also similar to an Olive-backed Pocket Mouse. It is more grey-brown to slate grey on the back, with a dark stripe along the backbone, pale grey underside, and a bicoloured tail. It occurs in grasslands and is very rare. The other pocket mouse species in Canada – the Great Basin Pocket Mouse – is very similar, but larger. It occurs only in south-central British Columbia, on the other side of the Rocky Mountains from the Olive-backed Pocket Mouse.

Olive-backed Pocket Mouse (*Perognathus fasciatus*)

SIZE

The following measurements apply to animals of reproductive age from the Canadian part of the range. Males and females are similar in size and appearance.

Total length: 115–143 mm; tail length: 51–69 mm; hind foot length: 15–19 mm; ear length: 7–8 mm.

Weight: 8.9–13.7 g; newborns, unknown.

RANGE

Endemic to central North America, these pocket mice are at the northern limits of their distribution in southwestern Manitoba, south-central Saskatchewan, and southeastern Alberta. They occur as far east as northwestern Nebraska, as far south as southern Colorado, and as far west as northeastern Utah.

ABUNDANCE

May be locally abundant in suitable habitat, but distribution is patchy throughout its range, as its presence is determined largely by soil type. This mouse is generally considered to be scarce in Canada.

ECOLOGY

In Canada, Olive-backed Pocket Mice typically occupy sandy to sandy-loam soils in arid grasslands with little shrub cover. They are sometimes found in sandy cultivated fields and are most common in recently disturbed areas such as field edges and road margins where weeds are abundant. They are nocturnal and sometimes crepuscular, and live in self-constructed underground burrows. In warm weather, they dig a network of burrows near the surface, with several entrances usually near a weed, sagebrush, or cactus. The nest and storage chamber is up to 30–45 cm below the surface. In autumn, the mice burrow around 2 m down to create a winter hibernaculum and larder chamber, which they plug when they enter hibernation. They drop into torpor for extended periods during the cold season, waking periodically to eat and eliminate. If the food supply fails during the summer, these little rodents remain in their burrows and allow their body temperature to drop to the ambient temperature.

This is called aestivation and may last for days or weeks. It is an energy-saving strategy that helps them survive until the plants set their seeds.

DIET

The principal diet of Olive-backed Pocket Mice is seeds of grasses and forbs. Many favoured seeds are weed species. Some insects may supplement the diet in summer. This rodent acquires all of its water from the digestion of the seeds it eats, and its kidneys produce highly concentrated urine, which further conserves water.

REPRODUCTION

The reproductive biology of this species is poorly understood. We are still uncertain whether there are two litters produced each summer, with a rest period in June, or simply a single litter over a prolonged reproductive period, with some females reproducing

Distribution of the Olive-backed Pocket Mouse (*Perognathus fasciatus*)

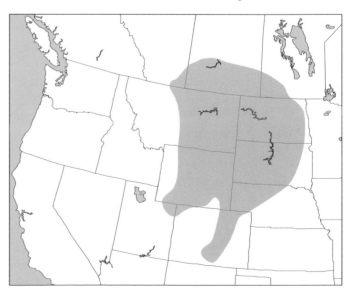

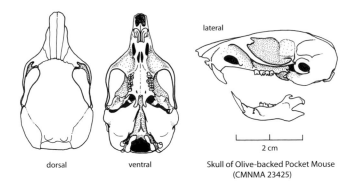

lateral

2 cm

Skull of Olive-backed Pocket Mouse
(CMNMA 23425)

dorsal   ventral

Mouse account for a schematic of a pocket mouse track. Burrow entrances are small, 2.5–3.0 cm in diameter, and may be plugged on the inside during the day. Small scuffed pits used for dust bathing may be found near the burrow entrances. Scat is normally deposited underground.

REFERENCES

Elbroch 2003; Gummer and Robertson 2003; Manning and Jones 1988; Smith, H.C., 1993; Wrigley 1980; Wrigley et al. 1991.

early and others later. The reproductive season begins when the snow melts in spring. An average litter of four (range 1–10) is born between mid-May and early June, following a gestation of about 30 days. Two litters may be possible for females that breed early, with the second born in late July or August. Males are in breeding condition from May to October. Juveniles are thought to become independent when they are about four weeks old and weigh around 6 g. It is thought that the young do not reach sexual maturity until the following spring.

BEHAVIOUR

Each pocket mouse aggressively defends its burrow system from other pocket mice and fighting occurs when adults meet, unless a female is in oestrus and willing to mate. To load the cheek pouches, a pocket mouse balances itself by crouching over its hind legs (as in the illustration) to free its front feet for food manipulation. Soil is sifted through the front toes and weed seeds are gathered and stuffed into the cheek pouches. A few quick swipes from back to front by a forefoot clear the seeds from a cheek pouch. Much time is spent grooming to keep the fur oil-free. Dust bathing is a favourite occupation undertaken with much energy. They roll on their backs and sides in the sand, then comb and scratch the fur diligently. The head is washed with the front feet, much like a cat does, by licking the wrist then rubbing it over the face. Cheek pouches are periodically everted with a front foot, cleaned, and then popped back into place with a contraction of the cheek muscles. These small rodents are very docile if captured and rarely attempt to bite.

VOCALIZATIONS
Unavailable or unknown.

SIGNS
Although the hind feet are somewhat enlarged, locomotion remains quadrupedal. At higher speeds, the animals gallop and at slower speeds they walk. A rapidly fleeing animal may make hops of one metre or more between strides, or make high evasive leaps and alter direction with each bounce. Tracks are rarely seen in snow, as Olive-backed Pocket Mice spend all winter in their underground nest chamber. All the pocket mice have similar and largely indistinguishable tracks. See the "Signs" section of the Great Basin Pocket

Great Basin Pocket Mouse

FRENCH NAME: **Souris-à-abajoues des pinèdes**
SCIENTIFIC NAME: *Perognathus parvus*

This charming little rodent extends into Canada only in south-central British Columbia.

DESCRIPTION
Great Basin Pocket Mice are darkest along their backbone, where the fur is dark olive grey. Their sides are a slightly lighter olive grey that is separated from the white underside and feet by a buffy yellowish band. Juveniles are grey on the upper surface and whitish on the underside. Fur-lined cheek pouches (also called "pockets") used to transport food lie on each side of the head. These open externally beside the mouth. The tail is longer than the body and is distinctly bicoloured: dark above and white below. Each ear contains a lobed antitragus (fleshy projection) at the base of the auditory opening. The summer moult that takes place from June to August takes about 31 days for each individual. It begins behind the ears and proceeds down and rearward and may at times appear as a line bisecting the body. The head moults around the midpoint of the moult. Lactating females moult later than adult males. Another moult is thought to occur during the winter while the animals are hibernating. The dental formula is incisors 1/1, canines 0/0, premolars 1/1, and molars 3/3, for a total of 20 teeth. The thin upper incisors have a distinct groove on their forward face.

SIMILAR SPECIES
The Great Basin Pocket Mouse is larger, though otherwise very similar to, the Olive-backed Pocket Mouse, but the two species are separated by the Rocky Mountains, with the Great Basin Pocket Mouse on the west and the Olive-backed Pocket Mouse on the east. The species most likely to be mistaken for a Great Basin Pocket Mouse is the Deer Mouse, which has larger ears, a shorter tail, no yellowish band between the dark back fur and the light belly fur, no cheek pouches, and only three teeth in the cheek

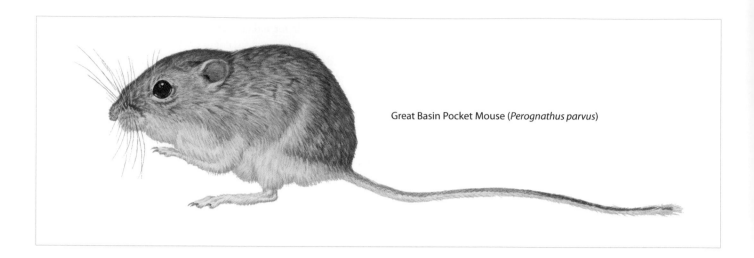

Great Basin Pocket Mouse (*Perognathus parvus*)

rows. Other species that resemble a pocket mouse, especially in their hopping movements, are jumping mice. Both the Meadow Jumping Mouse and the Western Jumping Mouse occur within the range of the Great Basin Pocket Mouse. They occupy moist meadows near water and are not found in the arid grasslands of the pocket mouse.

SIZE

The following measurements apply to animals of reproductive age from the Canadian part of the range. Males tend to be heavier than females.

Total length: 151–202 mm; tail length: 77–112 mm; hind foot length: 20–27 mm; ear length: 4–10 mm.

Weight: males 16.5–28.0 g, females 16–27 g; newborns, unknown, although three-day-olds weigh around 2.2 g.

RANGE

From the northern limit of its distribution in British Columbia, the Great Basin Pocket Mouse extends southward through central Washington and Oregon to eastern California and northern Arizona. The eastern limits of its range occur in Utah and southwestern Wyoming. In British Columbia, this rodent is found in three regions of the province: the Thompson River Valley from Ashcroft to Kamloops, the north Okanagan Valley from Okanagan Landing to Vernon, and from the southern Okanagan and Similkameen Valleys east to Midway and Grand Forks in the Kettle River Valley. Unsuitable habitat separates each of these populations.

ABUNDANCE

This species is listed on the British Columbia "Blue List," which means it is a species "of concern." Urban development and agriculture have destroyed or modified much of the original habitat of this mouse in the province and the remainder of the dry grasslands are fragmented and scattered. Several early collecting localities may no longer support populations due to urban growth and the expansion of agricultural irrigation. The Great Basin Pocket Mouse is being considered by COSEWIC for possible addition to one of their lists of terrestrial mammals in danger in Canada.

ECOLOGY

These animals are strictly nocturnal. They live in arid grasslands and forage only at night to avoid the daytime heat. By spending the hotter days in their cool, humid burrows they preserve precious moisture. Further water-saving abilities are discussed in the "Diet" section below. Sandy soils are preferred for ease of digging, but these pocket mice may also be found in somewhat rocky clay-till soils, as well as recently disturbed areas such as roadway embankments, recently burned regions, and overgrazed grasslands. Two kinds of burrows are constructed. Each home range will contain several short escape burrows, often with more than one entrance, as well as a more extensive permanent burrow with larder, nest, and latrine chambers. Entrances to the permanent burrow are often plugged on the inside during the day, while the escape burrows are normally left unplugged all the time. Nest chambers are globular and about 8 cm in diameter. Summer nests are 13–30 cm below the surface, while winter ones may be a metre deep. Size of home range is larger for males, but varies depending on food availability and density of neighbours. Although home ranges may be large (267–4005 m²), the main centre of activity is relatively small. Core areas in the home ranges are estimated to be around 6 m² for females and around 16 m² for males. A Great Basin Pocket Mouse can carry up to about 25% of its body weight in seeds in its pair of cheek pouches. This load is estimated to be a little less than its daily energy requirement. It is further estimated that each pocket mouse needs 60–90 warm days to accumulate sufficient food in its underground storage chamber to last through a winter. The mice rely on this stored food, an insulated nest below the frost-line, and periods of torpor to survive until spring. In the region of British Columbia where they are found, temperatures rarely fall below 5°C in the underground nest chamber, even during the winter. Pocket mice in these regions enter their hibernaculum in October or November depending on local temperatures and spend 60%–90% of the winter in torpor. In the Okanagan Valley, these rodents begin emerging from hibernation in late March or early April. Males emerge first and females about a month later. Torpor may also be used during dry periods in summer, when food is scarce, to conserve energy until the rains return. Relatively long lived for such a small rodent, captive animals have lived to be 4.5 years old and wild ones may live into their fourth year. Likely

Distribution of the Great Basin Pocket Mouse (*Perognathus parvus*)

days. Females in Canada typically have one or two and occasionally three litters annually of four to six (range 1–8) young. During dry years, one litter is common and some females do not produce any. Higher-elevation populations, where summers are shorter, experience a shorter breeding period and tend to have fewer litters than those in the valleys, where the growing season is longer. Females born in early spring may produce a litter of their own in late summer, if the warm season is long enough where they live. Juvenile males generally do not breed for the first time until after their first winter. The age of young at weaning and the distributional ecology of the weanlings is not known.

BEHAVIOUR

Each adult pocket mouse aggressively defends its exclusive burrow system. Burrow systems are not shared, although above-ground foraging areas may overlap. Rivals stand on their hind legs and bite and shove each other until one retreats. Presumably, females relax their territorial defence when they are ready to mate. This rodent crouches over its hind legs (see illustration) to balance itself so the front feet are free for food manipulation and digging. Food is gathered by sifting the soil with the forefeet and transferring the gathered seeds into the cheek pouches or by directly harvesting seeds by nipping the grass stems to place them within reach. Cheek pouches are stuffed using the forepaws and rapidly cleared using a forward swipe with the same paws. Although nocturnal, their activity is reduced on moonlight nights. Dust bathing is a favourite activity that removes excess oil from the fur.

VOCALIZATIONS

A single repeated sound, described as "que, que, que," is produced as a soft whine when an animal is handled roughly or as a shrill squeal during battles with rivals.

SIGNS

Although their hind feet are somewhat enlarged and their forefeet are small, gaits remain quadrupedal. They walk and gallop in a normal four-footed fashion, but will use a four-footed hop at times. Tracks during the winter are unlikely, as these rodents remain underground during the cold season. Both the hallux (big toe) on the hind foot and pollex (thumb) on the front foot are clawed but vestigial, and may not register in a footprint. The hallux is more

few live beyond two years old. Owls, mainly Burrowing Owls, Short-eared Owls, Western Screech-Owls, and Long-eared Owls, are their principal avian predators, while digging mammals such as American Badgers, Coyotes, and Striped Skunks are their main mammalian predators. Snakes, including the Western Rattlesnake, prey upon them as well.

DIET

Since seeds of a variety of plants make up the principal diet of this rodent, although they supplement this with green vegetation and arthropods during summer, Great Basin Pocket Mice are considered to be granivorous. Grass seeds are a favourite food, especially for winter storage. Underground larders have been discovered that contain up to 300 cm³ of seeds. A Great Basin Pocket Mouse needs at least 40–60 g of high-quality seeds to survive a winter. The bulk of their water requirements are met through the metabolism of their food. They also produce concentrated urine to conserve water.

REPRODUCTION

The mating season begins soon after the females emerge from their winter burrows. Depending on soil temperatures, this may be as early as May. Males emerge about a month before the females and are in breeding condition by the time the females emerge. They remain in breeding condition until around mid-August. Pregnant females have been captured in early September. The gestation period is 21–25

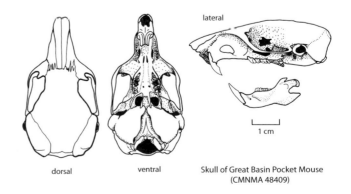

lateral

1 cm

dorsal   ventral   Skull of Great Basin Pocket Mouse
(CMNMA 48409)

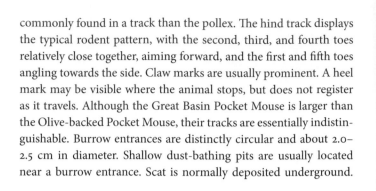

commonly found in a track than the pollex. The hind track displays the typical rodent pattern, with the second, third, and fourth toes relatively close together, aiming forward, and the first and fifth toes angling towards the side. Claw marks are usually prominent. A heel mark may be visible where the animal stops, but does not register as it travels. Although the Great Basin Pocket Mouse is larger than the Olive-backed Pocket Mouse, their tracks are essentially indistinguishable. Burrow entrances are distinctly circular and about 2.0–2.5 cm in diameter. Shallow dust-bathing pits are usually located near a burrow entrance. Scat is normally deposited underground.

REFERENCES
Cramer and Chapman 1990; Nagorsen 2005a; Vander Wall et al. 1998; Verts and Kirkland 1988.

stride length of a bound can vary from 7 to 100 cm or more

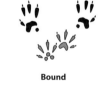

Bound

Left hind track
Length: 0.8–1.4 cm
Width: 0.6–1.2 cm

Left front track
Length: 0.6–1.0 cm
Width: 0.6–1.0 cm

Pocket Mice

FAMILY GEOMYIDAE
pocket gophers

Pocket gophers are fossorial rodents with large, external, fur-lined cheek pouches, long front claws, a short, thick, nearly naked tail, and small eyes and ears. Robust incisors are slightly protruding and have a rusty-orange anterior surface. This colour is caused by their iron-oxide content, which strengthens the enamel, rendering it harder and longer lasting.

Plains Pocket Gopher

FRENCH NAME: **Gaufre brun**

SCIENTIFIC NAME: *Geomys bursarius*

The range of these large pocket gophers extends into Canada only in a small portion of southern Manitoba along the Red River valley. Pocket gophers are named for their large cheek pouches or "pockets," which extend from the side of the mouth all the way back to their shoulders.

DESCRIPTION

Plains Pocket Gophers have thick, silky fur that is generally darker above and lighter below. Coat colour tends to match the colour of the moist soil in which it lives and is therefore highly variable, from buff to dark reddish brown and black. Most Canadian specimens are a deep mahogany brown to dark liver brown, often with white patches under the chin and on the forelegs. The feet are buffy white and the fur over the nose is also often whitish. Their large cheek pouches, which are furred on the inside, have large openings along the line of the lower jaw. The tail is short, pinkish, and sparsely furred with white hairs. Highly adapted for digging, their bodies are stocky, with short legs, and the front legs are thickly muscled, with long sharp claws. Ears and eyes are small and can be sealed to prevent dirt from getting inside. Long vibrissae extend from the upper lips, cheeks, and wrists to provide tactile information in the darkness of their tunnels. The tail also provides such information, as it is lashed about behind them as they back up with close to the same speed and facility as going forwards. Their hearing is poor and they are unable to localize a sound source during laboratory experiments. The dental formula is incisors 1/1, canines 0/0, premolars 1/1, and molars 3/3, for a total of 20 teeth. The skull of adult males is not only larger than that of adult females, but also tends to be more angular, with a higher sagittal crest.

SIMILAR SPECIES

Plains Pocket Gophers are most similar to Northern Pocket Gophers. Both live in underground burrow systems and construct mounds of excavated soil, and since neither is regularly seen above ground, especially in daylight, distinguishing them can be difficult. Where the two species overlap, the larger Plains Pocket Gophers are better competitors and take the preferred sandy and loamy areas, relegating the smaller Northern Pocket Gophers to the poorer soil types. The forward surface of the incisors has a useful characteristic that separates skulls of Plains from Northern Pocket Gophers. Plains Pocket Gopher teeth have a double groove running from the base to the tip of the tooth, while the Northern Pocket Gopher has a smooth incisor surface with no grooves. The mounds and tunnels of Plains Pocket Gophers may be confused with mole workings. The region of Manitoba where Plains Pocket Gophers occur also supports colonies of Star-nosed Moles. Moles are found in rich, moist to wet soils, where their aquatic invertebrate and earthworm prey abound, while gophers tend to occupy drier soils that are better drained. Mole hills are typically dome-shaped, while gopher mounds tend to be more sharply peaked and both higher and larger.

SIZE

Males are 10%–15% larger and up to 25% heavier than females. The following data applies to adult Canadian animals.
Total length, males: 285–298 mm; tail length: 85–94 mm; hind foot length: 36.5–38.5 mm; ear length: rudimentary.
Total length, females: 241–272 mm; tail length: 72–84 mm; hind foot length: 32.6–35.0 mm; ear length: rudimentary.
Weight: males, 226.5–343.0 g; females, 220.0–260.7 g.
Newborns: about 40 mm in length; weight around 5 g.

RANGE

These gophers reach the northern limits of their distribution in a small corridor along the Red River of southern Manitoba. They are widespread in the United States midwest from Minnesota to Indiana, and Mississippi through Texas to New Mexico, and north to Wyoming and North Dakota. Their Canadian range has slowly expanded north and eastward since their initial discovery in 1911 in Emerson, Manitoba, on the border between Manitoba and North Dakota. By 1991, they had expanded northward beyond the Roseau River as far as 6.6 km west of Rosa, Manitoba, and as far eastward as Gardenton, Manitoba; at that time they occupied an area of approximately 725 km². It is expected that their distribution will continue to expand in southern Manitoba, although much of the area surrounding the present range is unsuitable wetland, cultivated cropland, or woodland.

ABUNDANCE

Plains Pocket Gophers can be locally abundant in Canada, but their distribution is spotty as not all soil types are suitable and they avoid cultivated lands where ploughs destroy their burrow systems. They are fully protected in the Natural Grass Prairie Preserve near Tolstoi, Manitoba, but are otherwise treated as pests throughout the rest of their Canadian range and there are no restrictions on their control.

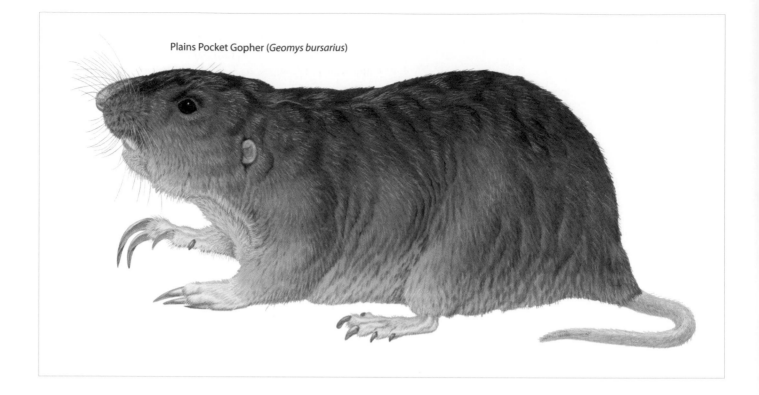

Plains Pocket Gopher (*Geomys bursarius*)

The Canadian population was roughly estimated at around 8000 in 1979, but surveying this species is very difficult due to their fossorial habits. Plains Pocket Gophers were classified by COSEWIC as "Rare" in 1979 based on a limited Canadian distribution. As their range continued to expand, this status was downgraded to "Vulnerable" in 1990, and in 1998 the species was de-listed altogether. Although numbers fluctuate seasonally, the population in Canada is thought to be stable or slowly increasing.

ECOLOGY

Like the closely related Northern Pocket Gopher, Plains Pocket Gophers are fossorial. They live in elaborate self-made underground burrow systems. The tunnels are constructed on two levels. Those near the surface (within 10–20 cm) are dug for the purpose of gathering food. The deeper tunnels (over a metre deep) typically contain chambers for nests, latrines, and food storage. The diameter of the tunnel equates to the girth of the animal, so larger animals create wider tunnels. Tunnel excavation is most intense in spring and autumn, with a lull during the summer. Porous, loose textured soils, usually with at least 50% sand, are preferred, but Plains Pocket Gophers are able to survive in somewhat more coarse-textured and compacted soils, although not with the same success as Northern Pocket Gophers. Land around and within areas of high food availability, such as fields of forage crops of Alfalfa and tame grasses, are favourite habitats. Snow melt and rising spring water levels are major mortality factors for pocket gophers in Canada. Flooded burrows drive them above ground, where they are vulnerable to predation. Although capable of swimming, these animals are poor swimmers, and are unlikely to survive in the turbulent, fast waters of a spring flood. Plains Pocket Gophers have an unexplained unequal sex ratio

with approximately three adult females to every adult male. Possibly, the males are more exposed to predation as they travel overland during the breeding season in search of females. They do not hibernate, instead occupying burrows below the frost-line, where they consume stored food. Adults tend to remain in the same home range for their whole lives. Most of the species dispersal is accomplished by the recently weaned juveniles as they search for a home range of their own. They travel above ground while dispersing, where they are exposed to many predators and may suffer a mortality rate as high as 90%. Primary predators are digging mammals such as skunks and badgers or animals like weasels, which are actually slim enough to travel into their burrow systems. Other predators, including Domestic Dogs and Cats, Coyotes, and foxes, may dig them out of surface tunnels, and owls and hawks may catch them above ground as they are foraging or dispersing. Bull Snakes are known to occasionally pursue them into their burrows. Most predation occurs when the gophers are at the mouth of a burrow excavating soil into a mound. During this activity, the gophers position themselves with their backs to the opening, gather surplus soil under their belly with their front legs, then scatter it with explosive kicks of their hind feet. They are surrounded by moist soil and their coat coloration tends to match the soil colour, providing them with a degree of camouflage that may foil some predators, but clearly not all. Their average lifespan is about two years, the maximum possible being around five years.

DIET

Plains Pocket Gophers are generalist herbivores. They consume a large number of grass species, as well as forbs such as dandelion and Alfalfa. The majority of their diet depends on below-ground portions – roots, bulbs, and tubers – but some green vegetation such as

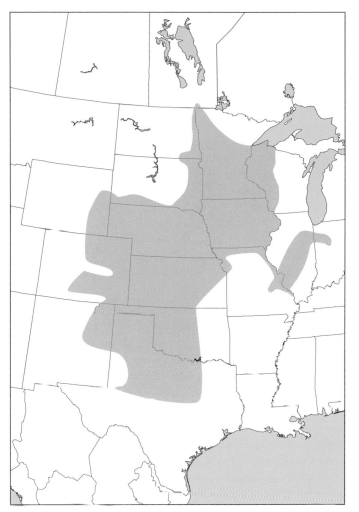

stems and leaves, as well as some fruits, acorns, walnuts, and seeds, are gathered. Like most rodents, they will also consume insects and the flesh of other mammals if the possibility presents itself. Their food choices vary with the season and with plant availability. If they venture above ground to forage, they prefer to do so at night and try to remain within an easy run to their burrow entrance, but will wander farther afield if food becomes scarce. Plains Pocket Gophers store food in large underground caches for later consumption, particularly during winter.

REPRODUCTION

Plains Pocket Gophers do not breed in captivity, so many details of their reproductive biology are unknown. Females produce a single litter annually, in the northern parts of the range, although two litters per year are possible farther south. The breeding period appears to be variable in different parts of the range. In the northern portions, it lasts from mid-April until early July, but its onset is difficult to determine as the animals are deep in their burrows under a snow cover often until late April. There is some evidence to suggest that pups may be born as early as January in Missouri and Kansas. Likely

females that come into early oestrus are bred by neighbouring males, while later oestrous females would have the opportunity to breed with more distant males that can travel overland as the weather moderates and snow disappears. Most females mate by mid-June. The extended breeding season allows time for the males to find all the fertile females, which may be widely scattered. The gestation period is uncertain and has been estimated from 28–51 days. Litter size varies from one to eight and averages three to four pups. Young are deaf, blind, and hairless at birth. Most newly weaned juveniles begin appearing above ground in June and July. Dispersing pups weigh 80–100 g. As many as a third of adult females do not breed in a year. Juveniles become sexually mature in the spring following their birth.

BEHAVIOUR

Pocket Gophers are highly territorial, vigorously defending their burrow systems from intrusion by other pocket gophers. Home ranges (which equate to the extent of the burrow system) of neighbouring individuals may be close, but they do not overlap or interconnect. How this is accomplished remains a mystery. Apparently, gophers are able to determine whether a burrow system is inhabited, as abandoned territories are quickly appropriated. Individuals are solitary, except during the mating and breeding season, when receptive females relax their aggressive territorial defence and allow the close proximity of adult males. Although these gophers were thought to be polygamous, recent evidence suggests that they may be serially monogamous, as males have been discovered attending pups along with the maternal female. Nursing females share their burrows with their pups and sometimes the paternal male, but the young are forcibly evicted soon after they are weaned when they begin to display competitive behaviour towards their mother and each other. In regions where both Plains and Northern Pocket Gophers occur, the larger Plains Pocket Gophers are more dominant and take the premium habitat, relegating the smaller Northern Pocket Gophers to regions with poorer soil conditions or driving them out altogether. Plains Pocket Gophers do not have a day/night cycle of activity, but are active throughout a 24-hour period, simply taking short rest breaks as needed. One summertime radio telemetry study shows that the animals are active for about a quarter to a third of each day and that they manage their movements to avoid the daily temperature extremes. Most above-ground food gathering takes place at night under cover of darkness.

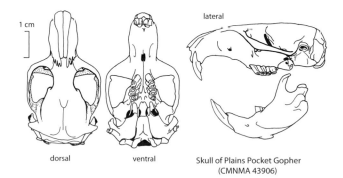

1 cm

lateral

dorsal   ventral   Skull of Plains Pocket Gopher
(CMNMA 43906)

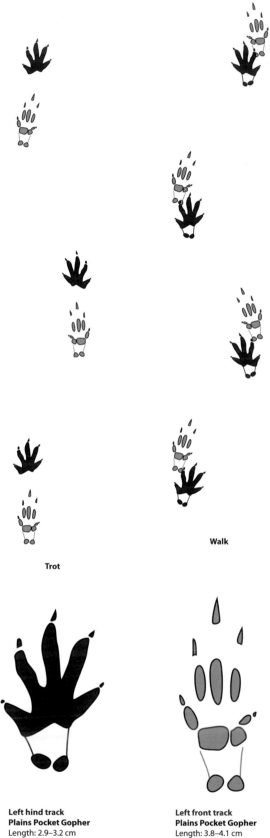

Trot

Walk

Left hind track
Plains Pocket Gopher
Length: 2.9–3.2 cm
Width: 2.9–3.2 cm
Northern Pocket Gopher
Length: 1.5–2.2 cm
Width: 1.0–1.7 cm

Left front track
Plains Pocket Gopher
Length: 3.8–4.1 cm
Width: 2.5 cm
Northern Pocket Gopher
Length: 2.2–3.3 cm
Width: 1.0–1.3 cm

Pocket Gopher

VOCALIZATIONS

These rodents clatter their teeth and hiss when captured, or if their tunnels are invaded by a rival. Juveniles still in the nest will squeak to solicit care from their mothers. Very little other information is available on intraspecific communication.

SIGNS

Dirt piles, created from surplus soil, accumulate around the territory of a Plains Pocket Gopher. Each one is deposited by the incumbent animal as it kicks waste soil out of a burrow into a fan-shaped mound, which typically is sharply peaked unless well weathered. The burrow is then backfilled to plug the opening, to prevent the entrance of a predator. Much of the damage that they do in hay fields is caused by the smothering action of the mounds and the subsequent damage to harvesting machinery as it strikes the pile of dirt. Extra-large mounds (up to a metre wide and 70 cm tall) likely contain a nest. Faeces are normally deposited in an underground latrine chamber, but some may end up above ground as the snow melts, or in the debris remaining after a predator has attempted to dig up a burrow. Scat tends to be filled with woody material and is smoothly ovoid-shaped with rounded ends (3–5 mm in diameter and 8–11 mm long). Most tracks radiate from a burrow entrance. Each foot has five toes – in the classic rodent pattern of toes 1 and 5 angled to the side and the three middle toes close together, aiming forward. The pollex (thumb) of the front foot is much reduced, but has a small claw and commonly registers in a track. Tracks of the Plains Pocket Gopher and the Northern Pocket Gopher are similar except for size. The long claws on the front feet are noticeable in the track. Another sign is "casts" – hollow tubes of packed earth that were pressed against the sides of snow tunnels during the winter as a means of disposing of excavation surpluses. As the snow melts in spring, these worm-like casts lying on the ground become temporarily visible until rain dissolves them.

REFERENCES

Benedix 1994; Cook F.R., and Muir 1984; Elbroch 2003; Heffner and Heffner 1990; Jones, J.K., et al. 1978; Krupa and Geluso 2000; Lowery 2006; Pitts and Choate 1990; Pitts et al. 1994; Wrigley et al. 1991; Zimmerman 1999.

Northern Pocket Gopher

FRENCH NAME: **Gaufre gris**
SCIENTIFIC NAME: *Thomomys talpoides*

Northern Pocket Gophers are stocky, chipmunk-sized, burrowing rodents that spend most of their life underground. They normally

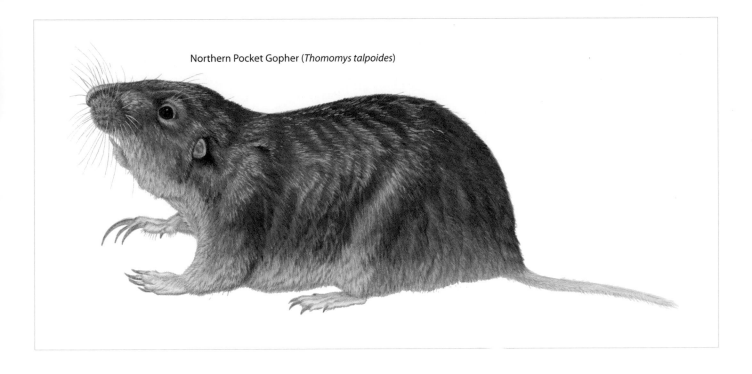

Northern Pocket Gopher (*Thomomys talpoides*)

venture out of their burrows only under cover of darkness. These animals are commonly called "moles," although they are not at all related to true moles.

DESCRIPTION

Northern Pocket Gophers are well adapted for a fossorial existence. Their eyes and external ears are small, and along with the nostrils can be tightly sealed off to protect them from dirt. Their shoulders and neck are heavily muscled for digging and their legs are short but powerful. Each front foot is armed with long, curved claws that make short work of loosening soil, and the toes and foot pads are surrounded by a fringe of stiff hairs that effectively increase the surface area of the foot. The lips can be closed behind the incisors so a pocket gopher can chew through a root or remove an obstructive rock without getting dirt in its mouth. The upper incisors have an orange pigmentation on their front surface, likely from iron in the enamel, which hardens the cutting surface. Both upper and lower incisors are regularly used as digging aids and, to replace this wear, grow at a rate of up to 1.22 mm per day; lower incisors wear more and grow about twice as quickly as upper incisors. A year's growth of a lower incisor could be as much as 45 cm. Sensory vibrissae (whiskers) extend well beyond the head and neck and out from the wrists to provide important tactile information as the animals move around in the subterranean darkness. The tail is short, sparsely furred, and pinkish and commonly has a white tip. The tail tip provides crucial sensory input as the animals cast it around behind them whenever they back up within their tunnel systems, which they regularly do. Large, fur-lined cheek pouches, also called "pockets," extend along either side of the head as far back as the shoulders. The openings, which are near the mouth, are exterior and not part of the oral cavity. Fur is silky and fine. Winter fur is longer and thicker than the summer pelage. Fur colour varies from steel grey, grey

brown, or dark brown to reddish or yellowish brown on the back, with a slightly lighter belly. White patches under the chin are common and some animals have variable amounts of additional white on the belly, throat, and forearms. Feet are usually buff or white. Melanism is common in some regions, although even these animals still tend to have the white chin patch. Albino and white animals are regularly reported. There are two annual moults, one in spring and another in autumn. Both are prolonged and animals are commonly seen with a bisecting moult line separating the new coat from the old. The dental formula is incisors 1/1, canines 0/0, premolars 1/1, and molars 3/3, for a total of 20 teeth.

SIMILAR SPECIES

In most parts of the Canadian plains, the only pocket gopher around is the Northern Pocket Gopher. The exception occurs in the Roseau River valley in southern Manitoba, where both of our pocket gopher species, the Northern and the Plains, co-occur. Despite the size difference (Plains Pocket Gophers are substantially larger), distinguishing the two species can be difficult as neither tends to appears above ground in daylight. Where the two species overlap, the Plains Pocket Gophers are better competitors and take the preferred sandy and loamy areas, relegating the smaller Northern Pocket Gophers to the poorer soil types. Looking at a skull from the front, the front surface of the incisors is diagnostic for separating Plains from Northern Pocket Gophers. The Plains teeth have a double groove running from the insertion to the tip of the tooth, while the Northern has a smooth incisor surface with no grooves. The mounds and tunnels of Northern Pocket Gophers are commonly confused with mole workings, but there are no moles living within the Canadian range of the Northern Pocket Gopher, except in southeastern Manitoba, where the western boundary of Star-nosed Mole distribution overlaps the far eastern distribution of Northern Pocket Gophers, and in the Skagit Valley of British Columbia, where the Coast Mole co-occurs. Moles are found in rich, moist to wet soils, where their aquatic invertebrate and earthworm prey abound, while gophers tend to occupy drier soils. Mole hills are typically dome-shaped, while gopher mounds tend to be more sharply peaked.

SIZE

The following data applies to reproductive-aged Canadian animals during the summer. There is considerable variation in size between populations. Adult females grow until they are two years old; adult males continue to grow through their whole life.

Total length: 154–261 mm; tail length: 38.0–82.8 mm; hind foot length: 22–36 mm; ear length: 4–10 mm.

Weight: males 61–212 g; females 49.0–204.8 g; newborns, around 2.5–3.6 g.

RANGE

All 5 genera and 35 species of pocket gophers are limited to North and Central America. Northern Pocket Gophers are the most widespread of the group, occurring in Canada from southern Manitoba through central Saskatchewan and Alberta, to southern British Columbia, then southwards to eastern California and northern New Mexico. Their distribution can be spotty within this range, as they are limited by soil type and the occurrence of surface bedrock. Northern Pocket Gophers are spreading rapidly in southwestern Alberta, so the range in that region is somewhat uncertain.

ABUNDANCE

One subspecies (*Thomomys talpoides segregatus*) with a limited range and population size in southern British Columbia (on the east side of the Kootenay River north of Creston) is on the provincial "Red List" and is considered at "Low Risk" by COSEWIC. The taxonomy of this subspecies is currently being questioned and it may turn out to be just a local variant of a neighbouring subspecies from the southern Purcell Mountains, a more numerous population that is not considered "At Risk." Otherwise, Northern Pocket Gophers can be locally common and farmers and foresters often consider this rodent to be a pest. It commonly girdles trees and shrubs, including fruit trees, causing extensive damage, especially in heavy snow years. Its taste for Alfalfa can make it a serious pest in hay fields, where it can diminish the crop by 20% or more and its mounds can foul harvesting equipment. Trapping and rodenticides are the usual control methods.

ECOLOGY

Like most fossorial, tunnel-digging mammals, soil quality is highly important to these gophers' ability to dig, and the wrong soil type can limit their distribution. Generally, Northern Pocket Gophers prefer relatively porous, moist soil, but can flourish in sandy and clay soils as well. They tend to avoid overly rocky areas, although they can cope with a gravelly soil. Open forests or forest edges are often acceptable, but they rarely inhabit the floor of a dense forest, since there is little undergrowth to provide them with food. They are most common between 1200 and 2750 m in elevation, but can

Distribution of the Northern Pocket Gopher (*Thomomys talpoides*)

? presence uncertain

occur as high as 3750 m and as low as 900 m in suitable habitats. Most Northern Pocket Gophers live in open grasslands, parklands, open forests, and alpine meadows. They are also quick to inhabit disturbed habitats opened up by fires, logging, road building, or farming, but avoid fields tended by heavy machinery that compacts the soil and collapses their burrows. Hayfields are a favoured habitat. Their activities loosen and aerate the soil, generally promoting plant growth, and their burrow systems create havens and homes for many small vertebrates and invertebrates. Northern Pocket Gophers can swim a 300 m-wide river, although a wider or rapidly flowing river can provide an effective barrier to their dispersal. Should the river become bridged, pocket gophers can, and often do, cross to colonize unoccupied lands on the other side. Adults typically occupy a home range of about 125–175 m² and remain within this area for their whole lives unless abnormally dry or wet weather forces them to relocate. Weaned juveniles disperse overland or under the snow and may travel as far as 1 km (more commonly 200–300 m) in search of a territory of their own. Most dispersal of the species is accomplished by such juveniles. Overland travel is very risky for a pocket gopher and many are lost to predators during the process. Maximum lifespan is thought to be around three years. Northern Pocket Gophers do not hibernate; rather, they remain active through the winter but burrow deeper to get below the frost-line. Occasionally, during years when the earth was saturated just before freeze-up, they are found roaming on top of the snow looking for dry ground. A burrow system typically consists of two levels of tunnels: shallow feeding tunnels 10–45 cm below the surface and deeper tunnels up to 3 m deep, which contain food caches and a nearby nest of finely

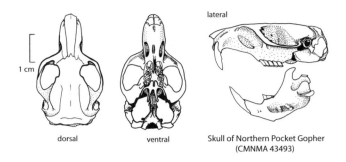

dorsal   ventral   lateral

1 cm

Skull of Northern Pocket Gopher
(CMNMA 43493)

shredded vegetation (up to 25 cm in diameter). Each adult maintains 45–400 m of tunnels. Food caches gathered over the summer provide them with much of their winter food, which they supplement with dried vegetation gathered from the ground surface under the protection of snow. Primary predators include owls, snakes, weasels, foxes, Coyotes, and American Badgers. Hawks, skunks, and Canada Lynx are occasional predators. Apart from predation, parasite load, cold, shortage of winter food, and spring flooding are significant causes of mortality.

DIET

Northern Pocket Gophers are herbivorous and eat a wide variety of plants. They prefer to eat stems and roots of forbs such as clover, dandelion, Alfalfa, and goldenrod along with some grasses, but their diet changes seasonally with plant availability and they will eat most plants within their range. Most foraging occurs above ground, usually within a metre or two of a burrow entrance, and the plants are quickly cut and pulled into the burrow to be consumed in safety. Sometimes plants are harvested from below by clipping the stem from the root and then pulling the plant into a tunnel. Roots, bulbs, and tubers are harvested below ground. What is not consumed immediately may be chewed into segments and carried in the cheek pouches to an underground cache. Each animal could have multiple caches, typically a main one near the nest chamber and several others in various stages of completion in the feeding tunnels, which will ultimately be moved down into the main cache. These caches of up to 3 kg of plant material provide winter food and are commonly composed of bulbs, corms, tubers, roots, and dried plant material. During the winter, gophers supplement their food stores by foraging for desiccated plants under the snow. They will also chew on shrubs and small trees that lie under a snow cover and can denude the bark up 1–2 m depending on snow depth. They are particularly hard on seedlings. Like most rodents, Northern Pocket Gophers will consume animal flesh, with the exception of insects, if it should become available.

REPRODUCTION

Females in Canada are thought to produce a single litter annually, although in rare locations during favourable years a second litter is suspected, especially when the earlier litter was small. Most litters are born in late April to late May and the long breeding season extends from late March through June. Gestation lasts 18–19 days. Litters range in size from three to ten young (five to seven on average). The altricial young are born hairless, deaf, and blind in an underground nest made of dried vegetation. Eyes and ears open at around 26 days old. The young start eating some solid food at around 17 days old and are fully furred and weaned by six to eight weeks old. They reach adult size at around 100 days old, but do not become sexually mature until the following spring.

BEHAVIOUR

Northern Pocket Gophers are largely solitary and vigorously defend exclusive burrow systems, except during the breeding season, when the females relax their aggression and allow males to approach. Should two adults meet outside of the breeding season, they will rise on hind legs and spar with open mouths and flailing front claws. The fight may escalate to lunging and biting, but usually one of the two animals will submit by rolling onto its back or side and allowing the victor to stand over it. When the dominant animal steps back, the subordinate animal retreats. Although aggressive towards their own species, Northern Pocket Gophers are indifferent to the occupation of their feeding tunnels by other creatures such as Deer Mice, toads, voles, salamanders, and invertebrates. Northern Pocket Gophers will venture above ground to forage, usually at night and rarely farther than a metre or so from their burrow entrance. They can dig at an amazing rate, producing a half-metre of tunnel in compacted clay soil in 15 minutes. They dig with teeth and front feet, then kick the loose soil behind with both hind feet. Soil may also be moved with front legs extended and head tipped down and to the side, using the chest to push. Pebbles and gravel-sized stones are usually discarded at the burrow entrance mound, while larger stones are avoided or undermined to sink them further into the earth. Burrow management takes a large part of their active time as new feeding tunnels must be dug, old ones blocked off, burrows opened and aired out on sunny days, surplus soil disposed of, and tunnels patrolled and defended. Waste, such as faeces, spoiled food, and old nesting materials is typically pushed into an underground chamber, which is then plugged. Activity continues throughout each 24-hour period, with intermittent rest breaks. As much as 2–3 hours out of 24 may be spent foraging and harvesting food at the surface. Burrow construction is most evident in spring just before breeding, and again in late summer and autumn until frost.

VOCALIZATIONS

These rodents are not very vocal. Young will squawk if handled and an aggressive animal will stand on its hind legs and squeak at a rival or predator.

SIGNS

The surest sign of pocket-gopher presence is the existence of mounds and snake-like "casts" – hollow tubes of packed earth that was pressed against the sides of snow tunnels during the winter as a means of disposing of excavation surpluses. As the snow melts in spring, these casts become temporarily visible until rain dissolves them. Most mounds are small (less than a metre in diameter and less than half a metre high) and are somewhat fan-shaped piles of earth that spread from the burrow entrances as the animal kicks the surplus soil out. The burrow entrance is commonly blocked after the excavations are complete. Inactive mounds may become ring-shaped as the blocked entrance sinks. Some mounds called "Mima mounds" are thought to be created by Northern Pocket Gophers in areas of the US Midwest, where the water table is high or winter snow melt causes low-lying burrows to flood. A group of gophers or a series of solitary ones raise the soil over many years into a super-mound, which may be 2 m high and 20–30 m in diameter. No other pocket gopher creates such massive mounds. Most tracks radiate from a burrow entrance. Each foot has five toes in the classic rodent pattern of toes 1 and 5 angled to the side and the three middle toes close together, aiming forward. The pollex (thumb) of the front foot is much reduced, but has a small claw and commonly registers in a track. See in the "Signs" section of the Plains Pocket Gopher account for the tracks of both gophers.

REFERENCES

Criddle 1930; Elbroch 2003; Friesen 1993; Koonz 1993; MacMahon 1999; Nagorsen 2005a; Proulx 2002, 2005; Proulx and Cole 1998, 2002; Proulx et al. 1995a, 1995b, 1996; Salt 2000a; Schowalter 2001; Smith, H.C., 1993; Verts and Carraway 1999.

FAMILY DIPODIDAE
jumping mice

Included in the family are birch mice, jerboas, and jumping mice. These mice are all adapted to some degree for jumping, with elongated hind feet and legs, short front legs, and a very long tail. The North American species further display fusion in the cervical vertebrae and enlarged auditory bullae. All four North American jumping mice have a deep groove on the anterior face of their upper incisors.

Woodland Jumping Mouse

FRENCH NAME: **Zapode des bois**
SCIENTIFIC NAME: *Napaeozapus insignis*

This colourful mouse is shy and nocturnal and, hence, rarely seen. Like all the jumping mice, it could easily be mistaken for a very athletic frog because of its hopping gait and small size.

DESCRIPTION
The Woodland Jumping Mouse is the most brightly coloured small mammal in eastern North America, apart from the Eastern Red Bat. Its flanks are a bright yellowish orange with some thin dark streaks caused by dark tips on the guard hairs. There is a broad, darker band of fur that runs along the back from the nose to the tail. The underside from chin to tail, including the legs and feet, is white. The hind legs and feet are elongated and the front legs and feet some-what reduced. The animal spends much of its time crouched over its large hind feet using the front paws to manipulate food. The tail is sparsely haired and longer than the head and body. It is dark above and white below, with an all-white tip that may be up to 5 cm in length. The dental formula is incisors 1/1, canines 0/0, premolars 0/0, and molars 3/3, for a total of 16 teeth. As with all the jumping mice, the front side of the upper incisors is distinctly grooved. A single annual moult occurs during late summer in Canadian parts of the range and is complete just before hibernation.

SIMILAR SPECIES
All four species of jumping mice are very similar, but only the Meadow Jumping Mouse occurs within the range of the Woodland Jumping Mouse. Meadow Jumping Mice are smaller, less colourful (flanks are a darker olive yellow) and the tip of their tail is dark. Both

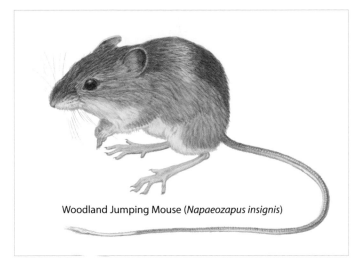

Woodland Jumping Mouse (*Napaeozapus insignis*)

usually contain a mix of conifers and deciduous trees, primarily spruce, fir, pine, hemlock, maple, beech, birch, and basswood. It can also survive in purely deciduous woodlands. They prefer moist, cool areas within the forest or forest margin with abundant cover, commonly found near streams or along the margins of lakes. These mice are relatively poor swimmers capable of travelling only short distances, but they can make short dives and will sometimes head for water to escape. They spend the day in a nest that is a hollow ball of dried grasses and leaves hidden in an underground burrow, in a debris pile, or in or under a fallen log. Their winter nest, in which they hibernate for up to seven months, is always underground. There are some indications that hibernacula in northern regions may be concentrated in favourable areas where drainage and soil type are most suitable. These conditions are most common along streams when slopes are formed, which may partly account for the density of populations along stream beds. During late summer and autumn, Woodland Jumping Mice gain around 7–8 g and sometimes up to 10 g of fat, which sustains them through hibernation. No food is stored for the winter. Adults typically depart to hibernate in late August–early September. Juveniles remain active for one to four weeks longer as they attempt to accumulate the necessary fat. It can be October before they too enter hibernation. A hibernating jumping mouse curls up on its side with its nose touching its belly, its hind feet on either side of its head, and its tail wrapped twice around. Winter mortality is high for those animals unable to store sufficient fat in autumn – a serious problem for many juveniles, as they lack experience and are especially vulnerable if snow comes early. Yearly population density varies in northern parts of the range depending on juvenile winter survival and the breeding success of females. In more southerly parts of the range, population density of this species remains fairly stable from one year to the next. Males emerge first in early May and are ready to mate when the females

share a similar movement pattern. An easy clue to identity is to take note of the habitat. Their names describe their preferred environment. The Meadow Jumping Mouse may occur in the woodland edges, in the absence of Woodland Jumping Mice, but the opposite situation is rare. Cranially, they are distinct, as the Meadow Jumping Mouse has four cheek teeth (molariform), while the Woodland Jumping Mouse has only three.

SIZE

The largest individuals appear to be female, but there is considerable overlap of size between the sexes, and it is not possible to distinguish gender by size in the field. The following measurements apply to adult animals in the Canadian part of the range.

Total length: 212–265 mm; tail length: 133–168 mm; hind foot length: 25–34 mm; ear length: 12–21 mm.

Weight: 19.0–32.6 g; newborns, about 1 g.

RANGE

All the jumping mice (four species) are endemic to North America, and all occur in Canada. The Woodland Jumping Mouse lives in eastern North America, from southern Labrador and southeastern Manitoba to northern Georgia. Increasingly higher elevations are occupied in southerly portions of the range, as the appropriate forest types necessary for the survival of this mouse occur higher and higher in the mountains. Many of these southerly populations have become isolated on mountain peaks.

ABUNDANCE

This mouse is rare and sporadic in the peripheries of its range, especially in Labrador and Manitoba, and is relatively uncommon throughout the rest of its range. It is most abundant in the debris and dense vegetation beside water, especially where there are many decomposing logs and stumps.

ECOLOGY

Most active during the dark hours, this rodent occasionally comes out to forage on heavily overcast days. It lives in woodlands that

Distribution of the Woodland Jumping Mouse (*Napaeozapus insignis*)

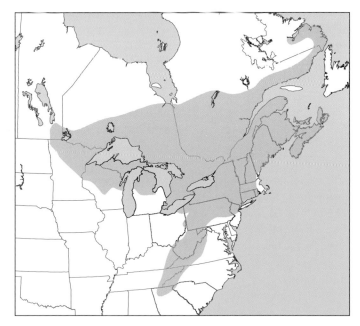

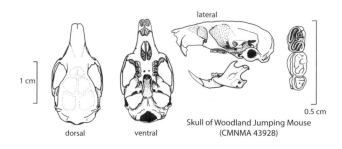

lateral

1 cm

0.5 cm

Skull of Woodland Jumping Mouse
(CMNMA 43928)

dorsal

ventral

**stride length of
a bound varies
from 15 to 122 cm**

emerge two weeks to a month later. The size of the males' home ranges vary from 0.4–3.6 ha and these tend to overlap the smaller home ranges of several females, whose home ranges vary from 0.4–2.6 ha. The maximum known lifespan in the wild is around four years, which is unusually long for a rodent of their size; there is no difference in longevity between males and females. Few survive more than two years. Most of the smaller predators, such as weasels, skunks, Bobcats, Domestic Cats, owls, snakes and likely even short-tailed shrews, take jumping mice.

DIET

Woodland Jumping Mice are considered to be omnivorous. They consume many seeds of grasses and forbs, fungi (various species of subterranean fungi called truffles), nuts and acorns, fleshy fruits such as raspberries, blackberries, and blueberries, leaves and roots of vegetation, and many insects such as Lepidoptera caterpillars, beetles, grubs, and centipedes. Their consumption of the hypogeous (underground fruiting) truffles helps the fungi to disperse its spores, both by uncovering portions (which they dig up to eat) and by scattering viable spores in their faeces. These fungi are highly beneficial to the health of the forest, as they infiltrate the roots of trees and improve their nutrient and mineral uptake.

REPRODUCTION

Females mate soon after they emerge in spring, so the date of birth of their litter varies based upon the timing of their emergence from hibernation, which is dependent on local conditions. In northern parts of the range (in Canada and the northern United States), a single litter each year is the norm, born in late June or early July. A second litter may occasionally occur following the failure of the first litter. Farther south, two litters are more common, the first born in June and the second in August. Gestation lasts for 20–25 days. Litters are made up of two to eight young (average four to five). Young are born hairless, deaf, and blind. They develop quickly and weigh around 9 g on day 26, when they are fully furred and their eyes finally open. Young begin eating solid food at around 30 days old and are weaned shortly thereafter, around 34 days old. They reach sexual maturity at around 35–40 days old, but in the north do not breed for the first time until the following spring.

BEHAVIOUR

Typically, a startled Woodland Jumping Mouse will take several bounds, then stop abruptly under cover, remaining motionless

4 3 2
5 1

Bound

furry
heel

Left hind track
Length: 1.1–2.2 cm
Width: 1.1–1.7 cm

Left front track
Length: 1.0–1.6 cm
Width: 1.0–1.6 cm

Jumping Mice

unless pursued. They are agile and able to climb small shrubs such as raspberry and blackberry, but do not climb trees. These rodents will dig their own burrows or utilize those of other small mammals. Adults rapidly thump their tails on the ground (called "tail drumming") if they are disturbed. This is likely detectable by neighbouring jumping mice and has been interpreted as a signal of anxiety. Captive studies show that they are generally tolerant of each other's presence and will share food and huddle together while sleeping, but females may drive males away when they are caring for a litter. Wild studies suggest that adult females exclude juveniles that are not their own from their home range by a combination of aggression and scent marking, pushing them into marginal areas when female density is high. After the females have entered hibernation, the juveniles relocate into the now empty, more favourable areas.

VOCALIZATIONS

Young squeak and make faint sucking sounds almost continuously while they are in the nest and will produce louder squeaks if disturbed. Adults emit a soft clucking sound and may squeal if disturbed.

SIGNS

Like all the jumping mice, the Woodland Jumping Mouse uses a quadrupedal walk at slow speeds and a quadrupedal hop at higher speeds. A walking trail is uncommonly found, as they usually bound. Hops of 0.6–1.2 m in length and up to 0.3–0.6 m in height are not uncommon, especially if the animal is fleeing. Leaps of up to 1.8 m have been reported. These mice may take to the water if spooked. Their tracks are not seen in winter, as they are hibernating. All toes are clawed and claw impressions are more prominent than with other mouse species. All four species of jumping mice produce very similar tracks, which are indistinguishable to species. Toe 1, the thumb (pollex) of the front track, is vestigial and rarely registers. The remaining four toes are long and slender. Toes 2 and 5 point to the side or even curve backwards and toes 3 and 4 aim more forward. The front track sometimes has a star-like appearance, with all the toes equally angled away from each other. The hind track is made by five clawed toes. Toe 1, the hallux (big toe), is much reduced in size, but usually registers. Toes 1 and 5 point to the side and toes 2, 3, and 4 point forward. Like the front toes, the hind toes are long and often toes 3 and 4 curve outwards. Nests are about the size of a grapefruit (10–15 cm in diameter), with a hollow centre. The entrance of an underground burrow is commonly blocked or covered, making it difficult to detect. Like all jumping mice, Woodland Jumping Mice do not create runways of their own, as their hopping gait frees them from the need to travel in defined, smooth pathways. Nevertheless, they will utilize the runways of other rodents, as well as those of larger mammals such as North American Beavers, Snowshoe Hares, and deer.

REFERENCES

Brannon 2005; Brower and Cade 1966; Collins and Cameron 1984; Kirkland and Kirkland 1979; Kurta 1995; Orrock et al. 2003; Ovaska and Herman 1986, 1988; Saunders 1988; Vickery 1979; Whitaker 1999b; Whitaker and Wrigley 1972; Wrigley 1972.

Meadow Jumping Mouse

FRENCH NAME: **Zapode des champs**
SCIENTIFIC NAME: *Zapus hudsonius*

These small mice are widespread across Canada, but rarely seen due to their shy and nocturnal habits.

DESCRIPTION

Meadow Jumping Mice are small rodents with large hind feet and a long tail. Their flanks are a yellowish olive brown with dark tips on the guard hairs which produce a streaky appearance. A broad, darker brown band runs along the back from nose to tail. The underside, including the legs, is white, sometimes with a faint buffy orange wash. The fur colours of the sides and belly are separated by a thin yellow or orange yellow band of fur. The sparsely furred tail is distinctly bicolour, dark above and white below. It is longer than the head-body length. Rare individuals display abnormal coloration such as all black, no black, white spots, and a white tail tip. The hind legs and feet are elongated and provide most of the propulsion. They commonly crouch over their hind feet to balance themselves while using their small front feet for food manipulation. The dental formula is incisors 1/1, canines 0/0, premolars 1/0, and molars 3/3, for a total of 18 teeth. Like all jumping mice, the front side of their upper incisors is distinctly grooved. The incisors of newborns are white, but gradually become orange, like those of the adults, by the time they are weaned.

SIMILAR SPECIES

There are four species of jumping mice: all occur in Canada and two overlap the distribution of the Meadow Jumping Mouse. Woodland Jumping Mice occur in eastern North America. They are larger, more colourful (flanks are a bright yellowish orange), and the tip of their tail is white. An easy clue to distinguishing the two is to take note of the habitat. Their names describe their preferred environment. The Meadow Jumping Mouse may occur in the woodland edges in the absence of Woodland Jumping Mice, but the opposite situation is

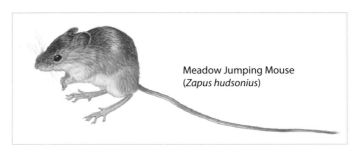

Meadow Jumping Mouse
(*Zapus hudsonius*)

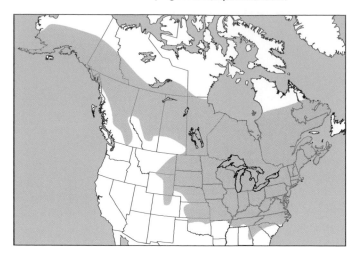

rare. Cranially, they are distinct, as the Meadow Jumping Mouse has four cheek teeth (molariform), while the Woodland Jumping Mouse has only three. Distinguishing the Western Jumping Mouse from the Meadow Jumping Mouse is more difficult. In general, the former is somewhat brighter on the flanks and a bit larger, although this is very difficult to discern in the field, especially as only one species is usually seen at a time, even when they share a habitat. Identification ultimately depends on a series of cranial characteristics. The skull of Western Jumping Mice is larger (incisive foramen length > 4.8 mm, lower toothrow length > 4.3 mm, zygomatic breadth > 12.0 mm). All share a similar movement pattern.

SIZE

The largest individuals appear to be female, but there is considerable overlap of size between the sexes and it is not possible to distinguish gender by size in the field. The following measurements apply to animals of reproductive age from the Canadian part of the range.
Total length: 180–237 mm; tail length: 106–149 mm; hind foot length: 27–36 mm; ear length: 10–19 mm.
Weight: 14–25 g; newborns, 0.7–0.9 g.

RANGE

This is the most widespread of the four species of jumping mice. Meadow Jumping Mice occur from Alaska to the northern Ungava Peninsula and southwards to Mississippi. They have reached Prince Edward Island, Manitoulin Island, and Akimiski and the Belcher Islands in James Bay and Hudson Bay respectively, but have not reached Vancouver Island or Newfoundland.

ABUNDANCE

Meadow Jumping Mice are generally common throughout the central portion of their range in areas of appropriate habitat. Around the perimeter of the range, especially in the north, they are more scarce. One subspecies, *Zapus hudsonius preblei*, which occupies a small range in Colorado and Wyoming was declared "Threatened" under the US Endangered Species Act in 2001.

ECOLOGY

Meadow Jumping Mice inhabit a range of meadow-type habitats, but are most abundant in moist, grassy, weedy meadows and in dense vegetation near water. Adequate cover is essential to their survival and they can live at some distance from water as long as the vegetation is dense. They will occupy the fringe vegetation on the edges of woodlands, but do not generally invade the forest. Dry grasslands are avoided. Largely nocturnal, they may become active in daylight if it is heavily overcast. Competition with the larger Meadow Voles can influence their distribution, as their numbers are greater when Meadow Voles are absent. Home ranges vary from about 0.08–1.10 ha and are similar in size for males and females. Home ranges of males show extensive overlap with those of females. These mice usually dig their own burrows, but may use those of other small mammals. Their summer nests are constructed in protected places, such as subterranean burrows, in clumps of vegetation, under logs, or in hollow logs. They look like a ball of vegetation, but have a hollow centre and usually a side entrance. In some cases, maternal nests may be lined with softer material. Summer nests are used only for a few weeks before a new one is constructed. Perhaps this is a strategy to reduce parasite infestation. Above-ground nests are used only in summer. Hibernation nests, and possibly also maternal ones, are underground. Hibernacula are likely plugged, both at the burrow access and at the nest chamber itself, and are excavated in areas with well-drained friable soil where digging is easy. There are indications that jumping mice may congregate at preferred hibernacula sites. A hibernation burrow may extend half a metre or more below ground surface and each mouse hibernates alone.

Meadow Jumping Mice are agile and can climb grass stems to harvest the seed heads. They will also reach up as high as they can, then cut the stem of a weed or grass to harvest the seeds. They are capable swimmers and can swim below the surface, but have stamina only for short swims. Towards the end of August, adults begin to accumulate body fat for winter. Some animals can double their weight over a 16-day period if food is abundant. In a normal year, the mice gain about a third of the mean adult body weight, or about 6 g. This fat sustains them during six to eight months of hibernation. Over-winter weight loss is around 35%–37% of total weight. The length of hibernation depends largely on local climate and the condition of the jumping mouse itself. Fat mice begin hibernating sooner than leaner animals. Typically, the juveniles of the later litters are less fat and must continue to feed after the adults have gone into hibernation. Mortality among this class of animals is high, as many cannot accumulate sufficient fat

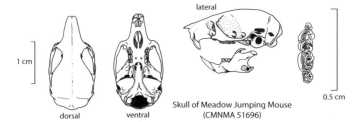

lateral

1 cm

0.5 cm

Skull of Meadow Jumping Mouse
(CMNMA 51696)

dorsal   ventral

reserves before snowfall to last them until spring, so they perish while hibernating. Fat adult males may begin hibernating in late August to early September and adult females follow two to four weeks later. They likely need the extra time to replace fat reserves they lost nursing their litter. Juveniles may delay until November. A hibernating Meadow Jumping Mouse tightly curls on its side with its hind feet beside its ears and its tail wrapped around its body. Males emerge first in the spring, in late April or early May, and are ready to mate when the females emerge around two weeks later. A three-year-old Meadow Jumping Mouse is old; few survive to be even one year old. The oldest captive animal lived five years. Predators are numerous and include most owls, as well as foxes, weasels, and Domestic Cats. Meadow Jumping Mice have even been found in the stomachs of moderate to large freshwater fishes, further evidence of their swimming abilities.

DIET

Seeds of grasses and forbs are the most important food of this species. The seed of choice varies as the plants ripen through the season. This mainstay is supplemented with a wide assortment of fungi (mainly various species of subterranean fungi, called truffles), nuts, acorns, many species of insects, and fruits such as raspberry, blackberry, blueberry, strawberry, and dogwood. For a few weeks in spring when seeds and fruit are scarce, animal matter (mainly insects and insect larva) may make up 50% of the diet.

REPRODUCTION

Breeding begins when the females emerge from hibernation. Females in warm regions of the country produce two litters annually, and some may produce three in years of abundant food. In the colder regions (including all but far western Canada), one litter per year is the norm. In Canada and the northern United States there are breeding peaks in June and August. The gestation period is 17–21 days and litter size varies from 2–9 (average 4–5). Pups are born blind, deaf, and hairless apart from the tips of their whiskers (vibrissae) showing through the skin. Ears open around day 20, they are fully furred by day 21, eyes open between days 22 and 25, and they begin to taste solid food about the same time. Weaning occurs soon after, and by 28 days old the juveniles are self-sufficient. By about day 60 the young reach a low adult weight and by day 90 attain a full adult weight. Some females born in an early litter may breed for the first time late in the same summer.

BEHAVIOUR

The behaviour of this species is poorly studied and not well known. Adult Meadow Jumping Mice are generally solitary. They will fight with each other, but are usually fairly tolerant towards other adults. Their typical escape behaviour is to make a series of long bounds (up to a metre long), then freeze for a moment under cover. If pursuit continues and there is sufficient cover, they may creep into the vegetation for a short distance and then jump again, often in a different direction; otherwise they will continue bounding, sometimes changing their direction as they go. Should an escape burrow be near at hand, they will dive into it. These mice are capable excavators, digging underground burrows for nests and hibernacula and uncovering subterranean fungi, which they probably detect by smell. They use runways of other small mammal species, but do not create any of their own.

VOCALIZATIONS

Newborns produce a high-pitched squeak during their first week that can be detected by humans at least a metre away. This call is supplemented by a fainter "suckling note," which is likely a solicitation call to their mother. Both these calls have been documented from a captive litter. A wild litter in an underground maternal nest would not likely be audible at all. Adults are relatively silent. Squeaking has been reported by adults and chirping sounds, like a small bird call, have been reported from two or more animals as they jump in close proximity to each other. The intent of this behaviour is unknown, but it does not appear to be aggressive. A faint clucking sound is emitted when the animals are excited.

SIGNS

Tracks of all four North American species of jumping mice are similar and not distinguishable to species. See the Woodland Jumping Mouse account ("Signs" section) for track information. The normal movement of Meadow Jumping Mice is either a quadrupedal walk or a quadrupedal hop (also called a bound). Normal hops are up to 15 cm in length, but if the mouse is pursued, a bound may be up to a metre, though more typically 45–60 cm in length. Tracks in snow are unlikely as this mouse hibernates during the cold season. Both winter and summer nests are about 8–10 cm across and 7–8 cm high, with a hollow inner chamber of about 5 cm high by 7.5 cm across. These are made of whatever dried vegetation is available but often include grass stems. Their small scat is deposited anywhere along their route and tends to break down quickly in such moist surroundings.

REFERENCES

Bain and Shenk 2002; Beauvais 2001; Cochran and Cochran 1999; Conner and Shenk 2003; Desrosiers et al. 2002; Elbroch 2003; Hoyle and Boonstra 1986; Jones, G.S., and Jones 1985; Kurta 1995; Nagorsen 2005a; Peirce and Peirce 2000b; Ryon 2001; Smith, H.C., 1993; Verts and Carraway 1998; Whitaker 1972; Wrigley 1972.

Western Jumping Mouse

FRENCH NAME: **Zapode de l'Ouest**
SCIENTIFIC NAME: *Zapus princeps*

One of the four similar jumping mice species that occur only in North America, the Western Jumping Mouse, like the others, is difficult to see due to its shy manner and nocturnal habits.

Western Jumping Mouse
(*Zapus princeps*)

DESCRIPTION

A wide, dark, olive-brown to brown stripe runs along the back from the nose to the base of the tail. The flanks are lighter olive brown to grey brown, usually washed with a yellow or orange that sometimes is very pale, verging on unnoticeable. A thin orange or yellow line separates the flank from the white or pale orange to yellow tinged belly. The tail is longer than the head-body length and is dark above and light below. The pinkish-white hind legs and feet are elongated and provide most of the propulsion, as the animals utilize a four-legged hopping gait. They commonly crouch over their hind feet to balance themselves, which frees up the small front legs and feet for food manipulation. The dental formula is incisors 1/1, canines 0/0, premolars 1/0, and molars 3/3, for a total of 18 teeth. Like all jumping mice, the front side of their upper incisors is distinctly grooved and dark orange. The upper incisors of newborns are white, but gradually become orange like those of the adults by the time they are weaned.

SIMILAR SPECIES

All four species of jumping mice are similar in appearance and all share a similar movement pattern. Two species, Meadow Jumping Mice and Pacific Jumping Mice, occur within the range of the Western Jumping Mouse in Canada. Distinguishing between the three is often very difficult. In general, Pacific Jumping Mice tend to be brighter than Western and Meadow Jumping Mice, but this varies with location and of course may be impossible to discern in the field. Individual guard hairs of Meadow Jumping Mice are very fine, with an average diameter of about 115 angstroms (one angstrom equals one ten millionth of a millimetre). Those of both Pacific and Western Jumping Mice are slightly coarser, with average diameters of around 140 μ. In the narrow zone region where Pacific and Western Jumping Mice overlap, the Pacific Jumping Mouse tends to be brighter in colour with a buffy underside, sometimes with buffy patches, and with a more distinct separation of colour between the back and the flanks, but these characteristics vary and are not always useful. DNA analysis has identified wild-born, fertile hybrids of Western and Pacific Jumping Mice in regions where their ranges overlap, imparting some doubt about the validity of the two

as distinct species. So far, no single feature or even series of cranial measurements are accurate in distinguishing these two species all the time. The overlap of Meadow and Western Jumping Mice is considerably more widespread and the two species may even occur together in the same habitat. Ultimately, the identification relies on cranial features. In general, the skull of the Western Jumping Mouse is larger than that of the Meadow Jumping Mouse (incisive foramen length > 4.8 mm, lower toothrow length > 4.3 mm, zygomatic breadth > 12.0 mm) and the outer margins of its incisive foramen tend to be straight, while those of the Meadow Jumping Mouse are curved inward.

SIZE

There appears to be no difference in size or appearance between adults of either sex. The following measurements apply to animals of reproductive age from the Canadian part of the range.
Total length: 209–274 mm; tail length: 122–176 mm; hind foot length: 27–35 mm; ear length: 10–18 mm.
Weight: 17.7–37.5 g; newborns, 0.7–0.9 g.

RANGE

Western Jumping Mice inhabit the plains and western mountains of North America from southern Yukon through most of British Columbia, southern Alberta, and Saskatchewan and southwestern Manitoba to northern New Mexico and California. Generally, these mice live in higher and higher elevations the farther south they occur, as their habitat zone moves up the mountain sides. They are replaced in the extreme southwestern portion of British Columbia by the Pacific Jumping Mouse and the only distribution overlap of these two species is along a narrow alpine zone running along the crest of the Cascade Range and the Coast Mountains. A small area of disjunct range occurs on the northwestern periphery of Winnipeg.

ABUNDANCE

Rarely seen and generally uncommon, these mice may be sporadically locally abundant in favourable habitats.

ECOLOGY

These rodents live in moist habitats in dense vegetation such as thickets, shrubbery, or grassy meadows near watercourses, in alpine or subalpine meadows, and in mature forests or recent burns where the vegetation is resurging. They are most abundant in productive habitats that support abundant forbs. Although primarily nocturnal, these mice may occasionally venture out during daylight. Western Jumping Mice spend seven to ten months in hibernation. They curl on their sides in a hibernation nest with their hind feet beside their head and their tails curled around their body. This round hollow nest, made of dried vegetation, is in a chamber of an underground burrow around half a metre or more beneath the ground surface. It has a single entrance and is plugged near the entrance while the occupant is in torpor. Fat is accumulated for about a month before hibernation, during a time when seeds are abundant and readily available. No food is stored for midwinter consumption, so the animals must survive solely on these fat reserves. Ten to fifteen grams

Distribution of the Western Jumping Mouse (*Zapus princeps*)

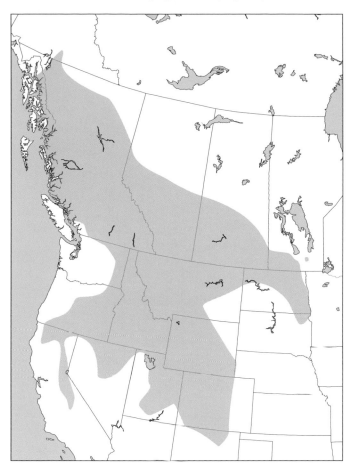

green vegetation, and subterranean fungi as well as arthropods (mainly spiders and insects) depending on season and availability. Arthropods may be a major component of their diet in spring before the majority of the seeds ripen. For about a month before hibernation they focus on high-calorie seeds of forbs and grasses in order to accumulate body fat.

REPRODUCTION

Breeding begins within a week of the female's emergence from hibernation. A single litter per year is the norm. Gestation lasts around 18 days and pups are nursed for 28–35 days. Litters average four to six young (range 2–7). Most Canadian litters are whelped in late May to early July. Juvenile males typically become sexually mature in the spring following their birth. In British Columbia, around 40% of yearling females breed in the season following their first hibernation, while 60% do not breed until their second year. During each reproductive season, around 30% of the adult females do not breed. Non-breeding females are thought to be those coming out of hibernation in poor condition, typically very thin. Since this situation is considerably more common among yearling females than among adult females, that single factor could account for the difference in breeding success between the two age classes.

BEHAVIOUR

Western Jumping Mice, like other jumping mice, use a series of bounds followed by a period of hidden immobility as their typical escape behaviour. They may move in zigzag leaps or in straight bounds of around 0.3–0.4 m in length. If there is sufficient vegetation, they will crawl through it for a short distance before resuming their bounding flight. When startled, leaps of up to 0.7 m long and 0.3 m high have been reported. Female Western Jumping Mice appear to be territorial, as their home ranges of around 0.10 ha (range 0.07–0.13 ha) do not overlap with those of other females. This does not seem to be the case with males. Their larger home ranges, which average 0.17 ha (range 0.03–0.33 ha), overlap extensively, not only with those of females but also with those of other males.

VOCALIZATIONS

Western Jumping Mice squeak when fighting and chatter their teeth at each other. They will thump their tails on the ground when agitated.

of body fat is necessary to survive the average hibernation. Juveniles typically enter their first hibernation with lower fat reserves than the adults and their winter mortality is significantly higher as a result (41%–70% for juveniles vs 16%–28% for adults). Most of the mice enter hibernation during September or early October or upon the first local snowfall, if that occurs earlier. Temperatures in their hibernacula rarely reach freezing (usually 2°C–6°C). As the soil warms following the disappearance of the snow pack, burrow temperatures of 8.0°C–9.5°C cue arousal. Emergence occurs in Canadian populations from early May to mid-July depending on local conditions and typically the males become active about a week before the females. Summer nests may be subterranean or constructed in or under logs and stumps, in dense vegetation, or under leaf litter. Western Jumping Mice are capable swimmers. They are relatively long-lived. Maximum lifespan is around six years, but if they survive their first winter, a lifespan of three to four years is very possible. Predators include several species of owls, Coyotes, American Martens, American Badgers, Bobcats, skunks, Northern Raccoons, several species of weasels (Ermine, Long-tailed, and Least), and snakes.

DIET

Western Jumping Mice depend on seeds of forbs and grasses for the bulk of their diet. In addition, they consume fruits and berries,

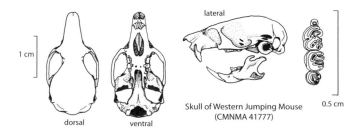

1 cm

lateral

Skull of Western Jumping Mouse (CMNMA 41777)

0.5 cm

dorsal   ventral

SIGNS

Feeding grounds may be identified by small untidy piles of grass stems about 10–12 cm long left by a mouse who has clipped the stems in order to reach and consume the seeds. This sign is likely common to all four species of jumping mice. Unlike other jumping mice, the Western Jumping Mouse does sometimes create runways, which are typically floored with grass clippings from foraging activities. See the Woodland Jumping Mouse account in its "Signs" section for a description of tracks. All the jumping mice make similar tracks, which are not distinguishable to species.

REFERENCES

Cranford, J.A., 1978, 1983a, 1983b; Falk and Millar 1987; Hart et al. 2004; Jameson and Peeters 2004; Jones, G.S., and Jones 1985; Jones, G.S., et al. 1978; Nagorsen 2005a; Smith, H.C., 1993; Verts and Carraway 1998; Wrigley et al. 1991.

Pacific Jumping Mouse

FRENCH NAME: **Zapode du Pacifique**

SCIENTIFIC NAME: *Zapus trinotatus*

As its name suggests, this rodent lives along the Pacific coast of North America, where it reaches the northern limit of its distribution in southwestern British Columbia. Very little is known about these animals in British Columbia. Most of the life-history data presented below is based on studies of specimens from farther south.

DESCRIPTION

This colourful mouse has yellowish to yellowish-orange sides and a broad band of dark brown to cinnamon brown along its back, which runs from nose to tail. The belly is white with a buffy tinge and often with buff patches at the throat, chest, and abdomen. A bright band of light orange buff or yellow separates the flank from the belly colours. The legs and feet are covered in fine, short, white hairs through which the pinkish skin shows. Hind legs and feet are elongated for hopping and the front legs and feet are reduced. This mouse spends much of its time balancing over its hind legs to free its front feet for food manipulation (as in the diagram). The long tail is bicoloured, dark above and light below, and is longer than the head-body length. As is the case for all the jumping mice, the tail is essential for balance in the air during long hops. The dental formula is incisors 1/1, canines 0/0, premolars 1/0, and molars 3/3, for a total of 18 teeth. All the jumping mice have a distinctly grooved, dark orange front side on the upper incisors. These incisors in newborns are white, but gradually become orange like those of the adults by the time they are weaned. There is one moult each year, which occurs during the summer.

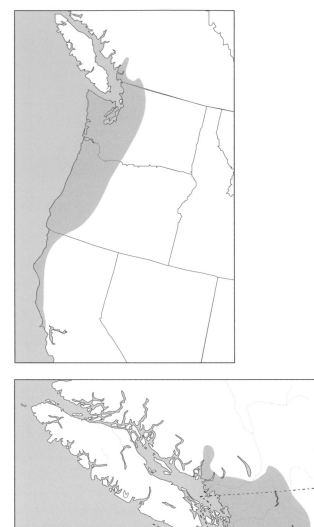

expanded view of Canadian range

SIMILAR SPECIES

Of the three other species of jumping mice, only the Western Jumping Mouse occurs within the range of the Pacific Jumping Mouse. The region of overlap in British Columbia occurs in a narrow alpine zone running along the crest of the Cascade Range and Coast Mountains. In general, the colour of the Pacific Jumping Mouse is brighter than that of the Western Jumping Mouse in that region and the Pacific form often has a buffy tinge to its belly hairs and buff patches on the abdomen and chest. Furthermore, the molars of the Pacific Jumping Mouse are somewhat larger. These characteristics are unfortunately variable and do not always satisfactorily distinguish the species. Even a series of skull measurements are not always useful for identification. Fertile hybrids of Western and Pacific Jumping Mice have been identified using DNA analysis in the wild, in regions where both occur, which imparts some doubt to the validity of the two as distinct species.

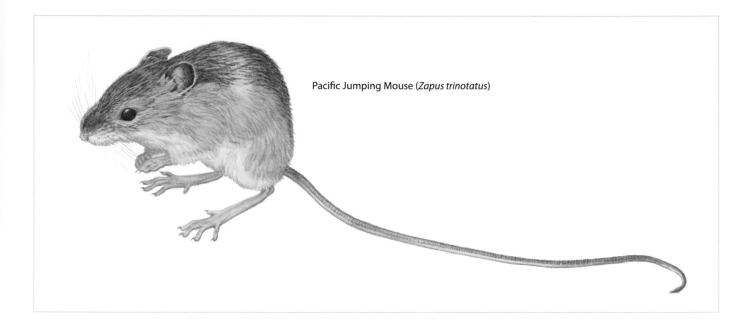

Pacific Jumping Mouse (*Zapus trinotatus*)

SIZE

There appears to be no difference in size or appearance between adults of either sex. The following measurements apply to animals of reproductive age from the Canadian part of the range.

Total length: 215–258 mm; tail length: 120–162 mm; hind foot length: 29–35 mm; ear length: 12–16 mm.

Weight: 17–36 g; newborns, 0.7–0.9 g.

RANGE

Pacific Jumping Mice occur in moist coastal forests from southwestern British Columbia to central California. The mountain crests of the Coast/Cascade/Sierra Nevada Ranges form a barrier to their eastward expansion and San Francisco Bay may limit their southward distribution.

ABUNDANCE

Pacific Jumping Mice are rarely seen, due to their shy, nocturnal habits, and although generally rare to scarce, they can be locally common, and even at times abundant, in favourable habitats.

ECOLOGY

Pacific Jumping Mice are usually nocturnal and crepuscular and are rarely seen in daylight unless disturbed. They live in a variety of moist habitats in dense forest and forest clearings with weedy undergrowth, in wet meadows and densely vegetated stream and marsh edges, and in alpine shrub and high alpine meadows. They are most abundant in regions where the yearly rainfall exceeds 30 cm. Although they do not make runways of their own, they will utilize those made by Mountain Beavers and Shrew-moles. In autumn, the mice begin to accumulate body fat and by late October most are in an underground burrow in torpor. They dig their hibernation burrows in loose, well-drained soil, well above the water level in a bank or mound, where they remain in hibernation for four to eight months. The nest chamber is up to 76 cm below the ground surface. The earliest specimen from British Columbia was collected on 8 March (Vancouver) and the latest specimen (Alta Lake) on 20 November. During hibernation the mice lie on their sides with their hind feet beside their head and curl their tails over the head and around their body. Warm soil temperatures in spring cause arousal. Many aspects of the biology of this species are poorly understood. For example, we know little of their population density, size of individual home ranges, social structure, or behaviour. Various species of owls are their main predators. Other predators, which likely take Pacific Jumping Mice when the opportunity arises, include skunks, small weasels, Bobcats, Domestic Cats, foxes, and snakes.

DIET

Pacific Jumping Mice are considered to be primarily granivorous as seeds of grasses compose more than 50% of their diet. These are supplemented with fleshy fruits (such as blackberry, thimbleberry, and huckleberry), seeds of forbs, insects, underground fungi (truffles), occasionally eggs of small birds, molluscs, and even small fishes (found washed up on shore).

REPRODUCTION

The breeding season begins soon after the females emerge from hibernation. In British Columbia, this occurs in May or June depending on local weather conditions. It is supposed that one litter per year is the norm. A litter of 4–8 pups is born after a gestation of 18–23 days. It is likely that yearling females breed later than adult females, so litters may be born from June to August. Young are born deaf, blind, and hairless (without even a hint of whiskers). Maternal nests are spherical and hollow, composed of woven dried vegetation (mostly grasses), and are commonly above ground in a clump of dense vegetation. One has even been discovered in a tree about 1.4 m above the ground. The young suckle for about a month. They reach sexual maturity just before entering hibernation and breed for the first time the following summer.

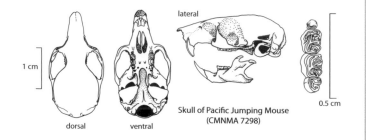

dorsal   ventral   Skull of Pacific Jumping Mouse
(CMNMA 7298)

1 cm

lateral

0.5 cm

BEHAVIOUR

If flushed, the typical escape behaviour of this mouse is to make three or four bounds of 0.9–1.8 m, changing direction with each bounce, then crouch motionless in hiding, unless further pursued. Variously described as passive or high strung and willing to bite, it appears that this mouse will behave aggressively if roughly handled. A report from Washington State describes a Pacific Jumping Mouse usurping the nest of a Dark-eyed Junco, eating the four eggs, and then creating a dome over the bird's nest before using it as its new home. Some of the hibernation burrows have been discovered with interconnecting passages, suggesting that neighbours may share hibernation nests.

VOCALIZATIONS

Pacific Jumping Mice will squeak when fighting and thump their tails against the substrate when agitated. Very young pups in the nest produce high-pitched squeals if left unattended. These sounds are sometimes loud enough to allow scientists to find the nest.

SIGNS

Nests are hollow balls of woven grasses and dried vegetation about 13–15 cm in diameter. Summer nests are loosely woven, while hibernation nests are more tightly woven. Burrow entrances are typically around 3–4 cm in diameter. Although Pacific Jumping Mice commonly dig their own burrows, they will use those of other species if available and suitable. Like other jumping mice, they commonly leave neatly cut sections of grass stems in spots where they foraged on the seed heads. All the jumping mice create similar signs and tracks. See the "Signs" section of the Woodland Jumping Mouse account for a full description of tracks.

REFERENCES

Jameson and Peeters 2004; Johnstone 1979; Jones, G.S., et al. 1978; Nagorsen 2005a; Smith, K.G., 1984; Verts and Carraway 1998.

FAMILY CRICETIDAE

lemmings, voles, rats, mice

Members of this family in Canada are classified into two subfamilies: the Arvicolinae and Neotominae.

SUBFAMILY ARVICOLINAE

lemmings, voles, muskrats

Members of this subfamily are characterized by their teeth. The crowns of the molars are composed of prisms of alternating triangles. They feed on plant material and have flat-topped molars to provide grinding surfaces and many have rootless molars that are ever-growing to compensate for the relentless wear caused by their diet. Muskrat, Heather Vole, and red-backed voles have rooted molars that have a finite size (i.e., they do not keep growing over the life of the animal). These stocky, short-legged rodents typically have a short muzzle, small eyes and ears, and a tail that is shorter than the head and body.

Northern Collared Lemming

also called Nearctic Collared Lemming, Varying Lemming, Hoofed Lemming

FRENCH NAME: **Lemming du Groenland**

SCIENTIFIC NAME: *Dicrostonyx groenlandicus,* formerly *Dicrostonyx torquatus*

This lemming is the most northerly rodent in the world. Like most other collared lemmings, it changes its coat colour to white in winter. Although North American lemming populations do undergo cyclical variations in population density, they do not experience the sporadic mass migrations of similar species in Scandinavia.

DESCRIPTION

The winter coat is highly insulative, being about twice as long and dense as the summer coat. Each white hair has a dark base,

Northern Collared Lemming (*Dicrostonyx groenlandicus*)

winter

summer

providing a greyish cast to the coat. The summer coat is more colourful and displays considerable variation across the range. A thin dark line runs along the middle of the back from the nose to the tail. It is most clearly seen on the shorter juvenile coat and may disappear on some adults. The short tail is dark above and white to buffy or orange below, with a long tuft of light-coloured hairs at its tip. Patches of dark to bright orange fur cover the ears, the muzzle where the vibrissae insert, and the sides of the neck at the shoulder. Tips of the belly hairs are silvery, buffy, or grey and sometimes are washed with dark orange, but the base of each hair is dark grey. A yellowish to orange band runs along the sides from throat to rump and separates the belly fur from the back fur. A "collar" or throat patch of dark to bright orange fur often extends down the midline of the belly for some distance (rarely more than half way). Most animals have hairs on their backs and upper sides that are banded with black and white, giving a grizzled grey appearance, but some animals, especially in the western Arctic, have a darker red wash over the shoulders and back that give the grizzled fur a brownish, almost purplish tint. These same lemmings often have a lighter rump. Most adults develop a pale indistinct band of fur that runs over the back of the neck just behind the ears. Juveniles are darker, browner, and less colourful than adults. Albino and melanin-deficient individuals are known, but are very rare. Like all the collared lemmings, this species exhibits a classic body shape well suited for burrowing and surviving in a cold Arctic climate. The body is stocky with a robust skull; the legs, ears, and tail are short to conserve heat; feet are well furred so the animal does not contact the cold substrate with its bare pads. As the photoperiod changes in autumn, collared lemmings develop thickened nails and cornified pads on their third and fourth front toes that act as shovels to assist in digging

burrows through hard-packed northern snowdrifts. The growth of these nails and pads diminishes in spring and they are gradually worn down, so the summer claws are normal. The dental formula is incisors 1/1, canines 0/0, premolars 0/0, and molars 3/3, for a total of 16 teeth. Adults undergo two moults annually in spring and autumn that are governed by lengthening and shortening photoperiods.

SIMILAR SPECIES

In the High Arctic islands, this lemming is the only small mammal. On the lower Arctic islands and the mainland, other lemmings co-occur. The Brown Lemming is grizzled brown on its head and neck and rusty brown on the back and hindquarters. It does not turn white in winter. Other similar rodents such as the Meadow Vole, Taiga Vole, Northern Bog Lemming, Tundra Vole, Singing Vole, Heather Vole, and Northern Red-backed Vole are smaller, longer tailed, and do not become white in winter. In Canada, the most similar small mammal to a Northern Collared Lemming is a Richardson's Collared Lemming. The two species occur together along a narrow band between their ranges in the vicinity of Chesterfield Inlet and the Thelon River and are not easily distinguished. Northern Collared Lemmings are typically brighter, with a grizzled black and white back, while Richardson's Collared Lemmings are grizzled black and brown in the same area and the orange patches on their coats are generally darker and duller. To distinguish vole skulls see appendix 2.

SIZE

The following measurements apply to Canadian animals of reproductive age. Males and females are similar in size and colouring. Weight is highest during winter and lowest in summer.

Distribution of the Northern Collared Lemming (*Dicrostonyx groenlandicus*)

Total length: 124–162 mm; tail length: 8–18 mm; hind foot length: 17–24 mm; ear length: 4–6 mm.
Weight: 50.5–118.4 g; newborns, 2.7–4.8 g.

RANGE

Northern Collared Lemmings occur from Alaska to northeast Greenland and south as far as Great Slave Lake. They have reached all the larger High Arctic islands and many of the smaller ones.

ABUNDANCE

Although reasonably common, their density varies from scarce to abundant, depending on the stage of the population cycle. This cycle of population expansion and decline normally occurs with a periodicity of three to five years.

ECOLOGY

Northern Collared Lemmings live in the tundra. In summer they prefer dry hummocks or rocky and often windswept ridges, where they dig numerous short, simple burrows or shelter under rocks. Most summer burrows have a single entrance and contain a nesting and latrine chamber and less than a metre of tunnel. During winter, they usually move to more protected, wetter areas with more vegetation and deeper snowdrifts. Snow burrows are more complex than summer burrows, with nest chambers and latrines, several entrances, and many metres of tunnels. Unlike all other small mammals that remain active under the snow, collared lemmings are so cold-adapted that they thrive under those conditions and are able to gain weight rather than lose it as the other species do. They are even capable of winter reproduction during years with adequate snow cover. Most Northern Collared Lemming populations are cyclic, experiencing dramatic numeric peaks and then crashes over about a four-year period. The reasons for these numeric swings are complex and may be variable from place to place. Most researchers feel that they are due to societal, behavioural, nutritional, and physiological effects caused by the lemmings themselves. But these reasons do not appear to regulate all populations. Some, especially those experiencing persistent, heavy predation, are less likely to be cyclic, while for other populations, heavy seasonal predation pressure may actually drive the cycle. Weather is likely a major factor as well. The cycles are not synchronous across the Arctic. These small mammals are

the cornerstone of the Arctic ecosystem. When lemming numbers are high, most Arctic predators, with the possible exception of Grey Wolves, tend to have higher reproductive success and lower juvenile mortality. When lemming numbers plummet, the predators suffer and refocus their foraging efforts towards other Arctic species, whose success then drops as well. The overall outcome of the lemming cycle is that the lemming's predators are unable to sustain any high population density themselves, as the poor lemming years prune their numbers back. Home-range size varies widely. Within each gender, larger adults are able to defend larger home ranges, and in general, males have considerably larger home ranges than do females. During lows in the population cycle, home ranges can be quite a bit larger than they are during the peaks, when adults of both sexes are squeezed into small home ranges and there is little difference in home-range size between the sexes. Home ranges of males vary from around 37,000 m² or more to less than 20 m², while those of females vary from 6000 m² or more to less than 20 m². Population density may vary from 10 individuals per hectare to 0.1 individual/ha during a cycle. Despite their body size, these small mammals are capable of extensive dispersion that accounts for their presence on many islands and for the genetic homogeneity of often even distant populations. Some of this dispersal occurs during winter when travel over sea ice is possible. Predators include Snowy and Long-eared Owls, jaegers, gulls, Ermine, Least Weasels, Arctic Foxes, and sometimes even Grey Wolves. Typically, only the Ermine and Least Weasel are resident predators and their major impact is felt during winter as they continue to hunt under the snow. Losses to predation are typically heaviest during summer when there are more predators around and when the lemmings are more exposed as they forage and search for mates above ground. Overall mortality due to predation may be as high as 50%.

DIET

Like most rodents, Northern Collared Lemmings are herbivorous. Their preferred food plants are shrubs, forbs, and sedges. They eat leaves and stems as well as flowers, roots, berries, and occasionally bark in summer. In winter, they focus mainly on leaves, buds, and bark of the recumbent Arctic willows that they can reach under the snow.

REPRODUCTION

Breeding typically begins in early spring when the ground is still snow covered. The season usually extends from early March until early September, but winter breeding commonly occurs, especially when the snow pack is substantial. Average litter size is four to five pups (range 1–7). The gestation period is 19–21 days and the average wild female produces up to three litters per year. The pups are born hairless, deaf, and blind in a hollow underground nest made of dried vegetation and insulated with whatever is available, such as eider down, Muskox wool (qiviut), moss, fluff from Arctic cotton grasses or shredded leaves. Their ears open and are functional by day 11 and their eyes open the following day. By day 15 they are fully furred and they are abruptly weaned when their mother abandons them in the maternal nest, between 15–20 days of age. The juveniles leave the

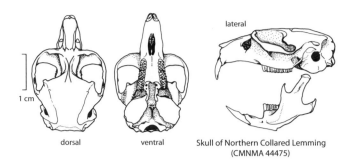

dorsal   ventral   Skull of Northern Collared Lemming
(CMNMA 44475)

1 cm

natal burrow up to 10 days later. Pups born late in the season moult from their grey-brown subadult coat into a white winter coat and their digging claws begin to grow before they are weaned. Young born earlier grow a dull brownish grey summer coat following their subadult coat, before moulting again in the autumn into the white winter pelage. Both sexes of juvenile generally become sexually mature around three months old.

BEHAVIOUR

As lemming numbers begin to increase, mature female lemmings start to abandon their home ranges to their latest litter and stake out a new territory for themselves some distance away. As adults, they stand a far better chance of acquiring and defending a new territory than do their callow and much smaller offspring. By providing their young with a territory, they ensure a higher juvenile survival rate. But when lemming numbers are low, the same females wean their young by simply moving to another burrow within their current territory. A low population density means that there are plenty of unoccupied territories nearby for the juveniles, so she does not have to sacrifice her space for them. Dispersing juveniles can travel 600 m or more in a single day, but most move less than 100 m to secure a territory of their own. Females defend their home range from other females and only allow a male's approach when in heat. Male home ranges overlap considerably with those of neighbouring males. Aggressive interactions ensure that larger females occupy the prime habitat, while smaller females are forced into less optimal areas. Natal burrows tend to be situated on the periphery of a female's range.

VOCALIZATIONS

Newborns produce high-pitched piping squeaks. A cornered adult will rise on its hind feet and squeal and attempt to bite its adversary, be it another lemming, an Arctic Fox, or even a human.

SIGNS

All the collared lemmings (*Dicrostonyx*) and the Brown Lemming (*Lemmus*) produce similar and indistinguishable sign, which are also very similar to that of large voles. The pollex (thumb) is vestigial and carries a diminutive, thick nail that rarely leaves an impression in a track. Toes 2 and 5 of the front feet angle to the side and toes 3 and 4 point forward. Claws do not always register. Tracks of the ribbed toes are long and slender and often connect to

the palm pad. Typical gaits are a trot, where the hind feet register directly over the front track (called a direct register trot), and a hop or bound, which is used when the animal is exposed or pursued. See the "Signs" section of the Meadow Vole account for a track schematic that applies to all the voles and lemmings. Scat is mouse-like and uncommonly seen above ground except in spring as the snow burrows melt and nesting and latrine chambers become visible. These droppings are invariably water-logged and swollen. Most summer faeces and urine are deposited in an underground latrine. This is thought to be a protective behaviour. Some raptors that can see ultraviolet wavelengths see urine as a bright colour, and in the barren, open Arctic environment where the vegetation is too short to hide the stains, the lemmings have learned to hide them below ground.

REFERENCES

Bety et al. 2002; Blackburn et al. 1998; Boonstra et al. 1996; Brooks 1993; Ehrich et al. 2001; Elbroch 2003; Gilg 2002; Gilg et al. 2006; Gruyer et al. 2008; Klein and Bay 1991; Nagy and Gower 1999; Predavec and Krebs 2000; Schmidt et al. 2002; Sittler et al. 2000; Wilson, D.J., and Bromley 2001; Wilson, D.J., et al. 1999.

Ungava Collared Lemming

also called Labrador Collared Lemming,
Ungava Varying Lemming

FRENCH NAME: **Lemming d'Ungava**
SCIENTIFIC NAME: *Dicrostonyx hudsonius*

This is one of three species of collared lemmings that are endemic to Canada.

DESCRIPTION

Ungava Collared Lemmings share the same body shape as other collared lemmings: a stout body, short legs, short tail, and short ears, and Toes 3 and 4 of the front feet develop thickened cornified pads and wider nails in winter to help dig into hard, wind-driven snow drifts. Like most other collared lemmings (genus *Dicrostonyx*), Ungava Collared Lemmings moult in autumn into a thick, long-haired, white winter coat. Each white hair has a dark base, giving the pelage a hint of a grey cast. The white coat is moulted in spring. The summer pelage is more colourful. A thin dark line runs along the back from nose to tail. The back and upper sides are a grizzled grey-brown. A pale yellow-orange line separated the buff-grey belly fur from the back colouring. The belly hairs are dark grey at their base, which imparts a distinct greyish cast to the fur. Yellowish-brown to rusty-brown patches typically occur on each cheek in the region where the vibrissae (whiskers) insert. A dull reddish to dark orange

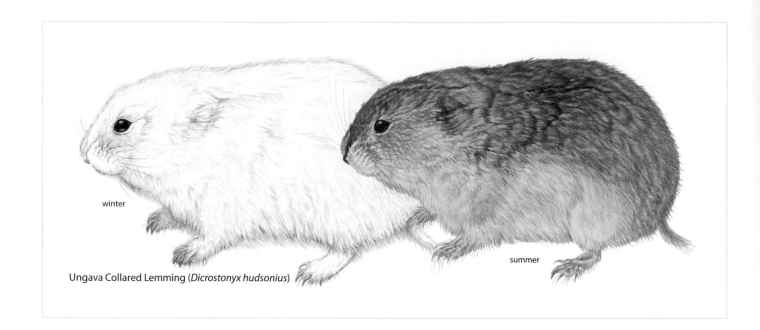

winter

Ungava Collared Lemming (*Dicrostonyx hudsonius*)

summer

patch of fur covers the ears. Typically, a similar dark orange patch (the collar) covers the throat and may attenuate and extend along the midline of the belly for some variable distance, but not usually more than halfway to the tail. A bright patch of orange fur often occurs on the outer portion of the upper front legs. Most adults develop an indistinct narrow pale band of fur that extends from the shoulder to the nape of the neck, running just behind the ears. The tail is dark above and lighter below, with a tuft of stiff white hairs at the tip. Juveniles are duller and greyer than the adults and the dark dorsal stripe is clearer on their shorter coat. The dental formula is incisors 1/1, canines 0/0, premolars 0/0, and molars 3/3, for a total of 16 teeth.

SIMILAR SPECIES
There are no other lemmings (*Dicrostonyx* or *Lemmus*) in Ungava. The white winter coat distinguishes this lemming from other rodents that share its range, such as Meadow Voles, Northern Bog Lemmings, Southern Red-backed Voles, and Heather Voles. Furthermore, none of these rodents have a thin, dark stripe running from nose to tail along their back. Appendix 2 provides information and illustrations to distinguish the skulls and teeth of all the voles and lemmings.

SIZE
These measurements are taken from adult animals. Males and females are similar in size and colouring.
Total length: 135–167 mm; tail length: 13–20 mm; hind foot length: 20–24 mm; ear length: 9 mm.
Weight: 55–112 g; newborns, likely around 4 g.

RANGE
This collared lemming is restricted to the northern tundra portion of the Ungava Peninsula in Labrador and Quebec and is also found offshore in Hudson Bay on the Belcher, King George, Sleeper, and Ottawa Islands as well as Smith Island, Long Island, Elsie Island, Christie Island, and Diggs Island. The species survived glaciation to the southeast of the last great ice sheet (the Wisconsinan Glaciation about 10,000 years ago), when it inhabited the eastern continent as far south as West Virginia. It is probable that, moving gradually northwards behind the retreating ice, Ungava Collared Lemmings from farther south eventually recolonized Ungava.

ABUNDANCE
Ungava Collared Lemmings may be scarce to common. Like all the collared lemmings, this species exhibits recurring population fluctuations that peak and crash periodically, but the length of time required for the cycle is unknown. These cycles are not necessarily synchronous throughout the range. But nearby locations (such as

Distribution of the Ungava Collared Lemming (*Dicrostonyx hudsonius*)

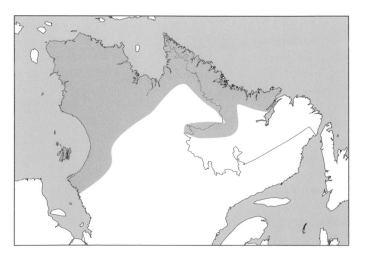

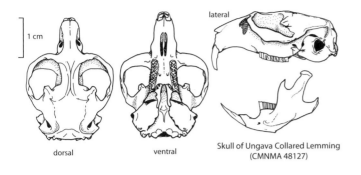

1 cm

dorsal   ventral

lateral

Skull of Ungava Collared Lemming
(CMNMA 48127)

VOCALIZATIONS
Unknown.

SIGNS
Tracks and other signs made by all the lemmings, both collared (*Dicrostonyx*) and brown (*Lemmus*) are similar and indistinguishable. See the "Signs" section of the Meadow Vole account for a track schematic that applies to all the voles and lemmings.

REFERENCES
Banfield 1974; Brooks 1993; Desrosiers et al. 2002; Engstrom 1999a; Lauriol et al. 2003.

neighbouring islands with little likelihood of gene exchange) may share a similar cycle.

ECOLOGY

The natural history of Ungava Collared Lemmings is not well known. They inhabit dry lichen-heath tundra on old beach terraces, sandy or rocky ridges, and dry alpine meadows above the timberline. Their habitat in Ungava (along the northern treeline and in alpine areas near the timberline) is commonly a mosaic of dwarf-shrub woodlands and tundra. They will travel through the woodlands to get to the tundra patches, especially during years of high population density. Like other collared lemmings, they burrow into soil and snow to create refuges. Summer burrows are fairly simple, less than a metre in length, and with one, or at most two, entrances. Winter burrows in snow may be more extensive. Each lemming has more than one burrow within its home range and each burrow contains a round nest constructed of dried vegetation. They spend most of their time in their burrows and when outside are either foraging or, if male, travelling in search of receptive females. These small rodents are prey for a wide variety of tundra birds and mammals. Snowy and Short-eared Owls, several species of jaegers and gulls, Rough-legged Hawks, Peregrines and Gyrfalcons, Arctic Foxes, weasels, and even Grey Wolves will kill and eat lemmings. Generally, an abundance of lemmings translates into a good year for the predators, as they can achieve high reproductive success and juvenile survival due to the easy hunting and plentiful food supply. Longevity is unknown.

DIET

Ungava Collared Lemmings are herbivorous. Although their diet is not well studied, burrows are usually found in the vicinity of food plants favoured by all the collared lemmings: willow, Arctic Avens, and blueberry as well as a variety of other Arctic forbs. Sedges and willow bark are known winter foods.

REPRODUCTION

This lemming is capable of breeding throughout the year, but fall and winter reproduction is most common during the expansion phase of the population cycle. Most reproductive details are unknown. Four to five pups per litter are thought to be average.

BEHAVIOUR

Unknown.

Ogilvie Mountains Collared Lemming

FRENCH NAME: **Lemming du Yukon**

SCIENTIFIC NAME: *Dicrostonyx nunatakensis*

Long thought to be a subspecies of the Northern Collared Lemming, this form has been granted full species status very recently, based on physical and distributional characteristics. Genetic studies remain necessary to verify this distinction.

DESCRIPTION

During summer, adults are pale grizzled greyish brown on their back and sides. Belly hairs are silvery or buffy at their tips and dark grey at their base. A narrow dark stripe runs down the centre of the back from forehead to base of tail. Adults have a rusty throat patch that may attenuate and extend some distance down the midline of the belly and may also extend onto the side of the chest and the upper part of the front leg. The extremely short tail is dark above and lighter below, with a white tip and a tuft of longer

summer

Ogilvie Mountains
Collared Lemming
(*Dicrostonyx nunatakensis*)

Distribution of the Ogilvie Mountains Collared Lemming
(*Dicrostonyx nunatakensis*)

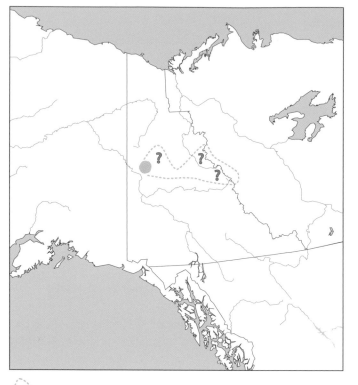

Total length: 110–129 mm; tail length: 11–12 mm; hind foot length: 15–21 mm; ear length: 3–6 mm.
Weight: unknown for adults and newborns.

RANGE

The distribution of this recently recognized species is not well understood and there are few specimens in museum collections. They have so far only been found on two peaks in the Ogilvie Mountains of northcentral Yukon, north of Dawson (Angelcomb Peak, 32 km S of Chapman Lake, and an adjacent peak 22.5 km S of Lomond Lake). Both these collecting localities are near a highway. Since this is a difficult region to survey beyond the road access, the possibility exists that the distribution may also extend into similar nearby habitat in the Wernecke and Selwyn Mountains. The Ogilvie Mountains Collared Lemming is one of only a handful of mammals whose range is wholly within Canada.

ABUNDANCE

Listed on the IUCN "Red List," where it is considered "Data Deficient," this species is likely vulnerable due to its limited distribution and specialized habitat requirements. The high-elevation tundra where it lives is probably very susceptible to the effects of global warming.

ECOLOGY

This is a very poorly studied rodent known from only eleven specimens housed at the Canadian Museum of Nature in Ottawa and a single specimen at the Royal Ontario Museum. These were found in rocky alpine tundra and talus considerably south of the High Arctic Tundra Zone. Youngman, the discoverer of the species, supposed that they represent a relict population that survived on isolated nunataks (peaks surrounded by glacial ice) at the southern limits of the ice sheet of a glaciation that occurred well over 10,000 years ago.

DIET

Unknown.

REPRODUCTION

Unknown.

BEHAVIOUR

Unknown.

whitish hairs at the tip. Hair around the ears of adults is rusty yellow and dark grey on juveniles and subadults. An indistinct band of lighter hairs extends around the neck from the throat to about the back of the ears and sometimes to the nape. Juveniles are greyer and slightly darker than adults. Claws are pale brown and eyes are black. Ears are very short and mostly hidden in the fur. All known specimens are in summer pelage. The winter fur colour of this lemming is unknown. Since this species is closely related to other collared lemmings, it is highly likely that Ogilvie Mountains Collared Lemmings moult twice each year and change colour to white in winter, but this has not been proved. The dental formula is incisors 1/1, canines 0/0, premolars 0/0, and molars 3/3, for a total of 16 teeth.

SIMILAR SPECIES

Northern Collared Lemmings and Richardson's Collared Lemmings are more colourful and do not occur within 400 km of the known distribution of Ogilvie Mountains Collared Lemmings. To distinguish vole and lemming skulls and teeth see appendix 2.

SIZE

Males and females are similar in size and colouring. The following measurements were taken from only four adult animals (CMNMA collection) and one large subadult (ROM collection).

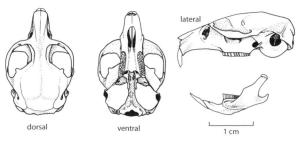

Skull of Ogilvie Mountains Collared Lemming
(CMNMA 29500)

VOCALIZATIONS
Unknown.

SIGNS
Front and hind feet have five toes; however, the pollex (thumb) is vestigial, with a short thick nail that does not show in a track. Although there are no specimens in winter pelage, it is likely that like all the other collared lemmings, the third and fourth toes on the forefeet develop paddle-like cornified extensions to their toes that act as shovels to dig through the hard windblown snowdrifts. These stop growing in late winter and wear away by the time spring arrives. See the "Signs" section of the Meadow Vole account for a track schematic that applies to all the voles and lemmings.

REFERENCES
Banfield 1974; Nagorsen 1998; Youngman 1967a, 1975.

Richardson's Collared Lemming
also called Richardson's Hoofed Lemming

FRENCH NAME: **Lemming de Richardson**

SCIENTIFIC NAME: *Dicrostonyx richardsoni*

No other northern rodents except collared lemmings (genus *Dicrostonyx*) develop a white winter pelage.

DESCRIPTION
Like all the collared lemmings (genus *Dicrostonyx*), Richardson's Collared Lemmings display many physical adaptations that suit them to the life of a burrowing mammal in a cold northern climate. They have a compact, stocky, torpedo-shaped body with short legs, a short tail, small eyes, and short, well furred ears that do not extend beyond even the shorter summer coat. Toes 3 and 4 on each front foot develop thickened cornified pads and larger claws during the winter that act as shovels to assist in digging into hardened, wind-packed snow drifts. They experience two annual moults. The spring moult begins on the head and proceeds rearwards. The autumn moult begins on the belly and moves towards the back. Hairs of the winter coat are about twice the length of those in the summer coat. Each winter hair is white for most of its length, but is dark grey at its base, providing a slight greyish cast to the otherwise pure white coat. The summer pelage is more colourful. A narrow dark line runs along the back from the nose to the base of the tail. This line is most apparent on the shorter-haired juveniles. The upper sides are grizzled dark reddish brown. The reddish colour is the result of a ruddy wash on the tip of the hairs. The belly and lower sides are buff-coloured, often washed with variable amounts of rufous, especially along the midline and the area on the sides where the belly hairs and the back hairs merge. All the belly hairs are dark grey at their base.

A rufous throat patch (collar) may extend up the sides of the neck and wash onto the top of the front legs. Mature adults often develop an indistinct pale band that extends from the shoulders to the nape, running just behind the ears. Fur over the ears is typically dark rufous. The short tail is dark above and buff below, with a white tuft of hairs at the tip. Feet are buff-coloured, sometimes with a darker brown upper surface, and the claws are light tan, often with a darker streak along the upper edge in winter. Juveniles are darker and less colourful than adults. The dental formula is incisors 1/1, canines 0/0, premolars 0/0, and molars 3/3, for a total of 16 teeth.

SIMILAR SPECIES
All the collared lemmings are similar, especially when in their white winter coat. The summer pelage is often a little more distinct. Ungava Collared Lemmings and Ogilvie Mountains Collared Lemmings do not occur within the known range of Richardson's Collared Lemmings. Northern Collared Lemmings occur along the northern border of the Richardson's range near Baker Lake and Chesterfield Inlet and can be quite difficult to distinguish. In general, the northern species has a grizzled black and light back with brighter highlights of reddish, orange, and white on their head, sides, and belly. Brown Lemmings that occur throughout the Richardson's range are generally found in more moist habitats. They are less grizzled on the back, considerably more rufous on the hind quarters, and lack the dark mid-dorsal stripe. Appendix 2 provides information and illustrations to distinguish the skulls and teeth of all the voles and lemmings.

SIZE
Males and females are similar in size and colouring.
Total length: 125–172 mm; tail length: 9–27 mm; hind foot length: 15–23 mm; ear length: 3 mm.
Weight: 64–120 g, newborns, around 4 g.

Distribution of the Richardson's Collared Lemming (*Dicrostonyx richardsoni*)

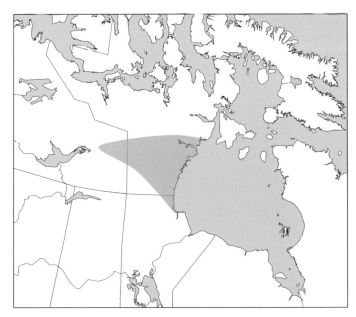

winter

summer

Richardson's Collared Lemming (*Dicrostonyx richardsoni*)

RANGE

Richardson's Collared Lemmings occur on the western side of Hudson Bay in extreme northeastern Manitoba, southeastern Northwest Territories, and southeastern Nunavut. They are one of three collared lemming species (genus *Dicrostonyx*) that are endemic to Canada.

ABUNDANCE

These lemmings are reasonably common, although density varies from scarce to abundant, depending on the stage of the population cycle.

ECOLOGY

Richardson's Collared Lemmings experience a three-to-four-year extreme population cycle. One study monitoring a 15 ha plot near Churchill, Manitoba, found 40 lemmings/ha as the population peaked, and approximately 1 lemming in 15 ha during low densities. These population fluctuations are localized and do not occur concurrently throughout the range, nor are they apparently determined by weather, predator abundance, or feed availability. Rate of reproductive success, rate of juvenile mortality, and higher rates of infanticide at higher densities have all been implicated as factors that stimulate the cycles. Richardson's Collared Lemmings prefer windswept, dry surroundings such as occur on sand, gravel, or rocky ridges, on raised beach ridges, and in drier sections of lichen-heath tundra. But during peak years, when lemming density is high, a variety of tundra habitats are inhabited, even marginal ones in wetlands and scrub woodlands. These lemmings live in burrows that they dig into snow, sand, gravel, peat, clay, and even mud. Summer burrows are simple and typically less than a metre in length, with one or two entrances. Maternal burrows may be somewhat more complex, as the female creates several entrances and, typically, a latrine chamber near the nest chamber. Winter burrow systems in the snow can be extensive. Each animal has several burrows within its home range. These are scattered to provide safe

refuge or a secure resting site. Females establish exclusive home ranges, but male ranges commonly overlap with those of several females and with other males. Home ranges of males are larger than those of females, except during the peak of population density, when high numbers of lemmings results in smaller territories for all. Home-range size varies from 0.18 to 1.77 ha for males and 0.07 to 0.42 ha for females. Unlike most other small rodents that live under the snow and remain active throughout the winter, collared lemmings do not lose weight during the cold season. Instead, they continue to grow and can even regularly and successfully reproduce under the snow. Like most animals that live all year in the north, they experience periods of prolonged darkness and daylight depending on the season and have adapted to that by adopting a 24-hour active period, rather than the more usual diurnal or nocturnal cycle. They take periodic rests as needed. Principal predators include Short-eared and Snowy Owls, Arctic Foxes, and Ermine. Other predators such as Rough-legged Hawks, Parasitic and Long-tailed Jaegers, and Least Weasels also take many lemmings, especially when lemming numbers are high. Captive lemmings have lived more than two years, but few in the wild survive longer than two seasons (around seven months on average).

DIET

These rodents are almost exclusively herbivorous. They eat leaves and stems as well as flowers, roots, berries, and occasionally bark. Their preferred food plants are shrubs and forbs such as willow, Arctic Avens, blueberry, and bearberry. Free water is consumed when available.

REPRODUCTION

Reproduction can take place at all times of the year, but fall and winter breeding typically only occur during periods of population increase. Furthermore, during this phase of the population cycle, most females will undergo a postpartum oestrus and will mate within 24 hours of whelping, so they are pregnant and nursing

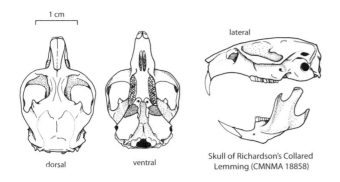

Skull of Richardson's Collared
Lemming (CMNMA 18858)

dorsal   ventral   lateral

1 cm

concurrently. This increase in litter production can rapidly enhance the numbers of lemmings. Under ideal conditions in captivity, one pair of lemmings produced 17 rapidly consecutive litters. Breeding drastically declines at the peak of population density and while the cycle is in the crash phase.

The gestation period averages 20–21 days and litter size varies from 1 to 8 and averages 3–4 pups. Newborns are blind, deaf, and hairless (apart from a few short whiskers), but they grow quickly. Their ears open and are functional by day 11 and their eyes open the following day. By day 15 they are fully furred, and they are abruptly weaned when their mother abandons them in the maternal nest between 16 and 22 days after their birth. She digs, or moves, to a new maternal burrow and soon whelps her next litter. Both sexes of juvenile generally become sexually mature around 3 months old, but young females can potentially breed as early as 25–30 days old, especially if they are winter born, as they are better nourished and grow larger than do the young born in summer. In captivity, males are known to assist the maternal female with the litter, but this does not appear to be the case in the wild, where females provide all the parental care.

BEHAVIOUR

Females are territorial and defend their territory from all other lemmings, except while in heat, when they will permit the approach of a male. Otherwise, the males are driven off. Males are non-territorial and occupy large home ranges, which commonly overlap with those of other males. Furthermore, males are quick to adjust the boundaries of their range to suit the location of nearby females. Mature males habitually make extended overland excursions, searching for and then regularly visiting the females in their area to check on their receptivity. One radio-tagged energetic male, during a period of low population density, several times travelled over 3 km within a 24-hour period while feeding and searching for receptive females. During periods of low population density, Richardson's Collared Lemmings are generally non-aggressive towards each other and, except for breeding, are most likely to actively avoid interactions. The adults become less tolerant of each other as the population density rises and home ranges shrink, so aggressive encounters and chases become more common. Lactating females will almost always chase an ardent male away from their territory, except during the postpartum oestrus, which typically occurs within 24 hours of whelping. Females invariably change burrows between litters, and

most maternal burrows are on the periphery of their range, perhaps as a predator avoidance strategy. Richardson's Collared Lemmings will commit infanticide, especially of an undefended litter. Although the culprit has never been observed, it is thought to be perpetrated by both males and females and to be more common at higher population densities. These rodents are very alert when foraging, as there is often little cover available in their northern habitat. They have good eyesight and can detect the movement of a predator up to 100 m away in daylight. Each lemming knows exactly where its nearest burrow entrance is located and immediately bolts for safety if threatened.

VOCALIZATIONS

Newborns persistently produce high-pitched squeaks. A cornered adult will rise on its hind feet, squeal, and attempt to bite its adversary, be it another lemming, an Arctic Fox, or even a human.

SIGNS

Tracks and other sign made by all the lemmings, both collared (*Dicrostonyx*) and brown (*Lemmus*) are similar and indistinguishable. See the "Signs" section of the Meadow Vole account for a track schematic that applies to all the voles and lemmings.

REFERENCES

Brooks 1993; Engstrom et al. 1993, 1999b; Gajda et al. 1993; Malcolm and Brooks 1993; Reiter and Andersen 2008; Scott, P.A., 1993; Shilton and Brooks 1989.

Sagebrush Vole

also called Sage Vole, Pallid Vole

FRENCH NAME: **Campagnol des armoises,** *also*
Campagnol des sauges

SCIENTIFIC NAME: *Lemmiscus curtatus,* formerly *Lagurus curtatus*

As its name implies, this small, nondescript vole is closely associated with sagebrush. Many aspects of its natural history remain obscure.

DESCRIPTION

Sagebrush Voles are greyish tan on the back and sides, with sides that are usually paler than the back and rump. Their overall sandy colour is made up of grizzled hairs that are grey-brown at their roots with buffy or black tips. Belly hairs are also grey-based, but have pale buffy to silver-grey tips. The bicoloured, short tail is brown above and pale grey or buff below, with a few longer hairs at the tip. Feet are buffy or light grey on the upper surface. The short, rounded ears protrude only slightly beyond the long fur. Adult males develop hip glands, which may be difficult to see, although the hair that grows over the gland is usually shorter and finer. The dental formula

Sagebrush Vole
(*Lemmiscus curtatus*)

is incisors 1/1, canines 0/0, premolars 0/0, and molars 3/3, for a total of 16 teeth. These voles have four pairs of mammae. There are two annual moults, one in spring and the other in autumn, and the new winter coat is somewhat darker than the summer coat being replaced.

SIMILAR SPECIES

Both Meadow Voles and Long-tailed Voles occur within the Sagebrush Vole range, but both are larger, darker, and have longer tails. Typically, the Sagebrush Vole is the only vole in its habitat in Canada. Skull and teeth characteristics, useful in discriminating voles, are illustrated in appendix 2.

SIZE

Males and females are similar in size and colouring. The following measurements apply to adult animals in the Canadian part of the range.
Total length: 110–135 mm; tail length: 16–23 mm; hind foot length: 16–19 mm; ear length: 9–11 mm.
Weight: 22.2–30.5 g; newborns, about 1.5–1.6 g.

RANGE

This is an endemic western North American species that occurs from southern Alberta and Saskatchewan to central Oregon, eastern California, and northern Colorado. Populations are scattered through the range, as Sagebrush Voles only occur in certain soil types.

ABUNDANCE

Sagebrush Vole populations experience density fluctuations that appear to be related to weather conditions. Warm winters, moist summers, and early autumn rains encourage population growth in these small mammals. Overall, this rodent is uncommon in Canada, but may be more common in more southerly portions of the range.

ECOLOGY

These voles occur in semi-arid prairie, rolling hills, dry streambeds, and coulees where sagebrush is plentiful. Loose, well-drained soil is essential, although the surface of the ground may be rocky. Sagebrush Voles can be active at any time of the day or night, but most

activity apparently takes place in the hours around sunrise and sunset, although this pattern may vary with season. Their external activity is more governed by wind conditions than light levels, as they rarely venture out of their burrows if it is windy. They do not hibernate, but remain active under the snow. Although Sagebrush Voles usually dig their own burrows, they will also readily take over and modify abandoned Northern Pocket Gopher burrows. Burrows lead downward to a level stretch just under the surface, then slope downward 8–46 cm to a level where horizontal tunnels are numerous. Tunnels are often floored with grass and sagebrush clippings. Nest chambers are usually 8–25 cm below the surface, have at least 2, and as many as 6 entrances, and are 10–20 cm by 10–25 cm in size, depending on the soil structure. Nests are constructed of stems and leaves of grasses and sometimes include feathers and paper and shredded sagebrush bark. Burrow systems commonly have multiple entrances and may be more than 4.5 m in total length. The home range size of Sagebrush Voles has so far been unreported. Average density of a population in Idaho ranged from 4–16 voles/ha over a 13-month period, with the peak in summer and the low in fall to early spring. Fleas associated with Sagebrush Voles have been responsible for several outbreaks of plague in North America. This infectious disease, caused by the bacterium *Yersinia pestis*, was inadvertently introduced to California from Asia in the early 1900s and has been reported in at least 76 different wild North American mammals. Fortunately, there have been few cases of transmission to humans (its most common form is called bubonic plague). Most cases are reported from the midwestern and southwestern United States. Owls are their primary predators, although predation by

Distribution of the Sagebrush Vole (*Lemmiscus curtatus*)

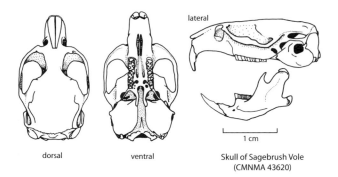

dorsal   ventral   Skull of Sagebrush Vole
(CMNMA 43620)

1 cm

snakes, weasels, Bobcats, and even shrikes has been reported. Likely domestic pets also capture some.

DIET

Sagebrush Voles are apparently strictly herbivorous, eating mainly forbs and grasses. They prefer to eat the tender leaves and stems, but will also consume the flowers and mature and green seeds of many forbs, as well as some fungi. They are not known to eat insects. While they rarely consume the leaves or bark of sagebrush, they will eat the flowers. Animals eating succulent vegetation do not need access to free water.

REPRODUCTION

This small rodent is capable of breeding year-round, although in the northern part of its range (including all of its Canadian range), breeding is normally restricted to March through early December. They may experience a breeding hiatus during a hot, dry summer. At least three litters may be produced in a breeding season. Gestation lasts an average of 25 days, and females experience a post-partum oestrus within 24 hours of parturition. Litters average four to six pups (range 1–11). Newborns are wrinkled, pink, and helpless. Their toes and claws are fully formed, but the toes are still connected by an interdigital membrane. By their seventh day, they are finely, but fully furred and their toes have separated. They respond to sounds by day 10, and begin eating solid food between 12 and 16 days old. Their eyes open at around 15 days old, and they can walk without wavering, but continue to attach themselves to a nipple and are still being dragged around by their mother. The pups are fully weaned by 17–21 days old and leave their natal nest to build one of their own. Growth rate of the pups is determined by the amount of milk they receive. Singleton pups grow faster than those from multiple litters and the larger litters have the smallest pups. Although females are thought to reach sexual maturity around 60 days old on average, there are reports of oestrous females as young as 30 days old. Males become sexually mature between 60 and 75 days old.

BEHAVIOUR

The clustering of burrows suggests that Sagebrush Voles are colonial, although it is possible that this social activity takes place more during winter than summer. Some researchers have found breeding pairs occupying a burrow system and suggest that family groups remain together as juveniles mature, but this has not yet been substantiated. Burrow systems may be periodically abandoned as the available forage is consumed. The voles move to nearby unused systems where the vegetation has been untouched. Apparently, the burrow systems may be repeatedly abandoned and reused as necessary. These voles do not store food, apart from the small piles gathered at a sheltered feeding station, or near the entrance of a burrow, where feeding can safely be undertaken. They must forage daily, winter and summer. Based on the degree of wounding seen during live capture studies, males are thought to fight with each other during the breeding season. They actively defend their territory from other mature males and mark it by scratching their flank glands with their hind feet, which spreads the odour around as they travel. The degree of communal behaviour is difficult to determine. Wild females are reported to block the entrances to the nest chamber where they leave their nursing young when they depart temporarily to mate during a post-partum oestrus. In one captive study, males were excluded from a natal nest, as they may kill and eat the young. In another, the male and female congenially shared a nest with the suckling young, and the male even provided occasional parental care by grooming the pups. Captive females are known to communally nurse litters born to other females within the colony. Sagebrush Voles are difficult to trap in conventional live traps, but can be successfully captured in specially designed live traps (Maser, 1967).

VOCALIZATIONS

Not reported.

SIGNS

Burrow entrances are 4–5 cm in diameter and usually located under cover of sagebrush, rocks, or debris. Burrows tend to occur in clusters, with 8–30 entrances hidden under the same piece of cover. Runways are 6–8 cm wide (wider than those of sympatric *Microtus*) and are typically indistinct unless the vegetation is thick. Unlike other small mammals, Sagebrush Voles do not carefully clear and clip all the vegetation in their runways. Some plants continue to grow in the runways, which may almost obstruct them in places. Most winter runways are in snow tunnels at the ground surface. Fresh scat is distinctly green, weathering to a brownish tan. Scat may be scattered along the runways, but is most often found in latrine sites just outside the burrow entrances or under cover along a runway. These piles can measure 10–20 cm long by 5–10 cm wide by 2.5–8.0 cm deep. Freshly clipped vegetation may be gathered into small piles at the burrow entrances or in protected spots along the runways. Short dead-end tunnels (up to 20 cm in length) frequently open off runways and may be used for escape, or as protected feeding stations. These can even occur in and under a dried cow-patty when other cover is scarce. Like other voles, Sagebrush Voles exhibit a four-toed front track, as the thumb is greatly reduced and rarely leaves an imprint. The hind track has five toe prints. Tracks are very similar to those reported for the Meadow Vole (see the Meadow Vole "Signs" section for further information).

REFERENCES
Carroll and Genoways 1980; Hill et al. 2001; Hofmann et al. 1989; Jameson and Peeters 2004; Jannett 1981a; Maser 1967; Mullican and Keller 1986, 1987; Smith, H.C., 1993; Verts and Carraway 1998; Zeveloff 1988.

Nearctic Brown Lemming

FRENCH NAME: **Lemming brun**

SCIENTIFIC NAME: *Lemmus trimucronatus,* formerly *Lemmus sibericus*

Although closely related to the famous mass-migrating Norwegian Lemmings (*Lemmus lemmus*), Nearctic Brown Lemmings do not migrate en masse during periods of superabundance. In fact, none of our North American lemmings exhibit this unusual behaviour.

DESCRIPTION

Like other lemmings, Nearctic Brown Lemmings have stocky, ovoid bodies, short legs, and a very short tail (< 20% of total length). Pelage colour and the extent of the colour vary geographically. In general, adults have a grizzled, brownish head, a band of light orange or buff fur around the lower sides, an even lighter belly, and a rump of a rust colour that may be as small as a discrete patch or that may extend to suffuse the entire back and rump. Guard hairs on the back are grey-brown to black at their base. Tips of the same hairs over the head and shoulders are a mix of dark brown and light grey to buff, resulting in a grizzled look, while those over the back and rump are tipped with tan to reddish orange. The cheeks and flanks are orange and the chin, throat, and belly are pale orange to buffy orange to light grey. Belly hairs have a grey base. The tail is usually bicoloured, darker above and lighter below with a group of long, stiff, pale hairs at the tip. These lemmings do not become white in

Nearctic Brown Lemming (*Lemmus trimucronatus*)

winter. Their pelage colour does not change appreciably from summer to winter, but winter fur is longer, especially over the back and rump, and is somewhat greyer. Younger animals tend to be darker and duller than adults. Ears are short and do not extend beyond the fur. Feet are buffy to dirty grey and claws are tan or brown. The dental formula is incisors 1/1, canines 0/0, premolars 0/0, and molars 3/3, for a total of 16 teeth. The anterior surface of the upper incisors is lightly grooved longitudinally. Juveniles undergo four moults in their first few months, terminating with a moult into an adult pelage in autumn. Adults moult into their winter coat in autumn and into their summer coat in spring.

SIMILAR SPECIES

This colourful and relatively easy to identify lemming cannot be confused with any other small rodent over most of its range. In eastern Manitoba, it co-occurs with Richardson's Collared Lemmings. These lemmings have a white winter coat, while Brown Lemmings remain brown all year round. Richardson's also develop a distinctively thickened front claw in winter. They are similar in size and colouring to Brown Lemmings, but their whole back is grizzled and they have a dark rufous wash (just on the very tips of the longer guard hairs) that is strongest over the shoulders. Most (except in winter pelage) have at least a suggestion of a dark line running from the nose to the rump, which is most distinct in juveniles. To distinguish the teeth and skull of Nearctic Brown Lemmings from those of other similar rodents see appendix 2.

SIZE

Males and females are similar in size and colouring. The following measurements apply to animals of reproductive age from the Canadian part of the range.

Total length: 122–170 mm; tail length: 8–29 mm; hind foot length: 13–26 mm; ear length: 9–16 mm.

Weight: 24.6–98.4 g; newborns, around 3.3 g.

RANGE

Nearctic Brown Lemmings occur as far north as Alaska and the lower islands of the Canadian Arctic Archipelago in the Northwest Territories and Nunavut. The southern limits of their distribution are in northeastern Manitoba, west-central Alberta, and the Rainbow Mountains north of Bella Coola, British Columbia. The current status of populations in northeastern Manitoba is uncertain. The population in west-central Alberta that was reported from Willmore Wilderness Park in 1985 has provided the only specimens from Alberta. This population may simply be disjunct, but it does put into question the absence of Brown Lemmings from the appropriate mountainous habitat between Alberta and the known boundary of distribution in adjoining British Columbia.

ABUNDANCE

Arctic populations of this species fluctuate on a three-to-five-year cycle and can go from superabundant to virtually non-existent over the course of a single year. In the High Arctic of North America, lemming populations can fluctuate dramatically from less than one

Distribution of the Nearctic Brown Lemming (*Lemmus trimucronatus*)

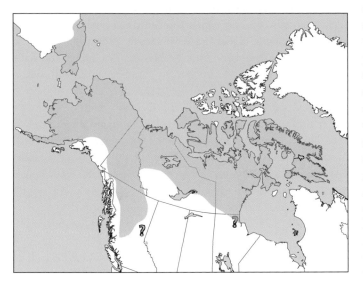

? presence uncertain

lemming per hectare to over 200/ha. The low phase of the cycle, when lemmings are difficult to find, can last two to four years. Anecdotal evidence suggests that large-scale irruptions do occur in alpine and subalpine populations farther south, but the regularity of such fluctuations is as yet unknown.

ECOLOGY

These lemmings occupy moist grass/sedge habitats in the Arctic, alpine, and subalpine zones of northwestern North America. They can be found in open meadows, dwarf shrub-sedge meadows, wetlands, and riparian habitats up to elevations of 1830 m. A radio-tracking study in Alaska revealed that male Brown Lemmings occupy larger home ranges than do females (0.01–2.85 ha for males as opposed to 0.02–1.68 ha for females). At intermediate densities, when populations are rapidly expanding, more movement of both adults and juveniles occurs and males of both age classes move greater distances than females. At the end of the breeding season in autumn, there tends to be more dispersal movement by all ages and sexes as the population adjusts to the higher density of animals. Although capable of long-distance travel (an individual was found on sea ice 16 km from shore in Alaska), Brown Lemmings are not as inclined to undertake long-distance movements over the snow or sea ice as are the closely related collared lemmings. So populations may be more likely to go extinct and not be repopulated by dispersal, especially in regions where their preferred wetland habitat is already scarce, such as in the central Canadian Arctic. Elsewhere in the range, genetic diversity is generally high despite the repeated genetic bottlenecks, as habitat is more widespread and dispersal between populations is more common. As one would expect based on preferred habitat, Nearctic Brown Lemmings are very capable and willing swimmers. They float fairly high in the water and have been seen swimming in meltwater ponds during the spring runoff.

Like other lemmings, Brown Lemmings dig burrow systems in soil in summer and in snow in winter. Summer burrows are often shallow tunnels that descend to a nest chamber. The depth of the burrows depends on local soil structure and the extent of the permafrost. Tunnels may descend up to 100 cm and nest chambers as deep as 60 cm if conditions permit, but may be only 5 cm deep in some locations where the permafrost is near the surface. Due to the cold regions where they live, the thermal properties of their nests are imperative for raising young and surviving a winter. Nests are globular, with a hollow centre. Temporary resting burrows in summer may have small nesting chambers with sparsely constructed globular nests, but winter nests and those used to rear young are much thicker-walled and more care is paid to the inner lining. Nests are constructed of plant material, often insulated with an inner lining of the lemmings' own moulted fur and are capable of increasing the insulative value of the lemmings' pelage by up to 46%. Predators include all the raptors and small-to-medium-sized carnivores within their range, such as jaegers, gulls, hawks, owls, foxes, Coyote, Grey Wolves, Ermine, and Least Weasels.

DIET

Nearctic Brown Lemmings are herbivorous, preferring to eat fresh vegetation all year round. In summer, they feed mainly on fresh green shoots and leaves of sedges and grasses. Their winter food is more restricted, so they must resort more to mosses, which they supplement with any frozen basal shoots of sedges and grasses they can find. Their harvesting activities in winter can create great amounts of clipped plant debris, which they commonly scatter along their runways. They do not appear to cache this material in any noticeable piles. One of the few mammals to consume large amounts of mosses (these plants are only about 25% digestible), Brown Lemmings spend large parts of their days foraging to acquire sufficient food in winter to keep warm. Their winter schedule is to forage for an hour or two, then rest for an hour or two, at roughly three-hour intervals, day and night.

REPRODUCTION

The onset of breeding is at least partly triggered by the growth of preferred plants, so most years, Nearctic Brown Lemmings begin breeding as the snow melts and sedges and grasses begin to produce fresh shoots and leaves. Chemicals in the plants apparently cue their

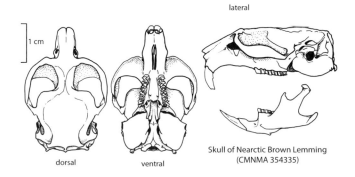

lateral

1 cm

dorsal   ventral

Skull of Nearctic Brown Lemming
(CMNMA 354335)

reproduction. Breeding under the snow may happen in the Arctic in some years, but does not take place during spring snowmelt (May and early June) or during snowpack formation (September and October). In most locations farther south, breeding ceases in October or earlier, depending on the onset of winter. Males do not assist in rearing the young. Adult females bear up to three litters annually and experience a postpartum oestrus shortly after parturition. Average litter size is around six to eight (range 3–12). Gestation lasts 21–23 days and the pups are born essentially hairless, pink, and blind. They develop very rapidly. Their eyes open when they are about 11 days old and they are weaned and leaving the nest by 16–17 days old. Juvenile females can breed as young as 3–4 weeks old, but are normally 5–6 weeks old at sexual maturity and may produce up to 2 litters in their first summer, if they were born in the first cohort of the season.

BEHAVIOUR

These small Arctic rodents are largely solitary. Adults maintain exclusive burrow systems and do not share a nest. The only enduring social bond is between a reproductive female and her pre-weaned young. Females dominate males in most social circumstances, except when they are in oestrus. Even so, females do manage to exercise some mate choice options, as they are more likely to mate with dominant males (whose status they can determine by their odour). Adult males appear to be decidedly intolerant of other males on their home ranges. Unlike collared lemmings (genus *Dicrostonyx*), Nearctic Brown Lemmings do not chatter their teeth during stressful situations.

VOCALIZATIONS

Like most of the lemmings, they squeak during agonistic encounters, but are otherwise relatively silent.

SIGNS

In their typical dense-grass and sedge habitat, the runways of these busy rodents are generally well worn and conspicuous and may persist for years after the population has crashed. Subnivean runways that have been littered with winter plant clippings may melt out as castings of the tunnel systems. Scat of the Nearctic Brown Lemming is greenish, rather than the more common brown or black of other voles and lemmings. Pellets are large, 5–10 mm in length, and capsule-shaped with blunt ends, and are generally deposited in scattered piles throughout the habitat. Winter nests (20–25 cm in diameter) are created near the base of the snowpack and at ground level and melt out in spring and summer as the snow disappears. These nests are made, used, and abandoned repeatedly throughout the winter, so nest counts do not form a reliable basis for estimates of population size. Sometimes an Ermine will capture and consume a lemming and then take over its nest, adding the fur of the deceased to the inner lining for added comfort and insulation. The front feet of a Brown Lemming have five toes, but toe 1 (the pollex) is much reduced and has a large, broad, flattened claw that rarely registers in a track. The other claws vary with season. Claws on toes 3–5 become larger and more recurved in winter. Hind feet have five toes, each with long curved claws that do not vary seasonally. All the collared lemmings (genus *Dicrostonyx*) and the Brown Lemmings (genus *Lemmus*) produce similar and indistinguishable sign that are also very similar to that of large voles. See the "Signs" section of the Meadow Vole account for a track schematic that applies to all the voles and lemmings.

REFERENCES

Banfield 1974; Banks et al. 1975, 1979; Batzli 1999; Casey 1981; Din 1981; Ehrich and Jorde 2005; Ehrich et al. 2001; Gruyer et al. 2008; Huck and Banks 1982; Jarrell and Fredga 1993; Krebs et al. 2002; Nagorsen 2005a; Negus and Berger 1998; Rogers, A.R., 1990; Slough and Jung 2007; Smith, D.A., and Foster 1957; Smith, H.C., 1993; Smith, H.C., and Edmonds 1985; Youngman 1975.

Rock Vole
also called Yellow-nosed Vole

FRENCH NAME: **Campagnol des rochers**
SCIENTIFIC NAME: *Microtus chrotorrhinus*

Rock Voles are medium-sized members of the *Microtus* genus, similar in size and proportion to the far more common Meadow Vole. As their name suggests, they are found in the vicinity of rock fields and talus slopes.

DESCRIPTION

Like all the voles in the genus *Microtus,* Rock Voles have a stocky body, a blunt head, and short, rounded ears that are a little longer than their fur. In Canada, their dorsal pelage is dark greyish to mahogany brown, with many black-tipped guard hairs and their

Rock Vole (*Microtus chrotorrhinus*)

Distribution of the Rock Vole (*Microtus chrotorrhinus*)

belly fur is dull grey, sometimes washed with white on the tips of the longer guard hairs. The fur on their snout from the nose to the eyes, and sometimes all the way to the ears, is a rusty orange. The colour is usually brightest where the whiskers insert on the upper lip and becomes less intense as it travels up the head. Their short tail is vaguely bicoloured, darker above and greyish below. Rock Voles within the Canadian portion of the range usually have light brown fur on the upper part of their feet. Juvenile pelage is generally darker and finer than that of adults. The dental formula is incisors 1/1, canines 0/0, premolars 0/0, and molars 3/3, for a total of 16 teeth. Each adult undergoes two annual moults, one in spring and early summer and one in autumn.

SIMILAR SPECIES

The primary species that may be confused with a Rock Vole in the wild is a Heather Vole, which occurs throughout most of the northern part of the Rock Vole range. Both may have an orange snout, but Heather Voles have smaller (usually < 19 mm), pinkish hind feet with white hairs on the upper surface, a slightly lighter, more yellowish-brown upper pelage and generally shorter tails (usually < 40 mm). Belly fur is typically lighter than that of Rock Voles, with more white-tipped guard hairs. Most Heather Voles in the east have a yellowish-brown to yellowish-orange snout that is not as rusty orange as that of Rock Voles. Some populations of Heather Voles are easier to distinguish as they do not develop a colourful snout at all. Juveniles of all populations of Heather Voles have darker pelage than the adults and a plain brown snout. Taiga Voles are similar, though considerably larger than Rock Voles, but the ranges of the two species do not overlap, as the Taiga Voles occur farther west. Meadow

Voles co-occur within the range of Rock Voles and, although they are similarly sized, lack the rufous snout of the Rock Vole and are found in grassy meadow habitats, rather than rock fields. The pattern of tooth enamel is a good characteristic to distinguish vole skulls. Appendix 2 outlines the skull and dental characters that distinguish voles.

SIZE

Males and females are similar in size and colouring. Weight fluctuates over the year and is lowest in mid–late winter and highest in late summer. The following measurements apply to adult animals in the Canadian part of the range.

Total length: 127–186 mm; tail length: 40–56 mm; hind foot length: 18.5–22.0 mm; ear length: 13–18 mm.

Weight: 19.0–57.9 g; newborns, about 3–4 g.

RANGE

Rock Voles are endemic to eastern North America, where they occur from northeastern Minnesota to southern Ungava and down through northern New Brunswick along the Appalachian Mountains. Two disjunct populations exist, one in the southeastern United States from North Carolina and Tennessee to northern West Virginia, and the other in northern Cape Breton Island. Because of its specialized habitat requirements (see in the "Ecology" section) this vole likely exists in isolated pockets throughout much of its range. Fossil evidence indicates that Rock Voles formerly occurred in some of the currently uninhabited areas bridging now disjunct populations. So this vole is a relict species, formerly more widely distributed, but why this is the case is unknown. It is entirely possible that the known range may expand as more sampling is done and more pockets of distribution are found, as has recently occurred in Labrador.

ABUNDANCE

Collection of this species is specialized as traps must be set between and under rocks in suitable habitat. The sampling difficulty and the isolated pockets of distribution could account for the scarcity of specimens in museum collections. Nevertheless, this vole is rarely abundant even in prime habitat.

ECOLOGY

Rock Voles are aptly named as they are found where rocks and talus slopes are conspicuous landscape features, especially where such rock is in close association with readily available water, usually a pond, bog, or a surface or subsurface stream. This cool, moist environment usually supports a profusion of mosses and lichens and an abundance of forbs, which provide both food and cover. Such sites may be found within a variety of open forest types (boreal, deciduous, coniferous, or mixed). Although most specimen records are from such moist habitats, Rock Voles have also been found in somewhat drier sites. The colonization ability of this vole has probably been underestimated in the past, as populations are found from time to time in unexpected habitats, such as disturbed areas that have been recently logged. Genetic studies support this dispersal capability, as many supposed isolated colonies remain relatively

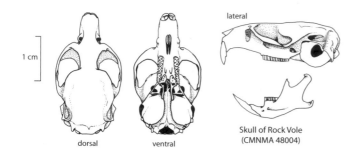

1 cm

dorsal   ventral

lateral

Skull of Rock Vole
(CMNMA 48004)

animals from early litters may breed later in the same breeding season, so at least females likely mature at around two to three months old. An adult male's testes shrink outside of the breeding season, but are not retracted abdominally.

BEHAVIOUR

Little is known of the behaviour of Rock Voles and they are difficult to maintain in captivity. Wild Rock Voles have been reported as being both mainly diurnal and active intermittently throughout the day and night. Whether their activity cycle varies with season or location is unknown, so this observed variation is unexplained. Captive females reportedly exclude males from their nest when young are present. Field observations of cut portions of plants in subsurface rock cavities occupied by Rock Voles suggest that they may cache food for later use.

VOCALIZATIONS

Unknown.

SIGNS

These voles leave few signs, as most of their runways are in thick vegetation or under rocks. Sometimes they attempt to pull whole plants into their burrows, leaving parts of the plant protruding above the opening if it is too large to fit. They appear to restrict both urination and defecation to latrine sites. One such site discovered deep in a rock crevice measured 500 mm in diameter and 240 mm in depth. Latrine sites are often on elevated rocks near the runway. During the preparation of trapped voles into museum specimens, it has been discovered that Rock Voles may have a number of pellets (up to 33) in their rectum awaiting deposit in such a latrine. Burrow entrances are likely comparable to those of the similarly sized Meadow Vole, which are about 2.5–3.5 cm in diameter. Snow tunnel entrances are around 1.9–2.2 cm in diameter. Similarly, Rock Vole runways are about the same dimension as those of Meadow Voles: 3.2–6.5 cm in width. In general, vole scat is straight-sided, with blunt ends and smooth surfaces, and is very regular in shape and size. Rock Vole scat is similar to that of Meadow Voles (0.2–0.3 cm in diameter and 0.4–0.8 cm in length). Tracks are very similar to those of other *Microtus* (see the "Signs" section of the Meadow Vole account).

REFERENCES

Christian and Daniels 1985; Cook 1984; Desrosiers et al. 2002; Dobbyn 1994; Elbroch 2003; French and Crowell 1985; Jannett et al. 2007; Jones, J.K., and Birney 1988; Kirkland and Knipe 1979; Kurta 1995; Lansing 2005; Martin, R.E., 1973; Merritt 1987; Nagorsen and Peterson 1981; Peterson 1966; Roscoe and Majka 1976.

diverse genetically. Rock Voles create subsurface runways within the rock or boulder field and under vegetative cover where they can move about undetected by aerial predators. Their burrow entrances are dug into soil trapped between or under the rocks. They are not as well protected from terrestrial predators and are recorded as prey of Bobcats and rattlesnakes (mainly in the United States, as there is only one rattlesnake that occurs within Rock Vole range in Canada – the Massasauga Rattlesnake along the north shore of Georgian Bay). Other likely predators include short-tailed shrews and the smaller weasels that can fit into their protected runways. Rock Voles remain active under the snow and do not hibernate. Southern Red-backed Voles commonly live in similar habitats and use the same runways as Rock Voles, without apparent conflict or competition. It is of interest that few of the far more common and abundant Meadow Voles occur in sites occupied by Rock Voles. Why this species separation occurs is unresolved, although competitive exclusion is suspected.

DIET

Rock Voles are primarily herbivorous, but become somewhat insectivorous during the summer. Bunchberry and Labrador Tea are major food plants, along with seeds, stems, and leaves of a variety of other forbs, mosses, and grasses. Insects, fungi, and some berries including those of raspberry, Bunchberry, Bakeapple, Blueberry, and Partridgeberry are consumed in season.

REPRODUCTION

The breeding season of this vole in Ontario and Quebec begins in early June and continues until late August, with females nursing their last litter into September. Females experience a postpartum oestrus, so are commonly pregnant and lactating concurrently, and two to three litters may be produced by each mature female annually. Gestation lasts 19–21 days. Litter size averages three to five pups (range 1–8). Larger, heavier females (presumably also older) tend to have larger litters than smaller, lighter females, and litter size, in general, tends to increase with increasing latitude. Pups are born hairless, apart from some short vibrissae (whiskers). Details of development and age at weaning are not known. Females become mature enough to breed when they achieve a total length of at least 127 mm and more often around 140 mm and a body weight of at least 20.5 g and more often 30 g. Males become fertile with descended testes (scrotal) when they are around 150 mm in total length and 30 g. The age of Rock Voles at sexual maturity is unknown, but

Long-tailed Vole

FRENCH NAME: **Campagnol à longue queue**

SCIENTIFIC NAME: *Microtus longicaudus*

This western vole is a medium-sized member of the vole genus *Microtus* and is distinguished from other *Microtus* by its longer than normal tail.

DESCRIPTION

Like other voles in the genus *Microtus*, Long-tailed Voles are stocky, with blunt heads, and short, rounded ears that protrude only slightly beyond the body fur. Upper pelage varies from greyish brown to reddish brown or dark brown and the belly is dark grey. Their tail is 33%–44% of their total length, more than twice the length of the hind foot and is vaguely bicoloured, dark brown above and greyish below. Interior populations may be slightly paler on the sides and belly with more distinctly bicoloured tails. On these animals, the top of the back stands out as a dark band. Feet are brown or grey on their upper surface. The dental formula is incisors 1/1, canines 0/0, premolars 0/0, and molars 3/3, for a total of 16 teeth. Each adult undergoes two annual moults, one in spring and early summer and one in autumn.

SIMILAR SPECIES

Seven other *Microtus* occur within the range of the Long-tailed Vole. These include Singing Voles, Creeping Voles, Tundra Voles, Montane Voles, Meadow Voles, Townsend's Voles, and Water Voles. Their relatively long tail is distinctive (> 50 mm) and distinguishes Long-tailed Voles from all others, except Townsend's Voles and Water Voles, which are also long-tailed, but are considerably larger (weight > 90 g and hind foot > 28 mm). Townsend's Voles typically have a lighter belly than Long-tailed Voles. Singing Voles are blonder than Long-tailed Voles, especially on their sides, and their very short tail is distinctly bicoloured and mostly white, with only the central portion of the upper surface brown. Creeping Voles are the smallest of the group (total length < 153 mm). Montane Voles are also smaller than Long-tailed Voles (total length < 165 mm) with a grizzled grey-brown pelage and a thin, distinctly bicoloured tail. Tundra Voles have a short, strongly bicoloured tail (length < 47 mm). Meadow Voles, although somewhat smaller, can be very similar and difficult to distinguish from Long-tailed Voles in the field, unless you get a close look at the tail. Heather Voles are small, with light brown to greyish-brown dorsal pelage and greyish to whitish belly fur. They have a short, distinctly bicoloured and sparsely furred tail. Populations in northern British Columbia have yellow fur on their snout, but those farther south may not. They can be distinguished from Long-tailed Voles mainly by their light colouring and tail length, < 46 mm. Skulls of all of the above can be distinguished using a variety of characters. See appendix 2.

SIZE

Males and females are similar in size and colouring. Weight fluctuates over the year and tends to be lowest in mid-to-late winter and highest in late summer. In Canada, animals along coastal British Columbia are larger and heavier, while those in Alberta and interior British Columbia are the smallest. The following measurements apply to adult animals in the Canadian part of the range.

Total length: 150–267 mm; tail length: 55–110 mm; hind foot length: 16–27 mm; ear length: 8–21 mm.

Weight: 26.9–85.0 g (average 32–40 g); newborns, 2.4 g on average.

RANGE

Long-tailed Voles occur in the western cordillera of North America from eastern Alaska, Yukon, and the western Northwest Territories to southern Arizona and eastward as far as South Dakota. Many populations in the eastern and southern boundaries of the range have become stranded at higher elevations by increasingly arid conditions over the last several thousand years. They now occupy coniferous forests at high elevations in those regions.

ABUNDANCE

This vole is generally rare and seldom achieves the high numbers characteristic of many other voles. Long-tailed Voles tend to

Long-tailed Vole (*Microtus longicaudus*)

Distribution of the Long-tailed Vole (*Microtus longicaudus*)

have been clearcut, once the regrowth has become advanced enough to provide abundant shelter and food, but not too tall to reduce light levels for the understorey plants. These conditions occur between approximately 5 and 20 years after logging. In Alberta they are most common in shrubby alpine meadows, while in British Columbia they are common in spruce-fir forests with many open glades. In both provinces, older clearcuts provide very favourable habitats, just as they do in Alaska. Most activity is nocturnal and Long-tailed Voles remain active all year round, but limit most of their winter activity to the snow layer, where they are protected from both predation and extreme cold. Long-tailed Voles are adapted for cool climates and are not able to survive for long when ambient air temperatures climb past 32°C, unless they have access to cool, underground burrows. In the southerly parts of their range they are restricted to higher, cooler elevations, but in British Columbia and northern areas where summer temperatures are more moderate, they may also occupy the valley lowlands. Occurrence of other voles seems to be a major factor in their distribution. Long-tailed Voles appear to compete poorly and may be displaced from their preferred habitats, even by much smaller species. The home ranges of males are considerably larger than those of females, which are in turn larger than those of juveniles. Long-tailed Voles nest in underground burrows or under fallen logs or stumps in warm weather, and in winter commonly construct nests within the snow layer. Generally, females have a longer life expectancy than males; the longest-lived Long-tailed Voles so far reported were two 13.5-month-old females from Alaska. Most Long-tailed Voles do not survive more than a few months and few survive their first winter. Potential predators include many raptors, Red Foxes, weasels, and Coyotes.

DIET

Like other voles, Long-tailed Voles are primarily herbivorous. During summer, they eat fresh vegetation such as forbs and grasses, berries, seeds, fungi, and insects. In winter, they eat primarily grasses and sedges, as well as roots and bulbs that they find under the snow. When winter food is scarce, they will chew on the bark of trees and shrubs and can cause the death of the tree if the trunk is girdled. Few Long-tailed Voles occur in agricultural lands, so they are not significant crop, orchard, or nursery pests. Their requirement for drinking water is poorly understood. Some populations appear to get their entire water requirement satisfied from their food, while other populations drink daily.

REPRODUCTION

There is limited and often conflicting information available on the reproductive biology of this species. In British Columbia, the breeding season extends from late March to November. Long-tailed Voles in southern Alaska and Yukon breed for three to four months from mid-May to mid-September, and those in southwestern Alberta have a 13-week reproductive season from late May to late August. Length of gestation is around 21 days and a postpartum oestrus does occur. Litter size averages five pups (range 2–7) in Canada and Alaska, but may be smaller in more southerly populations. Overwintered females are larger and bear larger

be more common in the mountains and along the west coast and more sparse on the grassland uplands. Density varies with habitat and degree of competition from other voles and mice. Densities as high as 120 Long-tailed Voles per ha have been reported (in New Mexico), but usually numbers vary from 1 to 40 per ha. Populations have a marked annual abundance cycle that peaks in August or September as the youngsters of the year are included. Highs and lows of population number vary from year to year, but it is not known if this fluctuation is cyclical, what the length of the cycle might be, or what drives it.

ECOLOGY

Long-tailed Voles live in a wide range of habitats, from moist and dry open meadows in grasslands, to alpine meadows, to scrubby woodlands, rocky slopes, bogs, and coniferous or mixed forests, from sea level to 2170 m in elevation. In Alaska, they are most abundant in scrubby woodlands and forest edges, including regions that

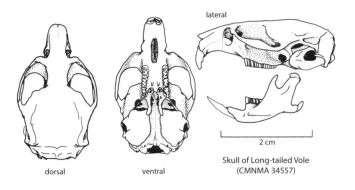

lateral

dorsal ventral

2 cm

Skull of Long-tailed Vole
(CMNMA 34557)

REFERENCES

Colvin 1973a, 1973b; Conley 1976; Conroy and Cook 2000; Cook et al. 2001; Elbroch 2003; Jameson and Peeters 2004; Loewen 1984; Nagorsen 2005a; Smith, H.C., 1993; Smolen and Keller 1987; Van Horne 1982; Verts and Carraway 1998.

litters than young-of-the-year females. Females in Alaska produce no more than two litters in a breeding season, but those in southwestern Alberta can produce three, sometimes up to four litters annually. Captive studies show that the pups, like other *Microtus,* are born hairless and blind. They are fully furred within 7 days, open their eyes at about 11 days old, and are weaned by 15 days old. A three-year study of wild Long-tailed Voles in Alaska did not discover any juveniles breeding during their first year, but a captive breeding study on voles from southwestern Alberta had juvenile females breeding at four to six weeks old and juvenile males at seven to nine weeks old.

BEHAVIOUR

Long-tailed Voles are considered to be relatively timid in their relations with other *Microtus* and prefer to avoid contact. Aggressive encounters between males may result in vocalization, chases, and wrestling, and the loser may suffer bite wounds to his rump.

VOCALIZATIONS

Young pups produce ultrasonic calls when disturbed or uncomfortable to attract their mother's attention. Adults sometimes squeak, especially during aggressive interactions.

SIGNS

These voles do not create distinctive runways like most other voles. If runways do develop, they are indistinct and hard to find. Like other *Microtus,* their front feet have five toes but toe 1 (the thumb or hallux) is greatly reduced and rarely leaves an impression in the track. The hind feet also have five toes, all of which register in the track. All toes have claws that may or may not register. Each front foot has a central cluster of three palm pads that are partially fused and two distinct heel pads that may register in the track. Each hind foot has four partially fused palm pads and two distinct heel pads, which may or may not register. A direct-register walk or trot (where the hind track is placed directly over the front track) are the two most common gaits, although hops and bounds may occur when the vole is threatened or exposed. Scat and tracks are similar to those of other *Microtus.* Distinguishing between them can be difficult to impossible. See the "Signs" section of the Meadow Vole account for a track schematic that applies to all the voles and lemmings.

Singing Vole
also called Hay Mouse, Alaska Vole

FRENCH NAME: **Campagnol chanteur**
SCIENTIFIC NAME: *Microtus miurus*

These voles are more social and vocal than other Canadian voles. They get their name from the high-pitched, pulsating warning trill usually delivered from a safe location, such as a burrow entrance. As an intruder moves through a colony, its path can be tracked by the trills that follow its progress.

DESCRIPTION

Upper back fur is yellowish brown to reddish brown, sides are lighter greyish or yellowish brown, and belly hairs are greyish at the base and buffy yellow or pale grey to whitish at the tips. Mature males have a brighter patch of fur covering their flank glands. The area on the upper lips where the long, vibrissae attach is commonly buffy or yellowish. The feet are greyish or dirty brown with rather large, narrow, brown claws. Their short, bristly tail is mainly buffy or white with a narrow line of brown in the centre of the upper surface. Their long, soft body fur gives this medium-sized vole a rounded body shape similar to that of a lemming. Ears are short and barely

Singing Vole (*Microtus miurus*)

protrude above the fur. A patch of fur directly behind each ear may be rufous or yellowish on some animals. The dental formula is incisors 1/1, canines 0/0, premolars 0/0, and molars 3/3, for a total of 16 teeth. Each adult undergoes two annual moults, one in spring and early summer and one in autumn.

SIMILAR SPECIES

Their "hay piles" (see in the "Ecology" section for more information) and the often rocky terrain of their colony sites resemble those of Collared Pikas, which share their range. Collared Pikas are considerably larger, with louder, single-note warning whistles. Several voles and lemmings could be confused with Singing Voles as they occur within the same range. These include Meadow Voles, Tundra Voles, Long-tailed Voles, Taiga Voles, Northern Bog Lemmings, Northern Collared Lemmings, and possibly Ogilvie Mountains Collared Lemmings. The overall yellowish pelage of most Singing Voles and their short strongly bicoloured tail distinguish them from the other small mammals. Only lemmings change to a white coat in winter. See appendix 2 for illustrated characteristics that identify their skulls.

SIZE

Males and females have similar colouring and are similar in size. The following measurements apply to adult animals in the Canadian part of the range.
Total length: 134–156 mm; tail length: 18–29 mm; hind foot length: 14–22 mm; ear length: 10–17 mm.
Weight: 28.0–51.3 g; newborns, 1.65–3.0 g (average 2.3 g).

RANGE

Singing Voles are northern rodents that occur only in North America, where they can be found in the Northwest Territories, Yukon, and Alaska. They occur only in suitable habitat, so their distribution tends to be spotty rather than continuous.

ABUNDANCE

Population size fluctuates between seasons and years, but even during peak years, these rodents tend to be localized in their distribution, so do not reach abundance levels that are common of the more diversified *Microtus* species.

ECOLOGY

In Yukon, Singing Voles may occur up to elevations of around 2000 m (throughout the alpine zone and into the subalpine zone). Most habitats where these voles are found are rocky, with nearby water, well-drained soils, and a diverse vegetation cover composed of grasses, forbs, mosses, lichens, and low or dwarf shrubs. Singing Voles are intermittently active throughout the day or night and they remain active under the snow during the winter. In a study of Alaskan animals, males and females occupied average home ranges of 1250 m² and 450 m² respectively, and range size was relatively stable regardless of vole density. The ranges of adult males overlap those of several adult females, but overlap between same-sex neighbours is much less evident. Resembling small pikas

Distribution of the Singing Vole (*Microtus miurus*)

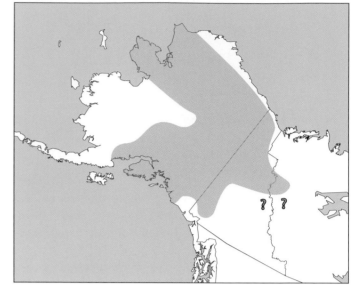

❓ possible but not verified range

in their posture and colonial behaviour, Singing Voles are similar to them in another important way: both gather and cache plants for winter consumption. These are collected into "hay piles" near a burrow entrance, where they are guarded from the pilfering of neighbours. When cured, the plants are carried underground for winter storage. Short, shallow burrows (10–20 cm below the surface) are dug that may reach a metre in length, but are usually less than 20 cm long. These terminate in a nesting and storage chamber that can exceed 30 cm in length. The roof of such a chamber may be only 5 cm below the ground surface. Although some early reports have native peoples in some parts of the Arctic raiding these caches for the stored roots and tubers, it is more likely those caches were the work of Tundra Voles (also called Root Voles) rather than Singing Voles, as Singing Vole caches are composed primarily of leaves and stems. During good years, vole densities reach annual peaks in late August with the recruitment of the young born that summer and are lowest in late winter–early spring. This cycle can vary dramatically from year to year, but the environmental factors that control these irregular density fluctuations remain unclear. In Alaska, Singing Voles can reach density highs of 47 per ha and lows of less than 4 per ha. Population fluctuations similarly occur in Yukon and likely also in the Northwest Territories. One female lived two winters and was around 20 months old when she disappeared, but mortality is high for most cohorts, except adults during the summer, and very few voles live through even one winter. Singing Voles are competent swimmers and can climb up into small shrubs as high as a metre and a half above the ground to harvest the leaves and shoots. Predators include Grey Jays, gulls, jaegers, owls, Brown Bears, Grey Wolves, foxes, Wolverines, and weasels.

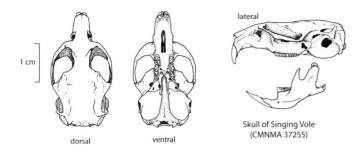

1 cm

dorsal   ventral

lateral

Skull of Singing Vole
(CMNMA 37255)

DIET

Singing Voles are primarily herbivorous. They consume forbs such as horsetails (*Equisetum* sp.), lupines, locoweed, and leaves of deciduous shrubs such as willow and blueberry. Horsetails and willow are preferred foods.

REPRODUCTION

Although an occasional litter may be born during any month of the year, the vast majority of breeding begins in late May before snowmelt is complete and continues until September, even after the first heavy snowfall of autumn. The mating system appears to be promiscuous, as both males and females mate with multiple partners throughout the breeding season. Females experience a postpartum oestrus, and up to three litters may be produced annually. Gestation averages 21 days. Litter size varies from four to twelve, but averages eight to nine pups. Births occur in an underground nest composed of dried leaves and plant stems. Young are born pink and hairless, with closed eyes and ear canals. Their eyes open when they are about 12 days old and they are weaned at around 20 days old. Although young males in a captive breeding colony became sexually active as young as 24 days old and young females as early as 41 days old, no juvenile females produced a first litter before they were at least 6 months of age. In the wild, early born females may breed late in the summer of their birth, but only large subadults and adults normally breed. Studies of most other *Microtus* species report that females mature earlier than males. Males' testes shrink and retract to an abdominal position outside of the breeding season.

BEHAVIOUR

Singing Voles are one of several *Microtus* species that are colonial. They live in loose association, but individual adults are generally solitary apart from mating pairs and females raising pups. Groups may nest communally during cold weather, presumably to share warmth. Territorial behaviour varies with season and gender, as adult males' active defence of their home range peaks early in the breeding season, while female defence peaks later in the summer. The reasons for this shifting territoriality are unclear. Singing Voles gather plants for winter food and by late summer the often large "hay piles" are common features of colony sites. Plants are cured in different ways depending on the species. Willow leaves and shoots, for example, are typically raised off the ground by placing them onto low-slung branches of nearby shrubs, while forbs are usually placed on rocks or well-drained slopes. It is primarily the immature animals

that cache winter food and they are the cohort that ordinarily survives the winter and reproduces the following season. Each vole may have multiple hay piles, and some especially large piles have been postulated as the cooperative work of several voles (likely immature siblings). Mature males develop prominent flank glands during the breeding season that recede over the winter. The odour from these glands is strong and has been described as similar to the resin of a Balsam Fir. Males scratch at the glands with their hind feet when sexually excited, presumably to spread the odour. Mature females sniff the male's flank gland, probably to determine his breeding condition and males sniff the perineal region of the female, likely for the same purpose. A researcher who had bred many different *Microtus* species in captivity made the particularly cogent remark that Singing Voles have an unusually gentle disposition compared to other similar voles.

VOCALIZATIONS

Singing Voles are aptly named. Their high-pitched, quavering trill, which appears to be used as a warning call, is distinctive. Since this call is especially evident during August when the hay piles are being created, it could also be interpreted as a territorial defence call. Quieter, single-note, bird-like squeaks may be heard within a colony.

SIGNS

Singing Voles are not particularly wary, so if you can find their hay piles or burrows, remain quietly nearby and one may appear. Burrows are about 2.5 cm in diameter and their colonial social system is usually evident by the proximity of multiple burrows. Occupied burrows are noticeable as the voles carry freshly excavated soil in their mouths to the burrow entrance, where it is deposited in a waste mound. By late summer, the presence of drying plants gathered into hay piles should be obvious. Especially large hay piles can exceed 45 cm in height and contain 30 litres of curing material. These piles may be stored under rock overhangs, in the lower branches of dwarf shrubs, or in the open. Although their runways are not as pronounced as those of most other *Microtus*, especially Meadow Voles, repeated passage near the burrows makes them most visible there. Toe 1 on the front feet (the pollex or thumb) is highly reduced, but toe 1 on the hind feet (the hallux or big toe) is prominent. Little data is available about Singing Vole tracks, but they are likely very similar to, and indistinguishable from, those left by Meadow Voles and Tundra Voles as they are similar in size. See the "Signs" section of the Meadow Vole account for a track schematic that applies to all the voles and lemmings. Singing Vole scat is comparable in size to that of Tundra Voles (blunt and oval-shaped, 0.2–0.3 cm in diameter, and 0.5–0.6 cm in length) and is common around the hay piles.

REFERENCES

Banfield 1974; Batzli and Henttonen 1993; Batzli and Lesieutre 1991; Elbroch 2003; Galindo and Krebs 1985; Gilbert et al. 1986; Krebs and Wingate 1985; Lidicker and Batzli 1999; Slough and Jung 2007; Youngman 1975.

Montane Vole

FRENCH NAME: **Campagnol montagnard**

SCIENTIFIC NAME: *Microtus montanus*

These medium-sized *Microtus* can become serious orchard pests, especially during winter, when they often resort to chewing bark and tree roots, which can weaken the tree and may cause its death if the trunk is girdled.

DESCRIPTION

The overall appearance of the dorsal fur is a grizzled, yellowish brown. Underfur hairs are dark grey, as are the bases of the guard hairs. Tips of the guard hairs are a mix of brown and black, while the middle section of each hair is a yellowish buff. This banded combination provides the "salt and pepper" look and yellowish-brown colour. Belly hairs are also grey at their base, but are either lighter grey or whitish at the tips. Fur colour on the sides is intermediate between the darker back and lighter belly fur. The medium-length tail (relative to other *Microtus*) is distinctly bicoloured, dark grey above and white or light greyish below. Feet are silvery or pale grey. Ears are short and protrude only slightly beyond the fur. Populations in the United States, especially those farther south, are generally darker. The dental formula is incisors 1/1, canines 0/0, premolars 0/0, and molars 3/3, for a total of 16 teeth. Each adult undergoes two annual moults, one in spring and early summer, the other in autumn. These voles continue to grow throughout their life, albeit at a reduced rate once they achieve sexual maturity, so the largest animals in any population are also the oldest. Montane Voles have four pairs of mammae. Both males and females develop hip glands as adults, but those of mature males are more pronounced. Other locations of glands used for scent marking include feet, chest, mouth, and the anogenital region.

SIMILAR SPECIES

The range of the Montane Vole in Canada overlaps that of Meadow Voles, Long-tailed Voles, and Heather Voles. Long-tailed Voles have tails that are > 50 mm in length and constitute 33%–44% of their total length. Meadow Voles do not have grizzled dorsal pelage and the upper surface of their feet is dark greyish brown. Heather Voles are lighter brown, with a thinner, sparsely furred tail and softer, finer fur. Appendix 2 provides an illustrated guide to identifying skull and dental characteristics of voles.

SIZE

Males and females have similar colouring and are similar in size. The following measurements apply to animals of reproductive age in the Canadian part of the range.

Total length: 120–165 mm; tail length: 29–51 mm; hind foot length: 15–22 mm; ear length: 9–15 mm.

Weight: 22.0–45.0 g; newborns, about 2.2 g.

RANGE

Montane Voles occur mainly in mountainous and intermontane regions of the western United States, but the northern limits of their distribution extend into the dry grasslands of south-central British Columbia as far north as the Chilcotin River Valley. The southern extent of their range occurs midway along the Arizona/New Mexico border and the eastern extent is in northeastern New Mexico. Montane Voles are found only at higher elevations in the southern United States, and populations there are often separated by the mountain valleys.

ABUNDANCE

Like many other *Microtus*, Montane Voles are subject to cyclic population fluctuations that, in their case, take two to five years (average three to four years) from peak to trough. Peak populations in British Columbia can reach densities of up to 186 voles per ha before declining to numbers averaging fewer than 60 voles per ha and as low as 1.4 voles per ha. A marsh in Utah recorded peak densities of up to 560 voles per ha.

ECOLOGY

British Columbia's Montane Voles inhabit the dry inland grasslands from 300–1190 m in elevation. They occur in grassland-steppe habitats dominated by hardy shrubs, in open grasslands and grassland thickets, and in brushy grasslands near open water. They reach their greatest densities in irrigated agricultural lands such as orchards, hay fields, and Alfalfa fields. Farther south, they range higher and higher into the mountains as the latitude decreases and temperatures rise in the lowlands. Montane Voles are not tolerant of ambient temperature above 31°C and will die if they cannot escape to a cooler burrow. During summer, they are active intermittently throughout the day or night, but peaks of activity occur in the morning and for a few hours before sunset and they rest in their burrows during the heat of the day. In winter, they are active under the snow. They construct extensive snow tunnels that are used mainly for foraging. An account from Colorado describes an early snowfall that later completely melted, revealing that Montane Voles had been busy under the snow. The ground had remained unfrozen during the previous three weeks after the snow fell and the vole tunnels were lined with

Montane Vole
(*Microtus montanus*)

Distribution of the Montane Vole (*Microtus montanus*)

clipped grasses and other plants and hollowed-out storage chambers were filled with roots. Such caches are not widely reported for this species, so may occur only under special circumstances. Nests of dried plant material are constructed above ground in dense vegetation or under woody debris, in nest chambers in underground burrows, and within the snow layer. Natal nests are usually situated in an underground burrow. Nests may be globular with hollow centres, cup-shaped, or simply a level platform. Both males and females dig underground burrows, but males may additionally make use of abandoned burrows of other mammals such as those of Northern Pocket Gophers. Burrows typically extend 20 cm or more in depth and may be 40 to over 200 cm in length. The survival and reproduction of Montane Voles is closely tied to climate. Maximum lifespan is about 18 months, but average lifespan depends on when birth occurs. During a summer with normal weather (most importantly, sufficient moisture for plant growth), animals born in the first cohort typically breed at four to five weeks old and live approximately six months. Those born in the second cohort become sexually mature at seven to eight weeks old and live about seven months. Pups born in the third cohort typically overwinter and do not become sexually mature until the following spring. This cohort is likely the long-lived one, with individuals living about 12 months on average. Pups born late in the season from a fourth or even fifth cohort are scarce and tend not to survive the winter. If they do, they also mature in spring.

During abnormally dry summers, pups born in the first or second cohort may not mature until the following spring and a third cohort is rare. During years of the population cycle when numbers are increasing, more of the third and fourth cohorts mature earlier and may breed over the winter. Although Montane Voles do not appear to climb, they can swim. Predators are legion and include numerous raptors as well as American Badgers, Coyotes, weasels, skunks, and rattlesnakes. Even gulls, shrikes, and magpies will take Montane Voles when they are plentiful.

DIET

Montane Voles are primarily herbivorous, consuming the leaves, stems, roots, and seeds of a wide variety of forbs, grasses, and sedges. They eat some fungi in spring before the new growth is available and some arthropods during the summer when they are abundant. When food is scarce in winter, Montane Voles will chew at the bark of trees and can cause their eventual death by girdling.

REPRODUCTION

Montane Voles have a highly variable and sometimes unusual reproductive strategy. Unlike many other mammals whose reproductive period is cued by increasing day length, Montane Vole breeding is triggered by chemical compounds in the plants that they eat. Since consumption of the first green shoots of spring is essential to activate the mating season, onset can vary by as much as a month from year to year, depending on when spring arrives. In British Columbia, breeding begins after the snowmelt and continues until late September. Sometimes breeding persists into the winter, typically when the population cycle is in the increase phase. Females exhibit postpartum oestrus and if well fed are capable of producing five or more litters each year, although wild females usually produce four or fewer litters annually. Gestation lasts about 21 days and litter size varies from 1–10 pups depending on factors such as the mother's age, population dynamics, and weather conditions. Larger litters are born to older females and during population expansion, while smaller litters are produced by young females and during population declines. Newborns are blind, deaf, toothless, and essentially hairless (apart from short vibrissae), with toes still fused together. They develop rapidly. Eyes open by 10 days and toes separate at 10–12 days old. Weaning begins around day 12, pups start eating solid food about day 13, and by day 15–16 they are fully weaned. Typically, the female then abandons the nest and moves to a new territory, where she prepares a new nest for the next litter. But when

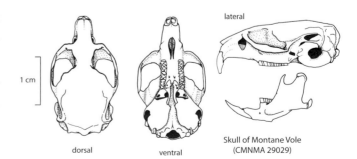

Skull of Montane Vole
(CMNMA 29029)

dorsal   ventral   lateral   1 cm

the population is at its peak and she is surrounded by other adult females, she stays on her home territory and, rather than driving her young away, allows them to remain with her, thereby forming an extended maternal family. Juvenile females in such a situation may have their own fertility suppressed. Should a nearby territory become available, the maternal female will abandon her extended family and establish a new nest in the unoccupied territory. Age at sexual maturity varies widely depending on weather and whether the pups were born early or late in the breeding season. For more information on sexual maturity of different age cohorts, see in the "Ecology" section.

BEHAVIOUR

Adult males and females are territorial with others of their same sex, but male territories normally overlap the territories of one or more females. Intense battles between males occur in spring as territorial ownership is determined. The males attack the rump and flanks of protagonists, often targeting the hip glands. Resident males vigilantly defend their territories by attacking and chasing intruding males away. Adult females are also territorial and maintain a home range exclusive of other females. They appear to do so more by mutual avoidance and more benign threat displays, rather than by direct confrontation like the males. Males create conspicuous scat piles in their runway system that they additionally mark with urine. They also mark their territories with secretions from their anal glands by anal dragging and from their prominent hip glands by raising their hindquarters and dragging them along the tunnel walls. Adults are monogamous when numbers are very low, and males especially are polygamous when densities are higher and more opportunities are available. Females commonly mate with the same male during consecutive heats, as long as that male remains available. Although captive males will provide some parental care of pups when they are confined within the same nest box as the female and litter, this situation is unlikely in the wild, as males normally occupy solitary nests and do not interact with the pups.

VOCALIZATIONS

Pups less than 15 days old produce ultrasonic distress calls in the 26–36 kHz range. These are not produced merely because their mother is absent, but only if they become stressed, typically from cold, hunger, or disturbance. Fighting males characteristically produce high-pitched squeaks.

SIGNS

Runways of Montane Voles tend to be conspicuous, with packed, well-worn surface trails 3–6 cm wide radiating from burrow entrances. Caches of plant cuttings are common along the runways and latrine sites regularly occur at intersections. Some of the larger latrine sites can become as large as 5 × 8 cm. Subnivean tunnels are 2.5–4.0 cm in width and up to 3 cm in height and can be extensive. Circular nests made of dried grasses and leaves are 12–24 cm in diameter and may be cup-shaped, flat platforms, or globular with a hollow centre that is 3.5–6.0 cm in diameter. Hollow nests may have more than one entrance. Cavities in

ball-shaped nests may be lined with finer material. These nests may be underground, in snow, or above ground under dense vegetation or woody debris. Tracks and scat are very similar to those of Meadow Voles.

REFERENCES

Benedict and Benedict 2001; Berger et al. 1997; Blake 2002; Drabek 1994; Elbroch 2003; Ferkin 2001; Jameson and Peeters 2004; Jannett 1978, 1981a, 1981b, 1984; Nagorsen 2005a; Negus et al. 1977; Negus et al. 1986; Rabon et al. 2001; Sera and Early 2003; Sullivan et al. 2003; Verts and Carraway 1998.

Prairie Vole

FRENCH NAME: **Campagnol des Prairies**
SCIENTIFIC NAME: *Microtus ochrogaster*

Since Prairie Voles commonly mate for life, they have been the lab animals of choice, for more than 25 years, of scientists studying the social and chemical basis of pair attachment. Monogamy is unusual behaviour in mammals, exhibited by only about 3% of the mammalian species.

DESCRIPTION

Pelage appears to be dark brown to almost black, but buff-tipped guard hairs intermix to give a grizzled appearance. Fur on the back is darkest. The sides are slightly lighter, and the belly is greyish with a light grey, buffy, or yellowish-orange wash caused by light-tipped guard hairs. Feet are brownish on their upper surface. Ears are rounded and short, and only the tips extend beyond the body fur. The sharply bicoloured tail is dark above and pale below, and is less than twice as long as the hind foot. Several colour variations occur rarely in the wild, including black, albino, yellow, and spotted. The dental formula is incisors 1/1, canines 0/0, premolars 0/0, and molars 3/3, for a total of 16 teeth. Winter pelage may be slightly darker and the belly fur more buffy. Prairie Voles have three pairs of mammae. They are reasonably tolerant to water restrictions, as they can produce concentrated urine and metabolize much of their water requirements from their food, especially when eating fresh, green vegetation.

SIMILAR SPECIES

In Canada, only one other vole occurs within the range of the Prairie Vole. Meadow Voles are larger, lack the grizzled pelage, and have either a dull grey or silvery grey belly. They have six small bumps (plantar tubercles) on the sole of each hind foot, while Prairie Voles have five. Meadow Voles have four pairs of mammae compared to three in the Prairie Vole. For skull diagnosis see appendix 2.

Prairie Vole
(*Microtus ochrogaster*)

SIZE

Males and females have similar colouring and are similar in size. The following measurements apply to adult animals in the Canadian part of the range. These voles are larger in the southern parts of their range.

Total length: 130–150 mm; tail length: 25–39 mm; hind foot length: 15.0–17.5 mm; ear length: 11–13 mm.

Weight: 26–38 g; newborns, about 3.5 g in the wild and 3 g in captivity.

RANGE

Prairie Voles occur in the prairie grasslands of central North America, from central Alberta to northeastern New Mexico and northern Texas, and from central Montana and eastern Manitoba to western West Virginia. Clearing forests for agriculture has allowed this vole to extend its range and abundance to the east, but many of the western populations have been negatively impacted by the reduction of native prairie due to ploughing and grazing. A disjunct population in eastern Texas and western Louisiana became extinct by the late 1980s. Current distribution is patchy.

ABUNDANCE

This small vole is sporadic in its density. Mostly it is scarce, to rare, in Canada, with occasional dramatic, but unpredictable population increases, which can make it locally common for a time. Natural tallgrass or shortgrass prairie habitats typically support low population densities with minor fluctuations, but the amplitude of fluctuation is more extreme in habitats providing more abundant food, such as the more artificial locations listed in the "Ecology" section. In either case, populations are highest in early autumn as the youngsters of the year are incorporated into the population. South of the border, maximum densities of 11 to 1060 per ha have been reported. Prairie Voles are listed as endangered in Michigan. Although a common agricultural pest south of the border, Prairie Voles are rarely numerous enough to be considered such in Canada.

ECOLOGY

As their name implies, these small, dark voles used to occur in natural tallgrass prairie, but now may also be found in other grassy locations such as road margins, railroad right-of-ways, fence rows,

old cemeteries, fallow fields, and hay fields. Regular ploughing discourages their occurrence. Their prairie habitat can vary from open grassland to grasslands dominated by shrubs, such as sagebrush, or upland grasslands mixed with aspen. Where Prairie Voles and Meadow Voles occur in the same area, Prairie Voles are found in drier sites with shorter and more varied vegetation. They live in colonies where their numerous burrows and runways provide concealment and protection, and a prime requirement of any habitat, is that it must provide sufficient and suitable cover for the creation of concealed runways. Prairie Vole populations are highly sensitive to moisture levels that support plant growth for food and cover, but too much water can flood burrows, drown nestlings, and force adults and juveniles to relocate. Although Prairie Voles can be active anytime of day or night, most of their summer activity takes place in darkness, and they are more active during the day in winter (usually under the snow), thereby avoiding the temperature extremes of each season. The home range size in natural populations is fairly stable at around 100 m² or 0.01 ha and pairs or communal groups (see "Behaviour" section) mutually defend their shared home range from intruders. Dispersal occurs mainly by juveniles. Average dispersal in a long-studied population in Illinois averaged 42 m or approximately four home ranges distant. Prairie Voles are capable swimmers. Although not regular climbers, they will climb more than half a metre into the lower branches of shrubs to chew the bark when food is scarce. Burrow length and depth varies with soil type, but may be up to several metres in length (maximum reported was 31 m long), although most are much shorter and are typically at least 5 cm beneath the surface. Nest chambers in cold regions such as Manitoba may be as deep as 92 cm. Burrows may be simple short passages terminating in a nest chamber, or more extensive, with multiple entrances and many interconnecting tunnels, dead ends, and food and nesting chambers. Food storage chambers and nesting chambers are up to

Distribution of the Prairie Vole (*Microtus ochrogaster*)

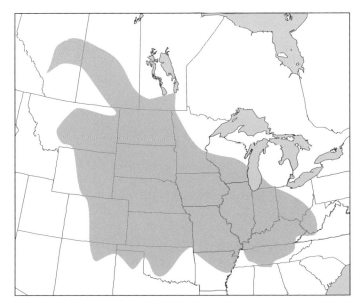

20 cm in diameter. Above-ground nests are usually located under cover of dense vegetation, fallen logs, or discarded construction materials, such as planks or metal sheeting. Predators are legion and include Coyotes, American Badgers, Bobcats, Northern Raccoons, Grasshopper Mice, short-tailed shrews, and all the raptors, shrikes, weasels, foxes, fox snakes, and rattlesnakes that live in the same region. Predation is so intense, combined with the effects of extreme weather, that the average life expectancy of a wild Prairie Vole is less than 10 weeks, and few live to be a year old, although captive animals can live up to 35 months. These voles rarely achieve sufficient numbers or do enough damage to be considered agricultural pests in Canada, although they are commonly considered as such in the United States.

DIET

During summer, Prairie Voles eat mostly vegetation, such as leaves, stems, roots, and seeds of grasses and forbs, and some insects, especially in the late summer. The basal portion of the grass stem is favoured, as are the seeds. Winter food includes the roots, rhizomes, tubers, and seeds of the same plants, as well as mosses, dried vegetation, and bark. Winter foraging under the snow is supplemented with cached items consisting primarily of seeds, dried vegetation, and underground plant parts. When available, they will add fruit and vegetable crops to their diet, and during winter they sometimes chew the bark of fruit and nut trees.

REPRODUCTION

These voles are capable of breeding all year round in the southern parts of their range, but in Canada reproductive activity occurs during the warm months from about May to September or October. Copulation stimulates ovulation, and the male deposits a fibrous copulatory plug in the vagina of the female when mating is concluded. Gestation requires around 21 days (range 20–23) and litters average 3–4 pups (range 1–8). Females undergo a postpartum oestrus and can be pregnant and lactating concurrently. They can produce several litters in a breeding season, but abnormally dry or hot weather can inhibit production. Both males and females in a pair bond provide parental care, although females restrict access to the newborn pups on the day of their birth. Newborns are hairless, toothless, blind, and deaf, with their toes partly fused together. Their incisors begin to emerge and their brown fur begins to appear within two days. Eyes and ears are open by day 10. Pups begin to eat solid food by day 12 and they are fully weaned by 21 days old. Juvenile females normally reach sexually maturity at about 30–40 days old and males at 35–45 days old. Juveniles are full grown by two months old. Testes regress in size during the winter months and may be retracted into the abdomen for around two months during the winter. Young may be born underground or in a well-concealed above-ground nest. Natal nests are made from dried grasses and are globular with a hollow centre. First oestrus of a juvenile female is induced by the direct sniffing of male urine on the genitalia. Only unfamiliar male urine can elicit this response, which thereby avoids mating or pair-bonding with close kin. Presence of a reproductive female can inhibit the reproduction of other females within the same communal group (see in the "Behaviour" section for an explanation of communal groups).

BEHAVIOUR

Prairie Voles are gregarious, highly social, and typically monogamous. Most partners remain together for life and share a home range and nest. In less than 20% of the cases will a vole accept another partner after the death or disappearance of their first mate. Males who lose their partner leave the pair's home range and wander the area, becoming itinerant males. Most females that lose their partner remain on the pair's home range and mate with itinerant males or possibly nearby resident males. A few such females may wander for a time before settling again. Most wandering males do not occupy a permanent nest. Each colony is composed of three types of groups: male-female pairs, single females (mostly survivors of a pair) and their litters, and communal groups usually composed of at least two adults of the same sex (average size in one study was 8 adults and groups of > 12 were common). The communal groups are typically composed of related adults, with the addition of occasional unrelated animals (possibly pair-bonded to one of the other adults in the group), and may be formed by a litter of a mated pair or a single female that stayed together after they were weaned. This is most likely to occur in late autumn as the last litter of the year remains to share a nest with their parents. It is possible that most male-female pairs would form the basis of a communal group (extended family) were it not for the high rate of predation of juveniles. Each communal group typically has one breeding pair and animals that are not reproductive within the group may assume some care of the pups born to the group, acting as helpers. The extent of pair bonding within a communal group beyond the dominant pair is still unknown. Burrow systems of communal groups are typically more extensive and complex than those of male-female pairs. About 55% of the breeding males in a colony are pair-bonded or part of a communal group. These males are termed "resident males." The remainder are wanderers that make frequent visits to the home ranges of pairs, communal groups, and single females searching for breeding opportunities. Sometimes even a pair-bonded female will mate with an itinerant male, if he happens along while she is still in oestrus and her mate is not around to drive him off. Resident males assist in raising the pups produced by their mate, whether they are all sired by him or not. Social groups display territorial behaviour, so there is little space overlap between the home ranges of adjacent groups. Resident males exhibit mate-guarding behaviour and will expel intruding strange males. A threat is signified by raising their forefeet, extending the head towards the threat, and chattering their teeth. They signal intent to fight by taking a fully upright stance from which they will lunge towards the rival. This may lead to boxing, wrestling, and biting and ultimately a chase as the dominant animal drives the other away.

VOCALIZATIONS

Prairie Vole pups produce ultrasonic distress calls centred at 32 kHz that attract either their sire or dam. Unlike most other voles, they

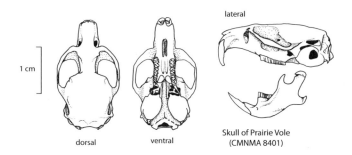

lateral

1 cm

dorsal   ventral

Skull of Prairie Vole
(CMNMA 8401)

Harvey and Barbour 1965; Jones and Birney 1988; Kurta 1995; Mankin and Getz 1994; McGuire et al. 2003; Pergams and Nyberg 2001; Pizzuto and Getz 1998; Poole and Matlack 2007; Rabon et al. 2001; Smith, H.C., 1993; Solomon and Jacquot 2002; Solomon et al. 2004; Soper 1961, 1964; Stalling 1990; Wrigley et al. 1991.

can call repeatedly from birth. Most other vole species only begin to call a few days after they are born. Reproductive males are also thought to use ultrasonic vocalizations when attempting to attract a mate.

SIGNS

Prairie Voles create a well-used system of runways within their home range, especially between burrow entrances or from a burrow to a favourite foraging area. Each main runway commonly has numerous less compacted side routes. About 5 cm in width, runways can be found by parting the long grass. They may be especially evident at the end of winter when the packed snow at the sides of the tunnels resists melting. These long, icy tubes snake across the brown litter after the surrounding snow has disappeared. Some main runways are so well used as to be floored with bare earth and sunken into the ground, while side runways may be only trampled grass and debris. Nests are elliptical and made of coarse, dried grasses with a finer inner lining of shredded vegetation. Whether constructed in a clump of vegetation above ground or in an underground chamber, they are around 18 cm in length, 15 cm in width, and 10 cm in depth, with a hollow centre. Underground nests are typically placed 25 cm or deeper below the ground surface. Active runways can be recognized by the presence of fresh clippings and scat and active burrows by the pile of fresh earth deposited in a throw mound at their mouths. Scat is similar in size, shape, and colour to that of Meadow Voles and is, in similar fashion, scattered throughout the home range, commonly accumulating in latrines, near nests, and at runway junctions. Burrow entrances are 2.5–3.5 cm in diameter and are often grouped at or near the base of a shrub. Like other voles, Prairie Voles rarely travel on the snow surface, especially if it is deep, preferring to burrow through the snow layer where it can afford them some cover and weather protection. When the snow melts, their droppings and plant clippings are concentrated along their runways and are usually evident. Prairie Vole tracks can be distinguished from those of Meadow Voles only under rare circumstances, when the track is pristine and very clear. Prairie Voles have five pads (palm and heel) on each hind foot, while Meadow Voles have six. For further information on vole tracks and scat see the Meadow Vole account.

REFERENCES

Blake 2002; Criddle 1926; Davis, W.H., and Kalisz 1992; Elbroch 2003; Getz and McGuire 2008; Getz et al. 1987, 1993, 1997, 2005;

Tundra Vole

also called Root Vole (Europe), Northern Vole

FRENCH NAME: **Campagnol nordique**
SCIENTIFIC NAME: *Microtus oeconomus*

The biology of this vole is complicated, because in Eurasia, where it has been relatively well studied and is known as the Root Vole, it is often dissimilar from that of our North American form. Unless otherwise stated, the information provided here applies to Tundra Voles from North America.

DESCRIPTION

Body fur is a golden brown, which is darker to rusty brown on the back and more golden or greyish on the sides and rump. Sometimes, especially in winter pelage, the sides are so light that the back fur stands out as a darker band. Belly fur is a greyish white often with a buffy wash. Juveniles are darker overall. Ears are rounded and barely show above the long pelage. The short tail is strongly bicoloured; dark grey above and greyish white below. Feet are greyish brown. Tundra Voles have four pairs of mammae. Both males and females develop visible hip glands during the breeding season, but they are larger on mature males. The dental formula is incisors 1/1, canines 0/0, premolars 0/0, and molars 3/3, for a total of 16 teeth. Tundra Voles experience two moults annually, one in spring and the other in autumn. The winter coat is long and silky and frequently lighter than the summer pelage. Feet in the winter pelage are edged with long, stiff, silvery hairs.

SIMILAR SPECIES

Meadow Voles share much of the Tundra Voles' range and may even use the same runways. Meadow Voles are more grey-brown on the sides and rump (without the golden sheen of the Tundra Vole), they have a dull grey belly, and their tail is not strongly bicoloured. Long-tailed Voles occur in parts of the Tundra Vole range and can be distinguished by their longer tail (> 50 mm and more than 33% of the animal's total length). Singing Voles also occur within the range of the Tundra Vole. The colonial Singing Voles are larger and generally have a much paler, yellowish pelage (especially on the sides of the body), and a very short tail. The more reddish Tundra Voles (especially in winter pelage) could be confused with Northern Red-backed Voles, whose tail has bristly hairs along its length and a long

Tundra Vole (*Microtus oeconomus*)

brush (10–12 mm) of darker hairs at its tip. Skull identification of the different vole species can be made using the teeth and cranial features illustrated in appendix 2.

SIZE

Males and females have similar colouring and are similar in size. The following measurements apply to adult animals in the Canadian part of the range.

Total length: 130–187 mm; tail length: 26–47 mm; hind foot length: 16–23 mm; ear length: 10–18 mm.

Weight: 20–69 g; newborns, unknown but likely in the 2.0 g range, based on other voles of similar size.

RANGE

Tundra Voles are found in northern North America from western Nunavut to Alaska and as far south as northern British Columbia. They have a Holarctic distribution and formerly occurred from northeastern Europe to the Bering Strait. The European distribution of Root Voles is becoming patchy and is no longer continuous.

ABUNDANCE

Most populations of this vole commonly experience marked fluctuations in density over a three-to-four-year cycle. During peaks of the cycle, summer densities can reach 70–80 voles per hectare. Some populations in the northern parts of the range may not cycle, as the short summers result in fewer litters, which prevent them from reaching high densities. The voles on the Alaskan islands are listed by the IUCN. Canadian populations are considered stable. In Europe, the Root Vole is declining in numbers and is being replaced by the Common Vole (*Microtus subterraneus*) in many areas. It is considered Endangered in some regions.

ECOLOGY

Tundra Voles are considered habitat specialists. They favour damp grass and sedge meadows where cover is abundant, but may also occur in more shrubby locations. Their range includes Arctic tundra and shrub tundra in the north, and alpine and subalpine areas

farther south, up to about 1550 m in elevation. Tundra Voles are largely nocturnal in summer and diurnal in winter, but may be active at any time of the day or night. During winter they tunnel through the snow, usually at ground level, to forage and move about under the protection of the snowpack. In summer, they burrow into the soil and create runways under cover of vegetation. The home ranges of males are larger than those of females and male ranges do not overlap, although they do overlap with those of one or more females. Most females' ranges overlap with those of other females, but typically only with one male. Female Tundra Voles tend to remain near their natal home range and it is likely that the females whose home ranges overlap are littermates, mothers and daughters, or otherwise closely related. These matrilineal females may sometimes inhabit a common area and form small colonies. The summer home range size of males was measured in a radio-telemetry study in the western Canadian Arctic at around 4000 m², while female home ranges were about 1000 m². Tundra Voles are willing and capable swimmers, as befits a wetland specialist and can even dive below the water surface for short distances. They climb readily to reach succulent shoots. Predators include many raptors, jaegers, gulls, Common Ravens, small weasels, foxes, Wolverines, and Coyotes. Tundra Voles' frequent aquatic excursions even allow some large predatory fishes such as Lake Trout to add them to their diet on occasion.

DIET

Tundra Voles eat mainly sedges and grasses, as well as shoots of shrubs and forbs and some seeds. Horsetails, which can be locally abundant in some areas, may be a major food source for some

North American distribution of the Tundra Vole (*Microtus oeconomus*)

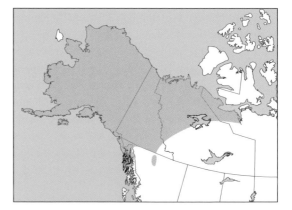

Eurasian distribution

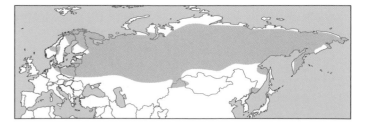

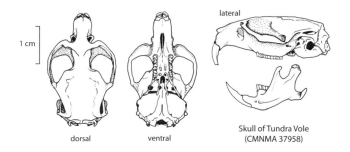

1 cm

lateral

dorsal ventral

Skull of Tundra Vole
(CMNMA 37958)

populations. Bird eggs and nestlings of ground-nesting small birds may be added to the summer diet when available. Winter food is made up predominately of dried plants and above-ground plant parts, as well as some mosses and lichens. While Tundra Voles in Europe are famous for caching rhizomes, tubers, and bulbs in autumn for winter consumption, it is not known to what extent North American voles do the same.

REPRODUCTION

The breeding season begins under the snow in mid-April to May and ends about late September. Gestation lasts 21–25 days and litters average 4–6 pups (known range in North America is 3–7). Females may produce up to four litters annually. Juveniles become sexually mature at around six weeks old. In early autumn the testes of mature males regress in size and are withdrawn from the scrotum into the abdomen until the following breeding season.

BEHAVIOUR

Little is known of the social structure of this vole. Male Tundra Voles are largely solitary in summer, but may nest communally in winter. Females may be solitary or communal in summer and nest communally in winter. During the summer breeding season, some female Tundra Voles share their home range with juvenile females that were probably successfully raised from earlier litters. It is possible that solitary females are so because they were unsuccessful in raising any female offspring to sexual maturity. In Eurasia, Root Vole females are considered to be highly territorial, while the home ranges of some males overlap. Agonistic behaviour typically begins with rearing to face an opponent and may proceed to boxing and wrestling if one vole doesn't retreat or submit. Deposition of scat in latrine sites along the runway is thought to be a means of olfactory communication, perhaps to denote territoriality or advertise reproductive condition.

VOCALIZATIONS

Older pups use ultrasonic distress calls to attract their mother's attention. Adults in some populations may use ultrasonic calls in the 30–35 kHz range during aggressive interactions.

SIGNS

Globular winter nests may be found in spring when the snow melts. These were constructed under the snow at about ground level. Both summer and winter nests are made of dried grasses and sedges, with a hollow centre lined with finely shredded vegetation. Nests are bulky and about 15 cm in diameter. Underground nests are rarely created in winter, likely due to the threat of flooding. Summer nests may be created in underground chambers or above ground in protected locations. Burrow entrances are often hidden, but may have a small throw mound of dirt to mark them. Most burrows are shallow due to the presence of permafrost. Tundra Voles use protected runways whenever possible and these may be under vegetation or follow recessed frost cracks or the sunken perimeter of frost polygons. Latrines are created just off the main runways. Runways in favoured foraging areas, such as along stream or pond margins, may be especially trampled to the point of being recessed into the soil. Small piles of clipped vegetation are scattered throughout the runway system. Scat is 5–10 mm in length and up to 3 mm in diameter. Tracks are similar to those of Meadow Voles.

REFERENCES

Andreassen et al. 1993; Banfield 1974; Batzli and Henttonen 1990; Bergman and Krebs 1993; Brunhoff et al. 2003; Kapusta et al. 1999; Krebs et al. 1995a; Lambin et al. 1992; Lance and Cook 1998; Macdonald and Barrett 1993; Nagorsen 2005a; Sealy 1982; Taitt and Krebs 1985.

Creeping Vole
also called Oregon Vole

FRENCH NAME: **Campagnol d'Oregon**
SCIENTIFIC NAME: *Microtus oregoni*

Most study of Creeping Voles in British Columbia has concentrated on populations from the lower Fraser River. The life history of these secretive voles in the more difficult-to-access alpine and subalpine regions of the Cascade Mountains remains virtually unknown.

DESCRIPTION

Creeping Voles are the smallest vole in their range. British Columbian animals average around 135 mm in total length and 21.5 g in weight. They have tiny dark eyes (< 4 mm in diameter) and a short, slightly bicoloured tail that is less than 30% of the vole's total length. Their back and sides are dark reddish brown and their belly fur is dark grey with a brownish wash. Body fur is short, thick and fine with the guard hairs barely protruding beyond the underfur, giving the plush coat a very shrew-like appearance. Fur on the upper parts of the feet is light brown. Ears are short and rounded and almost concealed by the fur. These voles have eight mammae. Hip glands on breeding males are reduced compared to other *Microtus* species. The dental formula is incisors 1/1, canines 0/0, premolars 0/0, and molars 3/3, for a total of 16 teeth. Each adult undergoes two annual moults, one in spring and early summer and one in autumn.

Creeping Vole
(*Microtus oregoni*)

SIMILAR SPECIES

This species occurs with four other voles in southwestern British Columbia. All are larger. Water Voles are the largest of the five; adults average 290 mm in total length and have a hind foot > 24 mm and weigh > 150 g. Townsend's Voles are also large. They average about 60 g and 192 mm in total length and have light belly fur. Long-tailed Voles average 182 mm in total length and their tail is > 50 mm long. Meadow Voles are the most similar to Creeping Voles, but average 147 mm in total length and 35 g in weight. Creeping Voles are darker than all the other voles, and their colour, along with their diminutive size, their fine, short fur, and their small eye, distinguishes them. Skulls of all the voles can be identified using teeth and cranial characters, as illustrated in appendix 2.

SIZE

Males and females are similar in size and colouring. The following measurements apply to adult animals in the Canadian part of the range.

Total length: 122–153 mm; tail length: 28–42 mm; hind foot length: 16–20 mm; ear length: 9–13 mm.

Weight: 10–31 g (average 21.5 g); newborns, 1.6–2.2 g.

RANGE

These small voles occur in western North America from southwestern British Columbia to northern California. The northernmost record is from Fishblue Lake, south of Lytton, on the east side of the Fraser River in British Columbia.

ABUNDANCE

Population numbers fluctuate seasonally, as these voles are most abundant in summer (average maximums, 32 voles per ha) and least numerous over the winter (average maximums, 7 voles per ha). The highest density was reported at 72 per ha from abandoned farmland. Some populations may cycle in abundance over a three-to-four-year period, but others seem to be relatively stable, showing only the seasonal variation. Even when there are abundance cycles, Creeping Vole maximum densities are considerably lower than those of many other vole species. This rodent is not considered of conservation concern in British Columbia.

ECOLOGY

Creeping Voles occupy several different habitat types in British Columbia. In the Fraser Valley, they are found in moist, coniferous forests, forest edges, riparian areas, recently logged forests, grassy fields, and abandoned agricultural lands. Old farm land and early successional stages of clearcut logging, especially those that were burned over, appear to provide ideal habitat with abundant forbs, grasses, and deciduous shrubs. As the young trees and shrubs grow, they begin to reduce the occurrence of the understorey plants, making these habitats less suitable. In the Cascade Mountains, Creeping Voles are found on eastern slopes in dry, subalpine forests and meadows up to 1980 m in elevation. The home range size of animals in British Columbia is unknown, but in nearby Oregon, males occupy home ranges of 0.05–0.38 ha and females 0.04–0.23 ha. These small voles dig their own burrows or use those of other small mammals. Many of their tunnels are shallow. Globular nests woven of dried grass stems are usually constructed in underground chambers, but are sometimes located above ground in well-protected sites such as under woody debris or inside a hollow log. Maximum longevity in the wild is 14–16 months, but very few live longer than 12 months and most survive only a few months. Predators include hawks and owls, Coyotes, Bobcats, Ermines, Long-tailed Weasels, and Western Spotted Skunks.

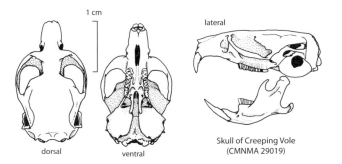

1 cm

lateral

dorsal

ventral

Skull of Creeping Vole
(CMNMA 29019)

DIET

The diet of this species has not been well studied. In summer, they consume mainly green vegetation (presumably forbs and grasses), fruit and berries, and some fungi, while in winter they eat the bark, buds, and twigs of shrubs and tree seedlings, and probably dried grasses.

REPRODUCTION

The breeding season usually extends from March to September in British Columbia, but Creeping Voles may begin earlier or continue breeding later, if the weather is unusually mild. Average litter size in British Columbia is three to four pups (range 1–5). Gestation takes about 23–24 days. Females may experience a postpartum oestrus. Newborns are hairless, blind, and deaf (their ear canals are fused shut). The pups develop rapidly. They begin to display visible fur at about 24 hours old. Incisors erupt and the pups can crawl weakly at five days old. By eight days old they are almost completely furred. Eyes and ears open and they begin to eat solid food by 10–11 days old and they are weaned at around 13 days old. Females are thought to produce four to five litters annually. Although captive breeding studies show that juvenile females become sexually mature at 22–24 days old and males as early as 42 days old, in the wild, few juvenile females breed in the same summer as they were born unless they are from an early litter.

BEHAVIOUR

Very little is known of the behaviour of wild Creeping Voles. Mating in captive animals is marked by lengthy periods of squeaking (10–15 minutes), which recur about every 30 minutes for up to 5 hours. Although each intromission is brief, multiple copulations are usual during that period.

VOCALIZATIONS

These voles are known to produce high-pitched squeaks during encounters. Otherwise, little has been reported concerning their vocalizations.

SIGNS

Although these small voles are probably not always as fossorial as once thought, they do create sometimes extensive burrow systems with small entrances (around 2 cm in diameter) in moist, friable soils, and sometimes their subsurface digging produces push-up ridges similar to those made by moles but in miniature. Some researchers think that these voles construct runways near the roots of grasses and seldom venture outside them, while others were unable to find runways or burrows but found that the voles entered traps set on the surface throughout their habitat. Clearly much is yet to be learned of this species. Tracks are similar to those of Meadow Voles, but about 30% smaller.

REFERENCES

Carraway and Verts 1985; Cowan and Arsenault 1954; Cowan and Guiguet 1956; Jameson and Peeters 2004; Nagorsen 2005a; Sullivan and Krebs 1981; Verts and Carraway 1998.

Distribution of the Creeping Vole (*Microtus oregoni*)

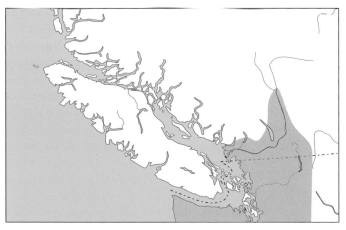

expanded view of Canadian range

Meadow Vole

also called Field Mouse

FRENCH NAME: **Campagnol des prés,** *also*
Campagnol des champs

SCIENTIFIC NAME: *Microtus pennsylvanicus*

Meadow Voles occur in all the provinces and territories of Canada. Theoretically, under perfect conditions with no deaths and abundant

Meadow Vole (*Microtus pennsylvanicus*)

food, a pair of Meadow Voles could produce a million descendants in a single year. However, countless causes of mortality exist in the wild and this level of fecundity is impossible.

DESCRIPTION

This is a medium- to large-sized vole depending on where in the country it lives and how much food is available. Generally, the smallest Meadow Voles occur in central North America and the largest are found on smaller, east coast islands (excluding the large islands such as Newfoundland, Cape Breton Island, and Prince Edward Island). This vole is also highly variable in coloration across the country. Back and sides vary from a rich, dark brown to olivaceous brown to reddish brown or even yellowish brown depending on geographic location. Typically, the upper back is darker than the sides, particularly in winter pelage. Belly hairs have grey bases with grey, white, or even buffy tips depending on the population. Fur on the upper parts of the feet varies from light grey to dark brown, again depending on the population. Juveniles and subadults within each population are generally darker than the adults. Colour variations such as blonde, and partial white (white hairs mixed with darker hairs in variable amounts) are rare. Although darker individuals occur in many populations, fully melanistic (all black) Meadow Voles are very rare, as are albino or all-white animals. The ears are short and rounded and barely protrude beyond the fur. Their somewhat bicoloured tail is dark brown above and varies from light to medium grey on the underside. The dental formula is incisors 1/1, canines 0/0, premolars 0/0, and molars 3/3, for a total of 16 teeth. Each adult undergoes two annual moults, one in spring and early summer and the other in autumn. Moult of animals in the northern parts of the range is delayed in spring and advanced in autumn. The winter coat is longer and thicker than the summer pelt.

SIMILAR SPECIES

Due to their widespread distribution, Meadow Voles occur within the ranges of most of the other *Microtus* in Canada, with the exception of Townsend's Voles. Prairie Voles are smaller, with dark salt

and pepper pelage. Long-tailed Voles have a longer tail, which is 33%–44% of their total length. Singing Voles are blonder than Meadow Voles, especially on their sides, and their very short tail is distinctly bicoloured and mostly white with brown only in the central portion of the upper surface. Tundra Voles are very similar to Meadow Voles, but have a strongly bicoloured tail. Creeping Voles and Woodland Voles are much smaller than Meadow Voles, with dark, reddish brown fur that is plush and shrew-like. Taiga Voles are larger, with distinctive rusty orange patches on each side of their snout. Rock Voles have similarly coloured patches on either side of their snout. Water Voles are distinctively larger. Montane Voles can be very similar to Meadow Voles. Where the two overlap in range in western Canada, Meadow Voles have dark brown fur on the upper surface of their feet, while Montane Voles have white feet and Meadow Voles lack the grizzled dorsal pelage of Montane Voles. Heather Voles are small voles with light brown to greyish brown dorsal pelage and greyish to whitish belly fur. They have a short, distinctly bicoloured and sparsely furred tail and lighter-coloured feet. Populations in northern British Columbia have yellowish fur on their snout, but those farther south may not. Appendix 2 provides an illustrated guide to vole teeth and skull characteristics.

SIZE

Males and females have similar colouring and are similar in size. The following measurements apply to adult animals in the Canadian part of the range and cover the diversity of size across the country. Weight fluctuates over the course of the year and is lowest in winter and highest in summer.

Total length: 120–204 mm; tail length: 21.0–61.5 mm; hind foot length: 14.5–26.0 mm; ear length: 9–20 mm.

Weight: 20.0–82.9 g; newborns, about 1.6–3.0 g.

RANGE

Meadow Voles have the largest range of any of the North American *Microtus* voles. They can be found from Alaska, Yukon, and the Northwest Territories to Mexico and from British Columbia to Newfoundland. Occurrence on both freshwater and marine islands is common, although they do not reach the marine islands of the Pacific, apart from Admiralty Island in the panhandle of southern Alaska. These voles are expanding their distribution southward in the midwestern United States in Kansas, Missouri, and Illinois and in the eastern United States in the state of Kentucky.

ABUNDANCE

Although some of the island populations may be in danger of decline, generally Meadow Voles are common and sometimes abundant, depending on the phase of the population cycle. Population densities as high as 600 voles per ha have been reported during population peaks in some habitats.

ECOLOGY

As their name suggests, Meadow Voles exhibit a strong preference for moist, grass/sedge meadows, but they can also survive, and sometimes flourish, in woodlands and open forests, especially at forest edges and clearings. Human-created habitat is often favoured

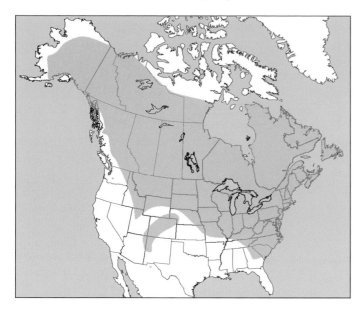

by these rodents. They can be common in hay fields and pasture lands, road margins, orchards, abandoned farmland, and early successional forests following clearcutting or forest fire. Other habitats such as bogs, fens, and alpine and subalpine meadows may support Meadow Vole populations. Like all the Canadian voles, Meadow Voles do not hibernate. They may be active at any time of the day or night. Their populations are usually subject to cyclical fluctuations of density that repeat over a two-to-five-year period. During years of super abundance, these small rodents can be jumping out from every step you take, or thronging in front of a harvester, whereas most other years they are rarely seen. Factors that may affect the periodicity of their population cycles include predation, food availability, disease, dispersal, climactic variation, physiological stress caused by crowding, or the consequences of genetic or behavioural make-up. Populations subject to heavy predation, or those on islands where dispersal is not possible, tend to have more stable population densities and appear to be subject only to the normal seasonal variations, which means they are more abundant in summer and autumn and decline in numbers over the winter. The home range size varies from 405–3480 m² (0.0405–0.3480 ha) for males and 160–3115 m² (0.016–0.3115 ha) for females. The extreme variation in home range size is affected not only by sex, but also by habitat variables such as food availability and terrain, as well as population density. Although often large, each animal may use only a small portion of their home range on a day-to-day basis. A radio-telemetry study found that males used about 192 m² and females about 69 m² during an average 24-hour period. During the breeding season, the activity range of mature males may vary daily depending on the location of oestrous females. Typically, a male's home range is two to three times the size of a female's and usually overlaps those of several females, as well as those of other males. Breeding females generally maintain exclusive home ranges. Home ranges are larger in summer than in winter.

Shallow, simple burrows or more complex multi-chambered and longer burrows may be dug during the summer for protection from predation, shelter, and temperature regulation. Nests are roughly oblong and globular with a hollow internal cavity. They are constructed of woven plant stems and lined with shredded vegetation. Most nests in summer are built in tussocks of vegetation, in shallow burrows, or under debris and are connected to a network of runways through the grasses. Winter nests are usually created at the ground surface under a snow layer and connected to runways through the snow. Predators are numerous and diverse and include virtually any small to medium-sized predators that occur within their range, including hawks, owls, shrikes, gulls, jaegers, snakes, foxes, weasels, Coyotes, Grey Wolves, Bobcats, Canada Lynx and Domestic Cats and Dogs. Short-tailed shrews are probably primary predators in eastern North America and Least Weasels are thought to be vole specialists. Meadow Voles are strong swimmers, buoyed by the air trapped in their fur, and they can even dive beneath the surface when pursued. Meadow Voles rarely climb. They can, however, sometimes access tender branches and trunks when they are buried in the snow. The height they can reach is governed by the height of the snowpack. Average longevity in the wild is about 2–3 months, but some lucky individuals may live 10–16 months. Captive voles can live more than 2.5 years. Infanticide can become a major cause of mortality of young, especially when densities are high.

DIET

Meadow Voles are eclectic herbivores, as would be expected of such a widespread species. They will eat leaves, stems, seeds, and fruit of most available species of grasses, sedges, and forbs, as well as insects, carrion, nuts, and some varieties of fungi when available. Winter foods include dried vegetation, roots, and evergreen herbs found under the snow and can include the bark of woody species. This diet is supplemented with roots, leaves, corms, and rhizomes cached along the runways under snow cover. High numbers of Meadow Voles over the winter can cause extensive damage to orchards, vineyards, and ornamental trees and shrubs by girdling the bark. Cannibalism does occur, especially of the young, mainly by adult males. Like many other rodents, Meadow Voles practise coprophagy, the reingestion of their own faeces. This becomes especially common when they have a low-quality diet, as a second pass through the digestive system allows the individual to extract further nutrients.

REPRODUCTION

In most of Canada, Meadow Voles begin breeding in early spring under the snow and continue until autumn. Winter breeding may

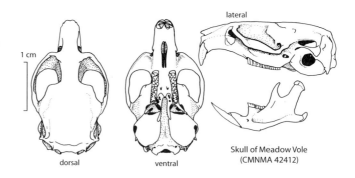

Skull of Meadow Vole
(CMNMA 42412)

occur at high population densities and is most common in the temperate portions of the range, but is rare in the north. Meadow Voles are polygamous and females generally raise their litter without assistance from the sire. Gestation lasts about 21 days. Multiple paternity does occur, most frequently in spring litters when the competing males are all of similar size, as they all survived the winter and are fully grown. Under these conditions, males are less successful at guarding their mated female from other males, so she is more likely to mate with more than one male. Litter size averages four to six (range 1–11) and is strongly correlated with the size of the mother, as larger females bear larger litters. A postpartum oestrus occurs about 55% of the time. Captive females are capable of producing up to 17 litters annually under ideal conditions in a laboratory. Wild females rarely survive long enough to produce more than two and at most three to four litters. Newborn pups are hairless, with closed eyes and ear canals. They can produce an audible squeak by day 4 and use only ultrasonic calls until then. Fur also begins to be visible by day 4 and the pups are largely fur-covered (except for the belly) by day 7. Eyes and ears open on the eighth day. Weaning occurs between 12 to 14 days old. Rarely do even 50% of a litter survive to weaning. Juvenile females can reach sexual maturity as early as 25 days old and males at 35–45 days old.

BEHAVIOUR

The social behaviour of Meadow Voles changes seasonally. During the breeding season, reproductively active females are generally territorial towards other females and maintain exclusive home ranges. But sometimes, especially early in the breeding season, two females (probably siblings) may share a nest and raise their pups communally. Female territoriality promotes dispersal of female young, reducing the possibility of interbreeding and exhaustion of food resources. Males do not defend their home ranges, choosing to avoid other males rather than risk a confrontation. But more is at stake when breeding opportunities are in jeopardy. If more than one male is attracted to the same oestrous female, scuffles and fighting may ensue. Such encounters can lead to wounding and possibly the establishment of temporary dominance hierarchies among the males. Dominance and the resulting access to females is clearly hotly contested during the breeding season, as many mature males bear bite marks during that time, especially when population densities are high and there is more competition. A commonly used threat display is a lunge with teeth bared, accompanied by a loud squeal. Adult females are generally dominant to males and a rejected male suitor will simply depart. Mate guarding does occur at times, as the dominant male will remain near the mated female to prevent other males from mating with her. During the non-breeding season (winter), these voles are more social, commonly sharing their nests, probably to conserve heat. As many as seven adult voles have been discovered in a single winter nest. Chemical signals are important means of communication for these small rodents. Latrine sites and traces of urine purposely deposited along the runways are likely signals of identity, proximity, home-range occupancy, breeding status, and in the case of breeding females, territory ownership.

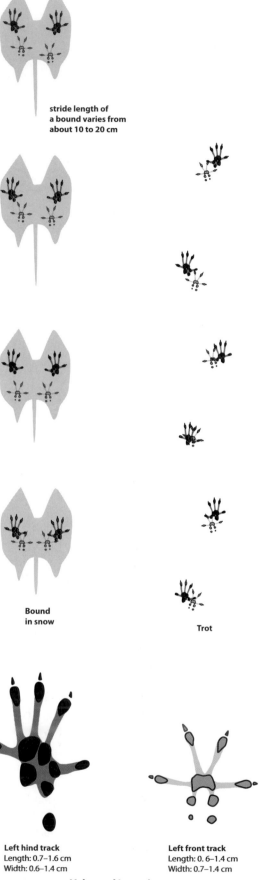

stride length of a bound varies from about 10 to 20 cm

Bound in snow

Trot

Left hind track
Length: 0.7–1.6 cm
Width: 0.6–1.4 cm

Left front track
Length: 0. 6–1.4 cm
Width: 0.7–1.4 cm

Voles and Lemmings

VOCALIZATIONS

The primary vocalization used by Meadow Voles is a squeak. It often accompanies an aggressive threat towards another Meadow Vole and is used by oestrous females to announce her condition to nearby males. Meadow Voles may squeak or squeal if grasped by a predator and they will also growl and tooth-chatter when threatened. Newborn pups emit four different ultrasonic vocalizations, usually in response to stressful situations.

SIGNS

Meadow Voles have six palm pads on each hind foot and five on each front foot. Other voles may have fewer pads, but rarely is a track clear enough to distinguish these minor differences. The tracks of all the voles are very similar, with some size variation between the smallest and the largest species. Each foot has five toes, but toe 1 (the thumb) on each front foot is greatly reduced and rarely leaves an impression in a track. The digital pads of voles are less bulbous than those of mice, and impressions of the long toes are often connected to the palm pads. Claws may or may not register. Hind tracks are larger than front tracks. These rodents typically travel in a direct-register trot (where the hind foot lands on top of the track left by the front foot on the same side), except in snow, where a bounding gait is more common. Tails frequently leave a drag mark in snow. The accompanying track schematic applies to all voles and lemmings. Above-ground nests of Meadow Voles are often sloppy-looking affairs of variable size, from 10–28 cm long by 8–23 cm wide by 4–10 cm high. Nests are 10–13 cm in diameter on average. The back of the nest is typically lower than the front. There may be one or several openings. The internal cavity is about 6–10 cm long by 5.0–9.5 cm wide by 3–8 cm in height. Maternity and winter nests have larger central hollows, as they are shared.

A maze of runways (3.0–6.5 cm wide) is created through the vegetation that can often be seen by parting the grass to observe the ground. Winter runway systems are most clear in spring when the earth that was pressed up onto the walls of the snow tunnels melts out as a casting of the tunnel system, or the trampled earthen base of the runway, appears as a depression in the grassy soil surface. Small piles of grass and leaf clippings are commonly scattered along the runways. Burrow openings, which are about 2.5–3.5 cm in diameter, may occur in snow, earth, or leaf litter. Most travel in winter occurs in runways within the snowpack, but voles will occasionally travel on the surface of the snow. Such trails often display how exposed and insecure they feel, as they will ordinarily run in a fairly straight line towards cover and may make repeated attempts to dig their way back down. Vole scat is tubular with blunt ends and is often deposited in a latrine site and along runways. Latrine sites are commonly located in dead-end runways or near a nest. Meadow Vole scat is 0.2–0.3 cm in diameter and 0.4–0.8 cm in length. Debarking of trees and branches occurs under the snow and the maximum height of the snowpack can sometimes be discerned by the height of the debarked section of tree trunk. Narrow grooves left by the small, paired incisors is a sign of small rodent activity.

REFERENCES

Banfield 1974; Boonstra et al. 1993; Brewer et al. 1993; Colvin 1973a and 1973b; Cranford and Johnson 1989; Dobson, F.S., and Myers 1989; Ebensperger et al. 2000; Elbroch 2003; Getz et al. 2001, 2006; Harper and Batzli 1996; Hill et al. 2001; Holt 1990; Innes and Millar 1990; Jones, E.N., 1990; Krupa and Haskins 1996; Kurta 1995; Madison and McShea 1987; McShea and Madison 1984; Nagorsen 2005a; Reich 1981; Smith, H.C., 1993; Storey et al. 1995; Wallace 2006; Weilert and Shump 1977; Youngman 1967b, 1975.

Woodland Vole
also called Pine Vole

FRENCH NAME: **Campagnol sylvestre**

SCIENTIFIC NAME: *Microtus pinetorum*

Woodland Voles are an often overlooked member of the fauna, as they are secretive and rarely appear above ground during daylight.

DESCRIPTION

These small rodents are semi-fossorial and have several adaptations for subterranean existence such as small eyes, short ears that are buried in their fur, and dense, soft, short fur that lacks the longer guard hairs, giving the pelage a plush, mole-, or shrew-like appearance. They have long whiskers on their snout to help them feel their way around in the darkness underground and their lips can close behind the incisors to prevent ingestion of soil while digging. Their short tail (approximately 20% of the length of the head and body) is somewhat bicoloured, dark brown above and greyish brown below. Fur on the back and sides is reddish brown. Belly fur is dark grey-based with paler tips, which gives the appearance of a light grey or buffy wash. Upper surfaces of the feet are medium brown. Albino, white, spotted, black, light brown, and orange individuals have been reported. The dental formula is incisors 1/1, canines 0/0, premolars

Woodland Vole
(*Microtus pinetorum*)

o/o, and molars 3/3, for a total of 16 teeth. These voles have only two pairs of mammae. Breeding males develop hip glands that secrete odoriferous compounds thought to stimulate oestrus in females.

SIMILAR SPECIES

No other eastern North American vole has the combined traits of a short tail (< 23 mm), small size, and plush, dark reddish fur. Southern Bog Lemmings could be mistaken for Woodland Voles, but they have even shorter tails, their grizzled pelage is coarser, their ears are more apparent and their eyes are larger. For skull diagnosis see appendix 2.

SIZE

Males and females have similar colouring and are similar in size. The following measurements apply to adult animals in the Canadian part of the range.

Total length: 107–131 mm; tail length: 14–23 mm; hind foot length: 15–19 mm; ear length: 9–10 mm.

Weight: 19–37 g; newborns, 1.9–3.2 g.

RANGE

Woodland Voles occur from southern Ontario, the Eastern Townships of Quebec, and southern Maine to Texas and Florida. They have reached some freshwater islands in the north of Michigan. They are at the northern limit of their distribution in Canada and have only been found in tracts of temperate deciduous forests. Most specimens are from southwestern Ontario, from northern Lambton County (Pinery Provincial Park) on Lake Huron to Halton County (Crawford Lake Conservation Area) north of Toronto. Only three localities have been reported in the Eastern Townships, all south of the St Lawrence River between Montreal and the Vermont border (Mount Pinnacle, Mont-St-Hilaire, and South Bolton).

ABUNDANCE

Like many other voles, populations of Woodland Vole are sometimes cyclic in density, although the length and extent of this cycle is unknown in Canadian populations. It is possible that, due to their rarity in Canada, Woodland Vole populations are relatively stable, exhibiting only seasonal fluctuations that reflect predation pressures, winter weather, and recruitment of the annual crop of young. Woodland Voles have become numerous enough in some parts of the United States at times to cause serious winter damage to orchards, as they will gnaw the bark around the base of fruit trees when other food is scarce. Their occurrence in Canada is limited primarily to hardwood forests, where they cause no noticeable agricultural damage. Woodland Voles are considered to be rare in Canada and are threatened by soil compaction and deforestation. They have been designated as species of "Special Concern" by COSEWIC since 2001, when their populations in Canada were deemed to be vulnerable.

ECOLOGY

In the southern United States, where they are more common, Woodland Voles inhabit a diverse assortment of habitats such as orchards, grassy fields, fence rows, bogs, and wet forests, but in Canada, at the

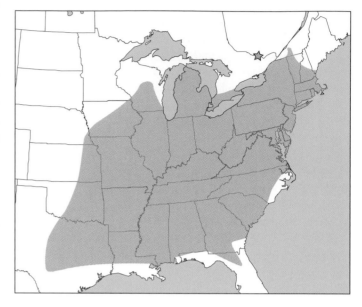

Distribution of the Woodland Vole (*Microtus pinetorum*)

☆ extralimital records

northern limit of their range, they are found primarily in light or sandy soils and thick leaf litter under upland hardwood forests of oak, beech, and maple. Either light soil or a deep humus layer is a primary habitat requirement. Woodland Voles do not hibernate and may be active any time of the day or night, but are most active in darkness, especially in summer, when daytime temperatures are high. Nests of woven grass and leaves are usually built in underground chambers or under woody debris. These are hollow and the interior sleeping space is lined with finely shredded vegetation and has one or more entrances. A nest, including a maternity nest, may be shared with other members of the extended family group and may house the litters of more than one female. Their inconspicuous tunnels are hidden under leaf litter, tree roots, and vegetation. Most of their foraging runways are 3–6 cm beneath the surface, but other tunnels may travel up to 30 cm down. Burrow depth is likely governed in large part by the depth of the sand or loam layer. Woodland Voles are capable swimmers, but poor climbers and jumpers, and are uneasy in the open. They are most comfortable when they can sidle up against an object like a fallen log, earthen bank, or preferably their own tunnel wall (a condition referred to as thigmotropism that is fairly common among small mammals). These voles can be drowned in their burrows by flood waters or heavy rains. Predators include Northern Raccoons, Northern Short-tailed Shrews, foxes, weasels, skunks, Bobcats, North American Opossums, snakes, and a multitude of raptors, as well as Domestic Dogs and especially Domestic Cats in inhabited areas. Their maximum lifespan in the wild is at least 18 months, but ordinarily 50%–80% survive less than two months.

DIET

Woodland Voles eat stems, leaves, seeds, fruit, roots, and tubers of grasses and forbs, and, to a lesser extent, carrion, insects, nuts, and

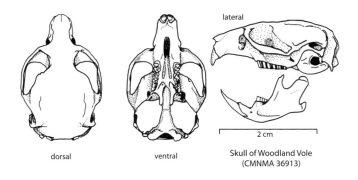

lateral

dorsal   ventral   Skull of Woodland Vole
(CMNMA 36913)

2 cm

fungi when available. They do forage above ground, but will also forage underground, where they can have a deleterious effect on tree seedling survival by consumption of the roots. They are occasionally cannibalistic. Winter food caches of roots, tubers, corms, and leaves are stored in underground burrow systems and supplement what can be found under the snow. Some caches have been found that were composed of almost 4 litres of material. The tuberous roots of Squirrel-corn (*Dicentra canadensis*) are included in the winter diet in Ontario, as they will remain fresh for the whole winter when stored around 10 cm underground. If winter food is scarce, these rodents frequently resort to chewing root bark below the snow and whatever tree bark they can reach within the snow layer. Woodland voles are commonly coprophagic. Reingestion of faeces typically occurs during resting phases before or after a feeding bout and is especially important for Woodland Voles on a low-quality diet. First-pass faeces still contain essential nutrients that can be extracted by a second pass through the digestive tract.

REPRODUCTION

There have been no studies to date that focus on the reproductive biology of this species in Canada, so most of what follows was learned through study of voles from the United States. The breeding season lasts from January to October in New York State, with a production peak from April to August. Likely the Canadian breeding season is similar. The timing and success of breeding varies from year to year, as these factors are subject to environmental variables. Woodland Voles appear to be monogamous. Like other voles and lemmings, both physical contact and chemical cues from an unrelated male are necessary to induce oestrus. Copulation lasts only a few seconds. Gestation takes 20–25 days, and most females experience a postpartum oestrus within a day of parturition. Although captive females can bear a litter each month all year round, it is unlikely that wild females in Ontario and Quebec could produce more than four litters annually, each consisting of two to three pups on average (range 1–6). The average litter size is greater in southern parts of the range. Newborns have a fuzz of hairs on their backs and well-developed vibrissae (whiskers), but they are toothless, deaf, and blind. They quickly anchor themselves to a nipple and are difficult to dislodge; ordinarily their mother must force them to release their tenacious hold before she leaves the nest. This behaviour effectively limits the number of pups that can survive until weaning to four, as each female has only four mammae. Captive females can sometimes wean up to five pups, but it is not known if this is possible in the wild. Incisors begin to erupt around five days old and the eyes and ears open between days 9 and 13. Weaning occurs during the third week, between days 17 and 21. Males become sexually mature at 6–8 weeks old and females at 10–12 weeks old.

BEHAVIOUR

These small voles occur in localized scattered colonies, and unless they are abundant, their presence can easily go unnoticed. Their social system consists of what are thought to be family units composed of a single breeding female, one or more breeding males, and their off-spring. Each family group occupies a shared home range of from 3 m in diameter in a prime New York apple orchard, to 148 m in diameter in a New England upland hardwood forest. Woodland Voles occupying grasslands have even larger home ranges. There is thought to be no overlap between the burrows of adjoining family groups. No home range sizes have been measured in Canadian populations. The family units share burrow systems, food, and nests. Woodland Voles are called cooperative breeders, as maternal females raise litters communally and continue to share their maternity nests with juveniles and adults of both sexes from their extended families. All the voles will groom and retrieve the pups and huddle over them to keep them warm. Nests have been found containing up to three litters of obviously different ages and, since none of the litters was yet weaned, they could not have been produced by the same female. Nevertheless, most family units are thought to contain only one breeding female (called a reproductive prime) and reproduction by other females is suppressed. Such suppression is accomplished in part by same-sex aggression. Mothers will lunge, bite or threaten a daughter, and fathers act similarly towards sons. The offender may also be grabbed and dragged around. This behaviour is called tugging, and it may have a threat or dominance component. Dispersal is undertaken by adults seeking breeding opportunities. Their goal is to find and join another group that is lacking a reproductive prime of their gender. Unlike most rodents that sit up on their haunches and use their front paws to hold and manipulate their food, Woodland Voles stand on all fours and press their food against the ground to stabilize it as they eat. They commonly approach a food plant from below where they can harvest roots, tubers, stems, and leaves without exposing themselves to the dangers above ground. Not especially powerful diggers and with no foreleg digging adaptations, these voles are nevertheless able to disappear from view in seconds through light soil or leaf litter if threatened. To dig their tunnels they use shovelling movements of their head and neck to loosen the soil and then they may engage their incisors (to chew through packed soil or roots) and their front feet to further loosen the earth. The waste soil is shoved behind them with a sideways movement of the forefeet, then flung farther back with the hind feet. The voles gather the waste soil and move it to the burrow entrance using their head as a bulldozer. Woodland Voles are difficult to trap due to their subterranean and subnivean habits.

VOCALIZATIONS

Pups can emit quiet squeaks from birth. Adults use a harsh chattering call to signal alarm and when fighting.

SIGNS

Underground tunnels and push-ups (tunnels that are just sub-surface) can be difficult to distinguish from those of moles. Above-ground runways are scarce and cryptic. Burrow openings are 3.0–3.5 cm in width and 2.5–3.0 cm in height. Globular hollow nests are 15–25 cm in diameter. Smaller simpler nests are likely used by one or only a few voles, while the larger nests are used communally by several. Burrow entrances often have a small pile of waste soil at their mouth, but this is usually hidden under the leaf litter. This throw mound may be 10 cm or more in diameter and 5.0–7.6 cm in height. Tracks are similar to those of Meadow Voles (see the "Signs" section of the Meadow Vole account) but uncommon, as most activity takes place below ground.

REFERENCES

Cranford and Johnson 1989; Davis, W.B., and Schmidly 1994; Marfori et al. 1997; McGuire and Sullivan 2001; Miller, D.H., and Getz 1969; Powell and Fried 1992; Raynor 1960; Ross, P.D., 1998; Schreiber and Swihart 2009; Smolen 1981; Solomon 1994; Solomon et al. 1998.

North American Water Vole

also called Richardson's Vole, Water Rat

FRENCH NAME: **Campagnol de Richardson**

SCIENTIFIC NAME: *Microtus richardsoni*, formerly *Arvicola richardsoni*

The North American Water Vole is the largest of the *Microtus* and possibly the most unique. It lives in small colonies that are typically widely separated from other colonies, it inhabits very specific habitat, and is semiaquatic.

DESCRIPTION

North American Water Voles are large voles. Their dorsal pelage is greyish brown to dark reddish brown and their underside is pale silvery to medium grey. Their long (>50 mm), well furred, thick tail is dark brown on the upper surface and pale to medium grey on the underside. Upper parts of the feet are dark brown. Both sexes have prominent hip glands during the breeding season marked by patches of greasy hairs over the glandular tissue. The glands regress outside the breeding season. The dental formula is incisors 1/1, canines 0/0, premolars 0/0, and molars 3/3, for a total of 16 teeth. Water Voles have four pairs of mammae.

SIMILAR SPECIES

The North American Water Vole distribution overlaps that of the Creeping Vole, Meadow Vole, and Long-tailed Vole. Apart from Common Muskrats, no other North American vole has a hind-foot measurement of more than 24 mm; large size (weight of adults > 105 g) and conspicuous hip glands (during the breeding season).

These physical traits combined with their mountainous habitat and colonial nature set Water Voles apart from all other small mammals. Appendix 2 illustrates all the diagnostic skull and teeth characteristics of Canadian voles.

SIZE

Males and females have similar colouring and are similar in size. The following measurements apply to adult animals in the Canadian part of the range. Body weights are highest in mid-summer and lowest in early spring.

Total length: 177–279 mm; tail length: 54–98 mm; hind foot length: 24–34 mm; ear length: 12–20 mm.

Weight: 118–190 g; newborns, 5 g.

RANGE

North American Water Vole distribution is divided into two portions, both in western North America. The easternmost and largest segment extends along the Rocky Mountains from east-central British Columbia and west-central Alberta to central Utah. The western segment extends along the Coastal Range from southwestern British Columbia to southern Oregon.

ABUNDANCE

North American Water Voles typically occur in remote mountainous terrain where (at least in Canada) they are little impacted

Distribution of the North American Water Vole (*Microtus richardsoni*)

North American Water Vole
(*Microtus richardsoni*)

by human activity. Their highly scattered populations tend not to reach the excessive densities of other *Microtus* species. Rather than experiencing years of superabundance, most Water Vole populations instead undergo an annual abundance cycle where numbers grow over the summer as the young are born and drop over the winter as weaker animals die. Largely depending on environmental conditions, the peak and trough of the annual density cycle can vary in amplitude from year to year. Densities during the summer breeding season may vary from 0.2 to 32.5 voles per ha in areas of favourable habitat in Canada. Occasional irruptions do occur on an irregular basis. In Alberta Water Voles are considered to be uncommon, while in British Columbia they are occasionally somewhat common in appropriate habitat and other times uncommon depending on their phase of the population cycle. North American Water Voles in Wyoming are considered to be "Critically Endangered."

ECOLOGY

These large western voles usually inhabit the alpine and subalpine zone from about 1340–2320 m in elevation in Canada. They will occasionally invade lower elevations (> 183 m elevation) during periods of high population density. Water Voles are unusual among the *Microtus* as they are habitat specialists. They prefer to live in meadows of mixed grasses and forbs or dwarf alder/willow thickets beside cold, moss-sided mountain streams (either glacier or spring-fed) where there is sufficient depth of soil along the banks to enable them to dig their burrows. Consequently, their occurrence is spotty, as colonies in favourable locations are often widely separated by stretches of unsuitable stream bed or dense forest. They may also inhabit the edges of still mountain lakes, ponds, and marshes. Radio-tracking of voles in a study in Alberta found that, on average,

males occupy home ranges of 770 m² and females 222 m². Adult females tend to remain in the same area for the whole breeding season and they maintain largely exclusive home ranges, both summer and winter, that overlap little with those of neighbouring females. Males attempt to have a summer home range that overlaps those of several females and they diligently travel between them to ensure they are nearby when a female is ready to breed. During the winter, males occupy even larger home ranges that overlap significantly with those of both females and males.

North American Water Voles are semiaquatic and very competent in swift waters. They can swim against the current and will dive beneath the water's surface in still pools. Most burrow systems have both above-ground and underwater entrances. Few Water Voles venture further than 10 m away from water, and individuals typically run for the water if threatened. Streams are commonly used avenues for daily movement and dispersal. Large unbroken tracts of forest appear to be barriers to dispersal. Most dispersal is accomplished by juveniles, usually in August of their natal year. Most young males (50%–100%) and up to 55% of the young females disperse. The remaining juveniles overwinter in their natal colony and by the following spring occupy territories left vacant by the death of an incumbent. Although North American Water Voles might be seen at any time, they are most active in darkness. During warmer weather they live in underground burrows. Their burrow system is complex, with branching tunnels 1–3 m in length and an assortment of nesting and food storage chambers. Most voles also dig some simple, short dead-end tunnels near foraging areas that are used as bolt holes or safe feeding sites. Sometimes summer nests are placed under woody debris, but most are underground. Snow lasts seven to eight months at the elevations inhabited by these voles. As they remain active all winter, they may continue to use their underground tunnels and nesting chambers, but they also dig a network of runways at ground level by burrowing through the snow pack. Sometimes new nests are constructed in the snow pack. These may be occupied throughout the winter, or only during the period of snow melt and flood. Predators include raptors, Coyotes, foxes, and weasels. Winters are harsh in the mountains and while some individuals may live through two winters, most Water Voles are lucky to survive one.

DIET

Leaves and some stems of various species of forbs supply most of the North American Water Vole's summer food, although grasses and sedges are also eaten, as are insects, seeds, and dwarf willow buds and twigs. Roots, corms, rhizomes, and tubers of forbs and grasses are especially important winter foods, but are likely consumed all year round. Although these voles will gather and store plant leaves and stems in their burrows during the summer, these are probably for consumption in the near future, as there is no evidence of winter food caches.

REPRODUCTION

In Alberta, the breeding season of this vole occurs in the snow-free period between May and early September and peaks from June to August. The first young typically begin appearing above ground in early July. Although unknown, the season in British Columbia is likely similar in duration and timing. Gestation lasts around 22 days and litter size in Canada averages 5–6 pups (range 2–10). In Canadian populations, females probably produce one to two litters annually. Parturient females experience a postpartum oestrus, and are capable of being pregnant and lactating concurrently. Like other *Microtus,* Water Vole newborns are pink and almost hairless as well as toothless, deaf, and blind. Their incisors are erupted by day 6 and eyes open by day 12. Captive pups swim voluntarily by day 17 and are weaned by day 21. The juveniles are usually on their own by the time they are 40 days old. Some juveniles are capable of breeding in the summer of their birth (around 25%), but most do not begin reproducing until the next breeding season.

BEHAVIOUR

The behaviour and social organization of this vole are poorly understood. Small groups of approximately 8–40 animals are scattered along mountainous waterways. When population density is high, the mature males focus their breeding efforts on a small, select cluster of females, but when there are few females, the males travel more widely and compete for the more limited breeding opportunities. Overwintered males often show evidence of bite injuries, suggesting that conflicts, likely for females, location, or dominance, take place under the snow. Overwintered females show wounding only when populations are dense and numbers are high, because then they are more likely to need to fight with other females to assert their right to a home range. Drum marking is a threat behaviour used by resident males when a strange male intruder is encountered. They rake the sole of alternating hind feet over the appropriate flank gland and then stomp the same foot several times, presumably to scent-mark the ground and perhaps at the same time deliver an audio or even vibrational signal. Another agonistic display involves the tail. It is held rigidly at about 45° and rapidly quivered from side to side. Again this display appears to be restricted to resident adult males.

VOCALIZATIONS

The vocalizations of Water Voles are not reported.

SIGNS

Scat is larger than that of other voles (about 3 mm in diameter by 6 mm long) and is deposited along the runways and often in or near the water. North American Water Voles do not create latrine sites, as do many other voles. When they are present along a stream bed they leave well-trampled, wide runways (5–7 cm in width) that commonly parallel the water and criss-cross the stream and its banks. These can become worn down to bare soil from long use by successive generations. Water Voles may also create small trampled clearings along the water's edge where they haul-out to groom and rest. These are not usually connected to runways and are often T- or Y-shaped. Clipped vegetation may be scattered along the runways or piled at a burrow entrance. Burrow systems can riddle the banks, as ideal locations may be occupied for generations. Burrow entrances are large, about 8 cm in diameter, and some may be underwater to provide hidden access to the burrow system. Underground nest chambers are about 9.5 cm in height, 15 cm in length, and 11 cm wide and are filled with globular nests constructed of vegetation. Newly excavated soil outside a burrow entrance gives evidence of an occupied burrow. Foraging tunnels are dug just below the root zone (4–6 cm of the ground surface) so plants can be harvested from below and their tangled roots support the tunnels and prevent them from caving in. Tracks are similar to, although somewhat larger than, those of Meadow Voles. See the "Signs" section of the Meadow Vole account for a track schematic that applies to all the voles and lemmings. Like other voles, North American Water Voles rarely travel on the surface of the snow once there is sufficient build-up to permit tunnel construction through the snowpack.

REFERENCES

Anderson, P.K., et al. 1976; Klaus 2003; Klaus et al. 2001; Ludwig 1984a, 1984b, 1988; Ludwig and Anderson 2009; Nagorsen 2005a; Smith, H.C., 1993; Verts and Carraway 1998.

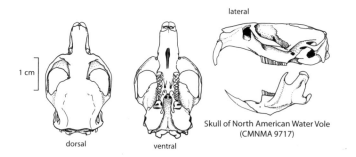

lateral

1 cm

Skull of North American Water Vole
(CMNMA 9717)

dorsal

ventral

Townsend's Vole

FRENCH NAME: **Campagnol de Townsend**

SCIENTIFIC NAME: *Microtus townsendii*

Large numbers of overwintering raptors in the Lower Mainland of British Columbia and on Vancouver Island depend on this vole for much of their food.

DESCRIPTION

These voles have a dark brown dorsal pelage. Belly hairs have a grey base and may be grey, white, or buffy at their tips. Many individuals have an elongated white patch under their throat. Feet are dark brown to medium brown, with dark brown to tan claws. Their tail is relatively long and indistinctly bicoloured, dark brown on the upper surface and grey-brown on the underside. Fur length is relatively short and ears protrude conspicuously beyond the hairs. Adult males and females have discernable hip glands, which are largest and most notable on the larger males, and most evident during the breeding season. These glands exude a greasy secretion that darkens and mats the hairs over the gland. The dental formula is incisors 1/1, canines 0/0, premolars 0/0, and molars 3/3, for a total of 16 teeth. These voles have three pairs of mammae.

SIMILAR SPECIES

Creeping Voles and Long-tailed Voles occur in the same range as Townsend's Voles in the lower Fraser Valley. Creeping Voles are the smallest of the group (total length < 153 mm) with small eyes that are < 4 mm in diameter and plush, shrew-like fur. Long-tailed Voles are somewhat smaller than Townsend's Voles and have a long tail that is 33%–44% of their total length. Townsend's Voles are the largest of the group, typically with a lighter belly than Long-tailed Voles. They are the only vole on Vancouver Island. For skull and teeth diagnoses, see appendix 2.

SIZE

Males and females have similar colouring and are similar in size. The following measurements apply to adult animals in the Canadian part of the range.

Total length: 155–235 mm; tail length: 42–75 mm; hind foot length: 19–29 mm; ear length: 10–18 mm.

Townsend's Vole
(*Microtus townsendii*)

Weight: 40–103 g; newborns, unreported despite several studies of captive breeding.

RANGE

A North American Pacific coast species, Townsend's Voles occur from northern Vancouver Island to northern California, west of the Cascade Range. They are found on many marine islands in both Canada and the United States. The most northerly population occurs on Triangle Island, 40 km offshore of Cape Scott on the northwestern tip of Vancouver Island. Not all the islands off Vancouver Island and the southern mainland are occupied by Townsend's Voles. Perhaps those unoccupied had insufficient grasslands to support viable populations.

ABUNDANCE

The Triangle Island subspecies is on British Columbia's "Provincial Red List" and is considered by the IUCN to be "Conservation Dependant." The whole island is a protected reserve, but the endemic vole population remains vulnerable to natural, accidental, or intentional introductions of predators such as American Mink or rats. Most other island populations are relatively secure, as are those on the mainland. Like many other *Microtus,* Townsend's Vole populations are subject to fluctuations of abundance both annually and over a several-year cycle. Annually, numbers are lowest in late spring and climb over the summer with the recruitment of the young until the population peaks in autumn. Most winters see gradual losses of juveniles and adults. Year-to-year cycles of abundance are irregular. Able to breed during mild winters, Townsend's Voles are quick to increase their numbers if food is plentiful. Numbers drop if the winter is cold and wet, if predation is high, or if food is in short supply. Townsend's Voles can reach the highest average densities of any of the North American *Microtus.* During periods of superabundance, which are uncommon and irregular, numbers can swing up to 525–880 voles per ha, but average annual fluctuations are more commonly 94–239 per ha.

ECOLOGY

Townsend's Voles live in open habitats such as marshes, bogs, grasslands, wet meadows, salt marshes, hay fields, riparian areas, irrigated fields, and alpine meadows. They may sometimes be found at the edges of open forests adjacent to grassy areas. On the mainland of southwestern British Columbia they are restricted to elevations below 300 m, but on Vancouver Island they have been found up to 1860 m in elevation in heights that experience considerable winter snow. These voles are active throughout the year and any time of the day or night. Dispersal may also reduce population density as animals leave their natal area to inhabit often less favourable habitats. Dispersal of both adults and juveniles occurs more commonly during upswings and peaks of population density. Males tend to disperse farther than do females and some females remain to share their mother's home range with her. The home range size for males measured by radio-telemetry averaged around 198–219 m², while those of females averaged 94–152 m² in a study near Ladner, British Columbia. The size of home ranges varies with the amount of food

Distribution of the Townsend's Vole (*Microtus townsendii*)

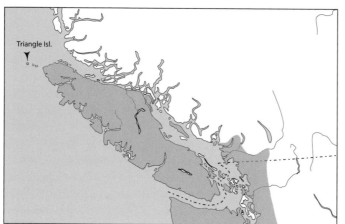

Triangle Isl.

expanded view of Canadian range

and cover. Those animals that are well fed require smaller home ranges. Townsend's Voles are good swimmers and divers and readily enter the water. They often successfully inhabit moist meadows that are partly flooded most winters. Predators are legion as this vole forms the basis of the diet of many. A wide variety of owls and hawks as well as shrikes, Great Blue Herons, Northern Raccoons, weasels, Coyotes, skunks, Bobcats, snakes, and Domestic Cats prey upon Townsend's Voles. Few Townsend's Voles live more than a few months, but the maximum possible lifespan in the wild is not known.

DIET

These voles are herbivorous, consuming leaves, stems, bulbs, and roots of a variety of grasses, sedges, and forbs. Agricultural crops such as clover and Alfalfa will be eaten if available. Apart from populations at elevation on Vancouver Island, most Canadian Townsend's Voles can readily find fresh food throughout the winter. Nevertheless, some populations will gather caches of roots and bulbs, which they consume through the winter along with fresh food. They will sometimes chew tree bark if food is scarce and may girdle trees up to 19 cm in diameter. Most damage is done to seedlings. The alpine

populations on Vancouver Island spend part of the winter under a snowpack. They supplement what plants they can harvest with piles of cuttings that they gathered onto their runways before the snow and with cached bulbs and roots that they store underground or in protected above-ground locations. A winter cache by a single Townsend's Vole living in similar circumstances in Washington contained 13 L of mint bulbs.

REPRODUCTION

The onset and length of the breeding season depends on winter weather conditions and population abundance. In normal years, breeding begins in January and continues until late autumn. In some years, when winters are mild and food is abundant, some females continue to breed all through the winter. Gestation lasts 21–24 days. The average litter size in British Columbia is four to five pups (range 1–9) and larger females tend to have larger litters. Like all *Microtus*, females may experience a postpartum oestrus and are capable of lactating while pregnant. With their short gestation and postpartum oestrus, Townsend's Vole females have the potential of producing numerous litters over the course of their long breeding season, but few survive that long. Like other *Microtus*, Townsend's Vole pups are born pink, hairless, and blind. They begin eating solid food by 15 days old and are weaned by 17 days old. The age at sexual maturity of juvenile females depends on their body weight and averages around 100 days old, but may be as early as 20 days old in very well fed individuals. Wild females need to be > 18 g before they will enter their first oestrus. Males in spring and summer that are > 30 g are sexually mature.

BEHAVIOUR

Adults communicate by scent marking with scat and secretions from their hip glands. They use a characteristic figure-eight movement of the hips to rub their hip glands against the substrate to deposit the scent. Adults of both sexes exclude individuals of the same sex from their home ranges and may fight to do so, especially early in the breeding season. Animals typically stand on their hind legs and box with each other before battles escalate into wrestling and biting. Such fighting may result in bite wounds in the vicinity of the hip glands and rump, especially of the males. A dominant animal may pounce on a subordinate to drive it away. When population density is low, as is commonly the case in spring, each male home range overlaps with the home range of only a single female and the

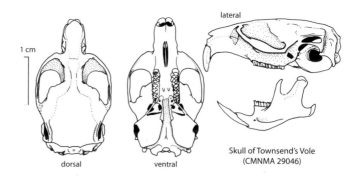

1 cm

lateral

dorsal   ventral

Skull of Townsend's Vole
(CMNMA 29046)

pair appears to be monogamous. In years of high density, or later in the season as density increases with the influx of juveniles, male home ranges begin to overlap with those of multiple females and the mating system becomes more polygynous. Female home ranges tend to become smaller and overlap more as the breeding season progresses and closely related females (mothers and daughters or female siblings) may communally share a home range and then tend to breed synchronously. Communal nursing of litters may occur in such cases. Townsend's Voles can be difficult to capture as they avoid traps, especially those newly placed. They may be more easily caught by hand when a field is being mowed or ploughed, or by overturning hay bales or other sources of cover. Alternatively, a baited but unset trap needs to remain in place for at least a week and, better, two weeks, before being set, to allow the resident vole to become accustomed to it.

VOCALIZATIONS

Little is reported on the vocalizations of this vole. Individuals may squeal when fighting or accosting an intruder on their territory.

SIGNS

Runways can be used for many years by successive generations and may be worn down 50 mm or more into the bare ground. The runways are used most during autumn, winter, and spring and may be partially abandoned during summer when the vegetation becomes rank and thick enough to provide complete concealment. Scat may be found along the runways and is often concentrated in latrine sites, which tend to occur at runway intersections. These latrines may reach 18 cm in length, 8 cm in width, and 13 cm in height, with a sloping ramp leading to the top. Fresh plant cuttings are piled along the runways and near the burrow entrances. Underground burrow systems are dug and used during dry seasons. Burrows may have an underwater entrance. Woven grass nests are constructed in underground chambers or, on or above ground, in hummocks of vegetation when the ground is too wet and burrows are flooded. Scat and tracks are similar to those of Meadow Voles. See the Meadow Vole account for more information.

REFERENCES

Campbell and Summers 1997; Krebs and Boonstra 1978; Krebs et al. 1977; Lambin 1994; Lambin and Krebs 1991; Nagorsen 2005a; Smith, H.C., 1993; Taitt and Krebs 1985; Verts and Carraway 1998.

Taiga Vole

also called Yellow-cheeked Vole, Chestnut-cheeked Vole

FRENCH NAME: **Campagnol à joues jaunes,** *also* **Campagnol Taiga**

SCIENTIFIC NAME: *Microtus xanthognathus*

This distinctively marked northern vole is the second largest of the *Microtus;* only the Water Vole is larger.

DESCRIPTION

Adults and subadults have a prominent rusty orange patch of fur on either side of their snout, on their upper lips where the whiskers insert. A dark stripe down the top of the nose separates the two colourful spots. Dorsal pelage is dark brown to yellowish brown and is grizzled with black-tipped guard hairs. Belly hairs are dark grey–based with lighter grey or white tips. The tail is moderately bicoloured, with a dark brown upper surface and a grey to pale grey underside. Feet are dark brown to tan, with dark brown to tan claws. Ears are short and almost concealed by the long fur. Juvenile pelage is generally darker and finer than that of adults. Adults of both sexes have noticeable hip glands during the breeding season. These glands secrete an oily substance that stains and mats the fur over them. The glands stop secreting and regress outside the breeding season and are then very difficult to detect. The dental formula is incisors 1/1, canines 0/0, premolars 0/0, and molars 3/3, for a total of 16 teeth.

SIMILAR SPECIES

No other vole within the range of the Taiga Vole attains its size. The smaller Rock Vole to the east has similar, but usually darker, rusty patches beside its nose, but the ranges of the two species do not overlap. Heather Voles may have a similar colourful snout, but are considerably smaller; no more than 164 mm in total length. Within the range of the Taiga Vole, the Singing Vole shares a colonial habit, but Singing Voles are found in rocky areas, whereas Taiga Voles occur most often in moist boreal forests near water. To identify skulls, see appendix 2.

Taiga Vole (*Microtus xanthognathus*)

SIZE

Males and females are similar in size and colouring. The following measurements apply to adult animals in the Canadian part of the range. The number of specimens (n) is listed after the measurement range, as records for this species are rare, and these may be the first published for ear length. All are wild-caught individuals in the collection of the Canadian Museum of Nature. Weight varies over the year and is lowest in late winter and highest in early to mid-summer. Total length: 183–209 mm (n=46); tail length: 35–50 mm (n=46); hind foot length: 16–27 mm (n=46); ear length: 17–21 mm (n=23). Weight: 82.2–119.5 g (n=45); newborns, averages 3.5 g (range 2.7–4.2 g [n=28], all born in a captive colony at the Canadian Museum of Nature in the 1970s).

RANGE

Taiga Voles are found only in North America and occur in taiga (boreal) forests from the western shore of Hudson Bay in northern Manitoba to central Alaska and from northern Yukon and the Northwest Territories to west-central Alberta. Their actual distribution is patchy and scattered within this broad range. The southern parts of the range (in Alberta, Saskatchewan, and Manitoba) have provided no recent specimens and the populations that may have existed in those areas are possibly extirpated.

ABUNDANCE

Like many *Microtus,* Taiga Vole populations may experience cyclic fluctuations of density. Irruptions are irregular and anecdotal evidence suggests they occur only about every 20 years or so in Yukon. Minimal data is available, but a three-year study in Alaska suggests that annual fluctuations of density do occur and populations can double in size from a low of 55 per ha to over 100 per ha from spring to autumn. Taiga Voles are unevenly distributed and generally considered to be uncommon.

ECOLOGY

These large voles are eclectic in their habitat choices, but must have good burrowing conditions and an abundant supply of rhizomes for winter food. They are commonly found in wet areas in association with sphagnum moss, but may also occupy dry grassy areas. Taiga Voles can become numerous where a burn or logging has decimated the forest and extensive regrowth has occurred, but they may also be found in boreal forests of primarily Black Spruce. Riparian forest edges are a prime habitat. Dispersal is an important part of the life cycle and individuals may move as far as 800 m away from their natal nest. Most dispersal occurs along water corridors. The voles will spread into marginal habitats during periods of high density, but colonies typically persist in regions of dense boreal forests. During summer, males appear to maintain home ranges exclusive of other males, but that overlap the ranges of several females. Female home ranges overlap with those of males and nearby females. The home range sizes of adult voles are approximately 625 m² and are about the same for males and females. Taiga Voles may be active above ground both day and night. They are very capable swimmers and do not hesitate to enter the water.

Distribution of the Taiga Vole (*Microtus xanthognathus*)

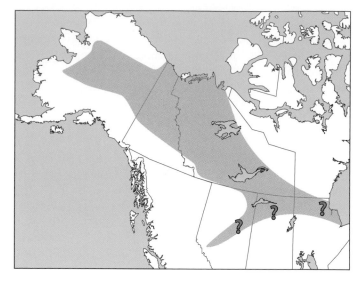

? distribution uncertain

These voles live in underground burrow systems that they dig for themselves or inherit from other Taiga Voles. Burrow systems may be elaborate, with many chambers and branching tunnels. Tunnels usually go down to bedrock or permafrost. Nest chambers may have multiple entrances, but food storage chambers have only one, typically leading directly from the nest chamber. Food caches and nest chambers are often 30 cm below the ground surface. Summer nests are constructed of dried grasses or sedges and are cup-shaped or globular with an inner hollow. They are usually located near or in the litter layer. Globular winter nests, also made of dried grasses or sedges, are typically located deeper underground in a nesting chamber that is tucked into a root or under a fallen log. Food cache chambers may be quite large: 15–30 cm high and 50–100 cm in diameter and contain more than 3 kg (dry weight) of stored food. This size is not unexpected as the contents provide 90%–95% of the food for a group of Taiga Voles for the eight to nine months of winter. The contents of the winter cache are gathered from mid-August until about mid-September. From October to December all the voles lose weight. Those adults still surviving usually die during this time. The surviving juveniles are considerably smaller, with commensurately smaller food requirements and are well suited to survive on a reduced winter food intake. Predators likely include Great Grey Owls and other raptors such as Northern Hawk Owls, Red-tailed Hawks, and Rough-legged Hawks and mammals such as American Martens, Grey Wolves, Red Foxes, Canada Lynx, Ermines, Least Weasels, and even Black Bears. Although some individuals may live two to three years in captivity, the maximum lifespan in the wild appears to be 18 months.

DIET

Stems, leaves, berries, and rhizomes of sedges, grasses, and forbs form the bulk of their diet. Horsetails, lichens, and leaves of short

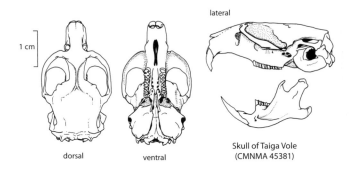

lateral

1 cm

dorsal   ventral

Skull of Taiga Vole
(CMNMA 45381)

males from the group in the spring and take over the group's burrow system and food cache. The subordinate males are then forced to relocate and establish a territory of their own.

VOCALIZATIONS

Taiga Voles are relatively vocal. An animal of any age and sex will produce a high, pulsating squeak as an alarm call. This alarm call may be taken up by other voles as the potential predator wanders through the colony and enters their home range, but ceases as the threat approaches to within 5–8 m. Males emit a low-pitched chirping call as they pursue an oestrous female.

SIGNS

Burrow entrances are 5–7 cm in diameter. Runways are about 5 cm wide and may be littered with small piles of plant cuttings. Some runways may be up to 10 m long. Some are tunnelled through the vegetation, especially sphagnum moss, and others are on the ground surface. Burrows may have throw mounds at their entrance that can measure 30–50 cm high and 200 cm in diameter. Summer nests are approximately 15 cm in diameter. Winter nests are larger as they are shared (25–30 cm in diameter). Latrine sites, which act as scent posts, are sometimes located at runway intersections. Like other voles, Taiga Voles rarely travel on the surface of the snow, preferring to travel along runways dug within the snow pack to benefit from the security and more moderate temperatures there. Tracks and scat are similar to those of other *Microtus*, but are somewhat larger. See the "Signs" section of the Meadow Vole account for more information.

REFERENCES

Carrière 1999; Conroy and Cook 1999; Guilday 1982; Smith, D.A., and Foster 1957; Smith, H.C., 1993; Wolff 1980, 1984; Wolff and Johnson 1979; Wolff and Lidicker 1980, 1981; Youngman 1975.

shrubs (such as dwarf willow) may also be eaten. Fresh parts of the plants are consumed in warm weather and rhizomes are cached for winter use. The voles supplement their winter caches by foraging under the snow for whatever edible plant material they can find.

REPRODUCTION

The breeding season lasts from around early May to August. The onset of breeding is closely tied to snowmelt, so is variable from year to year. Gestation takes 21 days and litter sizes range from 1–13 and average 8–9 pups. Like other *Microtus*, at least some females experience a postpartum oestrus and are capable of lactating and gestating concurrently. Few females produce more than two litters annually. Newborns are essentially hairless and blind. Their eyes open at 12–17 days old and they are weaned around 21 days old. Females raise their young without assistance from the sire, although the male may continue to nest with the female and her litter. Juveniles do not breed in the summer of their birth, waiting through their first winter until the following breeding season.

BEHAVIOUR

Taiga Voles are colonial, although this may be as much a factor of habitat as of sociality. During the breeding season, adult males become aggressive towards each other and defend a territory that includes the home ranges of several females. They then presumably sire all their litters. Females are not considered territorial, but it is possible that only the area immediately around a maternal nest is defended, but not the whole home range, as this differentiation has been observed in captive females. In late summer, breeding ceases, breeding territories break down, and groups of five to ten animals begin to form. These groups may be composed of some related individuals, but are mostly made up of completely unrelated animals. The group creates or expands a burrow, gathers food communally, and huddles together in the same winter nest, presumably for warmth. Scent marking is an important communication strategy for Taiga Voles. Reproductively active adults mark their runways, burrows, and food caches by scratching their hip glands with a hind foot then patting the foot onto the ground to spread the scent. They will also rub up against objects to directly place scent from their hip glands. Adult males have been seen to drag their anal region on the ground, presumably depositing scent from anal glands. Males will fight and are commonly wounded during dominance and territorial battles. The dominant male of a winter group will exclude other

Southern Red-backed Vole

also called Boreal Red-backed Vole

FRENCH NAME: **Campagnol de Gapper,** *also*
Campagnol-à-dos-roux de Gapper

SCIENTIFIC NAME: *Myodes gapperi,* formerly *Clethrionomys gapperi*

A common species in mature and second-growth forests, this vole is not one of the small mammals that try to move into dwellings in late autumn. The scientific name of this genus has recently been changed from *Clethrionomys* to *Myodes.*

DESCRIPTION

Most Southern Red-backed Voles have a silvery underside, greyish-brown sides that have an ocherous or yellowish wash, and a broad

Southern Red-backed Vole (*Myodes gapperi*)

dark phase

light phase

band of chestnut-coloured fur along their back from forehead to tail. Their thinly furred tail is weakly bicoloured, dark above and silvery-grey, buff, or slate grey below, usually with a black tip. Ears and eyes are noticeable. A second colour phase is possible in any population; this dark form (melanistic) is sooty black on their back, and their sides are greyish brown. Individuals with the dark pattern are more common in northern and eastern populations. For example, blackish animals are frequent in parts of central Ungava. Very rarely, an individual will be blond with no dark hairs at all, or partially albino. Young recently out of the nest have a short juvenile coat that is slate grey to silvery grey below and dark brownish above with rufous hairs along the back. They undergo a subadult moult that is complete at about one month of age, when they develop a slightly darker version of the adult pelage, but with adult-coloured (silvery) belly fur. The fully adult pelage is acquired between the ages of 40 and 120 days old depending on the condition of the animal and the elevation. Voles in better condition moult earlier, as do those from higher elevations, where the growing season is short. There are two annual moults, in autumn and in spring. Moult in pregnant or nursing females is delayed. Winter fur is longer, denser, and brighter than the summer pelt. The dental formula is incisors 1/1, canines 0/0, premolars 0/0, and molars 3/3, for a total of 16 teeth.

SIMILAR SPECIES

The reddish band of colour on the back of most red-backed voles is distinctive and clearly distinguishes them from other small rodents. In the northwestern part of their range, Southern Red-backed Voles co-occur with Northern Red-backed Voles. Distinguishing these two species is difficult. Adult Northern Red-backed Voles have a more densely furred tail that is thicker and yellowish on the underside and they are typically brighter rufous on the back. However, no single character distinguishes all Northern from Southern Red-backed Voles. To distinguish vole skulls see appendix 2.

SIZE

Size varies considerably across the range, with the smallest forms in the prairies and the largest in the northwest. Males and females are similar in size and colouring. Weight fluctuates over the year and is lowest in midwinter and highest in late spring. The following measurements apply to adult animals in the Canadian part of the range and cover the variation in size.

Total length: 110–164 mm; tail length: 22–58 mm; hind foot length: 14–22 mm; ear length: 9–19 mm.

Weight: 13–45 g; newborns, 1.7–2.3 g.

RANGE

The northern limit of distribution of Southern Red-backed Voles is northern Labrador and Ungava to southern Nunavut and the Northwest Territories and extreme southeastern Yukon. They occupy most of the remainder of southern Canada, except southern Ontario and the dry prairies of Saskatchewan and Alberta. They inhabit the Rocky Mountains southwards to Arizona and New Mexico, and the Appalachians southwards to far northern Georgia. In the mid-continent, they occur as far southwards as the northern border of Iowa. There are no Southern Red-backed Voles on certain large marine islands such as the Queen Charlottes (Haida Gwaii),

Distribution of the Southern Red-backed Vole (*Myodes gapperi*)

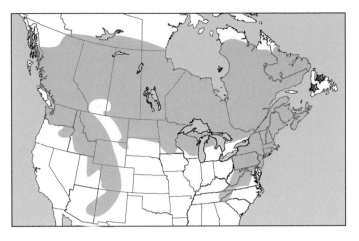

 introduced

Vancouver Island, and Anticosti, but they have reached Akimiski Island in James Bay, Cape Breton Island and Prince Edward Island, and many freshwater islands, including Manitoulin Island. First discovered in Newfoundland in 1999 after either an accidental or an intentional introduction, they now occur around Little Grand Lake (1999) and Stephenville (2000). Southern Red-backed Voles introduced for scientific studies onto Camel Island in Notre Dame Bay, Newfoundland, in the 1960s may still survive.

ABUNDANCE

These are common and often abundant rodents.

ECOLOGY

Southern Red-backed Voles live in moist habitats, preferring mixed forests with well-developed leaf litter and many rotting stumps and logs. But abandoned fields, stone fences, bogs, rockslides, meadows, prairie grasslands, coniferous forests, and even tundra will also support populations, as long as there is water nearby. Not usually an opportunistic species, Southern Red-backed Voles are most abundant in mature forests and typically do not exploit newly disturbed habitats, such as occur after clearcutting or fire. They will move into these habitats after a few years when rampant growth provides more cover. These rodents remain active all year, surviving in tunnels beneath the snow during the winter. They may be active in both darkness and daylight, but in summer tend to be more active at night. Population density reaches a peak in late autumn at the end of the breeding season and is at its lowest in early spring, especially after a hard winter. Winters with cold temperatures but limited snow or rapid spring thaws that result in flooded burrows, increase cold-season mortality substantially. Numbers rise and fall seasonally, and some populations do experience cyclic fluctuations possibly related to food abundance, but in general red-backed voles tend not to undergo cyclic population explosions as do other small mammals such as Meadow Voles or lemmings. They are capable swimmers and readily enter the water. Home ranges vary from 0.01

to 0.50 ha. Maximum longevity is about 20 months, but most live less than 10–12 months and very few survive more than a single winter. Hawks, owls, and small weasels (up to about American Marten size) are significant predators of Southern Red-backed Voles.

DIET

Diet varies seasonally, and Southern Red-backed Voles are opportunistic feeders with a taste for a wide variety of foods. Leaves, seeds, nuts, acorns, fruits and berries, mosses, lichens, and ferns are primary foods in summer. Insects and other invertebrates may supplement their diet, but are not a major component. They may consume large amounts of subterranean fungi (truffles) in summer and fall. During the winter they mainly eat the seeds, roots, bark, and perennial plants they can find under the snow. These voles drink a relatively large amount of water each day. This requirement plays a major role in their habitat selection.

REPRODUCTION

Southern Red-backed Voles are one of the more prolific rodent species. Their breeding season is long, from April until early October. Early spring breeding may occur under a deep snow cover in northern or subalpine areas. Gestation lasts 17–19 days and the average litter is 4–6 pups (range 1–10). Within this litter-size range, larger litters tend to occur in northern parts of the distribution. Well-fed females and older females also tend to have larger litters. Pups are born deaf, blind, and hairless (except for well-developed vibrissae). They grow quickly. By day 7–8 they have a full coat of short hairs (called the nestling coat). Eyes are open and they are eating solid food by day 15. Weaning begins at about 12 days old and is complete around day 17–20, when the youngsters weigh 11–12 grams. Meanwhile, their mother may have bred again soon after their birth and been both lactating and pregnant. She could whelp her new litter soon after the previous one is weaned. If a new litter is not imminent, the female may continue to suckle the latest litter until the pups are 28–30 days old. Age of sexual maturity varies from 2–4 months (3 months old on average). Ten litters per year are theoretically possible for an adult female, but few manage to rear more than two to three, and rarely four, litters annually. Young females born in a spring litter may produce one to two litters before their first winter.

BEHAVIOUR

Adult red-backed voles are generally solitary. Adult female–male relationships typically occur only when the female is in heat and

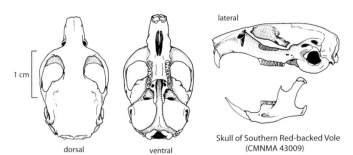

lateral

1 cm

Skull of Southern Red-backed Vole (CMNMA 43009)

dorsal   ventral

tend to be generally amicable, with the female holding the dominant position. Many females drive the males away after mating, but some are more tolerant of their mate's presence. There are reports of fathers congenially remaining nearby and interacting with the rapidly growing pups in the natal nest. Nevertheless, males do not contribute to the care of the young; they merely rest in contact with them, perhaps providing some warmth. Adults of the same gender mostly avoid contact and, when they do meet, often react with aggression towards each other and one is usually chased away. A dominant vole stretches upwards in the forequarters, while a submissive animal lowers its head and body and often closes its eyes. When fighting, the voles rise on their hind feet and pummel their opponent with their front feet. Physical contact may then evolve into a tumbling ball of voles. The victor chases the vanquished from the field and then may scent-mark the area by scratching one or both of its flank glands (to spread the scent onto the feet). The home ranges of males typically overlap extensively with the smaller home ranges of several females. Red-backed voles will also exclude other species, such as Woodland Jumping Mice and possibly Deer Mice and Meadow Voles, from their woodland territories. These voles are capable climbers and sometimes forage in trees and occasionally build their nests in tree holes. Most nests are constructed nearer the ground in natural cavities in deadfall, in abandoned subterranean or subnivean burrows created by other species, or in abandoned nests of other species.

VOCALIZATIONS

Southern Red-backed Voles are not especially vocal. A subordinate vole may produce a high-pitched squeal towards a dominant vole (for example, a juvenile encountering an adult). A similarly high-pitched, but different-sounding squeak is produced by a dominant animal as a threat. They sometimes squeal and squeak when fighting or when captured and they chatter their teeth when they are anxious.

SIGNS

Tracks, nests, and scat of the three red-backed vole species that occur in Canada are indistinguishable. Nests are spherical and hollow, 7.5–10.0 cm in diameter, typically woven from grasses, and lined with shredded dried leaves or moss. They may be found occasionally in trees, but are more commonly constructed under cover (in deadfall, under leaf litter, in dense vegetation) at ground surface and especially under snow. Tracks follow the classic rodent pattern of five toes, with toe 1 (the pollex) on the front feet highly reduced. On the hind foot, toes 1 and 5 angle to the side and toes 2, 3, and 4 cluster together and point forward. On the front foot, toe 1 rarely registers and toes 2 and 5 angle to the side, while toes 3 and 4 aim more forward. Front tracks are smaller than hind. Claws may or may not register depending on the substrate. Typical gait is a trot, but they will speed up to a lope or a bound when exposed or threatened. Red-back voles create runways or use those of other small mammals. They often create small holes in leaf litter or snow (diameter 2.5–3.2 cm). Scat is oval (circular in cross section) and about 0.2 cm in diameter and 0.6–0.8 cm in length. Droppings are dark brown or black. See the Meadow Vole account for a track schematic that applies to all voles.

REFERENCES

Banfield 1974; Bowman and Curran 2000; Choate et al. 1994; Elbroch 2003; Elias et al. 2006; Gliwicz and Glowacka 2000; Hearn et al. 2006; Innes and Millar 1993; Jung et al. 2006a; Kurta 1995; Mahoney 2001; McGuire 1997; Merritt 1981; Merritt and Zegers 1991; Mihok 1976; Nagorsen 2005a; Ovaska and Herman 1986; Sare et al. 2005; Slough and Jung 2007; Smith, H.C., 1993; Sopher 2000; Stewart, C.A., 1991; Verts and Carraway 1998; Vickery 1979; Wilson, D.E., and Reeder 2005.

Bank Vole
also called European Bank Vole

FRENCH NAME: **Campagnol roussâtre**
SCIENTIFIC NAME: *Myodes glareolus*, formerly *Clethrionomys glareolus*

Nine of these European rodents were introduced onto an island in Notre Dame Bay, Newfoundland, in the 1960s for scientific study. They rapidly took over the whole island, but have not reached the main island. Very little work has been conducted on this introduced population, so most of the information presented here is derived from studies on Eurasian populations. The scientific name of this genus has recently been changed from *Clethrionomys* to *Myodes*.

DESCRIPTION

This section describes animals from the introduced population in Newfoundland and does not include the full diversity of the species in Eurasia. Like the other voles in the *Myodes* group, the Bank Vole has a broad band of ruddy-coloured fur along its back from forehead to base of tail. Its sides are greyish brown, sometimes with a rufous wash, especially higher on the flanks nearer the back. The

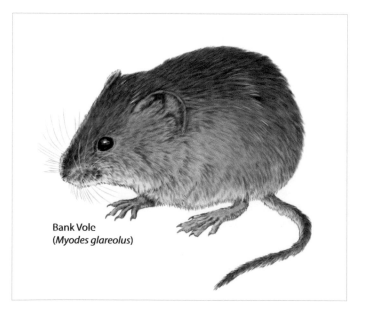

Bank Vole
(*Myodes glareolus*)

Distribution of the Bank Vole *(Myodes glareolus)*

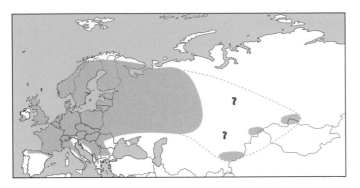

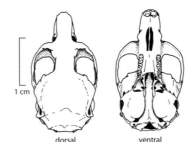
introduction   possible range

belly hairs are buffy at the tip and dark grey at their base. The feet are greyish white. The short tail is distinctly bicoloured, dark brown to brown above and buffy below. Ears and eyes are noticeable. Young, recently out of the nest, have a short juvenile coat that is slate grey below and dark greyish brown above, with some rufous hairs along the back. There are two annual moults, in autumn and in spring. Winter fur is longer, denser, and brighter than the summer pelt. The dental formula is incisors 1/1, canines 0/0, premolars 0/0, and molars 3/3, for a total of 16 teeth.

SIMILAR SPECIES

Although this vole is very similar to both the native Northern and Southern Red-backed Voles, its isolated location helps in the identification. No other species of red-backed voles occur on Yellow Fox Island, Newfoundland. Bank Voles tend to have longer tails than our native red-backed voles. To distinguish vole skulls see appendix 2.

SIZE

The following measurements apply to adult animals in the European part of the range and include several Canadian specimens.
Total length: 136–160 mm; tail length: 33–72 mm; hind foot length: 16–20 mm.
Weight: 14–40 g; newborns, around 2 g.

RANGE

The Bank Vole is a Eurasian species native to Europe, Scandinavia, and western Russia. It was introduced in 1967 onto Yellow Fox Island, a 57-ha island south of Fogo Island, Newfoundland, and fairly rapidly became abundant. A survey in 2004 found no Bank Voles on Yellow Fox Island, so it is possible that the colony has died out.

ABUNDANCE

These voles appear to be rare and possibly extinct on Yellow Fox Island, Newfoundland, the only Canadian introduction site. In Eurasia, where they are native, populations may be cyclically abundant or stable at relatively low numbers.

ECOLOGY

Bank Voles prefer to live in mixed deciduous or dense coniferous forests, but can survive and even flourish in scrubland, park land, hedges, and banks especially, where there is dense growth of forbs for food and cover. They will sometimes invade dwellings, especially during winter. Bank Voles are active all year and during both daylight and darkness, but tend to be more nocturnal during summer. They live in burrows in the leaf litter and under cover of vegetation and forest debris, and can be difficult to see at times. They are also agile climbers and sometimes make themselves a bit more visible as they scramble about in shrubs and over fallen logs. Home ranges vary from 0.05 to 0.73 ha and those of males are larger than those of females. Maximum lifespan in the wild is around 18 months and few survive more than one winter. In captivity they may live 40 months. Predators include a wide variety of hawks and owls, as well as Red Foxes and most of the smaller weasels. The population on Yellow Fox Island has avian, but no mammalian, predators.

DIET

These rodents are almost entirely herbivorous. They eat fruits, seeds, leaves and buds, fungi, grass, and some insects and worms during the summer and roots, bark, moss, and even dead leaves in winter. Food stores are collected by some northern populations, but it is not

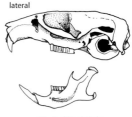

lateral

Skull of Bank Vole
(CMNMA 52297)

1 cm

dorsal   ventral

known whether this is the habit of the Yellow Fox Island population. Coprophagy occurs in this species.

REPRODUCTION

The breeding season extends from April to September/October. Females may mate with more than one male, so multiple paternity is possible. Onset of breeding may be delayed when populations are dense and may occur up to a month early in sparse populations when food is abundant. Each mature female is capable of producing four to six litters annually and can be pregnant and lactating simultaneously. When voles are present in high densities, some females do not breed during spring. Gestation lasts about 16–18 days (it may be extended to 19–22 days if the female is also lactating) and litters are typically 3–5 pups. Females provide all of the parental care and pups are weaned at around 14 days old. Sexual maturity of juveniles may occur as early as 35 days old in very well nourished populations, but the presence of a mature female can suppress the sexual maturity of young, especially in regions with high population density. Early-born juvenile females may become sexually mature and produce one to two litters in their first summer. Late-born juveniles do not mature until the following spring.

BEHAVIOUR

Outside the breeding season, these voles are gregarious and may even share nests and burrows. During the breeding season, mature females establish and aggressively defend exclusive territories, and only when they are in heat will they allow the approach of a male. Breeding male Bank Voles also defend territories that are larger than those of females. They fight with neighbouring males and thereby establish a local dominance hierarchy. Fighting males roll around biting each other and the winner chases the loser away. The dominance hierarchy is established early in the breeding season, but may not last for the whole period, especially if a dominant male dies. Males use urine to scent-mark their territories. Sexually mature males deposit urine trails, while females and immature males leave puddles. Dominant males scent-mark more often than do subordinate males and their urine carries chemical signals that indicate their status. Females can detect these odours and prefer to mate with dominant males, so dominant males sire more pups than subordinate males. The urine of Bank Voles is brightly visible in the ultraviolet range and may be used as a cue by raptors as they can detect these wavelengths of light. The urine of mature males is most reflective, perhaps making them more vulnerable to that manner of predation. Infanticide is commonly practised by adults of both sexes when they encounter an unattended nest with neonates. Between 25% and 67% of the time the stranger will kill all the pups in the nest. Voles unrelated to the juveniles are more likely to commit infanticide, so offspring survival is enhanced when related females cluster their territories together.

VOCALIZATIONS

Males squeal when fighting and a subordinate male will squeal if his burrow is invaded by a dominant male. Tooth chattering occurs when a vole is in an anxious state. Pups squeak if disturbed or hungry. Ultrasonic frequencies (beyond human hearing) are used by pups and their mothers to call to each other and are produced by males when they are mating.

SIGNS

See the "Signs" section of the Southern Red-backed Vole for a description of nests and scat that is applicable to all three species of red-backed voles in Canada. See the "Signs" section of the Meadow Vole account for a track schematic that applies to all voles and lemmings.

REFERENCES

Alibhai 1985; Eccard and Ylönen 2001; Horne and Ylönen 1996; Klemme et al. 2006; Koivula et al. 1999; Lee and Houston 1993; Macdonald and Barrett 1993; Mahoney 2001; Mappes and Koskela 2004; Mappes et al. 1995; Prévot-Julliard et al. 1999; Rozenfeld and Rasmont 1991; Rozenfeld et al. 1987; Stenseth 1985; Ylönen and Viitala 1991; Ylönen et al. 1997a, 1997b.

Northern Red-backed Vole
also called Ruddy Vole, Red Vole (in Europe)

FRENCH NAME: **Campagnol boreal**
SCIENTIFIC NAME: *Myodes rutilus*, formerly *Clethrionomys rutilus*

The Northern Red-backed Vole is, as its name suggests, a northern species that occurs in both North America and Eurasia. The scientific name of this genus has recently changed from *Clethrionomys* to *Myodes*.

DESCRIPTION

The coat colour of these small voles is quite variable. In general, they have a broad rufous coloured band of fur running along their back

Northern Red-backed Vole
(*Myodes rutilus*)

Global distribution of the Northern Red-backed Vole (*Myodes rutilus*)

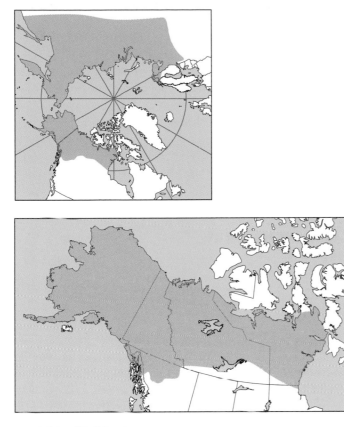

expanded view of North American range

from forehead to tail. Some of the ruddy colour washes down onto the yellowish to grey-brown sides. Belly hairs are tipped with silver, pale grey, or buff and the base of each hair is dark grey. Tops of the feet may be greyish, buff, or silver. The underside of the tail is silvery, buffy yellow, or light grey and lighter than the upper surface. The short, thick tail is covered with bristly hairs and the blunt end of the fleshy part of the tail has a long (10–12 mm) brush of dark hairs. Ears and eyes are noticeable. Two colour phases are possible (as with the Southern Red-backed Vole). The red phase is far more common than the dark phase, which has a sooty grey-black dorsal band, grey-brown sides, and a grey to silvery belly. Silky winter pelage is longer, denser, and brighter than summer pelage. Juveniles are duller and darker than adults. The dental formula is incisors 1/1, canines 0/0, premolars 0/0, and molars 3/3, for a total of 16 teeth.

SIMILAR SPECIES

The reddish band of colour on the back of most red-backed voles is distinctive and clearly distinguishes them from other small rodents. It is possible that there may be narrow zones of overlap along the southern border of the Northern Red-backed Vole range and the northern border of the Southern Red-backed Vole range. Distinguishing these two species can be difficult. Adult Southern Red-backed Voles have a more sparsely furred tail that is thinner and their fur colour is typically somewhat duller. However, no single characteristic distinguishes all Northern from Southern Red-backed

Voles. For an illustrated guide to vole skull and tooth identification see appendix 2.

SIZE

Males and females are similar in size and colouring. Weight fluctuates over the year and is lowest in midwinter and highest in late spring for males and early summer for females. The following measurements apply to adult animals in the Canadian part of the range.
Total length: 118–166 mm; tail length: 21–44 mm; hind foot length: 15–22 mm; ear length: 11–17 mm.
Weight: 18–35 g; newborns, about 1.7 g.

RANGE

The range of this northern species includes most of mainland Northwest Territories, Nunavut, Yukon, and Alaska as well as northwestern British Columbia and extreme northeastern Manitoba. Across the Bering Strait, Northern Red-backed Voles occur across northern Russia to Scandinavia.

ABUNDANCE

Generally, numbers of these voles are moderately low throughout the range, but they may become abundant at times in some regions.

ECOLOGY

Northern Red-backed Voles occupy a broad niche. They are found in a variety of habitat types such as stunted coniferous/birch forests typical along the treeline, dense boreal forest, bogs, well-drained slope forests, river and creek beds, clearcuts, regenerating burns, and even tundra. Active all year, these small rodents dig and use extensive snow tunnels to forage and travel undetected and protected during winter. Although they are nocturnal or crepuscular by choice, the long hours of summer daylight in the north requires that some foraging must occur under bright conditions. Extent of their home range is thought to vary with food availability. A study undertaken in a region in Scandinavia measured average home ranges of females at 0.42–0.65 ha. Home ranges of males were not measurable (as they extended beyond the 6.7 ha study area), but were clearly much larger than those of females. A female home range in a more productive environment could be as small as 0.05 ha. Population sizes of this vole fluctuate in an irregular manner. Usually vole numbers are low, but unpredictable outbreaks do occur, for reasons as yet poorly understood. Populations have been artificially enhanced in some regions by supplementing the natural food sources, so it is thought that this species is food limited, which is, in turn, likely

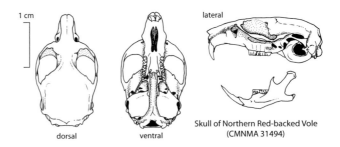

1 cm

dorsal   ventral

lateral

Skull of Northern Red-backed Vole
(CMNMA 31494)

weather-related. Maximum lifespan in the wild is around 14 months, although very few survive a single winter, as early winter mortality is high especially among late-born juveniles. Small weasels such as Least Weasels, Ermines, and Long-tailed Weasels are primary mammalian predators. Hawks and owls are the vole's primary avian predators.

DIET

The diet of Northern Red-backed Voles is poorly studied. It appears that they consume primarily plant material: mainly a variety of berries such as blueberries, cranberries, and crowberries and seeds supplemented with mosses, lichens, mushrooms, leaves, and some arthropod invertebrates.

REPRODUCTION

The breeding season extends from spring (May–June) until late summer or early autumn (August–September) and varies in length from year to year, depending on the local weather conditions. More northerly populations experience shorter breeding seasons than those farther south, and an early winter onset anywhere in the range will shorten the breeding season. The first litter of the year is usually conceived under the snow and is born in May or June. Gestation lasts 17–20 days. Litters average 4–6 pups (range 1–11). Juveniles normally reach sexual maturity around three months old, but this maturity may be delayed during years of high population density. Under those circumstances, young-of-the-year maturity is delayed in direct proportion to density. In years of peak numbers, juveniles may not breed until the following spring and even mature females show depressed reproduction. Late summer–born voles do not become sexually mature until the following spring. Mature females may have 2–3 and rarely 4 litters per year, and young females born in spring may have 1–2 litters before winter.

BEHAVIOUR

Unlike Southern Red-backed Voles, Northern Red-backed Voles often attempt to overwinter in buildings. Females defend exclusive territories and likely permit the presence of males only when they are in heat. Territories of males overlap those of several females and are much larger than those of the females. Territory size decreases in autumn. Neighbouring females display mutual avoidance. Males exercise a combination of primarily mutual avoidance and occasional aggression towards each other. These behaviours are suppressed in winter, as animals in some populations appear to congregate and possibly even huddle together for warmth in favourable habitats.

VOCALIZATIONS

Vocalizations are not discussed in the literature; however, it is likely that they are similar to those of the closely related Southern Red-backed Vole.

SIGNS

Signs are indistinguishable from Southern Red-backed Voles. See the "Signs" section of the Southern Red-backed Vole account for more information and the "Signs" section of the Meadow Vole account for a track schematic that applies to all voles and lemmings.

REFERENCES

Banfield 1974; Carrier and Krebs 2002; Gilbert, B.S., and Krebs 1991; Macdonald and Barrett 1993; McPhee 1984; Mihok 1976; Nagorsen 2005a; Novikov and Moshkin 1998; Soper 1961; Viitala 1987; West, S.D., 1977, 1982; Whitney 1976; Wilson, D.E., and Reeder 2005; Youngman 1975; Zuercher et al. 1999.

Common Muskrat
also called Muskrat

FRENCH NAME: **Rat-musqué commun**
SCIENTIFIC NAME: *Ondatra zibethicus*

These prolific rodents are North America's most important fur-bearers. They are also the largest and most aquatic of the voles on the continent. While Common Muskrats and North American Beavers are both rodents, they are not closely related, although both have evolved similar physical and behavioural adaptations to a semiaquatic lifestyle.

DESCRIPTION

A Muskrat's coat is composed of exceptionally dense, woolly underfur through which project the lustrous and much longer guard hairs. Underfur hairs are bicoloured, dark grey along most of the hair shaft, and creamy to buffy or pale grey at the tips. This underfur traps a layer of air next to the skin and is essentially waterproof. Most of the coat colour is derived from the guard hairs. Pelage colour is highly variable, although most individuals in an area tend to be similarly coloured. Coat colour on the back varies from dark chocolate brown to mahogany and chestnut brown, grey-brown, pale rufous, and yellowish brown, to pale brown. Colour mutations such as black, blonde, grey, and albino are rare. Fur colour is usually darkest along

Common Muskrat
(*Ondatra zibethicus*)

Distribution of the Common Muskrat (*Ondatra zibethicus*)

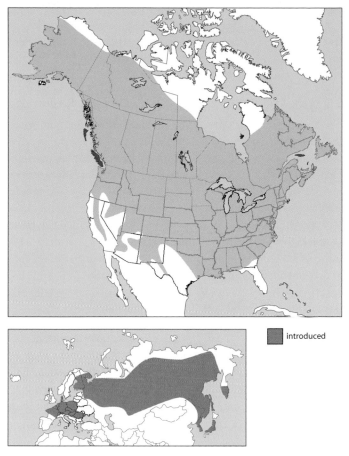

introduced

introduced range in Eurasia

the spine and top of the head, as that is where the most guard hairs are concentrated. The belly is consistently paler and takes on a wash that varies with the colour of the guard hairs, but the number of guard hairs is much reduced on the underside, so colour saturation is not pronounced and most of the colour is provided by the underfur. Juveniles are typically darker and greyer than adults. Muskrats have a long (about 50% of their total length), scaled, blackish tail that is laterally flattened and sparsely furred except for a ridge of stiff black hairs on the upper and lower keels. Older animals tend to have fewer hairs on the tail. Feet are usually dark brown or grey with stout, creamy, or light brown claws. Each hind toe and the edges of the large hind feet are surrounded by a stiff fringe of hairs that add considerably to the foot's surface area, no doubt to increase swimming capability. The very slight webbing between the hind toes contributes little to propulsion compared to the fringe. The front feet are small and not used while swimming, but are employed for terrestrial locomotion, digging and food handling. Their eyes are small and ears barely extend beyond the fur. Furred lips can close behind the incisors to prevent accidental ingestion of water while gnawing or transporting food underwater. All adults develop a pair of musk glands (also called anal glands) alongside the anus. These are most enlarged during the breeding season and produce a secretion used to mark home range. The dental formula is incisors 1/1, canines 0/0,

premolars 0/0, and molars 3/3, for a total of 16 teeth. Adult Muskrats undergo a single prolonged annual moult that extends through the summer and is completed in early autumn. Muskrats have six mammae.

SIMILAR SPECIES

This large, semiaquatic rodent is difficult to mistake. Adult North American Beavers are similarly aquatic, but considerably larger, with broad flat tails. Although juvenile Beavers may be the size of a Muskrat, they already have a Beaver's characteristic flat tail. The introduced Nutria is also larger than a Muskrat, with a larger head, a tail that is round in cross-section, and usually a white patch under the chin. Of the large aquatic North American rodents, only Nutria and Muskrats scull with their tails as they float at the surface. Distinguishing skull and teeth characteristics are illustrated in appendix 2.

SIZE

Males and females are similar in size and colouring. The following measurements apply to animals of reproductive age from the Canadian part of the range.

Total length: 448–685 mm; tail length: 195–295 mm; hind foot length: 63–90 mm; ear length: 19–26 mm.

Weight: 700–1770 g (average 1100–1300 g); newborns, around 15–22 g.

RANGE

The Common Muskrat is a widespread species indigenous to North America. It occurs naturally from Newfoundland to British Columbia and from Alaska and northern mainland Northwest Territories and Nunavut to Texas. Muskrats have been introduced to Europe and Asia, southern South America, as well as onto the following British Columbia islands: Graham Island, Vancouver Island, Pender Island, and 11 additional smaller islands between the mainland and Vancouver Island. Introductions have also occurred onto Anticosti Island and the Grand Manan archipelago. Most introductions originate as inadvertent escapees from fur farms, followed by naturalization and often rapid range expansion (as in Europe). Those in British Columbia were intentionally released with the aim of creating trappable populations and have spread extensively from the original introduction sites. Populations introduced into the British Isles in the early 1900s were eradicated within 40 years.

ABUNDANCE

Muskrats are mostly common in suitable habitats, but may be scarce in some parts of the range such as in south coastal British Columbia, the central Prairies, and the far north. Populations are sometimes subject to dramatic fluctuations in density, usually in the range of up to a threefold increase. Analysis of annual fur returns of the Hudson Bay Company suggests that populations in boreal and mountainous regions of Canada peak every eight to ten years and in the subarctic-Arctic region every three to five years, although the regularity of these peaks is subject to some controversy. Mink population peaks (in western and northern Canada) tend to follow those of Muskrat, with a one-to-three-year lag similar to those of Canada Lynx and

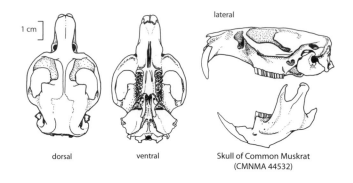

1 cm

lateral

dorsal ventral Skull of Common Muskrat
(CMNMA 44532)

Snowshoe Hare. In eastern Canada, no such relationship between Mink and Muskrat population cycles has been discovered. Perhaps prey diversity is lower in western and northern Canada, forcing Mink in those regions to specialize more on Muskrats. Densities of Muskrats in Canada range from 7 to 60 animals per ha in suitable habitat. Despite centuries of trapping pressure, Common Muskrats in Canada are thought to be as numerous now as they were before European settlement.

ECOLOGY

Almost any permanent waterway that supports emergent aquatic vegetation could sustain a Muskrat. They avoid rapidly rushing mountain streams or seasonally transient water and prefer slowly moving streams, rivers, ponds, lakes, and especially marshes where the water is deep enough not to freeze right to the bottom in winter. Although they live primarily in fresh water, Muskrats will occupy brackish estuaries and have been discovered swimming in saltwater at some distance from shore. When swimming at the surface, their front feet are tucked up under the chin and all the propulsion is provided by powerful, alternate strokes of the hind feet. The tail stretches out behind, moving from side to side acting as a rudder. When swimming underwater, the tail is skulled rapidly from side to side and provides some forward impulsion as well. Top forward swimming speed is about 5 km/h (both at the surface and submerged), and they can swim backwards when required. Average swimming speeds are 2–3 km/h. Muskrats can swim 90 m underwater and hold their breath for up to 20 minutes if necessary, but the extent of most dives is only a minute or two. Like many other aquatic mammals, Common Muskrats experience bradycardia (reduced heart rate) when diving. A laboratory test showed one animal's heart rate slowed to 54 beats per minute from the regular 310 beats per minute after only 2 seconds submerged. Common Muskrats are also highly resistant to the effects of carbon-dioxide build-up in their blood and tissues and can tolerate CO_2 levels much higher than most mammals. This ability is important both for diving and for their communal existence in winter lodges and burrows, where CO_2 levels tend to build up. Muskrats typically forage within 60 m of their burrow or lodge, so home ranges tend to be small. Normally the larger, older animals are more dominant and occupy the preferred sites. Muskrats may be active any time of the day or night, but appear to have two main activity peaks in each 24-hour period, one in late afternoon and

another in late evening. They are more likely to be out foraging in daylight on rainy days.

Muskrats do not hibernate, remaining active throughout the winter beneath the ice. Depending on the environment, they will den-up in a bank burrow or in lodges they build out of mud and available emergent aquatic vegetation (commonly cattails). These lodges are constructed on some kind of solid base, such as a raft of vegetation, a mud bar, rocks, snags, stumps, or collapsed and decaying lodges. The Muskrat mounds up a pile of mud and vegetation (not including any woody vegetation) and then hollows out the centre. Summer lodges are smaller, with only one central chamber. The larger winter lodges may have two or three chambers. The smaller lodges may be used as feeding shelters during the winter. Each lodge has at least one plunge hole with an underwater exit and larger lodges may have several. All lodges require ongoing maintenance during the warm season as they gradually decompose or dissolve in the weather. Bank burrows are preferred winter accommodation, if they are available, as they provide superior weather and predator protection. Push-ups are created over a 10–15 cm wide plunge hole that is chewed through the ice soon after freeze-up. Wet vegetation and mud is pushed up through the hole and may freeze into a dome over the hole, with an airspace below it to provide a single animal room to haul-out onto the ice to feed or rest. The protected ice surface under the push-up quickly becomes covered with feeding debris. The conical dome protects the Muskrat from predators and protects the plunge hole from being snow-filled. During severe winters with little snow, the plunge holes in push-ups are prone to freezing solid, severely limiting the foraging range of the Muskrats, leading to food shortages and starvation. Some push-ups do not develop into a dome, but instead become a volcano-like mound of feeding debris with a plunge hole through the centre. Both push-ups and lodges are vulnerable to consumption by ungulates, especially Wapiti in the west and Caribou in the north. These animals will paw open the structure and eat the aquatic plants used in the construction. Such activities may expose the central cavity, making the structure useless to the inhabitants unless rapid repairs are undertaken. Dispersing Muskrats may dig short tunnels into a stream or river bank for temporary shelter, but burrows in prime areas tend to evolve into complex burrow systems as they are enhanced and renovated by successive generations. Although they do occasionally make extended overland treks, moving slowly and awkwardly on land, most Muskrats disperse along waterways and most juveniles remain near their natal home range, unless forced to move farther away due to crowding or lack of food. Predators include hawks, eagles, owls, foxes, American Mink, Coyotes, Domestic Dogs, Northern Raccoons, American Badgers, Bobcats, Fishers, Northern River Otters, and likely Grey Wolves, Black Bears, and Canada Lynx, but their major predators are humans. Millions of Muskrats are trapped annually around the world for the fur market. Maximum longevity in the wild is three to four years, but few adults survive a second winter.

DIET

A Muskrat's diet is primarily composed of leaves, rhizomes, and basal portions of the stems of emergent aquatic plants such as cattails, bulrushes, sedges, horsetails, water lilies, arrowheads, and Sweet Flag. They are particularly fond of corn, apples, and carrots, which are successfully used as bait in traps. Some animal flesh in the form of fishes, frogs, and invertebrates like crayfish and clams is also consumed. They will also scavenge the flesh of dead Muskrats. Muskrats are voracious eaters. Captive animals will consume 25%–30% of their body weight daily, depending on the kind of food provided. When numbers are especially high, wild populations can cause what is called an "eat-out," where they literally mow down the available vegetation in their foraging efforts. Widespread dispersal and "die-outs" as a result of starvation or disease are common consequences of this situation.

REPRODUCTION

Male testes are enlarged from late February until late August. Mating occurs in the water while the animals are submerged, or as they float at the surface. Mating activity begins soon after the ice breaks up (March–June depending on latitude) and lasts until about mid-August. In lowland British Columbia, where ice is not common in winter, breeding begins in April. Young are born in a bank burrow or lodge from early May to early September after a gestation of 28–30 days. These large rodents are very prolific. Litter size averages five to nine (range 1–14). On average, females in Canada produce two litters annually (only one in the far north and infrequently three in the south) although they can reproduce year round in warmer parts of the United States. Northern Muskrats tend to produce larger litters than their more southerly compatriots. A postpartum oestrus is possible and females can gestate and lactate concurrently. Young are born nearly hairless, blind, and helpless. By five days old, the kits are covered in a coarse coat of grey hairs. By 14 days old, their eyes open and they are capable of limited swimming. They are weaned by the time they are four weeks old. The offspring of early litters disperse after they are weaned, but youngsters from the last litter of the season spend the winter with their mothers and do not disperse until the following spring. In northern North America, sexual maturity of both sexes of juveniles usually occurs the following spring, when they are yearlings. Yearling females typically reach oestrus later than do mature females and they tend to have smaller litters (averaging four to five kits). Early-born juveniles have more time to achieve a larger body size before the cold season sets in, so they tend to survive their first winter better than later-born kits. The following spring, these same larger-bodied animals have a competitive advantage in the contests for prime breeding sites, and as a result produce more offspring on average. For information on parental care and monogamy versus polygamy in Muskrats see the "Behaviour" section below.

BEHAVIOUR

Muskrats tend to be antisocial during the breeding season. They mark their home ranges with a yellowish, musky-smelling secretion from anal glands and defend this territory from other Muskrats of the same sex. Muskrats are believed to be largely monogamous, and although the females provide all of the primary care, males do enhance the survival of their weanlings by carrying food to the nest, by allowing the juveniles to follow while they forage, by being

tail drag

Walk

Left hind track
Length: 3.8–7.0 cm
Width: 3.8–6.4 cm

Left front track
Length: 2.2–3.8 cm
Width: 2.4–3.8 cm

Common Muskrat

vigilant around the lodge or burrow, and by conducting ongoing burrow and lodge maintenance. Some males will take the opportunity to become polygynous if it is possible, but do not assist the second female with her litter. Male movements often extend over the territory of more than one female, but the overlap is most extensive over the territory of their primary female. Fighting is most common in spring and early summer as adults and yearlings of both sexes compete for the prime breeding locations. Injuries are common and even death may result from such battles. Sociality increases during the winter as adult females overwinter with surviving offspring from their last litter and one or more adult males. These animals share a winter lodge and regularly huddle together to conserve heat. Individuals in these groups regularly engage in communal grooming of wet nest mates, which likely maintains the coat's waterproofing and reduces heat loss due to wetting. A further advantage of sharing a winter lodge with multiple animals is that more frequent use may prevent the plunge holes from freezing over during extreme cold. Muskrats are known on occasion to use active beaver lodges. Although they may excavate their own quarters within the lodge so as not to interfere with the more aggressive and larger North American Beavers, recent evidence from a video camera placed in a beaver lodge over the winter showed two Muskrats sharing the single chamber with the Beaver family, until an American Mink came through one day and killed one of the Muskrats. There are no reports of Muskrats using abandoned beaver lodges. Common Muskrats can be ferocious fighters if they are barred from an escape route to deep water and many an inexperienced Domestic Dog has learned the hard way to avoid such encounters.

VOCALIZATIONS

Muskrats produce four different vocal sounds. A whining-growl is emitted by disturbed animals; a chattering is produced by clacking the incisors together when they are threatened; and a squeak and a high-pitched "n-n-n-n" are produced during social encounters.

SIGNS

Lodges range in size from 0.8–2.5 m long, 0.5–2.4 m wide, and 0.3–1.8 m high. During winter, large lodges are used communally while smaller lodges are used as feeding shelters so the animals can climb out of the chilly water to eat. Most foraging occurs near a feeding shelter or a push-up. See in the "Ecology" section above for a discussion on lodge and push-up construction. Marking stations (also called scent posts) occur on high points of land, at the bases of lodges, around bank burrows, on logs and stumps, and usually include a latrine site. These are regularly marked with scat and anal-gland secretions. If such high sites are scarce, the Muskrats may even mound up a dirt pile to use instead. Their dark green or blackish scat comes in two forms: hard scat is shaped like an olive pit and is 0.5–0.8 cm in diameter by 1.0–2.5 cm long; soft scat is 1.0–2.5 cm in diameter and 2.2–4.4 cm long. Scat consistency varies with the moisture content of the diet. Hard scat typically forms into discrete pellets, while softer scat may be clumped pellets stuck together and the softest scat may be amorphous blobs. Runways are 10–15 cm in width on land or through floating vegetation. Muskrats may dig channels to connect ponds or to connect their lodge or burrow to the receding water. These are typically 20–30 cm wide; about half the width of similar, but deeper, North American Beaver channels. Burrow entrances that may become visible in low water are 9–15 cm in diameter. Most burrows are excavated on slopes and banks near a water source, but some may be found on relatively flat ground or under a tree root at the water's edge.

Muskrats have five toes on each foot, and although toe 1 (the thumb) on each front foot is highly reduced, it has a small claw that can, under ideal conditions, leave a noticeable impression in a track. Imprints of the front toes commonly connect to the palm pads. Each front foot has two somewhat bulbous heel pads that can leave a distinct double bump at the posterior edge of the track. The hair fringe around the hind toes broadens and blurs their track and can give the impression of extensive webbing between the toes. The two heel pads on the hind feet rarely register in a track. Apart from the claw on toe 1 of the front feet, most tracks register thick claw impressions. Tail-drag marks are a common component of a Muskrat track. Feeding signs may be found at the base of wetland plants. The cuts are angled, sometimes leaving incisor marks, and the stems are transported to a favourite feeding station either in the marsh or at its edge, where the remains are discarded. Some feeding stations are littered with the remains of crayfish or clams. Muskrats nibble the edge of a freshly procured clam or mussel and then discard it onto the feeding station. The flesh is retrieved after the mollusc dies and gapes open, leaving a relatively pristine shell.

REFERENCES

Banfield 1974; Campbell, K.L., and MacArthur 1996; Chubbs and Phillips 1993a; Elbroch 2003; Erb et al. 2000; Jameson and Peeters 2004; Kurta 1995; Long, J.L., 2003; MacArthur 1992; MacArthur and Humphries 1999; MacArthur et al. 1997; Macdonald and Barrett 1993; Marinelli and Messier 1993, 1995; Marinelli et al. 1997; McKinstry et al. 1997; Messier and Virgl 1992; Nagorsen 2005a; Slough and Jung 2007; Verts and Carraway 1998; Viljugrein et al. 2001; Virgl and Messier 1992.

Heather Vole

also called Heath Vole, Phenacomys Vole

FRENCH NAME: **Campagnol des bruyères**
SCIENTIFIC NAME: *Phenacomys intermedius*

The taxonomy of this enigmatic small vole has been contentious for over 100 years, especially the distinction between the eastern and western forms. Despite the widespread geographic range of the two forms, they are not common in museum collections, as Heather Voles are not only scarce but also difficult to trap, making

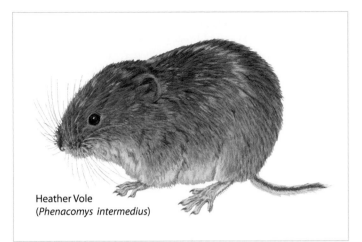

Heather Vole
(*Phenacomys intermedius*)

their study problematic. Absent the necessary specimens for a thorough review of geographic variation, a rigorous genetic study is necessary to resolve the taxonomy. Since this has not yet been conducted and there are no clear and effective morphological characteristics so far identified to distinguish the two forms, apart from the tendency for the eastern form to have a somewhat rusty or pale yellowish-brown snout, the two forms are treated as a single species herein.

DESCRIPTION

Dorsal pelage is greyish brown, sometimes with yellowish-brown undertones or slightly reddish highlights. The underside is white or pale grey, as are the upper surfaces of the feet. Feet look pinkish, as the skin beneath shows through the fur. Ears are short and almost concealed by the long, silky fur. Heather Voles have a distinctly bicoloured tail that is dark brown above and white or pale grey below. It is short (< 30% of the total length) and sparsely furred. Most adults of the eastern form have a yellowish-brown to orange snout, although some populations do not. The extent of this colouring is variable, as it may extend to the eyes or be limited to the fur around the nose. Adults and juveniles of the western form do not have a paler or differently coloured snout. Juveniles are darker, with shorter fur, and in the eastern form they often do not show the rusty nose colour of the adults and many subadults. The dental formula is incisors 1/1, canines 0/0, premolars 0/0, and molars 3/3, for a total of 16 teeth. Heather Voles have four pairs of mammae.

SIMILAR SPECIES

This vole can be easily mistaken for many of the *Microtus* voles, especially the juveniles. Heather Voles have softer, finer fur, a short tail that is sharply bicoloured and sparsely furred and pinkish feet with white or light grey fur on their dorsal surfaces. Meadow Voles do not have a distinctly bicoloured tail and the fur on the upper surface of their feet is usually a dirty brown colour. Montane Voles have longer tails (usually 40–50 mm in length) and tend to be darker with coarser fur. Long-tailed Voles are also darker and have a longer tail (usually > 60 mm in length). In the east, Rock Voles could be confused with Heather Voles. Both may have colourful snouts, but

the Rock Vole's snout is more rusty orange, while that of the Heather Vole in the same geographical region is more yellowish brown, or sometimes the same brown as the rest of the pelage. Tail length of the Rock Vole is generally > 45 mm, while the same measurement for Heather Voles is usually < 40 mm, except for the largest of the individuals. Belly fur on the Heather Vole is usually whiter than that of the Rock Vole and the hind-foot length of Heather Voles is usually < 19 mm. Southern Red-backed Voles, while sometimes similar to Heather Voles, do not have a sharply bicoloured tail and tend to be more reddish brown on the back, especially in winter pelage. Northern Red-backed Voles are even more reddish, and although they have a bicoloured tail, the upper surface near the base often retains the reddish colour of the back before changing to black for the remainder of the tail. Their tail is also covered in bristly hairs and has a full black tip. Bog Lemmings are darker and greyer in colour and have very short tails and grooved incisors. To identify vole skulls and teeth, see appendix 2.

SIZE

Males and females have similar colouring and are similar in size. The following measurements apply to adult animals in the Canadian part of the range.

Total length: 114–164 mm; tail length: 21–46 mm; hind foot length: 15–22 mm; ear length: 11–20 mm.

Weight: 17.5–50.0 g; newborns, about 1.9–2.4 g.

RANGE

This small vole has a widespread distribution from Labrador to Yukon and southwards to California and New Mexico. The eastern form is more widespread than the western and its range extends from Labrador through the northern part of the Prairie provinces to northern British Columbia and into southern Yukon, Northwest Territories, and Nunavut. The western form is more montane in its habitat and is found from the foothills of the Rocky Mountains in southwestern Alberta, diagonally across British Columbia to the northwestern corner, and southward to northern California and New Mexico.

ABUNDANCE

Although this species occupies a widespread range, populations are very scattered. There are rare reports of unusual abundances of Heather Voles, so populations may experience irregular "population explosions," but the frequency and amplitude of such irruptions are unknown. Heather Voles are generally rare and are rarely trapped.

ECOLOGY

Heather Voles are found primarily in coniferous and mixed forests in the boreal zone of North America, where they occupy a wide variety of habitats. They may be found in dry or moist forested sites, montane meadows, rocky areas with heavy moss cover, alpine tundra, or successional forests following logging or fire. Most sites have a dense understorey of shrubs and forbs for food and cover. In the west, Heather Voles commonly occur in alpine regions at or above the treeline.

Distribution of the Heather Vole (*Phenacomys intermedius*)

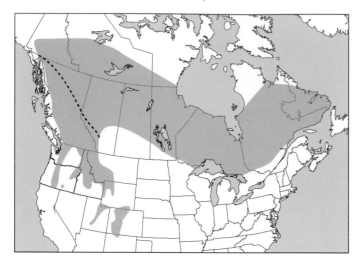

⟍ approximate demarcation line
⟍ between eastern and western forms

They may be active day or night, but are most active in darkness. They do not hibernate and do not move into human homes in the autumn. Heather Voles regularly create caches of food both summer and winter. Plant cuttings of stems and leaves are harvested at night during the warm months and piled near the burrow entrance, presumably to provide secure and convenient daytime food. Stockpiles of twigs, gathered and piled near the burrows in August and September, provide winter food. Winter harvesting continues under the snow. Burrows are generally short, less than a metre in length, and are limited to the environs of the nest. Like other voles, their winter activity is largely restricted to runways under the snow. Estimates of home range size are unavailable, but one older male was recaptured almost 550 m away from his first capture site 16 days earlier. Weasels, such as the American Marten and Long-tailed Weasel are primary mammalian predators. The smaller weasels, such as the Ermine and Least Weasel, are also likely predators. Many diurnal and nocturnal raptors take Heather Voles. Analysis of scat and raptor pellets often shows this vole to be more common than our trapping success would suggest. However, Heather Voles are thought to be comparatively unwary, which may account for the relatively high numbers that are caught and eaten by predators. Circumstantial evidence (remains found in fish stomachs) suggests that Heather Voles will take to the water and swim at times, but captives try to avoid immersion in water. Heather Voles do not climb shrubs and trees, but will chew what bark they can reach within the snowpack. Although some wild Heather Voles may live two, or perhaps even three winters, most are lucky to survive one winter.

DIET

Heather Voles appear to be entirely herbivorous. Their diet has largely been determined by analysis of the food piles mentioned in the "Ecology" section. Summer food is primarily composed of stems and leaves of forbs and shrubs, but may also include seeds, grasses, fungi, lichens, and fruits. Winter food is chiefly made up of twigs of a wide variety of broad-leaved shrubs and trees (such as willow, birch, maple, heather, and blueberry), as available. Coniferous twigs are not harvested in either season. Heather Voles are able to consume and digest the leaves and bark of the Sheep Laurel and Western Bog-laurel, species usually considered to be toxic to vertebrates.

REPRODUCTION

Heather Voles breed from May to August or September, but their breeding season may be shorter at high elevations. Females may produce up to three litters annually. Gestation takes 19–24 days. Litters average 3–5 pups (range 2–7 in Canadian animals). Males do not assist in rearing the young. Females that overwintered are larger and have larger litters (average of 6) than young of the year (average of 4). The following ontogeny data has been gathered from captive litters. Newborn pups are pink, hairless, toothless, and blind. By the end of their first day, darker pigmentation is visible in the skin on the back and tiny whiskers begin to erupt. They can vocalize by day 2 and crawl by day 8. They have a covering of short, satiny fur by 6 days old. Incisors begin to erupt as early as day 7. Their eyes open at 13–15 days old. They begin eating solid food about that same time and weaning begins. Pups are fully weaned by 19 days old. Some females are capable of breeding by as young as 4 weeks old. Juvenile males and females born early in the breeding season mature sufficiently to breed later in the season, but those born late in the season do not breed until the following spring.

BEHAVIOUR

Heather Voles are well known as docile captives that rarely struggle or bite. As sightings of wild voles are so scarce, much of what we know of Heather Vole behaviour is deduced from brief observations in the wild or from sign, or derived from the study of captive animals. Maintaining a long-term captive colony is very difficult, as many individuals refuse to eat and mortality, especially of infants, is high. Adults are solitary during the breeding season, apart from females with pups and, briefly, mating pairs. Reproductive females defend the area immediately around their nesting site and males are aggressive towards each other during the breeding season. Battles among captive males lead to bite wounds to rump and tail, but such wounds are not seen in wild-caught individuals. Aggression

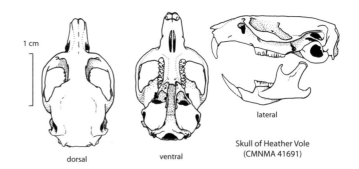

1 cm

dorsal   ventral

lateral

Skull of Heather Vole
(CMNMA 41691)

subsides during the cold season and family groups may huddle together in communal winter nests.

VOCALIZATIONS

The only report of sounds produced by this species is of squeaks produced by the young.

SIGNS

Heather Vole sign is rare and very similar to that left by other voles, but runways are usually short (less than a metre in length), often unobtrusive, and located only immediately around the nest. Heather Voles may use the runways created by other small mammals. In most parts of the Heather Vole range, only Heather Voles create caches of cut twigs near their burrow openings (the Red Tree Vole, *Arborimus longicaudus,* found in coastal Oregon and northern California, is the only other North American vole that creates twig caches). Winter twig caches of Heather Voles may exceed 1200 twigs from 8–125 mm in length and will not include twigs of coniferous species. Winter caches are placed near burrows, usually in the shelter of a log or standing trunk, under rocks, or under the branches of fallen trees. Most of the twigs will be stripped of bark and a pile of scat will be nearby. Summer and winter nests constructed by Heather Voles are different. Winter nests are about 15 cm in diameter and made of leaves, twigs and grasses with a soft interior lining of shredded vegetation. They may be weatherproofed with an exterior coating of lichens. These are placed either below ground, or at ground level beneath the snow, usually at the base of a tree, stump or rock. Latrine sites are located outside the nest and can accumulate into large piles that are revealed during snowmelt. Summer nests are smaller (around 10 cm in diameter) and may be located in or under a hollow log or a decaying stump, under a rock or in a shallow chamber just beneath the ground surface. They are constructed of mosses and grasses with shredded leaves and plant down commonly used as a liner. Typically, a latrine is situated in a chamber next to the nest chamber or near the mouth of the burrow. Scat is similar to that of other voles. Burrow openings may occur in snow or earth, but are usually obscured by vegetation. Tracks are likely similar and indistinguishable from those of other voles.

REFERENCES

Desrosiers et al. 2002; Dobbyn 1994; Foster 1961; George, S.B., 1999e; Jameson and Peeters 2004; Jannett and Oehlenschlager 1997; MacDonald, S.O., et al. 2004; McAllister and Hoffmann 1988; Nagorsen 1987b, 2005a; Simon et al. 1998; Smith, H.C., 1993; Verts and Carraway 1998.

Northern Bog Lemming
also called Northern Lemming Vole

FRENCH NAME: **Lemming des tourbières,** *also* **Campagnol-lemming boréal**

SCIENTIFIC NAME: *Synaptomys borealis*

These small, enigmatic rodents are poorly understood and rarely captured, even though they have a widespread distribution.

DESCRIPTION

Pelage on the back and sides is greyish to reddish brown depending on geographical location, with many dark-tipped guard hairs scattered throughout, which provide a grizzled appearance. Belly fur is dark grey at the base and the tips are buffy to silvery white. Feet are brown on their upper surface. The short tail (< 20% of the total length) is bicoloured, dark above and whitish below with a number of stiff hairs that project from the tip. Ears extend slightly beyond the hair in summer, but not in winter. Mature males develop a pair of lateral glands, one on each hip. Their position is usually marked by small patches of creamy white fur. The dental formula is incisors 1/1, canines 0/0, premolars 0/0, and molars 3/3, for a total of 16 teeth. The anterior face of each upper incisor has a single, shallow, longitudinal groove near the outer edge. They have eight mammae.

SIMILAR SPECIES

Their very short, bicoloured tail, dark reddish to greyish brown, grizzled upper pelage, grey belly, and grooved incisors separate the bog lemmings (genus *Synaptomys*) from other small mammals within their range. Unfortunately, Northern and Southern Bog Lemmings are virtually indistinguishable externally and require a close examination of the lower teeth to discriminate between them. See appendix 2 for skull and teeth characters that distinguish the two. Female Northern Bog Lemmings have two pairs of pectoral (chest) mammae and two inguinal (between the hind legs) pairs, so there are eight mammae in four pairs. Southern Bog Lemmings have the same double pair of inguinal mammae, but only one pair of pectoral, so they have only six mammae in three pairs. These are normally visible only on lactating females, making this distinguishing

Northern Bog Lemming
(*Synaptomys borealis*)

Distribution of the Northern Bog Lemming (*Synaptomys borealis*)

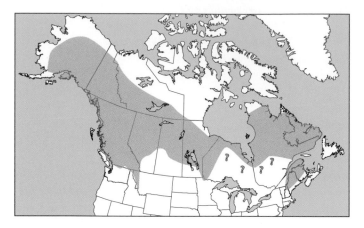

? occurrence possible but unsubstantiated

feature usable only with one sex and then only during the breeding season. Both Meadow Voles and Heather Voles lack the grooved incisors and have longer tails, and Meadow Voles are much larger than Northern Bog Lemmings.

SIZE
Males and females are similar in size and colouring. The following measurements apply to animals of reproductive age from the Canadian part of the range.
Total length: 102–150 mm; tail length: 15–30 mm; hind foot length: 14–22 mm; ear length: 11–15 mm.
Weight: 21.7–48.0 g; newborns, unknown.

RANGE
Northern Bog Lemmings are widely distributed across northern North America, from Labrador to Alaska and western Montana to New Brunswick. The known distribution in eastern Canada may be misleading, as this species probably occurs across most of northern Ontario and Quebec, but that area is difficult to access and specimens have not yet been collected there.

ABUNDANCE
Generally rare and populations are highly scattered in isolated pockets, especially in the east.

ECOLOGY
Little is known of the natural history of this boreal species. Bog lemmings are usually found in cool sphagnum bogs at low elevations, but may also occur in a wide variety of habitats such as moist spruce-horsetail forests, dry spruce-lichen forests, spruce-fir forests, hemlock-beech forests, mossy streamsides, dry larch parklands, wet and dry subalpine meadows, tundra and alpine tundra, dry grassy fields, and even dry sagebrush hills. The drier locations tend to be either in more northerly latitudes or at middle to upper elevations. They have been found from near sea level to 2300 m in elevation. Northern Bog Lemmings are active intermittently throughout the

day and night: under the vegetation in summer and under the snow in winter. They construct spherical nests of woven grass stems and sometimes other common herbaceous plants, either underground or in dense cover in summer and under the snowpack in winter. Predators likely include many northern hawks and owls as well as small northern carnivores such as weasels, American Martens, and foxes.

DIET
Scanty information is available on the diet of these small mammals. Most is derived from the examination of small piles of clipped plants gathered at burrow entrances in summer. These are composed mainly of grasses and sedges, but also include forbs such as hawkweed, Western Bog-laurel (a species normally considered toxic to vertebrates), Fan-leafed Cinquefoil, and saxifrage.

REPRODUCTION
Northern Bog Lemmings quickly die in captivity, so little is known of their reproductive biology. Based on museum specimens of pregnant females, embryo numbers average four to five (range 2–9). Reproductive females have been captured from mid-May to mid-September, so several litters annually are possible and the breeding season appears to extend through the length of the warm season.

BEHAVIOUR
It is reported that Northern Bog Lemmings will use the runways of Meadow Voles where their ranges overlap, but they also create extensive runway networks of their own, typically originating at burrow entrances. Bog lemmings appear to be more susceptible to capture in a pitfall trap than a snap trap and capture is much more likely if the trap is carefully placed in a runway; in other words, not along a fixed line or grid as is commonly the practice in many small mammal surveys.

VOCALIZATIONS
Unreported.

SIGNS
Bog lemmings leave tracks and trails that are very similar to those of Meadow Voles. See the "Signs" section of the Meadow Vole account for a track schematic that applies to all the voles and lemmings. Scat is 0.2–0.5 cm in diameter and 0.6–1.1 cm in length and ovoid with blunt ends, like that of voles. It is tan to dark brown and may

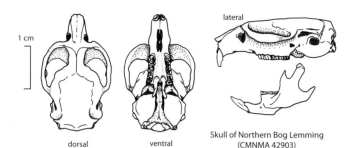

lateral

1 cm

dorsal   ventral

Skull of Northern Bog Lemming
(CMNMA 42903)

be scattered throughout the runway system, or accumulated above ground in winter latrine chambers that may be found during spring melt. Runways may honeycomb a boggy area.

REFERENCES

Desrosiers et al. 2002; Dobbyn 1994; Elbroch 2003; Groves and Yensen 1989; Jarrell and Fredga 1993; Kurta 1995; Nagorsen 2005a; Peterson, R.L., 1966; Prescott and Richard 2004; Salt 2000b; Smith, D.A., and Foster 1957; Smith, H.C., 1993; Wilson, C., et al. 1980; Youngman 1975.

Southern Bog Lemming

also called Southern Lemming Mouse, Southern Lemming Vole

FRENCH NAME: **Lemming de Cooper,** *also*
Campagnol-lemming de Cooper

SCIENTIFIC NAME: *Synaptomys cooperi*

Although somewhat better known than its close relative the Northern Bog Lemming, the Southern Bog Lemming is nevertheless still poorly understood and not well studied.

DESCRIPTION

Pelage on the back and sides is greyish to reddish brown depending on geographical location, with many dark-tipped guard hairs scattered throughout that provide a grizzled appearance. Belly fur is dark grey at the base and the tips are buffy to silvery white. Feet are brown on their upper surface with tan-coloured claws. The short tail (< 20% of the total length) is bicoloured, dark above and whitish below, with a number of stiff hairs that project from the tip. Ears barely extend beyond the hair in summer, but are completely hidden in winter. Juvenile pelage is darker and greyer than that of adults.

Southern Bog Lemming
(*Synaptomys cooperi*)

Adult males develop hip glands and their location may be highlighted on older males by the presence of white hairs that grow over the glands. These tufts of light hairs are not usually as prominent on this species as they are on Northern Bog Lemmings. The dental formula is incisors 1/1, canines 0/0, premolars 0/0, and molars 3/3, for a total of 16 teeth. The forward face of each upper incisor has a single, shallow, longitudinal groove near the outer edge. These small rodents normally have six mammae, although some may have fewer. Winter fur is moulted in early summer and summer fur is moulted in late autumn. Winter pelage is longer and generally somewhat greyer than summer pelage.

SIMILAR SPECIES

See the "Similar Species" section of the Northern Bog Lemming for distinguishing characteristics between these two very similar species. Meadow Voles, Woodland Voles, and Heather Voles have longer tails and do not have grooved incisors. See appendix 2 for skull and teeth characters that distinguish the voles and lemmings.

SIZE

Males and females are similar in size and colouring. The following measurements apply to animals of reproductive age from the Canadian part of the range.

Total length: 108–128 mm; tail length: 13.0–26.5 mm; hind foot length: 17.0–20.5 mm; ear length: 7–18 mm.

Weight: 20–30 g; newborns, 3.1–4.3 g.

RANGE

Southern Bog Lemmings occur in appropriate habitat from eastern Manitoba, northern Ontario, northern Quebec, and Nova Scotia throughout the New England states to southern Virginia and westward to southwestern Oklahoma. A small mammal survey in a previously unstudied part of northern Quebec found Southern Bog Lemmings along the Eastmain River and at Lac Boyd, both locations considerably north of the previous most northerly known specimen. Clearly, the northerly distribution limits of this species remain uncertain. Occurrence of bog lemmings within their range is erratic and patchy.

ABUNDANCE

The density of this bog lemming is highly variable. Estimates from areas of known occurrence range from 12 to 51 per ha. Some populations are reported to experience annual abundance cycles, while others apparently undergo multi-annual cycles, but the length and amplitude of this cycle is not known. These rodents are rare to uncommon in the eastern parts of the range and somewhat more numerous in the west, where they can become locally common at times.

ECOLOGY

The prime habitat for Southern Bog Lemmings is in moist, grassy areas around sphagnum bogs, swamps, and stream edges, but these rodents can inhabit a wide range of less preferred habitats, such as shrubby grasslands, mixed forests, wet meadows, pasture lands, fallow fields, woodland clearings, power-line rights-of-way, and even

Distribution of the Southern Bog Lemming (*Synaptomys cooperi*)

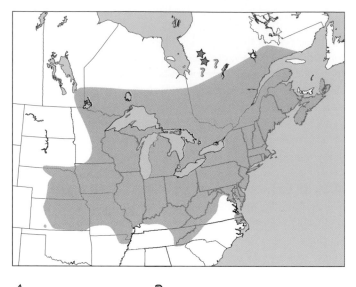

⭐ extralimital occurrences   ❓ occurrence likely but unsubstantiated

clearcuts. It is possible that their occupation of less preferred habitats may be the result of competition and exclusion by other small rodents, particularly voles in the genus *Microtus*. These more dominant voles may be forcing the bog lemmings to inhabit terrain with poor cover and less nutritious food. Some populations undertake a seasonal shift, moving to drier neighbouring habitats in winter. Their home range has been estimated at 0.06–0.32 ha, and females occupy smaller home ranges than do males. Bog lemmings may be active intermittently throughout the day and night both summer and winter, but most spring and summer activity occurs in darkness, unless cover is abundant. During the warm season they travel in a system of runways protected by vegetation and in winter they dig extensive tunnels at ground level under the snow, where they are protected from inclement weather and predators. Nests are globular, with a hollow centre and one or more entrances, and are constructed of whatever vegetation is readily available (typically stems and leaves of grasses and sedges). Winter nests are constructed at ground level in a snow chamber and summer nests are mainly placed underground in shallow burrows. Winter nests are about 20 cm in diameter and summer nests are smaller, 9–16 cm in diameter. Southern Bog Lemmings are capable and buoyant swimmers and they climb well. They are also efficient diggers, well able to produce their own burrow systems, although they are also quite willing to make use of tunnels and burrows created by other species. Direct reports of predation are few, but such predators as hawks, owls, snakes, weasels, foxes, Bobcats, Northern Raccoons, and Domestic Cats are known, and short-tailed shrews are probable. Captive animals may live more than 2 years, but likely few wild bog lemmings exceed 12 months old.

DIET

Primarily herbivorous, their diet is largely composed of fresh leaves, stems, basal rootstocks, and shoots of grasses and sedges, with the seasonal addition of mosses, fruits, fungi, bark, and roots.

REPRODUCTION

In southern parts of their range, these rodents are capable of breeding all year round, even under the snow, but it is unlikely that they do so in Canada, where breeding is probably restricted to the warm months from May to September. Based on a few captive females, the gestation period is 21–23 days, followed by a postpartum oestrus. Average litter size is three (range 1–8). Females are capable of producing multiple litters in a breeding season (possibly three to four), but the maximum annual productivity of a wild Southern Bog Lemming is unknown. Pups are born with a light covering of fine hairs on their backs and heads and short whiskers (vibrissae), but with their eyes and ears fused shut. By seven days old they are well furred. Lower incisors begin erupting around 7 days old and eyes open at 10–11 days old. They begin taking solid food by around day 16 and are fully weaned by the time they are 21 days old. Sexual maturity of both sexes occurs at around two months old.

BEHAVIOUR

Southern Bog Lemmings will share runway systems with other voles, especially Prairie Voles. Although from a small sample size of only three adult females, one in-depth study undertaken in New Jersey showed that these Southern Bog Lemming females established home ranges that did not overlap those of nearby females, suggesting some degree of territoriality. In the same study, male home ranges tended to overlap slightly with those of other males and, more broadly, with those of neighbouring females. There are some indications that bog lemmings cache food underground, but as yet this has not been proved.

VOCALIZATIONS

Southern Bog Lemmings are relatively vocal. Harsh, chattering notes (rarely squeaky) that appear to express threat, anger, or irritation are emitted when new animals are nearby, when captive animals are objecting to being handled, or when indicating alarm. Courtship calls appear to be produced mainly by the male, who emits a mellow, high-pitched, low-volume "tew-tew-tew-tew" call that, again, is not squeaky. Some males may emit a slow, deliberate, clicking note in the presence of a potential mate. Maternal females direct low-volume short calls towards their offspring and pups produce insistent piping calls to indicate distress.

SIGNS

Bog lemmings leave tracks and trails very similar to those of Meadow Voles. See the "Signs" section in the Meadow Vole account

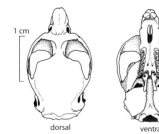

1 cm

dorsal   ventral

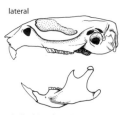

lateral

Skull of Southern Bog Lemming
(CMNMA 47948)

for more information. Scat is 0.2–0.5 cm in diameter and 0.6–1.1 cm in length and ovoid with blunt ends, like that of voles. Fresh scat is light green, but older droppings may be tan to bluish grey. Scat may be scattered throughout the runway system, piled in discrete latrines along the runways, or accumulated above ground in winter latrine chambers that may be found during spring melt. When these rodents occupy a habitat with heavy cover, their numerous runways can be obvious and cluttered with scat and piles of plant clippings. In a dense sphagnum bog, the moss must sometimes be lifted to reveal the runways. They may concentrate their feeding in one area for a time, the result being the accumulation of hundreds of cuttings that can carpet the runways. Bog lemmings tend to rise on their hind legs to make an initial cutting about 7–9 cm from the ground. The stem or leaf is then manipulated with the front feet, pulled down, and further cut into segments. Other times, the plant will be clipped at ground level and pulled into a hidden runway in the moss. Clippings may be accumulated at and into a burrow entrance or nest site or at random sites along the runways. Most clippings in a pile are almost the same length (3.0–7.5 cm long) and piles can have hundreds of clippings. Gathering bundles of full-length leaves of grasses or sedges appear to be a bog lemming characteristic. These leaves are cut at their base and carried to a spot along a runway or to a burrow entrance, where after numerous trips a bundle of several dozen similar leaves is accumulated. Some of these leaves may be up to 45 cm in length. Runways may be shallowly subsurface under the leaf litter or sphagnum moss or they may be on the ground surface under cover of vegetation. Runways are about 4 cm in width and can be extensive. Burrow entrances are 3.5–4.0 cm in diameter and typically circular. These may be found in the sphagnum or in the soil and are typically hidden under cover. See in the "Ecology" section above for more information on nest size and location.

REFERENCES

Brown, L.N., 1997; Choate et al. 1994; Conner 1959; Danielson and Swihart 1987; Desrosiers et al. 2002; Dobbyn 1994; Elbroch 2003; Fortin et al. 2004; Jarrell and Fredga 1993; Jones, J.K., and Birney 1988; Krupa and Haskins 1996; Kurta 1995; Linzey 1984; Long, C.A., and Long 1988; Prescott and Richard 2004; Sheffield 1998; Wilson, G.M., and Choate 1997.

SUBFAMILY NEOTOMINAE

New World rats and mice

Most members of this subfamily, and all those in Canada, have large ears, prominent eyes, long tails, and pointed muzzles.

Bushy-tailed Woodrat

also called Bushy-tailed Wood Rat, Bushy-tailed Packrat, Packrat

FRENCH NAME: **Néotoma à queue touffue**

SCIENTIFIC NAME: *Neotoma cinerea*

This western woodrat is named for the long hairs on its beautiful, heavily-furred, squirrel-like tail. It is the only native packrat in Canada and is the largest, most thickly furred of the woodrat group.

DESCRIPTION

These rodents are silver-grey to dark grey to brown on the back and sides and white or, rarely, light buff on the belly, throat, chin, and feet. Most adults have numerous black-tipped guard hairs on their back and rump that gives them a grizzled appearance. Juveniles and subadults are generally greyer than adults. The sharply bicoloured tail is grizzled grey above and white below. Tail hairs are around 30 mm in length. Stiff whiskers (vibrissae) are noticeably long and may reach 100 mm in length, some easily extending to the shoulders. These help the woodrat manoeuvre in the dark, rocky crevices where they live and are usually black, but sometimes mixed grey and black. Ears are large, often with a narrow rim of white fur. Adult male Bushy-tailed Woodrats develop a dermal thickening of sebaceous glands along the midline of their belly. Secretions from these glands stain the males' chest and belly fur a brownish or yellowish colour and impart a characteristic musky odour. Some females develop smaller versions of these glands while nursing. The glands in juvenile and most subadult males, as well as non-breeding females, are non-secretory. Feet are well furred on the upper surface, but the soles remain bare, even in winter. Claws are curved and sharp, but largely hidden in the fur. The dental formula is incisors 1/1, canines 0/0, premolars 0/0, and molars 3/3, for a total of 16 teeth. Juveniles undergo two or three moults during their first year. They have shorter tail fur than adults and may retain some white hairs on their tail tip through their first two moults, which can serve to distinguish yearlings from older adults. Adults have a single annual moult that occurs during the summer. Fur is long and silky. Bushy-tailed

Bushy-tailed Woodrat (*Neotoma cinerea*)

Woodrats have well-developed senses of smell and hearing and large eyes to help them see in low light conditions.

SIMILAR SPECIES

Although this rodent could be mistaken for a squirrel thanks to a bushy tail, its large eyes and nearly naked ears, pelage coloration, long, bicoloured tail, and long vibrissae distinguish it from any other mammal within its range.

SIZE

The following measurements apply to adult animals from the Canadian portion of the range. Adult males may be as much as 25% larger on average than adult females and both sexes become larger in the northern parts of the range.

Males: total length: 300–468 mm; tail length: 105–215 mm; hind foot length: 37–57 mm; ear length: 27–40 mm.

Females: total length: 318–412 mm; tail length: 112–190 mm; hind foot length: 37–49 mm; ear length: 23–40 mm.

Weight: males, 374.7–456.0 g; females, 263.0–394.6 g; newborn females average 14 g, while males average 15 g.

RANGE

All of the packrats are restricted to North America. Bushy-tailed Woodrats occur in mountainous or rocky terrain from southern Yukon and western Northwest Territories, through mainland British Columbia and western Alberta, to central North Dakota and southwards to central California and northern Arizona and New Mexico. They do not occur on any of the BC coastal islands. Distribution within this broad range is spotty, as Bushy-tailed Woodrats largely occur in discrete habitat patches (rocky outcrops). A disjunct area of distribution has recently been discovered (2002) in eastern Alberta along the South Saskatchewan River about 15 km from the Saskatchewan border. The occurrence of this population suggests that some of the older reports of extralimital individuals discovered in Alberta, largely ascribed to human accidental transport, may

in fact be representatives of more local undiscovered populations. There are reports of woodrat sign from the Milk River area of Grasslands National Park in southern Saskatchewan, but as yet neither a sighting nor a specimen has been procured.

ABUNDANCE

Bushy-tailed Woodrats remain relatively common in mountainous terrain in both Alberta and British Columbia. Populations fluctuate in density both seasonally and from year to year. They peak as the young are recruited in summer and autumn and drop as the weak (largely the juveniles and the old) die over the winter. Anecdotal reports of abnormally high population densities are occasionally received, but reasons for such sporadic abundance are so far unknown.

ECOLOGY

These woodrats occur in a wide variety of habitats from sea level to 2380 m in elevation in British Columbia and to 3600 m in Alberta, and from arid grassland to coniferous and mixed forests and mountainous, open alpine slopes. They usually den in rock crevices or caves, especially in cliffs, rocky outcrops and boulder or scree fields, and will commonly invade isolated, vacant buildings and abandoned mineshafts. The number of suitable natural den sites may limit population density in some areas, which may account for their willingness to occupy human-created alternatives. Radio-tracking of animals in the Rocky Mountains of Alberta found that males occupy home ranges of 1.6–11.23 ha (average 6.12 ha), while female home ranges were 0.13–10.44 ha (average 3.56 ha). No doubt, home range size is highly variable, depending on the all-important factors of den site and food availability. Bushy-tailed Woodrats are active at night and all year round, but are most active during nights of intermediate levels of light and less active under bright moonlight or on very dark nights. Perhaps the risk of predation is higher then. They forage and collect plants under cover of darkness, usually travelling less than 100 m from the protection of

Distribution of the Bushy-tailed Woodrat (*Neotoma cinerea*)

⅞ presence possible but uncertain

their den. Nursing females may travel much farther while foraging (a female was seen to venture 470 m), likely due to the increased energetic demands of lactation. The plants are carried back to the den and usually laid out to dry before being moved to a storage area for later consumption. Food is gathered throughout the summer, but the animals increase their pace in late summer and early autumn in order to put enough aside for winter. A large winter hoard can weigh 50 kg. As the woodrats forage, they also collect bones, sticks, and other plant material, which are added to a pile of interlocking sticks usually wedged into a corner or crevice at the den site. Only rarely are these stick nests created in the open. Within this mass of sticks and other vegetation, they build one or more cup-shaped or round, hollow nests made of shredded, fibrous plant material and grasses. Packrat den sites, especially those used by multiple generations, may develop large middens near the stick nest made up of discarded plant material, sticks, bones, and other accumulated oddities such as owl pellets, and native and modern artefacts that the animal has dragged home. The woodrat will often use these middens as a latrine for both scat and urine. Their nests commonly become infested with parasites such as chiggers, ticks, mites, lice, warble-fly larvae, and fleas. Some of these parasites can transmit fatal diseases to the woodrats, such as spotted fever or bubonic plague. High latitude or high elevation–adapted Bushy-tailed Woodrats cannot tolerate high temperatures. Their thick fur adapts them well to cold, but can be a liability when temperatures

rise. They prefer to den in cool, deep crevices or caves that moderate the daytime heat and they are active at night when temperatures are coolest. These same den sites also moderate the bitter winter cold. Bushy-tailed Woodrats are agile climbers, easily able to scale a tall tree or a near featureless cliff face. Predators include the Long-tailed Weasel, American Marten, Bobcat, Coyote, Great Horned Owl, and Red-tailed Hawk. The maximum attainable lifespan in the wild is at least three years, but few survive that long. In most populations, fewer than 35% survive their first year.

DIET

Bushy-tailed Woodrats are strictly herbivores, and although they prefer foliage, they will also consume seeds, fruits, bark, coniferous needles, and fungi. Across their range, they have a varied diet, but regionally they eat around 70% of the available plants in their habitat and tend to focus on a few preferred species. They appear universally to avoid sagebrush. Woodrats generally eat a low-quality diet, so must consume more of it to get the nutrients they need. They have an enlarged and efficient caecum (fermentation chamber in the hind gut) that produces soft pellets that the animals reingest. Coprophagy thereby enables them to reprocess their food to extract the maximum nutrition. Although woodrats often inhabit arid environments, they are not especially physiologically efficient in water conservation. Most of their water comes from the food they eat in summer and from snow in winter and large amounts of urine are voided daily.

REPRODUCTION

Onset of the breeding season is closely tied to the spring green-up and so varies from north to south, with elevation, and from year to year with local climate. Breeding generally occurs from May to August. Adults breed earlier in the season than yearlings. Females remain in heat for 5–7 days and gestation lasts about 30 days. Males are not known to provide any parental care to the pups. The average litter size in Alberta is 3–4 pups (range 3–6), and although females experience a postpartum oestrus 24 to 48 hours after giving birth, only 30%–62% of the females in Canada will produce a second litter. Up to three litters are possible farther south. Pups are very tenacious in their grip on a nipple and may get dragged along with the female as she leaves the nest. They are blind and nearly hairless at birth. By day 15 the young are fully furred, their eyes are open, and they are beginning to eat solid food. Weaning usually occurs by day 25, but

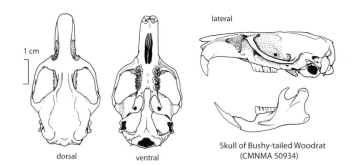

Skull of Bushy-tailed Woodrat
(CMNMA 50934)

1 cm

dorsal   ventral   lateral

**stride length
of a bound
can vary from
10 to 20 cm**

Bound

Left hind track
Length: 1.5–3.2 cm
Width: 1.4–2.5 cm

Left front track
Length: 1.4–2.4 cm
Width: 1.4–2.5 cm

Bushy-tailed Woodrat

some youngsters may be indulged until they are 30 or more days old. Young become sexually mature as yearlings, but the success of the younger females depends on their weight, which is directly influenced by food availability. Most yearling females have fewer pups per litter and generally lighter pups than older females. The same factor influences the number of litters produced in a season, as well-fed females are more likely to have a second litter.

BEHAVIOUR

Woodrats are widely known as packrats because they persist in gathering assorted items at their den site and are attracted especially to unusual or shiny objects. Ordinarily they collect natural items, but will also carry away artefacts such as clothing, jewellery, pieces of plastic, and tin can lids; perhaps abandoning the bone, stick, or rock they were already carrying at the time to make room in their mouth for the new article. They are capable of carrying a stick that is up to 1 m in length and 3 cm thick. Front paws are used to manipulate food and objects, and woodrats will sit on their haunches as they eat. They are very proficient at moving about within their home range as they memorize routes and can find their way quickly at low light levels. Aside from females raising young, or mating pairs, Bushy-tailed Woodrats are usually solitary, one to a den. Sociality varies across the range, but in Canada most outcrops will house a single male and one or more females. Woodrats commonly urinate and defecate on their middens, where dried, crystalline urine, deposited out of the weather (called amberat), can accumulate over many years. It becomes yellowish, dark brown, or black and rock hard, preserving pollen, bones, and plant fragments that have proved invaluable in paleontological studies of the changing local climate, flora, and fauna. Some such middens have been radiocarbon dated to over 40,000 years old. Scent marking is an important communication tool for woodrats. Both sexes mark their home ranges with urine. Their marking posts tend to be used repeatedly, even by successive generations. Amberat up to several centimetres thick can form at a traditional urine post, where it can easily be mistaken for a yellowish or amber coloured mineral. Males use an additional scent marking. They rub their underside on the rocks as they pass to deposit secretions from their belly glands. Females, especially oestrous females, are very attracted by this odour. Adults are loyal to their home range and rarely leave it once it has been established, perhaps because suitable den sites are so often rare. Adult males aggressively defend their home range from other males and attempt to overlap the home ranges of several females. Tooth chattering and foot thumping may be aggressive displays. Females are somewhat less territorial and will sometimes share their home range, and even their den, with daughters, although non-related females are excluded by residents. Juvenile females commonly settle near their birth site (if an all-important den site is available), but male offspring must either appropriate the territory of the dominant male (should he die), or find an unoccupied den within the colony. Failing that, they are forced to look for another colony in hopes of doing the same there. Most juvenile males disperse beyond their natal colony. Mortality of these displaced juvenile males is likely very high as there may be some distance between colonies.

VOCALIZATIONS
Apparently none or unreported.

SIGNS
Packrat urine contains a high level of insoluble calcium, which persists long after the organic material has been washed away and leaves distinctive white streaky stains on the rocks. These streaks can often be seen from a distance and are distinguishable from bird droppings, as they tend to follow the rock ledges and crevices, while birds concentrate their more spotty streaks at a roost site or a high point. Fresh urine has a distinctly musky odour that has been described as somewhat skunky. Bushy-tailed Woodrats have five toes on each foot, but toe 1 (the thumb) on the front foot is greatly reduced, and if it leaves an impression at all, it is just a bump beside the main pad rather than a toe print. Although all toes, except toe 1 on the front foot, have curved claws, these tend not to register in a track. A bound is the common gait when they are travelling across exposed areas. Substantial runs (6–10 cm in width) may be created from the den to favourite foraging areas. These runs tend to follow cover and often contain scat. Stick nests are easily recognized by their disorderly, haphazard appearance and by the scat that accumulates at the entrance and in the immediate vicinity. Not all Bushy-tailed Woodrats build an elaborate stick nest, instead choosing to use a burrow system or existing cavities in rocks or buildings. Nevertheless, plenty of scat will be scattered at the burrow or crevice entrance and along the runways to help identify the inhabitant. Burrow openings typically measure 7–12 cm in diameter and may be horizontally oblong, rather than round. Scat is usually roughly ovoid, sometimes slightly pointed at one end, 0.3–0.5 cm in diameter and 0.8–1.6 cm in length. Woodrats are competent climbers and can create a litter of nip twigs as they harvest fruit or seeds of shrubs or trees.

REFERENCES
Elbroch 2003; Hebda et al. 1990; Hickling et al. 1991; Jameson and Peeters 2004; Lausen 2002; Morton and Pereyra 2008; Moses and Millar 1992, 1994; Nagorsen 2005a; Slough and Jung 2007; Smith, F.A., 1997; Smith, H.C., 1993; Topping and Millar 1996a, 1996b, 1996c; Topping et al. 1999; Verts and Carraway 1998; Youngman 1975.

Northern Grasshopper Mouse

FRENCH NAME: **Souris-à-sauterelles boréale**

SCIENTIFIC NAME: *Onychomys leucogaster*

These stocky, short-tailed, and pugnacious mice are unusual as they are predatory rodents that hunt and kill large insects as well as other small rodents. There are three species of grasshopper mice (all in North America), one of which reaches Canada.

Northern Grasshopper Mouse
(*Onychomys leucogaster*)

DESCRIPTION
Their pelage is distinctly bicoloured. Back, upper head, and upper sides are light brown or pinkish cinnamon and the belly, feet, cheeks, chin, and lower sides are white. The back and top of the head are usually darker than the sides of the body, but older adults tend to look more orange in summer as the dark tips of the guard hairs from the last moult have been worn away. There are two maturational moults. The juvenile pelage is dark grey and white and the subadult pelage, which is assumed around three or four months old, is a darker grey brown with none of the cinnamon highlights of the adult and the belly fur tends to be slightly greyish rather than a clear white. Melanistic individuals occur rarely. The short, thick tail is also bicoloured, dark above and white below. Often the tip is all white. Ears are darker than the rest of the body, are often rimmed with short silvery hairs, and may have a small tuft of white hairs at the forward base. Eyes are large and dark. Front feet are large with long, flexible toes and long, curved claws. Northern Grasshopper Mice have keen senses of hearing, sight, and smell. The dental formula is incisors 1/1, canines 0/0, premolars 0/0, and molars 3/3, for a total of 16 teeth. Their incisors are not typically broad like those of other rodents, but are narrow and more piercing. Molars are high cusped, which provides ample cutting surfaces to shear flesh or crack tough insect chitin. They have six pairs of mammae. Adults undergo a single annual moult, which occurs from mid-July through September. Subadults moult into adult coat by September if they were born early in the season, or as late as midwinter if they were born late. Breeding females in either subadult or adult pelages moult later than males of the same age.

SIMILAR SPECIES
Northern Grasshopper Mice look like stocky Deer Mice, but have shorter, thicker tails (< 60% of their body length) and larger front feet.

SIZE
Males and females are similar in size and colouring. The following measurements apply to adult animals in the Canadian part of the range.
Total length: 125–155 mm; tail length: 34–42 mm; hind foot length: 19–22 mm; ear length: 14–17 mm.
Weight: 24.7–42.9 g; newborns, about 2 g.

RANGE

Northern Grasshopper Mice occur in western North America from the southern prairie provinces to northeastern Mexico, west to Oregon and California, and east to Minnesota and Iowa. Their populations are scattered throughout this range as they are generally restricted to sandy or dusty soils.

ABUNDANCE

These small predators are normally found in relatively low densities except in rare, localized instances.

ECOLOGY

Northern Grasshopper Mice occur in short-grass prairie, semi-stabilized sand dunes, sagebrush desert, and mixed grasslands. They appear to be limited to areas of sandy or dusty soils where they can dust bathe to remove excess oils from their coats. These rodents are largely nocturnal and are active all year round; they do not hibernate. Their home range is unusually large for a small mammal and likely reflects their predatory habits. Estimates of home range size, based on radio-tracking data in Colorado, show that male home ranges there are largest in summer, when they average 2.5 ha. Winter home ranges shrunk to 1.38 ha. Home range sizes of females showed the same trend, but were smaller than those of males (1.42 ha in summer and 1.0 ha in winter). Home ranges of males overlapped the home ranges of several females. Although these rodents may use abandoned burrows of other

Distribution of the Northern Grasshopper Mouse (*Onychomys leucogaster*)

species, especially for escape, they tend to dig their own nesting burrows. These are fairly short (about 48 cm long on average) and shallow (average of 12 cm at the deepest point), with a nest chamber that is about 12 cm long by 9 cm wide by 7 cm high. Within this chamber an oval pad of grasses is constructed; a cache of seeds may be placed nearby. Apart from nesting burrows, other common burrow types are scattered throughout the home range. These include: escape burrows (short dead-end tunnels used in emergencies); latrine burrows (very short dead-end tunnels with a scat pile at the end); cache burrows (short tunnels with a buried seed cache at the end), and signpost burrows, which are near the periphery of the home range (around 3 cm long – really just a pit), where the animals take a sand bath in the waste soil and anoint the pit with scent from their anal glands. These mice have been observed swimming and will use water as an avenue for escape. Although captives can survive to be five years old, it is highly doubtful that wild mice could survive that long. Predators include hawks, owls, foxes, Coyotes, and Badgers.

DIET

Northern Grasshopper Mice are largely carnivorous and insectivorous. They catch and eat 75%–100% arthropods, when available, including grasshoppers, beetles, spiders, and moth and butterfly larvae. But they are strong and agile hunters and can take down lizards and other rodents their equal in size and larger. They eat mainly arthropods in warm seasons and switch to vertebrates during colder weather. Their diet includes some seeds, which are only eaten when other food is scarce.

REPRODUCTION

Much of the following reproductive information is derived from captive animals, as Northern Grasshopper Mice breed readily in captivity. Captives will reproduce throughout the year and produce up to 12 litters annually, but in the wild, breeding occurs during the warm season and the length of the season likely varies depending on local weather. Breeding normally extends from May to August in the Canadian parts of the range, where wild females likely produce three or at most four litters per year (but this is not substantiated). Length of gestation of non-lactating females averages about 32 days (27–46 days have been reported), but lactating females have a longer gestation period, which can vary from 33–47 days. The gestation period for this species is clearly widely variable, the reasons for which are not well understood. Females experience a postpartum oestrus that occurs within a few hours of parturition and lasts for up to two days. Litters average 3–4 pups (range 1–6), with larger females tending to have larger litters. A Northern Grasshopper Mouse male takes a more active interest in his pups than do most rodents. He defends the nest site and assists in raising the litter by providing food to the female while she is lactating. Newborns are pink and hairless, apart from fine white vibrissae. Their toes are fused together and their eyes and ears are sealed shut, but they can be quite vocal – peeping to express distress. Their ear pinnae unfold by day 3, and by day 8 they have dark grey fur on their backs and the beginning of fine white hairs

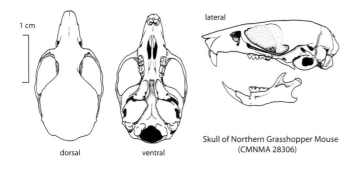

dorsal   ventral   lateral

Skull of Northern Grasshopper Mouse
(CMNMA 28306)

on their bellies and the first incisors start to erupt through the gums. By around 10 days old they begin to take some solid food. The ear canals open by day 14 and the eyes open by around day 19 or 20. Most pups are fully weaned by 23 days old. Juvenile females become sexually mature no earlier than 3 months old and males no earlier than 4 months old. Most do not breed until the following spring.

BEHAVIOUR

This tiny predator echoes the ecology and many of the habits of bigger, better-known predators. Like Grey Wolves and Coyotes, they can pull down prey larger than themselves, they occupy large, well-defended home ranges, they have developed a good long-distance means of communication, and they display a complex courtship, shared parental care, and tight social bonds. Unlike many of their larger analogues, Northern Grasshopper Mice do not hunt cooperatively. Nevertheless, they are formidable predators able to subdue prey up to four times their own weight. They usually chase their victim and pounce onto its back, seizing it with their forepaws before biting into the back of the head or neck to kill it, but they have the intelligence to learn to adapt their attack to suit the prey. For example, a naive grasshopper mouse will attack a large darkling ground beetle from behind as usual and then be deterred by a noxious spray into its face that will send it scurrying to rub itself in the sand. This is sufficiently discouraging that only when very hungry will a mouse attempt such prey again. But with experience, it will learn to approach the beetle from the front, jam the offending rear end into the dirt, and eat it from the head down. If their opponent is another grasshopper mouse, an attack is usually directed towards the rump and head, and only if the opponent persists will a killing bite be attempted. An attack will be temporarily abandoned when the subordinate animal assumes a submissive posture by lying on its side with its front paws tucked in, its ears laid back, and its eyes closed. This pose will then provide the subordinate a few moments to beat a hasty retreat before the dominant animal resumes its attack. Dominant animals will kill an intruder that does not withdraw with alacrity, but dominance is directly related to geography. An animal is only dominant on its own home range. Males are highly territorial, marking their home range with secretions from anal glands, especially in dust-bathing pits around the periphery. Their howling call may be used to space themselves by helping them locate other males, and it is likely that each animal's call can be individually identified. Males will not permit the entry of an adult male grasshopper mouse onto their home range. Females live on the home range of a male and do not establish or defend territories, apart from attacking any unknown female intruder.

As with most small mammals living in arid regions, burrows are essential to their survival, as they can moderate the heat of the day and maintain a higher humidity than the outside air, while providing shelter and protection from predators. Pairs work together to dig a nesting burrow, with the male doing most of the digging and the female removing the waste soil. The other burrow types (see in the "Ecology" section above) are dug by either mouse. A digging mouse stands on its hind legs, and using its forefeet together scoops sand under its belly. Then it tips forward to rest on all four legs and, using each hind leg alternately, kicks the sand to the rear, sometimes kicking so vigorously that it balances only on its front legs. The waste sand that accumulates at the burrow entrance is then scattered so that the burrow entrance is not raised above the soil surface and there is no noticeable throw pile.

Courtship between Northern Grasshopper Mice is considered to be the most complex of all the rodents. A courting male approaches an oestrous female by first circling her so that each animal can smell the other in safety. They sniff each other's anogenital and nasal regions and follow each other around for some distance, at times stopping to rise on their haunches and "kiss" (naso-naso sniffing). Then the pair rises, facing each other and holding each other's forelegs to press their bellies together. Females sometimes perform a backward somersault at this stage. This phase is followed by a period of mutual grooming and "neck-nibbling" and finally the female assumes a four-footed stand with her ears back, her tail to one side and her back arched downwards (this posture is called lordosis). The male will then mount. The pair maintains up to a one-minute copulatory lock, typically lying on their sides. The entire sequence can take up to three hours. Females are probably monogamous, but males may be polygamous. Nonetheless a strong pair bond develops between the partners. Normal territoriality may relax somewhat during the non-breeding season as more than two mice of the same sex may huddle together in one nest to conserve energy. Playing is an important part of the physical development of an infant. By pouncing, wrestling, and grappling with each other, they learn and practise important motor skills. Pups from smaller litters (3–4 pups) tend to play more with each other (perhaps they are better fed and have more energy to burn) and become more proficient hunters faster than do pups from larger litters (5–6 pups). Parents begin leaving insect prey around the nest site by the time the pups are 10–16 days old, perhaps to precondition them to viewing certain species as potential prey. Northern Grasshopper Mice are described by field researchers as self-assured and not timid like most other mice. They are reported to "walk with a swagger" rather than the usual rodent scurry. One captured adult attacked a gloved finger and gave it a shake before releasing it to attack yet another finger.

VOCALIZATIONS

Infants produce two calls. One is a distress call, a series of peeping chirps (4–6 per second at an approximate frequency of 8 kHz) and the other is a comfort call (also called a contentment vocalization) of a half-second-long, single tone in the 7–9 kHz frequency range, produced while suckling. Adults produce four calls: a single, sharp, high-pitched "chit," made during agonistic encounters; a soft series of "chirps" (5–7 per second) emitted by males during copulation; a low-volume pure tone lasting about 0.8 seconds in the 8–9 kHz range produced by either partner towards its mate, which may be analogous to the infant comfort call; and, lastly, their most famous vocalization, the loud, piercing "howl," so named because they often tip their nose into the air and open their mouth in the same posture assumed by a howling Grey Wolf. They may also rise on their hind feet, using their tail as a third leg for balance to make this call, but usually they merely stretch their head forward and upwards, half close their eyes, and flatten their ears before opening their mouth to howl. Both sexes will howl, but most calls are produced by males. Some animals will commonly climb to an elevated position before howling, and the call of one animal can prompt a flurry of howls from neighbouring territories. The sound produced, which appears to be used as a contact call, is a smooth, prolonged, high-pitched, piercing note (mainly around 12 kHz, but with harmonics extending to 64 kHz and possibly beyond) lasting about a second. The pitch of the howl call depends on the body size of the caller: larger animals produce a lower-pitched howl than do smaller ones, just as with wolves. A human can hear this loud, shrill call over at least 30 m on a still night, but the higher harmonics are ultrasonic and beyond the sharpest human hearing. The most commonly produced calls, those used in close-contact social situations, are in the ultrasonic range from 30 to 70 kHz. This high frequency (raptors can only detect sounds below 25 kHz) lends them nicely to this context, as aerial predators are unable to hear them.

SIGNS

Grasshopper Mice have the typical rodent pattern of five toes on their hind feet and four discernable toes on their front feet (toe 1, the thumb, is highly reduced and does not register in a track). The front toes may spread into a star pattern, or the three longer toes (toes 2, 3, and 4) may group together, aiming forward, with the outside toe (5) angled slightly to the side. The three middle toes on the hind feet usually group together, aiming forward, with the outside toes (1 and 5) angling to the sides. Distinguishable tracks of grasshopper mice are rarely observed in the wild. Burrow entrances are taller than wide and are around 6.7 cm high by 4.0–4.7 cm wide. Most nesting burrows have multiple entrances, but usually all (if the animals are inside), or all except one (if they are out hunting), are commonly plugged. Burrow entrances can be hard to find as they do not have a throw pile at their mouth. Dust-bathing pits may be found along territorial boundaries, which when fresh may retain some odour of secretions from the anal glands, said to smell musky and skunk-like. Vertebrate kill sites are normally just a few bones and pieces of skin. Insect kill sites may just be a pile of legs and other inedible parts, such as beetle wing covers. Scat, deposited underground, is not usually found. Northern Grasshopper Mice may be easiest to find by scouting an area from dusk to a couple of hours after dusk, when the animals first emerge from burrows to begin their nightly hunt. Listen for the high-pitched howl call, as it is commonly produced at that time.

Walk

Left hind track
Length: 13.0–14.0 cm*
Width: 11.0–12.0 cm*

Left front track
Length: 9.0–10.0 cm*
Width: 10.0–11.0 cm*

*approximate
Northern Grasshopper Mouse

REFERENCES

Eisner and Meinwald 1966; Elbroch 2003; Engstrom and Choate 1979; Finley, T.G., and Sikes 2004; Goheen et al. 2003; Hafner and Hafner 1979; Kemble 1984; Langley 1983; McCarty 1978; Pellis and Pellis 1992; Riddle 1999; Ruffer 1965a, 1965b, 1966, 1968; Russell and Finley 1954; Slobodchikoff et al. 1987; Smith, H.C., 1993; Stapp 1999; Verts and Carraway 1998; Wrigley et al. 1991.

Keen's Mouse

also called Cascade Deer Mouse, Sitka Mouse, Northwestern Deer Mouse

FRENCH NAME: **Souris de Keen**

SCIENTIFIC NAME: *Peromyscus keeni*

This large *Peromyscus* has only recently been recognized (1988) as a distinct species and much work remains to be done to determine its ecology and the full extent of its distribution. Populations formerly classified as subspecies of Deer Mouse (*Peromyscus maniculatus sitkensis, P. m. oreas, P. m. keeni,* and *P. m. algidus* as well as a number of island subspecies) have been reclassified now as Keen's Mice. While chromosomally distinct from Deer Mice, there are, unfortunately, no completely reliable physical traits that will distinguish all Keen's Mice from Deer Mice in the field.

DESCRIPTION

Fur colour and body size vary considerably across the range, and this mouse is inclined towards gigantism in isolated island populations. Generally, the dorsal fur is brown, reddish-brown, or dark grey-brown with a darker brown band of colour running along the spine from the head to the rump, but some island populations are

Keen's Mouse (*Peromyscus keeni*)

dark brown with a blackish dorsal band. Belly and chest hairs are mostly white or pale grey at their tips with dark grey roots, but some island populations have buffy to reddish-brown tips on their belly and chest fur. Some animals have a small patch of hairs under their chin that are white all the way to the root. Hairs on the upper surface of the front feet are white, but on the same part of the hind feet the fur may be white, pale silvery grey, or dark brown or even partially dark brown with white toes on a few animals. The large eyes are black, bulging, and luminous. Vibrissae are long and constantly in motion when the mouse is active. Ears are long and mostly covered with a thin layer of fine, very short, dark hairs. The long tail is sharply bicoloured, dark brown above and white below, with a tuft of stiff hairs at the tip that are 4–5 mm in length. Juveniles undergo two maturational moults before achieving their adult pelage, after which they moult once annually in late summer. Juveniles are usually dark grey above and grey to silver-grey below, while subadults are dark brown above and silver-grey to white below. The dental formula is incisors 1/1, canines 0/0, premolars 0/0, and molars 3/3, for a total of 16 teeth. Individuals have three pairs of mammae.

SIMILAR SPECIES

Deer Mice closely resemble Keen's Mice. The two species coexist in southwestern British Columbia, along the eastern slopes of the Cascade and Coast Mountains, and on some coastal islands. The following formula using diastema length (the gap between the back of the incisors and the front edge of the molars) and tail length separates the two species in the southern part of their range, including Washington State and Vancouver Island and the surrounding islands: 4.4 × length of tail (in mm) + length of mandibular diastema (in mm). Animals with a score of ≥ 436 are Keen's Mice, while those with scores of < 436 are Deer Mice. The efficacy of this formula, when applied to inland and more northerly populations has not yet been assessed, and the formula is ineffective in identifying living animals. See the "Similar Species" section in the Deer Mouse account for more information.

SIZE

The following measurements apply to animals of reproductive age from the Canadian portion of the range. Females and males are similar in size and appearance. Physical size and tail length vary considerably across the range. The largest individuals are from Triangle Island (off the northwest tip of Vancouver Island).
Total length: 140–263 mm; tail length: 70–126 mm; hind foot length: 19–32 mm; ear length: 12–26 mm.
Weight: 15.0–52.4 g; newborns, about 1.6–3.4 g.

RANGE

This largely coastal, western species occurs from southern Yukon to western Washington State. Most of its range is in British Columbia, where it is found all along the coastal mainland as well as on about 80 islands, including Vancouver Island, the Queen Charlotte Islands, Table Island, the Scott Islands, islands in Queen Charlotte Strait, Balaklava and Malcolm Islands (where it coexists with Deer Mice), the Moore Islands, the Goose Islands, and many of the coastal

Distribution of the Keen's Mouse (*Peromyscus keeni*)

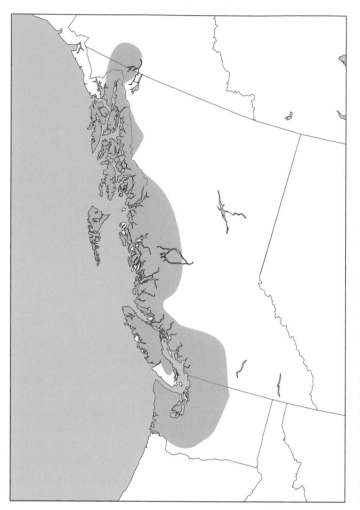

range has been very roughly estimated, in a low–medium density island population, at around 2600 m² (about a quarter of a hectare) and does not appear to vary by sex. Dispersal distances measured in the same population (in southeastern Alaska) vary from 248–381 m. As do the other two Canadian *Peromyscus* species, Keen's Mice are quick to live commensally with humans, especially in the cold season. Their invasion of homes, warehouses, and farm buildings is sufficient to provide them with pest status at times. Both Keen's Mice and Deer Mice occur in southwestern British Columbia. Persistent reports describe their jumping ability. Thought to be more arboreal than Deer Mice due to their longer tail, they can climb up to 15 m into a tree. Predators include a variety of raptors and mammalian carnivores such as the Coyote, Bobcat, Domestic Cat, and weasels. Some island populations have been significantly reduced following the introduction of Norway or Roof Rats either due to direct predation by the rats or through competition for food.

DIET

Keen's Mice, like other *Peromyscus*, consume a diverse omnivorous diet that varies seasonally and geographically. Generally, they eat seeds, green vegetation, fruits and berries, terrestrial invertebrates such as insects and spiders, and fungi. Some island populations specialize in consuming seabird eggs, possibly live nestlings, and probably carrion (carcasses of chicks and adults) when they are abundant. The mice primarily predate small seabirds that nest in ground burrows (such as Cassin's and Rhinoceros Auklets). Marine invertebrates, such as amphipods are a primary food source for Keen's Mice that forage in the intertidal zone. Although many coastal populations have access to salmon carcasses during the annual salmon spawning runs, there are no indications that Keen's Mice take advantage of this plenty by consuming salmon meat.

islands in Queen Charlotte Sound off Bella Coola. Keen's Mice also occupy many of the islands in the Alaska Panhandle (the Alexander Archipelago).

ABUNDANCE

These mice are relatively abundant and often are the dominant small mammal in their location. Estimates of densities of 0.30 animals per ha up to 96 per ha have been reported, but populations fluctuate drastically from year to year and from one habitat to another, probably due in large part to winter weather conditions and food abundance.

ECOLOGY

Like other mice in the genus *Peromyscus*, Keen's Mice are environmental generalists. They occupy a diversity of habitats from sea level to 2010 m in elevation. The zone where they live largely consists of temperate rainforest, where the climate is both moist and mild. The mice are found in coniferous forests, alder thickets, shrubby fields, subalpine and alpine meadows, and even heaps of driftwood pushed up above the intertidal zone. They can flourish in old-growth forests, floodplains, regenerating forests, and clearcuts. Average home

REPRODUCTION

These mice are difficult to breed in captivity, so many details of their reproduction are unknown. In the wild, most breeding occurs between late March and June, with the last litter weaned by the end of July to mid-August, although in coastal regions the breeding season may be somewhat more prolonged. Gestation lasts 23–25 days. Litter size averages 4–6 (range 3–9) and females produce several litters annually. Neonates are altricial, born essentially hairless and pink, with closed eyes and ears. Their eyes open in 13–21 days and

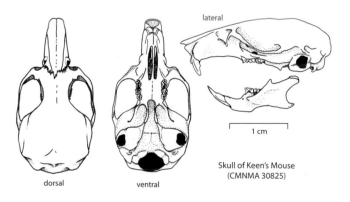

Skull of Keen's Mouse
(CMNMA 30825)

dorsal   ventral

they begin eating solid food about the same time. Young of the year are capable of breeding as early as 30 days old, but unless they are born early in the season, most do not breed for the first time until the following spring.

BEHAVIOUR

The social system of this rodent has not yet been described, but it is likely similar to that of other *Peromyscus* species.

VOCALIZATIONS

Not reported.

SIGNS

Tracks and sign of the three Canadian *Peromyscus* species are indistinguishable. See the "Signs" section of the Deer Mouse for further information.

REFERENCES

Allard and Greenbaum 1988; Chirhart et al. 2001; Cowan and Guiguet 1956; Drever et al. 2000; Hanley and Barnard 1999; Hanley et al. 1999; Hogan et al. 1993; Lucid and Cook 2004; Nagorsen 2005a; Reid, F.A. 2006; Slough and Jung 2007; Smith, W.P. and Nichols 2004; Wilson, D.E. and Reeder 2005; Youngman 1975; Zheng et al. 2003.

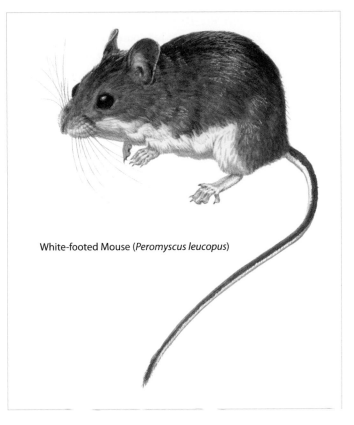

White-footed Mouse (*Peromyscus leucopus*)

White-footed Mouse

also called Wood Mouse, White-footed Deermouse

FRENCH NAME: **Souris à pattes blanches**

SCIENTIFIC NAME: *Peromyscus leucopus*

These mice commonly attempt to take up winter residence in homes, cottages, and outbuildings in wooded areas.

DESCRIPTION

White-footed Mice are large-eared, with orange-brown to yellowish-brown dorsal fur and white bellies and feet. Their white underside is sharply demarked from the coloured fur. Most animals have a wide, clearly defined, darker brown band of colour that runs along the spine from the top of the head to the tail. The dark eyes are large and luminous. Ears are largely covered by very short dark hairs. Tails are sparsely haired (some scales are visible) and bicoloured, brown above and creamy white below, and the demarcation between the dark and light fur is normally somewhat diffuse (some longer dark hairs extend over the separation). Tail length is very roughly about half of total length, but this varies across the range. Tail tip has a short tuft of hairs that are 2–3 mm in length. Juveniles (up to about 40 days old) are grey on the dorsum and white on the underside. Subadults have a white belly, and their back and sides are a

mix of grey and brown with the first signs of colour appearing low on the sides. Subsequent annual moults become brighter. Moulting begins in early autumn and is complete by late autumn. Albinos are known but very rare. The dental formula is incisors 1/1, canines 0/0, premolars 0/0, and molars 3/3, for a total of 16 teeth. White-footed Mice have three pairs of mammae.

SIMILAR SPECIES

Deer Mice are very similar to White-footed Mice and distinguishing between them can be a challenge. Within most regions of overlap, it is often possible to discern some reasonably clear morphological distinctions between the two species that can assist in identification, but these are both widespread species with considerable size and colour variation across their range. Attempting to outline characteristics that clearly separate the two all across their widely overlapping ranges is next to impossible. Juveniles are especially difficult to tell apart, as they are grey and white in both species. In the eastern Canadian parts of their range, adult Deer Mice tend to be drabber grey-brown on their dorsal surface, whereas adult White-footed Mice are typically more brightly coloured. This is largely true in the western part of the range as well, but is a less useful trait there as many Deer Mice are brighter than the norm in the east and approach White-footed Mice in colour. In general, adult Deer Mice lack a distinct mid-dorsal stripe, but this is definitely not the case in all populations. For example, Deer Mice from Grand Manan Island, New Brunswick, usually have a sharply defined mid-dorsal stripe, as do many from other parts of the range. Not all White-footed Mice have such a clearly defined stripe, although most do. Deer Mice classically have a more densely furred and more sharply bicoloured tail than do White-footed Mice, but again this is not a universally reliable trait. Tail length of eastern Deer Mice is generally longer than the head-body length, while in the west it is shorter. Deer Mouse belly fur is grey at its base and along most of its shaft and white or silvery-grey only at the tip. The belly fur of White-footed Mice may be white all the way to the root or white at least three-quarters of its length, with only a short section of grey right near the base. In most populations of White-footed Mice, the chin and throat fur is white all the way to its base and this white patch extends more than 1 cm from the mouth. Deer Mice may have a white chin patch, but it extends less than 1 cm from the mouth. Distinguishing the juveniles is usually impossible without a cleaned skull in hand. Some juvenile Deer Mice have greyish belly fur, while White-footed Mice juveniles always have white belly fur, but this characteristic is rarely useful, as many or most juvenile Deer Mice also have white belly fur. There is a tooth characteristic that may help identify the species, the use of which was first identified by Guilday and Handley in 1967. The anterior edge of the first molar of Deer Mice is angled inward, while on White-footed Mice it is more symmetrical and rounded. Both upper and lower first molars display this characteristic, but it is most clear on the lower molar. Rich et al. (1996) provide a discriminant function of 11 or 12 cranial characters that when correctly calculated will identify 94% (using the 11-character set) and 100% (12-character set) of the skulls of adults, subadults, and juveniles of both species from the eastern parts of the range. There is no reliable, fast, and easy way to differentiate between the two species in the field, although, as explained earlier, with experience, adults of many regional populations can be distinguished most of the time. Genetic analysis remains the only 100% reliable diagnostic technique for live or whole animals or for museum specimens without a skull for corroboration.

SIZE

The following measurements apply to animals of reproductive age from the Canadian portion of the range. Females and males are similar in size and appearance.

Total length: 161–200 mm; tail length: 71–100 mm; hind foot length: 19–22 mm; ear length: 16–20 mm.
Weight: 16.7–30.0 g; newborns, about 1 g.

RANGE

White-footed Mice reach the northern limits of their distribution in mainland Nova Scotia, eastern Quebec and southern Ontario, and extreme southeastern Manitoba. They extend as far south as southern Mexico and as far west as Arizona. The population in southern Nova Scotia is isolated from others by at least 200 km. Some recent specimens from the Gaspé region of Quebec have been identified as White-footed Mice using DNA testing, but the full extent of their range in that area remains unclear. Earlier reports and distribution maps to the contrary, no White-footed Mice have been discovered in Alberta or Saskatchewan and only one specimen is known from Manitoba.

ABUNDANCE

This species can be one of the most abundant small mammals in suitable habitat within its range, but population density can vary significantly due to factors such as weather and food availability. Numbers vary seasonally and annually. Most populations reach a peak in August, as the spring juveniles produce their litters along with the surviving adult females. Numbers plummet in September. Like many other mammals, White-footed Mice are generally less abundant at the peripheries of their range.

ECOLOGY

White-footed Mice are a widespread species and occupy many different habitats throughout their range. In eastern Canada they are most abundant in dry or upland mixed deciduous/coniferous forests, but may also be found in other habitats, such as regenerating old fields and brushy woodlands, and in agricultural areas, especially in fence lines and hedgerows between croplands. In the West they mostly live in protected woodlands in valleys and along rivers and streams. One of the major factors contributing to White-footed Mouse survival is the availability of abundant understorey vegetation that provides both food and cover. These small mammals are primarily nocturnal and crepuscular and are active throughout the year. Some winter activity may take place during the daylight in tunnels in the snowpack. Home ranges vary considerably by season, habitat, food availability, population density, and age of the individual. They are generally largest during the breeding season and smallest during the winter, with male home ranges larger than female. A typical home range is around 0.1 ha and few established mice shift their home range over their lifetime. Most juveniles disperse soon after they are weaned. Female juveniles generally travel only as far as they must to find a vacant territory and will remain

Distribution of the White-footed Mouse (*Peromyscus leucopus*)

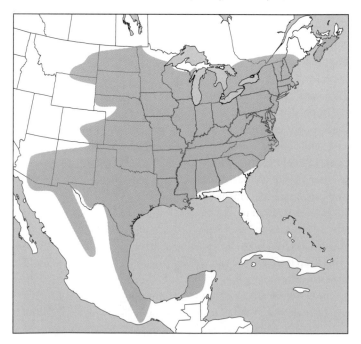

on their natal range if their mother permits it, or if she no longer continues to occupy it. Juvenile males will not settle on their natal home range, tending to disperse farther than juvenile females. Most juvenile dispersal is probably less than 100 m. Maximum known dispersal, by an adult (female) from a population at peak density that was suffering a poor mast year, is an astonishing 14.73 km. This extraordinary effort took no longer than four weeks.

Nests are important for resting, rearing young, and overwintering. These spherical constructions are made of plant fibres such as shredded bark, leaves, and grass stems and are insulated with moss, hair, feathers, fluff from milkweeds and cotton grass, or even shredded fabric. Nest sites may be above ground in a bird box, a tree cavity, under peeling bark, or in a roofed-over bird nest, but may also occur on or below the ground in rock piles, logs, and stumps and in ground burrows of other species such as Woodchucks. Although White-footed Mice do not hibernate, some do make energy savings by resorting to short, sometimes daily periods of torpor when temperatures are excessively cold. Torpor usually lasts at least three hours, during which the heart rate and body temperature may drop from an average 700 beats per minute and 35°C to about 60 beats per minute and 17°C. White-footed Mice and their closely related sibling species, Deer Mice, are very similar ecologically. In many regions where both occur, their diets are similar, their populations fluctuate in synchrony, their breeding seasons are the same length, and they are intra- and interspecifically territorial within each sex. White-footed Mice tend to be more woodland-adapted, while Deer Mice tend to be more grassland-adapted in many regions where their ranges overlap, but this general statement does not hold true in all areas, as there are many wooded habitats where both occur. In the Great Lakes region and parts of New England where extensive clearcutting occurred in the 1800s–1900s, many forests are beginning to regenerate, and White-footed Mice are returning to areas previously taken over by Deer Mice. Deer Mice are better suited to withstand severe winter weather and so extend farther north and westwards than do White-footed Mice. White-footed Mice are more three-dimensional in their foraging, commonly climbing trees and shrubs to harvest fruits and nuts. They will readily inhabit nest boxes placed in trees for birds.

In some parts of their range, White-footed Mice and Deer Mice are important intermediary hosts of the ticks that transmit the Lyme disease spirochete. The larval and nymph stages of the tick feed on mice and adult ticks feed on deer. Humans can contract the disease from an infected tick of any life stage. The two *Peromyscus* also serve as reservoirs of Hantavirus throughout their range. Hantavirus can be transmitted to humans, in whom it may cause a severe respiratory illness that can prove fatal. Rodents shed the virus in droppings, urine, and saliva. Transmission to humans typically occurs when the virus is inhaled. Crushed faeces, disturbed nesting material, or stirred-up fresh urine can cause tiny droplets of the virus to become airborne. Preventing mice from inhabiting human living spaces is a primary preventative measure.

White-footed Mice are agile climbers, so each home range has a significant vertical component, and they are very capable swimmers,

as evidenced by their repeated dispersal to islands. Captives may survive several years, but the maximum lifespan in the wild is about three years and few live longer than six months. Most populations undergo an almost complete turnover annually, as around 98% do not live beyond their first year. Most small to medium-sized carnivores will hunt White-footed Mice, including weasels, short-tailed shrews, foxes, American Badgers, Coyotes, and Domestic Cats and Dogs, as well as a legion of snakes and owls.

DIET

The diet of these omnivorous mice includes seeds and nuts, fruit, insects, green vegetation, and fungi, and varies seasonally. Cannibalism, especially of pups, may occur. Food hoarding is an important winter survival strategy in the northern parts of the range. The mice begin bringing food in their cheek pouches to cache sites in late summer and continue until snowfall. Food caches are placed in almost any protected site, such as in bird nests, in or under logs, in tree cavities, under loose bark, in underground chambers, or in or behind a piece of furniture or behind a hay bale in a barn.

REPRODUCTION

The breeding season of this species in Canada extends from March to October, with a midsummer hiatus or slowdown. Sexually mature females in the northern part of the range can produce four or more litters annually, while those farther south may produce even more as they continue to breed all year round. Females experience a postpartum oestrus and may mate again within a short time of giving birth, unless they are entering the midsummer hiatus or are reaching the end of the breeding season. Multiple paternities within a single litter do occur. Gestation lasts 22–24 days if the female is not concurrently lactating, and can last up to 37 days if she is still nursing a previous litter. Litters average 4–6 pups (range 2–7) in northern parts of the range. In the northern portions of the range, larger, typically also older, females have larger litters than smaller, usually first-time mothers. Most parental care is provided by the mother, although sires may cohabit and provide some care. See the "Behaviour" section below for more details on paternal care. Pups are born hairless and blind but develop very rapidly. Their eyes open when they are around 12 days old, and they are weaned at 3–4 weeks old. Lactating females require about 25% more food than do males or non-reproductive females. Females reach sexual maturity around 44 days old.

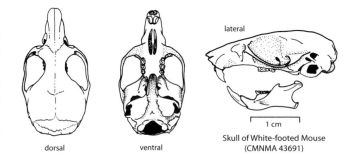

dorsal ventral lateral

Skull of White-footed Mouse
(CMNMA 43691)

1 cm

BEHAVIOUR

White-footed Mice are largely solitary in the warm season apart from mating pairs and lactating females. Adult females are highly aggressive towards each other (and also towards female Deer Mice), especially during the breeding season, which results in home ranges that are distinct and separated. Male home ranges also overlap little with those of other males (including male Deer Mice), but commonly overlap those of females. Intruders are attacked and chased away. Residents and larger-bodied animals tend to dominate in the majority of encounters. Urine and faeces appear to be used to mark territory and can repulse intruders without the physical presence of the resident. During periods of low population density, males may desert their territories to travel in search of females. Most pups are born to solitary females. Female-female aggression may be relaxed on occasion, as two females or even more (likely siblings or mother and daughters) have been known to share a nest and raise their litters together, even to the extent of nursing each other's pups.

A male White-footed Mouse may cohabit with a maternal female, although she will likely deny him access to her very young pups (less than four days old). Thereafter, he may share the nest, retrieve the young if they are dragged out of the nest and huddle over them to protect them and keep them warm. Wild paternal males have been observed foraging with their weaned offspring, thereby providing them with some food-finding experience. The extent of paternal care in the wild is not known and may vary from population to population. Females do not normally associate with weaned juveniles. Transient males and females will kill and sometimes eat unattended pups that they encounter. The exceptions are maternal females with their own litters and territorial males encountering litters on their home range (as these could likely be their own offspring). White-footed Mice tend to use the tops of distinct structural features such as rocks, logs, branches, or rock fences for their travel routes. These are used regardless of the amount of available lunar illumination, but the mice prefer softly padded substrates (moss-covered) where the sound of their passage is muffled.

VOCALIZATIONS

Pups up to about 10–12 days old use ultrasonic calls to express distress and solicit parental attention. Adults use a high-pitched vocalization that is similar to a bird's trilling and may continue to utilize ultrasonic frequencies.

SIGNS

Tracks and sign are indistinguishable from Deer Mouse sign. See the "Signs" section in the Deer Mouse account for more information.

REFERENCES

Anderson, C.S., and Meikle 2006; Barnum et al. 1992; Brown, L.N., 1997; Bruseo et al. 1999; Choate et al. 1994; Desrosiers et al. 2002; Dobbyn 1994; Elbroch 2003; Guilday and Handley 1967; Havelka and Millar 2004; Jacquot and Vessey 1994; Jannett et al. 2007; Kurta 1995; Lackey, J.A., 1999; Lackey, J.A., et al. 1985; Long, C.A., 1996; Maier 2002; McMillan and Kaufman 1995; McMillan et al. 1997; Myers and Lundrigan 2001; Prescott and Richard 2004; Reid, F.A., 1997, 2006; Rich et al. 1996; Schug et al. 1991, 1992; Shipp-Pennock et al. 2005; Soper 1961; Stafford 1993; Terman and Terman 1999; Tessier et al. 2004; Weston et al. 2005; Wolff 1993, 1996; Wolff and Cicirello 1990; Xia and Millar 1991.

Deer Mouse

also called North American Deer Mouse

FRENCH NAME: **Souris sylvestre**
SCIENTIFIC NAME: *Peromyscus maniculatus*

This is the most common and best known of the three Canadian *Peromyscus* species and the one with the largest distribution. These mice commonly attempt to invade homes, cottages, and outbuildings as the cold season approaches. They are easy to trap and remove. Sunflower seeds, peanut butter, or a mix of peanut butter and rolled oats or hulled sunflower seeds makes irresistible bait.

DESCRIPTION

The Deer Mouse is a widespread and highly variable species. To complicate identification, their colour, size, and body proportions vary in different environments. Dry grassland forms tend to be paler and yellowish brown, while dense forest forms are more colourful or darker. Those in woodlands that climb to forage in trees and shrubs have longer tails than those that spend most of their time on the ground. Furthermore, many populations have evolved in isolation on islands and may be quite different in appearance from nearby mainland forms. The adult colour pattern is some variation of brown above (grey-brown, orange-brown, yellowish brown, or reddish brown) and sharply delineated white or silvery grey below. A band of darker fur runs along the spine from the head to the rump

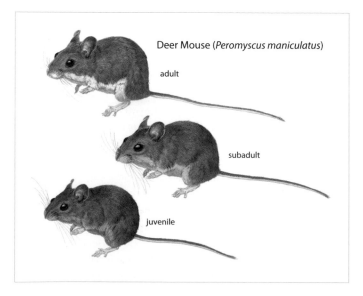

Deer Mouse (*Peromyscus maniculatus*)
adult
subadult
juvenile

Deer Mouse (*Peromyscus maniculatus*)
prairie form

of most animals. This band varies in width and colour saturation across the range and even within populations, but typically blends with the lighter fur on the sides rather than creating a distinct strip. A small patch of hairs that are white all the way to their root occurs under the chin. Hairs on the belly are generally only white or pale grey at their tips, with the remainder of the hair shaft being grey. Bulging black eyes are large and luminous. Ears are large and mostly covered in very short dark hairs. Long vibrissae on the upper lips are almost constantly in motion. The mice's semi-prehensile tail is sharply bicoloured and moderately furred, with a small tuft of stiff, 4–5 mm long, hairs at its tip. Pups go through two maturational moults before they achieve adult pelage, after which they experience a single annual moult that begins in late summer and is complete by early to mid-autumn. Juveniles are dark grey above and grey to white below with white feet, while subadults are grey brown above and pale grey to white below with white feet. Albinistic, melanistic, blonde, and even hairless specimens are reported. The dental formula is incisors 1/1, canines 0/0, premolars 0/0, and molars 3/3, for a total of 16 teeth. These mice have three pairs of mammae.

SIMILAR SPECIES

White-footed Mice overlap the range of Deer Mice in the southeastern quarter of the continent. They are very similar to Deer Mice and the two species can be difficult to distinguish, especially when they are in juvenile pelage. See the "Similar Species" section in the White-footed Mouse account for more details. Keen's Mice overlap the distribution of Deer Mice on Vancouver Island (and the associated islands of Balaklava and Malcolm), the coastal southern mainland, and the eastern slopes of the Cascade and Coast Mountains. As are White-footed Mice and Deer Mice, Keen's Mice and Deer Mice are genetically distinct, but there are no field characteristics that reliably separate the two. In general, Keen's Mice are somewhat larger, with a slightly longer tail. Any with tails longer than 100 mm will be Keen's Mice, as most Deer Mice in those areas have tails that are

< 98 mm long. There is some overlap in tail measurements, so this characteristic is not useful in distinguishing shorter-tailed Keen's Mice. Juveniles are especially similar and can only be distinguished genetically. See the Keen's Mouse account for further information. Northern Grasshopper Mice are larger and stockier than Deer Mice, but similar in pelage colouration. They have larger front feet and a thicker, shorter tail (< 42 mm long) that often has a white tip. Western Harvest Mice could be mistaken for small Deer Mice, but any Deer Mouse small enough to be confused would be in juvenile pelage (dark grey above, white below) and not the light grey-brown of a Western Harvest Mouse. Unlike the Western Harvest Mouse, Deer Mice do not have grooved incisors. House Mice are similar in shape to a Deer Mouse, but they are dark grey to grey brown overall with, at most, a slight lightening of colour on the underside.

SIZE

The following measurements apply to animals of reproductive age from the Canadian portion of the range and cover the variation in size and proportion across the country. Females and males are similar in size and appearance.
Total length: 120–209 mm; tail length: 42–110 mm; hind foot length: 15–28 mm; ear length: 10–22 mm.
Weight: 15.0–41.8 g; newborns, about 1.0–2.2 g.

RANGE

This widespread species occurs from the Northwest Territories to Mexico and from Labrador to British Columbia. Deer Mice have colonized many freshwater and marine islands including Anticosti Island, the Magdalene Islands, Prince Edward Island, Manitoulin Island, Vancouver Island, and 97 smaller BC islands around Vancouver Island, especially between Vancouver Island and the mainland. They do not occur on the Queen Charlotte Islands and many of the coastal islands of central and northern British Columbia, which are occupied instead by Keen's Mouse. Deer Mice have not naturally reached Newfoundland, but there is a population in southwestern Newfoundland that was presumably accidentally introduced.

ABUNDANCE

Deer Mice are one of the most abundant small mammal species in North America. Populations fluctuate widely in density, due to factors such as the severity of winter weather, food availability, and amount of summer rainfall. Generally in northern populations numbers peak near the end of the summer as the new crop of youngsters is recruited, and there is a gradual attrition during winter and early spring, such that the population density is lowest in late spring and early summer. Population densities can vary from 0.3/ha to 70/ha.

ECOLOGY

Across the continent, Deer Mice occur in a wide variety of habitats from sea level to at least 4300 m in elevation (2200 m in Canada). They can be found in grasslands, parklands, wetlands, deserts, shrubby woodlands, mature woodlands from coniferous to deciduous, regenerating fields, clearcuts, burns, alpine meadows, rocky mountainsides, and even alpine and some Arctic tundra. In short,

Distribution of the Deer Mouse (*Peromyscus maniculatus*)

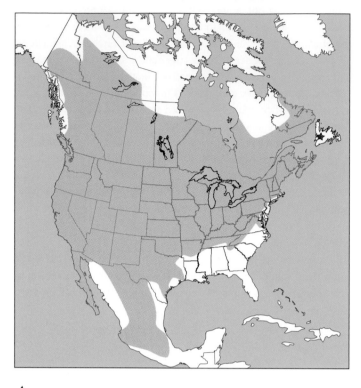

★ introduced

they are extremely adaptable, and are even able to survive and flourish in patchy and marginal urban environments. Deer Mice are primarily crepuscular and nocturnal and even in the north, where summer day length is prolonged, they shorten their activity period to restrict their foraging to the limited dark hours. Although Deer Mice remain active throughout the year, they commonly resort to short (4–9 hour) periods of torpor during daylight hours to conserve energy during cold periods. While torpid, their body temperature may fall to 13°C–17°C from a normal temperature of 37°C–38°C. Where winters are snowy, they tunnel through the snowpack or use tunnels created by other small mammals, but they do not restrict themselves to the snowpack, as they will sometimes travel on top of the snow while foraging. Deer Mice dig burrows that typically have one to three entrances, all leading to a single tunnel that slopes steeply to a nest chamber. Deer Mice living in grasslands may be forced to dig such shelter and escape burrows for themselves unless they can find nest sites in crevices between rocks or in the burrows of other species. Woodland inhabitants generally have a wider nest site selection, because they can not only easily burrow into the soft forest litter or find cavities in rock piles, but can also utilize tree cavities and spaces in or under logs or other forest debris. Deer Mice make use of little-used or abandoned burrows dug by other species, such as pocket gophers, ground squirrels, chipmunks, American Badgers, prairie dogs, or Woodchucks. Deer Mice are quick to live commensally with humans, and the species has probably benefited from settlement and agricultural practices since European colonization of North America. Their hollow,

globular nests are about 7–12 cm in diameter, and are constructed of grass stems and leaves with linings of shredded vegetation, plant fluff, fur, or feathers. Nests may be littered with droppings and seed casings. Most foraging takes place within 100 m of the nest site. Home range size varies with habitat, food availability, population density, sex, and age from an estimated 0.1 to 1.2 ha. Males generally have larger home ranges than females (around twice the size). Male home ranges overlap extensively with those of both neighbouring males and females, but females defend their home ranges during the breeding season, excluding other females. Apart from dispersing juveniles or adults forced to relocate due to lack of food, most Deer Mice are sedentary, remaining on the same home range for their entire life. Nevertheless, homing experiments show that displaced mice can find their way back to their home ranges from distances of 1980 m, or almost 2 km, in less than four days. It is unlikely that those homing mice were at all familiar with the route, but how they navigated remains unclear. Maximum dispersal distance, so far reported, is by a subadult male that, in less than 16 days, travelled 1768 m. Most dispersal is over a considerably shorter distance, but longer movements of 1000 m or more by subadults or adults may not be uncommon. Generally, juvenile females remain near their birth range if they can find a nearby home range and juvenile males disperse farther afield. Hantavirus, a potentially fatal pulmonary disease in humans, is normally contracted by inhalation of contaminated, pulverized faeces and the dried urine or saliva of Deer Mice or, secondarily, of White-footed Mice, two common reservoirs of the disease. The only cases of Hantavirus in Canada have been from the west (Manitoba, Saskatchewan, Alberta, and British Columbia) and have all been ascribed to Deer Mice. The same two species of mice act as reservoirs for Lyme disease (tick-borne) and Plague (flea-borne). Although they can live six to eight years in captivity, very few wild Deer Mice survive more than a year and a half, and an estimated 99% live less than a single year. Ground-dwelling Deer Mice are often less than agile climbers, although they will forage in low shrubs, but the longer-tailed woodland forms are able to scale the trunks of both large and small trees and run along branches, the tops of fences, power cables, clothes lines, and chair rails. They are also willing and capable swimmers in both fresh- and saltwater. By swimming or travelling over ice, Deer Mice have reached and colonized many near-shore islands. These small rodents are a staple in the diet of a host of predators across the continent. Owls, weasels, and foxes are primary predators, but many other animals consume Deer Mice when they can, including skunks, Coyotes, Bobcats, and snakes as well as Domestic Cats and Dogs and other small mammals such as short-tailed shrews and grasshopper mice.

DIET

The Deer Mouse diet varies with season. They mainly eat seeds, nuts, fruit, green vegetation, spiders, and insects. Slugs, snails, fungi, and bird eggs and nestlings may be added to the diet if available. An important part of their winter food is composed of seeds and nuts that have been previously cached, but Deer Mice continue to forage throughout the winter, searching for fresh vegetation, seeds, and even dormant insects under the snow. Opportunistic in their food

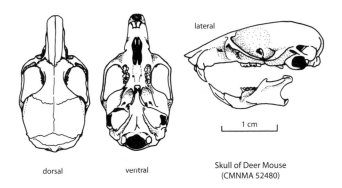

dorsal   ventral

lateral

1 cm

Skull of Deer Mouse
(CMNMA 52480)

choices, Deer Mice will readily consume most of the food stuffs in a modern kitchen along with stray crumbs that linger under furniture and the contents of a compost pile. They are not above helping themselves directly from the pet food bowl if it is left unattended at night. Access to an open bag of wild bird seed stored in a quiet corner of the shed or garage provides an overwhelmingly appealing opportunity. Small caches will begin appearing throughout the building, in all imaginable locations. Clearly, Deer Mice are more comfortable if their eggs are not all in one basket. Infanticide and sometimes cannibalism of unprotected young, by unrelated females and males, is not uncommon.

REPRODUCTION

The timing and duration of the breeding season depends on several external factors, primarily weather conditions, food availability, and day length, and can vary from year to year in the same population. In regions with mild winters (such as the Gulf Islands) breeding may begin in April and extend to December or even January. In colder areas with heavy snow, the season typically begins in May and lasts until September or October. High-elevation or high-latitude populations may breed only from June to August. Typically, breeding coincides with the local period of vegetation growth and insect activity and hence food abundance. During years of low food availability, fertility of both male and female mice is restricted. Gestation normally lasts 22–23 days. Litter size averages 3–6 (range 1–8). Females may experience a postpartum oestrus and breed again shortly after parturition. The breeding system is polygamous, with one to four litters produced annually in Canadian populations, depending on the length of the breeding season. Some litters may be multiply sired. Pups are altricial, born pink and helpless, with eyes and ears fused shut. They begin to show hair growth by their second day, their ear pinnae unfold on day 3, their eyes open by day 15, and they are weaned by 25 days old. A female typically drives any remaining pups away after five weeks to devote her attention to the next litter. Males become sexually mature as early as 40–60 days old. Subadult females may breed as young as 30 days old, so some females from early litters could breed in the summer of their birth if the breeding season is long enough, although most young Deer Mice in northern or alpine locations do not breed for the first time until the following spring.

BEHAVIOUR

Deer Mice are inveterate hoarders, stuffing their small internal cheek pouches with seeds and nuts and caching them in almost any protected container, including tree cavities, abandoned bird nests, underground chambers, behind peeling bark, in unused drawers, shoes, and dishes and inside mattresses, sofas, and even in insulation within walls. Creation of this type of large food cache is called larder-hoarding. Sometimes a small number of seeds may be "scatter-hoarded," meaning that only a few seeds are stored or buried in many locations. This latter form of hoarding is less common, but may play a major role in the dispersal of food trees. Individual mice use memory and scent to relocate caches. Adult females are usually territorial during the breeding season, aggressively excluding other females from their home range, but an instance of successful group nesting of five wild, breeding females has been reported. Only two of these animals were related. Males may remain near an oestrous female for a short time, but typically move on once she drops out of heat. Apart from mating pairs and maternal females with litters, adults are generally solitary during the warm months. Outside the breeding season, this social system breaks down and up to a dozen mice (typically 2–5) may share a nest, where they huddle together to conserve heat. Deer Mice regularly move to new nest sites. Breeding females will sometimes move their litter to a new nest site or may themselves relocate just before the birth of their next litter, leaving the original nest to offspring from previous litters. A mated male will not kill unattended pups found within his home range (who could very well be his offspring), but dispersing or unmated males will commonly kill and eat unattended pups they encounter. Resident females almost always kill unfamiliar pups found on their home range, as these would likely have been produced by an intruding and potentially competitive female. Deer Mice make an effort to travel on routes that reduce their risk of making noise that could cue a nocturnal predator to their position. They choose to move along mossy logs, rocks, and wet leaf litter rather than on dry crinkly leaf litter.

VOCALIZATIONS

Normally silent, Deer Mice will produce a high-pitched "chit" or a "squeal" when defending a nest with pups, accosting an intruder, fighting, or chasing an intruder away. Squeals are also used by adults and young (higher pitched) when in distress. Infants likely also use ultrasonic calls to solicit attention from their mother.

SIGNS

Trails in the snow may radiate from a single access hole, commonly exiting alongside the trunk of a small shrub or tree, or even a stick, where the mouse need only enlarge the natural space that develops between the wood and the snow. All the mice in the genus *Peromyscus* (including the Keen's Mouse, Deer Mouse, and White-footed Mouse in Canada) produce similar signs and tracks. The following description applies equally to all three species. Front feet have five toes, but the inside toe (toe 1) on each front foot is highly reduced and does not normally register in a track. The hind feet also have five toes. Toes 1 and 5 project to the side, while toes 2, 3, and 4 group together and point to the front. Claws and heel pads may or may

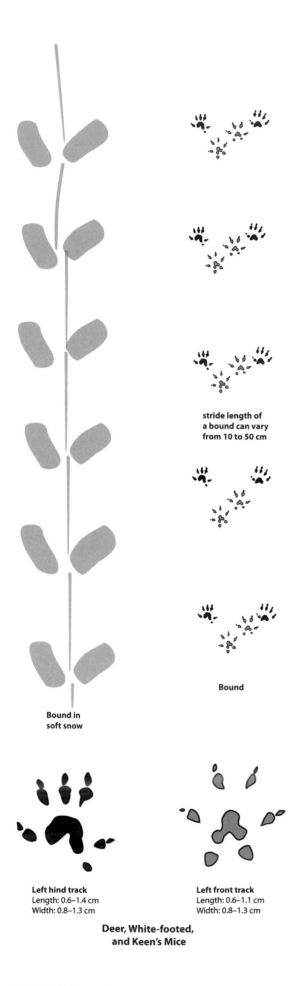

stride length of
a bound can vary
from 10 to 50 cm

Bound

**Bound in
soft snow**

Left hind track
Length: 0.6–1.4 cm
Width: 0.8–1.3 cm

Left front track
Length: 0.6–1.1 cm
Width: 0.8–1.3 cm

**Deer, White-footed,
and Keen's Mice**

not register, depending on the substrate. The typical gait of these mice, when in the open, is a bound. Tail drag is uncommon except in soft snow.

REFERENCES

Bowman et al. 1999; Brown, L.N., 1997; Choate et al. 1994; Davis, H., 1990; Dunmire 1960; Gilbert, B.S., et al. 1986; Gilbert, B.S., and Krebs 1991; Gould and Pruitt 1969; Hjertaas and Hjertaas 1990; Houseknecht 1968; Kalcounis-Ruepell et al. 2002; Lindquist et al. 2003; McAdam and Millar 1999; Millar and Derrickson 1992; Millar and Teferi 1993; Nagorsen 2005a; Perrigo 1990; Pierce and Vogt 1993; Prescott and Richard 2004; Rehmeier et al. 2004; Reid, F.A., 2006; Ribble and Millar 1996; Roche et al. 1999; Salt 2000a; Schmidly 2004; Sharpe and Millar 1990; Slough and Jung 2007; Sullivan et al. 2004; Teferi and Millar 1993; Tessier et al. 2004; Vander Wall 2000; Vander Wall and Longland 1999; Vander Wall et al. 2001; Walters and Miller 2001; Weston et al. 2005; Wolff 1996; Wolff and Cicirello 1990; Yates et al. 2002; Youngman 1975.

Western Harvest Mouse

FRENCH NAME: **Souris-moissonneuse de l'Ouest**
SCIENTIFIC NAME: *Reithrodontomys megalotis*

This is one of the smallest mice in Canada, with an average adult weight of only 11 g.

DESCRIPTION

The coat colour of these small mice varies geographically. In Canada, their short pelage is grey-brown above and whitish below. A diffuse, yellowish-brown band runs between these areas from cheek to rump. A distinctively dark stripe, most noticeable from above, runs along the midline of the back. Their long tail is about as long as the head and body, and is bicoloured (although not sharply so), brown above and whitish below. Ears are large and a pale cinnamon colour, with a covering of fine dark hairs through which the pinkish-tan skin of the pinnae shows. Feet are whitish on the upper surfaces. The dental formula is incisors 1/1, canines 0/0, premolars 0/0, and molars 3/3, for a total of 16 teeth. Upper incisors are grooved on their anterior face.

SIMILAR SPECIES

Western Harvest Mice could easily be mistaken for small Deer Mice. Any Deer Mouse equal in size to a Western Harvest Mouse would be a juvenile and these can be distinguished by their dark grey upper parts. House Mice are also larger but are uniformly dark grey–coloured, including their belly. Western Harvest Mice have a dark dorsal stripe that both Deer Mice and House Mice lack and neither of these species has grooved incisors.

Western Harvest Mouse (*Reithrodontomys megalotis*)

SIZE

The following measurements apply to animals of reproductive age from the Canadian part of the range. Males and females are similar in size and appearance.

Total length: 116–151 mm; tail length: 54–86 mm; hind foot length: 12–19 mm; ear length: 12–17 mm.

Weight: 8.0–16.2 g; newborns, 1.0–1.5 g.

RANGE

Globally this is an endemic western North American species that occurs from British Columbia and Alberta to southern Mexico. Western Harvest Mice are at the northern limits of their range in southwestern Canada. A finger of distribution extends into southeastern Alberta, and another into southcentral British Columbia. Although the species has not yet been discovered in Saskatchewan, it may well one day be found in the far southwestern part of the province. In British Columbia, it occurs throughout most of the Okanagan Valley as far north as Vernon and in the Similkameen River valley as far north as Keremeos. In Alberta, the most northerly records are from the protected grasslands of the Suffield National Wildlife Reserve. Despite numerous efforts, no other specimens have been found farther south since the 1960s, suggesting that the Suffield populations may be severed from the range, and isolated from the Montana populations by at least 150 km. Western Harvest Mouse distribution has been expanding slowly eastward in Illinois and Indiana as they invade moist or grassy fields that are left fallow.

ABUNDANCE

Western Harvest Mice are extremely rare in Alberta and rare in British Columbia. They are designated to be of "Special Concern" in British Columbia and "Endangered" in Alberta by COSEWIC, and the BC provincial government has placed the BC population

on the province's "Blue List." Density in appropriate habitat in British Columbia varies from 4.7 to 80 per ha, depending on habitat quality. In addition, dramatic seasonal fluctuations have been noted in British Columbia, where populations have been reported to peak during autumn and winter and decline sharply in midsummer. The 80 per ha estimate was in December in an irrigated field where the average annual density was 29.2 per ha. There are no indications of multi-year population cycles, although densities do vary from year to year. There are no density estimates available for Alberta populations.

ECOLOGY

In British Columbia, Western Harvest Mice are confined to valley bottoms or south facing slopes, where they occur primarily in shrubby grasslands with abundant tall grasses. In Alberta, they occur in flat or gently undulating prairie grasslands with sandy soils and dense grasses and shrubs. Suitable habitat in both provinces is diminishing with the advance of urban growth and agriculture. The primary zone of current distribution in Alberta is in a protected area of virgin prairie. Availability of cover is crucial to their survival, so grazing, and especially overgrazing, has a decidedly negative impact on their distribution. Grass fires can also reduce habitat for these mice, as they destroy both cover and food, although the

Distribution of the Western Harvest Mouse (*Reithrodontomys megalotis*)

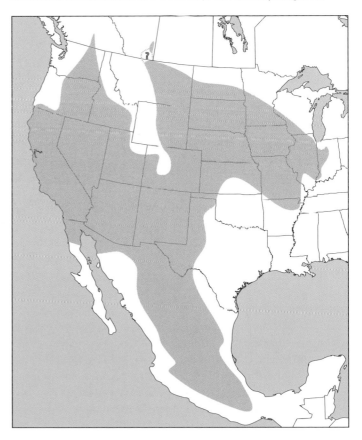

? current distribution uncertain

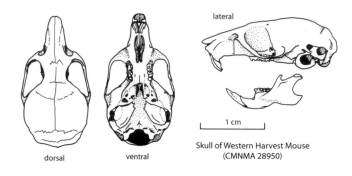

dorsal   ventral

lateral

1 cm

Skull of Western Harvest Mouse
(CMNMA 28950)

stride length of
a bound varies
from 5 to 23 cm

Bound

Left hind track
Length: 0.6–1.3 cm
Width: 0.6–0.8 cm

Left front track
Length: 0.6–1.0 cm
Width: 0.6–0.8 cm

Western Harvest Mouse

long-term effects of fire are likely more positive as vegetative growth is enhanced. Western Harvest Mice have not been studied comprehensively in Canada, so most of the biology reported herein is from studies undertaken in the United States. These mice are strongly nocturnal and most active on moonless and rainy nights. Individuals have been captured in all months of the year in British Columbia. Laboratory experiments with starved and chilled animals prove that they can enter torpor when stressed. This is possibly an important survival capability, especially in their Alberta range, where winters can be formidable; however, their ability to hibernate is not proved. No home-range estimates exist for Canadian populations, but in the American southwest home ranges are estimated from 0.44 to 1.12 ha. Dispersal is generally < 300 m, but during periods of high population density this can increase to as much as 3200 m. Males generally disperse farther than females. These mice are poor swimmers, but excellent climbers, usually building their nests in the lower branches of shrubs and capable of clinging to the outer twig tips to harvest flowers. As the vegetation grows and fruits over the summer, harvest mice spend more time climbing to forage and less time on the ground. Predators include owls, snakes, and small carnivores such as weasels, Coyotes, and foxes. Although they may live longer in captivity, the oldest known wild individual was 18 months old. Likely few survive more than six months.

DIET

These little mice eat the leaves, stems, and seeds of grasses and other herbaceous vegetation and will consume insects, particularly grasshoppers and caterpillars, that they encounter as they forage. They readily climb to collect grass seeds or to eat the flowers of shrubs in season. Despite their name, there is no evidence of food caching by this species in Canada.

REPRODUCTION

Although this mouse is capable of breeding throughout the year in warmer climates, the breeding season in Canada extends from March to November in British Columbia and is very likely considerably shorter in Alberta. Litter size ranges from one to seven and averages three. The gestation period lasts about 23 days and a postpartum oestrus occurs soon after parturition. Females may produce up to four to five litters annually in British Columbia, but this estimate remains unsubstantiated. Pups are born blind, deaf, and hairless except for short, fine vibrissae. Lower incisors begin to erupt at 4 days old. Their backs are darkly furred by the time they are 7 days old. By 10 days old, the auditory meatus has opened and they can hear. Their eyes open by day 11–12. They begin eating solid food at about the same time and are fully weaned by day 19–21. Both males and females reach sexual maturity in about four months. Most juveniles born late in the season in Canada do not reproduce until the following spring.

BEHAVIOUR

Very little is known of the behaviour of this small mouse. Their social structure is unknown; however, captive males appear to establish a strong dominance hierarchy and will breed with multiple females. Western Harvest Mice rarely develop or maintain their own runways, instead using those of other small mammals such as voles. They do not dig their own burrows either, choosing to take over abandoned burrows of voles or other small mammals.

VOCALIZATIONS

Newborns can produce a shrill screech if disturbed or stressed.

SIGNS

Nests are spherical or cup-shaped and about 7.5–12.5 cm in diameter. They are commonly placed above the ground (up to a metre high) in a tangle of vegetation or a shrub, or on the ground under debris, such as a board. Sometimes nests are built underground in a nest chamber. Nests are composed of coarsely woven grasses and other fibrous plant material, with a soft inner lining, often of plant down or dandelion fluff. Each animal constructs and maintains several nests. Nests of Marsh Wrens are sometimes taken over and roofed, but this has not been noted in Canada. In parts of the United States (Texas and Missouri), harvest mice have been found nesting in discarded aluminum drink cans tossed onto grassy road verges. Their small size allows them to comfortably enter through the pop tab, and they find the can nicely secure as well as waterproof. This behaviour has not yet been reported in Canada. Front feet have five toes, with toe 1 (the thumb) highly reduced and not registering in a track. Hind feet have five toes, with toe 1 (the hallux or big toe) registering farther back than is usual for small rodents. This hind toe 1 placement is distinctive for the genus. The usual gait of these mice is a bound.

REFERENCES

Clark, B.K., and Kaufman 1990; Cummins and Slade 2007; Elbroch 2003; Leibacher and Whitaker 1998; Lindgren 2007, Nagorsen 2005a; Peralta-García et al. 2007; Pigage and Pigage 1994; Pitts 1994; Skupski 1995; Smith, H.C., 1993; Verts and Carraway 1998; Webster and Jones 1982.

FAMILY MURIDAE
Old World rats and mice

This is the largest family of rodents, but only three occur in Canada. Two species of rats and a mouse were accidentally introduced from Europe around the time of European colonization, and are now widespread pests across North America. These species are commensal with humans, and likely first arrived onboard ships, as stowaways in the food cargo.

House Mouse

FRENCH NAME: **Souris domestique**
SCIENTIFIC NAME: *Mus musculus*

By stowing away on board ships or in shipments of foodstuffs travelling by air and land, this small rodent has been inadvertently introduced by humans to all continents, except Antarctica. House Mice are called commensal mammals, as they usually live in close association with humans.

DESCRIPTION

Wild forms are typically brownish grey with a slightly lighter grey to buffy belly, but the primary fur colour may vary from light brown to black, sometimes with white markings. Foot colour is similar to that of the belly, but with a pinkish undertone. The domestic form may be albino (like the standard lab mouse) or any number of fancy colour variations from fawn to black to spotted. The long tail (about the same length as the head-body) is uniformly coloured and sparsely haired, with obvious cross-sectional scales. The ears are hairless and relatively large. The dental formula is incisors 1/1, canines 0/0, premolars 0/0, and molars 3/3, for a total of 16 teeth. The tips of the upper incisors are notched in side view.

SIMILAR SPECIES

House Mice could be confused with Deer or White-footed Mice, except both of those species have a white belly and feet with a bicoloured tail and juveniles are greyish on their upper surface, while adults are brown to yellowish brown. Western Harvest Mice occur within House Mouse range in Canada in a very limited portion of south-central British Columbia. They are smaller than House Mice (adults weigh less than 15g), with a sparsely furred tail, whitish

House Mouse (*Mus musculus*)

do not depend directly on humans for food. Permanently feral populations are more common in warmer regions. Home range size is highly variable and has been recorded from 4 m² to 1–2 km² and is largest for dominant males. Their home range includes the entire group territory, while the home range of each of the females and juveniles in the group occupies a portion of the larger group territory. Dispersal movements of over two km have been reported, but for the most part these mice are relatively sedentary, especially those in buildings. Most dispersal is undertaken by subadults, although adults may move between temporary habitats on a seasonal basis. For example, some groups may live in or near a grain field during the summer and move into a house, barn, or granary for the winter. Outdoor House Mice dig short, but frequently complex, burrows with several entrances and nesting chambers. Their nests are typically constructed of dried vegetation. Indoor mice use a wide variety of available materials, such as shredded cardboard or paper, rags, insulation, and even stuffing from toys or mattresses, to construct their nests. These are built in quiet areas, such as inside little-used furniture or boxes, inside the walls and under the floor boards. Predators vary depending on the colony's location. Outdoor mice are captured by snakes, hawks, owls, weasels, Coyotes, foxes, and Domestic Cats. Indoor mice must elude Domestic Cats and Dogs, as well as the trapping and poisoning efforts of humans. These mice are agile climbers and jump well. They are also capable swimmers. Although captive House Mice can survive several years (up to 6 years in exceptional cases), few live beyond 12–18 months in the wild. Females live longer than males.

under parts, and a distinct longitudinal groove on the anterior face of each upper incisor.

SIZE
Total length: 142–197 mm; tail length: 64–95 mm; hind foot length: 15–23 mm; ear length: 11–18 mm.
Weight: 17–27 g; newborns, around 1 g.

RANGE
Thought to have originated in an area southeast of the Caspian Sea encompassing parts of modern-day Iran, Afghanistan, and Russia, this rodent now lives on most inhabited oceanic islands and all continents, with the exception of Antarctica. At least three of the sub-Antarctic islands (Macquarie, Marion, and South Georgia Islands) support introduced feral populations.

ABUNDANCE
In times past, "plague proportions" of these rodents have been reported in North America. An outbreak in California in 1926–7 was estimated to have reached over 170,000 mice per hectare. Feral populations typically range from 50 to 100 per ha, but may reach as many as 700 per ha or as few as less than 1 per ha. Populations living in buildings can reach an equivalent density of 1500 per ha. Although no North American estimates of overall population density exist, there is no doubt that this non-native mouse is generally common and frequently abundant in urban and agricultural regions.

ECOLOGY
These rodents are active all year, mainly at night, and are highly adaptable. Although they tend to be most abundant around buildings, they may also form colonies in agricultural land, where they

DIET
House Mice are omnivorous. Feral populations rely on seeds (especially of cereal grains), vegetation, and insects for the bulk of their

Distibution of the House Mouse (*Mus musculus*)

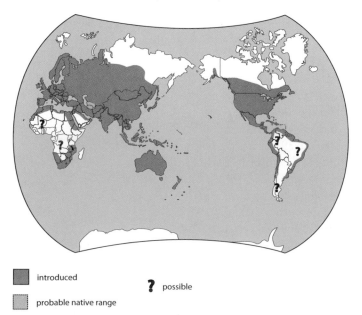

■ introduced
■ probable native range
? possible

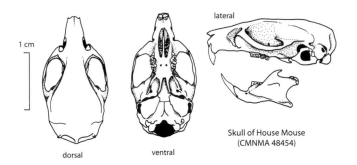

1 cm

dorsal

ventral

lateral

Skull of House Mouse
(CMNMA 48454)

food. Commensal populations add garbage, carrion, and a wide variety of human foodstuffs to that list. A pet's unattended food bowl is a favourite feeding station. Infanticide and cannibalism occur, especially in protein-deficient populations. These rodents can survive without drinking water if the moisture content of their food is greater than 15%–16%.

REPRODUCTION

House Mice can breed successfully at ambient temperatures between –20°C and +32°C, the widest range of any known mammal. The length of the breeding season in Canada is not known, but given the just mentioned temperature range, House Mice in most parts of Canada likely breed all year round, with perhaps a short winter hiatus in colder areas. Populations in heated buildings are unaffected by outside temperatures, and thus can breed all year round. The gestation period is 18–21 days and females experience a postpartum oestrus within 24 hours of whelping, so they may be simultaneously pregnant and nursing. Females typically produce 5–10 litters per year. Average litter size is 4–8 (range 1–14). Pups are altricial, born pink and naked with closed ears and eyes. They are furred by 10 days old, their eyes open by day 14, they begin wandering outside the nest shortly afterwards, and are weaned by 21–28 days old. Sexual maturity occurs around 35 days old in indoor and temperate regions and may require more than 60 days in especially cold regions.

BEHAVIOUR

House Mice are social rodents, typically living in relatively stable groups composed of a dominant male and several adult females and their young. Some groups may also include subordinate males. The dominant male urine-marks the boundaries of the group territory and all the adults of the group jointly defend it from other House Mice. Normally, each female defends her own burrow/nest site and keeps other mice away from her pups. Dominant males sometimes occupy the nest of a female and her litter, but his parental care is limited to occasional grooming of the pups. In captivity, related females (i.e., sisters or mother and daughter) may mutually nurse each other's litters, thereby enhancing the survival of both litters, but it is unknown whether this occurs in the wild. In high-density situations, surplus animals occupy the marginal peripheries of established group ranges in a subordinate, non-territorial, and non-breeding state. Inbreeding is common, especially

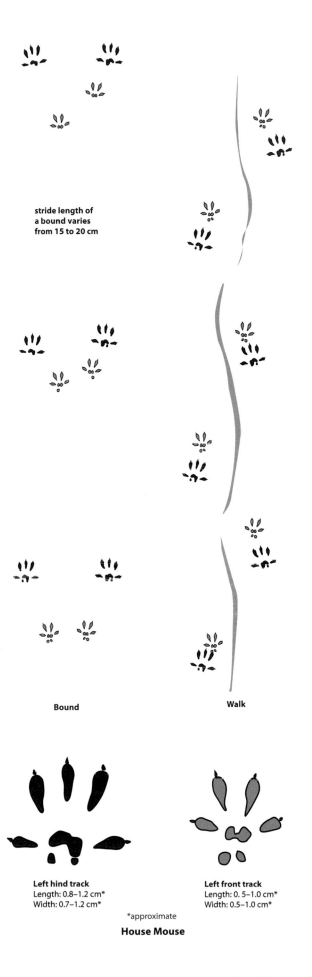

stride length of
a bound varies
from 15 to 20 cm

Bound

Walk

Left hind track
Length: 0.8–1.2 cm*
Width: 0.7–1.2 cm*

Left front track
Length: 0. 5–1.0 cm*
Width: 0.5–1.0 cm*

*approximate

House Mouse

in rapidly growing indoor populations. Like their cousins the Norway and Roof Rats, House Mice are positively thigmotropic – that is, they prefer to remain in contact with their surroundings by touch. This behaviour often leads to the creation of a dark smudge along an obstacle as they sidle along their way. They tend to travel on well-defined routes along walls, curbs, and other objects and will continue to follow the same route even after the obstacle has been removed. Whiskers are especially important to help them orient in the dark.

VOCALIZATIONS

House Mice sometimes produce a high-pitched squeak when interacting. Subordinates may rear onto their hind legs and squeal if attacked by a dominant animal. Communication between females and their litters is primarily confined to the ultrasonic levels and cannot be detected by humans. House Mice have a hearing range of 15–50 kHz, which extends considerably beyond our upper limit of 20 kHz.

SIGNS

A building infested with House Mice exudes a characteristic musky smell that is unmistakable once recognized. The odour is caused by urine and scent-gland excretions. Heavy infestations are often associated with "urinating pillars" – accumulations of crystallized urine, scat, and dirt up to 4 cm high and 1 cm wide that are created over many years by dominant males marking their territory. Greasy, dirty smudges may be visible at mouse-height at the base of walls along well-travelled routes. Hind feet have five toes. Toes 1 and 5 angle outwards and toes 2, 3, and 4 group together and aim more or less forward. Forefeet have four toes that register in a track. The first toe, the thumb, is highly vestigial and does not register. All four of the front toes typically angle about equally away from each other, producing a star-like pattern. Tracks toe outwards slightly, often with an undulating tail drag mark between, especially at slower gaits. All toes except the thumb are clawed and claws may or may not register in the track. Cylindrical, black, oval scat is about 6mm long by 2.0–2.5 mm in diameter and may have blunt or pointed ends depending on the diet. These are similar in colour, shape, and size to bat droppings, but do not crumble nearly as easily. House Mice regularly discard about one-third of each grain seed that they consume. The so-called kibbled grain that remains is a diagnostic sign of House Mice.

REFERENCES

Avenant and Smith 2004; Berry et al. 1979; Kurta 1995; Long, J.L., 2003; Macdonald and Barrett 1993; Mikesic and Drickamer 1992; Murie and Elbroch 2005; Nagorsen 2005a; Verts and Carraway 1998.

Norway Rat

also called Brown Rat, Sewer Rat, Wharf Rat, Common Rat

FRENCH NAME: **Rat surmulot**

SCIENTIFIC NAME: *Rattus norvegicus*

Very few humans in their lifetime will avoid a sighting of this non-native, urban, and farmland pest as they prefer to live commensally with humans. By stowing away in cargo, Norway Rats have achieved an almost global distribution.

DESCRIPTION

Norway Rats are stocky rodents with long, almost hairless tails. Their small black eyes display a pale green eye shine when captured in a spotlight. They are typically dull brown on their upper surface, but vary from white or pale reddish brown to almost black and sometimes even spotted. Their undersides are usually a paler version of the back colour, varying from white through buffy to silver or greyish brown. Like the coat, the tail is dark above and paler below. Fur on the feet matches the belly colour. Albino lab rats and multicoloured and diversely shaded so-called fancy rats bred for the pet trade are colour variants of the wild Norway Rat. Ears are lightly furred on the back side and barely reach the corner of the eye when gently pushed forward. Although Norway Rats have relatively poor vision (they can see only around 1–15 m and only in black and white), they make up for this by having acute senses of smell, touch, taste, and hearing. Their lips can close behind the incisors to prevent the shavings of whatever they are chewing on from falling into the throat. These same incisors grow up to 13 cm each year. If the rats do not wear these teeth down in their daily activities, they will grind

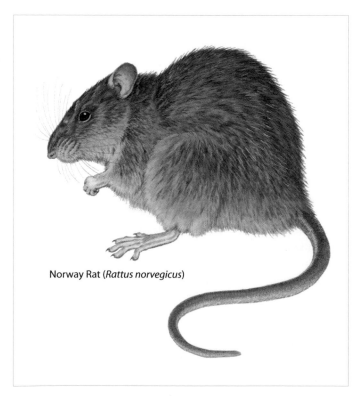

Norway Rat (*Rattus norvegicus*)

Distribution of the Norway Rat (*Rattus norvegicus*)

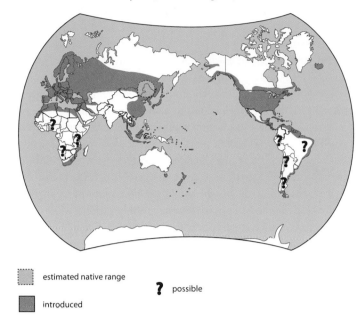

- estimated native range
- **?** possible
- introduced

their invasion by ship of more remote and smaller oceanic islands. Alberta has supported various detection and eradication programs since the 1950s, when Norway Rats first started to invade the province. With much expense and ongoing diligence, the province has essentially remained rat-free. Although difficult to carry out, this type of program (only possible in a land-locked province) has likely saved Alberta multimillions or even billions of dollars over the years in lost or spoiled crops. The range of this rat species in Canada is always in flux as new animals are introduced and reintroduced. Frequently, especially in the north, populations die out due to winter temperatures, but then may be reintroduced the following spring as new supplies arrive.

ABUNDANCE
These are common and often abundant rodents.

ECOLOGY
Norway Rats are superb survivors and consequently are highly invasive. This species is widespread and omnivorous, and has caused or contributed to the extinction or decline of many native mammals, birds, reptiles, amphibians, and even invertebrates by either predation or competition. Both Roof and Norway Rats carry many diseases that affect humans and domestic animals. They are crucial for the transmission of Bubonic Plague, also called the Black Death or simply the Plague. This disease is caused by the bacteria *Yersinia pestis*, which is carried by fleas that live on the rats. The disease is transmitted to humans by the bite of an infected flea. It can also be transmitted to wild rodents in which it can exist without necessarily killing its host. From this disease reservoir it can then spread and infect humans. Fortunately, there are effective remedies for this disease now if it is properly diagnosed in the early stages. Other rat-borne diseases include murine typhus, salmonella, trichinosis, rat-bite fever, rabies, and leptospirosis. Norway Rats qualify as the greatest mammalian pest of all time, as they eat or foul over a billion dollars' worth of crops globally every year. When fires, damage caused by their chewing on electrical cables, and the cost of pest control are taken into account, the costs climb to several billions. Norway Rats are largely nocturnal, but may occasionally be seen in daylight. They dig often complex burrow systems, primarily in soil, but also in old garbage in dumps, the insulation in homes, hay piles, construction waste, wood piles, and many other compacted substrates. Most burrows are fairly simple, with a short tunnel leading to a nest chamber. A well-made burrow will typically have a second concealed entrance. These rodents are nearly always found near human settlement. Even seemingly isolated populations are likely hold-outs still surviving after a homestead has been removed, or a dump abandoned and covered over. While capable climbers, they are not especially agile. Home range size varies with food abundance and population density and may be 12 m² to more than 5.8 ha. Single individuals may travel 3–4 km each night and will commonly cross open areas as large as 500 m wide. They are proficient swimmers and can readily cross 600 m of relatively calm open water. Their primary urban predators are humans, through their pest-control measures, although Domestic Cats and Dogs take some, especially the smaller

their upper incisors against the opposing lower incisors to accomplish the same result. The dental formula is incisors 1/1, canines 0/0, premolars 0/0, and molars 3/3, for a total of 16 teeth.

SIMILAR SPECIES
Roof Rats may be similar in coloration to a Norway Rat, but they have larger ears (longer than 20 mm) that are leathery and virtually hairless on the outer surface of the pinnae and generally a longer tail in proportion to the body length. The ears of a Roof Rat cover the eyes when folded forward, while those of a Norway Rat do not. A Norway Rat's tail is shorter than the head-body length (70%–90%) while that of a Roof Rat is longer (100%–160%). Generally, if the tail is folded along the body, it will exceed the snout on a Roof Rat, but not even reach the nose of a Norway Rat.

SIZE
Adult males tend to be somewhat larger and heavier than adult females.
Total length: 315–480 mm; tail length: 122–219 mm; hind foot length: 34–46 mm; ear length: 15–23 mm.
Weight: 180–463 g; newborns, around 6 g.

RANGE
Likely originating in northeastern China and southeastern Russia, Norway Rats travelled overland to reach eastern Europe by the late 1600s. This migration was made possible by the rat's association with human habitations. Within 100 years, they had infiltrated all of Europe and the British Isles. They reached the east coast of North America around 1775 and the west coast in the 1850s as stowaways on board sailing vessels. Now they occur on every continent except Antarctica, and on many of the larger oceanic and freshwater islands. They continue to stow away in cargo, and there is an ongoing risk of

young. Principal causes of mortality are poison, traps, and vehicles. Few wild rats live beyond one year, although pet rats may have a maximum longevity of almost five years (average 2–3 years).

DIET

These omnivorous rodents are quite catholic in their dietary choices. In the wild, they consume vegetation, carrion, insects, and whatever small mammals and birds they can find and catch. They are especially fond of bird eggs. In urban or farmyard settings they happily add fresh or stored grains, garbage, road-killed carrion, and small domestic birds such as chicks and ducklings and sometimes even adult barnyard fowl. Even seed, discarded beneath a birdfeeder, can be attractive to a rat. Given a choice, rats will select foods with high protein and fat contents. An adult Norway Rat needs to drink about 15–30 ml of water each day.

REPRODUCTION

If sheltered in buildings, Norway Rats can breed all year round, but outdoor populations in most of Canada tend to have about an eight-month breeding season from mid-March to mid-November. The exception occurs in coastal British Columbia, where the more moderate climate allows even outdoor rats to continue breeding through the winter. Mating in this species is promiscuous. Females will mate repeatedly during their oestrous period with many males and males routinely mate with as many of the oestrous females in their area as they can gain access to. Litter size may be 1–22 pups, but typically is 6–9. Gestation lasts 21–24 days. Young females have smaller litters than larger, more mature females. Mature females may have as many as 12 litters per year, but average 4–6 annually. They may enter oestrus and breed again shortly after the birth of their litter, so they can be concurrently lactating and gestating. Newborns are altricial and require intensive maternal care. They develop a hair coat and their ears open at around one week old. The eyes open by 14–17 days old and soon after that they begin to eat solid food and venture out of the nest following their mother. Pups are weaned when they are around one month old. If their mother becomes pregnant soon after their birth, and consequently has a new litter, they will disperse then to territories of their own. If she does not have a new litter, she may share her nest with at least the daughters until they are almost full-grown, but young males are forced out of the maternal burrow soon after they are weaned. Wild juveniles become sexually mature at around 3 months old. Captive juvenile females may be so well nourished that they can become fertile by as young as 3–4 weeks old.

Under similar circumstances, juvenile males may start producing sperm by as young as 5–6 weeks old.

BEHAVIOUR

Norway Rats are social animals. Although mature rats often maintain exclusive nesting burrows, neighbouring animals do form a hierarchy and a tightly knit colony. There appear to be two hierarchical systems, one for males and another for females. Typically, the dominant male and his mate will take the prime burrow within the group's territory. Mated pairs may share a burrow for a while and, as noted above, females may harbour their latest litter until it is almost full-grown. Additionally, some females permit closer than usual association with grown daughters. Social groups are highly territorial, defending their burrows from other groups. Most foraging occurs within the group territory, but during times of scarcity, rats will travel over 3 km to find food. Young rats learn the idiosyncrasies of their environment and where to find food by following their mothers, which they do for a few weeks around weaning time. Juvenile females tend to assume the social status of their mother. Female rats are diligent, caring mothers who can be highly protective of their young. In captivity, females in the same social group commonly combine their litters and nurse the pups indiscriminately. Many rats suffer wounds, sometimes severe enough to cause death, during hierarchy battles. These wounds are taken mainly on the face and hindquarters, primarily around the base of the tail. Norway Rats are also quick to cannibalize another rat caught in a trap.

VOCALIZATIONS

Most of this discussion is based on sounds produced by pet animals. The "peep" is a soft chirp used as a comfort sound, typically during grooming. Louder "chirrups" are produced in the same context. Short squeaks are produced during mild social interactions and aggressive encounters. Louder "long squeaks" are delivered as a protest during tense social encounters. An injured rat, or two rats fighting, will produce loud shrieks. A rat backed into a corner by an opponent may produce a soft hiss. Noises made by the teeth are also used to communicate. For example, a contented rat may quietly grind its lower incisors against the uppers, producing a soft repetitive sound. Teeth can also deliver louder sounds as they chatter against each other, a sound usually heard during highly stressful agonistic encounters. Hearing extends into the ultrasonic ranges (20–70 kHz and possibly to 100 kHz), and juveniles still in the nest vocalize at ultrasonic frequencies. Subordinate rats use ultrasonic vocalizations when confronted nose-to-nose by a dominant animal. Adult rats also use ultrasonic frequencies for certain loud aggressive calls and for a call referred to as "rat laughter." The latter call has been recorded from pet animals that were playing and being tickled. Norway Rats are suspected of using echolocation to navigate in the dark.

SIGNS

Norway Rats are describes as thigmotrophic, which means they prefer to maintain contact with their surroundings as they move

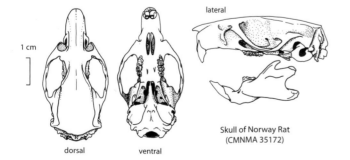

1 cm

lateral

Skull of Norway Rat
(CMNMA 35172)

dorsal   ventral

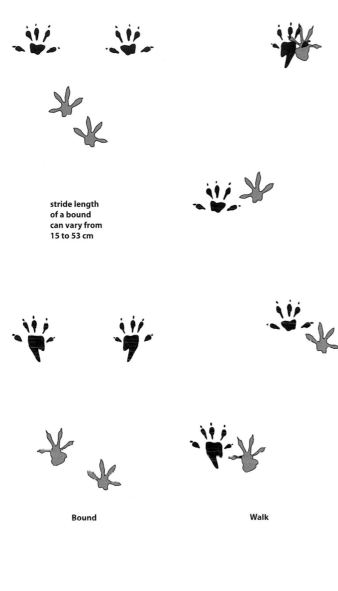

stride length
of a bound
can vary from
15 to 53 cm

Bound

Walk

Left hind track
Length: 2.5–3.2 cm
Width: 2.2–2.5 cm

Left front track
Length: 1.9–2.2 cm
Width: 1.3–2.5 cm

Norway and Roof Rats

about. Well-travelled runways often occur along obstacles such as walls, curbs, and pipes, and these can be recognized by the grease slick that parallels the route where the rats have leaned into the support. Trails are commonly marked with urine. Very well-used trails running over an open space, for example, along an attic beam, may develop pissicles of crystallized urine (they look like small stalactites) that gradually form due to the traditional use of the spot for a urine mark. An ultra-violet light causes rat urine to glow, making it easy to find in the dark. Scat is generally uniform in thickness, with blunt ends (0.3–0.6 cm in diameter and 0.6–2.4 cm long), and is scattered about the territory. It tends to accumulate near nest sites and favourite feeding spots, and along runways. The scat of Norway Rats and Roof Rats can usually be separated by dividing the diameter by the length. If the resulting number is between 0.42 and 0.56, then it was produced by a Norway Rat. Roof Rat droppings fall between 0.31 and 0.37 for the same calculation. Although Norway Rats have a long heavy tail, drag marks are not common, except in deep snow. Burrow openings are roughly circular and 5–8 cm in diameter. Earthen burrows may show a fan-shaped mound of discarded soil at their mouth. Norway Rats display the typical rodent track pattern of four toes on the front foot and five on the hind. The pollex (thumb) is highly reduced and does not leave an impression in a track. The toe impressions in a track commonly connect with the foot pad impressions. The heel pads on both front and hind feet may not register.

REFERENCES

Bertram and Nagorsen 1995; Drever and Harestad 1998; Jackson 1982; Long, J.L., 2003; Macdonald and Barrett 1993; McGuire et al. 2005; Monnella et al. 2004; Nagorsen 2005a; Storer and Davis 1953; Sullivan, R., 2004.

Roof Rat
also called Black Rat, House Rat, Ship Rat

FRENCH NAME: **Rat noir**
SCIENTIFIC NAME: *Rattus rattus*

This rat arrived in the New World before the Norway Rat and was fairly widely dispersed around the port cities. In the cooler parts of North America, Roof Rats were substantially displaced by the larger, more aggressive Norway Rat. There are no permanent populations on the Canadian east coast.

DESCRIPTION

Upper parts are typically brown to dark grey or black. The underside is white, buffy, yellowish, or dark grey, usually at least slightly lighter than the upper surface. Belly hairs are grey at their base.

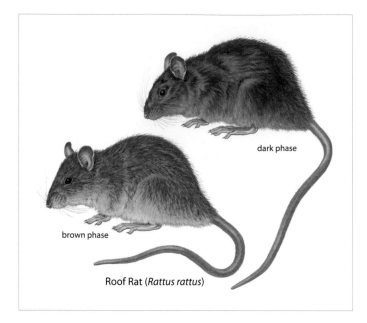

dark phase

brown phase

Roof Rat (*Rattus rattus*)

Two common colour variations are illustrated here. Some dark-backed animals may have white or yellowish bellies. Ears are long, tan-coloured, and hairless. When pressed forward, they cover the small dark eyes. The long, uniformly coloured tail appears hairless as the many thin, short hairs are too small to be seen easily, but their bristly texture can be felt if one rubs a finger along the tail towards the base. Although the tail length is shorter than the head-body length at birth, it grows to more than half the total length before the young leave the maternal nest. The feet are grey to brown on their upper surface and often pinkish below. Juveniles have shorter coats than adults and are generally greyer and somewhat lighter. Eye shine is reddish. Two glands lying on either side, just inside the anus, produce an unpleasant musky odour. The dental formula is incisors 1/1, canines 0/0, premolars 0/0, and molars 3/3, for a total of 16 teeth.

SIMILAR SPECIES
Roof Rats and Norway Rats may easily be confused, as colour is not necessarily a good distinguishing characteristic. Immatures can be especially difficult to differentiate. See the "Similar Species" section of the Norway Rat for traits that are useful in distinguishing the two species. The darker colour phase Roof Rats are physically similar, but much larger than House Mice.

SIZE
Adult males tend to be somewhat larger and heavier than adult females.
Total length: 325–455 mm; tail length: 160–255 mm; hind foot length: 30–42 mm; ear length: 17–27 mm.
Weight: 115–350 g (average 150–200 g); newborns, around 4.5 g.

RANGE
Originally from southeast Asia and possibly India, the extent of the native range of this species is somewhat uncertain. Some authors suggest that it may have been native as far north as southern China.

Roof rats have stowed away in cargo holds and been unintentionally carried on board ships to many temperate and tropical seaports around the world. Most Canadian records come from seaports, and no feral populations currently exist in Canada outside of British Columbia (southwestern mainland, southern Vancouver Island, and the Queen Charlotte Islands).

ABUNDANCE
Roof Rats may be locally abundant. Their populations are commonly eliminated or much reduced upon the arrival of Norway Rats.

ECOLOGY
Usually nocturnal, these rats may occasionally be seen foraging during the daytime, especially when population density is high. Roof Rats are very agile and graceful climbers. They use their long tail in a semi-prehensile manner by wrapping it around objects to improve stability, and it constantly flips from side to side as they travel on narrow surfaces to improve balance. They easily run along thin power lines, phone lines, and tree branches. Roof Rats typically move up when Norway Rats move in. The Norway Rats take over the prime ground- or basement-level regions and the Roof Rats are relegated to the attics and roofs. Wild populations of Roof Rats may be found in forested locations some distance from human habitation, where they regularly construct tree nests and sometimes ground burrows when there is no competition with Norway Rats. Although there are no estimates of population density or home range size for Canadian populations, Roof Rats in California, in a riparian habitat, had home ranges of 0.20 ha on average for adult males and 0.16 ha on average for adult females. During food shortages, or if a distant abundant source becomes available, Roof Rats will travel over 150 m to reach food. This rodent can become a serious pest in fruit-growing and other agricultural regions. A severe infestation may see a density of as many as 52 rats per ha. It, like the Norway Rat, is a carrier of disease, except for the transmittal of *Lepospira interogans,* the bacteria that causes a severe form of leptospirosis in humans, which is carried only by Norway Rats. See in the "Ecology" section of the Norway Rat for a list of the major diseases transmitted by both species. In broad terms, Roof Rats do similar damage to that of Norway Rats. They consume grain and foodstuffs and their resulting droppings and urine spoil even more. They chew electrical wiring, causing power failures and fires. Colonies in homes tear up and redistribute insulation. Tree crops such as nuts, fruit, and avocados, and vegetable crops, sugarcane, and seedlings are vulnerable to their depredations. The primary predators of Roof Rats are humans through pest-control activities. Domestic Dogs and Cats and Barn Owls will kill some. Maximum lifespan in the wild is thought to be about 18 months, but few exceed 12 months.

DIET
Roof Rats prefer to eat vegetation, grain, fruit, and nuts, but they are omnivorous and will readily consume birds and bird eggs, invertebrates, and, of course, human garbage. Pet food and livestock feed are common food choices. Daily water intake is usually necessary,

Distribution of the Roof Rat (*Rattus rattus*)

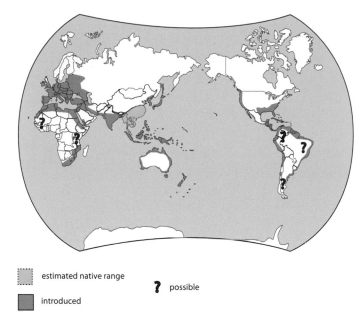

- estimated native range
- introduced
- **?** possible

although high-moisture-content fresh foods may significantly reduce the demand.

REPRODUCTION

Larger, older females tend to have larger litters than younger, smaller females (average 7–8 versus 5–6). Gestation lasts 21–22 days, but extends to 23–29 days for lactating females. Many females experience a postpartum oestrus about 24 hours after whelping, so may be lactating and pregnant concurrently. On average, a female Roof Rat produces 4–6 litters annually. Roof Rats in British Columbia appear to breed during the summer months from May to August. Elsewhere, in warmer regions, they may breed all year round. Newborns are altricial; they are essentially hairless, deaf, and blind. Pups grow quickly, being fully furred with fine short hairs by the time they are 8 days old. Their eyes open between 10 and 16 days old, and they begin eating solid food at 15–18 days old. Exploration outside the nest begins at 17–23 days old and the pups are weaned soon thereafter if their mother is already pregnant with her next litter, or as late as 35 days old if she is not. Much of the early time outside the nest is spent following their mother around learning the extent of the group territory and where to find food. Unlike many other rodents whose testes regress into the abdomen outside of the prime breeding season, those of an adult male Roof Rat remain scrotal throughout the year. Juveniles become sexually mature at around 3–4 months old.

BEHAVIOUR

Roof Rats are social animals. Because of difficulties of access, there are very few behavioural ecology studies of free-living Roof Rats. Most material presented here is derived from the study of captive colonies. Within each group, the dominant rat is an adult male. Below him is a single or a small group of females who are dominant to all other group members, both females and males. Internal hierarchy is often determined by fighting, which, depending on the size and determination of the individuals, may entail boxing and clutching with forefeet, vertical jumping, kicking, wrestling, and biting. The loser is typically chased off. Wounds to the tail and hindquarters are most common. Once the hierarchy is established, the subordinates withdraw, assuming a low stretched posture with ears folded back to signal their submission, or they approach the more dominant animal to sniff the corner of its mouth to acknowledge its superior status. In this manner, the status of the dominant animal is upheld without the need for potentially damaging fights. The dominant male of each group scent marks the territory by rubbing the scent glands on his cheeks and belly against the substrate. Each group aggressively defends their joint territory, but males tend to leave territorial defence against alien females to the females within his group, as male aggression towards females is inhibited. Only occasionally does a rat from another group manage to gain acceptance, and then only with a determined effort and much fighting. Threat is displayed by erecting the hair along the back while stretching the hind legs so that the hindquarters appear higher than the forequarters. Tooth chattering is used when a rat is uncertain, perhaps as a mild alarm signal. These rats are normally very vigilant, rising on their hind legs and sniffing the air if at all suspicious. A startled or highly alarmed animal will leap vertically into the air, commonly causing a chain reaction among the other group members present. This behaviour is thought to be evasive and can appear quite comical to an observer. Roof Rats prefer to create their bowl-shaped nests off the ground and will use trees, tree cavities, and their favourite location – within the walls of buildings. Maternal females carefully cover their young and block the nest entrance (typically with cut twigs) when they leave, presumably to prevent the young from falling out and perhaps to prevent other rats from gaining access. Food may be hoarded for later consumption, most often by females with pups. Such caches are commonly found in a wood pile or attic, or behind piled boxes. Roof Rats are highly neophobic (even more so than Norway Rats), being very suspicious of new items and foods in their environment, which they have a strong tendency to avoid.

VOCALIZATIONS

Fighting or wounded rats squeal loudly. Shrill squeaking is also common during battles. A long squeal is used as part of the appeasement

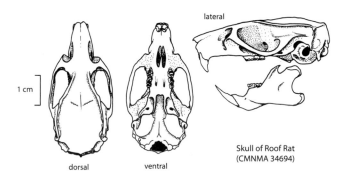

lateral

1 cm

Skull of Roof Rat
(CMNMA 34694)

dorsal   ventral

display of a subordinate rat towards a superior. This vocalization likely contains an ultrasonic component.

SIGNS

Roof Rats have a high tail carriage when travelling quickly, so tail drag marks only occur at slow speeds. The preference of this species for a warm climatic zone means there are few tracks in snow. Like Norway Rats, Roof Rats are thigmotrophic, which means they like to stay close enough to their environment to be able to sustain bodily contact. Favoured runways are consequently along walls, over beams near a pipe, or along debris piles. This behaviour commonly leaves dirty, greasy smudges where the rats frequently touch. Colonies in homes may be detected by simply listening for the sounds of the animals moving about in the ceiling and walls at night. Scat is dropped at random and urine appears not to be used for marking (as it is by Norway Rats). Roof Rat scat is elongated and typically pointed at both ends (0.2–0.3 cm in diameter and 1.0–1.2 cm in length). It is generally smaller than that of Norway Rats. The Norway Rats account ("Signs" section) provides an identification formula to differentiate scat of the two species. Tracks are similar to those of Norway Rats; see the Norway Rat account ("Signs" section) for track dimensions and a diagram.

REFERENCES

Bertram and Nagorsen 1995; Ewer 1971; Long, J.L., 2003; Macdonald and Barrett 1993; Storer and Davis 1953; Verts and Carraway 1998.

FAMILY ERETHIZONTIDAE
New World porcupines

Only one of the 10 species in this family occurs in North America. All have modified hairs that are quill-like and all are adapted to an arboreal lifestyle.

North American Porcupine

FRENCH NAME: **Porc-épic d'Amérique**
SCIENTIFIC NAME: *Erethizon dorsatum*

This large, prickly rodent is unique in North America. Its closest relative is a Central American species (genus *Coendou*). North American Porcupines have developed a physical defence that is so effective that they can afford to become ponderous and relatively unconcerned about predators. In fact, the two major natural causes of mortality among Porcupines are not predation or disease, but rather winter stress (usually starvation) and falling out of trees.

DESCRIPTION

Porcupine hair comes in four different forms (one more than most mammals): woolly underfur, longer guard hairs, sensory whiskers (vibrissae), and quills. Underfur is short and grey to dark brown or black. Guard hairs vary in length and may be one colour or banded depending on the individual and where on the body they occur. Winter guard hairs can be especially long, giving the animal a haloed-look in backlighting. Most Porcupines are an overall dark brown, but many western animals have long yellowish-tipped guard hairs on their upper surface that give them a blonde appearance. Hair colour in Porcupines varies from brown and black to albino. Most quills are banded, with a lighter base and a dark tip, but some are uniformly dark or light. There is a gradation between guard hairs and quills, with some guard hairs being thicker and quill-like, but not stiff enough to act as quills. When a Porcupine is threatened, it presents the erected quills on its rump and lower back to a potential predator. The banding on those quills creates a dramatic rosette of black and white, a classic warning coloration. The quills contain a fluorescent material that brightens the colours at night to ensure that the warning is visible to prospective predators. The rosette area, which covers about 25% of the dorsal surface, has a high density of large quills, but little underfur and guard hairs to interfere with the display. To avoid excessive heat

North American Porcupine
(*Erethizon dorsatum*)

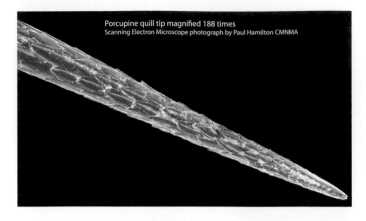

Porcupine quill tip magnified 188 times
Scanning Electron Microscope photograph by Paul Hamilton CMNMA

SIZE

Second largest of the North American rodents, Porcupines are exceeded in size only by North American Beavers. Weight fluctuates depending on the time of year and gender. Most adults are heaviest in late summer. Adult males then tend to lose some weight during the rut. Males are around 5%–20% heavier than females, and older males are somewhat larger than females, as they continue to grow longer than females. The following measurements apply to animals of reproductive age from the Canadian part of the range.

Total length: 610–932 mm; tail length: 170–330 mm; hind foot length: 80–126 mm; ear length: 15–40 mm.

Weight: 4.3–12.75 kg; average body weights in the east range from 5–7 kg and in the west and north from 7–9 kg; newborns, 340–640 g.

RANGE

North American Porcupines are widely distributed across North America, except on islands and in Arctic or alpine tundra, dry desert areas without trees, and intensively farmed lands that lack woodlots.

ABUNDANCE

North American Porcupines are relatively common, although population density does fluctuate depending on a variety of environmental factors, such as weather, predation, and habitat destruction. They are generally less common in dry regions and in the far northern and southern parts of the range. Porcupines do not endear themselves to foresters, thanks to their destructive consumption of trees. Porcupine controls are required only rarely, however, as most damage is localized and largely insignificant when viewed in the larger context.

ECOLOGY

Although Porcupines are most abundant in mixed coniferous/deciduous forests with abundant herbaceous understories, they may also occur in dry grasslands, and even deserts and tundra where riparian or protected areas can support shrubby trees. Most Porcupine activity is nocturnal and crepuscular, but many animals will rest in their feeding trees during the daytime in winter, especially

loss from this area in winter, the animals simultaneously erect the few hairs (both underfur and guard hairs) that remain and flatten the quills, like shingles, to maintain a boundary layer of warm air. Quills vary in length and width, depending on where on the body they grow. The longest, on the rump, can be up to 7.5 cm long and 2–3 mm wide, while those on the cheeks are less than 1 cm long and less than 1 mm wide. Quills occur on the upper surface of the body from the snout to the end of the tail. There are no quills on the belly, nose, inner sides of the limbs, soles of the feet, or underside of the tail. There are an estimated 30,000+ quills on each Porcupine, so a few hundred lost in a predator encounter are hardly missed. Unlike guard hairs, which will only regrow in the next moult, missing quills begin to regrow within a few weeks of being lost.

Porcupines have a small head and small, dark eyes. They can close their lips behind their incisors so they can gnaw without filling their mouth with unwanted wood shavings. Their short ears protrude only slightly beyond the summer fur and are completely hidden in the longer winter coat. Claws on all four feet are robust, long, hooked, sharp, and normally black or dark brown. Their dental formula is incisors 1/1, canines 0/0, premolars 1/1, and molars 3/3, for a total of 20 teeth. The anterior surfaces of all four incisors of adults are heavily pigmented a dark orange. This is caused by an iron inclusion in the enamel that hardens that part of the tooth and results in the chisel-like shape of the incisors. See in the "Description" section of the North American Beaver for an explanation of how this happens. Pups are born with white incisors. As they age the incisor roots absorb iron from their diet and a yellowish-orange colour begins to appear at the base of the growing tooth. By the time a juvenile is weaned, the whole front face of each incisor is coloured and the colouring continues to intensify for some time afterwards. Adult Porcupines experience two annual moults, one in spring and another in autumn.

SIMILAR SPECIES

Due to their quills, large size, and arboreal habits, this species cannot be confused with any other North American mammal. Beaver skulls are similar to those of Porcupines, but they lack the large infra-orbital foramina that are so prominent on a Porcupine skull. The accompanying figure illustrates this feature.

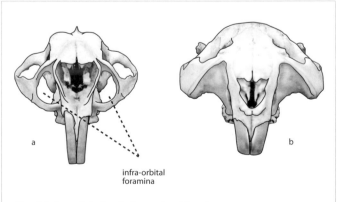

infra-orbital foramina

Frontal view of skulls of a Porcupine (a) and a North American Beaver (b)

Distribution of the North American Porcupine (*Erethizon dorsatum*)

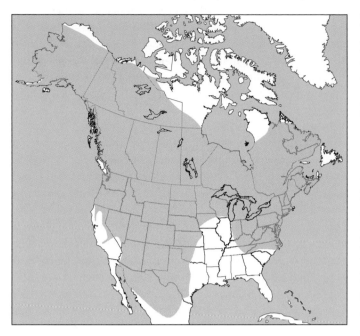

if it is sunny and they can absorb some solar heat. Porcupines are active all year round, even during very cold periods. They are known to reduce their core body temperature by as much as 5°C during very cold weather, most likely as an energy-saving strategy. Most of their time is spent high in a tree, resting or feeding. They do travel along the ground at times, especially during the summer while foraging and in the winter when moving from one feeding tree to another or back and forth to their den. Dens in caves, cracks in a rock face, natural openings in the root ball of a stump, in or under logs, in holes in living trees or snags, or in culverts and abandoned buildings may be used extensively after temperatures drop below freezing, especially during inclement weather and on clear, cold nights. Extent of den use varies across the range, largely depending on the severity of the winter. Dens may also be used in summer to avoid biting insects and predators. Porcupines will sometimes use a pre-existing abandoned burrow dug by another species, such as a Coyote or American Badger, but they do not dig their own den. In regions with abundant den sites, an individual animal will probably use more than one. Where den sites are scarce, more than one Porcupine may use the same den at the same time, usually a male and female pair. Home ranges are highly variable in size. Females are very territorial and exclude other females from their home ranges. Male home ranges frequently overlap, sometimes extensively with known male neighbours, but unknown (immigrant) males are forcibly ejected. Males tend to expand their home ranges as they mature, and the most dominant males have the largest home ranges, which hence overlap those of more females. Summer home ranges, especially of adult males, are considerably larger than winter home ranges. Reported home-range sizes in the east vary from 1.5 to 59 ha and in the west from 2.4 to over 100 ha. Maximum dispersal is so far known to be around 10 km for an adult searching for a new home

range. Recent evidence indicates that juvenile and yearling females generally disperse farther than the same age classes of males, which tend to remain on or near their natal site. Such female-biased dispersion is rare in mammals. Densities can reach up to 10/km², but are highly variable depending on habitat quality, predation levels, and phase of the population cycle.

A Porcupine quill is a very effective weapon. The tip has shallow, rear-projecting barbs that prevent it from moving backwards once it is imbedded. As the victim's muscle contracts, it drives the sharply pointed quill forward; the barbs then prevent the quill from returning to its original location when the muscle relaxes. In this slow, but inexorable way, the quill moves through the body unless prevented by bone, which will usually only divert its progress but may sometimes trap the quill. A quill lying parallel to a large bone may remain in situ alongside the bone for years. Quills have been known to perforate the gut and kill humans and predators that eat them by mistake. A quill was observed to travel a millimetre per hour through the muscle of a human arm. During normal activities, a Porcupine holds its quills relaxed and flattened against its skin. Each quill is attached to the skin by a thin collar of tissue at its base. When threatened, a Porcupine presents its back, erects its quills and lashes out with its tail. Loosely set quills will occasionally fall out during this flailing of the tail, but Porcupines cannot throw their quills. An elevated quill stretches the anchoring collar so that even a slight downward pressure on the quill tip breaks the collar and leaves the quill imbedded in the predator. In this simple and highly effective manner, a Porcupine cannot become even temporarily stuck to its victim. During fights or falls, Porcupines commonly become impaled by their own or another's quills. Fortunately, the quills have natural antiseptic properties that allow them to travel through flesh without creating an infection. Often the first indication of Porcupines in the area is the presence of quills sticking out of your dog's face. Unless there are only a few, the quills are best removed by a veterinarian for two reasons. First, the extraction is very painful and unless the dog is sedated, you can expect some serious opposition to the process. Second, a dog can get quills through its tongue and lips and even down its throat as well as in its ears, nose, chest, and shoulders. Every quill must be found and extracted to prevent the damage they may cause as they move within the body. The practice of cutting the end off the quill to release the internal pressure and make it easier to extract is ineffective. All this accomplishes is to provide less of the quill to hang on to as you try to pull it out. If you have to extract a quill, a pair of pliers and a quick yank work best.

Porcupines are buoyant and proficient swimmers and readily forage along shorelines for emergent aquatic plants. Their hearing and sense of smell is good, but their vision is thought to be myopic. They are slow and careful climbers, well adapted for an arboreal existence. They climb up and down a tree trunk head-up, bracing with the tail to prevent backsliding. Powerful thigh muscles and strong claws provide propulsion and grip. Nevertheless, accidents commonly do happen and many animals are killed or injured in falls. Eight to ten years is generally considered old age for a Porcupine, but they can survive up to 21 years in the wild. Although very resistant to predation, Porcupines are still vulnerable to a few predators that can learn to overcome their defence. These include mainly Fishers, Wolverines, Cougars, Coyotes, Red Foxes, and Bobcats. Of these, Fishers are by far the most successful. Predation mortality increases during late winter when hungry Porcupines spend increasing amounts of time searching for food on the ground, where they are more vulnerable. Winter mortality of Porcupines, especially of the smaller and more vulnerable juveniles, can be severe, particularly during exceptionally cold or snowy years. Vehicular traffic is a major cause of mortality in many areas.

DIET

Strictly herbivorous, Porcupines eat bark, leaves, fruits and nuts of trees and shrubs, as well as grasses, wetland plants, herbaceous forbs, fungi, and even some agricultural crops such as corn. There is a major shift in food choices between summer and winter. This is due not only to availability, but also in response to chemical controls being exerted by the plants. Preferred winter food trees are coniferous, mainly hemlock, pine, fir, and spruce. These are rarely consumed in summer, when deciduous trees such as poplar are preferred. Winter is a tough time for Porcupines, as they must subsist on low-quality and insufficient food during a period of high energy demands due to cold temperatures and travel in deep snow. Most adult females are pregnant during winter and must further support the growth of their foetus. Porcupines eat diligently all summer to gain weight so they can enter the winter in top shape with plenty of body fat. Individuals commonly crouch on their haunches or grip a tree branch with their hind claws to free up their front feet for food handling. Porcupines have an enormous caecum, located at the junction of the small and large intestine, which contains bacteria that ferment cellulose, making it digestible. Unlike many rodents, Porcupines do not appear to practise coprophagy (reingestion of faeces). Their diet is often deficient in salt, especially in the spring. Winter road salt can create a fatal attraction as Porcupines seek out the salty road margins, only to become vehicle casualties. Porcupines are notorious for chewing objects such as plywood, axe handles, outhouses, boat paddles, steering wheels, tires, and shoes – essentially anything impregnated with salt from human perspiration or urine – to satisfy their salt craving.

REPRODUCTION

The breeding season lasts about two to three months between September and November. Gestation is thought to take around 210 days, and a single pup is born annually, in April, May, or June. This gestation period is exceptionally long for a rodent, probably because the growth of the foetus is very slow as it corresponds to the season of reduced food availability for the mother. If the first mating is sterile, the female may cycle into a second heat, sometimes as late as December or even January. This accounts for the occasional pup being born as late as August. Parturition typically occurs on the ground in a protected fissure in rocks or at the base of a feeding tree in a crevice among the roots. Newborns are precocial, as they are born with their eyes open, incisors and premolars already erupted, and quills present and functional after they dry (which takes about

30 minutes to a few hours depending on the relative humidity). Pups are born with black fur and only exhibit adult coloration the following spring after their first moult. They begin eating solid food about one week after birth, but continue to nurse for approximately three to four more months. Pups remain with their mothers until the autumn breeding season, when they disperse. Juveniles become sexually mature at 18–25 months old.

BEHAVIOUR

Apart from brief pairing during the breeding season, maternal females with nursing pups, and groups of animals (up to eight so far reported) that may share a winter den, Porcupines are generally solitary. During the breeding season, pre-oestrous females advertise their condition by calling and urine marking. Numerous eager males converge on such females, fighting among themselves to establish dominance and, along with it, access to the breeding female. Typically the largest male wins alpha status and guards the female until she becomes receptive and mates with him. Although the courtship may take place in the trees, mating occurs only on the ground, and is always at the discretion of the female. She must raise her tail over her back to present her quill-less underside to the male before he can mount. An unusual aspect of Porcupine mating behaviour is male urine hosing. As he courts the female, he might rear onto his hind legs and spray her with squirts of urine. An unreceptive female will move away. A receptive one will not object.

Pups up to about 6 weeks old are unable to climb large trees. Their mothers leave them on the ground (often hidden in the hollow base of a tree or in a root junction) while she rests and feeds up the same tree. She will travel down to the pup to nurse it. The pups do not have the normal pungent odour of adults, but should a predator chance upon them, they do have a full complement of small, but effective quills with which to defend themselves. Their best defence, however, is concealment. Maternal females will not defend their young. For obvious reasons, Porcupines are very circumspect in grooming and their grooming is much less thorough than with most other mammals. They are also adept at quill extraction, using both their front teeth (incisors) and their very dexterous front feet. An experienced predator will attempt to attack a Porcupine's nose, where there are no quills. Eventually, after numerous wounds and much harassment, the Porcupine will weaken. It then gets flipped over and the kill is made. Inexperienced or impatient predators usually end up with a face-full of quills. If the beleaguered Porcupine is

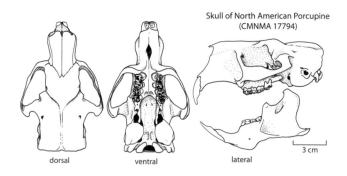

Skull of North American Porcupine
(CMNMA 17794)

3 cm

dorsal   ventral   lateral

**Walk in
deep snow**

Walk

Left hind track
Length: 2.9–4.1 cm
Width: 2.2–3.2 cm

Left front track
Length: 2.5–4.3 cm
Width: 2.4–3.8 cm

Porcupine

on the ground without anywhere to hide its head, it is more likely to be killed. Animals in the trees are safer, as they can cling to the end of a branch with their backs to the trunk, and using their bulk to effectively block the branch, they can usually prevent the predator from reaching their head. A Porcupine wedged into a den is virtually impregnable. Porcupines have an additional little-known defence: they are capable of producing a characteristic acrid odour, the smell of which has been likened to that of a goat or a strong cheese. It is excreted by skin glands at the base of the rosette and can make eyes water and a nose run when contained in an enclosed area like a den. The scent is emitted when the animal becomes alarmed and may be sufficient alone to deter a potential predator. It is usually the first line of defence and likely reinforces the warning presented by the rosette pattern, the quill rustling, tail flailing, and tooth chattering of a disturbed Porcupine.

VOCALIZATIONS

The vocal repertoire of Porcupines is impressive, varying from grunts and moans to tooth chattering, snorts, sobs, mews, whines, owl-like hoos, barks, child-like cries, and even screams and shrieks. Most calls are related to reproductive activities or battles. Some of the louder sounds can carry over 200 m. Since most of this vocalization takes place at night, it can be disconcerting to nearby humans, who may be unaware that their otherwise quiet neighbour could be producing such sounds.

SIGNS

Porcupines leave an abundance of signs of their presence. During spring and summer, the most visible are their nip-twigs. These are the remains of terminal twigs with leaves stripped to the petioles or fruit removed (e.g., oak acorns) then discarded onto the forest floor. Nip-twigs can litter the base of a feeding tree and may remain caught up in the foliage after the rest of the leaves have fallen. Twigs nipped by a Porcupine are cut at angles of at least 45° and often even more obliquely. A common sign is large areas of stripped bark on their chosen food trees. Dripping sap or indications of wound healing by the tree will help one judge the age of the scar. Fresh scars are brighter in colour and slowly fade to grey or brown over the course of several years. Porcupine digs (where they unearthed a truffle, an underground fungi) may be conical if the smell oriented them directly over the prize or widely messy as if the animal had to search. A clear impression of the removed bulbous fungi may remain at the bottom of the crater. Front feet have four toes, while hind feet have five. The wide foot pads have a pebbled texture, likely to improve grip while climbing. This pebbling will show in a clear track. Tracks in the snow are typified by a furrow ploughed by the body of this short-legged animal with the tracks punching through the bottom, long, curved foot-drag marks and a tail drag between. Tracks in mud or dust can be clearer, as long as the tail drag does not obscure them. Impressions of the claws are usually noticeable in tracks and the claws are large and long enough that they prevent the toe pads of either the front or hind feet from even hitting the ground. Most tracks, except those in a soft substrate such as snow or mud, show only the flat heel pad with the claw marks well beyond. Front tracks are considerably smaller than hind. Gait rarely increases beyond a slow walk where the hind track oversteps or registers on top of the front track. Tail drag is typical. Scat varies with the dryness of the food. The dull tan to brown winter scat is dry and may be a blunt-ended, curved cylinder up to 2.5 cm long, often with a groove on the inner surface, or a chain of pellets held together by narrow strands of fibrous material. Sometimes winter scat is shorter and, although rougher and more variable in size, can resemble deer droppings. The moister summer scat is shiny and blackish, more irregular, and often pointed at the ends. Good winter den sites can be hard to find and may be used for many years by generations of Porcupines. The entrances to these sites can accumulate flowing rivers of scat of such impressive dimensions that the animals must dig through it to gain access. Runways in soil or snow range in width from 15 to 23 cm. These commonly spread from a den or favoured feeding site and may be identified by scat or dropped quills.

REFERENCES

Berteaux et al. 2005; Comtois and Berteaux 2005; DeMatteo and Harlow 1997; Griesemer et al. 1996; Li et al. 1997; List et al. 1999; Morin et al. 2005; Nagorsen 2005a; Roze 1989, 1990, 2002, 2006; Smith, H.C., 1993; Somers and Thiel 2008; Sweitzer 1996; Sweitzer and Berger 1997, 1998; Woods 1973; Youngman 1975; Zimmerling 2005.

FAMILY MYOCASTORIDAE
nutria

There is only one species in this family, the South American Nutria, which has been introduced into various parts of North America as a farmed fur-bearer.

Nutria
also called Coypu

FRENCH NAME: **Ragondin**

SCIENTIFIC NAME: *Myocastor coypus*

These large, non-native rodents were widely fur-farmed in the early twentieth century and would regularly escape. Deliberate introductions occurred in some southern states where Water Hyacinth, a favourite food of Nutria, was clogging the waterways. Feral colonies can survive for years without support from humans in warmer regions of Canada such as southern Ontario and especially southern British Columbia, but most of these have died off or been purposely eradicated. Nutria continue to be considered a pest species in the southern and northwestern United States.

DESCRIPTION

Nutria are yellowish brown to dark brown, with reddish highlights depending on their subspecies, and they often have a white chin and white hairs around the nose pad and a darker head. Being descendents of farmed animals, many feral Nutria in North America are colour variants such as blonde, white, or black. Vibrissae are normally long and white. Their coat is composed of a soft, dense, short layer of underfur hairs, overtopped by two lengths of longer, coarser guard hairs. The longest of the guard hairs (up to 7.5 cm long) are most prominent on the back and rump. Unlike most other fur-bearers, the belly fur of Nutria is the most valuable part of the pelt, as the underfur is denser there. Nutria are highly adapted to an aquatic lifestyle. Their large triangular head has the ears, eyes, and nose set high on their heads, so that they are above the water surface as the animal floats. Nutria most often swim with the top of the head and much of the back exposed and the tail floating behind. Their lips close behind the incisors, so they can carry items when swimming and even gnaw while submerged without filling their mouths with water. Each eye is protected by a transparent nictitating membrane, and ears and nostrils have valves that can be closed when diving. Large hind feet provide the propulsion when swimming. The tail is long, round in cross-section, and sparsely furred. Tails are vulnerable to frostbite in freezing climates. Adults of both sexes have

Nutria (*Myocastor coypus*)

Total length: average 100 cm; tail length: 34.0–40.5 cm; hind foot length: 12.0–15.0 cm; ear length: 2.5–3.0 cm.
Weight: average about 5.5 kg, maximum female 8.2 kg, maximum male 9.1 kg; newborns, 170–332 g (average 225 g).

RANGE

Nutria are native to South America, where they occur in the southern portion of the continent from central Bolivia to Tierra del Fuego. Feral populations exist in Europe, Japan, North America, northern Asia, the Middle East, and Africa. Although colonies have existed in the Ottawa River basin and along the Canadian side of the Great Lakes in the past, there have been no new Ontario or Quebec records since the early 1960s. Populations in British Columbia were formerly more common, but apparently still survive in localized areas such as Salt Spring Island. There remains an ongoing risk of reinvasion into southwestern British Columbia by feral populations from northern Washington State. Cold winter temperatures and especially ice-bound waters limit expansion into the centre of the continent.

ABUNDANCE

Abundant in Louisiana and parts of eastern Texas, feral Nutria there support a lucrative fur industry. In the Pacific Northwest and Chesapeake Bay states, Nutria are considered a pest species, as they defoliate waterways and undermine shorelines and river banks. Population density is typically highest in autumn and lowest in spring. This species is rare and localized in Canada and may be eradicated in the future.

anal glands that secrete an oily substance used to delineate the home range. They also spread the oil from these glands, as well as from oil glands at the base of the whiskers, onto their fur when grooming to condition the fur and make it more waterproof. Their dental formula is incisors 1/1, canines 0/0, premolars 1/1, and molars 3/3, for a total of 20 teeth. The upper toothrows converge towards the incisors and the looping pattern on the grinding surface of the cheek teeth is distinctive. Incisors are broad, large, and ever growing, and both upper and lower incisors are pigmented with orange (caused by iron in the enamel) on the anterior surfaces. The iron hardens the area so that the softer material behind wears away faster. This results in the tooth assuming a chisel shape that is so effective for gnawing. Kits are born with white incisors, which gradually become orange. Nutria have four or five pairs of mammae that, amazingly, are located along the midline of the back on either side of the spine, perhaps so they can suckle while floating.

SIMILAR SPECIES

Nutria look and act like "giant" Muskrats, but have a pointed tail that is round in cross-section, and most have a white chin. Adult North American Beavers are considerably larger than Nutria, but beaver kits or yearlings could be mistaken for Nutria. All ages of beavers can be distinguished from Nutria by their flattened tail and lack of a white chin.

SIZE

Males are generally about 5% and up to 15% heavier than females.

ECOLOGY

Like North American Beavers and Muskrats, Nutria are highly aquatic, living alongside coasts, rivers, lakes, ponds, and marshes. They inhabit fresh, brackish and even saltwater wetlands, but appear to prefer freshwater habitats when a choice is available. Most of their time is spent within 10 m of the shore. Although Nutria may be active anytime, most activity takes place in darkness or in the pre-dawn and post-sunset hours. They do not hibernate. Populations may become more diurnal during cold temperatures to recover feeding time spent huddling together during the colder pre-dawn hours. Nutria can remain underwater for at least 10 minutes if necessary, although most dives last less than one minute. They tip their head to chew, much like beavers, gnawing with only one pair of incisors at a time. Nutria dig burrow systems in riverbanks and coastal shorelines. These may entail a simple tunnel with one or two entrances and a single nest chamber, or a complex group of tunnels with multiple entrances and chambers extending up to 15 m from the shore. Entrances may be above or below the water level. Nesting chambers are generally spherical and contain a pad of loose vegetation. Complex and extensive burrow systems are created and used by family groups and typically evolve from a simple burrow created by a founding pair. Burrows are more commonly constructed and used where winters are cool or cold. In warm regions, outside beds under overhangs or dense vegetation are created for resting and even producing and raising kits. During severe winter weather, Nutria may seek non-traditional protection such as hay or straw piles, barns, or outbuildings. Home range size varies with habitat quality, season, age,

Distribution of the Nutria (*Myocastor coypus*)

Salt Spring Island

☐ introduced

population density, and sex. In general, adult males occupy larger home ranges than females or juveniles. Male home ranges have been measured from 5.7 to 93.9 ha, while female ranges vary from 2.5 to 46.3 ha. Adults generally remain on their home range for life, but occasionally an adult male, or more rarely a female, will undertake a long-distance relocation of > 4 km. Severe drought or flooding can prompt a mass dispersal of animals of all ages. Most dispersal is by subadults. The farthest recorded dispersal movement so far was by a male who travelled 29.6 km in 187 days. Nutria can climb steep riverbanks, small trees, and over wire fences. Potential maximum lifespan is around 6.5 years, but typically less than 10% of a wild population survive to 3 years old, and the percentage is even lower if the population is subjected to trapping pressure or extreme climate. Some captive animals are claimed to have lived 12 years, but this is not well substantiated. Predators likely include most of the medium-sized to large predators in any given area, including American Alligators, snakes, Cougars, large fishes, and Domestic Dogs and Cats.

DIET

As a successful and widely introduced species, Nutria are clearly predisposed to accept and thrive on a wide variety of seasonally available food plants. They are considered to be opportunistic foragers.

Emergent aquatic plants dominate the diet when available. Typically they feed primarily on above-ground portions in summer, and on roots and rhizomes in winter. Bark and roots of woody vegetation may also be consumed in winter. Invertebrates associated with the plants may be ingested inadvertently and could contribute some fat and protein to the diet. Nutria will also eat freshwater mussels when they are available. Crops such as rice, sugar cane, sugar beets, and corn are commonly raided by Nutria in the southern United States. Nutria have a large caecum (near the end of the digestive tract) that produces large, soft pellets that are reingested, usually while the animals rest during the day. They recycle the food to extract the nutrients made available by fermentation, which takes place in the caecum. Food is manipulated by the front feet as the animal rests on its haunches. Nutria are estimated to eat about 25% of their body weight in a day, but lactating females must increase this to about 41% to support their milk production.

REPRODUCTION

Breeding may occur during any month of the year. Average litter size is 3–6 (range 1–13), tends to decline during cold or dry periods, and is generally greater in older, and hence larger, females. When conditions are good, a feral female will have 2–3 litters each year. Nutria have a relatively long gestation for their size. Pregnancy lasts 4.0–4.5 months and females may experience a postpartum oestrus within two days of parturition. Males do not provide any parental care. Young are precocial, born with a covering of soft fur, with their eyes and ears open, and some teeth already erupted, and they can swim soon after their birth. They begin to eat solid food within a few days of birth, but are not normally weaned until they are 5–8 weeks old. Females generally become sexually mature around 5–6 months old, but may delay their first breeding if they are born near the onset of a cold season. Males become sexually mature at 5–6 months old. Like many other rodents, both sexes begin reproducing before they reach adult size.

BEHAVIOUR

Nutria live in clans of related families, typically with a dominant pair and an assortment of subordinate males and females and offspring. Except when she is in heat, the alpha female dominates the alpha male. The entire group marks and defends the group territory and may feed or sunbathe together. Adult male Nutria patrol their large home ranges throughout the year, and hence are more prone to predation and hypothermia in cold waters. Females are more gregarious and spend more time huddling together out of the water when temperatures drop, to share heat and save energy. Courtship

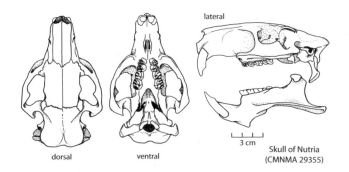

lateral

3 cm

dorsal   ventral

Skull of Nutria (CMNMA 29355)

stride length of a bound varies from 25 to 76 cm

Bound

Walk

Left hind track
Length: 6.5–15.0 cm
Width: 5.0–10.0 cm

Left front track
Length: 4.0–7.5 cm
Width: 3.5–6.5 cm

Nutria

may be a playful affair, especially if the female is not quite ready to mate. The partners chase each other around, calling as they move. Males may spray the female with urine or semen during the courtship activities. Young are born in a den, either underground or on the surface under dense cover. Apart from suckling the kits, mothers do not provide much more in the way of parental care. They do not groom the young or protect them from predators.

Scent marking is a major form of communication. Anal glands are everted and rubbed on objects around the home range, especially onto the ground where the animals enter or exit the water. A male will also back up to an object, raise its tail, and squirt urine onto it in much the same manner as a Domestic Cat urine marks. The common feeding technique is to dive down, clip a plant at the soil level, and swim with it to a feeding platform, where it is consumed and the debris discarded to add to the heap.

VOCALIZATIONS
Most calls are made in social contexts. Dominant animals produce a "mooing" call when driving away a subordinate. Maternal females call their offspring with a soft "maaw" contact call. Adults may make a loud "maawk" call during courtship.

SIGNS
Nutria tend to be shy and nocturnal, and so are rarely seen. The easiest way to know if they are around is to search for sign. Tracks may be found in mud or snow. Toes on the front feet are not webbed and the thumb (toe 1) is reduced, although its claw often leaves an impression in the track. The front tracks are larger, but otherwise very similar to those of Muskrat. The five-toed hind feet are webbed, except between the fourth and fifth toes, and the webbing sometimes leaves an impression in the track. The non-webbed fifth toe is used for grooming. Claws are prominent, sharp, and thick and usually register in a track. Nutria may leave a tail drag mark. Their tubular, blunt-ended, dark brown or dark green scat is large, approximately 5–13 mm in diameter by 20–50 mm long, with characteristic, evenly spaced, fine grooves that run longitudinally, making the scat readily distinguishable from that of Muskrats or North American Beavers. Most faeces are deposited in water as the animal forages and swims, and may sometimes be seen floating in areas frequented by Nutria. On land, scat may be deposited on grassy clumps or raised surfaces that may be used repeatedly.

Burrows can riddle and undermine the shoreline. Burrow entrances are usually wider than tall (17.0–25.5 cm high by 20–30 cm wide) and are usually connected by a well-worn runway to the water. Feeding platforms composed of plant debris are around 50–75 cm in diameter and are typically at least 15 cm above the water level. These may occur in the water, on shore at the water's edge, or on stationary objects such as a stump or a dock. Not all populations create feeding platforms. Emergent feeding platforms in shallow water can be mistaken for Muskrat houses or feeding platforms. A survey of the area to discover scat or tracks may be necessary to allow a proper identification.

REFERENCES
Bounds et al. 2003; Dobbyn 1994; Elbroch 2003; Woods et al. 1992.

Order Lagomorpha
pikas, hares, rabbits

American Pika *(Ochotona princeps)*

Photo: Brian Lasenby/iStockphoto

Of the 13 genera and 92 species of lagomorphs in the world, 4 genera and 9 species are found in Canada. Two of these species were introduced, the European Rabbit and the European Hare. Lagomorphs are native to every continent except Antarctica and Australia and to many of the oceanic islands such as Madagascar, New Zealand, and the West Indies. They are among the most frequently intentionally introduced species, usually as a potential food for human needs.

There are two main groups of lagomorphs: family Ochotonidae (the pikas) and family Leporidae (the rabbits and hares). Examples of both of these families are present in Canada. All lagomorphs are terrestrial and mainly grazers, although during hard times such as drought or heavy snow they will browse as well, chewing bark off trees and shrubs, and will even become opportunistic scavengers consuming carrion.

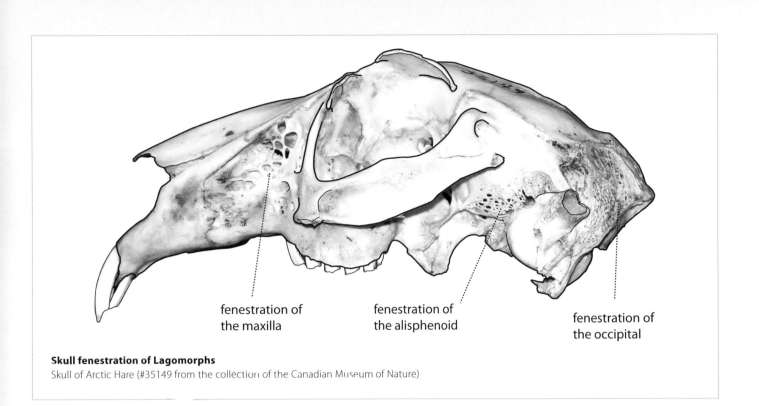

fenestration of
the maxilla

fenestration of
the alisphenoid

fenestration of
the occipital

Skull fenestration of Lagomorphs
Skull of Arctic Hare (#35149 from the collection of the Canadian Museum of Nature)

ANATOMY

Members of this order are identified primarily by one anatomical trait. Their upper incisor teeth are paired, one immediately in front of the other. The anterior tooth of each pair is predominant and has a longitudinal groove on the front surface. The posterior incisor is peg-like and much smaller. All of the incisors and cheek teeth grow continuously through the life of the animal, although the growth rate is most rapid in the anterior incisors and much slower in the cheek teeth and posterior incisors. Most lagomorphs have a red eye shine. The Leporidae have a very short, furred tail that resembles a powder puff. Members of the family Ochotonidae have short tails that are wholly within the flesh of the rump and are not visible externally. A further anatomical distinction of lagomorphs is the location of the males' testes, which lie in front of the penis as they do in marsupials rather than behind as they do in all other mammals. Lagomorph skulls are distinctive not only for their characteristic incisors but also for their sometimes elaborate fenestration. These lattice-like openings on the maxillary bones and sometimes portions of the parietal, squamosal, alisphenoid, and occipital bones reduce skull weight and allow the animal's centre of gravity to migrate rearward, closer to the main motivational power source, the hind legs.

FOSSIL RECORD

The earliest known lagomorph fossils are from China and date back to the late Palaeocene epoch, about 56 million years ago. The branch of the order that evolved into our modern rabbits and hares lived on the steppes and open grasslands of Asia where its members made a very choice meal for local predators. The long neck, ears, and legs and the large eyes on the side of an upright head for 360-degree vision were well suited to predator detection and evasion. Pikas evolved more for talus slopes and rocky areas where escape was less a matter of running and more a matter of hiding or scuttling into a burrow, hence their smaller, more compact bodies and shorter ears and legs.

ECOLOGY

Pikas (pronounced *pee-ka*) are small guinea-pig-like animals with rounded ears and no visible tail. Their fur is long and soft, and they are diurnal. There are 26 species of pikas around the world. The two North American species are very similar and occupy rocky, montane areas. None of the lagomorphs hibernate, but pikas surmount the long severe winters by storing food during the summer for consumption during the winter. Their short legs are poorly suited for travel in deep snow, so they spend most of the winter under the snow in the subnivean spaces. Rabbits and hares remain active above the snow.

Rabbits in general are smaller than hares, prefer a more wooded environment, are a little more gregarious, and give birth to altricial young. These blind, hairless babies (kittens) are helpless and require significant maternal care and protection. Hares, in general, have longer legs, are larger, usually have black ear tips, prefer more open spaces, are solitary, and give birth to precocial young. Baby hares (leverets) are born furred and able to hide and run, but they still, of course, need their mother's milk. Females of both rabbits and hares are somewhat larger than the males. Rabbits and hares are mainly nocturnal (active during the night), except those that live in the High Arctic, where they are intermittently active throughout the 24-hour period.

None of our native rabbits or hares digs underground burrows. Instead, they prefer to rest in a "form" (also called a scrape) when they are not active. These are oval-shaped depressions clawed out of the earth or snow, where the animal can crouch, unmoving, and alternately doze and wake to check for predators.

COPROPHAGY

Lagomorphs (and some rodents) have an unusual digestive strategy for coping with their difficult-to-digest, herbaceous diet. Food normally passes through the digestive tract at least twice. There is a valve at the end of the small intestine that diverts the smaller food particles into the caecum, a sac-like expansion of the large intestine filled with cellulose-digesting bacteria. The larger particles continue through the intestine. The fermented content of the caecum is excreted during the day while the animal is resting. These large, moist pellets are reingested directly from the anus. This is called coprophagy, or the eating of faecal material, and in lagomorphs it ensures that the minerals and nutrients made available by the bacteria in the caecum can be absorbed the next time around. It was originally thought that the familiar hard, round pellets were the product of the second pass through the digestive tract and that they were dropped one at a time as they were produced. New research indicates that during the day, as the animal rests, all faeces, both soft and hard, are reingested. The hard pellets produced at night when the animal is active are normally discarded, but even they are reingested during periods of starvation. Each time the hard pellets are eaten, the food particles are chewed into smaller fragments, allowing further action by the caecum and hence more food value to be extracted.

REPRODUCTION

"Breeding like rabbits" is an expression with more than a little truth behind it. Always a prime prey item, hares and rabbits can replenish their numbers quickly because they have a short gestation period, a large potential litter size, the ability to produce several litters in a season, and the capability to mature quickly and begin breeding at a young age. This phenomenon is referred to as having a high reproductive potential. During the times that conditions are good and predators are scarce, their numbers can skyrocket to the point of causing serious environmental damage. The introduction of the European Rabbit into previously rabbit-less Australia is a case in point. The downside of this type of reproductive strategy occurs when burgeoning populations create such a strain on the environment and such stress on the individuals that disease or lack of food can cause a population crash. Cyclic population fluctuations are commonly observed in some species of lagomorphs.

All lagomorphs are reflex ovulators, which means that the females ovulate following the stimulation of courtship and copulation. Only under optimum conditions do all of the fertilized eggs develop completely. Frequently, especially if the female is under stress (for example, from high population density or lack of food), resorption of some of the developing embryos reduces litter size.

FAMILY OCHOTONIDAE
pikas

Pikas are small lagomorphs with round bodies, short legs, rounded ears, and no visible tail. The skull is flattened in lateral profile, with a single, oval-shaped fenestration in the rostral region.

Collared Pika
also called Northern Pika, Rock Rabbit

FRENCH NAME: **Pika à collier**

SCIENTIFIC NAME: *Ochotona collaris*

Many populations of this northern species are threatened by climatic change. As the weather warms, less snow falls during the winter, and the Collared Pikas cannot benefit from the insulating properties of a dense snowpack over their burrows. Ironically, many die of the cold because of global warming. Owing to the remote location of these animals, they are not as well known as are their more southerly and more accessible relatives, American Pikas.

DESCRIPTION

The Collared Pika is a greyer version of the closely related, but more southerly, American Pika. It is about the size of a guinea pig, and the fur on its back is grey with grizzled black and white tips. Sometimes in summer the face and shoulders of adults will develop a rufous wash. Belly fur is creamy white. There is an indistinct lighter-grey collar around its neck from its throat to its ear, which provides the common name. This is more visible in summer pelage. The winter pelage is also greyish. Although they appear to be tailless, picas do in fact have a short tail that does not show externally. The dental formula is incisors 2/1, canines 0/0, premolars 3/2, and molars 2/3, for a total of 26 teeth. Pikas have one fewer upper molar than do the rabbits and hares. The Collared Pika has a single moult each year while the American Pika undergoes two.

SIMILAR SPECIES

The two pika species do not overlap in distribution; one is found in the north (Collared Pika) and the other in the south (American Pika). One of the best distinguishing features of the Collared Pika is the light-buff patch of fur on the side of its face, over the cheek gland. The same spot on an American Pika is usually similar in colour to the rest of the pelage.

Collared Pika
(*Ochotona collaris*)

SIZE

The following measurements apply to Canadian animals of reproductive age. Females and males are about the same size.
Total length: 155–217 mm; tail length: 3–14 mm; hind foot length: 20–36 mm; ear length: 19–25 mm.
Weight: 71.9–177.0 g; newborns, unknown.

RANGE

Collared Pikas are northern animals found in the mountainous regions of southeastern Alaska, Yukon, western Northwest Territories, and a small portion of northwestern British Columbia.

ABUNDANCE

They are locally common in the appropriate habitats but are already showing signs of decline because of a warming trend attributed to global warming. In some regions, numbers have plummeted by 90% in spring, and these declines have been attributed to warmer-than-usual winters. As they are already limited to the sides and tops of mountains, pikas cannot move to cooler locations; they would have to cross the even warmer lowland valleys to get to another mountain.

ECOLOGY

Collared Pikas are diurnal, and although they are active all year, they spend most of the winter under the snow. It is not known whether they dig in soil to make a burrow or they penetrate so deeply into the rock crevices that there is no need to dig further to be well sheltered, but they do dig foraging tunnels in the snow during winter. The activity level of pikas is directly correlated with wind speed, as wind reduces the efficacy of their alarm calls and consequently makes the animals more vulnerable to predation.

They are agile and have acute vision and hearing. Their colonies are found in montane regions in broken rock fields, talus, and boulder fields with bordering vegetation. Most colonies are on mountain sides above the timberline, but solitary individuals and some small colonies have been found in wooded areas that are as low as sea level. Pikas are closely tied to the protection of the rocks and rarely venture more than 10 m into the vegetation around the talus patches to feed or collect vegetation. Most foraging occurs within 3 m of the rock field. Like most pikas, Collared Pikas are insatiable hoarders of winter food. They gather a wide variety of alpine plants during the growing season and place them onto either a single major hay pile or a number of smaller hay piles within their territory. Often the size and position of the hay piles depend on wind. If the habitat is very windy, none of the hay piles will be out in the open; they will be sheltered under an overhang or beside a boulder, and the ultimate size of the pile will depend on the size of the shelter. Location of the hay piles tends to be traditional, with new plant material being deposited over the remains from previous years. Haymaking begins in late June (males) and early July (females) and becomes an increasingly time-consuming activity as winter approaches. The dried vegetation provides a substantial and crucial portion of the winter diet, and each pika stashes 20–30 kg of hay annually. Hoary Marmot faecal pellets are commonly added to the hay piles along with their own caecal pellets. Ermine occasionally use the hay piles as marking sites by depositing their faeces on the top. Collared Pikas share their habitat with Hoary Marmots and Arctic Ground Squirrels. All three species consume many of the same plants with no signs of competition. They are aware of the intent of each other's alarm calls and will react to them regardless of the species that produces it. As a cold-adapted species, these small mammals are vulnerable to high temperatures and will disappear down into the rocks to stay cool during hot days. As temperatures climb in the north (average temperatures have risen by 2°C per decade since the 1960s in southern Yukon),

Distribution of the Collared Pika (*Ochotona collaris*)

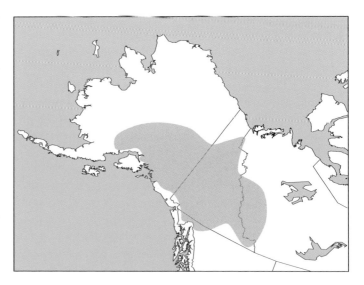

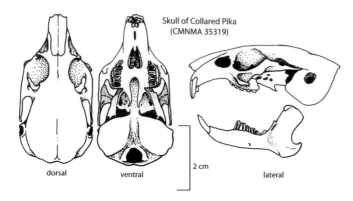

Skull of Collared Pika
(CMNMA 35319)

dorsal · ventral · 2 cm · lateral

unknown. They grow quickly and reach adult size in 40–50 days. The females experience a postpartum oestrus and breed again shortly after the birth of their first litter. Two litters per year are possible. The juveniles do not breed until the following spring.

BEHAVIOUR

Both sexes of Collared Pika are vocal and territorial. They establish and defend their territory by scent marking (using their cheek glands), by frequent calling, and by chasing and sometimes fighting with trespassers. These fights can result in bite wounds, particularly on the back and rump, usually on the intruder. When calling, pikas point their noses slightly upwards. They call frequently while haying, from their lookouts, and even while underground.

VOCALIZATIONS

The short call, described as a nasal bleat, a short sharp bark, or an *ank*, carries a considerable distance. It can have ventriloquistic properties, especially if the calling animal is down in the rocks. This species does not use a long call as described in studies of American Pika communications. Instead, Collared Pika males use a variation of the short call, repeated loudly and in series for the same purpose (to advertise their availability to female neighbours). A submissive squeak is emitted by an animal being chased or by juveniles towards an adult. One study describes an average day during hay-gathering season, when an adult pika was reported to call 217 times; about half of the calls were answered by a neighbour.

SIGNS

The front foot has five clawed digits, but the first toe on the inside (the pollex) is somewhat reduced and does not always register in a track. The hind foot has four toes. Tracks are uncommon, except in snow, and are more reminiscent of a rodent's than a lagomorph's. Like their southern relatives the American Pikas, Collared Pikas inhabit a burrow, the entrance of which can become visibly stained with urine and round faecal pellets during the course of the winter. Long-time favoured lookout sites can display conspicuous white stains, particularly in areas with low rainfall. These are caused by insoluble calcium in the urine, remaining after the rest of the urine has been washed away. Long-used latrine sites are commonly stained with bright-orange nitrophilous lichens that grow only on nitrogen-rich rocks or on rocks that are washed with nitrogen-rich faeces. Since the rocks in the region are generally nitrogen poor, these splashes of colour may indicate an old or active territory.

colonies may provide an early warning by beginning to disappear in warmer or lowland locations. Ermines can pursue pikas down into the broken rock fields and hence are their most dangerous predators. Larger species like American Martens, Coyotes, Golden Eagles, Red Foxes, and Canada Lynx capture migrants and the unwary. Since many of the small prey species in this part of the continent hibernate over the winter, pikas are important to the small carnivores at that time. Few Collared Pikas live more than three to four years.

DIET

Collared Pikas are herbivorous and will eat almost any edible plant. Their diet changes from summer to winter. They eat a wide variety of plants in summer and supplement their dried winter diet with whatever fresh vegetation they can access from their snow tunnels. Collared Pikas produce two types of faecal pellet. One looks like a shiny, black, narrow cylinder of toothpaste and is called a caecal pellet because it represents the contents of the caecum. The caecum is a pocket of the large intestine, located near the end of the digestive tract, which houses cellulose-digesting bacteria. Fine food particles are diverted into the caecum to allow the bacteria to work on the material. When it is eventually excreted, the pika eats the pellet in order to give its digestive tract an opportunity to extract the results of the bacterial digestion. This is called coprophagy, and it is practised by all of the lagomorphs. Having the ability to digest cellulose, even if it does require the food to pass twice or more through the digestive system, allows these small lagomorphs to survive and even thrive in marginal alpine habitats. The second type of pellet is a small, sticky, black pellet, which is not normally reingested; it is deposited around the animal's territory and eventually weathers to a golden-brown colour.

REPRODUCTION

Collared Pikas breed in May and early June and, after about a 30-day pregnancy, produce two to six young. Average litter size is unknown. Young pikas are born in an altricial state – blind, nearly hairless, and in need of much parental care. Age at weaning is

REFERENCES

Broadbooks 1965; Franken and Hik 2004; Hafner and Sullivan 1995; Hayes, A.R., and Huntly 2005; Hik 2002; Holmes, W.G., 1991; Kawamichi 1981; MacDonald, S.O., and Jones 1987; Morrison 2006; Morrison and Hik 2007; Morrison et al. 2004; Slough and Jung 2007; Smith, A.T., et al. 1990.

American Pika

also called Rocky Mountain Pika, Coney, Rock Rabbit,
Piping Hare, Whistling Hare, Mouse-Hare

FRENCH NAME: **Pika d'Amérique**
SCIENTIFIC NAME: *Ochotona princeps*

These guinea-pig-sized relatives of rabbits and hares prefer to live in cool temperatures, seeking higher elevations in warmer regions. They are restricted to mountain tops in the more southerly parts of their range. Their presence is usually heard before it is seen because they are alert and quick to call a general alarm if danger is spotted. There are two species of pikas in North America, and both occur in Canada.

DESCRIPTION

American Pikas are short legged and quite oval in appearance. They appear to be tailless but in fact have a length of tail vertebrae that is mostly enclosed in the flesh of the hind quarters and therefore does not show on the outside. Pelt colour is variable in summer from greyish with a rusty head and shoulders to rusty brown to a blackish brown in coastal forms. The belly is light grey in colour. The winter coat is greyer and considerably longer than the summer coat. Ears are moderately large and almost circular, with dark hairs inside and white hairs around the margin. The dental formula is incisors 2/1, canines 0/0, premolars 3/2, and molars 2/3, for a total of 26 teeth. Pikas have one fewer upper molar than do the rabbits and hares. American Pikas undergo two annual moults, in spring and autumn.

SIMILAR SPECIES

The American Pika lacks the light-grey collar that gives the Collared Pika its name and is generally larger and browner. The distributions of the two pika species do not overlap as there are at least 800 km between the closest extents of their ranges. The American Pika's

American Pika
(*Ochotona princeps*)

vocal repertoire includes the long call that the Collared Pika lacks (see "Vocalizations" section).

SIZE

The following measurements apply to Canadian animals of reproductive age. Females and males are about the same size.
Total length: 154–224 mm; tail length: 4–24 mm; hind foot length: 26–36 mm; ear length: 19–25 mm.
Weight: 81.7–190.0 g; newborns, usually 10–12 g.

RANGE

The species is distributed discontinuously in mountainous areas of western North America. They inhabit the Northern Rocky Mountains and the Coast Mountains in Canada, and the Sierra Nevada Mountains, the Cascade Mountains, the Wasatch Range, and the Southern Rocky Mountains in the United States. In the northern parts of the range, pikas can be found from sea level up to about 3000 m. The lower elevational limit becomes progressively higher as one travels south and the temperatures at sea level become warmer. Near the southern distributional limits, few pikas are found below 2500 m.

ABUNDANCE

This species is widely scattered and locally common in mountainous regions and is well adjusted to living in isolated genetic groups. However, the current warming trend is already subjecting the southern populations to significant summer temperature stress and decreased winter snowfall, resulting in serious population reductions and even local extinctions.

ECOLOGY

American Pikas live in cool habitats where the terrain is broken but fringed by vegetation. They prefer talus or broken rock fields, but, failing that, they have been found in boulder fields, mine tailings, or even piles of lumber or scrap metal. They shelter in broken terrain and, although not thought to dig burrows, they may expand existing cavities to enhance their living space. Pikas are diurnal, unlike the other lagomorphs, but if daytime temperatures become too warm, they retreat below ground during the hot parts of the day. By midsummer and through the autumn, pikas engage in an unusual activity called haying. They dart into the meadow near their territory, and instead of simply eating the plants, they carefully harvest them. Taking a mouth full, they return to their hay pile to deposit the plants. Pikas must make up to 13 such excursions per hour through the haying season in order to accumulate an average-sized hay pile. The plants dry in the hay pile and are later used by the owner as winter food. Each pika creates a hay pile or sometimes one major hay pile and one or more minor hay piles on its territory and defends them from other pikas. The heaps of vegetation are often used by biologists to assess the presence and size of a population. Pikas are active all year but spend winter under the snow. Although they dig tunnels through the snow to the meadows in order to forage for fresh vegetation, most pikas depend on their hay pile to see them through until spring. Major predators include weasels and Coyotes. The smaller species of weasels, Ermine and Long-tailed Weasel, are

Distribution of the American Pika (*Ochotona princeps*)

the most successful predators because they can follow the pikas into the rock talus. Coyotes catch only the dispersing juveniles. Occasionally marmots take nestlings, and bears will sometimes try to dig them out, usually with only limited success. Eagles and hawks probably take some pikas, but they are not major predators. For their size, American Pikas are long lived. Maximum age is around seven years. Mortality rates are highest in the first year and in the five-to-seven-year age group.

DIET

American Pikas are generalist herbivores with two types of diet. Their summer food is consumed immediately; their winter food is harvested in summer and autumn, dried, and stored for future use when fresh vegetation is scarce. Summer food is variable, depending on what is available and what is flowering, as pikas have a taste for the flower heads. The food plants they choose to dry in their hay piles often include species that they do not eat in a fresh state. These plants normally contain high levels of phenolic compounds, which make them distasteful or even toxic. Once the plants have been dried and stored for some time, the noxious chemicals begin to break down, and the pikas can then safely eat them. There is some evidence to suggest that the presence of these phenols actually enhances the preservation properties of the plant, making it more suitable for long-term storage. Like all of the lagomorphs, American Pikas produce two types of faecal pellets. The familiar hard, round pellet is dropped indiscriminately around the animal's territory or in latrine areas. The soft, black string from the caecum (the shape of toothpaste from a tube) is either reingested directly from the anus or added to the hay pile for future consumption.

REPRODUCTION

Most females produce two litters per year. Gestation is approximately 30 days. The first litter is born just as the snow melts and the alpine vegetation becomes available to the mother. Births occur as early as March in some low-elevation sites but can happen as late as June in the higher elevations. The females breed again shortly after the birth of the first litter; however, most do not wean more than one litter per season. Young in the first litter have the greatest chance of surviving their crucial first winter because they will have had more time to find a territory and create a hay pile large enough to last through the bad weather. Females invest a tremendous amount of energy feeding their first litter until it is weaned, and are normally so depleted that they cannot properly feed the second litter. The second litter usually only survives to weaning age if some accident or predation causes loss of the first litter. The average litter size is three (one to five is possible), and this is quite consistent throughout the range, regardless of the mother's age and whether it is the first or second litter of the year. The young are born in an altricial condition with eyes closed and only a light covering of dark hair, but their teeth are fully erupted. Their fur thickens quickly, and their eyes open at around nine days of age. Mothers spend much of the day feeding, and return to their young every two hours or so to allow them to nurse for about ten minutes. The young begin eating solid food at around twelve days of age and are about two-thirds of their adult size and fully weaned at three to four weeks old. Juveniles do not breed until the following spring.

BEHAVIOUR

Both males and females establish territories. Adult pikas become somewhat tolerant of neighbours, especially if they are of the opposite sex; however, they are very intolerant of strangers, especially if they are the same gender. This behavioural polarization by sex tends to result in a mosaic of territories within each colony, where individuals are surrounded by animals of the opposite gender. Females usually select a neighbouring male as their mate and often mate with the same male each year. Gathering of hay begins first by males in midsummer, followed by the females a week or two later. Often pikas

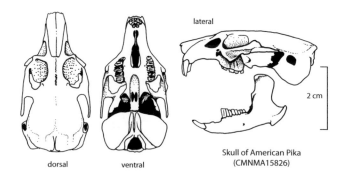

lateral

2 cm

dorsal　ventral

Skull of American Pika
(CMNMA15826)

Bound

Left hind track
Length: 1.6–3.0 cm
Width: 1.6–2.5 cm

Left front track
Length: 1.5–2.5 cm
Width: 1.7–2.2 cm

**American and
Collared Pikas**

will gather faecal pellets from other alpine species, such as marmots or ground squirrels, to add to their hay pile in addition to their own dried caecal pellets. About 30% of the day is spent on the surface, resting (surveying their territory), feeding, and gathering hay, or establishing and maintaining their territory by scent marking, vocalizing, and chasing intruders. The remainder of the day (typically the warmest portion) is spent out of sight within the rocks. Pikas scent mark by rubbing their cheek glands against rocks, and they appear to recognize individuals from their scent marks. Some colonies have developed a tradition of pilfering from neighbouring hay piles. In these colonies the incumbents may be forced to spend their resting time on or near their hay pile to defend it from thieving neighbours. After the juveniles have been weaned, their only chance for survival is to either occupy a vacant territory or create a new one between already existing territories. Once they have a territory, they must set about gathering a hay pile of their own. Those youngsters that cannot establish a nearby territory are forced to disperse away from the colony to search for appropriate habitat. Such juveniles are very vulnerable to predation, and most do not survive.

VOCALIZATIONS

American Pika whistles are either short or long calls. A single short call is used as a territorial call to deter trespassers. An alarm call is made up of a series of short calls delivered in rapid succession. The "chattering" long call is produced by adult males during the breeding season. Sometimes two males will perform a duet of long calls. Pikas also produce a submissive squeak, commonly heard when an animal is startled or when a juvenile is being chased by an adult.

SIGNS

Pikas have five toes on each front foot and four on each back foot. The first toe on the front foot (the pollex or thumb) is clawed and somewhat reduced, but not to the same extent as it is on rabbits and hares. The tracks of the two North American pika species are indistinguishable. Pika tracks are uncommon, except in snow, and are more reminiscent of rodent tracks than lagomorph tracks. Active hay piles may be found at the edges of talus slopes. Sometimes the hay piles are built in the lee of a large boulder and can be less obvious. Burrow entrances in the snow are frequently marked with urine stains, and accumulations of pellets in latrines can be found, especially after snow melt. Round faecal pellets are dark and sticky when fresh and gradually weather to a golden brown. They are small, usually between 2.5 and 3.5 mm in diameter. The favoured lookout site(s) of each pika can, with long usage, develop an accumulation of faeces and conspicuous white urine stains. Sharp alarm whistles may be heard; pikas are both vigilant and vocal.

REFERENCES

Beever et al. 2003; Broadbooks 1965; Dearling 1996, 1997a, 1997b; Elbroch 2003; Hafner and Sullivan 1995; Holmes, W.G., 1991; McKechnie et al. 1994; Smith, A.T., 1981; Smith, A.T., and Weston 1990; Smith, A.T., et al. 1990.

FAMILY LEPORIDAE
hares and rabbits

This family is characterized by long ears and long hind legs and feet. The skull is arched in lateral profile with prominent supraorbital processes and well-developed lattice-like fenestrations in the rostral, parietal, alisphenoid, and often occipital regions.

Snowshoe Hare

also called Varying Hare, Snowshoe Rabbit, Bush Hare

FRENCH NAME: **Lièvre d'Amérique**

SCIENTIFIC NAME: *Lepus americanus*

Over most of its range the Snowshoe Hare changes colour between the browns of summer and the white of winter. In areas where winter snowfall is uncommon, such as southwestern British Columbia and areas south, the summer and winter coats are both brown. Clearly, the colour change is for the purpose of camouflage, which is further enhanced by the gradual and patchy growth of the new hairs through the old coat; it blends with patches of snow and bare ground during early winter or the spring melt.

DESCRIPTION

This hare is the smallest of the three Canadian hares and the smallest of all 26 species of hares around the world. During winter the hind feet become thickly furred and snowshoe-like to help the animal travel on the surface of the snow. The eyes are large, luminous, and dark. The ears have black tips in both summer and winter pelage. Melanistic individuals are rare but have been reported in northern New York State, Maine, Prince Edward Island, and southern Quebec. The dental formula is incisors 2/1, canines 0/0, premolars 3/2, and molars 3/3, for a total of 28 teeth. In most parts of their range Snowshoe Hares undergo three moults each year: in early spring from white to brown, in summer from brown to brown, and in late fall from brown to white. Regional differences in the timing of the moult are directly related to differences in the duration and timing of the onset of snow cover. An abnormally early or late snowfall in an area may leave the animals alarmingly non-camouflaged for a time.

SIMILAR SPECIES

White-tailed Jackrabbits have longer ears (> 100 mm), and their adults are over 2500 g in weight. Arctic Hares are also considerably heavier (> 4000 g) as adults. The base of the white hair is white in Arctic Hares and White-tailed Jackrabbits, but dark in Snowshoe Hares. The introduced European Hare is the only one of Canada's hares to remain brown in winter. In summer, an adult European Hare can be distinguished from a Snowshoe Hare mainly by its much larger size (> 3000 g) and the black topside of its tail, which is held out and down from its body as it runs. All the other hares are usually found in open areas, while the Snowshoe Hare is a forest animal. Both Snowshoe Hares and Eastern Cottontails occur in parts of central and eastern Ontario and in southwestern British Columbia. The Eastern Cottontail has a reddish-orange patch of fur at the back of its neck, its hind legs are distinctly more delicate, its tail is larger and bounces as it runs, it does not turn white in winter (but neither does the Snowshoe Hare in southwestern British Columbia), and the tips of its ears are not black in any season. The skull of the two species can appear to be similar. When the top of the Snowshoe Hare's skull is viewed from above, it can be seen that the process over the eye socket, called the supraorbital process, has a wide scallop behind it and is clearly distinct. In the Eastern Cottontail the supraorbital processes fit tightly back against the skull and, in many examples, are actually fused to the skull. The brown form of a European Rabbit could be mistaken for a summer Snowshoe Hare. European Rabbits are generally more robust, with a brownish nape.

SIZE

The following measurements apply to animals of reproductive age from Canada. Adult females are larger than the males by 10%–40%.
Female: total length: 405–670 mm; tail length: 22–61 mm; hind foot length: 112–160 mm; ear length: 64–100 mm; weight: 1150–2311 g.
Male: total length: 381–620 mm; tail length: 18–60 mm; hind foot length: 110–153 mm; ear length: 58–95 mm; weight: 852–1730 g.
Newborns: weight 60–85 g, can drop to 40 g in especially large litters.

Distribution of the Snowshoe Hare (*Lepus americanus*)

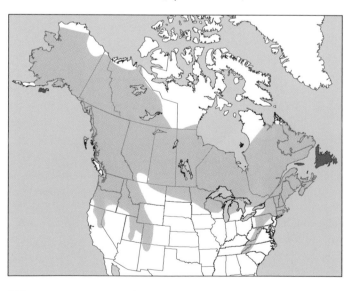

 introduced

Snowshoe Hare (*Lepus americanus*)

summer

winter

RANGE

Snowshoe Hares are found throughout Canada up to the treeline (except in a small part of treeless prairie in southeastern Alberta) and in deforested and heavily farmed southern Ontario. They have naturally inhabited some of our larger islands, such as Manitoulin Island and Prince Edward Island, but have not occupied Vancouver Island or the Queen Charlotte Islands. They have been introduced to many offshore islands in the North Atlantic such as Newfoundland, Anticosti Island, and several islands in the Grand Manan Archipelago.

ABUNDANCE

This is a widespread and common species. Numbers fluctuate periodically in the boreal forest portions of its range. Hunting by humans for food and recreation occurs in parts of the country. For example, in Newfoundland an estimated 1.8 million are killed annually.

ECOLOGY

Snowshoe Hares live in forests that typically have dense understorey vegetation. They are most abundant in successional forests following logging, forest fires, or other such wide-scale changes. Much research has been conducted over the years to attempt to explain the cyclical expansion and contraction of Snowshoe Hare numbers in the boreal forest regions of North America. The cycle takes usually between eight and eleven years to go from population peak, through decline, and back to population peak. Factors such as sunspot activity (which broadly affects climate), predator numbers, food availability, parasite load, stress, and disease have been investigated. Probably all of these factors play a role in the cycle, with predation and food availability being the most important. The number of hares per hectare can vary from 0.1 to 11–23 during the cycle. There is some evidence that the litter size varies with the population cycle, with females producing more leverets during the low phase than during the peak. Populations further south in other habitats undergo irregular and far less dramatic fluctuations or none at all. One major outcome of the "Snowshoe Hare cycle" is that predators are kept off balance, and most have been unable to specialize exclusively in this species. Canada Lynx has specialized to some degree in the Snowshoe Hare as a prey species, but it is forced to find other prey during periods of declining Snowshoe Hare numbers or to starve. In fact, Canada Lynx population numbers follow the hare cycle – with the peak and trough delayed by a year or two. A year or so after the hares have declined, the Canada Lynx population crashes, and thus the survival rate of the few remaining hares is enhanced; therefore, more will live to breed and rapidly produce more offspring to fuel the next population rebound. Home range size can vary from around 1.5 per ha to more than 12.0 per ha. Generally, the home ranges of males are larger and overlap portions of several females' home ranges. Snowshoe Hares are strong, buoyant, and willing swimmers, using hard kicks of the hind legs to provide propulsion in water as on land. They commonly travel to offshore islands or across rivers to feed,

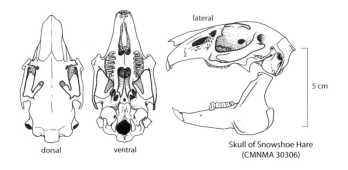

dorsal   ventral

lateral

5 cm

Skull of Snowshoe Hare
(CMNMA 30306)

and sometimes choose to escape a predator by water. This species is said to be capable of bursts of speed up to 50 km/h. Fewer than 20%, and some years as few as 2%, of leverets survive their first year of life, but once they reach adulthood, they can live to be five years old. The mortality rate of adults is also high. A captive Snowshoe Hare lived to be eight years old. Since Snowshoe Hares are widespread and high-quality prey items, they are hunted by many predators. The list includes almost every medium-sized mammalian carnivore, many of the smaller, and some of the larger ones. It also includes most of the larger raptors. The Canada Lynx, Bobcat, Coyote, Grey Wolf, Red Fox, Fisher, American Marten, American Mink, Long-tailed Weasel, Great Horned Owl, Great Grey Owl, Barred Owl, Snowy Owl, Goshawk, and Red-tailed Hawk are the major Snowshoe Hare predators. The Red Squirrel, Ermine, and several of the ground squirrel species are predators of the smaller leverets, as are sometimes Domestic Cats.

DIET

These hares eat a wide variety of woody and herbaceous plants and are known to occasionally feed in winter on carcasses of deer or other hares. Although mainly browsers of woody twigs and bark during winter, in some areas (particularly parts of the north) Snowshoe Hares will dig feeding craters to reach dried, standing plants buried in snow up to 36 cm deep. Snowshoe Hares reingest the soft form of faecal pellet that they produce, and sometimes even the hard form, in order to recycle the material through the digestive tract a second time. This is known as coprophagy and is common among rabbits and hares. The hard form of pellet, the one commonly associated with hares and rabbits, is the end result of the recycling process.

REPRODUCTION

The onset of the breeding season is directly related to day length and geography. Females enter their first oestrus of the year between late March (in the south) and early May (in the north). All of the females in a region come into oestrus at about the same time, with the first litter being born 35–40 days later, around mid-May to early July. Usually within 24 hours of parturition a female will experience a postpartum oestrus and mate again. She is capable of lactation and gestation concurrently. The litters within a region are born in distinct groups or cohorts, spaced apart by the 35–40

day gestation period. In the north of the range this can mean two, or perhaps three, litters per year depending on the length of the warm season. In warmer regions, where the season is longer, there can be as many as four litters. Average litter size is two to five leverets, but up to nine is known. Survival of all the leverets in such a large litter is unlikely, given that a female hare possesses only eight nipples. The first litter of the year tends to be the smallest of the season, and the second litter the largest. Juvenile females do not breed until they are one year old. Approximately 25%–30% of litters are fathered by more than one male. Females construct a shallow saucer-like depression, sometimes lined with dried grasses, for a nest. As with many prey species, the birthing process is quickly and quietly accomplished; an average-sized litter of four is delivered in around two minutes, including the production and consumption of the afterbirth. The female leaves once the birthing is over and spends little time in the vicinity of the nest thereafter. She will visit it only once a day to allow the young to nurse. Newborns are fully and thickly furred and able to weakly move about within 30 minutes. Their eyes open very soon after birth. By their fourth day they have scattered from the nest. They reassemble at the same time each evening, whereupon their mother will appear and allow them to nurse for five to ten minutes, after which all will scatter again. The young are nursed for 25–28 days, but those born in the last litter of the season may be allowed to nurse for up to twice that time. They begin eating solid food at seven to ten days old and usually are weaned and become fully independent at about 28 days old, at which time they are around 80% of their eventual adult size.

BEHAVIOUR

A solitary species, Snowshoe Hares are territorial, especially of others of their gender, and will maintain and defend a territory of varying size, depending on habitat and phase of the population cycle. Both genders defend their home ranges and exclude others of the same sex. Agonistic displays used in territorial defence include raising or lowering of ears and can escalate to chases and then kicking or biting. An ears-up posture indicates an inquisitive approach, while an ears-down posture indicates displeasure and aggression. Courtship is an energetic activity for both partners. Often several ardent bucks will follow a single doe, each kicking and drumming in an effort to gain her favour. "Boxing," as described in other lagomorphs, has not been observed in Snowshoe Hares. The solemn courtship procession soon escalates to a series of wild chases, with animals jumping and urinating over each other. The hares know every twist and turn of their carefully maintained runways, and every hidey-hole in their territory; hence they are reluctant to allow themselves to be driven by a predator into unknown regions and will usually attempt to circle back to stay in their known space. A dispersing juvenile may travel up to 10 km to find an unoccupied territory to appropriate, but most offspring settle closer to their natal territory. Like all of the Canadian leporids, Snowshoe Hares are nocturnal and active year round. They spend the day in their form (an oval, slightly depressed hollow that has been scratched out of the earth or snow), resting alertly, briefly snoozing, and occasionally grooming,

stride length of
a bound varies
from 20 to 183 cm

Bound in
deep snow

Bound in
light snow

Left hind track
Length: 8.3–15.2 cm
Width: 4.1–12.7 cm

Left front track
Length: 4.8–7.6 cm
Width: 2.9–5.7 cm

Snowshoe Hare

and they become more active at twilight. If they sense a disturbance, either with their acute senses or by feeling vibrations, they will freeze in their form or take flight along their runways. Hares of both sexes often warn others of impending danger by drumming their hind feet on the ground. These vibrations travel a considerable distance. Females are very solicitous of their newborn litters and will drum repeatedly to attract a potential predator away from the nest, risking their own lives in the process.

VOCALIZATIONS

Like most other lagomorphs, the Snowshoe Hare is usually quiet; for this reason, its chilling squeal when captured is often startling enough to make a predator or human drop it. Normally only low-volume grunts and snorts are used by females to call their leverets.

SIGNS

Snowshoe Hares spend a great deal of time preparing their escape routes. During the warm months they cut any vegetation that intrudes onto their chosen runways. These twigs and herbs are trimmed near the ground; the nipped ends are cleanly cut at a 45° angle (more or less) and can be quite visible to a careful observer. The runways are 10–20 cm wide. During the winter Snowshoe Hares spend a similar amount of energy in packing down the snow on these runways so they will have a clear, hard surface on which to make their escape. This is done by hopping up and down, making only a little forward progress with each hop; it looks like a dozen hares have passed. During protracted or heavy snowstorms they will come out even in daylight to perform this duty. If a clearing must be crossed, they hop across it, leaping about a metre per stride. When they cross back, they will carefully follow the first track, landing in the same spots, so that after a few passes they will have created islands of firm footing. As winter progresses and the snows deepen, the hares are faced with overhanging branches that need to be trimmed. A Snowshoe Hare's forms are smooth, oval scrapes in the soil or snow that closely match the size of the hare. Sometimes the ground surface takes an almost polished look in well-used forms. Snow forms tend to show some melting and refreezing on the surface from the body heat of the recumbent hare. Snowshoe Hares will also use appropriately sized abandoned burrows. They lie just inside the burrow entrance, probably as a protection from raptors. In winter they will do the same in short snow burrows that they have scratched out for themselves. During warmer weather the hares like to find a flat sandy area to enjoy a dust bath. These cleared, scuffed areas of varying size, but usually less than half a metre in diameter, are more commonly found in the Prairies. Fur may be seen in the vicinity. Faecal pellets in hard form (8–14 mm in diameter and often a bit flattened) are produced and dropped one at a time. If a group of pellets is found, the animal will have spent an extended period of time at that location. Urine tends to be orange or reddish in colour and is highly noticeable on snow. During winter Snowshoe Hares chew the bark and branches off trees up to the height of the snowpack and can leave a browse line similar to that made by White-tailed Deer, only lower (up to 50–60 cm above the ground or snow level). All of the rabbits, hares, and even pikas go beyond the cambium layer when they chew branches, taking some of the wood as well. This leaves a characteristically rough look to the chewed area on a standing tree, a fallen branch, or a raspberry cane. Snowshoe Hares will also peel bark from standing trees, leaving a ragged trunk with a smattering of deep chew marks and flapping bits of loose bark. Snowshoe Hares have five toes on each front foot. The first toe (on the inside of the foot) is greatly reduced but is clawed and sometimes registers in a track. The hind feet are heavily furred, especially in winter, and the four long toes can be very widely spread to provide a snowshoe effect in deep snow.

REFERENCES

Burton 2002; Cowan and Bell 1986; Criddle 1938; Elbroch 2003; Forsey et al. 1995; Gilbert, B.S., 1990; Graf 1985; Hodges 2000; Keith 1990; Krebs et al. 1995b; O'Donoghue and Bergman 1992; O'Donoghue and Boutin 1995; Rongstad and Tester 1971.

Arctic Hare

also called Polar Hare, Alpine Hare

FRENCH NAME: **Lièvre arctique**

SCIENTIFIC NAME: *Lepus arcticus*

This large, beautiful hare is the northernmost lagomorph in the world. It can survive, and indeed thrive, where winter temperatures of –40°C are common and light levels fluctuate from 24 hours of darkness during winter to 24 hours of daylight during summer.

DESCRIPTION

The colour of this hare's summer coat varies with latitude. On the mainland tundra it is grey on the back and throat, and white on the belly and tail. On the southern Arctic islands the summer coat is mainly white with pinkish-brown or grey patches on the face and back. The northern island hares tend to remain white all year. In winter all grow a thick, silky, white coat except for black ear tips, dark insides of the ears, and brownish-orange staining on the hairs of the bottom of the feet. Each body hair is white all the way to the base, unlike that of Snowshoe Hares which has a grey base. The leverets are born with a dark, grizzled-grey coat, which they shed and regrow to an almost fully white coat by the time they are eight weeks old. They retain a brown patch on the top of their head, which can be visible in good light. By the following spring they are indistinguishable from adults. The eyes are dark reddish-brown or yellow-brown, darker around the outside edge of the iris, with a somewhat oval pupil. The eyelashes are long and black, possibly serving as sunshades. The claws are long and strong, especially on the forefeet. As with all the hares, the dental formula is incisors 2/1, canines 0/0,

premolars 3/2, and molars 3/3, for a total of 28 teeth. The incisors of this species are angled forward more than of other hares, as an adaptation to feeding in snow.

Some scientists argue that the Alaskan Hare (*Lepus othus*) and the Arctic Hare (*Lepus arcticus*) are the same species as the Mountain Hare (*Lepus timidus*) of Europe and Asia. Recent genetic evidence suggests that each is distinct and should be considered a separate species.

SIMILAR SPECIES

This is the largest and the most northerly of the North American hares. The only other hare whose range overlaps that of the Arctic Hare is the much smaller Snowshoe Hare. The easiest way to distinguish these two species, apart from size, is habitat preference. The Arctic Hare prefers open, windswept locations while the Snowshoe Hare is a woodland species.

SIZE

The following measurements apply to Canadian animals of reproductive age. Body size increases with latitude; the largest animals are found on Greenland, and the smallest at the treeline. Females are slightly heavier than males.
Total length: 558–701 mm; tail length: 45–73 mm; hind foot length: 146–164 mm; ear length: 70–110 mm.
Weight: 2500–6300 g; newborns, around 100 g.

RANGE

As their name suggests, Arctic Hares are found above the treeline on the tundra around the coast of Greenland, and in Canada from Newfoundland to Yukon. They have been successfully introduced to a number of islands off Newfoundland, including Brunette, Valen,

Distribution of the Arctic Hare (*Lepus arcticus*)

Arctic Hare (*Lepus arcticus*)

juvenile coat and
summer coat
of adults in the
south of the range

winter coat of southern adults and
year-round coat of northern adults

Jude, Grey, Oderin, Emberley, and Kelly. Scatari Island, off the coast of Cape Breton, also had Arctic Hares introduced. The continuing presence of hares on these islands remains uncertain as trapping, hunting, poaching and the introduction of predators can have rapid and devastating effects on such insular populations.

ABUNDANCE

Numbers fluctuate, usually depending on weather and predation; thus Arctic Hares can be locally very abundant or totally absent in certain years. They are not considered to be threatened anywhere in Canada but are generally scarce in Newfoundland.

ECOLOGY

This animal is superbly adapted to a cold, barren, and windy environment. The northern populations find more appropriate habitats than do the southern, leading to greater numbers in the north. Like all of the lagomorphs, the Arctic Hare is active all through the year, even during the long, dark, bitter, Arctic winter. Its thick, insulating fur protects it from the cold, and it has developed many behaviours that help it to stay warm. (See the "Behaviour" section below for more details.) The Arctic Hare has been clocked running at speeds of almost 65 km/h, and it is a capable swimmer, readily crossing streams and rivers. Although the geographic ranges of the Arctic Hare and the Snowshoe Hare overlap slightly, there is little competition between the species. Each occupies a different habitat within the zone of overlap, and neither interacts with the other, although it is possible that a crash of the Snowshoe Hare population in Newfoundland could have an impact on the nearby Arctic Hares, simply by the diversion of

predators such as the Canada Lynx. Since Arctic Hares produce fewer offspring compared to Snowshoe Hares, any additional predation of the juveniles could potentially have a dramatic effect on the population. Arctic Hares are hunted by indigenous peoples, but their major predators are Grey Wolves, Arctic Foxes, raptors, and occasionally weasels and Canada Lynx. Most predation is upon the juveniles; only about 15% survive to breed the following year. The longevity of Arctic Hares is unknown but likely exceeds four years, once they have survived their first year.

DIET

Arctic Hares are herbivores. They consume a wide variety of low-growing Arctic plants and lichens. They particularly favour Arctic Willow (White Birch in Newfoundland) during the winter. They dig through hard-packed and ice-crusted snow to reach dried plants. Like all the hares, Arctic Hares will consume meat if they find it. Their water needs are met by eating snow in winter, spring, and autumn and by drinking standing water in summer. Like all of the lagomorphs, Arctic Hares cycle food through their digestive tract at least twice. The plant material travels through the digestive tract, and larger particles are diverted into the caecum, a pouch near the end of the intestine where bacteria breaks down some of the cellulose. Large, soft pellets produced by the caecum are reingested to allow the digestive system to absorb the available nutrients from this process. Soft pellets are excreted while the animal is resting, at a rate of about one every 30 minutes, and they are consumed directly from the anus. The distinctive hard "rabbit" pellet is formed after the maximum nutrition has been extracted, and it is dropped indiscriminately as the hare moves about.

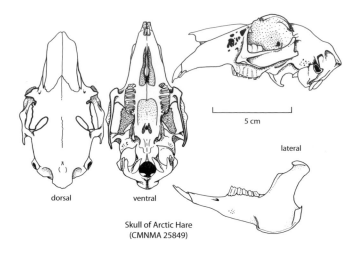

dorsal   ventral

5 cm

lateral

Skull of Arctic Hare
(CMNMA 25849)

REPRODUCTION

Initiation of the breeding season varies with latitude. In southern regions (on the northern mainland and in Newfoundland) it begins in early April, while in Greenland and the High Arctic islands it begins in early May. Ovulation is induced by copulation. After a gestation period of about 53 days, a litter of around four to six (the range is one to eight) leverets is born between late May and early July. In Newfoundland the average litter is three. Female Arctic Hares produce a single litter annually. The young are born fully furred and with their eyes open, in a shallow nest that is sometimes lined with grasses and their mother's fur. The female stays close to her litter for a while after the birth to protect it if necessary but soon leaves the leverets to their own devices. The newborns remain near the nest for the first few days, slinking away to hide if danger threatens. By freezing with their head on the ground and ears pinned back, they depend on their coloration to protect them from detection. The female spends her time far way from her offspring, returning for one-and-a-half to four minutes every 18–20 hours to allow them to nurse. Despite the absence of light-level cues during the 24 hours of summer daylight in the far north, Arctic Hare females still manage to coordinate a consistent 18–20 hour nursing cycle. How they do this is uncertain. As the young hares get older, they travel further from the nursing site between suckling sessions. The leverets begin eating vegetation in their first week and are usually weaned at around eight to nine weeks old. It is unknown, but likely, that the juveniles breed the following spring.

BEHAVIOUR

Arctic Hares are somewhat more gregarious than are their more southerly cousins. Litters may be seen together, usually while they are waiting for their mother to arrive to suckle them, and juveniles tend to group together after weaning. In winter in the northern potions of the range, herds of up to several hundred (but more commonly less than 100 animals) may gather. The animals in these herds display no signs of a rigid social dominance system. Spacing of about 1–2 m between hares is achieved by the aggression of each animal, to create and maintain individual distance. Some

males seem to be more dominant and can displace other hares from their feeding craters. As the herd moves, the hares tend to follow narrow traditional trails, which can become distinct after centuries of usage. Feeding and resting hares position themselves so that their backs are towards the wind. While resting during cold temperatures, the hares tuck their tails under, rock onto their hind feet, which are pushed forward to provide a base for the weight of the body, then with a characteristic rapid flurry of paddling, the front feet are tucked under the body and the hare tips forward to lie on its hind feet. With the ears pressed down into the fur on the back, the hare assumes a characteristic "resting sphere" posture, which is highly energy efficient. Arctic Hares occasionally scratch out shallow depressions in soil or snow, called forms. These could be in the open or in the lee of a boulder or other windbreak and are used while resting. Arctic Hares doze fitfully while resting, waking frequently to scan for predators. During winter Arctic Hares will also dig into a snowdrift to scratch out a shallow snow cave called a snow scrape, which they back into for shelter. These have an overhanging roof but tend to be less than 30 cm deep so that the resting hare can position its head at the opening to watch for wolves and foxes. Snow dens are similarly constructed but are considerably deeper and have a slightly enlarged terminal chamber. Both snow dens and scrapes are about 20 cm wide. Most snow dens are situated so that their openings face away from the prevailing winds. Oddly enough, snow den usage tends to take place in spring when temperatures are rising. It has been suggested that the hares construct and utilize the dens to provide shelter during the moult, when large patches of hair can be lost, thereby placing them under significant temperature stress during cold or windy days. During winter the hares can be extraordinarily tame and allow the approach of a human to the point of touching. In summer the hares are much more wary. Although it has not been previously reported, Dr David Gray has found that Arctic Hares, like other hares, possess chin glands. Males rub the underside of their chin on rocks and other prominent objects, leaving a distinctive, pungent odour that advertises their presence and gender. He has also discovered a peculiar form of penis display that is performed early in the breeding period. In spring, shortly after the onset of 24 hours of daylight, the males begin to extend their long, thin, dark penis in a quick, whip-like motion. The penis is uncoiled along the belly and out between the front legs in a movement that lasts about 1.5 seconds and is repeated up to 12 times. The males perform this display regardless of whether they are alone or in a group. Soon after the onset of the penis display the males begin sniffing other hares, looking for receptive females. This continual sniffing can lead to aggression by females that are not receptive. She will attempt to drive away her ardent suitor by facing him on her hind legs and batting at him with her front legs. Usually the male backs down immediately. If he is persistent, the fighting can escalate, sometimes resulting in serious injury. Many rebuffed males will quietly follow a chosen female, occasionally checking to see if she is becoming more receptive. Chases often follow as the female attempts to avoid advances. Eventually, as the females in a herd become receptive, a period of "mating madness" ensues, with many matings and attempted matings occurring over a few hours

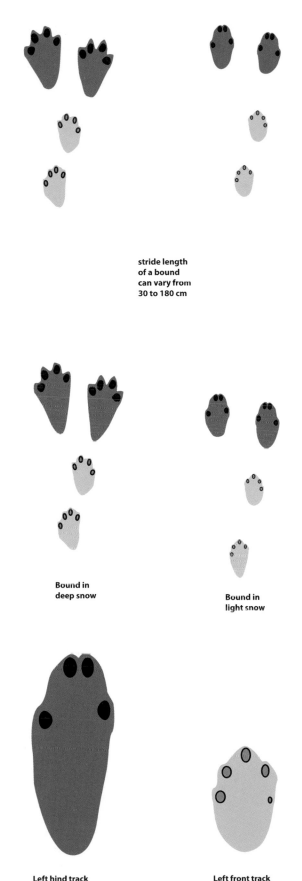

**stride length
of a bound
can vary from
30 to 180 cm**

**Bound in
deep snow**

**Bound in
light snow**

Left hind track
Length: 10.8–20.3 cm
Width: 6.0–12.0 cm

Left front track
Length: 8.0–9.5 cm
Width: 4.5–9.5 cm

Arctic Hare

in the late evening and early morning. If threatened by a predator, Arctic Hares will flee in a group or alone, often using a bipedal hop to rapidly travel uphill. They use their speed to outrun the predator.

VOCALIZATIONS

Arctic Hares are normally silent animals. As a female approaches her waiting (but motionless) young leverets, she will emit a series of low growls to call them to begin a nursing bout. Older leverets learn when to expect her and eagerly watch for her arrival, frequently running to meet her and then mobbing her before she has a chance to call them. Like all of the hares, Arctic Hares will sometimes produce a loud, startling scream when captured. They will also produce a fierce growl while being handled, in addition to biting and clawing.

SIGNS

Forms, snow scrapes, and snow dens are described in the "Behaviour" section above. Loose tufts of hair may be clinging to vegetation or snow in these areas because the hares commonly roll around to remove moulting hairs. Centuries of Arctic Hare use of common trails in the far north have left permanent depressions, which can be similar but narrower than those created by migrating Caribou. In the southern parts of their range Arctic Hares travel in the typical four-footed rabbit hop. In the far north an additional gait is added: the two-footed hop. When alarmed, the hares stand and sometimes bounce on their hind legs to provide themselves with a wider view as they assess the danger. They may then bound away, usually uphill. Considerable speed can be achieved with this bipedal bounce before the hares drop into their normal four-footed hop. The two-footed bounces can be up to 2.5 m apart. Each front foot has five toes, although the first toe (pollex or thumb) is reduced and often does not show in a track. Each hind foot has four toes. All toes are clawed. Arctic Hares have large, well-furred hind feet with long toes that can be spread like those of Snowshoe Hares to assist travel on deep snow. The heavily furred hind heel often does not register on a shallow substrate. Scat is similar to that of other large hares.

REFERENCES

Aniśkowicz et al. 1990; Best and Henry 1994; Fitzgerald and Keith 1990; Gray 1993; Larter 1999; Mercer et al. 1981; Parker 1977; Small et al. 1991; Waltari et al. 2004.

European Hare
*also called European Jackrabbit, Brown Hare, Common Hare
(The latter two names are used more often in Europe.)*

FRENCH NAME: **Lièvre d'Europe**

SCIENTIFIC NAME: *Lepus europaeus*

Nine European Hares (seven females and two males) were accidentally introduced near Brantford, Ontario, by a farmer in 1912. They

European Hare (*Lepus europaeus*)

have since spread through much of southern Ontario. There seems to be little competition between this introduced species and the two native lagomorphs, the Eastern Cottontail and the Snowshoe Hare, which share its range.

DESCRIPTION

This hare has long ears with black tips. Fur on the back is grizzled yellowish brown in colour, but the legs, shoulders, neck, and throat are rufous, and the haunches are greyish. The belly and inside of the upper legs are white. The tail is carried away and down from the body while the hare is running, which makes its black upper surface clearly visible. The eyes are yellowish brown. This hare does not change colour in winter. The dental formula is incisors 2/1, canines 0/0, premolars 3/2, and molars 3/3, for a total of 28 teeth.

SIMILAR SPECIES

The European Hare is larger than either the Snowshoe Hare or the Eastern Cottontail and is also larger than the introduced European Rabbit. The tail carriage and its black upper surface are good field identifiers, as are the open field habitat and the brown winter coat.

SIZE

The following measurements apply to Canadian animals of reproductive age. Males are on average about 5% heavier than the females. Weight varies seasonally, with the highest weights seen in mid-summer and the lowest in late winter.

Total length: 650–750 mm; tail length: 72–110 mm; hind foot length: 140–160 mm; ear length: 95–110 mm.

Weight: 3000–5500 g; newborns, around 100 g.

RANGE

As a result of several introductions into the United States and an introduction into southern Ontario, European Hares were found from Ontario to Michigan and throughout northern New York and adjoining states by around the 1930s. Since then the range has dwindled to southern Ontario, parts of eastern Ontario, and possibly a small portion of the adjoining northern New York State, where the current status of the animal is uncertain. There are also populations still existing on a few small islands in Boston Harbour. European Hares have also been introduced into Argentina and Chile, where they appear to be spreading. The Romans may have introduced European Hares into England and Ireland.

ABUNDANCE

These large hares can be common where their preferred habitat of open fields bordered by fencerows or woodlots prevails. Population numbers in Europe tend to fluctuate, and it seems that the North American populations do so as well.

ECOLOGY

European Hares in North America, as in Europe, are active all year. They are nocturnal, usually venturing out to feed after the sun has set, as it begins to get dark. This hare is a bit more active during daylight hours than are Canada's native hares. Long adapted to agricultural land in their native continent, European Hares in North America are most commonly seen in areas of mixed farming where

Distribution of the European Hare (*Lepus europaeus*)

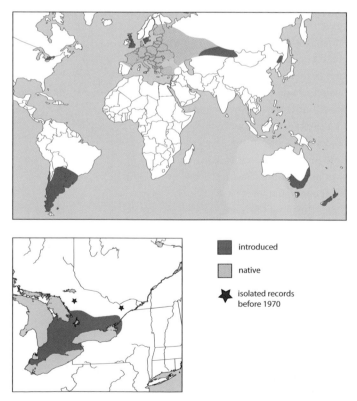

introduced

native

isolated records before 1970

expanded view of Canadian range

the fields are surrounded by fencerows and woodlots. When chased, these hares prefer to stay in the open and use their speed (around 70 km/h) to outrun the predator, but during foul weather they will retreat into the fencerows or the edges of the woodlots for shelter. Their "form" can be just a scrape in the soil or snow, or a somewhat more elaborate space pressed into the vegetation, with leaves or grass stems bending over the top. The hares back into this type of form so that only their head and shoulders are visible. Like the White-tailed Jackrabbit of western Canada, the European Hare will sit tight during a snowstorm and allow the snow to accumulate over and around it and provide some protection from the wind. The record longevity of this species in the wild is a 12.5 year-old hare from Poland. Numerous records exist for the five- to seven-year bracket; however, most hares die before they are two years old. Foxes, Coyotes, and Domestic Dogs are the most serious North American predators. Since these hares thrive in agricultural areas, they are vulnerable to agricultural chemicals and machinery.

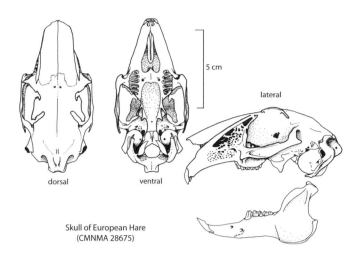

Skull of European Hare
(CMNMA 28675)

dorsal ventral lateral

5 cm

DIET

Like all hares, European Hares are herbivorous. They feed on grasses and forbs through the summer and add tree bark and woody shoots to their diet in winter. As they are closely associated with agricultural lands, European hares can cause crop damage. Two or three hares are estimated to eat as much vegetation as does one sheep. They are notorious for stripping the bark from grape vines and young fruit trees during the winter. These hares will dig through snow to reach dried grasses.

REPRODUCTION

In central Europe (the source of our European Hares) the breeding season lasts from December to October. The introduced North American population maintains this timing. Onset of the breeding season is governed by day length, not temperature. With a gestation period of around 40–42 days, this early breeding can result in leverets being born as early as January, which is a cold and snowy month in Ontario. No doubt in many years most, if not all, of the first litters perish from hypothermia; this is also often the case in colder parts of Europe, such as Austria. Females do not create a nest for their young; they usually scratch out a shallow depression in the soil into which the leverets are born. Litter size ranges from one to ten leverets, but most litters consist of two to four; the larger the litter, the lower the birth weight of each leveret. The first litter of the year tends to be the smallest. Leverets are born fully furred and with their eyes already open. They scatter from the nest site soon after birth but stay nearby. With little parental care and shelter, the babies must sometimes expend considerable energy to stay warm. Their mother returns to the nest area once each day to allow them to nurse for as little as five minutes. They begin eating solid food as early as their second week but continue to suckle until they are three to four weeks old. The last litter of the season may nurse for twice as long. Juvenile males are capable of breeding by the time they are six months old, and the females when they are seven to eight months old, but in Canada most juveniles do not breed until they are one year old. In Europe up to four litters may be produced by each female each year, but in Canada two litters per year is the norm. These births are usually not synchronous (all the females producing their leverets at about the same time), as they are with native hares. The females come into season again a few days after giving birth.

BEHAVIOUR

All parental care is provided by the female. The expression "mad as a March hare" originated with this animal and is related to courtship behaviour. The males throw caution to the wind during the breeding season; groups of them will follow a female and fight among themselves for access to her. Frequently seen, even in daylight, these males are so absorbed that they ignore possible dangers. Hares reported to be "boxing" during courtship were once thought to be two males fighting but are actually a female and a male, the female attempting to discourage the overly ardent male. Boxing hares may also be seen during winter when they are competing for a preferred food source. Although the species is often described as asocial, this is not always the case. There are reports of large herds in Europe looking for food during the winter, and frequent sightings of small groups at any time of year. The species has a whole suite of threatening, appeasement, and communication behaviours that they use to establish and maintain contact. Many of these behaviours are very subtle, such as raising the ears (show of interest), lowering them (keep away), stamping with the front feet towards a conspecific (intention to stand and fight), or drumming with the hind (warning of a predator). The sophistication of their communications strongly suggests a species that lives in groups. An adult occupies a home range of around 300 ha, which it will share with other hares without any signs of territoriality.

VOCALIZATIONS

The female produces a low "guttural" call to draw her offspring to nurse. This species, as with all hares, also emits a chilling squeal when wounded or frightened.

SIGNS

The droppings are generally larger, flatter, and more fibrous than those of the Eastern Cottontail, the Snowshoe Hare, and the European

Rabbit, but, depending on what each is eating at the time, it may be impossible to distinguish between them. Each front foot has five toes, and each hind foot has four. Their tracks are similar to those of other large hares such as White-tailed Jackrabbits and Arctic Hares. The only native lagomorphs that occur within the Canadian range of European Hares are Eastern Cottontails and Snowshoe Hares. Neither of these species occurs in the open field habitat preferred by European Hares. Cottontails are found in shelter belts and around buildings, while Snowshoe Hares are woodland species.

REFERENCES

Cowan and Bell 1986; Hackländer et al. 2002; Kurta 1995; Long, J.L., 2003; Macdonald and Barrett 1993; Schneider 1981.

White-tailed Jackrabbit

also called Prairie Hare, Prairie Jackrabbit

FRENCH NAME: **Lièvre de Townsend**

SCIENTIFIC NAME: *Lepus townsendii*

This hare has yellow eyes with a black pupil and is the only Canadian hare or rabbit, with the exception of the introduced European Hare, to have this eye colour.

DESCRIPTION

The White-tailed Jackrabbit is the only Canadian grassland hare. The most useful field characteristic that distinguishes this species is the colour of the top of its tail. As its name describes, this species has a completely white tail in both summer and winter pelage. The summer coat is yellowish-brown on the back, becoming yellowish-grey on the haunches, and whitish on the belly, chin, and inside of the legs. In the northern parts of the range, following a moult in late autumn, the winter coat grows in mostly white, sometimes with a buffy tinge to the ears, face, back, and feet. The tips of the ears remain black in both pelages. The dental formula is incisors 2/1, canines 0/0, premolars 3/2, and molars 3/3, for a total of 28 teeth.

SIMILAR SPECIES

The only other Canadian grassland lagomorph is the Mountain Cottontail. It is considerably smaller than the White-tailed Jackrabbit and has dark eyes. Mountain Cottontails tend to be found in areas that have more sagebrush and cover and rarely in the open habitats preferred by the White-tailed Jackrabbit.

SIZE

The following measurements apply to animals of reproductive age in the Canadian portion of the range. Females are larger than males by 5%–35%. The difference is mainly seen in weight. Weight varies

seasonally, with the highest weights being seen in mid-summer and the lowest in late winter.

Total length: 510–650 mm; tail length: 70–113 mm; hind foot length: 137–172 mm; ear length: 95–137 mm.

Weight: females, 2900–4100 g; males, 2800–3400 g, newborns, around 90 g.

RANGE

These hares are widely distributed in the prairie grasslands of western and central North America. Their range is slowly expanding northwards and eastwards as the forest is converted into more open farmland. Their range is shrinking in the southwest as increasingly arid conditions and the loss of native prairie habitat to agriculture make that environment less suitable for them and more suitable to their southerly competitor, the Black-tailed Jackrabbit. The population of White-tailed Jackrabbits in southern British Columbia was high enough in the 1920s and 1930s for the species to be considered a pest, but it has declined over the last 50 years to the point that now the White-tailed Jackrabbit may be extirpated (locally extinct). The first White-tailed Jackrabbit in Manitoba was captured in 1881 near Boissevain (southwest Manitoba), and the species has since expanded into all of southern Manitoba.

ABUNDANCE

White-tailed Jackrabbits are generally common in the appropriate habitat, although numbers fluctuate, usually owing to climatic conditions. The population in southern British Columbia is likely extirpated, and the occasional recent sightings are probably animals immigrating from nearby populations in Washington. These dispersing individuals probably do not survive long in British Columbia, because agricultural expansion in the Okanagan and Similkameen

Distribution of the White-tailed Jackrabbit (*Lepus townsendii*)

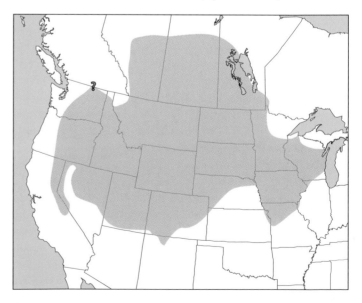

? current occurrence uncertain

White-tailed Jackrabbit (*Lepus townsendii*)

summer

winter

valleys has diminished the natural grasslands to the extent that they are too small and fragmented to support a viable population.

ECOLOGY

White-tailed Jackrabbits are active all year. They are primarily nocturnal and are most active during the first few hours of darkness. During an activity period they typically travel 0.4–1.0 km. Daylight hours are spent resting in shallow, scratched-out depressions in soil or snow, called forms, which are usually located at the base of a shrub or rock or even in a dip just below the crest of a hill, where the hare can rest with its eyes just at the ridgeline. (See the "Signs" section below for more details.) White-tailed Jackrabbits have robust and effective claws on all four feet. They use the front ones to dig their forms or to dig through the snow to find forage. They do not dig underground burrows. Claws on the hind feet provide traction during high-speed escape manoeuvres. Sight, hearing, and smell are all acute and are used to detect potential predators from a distance. They are capable of speeds up to 55 km/h when pursued and are able to make rapid turns at high speeds. They will sometimes swim, especially to escape a predator, paddling with their front feet and kicking out with their hind. Golden Eagles and the larger hawks and owls, along with Coyotes, Bobcats, Cougars, Grey Wolves, Red Fox, and American Badgers, hunt White-tailed Jackrabbits. Northern Raccoons, Striped Skunks, and rattlesnakes probably take the leverets. Maximum longevity is around eight years in the wild, but few live longer than one year.

DIET

These hares are herbivorous and feed primarily on grasses and forbs, adding bark and buds from shrubs to their diet during the winter.

Alfalfa is a favourite food, and even harvested fields can attract them. Like all of Canada's rabbits and hares, the White-tailed Jackrabbit ingests the soft form of pellet that it produces, in order to recycle more nutrients. The hard form represents the final stage in the hare's digestion. These round pellets are produced and dropped one at a time.

REPRODUCTION

The breeding season begins in March in Canada (earlier in warmer areas) and can extend until late July. It can be delayed by persistent snow cover. Environmental factors play a major role in the reproductive success. A wet summer can greatly reduce the number of leverets, as can a particularly hard winter with deep snow or prolonged, very cold temperatures. Generally one or two litters are produced each summer by White-tailed Jackrabbits in Canada. This may increase to as much as four litters in the more southerly parts of the range in the United States. Like many lagomorphs,

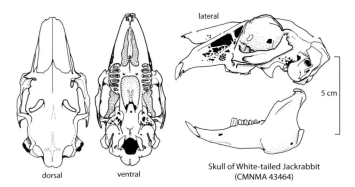

dorsal

ventral

lateral

5 cm

Skull of White-tailed Jackrabbit
(CMNMA 43464)

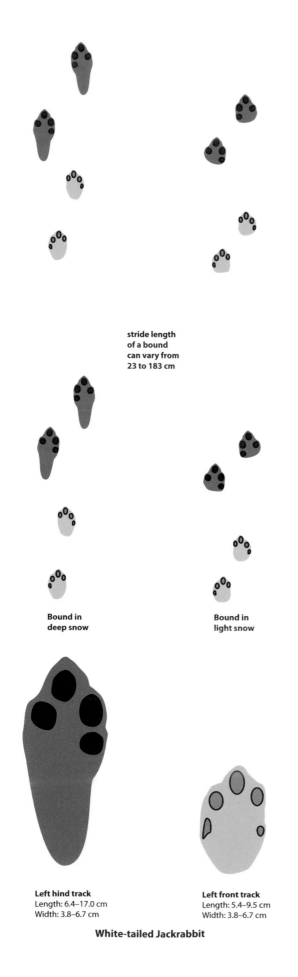

stride length
of a bound
can vary from
23 to 183 cm

**Bound in
deep snow**

**Bound in
light snow**

Left hind track
Length: 6.4–17.0 cm
Width: 3.8–6.7 cm

Left front track
Length: 5.4–9.5 cm
Width: 3.8–6.7 cm

White-tailed Jackrabbit

White-tailed Jackrabbits are synchronous breeders, which means that all of the females produce their litters at almost the same time. The glut of leverets ensures that some will survive predation and live to become adults. The second annual litter is usually the largest for each female, followed by the first and then the third if there are more than two litters in a season. Litter size varies from one to eleven leverets, but four to five is more common. Survival of all the leverets in the larger litters is unlikely, given that the females possess only eight nipples. The gestation period has been estimated at 42 days. Usually within 24 hours of giving birth, the females will breed again. Since the females are polygamous (will breed with more than one male), it is possible that a litter may have more than one father. Females merely use a shallow scrape for a nest or sometimes just drop the newborns onto the bare ground. The leverets are born with a thick coat, their eyes open, and their incisors already erupted. They are capable of movement shortly after birth and probably spend less than 24 hours at the nest site before scattering. Their mother returns to the birthing area once, or sometimes twice, each night for less than five minutes to allow them to nurse. The leverets begin eating solid food at around two weeks old and are fully weaned at around four to five weeks old, being about 35% of their adult size. Juveniles do not breed until the following spring.

BEHAVIOUR

Often called the least social of the North American hares, White-tailed Jackrabbits exhibit almost no territorial behaviour. Their home range is estimated at around 90 ha, with males having larger ranges than do females. The only social behaviour normally reported is that of courtship. One or more males will pursue a female by circling, chasing, and jumping over each other, the jumping and chasing often involving urination. These activities can last from five to twenty minutes and usually conclude with one of the males copulating with the female. If multiple males are involved, some fighting may ensue. This can escalate from charging to butting to biting to kicking with the hind feet if one of the males does not back down. Females can also become aggressive during the breeding season, becoming intolerant of nearby females. White-tailed Jackrabbits are normally nocturnal, spending the day resting alertly and intermittently dozing and grooming in a form scratched out of the soil or snow. When exposed to a cold wind, the hares typically turn their backs to it rather than seek shelter. Only the most severe winter weather will drive them to penetrate slightly into sheltering woodlands. If danger approaches, they usually freeze tightly in their form until almost stepped on, whereupon they will burst away at high speed, in bounds of 3–5 m long, and zigzagging if pursued. The startling manner of their departure can leave an unsuspecting human gasping. In winter a running jackrabbit will avoid snow if possible because it can be bogged down if the snow is deeper than 25 cm. Their preferred runways in winter are often over windblown ridges with little snow cover. White-tailed Jackrabbits may become habituated to the presence of humans if they are undisturbed. Although they are ordinarily solitary, groups may be observed during the mating season when up to five males can vie for the favour of a female. Occasionally groups are reported during winter at abundant food sources. In these loose groupings of as many as 100 or more, the jackrabbits are not known to interact with each other in either a social or a territorial manner. Drumming with the hind feet is used to communicate with other hares, usually about the presence of a threat.

VOCALIZATIONS

As with all of Canada's lagomorphs, the White-tailed Jackrabbit rarely utters any sounds, apart from a loud series of three or four squeals (screams) in quick succession when captured or wounded. Newborns that are still in the nest will emit similar but shriller squeals if disturbed. Females are known to produce a low-pitched growl towards males that approach too closely.

SIGNS

White-tailed Jackrabbit forms are oval shaped, around 35–60 cm long by 20–30 cm wide, and can be up to 20 cm deep; sometimes they have soil or snow piled at the front. They are usually scratched out at the base of a boulder or under a shrub but can also be found in the open, often on a small rise. Most often the hares choose a location for their form that is sheltered but open so that they can see an approaching predator. During a snowstorm hares will rest in a deep form and allow the snow to pile up around and over them; this is possibly the source of old reports of hares burrowing in the snow. They commonly dig feeding craters in the snow to reach the dried grasses buried below. White-tailed Jackrabbits have five toes on each front foot. The first toe on the inside of the foot (the pollex or thumb) is greatly reduced, but it is clawed and sometimes registers in a track. The hind feet have four toes.

REFERENCES

Blackburn 1968; Brunton 1981; Dunn et al. 1982; Lim 1987; Nagorsen 2005a; Rogowitz 1997; Rogowitz and Wolfe 1991; Wrigley 1979.

European Rabbit

also called Domestic Rabbit, Coney

FRENCH NAME: **Lapin de garenne**
SCIENTIFIC NAME: *Oryctolagus cuniculus*

Unlike our native species, this rabbit burrows and creates extensive underground warrens. It has been introduced to the Canadian mainland and many islands and is a frequent escapee in suburban regions. This is the rabbit that has virtually overrun Australia and Tasmania following its importation there. Fortunately its survival in Canada is less assured, so the introductions here have not been as invasive. Both domestic and wild phases of the European Rabbit are present in Canada.

European Rabbit (*Oryctolagus cuniculus*)

wild phase

DESCRIPTION

The wild phase is usually brownish grey in colour with a grizzled appearance. The tail is dark above and white below, but as it is usually held against the body, mainly the white shows. The tips of the ears are brown, not black. The legs and feet usually have a reddish cast, the belly is creamy white to buffy grey, and the throat is grey brown. There is a rufous patch on the nape of the neck. The domestic phase is highly variable in colour and is frequently more than one colour (white with black spots, for example). The size of the domestic phase varies from dwarf to giant. The dental formula is incisors 2/1, canines 0/0, premolars 3/2, and molars 3/3, for a total of 28 teeth.

SIMILAR SPECIES

The wild European Rabbit is usually larger than our native cottontails but otherwise is very similar, especially to the Eastern Cottontail, with subtle differences that can be difficult to detect. Cottontails have proportionately larger hind legs, hopping with their backs arched and their rumps high as a result. The European Rabbit with its shorter hind legs hops with its back flatter and rump lower. The ears of the European Rabbit are larger than those of the cottontails, and the membrane at the tip is considerably thicker. Generally, the head of the European Rabbit is heavier and broader than that of the Eastern Cottontail.

SIZE

Domestic phases are highly variable in size. The following measurements apply to the wild phase since longstanding colonies, even if founded by domestic phases, eventually revert to a wild phase.
Total length: 450–600 mm; tail length: 40–80 mm; hind foot length: 75–95 mm; ear length: 65–100 mm;
Weight: 1200–2300 g; newborns, 45–75 g.

Canadian distribution of the European Rabbit (*Oryctolagus cuniculus*)

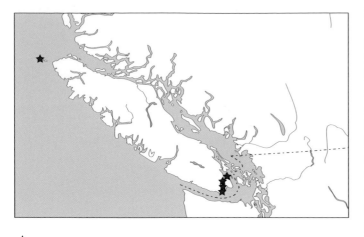

★ introduced

RANGE

As its name implies, the European Rabbit is native to the Old World (specifically Spain, Portugal, and parts of extreme northern Africa) and continues to be found there today, albeit in very reduced numbers. It is a popular small game animal and has been commonly introduced around the world. Its global distribution is continually changing as populations are introduced or relocated and they expand or die; therefore, no effort has been made to present a map of its worldwide distribution. Domestic phases are frequently seen in suburban areas. These escaped pets rarely survive their first winter and so have not produced self-sustaining wild populations in most parts of continental Canada. Deliberate introductions of both wild and domestic phases to temperate, offshore islands have been moderately more successful. Following introductions in the early 1900s, wild populations still persist in the west on Triangle Island, off the north coast of Vancouver Island, as well as on part of southern Vancouver Island around Victoria. A small population of the domestic phase has survived off the east coast in Anchorage Provincial Park on Grand Manan Island for at least 30 years, but these animals are protected and possibly fed by humans.

ABUNDANCE

The European Rabbit is often considered to be an agricultural and ecological pest in areas in which it has been introduced, and it is commonly subjected to eradication programs. Consequently the long-term continuance of many wild colonies is uncertain and largely determined by the forbearance of their human neighbours. The presently sparse populations on Vancouver Island, where the climate is mild, are likely to survive in the long run and are slowly expanding their range. Ironically, the IUCN has recently declared that in its native range the species is "near threatened" owing to over-hunting, habitat loss, eradication programs, and the effects of two diseases known to be lethal to the species: myxomatosis and rabbit haemorrhagic

virus. Native populations are currently estimated to be around 5% of their pre-1950s levels.

ECOLOGY

There are few studies of European Rabbits in North America. The following information is based either on studies made in Europe or on those conducted in San Juan Island, Washington, where the climate and environment is similar to that of southern Vancouver Island. European Rabbits are mainly diurnal if there is no human interference. They prefer to forage in the morning and afternoon and will retreat to their burrows to avoid the temperature extremes during both daytime and night-time. They prefer open woodlands and meadow or grassland areas where the soil is loose and easy to dig. Groups of up to 20 adults occupy interconnected underground burrow systems called warrens. These can be dug during years of occupation. Resting and the rearing of young usually occur underground. At the northernmost parts of their European range, European Rabbits behave more like hares in that they use shallow forms rather than underground burrows. This behaviour may also be seen in parts of Canada. Once established, their regular digging and grazing maintains the habitat at its optimum for their own needs, at the expense of local fauna. Eventually a smooth sward of short grasses surrounds each warren as the rabbits destroy all the taller plants to permit a clear view of approaching predators. Unlike our native lagomorphs, the European Rabbit is vulnerable to a disease called myxomatosis. This South American virus was introduced into Europe and Australia to control European Rabbits in the mid-1900s. Initially it killed 99% of the rabbits that it infected, but in recent years its virulence has declined and the rabbit populations are becoming more resistant. Most mid- to large-sized raptors and carnivores hunt European Rabbits. In North America the Coyote, Canada Lynx, fox, weasel, human, and Domestic Dog and Cat are the principal predators. Many populations on islands are largely predator free. Lifespan of the domestic phase in captivity is around 6–8 years, but some are known to live to 15 years old. In the wild the maximum known age is nine years, but few survive beyond their first year.

DIET

Strictly herbivorous, this species eats a wide variety of plants including agricultural crops and garden flowers and vegetables. Bark and buds of woody species, along with dried grasses, are the principal winter foods.

REPRODUCTION

In Europe the breeding season lasts all year, with a peak of litter production between February and August. Each female can have three to seven litters annually. Three litters per year is the average on San Juan Island, Washington, with the first being born in mid-April, the second in mid-May, and the third in mid-June. Most females in a colony will give birth at around the same time. Resorption of embryos is common if the mother is placed under strain because of poor weather conditions, food shortages, or periods of social stress. After a pregnancy of 28–30 days the does give birth in an underground

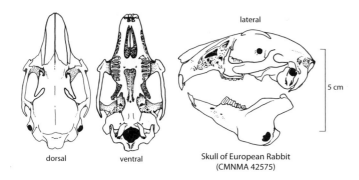

dorsal   ventral   Skull of European Rabbit
(CMNMA 42575)

burrow. They construct a birthing nest of moss and grasses that they line with their own belly and chest fur. Low-status females, who do not have access to the main warren, will dig a burrow of 1–2 m in length to house their nest. They conceal the entrance of this solitary burrow with soil whenever they leave it. Females produce litters of 3–12 (average 5–7) kittens that are born deaf, blind, and hairless. Their ears and eyes open by the tenth day. They are nursed only once during each 24-hour period, usually at night, and only for about 3–5 minutes. Each female has six nipples, so competition for access can become fierce among the larger litters. For the first 10 or so days the doe deposits a few faecal pellets as she leaves. The kittens begin to chew on these by around the eleventh day. It is thought that, while the pellets contain little nutritional value, they could serve to introduce the all-important gut flora into the kittens' digestive tracts. They begin eating solid food a few days later, at around 13 days, and are weaned as early as 25 days if their mother is pregnant with another litter, or around 31 days, if she is not. Since the does usually come into season again very shortly after giving birth, most are pregnant and nursing at the same time. Juvenile females are sufficiently mature at three-and-one-half months old to become pregnant, and the young bucks are fertile by the time they are four months old. Young born earlier in the year are more likely to survive their first winter.

BEHAVIOUR

Unlike Canada's native rabbits and hares, European Rabbits are gregarious. They form pairs at low density and groups as the density increases. Groups can number up to 20 adults, usually with a dominant male, several subordinate males, and a number of females. The females develop a hierarchical ranking within the colony, and the dominant doe produces her litter in the prime central burrow of the warren. A nesting burrow is called a breeding stop. The offspring of a dominant doe tend to achieve a high ranking as adults. The does with the lowest status are relegated to the margins of the warren and frequently must dig their own breeding stop. During periods of high rabbit density, competition for breeding stops can lead to infanticide, because the higher-ranking females will kill the nestlings of subordinates in order to take over their stop. Male European Rabbits are the only male lagomorphs known to exhibit any parental behaviour. After the juveniles have been weaned and are on their own, the males will protect them from attacks by adult females who are intolerant of any but their own young. In fact, the adult males will defend any of the colony's juveniles, regardless of whether or not they are genetically related. All members of the colony will defend the group territory, usually by chasing away interlopers. Several groups may share a feeding area. Much of the communication within the group is by odour. Glands in the anus, chin, toes, eyes, and groin all produce odoriferous substances that are used to mark territory, determine reproductive condition, and establish dominance. Dominant males possess the largest and most active glands, and they perform the majority of the scent marking. They are also the main contributors to the "dung hills" or communal latrines, which serve to advertise the group presence. Courtship is usually initiated by the male and involves sometimes extended periods of circling, chasing, anogenital sniffing, attempted mounting, and finally copulation. A rejected male may "guard" the female for some time while he waits for her to become more receptive.

VOCALIZATIONS

This species produces the typical lagomorph squealing distress call when captured or wounded. They will also thump the ground with their hind feet when alarmed.

SIGNS

Extensive underground warrens may be seen with as many as 40 entrances and the accompanying mounds of excavated soil. The openings vary from 10 to 50 cm in diameter. In low-density populations, solitary burrows can be found. Usually the warren area is worn bare, or the grass is turf-like from frequent grazing. This species, unlike Canada's native rabbits and hares, often uses a latrine area to deposit round faecal pellets. These group latrines tend to be located in bare open areas; accumulation of droppings into dung hills can be quite noticeable. Pellets are also scattered randomly around the warren and along favoured runways. The round pellets vary in size from 7 to 12 mm in diameter, depending on the plant being eaten, and are typically dark greenish brown to blackish in colour. Tracks are similar to those of the native Eastern Cottontail.

REFERENCES

Cowan, D.P., 1987; Flux and Fullagar 1992; Hudson et al. 1996; Künkele 1992; Macdonald and Barrett 1993; Sneddon 1991; Stevens, W.F., and Weisbrod 1981.

Eastern Cottontail

also called Cottontail Rabbit

FRENCH NAME: **Lapin à queue blanche**
SCIENTIFIC NAME: *Sylvilagus floridanus*

The Eastern Cottontail has widespread distribution in North and South America, which reaches its northern limits in southern and

Eastern Cottontail
(*Sylvilagus floridanus*)

to survive the winter in central Canada where the majority of the Eastern Cottontails are found. The occasional escaped Domestic Rabbit is usually a different colour – white, black, or even piebald – and is either much larger or, in the case of the dwarf forms, much smaller than the native rabbit. Occasionally a brown Domestic Rabbit may resemble an Eastern Cottontail. The native cottontails have proportionately larger hind legs and hop with their back arched and their rump high as a result. The European Rabbit, with its shorter hind legs, hops with its back flatter and the rump lower. See the "Description" section on the Snowshoe Hare for ways to distinguish the skulls of the Snowshoe Hare from the Eastern Cottontail. The New England Cottontail (*Sylvilagus transitionalis*) is not known in Canada; it is found just south of the border of Quebec in northern New York, Vermont, and New Hampshire. It can be distinguished from the Eastern Cottontail by a small black spot between the ears, by the lack of any white spot on the forehead, and by the distinct dark border on the leading edge of the ears.

eastern Ontario, southeastern Quebec, and southern Manitoba and Saskatchewan. This species is capable of prolific reproduction. It has been estimated that, assuming no mortality, a pair of Cottontail Rabbits could produce 350,000 descendants in five years.

DESCRIPTION

The Eastern Cottontail does not moult into a white winter coat. The summer pelage is generally brownish in colour, and the winter pelage is greyish, but both remain dark with grizzled tips. The tips of the ears are usually the same colour as the rest of the ears, although on some rabbits there may be some dark fur on the rim of each ear that can spread into a larger dark-brown area near the tip. There is a large rusty-brown patch on the nape of the neck, and frequently a small white spot on the forehead. The short tail is grizzled brown on the upper surface but is held up against the back, so only the fluffy white underside is visible. Since the tail is large and heavy, it may bounce as the animal hops. The belly fur is white with a grey base, and the legs and feet are reddish or buffy brown. Eyeshine is bright and reddish. The dental formula is incisors 2/1, canines 0/0, premolars 3/2, and molars 3/3, for a total of 28 teeth.

SIMILAR SPECIES

In Canada there are three lagomorphs that could be confused with the Eastern Cottontail. The Snowshoe Hare in summer pelage can sometimes resemble a large Eastern Cottontail. The hare has larger hind feet and longer legs, lacks a rusty-brown patch on the back of the neck, and has black tips on the ears and a smaller tail. The hare will only be seen in woodlands. The only cottontail in the prairie region of eastern Alberta and western Saskatchewan and in the dry Okanagan Valley of central British Columbia is the Mountain Cottontail; it resembles a small, light-coloured Eastern Cottontail, but their distributions do not overlap. The European Rabbit is often difficult to distinguish from an Eastern Cottontail; however, the European Rabbit and its variant, the Domestic Rabbit, are unable

SIZE

The following measurements apply to Canadian animals of reproductive age. Females are only about 1%–2% larger than males.
Total length: 370–490 mm; tail length: 35–70 mm; hind foot length: 80–110 mm; ear length: 53–76 mm.
Weight: 900–2000 g; newborns, 25–45 g; newborns from small litters are generally heavier than those from large litters.

RANGE

This species was absent from Canada at the time of European settlement. However, the environment became more favourable for Eastern Cottontails as land was cleared for agriculture, the prairies were planted with hedgerows, and human habitation increased. The first cottontails in southern Ontario were seen in the 1860s, and they had rapidly spread north to Ottawa and eastern Quebec by the 1930s. In Manitoba the earliest records date from 1912, and Eastern Cottontails had spread north to Dauphin by 1940 and west to Estevan in Saskatchewan by 1970. This rabbit has also been the subject of many introductions in eastern and western North American and Europe. Following population crashes of European Rabbits in Europe during the 1950s because of myxomatosis, Eastern Cottontails (which are immune) were introduced by sportsmen into Italy, northern Spain, and France, where they continue to flourish.

ABUNDANCE

Numbers fluctuate, but this is a common species in Canada. Most provinces within its range offer a hunting season.

ECOLOGY

Eastern Cottontails are very versatile in their habitat requirements, which may explain their movement northwards as humans cleared land for agriculture. They prefer a mixed farmland and hedgerow habitat, but they may also be found in swamps, open woodlands, and even grasslands, urban lawns, and gardens. This species is particularly good at taking advantage of patchy habitats, which so many other species find difficult to inhabit. In Canada these

Distribution of the Eastern Cottontail (*Sylvilagus floridanus*)

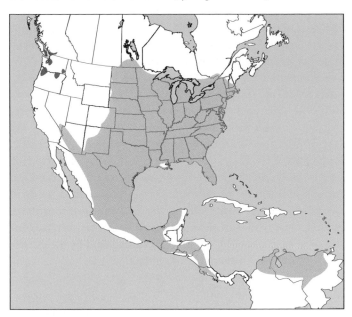

◼ introduced

range of European Introductions

rabbits are most common around human habitations where their major predators – Red Foxes, Coyotes, and raptors – are discouraged from hunting. Cottontails are hunted by most of the mammalian and avian predators within their range, including Domestic Dogs and Cats. They are particularly vulnerable in a snowy climate because their brown coat stands out against the white background and they can founder easily in deeper snow. They are mainly crepuscular but will venture out in late afternoon or early morning, especially if it is overcast. When they are inactive, their time is spent resting in a form scratched out of the soil or snow, although hollow logs and abandoned Woodchuck burrows are also used for shelter, especially in winter. They create the form in order to be sheltered overhead by adjacent vegetation, and favourite forms are reused repeatedly. Owing to high juvenile mortality, the usual life expectancy is less than a year, but some live to four or five, and a few to seven, years of age. These rabbits are active all year. Home range size is extremely variable, depending mainly on availability of food, cover, and season, and can be 1–3 ha. Reproductively active males tend to have the largest home ranges that overlap those of several females.

DIET

Eastern Cottontails are strictly herbivorous and consume a wide variety of plants depending on the geographical area. The summer diet consists mainly of grasses and forbs such as dandelion, clover, and Alfalfa, and often of garden vegetables and flowers if they are available. In winter they switch to bark and twigs of raspberry, apple, sumac, and many other woody species. As with all of the lagomorphs, Eastern Cottontails produce two types of droppings: a small, hard, round, brownish pellet and a much larger, soft, greenish pellet. The hard pellet is dropped indiscriminately as the rabbit moves about, and the soft pellet is reingested directly from the anus. See the introduction to the lagomorphs for an explanation of this peculiar behaviour called coprophagy.

REPRODUCTION

In the northern parts of the range Eastern Cottontails begin breeding in mid-March, with the first litter appearing in mid April after a gestation period of around 28 days. Inclement weather can delay the breeding season, and fine weather can hasten it, as happens in the introduced populations on the west coast of the United States and British Columbia. There the rabbits begin breeding in mid-February. A female constructs a nest before giving birth. She digs a slanting hole about 145 mm long, 120 mm wide, and 125 mm deep, which she lines with grass and dried leaves and then some of her own fur. She similarly constructs a covering of the same materials, which she drapes over the nest to protect and camouflage her young. At birth the kittens are blind, deaf, and usually pink skinned with a short coat of very fine hairs. Litter size, which usually numbers between three and six kittens, varies with the age and health of the mother and also with the season. The females can be quite protective of the nest for the first several days. They stay nearby but only visit briefly at dawn and dusk to allow the kittens to nurse, before covering them and leaving. In a week their eyes and ears have opened, and they are fully and darkly furred. The kittens leave the nest at around two weeks of age when they scatter and hide nearby. Their mother will visit each in turn once a night to allow them to nurse. They begin to eat solid food shortly after leaving the nest but continue to suckle until they are about four to five weeks old. The females normally experience a post-partum oestrus and will breed again shortly after giving birth. Up to eight litters per year may be produced in warm regions. In Canada, where the season is shorter, three litters

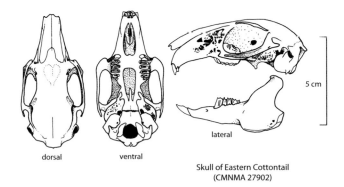

5 cm

lateral

dorsal   ventral

Skull of Eastern Cottontail
(CMNMA 27902)

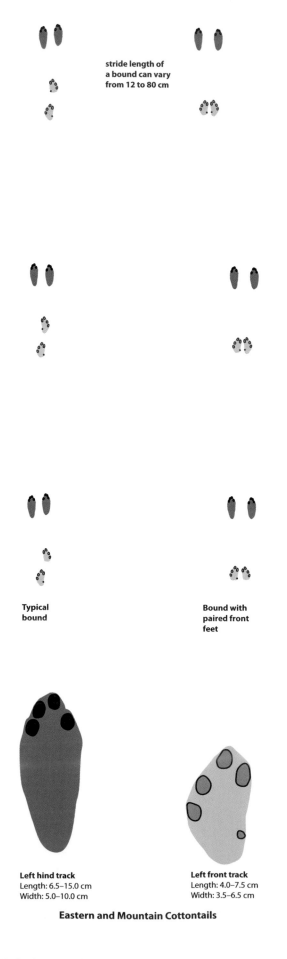

stride length of a bound can vary from 12 to 80 cm

Typical bound

Bound with paired front feet

Left hind track
Length: 6.5–15.0 cm
Width: 5.0–10.0 cm

Left front track
Length: 4.0–7.5 cm
Width: 3.5–6.5 cm

Eastern and Mountain Cottontails

appear to be the norm. The number of litters produced annually in the introduced British Columbia populations is uncertain but may be higher than the Canadian average, owing to the warmer climate. Female juveniles are capable of breeding at six months of age, and males at eight months. In northern parts of the range this effectively means that they breed for the first time as yearlings, because the warm season is over before they reach maturity. Occasionally nests with 11–15 young are found, leading scientists to conclude that more than one female is utilizing the same nest.

BEHAVIOUR

Eastern Cottontails use two techniques to escape danger. When there is sufficient cover and warning, the rabbit may slink away through the undergrowth with ears flattened and body close to the ground. Otherwise they will freeze until the last possible moment before flushing and making a rapid zigzagging dash for cover. They are well known for circling back to familiar trackways when pursued. Cottontails are not territorial, except during the breeding season. Bucks fight each other throughout the year to achieve dominance. In this manner a somewhat stable dominance hierarchy is formed. When the does come into oestrus, the dominant males devote their energy to breeding, without having to continually battle for access. The courtship display of the Eastern Cottontail is similar to that of other rabbits and hares. It begins with the buck and doe facing each other. Usually the male moves towards the female, and she threatens him. If he persists, she may rise on her hind legs and "box" at him with her forelegs. He will likely then dash towards and around her while attempting to spray her with his urine. An unreceptive female will usually retreat at this point. A receptive female will stay, and what follows is a sometimes lengthy period (up to seven minutes) of jumps over each other, charges, chases, and attempted mounts, usually concluding with the female submitting and allowing copulation. This species, like all of the lagomorphs, is polygamous, meaning that both sexes will mate with more than one partner. The males provide no parental care.

VOCALIZATIONS

Three types of vocalizations from Eastern Cottontails have been noted: a loud, chilling scream given by a captured or wounded rabbit, a squeal produced by one of a copulating pair (which one is unknown), and a low growl emitted by a female defending her nest.

SIGNS

The front feet have five toes, although the first digit is reduced and does not always register on the track, and the rear feet have four toes. The feet are well furred throughout the year, especially in winter, but the digital pads often show in a good track. Cottontail tracks can sometimes be confused with squirrel tracks. Straight-ahead-pointing hind tracks may be seen. In squirrels the hind tracks toe out slightly from the centre line. Many animals will chew on bark during the winter, but only rabbits, hares, and pikas regularly over-chew the cambium layer and take some of the wood as well. This leaves a rough-looking mark that holds the incisor shape. Small twigs cut by cottontails, as well as by Snowshoe

Hares, display a clean-cut end at a 45° angle. Forms used by this rabbit are usually about 150–200 mm long by 100–150 mm wide and are tucked under overhanging vegetation. Dust bathing is commonly performed by this species, resulting in circular, bare, slightly depressed patches of earth. In more wooded areas, where there is little bare ground, cottontails will utilize wood dust from punky, long-dead trees as an alternative. Scat varies considerably in size from 5 to 11 mm and in colour from dark greenish to light golden brown depending on the diet. Cottontails tend to have rounder and smaller droppings than do hares. Eastern Cottontails, like Snowshoe Hares, create runways over time and usage; these have rounded edges and take on a tunnel-like appearance when in dense vegetation.

REFERENCES

Chapman et al. 1980; Elbroch 2003; Keith and Bloomer 1993; Litvaitis et al. 1992; Nagorsen 2005a; Olcott and Barry 2000; Wrigley 1979.

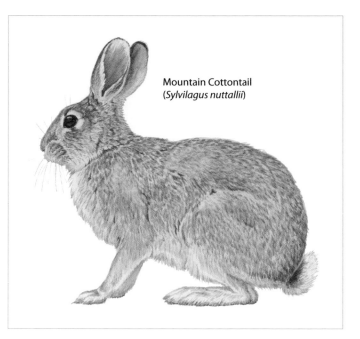

Mountain Cottontail
(*Sylvilagus nuttallii*)

Mountain Cottontail

also called Nuttall's Cottontail

FRENCH NAME: **Lapin de Nuttall**

SCIENTIFIC NAME: *Sylvilagus nuttallii*

This little western cottontail is famous for its very unrabbit-like habit of climbing trees, which it manages to do in spite of the lack of appropriate physical adaptations.

DESCRIPTION

Mountain Cottontails are the smallest rabbits in Canada. They are relatively short eared compared to the country's other rabbits and hares. The back is a grizzled brown in colour, and the rump and sides are distinctly grey. They have a pale rufous to brownish patch of fur at the nape. Throat, chest, and belly are creamy white to white. The tail is dark on the upper surface and white on the underside, but it is held up against the body, so mainly the white underside is visible. The tips of the ears are rounded and black. The dental formula is incisors 2/1, canines 0/0, premolars 3/2, and molars 3/3, for a total of 28 teeth. Mountain Cottontails moult twice annually, in spring and autumn; both pelages are similarly coloured, but the winter coat is thicker and longer.

SIMILAR SPECIES

Its small size, compact shape, and greyish rump serve to distinguish this species from the other lagomorphs in the same area – the Snowshoe Hare and the White-tailed Jackrabbit. These latter species turn white in winter, while the Mountain Cottontail retains the browns and greys of its summer pelage.

SIZE

The following measurements apply to animals of reproductive age from the Canadian part of the range. Females are about 4% larger than the males.

Total length: 338–430 mm; tail length: 33–69 mm; hind foot length: 88–110 mm; ear length: 54–65 mm.

Weight: 678–1500 g; newborns, unknown but, based on the weights of other cottontails, probably 20–30 g.

RANGE

The first reported Mountain Cottontail in Canada was discovered in 1909 in the Cypress Hills of southwestern Saskatchewan. Since then these rabbits have spread north and are found in most of southwestern Saskatchewan and southeastern Alberta. The first such rabbit in British Columbia was reported in 1939 near Osoyoos, at the south end of the Okanagan Valley. By 1993 they had moved up the Okanagan Valley to 18 km north of the town of Summerland on Okanagan Lake and up the Similkameen River to about 13 km west of the town of Keremeos. There is little suitable grassland habitat for the Mountain Cottontail elsewhere in British Columbia, so its range there is unlikely to expand further.

ABUNDANCE

Mountain Cottontails in British Columbia are on the provincial "Blue List" and are also classified by COSEWIC as being of "Special Concern." Therefore, hunting of this cottontail in the province is prohibited. Modern land use may be a threat to the ongoing health of the population in British Columbia. It is thought, however, that human agriculture may have assisted the species in advancing northward. In the Alberta and Saskatchewan part of the range, Mountain Cottontails can be locally common, but numbers fluctuate from year to year mainly as a result of dry weather. Dry summers result

Distribution of the Mountain Cottontail (*Sylvilagus nuttallii*)

expanded view of Canadian range

in less plant growth, which likely contributes to a shorter breeding season and poor survival of juveniles.

ECOLOGY

Rocky sagebrush-dominated grasslands with at least 30% vegetative cover are the preferred habitat of Mountain Cottontails in Canada. They favour rugged terrain such as ravines and coulees with brushy shrubs for cover and food and are rarely found in open, coverless prairie or cultivated fields. Old homesteads often provide suitable habitats. Population densities of 0.2 to 0.4 animals per ha have been reported in southern British Columbia. Home ranges are small (1–4 ha) and often overlap, especially during the fall and winter when the animals concentrate in areas with the best food and cover. This rabbit is considered to be crepuscular as most activity takes place in the second hour after sunset and during the hour before sunrise. It rarely ventures far from cover. Mountain Cottontails are known to survive in the wild for more than five years, but most die by predation, disease, starvation, or cold in their first year. Any mammalian or avian predator of sufficient size will take a Mountain Cottontail, including Canada Lynx, Bobcat, Red Fox, Coyote, and many of the larger raptors. The juveniles are especially vulnerable because their smaller size and often less wary attitude can make them easy prey.

DIET

Mountain Cottontails are strictly vegetarian and consume most plants available to them. They eat mainly grasses in the spring, add forbs during early summer, add juniper and sagebrush in mid-summer during the dry period if they are available, and then go back to mainly grasses in autumn after the rains. Sagebrush bark and twigs, along with dried grasses, are winter staples. Like all of our lagomorphs, Mountain Cottontails produce two kinds of faecal pellets. The large, soft, greenish ones are made up of food that has gone through the digestive tract once and spent time in the caecum, a sac in which some of the cellulose is broken down by bacterial action. These soft pellets are reingested so the animal can absorb more minerals and nutrients made available by the cellulose digesting bacteria. Food that has passed through the system at least twice forms into the familiar, round droppings that are deposited one at a time.

REPRODUCTION

Like other cottontails, Mountain Cottontails are polygamous; bucks and does mate with multiple partners. The breeding season begins in March in British Columbia and about a month later on the prairies. All of the does come into oestrus at about the same time in the spring, so all of the litters are born synchronously. Conception usually recurs almost immediately after parturition, and the females in any area tend to maintain this synchronicity of birth throughout the breeding season. Pregnancy is short, about 26–28 days, and two to seven kittens (usually four to five) are born into a nest made by the female. She digs a cup-like hole and lines it with dried grasses and then her own fur. She also makes a lid out of similar materials and small sticks, which she leaves on top. The kittens are born in an altricial state: blind, deaf, and more or less hairless. They grow quickly and are weaned after about 15 days. In British Columbia two or three litters are born each year, depending mainly on rainfall and hence vegetation growth. In the Alberta-Saskatchewan population the annual average is two litters, with three occasionally in a good year. Further south, where the breeding season is longer, some of the female juveniles from the early litter mature sufficiently to produce a litter in their first year. This is unlikely in Canada, given the short length of the breeding season.

BEHAVIOUR

Mountain Cottontails are relatively solitary and mostly nocturnal. Rarely is more than a single one seen, except during the breeding season or in areas supplying abundant food and cover. They are not reported to dig burrows but will readily utilize natural crevices in rocky areas or appropriately sized abandoned burrows dug by other

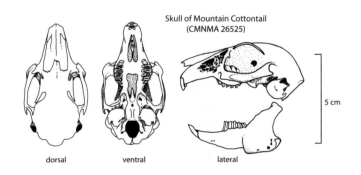

Skull of Mountain Cottontail
(CMNMA 26525)

dorsal   ventral   lateral

5 cm

species. Like Canada's other rabbits, Mountain Cottontails scratch out a shallow form under cover of vegetation, in which to rest during the day. If cover is sparse, they rest in burrow entrances or rock crevices that provide a suitable view, allowing them to maintain a predator watch. If disturbed, they will bound away for several metres (5–15 m) before darting into cover. Once hidden, they pause with ears erect, alert for further pursuit. They are reluctant to travel too far from their home range. Mountain Cottontails are known for their tree-climbing ability, but they rarely climb in Canada. In the more southerly portions of the range, where it is arid and Western Juniper is common, they will climb as high as 3 m into the branches of the juniper in search of drops of dew or succulent buds and leaves to supplement their water intake. This tree species is only found in the arid centre of British Columbia. Mountain Cottontails follow the general rabbit courtship behaviours of bucks facing does, bucks dashing towards does, and bucks and does jumping over and around each other, followed by the reproductive chase and then copulation. The majority of these incidents take place at night, although bucks have been seen quietly attending does during the day; these males are presumably waiting for the doe to become receptive.

VOCALIZATIONS

Mountain Cottontail vocalizations are unknown.

SIGNS

Small forms with overhanging vegetation in a sagebrush setting are likely made by Mountain Cottontails. The tracks are very similar to those of the Eastern Cottontail (see the "Signs" section of the latter for a track graphic that applies to both species). The Mountain Cottontail has five toes on each front foot. The first toe (on the inside of the foot) is greatly reduced but is clawed and sometimes registers in a track. Each hind foot has four toes. Scat is the familiar small, round pellets that are variable in colour from greenish to yellowish to brown, depending on the forage. There is considerable size variation in these droppings. Some are large enough to be confused with those of the two resident hares (Snowshoe Hare and White-tailed Jackrabbit). The smaller ones frequently fall below the size variation produced by these two hares. Generally, if they are less than 1 cm in width, they are from the Mountain Cottontail. Twigs nipped by Mountain Cottontails, like other lagomorphs, display a classic, cleanly cut, 45° angle. When they chew bark, they bite into the wood beyond the bark and leave a typically ragged wood surface with many roughened chew marks.

REFERENCES

Carter et al. 1993; Chapman 1975; Chapman and Flux 1990; Sullivan et al. 1989; Verts and Gehman 1991; Verts et al. 1984.

Order Soricomorpha

shrews, moles

Shrew (*Sorex* sp.)

Photo: Daniel Cadieux

This order comprises 4 families, 45 genera, and 429 species and has members on all continents except Australia and Antarctica. In Canada there are representatives of two families, the Soricidae (shrews) and the Talpidae (moles), with a total of 25 species.

FOSSIL RECORD

Their ancestors date back to the late Cretaceous in North America and Asia, about 90 million years ago.

ANATOMY

Many physical characteristics of the order Soricomorpha, such as a generalized body shape, a shared exit in some species for the genital and urinary tracts, the lack of a true scrotum, a small and smooth cranial cavity, and the long toothy snout, resemble characteristics ascribed to early mammals. So shrews and moles are considered to be primitive in many ways. Despite these primitive traits, however, some of the species show some highly specialized characteristics. The neurotoxic saliva of the Northern Short-tailed Shrew and the spade-like front feet of the moles are two examples. Shrew teeth are anything but primitive. Incisors are adapted into pincer-like grasping tools, and the teeth between the incisors and the premolars are modified to such an extent that they are no longer differentiated, becoming a graduated series of similar-looking simple teeth called unicuspids. It is not clear how these unicuspids relate to the normal dental formula of incisors, canines, and premolars. Like most mammals, shrews and moles grow a preliminary set of milk teeth before they develop their full adult dentition, but most members of this order lose these deciduous teeth in utero, and their permanent teeth emerge before they are weaned. The Shrew-mole is the only Canadian exception; it is born with a set of milk teeth that are similar to the adult or permanent teeth and which are replaced by permanent teeth around weaning. All North American shrews have a dark rusty-brown stain on the cusp of all or most of their teeth. This iron inclusion in the enamel hardens it and prevents premature wear, an important consideration for an animal that chews on hard insect chitin. None of the moles display this characteristic, likely because they have a softer diet. The fur is soft and often shiny.

ECOLOGY

Canadian representatives of this order are terrestrial, fossorial, or semiaquatic and are most active in darkness. They are small-bodied with minute eyes, small or hardly noticeable ears, five digits on each foot, and short soft fur. They often have a bad taste, which sometimes determines the predators that will eat them. For example, shrews and moles are often taken by owls, which have little or no sense of smell or taste, but less commonly by canines or felines, which have more discriminating palates. These latter species may catch and kill shrews and moles but normally only consume them if they are starving. These small mammals are often quite common but are usually so secretive that they are rarely seen. In captivity, shrews and moles are difficult to maintain for any length of time, and they will rarely breed; the Least Shrew is a notable exception.

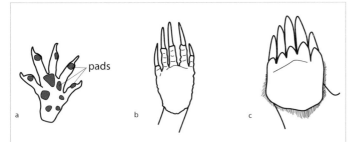

Soricomorph front feet (view of sole)
(a) generalized foot of shrew
(b) foot of a Shrew-mole which is partially adapted for digging
(c) foot of a Townsend's Mole which is highly adapted for digging

FAMILY SORICIDAE
shrews

Shrews are the size of mice or smaller. Their skulls lack zygomatic arches (the arc of bone that forms the outer edge of the eye socket). All shrews in North America have rusty stains on the cusps of most teeth, especially incisors and unicuspids. Shrews maintain an exceedingly high metabolic rate. Their heart rates can reach 1200 beats per minute, and respirations 750 per minute, when active.

Northern Short-tailed Shrew (*Blarina brevicauda*)

Northern Short-tailed Shrew
also called Short-tailed Shrew

FRENCH NAME: **Grande musaraigne**
SCIENTIFIC NAME: *Blarina brevicauda*

The mouse-sized Northern Short-tailed Shrew is the largest of the North American shrews. It has the distinction of having a venomous saliva with which it can paralyse its prey. Using the advantages of this neurotoxin, the Northern Short-tailed Shrew can subdue mice and invertebrate prey that are equal in size or even larger than itself. All three species of shrews in the North American genus *Blarina*, in company with the two European shrews in the genus *Neomys*, are the only shrews known to produce venomous salivary gland proteins.

DESCRIPTION

Northern Short-tailed Shrews have a slate-grey to blackish dorsal pelage, often suffused with a brownish wash that is caused by coloured tips on the hairs. The underside may be slightly paler than the back and sides. The coat is dense, short, and soft, appearing plush and velvety in texture. The snout is long and pointed but tapers more quickly than does that of other, smaller shrews. Northern Short-tailed Shrews have minute eyes, short ears that are totally hidden by their fur, a thick body, and short legs. Their short tail is well furred and faintly bicoloured. Their feet are relatively large and are thinly covered with buffy brown hairs, through which the pink skin is evident. These shrews appear somewhat mole-like because of their large size, the apparent lack of ear pinnae, and their extensive burrowing habits. This species has six mammae. Adults possess three scent glands, one on each flank and another on the chest or abdomen, which become larger on males during the breeding

season and smaller on females during pregnancy or lactation. (For more information see the "Behaviour" section below.) The teeth, skull, and skeleton are decidedly more robust than those of other North American shrews. Their eyesight and odour detection are thought to be rudimentary, and these shrews likely navigate and find prey using a combination of tactile vibrissae (whiskers), hearing, and echolocation. The dental formula is incisors 1/1, unicuspids 5/1, premolars 1/1, and molars 3/3, for a total of 32 teeth. The tips of all the teeth are stained a dark reddish-brown, caused by iron in the enamel. The iron is thought to strengthen the area and make the tooth more resistant to the abrasion and wear caused by grit and hard insect exoskeletons. The second upper unicuspid is the largest of the five, and the first unicuspid is the next largest. The third and fourth upper unicuspids are reduced, and the fifth is even smaller and is tucked into the premolar, rendering it invisible from the outside. Northern Short-tailed Shrews experience two annual moults. The spring moult begins in February and results in a somewhat paler, thinner, and shorter summer coat. The autumn moult takes place in October and November. It is impossible to externally determine the sex of individuals unless the animal is either a breeding male with well-developed side glands or a lactating female with noticeable nipples. Otherwise, gender determination can only be resolved by an internal examination.

Their poison is produced in modified salivary glands in the lower jaw and is presumably delivered along a shallow groove on the outside of each lower incisor. It is not injected but rather "chewed" into a wound. Introduced in sufficient quantities, it can cause respiratory failure in small mammals and paralysis in invertebrates. The toxin is thought to be mostly used to paralyse invertebrates so that they will remain alive longer in a food cache. A paralysed mealworm can live up to fifteen days. Northern Short-tailed Shrews will readily bite a human, but few of these bites break the skin. Effects from bites that do break the skin are variable, ranging from no effect at all to burning pain and sometimes swelling at the site that may last for several days. It is not known whether the toxin resides in the saliva at all times or, as in the case of a venomous snake, it is secreted at the time of the bite.

SIMILAR SPECIES

Their large size and more robust skull distinguish Northern Short-tailed Shrews from other North American shrews found in the same area. North American Water Shrews are almost as large as Northern

Short-tailed Shrews but have paler belly fur, a longer snout, and a considerably longer, distinctly bicoloured tail. While mole-like in appearance, Northern Short-tailed Shrews lack the large claws and paddle-shaped front feet that are characteristic of moles. Appendix 1 provides an illustrated identification guide to the skulls of Canadian shrews.

SIZE

The following measurements apply to animals of reproductive age in the Canadian portion of the range. Males and females are similar in size and appearance.
Total length: 90–145 mm; tail length: 17–35 mm; hind foot length: 13.5–18.0 mm.
Weight: 16.0–28.6 g; newborns, 0.8–1.0 g.

RANGE

The Northern Short-tailed Shrew is an eastern to central North American species that occurs from southern Saskatchewan to James Bay and the Maritimes, southward to western Nebraska, and along the Appalachian Mountains to central Georgia. Disjunct populations exist in southern Georgia and adjacent Alabama, and in southeastern Virginia and adjacent North Carolina.

ABUNDANCE

These shrews are generally common throughout their range, but like most small mammals they are careful to stay out of sight of their predators and so are rarely seen. Population density varies seasonally; it is highest in mid-summer as the juveniles are recruited into the population, and lowest in late winter and early spring after the winter attrition. Numbers within some, but not all, populations can also fluctuate dramatically from one year to the next. Population density estimates range from 0.75 to 121.00 per ha in suitable habitat, and moist locations support higher densities than do drier locations. Population crashes are rare, but have been reported, and recovery may take years.

ECOLOGY

This adaptable shrew can successfully occupy a wide variety of habitats from dense, swampy forests, bogs, and salt marshes to open, cultivated grasslands. In the east, they are most abundant in deciduous or mixed woodlands with thick leaf litter and profuse deadfall. In the west, they are most plentiful in tall, dense grasslands, which provide good cover. Northern Short-tailed Shrews readily use tunnels and runways created by other small mammals, but they will also dig their own tunnels through surface debris, leaf litter, and soil, digging as deep as 55 cm below the ground level. They can tunnel through loose soil at a rate of about 2.5 cm per minute but require frequent nap breaks. Digging is done with the front feet, and excess soil is kicked backwards with the hind feet. If necessary, the shrew will do a sideways somersault to reverse direction and push the loose soil along the tunnel with its snout, like a bulldozer. Most tunnels and runways run parallel to the ground surface within the top 10 cm, and these can honeycomb an area. Home range size varies with habitat and population density from about 0.2 to 0.8 per ha.

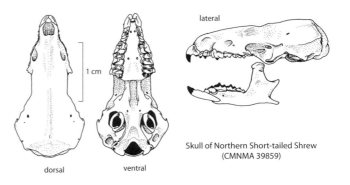

Skull of Northern Short-tailed Shrew
(CMNMA 39859)

Males occupy larger home ranges than do females and spend more time travelling around looking for prospective mates. Overlap of home ranges is common. Although they are active intermittently throughout a 24-hour period, most activity occurs at night or in early morning before full daylight. As with other North American shrews, Northern Short-tailed Shrews have a high metabolism and must eat frequently. The average activity phase is composed of about a five-minute foraging period followed by a twenty-five-minute rest. These shrews do not hibernate and are active under the snow throughout the winter, although they will often limit their activities to save energy, relying on stored food rather than foraging. Nests are typically constructed of locally available plant leaves and stems, and sometimes mammal hair, worked into a globular mass with a hollow centre. These are usually located under stumps, rocks, and fallen logs for protection and may be placed in underground burrows. Those built by breeding females are larger than the more common resting nests. Although these shrews spend most of their time on or below the surface, they are agile enough to climb rocks and trees and are capable of swimming. By late October and early November, population turnover is almost complete; most of the breeding adults that lived through the previous winter are dead, and only the young of the year remain. The maximum lifespan in captivity is 33 months; in the wild, very few live more than 14–17 months, and, depending on location, only 6%–11% of animals survive more than 12 months. Primary predators include owls, hawks, and snakes, and occasionally other predatory birds such as herons, shrikes, and grackles will take short-tailed shrews. Mammalian predators including Domestic Cats, weasels, skunks, Bobcats, foxes, and American Badgers kill many of these shrews but commonly discard the bodies, presumably because of their distasteful flavour.

DIET

Earthworms are the preferred prey of this shrew, to which are added a wide variety of invertebrates such as snails, slugs, insects (both adult and larval), spiders, sow bugs, and centipedes. Small vertebrates such as mice, shrews, small snakes, frogs, hatchling turtles, and nestling birds may also be included in the diet when they can be captured. Small mammals (mainly mice and other species of shrews) may be significant food sources during cold weather when invertebrates become scarce. Some plant material such as seeds and fungi will readily be eaten, especially in winter when other food is harder to find. Their penchant for sunflower seeds and cracked corn is well

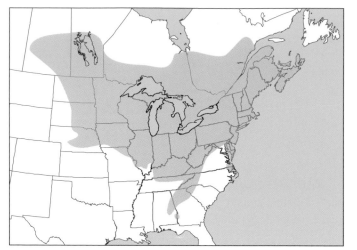

known to many bird-feeder providers. Carrion will be readily added to their diet. Unlike many other shrew species, Northern Short-tailed Shrews are apparently not cannibalistic. Individuals eat at least their body mass in food daily. Considerable free water is drunk every day.

REPRODUCTION

The breeding season of the Northern Short-tailed Shrew begins as early as February when the males' testes begin to enlarge, but the females do not start breeding until March; the first litters arrive in early April. Ovulation is stimulated by copulation that is prolonged. A pair may remain locked for up to 25 minutes, but most locks last around four to five minutes. During a lock the male dismounts. If the female then begins to walk around, she tows the passive male backwards. Multiple matings typically occur, and a female will remain in heat for two to four days. Females may have up to four litters annually and will continue to produce young until late September. Gestation lasts 21–22 days. The young are born in a bulky nest made by their mother (see the "Ecology" and "Signs" sections for more nest information). Litter size averages four to six (ranging from three to ten). Young are born in a highly altricial state. Neonates are pink, hairless, and toothless, with eyes and ears closed. They are about the size of a honeybee. They develop rapidly and are about half of their adult weight by the time they are a month old. Weaning occurs when they are around 25 days old, shortly after their eyes finally open. Females do not experience a postpartum oestrus. Instead, their next heat is delayed until the young are weaned. If she loses her litter before it is weaned, she may soon cycle again. Once the young are weaned, her interactions with them cease. Some juvenile females can become sexually mature by two months old, and males by three months old, under ideal conditions. Although some early-born young may breed in the year of their birth, most do not start breeding until the following spring.

BEHAVIOUR

Most adult short-tailed shrews are solitary, but the degree of social intolerance varies. Males and older animals are generally less sociable than are females and younger individuals. Males strongly curb their aggression when courting females, from whom they will passively accept even physical attacks. Most intraspecific behaviour is related to confrontation, attack, combat, or retreat from combat. Threat displays and aggressive calls play a large role in their interactions. They commonly bite or threaten to bite each other during confrontations but appear rarely to suffer from the effects of the toxic saliva. Northern Short-tailed Shrews are reported to occasionally attack animals that are much larger than themselves such as rabbits, young Snowshoe Hares, and snakes. One was observed tunnelling and jumping through snow to repeatedly bite the belly and legs of a large, very agitated, hunting dog that had unwittingly disturbed it. They will readily bite an incautious human finger. The secretions from their various glands are likely used to release a musky odour when the animals are agitated or courting and possibly to strongly flavour their flesh to make it less palatable to most mammalian predators. They do not appear to be territorial and do not seem to use their musk to mark their home ranges. The female retrieving her young will, depending on the size and mobility of the infant, either carry it in her mouth or allow it to scurry behind her while clinging to her rump fur or tail. This latter behaviour is called caravanning, but, unlike some shrews that will travel with their whole litter in this manner, Northern Short-tailed Shrews restrict this activity to a single young at a time. One of the important winter survival strategies of Northern Short-tailed Shrews is the creation of food caches for later consumption. Paralysed arthropod prey are hoarded in small underground chambers that radiate from major burrows, or sometimes in a nest. Caches of sunflower seeds and other seeds may also be gathered. Most hoarding takes place in autumn and winter, but will also occur whenever there is an overabundance of food beyond the shrew's immediate need.

VOCALIZATIONS

Northern Short-tailed Shrews are fairly vocal animals. Two aggressive individuals, upon meeting, may shriek loudly at each other, and if one, or both, does not retreat, the two may begin to produce shrill, chattering calls that begin with a loud high-pitched squeal and end with a series of rapid, short, buzzing notes. Another sound produced during aggressive encounters is a loud "chirp." The chirp call and the previously mentioned loud "buzz" call are both emitted with an open mouth and sometimes accompanied by stamping front feet. Breeding males produce a clicking call as they pursue a female. A receptive female will also click to attract a male. The same clicking call is used by distressed young and by adults exploring in strange surroundings. Click calls along with "put" and "twitter" calls are low-volume vocalizations, primarily used as contact calls and to advertise presence to nearby shrews. Ultrasonic frequencies in the 30–55 kHz range are used to assist navigation, especially in unfamiliar surroundings.

SIGNS

When tunnelling just below the ground surface, Northern Short-tailed Shrews can create pushed-up ridges similar to those of moles but smaller. Burrow entrances and tunnels are 2.5–3.2 cm in diameter and characteristically oval in cross-section, being wider

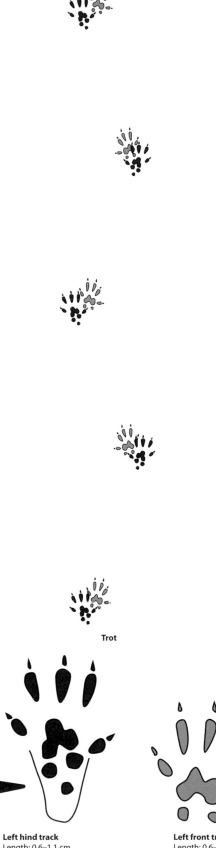

than they are high. Nests range from 12 to 15 cm in diameter. Scat rarely fouls a nest, as these shrews normally deposit it along runways, at burrow entrances, or just outside the nest entrance. Scat is variable in shape and size depending on diet. A diet of earthworms and slugs results in such soft faeces that they are more of a squirt than a scat. Harder foods result in droppings that are 0.2–0.5 cm in diameter and 0.6–1.7 cm in length. Scat may be twisted or straight and have one blunt and one tapered or pointed end, two blunt ends, or two tapered ends. Front feet have five toes, with the three middle toes aiming ahead and the terminal toe pads in a curved line, as the third toe is longer than the second and fourth toes. The first and fifth toes are angled to each side. The hind feet also have five toes arranged in a similar pattern, but the three middle toes are equal in length, so the toe pads at their tips form a straight line. All toes are clawed, and claws register fairly reliably in the track. A typical trotting gait leaves a splay-footed trail because the impressions from the front feet tend to toe in, while those from the hind feet tend to toe out. Tracks of Short-tailed Shrew differ from those of rodents by the fact that the shrews have five toes on their front feet and the rodents have four. They differ from other shrews (genus *Sorex*) mainly by size, although some of the larger shrews such as Water Shrews produce tracks that are almost as large as those of Short-tailed Shrews.

REFERENCES

Antipas et al. 1990; Benedict, R.A., 1999; Brant and Orti 2001, 2003; Dobbyn 1994; Elbroch 2003; George, S.B., 1999d; George, S.B., et al. 1984; Getz 1994; Getz et al. 1992, 2004; Long, C.A., 2008; Merritt and Adamerovich 1991; Olsen 1969; Reid, F.A., 2006; Sparks and Choate 1995; Standing et al. 2000; van Zyll de Jong 1983a; Whitaker and Hamilton 1998.

Trot

furry heel that may register

Left hind track
Length: 0.6–1.1 cm
Width: 0.5–0.9 cm

Left front track
Length: 0.6–1.0 cm
Width: 0.6–0.9 cm

Northern Short-tailed Shrew

Least Shrew

FRENCH NAME: **Petite musaraigne**
SCIENTIFIC NAME: *Cryptotis parvus*

Least Shrews are very rare in Canada and only occur along the northern shore of Lake Erie on the Long Point peninsula. Like other shrews, they usually move about in underground tunnels or under cover of vegetation and so are rarely seen.

DESCRIPTION

These shrews are sepia brown in colour on their back and sides in summer and brownish-grey to slate-grey in winter. Their belly is paler, usually silvery grey. Like other North American shrews, they have small eyes, a long body, short legs, and a long pointed snout. Their very short ears are completely hidden by fur. Least Shrews have exceptionally short tails, which are bicoloured, brown above

Least Shrew
(*Cryptotis parvus*)

and buffy white below, with a tiny tuft of brown hairs at the tip. Their feet are white to light grey brown. Their eyesight is thought to be weak. The dental formula is incisors 1/1, unicuspids 4/1, premolars 1/1, and molars 3/3, for a total of 30 teeth, which are two fewer than those of any other North American shrew. Least Shrews have three readily apparent upper unicuspids, which form a gradually declining series from the largest first to the smallest third. The tiny, fourth upper unicuspid is barely visible from an outside view because it is at least partially, if not wholly, tucked in behind the premolar. Least Shrews experience two annual moults, in spring and autumn.

SIMILAR SPECIES

Northern Short-tailed Shrews are similar to Least Shrews but are much larger, have less clearly bicoloured tails, and have five upper unicuspids. Other shrews that occur within the very small Canadian range of the Least Shrew have conspicuously larger ears and much longer tails. They also all have five, clearly apparent, upper unicuspids. Appendix 1 provides an illustrated identification guide to Canadian shrew skulls.

SIZE

As there are so few Canadian specimens, the following measurements apply to animals of reproductive age throughout the range. Males and females are similar in size and appearance.
Total length: 75–92 mm; tail length: 13–23 mm; hind foot length: 10.0–12.5 mm.
Weight: 4.4–7.5 g; newborns, 0.3–0.4 g.

RANGE

Least Shrews occur in eastern, central, and southern United States and in Mexico and Central America to western Panama. The very localized Canadian portion of the distribution is restricted to the peninsula of Long Point, Ontario, along the northern Lake Erie shore. Specimens were collected there in 1927, 1938, and 1941. An additional specimen has been reported near Hamilton, Ontario. No Least Shrews have been collected in Wisconsin since 1944, and the species may be extirpated in that state.

ABUNDANCE

Populations of this shrew may fluctuate to the extent that in some years none may be found, while in other years they may be locally abundant. While relatively common in southern parts of their range,

this shrew is considered to be very rare in Canada, if a population still exists. A large part of its Canadian range is within federally or regionally protected habitat. It is possible that more Least Shrews exist in grassy, dune habitat along northern Lake Erie, but this habitat is becoming increasingly rare owing to urbanization. It is also possible that this shrew is extirpated from Canada, as an extensive and specific search in 1986 failed to find any.

ECOLOGY

The biology of this shrew is poorly understood. Least Shrews generally live in dry, grassy areas and weedy abandoned or fallow fields, but they have also been found occasionally in marshes (both fresh and salt) and in woodlands. Owing to their high metabolic requirements, a characteristic of all shrews, Least Shrews must eat often. Their normal activity pattern is to forage for a short time (typically a few minutes) and then take a prolonged rest (typically around thirty minutes) to sleep and digest their meal. They may be active at any time of the day and night but are most active at night, and they remain active throughout the winter. Their hollow, spherical nests are constructed of plant stems and shredded leaves and usually placed in a protected site such as under a stump, a fallen log,

Distribution of the Least Shrew (*Cryptotis parvus*)

⭐ Canadian range

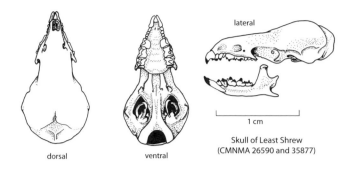

dorsal　　　ventral

lateral

1 cm

Skull of Least Shrew
(CMNMA 26590 and 35877)

or a rock or in a shallow, underground chamber. Nests are about 10–17 cm in diameter. Least Shrews often use runways and tunnel systems created by other small mammals but will also create their own runways and excavate tunnels for themselves in leaf litter or loose soil. Their burrows are seldom deeper than 20 cm beneath the ground surface. There is scant data available on home range size, but one female was estimated to occupy 0.23 ha, and a male 0.17 ha. They are willing and capable swimmers. Owls are thought to be their primary predators, although hawks, snakes, and mammals such as foxes and Domestic Dogs and Cats are known to kill them when the opportunity presents itself. Mammalian predators will kill but rarely eat these small shrews because of their pungent odour and subsequently their distasteful flavour. The maximum lifespan of wild Least Shrews is unknown, but captives can live 21 months. In very artificial conditions in a laboratory the maximum running speed of this shrew was measured at around 15 km/h. While this cannot, in any way, be equated to the speed at which Least Shrews can run in the cluttered environment of their own runways and tunnels, it still gives some indication of the velocity that their short legs can produce.

DIET

Like most other shrews, Least Shrews consume a wide variety of small invertebrates, such as insects, earthworms, slugs, spiders, centipedes, snails, sow bugs, and millipedes. The bulk of their insect prey is composed of crickets, grasshoppers, beetle larvae and adults, and moth and butterfly caterpillars. As captives, Least Shrews will eat their weight, or more, in food daily; in the wild, where they are more active, they very likely exceed this amount. Captives will become cannibalistic if food runs short. Free water is drunk frequently.

REPRODUCTION

The breeding season extends from around March to November but may last all year in some southerly portions of the range. Females are receptive to mating (in heat) for about 24 hours. A short copulatory lock of a few seconds is common, but a lock may persist for up to 40 minutes, during which the male may be dragged around backwards behind the female. Gestation takes 21–23 days, and several litters are likely produced annually. Females experience a postpartum oestrus and may breed again within one to four days of parturition. Litter size averages five (ranging from two to eight).

Both parents may provide parental care. Neonates are toothless, blind, deaf, and virtually hairless (stubs of vibrissae are visible). Eyes open around their fourteenth day, and the shrews have a full, albeit short, hair coat by the time they are 20 days old. Before the young are fully weaned, their parents bring insects back to the nest for them to sample. They are weaned at around three weeks old and are adult length (but not as heavy as adults) at about 30 days old. Captive young are capable of breeding for the first time at as early as five to six weeks old, but age at sexual maturity in the wild is unknown.

BEHAVIOUR

Least Shrews are thought to be the most social of the North American shrews. At least 31 adults and subadults have been discovered in a single nest. Reports of multiple adults occupying a nest are common, and these same communal animals will also share tunnel and runway construction and use without apparent conflict. Pairs are thought to nest together and cooperate in rearing the young. Like most shrews, Least Shrews will cache excess food for future use. Although Least Shrews do produce the pungent odour so distinctive of shrews, the purpose of the glandular secretions remains to be discovered. A gland behind the ear of the female is apparently of great interest to males. This gland is thought to produce a secretion when the female is pregnant and may provide males with an indication of the status of the female, thereby allowing them to redirect their attention away from "unavailable" (already pregnant) females. A female rebuffing an ardent male will raise her head and one foreleg in an aggressive posture, and the male will crouch down or lie on his side or back in a submissive posture. Infanticide of her own litter by a stressed maternal female is known to occur in captivity.

VOCALIZATIONS

Adults and young produce "twitter," "put," and "click" calls as well as a "shriek" and a "buzz" call. All are fairly low volume, especially the twitter and put calls, and even the loudest, the shriek, is only audible by humans within 50 cm. This call has been likened to the distant call of a flicker. Receptive females will produce click calls to attract a mate. Least Shrews use echolocation to navigate.

SIGNS

Burrow openings are 13–20 mm high by 18–25 mm wide, and runways are about 8–15 mm wide. Faeces are not deposited inside the nest. Instead, latrine sites are located nearby or just outside the entrance. Like moles and short-tailed shrews, Least Shrews will create pushed-up ridges of loose soil as they tunnel just beneath the ground surface. These are considerably smaller than those produced by short-tailed shrews and very much smaller than those made by moles.

REFERENCES

Backlund 1995; Brown, L.N., 1997; Formanowicz et al. 1989; Long, C.A., 2008; Punzo and Chavez 2000; Reid, F.A., 2006; Schmidly 2004; van Zyll de Jong 1983a; Whitaker and Hamilton 1998.

Arctic Shrew

also called Black-backed Shrew, Saddle-backed Shrew

FRENCH NAME: **Musaraigne nordique**

SCIENTIFIC NAME: *Sorex arcticus*

Arctic Shrews are one of our more colourful native shrews. A disjunct maritime population, once thought to be a subspecies of the Arctic Shrew, has recently been elevated to full species status. See the Maritime Shrew account for more information on this new species.

DESCRIPTION

Adults are dark chocolate brown to blackish in colour on their backs; their sides are lighter brown, sometimes with a rusty tinge around the shoulders and neck; and their bellies are an even paler grey brown to whitish colour. The demarcation between each colour is sharply distinct. The tail is faintly bicoloured, darker above and paler below, and the upper parts of the feet are a dirty buffy brown. Juveniles are generally browner than adults and not as distinctly tricoloured. The dental formula is incisors 1/1, unicuspids 5/1, premolars 1/1, and molars 3/3, for a total of 32 teeth. There are five readily apparent upper unicuspids, and the third is larger than the fourth. Adults have two annual moults, one in June (which may continue until late summer) and the other from mid-September until mid-October. Juveniles moult into adult pelage in their first autumn and then retain the distinctive tricoloured pelage for the remainder of their lives.

SIMILAR SPECIES

Tundra Shrews are almost as colourful as are Arctic Shrews, and their pelage patterns are similar though they are generally browner overall with lighter bellies. Although the ranges of the two species are not known to overlap, it is possible that this is due to lack of sampling rather than actual distributional distinction, so the differences are herein noted. The saddle-back pelage of the Tundra Shrew is brown, rather than the very dark brown or blackish fur of the Arctic Shrew in the same region. Tundra Shrews share the unicuspid traits of the Arctic Shrews, but they have shorter tails (≤ 36 mm in length), and their skull length is shorter (≤ 18.4 mm). Dusky Shrews are slightly smaller, and where their range overlaps with Arctic Shrews, their pelage is mainly grey-brown above and buffy-white below. They show no sign of the dark, saddle-back

Arctic Shrew (*Sorex arcticus*)

Distribution of the Arctic Shrew (*Sorex arcticus*)

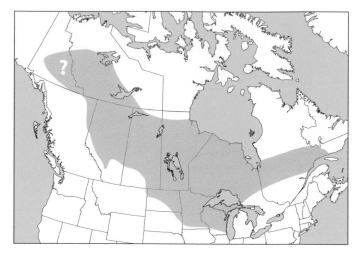

? range in this area uncertain

pelage pattern of the Arctic and Tundra Shrews. Cinereous Shrews can be somewhat tricoloured in some regions, but they are generally brown to dark brown above and greyish-white on the belly. They are smaller (weight ≤ 6 g). Their skull is smaller (length ≤ 17 mm), but they share the trait of the third unicuspid being larger than the fourth. See appendix 1 for figures and comments on shrew skull identification.

SIZE

The following measurements apply to animals of reproductive age from the Canadian part of the range. Males and females are similar in size and appearance.

Total length: 97–125 mm; tail length: 36–48 mm; hind foot length: 11–19 mm; ear length: 4–9 mm.
Weight: 5.1–13.0 g; newborns, unknown.

RANGE

This shrew occurs in northern North America, mainly in the boreal forest zone from eastern Quebec to southern Yukon and southward to southern Minnesota and Wisconsin. Populations are commonly localized in small patches of suitable habitat. As only one specimen exists from Yukon (from the western edge of the range map), there is some doubt as to the distribution of the species in that region.

ABUNDANCE

Abundance fluctuates both annually and seasonally. A normal seasonal population cycle starts with low numbers in the spring, which slowly increase until August or September as the young are recruited, followed by a major decline by late autumn and then a slow attrition over the winter. The generational period is about six months, and an estimated 80% of juveniles die before reaching sexual maturity. Mortality is especially high in the first month of life. Populations can also fluctuate annually, probably owing to weather

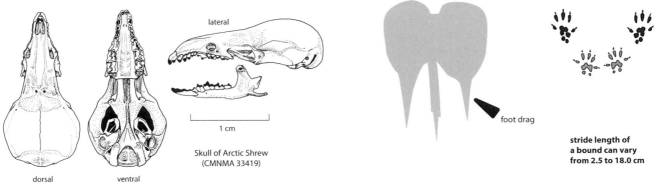

Skull of Arctic Shrew
(CMNMA 33419)

dorsal

ventral

lateral

1 cm

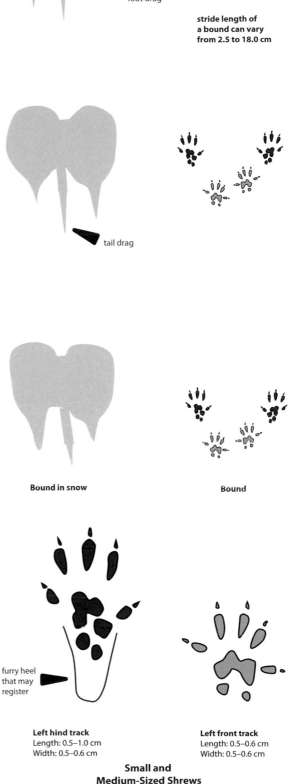

foot drag

stride length of
a bound can vary
from 2.5 to 18.0 cm

tail drag

Bound in snow

Bound

furry heel
that may
register

Left hind track
Length: 0.5–1.0 cm
Width: 0.5–0.6 cm

Left front track
Length: 0.5–0.6 cm
Width: 0.5–0.6 cm

**Small and
Medium-Sized Shrews**

conditions and food availability. These shrews are nowhere abundant even during population peaks. A long-term study in suitable habitat in Manitoba found that densities can vary from less than 1 per ha in low years to about 10 per ha in peak years.

ECOLOGY

Largely a species of the boreal forest region, these shrews favour moist (mesic) locations such as grass or sedge marshes, bogs, muskeg, damp meadows, soggy spruce forests, and shrubby alder or willow thickets or thick grass or sedge growth along the edges of watercourses and ponds. However, they may also be found in high numbers in dryer woodland-edge habitat, grassy meadows, and upland coniferous forests along streams and ponds. Arctic Shrews are active all year round and may be active at any time during a 24-hour period, although most activity takes place in darkness. Their normal routine is a short burst of activity (usually hunting), followed by a rest to digest. When resting, the shrew lies on its side or curls into a ball with its back up and its head tucked under its body. Home range size is around 0.6–0.7 ha and does not appear to vary greatly by sex or season. Dispersal movements of more than 0.6 km have been reported. Arctic Shrews dig using their front feet and use pushing movements of the head to move soil or litter. Burrows and nests remain undescribed, although it is almost certain that they are created and used. Maximum lifespan is around 18 months. Predators include most owls and sometimes hawks, weasels, and foxes. Most mammalian predators consider shrews to be fit only as starvation food, likely because of their foul taste and smell, and may leave the carcass uneaten.

DIET

Arctic Shrews are almost entirely insectivorous. Their diet varies seasonally and with availability. They eat about 8.2 grams of insects during a 24-hour period, approximately their own weight daily. Their primary diet is composed of sawfly pupa (in the cocoon phase), moth and butterfly caterpillars and adults, and grasshoppers, to which they will add earthworms, spiders, flies, and beetles as well as carrion. Non-nutritious insect parts, such as legs, wings, and wing coverings (elytra) are normally discarded. Arctic Shrews are major predators of the Larch Sawfly, a forest insect pest. A single shrew can consume an estimated 123 cocoons in a day. Arctic Shrews, like other shrews, need daily access to drinking water.

REPRODUCTION

Onset of the breeding season varies with geographical location. In the southern parts of the range (Wisconsin), breeding begins under the snow in late winter and extends until mid-summer. Farther north in Manitoba and British Columbia breeding begins in early May and extends to mid-August, while in Alberta it extends from late May to August. Farther north the season is likely even shorter. Litter size averages six to eight (ranging from five to eleven). A maximum of three litters may be produced annually. The length of the gestation period and the age at weaning are unknown. Like all shrews, the young are born altricial, with little hair and eyes sealed closed but with their adult teeth already erupted. The first young begin appearing on their own in June or July depending on the breeding season of the population. Juveniles reach sexual maturity at around four to six months old, and juveniles from early litters may themselves reproduce before the end of the breeding season in some populations. Breeding in their first year is rare in the southern parts of the range but is more common farther north.

BEHAVIOUR

Arctic Shrews are generally solitary, apart from mating pairs and females with dependent litters. Sometimes littermates may forage together for a time after weaning. Despite their small size, Arctic Shrews are formidable predators. They can detect their prey from a distance either visually or by sound, and they can pounce from several tens of centimetres away. They can also detect prey in darkness by touch. They hunt, not only on the ground but also in underground burrows and up in the vegetation. Two subadults have been reported to be hunting grasshoppers in sedges and grasses up to 31 cm above the ground. The animals saw the grasshoppers resting above on the vegetation and slowly climbed nearby plant stems, before jumping about 25–30 cm across onto the insects, grasping them with their feet and teeth. During 15 minutes of hunting the two shrews had pounced 37 times and killed and eaten 33 grasshoppers. Cool morning temperatures (6°C) undoubtedly enhanced their success, as the insects were rendered lethargic and slow to react while the shrews were unaffected. As the air temperature rose and the grasshoppers warmed up and began to make more effective escape manoeuvres, the shrews' success diminished and they moved on. This observation illustrates not only the complex hunting behaviour and visual ability of Arctic Shrews but also their voracious appetite. Arctic Shrews must fill their stomachs around 11 times in a 24-hour day in order to eat sufficiently to meet daily energy requirements. A shrew that has not eaten for three hours is starving and could die. Like many other shrews, Arctic Shrews regularly cache surplus food for later need. A food cache could mean the difference between life and death if hunting is unsuccessful, and it is probably also important on very wet days when foraging could result in soaked fur and hypothermia. Grooming is most evident following feeding and consists of several rapid swipes with the forepaws to the snout area. Water is lapped up in a manner similar to that of Domestic Dogs and Cats.

VOCALIZATIONS

Arctic Shrews are generally silent. Although echolocation ability has not yet been proven, these shrews emit an almost continuous low-volume, high-pitched, twittering squeak as they run along, which may be the audible portion of an echolocation pulse similar to that used for navigation by other shrew species.

SIGNS

Signs of all the smaller shrews are similar and often indistinguishable. Each front foot has five toes, four palm pads, and two heel pads. Each hind foot also has five toes, four palm pads and two heel pads. Not all of the pads register reliably in a track. Claws usually register. The middle toe (the third toe) on each front foot is longer than the others, so the front track is arched at the leading edge. The three middle toes on the hind feet (the second, third, and fourth toes) are roughly equal in length, so the hind track has a straight leading edge. Tracks of the largest shrews in Canada, Northern Short-tailed Shrews, are discussed separately in their own account. The smallest shrews, including Cinereous Shrews, Prairie Shrews, Barren Ground Shrews, Long-tailed Shrews, Trowbridge's Shrews, Least Shrews, and North American Pygmy Shrews, leave trails with short stride lengths (the bounding tracks are 2.5–9.0 cm apart) and narrow trail widths (1.9–2.5 cm). The medium-sized shrews, such as Arctic Shrews, Tundra Shrews, Maritime Shrews, Smoky Shrews, Water Shrews, Pacific Water Shrews, Dusky Shrews, and Vagrant Shrews, have longer stride lengths (the bounding tracks are 2.5–18.0 cm apart) and wider trail widths (2.2–3.3 cm). The tracks of shrews can usually be distinguished from those of rodents by their width (up to 3.3 cm) and by their five-toed front feet; rodent tracks show only four front toes on each foot because the first toe is highly reduced. Tail drag in soft snow is common, but snow tracks are rare as most winter activity takes place in tunnels under the snow. These small animals create small burrows in forest litter, moss, soil, and snow. The diameter of burrow entrances is less than 2.5 cm. Cinereous Shrew burrows are 1.6–1.7 cm in diameter. Burrows may have an accumulation of scat at their entrance either directly deposited there by the shrew or kicked out of the burrow during housekeeping. Shrews commonly use runways and burrows of other species. Shrew runways are 1.6–3.8 cm in width. Scat appearance varies with diet: shrews eating entirely soft-bodied prey, such as earthworms or slugs, produce a liquid squirt of faecal matter, but those eating hard-bodied insects produce small, tubular, black, sometimes twisted, scat with tapered ends that may contain undigested, often shiny bits of insect chitin. Scat is typically 0.2–0.3 cm in diameter and 0.5–0.6 cm in length.

REFERENCES

Buckner 1964, 1966, 1970; Desrosiers et al. 2002; Dobbyn 1994; Elbroch 2003; Jannett and Huber 1994; Kirkland 1999a; Kirkland and Schmidt 1996; Nagorsen 1996; Perry, N.D., et al. 2004; Prescott and Richard 2004; Salt 2005; Smith, H.C., 1993; Soper 1961; van Zyll de Jong 1983a.

Pacific Water Shrew

also called Marsh Shrew, Bendire Shrew

FRENCH NAME: **Musaraigne de Bendire**

SCIENTIFIC NAME: *Sorex bendirii*

The Pacific Water Shrew is the largest shrew in British Columbia and is the largest of the shrews in the genus *Sorex* in North America. It is similar in size to a medium-sized mouse, such as a House Mouse. The Pacific Water Shrew and the North American Water Shrew are similar, both physically and genetically, and are considered to be closely related sister species.

DESCRIPTION

Pacific Water Shrews are very simply coloured. Their back and sides are dark brown or blackish, and their belly is, at best, only slightly lighter and often the same colour as the upper surfaces. Fur is dense and plush with no protruding guard hairs. Their long tails are uniformly dark brown above and below. The upper surfaces of the feet are dark brown, and the hind feet are surrounded by a sparse fringe of short, stiff hairs (at most, 1 mm in length); these are commonly worn off in older individuals. This fringe enhances the surface area of the foot and contributes to the shrews' swimming capabilities. Their vision, at least above water, is poor. The dental formula is incisors 1/1, unicuspids 5/1, premolars 1/1, and molars 3/3, for a total of 32 teeth. The third unicuspid is slightly, or distinctly, smaller than the fourth. The upper incisors have a large medial tine that is positioned wholly within the pigmented zone. These skull characteristics are fully illustrated in appendix 1. Adult Pacific Water Shrews experience two annual moults, one in spring and the other in autumn. Summer pelage is browner than winter pelage.

SIMILAR SPECIES

Generally, their large size (> 130 mm in total length), uniformly dark pelage, and fringe of hairs on the hind feet distinguish this shrew. North American Water Shrews are similar to Pacific Water Shrews and have similar lifestyles and adaptations. The fringe on the feet of North American Water Shrews is longer (about 1.25 mm) and does not disappear with age. Most North American Water Shrews have paler, whitish to light-brownish bellies (especially in winter) and a distinctly bicoloured tail. Although these characteristics are useful if the specimen is in the hand, reliable identification of a live individual that is rapidly moving around in the wild is difficult, especially as it is uncommon to see the underside of such an animal. Appendix 1 provides an illustrated identification guide to Canadian shrew skulls.

SIZE

The following measurements apply to animals of reproductive age from the Canadian part of the range. Males and females are similar in size and appearance.

Total length: 137–176 mm; tail length: 61–81 mm; hind foot length: 17–21 mm; ear length: 7–9 mm.

Weight: 10.0–17.2 g; newborns, unknown.

RANGE

Pacific Water Shrews are limited to coastal, western North America where they reach the northern limits of their distribution in extreme southwestern British Columbia. They extend as far north as the north side of the lower Fraser Valley and eastwards to the Chilliwack River and Harrison Lake. They range through coastal Washington and Oregon into northern California, where their southern distributional limit occurs north of San Francisco. British Columbia populations on the southern side of the Fraser River are highly fragmented due to urbanization.

Distribution of the Pacific Water Shrew (*Sorex bendirii*)

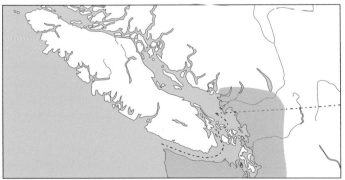

expanded view of Canadian range

Pacific Water Shrew
(*Sorex bendirii*)

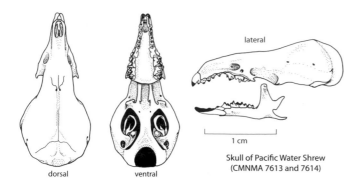

dorsal   ventral   lateral

1 cm

Skull of Pacific Water Shrew
(CMNMA 7613 and 7614)

ABUNDANCE

This water shrew is restricted to riparian areas in the Pacific Northwest where it is generally uncommon and localized in suitable habitat. This habitat is under threat in British Columbia from urban and agricultural development, which prompted COSEWIC (in 2005) to enhance the status of Pacific Water Shrews from "Threatened" to "Endangered." Provincially, this shrew is designated as "At Risk" under the British Columbia Forest and Range Practices Act, but this status provides only minimal protection from habitat destruction. Pacific Water Shrews are rare in Canada.

ECOLOGY

The ecology of Pacific Water Shrews is poorly understood. In general, they live in moist forests, alder thickets, grasses, sedges, or other wetland vegetation alongside ponds, marshes, and slow-moving streams. In British Columbia their ideal habitat is thought to be in lowland mature forests (mixed, coniferous, or deciduous) within 60 m of wetlands, with an upper elevational limit of around 850 m. Other habitats in British Columbia in which specimens have been collected include grassy areas alongside ditches and sloughs, and in beach debris near a freshwater spring. Sometimes individuals have been captured some distance from water. This is most common during the winter rainy period when animals may be found as far as one kilometre from their traditional habitats. Many of these shrews are young of the year that may be taking advantage of the wet conditions to disperse. Two breeding nests have been reported that can reliably be ascribed to the Pacific Water Shrew. One was constructed of shredded bark and was located, in late March, under some loose bark of a fallen Douglas fir that spanned a small stream; it was occupied by five juveniles. The second was located, in mid-June near Surrey, British Columbia, in a cavity in a fallen Red Alder snag that was lying on the ground near a creek. The hollow, spherical nest was composed of dried grasses and mosses, interwoven with strips of soft Paper Birch bark. Six partly furred juveniles were found within. The Pacific Water Shrew is a good swimmer, both at the surface and under water, and commonly hunts aquatic prey. Its large, semi-fringed feet allow it to run over the surface of the water for several seconds before it loses speed and must dive below. The maximum lifespan is unknown. Predators include owls, Domestic Cats, fish, and probably also Giant Pacific Salamanders. Like other species in the genus *Sorex,* Pacific Water Shrews have a pungent odour that makes their flesh distasteful to most predators.

DIET

The known prey of Pacific Water Shrews is all invertebrate, although it is possible that they may occasionally take some vertebrates such as minnows and small salamanders. Invertebrate prey includes snails, slugs, spiders, sow bugs, termites, terrestrial and aquatic insect larvae and nymphs, and earthworms. Most of their diet comprises soft-bodied species. This shrew hunts under water, at the surface, and on land.

REPRODUCTION

There is little information available concerning the breeding habits of this shrew in Canada. In Oregon the breeding season extends from February to August. The litter size of two pregnant females ranged from five to seven. The number of litters produced annually is unknown. Sexual maturity of males, and likely also of females, occurs at about 10 months old, after they have survived their first winter.

BEHAVIOUR

Pacific Water Shrews are frenetic hunters. Like other shrews, they have voracious appetites and must hunt regularly, about every hour or two. They are active periodically throughout a 24-hour period as they hunt for a meal, and then they rest for a while to digest it. They swim and dive with ease and agility, using alternate strokes of each hind foot for their primary underwater propulsion and all four feet (in a doggy-paddle) when they are at the surface. Each water-based hunt is brief. The maximum recorded swim is 3.5 minutes, and most are less than 2 minutes. Air trapped in the fur gives the shrew a silvery appearance when it is under water. All prey, whether aquatic or terrestrial, is consumed on land. Once its catch has been consumed, the shrew immediately grooms to remove water from its fur. Most grooming is performed with the hind feet and the mouth and tongue or by squeezing through dense vegetation to rub off the excess moisture. Terrestrial hunting usually entails rapid scurrying about with wiggling nose and vibrissae, searching for prey largely by odour and touch. Excess prey may be cached for later use.

VOCALIZATIONS

A shrill twittering call may be produced when a shrew is alarmed or during aggressive encounters between shrews.

SIGNS

The signs are unknown.

REFERENCES

Galindo-Leal and Zuleta 1997; Jameson and Peeters 2004; Nagorsen 1996, 2005b; Pattie 1973, 1999; Reid, F.A., 2006; Ryder and Campbell 2007; van Zyll de Jong 1983a; Verts and Carraway 1998.

Cinereous Shrew

also called Common Shrew, Masked Shrew

FRENCH NAME: **Musaraigne cendrée**

SCIENTIFIC NAME: *Sorex cinereus*

The Cinereous Shrew is the most common and widespread of the North American shrews. Despite their abundance, Cinereous Shrews, like other shrews, are rarely seen, owing to their habits of tunnelling through leaf litter and staying under cover to protect themselves from predators.

DESCRIPTION

Cinereous Shrews have small eyes, short ears that barely protrude beyond their plush fur, short legs, and (their most distinguishing trait apart from their small size) a long, pointed snout that is tipped with a highly mobile nose. This small shrew is typically dark greyish brown to dark brown on the dorsum and usually silvery grey to buffy grey beneath, but the pelage colour varies across the continent. Yellowish or reddish-brown individuals with buffy bellies occur, and in some regions there is a tendency towards lighter coats (interior British Columbia) or an indistinct tricolour pattern with paler sides. Vaguely tricoloured individuals are known from southern Quebec, central Ontario, Manitoba, central British Columbia, and Yukon. Tails are dark above and lighter below, often with a dark tip. Winter pelage tends to be darker, greyer, thicker, and longer (4.4 mm versus 3.6 mm). The dental formula is incisors 1/1, unicuspids 5/1, premolars 1/1, and molars 3/3, for a total of 32 teeth. There are five readily apparent upper unicuspids that form an almost evenly graduated series. The first is largest, and the fifth is smallest and can easily be overlooked. The fourth is generally smaller than or, at most, equal to the third. Older adults, towards the end of their second summer, commonly have teeth so badly worn that all of the red pigment has gone. Adults have two annual moults, one in April to July, depending on sex and latitude, and the other from mid-September to mid-October.

SIMILAR SPECIES

Cinereous Shrews are widespread and overlap the ranges of many other shrew species. Within Canada they are most similar to Prairie Shrews. They differ from Prairie Shrews by their longer, more tufted tails, dark tail tuft, and overall darker colour. The most reliable method of distinguishing these two shrews is by comparing the length of the upper unicuspid toothrow, which is < 2.3 mm in Prairie Shrews and longer in Cinereous Shrews. Arctic, Tundra, and Maritime Shrews are larger (> 6.0 g) than Cinereous Shrews and more strongly tricoloured. The two water shrews, the Pacific Water Shrew and the North American Water Shrew, are both much larger (> 8.5 g). Smoky Shrews are also larger (> 5.8 g) and are darker all around, lacking the paler belly of the Cinereous Shrew. Dusky and Vagrant Shrews are slightly larger than Cinereous Shrews but are similarly coloured. The best way to distinguish them is to look at the teeth. The third upper unicuspid is distinctly smaller than the fourth on both the Dusky and the Vagrant Shrews. North American Pygmy Shrews are even smaller than Cinereous Shrews and have a compressed upper unicuspid toothrow, with only three conspicuous unicuspids because the third and fifth unicuspids are highly reduced and compressed. The Least Shrew (known in Canada only from Long Point, Ontario) has a much shorter tail (≤ 12.5 mm) and is somewhat smaller (≤ 92 mm in total length) than the Cinereous Shrew. Barren Ground Shrews have shorter tails (≤ 31 mm) than do Cinereous Shrews, and the lighter fur on their belly extends high up the sides of the body. Long-tailed Shrews are grey rather than brown and have a distinctly bicoloured and longer tail (≥ 45 mm). Merriam's and Preble's Shrews are found in drier habitat than are Cinereous Shrews. A newly described species, the Olympic Shrew, is limited in Canada to a small region south of the Fraser River in extreme southwestern British Columbia. It has very small medial tines on the inside edges of its upper incisors (anterior view) and very pale reddish pigmentation on the same teeth, while Cinereous Shrews have distinct medial tines that occur within the extent of the dark reddish pigmentation on their upper incisors. Appendix 1 offers illustrations of these and other characteristics to assist in shrew skull identification. Cinereous Shrews also have a somewhat thicker tail (about 2 mm thick) with a distinct dark tuft at the tip, which is lacking in Olympic Shrews.

SIZE

The following measurements apply to animals of reproductive age from the Canadian part of the range. Males and females are similar in size and appearance. Size varies with latitude, the largest animals being in the most northerly portions of the range.

Total length: 75–125 mm; tail length: 28–50 mm; hind foot length: 8–14 mm; ear length: 6–10 mm.

Weight: 2.2–8.0 g (average 3.6–4.6 g); newborns, 0.2–0.3 g.

RANGE

Cinereous Shrews have the widest distribution of any North American Shrew. They occur across the northern portion of the continent from Alaska to northern Ungava and from northern Washington State to the Appalachian Mountains in southern Tennessee. They have reached only a few near-shore islands such as Prince Edward Island, Manitoulin Island, and a few coastal islands in British Columbia but have not reached Vancouver Island, the Queen Charlotte Islands, the Madeleine Islands, or Anticosti Island. Cinereous Shrews were introduced to Newfoundland in the 1950s as

Cinereous Shrew
(*Sorex cinereus*)

Distribution of the Cinereous Shrew (*Sorex cinereus*)

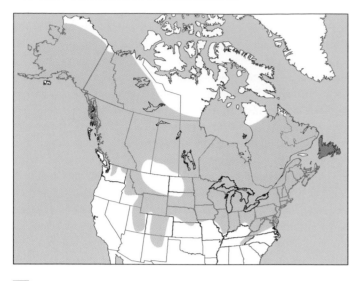

 introduced

a biological control against Larch Sawfly outbreaks and have since spread into appropriate habitat across the island. They are the only Newfoundland shrew. In the southern limits of the range, Cinereous Shrews are restricted to the higher cooler elevations. In the northern United States, the central plains have experienced a cooling, mesic (moist) trend since the 1960s, and Cinereous Shrews have been slowly expanding southward into southern Nebraska and north central Kansas.

ABUNDANCE

Generally common in appropriate habitat, Cinereous Shrews are frequently the dominant shrew in a wide variety of communities. Population densities of 0–27 per ha have been estimated in suitable habitats, and estimates from one year to another at the same location can easily fluctuate fivefold or more. Abundance normally increases as the cohort of young are added to the population over the course of the summer; therefore, population numbers are highest in August and September, and lowest in early spring after the winter attrition. Population size can also vary from place to place, possibly as a result of food availability. These little mammals are subject to intermittent population eruptions, during which numbers explode as subadults breed in the year in which they were born and the breeding season for all the shrews is prolonged. What triggers this phenomenon is unknown, although food abundance and a moist summer are suspected factors.

ECOLOGY

Cinereous Shrews are the most adaptable of all the North American shrews. They occupy a wide range of habitats from sea level to 2288 m in elevation. Cinereous Shrews may be found in damp grasslands, salt marshes, dense and open forests of various sorts (deciduous, coniferous, and mixed) up to the timberline, as well as in thickets, woodlands, and grassy meadows alongside streams,

ponds, and wetlands. Cinereous Shrews may also occupy disturbed sites (clearcuts, burns, cleared lands, stabilized talus), but they are often most abundant in damp, mossy woodlands. Primary requirements of their environment appear to be moisture and cover. These small mammals have a high metabolism and must eat frequently, at least every few hours. Consequently they may be active at any time during the day or night, although they appear to be most active during darkness. They alternate a foraging period with a subsequent rest period, during which they may exhibit a depressed metabolic rate. An active period may last a few minutes to over an hour, and a rest period follows as the animal digests its meal. Like all shrews, Cinereous Shrews do not hibernate, remaining active under the snow throughout the winter. Delicate, hollow, globular nests of grasses and leaves are constructed in protected locations such as in a stump, beneath a log or other debris, or in a shallow burrow. These are 4–6 cm in diameter, and the hollow cavity is 2–3 cm in diameter. Rare individuals may survive 24–30 months, but few even reach 16 months old, and fewer than half of the yearly cohort of juveniles survives more than 5 months. By autumn the populations are dominated by young of the year that overwinter as subadults. They become sexually mature in late winter, then breed, and usually die before the next autumn. The occasional dead shrew found in the woods in autumn is likely such an adult that has died essentially of old age. Predators include owls, hawks, herons, shrikes, snakes, weasels, foxes, Bobcats, larger shrews, and Domestic Cats. Even fishes have been discovered with shrews in their stomachs.

DIET

Cinereous Shrews are considered generalist insectivores. They eat a wide variety of invertebrates including insect larvae and pupae, spiders, grasshoppers, ants, sow bugs, butterflies, beetles, slugs, tiny snails, crickets, earthworms, and centipedes. Prey varies with season and availability, and during winter most prey is composed of dormant insects and pupae. Mycorrhizal fungi (truffles) and some seeds may be consumed, particularly in winter when more preferred food is scarce. Carrion will be consumed when available, and some vertebrate prey may also be taken (such as small salamanders and the eggs and nestlings of small birds). To fuel its rapid metabolism a Cinereous Shrew eats about its own weight in food each day, and pregnant or lactating females may triple that amount. These small mammals are important insect predators throughout the boreal zone because they consume large numbers of sawfly cocoons and budworm larvae, which are serious pests of many coniferous trees. Predation by Cinereous Shrews of 14%–88% of the annual sawfly cocoon production has been reported.

REPRODUCTION

Across their wide range, Cinereous Shrews exhibit considerable reproductive variation, probably as a result of weather and food availability. They typically breed between May and September, but some populations may experience a breeding hiatus or slowdown in mid-summer during dry years; some populations, in drier habitats, only breed from May to July. Onset and termination of reproduction

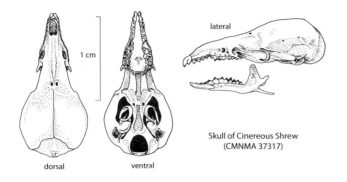

lateral

dorsal ventral

Skull of Cinereous Shrew
(CMNMA 37317)

can vary from year to year, even in the same location. Reproduction may begin as early as late March in some places and extend into November in others, where weather permits. A well-fed insular population in Nova Scotia is thought to experience limited breeding throughout the winter. Litter size ranges from one to twelve and averages five to seven. Females can produce up to six litters a year, but most generate one to three litters annually, depending on how long they live. Length of gestation is unknown. The incredibly tiny newborns are dark pink, and their eyes and ears are fused shut. Their abdominal organs (liver, stomach, and intestines) are visible through their skin, and the toes are fused together and have no claws. Vibrissae (whiskers) and claws begin to appear at around 2 days old, fine hairs begin to emerge from the skin at around 8 days old, and incisors erupt on days 13–14. The toes are completely free by 15–16 days old. Ears open at 14–17 days old, and eyes at 17–18 days old. Juveniles leave the nest at around 27 days old, when they are almost adult sized. Although young become sexually mature at around two months old, most do not breed until the following spring. However, during favourable years or during a population eruption, some subadults that are recently out of the nest may breed successfully, albeit with smaller than average litters (ranging from one to five).

BEHAVIOUR

Cinereous Shrews are versatile hunters. They normally scurry along runways created by other small mammals as they forage under cover of vegetation or leaf litter. The shrews will also dig tunnels in the leaf litter for themselves. They may climb trees and shrubs to raid bird nests, and beach-dwelling populations will forage for amphipods and kelp flies in the intertidal zone. A foraging shrew darts along a runway, poking its nose into cracks and crevices, all the while twittering quietly to itself. While it appears that most prey is detected by smell, touch, or hearing, Cinereous Shrews are also quite capable of displaying good visual detection despite their small eyes. Two adults were observed hunting butterflies one cool morning in Manitoba. The butterflies had gathered on a mudflat and were semi-torpid from the cold. Five juvenile shrews decamped under a log out of the way of the hunters. An adult would dash out from under the log, jump through the air onto a butterfly, and dart back under the log to share the prey. Within seconds all that remained of each insect were the discarded wings. The hunters ranged farther and farther as the nearby prey was depleted, dashing between small bits of cover until

they ran out of cover or were close enough to attack. They charged directly at their chosen prey from about 1–6 m away, launching themselves about 15 cm into the air from about 45 cm away, to land with all four feet right on the butterfly, sometimes as it was taking off or already in flight. After about an hour of hunting, the animals left, and the wings of 134 butterflies remained scattered about under the log. This observation serves to illustrate the shrews' hunting abilities, daytime visual acuity, and parental care (likely at least one of the adults was a parent). Most of the year, adults are thought to be solitary, and two shrews encountering one another may rear onto their hind legs and squeak in an aggressive display. This behaviour can escalate into a shoving match until one of the animals retreats or is pushed over and then retreats. Scent glands on their sides enlarge during the breeding season and secrete an oily substance that possibly serves as a sexual attractant or may be used for scent marking. Small aggregations, sometimes seen during summers of peak population years, are thought to be composed of a few (perhaps three or four) rapidly moving, squeaking, frenzied males that are pursuing an oestrous female.

VOCALIZATIONS

Young in the nest chirp or squeak when disturbed. Adults and sub-adults will emit a similar but deeper pitched rapid series of staccato squeaks when disturbed, agitated, or encountering another shrew. A foraging Cinereous Shrew commonly twitters faintly to itself as it travels. It produces brief, high-frequency echolocation pulses through its mouth, which are likely used to navigate. Most of the frequencies in these pulses are in the 30–60 kHz range, which is beyond the upper threshold of human hearing (which occurs at around 20 kHz). It is possible that the faint twittering sound is all that we can detect of these echolocation pulses.

SIGNS

See the Arctic Shrew account for signs and tracks of this small shrew.

REFERENCES

Bellocq and Smith 2003; Bellocq et al. 1994; Cawthorn 1994; Foresman and Long 1998; Frey 1992; Innes et al. 1990; Kurta 1995; Long, C.A., 2008; Maier et al. 2006; Matlack et al. 2001; McCay et al 1998; Merritt 1987; Moore, J.W., and Kenagy 2004; Nagorsen 1996; O'Donnell et al. 2005; Reid, F.A., 2006; Rickart and Heaney 2001; Rinehart-Whitt and Pagels 2000; Simon et al. 1998; Smith, H.C., 1993; Teferi and Herman 1995; Teferi et al. 1992; van Zyll de Jong 1983a, 1999a; Verts and Carraway 1998; Whidden et al. 2002; Whitaker 2004; Whitaker and Hamilton 1998; Youngman 1975.

Long-tailed Shrew

also called Rock Shrew, Gaspé Shrew

FRENCH NAME: **Musaraigne à longue queue,** *also*
Musaraigne longicaude, Musaraigne de Gaspé

SCIENTIFIC NAME: *Sorex dispar*

This species is composed of two taxa that were formerly considered to be separate species: the smaller Gaspé Shrew (*Sorex gaspensis*) and the larger Long-tailed Shrew (*Sorex dispar*). Since the scientific name of the Long-tailed Shrew is older, it takes precedence as the scientific name of the newly merged species.

DESCRIPTION

Long-tailed Shrews are slender, slate-grey shrews with long, fairly thick tails that comprise 50% of the total length. Pelage on the back and belly is usually the same colour during both winter and summer, but some individuals may be slightly paler on the underside during summer. Feet are pale. Tails are indistinctly bicoloured, darker above and lighter below. Tail fur commonly wears off as the animals age, so that older adults may have naked tails. Vibrissae (whiskers) are long. The dental formula is incisors 1/1, unicuspids 5/1, premolars 1/1, and molars 3/3, for a total of 32 teeth. Only four of the upper unicuspids are readily apparent, the fifth being highly reduced. The fourth upper unicuspid is noticeably smaller than the third. Adults have two annual moults (spring and autumn), and winter fur is longer. Long-tailed Shrews exhibit a size cline whereby animals in the southern parts of the range are the largest, while those at the northern limits of the distribution are the smallest. Between these extremes, size appears to diminish gradually from south to north. This reduction in size northwards is unusual and an exception to Bergmann's rule, which holds that among warm-blooded vertebrates body size increases with increasing latitude.

SIMILAR SPECIES

Smoky Shrews in their grey winter pelage may be confused with Long-tailed Shrews, but Smoky Shrews typically have a shorter, narrower, and distinctly bicoloured tail (usually < 40% of total length), their belly fur is lighter in colour than is their dorsal fur, and their dorsal fur is paler than that of the Long-tailed Shrew. North American Water Shrews may occur in similar rocky stream beds, but they are larger, have a distinctly lighter belly, and have noticeable fringes of hairs around their hind feet. Cinereous Shrews commonly have a dark tail tip, and both Cinereous and North American Pygmy Shrews have brown dorsal pelage, a distinctly paler underside, and a

Long-tailed Shrew (*Sorex dispar*)

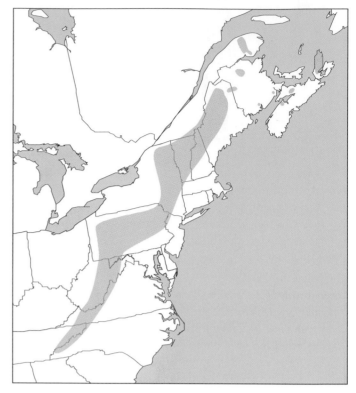

shorter tail. Appendix 1 provides an illustrated identification guide to Canadian shrew skulls.

SIZE

The following measurements apply to animals of reproductive age from the Canadian part of the range. Males and females are similar in size and appearance.

Total length: 95–136 mm; tail length: 45–67 mm; hind foot length: 11–15 mm.

Weight: 2.3–8.3 g; newborns, unknown.

RANGE

The Long-tailed Shrew is an Appalachian species, occurring in mountainous locations from Gaspé and Cape Breton to North Carolina and Tennessee. Although the American distribution appears continuous, this shrew is usually found in isolated pockets of suitable habitat along mountainsides. The patchy Canadian distribution more clearly reflects that reality. Specimens are scarce because of capture difficulty, but it is probable that this species is more widespread and abundant than is currently known.

ABUNDANCE

Uncommon to rare throughout their range, Long-tailed Shrews are considered to be rare to very rare in Canada. Owing to their largely subterranean existence, capture of this shrew is difficult, requiring traps to be set at arm's length, or more, into rock crevices. One extraordinary collection of 65 animals trapped in a single field season at

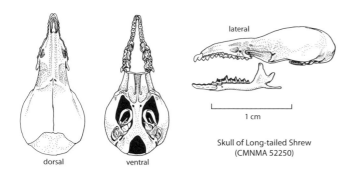

dorsal ventral lateral

1 cm

Skull of Long-tailed Shrew
(CMNMA 52250)

REFERENCES

Desrosiers et al. 2002; French and Crowell 1985; Kirkland 1981, 1999a, 1999b; Kirkland and Van Deusen 1979; Kirkland et al. 1979; Kurta 1995; McAlpine et al. 2004; Merritt 1987; Prescott and Richard 2004; Reid, F.A., 2006; Rhymer et al. 2004; Scott, F.W., and van Zyll de Jong 1989; Shafer and Stewart 2006, 2007; Shafer et al. 2008; van Zyll de Jong 1983a; Whitaker and French 1984; Woolaver et al. 1998.

Mount Carleton, New Brunswick, in 1980 suggests that population numbers may fluctuate from year to year.

ECOLOGY

The prime habitat for Long-tailed Shrews is cool, moist, moss-covered rock fields or talus slopes in boreal or mixed forests with a rapidly flowing stream nearby. It may also include rocky stream beds, rock slides, and man-made talus, such as roadway or railway embankments and open-pit mines. In all cases, moisture, and the crevices created by a jumble of rock, is a habitat requirement. Long-tailed Shrews are almost never seen, because most of their activity occurs well beneath the rock surface where they tunnel and forage in the humus and debris that accumulates between the rocks. Their long tail is thought to help them balance as they climb and manoeuvre among the rocks, and long vibrissae provide essential tactile information so that they can find their way or detect prey in darkness or dim light. The maximum lifespan is thought to be 14–17 months, similar to that of other shrews. Northern Short-tailed Shrews are common throughout the range of Long-tailed Shrews and are likely predators of Long-tailed Shrews.

DIET

Small invertebrates such as beetles, grasshoppers, flies, spiders, and centipedes make up the bulk of their diet. Plant material may also be consumed.

REPRODUCTION

The breeding season appears to last from mid-April (in Pennsylvania) to August, and known litter sizes range from two to six, based on embryo counts from captured pregnant females. Other reproductive details remain unknown.

BEHAVIOUR

Unknown.

VOCALIZATIONS

Unknown.

SIGNS

As this shrew spends all of its life in rock crevices well below the surface, it leaves no apparent signs above ground.

Smoky Shrew

FRENCH NAME: **Musaraigne fuligineuse**

SCIENTIFIC NAME: *Sorex fumeus*

The Smoky Shrew can be common in suitable habitat, and its range appears to be expanding in the northwestern portions, but numbers have declined drastically on Prince Edward Island.

DESCRIPTION

Pelage of this shrew varies seasonally and is brownish in summer and dark grey to almost black in winter. The underside is often slightly lighter than are the upper parts, being brownish yellow in summer and grey to silver grey in winter. The feet are similar in coloration to the underside. The tail is bicoloured, dark above and yellowish buff below. Both sexes possess side glands, but these are only evident on immature males and on adults of both sexes. The side glands become enlarged during the breeding season, and those of adult males produce an odoriferous secretion. The tail of both sexes may swell when they are in reproductive condition, sometimes to the extent that a noticeable constriction forms at the base of the tail. The dental formula is incisors 1/1, unicuspids 5/1, premolars 1/1, and molars 3/3, for a total of 32 teeth. There are five readily apparent upper unicuspids, and the third is larger than

Smoky Shrew (*Sorex fumeus*)

the fourth. Adults have two annual moults: the spring moult is complete by mid-July, and the autumn moult by early November. These shrews appear to lose hairs from all over the body during a moult, so there is no distinct moult line, as may be seen in many other mammals.

SIMILAR SPECIES

The range of the Smoky Shrew overlaps with the ranges of Arctic Shrews, North American Water Shrews, Cinereous Shrews, Long-tailed Shrews, North American Pygmy Shrews, Least Shrews, and Northern Short-tailed Shrews. Similar in size to Arctic and North American Water Shrews, Smoky Shrews are not distinctly tricoloured as are Arctic Shrews, nor light-bellied with a fringe of hairs around the feet as are water shrews. Long-tailed Shrews do not have a paler belly, and their tail is longer (50–64 mm) and thicker than that of Smoky Shrews. Cinereous Shrews are smaller than Smoky Shrews, usually with a more distinctly pale belly. North American Pygmy Shrews and Least Shrews are much smaller than Smoky Shrews, and Northern Short-tailed Shrews are much larger. Appendix 1 provides an illustrated identification guide to Canadian shrew skulls.

SIZE

The following measurements apply to animals of reproductive age from the Canadian portion of the range. Males and females are similar in size and appearance.
Total length: 104–125 mm; tail length: 42–54 mm; hind foot length: 12–15 mm.
Weight: 5.8–8.0 g; newborns: unknown.

RANGE

Smoky Shrews occur in eastern North America from extreme northeastern Minnesota, through south-central Ontario and southern Quebec to the Maritimes (excluding Cape Breton and Newfoundland). Occurrence on Prince Edward Island has recently been verified, but these shrews are very rare on the island. Their range extends southwards along the Appalachian Mountains to northern Georgia. In the southern extremes of the range Smoky Shrews are found mainly at the higher, cooler elevations.

ABUNDANCE

Smoky Shrews can be common in parts of the range in optimum habitat. Populations fluctuate seasonally (they are highest in the autumn) and from year to year. In late summer of peak years Smoky Shrew densities of 143 per ha may be achieved, but most populations fall within the 12–35 per ha range. Winter mortality can be heavy. These shrews tend to be rare in regions where Northern Short-tailed Shrews and Cinereous Shrews are common.

ECOLOGY

Smoky Shrews are primarily denizens of moist, cool forests in the Great Lakes–St Lawrence, Acadian, and Appalachian regions, where they occupy the thick humus layer beneath coniferous, deciduous, or mixed woodlands with abundant fallen woody debris or moss-covered boulders. They may also occur in sphagnum bogs, margins

Distribution of the Smoky Shrew (*Sorex fumeus*)

of freshwater wetlands, damp talus slopes, and rocky outcrops, and sometimes in dry forests and grasslands. Their home-range size is unknown. It is also not known whether or not they dig their own burrows, but Smoky Shrews regularly use tunnels created by other small mammal species such as red-backed voles, bog lemmings, short-tailed shrews, or moles. Spherical, hollow nests (about 6–8 cm in diameter) are constructed of shredded leaves and animal fur and placed in the underground labyrinth of shallow tunnels, commonly under rocks or fallen logs. Nests are usually 10–23 cm below the ground surface and are the hub or focus of many tunnels that radiate from them. Smoky Shrews may be active at any time of the day or night and do not hibernate. They tunnel through the snow pack in winter, foraging near the ground as they search for dormant insect prey. Like all shrews, Smoky Shrews have a voracious appetite and must eat regularly. Their normal activity pattern is to hunt until sated, and then rest to digest their meal. The length of the feeding and rest intervals is unknown. Maximum longevity is thought to be 14–17 months. Predators include owls, snakes, weasels, foxes, Northern Short-tailed Shrews, Bobcats, and Domestic Cats. Most mammalian predators (except other shrews) find shrews to be distasteful, and though they may kill them, they generally abandon the carcasses unless they are very hungry.

DIET

As much as 80% of the diet of Smoky Shrews is composed of a wide variety of leaf litter insects (adults, larvae, and pupae), to which they add other invertebrates such as earthworms, sow bugs, snails, centipedes, millipedes, and spiders, when available. Small amounts of plant material and fungal spores have been found in their digestive tracts. Smoky Shrews will occasionally take small vertebrates, such as small salamanders (which they kill with a neck

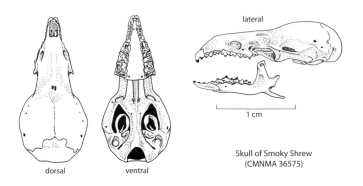

lateral

1 cm

dorsal ventral

Skull of Smoky Shrew
(CMNMA 36575)

bite), and they will also feed on carrion. During winter this shrew primarily forages on dormant insects and insect pupae. Captive Smoky Shrews consume up to about half of their own weight daily, and this is likely also the case among wild individuals. Foraging activity in dry sites increases during rainfall and shortly afterwards, while the same activity in the more mesic (moist), preferred locations is unchanged by rainfall.

REPRODUCTION

Much of the reproductive biology of this shrew remains unknown. The breeding season in Canada likely lasts from late March or early April until late July or early August, and two or possibly as many as three litters are possible annually if the female survives the whole summer. Males are reproductively mature a few weeks earlier than are females. The first pregnancies occur well into April. Gestation is thought to last slightly less than three weeks, and litter size averages four to five (ranging from two to seven). Females experience a postpartum oestrus and may mate soon after parturition. As is the case with all shrews, newborns are altricial, born pink and helpless. They are likely weaned just before the birth of the next litter, at about 18–20 days old. The weaned young often remain in their natal nest until they are adult size. Presumably their mother departs to create another nest for her new litter. These shrews do not breed in the year of their birth. Young of the year spend their first winter as immatures and breed for the first time during the following warm season. Most adults die by late autumn; only immatures overwinter, making the generational period about 12 months.

BEHAVIOUR

Very little is known of the social behaviour of Smoky Shrews. Although this shrew will kill a small salamander by holding it down with its feet, while delivering a lethal neck bite, and a captive individual was ferocious enough to attack a Deer Mouse, it was unable to kill this much larger mouse. During foraging, their noses are constantly twitching, searching for scent, and their vibrissae are held stiffly outstretched to afford them tactile information.

VOCALIZATIONS

A high-pitched grating sound is produced when these shrews are alarmed or disturbed. If highly disturbed, they may flip onto their back and produce this sound, while waving their legs in the air. Foraging Smoky Shrews constantly emit a low-volume "twitter,"

possibly the audible portion of a quiet echolocation call that they emit to assist navigation.

SIGNS

Signs of most shrews are similar. See the "Signs" section of the Arctic Shrew account for a generalized discussion of shrew signs.

REFERENCES

Brannon 2000, 2002a, 2002b; Desroches and Picard 2004; Dobbyn 1994; Hartling and Silva 2004; Jannett and Oehlenschlager 1994; Kirkland and Schmidt 1982; Kurta 1995; Merritt 1987; Owen 1984; Whitaker 1999a; Whitaker and French 1984; Whitaker and Hamilton 1998; van Zyll de Jong 1983a.

Prairie Shrew
also called Hayden's Shrew

FRENCH NAME: **Musaraigne des steppes**
SCIENTIFIC NAME: *Sorex haydeni*

Formerly thought to be a subspecies of the Cinereous Shrew, the Prairie Shrew has been recognized as a full species since 1980. Very little is known about its natural history.

DESCRIPTION

Prairie Shrews are small, brown to dark brown shrews with a greyish belly. Their tail is uniformly coloured but is slightly paler than the body fur and lacks a dark terminal tuft. Ears barely protrude beyond the fur. Their eyesight is reputed to be poor, and most prey is detected by sound, smell, or touch. Their side glands become enlarged during the breeding season and emit a pungent odour, likely used somehow as a sexual signal. The dental formula is incisors 1/1, unicuspids 5/1, premolars 1/1, and molars 3/3, for a total of 32 teeth. Only four of the upper unicuspids are easily visible in lateral view because the fifth is small and tucked in between the fourth unicuspid and the premolar. The fourth upper unicuspid is noticeably smaller than the third. On most Prairie Shrews the reddish pigment on the lower incisor is divided into two sections, one large one that

Prairie Shrew
(*Sorex haydeni*)

includes the tip and the first two cusps and a smaller second one that includes just the tip of the third cusp.

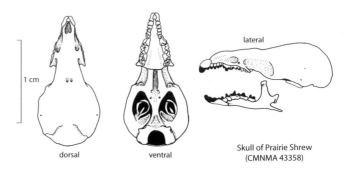

Skull of Prairie Shrew
(CMNMA 43358)

dorsal ventral lateral

1 cm

SIMILAR SPECIES

The Prairie Shrew is similar to the closely related Cinereous Shrew but is slightly smaller with a shorter tail and usually paler fur. The underside of the tail and the terminal tuft at the tail tip of the Cinereous Shrew is dark. Nevertheless, distinguishing between the two is very difficult in the field because a definitive diagnosis usually requires an examination of the skull. Cinereous Shrews have an upper unicuspid row length of > 2.2 mm, while the same in Prairie Shrews is < 2.2 mm. Barren Ground Shrews are tricoloured with three areas of red pigmentation on each upper incisor. Vagrant Shrews and Dusky Shrews are larger, with the third upper unicuspid being smaller than the fourth. North American Pygmy Shrews and Prairie Shrews are similar in size. They can be distinguished by their reduced upper unicuspid toothrow; North American Pygmy Shrews appear to have only three upper unicuspids because the third and fifth are highly reduced and often unnoticed. See Appendix 1 for figures that illustrate shrew skull characteristics.

SIZE

The following measurements apply to animals of reproductive age in the Canadian portion of the range. Males and females are similar in size and appearance.
Total length: 77–101 mm; tail length: 25–38 mm; hind foot length: 9–12 mm; ear length: 7–8 mm.
Weight: average 2.0–5.0 g; newborns, unknown.

RANGE

The Prairie Shrew lives on the northern plains, from southern Manitoba, Saskatchewan, and Alberta to northeastern Wyoming and northwestern Iowa. One specimen was captured in Beaverlodge, Alberta, in 1947. Habitat in the Peace River–Grimshaw area could support a population of Prairie Shrews, but their current status in that region is unknown.

ABUNDANCE

These small shrews are generally considered to be uncommon to rare throughout their range, although they can become locally abundant in suitable habitat in some years. Recent work in Alberta found these shrews to be common in appropriate habitat. It is likely that Prairie Shrews exhibit seasonal population cycles similar to those of other comparable shrews. Their populations are lowest in early spring, rise during the breeding season, and are highest in late summer after all the young of the year are recruited. The breeding adults die in late summer, and there is a slow attrition of the remaining subadults over the course of the winter.

ECOLOGY

Although its primary habitat is open prairie grasslands, the Prairie Shrew may also occur in sparsely treed aspen parklands and sometimes even in prairie wetlands and dry mixed or pine forests around the prairie edges. Prairie Shrews dig tunnels through the leaf litter to get around undetected. Their cup-shaped nests (60–100 mm in diameter) are constructed in underground tunnels, usually protected in rock crevices or beneath logs or rocks. Where they co-occur with Cinereous Shrews, the two species segregate by micro-habitat, the Cinereous Shrews occupying the more treed portions and the Prairie Shrews the more open prairie. Predators primarily include owls but may also include snakes, shrikes, foxes, Coyotes, and Domestic Cats. Owing to their bad taste, actual consumption of shrews killed by mammalian predators is uncommon, unless the predator is starving.

DIET

Thought to consume similar food to that of other shrews, Prairie Shrews eat mainly small invertebrates such as beetles, spiders, earthworms, grasshoppers, and insect larvae but will also consume carrion, seeds, and perhaps small vertebrates.

REPRODUCTION

These small mammals breed from early spring until July, when heat causes the insects to become scarce, thereby curtailing further breeding. A nest with eight young found in Manitoba on 14 October

Distribution of the Prairie Shrew (*Sorex haydeni*)

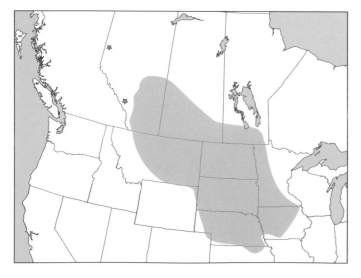

 extralimital records

1924 suggests that, in some years, breeding may resume after the late summer or early autumn rains. Females can produce 2–3 litters annually if they live through the whole summer. Litter size averages five to six (ranging from four to ten), and gestation probably is similar to that of the closely related Cinereous Shrew, 19–22 days. Young are weaned at about 20 days old and leave the nest by the age of 30 days old, when they are virtually adult size. Young of the year typically do not breed until the following spring.

BEHAVIOUR

Very little is known about the behaviour of Prairie Shrews. They are thought to be solitary after they reach 30–35 days old. Before that, they will play with littermates, and older young will "caravan" as they move from place to place. This interesting behaviour, which involves each animal placing its nose into the rump fur of the individual ahead, allows a female to safely move her entire litter. The animals form a long winding line, as they travel, with the maternal female in the lead. Excess food is cached for later use.

VOCALIZATIONS

Unknown.

SIGNS

Signs of all the smaller shrews are essentially indistinguishable. See the "Signs" section of the Arctic Shrew for a discussion of generalized shrew signs.

REFERENCES

Brunet et al. 2002; Clark and Stromberg 1987; Demboski and Cook 2003; Engley and Norton 2001; Frey and Moore 1990; George, S.B., 1999a; Petersen, S.D., and Roberts 1999; Smith, H.C., 1993; Stewart, D.T., et al. 1993; van Zyll de Jong 1980; Volobouev and van Zyll de Jong 1994; Whidden et al. 2002.

North American Pygmy Shrew

FRENCH NAME. **Musaraigne pygmée,** *also*
Musaraigne de Hoy

SCIENTIFIC NAME: *Sorex hoyi,* formerly *Microsorex hoyi*

Some adult North American Pygmy Shrews from Canada may weigh 2.1 g. Smaller forms, from more southerly parts of the range in the United States, average 2.0 g as adults. However, despite their small size, Pygmy Shrews are not small enough to claim the title of the world's smallest mammal. This distinction belongs to another shrew, the Alaska Tiny Shrew (*Sorex yukonicus*), which weighs 1.5–1.7 g as an adult; this diminutive shrew occurs in central and western Alaska and has not yet been discovered in Canada.

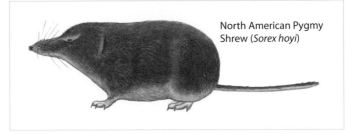

North American Pygmy Shrew (*Sorex hoyi*)

DESCRIPTION

North American Pygmy Shrews are small, long-bodied, short-legged animals with tiny eyes, short ears that barely protrude beyond their plush fur, and a long, pointed snout. North American Pygmy Shrews are reddish brown to greyish brown in colour with lighter, greyish to buffy-brown underparts and a relatively short, indistinctly bicoloured tail that is dark brown above and paler brown below. Their tail may have a darker tip. Winter pelage is longer and greyer than the more brownish summer pelage, and the underside is often whitish rather than the buffy brown of summer. The odoriferous secretion from well-developed flank glands (also called lateral or side glands) of this species is especially strong compared to that of many other shrews. The glands on adult males are visible from April to August. The dental formula is incisors 1/1, unicuspids 5/1, premolars 1/1, and molars 3/3, for a total of 32 teeth. There are three readily apparent upper unicuspids in each toothrow, as the third and fifth are highly reduced and easily overlooked. These reduced unicuspids are diagnostic of the species. The fifth unicuspids may be seen from a lateral view of the skull with magnification, but the even smaller, disc-shaped, third unicuspids can only be viewed from the inside edge of the toothrow, also with magnification. The North American Pygmy Shrew is the only Canadian shrew with just three readily apparent upper unicuspids. The skull characteristics of Pygmy Shrews are illustrated in appendix 1.

Recent DNA work indicates that the eastern and western forms of the North American Pygmy Shrew may be sufficiently distinct to justify species status for each, with the separation being somewhere between western Ontario and central Alberta. Until further work is done to delineate the contact zone, should there be one, and to verify that no hybridization occurs, most mammalogists continue to support the single species model as is herein reported; however, this may change as more information becomes available.

SIMILAR SPECIES

North American Pygmy Shrews closely resemble Cinereous Shrews but have shorter tails and are smaller and slighter. An examination of the upper toothrows is usually necessary to confirm identification. Cinereous Shrews have five upper unicuspids, although the fifth one is small and may be difficult to see. North American Pygmy Shrews also have five unicuspids in their upper toothrow, but the third and fifth are very small and may not be visible without magnification. Furthermore, the third unicuspid cannot be seen from the outside (lateral view) even with magnification. So North American Pygmy Shrews have only three readily apparent unicuspids in their upper

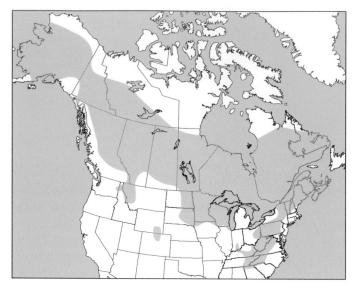

although they may become locally common at times. In most parts of the eastern range, they are rare to very rare. Estimates of population density range from 0.5 to 1.2 per ha.

ECOLOGY

These very small shrews occur in a variety of habitats. Throughout their range they are most abundant in grassy clearings in boreal forests, and they prefer moist areas over dry ones. In Canada they may be found up to about 1640 m in elevation in moist woodlands, marshes, bogs, and other riparian areas as well as in parklands (with grasses, shrubs, and scattered trees) and even sand dunes. North American Pygmy Shrews are active all year round. They make extensive use of tunnels in the snowpack, where they can travel widely while protected from the extreme cold of the winter and from most predators (apart from other shrews). Some of these tunnels may be self dug, but mostly they use tunnels that have been created by other small mammals or they take advantage of their small size to slip through the subnivean spaces where the snow melts back from the ground and vegetation. During snow-free seasons they create tiny burrows in the leaf litter and loose soil. Networks of tunnels have been found in protected areas such as under stumps and beneath fallen logs. North American Pygmy Shrews have a rapid metabolism and must eat frequently. Their normal activity pattern is a few minutes of foraging and eating followed by an extended period of resting. They are active for around four hours a day in total, with an activity peak during darkness. North American Pygmy Shrews are capable swimmers and climbers. Their maximum lifespan is about 16.0–17.5 months, but fewer than about 10% of individuals live longer than a year. Reported predators include garter snakes, hawks, and Domestic Cats, but the list is probably much longer and likely includes larger shrews, owls, and small weasels.

toothrow, while Cinereous Shrews have four and a much smaller fifth. Prairie Shrews have an upper toothrow pattern similar to that of Cinereous Shrews. Barren Ground Shrews have distinctive light sides and an upper toothrow pattern similar to that of Cinereous Shrews. All other shrews found within the North American Pygmy Shrew range are considerably and distinctively larger than North American Pygmy Shrews, and no others have only three readily apparent upper unicuspids. See appendix 1 for a fully illustrated guide to Canadian shrew skull identification.

SIZE

The following measurements apply to animals of reproductive age from the Canadian part of the range. Males and females are similar in size and appearance.
Total length: 62–104 mm; tail length: 22–35 mm; hind foot length: 6–12 mm; ear length: 5–9 mm.
Weight: 2.1–5.3 g; newborns, unknown.

RANGE

The geographic range of this shrew is extensive. North American Pygmy Shrews occur from Alaska, Yukon, and Northwest Territories to Labrador, and south through British Columbia to northern Idaho and Montana. Areas of dry prairie are avoided, so the distribution swings up into southern Alberta and Saskatchewan and from there drops down to northern Iowa and Illinois. An arm of distribution extends along the Appalachian Mountains to northern Georgia and Alabama. Disjunct populations occur in southeastern Wyoming and adjoining northern Colorado, in the northeastern corner of Montana, and in southern Illinois near the Indiana-Tennessee border.

ABUNDANCE

North American Pygmy Shrews are generally considered to range from rare to uncommon across the western part of the continent,

DIET

Their diet is mainly composed of invertebrates that generally are less than 5 mm long, such as insect larvae and adults, ants, isopods, and spiders. Captive North American Pygmy Shrews kill and eat grasshoppers, sawfly larvae, house flies, crane flies, and carrion. None of the studies found that slugs, earthworms, or snails were eaten. Perhaps these are too large for this tiny shrew to subdue. They will eat tree seeds, such as those from Jack Pine cones, in times of need and especially during winter when invertebrates are less abundant.

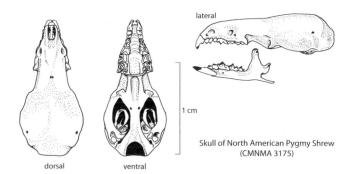

lateral

1 cm

Skull of North American Pygmy Shrew
(CMNMA 3175)

dorsal   ventral

REPRODUCTION

Despite its widespread distribution, this shrew has been poorly studied, and much of its reproductive biology remains unknown.

BEHAVIOUR

Pygmy Shrews rise and sit on their haunches to scan their surroundings. They also sit in this fashion to free their forepaws for food manipulation. When they run, their tail is extended horizontally with a slight upward curve. They set a frenetic pace as they scurry about wiggling and poking their long nose under leaves and into crevices, sniffing for food. They are timid and react to disturbance by briefly freezing, all aquiver, before darting away. The musky odour from their lateral glands appears to increase if the shrews are agitated.

VOCALIZATIONS

A few vocalizations have been reported. One is described as a weak, high-pitched purr, another as a high-pitched whisper, a third as a high-pitched whistle, and a fourth as a short, sharp squeak. It is not known whether this species uses ultrasonic frequencies to echolocate, but it is possible that some of the sounds reported are audible, lower-frequency components of just such vocalizations.

SIGNS

Pygmy Shrews are so tiny that their burrow openings dug into dirt can be mistaken for those made by a large earthworm. Most shrew signs are very difficult to ascribe to a species as they are very small and quite similar. See the "Signs" section in the Arctic Shrew account for generalized information on shrew signs.

REFERENCES

Backlund 1995; Foresman 1999; Hendricks 2001; Jung et al. 2007; King et al. 2000; Laerm et al. 1994, 1996; Long, C.A., 1974, 2008; Nagorsen 1996; Peirce and Peirce 2000a; Prescott and Richard 2004; Prince 1940; Roscoe and Majka 1976; Schowalter 2002; Slough and Jung 2007; Smith, H.C., 1993; Stewart, D.T., et al. 2003; Whitaker and French 1984; Whitaker and Hamilton 1998; van Zyll de Jong 1983a; Youngman 1975.

Maritime Shrew

FRENCH NAME: **Musaraigne des Maritimes**

SCIENTIFIC NAME: *Sorex maritimensis*

Until recently the Maritime Shrew was considered to be a subspecies of the Arctic Shrew. It was elevated to full species status in 2002, but much of its natural history is still unknown.

DESCRIPTION

Adult Maritime Shrews are tricoloured. Their backs are brown to chocolate brown, their sides are brown to yellow brown, and their

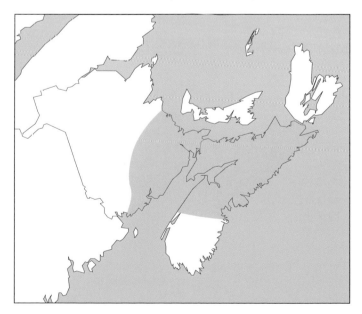

Maritime Shrew
(*Sorex maritimensis*)

bellies are yellowish to buffy grey. Their tails are somewhat bicoloured, being dark above and brown below. The upper surfaces of the feet are a dirty yellowish brown. Subadults and juveniles are less obviously tricoloured. Their backs are lighter than those of adults, and their bellies are somewhat darker. The dental formula is incisors 1/1, unicuspids 5/1, premolars 1/1, and molars 3/3, for a total of 32 teeth. There are five readily apparent upper unicuspids, and the third is larger than the fourth. Adults have two annual moults, one in June, which may continue until late summer, and the other in mid-September until mid-October.

SIMILAR SPECIES

Within their limited range only the Smoky Shrew matches the Maritime Shrew in size. Smoky Shrews have brownish upper parts, and their belly fur is only slightly lighter. Cinereous Shrews, Long-tailed Shrews, and North American Pygmy Shrews are smaller and do not have tricoloured pelage. North American Water Shrews are considerably larger and are dark grey to blackish above, with a whitish or silver-grey belly. Appendix 1 provides an illustrated identification guide to Canadian shrew skulls.

Distribution of the Maritime Shrew (*Sorex maritimensis*)

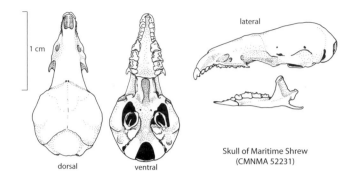

Skull of Maritime Shrew
(CMNMA 52231)

dorsal   ventral

lateral

1 cm

REFERENCES

Perry, N.D., et al. 2004; Stewart, D.T., et al. 2002; van Zyll de Jong 1983a, 1983b.

SIZE

The following measurements apply to animals of reproductive age. Males and females are similar in size and appearance.

Total length: 105–115 mm; tail length: 38–45 mm; hind foot length: 12.5–14.5 mm; ear length: 6–8 mm.

Weight: 6.5–10.5 g; newborns, unknown.

RANGE

Maritime Shrews are endemic to Canada. They occur in central Nova Scotia and southeastern New Brunswick.

ABUNDANCE

This species occupies a restricted habitat within a limited range, so it is considered to be rare; however, at times it can be somewhat more abundant locally, in suitable habitat.

ECOLOGY

These shrews favour moist (mesic) locations such as grass or sedge marsh edges, damp meadows, moist alder thickets, and damp, grassy banks of ditches, railway embankments, dikes, and streams. Maritime Shrews may be active at any time during a 24-hour period. Their usual routine is made up of a short burst of activity, usually foraging, followed by a rest to digest what they have caught. They dig tunnels through the snow near ground level or use those dug by other small mammals, and remain fully active all winter.

DIET

Unknown.

REPRODUCTION

Unknown.

BEHAVIOUR

Unknown.

VOCALIZATIONS

Unknown.

SIGNS

The signs of most of the smaller shrews are similar and often indistinguishable. See the "Signs" section of the Arctic Shrew account for more information.

Merriam's Shrew

FRENCH NAME: **Musaraigne de Merriam**

SCIENTIFIC NAME: *Sorex merriami*

There is only a single record of Merriam's Shrew in Canada. It was taken in 1996 in a pitfall trap set to survey invertebrates in the Kilpoola Lake region of the Okanagan Valley, in extreme southern British Columbia. Most of the following information is based on studies of Merriam's Shrews in the western United States.

DESCRIPTION

Merriam's Shrews are pale medium brown on their dorsum, becoming paler on their sides, and silvery-white to pinkish-buff on their underside, with pinkish-white feet. Their winter pelage is more greyish, with a clear white belly. They have a distinctly bicoloured tail, which is dark grey brown above and white below in both winter and summer. Adults, especially males, develop prominent side glands (3 mm wide by 7 mm long) during the early breeding season. These glands become hairless and exude a pungent secretion that is thought to be responsible for their strong "shrew" odour. Like other shrews, they have a long, pointed snout, small eyes, and plush fur. Their fur is relatively short, and their ears are somewhat more conspicuous than those of most other shrews. Their dental formula is incisors 1/1, unicuspids 5/1, premolars 1/1, and molars 3/3, for a total of 32 teeth. There are four readily apparent upper unicuspids, as the fifth is highly reduced and tucked in between the fourth unicuspid and the premolar. The second unicuspid is the largest, and the third is larger than the fourth. The upper incisors lack a medial tine. Each mandible has a post-mandibular foramen on the ramus. Appendix 1, "Identification of Shrew Skulls," illustrates all of the above-mentioned characteristics. Merriam's Shrews experience two moults annually, one in spring from March to June and the other in autumn from late September to early November.

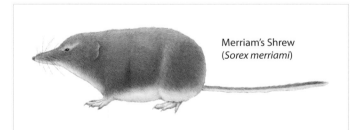

Merriam's Shrew
(*Sorex merriami*)

Distribution of the Merriam's Shrew (*Sorex merriami*)

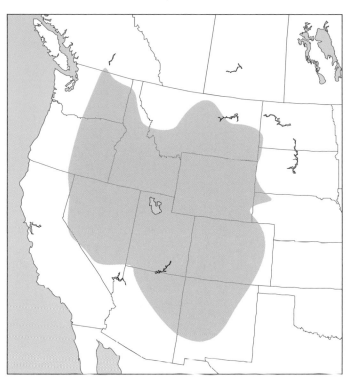

ABUNDANCE

This shrew is generally rare and does not appear to become more abundant at any particular location.

ECOLOGY

Merriam's Shrews tend to occupy drier sites than do most other shrews, where the principal vegetation is sagebrush and grasses. However, they have also been found in wet meadows, disturbed roadsides where the grasses are a mix of native and domestic, and grassy clearings in Ponderosa Pine forests. The habitat in which the Canadian specimen was discovered was dominated by grasses with sparse sagebrush and forbs and a stand of Ponderosa Pine within 75 m. The nearest water was 0.5 km away. Merriam's Shrews are reportedly active at any time of the day or night and throughout the year. Owls are their only known predators, although snakes, foxes, and Coyotes are also likely predators.

DIET

The diet of Merriam's Shrews, like most shrews, is mostly composed of invertebrates. They are reported to eat spiders, caterpillars, beetles, cave crickets, and wasps and likely will consume any invertebrate that is small enough to catch, as well as available carrion.

REPRODUCTION

No reproduction information from Canada is available for this shrew, and very little exists from elsewhere in its range. The breeding season in Washington State is thought to last from early April to October and may include a mid-summer hiatus during the hot, dry season. Litter size based on embryo counts is five to seven.

BEHAVIOUR

It is not known whether Merriam's Shrews create their own runways, but they will readily use those of other small mammals, notably those created by grassland voles.

VOCALIZATIONS

Unknown.

SIGNS

Unknown.

SIMILAR SPECIES

Few other shrews in Canada can survive in the dry sagebrush grasslands inhabited by this species. Preble's Shrews are also arid adapted. They are smaller than Merriam's Shrews and usually darker. They have a tine on the medial edge of their upper incisors, and their skull is shorter ($\leq$ 14.8 mm in length compared to 15.1–17.1 mm for Merriam's Shrews). Dusky Shrews are darker, and their tails are generally longer and not as strongly bicoloured; their upper incisors have a medial tine, their five unicuspids are easily apparent, and the medial edge of the incisor is straight in anterior view. These skull characteristics are illustrated in appendix 1.

SIZE

As there has been only one Canadian specimen to date, the following measurements were gathered from specimens throughout the range. Males and females are similar in size and appearance.
Total length: 88–107 mm; tail length: 30–43 mm; hind foot length: 11–14 mm; ear length: 8–9 mm.
Weight: 4.4–6.5 g; newborns, unknown.

RANGE

This is a western North American shrew, whose range extends from southern British Columbia southwards to eastern California, southern Arizona, and New Mexico and eastwards to western North Dakota and Nebraska. Populations are spotty within this wide range, and the actual distribution may be less continuous than presented.

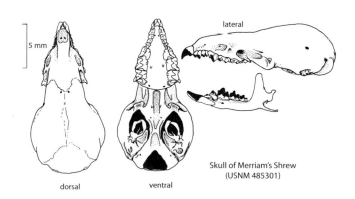

5 mm

lateral

dorsal

ventral

Skull of Merriam's Shrew
(USNM 485301)

REFERENCES

Armstrong and Jones 1971; Benedict, R.A., et al. 1999; Hafner and Stahlecker 2002; Jameson and Peeters 2004; Mullican 1994; Nagorsen et al. 2001; Reid, F.A., 2006; Verts and Carraway 1998.

Distribution of the Dusky Shrew (*Sorex monticolus*)

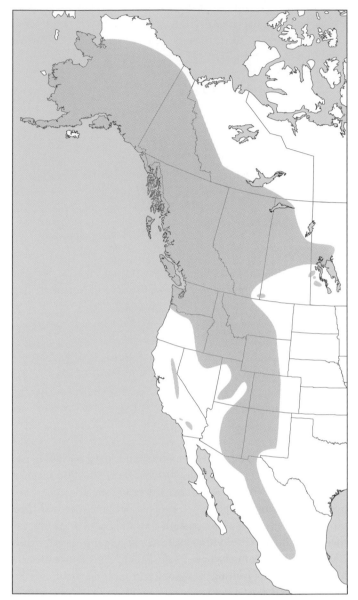

Dusky Shrew

also called Montane Shrew

FRENCH NAME: **Musaraigne sombre,** *also*
Musaraigne montane

SCIENTIFIC NAME: *Sorex monticolus*

Much remains to be learned about the Dusky Shrew, especially with regard to its ecology and behaviour.

DESCRIPTION

Dusky Shrews are very dark grey or greyish brown to brown (often with reddish or yellowish overtones) on their back and sides (sides may be slightly paler), with a somewhat lighter belly of silver to smoky grey to buffy brown. Interior populations in British Columbia tend to be brownish, but coastal populations may be almost black on the upper surfaces, especially in winter when the pelage is darker and greyer than in summer. Their long tail is bicoloured, dark brown above with a pinkish-grey or pale brown underside, and does not have a black tip. The upper surfaces of the feet are pale brown or grey brown. During the breeding season all adult males and about 30% of the adult females develop lateral glands that exude strong smelling secretions. The old adults present in the population at the end of the breeding season can be distinguished from the young of the year by their relatively darker pelage, the hairlessness of their feet and tails, their larger size and greater weight, and their well-worn teeth. The dental formula is incisors 1/1, unicuspids 5/1, premolars 1/1, and molars 3/3, for a total of 32 teeth. The five upper unicuspids are easily visible in lateral view, and the third is noticeably smaller than the fourth. Reddish pigment on the incisors extends above the medial tine. Adult Dusky Shrews experience two annual moults. The spring moult of females takes place in March and early April, and of males from late May to August. Both sexes, including juveniles, grow their longer, darker winter pelage in September and October.

Dusky Shrew
(*Sorex monticolus*)

SIMILAR SPECIES

Vagrant Shrews are slightly smaller but are otherwise similar to Dusky Shrews, and the two species may be impossible to tell apart in the field. Vagrant Shrews have a shorter tail, especially in areas of sympatry in mainland British Columbia and Vancouver Island, but overlap does occur in most measurements. East of the Cascade Range, Dusky Shrews and Vagrant Shrews have similar tail lengths; therefore, skull characters must be used to discriminate the species. With a skull in hand, the following characteristics distinguish the two species: the medial tines on the inside margin of the upper incisors (frontal view) of Vagrant Shrews are smaller and usually situated above the upper pigment limit or, if within the pigmented area, they are separated from the pigment by a lighter region; the medial tines of Dusky Shrews are usually larger and located wholly within the pigmented region.

Trowbridge's Shrews are larger, dark grey above and white below, and have whitish feet. The medial edges of their upper incisors are curved, and the medial tine is above the pigmented region. Cinereous Shrews, Arctic Shrews, and Tundra Shrews all have a third unicuspid that is larger or equal in size to the fourth. All of the mentioned skull characteristics are illustrated in appendix 1, "Identification of Shrew Skulls."

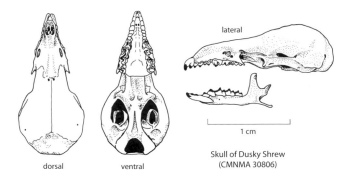

Skull of Dusky Shrew
(CMNMA 30806)

dorsal   ventral   lateral

1 cm

SIZE

The following measurements apply to animals of reproductive age in the Canadian portion of the range. Males and females are similar in size and appearance.
Total length: 93–140 mm; tail length: 35–68 mm; hind foot length: 10–18 mm; ear length: 4–10 mm.
Weight: typically 6–7 g (range 3.7–13.0 g); newborns, unknown.

RANGE

Dusky Shrews have a widespread western North American distribution. They occur from northern Alaska, Yukon, and Northwest Territories to northern Mexico. Their most easterly populations are in Manitoba. Populations in western Canada, the American west, and Mexico can be highly fragmented because they are largely restricted to isolated pockets of boreal or montane habitat, except in the states of the Pacific Northwest. Populations exist on many Pacific offshore islands, including the Queen Charlotte Islands, Vancouver Island, and many islands in the Alexander Archipelago of southern Alaska.

ABUNDANCE

Population density fluctuates seasonally. Generally, numbers are lowest in early spring and gradually climb over the summer as the young are recruited (that is, become independent and join the adult population). Numbers plummet in late autumn and then decline more slowly over the winter. Roughly 4% of juveniles will survive their first winter to breed in the following spring. Population densities of around 5–12 per ha have been reported from suitable habitat in British Columbia during the period of peak numbers in late summer. In general, the Dusky Shrew is uncommon.

ECOLOGY

Dusky Shrews are tolerant of a wide range of montane and boreal habitats but are most common in riparian areas. They have been found in grasslands, northern bogs, damp meadows, alpine tundra, tundra-taiga transition, boreal coniferous forests, temperate mixed, deciduous, and coniferous forests, and dense coastal forests. Moist conditions and the availability of dense undergrowth for cover are important for the Dusky Shrew, and this need may require them to slightly shift their habitat seasonally in order to take advantage of local microclimatic changes. Dusky Shrews typically inhabit the surface layer of the leaf litter and do not create burrows. The only nest found so far was a hollow globular structure constructed of plant stems, sited under woody debris at ground level. It was a maternity nest complete with young and was about 12 cm in diameter. As with other shrews, Dusky Shrews are active throughout the year and do not hibernate. They need to eat frequently and so

are likely to be active in spurts throughout the day or night. Their home-range size varies with the season. Winter ranges are smaller than breeding ranges. A study undertaken near Maple Ridge, British Columbia, on an open forty-year-old forest site found that breeding adults (summer) occupied home ranges that averaged 4020 m², while non-breeding animals (winter) had home ranges that averaged 1227 m². Home ranges of breeding males are larger than those of breeding females, to the extent that an individual male's home range may overlap the home ranges of as many as five breeding females. Individual shrews tend to remain on a home range for their entire life. Life expectancy is about 16–18 months, similar to that of other shrews. The population turnover is virtually 100% as very few, if any, breeding adults are thought to survive a second winter; only young of the year overwinter and reproduce the following spring. Likely predators include owls and possibly hawks as well as some mammalian carnivores that may kill but rarely consume shrews, probably owing to their foul taste.

DIET

The diet of this shrew has not been well studied. It appears to be a generalist, consuming mostly invertebrates but, at times, also conifer seeds, fungi, and lichens. Its large teeth suggest that its prey is predominantly hard-bodied insects, but one study in Oregon found that a large part of the Dusky Shrew's diet comprises insect larvae, which are soft bodied. It is probable that, like other shrews, Dusky Shrews are opportunistic predators capable of killing a wide variety of prey and that their diet varies seasonally and with prey availability.

REPRODUCTION

The breeding season begins once the spring moult is over, and onset varies with locality. In southern British Columbia most females begin breeding in mid-March to mid-April, but the populations further north may delay until May or even June. The season likely ends in August or early September. Females may have one litter or, at most, two litters annually in Canada, depending on the local length of the warm season and the aridity of the area. Dryer-region shrews have fewer litters and may only breed in early spring when local conditions are moist. The length of gestation is unknown. Litter size varies from two to eight young and is typically four to five. Females may undergo a postpartum oestrus. Newborns are altricial, being born hairless, pink, toothless, and blind. They grow quickly; their permanent teeth are fully erupted and they are weaned by the time they are three weeks old. In the Canadian parts of the range young females do not breed in the summer in which they were born.

BEHAVIOUR

Once the juveniles are independent, they attempt to establish and defend a territory, which they then occupy during the following winter. The territories likely allow the juveniles unobstructed access to winter food and enhance their likelihood of winter survival. The territoriality of the juveniles extends to other sympatric shrew species (such as Vagrant Shrews). During the breeding season side glands develop, especially on males, which emit a distinctive odour that is likely used to scent mark. The males, and possibly also the non-breeding females, roam more widely during the breeding season and do not appear to be territorial.

VOCALIZATIONS

Unknown.

SIGNS

It is almost impossible to distinguish shrew signs by species. See the "Signs" section in the Arctic Shrew account for a general discussion of shrew signs.

REFERENCES

Carraway 1990; Carraway and Verts 1999b; George, S.B., 1999b; Hawes 1977; Jameson and Peeters 2004; Nagorsen 1996; Reid, F.A., 2006; Smith, H.C., 1993; Smith, M.E., and Belk 1996; van Zyll de Jong 1983a; Verts and Carraway 1998; Woodward, S.M., 1994; Wrigley et al. 1979.

North American Water Shrew

also called Water Shrew

FRENCH NAME: **Musaraigne palustre,** *also* **Musaraigne aquatique**

SCIENTIFIC NAME: *Sorex palustris*

North American Water Shrews are the world's smallest mammalian divers. Amazingly, along with the similarly semiaquatic Star-nosed Mole, these shrews can actually smell underwater. They do this by expiring a bubble of air out of each nostril so that the bubbles contact the object they wish to investigate. The bubbles are then quickly re-inhaled to deliver the air-borne odour molecules to nasal scent detectors. This very rapid behaviour is detected by high-speed cameras.

DESCRIPTION

This large shrew is distinctly bicoloured, dark almost blue black above in winter and dark grey to grey brown in summer, with a white or silver-grey belly in both pelages. Some individuals may

North American Water Shrew (*Sorex palustris*)

have brownish bellies. The throat region is often the palest part of the underside. The front feet are pinkish white, and the large hind feet are pinkish brown with pinkish-white toes. A fringe of stiff hairs surrounds each front and hind foot, including the toes, and the hind toes are slightly webbed at their bases. The maximum length of the stiff fringe hairs on the edges of the hind foot, where the hairs are longest, is about 1.25 mm. Their long tail is distinctly bicoloured, dark above and whitish below. Their eyes are minute, their ears are just visible in the dense plush fur, and their snout is long and pointed. Mature males develop small (approximately 8 mm long), oval side glands during the breeding season. These glands are covered in whitish hairs and exude an oily, pungent secretion (some say nauseating), especially if the animal is excited or frightened. Side glands (also called flank glands) are not evident on immature males or females. The dental formula is incisors 1/1, unicuspids 5/1, premolars 1/1, and molars 3/3, for a total of 32 teeth. There are four readily apparent upper unicuspids, and the fifth is highly reduced. The third upper unicuspid is slightly larger than the fourth. Upper incisors usually have indistinct medial tines wholly within the reddish pigmentation, but these tines may be absent on some individuals. These shrews moult twice each year, in spring during May and June and in autumn from mid-August until late September.

SIMILAR SPECIES

The similar Pacific Water Shrew is dark on its back, sides, and belly, or only slightly lighter on the belly, and has a uniformly dark tail both above and below, while the North American Water Shrew is markedly lighter on the belly and has a distinctly bicoloured tail that is dark above and whitish below. Arctic and Maritime Shrews are smaller than North American Water Shrews, with shorter tails (≤ 48 mm), and are usually tricoloured with a darker back, lighter sides, and an even lighter belly. Appendix 1 provides an illustrated identification guide to Canadian shrew skulls.

SIZE

The following measurements apply to animals of reproductive age from the Canadian part of the range. Males and females are similar in size and appearance, except during the breeding season when sexually active males gain weight and become considerably heavier (at 13.0–17.9 g) than females (at 7.5–12.0 g).

Total length: 130–179 mm; tail length: 61–89 mm; hind foot length: 16–28 mm; ear length: 3–9 mm.

Weight: 7.5–17.9 g; newborns, unknown.

RANGE

North American Water Shrews are widespread within forested regions of North America, where they occur from central Alaska to northern Labrador, along the Appalachian Mountains to northern Georgia, and in western montane forests to California, central Arizona, and northeastern New Mexico. North American Water Shrews do not occur on any Pacific coast islands, apart from Vancouver Island. They have colonized some of the large islands in the eastern part of their range, including Prince Edward Island, Manitoulin Island, and Cape Breton Island.

ABUNDANCE

North American Water Shrews are generally considered to be rare to uncommon. They appear to be extirpated from Prince Edward Island; despite numerous attempts, no specimens have been captured in over 20 years.

ECOLOGY

As their name suggests, these shrews are rarely found far from water. They are most abundant around open water such as ponds, streams, and wetlands in forested areas, where adequate cover such as overhanging banks, dense vegetation, logs, and boulders is available along the shore. They are not dependant on the availability of open water; some sparse, largely terrestrial populations occur in moist habitats with no open water or with only transient seasonal water, such as bogs or damp, mossy, tamarack or spruce forests. North American Water Shrews spend most of their time on land, where they nest in dry areas and travel in grassy or muddy runways along the water's edge. Nests are globular and hollow and about 8 cm in diameter. They are constructed of leaves, shredded vegetation, and sometimes twigs and are usually found under fallen logs or in cavities in hollow logs. One enterprising water shrew built a twig-and-leaf nest inside a Beaver lodge just above the water line. Burrow use is poorly understood, but burrows that have been reported are mainly streamside in banks with entrances above the waterline. These are approximately 10–12 cm in length and were dug with the front feet, while the hind feet kicked backwards and moved the loosened soil out of the tunnel entrance.

These shrews are strong swimmers, quite capable of mastering cold, rapidly flowing mountain streams and agile enough to manoeuvre in vegetation-clogged marshes. Most North American Water Shrews hunt primarily in the water, although some hunting also takes place on land. Their dense vibrissae (whiskers) function as touch organs to detect not only their surroundings and the shape and texture of prey but also, and more important, any water currents caused by prey movement. Up to 85% of their brain's sensory cortex is devoted to processing information from touch, with only 8.5% devoted to vision and 6.5% to hearing. Within the touch percentage, 70% of that serves the vibrissae, and the remaining 30% serves all the rest of the body. Since most prey is detected by touch, these shrews can hunt successfully in total darkness. Their incredibly fast reflexes and high swimming speeds allow them to pursue and capture small fishes. They can visually detect swimming fish from 12 to 15 cm above the water. All four feet are used for propulsion when

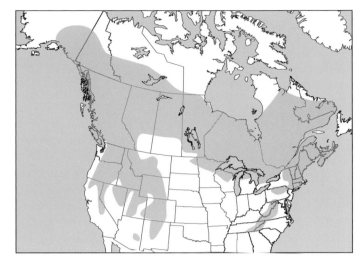

Distribution of the North American Water Shrew (*Sorex palustris*)

swimming, but the larger hind feet provide the primary thrust. The fringe of stiff hairs that surrounds each foot folds back as the foot is drawn forward and then flares out to provide additional surface area when the foot is pushed back against the water.

Like other shrews, North American Water Shrews must eat regularly to fuel their rapid metabolism. As a consequence, they alternately forage (for an average of 30 minutes) and rest (for an average of 60 minutes) throughout a 24-hour period, but they are most active in the few hours before sunrise and after sunset. North American Water Shrews are active through the year and do not hibernate. They continue to hunt in the water, even when it is ice covered. Their thick pelage traps a silvery layer of air near the skin, which insulates the shrews and slows the rate of cooling by about one-half the cooling rate of wet fur. The same air layer enhances their buoyancy, forcing them to paddle hard to dive or swim beneath the surface. After about a minute in the water the fur begins to get wet as the air layer is gradually lost, so most dives are short, especially in winter. Upon emergence and shaking itself, the shrew alternately consumes its catch and stops to groom with rapid strokes of its hind feet. If the swim was very brief, the shrew will emerge with its fur almost dry. During cold weather water shrews can enter daily torpor as an energy-saving measure. Furthermore, their nasal passages are able to warm incoming cold air and cool outgoing warm air to conserve body heat. Very few studies have focused on home-range size, mainly owing to the recapture difficulty of these rare animals. Two shrews in Manitoba reportedly occupied home ranges of 0.2–0.3 ha, and two from Ontario had estimated home ranges of 0.5 ha and 0.8 ha. Maximum lifespan for this water shrew is around 18 months. North American Water Shrews are captured by terrestrial predators such as snakes, weasels, hawks, and owls and by aquatic predators such as trout, bass, walleye, water snakes, and possibly large frogs.

DIET

North American Water Shrews eat primarily aquatic insects, especially juvenile forms of craneflies, caddisflies, stoneflies, and mayflies

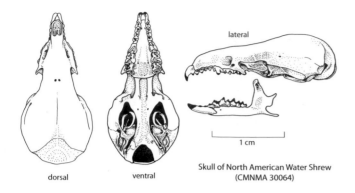

dorsal · ventral · lateral

Skull of North American Water Shrew
(CMNMA 30064)

as well as invertebrates such as crayfish, slugs, snails, spiders, and leaches. Occasionally carrion and aquatic vertebrates such as fishes (up to 8 cm long), small frogs, and salamanders may be taken if available. Terrestrial populations eat mainly insects, including beetles, moth and butterfly caterpillars, and sawfly pupae and larvae, along with invertebrates such as slugs, snails, and earthworms. A captive North American Water Shrew consumes the equivalent of about 95% of its own weight daily, and free water is drunk regularly.

REPRODUCTION

The breeding season of this shrew begins earlier than that of most other North American shrews. Females may become pregnant as early as February, and lactating females can be found until late August. Females experience a postpartum oestrus and are capable of lactation and gestation concurrently. Adult females may have two to three litters annually, with an average of five to six young per litter (ranging from three to ten). The gestation period of this species is unknown but is likely to be around three weeks, the normal for other shrews. Young first emerge from the nest at about three weeks old and start diving to capture their own food about a week later. Juveniles become sexually mature at about nine months old, so most do not breed for the first time until the following breeding season.

BEHAVIOUR

Adult North American Water Shrews are generally solitary, apart from breeding pairs and maternal females. When individuals meet, they normally behave aggressively, beginning by squeaking at each other. If one does not immediately retreat, one or both animals will rear up onto hind legs and continue vocalizing. Should neither retreat at this stage, the two will attempt to slash at each other's head with their teeth. Once physical contact has been made, the two will quickly drop to all fours into a rolling ball of rapidly fighting shrews until one breaks away and flees. Injuries to the head and tail often result from such severe battles, and death of one of the combatants is possible. Such physical attacks are reported in crowded captive situations, but their frequency in the wild is not known. Like many other shrews, North American Water Shrews hunt and kill beyond their immediate needs and then cache the excess food for future use. Large prey is torn into bite-sized chunks with quick upward jerks of the head while the shrew is standing on the carcass, with its front feet holding it down. Smaller prey is manipulated with the front feet

as the shrew sits on its haunches. These shrews commonly use water as escape terrain to elude potential land predators by either diving or running over the surface. With sufficient speed they can travel some distance on the water surface, presumably by the action of the fringe of hairs on their feet that holds air bubbles, allowing them to "skate" using the surface tension. Eventually, if they do not reach land, they lose momentum and must swim or dive.

VOCALIZATIONS

Although an echolocation ability has not yet been proven, these shrews emit an almost continuous low-volume, high-pitched squeak as they run, which may be the audible portion of an echolocation pulse similar to that used for navigation by other shrew species. Loud squeaks are used as a signal of aggressive intent, but the full extent of North American Water Shrew vocalizations has yet to be discovered.

SIGNS

The tracks and other signs of the shrews are similar. Although the North American Water Shrew is one of the larger shrews, its signs are rarely seen because it spends so much of its time in the water or along the shore, where wave action can quickly obliterate them. Burrow entrances are likely around 2.5 cm in diameter and may be camouflaged by sticks or soil. See the "Signs" section of the Arctic Shrew for more generalized shrew sign information.

REFERENCES

Beneski and Stinson 1987; Buckner 1968, 1970; Catania 2006, 2008; Cook, J.A., et al. 1997; Dobbyn 1994; Hartling and Silva 2004; Jameson and Peeters 2004; Laerm et al. 1995; Long, C.A., 2008; Nagorsen 1996; Prescott and Richard 2004; Sealy 2002; Slough and Jung 2007; Smith, H.C., 1993; Soper 1961; van Zyll de Jong 1983a; Verts and Carraway 1998; Whitaker and French 1984; Whitaker and Hamilton 1998; Wrigley et al. 1979; Youngman 1975.

Preble's Shrew

FRENCH NAME: **Musaraigne de Preble**

SCIENTIFIC NAME: *Sorex preblei*

The small Preble's Shrew was first discovered in Canada in the Okanagan Valley of southern British Columbia in late 1994 or early 1995 when one fell into a pitfall trap that had been set to take an inventory of the invertebrate fauna of the area. Since then there have been two further records (also from pitfall traps) from nearby sites. As there are so few Canadian specimens, most of the following information is based on studies of Preble's Shrews made in the western United States.

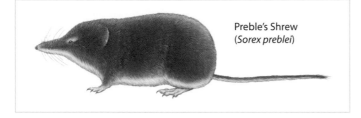

Preble's Shrew
(*Sorex preblei*)

DESCRIPTION

Preble's Shrews are small, dark brown to dark grey shrews with silvery-grey belly fur that extends up onto the sides. Their tail is somewhat bicoloured, brown above and lighter brown below, and darkens towards the tip. Their feet are buffy brown in colour. They have long pointed snouts, plush fur, and small eyes. Their ears are somewhat more conspicuous than those of others in the genus. Their dental formula is incisors 1/1, unicuspids 5/1, premolars 1/1, and molars 3/3, for a total of 32 teeth. The third upper unicuspid is equal to or larger than the fourth. The upper incisors have a medial tine that is relatively long, is pointed, and falls wholly within the pigmented region. The skull is short ($\leq$ 14.8 mm in total length), and the maxillary toothrow is $\leq$ 4.2 mm in length. Appendix 1 contains illustrations of these and other Preble's Shrew skull characteristics.

SIMILAR SPECIES

Cinereous Shrews, Merriam's Shrews, Dusky Shrews, North American Water Shrews, and Vagrant Shrews occur within the range of Preble's Shrews in Canada. North American Water Shrews are much larger (> 7.5 g), dark grey above and whitish below, have fringed hind feet, and are found in wet habitats. Vagrant Shrews are also larger (> 4.0 g), and their sides are about the same colour as their back, or only slightly lighter, and their third upper unicuspid is noticeably smaller than their fourth. Differentiation of Preble's Shrews from Cinereous and Merriam's Shrews requires examination of the skull. Appendix 1 illustrates the skull characteristics that distinguish Preble's Shrews.

SIZE

As there are so few Canadian specimens, the following measurements were gathered from specimens throughout the range. Males and females are similar in size and appearance.
Total length: 77–95 mm; tail length: 28–38 mm; hind foot length: 7–11 mm; ear length: 8–11 mm.
Weight: 2.1–4.1 g; newborns, unknown.

RANGE

Preble's Shrews occur in western North America from the Okanagan Valley of southern British Columbia to northern Nevada and Utah and from central Oregon to eastern Montana. Populations in southern British Columbia are separated from the nearest known populations in central Washington by about 170 km. Within their range Preble's Shrews appear to occur in disjunct pockets, but this could very well reflect unequal sampling effort rather than actual distribution. The three British Columbia records came from two separate sites along Vaseux Creek and another at Mount Kobau. Preble's Shrews are known to occur in northern Montana, not far south of the Alberta border, so it is possible that they may also occur in southern Alberta.

ABUNDANCE

Preble's Shrews are generally considered to be rare to uncommon and are very rare in Canada. The United States considers the Preble's Shrew to be a species of concern.

ECOLOGY

Apart from the habitat around collection sites, very little is known of the biology of this rare shrew. Throughout its range Preble's Shrew occupies a variety of arid to semi-arid grassland or shrub habitats that are typically dominated by sagebrush. However, this shrew has also been found in parklands of intermixed grass, sagebrush, and trees such as aspen, Douglas fir, or pines, as well as around marshes and in wet alkaline habitat. In British Columbia they occur in two somewhat different ecozones. The more montane site (specimen collected in 1998) on the slopes of Mount Kobau, elevation 1724 m, was a moderately grazed grassland patch surrounded by scattered stands of Douglas fir. The grassy clearing had an 80% shrub cover, largely of Big Sagebrush, Snowberry, and Antelope Bush, and standing water was over 2.3 km away. The sites along the Vaseux Creek (specimens captured in 2005 and 2006) were lightly grazed by wild species, although one site had been used as pasture for cattle and horses until 1996. Both sites were on terraces above the creek (elevation 343 m and 452 m) and were predominately grassland with 30%–40% coverage of 2.0–2.5 m high Antelope Bush. Scattered stands of Ponderosa Pine occurred nearby, and standing water was 350–900 m away. Preble's Shrews are active all year round. Predators have not been reported but probably include owls and snakes.

Distribution of the Preble's Shrew (*Sorex preblei*)

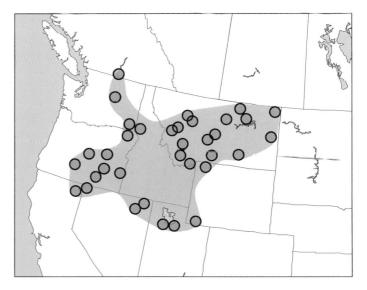

⬤ collection sites

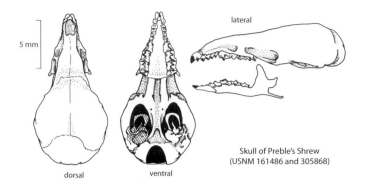

5 mm

dorsal · ventral · lateral

Skull of Preble's Shrew
(USNM 161486 and 305868)

DIET

Their food choices are unknown, although like other shrews it is likely that Preble's Shrews consume a variety of invertebrates. Examinations of the bite forces that exist between the upper and lower incisors predict that Preble's Shrews mostly eat soft-bodied prey (for example, invertebrates such as caterpillars, insect pupae, and spiders).

REPRODUCTION

Little is known of the reproductive biology of this small shrew. Based on embryo counts from five adult females, litter size is three to six. At least two litters are possible annually, and offspring from the earlier litters may become sexually mature in the year they are born. The onset and length of the breeding season are unknown.

BEHAVIOUR
Unknown.

VOCALIZATIONS
Unknown.

SIGNS
Unknown.

REFERENCES
Carraway and Verts 1999a; Cornely et al. 1992; Gitzen et al. 2009; Hendricks and Roedel 2002; Kirkland et al. 1997; Nagorsen et al. 2001; van Zyll de Jong 1983a; Verts and Carraway 1998.

Olympic Shrew

FRENCH NAME: **Musaraigne de Rohwer**
SCIENTIFIC NAME: *Sorex rohweri*

The Olympic Shrew, a very recently described species, was formerly considered to be a Cinereous Shrew until genetic and morphological evidence was presented that indicated it should be considered a full species.

DESCRIPTION

The Olympic Shrew is small with a long pointed snout, tiny eyes, short ears that are barely visible beyond the plush fur, and short legs. Fur on the back and sides is dark brown, and the belly is a somewhat paler tint of the same hue. The tail is bicoloured, dark above and buff below, and uniformly tapered from base to tip. Its feet are buffy brown, and the hair cover on the tail and feet is somewhat sparse. The dental formula is incisors 1/1, unicuspids 5/1, premolars 1/1, and molars 3/3, for a total of 32 teeth. There are five upper unicuspid teeth in each toothrow. The first and second unicuspids are the largest and about equal in size. The third and fourth unicuspids are also similar to each other in size but are smaller than the first pair, and the third may appear slightly smaller than the fourth. The fifth unicuspid is highly reduced and may be difficult to see, especially if gum tissue continues to adhere to the area. Medial tines (on the inside edge of the incisors when viewed from the front) are very small and located at or above the top margin of the pigmentation; they either are unpigmented or have only light pigmentation. The inside margins of the upper incisors diverge at about a 45° angle below the medial tines. Appendix 1 illustrates these skull characteristics.

SIMILAR SPECIES

Most other shrews that occur within the distribution of the Olympic Shrew can be differentiated by external physical differences. Trowbridge's Shrews are larger and dark grey or blackish with a long, sharply bicoloured tail (their total length averages 118 mm, and the tail length averages 55 mm). Both water shrews, the Pacific Water Shrew and the North American Water Shrew, are considerably larger (> 8.5 g). Dusky Shrews are also larger with a longer tail (their total length averages 117 mm, and the tail length averages 51 mm). Furthermore, the medial tines on their upper incisors are located wholly below the upper margin of the pigment and are well pigmented, and their third upper unicuspid is noticeably smaller than the fourth. Vagrant Shrews and Cinereous Shrews can be so similar

Olympic Shrew
(*Sorex rohweri*)

Distribution of the Olympic Shrew (*Sorex rohweri*)

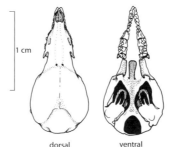
○ collection sites currently outside the known range

to the Olympic Shrew that skull features must be used to distinguish them. The medial tines on Vagrant Shrews are typically located well above the pigmented region of the cusp and are not pigmented; also, the interior margin of the cusp below the medial tine is only slightly splayed outwards at about a 15° angle, and the third upper unicuspid is distinctly smaller than the fourth. Cinereous Shrews have parallel, vertical margins to the medial edge of their upper incisors, and the medial tines are large, darkly pigmented, and located at the upper edge but within the pigmentation zone of the cusp. See appendix 1 for a fully illustrated guide to Canadian shrew skull identification.

SIZE
The following measurements apply to animals from the Canadian portion of the range. Males and females are similar in size and appearance.
Total length: 91–109 mm; tail length: 32–50 mm; hind foot length: 8–13 mm.
Weight: 2.5–6.0 g; newborns, unknown.

RANGE
This small shrew has a limited distribution in western North America. It occurs in three disjunct populations: the Olympic Peninsula in northwestern Washington State; the Cascade Mountains in southwestern Washington State; and, in the northern limit of its range, extreme southwestern British Columbia, south of the Fraser River. The species was first recognized in 2007, and the extent of its

range in British Columbia is still uncertain. There are 21 specimens of Olympic Shrews known from British Columbia, from eight localities. Eight of these specimens are from Burns Bog, located 25 km southeast of downtown Vancouver. Burns Bog is the most westerly locality, while the east side of Chilliwack Lake, roughly 100 km east of Burns Bog, is the most easterly.

ABUNDANCE
Olympic Shrews are rare and have a restricted distribution in Canada in a region where the habitat is highly fragmented by urban and agricultural development. Many of the recent specimens are from a 3000 ha patch of natural bog and forest (Burns Bog) that is poorly protected; it is encroached upon by highways and, at its south end, by a garbage dump for the City of Vancouver. The isolation of the disjunct populations makes the conservation of this shrew in North America, and especially in British Columbia, a concern. Some populations on the Olympic Peninsula that have been studied for nearly 20 years have fluctuated widely in density from year to year, but the cause of this variation is unknown.

ECOLOGY
Most of the biology of this species remains to be discovered. Olympic Shrews occupy a diversity of forested or forest-edge habitats, from about sea level to possibly as high as 1400 m in elevation, although most of the Canadian and all of the American records locate them below 700 m. Canadian specimens have been captured in coniferous or mixed forests and forest edges. The Burns Bog locality is a domed peat bog that is variously forested with coniferous trees and shrubs and interspersed with open areas. The American specimens occur in moist old growth and second-growth forests.

DIET
Unknown.

REPRODUCTION
Unknown.

BEHAVIOUR
Unknown.

VOCALIZATIONS
Unknown.

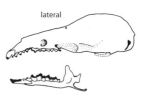
Skull of Olympic Shrew
(CMNMA 55048)

1 cm

dorsal   ventral   lateral

SIGNS

The signs of the Olympic Shrew are unknown but are likely to be similar to those of other small shrews. See the "Signs" section of the Arctic Shrew account for more information.

REFERENCES

Nagorsen and Panter 2009; Rausch et al. 2007.

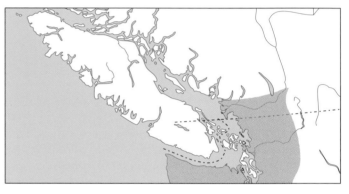

expanded view of Canadian range

Trowbridge's Shrew

FRENCH NAME: **Musaraigne de Trowbridge**

SCIENTIFIC NAME: *Sorex trowbridgii*

The medium-sized, Trowbridge's Shrew reaches the northern limit of its distribution in extreme southwestern British Columbia.

DESCRIPTION

Trowbridge's Shrews have long, sharply bicoloured tails that are dark grey brown or black above and white below. Their tail loses hair as they age, to the point where older animals have naked tails. Pelage on the back and sides is medium to dark sooty grey (winter) or brownish grey (summer), and the underside is similarly coloured but typically slightly paler. Juveniles are browner than adults until their first moult. Fur on the feet is sparse and whitish. Their feet have a pinkish tint as the skin shows through the fur. The dental formula is incisors 1/1, unicuspids 5/1, premolars 1/1, and molars 3/3, for a total of 32 teeth. There are five upper unicuspids, and the third is smaller than the fourth; the fifth unicuspid is very small and not easily apparent. The medial edges of the upper incisors are recurved in the anterior aspect, and each incisor has a small medial tine that is located at or slightly above the line of pigmentation. These features are illustrated in appendix 1. Age class can be determined by tooth wear; juveniles exhibit little wear, whereas animals of one year or older have heavily worn dentition. These shrews experience two annual moults: the spring moult takes place during late May and June, and the autumn moult occurs from late August until mid-November. The winter coat is longer, thicker, greyer, and often paler than the summer coat. Very few individuals live long enough to moult into winter pelage a second time.

SIMILAR SPECIES

Cinereous Shrews, Vagrant Shrews, and Dusky Shrews occur within the range of Trowbridge's Shrews. They are clearly distinguished from Trowbridge's Shrews by the lack of a sharply bicoloured tail. The Cinereous Shrew is smaller than the Trowbridge's Shrew, and its third upper unicuspid is larger than its fourth. Vagrant and Dusky Shrews have a fairly straight medial edge to their upper incisor. Skull characteristics that are useful in identifying shrew species are illustrated in appendix 1.

SIZE

The following measurements apply to animals of reproductive age from the Canadian part of the range. Males and females are similar in size and appearance.

Total length: 101–128 mm; tail length: 41–62 mm; hind foot length: 10–15 mm; ear length: 5–8 mm.
Weight: 3.8–8.0 g; newborns, unknown.

RANGE

Trowbridge's Shrews occur in western North America from southwestern British Columbia through coastal Washington State and

Trowbridge's Shrew
(*Sorex trowbridgii*)

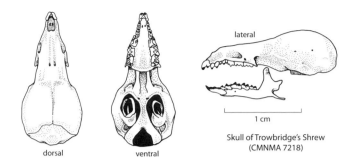

dorsal   ventral

lateral

1 cm

Skull of Trowbridge's Shrew
(CMNMA 7218)

Oregon into California. The only known island population occurs on Destruction Island, about 5 km off the west coast of Washington's Olympic Peninsula. In British Columbia the bulk of the range is south of the Fraser River, although there are three specimen records from the north side of the Fraser Valley (Point Grey, Maple Ridge, and the south side of Harrison Lake).

ABUNDANCE

These shrews are on the British Columbia "Blue List," although recent survey efforts have found that numbers may be healthier than previously thought. Population density fluctuates, peaking in late summer and early autumn as the juveniles are recruited, and dropping abruptly in early winter as the adults die. Numbers fall to their lowest in late winter and early spring following winter attrition. There are no population density figures available for this species, but it is currently the most abundant shrew in the lower Fraser Valley. Much of the range of Trowbridge's Shrews in British Columbia falls within the lower Fraser Valley. This area is undergoing rapid urbanization, clearcut logging, or intensive farming, which may place localized populations at risk in future and is already extensively fragmenting their habitat.

ECOLOGY

Trowbridge's Shrew prefers habitats with a deep organic layer and abundant cover, and it tends to occupy dryer sites than do most other shrews. When it occurs in riparian situations, it is not usually found right beside the water but on dryer nearby slopes. Coastal populations in British Columbia are most common in dry mixed or coniferous forests, but they can also be found in some recent clearcuts and in moister locations, such as wet forests and shrubby ravines. Trowbridge's Shrews, being strong diggers, mostly forage just beneath the surface. They are active intermittently at any time of the day or night but are most active in darkness. Each approximately one-hour-long activity cycle consists of a short active phase (mostly spent foraging) followed by a quiescent resting or digesting phase. About 40% of each 24-hour period is spent in the active phase. Like other shrews, they remain active throughout the winter and do not hibernate. Trowbridge's Shrews are strong and energetic burrowers, and their extensive tunnel systems are commonly used by shrews, like Vagrant and Dusky Shrews, that are less capable diggers. This digging capacity may explain the Trowbridge's Shrews' avoidance of water-saturated soils where there is a high water table. Trowbridge's Shrews are capable climbers and can forage above ground in the undergrowth and even into trees. Nests are placed in or under woody debris or in shallow burrows. Maximum lifespan is about 18 months. Most, if not all, of the breeding adults die by late autumn, and only the young of the year remain to overwinter and breed the following spring. There is no information available concerning their home-range size or territoriality. Owls, especially Barn Owls, are their primary predators in British Columbia. Giant Pacific Salamanders are reported to take Trowbridge's Shrews, and it is likely that some snakes do also. Domestic Cats sometimes kill Trowbridge's Shrews but rarely eat them, presumably owing to the disagreeable odour from the flank glands.

DIET

Trowbridge's Shrews are thought to be generalist foragers, taking a wide range of invertebrates and even some plant material, especially seeds of Douglas fir and some fungi. Beetles and other larval and adult insects as well as snails, slugs, spiders, and centipedes compose their main prey.

REPRODUCTION

Information about their reproduction is sparse. The breeding season begins earlier than it does for most other shrews in North America. Trowbridge's Shrews in British Columbia appear to breed from March until about early June, and females may have two litters annually. Further south, breeding may commence more than a month earlier, and females may be capable of producing three litters annually. Litter size is typically four to five (ranging from three to six). Females may experience a postpartum oestrus and can gestate and lactate concurrently. Newborns have not been described, and their development is unknown because it is difficult to breed Trowbridge's Shrews in captivity. In Canada, and possibly throughout the range, the young of the year do not breed until the following spring. Juvenile females become sexually mature about two weeks before the males do.

BEHAVIOUR

The social structure and behaviour of this shrew is very poorly known. Hoarding or food-caching behaviour has been noted in captives.

VOCALIZATIONS

Unknown.

SIGNS

The signs of all the shrews are difficult to find and largely impossible to ascribe to particular species. See the "Signs" section in the Arctic Shrew account for a general description of shrew signs.

REFERENCES

Carraway 1987; George, S.B., 1989, 1999c; Jameson and Peeters 2004; Lee 1995; Nagorsen 1996; Reid, F.A., 2006; van Zyll de Jong 1983a; Verts and Carraway 1998.

Tundra Shrew

FRENCH NAME: **Musaraigne de toundra**

SCIENTIFIC NAME: *Sorex tundrensis*

Tundra Shrews have had a confusing taxonomic history. They were considered to be a subspecies of the Arctic Shrew until the mid-1970s, when they were granted full species status and were thought to occur not only in North America but also across the Bering Strait into northern Eurasia. Genetic work in the early 1990s showed that there are substantial differences between the similar-looking Eurasian and North American forms, sufficient to consider each a distinct species.

DESCRIPTION

Tundra Shrews are medium-sized shrews with tricoloured summer pelage. This coat is brown on the back, with sharply contrasting pale-brown sides and a pale greyish-white belly. The dark dorsal band (the saddle) is paler, and the demarcation between the back and sides is less distinct on juveniles and subadults. The thick winter pelage is bicoloured, with the greyish-white sides and belly contrasting with the brown to reddish-brown back. The tail darkens towards the tip and is indistinctly bicoloured, brown above and buffy brown below in both seasons. The feet are buffy brown. The dental formula is incisors 1/1, unicuspids 5/1, premolars 1/1, and molars 3/3, for a total of 32 teeth. There are five readily apparent upper unicuspids in a fairly graduated series from the largest first unicuspid to the smallest fifth unicuspid except that the third unicuspid is considerably larger than the fourth. Adults have two annual moults. The spring moult begins in April or May and is complete by sometime in June. Winter fur begins growing in late August to September, and the pelage is complete by early to mid-October. Juveniles moult into adult pelage in their first autumn, and late-born juveniles moult later than do older juveniles.

SIMILAR SPECIES

Although the ranges of the Tundra and Arctic Shrews are not known to overlap, it is possible that this perception is caused by a lack of sampling rather than by an actual distributional distinction; therefore, the differences between the two are herein noted. The saddle-back pelage of the Tundra Shrew is brown, rather than the very dark brown or blackish of the Arctic Shrew in the same region. Tundra Shrews are almost as colourful as are Arctic Shrews, and their pelage patterns are similar, although Tundra Shrews are generally browner overall with lighter bellies. Tundra Shrews share the unicuspid traits

Tundra Shrew
(*Sorex tundrensis*)

Distribution of the Tundra Shrew (*Sorex tundrensis*)

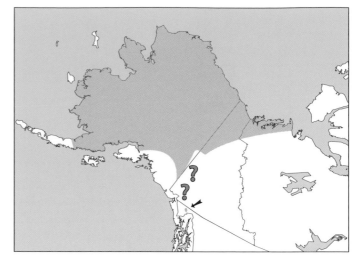

? presence uncertain

of the Arctic Shrews, but they have shorter tails (≤ 36 mm in length) and their skull length is shorter (16.8–18.4 mm vs 18.5–19.7 mm for Arctic Shrews). Juvenile Tundra Shrews resemble juvenile Arctic Shrews but have shorter tails and greyish, rather than brownish, underparts. Barren Ground Shrews are also similar in coloration to Tundra Shrews. The pale-coloured fur on their sides extends higher, and they are much smaller than Tundra Shrews (≤ 5.2 g in weight, and ≤ 15.4 mm in skull length). Skull characters that are useful in shrew identification are fully illustrated in appendix 1.

SIZE

As the number of Canadian specimens is limited, the following measurements apply to animals of reproductive age throughout the range. Males and females are similar in size and appearance. Total length: 83–120 mm; tail length: 22–37 mm; hind foot length: 11–15 mm; ear length: 4–10 mm.
Weight: 4.9–14.0 g; newborns, unknown.

RANGE

These northern shrews occur only in Northwest Territories, Yukon, Alaska, and extreme northwestern British Columbia. The British Columbia population is found about 300 km south of the closest known populations in Alaska. This disjunction may be real or it may be caused by a lack of successful sampling in the intervening region.

ABUNDANCE

Considered very rare in British Columbia, Tundra Shrews are on the provincial "Red List." Elsewhere throughout their range they are generally more common, but wide fluctuations of abundance do occur.

ECOLOGY

The biology of Tundra Shrews is poorly understood. They are most abundant in well-drained regions of dense vegetation in taiga

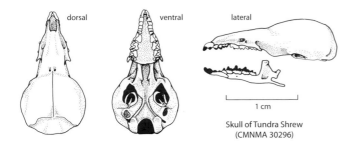

dorsal　　ventral　　lateral

1 cm

Skull of Tundra Shrew
(CMNMA 30296)

forests, and in alpine and arctic tundra. Shrubs like alder and Labrador Tea, and dwarf trees, such as dwarf birch and dwarf willow, along with grasses and forbs are common vegetative components of their habitat. In lesser numbers they may also be found in wetter habitats. These shrews cannot survive without eating every few hours, so they are intermittently active throughout each 24-hour period and, like other shrews, probably alternate a short foraging excursion of around 30 minutes with a longer resting or digesting period of roughly 60 minutes. They are thought to use runways created by other small mammals. During winter they remain active under the snow.

DIET
Information on the diet of Tundra Shrews is scanty. Specimens from Alaska have been found with insects, earthworms, and plant remains (grass florets) in their stomachs.

REPRODUCTION
The breeding season (in Arctic populations) likely begins in May and extends until September. Females have been reported to be pregnant from May until early September. Litter size, based on embryo counts, ranges from eight to twelve. Females may experience a postpartum oestrus, as concurrently lactating and pregnant individuals have been reported. Females are capable of producing up to two litters annually, and early-born females may mature sufficiently to breed late in their first summer.

BEHAVIOUR
Unknown.

VOCALIZATIONS
Unknown.

SIGNS
The Tundra Shrew's signs are unknown but are likely similar to generalized shrew signs. See the "Signs" section of the Arctic Shrew account for more information.

REFERENCES
Junge et al. 1983; Nagorsen 1996; Peirce and Peirce 2000a; Rausch and Rausch 1993; Reid, F.A., 2006; Slough and Jung 2007; van Zyll de Jong 1983a, 1999b; Youngman 1975.

Barren Ground Shrew

FRENCH NAME: **Musaraigne de Beaufort,** *also* **Musaraigne de Béringie**

SCIENTIFIC NAME: *Sorex ugyunak*

The biology of the Barren Ground Shrew is virtually unknown.

DESCRIPTION
This small shrew is distinctively coloured. It has a light- to medium-brown stripe running from its nose, along its back to the top of its tail, while the sides, cheeks, throat, belly, and feet are whitish. There is a fairly sharp demarcation high on the sides between the dark upper and paler lower fur. The tail is bicoloured like the body, darker above and whitish below. Juveniles are darker than adults, with a less sharply defined delineation between the dark and light fur. The dental formula is incisors 1/1, unicuspids 5/1, premolars 1/1, and molars 3/3, for a total of 32 teeth. Adults have two annual moults. The exact timing of the spring moult is unknown, but it is probably complete by early June, and the autumn moult is complete by mid-October.

SIMILAR SPECIES
The only other shrew of similar size that occurs within the range of the Barren Ground Shrew is the Cinereous Shrew. Cinereous Shrews are darker overall. Their paler belly fur does not extend up the sides of the body, and the belly colour merges with the upper coloration without leaving a distinct demarcation line. The other shrew that lives in the tundra is the Tundra Shrew. This shrew is considerably larger but shares a similar colour pattern to that of the Barren Ground Shrew, although the line between the dark and light fur is considerably lower on the sides of the Tundra Shrew. Appendix 1 provides an illustrated identification guide to Canadian shrew skulls.

SIZE
The following measurements apply to animals of reproductive age from the Canadian part of the range. Males and females are similar in size and appearance.
Total length: 74–103 mm; tail length: 22–31 mm; hind foot length: 10.0–13.5 mm.
Weight: 2.9–5.2 g; newborns, unknown.

Barren Ground Shrew
(*Sorex ugyunak*)

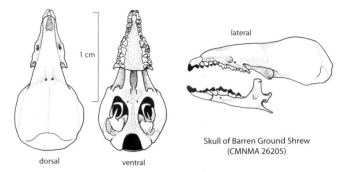

Skull of Barren Ground Shrew
(CMNMA 26205)

dorsal · ventral · lateral · 1 cm

RANGE

This northern shrew occurs in the tundra zone from the northwestern shore of Hudson Bay along the northern mainland of Nunavut, Northwest Territories, and Yukon and into the northern coastal plain of Alaska to Point Barrow. It was formerly thought to reach across the Bering Sea to the eastern tip of Siberia, but this has proven not to be the case, because the Russian shrew has been identified as *Sorex portenkoi* rather than *Sorex ugyunak*.

ABUNDANCE

The abundance of the Barren Ground Shrew is unknown.

ECOLOGY

Barren Ground Shrews occur north of the treeline. They are primarily collected in moist grass or sedge meadows and thickets of dwarf birch and dwarf willow. They have also been found among the rubble of old, abandoned Inuit winter homes.

DIET

It is unknown, but likely, that their diet, like that of other shrews, is mainly composed of invertebrates. When they can slip between the rocks composing the cairn, Barren Ground Shrews will raid Inuit food caches to consume the meat within. Likely they will also eat naturally occurring carrion that they find.

REPRODUCTION

Unknown.

BEHAVIOUR

Unknown.

VOCALIZATIONS

Unknown.

SIGNS

Its signs are probably similar to those of the Arctic Shrew.

REFERENCES

van Zyll de Jong 1983a; Zaitsev 1988.

Vagrant Shrew

also called Wandering Shrew, Salt Marsh Shrew

FRENCH NAME: **Musaraigne errante**

SCIENTIFIC NAME: *Sorex vagrans*

Vagrant Shrews are small, carnivorous mammals that employ a crude form of echolocation to help them find their way in the dark.

DESCRIPTION

The dorsal pelage colour of inland Vagrant Shrews is medium to dark brown or grey, but in some coastal populations it is almost black. The undersides are lighter, varying from light brown to grey. Juveniles have a distinctly bicoloured tail that is dark brown above and whitish below, but adults have a tail that is less distinctly bicoloured, being light to medium brown above and whitish to buffy below. Both males and females possess flank glands, but the glands on reproductively active males are the largest, most apparent, and most odoriferous. An experienced field biologist can accurately gender-identify a breeding male merely by its odour. The dental formula is incisors 1/1, unicuspids 5/1, premolars 1/1, and molars 3/3, for a total of 32 teeth. The five upper unicuspids are easily visible in lateral view, and the third is noticeably smaller than the fourth. Vagrant Shrews have a small medial tine on the inside edge of each upper incisor, which is positioned at or slightly above the pigmented portion of the tooth. This character is only useful for identification of younger animals because the pigmented portions of the teeth gradually wear away; the dentition of older adults is entirely white, and the medial tine may have been worn away. The medial edges of the incisors appear straight or only slightly curved when viewed from the front. Adults experience two annual moults. A spring

Distribution of the Barren Ground Shrew (*Sorex ugyunak*)

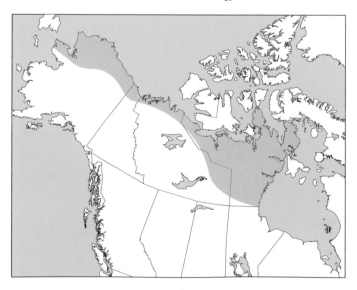

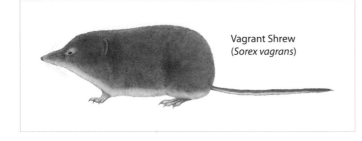

Vagrant Shrew
(*Sorex vagrans*)

at their peak. Other work in California found densities of up to 58.2 per ha, also in late summer. Population density drops drastically in autumn.

ECOLOGY

Vagrant Shrews are most common in moist grassy areas and open habitat that is scattered with shrubs and groups of deciduous or coniferous trees. They are rarely found in dense or closed forests or in arid regions, but they are flexible in their requirements and are known to occur in a variety of habitats. These include alpine tundra and high meadows, salt and fresh water marshes, forest clearings,

moult occurs in late winter and is usually complete by late March. An autumn moult begins in September, progressing from the snout rearward and from the rump forward at the same time. The winter coat is usually complete by early October and is longer and slightly darker than the summer coat.

SIMILAR SPECIES

The third upper unicuspid is equal in size to, or slightly larger than, the fourth in the Cinereous Shrew. Both the Dusky Shrew and the Trowbridge's Shrew share the Vagrant Shrew's characteristic of having a smaller third unicuspid. Trowbridge's Shrews have greyer fur and distinctly bicoloured tails that are usually longer than 50 mm. They also have a very curved medial edge to the upper incisor (in frontal view), while the same character of Vagrant and Dusky Shrews is relatively straight. Dusky Shrews usually have longer skull lengths (> 17.2 mm for Dusky Shrews versus < 17.2 mm for Vagrant Shrews), but there is considerable overlap because many smaller Dusky Shrews have skulls that are < 17.2 mm long. Another character that is useful in discriminating the two species is the medial tine on the upper incisors. Dusky Shrews have a larger medial tine, which is positioned within the pigmented region. Distinguishing live Dusky Shrews from Vagrant Shrews in coastal British Columbia is especially difficult. Appendix 1 provides illustrations of distinguishing skull characteristics.

SIZE

The following measurements apply to animals of reproductive age in the Canadian portion of the range. Males and females are similar in size and appearance.
Total length: 85–126 mm; tail length: 32–58 mm; hind foot length: 9–15 mm; ear length: 4–10 mm.
Weight: average 5–8 g, ranging from 4 to 10 g; newborns, 0.35–0.50 g.

RANGE

Vagrant Shrews are a western North American species that occurs from Vancouver Island and south-central British Columbia to extreme southwestern Alberta, and southward to central Arizona and central California. A very disjunct, relict population occurs in central Mexico along the southern end of the Mexican Plateau.

ABUNDANCE

This shrew is relatively common in suitable habitat within its range. A British Columbia study put the Vagrant Shrew population at around 12 per ha in the latter half of the summer when densities are

Distribution of the Vagrant Shrew (*Sorex vagrans*)

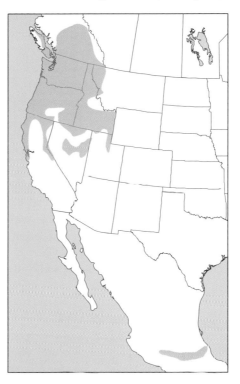

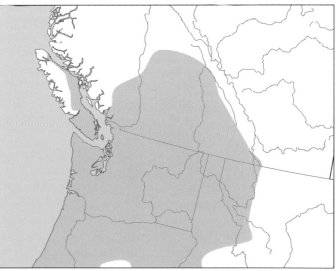

expanded view of Canadian range

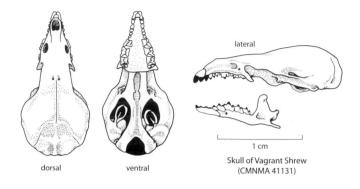

dorsal   ventral

lateral

1 cm

Skull of Vagrant Shrew
(CMNMA 41131)

REPRODUCTION

The breeding season begins as early as March in warm lowland sites in British Columbia, and as late as late April in cooler or upland regions, and extends until August or September. Breeding begins earlier in warmer, more southerly parts of the range. Gestation lasts about 20 days, and litter size averages four to six (ranging from one to nine). Females experience a postpartum oestrus and may produce up to three litters annually. They continue to reproduce even into old age. Young are altricial, being born helpless, with no hair, and with their eyes fused shut. They begin to show fur on their dark pinkish-red skin by two weeks, and their eyes open by their third week. By the time they are 25 days old, they are weaned, but they usually continue to use their natal nest until they are about five weeks old. Early born females may breed in the year in which they were born in some populations, but breeding may be delayed until the following spring in others where the breeding season is shorter owing to local climatic conditions. Males do not breed in the year in which they were born; they wait until the following breeding season, when they are around 7–11 months old.

BEHAVIOUR

These shrews are sometimes territorial. Newly solitary subadults establish and defend a home range from other subadults throughout their first autumn and winter. Since most of the animals in a winter population are subadults, this action may help ensure that they have adequate space to find sufficient winter food. The territories break down in spring with the onset of the breeding season because breeding individuals have overlapping home ranges. Captive Vagrant Shrews tend to be solitary and actively avoid each other. Individuals often defend their nest site from intruders, with vocalizations and short charges, but physical battles are rare. Nevertheless, like most shrews, Vagrant Shrews can be irascible and are known to attack animals larger than themselves. Three different kinds of nests are constructed. Warm weather resting nests are round or oval and shallowly cup-shaped (2–8 cm in diameter). Natal nests may be considerably larger and are domed, with a hollow centre, although they commonly lose the ceiling of the dome as the infants grow; they are 6–24 cm in diameter and 4–6 cm high. Cold weather resting nests are intermediate in size and are also domed. Nests are composed of dried leaves and grass stems and commonly incorporate moss, fur, and feather linings. They are placed under cover, such as woody debris or rocks. As do many other shrews, Vagrant Shrews will cache surplus food items for later consumption. Grooming activities are usually both rapid and brief, typically restricted to anogenital licking, scratching with the hind feet, and quick swipes of the snout with the front feet. Their long flexible snout is constantly twitching up and down and from side to side as the animals forage. Clearly, prey is regularly detected by scent.

VOCALIZATIONS

Vagrant Shrews are capable of a crude form of echolocation using double pulses in a relatively low frequency (18–60 kHz). Their short range capabilities imply that predator or prey detection is not likely but that echolocation is used to provide them with an impression of

early successional clearcuts and burns (< 10 years old), and open coniferous forests. In British Columbia most Vagrant Shrews live in the lowlands below 400 m elevation, but specimen records indicate that they exist in areas of up to 2133 m elevation. Their home ranges are largest during the early part of the breeding season. A study, conducted in British Columbia, found that breeding males occupy average home ranges of 4343 m², and breeding females 2233 m², while non-breeding males and females occupy around 1039 m². It also found that both females and males tend to remain on the same home range throughout their lives but that males are more likely to change their home range during the breeding season as they search for oestrous females. A similar study in Washington State, using the minimum area method, found adult males and females occupying an average of 316 m² in April and May. Home-range size likely varies depending on the habitat, food availability, and population density. Vagrant Shrews are active throughout the year and do not hibernate. Their high metabolism demands that most shrews are intermittently active day and night, although most activity takes place at night. Their normal activity pattern is a period of foraging followed by a rest to digest their meal. Vagrant Shrews are capable climbers, using their tail for balance in a semi-prehensile way by partially wrapping it around objects. Average life expectancy is 6–7 months, with a small percentage living 16–17 months and a very few living 24–26 months. Most breeding adults die by the autumn. Predators include mainly a variety of owls and snakes and occasionally mammalian carnivores such as Bobcats and Domestic Cats. The flesh of Vagrant Shrews, like that of most shrews, is strongly flavoured and generally distasteful to mammalian predators. Although they may be pursued and killed, especially by Domestic Cats, shrews in general are only eaten if the predator is desperately hungry.

DIET

These largely carnivorous small mammals eat a wide range of invertebrates, including earthworms, spiders, crickets, caterpillars, slugs, snails, beetles, centipedes, craneflies, grasshoppers, bees and wasps, carrion, and even some fungi and plant material such as flowers and seeds. Very small vertebrates, such as salamanders and small frogs, may occasionally be taken. Vagrant Shrews must eat at least every two to three hours or they risk death by starvation. In captivity, Vagrant Shrews eat more than 1.7 times their own weight in food each day, and they need access to drinking water.

their environment to help them get around quickly, expeditiously, and quietly. These shrews also produce squeaks during interactions.

SIGNS

It is almost impossible to distinguish shrew signs by species. See the "Signs" section in the Arctic Shrew account for a general discussion of shrew signs.

REFERENCES

Buchler 1976; Carraway 1987, 1990; Carraway and Verts 1999b; Findley, J.S., 1999; Foresman and Long 1998; Gillihan and Foresman 2004; Hawes 1977; Jameson and Peeters 2004; Nagorsen 1996; Reid, F.A., 2006; Smith, H.C., 1993; van Zyll de Jong 1983a; Verts and Carraway 1998; Woodward, S.M. 1994.

FAMILY TALPIDAE
moles

Moles are larger than shrews, although the Shrew-mole is an exception. Moles have complete zygomatic arches that are very thin and fragile. Their fur sticks straight out from the body and can lie either forwards or backwards, allowing the animals to move easily in any direction in tight spaces.

Star-nosed Mole

FRENCH NAME: **Taupe à nez étoilé,** *also* **Condylure à nez étoilé**

SCIENTIFIC NAME: *Condylura cristata*

The unique Star-nosed Mole is easily recognized by the twenty-two short, pink tentacles (also called nasal rays) that extend outwards from the tip of its snout. The nasal rays are covered by tens of thousands of touch-sensitive receptors, called Eimer's organs (which are only found in moles), making the star the most highly sensitive and developed touch organ among mammals, far surpassing the abilities of the human hand. As the mole forages or travels through its tunnels, the rays are constantly sweeping the air or water and touching the substrate, likely providing navigational cues as well as detecting prey. The Star-nosed Mole is the only semiaquatic mole in North America.

DESCRIPTION

The Star-nosed Mole is a dark mole with a long tail that is almost half of its total length (about as long as the body without the head). Adults are dark greyish black or brownish black with a slightly paler belly. The wrists are pale tan. Worn pelages show more of the underfur and are slightly paler. Juveniles, until their autumn moult, are even darker than adults. The pelage is composed of short guard hairs and underfur and of some guard hairs that are distinctly longer. This makes their coat different from that of the more fossorial moles, which lacks underfur and is very plush and velvety because all of the guard hairs are similar in length. Nevertheless, as with other moles, each guard hair can bend forwards or backwards with ease, allowing Star-nosed Moles uninhibited movement in either direction while they are underground. The scaly tail is sparsely furred with dark hairs, constricted at the base, and tapered towards the tip. It varies in diameter through the year, usually becoming enormously swollen with fat deposits in winter and early spring, possibly as an energy

Star-nosed Mole
(*Condylura cristata*)

Total length: 162–238 mm; tail length: 64–92 mm; hind foot length: 15–32 mm.
Weight: 50–70 g; newborns, about 1.5 g.

RANGE

Star-nosed Moles are an eastern North American species that occurs further north than do any other of the Canadian moles. They are found from northern Quebec and Labrador southwards to the Okefenokee Swamp of southeastern Georgia. The range extends westward from Nova Scotia to southeastern Manitoba and northern Minnesota. Isolated populations occur along coastal Georgia and possibly in Riding Mountain National Park in southwestern Manitoba and in the northeastern corner of North Dakota. The Star-nosed Mole has not occupied large offshore islands such as Prince Edward Island, Newfoundland, Anticosti Island, or Manitoulin Island. Owing to its habitat preferences (wet soils), the occurrence of this mole tends to be patchy throughout its range.

ABUNDANCE

Star-nosed Moles can be common in suitable habitat. Densities of 25 and 41 individuals per ha have been reported, although population densities and seasonal or annual variation of densities across the range are poorly understood.

ECOLOGY

This semiaquatic mole is equally at home in water or on land. It usually occupies wet, sometimes water-logged soils, often alongside

reservoir to help withstand the rigours of the early breeding season. The eyes are small but larger than those of other Canadian moles. However, given that the optic nerves are minute, the eyes are likely only useful for light-intensity discrimination. The ear openings are also larger and lack external pinnae (ear flaps), but they are well protected from soil and debris by the surrounding fur. Fossorial adaptations include short legs, a stocky body, and large, spade-like, scaly forefeet that are as broad as they are long and that are armed with pale tan-coloured, broad, flat claws for digging. The hind feet are smaller but also scaly. All feet have five toes. As with most other moles, the forearm articulates with the collar bone, rather than with the shoulder blade as is common in other mammals. This results in a more forward placement of the limb and in its rotation into an ideal digging position. Sensory vibrissae project from the eyebrows, wrists, and snout. Nostrils are at the tip of the snout, inside the ring of nasal rays, and open towards the front. Glandular secretions are most pronounced during the breeding season and can produce a golden stain on the fur on the top of the head, chin, wrists, and genital area. The dental formula is incisors 3/3, canines 1/1, premolars 4/4, and molars 3/3, for a total of 44 teeth. The first upper incisors protrude forwards, the second are minute, and the third look like canines and are larger than the next tooth in the toothrow, which is the actual canine. The auditory bullae are incomplete. They have eight mammae. Star-nosed Moles experience two annual moults: the spring moult occurs sometime within the three months of May to July; the autumn moult occurs in September or October. As with other mole species, it is impossible to externally determine the sex of individuals outside of the breeding season.

SIMILAR SPECIES

The fringe of nasal rays encircling the tip of the snout distinguishes this species from any other mole and is unique among mammals. The tail of the Star-nosed Mole is also significantly longer than that of fossorial mole species. The skull can be distinguished from that of other Canadian moles by the forward projection of the first upper incisors, which are tweezer-like in appearance and function (they appear to be an adaptation for grasping minute invertebrate prey that populate the moist environment of the Star-nosed Mole).

SIZE

The following measurements apply to animals of reproductive age from the Canadian part of the range. Males and females are similar in size and appearance, which is unusual in moles. Size increases with latitude.

Distribution of the Star-nosed Mole (*Condylura cristata*)

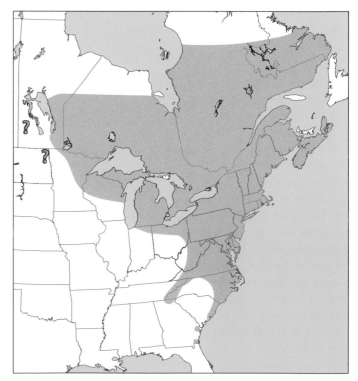

? possible distribution

waterways, seeps, and even seasonal rills, but it can also survive in dryer upland sites and seasonally wet locations. Wet sedge meadows, fields, woods, and swamp edges are its prime habitat. Although Star-nosed Moles typically occupy lowland habitats, they have been reported at elevations between 760 m and 1676 m at the southern edge of their range. Star-nosed Moles are intermittently active day and night with roughly a 4 hour activity cycle followed by about a 4 hour rest period. They have a higher rate of metabolism than do other Canadian moles (with the exception of the Shrew-mole) and must eat frequently; they can easily consume their body weight in prey items in a single day. Their nests are circular wads of leaves and grass about 15 cm in diameter, which are placed in chambers (about 13 cm wide by 10 cm high) that are commonly situated under stumps and tree roots and 7–25 cm below the ground surface. Maternal nests are larger and more elaborate than resting nests, and their location is carefully chosen so that they remain dry and above the flood line while the vulnerable young are resident. Tunnels are dug from just below the surface to around 60 cm in depth, although Star-nosed Moles also construct many surface runways that are generally hidden by thick thatches of vegetation. The shallow summer tunnels are often marked by a pushed-up ridge of soil as the humus-rich soil is compressed against the walls and roof of the tunnel by the mole as it digs. Molehills associated with surface tunnels are rare. Deeper tunnels usually require the creation of a molehill because the more mineralized soil at those depths cannot be compressed and the waste must be shoved to the surface. The deeper tunnels are used in winter, when earthworms and other invertebrates move down below the frost line. However, Star-nosed Moles occasionally travel to the surface in winter and have been observed tunnelling through snow. Tunnels produced by this mole are 3–5 cm in diameter, and entrances are commonly underwater in order to provide safe and convenient access to their hunting grounds, in both summer and winter.

The Star-nosed Moles are very competent swimmers and accomplished divers, using their broad forefeet as oars. When they are swimming, as when walking, the diagonally opposite front and hind limbs move together. Star-nosed Moles continue to forage in the water (under the ice) during winter, particularly in northern regions when terrestrial invertebrates are hibernating and hence inaccessible. Dives are generally less than ten seconds long but can last almost a minute under exceptional circumstances, and depths of 50–60 cm are easily within reach. A recent discovery has revealed that Star-nosed Moles, along with semiaquatic North American Water Shrews (*Sorex palustris*), can actually smell underwater. They accomplish this by rapidly expiring and re-inhaling bubbles of air via the nostrils (that is, sniffing) onto objects that they wish to investigate. Odour molecules that diffuse into the bubble are subsequently detected by nasal scent receptors, thus aiding their underwater foraging success. This mole can become a pest on golf courses and lawns but rarely impacts agriculture. Owls are their primary predators, but hawks, skunks, weasels, Domestic Cats and Dogs, and sometimes large fishes and bullfrogs will take them. Their mortality rates, average longevity, and maximum lifespan remain unknown.

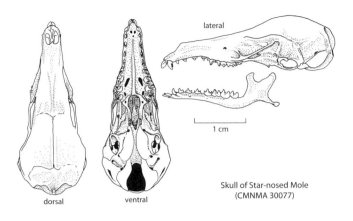

Skull of Star-nosed Mole
(CMNMA 30077)

DIET

Star-nosed Moles, with access to water, prefer to hunt in the muddy bottom where they prey upon aquatic invertebrates such as worms, leeches, and larvae of caddis fly, midge, and water beetle, as well as crustaceans and molluscs. Small fish and frogs are taken occasionally. When hunting on land, Star-nosed Moles predominantly target the small invertebrates and earthworms that populate their tunnel systems. Ever adaptable, these moles can occupy areas with only seasonal water by altering their diet to focus on whatever terrestrial invertebrates are abundant and available.

REPRODUCTION

Breeding is latitude dependent and mainly takes place in March and April. Males achieve breeding condition (enlarged testes) in late February or early March, and most return to a non-breeding condition (the testes shrink) in late May or early June. However, some males remain in breeding condition into mid-June and possibly later. Pregnant females may be found from mid-March until July, but most pregnancies occur in late April to mid-June. A single litter is produced annually. Litter size ranges from two to seven and averages five. The length of gestation is uncertain but is thought to last about 45 days. The young are born pink and virtually hairless, except for the stubs of their sensory vibrissae. Their eyes and ear canals are still closed, but nasal rays are already visible although they are still attached lengthwise along the side of the snout. Milk dentition is resorbed sometime after birth, without ever erupting through the gums, so the young are born toothless. The permanent dentition emerges from the gums before the young moles leave the nest. The juveniles are fully furred by two weeks, and they are weaned and independent when they are about a month old. By autumn, they are fully adult in size and weight but can still be differentiated from the adults by their unworn dentition. Juveniles become sexually mature in the late winter or in the early spring of the following year (at about ten months old), in time to participate in the next breeding season.

BEHAVIOUR

The behaviour and social organization of Star-nosed Moles is very poorly understood. These moles tend to occupy small areas and share their runways and tunnels with conspecifics. It has been speculated from anecdotal observations that they may even be colonial or semi-colonial. Star-nosed Moles are less fossorial than are most other North American moles (except for the Shrew-mole). They commonly forage in surface tunnels during the darkness and can rise onto their hind feet, using their tail as a brace – an action that is beyond most moles. When walking, they use the inside edge of their front feet rather than the palms. The sensitive tentacles are folded together for protection when the animal is digging and eating. Large prey is bitten into segments and held down with the front feet. Smaller prey is swallowed whole. When sleeping, like other moles, the Star-nosed Mole reclines on its belly and tucks its head under its chest.

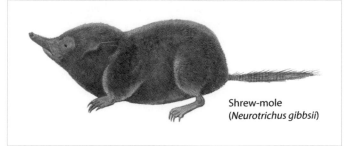

Shrew-mole
(*Neurotrichus gibbsii*)

VOCALIZATIONS

Moles are thought to be generally silent, but Star-nosed Moles will vigorously squeak while they are being handled or when they come in contact with other conspecifics. Nestlings produce shrill cries when touched. Audible squeaks and whirring sounds are occasionally emitted when the nose touches water, though the function of this vocalization remains unknown.

SIGNS

A shallow U-shaped surface runway or burrow found in muck or very wet soil likely has been made by a Star-nosed Mole. Burrow entrances are 3–5 cm in diameter and are slightly wider than they are high. Scat varies with diet. A diet wholly of earthworms results in a liquid squirt of faeces, but when the diet is harder and more diverse, a more solid, tubular scat is produced. These are 0.2–0.5 cm in diameter by 1.0–2.5 cm in length and are commonly deposited in latrine sites that may be above ground (commonly under cover) or at the mouth of burrows. Molehills vary in size. Most are 10–30 cm in diameter and 5–10 cm in height, but exceptionally large hills can be 60 cm in diameter and 15 cm high. The contour of a fresh hill is usually very irregular because the moist earth usually clumps together as the mole works it; erosion smoothes the hills over time until they gradually disappear. Surface ridges produced by this species tend to be less than 5 cm in width.

REFERENCES

Catania 1999, 2006; Catania and Kaas 1996; Dobbyn 1994; Gould et al. 1993; Grand et al. 1998; Hartman, G.D., 1999; Kurta 1995; Long, C.A., 2008; McIntyre et al. 2002; Norris and Kilpatrick 2007; Taylor 2003; van Zyll de Jong 1983a; Whitaker and Hamilton 1998.

Shrew-mole

FRENCH NAME: **Taupe naine**

SCIENTIFIC NAME: *Neurotrichus gibbsii*

The peculiar, tiny Shrew-mole has many shrew-like features, including a long, pointed snout with a pinkish nasal area and front feet that are only moderately adapted for digging. It also has a double coat (with both underfur and guard hairs) that is directed to the posterior like that of shrews, but it has the heavy skull and dentition of the moles. It is clearly a mole because its skull has complete zygomatic arches, which are always lacking in shrews.

DESCRIPTION

The fur colour ranges from dark greyish brown to dark sooty grey to blue black. The belly fur is the same colour as that on the back and sides. The fur has underfur and guard hairs. The sparsely haired, scaled tail is about one-third to one-half of the mole's total length, constricted at the base, and thickened with fat along the middle of its length. The eyes are minute, largely hidden in the fur. Their lack of reaction to light stimuli suggests that Shrew-moles are blind. Their ears have no external flap, only a slit that opens through the skin of the neck and is protected by fur. Two pairs of long vibrissae (whiskers) extend from about 5 mm in front of each ear, and several more, shorter vibrissae protrude from the snout. The thickened hairs on the tail also perform a sensory function. The long, tapering, sparsely furred snout has two nostrils at its reddish tip, each opening on the side. The feet are large. The front feet are slightly flattened for digging but are still longer than they are wide. The middle three claws on both front and hind feet are longer than the outer claws, and all claws are robust and long. Shrew-moles curl their front claws under their foot and walk on the backs of the claws. Male Shrew-moles have a pair of genital glands called ampullary glands, also found on shrews but not present on any other North American mole. The dental formula is incisors 3/3, canines 1/1, premolars 2/2, and molars 3/3, for a total of 36 teeth. The middle two upper incisors are broad and flattened, the second and third incisors are small, and the canines are larger than the third incisors. The canines and first premolars on the upper toothrow are flattened laterally. Auditory bullae are incomplete. Shrew-moles have eight mammae.

SIMILAR SPECIES

Shrew-moles are most easily mistaken for shrews. However, they are distinguished from shrews by their rudimentary and difficult-to-see eyes; their sparsely furred, thick tail with its basal constriction; their slightly flattened front feet; their long claws; and their uniformly coloured body and tail. The skulls of Shrew-moles are differentiated from those of shrews by their complete zygomatic arches, lack of red pigmentation on the teeth, and presence of 36 teeth. Shrew-moles are much smaller than the two mole species that share their range.

Skulls of Coast Moles and Townsend's Moles have 44 teeth, while Shrew-moles have only 36 teeth.

SIZE

The following measurements apply to animals of reproductive age. Males and females are similar in size and appearance.

Total length: 98–126 mm; tail length: 29–50 mm; hind foot length: 14–19 mm.

Weight: 8.0–14.5 g; newborns, 0.5–0.6 g.

RANGE

This mole is restricted to coastal western North America, from southern British Columbia to central California. Almost all specimens are taken from west of the Cascade mountain range. A single exception from British Columbia was taken in 1959 from a wet ravine southwest of Princeton, on the east side of the Cascade Mountains.

ABUNDANCE

Although generally considered uncommon in British Columbia, even in suitable habitat, these moles are thought to be relatively secure in Canada. An isolated population in Burns Bog on the Lower Mainland may be threatened by increasing urbanization. In favourable areas in Washington State, average densities are in the range of 12–15 per ha. These animals are too rarely trapped to provide data on annual or seasonal population fluctuations. Shrew-moles are on the

Distribution of the Shrew-mole (*Neurotrichus gibbsii*)

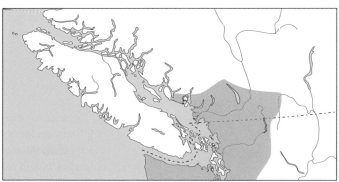

expanded view of Canadian range

British Columbia "Blue List," which means that they are considered to be vulnerable, owing to their restricted range and loss of habitat.

ECOLOGY

Shrew-moles are most abundant in moist ravines with deep, loose soils that are topped with a heavy layer of humus and leaf litter. Shrew-moles may be active at any time of the day or night, and, like other insectivores, they remain active throughout the winter. Unlike most other moles, they frequently forage above ground. A typical activity cycle involves a short foraging period followed by a more lengthy rest period. Shrew-moles dig two types of burrow systems. The shallow burrows are within about 2 cm of the ground surface and are about 3.5–4.0 cm wide. These systems are probably dug as the animals forage and generally are troughs in the soil, topped with leaf litter. They develop into extensive networks and are regularly travelled as the animals search for invertebrates that have fallen into them. The moles create nesting chambers around 13 cm wide by 8 cm high along the shallow tunnels, and they poke a small ventilation hole (using their snout) into the roof of this chamber. Only three nests of wild Shrew-moles have been described. Two were constructed of shredded leaves, and the other entirely of punky wood (this nest occurred in a tree stump, the interior of which was composed extensively of such material). Deep burrows are less common. These are no more than 2.5 cm in width and occur at depths down to about 30 cm. Within the deep burrow system of branching and intersecting tunnels the moles excavate another chamber at about the water level. This chamber will likely have a muddy floor. Shrew-moles are deliberate but capable climbers, and they are even known to nest above ground under certain conditions. The two known elevated nests were both in old stumps (up to 1.25 m above the ground), which allowed the moles to tunnel through the middle of the punky wood to reach a nest chamber. They are good swimmers and will readily take to water, with an undulating swimming motion as they alternately stroke with both legs on the same side. Owls are probably the most significant predator of Shrew-moles. Other predators include snakes, hawks, and Northern Raccoon and probably Domestic Cat.

DIET

Their favourite prey is earthworms, to which they add many other available invertebrates such as sow bugs, centipedes, slugs, flies, beetles, grasshoppers, snails, carrion, and sometimes vegetation. These moles have voracious appetites and in captivity readily consume tree seeds and fungi. A captive, 10 g animal ate a 1.3 g earthworm in ten seconds, and another consumed 4.7 g of earthworms in only two hours.

REPRODUCTION

Breeding information on this species is scarce, especially from British Columbia. Pregnant females have only been captured in British Columbia from April to mid-June, suggesting a spring breeding season from March to late June. In nearby Washington State the breeding season extends from late February to late September. The length of gestation is unknown but lasts at least 15 days. Litter size averages three (ranging from one to four). Neonates are hairless, toothless,

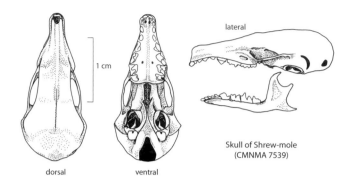

dorsal   ventral   Skull of Shrew-mole
                                      (CMNMA 7539)

blind, and deaf, with transparent, pink skin and no claws. The age at weaning is unknown. The young are not thought to become sexually mature until the following breeding season. Females probably produce only a single litter annually in British Columbia, although two or more may be possible in other parts of the range.

BEHAVIOUR

Shrew-moles are blind and have a poor sense of smell. They find their food and their way mainly by touch. As they travel, they rapidly tap their snout against the tunnel walls, roof, and ground surface, turning their head to extend their reach to the sides. This mole is remarkably agile and can scurry away very quickly if startled.

VOCALIZATIONS

Vocalizations that are audible to humans are rare. Some researchers have reported a faint twittering or a high, musical chittering that can be detected several feet away.

SIGNS

Their scat shape and size depend on diet. A mole that is eating primarily earthworms will produce a squirt of liquid brown faeces. If the diet is varied and includes harder-bodied prey, the faeces are black and shiny, irregular in shape, and about 4 mm in length and less than 1 mm wide. They disintegrate rapidly and are deposited randomly by the moles, with no apparent effort being made to confine them to specific areas. Runways may be found occasionally, usually on hard-packed soil between areas of cover. These are 1.5–2.5 cm wide.

REFERENCES

Carraway and Verts 1991a; Dalquest and Orcutt 1942; Elbroch 2003; Jameson and Peeters 2004; Kremsater et al. 1993; Moon and Leonard 2001; Nagorsen 1996; van Zyll de Jong 1983a; Verts and Carraway 1998.

Hairy-tailed Mole
also called Brewer's Mole

FRENCH NAME: **Taupe à queue velue**

SCIENTIFIC NAME: *Parascalops breweri*

Hairy-tailed Moles are medium-sized moles that are usually found in light soils, and their molehills are smaller than those of other moles in the same part of eastern North America.

DESCRIPTION

The dense, velvety body fur on this mole is generally metallic black to slate grey in colour. The belly may be slightly lighter. White spots on the chest and the belly are not uncommon. The colour of an individual may depend upon the angle of view because the coat can appear much lighter from some angles and much darker from others. Vibrissae extend from the snout and the top of the head and in a dense fringe around the edge of each forefoot, providing tactile information. Younger animals have darkly furred feet, tail, and snout, which become grizzled and then white as they age. The well-furred tail is short, fleshy, and slightly constricted at the base; it is less than 25% of the mole's total length. The legs are short. The feet of older animals appear pinkish because the skin is visible through the sparse white hairs. The mobile snout tip is pinkish to reddish and almost hairless, with nostrils that are laterally placed. The eyes are tiny (< 1 mm in diameter) and probably only detect dark and light. The ear holes are large and oval and lack external flaps. Like those of most other North American moles (except Shrew-moles), the forelimbs are rotated so that they protrude perpendicular to the shoulder, placing the broad spade-like front feet into an optimal digging position. Whitish claws on the forefeet are long and robust, while those on the hind feet are much weaker. Adults are well endowed with numerous large glands on the top of the head, snout, chin, throat, belly, and wrist. These glands produce a yellowish, pungent secretion, particularly during the breeding season, and impart a characteristic odour, especially to males. Fur from the underside of the jaw to the base of the tail in males, and under the jaw and throat of the females, remains stained yellowish brown to golden for some time after the end of the breeding season. The dental formula is incisors 3/3, canines 1/1, premolars 4/4, and molars 3/3, for a total of 44 teeth. Hairy-tailed Moles experience two annual moults. The summer coat grows in spring, from late March to the end of May, and the winter coat replaces it in autumn, between early

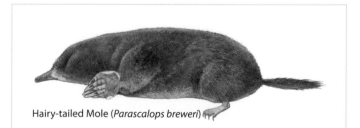

Hairy-tailed Mole (*Parascalops breweri*)

September and mid-October. During summer the juveniles have much shorter and slightly greyer fur than do adults. These moles have eight mammae.

SIMILAR SPECIES

Mole skulls are readily distinguished from shrew skulls by their thin, but complete, zygomatic arches and by the complete (in most moles) or almost complete (in the Hairy-tailed Mole and the Shrew-mole) auditory bullae. Hairy-tailed Moles are the smallest of the eastern North American moles and can be distinguished from other moles by their laterally placed nostrils and very short, hairy tail.

SIZE

The following measurements apply to animals of reproductive age from the Canadian part of the range. Males are somewhat heavier than females.
Total length: 138–170 mm; tail length: 25–33 mm; hind foot length: 15–20 mm.
Weight: 40–65 g; newborns, unknown.

RANGE

This eastern North American mole occurs from southern portions of Quebec and Ontario, through much of the Appalachian region of the eastern United States, to the northwest corner of Georgia in the south and to Ohio in the west. A single specimen has been reported from extreme southwestern New Brunswick (Charlotte County in 1884), but there have been no subsequent reports from New Brunswick.

Distribution of the Hairy-tailed Mole (*Parascalops breweri*)

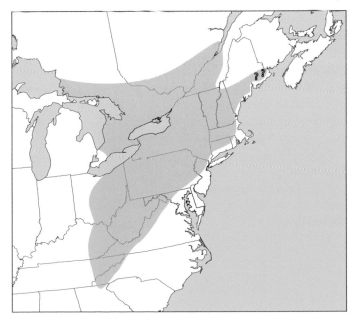

? presence uncertain

ABUNDANCE

This mole may be locally abundant, but it is patchy in its distribution, largely owing to the limitations of suitable habitat. There are no density figures available from Canadian populations, but in the United States the density estimates range from 3 to 30 per ha.

ECOLOGY

Hairy-tailed Moles occur from sea level up to 3000 m elevation (in the Appalachians). Prior to their westward range expansion, they were inhabitants of the eastern temperate, deciduous forests. Now they also occupy pasturelands, regenerating scrublands, and second-growth deciduous or mixed forests west of the Appalachians. They are most abundant in moist, well-drained, light, sandy or loam soils that are well mixed with humus. Heavy clay, wet, or hard dry soils are avoided. Hairy-tailed Moles dig tunnel systems at two different depths, some of which may be used for years. A wide-ranging, many-branched, shallow system (0–20 cm deep), mostly near the soil surface, supports all of the mole's warm-season activity and foraging. Tunnels on main routes are smooth sided and slightly larger, while single-use foraging tunnels are rougher sided and narrower. Surface access tunnels are normally plugged when not in immediate use. Occasional enlargements (about 7–13 cm in diameter) along tunnels at the 10–20 cm depth likely serve as nesting chambers and may be stuffed with spherical nests of dried leaves and grass stems. Larger maternity nests are constructed of shredded, dried leaves in spherical chambers about 16 cm in diameter, the roof of which is about 25 cm below the ground surface – close enough to receive some heat from the sun but deep enough to prevent predation or damage from above. Nest chambers tend to have multiple entrances. A less-extensive, deeper tunnel system at depths of 25–45 cm (in exceptional cases, down to about 55 cm) is used during winter after the frost has entered the ground and the worms have moved deeper. Winter nests are oval with a hollow central core. One such reported nest had external dimensions of 20 cm by 15 cm and was constructed of densely packed leaves and grasses. The home range in winter is estimated at 15–24 m³. The summer range is thought to be larger but remains unreported. Hairy-tailed Moles are capable swimmers. Like other small insectivores, they must eat frequently. Foraging periods alternate with rest periods throughout a 24 hour period, and they remain active all winter. They often forage above ground during the night, which makes them vulnerable to nocturnal predators such as foxes, owls, and Domestic Cats and Dogs. Maximum lifespan is thought to be around four years, by which time the teeth have usually been worn down to gum level.

DIET

This mole, like most others, feeds on soil invertebrates. Earthworms are their primary prey, followed by insects, both larval and adult forms, and then by other invertebrates such as centipedes, millipedes, snails, slugs, and sow bugs. Hairy-tailed Moles are prodigious eaters. If provided an unlimited number of earthworms, a captive can eat more than 300% of its own weight in 24 hours, but most can be maintained on considerably less. Another captive enthusiastically consumed a torpid Wood Frog, and it is possible that wild moles

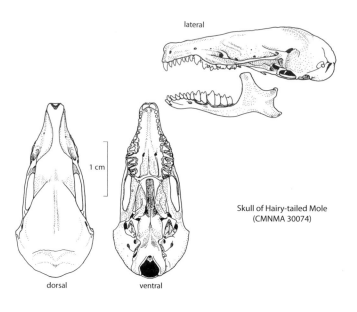

lateral

1 cm

Skull of Hairy-tailed Mole
(CMNMA 30074)

dorsal

ventral

also eat and even hunt Wood Frogs. Captives will drink frequently, but it is unknown whether wild moles do the same.

REPRODUCTION

Mating occurs in late March and early April. Females produce a single litter annually of an average of four young (up to eight have been reported). The gestation period is unknown but is likely in the range of four to six weeks. Litters are produced in late April or May. Young are born blind and hairless, apart from short vibrissae on the lips and a scattering of short hairs on the face. Their skin is whitish and wrinkled. They grow quickly and are hunting for themselves by the time they are about four weeks old. Juveniles become sexually mature the following spring at around 10 months old. As with other moles, the size of the testes varies dramatically through the year. During the breeding season they measure around 12 mm long by 7 mm wide, but they diminish quickly in May to 3 mm by 2 mm for the remainder of the year. Moles do not have a scrotum, as the testes are abdominal; they rest near the bladder for most of the year and descend ventrally to rest in an extension of the abdominal cavity during the breeding season. The female's vaginal opening is covered by skin throughout the rest of the year but opens from April to September to allow mating and parturition.

BEHAVIOUR

Adult Hairy-tailed Moles are generally solitary, especially in winter, but during the warm seasons they appear to be more tolerant of each other than are most other moles. Males associate without aggression during spring, and all ages and sexes use the same tunnels by late summer. In spring the females move from their deep winter tunnels into the more shallow tunnels above, but the males commonly abandon their tunnels altogether to search for receptive females. Like other moles, Hairy-tailed Moles cannot rise on their fore limbs. They rest on their chest and the inside edges (now the bottom edges) of their front feet. When travelling through a tunnel, they press all four feet against the sides so their body is suspended. Diagonal legs (that is, the

right front and the left hind in synchrony, followed by the left front and the right hind in synchrony) are moved together to maintain the body's suspension. Speeds of around 0.5 km/h can be achieved on the surface, and somewhat faster speeds are probable within the tunnels. When sleeping, a Hairy-tailed Mole, like other moles, tucks its head under its body. Prey is probably detected by using a combination of hearing, smelling, and touching. As the mole forages, its nose is constantly in motion, twitching from side to side and up and down and always sniffing. An earthworm can be detected, apparently by smelling, from about 5–6 cm away. Small prey is grasped in the mouth and swallowed whole. Larger prey is held down by the front feet and bitten into fairly large pieces, which are swallowed quickly without much chewing. Earthworms are consumed in a fairly stereotypic manner. Either end of the worm is grasped in the teeth, and both front feet are brought forward and tipped under the chin. The worm is squeezed between the index and middle claws, and jerking movements pull the worm through the claws so that segments can be bitten off and eaten. The squeezing action strips the slime and soil away from both the exterior and the guts of the worm.

VOCALIZATIONS

Hairy-tailed Moles appear to be fairly silent. Captives have been heard producing weak, high-pitched twitters while they are investigating their environment. A captured animal produced harsh, guttural squeaks while being held in the hand. Another captive, as it was awakening, was overheard emitting quiet, rhythmic squeaks at about one-second intervals.

SIGNS

Scat is about 15 mm long and 5 mm in diameter and is commonly deposited in latrine piles at tunnel entrances. Tunnels are somewhat oblong in cross-section, being about 37–50 mm wide by 25–40 mm high. Most Hairy-tailed Mole hills are around 15 cm in diameter and 7–8 cm high. Occasionally, larger hills are created in spring when a maternity chamber is being excavated. Most molehill creation occurs in autumn as the moles renovate and excavate their deeper tunnel systems in preparation for winter occupation. Hairy-tailed Moles tend to create molehills that are smaller than those of other moles in the same range. Although these moles will sometimes raise surface ridges when they dig just under the surface, for the most part these are hidden by the leaf litter; when visible, they are not as prominent as those of other moles in the same area.

REFERENCES

Dobbyn 1994; Elbroch 2003; Hallett 1978, 1999; Hecnar and Hecnar 1996; Hickman 1983; Jensen 1986; Kurta 1995; Peterson 1966; Whitaker and Hamilton 1998; van Zyll de Jong 1983a.

Eastern Mole

also called Prairie Mole

FRENCH NAME: **Taupe à queue glabre**

SCIENTIFIC NAME: *Scalopus aquaticus*

The Eastern Mole reaches the northern limit of its distribution in the Point Pelee area, where it is common. The fossorial nature of moles makes them difficult to study, and this species is no exception.

DESCRIPTION

Eastern Moles are greyish or brownish black, often with a silvery sheen visible from certain angles. The tail is short (about one-fifth of the mole's total length), basally constricted, scaled, and sparsely haired. The long, pointed snout is naked at its tip, where the nostrils open upwards. Fur on the tops of the feet is dirty white. A burrowing mammal, the Eastern Mole has many fossorial adaptations. The legs are short, and the body is stocky and muscular. The eyes are minute, often with fused eyelids, and are completely hidden by fur. The ears are merely holes protected by fur, with no external flaps. The shoulder girdle is modified so that the front limbs are rotated into an ideal digging position. The front feet are spade like, broader than long, and are equipped with large, flat claws for digging. Both the front and the hind toes are webbed to the claws. The dental formula is incisors 3/2 (sometimes 3/3), canines 1/0 (sometimes 1/1), premolars 3/3, and molars 3/3, for a total of 36 (or 38) teeth. The auditory bullae are complete. There are six mammae.

SIMILAR SPECIES

Within Canada only one other mole occurs in the range of the Eastern Mole. Star-nosed Moles are easily distinguished by the 22 pink tentacles they have on their noses and by their moist habitat preferences.

SIZE

The following measurements apply to animals of reproductive age from northern parts of the range, because there are so few Canadian specimens. Males are somewhat larger and heavier than females.
Total length: 173–200 mm; tail length: 28–38 mm; hind foot length: 20–27 mm.
Weight: 75–140 g; newborns, around 5 g.

Eastern Mole
(*Scalopus aquaticus*)

RANGE

Although this mole barely reaches Canada, it is the most widespread mole in North America. Its range extends throughout most of the eastern and central United States (except in the Appalachian Mountains) from Cape Cod to southeastern Wyoming, and southwards to northeastern Mexico in the State of Tamaulipas. There is an isolated record from Presidio County, Texas, and an isolated population south of that in the State of Coahuila, Mexico. Both of these populations are very small and possibly extinct. The Canadian distribution occupies about 350 km², in southern Essex County and the adjacent Kent County in extreme southwestern Ontario. Less than about 4% of the suitable habitat in this area remains natural, as most of the rest has been either urbanized or used for intensive agriculture, both of which are generally incompatible with mole habitation. About three-quarters of the remaining natural habitat occur within the boundaries of Point Pelee National Park, where the moles are legally protected. Outside the park the range appears to have expanded slightly to the east and west since the 1970s, but it is now constrained by unsuitable soil types. Some populations outside the park are protected in conservation areas.

ABUNDANCE

Mostly considered to be common to very common in suitable habitat, Eastern Moles have been designated to be of "Special Concern" by COSEWIC since 2010, largely because of their restricted and fragmented range in Canada. The Ontario Ministry of Natural Resources also lists the species as "Vulnerable." Populations in Point Pelee can display dramatic annual fluctuations based on the number of "pushups" (molehills), but over a thirteen-year survey the numbers appear to be relatively stable in the park. Studies in the United States reported estimated densities of 2–5 per ha, with most populations in the lower end of the range (2–12 per ha). Based on the available natural habitat and densities in the 2–12 per ha range, there are perhaps 2000–13,000 Eastern Moles in Canada.

ECOLOGY

Eastern Moles in Ontario live in meadows, pastures, wooded or bushy hedgerows, and open woodlands, with soft, moist, but well-drained soils. Some occur in gardens and lawns. They avoid soils that are gravelly, stony, hard, or dry, and heavy sand or clay. Although apparently not the case in the United States, moles in Canada do not penetrate more than a few metres into regularly cultivated croplands. Pastures, hayfields, and abandoned fields are the only agricultural lands that this mole finds suitable in this country. Two types of tunnel systems are constructed: shallow and deep. The shallow tunnels create a multi-branching gallery; the soil is loosened with the front feet and then compacted against the tunnel roof and walls with the body. Construction of these tunnels commonly raises a ridge of soil at the surface. The shallow system is occupied during warm to cool weather. Under the right moist conditions an Eastern Mole can dig shallow tunnels at the rate of 5 m per hour. A deeper system (10–60 cm below the surface) is used during cold or dry weather, for nesting, and for rearing young. Digging these deeper tunnels in the more mineralized soil below the humus layer means that since the soil cannot be compacted,

Distribution of the Eastern Mole (*Scalopus aquaticus*)

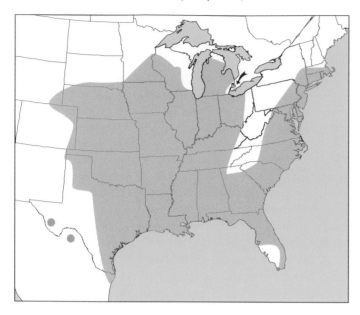

it must be shoved into abandoned tunnels or pushed to the surface to form molehills. Nest chambers (10–20 cm in diameter) are excavated in the deeper tunnels and usually have one or several entrances, including one that rises from below. Some nest chambers may contain a hollow, spherical nest made of grasses, leaves, and fine roots (130–150 mm in diameter), but others are devoid or almost devoid of nesting material. It is thought that only winter and maternal nests are regularly constructed and that males and non-reproductive females do not need an insulated nest in warm weather. Nests are often within 30 cm of the surface (ranging 8–46 cm below) and may be placed under a stump or a rock. Individual moles tend to have two or more nests scattered within their home range.

Home ranges, reported in the case of 12 moles in Kentucky, vary from about 1000 to 3400 m² for females and from 3000 to 18,000 m² for males. Male home ranges are typically at least three times the size of female home ranges in any given area. Eastern Moles are active all year round. Their high metabolism forces them to forage intermittently, both day and night. About 3.0–6.5 hours is the usual time period between bouts of activity. An active period is followed by a rest period when they sleep and digest. They are most active around dusk and dawn. Although quite capable swimmers, these moles are not as aquatic as their webbed feet and scientific name imply. They are highly fossorial but will travel on the surface at times. These above-ground individuals are likely to be dispersing juveniles, individuals retreating from a flood, or males seeking receptive females. Eastern Moles have few natural predators, but they are vulnerable on the surface, where they sometimes fall prey to owls, hawks, foxes, Coyotes, weasels, and occasionally Domestic Dogs and Cats. The mammalian predators may also dig them out of shallow foraging tunnels. Unless they are starving, many mammalian predators kill but then abandon the moles uneaten, likely owing to their rank odour. The maximum lifespan is probably six to seven years, but an annual mortality rate of more than 50% suggests that very few reach that age.

DIET

Eastern Moles mainly eat invertebrates, but they add plant material (primarily seeds but sometimes garden vegetables such as potatoes, corn, and tomatoes) during some seasons. Earthworms and insects make up most of their diet. Beetles (both adult and larval forms) and ants (in the pupal stage) are favoured insect prey, but many other insects and invertebrates are taken when they are available. Centipedes, spiders, and snails have been reported. Diet varies with availability, and these moles will switch to less preferred prey if necessary. Captives typically consume daily about 31%–55% of their body weight in food.

REPRODUCTION

In the northern portions of the range most Eastern Moles breed in late March and early April, and young are born in late April to early May. The length of gestation is uncertain; it has been reported to last as little as 28 days and as long as 45 days. A single litter is produced annually, and its size averages four (ranging from one to five). The hairless (apart from short stubs of vibrissae on the snout, face, and jaw) and toothless young are relatively large at birth and develop a short, fuzzy pelage in 7–10 days. Their milk teeth are shed in utero. Young are thought to be weaned and leave the maternal nest some time after they are four weeks old, but they may share their mother's tunnel system until they are capable of foraging on their own. Before autumn they develop their own tunnel systems. Juveniles become sexually mature in the spring following their birth, at around 10 months old.

BEHAVIOUR

Individual ranges tend to overlap, but adults rarely share the same tunnel system. Generally they occupy exclusive tunnel systems and defend them from intruders. This territoriality breaks down somewhat during the breeding season, when males invade the tunnel systems of neighbouring females, in search of a mate. Earthworms are eaten in a stereotypical manner. The worm is grasped at either end and squeezed between the front paws. Rapid jerks of the head pull a section of the worm through the paws, which is then bitten off and consumed. This process is repeated until the worm has been completely eaten. It removes the surface slime and dirt and expels the gut contents to reduce tooth wear. Large insects are smashed against the tunnel walls before being held down by a front paw to allow examination. They are then ripped into pieces and consumed. Smaller insects

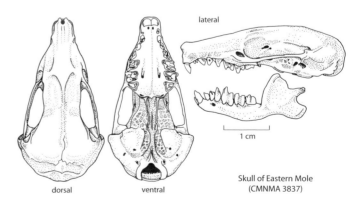

lateral

1 cm

Skull of Eastern Mole
(CMNMA 3837)

dorsal   ventral

are consumed whole. Like most moles, Eastern Moles explore their environment and navigate primarily by touch. Their sense of smell is weak and useful only within a few centimetres, but their hearing is well developed and likely helps them find prey.

VOCALIZATIONS
Vocalizations of the Eastern Mole are unknown or unreported.

SIGNS
Near-surface tunnels are 3.2–3.8 cm in diameter. Deeper, permanent tunnels have larger diameters (5.5 cm). Molehills tend to be roughly circular and up to 30 cm in diameter and 15 cm high. Mounds may be conical when fresh but are quickly rounded by erosion and become flattened over time. Digging and production of molehills is most noticeable when the ground is soft in spring or after a soaking rain. The hills tend to be larger, wider, and more flattened than those of Star-nosed Moles. Moles eating mostly earthworms produce a liquid or almost liquid squirt of faeces, but those eating mainly harder-bodied prey will produce blackish or dark brown, tubular scat that is 0.2–0.5 cm in diameter and 1.0–2.5 cm in length. This mole is not known to deposit scat in a latrine site; it is instead left throughout the tunnel system.

REFERENCES
Davis, F.W., and Choate 1993; Elbroch 2003; Hartman, G.D., 1995; Hartman, G.D., and Gottschang 1983; Hartman, G.D., et al. 2000; Harvey 1976; Hickman 1984; Jones, J.K., et al. 1978; Kurta 1995; Long, C.A., 2008; Pearce and Kirk 2010; Schmidly 2004; van Zyll de Jong 1983a; Waldron 1998; Waldron et al. 2000; Whitaker and Hamilton 1998; Yates 1999; Yates and Schmidly 1978.

Coast Mole
also called Pacific Mole

FRENCH NAME: **Taupe du Pacifique**

SCIENTIFIC NAME: *Scapanus orarius*

Coast Moles are common in the Lower Mainland of British Columbia. Their molehills are plentiful in winter and may be seen on golf courses, along road margins, in fields and pastures, and on lawns.

DESCRIPTION
The Coast Mole is a small mole with soft, short fur that is blackish brown to grey in colour. The belly fur is the same colour as the rest of the body. Their long, pointed snout is almost bare for about half the distance from the nose to the eyes, and the nostrils open at the tip of the snout, facing forwards. Like most other moles, apart from Shrew-moles, Coast Moles have many fossorial adaptations. Their front feet are large and spade-like with robust, flattened, white claws.

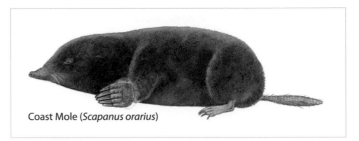
Coast Mole (*Scapanus orarius*)

The articulation of the humerus with the clavicle is modified so that the forelimbs stick out to the side to permit efficient digging. Their plush, velvety pelage lacks a distinct underfur, and their body hairs are a similar length (apart from the vibrissae). The eyes are tiny, and the ear openings are hidden in the fur and have no external flap. The dental formula is incisors 3/3, canines 1/1, premolars 4/4, and molars 3/3, for a total of 44 teeth. The first upper incisor is broad and longer than the other incisors in the toothrow, which are conical and similar to each other in length. As with other North American moles, except the Shrew-mole, the auditory bullae are complete. Coast Moles experience two moults annually, and their summer pelage is often paler than the winter pelage. The spring moult begins in March and is complete in all animals by the middle of May. The autumn moult can begin as early as mid-August in some individuals, but most grow their winter coat during October.

SIMILAR SPECIES
Apart from Shrew-moles that are much smaller, the only other mole that occurs within the range of the Coast Mole is the Townsend's Mole. These two species are similar and are distinguished by size. Townsend's Moles are larger; their adult total length is > 175 mm, the hind foot is > 24 mm, their skull is longer (> 37 mm, versus < 35 mm for the Coast Mole), and their weight is > 90 g.

SIZE
The following measurements apply to animals of reproductive age from the Canadian part of the range. Males are somewhat heavier than females.
Total length: 145–181 mm; tail length: 28–41 mm; hind foot length: 18–24 mm.
Weight: males, 46.0–78.1 g; females, 45.6–66.9 g; newborns, unknown.

RANGE
The Coast Mole occurs in western North America from southern British Columbia to northern California and eastwards as far as west-central Idaho. In British Columbia it is restricted to the southern Fraser Valley area, west of Hope.

ABUNDANCE
Coast Moles can be abundant in the Lower Mainland of British Columbia, where population densities range from less than 1 per ha to as many as 13 per ha in ideal habitats. In some parts of the range this mole is numerous enough to be considered a garden, crop, and golf-course pest, although its damage is largely cosmetic – mainly

Distribution of the Coast Mole (*Scapanus orarius*)

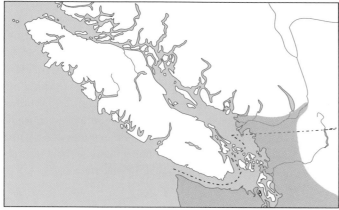

expanded view of Canadian range

molehills and root damage as they dig. Further damage to bulbs, corms, roots, and above-ground vegetation is done by the voles and other rodents that use the mole tunnels as runways.

ECOLOGY

Although Coast Moles are most abundant in moist, well-drained soils, they also occur in clays, glacial tills, sandy or gravelly soils, and river deposits as long as those soils support earthworms. The Coast Mole can live in a wider variety of soil and habitat types than can the Townsend's Mole and consequently has a wider range. Coast Moles are common in moist, cultivated soil and pasturelands, but they also occur in forested areas. They are highly subterranean in their habits and dig several types of underground tunnels: surface, shallow, and deep. Surface tunnels are created just below the ground surface and generally appear as raised ridges of loose soil. The waste soil is typically compacted against the side of the tunnel, so no molehills are created. These tunnels are made, and often used only once, while foraging, dispersing, or searching for a mate. Most hunting activity takes place in shallow tunnels, usually 7–20 cm below ground level. During dry periods or when the surface soil becomes frozen, deeper tunnels may be dug 1–2 m below the surface as the moles follow the worms. Small enlarged areas (10 cm by 8 cm) are scattered throughout the tunnel systems, likely for use as resting or nesting chambers. Construction of the shallow and the deeper tunnels usually

requires excavation of the waste soil, and this is kicked or shoved to the surface through voiding holes, where it forms mounds or molehills. The presence of molehills is most pronounced during winter, because the soil is more easily worked during the wet season, and this is the time of year in which populations may expand into new areas. From mid-autumn until early spring individual moles create exclusive shallow tunnel systems called encampments. A single Coast Mole, from October to March, can create 200–400 molehills. Breeding females construct nursery nests at shallow depths (around 15 cm below the ground surface) where they are protected (typically by tree roots, brush, or fallen woody debris) but are still warmed by the sun. The chamber is about 20 cm in diameter and is lined with dried grasses. Nests typically have several entrances. There is usually no molehill associated with the construction of this chamber, unlike the situation with Townsend's Mole.

Coast Moles are active all year and do not hibernate. Foraging takes place at any time of the day or night, followed by a rest period. A radioisotope-tagged individual that was followed around for over 50 hours displayed a two-to-four-hour activity period followed by a three-to-four-hour rest. The home range is 0.03–0.4 ha, and individuals may travel 2–30 m during an activity phase. Almost all of their time is spent underground with one major exception. Juveniles, dispersing from their natal nest, may travel for some distance on the surface, where they are vulnerable to predation. Adults are capable swimmers and can survive a flood by swimming as long as they are not caught deep in their burrows. Coast Moles are estimated to survive three to four years in the wild. They do not have any consistent predators, but owls and Domestic Cats and Dogs kill dispersing juveniles, and snakes may travel through the tunnel systems, where they could take nestlings. Probably the greatest predator of Coast Moles is humans: large numbers are killed each year as pests.

DIET

The favourite prey of this mole is earthworms. It will also eat other invertebrates that it encounters, such as slugs, larval and adult insects, centipedes, and millipedes, as well as carrion if it is available. In captivity this mole can consume almost double its body weight in earthworms in a 24-hour period.

REPRODUCTION

Little is known of the reproduction of these highly fossorial animals. Mating has never been witnessed. In British Columbia, Coast Moles

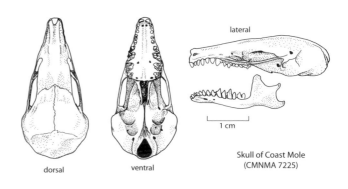

dorsal

ventral

lateral

1 cm

Skull of Coast Mole
(CMNMA 7225)

breed from January until early March. By May all of the females have given birth. The length of gestation is unknown. The litter size of mature females is almost always four young, but first-time breeders usually produce only two. A single litter is produced annually. Although neonates have not been observed, they are presumed to be hairless and toothless, as are most other newborn moles. The age of the young at weaning is also unknown. Juveniles breed for the first time at 9–10 months old, in the breeding season following their birth.

BEHAVIOUR

The behaviour of these moles is very poorly understood, largely owing to their highly subterranean lifestyle. They appear to be solitary during winter, but the extensive winter encampments become interconnected during the breeding season as males dig their way in, looking for receptive females. The moles tend not to intrude otherwise on occupied tunnel systems but will take over abandoned tunnels.

VOCALIZATIONS

No vocalizations have been reported.

SIGNS

The conical- or dome-shaped molehills formed by Coast moles are smaller than those created by Townsend's Moles. They average 29–30 cm in diameter and 10–13 cm in height, while those of Townsend's Mole are 40–50 cm in diameter and around 15–20 cm in height. Coast Mole shallow tunnels are 3.5–4.0 cm in diameter, while those of Townsend's Mole are 4.5–5.5 cm in diameter. Surface runways may be created under snow.

REFERENCES

Gitzen and West 2000; Jameson and Peeters 2004; Nagorsen 1996; Schaefer 1982; Sheehan and Galindo-Leal 1997; van Zyll de Jong 1983a; Verts and Carraway 1998; Yensen et al. 1986.

Townsend's Mole

FRENCH NAME: **Taupe de Townsend**

SCIENTIFIC NAME: *Scapanus townsendii*

The Townsend's Mole is the largest mole in North America and the rarest mole in British Columbia.

DESCRIPTION

Like most other moles, the Townsend's Mole is stocky, with plush fur, short legs and tail, and a long, pointed snout. The eyes are minute, and the ears are simply a hole protected by fur and lacking an external flap. The eyes are thought to detect only light intensity. Probably most of the sensory input is tactile, through the vibrissae

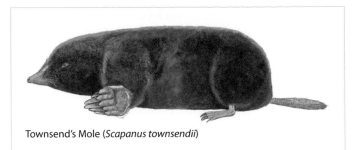

Townsend's Mole (*Scapanus townsendii*)

(sensory whiskers). The forefeet are broad, and the whole arm is rotated so that the forefeet stick out to the side of the shoulder in an optimal digging position. The claws on the forefeet are broad and long. The tail, the feet, and the tip of the snout are pinkish owing to a sparse covering of white hairs. The thick tail is constricted at its base. The tail hairs are thought to provide sensory input during manoeuvres. Some Townsend's Moles have irregular white markings on their body, and rarely are all white, but none of these colour variations is known from the Canadian populations. The dental formula is incisors 3/3, canines 1/1, premolars 4/4, and molars 3/3, for a total of 44 teeth. None of the mole species has pigmented teeth. Townsend's Moles experience two moults each year. The winter coat is almost black, and the summer coat is more greyish. There are eight mammae.

SIMILAR SPECIES

Apart from Shrew-moles, which are much smaller and have only 36 teeth, the only other mole that occurs within the range of the Townsend's Mole is the Coast Mole. These two species are similar and are distinguished mainly by size. Coast Moles are smaller. Their adult total length is usually < 175 mm; the hind foot is < 24 mm; their skull is shorter (< 35 mm, versus > 37 mm for the Townsend's Mole); and their weight is < 80 g. They also make smaller molehills (see the "Signs" section below for more information). Juvenile Townsend's Moles may be similar in size to Coast Moles, but their front feet are broader and more heavily clawed.

SIZE

The following measurements apply to animals of reproductive age from the Canadian part of the range. Males are somewhat heavier than females.
Total length: 179–237 mm; tail length: 31–45 mm; hind foot length: 23–29 mm.
Weight: males, 121–164 g; females, 96–122 g; newborns, around 5 g.

RANGE

This large, western mole occurs along coastal parts of California, Oregon, and Washington State and reaches the northern limit of its distribution several kilometres into southern British Columbia in the Abbotsford-Huntingdon area. A region of about 30 km² of Townsend's Mole habitat exists in the Fraser Valley. All but about 13 km² has undergone urban development, mainly housing.

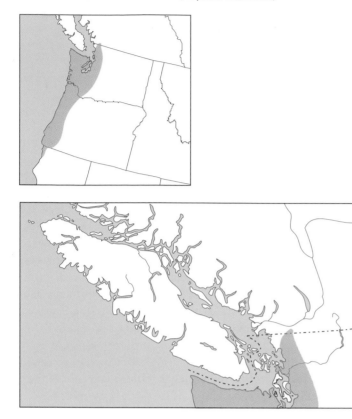

Distribution of the Townsend's Mole (*Scapanus townsendii*)

expanded view of Canadian range

ABUNDANCE

The Townsend's Mole is considered to be "Endangered" in Canada by COSEWIC, largely owing to a very limited Canadian distribution. It is on the British Columbia provincial "Red List," where it is also acknowledged as "Endangered." The primary threat to the species in Canada is habitat degradation because of the intensification of agricultural practices and the rapidly increasing urbanization. Most of the land currently occupied by this species is privately owned or is protected for agricultural development, but when such development results in intensive farming, it generally proves lethal to the moles. These animals are considered to be pests by homeowners, and kill trapping is permitted in residential areas where lawns are damaged. Population densities of 0.42–12.0 moles per ha have been estimated throughout the range. The density of Canadian populations, at the northern edge of the species' range, is most likely low. A very crude estimate of around 700 Townsend's Moles in Canada (including juveniles) was made in 2003, and the population appears to be stable.

ECOLOGY

All Townsend's Moles in British Columbia live in silty, loam soils in lowland meadows, farmlands, pastures, hayfields, and lawns. Densities are highest in manured pastures and hayfields, where the enhanced nutrients support abundant earthworms. There are some higher-elevation populations in Washington State and

Oregon, but none occurs in Canada. Townsend's Moles will live in sandy soils and open forests in other parts of the range but appear to be restricted to their preferred soil type in Canada. Although Townsend's Moles favour moister soils than do Coast Moles, their habitat and food requirements are otherwise similar, so the reason for the limited range of the Townsend's Mole is not clear. In areas where Townsend's Moles are removed by trapping, Coast Moles may move in to the vacated habitat, and Townsend's Moles seem unable to return.

Although details of the activity cycle of this mole are unavailable, the Townsend's Mole is reported to be active throughout a 24-hour period, probably in a cycle similar to that of its close relative, the Coast Mole. Coast Moles forage for two to four hours and then rest for three to four hours. Neither species hibernates; instead, they remain active all winter but may occupy deeper tunnels depending on the depth of the frost and the earthworms.

Townsend's Moles dig numerous tunnels at essentially three different depths. Surface tunnels lie within 10 cm of the surface and are used primarily for foraging because most earthworms occur in the top 7.5 cm of the soil and humus. Such shallow tunnels may be used only once. Dispersing juveniles and males searching for mates may also dig shallow tunnels. Waste soil from the digging is pressed against the sides of the tunnels, and molehills are not usually needed to dispose of the surplus. A more permanent system of intersecting tunnels occurs at a depth of 10–20 cm and is used for foraging, moving around the home range, resting, and raising young. Such tunnels may also be dug by males seeking access to the tunnel systems of neighbouring females. Occasionally deeper tunnels are dug, usually to follow the worms downwards during the dry or cold season, but also to travel underneath roadbeds and other buried obstructions. The two deeper tunnel systems are typically associated with the surface evacuation of waste soil (molehills). Townsend's Mole tunnels are commonly high in carbon dioxide and low in oxygen. This rarefied environment could place severe stress on these hard-working animals that have the ability to move up to 20 times their body weight in soil in 15 minutes. As do other moles, Townsend's Moles have likely adapted to this situation by having more powerful lungs and more blood haemoglobin to aid their bodies in oxygen intake and distribution. The tunnel system acts like a pitfall trap for invertebrates, and the moles forage as they travel along them. The home-range size varies from a low of 0.003 ha in prime habitat to 1.04 ha in poor habitat.

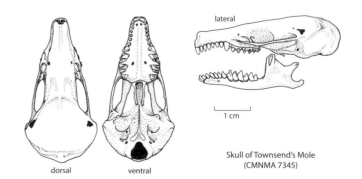

lateral

dorsal   ventral

Skull of Townsend's Mole
(CMNMA 7345)

Townsend's Moles are capable swimmers, able to cross water-filled ditches and small streams. Their fossorial lifestyle protects them from most predators, but they become vulnerable if they travel above ground. Males searching for females, juveniles dispersing, or moles evading a flood sometimes are forced to the surface, where they may face small mammal predators, hawks, or owls. Juveniles can travel over 800 m above ground when dispersing. Pest-control measures and domestic pets probably cause the highest mortality in Canadian populations. Some dispersing individuals are struck by motor vehicles while crossing roadways.

DIET

The principal food of this large mole is earthworms, although it will also eat a wide variety of soil-dwelling invertebrates as they are encountered, including centipedes, millipedes, slugs, and insects. Some plant material including bulbs, root vegetables such as parsnip, carrots, and potatoes, and grass roots may be eaten at times, as are small mammals like shrews and mice. Small mammals are common cohabitants of the mole tunnels and could be encountered as carrion. Free water is not necessary, as their diet is high in moisture. A Townsend's Mole typically eats 33%–66% of its weight daily.

REPRODUCTION

Breeding takes place from early December to late February, and in Canadian populations it likely peaks in late February. Young are born in late March to early April in an underground nest that has been constructed by their mother shortly before the birth of the litter. Gestation lasts four to six weeks. Females produce a single litter annually. Litters average three young (ranging from one to four). Neonates are pink and hairless. They remain naked until around 22 days old, when their hair begins to grow. The coat is complete about eight days later. Young remain in the nest until they are 30–36 days old, at which time they disperse (usually in May and June). Juveniles breed for the first time at around 10–11 months old, during the next breeding season.

BEHAVIOUR

These moles maintain exclusive home ranges and will aggressively repel the approach of conspecifics outside the breeding season. This territoriality breaks down somewhat during the breeding season when males spend much time and energy tunnelling in search of the encampments (tunnel systems) of receptive females. Townsend's Moles have strong homing instincts, which are perhaps useful after the flooding that may occur annually. Moles that are displaced at a distance of 100–200 m typically reoccupy their former territories when the flood abates, and artificially relocated animals have returned from distances of up to 450 m, even when there are canals, rivers, or roadways to cross. During the breeding season maternal females dig larger than usual chambers in which to deliver and raise their young. These nesting chambers are 15–20 cm below the surface and are roughly globular, with a height of about 15 cm and a diameter of 23 cm. Nest chambers are close enough to the surface to be warmed by the sun and are sometimes trampled by cattle. The large size of the chamber usually means that the female has created a larger than usual mound, or sometimes a cluster of several normal-sized mounds, to cope with the waste soil. Maternal nest chambers have several entrances (3–11 or more) that connect to either the adjacent tunnel system or to a bolt-hole that heads down from the bottom of the chamber and connects to one of the tunnels in the deeper tunnel system. Maternity chambers are usually dug in an elevated part of the encampment, presumably for flood protection. Within this chamber the female constructs a thick spherical nest of grass stems and leaves. The inner hollow is lined with fine, dry grasses, and the exterior, comprising about 75% of the nest, is made of coarse green grasses. New, fresh, moist material is added to the outside regularly. The use of freshly cut vegetation is thought to be a deliberate attempt to create heat from the fermentation of the vegetation in order to prevent the young from becoming chilled during their mother's absence. Breeding nests may be reused for several seasons.

VOCALIZATIONS

No vocalizations have been reported.

SIGNS

The size of molehills varies. A smaller than usual mound may indicate a repair to an existing tunnel, whereas larger hills likely indicate new tunnel construction. Typical molehills are 43 cm in diameter and 17 cm in height. The mounds start as conical shaped and erode into more dome shaped. Maternal nest chamber mounds are 70–130 cm in diameter and 30–45 cm in height. Tunnel diameter is 4.5–5.5 cm (versus 3.5–4.0 cm for Coast Moles).

REFERENCES

Carraway and Verts 1991b; Carraway et al. 1993; Giger 1973; Jameson and Peeters 2004; Nagorsen 1996; Reid, F.A., 2006; Schaefer 2003; Sheehan and Galindo-Leal 1997; van Zyll de Jong 1983a; Verts and Carraway 1998.

Order Chiroptera
bats

Bat (genus *Myotis*)

Photo: Carol Giffen

Bats – the only mammals capable of powered flight – are among the most unique and interesting of mammals. They are also among the most successful; approximately one in every five species of mammal is a bat. Bats have numerous adaptations not found in other mammals, even apart from the obvious ones related to flight.

There are 18 families, 202 genera, and 1116 species of bats in the world. In Canada, the 8 genera and 20 species of bats belong to one family, the plain-nosed bats (Vespertilionidae). There is one record of an additional family of free-tailed bats, Molossidae, which appears to be an accidental record; the species is not considered to be a regular part of the bat fauna.

DISTRIBUTION

Flight has allowed bats to reach and inhabit every continent except Antarctica, and most oceanic islands. Often the only native mammals on oceanic islands are bats. In Canada, bats inhabit every ecozone except the tundra. Their northern limit is determined by the treeline. Occasional records from north of the treeline appear to be accidental. Because they can cover a tremendous distance by flight, accidental occurrences of bats, like birds, are fairly common. All of Canadian species, including the Big Free-tailed Bat, are New World species not found outside of North and South America.

FOSSIL RECORD

The oldest known bat dates back over 50 million years to the early Eocene of Wyoming, Germany, Pakistan, and Australia. The first known fossil bats have fully functional wings and could echolocate, making them generally similar to modern bats. Clearly bat evolution extends back before the Eocene. The ancestors of bats were probably small, insectivorous, shrew-like mammals, living in trees. The origin of Chiroptera appears to have been in Africa.

FOOD

Our Canadian species are all insectivorous but in other parts of the world, other bat species eat fruit, pollen and nectar, leaves, fish, frogs, lizards, scorpions, small birds, and mammals – including other bats and, yes, even blood. Since they are mammals, all bats drink mother's milk as infants. Bats use a tremendous amount of energy flying and must stoke the internal engines on a regular basis unless in hibernation. An insect-eating bat that is non-reproductive will consume the equivalent of about 50% of its body weight every night. Nursing females eat even more, up to 110% of their body weight each night. Insect eating bats have a very rapid digestion provided they can find a warm spot to rest. Most can fill their stomachs and process their meals several times a night. Many of our bats (if not all of them) have bacteria in their gut that produce chitinase – an enzyme that breaks down chitin, the main component of the hard shell of an insect body. How much energy this provides to the bats remains unknown. Bats hunting insects usually either catch them in the air, called hawking, or take prey from surfaces (ground or foliage), called gleaning. It remains to be seen how many species use both hawking and gleaning when hunting, but we know that at least some Canadian bats do. Most insectivorous bats forage twice a night. The first feeding period begins around dark and may last for up to two hours. The second may occur just before dawn. The actual time spent foraging will depend upon the bat's situation. For example, females nursing young may return repeatedly to the roost to feed the young, while adult males may forage only once or twice a night.

SURVIVING WINTER

As eaters of insects, Canadian bats face a severe food shortage during winter. Some species migrate south for the winter, returning to Canada the following spring. Other species migrate to hibernation sites (usually underground in caves or abandoned mines). Some, for example Big Brown Bats, often hibernate in buildings and may not move appreciably between summer and winter ranges. During hibernation, bats require temperatures just above freezing and most need high humidity. Migrations, whether to warm wintering areas or to hibernation sites, may cover tens to hundreds of kilometres. Migrations to hibernacula may involve flights to the north (east, south, or west).

During hibernation, each bat species appears to require specific conditions of temperature and humidity. Biologists often find several species hibernating in one cave, each occupying different locations. The only energy available to hibernating bats is the fat stored in their bodies. Waking up from hibernation is expensive, and each time a Little Brown Myotis does so it uses the energy that would see it through 60 days of hibernation. We presume that the spring wake-up call for hibernating bats is the result of an empty fuel tank. White-nose syndrome, caused by a cold-hardy fungus that is externally evident around the muzzle and sometimes the wings, is prevalent in the northeastern United States and is already beginning to spread into bats in Ontario. This fungus affects hibernating bats, apparently causing them to awaken to fly around or attempt to forage. More than 90% of the bats in some cave hibernacula have been infected and subsequently died (in the United States), and between 2006 and 2009 more than half a million bats (and possibly as many as a million) have died as a result. Some Canadian bat populations, especially those along the border, may migrate to hibernacula in the United States, where they become infected and die. This disease does not appear to affect humans.

Studies involving the recovery of banded bats suggest that young learn the locations of hibernation sites by following adults. In August, many of the hibernating species can be found visiting caves and mines before any individuals begin hibernating there. August visits to future hibernation sites where the bats swirl around the entrance and fly through the site are called "swarming." We still do not know how bats find and assess potential new hibernacula.

ECHOLOCATION

All Canadian bats are nocturnal and use echolocation, or biosonar, to orient themselves in the dark. Using echolocation, bats avoid colliding with obstacles and also can detect and track their insect prey in total darkness. Echolocating bats emit pulses of sounds and listen for returning echoes. These pulses can be emitted through the nostrils or open mouth. The bat registers each outgoing pulse in its brain for comparison to the returning echoes. Differences in timing and frequency provide the bat with information about its surroundings or targets. Although most bats, including most Canadian species, produce echolocation calls well above the threshold of human hearing (around 20 kHz), some species including Spotted Bats produce lower frequency echolocation signals (around 10 kHz) that are readily audible to many people. Higher frequency sounds have shorter wavelengths, providing the bats with more details about their targets. Bats vary the rates at which they produce echolocation signals. A cruising Little Brown Myotis emits an echolocation pulse about 20 times a second when searching for prey. In the same situation, Hoary Bats produce about four pulses a second. When attacking a flying target, both species increase their call output to about 200 per second. Many insects can hear the echolocation calls of bats and avoid the bat's

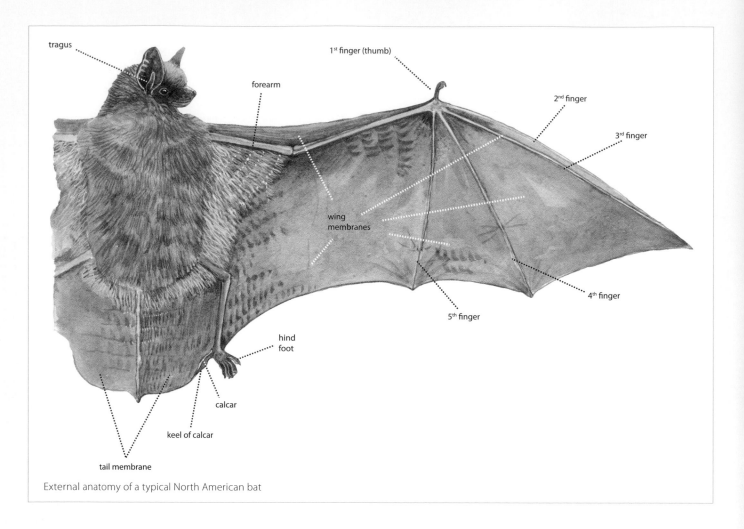

External anatomy of a typical North American bat

attacks. Some species of Tiger Moths (family Arctiidae) also produce clicks that interfere with bats' attacks.

Just because a bat is echolocating does not mean that it always makes the effort to interpret what it is hearing, especially in familiar surroundings where it has never encountered an obstacle in the past. Bat scientists use this characteristic to catch bats by placing a difficult to detect trap, called a harp trap, at the entrance to the roost or on a commonly used flight path. The trap consists of two offset rows of very thin plastic wires, whence its name. The bats run into the wires and slide down them into a cloth trough. From there they can quickly be caught, banded, measured, and released unharmed. A few of the more manoeuvrable and slower flying species are able to detect and avoid these traps.

Not all bats echolocate. Most of the flying foxes (fruit-eating bats of the Old World tropics) do not echolocate. The exception is some species of dog-faced bats (genus *Rousettus*). Other echolocating mammals include toothed whales and some shrews and tenrecs.

ENERGETICS

Like most mammals, a bat's metabolism varies depending upon its activity level. A flying Little Brown Myotis has a heart rate of over 1100 beats per minute, but after it lands the rate falls to about 200 beats per minute. When the same bat is hibernating its heart rate drops to 5 beats per minute. Other Canadian bat species appear to have similar ranges of heart rates based upon activity. Our temperate zone bats are heterothermic, which means their body temperature will drop as the ambient temperature drops. By carefully selecting a cool roost, a bat can go into torpor, a form of suspended animation, to save energy and wait until the weather improves and its insect prey become active again. At lower temperatures their lower metabolic rates allow them to conserve energy. Extended periods of lower body temperature constitute hibernation. Bats have a unique set of enzymes that allow the body's necessary chemical reactions to continue even at temperatures below which the normal enzymes cease functioning.

Juvenile bats need a fairly warm ambient temperature to maintain active growth. Their mothers move them around within the day roost to ensure that they stay warm but do not get overheated. Body temperatures of above 40°C will harm or even kill a bat. The adult males rarely occupy the nursery colonies. They prefer a cooler roost where they can allow their body temperature to drop so they can save energy and increase their fat reserves.

Flying bats generate large amounts of heat and apparently their wings act as heat radiators. In the wings, direct connections between small arteries and small veins allow rapid movement of blood without the loss of pressure that occurs when blood moves through a capillary bed. Heated blood quickly travels between the membranes and voids its heat into the air. We have yet to find out the details of the thermostats that control the movement of hot blood into a bat's wings. Flying requires high rates of energy expenditure usually fuelled by very rapid digestion.

LIFE SPAN

For their small size, bats live surprisingly long lives. The current record is 30+ years for two Little Brown Myotis that were first banded as adults and recaptured and re-released 30 years later. Many young bats probably do not survive their first year, reflecting perhaps their inexperience at foraging and their inability to accumulate enough body fat to see them through the winter. Banding studies suggest that Big Brown Bats do not live as long as Little Brown Myotis, up to 20 years as opposed to over 30.

HUMAN HEALTH

RABIES: Like other mammals, bats contract and are killed by rabies, a viral disease. The rabies virus often accumulates in the saliva of an infected animal, so bites from rabid animals can expose others to the disease. For all intents and purposes, rabies in humans is always fatal once a victim shows clinical symptoms of the disease. This makes it imperative that anyone who is bitten by a bat receive post-exposure vaccinations for rabies. To further minimize the risk, the area of the bite should be immediately and thoroughly washed with soap and water. It is important to realize that people do not always know that they have been bitten by a bat. While the larger species such as Hoary and Big Brown Bats can deliver painful bites, smaller species do not bite as hard. The site of the bite may be two small pin pricks caused by the upper canine teeth. Such wounds, like those inflicted when sewing, can be difficult to find once the drops of blood are washed away and may be overlooked or considered to be insignificant.

Different species of bats are associated with different varieties or strains of rabies. One strain, found in Silver-haired Bats and Eastern Pipistrelles, seems to be particularly virulent, and has accounted for most of the human deaths from rabies in Canada and the United States that have occurred since 1980. Neither of these bats occurs often in buildings, leaving it unclear just how people are exposed to the strain of rabies.

In Canada, bats, foxes, skunks, and Northern Raccoons are the main rabies vectors. In the past, before widespread use of rabies vaccines in domestic animals, Dogs and Cats were the main vectors. The disease was unwittingly brought to North America in the 1700s by European settlers in their domestic animals.

HISTOPLASMOSIS: This fungal disease enters through the lungs and has been known as "the curse of the mummy's tomb" and also "cave disease." The colourful name comes from Egypt, where the old tombs often supported colonies of bats. The spores of *Histoplasma capsulatum* are often found in the nitrogen-rich droppings of birds and bats, and they become suspended in the air following crushing of the dried faeces or disturbing of the soil. The symptoms can vary from those of a light chest cold or a dry cough to more serious tuberculosis-like effects. Blindness by macular degeneration occasionally

results. Sometimes the disease lies dormant in its human host and can manifest at a later date, when it can be very difficult to trace back to its source. The disease can be fatal, although most exposed people do not show any symptoms and do not need treatment. It is treatable in its more severe forms. In Canada bats frequently use caves as hibernation sites, but hibernating bats do not deposit faeces. The danger is more acute in buildings that support long-time summer roost sites. There have been a few outbreaks of histoplasmosis in Canada and most are restricted to the St Lawrence River Valley. Children, the elderly, and those with respiratory disease, HIV, or other auto-immune diseases are most susceptible. Precautions such as wearing a face mask with a HEPA filter rated to 2 microns or better when working around poultry or bats can help, although it is frequently impossible to avoid contact with the fungal spores. Poultry farmers, home renovators, construction workers, and bat scientists need to be aware of the potential of contracting this disease during the course of their work. Since this is such an uncommon disease in Canada, it often goes undiagnosed by medical personnel.

REPRODUCTION

Canadian bats have a low reproductive potential. Although they are long lived, most produce a single offspring per litter and one litter per year. Eastern Red Bats sometimes have litters of up to four, although two or three are more common. Plain-nosed bats of temperate regions typically mate in the late summer or autumn. Females store sperm in their uterus and become pregnant when they leave hibernation the following spring. Known as delayed fertilization, this phenomenon ensures that all of the offspring are born after the female leaves hibernation and insects are again abundant. Should the spring turn cold after the female has become active and pregnant, periods of torpor will reduce her energy consumption and prolong pregnancy. As the weather improves and food becomes more available, the female will arouse from torpor and foetal development resumes. The normal gestation period is around 50–60 days. Baby bats are large, usually around 25% and sometimes up to 30% of the mother's weight. Females usually reverse their normal position to hang upright while giving birth, with the tail membranes cupped under their body to catch the newborns. The young are born rump first, apparently to reduce the risk of wings becoming tangled in the birth canal. Deciduous milk-teeth are shaped to assist the neonates in attaching to their mother's nipple or fur. They will have grown their full adult dentition by the time they are weaned and begin flying, which for Little Brown Myotis may be by age three weeks. Young bats grow rapidly and nurse until they reach adult linear dimensions. Females of colonial species can recognize the call of their own offspring. As she approaches the nursery colony, she vocalizes, most of the juveniles respond, and she isolates the call of her own baby from the cacophony. A final sniff will confirm the pup's identity by odour and the female will allow it to suckle. Female North American bats are viewed as largely philopatric (they return to the maternity site where they were born to bear their own offspring) and males disperse when they mature. Recent genetic investigation in some species is proving that this is not an entirely strict practice, as some female dispersal is inferred from the mtDNA evidence.

CONSERVATION

Bats are very sensitive to disturbances to their roost sites. Even efforts to study them can lead to serious declines in their numbers. Females in nursery colonies and hibernating bats are particularly sensitive to disturbance and conservation measures must focus on reducing population disturbance to any roosting bats. For most Canadian species we lack details of population size and status. Unlike in the United States, where several species of bats are "Endangered," none are so listed in Canada. Pallid Bats are thought to be "Threatened" and Spotted Bats are of "Special Concern" in Canada because of their small population sizes. Migrating bats appear to be vulnerable to collision with wind-energy-producing turbines, especially those located on wooded ridges.

CONTROL

Colonies of bats in buildings are often unwelcome. To convince bats to move elsewhere requires closing off the routes they use to get into the roost space (often the attic or eaves of a building). Bats lack the stout claws and teeth necessary to create openings into buildings so that lightweight materials (spray foam insulation, screening) will exclude them. Wait until they leave for the night or preferably for the season and then block off the holes they use to get into the roost. Do not block access when the young are present (June and July). A nearby bat house might present an attractive alternative. Try putting it near the former access hole. Please read the "Human Health" section above before entering a bat colony, as some precautions are recommended.

BAT HOUSES

Insect control, environmental concerns, and just pure enjoyment of their presence encourage many people to have bats living nearby. By providing a secure roost in the form of a well-constructed and thoughtfully positioned bat house you might be able to attract a colony. A warm, dry, draught-free roost in an undisturbed location is the key. In most parts of Canada bat houses need to be painted black to absorb more heat from the sun's rays. The only part of Canada where the paint colour should be a medium tone is in the hotter parts of south-central British Columbia. Placement of the house can also increase the solar gain. If your average July temperatures are less than 26°C you will need to find a spot that gets 10 or more hours of direct sunlight. The necessary amount of direct sunlight drops to about 8 hours for July averages of 32°C and down to 6 hours if the July average is around 38°C. The house should be 3.5–6.5 m from the ground or higher and not closer than 6–7 m to the nearest trees. The chances of attracting bats are greater if the house can be located near open water. Most bats will change their roost site if given the opportunity, so multiple houses will increase the chances of a successful colony. Groups of three in close proximity but with different exposures work best. The females will move their young into the coolest site on a warm day and into the warmest on a cool day. They will also arbitrarily move for no known reason. Perhaps parasite avoidance is the goal. Larger constructions work better than smaller, so build the biggest you can for your location. Most commercially available houses are inadequate. The best house plans to date come from Bat Conservation International, an organization that continually researches the most successful designs. They also make an effort to create economical and efficient plans; for example, their most popular design calls for a single sheet of 1.25 cm (½ inch) plywood to build two bat houses.

Recommended reading and a good source for plans and placement: Tuttle, M.D., Kizer, M., and Kizer, S., 2005. *The Bat House Builder's Handbook.* Published by Bat Conservation International. Distributed by the University of Texas Press, PO Box 7819, Austin, TX 78713-7819. This publication is periodically revised to take into account the most recent successes.

VISION

All bats have fully functional eyes and see well. Bats' eyes appear to contain only rods so their view of the world is black and white. Apart from the tropical flying foxes, bats lack a *tapetum lucidum*, the reflective layer in the retina responsible for night-time eye-shine. This reflective layer magnifies the eyes' ability to detect and use low-level illumination.

ROOSTS

Roosts, protected places to spend the day (or the winter), are vital to bats. Bats are secretive about their roosts and even today we do not know where over half of the world's bat species spend the day. Bat roosts are labelled according to the role they play in the lives of bats. They may be in foliage, in crevices (such as those in rocks, trees, or buildings), or in hollows (also in buildings, rocks, and trees). Appropriate roost selection is learned and many young bats lose their lives when they make inappropriate choices.

DAY ROOSTS: Used to rest and sleep during the day, safely protected from predation. Selection of a day roost site usually varies with gender. Females select a day roost that will sustain enough heat during the cooler night-time to keep her young warm and secure while she is away foraging. The warmer environment also hastens the growth of the young. Males and non-reproducing females typically prefer a cooler day roost so that they can drop their core body temperature and save energy.

NIGHT ROOSTS: Used to rest, eat, and digest between foraging flights. These sites need to be fairly warm to speed up digestion so the bat can excrete the wastes and return to its foraging. A warm bat can pass a meal through its entire digestive tract in about 20–30 minutes, considerably reducing its weight.

HIBERNATION ROOSTS are discussed in the "Surviving Winter" section above.

Most roosting bats hang suspended from their hind claws. This allows their wings to be unencumbered and immediately ready for flight, and, incidentally, places the teeth into a strategically defensive position. Bats have a hind foot structure that locks a tendon into place as the animal places downward weight on its foot, so the claws can bear the weight for prolonged periods without expending any muscular energy. This lock is disengaged when the animal's weight is removed as it spreads its wings for take-off. Another adaptation to hanging upside down is the length of the toes. All are the same length so that each absorbs the animal's weight equally. Bats are clean creatures and spend a part of the daytime roosting phase grooming. If they need to urinate or defecate while roosting, they flip upside down (right-side up to us) and let gravity do the rest.

FLIGHT

Bats are the only mammals capable of flapping flight. Their wings are composed of two elastic layers of skin (patagium) stretched between elongated hand and finger bones. The hind legs are rotated so that the knees point outward. The uropatagium, the portion of the wing membrane between the hind legs, encloses the tail. The calcar is a cartilaginous spur that projects towards the tail from the ankle joint and helps to extend the uropatagium. In flight, the portion of the wings from the fifth fingers out provides thrust and the portion between the fifth fingers provides lift. Most of the wing movement takes place at the "hand" portion of the wing, but the hind legs do travel up and down as well. So technically bats fly with all four legs. Only most flying foxes and some fossil bats have claws on their second fingers. The rest of the bats have retained a claw only on the thumb and a few bats lack even thumb claws. The thumb is also the only digit not elongated to support the wing membrane. A bat's wing membranes heal quickly if torn or punctured.

In the wild, flight speeds of Canadian species of bats vary from 5 metres per second (35 km/h) for Little Brown Myotis to 10 m/s (70 km/h) for Hoary Bats. The largest bats in the world are the flying foxes, with wingspans of nearly two metres. At the other extreme are the butterfly bats of Asia, with wingspans of around 20 cm. Our Canadian species have wingspans in the 25–45 cm range.

Most bats need to drink water and all the Canadian species drink while on the wing. They fly low over a water body and dip their mouths into the water. This can be a very tricky manoeuvre, especially for the less coordinated juveniles, and can sometimes lead to a bat being forced to row itself to shore using its wings as oars. Provided a hungry fish is not lurking below and the water is not too cold, the bat will usually make it to shore and live to try again.

FAMILY MOLOSSIDAE
free-tailed bats

A tropical and subtropical family of bats that occurs in the Old World, as well as the southern United States, Mexico, and South America. These bats are medium-sized, insectivorous, and distinguished by the length of their tail, which extends beyond the tail membrane. The only individual of this family to be found in Canada was undoubtedly an accidental occurrence.

Big Free-tailed Bat

FRENCH NAME: **Grand Molosse**

SCIENTIFIC NAME: *Nyctinomops macrotis*, formerly *Tadarida macrotis*

This unusual looking bat has only been recorded once in Canada. A specimen was collected in November 1938 at Essondale, near New Westminister, British Columbia. No others have been reported since then, and that specimen is considered an accidental occurrence. Another free-tailed bat species, the Brazilian Free-tailed Bat, also called the Mexican Free-tailed Bat, has portions of its distribution quite near Canada, especially in the east. It is smaller and exists in much larger numbers than the Big Free-tailed Bat. It is perhaps more likely to occur accidentally in Canada in the future, especially given the effects of global warming. The Free-tailed Bats are members of the family Molossidae. Our other Canadian bat species belong to the family Vespertilionidae. No colour illustration is provided for this species.

DESCRIPTION

The Big Free-tailed Bat, the largest of the six species of free-tailed bats in North America, is of similar size to the Hoary Bat. The name derives from the at least 2.5 cm of tail vertebrae which protrude beyond the edge of the tail membrane. The wings are long, narrow, and tapered at the tips. Across the distribution of the species, fur colour varies from pale reddish brown to a dark brown (almost black). Each hair has a whitish base. The short, glossy fur is only slightly lighter on the belly. Young of the year are darker than adults. The upper lips of this species are heavily creased by a series of vertical wrinkles. The large, broad, forward flattened ears are joined at their base across the forehead. The tragus is small, actually just a bump at the base of the ear. The dental formula is incisors 1/2, canines 1/1, premolars 2/2, and molars 3/3, for a total of 30 teeth.

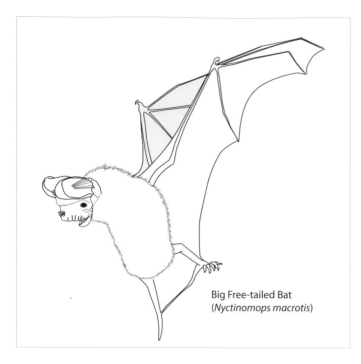

Big Free-tailed Bat
(*Nyctinomops macrotis*)

flying insects. The speculation about how they end up with flightless insects, such as crickets and grasshoppers, in their stomachs supposes that they cruise cliff edges and snatch them from the rocks. These bats like to roost in crevices in cliff faces during the daytime and are quite adept at crawling around on the rocks. Perhaps crickets and grasshoppers, which utilize the same resting places, are captured during daylight hours, becoming the equivalent to our midnight snack. Other day roost sites include buildings, caves, and occasionally holes in trees. Emergence from the day roost to hunt does not take place until it is fully dark. Unlike other free-tailed bats, this species can jump up into flight. Most of the other members of this family, and indeed most other North American bats, need to drop some distance before achieving sufficient speed to fly. The southern United States animals migrate south to Mexico for the winter. It is possible that some become torpid for part of the winter, but likely most go to locations in Mexico that are warm enough to allow them to continue with their normal activities. They are powerful flyers with great endurance. Calculations based on wing shape suggest a possible flight speed of at least 40 km/h.

DIET
Big Free-tailed Bats feed mainly on large moths, but are also known to consume crickets, grasshoppers, flying ants, stink bugs, froghoppers, leafhoppers, and beetles.

SIMILAR SPECIES
The Brazilian Free-tailed Bat (*Tadarida brasiliensis*) looks very similar, but is smaller – forearm < 46 mm.

SIZE
The following measurements are from Big Free-tailed Bats in Texas and Mexico. Males are slightly larger than females, especially in the total length measurement, although females are generally heavier.
Total length: 145–160 mm (males); 120–139 mm (females); tail length: 40–57 mm; hind foot length: 7–11 mm; forearm length: 58–63 mm; ear length: 25–32 mm.
Weight: 22–30g; newborns, about 4.5 g.
Wingspan: 42–44 cm.

RANGE
This species lives in the southern United States and the Caribbean and southward into South America. The northern part of its range is occupied only in summer. Records of this species outside its normal range are common, as this bat is a strong flier and undertakes a seasonal migration. Most of these are of misplaced animals in autumn, as was the only Canadian record.

ABUNDANCE
Although populations tend to fluctuate widely from year to year, this bat is generally considered rare throughout its range. It is listed as a "Species of Special Concern" by several states, but is not on the United States federal listing.

ECOLOGY
The biology of this rarely seen bat is poorly understood. Big Free-tailed Bats occupy rocky, arid habitats usually below about 1800 m in elevation. These bats utilize echolocation to find and capture

Distribution of the Big Free-tailed Bat (*Nyctinomops macrotis*)

⭐ extralimital records   ❔ presence probable but unconfirmed

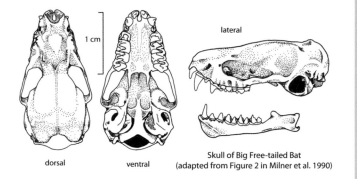

1 cm

lateral

dorsal   ventral

Skull of Big Free-tailed Bat
(adapted from Figure 2 in Milner et al. 1990)

REPRODUCTION

There is no evidence of sperm in the reproductive tracts of winter females, so it is presumed that mating takes place in the early spring. A single pup is born in late spring or early summer. Growth is rapid and the juveniles are flying by late August, and are almost indistinguishable from the adults by autumn.

BEHAVIOUR

Females form maternity colonies in late spring. The oldest known maternity colony in Texas supports around 130–150 adult females. Males are rarely found around the maternity sites and are probably roosting alone. Females returning to the roost site undertake some ritualized entry behaviour. They make several passes before attempting to land. If the roost is in a cliff face (as is the one in Texas mentioned above), the bats will make a steep dive towards the crevice with their wings partly folded before pulling up steeply and flying into the crevice. Often the steep dives are aborted before entry is achieved.

VOCALIZATIONS

The echolocation call of this species is audible to humans, as part of the call uses frequencies that fall below 20 kHz, the upper threshold of human audio sensitivity. It sounds like a loud series of chirps, sometimes getting closer together as they continue.

SIGNS

Maternity roosts are frequently used for many years, and consequently guano can accumulate into visible heaps below them. There are only a few known breeding sites in the southwestern United States, and there is little likelihood of seeing one of these animals in Canada.

REFERENCES

Fenton and Bell 1981; Milner et al. 1990; Parish and Jones 1999; van Zyll de Jong 1985.

FAMILY VESPERTILIONIDAE
plain-nosed bats

This widespread group of bats is distinguished by several characteristics: presence of a tragus, a long tail completely enclosed in tail membrane, a plain muzzle with no skin outgrowths, a relatively large thumb, lack of a postorbital process on the skull, and a well-developed calcar.

Pallid Bat

FRENCH NAME: **Chauve-souris blonde**
SCIENTIFIC NAME: *Antrozous pallidus*

This large, social bat is our only Canadian bat species that occasionally eats vertebrates, although its main prey is large insects.

DESCRIPTION

The Pallid Bat is light yellowish brown on the back and has an even paler belly. The fur is short, with a lighter base and darker tip. Wing membranes are a dark slate grey and the ears are greyish tan. It is a large bat, by Canadian standards, with oversized ears, eyes and hind feet. The ears are not joined at their base across the forehead and the series of horizontal pleats allow the bat to furl them over its back into the "rams-head" posture. The dental formula is incisors 1/2, canines 1/1, premolars 1/2, and molars 3/3, for a total of 28 teeth. A large bat, dipping and rising, and even briefly hovering as it forages near ground level, is likely a Pallid Bat. These bats have a noticeably slow flapping rate of only about 10–12 beats per minute. Pallid Bats have a musky skunk-like odour that intensifies if the animals are disturbed. The odour is emitted in tiny droplets by glands on either side of the nose and it is detectable from several metres away.

SIMILAR SPECIES

Townsend's Big-eared Bats are similar to Pallid Bats, but smaller. They have characteristically raised glandular lumps on the sides of the snout which the Pallid Bat lacks.

SIZE

The following measurements apply to animals of reproductive age from the Canadian portion of the range. Females and males are similar in appearance. Females are slightly larger than males. This is most clearly shown in forearm length and body weight.

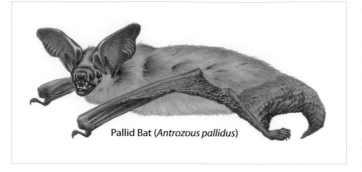

Pallid Bat (*Antrozous pallidus*)

Total length: 102–135 mm; tail length: 38–55 mm; hind foot length: 9–17 mm; ear length: 26–33 mm; forearm length: 48–57 mm (any forearm measurement over 54 mm is likely from a female).
Weight: male, 12–24 g; female, 14–28 g; newborns, about 3 g.
Wingspan: 31–37 cm.

RANGE

A western North American species, the Pallid Bat reaches the northern limit of its distribution in the Okanagan Valley of south-central British Columbia. Further surveys may show that the range is more extensive in southern British Columbia.

ABUNDANCE

This species has been designated "Threatened" by COSEWIC since 2002 and is on the "Red List" of endangered animals in British Columbia, where it is considered to be "at risk of extinction." Several specimens have been collected from southern British Columbia. Others have been heard in the same area with the aid of bat detectors, but the Canadian population size is unknown. Only 43 confirmed records have been made in the province since the bat was first detected there in 1931. All of these are in an area undergoing rapid urbanization and agricultural development. The major threat to this species in Canada is, hence, habitat loss.

ECOLOGY

The Pallid Bat is a desert-adapted species that reaches Canada in the open, dry regions of south-central British Columbia, where it occurs near rocky outcrops and water in habitat typically dominated by sagebrush. Pallid Bats are gregarious, roosting in colonies of 20 or more animals, mainly in rock crevices and buildings, but occasionally in caves, mines, stone piles, and tree cavities. The largest groups are maternity colonies of females and their young, which can comprise over 200 animals. These bats rarely venture out of their day roosts until it is well and truly dark. Colony emergence on warm nights is then fairly rapid, with each animal giving the "directive call" (see the "Vocalizations" section below) and usually voiding urine as it exits the roost site. On cool nights, emergence is more variable, with some animals choosing to remain at roost. Pallid Bats commonly undertake two foraging periods separated by a night roosting interval. The length of the night roosting interval varies with the season – longer when cool, shorter when warm. Nursing females spend only a short time at the night roost, choosing instead

to return to their pups to nurse them. The preferred night roosts are temperature-stable and maintain their heat when the outside temperature drops. The roosting bats thereby have enough ambient warmth to help them maintain their body heat while they digest their stomach contents. Clustering adds to the ambient heat and reduces the energy requirements of the individual bats. Rather than select a warm day roost, males at the northern limits of the distribution are more likely to select a cooler place to spend the daylight hours, so they can drop into shallow torpor during the cooler hours and conserve energy and water. Pallid Bats often carry large prey to the night roost to cull (pull off indigestible portions such as wings, legs, and heads) before eating. Softer or smaller food items are consumed on the wing. When outside temperatures are cold, especially towards the morning, the bats will forgo the night roost and head back to the day roost, skipping the second foraging period, when insect numbers would likely be very low anyway. Foraging takes place in open areas with low vegetation. Hunting Pallid Bats listen for the rustling sounds produced by their prey. During the approach and attack, they triangulate on those noises and also echolocate. Echolocation is also used in general orientation, and to capture flying prey. Pallid Bats produce concentrated urine and can survive on wet prey for an extended period, but prefer to drink water when possible. These bats hibernate in clusters or singly and are thought to travel only short distances between their summer and winter

Distribution of the Pallid Bat (*Antrozous pallidus*)

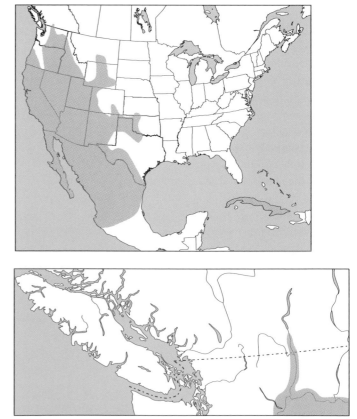

expanded view of Canadian range

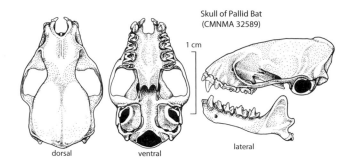

Skull of Pallid Bat
(CMNMA 32589)

1 cm

dorsal   ventral   lateral

sites. In northern regions such as Canada, the hibernacula may be farther south and require a longer migration. No winter hibernacula have been found in British Columbia. This is a long-lived species with known ages of more than nine years.

DIET

Pallid Bats are considered to be gleaners – bats that take prey from surfaces, the ground or vegetation, but they also take flying prey. The Canadian population preys mainly on beetles and moths. More data is available regarding the food habits of this species from populations in the United States. Large insects (2–7 cm) such as crickets, beetles, ant lions, grasshoppers, cicadas, katydids, moths, and flies are common prey items. This species also takes spiders, scorpions, lizards, small rodents, and possibly other smaller bats. Individual Pallid Bats may specialize in certain types of prey, reflecting perhaps the prey choices of their mothers or other group members. In some areas their preferred prey are poisonous (scorpions) or large and strong (Jerusalem Crickets). In arid regions of the southern United States, Pallid Bats frequently visit desert flowers (various cactae and agavae) to harvest the nectar-feeding insects that gather there. In the process, their heads become dusted with pollen and they inadvertently act as pollinators as they visit other flowers.

REPRODUCTION

Mating takes place from October to February, but likely in October and November for the Canadian populations. Sperm is stored internally by the female and fertilization is delayed until spring. Gestation lasts 53–71 days depending on the temperature. Most young are born over a two-week period in May or June, probably late June in northern regions. Yearling females tend to produce a single pup, while most adult females have twins, rarely triplets. Newborns are blind, deaf, and pink. They quickly attach (sometimes with their mother's assistance) to a nipple and are enfolded in her wing membranes. She will not carry her pups while she forages, rather choosing to leave them at the day roost while she hunts. Females will only allow their own young to nurse. The juveniles become volant in their fifth to sixth week and are usually weaned between 6 and 8 weeks old. Females become fertile in their first year, males in their second.

BEHAVIOUR

Pallid Bats display a well-developed and comfortable ability to crawl on horizontal surfaces. They are a social species that roost communally, occasionally tolerating other bat species in their clusters. Pups left while their mothers forage usually cluster together in the warmest part of the roost, often under the attendance of a single adult female "babysitter" whose actual role is unknown. Newly volant pups follow their mothers on her foraging excursions for about a month, presumably learning how to find prey and where to roost. Adult Pallid Bats also show abilities to learn new hunting skills, foraging locations or roost sites, and they are very vocal. They communicate with each other using a variety of sounds (see "Vocalizations" section below). All their vocalisations, both ultrasonic and subsonic, are emitted through an open mouth. These bats shift both day and night roosts fairly often. During the summer, they join together in groups, uttering the directive call and swirling around the roosting sites, eventually choosing one. This behaviour, known as swarming, can go on for up to 45 minutes and is thought to aid the young in finding roosts and familiarizing them with the roost area.

VOCALIZATIONS

Pallid Bats communicate with each other using a variety of subsonic vocalizations that are audible to humans. *Directive call* – described as a single call or series of calls sounding like a power line with a short circuit; used to call other bats, the directive call of young Pallid Bats (also called the isolation call) sounds more like a chirp and does not resemble the adult's until the pup is about 12 days old; pups use the isolation call to attract their mother. *Intimidation call* – described as a loud insect-like buzz; used to threaten enemies or a persistent youngster wanting to be nursed. *Squabble call* – described as a series of high pitched, dry, rasping, thin, double notes resembling the call of an Anna's Hummingbird; used to express mild irritation and serves to space bats within the roost. *Plaintive notes* are loud double note calls that are harsh and guttural and emitted by bats in pain such as labour. *Contentment notes* are a chittering emitted when bats are at ease and in physical contact. The ultrasonic echolocation call sweeps from 70 to 25 kHz and is beyond the detectable human range.

SIGNS

These bats often return with large prey items to the night roost, where they can cull and eat at their leisure. Depending on the size of the colony and the age of the roost site, this behaviour results in small to sometimes enormous heaps of disconnected legs and wings and other discarded bits included with the bats own mouse-like droppings, consisting mainly of insect parts. Their habit of vocalizing as they leave and enter their roosts will often give away their presence. They also have a characteristic musky odour which intensifies as they are handled and which lingers around a colony site. Most day roosts will display urine staining at the egress.

REFERENCES

Beck and Rudd 1960; Bell 1982; Cockrum 1973; Davis, R., 1969; Gaudet and Fenton 1984; Grindal et al. 1991; Hayward and Davis 1964; Hermanson and O'Shea 1983; Herrera et al. 1993; Johnston, D.S., and Fenton 2001; Orr 1954; O'Shea and Vaughan 1977; Rambaldini and Brigham 2008; Trune and Slobodchikoff 1976; Vaughan and O'Shea 1976; Willis, C.K.R., and Bast 2000.

Townsend's Big-eared Bat

also called Western Big-eared Bat, Lump-nosed Bat

FRENCH NAME: **Oreillard de Townsend**

SCIENTIFIC NAME: *Corynorhinus townsendii,* formerly *Plecotus townsendii*

Enormous brown ears, a highly variable flight pattern, and an overall brownish colouring distinguish this rare Canadian bat.

DESCRIPTION

A medium-sized bat, this species is characterized by its large ears and by the two raised glandular lumps along either side of its snout. The ears are easily half the length of the body (not counting the tail) and are joined at their base across the forehead. The tragus is about a third the length of the ear. Fur on the back is pale brown or dark brown at the tip with a greyish base. The darker form is found along the western coast, while the lighter form occurs inland. The belly fur is similar to the back colour, but slightly paler. The calcar is not keeled and extends almost halfway from the ankle to the tail along the edge of the uropatagium (tail flight membrane). The dental formula is incisors 2/3, canines 1/1, premolars 2/3, and molars 3/3, for a total of 36 teeth.

SIMILAR SPECIES

Pallid Bats are similar to Townsend's Big-eared Bats, but are larger. They also lack the glandular lumps on the size of the snout which characterize the Townsend's Big-eared Bat.

SIZE

The following measurements apply to animals of reproductive age from the Canadian portion of the range. Females and males are similar in size and appearance.

Total length: 83–113 mm; tail length: 38–57 mm; hind foot length: 9–12 mm; forearm length: 39–45 mm; ear length: 27–40 mm; tragus length: 10–15 mm.

Weight: 7–11 g; newborns, about 2.5 g.

Wingspan: 23–31 cm.

Townsend's Big-eared Bat
(*Corynorhinus townsendii*)

Distribution of Townsend's Big-eared Bat (*Corynorhinus townsendii*)

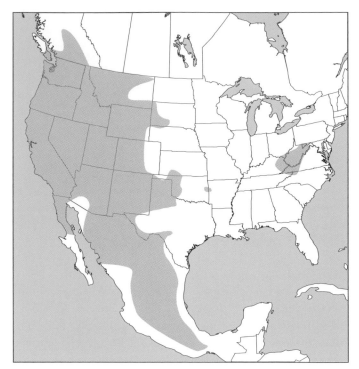

RANGE

This species occurs widely in western North America, from British Columbia to southern Mexico, with several disjunct populations in limestone karst areas of the eastern United States. The northern limit of its distribution is in south-central British Columbia and southern Vancouver Island.

ABUNDANCE

The extent of this species in Canada is poorly known and it is rare. About 16 Canadian hibernation sites have been discovered to date, but the total number of wintering bats in these hibernacula is fewer than 100 animals. There are only a few Canadian nursery colonies known, totalling around 350 bats. Elsewhere the species is generally uncommon and thought to be diminishing in numbers. The disjunct populations in the Ozark and central Appalachian regions of the eastern United States are considered "Endangered." This species is on the "Blue List" of endangered animals in British Columbia, where it is considered "at risk."

ECOLOGY

Townsend's Big-eared Bats are nimble flyers, capable of slow flight and occasional hovering. They readily detect and avoid the net traps set by biologists to catch birds in the daytime and bats at night. These bats are versatile insofar as their habitat preferences are concerned. They have been found in humid coastal forests as well as arid scrubland, open dry forests, and even grasslands. They appear to forage over wetlands, forest edges, and open woodlands, usually within a couple of kilometres of the day roost. They are closely associated with caves and, secondarily, abandoned mines, which they use both summer and winter, although old buildings, bridges, culverts, and large hollow trees are also used in the summer. The hibernation and warm weather locations are usually within 65 km of each other. In summer, males tend to be solitary, although bachelor roosts with multiple bats have been found. Townsend's Big-eared Bats prefer an open roost that allows them to fly to their roosting spot. They hang by their hind feet from horizontal or vertical surfaces with clear flight paths, never in cracks or crevices, and they rarely emerge to hunt until it is fully dark. Females will often fly within the nursery roost for some time before emerging. Nursery roosts are chosen that have a stable high temperature to allow rapid development of the pups. The females are very loyal to the nursery sites and will return year after year, as long as they are undisturbed and the conditions remain adequate. Nursery colonies are very sensitive to human disturbance. A single visit by a human can cause the females to abandon the site. During lactation, the females will forage and return to nurse their pups as many as four separate times each night. Males and non-nursing females spend the whole night away from their day roost, using temporary night roosts to hang up for short periods to cull prey and digest before resuming the hunt.

These bats tend to hibernate alone, although small clusters form from time to time and large clusters have been recorded. Caves and mines are the only known hibernation sites. This species prefers a cool hibernation site compared with many other bats, and can even tolerate short periods of below-freezing temperatures. They frequently arouse and fly around through the winter and commonly relocate within the cave or even between caves in an effort to find the right temperature regime. They appear not to eat during these times. In Canada, they will hibernate, depending on food availability and temperature, from the middle of September until the end of May. During deep torpor, the bats curl their ears back into the "ram's head" position and fold their wings around their bodies. Usually by spring they will have lost half or even more of their body weight. Human disturbances could tip the balance, during this already precarious time, by causing additional arousals that result in depletion of fat reserves before warm weather arrives and insects become available again. This degree of fat loss can result in the animal's death.

Colonies in buildings can be susceptible to rat predation but, like all our Canadian species, most mortality occurs over the winter as a result of insufficient fat reserves to last the bats until warm weather arrives. Yearlings are most susceptible to this type of mortality, and only about 40%–50% survive their first winter. This species is very vulnerable to human disturbances at both the maternity and the hibernation sites. Maximum lifespan is at least 16 years.

DIET

Moths in the 3–10 mm size range are the main food, but other flying insects such as beetles, flies, lacewings, and sawflies are also part of their diet. Although most of the big-eared bats are gleaners, the Townsend's Big-eared Bat likely catches most of its prey in flight. They typically forage near the vegetation surface either in the canopy or close to the ground, demonstrating that they are capable of hunting in both "cluttered" and open areas.

REPRODUCTION

Mating occurs in late autumn and over the winter, often while the female is torpid. The sperm is stored internally by the female and ovulation and fertilization of the egg takes place in the spring. Development of a single foetus takes 56–100 days depending on the temperature and how much torpor the mother experiences during gestation. Males do not breed in their first year, but yearling females will, although their first pup is born later in the season than those of the mature females. In British Columbia, most of the pups are born between mid-June and mid-July. Newborns are blind, deaf, and hairless, but develop quickly. If temperatures remain warm, they can be capable of flight in as little as three weeks, and are nearly adult-sized by five to six weeks old, when they are weaned.

BEHAVIOUR

A male Townsend's Big-eared Bat intent on mating approaches a female from the front while twittering, then proceeds to clasp her with his wings and spends more than a minute rubbing scent from his nose glands onto her fur. The females tend to remain indifferent to the attention, which may explain why so many copulations occur while they are torpid. Females do not carry pups while they hunt; however, they are known to transport the non-volant young to new maternity sites, especially following a disturbance by humans. Some researchers believe that this species commonly changes roost locations within a known area. This applies to night roosts, day roosts, maternity roosts, and hibernation roosts. They argue that the bats are faithful more to the area than to the specific roost site. However, the likelihood of finding a suitable nearby alternate nursery location, especially in the northern part of the range, is probably fairly low.

VOCALIZATIONS

Re-entry into a roost is often accompanied by a high-pitched vocalization audible to some humans. The echolocation call of this species sweeps from 90 to 20 kHz and it is emitted either through the nose or the open mouth. Most of the call is beyond the sensitivity range of human hearing. Breeding is accompanied by ritualistic vocalizations produced by both sexes.

SIGNS

Small clusters of culled moth wings or sometimes other insect parts, along with rice grain–sized faecal pellets, could be attributed to the

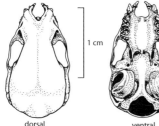

dorsal · ventral

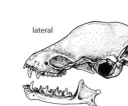

lateral

Skull of Townsend's Big-eared Bat
(CMNMA 27311)

night-time activities of this species. Long-standing roost locations (except hibernation roosts) will develop sometimes substantial guano piles made up of droppings and discarded insect parts – mainly moth wings. Hibernating bats do not produce faeces. Similar roosts of the Pallid Bat can be distinguished from this species by odour. (See the account of the Pallid Bat for a discussion of this odour.)

REFERENCES
Clark, B.S., et al. 1993, 2002; Fellers 2000; Fenton 1969; Humphrey and Kunz 1976; Kunz and Martin 1982; Lacki et al. 1994; López-González and Torres-Morales 2004; Mazurek 2004; Paradiso and Greenhall 1967; Pearson et al. 1952; Sherwin et al. 2000; Weyandt et al. 2005.

Big Brown Bat

FRENCH NAME: **Sérotine brune**

SCIENTIFIC NAME: *Eptesicus fuscus*

The Big Brown Bat is one of the most widely distributed mammals in North America. It is the biggest of our brownish-coloured bats, but not the largest Canadian bat species. The Hoary Bat enjoys that distinction.

DESCRIPTION
In Canada, this bat is usually reddish brown to dark brown on the back, with a distinctly lighter belly. The basal portion of each hair is darker and the hairs on the back are up to 12 mm long. The naked ears, snout, and wing membranes are black. The ears are short, barely reaching the nostrils when folded forward. The tragus is short

Big Brown Bat (*Eptesicus fuscus*)

and blunt and the calcar is keeled. Two caudal vertebrae protrude beyond the tail membranes. Fur on the back extends onto the tail membrane about one-third of the way down. Generally, a young of the year can be distinguished from an adult by its pelage, which is duller, darker, shorter, and has less distinction between the back and belly colour. The dental formula is incisors 2/3, canines 1/1, premolars 1/2, and molars 3/3, for a total of 32 teeth.

SIMILAR SPECIES
The only larger Canadian species is the Hoary Bat, which has silver-tipped hairs and is almost twice as heavy. In areas where the Little Brown Myotis co-occurs, identification of this species is usually based on size. In flight a Big Brown Bat has a much slower flapping rate than a Little Brown Myotis and is noticeably larger (almost twice as heavy).

SIZE
The following measurements apply to animals of reproductive age from the Canadian portion of the range. Females and males are similar in appearance. Females are slightly larger than males.
Total length: 93–130 mm; tail length: 37–59 mm; hind foot length: 10–15 mm; forearm length: 41–52 mm; ear length: 13–20 mm.
Weight: 15.0–29.6 g; newborns, around 3.3 g.
Wingspan: 32–39 cm.

RANGE
Big Brown Bats are widespread in southern Canada from coast to coast. Their range extends southward to northern South America and includes many islands, such as the West Indies and Vancouver Island. Isolated reports exist from central Alaska, Morris Lake, Yukon, and southwestern Northwest Territories in the Nahanni National Park region.

ABUNDANCE
This is a common species throughout most of its distribution, although it is uncommon to rare at the northern limits.

ECOLOGY
Big Brown Bats are generalists in many ways. They occupy and forage in many different habitats, from forests and agricultural regions to urban areas. They hibernate in caves, mines, and deep rock crevices, as well as heated buildings, and even in tree hollows in warmer regions. They tolerate a wider range of temperature and humidity during hibernation than other species, which allows them to make use of more marginal sites, such as within the walls of heated buildings. Cuban and other more southerly populations do not hibernate. The ability to adapt to, and even exploit, human activity is likely the reason this species is more numerous now than in historical records. Individual bats in summer roost in tree cavities, under loose bark, in rock crevices, and in buildings. In eastern Canada, maternity colonies are often in man-made structures, while those further west are more often found in tree hollows and crevices in rocks. Colony size of maternity sites can vary from 5 to 700 individuals, but typically contain fewer than 100 females and sometimes a small number of

Distribution of the Big Brown Bat (*Eptesicus fuscus*)

★ extralimital record

males. Summer colonies begin to disperse in August, but most Big Browns do not enter hibernation until November. They spend the time fattening up for the winter. Distance between summer roost and hibernation site is usually not more than 80 km. Many bats hibernate singly, but small clusters are common, usually of males. Big Brown Bats arouse from hibernation frequently and move about, presumably seeking a more comfortable location or perhaps to find water to drink. There is evidence that they may even change hibernacula during the course of the winter. Those that choose to hibernate in buildings are particularly vulnerable to drastic changes in temperature, especially during major outside temperature changes (warmer or colder). They may find their way into the interior of the building as they seek food or a warmer hibernation site.

Weather permitting, Big Brown Bats begin foraging soon after sundown and continue through the night. Usually the first stop upon emergence will be to a water source to replenish the liquid lost during the heat of the day. Rainy nights or temperatures below 15°C will delay emergence. Females with young to feed will likely hunt on these, and even cooler, nights, but the males may forgo the hunt altogether and just spend the time in torpor. Use of torpor as an energy-saving strategy is different for male and female Big Brown Bats. Males regularly go into torpor, both a daily light torpor, as well as a deeper torpor that can last for several days. Pregnant females will frequently enter a light daily torpor when foraging conditions

are poor, but nursing females rarely do. They must remain warm to continue producing the milk needed to feed their pups. Big Brown Bats eat a wide range of insects, apparently reflecting the prey available. When they feed on agricultural pests, they may be exposed to pesticides that accumulate in their bodies. These poisons are transmitted to the young through milk. High concentrations can cause death, but little is known about how lower concentrations may affect reproduction or behaviour. Predators include snakes, owls, American Kestrels, and even grackles, weasels, Northern Raccoons, Domestic Cats, rats, and bullfrogs, but there is no evidence that predation has a significant impact on populations of these (or other) bats. Postnatal mortality is 7%–10% and many subadults do not survive their first winter, usually because they fail to store sufficient fat reserves for their first hibernation. Accidents and bad weather also take their toll. Once adult, a Big Brown Bat can expect a fairly lengthy life. Males have been known to live more than 20 years. The lifespan of females is probably shorter due to reproductive demands.

DIET

Big Brown Bats have large teeth and powerful jaws and can easily crunch through the tough exoskeleton of small and medium-sized beetles. However, the Big Brown is flexible and eats a range of insects from beetles to moths to caddisflies. In southern Ontario, lactating females frequently eat June bugs. The teeth of Big Brown Bats are often worn down, apparently from their diet of hard-bodied insects. Worn teeth, however, do not appear to affect food preferences or hunting abilities. Nursing females with two fast-growing pups have tremendous energy demands to satisfy and will consume their own weight in insects each night – around 17.2 g. A maternity colony of 150 adult Big Brown Bats might eat around 2500 grams of insects a night. In an agricultural setting, this could be the equivalent of 600,000 cucumber beetles, 194,000 beetles (mainly June bugs and click beetles), 158,000 leafhoppers, and 335,000 stink bugs per year. It remains to be determined if these bats have any significant impact on populations of pest insects, but it is possible that they may be agents of some degree of biological control.

REPRODUCTION

Mating usually occurs in autumn before hibernation and the females store the sperm in their uteri until they ovulate in the spring. There are some recorded copulations over the winter, typically with torpid females, and more in the spring as the bats come out of hibernation. Females form maternity colonies ranging from 5 to over 500 animals soon after emergence from hibernation. Gestation lasts about 60 days depending on the weather, and in Canada most of the pups are born between early June and mid-July. In western Canada, females bear a single pup 85% of the time, while in eastern Canada the females produce twins 80% of the time. Multiple paternities are common, as one study found that about 50% of twin litters had two sires. The foetal sex ratio is essentially 50:50. Newborns weigh around 3.3 g, so total litter weight for twins is about 40% of the female's normal body weight. They are born blind, deaf, and hairless, but their eyes and ears open within a few hours of birth. Females prefer a relatively cool maternity roost temperature and they will

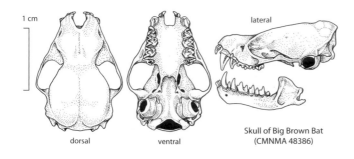

1 cm

lateral

dorsal

ventral

Skull of Big Brown Bat
(CMNMA 48386)

relocate if it climbs to 33°C–35°C. Juveniles begin to fly as early as 21 days old. Their wings are close to adult size but their weight is only 75% of an adult's weight. This low ratio of weight to wing size gives them an advantage as they learn to fly and hunt. By around 28 days old most juveniles are foraging for themselves. A cold, rainy spring will delay the development of the young and could add a week or more to these times. Males become sexually mature in their first autumn, but most females do not reproduce in their first spring.

BEHAVIOUR

Big Brown Bats can be aggressive in defending space. Generally they are defending their foraging territory and will attempt to exclude other bats and sometimes even birds, such as nighthawks which are four times their size. In southern Ontario, maternity colonies are usually located within 5 km of foraging areas, and within 7–10 km in south-central British Columbia. A lactating female must forage more than a male to satisfy her body's energy demands. She is on a tight schedule of hunting, digesting, producing milk, nursing, and excreting urine and faeces to reduce her weight, then beginning the cycle again. She will repeat this pattern between three and five times per night. Most adult Big Brown Bats forage for less than two hours over the course of a night. Males are usually solitary during the summer, but sometimes roost with the females, or form small all-male colonies. Big Brown Bats display strong roost fidelity, especially to their maternity and hibernation sites in buildings. In the west, when the maternity colonies are also in tree hollows or rock crevices, the bats tend to display more of a regional fidelity, switching roost sites often, but remaining in the same general area and continuing to use the same foraging areas. Females will retrieve their young if they fall from the roost. They do not carry their young while foraging, but will transport the non-volant pups if they change roost sites. Newly volant juveniles continue to roost with their mothers and follow them on their early foraging flights. As they get older, they join conspecifics and gradually blend with the adults.

VOCALIZATIONS

A foraging adult emits ultrasonic echolocation pulses of 5–10 milliseconds' duration, sweeping from about 50 to around 25 kHz. Their echolocation calls are similar to those of Silver-haired Bats. There is evidence of individual- and age- or gender-specific information encoded in echolocation signals, reflecting the reality that these signals also serve a communication function. Furthermore, the bats can alter their calls depending on context. For example, during the reproductive season, foraging calls are not distinguishable by gender, but those produced from roosts where mating activity is likely are gender dimorphic. Newborns produce an isolation call whenever they are separated from their mother. This call is detectable by humans and sounds like a chirping bird. Their mothers can identify them by these calls and will only nurse their own offspring.

SIGNS

Usually the first bat seen in the spring in eastern Canada is a Big Brown Bat, as it can tolerate the cool conditions and often hibernates in buildings. Signs of a maternity colony are similar to those of other bats – rice grain–sized dark faeces and stains at the egress. The large size, slow flapping, but rapid flight speed of this species can serve to distinguish it from other flying bats.

REFERENCES

Agosta and Morton 2003; Betts 1998c; Brigham 1987, 1991; Brigham and Brigham 1989; Brigham and Fenton 1986, 1991; Burnett 1983; Charbonneau et al. 2011; Davis, W.H., 1986; Grilliot et al. 2009; Grinevitch et al. 1995; Hamilton, I.M., and Barclay 1998a, 1998b; Kazial and Masters 2004; Keeler and Studier 1992; Kurta and Baker 1990; Kurta et al. 1990; Laborda and Cartwright 1993; Lausen and Barclay 2002, 2006; McAlpine et al. 2002a; Mensing-Solick and Barclay 2003; Rasmuson and Barclay 1992; Schowalter and Gunson 1979; Slough and Jung 2007; van Zyll de Jong 1985; Vonhof and Hobson 2001; Vonhof et al. 2008; Whitaker 1995; Whitaker and Gummer 1992; Whitaker et al. 1997a; Wilkinson, L.C., and Barclay 1997.

Spotted Bat

also called Pinto Bat

FRENCH NAME: **Oreillard maculé**

SCIENTIFIC NAME: *Euderma maculatum*

This is Canada's most flamboyant bat. Its spectacular ears and flashy pelage make it very recognizable. It also is the only bat species in Canada with an echolocation call that is audible to a human ear.

DESCRIPTION

The long silky fur of this bat is white on the belly and black on the back. What is most noticeable, and what gives this species its common name, are the three large, white spots on its back, one on each shoulder and a larger one on the rump. There are additional small white patches behind each ear. The pinkish-grey ears are enormous and are joined at their base across the forehead. They each have a series of horizontal pleats that allows the bat to fold them into a "rams-head." The bat only fully extends its ears while in flight or just before taking off; otherwise they are furled over the back. The flight membranes are pinkish red in live specimens, greyish if

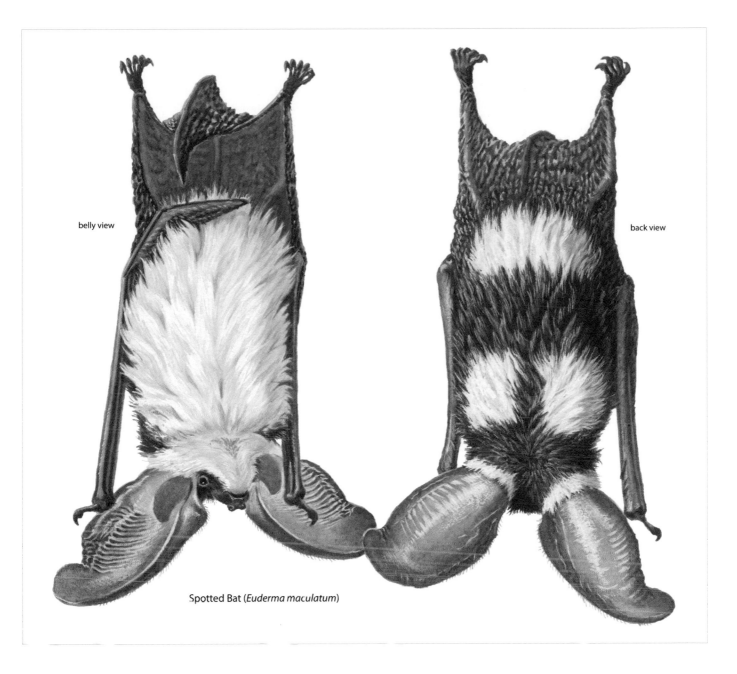

belly view

back view

Spotted Bat (*Euderma maculatum*)

the animal is dead. The last tail vertebra protrudes beyond the tail membrane. The calcar is not keeled. The dental formula is incisors 2/3, canines 1/1, premolars 2/2, and molars 3/3, for a total of 34 teeth. The high-pitched squeaks of its echolocation call can be heard over 250 m away.

SIMILAR SPECIES

Thanks to its dramatic fur and ears, there are no other bats that can be confused with this species.

SIZE

The following measurements are from Canadian and American specimens of reproductive age. Males and females are similar in appearance. Females are slightly larger than males. This size difference is most apparent in the forearm measurement.

Total length: 107–125 mm; tail length: 47–50 mm; hind foot length: 9–11 mm; forearm length: 48–53 mm (male average 50 mm, female average 52 mm); ear length: 34–41 mm.
Weight: 16.0–21.4 g; newborns, about 2.6–4.0 g.
Wingspan: about 34–35 cm.

RANGE

Spotted Bats were first detected in Canada in 1979 by Brock Fenton, C.G. van Zyll de Jong, and colleagues. A voucher specimen was collected at that time and deposited at the Royal British Columbia Museum. This remains the sole specimen of this species from Canada. Since 1979, extensive acoustical surveying has been undertaken to gather more information about the extent of its distribution and population size. The species is restricted to arid regions of south-central British Columbia. It occurs at elevations usually below

Distribution of the Spotted Bat (*Euderma maculatum*)

expanded view of Canadian range

ABUNDANCE

Current estimates in British Columbia suggest a population size of less than 1000 animals. It is possible that other populations exist that have not yet been found, since this is a rare species with localized populations and a patchy distribution. Spotted Bats are listed as a species of "Special Concern" by COSEWIC and are protected in British Columbia by the BC Wildlife Act.

ECOLOGY

Spotted bats occupy a variety of dry forested habitats, but are observed most often in arid locations, where they forage over open terrain, preferring forest openings, meadows, wetlands, and agricultural fields. Cliffs are usually evident nearby, along with a permanent water source. In Canada this species occupies grassland and open coniferous forests in dry river valleys. These bats prefer to roost in rock crevices on a vertical cliff face. They emerge from their roost when it is completely dark. The hunting grounds are usually within 10 km of the roost site. However, in one US location, the bats travel more than 35 km between the day roost and their foraging site. A strong, rapid flyer, this bat reaches flight speeds of about 50 km/h while commuting. Foraging speeds are considerably slower, in the 20 km/h range. Commuting bats fly in straight lines. Foraging bats travel in large elliptical patterns, usually at least 10 m off the ground (range 5–30 m). They will hunt while commuting, however, if the opportunity presents. Individual Spotted Bats appear to establish a foraging path that they will follow for several days, predictably visiting the same place at the same time night after night. Spotted Bats are moth specialists. The low frequency of their echolocation calls escapes the detection of their moth prey until the bat is 2 m away or closer, at which point the moths often enter a steep dive in an effort to avoid capture. Spotted Bats are fast manoeuvrable fliers, successful up to 88% of the time. When the hunting is good, they attack prey on average about every 45 seconds, a more common average being every 2 minutes. This is a considerably slower rate than for most insectivorous bats. Foraging takes place on the wing, and most Spotted Bats fly continuously from the time they leave their day roost until they return. Use of night roosts is only known in one population in Arizona that travels a long distance to its foraging site. In a study in the Okanagan Valley, most of the bats foraged 5 to 8 hours per night. Moonlight, cool temperatures, and light rain do not impede hunting, although they will roost-up to wait out a hard rain. Spotted Bats appear to migrate from higher elevations to lower ones in the autumn and at least some populations disappear during the colder periods, but where they go and whether or not they hibernate is unknown. Canadian populations are not detectable from November to March. There is a single report of a small cluster of four Spotted Bats seen hibernating in a Utah cave.

DIET

Moths around 10 mm or larger are the major food of Spotted Bats. The bats will usually cull the head, wings, legs, and antennae from the prey before consuming it. All of this is performed in flight.

900 m, in the dry interior valleys of the Okanagan, Similkameen, Thompson, Fraser, and Chilcotin Rivers. The most northerly record is near Macalister on the Fraser River. This is a western North American species with a range extending from southern British Columbia to central Mexico.

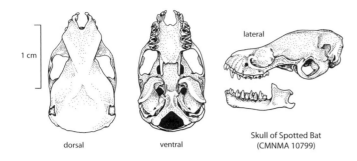

1 cm

dorsal ventral lateral

Skull of Spotted Bat
(CMNMA 10799)

Silver-haired Bat

also called Black Bat, Silver Bat, Silvery-haired Bat

FRENCH NAME: Chauve-souris argentée

SCIENTIFIC NAME: *Lasionycteris noctivagans*

Although rarely seen except during migration, this medium-sized bat is easily identified by its dark, but silver-tipped fur. It was previously thought to be solitary, but loose groups will migrate together and reproductive females will form nursery colonies.

REPRODUCTION

Reproductive information is very sparse for this species. Timing of mating is unknown, although likely similar to that of other Canadian species – autumn mating and delayed implantation until spring. Spotted Bats do not appear to form nursery colonies. Pregnant females have been captured up to the end of June, and lactating females have been captured from early June through to mid-August. In Canada, pups are probably born in late June or early July, nurse until mid-August, and are flying by mid-August. More southerly populations are likely born earlier. The pups are born naked, with their eyes closed, and their ears folded and not fully developed. Evidence indicates there is one offspring per year. Age of sexual maturity is unknown.

BEHAVIOUR

Females are very solicitous of their newborns. They enfold them in their wings and carry them on short flights between roosts, but not while foraging. While foraging, individual bats maintain a distance from each other that they enforce with agonistic displays. The bats will avoid an area where they can hear another calling, so the echolocation call provides a communication function as well as the more usual prey detection function. This species seems to exhibit roost loyalty, although this can be difficult to measure in a species that roosts in cliffs. After foraging, radio-tagged individuals returned to the same part of the cliff they had left at dusk.

VOCALIZATIONS

The echolocation calls of Spotted Bats typically sweep from 15 to 9 kHz. The call is loud and sounds like a series of high-pitched squeaks or clicks.

SIGNS

These bats rarely come into contact with humans. Although they do cull their prey, it is performed on-the-wing and no night roosts are used, so there are no telltale accumulations to suggest their presence. The best way to detect them is to listen for their calls at night.

REFERENCES

Best 1988; Cannings et al. 1999; Easterla 1965, 1971; Fenton et al. 1980; Fullard and Dawson 1997; Leonard and Fenton 1983; Nagorsen 2004; Rabe et al. 1998; Storz 1995; Wai-Ping and Fenton 1989; Watkins, L.C., 1977; Woodsworth et al. 1981; van Zyll de Jong 1985.

DESCRIPTION

Flight membranes, ears, and fur of this species are black. The hairs on the back and the belly are tipped with white, giving the bat a frosted appearance. This frosting is more conspicuous in younger animals. Old adults sometimes show very little frosting or the frosting has a yellowish tinge. The upper surface of the tail membrane is lightly furred halfway down to the tail. The dental formula is incisors 2/3, canines 1/1, premolars 2/3, and molars 3/3, for a total of 36 teeth.

SIMILAR SPECIES

The overall dark colour, the frosted hairs, and the furring on the upper surface of the tail membrane serve to distinguish this species from all other Canadian bats.

SIZE

The following measurements apply to animals of reproductive age from the Canadian portion of the range. There is very little size difference between the sexes, apart from weight in autumn, when females tend to be heavier as the bats enter hibernation.

Silver-haired Bat
(*Lasionycteris noctivagans*)

Distribution of the Silver-haired Bat (*Lasionycteris noctivagans*)

Total length: 80–113 mm; tail length: 25–48 mm; hind foot length: 7–12 mm; forearm length: 36–45 mm; ear length: 10–18 mm; tragus length: 5–9 mm.
Weight: 5.7–16.7 g; newborns, about 1.9 g.
Wingspan: 27–31 cm.

RANGE

This species is widespread across most of central North America. Since it is migratory, most of the northern portion of its range is deserted each winter. This region is mainly occupied by reproducing females each summer. Few males travel that far north, although they do move northward, since the southern portions of the range tend to be deserted in the spring and over the summer.

ABUNDANCE

The Silver-haired Bat is common to rare, depending on the season and the region. Population density is highly variable from one year to the next. These fluctuations are probably related mainly to temperature and the effect that has on reproductive success.

ECOLOGY

Silver-haired Bats have not been found hibernating in Canada and appear to emigrate south for the winter, although in the moderate climate of extreme southwestern British Columbia specimens have been collected in all seasons. Silver-haired Bats are highly adaptable to diverse habitats, but always near trees, especially old growth forests. Summer roost sites are found in tree cavities or under loose bark. Maternity colonies of 10–30 females have been found in tree cavities of tall, large diameter trees. Silver-haired Bats emerge from their day roosts shortly after sunset to forage for small flying insects in the treetops and over water. They tend to have two foraging cycles per night, separated by a night roosting period to digest, rest, and nurse pups. Their foraging flight is described as "erratic," as the bat twists and dips and glides in pursuit of its prey. Few hibernating individuals have been found north of the –6.7°C mean daily January isotherm, which goes from extreme southern British Columbia down the west coast to southern California, then angles across the continent to about Cape Cod. Small groups of bats have been sighted migrating and occasionally they get blown off course and turn up in unlikely places, like Bermuda or aboard ships at sea. Some, mainly juveniles, will also attempt to hibernate in regions that are too cold, and usually perish. Most of the winter records of this species north of their usual winter range are juveniles. Typically, Silver-haired Bats are thought to fly south of the –6.7°C mean daily January isotherm, where they have been found wintering in tree cavities, behind loose bark, in buildings, or in caves, mines, and rock crevices. They are generally solitary, but pairs and small mixed gender clusters of up to five or six have been found. Rarely are the bats in contact with each other in these communal situations. While torpid Silver-haired Bats have been found in these sites, it is not known if these bats were hibernating or in more temporary torpor. Remains of Silver-haired Bats have been discovered in owl pellets, and there are records of predation by skunks. This mortality is slight, however, compared to the threats of migration and possibly hibernation. It is possible that human forestry practices are a significant threat, as we select the biggest and tallest trees, which are also the favoured roosting trees.

DIET

In keeping with its wide distribution and twice yearly migrations, this species is not a specialist hunter. Rather, it is opportunistic and will forage on whatever insects are locally abundant. Main prey are beetles and moths with the addition of flies, leafhoppers, cicadas, flying ants and termites, true bugs, midges, and occasional oddities such as spiders and bees.

REPRODUCTION

As is typical of our northern bats, mating probably takes place in the autumn, perhaps during migration, and the females store the sperm internally. Fertilization occurs when the female ovulates in the spring. Gestation is generally in the 50–60 day range. Usually a litter consists of two pups or occasionally one. The sex ratio at birth is 1:1 males and females. In Canada, most pups are born in June or

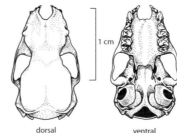

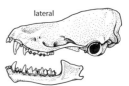

dorsal   ventral

Skull of Silver-haired Bat
(CMNMA 36675)

early July, although a pregnant female was collected in early August in northern Saskatchewan. Because this is a migratory species, the young have a little more time to gain weight, so the species is not as constrained with regard to birth season as are non-migratory bats. Nonetheless, the chance of survival of these late pups is probably poor. The females assume a head-up posture when birth is imminent and deliver the pup into their cupped tail membrane. Newborns are deaf, blind, and hairless. Their skin is either pink, black, or mottled pink and black. The young grow rapidly and are flying and weaned in about four weeks. Most become sexually mature in their first summer.

BEHAVIOUR

These bats migrate in small groups and individually. Migrants select older trees with furrowed bark for day roosting as well as the sides of buildings, in holes, and crevices in trees. Maternity colonies are invariably located in tree cavities (often abandoned woodpecker holes) in tall trees that extend above the canopy level, are not too decayed, and typically have either south-facing aspects or absorb solar radiation all day. Like other tree-roosting bats, Silver-haired Bats change roost sites regularly, and in southern British Columbia return to the same tree roosts year after year. Often whole maternity colonies will relocate together. Genetic testing has shown that most of the bats in these colonies are related along the maternal line.

VOCALIZATIONS

The echolocation call of this species is very similar to that of the Big Brown Bat, an equally widespread species. Since the two occupy similar habitats, distinguishing them by their echolocation calls can be challenging. Silver-haired Bat echolocation calls sweep from 35 down to 25 kHz. The "isolation call" of the young is audible to humans and sounds like a high pitched chirp. A disturbed colony will produce an insect like buzz of annoyance.

SIGNS

This species is rarely seen, except during migration, when individuals will occasionally select rather conspicuous roost sites, like the side of a building or the outside trunk of a tree. Provided they are not torpid, they are quite comfortable flying in daylight to find a quieter spot if disturbed. Silver-haired Bats are often the earliest bats in an area to begin foraging in the evening.

REFERENCES

Betts 1998a, 1998b, 1998c, 2000; Brigham 1995; Clark, M.K., 1993; Dunbar 2007; Kunz 1982; Kurta and Stewart 1990; Mattson, T.A., et al. 1996; Parsons et al. 1986; Reith 1980; Vonhof and Gwilliam 2000.

Eastern Red Bat

also called Red Bat

FRENCH NAME: **Chauve-souris rousse**

SCIENTIFIC NAME: *Lasiurus borealis*

The Eastern Red Bat and the Western Red Bat used to be considered subspecies of Red Bats. Canadian forms of this bat in the east and west have been confirmed through DNA testing to be the same species, Eastern Red Bat. The Western Red Bat (*Lasiurus blossevillii*) occurs from South America through Mexico as far north as northern California.

DESCRIPTION

The Eastern Red Bat has reddish-orange fur tipped with black or white on its back and slightly paler hairs on the belly. Below each shoulder is a whitish patch that fades out under the chin. The wing membranes are dark and the bare portions of the ears and snout are reddish tan. The upper side of the tail membrane is heavily furred and the tail is long and held straight out in flight. Males are more brightly coloured than females. The wings are long and pointed and the wing membranes are black. These same membranes are furred both above and below for a short distance from the body. The calcar is indistinctly keeled and less than half as long as the distance from heel to tail tip. The dental formula is incisors 1/3, canines 1/1, premolars 2/2, and molars 3/3, for a total of 32 teeth. The tiny first upper premolar is tucked in between the canine and the second premolar and is easily overlooked.

Distribution of the Eastern Red Bat (*Lasiurus borealis*)

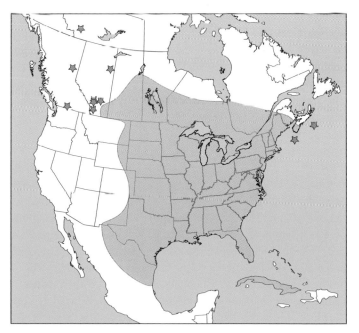

⭐ extralimital records

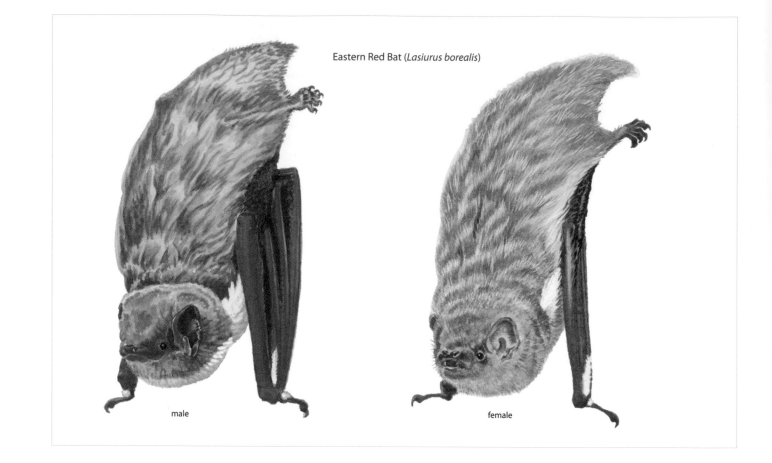

Eastern Red Bat (*Lasiurus borealis*)

male

female

SIMILAR SPECIES

Its bright reddish pelage separates this species from all other Canadian bats.

SIZE

The following measurements apply to animals of reproductive age from the Canadian portion of the range. Females are about 8% larger than males.

Total length: 87–120 mm; tail length: 36–65 mm; hind foot length: 7–10 mm; ear length: 10–13 mm; forearm length: 36–42 mm.

Weight: 10.0–17.4 g; newborns, about 0.5 g.

Wingspan: 28–33 cm.

RANGE

In Canada, Eastern Red Bats regularly occur from the southern Maritimes to southern Alberta, but the species is widespread across the central and eastern United States and Mexico. Increasing numbers of extralimital records in the west and northwest suggest the early stages of a range expansion, perhaps due to climate warming in that region. A female at the CMN has long been the only specimen of a red bat from British Columbia and until recently was identified as the only known Canadian specimen of the Western Red Bat (*Lasiurus blossevillii*). Recent genetic testing found that it is an Eastern Red Bat, eliminating the Western Red Bat from the Canadian faunal list.

ABUNDANCE

Can be locally common in Canada, but generally are rarely seen. This species has been considered abundant in the United States, but recent evidence suggests that numbers are declining by as much as 85% in some areas. Eastern Red Bats are especially vulnerable to wind turbines, and large numbers are killed at such developments.

ECOLOGY

The Eastern Red Bat is a solitary, tree-roosting species that prefers a mixed hardwood forest, where it roosts from ground level up to the highest canopy, depending on climatic conditions and sun position. It prefers to hang, sometimes from one foot, from small branches near the outside of the canopy, where it looks like a dried, partly rolled-up leaf. As it slowly twists, it resembles the movement of a dried leaf. Occasionally, red bats will roost on the trunk of trees, in the leaf litter, in dense grasses, and under the shingles of buildings. When temperatures drop, the bat curls its long tail up to its head like a blanket. The wings and ears are tucked under this "blanket," so all that is visible is red fur. Preferred roosting sites provide cover from the sides and above, but have an open flight path below. Like the other Canadian tree-roosting species, it switches roosts frequently, but stays within a known area. Eastern Red Bats forage in clearings at treetop level down to almost ground level. They fly either a straight path or in a large circle, broken by many dips and darts as they pursue their prey. They are often faithful to their foraging

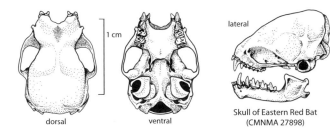

dorsal   ventral   lateral

Skull of Eastern Red Bat
(CMNMA 27898)

area and will return night after night to the same one, as long as the insects remain abundant. Home ranges have been determined, by radio-telemetry, to be 30–450 ha and are largest in unbroken forested habitat and smallest in highly fragmented agricultural landscapes. Eastern Red Bats travel around 0.4–5.5 km each night while foraging. Most nights are broken into two foraging periods, with a resting/nursing period between. Prey is typically captured in the cupped tail membrane and transferred to the mouth in flight. In Canada, this species emerges to begin foraging about a half hour after sunset. In the more southerly latitudes, it is known to emerge late, between one to two hours after sunset, long after most of the other species, except the Hoary Bat, are already hunting. The shortness of the northern nights likely causes this timing difference.

This species is considered to be highly migratory and, although usually solitary, is known to migrate in groups. During migration, individual bats are sometimes blown off course and have turned up in Bermuda, many of the Caribbean islands, and even on ships at sea. Most spend the winter in the southeastern United States from Delaware and the Ohio River southward. They travel to parts of the continent where the climate is sufficiently mild that they can hibernate under bark, in leaf litter, or in trees, without the risk of exposure to temperatures falling below 0°C. They will arouse and feed on warm days. The average foraging speed of Eastern Red Bats, determined by Doppler radar, is around 24 km/h.

DIET

This bat hunts insects in the 5–20 mm range and has a very diverse diet. It prefers moths when they are available, but also eats beetles, lacewings, flies, flying ants and termites, and surprisingly also many ground-dwelling species, such as crickets, cicadas, and ground beetles. Eastern Red Bats will exploit the gatherings of moths around artificial light.

REPRODUCTION

Breeding takes place in late summer and early autumn and fertilization is delayed until spring. Up to five pups are born (average of three) after a pregnancy of between 80 to 90 days. The same male may not father all of the litter. Newborns are small compared to other Canadian species. Based on an average adult female body weight of 13 g, a litter of five pups each weighing 0.5 g make the litter weight only 20% of the females' weight. Most births occur in June in Canada. Newborns cling to their mother with teeth, thumb-claws, and hind feet. As they grow larger, they clasp their mother with their wings and grasp a branch with one or both hind feet, creating a tight clump of

bats. The newborns are hairless, deaf, and blind, but like most bats, they develop quickly. By three weeks old, they are densely furred and fully half the weight of their mother. It is thought they begin to fly shortly after their third week and are weaned a week or two later.

BEHAVIOUR

Eastern Red Bats will initiate copulation in flight and then flutter to the ground, while still mating. They probably also mate in tree roosts. Groups, sometimes quite large, have been seen during migration, but otherwise, apart from females with young, they are solitary. Males and females migrate north at different times and to different summer ranges. Females undertake their northward and southward migrations earlier than do the males. The males are more common in summer in the west, and the females more common in the east and north. Females do not usually carry their young while foraging, but will transport them to new roosts if they are not yet too heavy. Whole families are sometimes found on the ground after the mother becomes dislodged from her roost and is unable to fly due to the weight of clinging pups. Many species of moths have bat-detecting ears, allowing them to detect and avoid foraging bats. Moths with ears are able to evade 60% of the first attacks by red bats, but are taken 90% of the time on second attacks that occur within 10 seconds of the first. As the moths undertake an evasive flight pattern to avoid the bats, they choose one of several stereotypic manoeuvres. The bats learn these patterns and are usually able to predict the moth's location and capture it during an immediate second attack. Arctiid moths like the Painted Lichen Moth use clicks to warn attacking red bats of their bad taste.

VOCALIZATIONS

The echolocation calls of this species fall in the range from 30 to 70 kHz. These bats can alter their echolocation call to suit the foraging location. Long calls are used in open areas to detect prey at a distance, and shorter calls in tighter locations, like clearings. When foraging in proximity to other bats, such as around an artificial light source, the bats occasionally produce vocalizations that can be detected by humans. These seem to occur during close encounters and appear to indicate irritation or perhaps antagonism. Young left on tree foliage roosts while their mothers forage are normally silent.

SIGNS

Few signs will be found, as this bat rarely roosts in or on buildings and is difficult to find on its tree roost. The long pointed wings, direct rapid flight pattern, and tail straight out behind identify a flying red bat.

REFERENCES

Baker, R.J., et al. 1988; Barclay 1984; Cannings et al. 1999; Cryan 2003; Fenton et al. 1980; Hickey and Fenton 1990; Hickey et al. 1996; Hutchinson and Lacki 2001; Mager and Nelson 2001; Mormann and Robbins 2007; Reddy and Fenton 2003; Salcedo et al. 1995; Saugey et al. 1998; Saunders 1990; Schmidt-French et al. 2006; Shump and Shump 1982a; Spradling et al. 2003; Walters et al. 2007; Whitaker et al. 1997b; Winhold et al. 2008.

Hoary Bat

FRENCH NAME: **Chauve-souris cendrée**

SCIENTIFIC NAME: *Lasiurus cinereus*

This is the largest of the Canadian bats. It has long slim wings, so it is a strong, speedy, but not very manoeuvrable flyer.

DESCRIPTION

This bat has distinctively coloured fur. On its back, the hairs are long and silky and a mix of dark brown/black and grey with distinct bands of light and dark, giving the animal a "hoary" appearance. The tail membrane is heavily furred on the upper surface. Short, brown hairs are present on the upper and lower wing membranes to the elbow and knee. Around the head the hairs are yellowish brown. The muzzle is dark, as are the wing membranes. There are small patches of fuzzy, white or creamy yellow hairs on both upper and lower sides of the wings over the elbow and wrist joints. The calcar is long and narrowly, but noticeably, keeled. The ears are short and rounded, with a black rim and short yellow hairs inside. The belly is creamy white with a yellowish brown throat. The tiny first upper premolar is tucked in between the canine and the second premolar and is easily overlooked. The dental formula is incisors 1/3, canines 1/1, premolars 2/2, and molars 3/3, for a total of 32 teeth.

SIMILAR SPECIES

Fur colour and size easily distinguish this bat from other Canadian bats. In flight the large size; narrow, pointed wings; slow wing beat; and fast, direct flight path help to identify this species. The Hoary Bat, along with other bat species, commonly feeds among insects attracted to artificial lights.

SIZE

The following measurements apply to animals of reproductive age from the Canadian portion of the range. Females and males are similar in appearance. Females are about 4% larger than males, but average 20%–25% heavier.

Total length: 99–143 mm; tail length: 40–64 mm; hind foot length: 9–14 mm; ear length: 13–20 mm; forearm length: 54–58 mm.

Weight: 25.0–35.7 g; newborns, about 4.7 g.

Wingspan: 34–41 cm.

RANGE

This widespread species occurs in North, Central, and South America. The North American population moves north for the summer and south for the winter. In South America, it does the opposite. There are resident, non-migrant populations of Hoary Bats on the archipelagos of Hawaii (the only native land mammal) and the Galapagos, and strays have turned up on Bermuda, Iceland, Southampton Island in Hudson Bay, Prince Edward Island, Newfoundland, and even the Orkney Islands in Scotland.

ABUNDANCE

Although rarely captured, this bat is often encountered by biologists listening with bat detectors. There is no data about its relative abundance.

ECOLOGY

Hoary Bats prefer to roost in trees, usually high in the branches concealed by leaves, but occasionally in hollows in the tree or fissures in the bark. Most roosts have a clear drop to allow easy take-offs and landings. Maternity roosts are usually oriented to provide enhanced solar radiation and shelter from wind. The roosting bats are invisible from above but sometimes can be seen from the ground, although their coloration and lack of daytime activity makes them very difficult to find. A solitary species except during migration, the males roost alone and the females with their offspring. The ranges of adult males and females appear to differ, with more females in the east of the continent and more males in the west. Females migrate north earlier than males. Most Hoary Bats depart for the wintering grounds between mid-August and October. They are presumed to fly south of a line that runs from Chesapeake Bay on the east coast along the coast to central Mexico and then up the west coast to San Francisco. Strong flyers, it is thought that these bats typically migrate far enough south that hibernation is not necessary. It is possible that some Hoary Bats hibernate in cooler parts of their southern wintering grounds, but no hibernating Hoary Bats have been found in Canada. The extensive, thick fur covering these bats provides exceptional insulation, allowing them to forage in temperatures far below the tolerances of other species. Hoary Bats will hunt at temperatures as low as freezing (0°C), although usually temperatures below 13°C send them into torpor. This species typically emerges to forage well after sunset;

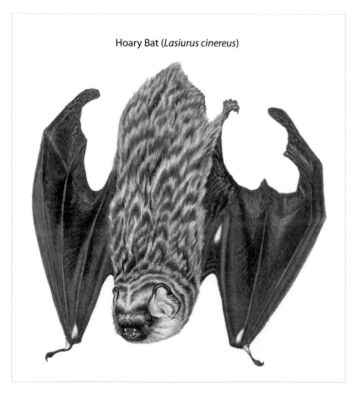

Hoary Bat (*Lasiurus cinereus*)

Distribution of the Hoary Bat (*Lasiurus cinereus*)

⬟ extralimital records

however, individuals have been seen foraging on warm winter afternoons. Hoary Bats forage high in clearings or over water. They are well known to take advantage of the rich concentration of moths that are attracted to street lights. Foraging Hoary Bats travel at about 28 km/h. Little is known about the predators of Hoary Bats. Likely owls and occasionally hawks will take a few, but this is probably not significant. These bats have been found accidentally impaled and entangled on barbed wire fences, and also found dead below microwave and TV towers, power lines, and wind turbines. Most of the collision mortality occurs during migration.

DIET

Hoary Bats eat a variety of insects, including many kinds of moths, and also beetles, flies, termites, dragonflies, and wasps. The preferred size ranges from 10 to 25 mm. Its robust skull and teeth make short work of even the larger dragonflies and beetles. Hoary Bats have been observed overtaking their moth prey from behind,

engulfing and swallowing the abdomen and thorax and allowing the severed head and wings to flutter to the ground. The Hoary Bat has an attack rate of around two to six times per minute, slow compared to smaller species, but it makes up for that by capturing larger prey items. The attack rate varies with prey density. Adult males consume an estimated 57% of their body weight each night, lactating females over 100%. The relatively low-frequency echolocation call of this species is well suited to detecting large insects at distances of 10–20 m.

During their first two weeks of hunting, juveniles have a different diet from the adults. Because they have full-sized wings, but a lower body weight, they are more manoeuvrable. Like other bats with low-frequency echolocation calls, it is likely that the juvenile call is a higher frequency until they are fully mature. The combination of higher frequency and greater manoeuvrability allows them to detect and capture smaller-sized insects like chironomids that are plentiful and easier for them to manage. The juveniles are not disadvantaged by having to compete with the more competent adults for the same resource. An added bonus to this behaviour is that the juveniles can usually find this prey nearby and can avoid the sometimes lengthy commutes required of their mothers.

REPRODUCTION

Mating takes place in the autumn, probably just before or during migration. The female stores the sperm internally and fertilization takes place when she ovulates in the spring. The length of pregnancy is unknown. The usual litter size is two, with a range of one to four, born in mid to late June in northern parts of the range. Pups can be born as early as mid-May in more southerly areas. Newborns have a fine silvery fur on their backs and their ears and eyes are closed. The females wrap their offspring in their wings during the day, but leave them at the roost when they forage at night. For the first couple of weeks after the pups are born, their mother reduces her foraging time in order to return to the nursery roost to warm up the babies. Juveniles are slow developing, partly due to the cool temperatures to which they are exposed during the night while their mother is hunting. They often drop into torpor for some of that time, which slows their growth. Their ears open at about three days old and their eyes open at about twelve days old. Although they begin to fly at around four weeks old, they are not weaned for another three weeks. Whether the mother leads her offspring on their first migration is unknown, but likely.

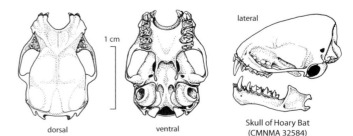

1 cm

dorsal   ventral

lateral

Skull of Hoary Bat
(CMNMA 32584)

BEHAVIOUR

Hoary Bats generally emerge about an hour after sunset, slightly later than some other Canadian bats. Foraging time varies from one to two hours where the bats hunt at lights, to the whole night in areas where they do not. For females, the length of the foraging period varies, depending on the phase of the reproductive cycle. Providing milk to young requires more energy than pregnancy, and those with multiple pups must produce even more milk. Some lactating females emerge at dusk and will hunt late-flying dragonflies and other large daytime insects not usually accessible to a night-flying bat. When prey is scarce or temperatures fall below 13°C, these bats usually drop into torpor to save energy. Hoary Bats are unusual in that both females and their young will regularly become torpid even during the nursing period. Large flocks of Hoary Bats have been observed in late summer and autumn, but it is uncertain whether these are mating or migrating flocks, or perhaps both. Flocks are also seen in spring, which again suggests a migration. Hoary Bats are at times aggressive in defence of their foraging territory and have been known to attack and drive away other smaller bats as well as others of their own species.

VOCALIZATIONS

The echolocation calls sweep from about 40 to 20 kHz and is easily identified using a bat detector. Pups are known to occasionally produce a high-pitched chirping "isolation call" when left alone. This isolation call falls at the upper audible limit for humans and can be detected by some people.

SIGNS

Because this species does not roost in buildings or caves, it is infrequently seen. The best way to discover it is to use a bat detector, which will help to identify the echolocation call.

REFERENCES

Barclay 1987, 1989; Bouchard et al. 2001; Cryan 2003; Fenton et al. 1983; Hickey and Fenton 1996; Hickey et al. 1996; Hill and Yalden 1990; Johnson, G.D., et al. 2003; Koehler and Barclay 2000; Maunder 1988; McAlpine et al. 2002b; Rolseth et al. 1994; Salcedo et al. 1995; Sealy 1978; Shump and Shump 1982b; Willis, C.K.R., and Brigham 2005.

California Myotis

also called California Bat

FRENCH NAME: **Vespertilion de Californie**

SCIENTIFIC NAME: *Myotis californicus*

This little western bat is notorious for having fur colour that is variable in different parts of its range. Within its relatively small

California Myotis (*Myotis californicus*)

Canadian distribution, it varies from blackish brown to a lighter reddish brown. Pelage colours in the southern parts of the range are even more bewildering in their variety.

DESCRIPTION

The California Myotis is not an easy bat to identify, either in flight or in the hand. In the field, look for a small, slow-flying bat making numerous dips and darts after its prey. It is similar to the Little Brown Myotis, but smaller. The ears, snout, and wing membranes are blackish. The length of the bare area on the snout is roughly equal to the width of the nostrils. The ears are fairly long, extending beyond the nose when gently pressed forward, but are less than 15 mm in length. The hind feet are small. The calcar has a distinct keel and its length is less than half the distance from the heel to the tip of the tail. The fur colour is variable, but dull with no sheen. Coastal bats have a dark brown back and a slightly lighter belly. Inland populations are reddish brown, also with a paler belly. The dental formula is incisors 2/3, canines 1/1, premolars 3/3, and molars 3/3, for a total of 38 teeth.

SIMILAR SPECIES

In British Columbia, there are two other mouse-eared bats (genus *Myotis*) with a keeled calcar, and similarly short ears. The Long-legged Myotis is much bigger and has larger feet, while the Western Small-footed Myotis usually has yellowish-brown fur and a distinct black mask and flight membranes. Western Small-footed Myotis also have a larger area of bare skin on its snout than California Myotis (see the figure in the "Similar Species" section of the Western Small-footed Myotis account).

SIZE

The following measurements apply to animals of reproductive age from the Canadian portion of the range. Females and males are similar in appearance. Females are slightly larger than males.

Total length: 74–95 mm; tail length: 34–41 mm; hind foot length: 5–8 mm; forearm length: 32–35 mm; ear length: 11–15 mm.

Weight: 3.3–5.4 g; newborns, unknown.

Wingspan: 22–23 cm.

Distribution of the California Myotis (*Myotis californicus*)

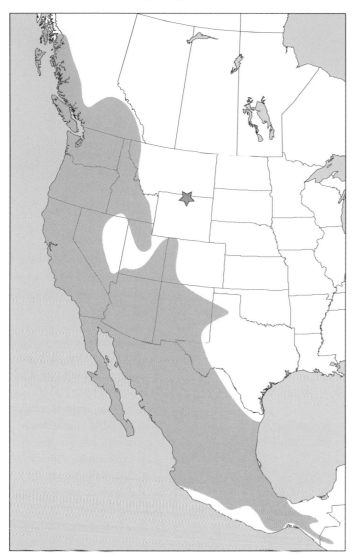

⭐ extralimital record

RANGE

The California Myotis occur from the Alaska Panhandle to southern Mexico, but can be scarce in parts of the distribution. Canadian animals live in British Columbia, from close to the Alberta border to the Gulf and Queen Charlotte Islands.

ABUNDANCE

It is considered common in British Columbia within its range.

ECOLOGY

In British Columbia, these versatile small bats are found from interior arid grasslands, to coastal rainforests, to montane forests up to about a 1300 m elevation. In the drier interior, they are rarely far from water. Although their kidneys are adapted to conserve fluids and they may not need as much water as other bats, their insect prey is more concentrated near water. California Myotis are slow,

but manoeuvrable flyers. They emerge for the nightly hunt around sunset and forage over water to within a metre of the surface, along the edges of the forest canopy, and well above ground level in open areas. Like most of the Myotis bats, their nightly foraging is bimodal. They have two foraging sessions, with a roosting/resting/nursing break in the middle, in order to take advantage of peaks of insect activity. On cool nights, when temperatures drop below 15°C, these bats hunt for about 4–5 hours starting near sunset and then call it a night. At those temperatures, few insects are flying anyway and the dawn peak is either reduced or non-existent. During the summer, singles or small groups (rarely more than 50) roost under tree bark, in cracks and holes in tree trunks, in narrow crevices on rocky hillsides, under bridges, or in buildings. Small maternity colonies have been found in the same types of places. There are no known hibernation sites in the province. Winter specimens have been collected, however, so it is likely that, like their more southerly relatives, at least some hibernate in buildings or caves and old mines near their summer ranges. California Myotis are known not only to come out of torpor, but also to forage on warmer winter days. Individuals live as long as 15 years, but the average lifespan is much shorter.

DIET

In Canada, this species hunts mainly caddisflies, with moths, flies, and beetles as secondary prey choices. In other more southerly parts of its range, the California Myotis prefers moths and flies.

REPRODUCTION

There are few details available concerning reproduction of this species. In Canada, mating occurs in the autumn, with fertilization delayed until spring. A single pup is born between late June and early July and is able to fly in about four weeks.

BEHAVIOUR

This bat's diet varies from place to place, reflecting insect availability and the presence of other bats. When California Myotis and Western Small-footed Myotis occur together, they partition the food resources. The former feeds mainly over water, the latter over adjacent rocky slopes. As a consequence, their food may be different as different insects are found over each habitat. Tree-roosting females regularly change their roost sites even when their offspring are unable to fly, possibly to avoid parasites or perhaps in search of better thermal conditions. They will carry their offspring to the new roost, but not while foraging. Large dead trees that extend above the canopy and are near water and foraging areas are

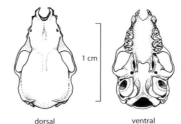

dorsal   ventral   1 cm

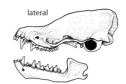

lateral

Skull of California Myotis
(CMNMA 42838)

preferred maternity roost sites. Although they move frequently, the females still remain within the neighbourhood of their foraging areas. The frequency and duration of the echolocation call of this species indicates that insects are detected at close range and the bat has less than a metre to alter its flight path to intercept its prey. The flight pattern of the California Myotis reflects this strategy. It is slow, but very manoeuvrable in the air, and while foraging the bats can be seen making abrupt, and often dramatic, changes in direction in pursuit of insects.

VOCALIZATIONS

The echolocation call sweeps downward from 82 to 45 kHz. If the bats detect another bat on a possible collision course, they will add a lower-frequency "honk" to the end of the call. The whole call, including the "honk," is beyond the frequencies detectable by humans.

SIGNS

California Myotis have rarely been found roosting in buildings in large enough groups to be noticeable. Their rice grain–sized droppings might be found below a roost. Their flight is slow and erratic, with many dips and diversions as they seek and pursue their flying prey. They often will forage over water.

REFERENCES

Barclay and Brigham 2001; Bogan 1999a; Brigham et al. 1997; Duke et al. 1979; Fenton and Bell 1979; Fenton et al. 1980; Gannon, W.L., et al. 2001; Hayward and Davis 1964; Simpson 1993; van Zyll de Jong et al. 1980.

Western Small-footed Myotis

also called Western Small-footed Bat

FRENCH NAME: **Vespertilion pygmée**

SCIENTIFIC NAME: *Myotis ciliolabrum*

The Western Small-footed Myotis is our smallest western bat.

DESCRIPTION

This beautiful little bat has pale yellow brown to orange brown fur with a paler buff belly. The black wing membranes, ears, face, and snout provide a lovely contrast to the fur. Tail vertebrae extend beyond the end of the uropatagium. There is a distinct keel on the calcar. The dental formula is incisors 2/3, canines 1/1, premolars 3/3 and molars 3/3, for a total of 38 teeth.

SIMILAR SPECIES

While there is little doubt that the Western Small-footed Myotis and the California Myotis are distinct species, it can be difficult to distinguish between them in the field. Both are small (forearm less than

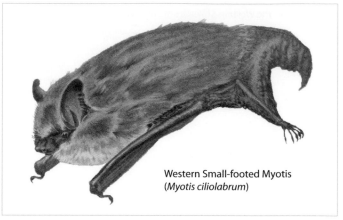

Western Small-footed Myotis
(*Myotis ciliolabrum*)

36 mm) and small-footed (hind foot less than 9 mm) and have a keeled calcar. The California Myotis is quite variable in its fur colour and can sometimes appear very similar to the Western Small-footed Myotis, which is usually paler, with more contrast between the fur colour and the wing membranes. It also has a larger bare patch on its snout. The fur begins up the snout about 1.5 times the distance between the nostrils (see figure on following page). The tail vertebrae of the California Myotis do not extend beyond the tail membrane. These characteristics separate the two species in Canada, but due to their variability are not as effective throughout the range, notably in the southwestern United States.

SIZE

The following measurements apply to animals of reproductive age from the Canadian portion of the range. Females and males are similar in appearance. Generally females are slightly larger than males.
Total length: 76–91 mm; tail length: 31–43 mm; hind foot length: 6–9 mm; forearm length: 30–33 mm; ear length: 13–17 mm.
Weight: 2.8–7.1 g; newborns, from 1.1–1.6 g.
Wingspan: 21–25 cm.

RANGE

This is a widespread western species that is found from south-central British Columbia, southern Alberta, and Saskatchewan down to central Mexico. It is not found in the wetter coastal regions. In British Columbia it is found up to about 850 m in altitude.

ABUNDANCE

Western Small-footed Myotis populations in both Alberta and British Columbia are provincially considered either "At Risk" or "Vulnerable." The US Office of Endangered Species lists the species as "Of Special Concern," which means that it is being considered for endangered status pending further information.

ECOLOGY

Western Small-footed Myotis are associated with badlands, dry grasslands, and arid valleys. These bats are slow, manoeuvrable flyers that can often detect and avoid nets. They are clearly capable

Distribution of the Western Small-footed Myotis (*Myotis ciliolabrum*)

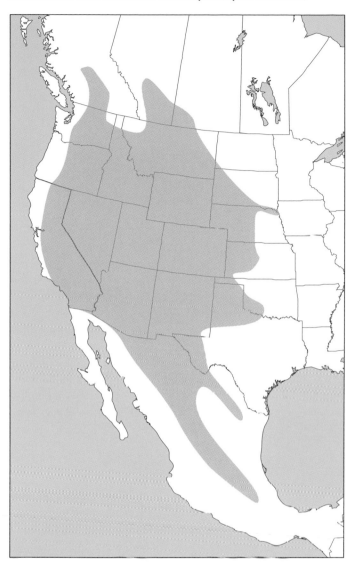

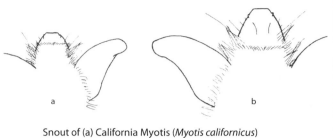

Snout of (a) California Myotis (*Myotis californicus*)
and (b) Western Small-footed Myotis (*Myotis ciliolabrum*)

DIET

Their preferred diet consists mainly of small moths, but also includes small flies, true bugs, and beetles. During the annual caddisfly emergence in the Okanagan Valley, these bats, along with other local species, will concentrate on this temporarily abundant prey.

REPRODUCTION

The timing of mating is not known, although circumstantial evidence, based on testes size of males in Alberta, suggests that this species follows the norm for our northern bats. Likely they mate in the autumn and early winter and the female stores the sperm internally until she ovulates in the spring. A single pup is born between mid-June and late July. Birthing dates vary yearly, depending on climatic conditions. The young likely do not become sexually mature as yearlings.

BEHAVIOUR

Reproductive females appear to form small nursery colonies. Little information is available on the location of maternity colonies in Canada; however, they are thought to be situated in similar locations to the day roosts listed above. Occasional use of abandoned buildings has been recorded among US populations. Like other northern bats, males are thought not to have any role in rearing the offspring and tend to be solitary. In areas where California Myotis and Western Small-footed Myotis occur together, they partition their foraging areas. California Myotis hunt in the preferred sites over the water and along shorelines, while Western Small-footed Myotis are relegated to the rockier, less productive areas. However, no aggressive interactions have been observed.

VOCALIZATIONS

The echolocation call of this species is highly variable geographically, but is generally within the 110–40 kHz range.

of detecting and capturing smaller insects. Emerging shortly after sunset, they usually take a drink at a nearby water source before setting off to hunt. They pursue small flying insects along cliffs and rocky slopes, over open water, and around trees from 1 m above the ground or water surface up to treetop height. Normally hunting and feeding rapidly, they then retire to a night roost or the maternity site, for a rest or to nurse their offspring before heading out again for a second feeding bout a few hours before dawn.

Western Small-footed Myotis are presumed to hibernate in the general area of their summer range mostly singly, although a small group of four individuals has been recorded in British Columbia. All known hibernation sites are in caves and old mines on the ceilings or sometimes in crevices in the ceilings, walls, or floors. Day roosts are reported in small crevices in hot, rocky areas and occasionally also under loose bark, in caves, and in abandoned swallow nests. The bats use buildings, bridges, caves, and old mines as night roosts.

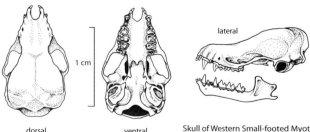

dorsal   ventral   Skull of Western Small-footed Myotis
(CMNMA 48336)

1 cm

lateral

This species rarely leaves indications of its presence in places where humans will encounter them.

REFERENCES
Constantine 1998; Gannon, W.L., et al. 2001; Garcia et al. 1995; Holloway and Barclay 2001; López-Wilchis et al. 1994; Nagorsen and Brigham 1993; Rodriguez and Ammerman 2004; Schowalter and Allen 1981.

Distribution of the Long-eared Myotis (*Myotis evotis*)

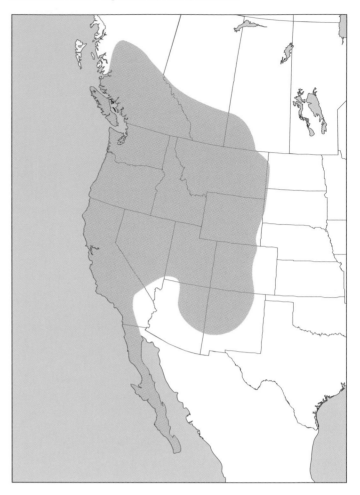

Long-eared Myotis

also called Western Long-eared Bat, Long-eared Bat

FRENCH NAME: **Vespertilion à longues oreilles**

SCIENTIFIC NAME: *Myotis evotis*

This bat is extremely hardy and is often the only bat flying at cool, higher elevations in western Canada.

DESCRIPTION

The long ears on this species extend at least 5 mm beyond the tip of the nose, when gently pressed forward. There is sometimes a fringe of very fine hairs on the trailing edge of the tail membrane, which can be seen under magnification. The fur is long and soft (8–10 mm on the back). The paler inland subspecies has a rich golden tip on each dark-based hair. The darker coastal subspecies has the same black base of each hair, but with a dark brown tip. The wings, ears, and naked snout are blackish. Both subspecies have a dark spot at the front of each shoulder, but it can be indistinct in the darker form. The calcar extends about half the distance along the trailing edge of the tail membrane and is usually not keeled although some individuals have a rudimentary keel with a minute lobule at the end. The dental formula, like all of the Myotis, is incisors 2/3, canines 1/1, premolars 3/3, and molars 3/3, for a total

of 38 teeth. A medium-sized bat with a very slow and erratic flight pattern, especially one flying on cool nights at higher altitudes, is likely a Long-eared Myotis.

SIMILAR SPECIES

This species can be difficult to distinguish from Keen's Myotis in the area where their ranges overlap and there is some debate whether Keen's Myotis should be considered as a distinct species, or as a subspecies of Long-eared Myotis.

SIZE

The following measurements apply to animals of reproductive age from the Canadian part of the range. Females are slightly larger than males.
Total length: 80–113 mm; tail length: 34–49 mm; hind foot length: 8–12 mm; forearm length: 36–41 mm; ear length: 18–22 mm.
Weight: 4.2–10.7 g; newborns, about 1.25 g.
Wingspan: 25–29 cm.

RANGE

This bat has a widespread western North American distribution, extending from the southern tip of Baja California to central British Columbia and over to the Saskatchewan-Montana border.

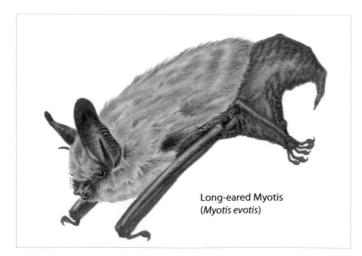

Long-eared Myotis
(*Myotis evotis*)

ABUNDANCE

Although widespread and found in many habitat types, this species is generally considered uncommon.

ECOLOGY

In altitudes ranging from sea level to 2830 m, Long-eared Myotis live in a variety of habitats from dry forest to subalpine forest to semiarid shrubland (especially if there are broken rock outcrops nearby). They are consistently found at higher elevations and are often the only bats seen there. They forage in and around trees and along water edges at variable heights, depending on the height of the canopy. Bats with attached radio transmitters were tracked to the same general foraging areas night after night. This species uses a low-frequency, low-volume echolocation call that, while almost inaudible to moths, only leaves the bat with a very short distance to detect and react to the prey. Likely the rate of prey capture is reduced, as many potential meals escape or go unnoticed. This species typically emerges for the nightly hunt once it is dark, at least half an hour after sunset. Male Long-eared Myotis follow the typical bimodal foraging pattern of most northern insectivorous bats. They forage for a few hours soon after dusk, and again for a few hours before dawn, and rest between. Reproductive Long-eared Myotis females do things a little differently. Especially those at the northern limits of the distribution will forage almost all night, regardless of the temperatures, and spend as little as 10% of the night roosting. Aerial insect activity drops drastically in the middle of the night and most insectivorous bats head for their night roost at that time. The female Long-eared Myotis (most of whom are lactating and have high energy demands) take advantage of the peaks of insect activity to do some aerial "hawking," but during the lull they switch to "gleaning." They hunt close to the vegetation and snatch insects as they move about on vegetation surfaces or the ground. Rather than using echolocation, these bats home in using passive sounds generated by the prey. Most bats specialize in one form of hunting or the other, but Long-eared Myotis do both. The slow manoeuvrable flight pattern, with occasional periods of hovering, required for this dual style of hunting is highly costly in energy. The lower frequency of insect capture, along with the occupation of marginal high-elevation habitats, probably accounts for the long foraging periods and the continuing use of torpor by the females, even during pregnancy and lactation. No winter records exist for Canada. A few have been found hibernating in caves and old mines in the United States. Favoured summer day roosts include rock crevices, under loose bark on snags and large stumps, in tree cavities, and in buildings, caves, and old mines. Pregnant and lactating females prefer to use sites that have a stable, high ambient temperature, or where they can cluster in numbers sufficiently large to raise the temperature of the roost with body heat. Rock crevices often satisfy these requirements. Some concern has been raised for those individuals that choose to roost in a recently man-made habitat – e.g., under the bark of large stumps. These individuals are close to the ground and become vulnerable to predators ranging from bears to Northern Raccoons and rodents, as well as snakes. The record of longevity for this species is of a male captured 22 years after it was first banded. Likely the average age is considerably shorter.

DIET

This species was once considered a moth specialist. The story is rather more complex. Many of the earlier studies discovered a distinct preponderance of moths in the stomach contents of captured bats, along with beetles and many other types of insects. It appears that aerial captures of flying insects are based on abundance and size of prey, but moths are usually captured while gleaning, as they are easier to hear and capture as they rest or move about on the vegetation. So this species is a generalist whose hunting style makes some kinds of moths preferred because they are more easily captured.

REPRODUCTION

Although the period of mating is unknown, circumstantial evidence based on testes size suggests that, as with most of our other bats, mating takes place in late autumn and early winter. Females store the sperm internally and the egg is fertilized when she ovulates in spring. A single hairless pup is born in late June to mid-July and is weaned and flying about 4–5 weeks later. By autumn, the juveniles are essentially indistinguishable from the adults.

BEHAVIOUR

Most males and non-reproductive females are solitary. Few maternity colonies have been found. All have been small – up to 30 adults, but more commonly around 5 adults, and occasionally a few males are included. Most maternity colonies of any size have been in buildings. Often females with young will roost alone. Female Long-eared Myotis are unusual because they continue to use torpor extensively as an energy-saving strategy, even while pregnant and nursing young.

VOICE

The echolocation call of this species is weak and short (less than 2 ms duration) and sweeps from 100 to 25 kHz.

SIGNS

Long-eared Myotis will occasionally hang up and cull their prey before eating it. Small clusters of moth wings or other insect pieces, along with some rice-sized droppings, might be found under a night roost. These bats rarely use the same night roost every night, so culled remains would only be found on an irregular basis. They are known to hunt around homes that use exterior lighting, since

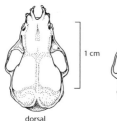

dorsal

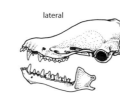

ventral

lateral

Skull of Long-eared Myotis
(CMNMA 48309)

1 cm

the lights attract moths. They capture the moths in flight or pluck them off the walls.

REFERENCES

Barclay 1991; Chruszcz and Barclay 2002, 2003; Fuare and Barclay 1992; Manning 1993; Manning and Jones 1989; Rancourt et al. 2005; Solick and Barclay 2006, 2007; Tuttle and Stevenson 1982; Verts and Carraway 1998; Vonhof and Barclay 1997; Waldien and Hayes 2001; Waldien et al. 2000.

Keen's Myotis (*Myotis keenii*)

Keen's Myotis

also called Keen's Bat, Keen's Long-eared Bat

FRENCH NAME: **Vespertilion de Keen**

SCIENTIFIC NAME: *Myotis keenii*

It is very possible, based on physical and mtDNA similarities, that the Keen's Myotis and the Long-eared Myotis will be determined in future to be different variations of the same species: one darker coastal form and one paler interior form. Because conclusive evidence for this hypothesis has not yet been produced, the two are treated herein as distinct, but closely related, species.

DESCRIPTION

The Keen's Myotis has dark brown, glossy fur on the back with indistinct dark spots at the shoulder. Fur on the underside is lighter, usually buffy brown. The flight membranes and ears are dark brown, but not black. Like all the Myotis, the ears are leathery and hairless. They extend 2–9 mm beyond the nose when gently pressed forward and the tragus is long (6–12 mm), slender, and pointed. The trailing edge of the tail membrane has a short fringe of fine hairs that are only visible with magnification. The cartilaginous spur on the heel bone (the calcar) extends along the trailing edge of the tail membrane about half the distance from the foot to the tail and is not distinctly keeled. The dental formula, as with all the Myotis, is incisors 2/3, canines 1/1, premolars 3/3, and molars 3/3, for a total of 38 teeth. The flight speed of foraging Keen's Myotis is reported to be rather slow.

SIMILAR SPECIES

Keen's Myotis and the dark brown coastal forms of Long-eared Myotis are so similar that they cannot be reliably distinguished in the field. Keen's Myotis are slightly smaller and on average have shorter ears and forearms, but these measurements overlap extensively between the two species. Even the more reliable skull measurements (van Zyll de Jong and Nagorsen 1994) show some overlap. Accurate identification requires a combination of skull and dental measurements placed through a statistical process called multivariate analysis.

SIZE

The following measurements apply to animals of reproductive age from the Canadian part of the range. Males and females are similar in size and appearance.

Total length: 63–94 mm; tail length: 32–44 mm; hind foot length: 8–10 mm; forearm length: 33.8–39.5 mm; ear length: 13–20 mm.

Weight: 3.8–6.7 g; newborns, about 1.0–1.25 g.

Wingspan: 22.4–26.2 cm.

RANGE

This bat has one of the most restricted ranges of any North American chiropteran, and is the only bat confined to the North American Pacific coast. Due to the difficulty of identification of this species, its distribution is not well defined. Keen's Myotis appears to be mainly limited to the coastal forest areas of southeastern Alaska, western British Columbia, including Vancouver Island, and western Washington State. A few recent records indicate that this bat will also venture inland, as evidenced by the capture of individuals along Nine Mile Creek in the Skeena Mountains. As there are so few distributional records, the range depicted is, by necessity, an approximation and no doubt a generalization. Since this species has been captured on many coastal islands (Whidbey and San Juan Islands off Washington; Chichagof, Prince of Wales, Revillagigedo, and Wrangel Islands in southeast Alaska; and Vancouver and Denman Islands, British Columbia), all the major islands have been included in the distribution, although specimens have not been collected from all of them.

ABUNDANCE

This is an uncommon and potentially endangered species throughout its range. The rarity and lack of ecological information has prompted the province of British Columbia to place it on their "Red List" of species under consideration as threatened or endangered. COSEWIC considered it a species of "Special Concern" in 1988 and in 2003 designated it "Data Deficient," as much of its biology and even taxonomy remains uncertain, and so few records were available that it was impossible to determine its population size

Distribution of Keen's Myotis (*Myotis keenii*)

or population trends. Both of the known maternity colonies are in protected areas, as are several of the hibernacula. Nonetheless, the dependence of this species on old-growth rainforest makes it vulnerable to the extensive clearcut logging practices still under way on the west coast.

ECOLOGY

The primary foraging habitat of Keen's Myotis is near water and in all ages of rainforest from near ground level to the canopy. But they are by no means limited to that habitat. They have also been captured in estuaries, over rivers and lakes, and in urban settings. Roost requirements of this species are not clearly understood, but they appear to use hollow trees, snags, rock crevices, cliff faces, caves, bridges, and buildings. The only known hibernacula (there are likely many others) are on northern Vancouver Island. Eight high and mid-elevation caves (above 550 m) in three areas (White Ridge near Gold River, Hankin Range east of Nimpkish Lake, and Weymer Creek near Tahsis) are used by Keen's Myotis and other bats as hibernation sites. About 100 m inside, these caves provide 100% humidity and stable winter temperatures of 2.4°C–4.0°C. Two maternity colonies have been discovered. One, on Gandl K'in Gwaayaay (Hot Spring Island), Haida Gwaii (Queen Charlotte Islands), has been known

for 40 years. It is in an unusual hydrothermal site that is shared with Little Brown Myotis. At this site, Keen's females tend to roost alone or in very small groups of conspecifics. They use narrower crevices than do Little Brown Myotis, so the two species do not roost together. About 40 female Keen's Myotis raise their young in heated crevices, under boulders, and in a small cave near the hot springs. Although non-reproductive bats appear to roost separately from the maternity colonies early in the season, they may join the colony later in the summer. In another maternity colony, near Tahsis, Vancouver Island, the bats were roosting in a Lodgepole Pine snag. It is likely that most breeding colonies occur in large snags as these locations provide large cavities that can hold clusters of bats to raise the internal temperature with their body heat and thick wooden sides to retain the heat. Such a microclimate provides stable conditions for the growth of the young. Male Keen's Myotis are less choosy about their roost sites. They will spend the day under rocks, in rocky crevices, in stumps, or under loose bark as well as in tree cavities, and most roost alone.

Keen's Myotis are adapted to foraging in the dense vegetation of the coastal rainforest with short, broad wings for slow flight speeds and manoeuvrability, and a low-intensity, high-frequency echolocation call most suited to the cluttered environment. Their ability to glean prey from the vegetation means they can continue to find food in rain or cool temperatures that prevent insects from flying. This flexibility is probably crucial for their survival, as such conditions are common in the Pacific Northwest. Banding data shows that these bats live more than 13 years, but their maximum lifespan is unknown. Keen's Myotis may be vulnerable to terrestrial predators as they do forage near the ground at times. Domestic Cats have contributed several specimens to museum collections. Keen's Deer Mice sign and skeletons have been found in the hibernation caves. These mice could be predating the torpid bats during the winter.

DIET

There is not a lot of information available concerning the diet of Keen's Myotis. Faecal and stomach analyses show that they eat a combination of flying insects and non-flying spiders, so they clearly both hawk and glean. Since bat diets typically vary with prey availability and abundance, it is expected that the kinds of flying insects and spiders, and their proportion in the diet, likely vary with locality and season. Adverse weather (e.g., periods of cool, wet weather) that negatively impact the hunting efficiency of the more aerial hunting bats, appear not to affect this species, since they can readily switch to a gleaning hunting style to make up any shortfall in prey.

REPRODUCTION

While the mating period is unknown, it is probable that like other North American bats, Keen's Myotis mate in the autumn, just before hibernation or during the hibernation period. Fertilized females then store the sperm internally and ovulation and fertilization occur in spring, after they arouse from hibernation. Length of gestation is unknown, but probably similar to that of other Myotis species (40–60 days), after which a single large pup is produced. It is uncertain whether females bear young annually. The young are born in

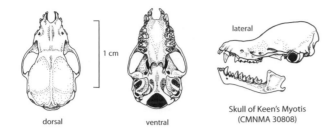

dorsal

ventral

lateral

Skull of Keen's Myotis
(CMNMA 30808)

Eastern Small-footed Myotis

also called Eastern Small-footed Bat, Least Bat

FRENCH NAME: **Vespertilion de Leib**

SCIENTIFIC NAME: *Myotis leibii*

The Eastern Small-footed Myotis is the smallest bat in eastern Canada.

June or July. The hot springs maternity site is used from late May until around the middle of August by females and young, and until early September by some of the late-departing juveniles. Young require about three weeks to become volant, but continue to suckle for another two weeks until they are proficient fliers and able to feed themselves. Age at sexual maturity is unknown. Some juveniles are thought to become sexually mature in time for the autumn breeding season, but probably most do not breed for the first time until their second autumn.

BEHAVIOUR

Nightly foraging is undertaken in several short bursts, with roosting periods between to digest the catch. The bats leave their day roosts about 20–30 minutes after sunset and return about two hours before sunrise. Like other bats that hibernate, Keen's Myotis practice a behaviour called "swarming," during which adults and juveniles congregate at night at a likely hibernaculum, and fly around the entrance and through the cave, to familiarize themselves with the site and possibly assess its potential. Swarming occurs from early August until early September. It is likely that, like other Myotis, swarming adult Keen's Myotis breed towards the end of the swarming period. Keen's Myotis hibernate singly or in small clusters.

VOCALIZATIONS

The echolocation call of a Keen's Myotis falls between 110 and 38 kHz and lasts less than 3 milliseconds. It is very similar to that of the Long-eared Myotis. This call is well above the 20 kHz threshold for human hearing.

SIGNS

Scat of all the Myotis are similar and may occur on the floor below roosting sites. Scat is rare in hibernacula, as the bats are usually torpid and not excreting. Faeces are about the size of rice grains, have tapered ends, and are blackish.

REFERENCES

Boland et al. 2009; Burles et al. 2008, 2009; Burles and Nagorsen 2003; Nagorsen and Brigham 1993; Parker and Cook 1996; van Zyll de Jong 1985; van Zyll de Jong and Nagorsen 1994.

DESCRIPTION

This small bat has thick, silky fur that is yellowish tan to dark brown on the back and slightly lighter on the belly. The shiny tips of the longest guard hairs give the bat a golden sheen. Sometimes the longer hairs become worn as moulting approaches and no longer display the sheen. Ears, snout, and wing membranes are blackish. The ears are less than 15 mm from tip to notch and reach, or slightly exceed, the tip of the snout when gently pressed forward. The tragus is slender, tapers to a dull point, and is about half as long as the ear. The calcar is obviously keeled. The dental formula is incisors 2/3, canines 1/1, premolars 3/3, and molars 3/3, for a total of 38 teeth. Their slow, somewhat erratic flight pattern can, with practice, be used to identify the species.

SIMILAR SPECIES

The best characteristic to separate this species from other small eastern bats is the foot size. The feet of an Eastern Small-footed Myotis are less than 8.5 mm long. The species closest in size and appearance is the Eastern Pipistrelle, which is bigger, with larger feet and tricoloured fur on the belly.

SIZE

The following measurements apply to animals of reproductive age from the Canadian portion of the range. Females and males are similar in size and appearance.

Total length: 74–93 mm; tail length: 30–40 mm; hind foot length: 6.0–8.5 mm; forearm length: 30–34 mm; ear length: 11–14 mm.

Weight: 3.2–5.5 g; newborns, unknown.

Wingspan: 21–25 cm.

Eastern Small-footed Myotis
(*Myotis leibii*)

Distribution of the Eastern Small-footed Myotis (*Myotis leibii*)

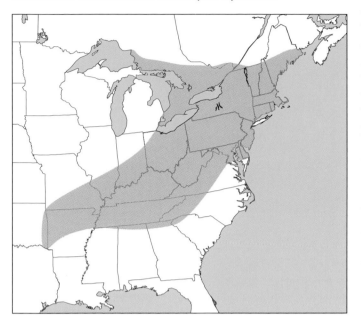

RANGE
This species occurs in eastern North America from central Ontario and southern Quebec, to central Maine southward to northern Alabama and westward into eastern Oklahoma.

ABUNDANCE
This species may be the rarest bat in eastern North America. This bat is listed as "Threatened" by the State of Pennsylvania.

ECOLOGY
Eastern Small-footed Myotis are most known from their hibernacula. They are rarely found in summer. These bats prefer hilly or mountainous terrain in deciduous or coniferous forests, although they are sometimes found in flat lands as well. They hunt over land and water at heights varying from 1–6 m above the surface. Little more is known about their habitat preferences. Despite their small size, these bats are extremely hardy and cold tolerant. They enter hibernation late and are one of the first to leave hibernacula in spring. In southern Ontario, for example, they arrive at hibernacula in late November and leave in early April. Their preferred hibernation roosts are near the entrances of caves and old mines, where humidity and temperatures are lower and more likely to fluctuate. They will only arouse to move to a warmer spot when temperatures fall below –9°C. They usually hibernate singly or in small groups. Hibernating Eastern Small-footed Myotis may be found hanging on the walls or roof of caves or under rocks on the floor. They occasionally arouse from hibernation and fly about the cave. Two bats banded in an Ontario cave were found less than 20 km away a few months later. This suggests that at least some Eastern Small-footed Myotis spend their lives in a relatively small area and do not undertake long migrations to the hibernacula.

In summer, their roost sites include buildings, bird nests, bridge expansion joints, behind doors and shutters on the outside of buildings, under rocks, in fissures in tree bark, and in caves and old mines. Based on known prey species, this bat mainly catches flying insects, but will also glean prey from the ground or vegetation at times. The lifespan of the Eastern Small-footed Myotis is thought to be shorter than many of our bat species (in the 6–12 year range), and northern females do not live as long as males, probably due to the stresses and demands of reproduction. The short lifespan, high female mortality, and low reproductive potential could account for the scarcity of this species.

DIET
The following information on the diet of the Eastern Small-footed Myotis is derived from a single paper published in 2007 (Moosman and Thomas) using data extracted from faeces analysis. Like other insectivorous bats, prey species vary with availability, which changes with season and climate. Moths, small beetles, and flies compose most of these bats' diet, with the addition of spiders and crickets.

REPRODUCTION
Very little information is available concerning the reproduction of this species. Based on a single report of a reproductively active male caught in the middle of September, it is thought that this species mates in the autumn before hibernation. It is highly likely that the female stores the sperm internally until spring, when she ovulates. Females produce a single pup between late May and early July, towards the latter period in more northerly areas. Few maternity colonies have been found; all have been in buildings and have consisted of 12–20 females.

BEHAVIOUR
Eastern Small-footed Myotis emerge to forage shortly after sunset. Their distinctive flight is slow and frequently erratic. They often use caves in summer as night roosts and can sometimes be seen flying in and out. Copulation takes place at the roost and is initiated by the male while the female remains passive. He bites the hair on the back of her neck and pulls her head back as he mounts.

VOCALIZATIONS
The echolocation call of this bat has only recently been recorded. It is a short high-frequency pulse that sweeps from about 85 to 45 kHz.

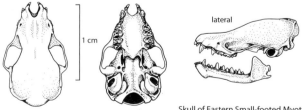

1 cm

dorsal ventral lateral

Skull of Eastern Small-footed Myotis
(CMNMA 17542)

SIGNS

Although rare, this species sometimes uses buildings during the summer for maternity colonies, but since they are usually occupied by small numbers of bats, they are rarely noticed. They also prefer to use the outside of buildings – behind doors and shutters – and so cause little disturbance.

REFERENCES

Bogan 1999b; Fenton 1972; Griffin et al. 1960; Hitchcock 1955, 1965; Moosman and Thomas 2007; Murray et al. 2001.

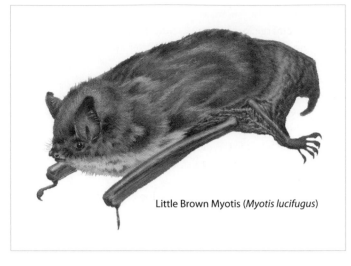

Little Brown Myotis (*Myotis lucifugus*)

Little Brown Myotis

also called Little Brown Bat

FRENCH NAME: **Vespertilion brune,** *also*
Petite chauve-souris brune

SCIENTIFIC NAME: *Myotis lucifugus*

The Little Brown Myotis is Canada's most ubiquitous and most studied bat species. These bats can be found from coast to coast, in all the provinces and territories with the exception of Nunavut, and on many of the offshore islands. When Canadians think "bat," this is usually the one that comes to mind.

DESCRIPTION

The fur colour of this species is variable. It can be yellowish brown, olive brown, rusty brown, dark brown, or almost black. The belly is usually noticeably lighter than the back. Juveniles are darker than the adults until August of their first year. In most populations, the adults have glossy fur because of the sheen on the outer third of the long guard hairs, which are about 9–10 mm in length on the back. Fur on the belly extends thinly onto the wing membranes to a line from the knee to about three-quarters down the humerus. The upper surface of the tail membrane is also lightly furred down to about the knees. The flight membranes and ears are dark brown. The calcar is not keeled. Hairs on the hind feet extend beyond the toes. The ears reach only to the nostrils when gently pressed forward. The dental formula is the same as for all the Myotis: incisors 2/3, canines 1/1, premolars 3/3, molars 3/3, for a total of 38 teeth.

SIMILAR SPECIES

In most parts of the range, this species is readily distinguished from most other small brown bats by its shorter ears and bluntly tipped tragus. Among the short-eared species, Long-legged Myotis have a small keel on the calcar, dense fur on the underside of the tail, and the belly fur is almost the same colour as the fur on the back. Eastern Small-footed Myotis, Western Small-footed Myotis, and California Myotis are smaller (forearm 30–35 mm) with a keeled calcar and hind feet < 8.5 mm in length; Yuma Myotis usually has shorter, dull fur with no gloss, and is slightly smaller (forearm < 37 mm), the skull length is also shorter, usually < 14 mm. Yuma Myotis and Little Brown Myotis in the south-central portions of British Columbia are very difficult to distinguish. Local experts suggest that the Yuma Myotis tends to be more passive and less aggressive than the Little Brown Myotis, when handled.

SIZE

The following measurements apply to animals of reproductive age from the Canadian portion of the range. Females and males are similar in appearance. Females are slightly larger than males.
Total length: 60–108 mm; tail length: 25–59 mm; hind foot length: 7–13 mm; forearm length: 33–41 mm; ear length: 12–16 mm.
Weight: 7–14 g, lowest in spring and highest in autumn, just before hibernation; newborns, about 2.5 g.
Wingspan: 22–27 cm.

RANGE

This is a widespread species occurring from Newfoundland, central Labrador, and south of Hudson Bay and across central Canada to central Alaska and the Queen Charlotte Islands. The most southerly extent of its range is a thin finger of distribution extending into the highlands of central Mexico. In many of the northern parts of its range, it is the only bat species, and in some of these same regions, it occurs where there are 24 hours of daylight during part of the summer. With the predicted rate of global climate change and warming of the northern parts of the range, Little Brown Myotis should move further north into Labrador and the Ungava Peninsula, to extend into all of Ontario, Manitoba, and Saskatchewan into the lower reaches of the Northwest Territories and Nunavut, the lower Yukon, and the southern half of Alaska by the year 2080.

ABUNDANCE

Little Brown Myotis are common throughout most of their range, becoming less common near the northern and southern limits.

ECOLOGY

These bats emerge for the nightly hunt around dusk. After drinking, they begin foraging around and over water or in the forest canopy and even along city streets and backyards, usually between 1 and 6 m above the ground or water level. Little Brown Myotis are very manoeuvrable, efficient hunters as they zig-zag around catching flying insects. Although gleaned spiders are included in their diet, most of their prey is captured by aerial "hawking." Often the prey is scooped up into the tail membrane and plucked from there to be eaten in flight. These bats regularly consume about one-half of their body weight each night, with nursing females eating even more, up to 110% of their body weight. A foraging bat with abundant insect prey can fill its stomach in less than 15 minutes. After a feeding session, the bats prefer to hang up in a warm night roost to speed the digestion and then lighten their load by excreting some of the waste (both solid and liquid). Under ideal warm conditions, the first urine and droppings can be produced within 35 minutes.

The Little Brown Myotis is considered to be mainly a cavity-roosting species, although there is one report of foliage roosting, and given their size and nondescript colouring, foliage roosting could be somewhat more common than so far recorded. Rock crevices may also be used for both day roosting and maternity roosts. Night roosts tend to be confined spaces that are warm or can be warmed by an accumulation of bats. Communal night roosts are used most often when the temperatures are cool. Nursing females do not use night roosts, choosing rather to return to the maternity colony. It is also warm there, and they can further reduce their weight by nursing their pup at the same time. This cycle is repeated two or more times each night. Maternity colonies form in late April and early May, just

Distribution of the Little Brown Myotis (*Myotis lucifugus*)

after the bats come out of hibernation. Colonies can become very large; over 1000 individuals roosting together is relatively common. Females select a variety of maternity roosts, usually in or around buildings, but tree cavities, exfoliating bark, cracks and crevices in cliffs, and even small caves or crevices heated by hot springs are used by these bats. The principal criterion for a maternity roost is one of temperature. It must be warm enough, preferably between 32°C and 36°C, for the rapid growth and development of the juveniles. Most maternity roosts are within a kilometre of water. Males and non-reproductive females typically roost away from the nursery roosts in cooler locations where they can allow their core temperature to drop so they can fall into a light torpor to preserve energy. However, in thermally challenging areas and especially in the northern portions of the range, the nursery colonies may be in the only warm roost site in the area. In these cases, the non-reproductive bats may opt to occupy a portion of the warm maternity roost during the day to save themselves the energy cost of arousing from their usual daily torpor. In late summer and autumn, Little Brown Myotis are occasionally found roosting in exposed locations. These are most often brick walls with a southerly or westerly exposure that gets late afternoon sun to warm the bats before they begin their evening foraging. Many of these bats are inexperienced juveniles who sometimes roost in inappropriate and often dangerously exposed locations.

Little Brown Myotis typically hibernate in caves and abandoned mines with favourable temperature and humidity regimes. Optimum temperatures for hibernation are between 1°C and 5°C with 70%–95% humidity. There is one record of these bats hibernating in an earthen basement in Prince Edward Island. Little Brown Myotis migrate up to 1000 km between hibernation sites and summer ranges. Most Canadian Little Brown Myotis enter hibernation some time after early September and become active again around mid-to-late April, when average daytime temperatures reach 10°C and insects become active. In the known hibernacula, a high percentage of the hibernating bats are male. Where the missing females hibernate is unknown, although some scientists suggest that they are just deeper in the cracks or cavities of the known hibernacula where they cannot be seen. In the West, a few hibernacula have been discovered, but certainly not enough to house the known summer populations. These bats are occasionally taken by predators such as owls, magpies, Domestic Cats, Northern Raccoons, and snakes, but the main cause of mortality is a failure to survive hibernation. This is most common among juveniles overwintering for the first time and is usually the result of insufficient fat accumulation in the autumn. Juvenile bats learning to fly and find safe roosts are also subject to high accidental mortality. If they survive their first winter, Little Brown Myotis commonly live beyond ten years and the current banding record is 31+ years for a male that was banded as an adult. A typical flight speed for this species is 18 km/h.

DIET

Little Brown Myotis eat a variety of insects, reflecting prey availability. Most of the prey is 3–10 mm long. Since most of their foraging is done over and near water, aquatic insects understandably make up a large part of the diet. Midges, mosquitoes, moths, spiders,

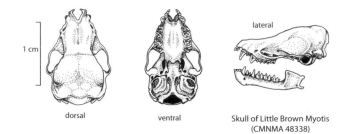

dorsal   ventral   lateral

Skull of Little Brown Myotis
(CMNMA 48338)

caddisflies, small beetles, termites, lacewings, crane flies, and leaf-hoppers are all consumed. Although these bats occasionally consume mosquitoes when they are abundant, these are not the usual prey because they are too small.

REPRODUCTION

Most mating takes place in late summer and autumn at the swarming sites (see a discussion on swarming in the "Behaviour" section, following). However, over the winter, some males, during their normal periodic arousals from hibernation, mate with nearby torpid animals, sometimes females. Females store sperm in the uterus until spring, when it is used to fertilize an egg released some time in April or early May. Gestation lasts 50–60 days depending on local temperatures, and a single pup is born mid-June to mid-July. The females right themselves just before the birth so that the young are born into the cupped tail membrane. Breech presentation is the norm. Newborns are blind and deaf with a fine, nearly invisible covering of silky hairs. Their ears and eyes open within hours of birth. Their birth weight of around 2.5 g is 25%–30% of their mother's normal weight. Pups cling to their mother during the day and are left behind, usually in large clusters, when she leaves to hunt. They are suckled for about 21 days, depending on roost temperatures. At around three weeks old they begin to fly and to eat insects. It is unknown whether they catch these first insects themselves, or whether their mothers deliver them. For a few days their stomachs contain both insects and milk. At around three weeks old they are weaned and become independent. By late autumn it is very difficult to distinguish young of the year from adults, although their pelage may still be a bit darker. The young males do not become sexually mature in their first autumn, but some young females are known to produce offspring the first spring after their birth. This happens more commonly in warmer, more southerly parts of the range. Generally, temperatures in Canada are cooler, and few of the young females mature early enough to mate in their first autumn.

BEHAVIOUR

The social behaviour of Little Brown Myotis has been well studied. Aside from echolocation calls, these bats produce ten other vocalizations whose purpose is specific. These are listed in the "Vocalizations" section, following. Although calls from mother and offspring are used to isolate the juvenile within the cluster, females use odour to confirm identity. The juveniles will nurse from any accommodating female, but the females will generally only allow their own pup to suckle. In southern Ontario in August (later in warmer areas), bats begin to gather at the hibernacula. They spend part of the night flying around the entrance to the roost and then through the site, a behaviour called swarming. In its initial stages, it appears to inform the juveniles about the location of the hibernation site. Later in August, the swarming takes on a reproductive significance. Adult males perch on the walls and ceiling of the hibernacula and advertise their presence by producing echolocation calls. Females fly around and select the male they wish for a mate, by landing beside him. Without further courtship, he mounts her from behind, grasps her neck fur in his teeth, and proceeds to copulate. Copulation can last from 3 to more than 20 minutes. Little Brown Myotis males will mate with as many females as they can, and most females will mate more than once with different males. Little Brown Myotis exhibit pronounced site loyalty. Hibernacula, maternity roosts, and night roosts are often used year after year by the same bats and their offspring.

VOCALIZATIONS

The echolocation call of this species is a high-intensity sweep from 80 to 35 kHz, with the majority of the energy focused around 40 to 45 kHz. If on a collision course with another bat, Little Brown Myotis will attach a lower frequency "honk" to the end of the call that sweeps from 40 to 25 kHz. Both types of call are beyond the audible range for humans. The social activities of this species have been well studied at the roost sites and much is known about the sub-sonic sounds produced during various social activities. These calls are detectable by humans. There are indications in some situations that high-frequency calls can reveal an individual's identity and age and even lactation condition, so may play a role in social communication.

COPULATION CALL – Of a low frequency, 5–12 kHz, it is produced by males while mating, but only if the female struggles. As she relaxes, the call frequency drops. Winter matings with torpid partners require no production of the copulation call.

ISOLATION CALL – Produced by juveniles to inform their mothers of their whereabouts within the colony. Also produced if the pup falls from its roost.

DOUBLE-NOTE CALL – Produced by females to alert their offspring of their whereabouts. Most commonly uttered as the female leaves or approaches the clustered juveniles, it prompts the juveniles to produce their isolation call. As one would expect, there is some evidence that both females and their young can distinguish each other's calls.

AGONISTIC CALLS – A series of four graduated calls used by all individuals to indicate levels of aggression.

SQUAWK CALL – Produced by all individuals to attract the attention of others. Commonly used during swarming or while entering and exiting the roost.

SIGNS

Night and maternity roosts used by a number of bats on a regular basis often accumulate mounds of droppings. Long and heavily used sites will also often develop pissicles, crystallized urine deposits in the form of a stalactite hanging from the rafters. Some pissicles contain droppings as well. A scattering of black rice grain–sized droppings probably indicates a temporary night roost. The access hole to a colony usually acquires a staining from faeces and urine over the course of time, making it easier to detect. Bat droppings are next to impossible to identify to species, without seeing the bat that produced it. However, huge bat colonies are uncommon in Canada and rarely are they of any other species besides the Little Brown Myotis.

REFERENCES
Barclay and Thomas 1979; Burles et al. 2008; Davy and Fraser 2007; Dubois and Monson 2007; Fenton 1999a, 2001; Fenton and Barclay 1980; Gould 1955; Humphries et al. 2002; Kazial et al. 2008; Lausen et al. 2008a; Nagorsen and Brigham 1993; Slough 2009; van Zyll de Jong 1985; Weller, T.J., et al. 2007; West and Swain 1999.

Northern Myotis

also called Northern Long-eared Myotis,
Northern Bat, Northern Long-eared Bat

FRENCH NAME: **Vespertilion nordique**
SCIENTIFIC NAME: *Myotis septentrionalis*

Northern Myotis are slow, but especially agile and manoeuvrable, flyers that often hover in their pursuit of insects resting on foliage. They can also catch flying insects, making them very versatile hunters.

DESCRIPTION
The Northern Myotis is a smallish bat with light- to medium-brown fur and darker brown ears, snout, and flight membranes. The ears are moderately long, extending about 5 mm beyond the nose when gently pressed forward. The tragus is long and pointed. The tail is also moderately long. The head and body length is less then 50 mm, although it can be up to 96 mm long when the tail length is included. The calcar is slightly keeled and extends along the full trailing edge of the tail membrane. The dental formula is incisors 2/3, canines 1/1, premolars 3/3, and molars 3/3, for a total of 38 teeth.

SIMILAR SPECIES
The Little Brown Myotis is the bat that is most similar to the Northern Myotis, but it can be distinguished by its shorter ears, which do not extend beyond the nose when gently pressed forward, and by its blunt rather than long and pointed tragus. The Western Long-eared Myotis has longer, darker ears, a dark shoulder spot, and a sparse fringe of hairs on the trailing edge of its tail membrane (sometimes only seen under magnification).

SIZE
The following measurements apply to animals of reproductive age from the Canadian portion of the range. Females and males are similar in appearance. Females are slightly larger than males.
Total length: 77–101 mm; tail length: 29–48 mm; hind foot length: 8–11 mm; forearm length: 34–40 mm; ear length: 14–19 mm.
Weight: 4.5–10.8 g; newborns, unknown.
Wingspan: 22–26 cm.

RANGE
The Northern Myotis is widespread in central North America from Newfoundland to British Columbia. It has recently been found in the only known Prince Edward Island bat hibernacula and joins the Little Brown Myotis and the Hoary Bat as the only three bat species proven to occur on Prince Edward Island.

ABUNDANCE
Northern Myotis are reasonably common in eastern North America, but uncommon or rare in the far western and northwestern portions of the range. In British Columbia this species is on the Provincial "Red List."

ECOLOGY
Preferred hibernation sites for Northern Myotis are caves or abandoned mines, which are regularly shared with many other bat species. An unusual hibernaculum on Prince Edward Island is perhaps an anomaly in the absence of natural caves or abandoned mines on the island. Instead, the bats were found hibernating in an earth and stone unheated basement of an occupied home. This species tends to be late entering hibernation. Depending on the environmental conditions, Northern Myotis will begin hibernation between September and November and emerge from hibernation some time between March and May. Longer hibernation times occur in colder regions and shorter times in warmer areas. These bats prefer to hibernate alone or in small clusters in deep crevices or drilled holes where they can be easily overlooked. This species is also known to relocate during the winter to other nearby hibernacula. In the two or three

Northern Myotis
(*Myotis septentrionalis*)

Distribution of the Northern Myotis (*Myotis septentrionalis*)

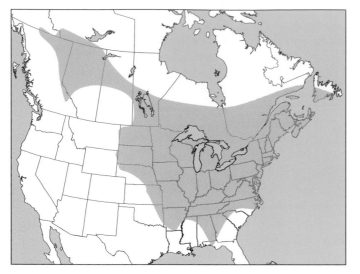

months before entering their winter torpor, Northern Myotis will gain between 40% and 45% of their normal body weight in fat to sustain them through the cold period.

In the summer the sexes roost separately and most males and non-breeding females roost alone, or in small groups of fewer than ten individuals. They choose trees, caves, or occasionally buildings for their roosts. Although barns and other buildings may be chosen as maternity sites, most breeding females select hollows or cavities in large trees. Depending on the cavity size, such colonies may include up to 60 adults, although most are much smaller. Colony size drops after the pups are born. Females will gather in cavities and crevices or under the loose bark of living or decaying trees. They switch roosts on average every few days and carry their non-volant offspring to the new site. Sometimes nursery colonies have been found in buildings. Typically they roost less than a 1 km from their foraging areas and they forage over an area of around 65 ha. These bats likely spend the summer not too far from their hibernation site. Movements of 56 km between summer and winter roosts have been documented. A genetic study in Ohio found that the females within subpopulations exhibit higher genetic similarities than do males, an indication that females are less likely to leave the area of their birth as they mature and that subadult males tend to disperse. Northern Myotis emerge around dusk to forage over small ponds, along paths and roadways, and under the forest canopy. They are slow and agile in flight and hunt in the very cluttered environment just above the understory vegetation, usually between 1 and 3 m from the ground. Once thought to hunt exclusively aerial prey (hawking), then exclusively prey resting on vegetation (gleaning), it has recently been shown that Northern Myotis use both approaches to capture prey. Longevity records for this species are few; however, at least one individual is known to have lived more than 18 years.

DIET
Northern Myotis feed mainly on flies, moths, beetles, caddisflies, lacewings, and leafhoppers as well as on non-flying insects such as spiders and caterpillars. They are opportunistic hunters and will prey on whatever insects are abundant at the time, but specialize in terrestrial insects that they catch in flight or by gleaning them from the vegetation. There are some indications that individuals prefer certain kinds of insects and will preferentially hunt for them while excluding others. Because of their habit of carrying their prey to a perch before consuming it, these bats are able to handle larger insects than other similar-sized species that feed on the wing.

REPRODUCTION
Very little information is available concerning the reproductive biology of this species. We do know that the bats begin swarming, flying through the hibernacula and around the entrances, in late summer and early autumn. Mating takes place at this time. The male mounts an accepting female from behind as she rests on a solid surface like a cave wall or a tree limb. Occasionally, he will grasp the fur at the back of her neck in his teeth. The sperm is likely stored internally by the female until she ovulates in the spring. A single pup is born between mid-May and mid-July, depending on the ambient temperatures, earlier in the warmer portions of the range and later in the colder areas. Births in any given area tend to be synchronous (within as few as six days). The young develop quickly and are indistinguishable from adults by their first autumn.

BEHAVIOUR
Northern Myotis fly slowly and frequently hover while attempting to localize sounds produced by potential prey as it moves about on the vegetation. They are capable of flying at speeds around 13 km/h when not foraging. The combination of a larger flight surface, which permits them to fly slowly and with more manoeuvrability, the ability to vary their echolocation frequency, and the low intensity of their echolocation calls allows these bats to forage successfully in cluttered areas not suited to most other species. These bats typically display a bimodal foraging strategy. Their first foraging period takes place within the first two hours after sunset, which coincides with a peak of insect activity. This is followed by a resting/roosting period of varying length, when the bats digest their meal and the females likely return to their pups to warm and nurse them. A second foraging period occurs during the few hours before sunrise and coincides with a lesser peak of insect activity. Nursing females may repeat the cycle more than twice on good nights. As with other species, this pattern likely changes during periods of cool, wet weather. Typically, the bats forgo the second, and possibly even the first, foraging period, choosing to remain

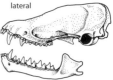

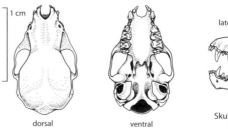

dorsal   ventral   Skull of Northern Myotis
(CMNMA 43405)

1 cm

dry and enter a light torpor instead of eating. Northern Myotis tend to return to the same hibernacula but not always in successive years. As they may change hibernation sites during the winter, it is perhaps more accurate to say that they are loyal to a group of hibernacula rather than to a single one. Hibernacula are usually shared with other species such as Little Brown Myotis, Eastern Small-footed Myotis, Eastern Pipistrelles, and Big Brown Bats. Northern Myotis make up a small percentage of the total number of bats and are often difficult to spot among the others as they roost deep in crevices, often out of sight. They tolerate cooler conditions than their more plentiful cousins, the Little Brown Myotis, and so are not usually found near that species.

VOCALIZATIONS

Northern Myotis use a broadband echolocation signal sweeping from 120 to 40 kHz, but with a very low intensity. An echolocating Northern Myotis is likely inaudible to a moth until it is less than 2 m away.

SIGNS

Northern Myotis are rarely encountered and deposit their droppings in roosts, where the faeces are unlikely to be found. The droppings of all our northern bats cannot be reliably identified to species. They and many other species "swarm" at their hibernacula in the late summer and autumn, but species are difficult to distinguish under those circumstances.

REFERENCES

Arnold, B.D., 2007; Broders et al. 2004; Brown, J.A., et al. 2007; Caceres and Barclay 2000; Fenton 1999b; Henderson and Broders 2008; Hubbs and Schowalter 2003; Jung et al. 2006b; Krochmal and Sparks 2007; Lausen et al. 2008b; Nagorsen and Brigham 1993; Owen et al. 2003; Ratcliffe and Dawson 2004; Sasse and Pekins 1996; Slough and Jung 2007; van Zyll de Jong 1985.

Fringed Myotis

also called Fringed Bat

FRENCH NAME: **Vespertilion à queue frangée**

SCIENTIFIC NAME: *Myotis thysanodes*

This rare western species is the largest of the Canadian Myotis. It reaches the northern limit of its distribution in southern British Columbia, where its life history is poorly known.

DESCRIPTION

The Fringed Myotis is a long-eared Myotis. The ears extend well beyond the nose when gently pressed forward. The ears and flight membranes are blackish and the fur is pale brown on the back and

Fringed Myotis (*Myotis thysanodes*)

cream-coloured on the belly. The base of each hair is dark. The calcar is long and not keeled. The distinguishing feature of this species is the easily visible fringe of stiff, short hairs on the trailing edge of the tail membrane. The flight membranes of this species are unusually thick and puncture resistant. The dental formula, as with all the Myotis, is incisors 2/3, canines 1/1, premolars 3/3, and molars 3/3, for a total of 38 teeth.

SIMILAR SPECIES

Fringed Myotis have the longest forearm of the Canadian Myotis, usually > 40 mm in the adults. Only the Long-legged Myotis and the Long-eared Myotis approach the size of the Fringed Myotis. The Long-legged Myotis has a keeled calcar and the underside of the flight membrane is furred to the knees and elbows. The Long-eared Myotis has longer ears and a sparse fringe of very fine hairs on the tail membrane, which are only visible under magnification.

SIZE

The following measurements apply to animals of reproductive age from the Canadian portion of the range. Females and males are similar in appearance. Females are slightly larger than males, especially in the forearm length, total length, and weight measurements.
Total length: 88–93 mm; tail length: 35–44 mm; hind foot length: 8–11 mm; forearm length: 40.0–44.5 mm; ear length: 18–20 mm.
Weight: 5.4–8.4 g; newborns, about 1.25 g.
Wingspan: 27–30 cm.

RANGE

Fringed Myotis can be found throughout much of western North American from south-central British Columbia to the central highlands of Mexico. A disjunct population is found in the Black Hills of South Dakota and Wyoming. The species reaches the northern limit of its distribution in Canada, and the Canadian portion of the total range is less than 5%. A single specimen from the Olympic Peninsula

Distribution of the Fringed Myotis (*Myotis thysanodes*)

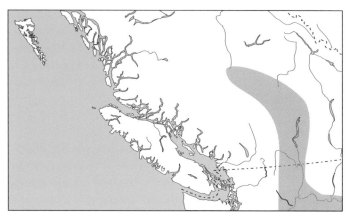

expanded view of Canadian range

of Washington suggests that a coastal population in southern British Columbia might be possible.

ABUNDANCE

It is likely that many Fringed Myotis are capable of evading capture in mist nets and harp traps in the same manner as other species that are extremely agile in flight and use low-intensity, but frequency-modulated echolocation calls. This may partly account for the low numbers of this species that have been captured in Canada. Fringed Bats are on the BC "Blue List" and are considered imperilled in the province. The species varies from vulnerable to critically imperilled in the United States, where it is listed as a federal Category 2 species. Fringed Myotis are generally rare in the northern portions of their range.

ECOLOGY

In Canada, this species is associated with grasslands and open forests in intermontane river valleys. There is very little information

available concerning the biology of this species in Canada. Most of the information provided in this account was garnered from studies on US populations.

Evening foraging generally begins shortly after sunset and coincides with the post-sunset peak of insect activity. Most Fringed Myotis, including the nursing females, do not return to the day roost until dawn. These bats are considered to be gleaners. They fly around the vegetation listening for insect sounds, and then either from flight, or by landing nearby, they pluck their prey from its perch. They are also capable of aerial hawking. Individual bats forage in an area of about 4 square km. Summer day roosts in Canada are unknown. Likely they are similar to those known from US populations – in caves and mine shafts, in large snags either under bark or in cavities and crevices, in rock crevices, or in buildings. Again in the United States, maternity colonies have been discovered in all these types of sites. So far in Canada, only buildings are known to be used as maternity sites. Fringed Myotis migrate to caves and mine shafts to hibernate. An extensive survey of BC underground hibernation sites within the known range of this species, found no hibernating Fringed Myotis, and there are no historical or museum records for Canada for the winter months. It is possible, perhaps likely, that these bats leave the province in search of suitable hibernacula. Fringed Myotis are very agile and manoeuvrable fliers and are capable of slow flight and short periods of hovering flight. This species is long lived. The longevity record is held by a banded animal that was recaptured at the same Oregon cave more than 18 years later.

DIET

Fringed Myotis eat a large variety of insects, including beetles, moths, lacewings, flies, spiders, and crickets. They appear to specialize on the harder-bodied beetles when available. Most of their prey is gleaned from the surface of the vegetation rather than in flight.

REPRODUCTION

The mating season of this species is not known. It occurs some time after the females leave the nursery colonies in early autumn and before they return the following spring. Females arrive at the maternity site in early to mid-April with sperm in their reproductive tract, but ovulation and fertilization does not take place for about another month. The gestation period is between 50 and 60 days, depending on the weather conditions, and a single pup is born between mid-June and mid-July. Indications are that the Canadian colonies reflect this timing. The birthing time appears to be synchronous in each maternity colony, usually within a two-week period. Only

1 cm

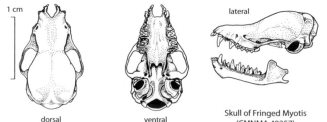

dorsal   ventral

lateral

Skull of Fringed Myotis
(CMNMA 48357)

one live birth has ever been witnessed and in this case the female gave birth in the head-down roosting position. The newborns are large at birth, approximately 22% of the weight of their mothers, and they are already around 54% of her total length. They are hairless, deaf, and blind. Typically, the ears unfurl and the eyes open within 24 hours of birth. The pups grow quickly and are capable of clumsy flight by the time they are 17 days old, and can fly as well as the adults by the time they are 21 days old. They are full grown by this time, but not as heavy as the adults. The male pups do not breed in their first year. Likely the young females do produce a pup the following spring.

BEHAVIOUR

Males are rarely found in the nursery colonies. They appear to roost alone or perhaps in loose groups, in cooler roosts. Maternity colonies, south of the border, have been discovered with over 1000 tightly packed bats. In Canada, only two maternity colonies have been found, both in attics and each containing fewer than 50 animals, including the young. In the United States, the species will also use tree roosts and rock crevices to raise young. It is possible that with more study we may find that they occupy similar locations in Canada. The maternity colonies in trees and rocks are small, with usually fewer than 30 animals. Females in these types of maternity roosts tend to change roost sites every few days. The whole colony will move to a new spot (usually not far from the old site) carrying the non-volant young with them. Roosts in buildings tend to be more permanent. Females, within 10 days of giving birth, become very secretive and will abandon a roost if disturbed. During the night, when most of the females are foraging, there are always a few adults that stay with the pups to act as guardians. These adults will retrieve any young pup that falls from the roost and emits the squeaky isolation call. Generally the older pups can climb back up without aid. Whether the females nurse any pup, or only their own, is unknown. The pups cluster tightly together, both during the night when their mothers are foraging and also through the day. The mother will fly in, allow a pup to attach to a nipple, and nurse. Then she will leave it with the pup group while she flies off to roost alone nearby. There is considerable movement within the roost during each 24-hour cycle as the bats select the appropriate temperature regime. Fringed Myotis in the United States display strong hibernation site fidelity.

VOCALIZATIONS

The echolocation call of the Fringed Myotis is of low intensity, ranging in frequency from 55 to 30 kHz. The isolation call emitted by the juveniles is audible to humans and sounds like a high squeaky buzz.

SIGNS

The usual indicators of a bat colony, black rice grain–sized droppings, stains at the egress locations, and noise, will alert one to the colony. There are no signs that can help to remotely identify the species, although the slow, manoeuvrable flight pattern can provide a clue.

REFERENCES

Cryan et al. 2001; Lacki and Baker 2007; Nagorsen 2004; Nagorsen and Brigham 1993; O'Farrell 1999; O'Farrell and Studier 1973, 1975; Weller and Zabel 2001.

Long-legged Myotis

also called Long-legged Bat

FRENCH NAME: **Vespertilion à longues pattes**
SCIENTIFIC NAME: *Myotis volans*

These hardy western bats are often out hunting in cool temperatures when other bats choose to remain at roost.

DESCRIPTION

Long-legged Myotis have short, rounded ears, a well-developed keel on the calcar, and belly fur that extends onto the flight membranes to the elbow and knee joints. The tail membrane is partly furred, as the hair on the back extends to cover about the top 25% of the membrane. The calcar is long, extending along almost the entire trailing edge of the tail membrane. The length of the tibia is longer than in other Myotis species, usually more than 18.5 mm (from knee to heel). This characteristic is less reliable for identification than others, as it changes depending on the age of the animal. Fur colour varies a lot in this species, from reddish brown to almost black. Generally, the darker individuals are found in the coastal regions. The belly fur is slightly lighter, but the colour difference is not as contrasting as it is in the Little Brown Myotis. Ears and flight membranes are blackish brown. Like all our Canadian Myotis The dental formula is incisors 2/3, canines 1/1, premolars 3/3, and molars 3/3, for a total of 38 teeth.

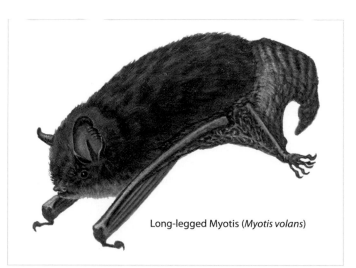

Long-legged Myotis (*Myotis volans*)

Distribution of the Long-legged Myotis (*Myotis volans*)

★ extralimital record

RANGE

Long-legged Myotis are widely distributed in the western portion of North America, from southeastern Alaska to central Mexico, and as far east as North Dakota.

ABUNDANCE

Long-legged Myotis are federally listed in the United States as a Category 2 species, which indicates that these bats are considered imperilled. Canadian populations appear to be stable, but by no means is this considered a common species.

ECOLOGY

These forest-dwelling bats are found in montane and coastal coniferous forests and in arid rangelands, especially along river valleys. During summer, Long-legged Myotis emerge from their day roosts around dusk. They forage in clearings, over water, along cliff faces, and within and above the forest canopy. Most fly high during the early evening and hunt lower later in the evening. Their flight is rapid and direct with no diversions to catch insects beyond their flight path. The impression is that these bats sense and select their prey at 5–10 metres' distance and focus on that prey to the exclusion of others that may be closer. They appear to be quite resistant to low temperatures and are able to fly and forage when other bats choose to remain in torpor. In some areas, the bats exhibit a bimodal pattern of foraging, with a peak of activity between 8 and 11 pm and another between 3 and 4 am. In other places, they are active all night. The summer home range size of adults from a continuously forested habitat in Idaho varied from 250 to 1000 ha, with a core use area of about 50 to 100 ha. The bats migrate to caves to hibernate in small clusters. Well-known hibernacula in Alberta, at Cadomin and Wapiabi Caves, host prehibernation swarming in late August and early September, and most Long-legged Myotis are hibernating by late September. There are no known hibernacula in British Columbia.

Banding data indicates that individuals can live at least 21 years.

SIMILAR SPECIES

Other similar short-eared species are the Little Brown Myotis and the Yuma Myotis, neither of which has a keeled calcar. Fur on their lower-wing membranes and upper-tail membranes is also less extensive. There are three species with prominently keeled calcars that may be confused with this species. The Western Small-footed Myotis and the California Myotis are significantly smaller – forearm < 35 mm, with much smaller hind feet. The Big Brown Bat is typically much larger – forearm 41–52 mm.

SIZE

The following measurements apply to animals of reproductive age from the Canadian portion of the range. Females and males are similar in appearance. Females are slightly larger than males.
Total length: 83–105 mm; tail length: 30–48 mm; hind foot length: 7–11 mm; forearm length: 36–44 mm; ear length: 9–16 mm.
Weight: 5.5–10.0 g; newborns, unknown.
Wingspan: 25–27 cm.

DIET

These bats are opportunistic hunters that will exploit whatever flying insects are locally abundant. Most of their diet is made up of moths, when there is a choice, but they also consume other soft-bodied insects such as termites, spiders, flies, beetles, leafhoppers, and lacewings.

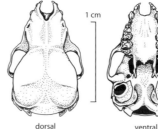

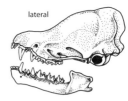

dorsal ventral

lateral

1 cm

Skull of Long-legged Myotis
(CMNMA 48359)

REPRODUCTION

Mating occurs in the autumn before hibernation and females store the sperm in their reproductive tracts over the winter. This sperm fertilizes their egg when they ovulate in the spring. Length of gestation is unknown. A single pup is the norm. In Canada, most of the young are born in late June and July. Long-legged Myotis appear to have an unusually long birthing season. Newborns have been recorded in August in California. There is little information available about the growth and development of the pups. Some young males, perhaps those born early in the season, are developed enough to breed in their first year. The age of first reproduction of females is unknown.

BEHAVIOUR

Females select specific types of day roosts to raise their pups. Most choose to roost under loose bark, or in cavities in snags that protrude beyond the forest canopy and are exposed to solar radiation throughout the daytime hours. Other choices for maternity roosts include crevices in cliff faces (such as in hoodoos in the Alberta Badlands) and buildings. The size of the group varies with the roost site. Generally, tree and cliff-crevice roosts are used by a single female and her pup, or occasionally by very small groups. This type of roost tends to be occupied for a few days, after which the female will move her pup to a new roost within the same general area. As is the trend with many of our Canadian Myotis bats, the larger maternity colonies are found in buildings and generally are more permanent. The largest Canadian colony of Long-legged Myotis so far discovered comprised 300 females in an old barn in central British Columbia. Males and non-reproductive females usually roost singly during the day. Caves, mine shafts, and the undersides of bridges are used as night roosts. Many bats of both genders will gather at the same night roost, but scatter over the surrounding area when they return to their day roosts.

VOCALIZATIONS

Their echolocation call is short but intense, sweeping from 89 to 40 kHz. If the bats detect another bat on a possible collision course, they will add a lower-frequency "honk" to the end of the call. The whole call, including the "honk," is still beyond the frequencies detectable by humans.

SIGNS

This bat might be identified in flight by its rapid direct flight pattern.

REFERENCES

Fenton and Bell 1979; Johnson, J.S., et al. 2007; Nagorsen and Brigham 1993; Ormsbee 1996; Ormsbee and McComb 1998; Schowalter 1980; Slough and Jung 2007; Verts and Carraway 1998; Warner and Czaplewski 1984.

Yuma Myotis
also called Yuma Bat

FRENCH NAME: **Vespertilion de Yuma**

SCIENTIFIC NAME: *Myotis yumanensis*

The Yuma Myotis occurs in British Columbia and regions to the south.

DESCRIPTION

The Yuma Myotis is one of the short-eared Myotis. The ears reach barely to the nostril when gently pressed forward. The fur on its back is short and dull, and varies from pale brown on those in the interior of British Columbia to almost black for those on the coast. The belly fur is lighter. Flight membranes, ears, and snout are dark brown. The calcar is not keeled. The dental formula, like that for all the Myotis, is incisors 2/3, canines 1/1, premolars 3/3, molars 3/3, for a total of 38 teeth.

SIMILAR SPECIES

Of the short-eared Myotis species, the only other without a keel on the calcar is the Little Brown Myotis. In parts of British Columbia it can be very difficult to discriminate between these two species. The ears and nose pad of the Yuma Myotis tend to be lighter than those of the Little Brown Myotis. The forearm measurement of the Yuma Myotis tends to be shorter, usually < 36 mm. The fur tends to be duller and shorter on the Yuma Myotis. The skull length of the Yuma Myotis is usually < 14 mm, while the same measurement on the Little Brown Myotis is usually > 14 mm. Especially in the Okanagan region, it is not unusual, even for experts, to have difficulty identifying the Yuma from the Little Brown Myotis, even though genetically the species are easily distinguished.

SIZE

The following measurements apply to animals of reproductive age from the Canadian portion of the range. Females and males are similar in size and appearance.

Yuma Myotis (*Myotis yumanensis*)

Distribution of the Yuma Myotis (*Myotis yumanensis*)

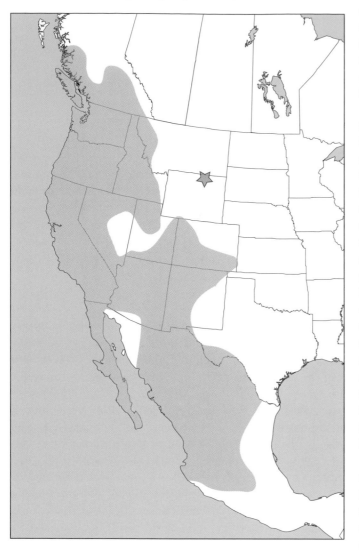

⭐ extralimital record

Total length: 60–99 mm; tail length: 29–45 mm; hind foot length: 7–13 mm; forearm length: 30–38 mm; ear length: 9–16 mm.
Weight: 4–8 g; newborns, about 1.4 g.
Wingspan: 21–25 cm.

RANGE

Yuma Myotis are western bats found in arid habitats near flowing water. They reach the northern limit of their distribution in southern British Columbia and the southern limit in the highlands of central Mexico.

ABUNDANCE

One of the largest known bat colonies in British Columbia is a maternity roost of Yuma Myotis in an old church in the Little Shuswap Indian Reserve near Squilax. In the early 1990s it numbered between 1500 and 2000 adults. Unfortunately the church burned down in 1994, and local volunteers immediately installed bat houses

to accommodate the pregnant females soon to arrive. About 600–1000 females remain at the site, using the bat houses. An additional colony of more than 600 bats resides in the attic and behind the store sign of the Squilax General Store. Yuma Myotis can be locally abundant, but tend not to be regionally common. These bats are classified as a "Species of Concern" in the United States and as protected wildlife in the state of Washington.

ECOLOGY

In British Columbia, these bats are found at low elevations (up to 730 m) in dry inland forests, arid grasslands, and coastal rainforests, never far from water. Yuma Myotis are more closely associated with water than any other Canadian bat and are one of the few species regularly seen foraging over salt marshes. These bats emerge soon after sunset and forage quickly and efficiently. If hunting is good, they can fill their stomachs in less than 15 minutes. They will then hang up in a nearby night roost such as in or on buildings, or under bridges, to digest and excrete the excess, before setting off on the hunt again. Large accumulations of guano are known from long-used night roosts. Females will usually return to the maternity roost to feed and warm their pup rather than select a separate night roost. Usually the foraging area and the maternity colonies are close together. The wing shape and size suggests that this species is capable of flying and foraging in cluttered environments. Most Canadian Yuma Myotis, however, seem to prefer to forage in open, uncluttered habitats over land and especially over flowing water. The only recorded longevity data for Yuma Myotis is 8.8 years; however, there is no reason to presume that the pattern of lifespan is different from that of other Myotis.

DIET

In Canada, Yuma Myotis forage opportunistically on mainly aquatic flying insects, changing their diet depending on what insects are available. Mayflies, caddisflies, flies such as mosquitoes, chironomids and midges, moths, and beetles are their main prey. Most of their prey items are relatively soft bodied.

REPRODUCTION

Mating occurs in the autumn. The sperm is stored internally by the female and fertilization is delayed until the following spring when she ovulates. The majority of births in Canada occur in June and early July. Unlike most of our Canadian bat species, which tend to begin the season considerably sooner in the warmer regions south of the border, Yuma Myotis births in the southern parts of the range are only

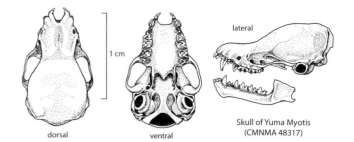

1 cm

lateral

dorsal   ventral

Skull of Yuma Myotis
(CMNMA 48317)

about a week earlier. Most females bear their first pup in the spring following their own birth. Some young males appear to become reproductively active in their first autumn, while others delay until the following autumn. A single pup is the norm. Newborn pups are about 20% of their mother's weight at birth. Older, heavier females tend to produce more male pups earlier in the season, while the younger, lighter females tend to produce more female pups a bit later in June. Within the whole colony this equalizes to an approximately similar number of male to female pups born each year.

BEHAVIOUR

In summer, while females will cluster together in a maternity roost, male Yuma Myotis are usually solitary and choose to roost in cooler locations than the maternity sites. They utilize buildings, crevices and cavities in trees and cliffs, the undersides of bridges, abandoned cliff swallow nests, caves, and mine shafts. Maternity colonies can be in similar sites; however, those in man-made locations are usually larger. Yuma Myotis are sensitive to disturbance at the roosts and will quickly abandon them. Frequent reports of female Yuma Myotis carrying non-volant young are likely mothers moving pups to quieter roosts after having been disturbed. Maternity colonies in British Columbia are vacated by late October. It is presumed that the bats migrate to caves and mine shafts to hibernate, but no Yuma Myotis have been found hibernating in British Columbia. Perhaps they have yet to be discovered, or they travel a bit farther south, to Washington or Oregon sites.

VOCALIZATIONS

The frequency range for an echolocating Yuma Myotis is a downward sweep from 110 to 45 kHz, with the majority of the energy focused around 50 to 55 kHz.

SIGNS

A small bat, seen slowly cruising back and forth with a steady wing stroke and not gliding, often in a circular or elliptical pattern only a few centimetres above flowing water, is likely of this species. The rice grain–sized droppings of this species cannot be distinguished from those of other Canadian bats.

REFERENCES

Brigham et al. 1992; Cockrum 1973; Harris, A.H., 1999; Milligan 1993; Weller, T.J., et al. 2007.

Evening Bat

FRENCH NAME: **Chauve-souris vespérale**

SCIENTIFIC NAME: *Nycticeius humeralis*

There is only one record of this species from Canada. A specimen was taken in Point Pelee in May of 1911. Mainly a southern species,

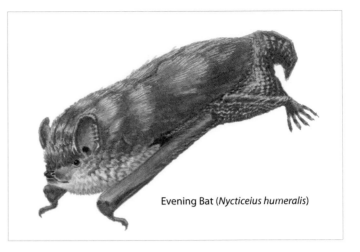

Evening Bat (*Nycticeius humeralis*)

this bat may be one of those poised to expand their range northward with the warming effects of climate change.

DESCRIPTION

A fairly undistinguished appearance can make this species difficult to identify. They have reddish to dull-brown fur on their back with a lighter, tawny belly area. The hairs on the back are often tipped with a lighter brown or greyish colour and have a dark base. Flight membranes, ears, and hairless portions of the face are blackish. The tragus is short and blunt and curved slightly forward. The calcar is not keeled. The most diagnostic trait of this species is found on the skull: the upper toothrow contains a single incisor tooth on each side. The dental formula is incisors 1/3, canines 1/1, premolars 1/2, and molars 3/3, for a total of 30 teeth.

SIMILAR SPECIES

The Big Brown Bat is superficially similar, but larger (forearm is > 41 mm), and has two incisors in its upper toothrow.

SIZE

The following measurements apply to animals from the US range. Females and males are similar in appearance. Females are slightly larger than males.
Total length: 86–104 mm; tail length: 33–40 mm; hind foot length: 8–10 mm; ear length: 11–14 mm; forearm length: 34–38 mm.
Weight: 6–12 g; newborns, 1.6–1.8 g.
Wingspan: 26–28 cm.

RANGE

Although only known from a single Ontario specimen, this bat reaches its northernmost distribution in the southern Great Lakes basin. It could potentially begin to appear more frequently in southern Ontario if average temperatures become a few degrees warmer as they are predicted to do as a result of global warming.

ABUNDANCE

Reasonably common in the southeastern United States, but rarely numerous, there does appear to be a trend towards a decline in the

Distribution of the Evening Bat (*Nycticeius humeralis*)

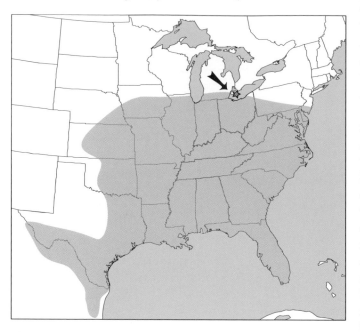

 Canadian specimen locality

number of Evening Bat maternity colonies over the past 15 years. This species is considered endangered in Indiana. The Evening Bat is very sensitive to human disturbance, especially while raising young. The presence of this species in Canada may be accidental.

ECOLOGY

Evening Bats are migratory and do not hibernate. Instead, they fly far enough south to remain active, where they roost in trees, usually in cavities, or under loose bark. They will enter torpor on cool days during the winter, but continue to forage on and off through the season. Preferred summer roosts are under bark, in tree cavities, and in buildings. As is often the case with tree-roosting bats, large snags and mature trees (typically more than 40 years old) are preferred sites. Males stay all year round in the southern portion of the species' distribution and only pregnant females migrate to the northern parts of the range. These females usually arrive at the summer maternity roosts in late April or early May. Evening Bats are efficient foragers. Studies of females have shown that, despite the demands of hungry offspring, most are able to eat enough in the two to three hours after they emerge in the evening to last them until the next night. The principle predators of Evening Bats are Domestic Cats, Northern Raccoons, and snakes, although humans are a more significant threat, as we cause the destruction of summer roost and maternity sites, which are often in buildings or in old-growth forest being clearcut. The accumulation of pesticides in the tissues could be playing a role in the species' decline. The average lifespan for this species is around two years. An Evening Bat is old if it lives five years. This short lifespan is quite different from many of our other bat species, particularly the Myotis bats, which can live 30 years or more.

DIET

The most common prey insects of the Evening Bat are beetles, notably the Spotted Cucumber Beetle, the adult form of the southern corn rootworm, a major agricultural pest. Leafhoppers, stink bugs, cicadas, moths, and chinch bugs are also consumed in huge numbers. A voracious eater, a single female Evening Bat, it has been conservatively estimated, eats at least 21,000 insects per year.

REPRODUCTION

Mating takes place in autumn, after the females join the males on the winter range. Sperm is internally stored by the female until she ovulates in the spring and only pregnant females migrate north. Within each maternity colony births are synchronous, most occurring within a six-day period around mid-June in the northern portion of the range. The length of the gestation period is unknown. Commonly, a litter consists of two pups, although litters of one and three occur rarely. Pups are born pink and blind, but within 24 hours their eyes are open, and within 9 days they are fully furred. Shortly after birth, the pups attach themselves to a nipple, sometimes with their mother's assistance, and are securely enclosed by her wing membranes. Females will retrieve pups if they fall from the roost. Newborn pups are very vocal if separated from their mother, and their calls are readily identified by their mothers, even in a large crowded colony. The females can then confirm their offspring's identity by odour, as she has already marked them by rubbing their bodies with her cheek glands. During their first week, the pups remain attached to their mother, except when she is foraging. After that the pups begin to crawl away for a while, but return to nurse. Females are prodigious milk producers. During the peak period of lactation, they secrete half their body weight in milk every day. Juveniles begin to fly at three weeks old and are weaned and already proficient hunters within four to five weeks.

BEHAVIOUR

Individuals rarely venture out of their roosts until it is truly dark, and return well before sunrise. Communal nursing has recently been discovered. Although the females are clearly able to distinguish their own offspring from all the others using sound and odour, they will allow unrelated juveniles to suckle about 20% of the time. A pup crawls up to a female and nudges her under the wing. If she is willing to nurse it, she lifts her wing to allow access. It has been proposed that this unusual bat behaviour could have evolved because successful hunters will have milk to share beyond the needs of their

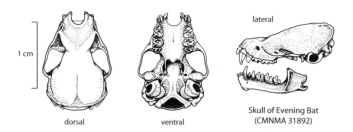

dorsal   ventral   Skull of Evening Bat
(CMNMA 31892)

own pups. If her milk supply were not drained each day, her production would drop before her own offspring were weaned. Because this species prefers prey that tends to cluster, it is not uncommon for some females to return to the roost after a less than fruitful hunt, while others return replete. Since milk production is diminished if the female is not well fed, her offspring will likely drain her supply before they are satisfied. These same juveniles will then go begging for more. Evening Bats can recognize that another bat was a more successful hunter, and will follow that bat on its next foraging excursion. So these bats can not only learn from each other, but also remember where the foraging was good. An experienced adult female will even follow a juvenile, if that juvenile was a successful hunter on its previous hunt. Newly volant juveniles follow adults to learn where to hunt and roost. Maternity colonies form cohesive groups from late April until late September or early October and then migrate together. Females display strong maternity-site fidelity, returning to their natal site year after year to produce their own offspring. Males are more solitary.

VOCALIZATIONS

The so-called isolation calls that the newborns produce allow their mothers to identify them from the other juveniles. These high-pitched squeaks can be detected by humans. Adult echolocation calls can only be heard with a properly set electronic bat detector that translates the ultrasonic sound down to a range humans can perceive.

SIGNS

Since Evening Bats have such a limited Canadian distribution any bat sign is unlikely to be from this species.

REFERENCES

Arnett and Hayes 2009; Clem 1992, 1993; Dowler et al. 1999; Hein et al. 2009; Kurta et al. 2005; Menzel, J.M., et al. 2001; Perry, R.W., and Thill 2008; Scherrer and Wilkinson 1993; Sparks et al. 2003; Watkins, L.C., 1972; Whitaker 1993; Whitaker and Clem 1992; Wilkinson, G.S., 1992a, 1992b; van Zyll de Jong 1985.

Eastern Pipistrelle

FRENCH NAME: **Pipistrelle de l'Est**

SCIENTIFIC NAME: *Pipistrellus subflavus*

A small, widespread species in eastern North America, the Eastern Pipistrelle reaches the northern limit of its distribution in eastern Canada. This species is noted for producing a litter of twins each year, whose combined birth weight can reach more than half the weight of their mother.

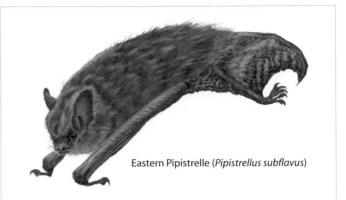

Eastern Pipistrelle (*Pipistrellus subflavus*)

DESCRIPTION

Pelage colour is variable, but usually yellowish to reddish brown on the back with a generally paler underside. A tricoloured hair pattern is diagnostic – dark at the base, lighter in the middle, and dark at the tip. The longer guard hairs are typically of one colour. Ears and face are brown and flight membranes are blackish, but the skin over the forearm and other wing bones is pinkish. The ears extend slightly past the tip of the snout when gently pressed forward. Total tooth number is a characteristic 34, with a small single premolar on the upper toothrow. The dental formula is incisors 2/3, canines 1/1, premolars 2/2, and molars 3/3, for a total of 34 teeth. The tragus is short, straight, and tapers to a rounded tip. This species is most commonly seen in hibernacula, where it usually roosts individually, and is often found covered in tiny water droplets that give it a frosted appearance. A roosting pipistrelle assumes a characteristic humped posture. A small eastern bat with a slow, erratic, fluttering flight pattern is likely a pipistrelle.

SIMILAR SPECIES

This species is distinguished from all other small Canadian bats by its tricoloured hairs.

SIZE

The following measurements apply to animals of reproductive age from the Canadian portion of the range. Females and males are similar in appearance. Females are slightly larger than males.
Total length: 74–98 mm; tail length: 30–46 mm; hind foot length: 8.0–10.5 mm; forearm length: 32–36 mm; ear length: 11.0–14.5 mm.
Weight: 6.0–7.9 g; newborns, about 1.4 g.
Wingspan: 20–26 cm.

RANGE

As their name suggests, Eastern Pipistrelles are found in eastern North America, from the southern Maritimes and eastern Quebec as far south as Honduras in Central America. It appears that the distribution of this species is closely tied to available hibernacula. More of these bats are being found in formerly vacant regions where there are no suitable natural cave sites, as the populations find acceptable abandoned mine shafts.

Distribution of the Eastern Pipistrelle (*Pipistrellus subflavus*)

⭐ extralimital record

ABUNDANCE

The Eastern Pipistrelle is regionally common in some places in its Canadian and US range, where suitable hibernacula and maternity sites are plentiful and temperatures are moderate.

ECOLOGY

The Eastern Pipistrelle forages most commonly over water and at forest edges. Maternity and hibernation roosts are usually less than a hundred kilometres from each other, as this bat is not a strong long-distance traveller. This species does not tolerate freezing temperatures and is usually the first to enter hibernation in autumn and the last to leave in the spring. It also prefers warmer hibernation conditions than most other bats – usually in the 6°C–9°C range, and so is found deeper in caves, where temperatures and humidity are higher and more consistent. Many caves that are acceptable hibernacula for other bats are too cold for Eastern Pipistrelles. Day roost sites in summer are in foliage, in tangles of Old Man's Beard lichen, or in tree cavities. The oldest known Eastern Pipistrelle is a male caught 15 years after he was first banded. The oldest known female was 10 years old. Average age is likely around four to six years for females and eight to ten years for males. Mortality is highest for juveniles during their first year. An Eastern Pipistrelle has been clocked with a flight speed of 18.7 km/h. Domestic Cats, Northern Raccoons, and Black Rat Snakes will occasionally prey upon a colony of Eastern Pipistrelles.

DIET

Eastern Pipistrelles hunt small flying insects, notably beetles, flies, moths, and leafhoppers. They particularly favour many agricultural pests, such as grain moths. Pipistrelles are thought to be among the most voracious of the insectivorous bats, consuming easily half their body weight in insects each night. Pregnant and especially nursing females may eat even more than their body weight per night.

REPRODUCTION

Mating occurs in autumn just before the bats enter the hibernacula and possibly occasionally again in spring at the time of ovulation. In Canada, most births take place during early July after a gestation of between 50 and 60 days, depending on the seasonal temperatures. Females typically produce two pups per litter, although commonly only one lives to become volant, especially in years with cool wet summers. Birth ratios tend to favour males, as does long-term survival, likely as a consequence of reproductive stress on the females. The weight of a full-term litter can be as much as 54% of the female's normal weight, but is more commonly in the 35%–40% range. This is the highest ratio recorded for any bat. Newborns are born naked and blind, but capable of emitting a loud clicking sound (called an isolation call) used to communicate with their mothers. The juveniles begin to fly at about three weeks of age and are usually weaned and foraging for themselves at four weeks old. At that time they are about 80% of their adult weight, but have almost fully grown wings.

BEHAVIOUR

Females gather into small maternity colonies of usually fewer than 20 adults. Males hunt and roost alone throughout the summer, using trees and occasionally caves for their night roosts. Preferred maternity sites are usually in the foliage of deciduous trees, often in clumps of dead leaves still attached to the branches, or in growths of Old Man's Beard lichen, and occasionally in buildings. Pipistrelles tend to hibernate alone. Roost-site loyalty is marked in this species, although the use of multiple roosts and movement between these roosts – even maternity roosts – is common. Females are faithful to traditional maternity roosting areas and young females will return to their natal area to produce their own pups. These bats begin their night of foraging shortly after sunset and usually show a second peak of activity just before dawn.

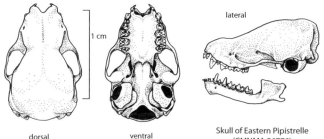

lateral

dorsal   ventral

Skull of Eastern Pipistrelle
(CMNMA 26776)

VOCALIZATIONS

This bat displays a characteristic ultrasonic echolocation signature that features a primary peak in the 20 kHz range and a major harmonic in the 40 kHz range. With two bat detectors, each set to one of these frequencies, a single echolocation call will produce a tonal signal simultaneously from both detectors. The newborn isolation call is a loud, clicking sound, easily detected by humans.

SIGNS

Eastern Pipistrelles rarely roost in buildings and then only in colonies of fewer than 50 individuals, so few people encounter them, and their maternity colonies are unlikely to create any health concerns. Signs of their presence are similar to those of other bat species: droppings the size of a small grain of rice might be found, as well as staining at colony access sites.

REFERENCES

Armstrong et al. 2006; Briggler and Prather 2003; Broders et al. 2001; Charbonneau et al. 2011; Fujita and Kunz 1984; Hooper et al. 2006; Hoying and Kunz 1998; Knowles 1992; Kurta and Teramino 1994; Kurta et al. 2007; MacDonald, K., et al. 1994; Perry, R.W., and Thill 2007; Unger and Kurta 1998; van Zyll de Jong 1985; Veilleux and Veilleux 2004; Veilleux et al. 2003; Whitaker 1998; Winchell and Kunz 1996.

Order Carnivora

felines, canines, bears, seals, walruses, weasels, skunks, raccoons

Red Fox (*Vulpus vulpus*)

Photo: Brian Tilson

The species included within the order Carnivora are native to all continents and oceans. Even the island continents of Australia and Antarctica, which have none of the terrestrial forms, support some of the aquatic forms around their coasts. There are 15 families, 126 genera, and 286 species of Carnivora. In Canada there are representatives from 9 families and 27 genera, with a total of 39 species.

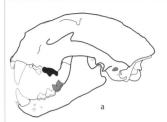
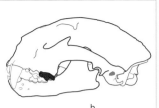

Cougar Carnassial
(a) and (b) show the left side of the skull with the mouth partly open (a) and almost closed (b), to show the action of the P4 (red) and the m1 (blue) as the mouth closes.

Harbour Seal (*Phoca vitulina*)   Photo: Mircea Costina

FOSSIL RECORD

The oldest known fossil of this order dates back to the Eocene epoch, around 60 million years ago.

ANATOMY

Gripping the prey is an integral part of being a predator; therefore, teeth are an important characteristic of the Carnivora. The presence of a large recurved canine tooth on each side of the upper and lower jaws is one of the main diagnostic traits of the order. Once the prey has been captured, other teeth help to cut the flesh into manageable pieces. The shearing action used by most modern carnivores to cut flesh is generated when the inner surface of the fourth upper premolar (P4) passes outside the outer surface of the first lower molar (m1). The P4 and m1 together make up the carnassial pair. As carnassials are most useful in species that specialize in flesh eating, they are reduced in omnivores such as raccoons and most bears (with the exception of the Polar Bear). The P4 and m1 in Sea Otters have converted to crushing surfaces for grinding up their shellfish prey, and carnassials in seals are entirely missing.

The divergence in size among this order is extreme. The smallest, the female Least Weasel, weighs an average of 40 g, while the largest, the male Southern Elephant Seal, weighs in at an average of 3700 kg. Many species within this order display marked sexual dimorphism. The males are larger than the females, sometimes even three or four times the weight.

ECOLOGY

As the name of the order indicates (*carnivora* is Latin for "flesh eating"), these mammals mainly specialize in eating flesh, although many are secondarily omnivorous, and one bear, the Giant Panda, is secondarily vegetarian. The carnivores have adapted to both terrestrial and aquatic environments in order to exploit their mainly vertebrate prey. Pinnipeds (the fin footed), including seals, sea lions, and walruses, evolved from terrestrial ancestors that over time achieved an almost completely aquatic existence. They are not, however, totally committed to the water, as are the whales. They still give birth on land or ice and leave the pups there for the first few weeks of their life (probably to avoid the aquatic predators) until they become proficient swimmers. Many of the terrestrial carnivores, such as otters, American Mink, and Polar Bears, are excellent swimmers and specialize in hunting aquatic prey. Although almost all of the Carnivora are flesh eating, there are many animals outside of this order that eat flesh. Bats, insectivores, and marsupials eat insects, fish, birds, frogs, or small mammals; whales eat fish, marine invertebrates, seals, and even other whales; and many rodents and primates are opportunistically flesh eating.

The predatory lifestyle, especially of those carnivores that specialize in eating other mammals larger than themselves, requires a special combination of mental and physical traits. In order to survive, the predator must be able to find and catch its prey. The senses must be keen; the mind must be active and able to quickly interpret very slight signals, sometimes in an abstract way, and make split-second decisions; and the body must be adapted to either stealth or speed, or both. It is little wonder that the development of the offspring requires considerable parental investment, especially when it comes to teaching the young how to hunt effectively. Members of this order also display a tendency to form social groups in order to increase the hunting potential. Working together, the group has a better chance of feeding all its members either by catching more prey or by catching larger individual prey animals.

CLASSIFICATION

There is some debate about the inclusion of seals and sea lions into this order. Some mammalogists prefer to provide a separate, albeit closely related, order for these animals, which they call the Pinnipedia.

FAMILY FELIDAE

cats

Cats have retractile claws, long sharp canines, short muzzles, and large lower-jaw muscles to deliver a powerful bite. They are penultimate predators and eat primarily flesh.

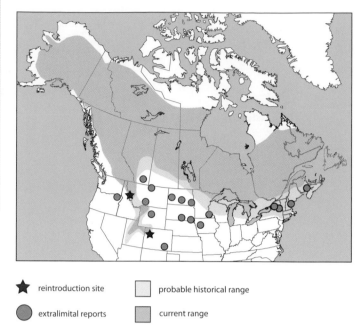

★ reintroduction site

● extralimital reports

▢ probable historical range

▢ current range

Canada Lynx

also called Canadian Lynx

FRENCH NAME: **Lynx du Canada**

SCIENTIFIC NAME: *Lynx canadensis,* formerly *Lynx lynx* and *Felis canadensis*

Canada Lynx are called the "grey ghosts" of the North, partly owing to the colour of their coat but also to their silent, stealthy habits. Actually seeing a wild Canada Lynx is rare, although their tracks may testify that they are around.

DESCRIPTION

A Canada Lynx's winter coat is composed of long, grizzled, grey or black guard hairs with a yellowish-brown or tawny-brown underfur. The combination of colours makes them look greyish to greyish-brown. Their shorter summer coat may be more reddish. The belly, throat, and chin in both pelages are buffy white. The pale throat fur flares out to the sides of the cheeks and throat to form a ruff that has a single or double pair of short drooping blackish stripes along the outward margins. Many individuals have faint spotting, especially on the legs and lower sides, which is most visible in the summer coat. The ears are dark brown to black on the back side, with white or silver-grey central spots and 4–5 cm long tufts of black hair that project from their tips. The nose pad is black or pink and black. Their large, yellow eyes have vertical, slit pupils and are surrounded by narrow white rims above and below. The hind legs are noticeably longer than the front legs. The paws are large and densely furred in winter, providing them with support on snow. Their claws are sharp and retractile. Young kittens are a fuzzy brown with blue eyes and often with noticeable small, dark spots on their sides and back. The dental formula is incisors 3/3, canines 1/1, premolars 2/2, and molars 1/1, for a total of 28 teeth. The canine teeth are long and sharp with shallow longitudinal grooves. The hearing and eyesight are acute. Scent detection is keen, but not as keen as that of canids.

Canada Lynx moult twice each year: they shed their winter pelt in spring, and their summer pelt in autumn. The eyeshine is bright yellowish-green.

SIMILAR SPECIES

Canada Lynx can easily be mistaken for Bobcats. In all parts of their overlapping range except Cape Breton Island, Bobcats are slightly smaller than Lynx, although this is very difficult to see in the field. The longer legs and denser fur of the Lynx usually make them look much larger. Canada Lynx have longer ear tufts, > 2.5 cm in length (usually 4–5 cm); heavily furred feet, especially on the underside between the pads; and a slightly shorter, plain-coloured tail with a full black tip all the way around; adults are more greyish with few well-defined spots (some indistinct dapples and sometimes dark spots on the light belly, at most). Bobcat ear tufts are < 2.5 cm in length. Animals in winter pelage can be the most difficult to tell apart as the Bobcat's longer winter fur often obscures its spots. In general, Bobcats are more red in colour, while Lynx are more grey, but these characteristics are too variable and seasonally affected to be reliable. Track differences between the two are often the best way to distinguish them. The well-furred feet of the Lynx make its tracks easily double the size of the Bobcat's tracks. Cougar cubs are spotted, and although similar in size to adult Bobcats, they lack the well-developed ruff of hair on the sides of their faces. Distinguishing Bobcat skulls from Lynx skulls can be difficult because no single character appears to be conclusive; their expression can vary individually and geographically. Most of the usable characters are visible on the ventral side of the skull. (See the accompanying figure.) Bobcat skulls are typically smaller than those of Canada Lynx but have relatively larger auditory bulla. The shape of the presphenoid bone is quite different between the two cats, narrower in the Bobcat

Canada Lynx (*Lynx canadensis*)

than in the Lynx. The relative position and orientation of the hypoglossal canal will identify some skulls to species. This canal may be confluent, or nearly confluent, with the jugal foramen in many Bobcats but is clearly distinct in Lynx skulls. All of these characters should be considered in concert as variation is considerable. While rare, hybrids are known between Bobcat and Canada Lynx. These animals tend to be intermediate in character and are fertile. Most occur in regions in which Bobcats are relatively common and Lynx are rare, and they appear to be the progeny of a male Bobcat and a female Lynx.

SIZE
Adult males are an average of 40%–55% heavier and 5%–10% longer than adult females. Canada Lynx at higher latitudes tend to be larger. The following measurements apply to adult animals in the Canadian part of the range.
Males: total length: 813–1050 mm; tail length: 95–120 mm; hind foot length: 228–240 mm; ear length: 74 mm; weight: 7.4–17.3 kg.
Females: total length: 810–950 mm; tail length: 86–114 mm; hind foot length: 203–235 mm; ear length: 74 mm; weight: 4.5–10.3 kg.
Newborns: weight: 150–211 g.

RANGE
This cat is closely tied to boreal forests and likely ranged from northern Canada and Alaska to about the 49th parallel, extending southwards in the Rocky Mountain region, following alpine coniferous forests, as far as Nevada. Persecution and habitat changes by

humans have extirpated Canada Lynx from Prince Edward Island, mainland Nova Scotia, and probably southern Alberta, Saskatchewan, and Manitoba. The northern range expansion of Bobcats in the last 100 years may have contributed to the decline of Lynx in the areas of shared range because Bobcats appear to be more aggressive and adaptable. There have been periodic winter mass migrations of Lynx out of the boreal forests and onto the prairie grasslands, where their presence is intermittent. One migration occurred in 1962–3, when Lynx were seen in western cities such as Edmonton, Calgary, and Winnipeg and invaded the grasslands as far south as Iowa and southwestern Wisconsin. Such mass movements are likely the result of high numbers of Lynx being forced to look elsewhere for food following a drastically declining hare population in the northern boreal forests. Since that time, Lynx have had several other peaks of abundance, in 1972–3, 1982–3, the very early 1990s, and the very late 1990s.

ABUNDANCE
Canada Lynx are easily trapped because they are highly curious, and this characteristic can be exploited by trappers to entice them. Trapping is licensed and regulated in Alaska and all provinces and territories, except Prince Edward Island, Nova Scotia, and New Brunswick. The provincial governments of Nova Scotia and New Brunswick both list Canada Lynx as "Endangered." Harvesting in most parts of northern Canada is thought to be sustainable, but careful regulation is required to reduce the allowable take when numbers dip. Canada Lynx have never been abundant in the

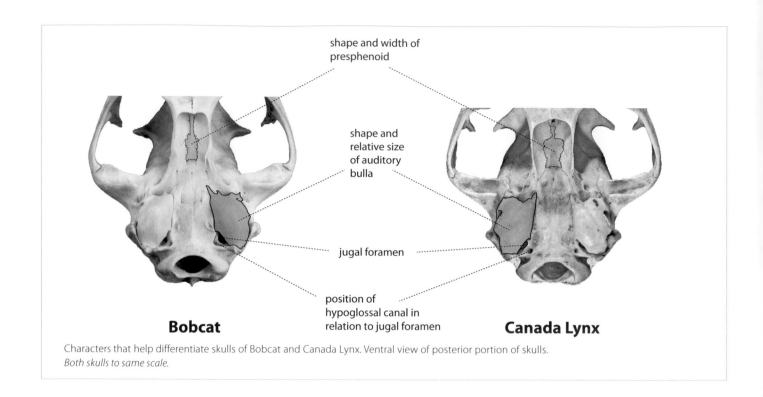

shape and width of presphenoid

shape and relative size of auditory bulla

jugal foramen

position of hypoglossal canal in relation to jugal foramen

Bobcat

Canada Lynx

Characters that help differentiate skulls of Bobcat and Canada Lynx. Ventral view of posterior portion of skulls. *Both skulls to same scale.*

continental United States and are now rare, and trapping is prohibited. Fewer than 200 reproductive adults are thought to remain, mostly in Maine, Minnesota, Montana, Washington State, and Wyoming where they are listed under the United States Endangered Species Act as "Threatened." Canada Lynx are negatively affected by road building and the human encroachment that inevitably follows. This is likely part of the reason for the disappearance of these cats at the southern limits of their range.

ECOLOGY

Canada Lynx generally inhabit densely forested wilderness. They prefer old-growth boreal forests with dense undergrowth and tangled windfalls, but they can manage well in dense mixed or even deciduous forests provided there is sufficient prey. Canada Lynx are specialized predators of Snowshoe Hares, and as such their population health is closely tied to that of the hare. Since the hares are often abundant in regenerating woodlands following logging and forest fires, these ecosystems can also support denser Lynx populations, at least for a time. In northern portions of the Lynx distribution (which includes most of Canada), Snowshoe Hare numbers undergo periodic crashes and peaks in roughly a 10-year cycle. Lynx density follows with about a one-year lag and varies from 2 to 45 Lynx per 100 km². At the peak of hare numbers, Lynx are prolific, and even the subadult females will bear litters, but in the year after hare numbers bottom out only a fraction of the adult female Lynx reproduce; even then they produce small litters, most of which do not survive their first winter. Predation by Lynx is one of several factors that drive the population crash of the Snowshoe Hares. It has been estimated that Canada Lynx catch and eat five times as many hares during the cyclic peak than during the low. Lynx can

also, upon rare occasions, have a major impact on ungulate populations, such as in Newfoundland in the 1960s when the introduced Snowshoe Hare populations crashed and Lynx turned their attention to Caribou calves. Virtually every calf was taken by a Lynx for a few years until the hares recovered. The Snowshoe Hare decline typically begins in mid-continent and, like a wave, spreads east and west, affecting more and more geographical area synchronously. The wave may be moderated or enhanced depending on the condition of the hare populations beyond the core area, so it can be amplified continent wide, as happened in 1962–3 (see the "Range" section above), or it can be diffused and more regional.

A Lynx's daily movement is closely tied to snow characteristics and prey density. When their primary prey is abundant, they may only travel 1–3 km each night to find sufficient food, but if hares are scarce, they could easily move 5–6 km. Since this distance is measured as a straight line from evening to morning locations and as Lynx normally travel in a zigzag path as they hunt, the distance actually travelled no doubt greatly exceeds the straight-line distance reported above. A snow tracking survey measured distances of 5–9 km for nightly hunting in winter. The average size of home range varies from 10 to 507 km² and is variable depending upon the sex of the animal, the season, and food availability. Typical home ranges are around 15–54 km². Males occupy home ranges that are at least twice the size of female home ranges in any geographic location. Although related females may permit some overlap of home ranges, generally Lynx of the same sex do not tolerate overlap. However, male ranges regularly overlap or even encompass those of several females. Summer ranges are typically larger than winter ranges, and individuals need larger home ranges when prey is scarce. Lynx show strong long-term fidelity to their home ranges, and tenure generally only changes following death or a severe food shortage that drives the animals away. Dispersal of subadults occurs in spring, and the distance they must travel to find a home range varies with prey density and availability. Female offspring may remain nearby, while male offspring tend to disperse further. The juveniles may move, settle for a while, and then move on if they find that a potential home range is already occupied or inadequate. Straight-line distances as far as 930 km have been reported. When hare numbers crash, resident Lynx often have no choice but to move on in search of regions where hares are still present. Long-distance movements of displaced adults can involve straight-line distances sometimes in excess of 500 km and up to a record 1530 km. Such dispersal of adults and subadults is significant in Lynx population dynamics because it can allow repopulation of regions in which Lynx have been extirpated owing to food shortages or over-harvesting, and it effectively homogenizes the gene pool.

Canada Lynx are passable swimmers and proficient tree climbers although they do all of their hunting on the ground. Most natural mortality is likely caused by starvation (up to 40% of the animals may die) following severe declines in the local hare populations, and unregulated trapping during such a decline can result in local extinction of Lynx. Otherwise, apart from the minor mortality caused by disease, accidental deaths, and predation by Cougars and Grey Wolves, the major cause of mortality is trapping. Canada Lynx can live for up to 20 years in captivity, but few survive more than 8 years in the wild. The oldest wild Lynx female so far reported came from Labrador and was trapped at the age of 14 years and 7 months. The oldest male died after a Cougar attack in Montana at the age of 16.

DIET

Canada Lynx are considered to be specialized predators of lagomorphs, primarily Snowshoe Hares, which constitute 35%–100% of their diet, depending on availability and season. Other prey are taken opportunistically and include Red Squirrels; mice and voles; flying squirrels; young of Caribou, White-tailed Deer, Moose, American Bison, and Thinhorn Sheep; smaller carnivores such as foxes and Northern Raccoons; ground squirrels; North American Beavers; Common Muskrats; ungulate carrion including domestic livestock; grouse, waterfowl, ptarmigans, and other birds; and fish. There are accounts of Lynx taking adult deer, Caribou, and Thinhorn Sheep during winter, but this apparently only occurs when hares are scarce. The Lynx (probably larger males) ambush the ungulates and jump on their backs, attempting to kill them with bites to the nape. These predation efforts may involve a lengthy chase, which is out of character with the Lynx's normal hunting activities. As just mentioned, prey switching may occur during a Snowshoe Hare decline and during summer and fall when fewer hares are caught and more alternate prey is consumed. Lynx depend on Snowshoe Hares and ungulate carrion more in winter, and when neither are available, the Lynx generally either move or starve. Like all felids, Canada Lynx are unable to convert beta-carotene into fat-soluble Vitamin A, so all of their Vitamin A needs are obtained from the kidney, liver, lungs, and adrenal glands of their prey.

REPRODUCTION

The breeding season varies geographically. It occurs in March and April in Newfoundland, Yukon, and Alaska and in April and May in Alberta. The bond of a male-female pair can persist for several days. If food is abundant, the breeding season may begin in February or even January. Gestation lasts 60–63 days, so kittens are born in May to July depending on location. The average litter size is three to four (ranging one to six). At most, females produce a single litter per year, and during years in which food is scarce many females will forgo a litter or they will mate and either fail to give birth or lose their litters soon after parturition. Young females producing their first litter tend to give birth later and have fewer kittens than do mature females, and even mature females produce smaller litters when there is a low food supply. During years of abundant food, 33%–79% of the yearling females and 73%–92% of the adult females breed. During lows in the food supply, 0%–10% of the yearling females and 0%–64% of the adult females breed. Most juvenile females breed for the first time as two-year-olds, except when food is abundant. Juvenile males generally become sexually mature as two-year-olds. Females rear the kittens without help from the sire. Young are born blind and deaf but are fully furred, with dark longitudinal stripes on their back and legs. Their eyes open when they are 10–12 days old. Their first milk teeth (deciduous) begin emerging from the gums at 11–14 days and are fully erupted by nine weeks. At two months old the kittens have lost their natal pelage pattern, but the stripes on the inner side of the forelegs are even more prominent. The permanent dentition begins to replace the milk teeth at around four months old and is completed by the time they are 34 weeks old. Kittens are weaned at 6–8 weeks old but usually remain with their mother on her home range through their first winter and disperse to find their own home range in early spring when they are around 10 months old.

BEHAVIOUR

Although Lynx may be seen at any time, they are most active in the hours before dawn and after dusk when their primary prey is also most active, and they will commonly hunt throughout the night, especially if they are unsuccessful. Tracking data shows that Lynx may be found in groups of two or three during the winter and that they appear to hunt cooperatively. These are likely family groups of a female with her last litter. Usually adult Lynx are solitary, except briefly during the breeding season. Lynx of both sexes mark their home ranges with excretions from their anal and cheek glands and

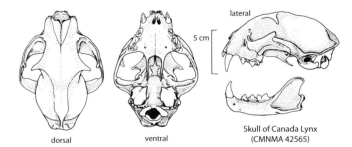

lateral

5 cm

Skull of Canada Lynx
(CMNMA 42565)

dorsal   ventral

with urine and scat. Urine is sprayed and scat is deposited at prominent locations and may accumulate into a latrine at preferred spots. The cats can then attempt to avoid interactions by paying close attention to these scent and visual markings. Nevertheless, males do occasionally battle each other, especially during the breeding season when their testosterone levels are high. Like other cats, Lynx hunt from ambush or by stalking. They prefer to follow well-used hare runways or to lie in ambush beside a runway. Generally, ambush is preferred when the cover is heavy, and stalking is used when the forest is more open. When hunting alternate prey, they search their home range until they detect a potential meal. Then they crouch and slink closer, using every bit of cover available. They will only pounce when they are close enough to almost assure a kill (within 6.5 m for Snowshoe Hares) as they are not built for a lengthy chase. Lynx can be remarkably tolerant of humans if they are undisturbed. Populations can thrive in heavily wooded and swampy habitats and even sometimes in partially cleared farmland at a short distance from human settlements. Maternal females rarely use underground dens. Most young are born in dens under dense brush piles, uprooted trees, or rocky overhangs or in hollow logs or very dense thickets. Females commonly move their dependent kittens to new den sites, some up to five times. A maternal den is used until the kittens have been weaned and is rarely reused unless appropriate den sites are uncommon. Solitary adults seek shelter under heavy evergreen boughs, in windfalls, under rocky overhangs or crevices, and in brush piles, and sometimes they curl up in a snowdrift. They rarely use the same site twice.

VOCALIZATIONS

Lynx vocalizations are similar to those of other mid-sized felids. They yowl (also called caterwaul), growl, hiss, grunt, meow, purr, and produce a host of quiet short calls (such as a bird-like chirp). Although normally fairly silent, both males and females commonly produce long wailing calls during the breeding season. Females produce a low drawn-out wail, similar to a mating call, when she is anxious about the safety of her kittens (for example, when they are being handled by a researcher). A family group of a maternal female and her kittens often make a series of five to ten short barking calls, presumably to locate each other.

SIGNS

Lynx have five toes on their front feet, but the first toe (the thumb or pollex) is positioned high up on the inside of the leg, and while it has a claw and is used for climbing and grasping prey, it does not touch the ground; therefore, it does not register in a track. Each hind foot has four toes. The front track is larger and rounder than the hind track. The palm pad on each foot is bilobed at the leading edge and trilobed at the rear margin. The bottom of a Lynx's foot, particularly in winter, is so well furred that only a small portion of the toe and palm pads actually contacts the ground, and the fur between the pads can blur the track. The fur acts as a snowshoe to enable the cat to travel over fluffy snow and, along with the long splayed toes, can enlarge the track to over 10 cm in width (close to the track size of an adult Cougar). Tracks on harder surfaces than snow, such as

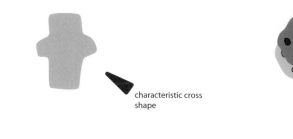

characteristic cross shape

Direct register walk in deep snow

Direct register walk

Left hind track
Length: 6.5–10.5 cm
Width: 15.5–12.7 cm

Left front track
Length: 6.0–10.8 cm
Width: 6.0–14.3 cm

Canada Lynx

silt or dust, are rarely wider than 7.6 cm. Retractable claws are only engaged on slippery surfaces or if the animal is travelling at high speed. Most hunting is conducted at a walk or a trot. When walking in snow, they typically register the hind track directly over the front track as an energy-saving measure. In fluffy snow the toes splay to such an extent that the overall track takes on a cross-like shape that is unique to this species. A stalking Lynx uses a shorter gait and may leave body marks or extensive drag marks in the snow as it crouches to reduce its body profile. Once close to its prey, a Lynx bounds in pursuit. Such bounds are 1.0–1.4 m in length. A hungry Lynx is quite capable of eating an entire Snowshoe Hare in one feeding, but if it does not, the remainder is cached for later consumption. Lynx caches are typically covered with snow or debris. When Lynx eat a hare, they usually leave the intestines and stomach, likely one or more feet and a flap of skin from along the back, which may include the tail. These remains may be cached but are only utilized if the Lynx is starving. Lynx scat is usually dark and segmented or has a beaded appearance. Kittens sharing their mother's range will cover their scat like Domestic Cats do, but adults always leave their scat as a visual and scent mark, usually depositing it in a prominent or elevated location. Scat is 1.3–2.4 cm in diameter by 7.5–25.5 cm in length. Resting beds may be sheltered in a brush pile or under a low-slung coniferous tree or out in the open. The cat curls up and creates an oval depression that is 38.0–53.5 cm in length by 25.5–38.0 cm in width. Hunting lays or lookouts tend to be long and narrow depressions that are made as the animal crouches in ambush near a game trail. They are 40.5–51.0 cm in length by 20.0–25.0 cm in width. Lynx sign is rare where Snowshoe Hare sign is absent.

REFERENCES

Anderson, E.M., and Lovallo 2003; Bayne et al. 2008; Beyer et al. 2001; Campbell, V., and Strobeck 2006; Carbyn 2003; Chubbs and Phillips 1993b; Elbroch 2003; Homyack et al. 2008; Hoving et al. 2005; Koble and Squires 2006; Lafond 1999; Mowat et al. 1996; Murray et al. 1995; Paragi et al. 1997; Poole 1995, 1997; Poole et al. 1996; Ranta et al. 1997; Rezendes 1992; Schwartz, M.K., et al. 2002; Slough 1999; Slough and Mowat 1996; Smith, H.C., 1993; Stephenson et al. 1991; Tumlison 1987.

Bobcat

also called Bay Lynx, Catamount, Red Lynx, Wildcat

FRENCH NAME: **Lynx roux,** *also*
Loup-cervier, Pichou

SCIENTIFIC NAME: *Lynx rufus,* also *Felis rufa*

Bobcats are about twice the size of a Domestic Cat. They are currently the most widespread North American felid, although their larger cousin, the Cougar, once held that honour.

DESCRIPTION

Bobcats are typically yellowish- to reddish-brown with numerous black spots and black-tipped guard hairs. The base colour varies across the continent. Cats in the dense forests of the Pacific Northwest and southwestern British Columbia tend to be dark reddish-brown; those in more arid central plains are typically pale yellowish- to greyish-brown; those in central Canada are usually a pale rusty brown, often with yellowish tones; and those in the Maritimes and southern Quebec tend to be a dark reddish-brown. Sometimes the spots on the upper sides and shoulders are roughly circular and vaguely paler in the centre. Their underside is paler, with black spots, and the legs are tawny brown to pale brown with black spots. There is often a darker band of black-tipped guard hairs running down the spine. Bobcats have a pinkish or blackish nose pad, white whiskers, and a black-striped forehead. The white ruff is an extension of the white throat. The long hairs flare out and droop downwards and are marked with two primary black stripes, the lower one usually being more distinct. The front of the muzzle, lips, and throat are white, but the rows of whisker insertions are black. The backs of the ears are black with a sometimes diffuse, central white spot. A tuft of longer dark hairs (less than 2.5 cm long) extends from the ear tips but may be absent during the moult. Hair on the insides of the ears is usually white. Their large golden eyes have a vertical elliptical pupil and are surrounded by narrow bands of white fur. The Bobcat is the only species in the genus *Lynx* that has rare melanistic individuals; all but one of those reported were from southern Florida, the other being from Saint John County, New Brunswick. A Bobcat's short tail looks like it has been bobbed (hence the English common name) and is often carried along the same line as the top of the back but with a jaunty upward bend at the tip. The upper surface of the tail is tawny with black, transverse bands and a black spot at the tip that does not extend all the way around the tip. The underside of the tail is white. The tail comprises 18%–22% of the head-body length. The dental formula is incisors 3/3, canines 1/1, premolars 2/2, and molars 1/1, for a total of 28 teeth. The canine teeth are long and sharp with shallow longitudinal grooves. The hearing and eyesight are acute. Felid scent detection is keen but not as keen as that of canids. The eyeshine is bright yellowish-green. Bobcats moult twice a year, in spring and autumn; the summer coat is more rufous, while the winter pelt is more greyish. Determining the sex of a Bobcat can be difficult and may require an examination of the internal organs, especially with regard to juveniles.

SIMILAR SPECIES

Bobcats can easily be mistaken for their close relative, the Canada Lynx. The "Similar Species" section of the Lynx account provides information on the ways to distinguish the two and includes a figure illustrating skull differences.

SIZE

Bobcats are sexually dimorphic. Males are up to 10% longer and 25%–80% heavier than females within each population. Size varies across the range, with the largest animals found in the northern and mountainous regions. The following measurements apply to adult animals in the Canadian part of the range.

Bobcat (*Lynx rufus*)

Males: total length: 736–1050 mm; tail length: 147–200 mm; hind foot length: 152–223 mm; ear length: 60–85 mm; weight: 10–16 kg, an exceptional individual in Minnesota weighed 17.6 kg; shoulder height: average 46 cm.

Females: total length: 724–925 mm; tail length: 127–180 mm; hind foot length: 146–180 mm; ear length: 60–76 mm; weight: average 7–9 kg, exceptional animals weighing up to 15.3 kg.

Newborns: weight: 280–340 g.

RANGE

Bobcats occur across much of southern Canada but have not reached the major offshore islands (Prince Edward Island, Newfoundland, and Vancouver Island). They reach midway up British Columbia to just beyond the fiftieth degree of latitude and as far south as central Mexico. Their range has extended northwards following the removal of the dense forests of southern Canada by clear-cutting, agriculture, and human settlement since the immigration of Europeans. Bobcats have been exterminated from much of the United States in the Great Lakes area as well as the state of Delaware by a combination of intensive agricultural practices, dense urbanization, persecution, and over-trapping. Likely the northern limits of the range are determined by snow depth, as Bobcats have difficulty in capturing prey in deep snow because they do not have large snowshoe-like feet like the Canada Lynx do. Bobcats invaded Cape Breton Island from mainland Nova Scotia in the 1950s, likely as a result of the construction of the Canso Causeway, which was completed in 1955. They appear to be displacing the native, less aggressive Canada Lynx on Cape Breton.

ABUNDANCE

Until the mid-1970s Bobcat pelts were considered to be second rate, and the animals were not heavily trapped. In 1975, with the ratification of the CITES (Convention on International Trade in Endangered Species) agreement and the subsequent trade embargo on spotted and striped cat skins from Africa and Asia, the Bobcat's soft, dense, winter pelt, especially the belly portion, took on a new lustre. Bobcats are now thought to be the most exploited wild felid on the planet. From the mid-1970s to the late 1980s an average of 94,000 Bobcat pelts were harvested annually in the United States and Canada. As prices for the fur dropped in the 1990s, fewer were trapped, and in 2002–3 trappers in North America harvested some 56,000 Bobcat pelts. Trapping is permitted in most states within the range, excluding Connecticut, Illinois, Indiana, Iowa, Maryland, New Hampshire, New Jersey, Ohio, and Rhode Island. In Canada trapping is permitted in all of the provinces except Quebec. The Bobcat population in Manitoba is considered "Vulnerable" although limited trapping is permitted. Trapping of Bobcats in Mexico is permitted in five states. Bobcats are considered to be abundant in the southeastern United States, and populations are stable or increasing in most jurisdictions. Human encroachment and habitat fragmentation are of concern in the northern reaches of the range, which includes all of their range in Canada.

ECOLOGY

Bobcats occupy a wide variety of habitats from swamps and arid scrub to grasslands, parklands, and temperate forests. Whatever the environment, it must provide several essentials including sufficient food, den sites, and adequate cover to both protect the cats from predation and allow them to stalk their prey unseen. Bobcats are active at any time of day or night but are most active around sunrise and sunset, which makes them primarily crepuscular. Not surprising, many of their favourite prey species are also most active at these times. Bobcats are somewhat more likely to move around during the day in winter to take advantage of any warmth at that time. Although their secretive ways make it difficult to study wild Bobcat ecology, it is thought that less than one in two predation attempts results in a kill and that adults are twice as likely as are subadults to hunt successfully. Bobcat numbers may fluctuate depending on the abundance of their preferred prey, but because these cats are generalist feeders that readily switch to other prey species, the fluctuations are rarely dramatic. Home ranges may vary with the season because Bobcats select their winter range based on ease of travel in snow, preferring low elevations, mature lowland coniferous forests that minimize snow beneath them, south-facing slopes, and windswept open or rocky areas. Winter home ranges may be larger than summer ranges. South-facing rocky ledges are particularly important for Bobcat survival in eastern forests as they provide an all-important sunning location in the cold months as well as resting and natal den sites during the spring and summer. Summer ranges are less habitat selective. Home ranges vary from 0.84 to 325.0 km² based on habitat, sex, season, and prey availability and tend to be larger in the northern parts of the distribution. Male home ranges on average are two to three times larger than those of females and may be up

Distribution of the Bobcat (*Lynx rufus*)

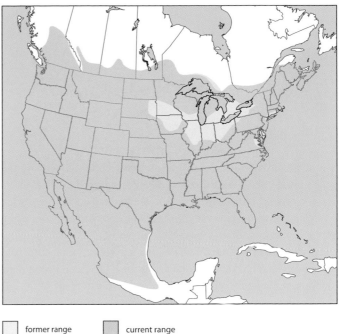

☐ former range ■ current range

to five times larger in some areas. Each male's home range typically overlaps the home ranges of several females. The males can thereby maximize their breeding opportunities as they are most likely to be able to detect the oestrous condition of those females nearby. Home ranges are not aggressively defended. Instead, transients and adjacent residents (see the "Behaviour" section for an explanation of transients and residents) usually respect the tenure of the incumbent. Neighbours generally avoid each other, and transients moving through the home ranges of residents are tolerated and allowed to simply pass through unmolested. There are rare reports of transients displacing residents, but most displaced residents voluntarily leave their home range owing to lack of prey. As residents die, their home ranges are taken over by lucky transients or by neighbouring residents.

Maternal dens are typically located in caves, under rock overhangs, in natural openings in a rock pile, under the roots of fallen trees, in hollow logs, in dense brush piles or thickets, or even in abandoned beaver lodges. Abandoned or little-used isolated buildings may also house a family. The female will sometimes line the den with moss or dried vegetation. Adult Bobcats den and rest during the day in similar sites as well as in tall dense grass. Non-maternal adults alternate their den and resting sites and rarely use the same one for more than a single day, except in winter. Bobcats have considerably less fur on their feet than do their close relatives, Canada Lynx. Lynx paws are about twice as good at distributing their weight over soft snow, and this gives Lynx an advantage in snowy climes that Bobcats cannot match. Juvenile dispersal occurs before the mother produces her next litter. The juveniles are at their most vulnerable as they search for a home range of their own. They may wander for a few days or up to ten months and over 175 km before finding a suitable location.

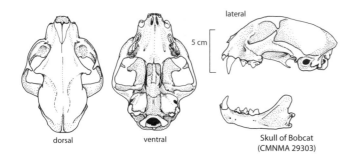

lateral

5 cm

dorsal   ventral   Skull of Bobcat
(CMNMA 29303)

Dispersal distances of 20–40 km are most common, with young males usually settling further away than do young females. Bobcats are good tree climbers, using trees to escape predators or to provide a resting site with a commanding view. Mortality may be caused by starvation, injury, disease, parasites, motor-vehicle collisions, trapping, shooting, poisoning, habitat loss, accidents, cannibalism, old age, and predation. Kittens are killed by owls, foxes, and adult male Bobcats, while adults and kittens are vulnerable to many of the larger carnivores such as Grey Wolves, Coyotes, Cougars, and Domestic Dogs. Bobcats do occasionally contract rabies but with nowhere near the frequency of the big three – Northern Raccoons, foxes, and skunks. Captive Bobcats have lived 32 years, but few in the wild survive beyond 16 years; the oldest known wild Bobcat was 23 years old.

DIET

Their size and strength make Bobcats best suited to prey upon animals in the 0.7–5.5 kg weight range, but they are generalists and do not turn their nose up at available smaller or larger prey. Rabbits and hares are their primary quarry. Adult males, and sometimes large adult females, will attack bedded, weak, injured, or bogged-down deer and Pronghorn that they encounter, and both males and females will take fawns. Older animals are generally larger and have more hunting experience and so are more able to bring down an adult ungulate. Males, being larger than females, eat more deer, especially in winter. Bobcats also eat many species of rodents including rats, mice, voles, Mountain Beavers, squirrels, North American Porcupines, and pocket gophers, as well as birds and bird eggs, Northern Raccoons, Domestic Sheep, Goats, Cats, and fowl, Bighorn Sheep, reptiles, beached fish, other Bobcats, and other smaller carnivores. Bobcats will feed on any large mammals that are victims of motor vehicles or sport hunters, and unspoiled carrion of all types is readily consumed. Even insects may become part of the diet at times when other food becomes scarce. Prey switching is common as prey abundance varies. Although Bobcats will drink water when it is available, they are capable of surviving for extended periods just on the water they ingest in the tissues of their prey.

REPRODUCTION

In the parts of their range that are subject to cold winters, the Bobcat breeding season occurs in early spring. Females are in heat for 5–10 days of their 45-day fertility cycle, and if they fail to become pregnant, they may undergo up to three cycles in a breeding season. Males mate with as many receptive females as they can find.

Following a gestation period that ranges from 50 to 70 days and averages 63 days, kittens are born mainly in May and June. One litter per year is the norm. Where winters are warm, breeding can occur at any time, and litters may be born in any month. Although rare in these warm regions, two litters annually are possible, but only if prey is abundant. Young are raised exclusively by the female. While mature females are capable of producing at least one litter every year, the actual pregnancy rate varies with the nutritional status of the female. When prey is abundant, up to 90% of the adult females will breed, but the rate drops as food becomes scarce and can be as low as 12.5% during years of drastic prey shortages. The litter size ranges from one to six but averages two to four. First-time breeders generally produce a litter of fewer kittens than that of a mature female. Females remain with their kittens for at least the first 48 hours after their birth and then leave the helpless young for a short time to drink and hunt. Kittens are born blind but fully furred. Each quickly establishes exclusive use of one nipple, which avoids future fights over milk supply. Suckling kittens purr and knead their mother's belly to enhance milk flow. Their eyes open 9–18 days after birth. Milk teeth begin to emerge from the gums at 11–14 days old, and the set is complete by the age of nine weeks. Adult or permanent teeth begin to erupt at 16–19 weeks old and have completely emerged from the gums by the time the kitten is 34 weeks old. Maternal females bring prey back to the kittens even before they are old enough to eat solid food. They smell it and play with it and in the process learn what is locally available. As they mature and start eating what she provides, she will deliver wounded or live prey so they can practise stalking, pouncing, and killing. Weaning occurs gradually by the end of the kitten's second month. Kittens venture after their mother on her nightly forays when they are three to five months old, following her upturned tail tip and white ear patches. She may change den sites if she is disturbed or if the den becomes fouled with faeces or rotting food. Young kittens are carried by their scruff, causing them to dangle limply. Older kittens are less likely to remain passive when she attempts to carry them in such a manner and can cause their mother considerable difficulty as she tries to relocate them. The youngsters are playful and learn many social and physical skills as they tumble around with each other in the safety of the den or at its mouth. Juveniles remain with their mother through their first winter. Some especially well-nourished young females may become sexually mature as early as 9–12 months old, but most become fertile as yearlings at 18–24 months old. If food is scarce, sexual maturity may be delayed until they are three years old. Males' sexual maturity depends on their body weight, but most are mature enough to have viable sperm in their second year.

BEHAVIOUR

Bobcats are largely solitary, and interactions are infrequent and brief. The only lengthy bonds that develop are between females and their kittens. A courting pair may consort for a few days, running, hunting, and playing together before mating. Females that are not ready to mate will repel an ardent male with teeth, claws, and snarls, but the males do not retaliate; they just back away to wait and continue to follow. When the female permits, the male mounts while grasping her nape in his teeth. Her copulatory cry is a low growl. Copulation typically lasts less than five minutes and may be repeated up to 16 times in 24 hours. Bobcat society is structured into three principal groups of animals: transients are juveniles that have not yet established a home range, or older adults that have been displaced; residents are adults with well-established home ranges; and kittens are juveniles of less than one year that are still dependent on their mother and who reside on her home range. Extralimital wanderers are likely to be transients as they may travel widely, looking for a suitable unoccupied home range. Like most of the felines, with the notable exception of Cheetahs, Bobcats are sprinters. They will not attack until they are within 10 m or less of their quarry, and if it cannot be captured within 3–18 m, depending on the prey, the pursuit is abandoned. Bobcats use two primary hunting techniques. They zigzag through their home range, searching until they detect prey, and then they stalk upwind, hiding behind every possible scrap of cover until they are close enough to pounce. They fix their gaze on their prey and slowly creep closer, usually when the animal has its head turned or eyes shut. Alternately, they may wait patiently in ambush until their unsuspecting prey draws near. Moving about in search of prey is most effective when prey is scarce or diverse, while ambush hunting is more successful when prey is abundant and predictable. The two strategies are also commonly used in concert, such as when a searching cat might crouch to rest in concealment beside a game trail. When hunting, Bobcats fan their highly sensitive whiskers to help them detect prey and obstacles. A high, arching pounce is used when hunting mice or voles in tall grass. A Bobcat can easily travel 8–10 km each night while hunting if prey is scarce, and males generally travel further than females because they spend more effort marking their larger home range. Prey is killed with bites to the nape and back of the skull or with a choke hold on the throat. If a kill is too large to eat in one feeding, the Bobcat will cache the remains. Sometimes they bury their kill if it is small, but most often they scrape dirt, snow, debris, and twigs over the remains and return to feed later. Typically, Bobcats rest near the cache to protect it.

During cold temperatures Bobcats sun themselves to gather solar heat, and they attempt to find resting and den sites that are sheltered from the wind. Adult Bobcats create scent marks by rubbing their mouth glands on objects, by depositing scat and spraying urine in prominent locations, and by digging scrapes upon which they may deposit urine and faeces. Anal gland secretions are dark brown in adult males and light yellow in adult females. Each scat produced is probably marked with this secretion as it is expressed from the anus, but the way in which these glands are used in scent communication is not well understood. Marking behaviours create visual or olfactory signs, or both, to signal home-range occupancy, den-site occupancy, sex, and reproductive condition. Fighting occurs occasionally, usually between males and during the breeding season when testosterone levels are especially high. Likely the males are competing for the attentions of an oestrous female. Such fights can be severe, causing injury and even death.

VOCALIZATIONS

Like most of the felines, Bobcats purr when they are contented; they emit various forms of "meow" calls (yowls, screams, squalls) to express varying levels of frustration, attraction, and demand; and they hiss, spit, snarl, and growl during agonistic encounters. A female that is caterwauling to attract a mate is considered to be producing a variant of the meow call. This call is easily heard by humans over a kilometre away on a calm night and is no doubt detectable by a male Bobcat from considerably further.

SIGNS

Secretive and stealthy, Bobcats are rarely seen. The best way to know whether they are around is to look for sign. Their front feet have five toes, but only four normally touch the ground. The fifth is high up on the inside of the foot and is used only to climb trees or grasp prey. The hind feet have four toes. Each of the eighteen toes has a retractile claw. These claws may be extended to ensure good traction when the animal is running or pouncing, but are not used during normal, slower locomotion. The front track is larger than the hind. Impressions of toe pads are teardrop shaped. The rear margin of both the front and the hind palm pads is trilobed and most noticeable on the hind foot. The same pad is bilobed on the leading edge. Palm pads can be measured in a crisp track to help distinguish between Cougar tracks and Bobcat tracks. The pads of Cougars are 4.0–7.3 cm in width, while those of Bobcats are 2.5–4.0 cm in width. Bobcats rarely run unless they are chasing prey or fleeing. Their normal pace is either a direct register walk (where the hind track on the same side registers directly over that of the front) or an overstep walk (where the hind track registers beyond the front track). They will occasionally break into a direct register trot for short distances. The direct register walk is the most common snow trail. Sit-downs are common features of Bobcat trails. The animals sit or crouch to rest and look or to listen for prey. If they remain still for longer periods, they usually produce a hunting lay. A Bobcat's hunting lay or lookout is the place in which it crouches in ambush to survey a game trail, thicket, or swamp margin. Hunting lays may be long and narrow or more widely packed down, depending on how long the cat stayed and whether it rotated. The cat rotates to get a different view. If a lookout is left in snow, the front feet leave imprints around the edge of the packed area, and frozen hairs may be found embedded in icy snow that was warmed by the cat's body heat.

Bobcats, like most other felids, like to scratch and stretch against tree roots and trunks. They rise on their hind legs and reach up with their front claws before kneading the wood and dragging their claws down. Bobcats avoid deep snow by spending more time under dense vegetation where the snow is shallower and by making use of game trails, ploughed roads, and snowmobile trails to get around. Scrapes are small mounds of dirt, debris, leaves, or twigs that a Bobcat creates with rearward digs of its hind feet. These may be simple swipes of the hind feet or more diligently produced digs that result in deep troughs. The parallel grooves are 15–51 cm in length and 7.5–19.0 cm in width. Scrapes often are squirted with urine, and sometimes scat is deposited on them as well. Resting beds measure 38–56 cm in length and 28–43 cm in width. Scat is a commonly used visual and scent mark. The cats form latrines of uncovered scat at prominent locations, such as the intersection of game trails, around a natal den site, or on a ledge favoured for resting and sunbathing.

Scat is typically dark blackish and of variable size and shape; it may be blunt ended and tubular or pointed and drawn out at one or both ends. Scat measures 1.0–2.5 cm in diameter and 7.5–23.0 cm in length. Bobcats at a deer kill typically begin eating at the rump near the spine, although in smaller prey they usually eat the brain first. When feeding on a deer, they bite away the hair to avoid eating it, and this discarded hair is frequently mixed with the debris that the cat drags over the kill to cover it, or it is left windblown around the carcass. A characteristic of Bobcat feeding is the amount of hair strewn around the carcass and the lack of broken long bones. Smaller animals like squirrels are often totally consumed, except for the tail and perhaps the stomach.

REFERENCES

Anderson, E.M., and Lovallo 2003; Benson et al. 2004; Elbroch 2003; Gipson and Kamler 2002; Hansen, K., 2007; Jacques and Jenks 2008; Janečka et al. 2007; Krebs et al. 2003; Larivière and Walton 1997; Lovallo and Anderson 1995, 1996; Lovallo et al. 1993; Matlack and Evans 1992; Rezendes 1992; Smith, H.C., 1993; Tischendorf and McAlpine 1995; Wigginton and Dobson 1999; Woelfl and Woelfl 1994b.

Cougar

also called Mountain Lion, Puma, Catamount, Panther

FRENCH NAME: **Couguar**

SCIENTIFIC NAME: *Puma concolor*, formerly *Felis concolor*

Cougars occupy the greatest latitudinal range (around 110 degrees) of any non-migratory terrestrial vertebrate besides humans, occurring from northern British Columbia to southern Argentina. The Cougar is the largest wild feline in Canada and the northern United States. The Jaguar, a larger feline, will occasionally reach the southern United States from Central and South America.

Cougar subspecies were reviewed in a paper published by Culver et al. in 2000. They concluded that, rather than there being 15 subspecies as previously recognized in North and Central America, all of the animals north of Nicaragua represent a single subspecies because they cannot be reliably separated by DNA or morphology. This widespread subspecies is called *Puma concolor couguar*, using the subspecies name formerly attached to the Eastern Cougar population. They further recognized five subspecies from Central and South America. This taxonomy is followed herein.

DESCRIPTION

Cougars and African Lions are the only large cats with relatively uniformly coloured coats (apart from melanistic Jaguars and Leopards). Some Cougars are more silver grey, greyish-brown, or reddish, but most are tawny brown on their back, sides, tail, and legs,

Overstep walk

Direct register walk

Left hind track
Length: 4.0–6.5 cm
Width: 3.0–7.0 cm

Left front track
Length: 4.0–6.5 cm
Width: 3.5–7.0 cm

Bobcat

Cougar (*Puma concolor*)

adult

cub

with a buffy, white, or cream-coloured underside. These slight colour variations can occur within the same population, and the basic coat colour can vary even among siblings. The backs of the ears and often the tail tip are darker brown or sometimes black, as is a patch on the upper lip where the whiskers insert. The gums are black, but the chin, throat, and front of the upper lips are white. Faint horizontal dark stripes may persist on the inside of the upper forelegs of younger adults beyond three years old. Although melanism (dark brown or black) is widely reported, dark specimens in collections are extremely rare, as are white or albino animals. It is likely that most of the large, dark cats reported in North America are animals that appear dark only as a consequence of lighting, rather than actual colour. The long, cylindrical tail is around one-third of the Cougar's total length and is normally carried drooping down close to the ground with the tip curling slightly upwards. Cougars have prominent, rounded ears and a relatively small head. Their feet appear to be oversized, and each toe has a sharp, retractile claw. The claw on the pollex (the front dewclaw) is the largest of all the claws, and although it cannot reach the ground, it is used for combat and for grasping prey. The hind legs are, proportionally, the longest of the cat family.

Adults of both sexes are similar in appearance, but males are significantly larger. The young have darker facial markings, often dark rings around the tail, and are marked with three irregular rows of black spots on their sides that remain distinct until they are three to four months old, when the spots rapidly fade. Their background coat colour is reddish-brown to greyish-brown. The spots have usually totally faded by the time the animal is around two years old, but some light-brown dapples may persist on the legs of some individuals until they are up to 30 months old. Cubs have blue eyes, which at 12–14 months old turn to greyish-brown or golden. A Cougar's large eyes have round pupils. Animals, in the parts of their range where there is a cold season, grow a long winter coat with fine, thick underfur. This is moulted in spring and replaced with a thinner, shorter summer coat. Despite the size differences between the sexes, it can be very difficult to ascribe gender in the field. Adults can sometimes be sexed when they are treed. Males have a black spot around the penile opening that is about 12 cm forward of the anus. A female's vulva is directly below the anus and often remains hidden by the tail. The dental formula is incisors 3/3, canines 1/1, premolars 3/2, and molars 1/1, for a total of 30 teeth. Molars in the upper toothrows may occasionally be highly reduced or even missing. Canines and carnassials are especially robust. Eyeshine is bright yellow.

SIMILAR SPECIES

There are no other large North American native cats that are uniformly coloured with a long tail and with hindquarters that are noticeably higher than the shoulders.

SIZE

Adult males are an average of 50% heavier and 20% longer than adult females. Cougars closer to the poles tend to be larger, and the smallest are near the equator. The following measurements apply to adult animals in the Canadian part of the range.

Males: total length: 201–230 cm; tail length: 73–85 cm; hind foot length: 25.4–29.0 cm; ear length: 7.0–9.2 cm; weight: 55–82 kg, an exceptional individual weighed 120 kg; shoulder height: 56.0–78.7 cm.

Females: total length: 182–206 cm; tail length: 60–82 cm; hind foot length: 22–27 cm; ear length: 6.4–9.2 cm; weight: 32–52 kg; shoulder height: 53.4–76.2 cm.

Newborns: weight: 400–500 g.

Distribution of the Cougar (*Puma concolor*)

current range

historical range

★ recent confirmed records of specimens or physical evidence

RANGE

These large cats formerly ranged from northern British Columbia to the Maritimes (except Prince Edward Island, Newfoundland, and Labrador) and southwards to southern Argentina and Chile. Hunting, poisoning, agriculture, logging, and human settlement have pushed them into mountainous or remote regions on both continents. In North America many populations in the east and mid-west are now isolated in pockets of appropriate habitat, and the largest contiguous portion of distribution is in the far west along the Rocky Mountain chain. The current North American range is about 50% of the historical range. At the peripheries of the range unsubstantiated sightings are commonly reported, but irrefutable physical evidence (specimens, tracks) is rare. For example, Cougars have been sporadically sighted in the southern half of the Yukon for decades, where they are often associated with sightings of Mule Deer, a natural Cougar prey species. The only specimen so far discovered was

found dead at Watson Lake in 2000, but successful breeding in the territory has not yet been documented.

Cougars are frequently displayed in game parks and road-side zoos, and they occasionally escape. Cubs, both wild and captive born, are sought after for the pet trade, especially in the United States and possibly southern Ontario and Quebec. These animals often escape or are intentionally released when they become too large and difficult to handle. Most of the previously captive Cougars are likely sterile owing to surgical neutering, probably done to improve their temperament, but if they are still fertile and can find a mate, there are no reproductive barriers to prevent breeding of such animals. It is probable that most, if not all, of the Cougar sightings in eastern North America (indicated with stars in the accompanying map) are of such former captives or their descendants. For example, the young male cougar shot in 1992 in the Abitibi region of Quebec, called the Abitibi Cougar, has proved to have DNA matching that of Cougars from South America, probably Chile, so it was likely a released or escaped captive.

ABUNDANCE

These cats are both shy and elusive and are rarely seen even when they are around. As a sometime predator of domestic livestock and a competitor of sport hunters, Cougars were shot on sight throughout North America and even killed for bounty in British Columbia and Alberta until the mid-1900s. Hunting is currently regulated in those two provinces and is prohibited in Saskatchewan, Manitoba, Ontario, New Brunswick, and Nova Scotia. Trapping for their pelts has never been economically viable and is not permitted in any North American jurisdiction. A previously recognized North American subspecies, the Eastern Cougar, was believed to be extinct by the mid-1900s, although persistent reports continue to suggest that a small population may survive in southwestern New Brunswick. This densely wooded region could support a small Cougar population, but it is unlikely to be a remnant of the so-called Eastern Cougar; rather, it could be derived from established exotic Cougars. This origin likely applies to any Cougars reported from the Maritimes, eastern Canada, and the eastern United States. The Florida Panther, a population formerly considered to be a subspecies (*Puma concolor coryi*), is the only North American population considered to be "Endangered." It has been protected by the United States Endangered Species Act since 1973. There are so few animals left of this dark-reddish group (30–50 individuals) that its extinction within the next two to four decades has been predicted. The major threats to their survival are clearcutting and draining of the Everglades, pollution, sport hunting, and vehicle collisions. Recent captive breeding efforts may prove beneficial. In 1976 there were an estimated 16,000 Cougars in Canada and the United States. The same population was estimated in 1997 at 29,000–30,000 animals. Roughly 20,000 Cougars were additionally thought to live in South America around the same time.

Cougars appear to be moving back into parts of their former historical range in some jurisdictions, such as Manitoba, Alberta, Texas, North Dakota, and South Dakota. They have also been increasingly infiltrating the northern peripheries of their range, even beyond the limits of their historical range, although possibly not with breeding populations. Cougars in the western states and provinces have been slowly, but steadily, increasing in numbers since the enactment of protective legislation in the 1960s and the expansion of prey populations. Cougars in some parts of the Pacific Northwest are defying this trend, and populations in Washington, Idaho, and southern British Columbia appear to be declining, despite increasing reports of human-Cougar interactions. Cougars in Central America are considered to be uncommon to rare, as are Cougars in many parts of South America. Primary threats to Cougars are human activities such as habitat fragmentation, logging, urbanization, and reduced prey resources owing to sport hunting.

ECOLOGY

Although it is possible to observe them in almost any habitat from sea level to 4000 m in North America, and up to 5000 m in the Andes of South America, Cougars are most commonly found in remote areas where the vegetative cover is dense or where the topography provides cover (such as in rocky, mountainous regions). Forests, rain forests, swamps, desert scrub, grasslands, and woodlands are all possible Cougar habitat. These large muscular and agile cats are highly competent tree climbers and can easily surmount rocky terrain in order to pull down animals larger than themselves. They are reluctant but capable swimmers. Cougars may be active at any time of day or night but are most active during low light levels (crepuscular and nocturnal). They live at low densities of 0.9–4.9 animals per 100 km², and there are typically two adult females to every adult male. The home ranges of these large cats vary by season and from year to year and are mainly dependant on prey availability, topography, and the presence of other Cougars. Most home ranges are largest in summer and autumn, and smallest in winter and spring when snow restricts movement. Some cats migrate between a winter range and a summer range, but most summer ranges are simply an enlargement of the winter range. Adult male Cougars are territorial, and their home ranges or territories usually do not to overlap. Adult females are not territorial, and their home ranges commonly overlap, sometimes wholly, with other females or with males. Nevertheless, the individual animals remain solitary, except when breeding. Males enhance their reproductive success by occupying larger home ranges than do females, typically 1.5–5.0 times larger. The home ranges of adult males vary from 140 to 1826 km² but average around 280 km². Adult females occupy 26–1717 km², but their home range averages around 140 km². Each male's home range typically overlaps those of several females. Female ranges, while smaller, must have suitable den locations and sufficient cover and especially food to raise their cubs. Maternal den sites may be in very dense brush, caves, overhangs, or natural spaces in rock crevices and may be reused. Non-maternal adults commonly rest in caves, under overhangs, in tree falls, or on tree branches.

Dispersing juveniles may travel considerable distances to find a home range that is not already occupied. Straight-line dispersal of occasional, exceptional cases may exceed 1000 km (a radio-collared female in Utah travelled over 1300 km during a year-long dispersal), but normal distances are in the 100–200 km range. Young males tend to disperse two to four times further than do young females. In regions where food is plentiful, dispersing females may settle adjacent to their mother's home range. Dispersal appears to be precipitated by the maternal female when she abandons the cub or prevents it from following her. Cougars can live longer than twenty years in captivity, but few wild individuals survive more than twelve years. As a summit predator, Cougars are not ordinarily subject to predation except as youngsters. Grey Wolves and Coyotes will attempt to kill any undefended cubs, and a pack of Grey Wolves is known to have killed an adult Cougar. Adult male Cougars will kill cubs. The major cause of death of North American Cougars is shooting by humans. These animals are killed by sport hunters, poachers, or protectors of livestock. Another significant cause of mortality is collision with motor vehicles, which may be the primary cause of death in fragmented populations near dense human habitations. Natural causes of death include snake bites, rockslides, lightning strikes, injuries brought about by prey or other Cougars, and post-partum complications. From 1890 to 2001 there were 98 attacks by Cougars on humans in North America, of which 17 were fatal. Cougar attacks have increased in the last few decades as people invade their space further and further. Vancouver Island has the dubious distinction of having the highest incidence of Cougar attacks on humans in the continent.

DIET

Highly carnivorous, Cougars are nevertheless very adaptable as far as prey choice is concerned and will even eat plant material and insects if food is scarce. Opportunistic hunters, they tend to eat what is available, including small mammals such as rabbits, hares, pikas, squirrels, and even mice and voles, as well as medium-sized prey such as North American Porcupine, waterfowl, Wild Turkey, North American Beaver, grouse, marmot, American Badger, and Coyote. Domestic Cat and Dog will be taken if they are encountered. However, Cougars prefer to kill large mammals such as Moose, Wapiti, Mountain Goat, Bighorn Sheep, and especially Mule and White-tailed Deer. Cougars kill proportionately more old, infirm, and young deer than are found in the population as a whole, likely because they are weaker and easier to catch. Domestic livestock may be predated in some regions, especially if regular wild prey is scarce. Typically, inexperienced juveniles resort to killing livestock. Cougars may sometimes become cannibalistic; large males will kill (and sometimes consume) cubs, females, and at times other males. Cougars will sometimes scavenge dead Cougars that they encounter, including their own relatives. An adult Cougar can hold up to 10 kg of meat in its stomach. On average, Cougars fast three days out of nine. Although they will return night after night to a kill, they rarely feed on a carcass that was not killed by themselves or another Cougar, unless they are starving. Cougars prefer to kill large prey but rarely take animals larger than 200–225 kg, such as mature cattle and horses; however, they readily hunt juveniles (foals, calves, and yearlings). Assuming a diet of only deer (Mule or White-tailed Deer), which is highly unlikely, an adult male would require 35 deer annually to survive, a non-maternal, adult female would need 25, and a maternal female with two to three yearling cubs would need 50–55. The actual rate of predation is likely higher in hot climates,

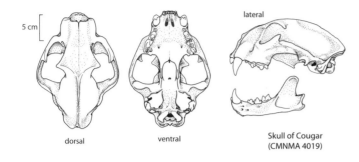

5 cm

dorsal   ventral   lateral

Skull of Cougar
(CMNMA 4019)

where spoilage is more rapid. Water is drunk from surface sources to supplement what is consumed in the tissues of prey.

REPRODUCTION

Although Cougars are polygamous, it is common for the same animals to mate year after year owing to the long-term stability of their home ranges and their lengthy lifespans. An adult male may breed with a number of different females that live within or around his territory, and a female may breed with more than one male during a single oestrous cycle. Multiple paternities in a single litter are theoretically possible. Females remain reproductively active to at least 12 years old, and males until they die. Although females may come into oestrus at any time of year, the majority of births occur between April and October in the Northern Hemisphere, except in the warmer parts of the Pacific Northwest, such as Vancouver Island, where births are equally as likely to occur in any month. Unmated females will cycle into another oestrus in about 21 days. Copulation is very brief, usually less than a minute, but may recur as frequently as nine times in an hour, and 50–70 times per day. Most females have a litter every 18–24 months. Gestation lasts 82–103 days (an average of 91 days). The litter size ranges from one to six, averaging two to three. Young Cougars are called cubs or kittens. Females rear their cubs without assistance from the sire. Cubs are born deaf and blind, with a full covering of densely spotted fur. The ears and eyes open within a few weeks. Milk teeth begin to emerge through the gums at 10–20 days old, and a full set is in place by about 50 days old. The permanent teeth start appearing around 5–6 months old and are fully emerged by 10–12 months old. Young are weaned by the time they are about three months old, although they may be brought to a carcass by their mother at as young as eight weeks old. Their distinct black spots fade rapidly when the cubs are 12–14 weeks old, about the time they start accompanying their mother on hunts. Young stay with their mother until they are one to two years old. Juveniles become sexually mature at around two years old but normally do not mate until they have established a home range. Cougars continue growing after sexual maturity, albeit very slowly, with females reaching physical maturity at around five to six years old, and males at around seven to nine years old.

BEHAVIOUR

Cougars are large, solitary predators. The only lengthy social bonds that occur are between a maternal female and her young (of one to two years old). A female in heat may be accompanied by an adult male for part of her oestrus (typically three to five days). A male will follow a female, intermittently attempting to mount. If she allows his approach, copulation will ensue; otherwise, she will hiss and snarl, and he will desist but continue to follow until she is receptive. Cougar society is divided into three classes. Transients are either young animals that have not yet established a home range or older adults that have been displaced. Residents are mature adults with well-established home ranges. Immatures are juveniles that are too young to leave their natal home range and still rely on their mother to feed them. Transients are the most likely to become problem cats because their inexperience and lack of hunting skills or their infirmities may lead them to prey on livestock or even humans. Residents are experienced hunters that usually focus on wild prey and tend to occupy the same home ranges for years. As is the habit of most cats, Cougars lick themselves clean with their rough tongue, ingesting some of their own hairs in the process. Cougars usually spring from ambush or stalk their prey, making a rush only when they are within 4–5 m. Large prey may be killed by a bite to the back of the neck or by a crushing bite to the throat. Successful hunters commonly cover a large kill with branches, debris, snow, or dirt and return to feed again in the following nights until it is all consumed, spoiled, or removed by other carnivores. A Cougar's normal nightly hunting pattern of travelling around 10 km in six bouts, averaging just over an hour each, is suspended while it feeds on a kill, and it will usually den-up at around 400 m (0–4200 m) from the cache site. Apart from the nights spent near a kill, Cougars rarely bed down in the same location on consecutive nights. Larder-hoarding of multiple carcasses at a single location has been occasionally reported. Territorial males patrol their home range and actively dispel male intruders. Such fights can mortally wound the combatants. The non-territorial adult females rarely fight another Cougar, except to defend cubs or possibly a kill. Annoyance or anger is expressed with a direct stare and flattened ears accompanied by a hiss or a growl. Cougars commonly resort to trees when pursued by dogs. This propensity is used by Cougar hunters and scientists alike to get access to a Cougar in order to shoot it or attach a tag or radio collar.

VOCALIZATIONS

Cougars are largely silent but can occasionally be quite vocal and have a fairly extensive repertoire. A female in heat will produce a throaty yowl or caterwaul to attract a male. Males respond with similar calls. Other reported vocalizations include chirps, peeps, whistles, squeaks, mews, snarls, growls, screams, spits, and hisses. Cubs call their mother with a loud chirping whistle. Contentment is expressed by purring. Two Cougars that are familiar with each other may express a "mra" call that is similar, though louder and deeper, to one used by a Domestic Cat for the same purpose.

SIGNS

Cougars are rarely seen in the wild, even in areas where there is a relatively high density. Finding signs is usually the only way to determine their presence. The front feet have five toes, with the inside toe (the pollex or thumb) being the smallest (but having the largest claw) and being set high up the leg. It does not strike the ground and so never leaves an impression. There are four relatively equal-sized toes on the hind feet. The front feet are slightly larger, rounder, and more asymmetrical than are the hind feet. A sure indication of a Cougar track is the combination of size, the three lobes separated by two indentations at the hind margin of the palm pads of both front and hind feet, and the double-lobed leading edge of the same pad. These features may only show under the right conditions. Another useful characteristic of Cougar tracks is the curved ridge between the impressions of the toes and the metacarpal pad, rather than the more pointed pyramidal mound that is common in canines. Feline tracks are often wider than they are long, especially if the toes are splayed. The cats travel with their claws hyper-retracted, so claw impressions rarely occur, except if the animal is running or loses its footing. Cougars do not waste energy by running unless they are pursued or are themselves chasing prey. The normal travelling pace is a walk where the hind foot registers partially or wholly over the front track. Although the heavily furred feet of a large Canada Lynx in winter may create a track whose outside dimensions can rival and even exceed those of a Cougar, the Canada Lynx pads are small and surrounded by a great deal of negative space and hair; Cougar tracks show much larger pads, with considerably less negative space and hair between them. Furthermore, the lighter Canada Lynx does not sink as far into the snow as does the substantially heavier Cougar. To distinguish Bobcat from Cougar, one can measure the width of the metacarpal pad in a crisp front track. Male Cougar pads have a width of 5.0–7.3 cm, and female pads 4.1–4.7 cm, while Bobcat pads are 2.4–4.0 cm wide. In deep snow a Cougar may leave a tail drag mark, which easily separates it from the other two native cats. Bounding strides during a prey approach can reach 6–12 m, but the typical walking stride is 50–81 cm.

Like Domestic Cats, Cougars enjoy scratching and stretching on a favourite tree, typically a leaning tree or a large exposed root. These actions can be identified by the height of the scratches and by the shed claw sheaths (lamella) that often persist at the site. Males, and on rare occasions females, make scrapes in prominent locations, especially along home-range boundaries. These are small piles of debris and earth pushed up by the hind feet, which measure 15–46 cm in length, 15–30 cm in width, and 3–5 cm in depth. Each scrape has two parallel troughs leading to it, made by the hind feet. Scrapes may indicate the direction of travel because the debris mound is always at the rear. Scat, and especially urine, may be associated with a scrape, typically on the debris pile. Scat is usually voluminous and may be folded over on itself. It is typically tubular and segmented, sometimes twisted, and contains little plant material. Ends are blunt or, at best, pointed only at the end that was the last to exit. The contents of the scat vary with the food eaten. A scat produced after a fresh kill will contain amorphous dark-grey or blackish material mixed with only a little hair, is typically smooth and loose, and may be twisted. A scat produced a few days later may contain bone chips and hair and will be tighter and more segmented. An even later scat could be primarily hair and very segmented. Scat usually accumulates around a kill site as the hunter spends time in the area. Scat may also be found under overhangs or

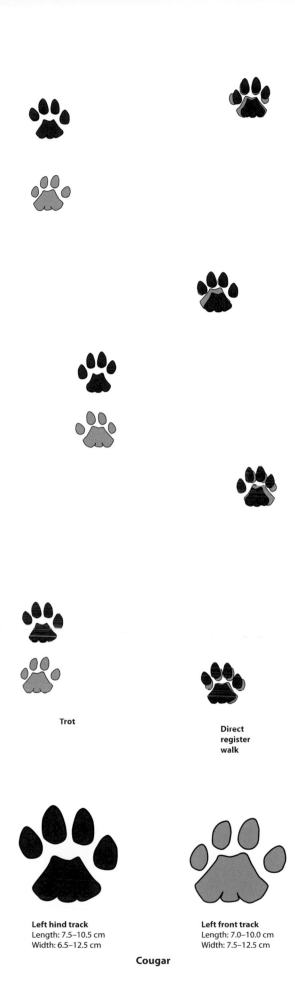

Trot

Direct register walk

Left hind track
Length: 7.5–10.5 cm
Width: 6.5–12.5 cm

Left front track
Length: 7.0–10.0 cm
Width: 7.5–12.5 cm

Cougar

in caves where the Cougar regularly rests. Cougar sometimes cover their scat but often leave it exposed as a scent post. A scrape to cover a scat is 50–90 cm in diameter. Scat is 2–4 cm in diameter and varies from 16.5 to 43.5 cm in length. Cougar beds measure 71–89 cm in length by 56–71 cm in width. Kill sites can be identified by a variety of sign. Broken necks are a common consequence of Cougar predation on large ungulates. If the bitten area on the neck of the prey is skinned back, the dual puncture marks of the large canine teeth should be visible, each having at least the diameter of a pencil. Drag marks are common, as the Cougar relocates its kill to an area with heavier cover where it can be cached and covered with debris, vegetation, and sometimes dirt or snow. The internal organs are often removed and may be consumed or cached separately nearby. Not all kills are covered. When the Cougar returns to feed, it may drag the kill to another spot and cache it again. The normal progression of the consumption of a Cougar kill over several days is first the internal organs (through an opening just behind the rib cage), then the shoulders and ribs, and finally the hindquarters. Viscera are normally discarded.

REFERENCES

Beier 1995, 1999; Bertrand et al. 2006; Boyd and Neale 1992; Bryant et al. 1996; Culver et al. 2000; Currier 1983; Elbroch 2003; Friesen 1999; Galentine and Swift 2007; Gau et al. 2001; Gay and Best 1996; Holt 1994; Hood and Neufeld 2004; Jung and Merchant 2005; Lambert et al. 2006; Logan and Sweanor 2000; Pierce and Bleich 2003; Pierce et al. 1999; Reid, F.A., 1997; Rezendes 1992; Ross and Jalkotzy 1992; Ross et al. 1995; Ross et al. 1997; Scott, F.W., 1998; Slough and Jung 2007; Smallwood 1993; Smith, H.C., 1993; Stoner et al. 2008; Thompson, M.J., and Stewart 1994; Tischendorf and Henderson 1994; Verts and Carraway 1998; Watkins 2005; Wilson, S.F., et al. 2004; Zakreski 1997.

FAMILY CANIDAE
dogs

Dogs have thin, often long legs, bushy tails, long muzzles, erect ears, and blunt, non-retractable claws. Most have 42 teeth. They are digitigrade (walk on their toes) and generally use speed to catch their prey, having detected it with their well-developed senses of smell and hearing.

Coyote
also called Brush Wolf, Prairie Wolf

FRENCH NAME: **Coyote**
SCIENTIFIC NAME: *Canis latrans*

Despite centuries of persecution and trapping, Coyotes are one of the few modern mammals whose range continues to expand. The name is typically pronounced *ky-OAT-ee* in the east and *KY-oat* or *KY-oot* in the west.

DESCRIPTION

Coyotes are lanky canids with long narrow muzzles, large erect triangular ears, and bushy tails. They are about the size of a Border Collie. Colour is tremendously variable in this species, both among individuals and between geographical populations. Coyotes from dry regions are generally more rufous or yellowish-grey, while forest and northern animals are usually darker and greyer and tend to have longer, coarser hair. Rufous sides, backs of ears, and outer legs are common in many eastern populations. In general, Canadian Coyotes have a grizzled grey or black back, and the fur colour on their sides ranges from rufous to yellowish-brown to grey. The belly and throat are normally paler than the rest of the body, and most individuals have a darker saddle over the shoulders and a somewhat darker tail tip of variable size. Blond and black animals do occur but are rare in Canada. Coyotes carry their tail low when running, usually holding it straight and at a downward angle. An elongate, dark patch of fur on the upper surface of the tail near the base covers the tail gland. Coyotes have yellow eyes with round pupils. Their eyesight and hearing are keen, and their sense of smell is exceptional. The dental formula is incisors 3/3, canines 1/1, premolars 4/4, and molars 2/3, for a total of 42 teeth. Fertile hybrids can occur between Grey Wolves, Red Wolves, and Domestic Dogs. Coyotes and Grey Wolves are known to hybridize from Minnesota east through southern Ontario and eastern Quebec to the extent that up to 100% of

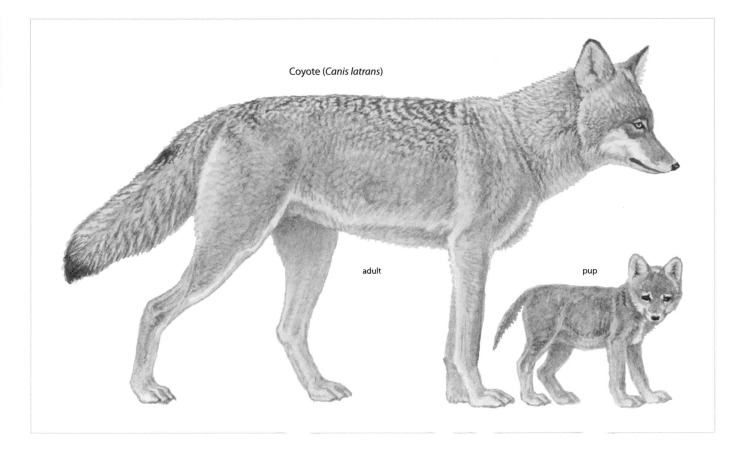

Coyote (*Canis latrans*)

adult

pup

wolves in some parts of this region (such as in central Ontario in the Algonquin Park area) have mitochondrial DNA genetic markers that are distinctive only to Coyotes. Interestingly, no Coyotes have been discovered with Wolf genetic markers, suggesting that, since mtDNA is solely transmitted by the female, the crosses have been exclusively the result of female Wolves mating with male Coyotes, followed by backcrosses to Wolves. The eyeshine is greenish-gold. Coyotes experience a prolonged moult that lasts from spring to autumn. As the old fur falls out, the new fur grows very slowly, such that the summer coat is shorter and thinner than the winter coat, but the colour remains the same.

SIMILAR SPECIES

In general, Coyotes run with their tail angled downward, wolves with their tails straight out behind, and Domestic Dogs anywhere from down to high over their backs. Wolves are generally larger than Coyotes, although this distinction can be difficult to determine in the field with no comparison to provide scale. The hind tracks of wolves are larger than those of the largest Coyotes. The skulls of Coyotes and Domestic Dogs can be distinguished by the ratio of palatal width (between the inner margins of the upper first molars) to toothrow length (from the anterior margin of the first premolar to the posterior margin of the last molar) with about 95% confidence. If the toothrow is ≥ 3.1 times the palatal width, the specimen is a Coyote. If the same ratio is < 2.7, the specimen is a Domestic Dog. The skulls of Grey Wolves can be distinguished from those of Coyotes by their greater zygomatic breadth and skull length, longer

more massive canines, and a relatively smaller braincase. Howling wolves produce a deeper call than do Coyotes.

SIZE

The following measurements and comments apply to animals of reproductive age from the Canadian part of the range. Size varies with geographic location. Coyotes in the northeast are larger (15%–20% heavier) than those in the west, and northern animals across the range are larger than more southerly individuals. Within a population, males are larger and around 10% heavier than females. Weight fluctuates considerably throughout the year depending on food availability and season and tends to be least in summer.

Males: total length: 1130–1600 mm; tail length: 295–460 mm; hind foot length: 180–220 mm; ear length: 55–67 mm; weight: 11.3–23.0 kg (maximum known wild weight is 33.9 kg).

Females: total length: 1093–1570 mm; tail length: 290–430 mm; hind foot length: 169–203 mm; ear length: 55–67 mm; weight: 11.3–17.9 kg (maximum known wild weight is 25.1 kg).

Newborns: weight: 225–275 g.

Adult Coyotes stand 50–66 cm high at the shoulder.

RANGE

Coyotes occur only in North America, where they are widespread from northern Panama to Alaska, Yukon, Nunavut, and the Northwest Territories, and their range continues to expand. Although some range expansion may have been the result of introductions by humans, notably in Florida and Georgia and perhaps parts of New

Distribution of the Coyote (*Canis latrans*)

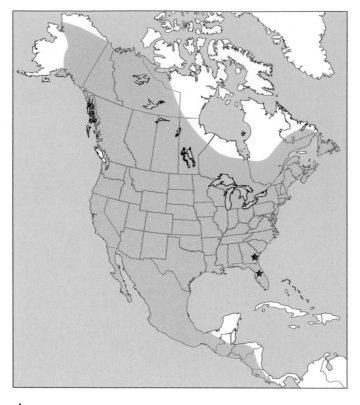

⭐ introduced

England, the remainder appears to have been unassisted. At the onset of European settlement in the New World, Coyotes were restricted to the aspen parklands and prairie grasslands of the Great Plains of central North America from the Canadian Prairies to central Mexico. In less than 200 years they have colonized much of the rest of the continent, including most of eastern North America, likely initially in response to the clearing of land by deforestation for logging and agriculture and later by the virtual decimation of the Grey Wolf, a predator and major competitor. These same habitat alterations also encouraged the range expansion of White-tailed Deer, providing prey for the Coyote in their expanded range, especially in the east. Coyotes arrived in Quebec in 1944, New Brunswick in the late 1950s, Nova Scotia in the 1970s, and Prince Edward Island in 1983 (long before the Confederation Bridge linked the island to the mainland). They were first documented on Newfoundland in 1985, coming ashore at the Port au Port Peninsula after crossing pack ice, possibly from Cape Breton. They arrived in Labrador in 1995 and the Magdalen Islands in 2001. Clearly their ability to cross ice bridges supports their range expansion. The first Coyote was reported in Yukon in 1907, near Whitehorse. Coyotes have not yet reached Vancouver Island, Anticosti Island, or the Queen Charlotte Islands.

ABUNDANCE

Coyotes are abundant throughout much of their range and are increasing in both distribution and numbers as humans continue to make environmental modifications, many of which encourage their increase. Coyotes remain scarce and only occasional sightings are reported in the far north (north of 65° latitude). Most North American jurisdictions permit licensed hunting and trapping of Coyotes, and some treat them as a nuisance species and allow hunting at any time of the year without the need of a licence.

ECOLOGY

This highly versatile canid is able to exploit a very wide range of habitats from prairie grassland to forest and from desert to mountains and tropical jungles. They readily occupy and can thrive in urban and rural environments that have been heavily modified by humans, but they do best in shrubby grasslands that support a diversity of habitats, including small woodlots, ravines, and thickets. Densities in Canada vary from 0.01 to 0.6 per km² in winter and are generally higher in summer as juveniles are recruited into the population.

In the absence of Grey Wolves and Cougars, Coyotes are now the top terrestrial predator in large parts of North America. They are adaptable in their hunting strategies, depending on the prey. Small mammals are pursued by solitary individuals and found largely by sound. The Coyote listens, determines the prey's location, and then pounces. Larger prey such as White-tailed Deer may be hunted in packs, but a single Coyote can take down an adult ungulate depending on its condition and the circumstances. For example, large prey weakened by nutritional stress or bogged down in deep snow is more likely to be killed by a single Coyote than is a healthy buck with antlers that it is willing to use or than is a female that is rigorously defending a fawn. The alpha pair leads and executes the kill if a pack is involved. Coyotes will cooperatively hunt prey. Sometimes one Coyote will chase a deer while the other waits; then the rested Coyote will take up the pursuit. This continues until the deer escapes or is exhausted and easier to kill. Coyotes have been clocked running at 48 km/h and can easily maintain a speed of 30 km/h for extended periods. A night of foraging may see them travel 5–30 km depending on the season and food availability. They are also very capable swimmers. Mainly active during low-light conditions around dusk and dawn, Coyotes may also be active throughout the day or night. They often adapt to high levels of hunting harassment by restricting their activity to the dark hours.

Coyotes remain active throughout the winter, although they may den-up during severe storms. Their home-range size is exceedingly variable and depends on the territorial behaviour of the alpha pair, food availability, and terrain. Across the country, home-range size has been estimated from around 10 km² to 190 km². Fidelity to the home range is strong and often persists for the animal's lifetime, although loss of a partner can result in a boundary shift, and loss of a neighbouring pack can result in a home-range expansion into that vacant territory.

Pups are usually born in underground dens or sometimes in hollow logs. Most dens are located on slopes or mounds or under overhangs. Although Coyotes mostly dig their own dens, they will sometimes modify a burrow that has been dug by another animal such as a fox, skunk, or Woodchuck. Favoured den sites are used for years. Once the young are large enough to follow their parents on the hunt, den use drops off and the pack often sleeps above ground, usually in a sheltered site such as a thicket or hollow or under an overhang, a fallen tree, or the spreading boughs of an evergreen. The family group breaks up in early autumn, and the juveniles hunt alone for small prey items or carrion during the winter. Mated pairs remain on their territory throughout the year.

Dispersal of the juveniles, yearlings, and non-breeding adults of subordinate status can occur at any time of the year but peaks in autumn and again during winter, just before the breeding season. The animals go looking for a mate and a territory and often travel more than 100 km, although many go no further than 2–4 km. Some juveniles, particularly females, may stay on their natal territories and assist their parents with the next litter. Coyotes are tireless and canny travellers that are capable of dispersal over long distances across substantial topographic obstacles such as mountain ranges, rivers, and sea ice. Even heavily urbanized regions appear to create no barriers to the dispersal of these highly mobile canids. The maximum reported dispersal distance is 440 km. Their wanderlust is likely one of the main reasons they have spread so quickly across eastern North America.

Like all mammals, Coyotes can contract and transmit rabies. They can also be afflicted with Canine Heartworm, tularaemia, and bubonic plague. Coyotes have few predators, but Grey Wolves and Cougars will kill them if they can, and Golden Eagles and American Badgers are known to take the occasional pup. Human hunting and trapping is probably the greatest cause of death. Starvation and roadkill may cause substantial mortality, especially among dispersing individuals. In captivity, Coyotes are known to live up to 21 years, but maximum longevity in the wild is 15.5 years, and few live beyond 6–8 years.

DIET

Although primarily carnivorous, Coyotes are both opportunistic and willing to consume almost anything that is digestible including insects, bird eggs, fruit, pet food, garbage, and carrion. Their primary prey includes small rodents (chiefly voles, mice, and squirrels), lagomorphs, and deer, but they will also kill birds, reptiles, amphibians, crustaceans, domestic pets (especially Cats and small Dogs), livestock, poultry, and smaller wild carnivores such as foxes, weasels, Bobcats, and Northern Raccoons. Most livestock are eaten as carrion, but Coyotes can take down and kill large wild and domestic ungulates. Fawns and calves of native ungulates are often particularly targeted. Coyotes are major predators of Domestic Sheep and their lambs.

REPRODUCTION

Canadian Coyotes mate between February and April, but precopulatory behaviour begins two to three months earlier. Females experience a single heat annually that lasts about two to five days. The percentage of females that breed varies, based on food availability and their individual physical condition. Typically 60%–90% of adult females and 0%–70% of yearlings produce a litter in any given year. Copulation terminates in a copulatory tie (similar to that of Domestic Dogs) that typically lasts 15–25 minutes. After a 58–63

day gestation period, one to ten pups are born, typically in late April or May. An average litter contains six pups. The litter size is directly related to the mother's nutritional status, and well-fed females have larger litters. The pups are born blind and helpless, with a fuzzy grey-brown coat. Their eyes open around 10–12 days after their birth, and their ears become erect around the same age. The pups begin to appear at the mouth of the den when they are about three weeks old, at which time they start to eat solid food (which has been regurgitated by parents and sometimes older siblings). Weaning occurs gradually and is usually complete by the time the pups are six to eight weeks old. Soon afterwards they are able to follow adults to large prey or carrion. Between the ages of eight and twelve weeks the pups follow their foraging parents, most commonly their mother, learning the skills that they will need in order to hunt for themselves. Juveniles are adult sized by the age of nine to ten months and may breed as yearlings if they are well nourished, but most commence breeding as two-year-olds.

BEHAVIOUR

The basic Coyote social unit is a pack consisting of two breeding adults, called the alpha pair, that share territorial maintenance. These animals may remain together for years, but not necessarily for life, although many do. Additional subordinate adults may join the dominant pair and their litter. These associate individuals are invariably offspring (mainly female) from earlier litters, and may eventually displace or replace one of the alpha pair. In areas where large ungulates are the main prey, Coyote packs can include up to ten individuals. If small prey is predominant, packs normally include just the alpha pair, their most recent litter, and sometimes one additional adult or subadult animal. Within each pack a dominance hierarchy is established, centred on the dominant pair, with the alpha male being the dominant individual and the alpha female the second most dominant. Normally only alpha females breed, although this can change during years of abundant food. Reports of excessively large litters may be cases in which multiple females share a male and produce their litters in the same den. A small percentage of Coyotes, either male or female, become nomadic individuals (called transients) that wander alone over large areas. They skirt occupied territories but stand ready to move in should a vacancy occur. Although single animals can establish and maintain a territory, pairs or packs hold most territories, using scent marking (urine and faeces) and vocal communication (mostly howling) to establish

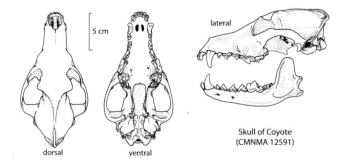

5 cm

lateral

dorsal   ventral

Skull of Coyote
(CMNMA 12591)

and maintain their boundaries, and using direct aggressive behaviour to enforce them. Typically the alpha individuals, especially the males, do most of the scent marking and territorial defence, often with the assistance of other pack members. They accost invading animals at a stiff-legged walk with head and tail held high, neck arched, and the fur on the nape erected. They snarl and stare with narrowed eyes and an open mouth that exposes their teeth. Physical battles are uncommon and rarely bloody, as most non-resident intruders beat a hasty retreat when confronted. The same aggressive postural display is used by dominant animals towards subordinate members of their own pack. The subordinates respond by crouching or rolling on their backs with tail wagging or curled under, ears flattened, and lips pulled back in a "grin." They may whine and urinate in submission. This submissive display usually preserves them from further aggression because it inhibits the dominant animal from attacking. Howling serves to advertise the presence and location of alpha animals that are willing to defend the territory, and plays a role in spacing packs across the landscape. It may also be used to coordinate social activities such as a group hunt. Only resident Coyotes howl.

Like many other carnivores, Coyotes kill more prey than they can eat when it is abundantly available. The excess is cached for future use by being buried in earth or snow, or, if it is composed of prey that is too large to bury, the carcass or carcasses will be visited repeatedly until consumed. Single Coyotes sometimes make hunting associations with other predators such as American Badgers and Grey Wolves. This temporary teamwork probably enhances both species' likelihood of catching prey.

All of the adults in a pack provide care to pups, delivering prey, grooming, guarding, and sometimes playing with them. Play is an important component of the muscular, neurological, intellectual, and social development of pups and one that they practise frequently, with every indication of enjoyment. Pre-copulatory behaviour includes enhanced scent marking by the male, increased play activity between the pair, increased genital sniffing by the male, and persistent following of the female by the male.

Most wild Coyotes are very wary and avoid humans. Attacks are rare and typically fall into two categories: unprovoked attacks by rabid animals, and assaults on unattended small children. Aggression and even attacks directed towards lone adult humans have been reported but are very rare. One fatality of an adult human has been attributed to Coyote predation. Attacks by healthy Coyotes are invariably associated with animals that have lost their fear of humans typically as a result of being fed, such as at garbage dumps or campgrounds or even at the pet food bowl on the back porch. Coyotes are elusive and difficult to capture because they are wary of people and avoid novel items in their environment (such as traps).

VOCALIZATIONS
Coyotes bark, wail, growl, whine, whimper, and squeal, but their most characteristic vocalization is their yelping or yelping and howling call that is most commonly heard at dusk and dawn. Howling rates are influenced by season; they are high during the breeding season and again in autumn when the juveniles disperse, but they are negatively affected by increased levels of moonlight (the opposite of prevailing folklore). Often the howling of one animal will trigger other nearby individuals' joining in, and neighbouring packs may reply. A siren, a barking dog, or even a thunderclap can stimulate a Coyote to howl. The howl starts with a series of high-pitched yips, followed by a prolonged howl, and may culminate in several sharp yaps. This characteristic call is the most common indicator that Coyotes are in the neighbourhood, although Coyotes in the east howl considerably less often than do those in the west.

SIGNS
Den entrances are 30–45 cm in diameter. Most dens have several entrances, and the main entrance will likely have a large throw mound of waste soil that has been kicked to the side where scat and tracks commonly accumulate. Since adults often bed down at the den entrance, hairs may also be evident. Coyotes typically travel at a ground-eating trot and often follow game trails, paths, snowmobile trails, and minor roadways. They avoid deep snow by using the ice of rivers and lakes and by selecting areas where snow depth is shallowest or the snow is packed down. Ordinarily they walk or trot when travelling in soft snow, but they may bound for short distances. Coyotes will tear at their prey during the chase, grabbing the neck, head, and flanks to slow the animal for a killing bite. A large ungulate kill site will have large patches of hide strewn around the carcass; any severing of the vertebral column will usually occur in the thoracic-lumbar region and sometimes just behind the skull; the nasal and maxillary bones will be chewed away; ribs, scapulae, and vertebrae will be chewed; ribs may be severed; and the limbs will be scattered, sometimes widely. Coyotes normally open a carcass from the rear and then proceed to the rib cage to remove and eat the prized, internal organs. The rumen is discarded early, and its location can often be used to determine the original kill site once the carcass or portions of it have been dragged around and scattered. The area around a fresh kill will probably have noticeable spots where the animals have rubbed their muzzle in the snow or on the vegetation to clean it. Wolves do the same. Domestic Dogs are typically less experienced hunters, and so their prey is often considerably more mutilated before it is finally dispatched. The dogs may consume none or very little of the meat because their pleasure is in the kill, not the eating. Truly feral dogs may be more experienced, and their kills can closely resemble those of Coyotes. Coyote faeces are highly variable. A meat diet produces dark, twisted, pointed scat that is 1.0–3.5 cm in diameter and 12.5–33.0 cm in length. Deer hair is a common component of scat, especially in winter in the east. A largely insect diet creates dark, crumbly scat in which fragments of insect chitin are readily visible. Fruit scat often takes on some of the fruit colours; for example, a diet of apples creates a pale to dark brown scat, while raspberries can produce a reddish scat. The components of a fruit scat are usually readily apparent because Coyotes digest vegetation poorly; therefore, the scat is replete with seeds and undigested portions of the fruit. Scat that is mostly fruit may be loose and amorphous or tubular with blunt ends. Coyotes use their faeces to mark their territory, so scat is commonly deposited in open areas where it is highly visible and may form small piles over

Trot

Direct register walk

Left hind track
Length: 5.5–7.5 cm
Width: 4.0–6.0 cm

Left front track
Length: 5.7–8.5 cm
Width: 3.8–6.5 cm

(these track measurements apply to western Coyotes; those in the east are about 10% larger)

Coyote

time. They even more regularly use urine to scent-mark, especially around the perimeter of their territory. Small dribbles are deposited on prominent landscape features, such as rocks, shrubs, fallen logs, tree trunks, or mounds. Urine marking may be accompanied by ground scratching with the hind feet. Urine is also used to scent-mark a carcass or portion of a carcass.

The front feet have five toes, but the first toe is reduced and raised on the inside of the front leg, where it cannot reach the ground and so does not normally register in a track. The hind feet have four toes. The front feet, and hence their tracks, are larger than the hind. All toes have fine, slightly curved claws that often leave an impression in a track. This is especially true of the two middle claws (on the third and fourth toes). Hind claws on the third and fourth toes commonly point inward towards each other, but on the front tracks those same claws are usually somewhat straighter. The claws on the outside toes may lie so close to the toe-pad impressions of the middle toes that they can be easily overlooked. The palm pads on both the front and the hind feet are triangular, and the negative space between the toe pads adjacent to the palm pads creates an X shape on a good track. Sometimes the palm pads on the hind feet do not fully register, leaving instead a central, oval impression similar in size to that of a toe pad. Eastern Coyotes, being larger as adults than western Coyotes, create comparably larger tracks (around 10% larger on average) and have a greater stride length. Beds are used for resting or sleeping. Summer beds are difficult to see, as the only indication is an oval of flattened vegetation. Furthermore, while the pups are young, the adults, especially the female, sleep with the pups in the den. Winter beds are easier to see. The Coyote may scratch up the snow before curling up with its tail over its nose. Snow beds are circular and about 35–58 cm in diameter. Most such beds are used only once. In winter, Coyotes will repeatedly use the same meandering trails, gradually producing well-defined, packed runways between hunting grounds that can resemble those made by deer.

REFERENCES
Atkinson and Shackleton 1991; Bekoff 1977; Bekoff and Gese 2003; Bender et al. 1996; Chubbs and Phillips 2002, 2005; Cluff 2006; Crête and Larivière 2003; Elbroch 2003; Fener et al. 2005; Gese 1998, 2001; Gese and Bekoff 2004; Gese and Ruff 1997, 1998; Gompper 2002; Kiliaan et al. 1991; Kitchen et al. 2000; Kuiken et al. 2003; Lehman et al. 1991; Long 2008; Minta et al. 1992; Muntz and Patterson 2004; Patterson, B.R., 1994; Patterson, B.R., and Messier 2001; Patterson, B.R., et al. 1999; Poulle et al. 1995; Reid, F.A., 1997; Rosatte 2002; Slough and Jung 2007; Smith, H.C., 1993; Smith, D.W., et al. 2001; Way 2007a, 2007b, 2007c; Way and Proietto 2004; Whitaker and Hamilton 1998; Woelfl and Woelfl 1994a.

Grey Wolf

also called Timber Wolf, Tundra Wolf, Polar Wolf, Arctic Wolf, Plains Wolf, Mexican Wolf

RENCH NAME: **Loup gris**

SCIENTIFIC NAME: *Canis lupus*

Although the species is called the Grey Wolf, the colour is variable, and there are Grey Wolves that are white, black, grey brown, orange brown, and pale to dark grey. The taxonomy of this species is controversial, and many refinements and possibly major changes may occur in the near future as more research, particularly genetic work, is conducted. A key area of research and vigorous debate in North America involves the taxonomic status of the small Eastern Wolf. As the outcome of this debate is inconclusive to date, the Eastern Wolf continues to be treated herein as a subspecies of the Grey Wolf (*Canis lupus lycaon*). The Eastern Wolf has been variously considered to be a subspecies of Grey Wolf; a hybrid between the Grey Wolf and the Coyote; a hybrid of the Red Wolf and the Grey Wolf; the same species as the Red Wolf; or, most likely, a new and distinct species (*Canis lycaon*) that is closely related to both the Red Wolf and the Coyote. Regardless, the Eastern Wolf is subject to considerable contemporary interbreeding with both Grey Wolves and Coyotes over much of its present range. Although muddying their genetic integrity, this hybridization may, in the long term, make Eastern Wolves more resilient. Eastern and Red

Wolves are thought to have evolved in North America as has the Coyote, while the Grey Wolf evolved in Eurasia and spread to western North America. Early humans tamed and eventually domesticated wolves (beginning more than 100,000 years ago, likely in eastern Asia), gradually selectively breeding them over millennia to create the multitude of shapes and sizes of Domestic Dogs. Genetic testing has verified that all Domestic Dogs descended from wolves in a direct line that involved no hybridization with other wild canids.

The Red Wolf (*Canis rufus*), not covered in this volume, is now endangered and restricted to the southeastern United States. However, it was thought to have occurred as far north as Maine and possibly into southeastern Quebec and Ontario prior to European settlement in North America.

DESCRIPTION

The Grey Wolf is the largest wild canid in the world. It is surpassed only by some breeds of Domestic Dogs. As mentioned above, Grey Wolves come in a variety of colours, not all of which are grey. Nevertheless, most Canadian wolves, except the all-white forms, have some grey in their coat, and all but the melanistic individuals have white hairs, at least around the mouth. The most common grizzled-grey form has long, black guard hairs (60–100 mm long) over its back, sides, and the upper side of the tail, with a paler belly, throat, legs, and underside of the tail. The belly and sides can vary from grey to fawn to rusty brown in colour. There is

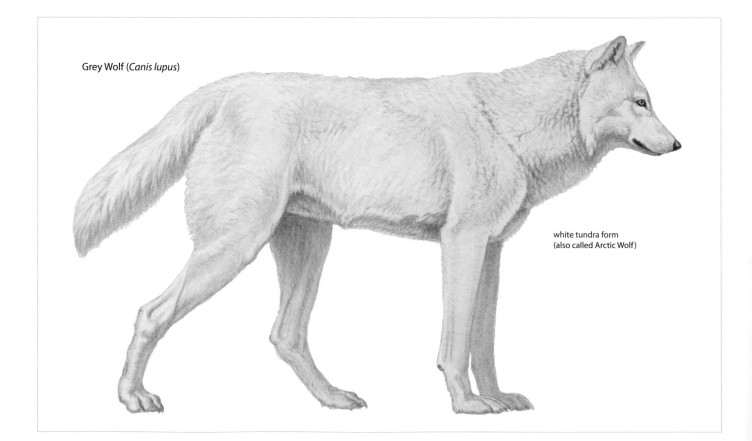

Grey Wolf (*Canis lupus*)

white tundra form
(also called Arctic Wolf)

a saddle of longer, coarser hairs that can reach 150 mm in length over the nape and shoulders. The nose pad is black, and the hairs inside the ears are whitish, while those on the backs of the ears are black or dark grey. The Eastern Wolf is smaller than the other Grey Wolves, has a fawn-coloured pelage with long, black guard hairs on the back and sides and a rust-coloured patch of fur behind each ear. All forms have yellow eyes with round pupils. A patch of stiff hairs (usually dark except on all-white individuals) surrounds the caudal gland on the upper side of the tail near the base. Grey Wolves are narrow in the chest, and their elbows press inward while their front feet turn outward. Their legs are long, and their feet are relatively large. They have mediocre eyesight but keen hearing and superlative olfactory abilities. Wolves can smell prey at distances of up to 2.4 km under suitable wind conditions. Dewclaws on the hind limbs are rare and appear to be an indicator of past cross-breeding with Domestic Dogs. Large jaw muscles and robust teeth allow for a powerful, bone-cracking bite. The dental formula is incisors 3/3, canines 1/1, premolars 4/4, and molars 2/3, for a total of 42 teeth. Fertile hybrids can occur with Coyotes, Red Wolves, and Domestic Dogs. Grey Wolves and Coyotes are known to hybridize from Minnesota east through southern Ontario and eastern Quebec to the extent that up to 100% of wolves in some parts of this region (such as in central Ontario in the Algonquin Park area) have mitochondrial DNA genetic markers distinctive only to Coyotes. Interestingly, no Coyotes have been discovered with Grey Wolf genetic markers, therefore suggesting that, since mtDNA is solely transmitted by the female, the crosses have been exclusively the result of female Grey Wolves mating with male Coyotes, followed by backcrosses to Grey Wolves. Wolves have one protracted moult that begins in late spring or early summer when the old coat is shed. The new hairs are already starting to develop by then; they remain short in summer but continue growing in autumn, to reach their full length in early to mid-winter. Pups begin to moult their fuzzy, dark natal coat at about a month old, exchanging it for one that is similar to their eventual adult pelage. The eyeshine is a bright yellow-green. Grey Wolves usually have ten mammae.

SIMILAR SPECIES

Some large northern breeds of Domestic Dog such as Huskies and Malamutes resemble Grey Wolves in appearance. Wolves have longer legs, larger feet, narrower chests, and tufts of fur that flare outward and droop down from below the ears to frame the head; they also carry their tail straight out behind when moving, not curled upward as do many Domestic Dogs. See the "Similar Species" section of the Coyote for characteristics that separate Coyotes from wolves. Some skulls of Domestic Dogs are as large as those of wolves, but the dogs usually have a larger braincase.

SIZE

The following measurements apply to animals of reproductive age from the Canadian part of the range. Size varies with geographic location, with a larger body size being more common in the northwest and far north. Eastern Wolves are the smallest of the group.

Within each population, the males are usually slightly larger than the females.

Males: total length: 1270–1850 mm; tail length: 405–540 mm; hind foot length: 280–310 mm; weight: 20–70 kg.
Females: total length: 1370–1585 mm; tail length: 400–490 mm; hind foot length: 240–300 mm; weight: 18–55 kg.
Newborns: weight: around 450 g.
Adults stand around 60–80 cm high at the shoulder.

RANGE

The Grey Wolf was once the most widely distributed mammal in the world, occurring throughout the Northern Hemisphere from central Mexico and southern India northward. Owing to persistent persecution by humans, many populations have been greatly reduced, threatened with extinction, or extirpated. In the Eastern Hemisphere, wolves have gone from most of Europe, northern Africa, and most of the southern Orient but are still widespread across northern Russia and China.

Grey Wolves historically occurred across Canada in all provinces except Prince Edward Island. They have been extirpated from Newfoundland, the Maritime Provinces and Gaspé, southern Ontario, the southern Prairie Provinces, and parts of the south-central Arctic (owing to a major crash of Caribou numbers). The Eastern Wolf occurs mainly in the Great Lakes and St Lawrence regions of Quebec and Ontario. As wolves are highly mobile, extralimital occurrences are common. Such wanderers could eventually re-establish populations in regions where wolves were formerly exterminated. The capture of a male wolf in 2002 (DNA tested as *Canis lupus lycaon*) from the Eastern Townships of southern Quebec was the first report from south of the St Lawrence River in over 100 years. Sporadic sightings and specimens continue to be reported from central Alberta, Wyoming, North and South Dakota, Idaho, and Washington State.

ABUNDANCE

Up until the middle of the twentieth century Grey Wolves were considered to be undesirable and were persecuted across their range. Hunters, trappers, livestock producers, and governmental support of bounties and predator-poisoning programs removed wolves from large areas of their original range. As environmental awareness increased in the 1970s, public interest started to focus more on wolf conservation, and the previous levels of persecution began to abate. Nevertheless, there remains considerable opposition and resistance to wolf conservation especially in areas where ranching and big-game hunting are prominent. There are an estimated 45,000–55,000 Grey Wolves in Canada, with an additional 6500–7000 in the United States (about 6000 of these are in Alaska), and likely less than 10 in Mexico. Populations in Greenland and Mexico are threatened with extinction and continue to be persecuted despite full legal protection. All Canadian jurisdictions with Grey Wolves regulate their hunting and trapping, and wolves are fully protected in national and provincial parks. The remainder of the world population includes around 115,000 animals, of which about 12,000 are in Europe and 70,000 are in Russia.

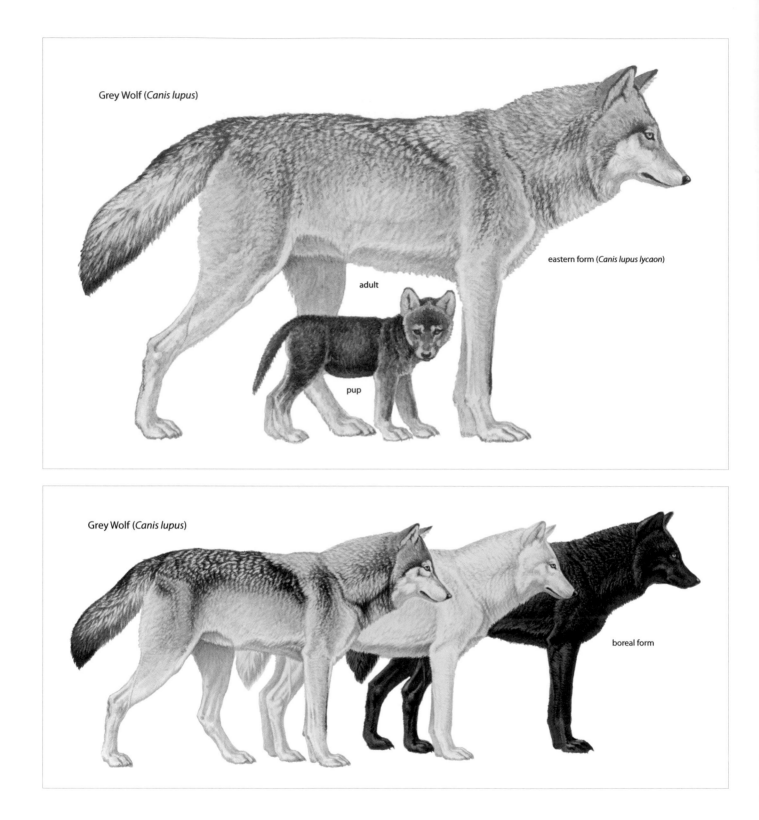

Grey Wolf (*Canis lupus*)

adult

pup

eastern form (*Canis lupus lycaon*)

Grey Wolf (*Canis lupus*)

boreal form

Population densities in North America depend on prey abundance and can vary from 0.5 to 0.00003 per km² in regions with stable wolf populations. The highest density represents a concentration of wolf packs in a deer yard in central Ontario, and the lowest density is on Ellesmere Island in the Canadian High Arctic. In 2001 there were an estimated 1300–2600 Eastern Wolves, and COSEWIC designated that population to be of "Special Concern" in 2001.

ECOLOGY

Grey Wolves are highly versatile in their habitat requirements, occupying almost all habitats in the Northern Hemisphere. Their primary need is food, and they are most numerous where large prey is abundant. Wolves can, and do, survive near urban areas as long as there is a sufficient prey base and their presence is tolerated by humans. Grey Wolves may be active at any time but often focus the bulk of their activity on the hours of darkness. Even Arctic Wolves, faced with 24 hours of daylight during the northern summer, are most active between 2200 and 0400 hours. Grey Wolves typically hunt in packs, and although they are not necessarily the biggest predator in their range, wolves usually assume the role of apex predator because those that are larger (Cougars and three species of bears in North America) are solitary, and most (except Polar Bears and the largest of the Brown Bears) will give way to a determined wolf pack. The size of a non-migratory pack's home range depends on prey density and can vary from 75 km² in rich habitat to more than 2500 km² in more barren areas such as the Arctic. Dispersing and lone animals wander more widely and can occupy home ranges of up to 13,000 km². These solitary individuals compose around 12% of a population. Although most packs occupy stable home ranges, some, especially in northern Canada and Alaska where Caribou are the primary prey, may be migratory, following their prey. These wolves travel between the Caribou's winter range in the forests and their summer calving grounds in the tundra but must cease their migrations when their pups are born, usually stopping somewhat south of the herds. Their annual home ranges may exceed 75,000 km². While feeding pups, these wolves may be forced to travel some distance to find food if local prey is scarce, sometimes even all the way to the Caribou calving grounds. Such hunting trips often last more than 24 hours and can take as long as two to three weeks, covering more than 400 km. In Algonquin Park, the centre for studies of Eastern Wolves, the packs in the eastern and central areas are migratory during some high-snow winters. Their primary prey, the White-tailed Deer, moves to deer yards outside the park, and the wolves follow. In so doing, these wolves remove themselves from the protections afforded by the park and become more vulnerable to hunting, trapping, and other forms of human-related mortality. Throughout North America the daily foraging distances can range from as little as a few kilometres to as much as 70 km or more.

Adult wolves usually eat meat at the kill site and deliver it in their stomachs to the hungry pups or the breeding female. The pups stimulate the adults to regurgitate this food by jumping up and nipping their muzzles. Larger pups may be fed with pieces of prey carried by mouth. Yearling and young adult wolves in the pack may disgorge food to the pups, but they also at times solicit regurgitation for themselves from the adults. Dens are only used in summer to raise offspring. These may be located in underground burrows, caves, rock crevices, under tree roots, in hollow logs, in abandoned beaver lodges after the pond has drained, or in other such sheltered locations. Pups have even been born in snow dens and under the low-spreading boughs of evergreens. Most dens have a southerly exposure to maximize solar heating, and favoured den sites may be reused year after year or intermittently over centuries. Sometimes dens are dug in autumn for use the following spring. Wolf packs typically maintain one or more rendezvous sites within their home range, and a pack will gather at a rendezvous site to care for larger pups and to loaf and socialize. Good rendezvous sites, usually well within the pack boundaries and with nearby water and a clear view, may be used repeatedly through the year for decades.

Grey Wolves are able to disperse over long distances and overcome substantial geographic obstacles such as mountain ranges and wide rivers. Most dispersal occurs in autumn or early winter, and relocating wolves are reported to travel 35 km or more in a day. Dispersal journeys of over 800 km have been recorded, but most average 50 km, although journeys of several hundred kilometres are

Global distribution of the Grey Wolf (*Canis lupus*)

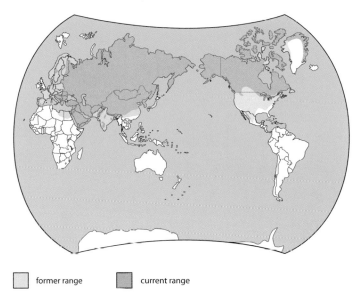

<table>
<tr><td>☐ former range</td><td>☐ current range</td></tr>
</table>

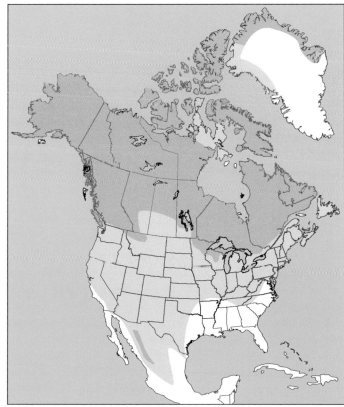

North American distribution

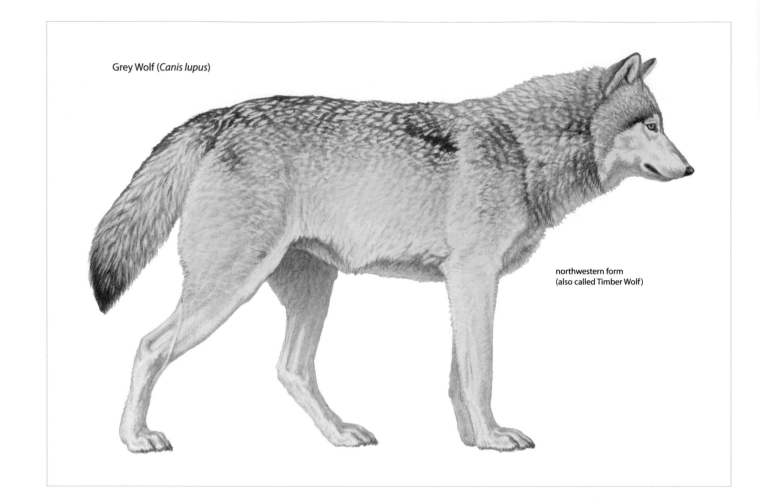

Grey Wolf (*Canis lupus*)

northwestern form
(also called Timber Wolf)

not uncommon. Long legs and powerful leg muscles allow the wolf to travel tirelessly at an average speed of 5–10 km/h. Top speed is close to 70 km/h. Grey Wolves are excellent swimmers and are quite capable of pursuing prey into the water or swimming across wide rivers or lakes. They can live up to 17 years in captivity and 15 years in the wild, but few live more than four to five years. Mortality is caused by human-related activities, even where wolves are legally protected, as well as by starvation, accidents (such as falling through thin ice), fights with other wolves, disease, and injuries caused by prey during predation attempts.

DIET

Wolves are eclectic and opportunistic carnivores that are willing to eat almost anything digestible, but they prefer large ungulates such as deer, mountain sheep, Moose, Wapiti, Muskox, Caribou, American Bison, and Mountain Goat when they are available. They will also take smaller prey like rodents, lagomorphs, smaller carnivores, birds and bird eggs, fish, and livestock, as well as carrion and garbage, and they will even consume fruit when it is seasonally abundant. Grey Wolves have an enormous stomach capacity, allowing them to eat up to 10 kg of meat at a single feeding. Much of this may later be regurgitated to feed pups or to be cached. On average a healthy wolf eats 2.5–6.3 kg of meat daily.

REPRODUCTION

Mating occurs between January and April depending on latitude. More northerly populations breed later in the season. Females are in oestrus once annually for 5–14 days. Copulation culminates with a copulatory lock, as is usual for canids. The base of the penis swells and locks into the vaginal sphincter in a tie that may continue for up to 30 minutes and during which ejaculation can occur many times. Gestation lasts around 62–63 days. Most Canadian wolves produce their litters in April or May. The litter size averages six pups (ranging from one to eleven). Females usually remain near their pups for at least their first three weeks and the alpha male and other pack members hunt and deliver food to her. Whelps are born with fuzzy dark fur. The flop-eared neonates are blind and deaf, easily chilled as they cannot regulate their body temperature, and must be stimulated by their mother's licking to eliminate bodily waste. Their eyes open, their ears stiffen, and they begin to wobble and stagger around the den by 5–15 days old. Most milk teeth have emerged and ear canals opened by around three weeks old. At three to four weeks old the pups begin to ingest solid food in the form of regurgitated meat, and weaning then begins gradually. They start to emerge from the den at around this time and may be seen playing near the entrance. Sometime before the pups are ten weeks old (possibly as early as three to four weeks old), their mother, carrying them one at a time in her mouth, transports them to a chosen rendezvous site that becomes their "nursery" area. The pack then focuses their activity on this site, bringing food to the pups, playing with them, and loafing. The pups romp and tussle over a hectare or more as they wait for food deliveries. They are rarely left alone at the rendezvous site; usually at least one adult attends them. Pups are fully weaned by eight to ten weeks old. Usually by September or October, at four to five months old, the young are old enough to travel with the pack, and the nursery site is abandoned. Juveniles have their full adult dentition by six to seven months old. They reach adult size at around twelve months old. A captive wolf may breed at as young as ten months old, but in the wild few reproduce for the first time before they are 22–46 months old. Wild wolves can continue breeding until they are at least ten years old.

BEHAVIOUR

Grey Wolves are pack animals, and a pack usually consists of an extended family. The most common ontogeny of a pack starts from the bonding of a male-female pair and then the gradual addition of their offspring from the last two to three litters. Outsiders may be accepted. In some areas up to 25% of packs include an unrelated immigrant, while in others such additions are rare. Grey Wolves typically mate for life, and their year-round pair bond ensures that there are always at least two experienced adults in a hunting unit. Most packs are composed of 4–12 individuals, but packs of up to 36 animals have been reported. The upper limit of a pack size is roughly, though not exclusively, related to prey size because the largest packs pursue the largest prey. Members of a pack are aligned in linear hierarchies by gender, with the alpha male and the alpha female being dominant in each hierarchy. Pack leadership may be assumed by either the alpha male or the alpha female, and it is commonly shared jointly, with each initiating activities in different areas: the alpha male in hunting and travelling, and the alpha female in pup care and protection. Usually only the alpha pair reproduces, although multiple litters (with up to three nursing females in a pack) have been reported. Typically, subordinate individuals are physically harassed by the alpha pair when they attempt to mate. The hierarchies are not entirely stable, and much jostling for position occurs during the year, especially each winter before the breeding season. If a member of the alpha pair dies, several outcomes are possible; usually an outsider assumes the vacant position, but one of the pack members could fill it or the pack could disintegrate. Pack cohesion is the result of the ties of affection that are reinforced during the annual courtship between the alpha pair and by their subsequent attention and care of their pups.

Grey Wolves are highly social animals with complex behaviours. They communicate with each other by using vocalizations (see the "Vocalizations" section following), scent marking (see the "Signs" section), and postural displays, often involving the face. During winter the pack tends to hunt together for larger prey, but during summer the animals are more likely to hunt singly, in pairs, or in small groups. Although a single, experienced wolf can take down a large ungulate, most such prey is brought down by the efforts of a pack of wolves after a chase that may extend up to 5 km. The wolves tend to select animals that are debilitated, young, old, or in poor condition because these are usually easier to catch. Most hunts are straightforward. The pack chases the prey until it is tired and then pulls it down. Large prey is attacked at the rump, but on smaller prey the head, flanks, or rump may be targeted. Studies of the predation ecology of wolves indicate that chases of White-tailed Deer and Moose are successful less than 10% of the time. Grey Wolves commonly kill more prey than they can immediately eat, as they are primarily hunters of large ungulates. They are also opportunistic and may kill a whole group of ungulates if they can catch them bogged down in snow. The surplus is often cached for future use by being buried in soil or snow. The meat may be eaten and then regurgitated at the cache site or larger portions may be separated from the carcass and carried away and buried. The wolves normally remain nearby or regularly visit the carcass until most of the meat has been devoured or cached. Wolf packs are typically territorial, but the migratory packs and the packs in regions with abundant prey may be less territorial and allow a certain degree of overlap with neighbouring packs. Grey Wolves scent-mark their home range with urine and scat and advertise their presence by howling. Intruders are dealt with harshly and may be killed. Juveniles remain with their natal pack at least through their first winter. Some pups remain as adults with their natal pack, helping care for future litters and trying to gradually increase their status in the hierarchy until they can become breeders. Other young adults, usually between one and three years old, choose to leave their pack to search for a mate and a territory in which to raise their own pups. Juveniles dominated by other pack members are most likely to leave. Aggressive behaviour by wild Grey Wolves towards humans is rare. Some attacks have involved rabid individuals, but most were made by animals that were habituated to humans either from foraging at a dump or from being fed. In Canada and Alaska since 1970 there have been 18 authenticated bite attacks by non-rabid Grey Wolves. Eight of these attacks were severe, and two resulted in human death. Wild wolves are generally shy of humans and avoid contact if at all possible. Grey Wolves will attack Domestic Dogs that invade their territory, because they view them much as they would an unknown wolf. Humans accompanied by dogs may be more at risk in wolf territory than those without.

VOCALIZATIONS

Grey Wolves bark, whine, yap, and growl, but their most famous vocalization is their howl. Its primordial sound has come to define the truly wild regions of the world. Howling by one pack may stimulate a howling response by other nearby packs, indicating that it may

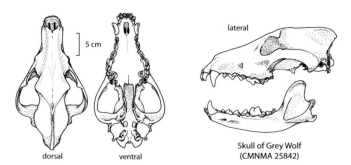

5 cm

lateral

dorsal   ventral

Skull of Grey Wolf
(CMNMA 25842)

Lope
A lope pattern of four tracks is 1.0 to 1.25 m long and the groups of tracks are 50 to 60 cm apart

Direct register trot

Left hind track
Length: 9.5–13.5 cm
Width: 6.7–11.5 cm

Left front track
Length: 9.5–14.5 cm
Width: 7.3–12.7 cm

Grey Wolf

serve a territorial function. Each wolf has a distinctive howl, so wolves can often identify the howling animal or animals at a distance, and packs likely recognize their neighbours by their howls. Usually, larger wolves howl at a deeper pitch than do smaller ones, so body size can be suggested at a distance. Howling is thought to help coordinate pack activities by assembling scattered pack members before a hunt, by calling absent pack members to a kill, or by helping synchronize hunting efforts by separated groups of the same pack. Howls can be heard for several kilometres under favourable conditions. Wolves will howl and bark if humans approach their den too closely. Barking, whining, and growling are used to indicate threat or alarm.

SIGNS

Scent-marking wolves deposit urine and scat as they travel through their territory. Urine is usually sprinkled on prominences in their landscape, from shrubs, trees, and branches to rocks, logs, and shallow mounds. Wolves leave about twice as many urine marks around the periphery of their range as they do in the core area. The alpha pair urine mark by lifting their legs, while other more subordinate pack members squat to urine mark. A female in oestrus often has small amounts of blood in her urine, which tints it orange or reddish. Since she comes into oestrus in late winter, this distinction may be visible in the snow. Invariably an oestrous female will be closely attended by a male, whose tracks will likely be evident alongside hers, and he will mark over her scent mark. Any urine and scat marking may be followed by vigorous scratching with the hind feet and sometimes also the front feet, creating a scrape measuring 10–20 cm long by 2.5–5.0 cm wide. Scat is deposited along travel routes and near den, rendezvous, and kill sites. The appearance of wolf scat varies with diet. Organ and meat scats are blackish and may be completely amorphous or elongated and pointed at one end and blunt at the other, while scat composed more of hair and bone is lighter or mottled in colour with blunt ends. Wolf scat measures 1.3–4.8 cm in diameter and 15.2–43.0 cm in length. A wolf's digestive juices are so powerful that even the bone fragments they consume are largely dissolved and are rarely evident in scat. Their front feet have five toes, but the first toe is rudimentary (although it has a well-developed claw) and is raised on the inside of the foreleg, where it does not reach the ground. Each hind foot has four toes. The front tracks are larger than the hind. All toes are clawed, and the claws (with the already noted exception of the claw on the front foot's first toe) usually register in a track. The front tracks especially tend to toe out. Although several gaits are used, the most common travelling gait is a trot. Wolves use both of the common canine trot patterns, either a direct register trot (as illustrated) or a side trot (as illustrated in the Coyote account). They commonly use a direct register walk (also illustrated in the Coyote account) and an overstep walk, where the hind foot registers beyond the front. Wolves try to take the easiest path and so will utilize trails and lesser-used roadways, as well as ski and snowmobile trails, ploughed roads, and ice-covered lakes and rivers when possible. Packs travel in single file in deep snow so that only one animal expends the energy necessary to break a trail.

Wolf tracks are most commonly found in groups although trails of solitary individuals do occur.

The kill sites of wolves can sometimes be distinguished from those of other canines, especially if the sites are discovered before scavengers have begun to scatter the clues. Wolves tend to attack their prey from the rear, then, as the animal is slowed down, another wolf will attempt to grab the nose while yet others in the pack will rip and slash the flanks and shoulders, attempting to get a grip to pull it down. Once the prey is down, the wolves enter at the abdomen, eating the organs first. The prey is then stripped of flesh, the bones are broken or the joints are separated, and the carcass is cut into pieces. Bones of even large prey can be disarticulated, and all but the largest may be entirely eaten. A hungry pack will consume everything except the stomach contents, the intestines, and some scattered pieces of hair and hide. Much of the meat may be consumed and later disgorged into a hidden cache (buried in soil or snow), after which the wolf can return and eat more. Pieces are also carried away and buried.

All rendezvous sites are occupied intermittently, but some are occupied long enough to become flattened and worn, and those used as a nursery for pups may be littered with chewed bones, antlers, and sticks. Worn or flattened ovals or circles in vegetation or snow may provide evidence of recent usage. Circular beds (made by a curled-up wolf) measure 56–69 cm in diameter, while oval beds (made by a more sprawling wolf) measure 66–76 cm long by 53–66 cm wide. Wolves may bed down in the open on knolls or hillsides where they have an unobstructed view, or under cover where they can be sheltered from the weather. Wolf dens are used only for raising suckling pups. The burrow opening is 51–76 cm high by 63–89 cm wide. A throw mound of waste soil is typically evident at the mouth, which may be littered with bones and scat if the den is in use. A tunnel leads to an underground chamber (1–2 m in diameter) where the pups are born.

REFERENCES

Bjorge and Gunson 1989; Boyd and Neale 1992; Carbyn 1983; Chambers 2006; Ciucci et al. 2003; Cook, S.J., et al. 1999; Darimont and Paquet 2002; Darimont et al. 2003; Elbroch 2003; Forbes and Theberge 1995; Frame et al. 2004; Gehring 1993; Grewal et al. 2004; Gunn et al. 2006; Heilhecker et al. 2007; Kuzyk et al. 2000, 2006; Kyle et al. 2006, 2007a; Lehman et al. 1991; Lohr and Ballard 1996; Marquard-Petersen 1998; McNay 2002; McNay and Mooney 2005; Mech 1994, 1999, 2000b, 2006, 2007; Mech and Adams 1999; Mech and Boitani 2004; Mech and Merrill 1998; Mech and Nelson 1990; Mech and Packard 1990; Mech et al. 1995, 1999; Miller, F.L., and Reintjes 1995; Musiani and Paquet 2004; Musiani et al. 2007; Nowak 2002; Person et al. 1996; Price et al. 2005; Riewe 1975; Samson 2001; Smith, H.C., 1993; Stahler et al. 2002; Theberge and Theberge 1998; Thiel 2006; Thiel et al. 1997, 1998; Via et al. 1997; Villemure and Jolicoeur 2004; Vors and Wilson 2006; Walton et al. 2001; Weaver et al. 1992; Wilson, P.J., et al. 2000, 2003.

Grey Fox

FRENCH NAME: **Renard gris**

SCIENTIFIC NAME: *Urocyon cinereoargenteus*

The small Grey Fox is extremely rare in Canada, occurring in southern and western Ontario and adjacent southern Manitoba, newly reported in extreme southwestern New Brunswick, and possibly occurring in the Eastern Townships of Quebec. The Grey Fox is unique among the North American canids as it is a very capable climber. It has been spotted resting in a crow's nest high in the canopy.

DESCRIPTION

Long, black-tipped guard hairs run along the spine and down the upper side of the tail, giving the Grey Fox a dark, dorsal stripe that extends from neck to tail tip. Otherwise, its upper pelage and rump are grizzled grey, which is caused by grey, white, and black banding on the guard hairs. The underfur is buffy tan or grey in colour. The lower sides, sides of the neck, and parts of the limbs are rusty orange brown. The inner ears, throat, chest, and belly are usually white (sometimes tan), and there may be white patches on the lower front legs and on the inner surfaces of the hind legs. The lower backs of the ears are rusty orange, and the tips are dark grey black. Most individuals have a blackish patch on either side of the muzzle, and black whiskers. The nose pad and lips are black, and the eyes are dark yellow with a bright bluish-white eyeshine. The claws are long, sharp, and curved. A long (around 110 mm) subcaudal musk gland occurs on the top of the tail, beginning near its base and running about one-third of the tail length. It is surrounded by short, stiff, black hairs that can be felt if the fur is brushed against the grain. Pups are dark grey at birth but are dark mahogany brown at about two-and-a-half months old when they begin to emerge from the den. The dental formula is incisors 3/3, canines 1/1, premolars 4/4, and molars 2/3, for a total of 42 teeth. A Grey Fox skull is distinctive. Prominent temporal ridges run from the orbits towards the back of the skull and form a U-shaped crest. Additionally, Grey Fox incisors are simple unlobed teeth, while Red Fox incisors are bilobed.

SIMILAR SPECIES

Red Foxes are longer legged, and only the cross-fox-colour phase (see the Red Fox account) could be confused with a Grey Fox. The Red Fox has a white tail tip in all colour phases, while the Grey Fox has a black tail tip. Grey Foxes are similar in colour pattern to some Swift Foxes, but the Swift Foxes are more sandy tan than grey, and if they exhibit any chestnut coloration on their sides, it is considerably more faded and not nearly as bright as that of a Grey Fox.

SIZE

The following measurements apply to animals of reproductive age from the North American range as there are few Canadian specimens. Males are slightly larger than females.

Total length: 825–1130 mm; tail length: 275–443 mm; hind foot length: 100–150 mm; ear length: 55–101 mm.

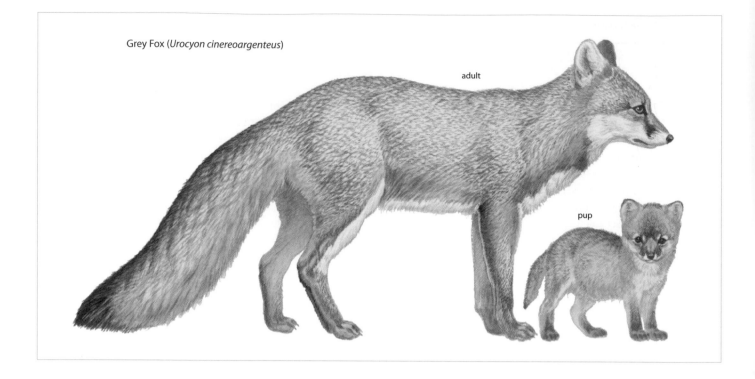

Grey Fox (*Urocyon cinereoargenteus*)

adult

pup

Weight: 2–7 kg (females average 3.9 kg; males average 4.1 kg); newborns: average 86 g.

RANGE

This fox occurs in southern North America from southern Canada and northern Oregon through the United States, Mexico, and Central America, and into northern South America in Columbia and Venezuela. It is at the northern limit of its range in southern Canada. Archaeological evidence suggests that Grey Foxes were almost as common as were Red Foxes in southern Ontario (north to Midland) prior to European colonization. There have been no confirmed records from Quebec in over 100 years, and the only previous report from that province was a sighting by an eleven-year-old boy made in the winter of 1893–4 in the Eastern Townships. An animal trapped near Lake Athabasca, northern Saskatchewan, in 1950 is considered to be an extralimital individual or perhaps a former captive. Nineteen of the last 21 confirmed reports from Canada over the last 20 years have been from southern Ontario. A recent trapping report from southwestern New Brunswick confirms Grey Foxes in that province for the first time, and a recent record from southern Manitoba shows that they still are found there occasionally. Records from the borders of northwestern Ontario and southeastern Manitoba are thought to consist of dispersing individuals from the American populations south of the border. There is one verified recent breeding record (a six-week-old pup was sighted), from Pelee Island in 1998. There is an earlier report (in 1952) of Grey Foxes breeding in the Kemptville, Ontario, area, but no sightings have been confirmed in the area since that time. The Grey Fox is thought to occur, albeit rarely, in St Lawrence Islands National Park. Over the past century (perhaps starting as long ago as the 1850s in parts of the range) Grey Foxes have been increasing in numbers, especially in the northern

states of Minnesota, Wisconsin, Michigan, New York, New Hampshire, and Maine, and have reappeared in Canada, not only reoccupying areas in which they were formerly extirpated but also extending into formerly unoccupied regions (southern New Brunswick). This slow expansion of their range northward is thought to be a response to the warming climate trends that began in the 1800s. The extent of Grey Fox distribution in Canada is unclear and changing and will require careful monitoring.

ABUNDANCE

The Grey Fox is common in parts of the United States (especially Minnesota, New York, Wisconsin, and Michigan) but is considered to be a "Threatened" species in Canada (COSEWIC 2002). There are likely fewer than 250 adults comprising the entire Canadian population, and proof of successful breeding in Canada is very limited. Grey Foxes were legally hunted in Canada, although less than 20 were taken annually (many in traps that were set for other species). Six to ten were shot each winter on Pelee Island, where the population is estimated at around 60 individuals and thought to include 10–15 breeding pairs. In 2008 the Ontario government designated Grey Foxes as "Threatened" in the province, and hunting and trapping of the species was banned. Quebec also does not permit hunting or trapping of the species in that province. Fewer than 10 are taken annually in southeastern Manitoba, where they fall victim to traps set for Red Foxes.

ECOLOGY

This small fox is closely associated with deciduous forest in the east and with rocky, broken terrain and shrubby woodlands in the west and mid-west. It may inhabit marshlands in the southeast United States, as it also does in Canada. Grey Foxes generally avoid open

Distribution of the Grey Fox (*Urocyon cinereoargenteus*)

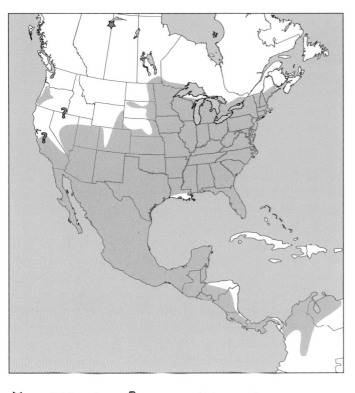

★ extralimital record   **?** presence possible but uncertain

fields and other bare areas but will take advantage of lightly wooded or well-vegetated riparian and old-field or successional habitats. The northern limit of distribution appears to be determined by winter temperatures, and possibly also snow depth, as this fox is generally considered to be a warm-adapted species. Grey Foxes are shy and generally active only under low-light conditions (both crepuscular and nocturnal), although they may be seen in daylight, especially during early morning or late afternoon. The Grey Fox is unique in North America for its adept tree climbing. Easily running up a sloping trunk and jumping from branch to branch, it can also climb straight up a trunk by grasping the bark with the front claws and pushing up with its hind legs. One fox was observed climbing 18 m in this fashion. A Grey Fox can rotate its front legs more than most other canids. This helps it to grasp the tree trunk with its claws. It can run down a slanted trunk head first but must back down a vertical trunk. Grey Foxes will climb trees to escape pursuit, to rest, and to forage. Populations in the United States vary from 0.4 to 2.1 foxes per km². The only density estimates for Canada are from Pelee Island where there are about 1.4 foxes per km². The home-range size is probably related to habitat quality and food availability. Home-range size in Canada is unknown. Estimates from U.S. populations vary from 30.0 ha to 27.6 km². Movement patterns vary seasonally, as females restrict their travel to a few hundred metres from the den site when they are suckling pups.

Pups are born in a secure den, usually located in a hollow tree, rock crevice, or underground burrow, although abandoned buildings

and debris piles may also serve. Natal den sites on the ground are typically concealed by dense vegetation and so are difficult to find. Underground dens may be modified burrows that were originally dug by other more fossorial mammals such as Woodchucks or American Badgers or they may be dug by the foxes. Dens in hollow trees can be up as high as 9 m. Dens are less frequently occupied during the remainder of the year because the foxes typically rest in dense vegetation during the day. Juveniles disperse in autumn, and males tend to travel further than do females. Dispersal distances vary between populations, and the previous maximum has been placed at around 80 km (by a female); a subadult male that was recently captured in New Brunswick likely travelled at least 135 km from the closest known sightings in Maine of a population north of Bangor. Predators include Golden Eagles, Domestic Dogs, Bobcats, and Coyotes, but predation is not thought to be a significant mortality factor. Grey Foxes are susceptible to rabies and especially to canine distemper but are resistant to mange and heartworm. Their mortality rate in the first year appears to be in the neighbourhood of 50%–90%, and few wild foxes live more than 4.5 years, although the maximum lifespan in captivity is 14–15 years. The chief cause of mortality (at least in the United States and possibly also in Canada) is human related: hunting, trapping, and motor-vehicle strikes.

DIET

Although primarily carnivorous, Grey Foxes are more omnivorous than are other North American canids; fruit and other plant material form a significant part of their diet in all seasons. Like other North American foxes, they are opportunistic foragers, willing to eat almost anything digestible, from small mammals such as mice, voles, shrews, squirrels, and rabbits to birds, insects, reptiles, acorns, fruit, carrion, poultry, and corn. Their diet necessarily varies with season and food availability; for example, small mammal prey is the most important part of their winter diet, probably because insects and plant material are less available, and fruit and nuts are their major summer and autumn food choice.

REPRODUCTION

Very little is known of the breeding habits of Grey Foxes in Canada. Like populations in the northern United States, Grey Foxes in Canada are thought to breed from mid-February to mid-March. Females produce a single litter annually that averages three to five pups (ranging from one to ten) after a gestation of about 59–60 days.

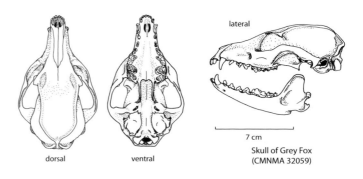

lateral

7 cm

Skull of Grey Fox
(CMNMA 32059)

dorsal   ventral

Straddle trot

Direct register trot

Left hind track
Length: 3.2–4.5 cm
Width: 2.4–3.8 cm

Left front track
Length: 3.5–5.0 cm
Width: 3.0–4.5 cm

Grey Fox

Their eyes open at 10–12 days old, and pups begin to appear above ground at 2.5–3.0 months old, at which time they begin to take solid food. Weaning is gradual and complete by the time the pups are about four months old. The pups accompany their mother as she forages when they are three to four months old, and they are foraging on their own by about four months old. Pups generally disperse from their natal home range in autumn although some delay until late winter. Juveniles become sexually mature at about 10 months old, and most breed in their first year.

BEHAVIOUR

The basic social unit is a mated pair and their pups. Grey Foxes are thought to be monogamous although this has not been confirmed. Members of a pair may remain together throughout the year or become solitary during the non-reproductive season. It is not known if they reconnect to breed the following year. Urine and faeces are deposited at scent posts for communication, but it is unclear whether these also represent territorial markings. For the most part, family groups maintain an exclusive core in the home range but will permit some overlap with adjacent family groups around the periphery. Several reports of two females and their litters sharing a maternal den suggest that, as with the Red Fox, polygamy occurs at times. Apparently both females suckle the combined litter. Family bonds are enhanced by mutual grooming, usually between adult pairs, and of juveniles by adults. It is unclear how much parental care is provided by the male, and the amount provided may vary from male to male. Some males do not provide prey to weaned pups, leaving that entirely to the female. In one case where his mate was killed, a male took over the care and feeding of their weaned pups.

VOCALIZATIONS

These foxes are normally silent, but they will communicate with each other by growls, screams, alarm barks, and long-distance-contact barks, which are produced most frequently at night during the breeding season. When greeting each other, they produce quiet cooing and mewing sounds.

SIGNS

The tracks of the Grey Fox are smaller, shorter, and broader than those of the Red Fox. They are similar to and are often mistaken for those of Domestic Cats, but the semi-retractable claws of the Grey Fox, unlike cats that sheath their claws, are often visible in a track. The most common gait of a Grey Fox is a straddle trot. This type of trot is often used by canids (and many ungulates), as the hind feet travel outside the front feet to avoid interference. Most canids use this gait only for short distances, but Grey Foxes maintain it for much longer and use it more frequently. Each front foot has five toes, but the first toe is highly reduced and raised on the inside of the front leg; it normally does not register in a track. The hind feet each have four toes. All of the toes on the front and rear feet have curved, sharp, semi-retractable claws. As with most canids, the negative space between the triangular palm pad and the outside toes forms an X. In the case of the Grey Fox, this X is broader and more flattened from top to bottom than that of other native canines and approaches an H rather than an X on the front tracks because the front tracks are larger and rounder than the hind tracks. Owing to their small foot size and relatively short legs compared to those of other foxes, Grey Foxes regularly take advantage of human-made pathways such as roads, snowmobile trails, cross-country ski trails, and farm tracks, especially when the snow is deep. They frequently use fallen logs, fences, walls, and other raised surfaces as travel corridors, and daytime rest sites are often above ground. Solitary trails are usual, except during the reproductive season. The scat of Grey Fox varies with diet and is impossible to distinguish from that of Red Fox. Both produce blackish meat scats and larger more tubular scat composed nearly entirely of fruit. Despite their body-size differences, contextual clues such as tracks are usually necessary to determine the species that produced the scat. See the Red Fox "Signs" section for scat descriptions and measurements, which are similar for both species. Like Red Foxes, Grey Foxes use urine and scat as communication tools and possibly territorial markers, and regularly mark elevated surfaces on their foraging rounds.

REFERENCES

Anderson, R.M., 1939; Chamberlain and Leopold 2000, 2002; Fedriani et al. 2000; Fuller and Cypher 2004; Jameson and Peeters 2004; Gerhardt and Gerhardt 1995; Greenberg and Pelton 1991; Judge and Haviernick 2002; McAlpine et al. 2008; Reid, F.A., 1997; Verts and Carraway 1998; Whitaker and Hamilton 1998.

Arctic Fox

also called White Fox, Polar Fox

FRENCH NAME: **Renard arctique**

SCIENTIFIC NAME: *Vulpes lagopus*, formerly *Alopex lagopus*

The Arctic Fox is the only member of the Canidae (Dog) family to undergo a seasonal colour change. It is also the wild carnivore with the largest known litter size (the Domestic Dog holds the record with up to 24 pups).

DESCRIPTION

This small canid is superbly adapted for the Arctic cold. To conserve heat, it has short ears, short legs, and a short muzzle. During cold weather and food shortage its metabolism drops to save energy. It also has a luxuriantly thick, long, and dense winter coat that even grows long between the toes pads and folds over so that the feet do not directly contact the cold substrate. Arctic Fox fur has the most insulative properties of all the terrestrial mammals. Although small, they do not begin to shiver until temperatures reach an astonishing –70°C. The winter fur is twice as long as the summer fur, and about 70% of the coat is composed of underfur. The longest hairs on the

Distribution of the Arctic Fox (*Vulpes lagopus*)

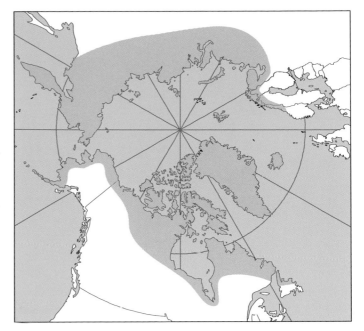

coat are on the parts of the body that are exposed to the elements when the fox is in a curled-up posture. Two colour morphs occur in Arctic Foxes. Foxes that are white in winter have a summer coat that is brownish dorsally and whitish to buff ventrally. In winter the "blue" form varies from blue black to dove grey, and in summer it is overall chocolate brown to the more common brownish (sometimes a dark bluish-grey) dorsally and white buff ventrally, in a pattern that is similar to that of the white form. There is considerable variation in each colour form, and most populations have individuals of both colours, although one colour or the other usually dominates. The tail is always one colour (not white tipped). White foxes are most common in the Canadian Arctic, while blue foxes are most common on islands in the North Atlantic and the Bering Sea, on the Aleutian Islands, and on Greenland. The eyes are golden yellow. Their feet are large for their size, to aid in locomotion on snow. Arctic Foxes have keen senses of sight, hearing, and smell. The dental formula is incisors 3/3, canines 1/1, premolars 4/4, and molars 2/3, for a total of 42 teeth. Sterile hybrids with Red Fox (typically a male silver-phase Red Fox crossed with a female blue-phase Arctic Fox) are commonly produced, using artificial insemination by the fur industry (notably in China), but hybridization between these two species does not occur in the wild. There are two annual moults, one in spring and the other in autumn. Arctic Foxes have 12–14 mammae.

SIMILAR SPECIES

Arctic Foxes are the only foxes in Arctic habitats, but along the southern border of their range they may overlap with Red Foxes. Red Foxes are larger, with longer legs and ears and a white-tipped tail even in the non-typical colour forms (see the Red Fox account for more details on colour variation). Red Foxes do not change colour in winter.

Arctic Fox (*Vulpes lagopus*)

winter — white phase — blue phase — summer — pup

SIZE

The following measurements apply to animals of reproductive age from the North American part of the range. Males are larger and 10%–20% heavier than females. Body weight is highest in autumn as fat is accumulated for insulation and a winter energy reserve.

Males: total length: 830–1100 mm; tail length: 280–425 mm; hind foot length: 130–161 mm; ear length: 55–67 mm; weight: 3.2–9.4 kg (average 3.5 kg).

Females: total length: 713–850 mm; tail length: 255–320 mm; hind foot length: 128–148 mm; weight: 1.4–3.3 kg (average 2.9 kg).

Newborns: weight: 55–90 g.

RANGE

These polar foxes occur in northern North America, Europe, and Russia and in Greenland and Iceland. They are highly nomadic and sometimes migratory, capable of travelling 2300 km annually, much of which is over sea ice. Extensive travel normally occurs only during spring, autumn, and winter, and not during the summer breeding season. Not surprising, Arctic Foxes have reached most polar islands, and tracks have been seen on the ice less than 150 km from the geographic North Pole. They commonly wander beyond their usual range, turning up in central Canada or floating on ice floes to Anticosti or Cape Breton Islands. Arctic Foxes now inhabit at least 450 of the Aleutian Islands, after being released from fur farms and then travelling from island to island by swimming or on winter ice. They are severely affecting seabird colonies on some of the Aleutian Islands, and efforts are underway to eradicate them from those areas. Red Foxes are thought to be such significant predators of Arctic Foxes that the southern limits of Arctic Fox distribution is largely determined by interspecific competition and predation by Red Foxes. In Europe, where Red Foxes are extending their range northwards, Arctic Foxes are becoming increasingly endangered on the mainland.

ABUNDANCE

Populations in North America and Iceland appear to be stable and secure. However, those in Norway, Finland, and Sweden crashed in the 1920s because of over-hunting and have not substantially recovered despite protection from further trapping. Populations on the Commander Islands, Russia, were decimated in the 1970s by over-hunting and by mange that was imported on Domestic Dogs, but they are recovering. The life of an Arctic Fox is often short. Most populations suffer about a 50% annual mortality rate of adults, and juvenile mortality is much higher. Arctic Fox populations appear to follow a three- to five-year density cycle: populations peak when food (primarily lemmings and voles) is abundant and bottom out a year or two after prey numbers crash. Fox numbers can decline by over 80% when food is limited, and different regions can be at different phases of the cycle. Many populations continue to be trapped for the fur trade, but this market is not as robust as it once was, so the number of foxes being taken has declined. There are likely around 100,000 Arctic Foxes being trapped annually, most in Canada, Greenland, and Iceland. Global population size is estimated at several hundred thousand. Anticipated changes in global climate are expected to have an effect (probably negative) on Arctic Fox abundance and distribution.

ECOLOGY

This fox is widespread across the Arctic in a variety of habitats from coastal to inland, and alpine to ice-covered marine, although they are most abundant near the coast. The home-range size varies, depending on food abundance. From the tracking of radio-collared individuals it has been learned that populations in North America and Greenland typically occupy home ranges of around 10–21 km². During summer the reproductive pairs are restricted to an area around the den site. Males forage over larger summer ranges than do their mates, who remain closer to the den and the pups. In winter many foxes (especially coastal foxes) occupy a larger range in the vicinity of their summer range, while others become transient in search of food. Winter movements are poorly documented and may be quite extensive. Some undertake seasonal migrations, which occur in late autumn or early winter and again in spring. Sporadic movements or periodic migrations are usually triggered by food shortages. The foxes are forced to abandon their home ranges and may travel almost 1000 km during a winter. They can easily average

> 24 km each day. Some follow Grey Wolf packs or Polar Bears in order to scavenge their kills, especially in winter and early spring. During years of low lemming and vole populations many foxes acquire up to half of their food requirements by foraging on sea ice. Access to marine food sources (mainly seal carrion and Ringed Seal pups) in winter depends on the extent and duration of the sea ice, which is increasingly threatened by global climate change. Occasionally, large numbers of foxes will invade the boreal forests along the southern margins of their range. These movements appear to be related to food shortages, particularly after population crashes of lemmings and voles.

Arctic Foxes are active all year round and prefer low-light conditions, but they are adaptable enough to accommodate the periods of 24-hour daylight and darkness that occur above the Arctic Circle. Reproductive dens are needed before the frost leaves the upper ground, so they are normally dug in permafrost-free areas in friable, often sandy soils on low mounds, hillsides, riverbanks, or eskers. Entrances on slopes face south whenever possible to take advantage of solar heating. Dens can vary from a simple burrow to complex burrow systems that may be used and renovated for centuries by generations of foxes. Natal dens are typically larger than non-reproductive dens, which tend to be simple holes used intermittently for shelter during both summer and winter. Natal dens with up to 100 entrances have been reported, and most have at least two access holes. Appropriate natal dens can be rare in some parts of the Arctic, and a good site is used repeatedly. Summer dens may provide winter shelter, and Arctic Foxes will also dig snow dens for winter shelter. Around settled areas Arctic Foxes may den in culverts and dig dens in road embankments. Rabies exists in most Arctic Fox populations and may become epidemic periodically, especially when numbers are high. Starvation is a serious mortality factor, especially during migrations and winter. Summer food shortages can result in pup starvation, but adults usually survive, though they may be compelled to abandon pups. Arctic Foxes are proficient and buoyant swimmers, able to swim for > 45 minutes and for distances > 2 km (this long-distance swim was interrupted halfway by a fifteen-minute rest on a convenient ice floe, so it is unknown whether the animal could have swum the estimated 2.4 km in a single effort). The usual maximum lifespan is around three to four years, with very rare individuals surviving to be nine to ten years old. Captives may reach 15 years old. Few pups survive their first winter. Primary predators of Arctic Foxes include Grey Wolves, Red Foxes, Wolverines, Snowy Owls, eagles, and Domestic Dogs.

DIET

Arctic Foxes are opportunistic predators and scavengers and will eat almost anything that is digestible, including plant material such as berries and seaweed, as well as molluscs, amphibians, human garbage, and insects, when other more preferred food sources are depleted. They will often specialize on one prey type when it is abundant. Arctic Foxes are the primary predators of most Arctic nesting birds and their eggs. Foxes in the vicinity of a large goose colony on Banks Island in the Canadian Arctic took an annual average of 900–1570 eggs per fox. Up to 97% of these were cached for future consumption and to provide crucial winter food. Despite the sometimes superabundance of eggs and nestlings that are available during the short Arctic summer, Arctic Foxes, especially inland populations, depend on small mammals such as lemmings and voles for the bulk of their diet. Birds and bird eggs, juvenile Arctic Hares, ground squirrels, fish, marine invertebrates, Ringed Seal pups, and the carcasses of marine and terrestrial mammals are added when available. Carrion is especially important in winter. Both adults and pups are cannibalistic at times, especially when food is scarce.

REPRODUCTION

Breeding typically occurs in March or April, and gestation lasts 50–55 days. Females enter oestrus only once each year and are in heat for three to five days. The timing of this heat varies with latitude, weather conditions, and the physical condition of the vixen. Well-fed foxes breed earlier, while poorly nourished animals breed late or not at all. Females begin each reproductive cycle with roughly the same number of embryos, but resorption of embryos, especially late in the pregnancy, is common if the female is nutritionally stressed. The litter size varies from year to year, ranging from three to eighteen pups, and is directly related to food availability; undernourished females produce smaller litters. Coastal populations tend to have small litters, averaging six (ranging from three to ten), while inland populations produce litters that average six to twelve. Despite this disparity of litter size, the breeding potential of the coastal and inland foxes remains more or less equal, as the inland populations are subject to greater fluctuations in food availability and may not reproduce every year. Most young are born in late spring (May–June). The pups are altricial, born blind and toothless, but with a full, albeit short, coat of woolly, dark blue-grey to grey-brown fur. Their eyes open at about 16 days old. The natal coat is shed by around eight weeks old, when the pups moult into the typical adult summer pelage. Their milk dentition erupts from the gums at three to four weeks old, and their adult teeth replace these by late summer. They emerge from the den at three to four weeks old to await the return of foraging parents and to play. They are weaned at six to seven weeks old and become independent around 12–14 weeks old. Pups are adult sized by 14–28 weeks old. Most leave their parents' territory by the time they are six months old. Female juveniles may become sexually mature in the year after their birth, at around 9–10 months old if they are well fed or a year or even two years later if not. Survival of juveniles is directly related to food availability, during both the parental-care period and their early independence when they must learn to hunt for themselves.

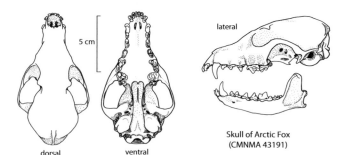

5 cm

lateral

dorsal   ventral

Skull of Arctic Fox
(CMNMA 43191)

BEHAVIOUR

Arctic Foxes are generally monogamous, and it is reported that pairs may remain together for life. As with most of the canids, their breeding systems can become more complex, especially in isolated populations. Plural breeding (multiple breeding females in a social group), polyandry (females breeding with more than one male), and multiple paternity (single litters with multiple sires) are reported. Although pairs may stay together over the winter, or part of the winter in rare cases, and groups may congregate around a rich food source such as a beached whale or a garbage dump, most foxes are solitary except during the breeding season. Rarely, females may share a den, resulting in erroneous reports of very large litters. Sometimes parents will move the pups to an alternate den, and there are reports of litters being split between two or even more den sites. Both parents raise the young, but the vixen is more diligent. Apart from suckling the litter, she delivers most of the solid food after the pups have been weaned. Once the pups are three to four weeks old and able to crawl to the burrow opening, they spend increasing amounts of time around the entrance, playing and awaiting their parents' return. They stalk and pounce on each other, chase one another, and tussle. As they get older, they begin to venture further from the den, looking for food caches or learning to hunt. The dog fox abandons the family in late July or early August, and the vixen leaves a few weeks later. By this time the pups are more or less self-sufficient and no longer dependent on their parents to provide food. The young usually disperse in late summer or autumn but may remain near the parental territory if food is abundant. If a female juvenile remains on her parents' territory beyond her first winter, she may act as a helper to the new litter; however, usually only the dominant female (her mother) breeds. Helpers do not provide much assistance to the parents apart from occasionally bringing food for the pups. Arctic Foxes are territorial during the summer while raising pups. Both partners will actively expel intruders, snarling, snapping, and growling, then chasing them away. Outside the summer reproductive season Arctic Foxes are highly mobile as they search for food, and are no longer territorial.

While naturally nervous, Arctic Foxes are not shy of humans if they are not molested or harassed. They are frequent scavengers at garbage dumps and will not hesitate to raid a food cache or run off with pilfered camping equipment, even if it is not edible. Like many other predators, Arctic Foxes, when provided with the opportunity, will kill more than they can eat and bury the excess for lean times in the future. This behaviour is most evident in summer when food is plentiful and sometimes superabundant. Fragments of goose eggs that had been cached during the previous summer have been detected in fox scat the following spring. Cached food is even moved at times, possibly to prevent thievery by other foxes. Food may be scatter-hoarded (small amounts of food stashed in many locations) or larder-hoarded (large amounts hidden in a single or only a few locations, typically near the den site). Arctic Foxes are not only diverse in their choice of prey but also clever and adaptable in how they acquire it. Fish are eaten as carrion or are caught alive. The foxes are reported to stir the water with their paw to attract the fish and then to snatch them when they come within reach. They

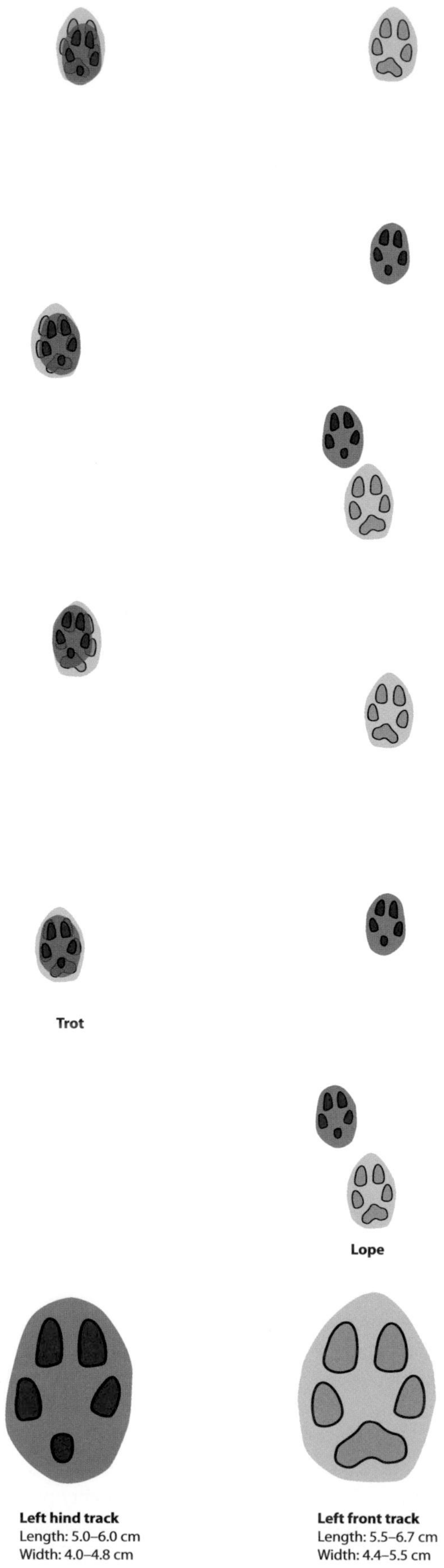

Trot

Lope

Left hind track
Length: 5.0–6.0 cm
Width: 4.0–4.8 cm

Left front track
Length: 5.5–6.7 cm
Width: 4.4–5.5 cm

Arctic Fox

will also patrol the coastline during low tide, searching for fishes that are confined to rock pools where they are easier to catch. They are smart enough to follow researchers on their rounds so that they can predate eggs or nestlings while the adults are distracted and off the nest owing to human activity. One observation of a Muskox charging a nesting goose included the note that a nearby fox was quick to take advantage of the disturbance to dash in and snatch an unguarded egg.

VOCALIZATIONS

A low chittering sound is used to call the pups. This call is commonly used by a female to attract the pups to suckle or by either parent to call them to accept prey. Arctic Foxes have a high-pitched bark rather like that of a small dog, heard most often during the summer breeding season. They bark in prolonged series (2–14 barks) as a signal of territorial occupancy or in the presence of potential enemies. Neighbouring foxes within earshot are usually stimulated to produce the same call in response, and there are sufficient differences between the barks of different animals that the foxes can distinguish the barks of members of their own social group from those of other foxes. Barks produced by juveniles are characteristically higher pitched, and juveniles bark out of excitement during many situations. Arctic Foxes also hiss, growl, and scream when in conflict, and "coo" at each other in greeting.

SIGNS

Sometimes, especially in dry tundra, the accumulation of food remains and excrement around a regularly active den enriches the soil sufficiently that the enhanced, darker green plant growth is visible from a distance and even from the air. Burrow entrances can be up to 34 cm in diameter, but most are around 15–25 cm in diameter. Arctic Foxes scent-mark by depositing scat along travel routes and on conspicuous or prominent locations such as a rock, a mound, or prey remains. Scat also accumulates around den sites, especially if pups are present. Scat varies in size and colour depending on diet. Most are 0.8–1.6 cm in diameter and 5.1–11.4 cm in length. Scat composed of flesh is black and can be amorphous or crumbly and blunt ended. Scat composed of bone and hair is paler, sometimes even whitish, and twisted with pointy ends. It may also be almost segmented looking or broken into short sections, each with a tapered, pointy end and a rounded, blunt end. Scat composed of berries or plant remains is usually tubular with little twisting and has a smooth surface, although intact berries can sometimes be seen within. Their diverse diet can mean that a scat may be a composite of several types of food. Urine stains may be visible, especially in snow. As with all canines, the urine of a male is deposited forward of the hind legs, while that of a female is commonly left just ahead or even between the rear tracks.

In cold or rainy weather Arctic Foxes rest in a perfect circle with their feet tucked under and their tail tip over their face and nose. They may dig a hollow into the snow to curl up in or, during a storm, allow the snow to drift around and over them. Snow beds are around 24–30 cm in diameter. The front feet have five toes, but only four of these register in a track. The first toe is elevated on the inner leg, with a claw that does not reach the ground. The hind feet have four toes. All toes are clawed, but the claws do not always register in a track. The front tracks are larger than the hind. The tracks of winter Arctic Foxes are often blurred by an abundance of hair. As with all canid tracks, the negative space between the toe pads of the two outside toes and the palm pad forms an X. Arctic Foxes have relatively large feet, so their tracks may seem excessively large for the size of the animal, especially when the tracks are further enlarged by the wealth of fur on the feet. Their most common travelling gait is a lope (canter).

REFERENCES

Anderson, C.G., 1999; Anthony 1997; Anthony et al. 2000; Audet et al. 2002; Bailey 1992; Ballard et al. 2000; Banfield 1974; Bantle and Alisauskas 1998; Carerau et al. 2008; Carmichael et al. 2007; Cypher 2003; Dobbyn 1994; Elbroch 2003; Frommolt et al. 1997, 2003; Kapel 1999; Kruchenkova et al. 2003; Macdonald and Barrett 1993; Møller Nielsen 1991; Murray and Larivière 2002; Pamperin et al. 2006; Peterson, R.L., 1966; Prescott and Richard 2004; Prestrud 1991; Reid, F.A., 2006; Roth 2002, 2003; Samelius and Alisauskas 2000; Samelius and Lee 1998; Sklepkovych and Montevecchi 1996; Smits and Slough 1993; Strand et al. 2000; Strub 1992; Tannerfeldt and Angerbjörn 1998; Tannerfeldt et al. 2002; Underwood and Reynolds 1980; Wrigley and Hatch 1976.

Swift Fox

FRENCH NAME: **Renard véloce**

SCIENTIFIC NAME: *Vulpes velox*

Swift Foxes are named for their ability to run at speeds of up to about 60 km/h. They are Canada's smallest foxes, about the size of a large house cat but with longer legs. Swift Foxes are closely related to Kit Foxes (*Vulpes macrotis*, which occur in the southwestern United States), and the two taxa are still considered by some to be subspecies. Skull and other morphological measurements, as well as some of the recent genetic work, support the species distinction of each, although some hybridization is known to occur in the wild along the thin line of overlap in the southwestern United States.

DESCRIPTION

This tiny western fox has a dark patch on either side of its muzzle between the nose and the eyes, and a black tip on its tail. The tail length averages 52% of the head-body length. The summer fur is short and coarse and is grizzled grey brown on the back and face, rusty to yellowish-tan on the sides and legs, and creamy white on the throat, belly, and inner legs. A band of colour extends from the

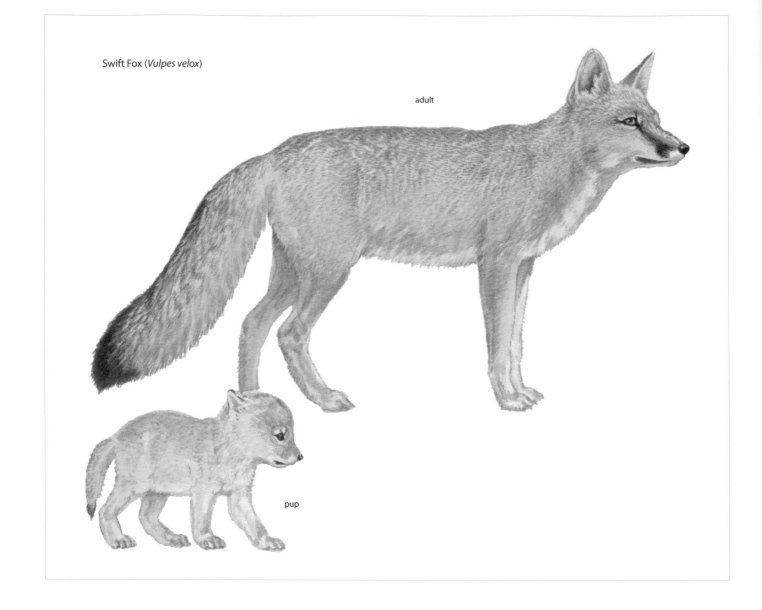

Swift Fox (*Vulpes velox*)

adult

pup

shoulder across the upper chest and separates the pale throat from the similarly pale belly. The colour of the upper side of the tail usually matches that of the back, while the underside of the tail is the same colour as the sides. The winter pelage is longer and denser and tends to be somewhat washed out and less reddish, but it retains the same colour pattern. Saturation of colour, especially on the sides, varies between individuals; some are pale tan, while others are much more reddish. The nasal pad is blackish, and the eyes are golden yellow. The dental formula is incisors 3/3, canines 1/1, premolars 4/4, and molars 2/3, for a total of 42 teeth. Swift Foxes experience two annual moults, one in spring and the other in autumn.

SIMILAR SPECIES

Their small size, light coloration, dark muzzle mark, and black tail tip distinguish Swift Foxes from all other western North American foxes, except the Kit Fox, which does not occur in Canada. Swift Foxes have pale legs compared to the dark legs of Red Foxes, and Red Foxes almost always have a white tail tip.

SIZE

The following measurements apply to animals of reproductive age from the North American range, as there are few Canadian specimens. Males are around 8% heavier than females and are slightly, but rarely noticeably, larger.

Males: total length: 735–880 mm; tail length: 240–340 mm; hind foot length: 114–135 mm; ear length: 56–75 mm; weight: 2.18–2.95 kg.
Females: total length: 680–850 mm; tail length: 225–340 mm; hind foot length: 110–128 mm; ear length: 58–68 mm; weight: 1.5–2.27 kg.
Newborns: weight: unknown, but about 200 g at two weeks old.
Shoulder height is about 30–32 cm.

RANGE

The Swift Fox is native to the Great Plains of central North America. The full extent of its historical range is difficult to determine, and it was likely patchy with disjunct populations in some areas. Before European settlement in the west it occurred from southern Alberta and Saskatchewan to Texas, as far west as the Rocky

Mountains, and as far east as Nebraska and possibly extreme western Iowa, Minnesota, and southwestern Manitoba, although these latter three localities have not been satisfactorily documented. The Swift Fox decline in the Canadian grasslands began in the late 1800s as native prairie was converted to cropland. These small foxes had been extirpated from Canada by the 1930s. They are less wary than are most other canids and will readily take poison bait or enter a trap. Therefore, their populations were severely affected by predator-eradication programs and by trapping for the fur trade. Reintroduction efforts from the 1980s and 1990s have proved successful to the extent that wild Swift Foxes have reinhabited a large portion of southern Alberta and Saskatchewan. Successful reintroduction sites in the Canadian Prairies were along the Alberta-Saskatchewan border south of the Cypress Hills and in both the east and the west block of Grasslands National Park, Saskatchewan. An introduction in the Milk River area south of Lethbridge, Alberta, failed, likely owing to rabies. This fox has also been successfully reintroduced into Montana, and efforts are currently underway to reintroduce them into currently unoccupied parts of South Dakota. Populations in the south-central United States are all located east of the Rocky Mountains, where they occupy about 40% of their former range.

ABUNDANCE

The current existence of Swift Foxes in Canada is largely thanks to a conservation-minded, expatriate British couple named Miles and Beryl Smeeton. They started with two pairs of Swift Foxes from Colorado in the early 1970s and began a private, successful captive breeding program on their ranch northwest of Calgary. Ten years later, with the help and approval of the Canadian Wildlife Service and the Alberta and Saskatchewan governments, and help from numerous non-governmental organizations, an initial release site was selected on the Lost River Ranch, south of Manyberries, Alberta. In the autumn of 1983 twenty foxes were released, followed by more over the next four years, including some in Grasslands National Park and others in the Nashlyn Community Pasture south of Consul, both in Saskatchewan. Wild-caught foxes from Colorado and Wyoming were also released during this time to bolster the gene pool and add to the numbers. Between 1983 and 1997 when the program ended, almost 1000 Swift Foxes were released in Alberta and Saskatchewan. Additionally, in 2004, 15 foxes from the captive breeding program were introduced into the Blood Tribe lands in southwestern Alberta. By the early 1990s there were an estimated 300 wild Swift Foxes in the Canadian Prairies. By 2005–6, an estimated 650 lived wild in Alberta and Saskatchewan, and an additional 515 existed in northern Montana. Alberta populations are classified provincially as "Endangered" and are fully protected. In Saskatchewan populations are fully protected by the Saskatchewan Wildlife Act. Both populations appear to be slowly increasing in numbers and distribution. Nevertheless, the Canadian numbers are low, and this fox may be negatively affected by global climate change as the prairies are tending to become hotter and dryer, and this trend is expected to continue. Such conditions are expected to impact prey numbers and availability. Swift Foxes are

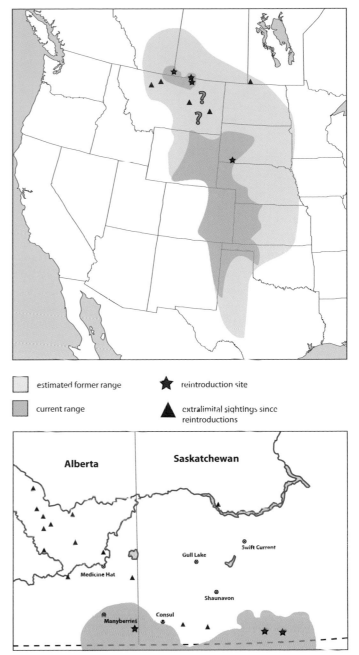

Distribution of the Swift Fox (*Vulpes velox*)

estimated former range   ★ reintroduction site

current range   ▲ extralimital sightings since reintroductions

expanded view of Canadian range

also threatened by poisoning campaigns directed towards Coyotes and by habitat degradation. Federally, COSEWIC designated the species as "Threatened" in 2009, upgrading it from "Endangered" in recognition of the successful reintroduction program that had returned the species to part of its former Canadian range. Within the southern extent of the current range Swift Foxes are sparse, endangered, or threatened and are carefully monitored by some states such as Texas, Nebraska, Montana, and South Dakota; they are more abundant in Colorado, Kansas, Wyoming, and parts of

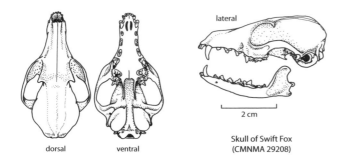

lateral

2 cm

Skull of Swift Fox
(CMNMA 29208)

dorsal   ventral

New Mexico and Oklahoma. Estimates of Swift Fox numbers in the United States range from 16,400 to 27,500 animals.

ECOLOGY

These small foxes are grassland animals, preferring open, sparsely vegetated, short grass or mixed-grass prairie on flat or gently rolling terrain that provides good visibility and optimum opportunities to detect and avoid predators. They avoid coulees and brushy or cultivated areas and are rarely seen near human settlements, although they may occupy pasture lands. Extensive cultivation and fragmentation of suitable habitat in Canada have resulted in a limited number of locations of sufficient size to support Swift Foxes. Most such remaining habitats occur in southeastern Alberta and adjacent southwestern Saskatchewan, and possibly into southwestern Manitoba; much of this is private, leased, or community pasture land. Unlike other North American canids, Swift Foxes use burrows year round. They are highly dependent on underground burrows, where they seek shelter from heat, cold, and predators. Unlike other foxes, Swift Foxes maintain multiple burrows throughout their home range and try not to wander too far from these sanctuaries. They relocate their primary den site at times, and it is probable that a heavy external parasite load is one of the prompts for such a move. The burrows may be dug by the foxes if the soil is friable or they may renovate those created by other fossorial mammals such as American Badgers and Richardson's Ground Squirrels. Whelping dens (also called natal dens) tend to be larger and more complex underground than are escape dens and have the most entrances, although simple escape dens usually have more than one entrance. Most whelping dens in Canada have been found on high points of ground, often on the brow of a hill. Within the underground den are tunnels (often laid out in a spiral pattern), and wider chambers for sleeping and caching food. Swift Foxes are largely nocturnal, emerging from their dens shortly after sunset to hunt intermittently throughout the night. They are rarely active during daylight, although they may spend extended time lounging or curled up at the den entrance on cold days, soaking up heat from the sun. Few foraging foxes venture more than 3 km from their den site, although they may travel several kilometres nightly within that 3 km radius. Hunting animals, in a radio-tracking study in Nebraska, averaged 5.7–18.5 km each night. The amount of travel varies with prey abundance and season, the longest distances being travelled in winter when prey abundance is reduced.

Home ranges are occupied year round, and an average home range in Canada is around 34 km². They are largest in winter. Home ranges further south have been measured at 7.6–32.3 km². The home-range size is probably largely influenced by habitat productivity and hence food availability. Most dispersal is undertaken by juveniles of four to five months old in late summer or autumn as they leave their natal home range. There is another smaller dispersal peak of juveniles in mid-winter, which coincides with the start of the breeding season. Dispersal distances are poorly known for wild juveniles in Canada. In Colorado juvenile males travel an average of 9–10 km, while females move an average of only 2 km. Many female juveniles remain near their natal range. Adults will also sometimes disperse and tend to be more successful than juveniles in this activity. Adult males disperse much more commonly than do females, usually following the death of a mate. This behaviour likely decreases the likelihood of a father mating with his daughter. Extraordinary dispersals of 70–190 km have been reported for the Swift Fox. Swift Foxes are small and vulnerable to many prairie predators. Predation is the principal cause of their mortality in Canada, and Coyotes are the foremost predators. They eagerly pursue and kill, but rarely consume, Swift Foxes. Other predators include Golden Eagles, American Badgers, and Bobcats. The larger owls (Great Horned and Snowy Owls) are also potential predators. A recently emerging predation threat may be the spread of Red Foxes into regions frequented by Canadian Swift Foxes. Red Foxes readily predate smaller foxes, as they do with Arctic Foxes in the north and Swift Foxes further south in North Dakota. Some Swift Foxes die in vehicular collisions, and this mortality factor is most significant for the naive juveniles, especially if natal dens are located near roadways. Swift Foxes experience a high annual mortality rate; around 54% of adults and 64% of juveniles in Canada die annually. Few wild Swift Foxes live more than two to four years, although a lucky few may live to six or seven years old. Captive individuals can live more than 12 years.

DIET

Swift Foxes are opportunistic predators that eat a varied diet, which includes small mammals, lagomorphs, birds, insects, carrion, and occasionally reptiles, amphibians, and fish. They also consume vegetation at times, typically grasses. Their diet varies with prey availability; mainly insects and small rodents are eaten in summer, and primarily rodents, lagomorphs, birds, and carrion are consumed in winter. The largest single prey item for Canadian foxes is the White-tailed Jackrabbit. Adults of this species weigh about as much as, or sometimes more than, an adult Swift Fox. Swift Foxes have a low requirement for drinking water as they probably fill much of their water needs through the prey they eat, and they reduce their water needs by remaining inactive during the heat of the day in a cool, humid den.

REPRODUCTION

Females have a single heat annually, typically coming into oestrus between mid- to late February and April in Canada, and most pups are born from mid-April to June following a gestation period of 49–55 days. Heavier (usually older) females have larger litters. The average litter size of the reintroduced population in Canada is four pups (ranging one to seven). Their eyes and ears open when they are about two weeks old, and pups are fully weaned by the time they are six weeks old. Juvenile males are sexually mature by the following late-winter breeding season, but not all yearling females are equally mature; many delay until they are two years old. Nevertheless, even in the harsh climate of the Canadian Prairies some juvenile females are able to gain sufficient weight over their first winter to enable them to become sexually mature and breed in their yearling year. Around 50% of juvenile females in Colorado, New Mexico, and Texas breed as yearlings.

BEHAVIOUR

Swift Foxes are typically monogamous, and mated pairs remain together on their home range all year. Members of a pair may share a den or use separate dens. Upon the death of a partner, the female usually remains on the pair's home range; the male is more likely to leave his home range unless an unpaired female that can become a replacement mate is nearby. The composition of pairs may change from year to year, but most pairs mate for life. A family grouping usually consists of two parents and their offspring, although groups of three adults, or a mated pair with a helper (probably a yearling female from the previous litter), have been reported, and one male was known to share a den with three females. Males assist females with the care and feeding of pups. Swift Foxes are apparently not territorial, and home ranges may overlap in habitat that has abundant prey. Each fox is a solitary hunter. Swift Foxes are rarely active above ground when it is windy, either in summer or in winter. They are generally considered to be the least aggressive of the North American canids. Like many other carnivores, Swift Foxes will engage in surplus killing if given the opportunity, after which they cache the excess for time of future need. Most food is cached in underground chambers in the den or buried just outside the den entrance.

VOCALIZATIONS

Swift Foxes growl and produce high-pitched yaps and a bubbling, chirping call when interacting. A high-pitched barking sequence consisting of 3–15 individual barks is used as a long-distance contact call by solitary animals and is most commonly heard during the breeding season. Such a call can elicit a responding call from any Swift Fox. The barking call is individually distinct, and neighbours and mates can very likely identify each other's calls.

SIGNS

Signs left by Kit and Swift Foxes are indistinguishable. Den entrances are about 19–20 cm in width and 20–22 cm in height and may be round, oval, or sometimes key-hole shaped with the widest opening at the top. The reason for this unusual shape is unknown, but it may possibly prevent predators, especially Coyotes, from gaining easy access. Swift Fox's burrow entrances may have no throw mound (which is sometimes the case with accessory openings in a large den site complex) or throw mounds that are 80–550 cm long by 30–100 cm wide. Long, thin throw mounds are characteristic of Swift Fox. Den sites with throw mounds can be recognized from a distance, owing to the size and elongated

Side trot

Direct register trot

Left hind track
Length: 2.9–4.1 cm
Width: 2.2–3.2 cm

Left front track
Length: 2.5–4.3 cm
Width: 2.4–3.8 cm

Swift Fox

shape of the waste soil. Whelping dens are the most complex of the burrow types used by Swift Foxes and have the most entrances and the largest throw mounds. Active entrances tend to be littered with scat, bones, feathers, and tracks. Burrows originally made by American Badgers can have the old crescent-shaped throw mound of the badger over-topped by a long, narrow throw mound that resulted from the renovations performed by a Swift Fox. The front feet have five toes, with the fifth lifted on the inside of the leg; it does not register in the track. The hind feet have four toes. Swift Foxes may have heavily furred soles, especially in winter, which can blur the track. The palm pads are fused into a single triangular pad on both front and hind feet but commonly register only partially in hind tracks. The front tracks are larger, and the toes are more widely spread. All toes are clawed, but claws may or may not register in a track. Tracks in loose sand are larger and more splayed in appearance. These small canids use a series of gaits including a walk, trot, lope (canter), and gallop. Their most common gait is a trot, either a direct register trot where the hind feet are planted directly onto the front track or a side trot where the body angles so that the front and hind feet travel on slightly offset paths. This common canine gait is also called two tracking, and it allows the hind feet to smoothly overstep the front feet without interfering or striking them. Scat varies with diet and may accumulate in latrine sites near or on prominent landscape features. Scat varies from 0.5–1.6 cm in diameter and may be 7.6–11.4 cm in length. It can be long and twisted with pointed ends, or dryer and segmented broken pieces, each with a blunt and a pointed end. Urine can have a somewhat skunky odour.

REFERENCES

Carbyn 1989, 1998; Cotterill 1997; Covell et al. 1996; Darden et al. 2003; Gedir 2009; Egoscue 1979; Herrero et al. 1991; Hines and Case 1991; Kamler et al. 2004a, 2004b; Kitchen et al. 2005; McGee et al. 2006; Mercure et al. 1993; Moehrenschlager and Moehrenschlager 2006; Olson and Lindzey 2002; Pruss 1999; Pruss et al. 2008.

Red Fox

FRENCH NAME: **Renard roux**
SCIENTIFIC NAME: *Vulpes vulpes*

Red Foxes are the most common and well-known of the Canadian foxes. Their typical flashy reddish coloration, with a long, white-tipped tail and dark lower legs, makes them immediately recognizable, and they often live in close proximity to human settlements. They are the most widespread mammal in the world (apart from humans), especially after numerous intentional and unintentional introductions and the global reduction in Grey Wolf distribution.

DESCRIPTION

The Red Fox is about the size of a small long-legged dog. There are three pelage phases in Red Foxes. The typical colour, as the name indicates, is rusty or orange red, usually mixed with black-tipped guard hairs. The chin, lower cheeks, chest, and belly are creamy white, the lower legs and feet are dark, a dark patch occurs on each side of the muzzle, and there is a bright white tip on the tail. The large prick ears are usually dark on the back and white on the inside. There is much variation of this colour pattern as some animals have a greyer hind end or a paler body than do others, some are almost yellow or greyish, and many have reduced dark areas. A cross fox is a colour variation that is predominantly greyish-brown (rarely reddish- or yellowish-brown), with a band of darker guard hairs along the spine that is crossed by another dark band running from shoulder to shoulder. A silver fox is pale grey to dark grey to almost black with varying degrees of white frosting caused by white-tipped guard hairs. Silver and cross-colour morphs are rare but can represent up to 10% and 25%, respectively, of the foxes in some northern areas. A single litter can contain pups of more than one colour. Although demand is currently low, Silver Foxes are still being raised in captivity to satisfy the fur trade. The long tail, which accounts for about 35% of the Red Fox's total length, is often mixed with numerous black guard hairs in the typical colour phase and is characteristically white tipped in all colour variations. A 2 cm long gland on the upper part of the tail near the base exudes a secretion that generates the characteristic "foxy" odour. The location of this subcaudal gland is normally marked by a patch of oily, dark fur.

Like all of the canids, Red Foxes have keen vision and hearing and a very acute sense of smell. Their eyes are yellow with a vertical pupil. A well-developed *tapetum lucidum* reflects light with a bright, pale greenish or white eyeshine. The cat-like vertical pupil allows a wide opening for maximum light reception while the fox is hunting in the dark, and a tight narrowing to block bright daylight. Long vibrissae (whiskers) project from the muzzle and from the front ankles. All toes have semi-retractable claws. The dental formula is incisors 3/3, canines 1/1, premolars 4/4, and molars 2/3, for a total of 42 teeth. Adult Red Foxes undergo a protracted moult during the summer, beginning in April and finishing by October. The fresh winter coat is long and lustrous, while the moulting, summer coat is often faded, short, and patchy. Pups are born with a dark natal coat. Between two and six weeks old, depending on the area, they moult into a sandy version of the adult pattern, which camouflages them as they frolic in the often sandy soil at the mouth of the den. There is considerable colour variation in this temporary coat from sandy to brighter red, and light or dark legs and tail. Reddish guard hairs soon begin to extend from the sandy fur, and by around 9–14 weeks old the pups have grown a coat similar in colour to that of an adult.

SIMILAR SPECIES

Swift Foxes, which occur only in the prairie grasslands of southern Alberta and Saskatchewan and southwards, are much smaller and tan or grey above with tawny brown sides and a black tail tip. Grey Foxes are also smaller than Red Foxes and are grizzled grey

Red Fox (*Vulpes vulpes*)

red phase

cross phase

silver phase

black phase

red phase
pup

on their back and head, with rusty sides. Arctic Foxes in summer pelage could be mistaken for small Red Foxes, but their coat colour is brown or blue grey rather than chestnut orange, and they do not have a white tail tip or dark legs. Coyotes are much larger and a pale greyish-tan colour. The Red Fox is the only wild North American canid with a white-tipped tail.

SIZE

The following measurements apply to animals of reproductive age from the Canadian range. Males on average are around 5% larger and may be up to 15%–25% heavier than females from the same region. These differences are very difficult, if not impossible, to distinguish in the field. There is considerable size variation across the wide range of this species, but the largest animals tend to be in the north.

Total length: 935–1110 mm; tail length: 350–455 mm; hind foot length: 155–174 mm; ear length: 89–102 mm.

Weight: males 3.8–7.0 kg; females 3.4–5.8 kg.

Newborns: weight: 85–115 g.

Shoulder height is about 35–40 cm.

RANGE

Red Foxes are widespread across the Northern Hemisphere in both Eurasia and North America, north of Mexico. In Canada they occur from Newfoundland to coastal British Columbia, north to the edge of the mainland, and have gradually island hopped since 1918 from Ungava Peninsula to Baffin Island and from there to Bylot, Devon, Somerset, Cornwallis, and southern Ellesmere Islands. Foxes were introduced to Vancouver Island probably as escapees from fur farms but remain rare and are now possibly extinct. Red Foxes in North America, prior to European settlement, probably occurred in boreal and hardwood forests north of the fortieth parallel. European Red Foxes, considered to be the same species as the North American foxes (although different subspecies), were introduced from Britain in the 1600s into the eastern and southeastern United States for the purpose of fox hunting. Introductions for sport hunting into Australia in the mid-1800s have resulted in the Red Fox becoming established across the island continent, except in the far north. Foxes have had a major impact on the decline and extinction of many native species in Australia. In the early 2000s, Red Foxes were detected on Tasmania, suggesting another surreptitious introduction, which

Global distribution of the Red Fox (*Vulpes vulpes*)

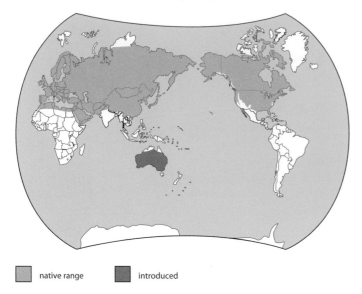

native range   introduced

North American distribution

is thought to have occurred in the late 1990s. Tasmania is taking serious and extensive measures to exterminate this threat before the foxes become established. Escaping from fur farms is the most common form of inadvertent introduction in North America.

ABUNDANCE

Red Foxes are relatively common throughout most of their North American range, although density tends to diminish in dryer, desert habitats in the southeast and in the north, especially in tundra regions. Density estimates within North America generally range from 0.1 to 1.0 per km² but may reach 3.0 to 10.0 per km² in limited but very productive areas. These foxes are probably more numerous now in North America than they were when the Europeans first arrived, thanks to forest clearing for agriculture and urbanization,

large predator-reduction programs, and introductions. Red Foxes are endangered on the islands of Japan and are considered to be threatened in the Sierra Nevada mountains of California.

ECOLOGY

The Red Fox is a highly adaptable, very successful generalist that is able to live in a wide variety of habitats, from open grassland and semi-desert to dense forests and Arctic tundra. Its preferred habitat is semi-open and especially woodland-edge habitat, where prey is abundant. Agricultural and urban regions are commonly occupied as long as prey is available, and the clearing of forests for agriculture has allowed Red Foxes to increase both their density and their range in eastern North America. Normally nocturnal, Red Foxes are sometimes seen during the daylight, especially as light levels drop during dawn and dusk or on heavily overcast days. Adults rearing pups are also more likely to be active in daylight. During winter, foxes may become more active in daylight because their normally nocturnal, small mammal prey is often active under the protecting blanket of snow regardless of light levels. Red Foxes are adaptable enough not to be inconvenienced by the 24 hours of light during the summer in the far north. Their home-range size varies with habitat, season, and prey abundance, from 4 to 35 km². While rearing pups, adults typically restrict their movements to an area of around 4–9 km² centred on the natal den. Daily travel while foraging can easily exceed 10 km. The family units begin to disintegrate in early autumn as the juveniles disperse from their parents' home range. Juvenile males tend to travel farther than do the females. They are known to disperse up to 250 km from their natal range.

Dens are used during the summer to raise pups. These may be dug by the adults or, more commonly, they may be renovated burrows initially dug by other foxes (such as Arctic Foxes in the north or Swift Foxes in the prairies) or by more fossorial mammals such as marmots (including Woodchucks), ground squirrels, or American Badgers. Most dens have multiple entrances and may be occupied for a single season or used by several generations. Adults may rest during the day in the den with the pups, especially when they are very young, but usually they find protected, above-ground resting places in both summer and winter, using thickets, dense vegetation, debris piles, rock overhangs, or low-hanging evergreen boughs for shelter. Predators in North America include larger canids (Coyotes and Grey Wolves), Cougars, Canada Lynx, Bobcats, and sometimes eagles (which take pups). More Red Foxes are killed by the fur trade, sport hunters, and motor-vehicle collisions than by all other predators. Foxes are primary vectors of rabies, as are Northern Raccoons, skunks, and to a lesser extent bats. Red Foxes may be afflicted by either the dumb form of rabies where they gradually succumb to progressive paralysis and in the process may lose their fear of humans, or the furious form where they attack objects or other animals including humans. They may also transmit many diseases and parasites including canine distemper, canine parvovirus, ringworm, and mange mites. Although captive foxes can live 15 years, few wild individuals live more than 3–4 years, with a rare individual living 7–12 years.

DIET

Omnivorous and highly flexible, Red Foxes consume almost anything edible including human garbage. Small mammals including mice, voles, and shrews are their preferred prey, to which they add birds and bird eggs, muskrats, squirrels, Woodchucks and other marmots, North American Opossums, rabbits and hares, Northern Raccoons, other smaller carnivores (even skunks on occasion), newborn Ringed Seals, invertebrates such as insects, crayfish and earthworms, fish, fruit, nuts and acorns, vegetation, reptiles, amphibians, and carrion as available. The diet varies with season and prey availability. Like many other carnivores, Red Foxes indulge in killing of excess prey beyond their immediate needs. This surplus is cached (usually carefully buried in earth or snow) for later consumption. Red Foxes are quick to exploit a windfall such as free-ranging poultry or colonially nesting waterfowl. Although it is rare, infanticide has been reported, perpetrated by strange adults who invaded the territory and snatched an unwary, above-ground youngster while its parents were away from the den.

REPRODUCTION

Vixens (females) are monoestrous and produce a single litter annually. Breeding occurs in February, March, or April. The onset of oestrus and the resulting parturition date vary with elevation and latitude and occur later in colder regions. Gestation lasts 52–53 days, and most pups in Canada are born from late March to May. The litter size averages four to eight (ranging from one to twelve). Well-nourished mothers produce larger litters. Males participate in raising the young by protecting the home range and delivering food to the vixen and pups. The degree of male parental care can vary between males, as some are considerably more diligent than others. Whelps are born with short, dark, grey-brown fur and are blind and deaf. By about two weeks they are well furred and large enough to maintain their own body temperature and no longer need the thermal blanket provided by their mother's presence. Their eyes open around 10–12 days old, and the pups begin to walk around this time. They start sampling solid food in their third week and begin to emerge from the den to play at its mouth at around four weeks old. Weaning begins at around five weeks, and the pups are gradually weaned by the age of eight weeks, around the time their milk dentition is complete. Their parents continue to bring prey back for the pups, but the pups also begin to explore the area around the den by 10 weeks old and may follow one of the adults on hunting expeditions. By 12 weeks old they are exploring farther afield and doing much of their own hunting, usually for small prey such as insects. The juveniles disperse from their parents' home range at the age of three to four months. Juvenile males become sexually mature around eight months old, and females around ten months old, in time for the next breeding season. Dispersing pups may travel together in small groups of two or three for a time, but as they are hunters of small, mostly scattered prey, they are soon forced to become solitary hunters.

BEHAVIOUR

Native peoples, trappers, farmers, and others living close to the land have long respected the cunning and adaptability of this curious and intelligent carnivore. Tales, fables, and legends that abound concerning the Red Fox and its continuing ability to thrive despite unrelenting persecution lend credence to its superb survival skills. Although usually shy, Red Foxes can become habituated to human activity, especially if undisturbed. Some individuals may reach such a level of acclimatization that they will continue their normal activity even under obvious human scrutiny.

As Red Foxes are generally solitary outside the breeding season, the onset of the annual mating period is announced by pairs running together and by their nocturnal barking. The female will clean out several dens on their shared territory and select one as the primary maternity den. One of the others will be put to use if the first is disturbed. Red Foxes are typically seasonally monogamous or polygynous (males mating with multiple females). Typically, pairs remain together through the breeding season. Vixens sequester themselves in the den for a few days before parturition and then for several days (up to 10 days in colder regions) afterwards to keep the pups warm. The dog fox (male) brings her food during that time. Partners may remain together over the winter if food is abundant, but usually they separate and may or may not reunite in the following late winter or early spring to breed again. When food is abundant, alternate arrangements beyond a male-female pair may be more successful for a season. "Helper" foxes, usually non-reproductive females from the previous litter, may remain on their natal territory and assist in catching prey for the new litter. During abundant prey years both the primary female and the helper may breed with the same male. Females tolerating a polygynous relationship with a common male are usually closely related. The females may occupy separate but nearby dens or share a den and suckle the combined litters. Red Foxes, on rare occasions, split a single litter between two or more dens. Females that have lost their mates may try to raise the pups alone but are rarely successful unless food is plentiful.

Pairs maintain exclusive territories, marked with urine, faeces, and anal gland secretions, and both sexes (particularly the males) aggressively defend it from interlopers. Vocalizing and threatening postures comprise most of the defence, which terminates in a chase to drive away the stranger. Actual physical battles are rare. The most common threat posture is a broadside display. The dominant animal presents its profile to its opponent. It erects its fur and arches its back and tail to appear larger and will advance in a graceful sideways gait while holding the broadside display. The subordinate animal

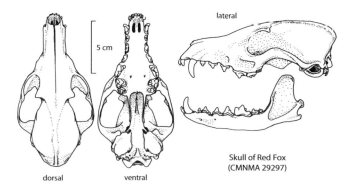

Skull of Red Fox
(CMNMA 29297)

lateral

5 cm

dorsal   ventral

Direct register trot

Side trot

Left hind track
Length: 4.0–6.5 cm
Width: 3.2–4.8 cm

Left front track
Length: 4.8–7.3 cm
Width: 3.5–5.5 cm

Red Fox

either runs away or crouches with ears back and sometimes beats its tail wildly and indulges in puppy-like begging actions towards the mouth of the dominant fox. Around 25 days old, littermates begin the serious business of fighting with their siblings for dominance. A dominance hierarchy, usually based on size, is firmly established amongst the pups over the following 10 days. This hierarchy can be extremely important because the higher-ranking pups can steal food from their lower-status siblings. Although the parents invariably present prey to the first pup that begs for it, that pup must then defend its food from its hungry littermates. If prey is scarce, the subordinate pups may starve as the parents make no effort to interfere with this, sometimes brutal, sibling rivalry. Once the hierarchy has been well established and entrenched, usually soon after they begin to appear above ground, aggression between the pups declines and they become more social and playful.

Red Foxes are dominant over other native foxes. They will readily chase down or dig out these smaller foxes and kill them. The newly created population of Swift Foxes (in the southern parts of Alberta and Saskatchewan) are threatened by an expanding population of Red Foxes in that region. Red Foxes are solitary hunters. Ever resourceful and opportunistic, they utilize different strategies for hunting different prey. Small mammals are normally hunted by sound. Their progress is tracked as the fox sits or stands in the "mousing" position – at high alert with the head elevated, eyes staring intently, and ears swivelling towards the sound. Once the prey's position and progress has been ascertained, capture is then accomplished with an arching leap that ends with the fox's forefeet pinning the prey to the ground. If the prey is larger or capable of flight, the fox will forgo the arching leap and lunge straight at it, attempting to grasp it in its jaws before it takes to the air or begins to run. Birds and larger mammals, such as squirrels, are stalked with belly almost to the ground as the fox slinks between clumps of cover until it is close enough for a lunge. Sometimes, if the fox misses on its first attempt, it will curl up for a "nap" while continuing to watch and listen for the prey to resume its former activity. With this patient, silent tactic the fox may garner a second chance. Red Foxes that forage in seabird colonies have learned to climb steep cliffs to reach the nests on ledges, and some foxes in Newfoundland have learned to travel on the top of the tuckamore (dwarfed and twisted firs caused by high, sheering winds). Red Foxes can detect a vole nest through almost a metre of snow. They likely hear the faint squeaks of the pups and know that the vulnerable pups cannot run away as they noisily dig down to expose the nest. Prey is killed either by crushing with the forefeet or by pressing the long, needle-sharp canines into the body, usually at the base of the neck. Red Foxes rarely shake their prey to kill it, like most other canids do. Excess food is normally buried. A hole, 7–10 cm deep, is carefully dug, no wider than the width of the two front paws, and the item is placed within. Usually each item of food is cached individually, but in some cases where prey is superabundant in a small area (such as a seabird colony) the caches may include dozens of items. The soil or snow is returned to the hole and tamped down with the nose. Surrounding litter or snow may then be scattered over the area with the nose to camouflage the dig site, even to the extent of smoothing their own tracks. Some experience is required to create a neat and well-hidden cache, but even pups of six weeks old will instinctively attempt to bury food. Recent caches are normally pinpointed by memory, but if the food is recovered up to several months later, as some may well be, the fox might remember the general area but then depend on its nose to find the actual site. Caches can become crucial to survival if a fox is injured, bad weather sets in, or prey becomes scarce. Foxes will steal each other's caches, but since these would ordinarily be stolen by a family member sharing the same territory, at least the food energy remains in the family.

Red Foxes can be playful. Pups tussle and chase their siblings and stalk their parents, and the adults will often join their fun. Adults also indulge in solitary play from time to time. Like cats, they sometimes toy with live or dead prey, exuberantly dancing around it and flinging it into the air before eating, caching, or, rarely but astonishingly, letting it go. Finding itself on slippery ice, a fox can quickly learn to enjoy sliding around and may adapt its normal foraging route to return to the ice for repeated sliding capers. Adults will also play with each other. Typically this activity involves mated pairs frisking around and chasing each other.

VOCALIZATIONS

Red Foxes are not normally loudly vocal, except during the spring breeding season when adults are pairing up and during the autumn dispersal period when the juveniles are competing for territory. A long-distance contact call sounds like a combination of a yapping bark and a scream, which may be repeated several times. Red Foxes may produce a barking call if they are startled or to warn other nearby foxes. Young pups (two to three weeks old) produce a low-volume warbling call to indicate distress or hunger. Adults call pups with a medium to low-volume chortling call. They also use a coughing bark to warn nearby pups of possible danger and send them scurrying for the den. Red Foxes may whine, growl, shriek, and scream during social interactions.

SIGNS

Red Fox urine has a decidedly skunky odour, which becomes especially strong during the breeding season. Urine scent posts can become so pungent that they can be detected by even our feeble noses. Urine marking is one of an adult Red Fox's most common behaviours, and foxes of both sexes mark throughout their territory as often as 70 times per hour. Only a few drops are released each time, but this means that a "normal" voluminous and noticeable urination is uncommon and is typically restricted to pups or dispersing juveniles that are leery of marking on an already occupied territory. Urine scent posts are often located on prominent landscape features such as rocks, fallen trees, shrubs, mounds, or stumps. Scat varies widely with diet, from 0.8–1.9 cm in diameter and 7.5–15.0 cm in length. Meat consumption produces a twisted, roughly tubular, blackish scat, often with long pointed ends. These scats may or may not include small bones and fur, depending on the diet. Scats composed of fruit and mast are tubular, with a smooth surface and blunt or slightly tapered ends. The composition of the scat is often easily visible.

Den entrances are about 15–30 cm wide and up to 40 cm high. The tunnels that connect to them are about 13 cm in diameter and can easily reach 14 metres in total length. Dens may be simple with only two entrances or more complex with several entrances. The main entrance usually has a conspicuous, large throw mound of excavated soil where scat and feeding debris accumulate. Additional entrances are often much more cryptic and smaller than the primary entrance. The front feet have five toes, but the first toe is highly reduced and raised on the inside of the leg, so it does not strike the ground or register in a track. The hind feet have four toes. The front tracks are larger and rounder than the hind. The palm pads on both front and hind feet are triangular, but they commonly register as roughly circular on the hind feet. The foot pads are usually well furred, especially during winter, which can blur the track and reduce the registration size of the toe and palm pads. Often the fur allows only the edge of the palm pads to touch the substrate, leaving a characteristic flat-line (hind feet) or chevron-shaped (front feet) impression in snow, a diagnostic trait of Red Foxes. As with all the North American canids, the negative space between the toes and the palm pads forms an X. All 18 toes are clawed, and apart from the first toes on the front feet, which do not touch the ground, the claws are semi-retractable and may or may not register in a track. The most common gait of a foraging Red Fox is a ground-eating trot where the hind feet register directly over the front track (see illustration). They commonly switch to a walk (also direct register but with shorter strides) in deep snow. Another common gait, used mainly when the footing is good, is a side trot (see illustration). In this gait the back angles to the side so that the hind feet travel outside the front feet in order that they can fully extend without striking the back of the front legs. Red Foxes are very agile and are able to jump up angled tree trunks; they are able to walk easily on the top of stone walls and logs. Trails are ordinarily single, except during the breeding season.

REFERENCES

Elbroch 2003; Finley, K., 1996; Henry 1996; Jameson and Peeters 2004; Larivière and Pasitschniak-Arts 1996; Pamperin et al. 2006; Reid, F.A., 1997; Schmidly 2004; Seidensticker 1999; Sklepkovych 1994; Sklepkovych and Montevecchi 1996; Vergara 2001a, 2001b; Verts and Carraway 1998; Whitaker and Hamilton 1998.

FAMILY URSIDAE

bears

The largest members of the order Carnivora are those of the family Ursidae. Male Brown and Polar Bears may exceed 600 kg. Bears are typically large bodied and heavy, with short powerful legs, small ears, and a very short tail. They are plantigrade; the whole sole of the foot strikes the ground, as it does with humans.

North American Black Bear

also called American Black Bear, Black Bear

FRENCH NAME: **Ours noir**

SCIENTIFIC NAME: *Ursus americanus*

The North American Black Bear is the most numerous and wide-spread of the three North American bears. It once roamed all the forested regions of North America and Mexico. This bear remains the most commonly seen bear in North America but only occupies approximately 62% of its former range.

DESCRIPTION

North American Black Bears have a wide variety of coat colours. Their uniformly coloured, coarse, dense fur is most commonly black in boreal, montane, and temperate rainforest populations in the east, north, and west coast, but some form of brown, chocolate-brown, reddish-brown (cinnamon), or even occasionally blond pelage is common in the open woodlands of the west. Brown animals may appear occasionally in the east. More than one colour phase can occur within the same litter. In southern Yukon some Black Bears are dark brown with yellow-tipped guard hairs on the back and rump. A creamy-white or pale-blond colour phase with a black nose and dark eyes (called a Kermode or Spirit Bear) and a bluish-grey phase (called a Glacier Bear) occur in coastal areas in the west. The muzzle area is usually light brown, except in the white, blond, and blue animals, and some individuals have a white blaze on their chest. Many black phase animals have brown spots on their eyebrows. The belly fur is the same colour as the rest of the animal. True albinos are rare but occur throughout the range. The feet are large and flat, and Black Bears, like all the bears, walk on the whole sole of their feet (plantigrade), as do humans. Their claws are large, curved, non-retractable, and dark brown or black

in colour, except on white or blond animals, which have white or pale-tan claws respectively. The claws are slightly longer on the front feet than on the hind. The eyes are blue at birth but soon turn dark brown. The tail is so short that it is rarely visible in the long rump fur. The prominent ears are rounded and well furred on both the front and the back sides. Distance vision is poor, but near vision is sharp, and their senses of smell and hearing are acute. Black Bears can perceive colour. The dental formula is incisors 3/3, canines 1/1, premolars 2–4 / 2–4, and molars 2/3, for a total of 34–42 teeth. The first three upper and lower premolars are usually rudimentary, and some may be absent altogether. These bears have six mammae. The eyeshine is reddish-orange.

SIMILAR SPECIES

Their stocky build, large size, round body, thick legs, and short tail make bears easily recognizable. No other North American bear is black, and Black Bears are the only bears that occur in eastern and central North America. Some confusion may arise in the west and northwest, where brown-phase Black Bears could be confused with young Brown Bears. Brown Bears are usually larger, have shorter more rounded ears, a high shoulder hump, and a heavier head with a distinct down curve from the forehead to the muzzle. Black Bears have a relatively straight head profile. Claw impressions are evident in the tracks of both the front and the hind feet of Black and Brown Bears. The key in comparing tracks is the relative distance of these impressions on the front feet compared to those on the hind feet. Both bear species have longer claws on the front feet, but those of the Brown Bear are much longer on the front than on the hind. See the track graphics of the two species for comparisons.

SIZE

The following measurements apply to animals of reproductive age from the northern part of the range (Canada and the northern United States). Males are larger than females by approximately 10% and heavier by 10%–200% or sometimes more. Body weight is greatest in autumn. Captive animals may exceed the provided weights.
Males: total length: 130–200 cm; tail length: 10–18 cm; hind foot length: 24–28 cm; ear length: 12–14 cm; weight: 60–226 kg (maximum reported weight in the wild is 370 kg).
Females: total length: 120–160 cm; tail length: 8.0–11.5 cm; hind foot length: 22–25 cm; ear length: 12–14 cm; weight: 40–136 kg (maximum reported weight in the wild is 236 kg).
Newborns: weight: 200–400 g.
Adults are 50–90 cm high at the shoulders.

RANGE

One of the most widely distributed mammals in Canada, North American Black Bears occur from coast to coast to coast. Many large oceanic islands are occupied, including Vancouver Island, the Queen Charlotte Islands, and Newfoundland. Notable exceptions are Prince Edward Island (where they were extirpated by 1937) and Anticosti Island (where they were extirpated by 1998). Manitoulin Island is the largest freshwater island that sustains a stable Black Bear population. The Spirit Bear or Kermode Bear occurs in

white phase (also called Kermode Bear)

grey phase (also called Glacier Bear)

cinnamon phase

black phase

cub

North American Black Bear (*Ursus americanus*)

the north-central coast of British Columbia from approximately the Burke Channel in the south to the Nass River in the north. The lighter-colour phases are most common on Princess Royal and Gribbel Islands, where about 10% of the bears are white or pale blond while the remainder are black. The blue phase, called a Glacier Bear, is occasionally seen in the extreme northwest corner of British Columbia and adjacent Alaska. Most of the coastal bears, including those on Vancouver Island and the Queen Charlotte archipelago are black. Black Bears in the mountains are often brown or cinnamon, and those in the east are black. Black Bears are the only bears in eastern North American forests, while in the west they are often sympatric with Brown Bears. As forests were converted to farmland in the 1800s and 1900s, Black Bears were eliminated from large swaths of former range, especially in the eastern and central United States. They have reclaimed some of this land in recent years as farmers abandoned marginal agricultural land and allowed it to revert to forest.

ABUNDANCE

Some populations of Black Bear are threatened, especially in the United States and Mexico, but Canadian Black Bears are generally secure. There are 120,000–160,000 Black Bears in British Columbia, about one-quarter of the roughly estimated 500,000 Black Bears in Canada. An additional 400,000–500,000 or more occur in the United States, up to half of these being in Alaska. Most jurisdictions within the Black Bear range sanction an annual hunt, and 40,000–50,000 are taken annually in Canada and the United States. In the parts of Canada with stable populations the densities vary from 0.05 to 0.62 bears per km². Black Bears are increasingly losing habitat to ruralization and expanding urban centres. Especially during low berry and mast years, the bears are attracted by gardens, outside pet food, garbage, and cooking odours and are often shot as nuisance animals. The illegal international trade in paws and gall bladders is an ongoing conservation concern for Black Bear populations.

ECOLOGY

This adaptable bear occurs in a wide range of forested habitats, from temperate hardwoods and swamp forests to coastal rainforest, open montane woodlands, dense boreal forests, and even open tundra above the treeline (Torngat Mountains in Labrador). North American Black Bears are active at any time of day or night. They generally manage their active periods to maximize their foraging opportunities but become most active in low light or darkness in areas where they may conflict with humans. During late summer when they forage for small berries, bears are most active in daylight because the light helps them find and harvest the fruit and determine its ripeness. They alternate activity periods of roughly four hours with short rest periods of about an hour during their active phase. Since

Distribution of the North American Black Bear (*Ursus americanus*)

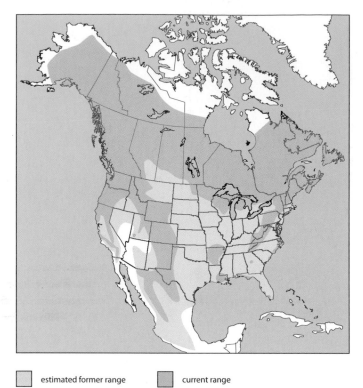

estimated former range	current range

their food all but disappears in winter, North American Black Bears have no alternative but to hibernate during that time. Most Black Bears in Canada hibernate for five to seven months depending on the length of the winter in their local area. They accumulate fat in late summer and early autumn to sustain them during this period of reduced metabolism and prolonged dormancy. The onset of hibernation varies with physical condition; fatter bears begin earlier than do leaner bears. Females that are pregnant or have cubs enter hibernation first, typically by early October. Most bears are hibernating by the end of October in Canada, although during heavy mast years many remain active (mainly the underweight subadults), eating acorns or beechnuts, until November despite freezing temperatures. Rarely, Black Bears in Canada may remain active until December. Most Black Bears in the northern parts of their range sleep heavily during hibernation, but females with cubs sleep lightly so that they can respond to the demands of their cubs. Black Bears emerge from winter dormancy in April or early May, and females with new cubs emerge last. Even though their body temperature does not drop appreciably (from 37°C–38°C to 31°C–35°C) because of their large body size, Black Bears are considered to be true hibernators by many scientists. Their metabolism drops by 30%, their heart rate slows from 40–50 beats per minute to around 8–10, and their oxygen usage drops by half. They do not eat, drink, urinate, or defecate while torpid, and their blood does not accumulate toxic levels of waste products as a result of the metabolism of their body fat, because the bears have a unique ability to recycle this waste into more nutrients. Furthermore, they lose little muscle tone because of the inactivity, and their bones do not lose calcium. They do drop

up to 30% of their body weight over a winter, and the weight loss of lactating females may be up to almost 44%. In early spring the best food that the bears can find are small green shoots, semi-dormant insects, or carrion if they are lucky. Those that run short of fat may awaken in spring with insufficient remaining energy in reserve to sustain them until food becomes plentiful again. Few bears actually die during hibernation, but spring starvation is a fate suffered by many, especially cubs and yearlings.

Winter dens may be in tree cavities and hollow logs if large trees are available. If not, or where snow can provide insulation, the bears find small caves, rock crevices, windfalls, or brush piles or they dig holes in the ground. Some bears even hibernate on open ground, although this is most common in the warmer southern United States. Older, more experienced animals are thought to select better and safer sites for hibernation. The bears typically drag in some vegetation to line the cavity, which is normally just large enough for them to curl up in. Entrances are at least partially, if not completely, plugged, especially in cold regions. A secure winter den is especially important for pregnant females or those with cubs. Some particularly favourable den sites are reused by either the same or different bears, but this is uncommon. Denning is preceded by a period of diminishing activity as the bear's digestive system changes for hibernation. This transition may last up to a month.

Male Black Bears establish ranges large enough to provide them with access to both mates and sufficient food. These vary in size (from 51 to 9500 km²) but average 60–150 km² in a forest setting and can encompass parts of the home ranges of up to 15 females. The home ranges of neighbouring males usually overlap to some degree. Female home ranges need only be large enough to provide sufficient food and den sites for her and her growing cubs and are normally 12%–50% of the size used by males in the same region. The larger home ranges quoted above derived from animals living on the barren grounds of northern Labrador. Some females defend a territory from strangers, allowing only her grown female offspring to share it. In other populations the home ranges of females overlap extensively, even with non-relatives. Home ranges may alter with the season, as the bears are known to travel up to 174 km to take advantage of temporary food abundances. After the intensive foraging activity in autumn that may take them well beyond their normal home range, most bears return to their summer range to den-up for winter. Some animals, mostly males, may leave their home range to travel some distance before selecting a den site, but these animals usually return the following summer. Dispersal of independent juveniles typically occurs in late spring or early summer, before the females breed. Most female offspring are permitted to establish a home range that overlaps somewhat with that of their mother, but male offspring disperse further before establishing a home range. Dispersal distances of 13–219 km have been reported, with an average of 61 km.

Black Bears can easily outdistance a human, having a top speed that is around 50–55 km/h, but this effort can only be sustained for short distances. Lean bears can run faster and further than can fatter bears, which overheat and fatigue more quickly. Even the largest males can climb trees easily, and Black Bears are capable swimmers, able to travel at least 1.5 km to reach islands. North American Black

Bears have few natural predators, but young may be killed by Bobcats, Coyotes, Grey Wolves, or other bears (both Brown and Black). Brown and Black Bears, particularly the larger adult males, may on rare occasions kill and consume adult (usually female) Black Bears that they chase down and kill or that they discover in hibernation. Most mortality of youngsters is caused by starvation, but in many, if not most, areas more than 90% of bears over 18 months old die from human-related causes, mainly shooting, trapping, and collisions with motor vehicles. Even in national parks, 80% of the mortality can be human related, primarily through motor-vehicle collisions and shooting or forced relocation of nuisance animals. Black Bears continue to suffer from a bad reputation, and many are shot unnecessarily just because they are close to human habitation. Captives can live 25–30 years, but few wild bears live more than 8–10 years. The oldest known wild bear is a 36-year-old female from Minnesota.

DIET

North American Black Bears spend up to seven months hibernating each year, so they must maximize the time remaining to rebuild their fat reserves. They have a tremendous appetite during the warm months and spend most of their time eating and sleeping. Their preferred foods are high in protein, fat, and sugar. Although classified as carnivores, Black Bears are functionally omnivorous, opportunistically consuming a wide variety of plant and animal foods as they are seasonally available. They eat insects, fruit and berries, fish (especially salmon where available), reptiles, and bird eggs; juveniles of Caribou, Wapiti, Moose, and deer; small mammals; fungi; carrion; and a wide variety of plant material such as grasses, the cambium layer of trees, nuts, and succulent aquatic vegetation. Most of the time, Black Bears ignore adult birds and large mammals, but they are certainly strong enough to take down an adult cow Moose, Caribou, or Wapiti and will do so under certain conditions, particularly when the ungulate has been weakened. Black Bears in northern Labrador are abnormal in their reliance on animal protein. In the absence of Brown Bears in the region, Black Bears have moved into the treeless, barren-ground habitat where they actively pursue and take down adult Caribou. Black Bears are also known to kill livestock in some regions. They can become significant predators of neonatal ungulates. Although they prefer meat and insects, the bulk of a Black Bear's diet is made up of plant material. Black Bears may be important distributors of fruit seeds, as many of the seeds benefit from a trip through the bear's digestive tract and are then deposited, in the scat, some distance from the harvesting site. Cannibalism of cubs and even adult females by male Black Bears is a significant mortality factor in some populations.

REPRODUCTION

Compared to most other mammals, bears have low reproductive rates. Female Black Bears enter oestrus approximately every two to four years between June and September, with most breeding taking place in June and July. An adult male seeks oestrous females, using scent. Once he finds her, he attends her for a few hours to a few days, during which time they may mate several times, before he moves on in search of another mate. Females may mate with more than one male, and multiple paternity within a litter is possible. In the east the inter-birth interval is around two years, while in the west it averages three years or more. Black Bears experience delayed implantation. The fertilized blastocyst floats around in the female's reproductive tract for several months before implanting on the uterus wall in late November or early December. Actual gestation, determined from the date of implantation, lasts 60–70 days. One to six cubs are born in January or early February while their mother is in hibernation. The average litter in western North America comprises two cubs, while in the east most females have three cubs. First-time mothers often have smaller litters. The nutritional status of the female in early winter is directly related to the size of the litter she will produce, as more fertilized blastocysts will implant and fewer foetuses will resorb in well-nourished mothers. When food is scarce, females may not enter oestrus at all that year, and if they do, their implantation rate will be lower and the resorption rate higher. Females in very poor condition may abandon cubs. Lactation suppresses oestrus, so a female that is suckling cubs will not breed that summer.

Newborns are sparsely furred, toothless, and blind, and their ear canals are closed. They are very small in relation to their ultimate adult size and grow slowly at first. To place the neonate size in perspective, they are similar in weight to a small rat and are approximately 1/250th the weight of their mother. Their eyes open at around six weeks old. The cubs nurse while their mother hibernates, and weigh 3–5 kg when she arouses in spring and they leave the den. Cubs begin to taste what their mother eats almost right away but do not actually consume it until their teeth have all erupted later in the spring. Weaning occurs gradually between July and September of their first summer. While some females may continue to allow yearling cubs to suckle, their milk production will have ceased. Cubs spend their first winter, and sometimes their second as well, hibernating with their mother, and she continues to support and protect them while teaching them to find food and avoid danger. She will drive them away when she is ready to breed again, at around 17–29 months old in the east and around 29 months old in the west and far north. Sexual maturity of females is achieved around two to eight years old, depending on their food intake, and they reach physical maturity at three to five years old. Males become sexually mature at three to four years old and physically mature between five and eight years old. As with most reproductive matters in this species, nutrition plays a major role in the age of both sexes at sexual maturity.

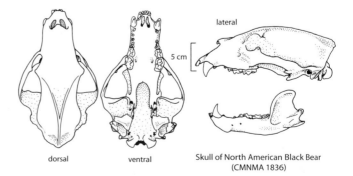

lateral

5 cm

dorsal

ventral

Skull of North American Black Bear
(CMNMA 1836)

Well-fed animals mature earlier. The testes of the male are scrotal only during the breeding season and retract into the abdomen for the rest of the year.

BEHAVIOUR

Males compete for access to oestrous females, and fighting is common, especially between individuals of similar size. Older males may sport battle scars on their head, neck, shoulders, and forelegs from the teeth or claws of their opponents. Males continue to grow in weight until they are 10–12 years old and will use their large size to dominate younger, smaller males without the need to fight. Generally the largest, strongest males in an area are dominant and perform a high percentage of the matings. Apart from a mother with cubs and the temporary mating bond, adult Black Bears are generally solitary. Aggregations around a rich food source, such as a garbage dump or spawning salmon, usually lack social cohesion because each bear fends for itself and individuals generally stay away from each other. Bears in long-lasting aggregations (such as at a garbage dump) may form dominance hierarchies, and the most dominant animals forage at the preferred locations and drive subordinates away from a rich food source. Gatherings tend to be avoided by mothers with cubs because the other bears may attack the vulnerable youngsters. Black Bears are intelligent and display considerable curiosity and problem-solving ability. A female Black Bear is a solicitous mother, even responding while in hibernation to vocalizations from her cubs pleading for warmth or milk, and she will move them to another winter den if disturbed. Once out of the den, the cubs follow her wherever she goes, learning the local geography and how to find food. They are able to remember and return to abundant food sources, in the appropriate season, even years later. A female continues to recognize her cubs for several years after they leave her, and possibly indefinitely. Captive females with newborn cubs will readily adopt an unrelated cub if it is placed with her in the den, and adoption of orphaned cubs has been reported in the wild. Young bears are fearful of unknown, larger bears and run from them as soon as they are detected.

Black Bears can rise on their hind legs, which gives them a visual advantage and at the same time increases their physical presence, making them appear more threatening. Their sweet tooth places Black Bears in regular conflict with honey and fruit producers across the country. They are clever and quickly become habituated to humans if food is provided, even inadvertently, as may happen at campgrounds where they learn to raid unattended coolers and cars with open windows. Some bears, even in the deep woods, learn to lie in wait at a canoe portage to tear open the campers' knapsacks while they have gone for another load. Bears that are accustomed to foraging around humans, such as at a landfill, a campground, an orchard, or at the roadside, can become dangerously aggressive, especially when food is scarce, but the vast majority are shy of humans and will retreat unless lured by food. Female Black Bears rarely attack people in defence of their cubs, preferring to send them up a tree for protection before retreating themselves. Some females will bluff charge before retreating. Occasionally, a defensive female will attack and may cause injury or death. Most Black Bear attacks

Walk

Left hind track
Length: 14.0–22.0 cm
Width: 9.0–15.0 cm

Left front track
Length: 10.0–20.0 cm
Width: 8.5–15.0 cm

**North American
Black Bear**

are clearly predatory, perpetrated by adult males. As a precaution, children should never be left unattended in the woods. Hibernating Black Bears should not be disturbed, as some can awaken aggressively in a matter of a few minutes. Despite millions of encounters, there were fewer than 36 human deaths in Canada and the United States that were attributed to Black Bears in the twentieth century, but the number of attacks appears to be increasing, likely owing to the increasing numbers of humans invading bear habitat for either recreation or habitation. If one is attacked, the best action is to fight back and make a lot of noise; playing dead is not a good option. If one suspects that a Black Bear is nearby, sufficient noise should be made to alert it so that it can retreat before being confronted.

VOCALIZATIONS

Black Bears are not especially vocal. Their most commonly used vocalization is a grunt or a huff that is used to communicate with cubs, playmates, or mates under amiable circumstances. A loud blowing sound will precede a bluff charge, and blowing combined with tooth chattering indicates a fearful animal. A short lunge followed by slapping an object or the ground with one front foot, or both, is a threatening, "move back" demand. They may also clatter their teeth when aggressive, and produce a loud, rumbling hum when contented. Cubs will scream or bawl to indicate distress, whine when approaching their mother, and produce a rumbling hum when contentedly nursing. Black Bears of all ages will bawl in pain, moan in fear, bellow in combat, and produce a deep-throated pulsing sound that could be mistaken for a growl when threatening another bear or a human.

SIGNS

Each foot has five toes, each with a large curved claw that usually leaves an impression in the track. The distance of claw impressions from the track is slightly further on the front foot than on the hind. The front feet have a separate, circular heel pad, located well behind the palm pad, that sometimes registers in a track. The hind heel pad is joined with the palm pad and almost always registers in a track. The first toe (thumb and big toe) is the smallest and is located on the inside of the track. The most common foraging gait is an over-step walk where the hind track precedes the front track, but Black Bears will also lope to traverse open ground or gallop when threatened. Large, ripped-up, or overturned logs, overturned rocks, large scats, well-worn paths through berry patches, and broken limbs and climbing marks on mast-bearing trees are common signs of Black Bear presence. Marked trees are typically found along a bear trail, especially at a ridge line with the marked portion facing the trail. These trees are scent marked and sometimes also scratch or bite marked, especially by adult males, although females will sometimes also mark trees, usually in late summer and autumn. The purpose of this mark is poorly understood but may be an advertisement of both presence and reproductive status. To rub a tree, the animal rises on its hind legs and rubs its back, shoulders, and especially the back of its head on the tree. Commonly, hairs are left, stuck in the bark. A bear passing a scent-marked tree will stop to sniff it and perhaps add its own scent. It is a misconception that the bears

attempt to advertise their size by reaching up as high as they can to mark or claw the spot. Both sexes use dribbles of urine to scent-mark their home range. This is accomplished by straddling a small shrub or sapling and urinating as they bend it and walk over it. Scat is deposited along travel routes and appears to be deposited without regard to location. Bear scat varies tremendously depending on diet, from amorphous patties (from a meat and sometimes fruit diet) to loose, broken tubular scat (from fruit, seeds, and nuts) to linked segmented tubes (from eating hide, hair, or roots). The fruit and seed components are often visible in the scat or stain it a distinctive colour, such as purple or blue from blueberries.

While resting, the bears use unremarkable beds in the forest that generally consist of shallow depressions of flattened leaf litter and are rarely reused. If there is snow on the ground, they may insulate the bed with a lining of vegetation, often torn from surrounding evergreen shrubs. Beds are typically oval and around 73–100 cm long by 56–84 cm wide. Black Bears commonly climb mast-bearing trees to harvest the fruit or nuts before they drop. Acorns are usually harvested from the ground, but if bears climb oak trees, the thick bark rarely shows claw marks. Apple trees can be severely damaged, with broken and chewed limbs and clawed bark. Beech trees, with their thin, easily damaged bark, show lasting effects of bear feeding. Claw marks and even partial footprints may be pressed into the grey bark, leaving a black scar. These scars widen as the tree grows and can present the impression that bears of extraordinary size have been scaling the trees. Serviceberry, hickory, shadbush, cherry, ash, and aspen may be subject to similar foraging. Other clawed mammals such as North American Porcupines, squirrels, and Northern Raccoons also climb these same trees, but their claw marks are much smaller. Another feeding sign of Black Bear in American Beech results in the creation of a "bear nest," which is not a real nest but an accumulation of beech twigs created as the bear sits in the canopy and pulls branches towards itself. The twigs are bitten off and discarded by simply dropping them once the beechnuts have been consumed. Black Bears climb tree trunks in two ways. They may "walk" up a tree, gripping with their front claws and walking up the trunk with their hind feet, or they will "jump" up the tree if they are in a hurry, by alternately gripping with the front feet and pushing off with both hind feet together. They always back down. Female bears sometimes use easily climbable trees (with furrowed bark, a leafy canopy, and large limbs upon which to rest) as "baby-sitter" trees. They leave the cubs safely there as they forage out of sight below. These trees may be surrounded by numerous scat from both the cubs and their mother, and marked trees are often nearby. If unattended cubs are found in a tree, cautiously leave the area immediately because their mother may be close at hand.

Black Bear kill sites typically involve fawns of Wapiti, Caribou, or deer; calves of Moose, Bison, or Domestic Cow; and sometimes adult or lambs of Domestic Sheep. These are usually dragged into cover to consume or cache. Black Bears may or may not cover the carcass, whereas Brown Bears almost always do. Although they often consume small prey completely, all bears try to avoid eating the hide of larger prey by inverting it as they feed. They usually pull the skin over the head, often leaving it rolled up. Most prey is killed by a bite to the neck or back. Cached prey should be inspected with care as bears are possessive of their prey and may be nearby.

REFERENCES

Auger et al. 2002; Benson and Chamberlain 2006; Boyd and Heger 2000; Chaulk et al. 2005; Côté 2005; Davis, H., and Harestad 1996; Davis, W.B., and Schmidly 1994; Dobbyn 1994; Elbroch 2003; Garshelis and Hellgren 1994; Harlow et al. 2001; Hebblewhite et al. 2003; Herrero and Higgins 2003; Jameson and Peeters 2004; Kamler et al. 2003b; Klenner and Kroeker 1990; Klinka and Reimchen 2009; Kovach and Powell 2003; Kurta 1995; Lamontagne 1998; Larivière et al. 1994; Long, C.A., 2008; Marshall and Ritland 2002; Mattson, D.J., et al. 1992; Pardy et al. 2004; Pelton 2000, 2003; Pelton and van Manen 1994; Powell et al. 1997; Prescott and Richard 2004; Reid, F.A., 2006; Reimchen 1998; Rogers, L.L., 1992, 1999; Samson and Huot 1995; Schenk and Kovacs 1995; Schenk et al. 1998; Smith, H.C., 1993; Veitch and Krizan 1996; Verts and Carraway 1998; Whitaker and Hamilton 1998; Zager and Beecham 2006.

Brown Bear

also called Grizzly Bear, Kodiak Bear

FRENCH NAME: **Ours brun**

SCIENTIFIC NAME: *Ursus arctos*

It is no surprise that native peoples in western North America hold the Brown Bear in high regard, both spiritually and culturally. Its rugged beauty, raw power, and massive size, as well as its aura of danger, make it a charismatic and provocative species.

The taxonomy of this species has been complicated because the skull shows substantial geographic variation, and it grows until a bear reaches its mid-teens. Natural scientists, even as late as 1981, perpetuated an old taxonomy by recognizing up to 77 different Brown Bear species in North America alone, based on minor differences of skull shape and tooth number. Current taxonomists follow the 1953 proposal by R.L. Rausch that the Eurasian Brown Bear and the North American Grizzly Bear are synonymous, and therefore the scientific name of the European form (*Ursus arctos,* first described by Linnaeus in 1758) has precedence. Herein, the common name Brown Bear is used. Brown Bears are closely related to Polar Bears, and fertile hybrids of the two have been produced in captivity for years, but only suspected in the wild. In 2006 a wild hybrid was shot in the Canadian Arctic, which was determined by DNA analysis to have been produced by a female Polar Bear that had mated with a male Brown Bear.

DESCRIPTION

The fur of the Brown Bear in North America is usually medium to dark brown but can vary from creamy white to blond, grey, silver, reddish-brown, and almost black. The full variation of coat colours may occur within any population. Often the back, neck, and top of the head appear sun-bleached and paler than the sides, belly, and legs. Animals often have creamy-white or silver-tipped hairs on their shoulders and back, giving them a grizzled appearance. Sometimes the muzzle is lighter brown than the rest of the head, and on some bears the hump region is darker. The shade of an individual bear's pelage can vary with the direction from which the light is striking it relative to the position of the observer, so that bears facing away from the light appear darker owing to reduced reflection. Claws on the front feet can be more than 8 cm long, are slightly curved, and about twice as long as those on the hind feet. The ears are short and rounded and heavily furred on both sides. The tail is very short and is rarely visible in the rump fur. Brown Bears have a dished head profile and a distinctively high, muscular hump on their shoulders, which is formed by the shoulder blades and powers the forelegs. Cubs may have a V-shaped, light-coloured chest blaze, or shoulder patches, in their first year that disappears thereafter. Brown Bears' keenest sense is that of smell, but their hearing and both near and distant vision are excellent. The dental formula is incisors 3/3, canines 1/1, premolars 2–4 / 2–4, and molars 2/3, for a total of 34–42 teeth. The coat is moulted once annually, usually beginning in spring, shortly after emergence from hibernation. Hairs are shed and regrow gradually throughout the summer, so that the full coat is in prime condition by late August.

SIMILAR SPECIES

Their massive size, round body, thick legs, and short tail make bears easily recognizable. Black Bears occur within almost the entire North American distribution of the Brown Bear except on the treeless tundra. Black Bears are commonly brown in the west, and some Brown Bears are blackish, so coat colour is not a diagnostic trait. Black Bears do not have the high shoulder hump of the Brown Bear, and they have larger ears and a smaller head with a relatively straight head profile. Claws on the front feet of Black Bears are more curved and are only slightly longer than those on the hind feet. Polar Bears are whitish, have a longer neck and legs, a smaller head, and a less prominent shoulder hump. Skulls of Brown Bears can be distinguished from those of Black Bears by molar measurements. The first lower molar has a crown length of > 20.4 mm and a crown width of > 10.5 mm for Brown Bears, which measure less for Black Bears. These measurements should work for any skull with permanent dentition.

SIZE

The following measurements apply to animals of reproductive age in North America. Males are approximately 1.2–2.2 times larger than females and can be more than twice as heavy, although on average they are about 1.5 times heavier. Their body weight is greatest in autumn. The largest, heaviest individuals in North America come from coastal regions of British Columbia and southern Alaska where they have access to salmon. Males take about fourteen years to reach 95% of their maximum weight and eight years to reach 95% of their skull length, while females take about nine

Brown Bear (*Ursus arctos*)

adult

cub

years and five years respectively. Captive animals may exceed the provided weights.

Males: total length: 170–230 cm; tail length: 7.0–14.5 cm; hind foot length: 24–40 cm; ear length: 11.0–13.5 cm; weight: typically 250–350 kg (maximum reported weight of a wild male is 640 kg).

Females: total length: 145–213 cm; tail length: 9–13 cm; hind foot length: 26–34 cm; ear length: 11–13 cm; weight: usually 80–225 kg (maximum reported weight of a wild female is 360 kg).

Newborns: weight: 340–680 g.

Adults are 80–150 cm high at the shoulders, and a large male can reach 3 m high when standing on his hind legs.

RANGE

Brown Bears once roamed most of central and western North America from Alaska, Yukon, Northwest Territories, and Nunavut to Mexico. They are now restricted to about half of their previous range, although they have recently been observed to be expanding northward to Victoria Island, Banks Island, and Melville Island. Recent observations in Wapusk National Park in northern Manitoba suggest an expansion eastward. The current distribution in Canada and Alaska is similar to the historical range (apart from the prairie area), but in the continental United States Brown Bears are found only in Idaho, Montana, Wyoming, and Washington State, and they have been extirpated from Mexico. A small number of Brown Bears may have lived in the Quebec-Labrador peninsula, but this evidence is ambivalent. The Brown Bear is a Northern Hemisphere species (occurring in both Eurasia and North America). It is widespread across Europe, Russia, and northern Asia.

ABUNDANCE

Many populations along the southern boundaries of the range in North America, Europe, and Asia are isolated in remnant wild areas and are threatened by their low numbers and by frequent contact with humans whose urban and agricultural development often completely surrounds these refuges. Being wide-ranging, opportunistic foragers, Brown Bears commonly come into contact with humans or human-altered landscapes, especially when natural foods are scarce and they are attracted to livestock, crops, and garbage. This places them at risk of being killed by sport hunters and poachers or of being declared nuisance bears and then shot or relocated. As surveying numbers and densities of these large, wide-ranging mammals is difficult and expensive, sustainable harvest limits can be overestimated. Low reproductive rates do not allow Brown Bear populations to rapidly recover from such mistakes. Poaching is also an issue because Brown Bears typically live in remote, underserviced regions where it can occur undetected. This is the case across their range but is especially true in eastern Russia where the bears are killed for their paws and especially their gall bladders, which are sold for Asian traditional medicine.

The greatest cause of Brown Bear population declines in North America is habitat degradation and fragmentation, which is caused by increasing urbanization, roadway creation, plantation forestry, and agriculture. Brown Bears were extirpated in North Africa by the mid-1800s and in Mexico and most of the southwestern United States during the twentieth century. Likely more than 100,000 Brown Bears inhabited North America before European settlement. Now there are 14,000–17,000 Brown Bears in British Columbia, and the average density in the province is estimated to be around 16 bears per 1000 km². Alberta estimated its population at 700–1000 animals in 2002. In 2004 the Northwest Territories estimated its population at 3500–4000. Yukon supports around 6000–7000 Brown Bears. Within the conterminous United States there are probably fewer than 1000 free-roaming Brown Bears, and Alaska supports an estimated 32,000. Density is dependant on food availability, terrain,

Distribution of the Brown Bear (*Ursus arctos*)

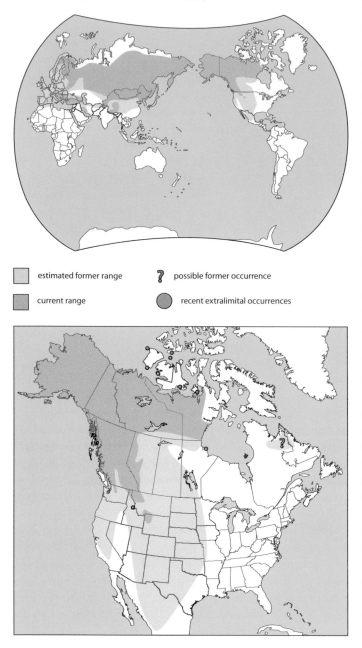

estimated former range

current range

? possible former occurrence

⬤ recent extralimital occurrences

Brown Bears occupy a variety of habitats including temperate rainforests, but they prefer open areas such as alpine meadows, river valleys, scrublands, and tundra. They hibernate for three to seven months depending on the severity of the winter in their area. There has been much debate as to whether bears are true hibernators or merely become dormant, called by some "carnivore lethargy." Most evidence shows that bears are indeed true hibernators. They do not eat, drink, urinate, or defecate, and their heart rate slows from a normal rate of 40–50 beats per minute to 8–10 beats per minute. Urea produced by the metabolism of stored fat is recycled into more nutrients. Owing to the bears' large size, their body temperature drops only 4°C–5°C from a normal of around 37°C. Since their muscles remain warm, they can arouse quickly, and females are able to respond to demands from their tiny cubs for warmth or milk. During hibernation the adults generally lose around 22%–24% of their body weight, but reproductive females lose about 40% because of the demands of lactation. Before hibernation Brown Bears dramatically increase their food consumption in an effort to add fat stores. Entrance into hibernation takes place between October and December, depending on the animal's nutritional condition, reproductive status, and the extent and severity of the local cold season. Pregnant females enter first, followed by adult males. Subadults and thin adults enter last.

Winter dens are usually located in mountainous, heavily forested areas, far from human disturbance, but open tundra and light forest can also provide den sites, as the bears will make use of the terrain available to them. Natural sites such as caves or rock crevices are sometimes chosen, but most dens are dug by the bear. A den may see single or multiple usage, and reused dens may shelter different bears in different years. Den-site selection and construction is experiential, and bears become more proficient as they age. Dens vary in shape and are dug in well-drained, often sloped sites, commonly but not always under trees or shrubs so that the roots can provide support for the roof. Some have short tunnels leading to a cavity; others are cavities dug immediately in from the ground surface. In either case, snow normally covers the entrance as Brown Bears prefer to den where snow cover is deep. The floor of the hibernation cavity may be covered with vegetation or left bare. Most maternal females do cover the ground. For cavity size see the "Signs" section below. Spring arousal generally occurs between March and May depending on the length of the cold season in the region. Males arouse first, and females with cubs emerge last. The bears usually find little to eat when they first emerge, and live largely on stored fat. Brown Bear activity varies with season, food availability, and human activity. In spring when they are hungry, they may forage day and night, while in summer they are often more active during low-light levels but will also readily forage during the day and at twilight, and bed down at night. Bears in regions with high human activity may become secretive and largely nocturnal.

The size of the home range is largely determined by terrain, food availability, and the bear's gender. Adult males have home ranges that are several times larger than those of females within the same area. The smallest home ranges are occupied in regions with reliable

and human interference and is generally highest in coastal areas with reliable, abundant salmon runs, at 175–550 bears per 1000 km². North American populations with the lowest densities occur in the far north of Alaska, Yukon, Northwest Territories, and Nunavut where there are typically < 10 bears per 1000 km². The IUCN estimated the worldwide population at more than 200,000 in 1998. Russia supports about 100,000 of those, and Canada is thought to have a relatively stable population of around 25,000 animals. There is a regulated hunt in Alaska and in all jurisdictions within the Brown Bear range in Canada, although the hunt in Alberta is currently suspended owing to low bear numbers. All Brown Bear populations in Canada are designated by COSEWIC as of "Special Concern," and the Prairie population as "Extirpated."

food, such as secure and abundant annual salmon runs. In that kind of environment the home ranges of males average 115–185 km², while females occupy 24–71 km². In the barren grounds of central Northwest Territories and Nunavut the home range of the adult male varies from 1154 to 8171 km², while that of the female varies from 670 to 2434 km². Their home ranges may be this large because some trail the migrating Caribou. Home ranges can overlap extensively, with no signs of territorial defence by either males or females. The size of the home range may increase during the breeding season as adult males and pre-oestrous females roam further, searching for mates. Adult females are especially faithful to their home range but, like any hungry Brown Bear, will travel more widely in the event of a widespread food failure, as is most common in autumn during low mast years. Adult males are especially aggressive towards subadult males, resulting in the dispersal of younger males away from their natal home range. Juvenile females typically disperse much shorter distances and often remain near or conjoining their mother's home range. During a 20-year study in southeastern British Columbia, three-year-old males dispersed an average of 30 km, while females travelled about 10 km (less than the average diameter of a female's home range in the region). The longest natal dispersal recorded over the 20 years was 67 km for a subadult male and 20 km for a subadult female. Brown Bears are excellent swimmers and are known to swim more than 15 km to offshore marine islands to feast on ground-burrowing seabirds. As apex predators, Brown Bears have few predators besides other Brown Bears and humans. Legal sport hunting, poaching, and shooting of nuisance animals results in 77%–88% of adult Brown Bear deaths in mountainous parts of the southern range (including British Columbia and Alberta). The oldest known wild bear was 29.5 years old when captured and released live, and the oldest known reproductive female produced cubs at 24.5 years old and successfully weaned them almost three years later. Brown Bears can survive 20–30 years in unhunted populations.

DIET

Brown Bears are omnivorous carnivores. Their diet is largely composed of vegetative matter: grasses, forbs, roots, tubers, bulbs, nuts, pine nuts, and fruit and berries as seasonally available. This menu is enhanced with fungi, insects, fish, small mammals, birds, carrion, and any large mammals they can catch, particularly ungulates such as Wapiti, Moose, deer, Caribou, Muskox, and mountain sheep. Brown Bears can reach speeds of 32–48 km/h for short distances. Although capable of taking down a healthy adult ungulate, they tend to concentrate on ungulate carrion, the very young, or adults weakened by old age, disease, malnutrition, or injury or mired in heavy snow. These large carnivores will also kill others of their own species as well as Black Bears. They adapt to human presence by scavenging in garbage dumps, orchards, bee hives, and crops and will kill penned or free-range livestock. Brown Bears are more carnivorous in regions with salmon runs or large numbers of ungulates. Seeds consumed by Brown Bears are often excreted in germinable condition, surrounded by natural fertilizer. Large quantities are thereby dispersed by the bears, making them among the most important dispersers of fleshy fruit seeds within their range.

REPRODUCTION

The breeding season in Canada occurs between early May and the end of July, with most mating taking place in June and July. Courtship usually takes several days as a male will attend a female, following her until she ovulates, which generally requires the prolonged presence of the male. Once the egg has been produced, the two may mate several times in the following days. The male may remain with the female for up to several weeks altogether in an attempt to sequester her and drive away other potential suitors. In areas of appropriate terrain (such as in the mountains) the male may herd the female to higher elevations and then remain below her to prevent her from descending towards other male bears. Copulation can take up to 60 minutes but usually lasts 10–25 minutes. Females are in oestrus for 10–30 days and may mate with several males during that time, so multiple paternity of a litter is possible. Brown Bears experience delayed implantation. The fertilized blastocysts float around in the female's uterus for approximately five months before implanting around late November. Active gestation then lasts about six to nine weeks, and the cubs are born in January or February while the female is hibernating.

The litter size can reach four cubs rarely but averages two to three in North America. Reports of litters of up to six may be the result of adoption or cub exchange between individual females. The litter size is directly related to the female's nutritional status, as well-fed females with a diet that is high in fats and protein produce larger litters. Litter size is also age related as young and old females have smaller litters than do those of prime breeding age (8–20 years old). Newborns are blind and helpless. Cubs grow quickly on the high-calorie (33% fat) milk produced by their mother. By about three months old, all of their milk teeth have fully emerged from the gums. Lactation lasts 1.5–2.5 years (an exceptionally long time for a carnivore), and juveniles may remain with their mothers beyond that until they are up to four years old. The cubs learn the many skills they will need to survive on their own, as well as the seasons and the locations of abundant food resources, during this lengthy period under their mother's care and protection. Her protection may be all that prevents the adult males of the region (or rarely neighbouring females) from killing the youngsters. The mother-cub association is typically severed when the female is ready to breed again, perhaps by the aggressive actions of the mother or by the presence of an adult male. Males play no role in raising the young. Sexual and physical maturity depends on the individual animal's nutritional condition.

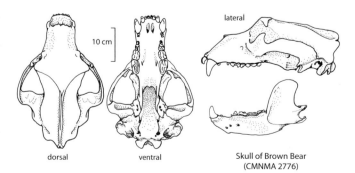

10 cm

dorsal　　　　ventral　　　　lateral

Skull of Brown Bear
(CMNMA 2776)

Male juveniles typically become sexually mature at 4.5–6.0 years old, but females may be up to 11 years old before they produce their first litter. Most North American females produce their first litters between the ages of five and eight years old and typically have a litter every two to four years thereafter, although the interbirth interval can extend to six years in some cases, particularly if the female is undernourished.

BEHAVIOUR

Apart from reproductive females with cubs and the temporary bond of mating pairs, adult Brown Bears are largely solitary. Subadults may associate with littermates for years after they leave their dam, but this solidarity does not usually last beyond their first breeding season. The solitary structure of Brown Bear society is accomplished mainly by mutual avoidance, but when abundant food draws the animals together, feeding aggregations can occur. Groups of bears foraging in close proximity, such as along a river during a salmon run, develop a dominance hierarchy, and the dominant animals take the preferred sites and achieve unrestricted access. The largest, most aggressive males top the rankings, and females with young are superior to subadults and barren females. A similar hierarchy develops among the breeding males of an area, and dominant males have access to oestrous females. Sometimes fights occur during dominance squabbles, and deaths of smaller animals have been reported, although fights usually only occur between similarly sized animals of the same sex. Fighting Brown Bears bite and swat at each other, trying to claw the neck and shoulders and to bite the head and neck of their opponent. Typically, a dominant animal will approach a subordinate head-on with the neck extended, ears back, and lips curled to display the canines. Subordinates remain sideways with their head low or turned away, and usually sit, lie down, or retreat by backing away before a fight can occur. Adult males are the most likely predators of cubs and females, so maternal females either avoid feeding aggregations altogether or only participate during daytime and try to maintain a distance from other bears for the safety of their cubs. Females are very protective of their cubs. Adoption in the wild has been reported. Cubs learn and remember the location of seasonal food abundances and may return to these sites as adults. Brown Bears may cache surplus food to hide it or to retard its spoilage. They pile dirt and debris over it and may lie over or near it in a protective manner, guarding it from intruders.

Most of the Brown Bear–human interactions in Canada occur in British Columbia because of the high numbers of both in the province. From 1960 to 1997 there were 41 attacks causing serious injury, of which eight resulted in a human fatality. An average of 45 problem bears are killed annually in British Columbia, and about the same number are captured and relocated. This latter solution is often less than satisfactory because many relocated bears eventually find their way back, and many others are killed by the bears already living in the relocation area. From 1960 to 1998 in Alberta there were 29 Brown Bear attacks that caused serious injuries or fatalities (12), and more than half of these attacks (18) occurred in Banff or Jasper National Parks. Females with cubs are

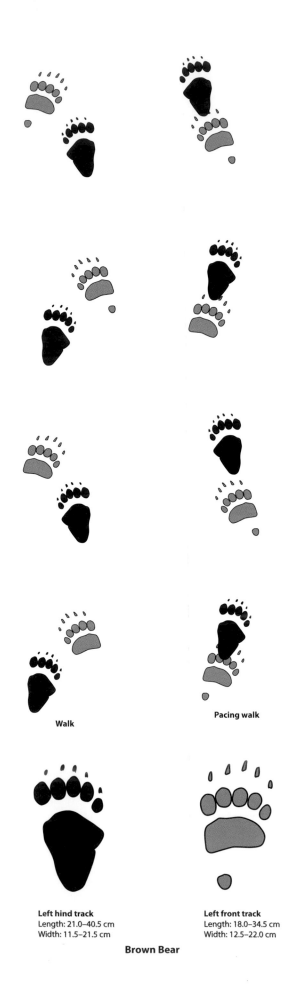

Walk

Pacing walk

Left hind track
Length: 21.0–40.5 cm
Width: 11.5–21.5 cm

Left front track
Length: 18.0–34.5 cm
Width: 12.5–22.0 cm

Brown Bear

the most likely Brown Bears to attack humans. One should never approach a female with cubs, a feeding bear, or a bear guarding carrion, because these animals may cause serious injury or death. Approaching too closely may provoke an attack. A bear about to attack may exhibit excessive salivation, clatter its teeth, growl or roar, rise on its hind legs, slap the ground with one or both front paws, press its ears back, and display its teeth. Alternately, it may attack with no warning at all. Most Brown Bear attacks in the daytime are protection related rather than predatory, and many are perpetrated by habituated bears that are used to finding food (usually garbage) in close proximity to humans.

VOCALIZATIONS

Distress calls are produced by cubs when they are hungry, cold, or separated from their siblings or mother. Adults in pain will also emit a distress call at times. Chuffs, huffs, and snorts are produced by anxious or stressed animals, and growls and roars are threat vocalizations. Chuffing is produced by a forceful exhalation combined with a popping sound by the cheeks and lips. Chuffs are also used in greeting, appeasement to a dominant, during courtship and mating, and between a female and her young. An anxious or threatened animal may also clatter its teeth.

SIGNS

All limbs have five, clawed digits. The first toe, the inside toe on each foot, is the smallest toe and does not reliably register in the front track. The claws on the front feet are much longer than those on the hind and leave impressions that are comparably further from the toe-pad impressions. Like other North American bears, Brown Bears walk with a plantigrade, pigeon-toed foot placement, especially with the front feet. The front feet have two pads, a palm pad that touches the ground with every step and a circular heel pad that sometimes strikes the ground, especially in soft substrate or if the animal is moving quickly. The hind-foot palm and the heel pads are connected to form a large sole. The most common gait while foraging is a pacing walk where both limbs on the same side move in synchrony. Cubs may climb trees, but adults usually stay on the ground. Beds may be constructed near foraging areas in dry, sheltered spots that afford the bear good visibility. They may register only as flattened, worn areas, but in cold weather the bear may tear limbs from nearby trees and shrubs and scrape them together to create a mattress. Beds may be used once or repeatedly. Most are oval in shape and measure 80–107 cm in length and 61–76 cm in width. In Yellowstone National Park, winter den chambers average 142 cm by 135 cm by 94 cm. Den entrances are 89–115 cm high by 89–115 cm wide. In the Canadian Arctic, den entrances are considerably smaller, 53–57 cm high by 58–63 cm wide, and the interior cavity averages 140 cm wide by 261 cm long by 82 cm high. The bears may start more than one den within their home range before deciding on one and fully excavating it. In open areas the den may be visible on a hillside as a large hole with a massive throw mound of fresh, waste soil on the down-slope side. Scat is variable in shape and colour depending on diet. It can be 3.0–7.3 cm in diameter and 18–51 cm in length. A mainly vegetation or fruit diet produces thick,

often crumbly or segmented, tubular scat with blunt ends. A meat diet produces blackish scat that may be ropey or amorphous. The character of the diet is usually visible in the scat (berry seeds, vegetation fragments, or fibres). While foraging for roots, bulbs, or other vegetation, Brown Bears can create large areas of disturbed soil and uprooted vegetation.

Large prey is killed by a bite to the head, neck, or back. Brown Bears may begin to feed by opening the abdomen, removing the stomach and intestines and consuming the internal organs, or by starting at the head and shoulders. Like Black Bears, Brown Bears avoid eating the fur and hide by inverting the skin as they feed, often leaving it beside the feeding site. They usually cache the remains of a kill or carrion by dragging it into cover and then scratching branches, debris, or soil over it. Feeding sites typically have plenty of scat around as well as bed sites where the bear rested between feedings. Brown Bears will remain near a kill to guard it and are extremely possessive of their food. Such a bear will readily attack any human who approaches too closely. Brown Bears, like Black Bears, commonly mark upright wood, usually trees, but also telephone poles, exterior cabin walls, and fence posts, to indicate their presence and possibly to scratch itchy, moulting spots. Such sites are usually located along well-used trails. Typically only a large adult male rubs, but other classes of bears visit the rubbing sites. He approaches a rubbing tree, dribbling urine, and then stands and rubs his back, rump, and the back of his head. Then he turns around and may bite or claw the tree before hugging it and rubbing his chest. These trees are likely used to advertise his presence and dominance. The bark on the rubbing trees often traps and holds hair, leaving a clue to the perpetrator's identity.

REFERENCES

Allan 2007; Bader 2000; Barnes and Smith 1993; Brady and Hamer 1992; Case and Buckland 1998; Churcher 1999; Clark, D., 2000; Clarkson and Liepins 1993; Craighead et al. 1995; Dahle and Swenson 2003a, 2003b; Doupé et al. 2007; Elbroch 2003; Gau et al. 2004; Gende and Quinn 2004; Gilbert, B.K., 1999; Green and Mattson 2003; Hall 1981; Hellgren 1998; Herrero and Higgins 1999, 2003; Kansas 2002; Klinka and Reimchen 2002; Kovach et al. 2006; Loring and Spiess 2007; Mattson, D.J., 2001; Mattson, D.J., et al. 1992, 1999; McLellan, B.N., and Hovey 2001; McLellan, B.N., et al. 1999; McLoughlin et al. 2002, 2003; Mowat and Heard 2006; Mowat et al. 2004; Munro et al. 2006; Nevin 2007; Pasitschniak-Arts 1993; Pasitschniak-Arts and Messier 2000; Proctor et al. 2004; Rausch 1953; Schwartz, C.C., et al. 2003; Servheen et al. 1998; Smith, H.C., 1993; Smith, M.E., and Follmann 1993; Struzik 2006; White, D., et al. 1998; Willson and Gendre 2004.

Polar Bear

also called White Bear, Ice Bear

FRENCH NAME: **Ours blanc,** *also* **Ours polaire**

SCIENTIFIC NAME: *Ursus maritimus*

Polar Bears are the largest of the living bears although the largest coastal Brown Bears that feed on salmon are a close match. Only the extinct Short-faced Bears (genus *Arctodus*) were larger than modern Polar Bears.

DESCRIPTION

Polar Bears have dark eyes and black skin, with a black nose pad, claws, foot pads and lips, and a pink and black tongue and gums. Scars appear as black marks. The cub's fur is white with a slight tawny cast that bleaches in the sun. The adult's coat tends to take on a creamy or yellowish tint. This coloration is thought to be the result of oils and stains, as individuals are whitest after a moult. Their white colour is the result of a complete lack of pigmentation in the hairs, and the fur can take on the yellow or orange hues of the rising or setting sun. Their claws are sharp and strongly curved to grip prey and ice. The ears are small, rounded, and densely furred. The tail is short and barely visible. Long hair grows profusely between the pads of the feet (especially in winter). This fur folds over and is walked on, shielding the pads from the cold substrate. Polar Bears have keen vision and hearing. Their acute sense of smell is likely as good as that of a Domestic Dog. The dental formula is incisors 3/3, canines 1/1, premolars 2–4 / 2–4, and molars 2/3, for a total of 34–42 teeth. Although Polar Bears evolved from Brown Bears about 150,000 years ago, their teeth have become significantly different, likely owing to their more carnivorous diet. Polar Bear teeth (apart from the canines, which are longer and larger in order to grip large prey) have higher, sharper cusps that are more suited to shearing meat and blubber than grinding the grasses and fruits that form a large part of the Brown Bear diet. Fur is moulted once annually, usually beginning in late May to July. Hairs are shed and regrow gradually throughout the summer so that the full coat is in prime condition by early winter. Polar Bears have a double coat made up of a very dense layer of underfur through which grow longer guard hairs. They can accumulate up to 10 cm of fat beneath the skin, depending on the body area (for example, the lower legs have little subcutaneous fat while the torso area tends to have the most). Polar Bears have four mammae in pairs at the top of the chest region, and females normally sit and recline on their rump to suckle.

The notion that individual Polar Bear guard hairs act like fibre optic cables, conducting ultraviolet light down through the dense underfur to the black skin beneath where it is absorbed, has been proved false. The fur instead absorbs the UV radiation, preventing it from reaching the skin.

The livers of Polar Bears are very high in Vitamin A, to the extent that most are toxic and unsafe for humans to eat. Consumption symptoms range from headache and drowsiness to skin peeling and possible death. Symptom severity varies with the amount ingested and the degree of toxicity of the individual liver.

The Brown Bear account discusses Polar Bear–Brown Bear hybrids.

SIMILAR SPECIES

These are the most northern of the bears. In parts of their range Polar Bears may overlap with Brown Bears, and in the Hudson Bay ecosystem with Black Bears. Although both of these bear species have light-colour morphs, these colours tend not to occur in the north, and so the only white bears in the region are Polar Bears. Polar Bears are usually larger, and the shoulder hump that is so prominent on

Polar Bear (*Ursus maritimus*)

adult

cub

Brown Bears is less evident in Polar Bears. Polar Bears have a longer neck and a narrower head than do Brown or Black Bears.

SIZE

The following measurements apply to animals of reproductive age in the North American part of the range. Adult males weigh about twice as much as adult females, and this size difference is clearly evident by the time the cubs are yearlings. Polar Bears are more sexually dimorphic than are other bears. Captive animals may exceed the following weights.

Males: total length: 200–255 cm; tail length: 12–19 cm; weight: 350–650 kg (an exceptional individual weighed an estimated 800 kg but was too heavy to weigh accurately); shoulder height: 1.0–1.7 m.

Females: total length: 185–215 cm; tail length: 10.0–14.0 cm; weight: 150–350 kg (exceptional individuals may weigh up to 450 kg).

Newborns: weight: 600–700 g.

Adults have hind foot lengths of 25–40 cm and ear lengths of 9–15 cm.

RANGE

Polar Bears occupy a circumpolar range. Their Canadian distribution includes ice and land from the northern tip of Newfoundland to the north of Yukon to the farthest extent of ice beyond the far northern Arctic islands. Polar Bears also inhabit the ice pack and land edges around Greenland, northern and western Alaska, and northern Russia to Svalbard (Norway). These bears are largely a species of the continental shelves where the combination of relatively shallow seas, high ocean productivity, and sea ice formation creates their preferred habitat. During the winter, when the ice pack drifts south, some Polar Bears travel with it and may reach the northern Bering Sea, coastal southern Labrador, and northern Newfoundland. Some bear populations, such as in the Beaufort Sea, may live entirely on the ice without needing to touch land during any part of their life cycle. Bears separated from the sea ice may wander several hundred kilometres inland. The bears seen farthest inland (ca 400 km) were a mother with two yearlings in Deline, Northwest Territories. Most of the central polar basin is very deep water and not good Polar Bear habitat even when ice covered.

ABUNDANCE

Polar Bears are powerful, dangerous, wide-ranging animals, living in remote and forbidding terrain, which makes them difficult and expensive to survey. Based on an estimate produced in 2002, there were around 15,000 Polar Bears shared by Canada and its immediate neighbours (Greenland and Alaska), about 2600 around Svalbard, and roughly 21,500–25,000 worldwide. Following years of over-hunting, mainly for the fur trade, an international agreement in 1973 prohibited all but subsistence hunting and some limited sport hunting. Abundance, at least in Canada, climbed for several decades thereafter, but some populations appear to be declining again. Canadian governments (federal, provincial, and territorial) monitor the number of bears in 13 different zones in order to identify how many can be taken yearly without damaging the health of each population. However, these estimates are an inexact science, and overestimations that led to over-hunting have severely reduced several populations, notably in the M'Clintock Channel and Viscount Melville Sound areas and likely also in the Kane Basin, Baffin Bay, and western Hudson Bay where the effects of climate warming are further stressing the populations.

As the top predator in the Arctic food web, Polar Bears are affected by a number of pollutants that biomagnify at higher trophic levels. Mercury and PCBs (polychlorinated biphenyls) appear to be most significant at present. The threat of climate warming, especially the earlier spring melt and the later autumn freeze-up, is already seriously affecting some Polar Bear populations in Canada, and this trend is expected to continue. The Arctic is warming faster than southern latitudes owing to the loss of white, highly reflective ice and snow and the subsequent increase in dark, open waters that absorb more solar radiation and heat. This effect exacerbates and may even accelerate the cycle of ice loss. Polar Bears depend on sea ice for feeding, breeding, and movement, and so modification of this substrate has significant ramifications to their survival. Many populations are suffering from ice that melts away progressively sooner each summer. Bears in the western Hudson Bay population have been well studied and were the first to show symptoms of the decreased survival of all age classes except prime-aged adults, as a result of this phenomenon. Ice break-up in Hudson Bay occurred three weeks sooner in 2004 than it did 20 years earlier in 1984. Increasingly early ice break-up reduces the bear's annual hunting period and increases the annual fasting period they must endure as they wait on shore for the new ice to form. Juveniles, subadults, and the oldest bears are most affected. More juveniles and subadults in this population are dying of starvation, even the adults are getting thinner, and reproductive rates have declined. Demographically, this means that the population is declining from both increased mortality and lower reproduction. More information on western Hudson Bay Polar Bears can be found in the "Ecology" section following. The effects of declining sea-ice

Distribution of the Polar Bear (*Ursus maritimus*)

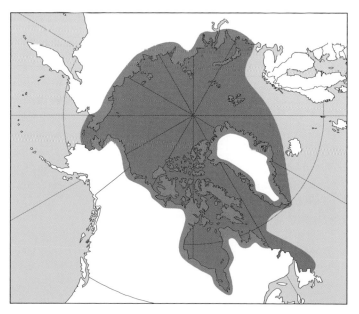

longevity on the western Hudson Bay bears is likely a prelude to what is happening to other Polar Bear populations (such as in Baffin Bay, Davis Strait, Foxe Basin, and southern Hudson Bay). If the warming trend continues as projected, an Arctic-wide summer melt could be seen by the year 2030 or sooner, and all Polar Bears would be forced to shore. Climate models indicate that global Polar Bear populations will suffer dramatic declines by the end of this century. Manitoba, Ontario, and Newfoundland and Labrador have listed Polar Bears as "Threatened" within their provinces, and federally COSEWIC has considered them to be of "Special Concern" since 1991. The IUCN placed Polar Bears on their "Red Data List" in the "Vulnerable" category in 2006.

ECOLOGY

The northern world of Polar Bears consists of months of daylight followed by comparable months of darkness. Most of their time is spent on sea ice, returning to land only when the ice disappears or when the pregnant females den up to produce their cubs. As soon as the youngsters are old enough, the family returns to the ice. Sea ice is a crucial part of the Polar Bear's ecosystem. Approximately 57% of Arctic productivity takes place in the lower 10–15 cm of sea ice, where the ice pores are filled with algae and other phyto- and zooplankton. In late spring as the ice melts, this plankton, which has been subjected to at least two months of full sunlight by then, is released into the water column where it feeds the food web, causing a prodigious burst of growth and activity in its predators. The fishes that feast on the plankton in turn feed the Beluga, Narwhals, and seals, which are in turn captured and eaten by Polar Bears. Ringed Seals, the Polar Bear's primary prey, are also dependent on sea ice because it harbours and supports their pups until they are old enough to swim. The peak feeding time in a Polar Bear's year is from May to July when the seal pups and moulting adults are most available on the ice. However, seals remain obtainable throughout the year while there is ice, so denning-up and hibernation have largely become part of a reproductive strategy to provide shelter to the neonates, rather than a foraging strategy to avoid periods of food scarcity as it is with other North American bears. Nevertheless, Polar Bears will eat as much as they can during the spring and summer in an effort to store crucial body fat to sustain them during the leaner times of the year. While on land they are usually fasting and resting to save energy. Unlike Brown and Black Bears that fast only during the winter denning period (and probably the few weeks before and after), Polar Bears can sustain long periods of fasting while remaining awake and are therefore able to exploit any feeding opportunities that may arise. While fasting, they maintain a near-normal body temperature, and their water and energy requirements are met by fat metabolism. Waste products of this metabolism are further recycled into nutrients and so do not remain in the blood stream to be extracted by the kidneys; therefore the fasting bears neither drink nor urinate. An example of the use of this fasting ability occurs in the western Hudson Bay region where the bears are typically landlocked by the melting sea ice from late July until the freeze in early November. During those approximately four months the bears are fasting. Pregnant females in that population must extend their fast by an additional four months because they den in autumn to await the birth of their cubs and cannot follow the other bears out onto the ice until the following spring. For the last two or three months of their extended fast they further stretch the limit of their resources by producing high-fat milk for their cubs.

Apart from pregnant females, Polar Bears remain active through most of the winter, only denning-up temporarily (usually on the sea ice) for part of the mid-winter if prey becomes scarce. Shelter dens in winter are dug into the snow and ice, but landlocked bears may dig shelter dens into the earth, not to escape the cold but to escape the summer heat. In late summer and autumn most pregnant females go inland where they find a snowdrift and dig a maternal den (usually not more than 16 km, and most often less than 8 km, from the coast, except in Hudson Bay where dens can be > 50 km inland). Their entry into maternal dens is largely determined by the availability of sufficient snow into which to dig, and may occur from late August to late November. In the Canadian Arctic most females begin hibernating in September or October. Earth hibernation dens are used only in the Hudson Bay area by pregnant females waiting for snow. They begin hibernating in an earth den and then dig a snow den in the drift that forms over the earth den. Some females in the Beaufort Sea region will den on drifting pack ice. Such locations are hazardous because the ice may travel over 1000 km during the denning period and be exposed to considerable stresses such as melting and refreezing, cracking, collapsing, and even overturning. Females hibernate and fast for four to eight months while birthing and rearing cubs and can lose as much as 50% (but more commonly around 43%) of their body weight during that time. Females and their new cubs generally emerge from maternal dens in late February to March and abandon the dens a couple of weeks later in late March or April, once the young are about three months old and sturdy enough to withstand the outside conditions. Polar Bears do not wander aimlessly. Radio-telemetry has show that most individuals, especially females, occupy the same general area from year to year and that females tend to display generalized den-site fidelity, choosing to dig another maternity den in the same general area as they did previously. Sizes of home ranges vary widely from region to region and from year to year, probably because ice conditions are different and variable. Home ranges vary from a few hundred square kilometres to several hundred thousand square kilometres. A startled Polar Bear can gallop at a speed of up to 40 km/h for a short distance, and most bears can sustain travel speeds of around 4 km/h for extended periods; so travel distances of > 50 km per day are not unusual. Still it comes as a surprise that some of these home ranges may exceed 550,000 km². Such home ranges are usually found in the Beaufort Sea, Barents Sea, or Greenland Sea where currents and wind move the ice pack. As befits an animal that lives on the ice, the Polar Bear is an excellent swimmer, capable of sustained speeds of around 6 km/h, and it can dive or swim completely submerged for up to two minutes. The longest known sea voyage was 687 km, but this mother lost her offspring during the trip. Polar Bears are often seen swimming at sea far from land, probably after their ice platform has melted. Propulsion in the water is provided by powerful paddling movements of the front legs, and the hind limbs trail behind. A travelling bear, even if it is just walking, can overheat fairly quickly in the Arctic summer, so it is common to see bears swimming in the ocean or dunking themselves in tundra ponds to cool off.

Polar Bears have no real predators apart from humans and other bears, although Grey Wolves may take the very occasional cub. Hunting is permitted in Greenland, Alaska, Canada, and eastern Russia (only Norway has a hunting ban) but is mostly limited to native subsistence harvesting. Cubs suffer a mortality rate of around 20%–40% during their first year. Most natural mortality beyond the cub stage occurs among subadults (two to five years old) and old bears and is usually caused by starvation because many individuals within these age classes are either inexperienced hunters or debilitated by age. Captive Polar Bears have lived more than 40 years, but wild males rarely survive longer than 25 years (28 is the record), and females past 30 years (32 is the record).

DIET

Polar Bears are more carnivorous and predacious than other bears. They are apex predators in their Arctic marine ecosystem, subsisting largely on Ringed Seals; to a lesser extent on Bearded, Harp, Harbour, and Hooded Seals; and to a much lesser extent on sea ducks, Walruses, and small whales (mainly Beluga and Narwhal). Larger whales and ungulates are eaten as carrion. Ordinarily the small whales and Walruses are taken only by the large adult males. The whales are most vulnerable while they are trapped in a small area of open water (called a savssat) within the ice pack. Polar Bears hunt Walrus mostly by startling them while they are basking. The herd stampedes towards the water, and sometimes a manageable-sized calf will trail behind within reach. The bears prefer to eat the fatty external blubber layer of their prey and often leave most of the meat for scavengers; for this reason, Arctic Foxes and gulls take note of a foraging bear. Polar Bears are highly opportunistic and are quick to exploit new food resources including garbage dumps. When marooned on land, these bears will sometimes eat berries, grasses, seaweeds that wash up on shore, and any carrion, bird eggs, or nestlings they encounter. Adult male Polar Bears will kill and consume smaller bears.

REPRODUCTION

Breeding occurs from March to June, with most mating occurring in April and May. Females appear to be induced ovulators. The release of eggs from the ovaries for fertilization is triggered only after sufficient stimulation; therefore, multiple copulations are necessary. Oestrous females have been observed to be consorting with more than one male during their receptive period, and multiple paternity of litters sometimes occurs. Pregnant females experience delayed implantation. The fertilized eggs float around in the uterus until cued by decreasing light levels to implant during the following autumn. Implantation will fail if the female becomes too thin to bear the energetic burden of pregnancy and nursing cubs. Following about 60 days of active gestation, cubs are born in November, December, or early January while their mother is hibernating. Like all newborn bears, the cubs are very small relative to the size of their parents. Neonates are about the size of a clenched fist. The litter size varies from one to three and is typically two. Neonates are blind and only lightly furred. They are entirely dependant on their mother to feed them and keep them warm, but they grow very quickly on her rich milk. Their eyes open at around six weeks old, and they weigh 10–22 kg at about three months old when they first emerge from the den. Females and offspring remain denned-up until sometime in March, when the female digs out and the cubs wander around the area, returning to the den periodically. After about 10–14 days the den is abandoned, and they travel to the sea ice where finally, after at least four months of fasting (four to eight months, depending on the population), she can hunt once more.

The cubs start tasting solid food and begin the long process of learning to hunt, but continue to nurse until their second year. Cubs remain with their mother for about 2.5 years before striking off on their own. Adoption is known to occur but appears to be rare. In the only recorded case, the female lost her two-cub litter in their first summer and adopted three cubs of similar age, which she nursed. Adult females breed about every three years. Young males become sexually mature at around three to four years old but are not large and strong enough to succeed in the male dominance battles to win a chance to mate until they are at least five and more likely six to seven years old. Males peak in reproductive success in their mid-teens. Females become sexually mature at around four to five years old and produce their first litter at five to six years old. Females reach physical maturity at around five years old, and males at around eight years old. Females are capable of breeding through their entire life although there is a marked decline after the age of 20.

BEHAVIOUR

Apart from females with cubs and the temporary pair bond of a mating pair, adult Polar Bears are largely solitary, and like most solitary animals their best and most effective communication sense is smell. Most important, scent is used by females to ensure that a mate is available when needed. A male bear crossing the trail of a female can discern whether or not she is near oestrus. If so, he will follow her for days; once paired, they will copulate multiple times and remain together for about two weeks. Males depart after this period and do not assist in raising the young. As pre-oestrous females wander, their enticing odour trail may attract several ardent males. Aggregations may occur around an abundant food source, such as a beached whale or a garbage dump, but these are uneasy and usually temporary. Aggregations may also occur along the shore when the entire population in the area is landlocked by melting sea ice. Siblings may stay together for a while after their mother leaves them, but eventually they wander alone. Females with cubs tend to avoid adult males and even the areas that they frequent, probably out of fear for the cubs' lives; if a large male approaches her and her cubs at a kill, she will usually abandon it. In western Hudson Bay where the bears gather on shore during the summer, pregnant females and those with cubs tend to go inland while the adult males gather and wait along the coast.

Starting in autumn, male Polar Bears will engage each other in "play fighting," which may be both practice and a precursor to the

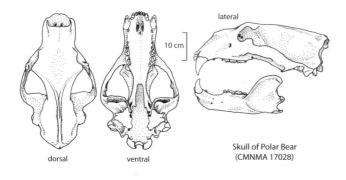

lateral

10 cm

dorsal   ventral

Skull of Polar Bear
(CMNMA 17028)

more meaningful dominance battles that take place in spring. These more dangerous competitions determine which bear will have access to an oestrous female. Teeth or bones may be broken, and wounds taken, and most adult males bear scars as mute testimony to their participation. Usually the wounds incurred are superficial, but these fights can result in the mortal injury or permanent disability (for example, blindness) of a combatant. Polar Bears are also competitive over kills, and larger bears will drive away smaller ones to take their kill, although sometimes not without a struggle. Polar Bears do not tend to cache food although they will engage in surplus killing under the rare conditions in which this is possible. Polar Bears are crafty and intelligent hunters, quite capable of learning by example and from experience. Cubs undergo an extended period (two to two-and-a-half years) of foraging tuition from their mother. They do not hunt during their first summer but carefully follow their mother, imitating her by sniffing where she sniffs and freezing when she freezes. They learn to lie unmoving behind her as she still-hunts or stalks, watching her intently. Although youngsters, especially siblings, may become distracted at times and begin to play, their mother's discipline usually settles them down again. Yearlings begin to hunt with limited success, but two-year-olds hunt more frequently and much more successfully, although most of their food is still provided by their mother. By the time the youngsters leave their mother early in their third year, they are large enough to break through a seal lair and are hopefully proficient enough to feed themselves, although many will start by scavenging carcasses abandoned by older bears.

Seals are primarily captured by stalking or by quietly waiting near a breathing hole. The bears search the ice surface and are thought to perceive Ringed Seal lairs primarily by scent, although sound detection may play a role at times. The lairs are usually about 1 m beneath the surface and consist of a hollow within the ice and snow, containing a breathing hole that leads directly to the water. Adult seals use these lairs to haul-out in safety, and females whelp in the lairs. Once a lair has been detected, the bear creeps towards it and freezes nearby to determine if it is occupied. If it is not, the bear will wait until a seal enters, and then jump onto the lair to break the roof so that the seal can be grabbed before it escapes down the breathing hole. Very young pups are easy prey to such tactics as they are reluctant to swim, likely because they have not yet accumulated sufficient blubber to insulate them from the cold water. However, this very lack of fat makes them less than preferred prey, and a reasonably well-fed bear may leave the carcass uneaten. When a bear

sees a seal basking on the ice, it freezes to assess the situation and then steadily approaches, counting on its white camouflage to keep it undetected until it is close enough for a rushing charge. Such a stalk can also take place from open water or along a lead as the bear slowly and very quietly swims until it is close enough to claw its way onto the ice beside the unsuspecting seal. Particularly in summer, the bears hunt adult seals by still-hunting beside a breathing hole, a method also used by Inuit hunters. The bear may sit or stand (in a characteristic pose with all four feet close together), but most recline on their chest and belly with their nose near the hole, and their hind legs under them, ready for a rapid lunge. Despite lying motionless for up to an hour and sometimes much longer, the bear can move with lethal speed to grab the surfacing seal and flip it out onto the ice, where it is dispatched with a bite to the head or back of the neck. Still-hunting can become long and boring, and the bears sometimes fall asleep while waiting.

Polar Bears typically wash themselves after eating to remove the blood and greasy seal oil. In summer they swim and then lick their paws, forearms, and muzzle. In winter they roll around, rubbing their bodies on the snow before licking. Polar Bears commonly increase their vision range by sitting up or standing on their hind legs. As the largest predator in the Arctic, they view any moving creature as a potential meal, including humans. This danger is ameliorated by the sparse human population in the north and by the fact that the bears are clever and can quickly learn that humans are dangerous and not good prey. There have been only 20 well-documented Polar Bear attacks on humans in North America, of which five proved fatal. Certainly, vigilance and a firearm are essential when one is in Polar Bear country.

VOCALIZATIONS

Polar Bears are relatively silent but will communicate vocally at times. Cubs and subadults utter bleating distress calls if they are cold, hungry, or lonely. Females produce low-volume moaning calls to attract their young, as well as low-intensity, repetitive chuffing calls to their young cubs. Huffs, chuffs, and snorts are emitted by anxious or apprehensive animals, and fighting adults will hiss, growl, and roar.

SIGNS

Polar Bear tracks are surprisingly large, especially in winter when the track may be exaggerated by the wealth of hair on the sole (shown as a paler blue in the track diagram). Each foot has five toes. The first toe (on the inside of each foot) is the smallest and sometimes might not leave a clear impression in a front track. All toes are clawed, and the claws on the front feet are slightly longer than those on the hind feet. Claws may not register in a track. The palm pads on the front feet are curved on the leading edge and are much wider than they are long. The front feet have a separate heel pad, which may or may not register. The palm and heel pads on the hind feet are fused (as they are in all the North American bears), so the hind track is considerably larger than the front.

Polar Bears often dig circular pits or shallow dens in snow banks for beds or they will rest on the snow surface and let themselves be

Slow walk

Left hind track
Length: 21.5–38.0 cm
Width: 15.0–25.0 cm

Left front track
Length: 15.0–35.5 cm
Width: 15.0–25.5 cm

Polar Bear

completely covered by drifting or falling snow. These beds measure up to 1 m in diameter. In the absence of snow they will also dig into the ground to create pits or shallow dens for temporary shelter. Maternal dens are usually located on the leeward slopes of hills near the coast. They are typically made up of a single egress tunnel (80–300 cm long) leading to one or more rooms carved out of the snow, but they may also be dug into the earth if the snow is late in arriving. Some dens have more than one egress tunnel. Entrances are 50–90 cm high by 45–190 cm wide and are often blocked with snow. The top of the entrance is typically much narrower than the base. The tunnel is typically larger than the entrance and may be slanted so that the rooms are slightly higher than the entrance, perhaps to conserve heat. Tunnels are normally small enough that the female bear must crawl to get through. If the den has more than one room (up to four have been reported), the largest is around 0.8–2.0 m long by 1.5–3.0 m wide by 1.0 m high. Dens housing females with older cubs tend to be generally larger but not higher. Deep or shallow earth dens are usually occupied by females, while even shallower earth pits are created and used mostly by males. Shallow shelter dens in snow and ice are used by all bears during inclement winter weather. These dens share a similar entrance size to that of the deeper maternal dens but generally have a single simple room, or they are only a shallow pocket gouged out of the snow or earth.

Polar Bear scat varies with diet. Largely meat droppings are flat, dark, amorphous liquid splatters, sometimes containing seal hair or feathers depending on the source of the meat. Vegetation scat may contain berries or fibrous remains of grasses and resemble Brown Bear scat. This tubular scat can be up to 5 cm in diameter and up to 25 cm in length if unbroken. Usually they are crumbly or break into segments.

REFERENCES

Aars et al. 2009; Amstrup 2003; Amstrup et al. 2000; Atkinson and Ramsay 1995; Atkinson et al. 1996; Banfield 1974; Born 1995; Born et al. 1995; Clark, D.A., et al. 1997; DeMaster and Stirling 1981; Durner et al. 2003; Dyck and Daley 2002; Ferguson, S.H., et al. 2000; Goodyear 2003; Harington 1968; IUCN/SSC Polar Bear Specialist Group 2002; Lindqvist et al. 2010; Lunn et al. 2002, 2004; Mauritzen et al. 2001, 2003; Messier et al. 1994; Parks, E.K., et al. 2006; Ramsay and Stirling 1990; Regehr et al. 2007; Richardson, E.S., and Andriashek 2006; Rugh 1993; Smith, A.E., and Hill 1996; Stirling 1988, 1999; Stirling and Parkinson 2006; Struzik 2006; Taylor et al. 2006; Van de Velde et al. 2003; Wiig et al. 2003.

FAMILY OTARIIDAE

fur seals and sea lions

Members of the Otariidae family of seals are identified by their small external ear flaps. The males are significantly larger than the females. Otariid seals can walk on the land by rotating their hind flippers to point forward and then shuffling along, using all four feet. Usually the whole body is lifted off the ground, although this can be next to impossible for some of the largest males. Their coat includes an underfur that is often very well developed.

Northern Fur Seal

FRENCH NAME: **Otarie à fourrure,** *also* **Otarie des Pribilofs**
SCIENTIFIC NAME: *Callorhinus ursinus*

Northern Fur Seals have long been hunted for their valuable pelt. The American commercial harvest of animals from the Pribilof Islands of Alaska was so lucrative that the value of a couple of years of indiscriminate harvesting in the late 1860s more than matched the price paid in 1867 by the United States to Russia for the purchase of all of Alaska.

DESCRIPTION

These seals have a short snout, small head, large and long hind flippers, and no fur below the wrist on their front flippers. Both front flippers and hind flippers appear to be oversized, but the hind flippers are actually the largest of all the seals relative to their body size. Their external ear pinnae are quite long for a seal and tend to become hairless at their tip in older animals. Adult females are small, sleek creatures, while the considerably larger adult males develop huge shoulders and chest, a thick neck, and a stiff mane that extends around the body from the shoulders to the neck. As with other fur seals, their double coat is exceedingly dense and soft and made up of guard hairs and shorter underfur. This fur is so thick that over 46,000 hairs grow on each square centimetre of skin. It protects them from the cold waters in which they swim, by trapping a layer of air, which prevents water from touching the skin. Even a small amount of crude oil can destroy the insulating properties of the fur and result in death by hypothermia. Their fur appears black when wet. Most adult males dry to a rich, dark-brown colour, with a black belly and flippers. Some may be reddish-brown or even silvery brown but usually still have a dark underside, and some mature males develop a pale cluster of hairs on the top of their head. Occasionally a male will have a mane colour that is different from the colour of the rest of his body, such as a silver mane and a black body, or a yellowish or reddish mane and a brown body. Dry females are usually dark brown or dark greyish-brown above, with a silvery-grey or creamy-brown throat and belly, and dark flippers. Some females are reddish on their chest. Pups that are less than two or three months old are black, often with white areas on their belly and occasionally their throat. This coat is water permeable; therefore, although the pups are capable of swimming, they cannot remain in the water very long until their adult coat has grown in. Between two and three months old the pups moult their dark birth coat and grow a pelage similar to that of adult females. Young males grow an adult coat at around five years old as they become sexually mature. The long whiskers (vibrissae) on the face are black at birth and remain black until the animal begins to mature sexually, at which time white ones begin to grow, and for a time they sport a mixture of both black and white vibrissae. Adult Northern Fur Seals have all-white vibrissae. These seals continue to grow throughout their lives, although the growth tapers off as they become adult. This means that larger animals are usually older. The dental formula is incisors 3/2, canines 1/1, and postcanines 6/5, for a total of 36 teeth (all the teeth behind the canines are called the postcanine teeth as premolars and molars cannot be distinguished).

SIMILAR SPECIES

Northern Fur Seals are quite different in many ways from other seals and sea lions. Their short snout, small head, and dark colour set them apart. The head shape of a female Northern Fur Seal is distinctive with her arched Roman nose. Both Steller Sea Lions and California Sea Lions can occur within the range of Northern Fur Seals. Sea lions are easily distinguished by their lighter tan or yellowish colour, their longer snout, and larger body size. Even female Steller Sea Lions are larger than male Northern Fur Seals, and male Stellers are considerably larger. A female California Sea Lion is around the size of an adult male Northern Fur Seal. Even a Harbour Seal is longer than a female Northern Fur Seal, which is the smallest seal in Canadian waters.

SIZE

Males are significantly larger than females. A mature male is about 3.4 times heavier than is the female prior to mating, and 5.4 times heavier when defending a breeding territory. Weight fluctuates during the year. Females are heaviest just prior to breeding, and all adults are lightest at the end of winter. Most weight is gained in spring during the northward migration.
Mature males: total length: 1.5–2.0 m; weight: 100–200 kg.
Mature females: total length: 1.1–1.4 m (average 1.3 m); weight: 30–60 kg (average 40 kg).
Newborns: total length: about 60 cm; weight: males 5.5–6.2 kg, females 4.5–5.5 kg.

RANGE

Limited to the Northern Hemisphere, Northern Fur Seals occur in the North Pacific Ocean, the Sea of Okhotsk, and the Bering Sea.

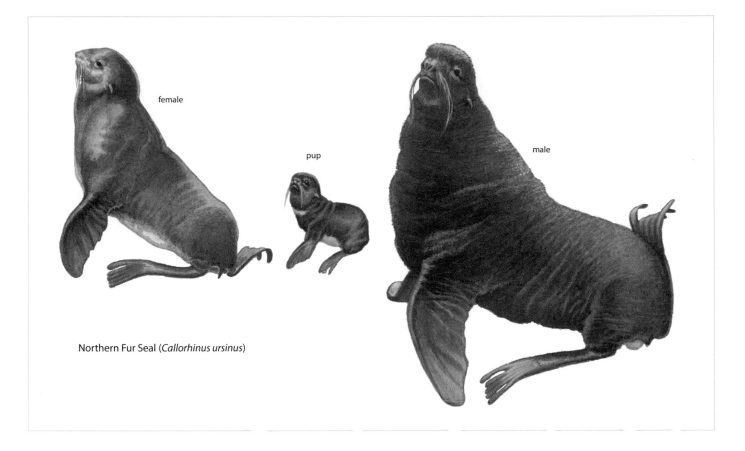

female

pup

male

Northern Fur Seal (*Callorhinus ursinus*)

There are seven known breeding locations. San Miguel Island, off the coast of southern California, is the most southerly. Robben Island (Russia), north of Japan, is the most westerly. Also in the western Pacific are the Russian colonies on the central Kuril Islands and Commander Islands. The last three colonies are on islands off the western coast of Alaska: St Paul, St George, and Bogoslof Islands, which host the large colonies in the eastern Pacific herd. The Bogoslof rookery was established in 1980. The southern population on San Miguel Island (United States) was created in the late 1950s and early 1960s following the introduction of animals born on the Pribilof and Commander Islands. By the late 1980s some of these animals had naturally established a small secondary breeding colony on the nearby South Farallon Islands. These southern animals are resident in the area all year round. Males from the other more northerly colonies spend August to May at sea, probably from the Sea of Okhotsk across to the North Pacific Ocean south of the Aleutian Islands, and over into the Gulf of Alaska. Females and juveniles from these same colonies migrate south in October. They fan out into the North Pacific Ocean, most ending up along the western and eastern margins, as far south as the United States–Mexico border in the east and the Japanese Honshu coast in the west. Distribution while at sea is patchy and directly related to food availability. Some recent archaeological evidence suggests that there once were rookeries scattered around the American west coast and a rookery on Vancouver Island (Benson Island). Presumably these rookeries were vulnerable to local aboriginal hunters since they are now extinct and are only known from archaeological sites. These remains might represent a closely related but distinct species. There are no modern breeding sites of Northern Fur Seals in Canada, but females and immatures do occur offshore in British Columbian waters year-round and occasionally wander inshore. They are most common from January to June. The most likely sighting locations are off the west coast of Vancouver Island and around the Queen Charlotte Islands.

ABUNDANCE

It is quite difficult to count these animals, not only because they are so dense in the rookeries but also because most of the immature animals (less than two years old) and some adults remain at sea during the breeding season, so their numbers can only be estimated. Nevertheless, attempts have been made over centuries to keep track of population size, mainly by counting the number of pups born. The current worldwide population is estimated at about 1.3 million animals. The Pribilof herd alone is estimated to have been twice this number before wholesale killing for skins began in 1786. Close to 50% of the Northern Fur Seals breed on the Pribilof Islands of St Paul, St George, and the closely associated Sea Lion Rock, where numbers have been declining since the 1950s and are now around 525,000. Other breeding sites have smaller colonies: Robben Island and central Kuril Islands around 200,000; the Commander Islands (225,000–250,000); Bogoslof Island in the Aleutian Island chain (10,000), colonized in 1980 from the Pribilof stock; and San Miguel Island in southern California (4300). The stock is divided into five herds named for their breeding islands: The Pribilof/Bogoslof

Distribution of the Northern Fur Seal (*Callorhinus ursinus*)

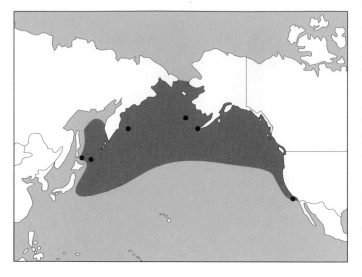

● breeding sites

herd, the Commander herd, the Kuril herd, the Robben herd, and the San Miguel herd. In 1991 the San Miguel herd may have begun a new rookery on the Farallon Islands, but its future is uncertain. The International Union for Conservation of Nature (IUCN "Red List") considers Northern Fur Seals to be "Vulnerable" because of habitat degradation from oil and gas exploration, heavy shipping traffic, depletion of fish stocks, high mortality owing to entrapment in discarded or lost fishing nets, and the continuation of harvesting. The Pribilof/Bogoslof herd appears to be suffering high juvenile and adult female mortality, possibly owing to a decline or change in the types of prey available to them in the Bering Sea or along their migration route in the North Pacific. Pup production dropped substantially during 1998–2004 and is continuing to slump. Unregulated hunting, which killed millions of animals in the late 1800s and early 1900s, resulted in the signing in 1911 of the North Pacific Fur Seal Convention by the United States, Japan, Russia, and Canada. This treaty, which lapsed in 1984, banned the killing of animals at sea and limited the harvest to immature males on land. Around the world, current hunting is primarily carried out by natives for subsistence, except in the Kuril and Commander herds of Russia, where a commercial hunt is still thought to be occurring. The Pribilof/Bogoslof herd is designated as "Depleted" under the U.S. Marine Mammal Protection Act. The species is protected in Canada by the 1993 Marine Mammal Regulations, which prohibit hunting except by indigenous peoples for subsistence use. In Canada, COSEWIC has designated the species as "Threatened."

ECOLOGY

Breeding rookeries are generally near the continental slope and are usually rocky. The exception occurs with the introduced population on San Miguel Island, where there are two rookeries, one in a rocky cove and the other on a sandy beach. Rookery sites are traditional, and Northern Fur Seals are very conservative when it comes

to adopting new breeding locations. These seals migrate annually between their winter feeding grounds and their breeding grounds. Adult males arrive on the rookeries in May and begin leaving in early August; all have left by late October or early November. The females and immatures begin returning in mid-June and leave in October or November. Pregnant females do not come ashore right away when they arrive at the breeding island. They continue to forage and only enter the rookery when parturition is imminent. Pups may leave the breeding islands before or after their mothers to migrate south and hunt for themselves. They will spend the next 22 months or so at sea, returning to land again as two- or three-year-olds at the breeding colony in which they were born. Pup mortality is high at around 50%–80% during their first year. The youngsters may lose up to half their body weight during their first winter. Male Northern Fur Seals have a higher natural mortality than do females, so the ratio of adult males to females is always lower. This factor is compounded by the harvest of immature males, which takes place next to the rookeries, where they commonly haul out so as not to disturb the females with their pups. Immature males that spend the summer near their natal rookery lose an estimated 20%–30% of their body weight owing to reduction of food intake. Presumably the learning of social and sexual skills during that time offsets the weight loss. Fur seals swim using their front flippers for propulsion. These seals rarely dive deeper than 200 m. Females may live for 26 years, but few live longer than four years. The maximum longevity of males is around 16–17 years. Predators include Steller Sea Lions, large sharks, and Killer Whales. Foxes may be seen around the breeding rookeries, scavenging dead pups, but they do not prey on them.

DIET

Northern Fur Seals take a wide variety (at least 75 different species) of near-shore and pelagic shoaling fish and squid, depending on availability. Most feeding occurs at night as the prey species take advantage of the darkness to move up the water column, thereby coming within easier reach of the seals. Walleye Pollock and squid are the most important prey during the summer while the population is in the Bering Sea, but fur seals are also known to feed to a lesser extent on sand lance, salmon, Capelin, Pacific Herring, mackerel, Pacific Hake, anchovy, Eulachon, rockfish, Pacific Whiting, lantern fish, and other fish and squid species. Males and females likely have different diets for at least part, if not most, of the year. Females tend to feed on inshore species throughout the year, whereas males spend the bulk of their year in the deep sea, feeding on more pelagic species. Most feeding dives of females are in the range of 20–140 m deep and last less than four minutes. Adult males, with their larger body size, can dive deeper and stay submerged longer. The maximum recorded dive depth of a male Northern Fur Seal is 230 m. The ability of the adults to partition their food resources and reduce feeding competition between the sexes is a crucial part of the ecology of the species.

REPRODUCTION

Males reach sexual maturity at four to five years old, but few begin breeding until they are between eight and nine years old and have

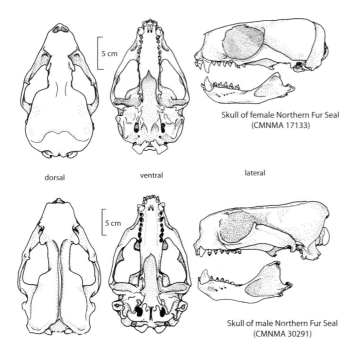

dorsal | ventral | lateral

Skull of female Northern Fur Seal
(CMNMA 17133)

5 cm

Skull of male Northern Fur Seal
(CMNMA 30291)

reached a body size that allows them to successfully contend for a portion of the rookery. Along with the space comes sole reproductive access to those females that choose to remain within it. Few males have the vitality to breed for more than two, or at most three, years between the ages of nine and fourteen years old. Females are three to five years old when they begin breeding, so they are four to six years old when they produce their first pup. Females typically produce a single pup each year, although after age 16 their reproduction rate diminishes. A female is receptive for less than 48 hours, sometime in late June or July, usually at about five to eight days after giving birth. The vast majority of copulations occur on land, but some have been observed at sea. The fertilized embryo floats around in the uterus for about four to five months before implanting in November. This delayed implantation permits parturition and mating to occur during the short period that the colony is all together on land. Gestation, including the delayed implantation phase, lasts about 358 days, so the pup is born in the following June or July, typically within two days of the female's arrival at the breeding site. Northern Fur Seal milk is composed of about 53% fat, and so the pups fill out fairly quickly. Weaning occurs abruptly when the pup is abandoned by its mother or the mother is abandoned by her pup, usually in late October. At this time the pup weighs around 15 kg.

BEHAVIOUR

Northern Fur Seals are hugely and densely gregarious during the three- to four-month breeding season and often solitary throughout the remainder of the year. During their months at sea they may occur in pairs or small groups, but many travel alone. A larger breeding colony may hold tens of thousands of tightly packed animals. As many as 70,000 animals can occupy a single kilometre of beach. Breeding males arrive at the traditional rookeries in May, claiming and battling for a portion of the space. The maximum recorded territory is 110 m². As the females begin arriving about a month later, each male endeavours to herd them onto his territory. Bulls are quite effective at restricting the movement of females and preventing them from moving about from territory to territory. Essentially the females are attracted to each other and form clusters in the middle of a bull's territory, where they are not bothered by other males. As the colony fills up, the space between female clusters becomes continuous, and bulls are scattered fairly evenly throughout. Only in depleted colonies does one see space between the territories. The bull's main role is to create an area of relative peace within the rookery, where the females can safely raise their pups. He vigorously defends his space and the females within it from other males by a combination of threat displays, vocalizations, and sometimes fighting.

Territorial bulls are tied to their territory and eat little while on land. During the breeding season they may lose as much as 25% of their body weight. Bachelor or "idle" adult males, without a territory, are relegated to peripheral areas and cluster around the rookery, intercepting the females as they attempt to put to sea. These females have probably already mated but must pass through this gauntlet each time they depart and return from feeding. Adult females are not territorial but react aggressively to a close approach by all other females, adult males, juveniles, and pups that are not her own. Any pup that approaches a female that is not its mother runs the risk of a nasty bite. Shortly after mating, when the pups are six to eight days old, the females leave them for the first time to feed. While their mothers are gone, the highly mobile pups frequently gather in large nursery aggregations. The females return about four or five days later and wander through the colony loudly calling. Many of the hungry pups respond, and she sorts through their voices to isolate the distinctive call of her pup. Counter-calling, the two reunite and verify identities by odour. Recent studies suggest that females and their pups can still recognize each other's voices at least four years later, which is the longest known individual recognition of calls between wild animals. She will remain on shore for a day or two to allow her pup to nurse, and then return to the sea to feed. Each successive trip to sea lasts longer (up to a maximum of eight to twelve consecutive days), but the suckling interval between sea voyages remains one or two days. Females may travel more than 400 km during such feeding excursions. Eventually, when the pup is around four months old, the female leaves for the last time and does not return to the pup.

Northern Fur Seals on land tend to overheat, especially on sunny days. Panting and waving their hind flippers in the air are means of venting excess heat. At sea, sleeping Northern Fur Seals assume a distinctive posture. Bobbing at the surface, they create an arch with one front flipper and both hind flippers held together out of the water, in a manner called jug handling. Likely they are raising these appendages to conserve body heat because flippers have little insulating fur, and heat loss is greater in water than in air. Fur seals are capable of resting one hemisphere of the brain at a time (called uni-hemispheric sleep), as has been found also in dolphins and porpoises.

VOCALIZATIONS

A Northern Fur Seal rookery is a noisy place. Whickering moans, grunts, and bawls resound around the area. Pups emit a distinctive bawling, mother-attraction call, and females produce a comparable pup-attraction call. These calls are similar, except that the adult call is a lower pitch. Advertisement roars of adult males are similar but are deeper yet. Adult males will produce this growling roar from the water as well as from land, and it can carry a considerable distance on a still day. Northern Fur Seals also produce clicking and bleating sounds underwater.

SIGNS

A noisy, smelly rookery that is occupied by thousands of fur seals is difficult to miss. Even when the animals are at sea, the hard-packed earth, smoothed rocks, and lack of vegetation may highlight the site. Their arched resting posture at sea is also a giveaway (further described in the "Behaviour" section above).

REFERENCES

Antonelis et al. 1997; Baird and Hanson 1997; Baker, J.D., and Donohue 2000; Baker, J.D., et al. 1994; Crockford et al. 2002; Gentry 1981; Insley 2000; Insley et al. 2008; Pyle et al. 2001; Reeves et al. 2002; Towell et al. 2006; Trites 1992; Trites and Bigg 1992, 1996; York and Scheffer 1997.

Steller Sea Lion

also called Northern Sea Lion, Steller's Sea Lion

FRENCH NAME: **Otarie de Steller,** *also* **Lion de mer de Steller**
SCIENTIFIC NAME: *Eumetopias jubatus*

Formerly listed as "Vulnerable" by COSEWIC, the protected Canadian populations of the Steller Sea Lion have recovered sufficiently since the 1970s that their status has been upgraded to that of "Special Concern."

DESCRIPTION

The largest of the world's five species of sea lions, Steller Sea Lions, like other sea lions, display considerable sexual dimorphism. Adult females are about half a metre shorter and weigh as little as one-fifth (average one-half to one-third) of the weight of adult males. Mature males develop a dense mane of long, coarse hairs on their massive shoulders, neck, and chest, similar to that of a male lion, hence the common name. The Latin *jubatus* means "mane." Dry animals are reddish-blond to light tan in colour, with contrasting black flippers and a chocolate-brown underside. Wet sea lions are darker. The flippers have short, black hairs on the upper surface and are hairless on the underside. Albino or white animals are rare. Females are typically a little paler than males. Pups are dark brown to black until they moult into a lighter brown coat at about the age of four to six months. Their pelage changes to the typical blond adult-female pelage after their next moult, at the end of their second year. Males begin to develop

their distinctive mane and thickened neck and shoulders at around four years old. Adult vibrissae (whiskers) on the snout are long and may be black, white, or a mix of both colours. Sea lions have small external ear flaps (pinnae). The dental formula is incisors 3/2, canines 1/1, and postcanines 5/5, for a total of 34 teeth (all the teeth behind the canines are called the postcanine teeth as the premolars and the molars cannot be distinguished). Steller Sea Lions moult once each year, typically taking about a month to do so during the summer. A newly grown pelage is lighter.

SIMILAR SPECIES

Northern Fur Seals and California Sea Lions both occur within the range of Steller Sea Lions. The fur seals are substantially smaller, have darker, more brownish fur, and a much smaller head. Mature male California Sea Lions are also darker brown than Steller Sea Lions. Female California Sea Lions are paler than the males and can be easily confused with female Steller Sea Lions, but few mature female California Sea Lions venture far north into the core of the Steller Sea Lions' distribution. Female California Sea Lions have a dog-like face with a distinct snout and a higher forehead. Female Steller Sea Lions have a wedge-shaped head as the forehead and the snout are on almost the same plane. Their darker colour, and the bulbous forehead of mature and even subadult male California Sea Lions, differentiates them from Steller Sea Lions.

SIZE

Like all of the sea lions, the adult male Steller Sea Lions are substantially larger and heavier than the adult females. Weight fluctuates during the year. The seals are heaviest just prior to breeding, and lightest at the end of winter.

Mature males: total length: average 2.7–3.1 m (maximum 3.25 m); weight: average 400–800 kg (fat males just prior to the breeding season may weigh over 1100 kg).

Mature females: total length: average 2.1–2.5 m (maximum 2.9 m); weight: average 200–300 kg (maximum 350 kg).

Newborns: total length: about 1 m; weight: 16–23 kg (males typically weigh more than females).

RANGE

Steller Sea Lions occupy the cool temperate and Subarctic continental shelf waters of the North Pacific Ocean from California to Alaska, across the Bering Sea, and along the coast of Russia to the waters off the tip of Hokkaido, Japan. DNA evidence suggests that three stocks exist: the eastern stock includes animals from southeastern Alaska southward to California; the western stock includes the animals from Prince William Sound to the tip of the Aleutian Archipelago; and the Asian stock includes the sea lions along the coast of Russia and northern Japan. The divide between the eastern and western stocks occurs just to the east of Cape Suckling, Alaska, near the top of the Alaska panhandle. There are 55–60 breeding sites and more than 300 haul-outs scattered throughout the range, mostly on coastal islands. Canada currently hosts five traditional breeding sites: the three Scott islands of Triangle, Beresford, and Maggot off the northwest tip of Vancouver Island; Cape St James, off the southern tip of Kunghit

Steller Sea Lion (*Eumetopias jubatus*)

male

female

pup

Island, Queen Charlotte Islands; and on the North Danger Rocks off the British Columbia coast near Banks Island. Rookeries on the Sea Otter Group of islands (Virgin and Pearl Rocks and possibly Watch Rock) were extirpated by hunting during the 1920s and 1930s. Over the past 10 years, increasing numbers of pups have been observed at some haul-outs. However, numbers of births are not yet high enough to classify these sites as rookeries.

ABUNDANCE

In the early 1900s, following fishermen's complaints that Steller Sea Lions were reducing their catches, a bounty was offered from British Columbia to California. Culling during this effort eliminated the rookeries on the Sea Otter Group of rocky reefs off the northwest tip of Vancouver Island. Practice bombings by the Canadian air force and navy in the 1940s were conducted using Steller Sea Lion breeding colonies as targets. Commercial harvests of pups were permitted in Alaska from 1959 to 1972. Predator-control efforts as well as commercial harvests in Canada stopped in 1970 when the Fisheries Act was enacted. Current harvesting is limited to native peoples' subsistence hunts, and this appears to be sustainable. Steller Sea Lions in Canadian waters are from the eastern stock (see the "Range" section). British Columbia places Steller Sea Lions on its "Red List," and they are considered to be of "Special Concern" by COSEWIC. Steller Sea Lions in United States territorial waters are from both the western and the eastern populations. In the United States the eastern population is considered to be "Threatened," and the western population to be "Endangered."

Abundance trends since the cessation of culling and harvesting are varied. The eastern stock (including the population in Canadian waters) is slowly increasing by around 3% annually. Historically the western population was much larger than the eastern population, and it accounted for roughly 90% of the total abundance. The decline of this population since the 1960s (by around 85% from the early 1970s to 2000) has left it severely depleted to the extent that it represents roughly half or less of the current numbers. There may be as many as 50,000 animals in the western stock, but likely less. This decline appears to have abated in most western rookeries since 2004. During the same time, the eastern population has slowly increased in numbers to around 45,000 animals. Between the 1950s and the 1970s the population worldwide was more or less stable, estimated at 240,000–300,000 animals. By 1989 the numbers had dropped to around 116,000; by 1994–5 to 97,500; and by 1999–2002 to 95,000 or less. There is evidence that sea lions, especially in the Gulf of Alaska and the Bering Sea, are nutritionally stressed, but it is not known if human overfishing of the fatty fishes such as herring is the cause or if oceanic changes that affect the fish species and abundance are causing the problem. It is very possible that both are factors. A recent theory holds that Killer Whales are consuming more seals than they were before, owing to the over-hunting and resulting scarcity of large whales, which were the Killer Whales' preferred prey in that region. This factor, combined with a weakened condition owing to nutritional stress, could mean that the sea lions are easier prey than they used to be. Considerably less information is available on the Asian stock, which is also declining and numbered around 12,000 in 2002.

ECOLOGY

There are three main types of terrestrial sites used by Steller Sea Lions. Rookeries are occupied during the breeding season

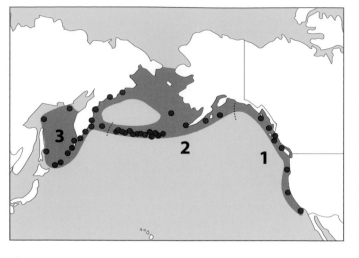

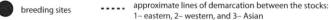

● breeding sites   ····· approximate lines of demarcation between the stocks: 1– eastern, 2– western, and 3– Asian

by reproductive adults and pups, and many are used in the off-breeding season as a convenient haul-out by a mixed group of animals. Year-round haul-out sites are used by a variety of age classes and both genders. Winter haul-out sites are mainly used in the non-breeding season, again by any of the animals in the population. Winter haul-outs in southern British Columbia may be shared with mature and juvenile male California Sea Lions. Most rookery sites have one or more bachelor haul-outs nearby that are occupied by mostly non-breeding adult and subadult males. Usage of these sites is generally traditional, especially the rookery locations, although new haul-outs are adopted from time to time, and on rare occasions haul-outs may become rookeries. Some sites have a documented history of over four centuries of consistent use. Most haul-outs and rookeries are on rocky beaches with significant wave action, but some are sandy. Although Steller Sea Lions do not migrate, they do undertake seasonal movements. Summer foraging areas are often in the neighbourhood of rookeries, while winter feeding areas are more widespread, with adult males generally moving northwards or remaining in colder northern waters, and females and juveniles moving southwards. Most foraging occurs between the edge of the continental shelf and the shore. Steller Sea Lions can dive to depths of over 325 m and remain submerged for over eight minutes. Even a ten-month-old juvenile can dive to over 200 m deep. The large adult males have the greatest capacity for both depth and length of time underwater. Most foraging dives are to the 15–50 m depth and last 1.5–2.5 minutes. Satellite monitoring shows that summer foraging trips are short, averaging 13–17 km, while winter trips last longer and average 28–153 km. During winter the sea lions may remain at sea for extended periods. Their top swimming speed is around 27 km/h, and they use their front flippers for propulsion in the water. Steller Sea Lions commonly lose their lives in fishing equipment, either in active nets or by entanglement in discarded ropes and nets. Probably a hundred, and likely many more, are shot annually

by British Columbia salmon farmers with licences to shoot seals that are threatening their pens. The animals are sensitive to disturbance while they are on land, which can cause stampedes to the water. Repeated disturbances may cause temporary abandonment of the area. Animals are easily startled and disturbed but can habituate to human activity, and large numbers may remain near large urban centres such as Vancouver and Victoria. Pups die from a variety of causes, including drowning. Pups that are washed into the ocean during storms are often unable to climb back out of the water onto the large rocks that often surround the haul-outs and rookeries. Pup mortality rates during the first three weeks of life vary from year to year, likely averaging around 30%–40%; they can be as low as 4% and as high as 60%. Females may live 22 years, and males about 14 years. The choice of islands for haul-outs and rookeries removes much of the threat of large terrestrial predators such as bears, and so the major predators of Steller Sea Lions are marine. These include Killer Whales and possibly large sharks.

DIET

Foraging generally occurs at night when the prey species rise up the water column in the darkness. Steller Sea Lions are opportunistic and catch and consume a wide variety of schooling fishes, squid, and octopus. Over 50 species have been recorded in their diet. Larger subadult and non-territorial males are also known to occasionally prey upon pups and juvenile Northern Fur Seals, California Sea Lions, Ringed Seals, Harbour Seals, and Sea Otters. Their principal prey are small to medium-sized shoaling fishes such as Pacific Herring, Hake, sand lance, salmon, Atka Mackerel, Eulachon, dogfish, Pacific Cod, Capelin, Walleye Pollock, and sardines. They also feed on bottom fishes such as sculpins, rockfish, flounder, and skate.

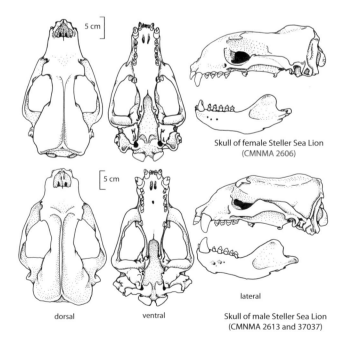

5 cm

Skull of female Steller Sea Lion (CMNMA 2606)

5 cm

dorsal   ventral   lateral

Skull of male Steller Sea Lion (CMNMA 2613 and 37037)

REPRODUCTION

Steller Sea Lions typically return to their natal rookery to breed, although there may be some exchange between neighbouring rookeries or even rookeries much further away. Mating occurs between late May and late July. The fertilized embryo floats around the female's reproductive tract for three to four months before implanting in late September or early October. Around 60%–75% of mature females produce a pup in any given year. Twins are rare. Premature pups are produced as early as February but are not known to survive. Full-term pups are born in mid-May to early July, with a peak in the third week of June. The gestation period, including the delayed implantation phase, takes around 354 days. Births at individual rookeries are synchronous, with 90% of pups born within a 25-day period. The most likely reason for the timing difference between rookeries is the nutritional status of the reproductive females. Females arrive on the rookery about three days before parturition. Growth is rapid during the first few weeks as the pups double their birth weight by the time they are seven weeks old. Pups are capable of swimming shortly after they are born but are very poor swimmers. Young Steller Sea Lions have a prolonged suckling period. Most are weaned near the end of their first year, but some may nurse into their second and even third year if their mother does not produce a new pup. Some females may be seen on the rookery with a new pup and a yearling, both suckling. Females come into oestrus and breed about 11–12 days after giving birth. They spend the bulk of their adult life simultaneously pregnant and lactating. Females become sexually mature between the ages of three and eight years old and so produce their first pups at the age of four to nine years old. They may breed into their early twenties. Males mature sexually between the ages of three and seven years old, but most are not physically large enough to hold a breeding territory until they are nine to thirteen years old.

BEHAVIOUR

Sexually mature sea lions begin returning to the rookeries in spring. Some of the females are accompanied by their yearling pups that are still suckling. Bulls arrive first in early May to compete for a territory. During the establishment and ongoing defence of their breeding territory, bulls face off against rivals by roaring, nodding their heads, and obliquely facing each other with mouths open and vibrissae aiming forward. Such threat displays establish the location of the territorial boundaries, which then tend to remain stable for the duration of the breeding season. Fighting occurs only occasionally when such displays are ineffective or insufficient. Bulls may establish an entirely terrestrial territory, one along the shore that is semiaquatic; less successful bulls may maintain wholly aquatic territories. Breeding territories are about 225 m² on average. Bulls typically stay on their territories for around two months, during which time they fast. The pregnant females begin arriving in the latter half of May and give birth to their pup within a few days of their arrival. Females gather on the males' territories and are usually later bred by the resident bull. As they prefer to give birth just above the high-tide line, the shoreline or semiaquatic territories are the most hotly contested. Each female aggressively defends a small space (about 5 m in diameter) to keep her newborn safe from other female sea lions and from disturbance caused by the bulky males patrolling and defending their borders. She uses threat displays similar to those of the bulls to defend her space from other females. Almost immediately after giving birth the mother will lift the pup into a suckling position. The two begin a vocal exchange about the same time. Each will then learn to recognize the other's smell and vocalizations. A female will aggressively repel any pup not her own that attempts to nurse. Maternal females remain on land for the first week after their pup is born and then begin alternating a day or two in foraging at sea, with a day on shore with their pup. At around two weeks old the pups begin to assemble away from the adults while their mothers are away foraging. The young are able to swim in inter-tidal areas and tidal pools by the time they are two weeks old and are sufficiently proficient to enter the open ocean at around four weeks old. Females often move their pups away from the breeding bulls to nearby sites at about that age. By the end of August most females and their pups are no longer at the rookeries.

VOCALIZATIONS

Steller Sea Lion rookeries tend to be noisy places. Territorial males wheeze and roar as part of their threat displays, in addition to producing a guttural sound in the air and underwater. Bellows and roars are produced by both males and females while they are on land. The pups' "attraction" call resembles the bleat of a sheep.

SIGNS

At sea, Steller Sea Lions are usually seen in small groups or as solitary animals, although up to 300 animals may feed together, porpoising and diving in apparent synchrony. Haul-outs and rookeries may contain a few to hundreds of individuals. Sleeping at sea, Steller Sea Lions characteristically float on their back with their flippers in the air to keep them out of the cold water.

REFERENCES

Ban and Trites 2007; Boltnev and Mathisen 1996; Byrnes and Hood 1994; Calkins et al. 1999; Gende et al. 2001; Kaplan et al. 2008; Loughlin et al. 1987, 2003; Maniscalco and Parker 2009; Matkin et al. 2002; Merrick et al. 1997; O'Daniel and Schneweis 1992; Olesiuk and Trites 2003; Pendleton et al. 2006; Pitcher et al. 2001; Rosen and Trites 2000; Schusterman 1981; Sease and York 2003; Sinclair and Zepplin 2002; Springer et al. 2003; Thomas and Thorne 2001; Trites and Donnelly 2003; Trites and Larkin 1996; Trujillo et al. 2004.

California Sea Lion

FRENCH NAME: **Otarie de Californie**

SCIENTIFIC NAME: *Zalophus californianus*

California Sea Lions are known for their noisy barking on the rookeries. They are usually the trained "seals" in aquariums, zoos, and circuses because they are intelligent, playful, and quick to learn. Some, mainly males, winter in Canadian waters along Vancouver Island.

DESCRIPTION

Immature and adult female California Sea Lions look similar. Right after moulting they are light grey to silver in colour but quickly fade to blond or light brown with a darker underside and rump, and blackish flippers. When males reach four or five years old, their fur darkens to a deep reddish or chocolate brown that does not fade. All California Sea Lions look much darker when they are wet, but this is especially true of females and immatures. Pups are dark brown at birth, but they fade to brown or even light brown in a few weeks. At four to six months old they moult their natal coat for a light-brown or silver coat similar to that of females. Mature males develop an exaggerated sagittal crest on the top of their skull, which gives them a distinctive bulging forehead. Often the fur on top of the bulge is noticeably pale blond. They also become increasingly thick through the neck, shoulders, and chest but do not develop a mane, as do Steller Seal Lion males. California Sea Lions are very slim in the hind end from the hips to the tail. The dental formula is incisors 3/2, canines 1/1, and postcanines 5–6 / 5–6, for a total of 34–38 teeth (all the teeth behind the canines are called the postcanine teeth as premolars and molars cannot be distinguished). They experience one moult each year. Adult males moult in January and February on their wintering grounds in the north, while females and immatures moult from early autumn into winter. Each animal's moult takes about one month to two months.

SIMILAR SPECIES

The ranges of Steller and California Sea Lions overlap in the waters from southern British Columbia southward to California. The larger Steller Sea Lion males lack the highly diagnostic bulging forehead of subadult and especially adult male California Sea Lions, and they are reddish-blond to dark tan, while male California Sea Lions are much darker. Females and immatures may be confused with Steller Sea Lion females and immatures. Steller Sea Lion females are larger with a wider snout and more robust head. The slim rear end of the California Sea Lions can often be distinguished if the animals are on land.

SIZE

Adult males are considerably heavier than adult females by a ratio of around 4:1.

Mature males: total length: up to 2.4 m; weight: up to 390 kg (average 375 kg).

Mature females: total length: up to 1.8 m; weight: up to 110 kg (average 94 kg).

Newborns: total length: 65–75 cm; weight: 6–9 kg (males are about 20% heavier than females at birth).

RANGE

There were three disjunct subspecies of California Sea Lion. The Japanese population (*Zalophus californianus japonicus*) is presumed

California Sea Lion (*Zalophus californianus*)

female

pup

male

to be extinct. The Galapagos population (*Zalophus californianus wollrebaeki*) is found only around the Galapagos Islands of Ecuador. The western coastal North American population (*Zalophus californianus californianus*) occurs from southern British Columbia to the coast of Mexico south of Baja California. California Sea Lions prefer coastal waters, and haul-outs and rookeries are normally on islands. The breeding islands include the Channel Islands off southern California, Guadalupe Island, and other islands off the coast of Baja California (Mexico), both on the Gulf of California side and along the Pacific side. After the breeding season large numbers of mature and subadult males and a few females migrate northward to winter from northern California to the shores of Vancouver Island. An increasing number of mature males are being sporadically sighted as far north as St Paul Island in the Bering Sea. These northern sightings occur mostly in spring and may be the result of the increasing populations in the south and the subsequent increasing competition for food, or of changes in environmental conditions, perhaps owing to climate warming. Females with pups remain in southern waters for the winter.

ABUNDANCE

Owing to their coastal habits, California Sea Lions have been within reach of aboriginal hunters for centuries. Europeans in North America killed them for food, and a bounty was paid on them from Oregon to California in the early 1900s because they were thought to compete with fishermen. There was a small commercial harvest from the 1920s to the 1940s off the coast of California. Since hunting was proscribed in 1970, numbers of California Sea Lions have increased at least fourfold. The number of pups born each year has almost tripled since 1975. In 2001 there were an estimated 200,000 California Sea Lions: 30,000 in the Galapagos Islands and 175,000 along the North American coast. Despite the mortality caused by pollution, entanglement in fishing gear, and disturbances at rookeries, their overall numbers continue to climb, although colonies in the Gulf of California have declined by about 20% in the first decade of the new millennium. These declines are thought to be caused by increasing temperatures.

ECOLOGY

Male California Sea Lions migrate during the winter to feeding areas off California, Oregon, Washington State, and British Columbia, but females with pups remain at the breeding colonies until the pups are weaned. Breeding colonies are occupied year-round but are used as haul-outs during the off-season. Trained California Sea Lions can swim at speeds of up to 35 km/h, a speed that can likely be matched by wild sea lions. Summer feeding dives are usually fairly shallow; females swim to depths of around 75 m and stay submerged for about four minutes. Winter dives are deeper and longer. The maximum recorded depth of a dive by a California Sea Lion is 536 m, and the maximum recorded duration is 12 minutes. Owing to the location of the breeding colonies, the North American population of California Sea Lions is affected in the years in which the tropical El Niño swings northward along the Central and North American coast. The warmer waters drive the normal cool-water fish species

non-breeding

breeding

★ isolated record

away, forcing the female sea lions to travel further and to forage longer in order to find sufficient food to support their milk production. As a result, their offspring grow more slowly as the pups are fed less frequently. The maximum lifespan of wild animals is unknown, but a captive California Sea Lion lived for 24 years. Killer Whales and Great White Sharks are their main predators.

DIET

Like other sea lions, California Sea Lions have a diverse diet. At least 100 different species have been identified to date. Their preferred foods are shoaling species such as anchovies, sardines, Whiting, Pacific and Jack Mackerel, and Market Squid. Rockfish of various species are also consumed with relish. They prefer to forage in areas with cool upwelling waters such as along the continental shelf or over seamounts, but they will also forage on the ocean bottom at times. Quick to learn, they have discovered that boats and fisherman often mean fish. They will approach and raid fishing lines and nets, sometimes unwittingly swallowing a hook with the fish or becoming entangled in the nets. They have also been discovered to prey upon newly fledged Common Murre chicks by swimming up from below and snatching an unsuspecting bird that is floating at the surface.

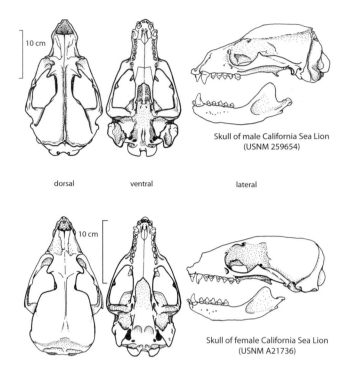

10 cm

dorsal   ventral   lateral

Skull of male California Sea Lion
(USNM 259654)

10 cm

Skull of female California Sea Lion
(USNM A21736)

REPRODUCTION

California Sea Lions do not reproduce in Canadian waters. All of the rookeries are far to the south along the coast of California and Mexico. Females give birth from late May to July after a gestation of around 340–345 days. Although delayed implantation has not been proved, it is thought that the inert but fertilized embryo floats around in the female's reproductive tract until it implants around mid-October and then begins its development. A single pup is produced. Imminent birth is signalled by a restless circling, nuzzling the perineal region, and vocalizing. The first sound that the pup hears is its mother's voice, and the first sea lion that it smells is its mother. She hears her pup's first bleats and learns its smell at the same time. From then on she will only allow her own pup to suckle, and the two can recognize each other in the crowded colony. Mothers remain on land with their pups for the first seven to ten days and then begin a foraging and nursing cycle of one to three days at sea followed by about a day on shore. Most pups are abruptly weaned and abandoned at around six months old, although some may be permitted to nurse for up to a year and occasionally even longer. Females enter oestrus and are willing to mate at about two to three weeks after they whelp. Rookeries further south in the Gulf of California are occupied earlier, pups are born earlier, the breeding season lasts longer, and females mate more than 30 days after the birth of their pups. The age at sexual maturity of either males or females is unknown.

BEHAVIOUR

California Sea Lions are social animals, and groups often rest closely packed together at favoured haul-out sites or float together on the ocean's surface in "rafts." They are sometimes seen porpoising, or jumping out of the water to breathe as they travel. Cooperative hunting is suspected. California Sea Lions are also known to play and "surf" in breaking waves. They breed on traditional island rookeries, and both adult males and adult females begin arriving at about the same time in late May. Adult males immediately start to establish breeding territories, which they aggressively defend from other males. They compete in ritualized displays involving belligerent open-mouthed threat postures, head-shaking, repetitive barking, lunges, and infrequently physical combat. Fights, mainly pushing and biting, rarely result in serious injury. All of the territories begin at the sea edge so that the males have uncontested access to the cool water on uncomfortably hot, sunny days. These territories are held for an average of 30 days (range 12–41 days), during which time the males fast. The average territory size on a rocky beach is 130 m² and may contain 10–15 adult females. Subadult males and adult males without a territory hang around the periphery of the rookery in the hope that when a territorial male is sleeping or otherwise occupied, they may have a chance to mate. Since the females are receptive about two to three weeks after giving birth, they are already into the foraging and nursing cycle (see the "Reproduction" section above) and have some opportunity to select the male of their choice, rather than simply mating with the territorial male on whose territory their pups were born. Many copulations occur in shallow water as the females arrive or depart on their foraging excursions. Males display limited courtship behaviour and do not systematically survey the reproductive condition of the females in their territories. Females solicit the attention of the male of their choice and conclude the mating when they have had enough. Older pups may accompany their mothers on short foraging trips, but those that remain congregate in large milling groups where they play and socialize. As each female returns from feeding, she calls to her pup, who recognizes her and replies. By repeatedly calling to each other, the two are eventually reunited, and after confirming its identity with a sniff, she permits the pup to nurse. As the pups get older, they play a more active role in these reunions, moving more quickly and for a greater distance towards their mother.

VOCALIZATIONS

Their main vocalization has been compared to a dog's bark. It is primarily a threat signal, and that of a mature male is the deepest, followed by that of immature males, and then of females. Maternal females produce a distinctive and individually identifiable pup-attraction call that they utter at birth and when returning to land after foraging at sea. A pup responds to its mother's call with a distinctive response call that allows its mother to find it in a crowd. Pups also produce a call that sounds like a bleating sheep. Mothers move their pups from place to place within the colony by walking away from it while uttering the pup-attraction call until it follows her. As the pups grow older, their mothers introduce them into the water by producing the pup-attraction call from the surf until the pups join them.

SIGNS

California Sea Lions, like Steller Sea Lions, often elevate one or more flippers (fore or hind) out of the water as they rest on the surface, probably to reduce heat loss to the cool water. Large groups may forage or "raft" together. This species is quick to take advantage of human activity and construction. They will adopt a moored pleasure boat, barge, or dock as a haul-out to the point of sinking them at times. Most travel is done below the surface, but this behaviour changes in areas in which shark attacks are common. They usually cluster into a tight group and swim at high speed in such regions, "porpoising" out of the water to breathe as they go.

REFERENCES

Antonelis et al. 1990; Boness et al. 1991; Garcia-Aguilar and Aurioles-Gamboa 2003; Garcia-Rodríguez and Aurioles-Gamboa 2004; Gisiner and Schusterman 1991; González-Suárez and Gerber 2008; Klimley 1994; Long, D.J., and Gilbert 1997; Odell 1981; Ono and Boness 1996; Peterson, R.S., and Bartholomew 1967; Reeves et al. 2002.

FAMILY ODOBENIDAE
walruses

The Walrus is the only living representative of the family Odobenidae. Walruses are large seals with distinctive long upper-canine teeth, hind flippers that rotate forwards, and no external tail or ears. They have very sparse fur.

Walrus

FRENCH NAME: **Morse**

SCIENTIFIC NAME: *Odobenus rosmarus*

Both male and female Walruses sport a pair of long upper canines. On adult males these can reach 65 cm in length. On females they are less robust and usually shorter. The ivory tusks of Walruses are one of the reasons they have been hunted commercially for at least 1000 years. Walruses have another anatomical distinction that is considerably less evident. Males have a bacula (penis bone) that is the largest of all the mammals. Lengths of 50–60 cm are possible in older males. This bone is wholly internal.

DESCRIPTION

Adult Walruses are typically cinnamon brown to medium brown in colour with a light covering of short (not more than 1 cm long), coarse, brown to reddish-blond hairs. Older males lose much of this hair, especially on the front end of the body, and may look completely hairless especially during the moult in June and July. Newborns appear grey to greyish-brown but quickly develop a tawny-brown coat with black flippers within one to two weeks. The blackish pigment in the skin fades as the animal ages, and so younger animals are darker. The lightest animals are usually old males. These bulls become quite pale when immersed in cold water because blood flow to the skin is restricted, leaving them with a ghostly greyish-white pallor. The same animals, basking in the sun, develop a rosy appearance, which is not sunburn but rather is caused by increased blood flow to the skin as their bodies work to shed excess heat. Females tend to retain their hair covering and so do not appear as pale as the males.

This species displays considerable sexual dimorphism. Adult females are considerably smaller than adult males, and bulls develop thickened shoulders, chest, and neck, and their neck

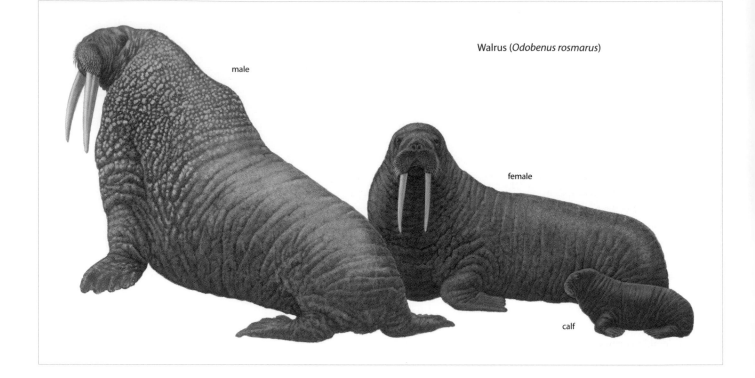

Walrus (*Odobenus rosmarus*)

male

female

calf

becomes increasingly warty with age. The skin on the neck and shoulders is 2–6 cm thick, and while thinner on the rest of the body, the skin is still substantially thicker than that of most mammals. The thickest skin is found on adult males. Sparsely haired, Walruses depend on fat deposits under their skin for insulation. Although they can rotate their hind flippers like fur seals and sea lions do so that they can "walk" on land, their large girth owing to the blubber makes them ponderous out of water. They utilize a combination of "walking" and undulating like hair seals to move on land or on ice floes. Their front flippers are squarish and about as wide as they are long. The hind flippers are somewhat triangular, widest at the tips of the toes and narrow at the insertion. The palms of the front flippers and the soles of the hind flippers are rough and thick skinned like a foot pad. During swimming, alternating hind flippers provide the propulsion, and front flippers are used mainly for steering, although they may be used as paddles during very slow manoeuvring. Adults have inflatable air bladders (called pharyngeal pouches) in their neck that serve as flotation devices so that they can rest (and perhaps even sleep) at sea. These are larger in adult males and additionally serve as resonating chambers for vocalizations. Most Walruses have a barely noticeable tail – just a bump at the end of the vertebrae. Compared to the body, the head appears undersized. The skull is grossly modified around the snout to accommodate the large tusks, making it roughly rectangular when viewed from above, below, or the side.

Their eyes are situated on the sides of the head, and although they can be protruded to a limited degree, Walruses have binocular vision in front only beyond about 2 m. They also have binocular vision both above their head and under their chin, simply by rolling their eyes in those directions. Like those of many seals, the eyes

of Walruses have a *tapetum lucidum* for maximum use of low-light levels. They also have cone cells in their retina, which indicates some degree of colour vision, the extent of which is unknown. There are no external ear flaps. Another peculiarity of a Walrus's head is the curving rows of stiff vibrissae that bristle from the large upper lip. These very coarse, translucent, brown to yellowish or even whitish whiskers are embedded in a thick pad of muscles and fat called a mystacial pad. The vibrissae are highly sensitive and can be moved independently by tiny muscles in the mystacial pad. The dental formula is incisors 0/0, canines 1/0, and postcanines 4/4, for a total of 18 teeth (all the teeth behind the canines are called the postcanine teeth as premolars and molars cannot be distinguished). Mature males moult in June and July, while females moult a little later, and their moult is more prolonged. Calves moult their birth coat in July of their first summer.

SIMILAR SPECIES

Their large size, unusual head shape, and distinctive tusks, which are present on all but calves under one year old, distinguish Walruses from all of the other seals that may occur in the same range. Of all the fin-footed mammals, also called pinnipeds (seals, Walruses, sea lions, and fur seals), only elephant seals are larger, and they have a more southerly distribution. No other pinniped has tusks.

SIZE

Adult males are larger and may be up to three times heavier than females, although they are commonly 1.5 times heavier. Adult males are lightest just after the breeding season in spring.

Pacific Walrus: total length: 270–380 cm (male), 225–297 cm (female); weight: 880–1900 kg (male), 580–1200 kg (female).

Atlantic Walrus: about 3% shorter and 10% lighter than its Pacific counterpart, although the female may actually be larger; Atlantic Walruses around Svalbard are almost as large as the Pacific form. Newborns: total length: 100–137 cm; weight: 45–77 kg (male), 33–85 kg (female).

RANGE

Walruses occur in the Northern Hemisphere. Their distribution is essentially pan-Arctic, where they are closely tied to the floating ice pack over the shallow waters of the continental shelf. Those in the North Pacific Ocean and Chukchi Sea are considered to be a different subspecies (*Odobenus rosmarus divergens*) than those in the North Atlantic and Arctic Ocean (*Odobenus rosmarus rosmarus*). The Atlantic subspecies is isolated into four disjunct populations: the eastern Canadian Arctic and the east coast of Greenland; western Greenland, Svalbard, and Franz Josef Land; the Kara Sea; and the Laptev Sea. The Laptev Sea population is intermediate in geography as well as in the physical and genetic characteristics of the two forms and has variously been placed in one or the other or as a distinct subspecies of its own. The current range of the Atlantic population has considerably shrunken from its pre-commercial hunting range when Walruses bred as far south as Sable Island, Nova Scotia, Prince Edward Island, and the Magdalen Islands in the Gulf of St Lawrence before being extirpated by the late 1600s. A population used Miscou Island, New Brunswick, at the mouth of the Baie des Chaleurs until the late 1700s. Post-glacial bones of Walrus continue to be dredged up around the Maritimes, including from the Gulf of St Lawrence and the Bay of Fundy. Walruses are rarely seen now even along the coast of Labrador, although wanderers occasionally still reach eastern and southern Newfoundland and the Gulf of St Lawrence. A juvenile was sighted in Halifax, Nova Scotia, in 2000. Historically, Pacific Walruses inhabited haul-outs in the Pribilof Islands, but those populations were hunted out, and few are seen in the area now.

ABUNDANCE

Walrus meat, blubber, bones, skin, and tusks have long attracted northern native hunters, and Walruses have been hunted throughout most of their range for thousands of years. Commercial hunting became extensive in the 1600s–1800s. Many local populations, especially in European waters and in the Gulf of St Lawrence (population at the time is estimated to have been about 100,000 animals), were exterminated. Commercial hunting of Walruses was banned in Canada in 1931 and in the United States in the 1940s. Walruses on Svalbard have been protected since 1952 and in the western Russian Arctic since 1956. Russia continues a commercial hunt of Pacific Walruses in the Bering and Chukchi seas within Russian territorial waters. Native subsistence hunters in Canada, Greenland, Russia, and Alaska also continue to take Walruses. In the 1990s Walrus populations in the Laptev Sea (around 4000–5000 animals), around Svalbard and east Greenland and Franz Josef Land (in the low thousands), and even in the Canadian Arctic (around 10,000–15,000 animals) were still relatively small, and their recovery is slow. In 2006, Atlantic Walruses in the Canadian Arctic were designated of "Special Concern" by COSEWIC. Pacific Walruses were extensively hunted into the early 1900s, but that population was larger to begin with and has recovered following U.S. hunting proscriptions. Pacific populations were estimated at around 200,000 in the 1980s but have declined since then because of reduced birth rates and increased juvenile mortality from unknown causes. Worldwide estimates of Walrus population size based on surveys conducted from 1975 to 1990 tentatively suggest an estimated 250,000 Walruses. A 2006 estimate for the global population is thought to be around 200,000. Owing to global warming, polar ice over much of the Walrus's range is thinning. The overall negative impact on Walrus populations remains to be seen, but abandoned calves are starting to be found in areas far from ice. Possibly they were left by their mother when lack of food prevented her from producing enough milk to continue feeding them.

ECOLOGY

Ice-loving habits and a unique feeding style (see the "Diet" section following) mean that Walruses inhabit shallow cold waters in coastal regions of the North Atlantic Ocean, Bering Sea, and Arctic Ocean. They can break through ice up to about 20 cm thick, but if it gets thicker, they move to areas of broken ice. They winter in regions where polynyas (areas of open water) are present and relatively shallow, and the surrounding ice is sufficiently thick to provide haul-outs. Walruses spend a great deal of time out of the water while moulting, rearing young, and resting. They prefer to haul out on ice, but when no ice is available, they use isolated rocky islands. The herds generally follow the edge of the ice pack as it advances

Distribution of the Walrus (*Odobenus rosmarus*)

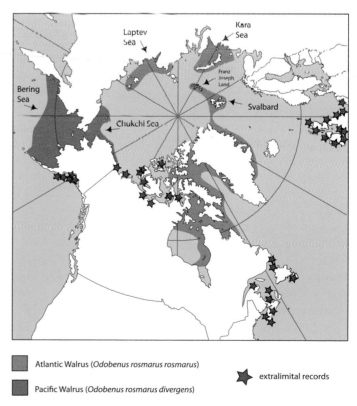

| | Atlantic Walrus (*Odobenus rosmarus rosmarus*) |
| | Pacific Walrus (*Odobenus rosmarus divergens*) |

⭐ extralimital records

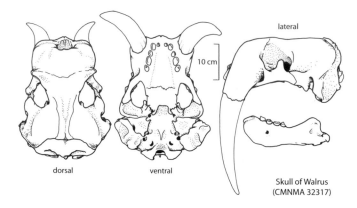

lateral

10 cm

dorsal   ventral

Skull of Walrus
(CMNMA 32317)

and recedes, resulting in a slow migration northwards in spring and a similar although often faster migration southwards in autumn. Some animals, especially those in the Pacific Ocean, may travel over 3000 km in a year. Pacific Walruses are segregated by gender during much of the annual cycle. Adult females, their dependent young, and most subadults remain near the sea ice throughout the year. Following the breeding season, from late spring until autumn, adult males form groups that may become quite large. They abandon the ice to feed and haul out at traditional coastal sites further to the south. Atlantic Walruses behave similarly, but the distances travelled are much shorter. While at sea, Walruses spend about 80% of their time underwater, likely feeding and diving back and forth to the sea floor. Predators include Polar Bears, Killer Whales, and humans. Walruses are known to fling themselves out of the water onto ice floes to escape a hungry pack of Killer Whales. Polar Bears are most dangerous to calves and immature Walruses. The bears typically charge a group that has hauled out on the ice, driving them into the water and then catching the young ones that are left behind in the rush. Human hunters have harpooned Walruses for as long as there have been humans in the North. Walruses may live 40 years, and it appears that females may live about five years longer than do males. Their normal swimming speed is around 10 km/h.

DIET

Unique among the pinnipeds, Walruses have a specialized diet that focuses on bottom-dwelling (benthic) invertebrates such as clams, whelks, cockles, and mussels, occasionally supplemented with marine worms, crabs, shrimp, fish, and even the odd seabird. The only fishes they eat are the slow-moving, sea-bottom species that they happen to encounter. To reach their prey, they must dive to the ocean floor, usually to depths of 10–50 m, remaining underwater for less than seven to eight minutes. The maximum known depth reached by a feeding Walrus is 133 m, and they can remain submerged for as long as 25 minutes. They swim slowly forward along the bottom with their head down, travelling with their tusks parallel to the substrate, perhaps acting as guides, and feeling for prey with their highly sensitive vibrissae. Owing to their eye placement and the width of the mystacial pads, it is impossible for Walruses to see what is in front of their nose (within 30 cm); furthermore, the dim light at their feeding depths often makes vision impractical, especially since much of their prey is buried in the

sediment. To capture a clam, they probably squirt water from their mouths to clear the sediment, increasing the turbidity of the water in the process. If they encounter a particularly dense concentration of clams, they will root around in a pig-like fashion, pushing the sediment about with their nose. The feeding furrows (around 40 cm wide), often dotted with various-sized pits (typically surrounded by broken bivalve mollusc shells turned so that the pale inside reflects light), that they create as they shuffle through a muddy substrate are distinctive. Normally only the meaty siphons and feet of the clams are consumed. Suction appears to be the mechanism used to quickly separate the flesh from the shell. The tongue acts like a piston in the oral cavity to create a vacuum or conversely to jet water out through the mouth to clear sediment. Suction also plays a role in the capture of slow-moving prey such as the occasional fish. Some so-called rogue Walruses will kill and eat seals.

REPRODUCTION

Walruses breed alongside and produce their calves on ice, and so, unlike many of the other pinnipeds, mature males cannot establish a territory on land. In addition, the unpredictable movement of the ice pack does not permit them to establish a territory on the ice. The bulls must take their show on the road. On the wintering grounds, from January to March in the Pacific, and February to April in the Atlantic, they follow clusters of females. The two subspecies, Pacific and Atlantic, appear to have slightly different breeding behaviour. In the Pacific, when females find a suitable ice floe upon which to rest, the dominant bulls surround the site and display, compete, and fight with each other in the water, each claiming and defending a small aquatic territory. Subordinate males are relegated to the periphery of the group, where they do not display. Some oestrous females select and join the male with whom they wish to mate; other females are forcibly pulled into the water by rutting males. In the latter cases, it is not known whether copulation then occurs or the females are able to elude the larger males underwater. In Atlantic populations where land-fast ice is more commonly used, adult males defend larger ice-edge aquatic territories in proximity to the females and can control access to the transient colony of females. These dominant males mate with all of the females within the herd for as long as the cows remain at that site or until they are displaced by another, more dominant bull. All Walruses mate in the water, usually beneath the surface. The fertilized embryo floats around in the female's uterus for four to five months and implants in about June or July. Gestation lasts for about 15–16 months from the time of mating, so most calves are born in April, May, or June while the females either are still on their northward migration or have just completed it. Calves are born on pack ice or, in the case of Atlantic Walruses, on pack or land-fast ice. Occasionally calves are born on land if ice is unavailable. Calves can swim almost from birth but may be carried on their mother's back when they are young. Cows bear a single calf every two to three years at most and undertake the longest lactation of any of the pinnipeds. Calves are nursed for 24 months, and possibly up to 30 months if they are not supplanted by a new calf. They may suckle on land, on ice, or in the water. They begin eating some solid food at around six months old, and weaning begins gradually around that time. Females breed for the first time at four to eleven years old but usually at five to six years old; so, most are six to seven years old when they produce their first calf. Males become sexually mature at seven to thirteen years old but are not large enough to compete for females until they reach their full physical maturity at about 15 years old.

BEHAVIOUR

Highly gregarious, Walruses usually travel in small groups, and they are exceptionally sociable while they are out of the water resting or moulting. They will pack into tight sprawling herds of several hundred up to several thousand individuals, each lying in close contact and some even lying on top of others at times. In a relaxed resting herd, individual animals lounge about in poses of somnolent abandon. If some animals are awake, they will quickly become alarmed and alert the herd to bolt for the water if a potential land predator or a human in a boat appears. Sometimes all of the animals are deeply asleep and can be closely approached. Females are equally gregarious during all seasons, but adult males during the breeding season typically travel alone, looking for female groups, but then congregate at other times of the year. Tusks play an important role in Walrus social activities. They are brandished as visual signals of size, age, or gender and are part of many threat displays. A tusk display with head raised and tusks parallel to the ground or water, pointing at an opponent, is a common threat posture. Tusks can become tools to poke a neighbour that is usurping one's space. Two males battling for dominance may bloody each other with their tusks, and tusks can come in handy as ice picks when a walrus is scrambling to scale a slippery ice floe. Walruses visually assess the tusk and body size of rivals, and smaller animals usually back down quickly from a superior foe. Combat is only necessary between equal-sized bulls. Young calves imprint on large moving objects, in almost all cases their mothers. Their mothers support this behaviour by isolating themselves from other Walruses before they give birth, and only rejoining other females in a "nursery herd" after their young have safely imprinted. The bond between mother and calf is strong during this period, and the two are inseparable. Mothers can become intensely protective of their calves and may charge or use their tusks against other Walruses or even humans that approach too closely. Mothers invariably move between their calves and another Walrus that might crush the smaller animal. Newly weaned females often remain in their mother's group, but young males leave to join an all-male group. The males may remain in the area, move to another region, or travel around, sometimes eventually returning to the region of their birth. Walruses exhibit strong rescue behaviour and will support wounded or sick individuals in the water and attempt to push them to the surface. Walruses are often very wary of hauling out on land and may swim near shore for a few hours, visually checking to make sure all is safe. They can push almost half of their body up above the water surface to get a better view

VOCALIZATIONS

On land Walruses tend to be noisy, with belches, growls, roars, snorts, coughs, barks, bellows, and grunts reverberating through the herd as the animals socialize. Walruses are able to hear in both air and water, and the sounds they produce are meant to be heard in one medium or the other and sometimes both. Most vocalizations are low frequency (in air, 1.0–8.0 kHz; in water, 0.2–8.0 kHz). During the breeding season, displaying males perform "songs," which are sequences of taps and knocks, punctuated with bell knocks (sounds like a ringing bell). Air sacs (also called pharyngeal pouches) that extend all the way over their shoulders are used to make the bell-knock sounds. Most of the song is performed underwater, but part of it is emitted as they float at the surface with their head submerged. Songs last for three to seven minutes in the Atlantic populations and around two to four minutes in the Pacific populations. Rutting whistles, emitted above the water at the end of the song, are the loudest sounds produced; they average around 120 dB. Songs are typically repeated, and breeding males may sing for more than 48 hours. The record to date is some 81 hours of continuous singing. The more mature males are the most proficient and persistent singers. These songs have been compared to the courtship songs of Humpback Whales in both their complexity and structure. By the end of the breeding season the dominant males often appear to be emaciated and exhausted because they commonly forgo feeding during that time, and the strain of continuous singing and defending their aquatic territories takes its toll. Females make soft, low-volume calls to their calves.

SIGNS

While on ice or land, resting Walruses can be noisy, and their vocalizations may be heard at some distance over the water. Adult-male breeding songs may be detectable kilometres away through the use of an underwater microphone (hydrophone). The pungent aroma and brownish stain of faeces and urine may remain for some time after a group of Walruses has hauled out on an ice floe.

REFERENCES

Born 2001, 2003; Born et al. 2003; Cronin et al. 1994; Dyke et al. 1999; Fay 1981, 1985; Fay et al. 1990, 1997; Jay and Hills 2005; Kastelein 2002; Kastelein et al. 1990, 1993, 1994, 1996; Kingsley 1998; Knudson 1998; Miller, R.E., 1997; Nowicki et al. 1997; Reeves et al. 2002; Rugh 1993; Sjare and Stirling 1996; Sjare et al. 2003; Sobey 2007; Stewart, B.E., and Burt 1994; Stewart, R.E.A., et al. 2003a; Stewart, R.E.A., 2008; Wiig and Gjertz 1996; Wright, B.S. 1951.

FAMILY PHOCIDAE
true seals

Members of the family Phocidae are also called hair seals. They lack external ear flaps, their hind flippers cannot be used to support the body on land, and as adults they lack a well-defined underfur.

Hooded Seal
also called Bladdernose Seal, Crested Seal

FRENCH NAME: **Phoque à capuchon**

SCIENTIFIC NAME: *Cystophora cristata*

The inflatable black nose skin (the hood) and the inflatable and shockingly extrudable red nasal sac (the bladder) of adult males give the Hooded Seal its common names. Hooded Seal pups are weaned only four days after they are born, which is the shortest known lactation period of any mammal.

DESCRIPTION

Juveniles shed their fluffy white "lanugo" (foetal) coat in utero and are born in the "blueback" phase with a slate-grey back and face and a light belly (as illustrated). During the next few annual moults they gradually develop a silver-grey base coat with increasingly dark-brown or black mottling, which is initially most noticeable on their back and upper sides. Adult Hooded Seals have a silvery-white base coat that is lighter ventrally, with black irregular mottling and a dark head. The male's flippers, both fore and hind, are black. Females trail some of their black-on-white mottling on to the upper portions of the flippers, but the digits are black. Adult females are typically lighter and have less mottling than do the males. Wet animals are considerably darker than dry ones. The "hood" of an adult male is an area of flexible black skin above the snout and the forehead that can be inflated into a balloon-like protuberance by closing the nostrils and blowing air into it. This hood has a transverse crease about two-thirds of the way up the snout, subdividing it into two unequal but connected parts that together can be inflated to about the size of a football. When deflated, the wrinkled hood droops over the end of the male's nose. The red nasal sac or bladder is actually an extensible portion of the nasal septum, which is inflated and balloons out of a nostril in an oblong bubble. The nasal sac can only be inflated if the hood is deflated or partly deflated. The hood and body size of males

gradually increases with age until they reach physical maturity at around 12 years old. The oldest, and hence the largest, males have the largest hoods and bladders. The dental formula is incisors 2/1, canines 1/1, and postcanines 5/5, for a total of 30 teeth (all the teeth behind the canines are called the postcanine teeth as premolars and molars cannot be distinguished). The annual moult takes place over one to two months between June and August.

SIMILAR SPECIES

Hooded Seals share their cold North Atlantic range with Harbour, Ringed, Harp, Bearded, and Grey Seals. The Hooded Seal adult coloration, with a dark head and dark mottling on a light background, easily distinguishes them from all the others except female Grey Seals. Female Grey Seals have a lighter head than do Hooded Seals, their snout is longer and more Roman nosed in appearance, and their background colour is usually slightly greyer. Juvenile Hooded Seals, with their dark colouring above and light below, are more difficult to identify. They may be confused with juvenile Harp Seals and even Juvenile Harbour Seals, both of which are typically more spotted. Subadults begin to show some of the adult colouring and may be confused with female Grey Seals, apart from the head shape.

SIZE

Adult males are noticeably larger and roughly 50% heavier than adult females. Weight fluctuates over the year, with animals at their leanest just after the moult and at their heaviest just prior to breeding.

Males: total length: up to 300 cm (average 240–260 cm); weight: 192–461 kg (average 300 kg).

Females: total length: up to 240 cm (average 200–220 cm); weight: 145–300 kg (average 200 kg).

Newborns: total length: 87–115 cm; weight: 11–30 kg (average 23–25 kg).

RANGE

Normally found in the Arctic and Subarctic waters of the North Atlantic Ocean region, Hooded Seals occur from the Gulf of St Lawrence and Nova Scotia to the Davis Strait, around southern Greenland to Iceland, and north to the island cluster of Svalbard. Whelping occurs in four principal areas: the Gulf of St Lawrence east of Newfoundland ("the Gulf"); off northern Newfoundland and southern Labrador ("the Front"); in central Davis Strait; and around Jan Mayen Island, north of Iceland ("the West Ice"). After breeding, adults from the Front and the Gulf populations follow the receding ice northwards and spend the moult on pack ice off southeastern Greenland. The Davis Strait population appears to moult north of Baffin Island. After moulting the seals are thought to head for deeper, ice-free waters in the North Atlantic. Juveniles swim out to sea after weaning and may spend the next several years at sea, presumably returning each summer to the moulting grounds. These juveniles and subadults are renowned for their wandering tendencies, which have been increasing since the mid-1990s. They are now common on Sable Island, Nova Scotia. Reports of their presence

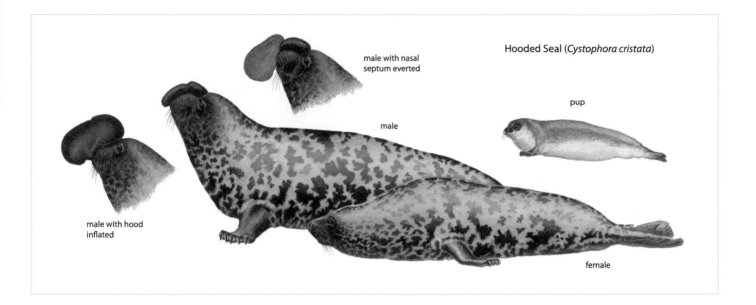

male with nasal septum everted

Hooded Seal (*Cystophora cristata*)

male

pup

male with hood inflated

female

have been received from as far away as the Bering and Beaufort Seas, the Canary Islands, the Caribbean, southern California, northern Portugal, and possibly the Mediterranean Sea.

ABUNDANCE

Humans have hunted Hooded Seals on their breeding grounds for over 170 years, primarily for the highly prized pelt of the pups. The Canadian commercial hunt has been greatly reduced since its peak in the 1970s, mainly owing to the 1983 European ban on seal products and the consequent collapse of the market. Since 1987, Canada has banned the killing of Hooded Seal pups for commercial purposes, although harvesting for personal use is still permitted under licence. A hunt for adult seals continues in Canada. Licences for harvesting up to 10,000 of these animals from the breeding grounds are granted annually, but owing to weather and other factors the actual take is considerably fewer. A so-called, subsistence hunt continues off Greenland, with much meat, oil, and pelts being exported to Denmark for sale. In this hunt, 4000–6000 animals die annually. Norway and Russia continue to hunt Hooded Seals on the West Ice breeding grounds each spring. Nevertheless, the number of seals killed annually is considerably fewer than that taken in the past, and Hooded Seal numbers are increasing. Surveys in 2005 placed the number of Hooded Seals at around 70,000 off Jan Mayen Island, approximately 500,000 off Newfoundland and in the Gulf of St Lawrence, and globally at up to 600,000.

ECOLOGY

Hooded Seals are capable of deep diving. Most foraging dives are to depths of 100–600 m, but dives of deeper than 1016 m have been recorded. The duration of most dives is less than 25 minutes, but dives of 52 minutes are known. While at sea, Hooded Seals spend an estimated 90% of their time foraging underwater. Adults begin gathering in one of the four breeding areas in late February. By late April most of the breeding adults have left the breeding grounds to migrate to traditional moulting grounds in regions with persistent

sea ice. Arriving there in June, the adults haul out in very loose aggregations to moult. This takes up to two months, and by the end of August most have widely dispersed again. Adults fast during both breeding and moulting. Many pups follow the adults to the moulting grounds, where they hunt while the adults moult. Other pups wander extensively, often beyond the normal range of the species. The maximum reported lifespan for both sexes is around 35 years, although it is expected that the larger males have a shorter lifespan than do the smaller females, as is the norm with other sexually dimorphic seals. Breeding on shifting and drifting pack ice can present some unique disadvantages, especially to the vulnerable pups. Should the weather turn windy or the ocean currents become strong, the ice pans can tilt, throwing pups into the water, where they may drown or be crushed between the floes. Late-born pups run an additional risk because their floating nursery may break up beneath them before they are mature enough to swim. Security from predation is a major advantage of breeding on such a transient surface, but nevertheless Polar Bears remain a threat at times. Human hunters are also significant predators. Greenland Sharks and Killer Whales may be occasional predators.

DIET

Pups begin feeding near the surface along the ice edge on crustaceans, squid, and small fishes, gradually learning to catch larger prey and becoming more physically capable of diving to greater depths. Adults forage along the edge of the continental shelf or in deep off shore (pelagic) waters where they hunt a wide variety of shoaling and bottom-dwelling fishes, as well as octopus, mussels, starfish, and shoaling squid and shrimp. Fish species captured include redfish, flounder, Atlantic Herring, Capelin, Atlantic and Polar Cod, Turbot (also called Greenland Halibut), and Blue Whiting.

REPRODUCTION

In North American waters, whelping and mating are highly synchronous and remarkably brief, occurring mainly in the middle of

Distribution of the Hooded Seal (*Cystophora cristata*)

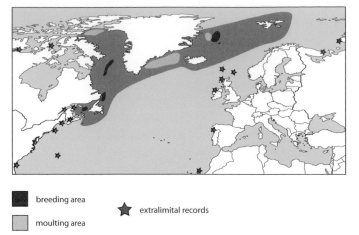

■ breeding area

▬ moulting area

★ extralimital records

March, with a few stragglers being born in early March or April. Females typically produce a single pup annually. Twins are unknown. At birth, pups already have a blubber layer of 2.5 cm thick. Even so, their skin, which is wrinkled and baggy, appears too large. Pups are born on the drifting sea ice and are suckled for about four days. A relatively late breeding season and the transient nature of the ice at that time of year may account for such a short lactation period. Approximately four to five hours out of each 24-hour period is spent nursing, and pups consume about 10–11 kg of rich milk each day. This milk is composed of an average of 58%–59% fat (the highest milk fat content recorded), and the pups nearly double their already considerable birth weight by the time they are weaned. At this point they look like inflated sausages. Weaning is abrupt as the females simply abandon the pups on the ice floe and return to sea to mate. The rotund pups then typically fast for up to two months, using their stored blubber for energy as they continue to develop. Often they will not begin swimming and hunting for themselves until they are forced into the water by disappearing ice. Mating occurs in the water. The fertilized egg then floats around in the female's reproductive tract for up to four months before implanting in the uterus wall and beginning to develop. The gestation period, including the delayed implantation phase, is about 12 months. Females become sexually mature between the ages of two and nine years old and typically produce their first pup between the ages of four and five years old. Males become sexually mature at around four to six years old but are unlikely to successfully compete for females until they are considerably older and larger.

BEHAVIOUR

Hooded Seals are generally solitary except while breeding and moulting. After a winter at sea the herd re-forms at the breeding grounds, and the females space themselves at about 50 m intervals. Some, if not most, pups return to the breeding colony of their birth to breed as adults. Each suckling female is attended by one or more rutting males that compete for proximity. Usually a large, older male occupies her ice pan, and less dominant males troll around it, popping up for a look once in a while as they search for an unattended

female. The dominant male must work tirelessly to maintain his exclusive position. He greets each potential rival with an aggressive display of his hood, nasal sac, or both. The bright red inflated nasal sac is bounced up and down in a highly noticeable, distinctive, and unmistakably threatening display towards rival males and even towards encroaching humans. These displays are usually sufficient to discourage the smaller males, but an equal-sized rival may haul out and perform his own counter display. Bloody battles using teeth and claws often result, and most mature males display battle scars on their head and chest. Having guarded and successfully mated with one female, a dominant male will often attempt to supplant the position of another attending male whose female is still nursing and has not yet mated. If he is successful, he will have another mating opportunity. Females are highly aggressive in defence of their suckling pups. They belligerently accost any ardent male that attempts to approach before the pup is weaned, and they are quick to rush a human who is bent on harming or trying to catch a pup. Hooded Seals can move surprisingly quickly on the ice, using their front claws like ice crampons for purchase and tobogganing, as well as humping along. Females returning to the water after abruptly weaning their pups may be pursued by several males, all vying for her favour. She makes them prove their worthiness by leading them on a chase over and under the floes.

VOCALIZATIONS

Females, repelling a male in defence of her pup, will growl, hiss, and roar. Fighting males also growl and roar at each other with open mouths, showing their startling pink and grey tongue and oral membranes as part of the display. Females produce soft growls (like an airy sigh) and guttural growls directed at displaying males. Attending males also use soft growls, moaning growls, and long growls as they react to a rival. Newly weaned pups, called weaners, emit a wailing moan as they call fruitlessly for their departed mother. An assortment of explosive loud snorts, soft grunts, and "humphs" also may be heard at the breeding colony from time to time. Underwater sounds recorded at a breeding colony include a commonly

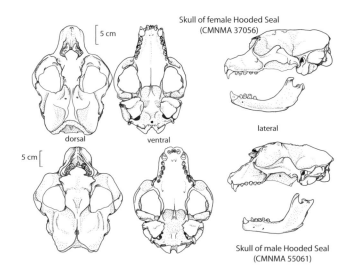

Skull of female Hooded Seal
(CMNMA 37056)

5 cm

dorsal   ventral   lateral

5 cm

Skull of male Hooded Seal
(CMNMA 55061)

produced wail similar to a pup wail, as well as clicks, trills, knocking, "beating," loud swallowing noises, and a sound reminiscent of blowing through pursed lips. Likely the bulk of the underwater sounds are produced by adult males. Although not strictly a vocalization, a commonly heard sound on the breeding grounds is that of whistling air, followed by a hollow "bloop," as the males collapse their inflated hood. The shaking of the extruded nasal sac creates a muffled but threatening rattling sound.

SIGNS

Blood stains and greyish disks of damp lanugo hairs that were in the embryonic fluids remain on the ice after the female and pup have gone to sea. The afterbirth is usually fully consumed by gulls. Ice floes that supported a female and pup often also have bright-red blood stains on their margins that are the result of battles between rival males fighting for the reproductive rights to the soon-to-be-receptive female.

REFERENCES

Ballard and Kovacs 1995; Campbell 1987a; Coltman et al. 2007; Folkow and Blix 1999; Folkow et al. 1996; Hammill 1993; Harris, D.E., et al. 2001; Iliff and Brinkley 2001; Kovacs 2002b; Kovacs and Lavigne 1986; Kovacs et al. 1996b; Lucas and Daoust 2002; Lucas et al. 2003; Lydersen et al. 1997; MacDonald, S.D., 1990; McAlpine et al. 1999b; Mignucci-Giannoni and Haddow 2002; Oftedal et al. 1993; Perry, E.A., and Stenson 1992; Potelov et al. 2000; Reeves and King 1981; Reeves et al. 2002; Stirling and Holst 2000.

Bearded Seal

also called Square Flipper Seal

FRENCH NAME: **Phoque barbu**

SCIENTIFIC NAME: *Erignathus barbatus*

The large, Arctic Bearded Seal is named for the long, luxuriant whiskers that partly obscure its mouth. The liver of Bearded Seals, like that of Polar Bears, can contain accumulated residues of mercury, DDT, DDE, and dieldrin, as well as a sufficiently high concentration of Vitamin A to be toxic if eaten by dogs or humans. Bearded Seals may harbour the nematode *Trichinella*, and so their flesh should be well cooked before consumption.

DESCRIPTION

Bearded Seals have a long body and a proportionately small head. The coat colour of adults is variable. Most are light to dark grey or brown, often darker along the back line and on the underside, and sometimes with irregular lighter patches anywhere on the body. Wet coats appear darker than dry ones. A rusty wash may colour the head or part of the head and all or part of the front flippers.

This is a ferrous stain, which results from feeding in the iron-rich sediments that are common in some regions. Lighter eyebrows and muzzle are not uncommon. Adult coats typically have no dark blotches. Juveniles have a similar pelage to that of adults, but they tend to have irregular dark blotches, which gradually diminish and eventually disappear with age. Newborns are often born with some of their white foetal coat (lanugo) still attached. The darker second coat is almost all grown in and can be seen beneath the longer white hairs that are soon lost. Most pups are dark blue grey or grey brown on their back, and lighter on the sides and belly. Their face is typically marked with light patches over each eye and on the muzzle around the dark nose. On some pups the light patches amalgamate to cover all of the face except the nose and a stripe along the midline of the forehead. The digits on the front flipper all end at about the same level, giving the appendage a squared-off appearance. Each "finger" terminates with a strong, sharp claw. All other seals have two mammary glands (one pair) except Bearded Seals and monk seals (genus *Monachus*), which have four (two pairs). The dense, creamy-white vibrissae that grow in rows along their upper lip are of varying lengths. Those below the nose are short, while further along the rows they become longer. The longest outside whiskers may be longer than the head and, when dry, curl at the tips. Bearded Seal whiskers are smooth, unlike those of other Arctic seals that have a beaded appearance and texture with regular thickened bumps separated by narrow necks all along their length. The 240–250 vibrissae are highly innervated and provide tactile information in the dark, benthic environment where the seals forage. The dental formula is incisors 3/2, canines 1/1, and postcanines 5/5, for a total of 34 teeth (all the teeth behind the canines are called the postcanine teeth as premolars and molars cannot be distinguished). Bearded Seals do not have a well-defined moult. They appear to shed hair almost constantly, more like humans do. However, there is a period in early summer when more hair is shed than usual.

SIMILAR SPECIES

Bearded Seals share the southern parts of their polar range with Grey, Harp, Hooded, and Harbour Seals in eastern Canada and with Spotted Seals in the western Arctic. The northern parts of their range are shared with Ringed Seals. The small head, square front flippers, long whiskers, large long body, and relatively unpatterned pelage distinguish this seal from other northern seals. Grey Seals are more heavily spotted; Harp Seals have a generally light background coat colour with a heavy dark saddle pattern on their back; Hooded Seals have a generally light coat with numerous dark patches; and Harbour and Spotted Seals are more heavily spotted, especially on their back. Ringed Seals are considerably smaller, and their dark back has a scattering of characteristic light-coloured rings.

SIZE

Adult females are slightly longer and may be significantly heavier than adult males. Weight fluctuates markedly over the year. Animals are heaviest in late winter and early spring and lightest just after the moult.

Males: total length: 210–255 cm; weight: 210–313 kg.

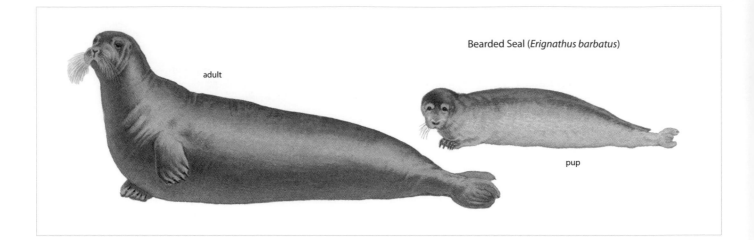

adult

Bearded Seal (*Erignathus barbatus*)

pup

Females: total length: 215–255 cm; weight: 210–425 kg.
Newborns: total length: 130–135 cm; weight: 33–34 kg.

RANGE

Bearded Seals are sparsely scattered in a patchy distribution through much of the Arctic and Subarctic waters. In the North Atlantic Ocean they occur as far south as the lower Labrador coast, the waters north of Iceland, and northern Norway. In the North Pacific Ocean they occur as far south as the Alaskan Peninsula, Kamchatka, and Sakhalin Island. Juveniles and subadults commonly wander beyond the normal range and are regularly seen off Newfoundland and Japan and occasionally as far south as Massachusetts.

ABUNDANCE

Although they are nowhere abundant, Bearded Seals are very roughly estimated to number 100,000–200,000 in the Canadian Arctic and perhaps as many as 500,000 worldwide. These seals have been hunted for centuries by native subsistence hunters around the Arctic. Normally not a serious problem, even this hunt can deplete a sedentary local population. It is estimated that between 3000 and 5000 animals are taken each year in Canadian waters by aboriginal hunters. A Russian commercial hunt during the 1950s and 1960s seriously depleted populations in the Okhotsk and Bering Seas, and quotas were instituted in the early 1970s. A population formerly around southern Labrador and the north shore of Newfoundland may have been exterminated by commercial sealers hunting Hooded and Harp Seals. Bearded Seals usually sink if they are killed in the water, and the sinking loss is estimated to be as high as 50%.

ECOLOGY

Bearded Seals prefer shallow open water and broken pack ice where breathing holes are relatively easy to maintain; they avoid thick shore-fast ice unless there are persistent, dependable leads (open areas and cracks in the ice). Shore haul-outs are only used if pack ice is unavailable. They keep their breathing holes open with regular use and by scratching the forming ice with their robust front claws. Thin new ice may be shattered with the head. They prefer to make short, shallow foraging dives. The deepest recorded dive was 480 m (and

this by a two-month-old pup), and the longest was 20–25 minutes in duration. Even one-week-old pups are able to dive in excess of 90 m and remain submerged for five minutes or more. Most foraging occurs in depths of less than 100 m, and dives are typically less than ten minutes in duration. Bearded Seals will travel into freshwater and have been observed as far as 50 km up the Nelson River (Manitoba) from Hudson Bay. Winters are spent around polynyas in areas of shifting pack ice where leads are common, or along the open water edge of the ice pack. Some coastal populations are thought to be more or less resident, while more pelagic populations tend to follow the receding ice northwards in summer and retreat southwards, in front of the advancing ice, in autumn. Polar Bears are the principal predators, and humans are the secondary predators. Greenland Sharks, Walruses (which only take pups), and Killer Whales may also take some. Bearded Seals usually sleep hauled out on the ice but are sometimes seen sleeping in the water, floating vertically with their heads above the surface. The maximum lifespan is 30+ years, but few live beyond 20 years.

DIET

Bearded Seals forage in shallow waters near the ocean floor where their diet consists mainly of shrimps, clams, and gastropods, such as whelks. They also consume crabs and other crustaceans, worms, octopus and other invertebrates, and small fishes like Polar Cod, Saffron Cod, Sand Lance, Capelin, sculpin, and flat fishes. They probably use their sensitive whiskers much like Walruses do to feel for food in the bottom sediments.

REPRODUCTION

A single pup is the norm, born on the ice between late March and late May depending on the latitude. In Canadian waters most pups are born between mid-April and late May. Females usually breed every year, but in some marginalized and poorly nourished populations they may only breed every other year. More northerly locations have a later breeding-period peak. Newborns already have a thin blubber layer, and unlike most other seals, Bearded Seal pups join their mothers in the water within hours of their birth, likely as a defence against Polar Bear predation. Pups may more than triple their birth weight during the 18–24 day lactation period. The milk contains about 50% fat content, and this does not fluctuate over the lactation period. Pups weigh about 110 kg at weaning, and weaning is not as abrupt for Bearded Seal pups as it is for most other seals. The pups begin foraging towards the end of the lactation period, so that when weaned, they are already skilled divers and well able to forage for themselves. Mating occurs in the water in May, around the time that the pups are weaned. Implantation occurs in late July or early August, following a period of about two months when the fertilized egg floats around in the uterus. The entire gestation period, including the two-month delayed implantation phase, lasts for about 11 months. Males become sexually mature at five to seven years old, while females mature at around four to six years old.

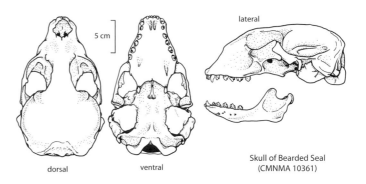

Skull of Bearded Seal
(CMNMA 10361)

dorsal   ventral   lateral

BEHAVIOUR

Apart from the breeding season, Bearded Seals are largely solitary. Small clusters may be seen in early summer during the moult. They prefer to be out of the water while they moult, but as appropriate ice floes are becoming scarce at this time of year, small clusters may form on the few remaining ice pans. Even when they are in groups, the animals are well spaced, and each faces the edge of the floe towards the open water. Bearded Seals rarely haul out on land unless they live in a region where the ice completely disappears in the summer (such as in the Okhotsk, Laptev, and White Seas). They typically sleep or rest on an ice floe at less than a body's length from the edge and with their nose near their breathing hole so that they can make a rapid escape should a predator approach. They are not especially wary of humans in boats and will often allow a fairly close approach. Nursing pups spend about half of their time hauled out, but their mothers rarely join them except to allow them to nurse. Reunions

between mother and pup are invariably preceded by a nose-to-nose greeting. Bearded Seals sometimes travel some distance up rivers, where they may become landlocked during freeze-up. During the breeding season, males may fight each other for access to receptive females or to defend their prime singing territory. (See the "Vocalizations" section for an explanation of this behaviour.) Many mature males bear series of parallel scars caused by the front claws of rivals. Mature males are thought to adopt one of two different breeding strategies. The dominant males occupy small exclusive singing territories that are strategically located to attract or intercept females; these are defended from all other males during the breeding season and are abandoned soon afterwards. Less dominant males occupy larger, mainly undefended territories that may overlap with others; they spend the breeding season travelling around within this area, singing at different locations. The length of song produced by territorial males is longer than that of roaming males, possibly reflecting the quality of the singing male and assisting the female's choice.

VOCALIZATIONS

The most notable call of Bearded Seals is created by a rutting male from late March until June. As the male sings, he spirals downwards, emitting bubbles. At the end of his song he surfaces in the centre of the bubble area. The song ranges over five octaves and is composed of a series of long (30–60 seconds each), distinctive, spiralling trills, each of which terminates with a moan. The males vocalize to both advertise their presence to females and deter other males from approaching. The sound may travel tens of kilometres during calm weather and will even reverberate through the ice or the hull of a boat (humans can hear it). The eerie song is considered a sign of spring in the Arctic. The male's stereotyped calling behaviour is well understood by seal hunters who use it to find the seals and then predict their surfacing location.

SIGNS

Like some whales and unlike other seals, Bearded Seals sometimes arch their back and lift their hind flippers out of the water at the beginning of a deep dive. Undisturbed seals characteristically swim with their head and back out of the water. A birthing site may be identified, after the departure of the mother and pup, by the presence of blood stains within 1 metre of open water and the 20–100 or more dark-grey disks (4–7 cm in diameter) composed primarily of lanugo hair shed in utero. Typically, gulls remove the placenta

Distribution of the Bearded Seal (*Erignathus barbatus*)

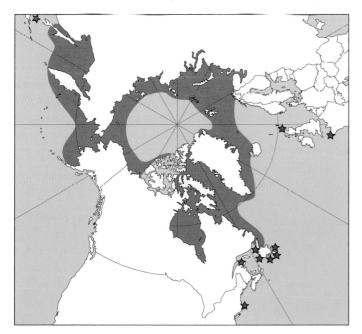

★ extralimital records

shortly after it has been produced. Hooded Seals also leave similar evidence at their birthing sites, but the lanugo disks are smaller.

REFERENCES
Andersen et al. 1999; Antonelis et al. 1994; Burns, J.J., 1981; Cleator 1996; Gjertz et al. 2000; Kingsley and Stirling 1991; Kovacs 2002a; Kovacs et al. 1996a; Krafft et al. 2000; Lydersen et al. 1994a; Reeves et al. 2002; Sardi and Merigo 2006; Terhune 1999; Van Parijs et al. 2001, 2003.

Grey Seal
also called Horse-head Seal

FRENCH NAME: **Phoque gris**

SCIENTIFIC NAME: *Halichoerus grypus*

The genus name of the Grey Seal, *Halichoerus,* is taken from Greek and means "sea pig." The trivial name, *grypus,* is Latin and means "hook nosed," referring to the Roman-nosed profile. Modern-day fishermen continue to regard this seal in similar, somewhat derogatory terms because it sometimes damages nets and is thought to compete for fish.

DESCRIPTION
Although the majority of Grey Seals are indeed greyish, there is considerable individual variation as well as colour differences between the sexes. Most fully mature bulls are almost completely black with only faint lighter blotches, mainly on the back. Most cows are the reverse, a creamy white interspersed with a few scattered black blotches also mainly on the back, although occasionally one is almost black all over. Some individuals have a rusty tinge to the fur on the head, neck, belly, and flippers. Whether this is a genetic trait or an environmental artefact is uncertain. Pups are born with a thick, silky, creamy-white coat. Some have a dark tinge at the tip of each hair, which produces a smoky appearance. This lanugo (foetal

pelage) is shed after one to three weeks and is rapidly replaced by a muted version of an adult-type coat with the appropriate gender distinctions already visible. The front flippers have long, slender claws that overhang the digits. The nostrils are nearly parallel slits that are slightly further apart at the top and are separated by a noticeable gap. Males are larger than females even at birth, and are considerably larger as adults. Mature males have a thicker, more wrinkled and scarred neck and larger chest than the sleeker females. The dental formula is incisors 3/2, canines 1/1, and postcanines 5–6 / 5, for a total of 34–36 teeth (all the teeth behind the canines are called the postcanine teeth as premolars and molars cannot be distinguished). Grey Seals undergo an annual moult. In North American populations this occurs over the course of about one to two months during late spring (May–June).

SIMILAR SPECIES
Harp, Hooded, Bearded, and Harbour Seals occur within the range of Grey Seals in Canada. The first three seals are usually associated with ice. Harp Seals are considerably smaller and have a distinctive dark saddle pattern and a dark face on a light body. Hooded Seals are similar in coloration to female Grey Seals. Male Hooded Seals are easily distinguished by their larger size and bulbous nose, but the smaller females and juveniles require a careful look. They have a dark head, their snout is shorter than that of female Grey Seals, and their dark blotches are larger with less distinct edges. Bearded Seals are larger than even male Grey Seals, and both sexes have long drooping whiskers. Harbour Seals are smaller than even female Grey Seals but could be easily confused with juvenile Grey Seals. In the eastern Atlantic, most Harbour Seals are dark phase, and so they are usually darker than a female Grey Seal. The head of a Harbour Seal is quite different from that of a Grey Seal. The snout is shorter, the forehead is rounder, and the nostrils are closer together and almost touching at the lower end. No other seal has the long, Roman-nosed profile of a Grey Seal.

SIZE
Grey Seals exhibit significant sexual dimorphism. Mature males are 1.5–3.0 times the weight of an adult female. Individuals in the western Atlantic (North America) are significantly larger than those in the eastern Atlantic (Europe). The following measurements apply to

Grey Seal (*Halichoerus grypus*)

female

male

pup

adult animals from the western Atlantic population. Weight fluctuates over the year, with animals being at their leanest just after the breeding season and during the moult.

Males: total length: 210–270 cm; weight: 235–381 kg.
Females: total length: 155–205 cm; weight: 150–260 kg.
Newborns: total length: 90–110 cm; weight: 16–18 kg (male), 14–16 kg (female).

RANGE

Grey Seals live in the North Atlantic Ocean. There are three distinct populations: two on the eastern side and one on the western side. In the western Atlantic Ocean, Grey Seals occur from Massachusetts to northern Labrador and the southern tip of Greenland, with breeding colonies on Sable Island, Îles de la Madeleine (Deadman Island), Bowen's Ledge, White Island, Hay Island, Amet Island, Point Michaud, and the sea ice of Georges Bay and Northumberland Strait. The Sable Island and Northumberland Strait locations are the largest. Other islands are used from time to time. In the eastern Atlantic Ocean one stock occurs from the British Isles to Iceland, Scandinavia, and southern Denmark, with colonies on the Faroe Islands, the Hebrides, North Rona Island, Orkney Islands, Shetland Islands, and Farne Islands. An isolated and much depleted herd in the Baltic Sea makes up the other eastern Atlantic stock.

ABUNDANCE

Following centuries of harvesting by indigenous peoples and settlers in the Gulf of St Lawrence area and along the eastern shore of Nova Scotia, Grey Seals were considered to be rare by the mid-1960s with an estimated population of only 5600 in all of eastern Canada. Since this low and despite government-sponsored culling programs and bounties, which only ended in 1987, numbers of Grey Seals have risen tremendously. The largest single breeding colony of Grey Seals on this side of the Atlantic is on Sable Island, off Nova Scotia. An estimated 115,000 animals comprised this colony in 1997, and it was increasing at a rate of approximately 13% per annum. A 2004 survey found that this rate of increase had dropped to 7%. The Gulf of St Lawrence supports about 60,000–70,000 animals, and this population is thought to be increasing at a rate of 3%–4%. The entire western Atlantic population was thought to contain around 192,000 animals in 1997. Currently their numbers are likely more than 200,000. In the eastern Atlantic Ocean, the Icelandic population contained around 11,000–12,000 animals in 1987, and there are about 110,000 in the British Isles, where the population is increasing at a rate of about 6% per annum. An estimated 3000 Grey Seals live in Norwegian waters. The Baltic population, which once numbered over 100,000 in the early 1900s, is now 2000–5000 and declining because of entanglements in fishing gear, poaching, and the extreme pollution of the waters. Climate change is likely to have a negative effect on those Grey Seals that whelp on the ice of Northumberland Strait because the ice could become thinner and less stable as the weather warms. Grey Seals are fully protected in U.S. waters except from hunting by indigenous peoples. In Canada, hunting is licensed outside of the breeding season, with up to six animals per hunter.

Distribution of the Grey Seal (*Halichoerus grypus*)

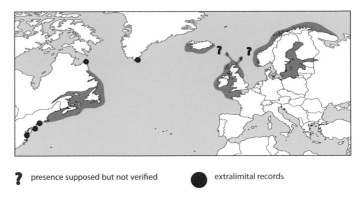

? presence supposed but not verified ● extralimital records

Regulations exist in different areas to restrict the hunt further. Sable Island is a preserve where hunting is forbidden.

ECOLOGY

Grey Seals occur in temperate and Subarctic coastal waters. They are one of the few seals that are able to whelp on both land and ice. Their colonies are typically located on rocky beaches, on islands or the mainland, and on transient sea ice if it is available. Haul-out sites are traditional and may be used primarily for breeding or resting, or both. Foraging typically takes place within 50–60 km of the haul-out at any time of day or night. Most foraging dives are shallow, down to 20–200 m, but dives of more than 300 m are known. Dives are usually short, four to ten minutes, but the longest known dive lasted around 30 minutes. From December to February the Canadian Grey Seals aggregate into two principal breeding colonies: Sable Island and the ice of the southern Gulf of St Lawrence. Breeding seals rarely forage. After the breeding season is over, they disperse widely along the Atlantic coast and around the Gulf, feeding extensively to replace reserves. During the May–June annual moult the seals come ashore and do not forage. The Sable Island seals commonly return to Sable Island to moult and then disperse again. In October–November the seals make their way back to the vicinity of the breeding colonies. From the northern Gulf of St Lawrence this could mean a migration of 350–800 km depending on where they were born. This migration takes six to ten days. Their average swimming speed while travelling is 3–4 km/h. While at sea most Grey Seals spend little time at the surface, remaining submerged for around 90% of the time. One of the main concerns of the fishing industry is the fact that Grey Seals are primary hosts of a parasitic worm called sealworm, which also invades the muscle tissue of fishes. Although this parasite is harmless to humans, many millions of dollars must be invested each year to remove the worms before the fish can be sold. With the exception of human hunting and possibly predation by sharks, Grey Seals are relatively predator free. They may live 40–50 years, and it is thought that more females reach old age than do males.

DIET

Grey Seals feed on a variety of coastal fishes, crustaceans, and cephalopods (squid and octopus), but the bulk of their diet is fish less

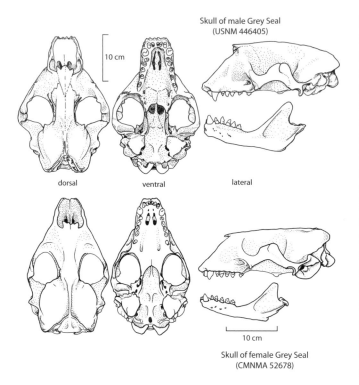

Skull of male Grey Seal
(USNM 446405)

10 cm

dorsal   ventral   lateral

10 cm

Skull of female Grey Seal
(CMNMA 52678)

than 40 cm long. They hunt in the lower levels of the water column near the bottom (benthic) and tend to take whatever prey is seasonally and locally abundant. Sand Eels and Sand Lance, Whiting, Atlantic Cod, Capelin, Haddock, Mackerel, Silver Hake, flatfish, skates, Atlantic Herring, and salmon are common prey, although few salmon are eaten in North America compared to Europe. The benthic flatfishes and skates are important year-round food, while the other fishes are eaten when their inshore migrations make them plentiful. Offshore populations around Sable Island eat mainly Sand Lance, Silver Hake, and squid in summer, and Sand Lance and Atlantic Cod in winter. Inshore populations on the Scotian Shelf eat mainly Atlantic Herring, Pollock, and Atlantic Cod in summer, and Atlantic Mackerel, Atlantic Cod, squid, and Atlantic Herring in winter. Although it is tempting to suppose that off Nova Scotia the predation by increasing numbers of Grey Seals on cod may have resulted in the collapse of the cod stocks, recent research shows that overfishing was the primary cause. Grey Seal predation may be playing a role in the slow recovery of these stocks, along with the high natural mortality rate of juvenile cod and the reduced reproductive rates of adult cod because of fewer fish and hence smaller schools.

REPRODUCTION

Females become sexually mature at four to nine years old (average age of five to six years old), and males at four to eight years old (average age of five to six years old). Despite their sexual capability, few young males can compete with the older bulls on the breeding grounds until they are closer to 11–12 years old. Virtually all adult females (around 90% or more) breed every year, producing a single pup. Births occur at different times in the different populations. Females whelp in September–November in the British Isles and in

late February–March in the Baltic Sea. In Canadian waters, land-breeding populations whelp in late December–late January, and ice-breeding populations in late February–early March. Pups suckle about every two to three hours and are nursed for 15–18 days. Their mother's milk is so rich that they can quadruple their weight during that time. The thick layer of blubber that they develop is essential to their survival, insulating them from the bitterly cold winter conditions, to which North American pups are exposed as they lie on the ice or on shore. Cows inevitably nuzzle their pups before or during a suckling session, probably to facilitate recognition by odour. Lactating females spend most of their time hauled out with their pup, but at least some ice-breeding females may forage nearby during the daytime. Nevertheless, almost all of the energy for milk production is derived from her stored fat, and she may expend as much as 40% of her body weight while nursing. Pups are weaned abruptly when they are abandoned by their mothers. They must then learn to swim before they can begin foraging for themselves. Typically they fast for up to four weeks before venturing into the water. At about the end of the lactation period the female comes into oestrus and mates. Mating usually takes place on land or ice but sometimes occurs in the water, perhaps more often than was once thought. The total gestation period is about 11–12 months, of which about 2.5–3.0 months are a delayed implantation phase where the fertilized embryo floats around in the uterus and does not grow; about 8.5–9.0 months are active gestation.

BEHAVIOUR

Although generally gregarious, Grey Seals tend to space themselves out on the haul-outs (5–10 m away from neighbours), so they avoid social contact, and most travel alone. Males adopt one of two strategies during the breeding season. Dominant males establish positions within and guard small scattered groups of breeding females. Other males roam through the colony, attempting to seize mating opportunities, or patrol the water just offshore with a similar intent. Sometimes a subdominant but mature male will guard and patrol an aquatic territory. Females are protective of their pups and will not permit the approach of other seals, especially ardent males who might crush the pup in their single-minded efforts to mate. They threaten to bite and actively chase these hopeful suitors away until about the time the pup is weaned. Once receptive, a female usually mates with the nearest territorial male and then may mate with other nearby territorial males or even with roaming younger males that have no territory. It also appears that some females select and breed with the same mate in successive years, regardless of whether he is a dominant bull. Females that did not bear a pup and so are not part of a dominant male's territory may approach a mature bull that is defending an aquatic territory off the breeding grounds and mate in the water. Similarly, many females that have a pup and are part of a dominant male's group may elude him to mate with other territorial males or with other males in the water. Males defend their territory and the associated females from other males, using a combination of open-mouth threats and chases, and will fight if their rival does not retreat. Larger, older males are generally more successful at defending a breeding territory for long periods and can expect to mate with about five to seven females in each breeding season. They spend a considerable part of their time challenging and chasing off other males, especially roving males, even to the point of interrupting their own breeding efforts. Since the males fast, or at best forage infrequently, while defending a breeding territory (during the worst of the winter weather), they may lose considerable weight over this period. Often towards the latter portion of the breeding season the exhausted dominant males retire from the field, and a surge of subordinate males come ashore in pursuit of the late-pupping females that have yet to mate. Their persistent attempts to breed distract and interfere with the cow's ability to suckle and protect her pup. Late-born pups, if they can avoid being crushed by the large males or being separated from their mothers by the relentless harassment, tend to be lower in weight at weaning and likely suffer a higher-than-average first-year mortality as a result. Grey Seals, especially females, display strong site fidelity. Females will usually return to their birth place to breed and will often whelp within metres of their pupping site of previous years, which was likely near their own birth site. Grey Seals exhibit excellent navigational abilities; they have been tracked swimming rapidly underwater directly between distant haul-outs at night. Grey Seal females in some colonies (including Sable Island) commonly foster an orphan, often in addition to their own pup.

VOCALIZATIONS

Grey Seals can be quite vocal when hauled out, especially if they are breeding. Seven distinct calls have been described that are common to all Grey Seals in the western Atlantic: a wail, a moan, a male roar, a female warble, an open-mouthed cough, a snort, and an infant cry. An additional call, so far only known from seals on Sable Island, resembles the falsetto part of a yodel. Pups produce the infant cry to solicit a suckling session from their mother. Since the nipples are often hidden under the female's body, she usually rolls and offers them when the pup produces this begging call. Underwater vocalizations of at least seven different types are also produced by Grey Seals during the breeding season. Most of these sounds vary in frequency from 100 to 3000 kHz. They have been described as ranging from clicks to guttural "rups" to growls and loud knocks that are similar to those produced by Walruses. Knock calls may be repeated two to four times. Some of the above-water vocalizations are also produced underwater (such as the roar, the growl, and the wail) and sound similar in either medium. Grey Seals produce more low-frequency underwater clicks in darkness or under ice cover, suggesting that the vocalization may have a navigational function.

SIGNS

Grey Seal colonies and haul-outs can often be identified from a distance by the scattered placement of the seals and the distance between the animals. Another interesting fact about Grey Seals is that they rarely swim while at the surface. They breathe and rest but do almost all of their swimming while submerged. They only swim with their heads out of the water when they slowly inspect a potential haul-out.

REFERENCES

Amos, B., et al. 1995; Asselin et al. 1993; Boness 1999; Bonner 1981; Bowen et al. 1993, 2003, 2007; Fu et al. 2001; Godsell 1991; Goulet et al. 2001; Hall, A., 2002; Hammill and Gosselin 1995; Hansen, S., and Lavigne 1997; Lesage and Hammill 2001; Lydersen and Kovacs 1999; Lydersen et al. 1994b; McCullogh and Boness 2000; Murie and Lavigne 1992; Perry, E.A., et al. 1998; Reeves et al. 2002; Rosing-Asvid et al. 2010; Schwartz, C.J., and Stobo 2000; Thompson, D., and Fedak 1993; Thompson, D., et al. 1991; Tinker et al. 1995; Watkins, J.F., 1990; Worthington-Wilmer et al. 1999; Yunker et al. 2005.

Northern Elephant Seal

FRENCH NAME: **Éléphant-de-mer boréal**
SCIENTIFIC NAME: *Mirounga angustirostris*

Northern Elephant Seals are the only known mammal to undertake an annual double migration. The mature males are the largest of the Northern Hemisphere seals. Among the pinnipeds, they are second in size only to Southern Elephant Seals, the largest seals in the world.

DESCRIPTION

This seal is named for the striking and awkward-looking nasal chamber of the mature male, which hangs down over its mouth like a short elephant trunk. The pendulous proboscis, which begins developing at puberty, is inflated and used as a resonation chamber for the sounds produced during the breeding season. Other sexually dimorphic traits developed by the mature males are their large size, massive neck and chest, and a thick, hairless, pinkish-grey callus on their throat and neck, which provides them some protection from the vicious bites of rival males. Males are generally a uniform dark-brown colour with a somewhat paler underside, while females and immatures are lighter buffy brown and sometimes blond, again usually with a slightly paler underside. Pups are born with a black coat. They moult into a light-silver coat after weaning, which fades to a buffy brown before they abandon the breeding beach where they were born to begin foraging for themselves. All are considerably darker when wet. The fore flippers are short but remarkably mobile with thick blunt claws. The hind flippers are U-shaped with the outer toes being considerably longer than the inner toes. There are no nails (claws) on the hind flippers. The annual moult for this seal is quite drastic, as the outer layer of epidermis is shed along with the hair. The skin and fur tend to come off in random clumps, lending a mottled, moth-eaten look to the animal because the newly exposed and as yet hairless skin is dark, while the old hair is sun bleached and lighter. The seals can look as though their skin is literally falling off. Furthermore, the newly exposed skin may develop low-level, but sometimes alarming-looking, infections that may bleed. Moulting animals spend extended periods of time on land and tend to

Northern Elephant Seal
(*Mirounga angustirostris*)

female

pup

male

have runny eyes, as tear production increases for sun protection. Many in this condition are mistaken as sick, when in fact they are largely healthy. Juveniles moult in April, adult females in May, sub-adult males in June, and adult males in July and August. The typical dental formula is incisors 2/1, canines 1/1, and postcanines 5/5, for a total of 30 teeth (all the teeth behind the canines are called the post-canine teeth as premolars and molars cannot be distinguished). The number of the peg-like postcanines may vary from three to seven, especially along the lower jaws.

SIMILAR SPECIES
Adult male Northern Elephant Seals are unmistakable, owing to both their size and the shape of their nose. Immatures and females at sea may be mistaken for Harbour Seals or Steller Sea Lions. The spotting and much smaller size distinguish the Harbour Seals, and the dark flippers, finer head, and smaller size identify the Steller Sea Lions. Moulting Northern Elephant Seals are very reluctant to venture into the water and will often allow a close approach. Moulting juveniles with their smaller size, ragged and mottled skin, and watery eyes may be mistaken for sick Harbour Seals. They may be mottled, but they are not spotted like Harbour Seals. Their hind flippers are also different. Harbour Seal hind flippers have toes all of similar length, giving a square look to the appendage, while Elephant Seal hind flippers are U-shaped. Northern Elephant Seals of all age classes and both sexes have a horizontal crease between the bottom of their nose and their upper lip line, which is only visible with a close view.

SIZE
A highly sexually dimorphic species, bulls are typically three to four times, and rarely up to ten times, the weight of cows. Weight fluctuates over the year, with animals being at their leanest just after the breeding season and heaviest just prior to breeding.
Mature males: total length: 360–450 cm (average 380 cm); weight: 914–2300 kg (average 1700 kg).
Mature females: total length: 280–300 cm (average 245 cm); weight: 400–900 kg (average 500–650 kg).
Newborns: total length: 125 cm; weight: average 35 kg.

RANGE
Northern Elephant Seals occur only in the North Pacific Ocean. Rookeries are found on offshore islands and remote coastal beaches in southern California, Baja California (Mexico), and recently on Shell Island, off Cape Arago (Oregon). Outside the breeding and moulting seasons, for eight to ten months annually, animals disperse widely. The majority travel northwards, with the larger males moving further into colder waters. They spend most of the year foraging in the Gulf of Alaska and around the Aleutian Islands. Females and immatures tend to move north, then veer to the west, and spend their time foraging in the deep waters of the central Pacific. Although they do occur regularly along the British Columbia coast, inshore sightings in Canadian waters are sparse because few people recognize them. Records of Elephant Seals around southern Vancouver Island are increasing every year, possibly owing to the growth in their population and our identification abilities. Each summer

Distribution of the Northern Elephant Seal (*Mirounga angustirostris*)

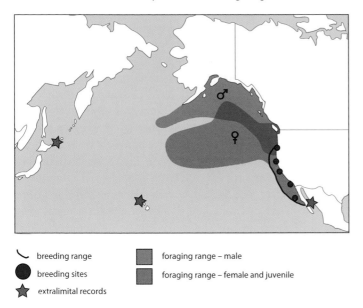

- breeding range
- breeding sites
- extralimital records
- foraging range – male
- foraging range – female and juvenile

more than an hour. An adult female has been recorded on a dive that lasted two hours. While these extremes are astounding, even the ordinary day-to-day diving abilities of these seals are formidable. While foraging, the animals are almost continuously diving to depths of 350–650 m on average and remain below the surface for 18–25 minutes per dive. The surface time is limited to only a few minutes between dives (approximately 90% of foraging time is spent underwater), and the animals rarely swim while at the surface. This species practices sexual segregation in foraging. Males tend to feed in more northerly coastal waters that are shallow enough for them to hunt for prey on the ocean floor. Females forage in deep mid-oceanic waters, where they hunt largely bioluminescent squid and fish species that migrate up the water column each night, to depths within the reach of these deep diving seals. All Northern Elephant Seals fast during the three- to four-week moult and while breeding. Some alpha males remain ashore for up to three months during the breeding season and lose around 35% of their body weight over that time. Breeding females lose up to 42% of their body weight during the one month that they remain ashore suckling their pup. Sandy beaches with a gentle slope are preferred for breeding.

Alternate sweeps of the hind flippers provide underwater propulsion, while the front flippers are pressed against the body. Elephant Seals travel about 100 km a day during migration. The top swimming speed so far recorded is 10.8 km/h. Pre-weaning mortality varies tremendously depending on both storm occurrences and crowding in the rookeries. Over-crowded rookeries permit little space between females, and so there is much squabbling, and pups are injured and killed by non-maternal females. The bulls are also forced into uncomfortable proximity, and they expend considerable additional effort in displaying and fighting with rival neighbours. Many pups that are unable to move aside are crushed by these oblivious behemoths as they manoeuvre. Great White Sharks and Killer Whales are the primary predators. Very few males live longer than 14 years, and females beyond 20 years.

some immature Northern Elephant Seals haul out along the British Columbia coast to moult. Wandering juveniles have been reported from Japan and the Hawaiian Islands. Adult and immature Northern Elephant Seals undertake prodigious migrations of up to 5250 km between their southern breeding range and northern foraging range. They make the round trip twice annually, as they return to the breeding colonies from their northern feeding grounds to moult, and then go back. With a yearly total of up to 21,000 km, these are the longest annual migrations and the only known double migration of any mammal. Although some Humpback and Grey Whales travel further between their summer and winter ranges (up to 8150 and 10,000 km, respectively), they only make one round trip annually.

ABUNDANCE

Northern Elephant seals were harvested for centuries for meat and oil by native subsistence hunters, both on their rookeries and as they migrated past the Strait of Juan de Fuca, British Columbia, but the intensive commercial seal hunting of the early and mid-1800s brought them to the verge of extinction by 1860. They were killed for the fine lubricating oil that could be extracted from their blubber. Thought to be extinct in 1884 after the Smithsonian Institute collected seven from the last known breeding herd of eight animals on Guadalupe Island, younger individuals, likely still at sea, were overlooked and saved the species. With strong Mexican, and then American, protection from hunting and harassment, those 10–20 animals became the progenitors of today's 150,000+ population. The small number of founders explains the abnormally low genetic diversity that is currently present in the population. Northern Elephant Seals are protected in Canadian waters and are not hunted.

ECOLOGY

Northern Elephant Seals are prodigious divers. Large males are capable of dives in excess of 1500 m deep and can stay under for

DIET

Depending on the ocean depth at which they are foraging, Northern Elephant Seals will take squid and fishes from the middle to lower depths if they can reach it. Shoaling squid are their primary prey, supplemented by fishes such as skates, rays, small sharks, rockfish, ratfish, flatfishes of various species, octopus, Pacific Lamprey, hagfish, Pacific Whiting, Northern Anchovies, Plainfin Midshipman, and Spotted Cusk Eel. Much of their deeper water prey is bioluminescent.

REPRODUCTION

The breeding season for this seal occurs from late November until early March, with a birthing peak in late January. Males come ashore at the rookeries in late November to begin their dominance struggles. Females begin arriving in mid-December until late January and whelp within six days of their arrival. Each female produces a single pup annually, following a gestation period of about 11 months. This includes a two- to three-month period of delayed implantation. Pups are suckled for about 21–28 days. During the last one to five

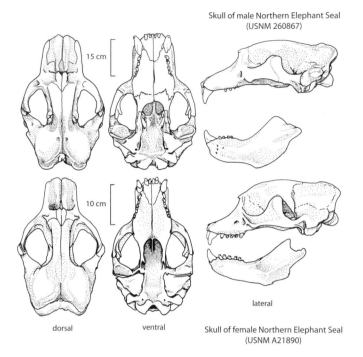

Skull of male Northern Elephant Seal
(USNM 260867)

15 cm

10 cm

dorsal   ventral

lateral

Skull of female Northern Elephant Seal
(USNM A21890)

days of lactation, the females come into oestrus and mate with dominant bulls. Pups are abruptly weaned when their mothers swim out to sea, abandoning them on the beach. Mating occurs on land. Newly weaned pups, called weaners, spend an additional four to six weeks on shore, fasting, moulting, and learning to swim in the tidal pools, before they venture into the ocean to forage. These weaners cluster together in dense groups on the beach (called weaner pods) and leave for the open ocean in small groups at about 14 weeks old. Their mothers may forage nearby after returning to sea, but most migrate north for a few weeks to two-and-a-half months before returning to haul out again on the breeding beaches to moult in May. Females become sexually mature at one to seven (average three to five) years old, producing their first pup the following year. The age of sexual maturity of males is six to seven years old, but few males younger than ten years old are large enough to compete for, and hold, a breeding territory.

BEHAVIOUR

Highly gregarious when breeding and moulting, Northern Elephant Seals are generally solitary while at sea. During the breeding season, males compete for access to females. Dominance hierarchies are determined with a combination of postural and vocal displays and occasionally by physical combat. Battles between alpha bulls are uncommon, but when they occur, they tend to be brutal. The bulls square off, chest to chest, and slash at each other with their canine teeth, and batter each other with powerful blows from head and neck, sometimes for hours. Although the canine teeth are not long enough to penetrate more than halfway through the neck callus, the pounding that each receives can break bones. They fight until one leaves or they are both too exhausted to continue. When the females come ashore to whelp, they are attracted to each other

and form groups. Each clump of females is typically attended by one alpha male and several subordinate males. The alpha males do tend to realize the highest rate of reproductive success, but many successful copulations are achieved by the lower-ranked peripheral males as well, particularly when the harem is large. Cows may be injured during mating, especially if they resist. The much larger males often bite the female's neck and rest their heavy head and neck on her during copulation. Injuries are more likely to occur as the already mated female leaves the protection of the harem and attempts to put to sea. She must pass through the gauntlet of subordinate males patrolling the periphery of an alpha male's harem. These ardent, and usually less than gentle, males rush to intercept any departing female. At least one in every 1000 females leaving a harem is killed each year by bites or head slams from these over-eager suitors, and many more are injured. Frustrated subadult males commonly mount weanlings of both sexes (weaners resemble cows at this age but are about one-third to one-half the size), and one in every 200–600 is killed annually by this misdirected behaviour. They will also attempt to mount yearlings, two-year-olds, dead Elephant Seals of all age classes, and even pups of other species of seals (usually causing their death).

Cows and their newborns vocalize and sniff one another for about an hour after the birth, learning to recognize each other's scent and vocalizations. Mothers attempt to remain in close contact with their pups during the whole nursing period, forgoing feeding to protect and nurture them. Cows become increasingly aggressive following the birth of their pups and attempt to prevent nearby females, often attracted by the newborn, from approaching, using a combination of postural and vocal threat displays and bites. The pups are endangered by these curious females that may bite them. A separation of as little as 2–3 m from its mother in a noisy, crowded colony can mean death for the pup from starvation or injury. Although most females nurse only their own pups and will violently drive strange pups away, adoption of orphans does occur, especially if the female has already lost her own pup. On the breeding colony, or later while moulting, seals exposed to the hot sun will flip sand onto their backs with their front flippers as a means of reducing heat absorption. Even newborn pups are capable of sand flipping.

VOCALIZATIONS

Northern Elephant Seal rookeries can be noisy as males snort and trumpet their threat vocalizations, pups bleat to be fed or squeal in distress, and females call their pups and squabble with each other over prime locations. Most calls are fairly loud and are usually repeated in an effort to be detected over the background din. The primary male threat vocalization is called a clap-threat call and makes use of their large nasal cavity to produce some impressive resonance and volume; the sound resembles a very loud hand clap. Snorts, gargles, grunts, belches, whimpers, bleats, squeaks, squeals, and the males' clap threats echo through a breeding colony. Low-frequency sounds that are transmitted through the earth to nearby seals are apparently used, especially at night, to transmit seismic signals of social dominance in noisy and crowded rookeries. Bulls especially appear to make use of this unusual mode of communication. Northern Elephant Seals are highly sensitive to underwater sounds, but this sensitivity seems to be related to prey detection rather than underwater communication.

SIGNS

In Canadian waters, adult male Northern Elephant Seals rarely haul out, but they are regularly seen resting vertically in the water with their heads above the surface, looking like a large deadhead. Unlike a real deadhead, these animals do not bob with the waves but slowly sink beneath the surface after several minutes and may not rise again for half an hour or more.

REFERENCES

Aurioles et al. 2006; Baird 1990; Campbell 1987b; Deutsch et al. 1990; Haley et al. 1994; Hindell 2002; Hodder et al. 1998; Hoelzel 1999; Hoelzel et al. 1999, 2002; Kastak and Schusterman 1999; Le Boeuf et al. 1992, 2000a; McGinnis and Schusterman 1981; Mesnick and Le Boeuf 1991; Mortenson and Follis 1997; Reeves et al. 2002; Reiter and Le Boeuf 1991; Rose et al. 1991; Sanvito et al. 2007; Stewart, B.S., 1997, 1999; Stewart, B.S., and DeLong 1995; Stewart, B.S., and Huber 1993.

Harp Seal

also called Saddleback Seal, Greenland Seal

FRENCH NAME: **Phoque de Groenland**

SCIENTIFIC NAME: *Pagophilus groenlandicus*, formerly, and still by many, *Phoca groenlandica*

The Harp Seal is the seal that is most involved in the controversial North Atlantic seal hunt. Harp Seals are considered to be the most numerous seal species in the Northern Hemisphere. Recently weaned pups are called beaters because of their awkward churning of the water as they learn to swim.

DESCRIPTION

The base coat of adult Harp Seals is variable and may be white to medium grey. Adults have a black head and a large black, roughly V shaped saddle (thought to resemble a harp in shape) on their back. The apex of the dark marking is situated at the upper back. From there it flares out along the sides to end just before the tail and hind flippers. Some adults do not develop a solid-dark saddle, remaining heavily spotted in the area instead. The full dark markings are usually a good indicator of sexual maturity. Fully mature males tend to display more contrast as they often have a lighter base coat and darker markings than do adult females. Pups are born with a dense, white, woolly birth coat (lanugo) and are called whitecoats. This white fur is lost at around three to four weeks old as the pups moult into their first juvenile pelage of silver grey, which is darker on the back, lighter on the underside, and spotted with scattered dark markings. With each annual moult the juvenile coat darkens

on the back and lightens on the underside so that the irregular dark spots become less distinct. Four-year-olds display a strong suggestion of the adult colour pattern, which continues to develop, especially in females, until the animal is physically mature. Harp Seals have a "double" coat, made up of longer guard hairs and some short woolly underfur. Although this arrangement offers some insulation, even underwater, the blubber layer just beneath the skin provides the bulk of the necessary insulation from cold polar waters. Like other true seals, Harp Seals have large, dark eyes that are adapted for dim light, with a large pupil that can be widely dilated and a light-reflecting *tapetum lucidum* layer at the back of the retina. They have good vision both underwater and in air. The dental formula is incisors 3/2, canines 1/1, and postcanines 5/5, for a total of 34 teeth (all the teeth behind the canines are called the postcanine teeth as premolars and molars cannot be distinguished). The annual moult takes place over a month or two in spring and early summer (April–June). Adult males moult first, followed by juveniles and non-breeders and finally by adult females.

SIMILAR SPECIES

Although Harp Seals share their range with Ringed, Harbour, Grey, Hooded, and Bearded Seals in the North Atlantic and with Bearded and Ringed Seals in the Russian Arctic Ocean, the adult pelage pattern makes them easily distinguishable. Young juveniles resemble Harbour Seals or female Grey Seals but are much less spotted than Harbour Seals and much smaller than female Grey Seals. Larger subadults are even less spotted than either Harbour or Grey Seals.

SIZE

Adult males are slightly longer and roughly 5% heavier than females, but this may be difficult to determine in the field. The following measurements apply to adult Canadian animals. Weight fluctuates through the year; adults are heaviest just before the breeding season and lightest in early summer after the moult.

Males: total length: 155–192 cm; weight: 98–140 kg, average 130–135 kg.

Females: total length: 140–191 cm; weight: 105–130 kg, average 115–120 kg.

Newborns: total length: 70–108 cm; weight: 7–12 kg.

RANGE

Harp Seals occur from the Gulf of St Lawrence and Nova Scotia to northern Greenland and across the North Atlantic to the Russian Arctic Ocean, as far east as the East Siberian Sea. There are three breeding populations: in the White Sea off northwestern Russia; in the Greenland Sea off Jan Mayen Island, north of Iceland; and in the Gulf of St Lawrence and southern Labrador. The Canadian breeding population occurs in two herds: the "Gulf" herd, which breeds in two locations, on the floating ice off northeastern Newfoundland and in the adjacent Gulf of St Lawrence; and the "Front" herd, which breeds on the floating ice off southern Labrador and northeastern Newfoundland. Harp Seals are migratory and follow the receding ice northward in late spring and summer. The reforming ice in autumn pushes them southward until they reach

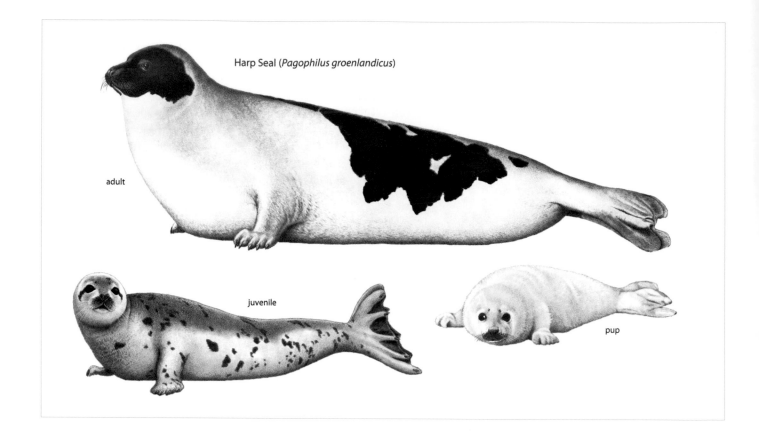

Harp Seal (*Pagophilus groenlandicus*)

adult

juvenile

pup

their breeding grounds in late winter. In some parts of the range this migration may be a 2500–3000 km round trip. Although not previously known as wanderers, Harp Seals, mainly juveniles, have been increasingly found outside their normal range since the early 1990s and for reasons unknown. Many of these wanderers are in poor physical condition. Winter vagrants occur as far south as Virginia in the eastern Atlantic Ocean, and France and Spain in the western Atlantic Ocean.

ABUNDANCE

Harp Seals are thought to number as many as 7 million, of which around 4.8 million occur in the western Atlantic Ocean off the coast of North America and Greenland. These estimates are based on 1998 surveys from the eastern North Atlantic, 1994 surveys from the western North Atlantic, and a 2001 report from an international joint working group. Fisheries officials placed the numbers at closer to 5.5 million in 2007. Harp Seals were heavily harvested in the western North Atlantic in the 1800s. During the First World War, hunting abated and numbers rebounded, only to be depleted again by overhunting in the decades that followed. A quota that was applied in 1972 and the crash of the market for seal products in the 1980s have led to another rapid recovery of seal numbers. The population has about tripled in the western North Atlantic since the 1970s. Currently the estimated population is growing and is thought to be only slightly more than the pre-exploitation population size. This rebound in numbers occurred at the same time as the collapse of Atlantic Cod, Capelin, and Atlantic Herring stocks in the North

Atlantic. It would be tempting to suggest that an overabundance of Harp Seals caused the decline of fish stocks were it not for the fact that fish stocks were also declining in regions not populated by Harp Seals and that the fish stocks sustained this number of seals in the past. Although the reasons for the collapse of the Atlantic Cod stock in the early 1990s are not fully understood, the likely cause was overfishing; however, predation by Harp and other seals may be retarding its recovery. Global warming and the resultant thinning of the ice sheets that support the breeding females and pups will likely result in the increased mortality of pups because the ice is more likely to deteriorate before they are able to swim.

ECOLOGY

Although generally considered to be associated with ice, Harp Seals are also quite comfortable and accustomed to spending time in the open ocean during warm seasons when ice is scarce. Young of the year, especially, tend not to follow the ice north during their first summer. Adults usually migrate north after breeding and moulting, in order to summer at the ice edge. Unlike many seals that fast during lactation, Harp Seal females may continue to forage during the nursing period, spending about half their time on the ice with their pups or at the surface of the water nearby, and the remainder in diving. Other Harp Seal females spend the whole period on the ice with their pup and do fast during the whole lactation period. Only the fattest of females are able to deliver high-energy milk to a pup for up to 15 days without replenishing themselves. The hind flippers, used alternately, provide most of the underwater propulsion, and

Distribution of the Harp Seal (*Pagophilus groenlandicus*)

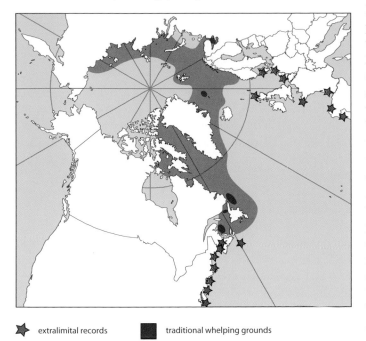

▲ extralimital records ■ traditional whelping grounds

the fore flippers are pressed tightly against the body unless needed for steering. At slow speeds the front flippers are also engaged and used like paddles. Harp Seals may live to be 35 years old or more. Their primary predators are Polar Bears, humans, and sharks. Killer Whales are occasional predators.

DIET

Diet varies seasonally, and many fish and crustacean species are taken. Consumption drops during spring and autumn migrations, spring whelping, and the annual moult in late spring to early summer. Amphipods form a large part of their spring diet. Polar and Saffron Cod are favourite summer prey. While these are also the favoured foods of co-occurring Ringed Seals, the two species divide the resource. Harps generally take larger fish and dive deeper to catch them than do the Ringed Seals. During winter and migration Harp Seals eat shoaling crustaceans (krill) and shoaling fishes such as Capelin, sand lance, and herring as well as benthic species such as Greenland Halibut. Except for very large fishes, most prey is eaten whole and head first. Harp Seals do occasionally take Atlantic Cod, but this fish comprises only a minor component of the diet (less than 4%). Although the seals can dive to depths that may exceed 400 m, most foraging takes place at depths of less than 90 m. Recently weaned pups feed in surface waters, catching krill, amphipods, and small fishes. Young animals typically make shallower dives than do adults and feed on smaller prey, gradually diving deeper and catching larger prey as their body size and expertise increase. Foraging dives are usually two to three minutes long, but adult Harp Seals are known for their ability to easily sustain dives for 10–18 minutes. Although Harp Seals gain much of their water requirements from their food, they also consume some fresh water in the form of snow and ice and likely also drink sea water.

REPRODUCTION

Reproductive females bear a single pup each year. Whelping takes place on floating pack ice from mid-February until early March in the northwestern Atlantic stocks, and a week or two later in the European stocks. There is some evidence to suggest that females can suppress the whelping process for a time until the formation of a suitable ice platform. Mating occurs in the water usually about two weeks after whelping, when the pup has been weaned. Following breeding, the fertilized egg floats around in the female's reproductive tract for up to four months before implanting in the wall of the uterus in late June to early August. Lactating female Harp Seals produce very rich milk for their pups. The composition of this milk is 50% fat, 37% water, 11% protein, and 2% ash, and as with other seals, the fat content increases and the water content drops as lactation progresses. The milk is greyish-white and the consistency of thick cream. Pups are nursed for about 10–15 days (average 13 days), during which time they develop a thick insulating blubber layer; they are abruptly weaned when their mothers abandon them on the ice. Weaned pups may voluntarily fast for several weeks before finally beginning to forage. Both males and females are four to seven years old at sexual maturity. Males do not become large enough to effectively compete for females until they are eight years old or older. Although they continue to forage intermittently, most rutting males spend the bulk of their time contending with other mature males and lose around 40 kg over the approximately six-week-long breeding period.

BEHAVIOUR

Highly gregarious, Harp Seals are rarely seen alone. They usually migrate, feed, moult, and breed in groups. Old males may become solitary or travel in small bachelor groups. Females with pups tend to space themselves out so that no pair is too close to another. Whelping usually occurs in rough, hummocky portions of the floes that provide some shelter to the pups. Adult males sometimes join the nursing females on the ice, escorting the pre-oestrous females until they become receptive. Males commonly fight each other with their teeth and front claws for access to a receptive female. Foraging maternal Harp Seals must relocate their pup on the floating and moving pack ice each time they return. They inevitably arrive at the closest haul-out to their pup, perhaps by recognizing the underwater topography of the floes. Identity is verified by nasal contact before the pup is permitted to suckle. Pups appear to recognize their

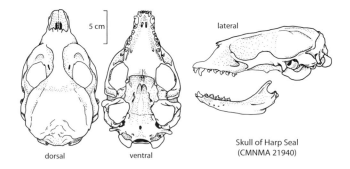

5 cm

lateral

dorsal   ventral

Skull of Harp Seal
(CMNMA 21940)

mother by sight as well as smell, as they rarely approach or solicit a strange female. Harp Seals maintain natural openings in the pack ice as breathing holes and, unlike most other seals, will share communal breathing holes. Harp Seals, especially pups, may "freeze" if approached. Maternal females are known to vigorously and aggressively defend their young.

VOCALIZATIONS

As expected for such a social marine mammal, Harp Seals are quite vocal. At least 27 different calls have been recorded that have been variously described as whistles, clicks, trills, chirps, warbles, grunts, dove-like cooing, a gull-like cry, a passerine-like call, knocking, and (the most common) squeaks and double grunts. Nineteen or more of these calls are produced underwater, and the remainder are produced in air. Underwater calls range in volume from about 100 to 180 dB and may be audible by the seals to distances of up to 80 km under quiet sea conditions. Aerial calls of adult seals are typically heard only during the breeding season. Distressed nursing females produce a roaring "urrrh" sound. Breeding males include an aerial call in their repertoire. This roaring sound is produced with the mouth and nostrils closed while the male is swimming at the surface. Pups will wail or bark to attract the attention of their mother. The breeding male's underwater vocal displays are varied. Some calls are directed towards rival males as part of territorial or threat displays, while others are meant to attract the attention of receptive females; as with many male singing birds, sometimes the same call can perform both functions.

SIGNS

Groups of Harp Seals often play at the water surface, and individuals float either with just the head above the water or with the back, side, or even belly out of the water.

REFERENCES

Bundy 2001; Chabot and Stenson 2000, 2002; Folkow et al. 2004; Hammill et al. 1995; Harris, D.E., et al. 2002; ICES Advisory Committee on Fisheries Management 2001; Kovacs 1995; Lacoste and Stenson 2000; Lawson and Hobson 2000; Lawson and Stenson 1997; Lucas and Daoust 2002; Lucas et al. 2003; Lydersen and Kovacs 1996; McAlpine and Walker 1990, 1999; McAlpine et al. 1999a; Nilssen et al. 2001; Pemberton et al. 1994; Potelov et al. 2000; Reeves et al. 2002; Renouf et al. 1990; Ronald and Healey 1981; Rossong and Terhune 2009; Serrano 2001; Shelton et al. 1996; Storeheier and Nordøy 2001; van Bree et al. 1997.

Spotted Seal
also called Largha Seal

FRENCH NAME: **Phoque tacheté**
SCIENTIFIC NAME: *Phoca largha*

Until about 1977, the Spotted Seal was thought to be a subspecies of the Harbour Seal. It is quite similar in appearance to a light-phase Harbour Seal but very different in its habits. Spotted Seals have not been well studied, partly because of their northern range and partly because of their location in the open ocean for part of the year and on poorly accessible pack ice for the remainder.

DESCRIPTION

Adult Spotted Seals have a light tan to pale silvery-grey base colour on the belly and often a somewhat darker base colour on their back. Brown and blackish spots are concentrated on the back and are fewer and more scattered on the flanks, belly, and throat. Their discrete dark spots are usually clearly defined and roughly oval in shape with the long axis of the spot matching the long axis of the body. The spots on the back may form faint pale rings in some areas, but the bold white rings as seen on some Harbour Seals do not occur on Spotted Seals. Otherwise Spotted Seals closely resemble light-phase Harbour Seals. Newborn Spotted Seals have a fluffy, thick, white lanugo coat, which they shed at two to four weeks after their birth. The dental formula is incisors 3/3, canines 1/1, and postcanines 5/4, for a total of 34 teeth (all the teeth behind the canines are called the postcanine teeth as premolars and molars cannot be distinguished). Spotted Seals undertake a single annual moult that takes place mainly in late spring during May and June.

SIMILAR SPECIES

Within Canadian waters this seal occurs only along the northern coast of Yukon and possibly the far western Northwest Territories during the summer. No Harbour Seals occur there, and

Distribution of the Spotted Seal (*Phoca largha*)

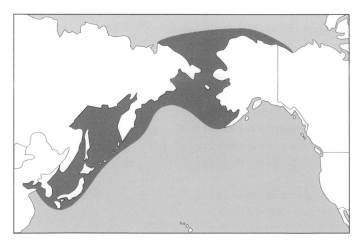

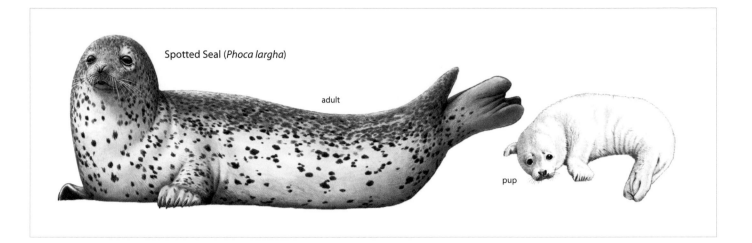

Spotted Seal (*Phoca largha*)

adult

pup

so Spotted Seals could only be confused with Ringed or Bearded Seals. Ringed Seals are readily distinguished by their dark pelage that is spotted with light rings. Bearded Seals are much longer and larger and have very distinct long vibrissae sprouting from their upper lip. For a description of ways to distinguish Spotted Seals from Harbour Seals, see the "Similar Species" section of the Harbour Seal account.

SIZE

Adult males are slightly longer and heavier than females, but this is almost impossible to determine in the field. The following measurements apply to adult animals from throughout the range as there are so few Canadian specimens.

Males: total length: 134–176 cm; weight: 85–130 kg.
Females: total length: 133–169 cm; weight: 66–114 kg and likely heavier.
Newborns: total length: 77–92 cm; weight: 7–12 kg.

RANGE

Spotted Seals occur only in the Northern Hemisphere in the North Pacific Ocean and adjoining Arctic Ocean. Their occurrence is seasonal; they migrate between more northerly summering grounds and more southerly wintering grounds. The large population that winters in the Bering Sea summers in the Chukchi and Beaufort Seas. Smaller populations that winter in the Sea of Okhotsk and further south around the coast of Japan likely do not migrate as far north. Breeding occurs on the receding spring ice while it is in the Bering Sea, Okhotsk Sea, Sea of Japan, and the Yellow Sea. At least half, if not more, of the herd resides on the western (Russian) side of the range. The few animals that reach Canadian waters can do so only in summer, when the ice has receded, and travel along the coast of northern Alaska to Yukon is possible.

ABUNDANCE

Along the eastern Russian and Alaskan coasts native peoples have traditionally killed a small number of these seals for subsistence purposes. This harvest continues today. Some small-scale commercial hunting occurred between the 1930s and the 1980s by the USSR and Japan in the western Bering Sea and the Sea of Okhotsk. Although the populations of this seal are poorly known, a very rough estimate of abundance placed their numbers at 235,000 to 270,000 in the 1980s. Only rare wanderers are seen in Canadian waters. Climate change, specifically global warming, which may reduce the ice extent and thickness and cause early break-up, could have a profound and negative affect on Spotted Seals in future.

ECOLOGY

In general, Spotted Seals winter at the south edge of the ice pack, breed on the receding ice pack, and summer in the open ocean. During summer, once the pups have been weaned, Spotted Seals do not continue to follow the retreating ice northwards as do most of the other ice-breeding seals. Instead, they resort to land-based haul-outs, hundreds of kilometres south of the permanent ice. In the eastern North Pacific they spend the long summer hours of daylight alternating between long foraging trips to sea and loafing at near-shore and coastal sites. Foraging trips may cover several hundred kilometres, and seals often stay at sea for days at a time. In the western North Pacific the seals congregate at river mouths during the time of salmon-spawning runs and abandon these locations in late summer after the runs are over. When the ice advances to their summering range, sometime during autumn, the seals abandon their land-based haul-outs in favour of those on the ice. They spend the winter and spring foraging at the ice edge in deep offshore continental-shelf waters and hauling out on the ice (from November or December to May or June). The ice front, called the fringe, is an area of broken and moving ice pans that provide ample breathing opportunities. Depending on air temperature, ice thickness, and water movement, this fringe may extend up to 350 km into the leading edge of the ice sheet in some years but normally extends 30–140 km. Those seals that winter in the Bering Sea follow the ice up into the Chukchi and Beaufort seas and haul out in northeastern Russia and northwestern Alaska. Southward movements begin in October, and most of these seals pass through the Bering Strait in November. As the ice continues to advance, they are pushed southwards into the

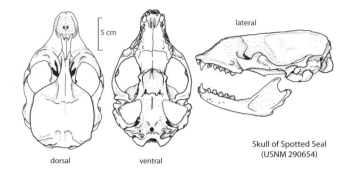

lateral

Skull of Spotted Seal
(USNM 290654)

dorsal   ventral

Bering Sea and the Sea of Okhotsk until about December when the ice stops advancing. Some of the weaned pups and juveniles from small southern breeding populations (the Sea of Japan and the Yellow Sea) show up on the shores of Hokkaido, Japan, during spring, having chosen to remain south rather than follow the retreating ice northwards. When hauled out, Spotted Seals, like Harbour Seals, are very vigilant to guard against land predators such as Polar Bear, Brown Bear, Walrus, Grey Wolf, and Arctic Fox, and the young are additionally vulnerable to Common Ravens, gulls, and eagles. While at sea they are preyed upon by Killer Whales and sharks. Males may live to 26 years old, while females may live up to 32 years.

DIET

Juveniles, subadults, and adults eat a wide variety of locally and seasonally available shoaling fishes such as herring, salmon, eelpout, Arctic Cod, Saffron Cod, Walleye Pollock, and Capelin, along with some fishes that live near or on the ocean bottom, like flatfishes, sculpins, and halibut. They supplement this diet with shrimp, crabs, and octopus that they catch at depths down to 300 m. Newly weaned pups do not dive this deeply, limiting themselves to more shallow water, where they catch amphipods, krill, and small crustaceans.

REPRODUCTION

Spotted Seals are thought to be monogamous, at least for the season. An adult male attends a breeding female on the ice as she suckles her pup. He joins her about a week after the pup has been born and mates with her when she comes into oestrus after the pup has been weaned. Mating is thought to take place in the water, as copulations have not been observed. Once they have mated, the male departs. Triads comprising a female, a pup, and an attending male can be seen scattered throughout the shifting ice pack from late January into mid-April. Pups are born when the ice pack is still stable in their area so that their weaning coincides with the seasonal breakup. More southerly populations breed earlier. The Yellow Sea population peaks in late January, the Sea of Japan population in February, and the Bering and Okhotsk Sea populations peak in early April. Pups do not swim for at least the first two to three weeks, and usually until weaning they remain exposed on the ice, protected by their dense, white birth coat and later by fat. They are weaned abruptly when they are abandoned by their mother at around four weeks old. At this point, they are the heaviest they will likely be during their

entire first year of life, about four times their birth weight. They are so fat and buoyant that they are poor divers for a while after weaning. The age of sexual maturity is unknown but is thought to be similar to that of Harbour Seals: three to six years old for females and four to five years old for males.

BEHAVIOUR

Apart from the already mentioned triad-forming behaviour during the breeding season, and their well-known preference for hauling out on ice, most of the behaviour of this species is unstudied.

VOCALIZATIONS

There are no records of the sounds made by this seal. It is possible that it is largely silent.

SIGNS

Although so far unrecorded, it is assumed that trails, if left on a sandy haul-out, would be similar to those created by Harbour Seals on a similar substrate (see the "Signs" section in the Harbour Seal account).

REFERENCES

Bigg 1981; Burns 2002; Lowry et al. 1998, 2000; O'Corry-Crowe and Westlake 1997; Reeves et al. 2002; Shaughnessy and Fay 1977; Sobolevsky 1996; Wada et al. 1992.

Harbour Seal

also called Common Seal

FRENCH NAME: **Phoque commun,** *also* **Loups-marin**
SCIENTIFIC NAME: *Phoca vitulina*

Harbour Seals are the most abundant and commonly seen seals in the Northern Hemisphere. They inhabit coastal regions of the North Pacific and North Atlantic Oceans and have the widest distribution of all the pinnipeds.

DESCRIPTION

The pelage colour and pattern are variable. Background colours vary from light tan to yellowish or yellowish-grey (light phase) to dark brown or black (dark phase), with many intermediate variations. All have dark blotches and spots on the upper surface, but these may be undetectable on very dark animals. Dark-phase individuals usually have light-coloured spots or rings on their belly and sometimes a mottling of lighter spots on their belly, flanks, and back. Light-phase individuals are typically lighter on the flanks and belly, while the back is covered in dark spots or blotches and sometimes a scattering of faint pale rings. Certain colour phases may dominate in some areas; for example, most animals in Ungava, the southern British

Harbour Seal (*Phoca vitulina*)

light-phase adult

dark-phase adult

light-phase pup

Columbia coast, and the Canadian Atlantic coast are dark phase or dark intermediate; on the northern British Columbia coast most are either light phase or are of lighter intermediate coloration. The eyes are large and close set, and the nostrils form a distinct V shape with the bases closer than the apexes. An adult coat is made up of three types of hairs: long, thick vibrissae (whiskers), which are commonly whitish and grow from mystacial pads on the upper lip and from spots over the eyes; coarse, flattened guard hairs, which may be up to 9 cm long; and shorter, finer underfur. Each hair follicle (apart from the vibrissae) produces a bundle that includes one guard hair and up to six underfur hairs. Newborns usually have shed their white lanugo or foetal coat in utero and have a spotted pelt similar to that of an adult. Premature pups and a small percentage of full-term pups, mainly in the northern parts of the range, may retain the lanugo coat for a day or two after birth. The dental formula is incisors 3/3, canines 1/1, and postcanines 5/4, for a total of 34 teeth (all the teeth behind the canines are called the postcanine teeth as premolars and molars cannot be distinguished). Moulting generally occurs during mid-summer to early autumn and may vary between regions. Most populations begin to moult within two to three months of their pupping season. Typically, juveniles begin the moult first, followed by adult females, and finally by adult males, but there may be considerable overlap between the age and gender classes in some regions. The full moult usually takes one to two months.

SIMILAR SPECIES

The most similar seal to a Harbour Seal is the closely related sister species, the Spotted Seal. It may be quite difficult to distinguish individuals of these two species because adult Spotted Seals have a similar pelt colour to that of light-phase Harbour Seals. Spotted Seals have a limited range in the North Pacific, but their distribution does overlap with Harbour Seals in that region. Usually Spotted Seals occupy a different habitat than Harbour Seals. Spotted Seals are typically found near floating sea ice, far out to sea, while Harbour Seals are more coastal in their habits and are rarely seen near pack ice in the North Pacific. Newborn Spotted Seals are born on the sea ice and are white, while newborn Harbour Seals are usually similar to adults in their pelage colour and are typically born on land. Other seals in the North Pacific that might be confused with Harbour Seals include Ribbon Seals, Ringed Seals, and Northern Fur Seals. Ribbon and Ringed Seals have distinctive dark and light markings and do not occur in Canadian waters in the Pacific; all are associated with pack ice further to the north, beyond the range of Harbour Seals. Female Northern Fur Seals might resemble the dark-phase Harbour Seals along the British Columbia coast. The fur seals have external ear flaps, and females have an arched head profile (Roman-nosed) and are considerably smaller than an adult Harbour Seal. Male Northern Fur Seals are substantially larger than Harbour Seals and have a distinctive domed head shape. Ringed, Grey, Hooded, Harp, and Bearded Seals overlap with Harbour Seals in the North Atlantic. Ringed, Hooded, and Harp Seals are usually associated with ice. Grey Seals are larger, with a bigger, longer, heavier snouted head that is not rounded like that of a Harbour Seal. Female Grey Seals are light coloured with large irregular dark splotches, while males are larger yet, with a dark body. Bearded Seals are much larger than Harbour Seals, with long distinctive whiskers (vibrissae).

SIZE

Adult males are slightly longer and heavier than females, but this may be very difficult to determine in the field. The following measurements apply to adult Canadian animals.

Distribution of the Harbour Seal (*Phoca vitulina*)

Males: total length: 150–190 cm, average 154 cm; weight: up to 170 kg, average 90 kg.
Females: total length: 140–170 cm, average 143 cm; weight: up to 130 kg, average 70 kg.
Newborns: total length: around 80 cm; weight: 8–12 kg.

RANGE

Harbour Seals occur only in the Northern Hemisphere. They inhabit temperate, Subarctic, and sometimes even Arctic coastal waters of the North Pacific and North Atlantic Oceans. On the western coast of North America, Harbour Seals occur from the Aleutian Islands of Alaska to Baja California. Along the eastern coast there are records from Florida and Ellesmere Island, but Harbour Seals occur commonly from Baffin Island, Greenland, and Iceland to New York State. The species is not strictly marine, although the vast majority live in salt water. Harbour Seals commonly move up river systems into freshwater drainages for a time. Along Hudson Bay, Harbour Seals may travel up to 240 km upriver. They were once seen in Lake Champlain, Lake Ontario, and as far up the Ottawa River as the city of Ottawa, although this has changed since the early 1900s, and they do not appear to travel as far inland in eastern Canada as they used to; they seldom even reach Montreal. Along the west coast, Harbour Seals are known to travel as far as 300 km inland. An unusual landlocked population in the Lacs des Loups Marin (Seal Lakes) region of the Ungava Peninsula of Quebec survives year-round in fresh water. Although some seals, notably juveniles, may sometimes travel great distances (over 1000 km), most adults are more sedentary and do not make long annual migrations; however, seasonal shifts and regional movements related to breeding or moulting are common. Some populations may undertake short seasonal migrations of a few hundred kilometres. Harbour Seals occur in many ice-free regions year-round. As they do not have long, robust front claws, Harbour Seals cannot excavate breathing holes in ice, and so they are reliant on permanent areas of open water in which to over-winter. Seals that inhabit areas such as the estuary of the St Lawrence River, which freezes during most winters, are forced to travel to areas with open water. Satellite tagging of seals in the St Lawrence estuary has shown that they move up to 520 km to find appropriate wintering conditions.

ABUNDANCE

Their coastal habits and relative abundance bring Harbour Seals into regular contact with humans, and native hunters across the Northern Hemisphere have pursued them for several hundreds and likely thousands of years. Subsistence hunting of Harbour Seals by aboriginal peoples is still practised in North America. In the early to mid-1900s (1913 to the 1970s in Canada), on both North American coasts – from California to British Columbia and Alaska, and from Massachusetts to the Maritimes – bounties were paid for Harbour Seals because of their supposed competition with commercial fishermen for fish. They have been partially protected since 1976 in Canada but may still be legally killed in order to protect fish farms in Canada, Norway, and the United Kingdom. The largest populations are in the North Pacific. Around 100,000 Harbour Seals were estimated on the British Columbia coast in 1990, and these numbers were growing at an estimated 12% per year, although this trend appears to be diminishing. Numbers in British Columbia are thought to be nearing the original levels before culling. Canadian Harbour Seals in the Atlantic were very roughly estimated in 2005 at 10,000 at least, excluding the Arctic and Labrador coastal seals that have not been recently surveyed. Bay of Fundy animals were estimated at about 3500 in 1994, and the Gulf of St Lawrence population was estimated at 4000–5000 in 2005. The entire eastern seaboard of North America may support as many as 100,000 Harbour Seals, as these animals are numerous along the New England coast. Harbour Seals are common on many parts of the eastern Canadian coast and are the most abundant marine mammals in British Columbia. The tiny freshwater Ungava population (estimated to be 100–600 animals, but likely closer to the lower number) was officially designated as of "Special Concern" by COSEWIC in 1996. However, the entire small range of this subspecies is on Crown land that is currently slated to be involved in the Grande Baleine hydro-electric project, which, although indefinitely postponed, has not yet been cancelled. Pervasive changes to their habitat could result if this project proceeds. COSEWIC, in its December 2007 meeting, ascribed the "Endangered" status to this imperilled population. The Greenland population has declined dramatically in the last 100 years, likely owing to overhunting, and is practically extirpated. The Japanese population is very small and declining due mainly to fishing equipment entanglement. Numbers in the Aleutian Archipelago have declined by an average of about 65%–70%. Nevertheless, there are an estimated 500,000 Harbour Seals living today. Climate change is anticipated to have a major positive impact on Harbour Seals in future as global warming and the resulting thinning of sea ice may make more ice-free habitat available to them.

ECOLOGY

Harbour Seals are coastal seals, hunting in near-shore waters (defined as within 30 km of the shore) and hauling out on rocky or sandy beaches. They are primarily marine but will travel for extended periods and, in the case of the seals that live in the "Seal Lakes" in Ungava, sometimes live year-round in fresh water. Unlike many other seals, Harbour Seals are not found in concentrations during at any time of the year. Breeding and pupping occur throughout the range. In their first year of life pups suffer a mortality rate of 20%–50%, and typically fewer than 50% survive to reproduce. Around Vancouver Island, where Killer Whales feed extensively on Harbour Seals, as few as 20% may survive to sexual maturity. Harbour Seals can swim at speeds of up to about 14 km/h, but they are awkward on land and rarely travel far from the water's edge. Although most foraging dives last less than four minutes, one radio-tagged seal stayed submerged for 13 minutes as it hid from hunting Killer Whales. Since they cannot rotate their hind flippers under themselves, they employ a distinctive caterpillar-like humping movement, using the front flippers for leverage to accomplish their terrestrial locomotion. The hind flippers are lifted and swung from side to side but do not provide propulsion. Aquatic movements are considerably more graceful. As with other pinnipeds, the hind limbs are the driving force (mostly with alternating strokes, but sometimes with simultaneous strokes), and the fore limbs are used for steering.

The home-range size of these seals is highly variable and ranges from 100 to 55,000 km². Harbour Seals are predators of farmed salmon on both east and west coasts. Besides shooting troublesome nuisance animals, the salmon farmers utilize seal "scarers" or "pingers" on their pens. These devices emit a loud underwater noise that is intended to discourage the seals from approaching. Avoidance of these acoustic devices may drive Harbour Seals from traditional haul-outs, feeding sites, and breeding sites, and exposure may result in deafness. Some evidence suggests that the sound does not exceed their pain threshold; in other words, the seals habituate to it and do not modify their behaviour. Other species such as Harbour Porpoises are also adversely affected by the noise makers. Harbour Seals inhabit some of the most industrialized and heavily polluted coastlines in the world. Increasing toxin loads may make the animals more susceptible to disease, starvation, and reproductive failure. The maximum lifespan is about 36 years for females and 31 years for males, but few reach these ages in the wild, where the oldest animals may reach about 30 years. Predators include Killer Whales and sharks. Pups are additionally predated by eagles, Steller Sea Lions (in the Pacific), ravens and gulls, and occasionally even Coyotes. An unknown number of Harbour Seals are shot annually by fishermen and fish farmers.

DIET

Harbour Seals are generalist foragers. They eat whatever pelagic or benthic prey is seasonally and locally available. Primary foods include hake, mackerel, herring, sardine, smelt, cod, salmon, sand lance, shad, Capelin, pollack, sculpins, greenling, anchovy, and a variety of flatfishes along with a host of other fish species. Cephalopods such as squid and octopus, as well as crab and shrimp, are also eaten. On the Pacific coast their tendency to follow salmon runs into the river mouths has put them into contention with salmon fishermen. At least 32 different species of fish and 18 different species of invertebrates are consumed on the Atlantic coast of Canada, and at least 27 fish species are caught and eaten by Harbour Seals along the Canadian Pacific coast. Although they have been accused of interfering with the Atlantic Herring weir fishery, it appears that Harbour Seals and herring move independently of each other; the seals do not follow and harass the herring. The landlocked population in Ungava eats freshwater fishes, mainly Brook Trout, Lake Trout, and Lake Whitefish. Small prey is eaten underwater, and larger fishes are brought to the surface before being chewed into smaller bite-sized pieces. Recent radio-tagging research shows that despite their coastal range Harbour Seals are capable of feeding to

depths of at least 500 m and may stay submerged for up to half an hour. Most dives are to depths of less than 50 m and last for four to six minutes or less. Newly weaned pups spend their first few months searching for their food near shore, where they feed on bottom-dwelling shrimps and other crustaceans that are easier for them to catch. As their body size increases, they are gradually able to dive deeper. Harbour Seals forage during both daylight and darkness. Their sensitive whiskers allow them to sense water and fish movements in the dark. They satisfy their body's need for fresh water from the food they eat.

REPRODUCTION

Female Harbour Seals reach reproductive maturity at around the ages of three to six years old, and males at four to five years old. Both require an additional three to four years to reach their full physical size. Within any given area, pupping lasts one to three months, with a peak during two weeks to a month within that period. In the western Atlantic most births occur in May–July (becoming later as one proceeds northwards). On the coasts of southern British Columbia and Washington State most pups are born in July and August. Along the northern British Columbia coast most births occur in May and June. The landlocked Ungava population produces pups very early, between mid-April and late May. The first hour or so after birth is a crucial time for the formation of the mother-pup bond, and so females isolate themselves before the birth. They almost always bear a single pup, usually on land between the high and low tide lines, although on Sable Island some females are known to travel overland for more than a kilometre to give birth beside inland freshwater ponds. In Alaska and Greenland some pups are born on icebergs calved from glaciers.

Newborns often swim almost immediately, especially if the tide is rising or if scavenging birds that are feeding or fighting over the afterbirth disturb or attack the pups. Although they are about the most precocial of the pinnipeds, newborns will often hang onto their mother's back with their front flippers while she swims. Mothers are solicitous of their pups and remain close to them, especially when hauled out on land. If danger threatens while they are together on shore, she will grab the pup in her mouth or with her fore flippers and lunge into the water. Despite her care, storms during the nursing period may separate mother from pup, dooming the pup unless the two can reunite or, as happens on rare occasions, a female who has lost her own pup fosters the foundling. Females with very young pups often spend much of their time hauled out, but they might leave their pups on shore for a short time while they forage nearby. During the

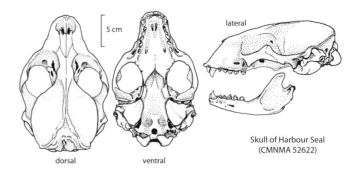

Skull of Harbour Seal
(CMNMA 52622)

5 cm

lateral

dorsal   ventral

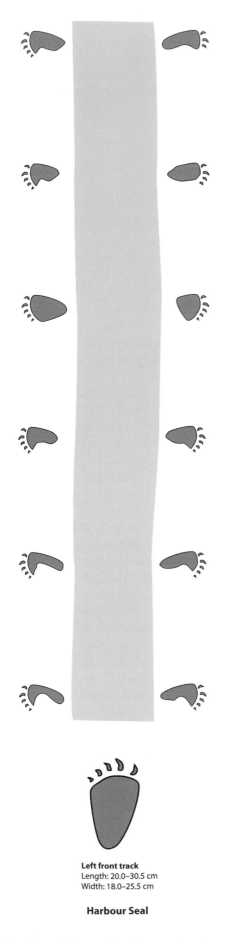

Left front track
Length: 20.0–30.5 cm
Width: 18.0–25.5 cm

Harbour Seal

mid to latter part of the nursing period, pups typically begin to follow their mother on her foraging, learning and sampling their mother's catch along the way. Pups are suckled for around three to four weeks, and sometimes as long as six weeks, after which they are abruptly weaned when they are abandoned by their mother. On average, pups undergo a fast of about 15–17 days, during which they survive on blubber, before they learn to become independent. They typically remain in the region of their birth for their whole life, although long-distance migrations do sometimes occur. Most females mate shortly after their pup has been weaned, from mid to late summer, and most mating occurs in the water. As with all pinnipeds, Harbour Seals experience a stage when the newly fertilized embryo floats around in the uterus but does not implant and begin to grow. In the case of Harbour Seals, this lasts for about 2.5–3.0 months. Their gestation period of 10.5–11.0 months includes this delayed-implantation stage. Around 85% of mature females bear a pup each year.

BEHAVIOUR

Harbour Seals typically forage alone but form groups when they haul out. In some regions, especially at traditional sites in the North Pacific, these groups can grow into the thousands, but most are 3–20 individuals of mixed age and sex. Females sometimes segregate while nursing, and newly weaned pups sometimes band together for a while. Hauled-out groups are evenly spaced with a clear area around each animal. The seals ensure this individual distance with threat displays including fore-flipper waving, fore-flipper clawing, head-butting, snorting and growling, and sometimes biting. Typically a larger animal will displace a smaller one, and more agonistic behaviour is evident in locations where space is limited. Harbour Seals may lie with their front and hind ends elevated into a characteristic "crescent" posture. While resting and sleeping, they rouse frequently to scan the area. Seldom can a Harbour Seal be surprised while it is on shore. Courtship and mating are rarely observed because they occur mainly in the water. As the oestrous females are widely and unpredictably dispersed, males make extensive use of underwater sounds to attract them and, at the same time, to deter rivals. One of the components of a male's aquatic courtship display consists of slapping the surface of the water with his flippers, either fore or hind. Aerial acrobatics are common among Scottish Harbour Seals. Displaying males arc out of the water, usually several times in succession. This dive display is unknown in North America. Males lose weight during the breeding season because they devote much of their time and energy to searching, displaying, and competing for females. Males try to establish their breeding territories in regions that are likely to be frequented by females, such as near a regular haul-out or along a well-travelled route. Females move freely between the territories, but should a rival male approach too closely, the males will lunge at each other with open mouths or wave their front flippers in a threat to strike with the claws. Breeding males with wounds on their neck and front flippers are not uncommon. Some Killer Whales on the west coast of North America catch and eat Harbour Seals. These whales, called transients, live in small, quiet groups and specialize in preying upon marine mammals. Residents, the other type of Killer Whale, live in noisy, social, extended groups and eat fish. Vocalizations of the two types of whale are somewhat different, and Harbour Seals in the west have learned to distinguish them. They will behave as usual when a resident pod is nearby but become secretive and silent if they hear a transient pod.

VOCALIZATIONS

Most reports of Harbour Seal vocalizations refer to animals on the east coast. Seven in-air vocalizations and one underwater vocalization have been described from that group. The in-air sounds are a short broadband bark produced by adult males; a short tonal honk, also produced by adult males; a grunt produced by adult and sub-adult males; a growl produced by both sexes and all age groups, including pups; a roar produced by both adult males and females; a moan produced by both adult males and females but predominantly by the males; and a pup attention call produced only by young pups. The first four calls and the sixth call are used to express aggressive threat intent. The fifth call is used by males during their courtship displays to both attract females and repel rival males. Females use the same roar to express aggressive intent. The last call is intended to attract the mother, and sounds like a sheep bleating. It is loud enough to be heard up to 1 km away on an abnormally calm day, but may only be heard 8 m away on a windy or rough day. Displaying male Harbour Seals produce a loud, pulsed, underwater sound described as an underwater roar that slowly builds in intensity and is followed by a pulsed expulsion of air. This call has been recorded at other Harbour Seal sites around the range and appears to be a commonly used mating call. The average rate of in-air calls is high among eastern Canadian Harbour Seals, perhaps owing, in part, to crowding at small and scattered haul-out sites.

SIGNS

Harbour Seals leave trails only on sandy or muddy haul-outs and usually below the high-tide line. On hard sand the front claws and part of the front flipper register in a track, with a hint of the body dragging between and smoothing the sand. On soft sand the trail includes the distinct drag marks of the body between the flipper marks and possibly also some drag marks from the claws. The hind feet do not create a track, as they are dragged behind the body. Both front flippers move together. Each has five claws and a flattened palm that can register in a track. The age classes of Harbour Seals can be estimated from the width of the trail. Juveniles (less than one year old) have trail widths of less than 32 cm from outside flipper to outside flipper; subadults (ages one to four years old) have trail widths of 32.1–56.0 cm; and adults have trail widths of more than 56.1 cm. Trails of seals can be distinguished from those of nesting sea turtles by the claw marks, because sea turtles do not have claws. Scat is rarely seen and is often missed. It is an amorphous patty-shaped mass found along a seal trail at a haul-out.

REFERENCES

Baird 2001a; Bigg 1981; Boness et al. 1992, 1994; Bowen and Harrison 1996; Bowen et al. 1999; Burns 2002; Colbourne and Terhune 1991; Deecke et al. 2002; Dehnhardt et al. 2001; Dubé et al. 2003; Frost et al. 2006; Härkönen and Harding 2001; Härkönen and Heide-Jørgensen 1990; Hastings et al. 2004; Hayes, S.A., et al. 2004; Jacobs, S.R., and Terhune 2000, 2002; Kovacs et al. 1990; Lesage et al. 1999, 2004; Muelbert and Bowen 1993; Neumann 1999; Reder et al. 2003; Reeves 1999a; Reeves et al. 2002; Reiman and Terhune 1993; Sjare et al. 2005; Small et al. 2008; Smith, R.J., 1997; Smith, R.J., et al. 1994, 2006; Teilmann and Dietz 1994; Terhune 1989; Van Parijs and Kovacs 2002; Van Parijs et al. 2000; Walker, B.G., and Bowen 1993; Westlake and O'Corry-Crowe 2002; Whitman and Payne 1990; Womble et al. 2007.

Ringed Seal

also called Hair Seal, Common Seal

FRENCH NAME: **Phoque annelé**

SCIENTIFIC NAME: *Pusa hispida*

Ringed Seals are the most common seals in Arctic waters, and they have long been a staple food for indigenous peoples in those regions. They are the smallest of the North American true seals (family Phocidae).

DESCRIPTION

These seals have a small head, small fore flippers, and a body that is well rounded by blubber. Adult Ringed Seals are silvery in colour on the underside and dark grey to black on the upper surface. The dark back and light belly grade into each other along the sides. The back and neck are speckled with numerous pale silver rings with dark centres that give them their common name. These rings become less distinct as they appear lower on the sides and onto the underside. The flippers are often dark above and lighter below but may be light silver on both surfaces. Ringed Seals have large dark eyes and fatty mystacial pads on each side of the muzzle that form the base for numerous rows of whiskers. The pups are born with a thick, fine, white lanugo coat, which they moult by around six to eight weeks old. Juveniles are silver grey above and light silver below, sometimes with traces of pale rings on the lower sides, until about one year old when they moult into a typical adult pattern. The dental formula is incisors 3/2, canines 1/1, and postcanines 5/5, for a total of 34 teeth (all the teeth behind the canines are called the postcanine teeth as premolars and molars cannot be distinguished). The annual moult takes place from late March until July. Juvenile seals moult early in that period, while breeding adults moult primarily in June and July.

SIMILAR SPECIES

Ringed Seals are the most ice adapted of the North American seals. They share parts of their habitat with Harp, Hooded, Grey, Bearded, Spotted, and Harbour Seals but are fairly easily distinguished from all but Spotted and Harbour Seals by their small size, rotund shape, and ringed pelage. Juvenile Harp Seals are similar in size to adult Ringed Seals, but they do not display the ringed pelage; they have a dark-spotted, pale-grey coat. Spotted Seals (which enter Canadian waters only in summer along the northern coast of Yukon)

Ringed Seal (*Pusa hispida*)

adult

pup

are larger, but slimmer, and spotted rather than ringed. They have a larger head and heavier snout. Harbour Seals also have a heavier head and are larger and slimmer but are rarely seen hauled out on ice, preferring ice-free habitats.

SIZE

The following measurements apply to adult Canadian animals. Mature males in most populations are slightly longer and heavier than mature females. Adults reach their full physical size at around eight to ten years old. The largest of the Ringed Seals occur in the Canadian Arctic. Their weight is greatest in winter and early spring.
Males: total length: 114–165cm; weight: 51–113 kg.
Females: total length: 114–149 cm; weight: 44–97 kg.
Newborns: total length: around 60–65 cm; weight: about 4.5–5.5 kg.

RANGE

Ringed Seals have a wide circumpolar distribution throughout the Arctic Ocean and the Baltic Sea and occur seasonally in Hudson Bay, James Bay, and the Bering and Okhotsk Seas. Two landlocked, freshwater populations occur in Lake Saimaa and nearby Lake Ladoga in Finland. They became separated about 8000–9000 years ago from the Baltic population following the last glaciation. Some Ringed Seals are seen from time to time in freshwater Nettilling Lake, Baffin Island, but that population is probably not restricted to the lake. This seal has a long history of becoming landlocked. The closely related Baikal Seals (*Phoca siberica*) in central Russia and the Caspian Seals (*Phoca caspica*) in eastern Russia have been separated in Lake Baikal and the Caspian Sea for long enough (at least 2–3 million years) to have evolved into similar but distinct species. Ringed Seals regularly wander beyond their usual distribution and have been found as far south as Portugal in the eastern Atlantic Ocean and Sable Island in the western Atlantic Ocean. In the Pacific Ocean, Ringed Seals have turned up in California and Northern Japan.

ABUNDANCE

Although no comprehensive worldwide surveys have been undertaken to suggest a total population size, it is roughly estimated that there may be as many as 2–4 million Ringed Seals alive today. As would be expected of a species that spends most of its time under the ice, Ringed Seals are very difficult to count. The Canadian population appears healthy and numerous despite centuries of subsistence hunting by northern aboriginal peoples for fur, flesh, and blubber. As northern animals that are closely tied to sea ice during all phases of their life history, Ringed Seals may prove to be practical indicators of climate change. Global warming, which is resulting in a noticeable diminution of the polar icefields, is already having an impact on breeding success and haul-out opportunities for these seals in some areas, such as western Hudson Bay, and may result in population reductions over time. The seals are vulnerable to warm weather during the breeding season. The accompanying rain, combined with their own body heat, can cause the slumping and collapse of the roofs of the ice lairs, subjecting the pups to exposure and increased predation. Early break-up of sea ice, which effectively causes early weaning, has a negative effect on pup growth and survival.

ECOLOGY

These small seals are closely associated with sea ice most of the year, and the species group, which includes Ringed Seals and the closely related Baikal Seals, is unique among seals in its use of snow lairs. These lairs are carved out of the snowdrifts and ice that form around a breathing and haul-out hole. Lairs are created by both male and female seals. They vary in size and shape and may contain several chambers. An average-sized birthing lair is roughly oval and around 2.5 m wide by 3.0 m long. These lairs serve to protect the pups from cold weather and likely from predators. Occasionally, pups are born in natural hollows in the ice surface without the shelter of a lair, usually in areas where snowfall is sparse. The seals use their stout, sharp front claws to scratch at the forming ice to keep their breathing holes open and also to dig their lairs. They can maintain an open breathing hole in ice as thick as 2.0–2.5 m, and most adults have several such holes, which may be kilometres apart. Polynyas (areas of ice-free water during winter that regularly form in about the same areas each year) are important feeding areas especially for juvenile Ringed Seals. Whelping occurs mostly on stable, thick, land-fast ice, but in some

Distribution of the Ringed Seal (*Pusa hispida*)

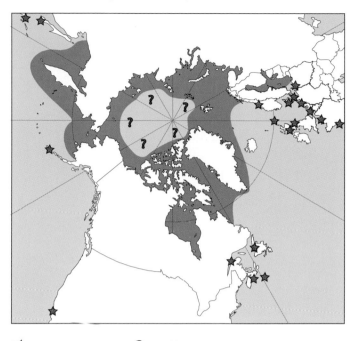

★ extralimital records   ? possible distribution

regions pack ice is also used. The maximum recorded age of a wild individual is 45 years, but few exceed 30 years old. Females appear to live longer than do males in some regions. Predators are numerous and include most Arctic carnivores: Common Raven, Glaucous Gull, Red Fox, Grey Wolf, Domestic Dog, Wolverine, Walrus, Killer Whale, and sharks and their primary predators – Polar Bear, Arctic Fox, and human. Polar Bears are by far the most significant of these.

DIET

Ringed Seals are somewhat opportunistic, and their diet varies with season. During autumn, winter, and spring they prey primarily on pelagic shoaling fishes such as Polar Cod, Arctic Cod, Saffron Cod, and Capelin. Shrimp such as *Crangon* species and *Pandalus* species, krill, amphipods, and other crustaceans are commonly eaten in summer. Younger seals may concentrate their feeding efforts on crustaceans all year-round. Most prey is relatively small, less than 20 cm in length, but considerably larger prey is taken occasionally. Much of the short Arctic summer is spent moulting. While moulting, the seals spend most of their time hauled out and fasting. Since most foraging occurs in the long periods of darkness during the polar winter, the way in which the seals find and capture their prey is unknown. Most foraging dives are 10–220 m deep, but the seals can dive to at least 500 m. Juveniles forage mostly in the top 50–100 m of the water column, while adults tend to forage at greater depths. The longest recorded dive of a radio-collared wild seal is 43 minutes, but few foraging dives exceed 8–9 minutes.

REPRODUCTION

Females become sexually mature between four and eight years old (average six to seven years old), and males between five and seven years old. Whelping takes place from mid-March to mid-April in Canadian waters. Mating occurs in April to early May, within a month of whelping and while the females are still lactating. Copulation occurs in the water. The almost year-long gestation includes a period of about 2.0–2.5 months of delayed implantation and around 8.5–9.0 months of active gestation. Females typically produce a single pup each year, but twins are known. Poorly nourished adult females may not ovulate in any given year. The pups nurse for three to seven weeks (usually around six weeks in the Canadian Arctic) and typically more than quadruple their weight during that time. Mothers will forage intermittently nearby but spend most of their time on the ice with their pups during the lactation period. Ringed Seal milk contains 38.1% fat, 9.9% protein, 2.3% lactose, and 1.0% ash. Weaning occurs abruptly when the females return to sea to forage, abandoning their pups on the ice. During the rut, mature males often develop a musky odour, which they may use to delineate their territory. The glands that produce the secretion are in their muzzle. The odour permeates their flesh, making it distasteful and generally avoided even by Polar Bears during that time of year. Rutting males, called tiggak (stinkers) by the Inuit, are considered to be unpalatable in that condition. Lairs used by rutting males become particularly pungent. Males are in rutting condition from about early March until the end of May.

BEHAVIOUR

Although Ringed Seals are largely solitary, groups may be seen at favourable haul-outs, near leads or ice margins where they bask in the sun, especially during the moult. Most of the time, individuals in these groups are well spread out with several metres or more between seals. Later in the summer, as the ice melts, seals may be seen in the water around the floating ice pans. Ringed Seals are rarely seen in ice-free regions, although if they do find themselves in open water during the moult, they are forced to haul out on shore. Hauled-out seals, whether on ice or on shore, are highly alert, looking up at least twice each minute to scan for Polar Bears. They rest on their stomach with their nose close to the breathing hole or lead so that they can make a rapid getaway. Very young pups can swim but are reluctant to do so, probably because they lack sufficient blubber to insulate them in the cold water. As they grow and accumulate a blubber layer, even while still in their white birth coat, they will readily swim and even dive and can flee with their mother through a breathing hole if the lair is attacked by a predator. Maternal females are very solicitous of their young. They will clasp young pups in their fore flippers or grasp them in their teeth to relocate them to a new breathing hole or lair. A native hunting technique involves tethering a pup to draw the female back to a lair, where she can be harpooned.

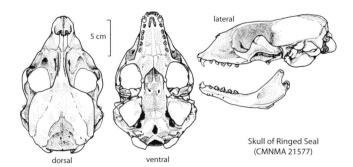

lateral

5 cm

Skull of Ringed Seal
(CMNMA 21577)

dorsal   ventral

Their northern distribution, early breeding season, the cold temperatures at that time of year, and the fact that the seals are laired up at that time accounts for the scarcity of the information regarding the breeding behaviour of Ringed Seals. Breeding occurs largely in fast-ice areas (ice attached to the shore). It is thought that adult males claim a territory beneath the ice that they patrol, looking for reproductive females. They exclude the subadults from this optimum ice type during this period, relegating them to the less stable, offshore ice. A mature male may accompany a nursing female for several days until she reaches oestrus and is willing to mate with him. When she cycles out of oestrus, he departs in search of other receptive females. Males defend underwater territories from early March until late May, thereby typically monopolizing the attention of around two or three reproductive females within their territory and excluding other adult and juvenile males. Breeding males will fight over territorial boundaries and are known to injure each other as well as juvenile males that blunder into their territory. The male's underwater advertisement displays contain a large audio component. It is also suggested that the scent from the muzzle glands (see the "Reproduction" section above) is deposited on the water surface of the breathing hole as a territorial marker. Breeding males eat very little during the breeding season.

VOCALIZATIONS
Vocalizations of Ringed Seals are poorly known. Low-frequency (0.4–16.0 kHz) barks, yelps, wails, and clicks have been recorded underwater. Some sounds made underwater are similar to those made in air. For example, pups will bark in both media to attract the attention of their mothers. Rutting males use low-frequency underwater sounds to advertise their presence as a territory holder.

SIGNS
Ringed Seals leave few clues to their presence. Detection of a lair usually requires a trained observer or a trained dog, although the lair of a rutting male may be more easily detected by its pungent odour. By smell, Arctic Foxes are able to distinguish the lairs containing newborns from those with older pups and will concentrate their efforts on the smaller young. They dig a small hole (10 cm in diameter) into the side of such a lair and attempt to capture the pup. The location of the now abandoned lair may then be detected by the occurrence of such a hole. Bigger pups will disappear down the breathing hole before the fox can dig through the lair wall, and so foxes do not waste their effort trying to capture them. Polar Bears also find lairs by odour, but signs of their predation are considerably more noticeable as they usually break in the roof of the lair to gain access.

REFERENCES
Born et al. 2002, 2004; Ferguson, S.H., et al. 2005; Frost and Lowry 1981; Furgal et al. 1996, 2002; Gjertz 1990; Hammill et al. 1991; Holst and Stirling 2002; Holst et al. 1999, 2001; Kelly and Wartzok 1996; Kingsley and Stirling 1991; Kunnasranta 2001; Lucas and McAlpine 2002; Lydersen and Hammill 1993; Lydersen et al. 1992; Palo 2003; Palo and Väinölä 2006; Reeves et al. 2002; Ryg et al. 1992; Smith, T.G., and Harwood 2001; Smith, T.G., et al. 1991; Stirling and Smith 2004; Wiig et al. 1999.

FAMILY MUSTELIDAE
weasels

Mustelidae is a very diverse family with members that are terrestrial, semiaquatic, fossorial, and arboreal. Most have a long, slender body with short legs and produce a pungent odour from their paired anal glands. The smallest, the Least Weasel, is often smaller than the mice and voles that are its primary prey. The largest is the Sea Otter, which may weigh up to 45 kg. Some mustelids, like the Wolverine and the badgers, are somewhat divergent and not as weasel-like as the others. Weasels have a short muzzle, long canines, and a relatively large braincase.

Sea Otter

FRENCH NAME: **Loutre de mer**
SCIENTIFIC NAME: *Enhydra lutris*

Sea Otters are the largest and most aquatic of the weasel family. They are also the smallest of the truly marine mammals. Once hunted almost to extinction for their beautiful, dense fur, the remaining Sea Otters have a genetic diversity that is very low owing to the severe bottleneck caused by the 99% population loss.

DESCRIPTION
Sea Otters have a long body, short limbs, and a thick neck. They are considered physically and ecologically intermediate between terrestrial and marine carnivores, with many adaptations that suit them to their aquatic environment. Their powerful forepaws are mitten-like with retractable claws. These are used for grooming and for grasping prey and manipulating it while eating. The hind limbs are posteriorly oriented and flipper-like, with digits that increase in length from the first to the fifth as with seals. Sea Otters have a luxuriant double coat composed of larger, longer, glossy guard hairs and shorter, interlocking underfur hairs, with a ratio of one guard hair to 70 underfur hairs. These hairs combine to hold a layer of air next to the skin so that water cannot penetrate. They also have specialized sebaceous glands that secrete oils, which enhance the water-repellent qualities of the fur. As their body does not have a subcutaneous fat layer as do other marine mammals, the thick fur and trapped air are their main defences against cold seawater. This air layer also provides buoyancy so that

Sea Otter (*Enhydra lutris*)

they can lounge and sleep at the surface. Sea Otters have the thickest fur of any mammal, varying from around 26,000 hairs per cm² on a hind foot to over 164,000 hairs per cm² on a foreleg. To put this into perspective, humans average around 100,000 hairs on the entire head, and Northern Fur Seals average 40,000–60,000 hairs per cm². Fur covers most of a Sea Otter's body except the nose pad, the ears, and the pads of the feet. Although the hind flippers are relatively sparsely furred, presumably to avoid turbulence, the otter is protected from excessive heat loss in these extremities by an extensive counter-current blood capillary network at their base that transfers heat from the warm blood entering the flipper to the cold blood returning from the flipper. The tail is about one-quarter of the total length, somewhat flattened dorso-ventrally, and equally thick from the base to almost the tip. It is mainly used to scull, which enables an otter to make minor surface movements as it floats on its back. Newborns have light-brown to yellowish, woolly natal fur. This is shed and replaced with dark juvenile fur over the course of around 14 days, by the age of 12–13 weeks. Juveniles moult into the lighter, adult pelage by about six months old. Adult fur is brown, ranging from pale brown or reddish-brown to very dark brown. The fur colour is similar on the back and belly, but the fur on the head, neck, chest, and sometimes even the front limbs and shoulders, becomes lighter with age. Most adults are at least grizzled on their faces. The ears are short and can be folded back and closed to keep water out. Sea Otter skeletons may be stained purple owing to absorption of polyhydroxynaphthoquinone, which they ingest in sea urchin shells.

The dental formula is incisors 3/2, canines 1/1, premolars 3/3, and molars 1/2, for a total of 32 teeth. Sea Otters have unusual teeth. They are the only weasel with just two lower incisors on each side of their lower jaw. These are spade shaped and protrude in order to scoop food out of a shell. Their molars have no cutting edges; instead, they are robust, broad, and flattened, with rounded crowns, suitable for grinding and breaking hard-shelled prey. In the field, gender can sometimes be determined, as they float on their backs, by the presence of penile and testicular bulges in males and a slightly swollen pair of abdominal mammae in females. Sea Otters replace hairs on an ongoing basis throughout the year, and so there is no noticeable moult and they are never unprotected from the cold seawater. Unlike most other weasels, Sea Otters have no anal glands.

SIMILAR SPECIES

Northern River Otters are smaller, with a tapered and much longer tail. Although they do forage along the marine coast, they do not venture far from shore. A raft of Sea Otters bobbing at the surface might resemble a group of very small seals, but only from a distance; otherwise their small, distinctive, floating body shape distinguishes them from the small seals that do not lounge at the surface.

SIZE

Sea Otters are moderately sexually dimorphic, as males are larger than females, averaging 34% heavier and around 8% longer. The following measurements apply to adult animals from the eastern North Pacific (Canadian and U.S. waters).

Males: total length: 126–148 cm; weight: 21.8–45.0 kg.
Females: total length: 107–140 cm; weight: 14.5–32.6 kg.
Newborns: weight: about 1.4–2.3 kg.

RANGE

The Sea Otter's original distribution was around the North Pacific Ocean from central Baja California, Mexico, to the island of Hokkaido, Japan. By 1911 only 13 small remnant colonies remained: on the Kuril Islands; Kamchatka; Commander Islands; five on various Aleutian Islands and the Alaska Peninsula; Kodiak Island; Prince William Sound; the Queen Charlotte Islands; central California; and the San Benito Islands. Several of these colonies subsequently declined and eventually disappeared, including the only remaining Canadian colony in the Queen Charlotte Islands, which had gone by 1929. Reintroductions of Alaskan otters began at Checleset Bay in 1969 and have successfully re-established Canadian populations along western Vancouver Island. By 2004, Sea Otters off Vancouver Island had expanded their range southwards as far as Vargas Island in Clayoquot Sound and northwards beyond Cape Scott to Hope Island. In 1989 a small colony was reported on the Goose Islands off mainland British Columbia, 125 km away from the then most northerly Vancouver Island populations. Although their origins

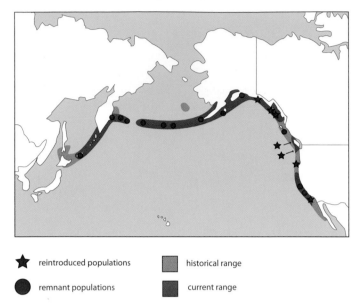

Distribution of the Sea Otter (*Enhydra lutris*)

★ reintroduced populations
● remnant populations
▮ historical range
▮ current range

are unknown, the otters in the Goose Islands genetically match the introduced otters of Vancouver Island, Washington State, and southeastern Alaska, all of which originated from Prince William Sound (Alaska) stock. Occasional wanderers have been seen off the Queen Charlotte Islands, and it is possible that Sea Otters may naturally repopulate that region in time. Current Sea Otter populations off British Columbia occupy an estimated 25%–33% of their historic Canadian range. Other successful reintroductions have re-established Sea Otters in southeastern Alaska and Washington State. Natural range expansion has re-established populations from the Gulf of Alaska westwards to northern Russia.

ABUNDANCE

From the 1740s until 1911, when an international treaty came into effect, an unregulated fur trade in their highly prized pelts reduced Sea Otter numbers from pre-exploitation levels of between 200,000–400,000 animals (some suggest up to 800,000) to less than 2000 and perhaps only a few hundred. By 2004 the reintroduced British Columbia populations were estimated to have reached over 3100 from an original introduction of 89 animals. These populations continue to grow. Reintroductions to Washington State and southeastern Alaska were equally successful, but several attempts in Oregon failed. The previously healthy, and increasing, Prince William Sound population declined following the Exxon Valdez oil spill in 1989 and has not increased appreciably since 1994, remaining at around 13,000 animals, less than pre-1989 levels. In California, Sea Otters have been subject to many anthropogenic factors, from pollution to gill nets, and increases have been balanced with losses. The population there numbered almost 3000 in 2004. The Aleutian remnant colonies prospered for many decades after 1911 and came to represent 80% of the world's Sea Otters by the 1960s. By the early 1980s there were an estimated 55,000–74,000 animals in that region. However, a steep decline began in the mid to late 1980s that saw the

numbers drop to a conservative estimate of 8742 otters by the year 2000. A similar, albeit less catastrophic, decline has been detected along the Alaskan Peninsula and as far as Kodiak Island. The reason for this decline is not clear, but many otters are believed to be falling prey to Killer Whales. Populations of Sea Otters in the Kuril Islands are declining, but those in the Commander Islands and the Kamchatka Peninsula remain stable. There were approximately 93,000 Sea Otters in 2003. Sea Otters were listed as "Endangered" by COSEWIC in 1978. The successful reintroductions were acknowledged in 1996, and this status was down listed to "Threatened." The continuing success of the introductions prompted COSEWIC in 2007 to down list the species further to "Special Concern." British Columbia designates Sea Otters as "Threatened," and it is illegal to kill, capture, or harass Canadian Sea Otters, or to destroy their habitat.

ECOLOGY

Sea Otters occur in shallow coastal marine waters. They forage at and near the ocean floor on rocky or soft substrates. Their maximum surface swimming speeds are 1.0–1.5 km/h, and underwater the animals can swim up to 5.0 km/h. The deepest dive, so far confirmed, was 97 m, but most dives are less than 40 m. Average dives last 50–60 seconds, but the maximum dive time recorded is 4.5 minutes. When Sea Otters are swimming, their hind flippers provide all the propulsion, and the front feet are tucked against the chest. While they are foraging, the front feet are engaged to grasp and manipulate underwater prey. Their skin is baggy and moves easily over the body. Sea Otters can tuck objects (typically food or rocks) under each front leg at the armpit for transport, especially from the ocean floor to the surface or vice versa. Sea Otters are awkward on land but can walk or run with a lumbering, bounding gait. Sea Otters are considered to be keystone species, as their foraging can dramatically and complexly alter the shallow ocean ecosystem. As they consume sea urchins that eat kelp, they alter the balance so that kelp is able to grow to profusion, and the shallow-water fish species that shelter and feed on the mature kelp proliferate. Once full-sized and abundant, the kelp reduces the force of the water currents and wave action, and so more invertebrates can survive.

The home-range size of Sea Otters varies with food abundance, season, gender, and geographical location and may be composed of several heavily used areas connected by travel corridors. It may also be composed of one area that is heavily used for an extended period, followed by a rapid long-distance relocation (60–100 km is not unusual), which can occur at any time of year. Males occupy two types of range during the year. Most of the year they share a large home range with other males, which may be more than 1000 ha, but during the breeding season each male defends an exclusive, smaller breeding territory that averages around 30–40 ha. The home ranges of females are typically smaller than those of males, except the small male-breeding territories, and females tend to be more sedentary. The areas of daily use by adults vary from around 7 to 1166 ha. There is little information available concerning maximum dispersal distances, but circumstantial evidence suggests that 125 km is not out of the question. Generally, juvenile females remain near their natal range, but juvenile males disperse further. Predators of the otters include those that kill them in the water – sharks, Killer Whales, and Bald Eagles, which can take only young pups – and those that attack them on land – Brown Bears and Coyotes. The greatest cause of natural mortality in northern populations is adverse weather conditions, which are typical in late winter. The animals, already weakened by winter, may find it difficult or impossible to forage during the storms and high seas and may die of starvation and starvation-related illnesses. Storms during the time that pups are small can separate them from their mother or negatively affect the female's ability to forage and then cause her milk to dry up. Either situation may lead to a pup's death by starvation, which on average kills around 15% of the pups annually.

Sea Otters are likely the most sensitive marine mammal to the effects of oil contamination. Oil quickly destroys their coat's ability to hold an insulating air layer and leads to hypothermia and loss of buoyancy. As little as 30% oil coverage will cause death. Smaller amounts may not kill immediately, but the ingestion of the oil during frantic grooming efforts can acutely or chronically poison the animal, and so even animals with small amounts of oil contamination may suffer internal damage. The long-term effects of an oil spill, as witnessed by the Exxon Valdez spill in Prince William Sound, can result in poor general health of the population and declining reproductive success and pup survival. Sea Otters in the southern parts of the range are subject to mortality owing to shark attacks, parasites, and disease, in addition to anthropogenic causes such as drowning in fishing nets. Females typically live longer than males. They can survive 15–20 years in the wild, while males live 10–15 years at most.

DIET

To produce sufficient heat to survive in cold waters, Sea Otters maintain a high basal metabolism: 2.4–3.2 times higher than would be expected of a similarly sized terrestrial mammal. They must eat at least 20%–25% of their body weight each day to fuel this energy cost. Food choices are variable depending on availability, season, age of the otter, and individual preference. In general, they eat benthic hard-shelled marine organisms like crabs, mussels, abalones, oysters, snails, bivalves, limpets, starfish, and sea urchins and softer-bodied prey such as squid, octopus, sea slugs, seabirds, sea cucumbers, fish eggs, and fishes of various species. They are known to occasionally eat kelp, perhaps secondarily as they forage for the small invertebrates that live in and on the kelp. Prey switching is a fact of life for Sea Otters, as foraging by an otter population will gradually change the inshore marine ecosystem and in turn alter the kind of prey available to them. Sea Otters forage equally in daylight and darkness, interspersing foraging bouts (which last an average of 2.0–2.5 hours) with rest periods (which last an average of 3 hours). In addition to the fresh water that they consume in their food, Sea Otters are able to drink sea water to top up their requirements.

REPRODUCTION

Breeding can occur at any time of year; however, in northern regions, including British Columbia, breeding activity peaks in October–December, and most pups arrive in May or June. In the southern part of the range, most are born from December to February. The total gestation period, including a variable period of delayed implantation (two to three months), lasts about six months. Copulation occurs in the water and normally lasts 15–30 minutes. The pair often forms a bond that keeps them together for a few days (one to four), and most of the females that do form such a bond mate with only that male. However, some females mate with more than one male, and sometimes a female will form a pair bond with two or more males during one oestrous period. During the bond period the female spends her time in the male's small breeding territory, and they typically mate several times. Males provide no parental care. Mature females produce a single pup at about one-year intervals, but rare births of twins have been observed. Most pups are born at sea, but occasionally females haul out to give birth – more often in regions without dense kelp beds. The precocial pups are born fully and thickly furred, with at least half of their milk teeth already erupted, and they can see, hear, and swim. Nevertheless, they are dependent upon their mother's care. Young pups ride and suckle on their mother's chest. Larger pups float alongside to nurse. Young can swim and dive proficiently by 14 weeks old. Their dive duration and depth increase as they grow. By 20–24 weeks old the pups can capture and break open their own prey. Weaning typically occurs gradually and is completed when the pups reach around six months old, at which time the pup is almost as large as its mother. In rare cases, nursing may continue until the pup is nine months old, or older. Pups may remain with their mother for up to a year after their birth, but most leave shortly after they are weaned. Normally the female will not enter oestrus and breed again until after her pup is weaned. Should she lose the pup before it is weaned, her milk flow stops, and she may cycle again and breed. Her new pup would then be born the following year. Sea Otters produce high-fat, low-lactose milk with a high fat-to-protein ratio, similar to that of whales and seals. Pups begin supplementing their mother's milk with solid food about a month after their birth and quickly learn that their mother will tolerate them literally taking the food out of her mouth. A female may adopt and nurse an orphaned pup, especially if she has recently lost her own. The majority of juveniles become sexually mature by three years old, but few males begin reproducing before they are fully physically and socially mature, at five to six years old. In populations where food is scarce, the sexual maturity of females may be delayed until they are five years old.

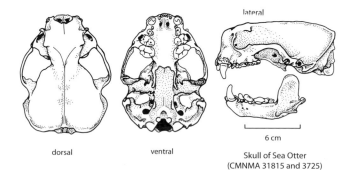

lateral

dorsal   ventral

6 cm

Skull of Sea Otter
(CMNMA 31815 and 3725)

BEHAVIOUR

Sea Otters are relatively social weasels. One individual rarely passes another without approaching for a sniff. Often the intruding otter performs a "head jerk," a rapid, ritualized, repeated, lateral jerk that is probably meant as a greeting. Although many Sea Otters spend their entire lives at sea, some populations traditionally haul out in groups at certain times of the year. They prefer shores with low relief where algae-covered rocks are visible at low tide, but they will use sand or pebble beaches. Excessive disturbance by humans or terrestrial predators causes them to haul out less often and can drive them away from traditional beaches. Sea Otters also congregate in rafts (large groups) to rest. Resting locations tend to be traditional, owing perhaps to a lack of current or to the presence of kelp fronds to serve as anchors. A resting Sea Otter floats on its back with the hind flippers and forelegs in the air, out of the cold water. A sleeping Sea Otter mostly floats on its back with its flippers on its belly or on the surface of the water. Occasionally, for periods of up to a minute, the otter might enter paradoxical sleep (also called REM sleep) when it can turn and float belly down. About 25%–30% of each 24-hour day is spent sleeping, typically in several episodes. When the animals become hungry, the raft disintegrates as each otter disperses to hunt alone. Most foraging occurs in the morning and evening. Heat-conservation behaviours include regular grooming to maintain the hair coat and preserve the vital insulating air layer; lifting of the flippers above the cold water during rest periods; and hauling out on land to rest and sunbathe. Grooming begins with rolling and somersaulting in the water, followed by vigorous rubbing of the entire body using its forepaws. The animal then blows air into the fur of the chest and belly, rubs its hind flippers together and finally spends some time rubbing and licking its face, nape, and chest.

Sea Otters segregate by sex in distinct and separate areas for most of the year, but during the breeding season mature males establish small solitary breeding territories along the shore near the rafts of females. These are actively defended from encroachment by other males. Although fighting is unusual, males patrol the boundaries with much kicking and splashing, warning other males away. When a female goes into heat, she normally selects the male with whom she wishes to mate and enters his breeding territory. Many males are rough with the females as they mate, and most females are lucky to escape a mating with only bloody wounds on her nose, because the males grip the female's nose and head in their jaws as they copulate. Female deaths have been reported owing to overly aggressive mating males. Females comfortably aggregate with other females and their pups most of the year except when they have small pups. Then they attempt to remain isolated and will aggressively attack other otters if they approach too closely, especially if inclement weather forces them to crowd together in a small, sheltered bay. Mothers develop a strong bond with their offspring and may even allow a pup from the previous year to take food from her, particularly if she does not have a new pup. Pups tend to adopt the food preferences of their mother. Sea Otters eat while at the surface. Their usual eating posture is to float on their back and use their chest as a table. A simple sideways rollover clears it for the next course. Sea Otters are ardent tool users. They regularly use rocks to help them crack open their prey. To get at the contents of a hard-shelled crab or bivalve, the otter floats on its back, places a rock on its chest, and, holding the prey in its forepaws, smashes it down onto the rock, pounding away until the flesh is revealed. The pounding behaviour may be innate, as pups have been observed pounding objects on their bare chests. Clearly the use of a rock anvil is learned. The rocks are also used while foraging, to batter a well-anchored abalone, mollusc, or oyster from its attachment site. Most rock tools are 6–15 cm in diameter and often flattened. Favoured rocks may be carried around for a while, tucked into the armpit pouch. Sometimes two hard-shelled food items may be whacked against each other in the absence of a rock anvil. Other food manipulation behaviours include holding a live crab securely on the chest with a hind flipper as another bit of food is consumed first; breaking the spines off sea urchins by rolling them between the front paws; and securing food on the chest with a kelp frond.

VOCALIZATIONS

Fourteen distinct vocalizations have been identified, many of low volume and low frequency. The loudest and most common vocalizations are produced by dependent pups. They emit a high-pitched call, reminiscent of a gull, when separated from their mother. It is called a baby cry and is the same call as the adult "scream," but much higher in pitch. When they resurface, mothers sometimes reply with the adult scream call. These calls can be detected over a kilometre away on a calm day. The scream call is also made by an otter in extreme distress and by a mother looking for her lost pup. There is sufficient variability in the pitch and tone of the scream calls to expect that females and pups can recognize each other's calls. Adults make cooing or grunting sounds when they are relaxed and content, such as when eating something especially satisfying or when grooming pups. Whining sounds are produced by mildly distressed or frustrated otters, such as a pup prevented from suckling or a male unable to reach a hauled-out, oestrous female. These calls may be interspersed with whimper calls. As the level of anxiety intensifies, whine calls may be followed by squeal calls. Squeal calls are high-pitched, loud calls that can be of two types: a "squeal whine"; or a "squeal scream," which is a nasal, distinctive, repetitive call used to communicate stress, discomfort, or medium-level distress. Both squeal calls are also used during the high intensity of courtship. High-pitched whistles, which resemble those made by dolphins in air, are emitted by very anxious animals. Growling and hissing sounds are commonly produced by captured wild otters, and likely signify fear or aggression. Hisses tend to be short and loud. Growls can vary in volume and intensity and may be interspersed with other calls such as hisses, squeals, and whimpers. Two versions of a "squeak" call have been reported. Both appear to be used during similar mildly stressful activities, and each individual otter seems to use only one type of squeak call. The last call, a bark, has only been reported once and was interpreted to be expressing high anxiety or frustration.

SIGNS

As a Sea Otter undertakes a deep dive, it rears up at the surface and rolls forward into a dive. Sea Otter rafts are commonly found in kelp beds when such are available. They haul out at times, most commonly on rocks, and so tracks are unreported. Most defecation occurs in the water, but occasionally scat may be found at haul-outs. Small, semi-liquid, amorphous scat (3–4 cm in diameter) is most common and is formed from a diet of clams and shrimp. A more uncommon, tubular scat (1.3–4.0 cm in diameter) may be produced from a diet of shellfish.

REFERENCES

Ballachey et al. 2003; Bodkin 2003; Brown, A.C., and Elias 2008; Doroff et al. 2003; Estes 1980; Estes and Duggins 1995; Estes et al. 1998, 2003; Hanson and Kusmer 2001; Kreuder et al. 2003; Lee et al. 2009; Lyamin et al. 2000; McShane et al. 1995; Nichol and Watson 2007; Ortiz 2001; Ralls et al. 1995, 1996; Raum-Suryan et al. 2004; Riedman and Estes 1990; Springer et al. 2003; Staedler and Riedman 1993; Watson et al. 1997; Wilson, D.E., et al. 1991; Zagrebelny 1998.

Wolverine
also called Glutton, Devil Bear, Skunk Bear

FRENCH NAME: **Carcajou, *also* Glouton**
SCIENTIFIC NAME: *Gulo gulo*

Wolverines are the hyenas of North America. They are scavengers with extremely powerful jaws and teeth that allow them to crush bones and feed on frozen carcasses, and, like the African hyenas, they are also capable predators. Wolverines are the largest living, terrestrial members of the weasel family (Mustelidae), exceeded in size only by Sea Otters, which are considered marine.

DESCRIPTION

Wolverines are powerfully muscled with a thick, short neck, a broad dog-like head, rounded close-set ears, and relatively short legs (although they are long compared to those of other weasels). They typically hold their head and tail lower than the level of their arched back. Their base coat colour is usually medium brown to black. The face, lower tail, legs, and back are generally darker than the forehead and lateral stripe. The forehead, occasionally including the upper neck, is usually a grizzled light brown to tan, and the lateral stripe tends to start as a reddish-brown and becomes a lighter cream as it broadens rearward. This stripe varies in length and width but typically extends from the shoulder or nape to the upper tail and may extend down the top of the tail. It is usually palest at the hips and root of the tail where it joins the stripe from the other side of the body. Irregular cream to orange patches of variable size and location often occur on the upper chest just ahead of the front legs or slightly higher, in the throat area. The extent, colour, and brightness of the lighter areas, including the lateral stripe, vary widely between animals. Some animals may be so dark that the lateral stripe is not apparent. Totally blond Wolverines are rare. The feet are oversized and have semi-retractable, robust, ivory-coloured claws that are used for climbing and digging. Wolverines have a shaggy double coat that comprises short (2–3 cm long), crimped, dense under-fur, overlaid with coarse, stiff guard hairs that are around 10 cm in length on most of the body and up to 20 cm long on the tail and rump. The tail makes up about 20% of the total length. Wolverines are thought to have an acute sense of smell, moderate hearing, and relatively poor eyesight. They have been known to approach within 45–55 m of a silent human before exhibiting alarm, if the wind does not carry the scent. Wolverines have a pair of walnut-sized anal glands (one on each side of the anus) that contain an odoriferous yellowish or brownish fluid, plantar glands (only on the hind feet), and a lower abdominal gland surrounded by orange fur. The dental formula is incisors 3/3, canines 1/1, premolars 4/4, and molars 1/2, for a total of 38 teeth. Their jaws are especially strong, and the teeth are large and powerful, particularly the carnassials (the upper fourth premolar and the lower first molar). Their bite force is similar to that of Spotted Hyenas. Unlike the smaller weasels, Wolverines can, and do, develop fat deposits upon which they depend when food is scarce. There is a single annual moult that extends from late spring or early summer to autumn.

SIMILAR SPECIES

Wolverines may be mistaken for an American Black Bear or Brown Bear cub, but their hunched back and rolling gait distinguish them. Wolverines have a bushy tail and a light side stripe that the bears lack. Wolverine tracks, made by their oversized feet, may be mistaken for those of small bears or large Domestic Dogs. Wolverine tracks show five toes on both hind and front feet, while canids show only four. Bear tracks show their five toes in a very shallow arc (almost a straight line) at the leading edge of the track, with claw impressions that extend considerably forward of the toe prints. Wolverine tracks are more circular with the toe impressions forming a rounded arc and the claw impressions being just in advance of the toes.

SIZE

On average, males are around 10% larger and 40%–60% heavier than females. The following measurements apply to adult animals in the North American portion of the range.
Males: total length: 938–1070 mm; tail length: 176–267 mm; hind foot length: 174–190 mm; ear length: 44–60 mm; weight: 11.3–20.7 kg on average, with exceptional animals up to 32.0 kg.
Females: total length: 865–932 mm; tail length: 161–250 mm; hind foot length: 155–165 mm; ear length: 38–56 mm; weight: 6.6–14.8 kg.
Newborns: weight: 85–100 g (average 90 g).

RANGE

The Wolverine is a Holarctic species now found in the boreal, sub-alpine, and tundra regions of North America and Eurasia. Before European settlement in North America, the Wolverine was found in all of Canada's ecozones. Its range has been greatly reduced in

Wolverine (*Gulo gulo*)

southern North America following the clearcutting of forests for timber and agriculture. Hunting, trapping, and the use of poison baits further reduced the populations, as did the removal of their ungulate prey base by over-hunting. The southern limits of their historical range are somewhat uncertain, as the species was likely scarce there to begin with, especially in the east. Remnant populations in the south are now fragmented and separated from the rest of the species distribution. Such disjunct populations may still exist in mountainous regions of Oregon, California, Vancouver Island, and the Saguenay region of Quebec. These small, localized populations are all declining and have an uncertain future. Wolverines may already be extirpated from Vancouver Island, and less than a handful of sightings have been reported in all of Quebec over the last 25 years. More than 40 sightings have been reported in Labrador since 1955, when the last specimen was acquired, suggesting that a much depleted population may still exist in northern Quebec and Labrador. Young animals regularly disperse 100 km or more from their natal home range, and some travel tremendous distances. Periodic sightings far to the south of the current range are likely these young adults.

ABUNDANCE

This species was probably never abundant, because large home ranges and relatively low reproductive potential result in naturally low numbers. The highest population densities in Canada occur in regions where large ungulates are common and carrion is abundant in winter, such as in the mountainous regions of Yukon, Northwest Territories, British Columbia, and northern Alberta. Montana, which boasts the highest density of Wolverines in North America,

reports 15.4 animals per 1000 km² in suitable habitat. By comparison, a portion of southwestern Yukon reports only 2.7 animals per 1000 km². British Columbia estimates 6.2 Wolverines per 1000 km² in high-quality habitat and less than 0.3 per 1000 km² in poor-quality habitat. In general, Wolverines are scarce to rare. The population in Canada from Yukon to Ontario is estimated at 15,000–19,000 animals. They are still legally trapped in many parts of Canada (the three territories and five western provinces) although the number of pelts produced in most jurisdictions is declining. Owing to their scavenging lifestyle, Wolverines are very susceptible to trapping as they are highly attracted to the bait; they are often caught in traps set for other species. In 1989, COSEWIC designated the Wolverine as "Vulnerable," and in 1996 the population east of Hudson Bay was determined to be "Endangered" and the western population was listed as of "Special Concern." The province of British Columbia lists the Vancouver Island population as "Critically Imperilled" and does not permit trapping of the populations there or in the southwestern mainland. The portion of the western population found in northern Ontario is classified by that province as "Threatened," and so trapping is restricted. Washington State, Oregon, California, Colorado, Idaho, and Wyoming provide Wolverines with protected status. Alaska and Montana remain the only American states in which trapping is still permitted. The IUCN classifies Wolverines as "Vulnerable" throughout their range.

ECOLOGY

Wolverines are very difficult to study because they occupy remote regions, are largely solitary, roam large home ranges, and hence

North American distribution of the Wolverine (*Gulo gulo*)

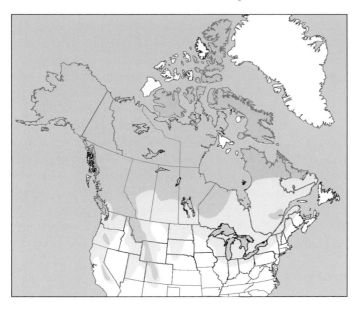

 current range   historical range – southern margin uncertain

Eurasian distribution

occur in low density. They are very versatile in their habitat requirements, and the home range of a single individual usually encompasses a variety of habitat types. Their current occupation of boreal,
alpine, and tundra zones is likely more a result of the availability of
food and a general movement away from human disturbance than
a direct habitat preference. In parts of western Canada and Alaska
(excluding Yukon) habitat preference may vary with season. Many
Wolverines prefer to summer at the higher elevations (alpine tundra)
and spend the winter at lower elevations (coniferous forests). These
preferences are thought to be related to presence of prey, as well as the
avoidance of excessive heat. Maternal females choose isolated areas
to have their kits and are especially sensitive at that time to human
proximity. Wolverines are active day and night and all year-round.
Their activity pattern is likely driven by hunger and prey availability rather than by the time of day, but they are most active at night.
Wolverines require large home ranges in which to find sufficient
food, owing to their scavenging lifestyle. The size of the home range
is inversely related to food availability and may fluctuate with season, age, sex, and the particular year. Males have the added difficulty
of locating females during the breeding season, and so their ranges
are usually considerably larger in order to overlap or encompass the
ranges of several (two to six) females. The home ranges of males are
typically hundreds of square kilometres (around 400–1500 km²) and

may extend up to an astounding 2000 km². Home ranges of adult
females are usually between 75 and 400 km² and become temporarily
smaller (average 70–100 km²) when they are nursing young. Ranges
of non-resident, subadult animals have been reported at up to 37,000
km², as they roam widely in search of a suitable home range. There is
evidence that both male and female subadults disperse widely.

Dens may be dug into snow or located in natural caves, in sheltered crevices in rockfalls, under tree roots, or in windfalls. Maternal females carefully select their birthing den, and a good site may
be reused year after year. The presence of deep snow and persistent
drifts is considered to be essential for denning females, as most natal
dens are dug into deep drifts with a main angled tunnel (about 25–35
cm in diameter) that may extend more than 30 m to ground level,
where the den itself may be sheltered by boulders or fallen trees. The
first metre or so of the tunnel drops almost vertically; then it levels
out and angles downwards only slightly. Natal dens often have only
a single entrance but may be composed of extensive interconnecting
tunnels, the length of which can exceed 54 m. Several beds (widened
chambers) may be present within one den. The beds do not usually
contain any bedding and become hard and icy from use. Females
use this natal den until the spring thaw forces them to switch den
sites. Secondary dens, selected after the young are born but before
they are weaned, may be located in natural rock caves or dug into
snow drifts like the natal dens.

Skilled tree climbers, Wolverines will readily retreat up trees if
threatened by Grey Wolves, and they sometimes cache food and rest
in trees. Their hind ankles rotate sufficiently to allow them to descend a tree at speed, head first. Reluctant but capable swimmers,
Wolverines are able to cross small waterways. Although their large
feet act to some degree as snowshoes, Wolverines living in areas
where the snow is deep but remains powder throughout the winter
often avoid the lower valleys in favour of the more windswept and
hard-packed snow of upland regions. Wolverines are too large and
powerful to be taken by any but the larger predators and then only
occasionally. Known deaths of Wolverines by Grey Wolf and Cougar
are documented. It is likely that bears and Golden Eagles (which
take only kits) cause some mortality. The maximum lifespan in the
wild is likely eight to ten years, but few live beyond four to six years.
The record lifespan in captivity is 17 years. The main natural cause
of mortality is starvation, followed by predation. About 1000–2000
are trapped each year in North America, about half of these from
Canada. There is a limited southern market for the skins, and so
most are used in the north for parkas, rugs, and taxidermy mounts.
Wolverine fur is preferred as trim for parka hoods because the
fur resists frosting by the wearer's breath, and any frost that does
accumulate can be easily knocked off as it tends not to form ice
chunks.

DIET

Wolverines are opportunistic carnivores that are primarily scavengers in winter and predators of small game in summer, and their diet
varies regionally depending on prey availability. They are too large
to survive for long on small game, too small to regularly kill large
game, and too slow to pursue fleet game. Therefore, they are generalists, and the bulk of their winter diet is carrion; they rely on the

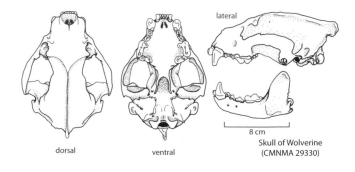

lateral

8 cm

Skull of Wolverine
(CMNMA 29330)

dorsal   ventral

carcasses of animals that have died of natural causes or were killed by other predators or humans. Furbearers caught in traps are welcome additions to this fare, and Wolverines will sometimes break into wilderness cabins or human caches in search of stored food. They can take down weakened larger animals like Moose, White-tailed Deer, or Caribou if conditions are favourable, such as when the animal is bogged down in snow. Their large feet allow Wolverines to travel more easily over snow than the larger ungulates can, and so they have an advantage under certain snow or ice conditions. Coastal Wolverines add whale, seal, and fish carrion to their menu. The main winter diet of carrion is supplemented with whatever else they can catch in the way of small game, including small rodents, hares, foxes, and North American Porcupines. In summer their diet is primarily composed of small game such as mice, voles, lemmings, pocket gophers, squirrels, hares, chipmunks, marmots, North American Beavers, wasp larvae, birds, and bird eggs, to which they occasionally add roots and fruit. Of course, carrion is not disdained in summer but is typically less available. If food is scarce, Wolverines may attempt to take down healthy large game animals such as Mountain Goats, Moose, deer, and Wapiti, but these attempts are rarely successful.

REPRODUCTION

The mating season occurs from late April to August. Some males mate with multiple females, but females usually mate with the male whose territory adjoins their own, and this mating may be repeated over the years unless there is a change in the territory-holding male. Most litters have a single sire. Implantation is delayed. Following fertilization, the eggs develop to the blastocyst stage; then development stops, and they float around in the uterus in a type of suspended animation for up to seven or eight months. This is followed by an active gestation of 30–40 days. The entire gestation period, including the delayed implantation phase, lasts seven to nine months. Litters average two to three young (ranging from one to four). Kits are born in late winter or early spring (February to April). Males provide no parental care. Kits are born with a short creamy-white coat, which they shed before emerging from the den. They are weaned at nine to ten weeks old. Kits remain in a den until they are around two months old. At that age they venture out of the den, and their mother begins to leave them at rendezvous sites while she hunts. She might also move them to a den site closer to a food cache. By around June (at 12 weeks old) they

begin travelling with their mother as she hunts, and a month later they begin to range alone and hunt for themselves, although they remain on their natal home range. Kits reach full adult size by the time they are eight to ten months old. Juveniles sometimes remain with their mother until they reach sexual maturity, but the young males especially tend to disperse when they are around one year old. Young males reach sexual maturity at two to five years old. Females may become sexually mature as young as 15 months old and breed for the first time as two-year-olds, but in most regions they do not mature before three to five years old. This may be a result of diet, primarily food availability, as well-fed females may mature earlier. Mature females typically do not breed every year or even every other year. Their reproductive potential is related to their individual nutritional levels and the availability of suitable rich, but isolated, habitat for raising kits. Kit survival is closely correlated to the availability of small game in their first summer.

BEHAVIOUR

Adult Wolverines are generally considered to be solitary, except briefly during the mating season and while females are raising kits. Males may accompany a female for several days while she is in heat. Like other carnivores that are solitary and occupy a large range, Wolverines rely on scent communication to keep track of each other. Wolverines mark mainly by urinating, biting trees or roots, defecating, scratching the ground, and rubbing their abdominal glands on the trees they climb. Scat and urine are deposited in prominent locations such as on rocks, trees, and fallen logs to mark home range and are sometimes added to the scat piles left by other animals, such as Coyotes and foxes. Planter glands on the hind feet are thought to leave a scent signal each time the animal places it onto the ground, a remarkably efficient means of marking for a wide-ranging species. Wolverines regularly mark a carcass to prevent other animals from eating it. The use of anal gland secretions in scent marking is still a matter of some contention. Many researchers feel that these secretions are used primarily when the animal is threatened or highly fearful (such as when it is attacked or caught in a trap) and that urine is the primary fluid used in scent marking. Considerable time and energy is devoted to scent marking, and Wolverines regularly diverge from their route to mark objects. They are believed to defend their home range from other Wolverines of the same gender, although the way in which this is accomplished is uncertain as most separation appears to be by mutual avoidance, and physical fighting has not been observed. Wolverines seem to have an insatiable need to keep moving. They typically travel 30–40 km while foraging (except for lactating females who stay close to their kits), at an average speed of about 8 km/h, but travel may be temporarily curtailed to take advantage of an abundant localized food source. Most immature females establish their home range next to, or even within, part of their natal range; if their mother dies or shifts her range, a young female from her last litter will take over her old home range. Some young females will disperse at greater distance, but young males are more likely to do so.

Kills are made with either a neck bite or, less commonly, a choke hold. Like other weasels, Wolverines cache surplus food. This may

Lope

Left hind track
Length: 9.0–15.2 cm
Width: 8.3–13.3 cm

Left front track
Length: 9.0–16.0 cm
Width: 9.0–13.3 cm

Wolverine

be buried under snow, ice, or soil, hidden under rocks, or stored up in a tree. Caches may be as simple as a single bone covered with a scuff of snow or soil or as complex as a series of small caches radiating from a central kill site. Most kills of large ungulates are made by Grey Wolves or other large predators (including humans), and the Wolverine caches the remaining hide and bones left after the killer has departed. Small game cached by Wolverines was likely killed by the Wolverine itself. Food caches may subsequently be important during food shortages or to allow females to provide solid food to young without leaving them alone for lengthy periods. Non-maternal Wolverines do sometimes kill kits. It is not certain whether the killers are male or female, but this behaviour may account for the isolated location of most natal den sites.

VOCALIZATIONS

Very little has been reported on Wolverine vocalizations. They are said to growl, snarl, and woof.

SIGNS

Like the other weasels, Wolverines have five toes on each foot, and the inside toe is reduced and may not register in a track or may register more lightly than do the other toes. A track where the small first toe does not register can easily be mistaken for a dog track. The soles of the feet are heavily furred in all of the negative spaces between the pads, which can obscure the tracks to some degree. Hind feet are slightly smaller than front feet and leave a smaller track unless the heel registers. All toes are clawed, and claws may or may not register. A loping gallop is the usual gait, and it can be maintained tirelessly over long distances. They may shift to a slower pace in deep snow. Their hunched appearance and rolling gait are unmistakable. Resting sites are commonly located on an elevated spot where a wide view is possible. These may occur at ground level or in a tree. Resting sites in snow are characterized by a compacted oval depression approximately 43–58 cm long by 35–43 cm wide. Long guard hairs or a latrine site may be nearby. Usually an escape tree is close to both resting sites and caches in case Grey Wolves appear while the Wolverine is resting or eating. Birthing dens are typically dug into a persistent snow drift and have an entrance opening that is 25–35 cm in diameter. Scat is commonly full of bone chips and is a simple tube that typically measures 1.0–2.5 cm in diameter and 7.6–20.0 cm in length. Wolverines roll on other animals' trails and scent posts and will "beat up" a sapling in a manner similar to that of a Fisher. They leave their mark as they bite and roll on the small tree, commonly breaking branches and defoliating parts or almost all of it.

REFERENCES

Banci 1994; Banci and Harestad 1990; Banfield 1974; Blomquist 1995; Blus et al. 1993; Copeland and Whitman 2003; Dalerum et al. 2007; Elbroch 2003; Hedmark et al. 2007; Heinrich and Biknevicius 1998; Inman et al. 2004; Krebs et al. 2004; Lee and Niptanatiak 1996; Lofroth et al. 2007; Macdonald and Barrett 1993; Magoun and Copeland 1998; Magoun et al. 2007; Moisan and Huot 1996; Pasitschniak-Arts and Larivière 1995; Persson 2005; Persson et al. 2003, 2006; Petersen, S., 1997; Samelius et al. 2002; Slough 2003, 2007; Smith,

H.C., 1993; Tomasik and Cook 2005; van Dijk et al. 2008; Vangen et al. 2001; van Zyll de Jong 1975; Verts and Carraway 1998; White, K.S., et al. 2002; Wright and Ernst 2004a, 2004b.

Northern River Otter
also called River Otter, Nearctic River Otter, North American River Otter, Fish Otter, Canadian Otter

FRENCH NAME: **Loutre du Canada, *also* Loutre de rivière**
SCIENTIFIC NAME: *Lontra canadensis,* formerly *Lutra canadensis*

Despite their name, the Northern River Otters are not restricted to fresh water. They commonly forage in coastal marine waters although they stay near the shore and do not cross the open ocean. Northern River Otters are intelligent, playful, and curious mammals that, as juveniles, can make a game out of almost any activity.

DESCRIPTION
This large, highly aquatic weasel has a long, tapering tail, short legs with webbed feet, a broad head and muzzle, a thick neck, and a long, supple body. The tail comprises at least one-third of the total length and has a greater surface area than all the limbs combined. It is flattened from side to side in the last third of its length and is used as a rudder and to add thrust when the animal is swimming. Their pelage is thick, lustrous, and pale chestnut to dark chocolate brown on the back and sides, paler on the underside, and silvery to golden under the chin and throat and sometimes upper chest. Very old individuals may develop white-tipped fur. Albinos are rare. River Otters do not have a layer of insulating fat under their skin. Instead, they have a dense layer of underfur hairs that interlock and trap air. This air layer prevents water from touching the skin. Guard hairs are short (averaging 23–24 mm in length), and fur (including both guard hairs and underfur) achieves an average density of 57,833 hairs per cm² in the mid-back area. More northerly animals have the longest and most dense coat. Whiskers (vibrissae) are very long and droop downwards. The fur of the Northern River Otter is considered to be the most durable of the North American furbearers and is used as a durability standard. The ears are short, rounded, and tight to the head, making them relatively inconspicuous. Secretions from paired anal glands and planter glands on the hind feet are used for scent marking. The dental formula is incisors 3/3, canines 1/1, premolars 4/3, and molars 1/2, for a total of 36 teeth.

SIMILAR SPECIES
American Minks, while similarly aquatic, are much smaller with a finer head. Sea Otters are larger than River Otters. They have light fur on their head and spend most of their time out at sea, and so are rarely seen on land. Although Northern River Otters will forage in coastal marine waters, they are rarely seen far from shore, spending much of their time foraging along the shore and in tide pools.

SIZE
Size may decrease slightly from north to south along the Pacific coast, but not from east to west of the continent. Males are larger than females, about 5% longer in body length and about 15% heavier on average. The following measurements apply to adult animals in the Canadian part of the range.
Males: total length: 1050–1300 mm; tail length: 368–500 mm; hind foot length: 110–135 mm; ear length: 23–25 mm; weight: 6.1–10.4 kg.
Females: total length: 900–1280 mm; tail length: 320–485 mm; hind foot length: 108–130 mm; ear length: 18–23 mm; weight: 6.1–10.0 kg.
Newborns: weight: about 130 g.

RANGE
Before European settlement in North America, Northern River Otters were widely distributed over most of the continent, except for some of the driest portions of the south and southwest. They were

Northern River Otter (*Lontra canadensis*)

Distribution of the Northern River Otter (*Lontra canadensis*)

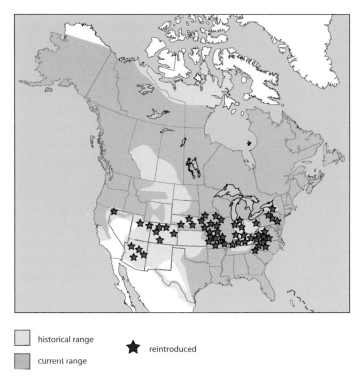

historical range

current range

★ reintroduced

food chain. As such, they have recently been recognized as useful indicator species to assess the environmental health of aquatic ecosystems. Population density fluctuates, in large degree because of the activities of North American Beavers that provide wetland habitat. If these beavers are trapped out in an area, the number of otters usually also drops.

ECOLOGY

These sleek aquatic weasels are active all year-round and may be seen during the day but are mainly nocturnal and crepuscular. They are somewhat more diurnal in winter. Northern River Otters occur near waterways, swamps, marshes, bogs, lake shores, and marine coasts. They are accomplished swimmers and divers, able to reach depths of 18–20 m, and remain submerged for up to four minutes as they hunt underwater. Their top swimming speed is 11 km/h. Depending on their swimming velocity, they may use two or four feet as well as their tail for propulsion, but the forelimbs are mainly used for turning. Northern River Otters are negatively buoyant and must continuously tread water or scull their tail to remain at the surface. Fishes are normally caught either by ambush or, less commonly, by direct pursuit. In the absence of fish, otters will roll or push over submerged rocks to find hiding prey. They will also stalk floating birds from below, seizing them and pulling them under. Although primarily found near water, some animals (mainly juveniles) may disperse up to 200 km, and even mountain ranges are not considered complete barriers. As a general rule, juvenile males disperse further than do juvenile females, many of whom remain near their natal home range. A top speed of 24–29 km/h can be achieved by running or by running and sliding (see the "Signs" section for an explanation of sliding). A family group's daily movements cover typically less than 10 km and vary with season, being shortest in winter. An otter's home-range size varies with food availability, weather, season, and the sex and age of the otter. Typically, the home range of a male will overlap the home ranges of several females, as well as those of other males. Lactating females have the smallest home ranges, and mature males generally have the largest. The ranges of both sexes shrink in winter. In northern Alberta, where home ranges are large as the seasons are severe and prey density is sparse, males occupy around 200 km² and females about 70 km². Marine habitats are usually richer, and so home ranges are smaller. In Alaska around the Prince William Sound area, males occupy around 20–40 km². The relative density of Northern River Otters may be as high as 1.25–3.9 otters per km of marine shoreline in prime habitat but is usually much less in freshwater habitats. They continue to hunt aquatic prey under the ice and make use of spring upwellings, swift-current areas, and openings in the ice created by other animals (primarily beavers and muskrats) to gain access to the water.

Northern River Otters use a variety of temporary dens as they move about their home range. These are typically created by other species, especially beavers and muskrats but also Woodchuck, Nutria, and foxes, or they are natural sites adapted by the otters in rock crevices, caves, logjams, undercut riverbanks, and cavities in tree roots or fallen woody debris. They rarely create their own burrows. Preferred den sites have a submerged entrance, such as in a beaver or muskrat

found as far south as the Rio Grande region of northern Mexico and as far north as Alaska and mainland Nunavut. By the mid-1970s the southern Canadian Prairies, Prince Edward Island, and most of the mid-western United States no longer supported Northern River Otters. Human settlement, overharvesting, and destruction of wetland habitats were largely responsible. This era was followed by several decades of reintroductions, which continue today. Current distribution is similar to the historical range, thanks to trapping controls as well as many successful reintroductions, but population densities are seriously reduced in many now-marginal areas, especially in the southern parts of the continent.

ABUNDANCE

Northern River Otters are rare to extirpated in southern Alberta and Saskatchewan and in Arizona, Colorado, Indiana, Iowa, Kansas, Kentucky, Nebraska, New Mexico, North Dakota, Ohio, Oklahoma, South Dakota, Tennessee, western Texas, Utah, and West Virginia. Reintroduction efforts in central Alberta, Pennsylvania, Arkansas, Missouri, and North Carolina have successfully sustained or re-established populations in those regions. Northern River Otters are still relatively common in appropriate habitats across most of Canada and Alaska, in the Great Lakes states, the Atlantic seaboard, Gulf of Mexico states, and the Pacific Northwest. Considered "Endangered" in Mexico by the IUCN, they are mainly extirpated there. This animal is an important North American furbearer, and 20,000–50,000 are harvested annually. Northern River Otters are sensitive to pollution and have disappeared from heavily polluted waterways. They readily accumulate heavy metals, organochlorines, and other chemicals owing to their position near the apex of the

bank burrow or an abandoned beaver lodge. Birthing dens may be some distance from water as the females appear to select locations that are above danger from flooding. During winter, snow cavities provide den sites. The nesting area is generally lined with dried vegetation, pieces of bark, and perhaps some animal hair. Otters are not subject to serious predation, but they are vulnerable to American Alligators, American Crocodiles (in the southern United States), and Killer Whales while swimming or more often to Bobcats, Cougars, Grey Wolves, Coyotes, and Domestic Dogs while on land. Most mortality is caused by humans, by trapping, accidental trapping in traps set for other species, vehicle collisions, illegal shooting, pollution, or wetland destruction. These otters can live to 13 years old in the wild and up to 25 years old in captivity.

DIET

Northern River Otters are carnivorous, eating mostly fishes (piscivorous), but they add crustaceans, frogs, snakes, turtles, molluscs, insects, birds, bird eggs, and the occasional mammal (mainly muskrat) to their diet when these are available. Their winter diet consists primarily of fish. They forage in fresh, salt, and brackish waters. The species of prey that they consume varies widely across the range as they are opportunistic hunters, eating whatever they can catch. Some otters favour certain fish species over others and will actively search for and select those species, despite their scarcity. Most of the fish they prey upon are slower-moving species or those that fatigue rapidly. They are quick to exploit an injured or weakened fish or those spawning in concentrations. Fish hatcheries can suffer huge losses if an otter gains access, because the small fish are densely packed and hence highly vulnerable to predation. Otters forage along the water's edge, searching places such as undercut banks, logjams, tidal pools, and under overhanging vegetation where fishes congregate or retreat for shelter. In some regions (such as Prince William Sound in Alaska) groups of otters have learned to coordinate their efforts, ensuring that all the individuals achieve a better quality diet, but cooperative hunting is not common for this species.

REPRODUCTION

Each mature female Northern River Otter comes into heat for 20–56 days, with peaks of receptivity about every six days, usually between December and April but occasionally as late as June. Otters in the northern, colder parts of the range may only reproduce every two years, but those in more southerly, warmer locations may breed annually. Males especially are promiscuous and seek more than one mate during each breeding season. A mating male grips the female at the nape of her neck with his teeth and may leave a noticeable wound. Copulation usually occurs in the water and may last 13–74 minutes. Northern River Otters are induced ovulators, and so the females must be sufficiently stimulated during intromission to trigger ovulation. Likely the longer copulations reported were multiple, with rest periods in between, while the animals remained locked. Following a successful mating, the fertilized eggs develop to the blastocyst stage and then enter a period of suspended animation as they float freely in the female's uterus for at least eight months. Once implantation occurs, active gestation takes 25–30 days. Active gestation is triggered by photoperiod. The entire pregnancy, including the delayed implantation phase, can last 10–12 months; many mature females are pregnant for almost all of their adult lives, as they have a post-partum heat and breed again soon after parturition.

Young (called kits or pups) are born as the temperatures moderate in spring, which, given the many climatic zones within the otter's vast geographic range, can be between February and May. Litters are normally composed of one to three kits (ranging from one to six). Younger females tend to have smaller litters. Newborns are sparsely furred, blind, and toothless but already have vibrissae that are about 5 mm in length, and well-formed claws. Their eyes open at 22–35 days old. The kits begin playing with each other at five to six weeks old, and they start eating solid food delivered by their mother at around nine weeks old. Weaning occurs at around 12 weeks old, but the females will continue to provide food to the kits until they are 37–38 weeks old, although the kits increasingly become capable of hunting for themselves as they mature. The youngsters go for their first swim at around eight weeks old. In the north, where mature females breed in consecutive years, youngsters disperse sometime from early winter of their first year to early winter of their second year, although occasionally one or more remain with their mother through the subsequent breeding season to help raise her next litter. In southern populations where mature females breed annually, the young disperse before the following breeding season, although again some may remain to act as "helpers." Juvenile females normally breed for the first time when they are 21–24 months old. Less common (and usually in warmer regions, including sometimes in southern British Columbia), a two-year-old will produce a litter, which means that she bred at 13–16 months old. Males in the northern parts of the range (including most of Canada, except southern British Columbia) become reproductively active in their second year, at around 21–22 months old. In southern parts of the continent they may be as young as eight months old when they produce viable sperm. Full physical maturity of both sexes is reached at three to four years old.

BEHAVIOUR

Northern River Otters are more social than most weasels, and they have a complex social system that varies across the range. Most social interactions take place between a maternal female and her young of the year, possibly with the addition of one or more helpers that may be members of a previous litter, unrelated adults, or yearlings. The

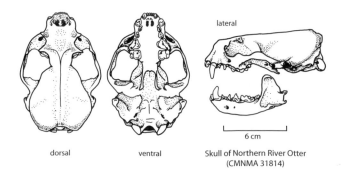

lateral

6 cm

dorsal ventral Skull of Northern River Otter
(CMNMA 31814)

helpers serve as companions to the kits but do not provide food or assume parental duties in the mother's absence. Apart from such "family" groups, adult otters tend to be solitary, except in some regions where "clans" composed largely of adult males (numbering 2–30) may associate to occupy a larger-than-normal home range and possibly engage in cooperative hunting. Males comprising the clans may or may not be related. Such clans are commoner in marine settings where higher-quality pelagic fishes can be caught by a coordinated group, but they may also occur in freshwater habitats. Even in regions where clans and large family groups are the norm, some adult and subadult otters are solitary. Most maternal females actively exclude mature males from their home range while they are caring for dependent kits. In some regions they also aggressively bar other mature females from their home range all year-round unless they are close relatives (mother, daughter, or littermate sister). At times, battles between mature females can be fierce to the point of death for one of the combatants. In other regions, females are much more social, excluding neither unrelated females nor males.

Scent marking is an important communication tool used by otters. A group of four planter glands on the sole of each hind foot probably leaves a scent signature any time the animal walks on land. River Otters, like many other weasels, have a pair of anal glands that produce a pungent exudate that is used for ritualized scent marking and may be released if the animal is in distress. Scent posts are maintained throughout each otter's home range and appear to function not as territorial markers but as a warning to neighbouring otters of each other's whereabouts so that contact can be avoided. An otter will regularly deviate from its path to refresh a scent post with scat and anal gland secretions. Fur maintenance is achieved mainly by rubbing and rolling in sand, snow, or vegetation, followed by vigorous scratching. This behaviour is commonly performed immediately upon emergence, to remove water from the fur. Mutual grooming may occur within "family" groups. Northern River Otters use few visual displays. They will make a threat face at one another, which consists of the ears being flattened back and the mouth agape with the teeth bared. The most evident visual display is called the latrine dance. The marking otter assumes a characteristic arched back, undulates its tail, alternately scratches, paws, and treads on the latrine site and then produces a stream of faeces. Other otters present will then advance to smell the faeces and repeat the dance.

VOCALIZATIONS

A low, purring grunt or a low-frequency chuckling may be emitted by otters at play or travelling in a group. Long-distance communication is accomplished using a high-pitched chirp. Their alarm call is a loud, explosive snort. A disturbed otter may bark or hiss, and a shrill whistle or scream is produced by an injured otter or one under extreme duress. While mating, females may caterwaul as they might also do if approached by an unwanted male.

SIGNS

The Northern River Otters will often walk, but when they want to move quickly, they actually use a variety of bounding gaits and can dramatically arch their back to extend stride length. Two bound

stride length of
a bound varies
from 15 to 75 cm

Bound

Lope

Left hind track
Length: 5.4–10.2 cm
Width: 5.4–9.5 cm

Left front track
Length: 5.4–8.3 cm
Width: 4.8–7.6 cm

Northern River Otter

variations are illustrated. In a walk the hind foot (on the same side) implants at the back edge of the front track. In a bound, all four feet usually leave distinctly individual tracks in a close group of four, but in one type of bound the hind tracks register directly over the front tracks, thereby creating pairs of offset tracks. The tail tip is usually held stiffly off the ground in a slight arch but may touch the ground, especially during a walk or in moderate to deep snow. Each foot leaves a five-toed track, with the inside toe (the first toe) on the front feet being somewhat reduced and smaller. The first toe on the hind foot usually angles inward. Both front and hind feet are webbed, but the webbing most frequently registers in a hind track. Claws may or may not leave an impression, and when they do, they typically register at or very near the tip of the toe and merge with the toe-pad impression. Lack of fur in the negative space between the toes is characteristic of an otter track. Unlike those of most other weasels, the front feet of an otter are smaller than their hind feet.

Sliding on snow, ice, and sometimes mud is also characteristic of Northern River Otters. They drop to their belly, tuck their front legs tightly against their chest, and push with their hind legs. Summer slides are usually seen on a muddy bank that slopes down to the water and are 15–30 cm or more in width, owing to the repetitive use. As the otters play on the slide, water from their coat makes the mud even more slippery. On snow and ice, sliding may be simply a game or it may be used as a way to travel quickly, while saving energy, as it is alternated with running or bounding. Such slides are typically short, around 3 m long and 15–25 cm in width, and are interspersed with tracks in a dot-dash pattern. A long slide on slippery ice could be up to 7.5 m long. Shore-edge haul-outs are worn, often muddy areas where the otters habitually emerge from the water. Invariably, a roll site will be nearby, distinguished by flattened vegetation, snow, or sand where the animal rolled to dry its coat. Roll sites are a fairly common sign if otters are around.

Latrine sites, also called scent posts or marking sites, usually occur nearby on rocks, logs, protruding hummocks, or along trails that bisect a narrow isthmus that juts into the water. Any place that protrudes above the water may be selected. These sites are used over extended periods and typically can be identified by the accumulation of fish scales from the dissolving scats and by the dead and dying vegetation that has been scalded by the acidic urine and scat. Trackers call such locations brown-outs. These sites may be 5–20 m across. Otter scat is dark, often blackish. Scats composed of crayfish, mammals, or birds may be somewhat tubular, but the more common fish scat is looser, varying from an amorphous patty to a squirt, and is defined by the fishy odour and preponderance of fish scales. Another material often found at marking sites is a gooey, whitish to yellowish substance that is believed to originate in the anal glands. It may be deposited at latrine sites and on scrapes. Scrapes are areas where the soil, snow, or vegetation is scratched into a shallow mound upon which is deposited scat or anal-gland secretions, or both. Otter dens are commonly located near a latrine site and typically have an entrance opening of about 20 cm. Mussels or clams predated by River Otters can sometimes be seen scattered along the shoreline. These commonly retain evidence of the otter's canine teeth where the shell was punctured. Trails of footprints

often include more than one animal, as otters are relatively sociable, and the young may stay with their mother for up to two years and sometimes longer.

REFERENCES

Blundell et al. 2002; Ceballos, G., 1999; Cote et al. 2008; Crait et al. 2006; Elbroch 2003; Gorman et al. 2006a, 2006b, 2008; Haskell 2006; Hodder and Rea 2006; Jameson and Peeters 2004; Johnson, S.A., and Berkley 1999; Koepfli and Wayne 1998; Kurta 1995; Larivière and Walton 1998; Melquist and Hornocker 1983; Melquist et al. 2003; Norment et al. 1999; Price and Aries 2007; Reid, D.G., et al. 1994a, 1994b; Rock et al. 1994; Scott, S.J., 1997; Serfass 1995; Smith, H.C., 1993; Stevens, S.S., and Serfass 2005; van Zyll de Jong 1972, 1987; Verts and Carraway 1998; Weisel et al. 2005; Youngman 1975.

American Marten

also called Canadian Sable, American Sable, Pine Marten

FRENCH NAME: **Martre d'Amérique**

SCIENTIFIC NAME: *Martes americana*

American Martens continue to play an important role in the North American fur trade, in which they are commonly called American or Canadian Sable to distinguish them from the European Sable, a similar and closely related weasel.

DESCRIPTION

This medium-sized weasel is a little larger than a mink and has the typical weasel shape: a long sleek body, short limbs, prominent round ears, and a pointed face. Its fur may vary in colour from pale gold to chestnut brown through light chocolate brown to almost black. Normal coloration is dark brown. The legs and tail are usually darker than the body. Winter pelage is longer, more luxuriant, brighter, and browner than the summer coat, which is shorter, thinner, and typically greyer and darker. A large cream, yellowish, or bright orange throat and chest patch is a diagnostic characteristic. Some martens do not develop the throat patch or have only a few small spots rather than a whole bib. Adults commonly develop lighter fur around their head, and white edges on their ears. Like other weasels, both sexes of American Marten have a pair of scent glands, one on each side of the anus. American Martens have a glandular area on their lower abdomen that exudes an oily secretion, which is used for scent marking and produces the characteristic musky "marten odour." Shorter, darkly stained hairs over the gland reveal its location. To deposit the scent they simply drag this area over logs or branches as they travel. American Martens are highly arboreal and have semi-retractable claws to help them climb and hang onto tree branches. Furthermore, their hind limbs can be rotated at the ankle

American Marten (*Martes americana*)

(like those of Grey Squirrels) to allow rapid head-first descent. Their long, bushy tail helps them balance in the treetops. Martens have large feet relative to their body size, which aid travel on snow. The dental formula is incisors 3/3, canines 1/1, premolars 4/4, and molars 1/2, for a total of 38 teeth. American Martens experience two moults each year, one in early summer and the other in autumn.

SIMILAR SPECIES

The American Marten is slightly larger and lighter in colour than the North American Mink. The mink lacks the characteristic large, orange bib of the marten although it often has a white chin patch of variable size. The mink is closely associated with water, unlike the marten, which is a forest dweller often seen up in a tree. As mink are semiaquatic, most sightings are of animals with wet spiky fur, not the dry, plush coat of a marten. A larger weasel, the Fisher, may be mistaken for an American Marten; the smaller, female Fisher especially, is close in size to a large male marten. Apart from being larger than American Martens, Fishers are so dark brown as to appear almost black, albeit usually with a lighter head.

SIZE

Size varies considerably across the range, with the smallest forms being in the south and the largest generally in the northwest. Males are considerably larger than females, about 15% longer in body length and up to 65% heavier. The following measurements apply to adult animals in the Canadian part of the range and cover the variation in size across the country.

Males: total length: 551–800 mm; tail length: 152–230 mm; hind foot length: 70–105 mm; ear length: 35–51 mm; weight: 700–1400 g.
Females: total length: 475–652 mm; tail length: 135–218 mm; hind foot length: 59–98 mm; ear length: 34–43 mm; weight: 400–900 g.
Newborns: weight: about 28 g.

RANGE

Previously found throughout the North American boreal forest region, American Martens no longer occupy many areas on the southern edges of the former range. They have been either driven out owing to habitat loss (largely because of logging) or eliminated by excessive fur trapping. They no longer occur in southern Ontario, Prince Edward Island, Anticosti Island, southern New Brunswick, and most of New England (except in northern parts of Maine, New Hampshire, and Vermont). Although American Martens have been exterminated in mainland Nova Scotia, recent reintroduction efforts in the southern mainland appear to have been at least partly successful. The population in northern Cape Breton is slowly declining and much reduced. Efforts are ongoing to bolster this remaining population with genetically similar animals from New Brunswick. Formerly found throughout Newfoundland, these weasels now occur in only a small area of the western part of the island. A second successful population has recently been established in the Terra Nova area, using relocated breeding stock captured from the more westerly Newfoundland population. There are some indications that martens are slowly beginning to repopulate suitable former habitat in central Saskatchewan, northeastern Minnesota, and adjacent Ontario.

ABUNDANCE

Before regulation, many American Marten populations were over harvested. There are likely 300–600 still surviving in Newfoundland, where they have been protected since 1934 and are federally listed as "Endangered." This population has apparently been more or less stable over the past 20 years. Captive breeding efforts are currently underway. Recent estimates from Nova Scotia suggest that there may be fewer than 50 animals remaining in Cape Breton, despite several reintroduction attempts. Earlier reintroductions in

Distribution of the American Marten (*Martes americana*)

- □ historical range
- ■ current range
- ★ introduced or reintroduced

Martens are bold and ferocious hunters. They are capable of catching and killing a Snowshoe Hare, which is considerably larger and heavier than they are. Martens are also known to kill Yellow-bellied Marmots up to a weight of 4 kg. Most carnivores that hunt larger prey do so with the help of a pack, but martens are solitary hunters. Although they are very skilled and agile tree climbers, capable of pursuing and catching arboreal species such as Red Squirrels and flying squirrels, most of their hunting takes place on the ground. Martens are small enough to hunt under the snow in the subnivean spaces created by woody debris, where small mammals can be plentiful. They also use these areas for den sites, as the protective blanket of snow moderates the outside conditions and provides a warmer environment. The size of their home range varies and is affected by factors such as climate, habitat quality, body size, and prey availability. The home-range size of males tends to be about twice that of females. A recent study in Labrador, using radio-tagged animals, discovered average home ranges of 45.0 km² for males and 27.6 km² for females. In more southerly locales, home ranges appear to be generally smaller; for example, in a study in Maine, males averaged 2.6 km² and females 2.0 km²; another study, in Ontario, showed that males averaged 5.2 km² and females 3.2 km². Martens in Newfoundland fit somewhere in the middle, as males occupy, on average, around 30.0 km² and females 15.0 km². The home ranges of male and female martens tend to have a small degree of overlap, likely imposed by the larger male so that he can more easily assess her reproductive condition. Again in Ontario, a study reported a density of 0.4 martens per km² of appropriate habitat during a period of prey scarcity, compared with 2.4 martens per km² when prey was abundant. Dispersal occurs among adults on occasion but is more common in subadults, typically in autumn as they leave their mother's territory in search of their own unoccupied space. Greater dispersal distances occur during years of low prey density. Adults that abandon their territory owing to prey scarcity are known to travel up to 80 km. Martens are willing and capable swimmers and can even swim underwater. Predators of martens include humans, Red Foxes, hawks, owls, eagles, Wolverines, Coyotes, Cougars, Canada Lynx, Bobcats, Grey Wolves, and Fishers. The maximum lifespan in the wild is about 11 years, but few live longer than four years. Captives may reach 15 years of age.

Digby County, Nova Scotia, appear to have been successful, as accidental captures in that area occur regularly; trapping specifically for marten is illegal in the province. Population numbers on mainland Nova Scotia are unknown but are likely low. In Saskatchewan, martens are slowly starting to reappear in the more southerly boreal forests after having been eliminated from that region for many years. Across the Canadian range, American Martens are legally trapped under provincially and territorially administered licences. Numbers are increasing or remaining stable in most jurisdictions in Canada. In the United States, martens are now rare over most of their range, except Alaska.

ECOLOGY

American Martens are most abundant in old-growth coniferous forests in the boreal zone but may also occur in mixed old-growth and sometimes mature second-growth forests within the boreal zone, especially in the east. Populations in the suboptimal habitats tend to be sparser with high predation mortality and are easily eradicated by trapping pressures or declining prey availability. Martens prefer woodlands with plentiful undergrowth and deadfall and tend to avoid open areas such as large clearings, burns, or clearcuts. They will sometimes hunt in open boulder fields and talus slopes where there are ample crevices between the rocks. Martens are active all year and may be out foraging at any time of the day or night. They are estimated to be active for as little as 16% of a 24-hour day in winter and for about 60% of a 24-hour day in summer. Since martens store very little fat, they generally must hunt every day. An average-sized marten needs an estimated three voles per day. Populations of martens are largely regulated by food availability. Well-fed females typically produce larger litters, and more juveniles survive and breed earlier when prey is plentiful. Adequate prey availability in winter is especially crucial. American

DIET

Largely carnivorous, American Martens consume a wide variety of prey including small rodents, tree squirrels, chipmunks, pikas, woodrats, Snowshoe Hares, Yellow-bellied Marmots, shrews, carrion, bird eggs, and birds up to about the size of grouse. Small mammals such as voles are the dominant prey. Martens will also eat amphibians, snakes, insects, earthworms, fish, and fruit and nuts in season. Martens can be flexible in their prey choices to adapt to fluctuations in prey numbers and can permanently adapt to regions with non-typical prey alternatives. On the Queen Charlotte Islands, for example, where there are no voles, the martens eat more birds, carrion, and fish; on Newfoundland, where voles are less available in winter, the martens switch to a wide variety of other prey choices.

REPRODUCTION

Timing of the breeding season varies across North America. It occurs early (late June to July) in Alaska, mid-June to mid-August in Alberta, and mid-July to early September in New York State. Most breeding in Nova Scotia takes place in July. The breeding season lasts about three to six weeks. Mating takes place on the ground or in the trees. Ovulation is likely induced, and the fertilized egg floats freely in the uterus (called delayed implantation) for 190–250 days. Implantation occurs about 27 days before the birth of the young, and births occur from mid-March to late April. The timing of implantation is governed by photoperiod. The entire gestation period, including the delayed implantation stage, ranges from 220 to 276 days. One litter at most is produced annually. Litter size varies from one to five and averages three kits. The young are typically born in a den in a tree cavity, large hollow log, or rock cavity. They are deaf and blind but already have a thin coat of fine yellowish hairs. This coat is moulted at around 21 days old, and they begin to grow the dark subadult coat that they will wear after weaning. Their ears open at around 26 days old, and their eyes open at 39 days old, at which time they are already eating solid food delivered by their mother. Weaning usually occurs when the kits are around six to seven weeks old. The juveniles then remain with their mother until they disperse in late summer or early autumn, at which time they are near adult length but not weight. A maternal female commonly uses more than one den site while raising young, and, once the kits are weaned, she moves them regularly. By following her around, they learn to find and catch prey, and by staying on her territory, they are protected from harassment by other territorial martens until they are large enough to secure their own territory. Well-nourished juveniles may breed for the first time as yearlings (around 15–16 months old), but in a population suffering from declining prey the juveniles may not breed until they are two years old (27–28 months old). The fertility rates of mature females are in large part governed by food. When prey is plentiful and the females are well fed, 90%–100% of them in a population will become pregnant annually; when prey is scarce, as few as 30% or even less may become pregnant, and their litter size also drops. Delayed implantation is an efficient strategy that allows a female to fertilize a large number of eggs (up to around nine in martens) long before there is any way of knowing how plentiful the prey will be when the young are born, eight or nine months later. Depending on her nutritional status at the implantation stage, which is only about 27 days before the young are born, more or fewer of the fertilized blastocystes will implant. Unless something drastic affects the prey populations during this short active gestation phase, female weasels will then produce the number of offspring that they are most likely to be able to successfully rear, given the amount of food available.

BEHAVIOUR

American Martens are solitary as adults, apart from mating pairs and females with kits. Adults can be very aggressive in defence of their territory, and territorial battles, especially between males, may lead to serious injury and even death. This aggression is only somewhat moderated between males and females during the breeding season. An oestrous female uses scent and sound to attract a mate. She urinates frequently and rubs her abdominal gland on rocks and branches to signal her location and receptivity. She also produces a chuckling call, which is also produced by mature males calling to females. Courtship in the wild is most commonly described as a chase by the male and attempted evasion by the female. Males can detect the reproductive state of the female (likely by scent) and will follow her, sometimes for several days, as she nears oestrus. A copulating male grasps the neck of the female with his teeth to hold her in position. Screams and growls, described by observers as similar to a cat fight, are common vocalizations during courtship and mating. Females determine the length of a mating and may bite the male as she departs. Females may mate several times with the same male if he is persistent and able to follow her, and may also be willing to mate on consecutive days with the same or different males. Males likely visit and mate with multiple females during the breeding season and are not known to provide any parental care to the young.

Martens find protected dens and resting sites throughout their home range to seek shelter and remain warm, especially in winter. They use a variety of locations, including hollow logs and stumps, tree cavities, rock crevices, rock and debris piles, natural cavities beneath the snow, in deadfall that is part of squirrel middens, excavations around tree roots, and in tree snags. They can even become habituated to human habitation if undisturbed, and are known to make use of under-utilized backwoods cabins as resting and possibly maternal dens. Severe or stormy weather often causes martens to den-up for the duration. An excited or anxious marten tends to erect the hairs on its tail, creating a "bushy" tail. Martens are very curious and quick to investigate unusual objects in their environment, including traps. This tendency makes them relatively easy to catch in a trap and, without appropriate controls, relatively easy to unwittingly eradicate.

VOCALIZATIONS

American Martens can be vocal at times. Huffs, pants, snarls, growls, chuckles, whines, screams, and "ccps" have been reported. During the breeding season adults will produce a throaty chuckling call, presumably to attract the attention of the opposite sex. A maternal female later uses the same call when calming or consoling her kits. Most of these calls have been noted from captive animals, so the repertoire of wild martens is likely wider.

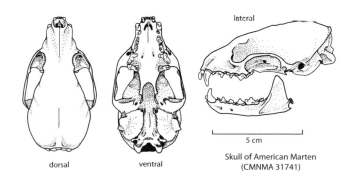

dorsal   ventral

lateral

5 cm

Skull of American Marten
(CMNMA 31741)

stride length of a bound can vary from 25 to 90 cm

Bound

Lope

Left hind track
Length: 3.8–7.0 cm
Width: 3.0–5.7 cm

Left front track
Length: 4.1–7.0 cm
Width: 3.3–6.7 cm

American Marten

SIGNS

Both the front and the hind feet have five toes, but the first toe on the inside of each foot (the thumb and the big toe) is much reduced and tends to register only in soft substrate. The first toe on a hind foot is further back than the comparable toe on the front foot, making a hind track more asymmetrical. The furred heel may leave an impression, especially in a soft substrate such as snow. In winter the spaces between the pads of the feet are heavily furred, which may blur the track. Claws may or may not register. A typical gait in snow is a lope in which the hind feet land in the track left by the front feet, which results in offset pairs of tracks with one side more advanced. They will also bound, especially when snow is deep. The stride length varies from 25 to 90 cm and occasionally up to 1.8 m. Other possible gaits are a walk (sometimes used when foraging) and a gallop. Fur and bone scats are long and thin and look twisted, bent, and very pointed at both ends. They may be long and folded over or arranged in a circle or semi-circle. Fruit and mast scats are tubular with little twisting and somewhat pointed ends. Scat varies depending on the diet, from 5.0 to 12.7 cm in length and from 0.5 to 1.6 cm in diameter. Droppings are commonly used as scent markings and deposited in prominent locations such as a distinctive rock or log. Several scats may be found together at such sites. Martens are energetic hunters and commonly criss-cross their own trail as they check potential prey sites. They will tunnel into soft snow. The diameter of a snow tunnel is about 7.5–8.5 cm. A marten jumping into snow from a tree may leave a full body impression. A body print without the tail varies from 20 to 35 cm long by 10 to 20 cm wide. Martens often use culverts to cross roadways. They commonly move large prey items to more secluded locations, such as under the snow, and trails with drag marks are not unusual. Entrances to snow tunnels are 7.6–8.3 cm in diameter. Distinguishing tracks made by American Martens from those made by Fishers can be difficult as they are similar, and a large male marten overlaps the size range of the smaller female Fisher. A clear track of a front foot is most useful in distinguishing the species. A high percentage of marten tracks do not register the small thumb (pollex), whereas the heavier Fishers usually do leave a thumb impression.

REFERENCES

Bissonette and Sherburne 1993; Bowman and Robitaille 1997; Bull and Heater 2001; Clark, T.W., et al. 1987; Drew and Bissonette 1997; Dumyahn et al. 2007; Foresman and Pearson 1999; Forsey et al. 1995; Fryxell et al. 1999; Gosse and Hearn 2005; Gosse et al. 2005; Grant and Hawley 1991; Henry and Ruggiero 1993; Holyan et al. 1998; Jameson and Peeters 2004; King and Powell 2007; Nagorsen et al. 1991; Powell, R.A., et al. 2003; Powell, T., et al. 2007; Price et al. 2005; Ruggiero and Henry 1993; Ruggiero et al. 1998; Smith, A.C., and Schaefer 2002; Smith, H.C., 1993; Swanson et al. 2006; Thompson, I.D., and Colgan 1987, 1994; Verts and Carraway 1998; Youngman 1975; Zielinski and Duncan 2004; Zielinski and Truex 1995; Zielinski et al. 2001.

Fisher

also called American Sable, Fisher Cat

FRENCH NAME: **Pékan**

SCIENTIFIC NAME: *Martes pennanti*

Probably named for their resemblance to a dark-phase European Polecat (sometimes called a fitchet), Fishers are now stuck with an inappropriate English name as they do not hunt or catch fish, although they will sometimes eat fish that they find washed ashore.

DESCRIPTION

Fishers display the typical weasel body shape with a long torso, short legs, large rounded ears, and a pointed face. They also have a long bushy tail that makes up about one-third of their total length. The ears are set close to the head and do not protrude as much as those of some of the other weasels. Their dense, luxurious fur is typically chocolate brown, with dark brown or black face, legs, tail, and often rump. Most animals develop considerable grizzled-looking, pale-brown highlights around the head, shoulders, and back, and around the rims of the ears. Most individuals have irregular and variably sized white or cream markings on their chest and lower abdomen. Seen in the forest, most Fishers appear essentially black. Arboreal adaptations include semi-retractable claws and ankle joints that rotate so that they can descend trees head-first. Fishers of both sexes have circular glands in the middle of the pads of the hind feet. These glands are visible as a patch of coarse hairs and emit a noticeable, distinct odour. They enlarge during the breeding season and are thought to deposit scent markings as the animal travels. The dental formula is incisors 3/3, canines 1/1, premolars 4/4, and molars 1/2, for a total of 38 teeth. Fishers experience a single yearly moult that begins in late summer and is complete by November or December. They have a pale-green eyeshine.

SIMILAR SPECIES

American Martens are slightly smaller, with more pronounced ears and a generally lighter, more reddish coat. The fur colour of the North American Mink is similar to that of a Fisher, but minks are considerably smaller and are usually associated with water, unlike Fishers, which are primarily forest animals.

SIZE

Size varies slightly across the range, with the smallest forms in the southeast and the largest generally in the northwest. Males are considerably larger than females, about 15%–25% longer in body length, and about twice as heavy on average. The following measurements apply to adult animals in the Canadian part of the range and cover the variation in size across the country.

Males: total length: 900–1200 mm; tail length: 368–422 mm; hind foot length: 100–135 mm; ear length: 40–59 mm; weight: 3500–5500 g on average, with exceptional animals being over 9000 g.

Females: total length: 750–950 mm; tail length: 340–380 mm; hind foot length: 89–120 mm; ear length: 40–48 mm; weight: 2000–2664 g.

Newborns: weight: 40–48 g.

RANGE

Fishers are endemic to North America. Historically, they occurred from Nova Scotia to British Columbia, northwards to extreme southern Yukon and Northwest Territories, southwards in the east along the Appalachians to Tennessee and in the west along the Rocky Mountain chain to southern California. They are most abundant in the eastern part of the continent, especially in Canada. Current distribution is much reduced, particularly in the United

Fisher (*Martes pennanti*)

Distribution of the Fisher (*Martes pennanti*)

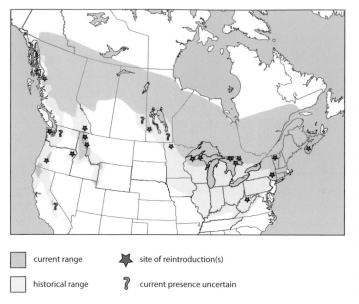

current range

historical range

site of reintroduction(s)

? current presence uncertain

per 1000 km², while similar estimates in eastern Canada can be 6–49 times higher, varying from 50 to 385 per 1000 km².

ECOLOGY

Fishers prefer forests with a diverse multi-layered canopy (mature trees, saplings, and shrubs), abundant large woody debris, large snags and cavity trees, and dense undergrowth, especially with a mix of woodland, wetland, and small clearings. Fishers in eastern North America do not appear to require any particular type of forest, as they occur in coniferous, mixed, and hardwood forests. Fishers in western North America (mainly in the Rocky Mountains) seem to have a more distinct relationship with old-growth coniferous forest habitat. Fishers regularly forage along forest edges and in riparian areas. While they may hunt in younger and more open forests, they favour older forests for resting and denning. Temporary or resting shelters may be in snags, hollow logs, tree cavities, abandoned raptor nests, marmot or Woodchuck burrows, brush piles, rock falls, and even dry abandoned beaver lodges. Maternal dens are usually in more protected tree cavities. In cold regions most Fishers rest and den under the snow during winter, typically in areas where coarse woody debris has created an air space. Fishers are active all year round and forage mainly in low-light conditions around dawn and dusk (crepuscular), although the energetic demands of cold weather or suckling young may require that they hunt at any time of the day or night. Most activity periods are about two to five hours long, followed by a rest period of at least 10 hours.

The home-range size varies widely and is smaller in winter. Typically, male home ranges are at least two or three times the size of female home ranges. Female home-range size and male winter home-range size is likely primarily determined by prey availability, while male summer home-range size is largely governed by their urge to breed as they may travel widely searching for oestrous females. In British Columbia, male home ranges average 74 km² in winter and 122 km² in summer, while female home ranges average 25 km² in winter and 33 km² in summer. The home-range size of Fishers living in the east tends to be smaller because prey is generally more abundant. In Gatineau Park, near Gatineau, Quebec, male Fishers occupy average home ranges of 9.2 km², and females 5.4 km², while in neighbouring eastern Ontario, males occupy 11.0 km², and females 2.1 km². On a typical winter day a Fisher covers around 5 km while foraging, but they are capable of travelling long distances in a short period of time. Translocated animals (released in summer) typically travel more than 50 km, and one exceptional male in British Columbia travelled at least 1055 km after being moved from the Chilcotin River area to a release site near Quesnel.

Population density is difficult to assess and is likely highly variable as Fishers occur naturally over large distances in relatively low numbers. Their density is also related to prey abundance, and, as Fishers rely heavily on Snowshoe Hares, their numbers appear to be affected by the approximately 10-year cycle of the hares. Recent studies indicate that Fisher densities are much higher in a mixed agricultural landscape than in traditional Fisher habitats such as Algonquin Park, likely as a response to more diversified and abundant prey in the settled areas. Fishers are primarily terrestrial but can climb

States and southern Canada, compared to the historical distribution. Many reintroductions have been undertaken across North America, partly to control the resulting North American Porcupine overabundance but also to re-establish the species for future harvesting. The accompanying range map is more provisional than most as the Fisher is rapidly re-establishing itself in many regions (especially in the east) but may be gradually losing ground elsewhere (especially in the far west).

ABUNDANCE

Fishers are generally considered uncommon to rare in most of their western and northern range, but they are becoming increasingly commoner in many parts of their former range in eastern North America. Over-trapping and habitat loss, mainly because of logging, are the two major reasons for Fisher decline and extirpation in eastern and southern North America. Predator poisoning (dead animals are poisoned and left lying around for the predators to eat) that was aimed at Grey Wolves, Cougars, and bears also played an unintended role in Fisher declines since Fishers are particularly susceptible to this form of predator control; carrion is a major component of their diet. The logging boom came to an end in the 1930s, and, in the decades that followed, the forests and abandoned farmland in eastern North America began to slowly regenerate. That and the suspension of predator poisoning, the closure of trapping seasons in many areas (by the 1930s and 1940s in the United States), and the assiduous management of trapping licences in those jurisdictions that continue to permit trapping have allowed Fisher numbers to slowly recover in parts of the range. Fishers in Pacific states and British Columbia continue to decline. In 2003 British Columbia placed Fishers on the provincial "Red List," and they may no longer be legally trapped in that province. Washington State declared Fishers to be "Endangered" in 1998. The density of Fishers in the spruce forests of north-central British Columbia varies from 8 to 11 animals

trees to reach den sites or to catch prey. The smaller females appear to be more adept in the trees. Both male and female juveniles disperse similar distances from their natal range, although males may disperse somewhat earlier than do females. Dispersal distances of 10–132 km have been reported. The distance that the juveniles must travel to find a suitable place to live varies. Distances are notably shorter in heavily trapped regions where vacant territories are common. Populations are normally made up of residents (usually adults) with well-established home ranges and transients (usually juveniles but sometimes adults that have abandoned their home ranges for a variety of reasons such as inadequate prey or injury). Transients generally live on the edge of starvation in less familiar and often marginal terrain. They are more vulnerable to trapping and death by starvation and more likely to come into conflict with humans. Fishers have a low reproductive output, which means that populations suffering from years of over-harvesting and habitat destruction take a long time to recover. They are capable swimmers, readily crossing rivers. Fishers have few natural predators although they are occasionally predated by Cougars, eagles, Wolverines, Bobcats, Canada Lynx, Coyotes, and other Fishers. Direct human-related mortality is primarily due to trapping or road kill. The maximum life expectancy in the wild is 10–12 years, but very few Fishers live longer than five years. Females appear to outlive males.

DIET

Fishers are generalist, opportunistic hunters and scavengers, and they consume a wide variety of animals and plants. They will eat virtually anything that they can catch and kill, including hares, rabbits, birds, bird eggs, smaller weasels, shrews, Porcupines, Northern Raccoons, Domestic Cats, and unsecured small livestock, especially chickens. Most of their prey is composed of Snowshoe Hares, small rodents, and squirrels. Amphibians, reptiles, invertebrates, stranded fish, and carrion are eaten when available, and ungulate carrion is especially important in winter. Fruit and nuts in season are also part of their diet but are typically only consumed when prey is scarce. Fishers are one of only a few carnivores that regularly kill and consume Porcupines, although these prickly mammals are by no means their preferred prey. Rarely do Porcupines comprise more than 10% of the total energy budget of a Fisher, although Porcupines may provide an important and even essential alternative when hare populations crash. Both male and female Fishers are capable of killing Porcupines, which are their largest living prey. Apart from Porcupines, Fishers favour prey that is smaller than they are. Larger food such as deer and moose are found as carrion, commonly after they have been killed by motor vehicles.

REPRODUCTION

The Fisher's mating season is short and lasts about two or three weeks in early April. Adult males produce sperm from mid-March through April, after which sperm production ceases. Each female has a six- to eight-day oestrous period. Fishers delay implantation. The fertilized egg develops into a blastocyst but then remains dormant until about mid-February when increasing photoperiod induces implantation onto the wall of the uterus, and active gestation

begins. This phase lasts about 35–40 days. Births occur from early March until early May (most from March to early April), and the female breeds at about three to nine days after parturition. Female Fishers are pregnant for 356–362 days each year, including the delayed implantation phase. They produce at most one litter annually. The nutritional condition of the females at the time of implantation determines the extent of her reproduction. If she is poorly fed, owing to, for example, prey scarcity or excessive competition, she may produce a small litter or fail to produce young altogether for the year. Litters average two to three kits (ranging from one to six). Newborns have a thin coat of fine hairs, and their eyes and ears are closed over. Development is relatively slow in the beginning as the kits only begin to crawl around the den at about three weeks old. Their silver-grey birth coat is thick by two weeks old, and they begin to moult and grow their juvenile chocolate-brown coat by three weeks old; it is complete by the time they are weaned at 10–12 weeks old. Their eyes open and they begin taking solid food by around seven weeks old. They are fully mobile by 10 weeks and are hunting with reasonable competence by four to five months old. The juveniles remain on their mother's territory until they are seven to eight months old, and females from the litter may cautiously continue to live on her territory into their first winter, by staying out of her way as much as possible. All parental care is provided by the mother. Both sexes may achieve sexual maturity at around one year old if they are well fed. Young females may bear their first litter as two year olds, but many delay until they are three, and most males do not become fully fertile until they are two.

BEHAVIOUR

Adult Fishers are solitary throughout most of the year. The obvious exceptions are the short liaisons of mating pairs and the longer interactions of maternal females and their dependent young. Fishers are active all year. They are secretive, and although they are active intermittently throughout the day and night, they are rarely seen. Their activity pattern appears to be driven by hunger rather than by time of day, but they tend to be most active around sunrise and sunset. Although in general they appear to avoid humans, Fishers can

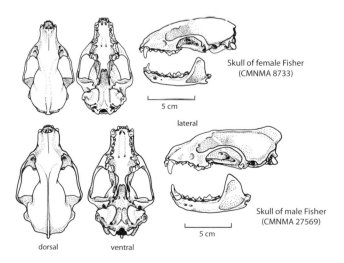

Skull of female Fisher
(CMNMA 8733)

5 cm

lateral

Skull of male Fisher
(CMNMA 27569)

5 cm

dorsal   ventral

Lope

Bound

Left hind track
Length: 5.0–8.0 cm
Width: 3.8–9.0 cm

Left front track
Length: 5.5–9.8 cm
Width: 4.8–10.8 cm

Fisher

become somewhat habituated to humans and will inhabit woodlands near low-density housing, roadways, and farms and may den under abandoned or little-used structures. They will even raid suet feeders left out for birds. Fishers use a zigzagging path as they search for small mammals, birds, and Snowshoe Hares, travelling back and forth between sites of suitable cover. They will also climb onto logs, snow banks, and rocks as vantage points to scan for prey.

Prey is dispatched with a bite to the neck. Larger prey such as hares and squirrels (with the obvious exception of Porcupines) are typically encircled (positioned at the Fisher's belly so that all four legs can grasp it) and killed with a bite to the back of the neck. This has been called a wraparound assist and is common among the weasels. Smaller prey is usually swallowed whole, and larger prey, such as hares and squirrels, are consumed in a single meal. A single Porcupine will feed a Fisher for two or three days. Porcupines are only attacked if they are on the ground. The Fisher circles the Porcupine, making repeated dives at the nose, which is free of quills. Eventually the severely bitten Porcupine succumbs to blood loss, shock, and exhaustion. Fishers learn the location of Porcupine denning sites within their home range and periodically travel directly to these locations without the usual zigzagging, hoping to surprise a Porcupine in a vulnerable position. The Porcupine's only escape is to find a spot that it can push into in order to protect its nose. Animals in single-entrance dens or in trees are unassailable as their noses are out of reach. Fishers, and other weasels, are known to attack and kill prey even when they are satiated. The extra prey is cached for future use. Although the likelihood of such an overabundance of prey occurring to a truly wild Fisher is relatively remote, still they cannot afford to overlook a predation opportunity. The home ranges of adults are defended as distinct territories, and those of neighbours of the same sex rarely overlap; however, those of males commonly overlap with those of several females, especially during the breeding season. Fishers, especially females with litters, aggressively defend their home range from other Fishers. Fights are common between mature males during the breeding season as they roam widely and encounter each other in their search for receptive females. Fishers mark their territory, using urine, faeces, and glandular secretions (from glands on their hind feet; see the "Description" section) on prominent objects such as stumps, logs, snow banks, and rocks. Urine and glandular marking is also used to indicate ownership of large carcasses.

VOCALIZATIONS
Fishers are relatively silent. Animals chatter when treed, produce a low chuckle when excited, and hiss and occasionally growl in threat.

SIGNS
Fishers have five toes on each foot. The inside toe (the first toe) on each foot is substantially smaller than the others and may not leave an impression in the track. Front feet are larger than hind feet and leave larger tracks. The negative space between the pads is filled with fur that may blur the impression of the foot. Claws may or may not register. Individual tracks that are less than 6.3 cm wide are likely those of a female Fisher or perhaps a male American Marten.

When tracking weasels, one must keep in mind the degree of sexual dimorphism. A large male American Marten is almost the size of a female Fisher. Distinguishing small Fisher tracks from American Marten tracks is difficult and often impossible, especially in snow. Both animals tend to travel in a lope (particularly in winter), leaving a distinctive paired track, with one foot slightly ahead of the other. Sometimes the habitat can provide a clue: Fishers tend to occupy more diverse habitats while martens are generally restricted to old-growth forests. Some Fisher tracks may occasionally look like Bobcat tracks. They can be differentiated by the claw impressions. Bobcat claws do not normally register in tracks as they are fully retractable. Also, Bobcats rarely use the distinctive lope pattern that is common in weasels. Fishers often do not consume the tails of their larger prey, particularly squirrels, Northern Raccoons, and Domestic Cats, commonly detaching and discarding them at the feeding site. Fisher feeding sites of large animals are distinguished by a complete lack of broken bones, as is common at wolf or Coyote sites. Fishers normally enter Porcupines through the skin of the stomach and consume the flesh and all but the largest bones, leaving the hide intact.

REFERENCES

Arthur et al. 1993; Aubry and Houston 1992; Aubry and Lewis 2003; Bowan et al. 2006; Dobbyn 1994; Earle and Kramm 1982; Garant and Crête 1997; Jameson and Peeters 2004; King and Powell 2007; Koen et al. 2007; Kohn et al. 1993; Lewis 2006; Powell 1981, 1993, 1994; Powell et al. 2003; Proulx 2006; Rezendes 1992; Slough and Jung 2007; Smith, H.C., 1993; Van Why and Giuliano 2001; Verts and Carraway 1998; Vinkey et al. 2006; Weir 2003; Weir and Corbould 2006, 2007; Weir et al. 2005; Zielinski and Duncan 2004; Zielinski and Truex 1995.

Ermine
also called Short-tailed Weasel, Stoat (Europe)

FRENCH NAME: **Belette hermine,** *also* **Hermine**

SCIENTIFIC NAME: *Mustela erminea*

Ermines are very curious and can be called up to a motionless observer who is making squeaking sounds. The thick, white winter fur of this weasel has been highly valued by Europeans and North American natives alike for hundreds of years.

DESCRIPTION

Like most of the weasels, Ermines are long and slender with short legs, an elongated neck, and large rounded ears. The typical summer pelage is monotone reddish-brown above but may vary from deep reddish-brown to light sandy brown. The throat, belly, and inner legs are usually white but may vary from white to creamy white to yellow. The back colour extends down the outside of the legs. Winter pelage is white (except for the black tail tip), sometimes washed with yellow, especially on the belly, rump, and tail. The colour change is thought to be for camouflage to help conceal the Ermine from predators. In regions with no winter snow cover, Ermines usually do not have a white winter pelage. However, Ermines on the Queen Charlotte Islands, for example, do have a white winter coat despite the total lack of snow. The tail is about 30%–44% of the head and body length, and the large black tail tip typically comprises about 25%–28% of the tail length in both summer and winter pelages. Ermines have a pair of anal glands, one gland lying on each side of the anus, which are under voluntary control. They contain a yellowish fluid that has a strong, unpleasant odour. Usually only a small amount is exuded at a time and is used for marking the home range. (For more information on the use of anal gland secretions see the "Behaviour" section below.) Ermines have an intense green eyeshine, and they also have slit, rather than rounded, pupils that are horizontal when contracted. Ermines have keen senses of smell, hearing, and sight. They are one of only a few non-primates known to have full colour vision in the red, yellow, blue, and green wavelengths (similar to that of humans). The dental formula is incisors 3/3, canines 1/1, premolars 3/3, and molars 1/2, for a total of 34 teeth. They undergo two moults, one in spring and the other in autumn. The spring moult begins on the head, moves along the back, and then goes down to the belly and legs. The autumn moult reverses that progression. Initiation of moult is governed by photoperiod but may be modified by temperature. For example, the spring moult may be delayed if winter temperatures, and hence snow, persist longer than usual.

SIMILAR SPECIES

The tail of the Ermine, like that of the Long-tailed Weasel, is black tipped. Ermines are generally smaller than Long-tailed Weasels, and their tail is less than 44% of their head and body length. Least Weasels are generally smaller than Ermines, their tail is much shorter (less then 25% of the head and body length), and it does not have a black tip. Size distinction among these small weasels is clouded by the fact that males are significantly larger than females. A male Least Weasel approaches the size of a female Ermine, and a large male Ermine is only slightly smaller than a small female Long-tailed Weasel. All three of the above-mentioned species are brownish above and whitish below in summer, and they change to all white in winter, except for the black tail tip of Ermines and Long-tailed Weasels.

SIZE

Size varies considerably across the range, with the smallest forms being in North America, northern Europe, and Siberia. The largest are in Britain, Ireland, and New Zealand (introduced from Britain). Within North America the largest are in the northwest (Alaska), and the smallest are in the east. Males are larger than females in all forms. North American males are about 10%–25% longer in body length and up to 80% heavier than females. The following measurements apply to adult animals in the Canadian part of the range and cover the variation in size across the country.

Ermine (*Mustela erminea*)

summer

winter

Males: total length: 300–355 mm; tail length: 69–98 mm; hind foot length: 36–51 mm; ear length: 17–26 mm; weight: 100.9–239.1 g. Females: total length: 218–290 mm; tail length: 45–91 mm; hind foot length: 25–45 mm; ear length: 11–19 mm; weight: 49.6–121.6 g. Newborns: weight: about 1.7–2.0 g.

RANGE

Ermines are circumboreal in distribution from North America to Europe and Russia. They were introduced from Britain to New Zealand in the 1880s to control imported European Rabbits. With their usual disregard for the arrangements of humans, the Ermines turned their attention to the nestlings, eggs, and even adults of many New Zealand birds as they were easier to catch. These birds, many of which are flightless and ground nesting, have all evolved without such predators and are poorly adapted to protect themselves from Ermines or the other introduced weasels such as the Least Weasel and the European Polecat (also called the European Ferret). It has been estimated that Ermines kill about 60% of the Kiwi born in New Zealand each year. Since another 35% are killed by introduced ferrets and rats, only around 5% of the hatchlings survive their first year.

ABUNDANCE

The population density of Ermines in North America is closely tied to prey abundance, mainly the numbers of voles and lemmings. When prey numbers decline, so too does the number of Ermines. This population cycle especially follows that of the prey in regions where one species is the primary prey (for example, in the Arctic where lemmings are the only winter food), but is less conclusive further south where Ermines can and do switch to other prey

when one species becomes unavailable. However, in years when all of the small rodents are scarce, such as after a major failure in the mast crop, Ermines usually become scarce as well. In favourable habitats in North America, Ermines can reach densities of up to 10 per km². The white winter pelts of Ermines and Long-tailed Weasels are valued by the fur trade, and both are called Ermine, but the Long-tailed Weasel pelt has a higher value owing to its larger size. Demand for such fur has declined in the last two decades, although trapping still takes place. The occurrence of Ermines on the Queen Charlotte Islands is becoming increasingly rare, and the subspecies there (*Mustela erminea haidarum*) was upgraded in 2001 by COSEWIC from "Special Concern" to "Threatened." The same subspecies receives special attention in British Columbia as it appears on the provincial "Red List."

ECOLOGY

Ermines are versatile in their habitat requirements, occupying successional forests, woodlands, parklands and forest edges, marshes, alpine meadows, farmlands, and riverbanks. They generally avoid dense woods and deserts and have not invaded the dry grasslands of the prairie region. Their long, thin shape allows them to move through the runways and burrows of their prey, but the price they pay is inefficiency in heat conservation; hence they must maintain a higher metabolism than do most other mammals in order to stay warm. A very well-insulated nest in which to rest is essential, and Ermines must eat regularly and frequently to fuel their internal furnace. They are said to be perpetually hungry. Ermines are active both day and night, foraging until successful and then resting for a few hours before hunger drives them out to hunt again.

North American distribution of the Ermine (*Mustela erminea*)

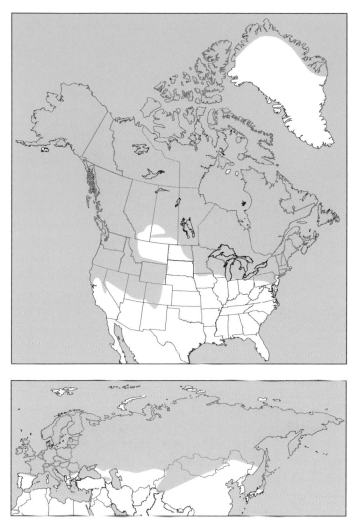

Eurasian distribution – New Zealand introduction is not illustrated

The principal cause of death is starvation, and only two or three consecutively missed meals are sufficient to kill an Ermine. They live their life on a tightrope as they attempt to balance survival and reproduction. At times a cache containing a single mouse can mean the difference between life and death, especially in winter when energy demands are highest or when a female is attempting to feed herself and her rapidly growing kits. Needless to say, Ermines are ferocious and relentless hunters, commonly going after prey that equals or exceeds their own size.

Like other mustelids (members of the weasel family), Ermines do not hibernate in winter. The distance travelled daily during foraging excursions averages around 0.5–1.3 km but may extend to as much as 8 km depending on hunting success and prey availability. Each home range includes several nests that are used in turn. These are usually near a favourite hunting area and are used while the animal hunts in that region. Dens are in hollow trees, rock crevices, prey burrows, and subnivean spaces. Most dens are appropriated from prey and further insulated with the fur of the

victim. The home-range size varies with habitat, prey availability, gender, and season. Home ranges of males are much larger than those of females and typically include parts of the home ranges of several nearby females. Male home ranges in Canada vary from 1.0–87.4 ha in the east to 123–205 ha in the west. Female ranges vary from 4–15 ha in the east to 66–95 ha in the west. Females exclude each other from their home ranges but simply avoid the males whose larger home ranges overlap theirs. Adult males will normally only venture onto the home range of a neighbouring male if he is not around. Juveniles dispersing from their natal home range can travel great distances. Young females tend to remain nearby, usually moving less than 5 or 6 km. Young males tend to travel further, more in the 8–23 km range. Longer distances are possible too. One young male moved 135 km between August and March, and a young female travelled 65 km in less than four weeks. Radio-tracked Ermines in Quebec travelled at speeds varying from 0.03 to 1.4 km/h depending on the gait. The average foraging speed is estimated to be around 0.25 km/h.

Ermines are agile in trees and have ankle joints that rotate to allow them to descend head-first. They will pursue prey into trees and also use trees as avenues of escape when threatened. They are capable swimmers. Ermines may live to a maximum of 10 years in captivity, but few in the wild exceed 1.5 years old, and most have only one or two opportunities to breed. Provided small livestock is secure, Ermines are useful rodent exterminators around a farm. Predators include hawks and owls, foxes, Coyotes, larger weasels, snakes, Domestic Cats, and humans. It is thought that the black tail tip is a mis-directional cue aimed towards predatory raptors. Presumably the bird will strike the prominent tail tip and miss grasping the animal.

DIET

Ermines are somewhat versatile in their prey choices and hunting abilities, and often the size difference between males and females from the same locality, or the general size increase from southeast to northwest, permits different prey choices. For example, the males in the northwest – northern Alberta, Northwest Territories, Yukon, and Alaska – will take Red Squirrels and young hares, while females in the same regions mainly eat lemmings and voles. In Ontario, Quebec, the Maritime Provinces, and the United States the Ermines of both sexes are smaller and kill primarily small rodents such as mice and voles. While small rodents provide most of the prey for Ermines worldwide, these adaptable small carnivores are able to modify their choices when prey is scarce or seasonally abundant. Birds, bird eggs, reptiles, amphibians, lagomorphs, and larger rodents such as chipmunks and small squirrels are added when they are available, and less palatable shrews, stranded fish, insects, and earthworms are eaten when the preferred prey is scarce. When the ground is snow covered and air temperatures are low, Ermines travel extensively in the subnivean runways of their small rodent prey, hunting mice, voles, and lemmings and appropriating their dens under the insulating blanket of snow. In temperate conditions male Ermines require about 32% of their body weight in food every day, while the smaller females need 23%–28%. This percentage climbs when temperatures

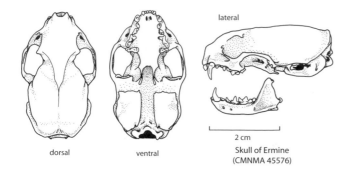

dorsal   ventral

lateral

2 cm

Skull of Ermine
(CMNMA 45576)

drop. Ermines need free water and will drink small amounts frequently during the day.

REPRODUCTION

Mating season occurs from April to early June. Females experience a post-partum oestrus at four to eight weeks after the birth of their litter. Like many other small weasels (with the notable exception of the Least Weasel), the Ermine reproduction cycle uses delayed implantation to schedule the breeding and birthing periods at appropriate times of the year. A fertilized egg develops into the blastocyst stage in around 14 days and then floats freely in the uterus for nine to ten months before implanting and undergoing active gestation for about 28 days. The entire gestation period therefore is about 10.5–11.5 months. Litters of up to 18 kits have been reported, but most litters average 4–8 kits. A single litter, at most, is produced by each female annually, born in spring (mostly in late April or May in Canada). The number of kits in a litter is strongly influenced by the nutritional status of the mother at the time of implantation. A well-fed female will typically have a larger litter than will an underfed female. Poorly nourished females may not have any blastocysts implant. Newborns are blind and deaf, with fine white hairs on the back only. Milk teeth appear during the third week, and eyes open at 28–42 days old (females first). Kits begin eating meat delivered by their mothers at around four to five weeks old. At birth, males and females are the same size, but sexual dimorphism becomes apparent by four to six weeks old. They begin to venture and play out of the nest at around five to six weeks old. Their permanent teeth begin to arrive at around 10 weeks old, and they begin to kill their own prey at 10–12 weeks old. Weaning occurs gradually and is completed by 12 weeks old or so, after the kits have been following their mother around for about two weeks as she hunts. The juveniles disperse by 16 weeks old. Young males leave sooner, usually in July or early August. Female juveniles achieve sexual maturity while they are still in the natal nest, usually by about two months old (some as early as three weeks old, despite the fact that they are still blind and deaf). A fortunate mature male that is accepted by the maternal female will then proceed to fertilize all of the precocious female young as well. This astonishing fact allows a single juvenile female leaving her natal nest to become an extremely capable colonizer as she has no need for another male to produce a litter. Male Ermines are highly efficient at finding oestrous females, and 99% of female Ermines are pregnant by June. Juvenile males become sexually and physically mature at

11–12 months old, just in time for the next breeding season, so they cannot inseminate their female littermates. Females reach physical maturity at around six months old.

BEHAVIOUR

Ermines can be playful, especially when they have recently fed. Juveniles tussle and play and in the process strengthen their bodies and learn many of the movements that they will need as adults. Adults engage in solitary play by bouncing off logs and other objects in a frenzy of twisting and contorting. Maternal females will also play with their young. Although males are usually dominant to females because of their greater size, this situation is reversed when the female is in oestrus; the male must solicit her for her attention. He is very persistent in this effort. Courtship is generally brief, and the male quickly grabs the female by her nape and may drag her around. She remains passive, as would a kit being carried in the same manner by its mother. Copulation lasts 2–20 minutes and may be repeated. Females will mate with multiple males, and males make concerted efforts to mate with as many females as they can find during the breeding season. The boundaries of home ranges are patrolled regularly and extensively scent-marked with secretions from the animals' anal glands. An anal drag is used to deposit anal gland secretions on the substrate. To accomplish this, the Ermine presses the anal area onto the ground and drags itself forward with its front legs. In moments of extreme distress an Ermine can voluntarily empty the anal glands, producing a "stink bomb" similar in odour to that of a skunk but without their range of ejection. They also have skin glands, especially on their belly, flanks, and cheeks, which they rub on objects as they move about their home range. Scat is most commonly deposited around the perimeter of the home range and at active den sites. The object is to mark the home range so that other Ermines will know that the area is occupied and keep away. Although Ermines are largely solitary as adults, they clearly keep track of their neighbours through their scent markings and thereby avoid open conflict. They can recognize the odour of a dominant animal and make efforts to avoid it. A dominant animal, encountering the scent marking of a subordinate, will cover the marking with its own. Although battles are generally avoided, they do occur, especially when weasel densities become high. Pugnacious fighters, Ermines are commonly injured during such encounters. Like other small weasels, Ermines perform an interesting "sham-death" behaviour. When subjected to mortal danger, such as a clearly superior predator like a Domestic Cat, they will freeze, appearing dead. If not killed, they will quickly revive and run away.

There are two types of weasels in Ermine society: residents and transients. Residents have a clearly established home range upon which they are always dominant to a transient. Transients are mostly unsettled juveniles looking for a home range, but they may also be former residents that have left their previous home range for a number of reasons, but mainly because of food shortage. Transients live a precarious existence as they search for a home range that is unoccupied and has sufficient prey. Ermines kill using a "wrap-around assist" technique. They jump on their prey, grasping it with all four legs and claws in an effort to subdue and contain its struggles so that they can deliver a killing neck bite. Caching food is common among the small and medium-sized weasels, as is surplus killing when prey is abundant. Ermines are quite capable of remembering the location of cached food and inevitably return for it. Females sometimes move their kits from one den to another by carrying each one separately, gripping it gently around the midsection. Once the kits are older and following her around as she hunts, she will take them to a variety of dens to rest. Like other weasels, Ermines are both bold and courageous, willing to face an animal many times their size. They will stand up to, and even bite, a human who is foolish enough to attempt to take a fresh kill.

VOCALIZATIONS

A trilling call is used by young and by subordinate adults that are attempting to placate a dominant individual. Shrieking is used in home-range defence; it is emitted while the Ermine is chasing off an intruder, but it may also be used defensively if the Ermine is cornered or feels itself to be severely threatened. Squeaking is a common vocalization that is used when complaining, fighting, or mating, or by youngsters while play-fighting. Kits squeak and chirp in response to disturbance, and they use an infantile begging cry to solicit nursing and to call their mother if they become lost while following her on a hunt. Females call to a male with a faint cooing call. A hissing sound is produced if a weasel is uneasy or mildly threatened.

SIGNS

Ermines have five toes on both front and hind feet, but the inside toe on each foot is much reduced and does not reliably register in a track. The front track is larger than the hind. The underside of the feet is entirely covered in hairs that are especially thick in winter; only the small pads are bare. Claws may or may not register. A lope is a typical gait used by these weasels, especially in snow. In this gait the hind feet land in the depression left by the front feet, and a trailing drag mark often leads into and out of each track depending on snow depth. The other common gait, especially on a solid ground surface, is a bound. In this gait the hind feet land near, or overstep, the front feet. A typical weasel trail is very winding and commonly crosses itself as the animal hunts under every nook and cranny and explores tree hollows, fallen debris, stumps, rock walls, and holes looking for prey. Distinguishing between trails of Least Weasels, Ermines, and Long-tailed Weasels is difficult and is best done by overall trail width and stride length, although there is a large margin of overlap. Trail width at a bound for Least Weasels varies from 1.7 to 3.5 cm, while that of Ermines is 4.5–6.5 cm and of Long-tailed Weasels is 4.8–7.3 cm. In regions with snow, Ermines spend most of the winter hunting under the snow, especially when it is very cold or windy. Scat is 2–8 cm in length and 0.3–0.8 cm in diameter, and typically twisted, drawn out, blackish, and usually pointed at each end. Scat may be found on prominent objects such as stumps, logs, or rocks, along trails, on prey runways, or in latrine sites near dens. Ermines typically carry their tails straight out behind their bodies or, at most, at a 45° angle.

foot drag on shorter strides

stride length on bounds can vary from 25 to 33 cm for a Least Weasel, 35 to 55 cm for an Ermine, and 38 to 64 cm for a Long-tailed Weasel

Bound on firm substrate

Lope in deep snow

Left hind track in shallow snow

Least Weasel:
Length: 0.8–2.2 cm
Width: 0.8–1.4 cm
Ermine:
Length: 1.1–2.5 cm
Width: 1.0–2.2 cm
Long-tailed Weasel:
Length: 1.9–3.8 cm
Width: 1.4–2.5 cm

Left front track in shallow snow

Least Weasel:
Length: 0.8–1.4 cm
Width: 1.0–1.4 cm
Ermine:
Length: 1.1–1.9 cm
Width: 1.1–2.2 cm
Long-tailed Weasel:
Length: 1.6–3.7 cm
Width: 1.9–3.0 cm

Least Weasel, Ermine, and Long-tailed Weasel

REFERENCES

Edie 2001; Edwards and Forbes 2003; Elbroch 2003; Jameson and Peeters 2004; Johnson, D.R., et al. 2000; King 1983; King and Powell 2007; Kurta 1995; Macdonald and Barrett 1993; McDonald and Case 1990; Robitaille and Raymond 1995; Samson and Raymond 1995a 1995b; Sittler 1995; Smith, H.C., 1993; Svendsen 2003; Tumanov and Sorina 1997; Youngman 1975.

Long-tailed Weasel

FRENCH NAME: **Belette à longue queue**

SCIENTIFIC NAME: *Mustela frenata*

Long-tailed Weasels are rarely seen but do become road kill at times as they are very unaware of vehicles, especially if they are in pursuit of prey.

DESCRIPTION

These weasels have a typical weasel shape with a long slender body, short legs, a pointed triangular head, and large rounded ears. In northern North America (including all of their Canadian range) the Long-tailed Weasel's pelage colour varies seasonally. Its summer coat is sandy to deep rust brown above, with white, cream, yellow, or orange underparts. The underside of the chin is usually white or cream coloured, even on those individuals with an orange belly. The brown back colour extends down the outside of the legs and sometimes the inside, often as far as the ankles, and may even cover the feet. More commonly, the feet and the lower legs are the colour of the belly fur. Some North American Long-tailed Weasels have an uncommon head coloration in summer pelage that makes them appear masked. The fur on their forehead and cheeks is light, while that on the snout and around the eyes is dark. There is considerable variation in the amount and intensity of this light and dark fur. Winter pelage is white, except for the tail tip, which remains black in both summer and winter pelages. In most regions with little or no winter snow cover the Long-tailed Weasels do not develop a white winter pelage, rather retaining the brown summer pattern all year round. The long tail is 44%–70% of the head and body length. Long-tailed Weasels have a pair of anal glands, one gland lying on each side of the anus, which are under voluntary control. These contain a foul smelling, oily fluid that is ordinarily used to mark the home range. The dental formula is incisors 3/3, canines 1/1, premolars 3/3, and molars 1/2, for a total of 34 teeth. Long-tailed Weasels undergo two moults, one in spring and the other in autumn. The spring moult takes 20–35 days and begins on the head and along the back and then progresses down to the belly. The autumn moult takes 37–70 days. The initiation of the moult is governed by photoperiod. The eyeshine is bright greenish to bluish. Scent detection and hearing are keen, and visual discernment is highly motion sensitive.

SIMILAR SPECIES

There are three North American weasels that are similar in both the brown summer pelage and the white winter pelage. Least Weasels are the smallest. They do not have a black tail tip as do Ermines

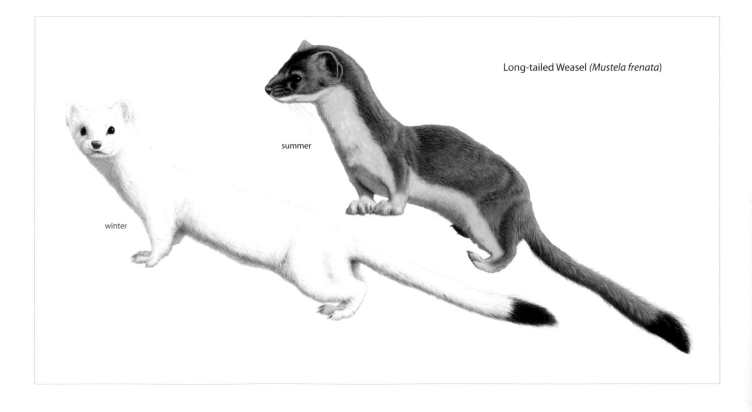

winter

summer

Long-tailed Weasel *(Mustela frenata)*

and Long-tailed Weasels. Ermines are next in size and occupy the middle ground between the larger Long-tailed Weasels and the smaller Least Weasels. Ermines' tails are also intermediate, longer than those of Least Weasels but shorter than those of Long-tailed Weasels. Least Weasel tails are less than 25% of the head and body length, Ermine tails are 30%–44% of the head and body length, and Long-tailed Weasel tails are 44%–70% of the head and body length. The marked sexual dimorphism of this group further confuses identification, as a large male Ermine is only a bit smaller than a small female Long-tailed Weasel. Proportional tail length and body size are the two best field identification clues for distinguishing Long-tailed Weasels from Ermines.

SIZE

Size varies considerably across the range, with the smallest forms in the eastern and southern regions and the largest in the west. Males are larger than females in all forms. North American males are about 15%–20% longer in total length and up to 50% heavier than females. The following measurements apply to adult animals in the Canadian part of the range and cover the variation in size across the country. Males: total length: 415–485 mm; tail length: 141–186 mm; hind foot length: 46–60 mm; ear length: 21–31 mm; weight: 199.3–423.7 g. Females: total length: 363–443 mm; tail length: 114–166 mm; hind foot length: 35–52 mm; ear length: 21–25 mm; weight: 154.3–242.6 g. Newborns: weight: about 3–4 g.

RANGE

Long-tailed Weasels are widely distributed in the New World from southern Canada to northern Bolivia, where they occupy all life zones except deserts and dense northern boreal forests. A recent report of a Long-tailed Weasel that was photographed on a suet feeder near Rimouski, Quebec, is indicated by a star on the range map.

ABUNDANCE

Generally rare or uncommon, these weasels normally occur at low densities, and populations are subject to periodic fluctuations, possibly owing to prey abundance. For example, populations in Alberta and Vermont appear to fluctuate in rhythm with the abundance cycle of Snowshoe Hares. Local populations are vulnerable to extinction, possibly as a result of habitat fragmentation combined with food shortages. The subspecies that occurs in the Prairie provinces is suspected to be rapidly disappearing and is currently under exhaustive survey. Alberta has designated it as "May Be at Risk."

ECOLOGY

This weasel is highly versatile and can survive in a wide variety of habitats. Grasslands, parklands, open coniferous forests, alpine meadows, scrub forests, tropical forests, agricultural lands, mixed forests, and deciduous forests all support Long-tailed Weasels. They are most common in open woodlands, forest edges, and marshland habitats and are absent from dense boreal forests and deserts. Of all the small North American weasels, this species is most comfortable around human settlement. Their long, slim, short-legged shape

allows them to enter the burrows of prey but is not very energy efficient. Long-tailed Weasels lose considerable heat because their ratio of surface area to weight is so disproportionate that they must maintain a higher than usual metabolic rate to compensate. As a consequence, they are always hungry. Captive adult Long-tailed Weasels consume 17%–33% of their body weight daily, and growing young can eat even more, up to 40% at times. The requirements of wild individuals are unknown but are likely higher. Daily activity patterns are governed by hunger and are typified by a period of foraging followed by a period of resting. Although Long-tailed Weasels may hunt in day or night, most of their activity occurs at night. They are active year-round and do not hibernate. Each adult weasel occupies a solitary home range, the size of which varies, mainly because of gender and prey availability. The ranges of males are usually considerably larger than those of females, partly because they are physically larger and need access to more food, but also so that their home range can overlap those of several females. The home-range size of males commonly increases during the breeding season to allow more contact with more females. Summer home ranges vary from 16 to 24 ha, and winter home ranges are 10 to 18 ha. If winter prey is scarce, however, ranges may expand to 44–59 ha for females and 120–240 ha for males. As weasels do not burrow, Long-tailed Weasels must find dens and burrows constructed by other animals, most often by their prey. Suitable nests are appropriated,

Distribution of the Long-tailed Weasel (*Mustela frenata*)

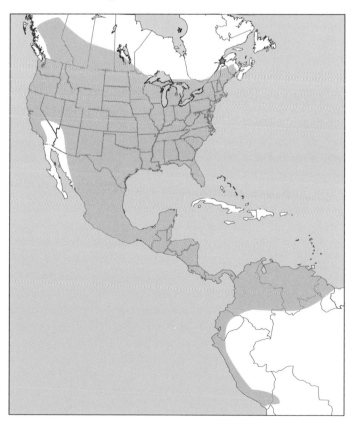

⭐ extralimital record

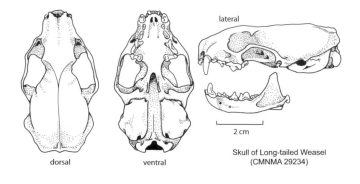

lateral

2 cm

Skull of Long-tailed Weasel
(CMNMA 29234)

dorsal   ventral

and an unlucky incumbent may be consumed and its fur used to further insulate the nest. Den sites include tree hollows, deep inside debris or wood piles, in rock piles, under tree roots, in stick nests in trees (formerly occupied by flying squirrels), in underground chambers, and in subnivean spaces where the snow protects them from extreme temperatures. Long-tailed Weasels are capable and willing swimmers. They are also very proficient tree climbers and able to rapidly descend head first as their hind feet are capable of enhanced rotation and so the claws can remain in contact with the trunk in that awkward-looking position. They will forage in trees and are often quick to retreat up a tree if threatened. Foxes and raptors, along with humans, are major predators of Long-tailed Weasels. Larger weasels, Bobcats, Coyotes, and Domestic Cats and Dogs also kill the occasional Long-tail. Although there is little information available on longevity, it is known that some Long-tailed Weasels may survive for at least three years in the wild. However, annual mortality is very high, and few weasels live more than one to two years.

DIET

Long-tailed Weasels are generalist predators that hunt and consume a wide variety of small prey and are capable of switching to alternative prey if the availability of preferred prey declines. Owing to their greater size, males can generally, and more frequently, take larger prey than can females. Diet varies across the range, owing to prey availability. Predominant prey in North America is composed of small and medium-sized rodents such as mice and voles, also squirrels and lagomorphs (rabbits, hares, and pikas), with the addition of seasonally available birds, bird eggs, snakes, lizards, and insects. Occasional prey includes shrews, moles, bats, smaller weasels, and carrion. Chickens, ducks, and other small domestic fowl are commonly killed by Long-tailed Weasels and may be the subject of surplus killing, as they are confined and vulnerable. Provided poultry is well secured, Long-tailed Weasels are beneficial around a farmyard for their rodent-control abilities. These weasels drink frequently and need about 25 ml of free water each day.

REPRODUCTION

Like many other weasels (with the notable exception of the Least Weasel), female Long-tailed Weasels exhibit a lengthy period of delayed implantation. The fertilized eggs develop to the blastocyst stage and then float around in the uterus in a form of suspended animation. Increasing day length provides the cue for implantation

of the blastocysts into the uterine wall, and once they are implanted, active gestation lasts around 26–27 days. The entire gestation period, including the delayed implantation phase, may take 205–337 days but averages 279 days. Each female bears, at most, a single litter annually and provides all of the parental care. Litters in North America are born from April to early May and average four to five kits (ranging from one to nine). The shrew-sized kits are born with a fine coat of white hairs. Both sexes are about the same size at birth, but by the time they are two weeks old, sexual dimorphism is already apparent, as the males are noticeably larger. Milk teeth begin to emerge at around 21 days old, and the kits start eating solid food at around 28–35 days old when their mother begins bringing prey to the nest. Weaning usually occurs gradually once the kits begin eating meat. Their eyes open at 35–37 days old, and their permanent teeth begin erupting at around 10 weeks old. The kits learn to play outside the nest at around five to six weeks old and begin to follow their mother around as she forages a week or two later. They are able to make their own kills at around 10–12 weeks old and are weaned between five and 12 weeks old.

Females become sexually mature as early as three to four months old and reach physical maturity by the age of six months. Their male littermates reach sexual maturity much later, at around 11–15 months old, at about the same time as they achieve their full physical maturity, so they are not likely to mate with their sisters. Individual females are in heat for three to four days, but there is considerable variation as to when oestrus may occur. Maternal females come into heat 65–104 days after the birth of their litter, as their post-partum oestrus is inhibited by lactation. If the whole litter is lost in the early stages, she may be ready to mate earlier, 39–71 days post-partum, but will still not produce the next litter until the following spring. Other females may come into season as early as mid-March. Female juveniles usually mate for the first time in their first summer while they still reside on their mother's range. Their mate is usually one of the males whose home range overlaps that of their mother's; however, owing to the rapid turnover of resident males, this male is unlikely to have been her own sire. To accommodate this lengthy breeding period, adult males are sexually active as early as February and until September in Canadian populations. Copulation is a lengthy affair for Long-tailed Weasels because the female will not ovulate unless sufficiently stimulated. The pair typically remains locked for two to three hours, with alternating periods of thrusting and resting. Females will mate with more than one male during a single oestrous period, so a litter could exhibit multiple paternity.

BEHAVIOUR

Scent is used to mark the home range. Long-tailed Weasels have numerous skin glands, especially on the belly, throat, and cheeks, which they rub onto branches and other prominent objects as they travel around their home range. They also mark using small amounts of the secretions from their anal glands. To lay this scent down, the weasel places its anus in direct contact with the substrate and scoots along, dragging itself by the fore legs. The contents of the anal glands can be expelled all at once by voluntary muscular contraction if the weasel is under threat. The acrid foul-smelling fluid is not directed, as it is with skunks, but the odour is often sufficiently unpleasant to dissuade a potential predator. Individual weasels patrol the boundaries of their home range, marking and sometimes actively excluding others of their sex. Apart from the activities associated with breeding, adults are solitary and generally avoid each other whenever they can. Courtship behaviour is often fairly cursory for Long-tailed Weasels. Once the male has determined from the odour of her urine that a female is in oestrus, he pursues her, and a scuffle ensues. He attempts to subdue her by gripping the skin at the back of her neck. Once she ceases struggling, he orients his body over hers, and the pair copulates. Sometimes an adult male will approach a maternal female with an offering of prey. If he can gain her favour, he may also be able to mate with her female offspring. A mother with young is quite fearless and will take on animals many times her size to protect them. She is so ferocious that she temporarily becomes dominant even to the larger adult males; in those circumstances they must tread carefully, and perhaps a fine meal makes a safe introduction. A pregnant Long-tailed Weasel hunts as long as she can fit into the burrows of her prey, and builds up a cache of food. When hunting is no longer possible, she retires to her chosen den and waits for the birth of her young, surviving on her cache. Once the kits are born, she is again svelte and able to hunt, which she now must do at an increased rate in order to produce sufficient milk to feed her rapidly growing young.

Prey is killed in one of several different ways depending on its size and location. For large prey such as rabbits and hares, the weasel runs alongside and bites several times wherever it can in order to slow the prey down and eventually disable it. Medium-sized prey such as squirrels and chipmunks being hunted above ground are killed with a bite to the nape that severs the spinal cord. Smaller rodents are subdued using a "wrap-around assist," where the weasel grasps the prey with all four feet to manoeuvre it for a neck bite. Underground prey is dealt with somewhat differently by necessity. Since there is insufficient room for a wrap-around assist and as the prey cannot claw the weasel with its hind feet, a slower suffocation method is used as the weasel grips the face or throat. Sometimes a Long-tailed Weasel will be mobbed and driven off by the communal efforts of a social prey species such as ground squirrels. Food caching is a common behaviour of Long-tailed Weasels. Surplus prey is stored where it is best preserved. During warm weather, underground locations are common. During cool and cold weather, above-ground and even arboreal sites (abandoned bird nests) have been found. Play among weasels is common. Youngsters wrestle and play-mount each other from the time they get out of the nest until they disperse. This activity is important to their neuromuscular development, ultimately making them better hunters. Adults, especially when well fed, sometimes engage in solitary play, bouncing off obstacles and twisting in the air.

VOCALIZATIONS

An anxious Long-tailed Weasel may produce a faint hiss as a low-intensity threat signal. A trill is commonly used while playing, hunting, or just calmly moving around. Mothers call their young using the trill. Young weasels squeak from birth if they are disturbed. This

squeak changes to a chirp as the kits get to be about a month old. A "zhzhzhzhp" sound is often emitted with an open, almost smiling mouth and is used socially between calm, alert, interested animals, but apparently, and for unknown reasons, only when they are in summer pelage. A squeal is used as a distress call. A screech (also called an explosive bark) is used as a defensive threat and is often accompanied by an open mouth gape and a sudden lunge.

SIGNS

Long-tailed Weasels have five toes on both front and hind feet, but the inside toe on each foot is reduced and does not always register in a track. Negative space between the pads is fully furred, especially in winter, which may distort the track dimensions. Claws may or may not register, but sometimes, especially if the weasel is sprinting, only the claws leave impressions. The front track is larger than the hind. Trails commonly begin or end at water. The most common gaits are a bound and a lope. In the bound each foot leaves an individual track, with the hind feet commonly overstepping the front. A loping weasel places the hind feet directly onto the track left by the front feet. This gait is typically used in snow and leaves a characteristic two-by-two track with drag marks leading into and out of it. Stride length varies from 38–64 cm for a bound to 25.4–114.3 cm for a lope. Distinguishing between the trails of the three smaller weasels (Least Weasel, Ermine, and Long-tailed Weasel) can be difficult. See the "Signs" section of the Ermine account for track illustrations and a discussion of the species' distinctions. Scat is typically black, twisted, with pointed ends and often looped around itself (0.5–1.0 cm in diameter by 2.0–8.3 cm in length). Latrines are regularly placed just outside an active den and may include food scraps as well as scat. While running, a Long-tailed Weasel raises its tail either parallel to its body or, at most, at a 45° angle, but never straight up at a 90° angle.

REFERENCES

Boulva 2009; Bowman, J.C., 1997; Cowan and Guiget 1956; Dekker 1993; Elbroch 2003; Gehring and Swihart 2004; Harestad 1990; Jardine 2004; Johnston, C., et al. 1993; King and Powell 2007; Kurta 1995; Sheffield and Thomas 1997; Smith, H.C., 1993; Weeks 1993; Wilson, T.M., and Carey 1996.

Black-footed Ferret

FRENCH NAME: **Putois d'Amérique**
SCIENTIFIC NAME: *Mustela nigripes*

The last Canadian specimen of a wild Black-footed Ferret dates from 1937, and the last colonies in the United States had dwindled to extinction by 1987. Realizing that the wild populations were doomed, 18 animals were captured from the last known colony

Black-footed Ferret (*Mustela nigripes*)

in Wyoming in 1986. A highly successful captive breeding program has since produced thousands of ferrets. From this secure source, reintroductions began in 1991, and by 2009 ferrets had been reintroduced to 19 locations within their former range in the United States and Canada and to an additional site in northern Mexico that may also have been part of their former range. There are a finite number of areas where ferret reintroductions can be undertaken with any hope of success, as their shelter and diet depend on the presence of prairie dogs. A prairie dog colony must exceed 10,000 individuals to sustain 10 breeding pairs of ferrets, the minimum determined for a viable, long-term colony. There is only one location in Canada that can satisfy these requirements – the area around and including Grasslands National Park in southwestern Saskatchewan – and reintroductions began there in autumn 2009.

DESCRIPTION

Black-footed Ferrets are dramatically marked weasels with the typical weasel form of a long slender body, short legs, large rounded ears, and a triangular head. Their legs, tail tip, mask, and nose pad are dark brown to black. Their upper body is sandy brown, often with a slightly darker saddle in the middle of the back caused by brown-tipped guard hairs. Their underside is buffy or yellowish. Muzzle, cheeks, and throat are creamy white. A pair of small, light eyebrow marks occurs within the dark mask. Sometimes the dark fur of the front legs extends onto the chest area between the legs. The front feet and claws are large. Their fur is short, less than a

centimetre long on the back, and does not change colour seasonally. Males have a dark, longitudinal pubic stripe, which may be visible when they sit or stand upright on their haunches in the alert position. Females sometimes have a faint version of that stripe, but it is usually absent. The dental formula is incisors 3/3, canines 1/1, premolars 3/3, and molars 1/2, for a total of 34 teeth. These weasels have a bright emerald-green eyeshine.

SIMILAR SPECIES

Their tawny body, blackish legs and tail tip, and dark facial mask make this species distinct. Some Long-tailed Weasels in the prairie region exhibit a facial "mask" that can be similar to that of a Black-footed Ferret, but on these animals the base body colour usually remains the ruddy brown of a normal Long-tailed Weasel, although some individuals may be a lighter brown. Both male and female Black-footed Ferrets are considerably larger than the largest Long-tailed Weasels. The North American weasel that is most similar in size to the Black-footed Ferret is the American Mink. Minks are dark chocolate brown and normally found near water.

SIZE

Adult Black-footed Ferret males are around 10% longer and up to 50% heavier than females. Owing to the shortage of data, the following measurements have been gathered from all known specimens. Males: total length: 490–662 mm; tail length: 107–140 mm; hind foot length: 56.0–63.5 mm; ear length: around 30–31 mm; weight: 588–1125 g.

Females: total length: 479–518 mm; tail length: 117–141 mm; hind foot length: 58–63 mm; ear length: unreported but likely similar to that of males; weight: 453–850 g.
Newborns: weight: about 7–10 g.

RANGE

Less than around 2% of their former habitat is currently available and suitable for ferret survival. Formerly occurring from southern Saskatchewan to Texas and possibly even northern Mexico, Black-footed Ferrets now occur in only a handful of carefully chosen locations where re-established colonies are derived from captive-bred stock. These are in the midwestern United States, southern Saskatchewan, and a large and highly promising region with hundreds of protected prairie dog towns just south of the Mexico–United States border, near the Mexican town of Janos. The Alberta specimen shown on the map as a light orange circle has an uncertain provenance and is thought by some to have been originally from southern Saskatchewan. Certainly there were no prairie dogs in the part of Alberta where it was captured.

ABUNDANCE

Although formerly widespread through the arid plains of central North America, Black-footed Ferrets were probably never numerous. The last known truly wild ferret was seen in 1987 (in Wyoming). There are no verified recent sightings in Canada, and the species was considered to be extirpated north of the 49th parallel until recent reintroductions. Unlike many other animals that were eradicated by over-hunting or over-trapping, Black-footed Ferrets were the unintended victims of extensive eradication of prairie dogs, their principal prey. European agricultural practices abhorred the disruption of the land surface caused by prairie dogs. As the Great Plains were settled and farmed, the dog towns succumbed to eradication programs. Prairie dogs are considered keystone grassland species, without which a host of other plant and animal species cannot survive. As the prairie dogs were poisoned, the habitat created by the dog towns disappeared, and so too did the ferrets. Once considered "the rarest animal in North America," the Black-footed Ferret has since lost this dubious title, thanks to the highly successful captive breeding program that produced over 3000 ferrets in the decade from 1990 to 2000. The American reintroduced colonies were created in the 1990s, and the Mexican colony in the early 2000s. Reintroductions began in early October 2009 in Canada onto private and public lands in the Grasslands National Park area of southern Saskatchewan, where Black-tailed Prairie Dog towns of sufficient size still exist. Thirty-four ferrets from captive colonies at the Metro Toronto and Calgary zoos, as well as some from the United States Fish and Wildlife Service, were released; around 20% were expected to survive their first winter and breed the following spring. Successful breeding in 2010 has been reported, but the extent is unknown. It will be several years before the success of the reintroduction is assured, and likely more releases will be necessary to bolster the population. The future of all these populations remains uncertain although at least two American colonies (in Montana and South Dakota) appear to be thriving. After a promising start, the Wyoming colony was seriously hit by an outbreak of Sylvatic Plague. There has never been a large market for Black-footed Ferret hides because the fur is short and does not become particularly plush or luxuriant in winter.

ECOLOGY

Although primarily nocturnal, Black-footed Ferrets may appear above ground at any hour (this is especially true of maternal females feeding young), and they are active all year-round. They live within prairie dog towns and are ideally suited to hunt prairie dogs. The ferrets do not dig their own burrows but can easily travel in the burrow systems of the prairie dogs and are large enough to kill a full-grown prairie dog underground. Ferrets are so flexible that they can easily reverse their direction in a tunnel by lying on their back, then bending and walking over their own hindquarters. A prairie dog in the same situation must back up or go forward until it reaches a wider section of tunnel before it can turn. Probably one of the main reasons that Black-footed Ferrets came to specialize on prairie dogs was to acquire their burrows. Ferrets make their dens in appropriated prairie dog burrows, and the extent of the connecting tunnels enables the ferrets to hunt and live primarily underground where exposure to the elements and predators is ameliorated. During summers the burrows are cooler and reduce heat and moisture loss, helping them to live in the arid short-grass prairie environments

Distribution of the Black-footed Ferret (*Mustela nigripes*)

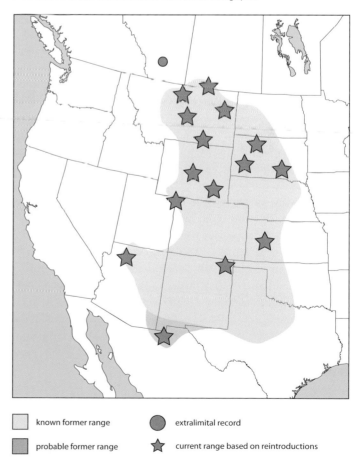

▢ known former range	⬤ extralimital record
▢ probable former range	★ current range based on reintroductions

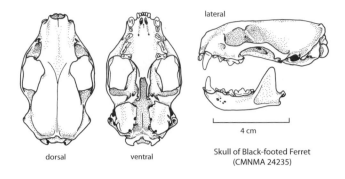

dorsal   ventral

lateral

4 cm

Skull of Black-footed Ferret
(CMNMA 24235)

where the dog towns occur. In winter the burrows are warmer and provide protection from wind and snow. Thriving Black-footed Ferret colonies are intimately tied to thriving prairie dog towns. The word colony is used in a general sense to describe a discrete group of Black-footed Ferrets that live within the geographical boundaries of a prairie dog town. As the prairie dogs declined, the ferret colonies became fragmented, isolated, and more vulnerable.

Each ferret occupies a home range within the prairie dog town that it defends from other ferrets of the same sex. Male home ranges are typically larger and tend to overlap those of several neighbouring females. There is limited information available on home-range size for Black-footed Ferrets, but ferret densities varied from one ferret per 30 ha to one ferret per 60 ha in the studies that were done in the few last remaining wild colonies. The longest dispersal distance known for a Black-footed Ferret is 7 km in a single night between prairie dog towns. Black-footed Ferrets are very susceptible to some communicable diseases. Two diseases in particular cause the most damage. Canine distemper, likely transmitted from Coyotes and American Badgers, is 100% lethal to ferrets. Sylvatic Plague (caused by the same organism that causes Bubonic Plague in humans) kills mainly prairie dogs but occasionally ferrets as well. Ultimately these two diseases could have as much impact as, or greater than, the more-than-century-long prairie dog eradication program because their effects on already decimated populations could threaten extinction. Predators of ferrets include Coyotes, owls, eagles, American Badgers, and possibly hawks. Juvenile ferrets are most vulnerable during the autumn when they disperse from their mother's care. They must travel above ground to find a home range of their own that can provide them with sufficient food and shelter, is not already occupied by another ferret, and yet is close enough to other ferrets to provide mating opportunities. Many are killed by predators, and those that are unable to find a suitable home range perish of cold and starvation. Of each year's batch of wild juveniles, 70%–80% or more die before they reach a year old. In general, these weasels do not survive long in the wild. The maximum lifespan of a wild Black-footed Ferret is around four to five years, but few live beyond three years.

DIET

Black-footed Ferrets are intimately tied to their prey. They are prairie dog specialists. Not only do prairie dogs make up around 90% of the ferret's diet, but their burrow systems provide shelter for the

ferrets as well as for many other species, some of which are then added to the ferret diet as secondary prey. These include mice, ground squirrels, and rabbits. Black-footed Ferrets do not need to drink free water in the wild. This is, at least in part, due to the protection they derive from living underground in the cool, moist burrows. Necessary moisture is acquired from their food, and they probably produce highly concentrated urine to conserve fluids.

REPRODUCTION

Black-footed Ferrets breed in March and April, although a female may come into a second oestrus if she does not become pregnant during her first heat. This may account for the few reports of breeding behaviour in early summer. Ovulation is induced, and so females require sufficient stimulation from mating in order to ovulate. Copulation can last several hours as a consequence. Unlike many other weasels, Black-footed Ferrets are not delayed implanters. After a normal period of development (around 12–14 days) the blastocyst implants into the uterine wall and begins developing into a fetus. Gestation lasts about 40–45 days, and the young are born in late April, May, or early June. Litter size averages three to five kits. Male Black-footed Ferrets provide no parental care. Kits are born deaf and blind with a sparse coat of fine white hairs and light brown skin around their muzzles. Newborns are not capable of full thermoregulation and must depend on their mother and huddling against each other to stay warm. The dark pattern on the legs is clearly visible by 33 days old. The young have most of their milk teeth and begin to eat solid food around 30 days. Their eyes open between 34 and 37 days, ears open soon after, and the young are walking around by 43 days old. Females must increase their energy intake by three to four levels of magnitude during the lactation period. Weaning usually occurs around the time that the kits become confident walkers. The permanent teeth begin replacing the milk teeth in the eighth week, and by 12 weeks old all the permanent teeth have emerged through the gums. Mothers bring their kits above ground for the first time when they are around 45–60 days old. By late August, at about three months old, the youngsters are essentially full sized. The males disperse first and tend to travel further than do the females. By late October all of the juveniles will have dispersed. If they survive the winter, they will participate in the next mating season.

BEHAVIOUR

Adult Black-footed Ferrets are solitary except during the breeding season. Courtship is difficult to observe in the wild as mating occurs underground where the pair is safer from predation. As reported from captive ferrets, courtship is perfunctory. An experienced male normally approaches an oestrous female and, without further formalities, grasps the nape of her neck in his teeth and proceeds to copulate. An inexperienced or reluctant male may need some help from the female. She may solicit his interest and make herself increasingly available until he responds. She approaches in a low crouch, emitting a chuckling call, and aligns her body alongside his. She will even crawl under him to the mating position if he fails to take appropriate action. Copulation may last several hours (from 0.5 to 4.0 hours) as the pair mates and rests in succession while remaining locked. The male may remain with the female for a few days while she is in oestrus, and the pair may mate several times. As the female transitions out of oestrus, she drives him out, and he returns to his regular haunts. A female will accept more than one male during her oestrous period, and so litters with multiple paternities are possible.

Like other weasels, Black-footed Ferrets cache surplus food. This is likely a very common occurrence as an adult prairie dog is far too large for a ferret to consume in a single feeding. Once their kits begin eating solid food, the maternal females will often move them to a new den near a kill rather than try to drag a full-grown prairie dog to the old den site. An alert prairie dog is a formidable opponent, and so most hunting occurs at night as the ferrets search underground for sleeping prairie dogs. These are wakened with a light nudge and then killed, using a throat hold that results in suffocation. The use of the choke hold is innate, but hunting success and efficiency increase with experience. Kits learn by watching their mother as she hunts. Black-footed Ferrets regularly mark the boundaries of their home range using the secretions from their anal glands. These glands are similar to those of skunks and are under voluntary control, but, unlike those of skunks, they serve mainly for scent marking rather than for defence. Their contents are deposited using a characteristic anal drag. The ferret drops its anogenital region to the ground and, pulling itself forward with its front legs, rapidly wiggles its pelvis from side to side, sometimes leaving a visible oily smear. A severely frightened or stressed ferret may emit a "stink bomb" as its tense muscles evacuate the entire contents of the anal glands. This may serve as a predator deterrent.

Scat and urine are commonly deposited in prominent areas and likely serve as both visual and olfactory signals that help the neighbouring animals know where each other's boundaries are so that contact can be avoided. The scent markings serve a further purpose during the breeding season by announcing the state of receptivity of the females. Ferrets often direct ritualized threat displays towards neighbours or intruders to enforce their dominance or territorial ownership without engaging in physical combat. These can take several forms: a bark or series of barks in conjunction with a rapid open- or closed-mouth lunge; a series of snake-head movements (the head and neck or whole upper body are rapidly whipped from side to side) as the ferret simultaneously and repeatedly stomps its front feet; and a flattened body with the feet underneath and the head lowered while hissing. A subordinate animal will retreat from such displays, avoiding possible injury from fighting. Kits are very playful with each other. They spend considerable time wrestling and jumping around, concurrently developing the strong muscles and lightning fast reflexes that they will need in order to hunt for themselves in future.

VOCALIZATIONS

A soft chuckling call is used by an oestrous female approaching an adult male or by friendly animals greeting each other. Newborns cheep soon after they are born if they are disturbed, hungry, or cold. Kits hiss, bark, and scream at each other when playing. Hissing by an adult ferret is a low-level threat vocalization. An explosive bark

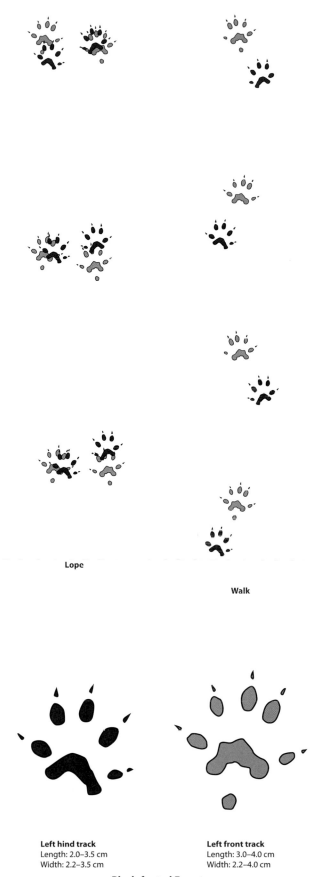

Lope

Walk

Left hind track
Length: 2.0–3.5 cm
Width: 2.2–3.5 cm

Left front track
Length: 3.0–4.0 cm
Width: 2.2–4.0 cm

Black-footed Ferret

often followed by a feigned lunge is the next level of threat. A serious threat is met with a succession of chattering barks (four to seven per second), often combined with rapid lunges. During a fight a shriek or scream may be emitted by the losing animal. This is usually accompanied by a musky odour, signalling the release of anal gland secretions.

SIGNS

These secretive, nocturnal, small carnivores are most easily detected by tracks and signs or by spotlighting after dark when their bright green eyeshine betrays their location. A characteristic trench-like indentation in the soil or snow at a prairie dog burrow entrance announces the active presence of a Black-footed Ferret. As the ferret removes loose soil, renovates a burrow for its own use, or removes the soil plug left by many prairie dogs as a barrier to predators, the waste earth is slowly dragged to the surface where it is pulled into a pile a short distance from the entrance. To move earth, ferrets walk backwards, dragging a small load of soil with their front feet. A narrow trench develops between the burrow entrance and the top of the excavated waste soil pile as the ferret repeatedly travels back and forth dragging loads to the heap, which typically develops a domed shape. By contrast, excavated soil worked by prairie dogs is energetically kicked out of the burrow mouth and accumulates into a fan-shaped throw mound. Trenching ferrets create their diggings all year-round, but most commonly from December to March. Most prairie dog digging occurs in the warm months. Although not well adapted for digging, a single ferret has been recorded moving more than 20 kg of soil in 109 minutes, which is the equivalent of a 90 kg human moving a metric ton of soil per hour – without a shovel. Most diggings are smaller, averaging 154 g of soil. Black-footed Ferrets, like the other weasels, have five toes on each foot, with the inside toe somewhat reduced. This toe does not reliably register in a track. The negative space between the toes and around the pads of each foot is filled with fur, which can distort the size of the track and influence its appearance. Claws may or may not register. The front feet are larger than the hind and leave slightly larger tracks. Ferrets travel above ground, usually in a lope where the hind feet register directly on top of the front tracks.

REFERENCES

Cully 1993; Dobson, A., and Lyles 2000; Harrington et al. 2003; Lockhart et al. 1998; Miller, B., et al. 1996; Miller, B.J., and Anderson 1993; Saskatchewan Department of Tourism and Renewable Resources 1978; Smith, H.C., 1993; Vargas and Anderson 1996, 1998; Wisely et al. 2008.

Least Weasel

also called Common Weasel (Europe), Snow Weasel (Russia and northern Europe)

FRENCH NAME: **Belette pygmée**

SCIENTIFIC NAME: *Mustela nivalis,* formerly *Mustela rixosa*

The Least Weasel is the smallest of the North American weasels and also the smallest member of the order Carnivora.

DESCRIPTION

Least Weasels have the typical weasel shape with a long slender body, a long neck, short legs, and large rounded ears. Their short tail is uniformly coloured and lacks a black tip although some individuals may have a few brown hairs on the tail tip in winter pelage. The pelage varies seasonally. The summer fur is about 10 mm in length, light brown to reddish to chocolate brown on the upper side, and creamy white to yellowish-white on the underside and legs. The belly fur is often marked with brown spots or blotches. The winter fur is white in northern populations (all of Canada) but remains bicoloured like the summer coat in more southerly populations. The white winter fur is about 15 mm in length. The skull is enlarged around the auditory bulla, indicating the importance of sound in the life of this weasel. Least Weasels can detect sounds from 52 Hz to at least 60.5 kHz with their best resolution being in the 1–16 kHz range. They are able to hear the ultrasonic calls of their prey, and they are also very sensitive to low-frequency sounds. Sight and olfaction are also well developed, but they hunt predominantly with the use of hearing. Least Weasels have paired anal glands that are used for marking and may have a defensive function. The dental formula is incisors 3/3, canines 1/1, premolars 3/3, and molars 1/2, for a total of 34 teeth. Least Weasels experience two moults each year, one in spring and the other in autumn. The spring moult begins on the head and upper back and moves rearwards and down the animal. The autumn moult is the reverse. Individuals complete the moult in about three weeks. The timing of the moult varies depending on photoperiod and is mediated by hormones. For example, breeding females moult later than do non-breeding females. In eastern Canada the winter pelage is typically complete by mid-November, and the summer pelage by March or April. Least Weasels have an intense green eyeshine, and they also have slit, rather than rounded, pupils, which are horizontal when contracted.

SIMILAR SPECIES

Least Weasels are most similar to Ermines and Long-tailed Weasels. A large male Least Weasel approaches the size of a small female Ermine. Least Weasels differ by not having a black tail tip, and the length of the tail is less than 25% of the head and body length. Even in white winter pelage the other two species retain their full black tail tip. The white hairs of North American Least Weasels fluoresce a dull lavender colour under ultra-violet light (those of Ermine and Long-tailed Weasels are bright white on the belly and lavender on the head and back).

Least Weasel (*Mustela nivalis*)

winter

summer

SIZE

Size varies considerably across the range, with the smallest forms being in North America, northern Europe, the Alps, and eastern Siberia. In all forms males are larger than females, especially in the larger European forms. In North America males are about 10%–15% longer in body length and up to 50% heavier than females. The following measurements apply to adult animals in the Canadian part of the range and cover the variation in size across the country.

Males: total length: 165–217 mm; tail length: 27–42 mm; hind foot length: 21–23 mm; ear length: 10–15 mm; weight: 28.2–64.2 g.
Females: total length: 153–188 mm; tail length: 23–33 mm; hind foot length: 18–29 mm; ear length: 10.0–14.5 mm; weight: 21.5–58.0 g.
Newborns: weight: about 1.1–1.7 g.

RANGE

Least Weasels are widely circumboreal in distribution. Their North American range stretches from Alaska to Ungava and southwards to Virginia along the Appalachian Mountains. This species is continuing to expand westwards and southwards in North America. A recent, extralimital report from Rimouski, Quebec, is indicated on the range map with a star. Eurasian Least Weasels (called Common Weasels) occupy the bulk of Europe (but not Ireland), Russia, and a strip of North Africa. Introduced from Britain into New Zealand, where they are seriously restricted by the lack of small native mammals, Common Weasels are not doing well, but they persist. These small weasels have also been introduced into the Azores, Crete, and Malta.

ABUNDANCE

Numbers vary regionally and annually. Least Weasels may be regularly common to rare, depending on location. They are quite rare in eastern Canada, less so in the west, and common in many parts of Europe and Asia. They may also be regionally or locally common or rare because their population density is often cyclical, following the population fluctuations of many of the weasel's rodent prey. Pelts of Least Weasels have almost no commercial value in North America, owing to their size, but some are killed in traps set for larger species.

ECOLOGY

These diminutive carnivores are very flexible in their habitat requirements. They can survive in coniferous forests, mixed forest, aspen parklands, grasslands, farmlands, cultivated fields, alpine meadows, semi-deserts, and Subarctic and Arctic tundra. Prey availability and cover are more crucial than is the vegetation type. Least Weasels specialize in capturing small mammals. They are small, short legged, and thin and can easily follow their prey along its grassy, earthen, or subnivean tunnels and into burrows. Further, the weasel's long neck allows it to carry prey in its mouth without stumbling over it. The price they pay for their shape and lack of body fat is the need to maintain a high metabolism in order to remain warm (up to six times the small mammal average during winter). This means that they must eat regularly, around nine to ten times daily, at a rate of at least a few grams per meal. A few consecutive missed meals can spell death by starvation, which is the major cause of mortality. They consume, on average, 40%–60% of their body weight each day, depending on the season (generally more in cold weather, less in warm weather). In northern regions, lemmings and voles are often the only available winter foods, which may be hunted to drastically low numbers, precipitating marked population fluctuations in first rodents and then Least Weasels. Ermine and Least Weasels are both small-mammal hunters. In regions of range overlap, when prey is plentiful, both survive without impact on the other. When prey becomes scarce, the size of the remaining available prey determines which weasel

North American distribution of the Least Weasel (*Mustela nivalis*)

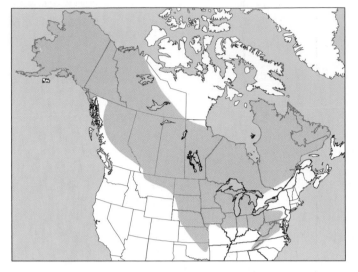

⭐ extralimital record

Eurasian distribution – New Zealand and Azores introductions are not illustrated

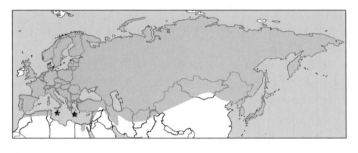

⭐ introduced

species retains the competitive advantage and which may become locally extinct. Least Weasels are well adapted to environments with wide fluctuations in food availability, as they can rapidly increase their numbers when prey is abundant. If they are locally depleted by a crash of prey numbers, they can expeditiously repopulate from neighbouring, unaffected regions through the immigration of dispersing juveniles. Juvenile females tend to remain closer to their natal home range (often within a few kilometres), but young males may travel many times that distance before settling.

Least Weasels do not burrow or create their own dens. Dens are the appropriated nests of prey species, often taken after the proprietor has been eaten, and then commonly lined with the plucked fur of the victim. Dens are not permanent, and each weasel uses several within its home range. Males occupy home ranges that vary from 0.6 to 26.2 ha and are considerably larger than those of females, which vary from 0.2 to 0.7 ha. Females hunt their rodent prey in their tunnels more often, and with greater success, than do the larger males; therefore, they can survive on smaller home ranges. Male home ranges tend not to overlap except when population density is very high, but each male's range normally overlaps one or more female ranges. The size of the home range is dependent on habitat,

population density, season, prey availability, and gender. Although Least Weasels may live to a maximum of ten years in captivity, very few wild individuals survive longer than three years, and most live a single year or less. Least Weasels will climb trees while foraging and are capable swimmers, albeit slow. Predators of this tiny weasel are numerous, ranging from hawks and owls to larger weasels, snakes, foxes, Domestic Cats and Dogs, and humans. The white winter pelage provides some camouflage against the snow. Although longer-tailed Ermine and Long-tailed Weasels have a black tail tip to misdirect potential predators, this bait would not be effective for the Least Weasels, because their tail is so short that a strike at a black tail tip would also capture the weasel's hind quarters.

DIET

Least Weasels are exclusively flesh eaters that primarily hunt small rodents such as mice, voles, and lemmings. Their prey choices vary seasonally and regionally. In general, they capture whatever is available, and when small rodents are scarce, they revert to alternatives such as bird eggs and nestlings, moles, chipmunks, rats, insects, lizards, frogs, earthworms, and carrion. Shrews are occasionally killed but are rarely eaten. Females tend to hunt for smaller prey than do the larger males. These weasels rarely predate domestic stock.

REPRODUCTION

Litter size, as well as the number of litters produced annually, is closely tied to food availability. In North America the average litter size is four to five (ranging from one to six in temperate areas, and up to 15 in Arctic regions during years of super prey abundance). Breeding may occur throughout the year but is concentrated in spring and late summer. Even in the Arctic, Least Weasels will breed under the snow when small rodents are abundant. Females remain in oestrus for about four days and may accept several males during that period. Two annual litters are the norm for adult females when prey is plentiful, but as many as three litters may be produced if food is very abundant. At such times females may experience a postpartum oestrus and mate again within five weeks of parturition, and so the second litter is born only four to five weeks after the previous one has dispersed. During periods of low prey availability, a single annual litter is usual, and some females (especially young females) may not reproduce at all. Unlike many other weasels, Least Weasels are not delayed implanters. Following a fairly normal development period of 10–12 days, the fertilized eggs develop into blastocysts, which then implant onto the uterus walls without undergoing any period of suspension. Active gestation following implantation lasts about 25 days, and the entire gestation period is 35–37 days. Kits are born blind, deaf, and hairless and completely dependent on their mother for food and warmth. By four days old they are covered in fine white hairs. From about six days old, young males are noticeably larger than young females in the same litter. By 11 days old, their milk teeth begin to erupt. By 18 days old, they are well furred. Their ears open at around 21–28 days old, and their eyes open by 26–30 days old. Their permanent teeth are fully erupted by 33 days old. Weaning begins around 32 days old and is completed by 42–56 days old. Their mother starts bringing meat to the nest to feed the young

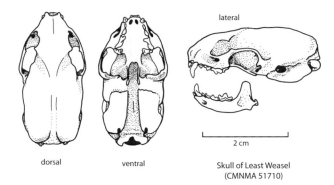

lateral

dorsal   ventral   Skull of Least Weasel
                                 (CMNMA 51710)

2 cm

when they are around 18 days old, and by 50 days old the young can kill their own small prey. Juveniles leave their mother's home range at eight to ten weeks old. Young females may become sexually mature as young as 12 weeks old if they are well nourished, but may not mature until the following spring if prey is scarce. Males are sexually mature at around three to five months old.

BEHAVIOUR

Like other weasels, Least Weasels are solitary except during the breeding season. They are driven by hunger to be active intermittently both day and night. A period of activity is followed by a few hours' rest. Foraging may be timed to coincide with a time that prey is most vulnerable. Least Weasels are stimulated to kill by movement and will continue killing moving prey until they are exhausted. Caching of surplus prey is a common behaviour of this weasel, and especially so of lactating females; they are reluctant to leave their young for extended hunting forays but must satisfy the energy demands of lactation. Since most prey is larger than can be consumed in one meal or even a day, some caching may always be necessary. Although the killing behaviour is innate, juveniles become proficient at an earlier age if they can watch their mothers hunt. Least Weasels typically use a "wrap-around assist" to capture small rodents that frequently exceed their own mass. They jump on the back of the prey and attempt to grasp it with all four feet in order to manoeuvre and contain it so that they can deliver a killing neck bite. A rolling ball of furiously struggling weasel and mouse may ensue until the prey is subdued.

Individual home ranges are defined by scent marking, using anal gland secretions, by dragging the anus over the ground or prominent objects. These secretions can also be voluntarily evacuated by a frightened weasel to deter a potential predator. This is called the dreaded "stink bomb" by experienced trappers. The weasels also rub their bodies on prominent objects to leave their body scent as a marker. Each sex will defend its home range from others of its gender. However, males, being larger, have generally free access to the home ranges of females, except in the later stages of their pregnancy and during lactation, when the increased defensiveness of the females will deter even a large male. During the breeding season, males travel widely in search of oestrous females. Females meanwhile continue to maintain their well-defined home range. By late summer the male home ranges are re-established, and some redistribution occurs

to accommodate juveniles. Courtship and mating is a rough-and-tumble exercise for Least Weasels. A female will only allow the male to approach closely if she is in heat. A scuffle and some fighting then result until the male can grasp the female by the back of the neck and then position her with his body and feet to permit intromission. Copulation may last up to 90 minutes. Males provide no parental care. Siblings tussle and mount each other, but once they become independent, they are intolerant of each other and disperse.

VOCALIZATIONS

Captive animals have been heard chirping, hissing, trilling, squealing, wailing, and squeaking. A hiss is a response to a low-intensity threat. A higher intensity, more pressing threat may be met with a series of loud, sharp chirps (like an explosive bark). A cornered Least Weasel may produce a loud, prolonged squeal or a plaintive wail. Quieter high-pitched trills are produced between friendly animals or by a mother calling her kits. Kits are especially vocal, producing high-pitched squeaks for the first month, which change to chirps after their eyes open.

SIGNS

Each foot has five toes with non-retractable claws and a fully furred underside except on the pads. The inside toe on each foot is reduced and does not register reliably. Claws may or may not leave an impression. The tracks are very small, not usually larger than those of their small rodent prey. The usual gait is the typical bounding lope favoured by all of the smaller weasels, with the hind feet registering directly onto the front tracks. Bounds are typically 10–32 cm apart and may be up to 76 cm apart. As with all the small weasels (including Ermines and Long-tailed Weasels), the trails are highly sinuous because the foraging weasel explores every nook and cranny, up into trees, and under the snow in search of prey. See the "Signs" section of the Ermine account for small weasel track illustrations. Snow tunnels are about 2.0–2.5 cm in diameter and can be quite lengthy, especially if the weasel is able to break into the subnivean tunnel systems of its prey. Scat is 20–50 mm long and 3–6 mm in diameter. As with other weasels, their scats are often twisted, usually have pointed drawn-out ends, and are blackish. Also like other weasels, they often create latrine sites near their dens. Kill sites may be difficult to discern as the weasel often carries the prey to a safer spot to consume it. Least Weasels start by eating the brain and then proceed rearwards. They commonly leave the front part of the cranium (which includes the toothrow), the feet, tail, and stomach. A well-fed Least Weasel may only consume the brain and then cache the rest. Caches may be found and are more common in winter; these are carefully hidden near the den site. A running Least Weasel usually carries its tail at a 90° angle to its body (straight up).

REFERENCES

Bellows et al. 1999; Boulva 2009; Cushing and Knight 1991; Elbroch 2003; Frey 1992; King and Powell 2007; Kurta 1995; Macdonald and Barrett 1993; Mock et al. 2001; Peterson, R.L., 1966; Sheffield and King 1994; Slough and Jung 2007; Smith, H.C., 1993; Svendsen 2003; Tumanov and Sorina 1996; Youngman 1975.

Sea Mink

FRENCH NAME: **Vison de mer**

SCIENTIFIC NAME: *Neovison macrodon,* formerly *Mustela macrodon*

A large mink, the Sea Mink, once lived on islands and possibly also coastal locations along the eastern coast of North America; it is known only from bones in archaeological sites. Sea Mink are thought to have become extinct perhaps as early as 1860 or as late as 1920. There is considerable debate about the validity of this species as some mammalogists continue to consider the Sea Mink to be a large subspecies of American Mink, rather than a unique species.

DESCRIPTION

Sea Minks were extinct before being described by science. A specimen that had been caught in 1894 on Campobello Island, New Brunswick, and then mounted, has been the subject of continuing dispute as to whether it is a Sea Mink or an American Mink. Owing to the uncertainty, it is herein treated as an American Mink, which means that there are no fur samples available with which to describe the Sea Mink's coat colour. Nevertheless, anecdotal reports from the early fur trade indicate that fur buyers and trappers selectively differentiated some east coast mink pelts as being larger and hence more valuable. The fur on these pelts was said to be more reddish and coarser than that of other mink pelts. The dental formula is incisors 3/3, canines 1/1, premolars 3/3, and molars 1/2, for a total of 34 teeth.

SIMILAR SPECIES

Based on cranial measurements, the height of the mandible, and the length of toothrows, male Sea Minks were on average about 20% larger, and female Sea Minks about 10% larger, than male American Minks from the same region. Sea Minks were not only larger but also more robust than American Minks.

SIZE

Like all the weasels, Sea Mink males were larger than females (from 10% to 100% larger based on measurements of bones). There are no measurements of whole animals available.

RANGE

Sea Minks formerly ranged along the east coast of Maine and nearshore islands, possibly as far north as New Brunswick and as far south as Massachusetts. Most of the bones have been found in shell middens along the central Maine coast from Casco Bay to Mount Desert Island. Sites to the north and south of this area contain single bones or only a few scattered bones, never a whole animal. The oldest bones that have been reliably radiocarbon dated were around 5100 years BP. Bones from the most northerly sites, in New Brunswick, are associated with stone tools made from rock called Kineo-Traveller Mountain porphyry (a very hard stone containing small feldspar crystals), known to occur only in north-central Maine, which was favoured by stone-tool makers of the region. There is a good chance that the Sea Mink bones as well as the stone were trade goods from further south. The possibility also exists that a similar origin may account for the few bones that have been found south of Maine in Massachusetts.

ABUNDANCE

The extinct Sea Mink is a classic example of the results of over-exploitation. They were hunted by indigenous people along the eastern seaboard for food and fur for centuries. Their larger pelts brought a high price in the European fur trade, and so Sea Mink were sought after in the 1800s by European and native trappers. Excessive trapping quickly decimated the already sparse population density of this mammal and was probably responsible for the extermination of the species.

ECOLOGY

This large, maritime mink inhabited rocky coasts on nearshore islands and so were likely strong swimmers. Like the American Mink, they were also probably solitary and primarily nocturnal and had a low population density. Persistent anecdotal reports indicate that they had a distinct and strong odour, as do all members of the weasel family.

DIET

Unknown.

REPRODUCTION

Unknown.

Distribution of the Sea Mink (*Neovison macrodon*)

? distribution uncertain

REFERENCES
Black et al. 1998; Campbell, R.R., 1988; Mead et al. 2000; Seaflon 2006, 2007; Waters and Ray 1961.

American Mink

FRENCH NAME: **Vison d'Amérique**

SCIENTIFIC NAME: *Neovison vision,* formerly *Mustela vison*

Their soft, lustrous fur has made the American Mink one of the most harvested fur-bearers in North America. American Mink have been raised in fur farms for close to 100 years, and about 30 million mink pelts are produced annually worldwide.

DESCRIPTION
Like most weasels, American Mink have a thin, elongated body with short legs and a pointed face. Their rounded ears are short and barely protrude beyond the fur. A long bushy tail is equal to about one-half of their head and body length. Their glossy fur is dark brown to dark reddish-brown, often blending to black at the tip of the tail. A small, irregular, white patch often occurs on the chin, and other white markings may be found on the throat, chest, and belly. The configuration of light markings is unique to each individual. White, tan, light-brown, and blond coats show up in wild populations infrequently. The coat colours of mink raised on fur farms are diverse and include white, black, white with black spots, pastel, sapphire, platinum, mahogany, pearl, and lavender. American Mink have paired anal glands that they use for scent marking. Their hearing is acute, and they can detect ultrasonic vocalizations produced in the 40 kHz range by many of their small rodent prey. American Mink are habitat generalists with a mix of terrestrial and aquatic adaptations. Aquatic adaptations include a streamlined body shape, partially webbed toes, a thick undercoat, and oily, water repellent guard hairs. Mink are additionally adapted for an arboreal lifestyle with a rotating ankle joint that allows them to rapidly descend a tree head-first. The dental formula is incisors 3/3, canines 1/1, premolars 3/3, and molars 1/2, for a total of 34 teeth. American Mink experience two annual moults. The dense, winter coat is shed in spring (April–May), and the shorter summer coat is fully grown by late June. The summer pelt is shed in autumn (September–October), and the full luxuriant winter coat is complete by November. Both coats are similarly coloured, but the winter coat is longer and denser to provide more insulation. Unlike most of the smaller weasels, mink do not turn white in winter.

SIMILAR SPECIES
Fishers and Northern River Otters are similarly coloured, but both are considerably larger than mink. Fishers are primarily forest weasels. Otters share an aquatic lifestyle with mink but have a long, thick, tapered tail and no white markings. Other small weasels (Least Weasels, Short-tailed Weasels, and Long-tailed Weasels) found in the mink's Canadian range share the same body shape but are smaller and bicoloured in summer (reddish-brown above and white to creamy yellow below) and turn all white in winter. Mink are most similar in size to American Martens, but martens usually have a longer, bushier tail and a prominent yellowish or orange throat patch and spend most of their time in forested habitats.

American Mink (*Neovison vison*)

SIZE

There is some variation in size across the country. Mink in the northwest are largest and may be more than 50% heavier than those in the east and southeast. Adult males are longer than females by about 10% and are 1.6 to 1.9 times their weight. Weight fluctuates throughout the year, especially for males. Animals are heaviest in late autumn and lightest in late summer. The following measurements apply to adults from within the Canadian range and encompass the variation from west to east.

Males: total length: 491–647 mm; tail length: 145–225 mm; hind foot length: 56–76 mm; ear length: 19–28 mm; weight: 548.5–1741.0 g.

Females: total length: 420–550 mm; tail length: 128–201 mm; hind foot length: 47–70 mm; ear length: 13–25 mm; weight: 424.7–856.8 g.

Newborns: weight: about 6–10 g. Sexual dimorphism is already apparent, with male neonates being larger than females.

RANGE

American Mink are endemic to North America and are found throughout the continent, except beyond the treeline in northern Canada and in the drier parts of the southwest and south. They are excellent swimmers and have populated many freshwater and nearshore marine islands. This species has been farmed for its fur since the 1920s, and escapees from such endeavours were accidentally introduced into Newfoundland in the 1930s, where they are now successfully established. An introduction onto Anticosti Island appears to have been less successful, as mink there are now believed to be extirpated. Deliberate introductions, or fur farm escapees, have populated the British Isles, Iceland, Scandinavia, and most of central Russia. South America supports further colonies, but their extent on that continent is uncertain.

ABUNDANCE

Although rare or uncommon in some areas such as Cape Breton Island and southern Florida, American Mink are considered to be generally common throughout most of their geographic range, in appropriate habitat.

ECOLOGY

American Mink are closely associated with water, although they may occasionally occur in drier habitats if prey is plentiful. They are highly adaptable, as shown by their rapid range expansion after introductions into Europe. There they successfully compete with the native European Mink, likely contributing to their decline in many areas of overlap. Further evidence of this same adaptability is the regular occurrence of American Mink in urbanized areas, often along heavily industrialized watercourses. American Mink forage in salt, fresh, and alkali waters. They are competent swimmers, but their underwater vision is poor. They can dive down as far as 5–6 m but tend to forage near shore in the vegetation. Most dives are 5–20 seconds long. They prefer to forage in calm water with few waves, and they avoid unproductive pebble or sandy beaches. When hunting in salt water, they commonly frequent tidal pools at low or mid tide when prey is confined and probably easier for them to catch. Mink are active all year, although the level of their activity decreases

Distribution of the American Mink (*Neovison vison*)

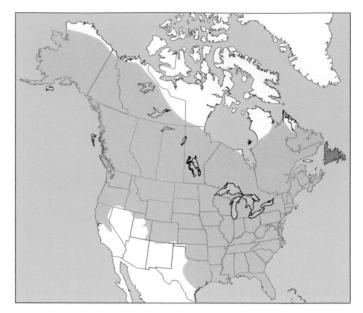

introduced

Eurasian distribution

during the cold season and especially during severe weather. Depending on prey availability, mink may travel up to 12 km daily while foraging.

Individual mink occupy home ranges of varying size, depending on the habitat and prey abundance. The home ranges of males may be 1.5 km² in prime habitat or over 16 km² in poorer areas with less prey, and they are up to two times larger than those of females. The home ranges of mink of the same sex tend not to overlap (this is thought to be more a matter of mutual avoidance than territoriality), but a male's home range usually includes portions of the home ranges of at least one female, and often several. Radio-tracked juveniles dispersing in autumn may travel up to 45 km before finding a suitable home range. Once they have established a home range, American Mink use a variety of resting and denning sites, often changing them nightly. Dens may be located in abandoned burrows created by other animals such as Woodchucks, marmots, or ground squirrels, in muskrat mounds and burrows (usually after the resident has been caught and eaten), in abandoned beaver lodges or bank burrows, in brush piles, in hollow logs, in rock crevices, in cavities under waterside trees, under bridges (usually in cracks in the bridge foundations), in tree cavities, and under the snow in natural cavities. Most dens have two to five entrances and are less than 2 m from water. Mink are capable tree climbers, able to jump from tree to tree and descend rapidly head-first. Their bounding speed averages 9–10 km/h, while their top swimming speed is around 2–3 km/h. Natural predators include hawks and owls, Coyotes, Red Foxes, Bobcats, Canada Lynx, alligators, and otters. Some mortality occurs through roadkill and drowning in fish traps, but by far the greatest human-caused mortality comes from trapping. Although about 90% of the economic demand for mink fur is satisfied by farmed animals, still around 400,000–700,000 wild mink are trapped each year in North America. Although captive animals can live up to eight years and wild individuals to six years, very few wild mink survive beyond three years of age.

DIET

American Mink are strictly carnivorous but consume a variety of prey when it is available. The bulk of their diet is composed of small, slow-moving fishes, frogs, crabs and crayfish, muskrats (typically captured by the larger males), and small mammals (rodents and shrews). They will also catch and eat rabbits and hares (typically captured by the larger males), birds and bird eggs, reptiles, insects, earthworms, and snails when the opportunity presents itself. They are major predators of the eggs of nesting waterfowl. Carrion, large or fast-moving fishes, and other smaller weasels may be added to the diet on rare occasions. Diet varies seasonally. For example, during summer 50% of a prairie mink's diet may be waterfowl. Mink can be serious pests to farmers for their depredations of poultry. They can also do immense harm to fry at fish farms and hatcheries. Mink can excrete indigestible parts of their dinner in as little time as one hour after eating, although, on average, they take around three hours to process food through their digestive tract.

REPRODUCTION

The mating season lasts about three weeks and occurs any time during February to May depending on the locality. In general, southerly populations breed earlier, whereas northern ones breed later. Mating is a vigorous activity for mink and usually begins, and may proceed, as a rough and tumble fight. The male eventually grips the female at the back of the neck, and recently mated females can often be identified by a bare patch of skin or a fresh wound at that site. Copulation is typically lengthy, lasting from ten minutes to over three hours, and averages over an hour. The male retains his grasp of the female's neck, and multiple copulations are typically interspersed with short rests, during which both animals lie together on their sides. Both sexes are polygamous, so a litter may have more than a single sire. American Mink, like most of the weasel family, exhibit delayed implantation. The fertilized eggs stop dividing at the blastocyst stage and float in the uterus for 8–45 days. As day length increases and temperatures warm, the blastocysts implant in the wall of the uterus, and active gestation commences. The total pregnancy period lasts 51 days on average (ranging from 40 to 75 days), and the active gestation portion is 28–32 days. The average implantation date in the same area may vary from year to year, depending on the local conditions.

Females produce, at most, a single litter per year. Young are born from April to July. The litter size averages four (ranging from one to eight). Older females tend to have larger litters. Kits are blind and deaf at birth and already have a thin, short coat of fine silvery hairs. Their first milk teeth appear at about 16 days after their birth; permanent teeth begin erupting at 44 days old and are fully emerged by 71 days old. Their eyes open at 21–25 days old, and weaning occurs around 32–35 days old. Breeding dens may be used for up to 40 days, after which the female and young use a variety of den and resting sites within the female's home range. The juveniles begin hunting when they are about eight weeks old but follow their mother closely during the early summer. They learn to fend for themselves after a month or so and disperse in autumn. Young females reach adult weight at around four months old and reproduce for the first time in the following breeding season, at about 12 months old. Males grow for a longer period, reaching adult weight at around 9–11 months old and sexual maturity at around 10 months old. They, too, are ready to breed as yearlings.

BEHAVIOUR

American Mink are generally nocturnal but may be out during the daytime, because if they are hungry, they can hunt whenever their prey is active. They are largely solitary, apart from pairings during the mating season and while females are raising kits. Sometimes young may hunt together before they disperse from the maternal home range. Mink mark their home range to identify their occupancy, as do others in the weasel family. They carefully deposit the secretions of their anal glands onto logs, stumps, branches, rocks, or the ground around the boundaries of their home range. The anal glands can be voluntarily emptied by tensing the muscles in that region (but only expelled to a distance of less than 30 cm) when the animal is under duress. The acrid, sulphurous, musky odour may have a defensive function similar to that made famous by skunks. Other glands (the proctodaeal glands that empty into the rectum) mark the faeces as they are excreted, and faeces are often deposited in prominent locations where they also serve to mark the home range. Mink typically follow the shoreline fairly closely as they forage, looking into the water from above or sticking their head underwater if reflection is a problem. Once prey has been located, they will then enter the water and swim in pursuit. Like most of the

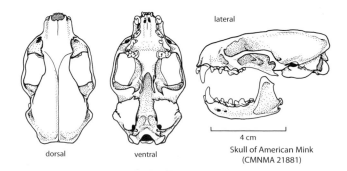

lateral

4 cm

Skull of American Mink
(CMNMA 21881)

dorsal   ventral

weasels, mink will kill and cache surplus prey for later consumption. Young mink can be quite playful, vocalizing, jumping on each other, and wrestling until the litter disperses in autumn. Even adults will occasionally create a snow or mud slide or playfully plunge into soft snow, creating short tunnels. Any mink, but especially juveniles, can be attracted towards a squeaking sound that can be deliberately produced by sucking against a hand or by pressing a squeaky toy.

VOCALIZATIONS

These weasels use a variety of vocalizations. They will scream, hiss, squeal, bark, and squeak. When alarmed or cornered, they raise their hackles and the fur on their tail, arch their back, hiss and emit high-pitched squeals, and sometimes empty their anal glands. The tail meanwhile waves rapidly back and forth, perhaps to distribute the strong odour. During the mating season either sex may produce a chuckling call.

SIGNS

Like all the weasels, American Mink have five toes on each front and hind foot. The inside toe on each foot is reduced in size and may not register in a track, especially a hind track. Claws sometimes register, depending on the substrate. Mink tracks are most commonly found along muddy stream beds and lake shores where they are left as the mink meander in search of food. Tracks tend to be stellate (star shaped) with all five toes about evenly spaced. The front tracks are larger than the hind. There is less fur in the space between the toes of mink than on other small weasels. Scats are commonly corkscrew shaped or twisted, dark, tapered or pointed at each end, 3–9 cm in length and 0.6–1.1 cm in diameter. When fresh, they have an unpleasant odour. Scats are used for marking home range, and so they are commonly deposited in conspicuous locations, such as on clumps of vegetation, fallen trees, stumps, or rocks usually near a waterway or marsh. Mink also regularly leave scat on muskrat scent posts and may create a latrine area near regular denning or resting sites. Mink scat can be easily mistaken for droppings produced by smaller or larger weasels. Unlike otter scat, which is primarily made up of fish bones, mink scat commonly contains fur or feathers along with fish bones. Fishers and American Martens regularly eat fruit (in season), which can be distinguished in their scat; mink do not consume plant material. Fisher and American Marten scat commonly contains the larger hairs of the bigger mammals that make up some of their diet, whereas mink scat usually contains parts of fishes, insects, or invertebrates (such as crayfish) or the fine fur of small mammals. Mink sometimes tunnel into the snow. The access opening to a snow tunnel is 4.8–5.4 cm in diameter.

**stride length of
a bound varies
from 28 to 66 cm**

Bound

Lope

Left hind track
Length: 2.0–4.5 cm
Width: 2.5–4.1 cm

Left front track
Length: 3.0–7.3 cm
Width: 2.2–4.5 cm

American Mink

REFERENCES

Arnold, T.W., and Fritzell 1990; Banfield 1974; Bonesi et al. 2000, 2006; Bonesi and Macdonald 2004; Elbroch 2003; Jameson and Peeters 2004; Larivière 1996, 1999, 2003; Macdonald and Barrett 1993; Mason 1994; Mech 2002; Price et al. 2005; Rust et al. 1965; Slough and Jung 2007; Smith, H.C., 1993; Stevens, R.T., and Kennedy 2006; Tumanov and Sorina 1996, 1997; Verts and Carraway 1998; Yamaguchi et al. 2003; Youngman 1975.

American Badger
also called Badger, North American Badger

FRENCH NAME: **Blaireau d'Amérique**

SCIENTIFIC NAME: *Taxidea taxus*

American Badgers are remarkably adapted for digging. Their legs and long front claws (4–5 cm) are robust and powerful, and their front feet are partially webbed for pushing dirt. They have a well-developed transparent membrane, called a nictitating membrane, which they can draw over their eyes to protect them from dirt. In friable soil a motivated badger can dig itself out of sight in around 90 seconds, making it one of the fastest digging mammals on the planet.

DESCRIPTION

This thick-bodied weasel has short muscular legs, long and stout front claws, a broad wedge-shaped head with a pointed nose, and a body that is somewhat flattened. The short, thick tail is well furred. American Badgers are grizzled grey on the back, often with a tan, yellow, or brownish undertone. Hairs on the back are long, giving the animal a shaggy look. The hairs on the sides are even longer than those on the back, adding to the animal's flattened appearance. This fringe can hide the short legs and make it seem as though the badger is flowing over the ground as it moves. The chin and throat are whitish, and their bellies are light buff to brownish-yellow, often with an elongated central white or pale buff marking. Distinctive black markings extend from the eyes to the muzzle, in a "badge" down the cheeks, and in front at the base of the ears. The forehead is dark brown. A prominent white stripe runs along the top of the head, usually from the nose to the nape, but the length is variable and it may extend as far as the shoulder in Canadian subspecies. The stripe may continue all the way to the base of the tail on some individuals in the American southwest. The legs and feet are black or dark brown with light tan to brown claws. The claws on the front feet grow faster than those on the hind to compensate for wear. The ears are large but close set with well-rounded, often black-rimmed, pinnae. Young are pale buff coloured, sometimes almost white. Badgers have tough, loosely connected skin, which helps them to manoeuvre underground, but it is not loose enough to allow them to "completely turn inside their skin to bite an attacker" as the old myth suggests. They have two pairs of scent glands, one pair on the abdomen and the other on either side of the anus. The musky-smelling secretions from the anal glands are potent but not really unpleasant. The dental formula is incisors 3/3, canines 1/1, premolars 3/3, and molars 1/2, for a total of 34 teeth. American Badgers moult twice a year, likely in spring and late autumn.

SIMILAR SPECIES

The distinctive and characteristic facial markings leave little doubt as to the identity of this mammal. Its ground works are also distinctive because of their size and extent.

SIZE

Males grow longer in body length than do females and are about 26% heavier on average. The following measurements apply to adult animals in the Canadian part of the range.

Males: total length: 730–880 mm; tail length: 110–195 mm; hind foot length: 110–128 mm; ear length: 41–55 mm; weight: average 8–10 kg, up to 13.6 kg.

Females: total length: 700–765 mm; tail length: 100–154 mm; hind foot length: 107–118 mm; ear length: 48–55 mm; weight: average 6.4 kg.

Newborns: weight: about 200–225 g.

American Badger
(*Taxidea taxus*)

RANGE

American Badgers occur only in North America. Their distribution has been highly affected by human activities. As an open country inhabitant, badgers were found mainly in the western part of the continent prior to European colonization. The extensive agricultural clearing that followed this settlement allowed badgers, and some of their primary prey species, to expand their ranges eastwards and northwards. American Badgers currently occur from southern Ontario to British Columbia and southwards to central Mexico.

ABUNDANCE

Badger fur was once prized for making shaving brushes, and their beautiful and durable winter pelt is still used in the fur trade, but less so now than in the past. Although American Badgers are more wide-spread now than they were previously, their numbers are declining in most jurisdictions. Much maligned, their digging can cause serious disruption of the ground surface, which can damage harvesting machinery. Farmers and ranchers are also concerned about their livestock injuring themselves in badger digs, a problem that is more possible than actual. Nevertheless badgers are considered to be nuisance animals in many regions and are frequently shot. They are also negatively affected by poisoning campaigns aimed at prey species such as ground squirrels and pocket gophers, which are also considered to be agricultural pests. Consequently, badgers are no longer common in most areas that are under intensive agriculture. By 2004 there were thought to be fewer than 240 badgers of reproductive age left in British Columbia, and the subspecies represented there (*Taxidea taxus jeffersonii*) has been considered "Endangered" by COSEWIC since 2000. The preferred badger habitat in British Columbia is highly fragmented, and much of what remains is rapidly being converted to cropland, orchards, or vineyards or is being invaded by trees and shrubs whose growth is encouraged by years of fire suppression. A protected population exists in Kootenay National Park, and British Columbia law prohibits shooting and poisoning. Badgers in Alberta are uncommon, as there were only an estimated 9000 animals in the province in 2000. In Saskatchewan they are also uncommon, with an estimated 13,000–26,000 in 1998. Manitoba, with an estimated 3000–5000 animals in 1998, has a stable population. Ontario badgers are rare, and there are perhaps up to 200 animals. Trapping is still permitted in Alberta, Saskatchewan, and Manitoba. Ontario badgers are protected and classified as "Endangered."

ECOLOGY

Badgers are most common in open grasslands and aspen parklands, but they also occur in semi-desert regions and open agricultural lands, as well as cleared land on golf courses, road margins, and clearcuts. Two requirements are listed for their survival: friable soil and prey availability. Badgers are exceptional diggers. They excavate their own underground dens (burrows), which are central to their life and provide a safe, temperature-moderated, and comfortable location for resting during daylight, storing food, raising young, and sleeping through the winter. Den sites are scattered throughout the home range and are used when the resident badger moves into

Distribution of the American Badger (*Taxidea taxus*)

the area to forage. If a burrow is not accessible, the badger simply digs another. Apart from nursing females, badgers rarely use a den site for more than a day at a time during the summer. Females are more reliant on a single burrow when their young are still suckling, but will move them to another burrow as necessary when the local prey becomes sparse. Suitable burrows are frequently reused and may remain intermittently active for years. Tunnels can be more than 3 m deep and 15 m long in a well-developed badger burrow. A maternal den tends to be more complex than a simple resting burrow, with more chambers and consequently a larger throw mound at the entrance. In cold regions badgers typically occupy a winter den for the whole cold season and block the burrow entrances before settling in for the winter. Abandoned badger burrows are used for shelter by a wide variety of small animals such as spiders, rabbits, insects, and salamanders and are used by "Endangered" Burrowing Owls for nest sites, by "Endangered" Swift Foxes for dens, and by snakes for hibernacula. Badger diggings aerate the soil, mix nutrients, and allow water to reach the lower levels.

During winter when their prey is hibernating, food is in short supply, and so American Badgers spend most of their time sleeping, emerging from their dens only on mild days. Although not true hibernators, they do drop into torpor for short periods during which their body temperature may drop 1°C–9°C. Torpor typically does not exceed 29 hours at a time, including 15 hours to drop into torpor and 6 hours to arouse. American Badgers are mainly crepuscular and nocturnal, preferring to remain in their underground burrows for most of the day and normally only venturing out in low light or darkness. Juveniles are exceptions as they are commonly active during daylight. The home ranges of all sex and age classes may overlap with those of neighbours, although the ranges of adult females tend to overlap the least. The home-range size varies with habitat, prey abundance, season, and sex, from 2 to 37,000 ha or more. Males generally occupy larger home ranges than do females. During the breeding season the home ranges of adult male badgers expand and in some cases may nearly triple in order to overlap the ranges of females. Before and after the breeding season and during the winter the home-range sizes of males and females are similar. Both sexes occupy a much-reduced home range during the winter. Home ranges in the East Kootenay region of British Columbia and the dry interior of Illinois are the largest recorded for the species, with breeding males averaging 30,000–37,000 ha and females averaging over 3400 ha. Likely the dry semi-desert habitat with sparse prey accounts for the extreme size of their home ranges as the animals must roam over vast distances to find enough to eat. The female range size is largely governed by prey availability, whereas the male range size is largely governed by female availability. In regions where prey is abundant, females have smaller ranges, and males do not have to travel far for breeding opportunities; so they too occupy smaller home ranges. Transient juveniles may reside temporarily on the home range of an adult without undue harassment.

Badgers are capable swimmers and are known to patrol shorelines from the water in search of amphibians or nesting waterfowl. Dispersing juveniles are known to travel long distances, sometimes through unsuitable terrain. Journeys of up to 110 km have been reported, which may involve crossing of rivers and highways. Such dispersal exacts a high cost and is the period of highest badger mortality because the naive juveniles face many hazards, from starvation to drowning to predators and motor vehicles. Like most of the weasel family, American Badgers are ferocious when cornered or attacked, and they can erect their hair to make themselves appear considerably larger and even more formidable. Few predators, except bears and Cougars, will attempt to kill an adult badger. Juveniles are additionally vulnerable to Coyotes, Grey Wolves, eagles, and adult badgers. However, human caused mortality can be high because many badgers are killed on roadways and as a result of farming activities. In the wild, badgers may live up to 14 years, but few survive more than four years.

DIET

American Badgers are carnivores that specialize in hunting small to medium-sized burrowing (fossorial) rodents such as ground squirrels, prairie dogs, pocket gophers, Woodchucks, and Yellow-bellied Marmots. Most often the badger digs these animals out of their underground burrows. A badger's menu varies with prey availability as the badger readily switches to alternatives when one favoured species becomes scarce. Like most other carnivores, badgers are opportunists and will supplement or alter their diet to take advantage of abundances of mice, voles, muskrats, rabbits, hares, reptiles, amphibians, fishes, birds, bird eggs, and insects. They will sometimes eat insignificant amounts of plant material such as berries, corn, and other grains, as well as herbs and grasses, but only when prey availability is low. Badgers readily consume carrion as it is encountered. Like most weasels, they will kill more than they can eat if the opportunity arises. The surplus is cached underground for future use.

REPRODUCTION

Mating takes place in summer and early autumn. Females are likely induced ovulators and require sufficient stimulation from mating to ovulate. Copulation may be lengthy (20 minutes or more), and females may accept the efforts of several males. Following fertilization, the ovum develops to the blastocyst stage and then stops growing as it floats freely in the uterus for around 6.5 months. This is called delayed implantation. Blastocysts implant in December–February, and a litter of one to five (average of two to three) kits is born in late March–early May following an active gestation phase of six weeks. The entire gestation period lasts seven to eight months. Northerly badgers (probably including all those in Canada) tend to have smaller than average litters (ranging from one to two). American Badgers are capable of producing a litter annually, but this is not normally the case for Canadian females. A 50% annual reproductive rate is the average, even in British Columbia, although some females manage to produce a litter every year, while others appear not to reproduce at all. It is likely that the nutritional status of the female determines her reproductive potential, and so well-fed animals breed more often and have larger litters. The sparsely furred and blind kits are born in an underground maternity den that is lined with dried grasses. Their eyes open at about one month old. Mothers nurse the kits for two to three months and begin to feed them solid food during the last month of suckling. Juveniles begin dispersing from the natal burrow at three to four months old. The majority of females mature sexually at about one year old, and a minority breed at four to five months old. Most males require about 14 months to become fully sexually potent, but it is thought that few make significant contributions to reproduction until they are around four years old.

BEHAVIOUR

Badgers are generally solitary apart from mating, raising young, and having short-term associations with siblings after the family breaks up. Family groups hunt together for about a month after the young are weaned and before they disperse. Male badgers will fight over an oestrous female, but otherwise aggression among neighbours or

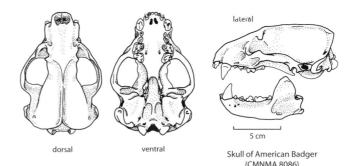

dorsal   ventral   lateral   Skull of American Badger
(CMNMA 8086)

5 cm

Trot

Walk

Left hind track
Length: 4.8–7.0 cm
Width: 3.5–5.1 cm

Left front track
Length: 7.3–9.8 cm
Width: 4.0–6.7 cm

American Badger

between intruders and residents is uncommon unless badger densities are unusually high, such as when their habitat is encroached by development. Hunting strategy depends in part on the prey. A badger hunting ground squirrels will locate an active burrow by smell, then may block all the burrow entrances but one, and excavate the burrow from the remaining open entrance until it captures the ground squirrel. When the ground squirrels are in hibernation, the badger does not bother to plug the entrances. In some cases a badger will plug an entrance to a ground squirrel burrow with rocks or other hard debris, which it carries or drags from locations close at hand. It may also use soil from a nearby mound, which it drags to the hole, leaving a characteristic ribbon-like trench of soil. Badgers hunting pocket gophers dig a series of test pits into the gopher's burrow system so that they can locate by smell the most recently used portion. This efficiency saves the badger from having to dig up the whole, often very extensive, burrow system. Badgers regularly patrol their own den sites, knowing that some hapless animals may be using the convenient protection. They are also known to ambush and chase prey from a hidden burrow entrance. One study found that over 180 litres of soil are moved on average at each predation site.

Like most weasels, Badgers are inveterate hoarders. They must consume on average the equivalent of two ground squirrels each day to maintain their condition. Most feeding occurs underground. The opportunity to kill surplus prey is rarely ignored, and the excess is cached safely for future use. Carcasses may be buried at the surface or stored in underground chambers. Each carcass is cached individually and is retrieved in the same order as it was cached. Caching is most common in autumn, although maternal females may engage in caching during the early summer. Small mammal prey is killed by grasping it around the thorax and clamping down with the teeth, causing massive internal injury. This is typically accomplished without puncturing the skin. Red-tailed Hawks and Coyotes may observe and sometimes follow a foraging American Badger in hope of capturing a distracted rodent fleeing from the badger's efforts. Some associations with Coyotes have been described as affectionate and cooperative, but it is still uncertain what the badger gains from the relationship, although it is probable that both predators hunt more effectively in association. Badgers are largely sedentary. Once they establish themselves in a home range, most remain there for life. Newly inhabited areas are usually colonized by dispersing juveniles. Badgers are relatively tolerant of human activity, perhaps owing to their nocturnal habits. Little is known about how badgers communicate using their scent glands. It has been noted that badgers investigate each den or dig site they encounter (usually their own) and defecate and urinate in the loose soil at the entrance. These deposits likely act as scent markings to alert other badgers of the site's occupancy. One researcher observed a badger rubbing its abdominal glands on the throw mound at a burrow entrance. Abdominal glands in males and anal glands in both sexes become prominent during summer.

VOCALIZATIONS
Little has been published about American Badger vocalizations. They will hiss, snarl, growl, and squeal if attacked or cornered or if

fighting with other badgers. They have also been heard to purr, puff, and grunt.

SIGNS

Badgers commonly dig new burrows and often renovate existing burrows; if a den is in use, there will likely be fresh signs of digging. Only hibernation and natal dens are occupied for longer than a day or two at a time. Burrow entrances are 15–27 cm in height by 18–30 cm in width and are typically wider than they are high. Badgers dig mainly on the sides of the tunnel; therefore, the excavations are oval, and claw marks at 3–5 cm apart may be visible along the walls. Badger burrows have characteristic, often enormous, throw mounds that are fan-shaped and tend to get hard packed and broaden over time. Tracks are often apparent in the loose soil at the burrow entrance. Day-use dens tend to have small throw mounds, while natal dens have larger throw mounds and characteristically have badger hairs mixed with the waste soil. Badgers also excavate tremendous amounts of soil as they dig up small mammalian prey. They may visit the same prey colony repeatedly over time and create increasing numbers of pits and throw mounds in their predation efforts. Scat can be deposited underground in a latrine chamber, buried in the throw mound, or deposited along travel routes and at the entrance of burrows. When used as a scent marking, scat is usually accompanied by a urine stain and deposited prominently on prey's throw mounds, or at the entrance of the badger's burrow. Most scat is weasel like – black, twisted, and pointed at each end – but occasionally it is tubular. Scat is typically 1–2 cm in diameter and 7.5–15.0 cm in length. Badgers can run quickly (faster than a human) using a low bounding lope, but their usual gait is either a leisurely waddle or a brisk trot. Badgers tend to trot while travelling, walk while foraging, and lope or gallop when alarmed. Tracks are characteristically pigeon toed, and the long front claws leave distinctive impressions. Each foot has five toes. The smallest is the first toe (on the inside of the track), which does not always register. The front track is decidedly larger than the rear track. Tracks in snow are often connected by drag marks.

REFERENCES

Apps et al. 2002; Brandt 1994; Casler and Murphy 2001; Cowan and Guiguet 1956; Elbroch 2003; Harlow et al. 1985; Hoodicoff 2003, 2006; Jameson and Peeters 2004; Jannett et al. 2007; *jeffersonii* Badger Recovery Team 2004; Kiliaan et al. 1991; Kurta 1995; Kyle et al. 2004; Lindzey 2003, Long, C.A., 1973; Michener 2000, 2004; Michener and Iwaniuk 2001; Minta 1993; Minta et al. 1992; Murie 1992; Newhouse and Kinley 2000; Rahme et al. 1995; Scobie 2002; Smith, H.C., 1993; Verts and Carraway 1998.

FAMILY MEPHITIDAE
skunks

Skunks have a distinctive black and white stripe pattern and are renowned for the potency of their anal gland contents. Long considered part of the weasel family, they were recently placed in a family of their own based on new genetic evidence.

Striped Skunk

FRENCH NAME: **Mouffette rayée**
SCIENTIFIC NAME: *Mephitis mephitis*

Few people are unfamiliar with the most important characteristic of the Striped Skunk: its ability to eject a repugnant and acrid liquid from its anal glands. The Latin word *mephitis* reflects this reality; it means "bad odour." The English name skunk is derived from the Algonquin name for the species.

DESCRIPTION

About the size of a Domestic Cat, the Striped Skunk has a black pelage with white stripes along the back that is classic "Don't come near me, I'm dangerous" coloration – easy to see and dramatic. There is considerable variation in the colour patterns and the extent of the white. Typically, Striped Skunks are black with a thin white stripe that runs from the nose to the forehead, and a broad white stripe that travels from the crown of the head to the shoulders, where it branches into two narrower dorsal stripes that run along the back just below the spine. The black bushy tail is commonly white tipped. Some skunks have a white chest patch, some have white stripes on the outside of their forelegs, and others have broken stripes. The dorsal stripes may be short or extend all the way to the tail and sometimes onto the tail. All-white or all-black skunks or those with dark brown replacing the black, or yellow replacing the white, are known to occur. A Striped Skunk's defensive armature consists of two large musk glands, each around 25 mm in diameter, which lie one on either side of the anus. These are surrounded by powerful muscles that, when contracted, shoot a yellowish musk up to 3.5 m through two nozzle-like ducts that emerge from the partially everted anus. These ducts can actually be aimed to some degree. Apart from the acrid and nauseating odour, musk can cause temporary blindness if it contacts the eyes. The ears are short and rounded. The nose pad is almost round and slightly bulbous, and the small black eyes lack a

Striped Skunk (*Mephitis mephitis*)

nictitating membrane. Striped Skunks are thought to have poor distance vision but excellent auditory, tactile, and olfactory senses. The dental formula is incisors 3/3, canines 1/1, premolars 3/3, and molars 1/2, for a total of 34 teeth. Their double coat is composed of long, silky guard hairs (38–76 mm) and shorter underfur (25–31mm). Adult Striped Skunks undergo a prolonged single moult each year over the summer. Juveniles do not moult until the summer after their birth. The eyeshine is amber.

NOTE: To neutralize the musk, mix in a well-ventilated area (preferably outside) the following: 1 quart of 3% hydrogen peroxide, ¼ cup of baking soda, and 1 teaspoon of liquid dish soap. Bathe the affected area with the mixture, wait a few minutes, and then rinse. Two or more applications may be necessary. Keep it away from eyes. This mixture is only recommended for skin and fur applications, not for clothes. Fabric is best washed in a diluted ammonia solution. The noxious odour will gradually dissipate over time, even if untreated.

SIMILAR SPECIES
The distinctive black-and-white markings leave little doubt concerning the identity of a skunk. There are two skunk species in Canada. The Striped Skunk is the only species in most of the country, but Western

Spotted Skunks may co-occur in a small region of southwestern British Columbia. Spotted Skunks are considerably smaller than Striped Skunks, and their white markings are broken rather than solid.

SIZE
Males are about 10% longer than females and about 20% heavier on average. There is much variation in size throughout the range. Weight varies through the year. Striped Skunks are heaviest in autumn and lightest in early spring. Skunks in urban areas are commonly heavier than their rural counterparts. The following measurements apply to adult animals in the Canadian part of the range.
Males: total length: 615–755 mm; tail length: 197–285 mm; hind foot length: 67–94 mm; ear length: 26–27 mm; weight: average 2–4 kg, but up to 5.5 kg.
Females: total length: 578–700 mm; tail length: 182–260 mm; hind foot length: 67–83 mm; ear length: 26–27 mm; weight: average 2.0–2.6 kg, but up to 4.4 kg.
Newborns: weight: about 25–40 g.

RANGE
All of the 13 species of skunks are endemic to the New World and most are tropical. Only the Striped Skunk and the Western Spotted

Distribution of the Striped Skunk (*Mephitis mephitis*)

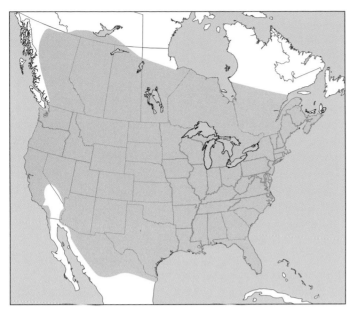

Skunk occur in Canada, although the Eastern Spotted Skunk is very close to the Ontario-Minnesota border and, with the effects of global warming, may be discovered in Canada in the future. Striped Skunks are well adapted to a cold climate and occur from southern Northwest Territories to northern Mexico and from coast to coast. Striped Skunks have not colonized most of the nearshore marine islands such as Vancouver Island, Cape Breton Island, and Newfoundland. Skunks were introduced into Prince Edward Island in the 1900s as farmed furbearers. Sufficient of these animals escaped or were released to create a sustaining and widespread Prince Edward Island population. Striped Skunks benefited from the forest cutting and agricultural intensification that followed European settlement in North America and were able to increase their population size and range. They have also become well adapted to urban life.

ABUNDANCE

These skunks are generally common, although in some regions the rabies-abatement programs have reduced (probably only temporarily) their numbers to rare.

ECOLOGY

Striped Skunks inhabit a wide variety of habitats from cities and towns to open grasslands, including woodlots, ravines, shelter belts, and farmlands. They prefer brush lands and agricultural lands where a diversity of forest, field, wetlands, and rocky outcrops provide abundant food and cover for den sites. Striped Skunks display their versatility by being comfortable, and even numerous, in highly urban settings and in areas of intensive agriculture. They are largely nocturnal, foraging from dusk to dawn. Striped Skunks den in stumps, hollow logs, rock, debris, and wood piles, under buildings and overhanging banks, and in ground burrows. They will dig their own simple burrows or renovate the more elaborate ones of other mammals, particularly Woodchucks, Badgers, Red Foxes,

and muskrats. Each burrow contains a nesting chamber of around 30–40 cm in diameter, which is about 1 m below the ground surface and is lined with dried vegetation. Dens are connected by 1.5–4.5 m of tunnel to one or more entrances. Each skunk maintains several burrows scattered throughout its home range that are used intermittently throughout the warmer months. Natal dens are typically underground, well protected in rock piles or under buildings. A single burrow is typically occupied throughout the whole winter, in regions where it is cold, and the entrance may be blocked with dried vegetation. Striped Skunks spend most of the winter sleeping in their dens and surviving on their fat. Their body temperature may drop a few degrees during this dormant period, and they may emerge to forage on mild days. Depending on the length of the cold period, these animals will lose 10%–55% of their body weight. Males are active at lower temperatures than are females and hence have more opportunities to forage and generally lose less weight.

During a night of foraging Striped Skunks may travel several kilometres, but they rarely go more than 1 km from their current den site. Except in winter, Striped Skunks forage for the whole night, returning to a den at around dawn. Females with dependent young often make one or more short forays out to forage, gradually extending the duration as the kits age. Population density averages 1.8–4.8 per km² but may reach 18.5 per km² in highly productive habitats. Home ranges vary with food availability and habitat. The average size ranges from 120 to 490 ha. The home ranges of neighbours, either male or female, tend to overlap, sometimes extensively. A natal den on a portion of the home range shared by adjoining females may be used by different females in different years. Striped Skunks are, along with Northern Raccoons and Red Foxes, one of the top three vectors for rabies in North America. They account for 20%–30% of Canadian cases annually. They also transmit and succumb to, among other things, leptospirosis, canine distemper, infectious canine hepatitis, pulmonary aspergillosis, histoplasmosis, and tularemia.

Great Horned Owls are the Striped Skunk's primary predator, but predation likely does not have a serious impact on skunk populations. These owls are large enough to kill the cat-sized, small carnivore, and, lacking odour acuity, they are unaffected by the skunk's defensive spray. Few other wild predators will attempt to take a skunk unless they are very, very hungry. Coyotes, Bobcats, Red Foxes, and Domestic Dogs occasionally kill them. Apparently prehistoric native cultures in North America enjoyed eating Striped Skunks, if the number of skunk bones in their middens is any indication. Probably the highest causes of mortality are disease and encounters with automobiles and farm machinery. Although skunks normally travel at a deliberate slow walk, they are capable of galloping at speeds of up to 16.5 km/h. Striped Skunks can swim reluctantly, but they are poor climbers. Captive skunks can live to 10 years old, but few wild Striped Skunks live more than four years, and almost none live beyond five to six years old.

DIET

Striped Skunks are omnivorous and highly insectivorous. They consume a wide variety of grubs, larvae, caterpillars, and adult insects,

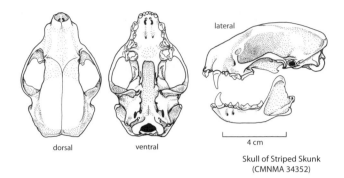

dorsal   ventral

lateral

4 cm

Skull of Striped Skunk
(CMNMA 34352)

worms, snails and small mammals, as well as amphibians (especially turtle eggs), reptiles, and ground nesting birds and their eggs in season. What they eat is largely determined by where they live and what is available. Their summer menu favours grubs, grasshoppers, beetles, crickets, and caterpillars, many of which are crop pests. They will dig out bumble bee nests and sometimes raid beehives, scratching at the entrance and catching the bees as they fly out. Occasionally a skunk will learn to raid a chicken coup. An energetic skunk foraging for grubs in a lawn can excavate what looks like a miniature minefield of shallow, cone-shaped pits around 3–6 cm in diameter. In spring, autumn, and winter skunks eat whatever they can find including plant material such as wild grapes, apples, grains, corn, and nuts, as well as small mammals such as mice, voles, shrews, and moles. Human garbage is an easy addition to their menu at any season. Of their diet, 80%–90% is thought to be of animal origin. Striped Skunks gain a tremendous amount of body fat in autumn to help them survive the lean winter season.

REPRODUCTION

Male skunks are polygamous and will attempt to mate with numerous females. Females are in oestrus for only two to three days each year and will vigorously repel any male attempting to mate outside of those days. If she does not breed during that first window of opportunity, she may undergo a second cycle about a month later. Females produce only one litter per year. Mating occurs during February or March (likely in the latter month in more northerly regions), and gestation lasts 59–77 days. Almost every female breeds annually, as the pregnancy rate is very high (92%–96%). Females breeding early in the breeding season may experience a short period of delayed implantation. Litter size ranges from one to ten but is typically five to seven. Young are usually born from mid-May to early June. Newborn kits are blind and deaf and have only sparse, fine fur, but their anal glands already contain musk. Their future colour pattern clearly shows on their wrinkled skin. The eyes open between 17 and 35 days old, at about the same time as the kits are capable of purposely discharging their musk. The pinnae unfurl and the auditory canal opens by 24–27 days old. Weaning occurs at six to eight weeks of age, as the youngsters gradually begin following their mother on her foraging expeditions. By November of the year in which they were born, most juveniles are adult sized. The majority of juveniles disperse when they are two to three months old, and few remain on their natal home range past five months old, although some females permit their young of the year to share their winter den. Both male and female juveniles become sexually mature at around 10 months old, although young females breeding for the first time tend to breed later than do adults and to produce smaller litters.

BEHAVIOUR

Striped Skunks are by nature placid and sluggish. In truth, they do not need to be highly alert and quick to react as they have one of the most potent chemical defences in the animal kingdom. Few predators, apart from some vengeful and slow-witted Domestic Dogs, or Great Horned Owls (which have a poor sense of smell), will take a second run at a skunk. Striped Skunks are normally reluctant to release their musk as they can only spray a limited number of times before the glands are emptied of their few teaspoons of musk. A skunk will first attempt to warn away the attacker by elevating and piloerecting all of its tail hairs to make itself appear larger and then drum with its front feet, hissing, growling, and arching its back. As it stomps, it usually retreats, shuffling backwards, although sometimes it may charge instead. When a skunk is about to spray, it bends around in a sideways U shape so that it can look behind to aim its rear end. The musk can be discharged over a 30°–45° arc either in a thin spray or as an atomized cloud of tiny droplets. Contrary to popular myth, a skunk hoisted by its tail can still spray. Skunks do not normally smell of their musk unless they have recently sprayed.

When a male skunk approaches a female, he attempts to do so from the rear as he wishes to sniff her to determine her reproductive receptivity. If she allows his approach, he will lick her and then mount after grasping her nape in his teeth. Copulation is rapid, taking only a few minutes. Males will do battle with each other for access to oestrous females, but they rarely discharge their musk during such conflicts. Clashes outside the breeding season are rare as most skunks are largely indifferent to one another. The exception is pregnant and lactating females, who continue to aggressively repel the attentions or presence of males near them and their offspring. Striped Skunks have devised an interesting method of predating an egg that is too large to be crushed in their mouth. They fling it between their hind legs with their front feet until it strikes something hard and cracks. When hunting small mammals (mice and voles), skunks are capable of locating a nest by smell. Following the initial attack on the nest, the skunk may then follow the individual scent trails of any escaping animals to capture them as well. During the warm months skunks shift den sites frequently and often spend the day above ground in a fencerow, in thick grass, or in a debris pile. Winter dens are always below ground and may be shared, for example by a maternal female and her last litter, but more commonly by a male and several females or sometimes by a unisex group of adults.

VOCALIZATIONS

Little has been reported concerning the variations and meanings of Striped Skunk vocalizations. They make a variety of sounds including hisses, growls, snarls, loud squeals, shrill screeches, soft cooing and churrings, tooth chattering, and bird-like twittering. Hisses,

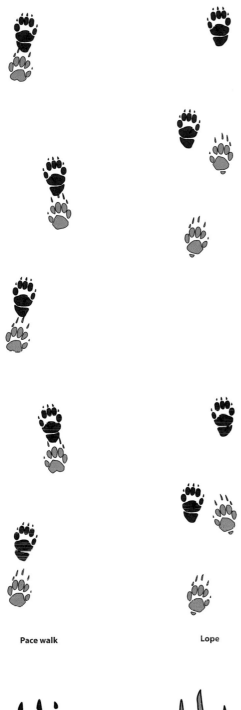

screeches, and squeals are often directed towards intruding skunks. Growls, hisses, snarls, and tooth chattering are aggressive sounds directed towards a possible predator, and cooing and twittering are produced by severely frightened animals.

SIGNS

Their typical foraging gait is a slow waddle (technically a pace, as the left legs advance together and then the right legs advance together), and both front and hind tracks appear very pigeon toed. The skunks are capable of a slow trot or a lope when travelling, or a gallop if alarmed. Striped Skunks have long, curved claws on their front feet and shorter, straight claws on their hind feet. The front claws reliably register in a track, but the shorter, hind claws may not. Each foot has five toes, with the first toe (the inside toe) being the smallest. Although the front track is wider than the hind, the hind track appears to be larger because of its extended heel pad. The smaller heel pad on the front foot often does not register in a track. Waterfowl nests predated by skunks typically have broken eggshells scattered up to 1 m away from the nest. The broken bits may be shattered, broken in half, or just punctured. When Striped Skunks predate nests of smaller birds, they may remove the eggs entirely. In neither case of nest depredation do the skunks dig up or otherwise damage the nest itself. During the warm months skunks often use day beds that are above ground, typically in a patch of dense vegetation. These are around 25–30 cm in diameter and appear as a circle of packed-down plant stems. Scat varies with diet but is typically smoothly tubular with blunt ends and is 5.0–12.7 cm long by 1.0–1.7 cm in diameter. Some scat is deposited in underground latrines, some gathers at the burrow entrances, and some is randomly scattered along the skunk's route. Skunks sometimes repeatedly deposit scat under overhangs or near prominent objects that they commonly pass in their nightly travels.

REFERENCES

Elbroch 2003; Greenwood et al. 1999; Jameson and Peeters 2004; Kurta 1995; Larivière 1998, 2001; Larivière and Messier 1996, 1997a, 1997b, 1998; Larivière et al. 2007; Laun 1962; Rosatte and Larivière 2003; Smith, H.C., 1993; Verts and Carraway 1998; Wade-Smith and Verts 1982; Walton and Larivière 1994.

Pace walk

Lope

Left hind track
Length: 3.3–5.1 cm
Width: 2.4–3.0 cm

Left front track
Length: 4.0–5.2 cm
Width: 2.5–3.0 cm

Striped Skunk

Western Spotted Skunk

also called Polecat, Civet Cat (it is neither a civet nor a cat)

FRENCH NAME: **Mouffette gracile,** *also* **Mouffette tachetée**
SCIENTIFIC NAME: *Spilogale gracilis*

Western Spotted Skunks are the smallest of the North American skunks, and they only reach Canada in far southwestern British

Western Spotted Skunk (*Spilogale gracilis*)

Columbia. Despite their bright pelage, these beautiful little skunks are rarely seen as they are almost entirely nocturnal and very shy.

DESCRIPTION

Western Spotted Skunks are about half the size of Striped Skunks and are more weasel like. They have short legs, a slender body, short rounded ears, and a small triangular head. Their short body fur is fine and soft, and the tail fur is much longer and coarser. Their curved front claws are around 7 mm long, while the almost straight hind claws are about half that length. The base colour of their pelage is glossy black. They have a small white patch in the middle of their forehead between their eyes and another in front of each ear, which may be confluent with the main dorsal stripes. Two or three pairs of dorsal stripes begin at the nape and behind the ears and travel parallel to the spine to approximately mid-back. Usually two stripes slash each flank across the hips, and one or two small white patches lie at the root of the tail, vestiges of a third stripe. The last vertical stripe may travel down the upper surface of the tail. Many of the stripes may be broken, providing a spotted appearance. Most Western Spotted Skunks have four to six segmented white stripes, and the pattern of white on each animal is unique. The last half or third of the tail is composed of many long white hairs that can be spread to form a large and noticeable white plume. On rare individuals the black fur is brownish or reddish, and the white is grey or yellowish.

Like Striped Skunks, Western Spotted Skunks have paired musk glands surrounded by muscular tissue that, when contracted, can propel the contents of the glands (yellowish foul-smelling musk) through nozzle-like ducts that project through the partially everted anus. The range (less than 2 m) of a Spotted Skunk spray is not as great as that of a Striped Skunk, but Western Spotted Skunk musk is considered to be somewhat more acrid and pungent. The dental formula is incisors 3/3, canines 1/1, premolars 3/3, and molars 1/2, for a total of 34 teeth.

SIMILAR SPECIES

Distinctive black-and-white pelage is characteristic of a skunk. Striped Skunks are larger, with an unbroken single pair of white dorsal stripes. Western Spotted Skunks are more agile and faster moving than Striped Skunks. The southwestern corner of British Columbia in the Puget Sound lowlands is the only part of Canada in which both Striped Skunks and Western Spotted Skunks may be found. Only the Striped Skunk and the Western Spotted Skunk occur in Canada, although the Eastern Spotted Skunk is very close to the Ontario-Minnesota border and, with the effects of global warming, may be discovered in Canada in the future. Eastern and Western Spotted Skunks are similar in appearance and cannot be safely identified by any single characteristic. The two species are separated by geography and are reproductively isolated; the

Distribution of the Western Spotted Skunk (*Spilogale gracilis*)

expanded view of Canadian range

eastern form breeds in March–April, and the western form breeds in September–October.

SIZE

Males are about 7%–10% heavier than females. There is much variation in size throughout the range, and the largest animals appear in the southwest. Weight varies through the year. Female Western Spotted Skunks are heaviest in autumn, while males are heaviest in spring. The following measurements apply to adult animals in both the Canadian part of the range and western Washington and Oregon that are of the same subspecies (*Spilogale gracilus latifrons*).
Males: total length: 370–453 mm; tail length: 110–162 mm; hind foot length: 44–54 mm; ear length: 22–32 mm; weight: 446–1200 g.
Females: total length: 368–457 mm; tail length: 104–150 mm; hind foot length: 41–52 mm; ear length: 22–30 mm; weight: average 269–965 g.
Newborns: weight: around 9–12 g.

RANGE

These small skunks reach the northern limit of their distribution in southwestern British Columbia. They occur west of the Continental Divide from British Columbia and southwestern Montana to Wyoming and then southwards to California, western Texas, and central Mexico.

ABUNDANCE

Although rarely seen, these small carnivores are not especially uncommon in appropriate habitat, except in California, where they are considered to be of "Special Concern." Trapping continues in most jurisdictions in the United States and Canada although the Canadian harvest is minimal.

ECOLOGY

Western Spotted Skunks occupy a variety of habitats from woodland to tallgrass prairie but seldom occur in low-lying deserts. In the Pacific Northwest they are widely distributed throughout upland coniferous forests especially in association with riparian habitats. In coastal areas they are most common along watercourses in thickets of alder, poplar, or salmonberry and along rocky stream beds. In more arid regions they occupy rocky creek and stream beds. In all cases, an abundance of cover is necessary. Western Spotted Skunks are more arid adapted and have an advantage over Striped Skunks in those areas. Most populations are disjunct, separated mainly by unsuitable habitat. These small skunks are almost entirely nocturnal and highly secretive. They spend the day in a den in a hollow log, a tree cavity, a rock crevice, inside or under a building, or in burrows constructed by other animals such as Mountain Beavers, woodrats, and ground squirrels. In areas that are intensively farmed, additional denning sites include hay and straw piles, culverts, grain elevators, woodpiles, and corncribs. Western Spotted Skunks are capable of digging their own burrows, but they commonly use abandoned burrows of other species if they are available. The actual sleeping chamber is typically lined with dried grasses. Underground burrows may be preferred den sites during hot weather. Each skunk may intermittently occupy several burrows scattered throughout its home range. They might range over a square kilometre to find food but will forage in smaller areas if prey is abundant. Average home ranges are 30–60 ha. The home ranges of neighbours frequently overlap because Western Spotted Skunks do not appear to defend their home ranges. In regions with cold winters these diminutive skunks will disappear into their underground burrows for four to eight months, where they remain inactive. It is not known whether they are true hibernators or they simply become dormant. They may emerge to forage on particularly warm winter nights. Agile climbers, Western Spotted Skunks will readily climb and den in

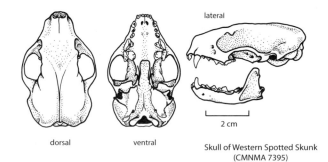

dorsal   ventral

lateral

2 cm

Skull of Western Spotted Skunk
(CMNMA 7395)

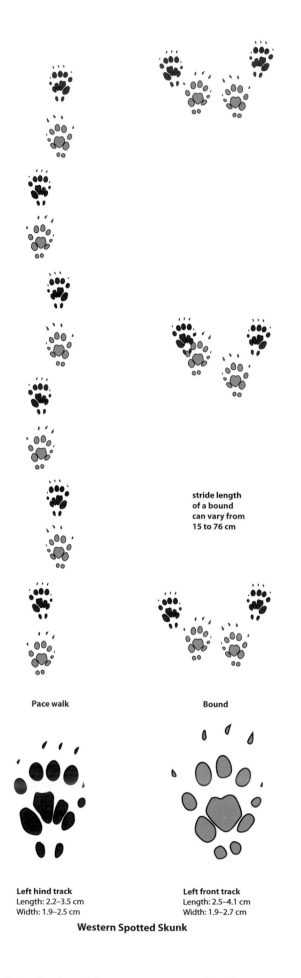

stride length of a bound can vary from 15 to 76 cm

Pace walk

Bound

Left hind track
Length: 2.2–3.5 cm
Width: 1.9–2.5 cm

Left front track
Length: 2.5–4.1 cm
Width: 1.9–2.7 cm

Western Spotted Skunk

trees and have been seen as high up as 7 m. They sometimes climb to escape predators and can expel musk from such a vantage point. Natural predators include Great Horned Owls and Bobcats, but most of their mortality is the result of vehicle collisions and shooting or trapping. Although cases of rabies have been reported in Western Spotted Skunks, the incidence is considerably less than that in Striped Skunks. While these skunks can survive up to ten years in captivity, few, if any, in the wild live longer than five years.

DIET

Western Spotted Skunks are omnivorous, but insects and small mammals constitute the majority of their diet. In addition, they consume lagomorphs, birds and bird eggs, carrion, amphibians, reptiles, and occasionally crops such as corn and fruit. Their diet varies with locality and season, but in general they are thought to be somewhat more carnivorous than other skunks.

REPRODUCTION

The breeding season extends from September to February, with a peak in September and October. Although two litters annually are possible in southerly populations, those skunks in Canada and the northern United States produce only one litter each year. Females that do not mate or whose eggs are not fertilized in their first cycle will come into heat again in around five days. Western Spotted Skunks experience an extended period of delayed implantation where the fertilized egg develops to the blastocyst stage and then becomes dormant, floating freely in the uterus. This phase lasts 200–220 days, followed by an active gestation phase of 28–31 days. The total gestation period is 230–250 days. Implantation into the wall of the uterus and continued growth of the embryo begins in April, and most litters are born in May. Litter size ranges from two to six and averages 3.8. Males provide no parental care. The kits are born blind, deaf, and toothless. Their bodies are sparsely covered in fine hairs, and their future colour pattern is already distinct in the pigmentation on their wrinkled skin. By 21 days old the kits are covered in short fur. Their eyes open after 30–32 days, and teeth begin erupting at around 35 days old. The first defensive movements (raising the tail) occur at 24 days old, and musk can be intentionally ejected at 46 days old. The young are weaned by the time they are two months old. Juveniles are almost adult sized by the time they are three months old. Most, but not all, juvenile males are sexually mature at four to five months old, but their actual contribution to their first breeding season is unknown and possibly minimal. Females breed for the first time at four to five months old, producing their first litter as yearlings.

BEHAVIOUR

Despite having a potent defensive armature, Western Spotted Skunks will only spray an attacker as a last resort. The musk glands have a finite capacity that must be utilized only in life-threatening situations, as they require some time to replenish. First, the skunks attempt to bluff their way clear. They hop towards the intruder and stamp their front feet on the ground. If the threat persists, they perform a unique handstand display (typically lasting less than five seconds) as they aim their anal sphincter with its chemical arsenal towards the adversary. During the handstand display their tail hairs are widely spread, creating a distinctive and very noticeable white plume on the black tail stock. If this clear warning is ignored, then the assailant will be sprayed. Although spotted skunks can spray from the handstand position, they normally drop to a more stable four-footed position and curve their bodies into a sideways U shape so that they can aim their musk glands and see their foe at the same time. In general, spotted skunks are more excitable than are Striped Skunks; they are more likely to spray with less provocation and more likely to exude musk during social interactions. They are also much more difficult to tame than Striped Skunks. Western Spotted Skunks have a peculiar egg-opening technique. First they straddle the egg and attempt to bite it open. If this is unsuccessful, as is likely for any large eggs, the animal will fling the egg between its hind legs, kicking it with a hind foot as it passes. The egg is then pursued and inspected for breakage. These actions may be repeated several times until the shell cracks against a solid object. Radio-tracking studies found that females may share dens, especially in winter. A male, intent on mating, first attempts to subdue a female by gripping her nape in his teeth. An unreceptive female can fight free of the male's attentions if she is determined.

VOCALIZATIONS

These small skunks do not normally reproduce in captivity, so captive colonies cannot be easily maintained. Consequently, much of their behaviour and especially their vocalizations are poorly known. Squeals, grunts, and bird-like twitters are the only vocalizations reported.

SIGNS

Den entrances are 8–13 cm wide by 7–14 cm high. The wide variation in size of the opening is likely due to the variety of species that originally created the burrow taken over by the skunk. Each foot of the Western Spotted Skunk has five toes, with the inside toe on each foot being the smallest. The front claws are longer than the hind claws and reliably register in a track. The front and hind tracks are similar in size. The inside toe on the hind track is further back than the other four toes, which form a fairly straight line. This characteristic distinguishes the track of the Western Spotted Skunk from that of the Striped Skunk. Spotted Skunks lope and bound when travelling or in the open, and walk when foraging and under cover. Although they can perform a regular diagonal walk at times, they mostly pace walk. Scat varies with diet but is typically tubular with smooth surfaces and blunt ends (0.6–1.9 cm in diameter).

REFERENCES

Carey and Kershner 1996; Crooks 1999; Elbroch 2003; Hattler et al. 2008; Jameson and Peeters 2004; Rosatte and Larivière 2003; Verts and Carraway 1998; Verts et al. 2001.

FAMILY PROCYONIDAE
raccoons

Most members of the Procyonidae family have alternate dark and light bands on their tail. They have long legs and plantigrade foot posture and are excellent climbers.

Northern Raccoon

FRENCH NAME: **Raton laveur**
SCIENTIFIC NAME: *Procyon lotor*

The Northern Raccoon, a medium-sized carnivore, is distinguished by its easily recognizable and striking markings. It has a dark mask over its eyes, and several prominent, dark rings around its light-coloured, bushy tail. Raccoons are commonly called masked bandits, referring both to their conspicuous facial markings and their penchant for raiding crops, gardens, and garbage cans.

DESCRIPTION

A Northern Raccoon's coat generally appears grizzled grey or grey brown in colour owing to the mix of long, banded guard hairs and shorter buffy underfur. The guard hairs on the back are typically black or dark brown, while those on the sides are light at the base but dark tipped. Guard hairs on the belly are usually all white. This makes the upper back look blackish, the sides greyish or brownish, and the belly buffy or greyish white. Albino or brown individuals are rare. Animals that are somewhat paler or darker than the norm are common because the individual extent of buff, brown, and black hairs varies. Geographical variation also occurs, with western-rainforest animals tending to be generally darker overall. A dark facial mask surrounds the eyes and is accentuated by a whitish band of fur along the upper margin and by the whitish muzzle below. The nose pad is black. The mask tends to fade as the animal ages. The prominent greyish ears are usually edged with white. The fur on the legs and tops of the feet is short and usually light grey. The hind legs are longer than the front legs, giving the animal a high-rumped look and an arched back. The long tail is buffy to greyish-white, encircled with four to seven (typically five) dark rings, and has a dark tip. Their senses of smell and hearing are acute, and although their eyes are highly adapted to low-light conditions, Northern Raccoons are thought to have poor distance vision. Their tactile sense, with extremely sensitive forepaws, is exceptional. Their dental formula is incisors 3/3, canines 1/1, premolars 4/4, and molars 2/2, for

Northern Raccoon (*Procyon lotor*)

a total of 40 teeth. Northern Raccoons moult twice a year, in spring and autumn. The spring moult occurs from mid-April to late June and may be patchy. The summer coat is shorter and sparser than the winter coat. The winter pelage requires six weeks to grow in and is usually complete by late November. Northern Raccoons in southern latitudes may not develop a dense winter coat. The eyeshine is bright yellow.

SIMILAR SPECIES

There are no other medium-sized mammals that resemble a Northern Raccoon except the Raccoon Dog (a canine), which only occurs in Eurasia.

SIZE

Adult males are an average of 10%–15% heavier (and can be up to 25%–30% heavier) and 5%–10% longer than adult females. Raccoons at higher latitudes tend to be larger and accumulate significant fat reserves before winter, and so they are heaviest in late autumn. They are lightest in early spring. The following measurements apply to adult animals in the northern part of the range.

Males: total length: 785–1050 mm; tail length: 180–405 mm; hind foot length: 106–138 mm; ear length: 48–72 mm; weight: average 6–11 kg (some autumn individuals may exceed 13.5 kg, and the heaviest on record is an exceptionally fat autumn male from Wisconsin that weighed 25.5 kg).

Females: total length: 740–890 mm; tail length: 184–340 mm; hind foot length: 95–129 mm; ear length: 58–62 mm; weight: 3.86–5.2 kg.

Newborns: weight: 62–98 g.

RANGE

Widespread throughout southern Canada, the United States, and Central America to central Panama, Northern Raccoons occur from coast to coast. Their range has extended northwards in recent decades (since around the 1920s), possibly owing to global warming and the enhanced growing season that results. At the turn of the 1900s, Northern Raccoons in the west extended about as far north as the Canada–United States border, and in the east into southeastern Ontario, southwestern Quebec, and perhaps southern New Brunswick. Within a century they had reached northwestern British Columbia and were found throughout the prairie grasslands as far north as the edge of the boreal forest. They occupied the entire Maritimes, except Cape Breton and Newfoundland, almost all of southern and central Ontario, and most of southern Quebec. Northern Raccoons have recently invaded Cape Breton Island via the causeway and are rapidly moving northwards to occupy the whole island. They have found their way onto most of the Gulf Islands and many other nearshore marine islands along the British Columbia coast and onto freshwater islands such as Manitoulin Island. Human water-management practices, such as those that occur with the establishment of cities and the expansion of agricultural irrigation in arid mid-continent regions, have allowed Northern Raccoons to expand their range into these previously inaccessible habitats. They have been introduced to the Queen Charlotte archipelago, several islands in southeast Alaska, Prince Edward Island, and Grand Manan Island, New Brunswick. Northern Raccoons on New Providence Island, Bahamas, are likely the result of introductions from nearby Florida. They have also been introduced into several parts of Europe, and populations are established in Germany, eastern Russia, southeastern Russia, and along the east side of the Caspian Sea. Initially protected in Europe, their numbers have dramatically increased in the absence of natural predators, and Northern Raccoons are now being actively culled as pests.

ABUNDANCE

Northern Raccoons are one of the most common, native, mid-sized mammals in southern Canada and the United States. They have been reported in densities as high as 244 per km² (in a Missouri marsh where they were not subjected to predation) and can reach more than 55 per km² in urban residential areas such as Toronto. In contrast, populations of 0.5 per km² have been reported from prairie grassland. Although population density may fluctuate dramatically, both seasonally and annually, owing to disease, food abundance, or hunting and trapping pressures, Northern Raccoon numbers appear to be stable or increasing in almost all jurisdictions, likely as a result of ever-increasing urbanization. There are an estimated 1,000,000 Northern Raccoons in Ontario alone. Annual trapping harvests fluctuate with pelt values. In the United States between 400,000 and 2,000,000 were trapped from the 1930s to the 1980s. Pelt prices rose during the 1970s, and over 5,000,000 were trapped annually until the early 1990s, when prices dropped again, as did the number of Northern Raccoons harvested. Canadian harvests closely followed these trends, but the take is only a fraction of that from the United States. Raccoons are rare only in the most northerly portions of their extended range and are abundant in the Maritimes and southern Ontario and Quebec. They remain somewhat uncommon in the southern Prairie provinces. Northern Raccoons are often considered pests in both urban and rural regions as they commonly forage in garbage cans and crops.

ECOLOGY

Northern Raccoons may be found in almost any North American habitat but are most abundant in urban areas and in riparian, swamp, marsh, and other wetland environments. They are primarily nocturnal, although they may at times be active during low light levels at dusk and dawn. Activity peaks between dusk and midnight. Adult males occupy home ranges of 16–2560 ha that generally overlap only slightly (< 10%) with those of neighbouring males. In regions such as the prairies, where food is in short supply, home ranges are large, and adult males aggressively defend them from other males to the extent that there is little overlap. Other regions that are more productive support higher Northern Raccoon densities, and males have smaller home ranges that are not so rigorously defended. In these areas the males tend to avoid male neighbours, and so interactions are few. Males in some regions form small groups of three to five individuals that share a home range for part of the year. The home ranges of adult females are roughly one-third the size of the local male home ranges (5–806 ha), and they usually overlap extensively with ranges of neighbouring females and males. Most females with overlapping ranges are related. The dispersal of juvenile males typically occurs before they are 12 months old. Those in northern populations wait until after the spring emergence from communal winter dens. They may travel as far as 275 km, but generally dispersal is less than 20 km. Juvenile females usually remain in their birth area, establishing a home range that either overlaps with or is included within the range of their mother.

In temperate climates with cold winters most Northern Raccoons will den-up and become dormant for weeks or months at a time to avoid the cold and the energy demands of travel in deep snow. They do not truly hibernate as their heart rate and body temperature do not drop; instead they live off their stored fat and save energy by remaining inactive. Consequently, the length of time each animal can spend in a winter den depends on the amount of fat they were able to accumulate before winter. Those with insufficient reserves either starve or are forced to forage in the snow, where they are exposed to hypothermia, increased predation risk, and the seasonal lack of food. Mortality of such individuals is understandably high. Raccoons do not normally create their dens; instead they make use of a variety of structures already present. Preferred winter dens are

Distribution of the Northern Raccoon (*Procyon lotor*)

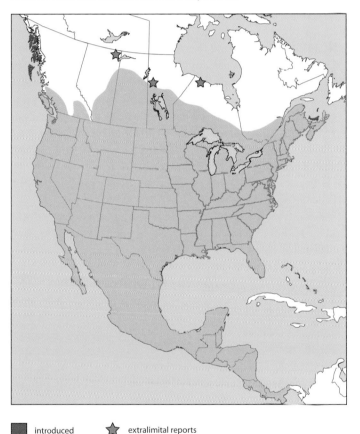

 introduced   ★ extralimital reports

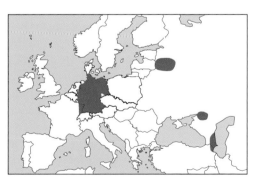

introduced range in Eurasia

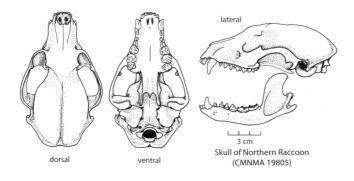

lateral

dorsal   ventral   3 cm
Skull of Northern Raccoon
(CMNMA 19805)

in tree cavities, as the wood provides some insulation value, but large rock crevices or buildings that can accommodate a great number of animals are also favoured. Communal nesting of as many as 23 animals improves each animal's thermal advantage. Such dens are especially important to juveniles, whose stored fat during their first winter can be significantly less than that of adults. Summer den sites are more variable and may include caves, rock crevices, brush piles, culverts, abandoned fox dens, abandoned buildings, attics, basements, barns, hay piles, mines, sewers, large tree limbs, muskrat houses, and even large squirrel dreys and abandoned bird nests. Apart from maternal dens, summer dens may be used repeatedly or just for a single day. Some individuals cycle between only a few dens, while others spend their daily rest period in a wide variety of den sites and rarely the same one for two days in a row. Particularly good dens, those that have multiple entrances and chambers, may be used frequently by a variety of individuals over the summer.

Raccoons are renowned for being able to jump or fall out of trees and drop without apparent injury. In one reported case, a pregnant female fell in excess of 9 m and not only was uninjured but successfully delivered her litter shortly afterwards. Northern Raccoons are capable and willing swimmers. They have been documented swimming up to 950 m to marine islands, and a mature female swam 1.2 km to escape a flood. Raccoons are very competent climbers with rear ankle joints that can be reversed to allow rapid head-first descents. They commonly adopt a bipedal posture to free their front paws for food manipulation. In some regions Northern Raccoon deaths are largely caused by human-related activities, primarily hunting, trapping, and motor vehicle collisions. Lightly hunted populations, such as those in most of Canada, suffer mortality from starvation (typically in late winter and early spring), disease, parasites, nuisance control, and motor vehicles. Juveniles, especially late-born juveniles, are most susceptible to starvation mortality.

Canine distemper and rabies are the diseases with the greatest impact on Northern Raccoons. Distemper mainly afflicts urban populations. Outbreaks are commonly followed by another viral disease, encephalitis, whose effects upon an already weakened animal can mimic the staggering symptoms of rabies. Foxes, skunks, and Northern Raccoons are the primary North American rabies vectors, and rabies can be a primary threat to Northern Raccoon populations, especially those in congested urban locations. Oral vaccine programs aimed at all three species are a major preventative measure for such outbreaks. As Northern Raccoons, skunks,

and foxes are so prevalent in both urban and rural areas, the common practice of inoculating Domestic Dogs and Cats for rabies and distemper helps to prevent the spread of the diseases to humans and livestock. Northern Raccoons occasionally act as reservoirs for other diseases such as leptospirosis, tularaemia, histoplasmosis, tuberculosis, and listeriosis, as well as various forms of encephalitis, including eastern equine and fox encephalitis. They can also transmit an intestinal roundworm called Raccoon Roundworm, which sometimes fatally affects humans and domestic animals. Raccoon predators include larger carnivores such as Bobcats, Coyotes, Grey Wolves, Cougars, Fishers, American Alligators, Domestic Dogs, and sometimes foxes and Great Horned Owls, which take primarily cubs and juveniles. Average longevity is less than two to three years in the wild, although approximately one in a hundred will live seven years, and exceptional animals may survive up to 16 years. Captives may live more than 17 years if properly fed.

DIET

Northern Raccoons are generalist omnivores that will consume almost anything that is organic. They are known to eat hundreds of different plant and animal species. The diversity of their diet has helped Northern Raccoons to make themselves at home in a wide range of habitats. Their intake varies with availability, season, and geographic location, and when food is abundant, they will become selective, eating only favourite foods. Raccoons are equally comfortable foraging along fresh or salt water. Slightly more than half of their diet is composed of plant material such as fruits, nuts, acorns, and seeds. Crops such as corn, melons, grapes, and other cultivated fruits, vegetables, and grains are readily eaten when available. The remainder of their food is dominated by invertebrate aquatic prey such as clams, crayfish, and crabs. Small mammals (including bats), fishes, reptiles and their eggs, and birds and their eggs are taken when opportunity provides. A wide variety of insects, spiders, snails, and slugs, as well as earthworms, caterpillars, and carrion, are also eaten. Rural Northern Raccoons will raid unsecured chicken coops, and their nest predation on wild passerines has been implicated in songbird decline in some regions. Northern Raccoons, introduced onto the Queen Charlotte Islands, are having a serious impact on colonies of burrow-nesting seabirds to the point of eradication. Nesting waterfowl on the prairies can be similarly affected. Urban Northern Raccoons will take advantage of human garbage and backyard gardens, as well as unattended pet food, lawn grubs and other miscellaneous food sources. Clearly Northern Raccoons are capable of learning to exploit new food sources, and these are then taught to successive generations. In most regions they eat more animal than plant food in spring. Summer and autumn is a time of abundant plant food, and Northern Raccoons then tend to eat more fruit, nuts, and various crops, especially corn. Depending on the location, Northern Raccoons in winter will either sleep for most of the time (in cold areas) or forage intermittently for both plant and animal food (in warmer areas). Even in colder regions the animals may come out to forage during warm spells. Northern animals generally bulk up during autumn, and over the course of the following winter the adults commonly drop 50% or more of their body weight. Northern Raccoons eat slowly and chew diligently to allow their well-developed salivary glands to begin the digestive process. Food typically takes 9–14 hours to travel the length of their short digestive tract.

REPRODUCTION

The breeding season lasts from early February to June, with an activity peak in March. Breeding begins earliest in the colder parts of the range and later in the warmer. A single litter is born annually to each female unless she fails to become pregnant or loses her first litter soon after parturition. Some females might then undergo a second oestrus and produce a late litter. Oestrus for each female lasts three to six days, and females will mate with more than one male during that time; therefore, multiple paternity is possible. Males spend the breeding season searching for oestrous females and will mate with multiple females if they can. They do not assist the female in raising the young (called cubs or kits). Adult males attempt to consort with a female for the full length of her oestrus if they can, and aggressively discourage the approach of other males. Copulation may last for an hour or more. Gestation typically takes 63–65 days (ranging 54–70), and litters average three to four cubs (ranging from one to seven). Larger litters are more common among northern females, perhaps owing to their bigger size. Young may be born in underground dens or in cavities in brush piles, but tree cavities are preferred. Urban Northern Raccoons additionally utilize attics, garages, sewers, and abandoned buildings. Most cubs are born in April in northern latitudes, and sometime later in warmer locations. Out-of-season litters may be produced by females who breed a second time after losing their first litter or by yearling females who are coming into oestrus a month or more after the mature females do. Such litters can be born up to six months late. The probability that late born offspring will survive their first year is low, especially in the north where winter starvation of juveniles is so severe.

Newborns are sparsely furred, and areas of darkly pigmented skin show where the facial mask and tail rings will develop. They are born blind and deaf. Their eyes and ear canals open when they are 18–30 days old, and they begin eating solid food at 10 or 11 weeks old. Their milk teeth begin to erupt at around 25 days old, and the set is complete by 65 days old. Permanent teeth begin erupting at around 65 days old and are fully emerged from the gums by about 105–120 days old, around the time that they stop nursing. At around the age of 10–12 weeks old, cubs begin travelling with their mother on her foraging trips, and after about 18 weeks old they forage independently. Both subadult females and subadult males can breed before they are one year old, and many do in some populations, but in northern parts of the range most breed for the first time when they are around 22 months old. Maturity is likely linked to nutritional status, with better-nourished juveniles breeding sooner.

BEHAVIOUR

Northern Raccoons are intelligent, bold, and adaptable. Ordinarily solitary when foraging, they may form aggregations around abundant food resources, but individuals in such circumstances tend to ignore each other and forage independently. Normally adult females are intolerant of adult males, so the more aggressive males do not take part in such gatherings. Typically, adult females associate with adult males only while they are in oestrus. In northern latitudes, aggregations often form again in winter in communal dens. These are usually segregated groups composed of adult males or of related adult females and their recent litters. Male Northern Raccoons in some regions form small social groups (usually three to five animals) that mutually defend a shared territory, rest together during the day, and forage as a "gang" at night. While members of such a group may be related, this is not exclusively the case. These groups tend to break apart during the breeding season, when male competition and aggression reach their peak, but may reform afterwards. Northern Raccoons have long been thought to "wash" their food. This behaviour is commonly seen in captive animals, but rarely seen in the wild. One theory contends that it occurs as a result of a lack of aquatic prey in the diet of captives. Wild Northern Raccoons commonly dabble in water, using their highly sensitive forepaws to find underwater prey by touch. Captive animals, denied this opportunity, satisfy their compulsion to dabble by placing other food into water, where they manipulate it, often staring vacantly into space. Although they do not actively hunt domestic pets, Northern Raccoons can be ferocious when cornered and are capable of deterring and even killing dogs much larger than themselves. Males contest access to oestrous females during the breeding season, and such encounters may leave their faces and ears scarred and nicked. Older males tend to develop a battered look. An angry Northern Raccoon lowers its head, raises its hackles and arches its back, lays its ears back, flares its facial ruff, and raises and thrashes its tail. It may also bare its teeth. This threatening physical display may be further enhanced with growling, snarling, screaming, or hissing vocalizations, the meaning of which is abundantly clear even to animals other than Northern Raccoons. Subordinate status is demonstrated by a general lowering of the head, body, and tail. Northern Raccoons are sometimes kept as pets, especially in the United States where exotic colours such as red, silver, blonde, and black are available. Many states and provinces now discourage or prohibit this practice owing to the disease risk. Northern Raccoons that are relocated for nuisance control should be moved at least 15 km away or they may find their way back.

VOCALIZATIONS

Northern Raccoons are highly vocal and are reported to produce 13 different sounds, seven of which involve females and young. Cubs calling for their mother use a high-pitched "trilling" call, also described as a quavering purr, which is similar to the call of a tree frog. Mothers calm their offspring with a quiet twittering or purring sound, to which the young respond with a contented "churr." A squeal call is used by distressed youngsters. Mothers grunt to warn their offspring. When females are leading their cubs on excursions, they call to them with a loud "chitter," and the cubs respond or call to her with a whistle. Females may produce shrill cries during mating, and as mentioned above, Northern Raccoons may snarl, growl, hiss, and scream during confrontations or if threatened or distressed.

Fast walk

Slow walk

Left hind track
Length: 5.4–10.2 cm
Width: 3.8–6.7 cm

Left front track
Length: 4.4–7.9 cm
Width: 3.8–8.3 cm

Northern Raccoon

SIGNS

Northern Raccoons have narrow feet with naked foot pads and non-retractable claws. Their long, finger-like front toes are highly dextrous and can be used to untie knots, open doors, unscrew jars, and lift the lids off garbage cans. Each foot has five toes and robust claws, which usually register as triangular dots ahead of each toe. The first toe (the big toe) on the hind foot is smallest, as is the first toe (the thumb) on the front foot. All toes are elongated and leave cigar-shaped impressions, which can look very much like a human hand, especially the impressions of the front feet. Impressions of the hind feet usually have the toes closer together than those of the front track. The hind heel may or may not register in a track, and full heel impressions are uncommon. Variation in width is largely determined by the amount of toe spread. The hind feet are larger and carry more weight, and so they tend to leave deeper tracks. Northern Raccoons typically walk when foraging, and both feet on the same side move together (a pace). In a slow walk the hind track on one side registers directly alongside or slightly behind the front track on the opposite side. In a faster walk the hind foot oversteps and registers beyond the front track of the opposite side, and the angle of each track set alternates. Both of these unusual track patterns are diagnostic for Northern Raccoons. Northern Raccoons will also gallop and bound in a more typical quadrupedal manner when startled or pursued or when chasing prey. When foraging on corn, Northern Raccoons typically knock the whole stalk over to get at the cobs. When foraging for ground bees and wasps, lawn grubs, and earthworms, they leave small digs that are very similar to those made by skunks (they are distinguished by hairs and tracks). Looking for insects, Northern Raccoons will pry old shingles off roofs and can quickly render the building unusable.

WARNING: Raccoon scat must be treated with respect as it may carry eggs of the Raccoon Roundworm, which can be fatal if inhaled or ingested. A mask and gloves should be worn when it is being handled. Children that consume soil are most at risk of contracting the parasite. The roundworm is likely as widely distributed as is the Northern Raccoon itself and may infect 3%–100% of a Northern Raccoon community.

Although Northern Raccoons do deposit some scat in random locations, they also often defecate in traditional communal latrine sites, which in urban regions could be on roofs, in debris in backyards, at the base of a fence, wall, or building, in the crook of a tree, or on top of a stump or stone wall. Over time the faeces are exposed to rain, and the tiny roundworm eggs are washed into the soil. Latrine sites in the wild are in locations such as under a ledge, on a large branch (up to 12 m off the ground), or at the base of a large tree, in brush piles and caves, along trails, on a fallen log, or at the base of a rock outcrop. Such accumulations usually indicate either a high-traffic area or a nearby den site. The purpose of latrines is not clear, but if defecation is an important marking activity, then multi-animal latrines could act as community bulletin boards. Scat can be blackish, brown, tan, or yellowish and may contain fruit or grain seeds, insect remains, or both. Most Northern Raccoon scat either is tubular, blunt at both ends, and breaks easily or is amorphous (plop-like) if the animal was eating fresh fruit. It is 0.8–3.0 cm in diameter and 9.0–17.8 cm in length. Runs are high-traffic trails that usually parallel a waterway. They may be 17.0–30.5 cm in width and typically contain numerous tracks.

REFERENCES

Barding and Nelson 2008; Belant 1992; Carleson 1991; Cowan and Guiguet 1956; Dobbyn 1994; Elbroch 2003; Gehrt 2003; Gehrt et al. 2008; Harfenist and Kaiser 1997; Hartman, L.H., and Eastman 1999; Hartman, L.H., et al. 1997; Jameson and Peeters 2004; Kamler et al. 2003a; Kurta 1995; Larivière 2004; Lotze and Anderson 1979; Lynch 1971; Macdonald and Barrett 1993; McClearn 1992; McCracken et al. 1986; Morris 1948; Page et al. 1998; Reid, F.A., 1997; Rezendes 1992; Rosatte 2000; Seidensticker 1999; Smith, H.C., 1993; Sutton 1964; Verts and Carraway 1998.

Order Perissodactyla
horses

Domestic Horse (*Equus caballus*)

Photo: Damian Lidgard

Worldwide there are three families, six genera, and 17 species in this order, which includes tapirs and rhinos as well as horses. The only member of this order that survives in the wild in Canada is the Domestic Horse, *Equus caballus,* an introduced species.

ORIGINS

The first perissodactyls appeared in Asia or Africa about 55 million years ago in the late Palaeocene epoch and spread into Europe and North America about 45 million years ago. The heyday of this order has passed, and the remaining species are mere remnants of this once widespread and diverse group. The evolutionary history of horses is one of the most studied and best understood histories of all the mammal groups.

Modern horses (genus *Equus*) evolved in North America, first showing up about four million years ago. Around 11,000 years ago, towards the end of the last great ice age, they vanished, along with Woolly Rhinos, Woolly Mammoths, and mastodons. The reason for their disappearance is not fully understood, although it was likely due to climate change and related vegetation changes. There has been some suggestion that over-hunting by humans contributed to the decline. Fortunately, before dying out in North America, some horses crossed over the Bering Isthmus into Asia, where they continued to survive and their descendants were domesticated by humans. Horses returned to North America with the help of humans, initially the Spanish conquistadores, about 8000 years later.

ECOLOGY

Rhinos and horses live in grassy plains or lightly wooded scrubland, while tapirs are found in humid tropical forests. All members of this order are herbivorous and are either browsers or grazers.

ANATOMY

This order represents the odd-toed ungulates. Their weight is concentrated on the third toe of the foot, and the remaining toes are reduced, as in tapirs and rhinos, or only exist as vestigial bones, as in horses. Tapirs display the most primitive condition with four toes on the front feet and three on the back. Rhinos have three toes on all feet, and horses have only one toe. All of the perissodactyls move about on the tips of their toes (digitigrade), an adaptation for speed. The nose horns of rhinos have no bony core; they are completely composed of compacted strands of keratinaceous material that is very similar to hair, much like the horn sheaths of the bison or the Muskox in the Bovid family of the order Artiodactyla.

Perissodactyls have several interesting adaptations that help them to chew and digest hard siliceous grasses. Their cheek teeth grow continuously, not wearing out until perissodactyls reach a very old age, and all have ridges and cusps that provide cutting edges for slicing the plant material into small bits. Unlike artiodactyls (the even-toed ungulates), which have a fermentation chamber in their forestomach (rumen) to predigest the tough cellulose, perissodactyls have developed a similar but less effective process. They have a caecum, an enlarged area between the large and small intestines where some fermentation takes place. The process is rapid and inefficient compared to that of artiodactyls, but the short fermentation does help to further break down some of the cellulose and thereby make the cell contents available for absorption by the intestine. Perissodactyls need to consume more food than do artiodactyls in order to gain the same nutritional benefit. As every farmer knows, a horse eats more than a cow does, but it can make do with poorer quality forage.

WILD HERDS

Four herds of horses living in Canada could be considered wild. Two in the Brittany Triangle of central British Columbia may have survived unassisted by humans for over 200 years. This group is currently under study. Attempts are being made to analyse the DNA of these animals to determine whether or not the horses that escaped from the conquistadores played a major role in their genetic background, thereby proving a very old ancestry. There is no doubt that these herds have existed in this location for many years. Another herd in southern Alberta is also a candidate for wild status. At the present time, the most certain wild herd is the protected one on Sable Island, Nova Scotia. A feral population has been established on the island since 1738. The herd has survived essentially unattended for almost 275 years.

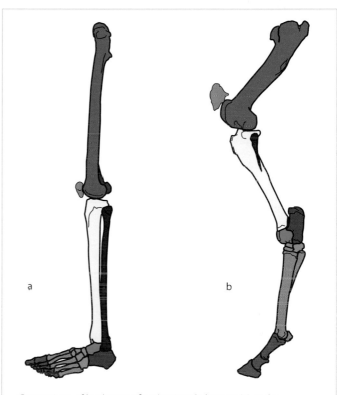

Comparison of leg bones of a plantigrade human (a) and a digitigrade horse (b), with equivalent bones similarly coloured.

FAMILY EQUIDAE

horses

Members of this family have tendon-locking mechanisms in both front and hind legs to allow extended periods of effortless standing. Horses can rest and even sleep while standing upright. This ability is likely an adaptation allowing them to more easily scan their surroundings for potential predators and then rapidly transition into running mode in order to escape.

Domestic Horse

FRENCH NAME: **Cheval**

SCIENTIFIC NAME: *Equus caballus*

Based on their first evidence in cave art, it appears that wild horses have been hunted for food by humans for at least 30,000 years. The first wild horses were domesticated around 3000 to 4000 years ago, probably in eastern Europe, making them the most recent of the large mammals to be domesticated for livestock by early humans. Like other ungulates, horses are engineering marvels with many locomotor adaptations. They can carry their own bodies at tremendous speeds and for great distances, even over broken terrain, and have the added distinction of being able to sustain this agility and stamina while carrying the extra weight of a human. The four herds of horses in Canada considered below are feral Domestic Horses that have survived unaided by humans for more than 200 years.

DESCRIPTION

Feral Domestic Horses on Sable Island are usually variations of reddish-brown (chestnut), dark brown, to brown red in colour with darker manes and tails (bay), and black with occasional blond with light manes and tails (palomino). In the western herds, coat colours are more varied and may be multicoloured such as in pintos, skewbalds, and Appaloosas, as well as white. All Domestic Horses in Canada, both feral and non-feral, undergo two moults each year. The colour changes little between the summer and the winter coats. The thick winter coat (see the colour illustration of the female) is a double coat with long guard hairs to shed rain and snow and with dense underfur to provide insulation; it is shed over the course of about 30–50 days, starting in early spring once the temperatures have risen above 6°C. Yearlings take longer to moult than do adults. The sleek summer coat (see the colour illustration of the male) is

made up only of guard hairs; it is shed in autumn as the new winter coat grows in, which is completed by early December. Foals begin to shed their birth coat two to three months after birth and develop a full winter coat by early winter. The hairs on their legs and head are normally shed first. Coat colour may change significantly between the birth coat and this first adult type coat. The dental formula is incisors 3/3, canines 0–1 / 0–1, premolars 3–4 / 3–4, and molars 3/3, for a total of 36–44 teeth. Males commonly have canines just behind the incisors. Females usually do not have canines, but if they do, these teeth typically are much smaller. The tiny first premolar, when present, is called a *wolf tooth*. Most foals are born toothless. When they are around a week old, the first of the milk teeth begins to break through the gums, and the full complement of 24 milk teeth is present by the age of 12 months. The adult teeth erupt in a prescribed order at specific ages, beginning with incisors and proceeding back to molars. Tooth eruption of both milk and adult teeth is commonly used to age a Domestic Horse until it is around five years old and has a full complement of adult teeth. Beyond five years old, tooth wear can be less precisely used to estimate age.

SIMILAR SPECIES

Surprisingly, the native mammal most commonly mistaken for a Domestic Horse is a non-antlered Moose. They are similar in size and can be similar in colouring. Moose have a pendulous dewlap and a tail so short as to be insignificant. They also have a large shoulder hump, no noticeable mane, and a dark body with pale legs. Non-antlered Wapiti/Elk are also sometimes mistaken for Domestic Horses. Their colour pattern is quite different from that of the Domestic Horse as they have a dark neck and legs and a lighter body. On Sable Island there are no other large mammals that could be mistaken for a feral Domestic Horse.

SIZE

Males are slightly to noticeably larger than females depending on the breed. Among feral forms, males tend to be only slightly larger than females.

Total length: 220–280 cm; tail length: 99–111 cm (without hair: 38–60 cm); ear length: 14–18 cm; weight: 200–360 kg; height at shoulder: 120–146 cm.

Feral newborns: weight: 25–30 kg.

RANGE

There are four main groups of feral Domestic Horses in Canada today: one on Sable Island (Nova Scotia), two in the Brittany Triangle (British Columbia's Chilcotin region), and one in the Siffleur Wilderness Area (southwestern Alberta). Animals in British Columbia and Alberta show evidence of Spanish blood, likely derived from animals brought to Mexico by the Spanish conquistadores, as some of them carry a rare DNA variant traced to a Spanish ancestry. The extent of this DNA within the British Columbia herds is still being tested because clearly some mixing of breeds has occurred, as evidenced by the variety of coat colours. Sable Island ponies are mainly descendants of animals abandoned during the forced relocation of the Acadians. There are many feral

male in
summer

Domestic Horse (*Equus caballus*)
Sable Island Horse

female in
winter

foal

Distribution of the Domestic Horse in North America (*Equus caballus*)

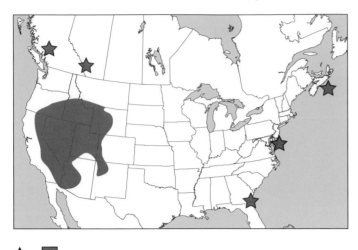

★ & ▓ introduced

herds of Domestic Horses around the world; the range map shows only those in North America.

ABUNDANCE

Domestic Horses have been transported all over the world by humans, and there are more now than there have ever been. Domestic Horses were derived from wild forms in Eurasia and the Middle East, of which there is only one that is not extinct – the Przewalski's Horse. Przewalski's Horses were eradicated in the wild in the late 1960s, but thanks to a diligent breeding and release program of captive zoo animals, there are now around 300 living wild in Mongolia again (in Hustai National Park and two other remote parks in the western Gobi Desert region). Sable Island Horses have been protected by the federal government since 1961 and may not be disturbed. The Siffleur herd is protected as long as it remains in the Wilderness Area, but feral herds in the Chilcotin area are unprotected and may be captured and sold for slaughter or domestication with impunity. Despite years of effort on the part of conservationists, Canada's western feral herds remain at risk. Determining population size is very difficult owing to the remote locations and shy habits of these animals. Estimates range from 140–200 to as high as 700 animals, divided into at least 14 bands in one of the Chilcotin herds, and around 50–60 animals in the other. An estimated 100 animals comprise the herd in the Siffleur Wilderness Area. Sable Island Horses have been carefully monitored for decades, and their numbers fluctuate, depending on winter weather conditions. Typically 165–360 animals are divided into 40–50 family bands.

ECOLOGY

Domestic Horses are highly adapted to grassland habitats, and the feral herds in western Canada also use nearby wooded parklands for shelter and cover. Sable Island Horses manage without the wooded shelter (there are no trees on Sable Island) by using the tall dunes as windbreaks. Horses' kidneys do not excel at concentrating urine, so horses must have regular access to water,

although some of the desert breeds can last up to four days between drinks, especially if their forage is succulent or dew covered in the mornings. Horses are quite capable of digging their own waterhole when water is scarce. Sable Island Horses commonly dig through the sand to find water. Winter stress and predation are the two major mortality factors in our western Canadian herds. These herds are faced with a full suite of predators including Grey Wolves, Coyotes, Cougars, and Brown Bears. Sable Island Horses have no predators. Weather conditions on the island, especially during winter, cause most of the mortality, compounded by the long-term effects of sand wearing down their teeth prematurely. Older Sable Island Horses typically have very worn teeth, which makes them vulnerable to winter starvation. The oldest Domestic Horse on record was 61 years old when it died. None of our feral Domestic Horses are likely to reach such ages, but females do generally live longer than do males.

DIET

Horses use hind gut fermentation in an enlarged caecum at the end of the digestive tract (see the introduction to the Order for a more detailed description of perissodactyl digestion). They have more rapid food-passage rates than do ruminants, but as they derive less nutrition from what they eat, they must consume more. A horse will graze for up to 15 hours out of every 24 in order to meet its energy requirement. This is opposed to ruminants such as cattle, which forage for about eight hours each day. Horses can, however, survive on lower-quality forage than can ruminants. They feed primarily on grasses but will alter their intake depending on availability. On Sable Island the horses consume primarily Beach Grass (also called Marram Grass) all year-round and add Beach Pea and Sandwort during spring through autumn. A variety of other forbs are consumed occasionally during the growing season, including the emergent aquatic vegetation growing in and around the freshwater ponds on the island. Horses paw through snow to uncover their winter forage.

REPRODUCTION

Male horses are capable of breeding at any time during the year, but their sex drive peaks strongly between April and June in the Northern Hemisphere. The age of sexual maturity varies. Females

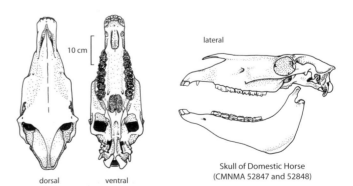

10 cm

lateral

dorsal   ventral

Skull of Domestic Horse
(CMNMA 52847 and 52848)

mature at between 10 and 24 months old, but very few yearlings and less than one-third of two-year-olds are able to carry a foal to term. Most conceive for the first time as three- to four-year-olds, about when they reach full size. Males are capable of mating when only 16 months old; however, they are not fully fertile until they are six years old. Semen production increases gradually and reaches significant levels by about two years old. Most feral males breed for the first time at five to six years old, when they are large enough to win a harem. Normally, a single foal is produced after a gestation period varying in length from 287 to 419 days. An average pregnancy lasts about 335 days. On Sable Island around 95% of foals are born between April and July, with most births occurring in May and June. Foals are nursed for up to two years but begin to eat solid food in their second week. How long they nurse depends on the fertility of their mother. If she breeds successfully during a post-partum oestrus (usually within 14–20 days after delivering her foal), a newborn will appear the following year, and the yearling is forcibly weaned so that she can suckle the foal. Some mares without a newborn will continue to tolerate the attentions of a yearling for up to another year, and rare females will allow a yearling to suckle occasionally while they are nursing a foal.

BEHAVIOUR

Horses are herd animals, and a basic feral herd consists of a dominant stallion and a harem of one or more unrelated mature females and their offspring that are up to the age of two to three years old. In a small proportion of herds (less than 10%) there may be one or more subordinate stallions distantly attached to the band and tolerated, to a limited degree, by the harem stallion. These subordinate stallions may cooperate with the harem stallion to defend the mares from other stallions. Multi-male bands tend to be short lived because most subordinates voluntarily move on after a few months or are driven away. Band size depends on many factors, including the number of mature males in the population and their individual fitness, as well as food availability and the habitat. The composition of adults within a band tends to be fairly stable over years, although adult mares sometimes leave to eventually join another band, and harem stallions may be deposed from time to time, especially as they reach old age. A harem stallion will attempt to coax a mare to stay but will not force her to remain. On Sable Island the band size is 2–10 animals, while in the Chilcotin area it may be as large as 15 animals. Males without a harem may form bachelor groups, which are made up of males of varying ages. The composition and the size of male herds vary. Sometimes two bachelor groups will temporarily amalgamate while the individual members interact. Lone stallions are seen occasionally. Horses are not territorial as they do not defend their grazing or watering areas. Harem stallions spend much of their time on the alert for possible danger or the approach of a rival. Battles between stallions for the right to a harem can be dramatic, involving biting, rearing, flailing with front hooves, and kicking with rear hooves, but most encounters between mature males are resolved after more benign mutual displays of neck arching, head tossing, and prancing. Aggression within a band is infrequent and limited to ear flattening and to

**Trot
(unshod)**

**Walk
(shod)**

Left hind track
Length: 10.0–13.5 cm
Width: 9.5–12.0 cm

Left front track
Length: 11.0–14.0 cm
Width: 10.0–13.5 cm

Domestic Horse

kicks and bites of a less serious nature. Typically, a harem stallion will not mate with females raised in his herd. These young mares leave their birth herd as they become sexually mature. Young males reaching sexual maturity are more forcibly ejected because their increasing sexual behaviour eventually provokes the herd stallion. Family bands are led by a dominant mare, and the stallion as a rule follows along at the back of the herd. During the breeding season, stallions test their mares to determine their willingness to mate. An unreceptive mare will move away from the male and even kick or bite him if he persists. If she is receptive, he will be allowed near, and the pair may indulge in some mutual grooming, and he might rest his head on her rump. Eventually she will lift her tail as a signal that she will permit him to mount. She may mate several times during her heat. Sneak matings between herd mares and subordinate males are very brief and tend to eliminate the more overt courtship signals. The subordinate males "tagging" along are looking for such an encounter while the harem stallion is distracted, but such opportunities are very rare. The vast majority of offspring are sired by harem stallions. Mares in a band with a single male tend to be healthier and less harassed than are mares in bands with multiple stallions. Foals are usually born at night, and although the mare stays close to her herd, she does not allow other horses near. Newborn foals exhibit a following instinct and will trail any large moving object, including other horses or humans. A feral mare prevents mistakes by driving other horses away from the foal or by standing between her foal and the rest of the herd until the foal is firmly imprinted onto her. Foals are normally up on their feet and nursing within an hour of their birth.

VOCALIZATIONS

Horses use a wide variety of sounds, body postures, movements, and facial expressions to communicate with each other. A whinny (or neigh) is their typical and most common vocalization. Neighs may serve to express anticipation or to locate the herd. Stallions during battle will scream, squeal, trumpet, snort, and grunt at their rival. Females squeal to drive away an ardent suitor and to defend their foals. Foals produce low-volume whickers to locate their dam and to tell her that they want to nurse. A male, approaching a female in season, may emit a giggling kind of neigh that ends in a grunt, as part of his courtship advance. Snorts are commonly used to signal alarm.

SIGNS

One of the ways that harem stallions advertise their presence is by creating large dung heaps known as "stallion piles." They may repeatedly defecate on the pile, presumably to increase its size and visibility. Of course, feral horses leave unshod tracks. The third toe forms the entire hoof, and the second and fourth toes are so reduced as to be retained only as remnants in the form of the "chestnuts" high on each leg and possibly the "ergot" on the fetlock. The front track is larger and rounder than is the hind track. The hard wall of the hoof leaves a deeper border to the track, while the softer cushioned base of the foot is inset with a triangular "frog" at the back of the track, pointing in the direction of travel. Family bands often travel in single file, creating a narrow rut in which each succeeding track obscures the one before. Scats are fibrous clumps to rounded pellets, whose size and degree of cohesion depends on the succulence of the diet. During winter, horses will sometimes gnaw bark for nourishment. Since horses have both upper and lower incisors (unlike all of the other Canadian hoofed mammals, which have only lower incisors), their scrape sign is distinctly two directional: the animal tips its head and engages both rows of teeth in a biting motion. Typically a double row of scrapes is produced, often with a strip of bark remaining between. The width of the incisor score is characteristic because horses have the largest front teeth of any of our ungulates.

REFERENCES

Bennett, D., and Hoffmann 1999; Berger 1986; Elbroch 2003; Findlay, A., 2005; Jenkins and Ashley 2003; Levine et al. 2000; Lucas et al. 1991; Turner and Morrison 2001.

Order Artiodactyla

deer, bison, sheep, and other even-toed ungulates

Muskox (*Ovibos moschatus*)

Photo: Michelle Valberg

Members of the order Artiodactyla are commonly called even-toed ungulates. The third and fourth toes have become dominant, and the remaining toes are reduced or vestigial or have disappeared altogether and, except in the hippos, are no longer weight bearing. Like the Perissodactyla, this order is digitigrade as an adaptation for speed. Artiodactyls are some of the best-known and most significant mammals to humans. Apart from Domestic Horses, most of our important domestic livestock are artiodactyls, as are many of our preferred prey species. Domestic Pigs, Cattle, Sheep, and Goats were derived from European forms about 8,000 to 10,000 years ago. As artiodactyls, both the domestic and the wild species, have often been introduced by humans into previously unoccupied areas, the natural distribution of some species can be difficult to piece together.

DISTRIBUTION

There are 10 families, 89 genera, and 240 species in this order. Artiodactyls are native on all of the continents except Australia and Antarctica and are missing from the faunas of most oceanic islands including larger ones like the West Indies, New Guinea, and New Zealand. Canada supports 3 families, which include 9 genera and 11 species.

FOSSIL RECORD

Artiodactyls first began to appear in the early Eocene epoch, around 54 million years ago. The order underwent a major radiation and increase in species numbers in the Miocene epoch, around 15 million years ago, which has persisted to modern times.

ANATOMY

Artiodactyls are divided into two main groups: the pigs and hippos, and the longer-legged ruminants. Pigs and hippos retain many of the primitive traits of this order: large canine teeth, a squat body with short legs, a simple stomach, and a tendency to be omnivorous. The other species of artiodactyls have longer legs and more complex teeth and are herbivorous and ruminant – they "chew their cud."

The stomach of a ruminant is usually divided into four chambers (three in camels): the first and second (the forestomachs) are the rumen and the reticulum, which store the gathered food and begin its processing using fermentation; the third and fourth (the hind stomachs) are the omasum and abomasum, which further break it down and absorb the nutrients. The animal regurgitates a bolus of food from the second stomach chamber (the reticulum), chews it more thoroughly, and swallows it once again either back into the first chamber (the rumen) if it needs more fermentation or into the third chamber (the omasum). The rumen contains billions of bacteria, protozoans, and yeasts that aid in fermentation and the breakdown of cellulose. These microorganisms are themselves digested in the later compartments and add to the nutrient uptake. The liquid bath that sustains the rumen organisms is composed mainly of saliva. Ruminants produce copious amounts of saliva, which they swallow with the food to replenish this broth. Such a sophisticated digestive system allows these animals to eat hard grasses and other difficult-to-digest material rapidly; they then retire to a safer location to complete the chewing and digestive process. Artiodactyls have sacrificed digestive speed for efficacy. Their plant food can take as long as 100 hours to digest. The majority of this time is taken by the fermentation process in the rumen and the reticulum. In a juvenile that is still drinking milk the rumen and the reticulum are bypassed altogether because the milk would begin to rot if it came under the action of the bacteria. The suckling action causes a reflex that closes off the forestomachs and shunts the milk directly to the omasum. In some species, the young are born with only an abomasum and require a further two to three months to complete the development of the remainder of the stomach sections, by which time they are beginning to eat plant material.

Ungulates in general, of both this order and the order that includes horses and rhinos, produce large offspring relative to the body weight of the female. The larger and more developed the newborn, the more quickly it will be able to keep up with its mother.

AVOIDING BEING EATEN

Being the preferred prey of many carnivores, artiodactyls depend on a keen sense of smell, good hearing, and, in most species, keen eyesight to help them detect predators. They also tend to gather into herds in order to benefit from additional eyes, ears, and noses. As mentioned earlier, most artiodactyls rely on speed to escape predation, which is a fine strategy for juveniles and adult animals. A newborn is less competent. There are two strategies for protecting newborns. In the follower strategy, used by some species that live in the open, the young are able to walk and then to run within hours of birth, and they instinctively follow their mother closely. The hider strategy, used by the majority of artiodactyls, involves the hiding of less mobile but well-camouflaged and almost odourless young by the mother while she grazes protectively nearby. These young instinctively flatten to the ground and keep silent for hours on end to avoid discovery.

HORNS AND ANTLERS

Artiodactyls have a monopoly on bony cranial ornamentation. The only other cranial ornamentation is found on perissodactyls. A rhino horn has no bony centre; it is composed entirely of compressed hair-like fibres. The dizzying array of horns and antlers produced by artiodactyls are often spectacular and are much sought after by human trophy hunters. There is a correlation between body size and the size of horns or antlers. Generally the bigger, older animals within each species have larger ornamentation. Horns continuously grow over the life of the animal and are not shed. They contain a bony core covered in the horn sheath, which is made up of compressed hair-like fibres. Both genders sport horns, although usually the male's horns are larger. Sometimes the older animals rub away their horn tips. This action can reduce the length of the horn and make it appear smaller. The diameter at the base of the horn is a good indicator of age if the tip has been abraded. Antlers are entirely bony, except while growing, during which time they have a thin but highly vascularized skin covering called velvet. They are the fastest growing structures in the animal kingdom. When the growth of the antler has been completed, the velvet dries up and falls off or is rubbed off. Among those species with antlers, it is almost always

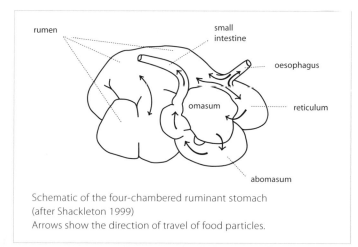

Schematic of the four-chambered ruminant stomach
(after Shackleton 1999)
Arrows show the direction of travel of food particles.

only the male that develops them. Caribou are the exception; in this species, both genders develop antlers, although the male versions are significantly larger. Antler size is usually an indication of both age and vigour. Large, healthy, mature males tend to produce the largest racks. The effects of old age, poor health, or poor nutrition will result in a reduction in antler size. Antlers are shed annually, to be regrown the following year. During the period in a year that males have no antlers they are called hummels.

Horns and antlers are important defensive weapons against predators, but the principle purpose of both horns and antlers is communication within the species. Size matters in the artiodactyls. Larger horns or antlers, combined with a massive body, convey messages of health, vigour, and dominance. The male with the largest ornamentation usually gains access to the most fertile females. This perception of dominance has an interesting corollary in Caribou. Female Caribou are the only female deer that sport antlers, albeit much smaller ones than those of the males. However, they do not shed them until much later. During the winter, when the pregnant female is struggling to supply food for herself and to help her previous year's offspring to forage as well, she still has antlers long after the males have lost theirs. With the status conferred by antlers, she can protect her hard-won feeding crater from the males who would otherwise assert their size dominance to push her out and take it for themselves.

FAMILY CERVIDAE
deer

Males in this family are antlered. Only in the exceptional case of Caribou are the females antlered. Deer have long skulls with a large vacuity anterior to the orbit. They also lack upper incisors.

Moose

FRENCH NAME: **Orignal**

SCIENTIFIC NAME: *Alces americanus,* formerly *Alces alces*

The Moose is the largest deer in the world. Adult males are more than 40% heavier than females, and a fully mature male of the largest subspecies (*Alces americanus gigas)* can weigh in at over 770 kg. His rack alone can weigh 35 kg. These antlers are the largest of all the living deer. They are exceeded only by a few extinct species, including the European Giant Deer or Irish Elk (*Megaloceros giganteus*), whose paired antlers are thought to weight up to 40 kg, and possibly by an extinct North American ancestor, the Giant Moose (*Alces latifrons*) with its massive beams.

DESCRIPTION

Rather homely looking with their large ears, long snout, droopy nostrils, and overhanging upper lip, these large deer are dark brown in colour, usually with a lighter grey-brown on the lower legs. The top of the pelage can become bleached to a reddish-brown. Moult begins in late May, and the winter coat is fully developed by late September. Mature males in new coat can be so dark brown as to appear black. Moose have very short tails. Females commonly have a small patch of white hair around their vulva that can often be seen from the air and thereby used during census taking. White and part-white Moose as well as true albinos are known but are rare. A small population near Fraser Lake, British Columbia, begin life as normal reddish-brown calves but start to turn white in their second year and become increasingly whitish-grey as they age. Moose have a double coat in both summer and winter, made up of guard hairs to shed rain and snow and of woolly underfur to provide insulation and protection from insects. Guard hairs can be as long as 25.5 cm on the top of the shoulder hump. Elsewhere on the body, they are shorter and vary in length, depending upon location. Those that make up the main part of the coat are up to about 8 cm long in the winter coat. Young are born with a reddish-brown coat

Moose (*Alces americanus*)

male in winter

male in summer with
developing antlers
still in velvet

female in
summer

calf

that has no spots. By their first autumn this coat has been replaced by one that is similar to their mother's. Adult males develop a pendulous dewlap on the underside of their throat; part of the dewlap is elongated into a "bell." Females also have a bell, but no dewlap, or one that is highly reduced. The shape and size of the bell is highly variable, and bells can be adversely affected by cold, and wholly or partly lost from frostbite. As with all of the deer except the Caribou, only males grow antlers. Antler growth begins in April, and these rapidly growing appendages remain in velvet until late August or early September. Demand for calcium and phosphorus is so high during antler growth that demineralization of the skeleton may occur, making the animals temporarily subject to osteoporosis. Antler size and body weight are correlated. The largest males usually carry the largest antlers. Antler size and body weight typically increase each year, reaching full size at seven to eleven years old. Thereafter, senescence reduces antler size until the death of the animal. Larger-antlered males drop their antlers first, sometimes as early as late November. Most males cast their antlers in December or early January, but subadult males can carry theirs until late February or even March. The dental formula is incisors 0/3, canines 0/1, premolars 3/3, and molars 3/3, for a total of 32 teeth. Calves are born with some of their milk teeth erupted (the incisors), and the remainder break through the gums and are functional by six weeks old. All of the permanent teeth are erupted by 20 months old.

SIMILAR SPECIES

At first glance, a non-antlered Moose can resemble a Domestic Horse or a non-antlered Wapiti/Elk. The body proportions and size are similar. Horses and Wapiti lack the pendulous dewlap and bell and the large shoulder hump. Horses have shorter legs and much longer tails. Wapiti have paler fur on their body and an even paler rump patch. Their legs and neck are darker, the reverse of the coloration of a Moose. Few Domestic Horses or Wapiti can travel through deep snow or bush as quickly and quietly as does a Moose. An antlered Moose is unmistakable.

SIZE

The following size discussion and data apply to North American adults. Size varies, depending upon location. The most northerly subspecies, *Alces americanus gigas*, is the largest, followed by *Alces americanus andersoni* (central and western) and *Alces americanus americanus* (eastern), which are mid-sized, and *Alces americanus shirasi* (southwestern), which is the smallest.

Males: total length: 224–345 cm or more; tail length: 9.5–17.8 cm or more; hind foot length: 76–85 cm or more; ear length: 24.8–25.1 cm; weight: 281–816 kg (average around 380–450 kg); height at shoulder: 169–192 cm.

Females: total length: 206–325 cm; tail length: 3.0–20.3 cm; hind foot length: 75–89 cm; ear length: unavailable but likely similar to males'; weight: 263–573 kg (average around 325–350 kg); height at shoulder: 165–184 cm.

Newborns: weight varies with gender. Males are about 2 kg heavier than females, and twins are on average about 3–4 kg lighter each than singletons. The average weight of a newborn male of the largest subspecies is 18 kg; of one of the mid-sized subspecies is 16 kg.

RANGE

Moose are found in northern forests in North America and eastern Asia. A closely related species, the European Moose (*Alces alces*) (often confusingly called the European Elk by many Europeans), has only recently been considered distinct by some researchers. They feel that the North American Moose occupies the northern portions of North America and eastern Russia, while the European Moose only occurs west of the Yenisey River in central Russia. Other scientists disagree and consider the two populations to be the same species. Successful introductions of Moose have been undertaken in Newfoundland, Anticosti Island, and the Cypress Hills on the Alberta-Saskatchewan border. They have been reintroduced into some areas where their populations had been extirpated, such as Cape Breton Island and the Adirondacks in northern New England. Their range is naturally expanding in some regions, such as the southwestern Rocky Mountains, the northern coast of British Columbia (recently detected on Pitt Island), the east coast of the United States, Labrador, and north of Lake Superior. There were Moose on Prince Edward Island at one time as evidenced by cast antlers that are occasionally found; when they became extinct on the island is uncertain. Western Canadian animals were introduced twice into New Zealand's South Island in the early 1900s, and a few may still survive there.

ABUNDANCE

North American Moose populations in the late 1920s were at all-time lows from over-hunting, especially in the east. Since then, steps have been taken to protect the herds and have been highly successful in most regions. Introduced Moose on the island of Newfoundland are even overabundant. Moose are highly sought-after game animals, by both sport and subsistence hunters, and are subject to intensive and careful management all across North America. Aerial surveys are the

Distribution of the Moose (*Alces americanus*)

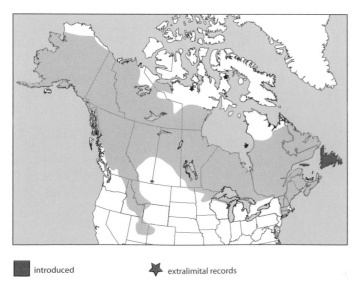

■ introduced ★ extralimital records

wildlife census technique of choice today. Current estimates suggest there may be as many as one million Moose roaming North America.

ECOLOGY

Although they can be active at any time of the day or night, Moose are weakly crepuscular most of the year, becoming most active at dusk and dawn. This pattern breaks down during the long darkness of winter and again at the peak of summer. These deer are key species in our northern forests and parklands, where they are most commonly found in rapidly regenerating areas and semi-open parklands. When forage is abundant, Moose populations are capable of rapidly increasing, especially because twinning can be common under these conditions. Cold-adapted, Moose seldom occur in regions where average winter temperatures are higher than −5°C and average summer temperatures climb above 15°C. Moose occupy a distinct home range varying from 2.2 km² to 17 km². They may use two or more portions of their home range during different seasons. Some Moose migrate from one seasonal range to another, while others remain in the same area year-round, even within a single population. Migrations can be as far as 170 km, but most are much shorter. Generally Moose migrate to better foraging areas, and most migration routes are traditional. Some individuals can occasionally disperse to new regions up to 250 km from their birthplace, which may then be successfully colonized. Still, most youngsters settle only a short distance away from the area in which they were born. Moose distribution can be influenced by that of White-tailed Deer. These smaller deer can carry the Brainworm or Meningeal Worm, to which they are partly immune, whereas Moose and other deer develop characteristic and fatal neurological symptoms. Therefore, their distribution tends not to overlap with regions of high White-tailed Deer density. The most frequent signs of an infection are an uncharacteristic lack of fear of humans, circling, and a loss of balance. Winter is a critical time for most Moose. Prolonged periods with deep snow can cause severe mortality and seriously diminish the nutritional health of the surviving animals. If a crust should form on the snow that will support predators but not the Moose, then more than usual can die by predation. Bulls will lose 12%–18% of their body weight during the rut because they generally cease eating. Consequently, dominant males often enter the winter in poor condition and are hard pressed to survive a difficult winter. Population fluctuations are typical for all of the above reasons and others, such as disease and hunting, but can be especially severe when several factors combine. Predators of Moose include the Grey Wolf, Brown Bear, Black Bear, Coyote, and Cougar. Most Moose populations are subject to several of these predators. Collisions with trains and motor vehicles cause serious mortality in some regions, especially during heavy-snow years when the cleared tracks and roadways are used by Moose as travel routes. Moose are also attracted to the roadway margins, either to lick salt in spring or to consume the successional plants that flourish there. The main natural causes of death are starvation and predation. Winter Ticks can infest a healthy Moose with little result, apart from discomfort and hair loss, but a Moose that is already weakened by starvation may succumb to a heavy infestation. Although they seldom travel faster than a walk, Moose are capable of speeds up to 55 km/h when alarmed. They are strong, capable swimmers and will even dive to reach the roots of aquatic vegetation. Moose can reach depths of 5.5 m and easily remain submerged for 30 seconds. Should they survive their first year, female Moose can live for 18–20 years, but few males live longer than 14 years. In most populations life expectancy for males averages 6–7 years and for females 7–8 years.

DIET

Moose are primarily browsers of woody vegetation, although they do seek out and consume many non-woody aquatic plants. They thrive best in areas that have been subjected to disturbances such as fire, logging, flooding, and glaciation and that are undergoing rapid regeneration by young woody plants. As a pioneering species to these quickly changing areas, Moose are quite adaptable in their dietary choices. They essentially select the most palatable and nutritious option available. In summer Moose prefer leaves and new shoots, and during winter they select the youngest woody stems they can, stretching as high as they can and sometimes straddling saplings to bend or break them over in order to reach the higher branches. In early spring and winter, if conditions are poor, they will also scrape bark off tree trunks with their lower incisors to reach the nutritious cambium layer. Preferred plants vary across the country: in the east Moose like Balsam Fir, Paper Birch, and Trembling Aspen. In the west and north they prefer Red Osier Dogwood and various willow species.

REPRODUCTION

Rut takes place from early September to early November. Well-nourished female yearlings (16–18 months old) may conceive, but most females breed when they are 28–30 months old. Poorly nourished females may not conceive until they are 40 months old, and unlike most cows that reproduce every year, an undernourished female may rest for a year between calves. Female Moose continue to produce calves even into old age; however, their most productive years are between 4 and 12 years old. Juvenile males of 16–18 months old are capable of breeding but rarely have an opportunity because they must battle for dominance with other bulls for the right to mate. They are not usually large enough to become successful in these encounters until they are over five years old. Cows also exercise considerable choice when it comes to mate selection and rarely are interested in a small bull. Most cows mate during their first heat in late September and early October, but should they fail to conceive, they will have a second or even third oestrous period at about 25-day intervals. Calves are born after a pregnancy of about 230 days, usually in May or June. Within each region the birthing peak is regular and occurs at nearly the same time each year, regardless of the current climate or the climate of the previous winter. Moose may be more susceptible to rapid climate change because of this factor. A single calf is most frequently produced, but twins are not uncommon under good conditions. Triplets are rare. The rate of twinning is a good indicator of population health and nutritional status and can vary from 5% to 67% of births. As parturition approaches, the cow drives her previous year's offspring away and seeks seclusion. The yearling tends not to go very far and frequently remains near the cow with her new calf. Births are rapid, usually requiring less than

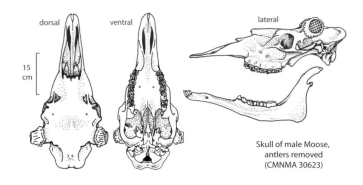

dorsal   ventral   lateral

15 cm

Skull of male Moose,
antlers removed
(CMNMA 30623)

20 minutes for singletons and 50 minutes for twins. The newborn is licked extensively and is soon standing. The calf may nurse as soon as one-and-a-half hours after its birth. It is unclear whether young Moose calves are left hidden by their mother while she forages or whether they follow her from birth. Typically, calves are nursed until the onset of rut but begin eating vegetation when they are about two weeks old. After weaning, they remain with their mother through their first winter to benefit from her foraging experience and protection. Moose have a higher reproductive output than do other large ungulates. Females reproduce earlier and have relatively large litters. The calves are born small in proportion to their mother but grow at a fast rate. The resulting faster-than-expected life cycle may be the result of adaptation to the unpredictable environmental conditions of the early successional habitats that they prefer.

BEHAVIOUR

Apart from cows with calves, Moose are largely solitary; however, in the open habitat in the far north their sociality is intermittent and thought to be relatively recent. In these non-forested regions Moose may gain some advantage by having several companions because the larger the group, the smaller is each animal's probability of being attacked by a predator. This sociality is uneasy because cows tend to be intolerant and aggressive towards each other. Bull Moose utilize two different mating strategies, depending on habitat. In a forested region, individual males search for pre-oestrous females with whom they will form a tending bond. She wanders about, calling plaintively, and all of the bulls within earshot converge on her location. If more than one male responds, they will compete for access. The bulls begin by circling each other, vocalizing with a moaning grunt, and walking in a peculiar stiff-legged swaying gait with eyeballs rolling, antlers rocking, and much thrashing of shrubbery. A nearby wallow may be urinated upon and then pawed and rolled into, the bull taking special care to saturate his dewlap and bell with urine-soaked mud. At some point in this process the smaller bull will normally retreat. If the animals are similarly sized and one bull does not yield, the battle escalates into an antler clash and some forceful shoving and antler wrestling, and each bull attempts to jab his opponent in the ribs. Broken ribs are not uncommon. Rarely, a fight may turn deadly, or antlers will lock, resulting in the death of both bulls. Typically the weaker

male will capitulate. The dominant bull will then accompany the female until she becomes receptive and allows him to mate. Despite the size differences, a cow will only allow a male of her choice to mate with her. She may remain receptive for 15 to 26 hours, and the bull will continue to tend her and breed repeatedly during that period. While he is tending a female, he may have to discourage other males from approaching since she may continue calling, and more battles may ensue. When she is no longer interested in mating, he departs, continuing his search for receptive females. In a more open habitat such as the tundra regions in the north, similar behaviours are used to establish dominance, but a dominant bull will have already conducted his dominance battles early in the rut before choosing an often traditional spot to advertise his presence. Receptive females travel around, viewing the prospective sires before making a choice. Characteristically, a group forms with the dominant bull in the middle, surrounded by cows, which are in turn surrounded by less dominant males. Cows are particularly attracted to the urine of a dominant bull, the odour of which may trigger ovulation. He will carefully urinate into a wallow to which the cows are immediately drawn. They, like the bulls, make an effort to drench their bells in the odoriferous mud. Both forest and open-country cows like to wallow in the urine-soaked mud, and battles for access can develop among them. Male Moose, when disturbed or threatened, may assume what is called a "horseshoe posture." The animals move their hind feet forward until they are just behind the front feet; when viewed from the side, the body and legs form an almost closed horseshoe shape. They may urinate on their tarsal glands (just below the hocks) at the same time. Although normally timid, Moose can be dangerous to approach and may lash out with their front legs. The cow-calf bond is particularly strong, especially during the first few weeks after birth, and females can be very aggressive in defence of their offspring. A cow will attempt and may succeed in holding off a wolf pack that is intent on killing her calf. She will also charge and trample an unwary human who approaches too closely. Bulls tend to be irascible and belligerent during the rut and are easily provoked into an attack; they are even known to contend with motor vehicles.

VOCALIZATIONS

Moose are most vocal during the rutting season. Pre-oestrous cows produce a loud, moaning or wailing call to attract a bull. This call is often mimicked by hunters or biologists to bring in a bull and may be audible by humans for more than 3 km (and presumably farther by other Moose, as they have keen hearing). During rut, bulls produce a loud grunting call that may perform two functions: advertising his presence to females and discouraging other males from intruding. Males in the open-habitat breeding groups will make a loud bark-like call, whose purpose is poorly understood. Both sexes will snort or produce a quiet appeasement whine during interactions, and cows produce soft mooing sounds during rut when they are interacting with a bull. Cows and calves emit low-volume grunts and snorts to help them locate each other, especially in a forested habitat, and calves and even yearlings will bawl if they cannot find their mother.

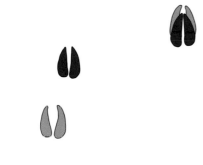

Trot

Walk

Left hind track
Length: 10.5–16.5 cm
Width: 9.0–12.7 cm

Left front track
Length: 11.0–18.0 cm
Width: 9.5–15.2 cm

Moose

SIGNS

Moose tracks are unique by their size and shape. The third and fourth toes create the characteristic long, pointed footprint. Dew-claws (the second and fifth toes) are positioned fairly low on the leg and sometimes register in a track (especially a front track) if the substrate is soft. The front track is larger than the hind, and since the front legs carry more of the body weight (which is especially true of antlered bulls), the toes tend to spread more, especially at the tips. The rear track is narrower with less space between the toes. The tracks of large males tend to be slightly blunter and broader than those of females and yearlings. Wallowing pits created by breeding males are shallow depressions with the vegetation scraped away by the front hooves. They vary in size but typically measure up to 3 m in length, 40–82 cm in width, and 9–15 cm in depth. If a bull has recently urinated into the pit, the odour will certainly be detectable, possibly from even 100 m away. Moose browse up to 2 m off the ground (some authors say up to 2.7 m), but the browsed twigs are torn off in a squared-cut, similar to that made by other deer. Moose, like other deer, rub their antlers against trees and shrubs. This tends to leave a bare section on the trunk that is highly visible. Unlike the rubs of deer, those of Moose are much higher on the tree, from 38 cm to 2.5 m from the ground. Moose will rub their orbital glands onto this bare area in order to leave both a visual and an olfactory signal, and the musky odour of a recent marking may still be detectable to a human. Incisor scrapes by Moose are typically performed on Red and Striped Maple, Willow, Trembling Aspen, Balsam Fir, and Mountain Ash but may be found on other species as well. These scrapes are made by the outer lower incisors (there are no upper incisors) with an upwards movement of the head. Scrapes can begin as low as 25 cm from the ground, and they can end as high as 2.4 m off the ground. The latter height was probably reached by an animal standing on snow. In addition to the usual incisor scrape, which is a clean scrape from top to bottom, Moose will also begin a scrape, then grab the piece of bark between their teeth and hard palate and pull it upwards, peeling off a strip. This type of sign is usually quite distinctive, owing to the shredded bits of bark left flapping at the top. Saplings that are broken over at about Moose chest height are a sign of winter feeding efforts. Summer scat is either a wet, flattened plop similar to a cow patty (up to around 28 cm in diameter) or an amorphous clump with suggestions of pellet margins, depending on how succulent the forage was. Winter scat takes the shape of a pellet with blunt ends. In spring and autumn, the pellets take a transitionary form, looking like they have been squeezed together and their shape distorted.

REFERENCES

Bowyer et al. 1998b, 1999, 2003; Chekchak et al. 1998; Chubbs and Schaefer 1997; Crichton 2002; Dumont and Crête 1996; Elbroch 2003; Franzmann 1981, 2000; Franzmann et al. 1978; Gaillard 2007; Keech et al. 2000; McLaren et al. 2004; Miller, B.K., and Litvaitis 1992; Molvar 1993; Molvar and Bowyer 1994; Price et al. 2005; Rezendes 1992; Schmitz and Nudds 1994; Stephenson and Van Ballenberghe 1995; Testa and Adams 1998; Timmerman 2003.

Wapiti

also called Elk, North American Elk

FRENCH NAME: **Cerf wapiti**

SCIENTIFIC NAME: *Cervus elaphus,* sometimes *Cervus canadensis*

Wapiti are currently considered to be part of the Red Deer (*Cervus elaphus*) complex. Red Deer are native across the Northern Hemisphere from the British Isles and Europe to Asia, Russia, and North America. They travelled over the Bering Isthmus around 250,000 years ago to North America, where they evolved for 200,000 years. Some migrated back to Siberia over another incarnation of the Bering Isthmus around 50,000 years ago. Wapiti and these closely related Siberian animals have a form and shape that is distinct among Red Deer. Other forms are smaller, have smaller antlers, and generally have redder body hair with a similar or lighter neck colour. Rutting males bugle in a lower register more suited to a forested environment. In the Old World, adult Red Deer are called stags and hinds, while adult Wapiti are commonly referred to as bulls and cows. Recent DNA work suggests that the New World/Siberian Wapiti might better be considered a separate species (*Cervus canadensis*), as has been the case in the past. Despite the genetic differences, both Red Deer and Wapiti can interbreed and produce fertile offspring, illustrating that, although distinct, the two are still closely related.

DESCRIPTION

Wapiti are large deer (second in size to Moose), with strongly contrasting coat colours, especially in winter. The neck, head, legs, and under portion of the belly are dark brown, and the rump patch and tail are creamy white. The body is light greyish-brown to reddish-brown in summer and brownish to golden tan in winter. A dark stripe rises up the back of the hind legs to border at least the lower margins of the rump patch. Some animals have a diffuse dark stripe along the spine. Hairs on the neck, especially on adult males, are longer than other body hair, giving a shaggy appearance. Newborn calves are orange-red with rows of creamy white spots and a barely distinguishable rump patch. The spots gradually fade, and the rump patch gets slightly paler over the summer. Calves grow a dark-brown juvenile coat in autumn, which lasts through their first winter. The following spring they grow their first adult coat and finally develop the characteristic large, bright rump patch. Adults undergo two moults each year. The spring moult in May and June often leaves the animals looking rather ragged for a time. The autumn moult in September is much less conspicuous. Like all other deer except Caribou, only male Wapiti develop antlers. These begin to grow each spring, and the velvet is rubbed off in late August and early September. Male calves are distinguishable by the 2–3 cm long "buttons," which can be seen only at close range. Yearlings usually grow a "spike"; however, a well-fed yearling's spike could have up to five tines. Two-year-olds produce small antlers with up to six, but usually three to four, points. Three-year-old males grow a heavier set, usually with four to five points. A pair of antlers produced by a mature bull typically weighs 7.0–15.5 kg; each antler usually has six

points. The largest antlers are grown by bulls seven to ten years old, beyond which age the antler size decreases. Antlers are shed in early spring. Mature bulls in good condition shed in late February or early March, younger males shed later, in April or even early May. The dental formula is incisors 0/3, canines 1/1, premolars 3/3, and molars 3/3, for a total of 34 teeth. Calves are born with a deciduous set of dentition with incisors, canines, premolars, and first molars already erupted. They achieve their full set of adult teeth between the age of 28 and 36 months. Upper canine teeth of adult Wapiti, called tusks, were of great importance to Plains Indians, who collected them for use as decoration and as a form of easily portable currency. Wapiti are the only North American deer that have erupted upper canine teeth. Caribou also may have upper tusks, but they are tiny and never break through the gums.

SIMILAR SPECIES

An antlered Wapiti is difficult to mistake. Antlerless Wapiti can resemble antlerless Moose or even Domestic Horses. Moose are larger and darker with no rump patch. Their coat colour is in the reverse pattern to that of Wapiti, as the legs are lighter than the body. Horses have long manes and tails and no rump patch.

SIZE

Wapiti exhibit substantial variation in weight depending on food availability and season. Within each population, adult males are larger and heavier than adult females.
Males: total length: 198.0–262.2 cm; tail length: 11–18 cm; hind foot length: 60.0–73.6 cm; ear length: 19–23 cm; weight: 178–500 kg; height at shoulder: around 130–155 cm.
Females: total length: 195–250 cm; tail length: 8–19 cm; hind foot length: 59.7–73.6 cm; ear length: 18–22 cm; weight: 160–360 kg; height at shoulder: around 127–154 cm.
Newborns: weight varies between 7 and 21 kg, largely determined by the nutritional status of their mother.

RANGE

When Europeans arrived in North America, Wapiti were found in areas ranging from southeastern Quebec and the Allegheny Mountains of New England, through the prairies and mountains to the coastal plains of western British Columbia, up to northern British Columbia, and down to California and Mexico. As settlers moved west, the Wapiti disappeared along the way. By the 1800s the eastern and southern forms had gone, and the western forms were in serious decline. The only large populations that remained were those native to mountainous regions in the west. Widespread reintroductions have occurred since then, and Wapiti now occupy more North American range than at any time since the mid-1800s. Wapiti and other varieties of Red Deer have been widely introduced around the world, primarily owing to their high value as game. Commercial production of venison has also led to the widespread use of tame (or perhaps more accurately, semi-tame) Red Deer and sometimes Wapiti in a farming context. Although there are many escapees from such establishments, these rarely result in viable wild populations.

male in summer with developing antlers still in velvet

male in winter

Wapiti (*Cervus elaphus*)

female in summer

calf

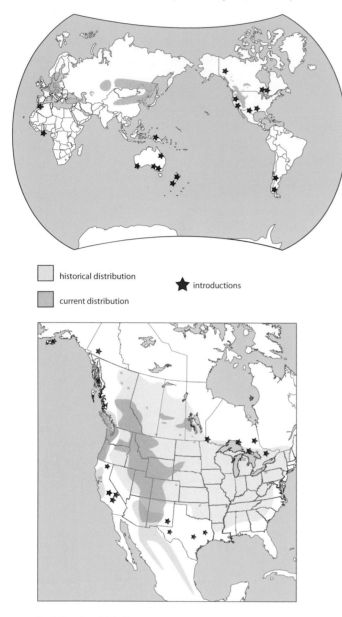

Global distribution of the Red Deer species complex (*Cervus elaphus*)

historical distribution

current distribution

★ introductions

North American distribution

ABUNDANCE

Before the arrival of Europeans, Wapiti were thought to number in the several millions in North America. They almost disappeared due to hunting, competition with livestock and introduction of their diseases, and habitat destruction or degradation. An estimated 40,000–60,000 remained by the early 1900s. This large deer is now extensively managed throughout its North American range, and there are more Wapiti now than there have been for close to two centuries. An estimated 800,000 wild Wapiti live in North America.

ECOLOGY

These deer are adaptable enough to thrive in dense rainforests, dry prairies, eastern deciduous forests, grassy valleys, and steep mountains. Wapiti avoid areas where snow is deeper than 30 cm. In forested regions they will move deeper into the forest, and in open regions they will move onto windblown slopes. Severe winters, especially those with heavy snowfall, present a serious hazard to Wapiti populations and can result in significant mortality, especially of bulls and calves, from starvation or increased predation. Mature bulls usually enter the winter in a depleted state because they do not feed during the rut. Winter mortality is so severe that the sex ratio of adult Wapiti is heavily skewed towards females, especially in those populations already affected by trophy hunting. Throughout the year Wapiti feed during the early morning and evening. They lie down to rest and chew their cud during midday and for a few hours after midnight. Feeding periods tend to be longer during winter, but otherwise the daily schedule remains the same. Preferred resting sites are just inside the forest edge, which provides protection. It also screens them from sight and provides a clear view in front to watch for approaching predators. Some Wapiti herds are migratory, travelling from higher-elevation summer ranges to lowland winter ranges. These migrations follow traditional routes and can be hundreds of kilometres long. Bulls lead the spring migration, as cows wait until their calves are old and strong enough to make the trip. Many populations do not undertake extensive migrations but rather move within the same general area, as weather and forage dictate. Within the same population, some may migrate and some may remain year round. Human civilization has had serious impacts on many herds by occupying some of the traditional seasonal range and cutting off access to crucial habitat. Some populations are being forced to abandon their traditional summer or winter ranges and are diminishing as a consequence. Wapiti rely on group vigilance to detect predators, and speed to escape them. Although there are no figures available regarding absolute speed capabilities, it is known that a Wapiti can outrun a Domestic Horse on the broken ground of the open prairie. Humans are the principal predators of Wapiti. Chief natural predators include Grey Wolves, Cougars, and Brown and Black Bears. Canada Lynx, Coyotes, and even Wolverines and Golden Eagles are secondary predators, mainly of young calves. In captivity Wapiti can live 20 years, but few wild individuals live that long, and bulls have shorter life expectancies than do cows. Even in unhunted populations, bulls rarely live more than 10 years.

DIET

Wapiti are flexible herbivores that both graze and browse, depending on the season and the availability of food. Grasses, sedges, forbs, and a wide variety of shrubs are consumed. Winter diet is composed mainly of grasses, sedges, and shrubs. Forbs are important in the spring and summer. If pressed, Wapiti will even eat horsetails, ferns, mushrooms, and pine needles. As might be expected, herds living in more open grassland habitats consume more grasses, while those in more forested regions eat more woody plants. While Wapiti can satisfy their water requirement in spring and early summer wholly from their food, they are often found near water as their preferred forage grows best there. Wapiti are also found in greater numbers in forest clearcuts and regrowth following a fire. They will happily graze on tame grasses in lawns, golf courses, and road margins if undisturbed. These large deer are ruminants; they quickly chew and swallow their forage and then retreat to and lie down in a safe spot, where they will regurgitate a bolus of compacted vegetation from their forestomach to chew more thoroughly.

REPRODUCTION

Well-fed yearlings (around 16 months old) are often capable of breeding, but sexual maturity may be delayed until the following year if their nutritional status is inadequate. Young males are usually unable to breed because older, more mature bulls drive them away from the receptive cows. Most bulls become breeders as four- or five-year-olds. In populations where hunting of mature males is intensive, yearlings and two-year-old males may become the main breeders. Most females conceive for the first time as two-year-olds and produce their first calves as three-year-olds. Rut takes place in late September and October. After a gestation of 247–262 days, a single calf is born in the following June. Underweight calves and those born later have less chance to bulk up for winter, and their survival chances are reduced. Twins occur rarely but are more common in larger, well-fed females. Cows typically produce a calf every year, but an underfed female may not come into season, thereby missing a year of breeding. Cows older than 12 years begin to display a declining pregnancy rate, until only about 20% of cows older than 16 years become pregnant. Most calves suckle for three to four months but begin nibbling on vegetation soon after they are born. Disturbance of young calves by humans can cause increased calf mortality.

BEHAVIOUR

Wapiti generally are social animals that practise gender segregation during most of the year. Maternal herds are made up of cows with their calves and yearlings and of a few older males that are younger than three years old. These groups can be large (up to 100 animals) on good range, but typically they are smaller (2–50 animals). A dominance hierarchy exists within the maternal herds, which are led by the dominant female. Usually the older, larger females are dominant. Male groups made up of similar-aged males are usually small. Solitary males are not uncommon during the summer because they spread out and sacrifice the safety of the herd for the chance to have exclusive access to a small area of top-quality forage. If they can increase their body and antler size sufficiently by using this risky strategy, they will have more opportunity to breed. Herd size tends to be larger in open habitat and smaller in forested regions. Large amalgamated herds will sometimes gather on good winter range. As pregnancy nears its conclusion, a cow will typically seek seclusion and stay alone with its new calf for as long as two to three weeks. The young quietly hide, motionless for extended periods during their first few weeks, depending for protection on being camouflaged and odourless. When the calves are older and stronger, they tag along with their mothers. They are drawn to each other to play and posture and are often left in the charge of a single female while the rest of the herd grazes. As rut approaches, Wapiti bulls seek maternal herds to join. The lead cow continues to determine the herd's movements and activity. The bull follows, guarding

and tending his "harem" of pre-oestrous cows. These bulls expend considerable effort maintaining their harem and eat very little during this period, sacrificing some of their winter survival chances in order to improve their breeding opportunities. Cows select the bull to mate with, rather than the reverse. If they are not impressed by one bull, they will leave his harem in search of another. They prefer to join a harem tended by a dominant bull as they will thereby be less bothered by other bulls, especially the overly hormonal, but inexperienced, yearlings. It is important that before winter arrives, the cows be left in peace to replenish their reserves that have been lost to a nursing calf. They can best do that under the protection of a dominant male. He benefits by being accepted by the cows as they become receptive to mating. Throughout the summer and especially as rut approaches, males rub the vegetation, bugle, spray urine, wallow, and spar with each other to determine dominance. These battles tend to be relatively harmless, although they can result in locked antlers on rare occasion. Once a bull has established himself with a harem, the battles take on a more violent tone, and losers can be seriously injured. A courting bull makes his intentions clear by approaching a cow with his head at shoulder height and his tongue flicking in and out, while uttering a low grunting call. She might simply walk away or turn and make submissive chewing motions. If she stands, he will mount. Wapiti are intelligent enough to become wary, and even nocturnal, if heavily hunted. Cows with calves, especially young calves, can be aggressive and are not safe to approach. Rutting bulls are known to be truculent, irritable, and prone to charge. In national parks and other areas where animals are accustomed to humans, Wapiti can become belligerent and may charge if disturbed. An angry Wapiti can severely injure and even kill a human. Antlered males fight by charging with head held low and tilted to present the antlers. Females and males without antlers, or with antlers still in velvet, rear up, and with heads held high and ears pinned back they flail with their sharp front hooves.

VOCALIZATIONS

Wapiti are vocal deer. The surprisingly high-pitched bugle of a rutting male ends with a long descending moan and is sometimes closely followed by a series of yelps. These grunting calls are low frequency and are usually accompanied by exaggerated pumping of the abdomen and sometimes by corresponding squirts of urine. The yelps do not carry as far as the higher-pitched bugle. While bugling, the male stands with his neck stretched and his mouth open. Because

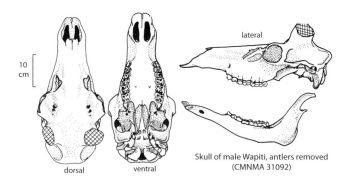

10 cm

lateral

dorsal   ventral

Skull of male Wapiti, antlers removed
(CMNMA 31092)

Trot

Walk

Left hind track
Length: 2.0–3.5 cm
Width: 2.2–3.5 cm

Left front track
Length: 3.0–4.0 cm
Width: 2.2–4.0 cm

Wapiti

of the seasonal timing of the rut, these calls are often accompanied by a vapour cloud. Females are said to be attracted to the loudest and most insistent bugler. A bull might attach the yelp call to the end of a bugle or perform just a series of yelps. Females rarely produce a bugle call and only in spring during the parturition season. It is usually emitted when the female is quietly resting or grazing with other females. Two males who are approaching each other in threat may curl their lips to expose the tusks, while producing a hissing sound and sometimes grinding their teeth. During mock fights the subordinate bull commonly emits a continuous series of high-pitched sparring squeaks. A short, high-pitched bleating contact call (also called a cohesion call) is emitted by individuals within a herd that is on the move, and probably serves to maintain herd unity. Cows often give a contact call to attract their calf to nurse. A courting bull utters quiet grunts as he approaches a cow. Faint cracking of hoof tendons allows animals to recognize cohorts moving about nearby in the dark, thereby preventing alarm and likely promoting herd cohesion. A calf in danger will call for its mother with a loud, high-pitched scream, which brings her and other nearby mothers at a run. Alarm squeals are prolonged high-pitched bleats typically produced by male yearlings or two-year-olds being chased away from the herd by a harem master. The animal assumes an alarm posture at the same time, with tail and head raised. Cows produce a different kind of alarm vocalization called an alarm bark, also produced while assuming the alarm posture and usually used to alert herd members to an approaching danger.

SIGNS

Depending on the amount of moisture in their forage, Wapiti scat can vary from an amorphous blob, like a small cow patty (12–15 cm in diameter), to a somewhat more solid blob showing faint indications of squashed pellet structure, to a discrete pellet with a characteristic acorn shape having a dimple at one end and often a small point at the other. These dry pellets are 1.1–1.7 cm in diameter and 1.3–2.5 cm in length. Wapiti do not deposit as many pellets during defecation as do Moose, and the pellets are smaller. Wapiti create two kinds of wallows. A rutting bull digs a muddy pit into which he urinates and wallows. These can usually be distinguished by odour. During summer both genders of Wapiti create a drier type of pit, with which to rub, scratch, and dust themselves to discourage biting insects. During their rest period Wapiti commonly utilize the same area day after day. Beds, which measure 1.0–1.3 m in length, tend to be used repeatedly. These are typically areas of flattened vegetation and sometimes bare earth. Feeding craters formed in snow caused by foraging Wapiti tend to be messy because the animals dig with front hooves and use their heads to shove the snow around. A whole meadow can be churned up by a feeding herd. Wapiti sign is commonly found on Trembling Aspen. The animals scrape the bark in a band as high as they can reach, exposing the bright-green cambium layer. Each year this damaged bark heals over with a black scar. In heavily used range, every aspen is scarred black from the ground to around 2 m up the trunk. Bull Wapiti typically leave signs of their antler rubs on the trunks of small and sometimes larger trees. Saplings may be rubbed clear of bark on one whole side from about 35 to

200 cm high. Larger trees tend to be rubbed in a smaller area from 58 to 152 cm up the trunk. The contents of the prominent preorbital gland beside each eye are smeared onto the bare portion of the wood to add an odoriferous component to the visual marker. Many fresh Wapiti rubs have some hairs deposited into the sap around the wound. Wapiti runways and migration trails are about the dimensions of a human trail and can usually be identified by the droppings deposited along the route. Wapiti make footprints with the third and fourth toes. The second and fifth toes are dewclaws raised up the back of the leg, which rarely register in a track, and the first toe is non-existent. The front track is larger than the rear track. The front legs carry more weight than the rear, and the tips of the hooves frequently splay at speed or on a soft substrate. The hind track usually overrides the front track at a walking pace. Their footprints tend to be larger and rounder than those of Mule and White-tailed Deer.

REFERENCES

Bauman et al. 2000; Bowyer and Kitchen 1987; Cannings et al. 1999; Cook, R.C., et al. 2001; Elbroch 2003; Feighny et al. 2006; Hebblewhite et al. 2006; Hunt 1980; Ludt et al. 2004; Peck and Peek 1991; Peek 2003; Polziehn and Strobeck 2002; Rosatte and Neuhold 2006; Shackleton 1999; Slough and Jung 2007; Wisdom and Cook 2000.

Fallow Deer

FRENCH NAME: **Daim d'Europe**

SCIENTIFIC NAME: *Dama dama, sometimes Cervus dama*

Fallow Deer is an Old World species that was introduced into Canada in about 1908 by the Duke of Devonshire. From his estate in England, they were brought to James Island, off the southeastern tip of Vancouver Island. These deer have been widely introduced around the globe as semi-tame park deer. There is considerable debate over the rightful genus name for the species. Some taxonomists favour *Cervus*; others favour *Dama*.

DESCRIPTION

Fallow Deer are the only deer in Canada whose adult pelage displays the spotted pattern normally associated only with juveniles. As this species has been selectively bred for over 1000 years, several colour variations exist. The most likely colour in Canada's population is of the wild type. In summer this coat is reddish, dotted with creamy white spots, some of which merge to form a whitish line midway along each side. A darker line may develop below the light line, and often a diffuse dark stripe runs along the spine from head to tail. The belly and lower legs are usually light tan to white, as is the throat region. The winter coat is darker and greyer, and the spots become less distinct or disappear. The tail is moderately sized for a deer and is dark above

and white below. A narrow, whitish rump patch surrounds the tail and has a dark outer border on either side, merging at the base of the tail. Other possible colour forms that might be seen are all white or light-tan (not true albino, because the animals have dark eyes; they are also called leucistic) and very dark (or melanistic), with some deer showing faint spotting and having either all dark or a lighter-brown belly. Piebald individuals may also occur. The "White Hart," so popular on signage, is actually a white male Fallow Deer. Fallow Deer undergo two moults each year, shedding their winter coat in spring and their summer coat in autumn. Bucks have a characteristic bulge on their throat over their larynx, like an Adam's apple, and a tuft of hair called a brush at the tip of their penis. The females have a comparable tuft of long hairs (up to 12 cm long) hanging just below their vulva. As in all deer, except Caribou, only males grow antlers. Fallow Deer antlers are unusually large for such a small deer and are unique in their form. A mature male will have a large brow tine, a large trey tine, and a cylindrical beam in the initial portion (like a Wapiti antler), which then flattens and flairs into a palmate rear portion, with several small points (called spellers) protruding from the edge (like a Moose antler). Male fawns begin to develop antler buds at around 6–7 months old. Before that time, the skulls of male and female fawns look similar. A yearling male produces a single spike. A two-year-old male's antlers look like a miniature version of a Wapiti antler with 3–5 points, as the flattened portion does not develop until the animal is mature. Antlers are shed in late winter and begin to regrow almost immediately. Velvet from the fully grown new antlers is shed in late summer, just before the beginning of the rut. The larger, heavier males grow larger antlers and shed them earlier than do other males. Once the males are past their prime, their antlers recede in size, and the body weight drops. Laboratory studies show that Fallow Deer see colour much as we do but are blind to red and green, which they probably see as black and shades of grey. This is likely the case for all ungulates including domestic species. The dental formula is incisors 0/3, canines 0/1, premolars 3/3, and molars 3/3, for a total of 32 teeth.

SIMILAR SPECIES

The only other deer likely to be confused with a Fallow Deer in the small area of British Columbia where Fallow Deer reside is the local native subspecies of Mule Deer, the Black-tailed Deer. This deer is slightly larger, and the male's antlers are smaller and not palmate in the outermost portion. The adult coat of a Black-tailed Deer is not spotted and is a pale greyish-tan, rather than the reddish of the Fallow Deer. Black-tailed Deer lack the black bordered rump patch of the Fallow Deer.

SIZE

Males are larger than females.
Males: total length: 140–164 cm; tail length: 16–20 cm; hind foot length: 38–42 cm; weight: 42–76 kg; height at shoulder: 76–89 cm.
Females: total length: 123–145 cm; tail length: 13–20 cm; hind foot length: 36–40 cm; weight: 34–43 kg; height at shoulder: 70–83 cm.
Newborns: weight: 2–4 kg.

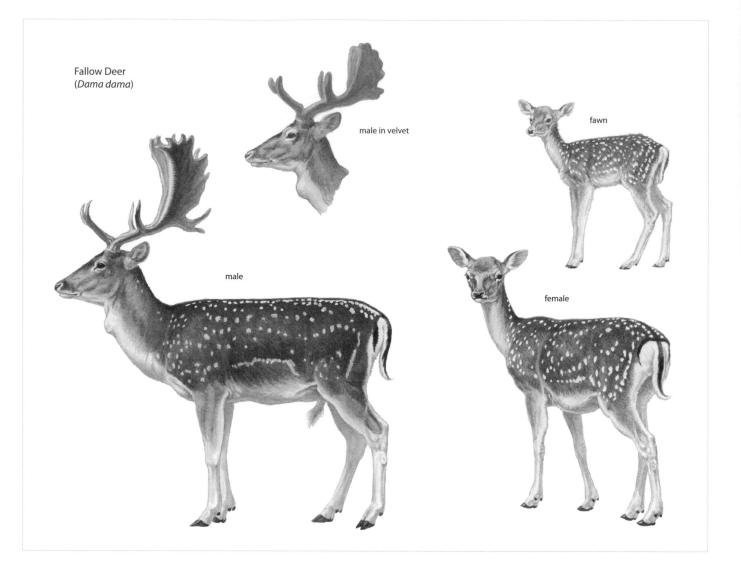

Fallow Deer
(*Dama dama*)

male in velvet

fawn

male

female

RANGE

Originally from the Mediterranean and Middle Eastern regions, Fallow Deer have been semi-domesticated for over a millennium and widely introduced around the world. The original Canadian introduction onto James Island, British Columbia, in 1908 was successful, and wandering animals soon swam to Sidney Island, where another viable colony became established. Fallow Deer have since been reported from D'Arcy Island and Mayne Island, where permanent populations are also thought to exist. Further North American introductions include Land Between the Lakes, a 70,000 ha national recreation area in Kentucky; and herds in Maryland, on Saint Simon's and Jekyll islands off the coast of Georgia, and in New York State, Alabama, Oklahoma, Texas, and California. Fallow Deer are also extensively farmed across Canada and the United States for venison and, to a lesser degree, for antlers.

ABUNDANCE

In Canada permanent wild populations of this introduced species currently exist only on four islands off the southeastern tip of Vancouver Island. Approximately 1000 animals occur in these four colonies. Recurrent reports of Fallow Deer in the lower British Columbia mainland are thought to refer to escapees from game farms, and as yet there are no records of these animals breeding in the wild. Hunting of Fallow Deer is permitted on some of the Gulf Islands.

ECOLOGY

Fallow Deer prefer to inhabit open mixed woodlands and adjacent meadows and grasslands. These deer are diurnal if undisturbed, but in Canada herds are most active between dusk and dawn, when they venture out into the open to graze. Daytime is spent mainly in the security of the trees. There are no predators of Canadian Fallow Deer apart from humans and possibly Domestic Dogs. Fallow Deer are proficient and willing swimmers. The majority of deaths are probably the result of hunting and old age. Longevity of Fallow Deer in British Columbia is unknown. However, in other regions of the globe Fallow Deer are known to live for 20 years in captivity; their lifespan in the wild is generally 10–12 years, although a few may live to 16 years old.

Distribution of the Fallow Deer (*Dama dama*)

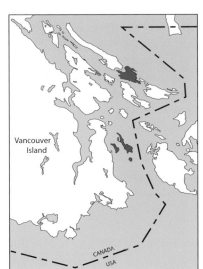

Vancouver Island

CANADA
USA

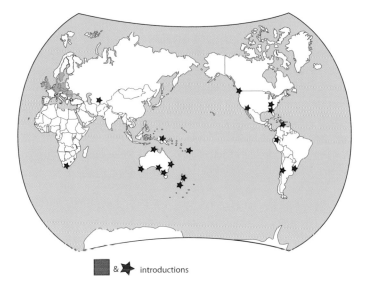

■ & ★➤ introductions

DIET

Fallow Deer are flexible feeders and will both graze and browse as conditions and forage permit. They consume a wide variety of vegetation, including grasses, forbs, and woody plants. On the Gulf Islands they are also known to forage on Common Eelgrass, which washes up onto the beaches.

REPRODUCTION

In the Gulf Islands rut occurs in October, with a peak around mid-October. Gestation is thought to last 180–220 days. Normally a single fawn is born in late May or June. Twinning has been recorded but is rare in Canada. Most does become sexually mature at 18 months old and produce their first fawn at the age of two years or sometimes three. Although the young males are also capable of breeding at 18 months old, they rarely have the opportunity because they are driven away by the more dominant, older males. Most bucks do not become breeders until they are five or six years old, but if hunting

has reduced the number of mature males, younger males are capable of filling the gap. Fawns are nursed for about four months and are weaned around the beginning of the rut.

BEHAVIOUR

In British Columbia adult males live in small groups of two to five animals, and females and calves form larger herds of five to twenty animals. In other parts of the world, herds can be as large as 150–175 animals. The nursery herds are loosely hierarchical and are usually led by the dominant doe. Bachelor herds break up at the early stages of rut, and the bucks travel back to the range of the maternal herds. Fallow Deer males curiously practise two different types of court-ship behaviour, perhaps depending on herd size and the level of competition. It is not known which form of courtship behaviour is followed by Canadian animals. Elsewhere around the world, Fallow bucks either collect a "harem," which they guard and tend, or they stake a small territory among other males and actively display and call to attract a female to them. This latter behaviour is called lekking and is rare among mammals, although common among birds. Activity on a Fallow Deer lekking ground can be quite chaotic. Some males will be vocalizing, others will be urinating onto their hind legs (rub-urinating) and into the mud of their scrape, while others will be bobbing their heads to draw attention to their armament. Antler-pushing contests break out from time to time. In the midst of all this, the females will be browsing for a mate and checking out the scrapes, while younger males will be dashing about trying to steal a mating. Large dominant males eat very little during the rut, whether they are tending a harem or displaying on the lekking ground, and can lose considerable body fat. This can negatively impact their survival chances if the following winter is severe. Does usually seek seclusion before giving birth. Fallow Deer fawns are "hiders" for the first one to two weeks of life: they flatten in the vegetation, remaining still for long periods while their mother grazes at a distance. She returns to nurse the fawn for only a few minutes at a time. Like Mule Deer, an alarmed Fallow Deer uses a bouncing gait called stotting. In other parts of the world this same gait has been called pronking. All four feet hit the ground at the same time, and the animal looks like it is on springs. A stotting deer is likely to have its tail high and even curved over its back to display the white underside. The gait, an accompanying alarm bark, and the flashing white tail are all alarm signals meant to warn others in the herd of a perceived danger.

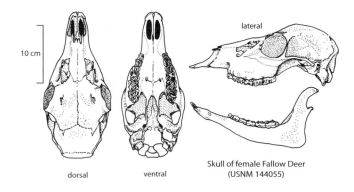

10 cm

lateral

dorsal   ventral

Skull of female Fallow Deer
(USNM 144055)

Trot

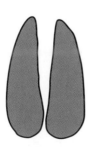

Left hind track
Length: 5.5–8.0 cm
Width: 3.0–5.0 cm

Left front track
Length: 6.0–9.0 cm
Width: 3.5–5.5 cm

Fallow Deer

VOCALIZATIONS

The most recognized sound produced by Fallow Deer is the belching groan heard during the rut. This sound is persistently repeated by dominant males as they occupy their "stands" or guard their harems. It is thought to fulfill several functions. It advertises the location of a buck to attract females and, at the same time, repels other males. The call's volume and repetition rate provides information of the size and vigour of the actor. A doe will bleat to encourage her hidden fawn to emerge and nurse, and the fawn learns to recognize its mother by her bleat. Fawns, in turn, will bleat to attract or locate their mother and can produce a wailing scream if in extreme distress. The barking alarm call is another well-known Fallow Deer sound; it is similar to a dog's bark and is used especially by females with fawns.

SIGNS

Fallow Deer's heavy browsing can create a distinct browse line on the trees and shrubs that may appear too high, but the deer will stand on their hind legs to reach the higher twigs. During the October rutting season, males makes scrapes on the ground, clearing patches of about 45–60 cm by 60–90 cm, onto which he rub-urinates. Even humans can detect the odour of the scrapes from a distance. Rutting bucks thrash nearby saplings, commonly leaving large bare areas of trunk. Playrings are a peculiar and mysterious sign left by Fallow Deer in Europe. They are large worn rings (up to 3 m in diameter) usually around a stump, and their purpose is unknown. Such rings have not been noticed in Canada. Droppings and footprints of Fallow Deer are indistinguishable from those of the native Mule Deer subspecies, the Black-tailed Deer (see Mule Deer signs).

REFERENCES

Long, J.L. 2003; Macdonald and Barrett 1993; Reby et al. 1998; Shackleton 1999; Sheats and Cason 2005.

Mule Deer

also called Black-tailed Deer, Columbian Black-tailed Deer, Sitka Black-tailed Deer, Rocky Mountain Mule Deer

FRENCH NAME: **Cerf mulet**

SCIENTIFIC NAME: *Odocoileus hemionus*

There are two main forms of this western species. Rocky Mountain Mule Deer are found inland, and Black-tailed Deer are found on the western coastal plains and rain forests.

DESCRIPTION

The Mule Deer is a large-eared deer with a stocky body and long, slender legs. Three subspecies occur in Canada: Rocky Mountain Mule Deer (*Odocoileus hemionus hemionus*) and two similar coastal subspecies of Black-tailed Deer, Columbian Black-tailed Deer (*Odocoileus*

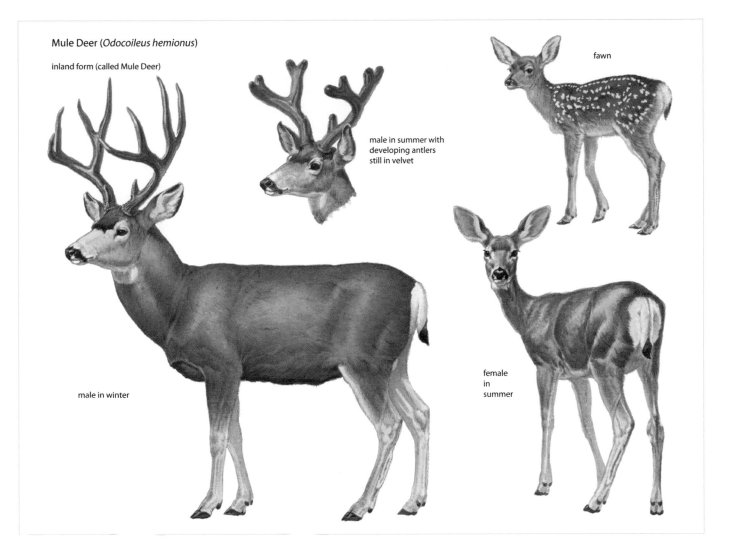

Mule Deer (*Odocoileus hemionus*)

inland form (called Mule Deer)

male in summer with developing antlers still in velvet

fawn

male in winter

female in summer

hemionus columbianus) and Sitka Black-tailed Deer (*Odocoileus hemionus sitkensis*). It is usually possible to distinguish between the interior form and the coastal forms, but it can be quite difficult, and usually impossible, to distinguish the two coastal subspecies in the field. Furthermore, there is considerable blending of subspecies by interbreeding in regions of overlap. The interior subspecies is generally larger and has a lighter, tan-grey winter coat, while the coastal subspecies are smaller with a darker, grey-brown winter coat. A distinct white rump patch is present in both summer and winter coats but is smaller in the two coastal subspecies. Only the tip of the tail of the Rocky Mountain Mule Deer is black, while the entire dorsal surface is black in the Sitka Black-tailed Deer, and almost all black in the Columbian Black-tailed Deer. The muzzle is blackish, the snout behind the muzzle tends to be lighter, and the forehead is dark. One white throat patch, and sometimes two, occurs. Adults moult twice yearly. In late spring the old winter coat is shed and replaced by a sleek reddish-brown summer coat, which has only sparse underfur. The spring moult of pregnant females is delayed, as it is with most ungulates. The warm-weather coat is shed in September, and by late November the full winter coat is complete with a thick, woolly undercoat and a greyish-brown overcoat of longer guard hairs. Both guard hairs

and underfur hairs on the winter coat are thicker, denser, and longer than those on the summer coat. Fawns' coats are spotted at birth and are darker brown than those of White-tailed Deer fawns, which tend to be lighter orange-red. They grow their first winter coat between July and October, at which time they lose their spots. Only males produce antlers. Male fawns commonly grow small button antlers in their first summer. The course of antler growth over the next two years depends on the nourishment that the young buck receives. A poorly fed yearling will grow a spike, a normally nourished youngster will develop a two-point forked antler, and a very well-nourished one may develop a three- or even four-point rack only slightly smaller than that of an adult. Normally, by two years old the antlers reach full size; however, if the diet is poor, a further year may be necessary before a full four-point rack has grown. Subsequent years' growth is most evident in the enhanced diameter of the beam, the increasing length of the antler, and sometimes in additional points, often by bifurcation. Apart from the usual walk, trot, lope, and gallop, Mule Deer have two additional and characteristic gaits. One is a stiff-legged bounce, called a stot, used by alarmed Mule Deer: all four feet hit the ground at the same time, and the animal bounds into the air for a distance of 3–5 m. A stotting (also called pronking) Mule Deer can quickly change or

Distribution of the Mule Deer (*Odocoileus hemionus*)

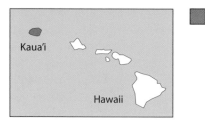

Kaua'i

Hawaii

introduced

display intermediate behaviour, gaits, and morphology and are probably fertile.

SIMILAR SPECIES

White-tailed Deer are the most likely species to be mistaken for Mule Deer. They have much smaller ears, lack the rump patch, and have a large tail that is brown above and white below. When they have been startled, they flip up and wave their tail as they run, a behaviour called flagging. Their antlers are simpler, lacking the double-forked tines (bifurcate) of the Mule Deer. White-tailed Deer do not stot, but they do make long leaping bounds when alarmed. White-tailed Deer fawns are reddish-orange in colour, while Mule Deer fawns are a darker reddish-brown, but both have white spots. In a few of the Gulf Islands, Black-tailed Deer co-occur with introduced Fallow Deer. Adult Fallow Deer are the only deer in North America whose coat retains the spotting normally found only on fawns. Distinguishing isolated fawns of Fallow and Black-tailed Deer in the field can be difficult, but Black-tailed Deer fawns tend to have darker foreheads. Since both species hide their fawns and graze away from their offspring, it is possible to see a fawn and imagine that it has been abandoned when in fact its mother is just out of sight.

SIZE

Mule Deer are quite variable in size, and males are 20%–58% heavier than females. Body weight varies seasonally; it is highest in October for all adults, and lowest in March for males and in April for females. The largest Mule Deer are found in the northern interior portion of the species distribution, and the smallest are found among the coastal forms.

Males: total length: 147.2–198.0 cm; tail length: 12.7–25.4 cm; hind foot length: 41.0–55.9 cm; ear length: 11.8–23.0 cm; weight: 45–131 kg; height at shoulder: up to 106 cm.

Females: total length: 127–175 cm; tail length: 15.0–21.6 cm; hind foot length: 40.0–50.8 cm; ear length: 11.8–22.0 cm; weight: 45–80 kg; height at shoulder: up to 100 cm.

Newborns: typical weight is 2–4 kg, although extreme weights of up to 5 kg have been reported. Male fawns are heavier than female fawns. Single fawns are typically heavier than either of a set of twins, and in a mixed gender set of twins the male foetus is up to 33% larger than the female foetus.

RANGE

The Mule Deer is strictly a New World species. It occurs over most of western North America from southern Alaska to Mexico, including offshore islands along the west coast, such as Vancouver Island and the Gulf Islands. Black-tailed Deer have been introduced to the island of Kaua'i in the Hawaiian chain, to the Queen Charlotte Islands, and to Kodiak Island and other nearby mainland regions of Alaska.

ABUNDANCE

Mule Deer are important game animals and are hunted in most Canadian jurisdictions where they occur. They are generally numerous

even reverse direction with each bounce, an exceptional advantage on rough and broken terrain, such as occurs throughout Mule Deer range. The second gait is a deliberate, stiff-legged walk commonly adopted if the animal is suspicious or uncertain. The dental formula is incisors 0/3, canines 0/1, premolars 3/3, and molars 3/3, for a total of 32 teeth. At birth, the milk teeth are composed only of deciduous lower canines and incisors. The deciduous premolars erupt soon after, and the permanent teeth slowly erupt over the course of the next 28 months, when the full permanent adult dentition is completed.

Hybrids have been seen of Mule Deer and the closely related White-tailed Deer but, despite extensive overlap of their distributions, these are rare. Some recent research suggests that the incidence may be much higher in certain areas. Hybrids typically involve a White-tailed Deer buck and a Mule Deer doe and hence are seen in association with Mule Deer herds. The hybrids typically

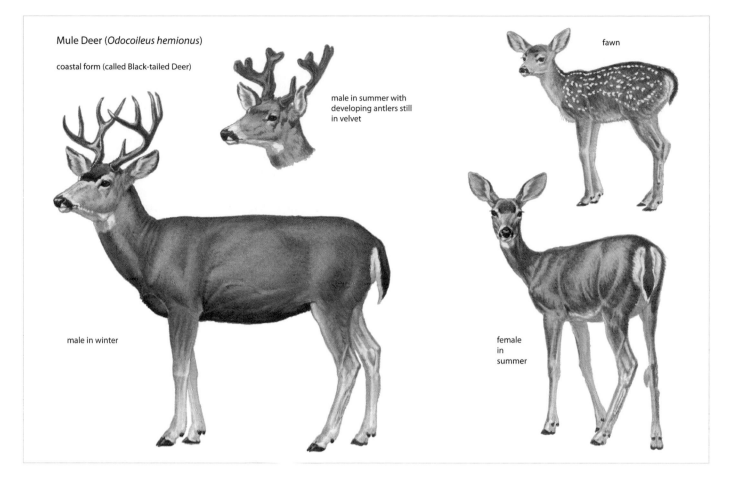

Mule Deer (*Odocoileus hemionus*)

coastal form (called Black-tailed Deer)

male in summer with
developing antlers still
in velvet

fawn

male in winter

female
in
summer

throughout the Canadian portion of their range, although populations are subject to occasional extensive die-offs during severe winters with heavy snowfall. Many wildlife managers believe that Mule Deer numbers are generally decreasing throughout North America (since about 1992), but reasons and widespread survey-based proof have not yet been provided.

ECOLOGY

Mule Deer are able to adapt to a wide range of habitat types, with the exception of extreme deserts, Arctic and tropical regions, and dense boreal forests. They are quick to take advantage of the new growth that follows a forest fire or logging operations. The interior subspecies is most commonly found in parklands or open forests. On the Prairies, Mule Deer are typically associated with coulees, whether forested or not, and with regions of dissected badlands or breaks, especially when these are intersected by agricultural lands. The two coastal subspecies are numerous where there are old-growth forests for shelter and near more open areas that are either natural or have been farmed, logged, or urbanized, and support plentiful new growth. The highest densities are recorded where forests were logged about five years previously and there is a superabundance of new growth. As the trees become denser (but are still low), deer numbers drop. Numbers rise again as the trees mature and their dense understorey provides good forage. Old-growth forests are most important during severe winters with deep snow accumulations. All Mule

Deer prefer to avoid snow that is deeper than about 30–40 cm, and migrations are often prompted by snowfall. When avoidance is not possible, the deer restrict themselves to an area (called a deer yard) where they can keep trails open in the deep snow by repeated use. Like most large North American herbivores, Mule Deer have a daily activity pattern that involves peaks of foraging around dawn and dusk and periods of resting around midday and in the middle of the night. Night-time activity varies, as some populations that are heavily hunted will forage only during times of darkness. During the winter, daytime foraging is common as more time must be spent eating and looking for food because of the low quality of the sparse forage. Some populations are sedentary, while others occupy seasonal ranges and migrate between them. In northern montane regions, populations are migratory in areas of high snowfall, generally moving to lower elevations in winter and to higher elevations in summer. During severe winters such as those experienced in the Prairies, Mule Deer select sheltered areas out of the wind and limit their movements in order to conserve energy. Deaths caused by starvation and predation become more numerous during extended winters with high snowfall. Winter mortality strikes heavily on fawns experiencing their first winter and on adult males entering the season already depleted from the rut. Sport hunting also takes a higher toll on adult males. As a consequence, the ratio of adult females to adult males in most populations is more than 2:1, even though there are slightly more male fawns born than female fawns. Heavily hunted populations may have a ratio of closer to 20:1. There are some suggestions that Mule Deer numbers may be declining, while White-tailed Deer numbers are increasing in some regions. These changes are complex, and the reasons vary from region to region. In some cases, predators attack Mule Deer more successfully; in other cases, human disturbance of the habitat by farming or logging benefits White-tailed Deer. In other regions, it seems that White-tailed Deer are repopulating areas where they were previously hunted out over a century ago. A parasite called a Brainworm or Meningeal Worm, which is carried and tolerated by White-tailed Deer, can be fatal to Mule Deer. A stotting Mule Deer is said to reach speeds of up to 60 km/h over a short distance. All Mule Deer are confident swimmers, and both of the coastal subspecies easily swim between the mainland and nearby offshore islands. Principal predators of Mule Deer include Coyote, Grey Wolf, and Cougar. Secondary predators, particularly of fawns, include Black and Brown Bears, Canada Lynx, Bobcat, Wolverine, Domestic Dog, and Golden Eagle. Unfortunately, it appears that Mule Deer do not change their behaviour near roadways when they are exposed to the ultrasonic sounds produced by deer whistles on motor vehicles. Although female Mule Deer may be capable of a lifespan of more than twenty years, and males only slightly less, the average lifespan of a wild deer is considerably shorter. Few females live longer than 10–12 years, and most males do not live more than 8 years.

DIET

These herbivores have a diverse diet that varies by season and location. They are primarily browsers but also consume forbs, grasses, mosses, lichens, mushrooms, and even algae. Mule Deer tend to select the most digestible and nutritious portion of the plant available. They occur in arid areas and are well adapted to extracting much of the moisture they require from their food. Still, some access to freestanding water is necessary, especially during the hot summer period.

REPRODUCTION

Although it is theoretically possible for a very well-nourished fawn to come into heat during her first autumn, few wild females are capable of this level of nutrition. The vast majority do not conceive until their second year. Similarly, most young males become fertile in their second year, although few have an opportunity to breed until they are at least three years old and have developed a rack that will enable them to compete with other males. Timing of the rut varies with location. In southern coastal British Columbia it

may begin in late October and last until mid-November. In more northerly coastal regions it may begin in early November and last until early December. Most Rocky Mountain Mule Deer begin the rut in mid-November and conclude in about mid-December. Onset of the rut is signalled by a noticeable thickening of the mature male's neck and by the males rubbing the velvet off their antlers. Does have an oestrous cycle of 20–29 days and come into heat and are receptive to a male for only 24–36 hours. Should conception not result, a doe will likely cycle again, perhaps resulting in a late fawn. Gestation lasts for an average of 200–205 days, and birth of a single fawn or twins occurs in the following June. Twins are common to well-nourished does, but most first-time breeders produce a singleton. Triplets are rare. Females typically isolate themselves just before the birth and establish a small summer home range, but some does remain with their small group of related does on a group range. Newborns and very young fawns use the camouflage provided by their spotted coat to hide, remaining still for hours at a time as their mothers graze. After a variable period spent hiding in the undergrowth (a few days to a few weeks), the fawns are strong enough to follow their mother. She will suckle her offspring all summer as they gradually consume more vegetation and less milk; by late summer they are fully weaned. The fawns remain with their mothers over at least their first winter.

BEHAVIOUR

Mule Deer tend to be gregarious, and especially the females and young are found in groups, although solitary animals are not uncommon. Group sizes are largest in winter and smallest in early summer when the fawns are being born. Herd size in all seasons is larger in open habitats and smaller in more dense vegetation, especially forests. Group living is clearly beneficial because the probability of being attacked is smaller, and when predator vigilance is shared, more time can be spent feeding. Mule Deer pay a price for this security, however, because forage must be shared and, as individuals can be aggressive towards each other, time may be spent dominating other group members. Consequently, herd membership is flexible as deer leave and join over time. Female herds during the summer are usually restricted to a female and her newest fawn and perhaps her offspring from the previous year. A large winter doe group may include several does and several generations. Male groups are often large, except during rut. Reproductive females select their range largely for its safety. They prefer a rugged habitat, which is less attractive to most predators, although this strategy is not as effective against Cougars as it is against Coyotes. In summer, males tend to select a habitat with higher-quality forage even though there can be a greater likelihood of predation, perhaps sacrificing safety in favour of better nutrition; this makes for a larger body and antler size and consequently more success on the breeding grounds. Mule Deer attacked by a single Coyote typically bunch together, stand their ground, and aggressively defend others in their herd. As long as each animal retains its nerve and does not flee, the whole herd stands a good chance of surviving such an assault. A solitary Coyote, especially if it is a juvenile, may be charged, chased, and kicked with the forelegs, and it runs the risk of being trampled

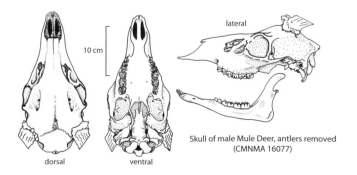

lateral

10 cm

dorsal   ventral

Skull of male Mule Deer, antlers removed
(CMNMA 16077)

and, even killed, if caught. In early autumn, females without fawns are generally in better physical condition since they have not been nursing a fawn all summer. They are more likely to produce a larger fawn the following spring, which by its size will have a better chance of survival to adulthood. Perhaps, somehow, the courting bucks recognize this, because they direct more of their courtship efforts towards the unattended does than towards those with fawns at foot. Before the rut, males spar with each other to determine their relative dominance status. Few of these battles result in serious injuries. Typically dominant males approach smaller males, inviting them to spar. As rut begins, the fights become more serious, and bucks attempting to disengage must retreat with alacrity to avoid being punctured by their superior rival, who will continue to pursue them until they are safely removed from the contested female. The longest, most ferocious battles occur between equal-sized bucks. Mule Deer frequently use scent to mark their home range and have several prominent scent glands. Rutting bucks will dig a scrape, then, using a characteristic hunched posture, position their hind legs under their bodies and urinate onto their metatarsal glands as they rub their legs together. This apparently enhances the musky odour from the glands and scents the urine, which then runs into the scrape. Females will also occasionally hock-rub but do not create scrapes. (Use of preorbital glands in scent marking is discussed in the "Signs" section below.) During the rut, males travel around looking for oestrous females. They may approach adult females slowly and calmly, with their tongue flicking in and out, while assuming a head-down, neck-stretched posture called a low stretch and emitting a soft buzzing call. If they are in a state of extreme excitation, they may rush at the doe, bleating with head lowered, then stop and strike the ground with the forefeet, while roaring. Both approaches usually result in the female crouching to urinate so that the male can sniff her urine to determine her reproductive state, using the lip-curl posture. A dominant buck will "tend" a pre-oestrous doe until she is receptive and allows him to mate, or he is displaced by a larger male. In Mule Deer, as in all deer, body and antler size determine dominance. The largest males perform most of the breeding. A receptive female allows the male to approach, lick her rear, rest his chin and neck on her rump, and then copulate.

VOCALIZATIONS

Males make a coughing grunt called a rut-snort, followed by a low-volume hissing sound, when confronting each other over access to a female. Males may produce a barking call when pursuing another male, and a loud coughing roar when they approach a female. Both sexes will produce whistling alarm snorts when startled or uncertain. Fawns bleat to call their mothers. Does produce a deeper blat to warn other deer away from their fawn. If captured by a predator or a human, a very young fawn will freeze and not make a sound, while a fawn older than about six days will produce a loud and startling squeal that attracts any nearby female. Courting males sometimes quietly approach a pre-oestrous female, emitting a low-volume buzzing sound. Alternately, he may rush at her while producing a long bleat.

stride length of a stot can vary from 2 to 5 m and be occasionally greater

Stot

Walk

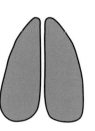

Left hind track
Length: 5.0–9.0 cm
Width: 3.8–6.0 cm

Left front track
Length: 5.7–10.2 cm
Width: 4.1–7.0 cm

Mule Deer

Mule Deer, like all browsing deer, nip twigs, leaving a rough squared-off end, and their preferred browse height is between about a human knee and hip. Antler rubs are made by bucks, usually in October and November. A buck will vigorously, or sometimes slowly and rhythmically, rub his antlers on a sapling, removing bark (typically between human knee and hip height) and creating a sight marking; he then adds a scent marking as he smears the secretion from his preorbital gland (just in front of the eye) over the raw surface. Most antler scrapes are smooth, but some have frayed bark at the top and bottom. During the rut, males produce scrapes that are often triangular in shape. These are created as the buck paws the earth with his forefeet, flinging vegetation and dirt backwards sometimes several metres. Then he will hock-rub and urinate onto the bare earth. Mule Deer urine often has a not unpleasant, piney smell. The does scout around looking for these scrapes, sniffing them intently. Preferred bedding sites may be reused over long periods of time, and each deer likely will have several within its home range. In summer, concentrations of biting flies can drive deer to run from one bed to another during a resting period. Beds of Mule and White-tailed Deer are similar in size, varying from 62 to 107 cm in length depending on the size of the individual and the length of use. Beds in snow are seen most frequently and, if carefully examined, may show the curved mark of the rump, a hind-leg print opposite, and marks where the front knees folded and dug into the snow at the top. Tracks are left by an imprint of the third and fourth toes. The second and fifth toes are dewclaws. These are slightly raised up the back of the legs and will leave an imprint in a soft substrate. Toes, especially those at the front, often splay apart at the tips depending on the animal's speed and whether it is travelling uphill or downhill. Track size varies from the smaller coastal subspecies to the larger inland form, and between the smaller female and larger male. Scat varies with diet. It can be like a small cow patty in summer or a mass of variably sized pellets, often with a dimple at one end and a small point at the other, in winter. Long thin pellets and short thick ones can appear in the same defecation mass. Unfortunately there are very few clues for distinguishing the Mule Deer sign from the White-tailed Deer sign because their habits are alike and their scat and tracks are similar. The primary difference in the track pattern is left by a stotting Mule Deer where all four feet strike the ground close together. White-tailed Deer do not use this gait. The illustrated walking track is common to both species. (See the White-tailed Deer account for illustrations of a trotting pattern and a deep-snow walking trail that are shared by both Mule Deer and White-tailed Deer.)

REFERENCES

Anderson, A.E., and Wallmo 1984; Bowyer et al. 1998a, 2001; Cathey et al. 1998; Collins and Urness 1982; Elbroch 2003; Geist 1999; Hoefs 2001; Hornbeck and Mahoney 2000; Kamler et al. 2001; Kay and Boe 1992; Kie and Czech 2000; Launchbaugh and Urness 1992; Lingle 1992; Lingle and Pellis 2002; Mackie et al. 2003; Main and Coblentz 1996; Margulis 1993; McCullough 1997; Rezendes 1992; Robinson et al. 2002; Romin and Dalton 1992; Shackleton 1999; Weckerly 1994; Wilkinson, J.A., and Duglass 2002.

White-tailed Deer

FRENCH NAME: Cerf de Virginie
SCIENTIFIC NAME: *Odocoileus virginianus*

The flashing wiggle-waggle of a white tail disappearing into the bushes is a familiar sight in North America, as the ever alert White-tailed Deer usually detects you before you see it. This animal is the most widespread, the most recognized, and the best understood of our North American deer. For many Canadians, the word deer conjures up this animal.

DESCRIPTION

White-tailed Deer are the smallest of our native deer, although they are similar in size to a small caribou (the northern Peary Caribou). They moult twice each year, and the winter and summer coats are different colours. The thick grey-brown winter coat is shed in spring. This moult begins at the neck before proceeding to the head and then farther down the body. A deer seen at this stage, with a skinny orange neck of summer fur and a fuzzy grey-brown head and body still in winter coat, can look very peculiar. The summer coat is a sleek orange-brown and is complete by May or early June. Healthy individuals finish moulting sooner than others. The

Distribution of the White-tailed Deer (*Odocoileus virginianus*)

⭐ extralimital record

★ introductions

White-tailed Deer
(*Odocoileus virginianus*)

male in summer with developing antlers still in velvet

fawn

female in summer

male in winter

winter coat begins to grow in August and is usually fully grown by early autumn. Winter hairs are long (5.0–7.5 cm on the back and sides), hollow, and densely packed, and their shafts are considerably thicker than those of the summer hairs. There is a sparse layer of shorter underfur beneath these guard hairs, which provides some additional insulation, but clearly the guard hairs retain most of the heat. The summer coat is composed only of short, fine guard hairs that are around 2–3 cm long. Melanism and albinism, as well as white animals and varying degrees of piebald colourings, have been reported, usually from localized areas (that is, where animals with recessive genes for unusual colours – but normal colouring – are most likely to breed with each other and produce unusually coloured offspring). There are conspicuous permanent white markings on the lower lip, throat, and chin, along the belly, inside the ears, around the muzzle and eyes, and over the tarsal glands (inside the hock joint, also called hock glands) and the metatarsal glands (outside of the hind leg, about two-thirds of the way down the leg between the hock and the hoof). The tip of the muzzle is black and leathery and is surrounded by black fur, with a strip of black fur extending down across the lower lip. The forehead, particularly of mature males, is often dark, and the top of the snout can have a greyish or even blackish cast. The upper surface of the large tail is brown or blackish-brown with a white border, and the underside is white. By raising it and erecting the hairs to display the bright

under-surface, and then waggling it to flash the white colour, the tail makes an effective alarm "flag." White-tails do not have a conspicuous white rump patch, but the size of the white flag is enhanced by the white regions of the hind end that are usually covered when the tail is down. Their eyeshine is bright pale yellow. Newborns have a bright reddish-orange coat with hundreds of small white spots. Some of these are arranged into two rows, one on either side of the backbone, and the remainder are randomly scattered on the sides and flanks. This coat is moulted at around three to four months old when the fawns grow a typical winter coat. Only males grow antlers. Antler size and shape are highly variable depending on age, nutrition, and genetics. Some early maturing, well-nourished fawns may produce a button antler by their first autumn, but most begin growing their first rack the following spring. Yearlings in Canada typically produce small simple antlers, although those that produced a button the previous year may grow a fairly respectable, albeit small, set. Usually the first full set of mature antlers is grown at two to three years old. The main beam bends forward, and four points on each antler is the norm. Growth continues, primarily of the beam length and diameter, until the buck is five to seven years old. Points may be added with age. Antlers begin growing in spring and are full grown and free of velvet in late August or September. They are usually cast in late December to early February. The dental formula is incisors 0/3, canines 0/1, premolars 3/3, and molars 3/3, for a total of

32 teeth. At birth the deciduous incisors are usually already erupted, followed closely by the incisiform canine and two premolars within the first four weeks. The third and last deciduous premolar erupts by 10 weeks old. There are no deciduous molars. By about 19 months old, all of the adult teeth have erupted and replaced the milk teeth.

For a discussion of hybridization between White-tailed and Mule Deer and the appearance of the hybrids, see the end of the "Description" section of the Mule Deer account.

SIMILAR SPECIES

The most similar species to White-tailed Deer is the closely related Mule Deer. In the regions in which both species co-occur, they are similar in size, although White-tails tend to be somewhat smaller. Some Mule Deer antlers have bifurcate (double) points unlike the simpler, branched antlers of White-tails. Mule Deer tails are smaller, have a black tip, and are surrounded by a white rump patch. Behaviourally, Mule Deer's reaction to alarm is quite different. They typically flee using a unique gait called a stot where all four feet strike the ground together, and the animal appears to bounce. White-tailed Deer bound in a more conventional way but elevate their tail to reveal the white underside, and they often wave it from side to side as they run away. Mule Deer fawns are usually a darker brownish-red, rather than the orange-red of White-tailed fawns, but are otherwise similar.

SIZE

Being so widespread and occupying so many different habitats, White-tailed Deer are extremely variable in size. Animals from the Florida Keys are among the smallest, and animals from the northern reaches of the distribution are among the largest. Males are 20%–55% heavier than females, depending on the location. The following measurements apply to adult Canadian animals.

Males: total length: 158–240 cm; tail length: 15.0–40.6 cm; hind foot length: 44.0–56.5 cm; ear length: 14–24 cm; height at shoulder: 90.0–106.7 cm; weight: 55–135 kg.

Females: total length: 151–200 cm; tail length: 17.0–35.6 cm; hind foot length: 41.0–52.1 cm; ear length: 13–17 cm; height at shoulder: 70–84 cm; weight: 50–80 kg.

Newborns: weight: 1.8–4.5 kg; singletons are generally larger than twins.

RANGE

Originating in North America over 3.5 million years ago, the genus *Odocoileus* probably split into the two modern species (White-tailed Deer and Mule Deer) around 3 million years ago. White-tailed Deer are widely distributed in North America and are found naturally in all Canadian provinces and territories except Prince Edward Island, Nunavut, and Newfoundland and Labrador. They occupy much of the United States and Mexico and even extend as far south as northern South America. There are some recent suggestions that the most southerly forms in Central and South America should be considered a different species. White-tails are slowly spreading northwards and westwards, possibly owing to human modification of the habitat by farming, but also likely owing to gradual range expansion following deglaciation. They have been introduced to Anticosti Island, Grand Manan Island, and Prince Edward Island as game animals.

ABUNDANCE

When Europeans arrived in North America there were an estimated 23–33 million White-tailed Deer. By 1800 there were about half of that number, and by 1900, following years of market hunting, habitat alteration, and livestock competition, approximately 350,000 remained. Restoration programs and hunting restrictions have returned the species to its former profusion. The White-tailed Deer is now considered a generally abundant species, prone to overabundance in favourable conditions. Numbers are closely monitored by wildlife managers, and, as the most popular large prey for sport hunters in North America, this species is carefully regulated in all Canadian and American jurisdictions where it occurs. Generally one deer per square kilometre is considered to be low density, while ten or more deer per square kilometre is considered high. Densities of fifty per square kilometre have been reported.

ECOLOGY

White-tailed Deer prefer areas of open forest interspersed with meadows, clearings, grasslands, and riparian flatlands. The forests provide shelter and protection, while the open areas provide the best forage. Many human activities favour White-tailed Deer, who are quick to take advantage of the new growth that follows clear-cutting forestry practices, forest fires, and the clearing for agriculture, roadways, railways, and power lines. Extensive logging of the climax coniferous forests of central Canada in the late 1900s opened much of that region to their use, and the creation of treed regions around homesteads in the open prairies have further encouraged White-tailed Deer. These deer are usually crepuscular but will venture out in daylight during overcast days. Many populations are migratory, moving between summer and winter ranges. These migrations typically cover 15–25 km but can be into the 100–200 km range. Migration is most common in northern and montane populations. Individuals tend to return to the same summer range year after year but are not considered to be territorial. Location of the winter range may be variable and largely depends on the depth of accumulated snow. Deer yards, areas where deer traditionally congregate during winters with heavy snowfall, are typically located within a coniferous forest where the trees block some of the snowfall. Within this area, which can include many hundreds of hectares, even hundreds of square kilometres, high deer density keeps the trails open, so movement is less restricted. Winter mortality can be very high in years of heavy snowfall or a late spring because malnourished deer starve or become easy prey. Winter survival is also reduced in regions of high deer density when the food becomes scarce, as the browse is overcropped, resulting in a clearly defined browse line. The size of the home range varies with the availability of forage, from around 60 to 500 ha, and tends to be larger in the northern regions. In those parts of the range that White-tailed Deer share with Mule Deer, White-tails generally prefer the gentler terrain; consequently, the two similar species occupy slightly different habitats and thereby reduce their

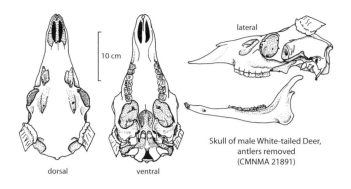

10 cm

dorsal   ventral

lateral

Skull of male White-tailed Deer,
antlers removed
(CMNMA 21891)

competition. Females usually do not disperse, remaining on their mother's home range and joining her band. About 50% of yearling males disperse, usually less than 12 km, but some may travel more than 200 km to a new home range. Most dispersal occurs during the autumn breeding season when the males are around 18 months old. White-tailed Deer commonly carry the Meningeal Worm (also called Brainworm) to which they are somewhat immune. Moose and other deer prove to be less resistant to this worm, which causes a number of often blatant neurological symptoms and usually leads to death. For this reason Moose tend not to occur in regions of high White-tailed Deer density. White-tailed Deer are exposed to a wide variety of potential predators owing to their extensive distribution. Depending on the locality, Grey Wolf, Coyote, Cougar, Black Bear, Bobcat, and Domestic Dog are the primary predators. Collisions with motor vehicles are a significant cause of mortality in many areas, as is hunting. White-tails are wary and typically either freeze or flee if they detect a threat. Does will defend their fawns, and a small proportion of deer of both sexes will act aggressively and have been known to stand off and even kill a predator up to wolf size. The average lifespan of a female is longer than that of a male, even in non-hunted populations. Part of the reason is that mature males normally enter the most difficult season of the year, winter, in a depleted state owing to the rigours of the rut. A severe or prolonged winter can cause tremendous mortality among this class of deer. Wild females remain reproductively active into old age. The oldest recorded wild doe was 19 years old. The oldest known wild male was 17 years old. Having survived their first winter, the average doe lives six years or less and the average buck lives three to four years or less. Top speeds of up to 80 km/h have been reported, but speeds of 50–60 km/h are usually cited as the typical top running speed. White-tailed Deer are strong swimmers and can jump over a 2 m fence from a standstill. Lyme disease is one of the major tick-borne diseases in North America. It is caused by a spirochaete bacterium (*Borrelia burgdorferi*) that is principally transmitted by the bite of a Black-legged Tick (*Ixodes scapularis*), formerly called a Deer Tick. This tick depends on White-tailed Deer for its reproduction, and White-footed Mice act as a reservoir for the spirochaete. The disease is rampant in northeastern and midwestern United States, and it is poised to become a serious problem in central and eastern Canada as our climate warms. The disease can infect domestic animals and humans and may cause a

multiplicity of long-term, deleterious health effects that are often difficult to diagnose.

DIET

These herbivores are very eclectic in their dietary selections, as befits such a widespread species. Their forage choices vary with location and season, but in general they require a high-quality diet and tend to improve their feed by choosing the most nutritious options available. In addition, White-tailed Deer have oversized salivary glands that produce copious amounts of saliva, enabling them to neutralize many of the secondary plant toxins in their food choices. Forbs, mast (such as fruit, acorns, and beech nuts), browse, and mushrooms make up the greatest portion of their food. In some regions around the Great Lakes, White-tails have learned to consume the carcasses of Alewives that wash ashore after spawning. The protein and fats of these fish are readily digestible, and the surplus occurs at a time (spring) when the deer are particularly in need of the nutrients. There have also been reports of deer eating insects and the nestlings of ground-nesting songbirds. They rarely consume sedges and ferns and select only fresh grasses; even then, they choose the most nutritious new growth or seeds. Their rumen capacity is too small to allow them to digest exclusively dried grasses (hay), and a White-tail will starve on such a diet. A starving deer is further challenged because the gradual reduction in food kills many of the rumen bacteria, and eating highly nutritious food after such deprivation can cause a deer to die with a full stomach. Typically, during winter deer's metabolism slows down, and they become less active in order to save energy. White-tailed Deer require regular access to freestanding water. Usage of mineral licks is highest in spring when their salt levels are most depleted. Consumption of agricultural crops can make White-tailed Deer a significant pest species.

REPRODUCTION

In regions where food is of high quality and plentiful, female fawns of only six months old may breed. In the northern portions of their range (including much of Canada) where feed is less plentiful and the season of plant growth is shorter, most females breed for the first time at the age of around 18 months old. A poorly nourished doe may require yet another year to reach sexual maturity. Once mature, does are capable of breeding every year, even into old age, provided they are properly nourished. Most male fawns also reach sexual maturity at around 18 months old but are typically prevented from breeding by older, larger bucks. When mature bucks are rare, as can occur from over-hunting of the older trophy males, the young bucks may conduct the majority of the breeding. In some cases of exceptional nutrition, as with the females, a male fawn of five to six months old may be fertile. A brief synchronous breeding period is crucial to infant survival in northern climates because births must coincide with the flush of new growth in the following spring. Rut of Canadian White-tails occurs from late October to mid-December, with a peak in mid-November. Its onset in any region is usually signalled by the swelling of the neck region of mature bucks. Following a pregnancy of around 200 days, the females give birth to the fawns after the worst of the early spring storms. Without the climatic constraints, White-tails farther south have a more prolonged rut, and those living near the equator have a continuous breeding season. First-time breeders usually produce a single fawn, while larger, mature does are more likely to produce twins. Triplets are rare in Canada but are most common among very well-nourished mature does. Quadruplets are very rare and have not been reported in a wild Canadian doe. It is possible for a litter of multiple fawns to be fathered by more than one male. Malnourished does suffer an increased possibility of stillbirths and are more likely to produce a single, often underweight fawn. They are also much more likely to abandon a newborn. Many poorly nourished does may not ovulate at all, resting for a year to build up some reserves before the energetically demanding production and lactation of offspring. Does generally will not tolerate a fawn that is not their own, but they have been known to occasionally adopt orphans. By two to three weeks old the fawns begin mouthing vegetation, and by four months old they are fully weaned. Both male and female fawns remain with their mothers during their first winter.

BEHAVIOUR

White-tailed Deer are somewhat gregarious, with two main types of groups. One is a family group made up of a matriarch and her fawns from previous generations. Her daughters typically stay with the group for two winters, leaving when their mothers give birth the following spring. Male offspring leave sooner, usually by their second autumn, to join a group of other males. It is thought that their mothers may force the young males to leave since they are just beginning to be sexually active. Sizes of both groups are smallest in summer and largest in winter, but fraternal groups break up completely during rut. Family groups remain together most of the year, with the exception of the early summer, when the mature does depart to a small nearby territory to bear their fawns. Does bond quickly with their newborn as they lick the infant dry and learn its scent. As fawns take longer to bond and learn to recognize their mother, the does are careful to make sure that they are the only large figure that the fawns see, until they are fully imprinted. They aggressively drive away any deer that intrude upon their fawning territory. Newborns stand and nurse for the first time shortly after their birth but do not accompany their mother for the first few weeks. She leaves them, and they hide, each separately. She must memorize their hiding spot and then only approach them briefly every three to four hours to allow them to suckle. These bedding sites are altered frequently by the doe. While they nurse, she licks their rear ends to encourage urination and defecation. By swallowing this material, she maintains the odour-free aspect of the area. Fawns are very susceptible to predators, and all too often even these precautions are insufficient. The fawns, for their part, lie still and make no sound during those two to three weeks, even though their mother spends most of her time out of sight. Females tend to remain closer to younger fawns to defend them, should they be detected by a predator. As they become older and more able to run away from a threat, she

may rejoin her doe group to feed while the fawns are hiding. By three to four weeks old fawns are strong enough to follow her at least part of the time, and thereafter she tends to hide twins together. By the time they are around eight weeks old, they and their mothers rejoin the family group. Fawns remain with their mothers through at least their first winter, learning the migration routes and seasonal ranges. Daughters tend to establish their own home range near that of their mother's, while sons wander more widely. Within male groups, White-tailed Deer maintain a dominance hierarchy that is determined by size. Among family groups, the hierarchy is typically determined by age. Rank is usually conveyed and accepted using the subtle body language of posture, staring, and movements, so that overt fighting can be avoided. Subordinates avoid eye contact and move aside when approached by a dominant animal. When rank is challenged, deer without hard antlers fight by pinning their ears back and striking with their sharp front hooves, either from a standing position or by rising on their hind legs. This form of aggression is typically practised against another deer, usually of the same gender, although a protective doe may attack a Coyote, a Domestic Dog, or even a human, in defence of her fawn. A buck with hard antlers will use them as weapons, rather than hooves. Apart from sparring, which occurs before rut begins or early in the rut and tends to be a relatively benign form of dominance activity, a true battle typically occurs between equal-sized rivals during rut, and frequently results in injury. The bucks sidle towards each other in a slight crouch with ears back and hackles raised. One will typically lunge and be met on the antlers of his rival. A brief pushing contest ensues as each buck twists his head, trying to unbalance his opponent. A loser is normally chased from the field by a winner who is intent on retribution, and he must depart with utmost speed to avoid being gored. Occasionally, both bucks die if their antlers lock. A rutting male spends much of his time marking. He noisily thrashes the bushes, peeling bark and depositing secretions from the glands on his forehead. He also digs and defends scrapes, which he marks with urine and secretions from hock glands. This odour is deposited as the deer hunches over with his hind legs under his belly and urinates while rubbing his hock (tarsal) glands together. The urine enhances the odour from the glands, as well as carrying the secretion down the legs to drip into the scrape. Bucks frequently investigate other bucks' scrapes and attempt to deposit their own urine and glandular secretions. Rutting males commonly display dark stains on the inside of their hind legs as a result of this activity. Fights are common around an active scrape, as the dominant bucks in the area gather to check it out and then challenge each other. Female deer visit the scrapes regularly, sniffing and sometimes leaving their own scent marks, usually limiting that to use of forehead glands and saliva. When a mature male finds an oestrous female, he "tends" her, remaining nearby, defending her from rivals, and testing her to determine if she is ready to stand for him. Does are receptive to mating for about 24 hours. Once she cycles out of oestrus and is no longer receptive, he departs in search of another receptive doe. Most rutting males are too focused to eat, although they drink copiously.

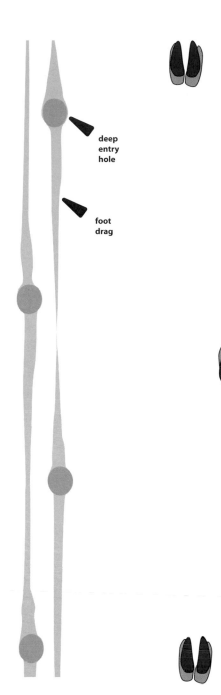

deep
entry
hole

foot
drag

**Walk in
deep snow**

Trot

Left hind track
Length: 4.8–8.9 cm
Width: 4.0–7.0 cm

Left front track
Length: 5.0–10.2 cm
Width: 4.0–7.6 cm

White-tailed Deer

VOCALIZATIONS

White-tails produce a variety of vocalizations, the most distinctive of which is the whistling alarm snort. This rapid exhalation is more commonly produced by does than by bucks and often follows foot stomping as a second, higher-intensity alarm reaction. Except for very young fawns that are less than a week old and normally silent, a deer in extreme distress will bawl. Any nearby doe will respond to a fawn's distress bawl. A young fawn walking around bawling is so unusual that it is generally assumed to have been abandoned and to be near death from starvation. Does approaching the spot where they have hidden a fawn will produce a low-volume grunt to alert the fawn and call it from hiding. The fawn in turn may either emit a care-soliciting mew or a louder, more demanding bleat. The volume of the call is proportional to the degree of need, and a very hungry fawn can bleat loudly enough to be heard by a human at 100 m distance. This loud solicitation call can be very compelling and may attract even non-related females. As the fawn suckles, it produces a characteristic nursing whine that may serve to identify it to its mother. Adults of both sexes may grunt at each other during dominance encounters. Rutting bucks use a similar grunt and add a snort and then a wheeze to the call to signal rising intensity and willingness to clash. A winner in such a confrontation will chase the loser away while uttering a coughing bark.

SIGNS

Deer do not have upper incisors; therefore, they cannot nip off the twig that they want to eat. They have to tear it off by pressing it between their hard palate and lower incisors. The rough and broken end is ripped perpendicular to the stem, rather than cleanly nipped at a 45° angle as done by rabbits, hares, and rodents. A browse line is a clear indication of the presence of deer. Higher densities result in higher and more clearly demarked browse lines. Preferred browse levels are between the height of a human knee and that of the hip. Higher browse lines up to and possibly higher than 1.8 m require the deer to stretch, stand on compacted snow, or rise up on their hind legs. Deer yards are regions of coniferous forest where deer congregate to escape deep snow. Some yards can be huge, such as the Loring Yard between Trout Creek and Parry Sound, Ontario. About 12,000–15,000 deer spend around eight weeks each winter in that yard, which occupies around 775 km². Most deer yards are quite a bit smaller but can easily encompass several tens of hectares. Realization that one is in a deer yard is usually gradual as the number of runs and amount of sign become increasingly obvious. Deer lie up in beds throughout the year, but in winter the sign left in the snow is clearer and more obvious. The animals lie to sleep or chew their cud and may reuse a bed or create a new one. The outline of the rump, and perhaps the print left by the front knees and the hind hock, may be seen. Early spring beds sometimes contain hair as the moult is beginning. Beds are commonly found under conifers where the animals can conserve body heat or in a sheltered spot with good visibility where they can keep watch while resting. Deer runs are trails frequently used by a single deer or a group of deer. They can be fairly obvious in long grass or snow and less obvious in leaf litter. Animals other than deer may choose to use a deer run. It is surprising how low the deer can crouch to get under obstacles along their runs. They prefer to duck under overhangs as low as 0.5 m off the ground and will normally choose not to go over an obstruction unless there is no alternative or they are running. Although females will occasionally produce a scrape, this sign is usually left by adult males more than two-and-a-half years old. Scrapes are made all year round but are most actively made and visited just before and during the early part of the rut. Typically, the buck will begin thrashing and mouthing an overhead branch along one of his runs, leaving some of his scent and saliva and commonly breaking the branch or peeling its bark. Then he will paw at the ground below with his front hooves to expose the earth. Some scrapes are very shallow, but heavily used scrapes can become quite sunken. He often finishes by hock-rubbing (see the "Behaviour" section). The odour of White-tailed Deer urine is described to be piney and not unpleasant. Antler rubs are another sign left by mature bucks. They rub their hard antlers on a sapling (usually from about human knee to hip height) to strip the bark and then use the peeled area of the trunk to deposit scent from the forehead glands and saliva. These rubs may be completely smooth, but most have ragged and frayed bark at the top and bottom. Many of the tree species chosen for rubs are aromatic, and very few dead trees are chosen. Incisor scrapes are signs left by deer of all ages. They scrape their lower incisors (since they do not have upper incisors) up a young tree to peel the bark away, leaving clear tooth marks. Typically, the top of each tooth scrape is a bit ragged where the deer had to tug the bark away from the tree. Often the large flattened incisiform canine (the outside tooth of the incisor group) is used for this purpose. The animal tilts its head to engage that tooth to the tree and then scrapes upwards at an angle. White-tail tracks and scat are indistinguishable from those of Mule Deer. (See the Mule Deer "Signs" section for descriptions and illustrations.)

REFERENCES

Alexy et al. 2001; Banfield 1974; DelGiudice et al. 2007; Elbroch 2003; Hoefs 2001; Hölzenbein and Marchinton 1992; Kennedy et al. 1995; Lingle 2002; Marchison et al. 1990; Miller, K.V., et al. 2003; Molina and Molinari 1999; Nelson and Mech 1990, 1992, 1993; Nixon et al. 1990, 1991; Oehler et al. 1995; Ozoga 1996; Pietz and Granfors 2000; Rezendes 1992; Schultz and Johnson 1992; Schwede et al. 1994a, 1994b; Shackleton 1999; Smith, W.P., 1991; Sorin 2004; Stafford 1993; Veitch 2001; Whittaker and Lindzey 1999.

Caribou
also called Reindeer

FRENCH NAME: **Caribou,** *also* **Renne**

SCIENTIFIC NAME: *Rangifer tarandus*

Caribou are unique among the world's deer species because both sexes, and even fawns by their first autumn, grow antlers. Adult males develop the largest antlers, but these are cast early, usually by mid-November, just after the rut. Females and juveniles retain their smaller antlers through the winter and therefore can use them to prevent the larger-bodied adult males from bullying them away from their winter food. The migration of hundreds of thousands of Caribou is one of the natural wonders of our north country.

DESCRIPTION

All Caribou, including Reindeer, belong to a single species, *Rangifer tarandus*. Reindeer is the name used to refer to both wild and domesticated European and Asian Caribou populations. Our North American Caribou are long-legged, large-hoofed deer. In order to fully discuss Caribou appearance, subspecies must be identified. As can be seen in the colour illustrations, there are three main forms of Caribou in Canada: Woodland Caribou (*Rangifer tarandus caribou*), two quite similar subspecies of Barren-ground Caribou (*Rangifer tarandus groenlandicus* and *Rangifer tarandus granti*), and Peary Caribou (*Rangifer tarandus pearyi*). Woodland Caribou are the largest and darkest of the subspecies. Their bodies are dark brown with a lighter neck, belly, and rump patch. A lighter-brown side stripe develops towards the end of the summer. Barren-ground Caribou are intermediate in size. Their head, neck, and side stripe are usually brighter than those of Woodland Caribou. Both of these forms produce white or piebald animals on rare occasions. Peary Caribou are the smallest and lightest-coloured of the North American Caribou. They are almost white in winter, sometimes with a hint of a side stripe, and have a greyish-brown saddle marking and white head, belly, and legs in summer. All Caribou have a white rim of hairs at the upper margin of their hooves, although this may not be apparent in the lighter Peary Caribou. Calves of all the subspecies are reddish-brown and tan in colour with an apparent rump patch and no spots. Cows usually drop their antlers in June at around calving time; new antlers begin growing almost immediately and remain in velvet until October. Bulls drop their antlers usually by mid-November; new antlers begin growing during the following March and remain in velvet until September. Bulls achieve their largest antlers at between four and nine years of age, after which time the antlers gradually become smaller until the animal's death. Caribou have small upper canines that rarely erupt through the gums. They and Wapiti are the only North American deer with canine teeth. The dental formula is incisors 0/3, canines 1/1, premolars 3/3, and molars 3/3, for a total of 34 teeth. Caribou undergo two moults each year. The spring moult can leave them looking shaggy as the thick winter underfur drops off. Typically, the older winter coat looks paler because of bleaching and breakage of the darker tips of the guard hairs. The autumn moult is less noticeable, and the full long winter coat is usually complete by October.

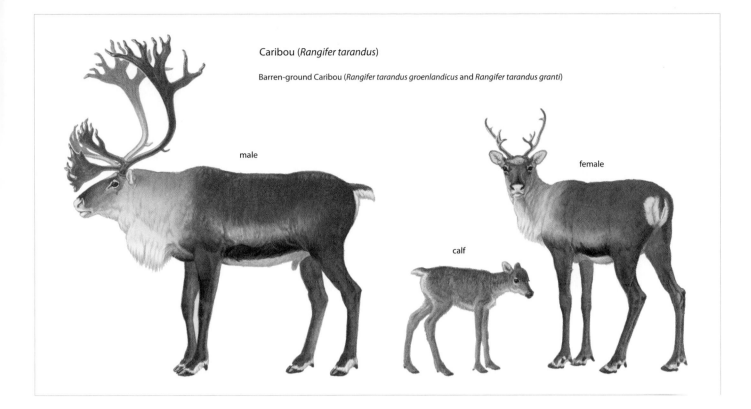

Caribou (*Rangifer tarandus*)

Barren-ground Caribou (*Rangifer tarandus groenlandicus* and *Rangifer tarandus granti*)

male

female

calf

SIMILAR SPECIES

These medium-sized deer are larger than Mule and White-tailed Deer and smaller than Moose and Wapiti. In the regions where other deer species occur with Caribou, body coloration will separate the species, as only Caribou have a lighter neck region combined with a darker body. Above the treeline, only Moose are found in regions that also support Caribou. Moose are considerably larger, with a dark neck and body and light legs. The most northerly Caribou subspecies, the Peary Caribou, is a ghostly white in winter and grey-brown and white in summer and is the only deer found that far north.

SIZE

In all of the subspecies, the males are larger and 10%–50% heavier than the females. The following size ranges include all four subspecies, although the smallest subspecies, Peary Caribou, is represented by only a few individuals. Body weight varies drastically, depending on season; for example, the late-winter weight of adult males can be as much as 35%–40% less than their pre-rut weight.

Males: total length: 167–298 cm; tail length: 11–22 cm; hind foot length: 46–66 cm; ear length: 10–16 cm; weight: 66–300 kg; height at shoulder: 87–158 cm.

Females: total length: 105.4–234.0 cm; tail length: 11–23 cm; hind foot length: 46–66 cm; ear length: 10–16 cm; weight: 51–156 kg; height at shoulder: around 80–139 cm.

Newborns: weight: around 6 kg.

RANGE

Caribou live in the northern regions of the Northern Hemisphere, both in North America and Eurasia. Until the early 1900s there

were Caribou in almost all parts of Canada except the Prairies. Since then, the southern boundary of their range has been pushed northwards, mainly by human habitation, lumbering practices, and the introduction of livestock and their diseases. Woodland Caribou occur in eastern and central North America; the two forms of Barren-ground Caribou occur on the Arctic mainland and lower Arctic islands; and Peary Caribou occur on the Queen Elizabeth Islands in the High Arctic and on Banks Island, western Victoria Island, Prince of Wales Island, Somerset Island, and the adjoining Boothia Peninsula in the lower Arctic islands. A subspecies unique to the Queen Charlotte Islands of British Columbia (*Rangifer tarandus dawsoni*) became extinct in 1908, and the Greenland subspecies (*Rangifer tarandus eogroenlandicus*) was extinct by 1900. Caribou currently living along the Greenland coast are the Barren-ground subspecies (*Rangifer tarandus groenlandicus*) that migrated over the ice from the lower Arctic islands sometime after the native subspecies became extinct. The subspecies distinctions can be blurred in regions of overlap, as intermediate forms are common. Now many of the remaining southerly populations (all of them Woodland Caribou) are isolated in small pockets of suitable habitat: the Slate Islands on the north shore of Lake Superior; in central Gaspé; north of Quebec City (the Charlevoix herd); south of Val-d'Or (the Val-d'Or herd); the Avalon Peninsula of Newfoundland; and the so-called Mountain Caribou in southern British Columbia and northern Idaho.

ABUNDANCE

In the early 1900s there were an estimated 1.8–2.5 million Caribou in North America. Recent estimates suggest that in the year 2000 this

Distribution of the Caribou (*Rangifer tarandus*)

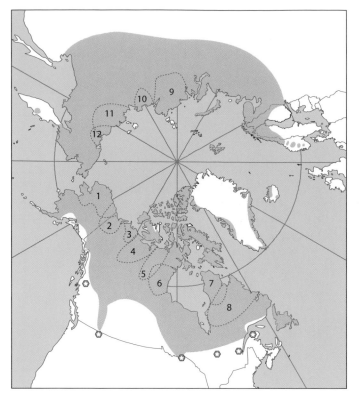

<svg><polygon/></svg> remnant populations

Approximate ranges of major migratory herds

1. Western Arctic herd – *R. t. granti*
2. Porcupine herd – *R. t. groenlandicus*
3. Bluenose herd – *R. t. groenlandicus*
4. Bathurst herd – *R. t. groenlandicus*
5. Beverly herd – *R. t. groenlandicus*
6. Qamanirjuaq herd – *R. t. groenlandicus*
   also called Kaminuriak herd

7. Leaf River herd – *R. t. caribou*
8. George River herd – *R. t. caribou*
9. Taimyr Peninsula herd – *R. t. tarandus*
10. Lena-Olenek herd – *R. t. tarandus*
11. Yana-Indigirka herd – *R. t. tarandus*
12. Sundrum herd – *R. t. tarandus*

number had increased to around 3.5–4.0 million. Caribou populations are naturally subject to cycles of increase and decline, and the Barren-ground herds appear to be at the bottom of the 30–50 year cycle. Globally, most herds are stable or in a slow decline. The exceptions are mainly those southerly herds that are most affected by humans, and the herds in the High Arctic, which are highly susceptible to severe winters, over-hunting, or predation. In the High Arctic, Peary Caribou live in conditions near the limit of the species' ability to survive, and their populations are prone to wide fluctuations in numbers and sometimes to local extirpations. Populations, especially in the western High Arctic, crashed in the late 1990s, and their future is uncertain. There were an estimated 50,000 Peary Caribou living on Canada's Arctic islands in the early 1960s. Estimates in 2001 placed numbers at less than 10,000, and some islands have lost almost 95% of their populations. There have not been so few Peary Caribou in recorded history. This subspecies was declared to be "Endangered" in 2004 by COSEWIC. It is possible that climate change is already a significant factor in their decline and will likely play an even greater role

in years to come. Warming climate is also having an impact on more southerly herds on the barren grounds as the amount of lichen growth is dropping, forcing the Caribou to eat more grasses during the winter. The long-term effect of this dietary change remains unresolved. The population of the Woodland Caribou isolated in Gaspé, Quebec, is classified as "Threatened," as are the Woodland Caribou living in the southern mountains of British Columbia. The Woodland Caribou population on Newfoundland numbered over 90,000 in the mid-1990s but has precipitously declined by around 60% since then and is now estimated at around 37,000 animals. Barren-ground Caribou that live near the Dolphin and Union Strait, Nunavut (herd 4 on the map), are listed as being of "Special Concern" because their numbers have declined substantially in recent years, from an estimated 100,000 in 2006 to around 32,000 in 2010. Nevertheless, Barren-ground Caribou are the most numerous of the subspecies with an estimated 1.4 million animals. The largest single herd is currently the George River herd of Woodland Caribou (herd 8 on the map), thought to be around 700,000 strong but declining. Long of major importance to the existence of northern peoples, Caribou remain an essential source of fat, protein, and hides for many communities. Ongoing diligent monitoring and management of these populations will be essential to maintain them at healthy levels that can sustain this harvesting demand.

ECOLOGY

Unlike all the other North American deer, Caribou have adapted to survive in regions of sparse and scattered forage, such as climax boreal forests and tundra. They have evolved a highly nomadic and sometimes migratory lifestyle, which allows the vegetation to recover between visits. The spectacular migrations of huge herds of Barren-ground and Woodland Caribou may be as lengthy as 1000 km each way. These deer travel from traditional tundra calving grounds to summering grounds, which are also on the tundra, and then to wintering grounds in the northern boreal forest. Fifty kilometres a day at speeds of 5–7 km/h is common for a migrating herd on its way to the calving grounds. All Caribou, even those that do not migrate, are on the move most of the time, pausing intermittently to rest, feed, and breed. A recent satellite tracking study of the George River herd of Woodland Caribou shows that these animals travel around 6000 km during the course of a year, of which about 2000 km are actual migration. These Woodland Caribou undertake the longest of the Caribou migrations and are similar in that way to the Barren-ground Caribou, rather than to other Woodland Caribou, most of which do not migrate or, at best, undertake only short migrations. Peary Caribou are generally nomadic rather than migratory. Caribou are superlatively adapted to cold and snow. All parts of their bodies are covered in fur, even their muzzle, lips, and inside the ears; in winter, long hairs hang down over the top of the hooves and even grow underneath the foot between the hooves. The main guard hairs are hollow and stiff, which, in combination with the dense underfur, traps a warm air layer around the body core. Blood moving into the legs is cooled by blood returning to the body in a counter-current heat exchanger of closely knotted blood vessels. The legs are thereby cooler than the body core, and less heat is lost to the environment. In addition, the nasal passages of Caribou are adapted for warming the cold air as it is inhaled

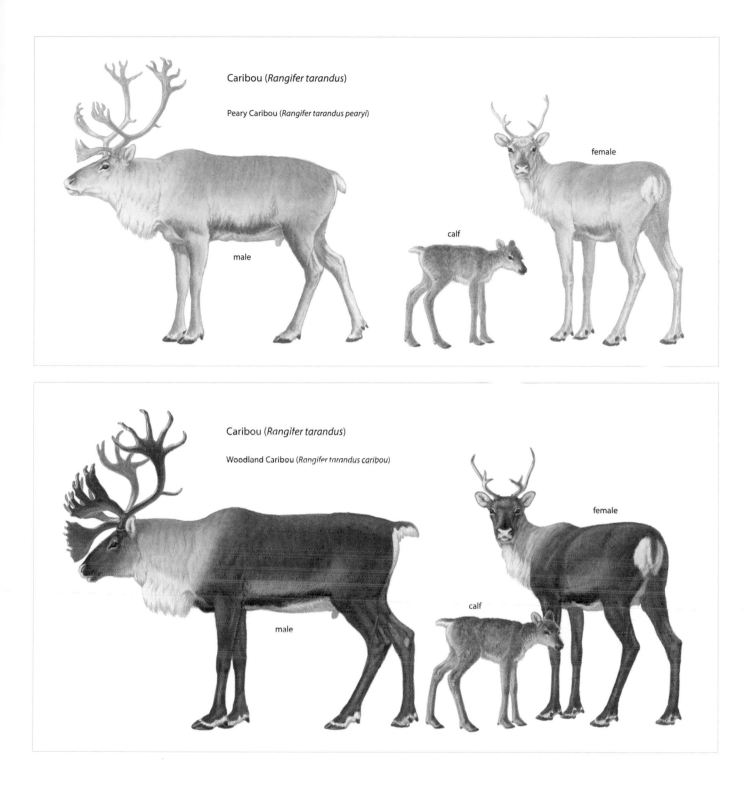

Caribou (*Rangifer tarandus*)

Peary Caribou (*Rangifer tarandus pearyi*)

female

calf

male

Caribou (*Rangifer tarandus*)

Woodland Caribou (*Rangifer tarandus caribou*)

female

calf

male

and for extracting the moisture and heat from it as it is exhaled. The water preserved this way reduces the waste of the body heat that is needed to convert snow to liquid inside the stomach. Despite all of these adaptations, winter puts considerable stress on Caribou, and heavy snow or a thick ice crust can result in significant mortality from starvation and predation. Populations in the High Arctic islands are suffering from recurrent winters of excessive snow and ice formation that makes their food harder to reach. These unusual weather conditions resulted in four major die-offs between 1973 and 1997. Climate change caused by global warming is thought to be a contributing factor. The mainland herds are also likely to be affected by global climate change. Overall herd range and migration routes are expected to alter as food availability and insect activity change with the changing temperature and precipitation regimes. Caribou activity is unaffected by light levels as they are equally active in darkness or daylight. Their activity pattern of feeding or travelling/resting varies seasonally with food availability and insect harassment rather than with light levels. Wildfires destroy the lichen growth, and Caribou make little winter

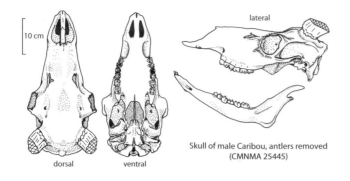

Skull of male Caribou, antlers removed
(CMNMA 25445)

dorsal   ventral

lateral

use of burn sites for many years after a fire. Grey Wolves are the main natural predators of Caribou. Cougars, Coyotes, Wolverines, Black Bears, Brown Bears, and Canada Lynx are also important predators in parts of the range, especially of calves. Golden Eagles will also kill calves. The sex ratio of calves at birth is close to 50:50, but more male calves suffer from predation than do females because they are more active and wander farther from the herd. Female Caribou live longer than do males. A wild female can live up to 16 years, but the average lifespan is 7–12 years. Due mainly to the rigours of the rut, few males live beyond 10 years, as they enter the winter in a depleted condition and are more prone to starvation and then predation. Caribou are confident and buoyant swimmers, capable of swimming speeds of up to nearly 10 km/h. They commonly use open water as escape terrain when pursued by predators, as none of these can swim as fast or as far. Their top land speed is around 60 km/h.

DIET

Caribou feed on a wide variety of grasses, sedges, forbs, and woody browse, but their enormous consumption of lichen in winter is unique. Most of these lichens are either ground dwelling and hence become snow covered, or hang from trees and become desiccated in the cold. During winter, Caribou dig feeding craters with their front hooves to reach fodder. Should the snow become too deep or thickly ice crusted, they will switch to the arboreal lichens and will include conifer needles and even moss into their diet when pressed. When green vegetation is available, they fastidiously select the most tender and succulent portions and will travel long distances in order to increase the quality of their forage. All Caribou are especially partial to mushrooms. In winter, Caribou tend to remain more sedentary to conserve energy, and they consume the lower-quality forage that is at hand. They commonly chew on discarded antlers in an effort to salvage the calcium.

REPRODUCTION

The rut occurs between October and early November, with the peak varying slightly between the different herds. Generally 80%–90% of the females in any given herd become receptive within the same ten-day period. The remaining females are usually thinner and tend to become receptive later, if at all. The birth of calves in the following late May to late June, after a gestation of about 225–235 days, is similarly synchronized and normally coincides with the resurgence of new spring vegetation at the calving grounds. Calves born late

have a lower likelihood of survival as they are more susceptible to predation. Inclement weather on the calving grounds can delay the peak of births by up to six days, indicating that the females may have some limited control of parturition. Most cows conceive for the first time at two-and-a-half years old, although some may conceive at as young as one-and-a-half years old if they are very well nourished. An exceptionally well-nourished female calf may even breed at five months old. While this is very uncommon, it does illustrate the potential of a Caribou population to rapidly recover its numbers on exceptional rangeland. Most yearling males are capable of breeding but are normally prevented from doing so through competition with older males. Most bulls are five to six years old before they begin breeding. Cows generally breed every year but will not conceive if they are too thin. This natural birth control enables them to take another year or more to accumulate sufficient body fat to sustain a healthy and successful pregnancy. Single calves are the norm and twins are rare. Populations, such as those in the High Arctic, which are prone to starvation during severe winters, may not conceive any calves for years at a time if the weather remains inclement. Females lie down to give birth. Newborns are precocial, as they are typically on their feet within an hour and nurse for the first time shortly thereafter. Most females with new calves remain near the birth site for the first few days to a week while the calf becomes strong enough to keep up with the herd. During that time the calves exhibit a partial hiding reflex but mainly follow their mothers. Calves are able to run as quickly as their mothers at around a week old. They are gradually weaned by early autumn, after nursing for about three to four months, but then remain near their mothers to learn the traditional grazing and migration routes until the following spring.

BEHAVIOUR

Caribou are gregarious deer and are rarely seen alone. Generally the herds are sexually segregated, with adult males being separate from the females with their calves and yearlings. Social tolerance is typical with much relocation between groups. Dominance interactions occur only when an essential commodity is at stake, such as a winter feeding crater, early spring greens, and milk (a calf attempting to steal milk from an unrelated cow), and when a mother protects her calf or a male vies for the right to mate. Normally the larger animal with larger antlers is superior; however, this does not apply when a cow is defending her calf. In that case, the fury of the defence can offset the lack of dominance. Herd size varies between the subspecies and at different times of the year. Peary Caribou and most Woodland Caribou form small herds all year round. Barren-ground Caribou and the two Ungava herds of Woodland Caribou may be seen in herds of several hundred thousand during migration; as they reach the wintering range, they spread out into smaller groups. Mature females are usually in the vanguard of any travelling herd, with males generally lagging behind. Moving Caribou are constantly accompanied by the sound of clacking foot tendons. This sound likely promotes herd unity during nocturnal feeding or migrations and allows resting animals to identify conspecifics moving about in the dark. Gregariousness is essential to large prey animals in an open habitat because it allows enhanced group vigilance for predator detection, while permitting more grazing time per individual. Females about to give birth forgo this protection and try to find a secluded spot away from other Caribou. While the new mothers are sedentary, their female yearlings become impatient to keep moving and leave their mothers for the first time to form small bands with other yearlings and barren females. Male yearlings usually leave their mothers earlier as all the females tend to begin the northward migration first, trailed by the males. Newborns imprint on the first moving object that they see. A new mother will imprint her calf early by standing in front of it and bobbing her head. She uses this motion to encourage the calf to stand for the first time and to then move towards her. As their calves become mobile, the females join into nursery bands that gradually roam more widely as the calves become stronger. In some subspecies like the Barren-ground Caribou these nursery bands can gather into huge and widespread "post-calving aggregations." These are thought to be formed as an attempt to escape biting insects, which can be a plague at that time of year, especially to the short-haired youngsters; many Caribou in a small area means fewer bites per individual. Caribou will also seek permanent snowbeds, windy hilltops, and dry rocky hillsides and will even submerge themselves in water during insect season so that they can rest more or less undisturbed. Persistent and unrelenting insect harassment can stampede whole herds. As the insects become less numerous or the weather becomes more windy, dispersing the insects, the herds spread out and concentrate on grazing to build fat for the winter. If a cow-calf pair becomes separated, the two call loudly to each other; if that fails, each member drops to the back of the moving herd and continues calling until they are reunited. By September the males begin to lose their velvet and begin sparring with each other and building up their neck muscles for the battles ahead. During the rut the battles become more intense, and injuries and even mortalities may occur. Locked antlers are also a rare possibility. Woodland bulls may collect a harem of cows, which are guarded until they are bred, but the tundra bulls concentrate on a single female at a time. The largest males fight to tend and court a female approaching oestrus. Other males chase each other and the females through the herd. As each female is bred, the male runs off in search of another. During this hectic time most of the adult males eat little, and the largest, most dominant males are too busy to eat at all. As the rut draws to a close, the larger males begin to lose their antlers first, and the younger fatter males, previously prevented from participating, begin to pester the already mated females. Caribou react to moving objects (such as approaching predators) and are less likely to take note of an immobile object. They have very acute olfactory abilities and will sniff above the snow or into the snow pack before expending energy necessary to dig out a feeding crater. Clearly, they can detect the presence of desiccated plants through several centimetres of snow. A deep feeding crater can become the object of considerable competition. Adult males attempt to force the smaller females, calves, and yearlings to yield one over; however, the small persistent antlers that these animals still carry help them to defend their hard-won food from the already antler-less larger males – a small set of antlers can trump a larger body. Protecting their feeding craters is especially important to pregnant females that

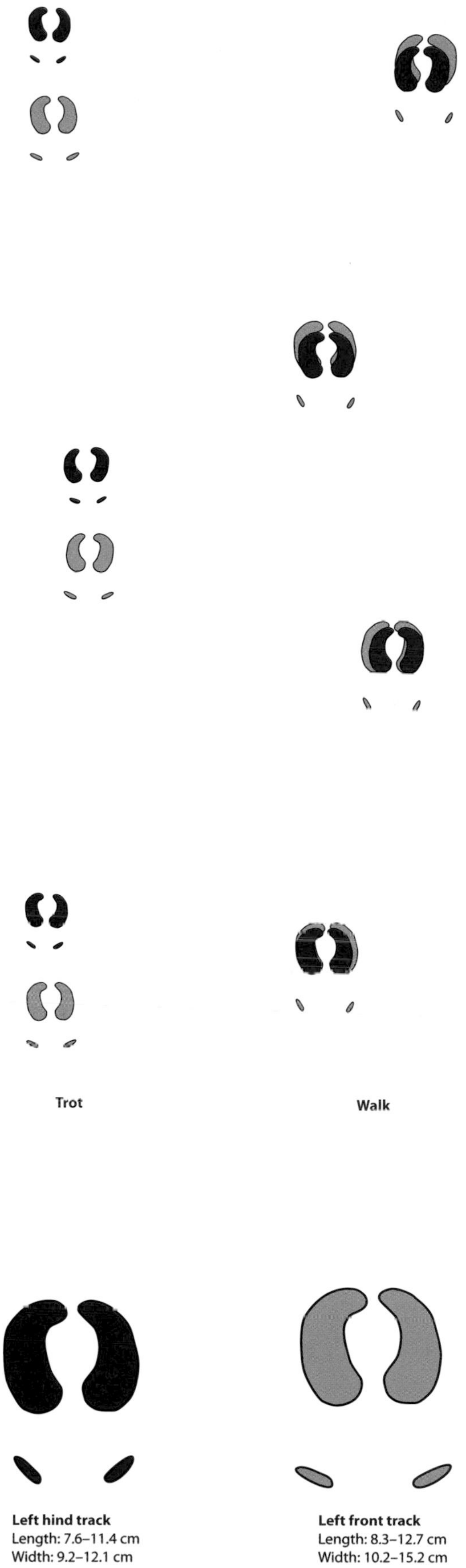

Trot

Walk

Left hind track
Length: 7.6–11.4 cm
Width: 9.2–12.1 cm

Left front track
Length: 8.3–12.7 cm
Width: 10.2–15.2 cm

Caribou

are developing a foetus and at the same time training and helping their latest calf to survive its first winter, and to yearlings that are coping with winter for the first time on their own.

VOCALIZATIONS

Cows grunt to call their calves. Calves bleat or bawl to call their mothers. The volume of these calls varies depending upon the anxiety level. Bulls hoot or bellow during the rut. A wheeze-snort call accompanies an excitation leap, which commonly results when a Caribou of either gender detects an unknown sight or sound. The excitation leap is thought to drive the hind feet into the ground, resulting in the deposition of a warning scent from the interdigital glands to complement the audio warning.

SIGNS

Caribou hooves are large in relation to their body. Their dewclaws are low on the legs and often register in a track, especially the front track. The third and fourth toes form the two hooves, and the second and fifth toes form the dewclaws. The front foot is slightly larger than the back foot. Male tracks are larger than the female's. The hooves are curved, producing an almost circular track, with a wide concave space between the toes. Registration of the dewclaws becomes more perpendicular to the line of travel the faster the animal is moving. The splay of the hooves also increases with increased speed. Trotting Caribou place their rear foot directly in front of the front foot, rather than off to the side slightly as is more common among other mammals. Hoof anatomy varies with season. In winter the inner foot pad of each hoof hardens and shrinks, leaving the sharp, hard outer rim to provide traction. At the same time, thick fur grows between the hooves to protect the foot from the cold substrate. Caribou migration trails may be reused for centuries, leaving deep ruts in the tundra. In autumn near rutting time, bull Caribou will flay saplings with their antlers if there are large enough trees present. In most parts of the Caribou range, trees are absent or shrubby and not substantial enough to provide a satisfying rub. Caribou scat varies with diet. Summer scat is softer, usually with compressed pellet shapes within a clump. Winter pellets are small and hard and are flat or concave on one end and pointed on the other, very similar to the scat of White-tailed Deer. They are around 1.0–1.5 cm in diameter and around 1.0–2.0 cm long.

REFERENCES

Banfield 1961; Bergerud 2000; Boan 2005; Boudreau et al. 2003; Cameron 1994; Chubbs 1993; Courtois et al. 2001; Crête and Desrosiers 1995; Cumming 1998; Edmonds 1998; Elbroch 2003; Farnell et al. 1998; Ferguson, S.H., and Elkie 2004; Ferguson, M.A.D., and Gauthier 1991; Gunn et al. 2000, 2006; Heard and Vagt 1998; Joly et al. 2003, 2007; Klein 1999; Larter and Nagy 2000; Mahoney and Schaefer 2002; Maier and White 1998; Miller, F.L., 2003; Miller, F.L., and Barry 2009; Post et al. 2003; Racey and Armstrong 2000; Rettie 2004; Rettie et al. 1998; Rezendes 1992; Russell 1998; Schaffer and Mahoney 2001; Shackleton 1999; Sharma et al. 2009; Snyder 2004; Stronen et al. 2007; Thomas et al. 1996.

FAMILY ANTILOCAPRIDAE
pronghorn

The Pronghorn is the only surviving member of a much more populous family. Fossil forms had four horns, spiral horns, and forked horns, and some were considerably larger than the Pronghorn. They all lived on the central Great Plains of North America, and although they resemble African antelopes in appearance and lifestyle, they are not closely related.

Pronghorn
also called Pronghorn Antelope

FRENCH NAME: **Antilocapre**
SCIENTIFIC NAME: *Antilocapra americana*

Thought to be the fastest terrestrial mammal in the world, Pronghorns can sprint at speeds of possibly as high as 95 km/h and can easily sustain speeds of 70 km/h for several kilometres. Not only are they fast, but they have tremendous stamina. None of their predators can approach these speeds, and hence only fawns that are less than three weeks old are vulnerable. We think that Pronghorns evolved their speed over two million years ago, when they shared the plains with the American Cheetah. This formidable predator became extinct only about 10,000 years ago – not long enough for Pronghorns to lose their, now unnecessary, fleetness.

DESCRIPTION

Described by one researcher as "bratwurst on chop sticks," Pronghorns have compact bodies with almost all of their muscle in the torso, which is supported on four strong, but decidedly spindly-looking, legs. An untutored eye watching them running through the vast open prairie would likely see a yellow-brown blur with no legs at all. Thin legs are more aerodynamic than those that are more heavily muscled because they encounter considerably less air resistance as they swing back and forth at high speed. At rest, Pronghorns are deceptively small, but they are packed with anatomical enhancements to allow them to run as fast as they do. Their windpipe is the diameter of a large vacuum hose to allow ample oxygen to the extra-large lungs. Their hearts are oversized to pump huge volumes of blood to the hard-working muscles, and their blood has a higher than usual haemoglobin count so that it can carry plenty of oxygen. They even have two pairs of countercurrent heat exchangers in their necks to cool overheated blood before it enters the heat-sensitive brain tissues.

Pronghorns have large black eyes and heavy black lashes that act as sunshades for both direct and reflected sunlight. They also have granules on the upper surface of their rectangular pupil that secondarily block additional sunlight. Their eyes protrude from the sides of the head so that they have almost a 360° field of vision (an excellent adaptation for an animal that lives in open grassland). There is a narrow blind spot directly behind the head. They are renowned for their excellent vision, which has been described as similar to our vision using binoculars with eight-times magnification. The tongue, the mucal membranes inside the mouth, and the hair on the top of the muzzle are black. The coarse pelage is coloured mainly in shades of rust and tan, with the belly, rump patch, neck bars, and patch around the ears in white. The top of the neck has a short stiff mane of somewhat darker hairs. There are sufficient individual differences in coloration and the extent of the white areas that most individuals can be recognized by a trained eye. Males of all ages have an additional black patch on the back of each cheek below their ears, which contains a gland used during the courtship display and for territorial marking. Each side of a rump patch also contains a gland, which emits an odour described as "reminiscent of buttered popcorn." This gland releases its odour when the stiff white hairs in the rump patch are erected as the animal becomes alarmed. Both sexes have horns, although those on females are simple spikes and considerably smaller (less than 7 cm in length), sometimes not even protruding through the skin. Male horns grow to about 30 cm, with a single forward-pointing "prong" and inwardly hooked tips. These are true horns because they grow over a permanent bony core. Unlike most horned ungulates, Pronghorns shed their horn sheaths annually. The old sheath is pushed off by the new sheath growing underneath. Bucks shed their horn sheath every autumn after the rut is over, and grow a full replacement set by the following spring. Females drop their horn sheaths at irregular times, usually during the summer. Pronghorn males achieve full horn growth by the age of two to three years old, which is unusually rapid. The dental formula is incisors 0/3, canines 0/1, premolars 3/3, and molars 3/3, for a total of 32 teeth.

SIMILAR SPECIES

Very deer-like, Pronghorns could be mistaken for either Mule or White-tailed Deer. Both the latter species are considerably larger and heavier. Male deer have large, light-coloured, multi-pronged antlers rather than the shorter, black, single-pronged, and hooked horns that give Pronghorns their English name. Female deer do not have antlers, while most female Pronghorns have small black horns.

SIZE

Males are slightly larger and heavier than females. Males are heaviest in late summer, and lightest in early spring and right after the rut. Females are heaviest in late autumn, and lightest in early spring after the fawns have been born.

Distribution of the Pronghorn (*Antilocapra americana*)

Males: total length: 134.4–149.0 cm; tail length: 8.5–14.4 cm; hind foot length: 39.2–45.1 cm; ear length: 12.0–14.5 cm; weight: 43.0–70.5 kg; height at shoulder: 87.3–94.8 cm (average 87.5 cm).
Females: total length: 128.3–152.9 cm; tail length: 7.7–11.0 cm; hind foot length: 39–44 cm; ear length: 13.0–14.2 cm; weight: 40.0–56.4 kg; height at shoulder: 83.0–91.4 cm (average 86 cm).
Newborns: usual weight: 3–4 kg.

RANGE

Pronghorns are an endemic North American ungulate. Historically, they inhabited the prairies from southwestern Manitoba to the foothills of the Rocky Mountains, and north as far as Edmonton and Saskatoon. In the United States and Mexico they reached northern California and the Baja peninsula in the west, northern Mexico in the south, and eastern South Dakota to central Texas in the east. Their current distribution in Canada is very similar to their historic distribution, except in the easterly extent (Pronghorns have not yet returned to Manitoba). In the United States and Mexico their current range is very spotty as Pronghorns are found mainly in rough grasslands and high-elevation deserts where farming is not possible. Pronghorns will move at any season to find better forage or water, sometimes for hundreds of kilometres.

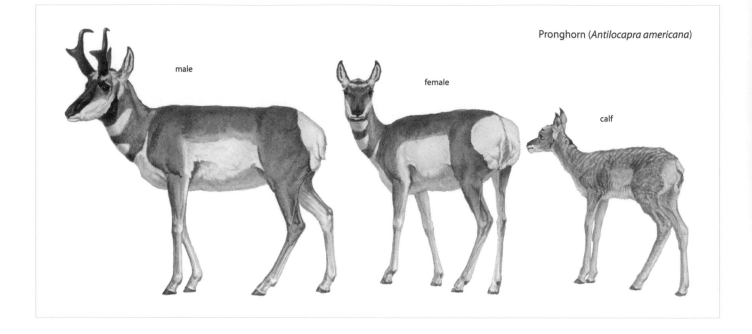

Pronghorn (*Antilocapra americana*)

male

female

calf

This travel is hindered by some types of impenetrable fencing (see the "Ecology" section below) that will permit movement in certain directions only. The range map for Pronghorns is consequently a generalized approximation and subject to even more variation than is normal for other mammals.

ABUNDANCE

Before the arrival of Europeans to North America there were an estimated 35 million Pronghorns roaming the prairie grasslands. By 1924, following years of fencing, ploughing, and hunting, there were an estimated 20,000 animals remaining. By 1980, following years of conservation efforts and controlled hunting in both Canada and the United States, numbers were back up to about 500,000. Currently, there are about one million, of which almost half occur in the state of Wyoming.

ECOLOGY

Pronghorns live in open grasslands where they can see a predator from a distance and can use their keen eyesight and blinding speed to avoid being captured. They do not require a source of water if the forage is succulent, but they do need access to water in hot dry weather and when forage is desiccated. Some populations migrate from low-snow areas where they spend the winter to moister areas where the vegetation is succulent in summer. They will also travel out of areas affected by drought. Movements and migrations are typically seasonal and based on the availability of water or the avoidance of severe weather or poor forage. Pronghorns prefer to crawl under fences, rather than jump them. Page-wire fences are impassable to them, and even barbed wire will stop them for a time. They will occasionally jump a small obstacle but are reluctant to risk jumping. In many regions Pronghorns are in peril from winters with heavy snow or summers with drought. As many of their traditional migration routes are now

blocked, they are unable to escape some regions and can perish in considerable numbers from starvation or thirst. Harem males tend to be most at risk during hard winters because they may have expended much energy during the rut and may not have had time to replenish their fat reserves before the snowfall. Male and female juveniles are equally likely to disperse. Some juveniles (of either sex) may become residents, establishing and occupying a home range very near their natal home range. Most of those that disperse are the young of the year. Dispersal typically occurs in autumn or early winter and varies in distance from a few kilometres to more than 200 km. The size of the home range varies with habitat, gender, season, and group size and has been reported to be from 0.2 km² to 2873 km². A recent study in South Dakota found that females there occupy home ranges that vary in size from 20 km² to 127 km². Most predation mortality occurs within the first 45 days of life when Coyotes, Golden Eagles, and sometimes Bobcats can catch and kill the fawns. Fawn mortality

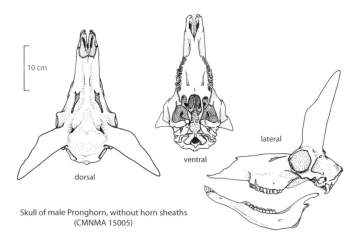

10 cm

dorsal

ventral

lateral

Skull of male Pronghorn, without horn sheaths (CMNMA 15005)

rates vary but can climb to 99% if there are a lot of Coyotes on the range. Assuming that a Pronghorn lives to become adult and the winters are not too deadly, a buck may live to be 11 years old, and a doe could survive to be 15 years old.

DIET

Pronghorns are grazing herbivores and use rumination to aid in the digestion of cellulose. They eat grasses and forbs and are very selective feeders, using their small muzzle and mobile lips to pick through the available forage in order to select either the portion or the type of plant that they prefer. Their choice of forage naturally changes with the season and with availability. Their digestive efficiency is high as they can subsist on very poor range if necessary. Pronghorns evolved in North America, and therefore many native plants that are poisonous to the introduced domestic livestock are relished with impunity.

REPRODUCTION

Rut lasts about three weeks from mid-September until early October. Gestation takes about 245–255 days, or around eight to eight-and-a-half months. Most commonly, twins are born. For about three weeks following their birth, the fawns are hidden by their mother and only rise when she calls them to suckle every three hours or so or if they are disturbed by a predator. Even a five-day-old fawn can sometimes outrun a human, and by about eight days old can outrun a Coyote, if the fawn detects it soon enough. The mother nurses her fawns for about 13 weeks, but by the fourth week she begins a gradual weaning process that forces the fawns to try solid food. Females generally become sexually mature at around 16 months and bear their first fawn at two years of age. They then produce every year thereafter. Rarely, a well-nourished female fawn will mate in her first autumn at five months old. It had always been thought that does would only mate once during each oestrous cycle; however, recent DNA evidence shows that a high percentage of the time (44% in one population), the father of each twin is different. Although males produce sperm by the time they are four months old, owing to the dominance hierarchy in males and the females' ability to choose their mate they rarely have an opportunity to breed until they are three years old and are able to defend a breeding territory and protect a small harem of fertile females.

BEHAVIOUR

Pronghorns typically are found in herds varying from a few animals in summer and early autumn to several hundred in winter or during migration. Harem males usually inhabit their future breeding grounds all summer in an effort to lay advance claim; therefore, they are solitary until late summer when they begin to gather females. Females become solitary just before and for about two weeks after giving birth. Summer bands are smallest and generally segregated by gender. "Bachelor bands" are made up of one- to three-year-old males. Females and their fawns, sometimes accompanied by a mature male, make up the other herd type. Summer herd structure tends to be highly dynamic as individuals leave and arrive regularly. After the rut, smaller groups coalesce into larger, mixed-sex herds that persist over the winter. Pronghorn social life is ruled by an almost perpetual struggle for dominance. Among both male and female bands this usually takes the form of subtle harassment where dominant individuals force subordinate individuals to get up and move, to relinquish a feeding site, or simply to move over or turn away. Interestingly, the social position of each year's crop of fawns is determined by their birth order, as the earliest born are naturally the largest. Although this size distinction vanishes as the fawns reach adulthood, the previous months of dominance skirmishes usually indelibly ingrain each animal's position in the hierarchy. Females and males each have a separate dominance hierarchy. As a result of this bullying, the subordinates are driven to the periphery of the herd, where they would have been most vulnerable to the now extinct American Cheetah. Today, with no predators of that class left, subordinate Pronghorns live as long as dominant ones. Among males this struggle also includes sparring bouts where the animals are essentially learning to fight. Once they are mature and on their harem territories, the skills learned in sparring are put to the test, and the fearsome battles during rut determine which buck will mate. The bucks fight to injure each other, and mortalities are not infrequent. Males will only do battle if they are in the presence of an oestrous female. Females actively compare harem males and travel back and forth between them until they select one. A female will only allow that one to mate when she is ready; she can easily foil his too early attempts by taking a single step forward. After mating with him, she may travel to another male and permit him to mate with her as well. She will only permit each male a single mating. Males are very busy during rut, courting and testing the readiness of females, attempting copulations, driving off rivals, and keeping their harem safely herded out of sight in a knoll or hollow. They typically take little time to feed. When a herd of Pronghorn take fright, they gallop off, bunch together into a rough ellipse, and tend to match their strides.

VOCALIZATIONS

Pronghorns communicate vocally using several different sounds. Adults of both sexes often produce an alarm snort, while simultaneously erecting their rump patch. A higher level of alarm is indicated with a louder, sneeze-like alarm snort. Territorial or harem bucks use a variation of the alarm snort in their snort-wheeze call. It starts with a higher-pitched and longer alarm snort, followed by a pause and then a series of 10–20 wheezing chugs, becoming softer at the end. These snort-wheeze calls are quite variable, and humans can often identify a buck by its call. Other bucks can likely also identify it. During rut, bucks will also emit a growling roar. The snort-wheeze is directed towards distant Pronghorns, male or female, while the roar is used at close range to call to the particular male or the female that the buck is currently chasing. Sometimes a buck will snort-wheeze to a female as he begins his courtship display. He may also produce a groaning whine as he approaches her. Mothers use a quiet, high-pitched grunt when calling to their fawns. Fawns are generally silent, especially when in hiding, but will produce a loud bleat if disturbed or grabbed by a predator or a human. Adults can emit a louder, even more startling version of this bleat under similar circumstances.

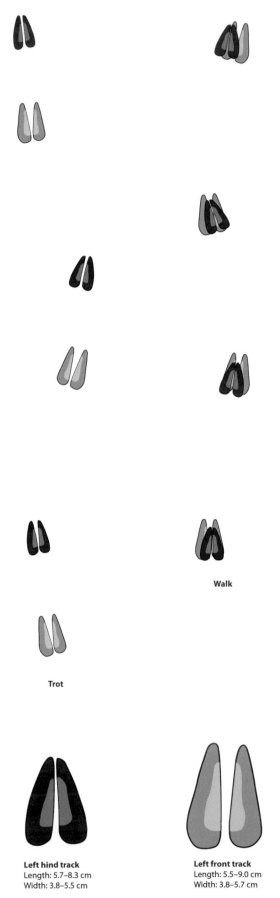

Walk

Trot

Left hind track
Length: 5.7–8.3 cm
Width: 3.8–5.5 cm

Left front track
Length: 5.5–9.0 cm
Width: 3.8–5.7 cm

Pronghorn

SIGNS

While a buck is guarding his harem in a hidden little knoll, he will scrape, then urinate, and defecate over any urine spot left by one of his females, presumably to hide her presence and reproductive condition. He scrapes the area with his front hooves and then, leaning forward, urinates, and finally steps forward and defecates onto the middle of his urine mark. Sometimes the faecal material ends up beside the scrape. During other times of the year, males use the scrape-urinate-defecate sequence to mark their presence. Their beds are Pronghorn-sized areas of flattened grass, used only once. Pronghorns often travel over their own well-worn trails, and these runs are most commonly seen leading to a water source. They can be confused with deer runs and could be used by both. Scat varies with diet. Winter pellets are 0.5–1.3 cm in diameter and 1.0–1.9 cm in length; they are found in clumps and, though usually individually distinct, can be joined like a knotted string. Summer droppings are looser and look like a bunch of soft pellets pressed together; they form a lump, but at least some individual pellet edges can usually be detected. Pronghorns do not have dewclaws. Tracks are an elongated heart shape made by two hooves. The inside toe (the second toe) is slightly smaller than the outside toe (the third toe). Soft pads on the back of each toe are bulbous and leave a rounded, flattened posterior edge to the track. The edges of the hoof at the front of the track are sharp and often are the only part of the hoof to contact the ground if the substrate is firm. Each hoof has a concave portion (shown in the diagram as a paler region) that may not impact the substrate; when it does, it often leaves a raised area that is not as deep as the rest of the track. The front hoof is larger than the hind. The split between the hooves varies in width, depending on the amount of splay. High speed or a soft substrate (such as mud or snow) can cause the toes to splay.

REFERENCES

Byers 1997, 2003a, 2003b; Carling et al. 2003; Elbroch 2003; Ferguson, T.A., 2003; Gainer 1995; Jacques and Jenks 2007; Jacques et al. 2009; Mitchell, C.D., and Maher 2006; O'Gara 1978, 1999; Wood 2002.

FAMILY BOVIDAE

bison, goats, muskoxen, sheep

These ungulates have permanent horns with a bony core and a horn sheath that is not shed and that continues to grow throughout the life of the animal.

American Bison

also called Buffalo

FRENCH NAME: **Bison**

SCIENTIFIC NAME: *Bison bison*

Bison are the heaviest terrestrial mammals native to North America. A large wild bull can weigh as much as 1100 kg or more. Captive bulls have been recorded up to 1724 kg. Males are bigger and heavier than females. This species is separated into two subspecies. The more northerly Wood Bison (*Bison bison athabascae*) tends to be larger and darker than the plains subspecies (*Bison bison bison*). Bison are inextricably linked with the culture and survival of many native peoples in western North America. Some of the earliest human artefacts (up to 11,000 years old) in North America are associated with bison kill sites.

DESCRIPTION

Bison have horns that grow slowly during the life of the animal and are never shed. Pelage is longer on the head and shoulders and shorter on the hind quarters. Most animals are shades of brown; rarely, pied, blue roan, and white individuals are seen. The massive and shaggy head, high-humped shoulders, and bushy forequarters, especially of the bulls, appear out of proportion with the slimmer, sleeker hindquarters. The short tail has a long, dark tassel of hair extending from its tip. Both sexes have a beard growing from their chin, which is usually fuller and longer in the mature males. The Wood Bison's hump is highest forward of the front legs, while the Plains Bison's hump is highest over, or even behind, the front legs. In general, Wood Bison have shorter hair on the back of the front legs (the chaps), along the bottom of the neck, between the horns, and in the beard and cape. The distinction between the colour of the longer shoulder hair (cape) and that of the shorter hair on the hindquarters is greater in the Plains Bison. Wood Bison usually appear to be dark brown on the hindquarters and blackish-brown on the cape, and the cape margin tends to gradually fade into the hindquarters, unlike the abrupt demarcation common to the Plains Bison. Bison have two seasonal moults, one in early spring and one in late summer. Nursing females usually moult later than do other bison. Calves are born with short reddish-orange fur, which moults to a dark-brown coat at around the age of two and a half months. The dental formula is incisors 0/3, canines 0/1, premolars 3/3, and molars 3/3, for a total of 32 teeth.

SIMILAR SPECIES

At a distance, Muskoxen can be mistaken for American Bison because of their overall large, dark appearance and herding behaviour. Silhouettes of the two grazers are quite different, however. Although Muskoxen have a shoulder hump, in general their outline is squarer due to their longer hair. They also are much smaller and are found only in the far north.

SIZE

Males are considerably larger and heavier than females (in both subspecies). The following measurements are derived from adult individuals.

Wood Bison

Males: total length: 245–386 cm; tail length: 35–57 cm; hind foot length: 58–71 cm; ear length: 12.5–18.0 cm; weight: 642–1179 kg (average 880 kg); height at shoulder: 168–201 cm.

Females: total length: 265–333 cm; tail length: 39–48 cm; hind foot length: 59–66 cm; weight: 493–567 kg (average 525 kg); height at shoulder: 155–172 cm

Plains Bison

Males: total length: 304–390 cm; tail length: 33–91 cm; hind foot length: 50–68 cm; weight: 544–1090 kg (average 769 kg); height at shoulder: 152–179 cm.

Females: total length: 213–318 cm; tail length: 30–51 cm; hind foot length: 50–53 cm; weight: 390–605 kg (average 425 kg); height at shoulder: 140–157 cm.

Newborns of both subspecies: weight: 15–25 kg.

RANGE

Bison evolved in Eurasia and travelled across a Bering Isthmus to North America over 130,000 years ago. Their continuing evolution into the most recent form included several now-extinct species. Modern American Bison were formerly widespread across North America from Alaska to northern Mexico and from New England and the southern Great Lakes to the Rocky Mountains. Today they survive in geographically isolated populations in parks and reserves that are scattered within the former range, and also on game farms and ranches within and even beyond their former range. A 2006 introduction of Wood Bison from Elk Island National Park into Lenskie Stolby National Park in the Russian republic of Sakha is too recent for comment on its success.

ABUNDANCE

Before Europeans came to North America with guns, there were an estimated 30–50 million American Bison roaming the grassland and northern parklands. The majority of these were the more southerly

American Bison (*Bison bison*)

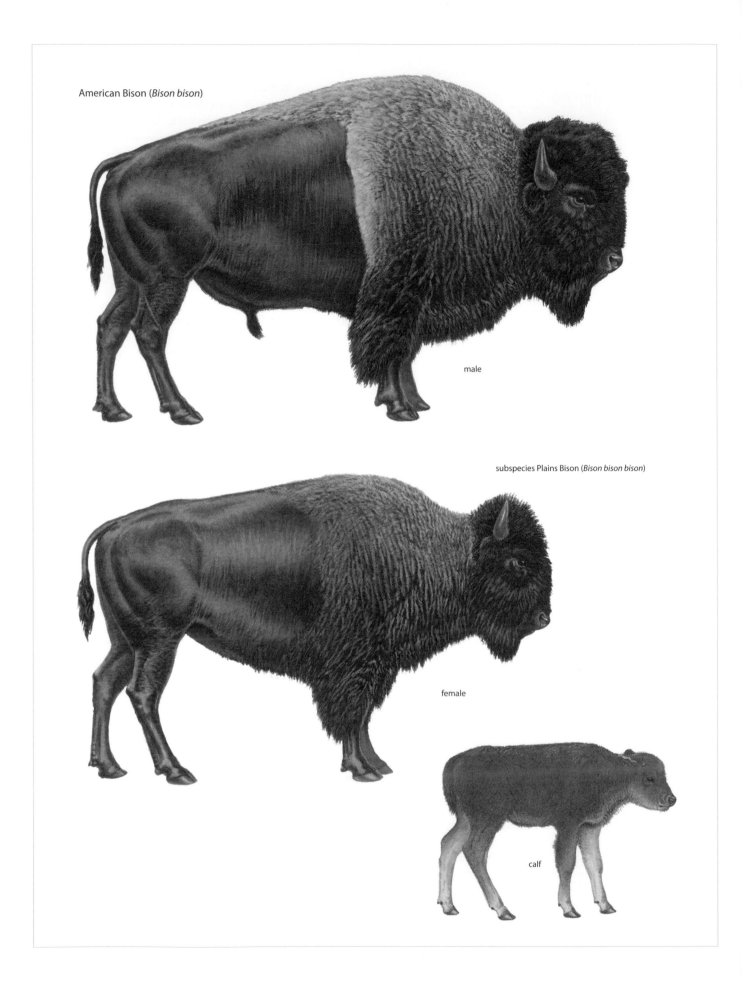

male

subspecies Plains Bison (*Bison bison bison*)

female

calf

Distribution of the American Bison (*Bison bison*)

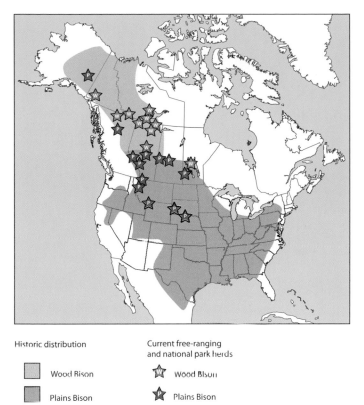

Historic distribution

Wood Bison

Plains Bison

Current free-ranging
and national park herds

Wood Bison

Plains Bison

Plains Bison. Wood Bison were already beginning to disappear from Alaska and western Yukon, perhaps partly owing to overhunting by natives using traditional tools. By the early 1900s, less than 2000 bison were left; numbers as low as 500 have been suggested for the remaining Plains Bison, and the rest were Wood Bison. Thanks to two wild herds (in Wood Buffalo National Park, Canada, and Yellowstone National Park, United States) and to a few farsighted private citizens and native bands who raised captive herds, both subspecies of American Bison have survived complete extermination. In the early 1900s the Canadian government embarked on a bison conservation strategy. They protected the wild bison that lived in what later became Wood Buffalo National Park on the border of Alberta and the Northwest Territories, and they created Buffalo Park near Wainwright, Alberta. On this southern reserve they then placed a small herd of Plains Bison that had been purchased from two Montana ranchers whose stock had originally been obtained from a native hunter. The bison thrived and by the 1920s were so numerous that they were overgrazing their range. To save the range, over 6500 young Plains Bison were shipped by train and river barge to Wood Buffalo National Park during the summers of 1925, 1926, and 1927. Opinions vary on the numbers that were released, but local estimates suggest that only about half survived the voyage. They brought with them the diseases brucellosis and bovine tuberculosis, which had been transmitted by Domestic Cattle. Unfortunately, the park was the home of the last remaining herd of around 2000 Wood Bison, which had never previously been exposed to these diseases. For many years it was assumed that the Wood Bison had disappeared because of disease and interbreeding with the more numerous Plains subspecies. In 1957 an isolated herd of about 200 animals was seen by air in a very remote corner of Wood Buffalo Park. It was possible that this Nyarling River herd might be pure-bred Wood Bison. Studies done later indicate that even this herd has some genes from the southern subspecies, but many of the animals still appear to be purebred Wood Bison. Some of the animals from this herd were brought south to form the nucleus of the current semi-wild herd in Elk Island National Park outside of Edmonton. Bison in Wood Buffalo National Park have slowly diminished in number from a high of 10,000–12,000 in the early 1970s to about 2200 in 1999. There has been a moderate increase since 1999, and the herd was estimated at almost 3500 in 2003.

By 2002 an estimated 600,000–720,000 Plains Bison lived in North America, 97% of which lived in private herds. In 2003, 50 bison from the Elk Island Plains Bison herd were released in Old Man on His Back Heritage and Conservation Area in southern Saskatchewan, and in 2006, 70 animals from the same Elk Island herd were reintroduced into Grasslands National Park in southwestern Saskatchewan. It is hoped that these two herds will become the first free-ranging herds of Plains Bison on the continent in more than a century. Approximately 3150 Wood Bison in six free-ranging, disease free herds existed in Canada in 2002. In 2005 one of these free-roaming herds, from the Slave River Lowlands, was discovered to harbour bovine tuberculosis, and the entire herd of 97 animals was slaughtered. An additional 1000 or more lived in five captive-breeding herds, and another approximately 4500, probably diseased and possibly hybridized, Wood Bison ranged free in Wood Buffalo National Park. Owing to some considerable recovery, Wood Bison were down listed from "Endangered" to "Threatened" by the IUCN

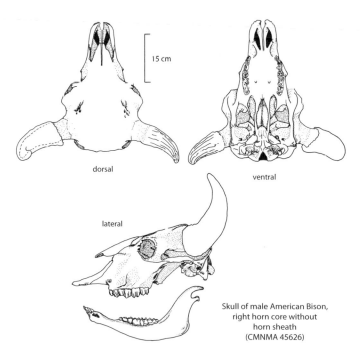

15 cm

dorsal

ventral

lateral

Skull of male American Bison,
right horn core without
horn sheath
(CMNMA 45626)

American Bison (*Bison bison*)

male

subspecies Wood Bison (*Bison bison athabascae*)

female

calf

in 1988. The Plains Bison is listed as "Conservation Dependent" in the 2000 version of the same list.

ECOLOGY

Bison are gregarious herbivores. They are mainly diurnal but will feed during moonlit nights, especially when there is snow cover, or during summer when grasses are lush. Length of active periods (mainly grazing) and resting periods (mainly ruminating) vary with the season. Generally, four to nine active periods occur each day. During the summer, around nine hours of each day are spent grazing. As forage quality drops in autumn, this period increases to eleven hours per day, despite the shorter day length. Bison generally use their heads to move snow away from their forage, rather than pawing with their front hooves as do all of our other ungulates. They also head into the wind during bad weather, unlike Domestic Cattle, which turn their backs to the wind. In times past, bison used traditional trails during their seasonal migrations. The distance of these travels has been recently estimated at about 300 to 450 km. They habitually travelled in single file, and their heavily worn trackways can still be discerned in areas of unplowed prairie. Most of our modern herds are confined and can undertake only limited migrations, but the herd in Wood Buffalo National Park still manages a jaunt of about 240 km down to the Peace River area each fall. Bison longevity is not well documented. The onset of old age is placed at 15 years. In the wild a few bison may live beyond 20 years. The oldest known was a cow that died when struck by a vehicle at the age of 27 years. Grey Wolves were the American Bison's chief predator, and still are for the free-ranging herd in Wood Buffalo National Park, although even a pack has difficulty taking down a lone adult bison. Most predation is of calves or unhealthy individuals. Wolves will sometimes stampede and chase a herd for several kilometres, and on one occasion in Wood Buffalo National Park they made a kill at over 16 km from the initial attack. Although bison are strong swimmers, spring and autumn floods in the Peace-Athabasca River Delta region of Wood Buffalo National Park have drowned thousands of Wood Bison over the years. They swim with their head and shoulders almost out of the water. Bison can achieve speeds of nearly 60 km/h for short distances and are quick to stampede if disturbed. Their hearing, vision, and sense of smell are acute and play an important role in detecting danger.

DIET

Bison are primarily grazers but will browse when graze is scarce, such as during the winter. They eat principally grasses and sedges in both summer and winter. Bison have an efficient digestive system and are able to thrive on lower-quality forage than can cattle. However, when they are both on good-quality forage, cattle gain more weight per day.

REPRODUCTION

The annual rut takes place between mid-July and mid-September. Female oestrus is synchronized within about a three-week period, so that most of the calves are born during the spring green-up when nutritious new growth is easily available to sustain the enormous energy demands of milk production. If a cow fails to conceive with her first mating, she may breed again 19–21 days later, producing a late calf. Gestation is 270–285 days, and most calves are born during mid-April to June. In northern areas both the rut and the calving season are towards the latter half of the times given. Typically, cows birth a single calf, producing two calves every three or four years. Twins are rare and have never been reported in Wood Bison. Birth weight depends on gender and the nutritional status and size of the cow. Smaller or poorly fed cows produce small calves. The calves are weaned at 7–12 months old. A cow will normally have her first calf at two to four years old. Male bison are capable of producing viable sperm by the time they are two years old, but bulls usually do not begin to breed until they are at least five, and more commonly at six, years old. This has more to do with their ability to compete with the bigger, older bulls in the herd than with their actual sexual maturity.

BEHAVIOUR

Migration herds were known to number in the hundreds of thousands, but during the rest of the year bison form smaller "bull" herds of up to 30 mature and subadult males, and larger "matriarchal" herds of up to 100–200 individuals; the latter herds are made up of mature females and their calves and sometimes a few bulls. In Plains Bison the groups tend to coalesce during the rut, and males establish a dominance hierarchy that allows a few of the larger bulls to conduct most of the breeding. Wood Bison generally form smaller groups than do Plains Bison, and during rut the groups become smaller still. Wood Bison appear to use a harem system where a small number of cows are guarded by a mature male. Herd size is typically larger in open habitat and smaller in parklands or partly forested areas. Often bulls, especially old bulls, live a solitary life. Rutting bulls compete for access to the females. They utilize a variety of threat displays, usually beginning with a broadside display, presumably to display their size, condition, and intent. These displays often move on to include pawing, wallowing, aggressive lunges, and head nodding and can escalate into butting, horn locking, shoving, hooking, and even head ramming – from a run. Bulls are sometimes injured and on rare occasions even killed, especially by hooking with the horns. Bulls can detect the future receptivity of a cow a few days before she ovulates, by a chemical odour in her urine and will "tend" that female until she becomes receptive and mates with him or he is ousted by a rival. Tending a pre-oestrous cow is not an easy task, because she is restless and frequently jumps away and careens through the herd, gathering adult bulls in her wake. When she settles, the trailing bulls quickly get down to resolving the question of which bull will continue to tend her. Before giving birth, the cows typically wander away from the herd, returning a day or two later with their calf. Cows are notorious for protecting their calves from perceived danger and hence are unsafe to approach at any time they have a calf at heel. Bulls will also charge intruders without warning, and both are very agile and astonishingly quick. Females also have a dominance hierarchy. Generally, older cows are more dominant, and during any given year female calves that are born earlier are usually larger and become more dominant over smaller female calves. This dominance is often retained into adulthood. Female calves tend to remain in their mother's herd

and continue to associate with her as they age. Male calves show no such extended relationships with their mothers. Cows generally take the lead when the herd is on the move.

VOCALIZATIONS

Grunts, roars, snorts, foot stamping and pawing, and tooth grinding are commonly used for communication. The use of these sounds increases during rut and herd movements. Calves and cows will bawl if separated.

SIGNS

Droppings are similar to those produced by Domestic Cattle. On succulent feed, a flat patty of about 30 cm in diameter is common. On drier feed, a layered, cylindrical dropping is more usual. Both sexes like to wallow, but wallowing is a favourite pastime for bulls during the rut. They churn up an area about 4–5 m in diameter, then use it for dust or mud baths, and can frequently be found resting in the middle of it. These hard-packed, depressed circles will gradually grass over if not used, but can often still be detected years later. Males also spend time "horning" nearby trees. The bark is worn and often stripped, sometimes to the point of killing the tree. Travelling Bison tend to follow behind each other in a line, etching their traditional trackways into the ground. Even captive herds create noticeable trails as they move from one part of their limited range to another. Where undisturbed by the plough or urbanization, some migratory trackways are still evident on the plains, even though they have not been used for several centuries. Bison tracks are large. The third and fourth toes register in a track, with the fourth toe being slightly larger. The second and fifth toes are dewclaws and are usually not seen in a track unless the footing is very soft, as in mud or snow. The front track is larger than the hind track. Bison tracks can easily be confused with those made by Domestic Cattle.

REFERENCES

Bradley and Wilmshurst 2005; Brookshier and Fairbanks 2003; Carbyn 1997; Dragon and Elkin 2001; Elbroch 2003; Fuller 2002; Gates et al. 2001; Komers et al. 1994; McMillan et al. 2000; Mitchell and Gates 2002; Reynolds, H.W., et al. 2003; Slough and Jung 2007; Stephenson et al. 2001; van Zyll de Jong 1986; van Zyll de Jong et al. 1995.

Slow walk

Left hind track
Length: 11.0–15.0 cm
Width: 10.0–14.0 cm

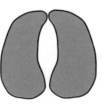

Left front track
Length: 11.5–16.5 cm
Width: 11.5–15.0 cm

American Bison

Mountain Goat

also called Rocky Mountain Goat, White Goat, Snow Goat

FRENCH NAME: **Chèvre de montagne**

SCIENTIFIC NAME: *Oreamnos americanus*

Female Mountain Goats are probably the most aggressive of all Canadian wild ungulates. They continually bicker amongst themselves to gain and maintain status, and they have dangerous horns to enforce their claims. Fortunately, few of their disputes reach the physical stage.

Mountain Goat (*Oreamnos americanus*)

male in winter

female in summer

kid

DESCRIPTION

Mountain Goats are medium-sized, white-haired ungulates with heavily muscled forequarters, a hump above the shoulder, thin, pointed black horns, and long pointed ears. Both sexes have beards, and the long hair on the backs of the upper front legs grows out to look like chaps, especially during winter. They have a typical double coat with long, white guard hairs for water repellence and a shorter, woolly, thick undercoat for insulation. Mountain Goats moult twice each year. During the spring moult, goats can look quite scruffy, as large patches of heavy winter hair hang loose. Pregnant and nursing females moult later than do males and juveniles. The short summer coat (2–5 cm long) is moulted in autumn, and the longer winter fur (often more than 20 cm long) is fully grown by late November or early December. Although adult males are larger than adult females, the sexes look similar, and detecting the size difference, especially of a solitary animal at a distance, is virtually impossible. The most obvious sign of gender comes from the horns. Male horns arc backwards in a gentle curve, and the circumference at their base is much larger than that of the eye. Female horns tend to be fairly straight, often with a backwards hook at the tip, and their circumference at the base is about the same size as the eye. Adult horns are usually 200–280 mm long, although the record is over 300 mm long. The black horn sheaths are smooth and shiny, except at the base where they are rougher and often ridged. Another useful way to determine gender, even of the kids, and especially at a distance, is to observe the urination posture. Females squat, while males stretch. Mountain Goats have short but powerful legs and hooves that are well designed for mountain climbing. The rough textured pad beneath the horny edge of the hoof protrudes beyond the hoof edge as well as at the back of the foot, creating a large cushion that helps them grip the rock. Although the coat is white or yellowish-white, the eyes, nose pad, lips, inside of the ears, tongue, horns, and hooves including dewclaws, are black. Both sexes have a black crescent shaped gland behind each horn, which is larger in males. These glands swell during rut and are thought to be used to scent-mark. Males use their sharp horns during battles for dominance over other males (see the "Behaviour" section). The majority of the horn strikes are aimed at the rear flanks. For protection, males develop an area of thickened skin called a dermal shield on the sides of their rump, which can reach 22 mm in thickness on older animals. Their dental formula is incisors 0/3, canines 0/1, premolars 3/3, and molars 3/3, for a total of 32 teeth. Kids are born with 18 milk teeth, and their full complement of adult teeth is achieved by the summer of the year they turn five.

SIMILAR SPECIES

The only other all-white ungulate in Canada is a subspecies of Thinhorn Sheep called Dall's Sheep. Its hair is shorter, especially in winter, and its longer, slenderer legs lack the chaps that are typical of Mountain Goats. Male Thinhorn Sheep have impressive curled light-brown horns, which are much larger than the thin, pointed black horns of the Mountain Goats. Female Thinhorn Sheep do have small horns that are similar to those of Mountain Goats in shape and size, but again the horns are light brown rather than black. Female Bighorn Sheep have similar horns to those of female Thinhorn Sheep, and lighter-coloured individuals can appear whitish from a distance, although their lighter rump patch remains distinct and they have a black tail. The characteristic shoulder hump and chaps of the Mountain Goat distinguishes it from all of the sheep.

SIZE

Males are heavier than females. Canadian goats tend to be on the larger end of the size range, and the more southerly animals are smaller. Greatest weight is at midsummer.

Male: total length: 125–183 cm; tail length: 9.0–20.3 cm; hind foot length: 33.0–38.9 cm; ear length: 11.0–14.9 cm; weight: 64.9–144.0 kg; shoulder height: 88–122 cm.

Female: total length: 132–194 cm; tail length: 7.3–15.2 cm; hind foot length: 30.5–37.5 cm; ear length: 9.5–14.5 cm; weight: 60.0–84.1 kg; shoulder height: 87–93 cm.

Newborns: weight: 2.5–3.5 kg.

RANGE

Mountain Goats occupy alpine and subalpine regions of western North America, primarily the Rocky Mountains and the coastal mountain ranges of British Columbia and Alaska. Fossil distribution of this mountain specialist is likely wider than the animal's present distribution. Fossils have been found on Vancouver Island, but no goats have been seen there in recorded history. Other native populations continue to exist in Alberta, Yukon, the Northwest Territories, Montana, Idaho, and Washington. Introduced populations have had mixed success. Successfully introduced herds occur in Colorado, Wyoming, South Dakota, Oregon, Utah, and Nevada.

ABUNDANCE

There are an estimated 75,000–110,000 Mountain Goats in North America. Almost 50,000 of these live in British Columbia. Most populations in Canada are considered to be stable, but almost

Distribution of the Mountain Goat (*Oreamnos americanus*)

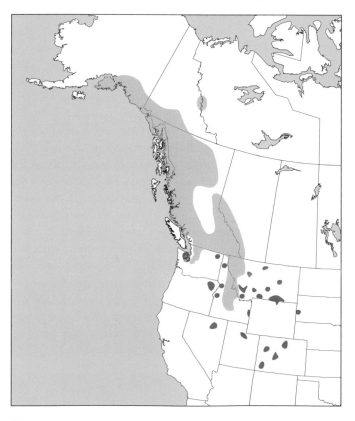

■ introductions

all, except those in national parks, are subject to a managed hunt. Mountain Goats are vulnerable to severe winter weather, especially heavy snow and ice. Populations in some areas have greatly diminished following recurring bad winters. Small, isolated populations (as are most in Alberta) are highly sensitive to eventual extinction, even without the added pressures of hunting.

ECOLOGY

Like all ruminants, Mountain Goats alternate feeding and resting bouts, and they typically have six to seven cycles per 24 hours. Activity peaks in early morning and during the evening, but goats are also active during the night and daylight hours. The active period of the cycle varies seasonally. During summer more time is spent feeding than resting, and during winter the reverse is true. In most parts of Canada, Mountain Goats undertake a seasonal altitudinal migration. Generally, they move higher in summer and lower in winter. Along the coast where snow accumulation can be particularly heavy, they may be driven all the way down to sea level. They prefer to graze in alpine meadows within a quick dash of some rugged escape terrain. Foraging in forested areas increases in autumn and winter. Their principal predators are Brown Bear, Grey Wolf, and Cougar. Occasional predators include Bobcat, Black Bear, Wolverine, Coyote, and Golden Eagle. Only kids and juveniles are vulnerable to eagles, which try to knock them off a cliff in order to feed below on the carcass. Avalanches and falls from cliffs are regular causes of mortality. Mountain Goats generally live 10–11 years. The oldest known male died at 15.5 years of age, and the female at 18 years.

DIET

These montane animals consume a wide range of herbaceous vegetation – grasses, forbs, lichens, twigs, and bark – depending on the season and availability. They are able to survive being temporarily trapped in a small area by a heavy snowfall, eating whatever vegetation is at hand. In many parts of their range the goats lack salt or other minerals, such as selenium, in their diet and will travel for many kilometres each summer, even through heavy timber, to reach a traditional mineral lick. Goats travelling in wooded areas are clearly uncomfortable and will often run.

REPRODUCTION

Rut occurs from late October to early December and normally peaks in mid-November. Most kids in a population are born within a two-week period during the following spring, indicating that female oestrus was synchronized. Based on the time from mid-rut to the birthing peak, the length of gestation seems to be around 190 days. Most births occur from mid-May to early June in Canada. Females find a safe birth site and stay alone for the birth and for a few days afterwards. The kids can follow their mothers over reasonably challenging terrain at two to three days old when they rejoin the nursery herd. Kids are essentially weaned at around four months old, but females will continue to sporadically suckle their kids for about a year. Milk production usually drops off early, and kids may begin foraging before they are even a week old. Females become sexually mature between the ages of two and six years old

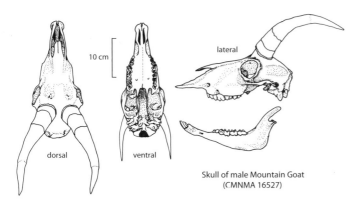

dorsal   ventral   lateral

10 cm

Skull of male Mountain Goat
(CMNMA 16527)

(usually between three and five); however, two-year-old females in introduced populations can produce their first kid, which means they mated and conceived when they were 17–19 months old. Although males become sexually mature at around two years old, they typically do not participate in the rut until they are at least five or six years old, when they are large enough to compete with the other adult males for dominance. Younger males may have an opportunity to mate occasionally, but many males may never mate. A litter usually comprises a single kid, but in rapidly increasing populations (such as recent introductions) twinning can occur up to one-third of the time. Triplets are possible but very rare. It is thought that larger litters occur more often when food is abundant, which is likely the case with introduced populations that have fresh range. In populations that are declining (possibly due to overuse of range) many of the nannies will produce kids only every other year. The same female can switch back and forth between annual and biennial reproduction. The yearly birth rate in most stable populations is usually less than 100%. A female's fertility is also known to drop off with increased age, with peak kid production taking place between the ages of five and ten years old. Older females are more likely to produce male kids, and younger females tend to produce more female kids.

BEHAVIOUR

Except during rut, adult males and females live in segregated groups. Male herds can be fairly large (up to 15 animals) in late spring but are usually small during the rest of the year. Mature males tend to be solitary or to join with two or three other males. Females and young live in nursery groups that can vary in size from two to 100 or more animals (but usually from two to ten), including yearlings and immature males. Most juvenile males leave the nursery groups when they are two or three years old, but some remain until they are four. Juvenile females typically remain with their nursery groups when they mature. Group size is largest in summer. Herds of over 100 animals have been counted on a rich alpine meadow, but this size is unusual. The bond between a nanny and her offspring (especially the kids) is very strong. Females will risk their lives to defend their beleaguered kid from a predator. Yearlings may continue to associate with their mother and may even periodically be allowed to suckle, especially if she does not have a new kid. This behaviour is common in populations where the

females reproduce only every other year. Should the nanny produce a kid in consecutive years, she will usually not permit the yearling to remain near her and her newborn. Female Mountain Goats are very aggressive towards each other throughout the year, while males are generally tolerant of each other, except during the rut. One study of a herd in Alberta showed that each adult female had three to four aggressive encounters per hour. These encounters usually consist of subtle body postures that indicate threat or superiority and rarely entail body contact. The dominance hierarchy for each year's cohort of females is apparently determined in their first summer as kids. Except during rut, adult females are the most aggressive and hence the most dominant class in Mountain Goat hierarchy. Adult males establish a similar hierarchy among themselves during the rut. Their struggles do occasionally lead to bloody battles, but fatalities are rare. A dominant male escorts an oestrous female and defends her from other males. Since females are aggressive and will strike a male if they are unwilling or unready to mate, males approach submissively and make no attempt to mate until they can fling their front leg and strike her haunch with no reaction. Since males can only defend a single female at a time, socially dominant, experienced males will actively select older, experienced females to escort and tend. This strategy appears to enable them breeding access to the most prolific segment of the female population, whose abilities to raise a kid successfully are greatest.

VOCALIZATIONS

Adults produce low-pitched grunts sometimes when they are threatening each other. They also bleat loudly at each other if they begin to actually fight. Separated kids and nannies will bleat loudly to locate each other. Kids also bleat if attacked or injured.

SIGNS

Mountain Goats commonly leave tufts of white hairs on their summer trails as they lose their winter coat. Winter hairs are long, often more than 20 cm in length. Mountain Goat tracks are somewhat indistinct as the edge of the hoof rarely contacts the substrate. The track is left by the fleshy pad under the hoof. As is usual for mammals, the front track is larger than the hind. Rutting males dig rutting pits into which they urinate. They then churn up the moist earth, pawing some onto their undersides. These pits are often visible for some time after the rut is over. Winter scat is 0.6–1.6 cm in diameter and 1.0–2.5 cm in length. These pellets usually have a blunt end and a pointed end and can easily last several seasons virtually unchanged, except in colour, as they slowly bleach out. Summer scat is more amorphous because the animals are eating vegetation that is more succulent. The elongated mass usually will show the edges of the compressed pellets and can be 10 cm in length. Mountain Goats, Bighorn Sheep, and Thinhorn Sheep create runs along the mountain slopes that can be visible even from a distance. Mountain Goats dig distinctive feeding craters in the winter snow. Using their forefeet, they push, pull, and scrape snow in all directions and then will enlarge the crater to reach more forage. The craters may become irregular in shape as they are enlarged.

Walk

Left hind track
Length: 10.5–16.5 cm
Width: 9.0–12.7 cm

Left front track
Length: 11.0–18.0 cm
Width: 9.5–15.2 cm

Mountain Goat

REFERENCES

Banfield 1974; Côté et al. 1997, 2000; Côté and Festa-Bianchet 2001, 2003; Dane 2002; Demarchi, M.W., et al. 2000; Elbroch 2003; Festa-Bianchet et al. 1994; Fournier and Festa-Bianchet 1995; Hamel et al. 2006; Hoefs and Nowlan 1998; Mainguy et al. 2008; Romeo and Lovari 1996; Shackleton 1999; Zettergreen 2006.

Muskox

also called Musk Ox

FRENCH NAME: **Boeuf musqué**
SCIENTIFIC NAME: *Ovibos moschatus*

Oomingmak (the bearded one) is the Inuit name for Muskoxen. The name certainly suits this shaggy beast, whose long dense fur keeps it warm through the bitter Arctic winter. While the shoulder hump and dark colouring make a Muskox appear large, in reality the top of a full grown bull's hump only reaches chest high on an average male human.

DESCRIPTION

Muskoxen have long, dark brown hair that hangs nearly to their hooves. Under the long guard hair they have a silky underfur called by its Inuit name of *qiviut*. Fabric made from spun qiviut has the warmth and weight of cashmere. The long outer guard hairs, which take about four years to reach full length, are shed continuously throughout the year, much like human hair is. The qiviut is shed each summer, giving the animals a ragged, wind-blown appearance for several weeks. The lower part of the legs, the muzzle, and a saddle-shaped patch on the back are not covered by the extra-long guard hairs but do have the fine underfur. These areas are a creamy-whitish colour. Only the hooves and horns are unprotected by fur. Over the shoulders, the fur is particularly thick and forms a distinct mane, which is especially prominent on mature males. Body hair hangs down the sides of the animal, creating a flowing skirt. This thick covering conceals a compact and deceptively agile body. Small ears and a short tail are hidden beneath the fur. Even the base of the horns is fur covered in winter, except in older males. Both sexes grow curved horns that have a thick base on the forehead. Males develop especially wide horn bases that can almost meet over the forehead, while females have narrower horn bases with a thick tuft of hair between them. A whitish band of woolly fur covers the horn bases of cows and young males up to the age of three years old. Even newborn calves have this white patch. Fully adult males lose the white band as their horn bases expand to fill the same space. The dental formula is incisors 0/3, canines 0/1, premolars 3/3, and molars 3/3, for a total of 32 teeth.

SIMILAR SPECIES

Bison look superficially similar to Muskoxen from a distance; however, the distributions of the two species do not overlap. Bison are found

Muskox (*Ovibos moschatus*)

female

male

calf

farther south than are Muskoxen, and they are considerably larger. In their Arctic habitat, Muskoxen are unique in size and colouring.

SIZE

Males are significantly larger and heavier than females, and the largest males occur in the southern portions of the range. Maximum and relative sizes of females remain consistent throughout the range. Males: total length: 200.7–260.0 cm; tail length: 7–14 cm; weight: 190–410 kg (captive males have reached 650 kg); shoulder height: 120–150 cm.

Females: total length: 190–240 cm; tail length: 6–12 cm; weight: 160–210 kg (captive females have reached 300 kg); shoulder height: 90–120 cm.

Newborns: weight: 10–14 kg.

RANGE

Although formerly found all around the Arctic, Muskoxen were exterminated in the Old World. Now their natural distribution is only in northern North America, mainly in Canada and Greenland. They live on many of the High Arctic islands (the Queen Elizabeth Islands) and the larger western Arctic islands, parts of the northern mainland of Nunavut, the Northwest Territories, and the North Slope of Alaska. They have gradually either naturally expanded into or been reintroduced into many former parts of their range, and in North America their current distribution may actually slightly exceed their historic distribution. They have been introduced extensively around the north into new areas, mainly in an effort to provide

potential food and revenue for indigenous peoples. Herds have been successfully introduced into western Greenland, the Ungava Peninsula of Quebec (from which they have since spread to northern Labrador), the Taimyr Peninsula and Wrangel Island in Russia, central Norway and Sweden, and many parts of western Alaska.

ABUNDANCE

Very vulnerable to human hunters, Muskoxen came perilously close to extinction in the early 1900s when their pelts supplied Europeans with warm carriage robes. Conservation measures taken by Canada in 1917 halted the slaughter, and Muskoxen have slowly repopulated the North. They now number around 100,000–120,000. Over 70% of the Canadian Muskoxen live on Banks and Victoria islands in the High Arctic.

ECOLOGY

Muskoxen live in cold northern regions that have low precipitation, long cold winters (eight to ten months' long), and short cool summers. Over most of their range, they endure a regime of approximately four months of complete darkness and four months of total daylight, with two months of lengthening or shortening days in between. Their long outer guard hairs protect them from rain and snow, and the long, fine underfur provides them with superb insulation. The inside of their hooves is recessed, so they walk only on the sharp horny outer edges that provide excellent traction on ice and snow. Muskoxen have adapted to their harsh Arctic habitat in several physiological ways as well as in the already mentioned physical ways. They have a relatively large rumen and a slow food-passage rate, which allows them to

Distribution of the Muskox (*Ovibos moschatus*)

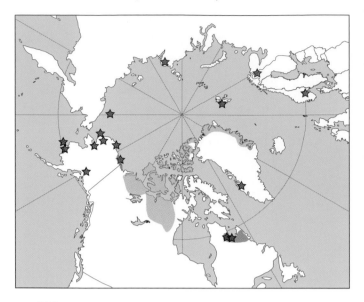

 & ▢ introductions

extract the most nutrition from tough, fibrous, and low-quality fodder. They have a low metabolic rate, especially when fasting, which allows them to maintain their body condition even if food is scarce. Their water requirements are similar to those of camels; therefore, they do not waste body heat by eating a lot of snow for moisture. Muskoxen are also very energy conservative during winter, moving only as far as necessary to find forage and spending much of their time huddled together, resting. Despite their many adaptations to cold weather, their winter survival and the following summer's reproductive success can be negatively affected by mid-winter thaws, sleet creating a thick ice crust over the forage, or heavy, late-winter snowfall. They are able to paw through moderate snow cover and thin ice to reach their forage. When the ice is too thick or the snow too deep, the vegetation exceeds their reach and the animals starve. Die-offs are common during such winters to the extent that the entire population on a High Arctic island can be eliminated. Most Muskoxen move only 10 km or less each day in the course of grazing, even during the summer. Although generally sedentary, they have been seen trekking across sea ice to reach other islands. Their principal predators are Grey Wolves, although attacks by Brown Bears and Polar Bears have been reported. Muskoxen are managed on a quota system, and human subsistence hunting is permitted in Canada, the United States (Alaska) and Greenland. The oldest known wild Muskox died at 24 years old, but few live beyond 10 years.

DIET

Muskoxen are ruminant herbivores that specialize in digesting high-fibrous plants. They eat sedges, Arctic Willow, and some grasses and assorted tundra forbs in the High Arctic islands. On the mainland, especially near the treeline, they add woody plants such as alder, crowberry, Labrador Tea, ground birch, and blueberry to their diet.

Preferred grazing areas on the mainland are usually windswept, where the snow cover is reduced or eliminated, thereby reducing the need to spend energy on digging feeding craters (see the "Behaviour" section). On the High Arctic islands, preferred grazing is in the lowlands where snow is less than 30 cm deep. Muskoxen appear to detect plants under the snow by smell. They sniff the surface of the snow carefully before digging and never dig where success is not assured.

REPRODUCTION

Rut occurs in August and September and is primarily determined by photoperiod, therefore being initiated earlier in more southerly locations and up to two to three weeks' later in more northerly areas. A cow's body weight in late summer determines whether she will enter oestrus and breed. If it is too low, she will not come into season. This physiological effect provides a crucial and early brake on reproductive effort when the cow would be unlikely to bring a healthy calf to term. Following an eight- to nine-month gestation, a single calf is born in April or May. Cows in a well-fed population may produce their first calf as two-year-olds, but most have their first calf in their third year. In poorly fed populations the cows may be five years old before they have their first calf. In productive areas well-nourished cows have a calf every year, and the calf is weaned in late autumn or early winter after nursing for about six to seven months. Lactating cows are able to become pregnant provided their body weight is high enough. In less productive regions where the cows reproduce only every two or more years, they may continue to nurse their calf until it is 15–18 months old, supplying it with essential additional calories through its first winter and enhancing its probability of survival. A particularly severe winter can stress the pregnant cows to the point that their fetuses are aborted or resorbed. As a result, for

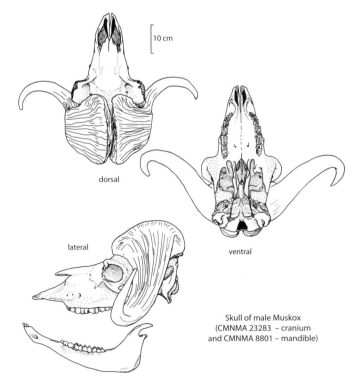

Skull of male Muskox
(CMNMA 23283 – cranium
and CMNMA 8801 – mandible)

some populations there are years when no calves are born. Healthy cows as old as 18 years continue to bear calves, showing no signs of age-related reproductive senescence. Newborns arrive with a thick woolly coat that insulates them to −30°C, provided there is little wind. They are able to stand and nurse within minutes of birth and quickly learn to use their mother as a mobile windbreak.

BEHAVIOUR

Muskoxen are social animals that are almost always seen in herds, except for the occasional solitary adult male (usually seen only in summer). Females and juveniles form herds that grow larger in winter and smaller in summer. Typically, a summer herd comprises 8–15 individuals, and a winter herd is made up of 10–20 animals. Bulls either band together in winter or join a mixed-herd. Like all ruminants, Muskoxen alternate periods of activity (mainly grazing) with periods of resting (mainly ruminating, also called "chewing their cud"). In a typical winter cycle the herd will spend 70 minutes grazing followed by 110 minutes of lying down. This timing is reversed during the short summer, as the animals spend longer grazing (100 minutes) and less time resting (50 minutes) while they race to consume enough succulent summer vegetation to grow fat before winter. As early as June, an adult bull will join a group of females, then proceed to spend most of his time defending his exclusive access to these females and smelling each in turn to check their state of oestrus. Harem bulls defend their position from other males with an escalating series of ritualized behaviours that ultimately lead to head-to-head clashes if the rival does not back away. The battles begin with gland rubbing (rubbing the preorbital gland against the inside of the foreleg), lateral displays (strutting in profile to enhance the appearance of size), horning (slashing the ground with the horns), roaring, and head swinging, and then move to head pushing and butting, and ultimately to charging and head-to-head clashing. The bulls back up from each other with heads swinging, then charge at speeds of up to 50 km/h to ram their heads together with a tremendous crash. The heavy helmet formed by the base of their horns protects them from the force of this collision. Still, many bulls are injured, and some die as a result of dominance battles. During winter considerable energy is expended in digging snow away from the fodder. The animals alternately dig with each front hoof until the plants are exposed, sometimes using their head to push away slabs of snow, and sometimes pounding with their chins to break up the snow slabs. Finally, the lips are used to clear the loose snow off the plants. Access to these feeding craters can be hotly contested as bigger or more dominant animals attempt to supplant a proprietor. The circle defence formation, for which these ungulates are famous, is a highly effective defence against wolves, their principal predator. As a wolf approaches, the cows and juveniles crowd into the adult male with their rumps touching. If the wolves remain in front of the Muskoxen, then a line is formed. If the wolf, or wolves, begins to circle the herd, the animals press against each other and form a rough circle, keeping their formidable horns oriented to the outside. Muskoxen will also stampede away from a potential threat, running at speeds of 40–50 km/h and often leaving young calves behind.

Walk

Left hind track
Length: 9.5–11.5 cm
Width: 10.2–11.5 cm

Left front track
Length: 10.8–14.0 cm
Width: 11.0–15.2 cm

Muskox

VOCALIZATIONS

The most dramatic sound produced by a Muskox is the call of a breeding bull. The bull stands with his mouth partly open and his tongue sticking out, making a deep rumbling roar described as not unlike that of the African Lion. Cows make a quieter, more guttural roar when calling their calves, and calves bleat when calling their mothers or playing. In addition, all of the animals in a herd will produce an alarm snort that is meant to warn others of approaching danger.

SIGNS

Muskox winter droppings are similar to those of Wapiti. They are clusters of oval to squarish fibrous pellets about 1.1–1.9 cm in diameter and 1.6–2.5 cm long. They are dark when fresh and bleach out to a yellowish-brown when weathered. Some have a somewhat pointed end. Summer droppings (found only in July and August in the High Arctic) are looser and clumped together owing to the more succulent forage. Other signs of range use by Muskoxen can be tufts of shed qiviut and even bleached skulls and bones from old wolf kills. Calcium is not recycled nearly as quickly in the Arctic as it is further south, so bone can sometimes remain on the tundra for hundreds of years. Winter signs include feeding craters and tracks. A Muskox feeding crater varies in size but can easily be as large as the Muskox. These snow craters usually have a leading edge. As the animals paw the snow, pushing it behind them, they produce a jumble of loose snow; it then bears the tracks where they stood, with a smooth, clean forward edge where the animal was last feeding. Muskoxen walk on the edges of their hooves and leave a track that is accentuated around the perimeter and concave in the interior. Each track is made by the third and fourth toe on each foot, and the third toe is slightly smaller than the fourth. Front tracks are larger than hind tracks. The negative space between the toes is much narrower than in Caribou.

REFERENCES

Adamczewski et al. 1997; Barr 1991; Chubbs and Brazil 2007; Clarkson and Liepins 1993; Elbroch 2003; Ferguson, M.A.D., and Gautier 1991; Frey et al. 2006; Gray 1987; Larter and Nagy 2001; Lent 1988; Lyberth et al. 2007; Mech 2000a; Reynolds, P.E., 2001; Rowell et al. 1993; Schaefer and Messier 1996; Slough and Jung 2007; White et al. 1997.

Bighorn Sheep

also called Mountain Sheep, Rocky Mountain Bighorn

FRENCH NAME: **Mouflon d'Amérique**

SCIENTIFIC NAME: *Ovis canadensis*

Bighorn males have the highest proportion of horn to body weight of all of the ungulates. About 8%–12% of an adult male's mass is made up of horns and the bony horn cores. Formerly there were thought to be two subspecies living in Canada: Rocky Mountain Bighorns in the British Columbia and Alberta Rockies and California Bighorns in the arid regions of south central British Columbia.

Recent evidence suggests that these subspecies are not distinct, and both should be considered the Rocky Mountain Bighorn subspecies.

DESCRIPTION

Bighorn Sheep are stocky ungulates with a light- to dark-brown coat and a white rump patch. Their short tail is black, and a thin dark line often runs from the top of the tail to the back, bisecting the rump patch. The muzzle, the insides of the ears, the backs of the legs, and the male's large visible scrotum are whitish. Bare skin on the nose is black. An annual spring moult can last up to two months, and nursing females complete their moult last. Both sexes have brown horns with distinct transverse ripples. Growth of the horns occurs only in the summer, and annual growth can be distinguished from the other transverse ripples by the deep groove created by the cessation of annual growth. The female's horns are short and curved back and to the side; their growth diminishes to a very slow rate around the age of one year, but annual rings can still be discerned (with close inspection) until the female is five to six years old. The male's horns continue to grow throughout his lifetime at a more rapid rate than do those of the female; they become very thick at their base and grow in a spiral out to the sides of the head. Although the length of a male's horns, judged by the degree of their "curl," is often used to age the animal, in reality there is too much variation between populations to make this valid. Age can only be accurately ascribed by counting annual growth rings. The tips of full-curl male horns are typically "broomed," which means they are worn or broken off during fights. Care must be exercised when aging a skull by the annual rings because several years can have been lost owing to brooming. The dental formula is incisors 0/3, canines 0/1, premolars 3/3, and molars 3/3, for a total of 32 teeth. Lambs are born with deciduous teeth, which are gradually replaced by permanent teeth until their full set of adult dentition is complete at four years old. The senses of Bighorn Sheep are acute.

SIMILAR SPECIES

The most similar species is the other mountain sheep species, Thinhorn Sheep. This more northerly sheep comes in two colours. The white form (called Dall's Sheep) is found only in northern Yukon and Alaska, far to the north of Bighorn range. The dark form (called Stone's Sheep), which has a similar pelage pattern to that of Bighorn Sheep, is found in northern British Columbia and southern Yukon, not far from the most northerly portions of the Bighorn range. The best way to distinguish the two species is by colour. Thinhorn Sheep are either charcoal grey to light grey and white, or all white, while Bighorn Sheep are brown to light brown and white. Thinhorn and Bighorn males have somewhat different horn shapes, which may be difficult to discern at a distance. Thinhorn Sheep horns are less robust and usually have a more sharply defined horn keel than that of Bighorn Sheep. They have a looser curl, so that the tips flair away from the animal's head, making the spread from tip to tip considerably wider than the spread of Bighorns. At a distance, Mountain Goats can resemble Bighorn Sheep. The Mountain Goats are pure white and longer haired, with short, sharp, black horns.

SIZE

Males are much larger and heavier than females, and northern sheep are sometimes larger and heavier than more southerly forms.

Males: total length: 132.1–195.6 cm; tail length: 8.3–15.2 cm; hind foot length: 35.6–48.3 cm; weight: 65–118 kg; height at shoulder: 81.3–111.8 cm.

Females: total length: 116.8–188.0 cm; tail length: 7.5–13.6 cm; hind foot length: 27.9–43.3 cm; weight: 43.0–82.5 kg; height at shoulder: 76–91 cm. Newborns: weight: 2.7–4.5 kg.

RANGE

Bighorn Sheep are widely distributed in western North America in the Rocky Mountains, the Cascade Range, and the Sierra Nevada Mountains. Their distribution was previously more extensive than it is today, especially in the United States, where they were almost extirpated by the mid 1900s. Numerous reintroductions have occurred since then, many of which have been successful. Scattered colonies currently exist throughout their former range, with the largest conterminous portion of range being in Canada.

ABUNDANCE

In 1996 there were an estimated 13,600 Bighorns in Canada, 3500 in Mexico, and a further 25,000 in the United States. Canadian numbers in 2002 were estimated at 13,700. Estimates of numbers before the 1900s vary from 500,000 to 4 million for both Bighorn and Thinhorn Sheep, and most scientists currently favour the lesser estimate. Probably about 75%–80% of that 500,000 were Bighorns.

Although rigorously managed, hunting is permitted of most populations, except those in national parks. Bighorns are considered "Vulnerable" and "At Risk" in British Columbia, and their numbers and distribution are carefully monitored. Bighorn Sheep were federally listed as "Endangered" species in the United States in 1998.

ECOLOGY

Bighorns are found most often in open shrub and grasslands located near or on rough slopes. This type of habitat provides good visibility and allows them to detect a potential threat at a distance. They are extremely competent and agile on the steep and broken terrain and use it to elude their predators. Rough escape terrain is crucial for all age and sex classes of sheep, and populations do not exist where such terrain is absent. Precipitous escape terrain is especially important for females with lambs or for females about to deliver. Bands of ewes with juveniles (called nursery herds) forage near these rocky safe zones, but male bands (called bachelor bands) may feed farther afield. In Canada, Bighorn Sheep occupy slopes from 300 m above sea level to those of the alpine and subalpine zones. Some sheep in south central British Columbia are resident all year, but generally most populations move downslope during autumn (October–November) and move gradually back up to the heights as the snows disappear in spring (May–June). They prefer to winter in regions with low snow accumulations and avoid snow deeper than 30 cm. Bighorns rarely enter dense forests, although they will shelter in the forest edges to avoid cold winter winds or to find shade during hot summer days. Prolonged fire suppression can encourage vegetation

Bighorn Sheep (*Ovis canadensis*)

female

lamb

male

male dark phase

Distribution of the Bighorn Sheep (*Ovis canadensis*)

☐ historical distribution

■ current distribution

growth, which eventually drives the sheep away because their visibility, and hence predator protection, is reduced. Bighorn Sheep feed during the day in relatively large herds. A typical day is spent alternately foraging and resting. Transmittal of diseases from livestock – mainly Domestic Sheep – has been the main factor in the historical decline of Bighorn Sheep and continues to be a serious threat today. In addition, they face competition from domestic livestock for forage, water, and space, as well as hunting, poaching, and increasing rates of human encroachment. Mortality rates are highest during a Bighorn's first year and vary between years and herds. The average first-year mortality within a herd may range from 40% to as high as 90% in exceptional circumstances. Causes of death include predation, inclement weather, human disturbance, poor maternal nutrition, and poor mothering. Coyotes and Cougars are the Bighorn's principal predators in Canada, and lambs and juveniles are most vulnerable. Sometimes a Cougar will learn to specialize on Bighorn Sheep. This can result in a very high hunting success that is known to cause alteration of sheep's traditional ranges or even local extinction.

Eagles prey on lambs occasionally, and Grey Wolves and bears will take adults. Die-offs of more than 50% of a population owing to disease infrequently occur. Diseases affect the adults as well as the juveniles and are compounded by the recurring stress of chronic Lungworm infections. Lungworm is a native North American nematode that is almost ubiquitous, especially where forage is poor. It weakens the animals, making them more susceptible to predation, disease, accidents, or starvation. If they do not die before they are two, Bighorns will likely live for 10–15 years. The oldest known wild Canadian Bighorn Sheep (a male and two females) were 20 years old.

DIET

Diet varies with season and location, and the animals are opportunistic in their forage choices. They eat whatever is most nutritious, palatable, and available at the time. In the Rocky Mountains, grasses, forbs, lichens, and sedges are the main food, and woody browse will be eaten when necessary. During winter the sheep must paw away the snow to reach the desiccated plants buried beneath. Consequently, the herds seek out south-facing sunny slopes and windswept open grasslands where snow cover is reduced. In most areas where the vegetation is succulent, Bighorn Sheep survive on the moisture from their food in summer and from snow in winter and do not need to drink freestanding water. In drier, hotter regions, access to water is necessary, especially during summer.

REPRODUCTION

Rut takes place in late autumn and early winter, usually November–December, with a peak from mid-November to mid-December. Some Okanagan herds may begin rutting as early as late October. Duration of the rut is longer at lower elevations and southern latitudes and shorter at higher elevations and more northerly latitudes. Following a gestation of 173–175 days the lambs are born in spring, when the vegetation is greening up to feed the nursing mother, and early enough to provide the lamb with the greatest possible amount of time to grow before winter arrives. If spring is late or winter is early, the lamb's survival chances are diminished. The peak of lambing takes place in mid-June, and the lambing period becomes shorter and more synchronized within each herd with increasing latitude. The most dominant rams can end up breeding about 50% of the local ewes. The remaining lambs are sired by younger subordinate males that fight and distract a dominant male and then chase the receptive female that he was tending, in order to quickly attempt a mating. Also, if two rams begin fighting over a receptive female, a third, smaller ram may sneak in and mate with her while the larger rams are occupied. A male's interest in a particular female lasts until she cycles out of oestrus, when he will leave to search for another receptive ewe. Males are sexually mature around 18 months old but rarely become sexually active until they are older and strong enough to compete with other rams, usually around seven to eight years old. Their reproductive period is generally limited to only a few years before they are deposed by younger, stronger rams. Females typically mate for the first time at the age of about 30 months, producing their first lambs in the spring that they turn three years old. Normally each female produces a single lamb each spring, and twins occur rarely. Ewes continue to reproduce even into old age. As her lambing time approaches, a ewe will seek steep and rugged terrain away from other sheep for a few days. Newborns are on their feet within 30 minutes and can follow their mother within 24 hours. Most lambs are nursed for three to five months before they are weaned.

BEHAVIOUR

Bighorns are social animals living in groups of 2–110 animals. In general, their society is segregated by age and gender for most of the year. Adult and subadult males older than two years old band together into small bachelor bands. Females with their lambs and one- to two-year-old offspring from previous years also band together into nursery herds of up to 25 animals. Nursery herds are matriarchal. Both herd types occupy home ranges that overlap to some degree. Male bands often venture farther from the rough escape terrain as they seek better-quality forage to help them grow large bodies and horns. Females graze near the escape terrain as their lambs are very vulnerable to predation. Membership in these herds can be irregular because there is some movement between neighbouring groups throughout the year, especially between male bands. Particularly fine grazing areas may be shared by several nearby herds. Herds are largest during autumn because both herd types will coalesce during the rut, usually on the home range of the females. Both male and female herds maintain a dominance hierarchy. Among females this is subtly accomplished and involves a great deal of displacement of subordinates by dominant ewes, usually by butting or threatening to butt. Social rank in females appears to be stable over time. Youngsters probably learn the location of mineral licks, migration routes, seasonal ranges, and good grazing areas from the older ewes. Most female lambs remain in their mother's group when they mature. Among males, horn size, body size, and fighting prowess are linked to dominance, and these factors vary over time. To determine individual fitness, males gather in pairs or small groups and clash. They square off and, rearing up, run on their hind legs, lunging down at the last moment to crash their heads together in an awesome and noisy spectacle of power and hard-headedness. The combined force of this collision has been estimated at around 900 kg. After clashing, the males often stand facing each other with their heads held high and their horns directed towards their opponent, perhaps in a horn threat. Males can become exhausted by these struggles, but their horns and a wealth of air spaces in the bone surrounding their brain absorb most of the shock, so injuries or deaths are rare. These battles can occur throughout the year, but most take place in autumn before the rut begins. Males use a "low-stretch" posture during courtship. He approaches a ewe with his head held low and stretched forward, his muzzle tipped up and tongue flicking, to which she usually responds by squatting and urinating. Using a posture called a lip curl, he then tastes the urine to determine whether she is coming into oestrus. Should she be near oestrus, he will likely begin to tend her by staying alongside her and driving away rivals. Occasionally a male will drive a female or a group of females away from the main herd and block them from returning until they become receptive and have been bred by him. This so-called blocking behaviour is relatively rare. Dominance mounting is common among male mountain sheep as they use this behaviour to reinforce their dominance. Juvenile males are subordinate to females. As they reach two to three years old, their larger body size provides them with social dominance over adult females. Ewes and their lambs learn to recognize each other very early, and many females refuse to suckle a lamb that is not their own. Lambs recognize their mother's voice, while ewes recognize their lambs by both their call and their odour.

VOCALIZATIONS

Bighorns produce similar vocalizations to those of Domestic Sheep but with more vibrato and in a lower register. Ewes and lambs bleat to locate each other. Females will emit an alarm bleat that their own and other lambs instantly recognize and respond to by scampering to their respective mothers. Males produce an undulating growl as they court a female.

SIGNS

Beds are used repeatedly and for extended periods for resting or chewing their cud. Typically, these are shallow depressions that are up to 25 cm deep and 63–101 cm long. Bighorn Sheep cursorily scrape their forefeet over the top of the bed each time they use it, to remove any loose stones, and over the years it gradually becomes a deeper depression. Scat is often found immediately adjacent to a bed. Extensive erosion sites on hillsides are favoured spots for resting and scratching and may have originated by individual sheep making beds or, where the hillside was disturbed, for example, by a bear digging out a marmot den. Depending on their extent, these

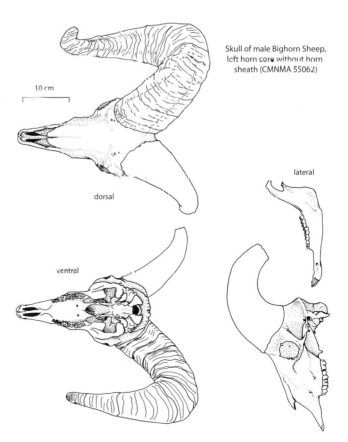

Skull of male Bighorn Sheep, left horn core without horn sheath (CMNMA 55062)

10 cm

dorsal

lateral

ventral

sites may be visible at a considerable distance. When feeding, the sheep spread out over the hillside and do not normally create a distinct trail. When travelling, however, Bighorn and Thinhorn Sheep, as well as Mountain Goats, follow behind each other in a line and create trails, especially on open talus slopes, which can sometimes be quite conspicuous and visible at a distance. Winter feeding craters made by both Bighorn and Thinhorn Sheep are distinctive. The animals scrape the snow downhill with their forelegs so that it mounds at the downslope edge of the crater. The sheep will forage in this crater and then move a short distance away and create another crater. Scat shape and contour varies with diet. Summer scat is more amorphous, although some suggestion of pellet shape, albeit compressed, remains detectable. Winter pellets are individually distinct and vary in length from 0.5 to 1.6 cm. Their shape is usually blunt at one end and roughly pointed at the other. Bighorns have four toes on each foot; the second and fifth toes are dewclaws, and only the third and fourth toes register in a track, unless the footing is very soft. Forefeet are slightly larger than hind feet. Hoof prints are more heart shaped and pointed than those of Mountain Goats.

REFERENCES

Bleich 1999; Coltman et al. 2002; Demarchi, R.A., et al. 2000a, 2000b; Elbroch 2003; Hass 1997; Krausman and Bowyer 2003; LeBlanc et al. 2001; Monello et al. 2001; Pelletier and Festa-Bianchet 2006; Pelletier et al. 2006; Shackleton 1999; Wehausen and Ramey 2000.

Trot

Direct register walk

Thinhorn Sheep

also called Dall's Sheep, White Sheep, Alaskan White Sheep, Stone's Sheep, Black Sheep, Fannin's Sheep

FRENCH NAME: **Mouflon de Dall**

SCIENTIFIC NAME: *Ovis dalli*

Thinhorn Sheep occur in two colour forms. The all-white version, Dall's Sheep (*Ovis dalli dalli*), is the more northerly of the two subspecies and is sometimes called the Alaskan White Sheep or just White Sheep. Stone's Sheep (*Ovis dalli stonei*), also called Black Sheep, has a variable grey to grey-brown body with a white rump patch and white underbelly and muzzle. This form occurs in the southern portion of the range. The area of intersection of the two subspecies is vague because there is a cline of colour intergrades between the darker sheep in the southeast and the white sheep in the northwest.

DESCRIPTION

Thinhorn Sheep are ungulates with short, stocky legs and short ears and tail. Both sexes have horns. Females have thin, blunt

Left hind track
Length: 5.2–8.3 cm
Width: 3.8–6.0 cm

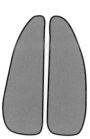

Left front track
Length: 5.4–8.6 cm
Width: 3.8–7.6 cm

Bighorn Sheep

Thinhorn Sheep (*Ovis dalli*)

female

male

lamb

northern white form – Dall's Sheep
(*Ovis dalli dalli*)

horns that curve back and away from the head. Their horns grow quickly until the animals are about a year old, and then the growth slows, continuing at a much reduced rate until they are five to six years old, when growth slows again, almost to a halt. Male's horns continue to grow at a rapid rate until they are four to five years old, then the growth slows. As a result, adult males have considerably larger horns than do females. The horns of both sexes cease to grow in winter, and a deep groove demarks the end of each growing season. The annual increments are easier to detect on male horns than on female horns, especially after they reach five years of age. Horns are amber or light brown in colour. Male's horns grow in a backwards and outwards curl, which then sweeps forward to complete an arc of 360° or more, depending on age. Thinhorn horns are not only less robust than Bighorn horns, but they also flare outwards to a greater degree and hence are wider from tip to tip. They typically are not broomed or worn at the tips and have a distinct ridge or keel on their outer edge. The northern subspecies has a completely white coat, although some individuals have a black tail tip. In the southeast of the range the Stone's Sheep is dark charcoal or grey-brown with a black tail and a white rump patch, belly, backs of legs, and head. In the darker forms the head and neck may also be a shade of grey with a white muzzle and white inside the ears. Between the northern and southern forms are Thinhorn Sheep of various intermediate shades and amounts of grey. Lambs are white or greyish, depending on the subspecies. A single moult occurs between March and July, with mature males completing it early, and females, juveniles, and animals in poor condition taking longer. In dark-coated animals, the bleached-out older coat is replaced by a glossier darker coat. Skin thickness is generally about 1 mm, except on the face and nose, where it is up to 6 mm thick on adult males. The dental formula is incisors 0/3,

canines 0/1, premolars 3/3, and molars 3/3, for a total of 32 teeth. Lambs are born with an almost completely erupted set of milk teeth, all of which are replaced by permanent adult teeth by the age of 40 months.

SIMILAR SPECIES

Although Bighorn Sheep are similar to Thinhorn Sheep, the more northerly all-white Dall's Sheep is relatively easy to distinguish just by colour and is more likely to be mistaken for a Mountain Goat – especially the female Dall's Sheep. However, unlike the mountain sheep, Mountain Goats have a beard, shorter legs with a fringe of chaps on the front legs, a shoulder hump and longer hair (especially in summer), and the short horns are black with sharp tips. In the more southeasterly portions of Stone's Sheep range the darker colour form more closely resembles a Bighorn. Coat colour in Bighorns is more brownish, and not black, grey, or grey-brown as in the Stone's Sheep. Mature Bighorn males have substantially more robust horns, which are more brownish and not amber coloured as in Thinhorn Sheep. Older adult male Bighorns usually have heavily broomed horns, whereas Thinhorn Sheep usually have complete or very lightly worn horn tips.

SIZE

Males are larger and around 40% heavier than females. Maximum annual weights are achieved in late autumn.
Males: total length: 132–180 cm; tail length: 7–13 cm; hind foot length: 37–46 cm; weight average: 82 kg (up to 129 kg in autumn); height at shoulder: 92–109 cm.
Females: total length: 132.5–162.0 cm; tail length: 7–10 cm; hind foot length: 32–41 cm; weight average: 49 kg (up to 61 kg); height at shoulder: 70–95 cm.
Newborns: weight: 3.2–4.1 kg.

Thinhorn Sheep (*Ovis dalli*)

female

lamb

southern dark form – Stone's Sheep
(*Ovis dalli stonei*)

male

RANGE

Thinhorn Sheep are found in northern North America from northern Alaska, through Yukon and the far western Northwest Territories, to north central British Columbia. Numerous introductions of both the Dall's and Stone's subspecies have occurred, usually to nearby unoccupied regions. Apart from these small introductions, Thinhorn Sheep are thought to occupy more or less the same range now as they did historically.

ABUNDANCE

Primarily owing to their remote range, but also to their rugged habitat, Thinhorn Sheep are not heavily hunted by sport hunters, nor have they been excessively affected by native subsistence hunters or railroad meat hunters. In 1997 there were an estimated 9000 Dall's Sheep in Yukon, 500 in British Columbia, 14,000 in the Northwest Territories, and more than 72,000 in Alaska. There were around 12,000 Stone's Sheep in British Columbia, and about 3000 in Yukon. The current population size of more than 100,000 Thinhorns is thought to be similar to historic levels.

ECOLOGY

Thinhorn Sheep are agile and capable on rough, rocky, mountainous terrain. They rarely graze far from this precipitous habitat, to which they can quickly run if threatened by predators. Most Thinhorn Sheep occupy distinct summer and winter ranges, but some populations do not appear to migrate. The distance between summer and winter habitats varies and can be up to 48 km, but most sheep do not migrate far and may only move a few hundred metres up or down the mountain. They also use different parts of their mountainous habitat according to the snowfall, generally in an effort to avoid deep snow. Although they are northern animals that are able to survive very cold temperatures, their short legs leave them vulnerable in even moderate snow depths; they tend to seek out windblown slopes with light or no snow cover, which usually can be found above 1500 m

in altitude. Owing to the lengthy winter season in their northern habitat, Thinhorn Sheep spend the majority of the year on their winter range: 240–303 days depending on the weather. In early spring they usually descend to lower altitudes to graze on early green vegetation, drifting gradually back to the higher elevations as spring travels up the mountain. They reach their alpine summer range by about July. In autumn they seek out yet a different habitat, searching for nutritious forage around the peaks or in the foothills around lakes and river deltas. Females, especially those with lambs, remain near escape terrain while foraging, although they may be forced by poor forage to move farther afield. Lambing grounds that supply both spring grazing and rough escape terrain with secluded lambing areas are traditionally used year after year. Male herds are commonly found grazing considerably farther away from the precipitous and rugged escape terrain. Mortality is, in large part, determined by weather. Seasons with heavy snow, ice, or thaws that then freeze to form a hard crust can cause high mortality and are particularly hard on lambs. Those lambs that do survive their first winter could live for 19–20 years, but sheep that are older than 12–13 years are rare. Maximum longevity is often determined by tooth wear as older animals commonly grind down or lose teeth and can no longer forage effectively. Lungworm, which can be so devastating to Bighorn Sheep farther south, has been detected in Thinhorn Sheep but does not cause the dramatic die-offs seen in Bighorns. Avalanches and accidental falls claim some Thinhorn Sheep each year. Predators of Thinhorn Sheep are primarily Grey Wolves and Coyotes. Golden Eagles will take small lambs. Other predators include Canada Lynx, Brown Bear, Black Bear, and Wolverine. Their primary limiting factor appears to be range quality, which is largely governed by weather.

DIET

Like all mountain sheep, Thinhorns eat mainly grasses, lichens, and some forbs but will browse some of the montane shrubs when other preferred forage is scarce. They are versatile in their dietary choices,

Distribution of the Thinhorn Sheep (*Ovis dalli*)

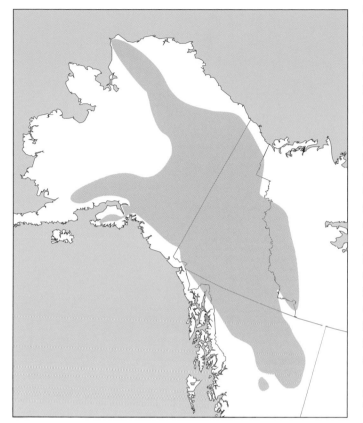

selecting the most nutritious and palatable option from whatever is available. Mineral licks, often characterized by moist, claylike soils, are important sources of salt and minerals for Thinhorns, and usage of traditional licks tends to be seasonal, primarily in late spring and early summer and again in early autumn, although some use occurs throughout the summer. The licks are so important that many populations govern their summer distribution according to their location. In the northern extent of their range severe weather conditions commonly affect plant growth, which can consequently have drastic impacts on sheep foraging. Starvation during late winter is a regular threat.

REPRODUCTION

Female Thinhorn Sheep are receptive to breeding for one to three days during each oestrous period. Onset of oestrus is largely governed by photoperiod, but the ewe's body condition and the presence of a ram have minor, but more immediate influences. Typically, the rut takes place in November and December, and if a ewe does not conceive during her first oestrus, she will cycle and become receptive again in 12–14 days. After a pregnancy of about 171–175 days, a single lamb is born the following May or June. Twins are very rare. Onset of lambing is timed to occur as the new spring vegetation becomes available for the nursing ewe, but as early as possible in the spring to allow sufficient time for the lamb to grow before facing its

first winter. The ewes leave the security of their herd before the birth and go to a rugged and secluded birthing site. They rejoin the herd a day or two after the lamb has been born. Most seclusion periods last less than 24 hours. Newborns stand within 32 minutes and begin suckling within three to four hours of birth, shortly after the placenta has been expelled. They are able to keep up with their mothers by the time they are 24 hours old. Lambs are nursed for three to five months, at which time they weigh about 30 kg. Weaning is a gradual process that is controlled by the ewe. The length of the nursing period varies from year to year. In a year when the lambs are born later, they might be allowed to suckle longer and hence get more milk during each bout of nursing, but then they will be weaned early, after only three months of nursing. This kind of flexibility allows the ewes to balance the needs of their offspring with the weather and their own nutritional status. Weaned lambs remain with their mothers until they are about a year old. Although well-nourished females are sometimes able to breed as young as 18 months old, most ewes are at least two-and-a-half years old, and more often three to four years old, before they breed for the first time. Males are capable of breeding as young as 18 months old but rarely have an opportunity because they are excluded by larger-horned, heavier, older rams. Females continue to bear offspring into old age, with no sign of reproductive senescence, even in very old ewes.

BEHAVIOUR

Thinhorn Sheep are very gregarious, but solitary rams commonly travel from nursery group to nursery group during rutting season. Herds are typically gender distinct. Maternal groups, composed of related ewes and their one- to two-year-old offspring, vary in size from four to twenty individuals. Ram groups (made up of males older than two to three years of age) are smaller, typically three to ten individuals. Herds of both types can be widely scattered while foraging, making their magnitude difficult to discern. During rut the two herd types coalesce, usually on the home range of the maternal group, and herd size becomes temporarily larger. Young sheep learn the traditional mineral licks and the seasonal and lambing ranges by following the members of the maternal group. Males go on to learn and inherit their home range from more senior rams after they move on to join a ram group. Young females usually remain with their maternal herd. Both male and female herds establish dominance relationships. The female dominance status is largely based on age, although body size does play a role, and male lambs are subordinate until they become larger than the ewes. Female dominance is largely intimidation based, and there are few overtly aggressive interactions because the older ewe usually wins by virtue of her long-established dominance and more aggressive horn threat. Dominance among males is based on horn and body size and thus is also largely age related since both these characteristics increase with age, up to about nine years old. Exhibitions of horns and body are common components of the displays that the males direct towards each other. If a non-physical display is insufficient to cause one of the protagonists to back down, as happens most frequently when the rams are similar in size, they will walk away for a few steps, then whirl around, rear up, and drop onto each other,

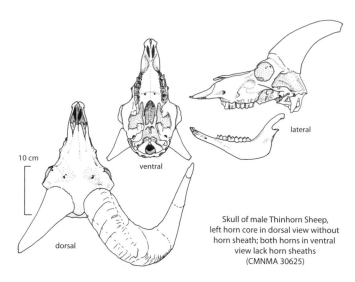

10 cm

ventral

lateral

dorsal

Skull of male Thinhorn Sheep,
left horn core in dorsal view without
horn sheath; both horns in ventral
view lack horn sheaths
(CMNMA 30625)

striking their horns together. The loser in these battles is treated, and will act, like a female. The winner or dominant male often nuzzles the subordinate's flanks and kicks him with a stiffened foreleg. The loser will accept the dominance displays of the larger ram without reacting and will even allow himself to be mounted by the dominant animal. The sound of clashes can ring through the mountains from spring through the autumn rut. Dall's Sheep clash considerably less often than do Stone's Sheep. Most of their battles end at the kicking phase. Fighting these dominance battles can be exhausting, and old males, despite their very large horns, lose their vigour and drop out of the hierarchy struggles. During rut the adult males travel from female to female, testing her readiness to mate by tasting her urine, presumably to analyse the hormone balance, using an action called a lip curl. If she nears receptivity, the male begins to accompany her and defends her from the approach of other males. A male tending a female draws the attention of nearby males, precipitating many battles to determine which will stay closest and have the best chance to mate. Usually the largest-horned or largest-bodied male wins this position.

VOCALIZATIONS

Like Bighorn Sheep, Thinhorns produce sounds that are similar to those of Domestic Sheep. Their vocalizations are lower pitched than Domestic Sheep's and higher pitched than Bighorns'. Lambs bleat to alert their mothers, and females produce an alarm bleat, to which their lambs instantly respond by running to them. Males displaying to each other and to females will emit a loud growl. Males searching for nursery groups will stop on a sheep trail and call with a loud "baa."

SIGNS

Thinhorns tend to use traditional areas as bedding sites. Not only are the resting spots scraped bare and slightly depressed (and may be confused with Mountain Goat beds), but the herd spends significant time around these bedding sites. Their eliminations fertilize the area, which will then stand out as greener and more lush than the surrounding vegetation. Erosion may follow from this usage, and

Walk

Left hind track
Length: 5.5–8.5 cm
Width: 5.0–6.5 cm

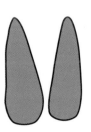

Left front track
Length: 5.5–9.0 cm
Width: 5.0–6.5 cm

Thinhorn Sheep

the resulting scar on the hillside can be visible for quite a distance. Alternately, the erosion may have been initiated by a Brown Bear digging for a ground squirrel or a marmot and then been taken over by the sheep. The small wall formed on the upslope side of these sites is favoured as a rubbing spot, where clinging hairs may be found. An uneroded bed measures 60–90 cm by 35–51 cm. The feeding craters that have been dug in the snow by sheep are distinct. The crater is small, and an intensively feeding sheep will feed from one crater and then move less than a metre to create another. They scrape the snow downhill with their forelegs, so it mounds on the downslope side of the crater. Mountain Goat craters are larger and messier, with snow thrown in all directions. Trackways or runs connect different feeding areas and can be visible on the mountainside from a considerable distance, especially if they occur on unvegetated slopes. Trackways of Mountain Goats, Bighorn Sheep, and Thinhorn Sheep can be very similar. Important mineral licks may be the focus of a spider web of converging runs. Thinhorn footprints are more pointed than those of either Mountain Goats or Bighorn Sheep. As with all ungulates, the first toe is missing. The third and fourth toes register in the track; the second and fifth toes are dewclaws. The front track is larger than the hind track. Winter scat is composed of discrete pellets of 0.6–1.1 cm in diameter that are commonly compressed slightly and may show a concave side. They are quite variable in shape: some are blunt at one end and sharply pointed at the other, while others are blunt at both ends. Summer scat is more amorphous; the edges of the individual pellets are usually still distinguishable, but several are compressed together into soft clumps.

REFERENCES

Bowyer and Leslie 1992; Bunnell 2005; Corti and Shackleton 2002; Elbroch 2003; Geist 1971; Hoefs 1991; Hoefs and Nowlan 1993; Krausman and Bowyer 2003; Nichols and Bunnell 1999; Rachlow and Bowyer 1991, 1994, 1998; Shackleton 1999; Walker, A.B.D., et al. 2006; Worley et al. 2004.

Order Cetacea

whales

Pacific White-sided Dolphin (*Lagenorhynchus obliquidens*)

Photo: Jamie Scarro

The species in this order are wholly aquatic. They inhabit all the world's oceans and some of its lakes and rivers. There are 10 families, 40 genera, and 84 species in this order. Of these, there are 8 families, 24 genera, and 37 species in Canadian marine waters. There likely are yet undiscovered species within this group, since, despite their size, they are difficult to see and identify and are often found only in the remotest areas of the oceans. The largest animals that have ever lived on Earth are members of this order and are still living in the world's oceans today.

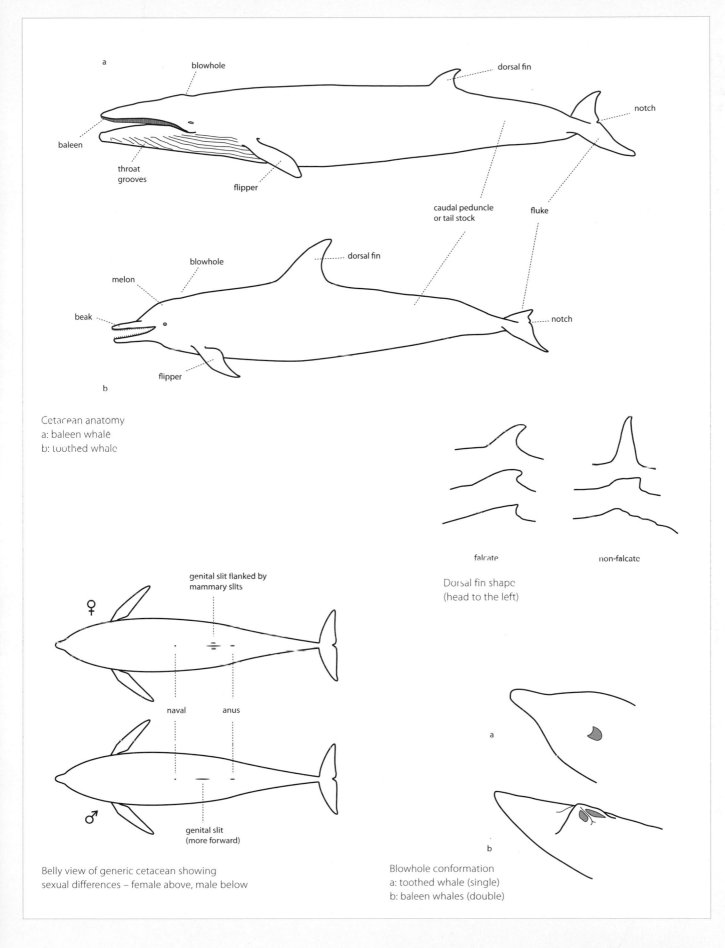

Cetacean anatomy
a: baleen whale
b: toothed whale

Belly view of generic cetacean showing
sexual differences – female above, male below

Dorsal fin shape
(head to the left)

falcate   non-falcate

Blowhole conformation
a: toothed whale (single)
b: baleen whales (double)

FOSSIL RECORD

Cetaceans evolved from four-legged, terrestrial ancestors that returned to the sea. The earliest known whale fossils date back to the Eocene, 55 million years ago.

TAXONOMY

There are two groups, or suborders, within the Cetacea, the toothed whales and the baleen whales. Both are carnivorous in the sense that they eat other animals. The toothed whales are the hunters. They capture and eat fish, squid including giant squid, birds, seals, and sometimes other whales. The baleen whales are truly the gentle giants; they sieve zooplankton and small fishes through their baleen plates. The largest animals on the planet eat some of the smallest.

Use of the terms whale, dolphin, and porpoise requires some explanation. Whale is used in this volume to include all cetaceans. Dolphins are those small whales that usually have a distinct beak and numerous conical sharp teeth. Porpoise refers to those small whales with blunt noses and flattened spade-shaped teeth. Canadian waters support nine species of dolphin and two species of porpoise.

ANATOMY

Whales are highly modified for living exclusively in the water. They are still air breathers like all the other mammals, but their nostrils have moved to the top of the head and have a direct connection with the lungs, which bypasses the mouth – with this arrangement a suckling infant or an adult swallowing cannot get liquid into the lungs. The toothed whales have a single nostril, while the baleen whales retain the more typically mammalian double nostril. Whales have lost their hind legs but retain the two small pelvis bones which serve to anchor the genitalia. Occasionally, individual adult whales will still have vestigial, non-functioning portions of some of the hind limb bones embedded in muscle. The front limbs form flippers – smooth, fleshy, rudder-like appendages used for steering that still contain the usual number of mammalian fingers (5), although extra length of the digits is common. Unlike most mammals, which have baby teeth and adult teeth, toothed whales have only a single generation of teeth that grow in before the calf is weaned. Baleen whales have teeth only as foetuses. These are shed before birth and short baleen plates are already present on newborns. Whales appear to have at least limited colour vision. They do not have tear ducts – which would be unnecessary underwater anyway – but these are replaced by glands that exude drops of grease to protect the eyes from salt and friction. Since sounds travel through the body to the inner ear, the normal external ear canal is generally unnecessary; it is often filled in and the structures of the inner ear are adapted to receive underwater sounds. Mammalian-style olfaction, or smelling, is based on detecting particles in the air, which is both difficult and pointless underwater. The toothed whales have a very limited olfactory capability and the baleen whales have no sense of smell at all. The skin of whales is smooth and the typical mammalian hair covering is reduced to a few bristles on the head and lips of a few species (all whales still have some hairs in the embryonic stages). To compensate for the lack of hair, there is a thick fibrous layer of blubber under the skin to provide insulation. In some species, particularly those living in very cold waters, this layer can be up to 50 cm thick. Whales also lack sweat and sebaceous glands in the skin. The tail is modified to form flukes – the main organ of propulsion. These flukes project horizontally, which distinguishes them from fishes, whose tail projects vertically. The flukes are composed of muscle and cartilaginous tissue, with no bony support. The whales' bodies are superbly shaped and streamlined for efficient hydrodynamics. The genitalia are all internal and there is no external ear. Because the water always supports them, their bones do not bear any weight, so they are spongy and serve as storage reservoirs for oil.

DIVING

Whales hold their breath while diving, so they are not normally subject to decompression sickness (the bends), as are humans, who continue to breathe under pressure while diving or mining deep below the earth's surface. There is some recent evidence that suggests that exposure to high-intensity underwater sonar, the sort used mainly for military purposes, can cause whales to suffer and die from the bends by forcing normally dissolved blood nitrogen to form gas bubbles in the bloodstream. Other adaptations of whales for diving include: valves to close off the nostrils; elastic, very extensible lungs that can inhale and expel large volumes of air in a short time; rigid areas in the lungs and bronchi that can safely contain the small amounts of air still left after the whale exhales before a dive; reduction in heart rate when not at the surface; elevated levels of oxygen-rich myoglobin in the muscles; increased oxygen-rich haemoglobin in the blood and additional blood volume; shunts that direct blood towards the brain and away from less important areas; a reduced metabolism while deep diving; and an ability to control the breathing reflex even when under serious carbon dioxide loading. Some whales, especially the species that hunt squid and other deep-dwelling prey, commonly dive for more than an hour. Records of 82 minutes and 120 minutes have been reported. Dives to depths over 2000 m have also been recorded. See the Sperm Whale and Northern Bottlenose Whale accounts for further information on the deep-diving abilities of these two species.

SLEEP

The dynamics of sleep in cetaceans is not well understood. It seems that some whales such as Bottlenose Dolphins, pilot whales, Harbour Porpoise, Amazon River Dolphins, and Belugas (all of the cetacean species examined to date) can rest half of their brain while the other half remains semi-alert to predators and retains control of breathing and movement. Behavioural observations suggest that likely all the Delphinidae and Monodontidae use this method. Whether this ability is shared by the larger whales is uncertain. Some species, like Right Whales, Sperm, Pygmy and Dwarf Sperm Whales, Grey Whales, and Humpbacks apparently can sleep relatively soundly while at the surface, making them susceptible to ship strikes. Their behaviour suggests that both hemispheres of their brains are asleep during this time. Other species are thought to sleep below the surface. How long whales need to sleep is not known and likely varies with the season and species.

REPRODUCTION

Whales usually produce only a single calf per pregnancy. Twins occur occasionally, at around the same rate as in humans – once in every 100 or so births. The large whales might have a single calf every three to five years. Even the smaller whales do not always produce a calf every year. Newborn whales are well developed and fairly large (between 5 and 15% of the weight of the mother depending on the species) after a gestation period of 10–12 months. The young are born backwards so they will not drown before their mother can guide them to the surface for their first breath. Because cetacean lips are fairly inflexible, the infant cannot grasp a nipple as firmly as do terrestrial mammals. Nursing occurs underwater, and the females swim slowly and often partly on their sides to make access as easy as possible. Muscles around the milk ducts contract as the calf grasps the nipple, squirting milk into the baby's mouth. Young whales can only hold their breath for about 30 seconds, so nursing must be rapid. The fat content of whale milk is as much as 20 times higher, and the protein as much as twice as high, as the average land mammal's milk. This rich diet allows the young to grow at a phenomenal rate. Blue Whale calves can double their already substantial birth weight in only seven days!

ECHOLOCATION

Like bats, which hunt at night when vision is limited, whales too have to deal with vision handicaps. No light at all penetrates below 390 m, and only 1% of the light even gets below 40 m. Under ideal conditions – clear water and bright sunlight – the range of underwater vision is only 17 m. Sonar or echolocation is the solution shared by both bats and whales.

It is difficult to determine the hearing range of a wild whale; however, experiments on captive Bottlenose Dolphins have pinpointed the audible range at between 1 and 150 kilohertz (kHz). Humans hear in the 1–20 kHz range. For more information on echolocation see the "Echolocation" section in the Bat (Order Chiroptera) introduction. As might be expected, the toothed whales are more accomplished at echolocation. They need it to catch elusive, fast-moving prey in conditions of poor visibility. The baleen whales seem to practise a more limited form of sound usage. While they commonly produce sound for communication, there is no substantiated evidence that they echolocate. They are able to produce sound in the ultrasonic range, but when and why they do so is unknown. The Blue Whale and other baleen whales also produce very low-frequency sounds that some scientists think are used to communicate over vast distances, possibly to locate dense swarms of zooplankton, and for navigation. The latter two uses could involve reception of returning echoes. Although echolocation is suspected in the baleen whales, it has not yet been proved.

SOCIAL LIFE

Most toothed whales travel in genetically related groups. These groups, called pods or herds, which can number into the hundreds at times for some species, are lead by a dominant animal, either female or male, again depending on the species. Most pods are made up of two to ten animals. Baleen whales are usually found singly, except during mating or while females are caring for young. Large numbers of baleen whales will gather at an attractive feeding site.

MASS STRANDING

Individual stranding of whales is entirely understandable. As an injured or dying animal becomes more debilitated it will float at the surface in order to breath. The likelihood of it being carried to shore by waves and tides is increased as it becomes more feeble.

The mass stranding phenomenon is a puzzling behaviour of toothed whales. Several animals (the record is 835) beach themselves at a time. The social bond between the animals of the pod has been cited as the main reason why large groups strand. They are either following an ill, injured, or disoriented leader or attempting to maintain contact with a pod member that is ill or injured. Often when live strandings are observed, attempts to return the animals to sea fail. Many, upon release, simply turn and head for shore once more. Although there has been much speculation and theorizing, scientists do not completely understand why cetaceans beach themselves.

Suborder Mysticiti

baleen whales

FAMILY BALAENIDAE

right and bowhead whales

These are thick-bodied baleen whales with highly arched rostrums and no throat grooves or dorsal fin. Their flippers are large and broad and their baleen is the longest of all the baleen whales. The head constitutes approximately 30%–40% of the body length in the three members of this family found in Canadian waters.

Whales in this suborder have keratinaceous plates called baleen that originate on the upper jaw and hang down along the sides of the mouth. The whale feeds by gulping a large mouthful of water, partly closing its mouth, and then pushing the water out through the baleen with its tongue. The baleen filters small organisms from the seawater, and then the tongue scrapes back against the baleen to collect the food. Baleen whales have small gullets for their size as their prey is tiny. Although the arterial blood vessels leaving the heart of an adult Blue Whale are large enough for a small child to crawl through, the throat opening is about as big around as a grapefruit (around 10 cm in diameter). Baleen whales have twin blowholes located centrally at the top of the head. These whales have not been proven to echolocate.

Bowhead Whale

also called Greenland Right Whale,
Greenland Whale, Rocknoser

FRENCH NAME: **Baleine boréale**

SCIENTIFIC NAME: *Balaena mysticetus*

Of all the baleen whales, Bowheads are most adapted to life under and near the ice. Huge heads and broad backs can smash through up to 60 cm of solid ice and they have a thick blubber layer to keep them warm in the cold water. Bowheads can hold their breath for over an hour, a useful ability when living under ice and travelling from breathing hole to breathing hole.

DESCRIPTION

Adults are black with variable white and black markings on their chins and usually a small amount of white around the genital slit. They often develop some white on the tail stock and flukes that can be visible from above. This white marking spreads with age and can serve as a handy, but not always reliable, indicator of maturity and age. Large subadults have noticeably less white on their tail stocks than full adults, but some adults remain all black. Newborns and very young calves are paler than adults, sometimes even a light brown. Albino or all-white animals are very rare. Skin can be up to 2.5 cm thick and can provide considerable protection during encounters with ice. Blubber

is normally 5.5 to 28 cm thick depending on the area of the body, but thicknesses of up to 50 cm have been reported. The enormous head is about 30%–40% of the total length of the body and has a highly arched mouthline and a narrow rostrum. From 230 to 360 black baleen plates hang on each side of the upper jaw, essentially where the teeth would normally be found. The longest baleen, in the middle of the row, can be over 4 m in length. This is the longest baleen in the world and is partly why Bowheads were hunted so assiduously in the past. The lower jaws bow outwards to accommodate the baleen. Each small eye is about the size of a billiard ball. Their dental formula is 0/0. They have large paddle-shaped flippers that taper at the tips and broad flukes with a distinct, deep, medial notch. They lack a dorsal fin. The flukes can be as wide as 6 m.

Bowheads have countercurrent heat exchangers in various parts of the body to conserve heat. These are intricate interweavings of blood vessels that allow outgoing blood to exchange its heat with cold incoming blood. A major one is present in the tail stock to warm the blood returning from the flukes. This countercurrent heat exchanger can be bypassed during periods of heat stress, so that the body can shunt large volumes of blood to the flukes, where that large surface area can be used to dump excess heat. Bowheads likely also have one at the base of their tongue to limit heat loss to the cold water while foraging. Their blood is less prone to becoming viscous at low temperatures and so can continue to flow even after being cooled. These adaptations, along with a thick blubber layer, allow this whale to live year-round in waters just around the freezing temperature. Injuries that penetrate the epidermis layer of the skin heal with a white scar. Many adults can be individually photo-identified based on the shape and size of the chin patch and accompanying black spots within the white along with the location of scars.

SIMILAR SPECIES

Only the closely related but smaller North Atlantic and North Pacific Right Whales are similar to Bowheads. Both species of Right

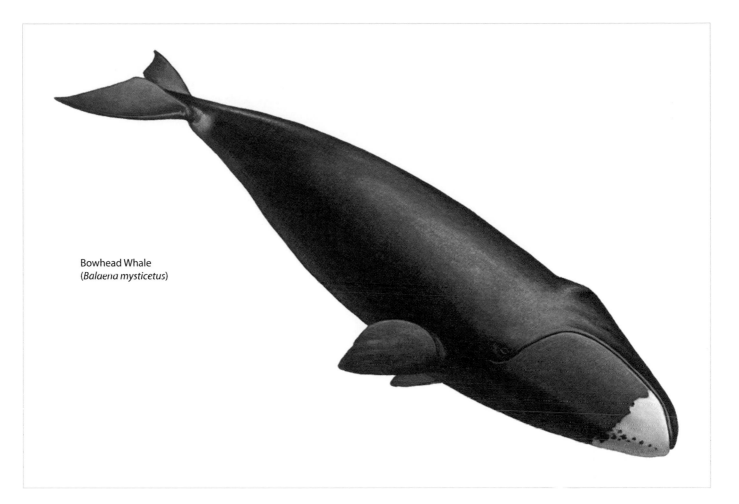

Bowhead Whale
(*Balaena mysticetus*)

Whales have yellowish-white callosities on their heads and chins and variable white splotches on their bellies. Usually when the Right Whales migrate northward, the Bowheads are doing the same; but the Bowheads stay farther north and close to the ice edge, while the Right Whales prefer the somewhat more southerly ice-free waters. Grey Whales are considerably smaller than Bowheads but also lack a dorsal fin. They do, however, have a series of "knuckles" that travel down the upper margin of the tailstock and simulate a series of shallow, blunt dorsal fins. Like Right Whales, Grey Whales move into the lower reaches of Bowhead range during the summer, but usually after the Bowheads have left to travel even farther north.

SIZE
Total length: males, 14–17 m; females, 15–18 m and possibly up to 20 m; newborns, 3.6–4.5 m.
Weight: males, up to about 80,000 kg; females, up to about 90,000 kg; newborns, around 900 kg.

RANGE
Bowheads are closely associated with ice all year. They winter at the unconsolidated southern edge of the pack ice or in large polynyas (semi-stable areas of open water within an ice pack). As the ice breaks up and retreats in the spring, the whales follow it northwards, usually the subadults, males and non-breeding females leading, with the females and calves following. The reverse occurs in autumn as the sea ice reforms, but all ages and genders depart together. There are thought to be four distinct stocks of Bowheads. The two eastern stocks include one that occupies the entire eastern Canadian Arctic to western Greenland. It summers in the eastern High Arctic, northern Hudson Bay, and the Foxe Basin and winters in the broken ice in Hudson Strait, Davis Strait, and Baffin Bay. The second eastern stock was formerly found from Spitsbergen Island to Iceland and eastern Greenland and is on the verge of extinction. Its summering and wintering areas are uncertain. There are two western stocks. The one found in the Sea of Okhotsk appears to remain in that sea all year round, possibly summering along the coast and wintering more offshore. The last and largest stock summers in the Beaufort and Chukchi Seas and winters in the slushy and broken ice of the western and central Bering Sea. Animals from this stock extend into Canadian waters from the Yukon/Alaska border to Amundsen Gulf during summer.

ABUNDANCE
Bowheads have been hunted for over 2000 years by aboriginal hunters around the north, without any harm to the stocks. The advent of commercial hunting in the 1700s changed all that. Whalers started with the Spitsbergen/Greenland stock, moved to the Davis Strait, and then to the Hudson Bay populations. Each group was systematically

Distribution of the Bowhead Whale (*Balaena mysticetus*)

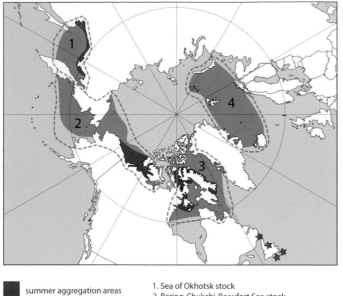

■ summer aggregation areas	1. Sea of Okhotsk stock
■ range	2. Bering-Chukchi-Beaufort Sea stock
★ extralimital records	3. Eastern Canada–West Greenland stock
	4. Spitsbergen Island–East Greenland stock

reduced until there were not enough remaining to sustain the harvest. The Bering Sea stock was next in the mid-1800s, followed soon after by the Sea of Okhotsk stock. The Soviets continued to harvest the Sea of Okhotsk population into the 1960s. Apart from the illegal hunting by the USSR, all commercial hunting of Bowheads has been banned since 1937. The Eastern Arctic stock was estimated at up to 6000–14,400 individuals in 2008, of which up to 2500–6000 are assumed to be mature animals. The Hudson Bay portion of that stock appears to be declining and currently numbers in the low hundreds at most, possibly 250–350 animals (formerly about 600); otherwise the entire stock is thought to be slowly increasing. The Spitsbergen/Greenland stock is thought to contain a few tens of individuals (formerly estimated at 25,000 to more than 100,000). The size of the Sea of Okhotsk population is unknown but small, perhaps around 300 animals (pre-whaling estimates are around 3000). Both of these stocks are considered critically endangered as neither has recovered from almost three centuries of commercial hunting. In the mid-2000s, the Bering Sea stock had an estimated 8000–13,500 Bowheads (formerly up to 23,000), of which around 2500–6000 were considered mature. The Bering stock is thought to be increasing at an annual rate of about 3.4%.

ECOLOGY

Individuals in some stocks migrate as far as 4000–5000 km, but others such as those in Davis Strait and the Sea of Okhotsk travel as little as 500 km between seasons. The Bering Sea Bowheads migrate from summering areas in the central and eastern Beaufort Sea and Amundsen Gulf to wintering areas in the Bering Sea. Migration routes are traditional and the passage of the whales is relatively predictable. Swimming speeds during migration can average up to

5–6 km/h, but most Bowhead movements are slow, averaging 2–5 km/h, although they can achieve speeds of up to 10 km/h for short distances. Bowheads usually occur in water less than 100 m deep, but have been recorded diving over 350 m deep and foraging in near shore trenches. Most dives are less than 16 m deep. Dives of more than 60 minutes have been recorded, but most last less than one minute. Harpooned Bowheads are said to be able to remain submerged for up to 80 minutes. Typically, the whales make a series of short surfacing dives (3–12) to reoxygenate, before making a deeper dive. They actually spend less than 8% of their time at the surface. Juveniles with shorter baleen tend to forage closer to shore in shallower waters, while adults feed more commonly offshore in deeper waters. Killer Whales are significant predators of Bowheads, both adults and calves, especially in the eastern Arctic, and most whales display the parallel scars and frayed fluke margins that typify a Killer Whale attack. Although Bowheads are superbly adapted to living under the ice, occasionally ice entrapment can present a mortality threat. Bowheads may be the longest living mammal. They likely live considerably more than a hundred years. One animal, using an aging technique based on eye tissue, was estimated to be over 200 years old. While this age seems extraordinary and possibly inaccurate, still, stone harpoon points have been found in three different whales taken in Alaska since 1992. These stone points were estimated to have been imbedded into those whales during failed aboriginal hunts at least 100–130 years previously and probably earlier.

DIET

Bowheads feed by filtering seawater through their baleen plates to strain out zooplankton. They eat a wide variety of small invertebrates, but their preferred prey is copepods, krill, and amphipods. These animals are the basis of the Arctic food chain. Up to 60 different species of small invertebrates have been found in the stomachs of whales killed in the Alaskan aboriginal hunt. A feeding whale swims with its mouth partly open for a while, then pushes its massive tongue up to force the water out. This tongue (up to 5 m long by 3 m wide) then scrapes the inside of the baleen, swiping the trapped zooplankton into a relatively small throat where they are swallowed. Bowheads forage in both daylight and darkness, as one would expect of an animal living in the north, where they would experience long periods of total darkness or 24 hours of daylight. Their zooplankton prey, which in more southerly regions moves up and down the water column depending on light conditions, does this only weakly in the Arctic. The depth at which the zooplankton is found is related more to season and ocean-floor topography than to light levels. In winter, most of the zooplankton retreats to great depths and is beyond the reach of the whales. Some feeding occurs in summer near the ice edge, but most of the year's feeding takes place in autumn and early winter when mature stages of the copepods are concentrated. Bowheads feed throughout the water column and are known to rise to the surface trailing mud plumes, as do bottom-feeding Grey Whales. Climatic change is likely to have serious impacts on water temperatures and currents in the Arctic, to the extent that the zooplankton productivity may decline.

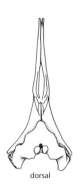

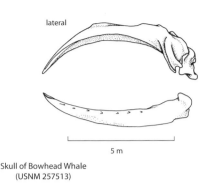

Skull of Bowhead Whale
(USNM 257513)

REPRODUCTION

The calving interval can be 3–7 years, but is usually every 3–4 years and is probably related to cyclic fluctuations in prey abundance. Calves are weaned between 9 and 15 months of age. Gestation is thought to be 13–14 months, after which a single calf is born, in April or May, usually on the spring migration north. Females become sexually mature at 12.5–14.0 m in length, perhaps by around 20–25 years old. This is very old for any mammal and likely reflects the slow growth of this species due to its harsh environment and relatively low food levels. Age at sexual maturity for males is unknown, but they are at least 12.5 m long before any significant sperm is found in their reproductive tract. Although sexual activity can take place throughout the year, most conceptions are thought to occur in late winter or early spring. Despite their extreme longevity, Bowheads appear to remain fecund to an old age and perhaps until death.

BEHAVIOUR

Bowheads are social animals. They tend to be found in loose associations with other Bowheads conducting a similar activity, such as feeding, migrating, or socializing. When conditions are favourable the whales feed, when the foraging is poor they socialize. These clumps of socializing animals can become quite boisterous with much pushing, chasing, lobtailing, and breaching. Courtship activities, including caressing and calling, often occur out of season, especially among subadults. Bowheads appear to have a similar courtship strategy to Right Whales. The female calls and males congregate around her, shoving and spinning, trying to gain the alpha position on either side of her. She rests on her back at the surface, rolling to one side or the other to invert herself for breathing. At each roll, the male on the appropriate side has a brief mating opportunity. Only the strongest males can hold the alpha position. Although the female has no choice of mate, she sets up a competition to ensure that only the best males may succeed. It is possible, as with Right Whales, that this type of activity is fairly common, but since the calves are all born in the same season, copulations outside the breeding period are not procreative.

Very young calves are known to ride the wake of their mother's back by staying closely above her, just about mid-back, where her back begins to slope down towards the tail. In this "piggyback" fashion they are able to keep up on migration with minimal effort. Since calves cannot hold their breath for as long as an adult, they are commonly left alone at the surface while their mother feeds below. As many as a dozen Bowheads have been seen feeding in echelon formation, similar to a skein of migrating geese, each whale offset a few metres from its neighbour's flipper. Juveniles and females with calves congregate in traditional shallow protected areas during summer. Other non-breeding adults also gather together into loose aggregations, but they traditionally summer in different parts of the range away from the nursery grounds. In the eastern Arctic, Bowheads are frequently harassed and injured and sometimes killed by Killer Whales. They have developed a healthy fear of the predators. When a pod of Killer Whales is detected, the Bowheads crowd into the shallows near shore or near the ice edge and will sometimes begin lobtailing. They will also display this type of behaviour if they are alarmed by shipping, aircraft, gunfire, or seismic explosions.

VOCALIZATIONS

Some scientists suggest that Bowheads use very loud (170–180 dB), low-frequency sounds (in the 100–1000 Hz range) over distances of as much as 2 km to determine the thickness and below-water geography of the pack ice. Circumstantial evidence seems to indicate that the whales somehow know whether the ice above them is thin enough for them to break, and a grounded iceberg, which cannot be swum under, is avoided long before it could possibly be detected by vision. It is possible that Bowheads possess a rudimentary form of echolocation. They are the most vocal of the large whales and maintain long-distance contact with each other, particularly during migration, by producing a contact call described as a low-frequency sweep termed a "J whoop" or by producing a broadband pulsed contact call. Bowheads produce highly variable and complex songs during the breeding season that slowly change from year to year. Two additional calls have been recorded, but their purpose is unknown. The "gunshot" sound is a loud short broadband call in the 600–1600 Hz range and the mechanical sounding "cr-unch" call is a pair of broadband signals in the 500–1600 Hz range. The latter call is emitted while socializing.

SIGNS

Bowheads are usually the only large whale associated with ice. The lack of a dorsal fin on a large black back, a "V"-shaped blow when viewed from front or back, a bump just before the blowhole and the white chin are good field signs. Breaching, flipper-slapping, and lobtailing are common aerial behaviours. Thrashing and splashing among groups on the surface is the usual sign of courtship or precourtship socializing.

REFERENCES

Blackwell et al. 2007; Cosens and Blouw 2003; Elsner et al. 2004a, 2004b; Fertl et al. 1999; Finley, K.J., 1990, 2001; George, J.C., et al. 1994, 1999, 2004; Heide-Jørgensen et al. 2006; Koski and Miller 2009; Krutzikowsky and Mate 2000; Ledwell et al. 2007; Mate et al. 2000; O'Hara et al. 2002; Reeves and Heide-Jørgensen 1996; Reeves et al. 2002; Richardson, W.J., et al. 1995; Rugh and Shelden 2002; Rugh et al. 1992; Würsig and Clark 1993; Würsig et al. 1999.

North Atlantic Right Whale

also called Black Right Whale

FRENCH NAME: **Baleine noire**

SCIENTIFIC NAME: *Eubalaena glacialis*

Over 3.5 million years ago, Panama emerged from the sea to split the Atlantic from the Pacific Ocean. Northern right whale populations in the two ocean basins were divided by increasingly shallow, warm seas in that area even before that, as long as 5–6 million years ago, and have evolved independently ever since. DNA evidence suggests that the North Atlantic Right Whale is distinct and a separate species from the North Pacific Right Whale. Physically, they are almost identical and cannot be distinguished. The colour illustration here applies to both.

DESCRIPTION

Northern right whales are large, black baleen whales with a broad back and no dorsal fin. They often have variable white splotches on their bellies. Their head and chin are adorned with a variety of callosities, which are areas of thickened skin, often with some sparse hairs, that become infested with whale lice within the calf's first few months. It takes several months for the callosities to develop into their full adult pattern. The whale lice provide the creamy-yellowish colour. Variations of placement and size of callosities, along with any distinctive white scarring, allows each whale to be identified individually. All the right whales have huge heads that constitute up to one-third of their total length. Flippers are large and broadly paddle shaped. Flukes are wide (about 40% of the total length) and deeply notched. The mouthline is highly arched and the head is large, with a broad chin. Their dental formula is 0/0. Between 200 and 270 baleen plates hang from each side of the upper palate. This baleen is long, up to 2.7 m at the middle of the row where the plates are longest.

SIMILAR SPECIES

The only other broad-backed black whale without a dorsal fin that may occur within the distribution of North Atlantic Right Whales is the Bowhead. Bowheads are larger, found most often near ice, and have no callosities. Usually Bowheads occur in North Atlantic Right Whale range during winter, and have moved farther north by the time the right whales reach the area in mid-summer.

SIZE

Adult females are about 10% larger than adult males.
Total length: males, up to 15.5 m; females, up to 17 m; newborns, 4.5–5.5 m.
Weight: up to 90 tonnes, or 90,000 kg; newborns, about 800–900 kg.

RANGE

Generally North Atlantic Right Whales migrate northwards in summer and southwards in late autumn. But each year some animals travel beyond the regular migration routes and are seen in regions normally outside the present distribution. Evidence suggests that these whales may be travelling to former feeding grounds that once

North Atlantic Right Whale (*Eubalaena glacialis*) and
North Pacific Right Whale (*Eubalaena japonica*)

Distribution of the North Atlantic Right Whale (*Eubalaena glacialis*)

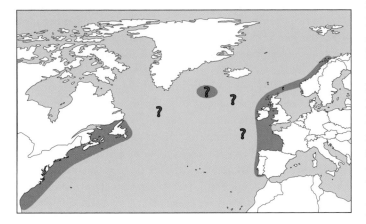

? may be part of range but usage is likely only occasional

supported large populations but were depopulated by whalers. In 1999, a male known from the Gulf of Maine made a return migration of at least 5700 km in no more than 117 days, to the Island of Sorpa off the coast of northern Norway. This area is thought to have been a traditional feeding ground. Before North Atlantic Right Whales were overexploited by whaling, they occurred all around the North Atlantic, from Florida to Labrador and Iceland, and over to Europe, and even down the east coast of Africa. Now there are only rare sightings in the eastern North Atlantic and most are likely wanderers from the west. Western North Atlantic sightings occur from Florida to Nova Scotia, with a few more northerly sightings, and as far east as the Cape Farewell whaling grounds southwest of Iceland. Most calves are born in the warm coastal waters of the southeastern United States, although some may be born during migration as far north as Cap Cod Bay and the Gulf of Maine. The Great South Channel, Massachusetts Bay, the Bay of Fundy, and the Scotian Shelf are other high-use areas. The whales travel 4000–5000 km each year between the wintering grounds off the southeastern United States and Gulf of Mexico and the summer feeding grounds in the Bay of Fundy and off the coast of New England. They leave the wintering/calving grounds in about April and arrive on the feeding grounds in June and July. They depart the summer feeding grounds in October and November and reach the wintering grounds in December and January. Seasonal occurrence of North Atlantic Right Whales in the Bay of Fundy was first reported in 1968, and since that time a whale-watching industry has grown around their presence, as this is one of the best regions to see these endangered animals. Historical whaling records suggest that there was another calving ground off northwestern Africa that was used by populations in the eastern Atlantic before that population was decimated by overhunting.

ABUNDANCE

There were fewer than 400 (likely 300–350) North Atlantic Right Whales left in 2007. If current trends of increased mortality (usually by human means) and reduced calving success continue, this population could be extinct in less than 200 years. The species is listed as "Endangered" by the World Conservation Union (IUCN) in their "Red Data Book." It bears mentioning that the extinction of North Atlantic Right Whales would also mean the end of three species of whale lice found nowhere else but on the callosities of these large whales.

ECOLOGY

Following centuries of exploitation, hunting of North Atlantic Right Whales was stopped in 1935. The population began to slowly rebound, but this slight increase halted in the 1980s and began to reverse in the 1990s. The mortality rate of reproductive-age females was rising and the calving interval was increasing, so calves were becoming more infrequent. Consequently, more whales were dying each year than were born. This, unfortunately, continues despite a bumper calf crop in 2000. Although calves are possibly vulnerable to Killer Whales and large sharks, these are not thought to be a serious mortality factor. Ship strikes and entanglement in fishing gear kill most Right Whales today. Ship strikes account for at least 35% of North Atlantic Right Whale mortality. The whales are particularly vulnerable to strikes for two reasons: first, their range occurs along some of the most heavily industrialized and busy coastlines in the world; and second, Right Whales "log" on the surface when they rest, with just a bit of their back, head, and blowhole awash. Their natural buoyancy allows them to hold this position with no effort, so they can sleep soundly without fear of drowning. It is possible for them to sleep through the approach of even a large vessel and by the time the noise wakes them, they cannot counteract their own buoyancy in time to avoid being hit. North Atlantic Right Whales have been studied and individually identified since 1935. At least one adult female with a calf, photo-identified in 1935, was seen for the last time in 1995 with a huge wound caused by a ship strike, which she probably did not survive. Assuming she was with her first calf and was ten years old in 1935, she would have been at least 70 years old when last seen. Maximum potential longevity is probably longer. There may be animals alive today that survived the heavy harvesting period of the early 1900s. These whales are slow moving, with migration speeds approaching 4.5 km/h and feeding speeds closer to 1.5 km/h.

DIET

Like all whales, Northern Right Whales are carnivorous. They are filter feeders that prey mainly on copepods, and sometimes krill and other zooplankton. Their prey is about the size of a flea, resulting in a predator/prey weight ratio of about 50 billion to 1. Zooplankton commonly moves up the water column during darkness and down during daylight. Right Whales feed both day and night and follow their prey up and down the water column. They swim slowly through the water with their mouths partly open (called skim feeding) and filter the small, free-swimming organisms through their long, fine baleen plates as they go. Although they will skim along the surface in some parts of their range, in Canadian waters, they usually dive to varying depths (10–200 m) to find concentrations of prey. Foraging typically occurs in areas of high productivity, such as upwellings and warm current/cold current contact zones, where

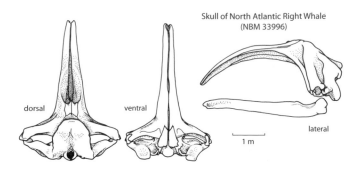

Skull of North Atlantic Right Whale
(NBM 33996)

dorsal

ventral

lateral

1 m

a high density of prey can be found. The bulk of the year's feeding takes place in higher-latitude feeding grounds from spring to late autumn. Right Whales are dependant on areas with high zooplankton density, which makes them vulnerable to climatic changes that affect marine water temperatures, currents, and hence the productivity of zooplankton.

REPRODUCTION

Females may reach sexual maturity as young as four years old and have their first calf at five, but most produce their first calf between the ages of eight and ten years old. The oldest known age of a female producing her first calf is 21 years old. About every three to eight years thereafter, another calf is produced. For unknown reasons, the average length of the calving interval in the North Atlantic Right Whale appears to be increasing (from 3.5 years to > 5 years). This results in fewer calves born each year, and although yearly calf counts are highly variable, on average they are lower than expected for the number of females. Females can be fertile for at least 30 years. Courtship occurs throughout the year, as do copulations, but most conceptions occur in early winter. A single calf is born 12–13 months later in the warm coastal waters of the southern calving grounds. The calf doubles its size within its first year and calves are weaned at between 8 and 17 months of age. Males become mature at around 13 m in length, but their age at that time is unknown. Penis lengths are about 2.3 m for an adult.

BEHAVIOUR

North Atlantic Right Whales are thought to learn experientially and to be very influenced by the location of feeding grounds and migration routes taught by their mother. A good feeding patch discovered one year is likely to be visited again the next year. Right Whales frequently perform aerial manoeuvres. They commonly breach, lobtail, and flipper-slap. The purpose of these actions is unknown, but they are loud. Perhaps there is a long-distance communication component, or they indicate aggression. Throughout the summer, in the Bay of Fundy, small courtship groups of Right Whales assemble, typically made up of a female surrounded by a number of males. The average size of these surface active groups is 4–5 animals, but groups of as many as 35–40 whales are known. More males gather around adult females than around subadult females. Some groups contain more than one female, the second female usually being a subadult at least three years old. These

gatherings may be accompanied by much flipper waving and swirling at the surface. Females in oestrus call to attract males. Usually several respond, jostling with each other and using their callosities as shoving and scraping weapons to maintain a position beside the inverted female, where they are in a favourable location to mate when she rights herself to take a breath. Typically the two alpha males will clasp her with their flippers, with one on each side. She rolls to one side or the other to expose her blowhole to breathe, about once a minute, so the male on that side has a short opportunity to attempt to mate with her. She rolls in either direction randomly, so each male has about an equal chance. The females do not actively select a mate, but instead, create a situation where the males compete with each other, so that only the strongest, most agile, and most persistent will be positioned to mate with her. Sperm competition is likely part of the contest as well, since Right Whale males have the largest testes of any mammal, with combined weights of 950 kg and more. Not only can a Right Whale male produce copious sperm, but he can repeatedly produce this sperm as he holds his position beside the female. Repeat mating by the same male enhances his chance of siring a calf. Not all surface active groups are conceptive, as groups have been reported in all months of the year and conceptions are common only in early winter. Furthermore, many surface active groups include immature individuals and fewer than 50% (where the participants can be determined) contain both a sexually mature male and female.

VOCALIZATIONS

Right Whales produce a variety of sounds, principally in the 500 Hz and lower range, although some up to 1500–2000 Hz have been recorded. The low-frequency vocalizations sound like moans, belches, burps, and groans. We think the sounds are used to maintain contact underwater, possibly to indicate a threat, for example, between males competing for access to a female, and probably by females to attract the attention of males. Solitary whales vocalize around 10 times per hour on average and groups average 60 calls per hour.

SIGNS

Right Whales have a distinctive V-shaped blow when viewed from in front or behind and a broad black back with no dorsal fin. Breaching, flipper slapping, and lobtailing are common, especially within mating groups. Fluke-ups occur before a deep dive. Blows easily reach 2 m in height.

REFERENCES

Caswell et al. 1999; Fujiwara and Caswell 2001; Hamilton, P.K., et al. 1995, 1998; International Whaling Commission 2001; Jacobsen et al. 2004; Kaliszewska et al. 2005; Kenney 2002; Knowlton et al. 1994, 2001; Kraus and Hatch 2001; Kraus and Rolland 2007; Kraus et al. 2001; Martin, A.R., and Walker 1997; Neave and Wright 1968; Nowacek et al. 2001; Parks, S.E., et al. 2005, 2007; Patrician et al. 2009; Reeves 2001; Rosenbaum et al. 2000; Smith, T.D., et al. 2006.

North Pacific Right Whale

FRENCH NAME: **Baleine du Japon**

SCIENTIFIC NAME: *Eubalaena japonica*

There are three species of right whale, two of which are present in Canadian waters. All are physically very similar, and can only be distinguished genetically or by geography. The colour illustration that represents both North Pacific Right Whales and North Atlantic Right Whales accompanies the species account of the North Atlantic Right Whale.

DESCRIPTION

The description and skull illustrations provided in the North Atlantic Right Whale account apply equally to the North Pacific Right Whale.

SIMILAR SPECIES

Few other whales in the North Pacific Ocean can be mistaken for a right whale. The most likely candidates are Bowheads and Grey Whales, as both are broad-backed and lack a dorsal fin. The Bowhead is larger, generally more northern in range, and has no callosities. Bowheads are pushed southwards during winter by the expansion of the northern ice sheet, then follow the receding ice northwards again. While some Bowheads may be seen in winter in the northern parts of the right whale range, the two species do not overlap, as the Bowheads have left before the right whales return in mid-summer. Grey Whales are considerably smaller and are mottled greyish, rather than black.

SIZE

Adult females are about 10% larger than adult males.
Total length: males, up to 16.5 m; females, up to 18–19 m; newborns, 4.5–5.5 m.
Weight: can reach over 100 tonnes, or 100,000 kg; newborns, about 800–900 kg.

RANGE

These large whales were once abundant, occurring from the South China Sea to the Bering Sea and down to the coast of California. The remaining population is divided into two stocks, one in the eastern Pacific, where rare sightings along the western shores of North America, from Alaska to California, continue to be reported. Somewhat more frequent sightings of individuals from the western Pacific stock are recorded from the Sea of Japan to the Bering Sea during summer. Occasional underwater calls continue to be recorded from the Kodiak Island region of the Gulf of Alaska within the summer feeding grounds of the eastern North Pacific stock. The location of the wintering/calving grounds is unknown.

ABUNDANCE

The North Pacific Right Whale is one of the most endangered whale species. There are likely even fewer of them than their sister species, the North Atlantic Right Whale, but no accurate estimates exist.

The eastern stock along the Alaska to California coast is the rarest. Any sighting is noteworthy, due to their scarcity. About two dozen animals were seen, in two separate sightings, during one month of surveying in 2004, and a calf was confirmed in 2002 for the first time in over a century. More recent surveys have found no right whales. The western stock is thought to be somewhat larger, as more sightings are reported. A worldwide ban on hunting this species was instituted in 1931, but illegal Soviet whaling following the Second World War and extending into the 1960s severely diminished an already struggling stock. North Pacific Right Whales are listed as "Endangered" by the IUCN. There is concern that the population size may be too small to permit recovery and that eventual extinction is inevitable.

ECOLOGY

There is little known of the life history of North Pacific Right Whales, although most of the information provided for the North Atlantic Right Whale likely also applies. One significant difference may be in the preference of winter grounds. The North Atlantic Right Whale winters near shore, where its presence has been noted. Such is not the case with North Pacific Right Whales, whose wintering grounds remain a mystery. North Pacific Right Whales are migratory, travelling between warmer southern waters in winter and cooler northern waters in summer.

DIET

Like all the right whales, North Pacific Right Whales are filter feeders that specialize in copepods, and secondarily on other zooplankton. They feed by skimming – swimming slowly through the water with their large mouths open, filtering the seawater, and trapping the copepods in the fine baleen. This activity can be conducted at the surface or at depth, depending on where the prey is concentrated. It is probable that all three of the right whale species, as well as the Bowhead, will be found capable of digesting the high-energy,

Distribution of the North Pacific Right Whale (*Eubalaena japonica*)

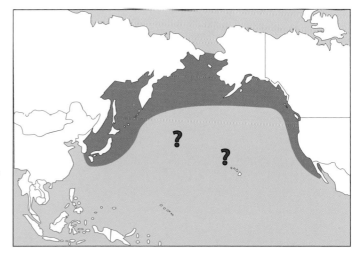

? may be part of range but usage is likely only occasional

wax esters available in copepods. Few other cetaceans apart from Minke Whales have this ability.

REPRODUCTION

Unknown, but probably similar to North Atlantic Right Whales.

BEHAVIOUR

Apart from the occasion report of breaching North Pacific Right Whales, very little behaviour has been observed, due to the scarcity of sightings.

VOCALIZATIONS

Vocalizations are similar to those described in the North Atlantic Right Whale account.

SIGNS

All the right whales produce a distinctive V-shaped blow and fluke-up before a deep dive.

REFERENCES

Brownell et al. 2001; Brueggeman et al. 1988; Goddard and Rugh 1998; Kenney 2002; Mellinger et al. 2004.

FAMILY BALAENOPTERIDAE
rorquals

These swift baleen whales have long, slim bodies, many throat grooves that extend to the chest and even the belly on some species, a small dorsal fin, and lower jaws that bow outward. Their baleen is comparatively short. Most of the members of this family have relatively small flippers. The exception is the Humpback, which has the longest flippers of all the cetaceans.

Minke Whale

also called Little Piked Whale, Lesser Rorqual, Sharpheaded Finner, Little Finner, Pike-headed Whale

FRENCH NAME: **Petit rorqual**

SCIENTIFIC NAME: *Balaenoptera acutorostrata*

Minke (pronounced mink-ee) Whales are the smallest of the North American rorquals. The only baleen whale smaller is the Pygmy Right Whale, not found in the northern hemisphere. Currently cetologists recognize three subspecies of Minke Whale: the North Atlantic (*Balaenoptera acutorostrata acutorostrata*), the North Pacific (*Balaenoptera acutorostrata scammoni*), and the Dwarf Minke Whale (*Balaenoptera acutorostrata* ssp.), a Southern Hemisphere form, which so far has no taxonomic status, and hence no scientific subspecific name. It is possible that, with more research, the southern form may be shown to be a species in its own right. The Dwarf Minke Whale shares Antarctic waters with the Antarctic Minke Whale (*Balaenoptera bonaerensis*), a similar but larger species.

DESCRIPTION

Minke Whales are counter-shaded. They are dark (dark grey or blackish) on their backs, with a white belly. Some, particularly in the North Atlantic, have one or more variable light-coloured chevrons behind the flippers. The dorsal fin is tall, usually highly falcate, and located about two-thirds of the way along the back. A band of white, about midway on the dark flippers, is a diagnostic feature of the species, and is bright enough to be visible even through the water when the whale nears the surface. Flukes are notched in the middle and their width is about one-fourth of the body length. Both North Pacific and North Atlantic subspecies have 230–360 short, creamy-white baleen plates hanging from each upper jaw. (The baleen of the

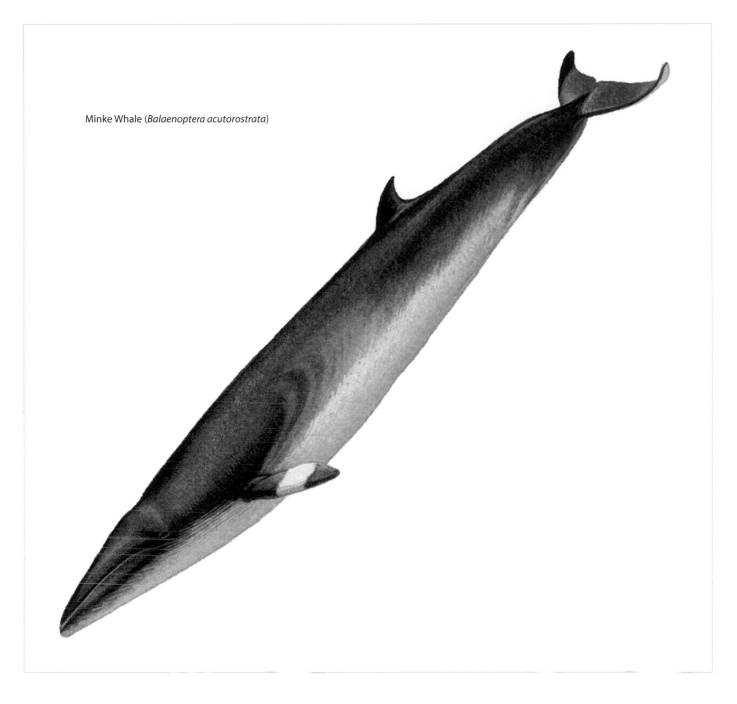

Minke Whale (*Balaenoptera acutorostrata*)

Dwarf Minke Whale is whitish near the snout and becomes greyish or brownish farther back.) The dental formula is 0/0. There are 50–70 grooves that run from the chin to just behind the flippers. The small head (about 20% of the body length) is triangular and narrow, both when viewed from above and in profile. There is a single, sharp ridge running down the middle of the head, starting at the blowhole and terminating at the tip of the snout. Some individuals have a light chevron on the back behind the head.

SIMILAR SPECIES

Small Sei and Fin Whales may be mistaken for Minke Whales at a distance. Both of these whales are generally considerably larger, do not have the white band on their flippers, and produce a pronounced blow. Some of the larger dolphins with large dark dorsal fins, such as the Atlantic White-sided Dolphin, the Striped Dolphin, the White-beaked Dolphin, the Bottlenose Dolphin, and both the Common Dolphins, could also be mistaken for a Minke Whale. The dolphins are considerably smaller, they have a pronounced beak, their dorsal fin in more centrally located on their backs, and, again, they lack the white band on their flippers. Any of the beaked whales, but especially True's, Blainville's, and Sowerby's Beaked Whales could easily be mistaken for Minke Whales. They all have a pronounced beak, are somewhat smaller, and have short, uniformly coloured flippers.

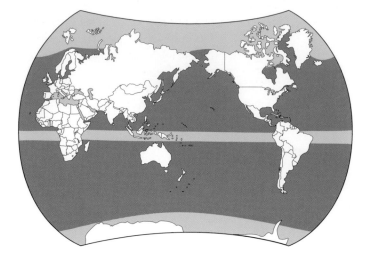

Distribution of the Minke Whale (*Balaenoptera acutorostrata*)

SIZE

The following measurements apply to animals of reproductive age from North American waters. Like all the rorquals, adult female Minke Whales are larger than adult males.

Total length: males, up to 9.8 m (average 7.8–8.2 m); females, up to 10.7 m (average 8.5–8.8 m); newborns, 2.4–2.8 m.

Weight: around 8 tonnes, but may be up to 10 tonnes; newborns, around 320 kg.

RANGE

Minke Whales are widespread in all the oceans except in a broad belt along the tropical waters at the equator. As is typical of baleen whales, they move to higher latitude feeding grounds during the warm months, as far as the ice edge in some areas, and most retreat to lower latitude breeding grounds during the cold season. Due to the difficulties of individually identifying these small, fast-moving whales, their migratory routes are poorly understood, as in fact is much of their behaviour and ecology. The map shows a wide distribution in Hudson Bay, but this reflects a few isolated records. Likely Minke Whales are rare in Hudson Bay, although they are more common in Hudson Strait.

ABUNDANCE

Minke Whales are the most abundant of the rorquals, although there are no worldwide population estimates available. Some stocks, such as those off West Greenland (estimated at 6300 whales) and in the northeast Atlantic (no population estimates available), are considered to be depleted. The population in the western North Pacific off Canada's west coast (estimated in the mid-hundreds) is thought to be naturally low. The population off the eastern coast of Canada is estimated at around 15,000 animals and is possibly the same stock as those whales off West Greenland. Minke Whales were not subjected to the commercial hunt until the 1960s, when the larger more lucrative species were becoming scarce. Norway (hunting in the northeastern Atlantic Ocean), Iceland (western

North Atlantic), and Japan (western North Pacific and Antarctic) continue to operate fleets that hunt Minke Whales, and a fairly large aboriginal hunt (139–170 animals per year from 1999 to 2002) continues in West Greenland. International trade in Minke Whale products is banned. This species was considered "Not at Risk" by COSEWIC in 2006.

ECOLOGY

Most Minke Whales remain on their Canadian summer feeding grounds from late March or early April to October or November, with peaks of abundance from July through September. It appears that a few animals stay north through each winter, as sightings exist for all months. Apparently females migrate to more northern regions, and arrive on the feeding grounds earlier in the season than do males. Although usually seen in coastal and continental shelf waters, migrating animals will travel over deep water. In those extraordinary studies where individual whales have been identified (using dorsal fin shape, coloration, and scarring) and studied for several years, it seems that the whales return to an area for several years running (at least nine years on one occasion), and most tend to remain in that area for the season, as long as prey continues to be available. While feeding at depth, the whales dive for around 4 minutes, and then spend about 20 seconds near the surface taking four to seven very rapid breaths. Travelling Minke Whales have a similar dive pattern. Whales feeding near the surface dive for a maximum of around 90 seconds and their surfacing periods are much more variable. In many regions Minke Whales are favoured prey of Killer Whales. Their only hope of escape is to swim quickly in a straight line for a long distance to outrun the Killer Whales and not to allow themselves to be surrounded. Minke Whales are known to beach themselves rather than be captured by Killer Whales. If the Minke cannot outdistance the Killer Whales, it will be repeatedly rammed and have its head forced underwater, until it is eventually battered to death or drowned. Minke Whales, like all the rorquals, are fast swimmers. They are capable of attaining speeds of 30–45 km/h and can sustain speeds of 15–25 km/h for over an hour. It is difficult to age baleen whales without a tooth to section in order to count the growth rings. Annual rings on the ear plugs are used instead. Their potential lifespan is estimated to be around 50 years.

DIET

Minke Whales tend to be flexible in their prey preferences and feeding style. They hunt small shoaling fishes such as herring, Capelin, sand lance, cod, mackerel, juvenile salmon, Pollock, Sprat, Haddock, and anchovy and krill, and even copepods and squid can be a major part of their diet, especially when fish are scarce. They focus on whatever fish species is locally abundant and are apparently more competent than the larger baleen whales at catching lower density prey. They feed by filtering prey through short plates of baleen that hang from the upper jaws and are considered to be primarily lunge feeders. When prey is concentrated near the top of the water column, they are known to lunge to the surface, filling their mouths and expanded throats with seawater and prey. Grooves running from their chin to their chest allow the throat to expand to several times its usual size. Their tongue then presses the water out and scrapes the food off the inside of the baleen, pushing it to the back of the throat, where it is swallowed. If prey is deeper, this same technique is accomplished below the surface. Foraging is usually done independently, even when several whales are exploiting the same food resource in the same area. Reports occasionally suggest temporary, but possibly cooperative, feeding when prey is dense, as two or up to six whales lunge to the surface at the same time, within a few metres of each other, sometimes for one lunge but other times for a more extended period and multiple lunges. Unlike the larger baleen whales, Minke Whales likely forage during their time in the lower latitudes (during the winter).With their small body size, it is unlikely they could store sufficient blubber to last through the whole winter without feeding. Most mammals are capable of only limited digestion of wax esters (a large component of the available energy in krill and copepods). Preliminary research suggests that Minke Whales (and Right Whales) can break down these wax esters and thereby extract and absorb most of the energy available from this process.

REPRODUCTION

Few calves have been seen in northern waters. It is thought that most females with calves remain in somewhat warmer waters until their calf is weaned at about five to six months old, before migrating to the high-latitude summer feeding grounds in colder waters. Calves have been sighted in Caribbean waters, but no near-shore calving grounds have been found. In their first season, the weanlings generally do not migrate as far north as do the adults. Female Minke Whales reach sexual maturity at around seven years old and males at around six years old. Most mature females breed each year and are capable of lactation and gestation concurrently. Mating occurs in late winter. After a pregnancy lasting around ten months, the single calf is born the following winter.

BEHAVIOUR

Unlike the other baleen whales, there are indications that Minke Whales may be somewhat territorial. Studies show that over the course of several years, the same animals return to the same small summer territory. No agonistic behaviour has ever been observed between Minke Whales. Perhaps the territorial boundaries are maintained using vocal communication. They tend to segregate more than do other rorquals, males being generally more pelagic, and females more coastal. Minke Whales adopt different feeding strategies to suit the local conditions. Their main methods are to look for, and then hunt below, the noisy sea bird aggregations that gather to feed on surface shoaling fishes; or they may themselves herd fishes into tight shoals at the surface and then lunge through the concentrated fish ball to capture a mouthful. Minke Whales sometimes hunt shoaling fish such as herring below the surface. Surface feeding activities can be dynamic and involve breaching and head slapping, but are more often relatively quiet.

VOCALIZATIONS

The predominant Minke Whale call in northern waters is a downsweep, lasting about 0.4 seconds, that starts at 100–200 Hz and terminates below 90 Hz. This call is emitted by feeding animals and is thought to allow spacing of individuals within the foraging area. Broadband pulses also called thump trains (200–400 Hz), high-frequency clicks (3–20 kHz), ratchet-like pulses (centred at 850 Hz), and grunt-like pulses (80–140 Hz) have also been reported from the wintering grounds. Apparently the downsweep is not produced in warmer waters. A "boing" sound, first recorded in 1964 from a US Navy submarine recording around Hawaii has recently been attributed to Minke Whales. The call consists of a low-frequency short pulse followed by a long, modulated call. The sound is estimated to be 150 decibels at source, which is not as loud as a Blue Whale (188 dB), Fin Whale (185 dB), or the loudest, a Sperm Whale (223 dB), but is still an amazing output from such a small whale. The purpose of this call is unknown, but may be related to breeding.

SIGNS

Minke Whales rarely have a visible blow, although if the viewer is nearby it may be heard. They will occasionally breach, usually onto their sides. Although they do arch their bodies considerably, they do not fluke up before sounding. The blowhole and dorsal fin appear simultaneously as they surface, but little time is spent at the surface and breaths are rapid. These factors, plus their relatively small size, can effectively hide them in a choppy sea. The bright-white band on the flippers may be perceptible through the water, especially when viewed from above, as from a large vessel. They are noted for being curious about ships and can suddenly appear nearby as they investigate.

REFERENCES

Abraham and Lim 1990; Dorsey et al. 1990; Edds-Walton 2000; Ford et al. 2005; Kuker et al. 2005; Lynas and Sylvestre 1988; Mansour et al. 2002; Mellinger et al. 2000; Mitchell, E.D., 1991; Murphy 1995; Nordøy 1995; Pattie and Webber 1992; Perrin and Brownell 2002; Rankin and Barlow 2005; Robinson and Tetley 2007; Stewart, B.S., and Leatherwood 1985; Waring et al. 2004.

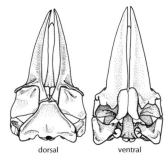

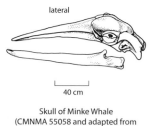

dorsal ventral lateral

40 cm

Skull of Minke Whale
(CMNMA 55058 and adapted from
Figure 4 in Stewart and Leatherwood 1985)

Sei Whale

FRENCH NAME: **Rorqual boréal**

SCIENTIFIC NAME: *Balaenoptera borealis*

Sei (*pronounced either say or sigh*) and Fin Whales are the swiftest of the great whales. They are capable of swimming at speeds of close to 50 km/h for short distances.

DESCRIPTION

Sei Whales are large rorquals. They are dark grey on their backs and somewhat lighter on the underside, often with a white patch under the chin that can extend to the umbilicus. The underside of the tail stock can be considerably lighter than the back. The body, especially the belly and tail stock, may have a scattering of light-coloured oval scars, presumably caused by bites from cookie-cutter sharks. (These small tropical and sub-tropical sharks specialize in removing a mouthful of living flesh from a large fish or whale.) There are 40–65 throat grooves that run from the chin to behind the flippers. Sei Whales have a single, sharp ridge running along the top of the head from the blowhole to the tip of the snout. There are between 300 and 410 dark baleen plates on each side of the upper jaw. The longest baleen plate in a row is less than 80 cm in length. The dental formula is 0/0. The dorsal fin is large, tall, and usually highly falcate, and the flukes are dark with a medial notch.

SIMILAR SPECIES

Sei Whales can be difficult to distinguish from Fin Whales. Although they are smaller than Fins, this can often be impossible to assess at sea. The most definitive difference, useful in the field, is the colour of the lower lips. Sei Whales may or may not have a white chin, but both lips are dark. Fin Whales have a dark lip on the left side and a white lip on the right side. Other more subtle differences include: Sei Whales tend to rise slowly at a shallow angle for a breath, so that their blowhole and dorsal fin are visible almost simultaneously; Fin Whales generally rise at a steeper angle, so their blowhole appears before their dorsal fin breaks the surface; Fin Whales arch their back quite significantly before sounding, whereas Sei Whales do not dive as deep and tend to just sink below the surface without "rounding out." These characteristics are only useful some of the time, as both Fin and Blue Whales will sometimes adopt the more leisurely approach of the Sei Whale and slowly sink below the surface without rounding out. Blue Whales, which are much larger than Sei Whales, have a smaller dorsal fin and a larger blow and also usually arch their backs before sounding. Minke Whales are much smaller, rarely have a noticeable blow, and have a distinct, bright, white band on each flipper. When viewing a carcass, take note of the throat grooves. Only Sei and Minke Whales have throat grooves that end before the umbilicus. All the other rorquals have longer throat grooves that terminate at or beyond the umbilicus.

SIZE

Like all the baleen whales, females are somewhat larger than males.

Total length: up to 18.6 m in the Pacific Ocean and 17.4 m in the Atlantic Ocean; newborns, around 4.4 m.
Weight: up to 30 tonnes; newborns, about 650 kg.

RANGE

This is a cosmopolitan species that tends to live in temperate waters, but is sometimes seen in sub-tropical and even tropical waters. This whale is renowned around the world for the unpredictability of its occurrence. It will avoid a region for years, then return in numbers for a time, before disappearing again. It has also been reported outside its regular range in large numbers for brief periods that coincide with a local profusion of plankton. Although these whales undertake a yearly migration from low latitudes in winter to higher latitudes in summer, the extent and route of these migrations is poorly understood.

ABUNDANCE

The tender flesh of this whale is very tasty by human standards, somewhere between pork and veal, it is said. These sleek, fast whales carry little blubber and were hunted mainly for meat and baleen. Like all the other whales in this fast-swimming genus, Sei Whales were more or less spared from the commercial hunt until the advent of steam and high-powered harpoons. By the 1880s they were being harvested in small numbers. As the larger whales, the Blue and Fin Whales, became scarcer, Sei Whales assumed a greater importance to whalers. Sei Whales were hunted at least until 1986, and their numbers in most areas, especially in the Southern Hemisphere and the eastern North Pacific, are considerably and probably even drastically reduced from pre-whaling times. Sei Whales are designated an "Endangered" species by IUCN. COSEWIC considers the Pacific population in Canadian waters to be "Endangered" and the Atlantic population to be "Data Deficient" because, although the whales do occur in Canadian waters in the North Atlantic, there is little information about how that population is faring since whaling ceased. It is thought that there are around 2000 animals in the western North Atlantic that feed in Canadian waters during the summer. In recent years no Sei Whales have

Distribution of the Sei Whale (*Balaenoptera borealis*)

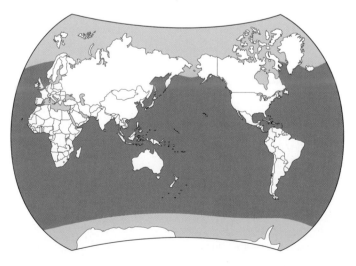

Sei Whale (*Balaenoptera borealis*)

been sighted in western Canadian waters, and only one sighting has been recorded in the whole eastern side of the North Pacific.

ECOLOGY

Most Sei Whales prefer water temperature between 8°C and 18°C, which may partly explain their unpredictable occurrence. If currents move, the whales may also. The intermediate baleen and flexible feeding style (see "Diet" section following) may allow Sei Whales to adapt to changing prey densities and oceanic conditions better than other whale species. Pregnant females tend to lead the migration to the summer feeding grounds, followed by males and non-breeding females. Juveniles and females with calves migrate last. Juveniles tend to leave the feeding grounds first to begin the southward migration and also do not travel as far north. From wintering grounds thought to be offshore along the south-central US east coast, the western North Atlantic population migrates up the edge of the continental shelf, passing the Maritimes in July and August and returning in September until November. In the eastern North Pacific, a similar migration peaked off the coast of British Columbia in July and then moved progressively farther offshore as the summer advanced. Although there are no records of Killer Whale predation upon Sei Whales, still Killer Whales are known to attack even larger baleen whales, so it is expected that they would attack a Sei Whale given the opportunity, especially a calf or debilitated adult. Sei Whales have not yet been reported entangled or drowned in fishing gear, probably due to their pelagic nature; however, offshore driftnet fisheries may present a future problem. Only one ship strike has been noted of this species; however, due to their deep-sea habitat, many strikes may go unreported. Sei Whales can live as long as 60 years.

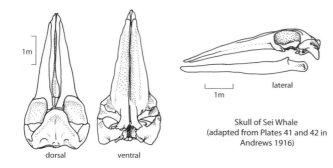

Skull of Sei Whale
(adapted from Plates 41 and 42 in
Andrews 1916)

1m

1m   lateral

dorsal   ventral

DIET

Sei Whales have the most diverse diet of the baleen whales. They eat small shoaling fishes of many species, as well as squid and krill; but they prefer copepods. Their baleen is the finest of the genus, as befits a whale that must filter tiny copepods. Sei Whales utilize two different feeding styles, largely determined by prey abundance. If copepods are plentiful, they will swim slowly through the dense cloud with their mouths partly open, sieving seawater and prey through their baleen plates, in a similar manner to Bowhead and Right Whales. This technique is called skimming. If fish, squid, or krill are available in numbers, the whales will lunge through the schools, gulping an enormous mouthful of seawater and prey. The throat grooves stretch open to accommodate an expanded capacity. Then, with mouth partly open, they use their massive tongue to press the water out through the baleen, and then to scrape the prey off the baleen to be swallowed. This feeding style is called lunging. Most feeding for the year takes place during migration and on the high-latitude summer feeding grounds.

REPRODUCTION

Breeding takes place in winter on the low-latitude wintering grounds. Pregnancy lasts around 10–12 months, and a single calf is born in early winter. Most females bear a calf every two to three years. In their first spring, the calves follow their mothers to higher-latitude feeding grounds, where they are weaned at around 6–7 months old. Growth of calves is rapid, as they double their length to about 9 m in those six or seven months. Sexual maturity apparently occurs between the ages of 5 and 15 years, but mostly around 8–10 years old. In the Northern Hemisphere, males are 12.8–13 m long and females 13.3–13.5 m long at that time.

BEHAVIOUR

Sei Whales are often found in the company of other Sei Whales, in small groups of up to six individuals; however, the make-up of these groupings, apart from cow/calf pairs, is variable. Associations appear to be transient, lasting less than 24 hours. Larger aggregations may occur on the feeding grounds, but whether the concentration of whales is due to their need for social contact, or merely a reaction to high prey density is unknown. Two typical respiration patterns have been observed. A single breath followed by a short dive of 45–90 seconds is the pattern used when making shallow, feeding dives. The whale is probably skimming near the surface (see "Diet" section).

After taking the single breath, they tend to roll 45–90° on the longitudinal axis, which probably keeps them within the copepod cloud. Deeper or larger prey requires deeper dives of up to 15 minutes, after which the whales remain near the surface and take several breaths before diving again. It is difficult, but possible, with much effort and luck, to photo-identify Sei Whales. Some individuals can be recognized based on dorsal fin shape, natural pigment patterns, and scars. Most young, and some adults, lack any distinctive markings, so unfortunately the whole population cannot be individually distinguished. Sei Whales are capable of tremendous short-term movement. A radio-tagged animal was tracked over 4000 km in 10 days at an average speed of 17 km/h. This degree of mobility suggests that Sei Whales are very well adapted to make long-range movements in response to food availability.

VOCALIZATIONS

The nature of Sei Whale vocalizations is unknown.

SIGNS

The blow of this whale reaches about 3 m high, in a thick column of condensation. Sei Whales tend to be unobtrusive at the surface. They surface quietly and smoothly, usually with blowhole and dorsal fin appearing at the same time, and they simply sink to re-submerge. This whale rarely breaches.

REFERENCES

Andrews 1916; Brown, S.G., 1977; Gambell 1985; Gregr 2003; Horwood 2002; Potter and Birchler 1999; Schilling et al. 1992.

Blue Whale

also called Great Blue Whale

FRENCH NAME: **Rorqual bleu**

SCIENTIFIC NAME: *Balaenoptera musculus*

This whale is the largest animal ever to have existed on our planet. Females, which are more massive than males, can weight up to 190–200 tonnes and exceed 33 m in length. The biggest Blue Whales live in the Southern Hemisphere near the Antarctic.

DESCRIPTION

Blue Whales are truly enormous. Adults are longer than two city buses placed end to end, and their heart is larger than a small family sedan. Their body is mottled in a blue-grey pattern that is highly variable and individually unique. Some animals are light enough to look aquamarine when near the surface. The underside is generally lighter than the back and can take on a yellowish or brownish cast from a thin coating of diatoms acquired in polar waters. Flippers have a light underside and are often whitish around their edges and at the tip.

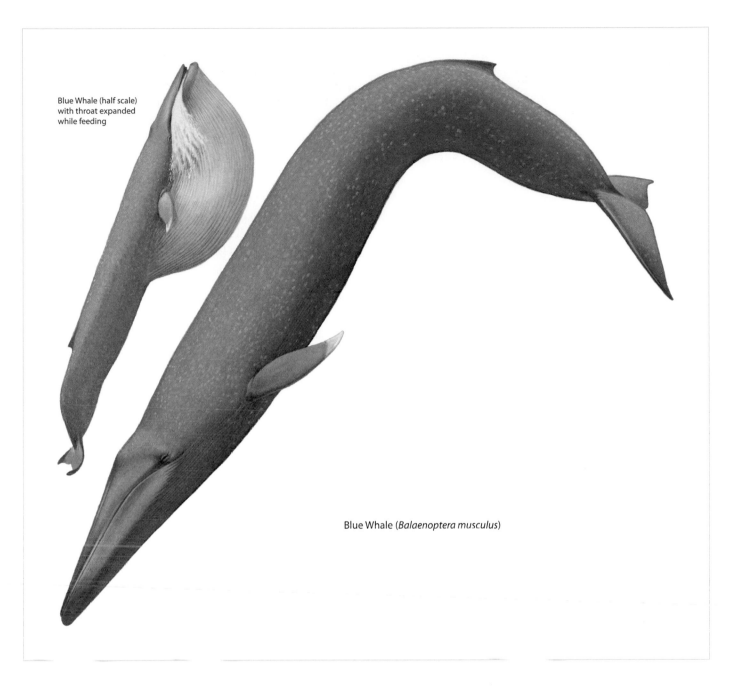

Blue Whale (half scale) with throat expanded while feeding

Blue Whale (*Balaenoptera musculus*)

Flukes are broad and usually uniformly grey above and below (occasionally the underside of the tail will have a patch of white, which can be used for individual recognition) with a medial notch and either a straight or slightly concave trailing edge. The dorsal fin is relatively small and quite variable in shape and size. It can vary from a bumpy nubbin to a falcate fin about 40 cm high. Regardless of its shape, it is placed well back on the body. Two rows of blackish baleen hang from the upper jaws. Each row is made up of 270–395 individual baleen plates, the longest of which (in the middle of the row) is no longer than 1 m. The dental formula is 0/0. A single ridge extends from the large fleshy splash guards in front of the blowholes to the tip of the snout. From above, the head is "U" shaped and flat. There are 55–88 throat grooves running from the chin to beyond the umbilicus that enable the throat to distend while feeding.

SIMILAR SPECIES

Blue Whales typically appear considerably paler than the other baleen whales, with the exception of the much smaller Grey Whale. Blue Whales are distinguished mainly by their size and pale coloration. Everything about them, except their dorsal fin, is larger. From the blow to the splash guard, to the head and shoulder, to the awesome length of time it takes for the animal to complete a roll at the surface, Blue Whales are just simply bigger. Fin Whales are commonly seen in the same areas and also sometimes in association with Blue Whales. They are almost as large with a similar, but shorter blow. Their dorsal fin is larger, and their coloration is darker with none of the mottling so dominant on the Blue Whales. Fins also have a white lower lip on the right side and a dark lower lip on the left side and they rarely fluke up.

Distribution of the Blue Whale (*Balaenoptera musculus*)

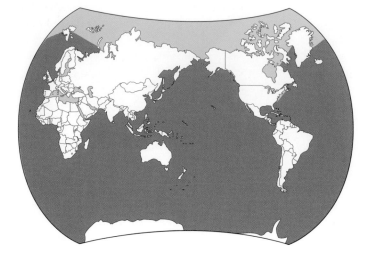

SIZE

As with all the baleen whales, females are somewhat larger than males.

The following measurements apply to adult Northern Hemisphere animals.

Total length: males, up to 27 m; females, up to 30 m; newborns, 6–8 m.

Weight: 80–160 tonnes. As it is impossible to weigh such a large creature, weight is necessarily an estimate; newborns, 2000–3000 kg (2–3 tonnes).

RANGE

Despite the numbers killed during the whaling era, Blue Whales remain a cosmopolitan species. They are separated into three subspecies. The largest in body size is found in the Antarctic, the next largest inhabits the Northern Hemisphere (both Pacific and Atlantic Oceans), and the smallest, also called the Pygmy Blue Whale, lives in the southern Indian and southwestern Pacific Oceans. Generally Blue Whales are believed to summer at higher latitudes and winter in warmer waters at lower latitudes, although some animals do remain in ice-free polar and cold temperate waters during the winter. They enter the Gulf of St Lawrence through the Cabot Strait in spring as the ice breaks up and may be seen along the North Shore from the estuary to the Strait of Belle Isle until late December, with a few staying until late January. Although individual whales appear to be largely faithful to their traditional summer feeding grounds, some (perhaps mainly the males), may spend a summer at another feeding ground. Southern Blue Whales are known to migrate distances of up to 8700 km between summer feeding grounds in the Antarctic and temperate to tropical wintering grounds near the equator.

ABUNDANCE

In the era of sailing ships, Blue Whales were generally too fast to be caught. Once steam ships, exploding harpoons, and deep-sea factory ships brought this fast-swimming giant within reach of whalers, the hunt was unrelenting. Unbridled killing of close to 400,000

Blue Whales by commercial whalers from 1880 to 1966 brought this species to the brink of extinction. It is considered an "Endangered" species by the World Conservation Union (IUCN) and, by international agreement, Blue Whales are no longer hunted. Pre-hunting worldwide estimates set numbers, very roughly, at around 350,000. Populations in the Antarctic remain highly depleted and sightings are rare. Abundance estimates for that region range from 710–1255 animals, where formerly there were close to 200,000. A few tentative estimates exist for some other populations: 600–1500 animals in the whole North Atlantic; 1500–2500 in the eastern North Pacific. There appears to be a trend towards more sightings of Blue Whales off California. This could be the result of an increase in numbers, but is more likely due to a gradual return of animals to a productive feeding area following the cessation of whaling. There are some indications that global numbers of Blue Whales may be slowly beginning to increase.

ECOLOGY

When feeding at depth, Blue Whale dives can be up to 36 minutes long, but most are less than 30 minutes long. When feeding near the surface, most of their dives are less than 10 minutes in duration. Following a deep dive, the whales generally remain near the surface and blow 8 to 15 times before undertaking another deep dive. Approximately 94% or more of their time is spent submerged. Few feeding dives are below 200 m in depth. Blue Whales swim at around 3–6 km/h when feeding, but can reach speeds of up to 35–40 km/h, if pursued. Unlike many baleen whales, Blues will feed while wintering in the low latitudes if they can find a productive feeding area, and some will remain in the higher latitudes feeding throughout the winter as well. Perhaps their massive size or reproductive condition makes demands that prevent them from fasting for the entire winter. Blue Whales often arrive in early spring to the summer feeding grounds. Wind or currents can move the early spring ice pack as it is breaking up and, with no breathing holes, trap the whales. These entrapments are relatively common in the Gulf of St Lawrence and off the southwestern coast of Newfoundland. At least 24 Blue Whales have died in this region by ice entrapment since 1976. Despite their size, even adult Blue Whales are vulnerable to Killer Whale attacks, and in areas with many Killer Whales, such as along the California coast, about 25% of the Blue Whales have scars they received during such encounters. Ship strikes can leave as many as 10% of a population scarred in a heavy traffic region and even cause mortalities. Estimated lifespan is thought to be about 80 years or more in unhunted populations. The longest ongoing study of a single known individual has lasted 34 years.

DIET

Blue Whales feed primarily on krill, a generic name used to describe many species of small swarming shrimp-like crustaceans. High concentrations of krill are most common in areas with cold current upwellings, and hence, that is also where Blue Whales are most common. Blue Whales are filter feeders that primarily use the lunge and gulp strategy. They lunge into a concentration of prey with their mouths open and gulp up to 40 tonnes of seawater and prey. Throat

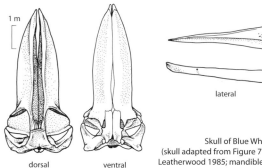

1 m

dorsal   ventral

lateral

Skull of Blue Whale
(skull adapted from Figure 7 in Yochem and
Leatherwood 1985; mandible CMNMA 55059)

grooves expand to allow such an enhanced throat capacity. Their massive tongue then presses the water out the slightly open mouth through the baleen, which filters out the small crustaceans. They will eat around 3.6 tonnes of krill each day (approximately 1.5 million calories). Blue Whales have a relatively small gullet (about 10 cm in diameter) for such a huge animal in keeping with the small prey they consume. Daytime feeding usually involves diving fairly deeply (at least 100 m) to find the krill as they retreat to depths to avoid daylight. Night-time feeding dives are usually much shallower, as krill rises up the water column during darkness. In polar regions, where extended daylight is the norm during summer, krill is abundant, but not usually accessible near the surface.

REPRODUCTION

Blue Whales in the Northern Hemisphere reach sexual maturity when they are 21–23 m in length for females and 20–21 m in length for males. Based on data from the whaling era, females are around five to six years old and males around eight to ten years old at that size. Mating takes place in late fall and throughout the winter. A single calf is born 10–12 months later. Calves are weaned and leave their mothers at the age of around six to eight months old when they are about 16 m long, which is more than double their birth length. They are eight to ten times their birth weight at that time. Drinking around 380 litres of their mother's rich milk each day, calves gain about 3.5 kg per hour. They will have followed their mother on their first migration and learned the location of some northern feeding grounds from her, before striking off on their own. Females produce a calf every two to three years, and it is not known if females undergo reproductive senescence in old age. No calving grounds have been discovered for Blue Whales, so it is assumed they give birth in the open ocean. The Costa Rica Dome, a large (about 2500 km²) region of upwelling currents in the Pacific Ocean, around 800 km west of Costa Rica, may be a calving ground for some eastern Pacific Blue Whales, as very young calves have been seen there. Another feeding and nursing area off the southern coast of Chile may also be an important calving area. Cows with older calves are regularly seen along the west coast of Mexico and in the Gulf of California.

BEHAVIOUR

Blue Whales are most often seen alone or in pairs; however, concentrations of 50 or more can be dispersed over a small area of high productivity, as the whales take advantage of a concentration of swarming krill. Male/female pairs are regularly seen in the Gulf of St Lawrence in late summer and autumn and some pairs will associate for as long as three weeks. If these pairs are approached by another Blue Whale, or even a Fin Whale, some very energetic surface displays can ensue. All three animals will race at the surface with shoulders high out of the water, almost breaching, and creating huge explosive bow wave splashes. This can go on for up to 15 minutes. Blue Whales will associate with Fin Whales and hybrids are known. Putative courtship behaviour has been reported for Blue Whales wintering in the Costa Rica Dome. The whales there often spend more time at the surface than usual and pairs and threesomes are commonly seen. A curious (possibly courtship) activity involving side-swimming and waving of flippers and part of a fluke out of the water have been noted. This is the only known breeding area for Blue Whales.

VOCALIZATIONS

Blue Whales have the second loudest known call in the animal kingdom. The song of a bull has been recorded at 188 dB, which is louder than a jumbo jet at full throttle from a few metres away. Recent recordings of male Sperm Whale clicks have surpassed this record by close to 45 dB. Most Blue Whale vocalizations appear to be low-frequency calls of less than 80Hz, the majority between 15 and 20 Hz. Such low, loud calls can carry for hundreds, if not thousands, of kilometres – an ideal mode of communication for a nomadic and widely distributed, but sparse, population. Some calls are prefaced by a higher-frequency component near 400Hz and other calls are in the 350 Hz range. While Blue Whales vocalize throughout the year, calling peaks in mid-summer and through the early winter, corresponding to the pre-breeding and breeding period. The gender of calling whales is suspected to be male. As with Humpback Whales, Blue Whale calls differ between geographic areas. All the North Atlantic animals share a call type, and eastern North Pacific and western North Pacific whales have recognizably different calls, both of which are distinct from the North Atlantic call.

SIGNS

The relatively small dorsal fin usually appears well after the blow, especially if the whale is surfacing after a deep dive. Flukes are often, but not always, raised before the whale sounds. The columnar blow is 6 to 10 or more metres high and is straight and narrow. Blue Whales raise their "shoulders" out of the water more than do the other rorquals and the fleshy splash guard in front of the dual blowholes is much more pronounced. Blue Whales seen underwater, as occasionally occurs before or after they surface, can appear startlingly turquoise blue. Depending on the viewing angle, they can look silvery grey as light reflects off their wet skin.

REFERENCES

Ahlborn et al. 2009; Baskin 1991; Bérubé and Aguilar 1998; Branch et al. 2007; Buchan et al. 2010; Lagerquist et al. 2000; McDonald et al. 2001; Rankin et al. 2006; Reeves et al. 2002; Reilly and Thayer 1990; Sears 2002; Sears and Larsen 2002; Sears et al. 1990; Stafford, K.M., 2003; Stafford, K.M., et al. 2001; Yochem and Leatherwood 1985.

Fin Whale

also called Finback Whale, Finner, Common Rorqual, Herring Whale, Razorback Whale

FRENCH NAME: **Rorqual commun**

SCIENTIFIC NAME: *Balaenoptera physalus*

Fin Whales and Blue Whales are very closely related. The two species diverged from a common ancestor about five million years ago, but are still known to hybridize occasionally. Fin and Sei Whales are the fastest of the great whales, capable of speeds close to 50 km/h over short distances.

DESCRIPTION

Fin Whales are large whales, second only to the Blue Whale in total length, although Right and Bowhead Whales are heavier. They are very sleek and streamlined in appearance, dark grey above and creamy white below. Most have a flash of light colour (called a "blaze") on the right side of the head that slants up from the upper lip and eye to behind the blowhole. Differences in colour, shape, and position of the blaze, along with size and shape of a light pattern of chevrons behind the head, and any distinctive scars and notches, help researchers identify individual whales. The undersides of the flukes are light with a dark grey border and the flippers are light on the underside as well. The dark dorsal fin is usually falcate, but can be variable in shape, which can also help to identify individual animals. The most distinctive feature of a Fin Whale is the asymmetrical lower lip colour. On the left side of the head the lower lip is dark, while on the right side it is creamy white. This pigmentation pattern is reflected in the colour of the baleen and the inside of the mouth. The baleen is all dark on the left side, but on the right side, the front third is white, while the remainder of the row is dark. The fringe fibres on the inside of all the baleen plates are yellowish-white. As they have no teeth, the whales' dental formula is 0/0. Each row of baleen consists of 260–480 individual baleen plates. There are 50–100 throat grooves that extend from the chin to beyond the umbilicus and allow the throat to expand to several times its normal capacity. The head has a sharp central ridge running from the blowhole to the tip of the snout. The flukes are broad, with a distinct notch in the middle. The tail stock is compressed laterally and a distinctive sharp ridge runs along the top of it from the dorsal fin to the flukes.

SIMILAR SPECIES

Fin Whales are very similar to Sei Whales, and a close view is usually necessary to distinguish the two. See the "Similar Species" section of the Sei Whale account for differences. Blue Whales are also similar to Fin Whales, but are a mottled blue-grey colour, have uniformly dark lower lips and a smaller dorsal fin, and often fluke up before sounding.

SIZE

Females are 5%–10% larger than males and Southern Hemisphere animals are larger than those in the north.

These measurements apply to Northern Hemisphere animals.
Total length: males, 19–21 m, up to 24 m; females, 20.0–22.5 m, up to 25 m; newborns, around 6 m.
Weight: males, average 55 tonnes, may be up to 85 tonnes; females, average 65 tonnes, may be up to 90 tonnes; newborns, about 1800–2000 kg.

RANGE

Fin Whales occur in all the major oceans and many of the seas. They are most common in temperate and sub-polar waters. Like most of the baleen whales, they generally undertake a seasonal migration from low-latitude warm waters, where they mate and calve over the winter, to higher-latitude cold waters, where they spend the summer feeding. Only about 75% of our Fin Whales travel to low-latitude wintering grounds each year. The remaining 25% linger in the northern feeding grounds, sometimes for the whole winter. Although they are most often found in coastal regions, they will also venture into deeper oceanic waters. Numbers of Fin Whales in the Gulf of St Lawrence peak from early June to October, and a similar peak has been noticed in the Atlantic. Off British Columbia, Fin Whales are common all year round, but it is uncertain if they are the same whales. The winter animals may have moved south from higher latitudes and replaced the summer residents that also moved southwards.

ABUNDANCE

Before the advent of steam-powered vessels and explosive harpoons, Fin Whales and the other fast rorquals were beyond the reach of whalers. This changed in the 1880s and Fin Whales now have the dubious distinction of having been one of the most heavily exploited whale species. Likely over a million were killed, mostly from the huge Antarctic population, which is still highly depleted, but possibly slowly starting to show signs of recovery. A moratorium on commercial whaling halted most of the hunting in 1985, and all commercial hunting stopped in 1989. Iceland has recently

Distribution of the Fin Whale (*Balaenoptera physalus*)

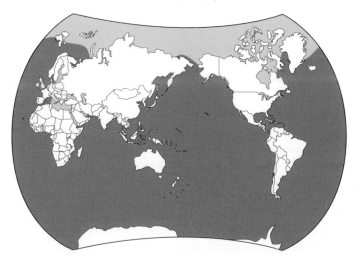

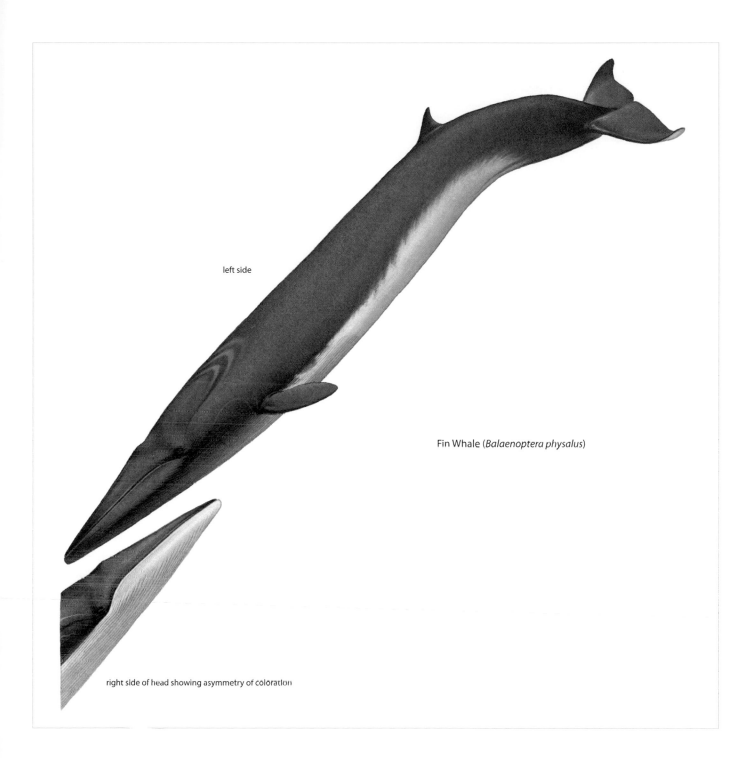

left side

Fin Whale (*Balaenoptera physalus*)

right side of head showing asymmetry of coloration

(2005) reinstituted a, so far small, Fin Whale commercial hunt (licences to kill nine issued in the 2005–6 season). Japan continues to take some Fin Whales for what they call research purposes. A small aboriginal hunt continues in West Greenland, where about 10–15 animals are taken per year. The status of Fin Whales in the North Atlantic and North Pacific is more favourable, as the populations seem to have at least partly recovered and may now even be considered abundant in some areas. The World Conservation Union (IUCN) considers Fin Whales to be "Endangered" internationally, although this status has recently been challenged for some of the northern populations.

ECOLOGY

Summer is spent intensively feeding and likely moving around considerably, to locate and exploit prey concentrations. The majority of the food eaten in a year is consumed in the four to five months spent on the northern feeding grounds. An adult whale is estimated to consume up to 1 tonne of food each day. Most feeding occurs in the early evening and at night, but some continues through the day. Most feeding dives are shallow (within 150 m of the surface), but Fin Whales have been recorded diving to at least 470 m. The longest recorded dive lasted almost 17 minutes, but usually dives are of three to ten minutes. Normal cruising speed is 10–15 km/h, but may reach

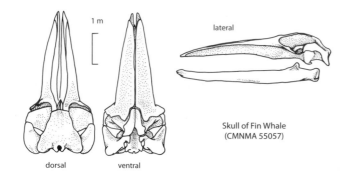

Skull of Fin Whale
(CMNMA 55057)

dorsal   ventral

up to 40–50 km/h for short bursts. During the winter, the whales calve and mate in warm tropical and subtropical waters. Since no defined calving grounds have been found, it is assumed that the whales are dispersed offshore. Given their ability for long-distance communication (see "Vocalizations" section below), finding a mate is not likely a problem, even if the whales are widely scattered. Fin Whales are probably only vulnerable to Killer Whales, which are known to kill even full-grown Blue Whales. In areas where Killer Whales are abundant, Fin Whales commonly display on their flukes, flippers, and flanks the distinctive parallel scars that indicate a wound caused by the bite of a Killer Whale. Their possible lifespan is at least 80–90, and perhaps as much as 100, years.

DIET

Fin Whales consume a diverse variety of prey, primarily depending on availability. They eat shoaling krill, squid, copepods, and fish. Their feeding style is called "lunging." They surge through a tight school of prey, taking an enormous mouthful of sea water and prey. Throat grooves expand and greatly enhance the amount of water that can be engulfed. As the tongue pushes the water out through the baleen plates, food is filtered out and can be scraped off the inside of the baleen and swallowed. The asymmetrical colour pattern on a Fin Whale's head has been the subject of considerable speculation. Many thought it may be related in some way to feeding, the suggestion being that the flash of white could be used to startle and herd prey into the necessary tight concentrations. Others have suggested that, by lunging through the schooled prey with the right side down, the whales can be less visible to prey as a consequence of their enhanced counter-coloration (light below/dark above). Recent studies do not support either of these theories.

REPRODUCTION

Mating takes place in winter, and after a gestation of 11–12 months, a single calf is born early the following winter, usually in October through January. Calves feed on their mother's rich milk for about six or eight months and approximately double their length. They swim with her during their first migration to the summer feeding grounds and are weaned over the summer. Their mothers then have a resting period of about five or six months before mating again in the winter. Females usually have a calf every two or three years. In the Northern Hemisphere females become sexually mature at around 18.5 m and males around 17.5 m in length, which typically correspond to ages of seven to eight years in females and six to seven years in males.

BEHAVIOUR

Fin Whales occur most commonly as singles or cow/calf pairs; however, given the distances that their vocalizations carry underwater (see "Vocalizations" section), it is possible that many judged to be alone could in fact be part of a group whose members are out of sight. Nevertheless, most Fin Whale groups are likely small, fewer than four animals, although unusually large groups of up to 14 have been reported in the Gulf of St Lawrence. Most pods are unstable, as animals move in and out of them frequently. Whales migrate north to Canadian waters in an age-structured migration pattern. Larger whales appear first ahead of smaller ones and larger whales travel farther north than smaller ones. Pregnant females remain closer to shore throughout the summer and probably do not travel as far north. They also leave for the south earlier than other whales, perhaps in order to arrive back at the wintering areas before giving birth. Nursing females and juveniles typically leave last. Calves learn the migration route and location of the summer feeding grounds from their mothers and tend to return to it year after year. Fin Whales are not renowned for their aerial displays; however, some populations, such as the one in the Mediterranean Sea, are known to breach.

VOCALIZATIONS

Pulsed vocalizations by Fin Whales in the 15–30 Hz range (average 20 Hz) were first recorded in 1951, although their source was not recognized until 1964. They are loud, low-frequency sounds (at the lower threshold of human hearing) of around 185 dB in volume. Long series (1–32.5 hours) are used for long-range communication by breeding males to achieve dominance over other males and to attract distant females. Calling males typically float about 50 m below the surface. Short series of similar loud, low-frequency pulses, usually lasting less than five minutes, are produced by all age groups and both genders. These are likely used to maintain contact with conspecifics and perhaps for long-distance communication. These shorter vocalizations are typical of Fin Whales on the high-latitude feeding grounds. Both types of low-frequency pulses are thought to carry hundreds of kilometres, although recent evidence suggests that increasing noise pollution in the world's oceans may be reducing the audibility range. How this may affect the whales is unknown.

SIGNS

Fin Whales rarely breach in Canadian waters and rarely raise their flukes when sounding. As the whales surface, the blowhole typically appears just before the dorsal fin breaks the water. Their large columnar blow is 6–8 m high. Fin Whales have a very sharp ridge running along the top of the tail stock, which is visible as the whale arches it back before sounding.

REFERENCES

Aguilar 2002; Charif et al. 2002; Clapham and Seipt 1991; Croll et al. 2001, 2002; Flinn et al. 2002; Gregr et al. 2000; Hain et al. 1992; Kopelman and Sadove 1995; Marini et al. 1996; Moore, S.E., et al. 1998; Panigada et al. 1999; Reeves et al. 2002; Tershy and Wiley 1992; Vikingsson 1997.

Humpback Whale

FRENCH NAME: **Rorqual à bosse**

SCIENTIFIC NAME: *Megaptera novaeangliae*

Because they are largely coastal in habitat and perform some spectacular aerial displays, Humpbacks are favoured by whalewatchers, making them one of the most familiar of the great whales. Humpbacks have long annual migrations. Some travel up to 16,500 km round trip each year. These distances are commonly topped by Grey Whales, Northern Elephant Seals, and perhaps by some Blue Whales.

DESCRIPTION

All Humpbacks have a black back, but the amount of white on the lower parts of the body can vary considerably. Individuals in the North Pacific tend to have a lot less white on their bellies than those in the North Atlantic; sometimes they are completely black with white only on the flippers. Humpbacks in the North Atlantic can be white and mottled white from their chin to their flukes and partway up their sides. Flippers in the North Pacific population tend to be mainly black with white edging, whereas North Atlantic Humpbacks tend to reverse this pattern, although much variation exists, both within a population and between populations. Flukes are broad, notched in the middle, and serrated on the trailing edge. The amount of white on the underside of the flukes is highly variable and, combined with scarring and fluke shape, has allowed whale researchers to photo-identify individual Humpbacks. The inordinately long flippers are up to one-third of the body length, and are one of the distinguishing features of the species. The leading edge of the flippers is scalloped and the knobs often contain barnacles. Dorsal fins are highly variable in shape. They can be a mere bump or a high falcate fin more like those of other whales, but all display the hump at the front base of the fin, which is most noticeable when the whale rounds its back

before a dive. The dental formula is 0/0. From 270–400 coarse baleen plates hang from each upper jaw. These are usually dark, although sometimes the first few are whitish or partly whitish. The throat has relatively few, but very deep grooves: 12–36 (average 28), as opposed to 38–100 for most other baleen whales. They extend from the chin all the way to the umbilicus. Another diagnostic characteristic of the species is the assortment of tubercles (rounded bumps) on the snout and chin, each of which contains a short sensory hair.

SIMILAR SPECIES

From a distance, a sounding Humpback can be similar to a Fin or Sei Whale, or even a Right or Grey Whale. Right Whales and Grey Whales do not have a dorsal fin and the underside of their flukes, as they fluke up before sounding, are dark. Fin and Sei Whales are larger and sleeker, and rarely fluke up before a dive. A closer sighting that allows a view of the rounded head bumps and long flippers is conclusive.

SIZE

Females are about 0.7–1.5 m longer than males.

Total length: males, up to 16 m long, average around 13 m; females, up to 17 m long, average around 14 m; newborns, 4–4.6 m.

Weights vary widely depending on the season. A very large Humpback can weigh 40,000 kg, but most average around 30,000 kg or 30 tonnes; newborns, around 680 kg.

RANGE

Humpback Whales are one of the more cosmopolitan of the cetaceans. They occur in all oceans, and have even been seen occasionally in the Mediterranean Sea. Although mainly coastal, they readily travel over deeper waters, especially during migration. Generally they spend the spring, summer, and autumn feeding at high latitudes, then migrate to winter mating and calving grounds in warmer

Humpback Whale
(*Megaptera novaeangliae*)

Distribution of the Humpback Whale (*Megaptera novaeangliae*)

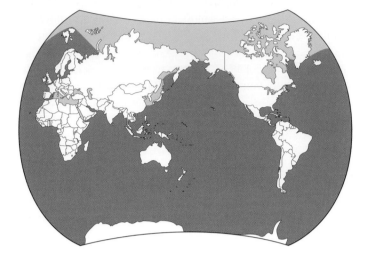

Humpbacks are highly migratory. They winter in tropical waters, where they generally do not feed. In late winter, they begin to travel to the feeding grounds. This migration can be up to 8000 km each way. A migrating Humpback covers about 100 km per day at an average rate of about 4–8 km/h. Cows with calves swim most slowly. The longest migrations occur in the Southern Hemisphere. Whales travelling to North American feeding grounds swim around 3000-7000 km each way. A Humpback, photographically identified in Alaska, was sighted 39 days later, and about 4440 km distant, in Hawaii. This is the fastest documented migration for this species. Most whales spend longer times resting along the route. Average migration speeds are 4 km/h, average foraging speeds are 2.2 km/h. Wintering and summering grounds are largely traditional, learned by a calf from its mother. Wintering grounds are often shared by whales from several different summering grounds. Since breeding occurs on the wintering grounds, whales from very diverse summer feeding grounds can be genetically related. Typically, Humpback dives vary with the season. In summer, most dives last less than 5 minutes and a 10-minute dive is unusually long. In winter, most dives are 10–15 minutes long and dives of up to 40 minutes have been recorded. These lengthy dives are not for foraging, as the whales rarely eat on the wintering grounds. The whales appear to be submerging to rest or to sing. Killer Whales are the Humpback's primary predator. Another cause of significant mortality is entanglement in fishing gear. A Humpback can live at least 50 years.

DIET

Humpbacks have coarse baleen, and so cannot capture very small prey. They focus primarily on krill and small schooling fishes such as herring, mackerel, sand lance, sardines, anchovies, and Capelin and sometimes somewhat larger fishes such as Pollock. Should these species not be available, they will resort to other shoaling species such as Banded Drum and Atlantic Croaker. They are termed "lunge" feeders because they lunge into a concentration of prey with their mouth open, capturing a large mouthful of water and prey. The expanding throat, enabled by the throat grooves, allows them to trap an enormous amount of seawater in each gulp. With lips partly open, the tongue pushes excess water out through the baleen, then scrapes the prey off the baleen and a mouthful is swallowed. Since this feeding method is only effective if the prey is concentrated, Humpbacks have a number of specialized methods of condensing their prey. The most interesting is their unique use of air bubbles. Individuals or coordinated groups will emit a curtain, cloud, or net of bubbles to herd or corral small schooling fishes or krill. A curtain is a line of bubbles, a cloud a dense surge of bubbles, and a net a cylindrical curtain of bubbles. The whales can then lunge through the concentrated prey to take their mouthful. The most dramatic example of this behaviour occurs when the lunge is made in a vertical manner through the middle of a bubble net and the whale concludes at the surface where we can see it. A coordinated group of lungers breaking the water together is an awesome sight. Some Humpbacks in the Gulf of Alaska have specialized in this cooperative technique. Bubble nets can be 45 m to 1.5 m in diameter depending on the prey – larger circles for

tropical or sub-tropical waters. The Arabian Sea population is aberrant in that it stays in tropical waters all year round. Two of the largest Northern Hemisphere wintering grounds appear to have been occupied by Humpbacks only recently. The Hawaiian and the Samana Bay (Caribbean Sea) calving grounds have apparently been used only since the early 1800s, perhaps in response to whaling efforts. Significant eastern North Pacific winter breeding grounds include the previously mentioned Hawaiian grounds, as well as a large area off the coast of California, another in the waters around Baja California, Mexico, and a fourth in the Philippine Sea south of Japan. The primary western North Atlantic breeding ground is the one mentioned above off Samana Bay, Dominican Republic. Over the last two decades, there has been a dramatic shift in Humpback distribution around British Columbia. Formerly commonly seen from shore during summer, they are now rare in inshore regions and may only be seen 20–50 km offshore along western Vancouver Island and around the Queen Charlotte Islands (Haida Gwaii).

ABUNDANCE

Like Grey Whales, Humpback Whales were vulnerable to whalers because of their inshore distribution on the northern feeding grounds and their large concentrations on the winter breeding grounds. Worldwide numbers were drastically reduced by around 90% by commercial whaling, mainly during the early 1900s. An international ban was agreed upon in 1966, after which time only the Soviet Union and Denmark (Greenland) continued to hunt illegally. Hunting finally halted in 1985, and today there is minimal hunting, most of it by aboriginal peoples. An estimated 8000–10,000 animals make up the population in the western North Atlantic (up to 12,000 in the whole North Atlantic) and around 10,000 live in the North Pacific. The Antarctic population is the largest, estimated to be at least 17,000 animals. All of the populations appear to be healthy and increasing. The World Conservation Union (IUCN) considers the Humpback Whale to be "Vulnerable." Canada (COSEWIC) considers the Pacific Humpbacks to be "Threatened" and the Atlantic population to be "Not at Risk."

fish and smaller circles for krill. Several whales may cooperate in the creation of the larger nets. The size and shape of the flippers and the tubercles along their leading edge assist the whales in conducting the tight turns and circles required for this method of foraging. Humpbacks are more manoeuvrable than any other large whale. Other foraging techniques use a sharp sound to startle a school that is near the surface and cause it to coalesce. Whales will mainly lobtail to create the noise, although breaches and flipper-slaps are used at times. Although they normally feed in the water column, Humpbacks are known to display a wide variety of foraging behaviours. In the Gulf of Maine off the Stellwagen Bank, a group of Humpbacks have learned to feed on fishes called sand lance, which normally burrow into the silty substrate. The whales will brush the bottom with the side of their head, their flippers or their flukes, to drive the fish up into the water where they can be engulfed. This feeding method is called prey flushing and tends to leave scuffed and sometimes bleeding marks on the jaws and flukes of the animals that practise it.

REPRODUCTION

As might be expected for a migratory whale, calving and breeding are seasonal. Breeding takes place on the wintering grounds. After a pregnancy of 11 to 12 months, a single calf is born, again in the warm waters of the wintering grounds. In the Southern Hemisphere, most calves are born in August. In the Northern Hemisphere most are born in February. Calves nurse for about a year, but begin to take solid food at around six months old. Usually calves leave their mothers following weaning, but some remain associated for up to two years, until she has her next calf. A female can become pregnant while nursing a calf. The calving interval is normally every two years (range 1–3 or more years). Calves grow very quickly while nursing on their mother's rich and highly nutritious milk. They drink at least 43 kg of milk every day and are between 8 and 10 m long when weaned. Both male and female Humpbacks in the eastern Atlantic populations become sexually mature around five to seven years old, while females in the western Pacific population are slower to mature, producing their first calves around eight to twelve years old.

BEHAVIOUR

The herd on the breeding ground breaks up in late winter to swim to their diverse, but traditional feeding grounds. Typically, cows with calves leave last. Calves travel with their mothers during their first migration and learn the route to and location of the summer feeding grounds from her. The social organization of Humpbacks, with the exception of mother/calf pairs, is generally fluid and any associations are temporary. No stable groups have been observed on the breeding grounds, but often whales will concentrate around a food resource and sometimes a group will learn to cooperatively exploit that resource. The same group may then stay together for an extended period during a season or re-amalgamate each summer on the feeding grounds for several feeding seasons. These cases are the exception, however, as most feeding Humpbacks forage alone or in constantly changing groups. On the northern summer feeding grounds, Humpbacks log (float) on the surface to sleep, at which time they are vulnerable to a stealthy approach by a small open boat.

Aboriginals around the world have taken advantage of this behaviour for centuries. Humpbacks on the wintering grounds rest below the surface for up to 40 minutes at a time. On the winter breeding grounds and sometimes on the northern feeding grounds before migration, groups of males are attracted to females, especially those without a calf. These males compete physically for dominance and hence proximity to a female. They jostle and lunge, sometimes to the extent of wounding each other with hits from flippers, flukes, or heads. Typically the largest male wins the spot closest to the female and becomes her "primary escort." He must continue to battle for position, however, as other adult males repeatedly challenge his preferred location. Copulation has not yet been witnessed in Humpback Whales, but likely the closest male would be the partner. Coalitions of two mature males on the breeding grounds are suspected. These males enter and leave a breeding group, compete together for access to the female, but show no aggression towards each other. Likely both of their chances of mating are increased due to their greater ability to compete, even against a larger male. Calves, like baby mammals everywhere, like to climb on their mothers and have been known to rest on her broad head for a ride. A female will defend her calf from Killer Whale attacks by attempting to shield the calf from the predators with her own body and by thrashing at them with her tail, flukes, and head. Humpbacks will come to the aid of

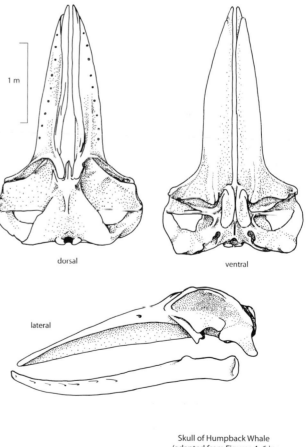

1 m

dorsal

ventral

lateral

Skull of Humpback Whale
(adapted from Figures 4–6 in
Winn and Reichley 1985)

other Humpbacks being attacked by Killer Whales, to join them in presenting a coordinated group defence.

VOCALIZATIONS

Humpback males are renowned for their complex and beautiful songs. Evidence suggests that the songs are a form of dominance competition between the males. Once thought to sing only on the wintering grounds, Humpbacks have recently been found to sing during the northwards and southwards migrations as well. Perhaps their hormone levels are still high (northwards) or rising (southwards) and they are trying to impress the other males before the next breeding season. Most of the song energy is between 200 and 2500 Hz and varies in intensity from 155 to 189 dB. The low frequencies are below our hearing and the high are equivalent to a human soprano's. So Humpbacks have a vocal range of over eight octaves. Songs are composed of repeated phrases generally produced in a fixed order which may go on for several hours. This order usually changes gradually as males add or remove phrases and males can be traced to specific wintering grounds by the composition of their song. Occasionally, a male from another wintering group will show up on the breeding grounds and provoke the locals into a major and rapid change. Mature females have been recorded producing what is called a "feeding sound." It is a highly stereotyped series of trumpeting calls that vary in frequency from 400–800 Hz. Humpback Whales of both genders are known to "trumpet" and forcefully (and loudly) exhale when attacked by Killer Whales, Pilot Whales, and probably False Killer Whales. Competing males will also "trumpet" at each other as part of their agonistic displays. Sounds called "grumbles," "snorts," "thwops," and "wops" are used to communicate between and within groups and "grunts," "groans," and "barks" enhance group cohesion. Sounds created by surface slaps (flipper or flukes) and breaches likely also serve a communication function as they increase in frequency as a group splits.

SIGNS

Breaching, lobtailing, spyhopping, flipper slapping, and fluke slapping are common Humpback aerial behaviours. These activities likely send percussive sounds through the water and air as a form of communication, but the activity is also known to dislodge bothersome external parasites. Fluke-ups normally occur before a deep dive. A Humpback blow is low and round, usually smaller than a Fin Whale's blow. The hump at the base of the dorsal fin is the most commonly seen distinguishing characteristic and the round bumps on the head and chin are indisputable if visible.

REFERENCES

Clapham 2002; Clapham and Mead 1999; Clapham et al. 1999; Craig et al. 2002; Darling and Bérubé 2001; Dunlop et al. 2008; Félix et al. 2006; Flórez-González et al. 1994; Ford and Reeves 2008; Gabriele et al. 1996, 2007; Hain et al. 1995; Katona and Beard 1990; Kingsley and Reeves 1998; Laerm et al. 1997; Lagerquist et al. 2008; Larsen and Hammond 2004; Mercado et al. 2003; Noad et al. 2000; Pack et al. 2002; Simão and Moreira 2005; Spitz et al. 2002; Stone, G.S., et al. 1990; Summers 2004; Weinrich 1995; Whitehead 1987; Winn and Reichley 1985.

FAMILY ESCHRICHTIIDAE
grey whales

The only living species in this family, the Grey Whale has two to four throat grooves, no dorsal fin, straight lower jaws, medium-length baleen, large, broad flippers, and a mottled appearance.

Grey Whale
also called Gray Whale (American spelling), Devilfish, Scrag Whale

FRENCH NAME: **Baleine grise**
SCIENTIFIC NAME: *Eschrichtius robustus*

Grey Whales have a unique feeding style that allows them to exploit the highly prized coastal habitat without competing with other whales. They make the second longest yearly migration of any mammal – up to 16,000–20,000 km annually, largely without feeding! Only Northern Elephant Seals travel farther.

DESCRIPTION

Grey Whales are large, mottled grey baleen whales with light-coloured callosities on various parts of their bodies, especially the head. The colour of the callosities is caused by a covering of barnacles and whale lice on the skin. Calves are darker than adults and although the areas of thickened skin are already visible where the callosities will later develop, their colonization has not yet taken place. A small number of short individual hairs occur on the front of the head, but are only visible on young calves. Wounds tend to heal with a white scar, which further adds to the light mottling. Albino or all-white animals are very rare. The mouth line arches upwards and about 130–180 coarse, short (up to 25 cm long) creamy yellowish baleen plates hang from the upper palate. These baleen plates are the fewest, shortest, and coarsest of all among the baleen whales. Although Grey Whales do not have the series of parallel throat grooves found in the rorquals, they do have two to seven (commonly three) deep creases that run between the lower jaws that allow them to expand their throat sufficiently to create some suction. A pair of blowholes is present at the top back of the head. Grey Whales do not have a dorsal fin, but there is a flattened bump at the top of the tail stock followed by a series (8–14) of smaller bumps trailing down to the flukes. The flukes have a

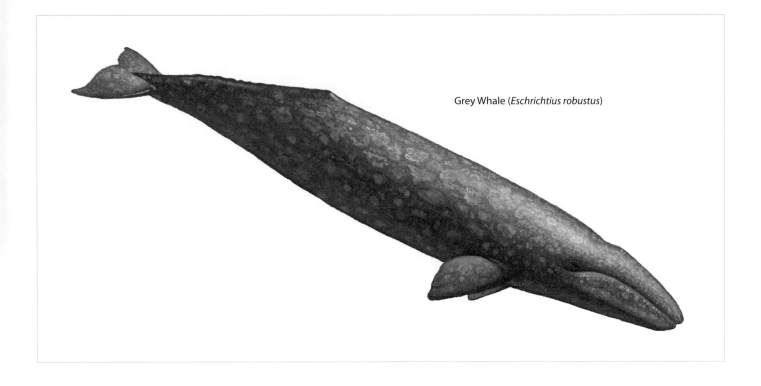

Grey Whale (*Eschrichtius robustus*)

convex trailing edge and distinct medial notch and are frequently scarred and battered. Flippers are large and paddle-shaped with pointed tips. As they have baleen rather than teeth, their dental formula is 0/0. The testes of adult males during breeding season are relatively large, averaging about 38 kg for a pair.

These filter feeders have a countercurrent heat-exchanger in their tongue to reduce the amount of body heat they lose when gulping, then filtering, large mouthfuls of cold seawater. This network of blood vessels allows the cold blood returning to the body to absorb heat from the warm blood entering the tongue. Blood flow to the tongue becomes cooled by this transfer and consequently the amount of heat lost to the water is drastically diminished. Likely all the filter-feeding whales that forage in cold waters will be found to have this energy-saving system.

SIMILAR SPECIES

Other large coastal whales such as Humpback, Pacific Right, and Bowhead Whales could be mistaken for Grey Whales. Apart from the Grey Whale, all the others are mainly black. Humpbacks have a dorsal fin, although it can be very small on some individuals, but their flippers are very long and their flukes sometimes are partly white beneath. Bowhead and Right Whales occur only in the northern parts of the Grey Whale range. Like the Grey Whale, neither has a dorsal fin, but both are considerably more rotund, and have smooth backs and highly arched lower jaw lines. In addition, Bowheads have a white chin and no callosities.

SIZE

Adult females are somewhat larger on average than adult males.
Total length: 11–16 m; newborns, 4.2–4.7 m.
Weight: 27,000–35,000 kg; newborns, 750–900 kg.

RANGE

Grey Whales formerly inhabited both the Atlantic and Pacific Oceans. The Atlantic population was likely hunted to extinction by the end of the seventeenth or early eighteenth century. Some argue that it was a small remnant population to begin with and was already declining due to habitat changes. The Pacific stock is divided into two populations, one on each side of the ocean. The eastern Pacific stock winters along the coast of California and Mexico and summers mainly in the Bering, Chukchi, and Beaufort Seas. Recent evidence of calls collected on subsurface hydrophones suggests that some animals may winter in the Bering Sea, perhaps as a response to sea ice reduction due to increased water temperatures. A few individuals summer along the North American coast from Washington to British Columbia. The main breeding lagoons are off the west coast of Baja. These include Magdalena Bay, Guerrero Negro Lagoon (also called Scammon's Lagoon), and Laguna Ojo de Libre. The western Pacific population winters along the coast of China, south of the Korean Peninsula, and summers mainly in the Okhotsk Sea near Sakhalin Island, Russia.

ABUNDANCE

There have been three separate harvesting peaks in the history of Grey Whale hunting. The first, in the early 1700s, eliminated the species in the Atlantic Ocean. The second, in the 1800s, reduced the eastern Pacific stock to a few thousand animals. The third peak, in the 1900s, almost decimated the western Pacific stock, leaving only a few hundred animals from a pre-whaling estimate of up to 10,000 animals. This stock was elevated from "Endangered" to "Critically Endangered" in the year 2000 by the World Conservation Union (IUCN) as numbers are not increasing and there are thought to be only 50–100 animals left. Unfortunately, poached Grey Whale meat

Distribution of the Grey Whale (*Eschrichtius robustus*)

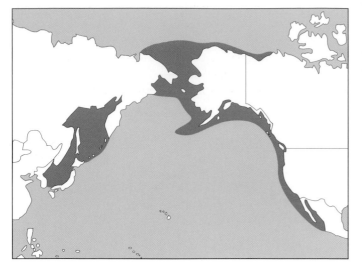

continues to be occasionally available for sale in Japan, despite a long-time ban on hunting. The fate of the eastern Pacific stock is more positive. After more than 50 years of partial protection from hunting, especially in the crucial breeding and calving lagoons, this stock has recovered in a spectacular fashion. Numbers are up to an estimated 26,000 animals, near pre-whaling estimates of around 30,000 animals. Their IUCN designation was reduced from "Endangered" to "Low Risk" in 1994. Limited hunting of this population occurs in the summer feeding area by aboriginal Russians. The future of this population is still somewhat uncertain as there has been unprecedented recent migration mortality, at least partly due to starvation, and a subsequent decline in calf production. Possibly this population has reached its current feeding capacity. Should these trends continue, we could see a decline in numbers of the eastern Pacific stock. In US waters, Grey Whales are protected by the Marine Mammal Protection Act, which restricts boats from approaching closer than 100 yards (91 m). British Columbia regulations require that any whale or group of whales not be observed for longer than 30 minutes and that vessels, both motorized and non-motorized, remain at least 100 m away.

ECOLOGY

These whales live in shallow waters 4–120 m deep, often in view of the coast. Their year is governed by their lengthy migration. Migration distances vary, but among the eastern Pacific population, can be 16,000–20,000 km for a round trip. This is not only one of the longest migrations by a cetacean, but is one of the longest for any mammal. Some Humpback Whales almost equal this feat during their annual migration. Migrating Grey Whales travel at a steady 5 km/h, although this picks up to about 8 km/h as they approach their destination. On average, they swim about 110 km each day at a speed of 6–7 km/h and the entire transit takes about two months. Migration routes are broad corridors that roughly follow the coastline. Cow/calf pairs, especially, tend to hug the coast. Most whales feed little during migration. There are a few profitable foraging areas along the

migration route that are visited by Grey Whales. These are regions where mysids and crab larvae form dense mats on the ocean floor or where plankton masses together in thick clouds. Clayquot Sound on the west-central coast of Vancouver Island and several mainland bays along the Queen Charlotte Strait are migratory stopping spots where feeding is commonly observed.

Again for the eastern Pacific population, the whales leave their northern summer feeding grounds around October/November. Pregnant females are the first to leave, followed by mature adults and finally by the juveniles. After one to two months south, the first echelon (males, juveniles, and females without calves) leaves the southern breeding grounds in January–February and reaches the feeding grounds in March–April. Often juveniles, many of which did not swim all the way down to Mexico, lead the northward movement. Once on the feeding grounds, the whales follow the receding pack ice northwards over the course of the summer. Cows with calves delay their northerly departure about two months, to allow the calves time to grow strong enough to travel and to accumulate a thick enough blubber layer to survive in the cold northern waters. Killer Whales are major predators of Grey Whales and many whales display whitish scars that are remnants of survived attacks. A typical Killer Whale tactic is to chase a baleen whale until it tires, then either drown it by holding it under water, or wait while it bleeds to death from a series of bite wounds. Cow/calf pairs are especially vulnerable during the northerly migration, as the young calf may not yet have the necessary stamina to swim far enough and fast enough to escape. Killer Whales commonly target the calves. Grey Whales occupy waters that are subject to intensive fishing efforts by humans. It is little wonder that they are sometimes injured or killed by collisions with powered boats or entanglement in fishing gear. Fishing gear entanglement not only drowns the whales, but has been implicated in many disfigurements including the complete loss of flukes on some Grey Whales. Constant swimming action eventually saws the ropes through the flesh at the base of the tail stock, causing necrosis and ultimately the loss of the flukes. In rare cases, when the mutilated whale survives the gradual amputation, it may be able to remain somewhat healthy, for a while, by drastically altering its swimming and diving style.

DIET

Grey Whales are primarily bottom feeders that specialize in catching and consuming tube-dwelling amphipod crustaceans of various species. These amphipods can form huge mats on the ocean floor with thousands, if not millions, of animals in a square metre. The whales feed by rolling on their side and sucking up a series of mouthfuls of bottom sediment before surfacing, trailing a characteristic mud plume and leaving behind several divots on the sea floor. Around 97% of feeding Grey Whales roll to the right (about 90% of humans are right handed). They also sometimes feed on swarms of free-swimming mysid shrimp. The small prey is filtered out as the tongue presses the water/prey mix through the baleen plates. Once the water is removed the tongue scrapes the prey off the baleen plates and it is swallowed. Aside from their primary prey, Grey Whales also ingest many other bottom-dwelling organisms that live in or on the sediment, such as polychaete tube worms, ghost shrimp, crab larvae, miscellaneous crustaceans, fish eggs and larvae, squid, bait fish, and even sponges. They do almost all their feeding for the year on the shallow Arctic feeding grounds from April to October. Very little food is eaten during migration and through the winter on the breeding grounds. Most whales lose about 25%–30% of their body weight during that three- to five-month period. Lactating females lose even more. To accumulate the energy necessary for the rest of the year, Grey Whales will consume more than a tonne of food each day during the polar summer's biological bloom (fostered by 24 hours of continuous daylight and the higher oxygen content of the cold water). The health of the eastern Pacific Grey Whale stock is dependant on the health of the northern amphipods. Should the waters of the Bering, Chukchi, and Beaufort Seas become only a few degrees warmer due to global warming, there could be a crash of the amphipod fauna and, hence, of the Grey Whale population. Alternately, if the ice is slower to recede in spring or early to return in autumn, the feeding opportunities for the whales are diminished. Current indications suggest that some or all of these possibilities may already be under way.

REPRODUCTION

Grey Whales are highly migratory and both breeding and calving are seasonal. Mating behaviour is observed at almost any time of the year, but conceptions appear to be restricted to a few weeks from late November to early January (most in late November and early December), while the animals are on the northern feeding grounds or during their southward migration. If a female does not conceive during her first oestrus, a second oestrus may occur 40 days later, when the animals are on their southern wintering grounds. Pregnancy lasts 13 months, so calves are born in the warm southern lagoons the following year, mostly in January. Occasionally, an early calf will be born towards the end of the southward migration. Females bear a single calf. Calving intervals range from one to four years, and average every two years. Typically, a female will mate in a year when she is not bearing a calf. Females that lose a very young calf may be capable of ovulating soon after, which may account for the few cows that produce calves on consecutive years. Females can lactate while pregnant. Calorie-rich milk (it is about 50% fat) produced by their mothers allows calves to gain up to 27 kg a day. Females fast for about half of the time they are suckling their calves. The calves begin taking solid food soon after they arrive on the northern feeding grounds, and are weaned and independent at around six to nine months old, just before the fall migration south. Females become sexually mature at around eight years old.

BEHAVIOUR

Grey Whales do not form lasting associations. Even calves do not stay with their mothers beyond weaning. Travel and feeding is usually solitary or in small, frequently changing groups. Large, loose aggregations do occur on the feeding grounds and in the breeding lagoons. The breeding season can be a boisterous time. Commonly, several males will court an oestrous female, with much splashing and physical contact but no aggression towards each other. Usually

births occur in isolation, but occasionally an adult female, who is not the mother, will assist with a newborn. Very young calves are lifted to the surface to breathe, and often ride on their mothers' backs until they are sturdy enough to swim on their own. As the calves grow stronger, they become quite rambunctious, jumping on and sliding off their mother's massive back. The mother/calf bond is very strong. Grey Whales have been called Devilfish by whalers, because of the aggressive defensive behaviour of females whose calves are threatened. This behaviour is also used towards Killer Whales attempting to prey upon a calf. The female will thrash her flukes, flippers, and head at attacking Killer Whales, and attempt to scrape her callosities against the sensitive skin of the predators. The mothers may even roll onto their backs at the surface, holding their calves above the water, in an effort to protect the calf and their own vulnerable underside from the Killer Whales. Grey Whales under attack sometimes try to flee into shallow water, where they can hide quietly amid the kelp, but running is often an unsuccessful tactic with a calf, as they are unable to swim quickly for long distances. Migration is relatively predictable, with animals passing certain points of land about the same time each year. Calves undertake their first northerly migration at their mother's side and learn the route from her. Commercial whale watching began with Grey Whales in 1955 and quickly spread to the east coast and then around the world. Careful whale watching is not too bothersome, unless boats approach the whales too closely or quickly, or from the front, causing them to alter course or slow down. These whales are curious and some even learn to be friendly to whale-watchers, to the point of allowing physical contact. We do not know if Grey Whales sleep. Rest periods tend to be brief during most of the year, except in the winter in the breeding grounds, when females near parturition or with newborns will spend up to an hour at a time floating near the surface, rhythmically lifting their heads for a breath. Possibly Grey Whales rest only one hemisphere of their brains at a time, as has been found in some of the dolphins.

VOCALIZATIONS

Grey Whales produce a whole series of sounds with frequencies from 100 Hz to 12 kHz, all of which fall within the range of human hearing. Most of the vocalizations have been described as moans, clicks, snorts, pops, and metallic knocks and bongs. Moans are frequently heard from migrating whales and series of metallic knocks are commonly heard on the summer feeding grounds and in the

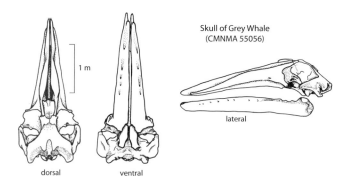

1 m

Skull of Grey Whale
(CMNMA 55056)

lateral

dorsal   ventral

breeding lagoons. Although none of the baleen whales are able to echolocate with high-frequency sounds as the toothed whales do, some, like the Grey Whale, are thought to use low-frequency sounds for a somewhat similar purpose; for example, to inform the whales of ocean floor topographic characteristics or perhaps to detect the presence of a large object (like another whale).

SIGNS

Grey Whales commonly fluke up before a deep dive. Their blow can be low and bushy (after a short dive) or larger and columnar (after a longer/deeper dive). Spyhopping and breaching are common. Grey Whale migrations are somewhat variable, but can be predicted with some degree of success. Frequently, local watchers can determine within a couple of weeks when sightings are likely.

REFERENCES

Baker, C.S., et al. 2002; Buckland and Breiwick 2002; Darling et al. 1998; Dunham and Duffus 2002; Fertl et al. 1999; Ford and Reeves 2008; Heckel et al. 2001; Heyning and Mead 1997; Jones, M.L., 1990; Jones, M.L., and Swartz 2002; Le Boeuf et al. 2000b; Mate 2005; Moore, S.E., and Clarke 2002; Moore, S.E., et al. 2001; Perryman and Lynn 2002; Perryman et al. 1999, 2002; Rugh et al. 2001; Stafford, K.M., et al. 2007; Stelle et al. 2008; Sumich et al. 2001; Swartz et al. 2006; Urbán-Ramírez et al. 2004; Weller, D.W., et al. 1999; Woodward, B.L., and Winn 2006.

Suborder Odontoceti

toothed whales

The toothed whales have a single offset blowhole (usually a conspicuous rounded melon housed in a concave depression on the dorsal surface of the skull), have teeth rather than baleen, are active hunters that prey on single relatively large prey items, and use echolocation to hunt and find their way around. Skulls of toothed whales are usually asymmetrical. The melon is thought to focus the echolocation sound waves, produced in the blowhole region.

FAMILY DELPHINIDAE

ocean dolphins

This family is the most diverse of the whale families. It comprises small- to medium-sized toothed whales. Males are generally larger than females, and most, but not all, have a prominent beak. Ocean dolphins tend to be quite vocal. Tooth number is highly variable, even within a species, but each tooth typically has a simple, sharp, conical shape. The flukes are always notched in the middle and the dorsal fin is near the middle of the back.

Long-beaked Common Dolphin

also called Long-beaked Saddleback Dolphin

FRENCH NAME: **Dauphin du Cap**
SCIENTIFIC NAME: *Delphinus capensis*

Before 1994, the Long-beaked Common Dolphin and the Short-beaked Common Dolphin were considered subspecies of the Common Dolphin, and neither was normally found in Canadian western waters. Until 1993, the Long-beaked form was known to extend only as far north as central California, but since then some have been seen in Canadian waters around Vancouver Island.

DESCRIPTION

The coloration of the Long-beaked Common Dolphin is very similar to, but more muted than, that of the closely related Short-beaked Common Dolphin. It has a light tan or ochre thoracic patch that flares out behind the eye. The back is dark, the sides are

Long-beaked Common Dolphin (*Delphinus capensis*)

male

dark greyish, and there is a darker V-shaped projection of the back colouring that drops down below the large dorsal fin. The dorsal fin and flippers are usually dark, but in older animals can become slightly paler, especially in the centre. A dark line, which begins along the beak, extends to the eye, and a similar, but broader black line runs from the chin or corner of the mouth to the flipper. Rarely does any of the white of the belly extend above this chin-to-flipper line. A flipper-to-anus dark stripe is fairly strongly developed, but hard to see in the field and usually only noticeable on stranded animals. The head of the Long-beaked Common Dolphin is narrower and less steeply inserted into the beak, giving an overall impression of a flatter head than the Short-beaked form. All the teeth are uniformly conical and the dental formula is 47-60/47-60, for a total of 188–240 teeth. Mature males develop a distinctive post-anal hump on the bottom of the tail-stock behind the anus. This can help identify gender in the field if the animal obligingly jumps clear of the water surface and a profile view is provided. The animal pictured here is a male.

SIMILAR SPECIES

Short-beaked Common Dolphins are very closely related and similar to Long-beaked Common Dolphins. Usually a close sighting in good lighting conditions is necessary in order to discern the subtle differences. A useful field characteristic to distinguish them is the bright white patch below the eye between the eye and the flipper on the Short-beaked form. This same area is dark on the Long-beaked form. The melon area of the head of the Short-beaked Common Dolphin is higher and meets the beak more steeply than on the Long-beaked form. Other dolphins, which are similar, are more readily distinguished, especially if one is granted a profile view in adequate light. Striped Dolphins and Pacific White-sided Dolphins both congregate in large boisterous groups, which also create considerable spray as they breach to breathe. Striped Dolphins are larger and occur mainly offshore. Pacific White-sided Dolphins share the inshore regions, are about the same size, but have a distinctly larger and bicoloured dorsal fin and a very short, hardly noticeable beak.

SIZE

Males are about 5% larger than females.
Total length: males, 200–254 cm; females, 193–222 cm; newborns, 80–100 cm.
Weight: males, up to 135 kg; newborns, unknown.

RANGE

The Long-beaked Common Dolphin is a widespread inshore species, which has many pockets of distribution that are separate from each other, called a disjunct distribution pattern. One of the larger pockets occurs along the west coast of North and South America from Peru to Vancouver Island. These animals move northward (or southward for those in the Southern Hemisphere) in the summer as the seas warm and retreat back to the Gulf of California and Central America during the winter. This species has only recently been seen in Canadian waters, and only during summers when warmer currents are present inshore.

Distribution of the Long-beaked Common Dolphin (*Delphinus capensis*)

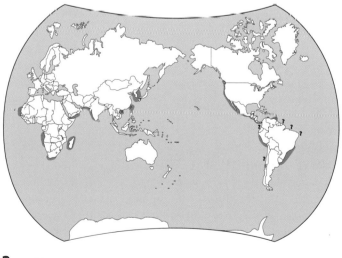

? possible range

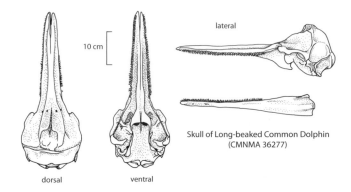

10 cm

lateral

dorsal

ventral

Skull of Long-beaked Common Dolphin
(CMNMA 36277)

ABUNDANCE

This is a rare species in Canadian waters, but as the temperatures of western inshore waters become warmer (as they are projected to do in future) due to global warming, increasing numbers of sightings are likely. Long-beaked Common Dolphin numbers around the world are unknown; however, the species is not considered to be under threat, although it is harvested in Peruvian waters. Estimates for the western North American population (from which no doubt our Canadian sightings derive) is at least a few thousand animals.

ECOLOGY

This is a near-shore species never found more than 100 km from land and usually much closer. Long-beaked Common Dolphins occur in warm temperate to tropical waters in a wide, but disjunct distribution. In areas where both the Long-beaked and Short-beaked forms occur, it has been suggested that in order to share the food resources, the Long-beaked Common Dolphins feed mainly on fish and the Short-beaked Common Dolphins concentrate on squid.

DIET

Like the Short-beaked Common Dolphin, this species consumes many species of small schooling fishes such as hake, pilchards, herring, sardines, and anchovies as well as schooling squid of various species.

REPRODUCTION

As a tropical species, their breeding season is probably not seasonally based but rather determined by the individual physiology and age of each female. Otherwise, the details of reproduction of this species are likely very similar to those indicated for the Short-beaked Common Dolphin.

BEHAVIOUR

Herd sizes of Long-beaked Common Dolphins vary from a single or a few, such as we see in Canadian waters, to herds of several thousand animals, as occur at times in tropical and subtropical coastal waters. Likely these large herds are amalgamations of many smaller herds congregating around a plentiful food resource. The average herd size off the California coast is about 180–200 animals. Rarely do the two Common Dolphin species form mixed herds, although

both species are known for associating with other cetacean species. Long-beaked Common Dolphins hunt cooperatively. One common strategy is for a group to herd fish into a tight "fish ball." As the other group members continue to herd the fish, individual animals can dash in to capture a meal. These small whales are active, energetic animals that spend considerable time on or near the surface of the water. They are eager bow-wave and stern-wave riders on powered vessels and often jump out of the water to breathe, especially when travelling.

VOCALIZATIONS

These dolphins are known to use whistles and clicks to communicate with each other. They also utilize echolocation to find and catch their prey and to navigate underwater.

SIGNS

Like many of the small dolphins, these animals are often seen in herds of rapidly surfacing animals creating considerable "splash." Due to the rarity of this species and their small herd size in Canadian waters, the surface "splash" and acrobatics of the species rather than large herd size may help observers to find and identify it.

REFERENCES

Ford 2005; Heyning 1999a; Heyning and Perrin 1994; Jefferson et al. 2009.

Short-beaked Common Dolphin

also called Common Dolphin, Short-beaked Saddleback Dolphin, Common Saddleback Dolphin, White-bellied Dolphin

FRENCH NAME: **Dauphin commun**

SCIENTIFIC NAME: *Delphinus delphis*

There is considerable current discussion about the taxonomic situation among the dolphins in this genus. It appears that there are two species, both of which are found in Canadian waters. Literature published before about the year 2000 should be scrutinized carefully, since most authors treated all the *Delphinus* as a single species until recently.

DESCRIPTION

The colour pattern on this small cetacean is quite elaborate, with a complex criss-cross of grey and white and tan on the sides. This pattern in the field makes it seem like the dark top of the back drops down below the dorsal fin to form a V-shaped dark point. The top of the back is dark grey or dark purplish brown, rapidly changing to blackish after death. The belly is white. There is a dark stripe from the top of the mouth to the eye and from the chin to the flipper. The tip of the beak is black. The blackish flippers project from the white

female

Short-beaked Common Dolphin (*Delphinus delphis*)

of the animal's lower body. A distinct, large, bright patch runs along the side from the eye to about half the length of the body, ending just below the dorsal fin. This "thoracic patch" can vary in colour from golden to tan to light grey. Mature males display a post-anal hump, a thickening of the tail stock just behind the anus, which from the side protrudes downwards, breaking the smooth line from the belly to the flukes. This hump can sometimes be seen in the field if the animals are breaching. The individual pictured here is a female, but see the male Long-beaked Common Dolphin illustration, which has a similar post-anal hump. The dorsal fin is large and curved. It is mainly dark, but the middle can be light coloured in adults. All the teeth are uniformly conical and the dental formula is 42–54/41–53, for a total of 166–214 teeth. The upper jaws usually have at least one more pair of teeth than the lower jaws.

SIMILAR SPECIES

The very similar Long-beaked Common Dolphin has recently been discovered in Canadian waters on the west coast. Likely any of the "Common Dolphins" (genus *Delphinus*) seen around Vancouver Island will be the Long-beaked form, as the only known western sighting of the Short-beaked form was a single stranded animal in 1953. If sea water temperatures continue to rise, there is potential for Short-beaked Common Dolphins to become more common, at least offshore around Vancouver Island. Field distinction of these two species where they co-occur can be tricky. Generally the melon size of the Long-beaked Common Dolphin is smaller, its beak is longer, and its colour pattern more subdued and generally darker. The area between the front flipper and the eye is greyish, rather than bright white, as in the Short-beaked form. All these characteristics can be very difficult to discern in the field, especially if the animal does not cooperate by jumping. The Striped Dolphin is very similar in size and shape to the Short-beaked Common Dolphin and also displays an eye stripe, but its eye stripe continues down the side all the way to the anus. Its "thoracic" patch is more diffuse and extends into a "blaze" along the shoulder. A good side view is usually required in order to discern the colour-pattern distinctions of these two species. In the North Atlantic, the Atlantic White-sided Dolphin with its yellowish side patch

could be confused with the Short-beaked Common Dolphin. This yellowish patch is beyond the dorsal fin on the rear of the animal, rather than the front, and the Atlantic White-sided Dolphin is larger, more robust, with a short, thick beak and a very distinct bright-white patch below the dorsal fin. In the North Pacific, the Pacific White-sided Dolphin is similarly gregarious, and even more acrobatic than the Short-beaked Common Dolphin. It, however, has an extremely short beak and does not exhibit the V-shaped pigmentation pattern on the side of its body below the dorsal fin.

SIZE

Males are about 5%–7% larger than females.

Total length: males, 172–223 cm; females, 164–215 cm; newborns, 80–100 cm.

Weight: males, up to 110 kg; newborns, around 10 kg.

RANGE

These dolphins are found in tropical and warm temperate waters, usually along continental shelves around the world. They prefer to

Distribution of the Short-beaked Common Dolphin (*Delphinus delphis*)

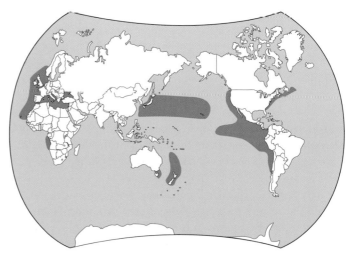

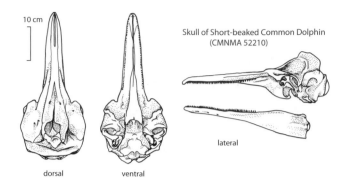

10 cm

Skull of Short-beaked Common Dolphin
(CMNMA 52210)

lateral

dorsal   ventral

forage along the shelf edge and over seamounts and are rarely, but occasionally seen in shallow coastal waters. Those on the northern and southern parts of the range will invade cooler waters as temperatures rise during summer. In these cases the dolphins tend to remain in the warmer surface waters except during the night when foraging. The small population off Nova Scotia and Newfoundland is present only in summer and spends the winter in the Chesapeake Bay and Gulf of Maine area. In the Pacific, the distribution does not reach Canadian waters, although a single individual of this species was stranded on Vancouver Island in 1953.

ABUNDANCE

Short-beaked Common Dolphins are disappearing from many parts of their range. The Mediterranean, northeastern Florida, and northern European stocks are seriously depleted and the Black Sea population may be verging on extinction. This species is vulnerable to drowning in fishing nets due to their surface habits and their pursuit of fish species also targeted by humans. Stocks in the Central Pacific and North Atlantic have been seriously threatened by purse-seining and drift net fisheries for squid and tuna. A recent United Nations moratorium on drift netting has reduced that threat, but large kills continue to be reported off Europe and South America in commercial trawl fisheries. Some populations, notably in the tropical eastern Pacific, are still healthy and are estimated to number in the millions. The migratory population along the eastern coast of North American has been estimated to number at least 10,000 animals.

ECOLOGY

Like most cetaceans, Short-beaked Common Dolphins are active both day and night. They are fast swimmers, capable of speeds up to 30 km/h. They prefer waters within the 7°C–23°C range, and in cooler waters will spend the bulk of their time in the warmer surface layers. In many parts of their range, most feeding occurs at night, when the deep-water prey species migrate up the water column to feed and rise to depths within the reach of the dolphins. Large loafing herds of dolphins break into smaller herds in the evening, in anticipation of this event. They undertake deep dives throughout the dark hours to reach their prey. They dive to at least 280 m and remain underwater for up to eight minutes at a time. Since most hunting and prey capture occurs in darkness, echolocation is essential. At daybreak, the smaller foraging herds reunite and daylight

hours are spent socializing, resting, and travelling. These are energetic animals that are rarely seen at rest.

Attempts have been made to maintain this species in captivity without much success, although births are known and one animal has lived for 22 years in captivity. Their known lifespan is around 25 years, but could be longer. Predation upon this species has not been observed except by humans, but likely sharks and Killer Whales take some.

DIET

Short-beaked Common Dolphins prefer to hunt schooling species of fish (between 10 and 30 cm long) and squid. In the western North Atlantic off the Canadian coast, mackerel, Butterfish, Common Squid, and herring are favourites.

REPRODUCTION

A single calf is born following a pregnancy lasting 10–11 months. Most calves in the northern hemisphere are born in May and June when the waters are warmer. Females produce a calf about every two years. Each calf is nursed for at least nine to ten months and attains sexual maturity between the ages of five and ten years, but this can be quite variable between different stocks and locations. In the central Pacific, females reach sexual maturity at around eight years old, males around ten years old.

BEHAVIOUR

Short-beaked Common Dolphins are very gregarious, almost always occurring in either nursery groups (females and their calves), mixed groups (adults and juveniles of both sexes), or male groups (adult and subadult males). Size and composition of the herds is subject to considerable fluctuation and is called a "fission-fusion" social structure. Herd size is variable, usually between 15 and 30 animals, but is known to exceed hundreds and even thousands of animals, and is generally largest in late summer. In Canadian waters, herds have been sighted with 5–40 individuals. Short-beaked Common Dolphins spend considerable time near the surface and produce a fair amount of surface "splash" as they breach to breathe. While travelling, the herd tends to synchronize their breathing and usually leap out of the water with each breath. The most common fishing strategies of this species are cooperative and highly coordinated. Colourful terms like carouseling (group circle-herding of the fish into a tight fish ball against the water surface into which individual dolphins can dash to grab a meal), line-abreast (a group of dolphins side-by-side, driving a school of fish and picking off stragglers), and wall formation (a line-abreast group driving the school towards a waiting group of dolphins) have been coined to describe some of their techniques. Generally, while foraging the dolphins will spread out perpendicularly to the direction of travel, presumably to increase the chances of detecting prey. Individual hunting behaviours – bubble-blowing, kerplunking, fish-whacking, and high-speed pursuits round out their hunting abilities. Bubble-blowing is used to separate fish from the school. Kerplunking (named for its sound) is the behaviour whereby the dolphin slaps its tail flukes onto the surface of the water. The loud sound and associated bubbles startle nearby prey into a tight school. Fish-whacking by Short-beaked Common Dolphin is similar to that described in the "Behaviour" section of the Bottlenose Dolphin. Bow riding is a favourite activity; Short-beaked Common Dolphins will approach a boat at speed in order to assume a preferred spot in the bow wave.

VOCALIZATIONS

Each dolphin produces a whistle unique to them, termed a signature whistle. Sonar (echolocation) is used by these dolphins to navigate and hunt underwater.

SIGNS

This species may gather into very large herds that travel at fairly high speeds, breaching in synchrony to breathe, although such large herds are more common in warmer waters. Bow riding is a favourite activity.

REFERENCES

Bearzi et al. 2003; Connor 2000; Ferrero and Walker 1995; Gaskin 1992b; Goold 2000; Heyning 1999b; Heyning and Perrin 1994; Kinze 1994; Murphy et al. 2006; Neumann 2001; Neumann and Orams 2003; Neumann et al. 2002; Reeves and Mead 1999; Reeves et al. 2002; Rohr et al. 2002.

Short-finned Pilot Whale
also called Blackfish, Pothead Whale, Pacific Pilot Whale

FRENCH NAME: **Globicéphale des Tropiques**

SCIENTIFIC NAME: *Globicephala macrorhynchus*

Both pilot whales, the Short-finned and the Long-finned, are commonly called Pothead Whales or Potheads due to their characteristic head shape. They can be very difficult to distinguish at sea, but fortunately in Canadian waters each occurs off a different coast: Short-finned Pilot Whales in the west and Long-finned Pilot Whales in the east.

DESCRIPTION

By about two years old the melon has grown sufficiently to overhang the tip of the upper jaw. There is usually only the merest hint of a beak, except in very small animals. The mouthline slopes upwards toward the eye. The large and long-based dorsal fin is distinctively placed forward on the back, well ahead of the midpoint. It is usually falcate and rounded at the tip and the leading edge tends to be longer and thicker in mature males. Flippers are usually about 1/6th of the body length and have a distinctive sickle shape. Flukes are notched at their midpoint and the trailing edge is slightly concave. Many animals exhibit a multi-toned grey saddle patch behind the dorsal fin, but most are black overall except for an anchor-shaped white chest patch, which usually extends onto the belly down to the genital slits. Calves are born light grey or cream coloured and darken early in their first year. The white chest patch is evident at

birth, but becomes more distinct as the body colour darkens. The dental formula is 7–9/7–9, for a total of 28–36 teeth.

SIMILAR SPECIES

The Short-finned and Long-finned Pilot Whales cannot be reliably distinguished at sea. In Canadian waters, the two species do not overlap in range. Only the Short-finned Pilot Whale occurs in the North Pacific Ocean and only the Long-finned Pilot Whale occurs in the North Atlantic. False Killer Whales are found in the same waters as Short-finned Pilot Whales. Both are dark-bodied and of similar size. The head area of the False Killer Whale is considerably less bulbous; its dorsal fin is narrower based and farther back on the back; its flippers have a pronounced "hump" at the middle of the leading edge; the body is more slender and it indulges in more dolphin-like behaviour than the pilot whales, commonly jumping out of the water, breaching, and bow riding. These distinctions can be subtle and often difficult to detect at sea.

SIZE

This species is sexually dimorphic, with males being longer and heavier than females.

Total length: males, average 550 cm long, but up to 700 cm; females, average 425 cm long, but up to 600 cm; newborns, 140–185 cm long. Weight: males, up to 3000 kg; females, up to 1500 kg; newborns, 37–84 kg.

RANGE

Short-finned Pilot Whales occur worldwide in tropical to warm temperate waters. All known records in Canadian waters have been from April to October, off the British Columbia coast. In the Pacific Ocean, the only pilot whales are Short-finned Pilot Whales, as the Long-finned species is extinct in that ocean. In other oceans in the warmer waters, both species occur and tend to be lumped together in surveys, as they are very difficult to distinguish at sea. Off the east coast of Canada, the ocean temperatures are too cold for Short-finned Pilot Whales, which only occur as far north as New Jersey.

Distribution of the Short-finned Pilot Whale (*Globicephala macrorhynchus*)

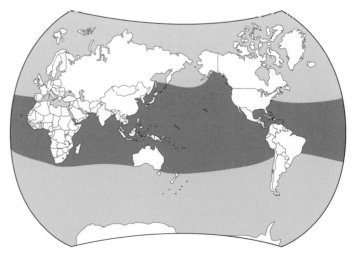

Short-finned Pilot Whale
(*Globicephala macrorhynchus*)

ABUNDANCE

Rare and seasonal in western Canadian marine waters, this whale is more common in warmer waters towards the middle of its range. Japanese, and possibly Caribbean, stocks are declining due to over-harvesting.

ECOLOGY

These whales tend to remain offshore during the late fall and winter and move inshore during the summer, following feeding or migrating squid. Most hunting occurs after dark when the deep-water prey migrates up the water column to feed. Captive animals have been trained to dive as deeply as 500 m and to remain submerged for 15 minutes. Mass strandings, sometimes involving hundreds of animals, are common for this species. In waters off Santa Catalina Island off the coast of California, Short-finned Pilot Whales were common until a 1982–3 incursion of El Niño warmed the waters and Risso's Dolphins moved in. Both species are squid eaters and the waters in dispute are rich feeding grounds. Encounters between the two species suggest that the smaller Risso's Dolphins were dominant and were able to drive out the larger Short-finned Pilot Whales. Like the closely related Long-finned Pilot Whale, the ratio of gender in calves is 1:1, while among adults the sex ratio is heavily biased towards females. Why the males suffer a higher mortality rate is a mystery. The degree of sexual dimorphism suggests that the males compete with each other for mates. Perhaps this competition increases their rate of mortality, although there is little evidence, in the form of strandings, to support this theory. Some males can become heavily scarred, presumably from the teeth of conspecifics. The lifespan of females appears to be almost 30% longer than that of males, which would certainly skew the ratio. Studies of populations around Hawaii and along the North American west coast, using techniques that identify individuals, shows they display a high degree of site fidelity, with the same animals appearing at the same time each year at the same locations.

While not common in oceanariums, this species has a long history in captivity dating back to 1935. It has bred and produced live calves, and captive individuals have lived longer than 30 years. Maximum known age of wild individuals is 62 years old for a female, 45 years for a male.

DIET

The diet of Short-finned Pilot Whales is dominated by squid, sometimes supplemented with octopus and fish. The major inshore and offshore movements of these whales are largely governed by similar movements of squid, and most prey is made up of deep-water species.

REPRODUCTION

Much of our information about reproduction in this species is derived from the study of animals captured in the Japanese drive fisheries. It is possible that the eastern Pacific stock (along our west coast) varies, in as yet unknown ways, from their Japanese counterparts.

The gestation period is about 15 months and the single calf is usually nursed for two to three years. The interval between calves is five to eight years, with younger females producing calves more frequently than older females. Females remain fertile until they are around 40 years old, after which they enter a period of reproductive senescence, but may continue to occasionally nurse their last calf for as long as 15 years. The breeding and calving season varies by location, and there is insufficient data available to comment on the reproductive timing of our North American stocks. Small calves

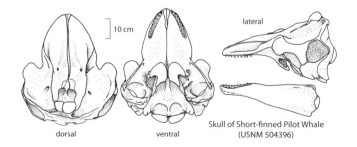

10 cm

lateral

Skull of Short-finned Pilot Whale
(USNM 504396)

dorsal   ventral

have been seen off the Atlantic coast of the United States in autumn. Females become sexually mature at around 9 years old, males at around 15 years old.

BEHAVIOUR

Pilot whales are highly social animals, usually seen in mixed herds of 15–50 individuals and occasionally in herds of up to several hundred. Probably the large herds are composed of several smaller groups spread over an expanse of ocean. Since most Canadian sightings are of single animals, it is possible that smaller groups and even solitary animals are more common on the fringes of the range. Apart from extensive vocal communication, Short-finned Pilot Whales also employ considerable non-vocal communication in the form of body posturing, spyhopping, lobtailing, and physical contact with other herd members. Adults have been seen disciplining juveniles and calves by striking them with their heads. Full breaches are uncommon with this species, but they do perform many other surface activities such as rolls and swimming upside down or on their sides. Dragging around inanimate objects, such as pieces of seaweed or plastic or even carcasses of dead sea lions may be play behaviour, but this is unclear at present. Many Short-finned Pilot Whales display scars that suggest tooth raking by other pilot whales, but since this activity has never been observed, it is uncertain whether the wounds are caused by herd mates or by pilot whales from other herds. The herds are matrilineal, with most females remaining in their natal herds and only the males sometimes moving to other herds. The small groups that make up the larger herds usually consist of an adult male, an older female, and several generations of her adult and juvenile descendents.

VOCALIZATIONS

Most of the vocalizations of this species recorded to date fall within the 2.2–10.8 kHz range. Like most of the social, toothed whales, this species communicates almost constantly with fellow herd members to coordinate travelling, hunting, and social activities.

SIGNS

These whales are very difficult to identify at sea. Look for a tight-knit herd of 15–50 large dark dolphins with dorsal fins that have a broad base and are placed forward on their body. Spyhopping is fairly common, but other aerial behaviours are uncommon. The flukes lift out of the water before a deep dive. Both this species and the Long-finned Pilot Whale characteristically create a crescent-shaped bow wave as they surface for a breath. This wave is the result of their extremely blunt head shape and is most visible from an airplane or a tall ship.

REFERENCES

Baird and Stacey 1993; Hoffman et al. 2004; Mintzer et al. 2008; Pyare, P.M., and Heinemann 1993; Reeves and Mead 1999; Rendell et al. 1999; Shane 1994, 1995b; Stacey and Baird 1993.

Long-finned Pilot Whale

also called Pothead Whale, Blackfish, Northern Pilot Whale

FRENCH NAME: **Globicéphale noir**

SCIENTIFIC NAME: *Globicephala melas*, previously *Globicephala melaena*

Both Long-finned and Short-finned Pilot Whales are also commonly called Pothead Whales, because of their characteristic head shape. In Canadian waters, the Long-finned Pilot Whale is found in the east and the Short-finned Pilot Whale in the west.

DESCRIPTION

The Long-finned Pilot Whale is a large, dark member of the dolphin family. It has a stocky body with a large bulbous melon that, in adults, overhangs the very short and barely noticeable beak. The line of the mouth slopes upwards toward the eye. A large, falcate and long-based dorsal fin is set forward of the mid-point of the back. The distinctive dorsal fin placement and shape are often the only identity clues provided at sea. The long, sickle-shaped flippers of this species are up to one-quarter of their total body length. Some of the whales have a light-coloured streak running back from the eye and a large, light saddle patch behind the dorsal fin. This coloration is more common among the southern stock; our North Atlantic stock does not usually display a distinct eye streak and rarely shows a distinct saddle patch. Most of the time, this saddle patch is an indistinct dark grey colour that does not stand out from the base colour. An "anchor"-shaped light grey or white patch occurs on the chest and is often connected to a pale stripe running down the belly that broadens and surrounds the genital area. The light areas on the belly are quite variable in size and shape. Albino or all-white animals have been reported, but are extremely rare. Calves are grey at birth and gradually darken as they age. The teeth of this species are conical and robust with a dental formula of 9–12/9–12, for a total of 36–48 teeth.

Distribution of the Long-finned Pilot Whale (*Globicephala melas*)

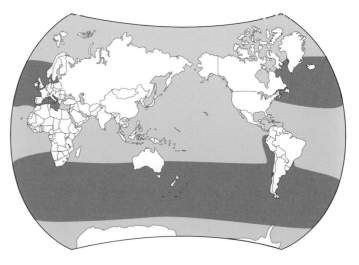

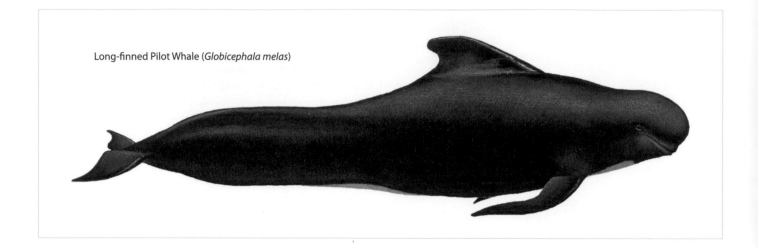

Long-finned Pilot Whale (*Globicephala melas*)

SIMILAR SPECIES

In the North Atlantic range of Long-finned Pilot Whales, there is only one other similar species, and that is the Risso's Dolphin. This whale has a large, dark dorsal fin but generally a much lighter body, tends to be seen in smaller herds, and is often more acrobatic in the air. It also has a squarish head, rather than the round bulbous head of the pilot whale, and is also considerably smaller than the Long-finned Pilot Whale, but this can be difficult to assess at sea. The most similar species is the very closely related Short-finned Pilot Whale. The main distinguishing features are often impossible to detect at sea; the Long-finned Pilot Whale has longer flippers, larger tail flukes, is somewhat larger, and has more teeth. The two species do not occupy the same waters off Canada's coasts, although that may change in future if the average sea temperatures rise and the Short-finned Pilot Whale is able to move northwards on the east coast. Currently, that species occurs only as far north as New Jersey. The False Killer Whale is also similar, but rarely reaches the cooler waters of the North Atlantic occupied by Long-finned Pilot Whales.

SIZE

This species is sexually dimorphic, with males achieving longer and heavier dimensions than females. The following measurements apply to eastern North Atlantic animals.

Total length: males, up to 6.3 m; females, up to 5.5 m; newborns, 160–200 cm.

Weight: males, up to 2500 kg; females, up to 1500 kg; newborns, around 75–100 kg.

RANGE

Long-finned Pilot Whales are a cold temperate species that mainly occur offshore, but will move inshore during warmer seasons. Their distribution is almost entirely determined by the presence of their prey. The seasonal movements inshore and into cooler northerly (and southerly in the southern hemisphere) waters as surface temperatures rise are driven by migrations of feeding squid and schooling mackerel and other fishes. They can be numerous in the Georges Bank, Scotian Shelf, outer Laurentian Channel, Grand Bank, the west coast of Newfoundland, and the Gulf of St Lawrence from early July to early December.

ABUNDANCE

Surveys tend to lump both pilot whale species together in their estimates, as discrimination is usually very difficult at sea. Pilot whales are considered abundant in parts of their range, especially in the North Atlantic, where most are the Long-finned species. Estimates of numbers vary. One survey suggests 400,000–500,000 Long-finned Pilot Whales, divided more or less equally between the North Atlantic population and the southern Antarctic population. Another survey estimates over 800,000 in the North Atlantic alone. The western North Atlantic stock (in waters off eastern Canada and western Greenland) has been estimated to number at least 10,000–20,000 animals. Drive fisheries off Newfoundland from 1947 to 1971 seriously depleted the local stocks, which have still not fully recovered. The only remaining drive fishery, undertaken since at least the 1500s, takes place in the Faeroe Islands north of Scotland. It appears to be sustainable, taking around 1500–2500 whales per year, with no apparent repercussions on the local stock. A North Pacific population, known only from archaeological records dating back to the 8th to the 12th centuries, is extinct, possibly through competition with the Short-finned Pilot Whale.

ECOLOGY

Most feeding takes place shortly after sunset until early morning, when the deep-water squid begin to move up the water column as light levels diminish. Most prey is taken at depths between 200 and 500 m, although pilot whales can dive deeper if necessary. Dives to depths around 650 m have been reported. The longest dive on record lasted almost 28 minutes, but most are 8–12 minutes long or less. Pilot whales have a long history of mass stranding, often in large numbers. It is possible that mass stranding is the highest single cause of mortality in the species. Killer Whales and sharks probably take some. Entanglement in fishing gear is a regular cause of mortality throughout their range and tends to be more common among juveniles. Human hunting was formerly common in many regions, but is now limited to a yearly drive fishery in the Faeroe Islands and

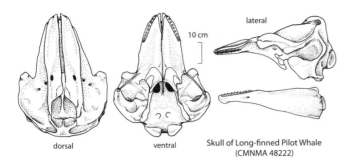

dorsal　　　　ventral　　　lateral

Skull of Long-finned Pilot Whale
(CMNMA 48222)

occasional catches in Greenland. These whales are strong swimmers and speeds of more than 16 km/h can be sustained for long periods. Males are known to live as long as 50 years and females up to 60 years old. Few females older than 40 produce calves; however, they likely continue to play an important social role in the pod. Although the birth ratio is equal for males and females, the ratio becomes skewed towards females as the animals mature. The reason for the higher rate of male mortality is unknown, but may be the result of fighting among the males. Intentional head-to-head collisions have been reported and males typically display more body, and particularly melon, scarring than do females. Long-finned Pilot Whales are rare in captivity. A growing threat to these whales is the increase in marine pollution, particularly the pervasive, albeit terrestrial in origin, organochlorides such as DDT and PCB and heavy metals such as mercury. Concentrations of these contaminants are increasing in their tissues and may eventually put an end to human consumption of their flesh.

DIET

Northern Long-finned Pilot Whales eat a variety of different prey, but focus on squid as their main food. Long-finned Squid and Short-finned Squid are the two principal western North Atlantic prey species. If squid becomes scarce, the whales readily divert their attention to whatever fishes are locally abundant. These whales are also known to eat shrimp, especially as juveniles. Southern populations of Long-finned Pilot Whales eat as much fish as they do squid. Suction is the usual method these whales use to capture their food.

REPRODUCTION

Females become sexually mature at around 6–9 years old while males take until they are closer to 12–14 years old. Males continue to grow until they are at least 20 years old, while females reach physical maturity between the ages of 12 and 15 years old. The breeding season in the western North Atlantic stretches from April to November, but mating peaks in April and May. Most births occur in late summer after a gestation period that has been estimated to be about 12–14 months. Calves nurse for at least two years, but weaning begins at around six months old, about the same time the teeth begin to erupt. Females produce a single calf about every four or five years and it is not uncommon for a female to become pregnant while she is still nursing her last calf. The sex ratio of calves is about 50:50.

BEHAVIOUR

Long-finned Pilot Whale herds in Canadian waters range from 2 to 135 individuals. In other parts of the range, herds may be as large as 200 whales. Herd size can be difficult to assess at times, as larger herds are often dispersed and broken into smaller close-knit groups of 10–20 individuals scattered over a large expanse of ocean. These smaller groups are thought to comprise a female and several generations of her offspring. In a very unusual exception to the more common mammalian practice of subadult males leaving their natal herd, adult male Long-finned Pilot Whales remain with their mother's herd for their whole life. There is a taboo against mating within the herd, so that mating only occurs when aggregations of more than a single herd temporarily coalesce. There are also rare accounts of small male-only herds travelling among the larger family herds. The strong social bonds within the herds allow them to be driven by boats into shore, where they can be slaughtered. The strength of this bond is noted in an ancient Faeroese expression, "The whales return to the blood." Hunters have observed that if any whales break from the main herd after the slaughter has begun and the sea contains the blood of their kin, they will rejoin their herd rather than take advantage of the opportunity to escape. This same powerful herding instinct is likely a significant factor in the rate of mass strandings noted for this species. Whole herds will follow a leader onto the beach to die, rather than be separated.

When foraging at depth, the whole herd dives in synchrony and is suspected of cooperatively hunting schooling squid hundreds of metres deep. As many as 1400 individual dives may be made each night. Travelling herds often advance in a "chorus line" of animals side by side, strung across almost a kilometre of ocean. This strategy probably enhances their chances of detecting prey. Most hunting takes place at night, when the deep-water prey moves up the water column. During daytime some hunting near the surface may be undertaken, but most of this time is spent resting, socializing, playing, and travelling. Pilot whales are commonly seen in association with other cetaceans.

VOCALIZATIONS

The social nature of this whale is reflected in its rich vocal repertoire. Simple sounds are produced when the herd is at rest, and more frequent and more complex sounds are produced while the whales are hunting or socializing. A variety of whistles ranging from 0.5 to 8.8 kHz, double clicks, and echolocation pulses have been recorded. This species is also capable of producing two entirely different audio signals simultaneously, and signature whistles (an individually recognizable whistle produced by an animal to identify itself to others in the herd) and dialects are suspected.

SIGNS

These whales are very difficult to identify at sea. Look for a tight-knit herd of 10–20 large dark dolphins with dorsal fins that have a broad base and are placed forward on their body. Spyhopping is fairly common, but other aerial behaviours are uncommon. Both this species and the Short-finned Pilot Whale characteristically create a crescent-shaped bow wave as they surface for a breath. This wave is the result of their extremely blunt head shape and is most visible from an airplane or a tall ship.

REFERENCES

Amos, W., et al. 1991; Andersen, L.W., and Siegismund 1994; Fertl et al. 1999; Fullard et al. 2000; Gannon, D.P., et al. 1997a, 1997b; Kingsley and Reeves 1998; Mate et al. 2005; Nawojchik et al. 2003; Nelson and Lien 1996; Ottensmeyer and Whitehead 2003; Pyare, P.M., and Heinemann 1993; Reeves 1999b; Reeves et al. 2002; Rendell et al. 1999; Werth 2000.

Risso's Dolphin

also called Grampus, Grey Grampus, White-headed Grampus

FRENCH NAME: **Dauphin de Risso**

SCIENTIFIC NAME: *Grampus griseus*

Like Sperm Whales, Risso's Dolphins have specialized in capturing squid, and, like Sperm Whales, they have no teeth in their upper jaw.

DESCRIPTION

Risso's Dolphin is a greyish-white, square-headed whale with no noticeable beak. Newborns are light bluish-grey, but soon darken to blackish. As the animals age, they become lighter, especially around the head, flanks, and belly. They have a large white patch between their flippers on the chest when they are born and gradually develop another in the middle of the belly below the dorsal fin. The skin of Risso's Dolphins scars easily. Adults develop widespread scarring, likely caused by the teeth of other Risso's Dolphins, along with scratches from squid beaks and tentacles and possibly as a result of bites by lampreys, Killer Whales, and sharks. These scars heal to a whitish colour and add to the overall light grey coloration of the adults. Some older adults look almost white, due partly to age and partly to scarring. The most distinctive feature of this species, aside from the extensive scarring on the body, is the forehead crease that begins at the tip of the upper lip and runs about one-third to one-half the way back towards the blowhole. No other cetacean has this crease. Unfortunately, this characteristic is very difficult to detect in the field. This medium-sized dolphin is fairly thick from the head to the dorsal fin and then noticeably slender in the tail stock. The dorsal fin is large, falcate, and, like the flukes and tips of the flippers, remains darker than the rest of the body. The dental formula is 0/2–7, for a total of 4–14 teeth. Most commonly each lower jaw has 3–4 teeth. These teeth are relatively large and usually become heavily worn in adults. Lack of teeth in the upper jaw and a reduction of numbers in the lower jaw has been suggested as an adaptation to eating squid. The bones of the face and jaws slope downward, providing room for the large melon. The angle can be visualized by following the downward angle of the mouth line.

SIMILAR SPECIES

The combination of a large, dark dorsal fin and a light body with extensive white scarring distinguishes this species from other whales within its range. Bottlenose Dolphins, False Killer Whales, Pacific White-sided Dolphins, and female and juvenile Killer Whales have a similar dorsal fin, which at a distance could be mistaken for Risso's Dolphin's. Bottlenose Dolphins have a noticeable beak and usually a darker grey body; False Killer Whales are dark-bodied; Killer Whales are much larger, with a distinct black and white body, and Pacific White-sided Dolphins are smaller, with a distinctive black and white body. Cuvier's Beaked Whales have a similar body colour, but are considerably larger.

SIZE

There appears to be little difference in size between sexes.
Total length: 250–385 cm; newborns, around 120–150 cm.
Weight: up to 400 kg; newborns, unknown.

Risso's Dolphin (*Grampus griseus*)

Distribution of the Risso's Dolphin (*Grampus griseus*)

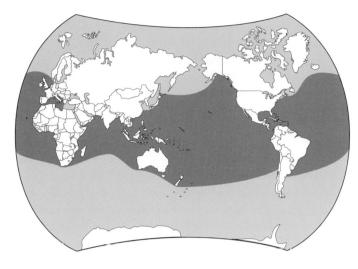

RANGE

Risso's Dolphins are found worldwide in warm and warm temperate seas with surface temperatures ranging from 4.5°C–28.0°C. Highest populations occur where temperatures fall within the range of 21°C–25°C. They are found most frequently along the edges of continental shelves, in waters averaging 1000 m deep, but herds will venture onto the continental shelves and into deeper oceanic waters at times. They are near the northern limits of their distribution in Canadian waters on both the east and west coasts, and strandings and sightings are rare, especially in the Atlantic.

ABUNDANCE

Although considered rare in Canadian waters, this dolphin can be seasonally common in some parts of its range. No worldwide population sizes are available; however, estimates of 16,000–30,000 along the US east coast, 32,000 along the US west coast, 175,000 in the eastern tropical Pacific, and 85,000 in the western North Pacific are suggested. This dolphin is still killed as by-catch in many types of fishing nets around the world and continues to be harvested in Japan, the Solomon Islands, Indonesia, and the Lesser Antilles, where it may be suffering local population declines as a result.

ECOLOGY

Surprisingly little is known about this species, partly because they are usually found in deeper waters and partly because they do not commonly strand. Risso's Dolphin numbers and distribution tend to fluctuate over the long term, likely as a result of environmental factors. Changes in the location of warm water currents can have a tremendous impact on their prey, hence on the dolphins. In the early 1980s around California, an incursion of El Niño brought warmer waters into the area. The number of resident Short-finned Pilot Whales dropped drastically and the number of Risso's Dolphins, formerly uncommon, increased dramatically. Possibly the lifestyle of Risso's Dolphins allows them to remain flexible and hence able to exploit such rapid habitat changes. Risso's Dolphins are most common along the edges of continental shelves where their squid prey

is found. The dolphins will follow the squid into shallower waters when they migrate to spawn.

Individuals have been held in captivity since at least 1957, and live births in captivity have been recorded since 1962. The longest an animal has lived in an oceanarium is 35 years, 8 months.

DIET

Risso's Dolphins primarily eat squid. Other cephalopods, such as octopus and cuttlefish, and some fish, such as anchovies, are taken occasionally.

REPRODUCTION

There is very little information available on the reproduction of this species. Sexual maturity of Risso's Dolphins is achieved around the time they reach 2.5–2.8 m in length for both sexes. Males are thought to be around 10–12 years old and females possibly around 10 years old at that length. The calving interval is unknown, but thought to be every two to three years at the shortest and likely much longer. Calving may be regionally seasonal, but young calves have been seen at all times of the year. The gestation period is thought to be 13–14 months. Females have successfully conceived and given birth in captivity. Male Risso's Dolphins are known to mate with female Bottlenose Dolphins to produce a hybrid calf. This occurs not only in captivity, but also in the wild.

BEHAVIOUR

These dolphins are highly social and are almost always seen in herds of 12–40 individuals. Herds in Canadian waters tend to be smaller than average, in the range of 2–30 animals. Amalgamated herds of up to 3000 have been recorded. Solitary individuals have been sighted, but not commonly. Within the larger herds, segregation occurs with clustering of females and calves, juveniles, and adults. Smaller herds have been reported that are made up of females, calves, and juveniles that are not yet sexually mature. It is theorized that the large mature males may rove between herds remaining long enough to inseminate the oestrous females before moving on. Risso's Dolphins are also frequently seen in mixed herds with other dolphins, especially Pacific White-sided Dolphins, pilot whales, Northern Right Whale Dolphins, and Dall's Porpoises, but also sometimes with the larger whales such as Sperm, Fin, and Grey Whales. They feed primarily at night when their squid prey migrates up the water column towards the surface. During the daytime, herds travel in a tight oval most of the time, but are also known to form tight lines and travel in echelon; their travelling speeds tend to be slow, 6 km/h or less. Likely these speeds reflect the fact that the animals are loafing or even sleeping. They are no doubt capable of considerably faster movement while hunting. Interactions between Risso's Dolphins and with other cetaceans can be quite physical. The animals rub, slap, splash, and even leap onto one another, which may account for many of the body scars the adults accumulate.

Although this species is not renowned for bow riding, a famous Risso's Dolphin named "Pelorus Jack" regularly escorted ships through Admiralty Bay, New Zealand, from 1888–1912 – 24 years of riding bow waves. They are known to "bow-ride" on the front wake

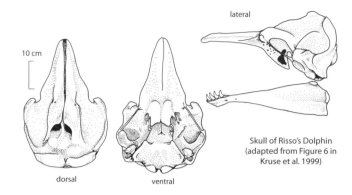

lateral

10 cm

dorsal   ventral

Skull of Risso's Dolphin
(adapted from Figure 6 in
Kruse et al. 1999)

of migrating large whales. Risso's Dolphins frequently leap clear of the water while travelling at speed, and at times will engage in considerable tail slapping and head bobbing and even leap partly out of the water to land with a crash on their sides. Tail slapping is usually an aggressive cetacean behaviour.

VOCALIZATIONS

Amazingly, despite the history of this species in captivity, the proof that Risso's Dolphins echolocate has only recently been provided. This same study shows that the angle of the sonar beam more or less matches the downward angle of the jaw, which measures 30°–40° off the horizontal. While it would seem likely that the unique vertical groove in the melon should have some impact on sound production, its function remains unproved. Most recorded sounds fall within the range of 2.2–13.4 kHz.

SIGNS

Tight herds of light-coloured dolphins with tall, dark dorsal fins are going to be Risso's Dolphins. Although this species is capable of creating considerable "splash" when it chooses, it is also able to surface fairly smoothly with hardly any spray.

REFERENCES

Amano and Miyazaki 2004; Baird and Stacey 1991a; Kruse et al. 1999; Lawson and Eddington 1998; Philips et al. 2003; Reeves and Mead 1999; Reeves et al. 2002; Rendell et al. 1999; Shane 1995a, 1995b; Shelden 1999; Shelden et al. 1995.

behaviour. Much of what is known is a result of the study of dead animals involved in two mass strandings in the 1970s.

DESCRIPTION

The bright white splotch on each side of this dolphin below the dorsal fin is the most visible diagnostic field characteristic. Identification can be confirmed if the yellowish caudal patch and the large, curved, and sharply pointed dorsal fin are also seen. On closer examination, the short beak, dark line from the eye to the beak, and faint grey line from the eye region to the flipper may be visible. The top of the body and flippers are black, the midline is grey and the belly is white. Demarcations between different coloured areas are sharply defined, without the smudgy appearance of many other species of small whales. Albino or all-white animals are very rare. The dental formula is 20–40/31–38, for a total of 102–156 small conical teeth.

SIMILAR SPECIES

Two other dolphins, the White-beaked Dolphin and the Short-beaked Common Dolphin, can be found in the same waters as the Atlantic White-sided Dolphin and may be confused with this species. The White-beaked Dolphin is of a similar size, with a similar dorsal fin, but has a lighter beak, a large black blotch on its side, and a lot of light greyish-white on its tail stock and sides. Its dark upper markings create a "saddle" around the dorsal fin. The Short-beaked Common Dolphin is slimmer and has a longer, more slender beak. Its dark upper markings dip down in a "V" shape rather than the straighter line of the White-sided Dolphin. The light patch on its side is farther forward, just behind the eye, rather than in the middle of the side, as in the White-sided Dolphin. Harbour Porpoises, which also occupy a similar range, are considerably smaller, with a short dorsal fin and a more coastal habitat preference.

SIZE

Males are about 10% longer and up to 45% heavier than females.
Total length: males, 220–282 cm; females, 200–243 cm; newborns, 100–120 cm.
Weight: males, 230 kg+; females, 180 kg+; newborns, around 20 kg.

Atlantic White-sided Dolphin

FRENCH NAME: **Dauphin à flancs blancs**

SCIENTIFIC NAME: *Lagenorhynchus acutus*

Despite the frequency of sightings of this species off the Canadian Atlantic coast, much remains to be learned of its ecology and

Distribution of the Atlantic White-sided Dolphin (*Lagenorhynchus acutus*)

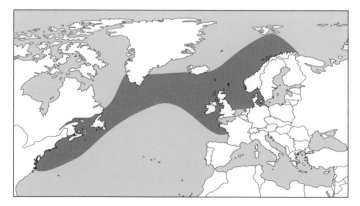

Atlantic White-sided Dolphin (*Lagenorhynchus acutus*)

RANGE

Atlantic White-sided Dolphins are found only in temperate and sub-polar waters of the North Atlantic, from about North Carolina to Greenland in the west and France to Svalbard in the east. Generally the populations travel north as the waters warm in the summer and then southwards as the temperatures drop in late autumn. They enter the Gulf of St Lawrence during the summer, moving as far west as the mouth of the Saguenay River.

ABUNDANCE

This dolphin can be locally common in season, but its total population size is unknown.

ECOLOGY

Atlantic White-sided Dolphins are found near land, for example, in the Bay of Fundy where depths are less than 300 m, as well as out at sea, for example, along the Scotian Shelf in waters 1000–2500 m deep. It is possible that, like the Bottlenose Dolphin, Atlantic White-sided Dolphins may be found to exist as two different populations with different habitats, one inshore and one offshore. These dolphins are tolerant of cool water temperatures, being most frequently found in waters between 7°C and 12°C. From the stranding data, it appears that the North American population moves north in the summer – June to September as far as Greenland – and move as far south as North Carolina as temperatures drop in the winter. They undertake dives to depths of up to 1000 m for periods of up to about four minutes. Most dives last less than a minute and are much shallower. Before the 1970s, White-sided Dolphins in US waters were found mainly offshore, and their close relatives, White-beaked Dolphins, were found in the shallower continental shelf waters. During the 1970s it seems there was a switch of habitat use between these two species.

These dolphins are frequent casualties of the fishing industry. They are caught in gill nets, drift nets, bottom trawls, and off-bottom trawls throughout their range. In US waters, an estimated 181 are drowned yearly in nets. Similar Canadian and European mortality

is unknown. Mass strandings of herds as large as 100 animals are not uncommon for this species. Reasons for these strandings are unknown, but historical records indicate that such strandings have occurred since antiquity. The travelling speed for this species depends on whether it is foraging at the same time and varies from around 3 km/h to around 45–50 km/h. The known lifespan is at least 22 years for males and 27 years for females. In all likelihood these ages will increase as more data becomes available.

DIET

Small schooling fish such as herring, mackerel, Whiting, Smelt, and Silver Hake and schooling squid are the principal prey of the Atlantic White-sided Dolphin. Shrimp are occasionally taken, especially by juveniles.

REPRODUCTION

Sexual maturity of both males and females is reached between 6 and 12 years old. Most births occur in June and July. Mature females produce a single calf about every two years, following a pregnancy of around 11 months. The calves nurse for about 18–24 months. Once weaned, the youngsters leave the breeding herd and join a juvenile herd or even a group of cetaceans of another species until they are mature.

BEHAVIOUR

Aggregations of hundreds of Atlantic White-sided Dolphins have been recorded at particularly favourable feeding sites, but most herds in Canadian waters contain 9–12 animals. Herd size changes around Newfoundland, where 50–60 animals are often seen together. Herd size is largest from August through October after the calves are born and the smaller herds begin to amalgamate, perhaps to socialize, breed, migrate, or feed. Herd members form strong bonds. The cohesiveness of these bonds, even in the large amalgamated herds, enables human hunters to drive them into shallow waters where they can be easily harpooned. Atlantic White-sided Dolphins often associate with other cetacean species, particularly

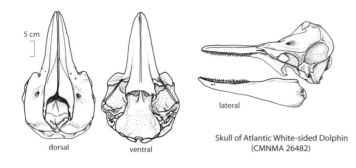

5 cm

dorsal   ventral   lateral

Skull of Atlantic White-sided Dolphin
(CMNMA 26482)

White-beaked Dolphin

FRENCH NAME: **Dauphin à nez blanc**

SCIENTIFIC NAME: *Lagenorhynchus albirostris*

White-beaked Dolphins share the dubious distinction, along with the St Lawrence River Belugas, of having the highest levels of toxic pollutants, such as PCBs and DDT, in their tissues of any of the whales.

DESCRIPTION

This is a robust, short-beaked dolphin whose distinct colour pattern allows for its identification in the field, given adequate lighting. Look for the large dorsal fin with a light tail stock behind, an irregularly shaped, black side patch behind the flippers, and a black saddle-shaped patch around the dorsal fin. Apart from the edges of the white chin patch, which are distinct, all the other regions of transition between dark and light pigmentation are smudged or blurry. The short beak, for which the species is named, is unfortunately not always white and so cannot act as a good distinguishing characteristic. The colour patterns in this species are highly variable and tend to be less apparent on juveniles. The lighter area behind the dorsal fin is usually a good discrimination feature even in the juveniles. White-beaked Dolphins have fewer, but larger teeth than their close relatives, Atlantic White-sided Dolphins. The dental formula is 23–28/22–28, for a total of 90–112 conical teeth.

SIMILAR SPECIES

Atlantic White-sided Dolphins are very similar in body shape and distribution to White-beaked Dolphins. There are subtle differences in head shape – the White-beaked Dolphin has a higher melon and a shorter beak – but the easiest differences to detect in the field are those of body pigmentation. The Atlantic White-sided Dolphin has a continuous black back from the top of its head to its tail, a distinct and discrete white patch in about the middle of its side, below the black back at about the midpoint of the body, and a tan or ochre patch on each side of its tail stock. The White-beaked Dolphin does not have a continuously black back. The dark pigmentation is broken behind the dorsal fin by a large whitish region that extends

the Northern Bottlenose Whale and Fin Whale. Bow-wave and stern-wake riding on powered vessels occurs, as does bow riding on the pressure wave produced by swimming large whales. Humpback Whales are less tolerant of this behaviour than other whales, and attempts will provoke these otherwise gentle animals into tail slashing, breaching, and even trumpet blows: all aggressive but otherwise harmless actions meant, no doubt, to discourage the actions of the dolphins. Aerial behaviours by Atlantic White-sided Dolphins are fairly common. Mainly these involve leaping out of the water in a smooth arc while taking a breath. Acrobatics are considerably less common, but occur most often in larger groups as individuals make higher leaps into the air with twists and turns included. Display swimming – two animals swimming side-by-side and upside-down at high speed – may be a precursor to mating. Lobtailing, when the dolphin pushes its tail up out of the water and slaps it forcibly down onto the surface, has been recorded by this species. Members of a herd will engage in cooperative fish herding. They circle a school of fish, driving them into a tight fish ball, then individual dolphins dart in to snatch a meal, while the others continue to herd the fish to prevent their escape. Herds of these dolphins have been seen swimming around large whales as they forage, likely taking advantage of and perhaps even assisting the fishing activities.

VOCALIZATIONS

A few "clicks" and squeals in the 1–24 kHz range have been recorded from wild Atlantic White-sided Dolphins, but otherwise little is known about the dolphin's echolocation and communication sounds.

SIGNS

This species will surface with just the dorsal fin and back exposed, with the whole body out of the water, or sometimes will perform aerial acrobatics.

REFERENCES

Fertl et al. 1999; Gowans and Whitehead 1995; Leopold and Couperus 1995; Mate et al. 1994; Palka et al. 1997; Weinrich 1996; Weinrich et al. 2001.

Distribution of the White-beaked Dolphin (*Lagenorhynchus albirostris*)

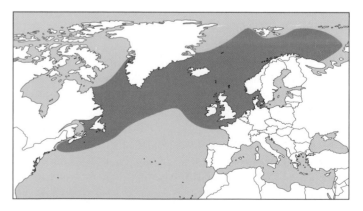

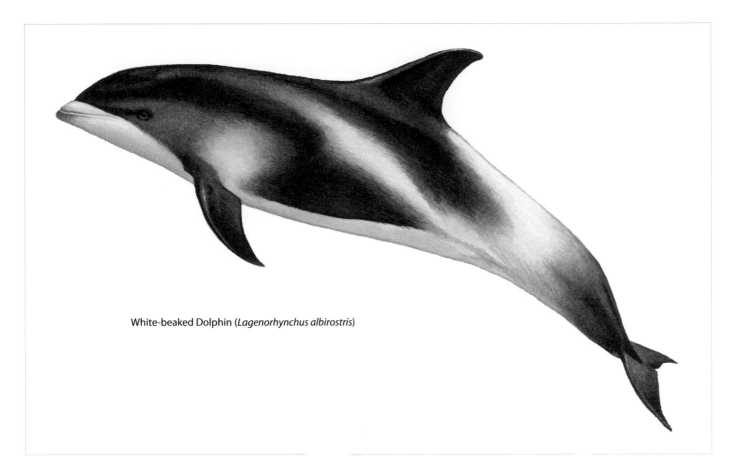

White-beaked Dolphin (*Lagenorhynchus albirostris*)

down onto the sides of the tail stock. Even when the animal shows only part of its back as it rolls at the surface to breathe, this light tail region is visible for a fraction of a second, unless the light is coming from behind. The Pacific White-sided Dolphin is also similar to both the dolphins already mentioned, but is only found in the Pacific Ocean.

SIZE
Males are larger and heavier than females, but insufficient data exists to specify by what percentage.
Total length: males, 251–310 cm long, but few grow beyond 300 cm; females, 174–248 cm; newborns, around 110–120 cm.
Weight: males, up to 354 kg; females, up to 310 kg; newborns, unknown.

RANGE
White-beaked Dolphins are found only in cold temperate and Subarctic waters of the North Atlantic. The population in the western North Atlantic spends the summer in the Davis Strait between Greenland and Baffin Island and along the Labrador coast, sometimes as far south as Newfoundland. Most winter in the Gulf of Maine, off Cape Cod. The population in the eastern North Atlantic is not as migratory, focusing their distribution in the North Sea and the seas around the north of the British Isles. There is some evidence that the distribution in the eastern North Atlantic has extended southwards since the 1980s to include the north coast of France.

ABUNDANCE
This species is more common in the eastern North Atlantic (off Europe) than in the western North Atlantic (off North America). The eastern population appears to be growing while the western one is declining. Total numbers have been suggested of perhaps as many as 100,000 animals.

ECOLOGY
White-beaked Dolphins are herd animals that have been seen in groups as large as 1500 individuals. Most herds are around five to eight animals and the very large herds are exceptional amalgamations of many subgroups. Pods of 10–35 and up to 100 animals have been seen on the north shore of the Gulf of St Lawrence. Herds tend to be of three types: mixed genders and ages, juveniles, and breeding-age females, with or without calves. This dolphin is tolerant of water temperatures between –1°C and 15°C.

Due to their presence in cold Subarctic waters, these dolphins can be vulnerable to ice entrapment in years when the pack ice is heavy and forms rapidly or is blown in quickly by the wind. They are regularly caught in ice off the coast of Newfoundland and Greenland. Mortality of entrapped animals is estimated at 55%. They are less subject to mass strandings than their close relative, the White-sided Dolphin, and most strandings are of smaller groups or single animals. No known incidences of predation are recorded, but likely Killer Whales and sharks take some. Although this species has been hunted in the past and continues to be taken

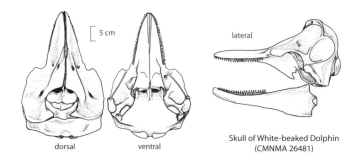

dorsal   ventral

lateral

Skull of White-beaked Dolphin
(CMNMA 26481)

occasionally, the major human pressures on it now are probably entrapment in fishing gear and marine pollution. Organochlorides and heavy metals in unprecedented high levels have been detected in the tissues of animals from Newfoundland; DDT and PCB levels were also high. These values compare with, and sometimes exceed, the levels found in the tissues of Belugas from the St Lawrence.

DIET

Schooling fishes like sand-eels, Capelin, Atlantic Cod, herring, Whiting, Haddock, Poor-cod, and Hake are the principal prey. Less frequently, some benthic crustaceans such as crab, as well as octopus and schooling squid, are also taken.

REPRODUCTION

Very little is known of the reproductive biology of White-beaked Dolphins. In the North American population, males are sexually mature by 13 years old, females by 16 years old. These ages are based on counting growth layers in the teeth of stranded animals and assume that each dental layer is equivalent to a year. In the European population, sexual maturity may be achieved at younger ages. Births occur from July to September in European waters and from May to October in North American waters. Length of gestation and lactation, and age at weaning, are unknown. Likely each female produces a single calf, but the time interval between pregnancies is also unknown. Weaned calves frequently leave the breeding herd and join juvenile herds until they reach breeding age.

BEHAVIOUR

As do many of the dolphins, this species will cooperatively herd schooling fish into a fish ball against the water surface. Individual dolphins then extract a meal, while the rest of the herd prevents the ball from dispersing. White-beaked Dolphins are known to hunt Capelin in association with Fin Whales. Whether this is an alliance that benefits either, or perhaps both, species, or is purely circumstantial due to the presence of the fish, is unknown. These dolphins have been seen taking Capelin out of the mouths of feeding Fin Whales.

White-beaked Dolphins frequently bow-ride and stern-wave ride on the wash from powered vessels. They are often acrobatic in the air and will leap almost straight out of the water, often turning to re-enter on their side or back.

VOCALIZATIONS

Like other dolphins, White-beaked Dolphins use whistles for communication between herdmates. These sounds fall within the range of human hearing. Echolocation clicks (sonar) fall mainly in the 59–75 kHz range with a secondary peak in the 200–250 kHz range. These are well beyond human hearing and in fact the secondary peak is a higher frequency than those known to be produced by any other dolphins. One of the proposed reasons for the high secondary peak is that sand eels, a group of fish species that are a primary prey in summer, do not have swim bladders and in addition possess very small ear bones. Both of these organs are highly reflective of sonar. The dolphins must compensate for the lack of sonar reflection by using highly directional, ultra-high-frequency sound in order to detect these fishes.

SIGNS

White-beaked Dolphins often bow-ride and breach. Frequently when they breach, they do not dive cleanly back into the water, but will turn on their side or even upside down in the air, returning to the water with much "splash."

REFERENCES

Hai et al. 1996; Kingsley and Reeves 1998; Lien et al. 2001; Northridge et al. 1997; Rasmussen and Miller 2002; Ree 1994; Reeves et al. 1999, 2002.

Pacific White-sided Dolphin

FRENCH NAME: **Dauphin de Gill**, *also* **Dauphin à flancs blancs de Pacifique**

SCIENTIFIC NAME: *Lagenorhynchus obliquidens*

Before 1984, these dolphins were rarely seen near land in British Columbia. Since then, they have become the region's most abundant inshore cetacean. Archaeological sites from the area dating back up to 4000 years contain many teeth from Pacific White-sided Dolphins, so clearly the species used to be common enough to attract the attention and be captured by the indigenous people using canoes and bone harpoons. After disappearing for a few centuries, they have returned.

DESCRIPTION

Pacific White-sided Dolphins have a very short, dark beak, a white belly, dark grey or black back and sides, with a large, light-grey patch above the flippers and a white stripe running from the upper back to the flukes, which broadens as it reaches the tail stock. The pigmentation pattern is variable and often sufficiently distinct to identify individuals. Their dorsal fin is large and recurved with a

Pacific White-sided Dolphin
(Lagenorhynchus obliquidens)

dark leading edge and a grey centre and trailing edge. The larger, older males display a very recurved dorsal fin, usually with many nicks and scars. The flippers are usually dark, but often have a greyish trailing edge. Colour variants occur infrequently. Melanistic and albinistic individuals have been noted. The dental formula is 23–36/25–35, for a total of 96–142 small conical teeth.

SIMILAR SPECIES

There are three other gregarious, acrobatic dolphins known to be in western Canadian waters. All of them are rare. They are the Short-beaked Common Dolphin, the Long-beaked Common Dolphin, and the Striped Dolphin. All three have prominent and noticeable beaks. The common dolphins both show a V-shaped dark pigment pattern on their sides. Striped Dolphins have a smaller, less recurved dorsal fin, and a dark strip along their sides running from the corner of the beak to the anus. Dall's Porpoises share the White-sided Dolphins' habit of racing along the water surface creating lots of "splash." If a closer look is possible, the smaller dorsal fin and more rotund body with a large distinct white patch that covers the belly and flanks will distinguish the porpoise. Furthermore, Dall's Porpoises rarely leap right out of the water as the dolphins commonly do.

SIZE

Adult males are slightly longer and heavier than females.
Total length: males, up to 250 cm; females, up to 236 cm; newborns, 90–105 cm. Calves in their first year grow to about 130 cm long, to about 140–180 cm in their second year, and to about 150–180 cm in their third year.
Weight: males, maximum weight 198 kg; females, maximum weight 145 kg; newborns, around 15–17 kg.

RANGE

Pacific White-sided Dolphins inhabit temperate and cold temperate waters of the North Pacific Ocean. They occur from the coast of Japan and the South China Sea across the Pacific to the Gulf of California and up to Alaska. In recorded history, the species has been considered an offshore species. This has changed along the coasts of British Columbia and Alaska since the mid-1980s, and the species is now commonly seen from shore in fjords and bays from approximately October through June.

ABUNDANCE

No reliable estimates of total population size are known, but the species is considered seasonally common in British Columbia waters. Numbers as high as 900,000 have been suggested for the world population.

ECOLOGY

Pacific White-sided Dolphins are often seen in large herds of around 500 or more animals. A herd of 6000 animals has been recorded, making this herd one of the largest of any cetacean. Such large aggregations are normally made up of several smaller herds that usually number in the range of 50–100 animals. The smaller herds will break up into even smaller subgroups when foraging. Some of these subgroups are made up of large, heavily scarred dolphins with exceptionally large, nicked, and hook-like dorsal fins. Probably made up of adult males, these small pods remain around the edge of the main herd and seem particularly shy and elusive, rarely approaching boats. Large amalgamated herds form in prime hunting areas, usually following spawning fish, or sometimes during migration or travel. In British Columbia, most

Distribution of the Pacific White-sided Dolphin (*Lagenorhynchus obliquidens*)

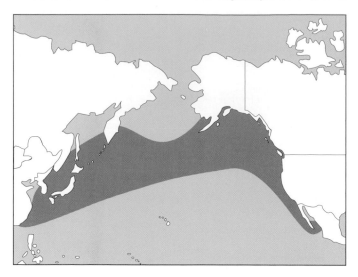

Pacific White-sided Dolphins are found in waters less than 600 m deep, although offshore herds are known to inhabit waters up to 1000 m deep. Even in the deeper waters, when the dolphins are foraging on deep-water prey, most dives are less than 20 seconds long. This suggests that the dolphins are waiting for their prey to move up the water column, rather than diving to great depths to find it. On average, inshore animals foraging in water less than 100 m deep stay submerged for 15 seconds. These dolphins prefer waters between 9°C and 17°C. The recently occupied inshore waters in British Columbia were 8.6°C before 1984, when the dolphins were not around, and since then have averaged 9.3°C. Likely some fish species have moved into the warmer waters, perhaps to spawn, and have drawn the dolphins into the area. Whether this change in water temperature is a result of climate change or is part of a normal cycle for the region is unknown.

Since 1994, the numbers of Pacific White-sided Dolphins in inshore waters have slowly declined in what appears to be a direct correlation to the increase in number of "pingers" on the farmed salmon pens, which are proliferating in the shallow waters. The "pingers" produce a high-volume underwater sound meant to discourage seals from hanging around the pens. Predation by Killer Whales has been recorded, and it is probable that sharks take some also. From 1970 to 1992, thousands of Pacific White-sided Dolphins were killed every year as by-catch by the Japanese and Korean high seas drift-net fishery for squid. A United Nations moratorium halted this fishery in 1993. Small numbers continue to be killed in Japanese waters in drive fisheries. Many more are probably caught and drowned in fishing nets across the Pacific, but go unreported.

Their maximum speed is around 26 km/h and their travelling speeds vary from 5 to 20 km/h. Although most dives are shallow and last less than 20 seconds, these dolphins are capable of staying submerged for as long as six minutes. Maximum known age is 46 for females and 42 for males; however, few animals survive beyond 30–35 years old.

DIET

Pacific White-sided Dolphins have very diverse and flexible prey preferences, depending on what is locally available. They eat at least 60 different fish species, as well as many species of squid, shrimp, and possibly jellyfish. Preferred prey, in Canadian waters, are fish that travel in schools such as herring, salmon, Northern Anchovy, Pacific Hake, and Horse Mackerel. These dolphins take fish within the range of 15–60 cm in length. The larger fish are usually broken up before swallowing and the head is often discarded. It has been estimated that Pacific White-sided Dolphins eat about 9% of their body weight per day.

REPRODUCTION

Both sexes reach sexual maturity between the ages of 7 and 12 years old. The calving season stretches from late January until August, with most births in Canadian waters occurring between June and August. The earliest known calf in Canada, to date, was born in April. The gestation period has been estimated at 11–12 months, and the length of the nursing period has been estimated at between eight and ten months, with the calves beginning to eat solid food by six months old.

BEHAVIOUR

Pacific White-sided Dolphins are mostly active during the day and at dawn and dusk, and rest or sleep at night. While sleeping, the members of the herd gather into a tight formation and swim slowly, breathe, and switch positions in synchrony. Captive animals will sleep for seven hours each night, although likely wild herds sleep less, due to the risks of predation. Cooperative hunting is a preferred strategy for this dolphin, both with others of their own species and with other cetacean species. Herding of schooling fish into a tight fish ball against the water surface from which the dolphins can pick a meal, while others continue to maintain the integrity of the ball, is a common foraging practice. There are many reports of Pacific White-sided Dolphins associating with other whales and even with California Sea Lions and Northern Fur Seals. Herds have been seen around foraging Humpbacks, jumping around and sometimes even onto the larger whale as they steal fish. In British Columbia waters, mixed herds with Northern Right Whale Dolphins have been seen, as well as a herd including Pacific White-sided Dolphins, Northern Right Whale Dolphins, and Short-finned Pilot Whales. For the most part, Pacific White-sided Dolphin herds in British Columbia

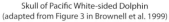

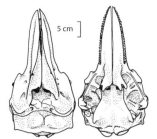

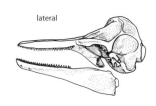

dorsal   ventral

Skull of Pacific White-sided Dolphin
(adapted from Figure 3 in Brownell et al. 1999)

waters tend to be single-species herds, probably due to the scarcity of the other dolphin species in the region. Mass strandings of this species have never been reported, although single strandings of sick, injured, or dead animals are not uncommon. Pacific White-sided Dolphins do well in captivity and are crowd favourites for their aerial performances. They are even known to reproduce in captivity.

VOCALIZATIONS

Like most dolphins, Pacific White-sided Dolphins use whistles and squeals to communicate with each other. These sounds can be detected by humans. They also use echolocation pulses to navigate and hunt underwater.

SIGNS

Swimming Pacific White-sided Dolphins commonly leap clear of the water when they take a breath. Often the white "splash" created with each jump is visible before the dolphins are, especially if a large herd is on the move. They are well known for their aerial displays of high leaps, belly flops, somersaults, and side flops. They appear to enjoy bow riding and will converge at high speed on a powered vessel travelling at the appropriate speed.

REFERENCES

Brownell et al. 1999; Dahlheim and Towell 1994; Fertl et al. 1999; Goley 1999; Heise 1997a, 1997b; Morton 2000; Stacey and Baird 1991b.

Northern Right Whale Dolphin

FRENCH NAME: **Dauphin à dos lisse**

SCIENTIFIC NAME: *Lissodelphis borealis*

These unusual-looking dolphins received their common name from early whalers who noted that, like right whales, they lacked a dorsal fin, allowing easy distinction, even from a distance.

DESCRIPTION

The most diagnostic feature of this species is the lack of a dorsal fin. Mainly black, Northern Right Whale Dolphins have a small white patch on their lower jaw and a larger patch on their chest, which narrows along the belly and extends all the way to the underside of the flukes. This white chest patch extends up the sides to the base of the flippers and onto the upper surface of the base of the flippers in some animals. Coloration of newborns is a muted grey version of the adult coloration, which gradually differentiates until they achieve full adult colours at about one year old. Another distinctive feature of this species is the abnormally long and slender tail stock. They have a short but well defined beak, with a shallow crease where it joins the melon. The flukes appear disproportionately

small. Northern Right Whale Dolphins have numerous fine, sharply pointed, conical teeth, with a dental formula of 37–54/37–54, for a total of 148–216. Typically there is a pair or two more in the lower jaw than the upper.

SIMILAR SPECIES

There are no other small cetaceans in the North Pacific that lack a dorsal fin. The low arc of the "porpoising" animals (they leap all the way out of the water in a long arc to take a breath) is reminiscent of a rapidly travelling fur seal or sea lion.

SIZE

Males are longer and heavier than females, but the size differences are not noticeable at sea.

Total length: males, 200–307 cm; females, 172–230 cm; newborns, 95–105 cm.

Weight: males, maximum 113 kg; females, unknown; newborns, unknown.

RANGE

These dolphins are endemic to the temperate North Pacific Ocean, with the area of highest density between the latitudes 30°N and 45°N.

ABUNDANCE

The Northern Right Whale Dolphin is at the northern limit of its distribution off the coast of British Columbia. Consequently, it is a rare Canadian cetacean, but is not particularly at risk in Canadian waters. In 2003 there were an estimated 20,000 Northern Right Whale Dolphins off the western coast of North America in the eastern Pacific. In the central and western Pacific Ocean during the late 1980s, an estimated 15,000–20,000 Northern Right Whale Dolphins died every year from drowning, after becoming entangled

Distribution of the Northern Right Whale Dolphin (*Lissodelphis borealis*)

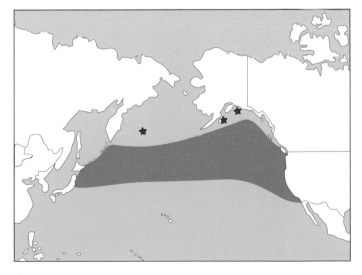

⭐ extralimital records

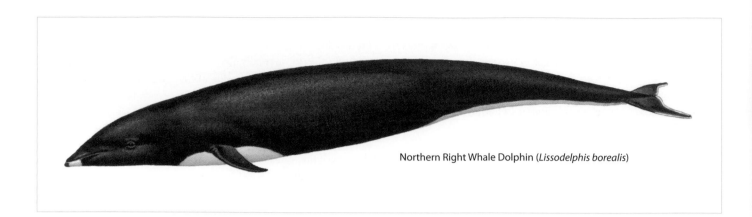

Northern Right Whale Dolphin (*Lissodelphis borealis*)

in drift nets set for squid. A United Nations moratorium was finally invoked in 1993 to halt the slaughter and the fishery was closed. That population was already depleted by this time, and is apparently still recovering, but its size is unknown. Many still die each year from entanglement in fishing gear throughout their range and a few hundred are taken annually in Japanese coastal waters for meat.

ECOLOGY
Northern Right Whale Dolphins inhabit waters with surface temperatures ranging from 8°C to 19°C. Calving occurs in the warmer waters of that range. Generally, the populations move inshore and southwards in late autumn and offshore and northwards in spring. A highly social species, the Northern Right Whale Dolphin is normally encountered in herds varying in number from 2 to 100 individuals. Amalgamated herds of as many as 2000 animals will gather in areas of high prey density. These dolphins are strong, speedy swimmers capable of speeds in excess of 40 km/h and of sustained speeds of 33 km/h over prolonged periods. An entire herd has been recorded diving for over six minutes. The expected lifespan is estimated at around 42 years. This species is not susceptible to mass stranding. Northern Right Whale Dolphins rarely survive for long in captivity. Although no natural predation has been observed, some likely fall prey to sharks and Killer Whales.

DIET
Northern Right Whale Dolphins eat deep-water fish and squid and they generally feed at night when their prey moves up the water column towards the surface, making it easier to reach. They may dive to depths in excess of 200 m to hunt. Some stranded specimens have had inshore fish remains in their stomachs, but it is thought that these were caught near the time of stranding and are probably not representative of the normal diet.

REPRODUCTION
Very little is known of the reproduction of this species and what we do know is mainly derived from study of animals killed in the North Pacific drift-net squid fishery. Both females and males become sexually mature at around 10 years old. The gestation period is around 12–13 months. The minimum calving interval is usually two years, but some females have been found to be pregnant and nursing at the same time. Most births in the central North Pacific occur in June, July, and August. The duration of nursing is unknown.

BEHAVIOUR
Configuration of a herd varies, but can usually be placed into one of four different patterns: (1) a tightly packed cluster with all animals surfacing independently; (2) scattered subgroups where each subgroup surfaces in unison; (3) a V-shaped chevron formation of tightly packed animals lined up side-by-side, each surfacing independently; (4) a "chorus line" of tightly packed animals lined up side-by-side, surfacing independently. Pattern 3 may be only an offset variation of pattern 4. Different herd formations are likely related to the activity of the herd at the time. Chevron and chorus-line formations have been noted in other night-hunting species when the herd was resting or sleeping during the day. Northern Right Whale Dolphins have a distinctive low arc when they jump out of the water for a breath (called "porpoising") while travelling at speed. Commonly part of mixed herds, this species is known to associate with at least 12 other species of marine mammals. They have been seen riding on the pressure waves of Grey and Fin Whales. Northern Right Whale Dolphins tend to be timid and usually avoid powered vessels. They are much more likely to bow-ride if in a mixed herd, especially if it includes Pacific White-sided Dolphins, who are eager bow-riders.

VOCALIZATIONS
Northern Right Whale Dolphin vocalizations have not been well studied. It appears that they have a similar repertoire to most

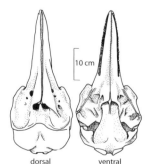

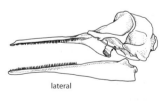

10 cm

dorsal   ventral

lateral

Skull of Northern Right Whale Dolphin
(adapted from Figure 5 in Jefferson et al. 1994)

dolphins, with the exception that they produce very few, if any, whistles. Burst-pulses (which occur in rapid series of 6–18 pulses) are the most common sounds that they generate, and a travelling herd is surrounded by these clicks, which are generally in the 14–24 kHz range. The clicks are thought to serve a communication function. Echolocation clicks are higher in frequency and are also produced in rapid series.

SIGNS

A large herd of Northern Right Whale Dolphins travelling along at high speed, porpoising in their characteristic low arc, and creating considerable "splash" is usually easy to identify, especially if lobtailing and belly and side flopping adds to the surface disturbance. They become harder to identify if the animals roll more smoothly at the surface to take a breath, causing barely a ripple. The absence of a dorsal fin as the animals roll at the surface is a giveaway, if the viewer is close enough to see that characteristic. Ordinarily the faster they are travelling, the more noticeable they become. Often this species travels with other small cetaceans, particularly Pacific White-sided Dolphins, and even with sea lions at times.

REFERENCES

Baird and Stacey 1991b; Barlow 2003; Chou et al. 1995; Ferrero and Walker 1993; Ferrero et al. 2002; Jefferson and Newcomer 1993; Jefferson et al. 1994; Leatherwood and Walker 1979; Rankin et al. 2007.

Killer Whale
also called Orca

FRENCH NAME: **Épaulard,** *also* **Orque**
SCIENTIFIC NAME: *Orcinus orca*

This whale is one of the most recognized and studied cetacean. Long-term research of pods off the west coast of North America and study of animals in captivity have helped to increase our understanding of the ecology and behaviour of this whale and to remove the former stigma attached to the species caused mostly by their occasionally ferocious feeding habits.

DESCRIPTION

Killer Whales are the largest of the dolphin family. They are strikingly black and white with broad, paddle-like black flippers and a very large black dorsal fin. This fin is somewhat falcate in females and juveniles, but is exceptionally tall and straight in the adult males. The dorsal fin on adult male Killer Whales is the largest among all the cetaceans, reaching heights of up to 1.8 m on some large individuals. Even the shorter dorsal fin of the females and juveniles is larger (up to 0.9 m tall) than in any other cetacean species. The majority of the body is black, with three clearly defined white areas.

The smallest of these is a patch behind the eye. The largest includes the whole chin and extends onto the chest, narrows to pass along the belly, flaring onto the sides, and ends around the anus. Another white area covers the underside of the flukes and an adjacent portion of the underside of the caudal peduncle. A variably shaped grey saddle occurs behind the dorsal fin. This saddle is barely noticeable at birth and becomes more distinct as the animal ages. The unique shape and colour of the saddle patch together with the shape of the dorsal fin and any nicks and scars allow scientists to recognize individual Killer Whales. Melanistic, and partly and almost fully, white individuals have been reported but are rare. Killer Whales have large curved conical teeth. The dental formula is 10–12/10–12, for a total of 40–48 teeth.

SIMILAR SPECIES

Its distinctive black and white markings, large dorsal fin, and large size make this whale relatively easy to identify. Other cetaceans that occur in the same waters and have large dorsal fins that may be confused at a distance are Risso's Dolphins and False Killer Whales. Risso's Dolphins are generally greyish-white, while False Killer Whales are a uniform black; neither displays the high-contrast pattern so diagnostic of Killer Whales.

SIZE

This species is highly sexually dimorphic, with males not only longer and heavier, but also with an easily visible gender-specific physical distinction – the size and shape of the dorsal fin.
Total length: males, up to 9 m, with dorsal fins up to about 1.8 m high; females, up to 7.7 m, with dorsal fins up to about 0.9 m high; newborns, 218–270 cm.
Weight: males, at least 5600 kg (highest known weight is 6600 kg); females, at least 3800 kg (highest known weight, 4700 kg); newborns, around 160 kg.

Distribution of the Killer Whale (*Orcinus orca*)

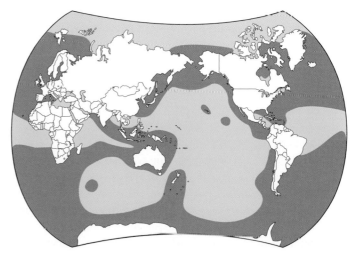

Killer Whale (*Orcinus orca*)

male

female

RANGE

Widespread around the globe, Killer Whales are able to live in both warm and cold waters, and are not limited by water depth, occupying both shallow and deep water. Concentrations occur in regions with abundant prey, especially at higher latitudes. In Canadian waters, Killer Whales occur in the Pacific, Atlantic, and Arctic Oceans as well as Hudson Bay, the Gulf of St Lawrence, and the mouth of the Bay of Fundy. Diminishing summer sea ice in the Arctic is allowing Killer Whales to extend the northwestern limits of their range and to occupy more of Hudson Bay. These whales are rarely seen in the western Canadian Arctic, despite extensive surveying efforts, and are rare on the east coast, apart from irregular occurrences off Newfoundland and Labrador.

ABUNDANCE

Japanese, Indonesians, Greenlanders, and Caribbeans continue to hunt Killer Whales. Chemical and noise pollution, increasing boat traffic, disappearance of prey, and even seal-pingers, used to frighten seals away from farmed-salmon pens, are having an effect on Killer Whale occurrence and abundance. Ironically, noise from whale-watcher boats can drive them out of traditional feeding and resting areas, although some populations seem to tolerate a large number of well-regulated and respectful whale-watch boats. The worldwide population has been estimated at around 95,000 whales, the vast majority living off Antarctica. The well-studied populations of residents and transients off the coasts of Washington and British Columbia number around 500–600 animals.

ECOLOGY

Killer Whales occupy a wide range of marine habitats both nearshore and offshore. They are tolerant of broad temperature and salinity variations, as long as they are relatively undisturbed by shipping activity, there is a sufficient prey base, and they are not subjected to excessive acoustic pollution that impedes their communication and prey detection or causes hearing impairment. Off the North American west coast, the Killer Whale population splits into three distinct and genetically separated communities that rarely if ever interbreed, but that occupy the same geographic range. These communities are made up of pods that share a social structure, dialect, and food preference. They are called "residents," "transients," and "offshores." Residents form pods of 3–59 animals all related on the maternal side, including the adult males. They are fish eaters and are frequently migratory or seasonal, following migrating, feeding, or spawning fish. Transients form small pods of two to six animals (most commonly three), also related maternally; typically, a female and her offspring. They are marine mammal and sea bird predators and also migrate as they follow their prey. The offshores are less well known, as they rarely travel near shore, and hence are poorly studied. They have only been seen to eat fish and their group size may be up to 50 animals. The genetic relationship within an offshore pod is uncertain, but is thought to be also matrilineal. All three groups have distinctly different dialects and some subtle physical differences, mainly in the shape of the dorsal fin and pigmentation of the saddle patch. In the western North Atlantic and Canadian Arctic waters, all the Killer Whales are thought to be one population.

Most pods are migratory, following the seasonal movements of their prey. While much remains to be learned of the distances, routes, and even wintering areas of most pods, an idea of the distances they are capable of travelling is indicated by the following observation. Three individuals in a transient pod of 17 Killer Whales observed attacking a Grey Whale and her calf in Monterey Bay, California, had last been positively identified more than 2600 km north in Glacier Bay, Alaska, less than three years previously. Killer Whales have been kept in captivity since 1964 and the first captive female to give birth in captivity produced a calf in 1977. The longest recorded dive to date lasted 17 minutes, but most dives are of less than 4 minutes. The maximum recorded dive depth of a wild Killer Whale is 264 m, but most dives are considerably shallower. Average speeds are 6–10 km/h, but bursts of at least 40 km/h can be achieved for short distances. As apex predators, Killer Whales are generally immune to predation except as calves, when they can be vulnerable to shark attacks. They do, however, die from many other causes such as starvation, disease, parasites, entanglement in fishing gear, boat collisions, and acute poisoning by chemical pollutants. The northern populations are vulnerable to ice entrapment. Among the well-studied west coast resident pods, mortality is greatest during the first six months, when approximately 37%–50% of the calves die and the maximum age of females is 80–90 years (average 30–46 years) and that of males 50–60 years (average 19–31 years). The oldest known wild individual is a female of 99 years.

DIET

Diet varies with population, geographical locations and season. Pods that specialize in fish typically consume many different species, but tend to concentrate on one preferred species or group of species. For example, off the coast of British Columbia, the resident pods are known to eat 22 species of fish and one species of squid, but 96% of the fish they eat are from the salmon group. There are some current concerns that the large-scale reduction in the salmon runs may be causing food shortages for some fish-eating Killer Whales, and starvation could become a significant mortality factor in future if salmon stocks continue to decline. A resident pod off the coast of Norway specializes in hunting herring. Other pods specialize in marine mammals and seabirds. Very little is known of the dietary habits of the offshore pods, but all indications suggest that they are primarily fish eaters. The transients off coastal British Columbia eat Harbour Seals over 50% of the time, but also kill and eat Northern

Elephant Seals, Steller Sea Lions, Dall's Porpoise, Harbour Porpoise, California Sea Lions, Pacific White-sided Dolphins, Grey Whales, Minke Whales, Sea and River Otters, and even swimming land mammals such as deer, as well as assorted sea birds. Birds account for about 6% of the kills, but because of their size form a very minor portion of the calories ingested. Killer Whales in the North Atlantic and Arctic Oceans have been observed preying on Bowhead Whales, Minke Whales, Belugas, Narwhals, White-beaked Dolphins, Humpback Whales, Long-finned Pilot Whales, Fin Whales, seals, seabirds, tuna, herring, and even fish discarded from fishing boats. Killer Whales learn the location of spawning fish or whelping seals and will time their arrival for peak season. They have also learned how to safely steal fish from longlines (sometimes ingesting the equipment at the same time). Small food items are swallowed whole and larger ones are ripped apart first. There are no recorded attacks of Killer Whales on humans.

REPRODUCTION

Mating is not known to occur within a pod, only between pods that are in the same "community." The mature males remain in their mothers' pod and only court females from other pods whenever two or more pods congregate. The gestation period is 15–18 months. Calving can occur year-round, but our resident pods tend to give birth between fall and spring, suggesting that most breeding may occur from spring to early summer. Killer Whale calves begin eating solid food very early and are weaned by the time they are about two years old. Among the resident pods, females can produce a calf every two years, but the calving interval is variable, from two to eleven years (average about five years). A female 51 years old is known to have produced a calf, but most females older than about 40 years are no longer reproductive. Females and males become sexually mature at about 12–17 years old. Births appear to occur near the surface and all members of the pod assist in ensuring that the neonate is able to reach the surface to breathe after its birth. Newborns are highly precocial and are able to swim well within about an hour or two.

BEHAVIOUR

Killer Whales are intelligent, social whales and their behaviour is complex and can change over time and in different habitats. They are cooperative hunters and occasionally develop new hunting techniques or discover new hunting opportunities, which they will then teach to future generations. The gruesome methods they use to kill

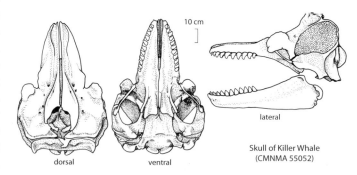

10 cm

dorsal   ventral   lateral

Skull of Killer Whale
(CMNMA 55052)

some of their prey have earned them their common name. Larger whales are well protected by blubber and are not easy for a pod of Killer Whales to kill. They must be rammed (to cause internal damage) or bitten (to bleed to death) and death tends to occur over a prolonged period. Killer Whales learn the timing and locations of prey aggregations such as fish and whale seasonal migrations, or seal breeding rookeries and haul-outs, and teach their young how and when to find them. One dramatic example of their specialized cooperative hunting techniques is called "wave-wash hunting," used to capture prey on floating ice flows. This practice, carried out by Antarctic pods, requires the careful coordination of several whales that charge the ice flow in synchrony to create a surge-wave large enough to wash over the flow and carry the prey (usually a seal) into the waiting mouth of pod members on the far side of the flow. The pod then shares the bounty. The basic Killer Whale social unit (among all the "communities" so far studied) is a matrilineal group consisting of one to three or four generations. A small pod may consist of a single matriline, while a larger one may be composed of a cluster of related matrilines. A pod hunts together, plays together, raises and teaches the young together, and usually even breathes in synchrony, with the senior female generally breathing first. Occasionally single males will roam alone for a time, but most males, contrary to the practices of most other mammals, remain in the pod in which they were born. In North American waters, most Killer Whale societies fit into one of three systems (transients, residents, and offshores), already discussed in the "Ecology" section above. Social systems in the southern hemisphere appear to blend these categories to varying degrees, illustrating the flexibility of these intelligent whales to adapt even their societies to different environments. Killer Whales often perform aerial acrobatics, including spy-hopping, breaching, lobtailing, and slapping the water surface with their flippers. Killer Whales, like humans, Gorillas, Bottlenose Dolphins, and False Killer Whales, are sufficiently cognitively aware to recognize their own reflection in a mirror.

VOCALIZATIONS

Killer Whales, like the other toothed whales, use a variety of clicks and pulses in the ultrasonic range and also produce many whistles, groans, screams, roars, and clicks that are audible to humans. Like the Bottlenose Dolphin and pilot whales, Killer Whales probably have more than one sound-producing structure, as they can produce two different sounds at the same time. Killer Whales use different echolocation strategies depending on their prey. Fishes generally do not hear in the ultrasonic range, so fish-eating Killer Whales freely use echolocation to hunt and while travelling. Mammal-eating whales produce only occasional ultrasonic pulses (which can be lost in the overall ambient noise), as their prey can hear them and will flee if they detect the echolocation sounds of a hunting pod. These hunters clearly are making an effort to curb their sound production, and consequently rely more on sounds produced by the prey, such as in swimming or breathing, and the environment, such as waves on a beach, rather than echolocation. Each Killer Whale "community" has a dialect and each pod within the "community" has an evolving variation of that dialect. By assessing a male's dialect, females can select a mate that is from the same "community," but is as different as possible in dialect from the dialect variations used by her pod, thereby assuring the least degree of genetic relatedness.

SIGNS

Pods of Killer Whales are usually unmistakable, due to their large size, distinctive coloration, and large dorsal fins (especially of mature males). Most Killer Whales seen from the whale-watching boats off the coast of British Columbia are residents. They congregate in large herds in predictable spots and have become very accustomed to boat traffic, often approaching or even showing off to a boat. Transients are encountered by chance, as their movements are generally less predictable. Seals, rapidly vacating the water or hiding in seaweed, have possibly detected nearby transients. Dolphins and even the great whales have been known to beach themselves to avoid the transients. Killer Whales seen off the east coast are likely mammal-eaters.

REFERENCES

Baird 2001b; Baird and Whitehead 2000; Baird et al. 2005; Dahlheim and Heyning 1999; Erbe 2002; Ford et al. 1998, 2005; Goley and Straley 1994; Heyning and Dahleim 1988; Jones, I.M., 2006; Nøttestad and Similä 2001; Olesiuk et al. 1990; Reeves and Mead 1999; Speckman and Sheffield 2001; Stacey and Baird 1997; Tosh et al. 2008; Visser et al. 2008; Vos et al. 2006.

False Killer Whale

also called Blackfish

FRENCH NAME: **Faux-orque**

SCIENTIFIC NAME: *Pseudorca crassidens*

This whale is commonly called Blackfish by west coast fishermen. The largest known mass stranding of cetaceans (835 animals) involved False Killer Whales.

DESCRIPTION

The False Killer Whale is a slender bulbous-nosed black whale. The melon overhangs the tip of the lower lip, especially in adult males, whose melon area is larger than the females'. Although the head area can be lighter than the rest of the body, in the field the animal looks uniformly dark. They are born dark with a lighter patch on the chest. This light patch often extends along the midline of the belly to the genital area. The lighter areas tend to darken as the animals age and are often grey or even dark grey in the adults. The dorsal fin is tall and falcate and located about the midpoint of the back. The flippers are characteristically humped at the middle of the leading edge. The dental formula is 7–11/8–12, for a total of 30–46 teeth. The teeth are heavy, conical, and curved, and often become heavily worn in older adults.

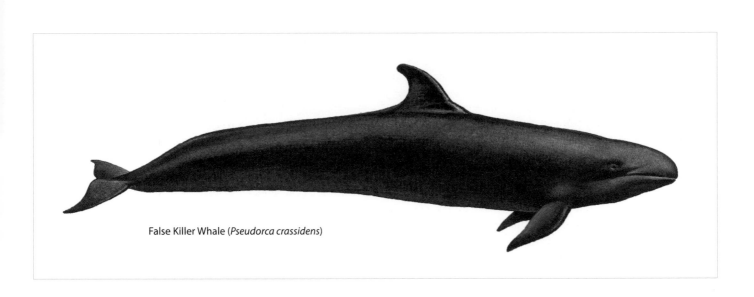

False Killer Whale (*Pseudorca crassidens*)

SIMILAR SPECIES

In Canadian waters, Risso's Dolphins and Short-finned Pilot Whales could be mistaken for False Killer Whales. Risso's Dolphins occur on both coasts and have a similar dorsal fin, but usually a light coloured body, and they are smaller with a squarish head. Short-finned Pilot Whales are similar in size and colouring, but have a more bulbous head. They only occur in Canadian waters off the west coast. The dorsal fin of the False Killer Whale is considerably farther back on its body than it is on the Short-finned Pilot Whale. Sometimes the dolphin-like behaviour of the False Killer Whale will help distinguish it from the pilot whales, which rarely breach. None of the similar species share the characteristic "hump" on the flippers and the resulting flipper shape of the False Killer Whale, although this characteristic can be difficult to detect in the field. Distinguishing between False Killer Whales and Short-finned Pilot Whales can be difficult, even under excellent viewing conditions.

SIZE

This species is sexually dimorphic and males are substantially longer and heavier than females.

Total length: males, 396–610 cm; females, 340–506 cm; newborns, 160–190 cm.

Weight: males, maximum 1300 kg; females, maximum unknown, but in excess of 775 kg; newborns, unknown.

RANGE

These whales have a widespread distribution, mainly in tropical and occasionally in temperate oceans, principally in offshore and deep waters. Most of our knowledge of their range is a result of strandings and their presence in mid-ocean regions is assumed, but largely unsubstantiated to date. Although False Killer Whales are occasionally found along the west coast of North America all the way to Alaska, they are rare north of Oregon and not known at all from eastern Canadian waters. All Canadian records have represented solitary individuals in the waters around Vancouver Island.

ABUNDANCE

This species is rare in Canadian waters, but is considerably more common in the main potions of its range. There are no estimates of worldwide population size.

ECOLOGY

The deep-water habitat of this whale makes study and even reliable sightings difficult; so much of our understanding is derived from captive animals or those that died in strandings. A study of three captive individuals showed that they have a relatively high metabolism, likely due to their elongated shape and the resulting high ratio of surface area to volume. This ratio results in substantially more loss of body heat than is usual for cetaceans and likely accounts, in large part, for their tropical distribution. Food passes through the gut in a surprisingly short time – 4½ hours or less. A resident population (of < 500 animals) around the Hawaiian Islands has been studied for about 30 years, and much of the known wild ecology and behaviour is derived from these studies. False Killer Whales are active both day and night. A pod may hunt together or forage individually and

Distribution of the False Killer Whale (*Pseudorca crassidens*)

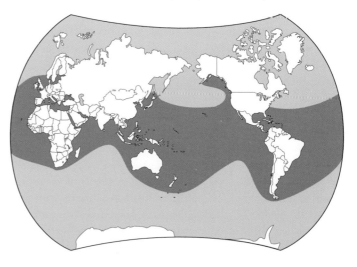

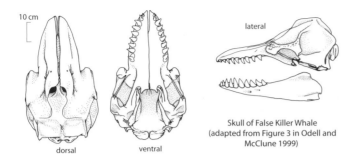

Skull of False Killer Whale
(adapted from Figure 3 in Odell and
McClune 1999)

then rest in synchrony. They are found in waters varying in surface temperature from 9°C to 31°C, but are most common in the warmer waters. Herd size varies considerably, from solitary animals to pods in excess of 800 individuals. Most sightings are of groups of 10–20 animals, but these are likely part of a large, scattered herd that could number into the hundreds. The oldest known female was 63 years old and the oldest known male 58 years old.

DIET

False Killer Whales are versatile predators that eat a wide variety of primarily squid and fish. Their large teeth and pack hunting behaviour allows them to catch and eat sizable fishes such as tuna, which they rip apart before swallowing. They have been known to kill and eat a Humpback Whale calf and several different dolphin and porpoise species that were caught in nets. They have been seen harassing a herd of Sperm Whales, presumably either to drive them away from a productive feeding area or to force them to regurgitate their stomach contents, which the False Killer Whales will then consume.

REPRODUCTION

Females achieve sexual maturity between the ages of 8 and 11 years old, males at around 18 years old. It appears that some populations in temperate waters reproduce seasonally, while in warmer waters calves may be born at any time of the year. Their gestation period is around 14–16 months. Calves nurse for about 18–24 months. The reproductive rate of False Killer Whales is very low. It has been estimated that females produce a calf about every seven years. Fertilization and successful pregnancy have occurred in captivity, as have crosses of captive female False Killer Whales with captive male Bottlenose Dolphins, which have produced live calves. There are no known wild crosses involving False Killer Whales. Females enter reproductive senescence around the age of 45, but continue to play an important social role in the herd.

BEHAVIOUR

False Killer Whales are very social animals, rarely seen alone. Their strong social bonds are presumed based on their tendency to mass-strand and the obvious reluctance of pod members to abandon a stranded individual in distress. They are often active near the surface and frequently bow-ride and breach. Sometimes when breaching an animal will turn in the air to land on its side with a great splash. False Killer Whales often associate with other cetaceans.

Up to 10 species are known to travel with them in mixed herds. Bottlenose Dolphins are the most common associates. A common foraging strategy has a travelling herd spread out in a wide line, up to several kilometres across, presumably to enhance their chances of encountering prey. Larger groups may break up into smaller subgroups that can be separated by 2–10 km. All travel in the same direction, foraging as they go. Group members typically dive in synchrony when hunting at depth and more randomly when foraging closer to the surface. The Hawaiian pods have displayed site fidelity and pod-member associations that extend over 21 and 15 years respectively, similar to those of Killer and pilot whales. Individuals have been seen to carry prey around for extended periods and to share food with other pod members. When a birth is imminent, most of the herd disperses around the pregnant female, possibly acting as sentinels in the event of danger. A small subgroup of adult-sized whales (8–10 animals) clusters around the mother so closely that they all touch each other. These whales are likely senior females. As soon as the infant is born, these adults help it to the surface and even lift it clear of the water in their efforts to keep it in the air long enough to breathe. After a few minutes of assistance, the neonate can surface and breathe on its own and within 25 minutes can swim well enough to join the main herd. Long-term recognition of individuals is hampered by the lack of distinctive individual colour patterns, but can be accomplished for some individuals using body scars, dorsal fin shape, and nicks and scars on the dorsal fin. Along with Killer Whales and Bottlenose Dolphins, False Killer Whales exhibit behaviour in front of a mirror that shows they recognize that they are looking at themselves. Such self-recognition is an important marker on the intelligence scale. False Killer Whales are not commonly held in captivity, due to their aggressive nature, although some captive births have been recorded and one animal has survived in captivity for more than 26 years. There are no known records of this species preying upon humans. They have been recorded swimming at speeds of 18 km/h and are likely capable of swimming even faster.

VOCALIZATIONS

False Killer Whales, as one would expect from such a social animal, are very vocal. They produce the usual range of whistles, clicks, and echolocation clicks and pulses in the frequency range of 1.3–8.3 kHz. Herds move about, constantly surrounded by sounds produced by one or more of the members.

SIGNS

The deep-water habitat of this species makes them difficult to see in the wild. Careful scrutiny and good visibility is required in order to separate this species from other similar dark whales such as Long-finned and Short-finned Pilot Whales.

REFERENCES

Baird 2008a; Delfour and Marten 2001; Kastelein et al. 2000; Notarbartolo-di-Sciara et al. 1997; Odell and McClune 1999; Palacios and Mate 1996; Reeves and Mead 1999; Rendell et al. 1999; Stacey and Baird 1991a; Stacey et al. 1994.

Striped Dolphin

also called Blue-white Dolphin

FRENCH NAME: **Dauphin bleu**

SCIENTIFIC NAME: *Stenella coeruleoalba*

Striped Dolphins are found offshore in deeper continental-shelf waters. They like warm temperatures and are only found in Canada in the summer. Rapidly swimming Striped Dolphins have been clocked at speeds of up to 60 km/h, making them among the fastest marine mammals.

DESCRIPTION

The Striped Dolphin is a small, beaked cetacean with an easily recognizable colour pattern, if seen in good light. The principal diagnostic marking is a dark blue-black stripe that runs from the eye to the anus on each side of the body. Above the stripe, the skin is light bluish-grey and below it is white, providing contrast to the dark stripe. The back and large dorsal fin are a bluish-black and the tail stock is a dark grey. Above the eye is a large, distinctive, light blaze that sweeps back and up to fade at about the level of the dorsal fin. The throat and belly are white and the small flippers are blue-black. A thin black line from the eye broadens as it reaches the front of the flipper. The bluish tones fade quickly after death to grey or black. The teeth are all conical and similar, with usually at least one pair more in the upper jaws; the total number is highly variable. The dental formula is 38–59/37–55, for a total of 150–228 teeth.

SIMILAR SPECIES

The two species of common dolphins (*Delphinus*) could be confused with Striped Dolphins, as all form large herds of small, acrobatic, beaked dolphins that cause a lot of "splash" as they breach to breathe. All three have a light patch above and behind the eye and can be found in similar regions. The two Common Dolphins lack the side stripe and their light side patch is distinct along its top margin and curves down towards the belly rather than up towards the dorsal fin, creating a distinctive V-shaped pigment pattern on their sides. Bottlenose Dolphins can also be seen in similar areas as Striped Dolphins, but they are a more uniform grey without the sharp demarcation of light and dark along the side stripe, and they are considerably larger than Striped Dolphins. In the Atlantic, the Atlantic White-sided Dolphin could be confused with the Striped Dolphin. This species is larger than the Striped Dolphin and the light patch on its side is in the middle of the body. It also tends to be seen in smaller herds. In the Pacific, the abundant Pacific White-sided Dolphin forms large boisterous herds of small dolphins that jump out of the water with each breath. They have an extremely short beak, unlike the Striped Dolphin, and a larger, more recurved and bi-coloured dorsal fin.

SIZE

Size varies among the different populations, with the Mediterranean being among the smallest and the western Pacific among the largest. Males are slightly larger than females, on average 2 cm longer.

Total length: males, up to 256 cm; females, up to 236 cm; newborns, around 90 100 cm.

Weight: males, up to 156 kg; females, up to 150 kg; newborns, 9–12 kg.

RANGE

The Striped Dolphin is frequently sighted in warm temperate and tropical seas, but it is clearly at its northern limits in Canadian waters and is rarely seen. These dolphins prefer waters which are above 15°C, and hence are usually only seen in Canadian waters in late July and August. On the east coast, most are sighted along the Scotia Shelf and the Sable Island Bank in years when the warm Gulf Stream

Striped Dolphin (*Stenella coeruleoalba*)

Distribution of the Striped Dolphin (*Stenella coeruleoalba*)

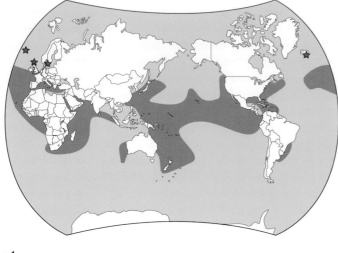

★ extralimital records

waters meander into the area. On the west coast records are even less common and are usually strandings or animals killed in fishing gear.

ABUNDANCE

This species is numerous over most of its range, although there has been a gradual, and now fairly serious, decline in the Japanese stocks caused by overharvesting, and there is some concern for the Mediterranean stocks, due to high incidence of net entanglements. Estimates of over 3 million worldwide have been suggested. Striped Dolphins are rare and seasonal in Canadian waters.

ECOLOGY

Striped Dolphins are usually found off the continental shelves along the continental slopes and in deeper oceanic waters. They are most common around the highly productive areas where currents converge. Most are found in areas where surface waters are between 15°C and 25°C, but they will tolerate waters as cool as 10°C for short periods. Many of their prey species have luminescent organs and are known to live at depths of 200–700 m. Striped Dolphins probably dive to those depths, but also likely hunt at night, when their prey moves up the water column to feed. Die-offs in the Mediterranean during the early 1990s have been attributed to morbillivirus, in the measles family of viruses. Considerable study of contaminant levels in this species has been undertaken in the last three decades, particularly in the Japanese and Mediterranean populations. Blubber PCB levels of animals from the Mediterranean Sea that died from the morbillivirus outbreak have recorded among the highest values ever for a mammal. This contaminant load could have contributed to a reduced resistance to the disease, which commonly occurs in cetaceans, but rarely with deleterious effects. Based on aging techniques that count annual deposition layers of the teeth, individuals of up to 58 years old have been discovered. Hunting of this species still occurs in Japan and some are taken for food in other parts of the range, but most

human-caused mortality is a result of incidental catches in fish nets. There is some natural mortality caused by sharks and Killer Whales and likely by False Killer Whales. This species is not successfully kept in captivity. Mass strandings are known, but are not common.

DIET

Schooling fish and squid are the primary prey, with some shrimp taken when abundant. They prefer fishes that are less than 13 cm long and squid less than 20 cm in mantle length. Food is swallowed whole, so the size of preferred prey and the size of their gullet are directly related. The soft-bodied squid are easier to swallow, and so can be larger.

REPRODUCTION

In the Mediterranean, the autumn calving season (September to October) matches the region's peak of productivity. In other parts of the range, such as the western Pacific, the births are spread out from early spring to mid-summer. The season of birthing in North American waters is probably late summer and early autumn, when the waters are warmest. A single calf is produced following a pregnancy of 12–13 months. Calves are suckled for about 15 months, but start to eat solid food at about 12 months old. The number of male newborns exceeds female newborns by about 40%. By the time the calves are weaned, the ratio is about even, suggesting that male calves, for reasons unknown, suffer a significantly higher mortality than females. Males reach sexual maturity between the ages of 7 and 15 years old, females between 5 and 13 years. Juveniles become sexually mature earlier, when the population density is low, and later, when the numbers are high: a rather neat solution to fluctuations in population size, usually caused by shortages of prey. Females produce a calf approximately every four years. This calving interval has dropped to less than three years in the heavily exploited Japanese populations.

BEHAVIOUR

Striped Dolphins are very social animals whose herds range in size from around 10 to more than 500 animals, averaging about 100. The strong bonds between individuals within a herd make them vulnerable to being driven into shallow water where they can be killed by human hunters. This type of harvesting has been practised for centuries in Japan, where whale meat is prized. The herds vary in composition; some contain only adults, others only juveniles, and yet others are mixed. Calves remain with their mother's herd for a year

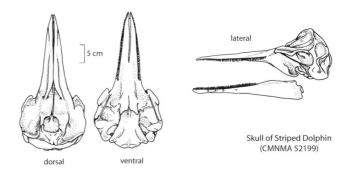

lateral

Skull of Striped Dolphin
(CMNMA 52199)

dorsal   ventral

5 cm

or two after weaning and then leave to join the juvenile herds. They rejoin either a non-breeding or breeding adult herd as young adults, although some females do so as subadults. Herds seen near a coast are likely made up of juveniles. These dolphins are active, speedy swimmers. Their average swimming speed is estimated at around 11 km/h and they are capable of sprinting at speeds of up to 60 km/h. They will bow-ride on powered vessels and large whales, and may school with other dolphin species, notably the Short-beaked Common Dolphin.

VOCALIZATIONS

The blow of this species is quiet and only detectable at short range. Like all the toothed whales, Striped Dolphins use echolocation to locate and catch prey. They produce echolocation clicks, pure tone whistles, burst-pulse sounds, and distress calls.

SIGNS

Herds of Striped Dolphins are usually quite conspicuous, as they create a considerable splash as they leap (breach) to breathe. They ride the bow wave of a boat or a large whale with every indication of enjoyment and are sometimes called "streakers" from their practice of darting away from the bow wave at high speed. Striped Dolphins perform a unique aerial movement, called "roto-tailing," which takes place during high arching leaps as the animal vigorously and rapidly rotates its tail several times before hitting the water.

REFERENCES

Aguilar 1991; Archer and Perrin 1999; Baird et al. 1993; Bloch et al. 1996; Dizon et al. 1994; Perrin et al. 1994; Reeves et al. 2002.

Bottlenose Dolphin

FRENCH NAME: **Grand Dauphin,** *also* **Dauphin à gros nez**
SCIENTIFIC NAME: *Tursiops truncatus*

When most people think "dolphin," this is the species they envisage. Not only are Bottlenose Dolphins widespread in marine waters off most temperate and tropical coasts, but they are popular and common in captivity and also gained much publicity through the 1960s hit television series *Flipper* and its 1990s remake as a series and movie.

DESCRIPTION

Bottlenose Dolphins are medium-sized dolphins. They are dark grey on the upper surface and light grey to pinkish-white on the belly, with a variable amount of intermediate grey between the two. There is subtle and varied shading on the head, which often resolves into a line from the corner of the eye to the flipper and other line markings at the base of the beak. These are only visible at close range. Albinos are

reported rarely. There is a short beak distinctly separated from the rest of the head by a crease. The dorsal fin is moderately large, curved, and positioned at about the midpoint of the back. The flukes are separated by a distinct notch. Teeth are uniformly conical. The dental formula is 20–26/18–24, for a total of 76–100 teeth. There are two recognized types of Bottlenose Dolphin: the inshore form and the offshore form. Offshore animals tend to be somewhat larger and darker, with smaller flippers and a more curved dorsal fin. Bottlenose Dolphins are energetic and commonly perform aerial acrobatics even without being prompted or trained by humans. They will converge at high speed to play in the bow wave of a boat. They are curious and adapt well to human presence, often being seen around fishing boats and nets.

SIMILAR SPECIES

Close to shore, the principal other small cetacean that could be confused with the Bottlenose Dolphin is the Harbour Porpoise, which is considerably smaller and has a simple small triangular dorsal fin. In Canadian offshore waters, there are three other dolphins that are similar to the Bottlenose Dolphin. Risso's Dolphins have a similar large dorsal fin, but their blunt forehead, white markings, and heavily scarred back serve to distinguish them. Striped Dolphins are also attracted to powered boats. They are smaller, with a distinctive pigment pattern on their sides and a longer, narrower beak. Short-beaked Common Dolphins display a distinct light patch on their sides and a black stripe from the eye to the beak. Bottlenose Dolphins can be surprisingly difficult to identify at sea, where they may appear dark with few identifying characteristics. Usually, a glimpse of the short beak and distinctive head shape along with the large dorsal fin is sufficient to identify the species.

SIZE

Males are larger and heavier than females.
Total length: males, 250–380 cm; females, 250–370 cm; newborns, 84–140 cm.
Weight: males, up to 500 kg; females, up to 260 kg; newborns, 14–21 kg.

Distribution of the Bottlenose Dolphin (*Tursiops truncatus*)

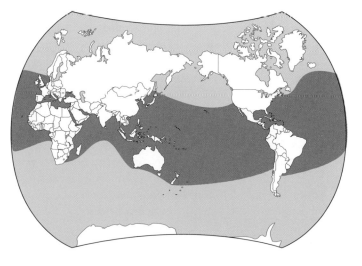

Bottlenose Dolphin (*Tursiops truncatus*)

RANGE

The Bottlenose Dolphin is a very widespread species found in all tropical and temperate seas around the world. In eastern North America, the inshore form is found within 5 km of the coast and will advance into river estuaries. The offshore form prefers deeper waters along the continental shelf up to Georges Bank and the Scotian Shelf. All the known Canadian sightings and strandings have likely been of the offshore form. On the west coast, the Pacific distribution does not reach Canadian waters. In North America, these whales are found in waters with surface temperature of 10°C–32°C. The Pacific populations have moved northward several times in the past few decades, following the warmer waters of the El Niño current, when it moved north. These expansions seem to have persisted even after the El Niño incursion receded.

ABUNDANCE

The size of the world population is unknown, but only limited areas such as the Black Sea suffer from severely depleted stocks. Otherwise, Bottlenose Dolphins are considered common.

ECOLOGY

Bottlenose Dolphins are active throughout the day and night, alternating sessions of feeding, socializing, travelling, and resting. As mentioned above, there are two types of Bottlenose Dolphins, the coastal or inshore populations and the pelagic or offshore populations. These groups eat different foods, and have slightly different appearances and ecology. The inshore population along the east coast of North America is further divided into resident and migratory stocks. Adults of the pelagic population are suspected of diving 500 m deep to catch deep-sea fishes. Newborns have only a limited ability to hold their breath. The blood-oxygen capacity of juveniles slowly increases as the animals mature, until, at around three years old, it reaches adult levels. Until then they cannot dive for as long or as deeply as do the adults. Females with calves are equally limited, as they will not dive beyond the capacity of their dependant calf unless it is in the care of another animal in the herd. Bottlenose Dolphins often associate with Yellowfin Tuna. Why they do this is

uncertain; possibly they are after the same prey, but the fact that they do has been exploited by fishermen for years. The dolphins, being air breathers, tend to be closer to the surface and readily visible. The fishermen encircle the herd in hopes that their nets will also capture the tuna that might be swimming below. How many dolphins die due to this by-catch is unknown, but the numbers are likely substantial. Many more thousands of these dolphins die each year from net entanglement, entanglement in monofilament fishing lines, pollution, and human hunting. Bottlenose Dolphins are vulnerable to shark attacks, especially calves and juveniles, and a high proportion of adults are swimming around with scars on their bodies that show they survived such an attack. Large numbers of these animals occasionally die from toxins associated with "red tide" algae, and in 1987–1988 as many as 50% of the US east coast migratory inshore animals were killed by a morbillivirus. Bottlenose Dolphins are long lived. The potential lifespan is probably 50 or more years for females and possibly slightly less for males. Normal travelling speeds are 1–4 km/h, but speeds of up to 19 km/h can be maintained for several kilometres and short bursts of speeds up to 30 km/h have been recorded. Commonly maintained in aquaria, Bottlenose Dolphins have bred in captivity since 1952.

DIET

These small whales are flexible in their feeding choices as they adapt to local conditions and prey. The inshore forms tend to concentrate on shallow-water fishes and invertebrates, while the offshore forms prefer deep-water fishes and squid. Most prey ranges from 10 to 40 cm long, but the dolphins must be cautious not to attempt to consume prey that is larger than the dolphin gullet opening of about 12 cm across, lest they choke to death. Bottlenose Dolphins are extremely agile and are capable of capturing fishes as small as 2 cm in length.

REPRODUCTION

Some populations display a synchronous breeding season, others a bimodal breeding season, and still others will mate at all times of year. The populations in the North Atlantic from Canada to at least as far south as North Carolina produce their calves in early summer, when the waters are warmer. A single calf is born after a gestation period of about 12 months. Females nurse their offspring for at least two years, and in some cases up to seven years, although the calves begin eating some solid food as early as their first year. Calves are weaned before the next calf is born. After weaning, the juvenile usually remains with its mother for several more years, learning to catch food and gaining experience with the surroundings. Juvenile females often join their mother's herd permanently. Females reproduce every 2–6 years. They achieve sexual maturity between the ages of 6 and 11, and have been known to produce calves until they are 45 years old. Age at sexual maturity for males is not clear, but given the species' sexual dimorphism and the vigorous competition for access to females, it is unlikely that they would have much opportunity to breed until they are full-sized, at 12–15 years old or older, regardless of whether they were fertile at an earlier age. In captivity, most breeding males are at least 20 years old. Orphaned calves are readily adopted, even by unrelated females, who are known to initiate lactation in order to nurse them.

BEHAVIOUR

Bottlenose Dolphins are gregarious animals usually found in herds, which vary in size depending on habitat. Inshore herds are commonly made up of two to fifteen individuals, although very large herds are known, while the offshore (pelagic) herds can number from ten to hundreds of individuals. These herds tend to amalgamate and break apart in a socio-behavioural pattern called "fission-fusion." Herds of related females may remain together as the nucleus of a herd for many years, during which they are visited for varying lengths of time by adult males or enhanced by juveniles produced by the females. Mixed-gender herds of subadults and all-male herds also occur. As with most mammals, occasionally one will find itself alone for a variety of reasons. These individuals will often seek out other species in order to satisfy their need for companionship. They may travel for a while with other cetaceans or even habituate to human swimmers.

Males adopt one of two different strategies in order to mate with females. They either form long-term pair bonds with specific females or they rove from female group to female group hoping to find a receptive female. Often two, or even three, males will form alliances that can last for more than 20 years. These partners coordinate their efforts to gain access to females. Many male partners are closely related on the maternal side. Bottlenose Dolphins display a high frequency of non-reproductive matings. They also are the only cetacean known to commit infanticide. Perhaps the females allow the males to mate in order to prevent the sexual frustration that could lead to this aberrant behaviour.

Males can be quite aggressive towards other dolphins, especially when competing for food or breeding opportunities. Bottlenose Dolphins spend a large part of their day engaging with others in their herd, and many types of aggressive, reconciliation, cooperative play, mating, and feeding behaviours have been recorded. They will toss objects (sometimes live objects) through the air to each other; they will play in each others' intentionally produced bubbles; and they will curiously pester other marine creatures such as crustaceans, sea

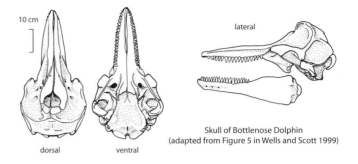

Skull of Bottlenose Dolphin
(adapted from Figure 5 in Wells and Scott 1999)

turtles, or molluscs. Play can also involve leaping out of the water and chasing other herd members of similar age. These dolphins are known for their ingenious hunting techniques, of which more is known of the inshore forms, due to their proximity to humans. They will drive their whole head up to the flippers into sand, in pursuit of sand-dwelling invertebrates; they will cooperatively drive schools of fish onto muddy shoals in a wave surge and partially beach themselves to catch the fish flopping on the land (called strand-feeding); they will fling fish up into the air with mighty blows from their flukes to stun or kill them and make them easy to catch (called fish whacking). These whales are very agile and perform a movement called "pinwheeling," which is similar to a swimmer's turn for a human. The dolphin, swimming on its side in pursuit of a fish, tucks its head and spins, rotating around the midpoint of the body while keeping its head in proximity to its prey, a particularly effective manoeuvre in shallow-water situations. They will often use natural and man-made barriers to enhance their hunting success. Driving fish along sand bars or sea walls is a common technique, as is circling a school and then darting into the fish ball to snatch an individual. They are quick to exploit human activities that might provide them with food, such as following fishing boats or hunting around nets. Bottlenose Dolphins are one of only a few cetaceans known to regularly body surf, often spectacularly leaping out of the wave just before the wave breaks. These intelligent mammals are one of only three cetacean species known to recognize their own reflection in a mirror. The others are the Killer Whale and the False Killer Whale.

VOCALIZATIONS

As befits an intensely social species, Bottlenose Dolphins are quite "noisy" animals. They produce three types of sounds: (1) short high-frequency echolocation clicks that are, for the most part, beyond the range of human hearing, (2) broadband pulsed sounds including barks, buzzes and cries, tail and body slaps, and (3) narrowband whistles. The latter two types of sounds are audible by humans, as are some of the lower-frequency components of the echolocation clicks. A sonar beam for echolocation is very directional, becoming narrower and more focused as the frequency increases. The dolphins therefore vary the frequency to suit the situation. They can also produce two different sounds simultaneously, suggesting they may have two different sound-production structures. Individuals also produce a whistle that is distinct to them, called a signature whistle.

SIGNS

Look for a large dorsal fin on a mid-sized grey dolphin in a moderate to large herd. The tail will occasionally rise out of the water when the animals dive, but only before a deep dive. Coastal sightings are common on both sides of the United States, south of 40°N latitude. Deeper-water sightings are less frequent. Aerial acrobatics and bow riding are frequent, along with tail slapping.

REFERENCES

Duffy-Echevarria et al. 2008; Duignan et al. 1996; Fertl et al. 1999; Maresh et al. 2004; Moore, S.E., and Ridgeway 1996; Noren et al. 2002; Pace 2000; Patterson, I.A.P., et al. 1998; Reeves and Mead 1999; Reeves and Read 2003; Reiss and Marino 2001; Ridgway et al. 1995; Rohr et al. 2002; Rossbach and Herzing 1997; Thayer et al. 2003; Weaver 2003; Wells and Scott 1999.

FAMILY MONODONTIDAE
belugas and narwhals

This family comprises two unusual species, the Narwhal and the Beluga. Both of these northern toothed whales have large melons and lack a dorsal fin. Most or all of their neck vertebrae are unfused, giving them the ability to turn their heads. The rear margins of the flukes of adults of both species become strongly convex, especially in males. Older animals also develop curled flippers.

Beluga or Beluga Whale
also called White Whale, Sea Canary

FRENCH NAME: **Béluga**

SCIENTIFIC NAME: *Delphinapterus leucas*

Unlike most whales, whose neck bones are fused together, the cervical vertebrae of Belugas are each distinct, making them one of the few whales that can actually turn its head.

DESCRIPTION

Belugas are stocky whales, with a short beak and a bulging forehead. The bulge, called a melon, is malleable and its shape can be altered by the whale. Their upper lip is cleft and their lips are more labile than is usual for whales. Calves are born a light brown, darkening soon after birth to dark, slate-grey, which sometimes develops a reddish-brown tinge. Their colour gradually fades to white by the time they are mature. Newborn Belugas are very similar in size and colouring to newborn Narwhals. Sometimes adults will retain some dark pigment on the back ridge and the trailing margins of the flippers and flukes. The adults will sometimes take on a distinct yellowish caste in spring and early summer before the moult, but are again bright white after they moult their old skin. Belugas are thick-skinned and lack a dorsal fin. Both of these characteristics are typical of polar whales that must sometimes break the sea ice with their bodies in order to reach air to breathe. Mature males develop a distinct upward curve to the flippers that increases with age, while the flippers of females remain straight. The tail flukes of older males become dramatically convex on the trailing edge. These characteristics can, under good conditions, be observed in the field, allowing gender identification of adults. Old males are called "flatbottoms" by some indigenous peoples as the

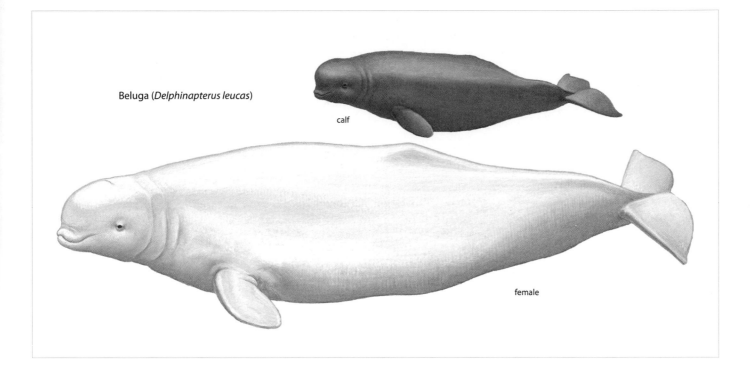

Beluga (*Delphinapterus leucas*)

calf

female

roll of blubber they develop low down along each side gives them a distinctly triangular cross-section. Belugas are born toothless and gradually grow thirty to forty teeth over the course of about four years. These teeth are quite uniform and cannot be classified into incisors, canines, premolars, and molars. The usual dental formula is 9/8, for a total of 34 teeth, with an additional rudimentary pair at the tip of the upper jaw that usually remain imbedded in the gums. Older Belugas, especially the males, commonly have teeth that are worn right down to the gum line.

SIMILAR SPECIES

Narwhals are the only other whales in the same area that could be confused with Belugas. Juveniles can be very similar, but Narwhal adults have a mottled dark and light skin that is quite distinct from the clear white of a Beluga. Belugas rarely raise their flukes out of the water, while Narwhals will before a deep dive.

SIZE

Overall size varies somewhat between populations. The smallest Belugas occupy Hudson Bay and the White Sea, medium-sized animals are found in the Gulf of St Lawrence and the eastern Canadian Arctic, and the largest animals are found off Greenland and in the Sea of Okhotsk. Adult males in some populations are about 10% longer than adult females and can be almost twice their weight. In other populations, they are about the same length, but usually become heavier when mature. The following measurements apply to animals of reproductive age from North American waters.

Total length: males, 4.0–5.5 m; females, 3.5–4.3 m; newborns, 1.2–1.8 m (average 1.6 m).

Weight: males, 1100–1600 kg; females, 700–1200 kg; newborns, 35–85 kg (average 79 kg).

RANGE

Belugas occur mainly in Arctic and Subarctic waters throughout the northern polar regions of North America and Eurasia, with the exception of ice-fast regions such as the pole. The two more southerly populations in the Gulf of St Lawrence and the Sea of Okhotsk survive in waters that border on cold-temperate and are mainly ice-covered in winter. Sightings of wanderers have been recorded far south of their usual habitat, mainly along the coast of Labrador and Newfoundland, but also along Prince Edward Island, in the Bay of Fundy, and along the New England coast as far south as New Jersey.

ABUNDANCE

The world population of Belugas in 2002 was estimated at about 100,000 animals, similar to 1992 levels. Of these, approximately 80,000 occur in Canadian waters. The small population of approximately 650–700 animals in the Gulf of St Lawrence is considered "Endangered" (since 1983) due partly to loss of habitat and past over-hunting (halted in 1979), but now mainly due to pollution. Contaminants such as PCBs, mercury, and pesticides accumulate in the Belugas' bodies over time and depress their general health, reduce reproductive success, jeopardize their immune systems, and make cancerous tumours common. Each generation becomes more contaminated as they start life with mother's milk laced with higher and higher concentrations of pollutants. Corpses that wash ashore are commonly treated as toxic waste. Another form of pollution currently of concern is increasing noise from boat and freighter traffic. Other "Endangered" populations include the southeast Baffin and the Ungava Bay stocks. Everywhere else, hunting by indigenous peoples is permitted. Stocks in the Canadian and Russian Arctic are generally stable. A few local populations are diminishing due to

Distribution of the Beluga (*Delphinapterus leucas*)

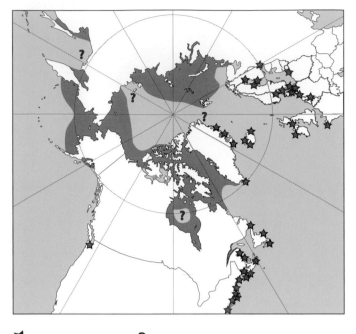

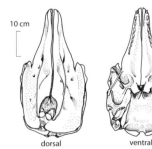 extralimital records   **?** possible distribution

unsustainable over-hunting. These include the stocks that summer in west Greenland, eastern Hudson Bay, and Cook Inlet, Alaska.

ECOLOGY

These medium-sized whales are generally found in coastal habitats, although they are equally at home diving to depths of over 800 m, or foraging in shallow river deltas. They will advance up large rivers into brackish or even fresh waters in pursuit of spawning fishes. Belugas are very proficient at manoeuvring in shallow waters and are able to swim backwards and upside down. Tidal stranding is not uncommon, but in cool northern climates these whales are usually capable of surviving until the next tide refloats them. Belugas have somewhat flexible lips, which they can purse to create an oval opening through which they rapidly draw seawater, effectively employing suction to catch prey. Belugas utilize several different hunting strategies. They will pursue schools of fish as a group; they will hunt individually in the water column; and they will dive to depths of more than 800 m to reach sea bottom, where they suck up benthic organisms, often ingesting mud and small rocks as well. The longest recorded dive is 23 minutes long, but most last less than 9 minutes. The longer and deeper dives are undertaken by the larger, older animals. Belugas are very loyal to their traditional summering and wintering waters and migrate as far as 1000 km between the two. Most stocks are known by their summering range, where they are easy to see near shore. The location and extent of wintering grounds is less well known. Belugas are adapted to live near and under sea ice. They can break through ice up to 20 cm thick in order to breathe, but heavy pack ice and landfast ice is a barrier, so the wintering grounds tend to be in polynyas (marine regions where currents maintain permanent ice-free conditions) or in areas of loose pack ice where

leads are available. Ice entrapment is a serious problem for Belugas. If a pod becomes trapped in an open lead that is isolated or freezing over (called a savsatt), they risk starvation, exhaustion, intensive hunting, or even drowning. Polar Bears can snatch a full-grown Beluga out of the water if it comes too near the ice edge. The fastest recorded swimming speed for a Beluga is 27.5 km/h, but most movement is undertaken at much more moderate speeds of 1–10 km/h. Migration travel is usually done at speeds of around 9–10 km/h. The maximum lifespan of Belugas is at least 35 years, although the average for the St Lawrence population is considerably less, around 22–23 years. Principal predators of Belugas are humans, Polar Bears, and Killer Whales. Secondary predators include Atlantic Walrus (which take calves occasionally) and possibly Great White Sharks in the Gulf of St Lawrence.

DIET

Belugas are toothed whales and they eat a diversity of prey species. Fish, squid, shrimp, and crab make up the bulk of their diet, which varies seasonally and by location, and also changes with the animal's age and hunting proficiency. The teeth are used only for capturing prey, which is then swallowed whole. The size of prey a Beluga can consume is determined by the size of its gullet. The opening is around 15–18 cm in diameter – large enough to swallow about a 4–5 kg fish comfortably. A dead Beluga found with a 9 kg Polar Cod wedged in its throat illustrates the hazards of gluttony.

REPRODUCTION

Breeding takes place between April and June while the Belugas are still offshore in their wintering grounds or are migrating between the winter and summer ranges. Timing varies geographically and is earlier in the south Alaskan populations, later in the Canadian Arctic and Greenland stocks. The gestation period is 14.0–14.5 months and a single calf is the norm. In the Canadian Arctic and in the Gulf of St Lawrence most births occur in July and early August. Females nurse their calves for 20–24 months, but solid food will supplement the milk after the first year. Females have a calf approximately every three years. Physical maturity is reached at seven to eight years old by females and nine to ten years old by males; however, sexual maturity occurs about two years earlier.

BEHAVIOUR

Belugas are very social animals, almost always occurring in groups. Males tend to form their own pods, as do females with calves and

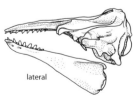

Skull of Beluga
(CMNMA 46732)

juveniles. All the groups converge into aggregations in summer in their traditional bays and estuaries. These assemblies can number in the thousands if they are undisturbed. The adults prefer to moult in the shallow, brackish warmer waters of the estuaries, and are sometimes seen actively rubbing their bodies on the sea bottom, presumably in an effort to scrape off the sloughing skin or perhaps to resolve an itch. These rubbing actions can be sufficiently vigorous as to leave permanent marks on the animal's underlying new skin. Calves are rarely born in the estuarine waters; they tend to be born farther offshore, but come into the estuary with their mothers soon after their birth. Cooperative behaviour by other females in the pod during a birth has been reported. They mill about the parturient female and assist the calf to the surface for its first breath (newborns are reported to be negatively buoyant – they sink) and help the mother carry it at the surface until it can swim for itself. Female Belugas are very solicitous of their calf and will allow it to swim in a tight formation just above her, where it can gain some hydrodynamic advantage to help it stay up. If the pod is attacked, the female will position her body between the threat and her calf. While both adult males and females will charge with an open mouth posture, only a female with a calf will actually strike. Male calves remain in their mother's pod until puberty, when they leave and join an all-male pod. Young adult females frequently stay in their mother's pod. Belugas often release air while under the surface, usually if the submerged animal is vocalizing. If pursued, Belugas are also known to exhale before surfacing, quietly breaking the water with only a small part of their head just around the blowhole, taking a quiet breath, and then smoothly submerging again. This strategy can make them all but invisible to a predator on the ice or in a boat, unless the sea is dead calm. Ordinarily the animals roll at the surface when breathing and part of their back is exposed.

VOCALIZATIONS

These whales have an extensive vocal repertoire and are almost always producing some kind of sound. Their common name of Sea Canary derives from the sounds that resonated through the wooden hulls of early Arctic explorers' ships. Sounds are produced either in the larynx or in the blowhole. Since Belugas use echolocation to find and capture prey and also to explore their environment, they are constantly producing a narrow beam of clicks and squeals and listening for the returning echo. In addition, they communicate with each other using trills, squawks, bell-like tones, jaw claps, and even a noise that sounds like the squeal of a rusty hinge. A pod of Belugas has been described as sounding like a string orchestra tuning up or like a crowd of children shouting in the distance.

SIGNS

A Beluga spends about three seconds at the water surface during each breath. Ordinarily, the top of the head breaks the surface, then submerges, as the back rolls into view. The tail does not rise out of the water. There is rarely a noticeable spout. Being curious animals, Belugas frequently spyhop (raise the head out of the water to achieve a better view) and will often approach a boat in areas where they are not hunted. Tail slapping is common.

REFERENCES

Béland et al. 1990; Belikov and Boltunov 2002; Brodie 1989; Brown Gladden et al. 1999; Curren and Lien 1998; Hamill et al. 2004; Heide-Jørgensen et al. 1998; Kingsley 1996; Martin, A.R., and Smith 1992, 1999; McAlpine et al. 1999; Prescott 1991; Reeves et al. 2002; Richard et al. 1998; Smith, T.G., et al. 1992; Stewart, B.E., and Burt 1994; Stewart, B.E., and Stewart 1989.

Narwhal

FRENCH NAME: **Narval**

SCIENTIFIC NAME: *Monodon monoceros*

Narwhals usually have only one visible tooth, but what a tooth it is. Found only on males, it erupts from the front of the left upper jaw, growing right through the lip and out in front of the animal for as much as 3 m. This unique tooth, with its gentle sinistral twist, may have contributed to the unicorn legend.

DESCRIPTION

Narwhals are medium-sized whales with large melons that often protrude beyond the lip and no dorsal fin or beak. Newborns are a solid grey colour. They gradually become darker grey/black and at around three years old begin to develop the characteristic mottling, as light patches grow and spread, starting on the belly and sides. Young adults are fairly evenly mottled. As they age, the white areas continue to spread, and in old adults, particularly males, the effect

Distribution of the Narwhal (*Monodon monoceros*)

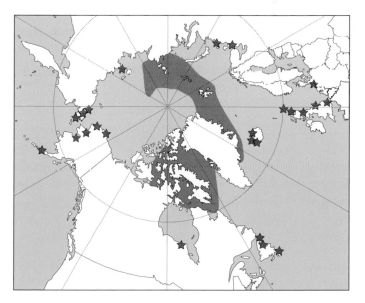

★ extralimital records

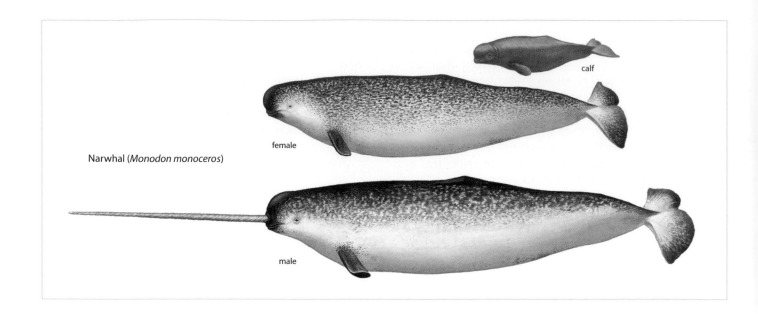

Narwhal (*Monodon monoceros*)

female

calf

male

is a light-coloured animal with the head, neck, midline of the back and belly, and edges of the flippers and flukes remaining dark. The flippers of older males curl upwards at the tip. Narwhal also display a unique tail fluke aspect that they share with Beluga. As the animals age, particularly the males, the leading edges become somewhat concave and the trailing edges more rounded, resulting in a tail contour that seems somehow reversed. Narwhal are essentially toothless in the usual sense. Their dental formula is 1/0, for a total of two teeth. In females, these teeth usually remain imbedded in the bones of the upper jaw, although females with single and double short tusks have been recorded. Males typically develop only the left tooth and the right remains imbedded in the skull, as with females. In rare instances when the right tusk also erupts, it is usually shorter than the left and displays the same left twist. Equally rarely, neither tooth will erupt. A large tusk (including the portion in the skull) can weigh up to 10 kg and be 3 m long. The tip of the tusk is frequently broken. This tusk likely continues to grow throughout the life of the animal, so the longest tusks belong to the oldest males. Usually only the outermost half-metre or so has a polished look, while the majority of the tusk may be darkened by algae growth. The tusk is a good gender indicator in the field, as part of it frequently extends out of the water as the males blow. Narwhals sometimes lift their tail flukes out of the water as they dive.

SIMILAR SPECIES
While newborn Narwhals are very similar to newborn Belugas, the adults are usually easy to distinguish by colour and by the presence of the tusk in males. Belugas rarely lift their flukes above the water, while Narwhals often do before diving.

SIZE
The following data apply to animals of reproductive age from North American waters. Adult males are about 10%–20% longer and up to 110% heavier than adult females.
Total length: males, 3.8–5.0 m; females, 3.7–4.1 m; newborns, around 1.6 m.

Weight: males, 1000–1600 kg; females, 700–1000 kg; newborns, about 80 kg.

RANGE
These whales have two populations. One occurs in Arctic waters to the east of Greenland, mainly around the eastern Queen Elizabeth Islands of Canada. The second population is found in Arctic waters to the west of Greenland, mainly in the Greenland Sea and northern Barents Sea north of Russia. Based on DNA evidence, there appears to be no contact between the two populations. Wanderers have been sighted off the coasts of Labrador and Newfoundland, in southern Hudson Bay, along the coast of Alaska and nearby Russia, around Iceland, and along the eastern coast of the British Isles.

ABUNDANCE
The North American population is estimated at between 40,000 and 60,000 animals, is thought to be declining, and has recently been designated as of "Special Concern." The Eurasian population is much smaller; its size was estimated at several thousand in 1974. Narwhals have been hunted for centuries in Canada and continue to be hunted by indigenous peoples (under a government-imposed quota). The current method of Narwhal hunting can be especially wasteful, as the whales are typically shot at with rifles from the ice edge. Many dead and dying animals sink before they can be retrieved and many others are needlessly wounded. The quota only applies to landed animals and the number killed-but-not-landed is unknown but likely high. Narwhals were even hunted commercially intermittently from the seventeenth century to recent times, but this kind of hunting is now prohibited. Unfortunately, Greenland has allowed unlimited hunting of Narwhals by indigenous peoples in the past and now allows a high quota. This overharvesting affects mainly the Canadian stock off western Greenland, which is rapidly declining. In Russian waters Narwhals are rare, have never been commercially hunted, and have been protected since 1983.

ECOLOGY

Narwhals are the most northerly of cetaceans and, not unexpectedly, the most closely linked to ice-covered waters. These whales are highly migratory, frequently travelling more than 1000 km along traditional migration routes between their offshore wintering areas in moving flow ice over deep waters and their inshore summering range in deep sounds and fjords. Their spring migration is frequently delayed by dense ice cover, but they wait at the ice edge, moving into the receding pack ice through leads and melt holes as they form. Movement of Narwhals into the inshore floating ice in spring is often prompted by the presence of Killer Whales in the deeper ice-free waters. Narwhals occupy their summer range through the summer and into early fall (August and September) and the offshore wintering areas from November through April. Their migration to the wintering region takes about four to six weeks. The spring migration back to the summering region can take three months, depending on ice break-up. Baffin Bay is an important wintering area for the majority of the Canadian Narwhal population. Generally the effect of global warming in the Arctic has been a reduction in the extent of sea-ice formation. This trend was reversed in the eastern Arctic from 1979 to 2001, likely due in part to shifts in the Gulf Current and to a decline in average winter temperatures. The area of open water in western Baffin Bay declined during that time and the ice became thicker. The Narwhals that winter there are dependent on moving ice to maintain openings for them to breathe. If a more permanent ice pack forms as a result of colder temperatures, they will be hard pressed to find another suitable area to overwinter and the Canadian stock could be threatened. But this trend appears to be reversing and ice cover in eastern Baffin Bay is actually diminishing.

Narwhals are known for their deep dives. Most can stay underwater for more than 15 minutes and dive to depths in excess of 1000 m, but a large adult can dive for more than 25 minutes and to depths of almost 1800 m. Narwhals are said to be capable of swimming up to 3 km under the ice between breathing holes. They can live for at least 25 years and may live more than 100 years. Humans are their main predator, but Polar Bears, Killer Whales, and Greenland Sharks take some. Narwhals are frequently caught in ice entrapments, where they can be vulnerable to predation. Many starve or drown if the entrapment is lengthy or the opening freezes over.

DIET

Fish, cephalopods (squid and octopus), and crustaceans (mainly shrimps) are the Narwhal's principal diet. Polar and Arctic Cod and Greenland Halibut, along with crustaceans and some squid, are the main inshore foods. In deeper waters, squid are the primary prey and some Greenland Halibut are added. The inshore feeding occurs mainly in summer, the offshore mainly in autumn and winter, and the majority of a year's calories are ingested during autumn and winter in the deeper offshore waters. Prey is probably captured by suction and swallowed whole.

REPRODUCTION

Mating takes place in late winter and early spring, with a peak of activity in April. During this time, the animals are still on their wintering grounds and are virtually inaccessible. Gestation lasts for around 15 months, with most births occurring in July and early August while the animals are in their inshore summering waters. Calves are nursed for at least a year, likely two. Most females raise a calf every three years. Males achieve sexual maturity at around eight to nine years old, but due to competition between males, it is unlikely that the young males with shorter tusks have any opportunities to mate, even if they are sexually mature. Females become sexually mature between five and seven years old.

BEHAVIOUR

Narwhal are highly social and will even cooperate to smash through ice to create a breathing hole. Ice up to 18 cm thick can be broken by Narwhals seeking air. Large aggregations of several hundred whales have been seen during the summer, but these are composed of smaller discrete groups of two to ten animals. The smaller groups are usually made up of Narwhal of similar ages or sexes, and the animals may be closely related. Mixed-gender groups are also seen at times. On the wintering grounds, the animals are often more scattered and even solitary, perhaps due to the unevenness of breathing and foraging locations. Resting groups of Narwhal often "log" on the surface. Part of their back is exposed and they appear to sleep for about 10 minutes before taking a series of breaths and then sleeping again. If the seas are rough, this activity takes place at some depth. Reaction to the presence of Killer Whales clearly shows that the Narwhals are afraid of the larger predators. They cluster at the surface or in shallow water near the noisy, surf zone and attempt to "hide" by remaining still and breathing quietly. Narwhal may even strand themselves in an effort to avoid the Killer Whale attacks.

One of the most intriguing questions about Narwhals is related to the tusk and its purpose. Current thinking is that the larger body size and long tusk of the fully mature and older males helps them to compete better for access to the breeding females. Based on the number of scars on the melon region of adult males, the tusk is probably also used as a weapon during head-on encounters. The noticeable sexual dimorphism (males are considerably larger than females and only males grow a tusk) further supports this argument, as does the discovery of a broken tusk tip imbedded in the melon of a dead male. The males are normally quite careful of their tusk as they are brittle. Any sparring that takes place tends to be done very gracefully and gently. When the animals are jostling to breathe in a confined opening in the ice, the tusked males are normally quite vigilant in their efforts to avoid harming their colleagues. Impalement by a

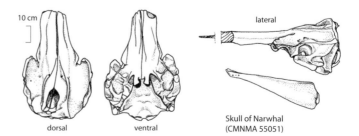

10 cm

dorsal   ventral

lateral

Skull of Narwhal
(CMNMA 55051)

tusk can be lethal, as evidenced by the occasional dead Narwhal, both male and female, found onshore with curious round puncture wounds. Whether these casualties were accidental or otherwise is unknown.

VOCALIZATIONS

Narwhal are very vocal animals. Their sounds can be heard through the hull of a boat and sometimes even above the surface. Most are narrow-band, repetitive clicks in the 24–60 kHz range. These may be emitted at a rate of 3–10 clicks per second, or as a click burst of 110–500 clicks per second. These click emission patterns are very similar to those produced by a hunting bat: slower click bursts are created as the bat flies around hunting for prey and click bursts provide it with detailed locational information once prey has been found and during its capture. It is postulated that Narwhal use their repetitive pulses in a similar way. Some of these clicks are quite loud, up to 218 dB, one of the loudest sounds created by a living organism and second only to a male Sperm Whale's clicks. Narwhal also produce whistles. Recognizably distinct whistles are thought to be created by individual animals as signature calls.

SIGNS

Narwhal are almost always found near ice except perhaps during August and September, when they are on their summering grounds. Blowing behaviour varies depending on whether the animal is breathing in open water or in a "lead." In open water the whale rolls at the surface for about three to four seconds with the blowhole and tusk tip (if it is a male) breaking first, followed by the back. If the whale is entering a deep dive, the tail will lift out of the water after the back disappears. In an ice crack or "lead", the whales often enter at a steeper angle and the whole head will emerge, followed by part of the back.

REFERENCES

Cosens and Dueck 1991; Garde et al. 2007; Heide-Jørgensen et al. 2002, 2003; Laidre and Heide-Jørgensen 2005; Laidre et al. 2002, 2006; Miller, L.A., et al. 1995; Palsbøll et al. 1997; Reeves and Tracey 1980; Stirling and Parkinson 2006.

FAMILY PHOCOENIDAE
porpoises

These small whales have small flippers, a melon, but not a prominent beak, and usually a dorsal fin that is located about midway along the back. (One species, the Finless Porpoise, has no dorsal fin.) The most diagnostic characteristic that identifies this family is tooth shape. Each tooth is compressed laterally and has a spatulate crown that forms a cutting edge.

Harbour Porpoise
also called Puffing Pig, Puffer, Common Porpoise

FRENCH NAME: **Marsouin commun,** *also* **Pourcil**
SCIENTIFIC NAME: *Phocoena phocoena*

This small cetacean is probably the one most commonly seen by Canadians, because it occurs near shore off both our east and west coasts, and is still numerous in most parts of its northern North American range.

DESCRIPTION

These thick-bodied porpoises are the smallest cetaceans in Canadian waters. They also occupy relatively cold waters, and so, to compensate, have a relatively thick blubber layer that averages 1.5–2.0 cm thick. The dorsal fin, flippers, and flukes are dark grey-black or blackish-brown, as is the back. The leading edge of the dorsal fin is lined with small, raised bumps known as tubercles. The chin and belly are white and the sides are greyish-brown, as the white underside diffuses into the darker back. Some animals have a black chin patch. A dark stripe runs from the corner of the mouth to the top of the flipper. The colour pattern is fairly simple, but with many subtle variations between individuals and also within the same individual as it ages. There is no noticeable beak. The dorsal fin is triangular in shape and may be slightly falcate in some animals. It is set just past the mid-body point. The dental formula is 21–29/20–29, for a total of 82–116 teeth. Porpoise teeth differ from dolphin teeth. The tips of dolphin teeth are simple sharp cones. Porpoise teeth are spade-shaped at the tip, which is flattened from side to side. These teeth actually can create a rough cutting edge, rather than serve simply for gripping as they do for dolphins.

Harbour Porpoise (*Phocoena phocoena*)

SIMILAR SPECIES

Harbour Porpoises might be confused with several species of coastal dolphins, but all are substantially larger and have a much larger, more falcate dorsal fin. In the west, Dall's Porpoises are small enough to be confused with Harbour Porpoises, but again the dorsal fin will distinguish them. It is similar in shape but dual coloured (black and white) in the Dall's Porpoise. Dall's Porpoises are also known to bow-ride, which the smaller Harbour Porpoises rarely do.

In the waters off southern British Columbia, a small number of female Dall's Porpoises are mating with male Harbour Porpoises and producing hybrid calves. These hybrids are intermediate in size and shape, and tend to be lighter grey than either parent, with variable dark grey or dark brown areas. They tend to associate with their maternal herd (Dall's Porpoises) and frequently bow-ride, unlike Harbour Porpoises.

SIZE

Females grow faster and achieve a larger size than males. The following dimensions apply to adult Harbour Porpoises in Canadian waters.
Total length: males, average of 144 cm, up to 170 cm; females, average of 155 cm, up to 190 cm; newborns, 60–90 cm.
Weight: males, up to 67 kg; females, up to 80 kg; newborns, 5.0–6.7 kg.

RANGE

Harbour Porpoises are found in temperate and Subarctic coastal waters over the continental shelves of the Northern Hemisphere. They are well adapted to cold waters and are seldom seen in waters warmer than 16°C. They occur in waters off British Columbia year-round. The eastern North American populations tend to migrate offshore and somewhat southerly in late autumn, but some can be sighted all year round off the east coast as well. They are readily seen from June to October in the Gulf of St Lawrence as far upstream as the mouth of the Saguenay River. The lower Bay of Fundy population migrates into the bay in July and leaves in September/October. In Newfoundland, Harbour Porpoises are commonly seen from May through July. There are substantial changes between years in the abundance and distribution of Harbour Porpoises in the Gulf of Maine and Bay of Fundy that appear to be related to prey distribution.

ABUNDANCE

Of all cetaceans, Harbour Porpoises are the most threatened by human activities. They inhabit some of the most heavily fished and industrialized coastlines in the world. It is little wonder that many thousands die every year from entanglement in fishing gear. The concentration of chemicals and heavy metals in their tissues is an ever-increasing threat. In some areas, the rate of death in bottom-set gill nets exceeded the rate of annual calf production. On the Atlantic coast since the early 1990s, this death rate has dropped dramatically, due in large part to the reduction to the groundfish fishery caused by the depletion of fish stocks. Harbour Porpoises remain fairly common in Canadian waters for now. Unfortunately, as fish stocks recover and fishing resumes, the death rate of Harbour Porpoises killed in entanglements is likely also to rise. They are rarely seen

Distribution of the Harbour Porpoise (*Phocoena phocoena*)

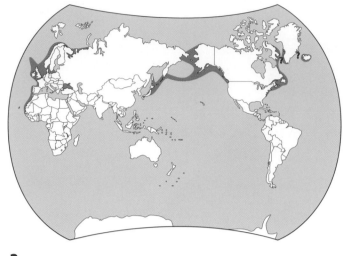

? presence likely seasonal and occasional

anymore in busy harbours, possibly due to the high noise levels. Populations in parts of Europe, particularly in the Black and Baltic Seas, which were heavily harvested in the past, have not recovered and are now threatened. Harbour Porpoises continue to be hunted in Greenland for meat. There is no limit to the take and the stock involved is likely from Labrador-Newfoundland. Recently, efforts have been made to attract the porpoise's attention to the presence of nets by studding them with acoustical "pingers." Although the animals can detect the netting with their echolocation, they may not use echolocation all the time, or may not be paying sufficient attention. The additional sensory stimulus does seem to help them avoid the potential death trap, at least for a while. Entrapment in Herring weirs in the Bay of Fundy area, formerly a major cause of mortality, has been largely ameliorated by a program of rapid detection, capture, and release, usually conducted by local fishermen and environmental agencies. High-amplitude Acoustic Harassment Devices (AHDs) attached to pens of farmed Atlantic Salmon are used primarily to deter Harbour and Grey Seals, but appear to be affecting Harbour Porpoises and driving them away from suitable habitat in the Bay of Fundy especially. Few porpoises will approach within 3.5 km of an AHD. Another source of potential acoustic harassment could include seismic exploration, especially if it occurs in feeding grounds or migration corridors. The Gulf of Maine/Bay of Fundy population was estimated at 89,700 animals following surveys in 1999, which is more than twice the number seen in 1991.

Harbour Porpoises are listed as "Vulnerable" by the World Conservation Union (IUCN) and are on appendix 2 of CITES.

ECOLOGY

These small cetaceans live in waters that vary in surface temperature from 4°C to 16°C and are usually less than 200 m in depth, although they may cross areas of deeper water during migration. Since most of their food comes from near the ocean floor and in the lower part of the water column, they are likely exposed to cooler temperatures

below the surface. Because of their small size, Harbour Porpoises cannot store a lot of food as blubber and so must remain near their food source so they can feed regularly and often. This is most crucial for lactating females, who expend tremendous amounts of energy producing high-fat milk. Calories are quickly extracted from food, as it takes only about 140 minutes for material to pass through the digestive tract. Brief dives of not more than five minutes are usually followed by several breaths and another dive. Most dives do not go below 50 m, but some can be as deep as 226 m. Feeding occurs around the clock during both daylight and darkness. Harbour Porpoises have been successfully rehabilitated and kept in captivity, but have never produced offspring in confinement. They typically travel more than 50 km in a day and have home ranges that may encompass enormous numbers of square kilometres. A pregnant female, with a dependant calf, was tagged in the Bay of Fundy in mid-July and travelled soon after to the Gulf of St Lawrence, where she and her calf spent the remainder of the summer. The maximum known age based on tooth layers is 24 years old in males and 23 years old in females, but few live into their teens. Killer Whales and sharks are serious predators of Harbour Porpoises, but probably most deaths occur following entanglements in fishing gear.

DIET

Harbour Porpoise prefer small schooling prey with an average length of 28–30 cm. Like many other small cetaceans, Harbour Porpoises are known to choke to death trying to swallow prey that is too large, and consequently lodges in their oesophagus. In the western Atlantic Ocean off Maine and the Maritimes, they eat mainly Atlantic Herring. The Fundy population adds some Silver Hake, while the Gulf of St Lawrence population adds Capelin, redfish, and mackerel. The Newfoundland population adds Capelin, sand lance, and Horned Lantern Fish to Atlantic Herring. Pacific porpoises have a more diverse diet. Market Squid, Blackbelly Eelpout, Pacific Herring, and Pacific Hake make up a large part of their food. Winter diets are more diverse, but not well understood. Calves begin to take some solid food by the end of the summer and often begin with euphausiid shrimps.

REPRODUCTION

Although relatively short-lived for cetaceans, Harbour Porpoises compensate by being very prolific. Some females mate as early as their second year, and produce a calf in their third year, but most produce their first calves in their fourth year, and then almost every year thereafter. Females spend the bulk of their adult life in the demanding situation of simultaneously nursing and being pregnant. In western waters, Harbour Porpoises give birth from May to September. Off the east coast of Canada, breeding is more seasonal. Mating occurs over the course of a few weeks in early summer, and after a gestation period of 10–11 months, most calves are born in May or early June. The arrival of the newborns is not timed for the period of warmest water as one might expect, but rather for the time when prey is most available to their mother to support her milk production. Around 37% of the weight of a newborn is skin and blubber, which is extremely important to protect the infant from cold waters.

Calves consume exclusively milk for their first three months (during which time they triple their birth weight) and then gradually begin to supplement that diet with shrimp and eventually fish. Harbour Porpoises are somewhat different from other cetaceans in that they nurse their calves for only 8–12 months and the young grow and achieve sexual maturity very quickly. During the breeding season the male testes are large, approximately 4% of body weight. Once breeding is over, they regress to a fraction of that size.

BEHAVIOUR

Although these small whales are often seen alone, they also form loosely connected groups of two to eight animals. Group size changes over the course of a summer, with the groups being smallest early and larger later. Most of the small groups (two individuals) are made up of a female and calf pair, but whether they form long-term social bonds is unknown. The genetics and social structure of the larger groupings is also unknown. Exclusive groups of juveniles, males, and females have been seen. Some naturally marked individuals have been seen returning to the same small cove over the course of several years. It is thought that Harbour Porpoises are not cooperative hunters; each appears to forage for itself. Large aggregations, which can be commonly seen in parts of the range, are likely drawn to an abundant food source. Harbour Porpoises travel many kilometres each day while foraging and marked individuals are known to travel long distances in a relatively short time (a migration of 300 km over the course of a month). Maximum swimming speed is about 22 km/h. These porpoises tend to be shy of boats and are difficult to study. Observations of captive animals indicate that they display complex social behaviour marked by frequent sexual and aggressive interactions.

VOCALIZATIONS

Harbour Porpoises echolocate and communicate using click-trains. They do not produce whistles. The clicks they create fall within two different frequency ranges, around 2 kHz and around 130 kHz. The lower frequency is audible to humans. The higher frequency is likely used for echolocation, the lower for social communication.

SIGNS

These small whales can be very difficult to spot, especially if the water is choppy. They roll quickly on the surface to breath and create little "splash." The small amount of their back that protrudes is shiny and dark and the small triangular dorsal fin provides a clue.

REFERENCES

Baird et al. 1998; Börjesson and Read 2003; Carlström 2005; Gaskin 1992a; Hult et al. 1980; Johnston, D.W., 2002; Kastelein et al. 1997; Kingsley and Reeves 1998; McLellan, W.A., et al. 2002; Neimanis et al. 2000; Olesiuk et al. 2002; Otani et al. 2000; Palka et al. 1996; Read 2001; Read and Hohn 1995; Read and Westgate 1997; Read et al. 1997; Richardson, S.F., et al. 2003; Teilmann et al. 2006; Trippel et al. 1999; Westgate and Read 1998; Willis, P.M., et al. 2004.

Dall's Porpoise

FRENCH NAME: **Marsouin de Dall**

SCIENTIFIC NAME: *Phocoenoides dalli*

Perhaps because of its size and stocky shape, this small porpoise is one of our fastest cetaceans. It is capable of achieving speeds in excess of 55 km/h, though it is clearly a sprinter, as such velocity can only be sustained for short distances.

DESCRIPTION

This small-headed, thick-bodied porpoise is striking, not only in its shape, but also in its black and white colouring. The body is basically black with two large oval side patches of white that join under the belly. The white patches are clearly defined and occupy a large area roughly in the middle of each side. The flukes and flippers are black and the flukes, and sometimes the flippers, have white on the trailing edges. The flippers are small and positioned forward on the body. The dorsal fin is wide-based, triangular in shape, and sometimes slightly falcate. It is black, with variable amounts of white on the upper portion or with a white centre. All-white, grey, or black animals are rare. The beak is poorly defined and hardly noticeable. Newborns display a muted version of the adult colouring, with dark greys and beiges or light grey rather than black and white. Their flippers and flukes are completely dark. Adult males have a thicker tail stalk than the adult females and also develop a pronounced hump just behind the anus. The trailing edges of the flukes of a fully mature male are strongly convex, unlike the more typical contour of a female. The animal pictured here is a female. The bulky body of these porpoises is not caused by blubber, but by muscle. They are heavily muscled through their trunk, especially in the lumbar region. These muscles, along with a larger than usual heart and an enhanced oxygen-carrying blood capacity, fuel the bursts of speed for which this species is renowned. The dental formula is 21–28/21–28, for a total of 84–112 small teeth. There are usually more teeth in the lower jaw than the upper. Dall's Porpoise teeth are not as spade-shaped at their tip as those of the Harbour Porpoise, but they are

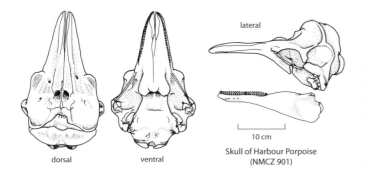

lateral

Skull of Harbour Porpoise
(NMCZ 901)

10 cm

dorsal   ventral

female

Dall's Porpoise (*Phocoenoides dalli*)

thicker at the crown than at the root, and erupt only slightly through the gums.

SIMILAR SPECIES

All the dolphins that might be confused with Dall's Porpoises are larger and have larger, more falcate dorsal fins. Harbour Porpoise are the only other small cetacean that shares the British Columbia waters and might be mistaken for Dall's Porpoise. A two-tone (black and white) dorsal fin distinguishes the Dall's Porpoise from the monotone dark dorsal fin of the Harbour Porpoise. Dall's Porpoises frequently bow-ride, much to the delight of whale watchers. Harbour Porpoises do not bow-ride.

In the waters off southern British Columbia, a small number of female Dall's Porpoises are mating with male Harbour Porpoises and producing hybrid calves. These hybrids are intermediate in size and shape between the two species, and tend to be lighter grey than either parent, with variable dark grey or dark brown areas. Their fertility is unknown. Most associate with their maternal herd (Dall's Porpoises) and frequently bow-ride, unlike Harbour Porpoises.

SIZE

Males are larger than females.
Total length: males, average 200 cm, up to 239 cm; females, average 172 cm, up to 205 cm; newborns, about 100 cm.

Weight: males, 160–200 kg; females, 100–170 kg; newborns, around 11–12 kg.

RANGE

Dall's porpoises are endemic to the cool temperate North Pacific Ocean. They occur from southern Alaska to southern California and from the sea of Okhotsk in northern Russia to about the middle of the most southerly island of the Japanese Archipelago (Honshū). They are seen year-round in the deeper inshore waters of British Columbia.

ABUNDANCE

Dall's Porpoises are generally considered abundant. There may be as many as 2,000,000 animals in the whole North Pacific. Approximately 130,000 live off the North American west coast. Thousands die every year in fishing gear and Japan continues to harvest close to 18,000 a year. Usually they are harpooned as they bow-ride. This species is considered to be common both offshore and in the deeper inshore waters of British Columbia.

ECOLOGY

These porpoises are found most commonly in cold waters between 2°C and 17°C and they tend to move on if surface temperatures rise above 17°C. Although there is little evidence of long migrations, the populations off the North American west coast do seem to show

Distribution of the Dall's Porpoise (*Phocoenoides dalli*)

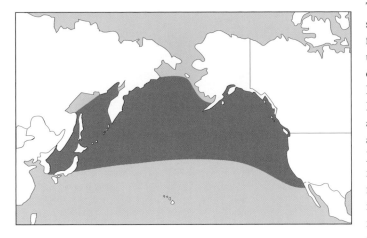

an inshore/southerly movement in winter and an offshore/northerly movement during summer. Some male Dall's Porpoises wander quite widely, while females tend to remain in a more restricted, albeit still quite extensive, home range. This species is difficult to maintain in captivity and most die within weeks of capture. Mammal-eating Killer Whales (transients) prey upon Dall's Porpoise, while the fish-eating Killer Whales (residents) interact more peacefully. Dall's Porpoises seem to be able to distinguish between the calls of the two types of Killer Whales and will behave accordingly, fleeing the "transients" while ignoring or even mingling with the "residents." Sharks may take a calf or a disabled animal, but probably cannot catch a healthy adult. Humans remain the Dall's Porpoises most serious predator, either directly by hunting or indirectly by entanglements in fishing gear. Dall's Porpoises typically are heavily infested with internal parasites, more so than most other small cetaceans. How this heavy parasite load affects them is unknown, but this species is not known to mass strand and even solitary individuals rarely strand. With their heavy muscle mass, a dead one likely sinks before washing ashore. The oldest known animal was 22 years old, but few live even into their teens.

DIET

These small cetaceans are opportunistic feeders. They eat small schooling fishes such as Pilchard and Walleye Pollock, and shoaling squid, and will feed on whatever is locally abundant. Most of their prey is less than 30 cm long, and generally less than 10 cm. Prey is swallowed whole. They feed during the day or night, depending on when prey is available. Many of their prey are deep-water species that rise towards the surface during darkness. If they hunt these species at night, the porpoises do not have to dive as deeply, or into the colder, deeper waters. Some schooling fish can be herded up against the water surface, where they can be taken fairly easily. This hunting technique is usually pursued during daylight. Seasonal occurrences of these porpoises may be related to prey abundance. Dall's Porpoises have a thinner blubber layer than most other cold-water cetaceans and so have to maintain a higher metabolism. They must eat frequently to get the calories they need to keep warm.

REPRODUCTION

Their offshore habits and speed make this porpoise very difficult to study. Much of our information on their reproduction is derived from the study of by-catch in the high seas drift-net fishery. Females usually produce their first calf when they are three to four years old. Males become sexually mature between 4.5 and 5.5 years old. Both genders achieve physical maturity at about seven years old. Pregnancy has been estimated to last 10–11 months and results in a single calf. Once they are sexually mature, most females produce a calf every year. Calving tends to be seasonal, peaking in June and July in the eastern Pacific off North America, although young calves may be seen from March to September. Calves begin eating solid food as early as three to four months old, but their age at weaning is uncertain. They may associate with their mother until she has her next calf. Since breeding occurs about a month after the birth, the females spend part of every year both nursing and pregnant.

BEHAVIOUR

Dall's Porpoises are most often seen in groups of two to twelve animals. Some of their foraging is done in coordination with the other animals in the herd. The herd will drive a school of fish up against the water's surface, where individual fishes can be plucked out of the fish ball. A herd of fast moving, zig-zagging porpoises, creating large amounts of "splash" as they breathe and change direction near the surface is likely feeding on such a fish ball. Other hunting strategies are more solitary, although the whole herd will hunt in unison. When foraging at depth for bottom-dwelling fishes and deepwater squid, each herd member will undertake long dives of two to four minutes interspersed with shallow dives and several breaths. How deep they dive is unknown. Occasionally, in areas of high prey density, aggregate herds of several thousand may form briefly while feeding. There is some segregation of the sexes during summer, as the pregnant females tend to travel farther north to calve and most of the males, subadults, and non-reproductive females remain farther south. Some mature males travel north with the females and presumably then are on hand to participate in breeding, as the females become receptive a month or so after the calves are born.

VOCALIZATIONS

The sound production of this species is poorly known. High-frequency sounds in the 120–160 kHz range have been recorded in the wild. These are likely used for echolocation. Captive animals are

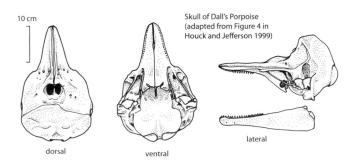

10 cm

Skull of Dall's Porpoise (adapted from Figure 4 in Houck and Jefferson 1999)

dorsal

ventral

lateral

known to produce low-frequency clicks that are audible to humans (< 20 kHz).

SIGNS

Dall's Porpoises are avid bow-riders and will converge on powered vehicles to enjoy this sport, darting back and forth in the wave with rapid, sometimes jerky movements. Mother-calf pairs rarely bow-ride. The optimum boat speed to attract bow-riders is between 26 and 32 km/h. These porpoises are not generally aerial and rarely show more than the top of their white side patch above the water's surface. During rapid swimming, they throw up a characteristic high spray of water called a "rooster tail" as they lunge to the surface for a breath. This "splash" will sometimes completely obscure the animal creating it. At slower speeds they are capable of rolling on the surface while taking a breath without creating any noticeable splash. On a slow roll, the top of the bent tail stock (but not the tail) appears out of the water just as the animal submerges.

REFERENCES

Amano et al. 1998; Baird et al. 1998; Escorza-Treviño and Dizon 2000; Ferrero and Walker 1999; Fertl et al. 1999; Houck and Jefferson 1999; Jefferson 1988, 1990; Law and Blake 1994; Walker, W.A., 1996.

FAMILY PHYSETERIDAE
sperm whales

Three species of toothed whales constitute this family and all occur in Canadian waters. Members of this family are recognized by the large melons on their foreheads that contain spermaceti oil. Their skulls are dished on the dorsal surface to contain the melon and a high occipital crest at the back of the skull forms the rear bulwark. All their teeth are confined to the lower jaw, which is under-slung and weak. Sperm Whales have a long rectangular head that constitutes about one-third of their total length, and a single blowhole located at the tip of the snout on the left side that is slanted towards the left so the spout is directed away from a blowing animal at a 45° angle. Pygmy and Dwarf Sperm Whales have smaller heads in proportion to their bodies, and they also have a single blowhole on the left side of the skull; however, it is located farther back, on the forehead.

Pygmy Sperm Whale
also called Pigmy Sperm Whale

FRENCH NAME: **Cachalot pygmée**
SCIENTIFIC NAME: *Kogia breviceps*

Pygmy and Dwarf Sperm Whales are very similar and are only distinguishable at sea under exceptional viewing conditions. They are rarely seen alive and most of our understanding of them is derived from the examination of stranded animals. Information published before 1966 should be carefully scrutinized, as both forms were considered to be the same species before that time.

DESCRIPTION

Pygmy Sperm Whales are small, but robust cetaceans with a distinctly shark-like appearance. They have a squared head (when viewed from the side), a small, under-slung lower jaw, and usually a distinct "false gill" marking behind the head. From above or below the head is thickly triangular in outline. Their bodies are counter-shaded, with dark grey above fading to whitish or pinkish-white on the underside. The flippers, dorsal fin, and flukes are dark grey. The dark front flippers stand out on the lighter sides,

Pygmy Sperm Whale (*Kogia breviceps*)

are set forward on the body, and can be tucked snugly against the body. The dorsal fin is small and falcate and positioned behind the animal's midpoint. The flukes have a concave trailing edge and a distinct medial notch. Like their larger cousins the Sperm Whales, Pygmy Sperm Whales have a spermaceti organ in their head filled with spongy tissue saturated in oil. Speculation abounds on the function of this organ, but most cetologists currently believe it serves some purpose in echolocation, probably to direct and focus the clicks. Their nasal passages have modified to the extent that the left opening extends to the outside of the animal and is used for breathing, while the right passage is blind and is likely used for vocalization. These whales have 12–16 teeth in each lower jaw (rarely 10 or 11) and usually no upper teeth. The dental formula is 0/12–16, for a total of 24–32 teeth.

SIMILAR SPECIES

The most similar species to the Pygmy Sperm Whale is the Dwarf Sperm Whale. These two whales are extremely difficult to distinguish at sea and most surveys lump them under the genus rather than attempt to determine the species. Dwarf Sperm Whales are typically smaller by around 45 cm and have a larger, more forward-placed dorsal fin. These characteristics are difficult to see at sea, but stranded individuals can be distinguished as follows: distance measured from the front of the snout to the anterior insertion of the dorsal fin is > 50% of the total body length for Pygmy Sperm Whales and < 50% for Dwarf Sperm Whales; dorsal fin height is < 5% of total body length for Pygmy Sperm Whales and > 5% for Dwarf Sperm Whales. Similar in size, Pygmy Sperm Whales are easily confused with the more common oceanic dolphins such as Bottlenose Dolphin, Atlantic White-sided Dolphin, and the White-beaked Dolphin, all of which also have a dark dorsal fin. Typically, the Pygmy Sperm Whale is quiet and unassuming at the surface. The dolphins tend to be more boisterous. Probably most sightings of Pygmy Sperm Whales are unrecognized.

SIZE

Adult males and females are about the same size.
Total length: around 250–380 cm; newborns, around 120–150 cm.
Weight: 318–450 kg; newborns, 23–25 kg.

RANGE

This whale is found in tropical and warm temperate waters worldwide, and reaches its northern limits in North America along the coasts of southern British Columbia and the Maritimes. Off our east coast, the presence of Pygmy Sperm Whales is closely linked to the warm Gulf Stream current. There are five strandings recorded from the east coast. One stranded on northeastern Vancouver Island in 2003, ending years of speculation concerning their presence off the west coast.

ABUNDANCE

Pygmy Sperm Whales are difficult to spot at sea and may be somewhat more abundant than sightings indicate; nevertheless, they are considered uncommon to rare throughout most of their distribution. Recent surveys indicate that the Gulf of Mexico may be a hot spot for *Kogia*, the most commonly seen cetaceans in the Gulf. Both the Pygmy and Dwarf Sperm Whales were lumped together in these surveys, due to the identification difficulties. There are no estimates of world population size, and although Pygmy Sperm Whales have been considered for "Endangered" status by the World Conservation Union (IUCN), this body has been forced to declare the species "Data Deficient" due to a lack of information.

Distribution of the Pygmy Sperm Whale (*Kogia breviceps*)

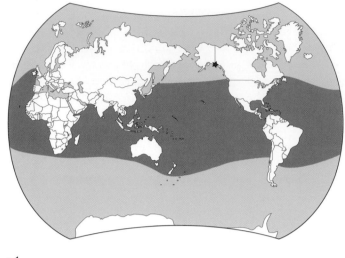

★ extralimital record

ECOLOGY

These whales are generally found in the deeper waters of the continental slope and open ocean, although juveniles are thought to remain closer to shore along the edge of the continental shelf and move farther offshore when they mature. Most sightings occur in waters that are 400–1000 m in depth and especially along the edge of the continental shelf, where cold upwellings concentrate their prey. The habit of this species of "logging" on the surface (see "Behaviour" section below) while resting makes them vulnerable to being hit by powered vessels, and many beach-cast animals show wounds that are likely the result of such collisions. Remains of Pygmy Sperm Whales have been found in the stomachs of Killer Whales and an attack by a Great White Shark has been reported. Likely shark attacks are fairly frequent and are probably a major cause of death. Stranded animals, brought to aquaria for rehabilitation, rarely thrive and generally die within days. Speeds of up to 11 km/h have been recorded from a radio-tracked individual, but the average travelling speed was about half that, or 5.5 km/h. Deeper dives lasted more than eight, and up to twelve, minutes. Shallow dives usually lasted two to five minutes. The deeper dives were made at night or on cloudy days and the shorter dives during clear days. Several animals have been discovered to have swallowed plastic wrap, which came to block the stomach or intestine, resulting in stranding and death. Both the Dwarf and Pygmy Sperm Whale are vulnerable to this phenomenon. Possibly the plastic film either acoustically or visually resembles their prey. This is of continuing and likely increasing conservation concern.

DIET

Their diet is principally composed of small to medium-sized squid, with the minor addition of some fishes, crustaceans, and molluscs. Few prey items in the stomachs of stranded animals exceeded 33 cm in length and most were less than 20 cm long. Their preferred prey inhabits the deeper parts of the continental shelf and along the continental slope at depths greater than 200 m to around 1000 m. Many of these species move upwards in the water column at night when light levels decline. Pygmy Sperm Whales foraging during the day must dive deeper than when they forage at night due to this daily migration. Prey is likely captured by suction. Bottom feeding is also suspected, as benthic crabs and fish have been found in their stomach. This further suggests a prodigious diving capacity for such a small whale.

REPRODUCTION

The mating and calving season for these whales is lengthy, from autumn to spring (October through April in the Northern Hemisphere and March through September in the Southern Hemisphere). The gestation period is thought to be around 11 months, following which a single calf is born. An unusual number of parturient or late-term females become stranded, and there are several records of births occurring on the beach. Calves are nursed for about a year and females can become pregnant while still lactating, which suggests that a calf may be born every year. Likely the usual calving interval is every two years. Females reach sexual maturity when they are around 2.5–2.7 m in length and males 2.7–3.0 m long. The testes of mature males are impressive in size: each one can weigh as much as 5 kg or more.

BEHAVIOUR

Pygmy Sperm Whales are often solitary or may be found in loose herds of two to six animals. This whale spends considerable time, especially in the afternoon, floating at the surface with its blowhole and part of its dorsal fin exposed, while the tail hangs down, a typical *Kogia* posture. Such "logging" Pygmy Sperm Whales can be very approachable and are suspected of being asleep. If startled, a logging whale may release a cloud of reddish-brown fluid to camouflage its escape. This behaviour, called "inking," typically occurs when the animal is stressed. The fluid (liquid faeces) is stored in the lower intestine and may leak out of the anus of stranded animals, making it appear that they are bleeding. The fluid is very sticky and smelly and is virtually impossible to wash out of stained clothing. Mass strandings of Pygmy Sperm Whales are very uncommon; only two have been recorded to date. The usual strandings involve a solitary whale or a mother-calf pair.

VOCALIZATIONS

These whales are not very vocal. Echolocation clicks have been recorded ranging in frequency from 60 kHz to over 200 kHz, with

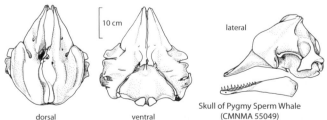

10 cm

lateral

dorsal   ventral   Skull of Pygmy Sperm Whale
(CMNMA 55049)

a dominant frequency of 120–130 kHz. Stranded animals or individuals in distress also produce audible squeals in the 1–2 kHz range. Indications are that the directionality of the echolocation clicks is considerably more focused than for most cetaceans. This quality of the sound stream is thought to be the result of the spermaceti organ in the head.

SIGNS

Pygmy Sperm Whales are usually inconspicuous at the surface. They have a small diffuse blow, rise slowly and quietly for a breath, and do not lift their flukes out of the water before a deep dive. Breaching has been reported, but generally they are timid and do not approach boats. They can sometimes be approached quite closely if discovered sleeping at the surface. All these characteristics apply also to its sister species, the Dwarf Sperm Whale.

REFERENCES

Bloodworth and Odell 2008; Caldwell and Caldwell 1989; Hückstädt and Antezana 2001; Lien et al. 2002; Long, D.J., 1991; Marten 2000; McAlpine 2002; McAlpine et al. 1997; Measures et al. 2004; Plön et al. 1999; Reeves et al. 2002; Santos et al. 2006; Scott, M.D., et al. 2001; Stamper et al. 2006.

Dwarf Sperm Whale

FRENCH NAME: **Cachalot nain**

SCIENTIFIC NAME: *Kogia sima*

Very little is known about this whale, as it is rarely seen alive. Most of our information is derived from the examination of stranded animals. To further complicate their study, another small whale, the Pygmy Sperm Whale, is very similar, and generally the two are indistinguishable at sea. Until 1966, both Dwarf and Pygmy Sperm Whales were considered to be the same species. The Dwarf Sperm Whale may be further divided into two species in future, as the North Atlantic population has been discovered to be significantly different genetically from the Indo-Pacific population.

DESCRIPTION

These small cetaceans are shark-like in both shape and colouring. They have a squarish head profile that is thickly triangular when viewed from above or below. The large dorsal fin, forward-placed flippers, and small flukes are dark grey-black. A medial notch occurs between the flukes. Their thick body is counter-shaded dark grey on the back, fading to a pale whitish or pinkish-white belly. An under-slung lower jaw and a more or less distinct "false gill" marking behind the head enhance the shark-like appearance. The falcate dorsal fin is placed about mid-body. Two, or sometimes more, short throat grooves have been noticed. Examination of more stranded

animals over time will validate whether this is a species characteristic, as throat grooves seem not to be present on Pygmy Sperm Whales. Seven to 13 pairs of thin sharp teeth (< 3 cm high) occur in each lower jaw, and up to three vestigial teeth, which do not erupt through the gums, in each upper jaw. The dental formula is 0–3/7–13, for a total of 14–32 teeth. Like Sperm Whales and Pygmy Sperm Whales, Dwarf Sperm Whales have a spongy, oil-filled spermaceti organ in their heads, probably used to focus their echolocation pulses. Dwarf Sperm Whales share a genus-specific characteristic with Pygmy Sperm Whales: a sac-like enlargement of the large intestine near the anus, which may contain up to 10–12 litres of a reddish-brown fluid (see "Behaviour" section below). This fluid can be forcibly expelled during times of stress or attack.

SIMILAR SPECIES

The Pygmy Sperm Whale is very similar to the Dwarf Sperm Whale, and the two species can rarely be distinguished at sea. The dorsal fin of the Pygmy Sperm Whale is smaller and set farther along the back than that of the Dwarf Sperm Whale, even though in body length and weight the Pygmy is larger than the Dwarf Sperm Whale. Adult Pygmy Sperms are generally more than 2.7 m long, while adult Dwarf Sperm Whales are less than 2.7 m long. The presence or absence of throat grooves should be noted, as indications suggest that they are found on Dwarf Sperm Whales, but not on Pygmy Sperm Whales. See the Pygmy Sperm Whale account for further information. The dark, falcate dorsal fin of the Dwarf Sperm Whale is very similar to that of many dolphins and likely many sightings are mistaken for dolphins.

SIZE

Both sexes reach similar sizes and weights as adults.
Total length: 190–270 cm; newborns, about 90–100 cm.
Weight: 136–272 kg and likely more; newborns, unknown.

Distribution of the Dwarf Sperm Whale (*Kogia sima*)

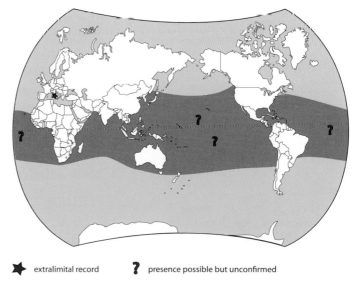

★ extralimital record   ？ presence possible but unconfirmed

Dwarf Sperm Whale (*Kogia sima*)

RANGE

Dwarf Sperm Whales occur worldwide in warm temperate and tropical marine waters. Their distribution is somewhat generalized, as the limits are based mainly on stranding records, which, of course, are coastal. Their oceanic presence is presumed, but the habitat of this whale tends to be somewhat more inshore than its sister species, the Pygmy Sperm Whale. It may be incorrect to assume their occurrence in deeper mid-oceanic waters. Strandings in Canadian waters have occurred on Sable Island in the Atlantic Ocean and Vancouver Island in the Pacific.

ABUNDANCE

There are no world population size estimates of Dwarf Sperm Whales. They are generally considered to be rare, although there are areas where they appear to be somewhat more abundant, such as the southern tip of Africa and the Gulf of Mexico. Difficulties of viewing and identification suggest that Dwarf Sperm Whales may be more common than sighting data imply. The World Conservation Union (IUCN) was forced to classify the species as "Data Deficient" after considering it for endangered-species status. Some animals are drowned each year in fishing gear, and some are harpooned or netted by shore-based small whaling operations around the world, but there is no existing commercial fishery for either the Pygmy or Dwarf Sperm Whales.

ECOLOGY

Dwarf Sperm Whales are thought to live near the edges of continental shelves and their slopes. There is some evidence that juveniles and pregnant or nursing females spend more time closer to shore on the continental shelf itself, feeding on inshore squid and consequently making shallower foraging dives. Non-reproductive adult females and mature males spend more time feeding in deeper continental-slope waters, capturing deeper-water species. A study from the Caribbean found that habitat usage varies with season. During summer the animals were found only in the deep (900–1600 m) regions, while in winter they occurred over both deep and slope (400–900 m) habitats. These variations are likely a result of seasonal movements of squid as the whales follow their favourite prey. An animal observed in daytime was found to dive for two to three minutes at a time, then rest on the surface for a minute or more in the characteristic "logging" posture (see the "Behaviour" section below for a description). Another foraging animal stayed underwater for almost 45 minutes. Dives of 25 minutes combined with surface intervals of up to three minutes are common. Dive times presumably vary with the depth of prey. During darkness many prey species migrate up the water column, becoming more accessible with shallower dives. Pursuit of the same prey during daylight would require considerably deeper and longer dives. The longevity of both *Kogia* species is unknown, due to the difficulties presented by their enamel-less teeth to the yearly layer count method commonly used for aging cetaceans. Shark predation is likely a major mortality factor and Killer Whales are also known predators. Examination of stranded animals has revealed that some are dying of digestive blockage from ingestion of plastic wrap or plastic bags.

DIET

These small whales, like the closely related Pygmy Sperm Whales, consume a variety of prey species, most of which are various kinds of squid, with some fishes and crustaceans added. Generally they capture whatever squids are locally abundant at the time. They are capable of relatively deep dives (more than 500 m), as their stomachs

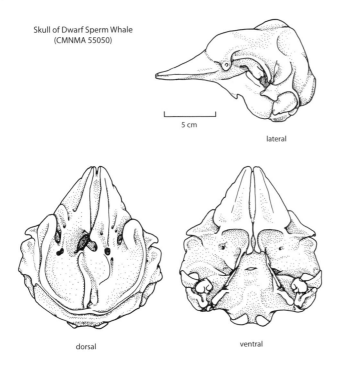

Skull of Dwarf Sperm Whale
(CMNMA 55050)

5 cm

lateral

dorsal

ventral

have been found to contain animals only found on the deep ocean floor. Suction may be used to capture prey.

REPRODUCTION

Both males and females appear to reach sexual maturity between the lengths of 210 and 220 cm, but their age at that time is unknown. A gestation period of 9.5 months has been estimated, followed by the birth of a single calf. A reproductive peak occurs in summer, December–January in the Southern Hemisphere and June–July in the Northern Hemisphere, with births occurring over a five-month period around those peaks. Females appear to be able to become pregnant while still lactating, which suggests that the birth interval is every one to two years. Females that breed soon after giving birth must wean their calf before the next is born. In that case, the calves are not likely nursed more than about eight to nine months, and probably less.

BEHAVIOUR

These small whales have been seen alone or in loosely formed groups of 2–12 animals. A long-term study from the Caribbean has found that group size is smaller in summer (1–8 animals) and larger in winter (1–12). The only exception to the loosely formed group is a cow-calf pair. The groups tend to be segregated by age or reproductive status (groups of females with calves, groups of immatures, and groups of adult males and non-reproductive adult females). Dwarf Sperm Whales, like their sister species, Pygmy Sperm Whales, have a habit of "logging" on the surface, often in the afternoon. The animals float with their blowhole and dorsal fin exposed and their flukes drooping down. As they can be very approachable while logging and appear startled if disturbed, it is presumed that they are sleeping. If surprised, attacked or stressed, a Dwarf Sperm Whale

may release a cloud of reddish-brown fluid to camouflage its escape, a behaviour called "inking." The fluid is stored in the lower intestine and may leak out of the anus of stranded animals, making it appear that they are bleeding. The fluid is very sticky and smelly and is virtually impossible to wash out of stained clothing. "Pugnacious" behaviour has been reported by a female Dwarf Sperm Whale who was accidentally captured in a purse seine along with her calf and a small herd of dolphins. As the fishermen were herding the whales out of the net, she rammed their boat several times, fortunately causing no harm to herself, the boat, or its occupants. Unless the animals are logging and hence unaware, Dwarf Sperm Whales are timid and difficult to approach and tend to avoid boats. A logging whale could be accidentally struck by a powered boat and some animals display wounds that could have been caused by such a collision.

VOCALIZATIONS

No recordings have been made of sounds produced by Dwarf Sperm Whales and they are generally considered to be not very vocal. Their vocalizations likely resemble those produced by their sister species, the Pygmy Sperm Whale.

SIGNS

Dwarf Sperm Whales are usually quiet and inconspicuous at the surface. They rise slowly and calmly, have a faint, usually unnoticed blow, and do not lift their flukes above the surface before diving. They are known to breach occasionally. If found logging on the surface, they can sometimes be approached quite closely. Unfortunately, all these characteristics are shared with the closely related, and very similar, Pygmy Sperm Whale.

REFERENCES

Birchler and Potter 1999; Caldwell and Caldwell 1989; Chivers et al. 2005; Dunphy-Daly et al. 2008; Lucas and Hooker 2000; McAlpine 2002; Nagorsen and Stewart 1983; Plön et al. 1999; Scott, M.D., and Cordaro 1987; Willis, P.M., and Baird 1998b.

Sperm Whale

FRENCH NAME: **Grand cachalot**

SCIENTIFIC NAME: *Physeter catodon*, formerly and sometimes still called *Physeter macrocephalus*

Sperm Whales are mammals best described in superlatives. They are the most massive of the toothed whales and, hence, the largest macropredators on the planet. They have the biggest brain of any animal, modern or extinct (averaging around 8 kg) and they display more

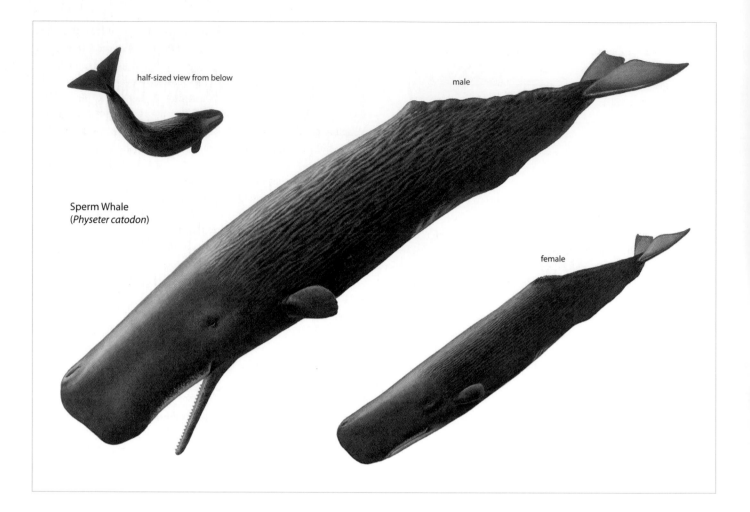

half-sized view from below

male

Sperm Whale
(*Physeter catodon*)

female

sexual dimorphism than any other whale. They produce the loudest sound of any biological organism that has ever been recorded. They also have the dubious distinction of being one of the cetaceans most hunted by humans. Over one million, and probably closer to two million, were killed worldwide in the years between the 1700s and the present.

DESCRIPTION

Their head shape and massive size make Sperm Whales distinct. The spermaceti organ takes up most of the bulging nose region, giving a Sperm Whale the largest nose on the planet. Head length typically makes up about one-quarter to one-third of the total length. This whale exhibits substantial sexual dimorphism, with males exceeding females in both length and weight. Adult males also have a more pronounced and bulbous head. Sperm Whales typically have visible teeth only in their lower jaws, although some have vestigial teeth in the upper jaws that rarely erupt through the gums. The usual dental formula is 0/20–26, for a total of 40–52 teeth. Teeth erupt around puberty. Their single nostril is at the tip of the snout on the left side, a highly unusual placement among cetaceans, most of whom have their blowhole towards the back of their heads. The offset nostril position directs the blow to the left and forward at a noticeable angle. Skin over the head is smooth and taut, as is usual for cetaceans, but behind the head it forms

into distinctive ripples. Most of the body is a dark grey colour, the lips are mottled white, and there are often irregularly shaped white blotches on the belly and flanks. Very rarely an animal will be unusually pale or even all white. The lower jaw is narrow, underslung, and when the mouth is shut, barely visible in profile. There are two to ten short, but deep, throat grooves. The dorsal fin is low, thick, and usually rounded, and the ridge behind it trailing down to the flukes is bumpy, especially on mature males. Around 75% of females and 30% of immature males have a light-coloured callous on their dorsal fin that does not occur on mature males. Flukes are triangular, with a more or less straight trailing edge and a medial notch. Flippers are short and paddle-shaped, and can be squeezed to fit snugly against the body.

SIMILAR SPECIES

Any of the large whales might be confused with a Sperm Whale at a distance. A closer view, especially of the head as the whale surfaces from a deep dive, or of the log-like body with its lumpy dorsal fin, as the animal rests on the surface, will distinguish this species from the other great whales. Northern Bottlenose Whales and Baird's Beaked Whales (North Pacific Ocean) are found in northern parts of the Sperm Whale's range and could be mistaken especially for a female or juvenile Sperm Whale. Both these whales have prominent beaks and dorsal fins that distinguish them from Sperm Whales.

SIZE

Adult males are about 50% longer, and typically around three times the weight of adult females.

Total length: males, 13–18 m, with the longest known at 18.3 m; females, 9.5–11.0 m, with the longest known at 12.5 m; newborns, about 4 m.

Weight: males, up to 57,000 kg; females, up to 24,000 kg, but rarely more than 15,000 kg; newborns, around 1000 kg.

RANGE

Sperm Whales inhabit deep marine waters and are most abundant near the edges of continental shelves and offshore banks and over submarine canyons and trenches, where currents and mixing of deeper waters with more surface waters create highly productive feeding areas. Males forage farther towards the poles than females, who tend to remain in warmer waters of roughly 15°C or warmer, year-round. No other mammal has such a large geographical separation between the ranges of males and females. Most Sperm Whales move towards the equator into warmer waters during the cold seasons. Males enter the Gulf of St Lawrence during the summer months and have been sighted in the mouth of the Bay of Fundy. They are reportedly common in deeper, offshore waters around the Maritimes.

ABUNDANCE

Sperm Whales have been hunted to some degree for centuries. A harvesting peak occurred in the mid-1800s, when their oil essentially lubricated the Industrial Revolution. Another major harvesting peak took place in the 1960s, when over 20,000 were taken each year. We again wanted them for oil (for soap and cosmetics as well as fine machine lubricants) and also for meat. Hundreds of thousands died from 1948 to 1978. All hunting ended in 1985, except by Japan, which continues to kill Sperm Whales in the western North Pacific, justified as research. Whale meat, and especially Sperm Whale meat, is considered a great delicacy by the Japanese

Distribution of the Sperm Whale (*Physeter catodon*)

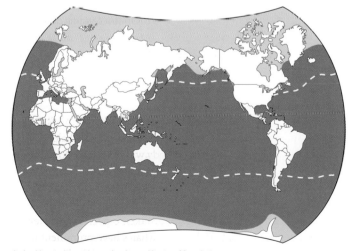

dashed line indicates normal poleward limits of female/immature range

and Koreans. Crude estimates set pre-harvested Sperm Whale populations at around 1,100,000 animals worldwide. This number is thought to have dropped about 30% by 1880 following the open-boat hunts. By 1985, when most of the modern explosive-harpoon whaling ceased, numbers had dropped by about 70% to less than 330,000 animals. Being slow reproducers, Sperm Whale numbers are slow to rebound and the estimated world population is not yet back up to 400,000.

ECOLOGY

Sperm Whales are the only deep-diving animals that prey upon large squid. Potential competitors are beaked whales and elephant seals, which can dive almost as deeply, but which typically capture smaller squid. Generally, the size of the Sperm Whale determines the size of its prey, with larger males consistently taking the largest squid. Dives vary considerably in both depth and duration, likely depending on prey availability. Young calves are not capable of deep or lengthy dives. Adult Sperm Whales can dive to depths in excess of 1000 m (the possible record is 3193 m) and stay submerged for up to 90 minutes and perhaps longer. These whales normally spend seven to ten minutes at the surface between dives, replenishing their oxygen reserves. They appear to forage throughout the day or night, taking rest breaks as necessary. Most prey is likely detected using echolocation, but since many of their preferred prey species are bioluminescent, there may be a visual component as well. Prey is likely captured by suction. Average travelling speed is around 4 km/h and the whales tend to stick to this speed during most activities. They have been clocked at speeds as high as 14 km/h while hunting. Female home ranges are about 1000–1500 km across. Males roam even more widely, but some males have appeared consistently in the same feeding area year after year. Killer Whales will attack pods of Sperm Whales, probably in an attempt to take a calf, although adults have been killed as well. Pilot whales and False Killer Whales have also been observed harassing and attacking Sperm Whale pods, although no fatalities are known in those cases. Large sharks may take the occasional calf. Maximum longevity could be over 100 years, but is at least 60–70 years.

DIET

Sperm Whales mostly eat medium to large squid. Medium to large deep-water fishes are a small, but regular part of their diet, except in high northern latitudes, where fish compose up to 98% of the diet of large males. Bottom organisms and debris are also commonly found in their stomachs, suggesting that prey capture may be accomplished using suction. Existence of throat grooves further supports the suction theory, as they would allow for rapid expansion to create the necessary vacuum. Sperm Whales have a large gullet for a whale and are capable of swallowing a medium-sized shark. Ambergris is a wax-like substance produced in the large intestine of 1% to 5% of Sperm Whales. It forms as roughly round, dark brown lumps that sometimes have concentric stratifications. Lumps usually weigh up to 10 kg, but masses of several hundred kilograms are known. Sometimes it is found floating or washed

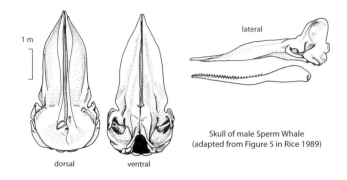

1 m

dorsal   ventral

lateral

Skull of male Sperm Whale
(adapted from Figure 5 in Rice 1989)

up on beaches, but most of the commercial supply was derived directly from the gut of animals killed by whalers. Ambergris was highly sought after as a binder by the perfume industry and has a pleasant musky odour.

REPRODUCTION

Mating activity in the Northern Hemisphere takes place from January through August, with most conceptions occurring in late spring. A single calf is born after a gestation of about 15 months. Typically a calf suckles for around two years, although this is highly variable. A 13-year-old juvenile male and a 7½-year-old juvenile female were found to have milk in their stomachs. Most calves start to consume solid food by, or even before, their second year, and weaning takes place gradually over several years. Juvenile females first conceive at around the age of nine and reach physical maturity around 30 years old. Younger adult females produce a calf about every four to six years. As they age, this interval increases. By the time a female is in her forties, the interval between births may be as long as 15 years. Females likely enter a phase of reproductive senescence as they reach their later years. Males gradually become sexually mature in their teens, but do not actively breed until their late twenties, when they begin to travel to tropical waters to search for receptive females. They achieve their full size (physical maturity) at roughly 40–50 years old.

BEHAVIOUR

Female Sperm Whales are highly social and form long-lasting stable associations, called units. They vary in composition from three to thirty whales and comprise individual mature females (usually genetically related) along with their offspring. Occasionally, an unrelated female will join a unit, and loose associations of several units do sometimes occur for short periods. Care of calves tends to be communal and there is some evidence that a calf might suckle from more than one female. Young calves cannot dive deeply and spend most of their time at the surface. Rather than leave them defenceless, the females will alternate dives, so that some remain at the surface with the youngsters. If a unit containing vulnerable individuals is attacked, the whales usually resort to one of two defensive strategies rather than simply diving to escape. In the first, the "marguerite" defence, the animals align their noses together with flukes outwards in a rosette pattern. In the second defence, called a "head-out" formation, unit members touch their tails together and

use their jaws to repel the attackers. In both cases, the juveniles and calves, being shorter, are tucked into the rosette, while the flukes or heads of the largest females stick out the farthest. Members of a unit are known to leave their companions to assist another who has been separated by an attack. Whalers used this strong social cohesion of group members and of females to their calves to "harvest" whole groups by carefully killing the animals in order so that uninjured adults remain in defence of the wounded juveniles and then of each other. Killer Whales adopt a similar strategy and may be able to mortally wound several members of the unit before they kill one and stop harrying the survivors as they settle down to feed.

Females typically remain in their maternal unit when they mature. Young males leave their mother's unit between the ages of 4 and 15 and often join a loosely knit, frequently changing bachelor group of similarly aged animals, which gradually moves to higher latitudes. As the males age, their group size shrinks and they are found at ever higher latitudes. The largest, fully mature males in their forties and older are usually solitary and live near the ice edges. The extent of roaming of adult males is uncertain and probably depends on the individual. Some tagged males have turned up thousands of kilometres away over the course of several years. Nevertheless, every mature male must migrate to the tropical haunts of the females in order to breed. While on the breeding grounds, the males roam between the units of females, presumably searching for those females that are receptive. He may spend only a few minutes or several hours or more with each group. These large breeding males tend to avoid each other, but fights do happen and most males carry scars from such encounters. Rare injuries, like broken jaws and teeth, are presumed to result from these battles. Mass stranding of either male or female groups are rare, but do occur. The largest involved 72 animals and must have been a staggering sight. Single strandings of dead, newborn, or moribund animals are more common, the waves and currents washing them ashore as they become increasingly feeble.

VOCALIZATIONS

The shape and structure of a Sperm Whale's snout is designed to focus and transmit sounds, and the larger the nose, the louder and deeper the sound. Most vocalizations fall within 5–25 kHz and do not include any whistles, as are common in other toothed whales. Clicks in the range of 223 dB have been recorded from large males, the loudest, naturally produced recorded sound. Clicks are arranged in different patterns for different purposes. "Usual" clicks are used to locate prey, "creaks" are rapid series of usual clicks used to track and capture prey, "trumpets" appear to signal a deep dive, but not all whales about to deep-dive will emit a trumpet. These click types are produced by both sexes and all age classes. "Clangs" are loud, lower-frequency ringing clicks produced primarily by adult males. They are transmitted without the usual directionality and so are widely heard and clearly meant for communication. Current theory supposes that the clangs are used to advertise presence to other Sperm Whales. The frequency and duration of the clang pulse provides information that would inform the receiver of the size of the sender. Another (perhaps smaller) male might be repelled by the sound, while females may be attracted to the clang of a large male. "Codas" are series of 3–20 clicks lasting from 0.2–5.0 seconds that appear to be used under social circumstances and can be unique to particular units or regions. "Chirrups" are uncommon sounds emitted, also in a social context, and are described as short bursts of resonant clicks. Their purpose is unknown. Sperm Whales can likely hear clangs over 60 km through the water, while usual clicks travel 16 km and creaks less than 6 km.

SIGNS

Compared to other large whales, a Sperm Whale blow is small and not particularly dense. It is directed towards the front and to the left. At the surface, Sperm Whales resemble large logs with a bulge on one end (the blowhole) and a bump on the other (the dorsal fin). The tail stock and flukes droop below the surface. Fluke-ups occur before a deep dive, and steep resurfacing at the end of a long dive often exposes part of the snout above the water. Breaches, lobtailing, and above-the-surface jaw clapping have been recorded. A pod of females and juveniles could be accompanied by a large breeding male during the late winter–early summer.

REFERENCES

Barlow and Taylor 2005; Cranford, T.W., 1999; Fertl et al. 1999; Fristrup and Harbison 2002; Goold et al. 2002; Madsen et al. 2002; McAlpine 1985; Møhl et al. 2003; Pitman et al. 2001; Reeves and Whitehead 1997; Rice 1989; Teloni 2005; Watkins, W.A., et al. 1999; Whitehead 2002, 2003; Whitehead and Weilgart 2000.

FAMILY ZIPHIIDAE
beaked whales

These deep-water inhabitants are poorly known and difficult to distinguish at sea. Some species, described from washed-up, dead animals, have never been seen alive. They are all primarily deep-diving squid-eaters that spend little time at the surface. Their skulls have a sturdy, often bulging, rear portion and a long, slender rostrum. Each beaked whale has a pair of throat grooves that almost converge under the jaw into a V-shaped groove, and all have "flipper pockets," depressions on the body into which the flippers can be tucked to present a very hydrodynamic contour, probably to reduce drag. The flippers are located closer to the head than is usual for most cetaceans. Dorsal fins are small and located at least two-thirds of the way along the back. All have well-developed melons, and most have a single pair of teeth (in the lower jaw). These teeth are larger in males and rarely erupt from the gums in females. Individuals in this family are reported to strand and mass strand in the same area and around the same time as naval sonar exercises are conducted. Such stranded beaked whales commonly display hemorrhaging around the ears or gaseous or fatty embolisms similar to those caused by decompression sickness.

Baird's Beaked Whale

also called North Pacific Bottlenose Whale, Giant Beaked Whale, Northern Giant Bottlenose Whale

FRENCH NAME: **Grande baleine-à-bec,** *also* **Baleine-à-bec de Baird**
SCIENTIFIC NAME: *Berardius bairdii*

The beaked whales are among the least known of all the whales, partly because of their deep-water habitat, but also due to their secretive habits. Baird's Beaked Whales are the largest of the beaked whales.

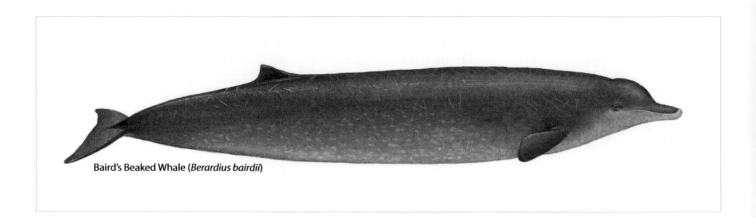
Baird's Beaked Whale (*Berardius bairdii*)

DESCRIPTION

They have a long streamlined body with a small dorsal fin set more than two-thirds of the way down the body. This swept-back dorsal fin has a straight or slightly falcate rear margin and typically a blunt tip. The flippers are small, blunt, and located forward on the body. The flukes have a nearly straight trailing edge with no notch, or at most a slight notch in the middle. Long throat creases (up to 70 cm in length) run from the chest to converge under the chin, but do not connect at their apex. Often there are two or more shorter creases between the longer outside pair. The principal body colour of adults is a mottled grey-black, often with a brownish tone. The belly is somewhat lighter, especially around the umbilicus and genital slits, and the throat and chin are often almost white. Frequently there are numerous whitish scars on the back and sides, especially on older males, that lighten the overall appearance. Juveniles are slate grey on both back and belly. The bulging melon on the forehead is somewhat larger on males. It slopes steeply, but smoothly, down to a long beak. There is often a visible crease where the melon slopes down at the back of the head just before the blowhole. The whitish lower jaws protrude beyond the upper jaws by about 10 cm, exposing a single tooth on each tip. Another smaller pair of teeth is behind, but they remain hidden in the mouth. Both males and females have the two pairs of erupted teeth in the lower jaw. The dental formula is 0/2, for a total of four teeth.

SIMILAR SPECIES

Minke Whales are similar in size, but have a more upright and falcate dorsal fin that is farther forward. The dorsal fin is visible at about the same time as the Minke Whale blows, unlike with a Baird's Beaked Whale, whose head is already submerged by the time the dorsal fin breaks the surface. Minke Whales have a white band on their flippers, tend to be found closer to shore, and are usually solitary. All the beaked whales can be difficult to identify at sea, and may be confused with each other. In the North Pacific, Stejneger's and Hubbs' Beaked Whales share similar distributions with Baird's Beaked Whales. Both these whales are about half the size of a Baird's Beaked Whale, with much less bulbous melons and shorter beaks.

SIZE

Females are somewhat larger than males.

Total length: males, up to 12 m, average around 10.5 m; females, up to 12.8 m, average around 11 m; newborns, around 450 cm.
Weight: males, more than 10,000 kg; females, more than 14,000 kg; newborns, unknown.

RANGE

This whale is endemic to the cool temperate North Pacific Ocean. It is a deep-water species rarely found in the shallow waters of the continental shelf. Most occur in areas where the water depth is between 1000 and 3000 m, which is most often found along the continental slopes and over the mid-ocean ridges. The populations in the western Pacific off Japan are migratory, spending their winters offshore, and move inshore and north each summer. The population off the coast of California moves offshore in winter and closer to shore in summer. These movements likely reflect prey movements. All the Canadian records are from spring and summer, May–September.

ABUNDANCE

Between 1950 and 1966, 25 Baird's Beaked Whales were taken in Canadian waters; after 1966 the fishery was closed. Japan is the only

Distribution of the Baird's Beaked Whale (*Berardius bairdii*)

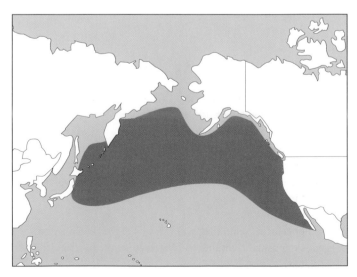

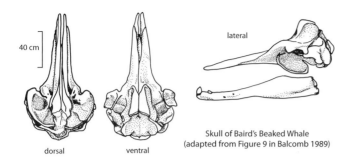

40 cm

dorsal   ventral   lateral

Skull of Baird's Beaked Whale
(adapted from Figure 9 in Balcomb 1989)

country that continues to hunt these whales. Between 50 and 60 are harpooned each year as they migrate past on their way to the summer feeding grounds in the Sea of Japan and the Sea of Okhotsk. There are no worldwide estimates of numbers. About 400 are thought to live off the west coast of the United States (not counting Alaska) and about 6000 in the waters off Japan. This species is uncommonly seen in Canadian waters, not because it is rare but because of its deep-water, open-ocean habitat.

ECOLOGY

Baird's Beaked Whales are capable of tolerating cold northern waters with floating ice, as well as warmer southerly waters, such as those around California. Prolonged dives of up to 67 minutes long have been recorded for this species. Most dives are about 20 minutes on average and usually less than five minutes are spent at the surface between deep dives. Dives of 1000 m are probably common, and this species may dive considerably deeper than that, perhaps to 2000 m or more. Although they can normally escape Killer Whales by diving, females with young that are unable to dive are likely vulnerable. White scars in parallel rows that match the spacing of Killer Whale teeth are not uncommon. This species tends to suffer from a heavy parasite load that affects especially the stomach, liver, kidneys, and blubber. The oldest known individuals were 84 years old for a male and 54 years old for a female. This is unusual among cetaceans, as normally the females live longer. Their longevity results in substantially more adult males in the populations than females, despite an initially equal birth ratio.

DIET

Deep-water fishes and squid are preferred prey. These species live between 600 and 3000 m down and do not rise in the water column during darkness, so the whales must swim to those depths to secure them. Average prey length is estimated at 36–52 cm. The proportions of fish to squid captured vary by locality and presumably by prey abundance. The stomachs of these whales commonly contain stones and gravel, most likely ingested incidentally during foraging. This suggests that they must, at least sometimes, dive all the way down to the ocean floor. Suction is suspected as a prey capture technique. Digestion in the beaked whales occurs at a faster rate than in most other cetaceans and they lack the more common forestomach, so food is deposited right into the stomach, where digestion begins immediately.

REPRODUCTION

The reproductive biology of this species is poorly understood. Sexual maturity is reached sometime between 10 and 15 years old, supposedly sooner for males and later for females. Males are at least 9.5 m long at sexual maturity, while females are at least 10 m long. The gestation period is thought to be around 17 months. In the Japanese populations, mating peaks in autumn and births peak in March and April. Females are thought to produce a single calf about every three years.

BEHAVIOUR

Pod size is usually 2–20, but can be up to 30–35 animals. They usually surface and breathe in synchrony. There appears to be some degree of segregation of the sexes, as many pods are made up exclusively of one gender or the other. Mass strandings are rare. There are only two records, one involving a pod of four and the other a pod of seven individuals. Although it is assumed that many of the scars on their bodies are caused by the teeth of other Baird's Beaked Whales, there are no reports of social interactions at the water surface where the whales are visible from boats. These whales sometimes spyhop, breach, and lobtail, and they tend to avoid powered boats.

VOCALIZATIONS

There are few sound recordings of this species, partly because there have been no captive-kept animals and partly due to the open-ocean, deep-water habitat of the species and the difficulties that presents to their study. What recordings we have indicate that the species echolocates, as we would expect, but not in the usual pattern. Their echolocation clicks are short and irregularly spaced, rather than regular and almost constant, as is more common. Most of the sounds fall within a 4–130 kHz range and have been described as pulses, clicks, and whistles.

SIGNS

The low, compact, round blow of this whale is difficult to detect at sea. The small dorsal fin is placed so far back on the body that on a normal rolling breath, the blowhole and head are already submerged before the dorsal fin appears. When surfacing after a deep dive, their re-entrant angle is so steep that the forehead and chin usually protrude out of the water. The white overshot lower jaw is a giveaway when visible. The flukes occasionally rise out of the water before a deep dive. This species is always seen in a group. Although the external teeth are small, they can be quite visible, as the light will often glint off their surfaces.

REFERENCES

Aurioles-Gamboa 1992; Balcomb 1989; Dawson et al. 1998; Kasuya et al. 1997; Mead 1984; Ohizumi et al. 2003; Reeves and Mitchell 1993; Reeves et al. 2002; Walker, W.A., et al. 2002.

Northern Bottlenose Whale

also called Bottle-nosed Whale, North Atlantic Bottlenose Whale

FRENCH NAME: **Baleine-à-bec commune,** *also*
Hyperoodon boréal

SCIENTIFIC NAME: *Hyperoodon ampullatus*

Northern Bottlenose Whales may undertake more deep dives than any other marine mammal yet known. They routinely dive 800–1000 m deep and spend a surprisingly short time on the surface reoxygenating.

DESCRIPTION

Young whales are black or dark brown on their backs and greyish-white on their bellies. Adults maintain this counter-shaded colouring pattern, but become somewhat lighter on their backs as they mature. White or yellowish-white diffuse spots are often present on the sides and belly. These increase in number and size with age and, especially on older females, can amalgamate on the belly. Mature males develop a white forehead and chin and sometimes the whole head becomes white. Rarely, on very old males, the dark areas can fade to a yellowish-white. Mature females sometimes develop a white band around their necks, but their head remains dark. The most distinctive feature of this species is the size of the melon. This is a sexually dimorphic trait. It is largest on old males and smallest on juveniles and females. But even the smaller melons are still large in comparison to those of other beaked whales. The animal pictured here is a mature male. Like all the beaked whales, the falcate to sub-falcate dorsal fin of the Northern Bottlenose Whale is far down on the back, about two-thirds of the way to the flukes. The flippers are small. Flukes do not display a medial notch on the trailing edge and sometimes the middle of the flukes, where the notch is normally found, protrudes beyond the trailing edge. There are two short throat grooves that almost converge under the chin. The beak on this whale is pronounced. The blowhole is wide and somewhat U-shaped, with the points facing forward. The dental formula is 0/1, for a total of two teeth. Usually only one tooth (about 5 cm long) erupts at the tip of each adult male's lower jaw (they remain embedded in the gums in females). In rare cases a second pair will also erupt behind the first pair, but these are always much smaller. Sometimes 10–20 vestigial teeth may be present in either upper or lower jaws, but these additional teeth remain in the gums.

SIMILAR SPECIES

If the bulbous head is not seen, the large, somewhat falcate, dorsal fin is reminiscent of a Minke Whale's. The blow and dorsal fin are visible at the same time on a Minke Whale, but because the dorsal fin is so far down the back on a Northern Bottlenose Whale, it will only just be breaking the surface as the head is submerging. A reasonably close view of the head region will eliminate all the other beaked whales found in the North Atlantic. Male Cuvier's Beaked Whales also have a whitish head, but they have a shorter beak, and a smaller melon. Sowerby's and Blainville's Beaked Whales are much smaller, have a very much flatter melon, and the adult male's teeth occur in the middle of the lower jaw rather than at the tip.

SIZE

Males are considerably larger than females.
Total length: males, average 8.5–9.0 m, maximum 9.8 m; females, average 7.5 m, maximum 8.7 m; newborns, about 3.6 m.
Weight: males, up to 7500 kg; females, around 5000–6000 kg; newborns, unknown.

RANGE

Northern Bottlenose Whales are an endemic North Atlantic species. They are found from Davis Strait between Baffin Island and Greenland down to Nova Scotia and across to Iceland, Spitsbergen, Norway, and the Bay of Biscay. Past records suggest that they sporadically found their way along the southern coast of Spain and into the Mediterranean Sea, but there are no recent records from these areas. Two regions of relative abundance occur in Canadian waters. The Davis Strait population is migratory and is found in the strait from early summer to late autumn. At least some of this population spend the winter in the Labrador Sea. A small population in the "Gully," a submarine canyon along the edge of the continental shelf, about 165 km northeast of Sable Island, is resident and found there winter and summer. Two other submarine canyons, the Shortland Canyon (50 km northeast of the Gully) and Haldimand Canyon (100 km

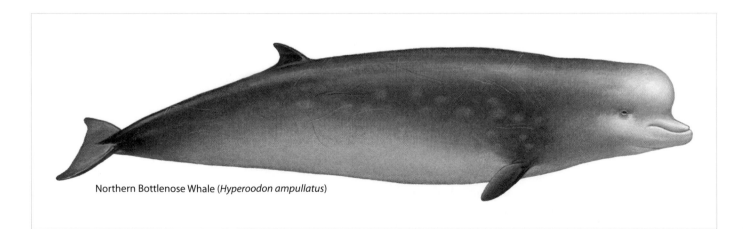

Northern Bottlenose Whale (*Hyperoodon ampullatus*)

Distribution of the Northern Bottlenose Whale (*Hyperoodon ampullatus*)

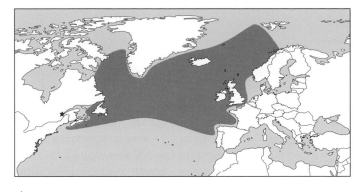

★ extralimital stranding

northeast of the Gully), are within the range of the Gully population. There have been four records of strandings in the Gulf of St Lawrence (two of these from the estuary area of the St Lawrence River), one from Trinity Bay, Newfoundland, and one from the Bay of Fundy.

ABUNDANCE

These whales have been harpooned for profit since the British discovered them in Frobisher Bay in 1852, when 28 were taken. The whalers hunted them for the spermaceti oil in their melon as well as for their blubber and sometimes meat. Since that time hunters, mainly from Norway, have killed thousands and seriously depleted the population. Canadians took 87 animals in a six-year period between 1962 and 1967 from the Gully off Sable Island, an estimated 30%–40% of that population. The whales in the Gully currently number about 130–230 animals. This population has been studied intensively for several years and much of the behavioural and ecological information presented in this account is derived from these studies. The Gully is the southernmost point where this species can be seen with any degree of reliability, and the area has been declared a "Whale Sanctuary" by the Department of Fisheries and Oceans. Shipping is asked to detour around it. The establishment of a more official "Marine Protected Area" under Canada's Oceans Act is presently being considered. Northern Bottlenose Whales were classified as "Provisional Protected Stock" in 1977 by the International Whaling Commission because of the depleted condition of the populations, and all harvesting was curtailed. There are no modern fisheries for Northern Bottlenose Whale. The intensive exploration and development of oil and gas fields in the region of the Gully could seriously threaten that population in future.

ECOLOGY

Northern Bottlenose Whales are largely deep-water animals seldom found in waters less than 1000 m deep. They are capable of dives lasting more than an hour, and based on fragments of starfish found in their stomachs, clearly reach the sea floor. The deepest recorded dive is 1453 m and the longest dive 70 minutes long. Whalers have reported dives of harpooned animals lasting two hours. Dives to depths of 800–1000 m are routine, and usually five minutes spent at the surface, blowing every 30–40 seconds, is sufficient to replenish the oxygen

reserves for another deep dive. Killer Whale tooth marks have been seen on Northern Bottlenose Whales, but most attacks are probably avoided by diving. Water temperature does not appear to be a concern for this species. They are found in waters varying in temperature from –2°C to 15°C. Each spring, as the whales move into the Davis Strait, they will wait at the edge of the sea ice for the pack to melt or shift. Some animals will even venture far into the ice. The population that inhabits The Gully has been verified as present from February through November and may prove (with additional surveying) to be year-round residents. The lifespan of this species is at least 37 years.

DIET

Not only is this species principally a squid eater, but it prefers squid of the genus *Gonatus*, and not just any *Gonatus*, but mainly females carrying egg masses, as they are less mobile than the males and juveniles, and easier to catch. These females lose their tentacles and drift in shoals at the 1000–2000 m depth, while their eggs develop, making them convenient prey. Other squid species, prawns, and bottom fishes such as redfish, Spiny Dogfish, Atlantic Herring, Greenland Halibut, and Ling are eaten in some locations during some seasons. Northern Bottlenose Whales have no functional teeth, so likely use suction to catch their prey.

REPRODUCTION

In the Gully population, mating probably occurs during the summer. The population off the coast of Labrador differs. Their mating season appears to be in early spring. Over the entire distribution of this whale, calves have been seen in all seasons. Pregnancy lasts about 12 months, followed by the birth of a single calf, so most calves seen in Canadian waters are born in spring or early summer. Calves are nursed for at least a year and probably longer. The shortest calving interval is two years, and probably most often longer. On average, females become sexually mature at around 11 years old and males between 7 and 11 years old.

BEHAVIOUR

Most groups of Northern Bottlenose Whales number one to four animals, but larger groups of up to 11 are seen from time to time. Pods are often segregated by sex, except during the mating season. Observations of male-male conflict in the Gully over the summer imply that the adult males either compete for dominance, or for territory, by head butting. There are also long-term associations of adult males, suggesting a coalition similar to that of Bottlenose Dolphins, where related males cooperate to drive away single males, ensuring their own shared access to females. It is possible to photo-identify these animals, as has been done by researchers studying the Gully population. They are curiously attracted to slow-moving or stationary boats, unlike all the other beaked whales, and have been known to circle around such a vessel for more than an hour. Whalers commonly took advantage of this curiosity to ensure a successful hunt.

VOCALIZATIONS

Northern Bottlenose Whales make a variety of clicks, chirps, squeals, and possibly whistles with frequencies between 3 and 16 kHz that

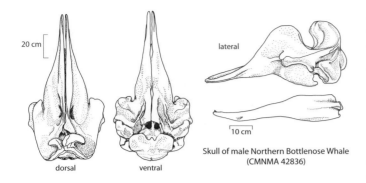

20 cm

lateral

10 cm

Skull of male Northern Bottlenose Whale
(CMNMA 42836)

dorsal

ventral

are within the range of human hearing. These may be produced on the surface as the whales socialize. They also produce unusually loud ultrasonic clicks in the 20–30 kHz range, which are presumably produced while foraging at depth and are used for echolocation.

SIGNS

The short, bushy blow of this species is inconspicuous at sea. Due to the placement of the dorsal fin far back on the animal, it is just breaking the surface as the head begins to submerge. Sometimes the flukes lift above the water when the animals dive, but usually they do not.

REFERENCES

Dawson et al. 1998; Gowans and Rendell 1999; Gowans and Whitehead 2001; Gowans et al. 2000; Hooker and Baird 1999b; Hooker and Whitehead 2002; Hooker et al. 2002a, 2002b; Mead 1989a; Reeves et al. 2002; Whitehead et al. 1997a, 1997b; Wimmer and Whitehead 2004.

Sowerby's Beaked Whale

also called North Atlantic Beaked Whale,
North Sea Beaked Whale

FRENCH NAME: **Baleine-à-bec de Sowerby**

SCIENTIFIC NAME: *Mesoplodon bidens*

Sowerby's Beaked Whale has recently been discovered to occur regularly in summer off the coast of Nova Scotia. Five *Mesoplodon* occur in Canadian waters. All are cryptic and very difficult to distinguish, even when lying on the beach, let alone when sighted at sea.

DESCRIPTION

Like all the mesoplodont beaked whales, Sowerby's Beaked Whales have a thick torso, a small head and tail stalk, and small flippers, flukes, and dorsal fin. They are dark grey on their backs. The colour gradually fades into lighter grey sides and a whitish belly region. The lower lip and chin are whitish and the beak, flippers, flukes,

and dorsal fin are dark grey. The lower lip protrudes slightly beyond the tip of the upper lip and there is a deep V-shaped groove on the throat. A dark grey-black, roughly oval patch surrounds each eye. Adults commonly have whitish oval scars on their belly and sides resulting from attacks by cookie-cutter sharks, and possibly other animals such as lamprey or even crustaceans. Males additionally sport numerous white linear scars on their backs, heads, and sides caused by tooth scrapes by other males. The melon is only slightly convex and definitely not bulbous. The long, slim, dark grey beak with a white chin and lower lip is a good identifying characteristic, but can be difficult to detect at sea. The flukes normally do not have a medial notch and the flippers are relatively long, about one-eighth of the body length. Both males and females typically have only two teeth, one in each lower jaw about one-third to one-half of the way back. These teeth only erupt from the gums and become visible in adult males. The dental formula is 0/1, for a total of two teeth.

SIMILAR SPECIES

In eastern Canadian waters, Sowerby's Beaked Whales share their range with True's, Blainville's, and Cuvier's beaked whales. Cuvier's Beaked Whale occurs throughout the eastern seaboard and is much larger, with a whitish head region. Of these three *Mesoplodon* beaked whales, Sowerby's Beaked Whale is the most cold-adapted and is found farthest north. True's Beaked Whale is the most warm-adapted and is found only occasionally as far north as Cape Breton, Nova Scotia, while Blainville's Beaked Whale falls somewhere between the two. Female and juvenile Sowerby's, Blainville's, and True's beaked whales are virtually indistinguishable. Teeth of adult male Sowerby's Beaked Whales erupt about in the middle of the lower jaw, but the jaw bone does not thicken or arch at the site, so the tooth sits about level with the eye. Adult male Blainville's Beaked Whales have teeth in about the same location on the lower jaw, but the site is strongly arched, so the tooth sits higher than the eye. Teeth of adult male True's Beaked Whales erupt at the tip of the lower jaw, as they do with male Cuvier's Beaked Whales. The habit of beaked whales of poking their heads out of the water when they surface for a breath may provide a clear view of the lower jaw under favourable viewing circumstances.

Distribution of the Sowerby's Beaked Whale (*Mesoplodon bidens*)

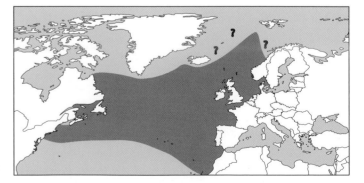

? possible range

Sowerby's Beaked Whale (*Mesoplodon bidens*)

SIZE

Adult males appear to be somewhat larger than adult females. Length at sexual maturity is thought to be approximately 4.7 m for both sexes.

Total length: males, up to 5.5 m; females, up to at least 5 m; newborns, around 2.4 m.

Adult weights for both males and females range between 1000 and 1300 kg; newborns, about 170 kg (based on a very limited sample size).

RANGE

This whale occurs in cool temperate waters of the North Atlantic from North America to Europe. The centre of highest density of the species is thought to be in the North Sea due to the higher number of strandings in Europe and Britain. Until 1999, there had been only 12 reports in North American waters (11 strandings and one reliable live sighting). Recent work in the Gully, a submarine canyon off Sable Island, Nova Scotia, has reported that a small population occurs there in summer. Animals from this population have been seen alive on at least four reported occasions to date. There have also been a few recent sightings in the waters off Newfoundland, so clearly there is a regular, albeit probably sparse, population in the western North Atlantic extending from around Cape Cod to Davis Strait. A single Canadian stranding has been reported from Sable Island, Nova Scotia.

ABUNDANCE

There are no abundance estimates for Sowerby's Beaked Whales. While there is no commercial hunting of this species, some die on occasion following entanglement in fishing gear. Like all the mesoplodont beaked whales, Sowerby's Beaked Whale has been considered for endangered status by the World Conservation Union (IUCN). Again, like the other mesoplodonts, the IUCN has been forced to declare its status "Data Deficient" due to a lack of information. Although the number of sightings and strandings of this normally shy whale likely do not reflect its true abundance, this species is still thought to be very rare in Canadian waters. It is considered to be of "Special Concern" by COSEWIC.

ECOLOGY

Most of the Canadian sightings have been made in cool seas that are 550–1500 m in depth. Across the whole range, these whales are mainly pelagic (found in deep waters), but sometimes venture into shallower seas such as the Baltic and the North Sea, and one uncertain record suggests the Mediterranean Sea. Sowerby's Beaked Whales have been recorded diving for 12 to at least 28 minutes, typically with a short, one-to-two minute period spent at the surface taking six to eight breaths before another deep dive. Like other deep-diving pelagic whales, Sowerby's Beaked Whales are likely vulnerable to loud, low- to mid-frequency underwater sounds such as are produced by the military for submarine detection or during seismic surveying for oil and gas deposits.

DIET

Examinations of the stomachs of stranded animals have found mainly squid remains and some unidentified fish. Other studies that detect nitrogen isotopes in the tissue and can determine if the food consumed originated in deep marine waters corroborate these results. Sowerby's Beaked Whales appear to consume primarily deep-water squid species, which they probably capture using suction.

REPRODUCTION

Nothing is known of the reproduction of this species apart from the size of newborns.

BEHAVIOUR

These whales apparently travel in small groups of two to ten animals that swim in close proximity to each other, surfacing and

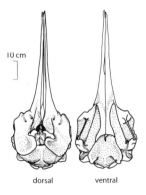

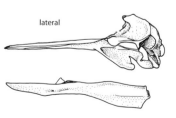

10 cm

dorsal   ventral

lateral

Skull of male Sowerby's Beaked Whale (adapted from Figure 15 in Mead 1989b and Figure 1 in True 1910)

breathing in synchrony. One North American sighting was of a small group of three adult males, but most sightings and mass strandings involve mixed herds of adults and calves. Mass strandings of up to six individuals are known from North America, but single strandings are more common. Only a few small mass strandings are known from the eastern side of the range, where by far most of the strandings are of single animals. The emerged pair of teeth of adult males is used during battles with other males. Because of their placement and height, only one tooth can be brought into play at a time, so the majority of the white scars on the bodies of older males are single, rather than pairs of parallel scars. Females and juveniles rarely have these long white scratches.

VOCALIZATIONS

Cow-like sounds have been reported from dying, stranded animals, but little else is known of the vocal communication of this species, although we assume they are able to echolocate.

SIGNS

Usually the first view of these whales is of the arched back and small rearward-placed dorsal fin. The blow is inconspicuous. When Sowerby's Beaked Whales surface for a breath, they poke their long beaks out of the water at a 30°–45° angle and then begin to roll. Typically the head is just disappearing as the dorsal fin emerges. Lobtailing has been observed, as has breaching. The flukes do not lift out of the water before a deep dive. Unfortunately, all these characteristics are also representative of other beaked whales in Canadian waters.

REFERENCES

Hooker and Baird 1999a; Ledwell et al. 2005; Lien and Barry 1990; Lien et al. 1990; Lucas and Hooker 2000; Mead 1989b; McAlpine and Rae 1999; Ostrom et al. 1993; True 1910.

whitish, the sides are greyish-white, the back is medium grey, and the "flipper pockets" are dark grey. Adult males are black, except on their melon and the front of their beak, which is white. The white colour reaches from the snout to the back of the tooth. The distinction of dark body and white markings on the head is retained after death. The same areas on females and juveniles are somewhat lighter than their base colour, but not as distinctly so as on the adult males. Most adults have oval whitish scars on their bodies. Males develop an extensive network of criss-crossed white scratches probably caused by the teeth of other males. There is no medial notch in the flukes. Teeth are only visible on adult males. One large tooth erupts slightly forward of the middle of each lower jaw and the jaw bone arches upwards right where the tooth erupts. These teeth angle forward and slightly inward towards each other, somewhat restricting the mouth opening. The same teeth in females and juveniles are much smaller and remain embedded in the gums. The dental formula is 0/1, for a total of two teeth.

SIMILAR SPECIES

All the mesoplodonts, especially the females and juveniles, look very similar, and even stranded animals require an expert to identify them. Chances of positively identifying them at sea are remote. Adult males are sometimes easier to distinguish. Only one other *Mesoplodon* species overlaps the Canadian distribution of Hubbs' Beaked Whale and that is Stejneger's Beaked Whale. The head of an adult male of this species is all black and lacks the white "cap" and partial white beak of Hubbs' Beaked Whale. The exposed teeth appear longer and are more noticeable, as they are white on a dark face. The melon of an adult male Hubbs' Beaked Whale is white, surrounded by a dark head, and resembles a white "beanie." The front portion of the beak is also white. The lower jaw of Stejneger's Beaked Whales is a medium grey and the top of the head and beak is dark. This subtle coloration difference is noticeable on a live animal,

Distribution of the Hubbs' Beaked Whale (*Mesoplodon carlhubbsi*)

Hubbs' Beaked Whale

also called Arch-beaked Whale

FRENCH NAME: **Baleine-à-bec de Moore**

SCIENTIFIC NAME: *Mesoplodon carlhubbsi*

This North Pacific endemic is one of the enigmatic mesoplodont (genus *Mesoplodon*) beaked whales. They are rarely seen alive and most of the information we have about them is derived from stranded animals. Hubbs' Beaked Whale is one of the better known of the beaked whales.

DESCRIPTION

Like the other *Mesoplodon*, Hubbs' Beaked Whales have a thick body, small appendages, a small head, forward-placed flippers, and a rearward-placed dorsal fin. In females and juveniles the belly is

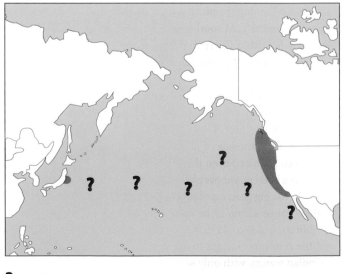

? possible range

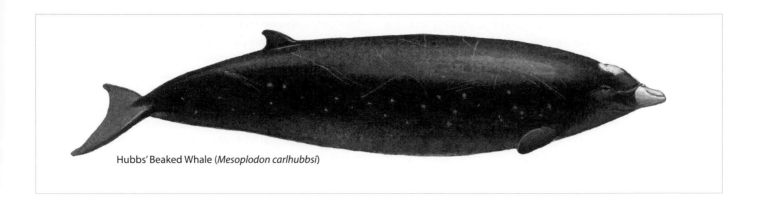

Hubbs' Beaked Whale (*Mesoplodon carlhubbsi*)

but the lighter colours rapidly darken after death, giving an overall blackish look. Larger beaked whales such as Cuvier's or even Baird's Beaked Whales may be mistaken for Hubbs' or Stejneger's Beaked Whales if seen from a distance. Cuvier's Beaked Whales have a light head area, a hardly noticeable beak, and a more prominent melon. Baird's Beaked Whales also have a more prominent melon, their beak is longer, and they are about twice the size of the Hubbs'. The dorsal fin of a Minke Whale is similar to that of Hubbs' or Stejneger's Beaked Whales. Minke Whales have a much larger head and a white band on their flippers, which is bright enough to be visible sometimes through the water.

SIZE
There is no known difference in size between males and females. Total length: maximum around 5.4 m; newborns, about 2.4 m. Weight: up to 1500 kg; newborns, unknown.

RANGE
Hubbs' Beaked Whales are known from only two disjunct parts of the North Pacific Ocean. The largest portion of the range is along the coast of North America from about the Queen Charlotte Islands to at least San Diego, California. This distribution more or less matches the confluence of the cold southerly Subarctic Current with the warm northerly California Current. The second, smaller area of occurrence is along the northeastern coast of the island of Honshū, Japan, where the cold southerly Oyashio Current meets the warm northerly Kuroshio Current. It is possible that the two regions of distribution are contiguous across the North Pacific, but there is no direct evidence of that yet. Seven strandings have been reported in Canada since 1963, all from Vancouver Island.

ABUNDANCE
There are no estimates of abundance for this species, as they are rarely seen alive and even then identification is chancy. Some concerns have been raised over how many die each year from entanglements in fishing gear, especially in gill nets off California.

Their range seems to be constrained to specific oceanographic conditions, which could make them vulnerable to changes in climate and the resulting changes to ocean currents. This species is rare in Canadian waters, with only seven known strandings from the British Columbia coast and no reliable sightings of living animals at sea.

ECOLOGY
Aside from the fact that these whales prefer deeper waters and likely can dive fairly deep for extended periods, little is known of their lifestyle.

DIET
Based on the stomach contents of stranded animals, Hubbs' Beaked Whales are primarily squid and fish eaters. Like the other beaked whales, they probably capture their food by suction.

REPRODUCTION
Calving is believed to take place in early to mid-summer. A single calf is born after a pregnancy lasting about 12 months. Length of the nursing period and the calving interval are unknown.

BEHAVIOUR
Hubbs' Beaked Whales are shy and avoid boats. The long scratches on the bodies of the males are attributed to other males. They aggressively use their pair of teeth during interactions to establish either territories or social dominance. It is not known, but is suspected, that this species undertakes seasonal migrations at least in some parts of the range.

VOCALIZATIONS
The only known recordings of sounds by this species were taken from two stranded small juveniles during rehabilitation. They produced ultrasonic clicks, centred near 1.77 kHz and bursts of three to eight pulses that ranged in frequency between 7 and 78 kHz. It is assumed that this species is capable of echolocation.

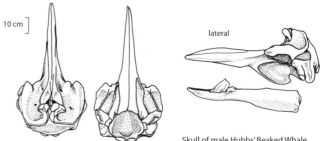

10 cm

dorsal   ventral

lateral

Skull of male Hubbs' Beaked Whale (adapted from Figure 15 in Mead 1989b)

This species is generally inconspicuous on the surface and has rarely been positively identified at sea. The white "beanie" and partial white beak of the male should be distinctive if seen at close range.

REFERENCES

Houston 1990b; Marten 2000; Mead 1989b; Reeves et al. 2002; Willis, P.M., and Baird 1998a.

Blainville's Beaked Whale

also called Dense-beaked Whale

FRENCH NAME: **Baleine-à-bec de Blainville**

SCIENTIFIC NAME: *Mesoplodon densirostris*

The densest bone currently known to science is found in the skull of this whale. It is more than 50% harder than elephant ivory. With the exception of one species (*Mesoplodon ginkgodens*, not found in Canadian waters), all the male mesoplodont beaked whales exhibit hardening of the bones of the beak and skull, but male Blainville's Beaked Whales win the prize for the world's hardest head.

DESCRIPTION

These whales have the typical beaked-whale body shape: thick torso and small head, tail stalk, and appendages. The flukes tend to be smooth along the trailing edge without a medial notch. A triangular or slightly falcate dorsal fin is placed about two-thirds of the way along the body. Blainville's Beaked Whales are countershaded with a dark grey back and a somewhat lighter grey belly, chin, and lower lip. The flippers, flukes, and dorsal fin are dark grey, as is a diffuse, relatively circular patch around each eye. Oval white marks are scattered over the body, which are scars resulting from attacks by cookie-cutter sharks or parasitic fishes or crustaceans. Adult males sport numerous additional, often parallel,

white scars that result from tooth scrapes inflicted by rival males. These scratches can accumulate on the older males to the extent that they appear to have a white head and cape. Male Blainville's Beaked Whales typically have a single pair of erupted teeth about halfway along the lower jaw. Both sexes display a thickening in the lower jaw bone at the site of the tooth, although it does not erupt through the gums in the females. This thickening is considerably more extreme in males. The dental formula is 0/1, for a total of two teeth. Occasionally, additional vestigial teeth may be present in the jaws of some individuals, which usually remain dormant and never emerge through the gums.

SIMILAR SPECIES

Similar in body shape and colouring to the other North Atlantic species of beaked whales, True's and Sowerby's Beaked Whales, the juveniles and adult females of all three species are easily confused. The thickening of the lower jaw in female Blainville's Beaked Whales can help distinguish this species from the others, but this feature is usually only visible in stranded animals. The development of the lower jaws and the placement of the pair of erupted teeth of adult males provide a potential means to identify them. See the "Similar Species" section of the Sowerby's Beaked Whale account for more information.

SIZE

Based on a limited sample, females appear to be larger than males.
Total length: males, at least 4.6 m; females, at least 4.6 m; newborns, around 2.0–2.5 m.
Weight: males, at least 800 kg; females, at least 1000 kg; newborns, around 60 kg.

RANGE

Found in tropical and warm temperate oceans around the world, this whale is found occasionally in Canadian waters of the North Atlantic Ocean and is probably near the northern limit of its distribution off the Maritimes. A single individual stranded at Peggy's Cove, Nova Scotia, in 1940 and another in Fourchu Bay, Nova Scotia, in 1968. Rare sightings are reported in deeper offshore waters. Populations do exist in the Pacific, but so far none is known to reach as far north as Canada.

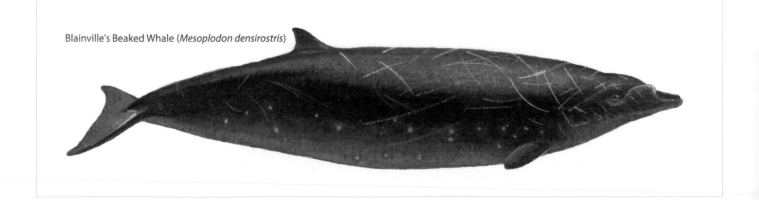

Blainville's Beaked Whale (*Mesoplodon densirostris*)

Distribution of the Blainville's Beaked Whale (*Mesoplodon densirostris*)

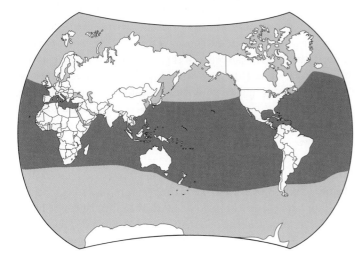

ABUNDANCE

Only three strandings are known from Canada, two from Nova Scotia and one from the New Brunswick side of the Bay of Fundy, and there have been no reliable live sightings in Canadian waters. Blainville's Beaked Whales are extremely rare off Canada's east coast. These whales are not common anywhere. Like all the mesoplodont beaked whales, Blainville's Beaked Whale has been considered for endangered status by the World Conservation Union (IUCN). Again, like the other mesoplodonts, the IUCN has been forced to declare its status as "Data Deficient" due to a lack of information.

ECOLOGY

Blainville's Beaked Whales appear to be the most pelagic (deep water) of an already pelagic group. More strandings are reported from islands than from the coasts of continents. Live sightings occur very rarely, although some have actually been photographed at sea near Hawaii and the Bahamas. These whales dive for extended periods, commonly 12–22 minutes, with a short period at the surface taking several breaths before diving again. The longest reported foraging dive is 83 minutes. The squid they eat do not rise in the water column at night, so the whales must dive down to them, likely to depths in excess of 1000 m. The maximum foraging depth reported for this species is 1599 m. Foraging takes place during both darkness and light and dive duration and depth is similar during either lighting condition. Mass stranding is not known by this species, although there is a record of several individuals stranding in the Bahamas in March of 2000 following an American naval exercise testing Low Frequency Active Sonar (LFAS), during which extremely loud, low-frequency sounds are produced (the ultimate purpose being to detect quiet diesel and nuclear-powered submarines at long distances). Reasons for the live strandings are poorly understood. The animals could have simply panicked at the unfamiliar noise and "stampeded" ashore, they could have been making every possible effort to get out of the water to avoid the sounds, or the sounds may have injured them and made them too feeble to avoid stranding. Some aspect of the hearing of beaked whales apparently makes them vulnerable to this kind of noise. Further evidence suggests that either the sound itself, or perhaps some actions by the whales when exposed to the sound, causes gaseous or fatty bubbles to form in the tissues (mainly the blood), which mirrors the symptoms of decompression sickness in humans and can prove fatal.

DIET

Like the other beaked whales, this species eats primarily deep-water squid and some deep-water fishes, which it likely captures using suction.

REPRODUCTION

Very little is known of the reproduction of this species. The minimum age of sexual maturity is thought to be nine years and the calving interval is suspected to be three to four years.

BEHAVIOUR

Herds of Blainville's Beaked Whales are generally small, usually three to seven individuals. Males of this species are thought to have such dense bone in their skulls because of male-male fighting that takes place, probably to gain access to females. Based on scarring patterns on adult males, scientists hypothesize that as two males approach each other, one turns upside down, so that each male's pair of teeth will contact the head and neck of its rival. They charge and rake each other with these tusks, leaving long linear slashes, which are deepest on the top of the head and trail out along the upper body. The scratches do not repigment as they heal, so whitish scars remain visible even years later. The whitish appearance of the head and top of the back of old males is the result of an accumulation of scars over years and may in itself become a visual signal of dominance. Females and juveniles are essentially scar free, apart from injuries caused by cookie-cutter sharks or parasitic fishes.

Long-term studies around Hawaii have shown that some Blainville's Beaked Whales (particularly females) display site fidelity to wintering and summering ranges that can extend over 15 years. Some of the females involved in these sighting also displayed long-term associations with other females that lasted at least nine years, although the individuals did not necessarily spend much of their time together. Males appear to exhibit female defence polygyny, whereby a mature male guards a female or small group of females and restricts the access of other males.

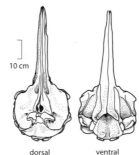

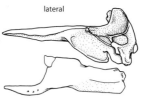

lateral

10 cm

dorsal   ventral

Skull of male Blainville's Beaked Whale (adapted from Figure 15 in Mead 1989b and Figure 2 in True 1910)

VOCALIZATIONS

The only sound-production information available relates to a stranded subadult male who emitted pulsed sounds in the 1–6 kHz frequency. These sounds, in the range audible to humans, were described as "whistles" and "chirps." The ability to echolocate is assumed in this species.

SIGNS

Like the other beaked whales, the blow of Blainville's Beaked Whale is inconspicuous. They usually poke their chin and head out of the water as they surface to take a breath. Flukes do not lift out of the water before a deep dive. Unfortunately, none of these signs are unique to Blainville's Beaked Whales.

REFERENCES

Baird et al. 2006; Baird et al. 2008b; Houston 1990a; Leatherwood et al. 1982; MacLeod 2002; McAlpine and Rae 1999; McSweeney et al. 2007; Mead 1989b; Reeves et al. 2002; Scott and Hebda 2004; True 1910.

True's Beaked Whale

FRENCH NAME: **Baleine-à-bec de True**

SCIENTIFIC NAME: *Mesoplodon mirus*

This beaked whale is very reclusive and there have only been two verifiable live sightings. Once in 1993, off the coast of North Carolina, when a pod of three were seen for almost 15 minutes, and again in 2001, when a boatload of fortunate ferry passengers in the Bay of Biscay off the north coast of Spain were treated to a series of unusual breaches by an adult male.

DESCRIPTION

True's Beaked Whale displays the typical beaked whale body shape, with a small head and appendages, a slightly falcate dorsal fin placed about two-thirds the way down the body, and flippers positioned more forward than is usual for cetaceans. The body is thick through the middle and narrows acutely at both ends. True's Beaked Whales have a small, but well-rounded melon, which slopes down to the blowhole, typically with a shallow crease where it meets the blowhole. The beak is distinct, and similar to that of a Bottlenose Dolphin. Coloration of males and females is similar. The back is a medium brownish-grey that fades to a whitish-grey on the under surface. The genital area is often surrounded by a diffuse white patch. The fins, flippers, and flukes are dark grey-black. There is often a dark line along the upper lip. A large, dark oval patch occurs over the eye. A narrow dark line runs along the spine from the back of the head to the dorsal fin. Beyond the dorsal fin, the tail stock flattens laterally and a distinct sharp dorsal ridge is formed from the back of the dorsal fin to the flukes. The trailing edge of the flukes typically does not have a medial notch. As is usual for this genus, only the adult males have exposed teeth. A single tooth emerges at the tip of each lower jaw. This small pair of teeth is exposed even when the mouth is closed, as the lower lip protrudes beyond the upper. The dental formula is 0/1, for a total of two teeth.

SIMILAR SPECIES

Most of the surveys along the Atlantic coast of North America lump all the sightings of beaked whales into one category, as identification to species is difficult to impossible at sea, even by experts. That being said, under ideal conditions, a positive identification of adult males can sometimes be made. See the "Similar Species" section of the Sowerby's Beaked Whale account for more information.

SIZE

Total length: adults appear to be around 5.0–5.5 m long; newborns are thought to be 2.0–2.5 m long.
Weight: at least 1000 kg, up to around 1400 kg; newborns, unknown.

RANGE

The distribution of this whale is patchy. The area of highest density appears to be in Gulf Stream waters off the coast of the central United States. Strandings are known from Nova Scotia (St Anns Bay, Victoria County) to the Bahamas. Another region of the North Atlantic where strandings and a sighting have occurred is from Ireland to the Canary Islands. Strandings on the southern coast of Africa and Australia indicate another population in the Indian Ocean. There are no known specimens or sightings from the Pacific Ocean. The

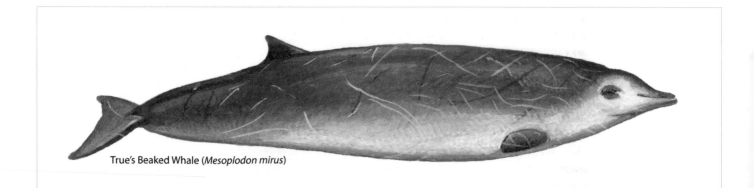

True's Beaked Whale (*Mesoplodon mirus*)

Distribution of the True's Beaked Whale (*Mesoplodon mirus*)

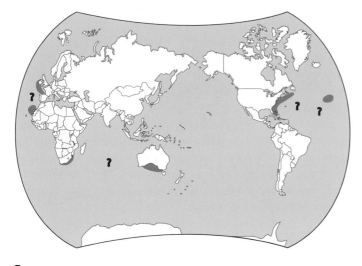

 possible range

continuity of the patches of distribution is unknown due to the lack of sightings at sea. The Northern and Southern Hemisphere populations are not likely contiguous, and it is possible that, with more samples, the two populations may prove to be distinct enough for each to be considered separate species.

ABUNDANCE

The worldwide population size of this species is entirely unknown. It has never been commercially exploited; however, a few are killed each year from entanglement in fishing gear. All the beaked whales have been considered by the International Union for the Conservation of Nature (IUCN) for inclusion in their "Red Book," but lack of information has prevented them from making a decision on the whales' status. True's Beaked Whales are therefore listed as being "Data Deficient" by the IUCN. While possibly more common than the occasional strandings and sightings suggest, this whale is nevertheless considered extremely rare in Canadian waters.

ECOLOGY

Both verifiable live sightings were made in waters that were deeper than 1000 m, along the slope of a continental shelf where it drops rapidly to depths in excess of 1500 m. This information, combined with locations of strandings, indicates that True's Beaked Whale is a deep-water species that prefers temperate water temperatures. Diet and similarity to other beaked whales further suggest that this whale is also a deep diver, probably to depths around, or even in excess of, 1000 m.

DIET

Based on examinations of the stomach contents of several stranded animals, this whale eats mainly deep-water squid, which do not rise in the water column during the night. The whales must therefore dive to considerable depths to capture them. It is very possible that further study of more specimens will add some deep-water fishes

to this list, as this is common for other species in the genus *Mesoplodon*. Like the other beaked whales, this species is thought to use suction to capture its prey.

REPRODUCTION

Almost nothing is known concerning the reproduction of this whale. The discovery of a near-term 2.2 m long foetus in an adult female, stranded in March along the coast of North Carolina, suggests an early spring birthing season in that region. A pregnant, lactating female who stranded with her 3.4 m long calf indicates that females can be pregnant and lactating concurrently.

BEHAVIOUR

Apart from the female with her calf mentioned above, mass strandings by this species are not known. The only pod ever reported was a small group of three animals thought to be two adult females and a juvenile. The other valid sighting was of a solitary male who breached 24 times in rapid succession, often flinging his whole body out of the water. Based on the number of paired scratches on the bodies of adult males and the noticeable lack of scratches on both adult females and juveniles, it is supposed that the males use their pair of exposed teeth to rake other males. These battles are likely over access to females. This species, like the other beaked whales, generally avoids powered vessels.

VOCALIZATIONS

No information on communication by this whale is available; however, their ability to echolocate is assumed.

SIGNS

The dark dorsal fin emerges just as the head is submerging and it stands out from the lighter surrounding body colour. The dolphin-like beak is visible at the beginning of each breath cycle as the head pokes out of the water and the dark oval patch surrounding the eye is visible for about a second. On some animals, the "lips" are defined by a dark line. The narrow band of darker pigment along the midline of the back that ends at the dorsal fin is usually visible as the whale surfaces. Behind the dorsal fin the body flattens laterally and a sharp dorsal ridge is sometimes noticeable depending on viewing conditions. These whales apparently do not expose their flukes before a deep dive, but are known to breach. Their diffuse blow is a columnar shape about the same height as the head length.

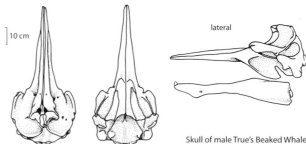

Skull of male True's Beaked Whale
(adapted from Figure 15 in Mead 1989b)

REFERENCES

Houston 1990d; MacLeod 2000; Mead 1989b; Reeves et al. 2002; Tove 1995; True 1910; Walker, D., et al. 2001.

Stejneger's Beaked Whale

also called Bering Sea Beaked Whale,
Sabre-toothed Beaked Whale

FRENCH NAME: **Baleine-à-bec de Stejneger**

SCIENTIFIC NAME: *Mesoplodon stejnegeri*

Like all the mesoplodont beaked whales (genus *Mesoplodon*), this species is very poorly known and rarely seen. Most of our information is derived from the examination of dead stranded animals.

DESCRIPTION

Stejneger's (pronounced sty-ne-gur's) Beaked Whale is a dark-bodied, small-headed whale whose flippers and flukes are small. The lower jaw, throat, and belly are often paler than the back. Adult females, juveniles, and younger adult males typically have a dark crown called a cranial cap, which dips down to encircle the eye on each side of the head, producing a "helmeted" appearance. Older adult females seem to develop a bright-white radiating pattern of concentric blotches on the underside of the flukes. Adult males, and some older females, are often dark all over. Adults usually have a variable number of white oval scars on their flanks that increase with age, and adult males can be covered in thin linear white scars caused by tooth raking. The falcate dorsal fin is placed about two-thirds of the way down the body. The flipper pockets are darker than the surrounding pigmentation on the lighter-bellied females. Like most of the beaked whales, there is usually no noticeable notch in the centre of the trailing edge of the flukes, although some animals exhibit a slight indentation at this point and rarely a notch is present. Only adult males have visible teeth. Each lower jaw displays a single tooth about in the middle of the jaw and the jawbone arches upwards at this tooth. The teeth are large and flattened and tilt towards each other, somewhat constricting the ability of the animal to open its mouth widely. The highest point of each tooth is on its leading edge. The same teeth in females and juveniles are much smaller and remain embedded in the gums. The dental formula is 0/1, for a total of two teeth.

SIMILAR SPECIES

Stejneger's Beaked Whale is the only small beaked whale in the northern North Pacific. In more temperate regions along the North American west coast, it shares the waters with similarly sized Hubbs' Beaked Whales. See the "Similar Species" section of the Hubbs' Beaked Whale account for more information.

SIZE

Females are slightly larger than males.
Total length: males, at least 5.37 m; females, 5.1–5.5 m; newborns, 2.1–2.3 m.
Weight: males, unknown; females, up to 1916 kg; newborns, estimated to be around 100 kg.

RANGE

Based almost exclusively on strandings, the distribution of Stejneger's Beaked Whale is assumed to be Subarctic and cold temperate waters of the North Pacific from the northern coast of Baja to Alaska and across the Bering Sea to the Sea of Japan. Three strandings are reported from Vancouver Island and two from the Queen Charlotte Islands (Haida Gwaii) since 1959.

ABUNDANCE

The population size of this species is totally unknown. The area of highest density is around the Aleutian Islands of Alaska, where most strandings occur. While it may not be as scarce as the few strandings and sightings suggest, it is undoubtedly a rare species in Canadian waters.

ECOLOGY

These whales appear to occur in cool, temperate, offshore waters where squid are plentiful. Likely, this is more related to the occurrence of prey than the habitat requirements of the whales. Since their prey is composed of species that do not undertake a nightly migration up the water column, the whales must dive to considerable depths to secure them, probably in excess of 1000 m. They have been seen in waters that vary from 730 to 1560 m deep. Like other beaked whales, they use suction to capture their food. There are some suggestions that these whales undertake annual migrations to warm temperate waters in winter. As the animals age, they tend to accumulate more oval scars caused by bites from cookie-cutter

Distribution of the Stejneger's Beaked Whale (*Mesoplodon stejnegeri*)

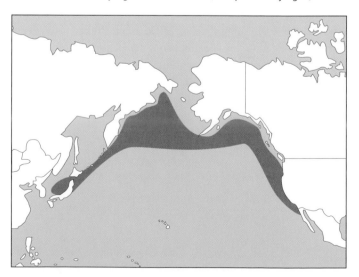

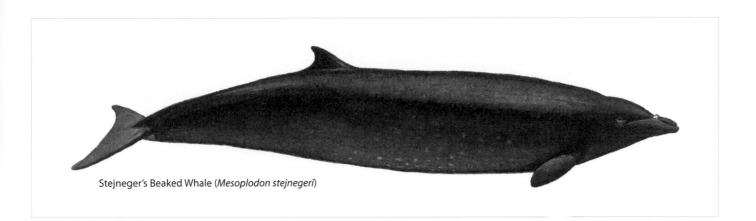

Stejneger's Beaked Whale (*Mesoplodon stejnegeri*)

sharks. These sharks are only found in temperate waters. There are no reliable longevity estimates.

DIET
Based on the stomach contents of stranded animals, Stejneger's Beaked Whales eat mainly deep-water squid and some deep-water fishes.

REPRODUCTION
Estimates based on the size of foetus in stranded females suggests that the young are born anywhere from April to September. The gestation period is thought to be about 12 months.

BEHAVIOUR
Stejneger's Beaked Whales are shy and avoid powered vessels. Herds of two or three up to 15 animals have been reported, as well as many solitary individuals. The animals in a herd swim close together and synchronize their surfacing, indicating that a group is highly cohesive. Small groups have been seen chasing schools of salmon in deep waters of the Sea of Japan. There is some sparse evidence that herds may separate along gender lines, based on a mass stranding of four animals that were all females. Other sightings involved herds of mixed ages and genders. Healed fractures of the rostrum and jaws of adult males, along with the often extensive white linear scars that accumulate on their bodies over time, suggest that they ram each other and drag their teeth over each other during fights. These battles are probably all about access to females, either through territorial defence or more likely during dominance interactions.

Their estimated cruising speed is 5.5–7.5 km/h, with a top speed of around 11 km/h.

VOCALIZATIONS
There are no reports of sounds produced by Stejneger's Beaked Whales. Like other toothed whales, it is presumed that they do echolocate.

SIGNS
The blow of this whale is short and diffuse and rarely seen under the open ocean conditions where this animal is found.

REFERENCES
Houston 1990c; Loughlin and Perez 1985; Mead 1989b; Reeves et al. 2002; True 1910; Walker, W.A., and Hanson 1999; Willis, P.M., and Baird 1998a.

Cuvier's Beaked Whale
also called Goose-beaked Whale

FRENCH NAME: **Baleine-à-bec de Cuvier**

SCIENTIFIC NAME: *Ziphius cavirostris*

These medium-sized whales are elusive and seldom seen despite their worldwide distribution.

DESCRIPTION
Cuvier's Beaked Whale has the thick body shape typical of beaked whales. The slightly falcate dorsal fin is placed about two-thirds of the way along the back. Small, blunt flippers are placed forward on the body. The flukes have a weak notch or often a straight trailing edge. Head shape distinguishes this whale from other beaked whales. Its beak is short and its mouthline curved upwards. The somewhat bulbous melon visually enhances the shortness of the head, making the animal appear foreshortened. A pair of short throat grooves starts below the eye and almost converges below the corner of the

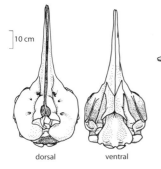

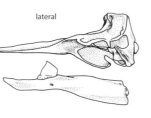

10 cm

lateral

dorsal   ventral

Skull of male Stejneger's Beaked Whale
(adapted from Figure 15 in Mead 1989b
and figures in True 1910)

Cuvier's Beaked Whale (*Ziphius cavirostris*)

mouth. Calves are dark above and lighter below. The general body colour of adults in the Atlantic is steel grey, while the Pacific animals are brownish-grey. There is a darkened area around the eye and some degree of lightening on the head usually encompasses at least the beak and chin. Adult males display considerably more white in the head area than do most adult females, often developing a whitish cape that can cover almost half of the back. This cape occasionally develops in adult females. The animal pictured here is a male. There is often a light chevron below the eye. Adults usually have numerous white scratches on their bodies that can enhance the light appearance. Males tend to accumulate more white scratches than do females. All adults have numerous oval, white scars scattered over their bodies, probably caused by bites from cookie-cutter sharks, lampreys, or even parasitic crustaceans. The body darkens quickly after death, so shape and size are better clues than colour in identifying a stranded animal. A conical tooth is present at the tip of each lower jaw. These erupt beyond the gums only in adult males and may remain exposed even when the mouth is closed. The dental formula is 0/1, for a total of two teeth.

SIMILAR SPECIES

All the beaked whales are very difficult to distinguish at sea. A detailed view is often necessary in order to make a valid identification. Generally the Cuvier's Beaked Whale adult males are easier to identify due to their lighter coloration. The juveniles and females could be confused with any of the *Mesoplodon* genus of beaked whales. In the Pacific Ocean, an adult male might be mistaken for the light-coloured, but much smaller Risso's Dolphin. The adult females are generally too dark for a Risso's Dolphin, but can be similar, although larger, than the Stejneger's or Hubbs' Beaked Whales. Both these beaked whales are smaller and have longer beaks than the Cuvier's Beaked Whale. In the Atlantic Ocean, both sexes could be mistaken for a Northern Bottlenose Whale, which is about 30% larger. Northern Bottlenose Whales have a hugely inflated melon and a longer, more defined beak. There are three beaked whales of the genus *Mesoplodon* in the Atlantic Ocean, True's, Sowerby's, and Blainville's. All are smaller than Cuvier's Beaked Whale and have longer beaks.

SIZE

Females are slightly longer and heavier than males.

Total average adult length: about 5.5–6.0 m, to a maximum of 7.5 m; newborns, about 270 cm.

Weight: maximum at least 5700 kg for males and around 6600 kg for females; newborns, unknown.

RANGE

Cuvier's Beaked Whales are the most widely distributed of the beaked whales, found throughout warm temperate, subtropical, and tropical waters in all the world's oceans. Strandings on Canada's east coast are rare. A single stranding has been reported from Sable Island, Nova Scotia. Strandings on the west coast are much more common, being reported from the Queen Charlotte Islands (Haida Gwaii) to Vancouver Island.

ABUNDANCE

The shyness and deep-water habits of this species make them very difficult to survey, as do the identification-at-sea difficulties. The population off the west coast of the United States has been estimated to number in the low thousands. Despite their seeming rarity, these are probably the most abundant of the beaked whales. There are occasional strandings in the Atlantic provinces and on the west coast of British Columbia. This whale is rare in Canadian waters, although this may be more a factor of visibility, than actual rarity.

Distribution of the Cuvier's Beaked Whale (*Ziphius cavirostris*)

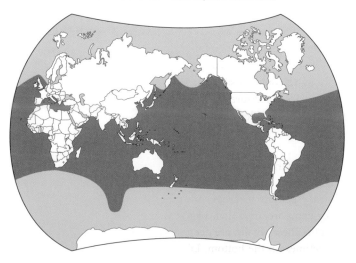

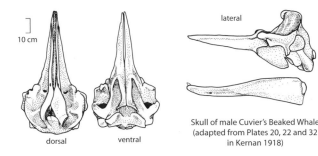

10 cm

dorsal   ventral   lateral

Skull of male Cuvier's Beaked Whale
(adapted from Plates 20, 22 and 32
in Kernan 1918)

It continues to be hunted along the coast of Japan in small numbers and is regularly drowned in deep-water fishing gear.

ECOLOGY

These are deep-water whales that are most often found where depths exceed 1000 m, particularly along the continental slopes, where there are steep gradients. They only approach the land in locations where deep water is near shore, and they prefer waters that are no colder than 10°C. There is some evidence that some populations are seasonal, moving into the higher-latitude waters as they warm in summer. Foraging dives are usually about 30–40 minutes long and, based on stomach contents, are likely at least 300–1000 m deep. The deepest recorded dive is 1484 m and the longest is 94.5 minutes. These whales rarely mass strand. Except under unusual circumstances, all the known strandings are of single animals. Recent mass strandings, apparently all of live animals – one in the Mediterranean Sea, another in the Caribbean, and three more off the Canary Islands – were all associated with military testing of Low Frequency Active Sonar (LFAS), during which extremely loud, low-frequency sounds are produced in order to detect quiet diesel and nuclear-powered submarines at long distances. The reasons for the strandings are still hypothetical. The animals could have panicked at the unfamiliar noise and "stampeded" ashore, they could have been making every possible effort to get out of the water to avoid the sounds, or the sounds may have injured their hearing and made them too feeble to avoid stranding. Some stranded individuals display internal ear injuries and others show symptoms similar to those suffered by humans with decompression sickness (the bends). There is some limited recent evidence that shipping noise from rapidly moving large cargo ships falls within the acoustic range of echolocating whales and may disrupt their foraging. The lifespan of Cuvier's Beaked Whales has been estimated at around 40 years, although it could possibly be as long as 60 years or more. Most travel takes place at a leisurely 5–6 km/h, although the whales are vigorous swimmers, capable of considerably faster speeds.

DIET

Their preferred prey is squid, with some crustaceans and some fish taken at times. Fish may be an important food source in some areas, especially in deeper waters (over 1000 m). Most individual prey items are more than 1 kg in weight and greater than 20 cm in length. These whales are deep divers that find their prey at depth, sometimes on the bottom. Like the other beaked whales, Cuvier's

Beaked Whales lack functional teeth and likely capture their prey using suction.

REPRODUCTION

There is very little information available concerning the reproduction of this species. Females apparently reach sexual maturity at around 5.8 m long and males around 5.5 m. A single calf is the norm and calves are born year-round.

BEHAVIOUR

Most pods are small, compact groups of two to five animals, usually including a single adult male. Larger pods of 10–15 may represent temporary assemblages of extended families. Solitary bulls are sometimes seen. Cuvier's Beaked Whales are shy, avoid boats, and so are rarely seen. Occasionally a pod may emerge from a deep dive near a vessel, apparently by mistake, take a series of shallow breaths, then sound, and not be seen again. Most adults sport numerous linear white scars likely made by the teeth of other Cuvier's Beaked Whales. Since only the adult males have erupted teeth, they must be the culprits. This suggests an active social life beneath the water surface, which is not yet supported by scientific observations.

VOCALIZATIONS

There are two known recordings made of Cuvier's Beaked Whale vocalizations, both recently. An animal tagged with a transitory acoustic monitor was heard to produce clicks and buzz sounds at the bottom of a series of deep foraging dives. Clicking was apparently used to detect prey, and the buzzes, a series of very rapid clicks, were likely used to facilitate capture of the prey. Other toothed whales, as well as many bats, produce similar vocalizations. The main energy of the click is within 29 and 51 kHz.

SIGNS

A deep dive is signalled by the whale lifting its tail flukes high out of the water in preparation for a nearly vertical descent. Returning to the surface after a deep dive is also almost vertical, so that the head and chin are often briefly exposed as the animals take their first breath. The head and chin are also commonly exposed as the animals roll at the surface for a breath, if they are swimming rapidly. The blow of this whale, like that of all the beaked whales, is low, diffuse, and difficult to spot at sea. If a clear, close view is provided, the blow can be seen to aim slightly forward and to the left. These whales will sometimes breach. When they roll at the surface to take a breath, the dorsal fin does not appear before the head and blowhole are submerged.

REFERENCES

Baird et al. 2006, 2008b; Dawson et al. 1998; Fiscus 1997; Frantzis 1998; Frantzis and Cebrian 1999; Houston 1991; Kernan 1918; Leatherwood et al. 1982; Lucas and Hooker 2000; MacLeod et al. 2003; McSweeney et al. 2007; Ohizumi and Kishiro 2003; Santos et al. 2001; Soto et al. 2006; Willis, P.M., and Baird 1998a; Zimmer et al. 2005.

Domestic Mammals

Lamb (*Ovis aries*)

Photo: Dan Barnes / iStockphoto

In Canada there are seven principal species of domesticated mammals. The horse is already covered in an earlier portion of the book as there are several wild populations. The others – goats, sheep, cattle, pigs, dogs, and cats – occasionally can survive for a while in the wild, but colonies of these animals have not proved to endure without some form of human intervention. The bones of domesticated animals are frequently unearthed in unusual places. For purposes of identification the skulls of these species are included in this section.

Domestic Cow

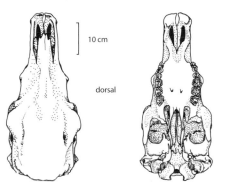

10 cm

dorsal

ventral

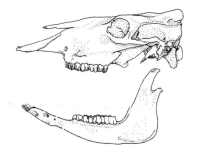

lateral

Skull of Domestic Cow
(CMNMA 75122)

Walk

Slow walk

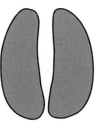

Left hind track
Length: 9.0–14.0 cm
Width: 8.5–11.5 cm

Left front track
Length: 10.0–15.0 cm
Width: 9.0–12.5 cm

Domestic Dog

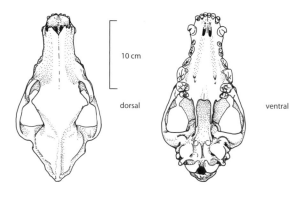

10 cm

dorsal

ventral

lateral

Skull of Domestic Dog
(CMNMA 41073)

Trot

Walk

Left hind track
Length: 11.0–15.0 cm
Width: 10.0–14.0 cm

Left front track
Length: 11.5–16.5 cm
Width: 11.5–15.0 cm

Domestic Goat

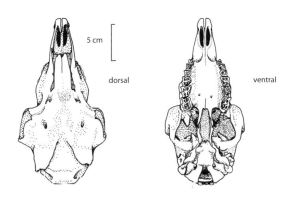

5 cm

dorsal

ventral

lateral

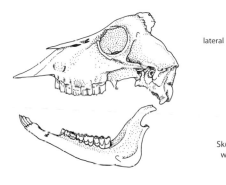

Skull of Domestic Goat,
without horn sheaths
(CMNMA 75478)

Walk

Left hind track
Length: 7.0–9.0 cm
Width: 6.0–8.0 cm

Left front track
Length: 8.0–11.0 cm
Width: 6.5–9.0 cm

Domestic Cat

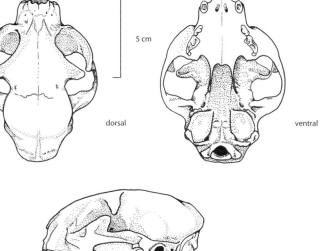

5 cm

dorsal

ventral

lateral

Skull of Domestic Cat
(CMNMA 51343)

Direct register walk

Direct register trot

Left hind track
Length: 4.0–6.5 cm
Width: 3.0–7.0 cm

Left front track
Length: 4.0–6.5 cm
Width: 3.5–7.0 cm

Domestic Sheep

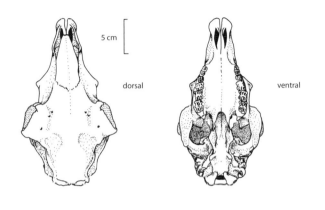

dorsal

ventral

5 cm

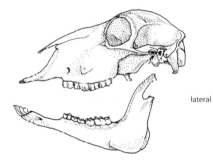

lateral

Skull of Domestic Sheep
(CMNMA 75326)

Walk

Left hind track
Length: 5.5–7.5 cm
Width: 5.0–6.0 cm

Left front track
Length: 6.0–8.0 cm
Width: 5.5–6.5 cm

Domestic Pig

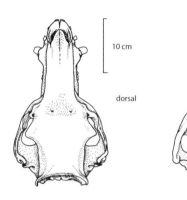

10 cm

dorsal

ventral

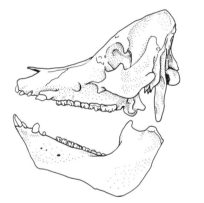

lateral

Skull of Domestic Pig
(CMNMA 41071)

Direct register trot

Direct register walk

Left hind track
Length: 4.8–6.5 cm
Width: 5.0–7.0 cm

Left front track
Length: 5.5–6.7 cm
Width: 5.5–7.5 cm

Appendices

Characteristics identified in bold type are diagnostic for the species.

Blarina brevicauda	**a)** 5 upper unicuspids, 4 clearly visible in lateral view, the 5th usually tucked in behind the 4th and the premolar; **b) skull length > 20 mm**; **c)** lateral edge of braincase sharply angled in dorsal view; **d)** height of coronoid process usually ≥ 6.0 mm **e)** mental foramen positioned beneath hypoconid of m1	
Cryptotis parvus	**a) 4 upper unicuspids, first 3 clearly visible in lateral view (U4 is vestigial and only visible in ventral view); b)** dentary length ≤ 7.1 mm; **c)** height of coronoid process ≤ 4.1 mm	
Sorex arcticus	**a)** 5 upper unicuspids, U3 ≥ U4; **b)** outline of anterior margin of U1 in occlusal view is rounded (as in *S. tundrensis*); **c)** skull length 18.8–19.5 mm; **d)** palatal length > 7.7 mm; **e)** pigment on i1 is broken into 2 or 3 segments; **f)** height of coronoid process > 4.6 mm; **g)** dentary length ≥ 7.9 mm	
Sorex bendirii	**a)** 5 upper unicuspids, U3 < U4; **b)** rostrum shallow and distinctly downcurved; **c)** skull length > 19 mm; **d)** length of unicuspid row >70% of the premolar-molar length; **e)** palatal length > 8.2 mm; **f)** each upper incisor has a medium-sized medial tine which is positioned wholly within the pigmented zone; **g)** length of i1 usually < 5.4 mm; **h)** band of pigment on ventomedial edge of i1 extends well back	
Sorex cinereus	**a)** 5 upper unicuspids in an almost graduated series, U3 ≥ U4; **b)** lacrimal foramen posterior to mesostyle of M1; **c)** skull length 14.7–17.0 mm; **d)** length of unicuspid row 2.2–2.6 mm; **e)** large medial tine on I1 wholly within pigmented zone; **f)** usually one distinct area of pigmentation on i1; **g)** length of coronoid-condyloid process usually ≤3.2 mm	

Characteristics identified in bold type are diagnostic for the species.

Sorex dispar	**a)** 5 upper unicuspids, U3 ≥ U4; **b)** posterior border of the infraorbital foramen behind the space between the M1 and M2; **c)** mental foramen level with the space between p4 and m1	
Sorex fumeus	**a)** 5 upper unicuspids, U3 ≥ U4; **b)** posterior border of infraorbital foramen ahead of space between the M1 and M2; **c)** mental foramen approximately level with the protoconid of m1; **d)** skull length > 17 mm	
Sorex haydeni	**a)** 5 upper unicuspids although the smallest (the 5th) may be difficult to see in lateral view; **b)** posterior border of lacrimal foramen above metastyle of M1; **c)** unicuspid toothrow length 1.86–2.20 mm; **d)** skull length ≤ 15.6 mm; **e)** usually 2 distinct areas of pigmentation on the i1; **f)** dentary length ≥ 6.2 mm; **g)** height of coronoid process usually ≥ 3.2 mm	
Sorex hoyi	**a)** 5 upper unicuspids, **3 clearly visible in lateral view (U3 and U5 are vestigial and only visible in ventral view); b)** alveolus of i1 extends beneath part of the paracone of m1 in labial view; **c)** height of coronoid process ≤ 4.1 mm; **d)** dentary length ≤ 6.1 mm	
Sorex maritimensis	**a)** 5 upper unicuspids, U3 ≥ U4; **b)** skull length 18.2–18.7 mm; **c)** height of coronoid process 4.1–4.4 mm; **d)** pigmentation on i1 broken into 2 or 3 segments; **e)** dentary length 7.5–7.85 mm	
Sorex merriami	**a)** 5 upper unicuspids although the smallest (the 5th) is difficult to see in lateral view, U3 > U4; **b)** no medial tine on the upper incisors; **c)** angle of the i1 relative to the horizontal ramus of the mandible is ≥ 13 degrees; **d)** only one area of pigmentation on i1; **e)** height of coronoid process ≥ 3.9 mm; **f)** dentary length ≥ 6.4 mm	

Characteristics identified in bold type are diagnostic for the species.

Sorex monticolus	**a)** 5 upper unicuspids, U3 < U4; **b)** small to large medial tine on upper incisors set well within the pigmented region; **c)** skull length < 19 mm; **d)** mental foramen usually beneath posterior half of m1; **e)** strip of pigment on ventromedial edge of i1 extends well back	
Sorex palustris	**a)** 5 upper unicuspids, U3 < U4; **b)** rostrum straight or downcurved only slightly; **c)** length of unicuspid row < 70% of premolar-molar length; **d)** palatal length < 8.2 mm; **e)** skull length > 19 mm; **f)** one area of pigmentation on i1; **g)** dentary length ≥ 8.0 mm; **h)** height of coronoid process ≥ 4.0 mm	
Sorex preblei	**a)** 5 upper unicuspids, U3 ≥ U4; **b)** maxillary toothrow length ≤ 4.2; **c)** 2 or 3 areas of pigmentation on i1; **d)** dentary length ≤ 6.6 mm; **e)** height of coronoid process ≤ 3.3 mm; **f)** maxillary breadth ≤ 4.2 mm; **g)** skull length ≤14.6 mm; **h)** tine on medial edge of upper incisor is long, acutely pointed, and set within the pigmented region	
Sorex rohweri	**a)** 5 upper unicuspids, U3 usually ≥ U4, rarely slightly shorter; **b) tine on medial edge of upper incisor small (barely projects above anterior face) and is located at the proximal edge or above the pigmented area;** **c)** large postmandibular foramen confluent with mandibular foramen; **d)** pigmentation on lower incisor broken into 2 isolated areas	
Sorex trowbridgii	**a)** 5 upper unicuspids, U3 < U4, U5 is vestigial; **b)** skull length < 19 mm; **c)** medial edges of upper incisors appear curved; **d)** small medial tines at upper edge of pigmented zone; **e)** inferior dentary foramen and postmandibular foramen usually confluent and large; **f)** mental foramen usually positioned beneath the anterior half of the m1	

Characteristics identified in bold type are diagnostic for the species.

Species	Characteristics	Illustrations
Sorex tundrensis	**a)** 5 upper unicuspids, U1=U2, U3≥U4, U3 to U5 diminish gradually; **b)** lacrimal foramen positioned over metacone of M1; **c)** skull length 16.8–18.5 mm; **d)** anterior margin of U1 is rounded in occlusal view (as in *S. arcticus*); **e)** one or two areas of pigmentation on i1; **f)** dentary length is ≤ 7.9 mm; **g)** height of coronoid process 3.9–4.1 mm	lateral view / dorsal view / occlusal view / labial side of mandible (lacrimal foramen, metacone)
Sorex ugyunak	**a)** 5 upper unicuspids, U3≥U4; **b)** length unicuspid row < 2.2 mm; **c)** upper unicuspids usually wider than long; **d)** skull length < 17 mm; **e)** only one area of pigmentation on i1; **f) two cusps on u1**; **g)** dentary length ≤ 6.9 mm; **h)** height of the coronoid process ≤ 3.2 mm	lateral view / dorsal view / occlusal views / labial side of mandible
Sorex vagrans	**a)** 5 upper unicuspids, **U3 noticeably smaller than U4**; **b)** small medial tine at top edge or above pigmentation and often separated by a pale gap from the pigmented region; **c)** medial edge of upper incisors appear straight in front view; **d)** skull length < 19 mm; **e)** only one area of pigmentation on i1; **f)** length of coronoid-condyloid process ≥ 3.1 mm	lateral view / dorsal view / frontal view / labial side of mandible

REFERENCES

Carraway, L.N. 1995. A key to recent Soricidae of the western United States and Canada based primarily on dentaries. *Occasional Papers of the Museum of Natural History of the University of Kansas* no. 175: 1–49.

Junge, J.A., and Hoffmann, R.S. 1981. An annotated key to the long-tailed shrews (genus *Sorex*) of the United States and Canada, with notes on Middle American *Sorex*. *Occasional Papers of the Museum of Natural History of the University of Kansas* no. 94: 1–48.

Nagorsen, D.W. 1996. *Opossums, shrews and moles of British Columbia*. Vol. 2 of the *The Mammals of British Columbia*. Royal BC Museum Handbook, Royal BC Museum, Victoria, BC.

Nagorsen, D.W., and Panter, N. 2009. Identification and status of the Olympic Shrew (*Sorex rohweri*) in British Columbia. *Northwestern Naturalist* 90(2): 117–129.

van Zyll de Jong, C.G. 1980. Systematic relationships of woodland and prairie forms of the Common Shrew, *Sorex cinereus cinereus* Kerr and *S. c. haydeni* Baird, in the Canadian prairie provinces. *Journal of Mammalogy* 61(1): 66–75.

– 1983a. *Marsupials and Insectivores*. Vol. 1 of the *Handbook of Canadian Mammals*. National Museums of Canada, Ottawa, ON.

– 1983b. A morphometric analysis of North American shrews of the *Sorex arcticus* group, with special consideration of the taxonomic status of *S. a. maritmensis*. *Naturaliste canadien* 110(4): 373–378.

– 1991. Speciation in the *Sorex cinereus* group. In J.S. Findley and T.L. Yates (eds), *The Biology of the Soricidae*, 65–73. Museum of Southwestern Biology, University of New Mexico, Albuquerque, NM.

Characteristics identified in bold type are diagnostic for the species.
Toothrows illustrate occlusal views of upper right (above) and lower left (below) with lingual to the top of the page and labial to the bottom.

genus *Dicrostonyx*	**a)** root of lower incisor shorter than that of most other voles, terminating between the mid-point and end of lower molar toothrow; **b)** labial and lingual re-entrant angles on upper and lower molars similar in depth Definitive identification of collared lemming skulls is difficult as pelage, distribution, and genetic characteristics are the primary features used to distinguish species.	
Dicrostonyx groenlandicus	**a)** M1 and M2 usually terminate with a narrow lingual and labial loop; **b)** m2 and m3 typically begin with similar narrow loops	
Dicrostonyx hudsonius	In general the molar patterns are simpler than those of other species in the genus. **a)** posterior margin of M1 and M2 usually have only a single labial terminal loop; **b)** sometimes the anterior margin of the m2 and usually of the m3 is simple (lacking the lingual and especially the labial loops)	
Dicrostonyx nunatakensis	Similar to *Dicrostonyx groenlandicus*	
Dicrostonyx richardsoni	Tends to be intermediate between the more complex molars of *Dicrostonyx groenlandicus* and *D. nunatakensis* and the simpler patterns found in *D. hudsonius* **a)** lingual terminal loops on the M1 and M2 and labial anterior loops on the m2 and m3 are usually vestigial and sometimes absent	
Lemmiscus curtatus	**a)** skull is flat in profile and may have a noticeable depression in the frontal region; **b)** large auditory bullae extend past the occiput; **c)** labial and lingual re-entrant angles of cheek teeth about equal (arrows) but angles are wider than those of most other voles; **d)** M3 has 3 loops of enamel on the lingual side; **e)** m3 has 3 enclosed triangles and a posterior closed loop	

Characteristics identified in bold type are diagnostic for the species.
Toothrows illustrate occlusal views of upper right (above) and lower left (below) with lingual to the top of the page and labial to the bottom.

Lemmus trimucronatus	**a)** skull robust; **b)** interorbital region narrow; **c)** zygomata flared; **d)** re-entrant angles disproportionately deep on labial side of upper cheek teeth and on lingual side of lower cheek teeth; **e)** lower molars have small closed triangles on the labial side (arrows); **f)** posterior palate terminates in simple tranverse shelf; **g)** toothrows widely divergent posteriorly	
genus *Microtus*	**a)** root of lower incisors originates within or near the condylar process and passes from the lingual (tongue side) to the labial side (lip side) of the molars between the bases of the m2 and m3; **b)** posterior edge of palate terminates between the last upper molars as a sloping medial ridge bordered by lateral pits; **c)** lingual and labial re-entrant angles of upper and lower molars about equal in depth (arrows)	
Microtus chrotorrhinus	**a)** M2 has 4 closed loops and lacks a posterior loop; **b)** 4-5 closed loops on the M3 (some individuals have the posterior loop confluent with the 4th loop, others have the 4th loop closed so the posterior loop forms a 5th closed loop as illustrated); **c)** m2 has 2 closed triangles and a posterior closed loop	
Microtus longicaudus	**a)** nasals usually hide incisors completely when viewed from above; **b)** incisive foramena taper gradually if at all; **c)** upper toothrow <7 mm long; **d)** 4 loops of enamel on lingual side of M3; **e)** M3 usually has 3 closed triangles; **f)** lingual side of M3 has 3 re-entrant angles while the labial side has 2; **g)** M2 has 3 closed triangles and a closed anterior loop	
Microtus miurus	**a)** skull long and narrow; **b)** zygomatic width about equal to half of greatest skull length; **c)** interorbital region narrow often with pronounced inter orbital crest and appears depressed in lateral view; **d)** M2 has 4 closed loops and no posterior loop; **e)** M3 has 4 closed loops and the lingual arm of the posterior loop forms an incipient 5th closed loop; **f)** m1 has 6 closed triangles between 2 terminal loops	
Microtus montanus	**a)** incisive foramina ≤ 5mm and constrict sharply to points posteriorly; **b)** upper incisors of adults protrude beyond the nasal bones when viewed from directly above; **c)** M3 has 3 closed triangles and a closed recurved posterior loop; **d)** m1 has 4 re-entrant angles on each side	

Characteristics identified in bold type are diagnostic for the species.
Toothrows illustrate occlusal views of upper right (above) and lower left (below) with lingual to the top of the page and labial to the bottom.

Microtus ochrogaster	**a) 2 closed triangles plus a closed anterior loop and 2 re-entrant angles (arrows) on each side of the M3**; **b)** m1 has 3 closed triangles and an anterior loop confluent with the 2 anterior-most open triangles (arrows); **c)** m3 has three transverse loops and no closed triangles	
Microtus oeconomus	**a)** incisive foramena gradually taper to a slit posteriorly; **b)** M2 has 3 closed triangles and an anterior loop and lacks a posterior loop; **c)** 4 closed triangles on m1, 5th triangle confluent with the anterior loop	
Microtus oregoni	**a)** M3 has 2 re-entrant angles (arrows) and 3 loops of enamel on the lingual side; **b)** m1 has 5 closed triangles, a closed crescent-shaped posterior loop, and a trefoil-shaped anterior loop	
Microtus pennsylvanicus	**a)** M2 has 5 closed loops (3 triangles and an anterior loop **plus a usually rounded, closed lingual posterior loop which may resemble an incipient 5th triangle in some individuals**); **b)** M3 usually has 3 re-entrant angles on either side but this trait is variable especially on the labial side, where a 4th incipient angle may occur; **c)** triangles (arrows) on m1 are narrow and usually sharply pointed, especially on lingual side; **d)** m1 has 4 and sometimes 5 lingual and 4 labial re-entrant angles	
Microtus pinetorum	**a) M3 has 3 transverse loops, no closed triangles, and 2 labial and 2 lingual re-entrant angles (arrows)**; **b)** m1 has 3–4 closed triangles, a posterior closed loop, the 1–2 open triangles (arrows) at the anterior end confluent with the anterior loop (one of the open anterior triangles is closed in some individuals)	
Microtus richardsoni	**a)** incisors protrude well beyond nasals when viewed from above; **b) incisive foramina are long and abruptly taper to narrow slits posteriorly**; **c)** upper toothrow > 7 mm; **d)** M3 has 3 lingual re-entrant angles, a crescent-shaped anterior loop, and a double-armed posterior loop; **e)** M2 lacks a posterior loop; **f)** m1 has 5 closed triangles, a closed posterior crescent, and a closed trefoil-shaped anterior loop; **g)** m2 has 5 closed loops	

Characteristics identified in bold type are diagnostic for the species.
Toothrows illustrate occlusal views of upper right (above) and lower left (below) with lingual to the top of the page and labial to the bottom.

Microtus townsendii	**a)** upper incisors protrude slightly beyond the nasal bones when viewed from above; **b)** incisive foramina taper gradually; **c)** M1 has 5 closed loops; **d)** M3 has 3 lingual re-entrant angles and 3 (one shallower than the others) labial re-entrant angles; **e)** M3 has 3 closed triangles (arrows), an anterior crescent, and a posterior loop; **f)** m2 has 4 closed triangles and a posterior loop	
Microtus xanthognathus	**a)** upper incisors protrude slightly beyond the nasal bones when viewed from above; **b)** adult upper toothrow length >7 mm; **c)** anterior loop of M3 is followed by 3 closed triangles (the 3rd is sometimes confluent with the posterior loop); **d)** M3 has 3 lingual and 2 labial re-entrant angles (arrows); **e)** anterior loop of m1 usually trefoil-shaped; **f)** lingual side of m1 has 3 deep and 1 shallow re-entrant angles while labial side has 3 deep re-entrant angles	
Phenacomys intermedius	**a)** posterior edge of palate with spinous process; **b)** roots of lower incisors extend beyond mid point of molar toothrow; **c)** lingual re-entrant angles of lower cheek teeth disproportionately large (more than halfway through the tooth); **d)** m1 has a variable number of closed triangles (3-7); **e)** lingual and labial re-entrant angles on upper cheek teeth about equal	
genus *Myodes*	**a)** posterior edge of palate a shelf (either complete * or incomplete **); **b)** lingual and labial re-entrant angles on upper and lower cheekteeth about equal in depth	
Myodes gapperi	**a)** posterior edge of palate complete laterally; **b)** posterior loop of M3 is usually closed; **c)** m2 typically has 4 closed loops (3-5)	
Myodes rutilus	**a)** posterior edge of palate usually incomplete laterally; **b)** posterior loop of M3 is usually closed; **c)** m2 has 3 closed loops	

Characteristics identified in bold type are diagnostic for the species.
Toothrows illustrate occlusal views of upper right (above) and lower left (below) with lingual to the top of the page and labial to the bottom.

Myodes glareolus	a) posterior edge of palate complete laterally; b) posterior loop of M3 is usually closed; c) m2 has 3-4 closed loops	
genus *Synaptomys*	a) upper incisors are grooved on anterior surface; b) rostral length <25% of total skull length; c) re-entrant angles disproportionately deep on labial side of upper cheek teeth and on lingual side of lower cheek teeth	
Synaptomys borealis	a) posterior edge of palate has a well-developed spinous process; b) **m3 has 3 closed loops**; c) mandibular molars lack closed triangles on the lingual side	
Synaptomys cooperi	a) posterior edge of palate with a weakly developed spinous process; b) **m3 has 4 closed triangles,** one of which is much smaller (labial side); c) each mandibular molar has a small closed triangle on the labial side	
Ondatra zibethicus	a) skull length > 50 mm; b) inter-orbital region strongly constricted; c) incisors protrude beyond nasals when viewed from above; d) incisive foramina long and narrow; e) maxillary toothrow > 14.0 mm; f) labial and lingual re-entrant angles similar in depth (arrows); g) m1 has 6 closed triangles, the anterior triangle confluent with the bi-lobed anterior loop	

REFERENCES

Engstrom, M.D. 1999a. Ungava Collared Lemming *Dicrostonyx hudsonius*. In D.E. Wilson and S. Ruff (eds), *The Smithsonian Book of North American Mammals*, 660–662. University of British Columbia Press, Vancouver, BC.

Guilday, J.E. 1982. Dental variation in *Microtus xanthognathus*, *M. chrotorrhinus*, and *M. pennsylvanicus* (Rodentia: Mammalia). *Annals of the Carnegie Museum of Natural History* 51(11): 211–230.

Nagorsen, D.W. 2002. *An Identification Manual to the Small Mammals of British Columbia*. Ministry of Sustainable Resource Management, Ministry of Water, Land and Air Protection, and the Royal British Columbia Museum.

Wallace, S.C. 2006. Differentiating *Microtus xanthognathus* and *Microtus pennsylvanicus* lower first molars using discriminant analysis of landmark data. *Journal of Mammalogy* 87(6): 1261–1269.

BRENDA CARTER

North American Opossum
Northern Short-tailed Shrew
Least Shrew
Arctic Shrew
Pacific Water Shrew
Cinereous Shrew
Long-tailed Shrew
Smoky Shrew
North American Pygmy Shrew
Dusky Shrew
North American Water Shrew
Trowbridge's Shrew
Tundra Shrew
Barren Ground Shrew
Star-nosed Mole
Shrew-mole
Hairy-tailed Mole
Eastern Mole
Coast Mole
Townsend's Mole
partial bat used to show anatomy

JULIUS CSOTONYI

Woodchuck (partial, with abnormal pelage)
Bank Vole
Ungava Collared Lemming (winter pelage)
Ogilvie Mountains Collared Lemming
Richardson's Collared Lemming
Prairie Shrew
Merriam's Shrew
Preble's Shrew
Olympic Shrew
Vagrant Shrew
Canada Lynx
Bobcat
Cougar
Harp Seal
Spotted Seal
Harbour Seal
Ringed Seal
Sea Otter
Wolverine
Northern River Otter
American Marten
Fisher
Ermine
Long-tailed Weasel
Black-footed Ferret
Least Weasel
American Mink
American Badger
Striped Skunk
Western Spotted Skunk
Northern Raccoon

PAUL GERAGHTY

Mountain Beaver
Black-tailed Prairie Dog
Northern Flying Squirrel
Southern Flying Squirrel
Woodchuck (typical pelage)
Yellow-bellied Marmot
Hoary Marmot
Vancouver Island Marmot
Eastern Grey Squirrel
Eastern Fox Squirrel
Columbian Ground Squirrel
Franklin's Ground Squirrel
Golden-mantled Ground Squirrel
Arctic Ground Squirrel
Richardson's Ground Squirrel
Cascade Mantled Ground Squirrel
Thirteen-lined Ground Squirrel
Yellow-pine Chipmunk
Least Chipmunk
Red-tailed Chipmunk
Eastern Chipmunk
Townsend's Chipmunk
Douglas' Squirrel
Red Squirrel
North American Beaver
Olive-backed Pocket Mouse
Great Basin Pocket Mouse
Ord's Kangaroo Rat
Plains Pocket Gopher
Northern Pocket Gopher
Meadow Jumping Mouse
Western Jumping Mouse
Pacific Jumping Mouse
Woodland Jumping Mouse
Southern Red-backed Vole
Northern Red-backed Vole
Northern Collared Lemming
Ungava Collared Lemming (summer pelage)
Sagebrush Vole
Nearctic Brown Lemming
Rock Vole
Long-tailed Vole
Singing Vole
Montane Vole
Prairie Vole
Tundra Vole
Creeping Vole
Meadow Vole
Woodland Vole
North American Water Vole
Townsend's Vole

Taiga Vole
Common Muskrat
Heather Vole
Northern Bog Lemming
Southern Bog Lemming
Western Harvest Mouse
Deer Mouse
Keen's Mouse
White-footed Mouse
Northern Grasshopper Mouse
Bushy-tailed Woodrat
Norway Rat
Roof Rat
House Mouse
North American Porcupine
Nutria
Collared Pika
American Pika
Snowshoe Hare
Arctic Hare
European Hare
White-tailed Jackrabbit
European Rabbit
Eastern Cottontail
Mountain Cottontail
Maritime Shrew
Pallid Bat
Townsend's Big-eared Bat
Big Brown Bat
Spotted Bat
Silver-haired Bat
Eastern Red Bat
Hoary Bat
California Myotis
Western Small-footed Myotis
Long-eared Myotis
Keen's Myotis
Eastern Small-footed Myotis
Little Brown Myotis
Northern Myotis
Fringed Myotis
Long-legged Myotis
Yuma Myotis
Evening Bat
Eastern Pipistrelle
Coyote
Grey Wolf
Grey Fox
Arctic Fox
Swift Fox
Red Fox
North American Black Bear

Brown Bear
Polar Bear
Northern Fur Seal
Steller Sea Lion
California Sea Lion
Walrus
Hooded Seal
Bearded Seal
Grey Seal
Northern Elephant Seal
Domestic Horse
Moose
Wapiti
Fallow Deer
Mule Deer
White-tailed Deer
Caribou
Pronghorn
American Bison
Mountain Goat
Muskox
Bighorn Sheep
Thinhorn Sheep
Bowhead Whale
North Atlantic Right Whale
Minke Whale
Sei Whale
Blue Whale
Fin Whale
Humpback Whale
Grey Whale
Long-beaked Common Dolphin
Short-beaked Common Dolphin
Long-finned Pilot Whale
Short-finned Pilot Whale
Risso's Dolphin
White-beaked Dolphin
Atlantic White-sided Dolphin
Pacific White-sided Dolphin
Northern Right Whale Dolphin
Killer Whale
False Killer Whale
Striped Dolphin
Bottlenose Dolphin
Beluga
Narwhal
Harbour Porpoise
Dall's Porpoise
Baird's Beaked Whale
Northern Bottlenose Whale
Sowerby's Beaked Whale
Hubbs' Beaked Whale

Blainville's Beaked Whale
True's Beaked Whale
Stejneger's Beaked Whale
Cuvier's Beaked Whale
Pygmy Sperm Whale
Dwarf Sperm Whale
Sperm Whale

Alaskan Indian Paintbrush: *Castilleja unalaschcensis*
alder: *Alnus* sp.
Alfalfa: *Medicago sativa*
American Alligator: *Alligator mississippiensis*
American Kestrel: *Falco sparverius*
anchovy: *Engraulis* sp.
Antelope Bush: *Purshia tridentata*
Arctic Cod: *Boreogadus saida*
Arctic Willow: *Salix arctica*
Atka Mackerel: *Pleurogrammus monopterygius*
Atlantic Cod: *Gadus morhua*
Atlantic Croaker: *Micropogonias undulatus*
Atlantic Herring: *Clupea harengus*
Atlantic Mackerel: *Scomber scombrus*
Bakeapple: *Rubus chamaemorus*
Balsam Fir: *Abies balsamea*
Banded Drum: *Larimus fasciatus*
Beach Grass: *Ammophila breviligulata*
Beach Pea: *Lathyrus maritimus*
Big Sagebrush: *Artemisia tridentata*
Bigleaf Maple: *Acer macrophyllum*
Black Spruce: *Picea mariana*
Blackbelly Eelpout: *Lycodopsis pacifica*
blackberry: *Rubus* sp.
Black Rat Snake: *Elaphe obsolete*
Black-tailed Jackrabbit: *Lepus californicus*
Blue Whiting: *Micromesistius poutassou*
blueberry: *Vaccinium* sp.
Brainworm or Meningeal Worm: *Parelaphostrongylus tenuis*
Brook Trout: *Salvelinus fontinalis*
Bunchberry: *Cornus canadensis*
Burrowing Owl: *Athene cunicularia*
Capelin: *Mallotus villosus*
Capybara: *Hydrochoerus hydrochaeris*
Cassin's Auklet: *Ptychoramphus aleuticus*
cod: several species in the genera *Gadus, Arctogadus,* and *Eleginus*
Common Murre: *Uria aalgae*
Cooper's Hawk: *Accipiter cooperi*
Crowberry: *Empetrum nigrum*
darkling ground beetles: family Tenebrionidae
dogfish: small sharks in the families Scyliorhinidae, Dalatiidae, and Squalidae
Douglas-fir: *Pseudotsuga menziesii*
Dwarf Birch: *Betula* sp.
Dwarf Willow: *Salix* sp.
eelpout: *Lycodes* sp.
Eulachon: *Thaleichtys pacificus*
euphausids (krill): *Thysanoessa* sp.
False Indian Hellebore: *Veratrum viride*
Fan-leafed Cinquefoil: *Potentilla flabellifolia*
ferret: several species in the genus *Mustela*
Finless Porpoise: *Neophocaena phocaenoides*

fir: *Abies* sp.
fleabane: family Asteraceae
flicker: *Colaptes* sp.
flounder: flatfishes in the families Bothidae and Pleuronectidae
Giant Pacific Salamander: *Dicamptodon tenebrosus*
Giant Panda: *Ailuropoda melanoleuca*
Glaucous Gull: *Larus hyperboreus*
Golden Eagle: *Aquila chrysaetos*
Gopher Snake: *Pituophis melanoleaucus*
Great Grey Owl: *Strix nebulosa*
Great Horned Owl: *Bubo virginianus*
Greenland Halibut: *Reinhardtius hippoglossoides;* see also turbot
Greenland Shark: *Somniosus microcephalus*
greenling: *Hexagrammos* sp.
Grey Jay: *Perisoreus canadensis*
Ground Birch: *Betula glandulosa*
Haddock: *Melanogrammus aeglefinus*
hagfish: *Epatretus* sp.
Hake: *Merluccius merluccius*
hawkweed: *Heiracium* sp.
hemlock: *Tsuga* sp.
herring: *Clupea* sp.
Horned Lantern Fish: *Diaphus splendidus*
horse mackerel: *Trachurus* sp.
horseweed: *Conyza* (also *Erigeron*) sp.
huckleberry: *Vaccinium* sp.
jack mackerel: *Trachurus* sp.
Jack Pine: *Pinus banksiana*
Jerusalem Cricket: *Stenopelmatus* sp.
krill: *Thysanoessa* spp.
Labrador Tea: *Ledum* sp. (now *Rhododendron* sp.)
Lake Trout: *Salvelinus namaycush*
Lake Whitefish: *Coregonus clupeaformis*
lantern fish: family Mycophidae
Larch Sawfly: *Pristiophora erichsonii*
Largemouth Bass: *Micropterus salmoides*
Ling: *Molva molva*
Lodgepole Pine: *Pinus contorta*
Long-eared Owl: *Asio otus*
Long-finned Squid: *Loligo pealei*
mackerel: *Scomber* sp.
Market Squid: *Loligo opalescens*
Marram Grass: *Ammophila breviligulat*
Marsh Wren: *Cistothorus palustris*
Massasauga Rattlesnake: *Sistrurus catenatus*
Mexican Free-tailed Bat: *Tadarida brasiliensis*
Mountain Ash: *Sorbus americana*
Mountain Plover: *Charadrius montanus*
mysid shrimp, also called opossum shrimp: a number of genera in the family Mysidae
Narrow-leafed Fireweed: *Chamerion angustifolium*
New England Cottontail: *Sylvilagus transitionalis*

Northern Anchovy: *Engraulis mordax*
Northern Goshawk: *Accipiter gentilis*
Northern Hawk Owl: *Strix ulula*
Old Man's Beard: *Usnea* sp.
Pacific Cod: *Gadus macrocephalus*
Pacific Hake: *Merluccius productus*
Pacific Lamprey: *Lampetra tridentata*
Pacific Mackerel: *Scomber japonicus*
Pacific Whiting: *Merluccius productus*
paintbrush: *Castilleja* sp.
Painted Lichen Moth: *Hypoprepia fucosa*
Paper Birch: *Betula papyrifera*
Partridgeberry: *Vaccinium vitis-ideae*
pilchard: several small members of the family Clupeidae
Plainfin Midshipman: *Porichthes notatus*
Polar Cod: *Arctogadus glacialis*
Pollock: *Pollachius virens*
Ponderosa Pine: *Pinus ponderosa*
Poor-cod: *Trisopterus minutus*
poplar: *Populus* sp.
porcelain crab: family Porcellanidae
Prairie Dog Weed: *Dyssodia papposa*
prickly pear cactus: *Opuntia* sp.
Przewalski's Horse: *Equus przewalskii*
Pygmy Right Whale: *Caperea marginata*
Raccoon Dog: *Nyctereutes procyonoides*
Raccoon Roundworm: *Baylisascaris procyonis*
raspberry: *Rubus* sp.
ratfish: *Hydrolagus* sp.
Red Alder: *Alnus rubra*
Red Oak: *Quercus rubra*
Red Osier Dogwood: *Cornus sericea/stolonifera*
Red Maple: *Acer rubrum*
redfish: *Sebastes* sp.
Red-tailed Hawk: *Buteo jamaicensis*
Rhinoceros Auklet: *Cerorhinca monocerata*
rockfish: *Sebastes* sp.
Rough-legged Hawk: *Buteo lagopus*
Saffron Cod: *Eleginus gracilus*
sagebrush: *Artemisia* sp.
salmon: *Salmo* sp. and *Oncorhynchus* sp.
sand-eel or sand lance: *Ammodytes* sp.
sandwort: several species in the genera *Honckenya, Arenaria, Minuartia,* and *Moehringia*
sardines: either *Sardinops* sp. or *Sardinella* sp.
saxifrage: *Saxifraga* sp.
sculpin: family Cottidae
Sealworm: *Pseudoterranova decipiens*
sedges: *Carex* sp. and *Eriophorum* sp.
Sheep Laurel: *Kalmia angustifolia*
Short-eared Owl: *Asio flammeus*
Short-finned Squid: *Illex illecebrosus*

Silver Hake: *Merluccius bilinearis*
skate: family Rajidae
Smelt: *Osmerus mordax*
Snowberry: *Symphoricarpos albus*
Southern Corn Rootworm: *Diabrotica undecimpunctata*
Southern Elephant Seal: *Mirounga leonina*
Spiny Dogfish: *Squalus acanthias*
Spotted Cucumber Beetle: *Diabrotica undecimpunctata*
Spotted Cusk Eel: *Otophidium taylori*
Spotted Hyena: *Crocuta crocuta*
Spotted Owl: *Strix occidentalis*
Sprat: *Clupea sprattus*
spruce: *Picea* sp.
Striped Maple: *Acer pensylvanicum*
tamarack: *Larix* sp.
thimbleberry: *Rubus* sp.
thistle: *Cirsium* sp.
threeawn: *Aristida* sp.
Trembling Aspen: *Populus tremuloides*
Turbot: *Reinhardtius hippoglossoides*
vetch: *Vicia* sp.
Vine Maple: *Acer circinatum*
Walleye Pollock: *Theragra chalcogramma*
Western Anemone: *Anemone occidentalis*
Western Bog-laurel: *Kalmia microphylla*
Western Hemlock: *Tsuga heterophylla*
Western Rattlesnake: *Crotalus viridis*
Western Redcedar: *Thuja plicata*
Western Screech Owl: *Otus kennicottii*
Western Swamp Laurel: *Kalmia microphylla*
Water Hyacinth: *Eichhornia* sp.
White Birch, also called Paper Birch: *Betula papyrifera*
White Oak: *Quercus alba*
Whitefish: *Coregonus clupeaformis*
Whiting: *Merlangius merlangus*
Wild Turkey: *Meleagris gallopavo*
willow: *Salix* sp.
Winter Tick: *Dermacentor albipictus*
Woolly Eriophyllum: *Eriophyllum lanatum*
Yellow Glacier-lily: *Erythronium grandiflorum*

Acknowledgments

Many people contributed to the creation of this volume. Micheline Beaulieu-Bouchard produced the lovely skull drawings. Paul Geraghty, Julius Csotonyi, and Brenda Carter created the magnificent colour paintings. See Appendix 3 for a complete listing of each artist's contribution. Alan McDonald produced the tooth illustrations used in Appendix 2. David Campbell did much of the reference acquisition and literature searches and provided images and specimens. Dr C.G. van Zyll de Jong oversaw the creation of about three-quarters of the colour paintings. A series of experts reviewed different portions of the text. They include Wendy McPeake and Drs M. Brock Fenton, Donald McAlpine, David Shackleton, Robert Wrigley, Francis Cook, C.R. Harington, Kevin Campbell, Andrew Trites, Judith Eger, Neil Woodman, George Feldhammer, and Andrew E. Derocher. I would like to thank a host of people who volunteered to read and comment on the text. These include Drs Erwin and Fenya Brodo, Dr Robert Holmes, Dr Steve Cumbaa, Michel Gosselin, Diana Boudreault, Sheila Edwards, and Jean Sabourin. The skilled team of Bruce Williams, Wendy McPeake, and Richard Martin shepherded the original idea into a solid product and Mark Graham carried it through to completion. Wendy further contributed to the initial project outline, helped develop the writing schedule, liaised between the museum and the publishers, negotiated and oversaw the publishing contract, and provided support and invaluable advice during the long creative process. Volunteers Jennifer Horne-McCaig, Jennifer Haughton, David Campbell, Dennis Bason, Lory Beaudoin, Patrika Novotny, and Sheila Edwards assisted in various capacities along the way. Sheila Edwards, Susan Goods, Kathleen Quinn, Anne Botman, and Russ Brooks supplied valuable advice. Susan Goods and Alex Tirabasso provided much-needed last-minute map-making support and Dominique Dufour carefully scanned the colour art. I would like to thank Don McAlpine and David Nagorsen for their gracious assistance over the years.

A group of donors supplied the funds necessary to complete the colour art. These include Roger and Gertrude Boudreault, Diana Boudreault, Nick and Carol Bélanger, Cecilia A. Bowden, Wendy Dion, Leo and Dorothy Flaherty, Celia Fothergill, Frank and Olive Healy, Sheila and Paul Jenkins, Cécile and Richard Julien, Elizabeth Rose MacSween, Edward Naughton, Carol Passfield, Douglas and Nancy Perkins, Gillian Perry and Manny Pittson, Susan Joan Perry, Gary Sealey, and Sylvia Torrance. Diana Boudreault and Cécile Julien developed, organized, and managed the fundraising campaign.

Museums that provided collection access include the Canadian Museum of Nature, the New Brunswick Museum, the Royal Saskatchewan Museum, and the United States National Museum (Smithsonian). I would also like to thank the CMN Library and particularly Mike Wayne and Ted Sypniewski for their long-standing and diligent support.

The *Canadian Field-Naturalist*, the premier Canadian natural history journal, published a great deal of the extremely valuable scientific data over the years upon which much of this volume is based. I am grateful to the Canadian Field-Naturalist Club for their support of the publication and to the journal's long-time editor Dr Francis Cook. Another major Canadian scientific journal, the *Canadian Journal of Zoology,* also deserves recognition for continuing to publish relevant and important Canadian natural history information.

Glossary

abomasum. The fourth and last portion of a ruminant stomach, where most of the nutrient absorption takes place.

aestivation. A period of torpor or dormancy in which metabolism is greatly reduced to save energy, but in response to heat or drought rather than cold.

agonistic. Pertaining to aggressive behaviour often related to fighting, or territorial or parental defence.

albinism. An uncommon condition caused by a genetic lack of the skin pigment called melanin. A full albino has white skin and fur and pink eyes. Partial albinos are called leucistic and may have variable spots or patches of white or can even be all white, but with pigmented eyes.

altricial. Refers to young born in a very underdeveloped state and requiring a prolonged period of development and parental care.

ambient. Surrounding.

anterior. Toward the head.

antitragus. Fleshy lobe at the exterior base of the ear opening, opposite the tragus.

arboreal. Inhabiting or frequenting trees.

arthropod. A member of the enormous and diverse group of invertebrates in the phylum Arthropoda, which includes insects, spiders, crustaceans, and related forms.

auditory bulla. Bone visible on the ventral side of the skull that contains the inner ear structures. Illustrated in Figure 1 (p. xv).

baculum. Penis bone.

baleen. Cornified epithelial plates that hang down from the upper jaws of baleen whales. Also called whalebone.

benthic. Relating to the sea or lake bottom.

bifurcate. Divided or forked into two branches.

bipedal. Moving on two legs (normally the hind legs).

breach. Leap from the water showing at least half of the body.

browser. Herbivorous mammals that feed on the leaves, twigs, shoots, and bark of woody vegetation.

caecum. A pouch at the junction of the small and large intestines, which is often the site of microbial fermentation in herbivores.

calcar. A boney or cartilaginous spur that projects from the ankles of many bats and from the wrists of flying squirrels to provide support for the flying or gliding membranes.

callosity. An area of thickened skin, often with a few hairs, found on the bodies of some whales. Most common in the head area, these callosities are usually inhabited by cynamid whale lice and barnacles.

canid. Member of the Canidae family, which includes dogs, wolves, Coyotes, jackals, and foxes.

canine. Tooth (sometimes called a fang) between the last incisor and the first premolar. Most flesh-eating mammals have a single, long, dagger-like canine on each side of both the upper and lower jaws. These four teeth pinch together to provide a grip on active prey. Also used to define a member of the Canidae family.

cartilaginous. Made up of cartilage, a flexible skeletal tissue. Skeletons of mammalian embryos are composed of cartilage, which gradually converts to bone as the animal ages.

carnivorous. Meat-eating; describes animals that consume mainly flesh.

castoreum. Brown, pasty, odoriferous secretion of the North American Beaver. Produced by the castor glands, located beside the anus.

caudal peduncle. Also called the tail stock; that portion of a whale's body between the dorsal fin and the flukes.

cetologist. Biologist who specializes in the study of whales.

cheek teeth. The teeth behind the canines – the premolars and molars.

cline. A graded series of characteristics within a species that vary typically with altitude, latitude, or sometimes longitude.

cohort. A group sharing a demographic characteristic; e.g., all the voles born in the first spring litter.

conspecifics. Individuals or populations of the same species.

conterminous. Having a common boundary; contiguous.

coprophagy. The ingestion of an individual's own faeces.

COSEWIC. Committee on the Status of Endangered Wildlife in Canada.

cosmopolitan. Found in all regions of the world.

crepuscular. Refers to animals active during the low light levels at dawn and dusk.

Cretaceous. Geologic period of time from roughly 145 to 65 million years ago.

deciduous. Not permanent. Meant to describe antlers, teeth, or leaves that fall off or are shed seasonally; deciduous teeth are also referred to as baby or milk teeth.

delayed fertilization. An adaptation of females of some species of hibernating bats whereby they breed in autumn but store the acquired sperm internally until fertilization occurs the following spring.

delayed implantation. An adaptation of many mustelids and several other northern mammals whereby implantation of an embryo onto the wall of the uterus, and its subsequent growth, is delayed for several months.

dentition. All of an animal's teeth; the arrangement and structure of an animal's teeth.

diastema. Refers to a gap between teeth, as between the incisors and cheek teeth of a rodent.

digitigrade. Walking on the tips of the toes or fingers.

disjunct. Refers to a population that is separated geographically from the main range of the species.

dispersal. The permanent emigration of individuals from a colony or population.

distribution. The geographical area occupied by members of the same species.

diurnal. Of or during the daytime. Refers to animals active during the daylight.

dog fox. Adult male fox.

dorsal. The back or upper surface.

drey. Leafy nest in the canopy of a tree, usually constructed on a twig framework, by tree squirrels.

echolocation. The generation of sounds and reception of returning echoes to locate obstacles and prey; sonar.

ectoparasite. External parasite such as a tick.

endemic. Restricted to a particular region or locality; e.g., Vancouver Island Marmots are endemic to Vancouver Island and Ungava Lemmings are endemic to northern Quebec.

endoparasite. Internal parasite such as a tapeworm.

Eocene. Geological epoch from about 56 to 34 million years ago; in the Paleocene Period.

esker. Elongated, often flat-topped, mounds of gravel and sand deposited by melting glaciers.

extinct. Refers to complete and irrevocable disappearance of all individuals of a taxa.

extirpated. Refers to disappearance of individuals of a taxa within a defined area but not from the global distribution; e.g., Canada Lynx is extirpated from Prince Edward Island but still occur on the mainland.

extralimital. Outside the usual range.

falcate. Recurved like a sickle.

fauna. A collective name for the animals of a particular area or region; e.g., the mammal fauna of the Arctic.

fenestrate. Having bone filled with a lattice of small openings.

feral. Refers to a former domestic species now living in the wild and no longer dependent on human husbandry.

flehmen. A drawing back of the lips, performed by (usually male) ungulates, felids, and canids after tasting or smelling something strange or exciting such as a mature female's urine. It is thought that the action enhances the ability of the nasal organs to analyse the odour.

fluke-up. Rise of the flukes out of the water prior to a dive.

foetus. An unborn or unhatched vertebrate once it has developed the rudimentary body plan of its species, prior to which it is called an embryo.

forb. Non-woody, broad-leafed flowering plants (not grasses).

fossorial. Adapted for underground living.

genera. Plural of genus.

genus. A taxonomic group of one or more closely related species.

gestation. Pregnancy.

gleaning. Form of hunting in which a bat plucks prey off surfaces rather than out of the air.

granivorous. Seed-eating; describes animals that consume primarily seeds.

gravid. Pregnant.

haemoglobin. Red blood cells that carry oxygen to the tissues throughout the body.

haul-out. Site on ice or land where marine mammals come out of the water to moult or rest.

hawking. Form of hunting in which a bat catches flying insects in flight.

herbivorous. Plant-eating; describes animals that consume mainly vegetation.

heterothermic. Describes small birds and mammals that are sometimes cold-blooded (poikilothermic) and sometimes hot-blooded (homeothermic). For example, the metabolism and body temperature of many bats are high only while the animal is active. However, both will drop to save energy if the bat chooses a cool roost site to rest.

hibernacula. Plural of hibernaculum.

hibernaculum. A winter shelter for a hibernating animal or group of animals.

hibernation. Long-term state of inactivity during which the metabolism and body temperature drop in order to save energy.

histoplasmosis. An infectious disease contracted by breathing the airborne fungal spores of *Histoplasmosis capsulatum*. It occurs in soil or in the faeces of infected animals and can become airborne if the soil is disturbed or the faeces are crushed.

home range. The area occupied by an animal during its normal day-to-day activities.

homeothermic. Describes animals whose body temperature remains at a relatively consistent temperature despite environmental fluctuations; warm-blooded.

hypogeous. Growing underground; commonly used to describe a group of fungi (commonly called truffles), whose fruiting bodies develop underground, many of which are favoured foods of squirrels, North American Porcupine, and humans.

hypothermia. State in which the body temperature is substantially below normal.

incisors. The front teeth in both upper and lower jaws. They precede the canines.

insectivorous. Insect-eating; describes animals that consume primarily insects and other invertebrates.

intraspecific. Within a species.

interspecific. Between species.

karst. An area of limestone bedrock characterized by sinks, caves, caverns, ravines, and underground streams.

labial. Of the lips; used to refer to the side of a tooth nearest the lips.

lactation. The process or action of producing milk and suckling an infant.

lagomorph. Member of the order Lagomorpha, which includes rabbits, hares, and pikas.

lanugo. Soft white foetal coat of seals.

larder-hoard. To hide many items of food in a single site.

leads. In the context of this volume refers to cracks or open areas in ice where seals or whales can reach the surface to take a breath.

lingual. Of the tongue; used to refer to the side of a tooth nearest the tongue.

lobtail. Thrashing of the flukes onto the water surface.

malocclusion. Imperfect or abnormal occlusion of teeth, which may result (in the case of mammals with ever-growing teeth) in excessive or unusual growth of teeth.

mammae. Mammary glands; breasts, udders, teats.

mammalogist. Scientist that studies mammals.

marsupium. A pouch composed of folds of abdominal skin that cover the mammae and protect the developing young.

mast. Tree and shrub nuts and seeds including beech nuts, acorns, coniferous cones, and maple keys.

matrilineal. Refers to a form of social organization in which the senior female (a grandmother, mother, or aunt) is the group leader and is generally closely related to the younger females of the group.

medial notch. Triangular gap at the middle of the trailing edge of a whale's flukes.

melanism. Abnormally dark colour produced by a hereditary defect that results in excessive amounts of melanin in the skin or fur of mammals.

molar. The back teeth in a toothrow (maximum of four). They are posterior to the premolars.

monoestrous. Having a single period of oestrus each year.

mtDNA. Mitochondrial DNA.

mycorrhizal fungi. Fungi whose thread roots (mycorrhiza) form close associations with plant roots, thereby improving the plant's ability to take up nutrients.

mystacial. Occurring beside and slightly below the nose, like a moustache.

natal. Related to an animal's birth.

Nearctic. A biogeographic area that includes North America, south to the Mexican plateau.

neonate. Newborn.

neophobic. Quick to notice and highly suspicious of new or relocated items in the environment.

nictitating membrane. Transparent membrane that can be slid over the eyeball to cover and protect the eye.

nocturnal. Of or during darkness; refers to animals active in darkness.

occlusal view. View from above, showing the biting surface.

olfaction. The act of smelling; the sense of smell.

olivaceous. Having a dusky green colour, with a tinge of yellow.

omasum. The third and smallest portion of a ruminant stomach. It receives food particles from the reticulum and presses and absorbs moisture from it before passing it into the abomasum.

omnivorous. Eats both plant and animal matter.

ontogeny. Development. Usually used to describe growth and development of juveniles from newborns to adults.

opportunistic. Taking advantage of circumstances.

pace. Both front and hind legs on the same side moving forward together. A walk or trot can be conduced in a pace.

Palaearctic. A biogeographic area that includes the cold and temperate regions of the Old World (Europe, North Africa, Russia, and Asia north of the Himalayas).

Palaeocene. A geologic epoch roughly spanning the period from 65 to 56 million years ago.

parturient. About to or in the process of giving birth.

parturition. Act of giving birth.

patagium. Flight membrane of bats or gliding membrane of flying squirrels or other mammalian or reptilian gliders.

phylogenetic. Relating to the evolutionary development and diversification of species or groups of organisms.

photoperiod. Duration of daylight.

piloerection. The erection or bristling of hairs or fur.

pinna. The fleshy outer or external ear.

pinnipeds. Fin-footed. Group name that includes all of the seals, fur seals, sea lions, and walrus.

piscivorous. Fish-eating; describes animals that consume mainly fish.

pitfall trap. A bucket dug into the earth so the lip is level with the ground surface. It may be filled with water to create a kill trap or may be stocked with food and bedding and covered with a plastic rain shield to act as a live trap. Especially useful for catching shrews.

placenta. The pad of vascular tissue formed by mammalian embryos that allows for the exchange of gases, nutrients, and waste products between the mother and the developing embryo. Marsupials such as the North American Opossum have only a rudimentary placenta.

placental. Relating to the placenta.

plantigrade. Refers to a mammal that walks on the soles of the feet, such as bears, raccoons, and humans.

Pleistocene. A geological epoch roughly encompassing the period from 1.8 million to 11,000 years ago.

poikilothermic. Having a body temperature that varies with that of the surroundings; cold-blooded.

polyandrous. Having more than one male mate per female.

polyoestrous. Having more than one heat (oestrous cycle) per year.

polygynous. Having more than one female mate per male.

polynya. An ice-free patch of open water that occurs in the same area every winter, year after year.

posterior. Toward the rear.

precocial. Refers to young born in a well-developed state, fully furred, with eyes open, and able to move about well soon after birth.

premolars. Teeth (up to four in each toothrow) that are between the canines and the molars.

preorbital gland. Gland whose outlet occurs anterior to the eye.

predation. The act of one animal capturing and consuming another.

prehensile. Adapted for grasping, as in the prehensile tail of the North American Opossum.

range. The geographic region where a species occurs.

recruitment. Increase in a population as progeny grow and become independent members.

refugia. Plural of refugium.

refugium. An area where environmental circumstances allow a species or community to survive despite extinction in surrounding areas.

resorption. The dissolution or disappearance of tissue by biochemical activity.

riparian. Pertaining to the lowlands alongside a river, stream, or lake.

reticulum. The second portion of the ruminant stomach. It filters out and returns large particles to the rumen for further processing.

rorqual. Group of baleen whales of the family Balaenopteridae, which include Blue, Fin, Sei, Minke, and Humpback Whales.

rostrum. The snout or nasal portion of a skull.

rumen. The first and largest portion of the four-chambered ruminant stomach. It stores food awaiting further chewing and supports millions of fermentation and cellulose-digesting bacteria that help break down food particles.

saltatorial. Adapted for leaping.

savannah. Grassland with scattered trees.

scat. Dung, droppings, faeces.

scatter-hoard. To hide small caches of food in many locations.

sexual dimorphism. Condition in which males and females of the same species are morphologically different.

siliceous. Containing or consisting of silica.

snap trap. A spring-loaded guillotine-type trap commonly used by householders to kill rodents.

spatulate. Having a broad, flattened end like a spatula or shovel.

speciation. An evolutionary process in which, over time, two are more species are produced in which only one formerly existed.

species. A group of animals, plants, minerals, fungi, or bacteria having certain common and permanent physical characteristics that clearly distinguish it from other groups.

spyhop. Slow raising of the head above the surface until the eye is above the water level.

subnivean. Living within or beneath the snowpack.

subspecies. Populations or groups within a species that are deemed to be distinct at some level but are still capable of interbreeding with individuals of other subspecies of the same species. This designation is recognized with a three-part scientific name, e.g., *Alces americanus gigas* or *Alces americanus andersoni*.

subterranean. Living under the ground surface.

systematics. The science of classification and organization of organisms hierarchically.

taiga. The swampy coniferous forest region that stretches across northern Canada, Alaska, and Siberia. Also called *boreal forest*.

tail stock. See *caudal peduncle*.

talus. A slope formed by the gradual accumulation of rocky debris at the base of a cliff or mountain.

tapetum lucidum. A layer in back of the eye that reflects light, causing the eye to glow when struck by a light in the dark. Assists in night vision. Found chiefly in nocturnal animals or those active in low light.

taxonomy. The process of defining the bounds of large or small groups of organisms and the investiture of names to these groups.

terrestrial. Living mainly on the ground.

Tertiary. A geological term that pertains to the third series of stratified formations. Includes the strata from the Eocene to the Pliocene epochs, around 56–1.8 million years ago.

torpor. A short-term (may be daily) state of inactivity during which metabolism and body temperature drop in order to save energy.

tragus. The flap of skin (sometimes prominent, other times almost obscure) that extends vertically from the base of a bat's ear. Thought to be involved with echolocation.

unicuspid. The simple conical teeth between the distinct incisor and the first recognizable premolar in the upper and lower toothrows of a shrew. Their origins as canines and premolars are uncertain.

ungulate. A group that includes all the mammals that are hoofed. Includes the members of the orders Artiodactyla and Perissodactyla.

uropatagium. Flight membrane of a bat; extends from the tail to the hind legs.

ventral. Of, on, or relating to the underside.

vertebrate. Animals having a backbone, including fishes, birds, reptiles, amphibians, and mammals.

vestigial. Organ or structure surviving in a degenerate, atrophied or imperfect form.

vibrissae. Whiskers. Long stiff hairs mainly on the face and head that provide tactile information.

volant. Capable of powered flight.

Wisconsinan. Last of the Ice Ages in the Pleistocene epoch; occurred roughly from 110,000 to 11,000 years ago.

xeric. Containing little moisture.

zooplankton. A collective term for small drifting or floating aquatic animals.

zygomatic arch. The bony arch that protects the eyeball on a mammal's skull, composed of parts of the squamosal and jugal bones. Illustrated in Figure 1 (page xv).

Bibliography

Aars, J., Marques, T.A., Buckland, S.T., Andersen, M., Belikov, S., Boltunov, A., and Wiig, Ø. 2009. Estimating the Barents Sea Polar Bear subpopulation size. *Marine Mammal Science* 25(1): 35–52.

Abraham, K.F. and Lim, B.K. 1990. First Minke Whale, *Balaenoptera acutorostrata*, record for James Bay. *Canadian Field-Naturalist* 104(2): 304–305.

Adamczewski, J.Z., Flood, P.F., and Gunn, A. 1997. Seasonal patterns in body composition and reproduction of female muskoxen (*Ovibos moschatus*). *Journal of Zoology* (London) 241: 245–269.

Agosta, S.J., and Morton, D. 2003. Diet of the Big Brown Bat, *Eptesicus fuscus*, from Pennsylvania and western Maryland. *Northeastern Naturalist* 10(1): 89–104.

Aguilar, A. 1991. Calving and early mortality in the western Mediterranean Striped Dolphin, *Stenella coeruleoalba*. *Canadian Journal of Zoology* 69(5): 1408–1412.

– 2002. Fin Whale *Balaenoptera physalus*. In W.F. Perrin, B. Wursig, and J.G.M. Thewissen (eds), *Encyclopedia of Marine Mammals*, 435–438. Academic Press, San Diego, CA.

Ahlborn, B.K., Blake, R.W., and Chan, K.H.S. 2009. Optimal fineness ratio for minimum drag in large whales. *Canadian Journal of Zoology* 87(2): 124–131.

Alexander, L.F., and Riddle, B.R. 2005. Phylogenetics of the New World rodent family Heteromyidae. *Journal of Mammalogy* 86(2): 366–379.

Alexy, K.J., Gassett, J.W., Osborn, D.A., and Miller, K.V. 2001. Remote monitoring of scraping behaviors of a wild population of White-tailed Deer. *Wildlife Society Bulletin* 29(3): 873–878.

Alibhai, S.K. 1985. Effects of diet on reproductive performance of the Bank Vole (*Clethrionomys glareolus*). *Journal of Zoology* (London) 205: 445–452.

Allard, M.W., and Greenbaum, I.F. 1988. Morphological variation and taxonomy of chromosomally differentiated *Peromyscus* from the Pacific Northwest. *Canadian Journal of Zoology* 66(12): 2734–2739.

Allan, N. 2007. Grizzly/Polar Bear hybrid. *Arctic Bulletin* 2(6): 6–7.

Amano, M., Yoshioka, M., Kuramochi, T., and Mori, K. 1998. Diurnal feeding by Dall's Porpoise *Phocoenoides dalli*. *Marine Mammal Science* 14(1): 130–135.

Amano, M., and Miyazaki, N. 2004. Composition of a school of Risso's Dolphins, *Grampus griseus*. *Marine Mammal Science* 20(1): 152–160.

Amos, B., Twiss, S., Pomeroy, P., and Anderson, S. 1995. Evidence for mate fidelity in the Gray Seal. *Science* 268(30 June): 1897–1899.

Amos, W., Barrett, J.A., and Dover, G.A. 1991. Breeding behaviour of pilot whales revealed by DNA fingerprinting. *Heredity* 67: 49–55.

Amstrup, S.C. 2003. Polar Bear (*Ursus maritimus*). In G.A. Feldhamer, B.C. Thompson, and J.A. Chapman (eds), *Wild Mammals of North America: Biology, Management, and Conservation*, 2nd ed., 587–610. Johns Hopkins University Press, Baltimore, MD.

Amstrup, S.C., Durner, G.M., Stirling, I., Lunn, N.J., and Messier, F. 2000. Movements and distribution of Polar Bears in the Beaufort Sea. *Canadian Journal of Zoology* 78(6): 948–966.

Andersen, L.W., and Siegismund, H.R. 1994. Genetic evidence for migration of males between schools of the Long-finned Pilot Whale *Globicephala melas*. *Marine Ecology Progress Series* 105: 1–7.

Andersen, M., Hjelset, A.M., Gjertz, I., Lydersen, C., and Gulliksen, B. 1999. Growth, age at sexual maturity and condition in Bearded Seals (*Erignathus barbatus*) from Svalbard, Norway. *Polar Biology* 21: 179–185.

Anderson, A.E., and Wallmo, O.C. 1984. *Odocoileus hemionus*. *Mammalian Species* no. 219: 1–9.

Anderson, C.G. 1999. Arctic Fox *Alopex lagopus*. In D.E. Wilson and S. Ruff (eds), *The Smithsonian Book of North American Mammals*, 146–148. UBC Press, Vancouver, BC.

Anderson, C.S., and Meikle, D.B. 2006. Annual changes in structural complexity of understory vegetation and relative abundance of *Peromyscus leucopus* in fragmented habitats. *Acta Theriologica* 51(1): 43–51.

Anderson, E.M., and Lovallo, M.J. 2003. Bobcat and Lynx (*Lynx rufus and Lynx canadensis*). In G.A. Feldhamer, B.C. Thompson, and J.A. Chapman (eds), *Wild Mammals of North America: Biology, Management, and Conservation*, 2nd ed., 758–786. Johns Hopkins University Press, Baltimore, MD.

Anderson, P.K., Whitney, P.H., and Huang, J.P. 1976. *Arvicola richardsoni*: Ecology and biochemical polymorphism in the Front Ranges of southern Alberta. *Acta Theriologica* 21(24–31): 425–468.

Anderson, R.M. 1939. Mammifères de la province de Québec. *Société Provancher 1939 Rapport Annuel*, 1939: 37–111.

Andreassen, H.P., Ims, R.A., Stenseth, N.C., and Yoccoz, N.G. 1993. Investigating space use by means of radiotelemetry and other methods: A methodological guide. In N.C. Stenseth and R.A. Ims (eds), *The Biology of Lemmings*, 588–618. Academic Press, San Diego, CA.

Andrews, R.C. 1916. The Sei Whale (*Balaenoptera borealis* Lesson). *Memoirs of the American Museum of Natural History*, new ser. 1(6): 291–388.

Aniśkowicz, B.T., Hamilton, H., Gray, D.R., and Downes, C. 1990. Nursing behaviour of Arctic Hares (*Lepus arcticus*), 643–664. In C.R. Harington (ed.), *Canada's Missing Dimension: Science and History in the Canadian Arctic Islands*, vol. 2, 588–618. Canadian Museum of Nature, Ottawa, ON.

Anthony, R.M. 1997. Home ranges and movements of Arctic Fox (*Alopex lagopus*) in western Alaska. *Arctic* 50(2): 147–157.

Anthony, R.M., Barten, N.L., and Seiser, P.E. 2000. Foods of Arctic Foxes (*Alopex lagopus*) during winter and spring in western Alaska. *Journal of Mammalogy* 81(3): 820–828.

Antipas, A.J., Madison, D.M., and Ferraro, J.S. 1990. Circadian rhythms in the Short-tailed Shrew, *Blarina brevicauda*. *Physiology and Behavior* 48: 255–260.

Antonelis, G.A., Melin, S.R., and Bukhtiyarov, Y.A. 1994. Early spring feeding habits of Bearded Seals (*Erignathus barbatus*) in the central Bering Sea, 1981. *Arctic* 47(1): 74–79.

Antonelis, G.A., Sinclair, E.H., Ream, R.R., and Robson, B.W. 1997. Inter-island variation in the diet of female Northern Fur Seals (*Callorhinus ursinus*) in the Bering Sea. *Journal of Zoology* (London) 242: 435–451.

Antonelis, G.A., Stewart, B.S., and Perryman, W.F. 1990. Foraging characteristics of female Northern Fur Seals (*Callorhinus ursinus*) and California Sea Lions (*Zalophus californianus*). *Canadian Journal of Zoology* 68(1): 150–158.

Apps, C.D., Newhouse, N.J., and Kinley, T.A. 2002. Habitat associations of American Badgers in southeastern British Columbia. *Canadian Journal of Zoology* 80(7): 1228–1239.

Arbogast, B.S., Browne, R.A., and Weigl, P.D. 2001. Evolutionary genetics and Pleistocene biogeography of North American tree squirrels (*Tamiasciurus*). *Journal of Mammalogy* 82(2): 302–319.

Archer, F.I., II, and Perrin, W.F. 1999. *Stenella coeruleoalba*. *Mammalian Species* no. 603: 1–9.

Arjo, W.M., Huenefeld, R.E., and Nolte, D.L. 2007. Mountain Beaver home ranges, habitat use, and population dynamics in Washington. *Canadian Journal of Zoology* 85(2): 328–337.

Armitage, K.B. 1991. Social and population dynamics of Yellow-bellied Marmots: Results from long-term research. *Annual Review of Ecology and Systematics* 22: 379–407.

– 2003. Marmots (*Marmota monax* and allies). In G.A. Feldhamer, B.C. Thompson, and J.A. Chapman (eds), *Wild Mammals of North America. Biology, Management, and Conservation*, 2nd ed., 188–210. Johns Hopkins University Press, Baltimore, MD.

Armstrong, D.M., Adams, R.A., and Taylor, K.E. 2006. New Records of the Eastern Pipistrelle (*Pipistrellus subflavus*) in Colorado. *Western North American Naturalist* 66(2): 268–269.

Armstrong, D.M., and Jones, J.K., Jr. 1971. *Sorex merriami*. *Mammalian Species* no. 2:1–2.

Arnett, E.B., and Hayes, J.P. 2009. Use of conifer snags as roosts by female bats in western Oregon. *Journal of Wildlife Management* 73(2): 214–224.

Arnold, B.D. 2007. Population structure and sex-biased dispersal in the forest dwelling vespertilionid bat, *Myotis septentrionalis*. *American Midland Naturalist* 157(2): 374–384.

Arnold, T.W., and Fritzell, E.K. 1990. Habitat use by male mink in relation to wetland characteristics and avian prey abundances. *Canadian Journal of Zoology* 68(10): 2205–2208.

Arthur, S.M., Paragi, T.F., and Krohn, W.B. 1993. Dispersal of juvenile Fishers in Maine. *Journal of Wildlife Management* 57(4): 868–874.

Asselin, S. Hammill, M.O., and Barrette, C. 1993. Underwater vocalizations of ice breeding Grey Seals. *Canadian Journal of Zoology* 71(11): 2211–2219.

Atkinson, K.T., and Shackleton, D.M. 1991. Coyote, *Canis latrans*, ecology in a rural-urban environment. *Canadian Field-Naturalist* 105(1): 49–54.

Atkinson, S.N., Cattet, M.R.L., Polischuk, S.C., and Ramsay, M.A. 1996. A case of offspring adoption in free-ranging Polar Bears (*Ursus maritimus*). *Arctic* 49(1): 94–96.

Atkinson, S.N., and Ramsay, M.A. 1995. The effects of prolonged fasting on the body composition and reproductive success of female Polar Bears (*Ursus maritimus*). *Functional Ecology* 9: 559–567.

Aubry, K.B., and Houston, D.B. 1992. Distribution and status of the Fisher (*Martes pennanti*) in Washington. *Northwestern Naturalist* 73: 69–79.

Aubry, K.B., and Lewis, J.C. 2003. Extirpation and reintroduction of Fishers (*Martes pennanti*) in Oregon: Implications for their conservation in the Pacific states. *Biological Conservation* 114(1): 79–90.

Audet, A.M., Robbins, C.B., and Larivière, S. 2002. *Alopex lagopus*. *Mammalian Species* no. 713:1–10.

Auger, J., Meyer, S.E., and Black, H.L. 2002. Are American Black Bears (*Ursus americanus*) legitimate seed dispersers for fleshy-fruited shrubs? *American Midland Naturalist* 147(2): 352–367.

Aurioles, D., Koch, P.L., and Le Boeuf, B.J. 2006. Differences in foraging location of Mexican and California elephant seals: Evidence from stable Isotopes in pups. *Marine Mammal Science* 22(2): 326–338.

Aurioles-Gamboa, D. 1992. Notes on a mass stranding of Baird's Beaked Whales in the Gulf of California, Mexico. *California Fish and Game* 78(3): 116–123.

Avenant, N.L., and Smith, V.R. 2004. Seasonal changes in age class structure and reproductive status of House Mice on Marion Island (sub-Antarctic). *Polar Biology* 27: 99–111.

Baack, J.K., and Switzer, P.V. 2000. Alarm calls affect foraging behavior in Eastern Chipmunks (*Tamias striatus*, Rodentia: Sciuridae). *Ethology* 106: 1057–1066.

Backlund, D.C. 1995. New records for the Dwarf Shrew, Pygmy Shrew, and Least Shrew in South Dakota. *Prairie Naturalist* 27(1): 63–64.

Bader, M. 2000. Distribution of Grizzly Bears in the U.S. northern Rockies. *Northwest Science* 74(4): 325–334.

Bailey, E.P. 1992. Red Foxes, *Vulpes vulpes*, as biological control agents for introduced Arctic Foxes, *Alopex lagopus*, on Alaskan islands. *Canadian Field-Naturalist* 106(2): 200–205.

Bain, M.R., and Shenk, T.M. 2002. Nests of Preble's Meadow Jumping Mouse (*Zapus hudsonius preblei*) in Douglas County, Colorado. *Southwestern Naturalist* 47(4): 630–633.

Baird, R.W. 1990. Elephant Seals around southern Vancouver Island. *Victoria Naturalist* 47(2): 6–7.

– 2001a. Status of Harbour Seals, *Phoca vitulina*, in Canada. *Canadian Field-Naturalist* 115(4): 663–675.

– 2001b. Status of Killer Whales, *Orcinus orca*, in Canada. *Canadian Field-Naturalist* 115(4): 676–701.

Baird, R.W., Gorgone, A.M., McSweeney, D.J., Webster, D.L., Salden, D.R., Deakos, M.H., Ligon, A.D., Schorr, G.S., Barlow, J., and Mahaffy, S.D. 2008a. False Killer Whales (*Pseudorca crassidens*) around the main Hawaiian Islands: Long-term site fidelity, inter-island movements, and association patterns. *Marine Mammal Science* 24(3): 591–612.

Baird, R.W., and Hanson, M.B. 1997. Status of the Northern Fur Seal, *Callorhinus ursinus*, in Canada. *Canadian Field-Naturalist* 111(2): 263–269.

Baird, R.W., Hanson, M.B., and Dill, L.M. 2005. Factors influencing the diving behaviour of fish-eating Killer Whales: Sex differences and diel and interannual variation in diving rates. *Canadian Journal of Zoology* 83(8): 257–267.

Baird, R.W., and Stacey, P.J. 1991a. Status of Risso's Dolphin, *Grampus griseus*, in Canada. *Canadian Field-Naturalist* 105(2): 233–242.

– 1991b. Status of the Northern Right Whale Dolphin, *Lissodelphis borealis*, in Canada. *Canadian Field-Naturalist* 105(2): 243–250.

– 1993. Sightings, strandings and incidental catches of Short-finned Pilot Whales, *Globicephala macrorhynchus*, off the British Columbia coast. *Report of the International Whaling Commission*, Special issue, 14: 475–479.

Baird, R.W., Stacey, P.J. and Whitehead, H. 1993. Status of the Striped Dolphin, *Stenella coeruleoalba*, in Canada. *Canadian Field-Naturalist* 107(4): 455–465.

Baird, R.W., Webster, D.L., McSweeney, D.J., Ligon, A.D., Schorr, G.S., and Barlow, J. 2006. Diving behaviour of Cuvier's (*Ziphius cavirostris*) and Blainville's (*Mesoplodon densirostris*) beaked whales in Hawai'i. *Canadian Journal of Zoology* 84(8): 1120–1128.

Baird, R.W., Webster, D.L., Schorr, G.S., McSweeney, D.J., and Barlow, J. 2008b. Diel variation in beaked whales diving behaviour. *Marine Mammal Science* 24(3): 630–642.

Baird, R.W., and Whitehead, H. 2000. Social organization of mammal-eating Killer Whales: Group stability and dispersal patterns. *Canadian Journal of Zoology* 78(12): 2096–2105.

Baird, R.W., Willis, P.M., Guenther, T.J., Wilson, P.J., and White, B.N. 1998. An inter generic hybrid in the family Phocoenidae. *Canadian Journal of Zoology* 76(1): 198–204.

Baker, C.S., Dalebout, M.L., and Lento, G.M. 2002. Gray Whale products sold in commercial markets along the Pacific coast of Japan. *Marine Mammal Science* 18(1): 295–300.

Baker, J.D., and Donohue, M.J. 2000. Ontogeny of swimming and diving in Northern Fur Seal (*Callorhinus ursinus*) pups. *Canadian Journal of Zoology* 78(1): 100–109.

Baker, J.D., Fowler, C.W., and Antonelis, G.A. 1994. Mass changes in fasting immature male Northern Fur Seals. *Canadian Journal of Zoology* 72(2): 326–329.

Baker, R.J., Patton, J.C., Genoways, H.H., and Bickham, J.W. 1988. Genic studies of *Lasiurus* (Chiroptera: Vespertilionidae). *Occasional Papers of the Museum, Texas Tech University* no. 117:1–15.

Balcomb, K.C. 1989. Baird's Beaked Whale *Berardius bairdii* Stejneger, 1883: Arnoux's Beaked Whale *Berardius arnuxii* Duvernoy, 1851. In S.H. Ridgway and R. Harrison (eds), *Handbook of Marine Mammals*, vol. 4, 261–288, Academic Press, San Diego, CA.

Ballachey, B.E., Bodkin, J.L., Howlin, S., Doroff, A.M., and Rebar, A.H. 2003. Correlates to survival of juvenile Sea Otters in Prince William Sound, Alaska, 1992–1993. *Canadian Journal of Zoology* 81(9): 1494–1510.

Ballard, K.A., and Kovacs, K.M. 1995. The acoustic repertoire of Hooded Seals (*Cystophora cristata*). *Canadian Journal of Zoology* 73(7): 1362–1374.

Ballard, W.B., Cronin, M.A., Rodrigues, R., Skoog, R.O., and Pollard, R.H. 2000. Arctic Fox, *Alopex lagopus*, den densities in the Prudhoe Bay oil field, Alaska. *Canadian Field-Naturalist* 114(3): 453–456.

Ban, S., and Trites, A.W. 2007. Quantification of terrestrial haul-out and rookery characteristics of Steller Sea Lions. *Marine Mammal Science* 23(3): 496–507.

Banci, V. 1994. Wolverine. *US Forest Service General Technical Report RM 1994* 254: 99–127.

Banci, V., and Harestad, A.S. 1990. Home range and habitat use of Wolverines *Gulo gulo* in Yukon, Canada. *Holarctic Ecology* 13: 195–200.

Banfield, A.W.F. 1961. A revision of the Reindeer and Caribou genus *Rangifer*. *Bulletin No. 177, Biological Series 66*. National Museum of Canada, Ottawa.

– 1974. *The Mammals of Canada*. National Museums of Canada, University of Toronto Press, Toronto, ON.

Banks, E.M., Brooks, R.J., and Schnell, J. 1975. A radiotracking study of home range and activity of the Brown Lemming (*Lemmus trimucronatus*). *Journal of Mammalogy* 56(4): 888–901.

Banks, E.M., Mankovich, N.J., and Huck, U.W. 1979. Female interspecific aggression in two species of lemmings. *Behavioral and Neural Biology* 26: 372–378.

Bantle, J.L., and Alisauskas, R.T. 1998. Spatial and temporal patterns in Arctic Fox diets at a large goose colony. *Arctic* 51(3): 231–236.

Barash, D.P. 1989. *Marmots: Social Behavior and Ecology*. Stanford University Press, Stanford, CA.

Barclay, R.M.R. 1984. Observations on the migration, ecology and behavior of bats at Delta Marsh, Manitoba. *Canadian Field-Naturalist* 98(3): 331–336.

– 1987. Foraging strategies of lactating and juvenile Hoary Bats (*Lasiurus cinereus*). *Bat Research News* 28(3–4): 31.

– 1989. The effect of reproductive condition on the foraging behavior of female Hoary Bats, *Lasiurus cinereus*. *Behavioral Ecology and Sociobiology* 24: 31–37.

– 1991. Population structure of temperate zone insectivorous bats in relation to foraging behaviour and energy demand. *Journal of Animal Ecology* 60: 165–178.

Barclay, R.M.R., and Brigham, R.M. 2001. Year-to-year reuse of tree-roosts by California Bats (*Myotis californicus*) in southern British Columbia. *American Midland Naturalist* 146(1): 80–85.

Barclay, R.M.R., and Thomas, D.W. 1979. Copulation call of *Myotis lucifugus*: A discrete situation-specific communication signal. *Journal of Mammalogy* 60(3): 632–634.

Barding, E.E., and Nelson, T.A. 2008. Raccoons use habitat edges in northern Illinois. *American Midland Naturalist* 159(2): 394–402.

Barlow, J. 2003. Preliminary estimates of the abundance of cetaceans along the U.S. west coast: 1991–2001. *Southwest Fisheries Science Center Administrative Report* LJ-03-03: 1–31.

Barlow, J., and Taylor, B.L. 2005. Estimates of Sperm Whales abundance in the northeastern temperate Pacific from a combined acoustic and visual survey. *Marine Mammal Science* 21(3): 429–445.

Barnes, D.M., and Mallik, A.U. 1996. Use of woody plants in construction of Beaver dams in northern Ontario. *Canadian Journal of Zoology* 74(9): 1781–1786.

Barnes, V.G., Jr, and Smith, R.B. 1993. Cub adoption by Brown Bears, *Ursus arctos middendorffi*, on Kodiak Island, Alaska. *Canadian Field-Naturalist* 107(3): 365–367.

Barnum, S.A., Manville, C.J., Tester, J.R., and Carmen, W.J. 1992. Path selection by *Peromyscus leucopus* in the presence and absence of vegetative cover. *Journal of Mammalogy* 73(4): 797–801.

Baron, D. 1979. Evidence of kangaroo rats near Burstall, Saskatchewan. *Blue Jay* 37(4): 240.

Barr, W. 1991. Back from the Brink: The road to Muskox conservation in the Northwest Territories. *Komatik Series* 3: 1–127.

Bartels, M.A., and Thompson, D.P. 1993. *Spermophilus lateralis*. *Mammalian Species* no. 440:1–8.

Baskin, Y. 1991. Blue behemoth bounds back. *Bioscience* 43(9): 603–605.

Batzli, G.O. 1999. Brown Lemming *Lemmus sibericus*. In D.E. Wilson and S. Ruff (eds), *The Smithsonian Book of North American Mammals*, 653–654. UBC Press, Vancouver, BC.

Batzli, G.O., and Henttonen, H. 1990. Demography and resource use by microtine rodents near Toolik Lake, Alaska, U.S.A. *Arctic and Alpine Research* 22(1): 51–64.

– 1993. Home range and social organization of the Singing Vole (*Microtus miurus*). *Journal of Mammalogy* 74(4): 868–878.

Batzli, G.O., and Lesieutre, C. 1991. The influence of high quality food on habitat use by Arctic microtine rodents. *Oikos* 60: 299–306.

Bauman, P.J., Jenks, J.A., and Roddy, D.E. 2000. Bull American Elk, *Cervus elaphus*, mortality resulting from locked antlers during spring sparring. *Canadian Field-Naturalist* 114(1): 144–147.

Baumgartner, L.L. 1939. Fox squirrel dens. *Journal of Mammalogy* 20: 456–465.

Bayne, E.M., Boutin, S., and Moses, R.A. 2008. Ecological factors influencing the spatial pattern of Canada Lynx relative to its southern range edge in Alberta, Canada. *Canadian Journal of Zoology* 86(10): 1189–1197.

Bearzi, G., Reeves, R.R., Notarbartolo de Sciara, G., Politi, E., Canadas, A., Frantzis, A., and Mussi, B. 2003. Ecology, status and conservation of Short-beaked Common Dolphins *Delphinus delphis* in the Mediterranean Sea. *Mammal Review* 33(3–4): 224–252.

Beauvais, G.P. 2001. Preble's Meadow Jumping Mouse (*Zapus hudsonius preblei*) in Wyoming. *Status Report, July 2001*. Prepared for the Wyoming Stockgrowers Association, Laramie, WY.

Beck, A.J., and Rudd, R.L. 1960. Nursery colonies in the Pallid Bat. *Journal of Mammalogy* 41(2): 266–267.

Becker, C.D., Boutin, S., and Larsen, K.W. 1998. Constraints on first reproduction in North American Red Squirrels. *Oikos* 81: 81–92.

Beever, E.A., Brussaed, P.F., and Berger, J. 2003. Patterns of apparent extirpation among isolated populations of pikas (*Ochotona princeps*) in the Great Basin. *Journal of Mammalogy* 84(1): 37–54.

Beier, P. 1995. Dispersal of juvenile Cougars in fragmented habitat. *Journal of Wildlife Management* 59(2): 228–237.

– 1999. Cougar *Puma concolor*. In D.E. Wilson and S. Ruff (eds), *The Smithsonian Book of North American Mammals*, 226–228. UBC Press, Vancouver, BC.

Bekoff, M. 1977. *Canis latrans*. *Mammalian Species* no. 79: 1–9.

Bekoff, M., and Gese, E.M. 2003. Coyote (*Canis latrans*). In G.A. Feldhamer, B.C. Thompson, and J.A. Chapman (eds), *Wild Mammals of North America: Biology, Management, and Conservation*, 2nd ed., 467–481. Johns Hopkins University Press, Baltimore, MD.

Béland, P., Faucher, A., and Corbeil, P. 1990. Observations on the birth of a Beluga Whale (*Delphinapterus leucas*) in the St. Lawrence Estuary, Quebec, Canada. *Canadian Journal of Zoology* 68(6): 1327–1329.

Belant, J.L. 1992. Homing of relocated Raccoons, *Procyon lotor*. *Canadian Field-Naturalist* 106(3): 382–384.

Belikov, S.E., and Boltunov, A.N. 2002. Distribution and migrations of cetaceans in the Russian Arctic according to observations from aerial ice reconnaissance. *NAMMCO Scientific Publications* 4: 69–86.

Bell, G.P. 1982. Behavioral and ecological aspects of gleaning by a desert insectivorous bat, *Antrozous pallidus* (Chiroptera: Vespertilionidae). *Behavioral Ecology and Sociobiology* 10: 217–223.

Bellocq, M.I., Bendell, J.F., and Innes, D.G.L. 1994. Diet of *Sorex cinereus*, the Masked Shrew, in relation to the abundance of Lepidoptera larvae in northern Ontario. *American Midland Naturalist* 132(1): 68–73.

Bellocq, M.I., and Smith, S.M. 2003. Population dynamics and foraging of *Sorex cinereus* (Masked Shrew) in the boreal forest of eastern Canada. *Annals Zool. Fennici* 40: 27–34.

Bellows, A.S., Pagels, J.F., and Mitchell, J.C. 1999. First record of the Least Weasel, *Mustela nivalis* (Carnivora: Mustelidae), from the coastal plain of Virginia. *Northeastern Naturalist* 6(3): 238–240.

Bender, D.J., Bayne, E.M., and Brigham, R.M. 1996. Lunar condition influences Coyote (*Canis latrans*) howling. *American Midland Naturalist* 136(2): 413–417.

Benedict, J.B., and Benedict, A.D. 2001. Subnivean root caching by a Montane Vole (*Microtus montanus nanus*), Colorado Front Range, USA. *Western North American Naturalist* 61(1): 241–244.

Benedict, R.A. 1999. Characteristics of a hybrid zone between two species of short-tail shrews (*Blarina*). *Journal of Mammalogy* 80(1): 135–141.

Benedict, R.A., Druecker, J.D., and Genoways, H.H. 1999. New records and habitat information for *Sorex merriami* in Nebraska. *Great Basin Naturalist* 59: 285–287.

Benedix, J.H., Jr. 1994. A predictable pattern of daily activity by the pocket gopher *Geomys bursarius*. *Animal Behaviour* 48: 501–509.

Beneski, J.T., and Stinson, D.W. 1987. *Sorex palustris*. *Mammalian Species* no. 296: 1–6.

Bennett, B. 1991. Big-eared burrowing beavers. *Victoria Naturalist* 47(5): 6–7.

Bennett, D., and Hoffmann, R.S. 1999. *Equus caballus*. *Mammalian Species* no. 628: 1–14.

Bennett, R. 1999. Status of the Red-tailed Chipmunk (*Tamias ruficaudus*) in Alberta. *Alberta Wildlife Status Report* no. 19: 1–12.

Bennett, R.P. 1999. Effects of food quality on growth and survival of juvenile Columbian Ground Squirrels (*Spermophilus columbianus*). *Canadian Journal of Zoology* 77(10): 1555–1561.

Benson, J.F., and Chamberlain, M.J. 2006. Cub adoption by a translocated Louisiana Black Bear. *Ursus* 17(2): 178–181.

Benson, J.F., Chamberlain, M.J., and Leopold, B.D. 2004. Land tenure and occupation of vacant home ranges by Bobcats (*Lynx rufus*). *Journal of Mammalogy* 85(5): 983–988.

Berger, J. 1986. *Wild Horses of the Great Basin: Social Competition and Population Size*. University of Chicago Press, Chicago, IL.

Berger, P.J., Negus, N., and Day, M. 1997. Recognition of kin and avoidance of inbreeding in the Montane Vole, *Microtus montanus*. *Journal of Mammalogy* 78(4): 1182–1186.

Bergerud, A.T. 2000. Caribou. In S. Demarais and P.R. Krausman (eds), *Ecology and Management of Large Mammals in North America*, 658–693. Prentice-Hall, Upper Saddle River, NJ.

Bergman, C.M., and Krebs, C.J. 1993. Diet overlap of collared lemmings and Tundra Voles at Pearce Point, Northwest Territories. *Canadian Journal of Zoology* 71(9): 1703–1709.

Berry, R.J., Bonner, W.N., and Peters, J. 1979. Natural selection in House Mice (*Mus musculus*) from South Georgia (South Atlantic Ocean). *Journal of Zoology* (London) 189: 385–398.

Berteaux, D., Klvana, I., and Trudeau, C. 2005. Spring-to-fall mass gain in a northern population of North American Porcupines. *Journal of Mammalogy* 86(3): 514–519.

Bertram, D.F., and Nagorsen, D.W. 1995. Introduced Rats, *Rattus* spp., on the Queen Charlotte Islands: Implications for seabird conservation. *Canadian Field-Naturalist* 109(1): 6–10.

Bertrand, A.-S., Kenn, S., Gallant, D., Tremblay, E., Vasseur, L., and Wissink, R. 2006. MtDNA analyses on hair samples confirm Cougar, *Puma concolor*, presence in southern New Brunswick, eastern Canada. *Canadian Field-Naturalist* 120(4): 438–442.

Bérubé, M., and Aguilar, A. 1998. A new hybrid between a Blue Whale, *Balaenoptera musculus*, and a Fin Whale, *B. physalus*: Frequency and implications of hybridization. *Marine Mammal Science* 14(1): 82–98.

Best, T.L. 1988. Morphologic variation in the Spotted Bat *Euderma maculatum*. *American Midland Naturalist* 119(2): 244–252.

– 1993. *Tamias ruficaudus*. *Mammalian Species* no. 452: 1–7.

Best, T.L., and Henry, T.H. 1994. *Lepus arcticus*. *Mammalian Species* no. 457: 1–9.

Betts, B.J. 1992. Spatial behaviour as a predictor of infanticidal individual in Columbian Ground Squirrels. *Northwestern Naturalist* 72: 85–91.

– 1998a. Variation in roost fidelity among reproductive female Silver-haired Bats in northeastern Oregon. *Northwestern Naturalist* 79(2): 59–63.

– 1998b. Roosts used by maternity colonies of Silver-haired Bats in northeastern Oregon. *Journal of Mammalogy* 79(2): 643–650.

– 1998c. Effects of interindividual variation in echolocation calls on identification of Big Brown and Silver-haired Bats. *Journal of Wildlife Management* 62(3): 1003–1010.

– 2000. Roosting behaviour of Silver-haired Bats (*Lasionycteris noctivagans*) and Big Brown Bats (*Eptesicus fuscus*) in northeast Oregon. In R.M.R. Barclay and R.M. Brigham (eds), *Bats and Forests Symposium*, 55–61. British Columbia Ministry of Forests, Victoria, BC.

Bêty, J., Gauthier, G., Korpimaki, E., and Giroux, J.F. 2002. Shared predators and indirect trophic interactions: Lemming cycles and arctic-nesting geese. *Journal of Animal Ecology* 71(1): 88–98.

Beyer, D.E., Jr, Roell, B.J., Hammill, J.H., and Earle, R.D. 2001. Records of Canada Lynx , *Lynx canadensis*, in the upper peninsula of Michigan, 1940 1997. *Canadian Field-Naturalist* 115(2): 234–240.

Bigg, M.A. 1981. Harbour Seal *Phoca vitulina* Linnaeus, 1758 and *Phoca largha* Pallas, 1811. In S.H. Ridgway and R. Harrison (eds), *Handbook of Marine Mammals*, vol. 2, *Seals*, 1–28. Academic Press, San Diego, CA.

Bihr, K.J., and Smith, R.J. 1998. Location, structure, and contents of burrows of *Spermophilus lateralis* and *Tamias minimus*, two ground-dwelling sciurids. *Southwestern Naturalist* 43(3): 352–362.

Birchler, B., and Potter, C.W. 1999. Dwarf Sperm Whale. In D.E. Wilson and S. Ruff (eds), *The Smithsonian Book of North American Mammals*, 302–303. UBC Press, Vancouver, BC.

Bissonette, J.A., and Sherburne, S.S. 1993. Subnivean access: The prey connection. *Proceedings of the International Union of Game Biologists Congress* 21(1): 225–228.

Bjorge, R.R., and Gunson, J.R. 1989. Wolf, *Canis lupus*, population characteristics and prey relationships near Simonette River, Alberta. *Canadian Field-Naturalist* 103(3): 327–334.

Black, D.W., Reading, J.E., and Savage, H.G. 1998. Archaeological records of the extinct Sea Mink, *Mustela macrodon* (Carnivora, Mustelidae), from Canada. *Canadian Field-Naturalist* 112(1): 45–49.

Blackburn, D.F. 1968. Courtship behavior among White-tailed and Black-tailed Jackrabbits. *Great Basin Naturalist* 33: 203–204.

Blackburn, G.S., Wilson, D.J., and Krebs, C.J. 1998. Dispersal of juvenile collared lemmings (*Dicrostonyx groenlandicus*) in a high-density population. *Canadian Journal of Zoology* 76(12): 2255–2261.

Blackwell, S.B., Richardson, W.J., Greene, C.R., Jr., and Streever, B. 2007. Bowhead Whale (*Balaena mysticetus*) migration and calling behaviour in the Alaskan Beaufort Sea, autumn 2001–04: An acoustic localization study. *Arctic* 60(3): 255–270.

Blake, B.H. 2002. Ultrasonic calling in isolated infant Prairie Voles (*Microtus ochrogaster*) and Montane Voles (*M. montanus*). *Journal of Mammalogy* 83(2): 536–545.

Bleich, V.C. 1999. Mountain sheep and Coyotes: Patterns of predator evasion in a mountain ungulate. *Journal of Mammalogy* 80(1): 283–289.

Bloch, D., Desportes, G., Petersen, A., and Sigurjøansson, J. 1996. Strandings of Striped Dolphins (*Stenella coeruleoalba*) in Iceland and the Faroe Islands and sightings in the northeast Atlantic, north of the 50°N latitude. *Marine Mammal Science* 12(1): 125–132.

Blomquist, L. 1995. Reproductive parameters of Wolverines (*Gulo gulo*) in captivity. *Annals Zoologica Fennici* 32: 441–444.

Bloodworth, B.E., and Odell, D.K. 2008. *Kogia brevicepts*. *Mammalian Species* no. 819:1–12.

Blumstein, D.T 1999. Alarm calling in three species of marmots. *Behaviour* 136: 731–757.

Blumstein, D.T., and Armitage, K.B. 1997. Alarm calling in Yellow-bellied Marmots: I. The meaning of situationally variable alarm calls. *Animal Behaviour* 53: 143–171.

Blumstein, D.T., Im, S., Nicodemus, A., and Zugmeyer, C. 2004. Yellow-bellied Marmots (*Marmota flaviventris*) hibernate socially. *Journal of Mammalogy* 85(1): 25–29.

Blundell, G.M., Ben-David, M., Groves, P., Bowyers, R.T., and Geffen, E. 2002. Characteristics of sex-biased dispersal and gene flow in coastal River Otters: Implications for natural recolonization of extirpated populations. *Molecular Ecology* 11: 289–303.

Blus, L.J., Fitzner, L., and Fitzner, R.E. 1993. Wolverine specimen from south-central Washington State. *Northwestern Naturalist* 74: 22.

Boan, J. 2005. Caribou continue to retreat. *ON Nature*, Spring: 10.

Bodkin, J.L. 2003. Sea Otter (*Enhydra lutris*). In G.A. Feldhamer, B.C. Thompson, and J.A. Chapman (eds), *Wild Mammals of North America: Biology, Management, and Conservation*, 2nd ed., 735–743. Johns Hopkins University Press, Baltimore, MD.

Bogan, M.A. 1999a. California Myotis. In D.E. Wilson and S. Ruff (eds), *The Smithsonian Book of North American Mammals*, 85–86. UBC Press, Vancouver, BC.

– 1999b. Eastern small-footed myotis. In Wilson and Ruff (eds), *Smithsonian Book of North American Mammals*, 93 94.

Boland, J.L., Hayes, J.P., Smith, W.P., and Huso, M.M. 2009. Selection of day-roosts by Keen's Myotis (*Myotis keenii*) at multiple spatial scales. *Journal of Mammalogy* 90(1): 222–234.

Boltnev, A.I., and Mathisen, O.A. 1996. Historical trends in abundance of Steller Sea Lions (*Eumetopias jubatus*) in the northwest Pacific Ocean. *Alaska Sea Grant Report* 96(1): 297–306.

Bonesi, L., Dunstone, N., and O'Connell, M. 2000. Winter selection of habitats within intertidal foraging areas by mink (*Mustela vison*). *Journal of Zoology* (London) 250: 419–424.

Bonesi, L., Harrington, L.A., Maran, T., Sidorovich, V.E., and Macdonald, D.W. 2006. Demography of three populations of American Mink *Mustela vison* in Europe. *Mammal Review* 36(1): 98–106.

Bonesi, L., and Macdonald, D.W. 2004. Evaluation of sign surveys as a way to estimate the relative abundance of American Mink (*Mustela vison*). *Journal of Zoology* (London) 262: 65–72.

Boness, D.J. 1999. Gray Seal *Halichoerus grypus*. In D.E. Wilson and S. Ruff (eds), *The Smithsonian Book of North American Mammals*, 211–212. UBC Press, Vancouver, BC.

Boness, D.J., Bowen, D., Iverson, S.J., and Oftedal, O.T. 1992. Influence of storms and maternal size on mother-pup separations and fostering in the Harbor Seal, *Phoca vitulina*. *Canadian Journal of Zoology* 70(8): 1640–1644.

Boness, D.J., Bowen, W.D., and Oftedal, O.T. 1994. Evidence of a maternal foraging cycle resembling that of otariid seals in a small phocid, the Harbor Seal. *Behavioral Ecology and Sociobiology* 34: 95–104.

Boness, D.J., Oftedal, O.T., and Ono, K.A. 1991. The effect of El Niño on pup development in the California Sea Lion (*Zalophus californianus*). I. Early postnatal growth. *Ecological Studies* 88: 173–179.

Bonner, W.N. 1981. Gray Seal *Halichoerus grypus* Fabricius, 1791. In S.H. Ridgway and R. Harrison (eds), *Handbook of Marine Mammals*, vol. 2, *Seals*, 111–144. Academic Press, San Diego, CA.

Boonstra, R., Krebs, C.J., and Kanter, M. 1990. Arctic Ground Squirrel predation on Collared Lemmings. *Canadian Journal of Zoology* 68(4): 757–760.

Boonstra, R., Krebs, C.J., and Kenney, A. 1996. Why lemmings have indoor plumbing in summer. *Canadian Journal of Zoology* 74(10): 1947–1949.

Boonstra, R., McColl, C.J., and Karels, T. 2001. Reproduction at all costs: The adaptive stress response of male Arctic Ground Squirrels. *Ecology* 82(7): 1930–1946.

Boonstra, R., Xia, X., and Pavone, L. 1993. Mating system of the Meadow Vole, *Microtus pennsylvanicus*. *Behavioral Ecology* 4(1): 83–89.

Börjesson, P., and Read, A.J. 2003. Variation in timing of conception between populations of the Harbor Porpoise. *Journal of Mammalogy* 84(3): 948–955.

Born, E.W. 1995. Status of the Polar Bear in Greenland 1993. *Occasional Papers of the IUCN Species Survival Commission* 10: 81–103.

– 2001. Reproduction in female Atlantic Walruses (*Odobenus rosmarus rosmarus*) from north-west Greenland. *Journal of Zoology* (London) 255: 165–174.

– 2003. Reproduction in male Atlantic Walruses (*Odobenus rosmarus rosmarus*) from the North Water (N Baffin Bay). *Marine Mammal Science* 19(4): 819–831.

Born, E.W., Rysgaard, S., Ehlmé, G., Sejr, M., Acquarone, M., and Levermann, N. 2003. Underwater observations of foraging free-living Atlantic Walruses (*Odobenus rosmarus rosmarus*) and estimates of their food consumption. *Polar Biology* 26: 348–357.

Born, E.W., Taylor, M.K., Rosing-Asvid, A., and Stirling, I. 1995. Summary status report for North Baffin Polar Bears. *Occasional Papers of the IUCN Species Survival Commission* 10: 175–176.

Born, E.W., Teilmann, J., Acquarone, M., and Riget, F.F. 2004. Habitat use of Ringed Seals (*Phoca hispida*) in the North Water Area (North Baffin Bay). *Arctic* 57(2): 129–142.

Born, E.W., Teilmann, J., and Riget, F.F. 2002. Haul-out activity of Ringed Seals (*Phoca hispida*) determined from satellite telemetry. *Marine Mammal Science* 18(1): 167–181.

Borrego, N., Ozgul, A., Armitage, K.B., Blumstein, D.T., and Oli, M.K. 2008. Spatio-temporal variation in survival of male Yellow-bellied Marmots. *Journal of Mammalogy* 89(2): 365–373.

Bouchard, S., Zigouris, J., and Fenton, M.B. 2001. Autumn mating and likely resorption of an embryo by a Hoary Bat, *Lasiurus cinereus* (Chiroptera: Vespertilionidae). *American Midland Naturalist* 145(1): 210–212.

Boudreau, S., Payette, S., Morneau, C., and Couturier, S. 2003. Recent decline of the George River Caribou herd as revealed by tree-ring analysis. *Arctic, Antarctic and Alpine Research* 35(2): 187–195.

Boulva, J. 2009. Présence de la belette pygmée (*Mustela nivalis*) et de la balette à longue queue (*Mustela frenata*) à Rimouski. *Naturalist Canadien* 133(2): 53–54.

Bounds, D.L., Sherfy, M.H., and Mollett, T.A. 2003. Nutria (*Myocastor coypus*). In G.A. Feldhamer, B.C. Thompson, and J.A. Chapman (eds), *Wild Mammals of North America: Biology, Management, and Conservation*, 2nd ed., 1119–1147. Johns Hopkins University Press, Baltimore, MD.

Boutin, S., Tooze, Z., and Price, K. 1993. Post-breeding dispersal by female Red Squirrels (*Tamiasciurus hudsonicus*): The effect of local vacancies. *Behavioral Ecology* 4: 151–155.

Bovet, J. 1990. Orientation strategies for long distance travel in terrestrial mammals, including humans. *Ethology Ecology and Evolution* 2: 117–126.

Bowan, J., Donovan, D., and Rosatte, R.C. 2006. Numerical response of Fishers to synchronous prey dynamics. *Journal of Mammalogy* 87(3): 480–484.

Bowen, W.D., Boness, D.J., and Iverson, S.J. 1999. Diving behaviour of lactating Harbour Seals and their pups during maternal foraging trips. *Canadian Journal of Zoology* 77(6): 978–988.

Bowen, W.D., and Harrison, G.D. 1996. Comparison of Harbour Seal diets in two inshore habitats of Atlantic Canada. *Canadian Journal of Zoology* 74(1): 125–135.

Bowen, W.D., Lawson, J.W., and Beck, B. 1993. Seasonal and geographic variation in the species composition and size of prey consumed by Grey Seals (*Halichoerus grypus*) on the Scotian Shelf. *Canadian Journal of Fisheries and Aquatic Sciences* 50(8): 1768–1778.

Bowen, W.D., McMillan, J. I., and Blanchard, W. 2007. Reduced population growth of Gray Seals at Sable Island: Evidence from pup production and age of primiparity. *Marine Mammal Science* 23(1): 48–64.

Bowen, W.D., McMillan, J., and Mohn R. 2003. Sustained exponential population growth of grey seals at Sable Island, Nova Scotia. *ICES Journal of Marine Science* 60: 1265–1274.

Bowman, J., and Curran, R.M. 2000. Partial albinism in a red-backed vole, *Clethrionomys gapperi*, from New Brunswick. *Northeastern Naturalist* 7(2): 181–182.

Bowman, J., Holloway, G.L., Malcolm, J.R., Middel, K.R., and Wilson, P.J. 2005. Northern range boundary dynamics of Southern Flying Squirrels: Evidence of an energetic bottleneck. *Canadian Journal of Zoology* 83(11): 1486–1494.

Bowman, J.C. 1997. An arboreal encounter between a Long-tailed Weasel, *Mustela frenata*, and three Red Squirrels, *Tamiasciurus hudsonicus*. *Canadian Field-Naturalist* 111(3): 480–481.

Bowman, J.C., Edwards, M., Sheppard, L.S., and Forbes, G.J. 1999. Record distance for a non-homing movement by a Deer Mouse, *Peromyscus maniculatus*. *Canadian Field-Naturalist* 113(2): 292–293.

Bowman, J.C., and Robitaille, J.-F. 1997. Winter habitat use of American Martens, *Martes americana* within second-growth forest in Ontario, Canada. *Wildlife Biology* 3: 97–105.

Bowyer, R.T., Kie, J.G., and Van Ballenberghe, V. 1998a. Habitat selection by neonatal Black-tailed Deer: Climate, forage, or risk of predation? *Journal of Mammalogy* 79(2): 415–425.

Bowyer, R.T., and Kitchen, D.W. 1987. Sex and age-class differences in vocalizations of Roosevelt Elk during rut. *American Field-Naturalist* 118(2): 225–235.

Bowyer, R.T., and Leslie, D.M., Jr. 1992. *Ovis dalli*. *Mammalian Species* no. 393: 1–7.

Bowyer, R.T., McCullough, D.R., and Belovsky, G.E. 2001. Causes and consequences of sociality in Mule Deer. *Alces* 37(2): 371–402.

Bowyer, R.T., Van Ballenberghe, V. and Kie, J.G. 1998b. Timing and synchrony of parturition in Alaskan Moose: Long-term versus proximal effects of climate. *Journal of Mammalogy* 79(4): 1332–1344.

Bowyer, R.T., Van Ballenberghe, V., and Kie, J.G. 2003. Moose (*Alces alces*). In G.A. Feldhamer, B.C. Thompson, and J.A. Chapman (eds), *Wild Mammals of North America: Biology, Management, and Conservation*, 2nd ed., 931–964. Johns Hopkins University Press, Baltimore, MD.

Bowyer, R.T., Van Ballenberghe, V., Kie, J.G., and Maier, J.A.K. 1999. Birth-site selection by Alaskan Moose: Maternal strategies for coping with a risky environment. *Journal of Mammalogy* 80(4): 1070–1083.

Boyd, D.K., and Heger, E.E. 2000. Predation of a denned Black Bear, *Ursus americanus*, by a Grizzly Bear, *U. arctos*. *Canadian Field-Naturalist* 114(3): 507–508.

Boyd, D.K., and Neale, G.K. 1992. An adult Cougar, *Felis concolor*, killed by Gray Wolves, *Canis lupus*, in Glacier National Park, Montana. *Canadian Field-Naturalist* 106(4): 524–525.

Bradley, M., and Wilmshurst, J. 2005. The fall and rise of Bison populations in Wood Buffalo National Park: 1971 to 2003. *Canadian Journal of Zoology* 83(9): 1195–1205.

Brady, K.M., and Armitage, K.B. 1999. Scent-marking in the Yellow-bellied Marmot (*Marmota flaviventris*). *Ethology Ecology and Evolution* 11: 35–47.

Brady, K.S., and Hamer, D. 1992. Use of a summit mating area by a pair of courting Grizzly Bears, *Ursus arctos*, in Waterton Lakes National Park, Alberta. *Canadian Field-Naturalist* 106(4): 519–520.

Branch, T.A., Abubaker, E.M.N., Mkango, S., and Butterworth, D.S. 2007. Separating southern Blue Whale subspecies based on length frequencies of sexually mature females. *Marine Mammal Science* 23(4): 803–833.

Brandt, D.A. 1994. Overwater foraging by a badger. *Prairie Naturalist* 26(2): 171.

Brannon, M.P. 2000. Niche relationships of two syntopic species of shrews, *Sorex fumeus* and *S. cinereus*, in the southern Appalachian Mountains. *Journal of Mammalogy* 81(4): 1053–1061.

– 2002a. Distribution of *Sorex cinereus* and *S. fumeus* on north- and south-facing slopes in the southern Appalachian Mountains. *Southeastern Naturalist* 1(3): 299–306.

– 2002b. Epigeal movement of the Smoky Shrew *Sorex fumeus* following precipitation in ridgetop and streamside habitats. *Acta Theriologica* 47(3): 363–368.

– 2005. Distribution and microhabitat of the Woodland Jumping Mouse, *Napaeozapus insignis*, and the White-footed Mouse, *Peromyscus leucopus*, in the southern Appalachians. *Southeastern Naturalist* 4(3): 479–486.

Brant, S.V., and Ortí, G. 2001. Molecular phylogeny of Short-tailed Shrews, *Blarina* (Insectivora: Soricidae). *Molecular Phylogenetics and Evolution* 22(2): 163–173.

– 2003. Phylogeography of the Northern Short-tailed Shrew, *Blarina brevicauda* (Insectivora: Soricidae): Past fragmentation and postglacial recolonization. *Molecular Ecology* 12: 1435–1449.

Brewer, S.R., Lucas, M.F., Mugnano, J.A., Peles, J.D., and Barrett, G.W. 1993. Inheritance of albinism in the Meadow Vole (*Microtus pennsylvanicus*). *American Midland Naturalist* 130(2): 393–396.

Briggler, J.T., and Prather, J.W. 2003. Seasonal use and selection of caves by the Eastern Pipistrelle Bat (*Pipistrellus subflavus*). *American Midland Naturalist* 149: 406–412.

Brigham, R.M. 1987. The significance of winter activity by the Big Brown Bat (*Eptesicus fuscus*): The influence of energy reserves. *Canadian Journal of Zoology* 65(5): 1240–1242.

– 1991. Flexibility in foraging and roosting behavior by the Big Brown Bat (*Eptesicus fuscus*). *Canadian Journal of Zoology* 69(1): 117–121.

– 1995. A winter record for the Silver-haired Bat in Saskatchewan. *Blue Jay* 53(3): 168.

Brigham, R.M., Aldridge, H.D.J.N., and Mackey, R.L. 1992. Variation in habitat use and prey selection by Yuma Bats, *Myotis yumanensis*. *Journal of Mammalogy* 73(3): 640–645.

Brigham, R.M., and Brigham, A.C. 1989. Evidence for association between mother bat and young during and after foraging. *American Midland Naturalist* 121: 205–207.

Brigham, R.M., and Fenton, M.B. 1986. The influence of roost closure on the roosting and foraging behavior of *Eptesicus fuscus* (Chiroptera: Vespertilionidae). *Canadian Journal of Zoology* 64(5): 1128–1133.

– 1991. Convergence in foraging strategies by two morphologically and phylogenetically distinct nocturnal aerial insectivores. *Journal of Zoology* (London) 223: 475–489.

Brigham, R.M., Vonhof, M.J., Barclay, R.M.R., and Gwilliam, J.C. 1997. Roosting behaviour and roost-site preferences of forest-dwelling California Bats (*Myotis californicus*). *Journal of Mammalogy* 78(4): 1231–1239.

Broadbooks, H.E. 1958. Life history and ecology of the chipmunk, *Eutamias amoenus*, in eastern Washington. *Miscellaneous Publications of the Museum of Zoology, University of Michigan* no. 103: 1–42.

– 1965. Ecology and distribution of the pikas of Washington and Alaska. *American Midland Naturalist* 73(2): 299–335.

Broders, H.G., Findlay, C.S., and Zheng, L. 2004. Effects of clutter on echolocation call structure of *Myotis septentrionalis* and *M. lucifugus*. *Journal of Mammalogy* 85(2): 273–281.

Broders, H.G., McAlpine, D.F., and Forbes, G.J. 2001. Status of the Eastern Pipistrelle (*Pipistrellus subflavus*)(Chiroptera: Vespertilionidae) in New Brunswick. *Northeastern Naturalist* 8(3): 331–336.

Brodie, P.F. 1989. The White Whale *Delphinapterus leucas* (Pallas, 1776). In S.H. Ridgway and R. Harrison (eds), *Handbook of Marine Mammals*, vol. 4, 119–144. Academic Press, San Diego, CA.

Brooks, R.J. 1993. Dynamics of home range in collared lemmings. In N.C. Stenseth and R.A. Ims (eds), *The Biology of Lemmings*, 355–386. Academic Press, San Diego, CA.

Brookshier, J.S., and Fairbanks, W.S. 2003. The nature and consequences of mother-daughter associations in naturally and forcibly weaned Bison. *Canadian Journal of Zoology* 81(3): 414–423.

Broussard, D.R., Michener, G.R., and Dobson, F.S. 2006. Age-specific resource investment strategies: Evidence from female Richardson's Ground Squirrels (*Spermophilus richardsonii*). *Journal of Zoology* 268: 389–394.

Brower, J.E., and Cade, T.J. 1966. Ecology and physiology of *Napaeozapus insignis* (Miller) and other woodland mice. *Ecology* 47(1): 46–63.

Brown, A.C., and Elias, E. 2008. Extralimital records of the Sea Otter (*Enhydra lutris*) in northern California. *Northwestern Naturalist* 89(1): 54–56.

Brown, J.A., McAlpine, D.F., and Curley, R. 2007. Northern Long-eared Bat, *Myotis septentrionalis* (Chiroptera: Vespertilionidae), on Prince Edward Island: First records of occurrence and over-wintering. *Canadian Field-Naturalist* 121(2): 208–209.

Brown, L.N. 1997. *A Guide to the Mammals of the Southeastern United States*. University of Tennessee Press, Knoxville.

Brown, S.G. 1977. Some results of Sei Whale marking in the Southern Hemisphere. *Report of the International Whaling Commission*, Special issue, 12: 349–356.

Brownell, R.L., Jr., Clapham, P.J., Miyashita, T., and Kasuya, T. 2001. Conservation status of North Pacific Right Whales. *Journal of Cetacean Research and Management*, Special issue, 2: 269–286.

Brownell, R.L., Jr, Walker, W.A., and Forney, K.A. 1999. Pacific White-sided Dolphin *Lagenorhynchus obliquidens* Gill, 1865. In S.H. Ridgeway and R. Harrison (eds), *Handbook of Marine Mammals*, vol. 6, 57–84. Academic Press, San Diego, CA.

Brown Gladden, J.G., Brodie, P.F., and Clayton, J.W. 1999. Mitochondrial DNA used to identify an extralimital Beluga Whale (*Delphinapterus leucas*) from Nova Scotia as originating from the St. Lawrence population. *Marine Mammal Science* 15(2): 556–558.

Brueggeman, J.J., Newby, T., and Grotefendt, R.A. 1988. Catch records of the twenty North Pacific Right Whales from two Alaska whaling stations, 1917–1939. *Arctic* 39(1): 43–46.

Brunet, A.K., Zink, R.M., Kramer, K.M., Blackwell-Rago, R.C., Farrell, S.L., Line, T.V., and Birney, E.C. 2002. Evidence of introgression between Masked Shrews (*Sorex cinereus*) and Prairie Shrews (*S. haydeni*), in Minnesota. *American Midland Naturalist* 147(1): 116–122.

Brunhoff, C., Galbreath, K.E., Fedorov, V.B., Cook, J.A., and Jaarola, M. 2003. Holarctic phylogeography of the Root Vole (*Microtus oeconomus*): Implications for late Quaternary biogeography of high latitudes. *Molecular Ecology* 12: 957–968.

Brunton, D.F. 1981. Nocturnal aggregations of White-tailed Jack-rabbits at Rimbey, Alberta. *Blue Jay* 39(2): 120–121.

Bruseo, J.A., Vessey, S.H., and Graham, J.S. 1999. Discrimination between *Peromyscus leucopus noveboracensis* and *Peromyscus maniculatus nubiterrae* in the field. *Acta Theriologica* 44(2): 151–160.

Bryant, A.A. 1996. Reproduction and persistence of Vancouver Island Marmots (*Marmota vancouverensis*) in natural and logged habitats. *Canadian Journal of Zoology* 74(4): 678–687.

– 2005. Reproduction rates of wild and captive Vancouver Island Marmots (*Marmota vancouverensis*). *Canadian Journal of Zoology* 83(5): 664–673.

Bryant, A.A., and Janz, D.W. 1996. Distribution and abundance of Vancouver Island Marmots (*Marmota vancouverensis*). *Canadian Journal of Zoology* 74(4): 667–677.

Bryant, A.A, and Page, R.E. 2005. Timing and causes of mortality in the endangered Vancouver Island Marmot (*Marmota vancouverensis*). *Canadian Journal of Zoology* 83(5): 674–682.

Bryant, H.N., Russell, A.P., Laroiya, R., and Powell, G.L. 1996. Claw retraction and protraction in the Carnivora: Skeletal microvariation in the phalanges of the Felidae. *Journal of Morphology* 229: 289–308.

Buchan, S.J., Rendell, L.E., and Hucke-Gaete, R. 2010. Preliminary recordings of Blue Whale (*Balaenoptera musculus*) vocalizations in the Gulf of Corcovado, northern Patagonia, Chile. *Marine Mammal Science* 26(2): 451–459.

Buchanan, J.B., Lundquist, R.W., and Aubry, K.B. 1990. Winter populations of Douglas' Squirrels in different-aged Douglas-fir forests. *Journal of Wildlife Management* 54(4): 577–581.

Buchler, E.R. 1976. The use of echolocation by the Wandering Shrew (*Sorex vagrans*). *Animal Behaviour* 24(4): 858–873.

Buck, C.L., and Barnes, B.M. 1999. Annual cycle of body composition and hibernation in free-living Arctic Ground Squirrels. *Journal of Mammalogy* 80(2): 430–442.

Buckland, S.T., and Breiwick, J.M. 2002. Estimated trends in abundance of eastern Pacific Gray Whales from shore counts (1967/68 to 1995/96). *Journal of Cetacean Research and Management* 4(1): 41–48.

Buckner, C.H. 1964. Metabolism, food capacity, and feeding behavior in four species of shrews. *Canadian Journal of Zoology* 42(2): 259–279.

– 1966. Populations and ecological relationships of shrews in tamarack bogs of southeastern Manitoba. *Journal of Mammalogy* 47(2): 181–194.

– 1968. Notes on the Water Shrew in bog habitats of southeastern Manitoba. *Blue Jay* 26(2): 95–96.

– 1970. Direct observations of shrew predation on insects and fish. *Blue Jay* 28(4): 171–172.

Bull, E.L., and Heater, T.W. 2001. Survival, causes of mortality, and reproduction in the American Marten in northeastern Oregon. *Northwestern Naturalist* 82: 1–6.

Bundy, A. 2001. Fishing on ecosystems: The interplay of fishing and predation in Newfoundland-Labrador. *Canadian Journal of Fisheries and Aquatic Sciences* 58(6): 1153–1167.

Bunnell, F.L. 2005. Thinhorn Sheep. *Wildlife Afield* 2(1): 22–30.

Burke da Silva, K., Kramer, D.L., and Weary, D.M. 1994. Context-specific alarm calls of the Eastern Chipmunk, *Tamias striatus. Canadian Journal of Zoology* 72(6): 1087–1092.

Burke da Silva, K., Mahan, C., and da Silva, J. 2002. The trill of the chase: Eastern Chipmunks call to warn kin. *Journal of Mammalogy* 83(2): 546–552.

Burles, D.W., Brigham, R.M., Ring, R.A., and Reimchen, T.E. 2008. Diet of two insectivorous bats, *Myotis lucifugus* and *Myotis keenii*, in relation to arthropod abundance in a temperate Pacific Northwest rainforest environment. *Canadian Journal of Zoology* 86(12): 1367–1375.

– 2009. Influence of weather on two insectivorous bats in a temperate Pacific Northwest rainforest. *Canadian Journal of Zoology* 87(2): 132–138.

Burles, D.W., and Nagorsen, D.W. 2003. Update COSEWIC Status Report on Keen's Long-eared Myotis *Myotis keenii*. Prepared for the Committee on the Status of Endangered Wildlife in Canada.

Burnett, C.D. 1983. Geographic and secondary sexual variation in the morphology of *Eptesicus fuscus. Annals of the Carnegie Museum* 52(7): 139–162.

Burns, J.J. 1981. Bearded Seal *Erignathus barbatus* Erxleben, 1777. In S.H. Ridgway and R. Harrison (eds), *Handbook of Marine Mammals*, vol. 2, *Seals*, 145–170. Academic Press, San Diego, CA.

– 2002. Harbor Seal and Spotted Seal *Phoca vitulina* and *Phoca largha*. In W.F. Perrin, B. Wursig, and H.G.T. Thewissen (eds), *Encyclopedia of Marine Mammals*, 552–560. Academic Press, San Diego, CA.

Burton, C. 2002. Microsatellite analysis of multiple paternity and male reproductive success in the promiscuous Snowshoe Hare. *Canadian Journal of Zoology* 80(11):1848–1856.

Byers, J.A. 1997. *American Pronghorn. Social Adaptations and the Ghosts of Predators Past.* University of Chicago Press, Chicago, IL.

– 2003a. *Built for Speed: A Year in the Life of Pronghorn.* Harvard University Press, Cambridge, MA.

– 2003b. Pronghorn (*Antilocapra americana*). In G.A. Feldhamer, B.C. Thompson, and J.A. Chapman (eds), *Wild Mammals of North America: Biology, Management, and Conservation*, 2nd ed., 998–1008. Johns Hopkins University Press, Baltimore, MD.

Byrnes, P.E., and Hood, W.R. 1994. First account of Steller Sea Lion (*Eumetopias jubatus*) predation on a California Sea Lion (*Zalophus californianus*). *Marine Mammal Science* 10(3): 381–383.

Byrom, A.E., and Krebs, C.J. 1999. Natal dispersal of juvenile Arctic Ground Squirrels in the boreal forest. *Canadian Journal of Zoology* 77(7): 1048–1059.

Caceres, M.C., and Barclay, R.M.R. 2000. *Myotis septentrionalis. Mammalian Species* no. 634: 1–4.

Caldwell, D.K., and Caldwell, M.C. 1989. Pygmy Sperm Whale *Kogia breviceps* (de Blainville, 1838): Dwarf Sperm Whale *Kogia simus* Owen, 1866. In S.H. Ridgway and R. Harrison (eds), *Handbook of Marine Mammals*, vol. 4, *River Dolphins and the Larger Toothed Whales*. Academic Press, San Diego, CA.

Calkins, D.G., McAllister, D.C., Pitcher, K.W., and Pendleton, G.W. 1999. Steller Sea Lion status and trend in southeast Alaska: 1979–1997. *Marine Mammal Science* 15(2): 462–477.

Cameron, R.D. 1994. Reproductive pauses by female Caribou. *Journal of Mammalogy* 75(1): 10–13.

Campbell, K.L., and MacArthur, R.A. 1996. Digestibility of animal tissue by Muskrats. *Journal of Mammalogy* 77(3): 755–760.

Campbell, R.R. 1987a. Status of the Hooded Seal, *Cystophora cristata*, in Canada. *Canadian Field-Naturalist* 101(2): 253–265.

– 1987b. Status of the Northern Elephant Seal, *Mirounga angustirostris*, in Canada. *Canadian Field-Naturalist* 101(2): 266–270.

– 1988. Status of the Sea Mink, *Mustela macrodon*, in Canada. *Canadian Field-Naturalist* 102(2): 304–306.

Campbell, R.W., and Summers, K.R. 1997. Vertebrates of Brooks Peninsula. *B.C. Parks Occasional Paper* 1997(5): 12.1–12.39.

Campbell, V., and Strobeck, C. 2006. Fine-scale genetic structure and dispersal in Canada Lynx (*Lynx canadensis*) within Alberta, Canada. *Canadian Journal of Zoology* 84(8): 1112–1119.

Cannings, S.G., Ramsay, L.R., Fraser, D.F., and Fraker, M.A. 1999. *Rare Amphibians, Reptiles and Mammals of British Columbia*. British Columbia Ministry of Environment, Lands and Parks Wildlife Branch and Resources Inventory Branch, Victoria, BC.

Carbyn, L.N. 1983. Wolf predation on Elk in Riding Mountain National Park, Manitoba. *Journal of Wildlife Management* 47(4): 963–976.

– 1989. Status of the Swift Fox in Saskatchewan. *Blue Jay* 47(1): 41–52.

– 1997. Unusual movement by Bison, *Bison bison*, in response to Wolf, *Canis lupus*, predation. *Canadian Field-Naturalist* 111(3): 461–462.

– 1998. Update COSEWIC status report on the Swift Fox *Vulpes velox* in Canada. *Committee on the Status of Endangered Wildlife in Canada*. Ottawa.

– 2003. *The Buffalo Wolf: Predators, Prey and the Politics of Nature.* Smithsonian Institution Press, Washington, DC.

Carerau, V., Girous, J.-F., Gautier, G., and Berteaux, D. 2008. Surviving on cached foods – the energetics of egg-caching by Arctic Foxes. *Canadian Journal of Zoology* 86(10): 1217–1223.

Carey, A.B. 1991. The biology of arboreal rodents in Douglas-fir forests. *United States Department of Agriculture Forestry Service General Technical Report* PNW-GTR-276: 1–46.

Carey, A.B., Colgan, W., III, Trappe, J.M., and Molina, R. 2002. Effects of forest management on truffle abundance and squirrel diets. *Northwest Science* 76(2): 148–157.

Carey, A.B., and Kershner, J.E. 1996. *Spilogale gracilis* in upland forests of western Washington and Oregon. *Northwestern Naturalist* 77: 29–34.

Carey, A.B., Wilson, T.M., Maguire, C.C., and Biswell, B.L. 1997. Dens of Northern Flying Squirrels in the Pacific Northwest. *Journal of Wildlife Management* 61(3): 684–699.

Carleson, M. 1991. Notes on Raccoons from Edam, Saskatchewan. *Blue Jay* 49(2): 101–103.

Carling, M.D., Wiseman, P.A., and Byers, J.A. 2003. Microsatellite analysis reveals multiple paternity in a population of wild Pronghorn Antelopes (*Antilocapra americana*). *Journal of Mammalogy* 84(4): 1237–1243.

Carlström, J. 2005. Diel variation in echolocation behaviour of wild Harbor Porpoises. *Marine Mammal Science* 21(1): 1–12.

Carmichael, L.E., Szor, G., Berteaux, D., Giroux, M.A., Cameron, C., and Strobeck, C. 2007. Free love in the far north: Plural breeding and polyandry of Arctic Foxes (*Alopex lagopus*) on Bylot Island, Nunavut. *Canadian Journal of Zoology* 85(3): 338–343.

Carraway, L.N. 1987. Analysis of characters for discriminating *Sorex trowbridgii* from sympatric *S. vagrans. Murrelet* 68(1): 29–30.

– 1990. A morphologic and morphometric analysis of the 'Sorex vagrans species complex' in the Pacific Coast Region. *Special Publications, Museum of Texas Tech University* no. 2: 1–76.

Carraway, L.N., Alexander, L.F., and Verts, B.J. 1993. *Scapanus townsendii. Mammalian Species* no. 434: 1–7.

Carraway, L.N., and Verts, B.J. 1985. *Microtus oregoni. Mammalian Species* no. 233: 1–6.

– 1991a. *Neurotrichus gibbsii. Mammalian Species* no. 387: 1–7.

– 1991b. Pattern and color aberrations in pelages of *Scapanus townsendii. Northwest Science* 65(1): 16–21.

– 1993. *Aplodontia rufa. Mammalian Species* no. 431: 1–10.

– 1999a. Records of reproduction in *Sorex preblei. Northwestern Naturalist* 80(3): 115–116.

– 1999b. Age-related fecundity in four taxa of western shrews (*Sorex* spp.). *American Midland Naturalist* 142(2): 424–426.

Carrier, P., and Krebs, C.J. 2002. Trophic effects of rainfall on *Clethrionomys rutilus* voles: An experimental test in a xeric boreal forest in the Yukon Territory. *Canadian Journal of Zoology* 80(5): 821–829.

Carrière, S. 1999. Small mammal survey in the Northwest Territories. *Wildlife and Fisheries, Dept. of Resources and Economic Development, Government of the Northwest Territories* Manuscript report no. 115: 1–22

Carroll, L.E., and Genoways, H.H. 1980. *Lagurus curtatus. Mammalian Species* no. 124: 1–6.

Carter, D., Harestad, A., and Bunnell, F.L. 1993. Status of Nuttall's Cottontail in British Columbia. *Wildlife Working Report* no. WR-56: 1–26. Wildlife Branch, British Columbia Ministry of Environment, Lands and Parks.

Case, R.L., and Buckland, L. 1998. Reproductive characteristics of Grizzly Bears in the Kugluktuk area, Northwest Territories, Canada. *Ursus* 10: 41–47.

Casey, T.M. 1981. Nest insulation: Energy savings to Brown Lemmings using a winter nest. *Oecologia* 50: 199–204.

Casimir, D.L., Moehrenschlager, A., and Barclay, R.M.R. 2007. Factors influencing reproduction in captive Vancouver Island Marmots: Implications for captive breeding and reintroduction programs. *Journal of Mammalogy* 88(6): 1412–1419.

Casler, B.R., and Murphy, R.K. 2001. Badger removes egg from island in prairie alkali lake. *Prairie Naturalist* 33(2): 109–110.

Caswell, H., Fujiwara, M., and Brault, S. 1999. Declining survival probability threatens the North Atlantic Right Whale. *Proceedings of the National Academy of Science* 96: 3308–3313.

Catania, K.C. 1999. A nose that looks like a hand and acts like an eye: The unusual mechanosensory system of the Star-nosed Mole. *Journal of Comparative Physiology A* 185: 367–372.

– 2006. Olfaction: Underwater 'sniffing' by semi-aquatic mammals. *Nature* 444: 1024–1025.

– 2008. No taming the shrew. *Natural History* 117(2): 56–60.

Catania, K.C., and Kaas, J.H. 1996. The unusual nose and brain of the Star-nosed Mole. *BioScience* 46(8): 578–586.

Cathey, J.C., Bickham, J.W., and Patton, J.C. 1998. Introgressive hybridization and nonconcordant evolutionary history of maternal and parental lineages in North American deer. *Evolution* 52(4): 1224–1229.

Cawthorn, J.M. 1994. A live-trapping study of two syntopic species of *Sorex, S. cinereus* and *S. fumeus,* in southwestern Pennsylvania. *Carnegie Museum of Natural History, Special Publication* 18: 39–43.

Ceballos-G., G. 1999. Northern River Otter / *Lontra canadensis.* In D.E. Wilson and S. Rue (eds), *The Smithsonian Book of North American Mammals,* 179–180. UBC Press, Vancouver, BC.

Chabot, D., and Stenson, G.B. 2000. Implantation date, growth rate, and allometric relationships in foetal Northwest Atlantic Harp Seals (*Phoca groenlandica*). *Canadian Journal of Zoology* 78(3): 501–505.

– 2002. Growth and seasonal fluctuations in size and condition of male Northwest Atlantic Harp Seals *Phoca groenlandica*: An analysis using sequential growth curves. *Marine Ecology Progress Series* 227: 25–42.

Chamberlain, M.J., and Leopold, B.D. 2000. Spatial use patterns, seasonal habitat selection, and interactions among adult Gray Foxes in Mississippi. *Journal of Wildlife Management* 64(3): 742–751.

– 2002. Movements and space use of Gray Foxes (*Urocyon cinereoargenteus*) following mate loss. *American Midland Naturalist* 147(2): 409–412.

Chambers, A. 2006. Wolf attack. *Canadian Geographic,* November/December. 38.

Chapman, J.A. 1975. *Sylvilagus nuttallii. Mammalian Species* no. 56: 1–3.

Chapman, J.A., and Flux, J.E.C. 1990. *Rabbits, Hares and Pikas: Status survey and conservation action plan.* IUCN/SSC Lagomorph Specialist Group, IUCN, Gland, Switzerland.

Chapman, J.A., Hockman, J.G., and Ojeda, M.M. 1980. *Sylvilagus floridanus. Mammalian Species* no. 136: 1–8.

Charbonneau, P., Julien, J.-R., and Tremblay, G. 2011. Premier inventaire de chiroptères sur l'île aux Basques. *Naturaliste Canadien* 135(1): 53–62.

Charif, R.A., Mellinger, D.K., Dunsmore, K.J., Fristrup, K.M., and Clark, C.W. 2002. Estimated source levels of Fin Whale (*Balaenoptera physalus*) vocalizations: Adjustments for surface interference. *Marine Mammal Science* 18(1): 81–98.

Chaulk, K., Bondrup-Nielsen, S., and Harrington, F. 2005. Black Bear, *Ursus americanus,* ecology on the northeast coast of Labrador. *Canadian Field-Naturalist* 119(2): 164–174.

Chekchak, T., Courtois, R., Ouellet, J.-P., Breton, L., and St-Onge, S. 1998. Caractéristiques des sites de mise bas de l'orignal (*Alces alces*). *Canadian Journal of Zoology* 76(9): 1663–1670.

Chirhart, S.E., Arianpour, R., Honeycutt, R.L., and Greenbaum, I.F. 2001. Mitochondrial DNA sequence variation and the specific identification of deer mice (*Peromyscus*) from Triangle Island, British Columbia, Canada. *Canadian Journal of Zoology* 79(12): 2257–2260.

Chivers, S.J., Leduc, R.G., Robertson, K.M., Barros, N.B., and Dizon, A.E. 2005. Genetic variation of *Kogia* spp. with preliminary evidence for two species of *Kogia sima. Marine Mammal Science* 21(4): 619–634.

Choate, J.R., Jones, J.K., Jr, and Jones, C. 1994. *Handbook of Mammals of the South-Central States.* Louisiana State University, Baton Rouge.

Chou, L.-S., Bright, A.M., and Yeh, S.-Y. 1995. Stomach contents of dolphins (*Delphinus delphis* and *Lissodelphis borealis*) from North Pacific Ocean. *Zoological Studies* 34(3): 206–210.

Christian, D.P., and Daniels, J.M. 1985. Distributional records of Rock Voles, *Microtus chrotorrhinus,* in northeastern Minnesota. *Canadian Field-Naturalist* 99(3): 356–359.

Chruszcz, B.J., and Barclay, R.M.R. 2002. Thermoregulatory ecology of a solitary bat, *Myotis evotis,* roosting in rock crevices. *Functional Ecology* 16: 18–26.

– 2003. Prolonged foraging bouts of a solitary gleaning/hawking bat, *Myotis evotis. Canadian Journal of Zoology* 81(5): 823–826.

Chubbs, T.E. 1993. Observations of calf-hiding behavior by female Woodland Caribou, *Rangifer tarandus caribou,* in east-central Newfoundland. *Canadian Field-Naturalist* 107(3): 368–369.

Chubbs, T.E., and Brazil, J. 2007. The occurrence of Muskoxen, *Ovibos moschatus,* in Labrador. *Canadian Field-Naturalist* 121(1): 81–84.

Chubbs, T.E., and Phillips, F.R. 1993a. Unusually high number of embryos in a Muskrat, *Ondatra zibethicus,* from central Labrador. *Canadian Field-Naturalist* 107(3): 363.

– 1993b. An apparent longevity record for Canada Lynx, *Lynx canadensis,* in Labrador. *Canadian Field-Naturalist* 107(3): 367–368.

– 1994. Long distance movement of a transplanted Beaver, *Castor canadensis,* in Labrador. *Canadian Field-Naturalist* 108(3): 366.

– 2002. First record of an Eastern Coyote, *Canis latrans,* in Labrador. *Canadian Field-Naturalist* 116(1): 127–129.

– 2005. Evidence of range expansion of Eastern Coyotes, *Canis latrans,* in Labrador. *Canadian Field-Naturalist* 119(3): 381–384.

Chubbs, T.E., and Schaefer, J.A. 1997. Population growth of Moose, *Alces alces,* in Labrador. *Canadian Field-Naturalist* 111(2): 238–242.

Churcher, C.S. 1999. Grizzly or Brown Bear *Ursus arctos.* In D.E. Wilson and S. Ruff (eds), *The Smithsonian Book of North American Mammals,* 160–163. UBC Press, Vancouver, BC.

Ciucci, P., Lucchini, V., Boitani, L., and Randi, E. 2003. Dewclaws in wolves as evidence of admixed ancestry with dogs. *Canadian Journal of Zoology* 81(12): 2077–2081.

Clapham, P.J. 2002. Humpback Whale (*Megaptera novaeangliae*). In W.F. Perrin, B. Wursig, and H.G.T. Thewissen (eds), *Encyclopedia of Marine Mammals,* 589–592. Academic Press, San Diego, CA.

Clapham, P.J., and Mead, J.G. 1999. *Megaptera novaeangliae. Mammalian Species* no. 604: 1–9.

Clapham, P.J., and Seipt, I.E. 1991. Resightings of independent Fin Whales, *Balaenoptera physalus,* on maternal summer ranges. *Journal of Mammalogy* 72(4): 788–790.

Clapham, P.J., Wetmore, S.E., Smith, T.D., and Mead, J.G. 1999. Length at birth and at independence in Humpback Whales. *Journal of Cetacean Research and Management* 1(2): 141–146.

Clark, B.K., and Kaufman, D.W. 1990. Short-term responses of small mammals to experimental fire in tallgrass prairie. *Canadian Journal of Zoology* 68(11): 2450–2454.

Clark, B.S., Clark, B.K., and Leslie, D.M., Jr. 2002. Seasonal variation in activity patterns of the endangered Ozark Big-eared Bats (*Corynorhinus townsendii ingens*). *Journal of Mammalogy* 83(2): 590–598.

Clark, B.S., Leslie, D.M., Jr, and Carter, T.S. 1993. Foraging activity of adult female Ozark Big-eared Bats (*Plecotus townsendii ingens*) in summer. *Journal of Mammalogy* 74(2): 422–427.

Clark, D. 2000. Recent reports of Grizzly Bears, *Ursus arctos*, in northern Manitoba. *Canadian Field-Naturalist* 114(4): 692–696.

Clark, D.A., Stirling, I., and Calvert, W. 1997. Distribution, characteristics, and use of earth dens and related excavations by Polar Bears on the western Hudson Bay lowlands. *Arctic* 50(2): 158–166.

Clark, M.K. 1993. A communal winter roost of Silver-haired Bats, *Lasionycteris noctivagans* (Chiroptera: Vespertilionidae). *Brimleyana* 19: 137–139.

Clark, T.W., Anderson, E., Douglas, C., and Strickland, M. 1987. *Martes americana*. *Mammalian Species* no. 289: 1–8.

Clark, T.W., and Stromberg, M.R. 1987. *Mammals in Wyoming*. University of Kansas Museum of Natural History, Lawrence.

Clarke, M.F., and Kramer, D.L. 1994. Scatter-hoarding by a larder-hoarding rodent: Intraspecific variation in the hoarding behaviour of the Eastern Chipmunk, *Tamias striatus*. *Animal Behaviour* 48: 299–308.

Clarkson, P.L., and Liepins, I.S. 1993. Grizzly Bear, *Ursus arctos*, predation on Muskox, *Ovibos moschatus*, calves near the Horton River, Northwest Territories. *Canadian Field-Naturalist* 107(1): 100–102.

Cleator, H.J. 1996. The status of the Bearded Seal, *Erignathus barbatus*, in Canada. *Canadian Field-Naturalist* 110(3): 501–510.

Clem, P.D. 1992. Seasonal population variation and emergence patterns in the evening bat, *Nycticeius humeralis*, at a west-central Indiana colony. *Proceedings of the Indiana Academy of Science* 101: 33–44.

– 1993. Foraging patterns and the use of temporary roosts in female Evening Bats, *Nycticeius humeralis*, at an Indiana maternity colony. *Proceedings of the Indiana Academy of Science* 102: 201–206.

Cluff, H.D. 2006. Extension of Coyote, *Canis latrans*, breeding range in the Northwest Territories, Canada. *Canadian Field-Naturalist* 120(1): 67–70.

Cochran, P.A., and Cochran, J.A. 1999. Predation on a Meadow Jumping Mouse, *Zapus hudsonius*, and a House Mouse, *Mus musculus*, by Brown Trout, *Salmo trutta*. *Canadian Field-Naturalist* 113(4): 684–685.

Cockrum, E.L. 1973. Additional longevity records for American bats. *Journal of the Arizona Academy of Science* 8: 108–110.

Colbourne, P.L., and Terhune, J.M. 1991. Harbour Seals (*Phoca vitulina*) do not follow herring movements in the Bay of Fundy, Canada. *Ophelia* 33(2): 105–112.

Collins, V.E., and Cameron, D.M. 1984. The effects of diet and photoperiod on hibernation in the Woodland Jumping Mouse, *Napaeozapus insignis* (Miller). *Canadian Journal of Zoology* 62(10): 1938–1945.

Collins, W.B., and Urness, P.J. 1982. Mule Deer and Elk responses to horsefly attacks. *Northwest Science* 56(4): 299–302.

Coltman, D.W., Festa-Bianchet, M., Jorgenson, J.T., and Strobeck, C. 2002. Age-dependent sexual selection in Bighorn Rams. *Proceedings of the Royal Society of London, B series* 269: 165–172.

Coltman, D.W., Stenson, G., Hammill, M.O., Haug, T., Davis, C.S., and Fulton, T.L. 2007. Panmictic population structure in the Hooded Seal (*Cystophora cristata*). *Molecular Ecology* 16(8): 1639–1648.

Colvin, D.V. 1973a. Agonistic behaviour in males of five species of voles *Microtus*. *Animal Behaviour* 21(2): 471–480.

– 1973b. Analysis of acoustic structure and function in ultrasounds of neonatal *Microtus*. *Behaviour* 44: 234–263.

Comtois, A., and Berteaux, D. 2005. Impacts of mosquitoes and black flies on defensive behaviour and microhabitat use of the North American Porcupine (*Erethizon dorsatum*) in southern Quebec. *Canadian Journal of Zoology* 83(5): 754–764.

Conley, W. 1976. Competition between *Microtus*: A behavioral hypothesis. *Ecology* 57: 224–237.

Conner, M.M., and Shenk, T.M. 2003. Distinguishing *Zapus hudsonius preblei* from *Zapus princeps princeps* by repeated cranial measurements. *Journal of Mammalogy* 84(4): 1456–1463.

Conner, P.F. 1959. The bog lemming *Synaptomys cooperi* in southern New Jersey. *Publications of the Museum – Michigan State University, Biological Series* 1(5): 161–248.

Connor, R.C. 2000. Group living in whales and dolphins. In J. Mann, R.C. Connor, P.L. Tyack, and H. Whitehead (eds), *Cetacean Societies: Field Studies of Dolphins and Whales*, 199–218. University of Chicago Press, Chicago, IL.

Conroy, C.J., and Cook, J.A. 1999. *Microtus xanthognathus*. *Mammalian Species* no. 627: 1–5.

– 2000. Phylogeography of a post-glacial colonizer: *Microbus longicaudus* (Rodentia: Muridae). *Molecular Ecology* 9: 165–175.

Constantine, D.G. 1998. An overlooked external character to differentiate *Myotis californicus* and *Myotis ciliolabrum* (Vespertilionidae). *Journal of Mammalogy* 79(2): 624–630.

Cook, F.R. 1984. *Introduction to Canadian Amphibians and Reptiles*. National Museums of Canada, Ottawa, ON.

Cook, F.R., and Muir, D. 1984. The Committee on the Status of Endangered Wildlife in Canada (COSEWIC): History and progress. *Canadian Field-Naturalist* 98(1): 63–70.

Cook, J.A., Bidlack, A.L., Conroy, C.J., Demboski, J.R., Fleming, M.A., Runck, A.M., Stone, K.D., and MacDonald, S.O. 2001. A phylogeographic perspective on endemism in the Alexander Archipelago of southeast Alaska. *Biological Conservation* 97: 215–227.

Cook, J.A., Conroy, C.J., and Herriges, J.D., Jr. 1997. Northern record of the Water Shrew, *Sorex palustris*, in Alaska. *Canadian Field-Naturalist* 111(4): 638–640.

Cook, R.C., Murray, D.L., Cook, J.G., Zager, P., and Monfort, S.L. 2001. Nutritional influences on breeding dynamics in Elk. *Canadian Journal of Zoology* 79(5): 845–853.

Cook, S.J., Norris, D.R., and Theberge, J.B. 1999. Spatial dynamics of a migratory wolf population in winter, south-central Ontario (1990–1995). *Canadian Journal of Zoology* 77(11): 1740–1750.

Copeland, J.P., and Whitman, J.S. 2003. Wolverine (*Gulo gulo*). In G.A. Feldhamer, B.C. Thompson, and J. A. Chapman (eds), *Wild Mammals of North America: Biology, Management, and Conservation*. 2nd ed., 672–682. Johns Hopkins University Press, Baltimore, MD.

Cork, S.J., and Kenagy, G.J. 1989. Rates of gut passage and retention of hypogeous fungal spores in two forest-dwelling rodents. *Journal of Mammalogy* 70(3): 512–519.

Cornely, J.E., Carraway, L.N., and Verts, B.J. 1992. *Sorex preblei*. *Mammalian Species* no. 416: 1–3.

Corti, P., and Shackleton, D.M. 2002. Relationship between predator-risk factors and sexual segregation in Dall's Sheep (*Ovis dalli dalli*). *Canadian Journal of Zoology* 80(12): 2108–2117.

Cosens, S.E., and Blouw, A. 2003. Size- and age-class segregation of Bowhead Whales summering in northern Foxe Basin a photogrammetric analysis. *Marine Mammal Science* 19(2): 284–296.

Cosens, S.E., and Dueck, L.P. 1991. Group size and activity patterns of Belugas (*Delphinapterus leucas*) and Narwhals (*Monodon monoceros*) during spring migration in Lancaster Sound. *Canadian Journal of Zoology* 69(6): 1630–1635.

Cote, D., Stewart, H.M.J., Gregory, R.S., Gosse, J., Reynolds, J.J., Stenson, G.B., and Miller, E.H. 2008. Prey selection by marine-coastal River Otters (*Lontra canadensis*) in Newfoundland, Canada. *Journal of Mammalogy* 89(4): 1001–1011.

– 2000. Dominance hierarchies in female Mountain Goats: Stability, aggressiveness and determination of rank. *Behaviour* 137: 1541–1566.

– 2005. Extirpation of a large black bear population by introduced white-tailed deer. *Conservation Biology* 19(5): 1668–1671.

Côté, S.D., and Festa-Bianchet, M. 2001. Offspring sex ratio in relation to maternal age and social rank in Mountain Goats (*Oreamnos americanus*). *Behavioral Ecology and Sociobiology* 49: 260–265.

– 2003. Mountain Goat (*Oreanmos americanus*). In G.A. Feldhamer, B.C. Thompson, and J.A. Chapman (eds), *Wild Mammals of North America: Biology, Management, and Conservation*, 2nd ed., 1061–1075. Johns Hopkins University Press, Baltimore, MD.

Côté, S.D., Peracino, A., and Simard, G. 1997. Wolf, *Canis lupus*, predation and maternal defensive behavior in Mountain Goats. *Canadian Field-Naturalist* 111(3): 389–392.

Cotterill, S.E. 1997. Status of the Swift Fox (*Vulpes velox*) in Alberta. *Wildlife Status Report* no. 7: 1–17. Alberta Environmental Protection, Wildlife Management Division, Edmonton, AB.

Cotton, C.L., and Parker, K.L. 2000a. Winter activity patterns of Northern Flying Squirrels in sub-boreal forests. *Canadian Journal of Zoology* 78(11): 1896–1901.

– 2000b. Winter habitat and nest trees used by Northern Flying Squirrels in sub-boreal forests. *Journal of Mammalogy* 81(4): 1071–1086.

Courtois, R., Ouellet, J.-P., Gingras, A., Dussault, C., and Banville, D. 2001. La situation du caribou forestier au Québec. *Naturaliste Canadien* 125(3): 53–63.

Covell, D.F., Miller, D.S., and Karasov, W.H. 1996. Cost of locomotion and daily energy expenditure by free-living Swift Foxes (*Vulpes velox*): A seasonal comparison. *Canadian Journal of Zoology* 74(2): 283–290.

Cowan, D.P. 1987. Aspects of the social organisation of the European Wild Rabbit (*Oryctolagus cuniculus*). *Ethology* 75: 197–210.

Cowan, D.P., and Bell, D.J. 1986. Leporid social behaviour and social organization. *Mammal Review* 16(3/4): 169–179.

Cowan, I. McT., and Arsenault, M.G. 1954. Reproduction and growth in the Creeping Vole, *Microtus oregoni serpens* Merriam. *Canadian Journal of Zoology* 32(3): 198–208.

Cowan, I. McT., and Guiguet, C.J. 1956. *The Mammals of British Columbia*. Handbook no. 11, British Columbia Provincial Museum, Victoria.

Craig, A.S., Herman, L.M., and Pack, A.A. 2002. Male mate choice and male-male competition coexist in the Humpback Whale (*Megaptera novaeangliae*). *Canadian Journal of Zoology* 80(4): 745–755.

Craighead, L., Paetkau, D., Reynolds, H.V., Vyse, E.R., and Strobeck, C. 1995. Microsatellite analysis of paternity and reproduction in Arctic Grizzly Bears. *Journal of Heredity* 86: 255–261.

Crait, J.R., Blundell, G.M., Ott, K.E., Herreman, J.K., and Ben-David, M. 2006. Late seasonal breeding of River Otters in Yellowstone National Park. *American Midland Naturalist* 156(1): 189–192.

Cramer, K.L., and Chapman, J.A. 1990. Reproduction of three species of pocket mice (*Perognathus*) in the Bonneville Basin, Utah. *Great Basin Naturalist* 50(4): 361–365.

Cranford, J.A. 1978. Hibernation in the Western Jumping Mouse (*Zapus princeps*). *Journal of Mammalogy* 59(3): 496–509.

– 1983a. Body temperature, heart rate and oxygen consumption of normothermic and heterothermic Western Jumping Mouse (*Zapus princeps*). *Comparative Biochemical Physiology* 74A(3): 595–599.

– 1983b. Ecological strategies of a small hibernator, the Western Jumping Mouse, *Zapus princeps*. *Canadian Journal of Zoology* 61(1): 232–240.

Cranford, J.A., and Johnson, E.O. 1989. Effects of coprophagy and diet quality on two microtine rodents (*Microtus pennsylvanicus* and *Microtus pinetorum*). *Journal of Mammalogy* 70(3): 494–502.

Cranford, T.W. 1999. The Sperm Whale's nose: Sexual selection on a grand scale? *Marine Mammal Science* 15(4): 1133–1157.

Crête, M., and Desrosiers, A. 1995. Range expansion of Coyotes, *Canis latrans*, threatens a remnant herd of Caribou, *Rangifer tarandus*, in southeastern Québec. *Canadian Field-Naturalist* 109(2): 227–235.

Crête, M., and Larivière, S. 2003. Estimating the costs of locomotion in snow for Coyotes. *Canadian Journal of Zoology* 81(11): 1808–1814.

Crichton, V. 2002. The horseshoe posture in Moose – a reaction to perceived threats. *Alces* 38: 109–111.

Criddle, S. 1926. The habits of *Microtus minor* in Manitoba. *Journal of Mammalogy* 7: 193–200.

– 1930. The Prairie Pocket Gopher, *Thomomys talpoides rufescens*. *Journal of Mammalogy* 11(3): 265–280.

– 1938. A study of the Snowshoe Rabbit. *Canadian Field-Naturalist* 52: 31–40.

– 1939. The Thirteen-striped Ground Squirrel in Manitoba. *Canadian Field-Naturalist* 53(1): 1–6.

Crockford, S.J., Frederick, S.G., and Wigen, R.J. 2002. The Cape Flattery Fur Seal: An extinct species of *Callorhinus* in the eastern North Pacific? *Canadian Journal of Archaeology* 26: 152–174.

Croll, D.A., Acevedo-Gutiérrez, A., Tershy, B.R., and Urbán-Ramírez, J. 2001. The diving behaviour of Blue and Fin Whales: Is dive duration shorter than expected based on oxygen stores? *Comparative Biochemistry and Physiology Part A* 129: 797–809.

Croll, D.A., Clark, C.W., Acevedo, A., Tershy, B., Flores, S., Gedamke, J., and Urban, J. 2002. Only male Fin Whales sing loud songs. *Nature* 417: 809.

Cronin, M.A., Hills, S., Born, E.W., and Patton, J.C. 1994. Mitochondrial DNA variation in Atlantic and Pacific Walruses. *Canadian Journal of Zoology* 72(6): 1035–1043.

Crooks, K. 1999. Western Spotted Skunk *Spilogale gracilis*. In D.E. Wilson and S. Ruff (eds), *The Smithsonian Book of North American Mammals*, 183–185. UBC Press, Vancouver, BC.

Cryan, P.M. 2003. Seasonal distribution of migratory tree bats (*Lasiurus* and *Lasionycteris*) in North America. *Journal of Mammalogy* 84(2): 579–93.

Cryan, P.M., Bogan, M.A., and Yanega, G.M. 2001. Roosting habits of four bat species in the Black Hills of South Dakota. *Acta Chiroptera* 3(1): 43–52.

Cully, J.F., Jr. 1993. Plague, prairie dogs, and Black-footed Ferrets. *U.S. Fish and Wildlife Service Biological Report* 13: 38–48.

Cully, J.F., Jr. and Williams, E.S. 2001. Interspecific comparisons of Sylvatic Plague in prairie dogs. *Journal of Mammalogy* 82(4): 894–905.

Culver, M., Johnson, W.E., Pecon-Slattery, J., and O'Brien, S.J. 2000. Genomic ancestry of the American Puma (*Puma concolor*). *Journal of Heredity* 91: 186–197.

Cumming, H.G. 1998. Status of Woodland Caribou in Ontario. *Rangifer*. Special issue, 10: 99–104.

Cummins, T., and Slade, N.A. 2007. Summer captures of *Reithrodontomys megalotis* in elevated traps. *Southwestern Naturalist* 52(1): 79–82.

Currah, R.S., Smreciu, E.A., Lehesvirta, T., Niemi, M., and Larsen, K.W. 2000. Fungi in the winter diets of Northern Flying Squirrels and Red Squirrels in the boreal mixedwood forest of northeastern Alberta. *Canadian Journal of Botany* 78: 1514–1520.

Curren, K., and Lien, J. 1998. Observations of White Whales, *Delphinapterus leucas*, in waters off Newfoundland and Labrador and in the Gulf of St Lawrence, 1979–1991. *Canadian Field Naturalist* 112(1): 28–31.

Currier, M.J.P. 1983. *Felis concolor*. *Mammalian Species* no. 200: 1–7.

Cushing, B.S., and Knight, F.M. 1991. Range extension and first reported female Least Weasel in Tennessee. *Journal of the Tennessee Academy of Science* 66(1): 12.

Cypher, B.L. 2003. Foxes (*Vulpes* species, *Urocyon* species, and *Alopex lagopus*). In G.A. Feldhamer, B.C. Thompson, and J. A. Chapman, (eds), *Wild Mammals of North America: Biology, Management, and Conservation*, 2nd ed., 511–546. Johns Hopkins University Press, Baltimore, MD.

Dahle, B., and Swenson, J.E. 2003a. Family breakup in Brown Bears: Are young forced to leave? *Journal of Mammalogy* 84(2): 536–540.

– 2003b. Seasonal range size in relation to reproduction strategies in Brown Bears *Ursus arctos*. *Journal of Animal Ecology* 72: 660–667.

Dahlheim, M.E., and Heyning, J.E. 1999. Killer Whale *Orcinus orca* (Linnaeus, 1758). In S.H. Ridgeway and R. Harrison (eds), *Handbook of Marine Mammals*, vol. 6, 281–322. Academic Press, San Diego, CA.

Dahlheim, M.E., and Towell, R.G. 1994. Occurrence and distribution of Pacific White-side Dolphins (*Lagenorhynchus obliquidens*) in southeastern Alaska, with notes on an attack by Killer Whales (*Orcinus orca*). *Marine Mammal Science* 10(4): 458–464.

Dalerum, F., Loxterman, J., Shults, B., Kunkel, K., and Cook, J.A. 2007. Sex-specific dispersal patterns of Wolverines: Insights from microsatellite markers. *Journal of Mammalogy* 88(3): 793–800.

Dalquest, W.W., and Orcutt, D.R. 1942. The biology of the Least Shrew-mole, *Neurotrichus gibbsii minor*. *American Midland Naturalist* 27: 387–401.

Dane, B. 2002. Retention of offspring in a wild population of ungulates. *Behaviour* 139: 1–21.

Danielson, B.J., and Swihart, R.K. 1987. Home range dynamics and activity patterns of *Microtus ochrogaster* and *Synaptomys cooperi* in syntopy. *Journal of Mammalogy* 68(1): 160–165.

Darden, S.K., Dabelsteen, T., and Pedersen, S.B. 2003. A potential tool for Swift Fox (*Vulpes velox*) conservation: Individuality of long-range barking sequences. *Journal of Mammalogy* 84(4): 1417–1427.

Darimont, C.T., and Paquet, P.C. 2002. The Gray Wolves, *Canis lupus*, of British Columbia's central and north coast: Distribution and conservation assessment. *Canadian Field-Naturalist* 116(3): 416–422.

Darimont, C.T., Reimchen, T.E., and Paquet, P.C. 2003. Foraging behaviour by Gray Wolves on salmon streams in coastal British Columbia. *Canadian Journal of Zoology* 81(2): 349–353.

Darling, J.D., and Bérubé, M. 2001. Interactions of singing Humpback Whales with other males. *Marine Mammal Science* 17(3): 570–584.

Darling, J.D., Keogh, K.E., and Steeves, T.E. 1998. Gray Whale (*Eschrichtius robustus*) habitat utilization and prey species off Vancouver Island, B.C. *Marine Mammal Science* 14(4): 692–720.

Davis, F.W., and Choate, J.R. 1993. Morphologic variation and age structure in a population of the Eastern Mole, *Scalopus aquaticus*. *Journal of Mammalogy* 74(4): 1014–1025.

Davis, H. 1990. On the longevity of a Deer Mouse, *Peromyscus maniculatus*: A Canadian record. *Canadian Field-Naturalist* 104(3): 476.

Davis, H., and Harestad, A.S. 1996. Cannibalism by Black Bears in the Nimpkish Valley, British Columbia. *Northwest Science* 70(2): 88–92.

Davis, R. 1969. Growth and development of young Pallid Bats, *Antrozous pallidus*. *Journal of Mammalogy* 50(4): 729–736.

Davis, W.B., and Schmidly, D.J. 1994. *Mammals of Texas*. University of Texas Press, Austin.

Davis, W.H. 1986. An *Eptesicus fuscus* lives 20 years. *Bat Research News* 27(3–4): 21.

Davis, W.H., and Kalisz, P.J. 1992. Burrow systems of the Prairie Vole, *Microtus ochrogaster*, in central Kentucky. *Journal of Mammalogy* 73(3): 582–585.

Davy, C.M., and Fraser, E.E. 2007. Observation of foliage-roosting in the Little Brown Bat, *Myotis lucifugus*. *Canadian Field-Naturalist* 121(4): 420.

Dawson, S., Barlow, J., and Ljungblad, D. 1998. Sounds recorded from Baird's Beaked Whale, *Berardius bairdii*. *Marine Mammal Science* 14(2): 335–344.

Dearling, M.D. 1996. Disparate determinants of summer and winter diet selection of a generalist herbivore, *Ochotona princeps*. *Oecologia* 108: 467–478.

– 1997a. The function of haypiles of pikas (*Ochotona princeps*). *Journal of Mammalogy* 78(4): 1156–1163.

– 1997b. The manipulation of plant toxins by a food-hoarding herbivore, *Ochotona princeps*. *Ecology* 78(3): 774–781.

Deecke, V.B., Slater, P.J.B., and Ford, J.K.B. 2002. Selective habituation shapes acoustic predator recognition in Harbour Seals. *Nature* 420(6912): 171–173.

Dehnhardt, G., Mauck, B., Hanke, W., and Bleckmann, H. 2001. Hydrodynamic trail-following in Harbour Seals (*Phoca vitulina*). *Science* 293 (6 July): 102–104.

Dekker, D. 1993. Tree-climbing by Long-tailed Weasel: An anti-predator strategy? *Blue Jay* 51(3): 179 –180.

Delfour, F., and Marten, K. 2001. Mirror image processing in three marine mammal species: Killer Whales (*Orcinus orca*), False Killer Whales (*Pseudorca crassidens*) and California Sea Lions (*Zalophus californianus*). *Behavioural Processes* 53: 181–190.

DelGiudice, G.D., Lenarz, M.S., and Castensen Powell, M. 2007. Age-specific fertility and fecundity in northern free-ranging White-tailed Deer: Evidence for reproductive senescence? *Journal of Mammalogy* 88(2): 427–435.

Demarchi, M.W., Johnson, S.R., and Searing, G.F. 2000. Distribution and abundance of Mountain Goats, *Oreamnos americanus*, in westcentral British Columbia. *Canadian Field-Naturalist* 114(2): 301–306.

Demarchi, R.A., Hartwig, C.L., and Demarchi, D.A. 2000a. Status of Rocky Mountain Bighorn Sheep in British Columbia. *Wildlife Bulletin* no. B-99: 1–56. British Columbia Ministry of Environment, Lands and Parks, Wildlife Branch.

– 2000b. Status of California Bighorn Sheep in British Columbia. *Wildlife Bulletin* no. B-98: 1–53. British Columbia Ministry of Environment, Lands and Parks, Wildlife Branch.

DeMaster, D.P., and Stirling, I. 1981. *Ursus maritimus. Mammalian Species* no. 145: 1–7.

DeMatteo, K.E., and Harlow, H.J. 1997. Thermoregulatory responses of the North American Porcupine (*Erethizon dorsatum bruneri*) to decreasing ambient temperature and increasing wind speed. *Comparative Biochemistry and Physiology* 116B(3): 339–346.

Demboski, J.R., and Cook, J.A. 2003. Phylogenetic diversification within the *Sorex cinereus* group (Soricidae). *Journal of Mammalogy* 84(1): 144–158.

Demboski, J.R., and Sullivan, J. 2003. Extensive mtDNA variation within the Yellow-Pine Chipmunk, *Tamias amoenus* (Rodentia: Sciuridae), and phylogeographic inferences for northwest North America. *Molecular Phylogenetics and Evolution* 26: 389–408.

Desroches, J.-F. and Picard, I. 2004. Extension de l'aire de distribution connue de la Musaraigne fuligineuse, *Sorex fumeus*, dans le nord-est du Québec. *Canadian Field-Naturalist* 118(3): 441–442.

Desrosiers, N., Morin, R., and Jutras, J. 2002. *Atlas des Micromammifères du Québec*. Société de la faune et des parcs du Québec. Direction du développement de la faune. Quebec, QC.

DeStefano, S., Koenen, K.K.G., Henner, C.M., and Strules, J. 2006. Transition to independence by subadult beavers (*Castor canadensis*) in an unexploited, exponentially growing population. *Journal of Zoology* 269: 434–441.

Deutsch, C.J., Haley, M.P., and Le Boeuf, B.J. 1990. Reproductive effort of male Northern Elephant Seals: Estimates from mass loss. *Canadian Journal of Zoology* 68(12): 2580–2593.

Devenport, L., Devenport, J., and Kokesh, C. 1999. The role of urine marking in the foraging behaviour of Least Chipmunks. *Animal Behaviour* 57: 557–563.

Din, N.A. 1981. The molt patterns of the Brown Lemming (*Lemmus trimucronatus*) at Baker Lake, Northwest Territories with a note on its hair morphology. *Pakistan Journal of Zoology* 13(1&2): 127–140.

Dizon, A.E., Perrin, W.F., and Akin, P.A. 1994. Stocks of dolphins (*Stenella* spp. and *Delphinus delphis*) in the eastern tropical Pacific: A phylogeographic classification. *NOAA Technical Report of the National Marine Fisheries Series* 119: 1–20.

Dobbyn, J. 1994. *Atlas of the Mammals of Ontario*. Federation of Ontario Naturalists, Don Mills, ON.

Dobson, A., and Lyles, A. 2000. Black-footed Ferret recovery. *Science* 288: 985–988.

Dobson, F.S., Chesser, R.K., Hoogland, J.L., Sugg, D.W., and Foltz, D.W. 2004. The influence of social breeding groups on effective population size in Black-tailed Prairie Dogs. *Journal of Mammalogy* 85(1): 58–66.

– 1997. Do Black-tailed Prairie Dogs minimize inbreeding? *Evolution* 51(3): 970–978.

Dobson, F.S., and Myers, P. 1989. The seasonal decline in litter size of Meadow Voles. *Journal of Mammalogy* 70(1): 142–152.

Dobson, F.S., and Oli, M.K. 2001. The demographic basis of population regulation in Columbian Ground Squirrels. *American Naturalist* 158(3): 236–247.

Doroff, A.M., Estes, J.A., Tinker, M.T., Burn, D.M., and Evans, T.J. 2003. Sea Otter population declines in the Aleutian Archipelago. *Journal of Mammalogy* 84(1): 55–64.

Dorsey, E.M., Stern, S.J., Hoelzel, A.R., and Jacobsen, J. 1990. Minke Whales (*Balaenoptera acutorostrata*) from the west coast of North America: Individual recognition and small-scale site fidelity. *Report of the International Whaling Commission*, Special issue, 12: 357–368.

Doucet, C.M., Adams, I.T., and Fryxell, J.M. 1994. Beaver dam and cache composition: Are woody species used differently? *Ecoscience* 1(3): 268–270.

Doupé, J.P., England, J.H., Furze, M., and Paetkau, D. 2007. Most northerly observation of a Grizzly Bear (*Ursus arctos*) in Canada: Photographic and DNA evidence from Melville Island, Northwest Territories. *Arctic* 60(3): 271–276.

Dowler, R.C., Dawkins, R.C., and Maxwell, T.C. 1999. Range extensions for the Evening Bat (*Nycticeius humeralis*) in west Texas. *Texas Journal of Science* 51(2): 193–195.

Drabek, C.M. 1994. Summer and autumn temporal activity of the Montane Vole (*Microtus montanus*) in the field. *Northwest Science* 68(3): 178–184.

Dragon, D.C., and Elkin, B.T. 2001. An overview of early anthrax outbreaks in northern Canada: Field reports of the Health of Animals Branch, Agriculture Canada, 1962–71. *Arctic* 54(1): 32–40.

Drever, M.C., Blight, L.K., Hobson, K.A., and Bertram, D.F. 2000. Predation on seabird eggs by Keen's Mice (*Peromyscus keeni*): Using stable isotopes to decipher the diet of a terrestrial omnivore on a remote offshore island. *Canadian Journal of Zoology* 78(11): 2010–2018.

Drever, M.C., and Harestad, A.S. 1998. Diets of Norway Rats *Rattus norvegicus*, on Langara Island, Queen Charlotte Islands, British Columbia: Implications for conservation of breeding seabirds. *Canadian Field-Naturalist* 112(4): 676–683.

Drew, G.S., and Bissonette, J.A. 1997. Winter activity patterns of American Martens (*Martes americana*): Rejection of the hypothesis of thermal-cost minimization. *Canadian Journal of Zoology* 75(5): 812–816.

Dubé, Y., Hammill, M.O., and Barrette, C. 2003. Pup development and timing of pupping in Harbour Seals (*Phoca vitulina*) in the St. Lawrence River estuary, Canada. *Canadian Journal of Zoology* 81(2): 188–194.

Dubois, J.E., and Monson, K.M. 2007. Recent distribution records of the Little Brown Bat, *Myotis lucifugus*, in Manitoba and northwestern Ontario. *Canadian Field-Naturalist* 121(1): 57–61.

Duffy-Echevarria, E.E., Connor, R.C., St Aubin, D.J., 2008. Observations of strand-feeding behavior by Bottlenose Dolphins (*Tursiops truncatus*) in Bull Creek, South Carolina. *Marine Mammal Science* 24(1): 202–206.

Duignan, P.J., House, C., Odell, D.K., Wells, R.S., Hansen, L.J., Walsh, M.T., St. Aubin, D.J., Rima, B.K., and Geraci, J.R. 1996. Morbillivirus infection in bottlenose Dolphins: Evidence for recurrent epizootics in the western Atlantic and Gulf of Mexico. *Marine Mammal Science* 12(4): 499–515.

Duke, S.D., Bateman, G.C., and Bateman, M.M. 1979. Longevity record for *Myotis californicus*. *Southwest Naturalist* 24(4): 693.

Dumont, A., and Crête, M. 1996. The Meningeal Worm, *Parelaphostrongylus tenuis*, a marginal limiting factor for Moose, *Alces alces*, in southern Québec. *Canadian Field-Naturalist* 110(3): 413–418.

Dumyahn, J.B., Zollner, P.A., and Gilbert, J.H., 2007. Winter home-range characteristics of American Marten (*Martes americana*) in northern Wisconsin. *American Midland Naturalist* 158(2): 382–394.

Dunbar, M.B. 2007. Thermal energetics of torpid Silver-haired Bats *Lasionycteris noctivagans*. *Acta Theriologica* 52(1): 65–68.

Dunham, J.S., and Duffus, D.A. 2002. Diet of Gray Whales (*Eschrichtius robustus*) in Clayoquot Sound, British Columbia, Canada. *Marine Mammal Science* 18(2): 419–437.

Dunlop, R.A., Cato, D.H., and Noad, M.J. 2008. Non-song acoustic communication in migrating Humpback Whales (*Megaptera novaeangliae*). *Marine Mammal Science* 24(3): 613–629.

Dunmire, W.W. 1960. An altitudinal survey of reproduction in *Peromyscus maniculatus*. *Ecology* 41(1): 174–182.

Dunn, J.P., Chapman, J.A., and Marsh, R.E. 1982. Jackrabbits (*Lepus californicus* and allies). In J.A. Chapman and C.A. Feldhamer (eds), *Wild Mammals of North America*, 124–145. Johns Hopkins University Press, Baltimore, MD.

Dunphy-Daly, M.M., Heithaus, M.R., and Claridge, D.E. 2008. Temporal variation in Dwarf Sperm Whale (*Kogia sima*) habitat use and group size off Great Abaco Island, Bahamas. *Marine Mammal Science* 24(1): 171–182.

Durner, G.M., Amstrup, S.C., and Fischbach, A.S. 2003. Habitat characteristics of Polar Bear terrestrial maternal den sites in northern Alaska. *Arctic* 56(1): 55–62.

Dyck, A.P., and MacArthur, R.A. 1993. Seasonal variation in the microclimate and gas composition of Beaver lodges in a boreal environment. *Journal of Mammalogy* 74(1): 180–188.

Dyck, M.G., and Daley, K.J. 2002. Cannibalism of a yearling Polar Bear (*Ursus maritimus*) at Churchill, Canada. *Arctic* 55(2): 190–192.

Dyke, A.S., Hooper, J., Harington, C.R., and Savelle, J.M. 1999. The Late Wisconsinan and Holocene record of Walrus (*Odobenus rosmarus*) from North America: A review with new data from Arctic and Atlantic Canada. *Arctic* 52(2): 160–181.

Earle, R.D., and Kramm, K.R. 1982. Correlation between Fisher and Porcupine abundance in Upper Michigan. *American Midland Naturalist* 107(2): 244–249.

Eason, P.K. 1998. Predation of a female House Finch, *Carpodacus mexicanus*, by a Gray Squirrel, *Sciurus carolinensis*. *Canadian Field-Naturalist* 112(4): 713–714.

Easterla, D.A. 1965. The Spotted Bat in Utah. *Journal of Mammalogy* 46(4): 664–668.

– 1971. Notes on young and adults of the Spotted Bat, *Euderma maculatum*. *Journal of Mammalogy* 52(2): 475–476.

Ebensperger, L.A., Botto-Mahan, C., and Tamarin, R.H. 2000. Nonparental infanticide in Meadow Voles, *Microtus pennsylvanicus*: The influence of nutritional benefits. *Ethology Ecology and Evolution* 12: 149–160.

Eccard, J.A., and Ylönen, H. 2001. Initiation of breeding after winter in Bank Voles: Effects of food and population density. *Canadian Journal of Zoology* 79(10): 1743–1753.

Edds-Walton, P.L. 2000. Vocalizations of Minke Whales *Balaenoptera acutorostrata* in the St. Lawrence estuary. *Bioacoustics* 11: 31–50.

Edie, A. 2001. Update COSEWIC Status Report on the Ermine *haidarum* subspecies *Mustela erminea haidarum* in Canada. Committee on the Status of Endangered Wildlife in Canada, Ottawa.

Edmonds, J. 1998. Status of Caribou in Alberta. *Rangifer*, Special issue, 10: 111–115.

Edwards, M.A., and Forbes, G.J. 2003. Food habits of Ermine, *Mustela erminea*, in a forested landscape. *Canadian Field-Naturalist* 117(2): 245–248.

Egoscue, H.J. 1979. *Vulpes velox*. *Mammalian Species* no. 122: 1–5.

Ehrich, D., and Jorde, P.E. 2005. High genetic variability despite high-amplitude population cycles in lemmings. *Journal of Mammalogy* 86(2): 380–385.

Ehrich, D., Krebs, C.J., Kenney, A.J., and Stenseth, N.C. 2001. Comparing the genetic population structure of two species of arctic lemmings: More local differentiation in *Lemmus trimucronatus* than in *Dicrostonyx groenlandicus*. *Oikos* 94: 143–150.

Eiler, K.C., and Banack, S.A. 2004. Variability in the alarm call of golden-mantled ground squirrels (*Spermophilus lateralis* and *S. saturatus*). *Journal of Mammalogy* 85(1): 43–50.

Eisner, T., and Meinwald, J. 1966. Defensive secretions of arthropods. *Science* 153: 1341–1350.

Elbroch, M. 2003. *Mammal Tracks & Sign*. Stackpole Books, Mechanicsburg, PA.

Elias, S.P., Witham, J.W., and Hunter, M.L., Jr. 2006. A cyclic red-backed vole (*Clethrionomys gapperi*) population and seedfall over 22 years in Maine. *Journal of Mammalogy* 87(3): 440–445.

Elliott, C.L., and Flinders, J.T. 1991. *Spermophilus columbianus*. *Mammalian Species* no. 372: 1–9.

Elsey, R.M., and Kinler, N. 1996. Range extension of the American Beaver, *Castor canadensis*, in Louisiana. *Southwest Naturalist* 41(1): 91–93.

Elsner, R., George, J.C., and O'Hara, T. 2004a. Vasomotor responses of isolated peripheral blood vessels from Bowhead Whales: Thermoregulatory implications. *Marine Mammal Science* 20(3): 546–553.

Elsner, R., Meiselman, H.J., and Baskurt, O.K. 2004b. Temperature–viscosity relations of Bowhead Whale blood: A possible mechanism for maintaining cold blood flow. *Marine Mammal Science* 20(2): 339–344.

Engley, L., and Norton, M. 2001. Distribution of selected small mammals in Alberta. *Alberta Species at Risk Report* no. 12: 1–75. Alberta Sustainable Resource Development, Fisheries and Wildlife Service.

Engstrom, M.D. 1999a. Ungava Collared Lemming *Dicrostonyx hudsonius*. In D.E. Wilson and S. Ruff (eds), *The Smithsonian Book of North American Mammals*, 660–662. UBC Press, Vancouver, BC.

– 1999b. Richardson's Collared Lemming *Dicrostonyx richardsoni*. In Wilson and Ruff (eds), *The Smithsonian Book of North American Mammals*, 662–664.

Engstrom, M.D., Baker, A.J., Eger, J.L., Boonstr, R., and Brooks, R.J. 1993. Chromosomal and mitochondrial DNA variation in four laboratory populations of collared lemmings (*Dicrostonyx*). *Canadian Journal of Zoology* 71(1): 42–48.

Engstrom, M.D., and Choate, J.R. 1979. Systematics of the Northern Grasshopper Mice (*Onychomys leucogaster*) on the Central Great Plains. *Journal of Mammalogy* 60(4): 723–739.

Erb, J., Stenseth, N.C., and Boyce, M.S. 2000. Geographic variation in population cycles of Canadian Muskrats (*Ondatra zibethicus*). *Canadian Journal of Zoology* 78(6): 1009–1016.

Erbe, C. 2002. Underwater noise of whale-watching boats and potential effects on Killer Whales (*Orcinus orca*), based on an acoustic impact model. *Marine Mammal Science* 18(2): 394–418.

Escorza-Treviño, S., and Dizon, A.E. 2000. Phylogeography, intraspecific structure and sex-biased dispersal of Dall's Porpoise *Phocoenoides dalli*, revealed by mitochondrial and microsatellite DNA analyses. *Molecular Ecology* 9: 1049–1060.

Estes, J.A. 1980. *Enhydra lutris*. *Mammalian Species* no. 133: 1–8.

Estes, J.A., and Duggins, D.O. 1995. Sea Otters and kelp forests in Alaska: Generality and variation in a community ecological paradigm. *Ecological Monographs* 65(1): 75–100.

Estes, J.A., Riedman, M.L., Staedler, M.M., Tinker, M.T., and Lyon, B.E. 2003. Individual variation in prey selection by Sea Otters: Patterns, causes and implications. *Journal of Animal Ecology* 72: 144–155.

Estes, J.A., Tinker, M.T., Williams, T.M., and Doak, D.F. 1998. Killer Whale predation on Sea Otters linking oceanic and nearshore ecosystems. *Science* 282: 473–476.

Ewer, R.F. 1971. The biology and behaviour of a free-living population of Black Rats (*Rattus rattus*). *Animal Behaviour Monographs* 4: 127–174.

Fairbanks, L., and Koprowski, J.L. 1992. Piscivory in fox squirrels. *Prairie Naturalist* 24(4): 283–284.

Falk, J.W., and Millar, J.S. 1987. Reproduction by female *Zapus princeps* in relation to age, size, and body fat. *Canadian Journal of Zoology* 65(3): 568–571.

Farnell, R., Florkiewicz, R., Kuzyk, G., and Egli, K. 1998. The status of *Rangifer tarandus caribou* in Yukon, Canada. *Rangifer*, Special issue, 10: 131–137.

Fay, F.H. 1981. Walrus *Odobenus rosmarus* (Linnaeus, 1758). In S.H. Ridgway and R. Harrison (eds), *Handbook of Marine Mammals*, vol. 1, *The Walrus, Sea Lions, Fur Seals and Sea Otter*. Academic Press, San Diego, CA.

– 1985. *Odobenus rosmarus*. *Mammalian Species* no. 238: 1–7.

Fay, F.H., Eberhardt, L.L., Kelly, B.P., Burns, J.J., and Quakenbush, L.T. 1997. Status of the Pacific Walrus population, 1950–1989. *Marine Mammal Science* 13(4): 537–565.

Fay, F.H., Sease, J.L., and Merrick, R.L. 1990. Predation on a Ringed Seal *Phoca hispida*, and a Black Guillemot, *Cepphus grylle*, by a Pacific Walrus, *Odobenus rosmarus divergens*. *Marine Mammal Science* 6(4): 348–350.

Fedriani, J.M., Fuller, T.K., Sauvajot, R.M., and York, E.C. 2000. Competition and intraguild predation among three sympatric carnivores. *Oecologia* 125: 258–270.

Feighny, J.A., Williamson, K.E., and Clarke, J.A. 2006. North American Elk bugle vocalizations: Male and female bugle call structure and context. *Journal of Mammalogy* 87(6): 1072–1077.

Feldhamer, G.A., Rochelle, J.A., and Rushton, C.D. 2003. Mountain Beaver (*Aplodontia rufa*). In G.A. Feldhamer, B.C. Thompson, and J.A. Chapman (eds), *Wild*

Mammals of North America: Biology, Management, and Conservation, 2nd ed., 179–187. Johns Hopkins University Press, Baltimore, MD.

Félix, F., Bearson, B., and Falconí, J. 2006. Epizoic barnacles removed from the skin of a HumpbackWhale after a period of intense surface activity. *Marine Mammal Science* 22(4): 979–984.

Fellers, G.M. 2000. Predation on *Corynorhinus townsendii* by *Rattus rattus*. *Southwestern Naturalist* 45(4): 524–527.

Fener, H.M., Ginsberg, J.R., Sanderson, E.W., and Gompper, M.E. 2005. Chronology of range expansion of the Coyote, *Canis latrans*, in New York. *Canadian Field-Naturalist* 119(1): 1–5.

Fenton, M.B. 1969. The carrying of young by females of three species of bats. *Canadian Journal of Zoology* 47(1): 158–159.

– 1972. Distribution and overwintering of *Myotis leibii* and *Eptesicus fuscus* (Chiroptera: Vespertilionidae) in Ontario. *Royal Ontario Museum, Life Sciences Occasional Papers* 21: 1–8.

– 1999a. Little Brown Bat. In D.E. Wilson and S. Ruff (eds), *The Smithsonian Book of North American Mammals*, 94–95. UBC Press, Vancouver, BC.

– 1999b. Northern long-eared myotis. In Wilson and Ruff (eds), *The Smithsonian Book of North American Mammals*, 96.

– 2001. *Bats*. Revised Edition. Checkmark Books, New York, NY.

Fenton, M.B., and Barclay, R.M.R. 1980. *Myotis lucifugus*. *Mammalian Species* no. 142: 1–8.

Fenton, M.B., and Bell, G.P. 1979. Echolocation and feeding behaviour in four species of *Myotis* (Chiroptera). *Canadian Journal of Zoology* 57(6): 1271–1277.

– 1981. Recognition of species of insectivorous bats by their echolocation calls. *Journal of Mammalogy* 62(2): 233–243.

Fenton, M.B., Merriam, H.G., and Holroyd, G.L. 1983. Bats of Kootenay, Glacier, and Mount Revelstoke national parks in Canada: Identification by echolocation calls, distribution, and biology. *Canadian Journal of Zoology* 61(11): 2503–2508.

Fenton, M.B., van Zyll de Jong, C.G., Bell, G.P., Campbell, D.B., and LaPlante, M. 1980. Distribution, parturition dates, and feeding of bats in south-central British Columbia. *Canadian Field-Naturalist* 94(4): 416–420.

Ferguson, M.A.D., and Gautier, L. 1991. Status and trends of *Rangifer tarandus* and *Ovibos moschatus* populations in Canada. *Rangifer* 12(3): 127–141.

Ferguson, S.H., and Elkie, P.C. 2004. Seasonal movement of Woodland Caribou (*Rangifer tarandus caribou*). *Journal of Zoology* (London) 262: 125–134.

Ferguson, S.H., Stirling, I., and McLoughlin, P. 2005. Climate change and Ringed Seal (*Phoca hispida*) recruitment in western Hudson Bay. *Marine Mammal Science* 21(1): 121–135.

Ferguson, S.H., Taylor, M.K., Rosing-Asvid, A., Born, E.W., and Messier, F. 2000. Relationships between denning of Polar Bears and conditions of sea ice. *Journal of Mammalogy* 81(4): 1118–1127.

Ferguson, T.A. 2003. Documenting Pronghorn Antelope, *Antilocapra americana*, in the Peace River Grasslands, Alberta. *Canadian Field-Naturalist* 117(4): 657–658.

Ferkin, M.H. 2001. Patterns of sexually distinct scents in *Microtus* spp. *Canadian Journal of Zoology* 79(9): 1621–1625.

Ferrero, R.C., Hobbs, R.C., and VanBlaricom, G.R. 2002. Indications of habitat use patterns among small cetaceans in the central North Pacific based on fisheries observer data. *Journal of Cetacean Research Management* 4(3): 311–321.

Ferrero, R.C., and Walker, W.A. 1993. Growth and reproduction of the Northern Right Whale Dolphin, *Lissodelphis borealis*, in the offshore waters of the North Pacific Ocean. *Canadian Journal of Zoology* 71(12): 2335–2344.

– 1995. Growth and reproduction of the Common Dolphin, *Delphinus delphis* Linnaeus, in the offshore waters of the North Pacific Ocean. *U.S. National Marine Fisheries Service Fisheries Bulletin* 93(3): 483–494.

– 1999. Age, growth, and reproductive patterns of Dall's Porpoise (*Phocoenoides dalli*) in the central North Pacific Ocean. *Marine Mammal Science* 15(2): 273–313.

Ferron, J. 1983. Scent marking by cheek rubbing in the Northern Flying Squirrel (*Glaucomys sabrinus*). *Canadian Journal of Zoology* 61(11): 2377–2380.

– 1996. How do Woodchucks (*Marmota monax*) cope with harsh winter conditions? *Journal of Mammalogy* 77(2): 412–416.

Ferron, J., and Ouellet, J.-P. 1991. Physical and behavioral postnatal development of Woodchucks (*Marmota monax*). *Canadian Journal of Zoology* 69(4): 1040–1047.

Fertl, D., Pusser, L.T., and Long, J.J. 1999. First record of an albino Bottlenose Dolphin (*Tursiops truncatus*) in the Gulf of Mexico, with a review of anomalously white cetaceans. *Marine Mammal Science* 15(1): 227–234.

Festa-Bianchet, M., and King, W.J. 1984. Behavior and dispersal of yearling Columbian Ground Squirrels. *Canadian Journal of Zoology* 62(2):161–167.

Festa-Bianchet, M., Urquhart, M., and Smith, K.G. 1994. Mountain Goat recruitment: Kid production and survival to breeding age. *Canadian Journal of Zoology* 72(1): 22–27.

Findlay, A. 2005. Mustang Valley. *Canadian Geographic*, March/April: 47–62.

Findley, J.S. 1999. Vagrant Shrew (*Sorex vagrans*). In D.E. Wilson and S. Ruff (eds), *The Smithsonian Book of North American Mammals*, 46–47. UBC Press, Vancouver, BC.

Finley, K. 1996. The Red Fox invasion and other changes in wildlife populations in west-central Saskatchewan since the 1960s. *Blue Jay* 54(4): 206–210.

Finley, K.J. 1990. Isabella Bay, Baffin Island: An important historical and present-day concentration area for the endangered Bowhead Whale (*Balaena mysticetus*) of the eastern Canadian Arctic. *Arctic* 43(2): 137–152.

– 2001. Natural history and conservation of the Greenland Whale, or Bowhead, in the northwest Arctic. *Arctic* 54(1): 55–76.

Finley, T.G., and Sikes, R.S. 2004. Ecological comparison of grasshopper mouse vocalizations. *Abstract from 84th Annual Meeting of the American Society of Mammalogists*. Humbolt State University, Arcata, CA.

Fiscus, C.H. 1997. Cephalopod beaks in a Cuvier's Beaked Whale (*Ziphius cavirostris*) from Amchitka Island, Alaska. *Marine Mammal Science* 13(3): 481–486.

Fish, F.E. 1993. Comparison of swimming kinematics between terrestrial and semiaquatic opossums. *Journal of Mammalogy* 74(2): 275–284.

Fitzgerald, S.M., and Keith, L.B. 1990. Intra- and inter-specific dominance relationships among Arctic and Snowshoe Hares. *Canadian Journal of Zoology* 68(3): 457–464.

Flinn, R.D., Trites, A.W., Gregr, E.J., and Perry, R.I. 2002. Diets of Fin, Sei, and Sperm Whales in British Columbia: An analysis of commercial whaling records, 1963–1967. *Marine Mammal Science* 18(3): 663–679.

Flórez-González, L., Capella, J.J., and Rosenbaum, H.C. 1994. Attack of Killer Whales (*Orcinus orca*) on Humpback Whales (*Megaptera novaeangliae*) on a South American Pacific breeding ground. *Marine Mammal Science* 10(2): 218–222.

Flux, J.E.C., and Fullagar, P.J. 1992. World distribution of the Rabbit *Oryctolagus cuniculus* on islands. *Mammal Review* 22(3/4): 151–205.

Folkow, L.P., and Blix, A.S. 1999. Diving behaviour of Hooded Seals (*Cystophora cristata*) in the Greenland and Norwegian Seas. *Polar Biology* 22: 61–74.

Folkow, L.P., Mårtensson, P.-E., and Blix, A.S. 1996. Annual distribution of Hooded Seals (*Cystophora cristata*) in the Greenland and Norwegian Seas. *Polar Biology* 16: 179–189.

Folkow, L.P., Nordøy, E.S., and Blix, A.S. 2004. Distribution and diving behaviour of Harp Seals (*Pagophilus groenlandicus*) from the Greenland Sea stock. *Polar Biology* 27: 281–298.

Forbes, G.J., and Theberge, J.B. 1995. Influences of a migratory deer herd on wolf movements and mortality in and near Algonquin Park, Ontario. *Canadian Circumpolar Institute Occasional Publication* 35: 303–313.

Ford, J.K.B. 2005. First records of Long-beaked Common Dolphins, *Delphinus capensis*, in Canadian waters. *Canadian Field-Naturalist* 119(1): 110–113.

Ford, J.K.B., Ellis, G.M., Barrett-Lennard, L.G., Morton, A.B., Palm, R.S., and Balcomb, K.C., III. 1998. Dietary specialization in two sympatric populations of Killer Whales (*Orcinus orca*) in coastal British Columbia and adjacent waters. *Canadian Journal of Zoology* 76(7): 1456–1471.

Ford, J.K.B., Ellis, G.M., Matkin, D.R., Balcomb, K.C., Briggs, D., and Morton, A.B. 2005. Killer Whale attacks on Minke Whales: Prey capture and antipredator tactics. *Marine Mammal Science* 21(4): 603–618.

Ford, J.K.B., and Reeves, R.R. 2008. Fight or flight: Antipredator strategies of baleen whales. *Mammal Review* 38(1): 50–86.

Foresman, K.R. 1999. Distribution of the Pygmy Shrew, *Sorex hoyi*, in Montana and Idaho. *Canadian Field-Naturalist* 113(4): 681–683.

Foresman, K.R., and Long, R.D. 1998. The reproductive cycles of the Vagrant Shrew (*Sorex vagrans*) and the Masked Shrew (*Sorex cinereus*) in Montana. *American Midland Naturalist* 139(1): 108–113.

Foresman, K.R., and Pearson, D.E. 1999. Activity patterns of American Martens, *Martes americana*, Snowshoe Hares, *Lepus americanus*, and Red Squirrels, *Tamiasciurus hudsonicus*, in westcentral Montana. *Canadian Field-Naturalist* 113(3): 386–689.

Formanowicz, D.R., Jr, Bradley, P.J., and Brodie, E.D., Jr. 1989. Food hoarding by the Least Shrew (*Cryptotis parva*): Intersexual and prey type effects. *American Midland Naturalist* 122(1): 26–33.

Forsey, O., Bissonette, J., Brazil, J., Curnew, K., Lemon, J., Mayo, L., Thompson, I., Bateman, L., and O'Driscoll, L. 1995. National recovery plan for the Newfoundland Marten. *Recovery of Nationally Endangered Wildlife Committee Report* no. 14: 1–29.

Fortin, C., Rousseau, J.-F. and Grimard, M.-J. 2004. Extension de l'aire de répartition du campagnol-lemming de Cooper (*Synaptomys cooperi*): Mentions les plus nordiques. *Naturaliste Canadien* 128(2): 35–37.

Foster, J.B. 1961. Life history of the Phenacomys Vole. *Journal of Mammalogy* 42(2): 181–198.

Fournier, F., and Festa-Bianchet, M. 1995. Social dominance in adult female Mountain Goats. *Animal Behaviour* 49: 1449–1459.

Frame, P.F., Hik, D.S., Cluff, H.D., and Paquet, P.C. 2004. Long foraging movement of a denning Tundra Wolf. *Arctic* 57(2): 196–203.

Franken, R.J., and Hik, D.S. 2004. Influence of habitat quality, patch size and connectivity on colonization and extinction dynamics of Collared Pikas *Ochotona collaris*. *Journal of Animal Ecology* 73: 889–896.

Frantzis, A. 1998. Does acoustic testing strand whales? *Nature* 392(6671): 29.

Frantzis, A., and Cebrian, D. 1999. A rare atypical mass stranding of Cuvier's Beaked Whales: Cause and implications for the species' biology. *European Research on Cetaceans* 12: 332–335.

Franzmann, A.W. 1981. *Alces alces*. *Mammalian Species* no. 154: 1–7.

– 2000. Moose. In S. Demarais and P.R. Krausman (eds), *Ecology and Management of Large Mammals in North America*, 578–600. Prentice Hall, Upper Saddle River, NJ.

Franzmann, A.W., LeResche, R.E., Rausch, R.A., and Oldemeyer, J.L. 1978. Alaskan Moose measurements and weights and measurement–weight relationships. *Canadian Journal of Zoology* 56(2): 298–306.

Frase, B.A., and Hoffmann, R.S. 1980. *Marmota flaviventris*. *Mammalian Species* no. 135: 1–8.

Frederiksen, J.K., and Slobodchikoff, C.N. 2007. Referential specificity in the alarm calls of the Black-tailed Prairie Dog. *Ethology Ecology and Evolution* 19: 87–99.

French, A.R. 2000. Interdependency of stored food and changes in body temperature during hibernation of the Eastern Chipmunk, *Tamias striatus*. *Journal of Mammalogy* 81(4): 979–985.

French, T.W., and Crowell, K.L. 1985. Distribution and status of the Yellow-nosed Vole and Rock Shrew in New York. *New York Fish and Game Journal* 32(1): 26–40.

Frey, J.K. 1992. Response of a mammalian faunal element to climatic changes. *Journal of Mammalogy* 73(1): 43–50.

Frey, J.K., and Moore, D.W. 1990. Status of Hayden's Shrew (*Sorex haydeni*) in Kansas. *Southwestern Naturalist* 35(1): 84–86.

Frey, R., Gebler, A., and Fritsch, G. 2006. Arctic roars – laryngeal anatomy and vocalization of the Muskox (*Ovibos moschatus* Zimmerman, 1780, Bovidae). *Journal of Zoology* 268: 433–448.

Friesen, V.C. 1993. Northern Pocket Gophers occupy new territory. *Blue Jay* 51(2): 122–3.

– 1999. Cougar sighting at Rosthern. *Blue Jay* 57(4): 187–188.

Fristrup, K.M., and Harbison, G.R. 2002. How do Sperm Whales catch squids? *Marine Mammal Science* 18(1): 42–54.

Frommolt, K.-H., Goltsman, M.E., and Macdonald, D.W. 2003. Barking foxes, *Alopex lagopus*: Field experiments in individual recognition in a territorial mammal. *Animal Behaviour* 65: 509–518.

Frommolt, K.-H., Kruchenkova, E.P., and Russig, H. 1997. Individuality of territorial barking in Arctic Foxes, *Alopex lagopus* (L., 1758). *Zeitchrisft fur Saugetierkunde* 62: 66–70 (Supplement 2).

Frost, K.J., and Lowry, L.F. 1981. Ringed, Baikal and Caspian Seals *Phoca hispida* Schreber, 1775, *Phoca siberica* Gmelin, 1788 and *Phoca caspica* Gmelin, 1788. In S.H. Ridgway and R. Harrison (eds), *Handbook of Marine Mammals*, vol. 2, *Seals*, 29–53. Academic Press, San Diego, CA.

Frost, K.J., Simpkins, M.A., Small, R.J., and Lowry, L.F. 2006. Development of diving by Harbor Seal pups in two regions of Alaska: Use of the water column. *Marine Mammal Science* 22(3): 617–643.

Fryxell, J.M., Falls, J.B., Falls, E.A., Brooks, R.J., Dix, L., and Strickland, M.A. 1999. Density dependence, prey dependence, and population dynamics of martens in Ontario. *Ecology* 80(4): 1311–1321.

Fu, C., Mohn, R., and Fanning, L.P. 2001. Why the Atlantic Cod (*Gadus morhua*) stock off eastern Nova Scotia has not recovered. *Canadian Journal of Fisheries and Aquatic Sciences* 58(8): 1613–1623.

Fuare, P.A., and Barclay, R.M.R. 1992. The sensory basis of prey detection by the Long-eared Bat, *Myotis evotis*, and the consequences of prey selection. *Animal Behaviour* 44: 31–39.

Fujita, M.S., and Kunz, T.H. 1984. *Pipistrellus subflavus*. *Mammalian Species* no. 228: 1–6.

Fujiwara, M., and Caswell, H. 2001. Demography of the endangered North Atlantic Right Whale. *Nature* 414(6863): 537–541.

Fullard, J.H., and Dawson, J.W. 1997. The echolocation calls of the Spotted Bat *Euderma maculatum* are relatively inaudible to moths. *Journal of Experimental Biology* 200: 129–137.

Fullard, K.J., Early, G., Heide-Jørgensen, M.P., Bloch, D., Rosing-Asvid, A., and Amos, W. 2000. Population structure of Long-finned Pilot Whales in the North Atlantic: A correlation with sea surface temperature. *Molecular Ecology* 9: 949–958.

Fuller, C.A., and Blaustein, A.R. 1990. An investigation of sibling recognition in a solitary sciurid, Townsend's Chipmunk, *Tamias townsendii*. *Behaviour* 112(1–2): 36–52.

Fuller, T.K., and Cypher, B.L. 2004. *Urocyon cinereoargenteus* (Schreber, 1775). In C. Sillero Zubiri, M. Hoffman, and D.W. Macdonald (eds), *Canids: Foxes, Wolves, Jackals and Dogs: Status Survey and Conservation Plan*, 92–97. IUCN, Gland, Switzerland.

Fuller, W.A. 2002. Canada and the 'Buffalo,' *Bison bison*: A tale of two herds. *Canadian Field-Naturalist* 116(1): 141–159.

Furgal, C.M., Innes, S., and Kovacs, K.M. 1996. Characteristics of Ringed Seal, *Phoca hispida*, subnivean structures and breeding habitat and their effects on predation. *Canadian Journal of Zoology* 74(5): 858–874.

– 2002. Inuit spring hunting techniques and local knowledge of the Ringed Seal in Arctic Bay (Ikpiarjuk), Nunavut. *Polar Research* 21(1): 1–16.

Gabriele, C.M., Straley, J.M., Herman, L.M., and Coleman, R.J. 1996. Fastest documented migration of a North Pacific Humpback Whale. *Marine Mammal Science* 12(3): 457–464.

Gabriele, C.M., Straley, J.M., and Neilson, J.L. 2007. Age at first calving of female Humpback Whales in southeastern Alaska. *Marine Mammal Science* 23(1): 226–239.

Gaillard, J.-M. 2007. Are Moose only a large deer? Some life history considerations. *Alces* 43: 1–11.

Gainer, R.S. 1995. Recent winter biology of Pronghorn near Hanna, Alberta. *Alberta Naturalist* 25(4): 66–69.

Gajda, A.M. Tuchscherer and Brooks, R.J. 1993. Paternal care in collared lemmings (*Dicrostonyx richardsoni*): Artifact or adaptation? *Arctic* 46(4): 312–315.

Galentine, S.P., and Swift, P.K. 2007. Intraspecific killing among Mountain Lions (*Puma concolor*). *American Midland Naturalist* 52(1): 161–164.

Galindo, C., and Krebs, C.J. 1985. Habitat use by Singing Voles and Tundra Voles in the southern Yukon. *Oecologia* 66: 430–436.

Galindo-Leal, C., and Zuleta, G. 1997. The distribution, habitat, and conservation status of the Pacific Water Shrew, *Sorex bendirii*, in British Columbia. *Canadian Field-Naturalist* 111(3): 422–428.

Gallant, D., Bérubé, C.H., Tremblay, E., and Vasseur, L. 2004. An extensive study of the foraging ecology of beavers (*Castor canadensis*) in relation to habitat quality. *Canadian Journal of Zoology* 82(5): 922–933.

Gambell, R. 1985. Sei Whale *Balaenoptera borealis* Lesson, 1828. In S.H. Ridgway and R. Harrison (eds), *Handbook of Marine Mammals*, vol. 3, 155–170. Academic Press, San Diego, CA.

Gannon, D.P., Read, A.J., Craddock, J.E., Fristrup, K.M., and Nicholas, J.R. 1997a. Feeding ecology of Long-finned Pilot Whales *Globicephala melas* in the western North Atlantic. *Marine Ecology Progress Series* 148: 1–10.

Gannon, D.P., Read, A.J., Craddock, J.E., and Mead, J.G. 1997b. Stomach contents of Long-finned Pilot Whales (*Globicephala melas*) stranded on the U.S. mid-Atlantic coast. *Marine Mammal Science* 13(3): 405–418.

Gannon, W.L., Sherwin, R.E., deCarvalho, T.N., and O'Farrell, M.J. 2001. Pinnae and echolocation call differences between *Myotis californicus* and *M. ciliolabrum* (Chiroptera: Vespertilionidae). *Acta Chiropterologica* 3(1): 77–91.

Garant, Y., and Crête, M. 1997. Fisher, *Martes pennanti*, home range characteristics in a high density untrapped population in southern Québec. *Canadian Field-Naturalist* 111(3): 359–364.

Garcia, P.F.J., Rasheed, S.A., and Holroyd, S.L. 1995. Status of the Western Small-footed Myotis in British Columbia. *Wildlife Working Report* no. WR-74: 1–14. British Columbia, Ministry of Environment, Lands and Parks, Wildlife Branch.

Garcia-Aguilar, M.C., and Aurioles-Gamboa, D. 2003. Breeding season on the California Sea Lion (*Zalophus californianus*) in the Gulf of California, Mexico. *Aquatic Mammals* 29(1): 67–76.

Garcia-Rodríguez, F.J., and Aurioles-Gamboa, D. 2004. Spatial and temporal variation in the diet of the California Sea Lion (*Zalophus californianus*) in the Gulf of California, Mexico. *Fisheries Bulletin* 102: 47–62.

Garde, E., Heide-Jørgensen, M.P., Hansen, S.H., Nachman, G., and Forchhammer, M.C. 2007. Age-specific growth and remarkable longevity in Narwhals (*Monodon monoceros*) from west Greenland as estimated by aspartic acid racemization. *Journal of Mammalogy* 88(1): 49–58.

Gardner, A.L., and Sunquist, M.E. 2003. Opossum (*Didelphis virginiana*). In G.A. Feldhamer, B.C. Thompson, and J.A. Chapman (eds), *Wild Mammals of North America: Biology, Management, and Conservation*, 2nd ed., 3–29. Johns Hopkins University Press, Baltimore, MD.

Garrison, T.E., and Best, T.L. 1990. *Dipodomys ordii. Mammalian Species* no. 353: 1–10.

Garshelis, D.L., and Hellgren, E.C. 1994. Variation in reproductive biology of male Black Bears. *Journal of Mammalogy* 75(1): 175–188.

Gaskin, D.E. 1992a. Status of the Harbour Porpoise, *Phocoena phocoena*, in Canada. *Canadian Field-Naturalist* 106(1): 36–54.

– 1992b. Status of the Common Dolphin, *Delphinus delphis*, in Canada. *Canadian Field-Naturalist* 106(1): 55–63.

Gates, C.C., Stephenson, R.O., Zimov, S., and Chapin, M.C. 2001. Wood Bison recovery: Restoring grazing systems in Canada, Alaska, and eastern Siberia. In B.D. Rutley (ed.), *Bison Are Back, 2000 Proceedings of the Second International Bison Conference*, 82–102. Edmonton, Alberta, Canada.

Gau, R.J., McLoughlin, P.D., Case, R.L., Cluff, H.D., Mulders, R., and Messier, F. 2004. Movements of subadult male Grizzly Bears, *Ursus arctos*, in the central Canadian Arctic. *Canadian Field-Naturalist* 118(2): 239–242.

Gau, R.J., Mulders, R., Lamb, T., and Gunn, L. 2001. Cougars (*Puma concolor*) in the Northwest Territories and Wood Buffalo National Park. *Arctic* 54(2): 185–187.

Gaudet, C.L., and Fenton, M.B. 1984. Observational learning in three species of insectivorous bats (Chiroptera). *Animal Behaviour* 32: 385–388.

Gay, S.W., and Best, T.L. 1996. Age-related variation in skulls of the Puma (*Puma concolor*). *Journal of Mammalogy* 77(1): 191–198.

Gedir, J.V. 2009. COSEWIC assessment and status report on the Swift Fox *Vulpes velox* in Canada. Committee on the Status of Endangered Wildlife in Canada, Ottawa.

Gehring, T.M. 1993. Adult Black Bear, *Ursus americanus*, displaced from a kill by a Wolf, *Canis lupus*, pack. *Canadian Field-Naturalist* 107(3): 373–374.

Gehring, T.M., and Swihart, R.K. 2004. Home range and movements of Long-tailed Weasels in a landscape fragmented by agriculture. *Journal of Mammalogy* 85(1): 79–86.

Gehrt, S.D. 2003. Raccoon (*Procyon lotor* and Allies). In G.A. Feldhamer, B.C. Thompson, and J.A. Chapman (eds). *Wild Mammals of North America: Biology, Management, and Conservation*, 2nd ed., 611–634. Johns Hopkins University Press, Baltimore, MD.

Gehrt, S.D., Gergits, W.F., and Fritzell, E.K. 2008. Behavioral and genetic aspects of male social groups in Raccoons. *Journal of Mammalogy* 89(6): 1473–1480.

Gehrt, S.D., Hubert, G.F., Jr, and Ellis, J.A. 2006. Extrinsic effects on long-term population trends of Virginia Opossums and Striped Skunks at a large spatial scale. *American Midland Naturalist* 155(1): 168–180.

Geist, V. 1971. *Mountain Sheep.* University of Chicago Press, Chicago, IL.

– 1999. *Deer of the World: Their Evolution, Behaviour, and Ecology.* Stackpole Books, Mechanicsburg, PA.

Gende, S.M., and Quinn, T.P. 2004. The relative importance of prey density and social dominance in determining energy intake by bears feeding on Pacific salmon. *Canadian Journal of Zoology* 82(1): 75–85.

Gende, S.M., Womble, J.N., Willson, M.F., and Marston, B.H. 2001. Cooperative foraging by Steller Sea Lions, *Eumetopias jubatus. Canadian Field-Naturalist* 115(2): 355–356.

Gentry, R.L. 1981. Northern Fur Seal *Callorhinus ursinus* (Linnaeus, 1758). In S.H. Ridgway and R. Harrison (eds), *Handbook of Marine Mammals*, vol. 1, *The Walrus, Sea Lions, Fur Seals and Sea Otter*, 143–160. Academic Press, San Diego, CA.

George, J.C., Bada, J., Zeh, J., Scott, L., Brown, S.E., O'Hara, T., and Suydam, R. 1999. Age and growth estimates of Bowhead Whales (*Balaena mysticetus*) via aspartic acid racemization. *Canadian Journal of Zoology* 77(4): 571–580.

George, J.C., Philo, L.M., Hazard, K., Withrow, D., Carroll, G.M., and Suydam, R. 1994. Frequency of Killer Whale (*Orcinus orca*) attacks and ship collisions based on scarring on Bowhead Whales (*Balaena mysticetus*) of the Bering-Chukchi-Beaufort Seas stock. *Arctic* 47(3): 247–255.

George, J.C., Zeh, J., Suydam, R., and Clark, C. 2004. Abundance and population trend (1978–2001) of western Arctic Bowhead Whales surveyed near Barrow Alaska. *Marine Mammal Science* 20(4): 755–773.

George, S.B. 1989. *Sorex trowbridgii. Mammalian Species* no. 337: 1–5.

– 1999a. Prairie Shrew *Sorex haydeni*. In D.E. Wilson and S. Ruff (eds), *The Smithsonian Book of North American Mammals*, 24–5. UBC Press, Vancouver, BC.

– 1999b. Montane Shrew *Sorex monticolus*. In Wilson and Ruff (eds), *The Smithsonian Book of North American Mammals*, 31–33.

– 1999c. Trowbridge's Shrew *Sorex trowbridgii*. In Wilson and Ruff (eds), *The Smithsonian Book of North American Mammals*, 42–44.

– 1999d. Northern Short-tailed Shrew, *Blarina brevicauda*. In Wilson and Ruff (eds), *The Smithsonian Book of North American Mammals*, 47–49.

– 1999e. Eastern Heather Vole *Phenacomys ungava*. In Wilson and Ruff (eds), *The Smithsonian Book of North American Mammals*, 618–619.

George, S.B., Choate, J.R., and Genoways, H.H. 1984. *Blarina brevicauda. Mammalian Species* no. 261: 1–9.

Gerhardt, F. 2005. Food pilfering in larder-hoarding Red Squirrels (*Tamiasciurus hudsonicus*). *Journal of Mammalogy* 86(1): 108–114.

Gerhardt, R.P., and Gerhardt, D. McA. 1995. Two Gray Fox litters share a den. *Southwestern Naturalist* 40(4): 419.

Gese, E.M. 1998. Response of neighboring Coyotes (*Canis latrans*) to social disruption in an adjacent pack. *Canadian Journal of Zoology* 76(10): 1960–1963.

– 2001. Territorial defense by Coyotes (*Canis latrans*) in Yellowstone National Park, Wyoming: Who, how, where, when, and why. *Canadian Journal of Zoology* 79(6): 980–987.

Gese, E.M., and Bekoff, M. 2004. Coyote, *Canis latrans* Say, 1823. In C. Sillero-Zubiri, M. Hoffmann, and D.W. Macdonald (eds), *Canids: Foxes, Wolves, Jackals and Dogs – 2004 Status Survey and Conservation Action Plan*, 81–87. Oxford University Press, Oxford, UK.

Gese, E.M., and Ruff, R.L. 1997. Scent-marking by Coyotes, *Canis latrans*: The influence of social and ecological factors. *Animal Behaviour* 54(5): 1155–1166.

– 1998. Howling by Coyotes (*Canis latrans*): Variation among social classes, seasons, and pack sizes. *Canadian Journal of Zoology* 76(6): 1037–1043.

Getz, L.L. 1994. Population dynamics of the Short-tailed Shrew, *Blarina brevicauda. Special Publication of the Carnegie Museum of Natural History* 18: 27–38.

Getz, L.L., Hofmann, J.E., Klatt, B.J., Verner, L., Cole, F.R., and Lindroth, R.L. 1987. Fourteen years of population fluctuations of *Microtus ochrogaster* and *M. pennsylvanicus* in east central Illinois. *Canadian Journal of Zoology* 65: 1317–1325.

Getz, L.L., Hofman, J.E., McGuire, B., and Dolan, T.W., III. 2001. Twenty-five years of population fluctuations of *Microtus ochrogaster* and *M. pennsylvanicus* in three habitats in east-central Illinois. *Journal of Mammalogy* 82(1): 22–34.

Getz, L.L., Hofmann, J.E., McGuire, B., and Oli, M.K. 2004. Population dynamics of the Northern Short-tailed Shrew, *Blarina brevicauda*: Insights from a 25-year study. *Canadian Journal of Zoology* 82(11): 1679–1686.

Getz, L.L., Larson, C.M., and Lindstrom, K.A. 1992 *Blarina brevicauda* as a predator on nestling voles. *Journal of Mammalogy* 73(3): 591–596.

Getz, L.L., and McGuire, B. 2008. Nestling survival and population fluctuations of the Prairie Vole *Microtus ochrogaster*. *American Midland Naturalist* 159(2): 413–420.

Getz, L.L., McGuire, B., and Carter, C.S. 2003. Social behavior, reproduction and demography of the Prairie Vole, *Microtus ochrogaster*. *Ethology Ecology and Evolution* 15: 105–118.

Getz, L.L., McGuire, B., Pizzuto, T., Hofmann, J.E., and Frase, B. 1993. Social organization of the Prairie Vole (*Microtus ochrogaster*). *Journal of Mammalogy* 74(1): 44–58.

Getz, L.L., Oli, M.K., Hofmann, J.E., and McGuire, B. 2005. Habitat-specific demography of sympatric vole populations over 25 years. *Journal of Mammalogy* 86(3): 561–568.

– 2006. Vole population fluctuations: Factors that initiate and determine intervals between them in *Microtus pennsylvanicus*. *Journal of Mammalogy* 87(5): 841–847.

Getz, L.L., Simms, L.E., McGuire, B., and Snarski, M.E. 1997. Factors affecting life expectancy of the Prairie Vole, *Microtus ochrogaster*. *Oikos* 80: 362–370.

Giger, R.D. 1973. Movements and homing in Townsend's Mole near Tillamook, Oregon. *Journal of Mammalogy* 54(3): 648–659.

Gilbert, B.K. 1999. Opportunities for social learning in bears. *Symposium of the Zoological Society of London* 72: 225–235.

Gilbert, B.S. 1990. Use of winter feeding craters by Snowshoe Hares. *Canadian Journal of Zoology* 68(7): 1600–1602.

Gilbert, B.S., Cichowski, D.B., Talarico, D., and Krebs, C.J. 1986. Summer activity patterns of three rodents in southwestern Yukon. *Arctic* 39(3): 204–207.

Gilbert, B.S., and Krebs, C.J. 1991. Population dynamics of *Clethrionomys* and *Peromyscus* in southwestern Yukon 1973–1989. *Holarctic Ecology* 14: 250–259.

Gilg, O. 2002. The summer decline of the collared lemming, *Dicrostonyx groenlandicus*, in high arctic Greenland. *Oikos* 99: 499–510.

Gilg, O., Sittler, B., Sabard, B., Hurstel, A., Sané, R., Delattre, P., and Hanski, I. 2006. Functional and numerical responses of four lemming predators in high arctic Greenland. *Oikos* 113(2): 193–216.

Gillihan, S.W., and Foresman, K.R. 2004. *Sorex vagrans*. *Mammalian Species* no. 744: 1–5.

Gillis, E.A., Hik, D.S., Boonstra, R., Karels, T.J., and Krebs, C.J. 2005a. Being high is better: Effects of elevation and habitat on Arctic Ground Squirrel demography. *Oikos* 108(2): 231–240.

Gillis, E.A., Morrison, S.F., Zazula, G.D., and Hik, D.S. 2005b. Evidence of selective caching by Arctic Ground Squirrels living in alpine meadows in the Yukon. *Arctic* 58(4): 354–360.

Gipson, P.S., and Kamler, J.F. 2001. Survival and home ranges of Opossums in northeastern Kansas. *Southwestern Naturalist* 46(2): 178–182.

– 2002. Bobcat killed by a Coyote. *Southwestern Naturalist* 47(3): 511–513.

Gisiner, R., and Schusterman, R.J. 1991. California Sea Lion pups play an active role in reunions with their mothers. *Animal Behaviour* 41: 364–366.

Gitzen, R.A., Bradley, J.E., Kroeger, M.R., and West, S.D. 2009. First record of Preble's Shrew in the Northern Columbia Basin, Washington. *Northwestern Naturalist* 90(1): 41–43.

Gitzen, R.A., and West, S.D. 2000. Occurrences of the Coast Mole (*Scapanus orarius orarius*) in the southern Washington Cascades. *Northwestern Naturalist* 81(2): 65–66.

Gjertz, I. 1990. Ringed Seal *Phoca hispida* fright behaviour caused by Walrus *Odobenus rosmarus*. *Polar Research* 8(2): 317–319.

Gjertz, I., Kovacs, K.M., Lydersen, C., and Wiig, Ø. 2000. Movements and diving of Bearded Seal (*Erignathus barbatus*) mothers and pups during lactation and postweaning. *Polar Biology* 23: 559–566.

Gleason, J.S., Hoffman, R.A., and Wendland, J.M. 2005. Beavers, *Castor canadensis*, feeding on salmon carcasses: Opportunistic use of a seasonally superabundant food source. *Canadian Field-Naturalist* 119(4): 591–593.

Gliwicz, J., and Glowacka, B. 2000. Differential responses of *Clethrionomys* species to forest disturbance in Europe and North America. *Canadian Journal of Zoology* 78(8): 1340–1348.

Goddard, P.D., and Rugh, D.J. 1998. A group of Right Whales seen in the Bering Sea in July 1996. *Marine Mammal Science* 14(2): 344–349.

Godsell, J. 1991. The relative influence of age and weight on the reproductive behaviour of male Grey Seals *Halichoerus grypus*. *Journal of Zoology* (London) 224: 537–551.

Goheen, J.R., Kaufman, G.A., and Kaufman, D.W. 2003. Effect of body size on reproductive characteristics of the Northern Grasshopper Mouse in north-central Kansas. *Southwestern Naturalist* 48(3): 427–431.

Goley, P.D. 1999. Behavioral aspects of sleep in Pacific White-sided Dolphins (*Lagenorhynchus obliquidens*, Gill 1865). *Marine Mammal Science* 15(4): 1054–1064.

Goley, P.D., and Straley, J.M. 1994. Attack on Gray Whales (*Eschrichtius robustus*) in Monterey Bay, California, by Killer Whales (*Orcinus orca*) previously identified in Glacier Bay, Alaska. *Canadian Journal of Zoology* 72(8): 1528–1530.

Gompper, M.E. 2002. The ecology of northeast Coyotes. *Working Paper of the Wildlife Conservation Society* no. 17: 1–46.

Gonzales, E.K. 2005. The distribution and habitat selection of introduced Eastern Grey Squirrels, *Sciurus carolinensis*, in British Columbia. *Canadian Field-Naturalist* 119(3): 343–350.

González-Suárez, M., and Gerber, L.R. 2008. Habitat preferences of California Sea Lions: Implications for conservation. *Journal of Mammalogy* 89(6): 1521–1528.

Goodyear, M.A. 2003. Extralimital sighting of a Polar Bear, *Ursus maritimus*, in northeast Saskatchewan. *Canadian Field Naturalist* 117(4): 648–649.

Goold, J.C. 2000. A diel pattern in vocal activity of Short-beaked Common Dolphins (*Delphinus delphis*). *Marine Mammal Science* 16(1): 240–244.

Goold, J.C., Whitehead, H., and Reid, R.J. 2002. North Atlantic Sperm Whale, *Physeter macrocephalus*, strandings on the coastlines of the British Isles and eastern Canada. *Canadian Field-Naturalist* 116(3): 371–388.

Gorman, T.A., Erb, J.D., McMillan, B.R., and Martin, D.J. 2006a. Space use and sociality of River Otters (*Lontra canadensis*) in Minnesota. *Journal of Mammalogy* 87(4): 740–747.

Gorman, T.A., Erb, J.D., McMillan, B.R., Martin, D.J., and Homyack, J.A. 2006b. Site characteristics of River Otter (*Lontra canadensis*) natal dens in Minnesota. *American Midland Naturalist* 156(1): 109–117.

Gorman, T.A., McMillan, B.R., Erb, J.D., Deperno, C.S., and Martin, D.J. 2008. Survival and cause-specific mortality of a protected population of River Otters in Minnesota. *American Midland Naturalist* 159(1): 98–109.

Gosse, J.W., Cox, R., and Avery, S.W. 2005. Home-range characteristics and habitat use by American Martens in eastern Newfoundland. *Journal of Mammalogy* 86(6): 1156–1163.

Gosse, J.W., and Hearn, B.J. 2005. Seasonal diets of Newfoundland Martens, *Martes americana atrata*. *Canadian Field-Naturalist* 119(1): 43–47.

Gould, E. 1955. The feeding efficiency of insectivorous bats. *Journal of Mammalogy* 36: 399–407.

Gould, E., McShea, W., and Grand, T. 1993. Function of the star in the Star-nosed Mole, *Condylura cristata*. *Journal of Mammalogy* 74(1): 108–116.

Gould, W.P., and Pruitt, W.O., Jr. 1969. First Newfoundland record of *Peromyscus*. *Canadian Journal of Zoology* 47(3): 469.

Goulet, A.-M., Hammill, M.O., and Barrette, C. 2001. Movements and diving of Grey Seal females (*Halichoerus grypus*) in the Gulf of St. Lawrence, Canada. *Polar Biology* 24: 432–439.

Gowans, S., and Rendell, L. 1999. Head-butting in Northern Bottlenose Whales (*Hyperoodon ampullatus*): A possible function for big heads? *Marine Mammal Science* 15(4): 1342–1350.

Gowans, S., and Whitehead, H. 1995. Distribution and habitat partitioning by small odontocetes in the Gully, a submarine canyon on the Scotian Shelf. *Canadian Journal of Zoology* 73(9): 1599–1608.

– 2001. Photographic identification of Northern Bottlenose Whales (*Hyperoodon ampullatus*): Sources of heterogeneity from natural marks. *Marine Mammal Science* 17(1): 76–93.

Gowans, S., Whitehead, H., Arch, J.K., and Hooker, K. 2000. Population size and residency patterns of Northern Bottlenose Whales (*Hyperoodon ampullatus*) using the Gully, Nova Scotia. *Journal of Cetacean Research* 2(3): 201–210.

Graf, R.P. 1985. Social organization of Snowshoe Hares. *Canadian Journal of Zoology* 63(3): 468–474.

Grand, T., Gould, E., and Montali, R. 1998. Structure of the proboscis and rays of the Star-nosed Mole *Condylura cristata*. *Journal of Mammalogy* 79(2): 492–501.

Grant, J., and Hawley, A. 1991. Some observations on the mating behaviour of captive American Pine Martens *Martes americana*. *Acta Theriologica* 41(4): 439–442.

Gray, D.R. 1967. Activity and behaviour in a colony of Hoary Marmots (*Marmota caligata*) in Manning Park, B.C. Honours B.Sc. thesis, University of Victoria, Victoria, BC.

– 1975. The Marmots of Spotted Nellie Ridge. *Nature Canada* 4(1): 3–8.

– 1987. *The Muskoxen of Polar Bear Pass*. Fitzhenry and Whiteside, Toronto, ON.

– 1993. Behavioural adaptations to arctic winter: Shelter seeking by Arctic Hare (*Lepus arcticus*). *Arctic* 46(4): 340–353.

Green, G.I., and Mattson, D.J. 2003. Tree rubbing by Yellowstone Grizzly Bears *Ursus arctos*. *Wildlife Biology* 9(1): 1–9.

Greenberg, C.H., and Pelton, M.R. 1991. Food habits of Gray Foxes (*Urocyon cinereoargenteus*) and Red Foxes (*Vulpes vulpes*) in east Tennessee. *Journal of the Tennessee Academy of Science* 66(2): 79–84.

Greene, E., and Meagher, T. 1998. Red Squirrels, *Tamiasciurus hudsonicus*, produce predator-class specific alarm calls. *Animal Behavior* 55: 511–518.

Greenwood, R.J., Sargeant, A.B., Piehl, J.L., Buhl, D.A., and Hanson, B.A. 1999. Foods and foraging of prairie Striped Skunks during the avian nesting season. *Wildlife Society Bulletin* 27(3): 823–832.

Gregr, E.J. 2003. COSEWIC assessment and status report on the Sei Whale *Balaenoptera borealis* in Canada. Committee on the Status of Endangered Wildlife in Canada. Ottawa.

Gregr, E.J., Nichol, L., Ford, J.K.B., Ellis, G., and Trites, A.W. 2000. Migration and population structure of northeastern Pacific whales off coastal British Columbia: An analysis of commercial whaling records from 1908–1967. *Marine Mammal Science* 16(4): 699–727.

Grewal, S.K., Wilson, P.J., Kung, T.K., Shami, K., Theberge, M.T., Theberge, J.B., and White, B.N. 2004. A genetic assessment of the Eastern Wolf (*Canis lycaon*) in Algonquin Provincial Park. *Journal of Mammalogy* 85(4): 625–632.

Griesemer, S.J., Fuller, T.K., and DeGraaf, R.M. 1996. Denning patterns of Porcupines, *Erethizon dorsatum. Canadian Field-Naturalist* 110(4): 634–637.

Griffin, D.R., Webster, F.A., and Michael, C.R. 1960. The echolocation of flying insects by bats. *Animal Behaviour* 8: 141–154.

Grilliot, M.E., Burnett, S.C., and Mendonça, M.T. 2009. Sexual dimorphism in Big Brown Bat (*Eptesicus fuscus*) ultrasonic vocalizations is context dependent. *Journal of Mammalogy* 90(1): 203–209.

Grindal, S.D., Collard, T.S., and Brigham, R.M. 1991. Evidence for a breeding population of Pallid Bats, *Antrozous pallidus* (Chiroptera: Vespertilionidae), in British Columbia. *Contributions to Natural Science* no.14: 1–4.

Grinevitch, L., Holroyd, S.L., and Barclay, R.M.R. 1995. Sex differences in the use of daily torpor and foraging time by Big Brown Bats (*Eptesicus fuscus*) during the reproductive season. *Journal of Zoology* (London) 235: 301–309.

Groves, C., and Yensen, E. 1989. Rediscovery of the Northern Bog Lemming (*Synaptomys borealis*) in Idaho. *Northwestern Naturalist* 70(1): 14–15.

Grubbs, S.E., and Krausman, P.R. 2009. Observations of Coyote–Cat interactions. *Journal of Wildlife Management* 73(5): 683–685.

Gruyer, N., Gauthier, G., and Berteaux, D. 2008. Cyclic dynamics of sympatric lemming populations on Bylot Island, Nunavut, Canada. *Canadian Journal of Zoology* 86(6): 910–917.

Guilday, J.E. 1982. Dental variation in *Microtus xanthognathus, M. chrotorrhinus,* and *M. pennsylvanicus* (Rodentia, Mammalia). *Annals of the Carnegie Museum* 51(11): 211–230.

Guilday, J.E., and Handley, C.O., Jr. 1967. A new *Peromyscus* (Rodentia: Cricetidae) from the Pleistocene of Maryland. *Annals of the Carnegie Museum* 39: 91–103.

Gummer, D.L., and Bender, D.J. 2006. COSEWIC assessment and update status report on the Ord's Kangaroo Rat *Dipodomys ordii* in Canada. Committee on the Status of Endangered Wildlife in Canada.

Gummer, D.L., and Robertson, S.E. 2003. Distribution of Ord's Kangaroo Rats in southeastern Alberta. *Alberta Species at Risk Report* no. 63: 1–16. Alberta Sustainable Resource Development, Fish and Wildlife Division.

Gunn, A., Miller, F.L., Barry, S.J., and Buchan, A. 2006. A near-total decline in Caribou on Prince of Wales, Somerset and Russell Islands, Canadian Arctic. *Arctic* 59(1): 1–13.

Gunn, A., Miller, F.L., and Nishi, J. 2000. Status of endangered and threatened caribou on Canada's arctic islands. *Rangifer*, Special issue, 12: 39–49.

Gyug, L.W. 2000. Status, distribution and biology of the Mountain Beaver, *Aplodontia rufa*, in Canada. *Canadian Field-Naturalist* 114(3): 476–490.

Haberman, C.G., and Fleharty, E.D. 1971. Natural history notes on Franklin's Ground Squirrel in Boone County, Nebraska. *Transactions of the Kansas Academy of Science* 74(1): 76–80.

Hackett, H.M., and Pagels, J.F. 2004. Nest site characteristics of the endangered Northern Flying Squirrel (*Glaucomys sabrinus coloratus*) in southwest Virginia. *American Midland Naturalist* 150(2): 321–331.

Hackländer, K., Arnold, W., and Ruf, T. 2002. Postnatal development and thermoregulation in the precocial European Hare (*Lepus europaeus*). *Journal of Comparative Physiology B* 172: 183–190.

Hackmann, N., Zamora, C.S., and Stauber, E. 1990. The white eye secretion in Aplodontia. *Chemical Signals in Vertebrates* 5: 139–146.

Hafner, D.J., and Stahlecker, D.W. 2002. Distribution of Merriam's Shrew (*Sorex preblei*) and the Dwarf Shrew (*Sorex nanus*), and new records for New Mexico. *Southwestern Naturalist* 47(1): 134–137.

Hafner, D.J., and Sullivan, R.M. 1995. Historical and ecological biogeography of Nearctic Pikas (Lagomorpha: Ochotonidae). *Journal of Mammalogy* 76(2): 302–321.

Hafner, M.S., and Hafner, D.J. 1979. Vocalizations of grasshopper mice (genus *Onychomys*). *Journal of Mammalogy* 60(1): 85–94.

Hai, D.J., Lien, J., Nelson, D., and Curren, K. 1996. A contribution to the biology of the White-beaked Dolphin, *Lagenorhynchus albirostris*, in waters off Newfoundland. *Canadian Field-Naturalist* 110(2): 278–287.

Hain, J.H.W., Ellis, S.L., Kenney, R.D., Clapham, P.J., Gray, B.K., Weinrich, M.T., and Babb, I.G. 1995. Apparent bottom feeding by Humpback Whales on Stellwagen Bank. *Marine Mammal Science* 11(4): 464–479.

Hain, J.H.W., Ratneswamy, M.J., Kenney, R.D., and Winn, H.E. 1992. The Fin Whale, *Balaenoptera physalus*, in waters of the northeastern United States continental shelf. *Report of the International Whaling Commission* 42: 653–669.

Haley, M.P., Deutsch, C.J., and Le Boeuf, B.J. 1994. Size, dominance and copulatory success in male Northern Elephant Seals *Mirounga angustirostris. Animal Behaviour* 48: 1249–1260.

Hall, A. 2002. Gray Seal *Halichoerus grypus*. In W.F. Perrin, B. Wursig, and H.G.T. Thewissen (eds), *Encyclopedia of Marine Mammals*, 522–524. Academic Press, San Diego, CA.

Hall, E.R. 1981. *The Mammals of North America*. 2nd ed. John Wiley and Sons, New York, NY.

Hallett, J.G. 1978. *Parascalops breweri. Mammalian Species* no. 98: 1–4.

– 1999. Hairy-tailed Mole *Parascalops breweri*. In D.E. Wilson and S. Ruff (eds), *The Smithsonian Book of North American Mammals*, 62–63. UBC Press, Vancouver, BC.

Halley, D.J., and Rosell, F. 2003. Population and distribution of European Beavers (*Castor fiber*). *Lutra* 46(2): 91–101 [*includes Palaearctic distribution of introduced North American Beaver*].

Hamel, S., Côté, S.D., Smith, K.G., and Festa-Bianchet, M. 2006. Population dynamics and harvest potential of Mountain Goat herds in Alberta. *Journal of Wildlife Management* 70(4): 1044–1053.

Hamilton, I.M., and Barclay, R.M.R. 1998a. Diets of juvenile, yearling, and adult Big Brown Bats (*Eptesicus fuscus*) in southeastern Alberta. *Journal of Mammalogy* 79(3): 764–771.

– 1998b. Ontogenetic influences on foraging and mass accumulation by Big Brown Bats (*Eptesicus fuscus*). *Journal of Animal Ecology* 67: 930–940.

Hamilton, P.K., Knowlton, A.R., Marx, M.K., and Kraus, S.D. 1998. Age structure and longevity in North Atlantic Right Whales *Eubalaena glacialis* and their relation to reproduction. *Marine Ecology Progress Series* 171: 285–292.

Hamilton, P.K., Marx, M.K., and Kraus, S.D. 1995. Weaning in North Atlantic Right Whales. *Marine Mammal Science* 11(3): 386–390.

Hamilton, W.J. 1934. The life history of the Rufescent Woodchuck. *Annals of the Carnegie Museum* 23: 87–118.

Hammill, M.O. 1993. Seasonal movements of Hooded Seals tagged in the Gulf of St. Lawrence, Canada. *Polar Biology* 13: 307–310.

Hammill, M.O., and Gosselin, J.F. 1995. Grey Seal (*Halichoerus grypus*) from the Northwest Atlantic: Female reproductive rates, age at first birth, and age of maturity in males. *Canadian Journal of Fisheries and Aquatic Sciences* 52(12): 2757–2761.

Hammill, M.O., Kingsley, M.C.S., Beck, G.G., and Smith, T.G. 1995. Growth and condition in the Northwest Atlantic Harp Seal. *Canadian Journal of Fisheries and Aquatic Sciences* 52(3): 478–488.

Hammill, M.O., Lesage, V., Gosselin, J.-F., Bourdages, H., de March, B.G.E., and Kingsley, M.C.S. 2004. Evidence for a decline in northern Quebec (Nunavik) Belugas. *Arctic* 57(2): 183–195.

Hammill, M.O., Lydersen, C., Ryg, M., and Smith, T.G. 1991. Lactation in the Ringed Seal (*Phoca hispida*). *Canadian Journal of Fisheries and Aquatic Sciences* 48(5): 2471–2476.

Hanley, T.A., and Barnard, J.C. 1999. Spatial variation in population dynamics of Sitka Mice in floodplain forests. *Journal of Mammalogy* 80(3): 866–879.

Hansen, K. 2007. *Bobcat: Master of Survival.* Oxford University Press, Oxford, UK.

Hansen, R.M. 1954. Molt patterns in ground squirrels. *Proceedings of the Utah Academy of Science, Arts and Letters* 31: 57–60.

Hansen, S., and Lavigne, D.M. 1997. Temperature effects on the breeding distribution of Grey Seals (*Halichoerus grypus*). *Physiological Zoology* 70(4): 436–443.

Hanson, D.K., and Kusmer, K.D. 2001. Sea Otter (*Enhydra lutris*) scarcity in the Strait of Georgia, British Columbia. *BAR International Series* 944: 58–56.

Harder, J.D., Stonerook, M.J., and Pondy, J. 1993. Gestation and placentation in two New World opossums: *Didelphis virginiana* and *Monodelphis domestica*. *Journal of Experimental Zoology* 266: 463–479.

Hare, J.F. 1998. Juvenile Richardson's Ground Squirrels, *Spermophilus richardsonii*, discriminate among individual alarm callers. *Animal Behaviour* 55: 451–460.

Hare, J.F., and Atkins, B.A. 2001. The squirrel that cried wolf: Reliability detection by juvenile Richardson's Ground Squirrels (*Spermophilus richardsonii*). *Behavioral Ecology and Sociobiology* 51(1): 108–112.

Hare, J.F., Todd, G., and Untereiner, W.A. 2004. Multiple mating results in multiple paternity in Richardson's Ground Squirrels, *Spermophilus richardsonii*. *Canadian Field-Naturalist* 118(1): 90–94.

Harestad, A.S. 1990. Mobbing of a Long-tailed Weasel, *Mustela frenata*, by Columbian Ground Squirrels, *Spermophilus columbianus*. *Canadian Field-Naturalist* 104(3): 483–484.

– 1991. Spatial behaviour of Townsend's Chipmunks and habitat structure. *Acta Theriologica* 36(3–4): 247–254.

Harfenist, A., and Kaiser, G.W. 1997. Effects of introduced predators on the nesting seabirds of the Queen Charlotte Islands. *Occasional Paper of the Canadian Wildlife Service* 93: 132–136.

Harington, C.R. 1968. Denning habits of the Polar Bear (*Ursus maritimus* Phipps). *Canadian Wildlife Service Report Series* 5: 1–30.

Härkönen, T., and Harding, K.C. 2001. Spatial structure of Harbour Seal populations and the implications thereof. *Canadian Journal of Zoology* 79(12): 2115–2127.

Härkönen, T., and Heide-Jørgensen, M.-P. 1990. Comparative life histories of east Atlantic and other Harbour Seal populations. *Ophelia* 32(3): 211–235.

Harlow, H.J., Lohuis, T., Beck, T.D.I., and Iaizzo, P.A. 2001. Muscle strength in overwintering bears. *Nature* 409(22 Feb.): 997.

Harlow, H.J., Miller, B., Ryder, T., and Ryder, L. 1985. Energy requirements for gestation and lactation in a delayed implanter, the American Badger. *Comparative Biochemistry and Physiology A* 82(4): 885–890.

Harper, S.J., and Batzli, G.O. 1996. Effects of predators on structure of the burrows of voles. *Journal of Mammalogy* 77(4): 1114–1121.

Harrington, L.A., Biggins, D.E., and Alldredge, A.W. 2003. Basal metabolism of the Black-footed Ferret (*Mustela nigripes*) and the Siberian Polecat (*Mustela eversmannii*). *Journal of Mammalogy* 84(2): 497–504.

Harris, A.H. 1999. Yuma myotis. In D.E. Wilson and S. Ruff (eds), *The Smithsonian Book of North American Mammals*, 103–104. UBC Press, Vancouver, BC.

Harris, D.E., Lelli, B., and Jakush, G. 2002. Harp Seal records from the southern Gulf of Maine: 1997–2001. *Northeastern Naturalist* 9(3): 331–340.

Harris, D.E., Lelli, B., Jakush, G., and Early, G. 2001. Hooded Seal (*Cystophora cristata*) records from the southern Gulf of Maine. *Northeastern Naturalist* 8(4): 427–434.

Harris, M.A., and Murie, J.O. 1984. Inheritance of nest sites in female Columbian Ground Squirrels. *Behavioral Ecology and Sociobiology* 15: 97–102.

Hart, E.B., Belk, M.C., Jordan, E., and Gonzalez, M.W. 2004. *Zapus princeps*. *Mammalian Species* no. 749: 1–7.

Hartling, L., and Silva, M. 2004. Abundance and species richness of shrews within forested habitats on Prince Edward Island. *American Midland Naturalist* 151(2): 399–407.

Hartman, G.D. 1995. Age determination, age structure, and longevity in the mole, *Scalopus aquaticus* (Mammalia: Insectivora). *Journal of Zoology* (London) 237(1): 107–122.

– 1999. Star-nosed Mole *Condylura cristata*. In D.E. Wilson and S. Ruff (eds), *The Smithsonian Book of North American Mammals*, 65–67. UBC Press, Vancouver, BC.

Hartman, G.D., and Gottschang, J.L. 1983. Notes on sex determination, neonates, and behaviour of the Eastern Mole, *Scalopus aquaticus*. *Journal of Mammalogy* 64(3): 539–540.

Hartman, G.D., Whitaker, J.O., Jr, and Munsee, J.R. 2000. Diet of the mole *Scalopus aquaticus*, from the Coastal Plain Region of South Carolina. *American Midland Naturalist* 144(2): 342–351.

Hartman, L.H., and Eastman, D.S. 1999. Distribution of introduced Raccoons *Procyon lotor* on the Queen Charlotte Islands: Implications for burrow-nesting seabirds. *Biological Conservation* 88: 1–13.

Hartman, L.H., Gaston, A.J., and Eastman, D.S. 1997. Raccoon predation on Ancient Murrelets on East Limestone Island, British Columbia. *Journal of Wildlife Management* 61(2): 377–388.

Harvey, M.J. 1976. Home range, movements and diel activity of the Eastern Mole, *Scalopus aquaticus*. *American Midland Naturalist* 95: 436–445.

Harvey, M.J., and Barbour, R.W. 1965. Home range of *Microtus ochrogaster* as determined by a modified minimum area method. *Journal of Mammalogy* 46(3): 398–402.

Haskell, S.P. 2006. First record of a River Otter, *Lontra canadensis*, captured on the northeastern coast of Alaska. *Canadian Field-Naturalist* 120(2): 235–236.

Hass, C.C. 1997. Seasonality of births in Bighorn Sheep. *Journal of Mammalogy* 78(4): 1251–1260.

Hastings, K.K., Frost, K.J., Simpkins, M.A., Pendleton, G.W., Swain, U.G., and Small, R.J. 2004. Regional differences in diving behavior of Harbor Seals in the Gulf of Alaska. *Canadian Journal of Zoology* 82(11): 1755–1773.

Hattler, D.F., Nagorsen, D.W., and Beal, A.M. 2008. *Carnivores of British Columbia*. Vol. 5: The Mammals of British Columbia, Royal BC Museum Handbook, Royal BC Museum, Victoria, BC.

Havelka, M.A., and Millar, J.S. 2004. Maternal age drives seasonal variation in litter size of *Peromyscus leucopus*. *Journal of Mammalogy* 85(5): 940–947.

Hawes, M.L. 1977. Home range, territoriality, and ecological separation in sympatric shrews, *Sorex vagrans* and *Sorex obscurus*. *Journal of Mammalogy* 58(3): 354–367.

Hayes, A.R., and Huntly, N.J. 2005. Effects of wind on the behavior and call transmission of pikas (*Ochotona princeps*). *Journal of Mammalogy* 86(5): 974–981.

Hayes, J.P., Horvath, E.G., and Hounihan, P. 1995. Townsend's Chipmunk populations in Douglas-fir plantations and mature forests in the Oregon Coast Range. *Canadian Journal of Zoology* 73(1): 67–73.

Hayes, S.A., Costa, D.P., Harvey, J.T., and Le Boeuf, B.J. 2004. Aquatic mating strategies of the male Pacific Harbor Seal (*Phoca vitulina richardii*): Are males defending the hotspot? *Marine Mammal Science* 20(3): 639–656.

Hayward, B., and Davis, R. 1964. Flight speeds of western bats. *Journal of Mammalogy* 45(2): 236–241.

Hayward, G.D., and Rosentreter, R. 1994. Lichens as nesting material for Northern Flying Squirrels in the northern Rocky Mountains. *Journal of Mammalogy* 75(3): 663–673.

Heard, D.C. 1977. The behaviour of Vancouver Island Marmots (*Marmota vancouverensis*). M.Sc. thesis, University of British Columbia, Vancouver, BC.

Heard, D.C., and Vagt, K.L. 1998. Caribou in British Columbia: A 1996 status report. *Rangifer*, Special issue, 10: 117–124.

Hearn, B.J., Neville, J.T., Curran, W.J., and Snow, D.P. 2006. First record of the Southern Red-backed Vole, *Clethrionomys gapperi*, in Newfoundland: Implications for the endangered Newfoundland Marten, *Martes americana atrata*. *Canadian Field-Naturalist* 120(1): 50–56.

Hebda, R.J., Warner, B.G., and Cannings, R.A. 1990. Pollen, plant macrofossils, and insects from fossil woodrat (*Neotoma cinerea*) middens in British Columbia. *Géographie physique et Quaternaire* 44(2): 227–234.

Hebblewhite, M., Merrill, E.H., Morgantini, L.E., White, C.A., Allen, J.R., Bruns, E., Thurston, L., and Hurd, T.E. 2006. Is the migratory behavior of montane Elk herds in peril? The case of Alberta's Ya Ha Tinda Elk herd. *Wildlife Society Bulletin* 34(5): 1280–1294.

Hebblewhite, M., Percy, M., and Serrouya, R. 2003. Black Bear (*Ursus americanus*) survival and demography in the Bow Valley of Banff National Park, Alberta. *Biological Conservation* 112: 415–425.

Heckel, G., Reilly, S.B., Sumich, J.L., and Espejel, I. 2001. The influence of whale-watching on the behaviour of migrating Gray Whales (*Eschrichtius robustus*) in Todos Santos Bay and surrounding waters, Baja California, Mexico. *Journal of Cetacean Research and Management* 3(3): 227–237.

Hecnar, S.J., and Hecnar, D.R. 1996. Range extension of the Hairy-tailed Mole, *Parascalops breweri*, in northern Ontario. *Canadian Field-Naturalist* 110(4): 702–703.

Hedmark, E., Persson, J., Segerström, P., Landa, A., and Ellegren, H. 2007. Paternity and mating system in Wolverines *Gulo gulo*. *Wildlife Biology* 13(Suppl. 2): 13–30.

Heffner, R.S., and Heffner, H.E. 1990. Vestigial hearing in a fossorial mammal, the Pocket Gopher (*Geomys bursarius*). *Hearing Research* 46: 239–252.

Heilhecker, E., Thiel, R.P., and Hall, W., Jr. 2007. Wolf, *Canis lupus*, behavior in areas of frequent human activity. *Canadian Field-Naturalist* 121(3): 256–260.

Heide-Jørgensen, M.P., Dietz, R., Laidre, K.L., and Richard, P. 2002. Autumn movements, home ranges, and winter density of Narwhals (*Monodon monoceros*) tagged in Tremblay Sound, Baffin Island. *Polar Biology* 25: 331–341.

Heide-Jørgensen, M.P., Dietz, R., Laidre, K.L., Richard, P., Orr, J., and Schmidt, H.C. 2003. The migratory behaviour of Narwhals (*Monodon monoceros*). *Canadian Journal of Zoology* 81(7): 1298–1305.

Heide-Jørgensen, M.P., Lairde, K.L., Jensen, M.V., Dueck, L., and Postma, L.D. 2006. Dissolving stock discreteness with satellite tracking: Bowhead Whales in Baffin Bay. *Marine Mammal Science* 22(1): 34–45.

Heide-Jørgensen, M.P., Richard, P.R., and Rosing-Asvid, A. 1998. Dive patterns of Belugas (*Delphinapterus leucas*) in waters near eastern Devon Island. *Arctic* 51(1): 17–26.

Hein, C.D., Miller, K.V., and Castleberry, S.B. 2009. Evening Bat summer roost-site selection on a managed pine landscape. *Journal of Wildlife Management* 73(4): 511–517.

Heinrich, R.E., and Biknevicius, A.R. 1998. Skeletal allometry and interlimb scaling patters in mustelid carnivorans. *Journal of Morphology* 235: 121–134.

Heise, K. 1997a. Diet and feeding behaviour of Pacific White-sided Dolphins (*Lagenorhynchus obliquidens*) as revealed through the collection of prey fragments and stomach contents. *Report of the International Whaling Commission* 47: 807–813.

– 1997b. Life history and population parameters of Pacific White-sided Dolphins (*Lagenorhynchus obliquidens*). *Report of the International Whaling Commission* 47: 817–825.

Hellgren, E.C. 1998. Physiology of hibernation in bears. *Ursus* 10: 467–477.

Henderson, L.E., and Broders, H.G. 2008. Movements and resource selection of the Northern Long-eared Myotis (*Myotis septentrionalis*) in a forest-agriculture landscape. *Journal of Mammalogy* 89(4): 952–963.

Hendricks, P. 2001. A significant new record of the Pygmy Shrew, *Sorex hoyi*, on the Montana-Alberta border. *Canadian Field-Naturalist* 115(3): 513–514.

Hendricks, P., and Roedel, M. 2002. Preble's Shrew and Great Basin Pocket Mouse from the Centennial Valley Sandhills of Montana. *Northwestern Naturalist* 83(1): 31–34.

Henry, J.D. 1996. *Red Fox: The Catlike Canine*. Smithsonian Institution Press, Washington, DC.

Henry, S.E., and Ruggiero, L.F. 1993. Den use and kit development of marten in Wyoming. *Proceedings of the International Union of Game Biologists Congress* 21(1): 233–237.

Hermanson, J.W., and O'Shea, T.J. 1983. *Antrozous pallidus*. *Mammalian Species*. no. 213: 1–8.

Herrera, L.G., Fleming, T.H., and Findley, J.S. 1993. Geographic variation in carbon composition of the Pallid Bat, *Antrozous pallidus*, and its dietary implications. *Journal of Mammalogy* 74(3): 601–606.

Herrero, S., and Higgins, A. 1999. Human injuries inflicted by bears in British Columbia: 1960–97. *Ursus* 11: 201–218.

– 2003. Human injuries inflicted by bears in Alberta: 1960–1998. *Ursus* 14(1): 44–54.

Herrero, S., Mamo, C., Carbyn, L., and Scott-Brown, M. 1991. Swift Fox reintroduction into Canada. *Provincial Museum of Alberta Natural History Occasional Paper* no. 15: 246–252.

Heyning, J.E. 1999a. Long-beaked Saddleback Dolphin (*Delphinus capensis*). In D.E. Wilson and S. Ruff (eds), *The Smithsonian Book of North American Mammals*, 272–273. UBC Press, Vancouver, BC.

– 1999b. Short-beaked Saddleback Dolphin (*Delphinus delphis*). In Wilson and Ruff (eds), *The Smithsonian Book of North American Mammals*, 274–275.

Heyning, J.E., and Dahleim, M.E. 1988. *Orcinus orca*. *Mammalian Species* no. 304: 1–9.

Heyning, J.E., and Mead, J.G. 1997. Thermoregulation in the mouths of feeding Gray Whales. *Science* 278(5340): 1138–1139.

Heyning, J.E., and Perrin, W.F. 1994. Evidence for two species of Common Dolphins (genus *Delphinus*) from the eastern North Pacific. *Contributions in Science, Natural History Museum of Los Angeles County* no. 442: 1–35.

Hickey, M.B.C., Acharya, L., and Pennington, S. 1996. Resource partitioning by two species of Vespertilionid bats (*Lasiurus cinereus* and *Lasiurus borealis*) feeding around street lights. *Journal of Mammalogy* 77(2): 325–334.

Hickey, M.B.C., and Fenton, M.B. 1990. Foraging by red bats (*Lasiurus borealis*): Do intraspecific chases mean territoriality? *Canadian Journal of Zoology* 68(12): 2477–2482.

– 1996. Behavioural and thermoregulatory responses of female Hoary Bats, *Lasiurus cinereus* (Chiroptera: Vespertilionidae), to variations in prey availability. *Ecoscience* 3(4): 414–422.

Hickling, G.J., Millar, J.S., and Moses, R.A. 1991. Reproduction and nutrient reserves of Bushy-tailed Wood Rats (*Neotoma cinerea*). *Canadian Journal of Zoology* 69(12): 3088–3092.

Hickman, G.C. 1983. Burrow structure of the talpid mole *Parascalops breweri* from Oswego County, New York State. *Zeitschrift Säugetierkunde* 48(5): 265–269.

– 1984. An excavated burrow of *Scalopus aquaticus* from Florida, with comments on Nearctic talpid/geomyid burrow structure. *Säugetierkundliche Mitteilungen* 31(2/3): 243–249.

Hik, D.S. 2002. Mammalian herbivores in a dynamic alpine environment. In *Mountain Science Highlights*, proceedings of a symposium on 'The State of Ecological and Earth Sciences in Mountain Areas,' Banff, Alberta, Canada.

Hill, J.E., and Yalden, D.W. 1990. The status of the hoary bat, *Lasiurus cinereus*, as a British species. *Journal of Zoology* (London) 222: 694–697.

Hill, M.M.A., Powell, G.L., and Russell, A.P. 2001. Diet of the Prairie Rattlesnake, *Crotalus viridis viridis*, in southeastern Alberta. *Canadian Field Naturalist* 115(2): 241–246.

Hindell, M.A. 2002. Elephant Seals *Mirounga angustirostris* and *M. leonina*. In W.F. Perrin, B. Wursig, and H.G.T. Thewissen (eds), *Encyclopedia of Marine Mammals*, 370–373. Academic Press, San Diego, CA.

Hines, T.D., and Case, R.M. 1991. Diet, home range, movements, and activity periods of Swift Fox in Nebraska. *Prairie Naturalist* 23(3): 131–138.

Hitchcock, H.B. 1955. A summer colony of least bat, *Myotis subulatus leibii* (Audubon and Bachman). *Canadian Field-Naturalist* 69(2): 31.

– 1965. Twenty-three years of bat banding in Ontario and Quebec. *Canadian Field-Naturalist* 79(1): 4–14.

Hjertaas, D.G., and Hyertaas, P. 1990. Predation at Bank Swallow colonies near Katepwa Lake. *Blue Jay* 48(3): 162–165.

Hodder, D.P., and Rea, R.V. 2006. Winter habitat use by River Otters (*Lontra canadensis*) in the John Prince Research Forest, Fort St. James, British Columbia. *Wildlife Afield* 3(2): 111–116.

Hodder, J., Brown, R.F., and Cziesla, C. 1998. The Northern Elephant Seal in Oregon: A pupping range extension and onshore occurrence. *Marine Mammal Science* 14(4): 873–881.

Hodges, K.E. 2000. Ecology of Snowshoe Hares in southern boreal and montane forests. In L.F. Ruggiero, K.B. Aubry, S.W. Buskirk, G.M. Koehler, C.J. Krebs, K.S. McKelvey, and J.R. Squires (eds), *Ecology and Conservation of Lynx in the United States*, 163–206. University Press of Colorado, Boulder.

Hoefs, M. 1991. A longevity record for Dall Sheep, *Ovis dalli dalli*, Yukon Territory. *Canadian Field-Naturalist* 105(3): 397–398.

– 2001. Mule, *Odocoileus hemionus*, and White-tailed, *O. virginianus*, Deer in the Yukon. *Canadian Field-Naturalist* 115(2): 296–300.

Hoefs, M., and Nowlan, U. 1993. Minimum breeding age of Dall Sheep, *Ovis dalli dalli*, ewes. *Canadian Field-Naturalist* 107(2): 241–243.

– 1998. Triplets in Mountain Goats, *Oreamnos americanus*. *Canadian Field-Naturalist* 112(3): 539–540.

Hoelzel, A.R. 1999. Impact of population bottlenecks on genetic variation and the importance of life-history: A case study of the Northern Elephant Seal. *Biological Journal of the Linnean Society* 68: 23–39.

Hoelzel, A.R., Fleischer, R.C., Campagna, C., Le Boeuf, B.J., and Alvord, G. 2002. Impact of a population bottleneck on symmetry and genetic diversity in the Northern Elephant Seal. *Journal of Evolutionary Biology* 15: 567–575.

Hoelzel, A.R., Le Boeuf, B.J., Reiter, J., and Campagna, C. 1999. Alpha-male paternity in elephant seals. *Behavioral Ecology and Sociobiology* 46: 298–306.

Hoffman, B., Scheer, M., and Behr, I.P. 2004. Underwater behaviors of Short-finned Pilot Whales (*Globicephala macrorhynchus*) off Tenerife. *Mammalia* 68(2–3): 221–224.

Hofman, J.E., McGuire, B., and Pizzuto, T.M. 1989. Parental care in the Sagebrush Vole (*Lemmiscus curtatus*). *Journal of Mammalogy* 70(1): 162–165.

Hogan, K.M., Hedin, M.C., Koh, H.S., Davis, S.K., and Greenbaum, I.F. 1993. Systematic and taxonomic implications of karyotypic, electrophoretic, and mitochondrial-DNA variation in *Peromyscus* from the Pacific Northwest. *Journal of Mammalogy* 74(4): 819–831.

Holloway, G.L., and Barclay, R.M.R. 2001. *Myotis ciliolabrum*. *Mammalian Species* no. 670: 1–5.

Holloway, G.L., and Malcolm, J.R. 2007. Nest-tree use by Northern and Southern flying squirrels in central Ontario. *Journal of Mammalogy* 88(1): 226–233.

Holmes, D.J. 1990. Social and other correlates of scent marking in captive Virginia Opossums, *Didelphis virginiana* Kerr. *Chemical Signals in Vertebrates* 5: 451–458.

– 1991. Social behavior in captive Virginia Opossums, *Didelphis virginiana*. *Journal of Mammalogy* 72(2): 402–410.

– 1992. Sternal odors as cues for social discrimination by female Virginia Opossums, *Didelphis virginiana*. *Journal of Mammalogy* 73(2): 286–921.

Holmes, W.G. 1991. Predator risk affects foraging behaviour of pikas: Observational and experimental experience. *Animal Behaviour* 42: 111–119.

Holst, M., and Stirling, I. 2002. A comparison of Ringed Seal (*Phoca hispida*) biology on the east and west sides of the North Water Polynya, Baffin Bay. *Aquatic Mammals* 28(3): 221–230.

Holst, M., Stirling, I., and Calvert, W. 1999. Age structure and reproductive rates of Ringed Seals (*Phoca hispida*) on the northwestern coast of Hudson Bay in 1991 and 1992. *Marine Mammal Science* 15(4): 1357–1364.

Holst, M., Stirling, I., and Hobson, K.A. 2001. Diet of Ringed Seals (*Phoca hispida*) on the east and west sides of the North Water Polynya, northern Baffin Bay. *Marine Mammal Science* 17(4): 888–908.

Holt, D.W. 1990. 'Blond' color morph of Meadow Voles, *Microtus pennsylvanicus*, from Massachusetts. *Canadian Field-Naturalist* 104(4): 596–597.

– 1994. Larder hoarding in the Cougar, *Felis concolor*. *Canadian Field-Naturalist* 108(2): 240–241.

Holyan, J.A., Jones, L.L.C., and Raphael, M.G. 1998. American Marten use of cabins as resting sites in central Oregon. *Northwestern Naturalist* 79: 68–70.

Hölzenbein, S., and Marchinton, R.L. 1992. Spatial integration of maturing-male White-tailed Deer into the adult population. *Journal of Mammalogy* 73(2): 326–334.

Homyack, J.A., Vashon, J.H., Libby, C., Lindquist, E.L., Loch, S., McAlpine, D.F., Pilgrim, K.L., and Schwartz, M.K. 2008. Canada Lynx-Bobcat (*Lynx canadensis* × *L. rufus*) hybrids at the southern periphery of Lynx range in Maine, Minnesota and New Brunswick. *American Midland Naturalist* 159(2): 504–508.

Hood, G.A., and Neufeld, T. 2004. First record of Mountain Lions, *Puma concolor*, in Elk Island National Park, Alberta. *Canadian Field-Naturalist* 118(4): 605–607.

Hoodicoff, C.S. 2003. Ecology of the badger (*Taxidea taxus jeffersonii*) in the Thompson Region of British Columbia: Implications for conservation. M.Sc. thesis, University of Victoria, Victoria, BC.

– 2006. Badger prey ecology: The ecology of six small mammals found in British Columbia. *Wildlife Working Report* no. WR-109: 1–31. BC Ministry of the Environment, Ecosystems Branch.

Hoogland, J.L. 1995. *The Black-Tailed Prairie Dog: Social Life of a Burrowing Mammal*. University of Chicago Press, Chicago, IL.

– 1996. *Cynomys ludovicianus*. *Mammalian Species* no. 535: 1–10.

– 2001. Black-tailed, Gunnison's and Utah Prairie Dogs reproduce slowly. *Journal of Mammalogy* 82(4): 917–927.

Hooker, S.K., and Baird, R.W. 1999a. Observations of Sowerby's Beaked Whales, *Mesoplodon bidens*, in the Gully, Nova Scotia. *Canadian Field-Naturalist* 113(2): 273–277.

– 1999b. Deep-diving behaviour of the Northern Bottlenose Whale, *Hyperoodon ampullatus* (Cetacea: Ziphiidae). *Proceedings of the Royal Society of London, B Series* 266: 671–676.

Hooker, S.K., and Whitehead, H. 2002. Click characteristics of Northern Bottlenose Whales (*Hyperoodon ampullatus*). *Marine Mammal Science* 18(1): 69–80.

Hooker, S.K., Whitehead, H., and Gowans, S. 2002a. Ecosystem consideration in conservation planning: Energy demand of foraging Bottlenose Whales (*Hyperoodon ampullatus*) in a marine protected area. *Biological Conservation* 104: 51–58.

Hooker, S.K., Whitehead, H., Gowans, S., and Baird, R.W. 2002b. Fluctuations in distribution and patterns of individual range use of Northern Bottlenose Whales. *Marine Ecology Progress Series* 225: 287–297.

Hooper, S.R., Van Den Bussche, R.A., and Horáček, I. 2006. Generic status of the American pipistrelles (Vespertilionidae) with description of a new genus. *Journal of Mammalogy* 87(5): 981–992.

Hornbeck, G.E., and Mahoney, J.M. 2000. Introgressive hybridization of Mule Deer and White-tailed Deer in southwestern Alberta. *Wildlife Society Bulletin* 28(4): 1012–1015.

Horne, T.J., and Ylönen, H. 1996. Female Bank Voles (*Clethrionomys glareolus*) prefer dominant males; but what if there is no choice? *Behavioral Ecology and Sociobiology* 38(6): 401–405.

Horwood, J. 2002. Sei Whale *Balaenoptera borealis*. In W.F. Perrin, B. Wursig, and H.G.T. Thewissen (eds), *Encyclopedia of Marine Mammals*, 1069–1071. Academic Press, San Diego, CA.

Hossler, R.J., McAninch, J.B., and Harder, J.D. 1994. Maternal denning behavior and survival of juveniles in Opossums in southeastern New York. *Journal of Mammalogy* 75(1): 60–70.

Houck, W.J., and Jefferson, T.A. 1999. Dall's Porpoise *Phocoenoides dalli* (True, 1885). In S.H. Ridgway and R. Harrison (eds), *Handbook of Marine Mammals*, vol. 6, 443–472. Academic Press, San Diego, CA.

Houseknecht, C.R. 1968. Sonographic analysis of vocalizations of three species of mice. *Journal of Mammalogy* 49(3): 555–560.

Houston, J. 1990a. Status of Blainville's Beaked Whale, *Mesoplodon densirostris*, in Canada. *Canadian Field-Naturalist* 104(1): 117–120.

– 1990b. Status of Hubbs' Beaked Whale, *Mesoplodon carlhubbsi*, in Canada. *Canadian Field-Naturalist* 104(1): 121–124.

– 1990c. Status of Stejneger's Beaked Whale, *Mesoplodon stejnegeri*, in Canada. *Canadian Field-Naturalist* 104(1): 131–134.

– 1990d. Status of True's Beaked Whale, *Mesoplodon mirus*, in Canada. *Canadian Field-Naturalist* 104(1): 135–137.

– 1991. Status of Cuvier's Beaked Whale, *Ziphius cavirostris*, in Canada. *Canadian Field-Naturalist* 105(2): 215–218.

Hoving, C.L., Harrison, D.J., Krohn, W.B., Joseph, R.A., and O'Brien, M. 2005. Broad-scale predictors of Canada Lynx occurrence in eastern North America. *Journal of Wildlife Management* 69(2): 739–751.

Hoying, K.M., and Kunz, T.H. 1998. Variation in size at birth and post-natal growth in the insectivorous bat *Pipistrellus subflavus* (Chiroptera: Vespertilionidae). *Journal of Zoology* (London) 245: 15–27.

Hoyle, J.A., and Boonstra, R. 1986. Life history traits of the Meadow Jumping Mouse, *Zapus hudsonius*, in southern Ontario. *Canadian Field-Naturalist* 100(4): 537–544.

Hubbs, A.H., and Boonstra, R. 1998. Effects of food and predators on the home-range sizes of Arctic Ground Squirrels (*Spermophilus parryii*). *Canadian Journal of Zoology* 76(3): 592–596.

Hubbs, A.H., Karels, T., and Byrom, A. 1996. Tree-climbing by Arctic Ground Squirrels, *Spermophilus parryii*, in the southwestern Yukon Territory. *Canadian Field-Naturalist* 110(3): 533–534.

Hubbs, A., and Schowalter, T. 2003. Survey of bats in northeastern Alberta. *Alberta Species at Risk Report* no. 68: 1–18.

Huck, U., and Banks, E.M. 1982. Male dominance status, female choice and mating success in the Brown Lemming, *Lemmus trimucronatus*. *Animal Behaviour* 30(3): 665–675.

Hückstädt, L., and Antezana, T. 2001. An observation of parturition in a stranded *Kogia breviceps*. *Marine Mammal Science* 17(2): 362–365.

Hudson, R., Bilkó, Á., and Altbäcker, V. 1996. Nursing, weaning and the development of independent feeding in the rabbit (*Oryctolagus cuniculus*). *Zeitschrift für Säugetierkunde* 61: 39–48.

Hult, R.W., Dupey, S.E., and Badley, R.W. 1980. Mortalities associated with prey ingestion by small cetaceans. *Cetology* 38: 1–2.

Humphries, M.M., Thomas, D.W., and Speakman, J.R. 2002. Climate-mediated energetic constraints on the distribution of hibernating mammals. *Nature* 418: 313–316.

Humphrey, S.R., and Kunz, T.H. 1976. Ecology of a Pleistocene relict, the Western Big-Eared Bat (*Plecotus townsendii*), in the southern Great Plains. *Journal of Mammalogy* 57(3): 470–494.

Hunt, H.M. 1980. Autumn diet of Elk in three areas of Saskatchewan. *Blue Jay* 38(2): 125–129.

Hurly, T.A., and Robertson, R.J. 1990. Variation in the food hoarding behaviour of Red Squirrels. *Behavioral Ecology and Sociobiology* 26: 91–97.

Hutchinson, J.T., and Lacki, M.J. 2001. Possible microclimate benefits of roost site selection in the red bat, *Lasiurus borealis*, in mixed mesophytic forests of Kentucky. *Canadian Field-Naturalist* 115(2): 205–209.

ICES Advisory Committee on Fisheries Management. 2001. Report of the Joint ICES/ NAFO Working Group on Harp and Hooded Seals. ICES CM 2001/ACFM:08. International Council for the Exploration of the Sea, Advisory Committee on Fisheries Management, 2–6 October 2000, Copenhagen.

Iliff, M.J., and Brinkley, E.S. 2001. Observation of a yearling Hooded Seal (*Cystophora cristata*) at Back Bay National Wildlife Refuge, Virginia Beach. *Banisteria* 18: 38–40.

Inman, R.M., Wigglesworth, R.R., Inman, K.H., Schwartz, M.K., Brock, B.L., and Rieck, J.D. 2004. Wolverine makes extensive movements in the Greater Yellowstone Ecosystem. *Northwest Science* 78(3): 261–266.

Innes, D.G.L., Bendell, J.F., Naylor, B.J., and Smith, B.A. 1990. High densities of the Masked Shrew, *Sorex cinereus*, in Jack Pine plantations in northern Ontario. *American Midland Naturalist* 124(2): 330–341.

Innes, D.G.L., and Millar, J.S. 1990. Numbers of litters, litter size and survival in two species of microtines at two elevations. *Holarctic Ecology* 13: 207–216.

– 1993. Factors affecting litter size in *Clethrionomys gapperi*. *Annals Zoologica Fennica* 30: 239–245.

Inouye, D.W., Barr, B., Armitage, K.B., and Inouye, B.D. 2000. Climate change is affecting altitudinal migrants and hibernating species. *Proceedings of the National Academy of Science of the USA* 97(4): 1630–1633.

Insley, S.J. 2000. Long-term vocal recognition in the Northern Fur Seal. *Nature* 406(27 July): 404–405.

Insley, S.J., Robson, B.W., Yack, T., Ream, R.R., and Burgess, W.C. 2008. Acoustic determination of activity and flipper stroke rate in foraging Northern Fur Seal females. *Endangered Species Research* 4: 147–155.

International Whaling Commission. 2001. Report of the workshop on status and trends of western North Atlantic Right Whales. A worldwide comparison. *Journal of Cetacean Research and Management*, Special issue, 10: 1–33.

IUCN/SSC Polar Bear Specialist Group. 2002. Status of the Polar Bear. *Occasional Papers of the IUCN Species Survival Commission* 26: 21–35.

Iverson, S.L., and Turner, B.N. 1972. Natural history of a Manitoba population of Franklin's Ground Squirrels. *Canadian Field-Naturalist* 86(2): 145–149.

Jackson, W. B. 1982. Norway rat and allies. In J.A. Chapman and G.A. Feldhamer (eds), *Wild Mammals of North America: biology, management, and economics*, 1077–1088. The Johns Hopkins University Press, Baltimore, MD.

Jacobs, L.F., and Liman, E.R. 1991. Grey Squirrels remember the locations of buried nuts. *Animal Behaviour* 41: 103–110.

Jacobs, S.R., and Terhune, J.M. 2000. Harbor Seal (*Phoca vitulina*) numbers along the New Brunswick coast of the Bay of Fundy in autumn in relation to aquaculture. *Northeastern Naturalist* 7(3): 289–296.

– 2002. The effectiveness of acoustic harassment devices in the Bay of Fundy, Canada: Seal reactions and a noise exposure model. *Aquatic Mammals* 28(2): 147–158.

Jacobsen, K.-O., Marx, M., and Øien, N. 2004. Two-way trans-Atlantic migration of a North Atlantic Right Whale (*Eubalaena glacialis*). *Marine Mammal Science* 20(1): 161–166.

Jacques, C.N., and Jenks, J.A. 2007. Dispersal of yearling Pronghorns in western South Dakota. *Journal of Wildlife Management* 71(1): 177–182.

– 2008. Visual observation of Bobcat predation on an adult female Pronghorn in northwestern South Dakota. *American Midland Naturalist* 160(1): 259–261.

Jacques, C.N., Jenks, J.A., and Klaver, R.W. 2009. Seasonal movements and home-range use by female Pronghorns in sagebrush-steppe communities of western South Dakota. *Journal of Mammalogy* 90(2): 433–441.

Jacquot, J.J., and Vessey, S.H. 1994. Non-offspring nursing in the White-footed Mouse, *Peromyscus leucopus*. *Animal Behaviour* 48: 1238–1240.

Jameson, E.W., Jr, and Peeters, H.J. 2004. *Mammals of California*. University of California Press, Berkeley.

Janečka, J.E., Blankenship, T.L., Hirth, D.H., Kilpatrick, C.W., Tewes, M.E., and Grassman, L.I., Jr. 2007. Evidence for male-biased dispersal in Bobcats *Lynx rufus* using relatedness analysis. *Wildlife Biology* 13(1): 38–47.

Jannett, F.J., Jr. 1978. The density-dependant formation of extended maternal families of the Montane Vole, *Microtus montanus*, and the behavior of territorial males. *Behavioral Ecology and Sociobiology* 3: 245–263.

– 1981a. Scent mediation of intraspecific, interspecific, and intergeneric agonistic behaviour among sympatric species of voles (Microtinae). *Behavioral Ecology and Sociobiology* 8(4): 293–296.

– 1981b. Sex ratios in high-density populations of the Montane Vole, *Microtus montanus*, and the behaviour of territorial males. *Behavioral Ecology and Sociobiology* 8(4): 297–307.

– 1984. Reproduction of the Montane Vole, *Microtus montanus*, in subnivean populations. *Special Publications of the Carnegie Museum* 10: 215–224.

Jannett, F.J., Jr, Broschart, M.R., Grim, L.H., and Schaberl, J.P. 2007. Northerly range extensions of mammalian species in Minnesota. *American Midland Naturalist* 158(1): 168–176.

Jannett, F.J., Jr, and Huber, R.L. 1994. Range extension and first Holocene record of the Arctic Shrew, *Sorex arcticus*, from the Driftless Area, southeastern Minnesota. *Canadian Field-Naturalist* 108(2): 226–228.

Jannett, F.J., Jr, and Oehlenschlager, R.J. 1994. Range extension and first Minnesota records of the Smoky Shrew *Sorex fumeus*. *American Midland Naturalist* 131(2): 364–365.

– 1997. Range extension and unusual occurrences of the Heather Vole, *Phenacomys intermedius*, in Minnesota. *Canadian Field-Naturalist* 111(3): 459–461.

Jardine, C. 2004. Unusual 'masked' Long-tailed Weasel. *Blue Jay* 62(2): 116–117.

Jarrell, G.H., and Fredga, K. 1993. How many kinds of lemmings? A taxonomic overview. In N.C. Stenseth and R.A. Ims (eds), *The Biology of Lemmings*, 45–57. Linnean Society Symposium Series no. 15. Academic Press, San Diego, CA.

Jay, C.V., and Hills, S. 2005. Movements of Walruses radio-tagged in Bristol Bay, Alaska. *Arctic* 58(2): 192–202.

Jefferson, T.A. 1988. *Phocoenoides dalli*. *Mammalian Species* no. 319: 1–7.

– 1990. Status of Dall's porpoise, *Phocoenoides dalli*, in Canada. *Canadian Field-Naturalist* 104(1): 112–116.

Jefferson, T.A., Fertl, D., Bolaños-Jiménez, J., and Zerbini, A.N. 2009. Distribution of common dolphins (*Delphinus* spp.) in the western Atlantic Ocean: A critical re-examination. *Marine Biology* 156(6): 1109–1124.

Jefferson, T.A., and Newcomer, M.W. 1993. *Lissodelphis borealis*. *Mammalian Species* no. 425: 1–6.

Jefferson, T.A., Newcomer, M.W., Leatherwood, S., and van Waerebeek, K. 1994. Right Whale Dolphins *Lissodelphis borealis* (Peale, 1848) and *Lissodelphis peronii* (Lacépède, 1804). In S.H. Ridgway and R. Harrison (eds), *Handbook of Marine Mammals*, vol. 5, 335–362. Academic Press, San Diego, CA.

jeffersonii Badger Recovery Team 2004. National recovery strategy for American Badger, *jeffersonii* subspecies (*Taxidea taxus jeffersonii*). *Recovery of Nationally Endangered Wildlife (RENEW)*. Ottawa, ON.

Jenkins, S.H., and Ashley, M.C. 2003. Wild Horses (*Equus caballus* and allies). In G.A. Feldhamer, B.C. Thompson, and J.A. Chapman (eds), *Wild Mammals of North America: Biology, Management, and Conservation*, 2nd ed., 1148–1163. Johns Hopkins University Press, Baltimore, MA.

Jenkins, S.H., and Busher, P.E. 1979. *Castor canadensis. Mammalian Species* no. 120: 1–8.

Jensen, I.M. 1986. Foraging strategies of the mole (*Parascalops breweri* Bachman, 1842). I. The distribution of prey. *Canadian Journal of Zoology* 64(8): 1727–1733.

Johnson, D.R., Swanson, B.J., and Eger, J.L. 2000. Cyclic dynamics of eastern Canadian Ermine populations. *Canadian Journal of Zoology* 78(5): 835–839.

Johnson, G.D., Erickson, W.P., Strickland, M.D., Shepherd, M.F., Shepherd, D.A., and Sarappo, S.A. 2003. Mortality of bats at a large-scale wind power development at Buffalo Ridge, Minnesota. *American Midland Naturalist* 150(2): 332–342.

Johnson, J.S., Lacki, M.J., and Baker, M.D. 2007. Foraging ecology of Long-legged Myotis (*Myotis volans*) in north-central Idaho. *Journal of Mammalogy* 88(5): 1261–1270.

Johnson, S.A., and Berkley, K.A. 1999. Construction of a natal den by an introduced River Otter, *Lutra canadensis*, in Indiana. *Canadian Field-Naturalist* 113(2): 301–304.

Johnson, S.A., and Choromanski-Norris, J. 1992. Reduction in the eastern limit of the range of the Franklin's Ground Squirrel (*Spermophilus franklinii*). *American Midland Naturalist* 128(2): 325–331.

Johnston, C., Runge, W., and McFetridge, R. 1993. Status Report on the Prairie Long-tailed Weasel, *Mustela frenata longicauda*, in Canada. *Committee on the Status of Endangered Wildlife in Canada*, Ottawa.

Johnston, D.S., and Fenton, M.B. 2001. Individual and population-level variability in diets of Pallid Bats (*Antrozous pallidus*). *Journal of Mammalogy* 82(2): 362–373.

Johnston, D.W. 2002. The effect of acoustic harassment devices on Harbour Porpoises (*Phocoena phocoena*) in the Bay of Fundy, Canada. *Biological Conservation* 108: 113–118.

Johnstone, R. 1979. An unusual nest site of the Pacific Jumping Mouse with notes on its natural history. *Murrelet* 60(2): 72.

Joly, K., Cole, M.J., and Jandt, R.R. 2007. Diets of overwintering Caribou, *Rangifer tarandus*, track decadal changes in Arctic tundra vegetation. *Canadian Field-Naturalist* 121(4): 379–383.

Joly, K., Dale, B.W., Collins, W.B., and Adams, L.G. 2003. Winter habitat use by female Caribou in relation to wildland fires in interior Alaska. *Canadian Journal of Zoology* 81(7): 1192–1201.

Jones, E.N. 1990. Effects of forage availability on home range and population density of *Microtus pennsylvanicus. Journal of Mammalogy* 71(3): 382–389.

Jones, G.S., and Jones, D.B. 1985. Observations of intraspecific behavior of Meadow Jumping Mice, *Zapus hudsonius*, and escape behavior of a Western Jumping Mouse, *Zapus princeps*, in the wild. *Canadian Field-Naturalist* 99(3): 378–380.

Jones, G.S., Whitaker, J.O., Jr, and Maser, C. 1978. Food habits of jumping mice (*Zapus trinotatus* and *Zapus princeps*) in western North America. *Northwest Science* 52(1): 57–60.

Jones, I.M. 2006. A northeast Pacific offshore Killer Whale (*Orcinus orca*) feeding on a Pacific Halibut (*Hippoglossus stenolepis*). *Marine Mammal Science* 22(1): 198–200.

Jones, J.K., Jr, and Birney, E.C. 1988. *Handbook of Mammals of the North-Central States*. University of Minnesota Press, Minneapolis.

Jones, J.K., Jr, Choate, J.R., and Wilhelm, R.B. 1978. Notes on the distribution of three species of mammals in South Dakota. *Prairie Naturalist* 10(3): 65–70.

Jones, M.L. 1990. The reproductive cycle in Gray Whales based on photographic resightings of females on the breeding grounds from 1977–1982. *Report of the International Whaling Commission*, Special issue, 12: 177–182.

Jones, M.L., and Swartz, S.L. 2002. Gray Whale *Eschrichtius robustus*. In W.F. Perrin, B. Wursig, and H.G.T. Thewissen (eds), *Encyclopedia of Marine Mammals*, 524–536. Academic Press, San Diego, CA.

Judge, K.A., and Haviernick, M. 2002. Update COSEWIC status report on the Grey Fox *Urocyon cinereoargenteus* in Canada, In COSEWIC assessment and update status report on the Grey Fox *Urocyon cinereoargenteus interior* in Canada. *Committee on the Status of Endangered Wildlife in Canada*. Ottawa.

Jung, T.S., and Merchant, P.J. 2005. First confirmation of Cougar, *Puma concolor*, in the Yukon. *Canadian Field-Naturalist* 119(4): 580–581.

Jung, T.S., Pretzlaw, T.D., and Nagorsen, D.W. 2007. Northern Range extension of the Pygmy Shrew, *Sorex hoyi*, in the Yukon. *Canadian Field-Naturalist* 121(1): 94–95.

Jung, T.S., Runck, A.M., Nagorsen, D.W., Slough, B.G., and Powell, T. 2006a. First records of the Southern Red-backed Vole, *Myodes gapperi*, in the Yukon. *Canadian Field-Naturalist* 120(3): 331–334.

Jung, T.S., Slough, B.G., Nagorsen, D.W., Dewey, T.A., and Powell, T. 2006b. First records of the Northern Long-eared Bat, *Myotis septentrionalis*, in the Yukon Territory. *Canadian Field-Naturalist* 120(1): 39–42.

Junge, J.A., Hoffmann, R.S., and Derby, R.W. 1983. Relationships within the Holarctic *Sorex arcticus – Sorex tundrensis* complex. *Acta Theriologica* 28(21–31): 339–350.

Kalcounis-Ruepell, M.C., Millar, J.S., and Herdman, E.J. 2002. Beating the odds: Effects of weather on a short-season population of Deer Mice. *Canadian Journal of Zoology* 80(9): 1594–1601.

Kaliszewska, Z.A., Seger, J., Rowntree, V.J., Barco, S.G., Benegas, R., Best, P.B., Brown, M.W., Brownell, R.L., Jr, Carribero, A., Harcourt, R., Knowlton, A.R., Marshall-Tilas, K., Patenaude, N.J., Rivarola, M., Schaeff, C.M., Sironi, M., Smith, W.A., and Yamada, T.K. 2005. Population histories of right whales (Cetacea: *Eubalaena*) inferred from mitochondrial sequence diversities and divergences of their whale lice (Amphipoda: *Cyamus*). *Molecular Ecology* 14(11): 3439–3456.

Kamler, J.F., Ballard, W.B., Gese, E.M., Harrison, R.L., and Karki, S.M. 2004a. Dispersal characteristics of Swift Foxes. *Canadian Journal of Zoology* 82(12): 1837–1842.

Kamler, J.F., Ballard, W.B., Gese, E.M., Harrison, R.L., Karki, S., and Mote, K. 2004b. Adult male emigration and a female-based social organization in Swift Foxes, *Vulpes velox. Animal Behaviour* 67(4): 699–702.

Kamler, J.F., Ballard, W.B., Helliker, B.R., and Stiver, S. 2003a. Range expansion of Raccoons in western Utah and central Nevada. *Western North American Naturalist* 63(3): 406–408.

Kamler, J.F., Ballard, W.B., and Swepston, D.A. 2001. Range expansion of Mule Deer in the Texas Panhandle. *Southwestern Naturalist* 46(3): 378–379.

Kamler, J.F., Green, L.A., and Ballard, W.B. 2003b. Recent occurrence of Black Bears in the southwestern Great Plains. *Southwestern Naturalist* 48(2): 303–306.

Kansas, J.L. 2002. Status of the Grizzly Bear (*Ursus arctos*) in Alberta. *Wildlife Status Report* no. 37: 43. Alberta Sustainable Resource Development, Fish and Wildlife Division, and Alberta Conservation Association.

Kapel, C.M.O. 1999. Diet of Arctic Fox (*Alopex lagopus*) in Greenland. *Arctic* 52(3): 289–293.

Kaplan, C.C., White, G.C., and Noon, B.R. 2008. Neonatal survival of Steller Sea Lions (*Eumetopias jubatus*). *Marine Mammal Science* 24(3): 443–461.

Kapusta, J., Pachinger, K., and Marchlewska-Koj, A. 1999. Behavioural variation in two populations of Root Voles. *Acta Theriologica* 44(4): 337–343.

Karban, R., Karban, C., and Karban, J. 2007. Hay piles of the Mountain Beaver (*Aplodontia rufa*) delay plant decomposition. *Western North American Naturalist* 67(4): 618–621.

Karels, T., and Boonstra, R. 1999. The impact of predation on burrow use by Arctic Ground Squirrels in the boreal forest. *Proceedings of the Royal Society of London, B series* 266: 2117–2123.

Kastak, D., and Schusterman, R.J. 1999. In-air and underwater hearing sensitivity of a Northern Elephant Seal (*Mirounga angustirostris*). *Canadian Journal of Zoology* 77(11): 1751–1758.

Kastelein, R.A. 2002. Walrus *Odobenus rosmarus*. In W.F. Perrin, B. Wursig, and H.G.T. Thewissen (eds), *Encyclopedia of Marine Mammals*, 1294–1300. Academic Press, San Diego, CA.

Kastelein, R.A., Mosterd, J., Schooneman, N.M., and Wiepkema, P.R. 2000. Food consumption, growth, body dimensions, and respiration rates of captive False Killer Whales (*Pseudorca crassidens*). *Aquatic Mammals* 26(1): 33–44.

Kastelein, R.A., Mosterd, P., van Ligtenberg, C.L., and Verboom, W.C. 1996. Aerial hearing sensitivity tests with a male Pacific Walrus (*Odobenus rosmarus divergens*), in the free field and with headphones. *Aquatic Mammals* 22(2): 81–93.

Kastelein, R.A., Muller, M., and Terlouw, A. 1994. Oral suction of a Pacific Walrus (*Odobenus rosmarus divergens*) in air and under water. *Zeitschrift Säugetierkunde* 59: 105–115.

Kastelein, R.A., Nieuwstraten, S.H., and Verstegen, M.W.A. 1997. Passage of carmine red dye through the digestive tract of Harbour Porpoises (*Phocoena phocoena*). In A.J. Read, P.R. Wiepkema, and P.E. Nachtigall (eds), *The Biology of the Harbour Porpoise*, 265–278. De Spil Publishers, Woerden, The Netherlands.

Kastelein, R.A., Stevens, S., and Mosterd, P. 1990. The tactile sensitivity of the mystacial vibrissae of a Pacific Walrus (*Odobenus rosmarus divergens*). Part 2: Masking. *Aquatic Mammals* 16(2): 78–87.

Kastelein, R.A., Zweypfenning, R.C.V.J., Spekreijse, H., Dubbeldam, J.L., and Born, E.W. 1993. The anatomy of the Walrus head (*Odobenus rosmarus*). Part 3: The eyes and their function in Walrus ecology. *Aquatic Mammals* 19(2): 61–92.

Kasuya, T., Brownell, R.L., Jr, and Balcomb, K.C. 1997. Life history of Baird's Beaked Whales off the Pacific coast of Japan. *Report of the International Whaling Commission* 47: 969–979.

Katona, S.K., and Beard, J.A. 1990. Population size, migrations and feeding aggregations of the Humpback Whale (*Megaptera novaeangliae*) in the western North Atlantic Ocean. *Report of the International Whaling Commission*, Special issue, 12: 295–305.

Kawamichi, T. 1981. Vocalizations of *Ochotona* as a taxonomic character. In K. Myers and C.D. MacInnes (eds), *Proceeding of the World Lagomorph Conference, August 12–16, 1979*, 324–339. University of Guelph, Guelph, ON.

Kay, C.E., and Boe, E. 1992. Hybrids of White-tailed and Mule Deer in western Wyoming. *Great Basin Naturalist* 52(3): 290–292.

Kazial, K.A., and Masters, W.M. 2004. Female Big Brown Bats (*Eptesicus fuscus*), recognise sex from a caller's echolocation signals. *Animal Behaviour* 67(5): 855–863.

Kazial, K.A., Pacheco, S., and Zielinski, K.N. 2008. Information content of sonar calls of Little Brown Bats (*Myotis lucifugus*): Potential for communication. *Journal of Mammalogy* 89(1): 25–33.

Keech, M.A., Bowyer, R.T., Ver Hoef, J.M., Boertje, R.D., Dale, B.W., and Stephenson, T.R. 2000. Life-history consequences of maternal condition in Alaskan Moose. *Journal of Wildlife Management* 64: 450–462.

Keeler, J.O., and Studier, E.H. 1992. Nutrition in pregnant Big Brown Bats (*Eptesicus fuscus*) feeding on June beetles. *Journal of Mammalogy* 73(2): 426–430.

Keith, L.B. 1990. Dynamics of Snowshoe Hare populations. *Current Mammalogy* 2: 119–195.

Keith, L.B., and Bloomer, S.E.M. 1993. Differential mortality of sympatric Snowshoe Hares and Cottontail Rabbits in central Wisconsin. *Canadian Journal of Zoology* 71(8): 1694–1697.

Kelly, B.P., and Wartzok, D. 1996. Ringed Seal diving behavior in the breeding season. *Canadian Journal of Zoology* 74(8): 1547–1555.

Kemble, E.D. 1984. Effects of preweaning predatory or consummatory experience and litter size on cricket predation in Northern Grasshopper Mice *Onychomys leucogaster*. *Aggressive Behavior* 10: 55–58.

Kenagy, G.J., and Barnes, B.M. 1988. Seasonal reproductive patterns in four coexisting rodent species from the Cascade Mountains, Washington. *Journal of Mammalogy* 69(2): 274–292.

Kennedy, J.F., Jenks, J.A., Jones, R.L., and Jenkins, K.J. 1995. Characteristics of mineral licks used by White-tailed Deer (*Odocoileus virginianus*). *American Midland Naturalist* 134: 324–331.

Kenney, R.D. 2002. North Atlantic, North Pacific, and Southern Right Whales. In W.F. Perrin, B. Wursig, and H.G.T. Thewissen (eds), *Encyclopedia of Marine Mammals*, 806–813. Academic Press, San Diego, CA.

Kernan, J.D. 1918. The skull of *Ziphius cavirostris*. *Bulletin of the American Museum of Natural History* 38(11): 349–394.

Kie, J.G., and Czech, B. 2000. Mule and Black-tailed Deer. In S. Demarais and P.R. Krausman (eds), *Ecology and Management of Large Mammals in North America*, 629–657. Prentice Hall, Upper Saddle River, NJ.

Kilburn, K.S. 1997. Skeletal architecture of the forelimbs in kangaroo rats (Heteromyidae: *Dipodomys*): Adaptations for digging and food handling. *Special Publication of the Museum of Southwestern Biology* 3: 215–224.

Kiliaan, H.P.L., Mamo, C., and Paquet, P.C. 1991. A Coyote, *Canis latrans*, and Badger, *Taxidea taxus*, interaction near Cypress Hills Provincial Park, Alberta. *Canadian Field-Naturalist* 105(1): 122–123.

King, C.B., Wilson, G.M., and Sudman, P.D. 1999. New records of the Pygmy Shrew in southeast South Dakota. *Prairie Naturalist* 31(2): 115–117.

King, C.M. 1983. *Mustela erminea*. *Mammalian Species* no. 195: 1–8.

King, C.M., and Powell, R.A. 2007. *The Natural History of Weasels and Stoats*. 2nd ed. Oxford University Press, Oxford, UK.

Kingsley, M.C.S. 1996. Population index estimate for the Belugas of the St. Lawrence in 1995. *Canadian Technical Report of Fisheries and Aquatic Sciences* no. 2117: 1–38.

– 1998. Walrus, *Odobenus rosmarus*, in the Gulf and estuary of the St. Lawrence, 1992–1996. *Canadian Field-Naturalist* 112(1): 90–93.

Kingsley, M.C.S., and Reeves, R.R. 1998. Aerial surveys of cetaceans in the Gulf of St. Lawrence in 1995 and 1996. *Canadian Journal of Zoology* 76(8): 1529–1550.

Kingsley, M.C.S., and Stirling, I. 1991. Haul-out behaviour of Ringed and Bearded Seals in relation to defence against surface predators. *Canadian Journal of Zoology* 69(7): 1857–1861.

Kinze, C.C. 1994. *Marine Mammals of the North Atlantic*. Princeton University Press, Princton, NJ.

Kirkland, G.L., Jr. 1981. *Sorex dispar* and *Sorex gaspensis*. *Mammalian Species* no. 155: 1–4.

– 1999a. Arctic Shrew *Sorex arcticus*. In D.E. Wilson and S. Ruff (eds), *The Smithsonian Book of North American Mammals*, 15–16. UBC Press, Vancouver, BC.

– 1999b. Long-tailed Shrew *Sorex dispar*. In Wilson and Ruff (eds), *The Smithsonian Book of North American Mammals*, 21–22.

– 1999c. Gaspé Shrew *Sorex gaspensis*. In Wilson and Ruff (eds), *The Smithsonian Book of North American Mammals*, 24.

Kirkland, G.L., Jr, and Kirkland, C.J. 1979. Are small mammal hibernators K-selected? *Journal of Mammalogy* 60(1): 164–168.

Kirkland, G.L., Jr, and Knipe, C.M. 1979. The Rock Vole (*Microtus chrotorrhinus*) as a Transition Zone species. *Canadian Field-Naturalist* 93(3): 319–321.

Kirkland, G.L., Jr, Parmemter, R.R., and Skoog, R.E. 1997. A five-species assemblage of shrews from the sagebrush-steppe of Wyoming. *Journal of Mammalogy* 78(1): 83–89.

Kirkland, G.L., Jr, and Schmidt, D.F. 1982. Abundance, habitat, reproduction and morphology of forest-dwelling small mammals of Nova Scotia and southeastern New Brunswick. *Canadian Field-Naturalist* 96(2): 156–162.

– 1996. *Sorex arcticus*. *Mammalian Species* no. 524: 1–5.

Kirkland, G.L., Jr, Schmidt, D.F., and Kirkland, C.J. 1979. First record of the Long-tailed Shrew (*Sorex dispar*) in New Brunswick. *Canadian Field-Naturalist* 93(2): 195–198.

Kirkland, G.L., Jr, and Van Deusen, H.M. 1979. The shrews of the *Sorex dispar* group: *Sorex dispar* Batchelder and *Sorex gaspensis* Anthony and Goodwin. *American Museum Novitates* no. 2675: 1–21.

Kitchen, A.M., Gese, E.M., Karki, S.M., and Schauster, E.R. 2005. Spatial ecology of Swift Fox social groups: From group formation to mate loss. *Journal of Mammalogy* 86(3): 547–554.

Kitchen, A.M., Gese, E.M., and Schauster, E.R. 2000. Changes in Coyote activity patterns due to reduced exposure to human persecution. *Canadian Journal of Zoology* 78(5): 853–857.

Kiupel, M., Simmons, H.A., Fitzgerald, S.D., Wise, A., Sikarskie, J.G., Cooley, T.M., Hollamby, S.R., and Maes, R. 2002. West Nile Virus in Eastern Fox Squirrels (*Sciurus niger*). *Veterinary Pathology* 40: 703–707.

Kivett, V.K., Murie, J.O., and Steiner, A.L. 1976. A comparative study of scent-gland location and related behavior in some northwestern Nearctic ground squirrel species (Sciuridae): An evolutionary approach. *Canadian Journal of Zoology* 54(8): 1294–1306.

Klaus, M. 2003. The status, habitat, and response to grazing of Water Vole populations in the Big Horn Mountains of Wyoming, U.S.A. *Arctic, Antarctic and Alpine Research* 35(1): 100–109.

Klaus, M., Moore, R.E., and Vyse, E. 2001. Microgeographic variation in allozymes and mitochondrial DNA of *Microtus richardsoni*, the Water Vole, in the Beartooth Mountains of Montana and Wyoming, U.S.A. *Canadian Journal of Zoology* 79(7): 1286–1295.

Klein, D.R. 1999. The role of climate and insularity in establishment and persistence of *Rangifer tarandus* populations in the High Arctic. *Ecological Bulletin* 47: 96–104.

Klein, D.R., and Bay, C. 1991. Diet selection by vertebrate herbivores in the high arctic of Greenland. *Holarctic Ecology* 14: 152–155.

Klemme, I., Eccard, J.A., and Ylönen, H. 2006. Do female Bank Voles (*Clethrionomys glareolus*) mate multiply to improve on previous mates? *Behavioral Ecology and Sociobiology* 60(3): 415–421.

Klenner, W., and Kroeker, D.W. 1990. Denning behavior of Black Bears, *Ursus americanus*, in western Manitoba. *Canadian Field-Naturalist* 104(4): 540–544.

Klimley, A.P. 1994. The predatory behavior of the White Shark. *American Scientist* 82: 122–133.

Klinka, D.R., and Reimchen, T.E. 2002. Nocturnal and diurnal foraging behaviour of Brown Bears (*Ursus arctos*) on a salmon stream in coastal British Columbia. *Canadian Journal of Zoology* 80(8): 1317–1322.

– 2009. Darkness, twilight, and daylight foraging success of bears (*Ursus americanus*) on salmon in coastal British Columbia. *Journal of Mammalogy* 90(1): 144–149.

Knowles, B. 1992. Bat hibernacula on Lake Superior's north shore, Minnesota. *Canadian Field-Naturalist* 106(2): 252–254.

Knowlton, A.R., Kraus, S.D., and Kenny, R.D. 1994. Reproduction in North Atlantic Right Whales (*Eubalaena glacialis*). *Canadian Journal of Zoology* 72(7): 1297–1305.

– 2001. Mortality and serious injury of Northern Right Whales (*Eubalaena glacialis*) in the western North Atlantic Ocean. *Journal of Cetacean Research and Management*, Special issue, 2: 193–208.

Knudson, P. 1998. *The Nature of Walruses*. Greystone Books, Vancouver, BC.

Koble, J.A., and Squires, J.R. 2006. Longevity record for Canada Lynx, *Lynx canadensis*, in western Montana. *Western North American Naturalist* 66(4): 535–536.

Koehler, C.E., and Barclay, R.M.R. 2000. Post-natal growth and breeding biology of the Hoary Bat (*Lasiurus cinereus*). *Journal of Mammalogy* 81(1): 234–244.

Koen, E.L., Bowman, J., Findlay, C.S., and Zheng, L. 2007. Home range and population density of Fishers in eastern Ontario. *Journal of Wildlife Management* 71(5): 1484–1493.

Koepfli, K.-P., and Wayne, R.K. 1998. Phylogentic relationships of otters (Carnivora: Mustelidae) based on mitochondrial cytochrome *b* sequences. *Journal of Zoology* (London) 246: 401–416.

Koeppl, J.W., Hoffman, R.S., and Nadler, C.F. 1978. Pattern analysis of acoustical behaviour in four species of ground squirrels. *Journal of Mammalogy* 59(4): 677–696.

Koford, R.R. 1982. Mating system of a territorial tree squirrel (*Tamiasciurus douglasii*) in California. *Journal of Mammalogy* 63(2): 274–283.

Kohn, B.E., Payne, N.F., Ashbrenner, J.E., and Creed, W.A. 1993. The Fisher in Wisconsin. *Technical Bulletin, Department of Natural Resources* no. 183: 1–24

Koivula, M., Koskela, E., and Viitala, J. 1999. Sex and age-specific differences in ultraviolet reflectance of scent marks of Bank Voles (*Clethrionomys glareolus*). *Journal of Comparative Physiology A* 185(6): 561–564.

Komers, P.E., Messier, F., Flood, P.F., and Gates, C.C. 1994. Reproductive behavior of male Wood Bison in relation to progesterone level in females. *Journal of Mammalogy* 75(3): 757–765.

Koonz, W.H. 1993. Northern Pocket Gophers above ground in winter. *Blue Jay* 51(2): 124.

Kopelman, A.H., and Sadove, S.S. 1995. Ventilatory rate difference between surface-feeding and non surface feeding Fin Whales (*Balaenoptera physalus*) in the waters off eastern Long Island, New York, U.S.A., 1981–1987. *Marine Mammal Science* 11(2): 200–208.

Koprowski, J.L. 1992a. Do estrous female Gray Squirrels, *Sciurus carolinensis*, advertise their receptivity? *Canadian Field-Naturalist* 106(3): 392–394.

– 1992b. Removal of copulatory plugs by female tree squirrels. *Journal of Mammalogy* 73(3): 572–576.

– 1993. Behavioral tactics, dominance, and copulatory success among male fox squirrels. *Ethology Ecology and Evolution* 5:169–176.

– 1994a. *Sciurus carolinensis*. *Mammalian Species* no. 480: 1–9.

– 1994b. *Sciurus niger*. *Mammalian Species* no. 479: 1–9.

– 1996. Natal philopatry, communal nesting, and kinship in Fox Squirrels and Gray Squirrels. *Journal of Mammalogy* 77(4): 1006–1016.

Koski, W.R., and Miller, G.W. 2009. Habtat use by different size classes of Bowhead Whales in the central Beaufort Sea during late spring and autumn. *Arctic* 62(2): 137–150.

Kotliar, N.B., Baker, B.W., Whickler, A.D., and Plumb, G. 1999. A critical review of assumptions about the prairie dog as a keystone species. *Environmental Management* 24: 177–192.

Kovach, A.I., and Powell, R.A. 2003. Effects of body size on male mating tactics and paternity in Black Bears, *Ursus americanus*. *Canadian Journal of Zoology* 81(7): 1257–1268.

Kovach, S.D., Collins, G.H., Hinkes, M.T., and Denton, J.W. 2006. Reproduction and survival of Brown Bears in southwest Alaska. *Ursus* 17(1): 16–29.

Kovacs, K.M. 1995. Mother-pup reunions in Harp Seals, *Phoca groenlandica*: Cues for the relocation of pups. *Canadian Journal of Zoology* 73(5): 843–849.

– 2002a. Bearded Seal *Erignathus barbatus*. In W.F. Perrin, B. Wursig, and H.G.T. Thewissen (eds), *Encyclopedia of Marine Mammals*, 84–87. Academic Press, San Diego, CA.

– 2002b. Hooded Seal *Cystophora cristata*. In Perrin, Wursig, and Thewissen (eds), *Encyclopedia of Marine Mammals*, 580–582.

Kovacs, K.M., Jonas, K.M., and Welke, S.E. 1990. Sex and age segregation by *Phoca vitulina concolor* at haul-out sites during the breeding season in the Passamaquoddy Bay region, New Brunswick. *Marine Mammal Science* 6(3): 204–214.

Kovacs, K.M., and Lavigne, D. 1986. *Cystophora cristata*. *Mammalian Species* no. 258: 1–9.

Kovacs, K.M., Lydersen, C., and Gjertz, I. 1996a. Birth-site characteristics and prenatal molting in Bearded Seals (*Erignathus barbatus*). *Journal of Mammalogy* 77(4): 1085–1091.

Kovacs, K.M., Lydersen, C., Hammill, M., and Lavigne, D.M. 1996b. Reproductive effort of male Hooded Seals (*Cystophora cristata*): Estimates from mass loss. *Canadian Journal of Zoology* 74(8): 1521–1530.

Krafft, A.B., Lydersen, C., Kovacs, K.M., Gjertz, I., and Haug, T. 2000. Diving behaviour of lactating Bearded Seals (*Erignathus barbatus*) in the Svalbard area. *Canadian Journal of Zoology* 78(8): 1408–1418.

Kraus, S.D., Hamilton, P.K., Kenney, R.D., Knowlton, A.R., and Slay, C.K. 2001. Reproductive parameters of the North Atlantic Right Whale. *Journal of Cetacean Research and Management*, Special issue, 2: 231–236.

Kraus, S.D., and Hatch, J.J. 2001. Mating strategies in the North Atlantic Right Whale (*Eubalaena glacialis*). *Journal of Cetacean Research and Management*, Special issue, 2: 237 244.

Kraus, S.D., and Rolland, R.M. 2007. *The Urban Whale: North Atlantic Right Whales at the Crossroads*. Harvard University Press, Cambridge, MA.

Krausman, P.R., and Bowyer, R.T. 2003. Mountain Sheep (*Ovis canadensis* and *O. dalli*) In G.A. Feldhamer, B.C. Thompson, and J.A. Chapman (eds), *Wild Mammals of North America: Biology, Management, and Conservation*, 2nd ed., 1095–1115. Johns Hopkins University Press, Baltimore, MA.

Krebs, C.J., and Boonstra, R. 1978. Demography of the spring decline in populations of the vole, *Microtus townsendii*. *Journal of Animal Ecology* 47: 1007–1015.

Krebs, C.J., Boonstra, R., and Kenney, A.J. 1995a. Population dynamics of the Collared Lemming and the Tundra Vole at Pearce Point, Northwest Territories, Canada. *Oecologia* 103: 481–489.

Krebs, C.J., Boutin, S., Boonstra, R., Sinclair, A.R.E., Smith, J.N.M., Dale, M.R.T., Martin, K., and Turkington, R. 1995b. Impact of food and predation on the Snowshoe Hare Cycle. *Science* 269: 1112–1115.

Krebs, C.J., Halpin, Z.T., and Smith, J.N.M. 1977. Aggression, testosterone, and the spring decline in populations of the vole *Microtus townsendii*. *Canadian Journal of Zoology* 55(2): 430–437.

Krebs, C.J., Kenney, A.J., Gilbert, S., Danell, K., Angerbjörn, A., Erlinge, S., Bromley, R.G., Shank, C. and Carriere, S. 2002. Synchrony in lemming and vole populations in the Canadian Arctic. *Canadian Journal of Zoology* 80(8): 1323–1333.

Krebs, C.J., and Wingate, I. 1985. Population fluctuations in the small mammals of the Kluane Region, Yukon Territory. *Canadian Field-Naturalist* 99(1): 51–61.

Krebs, J., Lofroth, E., Copeland, J., Banci, V., Cooley, D., Golden, H., Magoun, A., Mulders, R., and Shults, B. 2004. Synthesis of survival rates and causes of mortality in North American Wolverines. *Journal of Wildlife Management* 68(3): 493–502.

Krebs, J.W., Williams, S.M., Smith, J.S., Rupprecht, C.E., and Childs, J.E. 2003. Rabies among infrequently reported mammalian carnivores in the United States, 1960–2000. *Journal of Wildlife Diseases* 39(2): 253–261.

Kremsater, L., Andrusiak, L., and Brunnell, F.L. 1993. Status of the Shrew-mole in British Columbia. *Wildlife Working Report* no. WR-55: 1–22.

Kreuder, C., Miller, M.A., Jessup, D.A., Lowenstine, L.J., Harris, M.D., Ames, J.A., Carpenter, T.E., Conrad, P.A., and Mazet, J.A.K. 2003. Patterns of mortality in Southern Sea Otters (*Enhydra lutris*) from 1998–2001. *Journal of Wildlife Diseases* 39(3): 495–509.

Krochmal, A.R., and Sparks, D.W. 2007. Timing of birth and estimation of age of juvenile *Myotis septentrionalis* and *Myotis lucifugus* in west-central Indiana. *Journal of Mammalogy* 88(3): 649–656.

Kruchenkova, E.P., Goltsman, M.E., and Frommolt, K.-K. 2003. Rhythm structure of Arctic Fox (*Alopex lagopus beringensis* and *A. l. semenovi*) serial barking: Age, sex, and context determinants. *Zoologicheskii Zhurnal* 82(4): 525–533.

Krupa, J.J., and Geluso, K.N. 2000. Matching the color of excavated soil: Cryptic coloration in the Plains Pocket Gopher (*Geomys bursarius*). *Journal of Mammalogy* 81(1): 86–96.

Krupa, J.J., and Haskins, K.E. 1996. Invasion of the Meadow Vole (*Microtus pennsylvanicus*) in southeastern Kentucky and its possible impact on the Southern Bog Lemming (*Synaptomys cooperi*). *American Midland Naturalist* 135(1): 14–22.

Kruse, S., Caldwell, D.K., and Caldwell, M.C. 1999. Risso's Dolphin *Grampus griseus* (G. Cuvier, 1812). In S.H. Ridgway and R.Harrison, (eds), *Handbook of Marine Mammals*, vol. 6, 183–212. Academic Press, San Diego, CA.

Krutzikowsky, G.K., and Mate, B.R. 2000. Dive and surfacing characteristics of Bowhead Whales (*Balaena mysticetus*) in the Beaufort and Chukchi seas. *Canadian Journal of Zoology* 78(7): 1182–1198.

Kuiken, T., Leighton, A., and Johnson, D. 2003. Grasshoppers in Coyote scats. *Blue Jay* 61(1): 51–55.

Kuker, K.J., Thomson, J.A., and Tscherter, U. 2005. Novel surface feeding tactics of Minke Whales, *Balaenoptera acutorostrata*, in the Saguenay–St. Lawrence National Marine Park. *Canadian Field-Naturalist* 119(2): 214–218.

Künkele, J. 1992. Infanticide in wild rabbits (*Oryctolagus cuniculus*). *Journal of Mammalogy* 73(2): 317–320.

Kunnasranta, M. 2001. Behavioural biology of two Ringed Seals (*Phoca hispida*) subspecies in the large European lakes Siamaa and Ladoga. PhD thesis, University of Joensuu, Finland.

Kunz, T.H. 1982. *Lasionycteris noctivagans*. *Mammalian Species* no. 172: 1–5.

Kunz, T.H., and Martin, R.A. 1982. *Plecotus townsendii*. *Mammalian Species* no. 175: 1–6.

Kurta, A. 1995. *Mammals of the Great Lakes Region*. University of Michigan Press, Ann Arbour.

Kurta, A., and Baker, R.H. 1990. *Eptesicus fuscus*. *Mammalian Species* no. 356: 1–10.

Kurta, A., Foster, R., Hough, E., and Winhold, L. 2005. The Evening Bat (*Nycticeius humeralis*) on the northern edge of its range – a maternity colony in Michigan. *American Midland Naturalist* 154(1): 264–267.

Kurta, A., Kunz, T.H., and Nagy, K.A. 1990. Energetics and water flux of free-ranging Big Brown Bats (*Eptesicus fuscus*) during pregnancy and lactation. *Journal of Mammalogy* 71(1): 59–65.

Kurta, A., and Stewart, M.E. 1990. Parturition in the Silver-haired Bat, *Lasionycteris noctivagans*, with a description of the neonates. *Canadian Field-Naturalist* 104(4): 598–600.

Kurta, A., and Teramino, J.A. 1994. A novel hibernaculum and noteworthy records of the Indiana Bat and Eastern Pipistrelle (Chiroptera: Vespertilionidae). *American Midland Naturalist* 132: 410–413.

Kurta, A., Winhold, L., Whitaker, J.O., Jr, and Foster, R. 2007. Range expansion and changing abundance of Eastern Pipistrelle (Chiroptera: Vespertilionidae) in the central Great Lakes Region. *American Midland Naturalist* 157(2): 404–411.

Kuzyk, G.W., Kneteman, J., and Schmiegelow, F.K.A. 2006. Pack size of wolves, *Canis lupus*, on Caribou, *Rangifer tarandus*, winter ranges in westcentral Alberta. *Canadian Field-Naturalist* 120(3): 313–318.

Kuzyk, G.W., Rohner, C., and Kneteman, J. 2000. Grizzly Bear defends Moose carcass from wolves in west-central Alberta. *Alberta Naturalist* 30(4): 75.

Kwiecinski, G.G. 1998. *Marmota monax*. *Mammalian Species* no. 591: 1–8.

Kyle, C.J., Johnson, A.R., Patterson, B.R., Wilson, P.J., Shami, K., Grewal, S.K., and White, B.N. 2006. Genetic nature of Eastern Wolves: Past, present and future. *Conservation Genetics* 7(2): 273–287.

Kyle, C.J., Johnson, A.R., Patterson, B.R., Wilson, P.J., and White, B.N. 2007a. The conspecific nature of Eastern and Red Wolves: Conservation and management implications. *Conservation Genetics* 9(3): 699–701.

Kyle, C.J., Karels, T.J., Davis, C.S., Mebs, S., Clark, B., Strobeck, C., and Hik, D.S. 2007b. Social structure and facultative mating systems of Hoary Marmots (*Marmota caligata*). *Molecular Ecology* 16(6): 1245–1255.

Kyle, C.J., Weir, R.D., Newhouse, N.J., Davis, H., and Strobeck, C. 2004. Genetic structure of sensitive and endangered northwestern badger populations (*Taxidea taxus taxus* and *T.t. jeffersonii*). *Journal of Mammalogy* 85(4): 633–639.

Laborda, J.A., and Cartwright, A. 1993. Initial emergence factors of the Big Brown Bat (*Eptesicus fuscus*). *Proceedings of the Indiana Academy of Science* 102: 273–277.

Lacey, E.A., Wieczorek, J.R., and Tucker, P.K. 1997. Male mating behaviour and patterns of sperm precedence in Arctic Ground Squirrels. *Animal Behaviour* 53: 767–79.

Lackey, J.A. 1999. White-footed Mouse *Peromyscus leucopus*. In D.E. Wilson and S. Ruff (eds), *The Smithsonian Book of North American Mammals*, 572–574. UBC Press, Vancouver, BC.

Lackey, J.A., Huckaby, D.G., and Ormiston, B.G. 1985. *Peromyscus leucopus*. *Mammalian Species* no. 247: 1–10.

Lacki, M.J., Adam, M.D., and Shoemaker, L.G. 1994. Observations on seasonal cycle, population patterns and roost selection in summer colonies of *Plecotus townsendii virginianus* in Kentucky. *American Midland Naturalist* 131: 34–42.

Lacki, M.J., and Baker, M.D. 2007. Day roosts of female Fringed Myotis (*Myotis thysanodes*) in xeric forests of the Pacific Northwest. *Journal of Mammalogy* 88(4): 967–973.

Lacoste, K.N., and Stenson, G.B. 2000. Winter distribution of Harp Seals (*Phoca groenlandica*) off eastern Newfoundland and southern Labrador. *Polar Biology* 23: 805–811.

Ladine, T.A. 1997. Activity patterns of co-occurring populations of Virginia Opossums, (*Didelphis virginiana*) and Raccoons (*Procyon lotor*). *Mammalia* 61(3): 345–354.

Laerm, J., Ford, W.M., Weinland, D.C., and Menzel, M.A. 1994. First records of the Pygmy Shrew, *Sorex hoyi* (Insectivora: Soricidae), in western Maryland and Pennsylvania. *Maryland Naturalist* 38(1–2): 23–27.

Laerm, J., Lepardo, L., Gaudin, T., Monteith N., and Szymczak, A. 1996. First records of the Pygmy Shrew, *Sorex hoyi winnemania* Preble (Insectivora: Soricidae), in Alabama. *Journal of the Alabama Academy of Science* 67(1): 43–48.

Laerm, J., Wenzel, F., Craddock, J.E., Weinard, D., McGurk, J., Harris, M.J., Early, G.A., Mead, J.G., Potter, C.W., and Barros, N.B. 1997. New prey species for northwestern Atlantic Humpback Whales. *Marine Mammal Science* 13(4): 705–711.

Laerm, J., Wharton, C.H., and Ford, W.M. 1995. First record of the Water Shrew, *Sorex palustris* Richardson (Insectivora: Soricidae), in Georgia with comments on its distribution and status in the southern Appalachians. *Brimleyana* 22: 47–51.

Lafond, R. 1999. La gestion du Lynx du Canada au Québec. Une histoire à suivre. *Naturaliste Canadien* 123(3): 26–31.

Lagerquist, B.A., Mate, B.R., Ortega-Ortiz, J.G., Winsor, M., and Urbán-Ramirez, J. 2008. Migratory movements and surfacing rates of Humpback Whales (*Megaptera novaeangliae*) satellite tagged at Socorro Island, Mexico. *Marine Mammal Science* 24(4): 815–830.

Lagerquist, B.A., Stafford, K.M., and Mate, B.R. 2000. Dive characteristics of satellite-monitored Blue Whales (*Balaenoptera musculus*) off the central California coast. *Marine Mammal Science* 16(2): 375–391.

Laidre, K.L., and Heide-Jørgensen, M.P. 2005. Winter feeding intensity of Narwhals (*Monodon monoceros*). *Marine Mammal Science* 21(1): 45–57.

Laidre, K.L., Heide-Jørgensen, M.P., and Dietz, R. 2002. Diving behaviour of Narwhals (*Monodon monoceros*) at two coastal localities in the Canadian High Arctic. *Canadian Journal of Zoology* 80(4): 624–635.

Laidre, K.L., Heide-Jørgensen, M.P., and Orr, J.R. 2006. Reactions of Narwhals, *Monodon monoceros*, to Killer Whale, *Orcinus orca*, attacks in the eastern Canadian Arctic. *Canadian Field-Naturalist* 120(4): 457–465.

Lair, H. 1990. The calls of the Red Squirrel: A contextual analysis of function. *Behaviour* 115(3–4): 245–281.

Lambert, C.M.S., Wielgus, R.B., Robinson, H.S., Katnik, D.D., Cruickshank, H.S., Clarke, R., and Almack, J. 2006. Cougar population dynamics and viability in the Pacific Northwest. *Journal of Wildlife Management* 70(1): 246–254.

Lambin, X. 1994. Sex ratio variation in relation to female philopatry in Townsend's Voles. *Journal of Animal Ecology* 63: 945–953.

Lambin, X., and Krebs, C.J. 1991. Spatial organization and mating system of *Microtus townsendii*. *Behavioral Ecology and Sociobiology* 28: 353–363.

Lambin, X., Krebs, C.J., and Scott, B. 1992. Spacing system of the Tundra Vole (*Microtus oeconomus*) during the breeding season in Canada's western Arctic. *Canadian Journal of Zoology* 70(10): 2068–2072.

Lamontagne, G. 1998. Le plan de gestion de l'ours noir au Québec (1998–2000). *Le Naturalist canadien* 122: 13–20.

Lance, E.W., and Cook, J.A. 1998. Biogeography of Tundra Voles (*Microtus oeconomus*) of Beringia and the southern coast of Alaska. *Journal of Mammalogy* 79(1): 53–65.

Langley, W.M. 1983. Relative importance of the distance senses in Grasshopper Mouse predatory behaviour. *Animal Behaviour* 31: 199–205.

Lansing, S.W. 2005. A range extension for the Rock Vole, *Microtus chrotorrhinus*, in Labrador. *Canadian Field-Naturalist* 119(3): 412–416.

Larivière, S. 1996. The American Mink, *Mustela vison* (Carnivora, Mustelidae), can climb trees. *Mammalia* 60(3): 485–486.

– 1998. The radiating mousing technique of the Striped Skunk. *Blue Jay* 56(4): 218–220.

– 1999. *Mustela vison*. *Mammalian Species* no. 608: 1–9.

– 2003. Mink (*Mustela vison*). In G.A. Feldhamer, B.C. Thompson, and J.A. Chapman (eds), *Wild Mammals of North America: Biology, Management, and Conservation*, 2nd ed., 662–671. Johns Hopkins University Press, Baltimore, MD.

– 2004. Range expansion of Raccoons in the Canadian prairies: Review of hypotheses. *Wildlife Society Bulletin* 32(3): 955–963.

Larivière, S., Howerter, D., and Messier, F. 2007. Influence of gender and den type on home range shape for Striped Skunks, *Mephitis mephitis*, in Saskatchewan. *Canadian Field-Naturalist* 121(3): 261–264.

Larivière, S., Huot, J., and Samson, C. 1994. Daily activity patterns of female Black Bears in a northern mixed-forest environment. *Journal of Mammalogy* 75(3): 613–620.

Larivière, S., and Messier, F. 1996. Aposematic behavior in the Striped Skunk, *Mephitis mephitis*. *Ethology* 102: 986–992.

– 1997a. Seasonal and daily activity patters of Striped Skunk (*Mephitis mephitis*) in the Canadian prairies. *Journal of Zoology* (London) 243: 255–262.

– 1997b. Characteristics of waterfowl nest depredation by the Striped Skunk (*Mephitis mephitis*): Can predators be identified from nest remains? *American Midland Naturalist* 137: 393–396.

– 1998. Denning ecology of the Striped Skunk in the Canadian prairies: Implications for waterfowl nest predation. *Journal of Applied Ecology* 35: 207–213.

– 2001. Space-use patterns by female Striped Skunks exposed to aggregations of simulated duck nests. *Canadian Journal of Zoology* 79(9): 1604–1608.

Larivière, S., and Pasitschniak-Arts, M. 1996. *Vulpes vulpes*. *Mammalian Species* no. 537: 1–11.

Larivière, S., and Walton, L.R. 1997. *Lynx rufus*. *Mammalian Species* no. 563: 1–8.

– 1998. *Lontra canadensis*. *Mammalian Species* no. 587: 1–9.

Larsen, E., and Hammond, P.S. 2004. Distribution and abundance of West Greenland Humpback Whales (*Megaptera novaeangliae*). *Journal of Zoology* (London) 263: 343–358.

Larsen, K.W., and Boutin, S. 1994. Movements, survival, and settlement of Red Squirrels (*Tamiasciurus hudsonicus*) offspring. *Ecology* 75(1): 214–223.

Larter, N.C. 1999. Seasonal changes in Arctic Hare, *Lepus arcticus*, diet composition and differential digestibility. *Canadian Field-Naturalist* 113(3): 481–486.

Larter, N.C., and Nagy, J.A. 2000. Calf production and overwinter survival estimates for Peary Caribou, *Rangifer tarandus pearyi*, on Banks Island, Northwest Territories. *Canadian Field-Naturalist* 114(4): 661–670.

– 2001. Overwinter changes in the urine chemistry of Muskoxen from Banks Island. *Journal of Wildlife Management* 65(2): 226–234.

Laun, H.C. 1962. Loud vocal sounds produced by Striped Skunk. *Journal of Mammalogy* 43(3): 432–433.

Launchbaugh, K.L., and Urness, P.J. 1992. Mushroom consumption (mycophagy) by North American cervids. *Great Basin Naturalist* 52(4): 321–327.

Lauriol, B., Deschamps, E., Carrier, L., Grimm, W., Morlan, R., and Talon, B. 2003. Cave infill and associated biotic remains as indicators of Holocene environment in Gatineau Park (Quebec, Canada). *Canadian Journal of Earth Sciences* 40(5): 789–803.

Lausen, C.L. 2002. An extralimital prairie population of Bushy-tailed Woodrats, *Neotoma cinerea*, along the South Saskatchewan River, Alberta, Canada. *Northwestern Naturalist* 83(3): 125–128.

Lausen, C.L., and Barclay, R.M.R. 2002. Roosting behaviour and roost selection of female Big Brown Bats (*Eptesicus fuscus*) roosting in rock crevices in southeastern Alberta. *Canadian Journal of Zoology* 80(6): 1069–1076.

– 2006. Winter bat activity in the Canadian prairies. *Canadian Journal of Zoology* 84(8): 1079–1086.

Lausen, C.L., Delisle, I., Barclay, R.M.R., and Strobeck, C. 2008a. Beyond mtDNA: Nuclear gene flow suggests taxonomic oversplitting in the Little Brown Bat (*Myotis lucifugus*). *Canadian Journal of Zoology* 86(7): 700–713.

Lausen, C.L., Jung, T.S., and Talerico, J.M. 2008b. Range extension of the Northern Long-eared Bat (*Myotis septentrionalis*) in the Yukon. *Northwestern Naturalist* 89(2): 115–117.

Lavers, A.J., Petersen, S.D., Stewart, D.T., and Herman, T.B. 2006. Delineating the range of a disjunct population of Southern Flying Squirrels (*Glaucomys volans*). *American Midland Naturalist* 155(1): 188–196.

Law, T.C., and Blake, R.W. 1994. Swimming behaviors and speeds of wild Dall's Porpoises (*Phocoenoides dalli*). *Marine Mammal Science* 10(2): 208–213.

Lawson, J.W., and Eddington, J.D. 1998. A first eastern Canadian stranding record for Risso's Dolphin, *Grampus griseus*. *Northeastern Naturalist* 5(3): 215–218.

Lawson, J.W., and Hobson, K.A. 2000. Diet of Harp Seals (*Pagophilus groenlandicus*) in nearshore northeast Newfoundland: Inferences from stable-carbon (δ^{13}C) and nitrogen (δ^{15}N) isotope analyses. *Marine Mammal Science* 16(3): 578–591.

Lawson, J.W., and Stenson, G.B. 1997. Diet of northwest Atlantic Harp Seals (*Phoca groenlandica*) in offshore areas. *Canadian Journal of Zoology* 75(12): 2095–2106.

Leatherwood, S., Reeves, R.R., Perrin, W.F., and Evans, W.E. 1982. Whales, dolphins, and porpoises of the eastern North Pacific and adjacent Arctic waters: A guide to their identification. *NOAA Technical Report NMFS Circular* 444: 1–244.

Leatherwood, S., and Walker, W.A. 1979. The Northern Right Whale Dolphin *Lissodelphis borealis* Peale in the eastern North Pacific. *Behavior of Marine Animals* 3: 85–141.

LeBlanc, M., Festa-Bianchet, M., and Jorgenson, J.T. 2001. Sexual size dimorphism in Bighorn Sheep (*Ovis canadensis*): Effects of population density. *Canadian Journal of Zoology* 79(9): 1661–1670.

Le Boeuf, B.J., Crocker, D., Costa, D.P., Blackwell, S.B., Webb, P.M., and Houser, D.S. 2000a. Foraging ecology of Northern Elephant Seals. *Ecological Monographs* 70(3): 353–382.

Le Boeuf, B.J., Naito, Y., Asaga, T., Crocker, D., and Costa, D.P. 1992. Swim speed in a female Northern Elephant Seal: Metabolic and foraging implications. *Canadian Journal of Zoology* 70(4): 786–795.

Le Boeuf, B.J., Pérez-Cortés, M.H., Urbán, R.J., Mate, B.R., and Ollervides, U.F. 2000b. High Gray Whale mortality and low recruitment in 1999: Potential causes and implications. *Journal of Cetacean Research and Management* 2(2): 85–99.

Ledwell, W., Benjamins, S., Lawson, J., and Huntington, J. 2007. The most southerly record of a stranded Bowhead Whale, *Balaena mysticetus*, from the western North Atlantic Ocean. *Arctic* 60(1): 17–22.

Ledwell, W., Lien, J., and Wakeham, D. 2005. A possible ship collision with a Sowerby's Beaked Whale that stranded in Conception Bay, Newfoundland. *Osprey* 36(1): 95–97.

Lee, J., and Niptanatiak, A. 1996. Observation of repeated use of a Wolverine, *Gulo gulo*, den site on the tundra of the Northwest Territories. *Canadian Field-Naturalist* 110(2): 349–350.

Lee, O.A., Olivier, P., Wolt, R., and Davis, R.W. 2009. Aggregations of Sea Otters (*Enhydra lutris kenyoni*) feeding on fish eggs and kelp in Prince William Sound, Alaska. *American Midland Naturalist* 161(2): 401–405.

Lee, S.D. 1995. Comparison of population characteristics of three species of shrews and the Shrew-mole in habitats with different amounts of coarse woody debris. *Acta Theriologica* 40(4): 415–424.

Lee, W.B., and Houston, D.C. 1993. The role of coprophagy in digestion in voles (*Microtus agrestis* and *Clethrionomys glareolus*). *Functional Ecology* 7: 427–432.

Lehman, N., Eisenhawer, A., Hansen, K., Mech, L.D., Peterson, R.O., Gogan, P.J.P., and Wayne, R.K. 1991. Introgression of Coyote mitochondrial DNA into sympatric North American Gray Wolf populations. *Evolution* 45(1): 104–119.

Lehmer, E.M., Van Horne, B., Kulbartz, B., and Florant, G.L. 2001. Facultative torpor in free-ranging Black-tailed Prairie Dogs (*Cynomys ludovicianus*). *Journal of Mammalogy* 82(2): 551–557.

Leibacher, B., and Whitaker, J.O., Jr. 1998. Distribution of the Western Harvest Mouse, *Reithrodontomys megalotis*, in Indiana. *Proceedings of the Indiana Academy of Science* 107: 167–170.

Lenihan, C., and Van Vuren, D. 1996. Growth and survival of juvenile Yellow-bellied Marmots (*Marmota flaviventris*). *Canadian Journal of Zoology* 74(2): 297–302.

Lent, P.C. 1988. *Ovibos moschatus*. *Mammalian Species* no. 302: 1–9.

Leonard, M.L., and Fenton, M.B. 1983. Habitat use by spotted bats (*Euderma maculatum*, Chiroptera:Vespertilionidae): Roosting and foraging behaviour. *Canadian Journal of Zoology* 61(9): 1487–1491.

Leopold, M.F., and Couperus, A.S. 1995. Sightings of Atlantic White-sided Dolphins *Lagenorhynchus acutus* near the south-eastern limit of the known range in the north-east Atlantic. *Lutra* 38(2): 77–80.

Lesage, V., and Hammill, M.O. 2001. The status of the Grey Seal *Halichoerus grypus*, in the northwest Atlantic. *Canadian Field-Naturalist* 115(4): 653–662.

Lesage, V., Hammill, M.O., and Kovacs, K.M. 1999. Functional classification of Harbor Seals (*Phoca vitulina*) dives using depth profiles, swimming velocity, and an index of foraging success. *Canadian Journal of Zoology* 77(1): 74–87.

– 2004. Long-distance movements of Harbour Seals (*Phoca vitulina*) from a seasonally ice-covered area, the St. Lawrence River estuary, Canada. *Canadian Journal of Zoology* 82(7): 1070–1081.

Leung, M.C., and Cheng, K.M. 1997. The distribution of the Cascade Mantled Ground Squirrel, *Spermophilus saturatus*, in British Columbia. *Canadian Field-Naturalist* 111(3): 365–375.

Levine, M.A., Bailey, G.N., Whitwell, K.E., and Jeffcott, L.B. 2000. Palaeopathology and horse domestication, 123–133. In G. Bailey, R. Charles and N. Winder (eds), *Human Ecodynamics and Environmental Archaeology*. Oxbow, Oxford, UK.

Lewis, J.C. 2006. *Implementation Plan for Reintroducing Fishers to Olympic National Park*. Washington Department of Fish and Wildlife, Olympia, WA.

Li, G., Roze, U., and Locke, D.C. 1997. Warning odor of the North American Porcupine (*Erethizon dorsatum*). *Journal of Chemical Ecology* 23(12): 2737–2754.

Lidicker, W.Z., Jr, and Batsli, G.O. 1999. Singing Vole (*Mictorus miurus*). In D.E. Wilson and S. Ruff (eds), *The Smithsonian Book of North American Mammals*, 632–633. UBC Press, Vancouver, BC.

Lien, J., and Barry, F. 1990. Status of Sowerby's Beaked Whale, *Mesoplodon bidens*, in Canada. *Canadian Field-Naturalist* 104(1): 125–130.

Lien, J., Barry, F., Breeck, K., and Zuschlag, U. 1990. Multiple strandings of Sowerby's Beaked Whales, *Mesoplodon bidens*, in Newfoundland. *Canadian Field-Naturalist* 104(3): 414–420.

Lien, J., Nelson, D., and Hai, D.J. 2001. Status of the White-beaked Dolphin, *Lagenorhynchus albirostris*, in Canada. *Canadian Field-Naturalist* 115(1): 118–126.

Lien, J., Whitney, H., Ledwell, W., Joy, J., Forzan, M., Conboy, G., Daoust, P.Y., and Sjaer, B. 2002. A new record of a Pigmy Sperm Whale (*Kogia breviceps*) from Newfoundland. *Osprey* 33(4): 104–108.

Lim, B.K. 1987. *Lepus townsendii*. *Mammalian Species* no. 288: 1–6.

Lindgren, P. 2007. COSEWIC assessment and update status report on the Western Harvest Mouse *Reithrodontomys megalotis megalotis* and *Reithrodontomys megalotis dychei* in Canada. Committee on the Status of Endangered Species in Canada.

Lindquist, E.S., Aquadro, C.F., McClearn, D., and McGowan, K.J. 2003. Field identification of the mice *Peromyscus leucopus noveboracensis* and *P. maniculatus gracilis* in central New York. *Canadian Field-Naturalist* 117(2): 184–189.

Lindqvist, C., Schuster, S.C., Sun, Y., Talbot, S.L., Qi, J., Ratan, A., Tomsho, L.P., Kasson, L., Zeyl, E., Aars, J., Miller, W., Ingólfsson, Ó., Bachmann, L., and Wiig, Ø. 2010. Complete mitochondrial genome of a Pleistocene jawbone unveils the origin of Polar Bear. *Proceedings of the National Academy of Sciences of the Untited States of America* 107(11): 5053–5057.

Lindzey, F.G. 2003. Badger (*Taxidea taxus*). In G.A. Feldhamer, B.C. Thompson, and J.A. Chapman (eds), *Wild Mammals of North America: Biology, Management, and Conservation*, 2nd ed., 683–691. Johns Hopkins University Press, Baltimore, MD.

Lingle, S. 1992. Escape gaits of White-tailed Deer, Mule Deer and their hybrids: Gaits observed and patterns of limb coordination. *Behaviour* 122(3–4): 153–181.

– 2002. Coyote predation and habitat segregation of White-tailed Deer and Mule Deer. *Ecology* 83(7): 2037–2048.

Lingle, S., and Pellis, S.M. 2002. Fight or flight? Antipredator behavior and the escalation of coyote encounters with deer. *Oecologia* 131: 154–164.

Linzey, A.V. 1984. Patterns of coexistence in *Synaptomys cooperi* and *Microtus pennsylvanicus*. *Ecology* 65(2): 382–393.

Lishak, R.S. 1982a. Vocalizations of nestling Gray Squirrels. *Journal of Mammalogy* 63(3): 446–452.

– 1982b. Gray Squirrel mating calls: A spectrographic and ontogenic analysis. *Journal of Mammalogy* 63(4): 661–663.

– 1984. Alarm vocalizations of adult Gray Squirrels. *Journal of Mammalogy* 65(3): 681–684.

List, R., Ceballos, G., and Pacheco, J. 1999. Status of the North American Porcupine (*Erethizon dorsatum*) in Mexico. *Southwestern Naturalist* 44(3): 400–404.

Litvaitis, J.A., Verbyla, D.L., and Litvaitis, M.K. 1992. A field method to differentiate New England and Eastern Cottontails. *Transactions of the Northeast Section of the Wildlife Society* 48: 11–14.

Livoreil, B., and Baudoin, C. 1996. Differences in food hoarding behaviour in two species of ground squirrels *Spermophilus tridecemlineatus* and *S. spilosoma*. *Ethology Ecology and Evolution* 8: 199–205.

Lloyd, J.E. 1972. Vocalization in *Marmota monax*. *Journal of Mammalogy* 53(1): 214–216.

Lockhart, M., Vargas, A., Martinari, P., and Gober, P. 1998. Black-footed Ferret (*Mustela nigripes*) recovery update. *Endangered Species Update* 15(6): 92–93.

Loew, S.S. 1999. Sex-biased dispersal in Eastern Chipmunks, *Tamias striatus*. *Evolutionary Ecology* 13: 557–577.

Loewen, V.A. 1984. Life-history of the Long-tailed Vole, *Microtus longicaudus* in southwestern Alberta. M.Sc. thesis, University of Western Ontario, London, ON.

Lofroth, E.C., Krebs, J.A., Harrower, W.L., and Lewis, D. 2007. Food habits of Wolverine *Gulo gulo* in montane ecosystems of British Columbia, Canada. *Wildlife Biology* 13(Suppl. 2): 31–37.

Logan, K.A., and Sweanor, L.L. 2000. Puma. In S. Demarais and P.R. Krausman (eds), *Ecology and Management of Large Mammals in North America*, 347–377. Prentice Hall, Upper Saddle River, NJ.

Lohr, C., and Ballard, W.B. 1996. Historical occurrence of wolves, *Canis lupus*, in the Maritime Provinces. *Canadian Field-Naturalist* 110(4): 607–610.

Lomolino, M.V., and Smith G.A. 2001. Dynamic biogeography of prairie dog (*Cynomys ludovicianus*) towns near the edge of their range. *Journal of Mammalogy* 82(4): 937–945.

Long, C.A. 1973. *Taxidea taxus*. *Mammalian Species* no. 26: 1–4.

– 1974. *Sorex hoyi* and *Sorex thompsoni*. *Mammalian Species* no. 33: 1–4.

– 1996. Ecological replacement of the Deer Mouse, *Peromyscus maniculatus*, by the White-footed Mouse, *P. leucopus*, in the Great Lakes Region. *Canadian Field-Naturalist* 110(2): 271–277.

– 2008. *The Wild Mammals of Wisconsin*. Pensoft Publishers, Sofia, Bulgaria.

Long, C.A., and Long, J.E. 1988. Southern Bog Lemming, *Synaptomys cooperi*, new to islands in Lake Michigan. *Canadian Field-Naturalist* 102(1): 64–65.

Long, D.J. 1991. Apparent predation by a White Shark *Carcharodon carcharias* on a Pygmy Sperm Whale *Kogia breviceps*. U.S. Fish and Wildlife Service, *Fishery Bulletin* 89(3): 538–540.

Long, D.J., and Gilbert, L. 1997. California Sea Lion predation on chicks of the Common Murre. *Journal of Field Ornithology* 68(1): 152–154.

Long, J.L. 2003. *Introduced Mammals of the World*. CSIRO Publishing, Collingwood, Australia.

López-González, C., and Torres-Morales, L. 2004. Use of abandoned mines by long-eared bats, genus *Corynorhinus* (Chiroptera: Vespertilionidae) in Durango, Mexico. *Journal of Mammalogy* 85(5): 989–994.

López-Wilchis, R., López–Ortega, G., and Owen, R.D. 1994. Noteworthy record of the Western Small-footed Myotis (Mammalia: Chiroptera: *Myotis ciliolabrum*). *Southwestern Naturalist* 39(2): 211–212.

Loring, S., and Spiess, A. 2007. Further documentation supporting the former existence of Grizzly Bears (*Ursus arctos*) in northern Quebec-Labrador. *Arctic* 60(1): 7–16.

Lotze, J.-H., and Anderson, S. 1979. *Procyon lotor*. *Mammalian Species* no. 269: 1–8.

Loughlin, T.R., and Perez, M.A. 1985. *Mesoplodon stejnegeri*. *Mammalian Species* no. 250: 1–6.

Loughlin, T.R., Perez, M.A., and Merrick, R.L. 1987. *Eumetopias jubatus*. *Mammalian Species* no. 283: 1–7.

Loughlin, T.R., Sterling, J.T., Merrick, R.L., Sease, J.L., and York, A.E. 2003. Diving behavior of immature Steller Sea Lions (*Eumetopias jubatus*). *Fisheries Bulletin* 101: 566–582.

Lovallo, M.J., and Anderson, E.M. 1995. Range shift by a female Bobcat (*Lynx rufus*) after removal of neighboring female. *American Midland Naturalist* 134: 409–412.

– 1996. Bobcat (*Lynx rufus*) home range size and habitat use in northwest Wisconsin. *American Midland Naturalist* 135: 241–252.

Lovallo, M.J., Gilbert, J.H., and Gehring, T.M. 1993. Bobcat, *Felis rufus*, dens in an abandoned Beaver, *Castor canadensis*, lodge. *Canadian Field Naturalist* 107(1): 108–109.

Lowery, J.C. 2006. *The Tracker's Field Guide: A Comprehensive Handbook for Animal Tracking in the United States*. FalconGuide, Globe Pequot Press, Guilford, CT.

Lowry, L.F., Burkanov, V.N., Frost, K.J., Simpkins, M.A., Davis, R., DeMaster, D.P., Suydam, R.S., and Springer, A. 2000. Habitat use and habitat selection by Spotted Seals (*Phoca largha*) in the Bering Sea. *Canadian Journal of Zoology* 78(11): 1959–1971.

Lowry, L.F., Frost, K.J., Davis, R., DeMaster, D.P., and Suydam, R.S. 1998. Movements and behavior of satellite-tagged Spotted Seals (*Phoca largha*) in the Bering and Chukchi Seas. *Polar Biology* 19: 221–230.

Lucas, Z.N., and Daoust, P.-Y. 2002. Large increases of Harp Seals (*Phoca groenlandica*) and Hooded Seals (*Cystophora cristata*) on Sable Island, Nova Scotia, since 1995. *Polar Biology* 25: 562–568.

Lucas, Z.N., Daoust, P.-Y., Conboy, G., and Brimacombe, M. 2003. Health status of Harp Seals (*Phoca groenlandica*) and Hooded Seals (*Cystophora cristata*) on Sable Island, Nova Scotia, Canada, concurrent with their expanding range. *Journal of Wildlife Diseases* 39(1): 16–28.

Lucas, Z.N., and Hooker, S.K. 2000. Cetacean strandings on Sable Island, Nova Scotia, 1970–1998. *Canadian Field-Naturalist* 114(1): 45–61.

Lucas, Z.N., and McAlpine, D.F. 2002. Extralimital occurrences of Ringed Seals, *Phoca hispida*, on Sable Island, Nova Scotia. *Canadian Field-Naturalist* 116(4): 607–610.

Lucas, Z.N., Raeside, J.I., and Betteridge, K.J. 1991. Non-invasive assessment of the incidence of pregnancy and pregnancy loss in the feral horses of Sable Island. *Journal of Reproduction and Fertility, Supplement* 44: 479–488.

Lucid, M.K., and Cook, J.A. 2004. Phylogeography of Keen's Mouse (*Peromyscus keeni*) in a naturally fragmented landscape. *Journal of Mammalogy* 85(6): 1149–1159.

Ludt, C.J., Schroeder, W., Rottmann, O., and Kuehn, R. 2004. Mitochondrial DNA phylogeography of Red Deer (*Cervus elaphus*). *Molecular Phylogenetics and Evolution* 31: 1064–1083.

Ludwig, D.R. 1984a. *Microtus richardsoni. Mammalian Species* no. 223: 1–6.

– 1984b. *Microtus richardsoni* microhabitat and life history. *Carnegie Museum of Natural History Special Publication* 10: 319–331.

– 1988. Reproduction and population dynamics of the Water Vole, *Microtus richardsoni. Journal of Mammalogy* 69(3): 532–541.

Ludwig, D.R., and Anderson, P.K. 2009. Metapopulation biology: *Microtus richardsoni* in the Rocky Mountain front range of Alberta. *Northwestern Naturalist* 90(1): 1–16.

Lunn, N.J., Atkinson, S., Branigan, M., Calvert, W., Clark, D., Doidge, B., Elliott, C., Nagy, J., Obbard, M., Otto, R., Stirling, I., Taylor, M., Vandal, D., and Wheatley, M. 2002. Polar Bear management in Canada 1997–2000. *Occasional Papers of the IUCN Species Survival Commission* 26: 41–52.

Lunn, N.J., Stirling, I., Andriashek, D., and Richardson, E. 2004. Selection of maternity dens by female Polar Bears in western Hudson Bay, Canada and the effects of human disturbance. *Polar Biology* 27: 350–356.

Lyamin, O.I., Oleksenko, A.I., Sevostiyanov, V.F., Nazarenko, E.A., and Mukhametov, L.M. 2000. Behavioral sleep in captive Sea Otters. *Aquatic Mammals* 26(2): 132–136.

Lyberth, B., Landa, A., Nagy, J., Loison, A., Olesen, C.R., Gunn, A., and Forchhammer, M.C. 2007. Muskoxen in the high Arctic – temporal and spatial differences in body size. *Journal of Zoology* 272: 227–234.

Lydersen, C., and Hammill, M.O. 1993. Diving in Ringed Seal (*Phoca hispida*) pups during the nursing period. *Canadian Journal of Zoology* 71(5): 991–996.

Lydersen, C., Hammill, M.O., and Kovacs, K.M. 1994a. Diving activity in nursing Bearded Seal (*Erignathus barbatus*) pups. *Canadian Journal of Zoology* 72(1): 96–103.

– 1994b. Activity of lactating ice-breeding Grey Seals, *Halichoerus grypus*, from the Gulf of St. Lawrence, Canada. *Animal Behaviour* 48: 1417–1425.

Lydersen, C., Hammill, M.O., and Ryg, M.S. 1992. Water flux and mass gain during lactation in free-living Ringed Seal (*Phoca hispida*). *Journal of Zoology* (London) 228: 361–369.

Lydersen, C., and Kovacs, K.M. 1996. Energetics of lactation in Harp Seals (*Phoca groenlandica*) from the Gulf of St. Lawrence, Canada. *Journal of Comparative Physiology B* 166: 295–304.

– 1999. Behaviour and energetics of ice-breeding, North Atlantic phocid seals during the lactation period. *Marine Ecology Progress Series* 187: 265–281.

Lydersen, C., Kovacs, K.M., and Hamill, M.O. 1997. Energetics during nursing and early postweaning fasting in Hooded Seal (*Cystophora cristata*) pups from the Gulf of St. Lawrence, Canada. *Journal of Comparative Physiology B* 167: 81–88.

Lynas, E.M., and Sylvestre, J.P. 1988. Feeding techniques and foraging strategies of Minke Whales (*Balaenoptera acutorostrata*) in the St. Lawrence River estuary. *Aquatic Mammals* 14(1): 21–32.

Lynch, G.M. 1971. Raccoon increases in Manitoba. *Journal of Mammalogy* 52(3): 621–2.

MacArthur, R.A. 1992. Foraging range and aerobic endurance of Muskrats diving under ice. *Journal of Mammalogy* 73(3): 565–569.

MacArthur, R.A., and Humphries, M.M. 1999. Postnatal development of thermoregulation in the semiaquatic Muskrat (*Ondatra zibethicus*). *Canadian Journal of Zoology* 77(10): 1521–1529.

MacArthur, R.A., Humphries, M.M., and Jeske, D. 1997. Huddling behavior and the foraging efficiency of Muskrats. *Journal of Mammalogy* 78(3): 850–858.

Macdonald, D.W., and Barrett, P. 1993. *Mammals of Europe*. Princeton Field Guides, Princeton University Press, Princeton, NJ.

MacDonald, I.M.V. 1992. Grey Squirrels discriminate red from green in a foraging situation. *Animal Behaviour* 43: 694–695.

MacDonald, K., Matsui, E., Stevens, R., and Fenton, M.B. 1994. Echolocation calls and field identification of the Eastern Pipistrelle (*Pipistrellus subflavus*: Chiroptera: Vespertilionidae), using ultrasonic bat detectors. *Journal of Mammalogy* 75(2): 462–465.

MacDonald, S.D. 1990. The whelping patch. *Nature Canada* 19(4): 23–27.

MacDonald, S.O., and Cook, J.A. 2009. *Recent Mammals of Alaska*. University of Alaska Press, Fairbanks, AK.

MacDonald, S.O., and Jones, C. 1987. *Ochotona collaris. Mammalian Species* no. 281: 1–4.

MacDonald, S.O., Runck, A.M., and Cook, J.A. 2004. The Heather Vole, genus *Phenacomys*, in Alaska. *Canadian Field-Naturalist* 118(3): 438–440.

MacHutchon, A.G., and Harestad, A.S. 1990. Vigilance behaviour and use of rocks by Columbian Ground Squirrels. *Canadian Journal of Zoology* 68(7): 1428–1433.

Mackie, R.J., Kie, J.G., Pac, D.F., and Hamlin, K.L. 2003. Mule Deer (*Odocoileus hemionus*). In G.A. Feldhamer, B.C. Thompson, and J.A. Chapman (eds), *Wild Mammals of North America: Biology, Management, and Conservation*, 2nd ed., 889–905. Johns Hopkins University Press, Baltimore, MD.

MacLeod, C. 2000. Review of the distribution of *Mesoplodon* species (order Cetacea, family Ziphiidae) in the North Atlantic. *Mammal Review* 30(1): 1–8.

MacLeod, C.D. 2002. Possible functions of the ultradense bone in the rostrum of Blainville's Beaked Whale (*Mesoplodon densirostris*). *Canadian Journal of Zoology* 80(1): 178–184.

MacLeod, C.D., Santos, M.B., and Pierce, G.J. 2003. Review of data on diets of beaked whales: Evidence of niche separation and geographic segregation. *Journal of the Marine Biological Association of the United Kingdom* 83: 651–665.

MacMahon, J.A. 1999. Northern Pocket Gopher. In D.E. Wilson and S. Ruff (eds), *The Smithsonian Book of North American Mammals*, 472–477. UBC Press, Vancouver, BC.

Madison, D.M., and McShea, W.J. 1987. Seasonal changes in reproductive tolerance, spacing, and social organization in Meadow Voles: A microtine model. *American Zoologist* 27: 899–908.

Madsen, P.T., Payne, R., Kristiansen, N.U., Wahlberg, M., Kerr, I., and Møhl, B. 2002. Sperm Whale sound production studied with ultrasonic time/depth-recording tags. *Journal of Experimental Biology* 205(13): 1899–1906.

Mager, K.J., and Nelson, T.A. 2001. Roost-site selection by Eastern Red Bats (*Lasiurus borealis*). *American Midland Naturalist* 145(1): 120–126.

Magoun, A.J., and Copeland, J.P. 1998. Characteristics of Wolverine reproductive den sites. *Journal of Wildlife Management* 62(4): 1313–1320.

Magoun, A.J., Ray, J.C., Johnson, D.S., Valkenburg, P., Dawson, F.N., and Bowman, J. 2007. Modeling Wolverine occurrence using aerial surveys of tracks in snow. *Journal of Wildlife Management* 71(7): 2221–2229.

Mahan, C.G., and Yahner, R.H. 1996. Effects of forest fragmentation on burrow-site selection by Eastern Chipmunk (*Tamias striatus*). *American Midland Naturalist* 136: 352–357.

Mahoney, S.P. 2001. The land mammals of insular Newfoundland. *Antigonish Review* no. 124: 87–94.

Mahoney, S.P., and Schaefer, J.A. 2002. Long-term changes in demography and migration of Newfoundland Caribou. *Journal of Mammalogy* 83(4): 957–963.

Maier, J.A.K., and White, R.G. 1998. Timing and synchrony of activity in Caribou. *Canadian Journal of Zoology* 76(11): 1999–2009.

Maier, T.J. 2002. Long-distance movements by female White-footed Mice, *Peromyscus leucopus*, in extensive mixed-wood forest. *Canadian Field-Naturalist* 116(1): 108–111.

Maier, T.J., and Doyle, K.L. 2006. Aggregations of Masked Shrews (*Sorex cinereus*): Density-related mating behavior? *Mammalia* 70(1): 86–89.

Main, M.B., and Coblentz, B.E. 1996. Sexual segregation in Rocky Mountain Mule Deer. *Journal of Wildlife Management* 60(3): 497–507.

Mainguy, J., Coté, S.D., Cardinal, E., and Houle, M. 2008. Mating tactics and mate choice in relation to age and social rank in male Mountain Goats. *Journal of Mammalogy* 89(3): 626–635.

Malcolm, J.R., and Brooks, R.J. 1993. The adaptive value of photoperiod-induced shape changes in the collared lemming. In N.C. Stenseth and R.A. Ims (eds), *The Biology of Lemmings*, 311–328. Academic Press, San Diego, CA.

Maniscalco, J.M., and Parker, P. 2009. A case of twinning and the care of two offspring of different ages in Steller Sea Lions. *Marine Mammal Science* 25(1): 206–213.

Mankin, P.C., and Getz, L.L. 1994. Burrow morphology as related to social organization of *Microtus ochrogaster*. *Journal of Mammalogy* 75(2): 492–499.

Manning, R.W. 1993. Systematics and evolutionary relationships of the Long-eared Myotis, *Myotis evotis* (Chiroptera: Vespertilionidae). *The Museum of Texas Tech University Special Publications* no. 37: 1–58.

Manning, R.W., and Jones, J.K., Jr. 1988. *Perognathus fasciatus*. *Mammalian Species* no. 303: 1–4.

– 1989. *Myotis evotis*. *Mammalian Species* no. 329: 1–5.

Mansour, A.A.H., McKay, D.W., Lien, J., Orr, J.C., Banoub, J.H., Øien, N., and Stenson, G. 2002. Determination of pregnancy status from blubber samples in Minke Whales (*Balaenoptera acutorostrata*). *Marine Mammal Science* 18(1): 112–120.

Mappes, T., and Koskela, E. 2004. Genetic basis of the trade-off between offspring number and quality in the Bank Vole. *Evolution* 58(3): 645–650.

Mappes, T., Ylönen, H., and Viitala, J. 1995. Higher reproductive success among kin groups of Bank Voles (*Clethrionomys glareolus*). *Ecology* 76(4): 1276–1282.

Marchison, R.L., Johansen, K.L., and Miller, K.V. 1990. Behavioural components of White-tailed Deer marking: Social and seasonal effects. *Chemical Signals in Vertebrates* 5: 295–301.

Maresh, J.L., Fish, F.E., Nowacek, D.P., Nowacek, S.M., and Wells, R.S. 2004. High performance turning capabilities during foraging by Bottlenose Dolphins (*Tursiops truncatus*). *Marine Mammal Science* 20(3): 498–509.

Marfori, M.A., Parker, P.G., Gregg, T.G., Vandenbergh, J.G., and Solomon, N.G. 1997. Using DNA fingerprinting to estimate relatedness within social groups of Pine Voles (*Microtus pinetorum*). *Journal of Mammalogy* 78(3): 715–724.

Margulis, S.W. 1993. Mate choice in Rocky Mountain Mule Deer bucks (*Odocoileus hemionus hemionus*): Is there a preference for does without fawns? *Ethology Ecology and Evolution* 5: 115–119.

Marin, L., Consiglio, C., Catalono, B., Valentini, T., and Villetti, G. 1996. Aerial behavior in Fin Whales (*Balaenoptera physalus*) in the Mediterranean Sea. *Marine Mammal Science* 12(3): 489–495.

Marinelli, L., and Messier, F. 1993. Space use and the social system of Muskrats. *Canadian Journal of Zoology* 71(5): 869–875.

– 1995. Parental-care strategies among Muskrats in a female-biased population. *Canadian Journal of Zoology* 73(8): 1503–1510.

Marinelli, L., Messier, F., and Plante, Y. 1997. Consequences of following a mixed reproductive strategy in Muskrats. *Journal of Mammalogy* 78(1): 163–172. Marini, L., Consiglio, C., Catalono, B., Valentini, T., and Villetti, G. 1996. Aerial behaviour in Fin Whales (*Balaenoptera physalus*) in the Mediterranean Sea. *Marine Mammal Science* 12(3): 489–495.

Marquard-Petersen, U. 1998. Food habits of Arctic Wolves in Greenland. *Journal of Mammalogy* 79(1): 236–244.

Marshall, H.D., and Ritland, K. 2002. Genetic diversity and differentiation of Kermode Bear populations. *Molecular Ecology* 11: 685–697.

Martell, A.M., and Milko, R.J. 1986. Seasonal diets of Vancouver Island Marmots, *Marmota vancouverensis*. *Canadian Field-Naturalist* 100(2): 241–245.

Marten, K. 2000. Ultrasonic analysis of Pygmy Sperm Whale (*Kogia breviceps*) and Hubbs' Beaked Whale (*Mesoplodon carlhubbsi*) clicks. *Aquatic Mammals* 26(1): 45–48.

Martin, A.R., and Smith, T.G. 1992. Deep diving in wild, free-ranging Beluga Whales, *Delphinapterus leucas*. *Canadian Journal of Fisheries and Aquatic Sciences* 49(3): 462–466.

– 1999. Strategy and capability of wild Belugas, *Delphinapterus leucas*, during deep, benthic diving. *Canadian Journal of Zoology* 77(11): 1783–1793.

Martin, A.R., and Walker, F.J. 1997. Sighting of a Right Whale (*Eubalaena glacialis*) with calf off s.w. Portugal. *Marine Mammal Science* 13(1): 139–140.

Martin, J.M., and Heske, E.J. 2005. Juvenile dispersal of Franklin's Ground Squirrel (*Spermophilus franklinii*) from a prairie "island." *American Midland Naturalist* 153(2): 444–449.

Martin, J.M., Heske, E.J., and Hofmann, J.E. 2003. Franklin's Ground Squirrel (*Spermophilus franklinii*) in Illinois: A declining prairie mammal? *American Midland Naturalist* 150(1): 130–138.

Martin, K.J., and Anthony, R.G. 1999. Movements of Northern Flying Squirrels in different-aged forest stands of western Oregon. *Journal of Wildlife Management* 63(1): 291–297.

Martin, R.L. 1973. Molting in the Rock Vole, *Microtus chrotorrhinus*. *Mammalia* 37: 342–347.

Maser, C. 1967. A life trap for microtines. *Murrelet* 48: 58.

Mason, G.A. 1994. The influence of weight, sex, birthdate, and maternal age on the growth of weanling mink. *Journal of Zoology* (London) 233: 203–214.

Mate, B.R. 2005. Gray Whales (*Eschrichtius robustus*). Oregon State University and Oregon Sea Grant ORESU G-05-002, 2-page pamphlet.

Mate, B.R., Krutzikowsky, G.K., and Winsor, M.H. 2000. Satellite-monitored movements of radio-tagged Bowhead Whales in the Beaufort and Chukchi seas during the late-summer feeding season and fall migration. *Canadian Journal of Zoology* 78(7): 1168–1181.

Mate, B.R., Lagerquist, B.A., Winsor, M., Geraci, J., and Prescott, J.H. 2005. Movements and dive habits of a satellite-monitored Longfinned Pilot Whale (*Globicephala melas*) in the northwest Atlantic. *Marine Mammal Science* 21(1): 136–144.

Mate, B.R., Stafford, K.M., Nawojchik, R., and Dunn, J.L. 1994. Movements and dive behavior of a satellite-monitored Atlantic White-sided Dolphin (*Lagenorhynchus acutus*) in the Gulf of Maine. *Marine Mammal Science* 10(1): 116–121.

Matkin, C.O., Lennard, L.B., and Ellis, G. 2002. Killer Whales and predation on Steller Sea Lions. *Alaska Sea Grant Report* 2002 (02-02): 61–66.

Matlack, C.R., and Evans, A.J. 1992. Diet and condition of Bobcats, *Lynx rufus*, in Nova Scotia during autumn and winter. *Canadian Journal of Zoology* 70(6): 1114–1119.

Matlack, R.S., Kaufman, D.W., and Charlton, R.E. 2001. First record of the Cinereus Shrew for Riley County, Kansas. *Prairie Naturalist* 33(2): 107–108.

Mattson, D.J. 2001. Myrmecophagy by Yellowstone Grizzly Bears. *Canadian Journal of Zoology* 79(5): 779–793.

Mattson, D.J., Green, G.I., and Swalley, R. 1999. Geophagy by Yellowstone Grizzly Bears. *Ursus* 11: 109–116.

Mattson, D.J., Knight, R.R., and Blanchard, B.M. 1992. Cannibalism and predation on Black Bears by Grizzly Bears in the Yellowstone ecosystem, 1975–1990. *Journal of Mammalogy* 73(2): 422–425.

Mattson, T.A., Buskirk, S.W., and Stanton, N.L. 1996. Roost sites of the Silver-haired Bay (*Lasionycteris noctivagans*) in the Black Hills, South Dakota. *Great Basin Naturalist* 56(3): 247–253.

Maunder, J.E. 1988. First Newfoundland record of the Hoary Bat, *Lasiurus cinereus*, with a discussion of other records of migratory tree bats in Atlantic Canada. *Canadian Field-Naturalist* 102(4): 726–728.

Mauritzen, M., Derocher, A.E., Pavlova, O., and Wiig, Ø. 2003. Female Polar Bears, *Ursus maritimus*, on the Barents Sea drift ice: Walking the treadmill. *Animal Behaviour* 66(1): 107–113.

Mauritzen, M., Derocher, A.E., and Wiig, Ø. 2001. Space-use strategies of female Polar Bears in a dynamic sea ice habitat. *Canadian Journal of Zoology* 79(9): 1704–1713.

Mazurek, M.J. 2004. A maternity roost of Townsend's Big-eared bats (*Corynorhinus townsendii*) in Coast Redwood basal hollows in northwestern California. *Northwestern Naturalist* 85: 60–62.

McAdam, A.G., and Boutin, S. 2004. Maternal effects and the response to selection in Red Squirrels. *Proceedings of the Royal Society of London, B series – Biological Sciences* 271: 75–79.

McAdam, A.G., and Millar, J.S. 1999. Breeding by young-of-the-year female Deer Mice: Why weight? *Ecoscience* 6(3): 400–405.

McAllister, J.A., and Hoffmann, R.S. 1988. *Phenacomys intermedius. Mammalian Species* no. 305: 1–8.

McAlpine, D.F. 1985. First records of the Sperm Whale (*Physeter macrocephalus*) from New Brunswick and the Bay of Fundy. *Naturaliste canadien* 112: 433–434.

– 2002. Pygmy and Dwarf Sperm Whales (*Kogia breviceps* and *K. sima*). In W.F. Perrin, B. Wursig, and H.G.T. Thewissen (eds), *Encyclopedia of Marine Mammals*, 1007–1009. Academic Press, San Diego, CA.

McAlpine, D.F., Cox, S.L., McCabe, D.A., and Schnare, J.-L. 2004. Occurrence of the Long-tailed Shrew (*Sorex dispar*) in the Nerepsis Hills, New Brunswick. *Northeastern Naturalist* 11(4): 383–386.

McAlpine, D.F., Kingsley, M.C.S., and Daoust, P.-Y. 1999. A lactating record-age St Lawrence Beluga (*Delphinapterus leucas*). *Marine Mammal Science* 15(3): 854–859.

McAlpine, D.[F.], Martin, J.D., and Libby, C. 2008. First occurrence of the Grey Fox, *Urocyon cinereoargenteus*, in New Brunswick: A climate-change mediated range expansion? *Canadian Field-Naturalist* 122(2): 169–171.

McAlpine, D.F., Muldoon, F., Forbes, G.J., Wandeler, A.I., Makepeace S., Broders, H.G., and Goltz, J.P. 2002a. Over-wintering and reproduction by the Big Brown Bat, *Eptesicus fuscus*, in New Brunswick. *Canadian Field-Naturalist* 116(4): 645–647.

McAlpine, D.F., Muldoon, F., and Wandeler, A.I. 2002b. First record of the Hoary Bat, *Lasiurus cinereus* (Chiroptera: Vespertilionidae), from Prince Edward Island. *Canadian Field-Naturalist* 116(1): 124–125.

McAlpine, D.F., Murison, L.D., and Hoberg, E.P. 1997. New records for the Pygmy Sperm Whale, *Kogia breviceps* (Physeteridae) from Atlantic Canada with notes on diet and parasites. *Marine Mammal Science* 13(4): 701–704.

McAlpine, D.F., and Rae, M. 1999. First confirmed reports of beaked whales, cf. *Mesoplodon bidens* and *M. densirostris* (Ziphiidae), from New Brunswick. *Canadian Field-Naturalist* 113(2): 293–295.

McAlpine, D.F., Stevick, P.T., and Murison, L.D. 1999a. Increase in extralimital occurrences of ice-breeding seals in the northern Gulf of Maine region: More seals or fewer fish? *Marine Mammal Science* 15(3): 906–911.

McAlpine, D.F., Stevick, P.T., Murison, L.D., and Turnbull, S.D. 1999b. Extralimital records of Hooded Seals (*Cystophora cristata*) from the Bay of Fundy and northern Gulf of Maine. *Northeastern Naturalist* 6(3): 225–230.

McAlpine, D.F., and Walker, R.J. 1990. Extralimital records of the Harp Seal, *Phoca groenlandica*, from the western North Atlantic: A review. *Marine Mammal Science* 6(3): 248–252.

– 1999. Additional extralimital records of the Harp Seal, *Phoca groenlandica*, from the Bay of Fundy, New Brunswick. *Canadian Field-Naturalist* 113(2): 290–292.

McCarty, R. 1978. *Onychomys leucogaster. Mammalian Species* no. 87: 1–6.

McCay, T.S., Menzel, M.A., Laerm, J., and Lepardo, L.T. 1998. Timing of parturition of three long-tailed shrews (*Sorex* spp.) in the southern Appalachians. *American Midland Naturalist* 139(2): 394–397.

McClearn, D. 1992. Locomotion, posture, and feeding behavior of Kinkajous, Coatis, and Raccoons. *Journal of Mammalogy* 73(2): 245–261.

McCracken, G.F., Gustin, M.K., and McKamey, M.I. 1986. Raccoons catch Mexican Free-tailed Bats "on the wing." *Bat Research News* 27(3–4): 21–22.

McCullogh, S., and Boness, D.J. 2000. Mother-pup vocal recognition in the Grey Seal (*Halichoerus grypus*) of Sable Island, Nova Scotia, Canada. *Journal of Zoology* (London) 251(4): 449–455.

McCullough, D.R. 1997. Breeding by female fawns in Black-tailed Deer. *Wildlife Society Bulletin* 25(2): 296–297.

McDonald, K.P., and Case, D.S. 1990. The status of Ermine (*Mustela erminea*) in Ohio. *Ohio Journal of Science* 90(1): 46–47.

McDonald, M.A., Calambokidis, J., Teranishi, A.M., and Hildebrand, J.A. 2001. The acoustic calls of Blue Whales off California with gender data. *Journal of the Acoustical Society of America* 109(4): 1728–1735.

McGee, B.K., Nicholson, K.L., Ballard, W.B., and Butler, M.J. 2006. Characteristics of Swift Fox dens in northwest Texas. *Western North American Naturalist* 66(2): 239–245.

McGinnis, S.M., and Schusterman, R.J. 1981. Northern Elephant Seal *Mirounga angustirostris* Gill, 1866. In S.H. Ridgway and R. Harrison (eds), *Handbook of Marine Mammals*, vol. 2, *Seals*, 329–349. Academic Press, San Diego, CA.

McGuire, B. 1997. Influence of father and pregnancy on maternal care in red-backed voles. *Journal of Mammalogy* 78(3): 839–849.

McGuire, B., Henyey, E., McCue, E., and Bemis, W.E. 2003. Parental behavior at parturition in Prairie Voles (*Microtus ochrogaster*). *Journal of Mammalogy* 84(2): 513–523.

McGuire, B., Pizzuto, T., Bemis, W.E., and Getz, L.L. 2006. General ecology of a rural population of Norway Rats (*Rattus norvegicus*) based on intensive live trapping. *American Midland Naturalist* 155(1): 221–236.

McGuire, B. and Sullivan, S. 2001. Suckling behavior of Pine Voles (*Microtus pinetorum*). *Journal of Mammalogy* 82(3): 690–699.

McIntyre, I.W., Campbell, K.L., and MacArthur, R.A. 2002. Body oxygen stores, aerobic dive limits, and diving behaviour of the Star-nosed Mole (*Condylura cristata*) and comparisons with non-aquatic talpids. *Journal of Experimental Biology* 205(1): 45–54.

McKechnie, A.M., Smith, A.T., and Peacock, M.M. 1994. Kleptoparasitism in pikas (*Ochotona princeps*): Theft of hay. *Journal of Mammalogy* 75(2): 488–491.

McKenzie, A. 1990. Seeking the mechanisms of hibernation. *Bioscience* 40(6): 425–427.

McKinstry, M.C., Karhu, R.R., and Anderson, S.H. 1997. Use of active Beaver, *Castor canadensis*, lodges by Muskrats, *Ondatra zibethicus*, in Wyoming. *Canadian Field-Naturalist* 111(2): 310–311.

McLaren, B.E., Roberts, B.A., Djan-Chékar, N., and Lewis, K.P. 2004. Effects of overabundant Moose on the Newfoundland landscape. *Alces* 40: 45–59.

McLean, I.G. 1978. Plugging of nest burrows by female *Spermophilus columbianus*. *Journal of Mammalogy* 59(2): 437–439.

McLellan, B.N., and Hovey, F.W. 2001. Natal dispersal of Grizzly Bears. *Canadian Journal of Zoology* 79(5): 838–844.

McLellan, B.N., Hovey, F.W., Mace, R.D., Woods, J.G., Carney, D.W., Gibeau, M.L., Wakkinen, W.L., and Kasworm, W.F. 1999. Rates and causes of Grizzly Bear mortality in the interior mountains of British Columbia, Alberta, Montana, Washington, and Idaho. *Journal of Wildlife Management* 63(3): 911–920.

McLellan, W.A., Koopman, H.N., Rommel, S.A., Read, A.J., Potter, C.W., Nicholas, J.R., Westgate, A.J., and Pabst, D.A. 2002. Ontogenetic allometry and body composition of Harbour Porpoises (*Phocoena phocoena*, L.) from the western North Atlantic. *Journal of Zoology* (London) 257: 457–471.

McLoughlin, P.D., Cluff, H.D., and Messier, F. 2002. Denning ecology of Barren-ground Grizzly Bears in the Central Arctic. *Journal of Mammalogy* 83(1): 188–198.

McLoughlin, P.D., Taylor, M.K., Cluff, H.D., Gau, R.J., Mulders, R., Case, R.L., Boutin, S., and Messier, F. 2003. Demography of Barren-ground Grizzly Bears. *Canadian Journal of Zoology* 81(2): 294–301.

McMaster, A. 2002. Flying squirrels in Bluebird nest boxes. *Blue Jay* 60(4): 225–226.

McMillan, B.R., Cottam, M.R., and Kaufman, D.W. 2000. Wallowing behavior of American Bison (*Bos bison*) in tallgrass prairie: An examination of alternative explanations. *American Midland Naturalist* 144: 159–167.

McMillan, B.R., and Kaufman, D.W. 1995. Travel path characteristics for free-living White-footed Mice (*Peromyscus leucopus*). *Canadian Journal of Zoology* 73(8): 1474–1478.

McMillan, B.R., Kaufman, G.A., and Kaufman, D.W. 1997. A case of senescence for the White-footed Mouse. *Southwestern Naturalist* 42(2): 236–237.

McNay, M.E. 2002. A case history of wolf-human encounters in Alaska and Canada. *Alaska Department of Fish and Game Wildlife Technical Bulletin* 13: 1–43.

McNay, M.E., and Mooney, P.W. 2005. Attempted predation of a child by a Gray Wolf, *Canis lupus*, near Icy Bay, Alaska. *Canadian Field-Naturalist* 119(2): 197–201.

McPhee, E.C. 1984. Ethological aspects of mutual exclusion in the parapatric species of *Clethrionomys, C. gapperi* and *C. rutilus. Acta Zoologica Fennica* 172: 71–73.

McShane, L.J., Estes, J.A., Riedman, M.L., and Staedler, M.M. 1995. Repertoire, structure, and individual variation of vocalizations in the Sea Otter. *Journal of Mammalogy* 76(2): 414–427.

McShea, W.J., and Madison, D.M. 1984. Communal nesting between reproductively active females in a spring population of *Microtus pennsylvanicus. Canadian Journal of Zoology* 62(3): 344–346.

McSweeney, D.J., Baird, R.W., and Mahaffy, S.D. 2007. Site fidelity, associations, and movements of Cuvier's (*Ziphius cavirostris*) and Blainville's (*Mesoplodon densirostris*) beaked whales off the island of Hawaii. *Marine Mammal Science* 23(3): 666–687.

Mead, J.G. 1984. Survey of reproductive data for the beaked whales (Ziphiidae). *Report of the International Whaling Commission* (special issue) 6: 91–96.

– 1989a. Bottlenose Whales (*Hyperoodon ampillatus*) (Forster 1770) and *Hyperoodon planifrons* (Flower 1982). In S.H. Ridgway and R. Harrison (eds), *Handbook of Marine Mammals*, vol. 4, 321–348. Academic Press, San Diego, CA.

– 1989b. Beaked whales of the genus *Mesoplodon*. In S.H. Ridgway and R. Harrison (eds), *Handbook of Marine Mammals*, vol. 4, 349–430. Academic Press, San Diego, CA.

Mead, J.I., Spiess, A.E., and Sobolik, K.D. 2000. Skeleton of extinct North American Sea Mink (*Mustela macrodon*). *Quaternary Research* 53: 247–262.

Measures, L., Roberge, B., and Sears, R. 2004. Stranding of a Pygmy Sperm Whale, *Kogia breviceps*, in the northern Gulf of St Lawrence, Canada. *Canadian Field-Naturalist* 118(4): 495–498.

Mech, L.D. 1994. Regular and homeward travel speeds of Arctic Wolves. *Journal of Mammalogy* 75(3): 741–742.

– 1999. Gray Wolf (*Canis lupus*). In D.E. Wilson and S. Ruff (eds), *The Smithsonian Book of North American Mammals*, 141–143. UBC Press, Vancouver, BC.

– 2000a. Lack of reproduction in Muskoxen and Arctic Hares caused by early winter? *Arctic* 53(1): 69–71.

– 2000b. Leadership in Wolf, *Canis lupus*, packs. *Canadian Field-Naturalist* 114(2): 259–264.

– 2003. Incidence of Mink, *Mustela vison*, and River Otter, *Lutra canadensis*, in a highly urbanized area. *Canadian Field-Naturalist* 117(1): 115–116.

– 2006. Estimated age structure of wolves in northeastern Minnesota. *Journal of Wildlife Management* 70(5): 1481–1483.

– 2007. Possible use of foresight, understanding, and planning by Wolves hunting Muskoxen. *Arctic* 60(2): 145–149.

Mech, L.D., and Adams, L.G. 1999. Killing of a Muskox, *Ovibos moschatus*, by two Wolves, *Canis lupus*, and subsequent caching. *Canadian Field-Naturalist* 113(4): 673–675.

Mech, L.D., and Boitani, L. 2004. Grey Wolf, *Canis lupus* (Linnaeus 1758). In C. Sillero-Zubiri, M. Hoffmann, and D.W. Macdonald (eds), *Canids: Foxes, Wolves, Jackals and Dogs; 2004 Status Survey and Conservation Action Plan*, 124–129. Oxford University Press, Oxford, UK.

Mech, L.D., Meier, T.J., Burch, J.W., and Adams, L.G. 1995. Patterns of prey selection by wolves in Denali National Park, Alaska. *Canadian Circumpolar Institute Occasional Publication* 35: 231–243.

Mech, L.D., and Merrill, S.B. 1998. Daily departure and return patterns of Wolves, *Canis lupus*, from a den at 80°N latitude. *Canadian Field-Naturalist* 112(3): 515–517.

Mech, L.D., and Nelson, M.E. 1990. Non-family wolf, *Canis lupus*, packs. *Canadian Field-Naturalist* 104(3): 482–483.

Mech, L.D., and Packard, J.M. 1990. Possible use of wolf, *Canis lupus*, den over several centuries. *Canadian Field-Naturalist* 104(3): 484–485.

Mech, L.D., Wolf, P.C., and Packard, J.M. 1999. Regurgitative food transfer among wild wolves. *Canadian Journal of Zoology* 77(8): 1192–1195.

Meier, P.T. 1992. Social organization of Woodchucks (*Marmota monax*). *Behavioral Ecology and Sociobiology* 31: 393–400.

Mellinger, D.K., Carson, C.D., and Clark, C.W. 2000. Characteristics of Minke Whale (*Balaenoptera acutorostrata*) pulse trains recorded near Puerto Rico. *Marine Mammal Science* 16(4): 739–756.

Mellinger, D.K., Stafford, K.M., Moore, S.E., Munger, L., and Fox, C.G. 2004. Detection of North Pacific Right Whale (*Eubalaena japonica*) calls in the Gulf of Alaska. *Marine Mammal Science* 20(4): 872–879.

Melquist, W.E., and Hornocker, M.G. 1983. Ecology of river otters in west central Idaho. *Wildlife Monographs* 83: 1–60.

Melquist, W.E., Polechla, P.J., Jr., and Toweill, D. 2003. River Otter (*Lontra canadensis*). In G. A. Feldhamer, B.C. Thompson, and J.A. Chapman (eds), *Wild Mammals*

of North America: Biology, Management, and Conservation, 2nd ed., 708–734. Johns Hopkins University Press, Baltimore, MD.

Mensing-Solick, Y.R., and Barclay, R.M.R. 2003. The effect of canine tooth wear on the diet of Big Brown Bats (*Eptesicus fuscus*). *Acta Chiropterologica* 5(1): 91–95.

Menzel, J.M., Ford, W.M., Edwards, J. W., and Menzel, M.A. 2004. Nest tree use by the endangered Virginia Northern Flying Squirrel in the central Appalachian Mountains. *American Midland Naturalist* 151(2): 355–368.

Menzel, M.A., Carter, T.C., Ford, W.M., and Chapman, B.R. 2001. Tree-roost characteristics of subadult and female adult Evening Bats (*Nycticeius humeralis*) in the upper coastal plain of South Carolina. *American Midland Naturalist* 145: 112–119.

Mercado, E., III, Herman, L.M., and Pack, A.A. 2003. Stereotypical sound patterns in Humpback Whale songs: Usage and function. *Aquatic Mammals* 29(1): 37–52.

Mercer, W.E., Hearn, B.J., and Finlay, C. 1981. Arctic Hare populations in insular Newfoundland. In K. Myers and C.D. MacInnes (eds), *Proceedings of the World Lagomorph Conference, 1979*, 450–468. University of Guelph, Guelph, ON.

Mercure, A., Ralls, K., Koepfli, K.P., and Wayne, R.K. 1993. Genetic subdivisions among small canids: Mitochondrial DNA differentiation of Swift, Kit, and Arctic Foxes. *Evolution* 47(5): 1313–1328.

Merrick, R.L., Chumbley, M.K., and Byrd, G.V. 1997. Diet diversity of Steller Sea Lions (*Eumetopias jubatus*) and their population decline in Alaska: A potential relationship. *Canadian Journal of Fisheries and Aquatic Sciences* 54(6): 1342–1348.

Merritt, J.F. 1981. *Clethrionomys gapperi. Mammalian Species* no. 146: 1–9.

– 1987. *Guide to the Mammals of Pennsylvania*. University of Pittsburgh Press for the Carnegie Museum of Natural History, Pittsburgh, PA.

Merritt, J.F., and Adamerovich, A. 1991. Winter thermoregulatory mechanisms of *Blarina brevicauda* as revealed by radiotelemetry. In J.S. Findley and T.L. Yates (eds), *The Biology of the Soricidae*, 47–64. University of New Mexico, Albuquerque, NM.

Merritt, J.F., and Zegers, D.A. 1991. Seasonal thermogenesis and body-mass dynamics of *Clethrionomys gapperi. Canadian Journal of Zoology* 69(11): 2771–2777.

Merritt, J.F., Zegers, D.A., and Rose, L.R. 2001. Seasonal thermogenesis of Southern Flying Squirrels (*Glaucomys volans*). *Journal of Mammalogy* 82(1): 51–64.

Mesnick, S.L., and Le Boeuf, B.J. 1991. Sexual behavior of male Northern Elephant Seals: II, Female response to potentially injurious encounters. *Behaviour* 117(3–4): 262–280.

Messier, F., Taylor, M.K., and Ramsay, M.A. 1994. Denning ecology of Polar Bears in the Canadian Arctic Archipelago. *Journal of Mammalogy* 75(2): 420–430.

Messier, F., and Virgl, J.A. 1992. Differential use of bank burrows and lodges by Muskrats, *Ondatra zibethicus*, in a northern marsh environment. *Canadian Journal of Zoology* 70(6): 1180–1184.

Michener, G.R. 1998. Sexual differences in reproduction effort of Richardson's Ground Squirrels. *Journal of Mammalogy* 79(1): 1–19.

– 2000. Caching of Richardson's Ground Squirrels by North American Badgers. *Journal of Mammalogy* 81(4): 1106–1117.

– 2002. Seasonal use of subterranean sleep and hibernation sites by adult female Richardson's Ground Squirrels. *Journal of Mammalogy* 83(4): 999–1012.

– 2004. Hunting techniques and tool use by North American Badgers preying on Richardson's Ground Squirrels. *Journal of Mammalogy* 85(5): 1019–1027.

– 2005. Limits on egg predation by Richardson's Ground Squirrels. *Canadian Journal of Zoology* 83(8): 1030–1037.

Michener, G.R., and Iwaniuk, A.N. 2001. Killing technique of North American Badgers preying on Richardson's Ground Squirrels. *Canadian Journal of Zoology* 79(11): 2109–2113.

Michener, G.R., and Koeppl, J.W. 1985. *Spermophilus richardsonii. Mammalian Species* no. 243: 1–8.

Michener, G.R., and McLean, I.G. 1996. Reproductive behaviour and operational sex ratio in Richardson's Ground Squirrels. *Animal Behaviour* 52: 743–758.

Mignucci-Giannoni, A.A., and Haddow, P. 2002. Wandering Hooded Seals. *Science* 295 (25 January): 627–628.

Mihok, S. 1976. Behavior of subarctic red-backed voles (*Clethrionomys gapperi athabascae*). *Canadian Journal of Zoology* 54(11): 1932–1945.

Mikesic, D.G., and Drickamer, L.C. 1992. Factors affecting home-range size in House Mice (*Mus musculus domesticus*) living in outdoor enclosures. *American Midland Naturalist* 127: 31–40.

Millar, J.S., and Derrickson, E.M. 1992. Group nesting in *Peromyscus maniculatus*. *Journal of Mammalogy* 73(2): 403–407.

Millar, J.S., and Teferi, T. 1993. Winter survival in northern *Peromyscus maniculatus*. *Canadian Journal of Zoology* 71(1): 125–129.

Miller, B., Reading, R.P., and Forrest, S. 1996. *Prairie Night. Black-footed Ferrets, and the Recovery of Endangered Species*. Smithsonian Institution Press, Washington, DC.

Miller, B.J., and Anderson, S.H. 1993. Ethology of the endangered Black-footed Ferret (*Mustela nigripes*). *Advances in Ethology* no. 31: 1–47.

Miller, B.K., and Litvaitis, J.A. 1992. Use of roadside salt licks by Moose, *Alces alces*, in northern New Hampshire. *Canadian Field-Naturalist* 106(1): 112–117.

Miller, D.H., and Getz, L.L. 1969. Life-history notes on *Microtus pinetorum* in central Connecticut. *Journal of Mammalogy* 50(4): 777–784.

Miller, F.L. 2003. Caribou (*Rangifer tarandus*). In G.A. Feldhamer, B.C. Thompson, and J.A. Chapman (eds), *Wild Mammals of North America: Biology, Management and Conservation*, 2nd ed., 965–997. Johns Hopkins University Press, Baltimore, MD.

Miller, F.L., and Barry, S.J. 2009. Long-term control of Peary Caribou numbers by unpredictable, exceptionally severe snow or ice conditions in a non-equilibrium grazing system. *Arctic* 62(2): 175–189.

Miller, F.L., and Reintjes, F.D. 1995. Wolf-sightings on the Canadian Arctic Islands. *Arctic* 48(4): 313–323.

Miller, K.V., Muller, L.I., and Demarais, S. 2003. White-tailed Deer (*Odocoileus virginianus*). In G.A. Feldhamer, B.C. Thompson, and J.A. Chapman (eds), *Wild Mammals of North America: Biology, Management, and Conservation*, 2nd ed., 906–930. Johns Hopkins University Press, Baltimore, MD.

Miller, L.A., Pristed, J., Møhl, B., and Surlykke, A. 1995. The click-sounds of Narwhals (*Monodon monoceros*) in Inglefield Bay, Northwest Greenland. *Marine Mammal Science* 11(4): 491–502.

Miller, R.F. 1997. New records and AMS radiocarbon dates on Quaternary Walrus (*Odobenus rosmarus*) from New Brunswick. *Géographie physique et Quaternaire* 51(1): 1–5.

Milligan, B.N. 1993. Carrying of young by female Yuma Bats (*Myotis yumanensis*). *Northwestern Naturalist* 74(2): 55–56.

Milner, J., Jones, C., and Jones, J.K., Jr. 1990. *Nyctinomops macrotis*. *Mammalian Species* no. 351: 1–4.

Minta, S.C. 1993. Sexual differences in spatio-temporal interaction among badgers. *Oecologia* 96: 402–409.

Minta, S.C., Minta, K.A., and Lott, D.F. 1992. Hunting associations between Badgers (*Taxidea taxus*) and Coyotes (*Canis latrans*). *Journal of Mammalogy* 73(4): 814–820.

Mintzer, V.J., Gannon, D.P., Barros, N.B., and Read, A.J. 2008. Stomach contents of mass-stranded Short-finned Pilot Whales (*Globicephala macrorhynchus*) from North Carolina. *Marine Mammal Science* 24(2): 290–302.

Mitchell, C.D., and Maher, C.R. 2006. Horn growth in male Pronghorns *Antilocapra americana*: Selection for precocial maturation in stochastic environments. *Acta Theriologica* 51(4): 405–409.

Mitchell, E.D., Jr. 1991. Winter records of the Minke Whale (*Balaenoptera acutorostrata* Lacépède 1804) in the southern North Atlantic. *Report of the International Whaling Commission* 41: 455–457.

Mitchell, J.A., and Gates, C.C. 2002. Status of the Wood Bison (*Bison bison athabascae*) in Alberta. *Wildlife Status Report* no. 38: 1–32. Alberta Sustainable Resource Development, Fish and Wildlife Division, and Alberta Conservation Association.

Mock, O.B., Sells, G.D., and Easterla, D.A. 2001. The first distributional record of the Least Weasel, *Mustela nivalis*, in northeastern Missouri. *Transactions of the Missouri Academy of Science* 35: 7–12.

Moehrenschlager, A., and Moehrenschlager, C. 2006. Population census of reintroduced Swift Fox (*Vulpes velox*) in Canada and Northern Montana, 2005/2006. Draft report to Alberta Sustainable Resource Development and the Alberta Conservation Association. Edmonton, Alberta.

Møhl, B., Wahlberg, M., Madsen, P.T., Heerford, A., and Lund, A. 2003. The monopulsed nature of Sperm Whale sonar clicks. *Journal of the Acoustical Society of America* 107: 638–648.

Moisan, M., and Huot, M. 1996. Le carcajou, une légende vivante? *Naturaliste Canadien* 120(1): 30–33.

Molina, M., and Molinari, J. 1999. Taxonomy of Venezuelan White-tailed Deer (*Odocoileus*, Cervidae, Mammalia), based on cranial and mandibular traits. *Canadian Journal of Zoology* 77(4): 632–645.

Møller Nielsen, S. 1991. Fishing Arctic Foxes *Alopex lagopus* on a rocky island in West Greenland. *Polar Research* 9(2): 211–113.

Molvar, E.M. 1993. Nursing by a yearling Moose, *Alces alces gigas*, in Alaska. *Canadian Field-Naturalist* 107(2): 233–235.

Molvar, E.M., and Bowyer, R.T. 1994. Costs and benefits of group living in a recently social ungulate: The Alaskan Moose. *Journal of Mammalogy* 75(3): 621–630.

Monello, R.J., Murray, D.L., and Cassirer, E.F. 2001. Ecological correlates of pneumonia epizootics in Bighorn Sheep herds. *Canadian Journal of Zoology* 79(8): 1423–1432.

Monnella, J.A., Blumburg, M.S., McClintock, M.K., and Moltz, H. 2004. Inter-litter competition and communal nursing among Norway Rats: Advantages of birth synchrony. *Behavioral Ecology and Sociobiology* 27(3): 183–190.

Moon, B.R., and Leonard, W.P. 2001. A semiarboreal nest of the American Shrew-mole, *Neurotrichus gibbsii*. *Northwestern Naturalist* 82(1): 26–27.

Moore, J.W., and Kenagy, G.J. 2004. Consumption of shrews, *Sorex* spp., by Arctic Grayling, *Thymallus arcticus*. *Canadian Field-Naturalist* 118(1): 111–114.

Moore, S.E., and Clarke, J.T. 2002. Potential impact of offshore human activities on Gray Whales (*Eschrichtius robustus*). *Journal of Cetacean Research and Management* 4(1): 19–25.

Moore, S.E., and Ridgeway, S.H. 1996. Patterns of dolphin sound production and ovulation. *Aquatic Mammals* 22(3): 175–184.

Moore, S.E., Stafford, K.M., Dahlheim, M.E., Fox, C.G., Braham, H.W., Polovina, J.J., and Bain, D.E. 1998. Seasonal variation in reception of Fin Whale calls at five geographic areas in the North Pacific. *Marine Mammal Science* 14(3): 617–627.

Moore, S.E., Urbán-Ramirez, J., Perryman, W.L., Gulland, F., Perez-Cortes Moreno, H., Wade, P.R., Rojas-Bracho, L., and Rowles, T. 2001. Are Gray Whales hitting "K" hard? *Marine Mammal Science* 17(4): 954–958.

Moosman, P.R., Jr., and Thomas, H.H. 2007. Food habits of Eastern Small-footed Bats (*Myotis leibii*) in New Hampshire. *American Midland Naturalist* 158(2): 354–360.

Morin, P., Berteaux, D., and Klvana, I. 2005. Hierarchical habitat selection by North American Porcupines in southern boreal forest. *Canadian Journal of Zoology* 83(10): 1333–1342.

Mormann, B.M., and Robbins, L.W. 2007. Winter roosting ecology of Eastern Red Bats in southwest Missouri. *Journal of Wildlife Management* 71(1): 213–217.

Morris, R.F. 1948. The land mammals of New Brunswick. *Journal of Mammalogy* 29(2): 165–176.

Morrison, S.F. 2006. Foraging ecology and population dynamics of Collared Pikas in southwestern Yukon. *Arctic* 59(1): 104–107.

Morrison, S.F., Barton, L., Caputa, P., and Hik, D.S. 2004. Forage selection by Collared Pikas, *Ochotona collaris*, under varying degrees of predator risk. *Canadian Journal of Zoology* 82(4): 533–540.

Morrison, S.F., and Hik, D.S. 2007. Demographic analysis of a declining pika *Ochotona collaris* population: Linking survival to broad-scale climate patterns via spring snowmelt patterns. *Journal of Animal Ecology* 76(5): 899–907.

Mortenson, J., and Follis, M. 1997. Northern Elephant Seal (*Mirounga angustirostris*) aggression on Harbor Seal (*Phoca vitulina*) pups. *Marine Mammal Science* 13(3): 526–530.

Morton, A. 2000. Occurrence, photo-identification, and prey of Pacific White-sided Dolphins (*Lagenorhynchus obliquidens*) in the Broughton Archipelago, Canada, 1984–1998. *Marine Mammal Science* 16(1): 80–93.

Morton, M.L., and Pereyra, M.E. 2008. Haying behavior in a rodent, the Bushy-tailed Woodrat (*Neotoma cinerea*). *Northwestern Naturalist* 89(2): 113–115.

Moses, R.A., and Millar, J.S. 1992. Behavioural asymmetries and cohesive mother-offspring sociality in Bushy-tailed Wood Rats. *Canadian Journal of Zoology* 70(3): 597–604.

– 1994. Philopatry and mother-daughter associations in Bushy-tailed Woodrats: Space use and reproductive success. *Behavioral Ecology and Sociobiology* 35: 131–140.

Mowat, G., and Heard, D.C. 2006. Major components of Grizzly Bear diet across North America. *Canadian Journal of Zoology* 84(3): 473–489.

Mowat, G., Heard, D.C., and Gaines, T. 2004. *Predicting Grizzly Bear (Ursus arctos) Densities in British Columbia Using a Multiple Regression Model.* B.C. Ministry of Water, Land, and Air Protection, Victoria, BC.

Mowat, G., Slough, B.G., and Boutin, S. 1996. Lynx recruitment during a Snowshoe Hare population peak and decline in southwest Yukon. *Journal of Wildlife Management* 60(2): 441–452.

Muelbert, M.M.C., and Bowen, W.D. 1993. Duration of lactation and postweaning changes in mass and body composition of Harbour Seal, *Phoca vitulina*, pups. *Canadian Journal of Zoology* 71(7): 1405–1414.

Mullican, T.R. 1994. First record of Merriam's Shrew from South Dakota. *Prairie Naturalist* 26(2): 173.

Mullican, T.R., and Keller, B.L. 1986. Ecology of the Sagebrush Vole (*Lemmiscus curtatus*) in southeastern Idaho. *Canadian Journal of Zoology* 64(6): 1218–1223.

– 1987. Burrows of the Sagebrush Vole (*Lemmiscus curtatus*) in southeastern Idaho. *Great Basin Naturalist* 47(2): 276–279.

Munro, R.H.M., Nielsen, S.E., Price, M.H., Stenhouse, G.B., and Boyce, M.S. 2006. Seasonal and diel patterns of Grizzly Bear diet and activity in west-central Alberta. *Journal of Mammalogy* 87(6): 1112–1121.

Muntz, E.M., and Patterson, B.R. 2004. Evidence for the use of vocalization to coordinate the killing of a White-tailed Deer, *Odocoileus virginianus*, by Coyotes, *Canis latrans*. *Canadian Field-Naturalist* 118(2): 278–280.

Murie, D.J., and Lavigne, D.M. 1992. Growth and feeding habits of Grey Seals (*Halichoerus grypus*) in the northwestern Gulf of St Lawrence, Canada. *Canadian Journal of Zoology* 70(8): 1604–1613.

Murie, J.O. 1973. Population characteristics and phenology of a Franklin Ground Squirrel (*Spermophilus franklinii*) colony in Alberta, Canada. *American Midland Naturalist* 90(2): 334–340.

– 1992. Predation by badgers on Columbian Ground Squirrels. *Journal of Mammalogy* 73(2): 385–394.

Murie, J.O., and Harris, M.A. 1988. Social interactions and dominance relationships between female and male Columbian Ground Squirrels. *Canadian Journal of Zoology* 66(6): 1414–1420.

Murie, J.O., and McLean, I.G. 1980. Copulatory plugs in ground squirrels. *Journal of Mammalogy* 61(2): 355–356.

Murie, O.J., and Elbroch, M. 2005. *A Field Guide to Animal Tracks.* Peterson Field Guides, 3rd ed. Houghton Mifflin, New York, NY.

Murphy, M.A. 1995. Occurrence and group characteristics of Minke Whales, *Balaenoptera acutorostrata*, in Massachusetts Bay and Cape Cod Bay. *US National Marine Fisheries Service Fisheries Bulletin* 93(3): 577–585.

Murphy, S., Herman, J.S., Pierce, G.J., Rogan, E., and Kitchener, A.C. 2006. Taxonomic status and geographical cranial variation of Common Dolphins (*Delphinus*) in the eastern North Atlantic. *Marine Mammal Science* 22(3): 573–599.

Murray, D.L., Boutin, S., O'Donoghue, M., and Nams, V.O. 1995. Hunting behaviour of a sympatric felid and canid in relation to vegetative cover. *Animal Behaviour* 50: 1203–1210.

Murray, D.L., and Larivière, S. 2002. The relationship between foot size of wild canids and regional snow conditions: Evidence for selection against a high footload? *Journal of Zoology* (London) 256: 289–299.

Murray, K.L., Britzke, E.R., and Robbins, L.W. 2001. Variation in search-phase calls of bats. *Journal of Mammalogy* 82(3): 728–737.

Musiani, M., Leonard, J.A., Cluff, H.D., Gates, C.C., Mariani, S., Paquet, P.C., Vilàs, C., and Wayne, R.K. 2007. Differentiation of tundra/taiga and boreal coniferous forest wolves: Genetics, coat colour, and association with migratory Caribou. *Molecular Ecology* 16(19): 4149–4170.

Musiani, M., and Paquet, P.C. 2004. The practices of wolf persecution, protection, and restoration in Canada and the Unites States. *BioScience* 54(1): 50–60.

Myers, P., and Lundrigan, B. 2001. Great Lakes *Peromyscus*: Is recent climate change affecting their distribution? *Michigan Academician* 33: 127–135.

Nagorsen, D.W. 1987a. *Marmota vancouverensis. Mammalian Species* no.270: 1–5.

– 1987b. Summer and winter food caches of the Heather Vole, *Phenacomys intermedius*, in Quetico Provincial Park, Ontario. *Canadian Field-Naturalist* 101(1): 82–85.

– 1996. *Opossums, Shrews, and Moles of British Columbia.* Vol. 2 of *The Mammals of British Columbia.* Royal BC Museum Handbook. Royal BC Museum, Victoria, BC.

– 1998. *Dicrostonyx nunatakensis* (Youngman 1967), Ogilvie Mountain Collared Lemming. In D.J. Hafner, E. Yensen, and G.L. Kirkland, Jr. (comps and eds), *North American Rodents: Status Survey and Conservation Action Plan*, 88. IUCN/SSC Rodent Specialist Group, IUCN, Gland, Switzerland, and Cambridge, UK.

– 2002. *An Identification Manual to the Small Mammals of British Columbia.* Published by the British Columbia Ministry of Sustainable Resource Management; Ministry of Water, Land, and Air Protection, Biodiversity Branch; and the Royal British Columbia Museum.

– 2004. *COSEWIC Assessment and Update Status Report on the Spotted Bat* Euderma maculatum *in Canada.* Committee on the Status of Endangered Wildlife in Canada, Ottawa.

– 2005a. *Rodents and Lagomorphs of British Columbia.* Vol. 4 of *The Mammals of British Columbia.* Royal BC Museum Handbook. Royal BC Museum, Victoria, BC.

– 2005b. *Update: COSEWIC Status Report on Pacific Water Shrew, Sorex bendirii.* Prepared for the Committee on the Status of Endangered Wildlife in Canada.

Nagorsen, D.W., and Brigham, R.M. 1993. *Bats of British Columbia.* Vol. 1 of *The Mammals of British Columbia.* Royal BC Museum Handbook. Royal BC Museum, Victoria, BC.

Nagorsen, D.W., Campbell, R.W., and Giannico, G.R. 1991. Winter food habits of Marten, *Martes americana*, on the Queen Charlotte Islands. *Canadian Field-Naturalist* 105(1): 55–59.

Nagorsen, D.W., and Panter, N. 2009. Identification and status of the Olympic Shrew (*Sorex rohweri*) in British Columbia. *Northwestern Naturalist* 90(2): 117–129.

Nagorsen, D.W., and Peterson, R.L. 1981. Distribution, abundance, and species diversity of small mammals in Quetico Provincial Park, Ontario. *Naturalist Canadien* 108(3): 209–218.

Nagorsen, D.W., Scudder, G.G.E., Huggard, D.J., Stewart, H., and Panter, N. 2001. Merriam's Shrew, *Sorex merriami*, and Preble's Shrew, *Sorex preblei*: Two new mammals for Canada. *Canadian Field-Naturalist* 115(1): 1–8.

Nagorsen, D.W., and Stewart, G.E. 1983. A Dwarf Sperm Whale (*Kogia simus*) from the Pacific coast of Canada. *Journal of Mammalogy* 64(3): 505–506.

Nagy, T.R., and Gower, B.A. 1999. Northern Collared Lemming (*Dicrostonyx groenlandicus*). In D.E. Wilson and S. Ruff (eds), *The Smithsonian Book of North American Mammals*, 659–660. UBC Press, Vancouver, BC.

Nawojchik, R., St Aubin, D.J., and Johnson, A. 2003. Movements and dive behavior of two stranded, rehabilitated Long-finned Pilot Whales (*Globicephala melas*) in the northwest Atlantic. *Marine Mammal Science* 19(1): 232–239.

Neave, D.J., and Wright, B.S. 1968. Seasonal migrations of the Harbor Porpoise (*Phocoena phocoena*) and other Cetacea in the Bay of Fundy. *Journal of Mammalogy* 49(2): 259–264.

Negus, N.C., and Berger, P.J. 1998. Reproductive strategies of *Dicrostonyx groenlandicus* and *Lemmus sibericus* in high-arctic tundra. *Canadian Journal of Zoology* 76(3): 390–399.

Negus, N.C., Berger, P.J., and Brown, B.W. 1986. Microtine population dynamics in a predictable environment. *Canadian Journal of Zoology* 64(3): 785–792.

Negus, N.C., Berger, P.J., and Forslund, L.G. 1977. Reproductive strategy of *Microtus montanus. Journal of Mammalogy* 58(2): 347–353.

Neimanis, A.S., Read, A.J., Foster, R.A., and Gaskin, D.E. 2000. Seasonal regression in testicular size and histology in Harbour Porpoises (*Phocoena phocoena*) from the Bay of Fundy and the Gulf of Maine. *Journal of Zoology* (London) 250: 221–229.

Nelson, D., and Lien, J. 1996. The status of the Long-finned Pilot Whale, *Globicephala melas*, in Canada. *Canadian Field-Naturalist* 110(3): 511–524.

Nelson, M.E., and Mech, L.D. 1990. Weights, productivity, and mortality of old White-tailed Deer. *Journal of Mammalogy* 71(4): 689–691.

– 1992. Dispersal in female White-tailed Deer. *Journal of Mammalogy* 73(4): 891–894.

– 1993. A single deer stands off three wolves. *American Midland Naturalist* 131: 207–208.

Nero, R.W. 1993. Northern Flying Squirrel and Red Bat caught on barbed-wire. *Blue Jay* 51(4): 215–215.

Nero, R.W., and Wrigley, R.E. 1977. Rare, endangered, and extinct wildlife in Manitoba. *Manitoba Nature*, summer, 4–37.

Neuhaus, P. 2000a. Weight comparisons and litter size manipulation in Columbian Ground Squirrels (*Spermophilus columbianus*) show evidence of costs of reproduction. *Behavioral Ecology and Sociobiology* 48: 75–83.

– 2000b. Timing of hibernation and molt in female Columbian Ground Squirrels. *Journal of Mammalogy* 81(2): 571–577.

Neuhaus, P., and Pelletier, N. 2001. Mortality in relation to season, age, sex, and reproduction in Columbian Ground Squirrels (*Spermophilus columbianus*). *Canadian Journal of Zoology* 79(3): 465–470.

Neumann, D.R. 1999. Agonistic behavior in Harbor Seals (*Phoca vitulina*) in relation to the availability of haul-out space. *Marine Mammal Science* 15(2): 507–525.

– 2001. The activity budget of free-ranging Common Dolphins (*Delphinus delphis*) in the northwestern Bay of Plenty, New Zealand. *Aquatic Mammals* 27(2): 121–136.

Neumann, D.R., and Orams, M.B. 2003. Feeding behaviours of Short-beaked Common Dolphins, *Delphinus delphis*, in New Zealand. *Aquatic Mammals* 29(1): 137–149.

Neumann, D.R., Russell, K., Orams, M.B., Baker, C.S., and Duignan, P. 2002. Identifying sexually mature, male Short-beaked Common Dolphins (*Delphinus delphis*) at sea, based on the presence of a postanal hump. *Aquatic Mammals* 28(2): 181–187.

Nevin, O. 2007. Bear facts. *Science* 317(5844): 1477.

Newhouse, N.J., and Kinley, T.A. 2000. *Update: COSEWIC Status Report on American Badger* (Taxidea taxus). Committee on the Status of Endangered Wildlife in Canada, Ottawa, ON.

Nichol, L.M., and Watson, J.C. 2007. *Update: COSEWIC Status Report on Sea Otter* Enhydra lutris. Committee on the Status of Endangered Wildlife in Canada.

Nichols, L., and Bunnell, F.L. 1999. Natural history of Thinhorn Sheep. In R. Valdez and P.R. Krausan (eds), *Mountain Sheep of North America*, 23–77. University of Arizona Press, Tucson, AZ.

Nilssen, K.T., Haug, T., and Lindblom, C. 2001. Diet of weaned pups and seasonal variations in body condition of juvenile Barents Sea Harp Seals *Phoca groenlandica*. *Marine Mammal Science* 17(4): 926–936.

Nixon, C.M., Brewer, P.A., and Hansen, L.P. 1990. White-tailed doe tolerates nursing by non-offspring. *Transactions of the Illinois State Academy of Science* 83(1/2): 104–106.

Nixon, C.M., Hansen, L.P., Brewer, P.A., and Chelsvig, J.E. 1991. Longevity of female White-tailed Deer on a refuge in Illinois. *Transactions of the Illinois State Academy of Science* 84(1/2): 84–91.

Nixon, C.M., Sullivan, J.B., and Koerkenmeier, R. 1994. Notes on the life history of Opossums in west-central Illinois. *Transactions of the Illinois State Academy of Science* 87(3/4): 187–193.

Noad, M.J., Cato, D.H., Bryden, M.M., Jenner, M.-N., and Jenner, K.C.S. 2000. Cultural revolution in whale songs. *Nature* 408(6812): 537.

Nolte, D.L., Epple, G., Campbell, D.L., and Mason, J.R. 1993. Response of Mountain Beaver (*Aplodontia rufa*) to conspecifics in their burrow system. *Northwest Science* 67(4): 251–255.

Nordøy, E.S. 1995. Do Minke Whales (*Balaenoptera acutorostrata*) digest wax esters? *British Journal of Nutrition* 74: 717–722.

Noren, S.R., Lacave, G., Wells, R.S., and Williams, T.M. 2002. The development of blood oxygen stores in Bottlenose Dolphins (*Tursiops truncatus*): Implications for diving capacity. *Journal of Zoology* (London) 258: 105–113.

Norment, C.J., Hall, A., and Hendricks, P. 1999. Important bird and mammal records in the Thelon River Valley, Northwest Territories: Range expansions and possible causes. *Canadian Field-Naturalist* 113(3): 375–385.

Norris, R.W., and Kilpatrick, C.W. 2007. A high elevation record of the Star-nosed Mole (*Condylura cristata*) in northeastern Vermont. *Canadian Field-Naturalist* 121(2): 206–207.

Northridge, S., Tasker, M., Webb, A., Camphuysen, K., and Leopold, M. 1997. White-beaked *Lagenorhynchus albirostris* and Atlantic White-sided Dolphin *L. acutus* distributions in northwest European and US North Atlantic waters. *Report of the International Whaling Commission* 47: 797–805.

Notarbartolo-di-Sciara, G., Barbaccia, G., and Azzellino, A. 1997. Birth at sea of a False Killer Whale, *Pseudorca crassidens*. *Marine Mammal Science* 13(3): 508–511.

Nøttestad, L., and Similä, T. 2001. Killer Whales attacking schooling fish: Why force herring from deep water to the surface? *Marine Mammal Science* 17(2): 343–352.

Novikov, E., and Moshkin, M. 1998. Sexual maturation, adrenocortical function, and population density of red-backed vole, *Clethrionomys rutilus* (Pall.). *Mammalia* 62(4): 529–540.

Nowacek, D.P., Johnson, M.P., Tyack, P.L., Shorter, K.A., McLellan, W.A., and Pabst, D.A. 2001. Buoyant balaenids: The ups and downs of buoyancy in right whales. *Proceedings of the Royal Society of London, B Series* 268: 1811–1816.

Nowak, R.M. 2002. The original status of wolves in eastern North America. *Southeastern Naturalist* 1(2): 95–130.

Nowicki, S.N., Stirling, I., and Sjare, B. 1997. Duration of stereotyped underwater vocal displays by male Atlantic Walruses in relation to aerobic dive limit. *Marine Mammal Science* 13(4): 566–575.

O'Corry-Crowe, G.M., and Westlake, R.L. 1997. Molecular investigations of Spotted Seals (*Phoca largha*) and Harbor Seals (*Phoca vitulina*), and their relationship in areas of sympatry. *Molecular Genetics of Marine Mammals Special Publication* 3: 291–304.

O'Daniel, D., and Schneeweis, J.C. 1992. Steller Sea Lion, *Eumetopias jubatus*, predation on Glaucous-winged Gulls, *Larus glaucescens*. *Canadian Field-Naturalist* 106(2): 268.

Odell, D.K. 1981. California Sea Lion *Zalophus californianus* (Lesson 1828). In S.H. Ridgway and R. Harrison (eds), *Handbook of Marine Mammals*, vol. 1, *The Walrus, Sea Lions, Fur Seals, and Sea Otter*, 67–97. Academic Press, San Diego, CA.

Odell, D.K., and McClune, K.M. 1999. False Killer Whale *Pseudorca crassidens* (Owen 1846). In S.H. Ridgway and R. Harrison (eds), *Handbook of Marine Mammals*, vol. 6, 213–243. Academic Press, San Diego, CA.

O'Donnell, R.P., Urling, E.P., Sato, C.L., and Hayes, M.P. 2005. First record of a Masked Shrew (*Sorex cinereus*) in the Willapa Hills, Washington. *Northwestern Naturalist* 86(3): 154–156.

O'Donoghue, M., and Bergman, C.M. 1992. Early movements and dispersal of juvenile Snowshoe Hares. *Canadian Journal of Zoology* 70(9): 1787–1791.

O'Donoghue, M., and Boutin, S. 1995. Does reproductive synchrony affect juvenile survival rates of northern mammals? *Oikos* 74: 115–121.

Oehler, M.W., Sr, Jenks, J.A., and Bowyer, R.T. 1995. Antler rubs by White-tailed Deer: The importance of trees in a prairie environment. *Canadian Journal of Zoology* 73(7): 1383–1386.

O'Farrell, M.J. 1999. Fringed Myotis (*Myotis thysanodes*). In D.E. Wilson and S. Ruff (eds), *The Smithsonian Book of North American Mammals*, 98–100. UBC Press, Vancouver, BC.

O'Farrell, M.J., and Studier, E.H. 1973. Reproduction, growth, and development in *Myotis thysanodes* and *M. lucifugus* (Chiroptera: Vespertilionidae). *Ecology* 54(1): 18–30.

– 1975. Population structure and emergence activity patterns in *Myotis thysanodes* and *M. lucifugus* (Chiroptera: Vespertilionidae) in northeastern New Mexico. *American Midland Naturalist* 93(2): 368–376.

Oftedal, O.T., Bowen, W.D., and Boness, D.J. 1993. Energy transfer by lactating Hooded Seals and nutrient deposition in their pups during the four days from birth to weaning. *Physiological Zoology* 66(3): 412–436.

O'Gara, B.W. 1978. *Antilocapra americana*. *Mammalian Species* no. 90: 1–7.

– 1999. Pronghorn (*Antilocapra americana*). In D.E. Wilson and S. Ruff (eds), *The Smithsonian Book of North American Mammals*, 339–341. UBC Press, Vancouver, BC.

O'Hara, T.M., George, J.C., Tarpley, R.J., Burek, K., and Suydam, R.S. 2002. Sexual maturation in male Bowhead Whales (*Balaena mysticetus*) of the Bering-Chukchi-Beaufort Seas stock. *Journal of Cetacean Research and Management* 4(2): 143–148.

Ohizumi, H., Isoda, T., Kishiro, T., and Kato, H. 2003. Feeding habits of Baird's Beaked Whale *Berardius bairdii*, in the western North Pacific and Sea of Okhotsk off Japan. *Fisheries Science* (Tokyo) 69(1): 11–20.

Ohizumi, H., and Kishiro, T. 2003. Stomach contents of a Cuvier's Beaked Whale (*Ziphius cavirostris*) stranded on the central Pacific coast of Japan. *Aquatic Mammals* 29(1): 99–103.

Olcott, S.P., and Barry, R.E. 2000. Environmental correlates of geographic variation in body size of the Eastern Cottontail (*Sylvilagus floridanus*). *Journal of Mammalogy* 81(4): 986–998.

Olesiuk, P.F., Bigg, M.A., and Ellis, G.M. 1990. Life history and population dynamics of resident Killer Whales (*Orcinus orca*) in the coastal waters of British Columbia and Washington State. *Report of the International Whaling Commission, Special Issue* 12: 209–243.

Olesiuk, P.F., Nichol, L.M., Sowden, M.J., and Ford, J.K.B. 2002. Effect of the sound generated by an acoustic harassment device on the relative abundance and distribution of Harbor Porpoises (*Phocoena phocoena*) in Retreat Passage, British Columbia. *Marine Mammal Science* 18(4): 843–862.

Olesiuk, P.F., and Trites, A.W. 2003. *COSEWIC Assessment and Update Status Report on the Steller Sea Lion* (Eumetopas jubatus) *in Canada*. Committee on the Status of Endangered Wildlife in Canada, Ottawa.

Oli, M.K., and Armitage, K.B. 2003. Sociality and individual fitness in Yellow-bellied Marmots: Insights from a long-term study (1962–2001). *Oecologia* 136: 543–550.

Olsen, R.W. 1969. Agonistic behavior of the Short-tailed Shrew (*Blarina brevicauda*). *Journal of Mammalogy* 50(3): 494–500.

Olson, T.L., and Lindzey, F.G. 2002. Swift Fox (*Vulpes velox*) home-range dispersion patterns in southeastern Wyoming. *Canadian Journal of Zoology* 80(11): 2024–2029.

Ono, K.A., and Boness, D.J. 1996. Sexual dimorphism in sea lion pups: Differential maternal investment, or sex-specific differences in energy allocation? *Behavioral Ecology and Sociobiology* 38: 31–41.

Ormsbee, P.C. 1996. Characteristics, use, and distribution of day roosts selected by female *Myotis volans* (Long-legged Myotis) in forested habitat of the central Oregon Cascades. In Robert M. R. Barclay and R. Mark Brigham (eds), *Bats and Forests Symposium, October 19–21, 1995, Victoria, British Columbia, Canada*, 124–131. Working Paper 23/1996. Research Branch, BC Ministry of Forests, Victoria, BC.

Ormsbee, P.C., and McComb, W.C. 1998. Selection of day roosts by female Long-legged Myotis in the central Oregon Cascade Range. *Journal of Wildlife Management* 62(2): 596–603.

Orr, R.T. 1954. Natural history of the Pallid Bat, *Antrozous pallidus* (LeConte). *Proceedings of the California Academy of Sciences* 28(4): 165–256.

Orrock, J.L., Farley, D., and Pagels, J.F. 2003. Does fungus consumption by the Woodland Jumping Mouse vary with habitat type or the abundance of other small mammals? *Canadian Journal of Zoology* 81(4): 753–756.

Ortiz, R.M. 2001. Osmoregulation in marine mammals. *Journal of Experimental Biology* 204: 1831–1844.

O'Shea, T.J., and Vaughan, T.A. 1977. Nocturnal and seasonal activities of the Pallid Bat, *Antrozous pallidus*. *Journal of Mammalogy* 58(3): 269–284.

Ostroff, A.C., and Finck, E.J. 2003. *Spermophilus franklinii*. *Mammalian Species* no. 724: 1–5.

Ostrom, P.H., Lien, J., and Macko, S.A. 1993. Evaluation of the diet of Sowerby's Beaked Whale, *Mesoplodon bidens*, based on isotopic comparisons among northwestern Atlantic cetaceans. *Canadian Journal of Zoology* 71(4): 858–861.

Otani, S., Naito, Y., Kato, A., and Kawamura, A. 2000. Diving behavior and swimming speed of a free-ranging Harbor Porpoise, *Phocoena phocoena*. *Marine Mammal Science* 16(4): 811–814.

Ottensmeyer, C.A., and Whitehead, H. 2003. Behavioural evidence for social units in Long-finned Pilot Whales. *Canadian Journal of Zoology* 81(8): 1327–1338.

Ovaska, K., and Herman, T.B. 1986. Fungal consumption by six species of small mammals in Nova Scotia. *Journal of Mammalogy* 67(1): 208–211.

– 1988. Life history characteristics and movements of the Woodland Jumping Mouse, *Napaeozapus insignis*, in Nova Scotia. *Canadian Journal of Zoology* 66(8): 1752–1762.

Owen, J.G. 1984. *Sorex fumeus*. *Mammalian Species* no. 215: 1–8.

Owen, S.F., Menzel, M.A., Ford, W.M., Chapman, B.R., Miller, K.V., Edwards, J.W., and Wood, P.B. 2003. Home-range size and habitat used by the Northern Myotis (*Myotis septentrionalis*). *American Midland Naturalist* 150(2): 352–359.

Ozoga, J.J. 1996. *Whitetail Spring*. Willow Creek Press, Minocqua, WI.

Pace, D.S. 2000. Fluke-made bubble rings as toys in Bottlenose Dolphin calves (*Tursiops truncatus*). *Aquatic Mammals* 26(1): 57–64.

Pack, A.A., Herman, L.M., Craig, A.S., Spitz, S.S., and Deakos, M.H. 2002. Penis extrusions by Humpback Whales (*Megaptera novaeangliae*). *Aquatic Mammals* 28(2): 131–146.

Page, L.K., Swihart, R.K., and Kazacos, K.R. 1998. Raccoon latrine structure and its potential role in transmission of *Baylisascaris procyonis* in vertebrates. *American Midland Naturalist* 140(1): 180–185.

Palacios, D.M., and Mate, B.R. 1996. Attack by False Killer Whales (*Pseudorca crassidens*) on Sperm Whales (*Physeter macrocephalus*) in the Galápagos Islands. *Marine Mammal Science* 12(4): 582–587.

Palka, D.L., Read, A.J., and Potter, C.W. 1997. Summary of knowledge of White-sided Dolphins (*Lagenorhynchus acutus*) from US and Canadian North Atlantic waters. *Report of the International Whaling Commission* 47: 729–734.

Palka, D.L., Read, A.J., Westgate, A.J., and Johnston, D.W. 1996. Summary of current knowledge of Harbour Porpoises in US and Canadian Atlantic waters. *Report of the International Whaling Commission* 46: 559–565.

Palo, J.U. 2003. Genetic diversity and phylogeography of landlocked seals. PhD thesis, University of Helsinki, Finland.

Palo, J.U., and Väinölä, R. 2006. The enigma of the landlocked Baikal and Caspian Seals addressed through phylogeny of phocine mitochondrial sequences. *Biological Journal of the Linnaean Society* 88(1): 61–72.

Palsbøll, P.J., Heide-Jørgensen, M.P., and Dietz, R. 1997. Population structure and seasonal movements of Narwhals, *Monodon monoceros*, determined from mtDNA analysis. *Heredity* 78(1997): 284–292.

Pamperin, N.J., Follmann, E.H., and Petersen, B. 2006. Interspecific killing of an Arctic Fox by a Red Fox at Prudhoe Bay, Alaska. *Arctic* 59(4): 361–364.

Panigada, S., Zanardelli, M., Canese, S., and Jahoda, M. 1999. How deep can baleen whales dive? *Marine Ecology Progress Series* 187: 300–311.

Paradiso, J.L., and Greenhall, A.M. 1967. Longevity records for American bats. *American Midland Naturalist* 78: 251–252.

Paragi, T.F., Johnson, W.N., and Katnik, D.D. 1997. Selection of post-fire seres by Lynx and Snowshoe Hares in the Alaskan taiga. *Northwestern Naturalist* 78: 77–86.

Pardy, C.K., Wohl, G.R., Ukrainetz, P.J., Sawers, A., Boyd, S.K., and Zernicke, R.F. 2004. Maintenance of bone mass and architecture in denning Black Bears (*Ursus americanus*). *Journal of Zoology* (London) 263: 359–364.

Parish, D.A., and Jones, C. 1999. Big free-tailed bat (*Nyctinomops macrotis*). In D.E. Wilson and S. Ruff (eds), *The Smithsonian Book of North American Mammals*, 130–131. UBC Press, Vancouver, BC.

Parker, D.I., and Cook, J.A. 1996. Keen's Long-eared Bat, *Myotis keenii*, confirmed in southeast Alaska. *Canadian Field-Naturalist* 110(4): 611–614.

Parker, G.R. 1977. Morphology, reproduction, diet, and behavior of the Arctic Hare (*Lepus arcticus monstrabilis*) on Axel Heiberg Island, Northwest Territories. *Canadian Field-Naturalist* 91(1): 8–18.

Parks, E.K., Derocher, A.E., and Lunn, N.J. 2006. Seasonal and annual movement patterns of Polar Bears on the sea ice of Hudson Bay. *Canadian Journal of Zoology* 84(9): 1281–1294.

Parks, S.E., Brown, M.W., Conger, L.A., Hamilton, P.K., Knowlton, A.R., Kraus, S.D., Slay, C.K., and Tyack, P.L. 2007. Occurrence, composition, and potential functions of North Atlantic Right Whale (*Eubalaena glacialis*) surface active groups. *Marine Mammal Science* 23(4): 868–887.

Parks, S.E., Hamilton, P.K., Kraus, S.D., and Tyack, P.L. 2005. The gunshot sound produced by male North Atlantic Right Whales (*Eubalaena glacialis*) and its potential function in reproductive advertisement. *Marine Mammal Science* 21(3): 458–475.

Parsons, H.J., Smith, D.A., and Whittam, R.F. 1986. Maternity colonies of Silver-haired Bats, *Lasionycteris noctivagans*, in Ontario and Saskatchewan. *Journal of Mammalogy* 67(3): 598–600.

Pasitschniak-Arts, M. 1993. *Ursus arctos*. *Mammalian Species* no. 439: 1–10.

Pasitschniak-Arts, M., and Larivière, S. 1995. *Gulo gulo*. *Mammalian Species* no. 499: 1–10.

Pasitschniak-Arts, M., and Messier, F. 2000. Brown (Grizzly) and Polar Bears. In S. Demarais and P.R. Krausman (eds), *Ecology and Management of Large Mammals in North America*, 409–428. Prentice Hall, Upper Saddle River, NJ.

Patrician, M.R., Biedron, I.S., Esch, H.C., Wenzel, F.W., Cooper, L.A., Hamilton, P.K., Glass, A.H., and Baumgartner, M.F. 2009. Evidence of a North Atlantic Right Whale calf (*Eubalaena glacialis*) born in northeastern U.S. waters. *Marine Mammal Science* 25(2): 462–477.

Patterson, B.R. 1994. Surplus killing of White-tailed Deer, *Odocoileus virginianus*, by Coyotes, *Canis latrans*, in Nova Scotia. *Canadian Field-Naturalist* 108(4): 484–487.

Patterson, B.R., Bondrup-Nielsen, S., and Messier, F. 1999. Activity patterns and daily movements of the eastern Coyote, *Canis latrans*, in Nova Scotia. *Canadian Field-Naturalist* 113(2): 251–257.

Patterson, B.R., and Messier, F. 2001. Social organization and space use of Coyotes in eastern Canada relative to prey distribution and abundance. *Journal of Mammalogy* 82(2): 463–477.

Patterson, I.A.P., Reid, R.J., Wilson, B., Grellier, K., Ross, H.M., and Thompson, P.M. 1998. Evidence for infanticide in Bottlenose Dolphins: An explanation for violent interactions with Harbour Porpoises? *Proceedings of the Royal Society of London* 265: 1167–1170.

Patterson, J.E.H., Patterson, S.J., and Malcolm, J.R. 2007. Cavity nest materials of Northern Flying Squirrels, *Glaucomys sabrinus*, and North American Red Squirrels, *Tamiasciurus hudsonicus*, in a secondary hardwood forest of southern Ontario. *Canadian Field-Naturalist* 121(3): 303–307.

Pattie, D. 1973. *Sorex bendirii. Mammalian Species* no. 27: 1–2.

– 1999. Marsh Shrew (*Sorex bendirii*). In D.E. Wilson and S. Ruff (eds), *The Smithsonian Book of North American Mammals*, 19–20. UBC Press, Vancouver, BC.

Pattie, D.L., and Webber, M. 1992. First record of the Minke Whale, *Balaenoptera acutorostrata*, in Manitoba waters. *Canadian Field-Naturalist* 106(2): 266–267.

Pauli, J.N. 2005. Evidence for long-distance swimming capabilities in Red Squirrels, *Tamiasciurus hudsonicus. Northeastern Naturalist* 12(2): 245–248.

Payne, P.M., and Heinemann, D.W. 1993. The distribution of pilot whales (*Globicephala* spp.) in shelf / shelf-edge and slope waters of the northeastern United States, 1978–1988. *Reports of the International Whaling Commission, Special Issue* no. 14: 51–68.

Pearce, J.L., and Kirk, D.A. 2010. COSEWIC assessment and status report on the Eastern Mole *Scalopus aquaticus* in Canada. Committee on the Status of Endangered Wildlife in Canada, Ottawa.

Pearson, O.P., Koford, M.R., and Pearson, A.K. 1952. Reproduction of the Lumpnosed Bat (*Corynorhinus rafinesquei*) in California. *Journal of Mammalogy* 33(3): 273–320.

Peck, V.R., and Peek, J.M. 1991. Elk, *Cervus elaphus*, habitat use related to prescribed fire, Tuchodi River, British Columbia. *Canadian Field-Naturalist* 105(3): 354–362.

Peek, J.M. 2003. Wapiti (*Cervus elaphus*). In G.A. Feldhamer, B.C. Thompson, and J.A. Chapman (eds), *Wild Mammals of North America: Biology, Management, and Conservation*, 2nd ed., 877–888. Johns Hopkins University Press, Baltimore, MD.

Peirce, K.N., and Peirce, J.M. 2000a. Range extensions for the Alaska Tiny Shrew and Pygmy Shrew in southwestern Alaska. *Northwestern Naturalist* 81(2): 67–68.

– 2000b. A range extension for the Meadow Jumping Mouse, *Zapus hudsonius*, in southwestern Alaska. *Canadian Field-Naturalist* 114(2): 311.

– 2005. Occurrence and distribution of small mammals on the Goodnews River, southwestern Alaska. *Northwestern Naturalist* 86(1): 20–24.

Pelletier, F., and Festa-Bianchet, M. 2006. Sexual selection and social rank in Bighorn rams. *Animal Behaviour* 71: 649–655.

Pelletier, F., Gendreau, Y., and Feder, C. 2006. Behavioural reactions of Bighorn Sheep (*Ovis canadensis*) to Cougar (*Puma concolor*) attacks. *Mammalia* 70(1–2): 160–162.

Pellis, S.M., MacDonald, N.L., and Michener, G.R. 1996. Lateral display as a combat tactic in Richardson's Ground Squirrels *Spermophilus richardsonii. Aggressive Behavior* 22: 119–134.

Pellis, S.M., and Pellis, V.C. 1992. Analysis of the targets and tactics of conspecific attack and predatory attack in Northern Grasshopper Mice *Onychomys leucogaster. Aggressive Behavior* 18: 301–316.

Pelton, M.R. 2000. Black Bear. In S. Demarais and P.R. Krausman (eds), *Ecology and Management of Large Mammals in North America*, 389–408. Prentice Hall, Upper Saddle River, NJ.

2003. Black Bear (*Ursus americanus*). In G.A. Feldhamer, B.C. Thompson, and J.A. Chapman (eds), *Wild Mammals of North America: Biology, Management, and Conservation*, 2nd ed., 547–555. Johns Hopkins University Press, Baltimore, MD.

Pelton, M.R., and van Manen, F. 1994. Distribution of Black Bears in North America. *Proceedings of the Eastern Workshop on Black Bear Management and Research* 12: 133–138.

Pemberton, D., Merdsoy, B., Gales, R., and Renouf, D. 1994. The interaction between offshore cod trawlers and Harp *Phoca groenlandica* and Hooded *Cystophora cristata* Seals off Newfoundland, Canada. *Biological Conservation* 68: 123–127.

Pendleton, G.W., Pitcher, K.W., Fritz, L.W., York, A.E., Raum-Suryan, K.L., Loughlin, T.R., Calkins, D.G., Hastings, K.K., and Gelatt, T.S. 2006. Survival of Steller Sea Lions in Alaska: A comparison of increasing and decreasing populations. *Canadian Journal of Zoology* 84(8): 1163–1172.

Peralta-García, A., Samaniego-Herrera, A., and Valdez Villavicencio, J.H. 2007. Western Harvest Mouse, *Reithrodontomys megalotis* (Rodentia: Muridae), on Magdalena Island, Mexico. *Southwestern Naturalist* 52(4): 595–597.

Pereira, M.E., Aines, J., and Scheckter, J.L. 2002. Tactics of heterothermy in Eastern Gray Squirrels (*Sciurus carolinensis*). *Journal of Mammalogy* 83(2): 467–477.

Pergams, O.R.W., and Nyberg, D. 2001. Museum collections of mammals corroborate the exceptional decline of prairie habitat in the Chicago region. *Journal of Mammalogy* 82(4): 984–992.

Perrigo, G. 1990. Food, sex, time, and effort in a small mammal: Energy allocation strategies for survival and reproduction. *Behaviour* 114(1–4): 191–205.

Perrin, W.F., and Brownell, R.L., Jr. 2002. Minke Whales: *Balaenoptera acutorostrata* and *B. bonaerensis*. In W.F. Perrin, B. Wursig, and H.G.T. Thewissen (eds), *Encyclopedia of Marine Mammals*, 750–754. Academic Press, San Diego, CA.

Perrin, W.F., Wilson, C.E., and Archer, F.I, II. 1994. Striped Dolphin *Stenella coeruleoalba* (Meyen 1833). In S.H. Ridgeway and R. Harrison (eds), *Handbook of Marine Mammals*, vol. 5, 129–159. Academic Press, San Diego, CA.

Perry, E.A., Boness, D.J., and Fleischer, R.C. 1998. DNA fingerprinting evidence of nonfilial nursing in Grey Seals. *Molecular Biology* 7: 81–85.

Perry, E.A., and Stenson, G.B. 1992. Observations on nursing behaviour of Hooded Seals, *Cystophora cristata. Behaviour* 122(1–2): 1–10.

Perry, N.D., Stewart, D.T., Madden, E.M., and Maier, T.J. 2004. New records for the Arctic Shrew, *Sorex arcticus*, and the newly recognized Maritime Shrew, *Sorex maritimensis. Canadian Field-Naturalist* 118(3): 400–404.

Perry, R.W., and Thill, R.E. 2007. Tree roosting by male and female Eastern Pipistrelles in a forested landscape. *Journal of Mammalogy* 88(4): 974–981.

– 2008. Diurnal roots of male Evening Bats (*Nycticeius humeralis*) in diversely managed pine-hardwood forests. *American Midland Naturalist* 160(2): 374–385.

Perryman, W.L., Donahue, M.A., Laake, J.L., and Martin, T.E. 1999. Diel variation in migrating rates of Eastern Pacific Gray Whales measured with thermal imaging sensors. *Marine Mammal Science* 15(2): 426–445.

Perryman, W.L., Donahue, M.A., Perkins, P.C., and Reilly, S.B. 2002. Gray Whale calf production, 1994–2000: Are observed fluctuations related to changes in seasonal ice cover? *Marine Mammal Science* 18(1): 121–144.

Perryman, W.L., and Lynn, M.S. 2002. Evaluation of nutritive condition and reproductive status of migrating Gray Whales (*Eschrichtius robustus*) based on analysis of photogrammetric data. *Journal of Cetacean Research and Management* 4(2): 155–164.

Person, D.K., Kirchhoff, M., Van Ballenberghe, V., Iverson, G.C., and Grossman, E. 1996. The Alexander Archipelago wolf: A conservation assessment. *US Dept of Agriculture Forestry Service Gen. Tech. Rep.* PNW-GTR-384: 1–42.

Persson, J. 2005. Female Wolverine (*Gulo gulo*) reproduction: Reproductive costs and winter food availability. *Canadian Journal of Zoology* 83(11): 1453–1459.

Persson, J., Landa, A., Andersen, R., and Segerström, P. 2006. Reproductive characteristics of female Wolverines (*Gulo gulo*) in Scandinavia. *Journal of Mammalogy* 87(1): 75–79.

Persson, J., Willebrand, T., Landa, A., Andersen, R., and Segerström, P. 2003. The role of intraspecific predation in the survival of juvenile Wolverines (*Gulo gulo*). *Wildlife Biology* 9(1): 21–28.

Petersen, S. 1997. Status of the Wolverine (*Gulo gulo*) in Alberta. *Wildlife Status Report no. 2.* Alberta Environmental Protection, Wildlife Management Division, Edmonton, AB.

Petersen, S.D., and Roberts, W. 1999. Notes on the abundance and distribution of Prairie Shrews (*Sorex haydeni*) in Alberta. *Alberta Naturalist* 29(2): 24–27.

Petersen, S.D., and Stewart, D.T. 2006. Phylogeography and conservation genetics of Southern Flying Squirrels (*Glaucomys volans*) from Nova Scotia. *Journal of Mammalogy* 87(1): 153–160.

Peterson, R.L. 1966. *The Mammals of Eastern Canada*. Oxford University Press, Toronto, ON.

Peterson, R.S., and Bartholomew, G.A. 1967. The natural history and behavior of the California Sea Lion. *American Society of Mammalogists, Special Publication* no. 1: 79.

Petterson, J.R. 1992. Yellow-bellied Marmot, *Marmota flaviventris*, predation on Pikas, *Ochotona princeps. Canadian Field-Naturalist* 106(1): 130–131.

Philips, J.D., Nachtigall, P.E., Au, W.W.L., Pawloski, J.L., and Roitblat, H.L. 2003. Echolocation in the Risso's Dolphin, *Grampus griseus. Journal of the Acoustical Society of America* 113(1): 605–616.

Pierce, B.M., and Bleich, V.C. 2003. Mountain Lion (*Puma concolor*). In G.A. Feldhamer, B.C. Thompson, and J.A. Chapman (eds), *Wild Mammals of North America: Biology, Management, and Conservation*, 2nd ed., 744–757. Johns Hopkins University Press, Baltimore, MD.

Pierce, B.M., Bleich, V.C., Wehausen, J.D., and Bowyer, R.T. 1999. Migratory patterns of Mountain Lions: Implications for social regulation and conservation. *Journal of Mammalogy* 80(3): 986–992.

Pierce, S.S., and Vogt, F.D. 1993. Winter acclimatization in *Peromyscus maniculatus gracilis, P. leucopus noveboracensis,* and *P.l. leucopus. Journal of Mammalogy* 74(3): 665–677.

Pietz, P.J., and Granfors, D.A. 2000. White-tailed Deer (*Odocoileus virginianus*) predation on grassland songbird nestlings. *American Midland Naturalist* 144: 419–422.

Pigage, J.C., and Pigage, H.K. 1994. The Western Harvest Mouse (*Reithrodontomys megalotis*) moves into northeastern Illinois. *Transactions of the Illinois State Academy of Science* 87(1 and 2): 47–50.

Pitcher, K.W., Burkanov, V.N., Calkins, D.G., Le Boeuf, B.J., Mamaev, E.G., Merrick, R.L., and Pendleton, G.W. 2001. Spatial and temporal variation in the timing of births of Steller Sea Lions. *Journal of Mammalogy* 82(4): 1047–1053.

Pitman, R.L., Ballance, L.T., Mesnick, S.I., and Chivers, S.J. 2001. Killer Whale predation on Sperm Whales: Observations and implications. *Marine Mammal Science* 17(3): 494–507.

Pitts, R.M. 1994. Unusual nesting behavior of harvest mice. *Prairie Naturalist* 26(4): 311.

Pitts, R.M., and Choate, J.R. 1990. Winter breeding by the Plains Pocket Gopher. *Prairie Naturalist* 22(4): 277.

Pitts, R.M., Choate, J.R., and Scharninghausen, J.J. 1994. Plains Pocket Gopher storing black walnuts. *Prairie Naturalist* 26(4): 312.

Pizzuto, T., and Getz, L.L. 1998. Female Prairie Voles (*Microtus ochrogaster*) fail to form a new pair after loss of mate. *Behavioural Processes* 43: 79–86.

Plön, S., Bernard, R.T.F., Klages, N.T.K., and Cockcroft, V.G. 1999. Stomach content analysis of Pygmy and Dwarf Sperm Whales and its ecological implications: Is there niche partitioning? *European Research on Cetaceans* 13: 336–339.

Polziehn, R.O., and Strobeck, C. 2002. A phylogenetic comparison of Red Deer and Wapiti using mitochondrial DNA. *Molecular Phylogenetics and Evolution* 22(3): 342–356.

Poole, K.G. 1995. Spatial organization of a Lynx population. *Canadian Journal of Zoology* 73(4): 632–641.

– 1997. Dispersal patterns of Lynx in the Northwest Territories. *Journal of Wildlife Management* 61(2): 497–505.

Poole, K.G., Gunn, A., Patterson, B.R., and Dumond, M. 2010. Sea ice and migration of the Dolphin and Union Caribou herd in the Canadian Arctic: An uncertain future. *Arctic* 63(4): 414–428.

Poole, K.G., Wakelyn, L.A., and Nicklen, P.N. 1996. Habitat selection by Lynx in the Northwest Territories. *Canadian Journal of Zoology* 74(5): 845–850.

Poole, M.W., and Matlack, R.S. 2007. Prairie Vole and other small mammals from the Texas Panhandle. *Southwestern Naturalist* 52(3): 442–445.

Post, E., Bøving, P.S., Pedersen, C., and MacArthur, M.A. 2003. Synchrony between Caribou calving and plant phenology in depredated and non-depredated populations. *Canadian Journal of Zoology* 81(10): 1709–1714.

Poszig, D., Apps, C.D., and Dibb, A. 2004. Predation of two Mule Deer, *Odocoileus hemionus*, by a Canada Lynx, *Lynx canadensis*, in the southern Canadian Rocky Mountains. *Canadian Field-Naturalist* 118(2): 191–194.

Potelov, V., Nilssen, K.T., Svetochev, V., and Haug, T. 2000. Feeding habits of Harp (*Phoca groenlandica*) and Hooded Seals (*Cystophora cristata*) during late winter, spring, and early summer in the Greenland Sea. *NAMMCO Scientific Publications* 2: 40–49.

Potter, C.W., and Birchler, B. 1999. Sei Whale (*Balaenoptera borealis*). In D.E. Wilson and S. Ruff (eds), *The Smithsonian Book of North American Mammals*, 248–249. UBC Press, Vancouver, BC.

Poulle, M.-L., Crête, M., and Huot, J. 1995. Seasonal variation in body mass and composition of eastern Coyotes. *Canadian Journal of Zoology* 73(9): 1625–1633.

Powell, R.A. 1981. *Martes pennanti. Mammalian Species* no. 156: 1–6.

– 1993. *The Fisher: Life History, Ecology, and Behavior*. 2nd ed., University of Minnesota Press, Minneapolis, MN.

– 1994. Effects of scale on habitat selection and foraging behavior in Fishers in winter. *Journal of Mammalogy* 75(2): 349–356.

Powell, R.A., Buskirk, S.W., and Zielinski, W.J. 2003. Fisher (*Martes pennanti*) and Marten (*Martes americana*). In G.A. Feldhamer, B.C. Thompson, and J.A. Chapman, (eds), *Wild Mammals of North America: Biology, Management, and Conservation*, 2nd ed., 635–649. Johns Hopkins University Press, Baltimore, MD.

Powell, R.A., and Fried, J.J. 1992. Helping by juvenile Pine Voles (*Microtus pinetorum*), growth and survival of younger siblings, and the evolution of Pine Vole sociality. *Behavioral Ecology* 3(4): 325–333.

Powell, R.A., Zimmerman, J.W., and Seaman, D.E. 1997. *Ecology and Behaviour of North American Black Bears*. Chapman and Hall, New York, NY.

Powell, T., Jung, T.S., and Clyde, K.J. 2007. Apparent predation of an American Water Shrew, *Sorex palustris*, by an American Marten, *Martes americana. Canadian Field-Naturalist* 121(4): 422–423.

Predavec, M., and Krebs, C.J. 2000. Microhabitat utilisation, home ranges, and movement patterns of the Collared Lemming (*Dicrostonyx groenlandicus*) in the central Canadian Arctic. *Canadian Journal of Zoology* 78(11): 1885–1890.

Prescott, J. 1991. The St Lawrence Beluga: A concerted effort to save an endangered species. *Occasional Paper of the Provincial Museum of Alberta* 15: 269–273.

Prescott, J., and Richard, P. 2004. *Mammifères du Québec et de l'est du Canada*. 2nd ed. Éditions Michel Quintin, Waterloo, QC.

Prestrud, P. 1991. Adaptations by the Arctic Fox (*Alopex lagopus*) to the polar winter. *Arctic* 44(2): 132–138.

Pretzlaw, T., Trudeau, C., Humphries, M. M., LaMontagne, J. M., and Boutin, S. 2006. Red Squirrels (*Tamiasciurus hudsonicus*) feeding on Spruce Bark Beetles (*Dendroctonus rufipennis*) in the Canadian boreal forest: Energetic and ecological implications. *Journal of Mammalogy* 87(5): 909–914.

Prévot-Julliard, A.-C., Henttinen, H., Yoccoz, N.G., and Stenseth, N.C. 1999. Delayed maturation in female Bank Voles: Optimal decision or social constraint? *Journal of Animal Ecology* 68(4): 684–697.

Price, K., and Boutin, S. 1993. Territorial bequeathal by Red Squirrel mothers. *Behavioral Ecology* 4(2): 144–150.

Price, M.H.H., and Aries, C.E. 2007. A River Otter's, *Lontra canadensis*, capture of a Double-crested Cormorant, *Phalacrocorax auritus*, in British Columbia's Gulf Island waters. *Canadian Field-Naturalist* 121(3): 325–326.

Price, M.H.H., Darimont, C.T., Winchester, N.N., and Paquet, P.C. 2005. Facts from faeces: Prey remains in Wolf, *Canis lupus*, faeces revise occurrence records for mammals of British Columbia's coastal archipelago. *Canadian Field-Naturalist* 119(2): 192–196.

Prince, L.A. 1940. Notes on the habits of the Pigmy Shrew (*Microsorex hoyi*) in captivity. *Canadian Field-Naturalist* 54: 97–100.

Proctor, M.F., McLellan, B.N., Strobeck, C., and Barclay, R.M.R. 2004. Gender-specific dispersal distances of Grizzly Bears estimated by genetic analysis. *Canadian Journal of Zoology* 82(7): 1108–1118.

Proulx, G. 2002. Reproductive characteristics of Northern Pocket Gophers, *Thomomys talpoides*, in Alberta alfalfa fields. *Canadian Field-Naturalist* 116(2): 319–321.

– 2005. Body weights of adult and juvenile Northern Pocket Gophers, *Thomomys talpoides*, in central Alberta alfalfa fields. *Canadian Field-Naturalist* 119(4): 551–559.

– 2006. Using forest inventory data to predict winter habitat use by Fisher *Martes pennanti* in British Columbia, Canada. *Acta Theriologica* 51(3): 275–282.

Proulx, G., Badry, M.J., Cole, P.J., Drescher, R.K., Kolenosky, A.J., and Pawlina, I.M. 1995a. Summer activity of Northern Pocket Gophers, *Thomomys talpoides*, in a simulated natural environment. *Canadian Field-Naturalist* 109(2): 210–215.

– 1995b. Summer above-ground movements of Northern Pocket Gophers, *Thomomys talpoides*, in Alberta alfalfa fields. *Canadian Field-Naturalist* 109(2): 256–258.

Proulx, G., and Cole, P.J. 1998. Identification of Northern Pocket Gopher, *Thomomys talpoides*, remains in Long-tailed Weasel, *Mustela frenata longicauda*, scats. *Canadian Field-Naturalist* 112(2): 345–346.

– 2002. Evidence of a second litter in Northern Pocket Gophers, *Thomomys talpoides. Canadian Field-Naturalist* 116(2): 322–323.

Proulx, G., Lounsbury, L., and Bryant, H.N. 1996. Northern Pocket Gophers, *Thomomys talpoides*, with white pelage from Alberta. *Canadian Field-Naturalist* 110(2): 331.

Pruss, S.D. 1999. Selection of natal dens by the Swift Fox (*Vulpes velox*) on the Canadian prairies. *Canadian Journal of Zoology* 77(4): 646–652.

Pruss, S.D., Fargey, P., and Moehrenschlager, A. 2008. Recovery strategy for the Swift Fox (*Vulpes velox*) in Canada. Prepared in consultation with the Canadian Swift Fox Recovery Team. Species at Risk Act Recovery Strategy Series. Parks Canada Agency.

Punzo, F., and Chavez, S. 2000. Effect of aging on the spatial learning and running speed in the shrew (*Cryptotis parva*). *Journal of Mammalogy* 84(3): 1112–1120.

Pyare, S., and Longland, W.S. 2001. Mechanisms of truffle detection by Northern Flying Squirrels. *Canadian Journal of Zoology* 79(6): 1007–1015.

Pyare, S., Smith, W.P., Nicholls, J.V., and Cook, J.A. 2002. Diets of Northern Flying Squirrels, *Glaucomys sabrinus*, in southeastern Alaska. *Canadian Field-Naturalist* 116(1): 98–103.

Pyle, P., Long, D.J., Schonewald, J., Jones, R.E., and Roletto, J. 2001. Historical and recent colonization of the South Farallon Islands, California, by Northern Fur Seals (*Callorhinus ursinus*). *Marine Mammal Science* 17(2): 397–402.

Rabe, M.J., Siders, M.S., Miller, C.R., and Snow, T.K. 1998. Long foraging distance for a Spotted Bat (*Euderma maculatum*) in northern Arizona. *Southwestern Naturalist* 43(2): 266–286.

Rabon, D.R., Jr., Sawrey, D.K., and Webster, W.D. 2001. Infant ultrasonic vocalizations and parental responses in two species of voles (*Microtus*). *Canadian Journal of Zoology* 79(5): 830–837.

Racey, G.D., and Armstrong, T. 2000. Woodland Caribou range occupancy in northwestern Ontario: Past and present. *Rangifer*, special issue no. 12: 173–184.

Rachlow, J.L., and Bowyer, R.T. 1991. Interannual variation in timing and synchrony of parturition in Dall's Sheep. *Journal of Mammalogy* 72(3): 487–492.

– 1994. Variability in maternal behavior by Dall's Sheep: Environmental tracking or adaptive behaviour? *Journal of Mammalogy* 75(2): 328–337.

– 1998. Habitat selection by Dall's Sheep (*Ovis dalli*): Maternal trade-offs. *Journal of Zoology* (London) 245: 457–465.

Rahme, A.H., Harestad, A.S., and Bunnell, F.L. 1995. Status of the Badger in British Columbia. *Wildlife Working Report*, no. WR-72, 1–51. BC Ministry of the Environment, Wildlife Branch.

Ralls, K., Eagle, T.C., and Siniff, D.B. 1996. Movement and spatial use patterns of California Sea Otters. *Canadian Journal of Zoology* 74(10): 1841–1849.

Ralls, K., Hatfield, B.B., and Siniff, D.B. 1995. Foraging patterns of California Sea Otters as indicated by telemetry. *Canadian Journal of Zoology* 73(3): 523–531.

Rambaldini, D.A., and Brigham, R.M. 2008. Torpor use by free-ranging Pallid Bats (*Antrozous pallidus*) at the northern extent of their range. *Journal of Mammalogy* 89(4): 933–941.

Ramsay, M.A., and Stirling, I. 1990. Fidelity of female Polar Bears to winter-den sites. *Journal of Mammalogy* 71(2): 233–236.

Rancourt, S.J., Rule, M.I., and O'Connell, M.A. 2005. Maternity roost site selection of Long-eared Myotis, *Myotis evotis*. *Journal of Mammalogy* 86(1): 77–84.

Rankin, S., and Barlow, J. 2005. Source of North Pacific "boing" sound attributed to Minke Whales. *Journal of the Acoustical Society of America* 188(5): 3346–3351.

Rankin, S., Barlow, J., and Stafford, K.M. 2006. Blue Whale (*Balaenoptera musculus*) sightings and recordings south of the Aleutian Islands. *Marine Mammal Science* 22(3): 708–713.

Rankin, S., Oswald, J., Barlow, J., and Lammers, M. 2007. Patterned burst-pulse vocalizations of the Northern Right Whale Dolphin, *Lissodelphis borealis*. *Journal of the Acoustical Society of America* 121(2): 1213–1218.

Ranta, E., Kaitala, V., and Lindström, J. 1997. Dynamics of Canadian Lynx populations in space and time. *Ecography* 20: 454–460.

Rasmuson, T.M., and Barclay R.M.R. 1992. Individual variation in the isolation calls of newborn Big Brown Bats (*Eptesicus fuscus*): Is variation genetic? *Canadian Journal of Zoology* 70(4): 698–702.

Rasmussen, M.H., and Miller, L.A. 2002. Whistles and clicks from White-sided Dolphins, *Lagenorhynchus albirostris*, recorded in Faxaflói Bay, Iceland. *Aquatic Mammals* 28(1): 78–89.

Ratcliffe, J.M., and Dawson, J.W. 2004. Behavioural flexibility: The Little Brown Bat, *Myotis lucifugus*, and the Northern Long-eared Bat, *Myotis septentrionalis*, both glean and hawk prey. *Animal Behaviour* 66(5): 847–856.

Raum-Suryan, K., Pitcher, K., and Lamy, R. 2004. Sea Otter, *Enhydra lutris*, sightings off Haida Gwaii / Queen Charlotte Islands, British Columbia, 1972–2002. *Canadian Field-Naturalist* 118(2): 270–272.

Rausch, R.L. 1953. On the status of some Arctic mammals. *Arctic* 6: 91–148.

Rausch, R.L., Feagin, J.E., and Rausch, V.R. 2007. *Sorex rohweri* sp. nov. (Mammalia, Soricidae) from northwestern North America. *Mammalian Biology (Zeit. Säugetier.)* 72(2): 93–105.

Rausch, V.R., and Rausch, R.L. 1993. Karyotypic characteristics of *Sorex tundrensis* Merriam (Mammalia: Soricidae), a Nearctic species of the *S. araneus* group. *Proceedings of the Biological Society of Washington* 106(2): 410–416.

Raynor, G.S. 1960. Three litters in a Pine Vole nest. *Journal of Mammalogy* 41(2): 275.

Raynor, L.S., and Armitage, K.B. 1991. Social behavior and space-use of young of ground-dwelling squirrel species with different levels of sociality. *Ethology Ecology and Evolution* 3: 185–205.

Read, A.J. 2001. Trends in the maternal investment of Harbour Porpoises are uncoupled from the dynamics of their primary prey. *Proceedings of the Royal Society of London, B Series* 268: 573–577.

Read, A.J., and Hohn, A.A. 1995. Life in the fast lane: The life history of Harbor Porpoises from the Gulf of Maine. *Marine Mammal Science* 11(4): 423–440.

Read, A.J., and Westgate, A.J. 1997. Monitoring the movements of Harbour Porpoises (*Phocoena phocoena*) with satellite telemetry. *Marine Biology* 130: 315–322.

Read, A.J., Wiepkema, P.R., and Nachtigall, P.E. 1997. The Harbour Porpoise (*Phocoena phocoena*). In A.J. Read, P.R. Wiepkema, and P.E. Nachtigall (eds), *The Biology of the Harbour Porpoise*, 3–6. De Spil Publishers, Woerden, The Netherlands.

Réale, D., McAdam, A.G., Boutin, S., and Berteaux, B. 2003. Genetic and plastic responses of a northern mammal to climate change. *Proceedings of the Royal Society of London, B Series* 270: 591–596.

Reby, D., Joachim, J., Lauga, J., Lek, S., and Aulagnier, S. 1998. Individuality in the groans of Fallow Deer (*Dama dama*) bucks. *Journal of Zoology* (London) 245: 79–84.

Reddy, E., and Fenton, M.B. 2003. Exploiting vulnerable prey: Moths and Red Bats (*Lasiurus borealis*; Vespertilionidae). *Canadian Journal of Zoology* 81(9): 1553–1560.

Reder, S., Lydersen, C., Arnold, W., and Kovacs, K.M. 2003. Haulout behaviour of High Arctic Harbour Seals (*Phoca vitulina vitulina*) in Svalbard, Norway. *Polar Biology* 27: 6–16.

Ree, V. 1994. Field identification of White-beaked Dolphin *Lagenorhynchus albirostris* and Atlantic White-sided Dolphin *L. acutus*. *Fauna* 47: 132–165.

Reeves, R.R. 1999a. Harbor Seal (*Phoca vitulina*). In D.E. Wilson and S. Ruff (eds), *The Smithsonian Book of North American Mammals*, 209–211. UBC Press, Vancouver, BC.

– 1999b. Long-finned Pilot Whale (*Globicephala melas*). In Wilson, and Ruff, *The Smithsonian Book of North American Mammals*, 286–287.

– 2001. Overview of catch history, historic abundance, and distribution of Right Whales in the western North Atlantic and in Cintra Bay, West Africa. *Journal of Cetacean Research and Management (Special Issue)* 2: 187–192.

Reeves, R.R., and Heide-Jørgensen, M.P. 1996. Recent status of Bowhead Whales, *Balaena mysticetus*, in the wintering grounds off West Greenland. *Polar Research* 15(2): 115–125.

Reeves, R.R., and King, J.K. 1981. Hooded Seal *Cystophora cristata* (Erxleben 1777). In S.H. Ridgway and R. Harrison (eds), *Handbook of Marine Mammals*, vol. 2, *Seals*, 171–194. Academic Press, San Diego, CA.

Reeves, R.R., and Mead, J.G. 1999. Marine mammals in captivity. In J.R. Twiss, Jr, and R.R. Reeves (eds), *Conservation and Management of Marine Mammals*, 412–436. Smithsonian Institute Press, Washington, DC.

Reeves, R.R., and Mitchell, E. 1993. Status of Baird's Beaked Whale, *Berardius bairdii*. *Canadian Field-Naturalist* 107(4): 509–523.

Reeves, R.R., and Read, A.J. 2003. Bottlenose Dolphin, Harbor Porpoise, Sperm Whale (*Tursiops truncatus, Phocoena phocoena*, and *Physeter macrocephalus*), and other toothed cetaceans. In G.A. Feldhamer, B.C. Thompson, and J.A. Chapman (eds), *Wild Mammals of North America*, 2nd ed., 327–424. Johns Hopkins University Press, Baltimore, MD.

Reeves, R.R., Smeenk, C., Kinze, C.C., Brownell, R.L., Jr., and Lien, J. 1999. White-beaked Dolphin *Lagenorhynchus albirostris* (Gray 1846). In S.H. Ridgeway and R. Harrison (eds), *Handbook of Marine Mammals*, vol. 6, 1–30. Academic Press, San Diego, CA.

Reeves, R.R., Stewart, B.S., Chapman, P.J., and Powell, J.A. 2002. *Guide to Marine Mammals of the World*. Alfred A. Knopf, New York, NY.

Reeves, R.R., and Tracey, S. 1980. *Monodon monoceros*. *Mammalian Species* no. 127: 1–7.

Reeves, R.R., and Whitehead, H. 1997. Status of the Sperm Whale, *Physeter macrocephalus*, in Canada. *Canadian Field-Naturalist* 111(2): 293–307.

Regehr, E.V., Lunn, N.J., Amstrup, S.C., and Stirling, I. 2007. Effects of earlier ice breakup on survival and population size of Polar Bears in western Hudson Bay. *Journal of Wildlife Management* 71(8): 2673–2683.

Rehmeier, R.L., Kaufman, G.A., and Kaufman, D.W. 2004. Long-distance movements of the Deer Mouse in tallgrass prairie. *Journal of Mammalogy* 85(3): 562–568.

Reich, L.M. 1981. *Microtus pennsylvanicus. Mammalian Species* no. 159: 1–8.

Reid, D.G., Code, T.E., Reid, A.C.H., and Herrero, S.M. 1994a. Food habits of the River Otter in a boreal ecosystem. *Canadian Journal of Zoology* 72(7): 1306–1313.

– 1994b. Spacing, movements, and habitat selection of the River Otter in boreal Alberta. *Canadian Journal of Zoology* 72(7): 1314–1324.

Reid, F.A. 1997. *A Field Guide to the Mammals of Central America and Southeast Mexico.* Oxford University Press, New York, NY.

– 2006. *A Field Guide to the Mammals of North America.* Peterson Field Guide Series, 4th ed. Houghton Mifflin, New York, NY.

Reilly, S.B., and Thayer, V.G. 1990. Blue Whale (*Balaenoptera musculus*) distribution in the eastern tropical Pacific. *Marine Mammal Science* 6(4): 265–277.

Reiman, A.J., and Terhune, J.M. 1993. The maximum range of vocal communication in air between a Harbor Seal (*Phoca vitulina*) pup and its mother. *Marine Mammal Science* 9(2): 182–189.

Reimchen, T.E. 1998. Nocturnal foraging behaviour of Black Bears, *Ursus americanus*, on Moresby Island, British Columbia. *Canadian Field-Naturalist* 112(3): 446–450.

Reiss, D., and Marino, L. 2001. Mirror self-recognition in the Bottlenose Dolphin: A case of cognitive convergence. *Proceedings of the National Academy of Science* 98(10): 5937–5942.

Reiter, J., and Le Boeuf, B.J. 1991. Life history consequences of variation in age at primiparity in Northern Elephant Seals. *Behavioral Ecology and Sociobiology* 28: 153–160.

Reiter, M.E., and Andersen, D.E. 2008. Trends in abundance of Collared Lemmings near Cape Churchill, Manitoba, Canada. *Journal of Mammalogy* 89(1): 138–144.

Reith, C.C.1980. Shifts in times of activity by *Lasionycteris noctivagans*. *Journal of Mammalogy* 61(1): 104–108.

Rendell, L.E., Matthews, J.N., Gill, A., Gordon, J.C.D., and Macdonald, D.W. 1999. Quantitative analysis of tonal calls from five odontocete species, examining interspecific and intraspecific variation. *Journal of Zoology* (London) 249: 403–410.

Renouf, D., Noseworthy, E., and Scott, M.C. 1990. Daily fresh water consumption by captive Harp Seals (*Phoca groenlandica*). *Marine Mammal Science* 6(3): 253–257.

Rettie, W.J. 2004. Morphology of female Woodland Caribou, *Rangifer tarandus caribou*, in Saskatchewan. *Canadian Field-Naturalist* 118(1): 119–121.

Rettie, W.J., Rock, T., and Messier, F. 1998. The status of Woodland Caribou in Saskatchewan. *Rangifer*, special issue no. 10: 105–110.

Reynolds, H.W., Gates, C.C., and Glaholt, R.D. 2003. Bison (*Bison bison*). In G.A. Feldhamer, B.C. Thompson, and J.A. Chapman (eds), *Wild Mammals of North America: Biology, Management, and Conservation*, 2nd ed., 1009–1060. Johns Hopkins University Press, Baltimore, MD.

Reynolds, P.E. 2001. Reproductive patterns of female Muskoxen in northeastern Alaska. *Alces* 37(2): 403–410.

Reynolds, P.S. 1993. Size, shape, and surface area of Beaver, *Castor canadensis*, a semiaquatic mammal. *Canadian Journal of Zoology* 71(5): 876–882.

Rezendes, P. 1992. *Tracking and the Art of Seeing: How to read animal tracks and sign.* Camden House Publishing, Charlotte, VT.

Rhymer, J.M., Barbay, J.M., and Givens, H.L. 2004. Taxonomic relationship between *Sorex dispar* and *S. gaspensis*: Inferences from mitochondrial DNA sequences. *Journal of Mammalogy* 85(2): 331–337.

Ribble, D.O., and Millar, J.S. 1996. The mating system of northern populations of *Peromyscus maniculatus* as revealed by radiotelemetry and DNA fingerprinting. *Ecoscience* 3(4): 423–428.

Rice, D.W. 1989. Sperm Whale *Physter macrocephalus* (Linnaeus 1758). In S.H. Ridgway and R. Harrison (eds), *Handbook of Marine Mammals*, vol. 4, *River Dolphins and the Larger Toothed Whales*, 177–233. Academic Press, San Diego, CA.

Rich, S.M., Kilpatrick, W., Shippee, J.L., and Crowell, K.L. 1996. Morphological differentiation and identification of *Peromyscus leucopus* and *P. maniculatus* in northeastern North America. *Journal of Mammalogy* 77(4): 985–991.

Richard, P.R., Heide-Jørgensen, M.P., and St Aubin, D. 1998. Fall movements of Belugas (*Delphinapterus leucas*) with satellite-linked transmitters in Lancaster Sound, Jones Sound, and Northern Baffin Bay. *Arctic* 51(1): 5–16.

Richardson, E.S., and Andriashek, D. 2006. Wolf (*Canis lupus*) predation of a Polar Bear (*Ursus maritimus*) cub on the sea ice off northwestern Banks Island, Northwest Territories, Canada. *Arctic* 59(3): 322–324.

Richardson, S.F., Stenson, G.B., and Hood, C. 2003. Growth of the Harbour Porpoise (*Phocoena phocoena*) in eastern Newfoundland, Canada. *NAMMCO Scientific Publications* 5: 211–222.

Richardson, W.J., Finley, K.J., Miller, G.W., Davis, R.A., and Koski, W.R. 1995. Feeding, social and migration behavior of Bowhead Whales, *Balaena mysticetus*, in Baffin Bay vs the Beaufort Sea: Regions with different amounts of human activity. *Marine Mammal Science* 11(1): 1–45.

Rickart, E.A., and Heaney, L.R. 2001. Shrews of the La Sal Mountains, southeastern Utah. *Western North American Naturalist* 61(1): 103–108.

Riddle, B.R. 1999. Northern Grasshopper Mouse (*Onychomys leucogaster*). In D.E. Wilson and S. Ruff (eds), *The Smithsonian Book of North American Mammals*, 588–590. UBC Press, Vancouver, BC.

Ridgway, S., Kamolnick, T., Reddy, M., Curry, C., and Tarpley, R.J. 1995. Orphaninduced lactation in *Tursiops* and analysis of collected milk. *Marine Mammal Science* 11(2): 172–182.

Riedman, M.L., and Estes, J.A. 1990. The Sea Otter (*Enhydra lutris*): Behavior, Ecology, and Natural History. *U.S. Fish and Wildlife Service, Biological Report* 90(14): 1–126.

Riewe, R.R. 1975. The High Arctic Wolf in Jones Sound region of the Canadian High Arctic. *Arctic* 28(3): 209–212.

Rinehart-Whitt, S.C., and Pagels, J.F. 2000. Seasonal pelage variation in the Masked Shrew *Sorex cinereus*: A demographic perspective. *Acta Theriologica* 45(1): 111–116.

Robinson, H.S., Wielgus, R.B., and Gwilliam, J.C. 2002. Cougar predation and population growth of sympatric Mule Deer and White-tailed Deer. *Canadian Journal of Zoology* 80(3): 556–568.

Robinson, K.P., and Tetley, M.J. 2007. Behavioural observations of foraging Minke Whales (*Balaenoptera acutorostrata*) in the outer Moray Firth, north-east Scotland. *Journal of the Marine Biological Association of the United Kingdom* 87: 85–86.

Robitaille, J.-F., and Raymond, M. 1995. Spacing patterns of Ermine, *Mustela erminea* L., in a Quebec agrosystem. *Canadian Journal of Zoology* 73(10): 1827–1834.

Roche, B.E., Schulte-Hostedde, A.I., and Brooks, R.J. 1999. Route choice by Deer Mice (*Peromyscus maniculatus*): Reducing the risk of auditory detection by predators. *American Midland Naturalist* 142: 194–197.

Rock, K.R., Rock, E.S., Bowyer, R.T., and Faro, J.B. 1994. Degree of association and use of a helper by coastal River Otters, *Lutra canadensis*, in Prince William Sound, Alaska. *Canadian Field-Naturalist* 108(3): 367–369.

Rodriguez, R.M., and Ammerman, L.K. 2004. Mitochondrial DNA divergence does not reflect morphological difference between *Myotis californicus* and *Myotis ciliolabrum*. *Journal of Mammalogy* 85(5): 842–851.

Rogers, A.R. 1990. Summer movement patterns of Arctic lemmings (*Lemmus sibericus* and *Dicrostonyx groenlandicus*). *Canadian Journal of Zoology* 68(12): 2513–2517.

Rogers, L.L. 1992. *Watchable Wildlife: The Black Bear.* Pamphlet prepared by the North Central Forest Experimental Station for the U.S. Forestry Service, United States Department of Agriculture.

– 1999. American Black Bear (*Ursus americanus*). In D.E. Wilson and S. Ruff (eds), *The Smithsonian Book of North American Mammals*, 157–160. UBC Press, Vancouver, BC.

Rogowitz, G.L. 1997. Locomotor and foraging activity of the White-tailed Jackrabbit (*Lepus townsendii*). *Journal of Mammalogy* 78(4): 1172–1181.

Rogowitz, G.L., and Wolfe, M.L. 1991. Intra-specific variation in life-history traits of the White-tailed Jackrabbit (*Lepus townsendii*). *Journal of Mammalogy* 72(4): 796–806.

Rohr, J.J., Fish, F.E., and Gilpatrick, J.W. 2002. Maximum swim speeds of captive and free-ranging delphinids: Critical analysis of extraordinary performance. *Marine Mammal Science* 18(1): 1–19.

Rolseth, S.L., Koehler, C.E., and Barclay, R.M.R. 1994. Differences in the diets of juvenile and adult Hoary Bats, *Lasiurus cinereus*. *Journal of Mammalogy* 75(2): 394–398.

Romeo, G., and Lovari, S. 1996. Summer activity rhythms of the Mountain Goat *Oreamnos americanus* (de Blainville 1816). *Mammalia* 60(3): 496–469.

Romin, L.A., and Dalton, L.B. 1992. Lack of response by Mule Deer to wildlife warning whistles. *Wildlife Society Bulletin* 20: 382–384.

Ronald, K., and Healey, P.J. 1981. Harp Seal *Pagophilus groenlandicus* (Erxleben 1777). In S.H. Ridgway and R. Harrison (eds), *Handbook of Marine Mammals*, vol. 2, *Seals*, 55–87. Academic Press, San Diego, CA.

Rongstad, O.J., and Tester, J.R. 1971. Behavior and maternal relations of young Snowshoe Hares. *Journal of Wildlife Management* 35(2): 338–346.

Rosatte, R.C. 2000. Management of Raccoons (*Procyon lotor*) in Ontario, Canada: Do human intervention and disease have significant impact on Raccoon populations? *Mammalia* 64(4): 369–390.

– 2002. Long distance movement by a Coyote, *Canis latrans*, and Red Fox, *Vulpes vulpes*, in Ontario: Implications for disease-spread. *Canadian Field-Naturalist* 116(1): 129–131.

Rosatte, R.[C.], and Larivière, S. 2003. Skunk (genera *Mephitis, Spilogale* and *Conepatus*). In G. A. Feldhamer, B.C. Thompson, and J. A. Chapman (eds), *Wild Mammals of North America: Biology, Management, and Conservation*, 2nd ed., 692–707. Johns Hopkins University Press, Baltimore, MD.

Rosatte, R.C., and Neuhold, J. 2006. Late-born Elk, *Cervus elaphus*, calf observed near Bancroft, Ontario. *Canadian Field-Naturalist* 120(2): 188–191.

Roscoe, B., and Majka, C. 1976. First records of the Rock Vole (*Microtus chrotorrhinus*) and the Gaspé Shrew (*Sorex gaspensis*) from Nova Scotia and a second record of the Thompson's Pygmy Shrew (*Microsorex thompsoni*) from Cape Breton Island. *Canadian Field-Naturalist* 90(4): 497–498.

Rose, N.A., Deutsch, C.J., and Le Boeuf, B.J. 1991. Sexual behavior of male Northern Elephant Seals: III, The mounting of weaned pups. *Behaviour* 119(3–4): 171–192.

Rosell, F., Bozsér, O., Collen, P., and Parker, H. 2005. Ecological impact of beavers *Castor fiber* and *Castor canadensis* and their ability to modify ecosystems. *Mammal Review* 35(3–4): 248–276.

Rosen, D.A.S., and Trites, A.W. 2000. Pollock and the decline of Steller Sea Lions: Testing the junk-food hypothesis. *Canadian Journal of Zoology* 78(7): 1243–1250.

Rosenbaum, H.C., Brownell, R.L., Jr., Brown, M.W., Schaeff, C., Portway, V., White, B.N., Malik, S., Pastene, L.A., Patenaude, N.J., Baker, C.S., Goto, M., Best, P.B., Clapham, P.J., Hamilton, P., Moore, M., Payne, R., Rowntree, V., Tynan, C.T., Bannister, J.I., and DeSalle, R. 2000. World-wide genetic differentiation of *Eubalaena*: Questioning the number of Right Whale species. *Molecular Ecology* 9: 1793–1802.

Rosing-Asvid, A., Teilmann, J., Dietz, R., and Tange Olsen, M. 2010. First confirmed record of Grey Seals in Greenland. *Arctic* 63(4): 471–473.

Ross, P.D. 1998. *COSEWIC Status Report on the Woodland Vole*, Microtus pinetorum. Committee on the Status of Endangered Wildlife in Canada, 1–44.

Ross, P.I., and Jalkotzy, M.G. 1992. Characteristics of a hunted population of Cougars in southwestern Alberta. *Journal of Wildlife Management* 56(2): 417–426.

Ross, P.I., Jalkotzy, M.G., and Daoust, P.-Y. 1995. Fatal trauma sustained by Cougars, *Felis concolor*, while attacking prey in southern Alberta. *Canadian Field-Naturalist* 109(2): 261–263.

Ross, P.I., Jalkotzy, M.G., and Festa-Bianchet, M. 1997. Cougar predation on Bighorn Sheep in southwestern Alberta during winter. *Canadian Journal of Zoology* 75(5): 771–775.

Rossbach, K.A., and Herzing, D.L. 1997. Underwater observations of benthic-feeding Bottlenose Dolphins (*Tursiops truncatus*) near Grand Bahama Island, Bahamas. *Marine Mammal Science* 13(3): 498–504.

Rossong, M.A., and Terhune, J.M. 2009. Source levels and communication range models for Harp Seal (*Pagophilus groenlandicus*) underwater calls in the Gulf of St Lawrence, Canada. *Canadian Journal of Zoology* 87(7): 609–617.

Roth, J.D. 2002. Temporal variability in Arctic Fox diet as reflected in stable-carbon isotopes: The importance of sea ice. *Oecologia* 133: 70–77.

– 2003. Variability in marine resources affects Arctic Fox population dynamics. *Journal of Animal Ecology* 72: 668–676.

Rowell, J.E., Pierson, R.A., and Flood, P.F. 1993. Endocrine changes and luteal morphology during pregnancy in Muskoxen (*Ovibos moschatus*). *Journal of Reproductive Fertility* 99: 7–13.

Roze, U. 1989. *The North American Porcupine*. Smithsonian Institution Press, Washington, DC.

– 1990. Antibiotic properties of Porcupine quills. *Journal of Chemical Ecology* 16(3): 725–734.

– 2002. A facilitative release mechanism for quills of the North American Porcupine (*Erethizon dorsatum*). *Journal of Mammalogy* 83(2): 381–385.

– 2006. Smart weapons. *Natural History* 115(2): 48–53.

Rozenfeld, F.M., Le Boulangé, E., and Rasmont, R. 1987. Urine marking by male Bank Voles (*Clethrionomys glareolus* Schreber, 1780; Microtidae, Rodentia) in relation to their social rank. *Canadian Journal of Zoology* 65(11): 2594–601.

Rozenfeld, F.M., and Rasmont, R. 1991. Odour cue recognition by dominant male Bank Voles, *Clethrionomys glareolus*. *Animal Behaviour* 41(5): 839–850.

Rue, L.L., III. 2002. *Beavers*. Voyageur Press, Stillwater, MN.

Ruffer, D.G. 1965a. Burrows and burrowing behavior of *Onychomys leucogaster*. *Journal of Mammalogy* 46(2): 241–247.

– 1965b. Sexual behaviour of the Northern Grasshopper Mouse (*Onychomys leucogaster*). *Animal Behaviour* 13(4): 447–452.

– 1966. Observations on the calls of the Grasshopper mouse (*Onychomys leucogaster*). *Ohio Journal of Science* 66(2): 219–220.

– 1968. Agonistic behavior of the northern grasshopper mouse (*Onychomys leucogaster breviauritus*). *Journal of Mammalogy* 49(3): 481–487.

Ruggiero, L.F., and Henry, S.E. 1993. Courtship and copulatory behavior of *Martes americana*. *Northwestern Naturalist* 74: 18–22.

Ruggiero, L.F., Pearson, D.E., and Henry, S.E. 1998. Characteristics of American Marten den sites in Wyoming. *Journal of Wildlife Management* 62(2): 663–673.

Rugh, D.J. 1993. A Polar Bear kills a Walrus calf. *Northwestern Naturalist* 74: 23–24.

Rugh, D.J., Miller, G.W., Withrow, D.E., and Koski, W.R. 1992. Calving intervals of Bowhead Whales established through photographic identifications. *Journal of Mammalogy* 73(3): 487–490.

Rugh, D.J., and Shelden, K.E.W. 2002. Bowhead Whale (*Balaena mysticetus*). In W.F. Perrin, B. Wursig, and H.G.T. Thewissen (eds), *Encyclopedia of Marine Mammals*, 129–131. Academic Press, San Diego, CA.

Rugh, D.J., Shelden, K.E.W., and Schulman-Janiger, A. 2001. Timing of Grey Whale southbound migration. *Journal of Cetacean Research and Management* 3(1): 31–39.

Russell, H.J. 1998. *The World of the Caribou*. Greystone Books, Vancouver, BC.

Russell, R.J., and Finley, E.S. 1954. Swimming ability of the grasshopper mouse. *Journal of Mammalogy* 35(1): 118.

Rust, C.C., Shackelford, R.M., and Meyer, R.K. 1965. Hormonal control of pelage cycles in the mink. *Journal of Mammalogy* 46: 549–565.

Ryder, G.R., and Campbell, R.W. 2007. First Pacific Water Shrew nest for British Columbia. *Wildlife Afield* 4(1): 74–75.

Ryg, M., Solberg, Y., Lydersen, C., and Smith, T.G. 1992. The scent of rutting male Ringed Seals (*Phoca hispida*). *Journal of Zoology* (London) 226: 681–689.

Ryon, T.R. 2001. Summer nests of Preble's Meadow Jumping Mouse. *Southwestern Naturalist* 46(3): 376–378.

Ryser, J. 1992. The mating system and male mating success of the Virginia Opossum (*Didelphis virginiana*) in Florida. *Journal of Zoology* (London) 228: 127–139.

Salcedo, H. de la Cueva, Fenton, M.B., Hickey, M.B.C., and Blake, R.W. 1995. Energetic consequences of flight speeds of foraging Red and Hoary Bats (*Lasiurus borealis* and *Lasiurus cinereus*; Chiroptera: Vespertilionidae). *Journal of Experimental Biology* 198: 2245–2251.

Salsbury, C.M., and Armitage, K.B. 1994. Home-range size and exploratory excursions of adult, male Yellow-bellied Marmots. *Journal of Mammalogy* 75(3): 648–656.

Salt, J.R. 2000a. Pocket gopher/mouse associations on the Milk River Grasslands. *Blue Jay* 58(3): 139–143.

– 2000b. A note on status and habitat of Northern Lemming Vole, *Synaptomys borealis*. *Alberta Naturalist* 30(3): 54–56.

– 2005. Habitat preferences of Arctic Shrew in central and southern Alberta. *Blue Jay* 63(2): 85–6.

Samelius, G., and Alisauskas, R.T. 2000. Foraging patterns of Arctic Foxes at a large Arctic goose colony. *Arctic* 53(3): 279–288.

Samelius, G., Alisauskas, R.T., Larivière, S., Bergman, C., Hendrickson, C.J., Phipps, K., and Wood, C. 2002. Foraging behaviours of Wolverines at a large Arctic goose colony. *Arctic* 55(2): 148–150.

Samelius, G., and Lee, M. 1998. Arctic Fox, *Alopex lagopus*, predation on Lesser Snow Geese, *Chen caerlescens*, and their eggs. *Canadian Field-Naturalist* 112(4): 700–701.

Samson, C. 2001. *COSEWIC Status Report on Eastern Grey Wolf, Loup gris de l'est* (*Canis lupus lycaon*). Committee on the Status of Endangered Wildlife in Canada, Ottawa.

Samson, C., and Huot, J. 1995. Reproductive biology of female Black Bears in relation to body mass in early winter. *Journal of Mammalogy* 76(1): 68–77.

Samson, C., and Raymond, M. 1995a. Daily activity pattern and time budget of stoats (*Mustela erminea*) during summer in southern Québec. *Mammalia* 59(4): 501–510.

– 1995b. Movement and habitat preference of radio tracked stoats, *Mustela erminea*, during summer in southern Québec. *Mammalia* 62(2): 165–174.

Samways, K.M., Poulin, R.G., and Brigham, R.M. 2004. Directional tree felling by Beavers (*Castor canadensis*). *Northwestern Naturalist* 85(2): 48–52.

Santos, M.B., Pierce, G.J., Herman, J., López, A., Guerra, A., Mente, E., and Clarke, M.R. 2001. Feeding ecology of Cuvier's Beaked Whale (*Ziphius cavirostris*): A review with new information on the diet of this species. *Journal of the Marine Biological Association of the United Kingdom* 81: 687–694.

Santos, M.B., Pierce, G.J., López, A., Reid, R.J., Ridoux, V., and Mente, E. 2006. Pygmy Sperm Whale *Kogia breviceps* in the northeast Atlantic: New information on stomach contents and strandings. *Marine Mammal Science* 22(3): 600–616.

Sanvito, S., Galimberti, F., and Miller, E.H. 2007. Having a big nose: Structure, ontogeny, and function of the Elephant Seal proboscis. *Canadian Journal of Zoology* 85(2): 207–220.

Sardi, K.A., and Merigo, C. 2006. *Erignathus barbatus* (Bearded Seal) vagrant in Massachusetts. *Northeastern Naturalist* 13(1): 39–42.

Sare, D.T.J., Millar, J.S., and Longstaffe, F.J. 2005. Moulting patterns in *Clethrionomys gapperi*. *Acta Theriologica* 50(4): 561–569.

Sargeant, A.B., Sovada, M.A., and Greenwood, R.J. 1987. Responses of three Prairie ground squirrel species, *Spermophilus franklinii, S. richardsonii,* and *S. tridecemlineatus,* to duck eggs. *Canadian Field-Naturalist* 101(1): 95–97.

Saskatchewan Department of Tourism and Renewable Resources. 1978. *COSEWIC Status Report on the Black-footed Ferret* Mustela nigripes *in Canada.* Committee on the Status of Endangered Wildlife in Canada, Ottawa.

Sasse, D.B., and Pekins, P.J. 1996. Summer roosting ecology of Northern Long-eared Bats (*Myotis septentrionalis*) in the White Mountain National Forest. *Bats and Forests Symposium, 19–21 October 1995, Victoria, British Columbia.* Published by the BC Ministry of Forests Research Program.

Saugey, D.A., Crump, B.G., Vaughn, R.I., and Heidt, G.A. 1998. Notes on the natural history of *Lasiurus borealis* in Arkansas. *Journal of the Arkansas Academy of Science* 52: 92–99.

Saunders, D.A. 1988. *Adirondack Mammals.* State University of New York, College of Environmental Science and Forestry.

Saunders, M.B. 1990. Fourth red bat found in Alberta. *Blue Jay* 48(1): 57–58.

Sayner, D.P. 1982. *Tamias striatus. Mammalian Species* no. 168: 1–8.

– 2003. *COSEWIC Status Report on the Townsend's Mole,* Scapanus townsendii, *in Canada.* Committee on the Status of Endangered Wildlife in Canada.

Schaefer, J.A., and Messier, F. 1996. Winter activity of Muskoxen in relation to foraging conditions. *Ecoscience* 3(2): 147–153.

Schaefer, J.A., and Mahoney, S.P. 2001. Antlers on female Caribou: Biogeography of the bones of contention. *Ecology* 82(12): 3556–3560.

Schaefer, V.H. 1982. Movements and diel activity of the Coast Mole *Scapanus orarius* True. *Canadian Journal of Zoology* 60(3): 480–482.

Scheibe, J.S., Smith, W.P., Bassham, J., and Magness, D. 2006. Locomotor performance and cost of transport in the Northern Flying Squirrels *Glaucomys sabrinus. Acta Theriologica* 51(2): 169–178.

Schenk, A., and Kovacs, K.M. 1995. Multiple mating between Black Bears revealed by DNA fingerprinting. *Animal Behaviour* 50(6): 1483–1490.

Schenk, A., Obbard, M.E., and Kovacs, K.M. 1998. Genetic relatedness and home-range overlap among female Black Bears (*Ursus americanus*) in northern Ontario, Canada. *Canadian Journal of Zoology* 76(8): 1511–1519.

Scherrer, J.A., and Wilkinson, G.S. 1993. Evening Bat isolation calls provide evidence for heritable signatures. *Animal Behaviour* 46: 847–860.

Schilling, M.R., Seipt, I., Weirich, M.T., Frohock, S.E., Kuhlberg, A.E., and Clapham, P.J. 1992. Behavior of individually identified Sei Whales *Balaenoptera borealis* during an episodic influx into the southern Gulf of Maine in 1986. *US National Marine Fisheries Service Fisheries Bulletin* 90(4): 749–755.

Schmidly, D.J. 2004. *The Mammals of Texas.* Rev. ed. University of Texas Press, Austin, TX.

Schmidt, N.M., Berg, T.B., and Jensen, T.S. 2002. The influence of body mass on daily movement patterns and home ranges of the Collared Lemming (*Dicrostonyx groenlandicus*). *Canadian Journal of Zoology* 80(1): 64–69.

Schmidt-French, B., Gillam, E., and Fenton, M.B. 2006. Vocalizations emitted during mother-young interactions by captive Eastern Red Bats *Lasiurus borealis* (Chiroptera: Vespertilionidae). *Acta Chiropterologica* 8(2): 477–484.

Schmitz, O.J., and Nudds, T.D. 1994. Parasite-mediated competition in deer and Moose: How strong is the effect of Meningeal Worm on Moose? *Ecological Applications* 4(1): 91–103.

Schneider, E. 1981. Studies on the social behaviour of the Brown Hare. In K. Myers (ed.), *Proceedings of the World Lagomorph Conference, 1979,* 340–348. University of Guelph, Guelph, ON.

Schowalter, D.B. 1980. Swarming, reproduction, and early hibernation of *Myotis lucifugus* and *M. volans* in Alberta, Canada. *Journal of Mammalogy* 61(2): 350–354.

– 2001. Distribution and abundance of small mammals of the Western Plains as determined from Great Horned Owl pellets. *Alberta Species at Risk Report,* no. 17. Alberta Sustainable Resource Development, Fish and Wildlife Service, Edmonton, Alberta.

– 2002. Records of Pygmy Shrew, Northern Bog Lemming, and Heather Vole from owl pellets from north-central Alberta. *Alberta Naturalist* 32(1): 72–73.

Schowalter, D.B., and Allen, A. 1981. Late summer activity of Small-footed, Long-eared and Big Brown Bats in Dinosaur Park, Alberta. *Blue Jay* 39(1): 50–53.

Schowalter, D.B., and Gunson, J.R. 1979. Reproductive biology of the Big Brown Bat (*Eptesicus fuscus*) in Alberta. *Canadian Field-Naturalist* 93(1): 48–54.

Schreiber, L.A., and Swihart, R.K. 2009. Selective feeding of Pine Voles on roots of tree seedlings. *Canadian Journal of Zoology* 87(2): 183–187.

Schug, M.D., Vessey, S.H., and Korytko, A.I. 1991. Longevity and survival in a population of White-footed Mice (*Peromyscus leucopus*). *Journal of Mammalogy* 72(2): 360–366.

Schug, M.D., Vessey, S.H., and Underwood, E.M. 1992. Paternal behavior in a natural population of White-footed Mouse (*Peromyscus leucopus*). *American Midland Naturalist* 127: 373–380.

Schulte, B.A. 1998. Scent marking and responses to male castor fluid by Beavers. *Journal of Mammalogy* 79(1): 191–203.

Schultz, S.R., and Johnson, M.K. 1992. Breeding by male White-tailed Deer fawns. *Journal of Mammalogy* 73(1): 148–150.

Schultze-Hostedde, A.I., and Millar, J.S. 2002. "Little chipmunk" Syndrome? Male body size and dominance in captive Yellow-pine Chipmunks (*Tamias amoenus*). *Ethology* 108: 127–137.

Schusterman, R.J. 1981. Steller Sea Lion *Eumetopias jubatus* (Schreber 1776). In S.H. Ridgway and R. Harrison (eds), *Handbook of Marine Mammals,* vol. 1, *The Walrus, Sea Lions, Fur Seals, and Sea Otter,* 119–140. Academic Press, San Diego, CA.

Schwagmeyer, P.L. 1995. Searching today for tomorrow's mates. *Animal Behaviour* 50: 759–767.

Schwagmeyer, P.L., and Parker, G.A. 1990. Male mate choice as predicted by sperm competition in Thirteen-lined Ground Squirrels. *Nature* 348: 62–64.

Schwartz, C.C., Miller, S.D., and Haroldson, M.A. 2003. Grizzly Bear (*Ursus arctos*). In G.A. Feldhamer, B.C. Thompson, and J. A. Chapman, (eds), *Wild Mammals of North America: Biology, Management, and Conservation,* 2nd ed., 556–586. Johns Hopkins University Press, Baltimore, MD.

Schwartz, C.J., and Stobo, W.T. 2000. Estimation of juvenile survival, adult survival, and age-specific pupping probabilities for the female Grey Seal (*Halichoerus grypus*) on Sable Island from capture-recapture data. *Canadian Journal of Fisheries and Aquatic Sciences* 57(2): 247–253.

Schwartz, M.K., Mills, L.S., McKelvey, K.S., Ruggiero, L.F., and Allendorf, F.W. 2002. DNA reveals high dispersal synchronizing the population dynamics of Canada Lynx. *Nature* 415: 520–522.

Schwartz, O.A., and Armitage, K.B. 2005. Weather influences on demography of the Yellow-bellied Marmot (*Marmota flaviventris*). *Journal of Zoology* (London) 265: 73–79.

Schwede, G., Hendrichs, H., and Wemmer, C. 1994a. Sibling relations in young White-tailed Deer fawns *Odocoileus virginianus. Mammalia* 58(2): 175–181.

– 1994b. Early mother-young relations in White-tailed Deer. *Journal of Mammalogy* 75(2): 438–445.

Scobie, D. 2002. Status of the American Badger (*Taxidea taxus*) in Alberta. *Wildlife Status Report* no. 43: 1–17. Alberta Sustainable Resource Development, Fish and Wildlife Division, and Alberta Conservation Association.

Scott, F.W. 1998. *Update COSEWIC Status Report on Cougar, Eastern Population* (Puma concolor couguar). Committee on the Status of Endangered Wildlife in Canada.

Scott, F.W., and Hebda, A.J. 2004. Annotated list of the mammals of Nova Scotia. *Proceedings of the Nova Scotia Institute of Science* 42 (2): 189–208.

Scott, F.W., and van Zyll de Jong, C.G. 1989. New Nova Scotia records of the Long-tailed Shrew, *Sorex dispar*, with comments on the taxonomic status of *Sorex dispar* and *Sorex gaspensis*. *Naturalist Canadien* 116(3): 145–154.

Scott, M.D., and Cordaro, J.G. 1987. Behavioral observations of the Dwarf Sperm Whale, *Kogia simus*. *Marine Mammal Science* 3(4): 353–354.

Scott, M.D., Hohn, A.A., Westgate, A.J., Nicolas, J.R., Whitaker, B.R., and Campbell, W.B. 2001. A note on the release and tracking of a rehabilitated Pygmy Sperm Whale (*Kogia breviceps*). *Journal of Cetacean Research and Management* 3(1): 87–94.

Scott, P.A. 1993. Relationship between the onset of winter and collared lemming abundance at Churchill, Manitoba, Canada: 1932–90. *Arctic* 46(4): 293–296.

Scott, S.J. 1997. Social organisation of marine coastal otters: Overview of a work in progress. *IUCN Otter Specialist Group Bulletin* 14(1): 26–29.

Seabloom, R.W., and Theisen, P.W. 1990. Breeding biology of the Black-tailed Prairie Dog in North Dakota. *Prairie Naturalist* 22(2): 63–74.

Seaflon, R.A. 2006. Morphological evidence supports the status of the extinct Sea Mink (Mammalia: Carnivora: Mustelidae) as a separate species. In *Carnivore Talks, American Society of Mammalogists Conference, 2006*, 19.

– 2007. Dental divergence supports species status of the extinct Sea Mink (Carnivora: Mustelidae: *Neovison macrodon*). *Journal of Mammalogy* 88(2): 371–383.

Sealy, S.G. 1978. Litter size and nursery sites of the Hoary Bat near Delta, Manitoba. *Blue Jay* 36(1): 51–52.

– 1982. Voles as a source of egg and nestling loss among nesting auklets. *Murrelet* 63: 9–14.

– 2002. Discovery of a "lost" specimen of the American Water Shrew from Churchill, MB. *Blue Jay* 60(2): 113–114.

Sears, R. 2002. Blue Whale *Balaenoptera musculus*. In W.F. Perrin, B. Wursig, and H.G.T. Thewissen (eds), *Encyclopedia of Marine Mammals*, 112–116. Academic Press, San Diego, CA.

Sears, R., and Larsen, F. 2002. Long range movements of a Blue Whale (*Balaenoptera musculus*) between the Gulf of St Lawrence and West Greenland. *Marine Mammal Science* 18(1): 281–285.

Sears, R., Williamson, J.M., Wenzel, F.W., Bérubé, M., Gendron, D., and Jones, P. 1990. Photographic identification of the Blue Whale (*Balaenoptera musculus*) in the Gulf of St Lawrence, Canada. *Report of the International Whaling Commission (Special Issue)* 12: 335–342.

Sease, J.L., and York, A.E. 2003. Seasonal distribution of Steller's Sea Lions at rookeries and haul-out sites in Alaska. *Marine Mammal Science* 19(4): 745–763.

Seidensticker, J. 1999. Red Fox *Vulpes vulpes*. In D.E. Wilson and S. Ruff (eds), *The Smithsonian Book of North American Mammals*, 150–152. UBC Press, Vancouver, BC.

Sera, W.E., and Early, C.N. 2003. *Microtus montanus*. *Mammalian Species* no. 716: 1–10.

Serfass, T.L. 1995. Cooperative foraging by North American River Otters, *Lutra canadensis*. *Canadian Field-Naturalist* 109(4): 458–459.

Serrano, A. 2001. New underwater and aerial vocalizations of captive Harp Seals (*Pagophilus groenlandicus*). *Canadian Journal of Zoology* 79(1): 75–81.

Servheen, C., Herrero, H., and Peyton, B. (comps). 1998. *Bears: Status Survey and Conservation Action Plan*. IUCN/SSC Bear and Polar Bear Specialist Groups, IUCN, Gland, Switzerland.

Seton, E.T. 1929. *Lives of Game Animals*, vol. 4, part 2, *Rodents etc*. Doubleday, Doran, Garden City, NJ.

Shackleton, D. 1999. *Hoofed Mammals of British Columbia*. Vol. 3 of the Mammals of British Columbia Series. Royal British Columbia Museum Handbook. UBC Press, Vancouver, BC.

Shafer, A.B.A., Scott, F.W., Petersen, S.D., Rhymer, J.M., and Stewart, D.T. 2008. Following the SINEs: A taxonomic revision of the Long-tailed Shrew complex, *Sorex dispar* and *S. gaspensis*. *Journal of Mammalogy* 89(6): 1421–1427.

Shafer, A.B.A., and Stewart, D.T. 2006. A disjunct population of *Sorex dispar* (Long-tailed Shrew) in Nova Scotia. *Northeastern Naturalist* 13(4): 603–608.

– 2007. Phylogenetic relationships among Nearctic shrews of the genus *Sorex* (Insectivora, Soricidae) inferred from combined cytochrome b and inter-SINE fingerprint data using Bayesian analysis. *Molecular Phylogenetics and Evolution* 44(1): 192–203.

Shane, S.H. 1994. Pilot whales carrying dead sea lions. *Mammalia* 58(3): 494–498.

– 1995a. Relationship between pilot whales and Risso's Dolphins at Santa Catalina Island, California, USA. *Marine Ecology Progress Series* 123: 5–11.

– 1995b. Behavior patterns of pilot whales and Risso's Dolphins off Santa Catalina Island, California. *Aquatic Mammals* 21(3): 195–197.

Sharma, S., Couturier, S., and Côté, S.D. 2009. Impacts of climate change on the seasonal distribution of migratory Caribou. *Global Climate Change* 15(10): 2549–2562.

Sharpe, S.T., and Millar, J.S. 1990. Relocation of nest sites by female Deer Mice, *Peromyscus maniculatus borealis*. *Canadian Journal of Zoology* 68(11): 2364–237.

Shaughnessy, P.D., and Fay, F.H. 1977. A review of the taxonomy and nomenclature of North Pacific Harbour Seals. *Journal of Zoology* (London) 182: 385–419.

Shaw, D.C., and Flick, C.J. 2002. Seasonal variation in vertical distribution of Douglas' Squirrel, *Tamiasciurus douglasii*, in an old-growth Douglas-fir and Western Hemlock forest in the morning. *Northwestern Naturalist* 83: 123–125.

Sheats, R., and Cason, K. 2005. Deer Speak. *University of Georgia Research Magazine*, summer.

Sheehan, S.T., and Galindo-Leal, C. 1997. Identifying Coast Moles, *Scapanus orarius*, and Townsend's Mole, *Scapanus townsendii*, from tunnel and mound size. *Canadian Field-Naturalist* 111(3): 463–465.

Sheffield, S.R. 1998. The Southern Bog Lemming in central Kansas. *Prairie Naturalist* 30(2): 129–132.

Sheffield, S.R., and King, C.M. 1994. *Mustela nivalis*. *Mammalian Species* no. 454: 1–10.

Sheffield, S.R., and Thomas, H.H. 1997. *Mustela frenata*. *Mammalian Species* no. 570: 1–9.

Shelden, K.E.W. 1999. Risso's Dolphin (*Grampus griseus*). In D.E. Wilson, and S. Ruff (eds), *The Smithsonian Book of North American Mammals*, 280–282. UBC Press, Vancouver, BC.

Shelden, K.E.W., Baldridge, A., and Withrow, D.E. 1995. Observations of Risso's Dolphin, *Grampus griseus*, with Gray Whales, *Eschrichtius robustus*. *Marine Mammal Science* 11(2): 231–240.

Shelton, P.A., Stenson, G.B., Sjare, B., and Warren, W.G. 1996. Model estimates of Harp Seal numbers-at-age for the Northwest Atlantic. *Northwest Atlantic Fisheries Organization Scientific Council Studies* 26: 1–14.

Sherwin, R.E., Stricklan, D., and Rogers, D.S. 2000. Roosting affinities of Townsend's Big-eared Bat (*Corynorhinus townsendii*) in northern Utah. *Journal of Mammalogy* 81(4): 939–947.

Shilton, C.M., and Brooks, R.J. 1989. Paternal care in captive Collared Lemmings (*Dicrostonyx richardsoni*) and its effect on development of the offspring. *Canadian Journal of Zoology* 67(11): 2740–2745.

Shipp-Pennock, M.A., Webster, W.D., and Freshwater, D.W. 2005. Systematics of the White-footed Mouse (*Peromyscus leucopus*) in the mid-Atlantic Region. *Journal of Mammalogy* 86(4): 803–813.

Shriner, W. 1998. Yellow-bellied Marmot and Golden-mantled Ground Squirrel responses to heterospecific alarm calls. *Animal Behaviour* 55: 529–536.

Shump, K.A., Jr., and Shump, A. 1982a. *Lasiurus borealis*. *Mammalian Species* no. 183: 1–6.

– 1982b. *Lasiurus cinereus*. *Mammalian Species* no. 185: 1–5.

Sidle, J.G., Johnson, D.H., and Euliss, B.R. 2001. Estimated areal extent of colonies of Black-tailed Prairie Dogs in the northern Great Plains. *Journal of Mammalogy* 82(4): 928–936.

Simão, S.M., and Moreira, S.C. 2005. Vocalizations of a female Humpback Whale in Arraial do Cabo (RJ, Brazil). *Marine Mammal Science* 21(1): 150–153.

Simon, N.P.P., Schwab, F.E., Baggs, E.M., and Cowan, I. McT. 1998. Distribution of small mammals among successional and mature forest types in western Labrador. *Canadian Field-Naturalist* 112(3): 441–445.

Simpson, M.R. 1990. Observation of an Arctic Ground Squirrel, *Spermophilus parryii*, – Short-tailed Weasel, *Mustela erminea*, interaction. *Canadian Field-Naturalist* 104(3): 473–474.

– 1993. *Myotis californicus*. *Mammalian Species* no. 428: 1–4.

– 1994. Possible selective disadvantage of a coat color mutant in the Arctic Ground Squirrel *Spermophilus parryii. American Midland Naturalist* 132: 199–201.

Sinclair, E.H., and Zepplin, T.K. 2002. Seasonal and spatial differences in diet in the western stock of Steller Sea Lion (*Eumetopias jubatus*). *Journal of Mammalogy* 83(4): 973–990.

Sinha Hikim, A.P., Woolf, A., Bartke, A., and Amador, A.G. 1992. Further observations on estrus and ovulation in Woodchucks (*Marmota monax*) in captivity. *Biology of Reproduction* 46: 10–16.

Sittler, B. 1995. Response of Stoats (*Mustela erminea*) to a fluctuating Lemming (*Dicrostonyx groenlandius*) population in North East Greenland: Preliminary results from a long-term study. *Annals Zoologici Fennici* 32: 79–92.

Sittler, B., Gilg, O., and Berg, T.B. 2000. Low abundance of King Eider nests during low lemming years in northeast Greenland. *Arctic* 53(1): 53–60.

Sjare, B., Lebeuf, M., and Veinot, G. 2005. *Harbor Seals in Newfoundland and Labrador: A Prelinary Summary of New Data on Aspects of Biology, Ecology and Contaminant Profiles*. Canadian Science Advisory Secretariat, Research Document 2005/030, 1–42.

Sjare, B., and Stirling, I. 1996. The breeding behaviour of Atlantic Walruses, *Odobenus rosmarus rosmarus*, in the Canadian High Arctic. *Canadian Journal of Zoology* 74(5): 897–911.

Sjare, B., Sterling, I., and Spencer, C. 2003. Structural variation in the songs of Atlantic Walruses breeding in the Canadian High Arctic. *Aquatic Mammals* 29(2): 297–318.

Sklepkovych, B. 1994. Arboreal foraging by Red Foxes, *Vulpes vulpes*, during winter food shortage. *Canadian Field-Naturalist* 108(4): 479–481.

Sklepkovych, B.O., and Montevecchi, W.A. 1996. Food availability and food hoarding behaviour by Red and Arctic Foxes. *Arctic* 49(3): 228–234.

Skupski, M.P. 1995. Population ecology of the Western Harvest Mouse *Reithrodontomys megalotis*: A long-term perspective. *Journal of Mammalogy* 76(2): 358–367.

Slobodchikoff, C.N., Vaughan, T.A., and Warner, R.M. 1987. How prey defenses affect a predator's net energetic profit. *Journal of Mammalogy* 68(3): 668–671.

Slough, B.G. 1999. Characteristics of Canada Lynx, *Lynx canadensis*, maternal dens and denning habitat. *Canadian Field-Naturalist* 113(4): 605–608.

– 2003. *COSEWIC Assessment and Update Status Report on the Wolverine* Gulo gulo *in Canada*. Committee on the Status of Endangered Wildlife in Canada, Ottawa, ON.

– 2007. Status of the Wolverine Gulo gulo in Canada. *Wildlife Biology* 13(suppl. 2): 76–82.

– 2009. Behavioral thermoregulation by a maternity colony of Little Brown Bats in Yukon. *Northwestern Naturalist* 90(1): 47–51.

Slough, B.G., and Jung, T.S. 2007. Diversity and distribution of the terrestrial mammals of the Yukon Territory: A review. *Canadian Field-Naturalist* 121(2): 119–127.

Slough, B.G., and Mowat, G. 1996. Lynx population dynamics in an untrapped refugium. *Journal of Wildlife Management* 60(4): 946–961.

Small, R.J., Boveng, P.L., Byrd, G.V., and Withrow, D.E. 2008. Harbor Seal population decline in the Aleutian Archipelago. *Marine Mammal Science* 24(4): 845–863.

Small, R.J., Keith, L.B., and Barta, R.M. 1991. Dispersion of introduced Arctic Hares (*Lepus arcticus*) on islands off Newfoundland's south coast. *Canadian Journal of Zoology* 69(10): 2618–2623.

Smallwood, K.S. 1993. Mountain Lion vocalizations and hunting behavior. *Southwestern Naturalist* 38(1): 65–67.

Smith, A.C., and Schaefer, J.A. 2002. Home-range size and habitat selection by American Marten (*Martes americana*) in Labrador. *Canadian Journal of Zoology* 80(9): 1602–1609.

Smith, A.E., and Hill, M.R.J. 1996. Polar Bear, *Ursus maritimus*, depredation of Canada Goose, *Branta canadensis*, nests. *Canadian Field Naturalist* 110(2): 339–340.

Smith, A.R. 2009. 36th annual Saskatchewan Christmas mammal count, 2008. *Blue Jay* 67(1): 42–49.

Smith, A.T. 1981. Territoriality and social behaviour of *Ochotona princeps*. In K. Myers and C.D. MacInnes (eds), *Proceeding of the World Lagomorph Conference, August 12–16, 1979*, 310–323. University of Guelph, Guelph, ON.

Smith, A.T., Formozov, N.A., Hoffmann, R.S., Changlin, Z., and Erbajeva, M.A. 1990. The Pikas. In J.A. Chapman and J.E.C. Flux (eds), *Rabbits, Hares, and Pikas: Status Survey and Conservation Action Plan*, 14–60. IUCN Press, Gland, Switzerland.

Smith, A.T., and Weston, M.L. 1990. *Ochotona princeps. Mammalian Species* no. 352: 1–8.

Smith, C.C. 1968. The adaptive nature of social organization in the genus of tree squirrels *Tamiasciurus. Ecological Monographs* 38(1): 31–63.

Smith, D.A., and Foster, J. B. 1957. Notes on small mammals of Churchill, Manitoba. *Journal of Mammalogy* 38(1): 98–115.

Smith, D.W., and Jenkins, S.H. 1997. Seasonal change in body mass and size of tail of Northern Beavers. *Journal of Mammalogy* 78(3): 869–876.

Smith, D.W., Murphy, K.M., and Monger, S. 2001. Killing of a Bison, *Bison bison*, calf by a Wolf, *Canis lupus*, and four Coyotes, *Canis latrans*, in Yellowstone National Park. *Canadian Field-Naturalist* 115(2): 343–345.

Smith, F.A. 1997. *Neotoma cinerea. Mammalian Species* no. 564: 1–8.

Smith, H.C. 1993. *Alberta Mammals: An Atlas and Guide*. Provincial Museum of Alberta, Edmonton, AB.

Smith, H.C., and Edmonds, E.J. 1985. The Brown Lemming, *Lemmus sibiricus*, in Alberta. *Canadian Field-Naturalist* 99(1): 99–100.

Smith, K.G. 1984. Dark-eyed Junco, *Junco hyemalis*, nest usurped by Pacific Jumping Mouse, *Zapus trinotatus. Canadian Field-Naturalist* 98(1): 47–48.

Smith, M.E., and Belk, M.C. 1996. *Sorex monticolus. Mammalian Species* no. 528: 1–5.

Smith, M.E., and Follmann, E.H. 1993. Grizzly Bear, *Ursus arctos*, predation of a denned adult Black Bear, *U. americanus. Canadian Field-Naturalist* 107(1): 97–99.

Smith, R.J. 1995. Harvest rates and escape speeds in two coexisting species of montane ground squirrels. *Journal of Mammalogy* 76(1): 189–195.

– 1997. Status of the Lacs des Loups Marins Harbour Seal, *Phoca vitulina mellonae*, in Canada. *Canadian Field-Naturalist* 111(2): 270–276.

Smith, R.J., Cox, T.M., and Westgate, A.J. 2006. Movements of Harbor Seals (*Phoca vitulina mellonae*) in Lacs des Loups Marins, Quebec. *Marine Mammal Science* 22(2): 480–485.

Smith, R.J., Lavigne, D.M., and Leonard, W.R. 1994. Subspecific status of the freshwater Harbor Seal (*Phoca vitulina mellonae*): A re-assessment. *Marine Mammal Science* 10(1): 105–110.

Smith, T.D., Barthelmess, K., and Reeves, R.R. 2006. Using historical records to relocate a long-forgotten summer feeding ground of North Atlantic Right Whales. *Marine Mammal Science* 22(3): 723–734.

Smith, T.G., Hammill, M.O., and Taugbøl, G. 1991. A review of the developmental, behavioural and physiological adaptations of the Ringed Seal, *Phoca hispida*, to life in the Arctic winter. *Arctic* 44(2): 124–131.

Smith, T.G., and Harwood, L.A. 2001. Observations of neonate Ringed Seals, *Phoca hispida*, after early break-up of sea-ice in Prince Albert Sound, Northwest Territories, Canada, spring 1998. *Polar Biology* 24: 215–219.

Smith, T.G., St Aubin, D.J., and Hammill, M.O. 1992. Rubbing behaviour of Belugas, *Delphinapterus leucas*, in a High Arctic estuary. *Canadian Journal of Zoology* 70(12): 2405–2409.

Smith, W.J., Smith, S.L., Oppenheimer, E.L., and DeVilla, J.G. 1977. Vocalizations of the Black-tailed Prairie Dog, *Cynomys ludovicianus. Animal Behaviour* 25: 152–164.

Smith, W.P. 1991. *Odocoileus virginianus. Mammalian Species* no. 388: 1–13.

Smith, W.P., and Nichols, J.V. 2004. Demography of two endemic forest-floor mammals of southeastern Alaskan temperate rain forest. *Journal of Mammalogy* 85(3): 540–551.

Smits, C.M.M., and Slough, B.G. 1993. Abundance and summer occupancy of Arctic Fox, *Alopex lagopus*, and Red Fox, *Vulpes vulpes*, dens in the northern Yukon Territory, 1984–1990. *Canadian Field-Naturalist* 107(1): 13–18.

Smolen, M.J. 1981. *Microtus pinetorum. Mammalian Species* no. 147: 1–7.

Smolen M.J., and Keller, B.L. 1987. *Microtus longicaudus. Mammalian Species* no. 271: 1–7.

Sneddon, I.A. 1991. Latrine use by the European Rabbit (*Oryctolagus cuniculus*). *Journal of Mammalogy* 72(4): 769–775.

Snyder, J. 2004. Survival on the hoof: Is time running out for BC's Mountain Caribou? *BC Naturalist* no. 5: 18–19, 26.

Sobey, D.G. 2007. An analysis of the historical records for the native mammalian fauna of Prince Edward Island. *Canadian Field-Naturalist* 121(4): 384–396.

Sobolevsky, Y.I. 1996. Distribution and seasonal feeding behavior of Spotted Seals (*Phoca largha*) in the Bering Sea. *Alaska Sea Grant Report* 96(1): 289–296.

Solick, D.I., and Barclay, R.M.R. 2006. Thermoregulation and roosting behaviour of reproductive and nonreproductive female Western Long-eared Bats (*Myotis evotis*) in the Rocky Mountains of Alberta. *Canadian Journal of Zoology* 84(4): 589–599.

– 2007. Geographic variation in the use of torpor and roosting behaviour of female Western Long-eared Bats. *Journal of Zoology* 272: 358–366.

Sollberger, D.E. 1940. Notes on the life history of the small eastern flying squirrel. *Journal of Mammalogy* 21: 282–293.

Sollberger, D.E. 1943. Notes on the breeding habits of the Eastern Flying Squirrel (*Glaucomys volans volans*). *Journal of Mammalogy* 24: 163–173.

Solomon, N.G. 1994. Eusociality in a microtone rodent. *Trends in Ecology and Evolution* 9(7): 264.

Solomon, N.G., and Jacquot, J.J. 2002. Characteristics of resident and wandering Prairie Voles, *Microtus ochrogaster*. *Canadian Journal of Zoology* 80(5): 951–955.

Solomon, N.G., Keane, B., Knoch, L.R., and Hogan, P.J. 2004. Multiple paternity in socially monogamous Prairie Voles (*Microtus ochrogaster*). *Canadian Journal of Zoology* 82(10): 1667–1671.

Solomon, N.G., Vandenbergh, J.G., and Sullivan, W.T., Jr. 1998. Social influences on intergroup transfer by Pine Voles (*Microtus pinetorum*). *Canadian Journal of Zoology* 76(12): 2131–2136.

Somers, M., and Thiel, R.P. 2008. Use of winter dens by Porcupines, *Erethizon dorsatum*, in Wisconsin. *Canadian Field-Naturalist* 122(1): 45–48.

Soper, J.D. 1961. The mammals of Manitoba. *Canadian Field-Naturalist* 75(4): 171–219.

– 1964. *The Mammals of Alberta*. Hamly Press, Edmonton, AB.

Sopher, L. 2000. The Boreal Red-backed Vole. *Newfoundland and Labrador Forestry News* 2(2): 3.

Sorin, A.B. 2004. Paternity assignment for White tailed Deer (*Odocoileus virginianus*): Mating across age classes and multiple paternity. *Journal of Mammalogy* 85(2): 356–362.

Soto, N.A., Johnson, M., Madsen, P.T., Tyack, P.L., Bocconcelli, A., and Borsani, J.F. 2006. Does intense ship noise disrupt foraging in deep-diving Cuvier's Beaked Whales (*Ziphius cavirostris*)? *Marine Mammal Science* 22(3): 690–699.

Sowls, L.K. 1948. The Franklin's Ground Squirrel, *Citellus franklinii* (Sabine), and its relationship to nesting ducks. *Journal of Mammalogy* 29: 113–137.

Sparks, D.W., and Choate, J.R. 1995. Attempted predation on a Short-tailed Shrew by a Common Grackle. *Kentucky Warbler* 71(3): 48.

Sparks, D.W., Simmons, M.T., Gummer, C.L., and Duchamp, J.E. 2003. Disturbance of roosting bats by woodpeckers and raccoons. *Northeastern Naturalist* 10(1): 105–108.

Speckman, S.G., and Sheffield, G. 2001. First record of an anomalously white Killer Whale, *Orcinus orca*, near St Lawrence Island, northern Bering Sea, Alaska. *Canadian Field-Naturalist* 115(3): 501–502.

Spitz, S.S., Herman, L.M., Pack, A.A., and Deakos, M.H. 2002. The relation of body size of male Humpback Whales to their social roles on the Hawaiian winter grounds. *Canadian Journal of Zoology* 80(11): 1938–1947.

Spradling, K.D., Stangl, F.B., Jr., and Cook, W.B. 2003. Evidence of a case of multiple paternity in the Red Bat (*Lasiurus borealis*) as indicated by DNA fingerprinting. *Occasional Papers of the Museum of Texas Tech University* no. 224: 1–10.

Springer, A.M., Estes, J.A., van Vliet, G.B., Williams, T.M., Doak, D.F., Danner, E.M., Forney, K.A., and Pfister, B. 2003. Sequential megafaunal collapse in the North Pacific Ocean: An ongoing legacy of industrial whaling? *Publications of the National Academy of Sciences* 100(21): 12223–12228.

Stacey, P.J., and Baird, R.W. 1991a. Status of the False Killer Whale, *Pseudorca crassidens*, in Canada. *Canadian Field Naturalist* 105(2): 189–197.

– 1991b. Status of the Pacific White-sided Dolphin, *Lagenorhynchus obliquidens*, in Canada. *Canadian Field Naturalist* 105(2): 219–232.

– 1993. Status of the Short-finned Pilot Whale, *Globicephala macrorhynchus*, in Canada. *Canadian Field Naturalist* 107(4): 481–489.

– 1997. Birth of a "resident" Killer Whale off Victoria, British Columbia, Canada. *Marine Mammal Science* 13(3): 504–508.

Stacey, P.J., Leatherwood, S., and Baird, R.W. 1994. *Pseudorca crassidens*. *Mammalian Species* no. 456: 1–6.

Staedler, M., and Riedman, M.L. 1993. Fatal mating injuries in female Sea Otters (*Enhydra lutris nereis*). *Mammalia* 57(1): 135–139.

Stafford, K.C., III. 1993. The epizootiology of Lyme disease. *Northeast Wildlife* 50: 181–189.

Stafford, K.M. 2003. Two types of Blue Whale calls recorded in the Gulf of Alaska. *Marine Mammal Science* 19(4): 682–693.

Stafford, K.M., Moore, S.E., Spillane, M., and Wiggins, S. 2007. Gray Whale calls recorded near Barrow, Alaska, throughout the winter of 2003–04. *Arctic* 60(2): 167–172.

Stafford, K.M., Nieukirk, S.L., and Fox, C.G. 2001. Geographic and seasonal variation on Blue Whale calls in the North Pacific. *Journal of Cetacean Research and Management* 3(1): 65–76.

Stahler, D.R., Smith, D.W., and Landis, R. 2002. The acceptance of a new breeding male into a wild wolf pack. *Canadian Journal of Zoology* 80(2): 360–365.

Stalling, D.T. 1990. *Microtus ochrogaster*. *Mammalian Species* no. 355: 1–9.

Stallman, E.L., and Holmes, W.G. 2002. Selective foraging and food distribution of high-elevation Yellow-bellied Marmots. *Journal of Mammalogy* 83(2): 576–584.

Stamper, M.A., Whitaker, B.R., and Schofield, T.D. 2006. Case study: Morbidity in a Pygmy Sperm Whale *Kogia breviceps* due to ocean-borne plastic. *Marine Mammal Science* 22(3): 719–722.

Standing, K.L., Herman, T.B., and Morrison, I.P. 2000. Predation of neonate Blanding's Turtles (*Emydoidea blandingii*) by Short-tailed Shrews (*Blarina brevicauda*). *Chelonian Conservation and Biology* 3(4): 658–660.

Stapp, P. 1992. Energetic influences on the life history of *Glaucomys volans*. *Journal of Mammalogy* 73(4): 914–920.

– 1999. Size and habitat characteristics of home ranges of Northern Grasshopper Mice (*Onychomys leucogaster*). *Southwestern Naturalist* 44(1): 101–105.

Stapp, P., and Mautz, W.W. 1991. Breeding habits and postnatal growth of the Southern Flying Squirrel (*Glaucomys volans*) in New Hampshire. *American Midland Naturalist* 126(1): 203–208.

Stapp, P., Pekins, P.J., and Mautz, W.W. 1991. Winter energy expenditure and the distribution of Southern Flying Squirrels. *Canadian Journal of Zoology* 69(10): 2548–2555.

Steele, M.A. 1999. *Tamiasciurus douglasii*. *Mammalian Species* no. 630: 1–8.

Steele, M.A., and Koprowski, J.L. 2002. *North American Tree Squirrels*. Smithsonian Books, Washington, DC.

Stelle, L.L., Megill, W.M., and Kinzel, M.R. 2008. Activity budget and diving behavior of Gray Whales (*Eschrichtius robustus*) in feeding grounds off coastal British Columbia. *Marine Mammal Science* 24(3): 462–478.

Stenseth, N.C. 1985. Geographic distribution of *Clethrionomys* species. *Annales Zoologici Fennici* 22: 215–219.

Stephenson, R.O., Gerlach, S.C., Guthrie, R.D., Harington, C.R., Mills, R.O., and Hare, G. 2001. Wood Bison in Late Holocene Alaska and adjacent Canada: Paleontological, archaeological, and historical records. *BAR International Series* 944: 124–158.

Stephenson, R.O., Grangaard, D.V., and Burch, J. 1991. Lynx, *Felis lynx*, predation on Red Foxes, *Vulpes vulpes*, Caribou, *Rangifer tarandus*, and Dall Sheep, *Ovis dalli*, in Alaska. *Canadian Field-Naturalist* 105(2): 255–262.

Stephenson, T.R., and Van Ballenberghe, V. 1995. Defense of one twin calf against wolves, *Canis lupus*, by a female Moose, *Alces alces*. *Canadian Field-Naturalist* 109(2): 251–253.

Stevens, R.T., and Kennedy, M.L. 2006. Geographic variation in body size of American Mink (*Mustela vison*). *Mammalia* 70(1/2): 145–152.

Stevens, S.D. 1998. High incidence of infanticide by lactating females in a population of Columbian Ground Squirrels (*Spermophilus columbianus*). *Canadian Journal of Zoology* 76(6): 1183–1187.

Stevens, S.D., Coffin, J., and Strobeck, C. 1997. Microsatellite loci in Columbian Ground Squirrels, *Spermophilus columbianus*. *Molecular Ecology* 6: 493–495.

Stevens, S.S., and Serfass, T.L. 2005. Sliding behavior in Nearctic River Otters: Locomotion or play? *Northeastern Naturalist* 12(2): 241–244.

Stevens, W.F., and Weisbrod, A.R. 1981. The biology of the European Rabbit on San Juan Island, Washington, USA. In K. Myers and C.D. MacInnes (eds), *Proceedings of the World Lagomorph Conference, 1979*, 870–879. University of Guelph, Guelph, ON.

Stewart, B.E., and Burt, P.M. 1994. Extralimital occurrences of Beluga, *Delphinapterus leucas*, and Walrus, *Odobenus rosmarus*, in Bathurst Inlet, Northwest Territories. *Canadian Field-Naturalist* 108(4): 488–490.

Stewart, B.E., and Stewart, R.E.A. 1989. *Delphinapterus leucas*. *Mammalian Species* no. 336: 1–8.

Stewart, B.S. 1997. Ontogeny of differential migration and sexual segregation in Northern Elephant Seals. *Journal of Mammalogy* 78(4): 1101–1116.

– 1999. Northern Elephant Seal, *Mirounga angustirostris*. In D.E. Wilson and S. Ruff (eds), *The Smithsonian Book of North American Mammals*, 217–218. UBC Press, Vancouver, BC.

Stewart, B.S., and DeLong, R.L. 1995. Double migrations of the Northern Elephant Seal, *Mirounga angustirostris*. *Journal of Mammalogy* 76(1): 196–205.

Stewart, B.S., and Huber, H.R. 1993. *Mirounga angustirostris*. *Mammalian Species* no. 449: 1–10.

Stewart, B.S., and Leatherwood, S. 1985. Minke Whale. In S.H. Ridgway and R.J. Harrison (eds), *Handbook of Marine Mammals*, vol. 3, 91–136. Academic Press, San Diego, CA.

Stewart, C.A. 1991. A note on the microdistribution of the red-backed vole, *Clethrionomys gapperi*, in the E.N. Huyck Preserve, New York. *Canadian Field-Naturalist* 105(2): 274–275.

Stewart, D.T., Baker, A.J., and Hindocha, S.P. 1993. Genetic differentiation and population structure in *Sorex haydeni* and *S. cinereus*. *Journal of Mammalogy* 74(1): 21–32.

Stewart, D.T., McPherson, M., Robichaud, J., and Fumagalli, L. 2003. Are there two species of Pygmy Shrews (*Sorex*)? Revisiting the question using DNA sequence data. *Canadian Field-Naturalist* 117(1): 82–88.

Stewart, D.T., Perry, N.D., and Fumagalli, L. 2002. The Maritime Shrew, *Sorex maritimensis* (Insectivora: Soricidae): A newly recognized Canadian endemic. *Canadian Journal of Zoology* 80(1): 94–99.

Stewart, R.E.A., et al. 2008. Redefining Walrus stocks in Canada. *Arctic* 61(3): 292–308.

Stewart, R.E.A., Outridge, P.M., and Stern, R.A. 2003. Walrus life-history movements reconstructed from lead isotopes in annual layers of teeth. *Marine Mammal Science* 19(4): 806–818.

Stirling, I. 1988. *Polar Bears*. University of Michigan Press, Ann Arbor, MI.

– 1999. Polar Bear *Ursus maritimus*. In D.E. Wilson and S. Ruff (eds), *The Smithsonian Book of North American Mammals*, 163–164. UBC Press, Vancouver, BC.

Stirling, I., and Holst, M. 2000. Observations of Hooded Seals, *Cystophora cristata*, in the northwestern Labrador Sea and southern Davis Strait in March–April 1998. *Canadian Field-Naturalist* 114(1): 147–149.

Stirling, I., and Parkinson, C.L. 2006. Possible effects of climate warming on selected populations of Polar Bears (*Ursus maritimus*) in the Canadian Arctic. *Arctic* 59(3): 261–275.

Stirling, I., and Smith, T.G. 2004. Implications of warm temperatures and an unusual rain event for the survival of Ringed Seals on the coast of southeastern Baffin Island. *Arctic* 57(1): 59–67.

Stone, G.S., Flórez-Gonzalez, L., and Katona, S. 1990. Whale migration record. *Nature* 346(6286): 705.

Stone, K.D., Heidt, G.A., Caster, P.T., and Kennedy, M.L. 1997. Using geographic information systems to determine home range of the Southern Flying Squirrel (*Glaucomys volans*). *American Midland Naturalist* 137(1): 106–111.

Stoner, D.C., Rieth, W.R., Wolfe, M.L., Mecham, M.B., and Neville, A. 2008. Long-distance dispersal of a female Cougar in a basin and range landscape. *Journal of Wildlife Management* 72(4): 933–939.

Storeheier, P.V., and Nordøy, E.S. 2001. Physiological effects of seawater intake in adult Harp Seals during phase 1 of fasting. *Comparative Biochemistry and Physiology, Part A* 128: 307–315.

Storer, T.I., and Davis, D.E. 1953. Studies of rat reproduction in San Francisco. *Journal of Mammalogy* 34(3): 365–373.

Storey, A.E., French, R.J., and Payne, R. 1995. Sperm competition and mate guarding in Meadow Voles (*Microtus pennsylvanicus*). *Ethology* 101: 265–279.

Storz, J.F. 1995. Local distribution and foraging behavior of the Spotted Bat (*Euderma maculatum*) in northwestern Colorado and adjacent Utah. *Great Basin Naturalist* 55(1): 78–83.

Strand, O., Landa, A., Linnell, J.D.C., Zimmerman, B., and Skogland, T. 2000. Social organization and parental behavior in the Arctic Fox. *Journal of Mammalogy* 81(1): 223–233.

Streubel, D.P., and Fitzgerald, J.P. 1978. *Spermophilus tridecemlineatus*. *Mammalian Species* no. 103: 1–5.

Stronen, A.V., Paquet, P., Herrero, S., Sharpe, S., and Waters, N. 2007. Translocation and recovery efforts for the Telkwa Caribou, *Rangifer tarantus caribou*, herd in west-central British Columbia, 1997–2005. *Canadian Field-Naturalist* 121(1): 155–163.

Strub, H. 1992. Swim by an Arctic Fox, *Alopex lagopus*, in Alexandra Fiord, Ellesmere Island, Northwest Territories. *Canadian Field-Naturalist* 106(4): 513–514.

Struzik, E. 2006. Pizzlies of the Arctic. *Canadian Geographic*, Nov/Dec: 40–41.

Stuart-Smith, A.K., and Boutin, S. 1994. Costs of escalated territorial defence in Red Squirrels. *Canadian Journal of Zoology* 72(6): 1162–1167.

Sullivan, R. 2004. *Rats*. Bloomsbury, New York.

Sullivan, T.P., Jones, B., and Sullivan, D.S. 1989. Population ecology and conservation of the Mountain Cottontail, *Sylvilagus nuttallii nuttallii*, in southern British Columbia. *Canadian Field-Naturalist* 103(3): 335–340.

Sullivan, T.P., and Krebs, C.J. 1981. *Microtus* population biology: Demography of *M. oregoni* in southwestern British Columbia. *Canadian Journal of Zoology* 59(11): 2092–2102.

Sullivan, T.P., Sullivan, D.S., and Hogue, E.J. 2003. Demography of Montane Voles in old field and orchard habitats in southern British Columbia. *Northwest Science* 77(3): 228–236.

– 2004. Population dynamics of Deer Mice, *Peromyscus maniculatus*, and Yellow-pine Chipmunks, *Tamias amoenus*, in old field and orchard habitats. *Canadian Field Naturalist* 118(3): 299–308.

Sumich, J.L., Goff, T., and W.L. Perryman 2001. Growth of two captive Gray Whale calves. *Aquatic Mammals* 27(3): 231–233.

Summers, A. 2004. As the whale turns. *Natural History* 113(5): 24–25.

Sun, C. 1997. Dispersal of young in Red Squirrels (*Tamiasciurus hudsonicus*). *American Midland Naturalist* 138: 252–259.

Sun, L., and Müller-Schwarze, D. 1997. Sibling recognition in the Beaver: A field test for phenotype matching. *Animal Behaviour* 54: 493–502.

Sun, L., Müller-Schwarze, D., and Schulte, B.A. 2000. Dispersal distance and effective population size of the Beaver. *Canadian Journal of Zoology* 78(3): 393–398.

Sutton, D.A. 1992. *Tamias amoenus*. *Mammalian Species* no. 390: 1–8.

– 1993. *Tamias townsendii*. *Mammalian Species* no. 435: 1–6.

Sutton, R.W. 1964. Range extension of Raccoon in Manitoba. *Journal of Mammalogy* 45(2): 311–312.

Svendsen, G.E. 2003. Weasels and Black-footed Ferret (*Mustela* species). In G.A. Feldhamer, B.C. Thompson, and J. A. Chapman (eds), *Wild Mammals of North America: Biology, Management, and Conservation*, 2nd ed., 650–661. Johns Hopkins University Press, Baltimore, MD.

Svendsen, G.E., and White, M.M. 1997. Body mass and first-time reproduction in female chipmunks (*Tamias striatus*). *Canadian Journal of Zoology* 75(11): 1891–1895.

Swanson, B.J., Peters, L.R., Kyle, C.J. 2006. Demographic and genetic evaluation of an American Marten reintroduction. *Journal of Mammalogy* 87(2): 272–280.

Swartz, S.L., Taylor, B.L., and Rugh, D.J. 2006. Gray Whale *Eschrichtius robustus* population and stock identity. *Mammal Review* 36(1): 66–84.

Sweitzer, R.A. 1996. Predation or starvation: Consequences of foraging decisions by Porcupines (*Erethizon dorsatum*). *Journal of Mammalogy* 77(4): 1068–1077.

Sweitzer, R.A., and Berger, J. 1997. Sexual dimorphism and evidence for intrasexual selection from quill impalements, injuries, and mate guarding in Porcupines (*Erethizon dorsatum*). *Canadian Journal of Zoology* 75(6): 847–854.

– 1998. Evidence for female-biased dispersal in North American Porcupines (*Erethizon dorsatum*). *Journal of Zoology* (London) 244: 159–166.

Swihart, R.K., and Picone, P.M. 1991. Arboreal foraging and palatability of tree leaves to Woodchucks. *American Midland Naturalist* 125: 372–374.

Sykes, J. 1996. Great Blue Heron eating a Richardson's Ground Squirrel. *Blue Jay* 54(3): 165–170.

Taitt, M.J., and Krebs, C.J. 1985. Population dynamics and cycles. In R.H. Tamarin (ed.), *Biology of the New World Microtus*, 567–620. Special Publication no. 8, American Society of Mammalogists.

Tannerfeldt, M., and Angerbjörn, A. 1998. Fluctuating resources and the evolution of litter size in the Arctic Fox. *Oikos* 83(3): 545–559.

Tannerfeldt, M., Elmhagen, B., and Angerbjörn, A. 2002. Exclusion by interference competition? The relationship between Red and Arctic Foxes. *Population Ecology* 132: 213–220.

Taulman, J.F. 1990a. Late summer activity patterns in Hoary Marmots. *Northwestern Naturalist* 71: 21–26.

– 1990b. Observations on scent marking in Hoary Marmots, *Marmota caligata*. *Canadian Field-Naturalist* 104(3): 479–482.

– 1998. Observations on quantitative cognition in Southern Flying Squirrels, *Glaucomys volans*. *Canadian Field-Naturalist* 112(2): 347–349.

Taulman, J.F., and Seaman, D.E. 2000. Assessing Southern Flying Squirrel, *Glaucomys volans*, habitat selection with kernel home range estimation and GIS. *Canadian Field-Naturalist* 114(4): 591–600.

Taylor, M.K., Laake, J., McLoughlin, P.D., Cluff, H.D., and Messier, F. 2006. Demographic parameters and harvest-explicit population viability analysis for Polar Bears in M'Clintock Channel, Nunavut, Canada. *Journal of Wildlife Management* 70(6): 1667–1673.

Taylor, P. 2003. Close encounters with a Star-nosed Mole. *Blue Jay* 61(4): 210–213.

Teferi, T., and Herman, T.B. 1995. Epigeal movement by *Sorex cinereus* on Bon Portage Island, Nova Scotia. *Journal of Mammalogy* 76(1): 137–140.

Teferi, T., Herman, T.B., and Stewart, D.T. 1992. Breeding biology of an insular population of the Masked Shrew (*Sorex cinereus* Kerr) in Nova Scotia. *Canadian Journal of Zoology* 70(1): 62–66.

Teferi, T., and Millar, J.S. 1993. Long distance homing by the Deer Mouse, *Peromyscus maniculatus*. *Canadian Field-Naturalist* 107(1): 109–111.

Teilmann, J., and Dietz, R. 1994. Status of Harbour Seal, *Phoca vitulina*, in Greenland. *Canadian Field-Naturalist* 108(2): 139–155.

Teilmann, J., Tougaard, J., Miller, L.A., Kirketerp, T., Hansen, K., and Brando, S. 2006. Reactions of captive Harbor Porpoises (*Phocoena phocoena*) to pinger-like sounds. *Marine Mammal Science* 22(2): 240–260.

Teloni, V. 2005. Patterns of sound production in diving Sperm Whales in the northwestern Mediterranean. *Marine Mammal Science* 21(3): 446–457.

Terhune, J.M. 1989. Underwater click hearing thresholds of a Harbour Seal, *Phoca vitulina*. *Aquatic Mammals* 15(1): 22–26.

– 1999. Pitch separation as a possible jamming-avoidance mechanism in underwater calls of Bearded Seals (*Erignathus barbatus*). *Canadian Journal of Zoology* 77(7): 1025–1034.

Terman, C.R., and Terman, J.R. 1999. Early summer reproductive hiatus in wild adult White-footed Mice. *Journal of Mammalogy* 80(4): 1251–1256.

Tershy, B.R., and Wiley, D.N. 1992. Asymmetrical pigmentation in the Fin Whale: A test of two feeding related hypotheses. *Marine Mammal Science* 8(3): 315–318.

Tessier, N., Noël, S., and Lapointe, F.-J. 2004. A new method to discriminate the Deer Mouse (*Peromyscus maniculatus*) from the White-footed Mouse (*Peromyscus leucopus*) using species-specific primers in multiplex PCR. *Canadian Journal of Zoology* 82(11): 1832–1835.

Testa, J.W., and Adams, G.P. 1998. Body condition and adjustments to reproductive effort in female Moose (*Alces alces*). *Journal of Mammalogy* 79(4): 1345–1354.

Thayer, V.G., Read, A.J., Friedlaender, A.S., Colby, D.R., Hohn, A.A., McLellan, W.A., Pabst, D.A., Dearolf, J.L., Bowles, N.I., Russell, J.R., and Rittmaster, K.A. 2003. Reproductive seasonality of western Atlantic Bottlenose Dolphins off North Carolina, USA. *Marine Mammal Science* 19(4): 617–629.

Theberge, J.B., and Theberge, M.T. 1998. *Wolf Country: Eleven years of tracking Algonquin wolves*. McClelland and Stewart, Toronto, ON.

Thiel, R.P. 2006. Conditions for sexual interactions between wild Grey Wolves, *Canis lupus*, and Coyotes, *Canis latrans*. *Canadian Field-Naturalist* 120(1): 27–30.

Thiel, R.P., Hall, W.H., and Schultz, R.N. 1997. Early den digging by wolves, *Canis lupus*, in Wisconsin. *Canadian Field-Naturalist* 111(3): 481–482.

Thiel, R.P., Merrill, S., and Mech, L.D. 1998. Tolerance by denning Wolves, *Canis lupus*, to human disturbance. *Canadian Field-Naturalist* 112(2): 340–342.

Thomas, D.C., Edmonds, E.J., and Brown, W.K. 1996. The diet of Woodland Caribou populations in west-central Alberta. *Rangifer,* special issue no. 9: 337–342.

Thomas, G.L., and Thorne, R.E. 2001. Night-time predation by Steller Sea Lions. *Nature* 411(28 June): 1013.

Thompson, D., and Fedak, M.A. 1993. Cardiac responses of Grey Seals during diving at sea. *Journal of Experimental Biology* 174: 139–164.

Thompson, D., Hammond, P.S., Nicholson, K.S., and Fedak, M.A. 1991. Movements, diving and foraging behaviour of Grey Seals (*Halichoerus grypus*). *Journal of Zoology* (London) 224: 223–232.

Thompson, I.D., and Colgan, P.W. 1987. Numerical responses of Martens to a food shortage in north-central Ontario. *Journal of Wildlife Management* 51: 824–835.

– 1994. Marten activity in uncut and logged boreal forests in Ontario. *Journal of Wildlife Management* 58(2): 280–288.

Thompson, M.J., and Stewart, W.C. 1994. Cougar(s), *Felis concolor*, with a kill for 27 days. *Canadian Field-Naturalist* 108(4): 497–498.

Timmerman, H.R. 2003. The status and management of Moose in North America, circa 2000–01. *Alces* 39: 131–151.

Tinker, M.T., Kovacs, K.M., and Hammill, M.O. 1995. The reproductive behaviour and energetics of male Grey Seals (*Halichoerus grypus*) breeding on a land–fast ice substrate. *Behavioral Ecology and Sociobiology* 36(3): 159–170.

Tischendorf, J.W., and Henderson, F.R. 1994. The Puma in the Central Mountains and Great Plains: A synopsis. *Blue Jay* 52(4): 218–223.

Tischendorf, J.W., and McAlpine, D.F. 1995. A melanistic Bobcat from outside Florida. *Florida Field Naturalist* 23(1): 13–14.

Tomasik, E., and Cook, J.A. 2005. Mitochondrial phylogeography and conservation genetics of Wolverine (*Gulo gulo*) of northwestern North America. *Journal of Mammalogy* 86(2): 386–396.

Topping, M.G., and Millar, J.S. 1996a. Spatial distribution in the Bushy-tailed Wood Rat (*Neotoma cinerea*) and its implications for the mating system. *Canadian Journal of Zoology* 74(3): 565–569.

– 1996b. Foraging movements of female Bushy-tailed Wood Rats (*Neotoma cinerea*). *Canadian Journal of Zoology* 74(5): 798–801.

– 1996c. Home range size of Bushy-tailed Woodrats, *Neotoma cinerea*, in southwestern Alberta. *Canadian Field-Naturalist* 110(2): 351–353.

Topping, M.G., Millar, J.S., and Goddard, J.A. 1999. The effects of moonlight on nocturnal activity in Bushy-tailed Wood Rats (*Neotoma cinerea*). *Canadian Journal of Zoology* 77(3): 480–485.

Tosh, C.A., De Bruyn, P.J.N., and Bester, M.N. 2008. Preliminary analysis of the social structure of Killer Whales, *Orcinus orca*, at subantarctic Marion Island. *Marine Mammal Science* 24(4): 929–940.

Tove, M. 1995. Live sighting of *Mesoplodon* cf. *M. mirus*, True's Beaked Whale. *Marine Mammal Science* 11(1): 80–85.

Towell, R.G., Ream, R.R., and York, A.E. 2006. Decline in Northern Fur Seal (*Callorhinus ursinus*) pup production on the Pribilof Islands. *Marine Mammal Science* 22(2): 486–491.

Trippel, E.A., Strong, M.B., Terhune, J.M., and Conway, J.D. 1999. Mitigation of Harbour Porpoise (*Phocoena phocoena*) by-catch in the gillnet fishery in the lower Bay of Fundy. *Canadian Journal of Fisheries and Aquatic Sciences* 56(1): 113–123.

Trites, A.W. 1992. Northern Fur Seals: Why have they declined? *Aquatic Mammals* 18(1): 3–18.

Trites, A.W., and Bigg, M.A. 1992. Changes in body growth of Northern Fur Seals from 1958 to 1974: Density effects or changes in the ecosystem? *Fisheries Oceanography* 1(2): 127–136.

– 1996. Physical growth of Northern Fur Seals (*Callorhinus ursinus*): Seasonal fluctuations and migratory influences. *Journal of Zoology* (London) 238: 459–482.

Trites, A.W., and Donnelly, C.P. 2003. The decline of Steller Sea Lions *Eumetopias jubatus* in Alaska: A review of the nutritional stress hypothesis. *Mammal Review* 33(1): 3–28.

Trites, A.W., and Larkin, P.A. 1996. Changes in abundance of Steller Sea Lions (*Eumetopias jubatus*) in Alaska from 1956 to 1992: How many were there? *Aquatic Mammals* 22(3): 153–166.

Trombulak, S.C. 1987. Life history of the Cascade Golden-mantled Ground Squirrel (*Spermophilus saturatus*). *Journal of Mammalogy* 68(3): 544–554.

– 1988. *Spermophilus saturatus*. *Mammalian Species* no. 322: 1–4.

True, F.W. 1910. An account of the beaked whales of the family Ziphiidae in the collection of the United States National Museum, with remarks on some species in other American museums. *Bulletin of the United States National Museum* 73: 1–89.

Trujillo, R.G., Loughlin, T.R., Gemmell, N.J., Patton, J.C., and Bickham, J.W. 2004. Variation in microsatellites and mtDNA across the range of the Steller Sea Lion, *Eumetopias jubatus*. *Journal of Mammalogy* 85(2): 338–346.

Trune, D.R., and Slobodchikoff, C.N. 1976. Social effects of roosting on the metabolism of the Pallid Bat (*Antrozous pallidus*). *Journal of Mammalogy* 57(4): 656–663.

Tumanov, I.L., and Sorina, E.A. 1996. Seasonal changes in energetics of nutrition in males of small mustelid species (Mustelidae). *Small Carnivore Conservation* 15: 6–9.

– 1997. Dynamics of the nutritional energetics in female mustelids (Mustelidae). *Small Carnivore Conservation* 17: 10–14.

Tumlison, R. 1987. *Felis lynx. Mammalian Species* no. 269: 1–8.

Turner, J.W., Jr., and Morrison, M.L. 2001. Influence of predation by Mountain Lions on numbers and survivorship of a feral Horse population. *Southwestern Naturalist* 46(2): 183–190.

Tuttle, M.D., and Stevenson, D. 1982. Growth and survival of bats. In T.H. Kunz (ed.), *Ecology of Bats*, 105–150. Plenum Press, New York, NY.

Underwood, L.S., and Reynolds, P. 1980. Photoperiod and fur lengths in the Arctic Fox (*Alopex lagopus* L.). *International Journal of Biometeorology* 24(1): 39–48.

Unger, C.A., and Kurta, A. 1998. Status of the Eastern Pipistrelle (Mammalia: Chiroptera) in Michigan. *Michigan Academician* 30: 423–437.

Urbán-Ramírez, J., Flores de Sahagún, V., Jones, M.L., Swartz, S.L., Mate, B., Gómez-Gallardo, A., and Guerrero-Ruíz, M. 2004. Gray Whales with loss of flukes adapt and survive. *Marine Mammal Science* 20(2): 335–338.

van Bree, P.J.H., Vedder, E.J., and T'Hart, L. 1997. Nogmaals over Zadelrobben *Phoca groenlandica* de kust van continentaal West-Europa. *Lutra* 40: 23–27.

Van Deelen, T.R., and Pletscher, D.H. 1996. Dispersal characteristics of two-year-old Beavers, *Castor canadensis*, in western Montana. *Canadian Field-Naturalist* 110(2): 318–321.

Vander Wall, S.B. 1991. Mechanisms of cache recovery by Yellow Pine Chipmunks. *Animal Behaviour* 41: 851–863.

– 1995. The effects of seed value on the caching behavior of Yellow Pine Chipmunks. *Oikos* 74: 533–537.

– 2000. The influence of environmental conditions on cache recovery and cache pilferage by Yellow Pine Chipmunks (*Tamias amoenus*) and Deer Mice (*Peromyscus maniculatus*). *Behavioral Ecology* 11(5): 544–549.

Vander Wall, S.B., and Longland, W.S. 1999. Cheek pouch capacities and loading rates of Deer Mice (*Peromyscus maniculatus*). *Great Basin Naturalist* 59(3): 278–280.

Vander Wall, S.B., Longland, W.S., Pyare, S., and Veech, J.A. 1998. Cheek pouch capacities and loading rates of heteromyid rodents. *Oecologia* 113: 21–28.

Vander Wall, S.B., Thayer, T.C., Hodge, J.S., Beck, M.J., and Roth, J.K. 2001. Scatter-hoarding behavior of Deer Mice (*Peromyscus maniculatus*). *Western North American Naturalist* 61(1): 109–113.

Van de Velde, F., Stirling, I., and Richardson, E. 2003. Polar Bear (*Ursus maritimus*) denning in the area of the Simpson Peninsula, Nunavut. *Arctic* 56(2): 191–197.

van Dijk, J., Andersen, T., May, R., Andersen, R., Andersen, R., and Landa, A. 2008. Foraging strategies of Wolverines within a predator guild. *Canadian Journal of Zoology* 86(9): 966–975.

Vangen, K.M., Persson, J., Landa, A., Andersen, R., and Segerström, P. 2001. Characteristics of dispersal in Wolverines. *Canadian Journal of Zoology* 79(9): 1641–1649.

Van Horne, B. 1982. Demography of the Longtail Vole *Microtus longicaudus* in seral stages of coastal coniferous forest, southeast Alaska. *Canadian Journal of Zoology* 60(7): 1690–1709.

Van Parijs, S.M., Janik, V.M., and Thompson, P.M. 2000. Display-area size, tenure length, and site fidelity in the aquatically mating male Harbour Seal, *Phoca vitulina*. *Canadian Journal of Zoology* 78(12): 2209–2217.

Van Parijs, S.M., and Kovacs, K.M. 2002. In-air and underwater vocalizations of eastern Canadian Harbour Seals, *Phoca vitulina. Canadian Journal of Zoology* 80(7): 1173–1179.

Van Parijs, S.M., Kovacs, K.M., and Lydersen, C. 2001. Spatial and temporal distribution of vocalising male Bearded Seals: Implications for male mating strategies. *Behaviour* 138: 905–922.

Van Parijs, S.M., Lydersen, C., and Kovacs, K.M. 2003. Vocalizations and movements suggest alternative mating tactics in male Bearded Seals. *Animal Behaviour* 65: 273–283.

van Staaden, M.J., Michener, G.R., and Chesser, R.K. 1996. Spatial analysis of microgeographic genetic structure in Richardson's Ground Squirrels. *Canadian Journal of Zoology* 74(7): 1187–1895.

Van Vuren, D., and Armitage, K.B. 1991. Duration of snow cover and its influence on life-history variation in Yellow-bellied Marmots. *Canadian Journal of Zoology* 69(7): 1755–1758.

Van Why, K.R., and Giuliano, W.M. 2001. Fall food habits and reproductive condition of Fishers, *Martes pennanti*, in Vermont. *Canadian Field-Naturalist* 115(1): 52–56.

van Zyll de Jong, C.G. 1972. A systematic review of the Nearctic and Neotropical river otters (genus *Lutra*, Mustelidae, Carnivora). *Life Sciences Contributions, Royal Ontario Museum* no. 80: 104.

– 1975. The distribution and abundance of the wolverine (*Gulo gulo*) in Canada. *Canadian Field-Naturalist* 89(4): 431–437.

– 1980. Systematic relationships of woodland and prairie forms of the Common Shrew, *Sorex cinereus cinereus* Kerr and *S.c. haydeni* Baird, in the Canadian prairie provinces. *Journal of Mammalogy* 61(1): 66–75.

– 1983a. *Marsupials and Insectivores.* Vol. 1 of *Handbook of Canadian Mammals.* National Museums of Canada, Ottawa, ON.

– 1983b. A morphometric analysis of North American shrews of the *Sorex arcticus* group, with special consideration of the taxonomic status of *S. a. maritimensis. Naturaliste canadien* 110(4): 373–378.

– 1985. *Bats.* Vol. 2 of *Handbook of Canadian Mammals.* National Museums of Canada, Ottawa, ON.

– 1986. A systematic study of recent bison, with particular consideration of the Wood Bison (*Bison bison athabascae* Rhoads 1898). *Publications in Natural Sciences* no. 6: 1–69. National Museum of Natural Sciences, Ottawa, Canada.

– 1987. A phylogenetic study of the Lutrinae (Carnivora; Mustelidac) using morphological data. *Canadian Journal of Zoology* 65(10): 2536–2544.

– 1999a. Cinereus Shrew (*Sorex cinereus*). In D.E. Wilson and S. Ruff (eds), *The Smithsonian Book of North American Mammals*, 20–21. UBC Press, Vancouver, BC.

– 1999b. Tundra Shrew (*Sorex tundrensis*). In Wilson and Ruff, *The Smithsonian Book of North American Mammals*, 44–45.

van Zyll de Jong, C.G., Fenton, M.B., and Woods, J.G. 1980. Occurrence of *Myotis californicus* at Revelstoke and a second record of *Myotis septentrionalis* for British Columbia. *Canadian Field-Naturalist* 94(4): 455–456.

van Zyll de Jong, C. G., Gates, C.C., Reynolds, H., and Olson, W. 1995. Phenotypic variation in remnant populations of North American Bison. *Journal of Mammalogy* 76: 391–405.

van Zyll de Jong, C.G., and Nagorsen, D.W. 1994. A review of the distribution and taxonomy of *Myotis keenii* and *Myotis evotis* in British Columbia and the adjacent United States. *Canadian Journal of Zoology* 72(6): 1069–1078.

Vargas, A., and Anderson, S.H. 1996. Growth and physical development of captive-raised Black-footed Ferrets (*Mustela nigripes*). *American Midland Naturalist* 135: 43–52.

– 1998. Ontogeny of Black-footed Ferret predatory behavior towards prairie dogs. *Canadian Journal of Zoology* 76(9): 1696–1704.

Vaughan, T.A., and O'Shea, T.J. 1976. Roosting ecology of the Pallid Bat, *Antrozous pallidus. Journal of Mammalogy* 57(1): 19–42.

Veilleux, J.P., and Veilleux, S.L. 2004. Intra-annual and interannual fidelity to summer roost areas by female Eastern Pipistrelles, *Pipistrellus subflavus. American Midland Naturalist* 152(1): 196–200.

Veilleux, J.P., Whitaker, J.O., Jr., and Veilleux, S.L. 2003. Tree-roosting ecology of reproductive female Eastern Pipistrelles, *Pipistrellus subflavus*, in Indiana. *Journal of Mammalogy* 84(3): 1068–1075.

Veitch, A.M. 2001. An unusual record of a White-tailed Deer, *Odocoileus virginianus*, in the Northwest Territories. *Canadian Field-Naturalist* 115(1): 172–175.

Veitch, A.M., and Krizan, P.K. 1996. Black Bear predation on vertebrates in northern Labrador. *Journal of Wildlife Research* 1(2): 193–194.

Vergara, V. 2001a. Comparison of parental roles in male and female Red Foxes, *Vulpes vulpes*, in southern Ontario. *Canadian Field-Naturalist* 115(1): 22–33.

– 2001b. Two cases of infanticide in a Red Fox, *Vulpes vulpes*, family in southern Ontario. *Canadian Field-Naturalist* 115(1): 170–173.

Vernes, K. 2001. Gliding performance of the Northern Flying Squirrel (*Glaucomys sabrinus*) in mature mixed forest of eastern Canada. *Journal of Mammalogy* 82(4): 1026–1033.

Vernes, K., Blois, S., and Bärlocher, F. 2004. Seasonal and yearly changes in consumption of hypogeous fungi by Northern Flying Squirrels and Red Squirrels in old-growth forest, New Brunswick. *Canadian Journal of Zoology* 82(1): 110–117.

Verts, B.J., and Carraway, L.N. 1998. *Land Mammals of Oregon.* University of California Press, Berkeley, CA.

– 1999. *Thomomys talpoides. Mammalian Species* no. 618: 1–11.

– 2001. *Tamias minimus. Mammalian Species* no. 653: 1–10.

Verts, B.J., Carraway, L.N., and Kinlaw, A. 2001. *Spilogale gracilis. Mammalian Species* no. 674: 1–10.

Verts, B.J., and Gehman, S.D. 1991. Activity and behaviour of free-living *Sylvilagus nuttallii. Northwest Science* 65(5): 231–237.

Verts, B.J., Gehman, S.D., and Hundertmark, K.J. 1984. *Sylvilagus nuttallii*: A semiarboreal lagomorph. *Journal of Mammalogy* 65(1): 131–135.

Verts, B.J., and Kirkland, G.L., Jr. 1988. *Perognathus parvus. Mammalian Species* no. 318: 1–8.

Vestal, B.M. 1991. Infanticide and cannibalism by male Thirteen-lined Ground Squirrels. *Animal Behaviour* 41: 1103–1104.

Via, C., Savolainen, P., Maldonado, J.E., Amorin, I.R., Rice, J.E., Honeycutt, R.L., Crandall, K.A., Lundeberg, J., and Wayne, R.K. 1997. Multiple and ancient origins of the Domestic Dog. *Science* 276(5319): 1687–1689.

Vickery, W.L. 1979. Food consumption and preferences in wild populations of *Clethrionomys gapperi* and *Napaeozapus insignis. Canadian Journal of Zoology* 57(8): 1536–1542.

Viitala, J. 1987. Social organization of *Clethrionomys rutilus* (Pall.) at Kilpisjärvi, Finnish Lapland. *Annales Zoologici Fennici* 24: 267–273.

Vikingsson, G.A. 1997. Feeding of Fin Whales (*Balaenoptera physalus*) off Iceland: Diurnal and seasonal variation and possible rates. *Journal of Northwest Atlantic Fisheries Science* 22: 77–89.

Viljugrein, H., Lingjaerde, O.C., Stenseth, N.C., and Boyce, M.S. 2001. Spatio-temporal patterns in Mink and Muskrat in Canada during a quarter century. *Journal of Animal Ecology* 70: 671–682.

Villa, L.J., Carey, A.B., Wilson, T.M., and Glos, K.E. 1999. Maturation and reproduction of Northern Flying Squirrels in Pacific Northwest forests. *United States Department of Agriculture, Forest Service, Pacific Northwest Research Station General Technical Report* PNW-Gtr-444: 1–49.

Villemure, M., and Jolicoeur, H. 2004. First confirmed occurrence of a Wolf, *Canis lupus*, south of the St Lawrence River in over 100 years. *Canadian Field-Naturalist* 118(4): 608–610.

Vinkey, R.S., Schwartz, M.K., McKelvey, K.S., Foresman, K.R., Pilgrim, K.L., Giddings, B.J., and LoForth, E.C. 2006. When reintroductions are augmentations: The genetic legacy of Fishers (*Martes pennanti*) in Montana. *Journal of Mammalogy* 87(2): 265–271.

Virgl, J.A., and Messier, F. 1992. The ontogeny of body composition and gut morphology in free-ranging Muskrats. *Canadian Journal of Zoology* 70(7): 1381–1388.

Vispo, C.R., and Bakken, G.S. 1993. The influence of thermal conditions on the surface activity of Thirteen-lined Ground Squirrels. *Ecology* 74(2): 377–389.

Visser, I.N., Smith, T.G., Bullock, I.D., Green, G.D., Carlsson, O.G.L., and Imberti, S. 2008. Antarctic peninsula Killer Whales (*Orcinus orca*) hunt seals and a penguin on floating ice. *Marine Mammal Science* 24(1): 225–234.

Volobouev, V.T., and van Zyll de Jong, C.G. 1994. Chromosome banding analysis of two shrews of the *cinereus* group: *Sorex haydeni* and *Sorex cinereus* (Insectivora, Soricidae). *Canadian Journal of Zoology* 72(5): 958–964.

Vonhof, M.J., and Barclay, R.M.R. 1997. Use of tree stumps by the Western Long-eared Bat. *Journal of Wildlife Management* 61(3): 674–684.

Vonhof, M.J., and Gwilliam, J.C. 2000. *A Summary of Bat Research in the Pend d'Oreille Valley in Southern British Columbia.* Report prepared for the Columbia Basin Fish and Wildlife Compensation Program in partnership with the British Columbia Ministry of Forests.

Vonhof, M.J., and Hobson, D. 2001. Survey of bats of central and northwestern Alberta. *Alberta Species at Risk Report* no. 4: 1–33.

Vonhof, M.J., Strobeck, C., and Fenton, M.B. 2008. Genetic variation and population structure in Big Brown Bats (*Eptesicus fuscus*): Is female dispersal important? *Journal of Mammalogy* 89(6): 1411–1420.

Vors, L.S., and Wilson, P.L. 2006. A new record size wolf, *Canis lupus*, group for Ontario. *Canadian Field-Naturalist* 120(3): 367–369.

Vos, D.J., Quakenbush, L.T., and Mahoney, B.A. 2006. Documentation of Sea Otters and birds as prey for Killer Whales. *Marine Mammal Science* 22(1): 201–205.

Wada, K., Hamanaka, T., Nakaoka, T., Tanahashi, K. 1992. Food and feeding habits of Kuril and Largha Seals in southeastern Hokkaido. *Mammalia* 56(4): 555–566.

Wade-Smith, J., and Verts, B.J. 1982. *Mephitis mephitis. Mammalian Species* no. 173: 1–7.

Wai-Ping, V., and Fenton, M.B. 1989. Ecology of Spotted Bat (*Euderma maculatum*) roosting and foraging behaviour. *Journal of Mammalogy* 70(3): 617–622.

Waldien, D.L., and Hayes, J.P. 2001. Activity areas of female Long-eared Myotis in coniferous forests in western Oregon. *Northwest Science* 75(3): 307–314.

Waldien, D.L., Hayes, J.P., and Arnett, E.B. 2000. Day-roosts of female Long-eared Myotis in western Oregon. *Journal of Wildlife Management* 64(3): 785–796.

Waldron, G. 1998. *COSEWIC Status Report on the Eastern Mole,* Scalopus aquaticus. Committee on the Status of Endangered Wildlife in Canada.

Waldron, G., Rodger, L., Mouland, G., and Lebedyk, D. 2000. Range, habitat, and population size of the Eastern Mole, *Scalopus aquaticus*, in Canada. *Canadian Field-Naturalist* 114(3): 351–358.

Walker, A.B.D., Parker, K.L., and Gillingham, M.P. 2006. Behaviour, habitat associations, and intrasexual differences of female Stone's Sheep. *Canadian Journal of Zoology* 84(8): 1187–1201.

Walker, B.G., and Bowen, W.D. 1993. Changes in body mass and feeding behaviour in male Harbour Seals, *Phoca vitulina*, in relation to female reproductive status. *Journal of Zoology* (London) 231: 423–436.

Walker, D., Diamond, J., and Stokes, J. 2001. True's Beaked Whale: A first confirmed live sighting for the eastern North Atlantic. *Orca* 2: 63–66.

Walker, W.A. 1996. Summer feeding habits of Dall's Porpoise, *Phocoenoides dalli*, in the southern Sea of Okhotsk. *Marine Mammal Science* 12(2): 167–181.

Walker, W.A., and Hanson, M.B. 1999. Biological observations on Stejneger's Beaked Whale, *Mesoplodon stejnegeri*, from strandings on Adak Island, Alaska. *Marine Mammal Science* 15(4): 1314–1329.

Walker, W.A., Mead, J.G., and Brownell, R.L., Jr. 2002. Diets of Baird's Beaked Whales, *Berardius bairdii*, in the southern Sea of Okhotsk and off the Pacific coast of Honshu, Japan. *Marine Mammal Science* 18(4): 902–919.

Wallace, S.C. 2006. Differentiating *Microtus xanthognathus* and *Microtus pennsylvanicus* lower first molars using discriminant analysis of landmark data. *Journal of Mammalogy* 87(6): 1261–1269.

Waltari, E., Demboski, J.R., Klein, D.R., and Cook, J.A. 2004. A molecular perspective on the historical biogeography of the northern high latitudes. *Journal of Mammalogy* 85(4): 591–600.

Walters, B.L., Ritzi, C.M., Sparks, D.W., and Whitaker, J.O., Jr. 2007. Foraging behavior of Eastern Red Bats (*Lasiurus borealis*) at an urban-rural interface. *American Midland Naturalist* 157(2): 365–373.

Walters, E.L., and Miller, E.H. 2001. Predation on nestling woodpeckers in British Columbia. *Canadian Field-Naturalist* 115(3): 413–419.

Walton, L.R., Cluff, H.D., Paquet, P.C., and Ramsay, M.A. 2001. Movement patterns of Barren-ground Wolves in the Central Canadian Arctic. *Journal of Mammalogy* 82(3): 867–876.

Walton, L.R., and Larivière, S. 1994. A Striped Skunk, *Mephitis mephitis*, repels two Coyotes, *Canis latrans*, without scenting. *Canadian Field-Naturalist* 108(4): 492–493.

Waring, G.T., Pace, R.M., Quintal, J.M., Fairfield, C.P., and Maze-Foley, K. (eds). 2004. Minke Whale (*Balaenoptera acutorostrata*): Canadian east coast stock. In *NOAA Technical Memorandum* NOAA–TM–NMFS–NE–182: 30–39. U.S. Atlantic and Gulf of Mexico Marine Mammal Stock Assessments, 2003, U.S. Department of Commerce.

Warkentin, K.J., Keeley, A.T.H., and Hare, J.F. 2001. Repetitive calls of juvenile Richardson's Ground Squirrels (*Spermophilus richardsonii*) communicate response urgency. *Canadian Journal of Zoology* 79(4): 569–573.

Warner, R.M., and Czaplewski, N.J. 1984. *Myotis volans. Mammalian Species* no. 224: 1–4.

Waters, J.H., and Ray, C.E. 1961. Former range of the Sea Mink. *Journal of Mammalogy* 42(3): 380–383.

Watkins, B. 2005. Cougars confirmed in Manitoba. *Wild Cat News* (an online publication) 1(1): 6–8.

Watkins, J.F. 1990. Observations of an aquatic Grey Seal (*Halichoerus grypus*) mating. *Journal of Zoology* (London) 222: 677–680.

Watkins, L.C. 1972. *Nycticeius humeralis. Mammalian Species* no. 23: 1–4.

– 1977. *Euderma maculatum. Mammalian Species* no. 77: 1–4.

Watkins, W.A., Daher, M.A., DiMarzio, N.A., Samuels, A., Wartzok, D., Fristrup, K.M., Gannon, D.P., Howey, P.W., and Maiefski, R.R. 1999. Sperm Whale surface activity from tracking by radio and satellite tags. *Marine Mammal Science* 15(4): 1158–1180.

Watson, J.C., Ellis, G.M., Smith, T.G., and Ford, J.K.B. 1997. Updated status of the Sea Otter, *Enhydra lutris*, in Canada. *Canadian Field-Naturalist* 111(2): 277–286.

Way, J.G. 2007a. A comparison of body mass of *Canis latrans* (Coyotes) between eastern and western North America. *Northeastern Naturalist* 14(1): 111–124.

– 2007b. Movements of transient Coyotes, *Canis latrans*, in urbanized eastern Massachusetts. *Canadian Field-Naturalist* 121(4): 364–369.

– 2007c. Social and play behavior in a wild Eastern Coyote, *Canis latrans*, pack. *Canadian Field-Naturalist* 121(4): 397–401.

Way, J.G., and Proietto, R.L. 2004. Record size female Coyote, *Canis latrans. Canadian Field-Naturalist* 119(1): 139–140.

Weaver, A. 2003. Conflict and reconciliation in captive Bottlenose Dolphins, *Tursiops truncatus. Marine Mammal Science* 19(4): 836–846.

Weaver, J.L., Arvidson, C., and Wood, P. 1992. Two Wolves, *Canis lupus*, killed by a Moose, *Alces alces*, in Jasper National Park, Alberta. *Canadian Field-Naturalist* 106(1): 126–127.

Webster, W.D., and Jones, J.K. 1982. *Reithrodontomys megalotis. Mammalian Species* no. 167: 1–5.

Weckerly, F.W. 1994. Selective feeding by Black-tailed Deer: Forage quality or abundance? *Journal of Mammalogy* 75(4): 905–913.

Weeks, H.P., Jr. 1993. Arboreal food caching by Long-tailed Weasels. *Prairie Naturalist* 25(1): 39–42.

Wehausen, J.D., and Ramey, R.R., II. 2000. Cranial morphometric and evolutionary relationships in the northern range of *Ovis canadensis. Journal of Mammalogy* 81(1): 145–161.

Weilert, N.G., and Shump, K.A., Jr. 1977. Physical parameters of *Microtus* nest construction. *Transactions of the Kansas Academy of Science* 79: 161–164.

Weinrich, M. 1995. Humpback Whale competitive groups observed on a high-latitude feeding ground. *Marine Mammal Science* 11(2): 251–254.

– 1996. Abandonment of an entangled conspecific by Atlantic White-sided Dolphins (*Lagenorhynchus acutus*). *Marine Mammal Science* 12(2): 293–296.

Weinrich, M.T., Belt, C.R., and Morin, D. 2001. Behavior and ecology of the Atlantic White-sided Dolphin (*Lagenorhynchus acutus*) in coastal New England waters. *Marine Mammal Science* 17(2): 231–248.

Weir, R.D. 2003. Status of the Fisher in British Columbia. *Wildlife Bulletin* no. B-105: 1–47. BC Ministry of Sustainable Resource Management, Conservation Data Centre, and Ministry of Water, Land and Air Protection, Biodiversity Branch.

Weir, R.D., and Corbould, F.B. 2006. Density of Fishers in the sub-boreal spruce biogeoclimatic zone of British Columbia. *Northwestern Naturalist* 87(2): 118–127.

– 2007. Factors affecting diurnal activity of Fishers in north-central British Columbia. *Journal of Mammalogy* 88(6): 1508–1514.

Weir, R.D., Harestad, A.S., and Wright, R.C. 2005. Winter diet of Fishers in British Columbia. *Northwestern Naturalist* 86: 12–19.

Weisel, J.W., Nagaswami, C., and Peterson, R.O. 2005. River Otter hair structure facilitates interlocking to impede penetration of water and allow trapping of air. *Canadian Journal of Zoology* 83(5): 649–655.

Weller, D.W., Würsig, B., Bradford, A.L., Burdin, A.M., Blokhin, S.A., Minakuchi, H., and Brownell, R.L., Jr. 1999. Gray Whales (*Eschrichtius robustus*) off Sakhalin Island, Russia: Seasonal and annual patterns of occurrence. *Marine Mammal Science* 15(4): 1208–1227.

Weller, T.J., Scott, S.A., Rodhouse, T.J., Ormsbee, P.C., and Zinck, J.M. 2007. Field identification of the cryptic vespertilionid bats, *Myotis lucifugus* and *M. yumanensis. Acta Chiropterologica* 9(1): 133–147.

Weller, T.J., and Zabel, C.J. 2001. Characteristics of Fringed Myotis day roosts in Northern California. *Journal of Wildlife Management* 65(3): 489–497.

Wells, R.S., and Scott, M.D. 1999. Bottlenose Dolphin *Tursiops truncatus* (Montagu 1821). In S.H. Ridgway and R. Harrison (eds), *Handbook of Marine Mammals*, vol. 6, 137–182. Academic Press, San Diego, CA.

Wells-Gosling, N. 1985. *Flying Squirrels: Gliders in the Dark.* Smithsonian Institution Press, Washington, DC.

Wells-Gosling, N., and Heaney, L.R. 1984. *Glaucomys sabrinus. Mammalian Species* no. 229: 1–8.

Werth, A. 2000. A kinematic study of suction feeding and associated behavior in the Long-finned Pilot Whale, *Globicephala melas* (Traill). *Marine Mammal Science* 16(2): 299–314.

West, E.W., and Swain, U. 1999. Surface activity and structure of a hydrothermally-heated maternity colony of the Little Brown Bat, *Myotis lucifugus*, in Alaska. *Canadian Field-Naturalist* 113(3): 425–429.

West, S.D. 1977. Midwinter aggregation in the Northern Red-backed Vole, *Clethrionomys rutilus. Canadian Journal of Zoology* 55(9): 1404–1409.

– 1982. Dynamics of colonization and abundance in central Alaskan populations of Northern Red-backed Vole, *Clethrionomys rutilus. Journal of Mammalogy* 63(1): 128–143.

Westgate, A.J., and Read, A.J. 1998. The application of new technology to the conservation of porpoises. *Marine Technology Society Journal* 32: 70–81.

Westlake, R.L., and O'Corry-Crowe, G.M. 2002. Macrogeographic structure and patterns of genetic diversity in Harbour Seals (*Phoca vitulina*) from Alaska and Japan. *Journal of Mammalogy* 83(4): 1111–1126.

Weston, J.L., Blake, B.H., Reynolds, J., and Dewey, M.J. 2005. Patterns of ultrasonic vocalizations in neonates of the genus *Peromyscus*. Abstract and poster presentation, 85th ASM Annual Meeting.

Weyandt, S.E., Van Den Bussche, R.A., Hamilton, M.J., and Leslie, D.M., Jr. 2005. Unraveling the effects of sex and dispersal: Ozark Big-eared Bat (*Corynorhinus townsendii ingens*) conservation genetics. *Journal of Mammalogy* 86(6): 1136–1143.

Wheatley, M. 1997. Beaver, *Castor canadensis*, home range size and pattern of use in the taiga of southeastern Manitoba: I, Seasonal variation. *Canadian Field-Naturalist* 111(2): 204–210.

Wheatley, M., Larsen, K.W., and Boutin, S. 2002. Does density reflect habitat quality for North American Red Squirrels during a spruce-cone failure? *Journal of Mammalogy* 83(3): 716–727.

Whidden, H.P., Ray, A.W., and Bowles, J.B. 2002. Identification and distribution of Masked and Hayden's Shrews (genus *Sorex*) in Iowa. *Journal of the Iowa Academy of Science* 109(1/2): 19–24.

Whitaker, J.O., Jr. 1972. *Zapus hudsonius. Mammalian Species* no. 11: 1–7.

– 1995. Food of the Big Brown Bat *Eptesicus fuscus* from maternity colonies in Indiana and Illinois. *American Midland Naturalist* 134: 346–360.

– 1998. Life history and roost switching in six summer colonies of Eastern Pipistrelles in buildings. *Journal of Mammalogy* 79(2): 651–659.

– 1999a. Smoky Shrew (*Sorex fumeus*). In D.E. Wilson and S. Ruff (eds), *The Smithsonian Book of North American Mammals*, 22–23. UBC Press, Vancouver, BC.

– 1999b. Woodland Jumping Mouse (*Napaeozapus insignis*). In Wilson and Ruff, *The Smithsonian Book of North American Mammals*, 665–6.

– 2004. *Sorex cinereus. Mammalian Species* no. 743: 1–9.

Whitaker, J.O., Jr., and Clem, P.D. 1992. Food of the Evening Bat *Nycticeius humeralis* from Indiana. *American Midland Naturalist* 127: 211–214.

Whitaker, J.O., Jr., and French, T.W. 1984. Food of six species of sympatric shrews from New Brunswick. *Canadian Journal of Zoology* 62(4): 622–626.

Whitaker, J.O., Jr., and Gummer, S.L. 1992. Hibernation of the Big Brown Bat, *Eptesicus fuscus*, in buildings. *Journal of Mammalogy* 73(2): 312–316.

– 1994. The status of the Evening Bat, *Nycticeius humeralis*, in Indiana. *Proceedings of the Indiana Academy of Science* 102: 283–291.

Whitaker, J.O., Jr., and Hamilton, W.J., Jr. 1998. *Mammals of the Eastern United States.* 3rd ed. Cornell University Press, Ithaca, NY.

Whitaker, J.O., Jr., McKenzie, R., Rakow, M., Leibacher, B., and Leibacher, P. 1997a. Seasonal flight counts in three Big Brown Bat (*Eptesicus fuscus*) colonies. *Proceedings of the Indiana Academy of Science* 106: 79–84.

Whitaker, J.O., Jr., Rose, R.K., and Padgett, T.M. 1997b. Food of the Red Bat *Lasiurus borealis* in winter in the Great Dismal Swamp, North Carolina and Virginia. *American Midland Naturalist* 137(2): 408–411.

Whitaker, J.O., Jr., and Wrigley, R.E. 1972. *Napaeozapus insignis. Mammalian Species* no. 14: 1–6.

White, D., Jr., Berardinelli, J.G., and Aune, K.E. 1998. Reproductive characteristics of the male Grizzly Bear in the continental United States. *Ursus* 10: 497–501.

White, K.S., Golden, H.N., Hundertmark, K.J., and Lee, G.R. 2002. Predation by Wolves, *Canis lupus*, on Wolverines, *Gulo gulo*, and an American Marten, *Martes americana*, in Alaska. *Canadian Field-Naturalist* 116(1): 132–134.

White, R.G., Rowell, J.E., and Hauer, W.E. 1997. The role of nutrition, body condition, and lactation on calving success in Muskoxen. *Journal of Zoology* (London) 243: 13–20.

Whitehead, H. 1987. Updated status of the Humpback Whale, *Megaptera novaeangliae*, in Canada. *Canadian Field-Naturalist* 101(2): 284–294.

– 2002. Sperm Whale (*Physeter macrocephalus*). In W.F. Perrin, B. Wursig, and H.G.T. Thewissen (eds), *Encyclopedia of Marine Mammals*, 1165–1172. Academic Press, San Diego, CA.

– 2003. *Sperm Whales: Social Evolution in the Ocean.* University of Chicago Press, Chicago, IL.

Whitehead, H., Faucher, A., Gowans, S., and McCarrey, S. 1997a. Status of the Northern Bottlenose Whale, *Hyperoodon ampullatus*, in the Gully, Nova Scotia. *Canadian Field-Naturalist* 111(2): 287–292.

Whitehead, H., Gowans, S., Faucher, A., and McCarrey, S. 1997b. Population analysis of Northern Bottlenose Whales in the Gully, Nova Scotia. *Marine Mammal Science* 13(2): 173–185.

Whitehead, H., and Weilgart, L. 2000. The Sperm Whale: Social females and roving males. In J. Mann, R.C. Connor, P.L. Tyack, and H. Whitehead (eds), *Cetacean Societies: Field Studies of Dolphins and Whales*, 154–172. University of Chicago Press, Chicago, IL.

Whitman, A.A., and Payne, P.M. 1990. Age of Harbour Seals, *Phoca vitulina concolor*, wintering in southern New England. *Canadian Field-Naturalist* 104(4): 579–582.

Whitney, P. 1976. Population ecology of two sympatric species of subarctic microtine rodents. *Ecological Monographs* 46: 85–104.

Whittaker, D.G., and Lindzey, F.G. 1999. Effect of Coyote predation on early fawn survival in sympatric deer species. *Wildlife Society Bulletin* 27(2): 256–262.

Wigginton, J.D., and Dobson, F.S. 1999. Environmental influences on geographic variation in body size of western Bobcats. *Canadian Journal of Zoology* 77(5): 802–813.

Wiig, Ø., Born, E.W., and Pedersen, L.T. 2003. Movements of female Polar Bears (*Ursus maritimus*) in the East Greenland pack ice. *Polar Biology* 26: 509–516.

Wiig, Ø., Derocher, A.E., and Belikov, S.E. 1999. Ringed Seal (*Phoca hispida*) breeding in the drifting pack ice of the Barents Sea. *Marine Mammal Science* 15(2): 595–598.

Wiig, Ø., and Gjertz, I. 1996. Body size of male Atlantic Walruses (*Odobenus rosmarus rosmarus*) from Svalbard. *Journal of Zoology* (London) 240: 495–499.

Wilkinson, G.S. 1992a. Information transfer at evening bat colonies. *Animal Behaviour* 44: 501–518.

– 1992b. Communal nursing in the evening bat, *Nycticeius humeralis*. *Behavioral Ecology and Sociobiology* 31: 225–235.

Wilkinson, J.A., and Duglass, J.F. 2002. Mule Deer group kills Coyote. *Western North American Naturalist* 62(2): 253.

Wilkinson, L.C., and Barclay, R.M.R. 1997. Differences in the foraging behaviour of male and female Big Brown Bats (*Eptesicus fuscus*) during the reproductive period. *Ecoscience* 4(3): 279–285.

Willis, C.K.R., and Bast, M.L. 2000. Status report for the Pallid Bat (*Antrozous pallidus*) in Canada. Accepted at the annual meeting of the Committee on the Status of Endangered Wildlife in Canada (COSEWIC), May.

Willis, C.K.R., and Brigham, R.M. 2005. Physiological and ecological aspects of roost selection by reproductive female Hoary Bats (*Lasiurus cinereus*). *Journal of Mammalogy* 86(1): 85–94.

Willis, P.M., and Baird, R.W. 1998a. Sightings and strandings of beaked whales on the west coast of Canada. *Aquatic Mammals* 24(1): 21–25.

– 1998b. Status of Dwarf Sperm Whale, *Kogia simus*, with special reference to Canada. *Canadian Field-Naturalist* 112(1): 114–125.

Willis, P.M., Crespi, B.J., Dill, L.M., Baird, R.W., and Hanson, M.B. 2004. Natural hybridization between Dall's Porpoises (*Phocoenoides dalli*) and Harbour Porpoises (*Phocoena phocoena*). *Canadian Journal of Zoology* 82(5): 828–834.

Willson, M.F., and Gende, S.M. 2004. Seed dispersal by Brown Bears, *Ursus arctos*, in southeastern Alaska. *Canadian Field-Naturalist* 118(4): 499–503.

Wilson, C., Johnson, R.E., and Reichel, J.D. 1980. New records of the Northern Bog Lemming in Washington. *Murrelet* 61(3): 104–106.

Wilson, D.E., Bogan, M.A., Brownell, R.L., Jr., Burdin, A.M., and Maminov, M.K. 1991. Geographic variation in Sea Otters, *Enhydra lutris. Journal of Mammalogy* 72(1): 22–36.

Wilson, D.E., and Reeder, D.M. (eds) 2005. *Mammal Species of the World: A Taxonomic and Geographic Reference.* Vols. 1 and 2. Johns Hopkins University Press, Baltimore, MD.

Wilson, D.J., and Bromley, R.G. 2001. Functional and numerical responses of predators to cyclic lemming abundance: Effects on loss of goose nests. *Canadian Journal of Zoology* 79(3): 525–532.

Wilson, D.J., Krebs, C.J., and Sinclair, A.R.E. 1999. Limitation of collared lemming populations during a population cycle. *Oikos* 87: 382–398.

Wilson, D.R., and Hare, J.F. 2004. Animal communication: Ground squirrel uses ultrasonic alarms. *Nature* 430(6999): 523.

– 2006. The adaptive utility of Richardson's Ground Squirrel (*Spermophilus richardsonii*) short-range ultrasonic alarm signals. *Canadian Journal of Zoology* 84(9): 1322–1330.

Wilson, G.M., and Choate, J.R. 1996. Continued westward dispersal of the Woodchuck in Kansas. *The Prairie Naturalist* 28(1): 21–22.

– 1997. Taxonomic status and biogeography of the Southern Bog Lemming, *Synaptomys cooperi*, on the Central Great Plains. *Journal of Mammalogy* 78(2): 444–458.

Wilson, P.J., Grewal, S., Lawford, I.D., Heal, J.N.M., Granacki, A.G., Pennock, D., Theberge, J.B., Theberge, M.T., Voigt, D.R., Waddell, W., Chambers, R.E., Paquet, P.C., Goulet, G., Cluff, D., and White, B.N. 2000. DNA profiles of the Eastern Canadian Wolf and the Red Wolf provide evidence for a common evolutionary history independent of the Gray Wolf. *Canadian Journal of Zoology* 78(12): 2156–2166.

Wilson, P.J., Grewal, S., McFadden, T., Chambers, R.C., and White, B.N. 2003. Mitochondrial DNA extracted from eastern North American wolves killed in the 1800s is not of Gray Wolf origin. *Canadian Journal of Zoology* 81(5): 936–940.

Wilson, S.F., Hahn, A., Gladders, A., Goh, K.L.M., and Shackleton, D.M. 2004. Morphology and population characteristics of Vancouver Island Cougars, *Puma concolor vancouverensis. Canadian Field-Naturalist* 118(2): 159–163.

Wilson, T.M., and Carey, A.B. 1996. Observations of weasels in second-growth Douglas-fir forests in the Puget Sound Trough, Washington. *Northwestern Naturalist* 77: 35–39.

Wimmer, T., and Whitehead, H. 2004. Movements and distribution of Northern Bottlenose Whales, *Hyperoodon ampulatus*, on the Scotian Slope and in adjacent waters. *Canadian Journal of Zoology* 82(11): 1782–1794.

Winchell, J.M., and Kunz, T.H. 1996. Day-roosting activity budgets of the Eastern Pipistrelle Bat, *Pipistrellus subflavus* (Chiroptera: Vespertilionidae). *Canadian Journal of Zoology* 74(3): 431–441.

Winhold, L., Kurta, A., and Foster, R. 2008. Long-term changes in an assemblage of North American bats: Are Eastern Red Bats declining? *Acta Chiropterologica* 10(2): 359–366. Winn, H.E., and Reichley, N.E. 1985. Humpback Whale *Megaptera novaeangliae* (Borowski 1781). In S.H. Ridgway and R. Harrison, Jr. (eds), *Handbook of Marine Mammals*, vol. 3, 241–274. Academic Press, San Diego, CA.

Winterrowd, M.F., Gergits, W.F., Laves, K.S., and Weigl, P.D. 2005. Relatedness within nest groups of the Southern Flying Squirrel using microsatellite and discriminant function analyses. *Journal of Mammalogy* 86(4): 841–846.

Wisdom, M.J., and Cook, J.G. 2000. North American Elk. In S. Demarais and P.R. Krausman (eds), *Ecology and Management of Large Mammals in North America*, 694–735. Prentice Hall, Upper Saddle River, NJ.

Wisely, S.M., Statham, M.J., and Fleischer, R.C. 2008. Pleistocene refugia and Holocene expansion of a grassland-dependent species, the Black-footed Ferret (*Mustela nigripes*). *Journal of Mammalogy* 89(1): 87–96.

Witt, J.W. 1992. Home range and density estimates for the Northern Flying Squirrel, *Glaucomys sabrinus*, in western Oregon. *Journal of Mammalogy* 73(4): 921–929.

Woelfl, M., and Woelfl, S. 1994a. Golden Eagles, *Aquila chrysaetos*, preying on a Coyote, *Canis latrans. Canadian Field-Naturalist* 108(4): 494–495.

– 1994b. Evidence of Bobcats, *Lynx rufus*, in southeastern Alberta. *Canadian Field Naturalist* 108(4): 495–496.

Wolff, J.O. 1980. Social organization of the Taiga Vole (*Microtus xanthognathus*). *The Biologist* 62(1–4): 34–45.

– 1984. Overwintering behavioral strategies in Taiga Voles (*Microtus xanthognathus*). *Carnegie Museum of Natural History, Special Publication* 10: 315–318.

– 1993. Reproductive success of solitary and communally nesting White-footed Mice and Deer Mice. *Behavioral Ecology* 5(2): 206–209.

– 1996. Population fluctuations of mast-eating rodents are correlated with production of acorns. *Journal of Mammalogy* 77(3): 850–856.

Wolff, J.O., and Cicirello, D.M. 1990. Comparative paternal and infanticidal behaviour of sympatric White-footed Mice (*Peromyscus leucopus noveboracensis*) and Deer Mice (*P. maniculatus nubiterrae*). *Behavioral Ecology* 2(1): 38–45.

Wolff, J.O., and Johnson, M.F. 1979. Scent marking in Taiga Voles, *Microtus xanthognathus*. *Journal of Mammalogy* 60(2): 400–404.

Wolff, J.O., and Lidicker, W.Z., Jr. 1980. Population ecology of the Taiga Vole, *Microtus xanthognathus*, in interior Alaska. *Canadian Journal of Zoology* 58(10): 1800–1812.

– 1981. Communal winter nesting and food sharing in Taiga Voles. *Behavioral Ecology and Sociobiology* 9: 237–240.

Womble, J.N., Gende, S.M., and Blundell, G.M. 2007. Dive behavior of a Harbor Seal (*Phoco vitulina richardii*) in the presence of transient Killer Whales (*Orcinus orca*) in Glacier Bay National Park, Alaska. *Marine Mammal Science* 23(1): 203–208.

Wood, W.F. 2002. 2-Pyrrolidinine, a putative alerting pheromone from rump glands of Pronghorn, *Antilocapra americana*. *Biochemical Systematics and Ecology* 30: 361–363.

Woods, C.A. 1973. *Erethizon dorsatum*. *Mammalian Species* no. 29: 1–6.

Woods, C.A., Contreras, L., Willner-Chapman, G., and Widden, H.P. 1992. *Myocastor coypus*. *Mammalian Species* no. 398: 1–8.

Woodsworth, G.C., Bell, G.P., and Fenton, M.B. 1981. Observations of the echolocation, feeding behaviour, and habitat use of *Euderma maculatum* (Chiroptera: Vespertilionidae) in south-central British Columbia. *Canadian Journal of Zoology* 59(6): 1099–1102.

Woodward, B.L., and Winn, J.P. 2006. Apparent lateralized behaviour in Gray Whales feeding off the central British Columbia coast. *Marine Mammal Science* 22(1): 64–73.

Woodward, S.M. 1990. Population density and home range characteristics of Woodchucks, *Marmota monax*, at expressway interchanges. *Canadian Field-Naturalist* 104(3): 421–428.

– 1994. Identification of *Sorex monticolus* and *Sorex vagrans* from British Columbia. *Northwest Science* 68(4): 277–284.

Woolaver, L.G., Elderkin, M.F., and Scott, F.W. 1998. *Sorex dispar* in Nova Scotia. *Northeastern Naturalist* 5(4): 323–330.

Worley, K., Strobeck, C., Arthur, S., Carey, J., Schwantje, H., and Veitch, A. 2004. Population genetic structure of North American Thinhorn Sheep (*Ovis dalli*). *Molecular Ecology* 13: 2545–2556.

Worthington-Wilmer, J., Allen, P.J., Pomeroy, P.P., Twiss, S.D., and Amos, W. 1999. Where have all the fathers gone? An extensive microsatellite analysis of paternity in the Grey Seal (*Halichoerus grypus*). *Molecular Ecology* 8: 1417–1429.

Wright, B.S. 1951. A Walrus in the Bay of Fundy; The first record. *Canadian Field-Naturalist* 65: 61–65.

Wright, J.D., and Ernst, J. 2004a. Effects of mid-winter snow depth on stand selection by Wolverines, *Gulo gulo luscus*, in the boreal forest. *Canadian Field-Naturalist* 118(1): 56–60.

– 2004b. Wolverine, *Gulo gulo luscus*, resting sites and caching behavior in the boreal forest. *Canadian Field-Naturalist* 118(1): 61–64.

Wrigley, R.E. 1972. Systematics and biology of the Woodland Jumping Mouse *Napaeozapus insignis*. *Illinois Biological Monographs* no. 47: 1–117.

– 1974. Ecological notes on the animals of the Churchill region of Hudson Bay. *Arctic* 27: 201–214.

– 1979. History of the mammal fauna of southern Manitoba. *Manitoba Nature* 20(1): 3–17.

– 1980. The mouse with pockets. *Manitoba Nature* 21(1): 16–17.

Wrigley, R.E., Drescher, H., and Drescher, S. 1973. First record of the Fox Squirrel in Canada. *Journal of Mammalogy* 54: 782–783.

Wrigley, R.E., Dubois, J.E., and Copeland, H.W.R. 1979. Habitat, abundance, and distribution of six species of shrews in Manitoba. *Journal of Mammalogy* 60(3): 505–520.

– 1991. Distribution and ecology of six rare species of prairie rodents in Manitoba. *Canadian Field-Naturalist* 105(1): 1–12.

Wrigley, R.E., and Hatch, D.R.M. 1976. Arctic Fox migrations in Manitoba. *Arctic* 29: 147–158.

Würsig, B., and Clark, C. 1993. Behavior. In J.J. Burns, J.J. Montague, and C.J. Cowles (eds), *The Bowhead Whale*, special publication no. 2, 157–199. Society for Marine Mammalogy, Lawrence, KS.

Würsig, B., Koski, W.R., and Richardson, W.J. 1999. Whale riding behavior: Assisted transport for Bowhead Whale calves during spring migration in the Alaskan Beaufort Sea. *Marine Mammal Science* 15(1): 204–210.

Xia, X., and Millar, J.S. 1991. Genetic evidence of promiscuity in *Peromyscus leucopus*. *Behavioral Ecology and Sociobiology* 28: 171–178.

Yamaguchi, N., Rushton, S., and Macdonald, D.W. 2003. Habitat preferences of feral American Mink in the Upper Thames. *Journal of Mammalogy* 84(4): 1356–1373.

Yamamoto, O., Moore, B., and Brand, L. 2001. Variation in the bark call of the Red Squirrel (*Tamiasciurus hudsonicus*). *Western North American Naturalist* 61(4): 395–402.

Yates, T.L. 1999. Eastern Mole (*Scalopus aquaticus*). In D.E. Wilson and S. Ruff (eds), *The Smithsonian Book of North American Mammals*, 63–64. UBC Press, Vancouver, BC.

Yates, T.L., Mills, J.N., Parmenter, C.A., Ksiazek, T.G., Parmenter, R.R., Vande Castle, J.R., Calisher, C.H., Nichol, S.T., Abbott, K.D., Young, J.C., Morrison, M.L., Beaty, B.J., Dunnum, J.L., Baker, R.J., Salazar-Bravo, J., and Peters, C.J. 2002. The ecology and evolutionary history of an emergent disease: Hantavirus Pulmonary Syndrome. *BioScience* 52(11): 989–998.

Yates, T.L., and Schmidly, D.J. 1978. *Scalopus aquaticus*. *Mammalian Species* no. 105: 1–4.

Yensen, E., Stephens, D.A., and Post, M. 1986. An additional Idaho mole record. *Murrelet* 76(3): 96.

Ylönen, H., Koskela, E., and Mappes, T. 1997a. Infanticide in the Bank Vole (*Clethrionomys glareolus*): Occurrence and the effect of familiarity on female infanticide. *Annales Zoologici Fennici* 34: 259–266.

Ylönen, H., Mappes, T., and Koskela, E. 1997b. Territorial behaviour and reproductive success of Bank Vole *Clethrionomys glareolus* females. *Journal of Animal Ecology* 66: 341–349.

Ylönen, H., and Viitala, J. 1991. Social overwintering and food distribution in the Bank Vole *Clewthrionomys glareolus*. *Ecography* 14(2): 131–137.

Yochem, P.K., and Leatherwood, S. 1985. Blue Whale *Balaenoptera musculus* (Linnaeus 1758). In S.H. Ridgway and R. Harrison (eds), *Handbook of Marine Mammals*, vol. 3, 193–240. Academic Press, San Diego, CA.

York, A.E., and Scheffer, V.B. 1997. Timing of implantation in the Northern Fur Seal *Callorhinus ursinus*. *Journal of Mammalogy* 78(2): 675–683.

Young, P.J. 1990. Structure, location, and availability of hibernacula of Columbian Ground Squirrels (*Spermophilus columbianus*). *American Midland Naturalist* 123: 357–364.

Youngman, P.M. 1967a. A new subspecies of varying lemming, *Dicrostonyx torquatus* (Pallas), from Yukon Territory (Mammalia, Rodentia). *Proceedings of the Biological Society of Washington* 80: 31–34.

– 1967b. Insular populations of the Meadow Vole, *Microtus pennsylvanicus*, from northeastern North America, with descriptions of two new subspecies. *Journal of Mammalogy* 48(4): 579–588.

– 1975. *Mammals of the Yukon Territory*. National Museums of Canada Publications in Zoology, no. 10.

Yunker, G.B., Hammill, M.O., Gosselin, J.-F., Dion, D.M., and Schreer, J.F. 2005. Foetal growth in north-west Atlantic Grey Seals (*Halichoerus grypus*). *Journal of Zoology* (London) 265: 411–419.

Zager, P., and Beecham, J. 2006. The role of American Black Bears and Brown Bears as predators on ungulates in North America. *Ursus* 17(2): 95–108.

Zagrebelny, S.V. 1998. Morphological characteristics of Sea Otter *Enhydra lutris* L. (Carnivora, Mustelidae) pelage and first age moult. *IUCN Otter Specialist Group Bulletin* 15(2): 93–102.

Zaitsev, M.V. 1988. On the nomenclature of Red-toothed Shrews of the genus *Sorex* in the fauna of the USSR. *Zool. Zhurnal* 67(12): 1878–1888.

Zakreski, D. 1997. Big cat startles farmer. *Saskatoon Star Phoenix*, 10 December.

Zervanos, S.M., and Salsbury, C.M. 2003. Seasonal body temperature fluctuations and energetic strategies in free-ranging eastern Woodchucks (*Marmota monax*). *Journal of Mammalogy* 84(1): 299–310.

Zettergreen, B. 2006. Golden Eagle attacks and kills yearling Mountain Goat. *Wildlife Afield* 3(1): 27–28.

Zeveloff, S.I. 1988. *Mammals of the Intermountain West*. University of Utah Press, Salt Lake City, UT.

Zheng, X., Arbogast, B.S., and Kenagy, G.J. 2003. Historical demography and genetic structure of sister species: Deer Mice (*Peromyscus*) in the North American temperate rain forest. *Molecular Ecology* 12: 711–724.

Zielinski, W.J., and Duncan, N.P. 2004. Diets of sympatric populations of American Martens (*Martes americana*) and Fishers (*Martes pennanti*) in California. *Journal of Mammalogy* 85(3): 470–477.

Zielinski, W.J., Slauson. K.M., Carroll, C.R., Kent, C.J., and Kudrna, D.G. 2001. Status of American Martens in coastal forests of the Pacific states. *Journal of Mammalogy* 82(2): 478–490.

Zielinski, W.J., and Truex, R.L. 1995. Distinguishing tracks of marten and fisher at track-plate stations. *Journal of Wildlife Management* 59(3): 571–579.

Zimmer, W.M.X., Johnson, M., Madesen, P.T., and Tyack, P. 2005. Echolocation clicks of free-ranging Cuvier's Beaked Whale (*Ziphius cavirostris*). *Journal of the Acoustical Society of America* 117: 3919–3927.

Zimmerling, T.N. 2005. The influence of thermal protection on winter den selection by Porcupines, *Erethizon dorsatum*, in second-growth conifer forests. *Canadian Field-Naturalist* 119(2): 159–163.

Zimmerman, E.G. 1999. Plains Pocket Gopher. In D.E. Wilson and S. Ruff (eds), *The Smithsonian Book of North American Mammals*, 485–486. UBC Press, Vancouver, BC.

Zuercher, G.L., Roby, D.D., and Rexstad, E.A. 1999. Seasonal changes in body mass, composition, and organs of Northern Red-backed Voles in interior Alaska. *Journal of Mammalogy* 80(2): 443–459.

Index